Fluid Mechanics for Civil and Environmental Engineers

Fluid Mechanics for Civil
and Environmental
Engineers

Fluid Mechanics for Civil and Environmental Engineers

Ahlam I. Shalaby

CRC Press
Taylor & Francis Group
Boca Raton London New York

CRC Press is an imprint of the
Taylor & Francis Group, an **informa** business

CRC Press
Taylor & Francis Group
6000 Broken Sound Parkway NW, Suite 300
Boca Raton, FL 33487-2742

© 2018 by Taylor & Francis Group, LLC
CRC Press is an imprint of Taylor & Francis Group, an Informa business

No claim to original U.S. Government works

Printed on acid-free paper

International Standard Book Number-13: 978-0-8493-3737-6 (Hardback)

Library of Congress Cataloging-in-Publication Data

Names: Shalaby, A. I. (Ahlam Ibrahim), 1957- author.
Title: Fluid mechanics for civil and environmental engineers / Ahlam I. Shalaby.
Description: Boca Raton : Taylor & Francis a CRC title, part of the Taylor & Francis imprint, a member of the Taylor & Francis Group, the academic division of T&F Informa, plc, [2018] | Includes bibliographical references and index.
Identifiers: LCCN 2017038418| ISBN 9780849337376 (hardback : acid-free paper) | ISBN 9781315156637 (ebook)
Subjects: LCSH: Fluid mechanics. | Hydrodynamics.
Classification: LCC TA357 .S425 2018 | DDC 620.1/06--dc23
LC record available at https://lccn.loc.gov/2017038418

Visit the Taylor & Francis Web site at
http://www.taylorandfrancis.com

and the CRC Press Web site at
http://www.crcpress.com

Visit the eResources: https://www.crcpress.com/9780849337376

Printed and bound in the United States of America by
Edwards Brothers Malloy on sustainably sourced paper

I dedicate this to God Almighty, Who made writing this textbook even a possibility, and its completion a reality.

In loving memory of my precious and devoted mother, Mrs. Elham Ali Osman Shalaby, who gave me inspiration to keep going even when it seemed impossible, and my dear and dedicated father, Dr. Ibrahim Mahmoud Shalaby, who gave me motivation to pursue my dreams.

I dedicate this to my loving and very special boys, Omar Mohsen Bagnied and Basem Mohsen Bagnied, who give me unconditional love, and give me a reason to be the best in everything that I do.

To all of my students: you are the reason why I wrote this textbook.

Dream Big, Take a Leap of Faith, and Throw Caution to the Wind.

Contents

Preface xxxiii
Acknowledgments xxxvii
Author xxxix
List of Symbols xli
List of Abbreviations xlix
Standards for Units, Unit Prefixes, and Fluid Properties li
Introduction to the Textbook lv

1. Introduction 1
  1.1   Introduction 1
  1.2   Fundamental Principles of Fluid Mechanics 1
  1.3   Types of Fluid Flow 2
  1.4   Dimensions and Units 3
      1.4.1   Primary and Secondary Dimensions 3
           1.4.1.1   Dimensions for the British Gravitational (BG) System 3
           1.4.1.2   Dimensions for the International System (SI) 5
      1.4.2   System of Units 6
           1.4.2.1   Units for the British Gravitational (BG) System 9
           1.4.2.2   Units for the International System (SI) 10
  1.5   Pressure Scales 12
  1.6   Standard Atmosphere 13
  1.7   Standard Reference for Altitude, Pressure, Latitude, and Temperature 15
      1.7.1   Standard Atmospheric Pressure for Fluid Properties 16
      1.7.2   Variation in Temperature for Fluid Properties 16
      1.7.3   Standard Temperature Assumed for the Reference Fluid 19
  1.8   Newton's Second Law of Motion 19
      1.8.1   Acceleration due to Gravity for the British Gravitational (BG) System 19
      1.8.2   Acceleration due to Gravity for the International System (SI) 20
  1.9   Dynamic Forces Acting on a Fluid Element 21
  1.10 Physical Properties of Fluids 21
      1.10.1 Mass Density 22
      1.10.2 Specific Gravity 25
      1.10.3 Specific Weight 27
      1.10.4 Viscosity 32
           1.10.4.1 Laminar versus Turbulent Flow 36
           1.10.4.2 Shear Stress Distribution/Profile for Laminar or Turbulent Pipe Flow 38
           1.10.4.3 Newton's Law of Viscosity for Laminar Flow 39
           1.10.4.4 Dynamic Viscosity versus Kinematic Viscosity 44
           1.10.4.5 Newton's Law of Viscosity: Theoretically Derived Expression for Shear Stress for Laminar Flow 45

          1.10.4.6   Empirically Derived Expression for Shear Stress
                     for Turbulent Flow 53
     1.10.5 Surface Tension 55
          1.10.5.1   Definition of Surface Tension 56
          1.10.5.2   The Formation of a Droplet in the Atmosphere 58
          1.10.5.3   The Formation of a Droplet on a Solid Surface and
                     Capillarity 60
          1.10.5.4   The Formation of a Droplet on a Solid Surface 61
          1.10.5.5   Capillarity 61
     1.10.6 Vapor Pressure 64
          1.10.6.1   Partial Pressure, Atmospheric Pressure, and Vapor
                     Pressure 64
          1.10.6.2   Evaporation, Condensation, Boiling Point, and Vapor
                     Pressure of a Liquid 66
          1.10.6.3   Cavitation 68
     1.10.7 Elasticity, Compressibility, or Bulk Modulus of Elasticity 70
          1.10.7.1   Definition of Bulk Modulus of Elasticity 71
          1.10.7.2   Ideal Gas Law: Definition of Gas Constant and
                     Molecular Weight 75
          1.10.7.3   First Law of Thermodynamics: Definition of Specific
                     Heat and Specific Heat Ratio 79
          1.10.7.4   Determination of Bulk Modulus of Elasticity for Gases ... 85
          1.10.7.5   Definition of the Sonic (Acoustic) Velocity for Fluids 89
          1.10.7.6   Definition of the Mach Number for Fluids: A Measure
                     of Compressibility 92
  End-of-Chapter Problems 93

2. Fluid Statics 115
  2.1   Introduction 115
  2.2   The Principles of Hydrostatics 116
     2.2.1 Pascal's Law: The Hydrostatic Pressure at a Point 117
     2.2.2 The Hydrostatic Pressure Equation: Variation of Pressure
           from Point to Point 118
     2.2.3 The Hydrostatic Pressure Head 121
     2.2.4 The Hydrostatic Pressure Distribution 123
     2.2.5 Application of the Principles of Hydrostatics 124
  2.3   Measurement of Hydrostatic Pressure at a Point 126
     2.3.1 Pressure Scales 126
          2.3.1.1   Standard Atmospheric Pressure 128
     2.3.2 Methods of Pressure Measurement 129
     2.3.3 Barometers 129
     2.3.4 Manometers 133
          2.3.4.1   Manometry 134
     2.3.5 Piezometer Columns/Tubes: Simplest Manometers 135
     2.3.6 Open U-Tube Manometers: Simple Manometers 139
          2.3.6.1   Single-Fluid Simple Manometer 140
          2.3.6.2   Multifluid Simple Manometer 148
     2.3.7 Differential Manometers 150

2.3.7.1 Single-Fluid Differential Manometer within
a Single Pipe 152
2.3.7.2 Single-Fluid Differential Manometer between
Two Pipes.. 154
2.3.7.3 Multifluid Differential Manometer between
Two Pipes.. 159
2.4 Hydrostatic Forces on Submerged Surfaces 161
2.4.1 The Variation of the Hydrostatic Pressure along a Surface 162
2.4.1.1 Variation of the Pressure Prism for a Surface
Submerged in a Gas 162
2.4.1.2 Variation of the Pressure Prism for a Surface
Submerged in a Liquid Open to the Atmosphere 162
2.4.1.3 Variation of the Pressure Prism for a Surface
Submerged in a Liquid, Enclosed, and Pressurized 164
2.4.2 Magnitude and Location of the Hydrostatic Force for
Plane Surfaces 165
2.4.2.1 Magnitude of the Resultant Hydrostatic Force on
Plane Surfaces 166
2.4.2.2 Location of the Resultant Hydrostatic Force on Plane
Surfaces.. 169
2.4.3 Planes Submerged in a Gas.. 173
2.4.4 Planes Submerged in a Liquid Open to the Atmosphere 174
2.4.4.1 Submerged ($h = H$) Horizontal Plane (Open)............. 174
2.4.4.2 Submerged ($h = y_o$) Vertical Plane (Open) 176
2.4.4.3 Submerged ($h = y_o \sin \alpha$) Sloping Plane (Open)........... 180
2.4.5 Planes Submerged in an Enclosed and Pressurized Liquid.......... 184
2.4.5.1 Submerged ($h = H$) Horizontal Plane (Enclosed) 184
2.4.5.2 Submerged ($h = 0$) Vertical Plane (Enclosed) 186
2.4.5.3 Submerged ($h = y_o$) Vertical Plane (Enclosed) 189
2.4.5.4 Submerged ($h = y_o \sin \alpha$) Sloping Plane (Enclosed) 193
2.4.6 Submerged Nonrectangular Planes................................... 196
2.4.6.1 Submerged ($h = H$) Horizontal Nonrectangular Plane.... 197
2.4.6.2 Submerged ($h = H$) Horizontal Circular Plane............ 197
2.4.6.3 Submerged ($h = H$) Horizontal Triangular Plane 199
2.4.6.4 Submerged ($h = y_o$) Vertical Nonrectangular Plane....... 200
2.4.6.5 Submerged ($h = y_o$) Vertical Circular Plane.............. 201
2.4.6.6 Submerged ($h = y_o$) Vertical Triangular Plane 204
2.4.6.7 Submerged ($h = y_o \sin \alpha$) Sloping Nonrectangular
Plane... 208
2.4.6.8 Submerged ($h = y_o \sin \alpha$) Sloping Circular Plane.......... 208
2.4.6.9 Submerged ($h = y_o \sin \alpha$) Sloping Triangular Plane 212
2.4.7 Magnitude and Location of the Hydrostatic Force for Curved
Surfaces.. 216
2.4.7.1 Magnitude and Location of the Horizontal Component
of Force on Curved Surfaces............................. 217
2.4.7.2 Magnitude and Location of the Vertical Component
of Force on Curved Surfaces............................. 219
2.4.7.3 Magnitude and Location of the Resultant Force
on Curved Surfaces..................................... 220

    2.4.8   Magnitude and Location of the Hydrostatic Force for Surfaces
             Submerged in a Multilayered Fluid 223
             2.4.8.1   Magnitude and Location of the Resultant Force Acting
                        in First Fluid 223
             2.4.8.2   Magnitude and Location of the Resultant Force Acting
                        in Second Fluid 225
             2.4.8.3   Magnitude and Location of the Resultant Force Acting
                        in Multilayer Fluid 227
  2.5   Buoyancy and Stability of a Floating or a Neutrally Buoyant Body 229
    2.5.1   Buoyancy and Archimedes Principle 230
             2.5.1.1   Magnitude and Location of the Buoyant Force 230
             2.5.1.2   Buoyancy of a Completely Submerged Sinking Body 231
             2.5.1.3   Buoyancy of a Completely Submerged Neutrally
                        Buoyant Body 233
             2.5.1.4   Buoyancy of a Partially Submerged Floating Body 235
             2.5.1.5   The Role of the Buoyant Force in Practical
                        Applications 239
    2.5.2   Stability of a Floating or a Neutrally Buoyant Body 240
             2.5.2.1   Vertical Stability of a Floating or Neutrally
                        Buoyant Body 240
             2.5.2.2   Rotational Stability of a Floating or Neutrally Buoyant
                        Body 241
             2.5.2.3   Rotational Stability of Neutrally Buoyant/Suspended
                        Bodies 241
             2.5.2.4   Rotational Stability of Floating Bodies 243
             2.5.2.5   Maximum Design Angle of Rotation for a Top-Heavy
                        Floating Body 246
             2.5.2.6   Metacentric Height for a Top-Heavy Floating Body 246
             2.5.2.7   Computation of the Metacentric Height and the
                        Resulting Moment (Restoring, Overturning, or Zero)
                        for a Top-Heavy Floating Body 248
  End-of-Chapter Problems 259

3. Continuity Equation 301
  3.1   Introduction 301
  3.2   Fluid Kinematics 302
  3.3   Types of Fluid Flow 302
    3.3.1   Internal versus External Flow 302
    3.3.2   Pressure versus Gravity Flow 303
    3.3.3   Real versus Ideal Flow 303
             3.3.3.1   Laminar versus Turbulent Flow 303
    3.3.4   Compressible versus Incompressible Flow 303
    3.3.5   Spatially Varied (Nonuniform) versus Spatially
             Uniform Flow 303
    3.3.6   Unsteady versus Steady Flow 304
    3.3.7   One-, Two-, and Three-Dimensional Flows 304
  3.4   Flow Visualization/Geometry 304
    3.4.1   Path Lines 305

3.4.2   Streamlines 305

3.4.3   Stream Tubes 306

3.5   Describing and Observing the Fluid Motion................................. 307

3.5.1   Definition of a Fluid System: The Lagrangian Point of View 308

3.5.1.1   Definition of the Fluid Particle Velocity in the Lagrangian Point of View 308

3.5.2   Definition of a Control Volume: The Eulerian Point of View 309

3.5.2.1   Definition of the Fluid Particle Velocity in the Eulerian Point of View 310

3.5.2.2   Fixed versus Moving Control Volume in the Eulerian Point of View 311

3.5.3   Deriving One View from the Other................................. 312

3.5.4   Application of the Lagrangian View versus Application of the Eulerian View... 312

3.6   The Physical Laws Governing Fluid in Motion 313

3.6.1   Statement of the Governing Physical Laws Assuming a Fluid System 314

3.6.1.1   Extensive and Intensive Fluid Properties 314

3.6.2   The Reynolds Transport Theorem 314

3.6.3   Statement of the Assumptions Made for the Type of Fluid Flow.... 317

3.6.3.1   The Assumption of Spatial Dimensionality of the Fluid Flow 318

3.6.3.2   The Modeling of Two-Dimensional Flows................ 319

3.6.3.3   The Modeling of One-Dimensional Flows................ 320

3.6.3.4   A Comparison of Velocity Profiles for Various Flow Types 323

3.7   Conservation of Mass: The Continuity Equation 324

3.7.1   The Continuity Equation 325

3.7.1.1   The Continuity Equation for a Fluid System 326

3.7.1.2   The Continuity Equation for a Control Volume 335

3.8   Measurement of the Volume Flowrate...................................... 348

End-of-Chapter Problems... 348

4. Energy Equation 363

4.1   Introduction... 363

4.2   Fluid Dynamics 363

4.3   Derivation of the Energy Equation 365

4.4   Conservation of Momentum Principle: Newton's Second Law of Motion.... 365

4.4.1   Newton's Second Law of Motion for a Fluid System................ 367

4.4.2   Newton's Second Law of Motion for a Control Volume............. 369

4.5   The Energy Equation Based on Newton's Second Law of Motion............ 371

4.5.1   The Energy Equation for a Fluid System 372

4.5.2   The Energy Equation for a Control Volume 373

4.5.2.1   Definition of the Terms in the Energy Equation 374

4.5.2.2   Practical Assumptions Made for the Energy Equation ... 375

4.5.2.3   Application of the Energy Equation for Pressure Flow versus Open Channel Flow 376

4.5.3   The Bernoulli Equation 376

4.5.3.1   Applications of the Bernoulli Equation 379

4.5.3.2   The Pitot-Static Tube: Dynamic Pressure Is Modeled
            by a Pressure Rise 380

4.5.3.3   Ideal Flow Meters, Ideal Gradual Channel
            Contractions, and Ideal Flow from a Tank: Dynamic
            Pressure Is Modeled by a Pressure Drop 382

4.5.3.4   The Energy Grade Line (EGL) and the Hydraulic Grade
            Line (HGL) 383

4.5.3.5   The Hydraulic Grade Line (HGL) and Cavitation 389

4.5.4   Application of the Energy Equation in Complement to the
        Momentum Equation 392

4.5.4.1   Application of the Governing Equations 392

4.5.4.2   Modeling Flow Resistance: A Subset Level Application
            of the Governing Equations 401

4.5.5   Application of the Energy and Momentum Equations for Real
        Internal Flow 403

4.5.5.1   Evaluation of the Major and Minor Head
            Loss Terms 403

4.5.5.2   Evaluation of the Pump and Turbine Head Terms 405

4.5.5.3   Applications of the Governing Equations for Real
            Internal Flow 406

4.5.5.4   Application of the Energy Equation for Real
            Pipe Flow 406

4.5.5.5   Application of the Energy Equation for Real Open
            Channel Flow 410

4.5.6   Application of the Energy and Momentum Equations for Ideal
        Internal Flow and Ideal Flow from a Tank 412

4.5.6.1   Evaluation of the Actual Discharge 413

4.5.6.2   Applications of the Governing Equations for Ideal
            Internal Flow and Ideal Flow from a Tank 414

4.5.6.3   Application of the Bernoulli Equation for Ideal
            Pipe Flow 415

4.5.6.4   Application of the Bernoulli Equation for Ideal Open
            Channel Flow 419

4.5.6.5   Applications of the Bernoulli Equation for Ideal
            Flow from a Tank 423

4.5.7   Application of the Energy and Momentum Equations for Ideal
        External Flow 442

4.5.7.1   Evaluation of the Drag Force 442

4.5.7.2   Applications of the Governing Equations for Ideal
            External Flow 444

4.5.7.3   Applications of the Bernoulli Equation for Ideal
            External Flow 444

4.5.8   Application of the Energy and Momentum Equations for a
        Hydraulic Jump 449

4.5.8.1   Applications of the Governing Equations for a
            Hydraulic Jump 450

4.5.8.2   Application of the Energy Equation for a
            Hydraulic Jump 451

4.6   Conservation of Energy Principle: First Law of Thermodynamics............ 452
     4.6.1   Total Energy 453
         4.6.1.1   Total Energy for a Fluid Flow System 454
         4.6.1.2   Dimensions and Units for Energy 455
     4.6.2   Energy Transfer by Heat... 455
     4.6.3   Energy Transfer by Work.. 456
4.7   The Energy Equation Based on the First Law of Thermodynamics........... 457
     4.7.1   The Energy Equation for a Fluid System 457
     4.7.2   The Energy Equation for a Control Volume 458
         4.7.2.1   The Energy Equation Expressed in Power Terms 460
         4.7.2.2   Pump and Turbine Losses Are Modeled by Their
                 Efficiencies... 461
         4.7.2.3   Pump, Turbine, and System Efficiencies................. 463
         4.7.2.4   The Energy Equation Expressed in Energy
                 Head Terms... 465
         4.7.2.5   Definition of the Terms in the Energy Equation 466
         4.7.2.6   Practical Assumptions Made for the Energy Equation ... 467
     4.7.3   Application of the Energy Equation in Complement to the
         Momentum Equation 467
     4.7.4   Application of the Energy Equation for Real Internal Flow
         with a Pump or a Turbine 468
         4.7.4.1   Dimensions and Units for the Pump and Turbine
                 Head Terms... 470
         4.7.4.2   Applications of the Energy Equation for Real
                 Internal Flow with a Pump 471
         4.7.4.3   Applications of the Energy Equation for Real
                 Internal Flow with a Turbine 486
End-of-Chapter Problems.. 513

5. Momentum Equation 555
5.1   Introduction... 555
5.2   Derivation of The Momentum Equation...................................... 556
     5.2.1   The Momentum Equation for a Fluid System: Differential Form
         of the Momentum Equation 556
         5.2.1.1   Differential Form of the Momentum Equation 558
         5.2.1.2   Application of the Differential Form of the Momentum
                 Equation 559
     5.2.2   The Momentum Equation for a Control Volume: Integral
         Form of the Momentum Equation 564
         5.2.2.1   Fixed versus Moving Control Volume in the
                 Eulerian (Integral) Point of View 565
         5.2.2.2   Application of the Integral Form of the Momentum
                 Equation 566
5.3   Application of the Energy Equation in Complement
     to the Momentum Equation 576
     5.3.1   Application of the Governing Equations 577
     5.3.2   Modeling Flow Resistance: A Subset Level Application of
         the Governing Equations 578

5.3.3 Application of the Energy and Momentum Equations for Real
Internal Flow 580
5.3.3.1 Evaluation of the Major and Minor Head Loss Terms.... 580
5.3.3.2 Evaluation of the Pump and Turbine Head Terms 583
5.3.3.3 Applications of the Governing Equations for Real
Internal Flow 583
5.3.3.4 Applications of the Governing Equations for Real
Internal Flow: Differential Fluid Element................. 584
5.3.3.5 Applications of the Governing Equations for Real
Internal Flow: Finite Control Volume 592
5.3.4 Application of the Energy and Momentum Equations for Ideal
Internal Flow and Ideal Flow from a Tank 598
5.3.4.1 Evaluation of the Actual Discharge....................... 598
5.3.4.2 Applications of the Governing Equations for Ideal
Internal Flow and Ideal Flow from a Tank 600
5.3.4.3 Applications of the Governing Equations for Ideal
Pipe Flow 601
5.3.4.4 Applications of the Governing Equations for Ideal
Open Channel Flow 602
5.3.4.5 Applications of the Governing Equations for Ideal
Flow from a Tank... 605
5.3.5 Application of the Energy and Momentum Equations for
Ideal External Flow 614
5.3.5.1 Evaluation of the Drag Force 614
5.3.5.2 Applications of the Governing Equations for Ideal
External Flow 616
5.3.6 Application of the Energy and Momentum Equations for a
Hydraulic Jump... 618
5.3.6.1 Applications of the Governing Equations for a
Hydraulic Jump 618
End-of-Chapter Problems... 620

6. Flow Resistance Equations 643
6.1 Introduction... 643
6.1.1 Modeling Flow Resistance: A Subset Level Application of the
Governing Equations 644
6.1.2 Derivation of the Flow Resistance Equations and the Drag
Coefficients 645
6.1.2.1 Empirical Evaluation of the Drag Coefficients and
Application of the Flow Resistance Equations........... 646
6.1.3 Modeling the Flow Resistance as a Loss in Pump and Turbine
Efficiency in Internal Flow.. 646
6.2 Types of Flow... 647
6.2.1 Internal Flow versus External Flow................................. 647
6.2.2 Pipe Flow versus Open Channel Flow.............................. 647
6.2.3 Real Flow versus Ideal Flow.. 648
6.2.4 Ideal Flow 649
6.2.4.1 Ignoring the Flow Resistance in Ideal Flow 649

6.2.4.2   Subsequently Modeling the Flow Resistance in
Ideal Flow 650

6.2.4.3   The Pitot-Static Tube: Dynamic Pressure Is Modeled
by a Pressure Rise 651

6.2.4.4   The Venturi Meter: Dynamic Pressure Is Modeled
by a Pressure Drop 653

6.2.5   Real Flow 654

6.2.5.1   Directly and Subsequently Modeling the Flow
Resistance in Real Flow................................ 654

6.2.5.2   Directly Modeling the Flow Resistance in Real Flow 655

6.2.5.3   Laminar Flow versus Turbulent Flow 656

6.2.5.4   The Velocity Profiles for Laminar and Turbulent
Internal Flows 657

6.2.5.5   Developing Flow versus Developed Flow 659

6.3   Modeling the Flow Resistance as a Drag Force in External Flow 659

6.3.1   Evaluation of the Drag Force 660

6.3.2   Application of the Bernoulli Equation: Derivation of the Ideal
Velocity 661

6.3.3   Application of the Momentum Equation and Dimensional
Analysis: Derivation of the Actual Velocity, Drag Force Equation,
and the Drag Coefficient... 661

6.3.4   Modeling the Drag Force in the Momentum Equation 662

6.3.4.1   Submerged Stationary Body in a Stationary Real or
Ideal Fluid 662

6.3.4.2   Submerged Moving Body in a Stationary
Ideal Fluid 663

6.3.4.3   Submerged Moving Body in a Stationary Real Fluid..... 664

6.3.5   Supplementing the Momentum Equation with Dimensional
Analysis: Derivation of the Drag Force and the Drag Coefficient ... 665

6.4   Modeling the Flow Resistance as a Major Head Loss in Internal Flow 666

6.4.1   Evaluation of the Major Head Loss................................. 667

6.4.2   Application of the Energy Equation: Derivation of the
Head Loss 668

6.4.3   Application of the Momentum Equations and Dimensional
Analysis: Derivation of the Pressure Drop, Shear Stress, and the
Drag Coefficient 668

6.4.4   Application of the Governing Equations to Derive the
Major Head Loss Equations for Laminar and Turbulent Flow 669

6.5   Laminar and Turbulent Internal Flow Characteristics 671

6.5.1   Determining the Variation of Shear Stress with Radial Distance
for Laminar and Turbulent Flow 672

6.5.2   The Variation of the Velocity with Time for Laminar and
Turbulent Flow 675

6.5.3   The Variation of the Velocity, Pressure Drop, and the Wall Shear
Stress with Length of Pipe for Laminar and Turbulent Flow:
Developing versus Developed Flow................................ 675

6.5.3.1   The Extent of Developing Flow in the Pipe Length....... 676

6.5.3.2   The Variation of the Velocity Profile with
Pipe Length.. 678

6.5.3.3   The Variation of the Pressure Drop and the
Wall Shear Stress with Pipe Length 679

6.5.3.4   Laminar versus Turbulent Flow 680

6.6   Derivation of the Major Head Loss Equation for Laminar Flow.............. 681

6.6.1   Application of the Integral Form of the Momentum
Equation: Deriving the Velocity Profile for Laminar Flow 681

6.6.2   Application of the Differential form of the Continuity Equation:
Deriving Poiseuille's Law... 683

6.6.3   Substituting Poiseuille's Law into the Integral Form of the Energy
Equation: Deriving the Major Head Loss Equation for Laminar
Flow (Poiseuille's Law) 684

6.6.4   Interpretation of Poiseuille's Law Expressed in Terms of the
Pressure Drop.. 685

6.7   Derivation of the Major Head Loss Equation for Turbulent Flow 686

6.7.1   Application of the Integral Momentum Equation: Deriving an
Expression for the Pressure Drop................................... 687

6.7.2   Application of the Differential Momentum Equation: Interpreting
the Friction Slope in Turbulent Flow 687

6.7.3   Application of Dimensional Analysis: Empirically Interpreting
Friction Slope (Empirically Deriving an Expression for Wall Shear
Stress) in Turbulent Flow 688

6.7.3.1   Derivation of an Empirical Expression for the Wall
Shear Stress as a Function of the Velocity 689

6.7.3.2   Derivation of the Chezy Equation and Evaluation of the
Chezy Coefficient, C..................................... 690

6.7.3.3   Application of the Chezy Equation....................... 690

6.7.4   Substituting the Chezy Equation into the Integral Form of the
Energy Equation: Deriving the Major Head Loss Equation for
Turbulent Flow 691

6.7.5   The Darcy–Weisbach Equation 692

6.7.5.1   Derivation of the Darcy–Weisbach Friction
Coefficient, f 692

6.7.5.2   Evaluation of the Darcy–Weisbach Friction
Coefficient, f 693

6.7.5.3   The Darcy–Weisbach Head Loss Equation 694

6.7.6   Manning's Equation 694

6.7.6.1   Derivation and Evaluation of the Manning's
Roughness Coefficient, n 695

6.7.6.2   Manning's Equation 695

6.7.6.3   Manning's Head Loss Equation 696

6.7.7   The Hazen–Williams Equation 696

6.7.7.1   The Hazen–Williams Equation 696

6.7.7.2   Evaluation of the Hazen–Williams Equation
Roughness Coefficient, C_h 697

6.7.7.3   The Hazen–Williams Head Loss
Equation 697

6.7.8   The Relationship between the Drag Coefficient, C_D; the Chezy
Coefficient, C; the Darcy–Weisbach Friction Factor, f; and
Manning's Roughness Coefficient, n................................ 698

6.7.9   A Comparison between Laminar and Turbulent Flow Using the
Darcy–Weisbach Head Loss Equation 698
     6.7.9.1   The Darcy–Weisbach Head Loss Equation and the
Reynolds Number, R for Noncircular Pipes 699
     6.7.9.2   Evaluating the Darcy–Weisbach Friction Coefficient,
f for Laminar Pipe Flow 700
     6.7.9.3   A Comparison between Laminar and Turbulent Flow
Using the Darcy–Weisbach Head Loss Equation 700
6.7.10   Determining the Velocity Profile for Turbulent Flow 702
     6.7.10.1   A Comparison between the Velocity Profiles for
Laminar Flow and Turbulent Flow 704
     6.7.10.2   The Role of the Boundary Roughness in Laminar Flow
and Turbulent Flow 705
6.7.11   Application of the Major Head Loss Equation for Open
Channel Flow 706
     6.7.11.1   Application of the Chezy Equation for Open
Channel Flow 708
6.8   Modeling the Flow Resistance as a Minor Head Loss in Pipe Flow 709
6.8.1   Evaluation of the Minor Head Loss 710
6.8.2   Application of the Energy Equation: Derivation of the
Head Loss 711
6.8.3   Application of the Momentum Equation and Dimensional
Analysis: Derivation of the Pressure Drop, Shear Stress, and the
Drag Coefficient 713
6.8.4   Analytical Derivation of the Minor Head Loss Equation due to
a Sudden Pipe Expansion 714
6.8.5   Empirical Derivation of the Minor Head Loss Equation due to
Pipe Components in General 715
     6.8.5.1   Application of the Energy Equation: Derivation of the
Head Loss 715
     6.8.5.2   Application of the Momentum Equation and
Dimensional Analysis: Derivation of the Pressure
Drop, Shear Stress, and the Drag Coefficient 716
     6.8.5.3   Evaluation of the Minor Head Loss
Coefficient, k 716
6.9   Modeling the Flow Resistance as a Loss in Flowrate in
Internal Flow 718
6.9.1   Evaluation of the Actual Discharge 719
6.9.2   Application of the Bernoulli Equation: Derivation of the Ideal
Velocity 720
6.9.3   Application of the Continuity Equation: Derivation of the Ideal
Discharge and the Actual Discharge 720
6.9.4   Application of the Momentum Equation and Dimensional
Analysis: Derivation of the Actual Velocity, Actual Area, Actual
Discharge, and the Discharge Coefficient 721
     6.9.4.1   Supplementing the Momentum Equation with
Dimensional Analysis: Derivation of the
Reduced/Actual Discharge Equation for
Pipe Flow 722

         6.9.4.2   Supplementing the Momentum Equation with
                   Dimensional Analysis: Derivation of the
                   Reduced/Actual Discharge Equation for Open
                   Channel Flow... 723
      6.9.5   A Comparison of the Velocity Profiles for Ideal and Real Flows.... 724
  6.10 Modeling the Flow Resistance as a Loss in Pump and Turbine Efficiency
        in Internal Flow... 725
      6.10.1 Evaluation of the Efficiency of Pumps and Turbines................ 726
         6.10.1.1 Supplementing the Momentum Equation with
                   Dimensional Analysis: Derivation of the Efficiency
                   of Pumps and Turbines................................. 726
  End-of-Chapter Problems.. 726

7. **Dimensional Analysis** 749
  7.1   Introduction... 749
      7.1.1   The Role of Dimensional Analysis in the Modeling of
              Fluid Flow 750
      7.1.2   The Role of Dimensional Analysis in the Empirical Modeling of
              Flow Resistance in Real Fluid Flow................................. 752
         7.1.2.1 Modeling of Flow Resistance in Real Fluid Flow........ 752
         7.1.2.2 Using Dimensional Analysis in the Empirical Modeling
                  of Flow Resistance in Real Fluid Flow.................. 753
      7.1.3   Flow Types and Dimensional Analysis............................. 755
      7.1.4   Internal Flow versus External Flow............................... 755
      7.1.5   Modeling Flow Resistance: A Subset Level Application of the
              Governing Equations 755
      7.1.6   Supplementing the Momentum Theory with Dimensional
              Analysis... 756
      7.1.7   Derivation of the Flow Resistance Equations and the Drag
              Coefficients 756
         7.1.7.1 Empirical Evaluation of the Drag Coefficients and
                  Application of the Flow Resistance Equations........... 757
      7.1.8   Derivation of the Efficiency of Pumps and Turbines 757
  7.2   Dimensional Analysis of Fluid Flow...................................... 757
      7.2.1   Dynamic Forces Acting on a Fluid Element 758
      7.2.2   Two-Dimensional Systems... 759
      7.2.3   Deriving a Functional Relationship/Dimensionless Numbers....... 759
      7.2.4   Main Pi Terms 761
      7.2.5   The Definition of New Pi Terms................................... 762
      7.2.6   Guidelines in the Derivation of the Flow Resistance
              Equations and the Drag Coefficients 762
      7.2.7   The Definition of the Drag Coefficient........................... 769
      7.2.8   Guidelines in the Derivation of the Efficiency of Pumps
              and Turbines... 770
      7.2.9   The Definition of the Pump (or Turbine) Efficiency 773
      7.2.10 Specific Guidelines and Summary in the Application of
              Dimensional Analysis for Example Problems and End-of-
              Chapter Problems.. 774

7.3   Modeling the Flow Reistance as a Drag Force in External Flow 776
    7.3.1   Evaluation of the Drag Force 776
    7.3.2   Application of the Bernoulli Equation: Derivation of the Ideal
        Velocity 777
    7.3.3   Application of the Momentum Equation and Dimensional
        Analysis: Derivation of the Actual Velocity, Drag Force
        Equation, and the Drag Coefficient 777
        7.3.3.1   Application of Dimensional Analysis: Derivation of the
            Drag Force and the Drag Coefficient 778
        7.3.3.2   Application of Dimensional Analysis to Derive the
            Drag Force and Drag Coefficient for More Specific
            Assumptions of External Flow 787
    7.3.4   Derivation of the Lift Force and the Lift Coefficient 791
7.4   Modeling the Flow Resistance as a Major Head Loss in Internal Flow 795
    7.4.1   Evaluation of the Major Head Loss 796
    7.4.2   Application of the Energy Equation: Derivation of the Head
        Loss 797
    7.4.3   Application of the Momentum Equations and Dimensional
        Analysis: Derivation of the Pressure Drop, Shear Stress,
        and the Drag Coefficient 797
    7.4.4   Derivation of the Major Head Loss Equation for Laminar Flow 799
    7.4.5   Derivation of the Major Head Loss Equation for Turbulent Flow ... 799
        7.4.5.1   Application of Dimensional Analysis: Derivation
            of the Wall Shear Stress and Drag Coefficient in
            Turbulent Flow 800
        7.4.5.2   Derivation of the Chezy Equation and Evaluation of
            the Chezy Coefficient, C 804
        7.4.5.3   Substituting the Chezy Equation into the Energy
            Equation: Deriving the Major Head Loss Equation
            for Turbulent Flow 805
    7.4.6   The Darcy–Weisbach Equation 806
        7.4.6.1   Derivation of the Darcy–Weisbach Friction
            Coefficient, f 806
        7.4.6.2   Evaluation of the Darcy–Weisbach Friction
            Coefficient, f 808
        7.4.6.3   The Darcy–Weisbach Head Loss Equation 808
        7.4.6.4   Evaluating the Darcy–Weisbach Friction
            Coefficient, f for Laminar Pipe Flow 809
    7.4.7   Manning's Equation 809
        7.4.7.1   Derivation and Evaluation of the Manning's
            Roughness Coefficient, n 809
        7.4.7.2   Manning's Equation 810
        7.4.7.3   Manning's Head Loss Equation 811
    7.4.8   Application of the Major Head Loss Equation for Open
        Channel Flow 811
        7.4.8.1   Interpretation of the Results of Dimensional Analysis
            for Open Channel Flow 813
        7.4.8.2   Application of the Chezy Equation for Open
            Channel Flow 814

7.4.8.3   Application of the Darcy–Weisbach Equation for
Open Channel Flow 815

7.4.8.4   Application of Manning's Equation for Open
Channel Flow.. 815

7.4.9   Application of Dimensional Analysis to Derive the Wall Shear
Stress and Drag Coefficient for More Specific Assumptions of
Internal Flow 816

7.5   Modeling the Flow Resistance as a Minor Head Loss in Pipe Flow 820

7.5.1   Evaluation of the Minor Head Loss 820

7.5.2   Application of the Energy Equation: Derivation of the
Head Loss 821

7.5.3   Application of the Momentum Equation and Dimensional
Analysis: Derivation of the Pressure Drop, Shear Stress, and the
Drag Coefficient 822

7.5.4   Analytical Derivation of the Minor Head Loss Equation
due to a Sudden Pipe Expansion 823

7.5.5   Empirical Derivation of the Minor Head Loss Equation
due to Pipe Components in General............................... 824

7.5.5.1   Application of Dimensional Analysis: Derivation
of the Pressure Drop and Drag Coefficient for Pipe
Components in General 824

7.5.5.2   Substituting the Pressure Drop into the Energy
Equation: Deriving the Minor Head Loss Equation for
Pipe Components in General 831

7.5.5.3   Evaluation of the Minor Head Loss Coefficient, k 831

7.5.5.4   Application of Dimensional Analysis to Derive the
Pressure Drop and Drag Coefficient for More Specific
Assumptions for Pipe Flow Components................. 833

7.6   Modeling the Flow Resistance as a Loss in Flowrate in Internal Flow........ 838

7.6.1   Evaluation of the Actual Discharge............................... 839

7.6.2   Application of the Bernoulli Equation: Derivation of the Ideal
Velocity 840

7.6.3   Application of the Continuity Equation: Derivation of the Ideal
Discharge and the Actual Discharge............................... 840

7.6.4   Application of the Momentum Equation and Dimensional
Analysis: Derivation of the Actual Velocity, Actual Area, Actual
Discharge, and the Discharge Coefficient.......................... 841

7.6.4.1   Application of Dimensional Analysis: Derivation
of the Reduced/Actual Discharge and Drag Coefficient
for Pipe Flow Measuring Devices 842

7.6.4.2   Application of Dimensional Analysis: Derivation of the
Reduced/Actual Discharge and Drag Coefficient for
Open Channel Flow-Measuring Devices 854

7.6.4.3   Application of Dimensional Analysis to Derive the
Reduced/Actual Discharge and Drag Coefficient for
More Specific Assumptions for Open Channel
Flow-Measuring Devices 861

7.7   Modeling the Flow Resistance as a Loss in Pump and Turbine
Efficiency in Internal Flow... 866

	7.7.1	Evaluation of the Efficiency of Pumps and Turbines 867
		7.7.1.1 Application of Dimensional Analysis: Derivation of the Efficiency of Pumps and Turbines 867
		7.7.1.2 Application of Dimensional Analysis to Derive the Efficiency of Pumps and Turbines for More Specific Flow Assumptions 878
	7.8	Experimental Formulation of Theoretical Equations 893
		End-of-Chapter Problems 901

8. Pipe Flow 919
  8.1 Introduction 919
  8.2 Application of the Eulerian (Integral) versus Lagrangian (Differential) Forms of the Governing Equations 920
      8.2.1 Eulerian (Integral) Approach for Pipe Flow Problems 920
      8.2.2 Lagrangian (Differential) Approach for Pipe Flow Problems 921
  8.3 Modeling Flow Resistance in Pipe Flow: A Subset Level Application of the Governing Equations 921
  8.4 Application of the Governing Equations in Pipe Flow 922
      8.4.1 Application of the Governing Equations for Real Pipe Flow 923
            8.4.1.1 Integral Approach for Real Pipe Flow Problems 923
            8.4.1.2 Evaluation of the Major and Minor Head Loss Terms 923
            8.4.1.3 Evaluation of the Pump and Turbine Head Terms 926
            8.4.1.4 Differential Approach for Real Pipe Flow Problems 926
            8.4.1.5 Applications of the Governing Equations for Real Pipe Flow Problems 927
      8.4.2 Application of the Governing Equations for Ideal Pipe Flow 927
            8.4.2.1 Evaluation of the Actual Discharge 928
            8.4.2.2 Applications of the Governing Equations for Ideal Pipe Flow 929
  8.5 Single Pipes: Major Head Loss in Real Pipe Flow 929
      8.5.1 Evaluation of the Major Head Loss Term in the Energy Equation 930
            8.5.1.1 Laminar Pipe Flow: Poiseuille's Law 930
            8.5.1.2 Turbulent Pipe Flow: The Chezy Equation 934
      8.5.2 Turbulent Pipe Flow Resistance Equations and Their Roughness Coefficients 937
            8.5.2.1 A Comparison of the Three Standard Empirical Flow Resistance Coefficients 938
            8.5.2.2 A Comparison between Manning's and Hazen–Williams Roughness Coefficients 939
            8.5.2.3 Turbulent Pipe Flow Resistance Equations 939
      8.5.3 The Darcy–Weisbach Friction Coefficient, f and the Darcy–Weisbach Equation 941
            8.5.3.1 Evaluation of the Darcy–Weisbach Friction Coefficient, f 942
            8.5.3.2 Application of the Darcy–Weisbach Equation 944

8.5.4   Manning's Roughness Coefficient, n and Manning's
Equation 952
       8.5.4.1   Application of Manning's Equation 954
8.5.5   The Hazen–Williams Roughness Coefficient, C_h and the
Hazen–Williams Equation 957
       8.5.5.1   Application of the Hazen–Williams Equation 959

8.6   Pipes with Components: Minor Head Losses and Reaction Forces
in Real Pipe Flow 962
   8.6.1   Evaluation of the Minor Head Loss Term in the
Energy Equation 963
       8.6.1.1   Evaluation of the Minor Head Loss 963
       8.6.1.2   Derivation of the Minor Head Loss Coefficient, k 965
       8.6.1.3   Evaluation of the Minor Head Loss Coefficient, k 965
   8.6.2   Alternative Modeling of the Minor Head Loss Term in the
Energy Equation 966
   8.6.3   Evaluation of the Minor Head Loss due to Pipe Components 967
   8.6.4   Minor Losses in Valves and Fittings (Tees, Unions, Elbows,
and Bends) 968
       8.6.4.1   Valves 969
       8.6.4.2   Fittings (Tees, Unions, Elbows, and Bends) 974
   8.6.5   Minor Losses in Entrances, Exits, Contractions, and Expansions 979
       8.6.5.1   Pipe Entrances 980
       8.6.5.2   Pipe Exists 983
       8.6.5.3   Sudden Pipe Expansions 984
       8.6.5.4   Sudden Pipe Contractions 989
       8.6.5.5   Gradual Pipe Expansions 990
       8.6.5.6   Gradual Pipe Contractions 995
   8.6.6   Reaction Forces on Pipe Components 998

8.7   Pipe Systems: Major and Minor Head Losses in Real Pipe Flow 1000
   8.7.1   Single Pipes 1001
   8.7.2   Pipes with Components 1001
   8.7.3   Pipes with a Pump or a Turbine 1004
   8.7.4   Pipes in Series 1009
   8.7.5   Pipes in Parallel 1017
   8.7.6   Branching Pipes 1030
       8.7.6.1   Branching Pipes Connected to Three Reservoirs 1031
       8.7.6.2   Branching Pipes Connected to a Water Supply
Source under Pressure 1044
   8.7.7   Pipes in a Loop 1055
   8.7.8   Pipe Networks 1063
       8.7.8.1   Continuity Principle 1063
       8.7.8.2   Energy Principle 1064
       8.7.8.3   Momentum Principle 1067
       8.7.8.4   Summary of Governing Equations for Pipe
Networks 1067

8.8   Pipe Flow Measurement and Control Devices: Actual Flowrate
in Ideal Flow Meters 1078
   8.8.1   Evaluation of the Actual Flowrate for Ideal Flow Meters
in Pipe Flow 1079

8.8.1.1   Derivation of the Discharge Coefficient, C_d 1080

8.8.1.2   Evaluation of the Discharge Coefficient, C_d 1081

8.8.2   Evaluation of the Actual Flowrate for a Pitot-Static Tube 1081

8.8.3   Evaluation of the Actual Flowrate for Orifice, Nozzle, and Venturi Meters 1086

8.8.3.1   Evaluation of the Actual Flowrate for an Orifice, a Nozzle, or a Venturi Meter 1088

8.8.3.2   Actual Flowrate for an Orifice Meter 1090

8.8.3.3   Actual Flowrate for a Nozzle Meter 1093

8.8.3.4   Actual Flowrate for a Venturi Meter.................... 1097

8.8.3.5   Evaluation of the Minor Head Loss due to an Orifice, a Nozzle, or a Venturi Meter 1100

End-of-Chapter Problems.. 1101

9. Open Channel Flow 1115

9.1   Introduction.. 1115

9.2   Application of the Eulerian (Integral) versus Lagrangian (Differential) Forms of the Governing Equations 1116

9.2.1   Eulerian (Integral) Approach for Open Channel Flow Problems.. 1117

9.2.2   Lagrangian (Differential) Approach for Open Channel Flow Problems.. 1118

9.3   The Occurrence of a Major Head Loss in Open Channel Flow 1118

9.4   Modeling Flow Resistance in Open Channel Flow: A Subset Level Application of the Governing Equations.................................... 1118

9.5   Application of the Governing Equations in Open Channel Flow 1119

9.5.1   Application of the Governing Equations for Real Open Channel Flow... 1120

9.5.1.1   Integral Approach for Real Open Channel Flow Problems.. 1120

9.5.1.2   Evaluation of the Major Head Loss Term 1120

9.5.1.3   Evaluation of the Pump and Turbine Head Terms 1122

9.5.1.4   Differential Approach for Real Open Channel Flow Problems.. 1122

9.5.1.5   Applications of the Governing Equations for Real Open Channel Flow 1123

9.5.2   Application of the Governing Equations for Ideal Open Channel Flow... 1123

9.5.2.1   Evaluation of the Actual Discharge..................... 1124

9.5.2.2   Applications of the Governing Equations for Ideal Open Channel Flow 1126

9.5.3   Application of the Governing Equations for a Hydraulic Jump 1126

9.6   Major Head Loss due to Flow Resistance in Real Open Channel Flow 1126

9.6.1   Uniform versus Nonuniform Open Channel Flow 1127

9.6.2   Evaluation of the Major Head Loss Term in the Energy Equation 1128

9.6.3   The Chezy Equation and Evaluation of the Chezy Coefficient, C 1128

9.6.4   Turbulent Channel Flow Resistance Equations and Their
         Roughness Coefficients 1129
9.6.5   Manning's Equation and Evaluation of Manning's Roughness
         Coefficient, n 1129
9.7   Flow Type (State) and Flow Regime 1131
9.7.1   The Role and Significance of Uniform Flow 1131
9.7.2   The Definition of Flow Regimes for Uniform Flow 1132
9.7.3   The Occurrence of Nonuniform Flow and Changes in the
         Flow Regime 1132
9.8   Transitions and Controls in Open Channel Flow 1132
9.8.1   The Definition of a Control in Open Channel Flow 1133
         9.8.1.1   The Occurrence of Critical Flow at Controls 1134
9.8.2   The Definition of a Flow-Measuring Device in Open
         Channel Flow 1135
9.9   Energy Concepts in Open Channel Flow 1135
9.9.1   Hydrostatic Pressure Distribution 1136
9.9.2   Deviation from a Hydrostatic Pressure Distribution 1137
         9.9.2.1   Sharp-Crested Weir and Free Overfall 1138
9.9.3   Specific Energy in Open Channel Flow 1139
9.9.4   Specific Energy for Rectangular Channel Cross Sections 1140
         9.9.4.1   The Specific Energy Curve for Rectangular Channel
                    Sections 1140
         9.9.4.2   Derivation of Critical Flow for Rectangular Channel
                    Sections 1141
         9.9.4.3   The Depth–Discharge Curve for Rectangular Cross
                    Sections 1143
9.9.5   Specific Energy for Nonrectangular Channel Cross Sections 1145
         9.9.5.1   Derivation of Critical Flow for Nonrectangular
                    Channel Sections 1146
9.9.6   Subcritical and Supercritical Flow 1147
9.9.7   Analysis of the Occurrence of Critical Flow at Controls 1148
         9.9.7.1   Critical Flow at Controls due to an
                    Abrupt/Maximum Vertical Constriction 1149
         9.9.7.2   Critical Flow at Controls due to an Abrupt/Maximum
                    Horizontal Constriction 1153
         9.9.7.3   Critical Flow at Controls due to a Change in
                    Channel Bottom Slope 1156
9.9.8   Analysis of the Occurrence of Uniform Flow as a
         Control 1159
9.10   Momentum Concepts in Open Channel Flow 1162
9.10.1   The Momentum Function 1162
         9.10.1.1   The Use of Controls in the Formation of a
                     Hydraulic Jump 1163
         9.10.1.2   Definition of the Momentum Function for
                     Rectangular Channel Sections 1164
         9.10.1.3   The Momentum Function Curve for Rectangular
                     Channel Sections 1166
         9.10.1.4   Definition of the Momentum Function for
                     Nonrectangular Channel Sections 1170

9.11 Geometric Properties of Some Common Channel Sections 1171
    9.11.1 Geometric Properties of Rectangular Channel Sections............. 1172
    9.11.2 Geometric Properties of Trapezoidal Channel Sections............. 1172
    9.11.3 Geometric Properties of Triangular Channel Sections 1174
    9.11.4 Geometric Properties of Partially Filled Circular
            Channel Sections... 1175
9.12 Flow Depth and Reaction Force for Short Channel Transitions in Open
     Channel Flow: Ideal Flow 1176
    9.12.1 Flow Depth for Gradual Channel Transitions: Not Controls 1176
            9.12.1.1 Flow Depth for a Gradual Upward Step 1177
            9.12.1.2 Flow Depth for a Gradual Decrease in Channel
                    Width.. 1180
    9.12.2 Flow Depth for Abrupt Channel Transitions: Controls............. 1187
            9.12.2.1 Flow Depth for an Abrupt Upward Step 1187
            9.12.2.2 Flow Depth for an Abrupt Decrease in
                    Channel Width 1192
            9.12.2.3 Flow Depth for Typical Flow-Measuring Devices 1196
    9.12.3 Reaction Force on Open Channel Flow Structures/Controls
            (Abrupt Flow Transitions and Flow-Measuring Devices) 1204
9.13 Flow Depth and Major Head Loss for a Hydraulic Jump in Open
     Channel Flow 1209
    9.13.1 Critical Flow at the Hydraulic Jump............................. 1210
    9.13.2 Derivation of the Hydraulic Jump Equations: Rectangular
            Channel Sections... 1211
    9.13.3 Numerical Solution for a Hydraulic Jump: Nonrectangular
            Channel Sections... 1214
    9.13.4 Computation of the Major Head Loss due to a Hydraulic Jump ... 1215
9.14 Flow Depth and Major Head Loss in Uniform Open Channel Flow:
     Real Flow 1221
9.15 Flow Depth and Major Head Loss in Nonuniform Open Channel
     Flow: Real Flow.. 1228
9.16 Actual Flowrate in Flow Measrurement and Control Devices in
     Open Channel Flow: Ideal Flow Meters 1253
    9.16.1 Flow-Measuring Devices in Open Channel Flow.................. 1255
            9.16.1.1 Ideal Flow Meters 1255
            9.16.1.2 Critical Depth Meters................................... 1255
    9.16.2 Controls Serving as Flow Measurement Devices in Open
            Channel Flow.. 1256
            9.16.2.1 Determination of the Depth–Discharge Relationship
                    for Controls Serving as Flow-Measuring Devices 1256
            9.16.2.2 Deviations from the Assumptions for Critical Flow
                    in Controls Serving as Flow-Measuring Devices 1257
            9.16.2.3 Applied Depth–Discharge Relationship for Ideal
                    versus Critical Flow Meters............................. 1258
    9.16.3 Actual Flowrate for Ideal Flow Meters 1258
    9.16.4 Actual Flowrate for Critical Flow Meters......................... 1259
    9.16.5 A Comparison between Ideal Flow Meters and Critical
            Depth Meters 1260
            9.16.5.1 Advantages of Critical Depth Meters................... 1261

                9.16.5.2   Disadvantages of Critical Depth Meters 1261
                9.16.5.3   Ideal Flow Meters: Typical Flow-Measuring
                           Devices 1261
    9.17   Evaluation of the Actual Flowrate for Ideal Flow Meters in Open
           Channel Flow 1262
           9.17.1   Evaluation of the Actual Flowrate for Ideal Flow Meters 1263
                9.17.1.1   Application of the Bernoulli Equation 1263
                9.17.1.2   Application of the Continuity Equation, the
                           Momentum Equation, and Dimensional Analysis 1264
                9.17.1.3   Derivation of the Discharge Coefficient, C_d 1265
                9.17.1.4   Evaluation of the Discharge Coefficient, C_d 1266
           9.17.2   Sluice Gates, Weirs, Spillways, Venturi Flumes, and Contracted
                    Openings in Open Channel Flow 1266
           9.17.3   Evaluation of the Actual Flowrate for a Sluice Gate 1267
           9.17.4   Evaluation of the Actual Flowrate for a Sharp-Crested Weir 1272
                9.17.4.1   Rectangular Sharp-Crested Weir 1275
                9.17.4.2   Triangular Sharp-Crested Weir 1278
                9.17.4.3   A Comparison between a Rectangular and a Triangular
                           Sharp-Crested Weir 1282
           9.17.5   Evaluation of the Actual Flowrate for a Spillway 1283
           9.17.6   Evaluation of the Actual Flowrate for a Broad-Crested Weir 1285
           9.17.7   Evaluation of the Actual Flowrate for the Parshall Flume
                    (Venturi Flume) 1291
           9.17.8   Evaluation of the Actual Flowrate for a Contracted
                    Opening 1294
    End-of-Chapter Problems 1294

10. **External Flow** 1305
    10.1   Introduction 1305
           10.1.1   Occurrence and Illustration of the Drag Force and
                    the Lift Force 1306
    10.2   Application of the Eulerian (Integral) versus Lagrangian
           (Differential) Forms of the Governing Equations 1308
    10.3   Modeling Flow Resistance in External Flow: A Subset Level
           Application of the Governing Equations 1308
    10.4   Application of the Governing Equations in External Flow 1309
           10.4.1   Application of the Governing Equations for Ideal
                    External Flow 1309
                10.4.1.1   Evaluation of the Drag Force 1310
    10.5   The Drag Force and the Lift Force in External Flow 1311
           10.5.1   Evaluation of the Drag Force in External Flow 1312
                10.5.1.1   Modeling the Drag Force in the Momentum
                           Equation 1313
                10.5.1.2   Derivation of the Drag Coefficient, C_D 1314
           10.5.2   Determination of the Drag Coefficient, C_D 1315
           10.5.3   Evaluation of the Drag Coefficient, C_D 1315
                10.5.3.1   The Role of the Velocity of Flow 1316
                10.5.3.2   The Role of the Shape of the Body 1316

10.5.3.3 Reducing the Total Drag Force by Optimally
Streamlining the Body 1319
10.5.3.4 The Occurrence of Flow Separation 1331
10.5.3.5 Reducing the Flow Separation (Pressure Drag).......... 1333
10.5.3.6 The Importance of the Reynolds Number, R 1333
10.5.3.7 Creeping Flow ($R \leq 1$) for Any Shape Body, and
Stokes Law for a Spherical Shaped Body 1338
10.5.3.8 Laminar and Turbulent Flow for Any Shape Body
except Round-Shaped Bodies........................... 1344
10.5.3.9 Laminar and Turbulent Flow for Round-Shaped
Bodies (Circular Cylinder or Sphere) 1357
10.5.3.10 Laminar and Turbulent Flow with Wave Action at
the Free Surface for Any Shape........................ 1362
10.5.3.11 The Importance of the Relative Surface Roughness 1362
10.5.3.12 The Importance of the Mach Number, M............... 1365
10.5.3.13 The Importance of the Froude Number, F.............. 1369
10.5.4 Evaluation of the Lift Force in External Flow....................... 1372
10.5.5 Evaluation of the Lift Coefficient 1375
10.5.5.1 The Role of the Shape of the Body and the
Angle of Attack.. 1375
10.5.5.2 Optimizing the Shape of an Airfoil and the
Angle of Attack.. 1376
10.5.5.3 Optimizing the Performance of an Airfoil 1382
10.5.5.4 Optimizing the Shape of an Airfoil by the Use
of Flaps 1384
10.5.5.5 Optimizing the Shape of an Airfoil by the
Aspect Ratio 1389
10.5.6 Estimating the Lift Force and Lift Coefficient for a
Hot-Air Balloon... 1392
End-of-Chapter Problems.. 1395

11. **Dynamic Similitude and Modeling** 1403
11.1 Introduction.. 1403
11.1.1 The Role of Dynamic Similitude in the Modeling of
Fluid Flow 1404
11.1.2 The Role of Dynamic Similitude in the Empirical Modeling
of Flow Resistance in Real Fluid Flow........................... 1406
11.1.2.1 Using Dynamic Similitude in the Empirical Modeling
of Flow Resistance in Real Fluid Flow.................. 1407
11.1.3 Developing and Applying the Laws of Dynamic Similitude
to Design Geometrically Scaled Physical Models of Real
Fluid Flow 1408
11.2 Primary Scale Ratios 1409
11.2.1 Geometric Similarity.. 1410
11.2.2 Kinematic Similarity 1410
11.2.3 Dynamic Similarity 1411
11.3 Interpretation of the Main π Terms 1412
11.3.1 The Euler Number 1413

11.3.2 The Froude Number 1414
11.3.3 The Reynolds Number.. 1414
11.3.4 The Cauchy Number 1415
11.3.5 The Weber Number... 1415
11.3.6 Implications of the Definitions of the Main π Terms................. 1416
11.4 Laws Governing Dynamic Similirity: Secondary/Similitude
Scale Ratios 1418
11.4.1 Similitude Scale Ratios for Physical Quantities 1419
11.4.2 Similitude Scale Ratios for Physical Properties of Fluids 1420
11.5 The Role and the Relative Importance of the Dynamic Forces in
Dynamic Similitude... 1421
11.6 Guidelines in the Application of the Laws Governing Dynamic
Similarity.. 1425
11.6.1 Definition of the Flow Resistance Prediction Equations
(or Equations for Any Other Physical Quantity)..................... 1426
11.6.2 Definition of the Dynamic Similarity Requirements................... 1426
11.6.3 "True Models" versus "Distorted Models" 1427
11.6.4 General Guidelines in the Application of "Distorted Models" 1428
11.7 Application of the Laws Governing Dynamic Similarity for Flow
Resistance Equations and Efficiency of Pumps and Turbines................. 1429
11.7.1 Specific Guidelines in the Application of
"Distorted Models"... 1430
11.7.1.1 Geometry Similarity Requirements....................... 1436
11.7.1.2 Relative Roughness Similarity Requirements............ 1436
11.7.1.3 "Pressure Model" Similarity Requirements.............. 1437
11.7.1.4 "Viscosity Model" Similarity Requirements 1438
11.7.1.5 "Elastic Model" Similarity Requirements............... 1439
11.7.1.6 "Gravity Model" Similarity Requirements 1439
11.7.1.7 "Surface Tension Model" Similarity Requirements...... 1440
11.7.1.8 "Viscosity Model" and "Elastic Model"
Similarity Requirements 1440
11.7.1.9 "Viscosity Model" and "Gravity Model"
Similarity Requirements 1441
11.7.1.10 "Viscosity Model" and "Surface Tension Model"
Similarity Requirements 1444
11.7.1.11 "Gravity Model," "Viscosity Model," and
"Surface Tension Model" Similarity
Requirements 1445
11.7.2 Application of the Similitude Scale Ratios for the Drag
Force in External Flow... 1446
11.7.2.1 Creeping Flow ($R \leq 1$) for Any Shape Body 1447
11.7.2.2 Laminar Flow ($R < 10,000$) for Any Shape Body
except Round-Shaped Bodies............................ 1451
11.7.2.3 Turbulent Flow ($R > 10,000$) for Any Shape Body
except Round-Shaped Bodies............................ 1459
11.7.2.4 Laminar and Turbulent Flow for Round-Shaped
Bodies (Circular Cylinder or Sphere) 1470
11.7.2.5 Laminar and Turbulent Flow with Wave Action
at the Free Surface for Any Shape Body 1479

11.7.3 Application of the Similitude Scale Ratios for the Major
Head Loss in Pipe Flow 1491
    11.7.3.1 Laminar Pipe Flow 1491
    11.7.3.2 Completely Turbulent Pipe Flow (Rough Pipes) 1495
    11.7.3.3 Transitional Pipe Flow................................. 1500
11.7.4 Application of the Similitude Scale Ratios for the Major
Head Loss in Open Channel Flow................................. 1504
    11.7.4.1 Turbulent Open Channel Flow 1505
11.7.5 Application of the Similitude Scale Ratios for the Minor
Head Loss in Pipe Flow 1510
    11.7.5.1 Turbulent Pipe Flow with Pipe Component............. 1511
11.7.6 Application of the Similitude Scale Ratios for the Actual
Discharge in Pipe Flow 1515
    11.7.6.1 Pipe Flow with a Flow-Measuring Device.............. 1516
11.7.7 Application of the Similitude Scale Ratios for the Actual
Discharge in Open Channel Flow 1520
    11.7.7.1 Open Channel Flow with Sluice Gate or Venturi
Meter 1521
    11.7.7.2 Open Channel Flow with Weir or Spillway
with Large Head....................................... 1525
    11.7.7.3 Open Channel Flow with Weir or Spillway with
Small Head 1529
11.7.8 Application of the Similitude Scale Ratios for the Efficiency
of Pumps and Turbines... 1535
    11.7.8.1 Similitude Scale Ratios for the Efficiency
of Pumps 1535
    11.7.8.2 Affinity Laws for the Efficiency of Homologous
Pumps 1537
    11.7.8.3 Modeling of Scaling Effects for the Efficiency of
Nonhomologous ("Distorted Models") Pumps
Using the Moody Equation 1539
    11.7.8.4 Similitude Scale Ratios for the Efficiency of
Turbines 1544
    11.7.8.5 Affinity Laws for the Efficiency of Homologous
Turbines 1545
    11.7.8.6 Modeling of Scaling Effects for the Efficiency of
Nonhomologous ("Distorted Models") Turbines
Using the Moody Equation 1546
End-of-Chapter Problems... 1550

Appendix A: Physical Properties of Common Fluids 1563

Appendix B: Geometric Properties of Common Shapes 1571

References.. 1575

Bibliography 1577

Index 1579

Preface

During my career in civil and environmental engineering for the past 31 years, I have had the fortunate opportunity to serve in the various capacities of water resources engineering, including teaching, research, and consulting. As such, throughout my professional career, I have been involved in the diverse aspects of water resources, including remote sensing, hydrologic modeling, hydraulic-open channel flow modeling, statistical modeling, computer modeling, and writing and publishing. My passion for each of these facets of water resources engineering has culminated in the writing of this fluid mechanics textbook. I will now share with you the reasons why I decided to take this long and arduous journey of writing this textbook. It was truly a labor of love.

My fascination with fluid flow phenomenon initially began while I was an undergraduate student taking my first course in fluid mechanics. As I started to teach the subject to my own undergraduate students, the analytical and empirical mathematical modeling of fluid flow began to pique my interest. In a dual effort to satisfy my ever-growing curiosity, and to continuously improve my fluid mechanics notes, I began to delve deeper into the subject. After researching, teaching, and applying fluid mechanics in both academia and in the industry, I recognized three basic problems that I sought to address in writing this textbook. First, it become evident that the subject is overflowing with theoretical and empirical concepts, which may initially seem difficult for the student to grasp and subsequently to apply to real-world practical fluid flow problems. Second, it became clear that the conventional solution approaches for many practical fluid flow problems are filled with tedious and lengthy trial-and-error procedures. In spite of the numerous suggestions to use programming and computer application tools in order to alleviate this problem, only the end-solution result is typically provided, without providing a step-by-step detailed solution to the problem. And, third, it became noticeable that while the currently available textbooks in fluid mechanics may be successfully adopted in the various disciplines in engineering in general, very few address the topic of open channel flow in the great detail that is of critical importance to the civil and environmental engineer. In order to address these three problems facing the study and application of fluid mechanics, I wrote this textbook with the hope of making a contribution to the learning of fluid mechanics by achieving the three objectives outlined below.

Upon learning a new concept in fluid mechanics, it is not always easy to for the student to grasp where it fits into the bigger scheme of the subject matter and how it is applied in the real world. Therefore, my first objective in writing this textbook was to provide an intuitive, detailed, and easy-to-understand approach in the presentation of the material, providing a clear distinction between the theoretical concepts versus the empirical concepts, and where and why they are applicable. Some of the important features of the textbook are mentioned herein. A clear distinction is made as to whether the Lagrangian (differential) versus the Eulerian (integral) approach is applicable, and why. The modeling of flow resistance, which is at the heart of all practical fluid mechanics problem solving, is highlighted as a major theme throughout this textbook and has an entire chapter dedicated to the topic, namely Chapter 6. The important role dimensional analysis plays in supplementing theory when modeling turbulent flow (as opposed to laminar flow) is highlighted as another major theme throughout this textbook, and also has an entire chapter dedicated to this topic, namely Chapter 7. The application of dimensional analysis to derive the flow resistance equations as well as theoretical equations is presented in a thorough and systematic approach, with a clear explanation as to how the variables are chosen. These are just a few key highlights of this textbook, which I have written with the student in mind and with the overall goal to present the study of fluid mechanics in a step-by-step, nonintimidating, yet precise manner.

In the practical application of the concepts to real-world fluid flow problems, the typical solution approaches are burdened with awkward and sometimes tiresome and lengthy trial-and-error procedures. Although many textbooks suggest the use of programming and computer application tools to alleviate these types of problems, only the end solution result is typically provided, without providing a step-by-step detailed solution to the problem. Thus, my second objective in writing this textbook was to facilitate and enhance the solution approach in the application of the fluid mechanics concepts. The conventional procedure is replaced with a detailed solution procedure using state-of -the art mathematical software. In particular, Mathcad was intentionally selected for application in this textbook, because it provides a very simple, elegant, yet very powerful mathematical software tool. Mathcad is used extensively throughout this textbook as a computational tool for both the derivation (integration, differentiation, evaluation, simplification, etc.) of the principles of fluid mechanics and the numerical and analytical solutions of empirical and analytical equations, including the solution of ordinary differential equations. Other types of programming and computer application tools may be used as an alternative to Mathcad. Alternatives to Mathcad are listed in the Introduction to the textbook.

While the currently available textbooks in fluid mechanics may be successfully adopted in the various disciplines in engineering in general, very few address the topic of open channel flow in great detail, as sought by the civil and environmental engineer. Therefore, my third objective in writing this textbook was to address the topic of open channel flow in great depth and detail for the civil and environmental engineer. A few of the important features of this textbook are mentioned herein. A clear distinction is made as to whether the Lagrangian (differential) versus the Eulerian (integral) approach is assumed in the application of the momentum equation; that is, the difference between using the Chezy equation, $S_f = (v^2/C^2 R_h)$ or Manning's equation, $v = (1/n)R_h^{2/3}S_f^{1/2}$ versus $\sum F_x = \rho Q(v_{2x} - v_{1x})$, respectively. Complete and detailed numerical solutions of the equations for uniform flow (Chezy and Manning's equations), and the numerical solutions of the ordinary differential equation for nonuniform flow $((dy/dx) = (S_o - S_f)/(1 - F^2))$ are presented and graphically illustrated using Mathcad. Furthermore, detailed and in-depth analytical and numerical solutions of the energy and momentum concepts in open channel flow for both rectangular and nonrectangular channels are presented and graphically illustrated using Mathcad.

In writing this textbook, accomplishing my first objective was like searching for clues to solve a mystery, or like discovering the missing pieces of a jigsaw puzzle. Each time I found a clue or a missing piece of the puzzle, it was truly a remarkable Eureka moment for me. It is my humble hope that this textbook has provided valuable learning and teaching material for all disciplines of engineering. Furthermore, I hope that this textbook has contributed to bringing the study of fluid mechanics into the twenty-first century, by applying state-of-the-art mathematical software to achieve efficient and detailed solution procedures. Moreover, I hope that this textbook has provided the necessary in-depth study of open channel flow that is essential for civil and environmental engineers. And, although the title of the textbook suggests that the intended audience is civil and environmental engineering students, this textbook may also be successfully adopted by other disciplines in engineering that teach fluid mechanics, such as mechanical, aerospace, nuclear, chemical, and agricultural engineering. It is my sincere hope that providing an insightful and easy-to-understand approach in the presentation and application of the concepts in fluid mechanics, coupled with the use of state-of-the-art mathematical software in order to elevate the learning and teaching experience will intrigue the curiosity of the student and professor in all engineering disciplines and incite you to share my passion for the subject of fluid mechanics.

Ahlam I. Shalaby
Springfield, VA

MATLAB® is a registered trademark of The MathWorks, Inc. For product information, please contact:

The MathWorks, Inc.
3 Apple Hill Drive
Natick, MA 01760-2098 USA Tel: 508 647 7000
Fax: 508-647-7001
E-mail: info@mathworks.com
Web: www.mathworks.com

Acknowledgments

First, I would like to thank Mr. Jonathan Plant, executive editor, CRC Press/Taylor & Francis Group, for being extremely patient and supportive throughout the many years it has taken me to complete the writing of this textbook. Jonathan: it was truly a pleasure working with you on this project. Thanks to Ms. Claudia Kisielewicz, editorial assistant, CRC Press/Taylor & Francis Group, for facilitating submission of the manuscript. Thanks to Mr. Hector Mojena III, editorial assistant, CRC Press/Taylor & Francis Group, for facilitating transmittal of the manuscript for production. Thanks to Mr. Richard Tressider, project editor, CRC Press/Taylor & Francis Group, for managing the production of the textbook. And thanks to Mr. Paul Beaney of Nova Techset for the editing, design, typesetting, proof trafficking and correction of this book.

Professor Shahram E. Zanganeh, my undergraduate professor, my colleague, and my dear friend: I would like to express to you my deep appreciation for spending that summer teaching me how to use Mathcad, and always being there to support and encourage me along the way. Without you and your patience and willingness to teach me, using Mathcad for this textbook would not have been possible. Thank you for our everlasting collegiality and friendship throughout the years.

Dr. Richard H. McCuen, my graduate professor and mentor: I would like to express to you my sincere thanks for teaching me how to write. Without you and your dedication as my graduate advisor, who was always willing to correct the numerous drafts of my master's thesis and doctoral dissertation, writing this textbook would not have been possible.

Dr. Subhash C. Mehrotra, my undergraduate fluid mechanics professor: I would like to thank you for making the subject of fluid mechanics so enjoyable and so easy to understand. After all these years, I hope that now you know that your teaching has inspired my enthusiasm and passion for the subject, culminating in writing a textbook.

Thanks to my very bright students: I wrote this book for you, and with you in mind. Thank you for your patience and for motivating me to write and rewrite the chapters, meanwhile helping me to debug them by teaching you from the numerous drafts of the textbook.

Joanne: yes, I am done with the book! Thanks for our special camaraderie and friendship, constant encouragement, and believing in me. Sandee: thanks for all our fun, our enduring friendship, and your constant support. Leigh: thanks for our true and everlasting friendship and your unwavering faith in me. Danielle: thanks for our amazing friendship and your support. Faiza and Hoda: thanks for our lifelong friendship and your encouragement. Deb and Mark: Thanks for our friendship and your constant encouragement.

Suma: thanks for being my "twin" and all of our fun. Amany: thanks for encouraging me to complete my "opus." Gila: thanks for always being there for me and encouraging me. Amira: thanks for your encouragement. Ameer: thanks for believing in me.

Omar: thank you for your unconditional love and unwavering support. Basem: thank you for your unconditional love and enduring support. Omar and Basoom: you are my true blessing in life, you fill my heart with love, and you fill my life with true meaning. And, I thank God Almighty, Who made writing this textbook even a possibility, and its completion a reality.

Author

Ahlam I. Shalaby is an associate professor of Civil and Environmental Engineering at Howard University, Washington, DC. She earned her BS in civil engineering from Howard University in 1979, and her MS and PhD in civil engineering from the University of Maryland, College Park, MD, in 1986. She has worked with Bechtel, NASA, and USDA/ARS in several academic and professional capacities. Since joining the faculty at HU in 1986, Dr. Shalaby has taught courses in fluid mechanics, open channel flow, hydraulic project research, water resources engineering, advanced hydrology, and probability and statistics. She has served as a hydrology consultant to the World Bank in Washington, DC, and as a disaster specialist for Dewberry in Fairfax, VA. She has also served as an associate expert on remote sensing and hydrological models for forecasting for the World Meteorological Organization, United Nations in Geneva, Switzerland. Dr. Shalaby has done extensive research with the U.S. Army Corps of Engineers CRREL on the statistical assessment of distributed snow process modeling using remote sensing/satellite data, and analysis of risk information derived from geospatial data for the HEC-FDA model. She has authored a number of technical publications in refereed journals and United Nations publications. She is the author of "Probable Maximum Flood Determination" in the hydrology textbook *Hydrologic Analysis and Design* by Dr. R. H. McCuen.

List of Symbols

Symbols

∇ = free surface of liquid is open to the atmosphere

English Symbols

a	acceleration
	opening or depth of flow at sluice gate
a_m	model scale acceleration
a_p	prototype scale acceleration
a_r	acceleration scale ratio
A	area
	cross-sectional area
	surface area
	planform area
A_a	actual cross-sectional area
A_c	critical cross-sectional channel area
A_i	ideal cross-sectional area
A_o	area of orifice opening
AR	aspect ratio
b	intensive fluid property
	channel width
	span of strut, etc.
b_t	channel width at channel transition
B	center of buoyancy
	extensive fluid property
	surface channel width
B_c	critical surface channel width
BG	British Gravitational (English system of units)
BM	distance between the center of buoyancy, B and the metacenter, M
c	specific heat for liquids
	sonic (acoustic) velocity
	chord length
c_m	model scale sonic velocity
c_p	ideal specific heat at constant pressure
	prototype scale sonic velocity
c_r	sonic velocity scale ratio
c_v	ideal specific heat at constant volume
cv	control volume
C	constant
	Chezy coefficient
	Cauchy number
C	Cauchy number
C_a	cavitation number

C_a cavitation number
C_c contraction coefficient
C_d discharge coefficient
C_D drag coefficient
               total drag coefficient
C_f friction drag coefficient
C_h Hazen–Williams roughness coefficient
C_H head coefficient
C_L lift coefficient
C_n nozzle discharge coefficient
C_o orifice discharge coefficient
C_p pressure (pressure recovery) coefficient
               pressure drag coefficient
C_P power coefficient
C_Q capacity coefficient
C_v velocity coefficient
               venturi discharge coefficient
dep dependent
D diameter
               thickness
D_h equivalent diameter (hydraulic diameter)
D_o orifice diameter
e total energy per unit mass
               total number of equations for a pipe network
E total energy
               specific energy
               Euler number
$\boldsymbol{E}$ Euler number
$\dot{E}$ time rate of change of total energy
E_c critical (minimum) specific energy
E_{min} minimum (critical) specific energy
E_t specific energy at channel transition
E_v bulk modulus of elasticity
E_{v_m} model scale bulk modulus of elasticity
E_{v_p} prototype scale bulk modulus of elasticity
E_{v_r} bulk modulus of elasticity scale ratio
EGL energy grade line
f Darcy–Weisbach friction factor
F force
               hydrostatic force
               resultant hydrostatic force
               Froude number
$\boldsymbol{F}$ Froude number
F_B buoyant force
F_D drag force
F_E elastic force
F_f friction force
               friction drag (surface drag)

F_G	gravitational force
F_H	horizontal component of resultant force
F_I	inertia force
F_L	lift force
F_m	model scale force
F_P	pressure force
	pressure drag (form drag)
F_p	prototype scale force
F_R	resultant force acting on a curved surface
F_r	force scale ratio
F_T	surface tension force
F_V	viscous force
	vertical component of resultant force
FT	momentum
g	acceleration due to gravity
g_r	acceleration due to gravity scale ratio
G	center of gravity
	weight flow rate
GB	distance between the center of gravity, G and the center of buoyancy, B
GM	metacentric height
Δh	drawdown
h	variable distance along the h-axis
	variable column of fluid
	enthalpy
	pressure head
h_a	acceleration head loss
h_{ca}	height to the center of area measured along the h-axis
h_e	total energy head loss
h_f	friction head loss
	head loss due to flow turbulence (as for a hydraulic jump)
	total head loss
$h_{f,maj}$	major friction head loss
	major head loss due to flow turbulence (as for a hydraulic jump)
$h_{f,min}$	minor friction head loss
$h_{f,total}$	total friction head loss
h_F	center of pressure for the resultant force, F measured along the h-axis
h_{F_H}	center of pressure for the resultant horizontal force, F_H measured along the h-axis
h_L	head loss (due to friction or turbulence)
h_m	manometer reading
h_p	energy head added to fluid by a pump
h_{pump}	energy head added to fluid by a pump
h_t	energy head removed from fluid by a turbine
$h_{turbine}$	energy head removed from fluid by a turbine
H	total energy head
	head of the weir or spillway
Hg	mercury
HGL	hydraulic grade line

i	total number of pipes in a pipe network
I_x	second moment of the area (moment of inertia) with respect to x-axis
I_{x-ca}	centroidal moment of inertia
I_y	moment of inertia with respect to y-axis
j	total number of junctions (nodes) in a pipe network
k	specific heat ratio
	adiabatic exponent
	minor head loss coefficient
ke	kinetic energy per unit mass
K	pipe constant
KE	kinetic energy
l	position
	total number of loops in a pipe network
L	distance
	length
	pipe or channel length
	chord length of strut, etc.
L_e	entrance length
L_{eq}	equivalent length of pipe
L_i	geometry of pipe, channel, pump, turbine, body, etc.
L_m	model scale length
L_p	prototype scale length
L_r	length scale ratio
L_w	length of horizontal crest on weir
m	molecular weight (molar mass)
	number of fundamental dimensions (mass or force, length, and time)
	model
	mass
M	mass
	moment
	metacenter
	momentum function
	Mach number
$\boldsymbol{M}$	Mach number
$\dot{M}$	mass flow rate
M_c	critical (minimum) momentum function
M_H	moment
M_m	model scale mass
M_p	prototype scale mass
M_r	mass scale ratio
Mv	impulse
n	Manning's roughness coefficient
	number of important physical quantities (variables)
N	number of moles of gas
p	pressure
	prototype
p_{atm}	atmospheric pressure
p_{ca}	pressure at the center of area

p_e	external pressure
p_g	gage air pressure
$p(gage)$	gage pressure
p_i	internal pressure
p_m	model scale pressure
p_p	prototype scale pressure
p_r	reference pressure
	pressure scale ratio
p_v	vapor pressure
p_{v_m}	model scale vapor pressure
p_{v_p}	prototype scale vapor pressure
p_{v_r}	vapor pressure scale ratio
pe	potential energy per unit mass
P	perimeter
	height of weir or spillway
	power
P_m	model scale power
P_p	power added to fluid by a pump
	prototype scale power
P_r	power scale ratio
P_t	power removed from fluid by a turbine
P_w	wet (wetted) perimeter
PE	potential energy
q	heat transfer per unit mass
	unit discharge
q_{max}	maximum unit discharge
q_t	unit discharge at channel transition
Q	discharge
	volume flow rate
	heat transfer
$\dot{Q}$	time rate of change of heat
Q_a	actual discharge
Q_i	ideal discharge
Q_m	model scale discharge
Q_p	prototype scale discharge
Q_r	discharge scale ratio
r	variable radius
	radius
	ratio
r_o	radius of pipe, tube, etc.
R	a gas constant
	resultant force
	reaction force
	Reynolds number
$\boldsymbol{R}$	Reynolds number
R_h	hydraulic radius
R_s	resultant force acting along a streamline, s
R_u	universal gas constant

s	specific gravity
	variable distance along s-axis
	streamline
s_m	model scale specific gravity
s_p	prototype scale specific gravity
s_r	specific gravity scale ratio
sys	system
S_a	acceleration slope
S_c	critical channel bottom slope
S_e	energy slope (slope of the energy grade line)
S_f	friction slope
S_o	channel bottom slope
SI	International System (metric system of units)
t	time
T	time
	temperature
	torque
T_m	model scale time
T_p	prototype scale time
T_r	time scale ratio
u	internal energy per unit mass
U	internal energy
v	velocity
	specific volume
v_a	actual velocity
v_{ave}	average velocity
v_c	critical velocity
v_{crp}	creeping velocity
v_i	ideal velocity
v_m	model scale velocity
v_{max}	maximum velocity
v_{oa}	actual velocity at orifice
v_{oi}	ideal velocity at orifice
v_p	prototype scale velocity
v_r	velocity scale ratio
V	volume
w	width
	work per unit mass
W	work
	work transfer
	weight (gravitational force)
	Weber number
$\mathbf{W}$	Weber number
$\dot{W}$	time rate of change of work
W_m	model scale work
W_p	prototype scale work
W_r	work scale ratio
x	variable distance along x-axis

x_{F_V}	center of pressure for the resultant vertical force, F_V measured along x-axis
y	variable distance along y-axis
	variable depth of flow in an open channel
y_c	critical depth of flow
y_{ca}	height to the center of area measured along y-axis
y_F	center of pressure for the resultant force, F measured along y-axis
y_{F_H}	center of pressure for the resultant horizontal force, F_H measured along y-axis
y_t	depth of flow at channel transition
z	variable distance along z-axis
	elevation
z_{ca}	height to the center of area measured along z-axis
z_F	center of pressure for the resultant force, F measured along z-axis

Greek Symbols

α (alpha)	angle
	angle of attack
γ (gamma)	specific weight
γ_m	model scale specific weight
γ_p	prototype scale specific weight
γ_r	specific weight scale ratio
γ_w	specific weight of water
Δ (delta)	change in
ε (epsilon)	absolute pipe, channel, or boundary roughness
ε_m	model scale absolute pipe, channel, or boundary roughness
ε_p	prototype scale absolute pipe, channel, or boundary roughness
ε_r	absolute pipe, channel, or boundary roughness scale ratio
η (eta)	efficiency
	thickness of the boundary layer
θ (theta)	temperature
	angle
	conical angle
λ (lambda)	model scale (model scale ratio)
μ (mu)	absolute (dynamic) viscosity
μ_m	model scale absolute (dynamic) viscosity
μ_p	prototype scale absolute (dynamic) viscosity
μ_r	absolute (dynamic) viscosity scale ratio
υ (nu)	kinematic viscosity
υ_m	model scale kinematic viscosity
υ_p	prototype scale kinematic viscosity
υ_r	kinematic viscosity scale ratio
π (pi)	dimensionless coefficients (pi terms)
ρ (rho)	density
ρ_m	model scale density
ρ_p	prototype scale density
ρ_r	density scale ratio

σ (sigma)	surface tension
σ_m	model scale surface tension
σ_p	prototype scale surface tension
σ_r	surface tension scale ratio
Σ (sigma)	summation
τ (tau)	shear stress
τ_w	wall shear stress
ϕ (phi)	angle
	unknown function of, or function of
ω (omega)	frequency
	angular speed
	rotational speed

List of Abbreviations

abs	absolute
ac	acre
atm	atmospheric
ave	average
Btu	British thermal unit
cal	calorie
cfm	cubic feet per minute
cfs	cubic feet per second
cm	centimeter
cP	centipoise
°C	degrees Celsius
d	day
fpm	feet per minute
fps	feet per second
ft	feet
°F	degrees Fahrenheit
g	gram
	gage
gal	gallon
gpd	gallons per day
gpm	gallons per minute
h	hour
ha	hectare
hr	hour
hp	horsepower
	550 ft-lb/sec
Hg	mercury
Hz	Hertz (cycles per second)
H_2O	water
in	inch
J	Joule
	N-m
kg	kilogram
km	kilometer
kW	kilowatt
kWh	kilowatt hr
°K	degrees Kelvin
lb	pound
ln	$\log_e$
log	$\log_{10}$
L	liter
m	meter
maj	major
max	maximum

mb	millibar
mech	mechanical
mgd	million gallons per day
mi	mile
min	minute
	minor
	minimum
mL	milliliter
mm	millimeter
mol	mole
mph	miles per hour
N	Newton
	kg-m/sec^2
oz	ounce
pcf	pounds per cubic foot
psf	pounds per square foot
psfa	pounds per square foot, absolute
psfg	pounds per square foot, gage
psi	pounds per square inch
psia	pounds per square inch, absolute
psig	pounds per square inch, gage
P	poise
Pa	Pascal
	N/m^2
rad	radians
rev	revolutions
rpm	revolutions per minute
rps	revolutions per second
°R	degrees Rankine
s	second
sec	second
slug	slug
	lb-sec^2/ft
sq	square
St	stoke
tsp	teaspoon
Tbsp	tablespoon
W	Watt
	N-m/sec

Standards for Units, Unit Prefixes, and Fluid Properties

BG (English) Units and SI (Metric) Units Conversion Table

Physical Quantity/ Property	Equivalence within BG (English) Units	BG (English) Units	Equal to	SI (Metric) Units	Equivalence within SI (Metric) Units
Geometrics					
Length, L	12 in = 1 ft	1 in	=	2.54 cm	1000 mm = 1 m
	3 ft = 1 yard	1 ft	=	0.3048 m	10 mm = 1 cm
	5280 ft = 1 mile	1 yard	=	0.9144 m	100 cm = 1 m
	6076.12 ft = 1 nautical mile	1 mile	=	1.60934 km	1000 m = 1 km
		1 nautical mile	=	1.852 km	
Area, A	43,560 ft^2 = 1 acre	1 in^2	=	6.4516 cm^2	hectare = ha
		1 ft^2	=	0.092903 m^2	10,000 m^2 = 1 ha
		1 acre	=	4046.86 m^2	
		1 acre	=	0.404686 ha	
Volume, V	1728 in^3 = 1 ft^3	1 in^3	=	16.3870 cm^3	milliliter = mL
	0.133681 ft^3 = 1USgal	1 ft^3	=	0.0283168 m^3	Liter = L
	3 tsp = 1 Tbsp	1 acre-ft	=	1233.48183755 m^3	1 cm^3 = 1 mL
	2 Tbsp = 1 oz	1 oz	=	29.5735 mL	1000 mL = 1 L
	8 oz = 1 cup	1 quart	=	0.946353 L	0.001 m^3 = 1 L
	2 cups = 1 pint	1 USgal	=	3.78541 L	
	2 pints = 1 quart	1 imperial gal	=	4.54609 L	
	4 quarts = 1 USgal				
	1.20095 USgal = 1 imperial gal				
Kinematics					
Time, T		1 sec	=	1 sec	
Frequency, ω		1 cycle/sec = sec^{-1}	=	1 cycle/sec = sec^{-1}	Hertz = Hz
					1 cycle/sec = 1 Hz = sec^{-1}
Velocity, v	ft/sec = fps	1 ft/sec	=	0.3048 m/sec	0.277778 m/sec = 1 km/hr
	mile/hr = mph	1 mph	=	1.60934 km/hr	
	1.46667 fps = 1mph	1 mph	=	0.44704 m/sec	
	1 nautical mph = 1 knot	1 knot	=	1.852 km/hr	
	1.15078 mph = 1 knot	1 knot	=	0.514444 m/sec	
	1.68781 fps = 1 knot				
Acceleration, a		1 ft/sec^2	=	0.3048 m/sec^2	

Continued

BG (English) Units and SI (Metric) Units Conversion Table

Physical Quantity/ Property	Equivalence within BG (English) Units	BG (English) Units	Equal to	SI (Metric) Units	Equivalence within SI (Metric) Units
Discharge, Q	$\text{ft}/^3\text{sec} = \text{cfs}$	1 cfs	$=$	$0.0283168\,\text{m}^3/\text{sec}$	$0.001\,\text{m}^3/\text{sec} = 1\,\text{L/sec}$
	gal/min = gpm	1 mgd	$=$	$0.0438126\,\text{m}^3/\text{sec}$	
	million gal/dy = mgd	1 USgpm	$=$	0.0630902 L/sec	
	448.831 USgpm = 1 cfs				
	1.547229 cfs = 1 mgd				
Dynamics					
Mass, M	$1\,\text{lb-sec}^2/\text{ft} = 1\,\text{slug}$	1 slug	$=$	14.59390 kg	
	1 slug of mass weighs 32.174 lb on earth				1 kg of mass weighs 9.80665 N on earth
Force, F	$\text{slug-ft/sec}^2 = 1\,\text{lb}$	1 lb	$=$	4.44822 N	Newton = N
					1000 N = 1 kN
	16 oz = 1 lb				$1\,\text{kg-m/sec}^2 = 1\,\text{N}$
					0.00001 N = 1 dyne
Weight, F_G	2000 lb = 1 US (short) ton	1 US ton	$=$	0.90718474 tonne	metric ton = tonne
	2240 lb = 1 British (long) ton	1 British ton	$=$	1.0160469 tonne	9800 N = 1 tonne
Pressure, p	$\text{lb/in}^2 = \text{psi}$	1 psi	$=$	$6894.76\,\text{N/m}^2$	pascal = Pa
	$\text{lb/ft}^2 = \text{psf}$	1 psf	$=$	$47.8803\,\text{N/m}^2$	$1\,\text{N/m}^2 = 1\,\text{Pa}$
	144psf = 1 psi	1 psi	$=$	$6.89476\,\text{kN/m}^2$	100,000 Pa = 1 bar
					millibar = mb
					0.001 bar = 1 mb
					1 kN = 1 kPa = 10 mb
Momentum, Mv or Impulse, FT		1 lb-sec	$=$	4.44822 N-sec	
Energy, E, Work, W, or Heat	777.649 ft-lb = 1 Btu	1 ft-lb	$=$	1.355818 N-m	joule = J
	1 Btu = heat required to raise 1 lb of water at 68°F by 1°F	1 ft-lb	$=$	3.76616×10^{-7} kWh	1 N-m = 1 J
		1 Btu	$=$	1054.350 J	$1\,\text{J} = 10^7\,\text{erg}$
		1 Btu	$=$	251.996 cal	watt-hr = Wh
					1000 watt-hr = 1 kWh
					3,600,000 J = 1 kWh
					4.1868 J = 1 cal
					1 cal = heat required to raise 1 g of water at 14.5°C by 1°C

Continued

BG (English) Units and SI (Metric) Units Conversion Table

Physical Quantity/ Property	Equivalence within BG (English) Units	BG (English) Units	Equal to	SI (Metric) Units	Equivalence within SI (Metric) Units
Power, P	550 ft-lb/sec = 1 hp	1 ft-lb/sec	=	1.355818 N-m/sec	watt = W
	2546.14 Btu/hr = 1 hp	1 hp	=	0.745700 kW	1 N-m/sec = 1 W
					1 J/sec = 1 W
					1000 W = 1 kW
Fluid Properties					
Density, ρ		1 slug/ft^3	=	515.379 kg/m^3	
Specific gravity, s		no units		no units	
Specific weight, γ	lb/ft^3 = pcf	1 lb/ft^3	=	157.0875 N/m^3	
Absolute (dynamic) viscosity, μ	1 lb-sec/ft^2 = 1 slug/(ft-sec)	1 lb-sec/ft^2	=	47.8803 N-sec/m^2	1 N-sec/m^2 = 1 kg/(m-sec)
					poise = P
					0.1 N-sec/m^2 = 1 P
					100 cP = 1 P
Kinematic viscosity, $v = \mu/\rho$		1 ft^2/sec	=	0.092903 m^2/sec	stoke = St
					10^{-4} m^2/sec = 1 St
Surface tension, σ		1 lb/ft	=	14.5939 N/m	
Vapor pressure, p_v		1 psia	=	6894.76 N/m^2 abs	
		1 psia	=	6.89476 kN/m^2 abs	
Bulk modulus of elasticity, E_v		1 lb/ft^2	=	47.8803 N/m^2	
Specific heat, c	ft-lb/(slug-°R) = ft^2/(sec^2-°R)	1 ft-lb/(slug-°R)	=	0.1672255 N-m/(kg-°K)	N-m/(kg-°K) = m^2/(sec^2-°K
Gas constant, R	ft-lb/(slug-°R) = ft^2/(sec^2-°R)	1 ft-lb/(slug-°R)	=	0.1672255 N-m/(kg-°K)	N-m/(kg-°K) = m^2/(sec^2- °K
Temperature, θ	Fahrenheit = F	°F	=	32 + (9/5)(°C)	Celsius = C
	Rankine = R	°R	=	(9/5)(°K)	Kelvin = K
	°F + 459.67° = °R				°C + 273.15° = °K

SI (Metric) Unit Standard Prefixes

Factor by Which Unit Is Multiplied	Prefix	Symbol
10^{18}	exa	E
10^{15}	peta	P
10^{12}	tera	T

Continued

SI (Metric) Unit Standard Prefixes

Factor by Which Unit Is Multiplied	Prefix	Symbol
10^9	giga	G
10^6	mega	M
10^3	kilo	k
10^2	hecto	h
10	deka	da
10^{-1}	deci	d
10^{-2}	centi	c
10^{-3}	milli	m
10^{-6}	micro	μ
10^{-9}	nano	n
10^{-12}	pico	p
10^{-15}	femto	f
10^{-18}	atto	a

Standard Physical Properties for Fluids

Fluid	Physical Property	Standard	BG (English) Units	SI (Metric) Units
Standard atmosphere	Pressure	Standard sea-level atmospheric pressure, p_{atm}	14.696 psia = 2116.2 psfa	101.325 kPa abs = 1013.25 millibars abs
	Pressure head	Mercury, Hg	29.92 in Hg	760 mm Hg = 760 torr
	Pressure head	Water, H_2O	33.91 ft H_2O	10.33 m H_2O
	Temperature	Standard sea-level atmospheric temperature, T	59°F	15°C
	Acceleration due to gravity	Standard acceleration due to gravity at sea level and 45° latitude, g	32.174 ft/sec^2 ~ 32.2 ft/sec^2	9.807 m/sec^2 ~ 9.81 m/sec^2
Reference fluid for liquids (water)	Pressure	Standard sea-level atmospheric pressure, p_{atm}	14.696 psia	101.325 kPa abs
	Temperature	Standard temperature, θ	39.2°F	4°C
	Density	Standard density, ρ	1.940 slug/ft^3	1000 kg/m^3
	Specific weight	Standard specific weight, $\gamma = \rho g$	62.417 lb/ft^3 ~ 62.4 lb/ft^3	9810 N/m^3
Fluids in general	Pressure	Standard sea-level atmospheric pressure, p_{atm}	14.696 psia	101.325 kPa abs
	Temperature	Room temperature, θ	68°F	20°C

Introduction to the Textbook

This introduction will give you a brief synopsis and a quick tour of the textbook, while revealing a bit of its contents, in order to demonstrate the benefits of adopting this textbook in your fluid mechanics course. This introduction will give you a quick impression of how the intuitive, detailed, and easy-to-understand presentation of the material will make adopting this textbook an excellent choice for all disciplines of engineering. This introduction will also give you a quick preview of how the mathematical software Mathcad has been applied throughout this textbook in order to provide efficient and detailed solutions to real-world problems in fluid mechanics. The use of state-of-the-art mathematical software for teaching and detailed problem solving makes this textbook an exceptional choice for learning and teaching fluid mechanics in the twenty-first century. Furthermore, this introduction will give you a glimpse at the in-depth and detailed coverage of the topic of open channel flow, which makes this textbook especially valuable and essential for civil and environmental engineers. This introduction will give you a quick glance at the chapters in this textbook, while revealing the insightful order and the meticulous presentation of the material. Each chapter presents the relevant concepts, which are applied in an abundance of real-world application example problems solved in complete detail using Mathcad throughout each chapter. Furthermore, each chapter ends with a wealth of end-of-chapter problems for the students to solve; their complete and detailed solutions using Mathcad are provided for the professor in a solutions manual. And, finally, this introduction ends with a short presentation of the history of fluid mechanics and a brief overview of practical applications of fluid mechanics. A glimpse into the historical developments made in field of fluid mechanics will quickly show how the principles, laws, and equations in the subject have been personified by naming them after the discovering scientist. It is hoped that making this connection between the concepts and the famous mathematicians, physicists, and engineers who discovered them will incite an interest in the student and teacher alike and bring the learning process of the subject full circle. Moreover, it is hoped that making the association between the study of fluid mechanics and practical problems in our everyday lives will motivate the student to learn and apply the concepts in order to attain practical solutions of real-life problems.

About the Textbook

Fluid Mechanics for Civil and Environmental Engineers is written for a first course in fluid mechanics. This textbook introduces the undergraduate student to the principles of fluid mechanics and their application to internal (pipe and open channel flow) and external (flow over body immersed in flow) fluid flow problems. While the title of the textbook suggests that the intended audience is civil and environmental engineering students, this textbook may also be successfully adopted by other disciplines in engineering that teach fluid mechanics, such as mechanical, aerospace, nuclear, chemical, and agricultural engineering. The topic of open channel flow is addressed in great depth and detail in the textbook; while open channel flow is of considerable importance in civil, environmental, and agricultural engineering, it may be of less importance in the other disciplines in engineering. Although this textbook is intended for use in an introductory course about fluid mechanics, the material provided in this textbook is more than ample enough to also be used for an intermediate course in fluid mechanics. Students who have completed courses in physics and dynamics and three semesters of calculus will be well prepared to easily understand and apply the material in this textbook.

Why Choose This Textbook?

After researching, teaching, and applying fluid mechanics for over 31 years, I recognized three basic problems that I sought to address in writing this textbook. First, it become apparent that the topic contains a wealth of theoretical and empirical concepts, which may initially seem difficult for the student to grasp, and subsequently, to apply to real-world practical fluid flow problems. Second, it became apparent that the conventional solution approaches for many practical fluid flow problems are filled with wearisome and lengthy trial-and-error procedures. In spite of the numerous suggestions to use programming and computer application tools in order to circumvent this problem, only the end solution result is typically provided, without providing a step-by-step detailed solution to the problem. And, third, it became noticeable that while the currently available textbooks in fluid mechanics may be successfully adopted in the various disciplines in engineering in general, very few address the topic of open channel flow in great detail, which is crucial to the civil and environmental engineer. As a result, this textbook has been written with the goal to address these three problems facing the study and the application of fluid mechanics. In order to achieve my goal, I identified a number of specific questions and concerns, which I sought to address by achieving three objectives. My first objective was to provide an intuitive, detailed, and easy-to-understand approach in the presentation of the material, providing a clear distinction between the theoretical concepts versus the empirical concepts, and where and why they are applicable. My second objective was to facilitate and enhance the solution approach in the application of the fluid mechanics concepts; the conventional procedure is replaced with a detailed solution procedure using state-of-the art mathematical software. And, my third objective was to address the topic of open channel flow in great depth and detail for the civil and environmental engineer. I have written this textbook with the student in mind, and with the overall goal to present the study of fluid mechanics in a step-by-step, nonintimidating, yet precise manner.

An Excellent Fluid Mechanics Textbook for All Engineers

In order to address the first problem that deals with the challenges in learning and applying the concepts in fluid mechanics, an intuitive, detailed, and easy-to-understand presentation of the material is provided, with a clear distinction between the theoretical concepts versus the empirical concepts, and where and why they are applicable in real-world problems. When learning the concepts for the first time, the student is faced with a number of questions and concerns regarding the appropriate application of the principles to real-world problems, which are carefully addressed in this textbook and highlighted as follows:

- What the applicable governing equations (continuity, energy, and momentum) are for a given fluid flow problem, and why the equations of motion (momentum and energy equations) are complementary equations.

- When to apply the Lagrangian (differential) versus the Eulerian (integral) approach in a given fluid flow problem, and why.
- When it is appropriate to assume real flow versus ideal flow in the application of the governing equations, and why (see Table 4.1).
- The difference between flow resistance (as for pipe or open channel flow) versus flow turbulence (as for a hydraulic jump) when modeling major head loss.
- Why the flow resistance is modeled theoretically for laminar flow, empirically for turbulent flow, and ignored for flow from a tank.
- How to model the flow resistance (major head loss, minor head loss, actual discharge, and drag force) in a given fluid flow problem, theoretically versus empirically using dimensional analysis, and why.
- The significance of the role of flow resistance (and the empirical modeling of flow resistance for turbulent flow using dimensional analysis) in fluid flow.
- The significance of the role of dimensional analysis in supplementing theory when modeling turbulent flow (as opposed to laminar flow).
- How to distinguish the differences and the relationship between the theoretical versus the empirical concepts.
- How to choose the appropriate variables when using dimensional analysis to derive empirical flow resistance equations, as well as theoretical equations.
- The significance of the relationship between the roles of dimensional analysis and dynamic similitude and modeling in fluid flow.
- Why it is important for the independent variables to remain a constant between the model and its prototype in the dynamic similitude and modeling of fluid flow.

In summary, the intuitive, detailed, and easy-to-understand presentation of the material makes this fluid mechanics textbook an excellent option for learning and teaching fluid mechanics in all engineering disciplines.

An Exceptional Fluid Mechanics Textbook for the Twenty-First Century

And, in order to address the second problem dealing with arduous conventional solutions and missing solution detail of suggested computer solutions, an efficient and detailed solution procedure using state-of-the-art mathematical software is provided in this textbook. In particular, application of the fluid mechanics concepts in this textbook is made significantly easier for both the student and professor by the use of the mathematical software tool Mathcad. Mathcad is used extensively throughout this textbook as a computational tool for both the derivation (integration, differentiation, evaluation, simplification, etc.) of the principles of fluid mechanics and the numerical and analytical solutions of empirical and analytical equations, including the solution of ordinary differential equations. This textbook provides an abundance of real-world application example problems that are solved using Mathcad in complete detail and presented throughout each chapter. This textbook also provides a wealth of end-of-chapter problems for the students to solve; their complete and detailed solutions using Mathcad are provided for the professor in a solutions manual. Other types of programming and computer application tools may be used as an alternative to Mathcad. Alternatives to Mathcad may include, but are not limited to, the following tools: generic mathematics software (Mathematica, MATLAB®, Maple, Derive, GAMS, Theorist,

TK Solver, etc.), generic software (Finite Difference Method, Finite Element Method, Boundary Element Method, Computational Fluid Mechanics, and Computational Fluid Dynamics), fluid mechanics applications software (e.g., Bentley Solutions), programmable scientific calculators (polynomial and equation solvers), spreadsheets, and programming languages. In summary, the detailed use of state-of-the-art mathematical software for teaching and problem solving makes this fluid mechanics textbook an exceptional option for learning and teaching fluid mechanics in the twenty-first century.

An Essential Fluid Mechanics Textbook for Civil and Environmental Engineers

And finally, in order to address the third problem dealing with traditional coverage of the topic of open channel flow, this topic is addressed in great depth and detail in this textbook in order to support the civil and environmental engineer. Highlights from Chapter 9, "Open Channel Flow," are as follows:

- Distinguish when to apply the Lagrangian (differential) versus the Eulerian (integral) approach in the application of the momentum equation; that is, the difference between using the Chezy equation, $S_f = (v^2/C^2 R_h)$ or Manning's equation, $v = (1/n)R_h^{2/3}S_f^{1/2}$ versus $\Sigma F_x = \rho Q(v_{2x} - v_{1x})$, respectively.
- Numerical solution of the Chezy and Manning's equations for the uniform channel depth of flow, y using Mathcad, as opposed to the conventional tedious and time-consuming trial-and-error solution approach.
- Numerical solution of the ordinary differential equation for nonuniform channel depth of flow, $(dy/dx) = (S_o - S_f)/(1 - F^2)$, for y versus x (surface water profile), using an ordinary differential equation solver in Mathcad, which uses the fourth-order Runge-Kutta fixed-step method.
- The formation of various nonuniform flow surface water profiles by controls such as a break in channel slope, a hydraulic drop, or a free overfall.
- Numerical solution for a hydraulic jump in nonrectangular channel sections using Mathcad.
- Detailed coverage of the energy and momentum concepts in open channel flow, with numerical solutions for various nonrectangular channel sections using Mathcad.
- Detailed coverage of flow measuring devices in open channel flow, with numerical solutions for various nonrectangular channel sections using Mathcad.

In summary, the in-depth and detailed coverage of open channel flow makes this fluid mechanics textbook an essential option for learning and teaching fluid mechanics for civil and environmental engineers.

Mathcad: The Mathematical Learning and Teaching Tool of Choice of the Textbook

The historically used conventional solution approach for most of the empirically-derived equations, and even for some of the analytical equations in fluid mechanics, requires cumbersome and sometimes tedious and time-consuming trial-and-error procedures.

Furthermore, the historical suggestions to use computer solutions typically provide only the end result, without providing a step-by-step detailed solution to the problem. The use of a mathematical software tool such as Mathcad (or an alternative such as those listed earlier) in this textbook either removes or significantly reduces this arduous task and provides a complete and detailed solution. Mathcad lends itself as an excellent learning and teaching tool for this purpose, because it allows the user to type and view the equations exactly as they appear on paper. In Mathcad, you do not need to write the equation in "code," as the user is required to do in MATLAB, for instance, which seems to be a bit more popular in the corporate world. Consequently, Mathcad is intentionally selected for application in this textbook, because it provides a very simple, elegant, yet very powerful mathematical software tool. Mathcad is used extensively throughout this textbook as a computational tool for both the derivation (integration, differentiation, evaluation, simplification, etc.) of the principles of fluid mechanics and the numerical and analytical solutions of empirical and analytical equations, including the solution of ordinary differential equations by using the fourth-order Runge-Kutta fixed-step method, for example. Using Mathcad in the derivation of the principles of fluid mechanics and for the analytical solution of analytical equations eliminates the need to look up mathematical procedures in mathematical hand-books and allows the student and professor to focus on the theory and its application. Furthermore, using Mathcad in the numerical solution of both empirical and analytical equations eliminates the need for the time-consuming and labor-intensive solution proce-dures and allows the student to focus on the concepts and efficiently obtain a solution to the problem. This allows the student to conduct open-ended studies of the fluid flow situa-tion (i.e., sensitivity analyses) and gain a deeper insight into the problem and its solutions. And, finally, using Mathcad provides an easy, quick, and powerful tool to graphically plot and illustrate the solution. Below are just a few examples of how Mathcad is utilized in this textbook to solve and graphically illustrate the solutions to the most seemingly complicated problems in fluid mechanics.

An example of using Mathcad to analytically integrate an equation is given in the derivation of the total ideal (theoretical) flow rate, Q_i over the triangular weir, given by Equation 9.306 as follows:

$$Q_i = \int dQ_i = \int_0^H 2\sqrt{2g}\tan\left(\frac{\theta}{2}\right)(H-y)y^{1/2}dy$$

Using Mathcad to integrate yields:

$$\int_0^H 2\sqrt{2g}\cdot\tan\left(\frac{\theta}{2}\right)\cdot(H-y)\cdot\sqrt{y}\,dy \rightarrow \frac{8}{15}\cdot H^{\frac{5}{2}}\cdot 2^{\frac{1}{2}}\cdot g^{\frac{1}{2}}\cdot\tan\left(\frac{1}{2}\cdot\theta\right)$$

Thus yielding Equation 9.307 as follows:

$$Q_i = \frac{8}{15}\sqrt{2g}\tan\left(\frac{\theta}{2}\right)H^{\frac{5}{2}}$$

An example of using Mathcad to numerically integrate is given in the numerical evaluation of the magnitude of the resultant hydrostatic force, F acting on the sloping triangular gate in Example Problem 2.35 as follows:

$$B := 5 \text{ ft} \qquad L := 8 \text{ ft} \qquad A := \frac{B \cdot L}{2} = 20 \text{ ft}^2 \qquad h_o := 6 \text{ ft} \qquad \gamma := 62.417 \frac{\text{lb}}{\text{ft}^3}$$

$$\phi := \text{atan}\left(\frac{\frac{B}{2}}{L}\right) = 17.354 \text{ deg} \qquad\qquad \alpha := 30 \text{ deg} \qquad\qquad y_o := \frac{h_o}{\sin(\alpha)} = 12 \text{ ft}$$

$$F := \int_0^L \gamma \cdot [(y_o + y) \cdot \sin(\alpha)] \cdot (2 \cdot y \cdot \tan(\phi)) dy = 10818.947 \text{ lb}$$

An example of using Mathcad to solve the flow resistance equation for nonuniform flow, which is an ordinary differential equation that is numerically solved for the depth of flow, y using an ordinary differential equation solver in Mathcad, "rkfixed," which uses the fourth-order Runge-Kutta fixed-step method, is given by Example Problem 9.16 as follows. The first-order ordinary differential equation initial value problem is defined below, and a plot of the nonuniform surface water profile is computed and plotted using Mathcad as shown below:

$$\frac{d}{dx} y(x) = f(x, y) = \frac{S_{o1} - \frac{v(y)^2}{C(y)^2 \cdot R_h(y)}}{1 - \left(\frac{v(y)}{v_v}\right)^2} \qquad\qquad y(x0) = y0 = y_c + 0.001$$

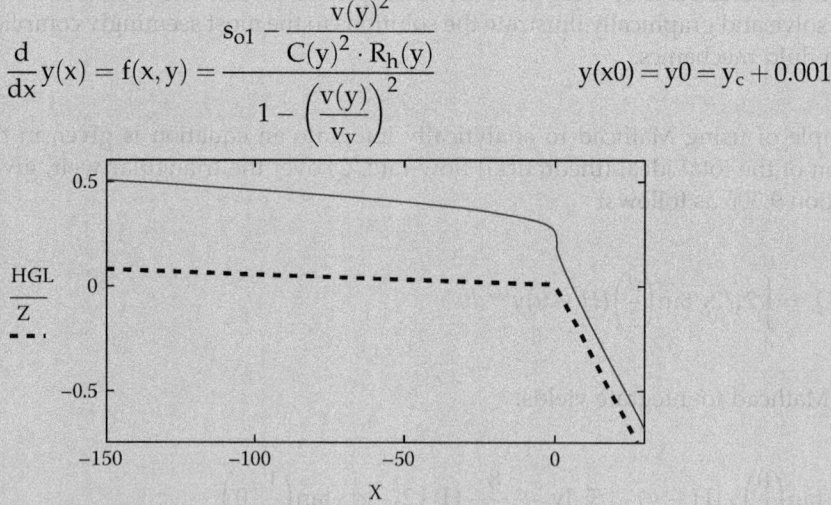

An example of using Mathcad to numerically solve a cubic equation for the depth of flow, y_2 and plot the specific energy curve is given by Example 9.3 as follows:

$$v_1 := 1 \frac{\text{m}}{\text{sec}} \qquad\qquad\qquad y_1 := 1.65 \text{ m} \qquad q := v_1 \cdot y_1 = 1.65 \frac{\text{m}^2}{\text{sec}}$$

$$g := 9.81 \frac{m}{\sec^2} \qquad y_c := \sqrt[3]{\frac{q^2}{g}} = 0.652 \text{ m} \qquad v_c := \sqrt{g \cdot y_c} = 2.53 \frac{m}{s}$$

$$E_c := \frac{3}{2} \cdot y_c = 0.978 \text{ m} \qquad \Delta z := 0.33 \text{ m}$$

Guess value: $\qquad y_2 := 0.5 \text{ m}$

Given

$$y_1 + \frac{v_1^2}{2 \cdot g} = y_2 + \frac{q^2}{2 \cdot g \cdot y_2^2} + \Delta z$$

$$y_2 := \text{Find}\,(y_2) = 0.373 \text{ m}$$

$$E(y) := y + \frac{q^2}{2 \cdot g \cdot y^2} \qquad a \equiv 0.1m \qquad b \equiv 5m \qquad h \equiv 0.1m \qquad y := a, a + h .. b$$

E-y curve

$$E(y), E_1, E_2, E_1, E_c, E_2$$

$$E(y)$$

Order of Presentation and Relevance of Material in the Textbook

With the student in mind, the chapters of the textbook are presented in an order that I hope will be intuitive to the learning process. Chapter 1 introduces the fundamental principles of fluid mechanics (statics, kinematics, and dynamics) and the physical properties of fluids. Chapter 2 presents the principles of fluid statics. Chapters 3 through 5 present the theory of fluid mechanics, which includes the principles of kinematics and dynamics. The theory of fluid mechanics is presented in both the Eulerian (integral) and the Lagrangian (differential) approaches, in the simplest, yet accurate and complete manner. Specifically, the principles of the conservation of mass, energy, and momentum are historically stated and applied for a fluid system (Lagrangian approach), and then rephrased and extended to a control volume (Eulerian approach) using the Reynolds transport theorem. The theory of fluid mechanics presented in Chapters 3 through 5 assumes that the fluid flow and the resulting flow resistance behave in a predictable/deterministic/analytical manner, as in the case of laminar flow. However, the theory of fluid mechanics is not completely adequate to model the case of the unpredictable/stochastic/empirical turbulent flow and the resulting flow resistance. As a result, there is a need to supplement the theory of fluid mechanics with dimensional analysis. As such, fluid dynamics also includes the topic of dimensional analysis, which is introduced in Chapter 7 and yields the resistance equations, which are introduced in Chapter 6. Chapter 6 presents the modeling of the flow resistance for turbulent flow by supplementing the theory of fluid mechanics with dimensional analysis. The modeling of flow resistance is at the heart of practical fluid mechanics problem solving. Chapter 7 presents the essential and invaluable tool of dimensional analysis that supplements the theory of fluid mechanics and yields an uncalibrated deterministic mathematical model of the stochastically behaving flow resistance for turbulent flow. Chapters 8 through 10 present the application of the theory of fluid mechanics, supplemented by dimensional analysis for internal and external flow. The historically documented calibrated mathematical models of the flow resistance equations are used in the application of the theory of fluid mechanics. And, finally, Chapter 11 presents the essential and invaluable tool of dynamic similitude (i.e., the laws governing dynamic similarity resulting from dimensional analysis), which provides a scaled physical model of the fluid flow situation that is used to calibrate the deterministic mathematical model of the stochastically behaving flow resistance for turbulent flow.

In this textbook, the theory of fluid mechanics is presented in both the Eulerian (integral) and the Lagrangian (differential) approaches. As a result, the applications of the resulting governing equations (continuity, energy, and momentum) throughout this textbook highlight to the student whether the Eulerian approach or the Lagrangian approach (or both) are appropriate and applicable to a given fluid flow problem.

In this textbook, Chapter 6, "Flow Resistance Equations," is entirely dedicated to the derivation of the flow resistance equations in internal and external flow (major head loss, minor head loss, actual discharge, and drag force), and the efficiency of a pump or turbine. Furthermore, this chapter highlights that a "subset-level" application of the appropriate governing equations is supplemented by dimensional analysis in order to derive flow resistance equations. The flow resistance equations are introduced and applied in their respective chapter in the textbook. However, Chapter 6, "Flow Resistance Equations," (in the middle of the textbook), emphasizes that the modeling of flow resistance, which is at the heart of practical fluid mechanics problem solving, is highlighted as a major theme throughout the textbook.

Because the topic of open channel flow is of significant importance to the civil and environmental engineer, Chapter 9, "Open Channel Flow," addresses this topic in great depth and detail. The analysis and design of and the occurrence of both uniform and nonuniform flow steady flow are presented with numerous practical real-world engineering applications and detailed solutions using Mathcad.

Because the topics of dimensional analysis (highlighted as another major theme throughout the textbook) and dynamic similitude are significantly related to one another, many fluid mechanics textbooks present these two topics in the same chapter. However, in this textbook, while the chapter on dimensional analysis (Chapter 7, which provides an uncalibrated mathematical model) precedes the chapters on the applications of the historically documented calibrated mathematical models (Chapters 8, 9, and 10), the chapter on dynamic similitude (Chapter 11, which designs a geometrically scaled physical model used to calibrate the mathematical model) is postponed until the very last. The reason for this is as follows: in order for the student to fully appreciate and understand how to design a geometrically scaled physical model of a given fluid flow situation, I believe that it is important for the student to first learn and understand how to apply the historically documented calibrated mathematical model of the fluid flow situation, as presented in Chapters 8, 9, and 10.

Content of Each Chapter in the Textbook

Each chapter begins with an introduction. Then, the principles and concepts of fluid mechanics are presented in a clear and complete manner. Mathcad is used in the derivation (integration, differentiation, evaluation, simplification, etc.) of the principles of fluid mechanics. A wealth of real-world application example problems are solved in complete detail and presented throughout each chapter. The example problems apply the resulting relevant governing equations and highlight the analysis and design of real-life fluid mechanics problems. The example problems are solved using Mathcad, and most examples are graphically illustrated, many of which use Mathcad to generate the graphics. Each chapter concludes with an abundance of end-of-chapter problems that also highlight the analysis and design of real-life fluid mechanics problems; their complete and detailed solutions using Mathcad are provided for the professor in a solutions manual. Both the British Gravitational (BG) (English) system and the International System (SI) (metric) system of units are used in the textbook. And, finally, because most practical engineering problems involve steady flow, the application of the principles of fluid mechanics in this textbook will be limited to steady fluid flow problems.

- Chapter 1, Introduction, introduces the three basic units of study in fluid mechanics (statics, kinematics, and dynamics), the fundamental principles of fluid mechanics (conservation of mass, momentum, and energy), dimensional analysis, the types of fluid flow, dimensions and units, Newton's second law of motion, dynamic forces acting on a fluid element, physical properties of fluids, and standard physical properties for fluids.

- Chapter 2, Fluid Statics, presents the principles of fluid statics/hydrostatics (Pascal's law and Newton's second law of motion), which yield the hydrostatic pressure equation that represents the scalar ordinary differential equation of fluid statics. The hydrostatic pressure equation is applied to measure the hydrostatic pressure at a point (using barometers and manometers), to determine the hydrostatic forces acting on a submerged surface (plane and curved surfaces), and to determine the role of the hydrostatic force in the buoyancy and stability of a floating or neutrally buoyant body.

- Chapter 3, Continuity Equation, presents the principles of fluid kinematics (conservation of mass), which yield the continuity equation. The conservation of mass is presented and applied in both the Eulerian (integral) and the Lagrangian (differential) approaches. The study of fluid kinematics considers the type of fluid flow, the flow visualization/geometry, and the description and observation of the fluid motion, and thus, the analysis of the velocity and acceleration of the fluid in motion.

- Chapter 4, Energy Equation, presents and applies the principles of fluid dynamics (conservation of momentum/Newton's second law of motion), which yield the equations of motion, known as the energy equation and the momentum equation. The conservation of momentum is presented in both the Eulerian (integral) and the Lagrangian (differential) approaches. Chapter 4 also presents and applies the principle of conservation of energy (the first law of thermodynamics), which yields the energy equation. The conservation of energy is presented in both the Eulerian (integral) and the Lagrangian (differential) approaches. While the energy equation that is derived based on the conservation of momentum principle allows for the modeling of unsteady flow, it does not directly consider the modeling of energy transfer by work (pumps, turbines, etc.). And while the energy equation that is derived based on the conservation of energy principle assumes steady flow, it directly considers the modeling of energy transfer by work (pumps, turbines, etc.).

- Chapter 5, Momentum Equation, presents the derivation of the momentum equation, which is based on the principles of fluid dynamics (conservation of momentum/Newton's second law of motion). The conservation of momentum is presented and applied in both the Eulerian (integral) and the Lagrangian (differential) approaches. The application of the governing equations (continuity, energy, and momentum) for real flow (internal), and ideal flow (flow measuring devices for internal flow, and external flow around objects) requires the modeling of flow resistance (frictional losses). The flow resistance in internal and external flow (major head loss, minor head loss, actual discharge, and drag force) is modeled by a "subset-level" application of the appropriate governing equations in Chapter 6.

- Chapter 6, Flow Resistance Equations, presents the derivation of the flow resistance equations in internal and external flow (major head loss, minor head loss, actual discharge, and drag force) and the efficiency of a pump or turbine. A "subset-level" application of the appropriate governing equations is supplemented by dimensional analysis in Chapter 7 to derive flow resistance equations when there is no theory to model the friction force (as in the case of turbulent internal and external flow). Chapter 6, "Flow Resistance Equations," emphasizes that the modeling of flow resistance, which is at the heart of practical fluid mechanics problem solving, is a major theme throughout the textbook.

- Chapter 7, Dimensional Analysis, presents the essential and invaluable tool of dimensional analysis (highlighted as another major theme throughout the textbook) that supplements the theory of fluid mechanics and yields an uncalibrated deterministic mathematical model of the stochastically behaving flow resistance for turbulent flow. A thorough and systematic approach to using dimensional analysis to derive empirical flow resistance equations, as well as theoretical equations, is presented; a clear explanation as to how the variables are chosen when applying dimensional analysis is presented and applied.

- Chapter 8, Pipe Flow, presents the application of the governing equations (continuity, energy, and momentum), and the results of dimensional analysis to pipe (pressure/closed conduit flow) flow. Applications of the governing equations for real pipe flow include the analysis and design of pipe systems (carrying water, oil, gasoline, chemicals, etc.) that include numerous arrangements of pipes, including single pipes, pipes with components, pipes with a pump or a turbine, pipes in series, pipes in parallel, branching pipes, pipes in a loop, and pipe networks. And, applications of the governing equations for ideal pipe flow include the analysis and design of pipe systems that include a pipe flow measuring device such as orifices, nozzles, or venturi meters, which are also known as ideal flow meters. A clear explanation of when to apply the Lagrangian (differential) versus the Eulerian (integral) approach in a given fluid flow problem is given, and why.

- Chapter 9, Open Channel Flow, presents the application of the governing equations (continuity, energy, and momentum), and the results of dimensional analysis to open channel flow, which occurs in natural rivers, streams, and channels and in artificial channels, canals, and waterways. Applications of the governing equations for real open channel flow include the analysis and design of uniform and nonuniform flow (generated by controls such as a break in channel slope, a hydraulic drop, or a free overfall). Applications of the governing equations for ideal open channel flow include the analysis and design of an open channel flow measuring device (such as sluice gates, weirs, spillways, venturi flumes, and contracted openings), which are also known as ideal flow meters, and the analysis and design of a gradual (vertical or horizontal) channel contraction/transition. And, applications of the governing equations for open channel flow include the analysis and design of a hydraulic jump. A clear explanation of when to apply the Lagrangian (differential) versus the Eulerian (integral) approach in a given fluid flow problem is given, and why.

- Chapter 10, External Flow, presents the application of the governing equations (energy and momentum), and the results of dimensional analysis to external flow. Applications of the governing equations for ideal external flow include the determination of the lift force, F_L and the drag force, F_D in the analysis and design of planes, cars, buildings, ships, submarines, and submerged pipes. The Eulerian (integral) approach is assumed in the application of the governing equations for external flow. An in-depth and very detailed study of how the drag and lift coefficients vary for a given flow situation is provided. In particular, a detailed study of how the drag coefficient, C_D varies with the velocity of flow, the shape of the body, and whether there is a wave action at the free surface is presented. And, a detailed study of how the lift coefficient, C_L varies with the shape of the body, the angle of attack, and the relative surface roughness is presented.

- Chapter 11, Dynamic Similitude and Modeling, presents the essential and invaluable tool of dynamic similitude (i.e., the laws governing dynamic similarity resulting from dimensional analysis), which provides a scaled physical model of the fluid flow situation that is used to calibrate the deterministic mathematical model of the stochastically behaving flow resistance for turbulent flow. A detailed set of guidelines for applying the laws governing dynamic similarity is given for both true models and distorted models. The reason why it is important for the independent variables to remain a constant between the model and its prototype in the dynamic similitude and modeling of fluid flow is highlighted and explained in detail.

- Appendix A, Physical Properties of Common Fluids, presents the physical properties of the standard atmosphere, water, and some common fluids in both BG and SI units.

- Appendix B, Geometric Properties of Common Shapes, presents the area and volume geometric properties of common shapes.

Solutions Manual for the Professor

A solutions manual is available for the professor. The solutions manual is available on a password-protected web site for the textbook, and includes the complete solutions and graphic illustrations to all of the end-of-chapter problems. The abundant end-of-chapter problems apply the relevant governing equations and highlight the analysis and design of real-life fluid mechanics problems. They are solved in great detail using Mathcad, and most solutions are graphically illustrated, many of which use Mathcad to generate the graphics.

History of Fluid Mechanics

A brief introduction to the great mathematicians, physicists, and engineers who discovered the principles, laws, and equations of fluid mechanics is presented below. The connection between the concepts and the famous scientists who discovered them will hopefully spark an interest in the student who is about to embark on the knowledge of fluid mechanics. As the student reads and learns from the textbook, he/she will quickly recognize that many of the principles, laws, and equations are named after the discovering scientist. As such, it is hoped that the following brief history of fluid mechanics will put in perspective the knowledge of fluid mechanics throughout the history of mankind, and hopefully bring the learning process of the subject full circle. Fluid mechanics has a long history and dates as far back as prehistoric times. The history of fluid mechanics includes both the art and the science of fluid mechanics. The art of fluid mechanics is archeologically documented as beginning in prehistoric times. Practical knowledge of fluid flow was demonstrated by ancient civilizations by the design and use of streamlined arrows, spears, boats, and ships, and in the solution of basic fluid flow problems such as water supply, irrigation, drainage, and flood protection. Although the mathematical knowledge of fluid mechanics began with the Greek mathematician Archimedes of Syracuse (287–212 BC), the science of fluid mechanics began in the seventeenth century with the Italian astronomer Galileo Galilei (1564–1642), who is known as the father of science. By the end of the eighteenth century, the basic equations of fluid mechanics theory had been developed. During the nineteenth century, dimensional analysis (derivation of empirical equations) was developed to supplement the fluid mechanics theory. While the science of fluid mechanics had greatly evolved by the beginning of the twentieth century, easily attained solutions to the equations were not as developed yet. For instance, the Navier-Stokes equation, which describes the motion of viscous flow, was too difficult to solve. A simplified application of the Navier-Stokes equation was achieved at the beginning of the twentieth century. However, methods to solve the basic equations of fluid mechanics supplemented by dimensional analysis were not developed until the latter half of the twentieth century. The second half of the twentieth century was a significant turning point for fluid flow research, with the development and application of experimental and numerical methods that continue to rapidly develop in the twenty-first century. Innovative tools used to solve numerical methods include mathematical software tools such as Mathcad, MATLAB, Mathematica, Maple, etc. Therefore, while fluid mechanics has a long history, fluid mechanics and its application to solve engineering problems are fairly recent in a historical context. And, finally, it is important to note that the study of fluid mechanics is essential in the various disciplines of engineering, including civil, environmental, mechanical, aerospace, nuclear, chemical, and agricultural engineering.

The Beginning of Fluid Mechanics

The study of fluid mechanics has its beginnings in the fields of mathematics and physics. Mathematical knowledge of fluid mechanics began with Archimedes of Syracuse

FIGURE I.1
Archimedes of Syracuse (287–212 BC). (Courtesy http://www.infoniac.com/hi-tech/famous-scientists-their-inventions-and-discoveries.html)

(287–212 BC) (see Figure I.1), who developed the fundamental principles of hydrostatics and dynamics. He applied the laws of buoyancy (known as the Archimedes Principle) to submerged and floating bodies. Sextus Julius Frontinus (40–103) was a Roman senator who documented the history and description of the water supply of the Roman aqueducts. Leonardo da Vinci (1452–1519) (see Figure I.2) was an Italian polymath who discovered the equation of conservation of mass for incompressible, one-dimensional, steady-state flow. da Vinci led the way to flow visualization, experimenting with waves, jets, hydraulic jumps, eddy formation, vortices, and drag on bodies.

Fluid Mechanics in the Seventeenth Century

The science of fluid mechanics began with Galileo Galilei (1564–1642) (see Figure I.3), who is known as the father of science and documented the science of motion: uniform and naturally accelerated motion, terminal velocity, and projectile motion. Benedetto Castelli (1578–1653) was an Italian mathematician who explained the dynamics of fluid flow in rivers and canals, and explained the continuity principle for fluid flow. Edme Mariotte (1620–1684) was a French physicist who discovered Mariotte's law (also known as Boyle's law) describing the inverse relationship between volume and pressures in gases. Mariotte built the first wind tunnel and tested models in it. Evangelista Torricelli (1608–1647) (see Figure I.4) was an Italian physicist who invented the barometer and deduced that the velocities of liquids are equal to the square root of the head, assuming ideal flow through an opening (orifice) in a tank (known as Torricelli's theorem). Blaise Pascal (1623–1662) (see Figure I.5) was a French mathematician who discovered the laws of the equilibrium of liquids (known as Pascal's law). Pascal also clarified concepts of pressure and vacuum by applying the work of Torricelli. Isaac Newton (1642–1727) (see Figure I.6) was a British physicist who is one of the most influential scientists in history. Newton formulated the three laws of motion and universal gravitation (known as Newton's first, second, and third laws of motion), and the law of viscosity of linear fluids (known as Newton's law of viscosity).

FIGURE I.2
Leonardo da Vinci (1452–1519). (Courtesy http://www.infoniac.com/hi-tech/famous-scientists-their-inventions-and-discoveries.html)

FIGURE I.3
Galileo Galilei (1564–1642). (Courtesy http://www.sciencekids.co.nz/pictures/scientists.html.)

FIGURE I.4
Evangelista Torricelli (1608–1647). (Courtesy http://www.infoniac.com/hi-tech/famous-scientists-their-inventions-and-discoveries.html)

FIGURE I.5
Blaise Pascal (1623–1662). (Courtesy http://www.gettyimages.com/)

FIGURE I.6
Isaac Newton (1642–1727). (Courtesy http://www.gettyimages.com/)

Fluid Mechanics in the Eighteenth Century

The theoretical science of fluid mechanics continued to grow in the eighteenth century, with a focus on inviscid (ideal, frictionless) flow. And by the end of the eighteenth century, the basic equations of fluid mechanics theory had been developed. Leonhard Euler (1707–1783) (see Figure I.7), who was a Swiss mathematician, was one of the most distinguished mathematicians of the eighteenth century, and one of the greatest in history. Euler developed the differential equation of motion for inviscid flow (known as Euler's equation of motion) and its integral form (known as the Bernoulli equation). The dimensionless pi term, the Euler number, is named after him. Daniel Bernoulli (1700–1782) (see Figure I.8) was a Swiss mathematician who discovered Bernoulli's principle (energy equation for inviscid flow), which is based on the conservation of energy. Henri Pitot (1695–1771) was a French hydraulic engineer who discovered that the height of the fluid column is proportional to the square of the velocity of the fluid at the depth of the inlet to the pitot tube, which he invented. Jean-Baptiste le Rond d'Alembert (1717–1783) was a French mathematician who is well known for D'Alembert's paradox, also known as the hydrodynamic paradox for incompressible and inviscid flow; the contradiction is that the drag force is zero on a body moving with a constant velocity relative to the fluid. Antoine-Laurent de Lavoisier (1743–1794) was a French biologist and chemist who discovered the law of the conservation of mass. Joseph-Louis Lagrange (1736–1813) was a French mathematician who developed the most comprehensive study of classical mechanics since Newton, and formed a basis for the development of mathematical physics in the nineteenth century.

Fluid Mechanics in the Nineteenth Century

While the theory of fluid mechanics continued to grow, the empirical science of fluid mechanics began to grow in the nineteenth century, with a focus on viscous and turbulent

FIGURE I.7
Leonhard Euler (1707–1783). (Courtesy http://www.infoniac.com/hi-tech/famous-scientists-their-inventions-and-discoveries.html)

flow and hydraulic engineering applications. During the nineteenth century, dimensional analysis (derivation of empirical equations using experimental methods) was developed to supplement the fluid mechanics theory. The theory of fluid mechanics discovered by the end of the eighteenth century assumes that the fluid flow and the resulting flow resistance behave in a predictable/deterministic/analytical manner, as in the case of laminar flow, or an ideal and frictionless flow. However, the theory of fluid mechanics is not

FIGURE I.8
Daniel Bernoulli (1700–1782). (Courtesy http://www.infoniac.com/hi-tech/famous-scientists-their-inventions-and-discoveries.html)

adequate to model the case of the unpredictable/stochastic/empirical turbulent flow and the resulting flow resistance. As a result, there was a need to supplement the theory of fluid mechanics with experimental studies, guided by dimensional analysis and dynamic similitude. Antoine de Chezy (1718–1798) was a French hydraulic engineer who is known for the Chezy equation (derived by supplementing the momentum equation with dimensional analysis). Henry Darcy (1803–1858) (see Figure I.9), a French hydraulic engineer, and Julius Weisbach (1806–1871), a German mathematician, are known for the Darcy-Weisbach frictional head loss equation. Jean Leonard Marie Poiseuille (1797–1869), a French physicist, and Gotthilf Heinrich Ludwig Hagen (1797–1884), a German civil engineer, independently, at the same time, experimentally derived and formulated Hagen-Poiseuille's law for laminar flow. William Froude (1810–1879) was an English hydrodynamicist who formulated reliable laws for the flow resistance on ships (hull speed equation) and for predicting their stability. Froude developed laws for the model testing of hulls, leading to the dimensionless pi term, the Froude number. Robert Manning (1816–1897), who was an Irish engineer, developed Manning's equation for turbulent flow. Rudolf Clausius (1822–1888) a German physicist, and William John Macquorn Rankine (1820–1872), a Scottish physicist, discovered the first law of thermodynamics (the law of the conservation of energy). Lord Rayleigh (1842–1919) was an English physicist who proposed dimensional analysis and is known for the Rayleigh method. Joseph Louis Francois Bertrand (1822–1900) was a French mathematician who proved the dimensional analysis procedure. Edgar Buckingham (1867–1940) (see Figure I.10) was an American physicist who formalized the dimensional analysis procedure, known as the Buckingham pi method. Osborne Reynolds (1842–1912) was a British physicist who studied and published experimental studies of the transition of laminar flow to turbulent flow, leading to the dimensionless pi term, the Reynolds number. Claude-Louis Navier (1785–1836), a French physicist, and George Gabriel Stokes (1819–1903), an Irish physicist, are known for the derivation of the one-dimensional Navier-Stokes equation, which is the

FIGURE I.9
Henry Darcy (1803–1858). (Courtesy https://www.google.com)

FIGURE I.10
Edgar Buckingham (1867–1940). (Courtesy http://www.infoniac.com/hi-tech/famous-scientists-their-inventions-and-discoveries.html.)

differential form of Newton's second law of motion (the law of the conservation of momentum) that describes the motion of viscous fluid. Stokes is also known for Stokes's law for the drag force on small spherical objects in viscous fluids.

Fluid Mechanics in the Twentieth Century

By the beginning of the twentieth century, the basic equations of fluid mechanics theory (continuity, energy, and momentum) and dimensional analysis (developed to supplement the fluid mechanics theory and yielding empirical flow resistance equations) had been developed by famous mathematicians, physicists, and engineers from all over the world. Thus, while the science of fluid mechanics had greatly evolved by the beginning of the new millennium, easily attaining solutions to the equations was not as developed yet. For instance, the Navier-Stokes equation, which describes the motion of viscous flow, was too difficult to solve. Furthermore, although, in general, the theoretical equations had analytical solutions, achieving their solutions was not possible due to the lack of solution methods. And, finally, achieving solutions to the empirical equations was also challenging due to the trial-and-error nature of the solution. A simplified application of the Navier-Stokes equation was achieved at the beginning of the twentieth century. However, in general, the solutions to the theoretical and empirical fluid mechanics equations required experimental, as did advanced numerical methods, which demand significant computational (computer) power that was not available until the second half of the twentieth century. In 1904, the German aerodynamic engineer Ludwig Prandtl (1875–1953) wrote and published one of the most important papers to be written on the subject of fluid mechanics. Prandtl developed the fundamental principles of subsonic aerodynamics, including transonic velocities, which form the basis of aeronautical engineering, in which he applied the Euler and

Bernoulli equations. Specifically, Prandtl demonstrated that a fluid flow with low viscosity (for instance, water and air flow) can be divided into two layers: (1) a thin viscous layer near the walls (boundary layer), where the friction effects are significant, and (2) a thin outer layer, where the flow is assumed to be inviscid (frictionless), and thus, the Euler and Bernoulli equations are applicable. As a result, boundary layer theory serves as the single most important solution approach in modern-day fluid mechanics. Theodor von Karman (1881–1963), a Hungarian-American mathematician; Paul Blasius (1883–1970), a German engineer; Johann Nikuradse (1894–1979), a German engineer; and other scientists further developed the boundary layer theory for application in hydraulic and aerodynamic engineering. The second half of the twentieth century was a significant turning point for fluid flow research and application, with the development and application of experimental and highly developed numerical methods to solve both theoretical and empirical equations that are easily implemented by the use of mathematical software tools. Innovative tools used to solve numerical methods include generic mathematical software tools (Mathcad, Mathematica, MATLAB, Maple, Derive, GAMS, Theorist, TK Solver, etc.), generic software (Finite Difference Method, Finite Element Method, Boundary Element Method, Computational Fluid Mechanics, and Computational Fluid Dynamics), fluid mechanics applications software (e.g., Bentley Solutions), programmable scientific calculators (polynomial and equation solvers), spreadsheets, and programming languages. The development of experimental fluid mechanics methods for laminar and turbulent flow includes flow visualization, laser- and phase-Doppler anemometry, and particle image velocimetery.

Fluid Mechanics in the Twenty-First Century

The development and application of experimental and highly developed numerical methods to solve fluid flow problems continue to rapidly develop and evolve in the twenty-first century. In addition to the continued growth of innovative software tools used to solve numerical methods listed above, there are considerable developments in the field of experimental fluid mechanics. These include the development of a number of measuring techniques to study fluid flow, including fast electronic components, lasers, integrated optics, numerous sensors, and micro-techniques. For instance, experimental research continues to grow in the areas of turbulent boundary-layer and pipe flow using thermal anemometry. Laser anemometry is applied for experimental unsteady and 3-dimensional open channel flow hydraulics. And open channel flow measurements are obtained using laser Doppler anemometry. The twenty-first century will include significant advances in flow optimization techniques in design, magneto-hydrodynamics, flow visualization and measurement techniques, solution of compressible and incompressible flow problems, and multiphase flows and solid-fluid integrations, to name a few.

Practical Applications of Fluid Mechanics

The association between the study of fluid mechanics and practical problems in our everyday lives will hopefully motivate the student to learn and apply the concepts in order to attain practical solutions of real-life problems. Practical applications of the principles of fluid mechanics are exemplified in the human body; our daily lives; and the various disciplines of engineering, including civil, environmental, mechanical, aerospace, nuclear, chemical, agricultural, and biomedical engineering. Because many functions of the human body (for

FIGURE I.11
Manometer. (Courtesy https://pixabay.com)

instance, our heart, lungs, kidneys, veins, blood vessels, etc.) are governed by the principles of fluid mechanics, advancements in biomedical engineering have applied these principles in order to artificially simulate these functions. The use of barometers and manometers (see Figure I.11) to measure the atmospheric pressure and pressure in our fresh water supply pipes is an example of the application of fluid statics. The analysis and design of boats, ships (see Figure I.12), and submarines (see Figure I.13) are examples of the application of the Archimedes Principle. The analysis and design of fresh water supply pipes, natural gas pipes (see Figure I.14), and oil supply pipes (see Figure I.15) are examples of pipes (pressure flow). The analysis and design of storm or sewer drainage pipes (see Figure I.16), rivers, streams, channels, canals (see Figure I.17), and dams (see Figure I.18) are examples of open channel (gravity) flow. The analysis and design of buildings (see Figure I.19), bridges, cars, trucks (see Figure I.20), airplanes (see Figure I.21), rockets, and submarines are examples of the application of the aerodynamic principles of drag and lift. The analysis and design of dams, reservoirs, and culverts are examples of the application of fluid statics

FIGURE I.12
Ship. (Courtesy https://pixabay.com)

FIGURE I.13
Submarine. (Courtesy https://pixabay.com)

FIGURE I.14
Gas pipes. (Courtesy https://pixabay.com)

FIGURE I.15
Oil supply pipeline. (Courtesy https://pixabay.com)

FIGURE I.16
Drainage pipe. (Courtesy https://pixabay.com)

FIGURE I.17
River. (Courtesy https://pixabay.com)

FIGURE I.18
Grand Coulee Dam. (Courtesy http://www.usbr.gov)

FIGURE I.19
Building. (Courtesy https://pixabay.com)

FIGURE I.20
Truck. (Courtesy https://pixabay.com)

FIGURE I.21
Airplane. (Courtesy https://pixabay.com)

FIGURE I.22
Wind turbines. (Courtesy https://pixabay.com)

and fluid dynamics. Furthermore, the analysis and design of numerous everyday appliances (refrigerators, heaters, air conditioners, hair dryers, fans, vacuum cleaners, etc.), pumps, turbines (see Figure I.22), and gas and oil pipe lines all demonstrate the application of the principles of fluid mechanics. In summary, there are endless examples in our daily lives that display the importance of fluid mechanics and how the various disciplines of engineering apply the principles of fluid mechanics in order to make our lives easier, more comfortable, and sustainable in an ever-changing climate and world.

In Summary

Fluid Mechanics for Civil and Environmental Engineers is written with the student in mind, and with the goal to present the study of fluid mechanics in a step-by-step, nonintimidating, yet precise manner. This introduction has presented you with a brief synopsis and a quick tour of the textbook, while revealing a bit of its contents, in order to demonstrate the benefits of adopting this textbook in your fluid mechanics course. The intuitive, detailed, and easy-to-understand presentation of the material makes this textbook an excellent option for learning and teaching fluid mechanics in all engineering disciplines. The detailed use of state-of-the-art mathematical software for teaching and problem solving makes this fluid mechanics textbook an exceptional option for learning and teaching fluid mechanics in the twenty-first century; Mathcad is highlighted as the mathematical learning and teaching tool of choice of the textbook. And, the in-depth and detailed coverage of the topic of open channel flow makes this textbook an essential option for learning and teaching fluid mechanics for civil and environmental engineers. A brief examination of the order of chapters and a quick glance of their contents are presented, along with the availability of a solutions manual for the professor. A quick review of the history of fluid mechanics clearly demonstrates the embodiment of the concepts of the subject matter, in anticipation that making the connection between the famous scientists and their discoveries will incite a passion for learning the subject matter. And, finally, making the association between the study of fluid mechanics and practical problems in our everyday lives will hopefully motivate the student to learn and apply the concepts in order to achieve practical solutions to real-life problems.

1

Introduction

1.1 Introduction

The study of fluid mechanics is important in numerous fields of engineering, including civil, environmental, agricultural, irrigation, mechanical, aerospace, nuclear, chemical, petroleum, biomedical, fire protection, and automotive engineering. The fundamental principles of fluid mechanics include three basic units of study: fluid statics, fluid kinematics, and fluid dynamics (Section 1.2). The physical properties/characteristics of a fluid system, along with the fluid kinematics and fluid dynamics, will determine the type of fluid flow (Section 1.3). The physical quantities of fluid flow (geometrics, kinematics, and dynamics) and the physical properties/characteristics of fluids (mass density, specific gravity, specific weight, viscosity, surface tension, vapor pressure, and bulk modulus) are expressed using four primary dimensions (force or mass, length, time, and temperature) and a specific system of units (metric or English) (Section 1.4). Most fluid properties vary with temperature and pressure, while the acceleration due to gravity varies with altitude and thus atmospheric pressure. As such, it is important to distinguish between two types of pressure scales (Section 1.5), define the conditions of standard atmosphere (Section 1.6), and define the standard reference for standard atmospheric pressure (Section 1.7). Furthermore, it is important to highlight Newton's second law of motion in the definition of the acceleration due to gravity (Section 1.8) and to note that the dynamic forces acting on a fluid element include those due to gravity, pressure, viscosity, elasticity, surface tension, and inertia (Section 1.9). And, finally, the physical properties of fluids are presented in Section 1.10.

1.2 Fundamental Principles of Fluid Mechanics

The fundamental principles of fluid mechanics can be subdivided into three units of study: fluid statics, fluid kinematics, and fluid dynamics. Fluid statics deals with fluids at rest, while fluid kinematics and fluid dynamics deal with fluids in motion. Fluid statics is based upon the principles of hydrostatics, which yield the hydrostatic pressure equation. Fluid kinematics is based upon the principle of conservation of mass, which yields the continuity equation. And fluid dynamics is based upon the principle of conservation of momentum (Newton's second law of motion), which yields the equations of motion, known as the energy equation and the momentum equation. The energy equation may alternatively be based on the principle of conservation of energy (the first law of thermodynamics). Furthermore, fluid dynamics also includes the topic of dimensional analysis, which yields the resistance equations. The continuity equation, the equations of motion, and the results of dimensional analysis (the resistance equations) are applied to internal flows, which include pressure (pipe) flow and gravity (open channel) flow. However, only the equations of motion and the results of dimensional analysis (the resistance equations) are applied to external flows, which include the flow around objects. For external flow, the Eulerian (integral) approach is assumed in the application of the governing equations. And, finally, depending upon the internal flow problem, one may use either the Eulerian (integral) approach or the Lagrangian (differential) approach, or both, in the application of the three governing equations (continuity, energy, and momentum). It is important to note that while the continuity and momentum equations may be applied using either the integral or the differential approach, application of the energy equation is useful only using the integral approach, as energy is work, which is defined over a distance, L.

Depending on the unit of study, certain principles and thus certain equations will be applicable in the solution of problems. For fluids in static equilibrium, the principles of hydrostatics and thus the pressure equation would be applicable. For fluids in motion, the principles of the conservation of mass and the conservation of momentum and thus the continuity equation and the equations of motion (energy equation and momentum equation), respectively, would be applicable. Furthermore, it may be noted that for fluid (internal flow) in motion, while the application of the continuity equation will always be necessary, the energy equation and the momentum equation play complementary roles in the analysis of a given flow situation. Additionally, the energy equation and the momentum equation play complementary roles for external flow. Complementary roles mean that when one of the two equations of motion breaks down, the other may be used to find the additional unknown quantity.

This book covers various topics within each of the three units of study. The fluid statics unit covers the topics of hydrostatic pressure, hydrostatic forces on submerged surfaces, and buoyancy and stability, which are introduced and applied in Chapter 2. The fluid kinematics unit covers the topic of the continuity equation, which is introduced in Chapter 3 and applied in the remaining chapters of the book (except for Chapter 10). The fluid dynamics unit covers the topics of the energy equation, the momentum equation, flow resistance, and dimensional analysis, which are introduced in Chapters 4 through 7, respectively, and are applied to pipe flow in Chapter 8, open channel flow in Chapter 9, external flow in Chapter 10, and dynamic similitude and modeling in Chapter 11.

1.3 Types of Fluid Flow

The characteristics/properties of a fluid system, along with the fluid kinematics and fluid dynamics, will determine the type of flow (see Chapter 3 for detailed discussion). First, a flow may be classified as internal flow (pipe or open channel) or external flow (flow around an object), depending upon the use of energy or work to move the fluid. Second, an internal flow may be classified as a pressure (pipe) flow or a gravity (open channel flow), depending upon whether a hydraulic gradient or gravity caused the flow. Third, a flow may be real (viscous) or ideal (inviscid), depending upon the value assumed for the fluid viscosity. A real flow may be subdivided into laminar or turbulent flow, depending upon the value of the Reynolds number, $R = \rho v L/\mu$. Furthermore, real fluids may be divided into Newtonian and non-Newtonian fluids. Fourth, a flow may be compressible (pressure flow) or incompressible (pressure flow or gravity flow), depending upon the spatial and/or temporal variation in the fluid density. Fifth, a flow may be spatially varied (nonuniform) or spatially uniform, depending upon the spatial variation in fluid velocity (convective acceleration). Sixth, a flow may be unsteady or steady, depending upon the temporal variation in the fluid velocity (local acceleration). And, seventh, a flow may be one-, two-, or three-dimensional, depending upon the assumption of spatial dimensionality.

1.4 Dimensions and Units

The physical quantities (geometrics, kinematics, and dynamics) of fluid flow and the physical properties (mass density, specific gravity, specific weight, viscosity, surface tension, vapor pressure, and bulk modulus) of fluids are expressed using four primary dimensions (force or mass, length, time, and temperature) and a specific system of units (metric or English). There are four fundamental/primary dimensions in the study of fluid mechanics, and they are: force, F or mass, M, length, L, time, T, and temperature, θ. In an engineering application, any equation that relates physical quantities comprising these dimensions must be dimensionally homogenous; this means that the dimensions of each term of the equation must be identical. Units are magnitudes that carry specific names for the primary dimensions and for combinations of dimensions. There are many systems of units, with two major systems in use: the British Gravitational (BG) (English) system and the International System (SI) (metric) system. The units for the primary dimensions in the English system, F, L, T, and θ, are pounds, feet, seconds, and Rankine or Fahrenheit, respectively. The units for the primary dimensions in the metric system, M, L, T, and θ, are kilogram, meter, seconds, and Kelvin or Celsius, respectively. Furthermore, secondary dimensions and units may be derived from the primary dimensions and units by applying their respective physical definitions.

1.4.1 Primary and Secondary Dimensions

The primary dimensions will depend upon the chosen system of units (BG or SI). As such, secondary dimensions for a given system of units may be derived by applying the corresponding physical definition.

1.4.1.1 Dimensions for the British Gravitational (BG) System

In the BG system of units, the fundamental/primary dimensions are F, L, T, and θ. Because a large number of problems in fluid mechanics require only the F, L, and T dimensions, the BG system is also known as the FLT system. The dimensions for M and all other secondary dimensions may be derived from the primary dimensions. Derivation of the M dimension (for the BG system) is done by applying Newton's second law of motion, $F = Ma$ as follows:

$$M = \frac{F}{a} = [FL^{-1}T^2] \tag{1.1}$$

Furthermore, derivation of secondary dimensions (such as for velocity, acceleration, pressure, etc.) in the FLT system (primary dimensions for the BG system) is done by applying their respective physical definitions. For instance, the derivation of the dimensions for the pressure, p in the FLT system is given as follows:

$$p = \frac{F}{A} = [FL^{-2}] \qquad (1.2)$$

The secondary dimensions in the *FLT* system for some common physical quantities are presented in Table 1.1 below.

TABLE 1.1

Dimensions and Units of Some Common Physical Quantities in Fluid Flow

Physical Quantity/Property	FLT System: Primary Dimensions for BG Units	MLT System: Primary Dimensions for SI Units	BG (English) Units	SI (Metric) Units
Geometrics				
Length, L	L	L	ft	m
Area, A	L^2	L^2	ft^2	m^2
Volume, V	L^3	L^3	ft^3	m^3
Kinematics				
Time, T	T	T	sec	sec
Frequency, ω	T^{-1}	T^{-1}	cycle/sec $=$ sec^{-1}	cycle/sec $=$ sec^{-1} $=$ Hz
Velocity, v	LT^{-1}	LT^{-1}	ft/sec (fps)	m/sec
Acceleration, a	LT^{-2}	LT^{-2}	ft/sec^2	m/sec^2
Discharge, Q	$L^3\,T^{-1}$	$L^3\,T^{-1}$	ft^3/sec (cfs)	m^3/sec
Dynamics				
Mass, M	$FL^{-1}T^2$	M	slug $=$ lb-sec^2/ft	kg
Force, F	F	MLT^{-2}	lb	N $=$ kg-m/sec^2
Pressure, p	FL^{-2}	$ML^{-1}T^{-2}$	lb/ft^2 (psf)	Pa $=$ N/m^2
Momentum, Mv or Impulse, FT	FT	MLT^{-1}	lb-sec	N-sec
Energy, E or Work, W	FL	ML^2T^{-2}	ft-lb	J $=$ N-m
Power, P	FLT^{-1}	ML^2T^{-3}	ft-lb/sec	W $=$ N-m/sec
Fluid Properties				
Density, ρ	$FL^{-4}\,T^2$	ML^{-3}	slug/ft^3	kg/m^3
Specific gravity, s	dimensionless	dimensionless	no units	no units
Specific weight, γ	FL^{-3}	$ML^{-2}T^{-2}$	lb/ft^3	N/m^3
Absolute (dynamic) viscosity, μ	$FL^{-2}\,T$	$ML^{-1}T^{-1}$	lb-sec/ft^2	N-sec/m^2
Kinematic viscosity, $v = \mu/\rho$	L^2T^{-1}	L^2T^{-1}	ft^2/sec	m^2/sec
Surface tension, σ	FL^{-1}	MT^{-2}	lb/ft	N/m
Vapor pressure, p_v	FL^{-2}	$ML^{-1}T^{-2}$	lb/ft^2 (psf)	Pa $=$ N/m^2
Bulk modulus of elasticity, E_v	FL^{-2}	$ML^{-1}T^{-2}$	lb/ft^2	N/m^2
Temperature, θ	θ	θ	°F	°C

Note the following abbreviations: Hertz (Hz), Newtons (N), Pascal (Pa), Joules (J), Watts (W), Fahrenheit (F), and Celsius (C).

1.4.1.2 Dimensions for the International System (SI)

In the SI system of units, the fundamental/primary dimensions are M, L, T, and θ. Because a large number of problems in fluid mechanics require only the M, L, and T dimensions, the SI system is also known as the MLT system. The dimensions for F and all other secondary dimensions may be derived from the primary dimensions. Derivation of the F dimension (for the SI system) is done by applying Newton's second law of motion, $F = Ma$ as follows:

$$F = Ma = [MLT^{-2}] \tag{1.3}$$

Furthermore, derivation of secondary dimensions (such as for velocity, acceleration, pressure, etc.) in the MLT system (primary units for the SI system) is done by applying their respective physical definitions. For instance, the derivation of the dimensions for the pressure, p in the MLT is given as follows:

$$p = \frac{F}{A} = [ML^{-1}T^{-2}] \tag{1.4}$$

The secondary dimensions in the MLT system for some common physical quantities are presented in Table 1.1 above.

Solution

(a) $v = \dfrac{x}{t} = [LT^{-1}]$

(b) $a = \dfrac{v}{t} = [LT^{-2}]$

(c) $Mv = [MLT^{-1}]$

(d) $P = \dfrac{Work}{t} = \dfrac{Fd}{t} = [MLT^{-2}LT^{-1}] = [ML^2T^{-3}]$

(e) $\gamma = \dfrac{W}{V} = [MLT^{-2}L^{-3}] = [ML^{-2}T^{-2}]$

EXAMPLE PROBLEM 1.3

Demonstrate that the drag force equation, $F_D = \dfrac{1}{2}C_D\rho v^2 A$ (see Chapter 10) is dimensionally homogenous, assuming (a) the BG system of units (FLT). (b) the SI system of units (MLT).

Solution

(a) $F_D = \dfrac{1}{2}C_D\rho v^2 A$

$[F] = [\,][\,][FL^{-4}T^2][L^2T^{-2}][L^2]$

$[F] = [F]$

(b) $F_D = \dfrac{1}{2}C_D\rho v^2 A$

$[MLT^{-2}] = [\,][\,][ML^{-3}][L^2T^{-2}][L^2]$

$[MLT^{-2}] = [MLT^{-2}]$

1.4.2 System of Units

The two commonly used systems of units are the British Gravitational (BG) (English) system and the International System (SI) (metric). As presented above, while the primary dimensions will depend upon the chosen system of units (BG or SI), the secondary dimensions for a given system of units are derived by applying the corresponding physical definition. Similarly, the secondary units corresponding to the secondary dimensions may be derived from the primary units by applying their respective physical definitions. A summary of the units for the English system (the primary dimensions F, L, and T, and the derived dimension M) and for the metric system (the primary dimensions M, L, and T, and the derived dimension F) are presented in Table 1.2 below. Additionally, a summary of the derived units for some common physical quantities are presented in Table 1.1 above. Furthermore, a conversion from the BG system to the SI system of units (or a conversion from the SI system of units

TABLE 1.2

Two Common Systems of Units: The BG (English) System and the SI (Metric) System

Dimensions	BG (English) Units	SI (Metric) Units
Force, F	Pound (lb)	Newton (N) = (kg-m/sec^2)
Mass, M	Slug = (lb-sec^2/ft)	Kilogram (kg)
Length, L	Foot (ft)	Meter (m)
Time, T	Second (s or sec)	Second (s or sec)
Temperature, θ		
• Absolute	Rankine (°R)	Kelvin (°K)
• Customary	Fahrenheit (°F)	Celsius (°C)

Note: The English unit for M, Slug and the metric unit for F, Newtons are derived based on Newton's second law of motion.

to the BG system of units) may be accomplished by applying the equivalence between the two system of units listed at the front of this textbook (BG [English] Unit and SI [metric] Unit Conversion Table). And, finally, one may note that because the SI system is based on a decimal relationship between the various units, there are standard prefixes that are used to form multiples and fractions of the various SI units, which are listed at the front of this textbook (SI [metric] Unit Standard Prefixes).

EXAMPLE PROBLEM 1.4

Define the variable and convert the units of the following quantities from the BG system of units to the SI system of units: (a) 3 knots (nautical mph). (b) 60 gallons. (c) 90 tons. (d) 4 Btus. (e) 30°R.

Mathcad Solution

The conversions from the BG system of units to the SI system of units for some common variables are given at the front of this textbook (and some are also available in Mathcad).

(a) The velocity is converted from BG to SI as follows:

$$1 \text{ knot} = 1.688 \frac{\text{ft}}{\text{s}} \qquad 1 \text{ ft} = 0.305 \text{ m}$$

$$3 \text{ knot } 1.688 \frac{\text{ft}}{\text{sec}} \frac{1}{1 \text{ knot}} \frac{0.305 \text{ m}}{1 \text{ ft}} = 1.545 \frac{\text{m}}{\text{sec}}$$

(b) The volume is converted from BG to SI as follows:

$$1 \text{ gal} = 0.134 \text{ ft}^3 \qquad (1 \text{ ft})^3 = 0.028 \text{ m}^3 \qquad 60 \text{ gal} \frac{0.134 \text{ ft}^3}{1 \text{ gal}} \frac{0.028 \text{ m}^3}{1 \text{ ft}^3} = 0.225 \text{ m}^3$$

(c) The weight is converted from BG to SI as follows:

$$1 \text{ ton} = 2 \times 10^3 \text{ lb} \qquad \text{lb} := 4.44 \text{ N} \qquad 90 \text{ ton} \frac{2000 \text{ lb}}{1 \text{ ton}} \frac{4.44 \text{ N}}{1 \text{ lb}} = 7.992 \times 10^5 \text{ N}$$

(d) The energy is converted from BG to SI as follows:

$$\text{Btu} := 777.649 \, \text{ft lb} \qquad 1 \, \text{ft lb} = 1.353 \, \text{J}$$

$$4 \, \text{Btu} \, \frac{777.649 \, \text{ft lb}}{1 \, \text{Btu}} \, \frac{1.353 \, \text{J}}{1 \, \text{ft lb}} = 4.209 \times 10^3 \, \text{J}$$

(e) The temperature is converted from BG to SI as follows. The conversion formula listed at the front of the textbook is as follows: $^\circ R = \dfrac{9}{5}(^\circ K)$. The direct computation using Mathcad is as follows:

$$30^\circ R = 16.667 \, \text{K}$$

EXAMPLE PROBLEM 1.5

Define the variable and convert the units of the following quantities from the SI system of units to the BG system of units: (a) 90 Pa. (b) 4 L. (c) 80 J. (d) 96 W. (e) 45°C.

Mathcad Solution

The conversions from the SI system of units to the BG system of units for some common variables are given at the front of this textbook (and some are also available in Mathcad).

(a) The pressure is converted from SI to BG as follows:

$$1 \, \text{Pa} = 1 \frac{\text{N}}{\text{m}^2} \qquad\qquad 1 \, \text{m} = 3.281 \, \text{ft} \qquad\qquad \text{N} := 0.225 \, \text{lb}$$

$$90 \, \text{Pa} \, 1 \frac{\text{N}}{\text{m}^2} \, \frac{1}{\text{Pa}} \, \frac{0.225 \, \text{lb}}{1 \, \text{N}} \, \frac{(1 \, \text{m})^2}{(3.281 \, \text{ft})^2} = 1.881 \frac{\text{lb}}{\text{ft}^2}$$

(b) The volume is converted from SI to BG as follows:

$$1 \, \text{L} = 1 \times 10^{-3} \, \text{m}^3 \qquad 1 \, \text{m} = 3.281 \, \text{ft} \qquad 4 \, \text{L} \, 1 \cdot 10^{-3} \frac{(1 \, \text{m})^3}{1 \, \text{L}} \, \frac{(3.281 \, \text{ft})^3}{(1 \, \text{m})^3} = 0.141 \, \text{ft}^3$$

(c) The energy (or work or quantity of heat) is convened from SI to BG as follows:

$$\text{J} := 1 \, \text{N m} \qquad\qquad \text{N} := 0.225 \, \text{lb} \qquad\qquad 1 \, \text{m} = 3.281 \, \text{ft}$$

$$80 \, \text{J} \, \frac{1 \, \text{N m}}{1 \, \text{J}} \, \frac{0.225 \, \text{lb}}{1 \, \text{N}} \, \frac{3.281 \, \text{ft}}{1 \, \text{m}} = 59.058 \, \text{ft lb}$$

(d) The power is converted from SI to BG as follows:

$$\text{W} := 1 \frac{\text{N m}}{\text{sec}} \qquad\qquad \text{N} := 0.225 \, \text{lb} \qquad\qquad 1 \, \text{m} = 3.281 \, \text{ft}$$

$$96 \, \text{W} \, 1 \frac{\text{N m}}{\text{sec}} \, \frac{1}{1 \, \text{W}} \, \frac{0.225 \, \text{lb}}{1 \, \text{N}} \, \frac{3.281 \, \text{ft}}{1 \, \text{m}} = 70.87 \frac{\text{ft lb}}{\text{sec}}$$

(e) The temperature is converted from SI to BG as follows. The conversion formula listed at the front of the textbook is as follows: $°F = 32 + \frac{9}{5}(°C)$. The direct computation using Mathcad is as follows:

$$45°C = 113°F$$

1.4.2.1 Units for the British Gravitational (BG) System

The units for the primary dimensions in the English system, F, L, T, and θ, are pounds, feet, seconds, and Rankine or Fahrenheit, respectively. The units for the derived dimensions for M and all other secondary dimensions may be derived from the primary units. Derivation of the units for derived dimensions for M in the English system is done by applying Newton's second law of motion, $F = Ma$ as follows:

$$M = \frac{F}{a} = [FL^{-1}T^2] = [lb - sec^2/ft] = [slug] \tag{1.5}$$

where a slug is the derived English unit of mass. As such, 1 slug accelerates at 1 foot per second squared (ft/sec^2) when acted upon a force of 1 pound as follows:

$$1\,slug = (1\,lb)/(1\,ft/sec^2) = 1\,lb - sec^2/ft \tag{1.6}$$

Alternatively stated, 1 pound is the force required to accelerate 1 slug of mass at a rate of 1 foot per second squared (ft/sec^2) as follows:

$$1\,lb = (1\,slug)(1\,ft/sec^2) \tag{1.7}$$

Furthermore, the derivation of all other secondary units (such as for velocity, acceleration, pressure, etc.) is done by applying their respective physical definitions. For instance, the derivation of the units for the pressure p in the English system is given as follows:

$$p = \frac{F}{A} = [lb/ft^2] = [psf] \tag{1.8}$$

or,

$$p = \frac{F}{A} = [lb/in^2] = [psi] \tag{1.9}$$

where psf is the abbreviation for the derived English unit, lb/ft^2 for pressure, p, and psi is the abbreviation for the derived English unit, lb/in^2 for pressure, p. In addition to the derived English units for mass, M and pressure, p, it is important to highlight the derived English units for power, P as follows:

$$P = \frac{W}{t} = \frac{[ft - lb]}{[sec]} \tag{1.10}$$

where a horsepower (*hp*) is defined as follows:

$$1 \, hp = 550 \, ft - lb/sec \tag{1.11}$$

Finally, one may note that the Fahrenheit scale is more commonly used than the Rankine scale to specify the primary dimension temperature, θ when using the English system. Furthermore, one may note that the customary/ordinary unit, Fahrenheit is related to the Rankine scale, which is an absolute scale, as follows:

$$°R = °F + 459.67° \tag{1.12}$$

where absolute temperature in $°R$ is measured above absolute zero.

EXAMPLE PROBLEM 1.6 (REFER TO EXAMPLE PROBLEM 1.1)

Based on their respective physical definitions, derive the BG units for the following variables: (a) velocity, v. (b) acceleration, a. (c) momentum, Mv. (d) power, $P = $ Work/time $= Fd/t$. (e) specific weight, $\gamma = $ Weight/Volume $= W/V$.

Solution

(a) $v = \dfrac{x}{t} = [LT^{-1}] = [lb/ft^2]$

(b) $a = \dfrac{v}{t} = [LT^{-2}] = [ft/sec^2]$

(c) $Mv = [FL^{-1}T^2LT^{-1}] = [FT] = [lb - sec]$

(d) $P = \dfrac{\text{Work}}{t} = \dfrac{Fd}{t} = [FLT^{-1}] = [ft - lb/sec]$

(e) $\gamma = \dfrac{W}{V} = [FL^{-3}] = [lb/ft^3]$

1.4.2.2 Units for the International System (SI)

The units for the primary dimensions in the metric system, M, L, T, and θ, are kilograms, meters, seconds, and Kelvin or Celsius, respectively. The units for the derived dimensions for F and all other secondary dimensions may be derived from the primary units. Derivation of the units for derived dimensions for F in the metric system is done by applying Newton's second law of motion, $F = Ma$ as follows:

$$F = Ma = [MLT^{-2}] = [kg - m/sec^2] = [N] \tag{1.13}$$

where a Newton (N) is the derived metric unit for force, F. As such, 1 Newton is the force required to accelerate 1 kilogram (*kg*) of mass at a rate of 1 meter per second squared (m/sec^2) as follows:

$$1 \, N = (1 \, kg)(1 \, m/sec^2) \tag{1.14}$$

A less commonly used derived metric unit for force, F is the dyne, which is defined as follows:

$$1\,dyne = (1\,g)(1\,cm/sec^2) = 10^{-5}N \qquad (1.15)$$

Furthermore, the derivation of all other secondary units (such as for velocity, acceleration, pressure, etc.) is done by applying their respective physical definitions. For instance, the derivation of the units for the pressure, p in the metric system is given as follows:

$$p = \frac{F}{A} = [N/m^2] = [Pa] \qquad (1.16)$$

where a Pascal (Pa) is the derived metric unit for pressure, p. Furthermore, a common unit for pressure, p in the metric system is the bar (b), which is defined as follows:

$$1\,N/m^2 = 1\,Pa = 10^{-5}\,bar \qquad (1.17)$$

$$1\,kN/m^2 = 1\,kPa = 10\,mb \qquad (1.18)$$

In addition to the derived metric units for force, F, and pressure, p, it is important to highlight the derived metric units for work, W and power, P, which are given, respectively, as follows. The metric unit for work, W is derived as follows:

$$W = (F)(d) = [N - m] = [J] \qquad (1.19)$$

where a Joule (J) is the derived metric unit for work, W. As such, 1 Joule is the work, W done when a force, F of 1 Newton is applied to displace a fluid element through a distance, d of 1 meter in the direction of the force as follows:

$$1J = (1\,N)(1\,m) \qquad (1.20)$$

The metric unit for power, P is derived as follows:

$$P = \frac{W}{t} = [J/sec] = [W] \qquad (1.21)$$

where a Watt (W) is the derived metric unit for power, P. As such, 1 Watt is the power, P done when work, W of 1 Joule is done over a period time, t of 1 second as follows:

$$1\,W = 1\,J/sec \qquad (1.22)$$

And, finally, one may note that the Celsius scale is more commonly used than the Kelvin scale to specify the primary dimension temperature, θ when using the metric system. Furthermore, one may note that although the customary/ordinary unit, Celsius (centigrade) is not actually part of the metric system of units, it is related to the Kelvin scale, which is an absolute scale, as follows:

$$^\circ K = {^\circ C} + 273.15^\circ \qquad (1.23)$$

where absolute temperature in $^\circ K$ is measured above absolute zero.

EXAMPLE PROBLEM 1.7 (REFER TO EXAMPLE PROBLEM 1.2)

Based on their respective physical definitions, derive the SI units for the following variables: (a) velocity, v. (b) acceleration, a. (c) momentum, Mv. (d) power, $P = $ Work/time $= Fd/t$. (e) specific weight, $\gamma = $ Weight/Volume $= W/V$.

Solution

(a) $v = \dfrac{x}{t} = [LT^{-1}] = [m/\text{sec}]$

(b) $a = \dfrac{v}{t} = [LT^{-2}] = [m/\text{sec}^2]$

(c) $Mv = [MLT^{-1}] = [kg - m/\text{sec}] = [N - \text{sec}]$

(d) $P = \dfrac{\text{Work}}{t} = \dfrac{Fd}{t} = [MLT^{-2}LT^{-1}] = [ML^2T^{-3}] = [kg - m^2/\text{sec}^{-3}] = [N - m/\text{sec}]$

$= [Watt]$

(e) $\gamma = \dfrac{W}{V} = [MLT^{-2}L^{-3}] = [ML^{-2}T^{-2}] = [kg/m^2 - \text{sec}^2] = [N/m^3]$

1.5 Pressure Scales

Most of the fluid properties vary with temperature and pressure, while the acceleration due to gravity, g varies with altitude (and thus atmospheric pressure) and latitude (distance north or south relative to the equator) on the earth. While the customary temperature scale (°F in English units or °C in metric units) is used in most engineering applications, there is a need to distinguish between the two types of pressure scales, which are absolute pressure and gage pressure. The gage pressure of a fluid is measured relative to the atmospheric pressure. Thus, the gage pressure (relative pressure) of a fluid at the atmospheric pressure registers as a zero pressure, while the gage pressure of a fluid above the atmospheric pressure registers as a positive pressure, and the gage pressure of a fluid below the atmospheric pressure registers as a negative pressure (suction or vacuum). The conditions in the atmosphere vary both temporally and spatially, and thus are a function of the weather, the time of year (seasons), the altitude, and the latitude. Therefore, the atmospheric pressure will vary with the elevation of the particular surface on earth and the weather conditions; thus, it is a moveable datum/scale (for the gage pressure), as illustrated in Figure 2.7. As such, it becomes important to establish a fixed datum relative to the moveable atmospheric pressure datum. This fixed datum is the absolute zero pressure scale, which occurs in a perfect vacuum such as in outer space. The moveable atmospheric pressure datum is measured relative to the fixed absolute zero pressure datum. Therefore, the variable atmospheric pressure above the fixed absolute zero is reported in absolute pressure units (*psi* [abs] or psia in the English system and *Pa* [abs] in the metric system). Furthermore, the absolute pressure of a fluid above the fixed absolute zero is computed as the sum of the atmospheric pressure and the gage pressure of the fluid (measured relative to the atmospheric pressure) as follows:

$$p_{abs} = p_{atm} + p(\text{gage}) \tag{1.24}$$

where the absolute pressure, p_{abs} is the pressure that is measured relative to absolute zero pressure. In the English system of units, the absolute pressure is reported as *psi* (abs) or *psia*, while in the metric system of units, the absolute pressure is reported as *Pa* (abs). Furthermore, in the metric system of units, the gage pressure is reported as *psi* (gage) or *psig*, while in the metric system of units, the gage pressure is reported as *Pa* (gage).

EXAMPLE PROBLEM 1.8

The atmospheric pressure at sea level is 14.6959 psi (abs). The gage pressure at a point in a tank of fluid is 300 psi (gage). (a) Determine the absolute pressure at the point in the tank of fluid.

Mathcad Solution

(a) The absolute pressure at the point in the tank of fluid is computed by applying Equation 1.24 as follows:

$$p_{abs} = p_{atm} + p(gage)$$

$$P_{atm} := 14.6959 \frac{lb}{in^2} \qquad\qquad P_{gage} := 300 \frac{lb}{in^2}$$

$$P_{abs} := P_{atm} + P_{gage} = 314.696 \frac{lb}{in^2}$$

EXAMPLE PROBLEM 1.9

The atmospheric pressure at sea level is 101.325 kPa (abs). The gage pressure at a point in a tank of fluid is 45 kPa (gage). (a) Determine the absolute pressure at the point in the tank of fluid.

Mathcad Solution

(a) The absolute pressure at the point in the tank of fluid is computed by applying Equation 1.24 as follows:

$$p_{abs} = p_{atm} + p(gage)$$

$$P_{atm} := 101.325 \, kPa \qquad\qquad P_{gage} := 45 \, kPa$$

$$P_{abs} := P_{atm} + P_{gage} = 1.463 \times 10^5 \, Pa$$

1.6 Standard Atmosphere

The atmosphere is made up of dry air (mostly nitrogen and oxygen) and water vapor (usually less than 3%). The conditions (physical properties) in the atmosphere include temperature, absolute pressure, density, specific weight, viscosity, speed of sound, and acceleration due to gravity. The conditions in the atmosphere vary both temporally and spatially, and thus are a function of the weather, the time of year (seasons), the altitude, and the latitude. However, in order to adopt a standard for practical applications, the International Civil Aviation Organization (ICAO 1993) has established the ICAO Standard Atmosphere, which serves as an approximation of average conditions in the atmosphere. As such, the physical properties (temperature, absolute pressure, density, specific weight, viscosity, speed of sound, and the acceleration due to gravity) for the ICAO Standard Atmosphere are presented as a function of elevation above sea level for altitudes (elevations) ranging from sea level to 100,000 *ft* above sea level, and 30 *km* above sea level, as seen in Table A.1 in Appendix A.

EXAMPLE PROBLEM 1.10

Using Table A.1 in Appendix A, (a) graphically illustrate the variation of the temperature of the standard atmosphere with altitude assuming the BG system of units; (b) graphically illustrate the variation of the pressure of the standard atmosphere with altitude assuming the SI system of units.

Mathcad Solution

(a) A graphical illustration of the variation of the temperature of the standard atmosphere with altitude assuming the BG system of units is given as follows:

$$
\text{Elev} := \begin{pmatrix} 0 \\ 5000 \\ 10000 \\ 15000 \\ 20000 \\ 25000 \\ 30000 \\ 35000 \\ 40000 \\ 45000 \\ 50000 \\ 60000 \\ 70000 \\ 80000 \\ 90000 \\ 100000 \end{pmatrix}
\qquad
\text{Temp} := \begin{pmatrix} 59 \\ 41.173 \\ 23.355 \\ 5.545 \\ -12.255 \\ -30.048 \\ -47.832 \\ -65.607 \\ -69.700 \\ -69.700 \\ -69.700 \\ -69.700 \\ -67.425 \\ -61.976 \\ -56.535 \\ -51.099 \end{pmatrix}
$$

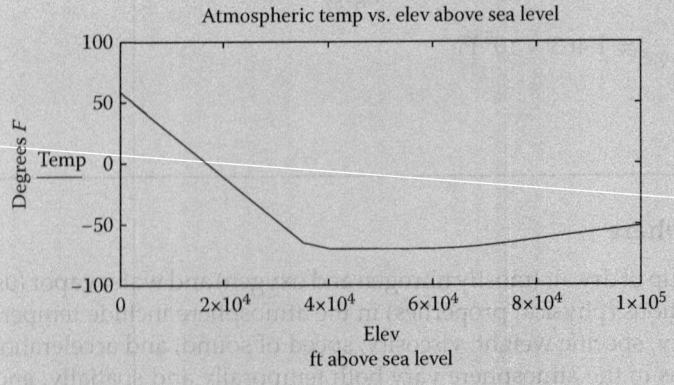

(b) A graphical illustration of the variation of the pressure of the standard atmosphere with altitude assuming the SI system of units is given as follows:

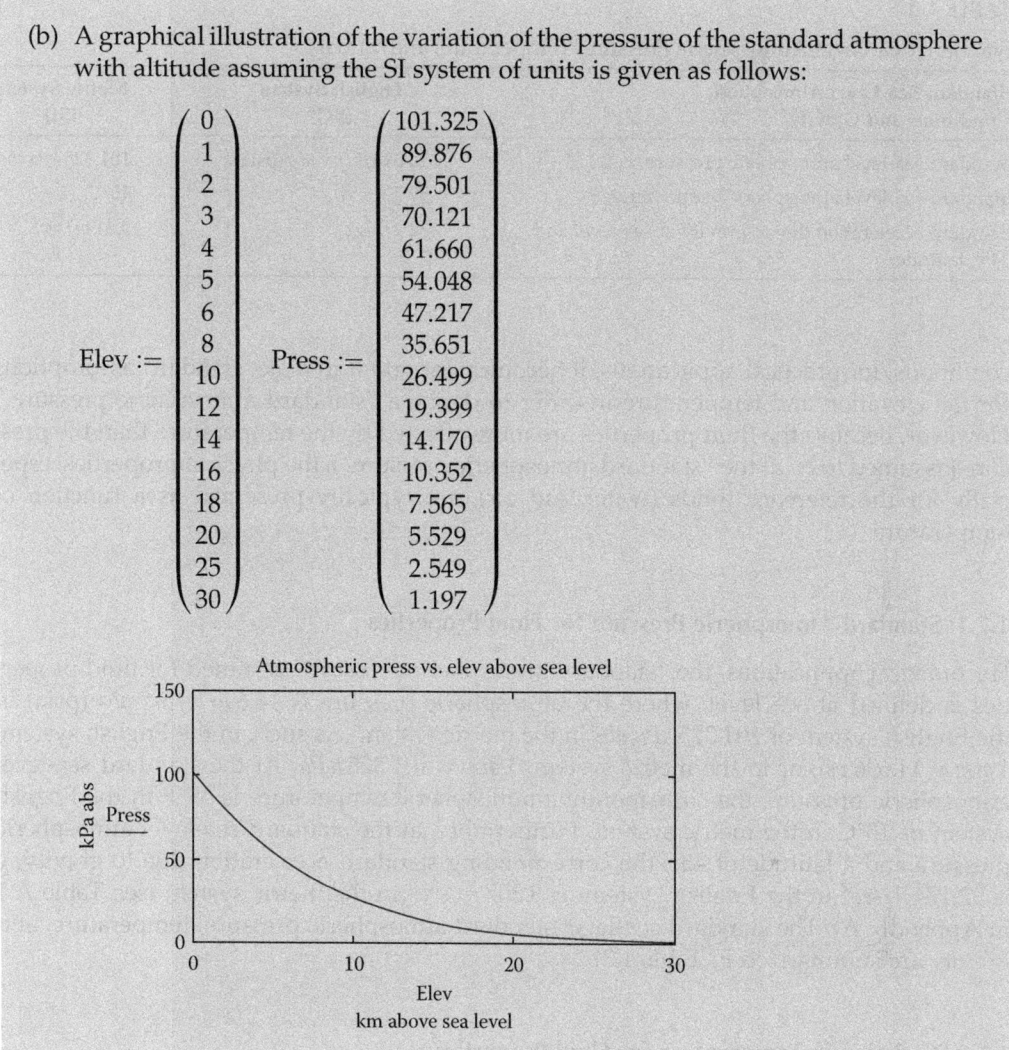

$$
\text{Elev} := \begin{pmatrix} 0 \\ 1 \\ 2 \\ 3 \\ 4 \\ 5 \\ 6 \\ 8 \\ 10 \\ 12 \\ 14 \\ 16 \\ 18 \\ 20 \\ 25 \\ 30 \end{pmatrix} \qquad \text{Press} := \begin{pmatrix} 101.325 \\ 89.876 \\ 79.501 \\ 70.121 \\ 61.660 \\ 54.048 \\ 47.217 \\ 35.651 \\ 26.499 \\ 19.399 \\ 14.170 \\ 10.352 \\ 7.565 \\ 5.529 \\ 2.549 \\ 1.197 \end{pmatrix}
$$

Atmospheric press vs. elev above sea level

$\frac{\text{Press}}{}$ kPa abs

Elev
km above sea level

1.7 Standard Reference for Altitude, Pressure, Latitude, and Temperature

There are seven major physical properties of fluids, given as follows: mass density, ρ; specific gravity, $s_f = (\rho_f / \rho_{ref\ fluid})$; specific weight, $\gamma = \rho g$; viscosity, μ; surface tension, σ; vapor pressure, p_v; and bulk modulus of elasticity, E_v, where g is the acceleration due to gravity. As noted above, most of the fluid properties vary with the temperature and pressure, while acceleration due to gravity, g varies with altitude (and thus atmospheric pressure) and latitude on the earth. It is important to note that the absolute pressure of a fluid, $p_{abs} = p_{atm} + p(gage)$, and thus most of its fluid properties, and the acceleration due to gravity, g, vary with the atmospheric pressure. However, because the atmospheric pressure will vary with the elevation of the particular surface on earth and the weather

TABLE 1.3

Standard Sea-Level Atmospheric Pressure, Temperature, and Gravity

Standard Sea-Level Atmospheric Conditions and Gravity	English System (BG)	Metric System (SI)
Standard sea-level atmospheric pressure, p_{atm}	14.696 lb/in^2 abs (psia)	101.325 kPa abs
Standard sea-level atmospheric temperature, θ	59°F	15°C
Standard acceleration due to gravity at sea level and 45° latitude, g	32.2 ft/sec^2	9.81 m/sec^2

conditions, for practical applications, it becomes important to make standard assumptions for the elevation and temperature in order to define a "standard atmospheric pressure." However, because the fluid properties are more affected by the temperature than the pressure (assumed to be at the "standard atmospheric pressure"), the physical properties, especially for the reference fluids (water and air), are typically presented as a function of temperature.

1.7.1 Standard Atmospheric Pressure for Fluid Properties

For practical applications, the "standard atmospheric pressure" assumed for fluid properties is defined at sea level, where the atmospheric pressure is 14.696 lb/in^2 abs (psia) in the English system or 101.325 kPa abs in the metric system. As such, in the English system, $1\,atm = 14.696\,psi$, or in the metric system, $1\,atm = 101.325\,kPa$. At the standard sea-level atmospheric pressure, the corresponding atmospheric temperature is 59°F in the English system or 15°C in the metric system. Furthermore, at the standard sea-level atmospheric pressure and a latitude of 45°, the corresponding standard acceleration due to gravity, g is 32.174 ft/sec^2 in the English system or 9.807 m/sec^2 in the metric system (see Table A.1 in Appendix A). The standard sea-level standard atmospheric pressure, temperature, and gravity are summarized in Table 1.3.

1.7.2 Variation in Temperature for Fluid Properties

The standard sea-level atmospheric pressure is assumed for fluid properties. However, because the fluid properties are more affected by the temperature than the pressure, the physical properties, especially for the reference fluids (water and air), are presented as a function of temperature. The reference fluid for liquids is water, while the reference fluid for gases is either air or hydrogen. As such, the physical properties for water and air at standard sea-level atmospheric pressure are presented as a function of temperature in Table A.2 in Appendix A and Table A.3 in Appendix A, respectively.

EXAMPLE PROBLEM 1.11

Using Table A.2 in Appendix A and Table A.3 in Appendix A, respectively: (a) Graphically illustrate the variation of the absolute viscosity of water with temperature assuming the BG system of units. (b) Graphically illustrate the variation of the density of air with temperature assuming the SI system of units.

Mathcad Solution

(a) A graphical illustration of the variation of the absolute viscosity of water with temperature assuming the BG system of units is given as follows:

$$
T := \begin{pmatrix} 32 \\ 40 \\ 50 \\ 60 \\ 70 \\ 80 \\ 90 \\ 100 \\ 110 \\ 120 \\ 130 \\ 140 \\ 150 \\ 160 \\ 170 \\ 180 \\ 190 \\ 200 \\ 212 \end{pmatrix} \qquad \mu := \begin{pmatrix} 37.46 \\ 32.29 \\ 27.35 \\ 23.59 \\ 20.50 \\ 17.99 \\ 15.95 \\ 14.24 \\ 12.84 \\ 11.68 \\ 10.69 \\ 9.81 \\ 9.05 \\ 8.38 \\ 7.80 \\ 7.26 \\ 6.78 \\ 6.37 \\ 5.93 \end{pmatrix}
$$

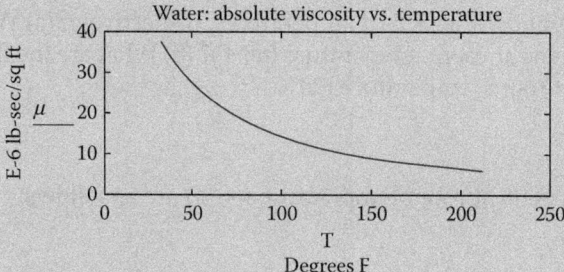

Water: absolute viscosity vs. temperature

(b) A graphical illustration of the variation of the density of air with temperature assuming the SI system of units is given as follows:

$$
T := \begin{pmatrix} -40 \\ -20 \\ 0 \\ 10 \\ 20 \\ 30 \\ 40 \\ 60 \\ 80 \\ 100 \\ 200 \end{pmatrix} \qquad \rho := \begin{pmatrix} 1.515 \\ 1.395 \\ 1.293 \\ 1.248 \\ 1.205 \\ 1.165 \\ 1.128 \\ 1.060 \\ 1.000 \\ 0.946 \\ 0.747 \end{pmatrix}
$$

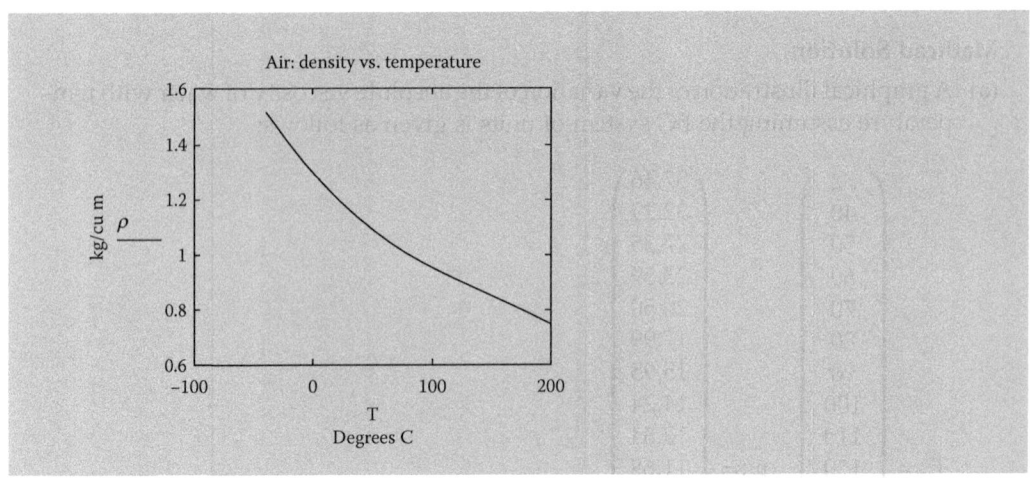

Furthermore, although the fluid properties of other common fluids will vary with temperature, for practical applications, room temperature (68°F or 20°C), in general, is assumed to be a standard temperature for fluids. As such, the physical properties for some common liquids and some common gases at standard sea-level atmospheric pressure are presented at room temperature (68°F or 20°C), in general, in Table A.4 in Appendix A and Table A.5 in Appendix A, respectively.

EXAMPLE PROBLEM 1.12

Using Table A.4 in Appendix A and Table A.5 in Appendix A, respectively: (a) What are the fluid properties for benzene at room temperature (68°F)? (b) What are the fluid properties for carbon dioxide at room temperature (20°C)?

Solution

(a) The fluid properties for benzene at room temperature (68°F) are as follows:
Density: $\rho = 1.70 \, slug/ft^3$
Specific gravity: $s = 0.876$
Absolute viscosity: $\mu = 14.37 \times 10^{-6} \, lb\text{-}sec/ft^2$
Surface tension: $\sigma = 0.0020 \, lb/ft$
Vapor pressure: $p_v = 1.45 \, psia$
Bulk modulus of elasticity: $E_v = 150{,}000 \, psi$
Specific heat: $c = 10{,}290 \, ft\text{-}lb/(slug\text{-}°R)$

(b) The fluid properties for carbon dioxide at room temperature (20°C) are as follows:
Molar mass: $m = 44.01 \, kg/kg\text{-}mol$
Density: $\rho = 1.84 \, kg/m^3$
Absolute viscosity: $\mu = 14.8 \times 10^{-6} \, N\text{-}sec/m^2$
Gas constant: $R = 188 \, N\text{-}m/(kg\text{-}K)$
Specific heat: $c_p = 858 \, N\text{-}m/(kg\text{-}K)$
Specific heat: $c_v = 670 \, N\text{-}m/(kg\text{-}K)$
Specific heat ratio: $k = \dfrac{c_p}{c_v} = 1.28$

1.7.3 Standard Temperature Assumed for the Reference Fluid

In the definition of specific gravity, $s_f = (\rho_f / \rho_{ref\ fluid})$, the reference fluid for liquids is water, while the reference fluid for gases is either air or hydrogen. For water, the standard temperature assumed by physicists (and assumed in this textbook) is 39.2°F or 4°C, and the pressure is assumed to be at the standard sea-level atmospheric pressure. It is important to note that while the standard temperature of 39.2°F or 4°C is assumed for the reference fluid, water, the fluid in question (and thus its corresponding s_f and ρ_f), may be at any given temperature. For instance, Table A.4 in Appendix A presents the specific gravity and density (and other physical properties) for some common liquids at standard sea-level atmospheric pressure and at room temperature (68°F or 20°C). Although the standard temperature of 39.2°F or 4°C is assumed for the reference fluid, water, the fluid in question (and thus its corresponding s_f and ρ_f), is given at room temperature (68°F or 20°C). Additionally, for water, another standard temperature assumed is 60°F or 15.56°C, which is sometimes used by engineers (not assumed in this textbook). However, for hydrogen or air, because there is no general agreement on a standard temperature and standard pressure for these reference fluids, these are specified for a given problem.

1.8 Newton's Second Law of Motion

Newton's second law of motion states that the sum of the forces in a given direction, s is equal to the product of the mass, M and acceleration in the given direction, a_s and is given as follows:

$$\sum_s F = M a_s \qquad (1.25)$$

The dynamic forces acting on a fluid element include those due to gravity, pressure, viscosity, elasticity, surface tension, and inertia. In particular, the weight, W or the gravitational force, F_G (force due to gravity, g) of a mass, M is given as follows:

$$F_G = Mg \qquad (1.26)$$

where $g =$ the acceleration due to gravity. Furthermore, because the primary BG units, $F, L,$ and T, are dependent on the value of g, which varies with location (altitude and latitude), it is referred to as a gravitational system of units. However, because the primary SI units, $M, L,$ and T, are independent of the value of g, it is referred to as an absolute system of units.

1.8.1 Acceleration due to Gravity for the British Gravitational (BG) System

The application of Newton's second law of motion for the BG system of units yields that 1 pound is the force required to accelerate 1 slug of mass at a rate of 1 foot per second squared (ft/sec^2), as follows:

$$1\,lb = (1\,slug)(1\,ft/sec^2) \qquad (1.27)$$

Furthermore, although the acceleration due to gravity, g varies with altitude (and thus, atmospheric pressure) and latitude on the earth, at the standard sea-level atmospheric pressure and a latitude of 45°, the corresponding standard acceleration due to gravity, g for the BG system is given as follows:

$$g = 32.174\,ft/sec^2 \cong 32.2\,ft/sec^2 \tag{1.28}$$

Thus, the application of Newton's second law of motion in order to define the gravitational force, F_G for the BG system yields the following:

$$F_G = Mg$$

$$F[lb] = M[slug]\,g[ft/sec^2]$$

$$32.2\,lb = (1\,slug)(32.2\,ft/sec^2) \tag{1.29}$$

Therefore, a mass of 1 slug weighs 32.2 lbs under a standard gravity, $g = 32.2\,ft/sec^2$.

EXAMPLE PROBLEM 1.13

A body has a mass of 4 slugs. (a) Determine the weight of the mass.

Mathcad Solution

(a) In order to determine the weight (gravitational force) of the mass, Newton's second law of motion is applied as follows:

$$slug := 1\,lb\,\frac{sec^2}{ft} \qquad M := 4\,slug \qquad g := 32.2\,\frac{ft}{sec^2} \qquad W := M\,g = 128.8\,lb$$

1.8.2 Acceleration due to Gravity for the International System (SI)

The application of Newton's second law of motion applied for the SI system of units yields that 1 Newton is the force required to accelerate 1 kilogram (kg) of mass at a rate of 1 meter per second squared (m/s^2) as follows:

$$1\,N = (1\,kg)(1\,m/sec^2) \tag{1.30}$$

Furthermore, although the acceleration due to gravity, g varies with altitude (and thus atmospheric pressure) and latitude on the earth, at the standard sea-level atmospheric pressure and a latitude of 45°, the corresponding standard acceleration due to gravity, g for the SI system is given as follows:

$$g = 9.807\,m/sec^2 \cong 9.81\,m/sec^2 \tag{1.31}$$

Thus, the application of Newton's second law of motion in order to define the gravitational force, F_G for the SI system yields the following:

$$F_G = Mg$$

$$F[N] = M[kg]\,g[m/sec^2]$$

$$9.81\,N = (1\,kg)(9.81\,m/sec^2) \tag{1.32}$$

Therefore, a mass of 1 kg weighs 9.81 N under a standard gravity, $g = 9.81\,m/sec^2$.

EXAMPLE PROBLEM 1.14

A body weighs 80 N. (a) Determine the mass of the body.

Mathcad Solution

(a) In order to determine the mass of the body, Newton's second law of motion is applied as follows:

$$N := kg \frac{m}{sec^2} \qquad W := 80 \, N \qquad g := 9.81 \frac{m}{sec^2} \qquad M := \frac{W}{g} = 8.155 \, kg$$

1.9 Dynamic Forces Acting on a Fluid Element

The dynamic forces acting on a fluid element include those due to gravity, F_G; pressure, F_P; viscosity, F_V; elasticity, F_E; surface tension, F_T; and inertia, F_I. The actual forces acting on a fluid element will depend upon whether the fluid is in static equilibrium or is in motion (fluid dynamics). As such, in the study of fluids in static equilibrium, the significant forces acting on a fluid element include those due to gravity, F_G and pressure, F_P. Additionally, other potentially significant forces for fluid statics may include those due to elasticity, F_E (compressible fluids) and surface tension, F_T (droplets, bubbles, thin films, and capillarity). However, the viscous force, F_V does not exist in a fluid in static equilibrium, regardless of the fluid viscosity, μ, and the inertia force, F_I does not come into play in fluid statics. In the study of fluids in motion (fluid dynamics), the significant forces acting on a fluid element include those due to gravity, F_G; pressure, F_P; viscosity, F_V; and inertia, F_I. Additionally, other potentially significant forces for fluid dynamics may include those due to elasticity, F_E (compressible fluids) and surface tension, F_T (sheets of fluids). The dynamic forces acting on a fluid element (F_G, F_P, F_V, F_E, F_T, and F_I) and the fluid properties (mass density, ρ; specific gravity, s; specific weight, γ; viscosity, μ; surface tension, σ; vapor pressure, p_v; and bulk modulus, E_v) are related by the corresponding physical definitions of the dynamic forces, as presented in Table 1.4 below. Table 1.4 also presents the significant fluid property associated with a given dynamic force and whether the force and fluid property is applied in fluid statics and fluid dynamics.

1.10 Physical Properties of Fluids

Each fluid has certain characteristics by which its physical condition may be described. Such characteristics are called properties, which are observable qualities of the fluid. These fluid properties are expressed in terms of secondary dimensions and secondary units, both of which are derived from the primary dimensions (F or M, L, and T) and units (English or metric) by applying their respective physical definitions, and are presented in Table 1.1 above. There are seven major fluid properties, given as follows: mass density, ρ; specific gravity, s; specific weight, γ; viscosity, μ; surface tension, σ; vapor pressure, p_v; and bulk modulus of elasticity, E_v. In the development of the fundamental principles of fluid mechanics, some fluid properties/characteristics play principal roles and others play only minor roles, while

TABLE 1.4

Relationship between the Dynamic Forces and the Fluid Properties

Dynamic Force	Physical Definition of Dynamic Force	Significant Fluid Property	Fluids Application	Fluids Application
Gravity force	$F_G = Mg = \gamma V$	$\gamma = \rho g$	Dynamics	Statics
Pressure force	$F_P = \Delta p A$	p_v	Dynamics	Statics
Laminar viscous force	$F_V = \tau A = \mu \dfrac{dv}{dy} A$	μ	Dynamics	
Turbulent viscous force	$F_V = \tau A = C_D \rho v^2 A$	ρ	Dynamics	
Elastic force	$F_E = E_v A$	E_v	Dynamics (compressible fluids)	Statics (compressible fluids)
Surface tension force	$F_T = \sigma L$	σ	Dynamics (sheets of fluids)	Statics (droplets, bubbles, thin films, and capillarity)
Inertia force	$F_I = Ma = \rho v^2 L^2$	ρ	Dynamics	

Note: Δp = fluid pressure difference, and C_D = drag coefficient.

some play no role. For instance, in fluid statics, the specific weight is the important fluid property. In both fluid statics and fluid flow (fluid dynamics), vapor pressure becomes important when very low pressures are involved; bulk modulus becomes important when compressible fluids are involved; furthermore, surface tension is important in droplets, bubbles, thin films, capillarity, or sheets of fluid. Viscosity, μ only becomes important when the fluid is no longer in static equilibrium, and thus, there is a fluid flow. And finally, in fluid flow, density and viscosity are the predominant properties. One may note that most fluid properties (especially the density for a gas/compressible flow) are a function of the temperature and pressure.

1.10.1 Mass Density

The mass density (or just density) is defined as mass, M per unit volume, V and is designated by ρ as follows:

$$\rho = \frac{M}{V} \tag{1.33}$$

One may note that density, ρ is considered to be an absolute fluid property, because it depends on mass, M (and thus characterizes the mass of the fluid system), which is independent of location (latitude and elevation above mean sea level). Furthermore, the density, ρ is a measure of the "heaviness" of the fluid and is a predominant fluid property in fluid flow/dynamics. The dimensions and units for the density, ρ in the English system are given as follows:

$$\rho = \frac{M}{V} = \frac{[FL^{-1}T^2]}{[L^3]} = \frac{[lb-sec^2/ft]}{[ft^3]} = [slug/ft^3] \tag{1.34}$$

The dimensions and units for the density, ρ in the metric system are given as follows:

$$\rho = \frac{M}{V} = \frac{[M]}{[L^3]} = [kg/m^3] \tag{1.35}$$

In general, the density of a fluid, ρ decreases with an increase in temperature, while it increases with an increase in pressure. Because the density of a gas is strongly dependent on both temperature and pressure, it is a variable, and gas flow is thus modeled as compressible flow (see Section 1.10.7). However, since the density of a liquid is more affected by the temperature than by the pressure, for practical changes in pressure, the density of a liquid is essentially constant and thus is accurately approximated as an incompressible flow. The mass density, ρ for water and air at standard sea-level atmospheric pressure are presented as a function of temperature in Table A.2 in Appendix A and Table A.3 in Appendix A, respectively. And, although the mass density, ρ for some other common liquids and some other common gases will vary with temperature, for practical applications, their mass density, ρ at standard sea-level atmospheric pressure at room temperature (68°F or 20°C), in general, are given in Table A.4 in Appendix A and Table A.5 in Appendix A, respectively.

For a given temperature, a change in density, ρ of a given fluid mass that is due to a change in pressure is a measure of the compressibility of the fluid (see Section 1.10.7). Although all fluids have a certain degree of compressibility, some fluids (liquids and gases) are assumed to be incompressible when the change in density due to a change in pressure is negligible. Therefore, for compressible fluids/flow, the fluid density is assumed to vary with pressure, while for incompressible fluids/flow, the fluid density is assumed to be a constant with respect to pressure. It is important to note that, although liquids can be compressed (resulting in small changes in density) at very high pressures, they are typically assumed to be incompressible for most practical engineering applications. Furthermore, it is important to note that, although gases are easily compressed (resulting in significant changes in density) at very high pressures, they are typically assumed to be incompressible, as indicated by the Mach number, M (discussed in Section 1.10.7) when the pressure variation is small in comparison to the absolute pressure (sum of atmospheric and gage pressure).

EXAMPLE PROBLEM 1.15

Using Table A.4 in Appendix A and Table A.5 in Appendix A, respectively, to read the fluid density: (a) Determine the volume of 8 slugs of kerosene at room temperature (68°F). (b) Determine the mass of 90 liters of methane at room temperature (20°C).

Mathcad Solution

(a) The volume of 8 slugs of kerosene at room temperature (68°F) is determined by applying Equation 1.33 as follows:

$$slug := 1\,lb\,\frac{sec^2}{ft} \qquad M := 8\,slug \qquad \rho := 1.57\,\frac{slug}{ft^3} \qquad V := \frac{M}{\rho} = 5.096\,ft^3$$

(b) The mass of 90 liters of methane at room temperature (20°C) is determined by applying Equation 1.33 as follows:

$$1\,L = 1 \times 10^{-3}\,m^3 \qquad V := 90\,L \qquad \rho := 0.668\,\frac{kg}{m^3} \qquad M := \rho V = 0.06\,kg$$

TABLE 1.5

Standard Density for Water at Standard Temperature and Standard Sea-Level
Atmospheric Pressure

Standard Density, $\rho = M/V$ for Water (Reference Fluid) at Standard Temperature and Standard Sea-Level Atmospheric Pressure	English System (BG)	Metric System (SI)
Standard temperature, θ	39.2°F	4°C
Standard sea-level atmospheric pressure, p_{atm}	14.696 lb/in^2 abs (psia)	101.325 kPa abs
Standard density for water, ρ (absolute property)	1.940 $slug/ft^3$	1000 kg/m^3

As noted above, pure/fresh water is considered to be the reference fluid for liquids. As such, the densities of water at the standard temperature and the standard sea-level atmospheric pressure are given for the English and metric systems, respectively, in Table 1.5.

The specific volume is defined as the volume, V occupied by a unit mass, M of fluid and is designated by v as follows:

$$v = \frac{V}{M} = \frac{1}{\rho} \tag{1.36}$$

where the specific volume, v is the reciprocal of the density, ρ. Therefore, the dimensions and units for the specific volume, v in the English system are given as follows:

$$v = \frac{V}{M} = \frac{[L^3]}{[FL^{-1}T^2]} = \frac{[ft^3]}{[lb-sec^2/ft]} = [ft^3/slug] \tag{1.37}$$

The dimensions and units for the specific volume, v in the metric system are given as follows:

$$v = \frac{V}{M} = \frac{[L^3]}{[M]} = [m^3/kg] \tag{1.38}$$

Finally, it may be noted that although the specific volume, v is not typically applied in fluid mechanics, it is commonly applied in the field of thermodynamics.

EXAMPLE PROBLEM 1.16

Refer to Example Problem 1.15 above: (a) Determine the specific volume of kerosene at room temperature (68°F). (b) Determine the specific volume of methane at room temperature (20°C).

Mathcad Solution

(a) The specific volume of kerosene at room temperature (68°F) is determined by applying Equation 1.36 as follows:

$$slug := 1\,lb\,\frac{sec^2}{ft} \qquad \rho := 1.57\,\frac{slug}{ft^3} \qquad v := \frac{1}{\rho} = 0.637\,\frac{ft^3}{slug}$$

(b) The specific volume of methane at room temperature (20°C) is determined by applying Equation 1.36 as follows:

$$\rho := 0.668 \, \frac{kg}{m^3} \qquad v := \frac{1}{\rho} = 1.497 \, \frac{m^3}{kg}$$

1.10.2 Specific Gravity

The mass density of a fluid, ρ is sometimes expressed as a dimensionless ratio called the specific gravity, which is defined as the ratio of the density of the fluid, ρ_f to the density of a reference fluid, $\rho_{ref\,fluid}$ at a specified temperature and pressure and is designated by s as follows:

$$s_f = \frac{\rho_f}{\rho_{ref\,fluid}} \tag{1.39}$$

For liquids, the reference fluid is water; therefore, the specific gravity for a liquid, s_{liquid} is given as follows:

$$s_{liquid} = \frac{\rho_{liquid}}{\rho_{water@standardtemp}} \tag{1.40}$$

As presented above, in the English system, the reference fluid for liquids is water at a standard temperature of 39.2°F and a standard sea-level atmospheric pressure of 14.696 psia, for which the density is $1.94 \, slug/ft^3$. Furthermore, in the metric system, the reference fluid for liquids is water at a standard temperature of 4°C and a standard sea-level atmospheric pressure of 101.325 kPa abs, for which the density is $1000 \, kg/m^3$. It is important to note that while the standard temperature of 39.2°F or 4°C is assumed for the reference fluid, water, the fluid in question (and thus its corresponding s_f and ρ_f), may be at any given temperature. The reference fluid for gases is either hydrogen or air at a specified temperature and pressure. Furthermore, because, unlike for water, there is no general agreement on a standard temperature and standard pressure for hydrogen or air, these are specified for a given problem.

It is important to note that because the density, ρ of a fluid varies with temperature and pressure, so will the corresponding specific gravity, s. Furthermore, similar to the density, ρ, the specific gravity, s of a fluid is considered to be an absolute fluid property because it depends on mass, M, which is independent of location (latitude and elevation above mean sea level). Furthermore, the specific gravity, s is a measure of the relative "heaviness" of the fluid with respect to water (or hydrogen or air in the case of gasses). As such, the specific gravity, s is primarily useful in identifying a particular fluid and may also be used to relate the property of one fluid to another. Although the specific gravity, s for some common liquids will vary with temperature, for practical applications, their specific gravity, s at standard sea-level atmospheric pressure at room temperature (68°F or 20°C), in general, are given in Table A.4 in Appendix A.

EXAMPLE PROBLEM 1.17

Using Table A.4 in Appendix A to read the fluid-specific gravity or fluid density and Equation 1.40: (a) Determine the specific gravity of water at room temperature (68°F). (b) Determine the specific gravity of crude oil at room temperature (68°F). (c) Determine the specific gravity of mercury at room temperature (68°F).

Mathcad Solution

(a) The specific gravity of water at room temperature (68°F) is determined from Table A.4 in Appendix A as follows:

$$S_{\text{water at 68 deg F}} = 0.998$$

Alternatively, it may be computed by applying Equation 1.40 as follows:

$$\text{slug} := 1 \, \text{lb} \frac{\sec^2}{\text{ft}} \qquad \rho_{w68F} := 1.936 \frac{\text{slug}}{\text{ft}^3} \qquad \rho_{w39.2F} := 1.94 \frac{\text{slug}}{\text{ft}^3}$$

$$S_{w68F} := \frac{\rho_{w68F}}{\rho_{w39.2F}} = 0.998$$

(b) The specific gravity of crude oil at room temperature (68°F) is determined from Table A.4 in Appendix A as follows:

$$S_{\text{crudeoil at 68 deg F}} = 0.856$$

Alternatively, it may be computed by applying Equation 1.40 as follows:

$$\rho_{co68F} := 1.66 \frac{\text{slug}}{\text{ft}^3} \qquad S_{co68F} := \frac{\rho_{co68F}}{\rho_{w39.2F}} = 0.856$$

(c) The specific gravity of mercury at room temperature (68°F) is determined from Table A.4 in Appendix A as follows:

$$S_{\text{Hg at 68 deg F}} = 13.557$$

Alternatively, it may be computed by applying Equation 1.40 as follows:

$$\rho_{Hg68F} := 26.3 \frac{\text{slug}}{\text{ft}^3} \qquad S_{Hg68F} := \frac{\rho_{Hg68F}}{\rho_{w39.2F}} = 13.557$$

EXAMPLE PROBLEM 1.18

Using Table A.4 in Appendix A to read the fluid specific gravity or fluid density and Equation 1.40: (a) Determine the specific gravity of water at room temperature (20°C). (b) Determine the specific gravity of crude oil at room temperature (20°C). (c) Determine the specific gravity of mercury at room temperature (20°C). (d) Refer to Example Problem 1.17 and compare the results.

Mathcad Solution

(a) The specific gravity of water at room temperature (20°C) is determined from Table A.4 in Appendix A as follows:

$$S_{\text{water at 20 deg C}} = 0.998$$

Alternatively, it may be computed by applying Equation 1.40 as follows:

$$\rho_{w20C} := 998 \frac{kg}{m^3} \qquad \rho_{w4C} := 1000 \frac{kg}{m^3}$$

$$s_{w20C} := \frac{\rho_{w20C}}{\rho_{w4C}} = 0.998$$

(b) The specific gravity of crude oil at room temperature (20°C) is determined from Table A.4 in Appendix A as follows:

$$S_{\text{crude oil at 20 deg C}} = 0.856$$

Alternatively, it may be computed by applying Equation 1.40 as follows:

$$\rho_{co20C} := 856 \frac{kg}{m^3} \qquad s_{co20C} := \frac{\rho_{co20C}}{\rho_{w4C}} = 0.856$$

(c) The specific gravity of mercury at room temperature (20°C) is determined from Table A.4 in Appendix A as follows:

$$S_{\text{Hg at 20 deg C}} = 13.55$$

Alternatively, it may be computed by applying Equation 1.40 as follows:

$$\rho_{Hg20C} := 13550 \frac{kg}{m^3} \qquad s_{Hg20C} := \frac{\rho_{Hg20C}}{\rho_{w4C}} = 13.55$$

(d) Thus, because the specific gravity is dimensionless, its magnitude for a given fluid will be the same regardless of whether the BG or SI system of units is assumed.

It is interesting to note Table A.5 in Appendix A, which presents physical properties of some common gases at standard sea-level atmospheric pressure, does not include the specific gravity, s for a given gas, as there is no general agreement on a standard temperature and standard pressure for the reference fluid for gas (hydrogen or air).

1.10.3 Specific Weight

The specific weight is defined as weight, W per unit volume, V and is designated by γ as follows:

$$\gamma = \frac{W}{V} \tag{1.41}$$

One may note that unlike the density, ρ, the specific weight, γ is not considered to be an absolute fluid property (rather it is a gravitational fluid property) because it depends on weight, $W = Mg$ (and thus characterizes the weight of the fluid system), which is dependent on location (latitude and elevation above mean sea level), as the acceleration due to gravity, g varies with location. Furthermore, the specific weight, γ is a measure of the "heaviness" of the fluid and is a predominant fluid property in fluid statics. The dimensions and units for the specific weight, γ in the English system are given as follows:

$$\gamma = \frac{W}{V} = \frac{[F]}{[L^3]} = [lb/ft^3] \tag{1.42}$$

The dimensions and units for the specific weight, γ in the metric system are given as follows:

$$\gamma = \frac{W}{V} = \frac{[MLT^{-2}]}{[L^3]} = \frac{[kg - m/sec^2]}{[m^3]} = [N/m^3] \tag{1.43}$$

Because the weight, W (gravitational force, F_G) is related to the mass, M by Newton's second law of motion, $W = Mg$, in which g is the acceleration due to the local force of gravity, the mass density, ρ (which characterizes the mass, M) and the specific weight, γ (which characterizes the weight, W) of a unit volume of fluid are related by a similar equation as follows:

$$\gamma = \frac{W}{V} = \frac{Mg}{V} = \rho g \tag{1.44}$$

The specific weight, γ is defined as a function of the mass density, ρ; therefore, similarly, in general, the specific weight, γ of a fluid decreases with an increase in temperature, while it increases with an increase in pressure. Because the specific weight, γ of a gas is strongly dependent on both temperature and pressure, it is a variable, and gas flow is thus modeled as compressible flow (see Section 1.10.7). However, since the specific weight of a liquid is more affected by the temperature than by the pressure, for practical changes in pressure, the specific weight of a liquid is essentially constant and thus is accurately approximated as an incompressible flow. The specific weight, γ for water and air at standard sea-level atmospheric pressure and standard acceleration due to gravity are presented as a function of temperature in Table A.2 in Appendix A and Table A.3 in Appendix A, respectively. The specific weight, γ for some other common liquids and some other common gases may be computed by applying Equation 1.44. Furthermore, although their specific weight, γ will vary with temperature, for practical applications, the corresponding mass densities, ρ (used in the definition of $\gamma = \rho g$) at standard sea-level atmospheric pressure at room temperature ($68°F$ or $20°C$), in general, are given in Table A.4 in Appendix A and Table A.5 in Appendix A, respectively.

Similar to the density, ρ for a given temperature, a change in specific weight, $\gamma = \rho g$ of a given fluid mass that is due to a change in pressure is a measure of the compressibility of the fluid (see Section 1.10.7). Although all fluids have a certain degree of compressibility, some fluids (liquids and gases) are assumed to be incompressible when the change in density due to a change in pressure is negligible. Therefore, for compressible fluids/flow, the fluid density is assumed to vary with pressure, while for incompressible fluids/flow, the fluid density is assumed to be a constant with respect to pressure. It is important to note that, although liquids can be compressed (resulting in small changes in density) at very high pressures, they are typically assumed to be incompressible for most practical engineering applications. Furthermore, it is important to note that, although gases are easily compressed (resulting in significant changes in density) at very high pressures, they are typically assumed to be incompressible (as indicated by the Mach number, M [discussed in Section 1.10.7] when the pressure variation is small in comparison to the absolute pressure [sum of atmospheric and gage pressure]).

EXAMPLE PROBLEM 1.19

Using Table A.2 in Appendix A and Table A.3 in Appendix A, respectively, to read the fluid specific weight: (a) Determine the volume of 13 lb of water at 100°F. (b) Determine the weight of 60 m^3 of air at 200°C.

Mathcad Solution

(a) The volume of 13 lb of water at 100°F is determined by applying Equation 1.41 as follows:

$$W := 13\,lb \qquad \gamma := 62.00\,\frac{lb}{ft^3} \qquad V := \frac{W}{\gamma} = 0.21\,ft^3$$

(b) The weight of 60 m^3 of air at 200°C is determined by applying Equation 1.41 as follows:

$$N := kg\,\frac{m}{sec^2} \qquad V := 60\,m^3 \qquad \gamma := 7.33\,\frac{N}{m^3} \qquad W := \gamma\,V = 439.8\,N$$

EXAMPLE PROBLEM 1.20

Using Table A.4 in Appendix A and Table A.5 in Appendix A, respectively, to read the fluid density: (a) Determine the volume of 56 lb of mercury at room temperature (68°F). (b) Determine the weight of 145 m^3 of water vapor at room temperature (20°C).

Mathcad Solution

(a) The specific weight of mercury at room temperature (68°F) is determined by applying Equation 1.44 as follows:

$$slug := 1\,lb\,\frac{sec^2}{ft} \qquad \rho := 26.3\,\frac{slug}{ft^3} \qquad g := 32.2\,\frac{ft}{sec^2} \qquad \gamma := \rho\,g = 846.86\,\frac{lb}{ft^3}$$

The volume of 56 lb of mercury at room temperature (68°F) is determined by applying Equation 1.41 as follows:

$$W := 56\,lb \qquad V := \frac{W}{\gamma} = 0.066\,ft^3$$

(b) The specific weight of water vapor at room temperature (20°C) is determined by applying Equation 1.44 as follows:

$$N := kg \frac{m}{sec^2} \qquad \rho := 0.747 \frac{kg}{m^3} \qquad g := 9.81 \frac{m}{sec^2} \qquad \gamma := \rho\, g = 7.328 \frac{N}{m^3}$$

The weight of 145 m³ of water vapor at room temperature (20°C) is determined by applying Equation 1.41 as follows:

$$V := 145\, m^3 \qquad W := \gamma\, V = 1.063 \times 10^3\, N$$

As discussed above, pure/fresh water is considered to be the reference fluid for liquids. Given the definition of the specific weight, $\gamma = \rho g$, the specific weight of water at the standard temperature, the standard sea-level atmospheric pressure, and the standard gravity are given for the English and metric systems, respectively, in Table 1.6 below.

And, finally, given the specific weight, $\gamma = \rho g$, the specific gravity, s may also be defined as the ratio of the specific weight of the fluid, γ_f to the specific weight of a reference fluid, $\gamma_{ref\ fluid}$ at a specified temperature and pressure as follows:

$$s_f = \frac{\gamma_f}{\gamma_{ref\ fluid}} = \frac{\rho_f g}{\rho_{ref\ fluid} g} = \frac{\rho_f}{\rho_{ref\ fluid}} \tag{1.45}$$

where (as stated above), for liquids, the reference fluid is water; therefore, the specific gravity, s_{liquid} for a liquid is given as follows:

$$s_{liquid} = \frac{\gamma_{liquid}}{\gamma_{water@\ standard\ temp}} = \frac{\rho_{liquid}}{\rho_{water@\ standard\ temp}} \tag{1.46}$$

As presented below, in the English system, the reference fluid for liquids is water at a standard temperature of 39.2°F, a standard sea-level atmospheric pressure of 14.696 *psia*, and a standard gravity of 32.174 *ft/sec²*, for which the density is 1.94 *slug/ft³*, and the specific weight is 62.417 *lb/ft³*. Furthermore, in the metric system, the reference fluid

TABLE 1.6

Standard Specific Weight for Water at Standard Temperature, Standard Sea-Level Atmospheric Pressure, and Standard Gravity

Standard Specific Weight, $\gamma = \rho g$ for Water (Reference Fluid) at Standard Temperature, Standard Sea-Level Atmospheric Pressure, and Standard Gravity	English System (BG)	Metric System (SI)
Standard temperature, θ	39.2°F	4°C
Standard sea-level atmospheric pressure, p_{atm}	14.696 *lb/in² abs* (psia)	101.325 *kPa abs*
Standard density for water, ρ (absolute property)	1.940 *slug/ft³*	1000 *kg/m³*
Standard gravity at sea level and 45° latitude, g	32.2 *ft/sec²*	9.81 *m/sec²*
Standard specific weight for water, γ	62.4 *lb/ft³*	9810 *N/m³*

for liquids is water at a standard temperature of 4°C, a standard sea-level atmospheric pressure of 101.325 *kPa abs*, and a standard gravity of 9.81 m/sec^2, for which the density is 1000 kg/m^3 and the specific weight is 9810 N/m^3. Additionally, since the mass density, ρ; the specific gravity, s; and the specific weight, γ are all interrelated, knowledge of one of these three fluid properties allows the calculation of the remaining two fluid properties as follows:

$$s_f = \frac{\gamma_f}{\gamma_{ref\ fluid}} = \frac{\rho_f g}{\rho_{ref\ fluid} g} \tag{1.47}$$

and, finally, since the acceleration due to gravity, g is a quantity that varies from one part of the earth to another (latitude) and with elevation above mean sea level (altitude), the mass density, ρ and the specific gravity, s are the more basic fluid properties; they are absolute since they depends on mass, which is independent of location, while the specific weight, $\gamma = \rho g$ depends on location.

EXAMPLE PROBLEM 1.21

A 90-gallon volume of saltwater weighs about 768 lb. (a) Determine the specific weight of the saltwater. (b) Determine the density of the saltwater. (c) Determine the specific gravity of the saltwater.

Mathcad Solution

(a) The specific weight of the saltwater is determined by applying Equation 1.41 as follows:

$$1\ \text{gal} = 0.134\ \text{ft}^3 \quad V := 90\ \text{gal} = 12.031\ \text{ft}^3 \quad W := 768\ \text{lb} \quad \gamma := \frac{W}{V} = 63.834\ \frac{\text{lb}}{\text{ft}^3}$$

(b) The density of the saltwater is determined by applying Equation 1.44 as follows:

$$\text{slug} := 1\ \text{lb}\frac{\text{sec}^2}{\text{ft}} \quad g := 32.2\frac{\text{ft}}{\text{sec}^2} \quad \rho := \frac{\gamma}{g} = 1.982\frac{\text{slug}}{\text{ft}^3}$$

(c) The specific gravity of the saltwater is determined by applying Equation 1.46 as follows:

$$\rho_{w39.2F} := 1.94\frac{\text{slug}}{\text{ft}^3} \qquad \gamma_{w39.2F} := \rho_{w39.2F}\ g = 62.468\frac{\text{lb}}{\text{ft}^3}$$

$$s := \frac{\gamma}{\gamma_{w39.2F}} = 1.022 \qquad s := \frac{\rho}{\rho_{w39.2F}} = 1.022$$

EXAMPLE PROBLEM 1.22

The specific gravity of a liquid is 1.6. (a) Determine the density of the liquid in SI units. (b) Determine the specific weight of the liquid in SI units.

Mathcad Solution

(a) The density of the liquid is determined by applying Equation 1.40 (or Equation 1.46) as follows:

$$s_{liquid} := 1.6 \qquad \rho_{w4C} := 1000\frac{kg}{m^3} \qquad \rho_{liquid} := s_{liquid}\rho_{w4C} = 1.6 \times 10^3 \frac{kg}{m^3}$$

(b) The specific weight of the liquid is determined by applying Equation 1.46 as follows:

$$N := kg\frac{m}{sec^2} \qquad g := 9.81\frac{m}{sec^2} \qquad \gamma_{w4C} := \rho_{w4C}g = 9.81 \times 10^3 \frac{N}{m^3}$$

$$\gamma_{liquid} := s_{liquid}\gamma_{w4C} = 1.57 \times 10^4 \frac{N}{m^3}$$

Alternatively, the specific weight of the liquid may be determined by applying Equation 1.44 as follows:

$$\gamma_{liquid} := \rho_{liquid}g = 1.57 \times 10^4 \frac{N}{m^3}$$

1.10.4 Viscosity

While the fluid viscosity, μ plays a principal role in the flow (fluid dynamics) of real (viscous) fluids, the viscosity, μ does not play an important role in (real) fluid statics, and, furthermore, the fluid viscosity, μ is assumed be equal to zero in the flow of ideal (inviscid) fluids. An ideal fluid is defined as a fluid in which there is no friction; thus, it is inviscid (its viscosity is zero). In (real) fluid statics, there are no shear forces, only pressure forces. However, in (real) fluid flow, there are shear forces, which create fluid friction, which in turn give rise to the fluid property viscosity. The viscosity is designated by μ, which is also referred to as the absolute or dynamic viscosity, and is probably the single most important fluid property in fluid dynamics. The dimensions and units for the absolute or dynamic viscosity, μ are based on its physical definition, which is given by Newton's law of viscosity, $\tau = \mu(dv/dy)$ (see Section 1.10.4.3). In addition to the definition of the absolute or dynamic viscosity, μ, the kinematic viscosity, $v = \mu/\rho$ is also defined and presented below. It is important to note that while Newton's law of viscosity provides a theoretical expression for the shear stress, τ (or friction/viscous force, $F_V = \tau A$), for laminar flow, one must resort to dimensional analysis in order to derive an empirical expression for the shear stress, τ (or friction/viscous force, $F_V = \tau A$, where $\tau_w = C_D \rho v^2$ [discussed below]) for turbulent flow. Furthermore, it is important to note that while the theoretical modeling of the friction/viscous force, F_V by Newton's law of viscosity results in a theoretical (parabolic) velocity distribution for laminar flow, the empirical modeling of the friction/friction force, F_V for turbulent flow results in an empirical velocity distribution (see Chapter 6).

Given that the fluid is in motion, the viscosity, μ of a fluid is a measure of the internal resistance ("fluidity" or "internal stickiness") to shear or angular deformation (caused by tangential or shearing forces) that is displayed by the fluid as a layer of fluid is moved relative to another layer of fluid or a solid boundary. It is important to note that the tangential or shearing forces, which oppose the motion of one fluid particle past another, create fluid friction,

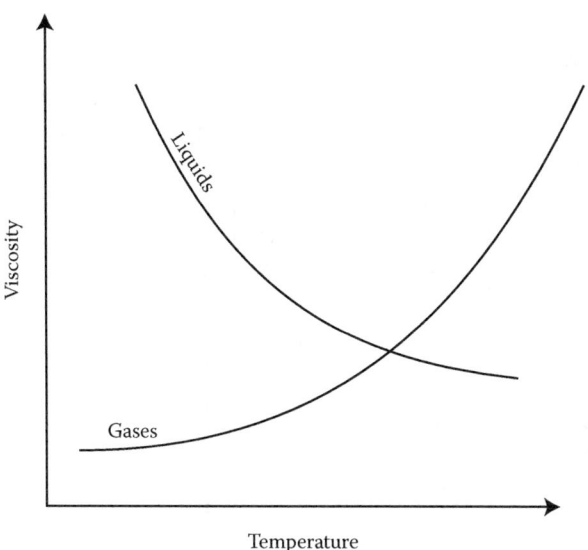

FIGURE 1.1
Variation of fluid (gases and liquids) viscosity (absolute, μ and kinematic, v) with temperature.

which results in flow resistance. The flow resistance, which is the shear stress, friction force, or drag force originating in the fluid viscosity, μ is modeled by the friction/viscous force, F_V in Newton's second law of motion, $\sum_s F = Ma_s$ (integral form of the momentum equation). There are two factors that produce viscosity, μ, and they are molecular cohesion and the rate of transfer of molecular momentum. Although both temperature and pressure govern these two factors, temperature is the most dominant and the effect of pressure is usually ignored. Furthermore, viscosity is very sensitive to temperature; it is much more sensitive to temperature than the mass density, ρ is. The viscosity for a liquid decreases with an increase in temperature, while the viscosity for a gas increases with an increase in temperature as illustrated in Figure 1.1. This is because for liquids, the cohesive forces predominate over the inertial forces caused by changes in momentum, while for gases, the reverse is true, due to the increased molecular activity in a gas (compared to a liquid). The absolute or dynamic viscosity, μ and the kinematic viscosity, v for water and air at standard sea-level atmospheric pressure are presented as a function of temperature in Table A.2 in Appendix A and Table A.3 in Appendix A, respectively. And, although the absolute or dynamic viscosity, μ (and the kinematic viscosity, v) for some common liquids and some common gases will vary with temperature, for practical applications, their absolute or dynamic viscosity, μ at standard sea-level atmospheric pressure at room temperature (68°F or 20°C), in general, are given in Table A.4 in Appendix A and Table A.5 in Appendix A, respectively.

EXAMPLE PROBLEM 1.23

Using Table A.2 in Appendix A and Table A.3 in Appendix A, respectively: (a) Graphically illustrate the variation of both the density and the absolute viscosity of water with temperature, assuming the BG system of units. (b) Graphically illustrate the variation of both the density and the absolute viscosity of air with temperature, assuming the SI system of units. (c) Compare the results.

Mathcad Solution

(a) A graphical illustration of the variation of both the density and the absolute viscosity of water with temperature, assuming the BG system of units, is given as follows:

$$
T := \begin{pmatrix} 32 \\ 40 \\ 50 \\ 60 \\ 70 \\ 80 \\ 90 \\ 100 \\ 110 \\ 120 \\ 130 \\ 140 \\ 150 \\ 160 \\ 170 \\ 180 \\ 190 \\ 200 \\ 212 \end{pmatrix} \qquad \rho := \begin{pmatrix} 1.940 \\ 1.940 \\ 1.940 \\ 1.938 \\ 1.936 \\ 1.934 \\ 1.931 \\ 1.927 \\ 1.923 \\ 1.918 \\ 1.913 \\ 1.908 \\ 1.902 \\ 1.896 \\ 1.890 \\ 1.883 \\ 1.876 \\ 1.868 \\ 1.860 \end{pmatrix} \qquad \mu := \begin{pmatrix} 37.46 \\ 32.29 \\ 27.35 \\ 23.59 \\ 20.50 \\ 17.99 \\ 15.95 \\ 14.24 \\ 12.84 \\ 11.68 \\ 10.69 \\ 9.81 \\ 9.05 \\ 8.38 \\ 7.80 \\ 7.26 \\ 6.78 \\ 6.37 \\ 5.93 \end{pmatrix}
$$

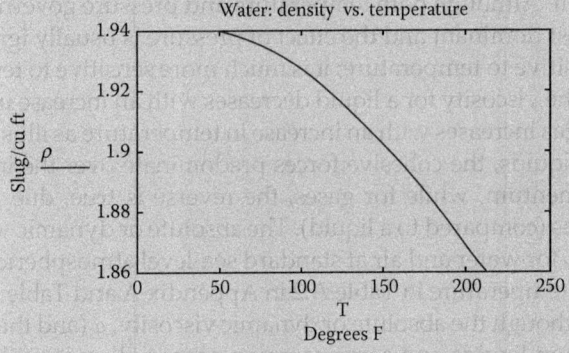

Water: density vs. temperature

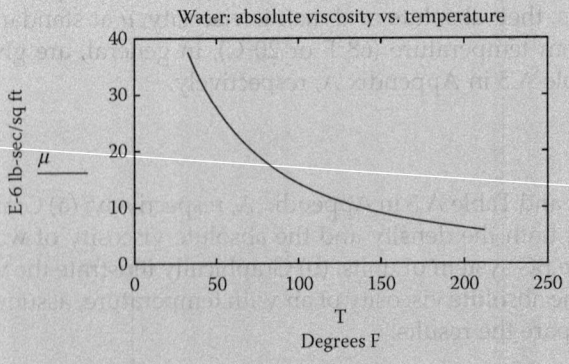

Water: absolute viscosity vs. temperature

(b) A graphical illustration of the variation of both the density and the absolute viscosity of air with temperature, assuming the SI system of units, is given as follows:

$$
T := \begin{pmatrix} -40 \\ -20 \\ 0 \\ 10 \\ 20 \\ 30 \\ 40 \\ 60 \\ 80 \\ 100 \\ 200 \end{pmatrix} \quad \rho := \begin{pmatrix} 1.515 \\ 1.395 \\ 1.293 \\ 1.248 \\ 1.205 \\ 1.165 \\ 1.128 \\ 1.060 \\ 1.000 \\ 0.946 \\ 0.747 \end{pmatrix} \quad \mu := \begin{pmatrix} 14.9 \\ 16.1 \\ 17.1 \\ 17.6 \\ 18.1 \\ 18.6 \\ 19.0 \\ 20.0 \\ 20.9 \\ 21.8 \\ 25.8 \end{pmatrix}
$$

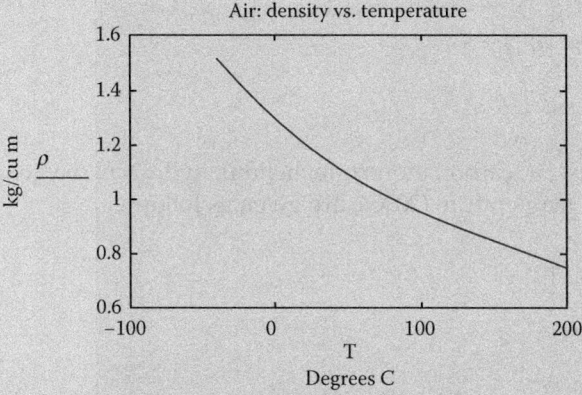

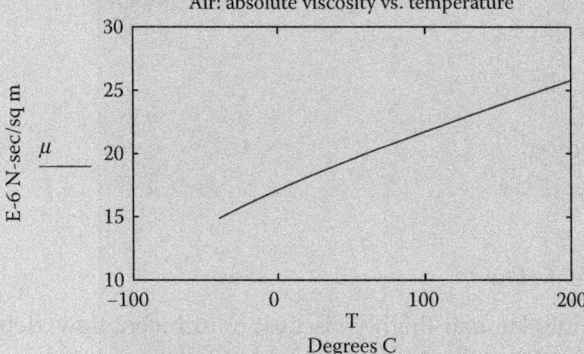

(c) Thus, both the density and the absolute viscosity for water (a liquid) decrease with an increase in temperature, as illustrated in (a). However, while the density of air (a gas) decreases with an increase in temperature, the absolute viscosity of air actually increases with an increase in temperature, as illustrated in (b).

EXAMPLE PROBLEM 1.24

Using Table A.4 in Appendix A and Table A.5 in Appendix A, respectively: (a) Read the absolute viscosity of benzene, crude oil, glycerin, mercury, SAE 30 oil, and fresh water, all at room temperature (68°F). (b) Read the absolute viscosity of air, carbon monoxide, helium, hydrogen, oxygen, and water vapor, all at room temperature (20°C).

Solution

(a) The absolute viscosities of benzene, crude oil, glycerin, mercury, SAE 30 oil, and fresh water, all at room temperature (68°F), are given as follows:

$$\mu_{benzene} = 14.37 \times 10^{-6} \, lb - sec/ft^2$$

$$\mu_{crudeoil} = 150 \times 10^{-6} \, lb - sec/ft^2$$

$$\mu_{glycerin} = 31{,}200 \times 10^{-6} \, lb - sec/ft^2$$

$$\mu_{Hg} = 33 \times 10^{-6} \, lb - sec/ft^2$$

$$\mu_{SAE30oil} = 9{,}200 \times 10^{-6} \, lb - sec/ft^2$$

$$\mu_{freshwater} = 21.0 \times 10^{-6} \, lb - sec/ft^2$$

(b) The absolute viscosities of air, carbon monoxide, helium, hydrogen, oxygen, and water vapor, all at room temperature (20°C), are given as follows:

$$\mu_{air} = 18.0 \times 10^{-6} \, N - sec/m^2$$

$$\mu_{CO} = 18.2 \times 10^{-6} \, N - sec/m^2$$

$$\mu_{He} = 19.7 \times 10^{-6} \, N - sec/m^2$$

$$\mu_{H_2} = 9.0 \times 10^{-6} \, N - sec/m^2$$

$$\mu_{O_2} = 20.0 \times 10^{-6} \, N - sec/m^2$$

$$\mu_{H_2O} = 10.1 \times 10^{-6} \, N - sec/m^2$$

1.10.4.1 Laminar versus Turbulent Flow

A real flow may be subdivided into laminar (highly viscous) or turbulent flow, depending upon the value of the Reynolds number, R, which is defined as follows:

$$R = \frac{F_I}{F_V} = \frac{\rho v^2 L^2}{\mu v L} = \frac{\rho v L}{\mu} \tag{1.48}$$

which is dimensionless, and where the Reynolds number, R expresses the relative importance of the fluid's inertia force, F_I compared to its viscosity force, F_V, and where v is the velocity of flow and L is the pipe diameter or a geometric/length flow variable. A flow is called laminar when $R < 2000$, transitional when $2000 < R < 4000$, and turbulent when $R > 4000$. As such, for laminar flow (small to moderate values for R), the viscous force, F_V is large enough to suppress the fluctuations in flow due to the significance of the inertia force, F_I; thus, the viscosity, μ plays a important role in the definition of the shear stress, τ. However, for turbulent flow (large values for R), the inertia force, F_I is large compared to the viscous force, F_V; thus, the fluid density, ρ plays a more important role (than the fluid viscosity, μ) in the definition of the shear stress, τ. Laminar flow is characterized by a smooth deterministic motion, while turbulent flow is characterized by random or chaotic motion. For instance, the flow of highly viscous fluids such as oils at a low velocity is typically laminar, while the flow of low viscosity fluids such as air at high velocities is typically turbulent. Additionally, a flow that alternates between laminar and turbulent is called transitional.

EXAMPLE PROBLEM 1.25

Water at 32°F flows at $4\,ft/sec$ in a pipe with a diameter of 3 ft. (a) Determine the Reynolds number assuming the BG system of units. (b) Determine the Reynolds number assuming the SI system of units. (c) Compare the results.

Mathcad Solution

(a) The fluid properties for water are read from Table A.2 in Appendix A. The Reynolds number, assuming the BG system of units, is determined by applying Equation 1.48 as follows:

$$slug := 1\,lb\,\frac{sec^2}{ft} \qquad \rho_{w32F} := 1.940\,\frac{slug}{ft^3} \qquad \mu_{w32F} := 37.46 \times 10^{-6}\,lb\,\frac{sec}{ft^2}$$

$$v := 4\,\frac{ft}{sec} \qquad D := 3\,ft \qquad R := \frac{\rho_{w32F}\,v\,D}{\mu_{w32F}} = 6.215 \times 10^5$$

(b) The fluid properties for water are read from Table A.2 in Appendix A. The Reynolds number, assuming the SI system of units, is determined by applying Equation 1.48 as follows:

$$N := kg\,\frac{m}{sec^2} \qquad \rho_{w0C} := 999.8\,\frac{kg}{m^3} \qquad \mu_{w0C} := 0.001781\,N\,\frac{sec}{m^2}$$

$$1\,ft = 0.305\,m \qquad v := 4\,\frac{ft}{sec}\,\frac{0.305\,m}{1\,ft} = 1.22\,\frac{m}{s} \qquad D := 3\,ft\,\frac{0.305\,m}{1\,ft} = 0.915\,m$$

$$R := \frac{\rho_{w0C}\,v\,D}{\mu_{w0C}} = 6.267 \times 10^5$$

(c) First, it is important to note that because the Reynolds number is dimensionless, its magnitude for a given flow situation will be the same regardless of whether the BG or SI system of units is assumed. As such, the difference between the results of (a) assuming; BG units ($R = 6.215 \times 10^5$) and (b) assuming; SI units ($R = 6.267 \times 10^5$) is only due to the round-off error in the conversion of units from the BG to the SI system of units. Thus, if the magnitudes of the converted variables were more precise (using more digits of precision), then one would expect the Rs to be identical in (a) and (b).

1.10.4.2 Shear Stress Distribution/Profile for Laminar or Turbulent Pipe Flow

As explained in detail in Chapter 6, application of Newton's second law of motion (integral form of the momentum equation) for a laminar or turbulent pipe flow yields a linear relationship between the shear stress, τ and the radial distance, r from the center of the pipe with a radius, r_o, which also describes a theoretical relationship between the pressure drop, Δp and the shear stress, τ, for both laminar and turbulent flow as follows:

$$\tau = \frac{r}{2}\frac{\Delta p}{L} \tag{1.49}$$

where L is the pipe length. Figure 1.2 illustrates laminar pipe flow; while Figure 1.2a illustrates the parabolic velocity distribution, Figure 1.2b illustrates that the shear stress, τ increases linearly with the radial distance, r to a maximum shear stress, τ_{max} at the pipe wall/boundary (or the wall shear stress, τ_w) where $r = r_o$, and is given as follows:

$$\tau_{max} = \tau_w = \frac{r_o}{2}\frac{\Delta p}{L} \tag{1.50}$$

and the shear stress, τ at the centerline of the pipe $r = 0$ is equal to zero. The linear relationship between the shear stress, τ and the radial distance, r may be illustrated by isolating the pressure drop, Δp over the length, L term from the expression for the variation of the shear stress, τ with the radial distance, r from the center of the pipe (Equation 1.49), and the expression for the maximum shear stress, τ_{max} (Equation 1.50), and equating as follows:

$$\frac{\Delta p}{L} = \frac{2\tau}{r} = \frac{2\tau_{max}}{r_o} \tag{1.51}$$

Thus:

$$\frac{\tau}{r} = \frac{\tau_{max}}{r_o} \tag{1.52}$$

$$\tau = \tau_{max}\frac{r}{r_o} \tag{1.53}$$

where the shear stress is a maximum at the pipe wall ($\tau_w = \tau_{max}$) and there is no shear stress at the centerline of the pipe.

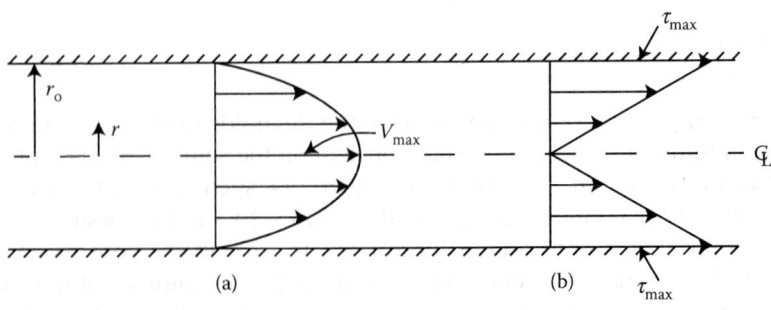

FIGURE 1.2
Laminar pipe flow. (a) Parabolic velocity distribution. (b) Linear shear stress distribution.

If the flow is laminar, then the friction/viscous force, $F_V = \tau A$ in Newton's second law of motion, $\sum_s F = Ma_s$ (integral form of the momentum equation) can be modeled theoretically by Newton's law of viscosity, $\tau = \mu(dv/dy)$ (discussed below). As such, a theoretical modeling of the friction force by Newton's law of viscosity for laminar flow results in a theoretical (parabolic) velocity distribution. However, if the flow is turbulent, then the friction/viscous force, F_V in Newton's second law of motion, $\sum_s F = Ma_s$ (integral form of the momentum equation) cannot be modeled theoretically. Rather, one must resort to dimensional analysis in order to derive an empirical relationship for the friction/viscous force, $F_V = \tau A$, where $\tau_w = C_D \rho v^2$ (discussed below). As such, an empirical modeling of the friction force for turbulent flow results in an empirical velocity distribution.

1.10.4.3 Newton's Law of Viscosity for Laminar Flow

The derivation of Newton's law of viscosity for laminar flow may be illustrated by an experimental procedure, using what is known as Couette flow, as follows. Figure 1.3 illustrates that the space between two very long parallel plates separated by distance, D is filled with a fluid of constant viscosity, μ. The lower plate is held stationary. However, a continuous tangential (shear) force, F_V is applied to the upper plate (with a cross-sectional area, A), thus moving the plate with a constant velocity, v_{plate} and continuously deforming the fluid at a rate of dv/dy. The fluid layers continuously deform due to the effect of shear stress. The shear stress of the fluid that is in contact with the moving plate is computed as follows:

$$\tau = \frac{F_V}{A} \tag{1.54}$$

Due to the "no-slip" condition, the fluid sticks to the solid boundaries, where the velocity of the fluid at the stationary plate is zero, increasing linearly to a maximum velocity where the fluid is in contact with the moving plate, and thus moves with the plate velocity, v_{plate}, as illustrated in Figure 1.3a. Thus, the velocity profile is given as follows:

$$v(y) = \frac{v_{plate}}{D} y \tag{1.55}$$

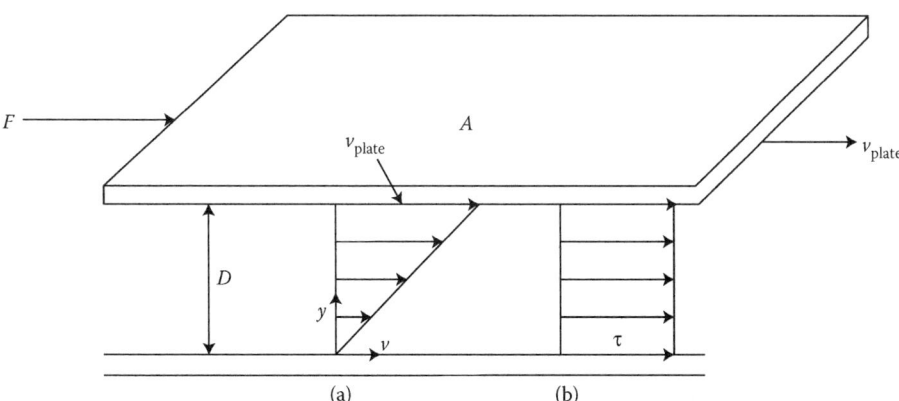

(a) (b)

FIGURE 1.3
Laminar Couette flow between two very long parallel plates. (a) Linear velocity distribution. (b) Constant shear stress distribution.

where the velocity gradient is given as follows:

$$\frac{dv}{dy} = \frac{v_{plate}}{D} \tag{1.56}$$

If the tangential (shear) force, F_V is increased, then the shear stress of the fluid that is in contact with the moving plate, $\tau = (F_V/A)$ will also increase; thus, the constant velocity of the plate, v_{plate} will also increase. Thus,

$$\tau = \frac{F_V}{A} \propto \frac{dv}{dy} \tag{1.57}$$

where the fluid is continuously deforming at a rate of dv/dy, and the constant of proportionality is defined as the absolute or dynamic viscosity, μ of the fluid; thus, Newton's law of viscosity is stated as follows:

$$\tau = \mu \frac{dv}{dy} \tag{1.58}$$

Furthermore, substituting the velocity gradient given by Equation 1.56 into Newton's law of viscosity, Equation 1.58, yields the shear stress distribution across the cross section and is determined as follows:

$$\tau = \mu \frac{dv}{dy} = \mu \frac{v_{plate}}{D} = constant \tag{1.59}$$

Thus, the shear stress distribution/profile is constant, as illustrated in Figure 1.3b.

Therefore, in the case of laminar flow, the friction/viscous force, F_V in Newton's second law of motion, $\sum_s F = Ma_s$ (integral form of the momentum equation) can be modeled theoretically by Newton's law of viscosity Equation 1.58), which states that when a shear stress, τ is applied to any fluid, the fluid will deform continuously at a rate of velocity, v so long as the shear stress is applied at the rate of shearing deformation, dv/dy (also called the velocity gradient), where μ is the viscosity, which is a measure of the resistance of the fluid flow. Therefore, based on Newton's law of viscosity, the dimensions and units for the absolute or dynamic viscosity, μ in the English system are given as follows:

$$\mu = \frac{\tau}{\dfrac{dv}{dy}} = \frac{[FL^{-2}]}{[LT^{-1}L^{-1}]} = [FL^{-2}T] = [lb - sec/ft^2] \tag{1.60}$$

The dimensions and units for the absolute or dynamic viscosity, μ in the metric system are given as follows:

$$\mu = \frac{\tau}{\dfrac{dv}{dy}} = \frac{[ML^{-1}T^{-2}]}{[LT^{-1}L^{-1}]} = \frac{[MLT^{-1}T]}{[L^2]} = [FTL^{-2}] = [N - sec/m^2] \tag{1.61}$$

Furthermore, one may note that a common unit for the dynamic viscosity, μ in the metric system is the poise (P), which is defined as follows:

$$1P = 10^{-1} N - sec/m^2 \tag{1.62}$$

and where the centipoise (*cP*) is defined as follows:

$$1cP = 10^{-2}P = 10^{-3}\,N - sec/m^2 = 1\,mN - sec/m^2 \tag{1.63}$$

where 1 *mN* represents 1 milli *N*.

EXAMPLE PROBLEM 1.26

The space between two very long parallel plates separated by distance 0.35 m is filled with glycerin at room temperature (20°C), as illustrated in Figure EP 1.26. The lower plate is held stationary, while a continuous tangential (shear) force, F_V is applied to the upper plate, thus moving the plate with a constant velocity of 0.25 m/sec. The dimensions of the upper plate are 2 m by 1 m. (a) Determine if the flow is laminar or turbulent. (b) Graphically illustrate the velocity profile at the cross section of flow. (c) Graphically illustrate the shear stress at the cross section of flow. (d) Determine the magnitude of the shear force required to keep the plate moving at the given constant velocity.

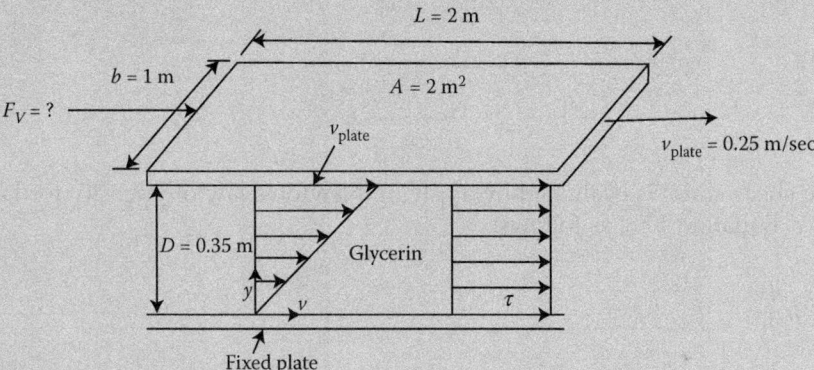

FIGURE EP 1.26
Laminar Couette flow of glycerin flows between two very long parallel plates.

Mathcad Solution

(a) The fluid properties for glycerin at room temperature (20°C) are read from Table A.4 in Appendix A. In order to determine if the flow is laminar or turbulent, the Reynolds number, *R* is computed from Equation 1.48 as follows:

$$N := kg\,\frac{m}{sec^2} \qquad\qquad \rho := 1258\,\frac{kg}{m^3} \qquad \mu := 1494 \times 10^{-3}\,N\,\frac{sec}{m^2}$$

$$v_{plate} := 0.25\,\frac{m}{sec} \qquad\qquad D := 0.35\,m$$

$$R := \frac{\rho\,v_{plate}\,D}{\mu} = 73.678$$

Thus, because $R = 73.678 < 2000$, the flow between the two long parallel plates is laminar.

(b) Because the flow between the two long parallel plates is laminar, the velocity profile is assumed to be linear and is evaluated by applying Equation 1.55 as follows:

$$v(y) = \frac{v_{plate}}{D} \, y$$

$$v(y) := \frac{v_{plate}}{D} \, y$$

A graphical illustration of the velocity profile at the cross section of flow is given as follows:

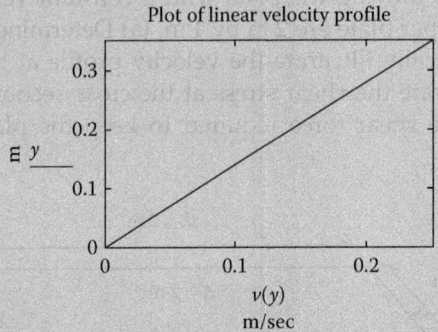

Plot of linear velocity profile

(c) The shear stress is evaluated by applying Newton's law of viscosity for laminar flow, Equation 1.58, as follows:

$$\tau = \mu \frac{dv}{dy}$$

$$v(y) := \frac{v_{plate}}{D} \, y \qquad\qquad \frac{d}{dy} v(y) \text{ simplify} \rightarrow \frac{0.714}{sec}$$

$$\tau(y) := \mu \frac{0.714}{sec}$$

A graphical illustration of the shear stress at the cross section of flow is given as follows:

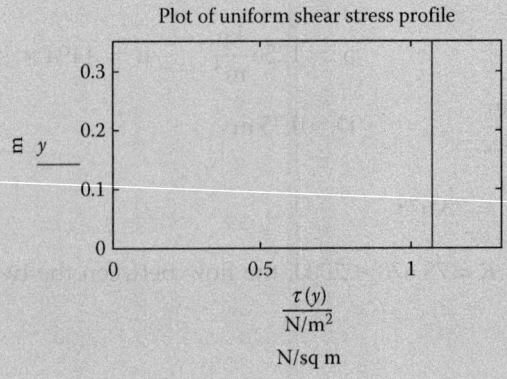

Plot of uniform shear stress profile

$$y := 0\,\text{m} \qquad\qquad \tau(y) = 1.067\,\frac{\text{N}}{\text{m}^2}$$

$$y := 0.35\,\text{m} \qquad\qquad \tau(y) = 1.067\,\frac{\text{N}}{\text{m}^2}$$

Note that the shear stress function, $\tau(y)$ was divided by N/m^2 in order to properly display the correct values for this function. Mathcad plots values in its default (base) units. The base unit for pressure (or shear stress) in the SI system of units is N/m^2, so Mathcad saves them in N/m^2. Thus, in order to properly display the values in N/m^2, one may divide the shear stress by N/m^2, which essentially removes the units and displays only the values.

(d) The magnitude of the shear force required to keep the plate moving at the given constant velocity is determined by applying Equation 1.54 as follows:

$$b := 1\,\text{m} \qquad\qquad L := 2\,\text{m} \qquad A := b\,L = 2\,\text{m}^2$$

$$\tau(0.35\,\text{m}) = 1.067\,\frac{\text{N}}{\text{m}^2} \qquad F_V := \tau(0.35\,\text{m})\,A = 2.133\,\text{N}$$

A plot of the shear stress, τ vs. the rate of shearing deformation, dv/dy (also called the velocity gradient) is given in Figure 1.4, where the viscosity, μ represents the slope of the linear or curvilinear function. This plot indicates that according to Newton's law of viscosity, the shear stress, τ increases with an increase in the rate of shearing deformation, dv/dy, where the viscosity, μ predicts the behavior of the shear stress, τ with respect to the rate of shearing deformation, dv/dy and thus describes the nature of the fluid. There are five general cases to consider on this plot. In the first case, where the plot yields a linear function

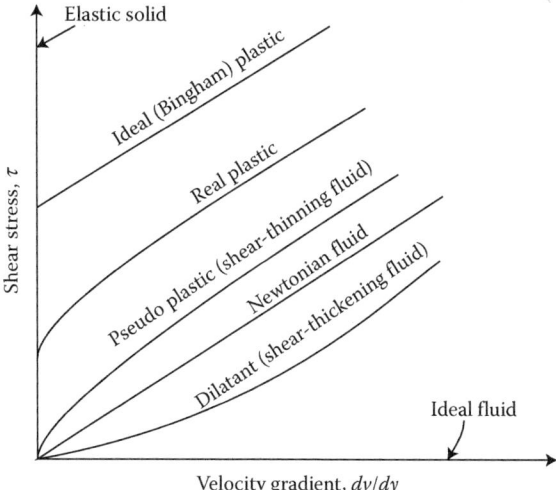

FIGURE 1.4
Relationship between the shear stress, τ and the velocity gradient, dv/dy for both Newtonian and non-Newtonian fluids.

with a zero-intercept, the constant slope represents a constant viscosity, μ regardless of the changes in the applied shear stress, τ and the resulting rate of shearing deformation, dv/dy, and these types of fluids are known as Newtonian fluids. Most common fluids, both liquids and gasses, are Newtonian and include water, oil, gasoline, alcohol, and air. In the second case, where the plot yields a curvilinear function, the variable slope represents a variable viscosity, μ, which changes as a function of the applied shear stress, τ and the resulting rate of shearing deformation, dv/dy, and these types of fluids are non-Newtonian fluids. These non-Newtonian fluids include real plastic, pseudoplastic (shear thinning: apparent viscosity decreases with increasing shear rate, e.g., colloidal suspensions and polymer solutions such as latex paint), and dilatant (shear thickening: apparent viscosity increases with increasing shear rate, e.g., water-corn starch mixture and water-sand mixture). In the third case, the ideal fluid for which the viscosity is zero plots along the horizontal axis, and in the fourth case, the true elastic solid (such as steel, for instance) plots along the vertical axis. And, finally, in the fifth case, the plot yields a linear function with a nonzero intercept. The nonzero intercept represents the initial shear stress, τ_o required to change the solid to a fluid, and the constant slope represents a constant viscosity, μ regardless of the changes in the applied shear stress, τ and the resulting rate of shearing deformation, dv/dy. These types of fluids are non-Newtonian fluids, and are known as ideal (Bingham) plastic (e.g., toothpaste, oil paint, and mayonnaise). Furthermore, while Newton's law of viscosity may represent non-Newtonian fluids as well as Newtonian fluids for the case of laminar flow, in this textbook, we will be concerned with its application for only Newtonian fluid flow, in which viscous action is strong (and thus, laminar/nonturbulent flow). Thus, for Newtonian fluids, the viscosity, μ of a given fluid is constant for a given temperature and a given pressure, and is independent of the rate of shearing deformation, dv/dy.

1.10.4.4 Dynamic Viscosity versus Kinematic Viscosity

Based on Newton's law of viscosity, the dimensions and units for the absolute or dynamic viscosity, μ in the English and metric systems were given above in Equations 1.60 and 1.61, respectively. However, because many of the equations in fluid mechanics include the combination of the ratio μ/ρ, this combination is called kinematic viscosity, v, as the force dimensions cancel out and the remaining units represent kinematics only. As such, the dimensions and units for the kinematic viscosity, v in the English system are given as follows:

$$v = \frac{\mu}{\rho} = \frac{[FL^{-2}T]}{[FL^{-4}T^2]} = [L^2 T^{-1}] = [ft^2/sec] \tag{1.64}$$

The dimensions and units for the kinematic viscosity, v in the metric system are given as follows:

$$v = \frac{\mu}{\rho} = \frac{[ML^{-1}T^{-1}]}{[ML^{-3}]} = [L^2 T^{-1}] = [m^2/sec] \tag{1.65}$$

Furthermore, one may note that a common unit for the dynamic kinematic viscosity, v in the metric system is the stoke (*St*), which is defined as follows:

$$1\,St = 10^{-4}\,m^2/sec = cm^2/sec \tag{1.66}$$

and where the centistoke (*cSt*) is defined as follows:

$$1\,cSt = 10^{-2}\,S = 10^{-6}\,m^2/sec \tag{1.67}$$

One may note that while the dynamic viscosity, μ was assumed to be mostly independent of pressure, the kinematic pressure, $v = \mu/\rho$ varies inversely with pressure (as it varies inversely with mass density, ρ [and density varies with pressure]). Furthermore, for liquids, both the dynamic and kinematic viscosities are practically independent of pressure, except at extremely high pressures. However, for gases, while the dynamic viscosity is practically independent of moderate to low pressure, the kinematic viscosity is inversely proportional to the pressure.

EXAMPLE PROBLEM 1.27

Using Table A.4 in Appendix A and Table A.5 in Appendix A, respectively: (a) Read the absolute viscosity and the density of carbon tetrachloride at room temperature (68°F), and compute the corresponding kinematic viscosity. (b) Read the absolute viscosity and density of nitrogen at room temperature (20°C), and compute the corresponding kinematic viscosity.

Mathcad Solution

(a) The kinematic viscosity of carbon tetrachloride at room temperature (68°F) is computed by applying Equation 1.64 as follows:

$$slug := 1\,lb\,\frac{sec^2}{ft} \qquad \mu := 20.35 \times 10^{-6}\,lb\,\frac{sec}{ft^2} \qquad \rho := 3.08\,\frac{slug}{ft^3}$$

$$v := \frac{\mu}{\rho} = 6.607 \times 10^{-6}\,\frac{ft^2}{sec}$$

(b) The kinematic viscosity of nitrogen at room temperature (20°C) is computed by applying Equation 1.65 as follows:

$$N := kg\,\frac{m}{sec^2} \qquad \mu := 17.6 \times 10^{-6}\,N\,\frac{sec}{m^2} \qquad \rho := 1.16\,\frac{kg}{m^3}$$

$$v := \frac{\mu}{\rho} = 1.517 \times 10^{-5}\,\frac{m^2}{sec}$$

1.10.4.5 Newton's Law of Viscosity: Theoretically Derived Expression for Shear Stress for Laminar Flow

Theoretically modeling the friction/viscous force, F_V in Newton's second law of motion, by Newton's law of viscosity, yields a theoretical (parabolic) expression for the velocity distribution for laminar flow (see Chapter 6 for details), as illustrated in Figure 1.2a for laminar pipe flow and Figure 1.5a for the laminar flow of a fluid over a solid boundary. This parabolic velocity profile is typical for laminar/viscous (nonturbulent) pipe flow or laminar/viscous (nonturbulent) flow along a solid boundary. The velocity of the fluid at the pipe walls or the stationary solid boundary is zero, as the fluid at the boundary surface sticks

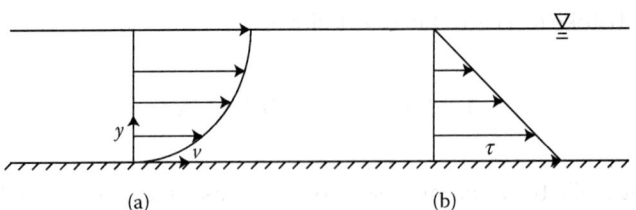

FIGURE 1.5
Laminar flow over a solid boundary. (a) Parabolic velocity distribution. (b) Linear shear stress distribution.

to the boundary and has the velocity of the pipe wall/boundary (this is known as the "no-slip" condition). Furthermore, the velocity at the centerline of the pipe or at the top of the fluid surface is a maximum. However, because the parabolic velocity profile for a pipe flow is a full parabola, while for a fluid flow over a solid boundary, it is only half a parabola, the parabolic velocity distributions will be addressed separately. And, similarly, the linear shear stress distribution for pipe flow depicts a full pipe cross section, while for a fluid flow over a solid boundary, it depicts only half, as illustrated in Figure 1.5b.

Assuming laminar pipe flow, the parabolic relationship between the velocity, v and the radial distance, r from the center of the pipe with a radius, r_o is derived in Chapter 6, and is given as follows:

$$v = \frac{\Delta p}{2\mu L}\left(\frac{r_o^2 - r^2}{2}\right) = \frac{\Delta p}{4\mu L}\left(r_o^2 - r^2\right) \tag{1.68}$$

which is expressed as a function of the pressure drop, Δp, with the maximum velocity, v_{max} at the center of the pipe, where $r = 0$, and $v = 0$ at $r = r_o$, as illustrated in Figure 1.2a. Thus, the expression for v_{max} for laminar flow is:

$$v_{max} = \frac{\Delta p r_o^2}{4\mu L} \tag{1.69}$$

where the pressure drop, Δp for laminar pipe flow is evaluated by Poiseuille's law (see Chapter 6) as follows:

$$\Delta p = \frac{128\mu QL}{\pi D^4} \tag{1.70}$$

where Q is the flow rate or discharge. Furthermore, substituting Equation 1.69 into Equation 1.68 yields the parabolic velocity distribution for laminar flow expressed as a function of v_{max} as follows:

$$v = v_{max}\left[1 - \left(\frac{r}{r_o}\right)^2\right] \tag{1.71}$$

where

$$v_{ave} = \frac{v_{max}}{2} = \frac{\Delta p r_o^2}{8\mu L} \tag{1.72}$$

Thus, alternatively, the parabolic velocity distribution for laminar flow may be presented as a function of v_{ave} as follows:

$$v = 2v_{ave}\left[1 - \left(\frac{r}{r_o}\right)^2\right]$$ (1.73)

For laminar pipe flow, the shear stress distribution in Figure 1.2b may be evaluated by applying either the results of Newton's second law of motion for either laminar or turbulent pipe flow (Equation 1.53), or Newton's law of viscosity for laminar flow (Equation 1.58). Newton's second law of motion for either laminar or turbulent flow above (see Chapter 6), given by Equation 1.53, illustrated that the velocity gradient and thus the shear stress decreases linearly with distance from the stationary pipe wall to center of the pipe, where there is no shear stress, as follows:

$$\tau = \tau_{max}\frac{r}{r_o}$$ (1.74)

where the shear stress is a maximum at the pipe wall ($\tau_w = \tau_{max}$) and there is no shear stress at the centerline of the pipe, and where the maximum shear stress is evaluated by Equation 1.50 (for either laminar or turbulent flow) as follows:

$$\tau_{max} = \tau_w = \frac{r_o}{2}\frac{\Delta p}{L}$$ (1.75)

and where for laminar pipe flow, the pressure drop, Δp may be evaluated from Poiseuille's law (Equation 1.70) as follows:

$$\Delta p = \frac{128\mu QL}{\pi D^4}$$ (1.76)

Alternately, Figure 1.2a illustrates that for the parabolic velocity profile for laminar pipe flow, the velocity gradient, $-dv/dr$ is a maximum at the pipe wall. Thus, Figure 1.2b illustrates that the resulting shear stress, as modeled by Newton's law of viscosity (Equation 1.58):

$$\tau = \mu\frac{dv}{dy} = \mu\left(-\frac{dv}{dr}\right)$$ (1.77)

is also a maximum at the pipe wall. Note that a negative sign was inserted in the derivative function for pipe flow because radial distance, r increases from the center of the pipe from 0 to the pipe radius, r_o. Furthermore, the velocity gradient and thus the shear stress decreases linearly with distance from the stationary pipe wall to the center of the pipe, where there is no shear stress.

EXAMPLE PROBLEM 1.28

Glycerin at room temperature (68°F) flows at a flow rate of 89 cfs in a 5-ft-diameter pipe for a length of 10 ft, as illustrated in Figure EP 1.28. (a) Determine if the flow is laminar or turbulent. (b) Graphically illustrate the velocity profile at the pipe cross section. (c) Graphically illustrate the shear stress at the pipe cross section.

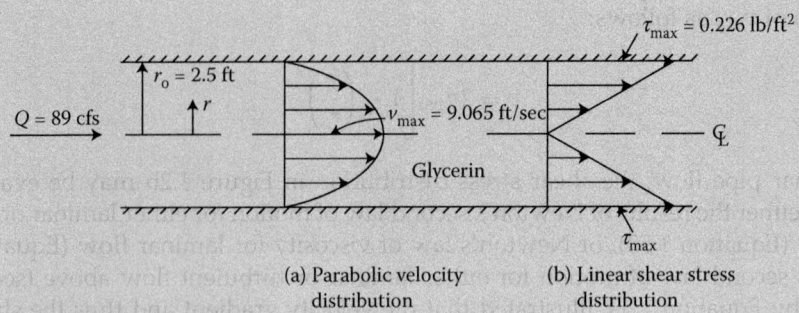

(a) Parabolic velocity   (b) Linear shear stress
distribution   distribution

FIGURE EP 1.28
Glycerin flows in a pipe.

Mathcad Solution

(a) The fluid properties for glycerin at room temperature (68°F) are read from Table A.4 in Appendix A. In order to determine if the flow is laminar or turbulent, the Reynolds number, R is computed from Equation 1.48 as follows:

$$slug := 1\,lb\,\frac{sec^2}{ft} \qquad \rho := 2.44\,\frac{slug}{ft^3} \qquad \mu := 31200 \times 10^{-6}\,lb\,\frac{sec}{ft^2}$$

$$Q := 89\,\frac{ft^3}{sec} \qquad D := 5\,ft \qquad A := \frac{\pi \cdot D^2}{4} = 19.635\,ft^2 \qquad v := \frac{Q}{A} = 4.533\,\frac{ft}{s}$$

$$R := \frac{\rho \cdot v \cdot D}{\mu} = 1.772 \times 10^3$$

Thus, because $R = 1.772 \times 10^3 < 2000$, the pipe flow is laminar.

(b) Because the pipe flow is laminar, the velocity profile is assumed to be parabolic and is evaluated by applying Equation 1.71 as follows:

$$v = v_{max}\left[1 - \left(\frac{r}{r_0}\right)^2\right]$$

where v_{max} is evaluated by applying Equation 1.69 as follows:

$$v_{max} = \frac{\Delta p\, r_0^2}{4\mu L}$$

and where Δp for laminar pipe flow is evaluated by applying Poiseuille's law (Equation 1.70) as follows:

$$\Delta p = \frac{128\mu Q L}{\pi D^4}$$

$$L := 10\,ft \qquad \Delta p := \frac{128 \cdot \mu \cdot Q \cdot L}{\pi \cdot D^4} = 1.807\,psf \qquad r_0 := \frac{D}{2} = 2.5\,ft$$

$$v_{max} := \frac{\Delta p \cdot r_0^2}{4 \cdot \mu \cdot L} = 9.065\,\frac{ft}{s} \qquad\qquad v(r) := v_{max}\left[1 - \left(\frac{r}{r_0}\right)^2\right]$$

A graphical illustration of the velocity profile at the pipe cross section is given as follows:

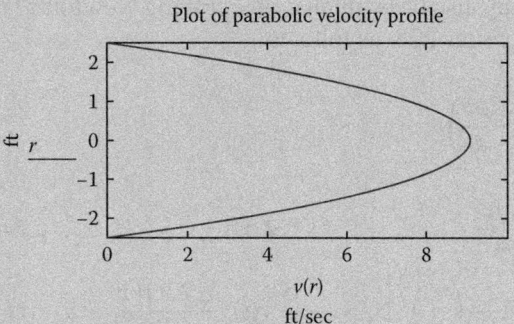

Plot of parabolic velocity profile

(c) The shear stress is evaluated by applying Newton's second law of motion for either laminar or turbulent flow (Equation 1.53) as follows:

$$\tau = \tau_{max} \frac{r}{r_o}$$

where τ_{max} is evaluated by applying Equation 1.50 as follows:

$$\tau_{max} = \tau_w = \frac{r_o}{2} \frac{\Delta p}{L}$$

and where Δp for laminar pipe flow was already evaluated by applying Poiseuille's law (Equation 1.70) in (b) above.

$$\tau_{max} := \frac{r_o}{2} \frac{\Delta p}{L} = 0.226 \, psf \qquad \tau(r) := \tau_{max} \frac{r}{r_o}$$

A graphical illustration of the shear stress at the pipe cross section is given as follows:

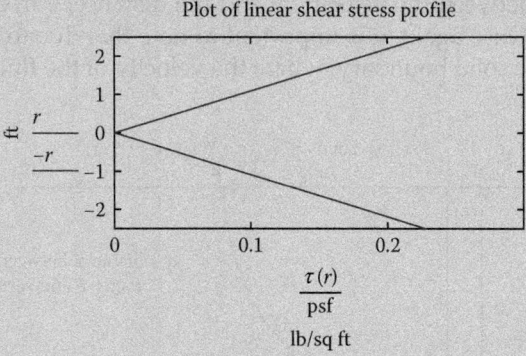

Plot of linear shear stress profile

Note that the shear stress function, $\tau(r)$ was divided by psf in order to properly display the correct values for this function. Mathcad plots values in its default (base) units. The base unit for pressure (or shear stress) in the BG system of units is psi, so Mathcad saves them in psi. Thus, in order to properly display the values in psf, one may divide the shear stress by psf, which essentially removes the units and displays only the values. Alternatively, because the pipe flow is laminar, the shear stress may also be evaluated by applying Newton's law of viscosity for laminar flow, Equation 1.58, as follows:

$$\tau = \mu \frac{dv}{dy} = \mu \left(-\frac{dv}{dr} \right)$$

$$v_{max} = 9.065 \, \frac{ft}{sec} \qquad r_o = 2.5 \, ft \qquad \mu := 31{,}200 \times 10^{-6} \, lb \, \frac{s}{ft^2}$$

$$v(r) := v_{max} \left[1 - \left(\frac{r}{r_o} \right)^2 \right] \qquad -\frac{d}{dr} v(r) \rightarrow \frac{2.901 \, r}{ft \, s} \qquad \tau(r) = \mu \, \frac{2.901 \, r}{ft \, s}$$

A graphical illustration of the shear stress at the pipe cross section is given as follows:

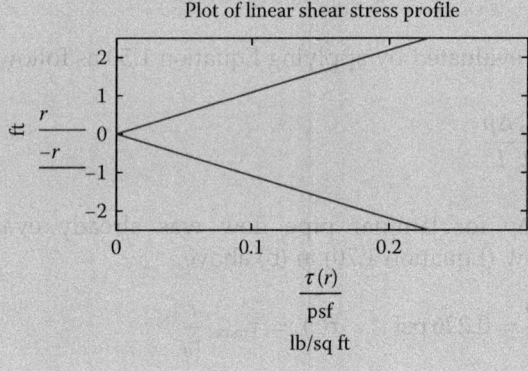

Plot of linear shear stress profile

Assuming laminar fluid flow over a solid boundary, as illustrated in Figure 1.6, the parabolic relationship between the velocity, v and the distance, y from the solid boundary is derived as follows. Once again, it is important to note that due to the "no-slip" condition, the fluid sticks to the solid boundary, where the velocity of the fluid at the stationary solid

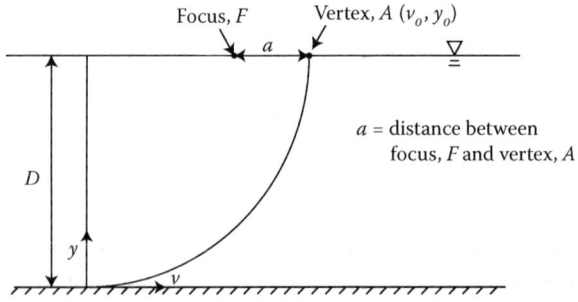

FIGURE 1.6
Parabolic relationship between the velocity, v and the distance, y from the solid boundary for laminar flow over a solid boundary.

boundary is zero, increasing parabolically to a maximum velocity. The general equation for a parabola that opens to the left, as illustrated in Figure 1.6, is defined as follows:

$$(y - y_o)^2 = -4a(v - v_o) \tag{1.78}$$

where $y_o = D$, $v_o = v_{max}$, and $y = 0$ at $v = 0$; thus, a is evaluated as follows:

$$(0 - D)^2 = -4a(0 - v_{max}) \tag{1.79}$$

$$D^2 = 4av_{max} \tag{1.80}$$

$$a = \frac{D^2}{4v_{max}} \tag{1.81}$$

Thus, the equation for the parabolic velocity distribution is given as follows:

$$(y - D)^2 = -\frac{D^2}{v_{max}}(v - v_{max}) \tag{1.82}$$

Isolating the velocity, v using Mathcad symbolic find yields the following:

Given

$$(y - D)^2 = -\frac{D^2}{v_{max}}(v - v_{max})$$

$$\text{Find}(v) \rightarrow -\frac{v_{max}y^2 - 2Dv_{max}y}{D^2} \text{ factor} \rightarrow \frac{v_{max}y(2D - y)}{D^2}$$

$$v = \frac{v_{max}}{D^2}y(2D - y) \tag{1.83}$$

Furthermore, the shear stress distribution for laminar fluid flow over a solid boundary may be evaluated by applying Newton's law of viscosity (Equation 1.58) as follows: $\tau = \mu(dv/dy)$.

Figure 1.5a illustrates that for the parabolic velocity profile for laminar flow over a solid boundary, the velocity gradient, dv/dy is a maximum at the solid boundary. Thus, Figure 1.5b illustrates that the resulting shear stress, τ is also a maximum at the boundary, decreasing linearly with distance from the stationary boundary to top of the fluid surface interface, where there is no shear stress.

EXAMPLE PROBLEM 1.29

Crude oil at room temperature (68°F) flows at a depth of 0.35 ft and a maximum velocity of 0.5 ft/sec over a solid boundary, as illustrated in Figure EP 1.29. (a) Determine if the flow is laminar or turbulent. (b) Graphically illustrate the velocity profile at the cross section of flow. (c) Graphically illustrate the shear stress at the cross section of flow.

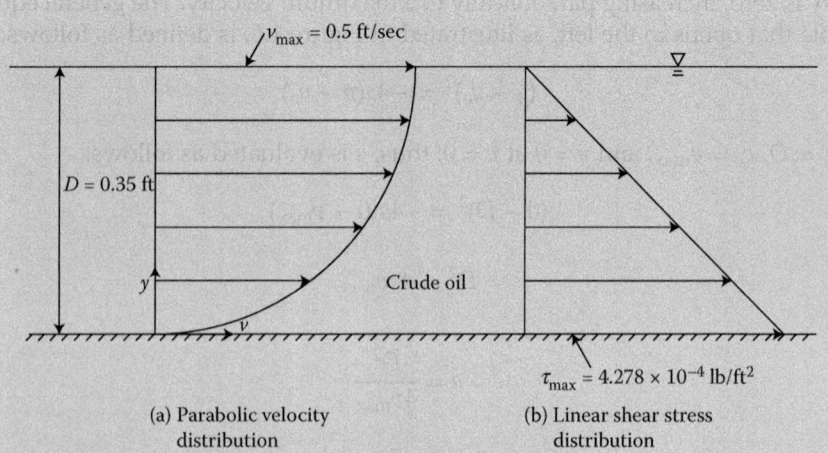

(a) Parabolic velocity
distribution

(b) Linear shear stress
distribution

FIGURE EP 1.29
Crude oil flows over a solid boundary.

Mathcad Solution

(a) The fluid properties for crude oil at room temperature (68°F) are read from Table A.4 in Appendix A. In order to determine if the flow is laminar or turbulent, the Reynolds number, R is computed from Equation 1.48 as follows:

$$\text{slug} := 1 \, \text{lb} \frac{\sec^2}{\text{ft}} \qquad \rho := 1.66 \, \frac{\text{slug}}{\text{ft}^3} \qquad \mu := 150 \times 10^{-6} \, \text{lb} \frac{\sec}{\text{ft}^2}$$

$$D := 0.35 \, \text{ft} \qquad v_{max} := 0.5 \, \frac{\text{ft}}{\sec} \qquad R := \frac{\rho \, v_{max} D}{\mu} = 1.937 \times 10^3$$

Thus, because $R = 1.937 \times 10^3 < 2000$, the flow over the solid boundary is laminar.

(b) Because the flow over the boundary is laminar, the velocity profile is assumed to be parabolic, and is evaluated by applying Equation 1.83 as follows:

$$v = \frac{v_{max}}{D^2} y(2D - y)$$

$$v(y) := \frac{v_{max}}{D^2} y(2D - y)$$

A graphical illustration of the velocity profile at the cross section of flow is given as follows:

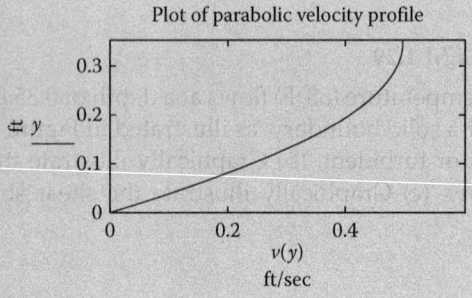

Plot of parabolic velocity profile

(c) The shear stress is evaluated by applying Newton's law of viscosity for laminar flow, Equation 1.58, as follows:

$$\tau = \mu \frac{dv}{dy}$$

$$v_{max} = 0.5 \frac{ft}{sec} \qquad\qquad D = 0.35\, ft \qquad\qquad \mu = 150 \times 10^{-6}\, lb\, \frac{sec}{ft^2}$$

$$v(y) := \frac{v_{max}}{D^2} y(2D - y) \qquad \frac{d}{dy} v(y) \text{ simplify} \rightarrow \frac{2.857\, ft - 8.163\, y}{ft\, sec}$$

$$\tau(y) := \mu \frac{2.857\, ft - 8.163\, y}{ft\, sec}$$

A graphical illustration of the shear stress at the cross section of flow is given as follows:

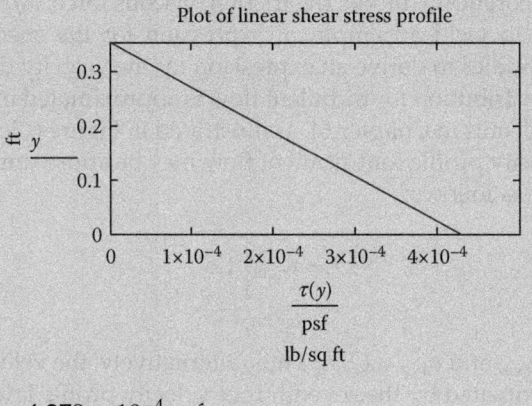

Plot of linear shear stress profile

$$y := 0\, ft \qquad \tau(y) = 4.278 \times 10^{-4}\, psf$$

1.10.4.6 Empirically Derived Expression for Shear Stress for Turbulent Flow

In the case of turbulent flow, the friction/viscous force, F_V in Newton's second law of motion, $\sum_s F = Ma_s$ (integral form of the momentum equation), cannot be modeled theoretically. As a result, one must resort to dimensional analysis in order to derive an empirical relationship for the friction/viscous force, $F_V = \tau A$, where, specifically, the wall shear stress, τ_w is derived as follows:

$$\tau_w = C_D \rho v^2 \tag{1.84}$$

where C_D is the drag coefficient that represents the flow resistance for turbulent flow and is a function of the Reynolds number, $R = (\rho v L / \mu)$ and the relative surface roughness, ε/L, where ε is the absolute surface roughness and L is the pipe diameter or a geometric/length flow variable. A comparison between the theoretical expression for the shear stress for laminar flow [i.e., Newton's law of viscosity, $\tau = \mu(dv/dy) = \mu(-dv/dr)$] and the empirical expression for the shear stress for turbulent flow, $\tau = C_D \rho v^2 (r/r_o)$ indicates the following: (1) the shear stress in laminar flow is linear in velocity, v, whereas for turbulent flow, it is nonlinear in velocity, v, and (2) the shear stress in laminar flow is a function of the fluid viscosity, μ (because the viscous force is large enough to suppress the fluctuations in flow due to the significance of the inertia force), whereas for turbulent flow, it is mostly a function of the fluid density, ρ (because the inertia force is large compared to the viscous force) and the relative surface roughness, ε/L, and to a lesser extent, a function of the fluid viscosity, μ.

The assumption made regarding the viscosity of the fluid will determine the resulting velocity profile for the flow. For an ideal flow through a pipe or a channel, the assumption of no viscosity results in a velocity profile with a rectangular distribution (see Figure 6.3a). However, for a real flow through a pipe or channel, modeling the viscosity results in the velocity of the fluid at the stationary solid boundary being zero, as the fluid at the boundary surface sticks to the boundary and has the velocity of the boundary (because of the "no-slip" condition). Furthermore, the deterministic nature of laminar flow results in the ability to theoretically model the friction/viscous force, F_V by Newton's law of viscosity, which in turn yields a theoretical (parabolic) expression for the velocity distribution for laminar flow, as illustrated in Figure 6.3b. However, the more stochastic nature of turbulent flow results in the inability to theoretically model the friction/viscous force, F_V. As a result, dimensional analysis is applied to yield an empirical expression for the shear stress, which in turn requires empirical studies to derive an expression for the velocity distribution for turbulent flow. The velocity distribution for turbulent flow is approximated by the seventh root velocity profile law (see details in Chapter 6), as illustrated in Figure 6.3c. Empirical studies have shown that the velocity profile for turbulent flow may be approximated by the seventh root velocity profile law as follows:

$$v = v_{\max}\left[1 - \frac{r}{r_o}\right]^{1/7} \tag{1.85}$$

where $v_{\max} = 1.22\, v_{\text{ave}}$, and $v_{\text{ave}} = Q/A$. Thus, alternatively, the velocity profile for turbulent flow may be approximated by the seventh root velocity profile law as a function of v_{ave} as follows:

$$v = 1.22 v_{\text{ave}}\left[1 - \frac{r}{r_o}\right]^{1/7} \tag{1.86}$$

Figure 1.2b illustrates that the shear stress, τ is a maximum at the pipe boundary, decreasing linearly with distance from the boundary to centerline of the pipe, where there is no shear stress (derived from Newton's second law of motion for either laminar or turbulent pipe flow in Chapter 6). Newton's second law of motion for either laminar or turbulent flow (see Chapter 6), given by Equation 1.53, illustrates that the velocity gradient and thus the shear stress decreases linearly with distance from the stationary pipe wall to the center of the pipe, where there is no shear stress, as follows:

$$\tau = \tau_{\max}\frac{r}{r_o} \tag{1.87}$$

where the shear stress is a maximum at the pipe wall ($\tau_w = \tau_{\max}$) and there is no shear stress at the centerline of the pipe, and where the maximum shear stress is evaluated by

Equation 1.50 (for either laminar or turbulent flow) as follows:

$$\tau_{\max} = \tau_w = \frac{r_o}{2}\frac{\Delta p}{L} \tag{1.88}$$

For laminar pipe flow, the pressure drop, Δp is theoretically evaluated from Poiseuille's law (Equation 1.70), while for turbulent flow, the pressure drop, Δp cannot be theoretically evaluated. Instead, the pressure drop, Δp for turbulent flow is indirectly and empirically determined by applying Equation 1.84 for the pipe wall shear stress, where the shear stress is a maximum ($\tau_w = \tau_{\max}$) as follows:

$$\tau_w = C_D \rho v^2 \tag{1.89}$$

Thus, substituting Equation 1.89 into Equation 1.87 yields the following expression for the shear stress for turbulent flow:

$$\tau = C_D \rho v^2 \frac{r}{r_o} \tag{1.90}$$

where C_D is a dimensionless drag coefficient that represents the flow resistance for turbulent flow. Finally, the drag coefficient, C_D is typically represented by more commonly used flow resistance factors/coefficients such as the Chezy coefficient, the Darcy–Weisbach friction factor, or Manning's roughness coefficient (see Chapter 6).

1.10.5 Surface Tension

While the more dominant forces in fluid statics and dynamics include those due to gravity, pressure, viscosity (fluid dynamics only), and inertia (fluid dynamics only), the surface tension force is generally negligible. However, surface tension, σ effects can be important in certain fluid situations. For instance, the fluid surface tension, σ plays an important role in the breakup of liquid jets; in the mechanics of the formation of droplets, bubbles, and thin films; in capillarity; in sheet flow; in the occurrence of osmosis; in the behavior of buoys; and in the interpretation of small hydraulic models of larger prototypes.

The effect of fluid surface tension, σ is displayed at the interface between a liquid and a gas; between two immiscible (not capable of being mixed) liquids; and between a liquid, gas, and solid. As such, the surface tension, σ is a liquid property (and not a gas property). As a result of the liquid property, σ, surface tension forces develop in the liquid surface, which cause the liquid surface to behave as a membrane/skin that is stretched over the fluid mass (gas or liquid). For instance, the surface tension, σ is responsible for: (1) the formation of a liquid membrane at the top of the water/liquid surface, which allows a small steel needle that is carefully placed on the surface to float and some insects to land and gently move across the surface; (2) the formation of a water droplet in the atmosphere or on a solid surface, or the formation of a spherical droplet of mercury, which can easily roll on a solid surface; (3) the formation of a soap bubble in the air; (4) the formation of a thin soap film of fluid within a thin plastic or steel frame prior to the formation of a soap bubble; and (5) the rise or fall of liquid in a capillary (small) tube or in pore (small) spaces/passages, or the intake of water by trees.

1.10.5.1 Definition of Surface Tension

Liquids are characterized by molecular attraction, including cohesion and adhesion. Molecular cohesion allows the liquid to resist tensile stress or surface tension, σ. Meanwhile, molecular adhesion allows the liquid to adhere to a solid surface as a result of the tensile stress or surface tension, σ. While a liquid surface, droplet, bubble, and thin film in the atmosphere involve a liquid–gas interface, a capillary rise or fall and a droplet on solid surface involve a liquid–gas–solid interface. Furthermore, one may note that while a liquid–gas interface deals with the molecular cohesive forces (which resist the surface tension, σ), a liquid–gas–solid interface deals with both the molecular cohesive forces (which resist the surface tension, σ) and adhesive forces (which are a result of the surface tension, σ). Therefore, depending upon the fluid situation, the role of the surface tension, σ will vary. However, regardless of the fluid situation, as a result of the liquid property, σ, surface tension forces develop in the liquid surface, which causes the liquid surface to behave as a membrane/skin that is stretched over the fluid mass (gas or liquid).

The simple case of the formation of a liquid membrane at the top of a liquid surface will be used to define the surface tension, σ. Figure 1.7a illustrates that the molecules in the interior of the liquid mass are surrounded by liquid molecules that are equally attracted to each other, which results in a balance of cohesive forces. However, the molecules along the liquid surface are not surrounded by liquid molecules at the top of the liquid surface and thus are subjected to an unbalanced net cohesive force toward the interior of the fluid mass. The unbalanced net inward cohesive force acting on the liquid molecules along the liquid surface causes the liquid surface to behave as a membrane/skin that is stretched over the air mass above it and results in a tensile force that acts in the plane of the stretched surface along any line in the surface. The tensile force, $F_T = \sigma L$ represents the intensity of the molecular attraction/cohesion per unit length along any line in the stretched liquid surface, which is the liquid surface tension, σ. Therefore, based on the definition of the tensile force (surface tension force), the dimensions and units for the surface tension, σ in the English system are given as follows:

$$\sigma = \frac{F_T}{L} = \frac{[F]}{[L]} = [lb/ft] \tag{1.91}$$

The dimensions and units for the surface tension, σ in the metric system are given as follows:

$$\sigma = \frac{F_T}{L} = \frac{[MLT^{-2}]}{[L]} = [N/m] \tag{1.92}$$

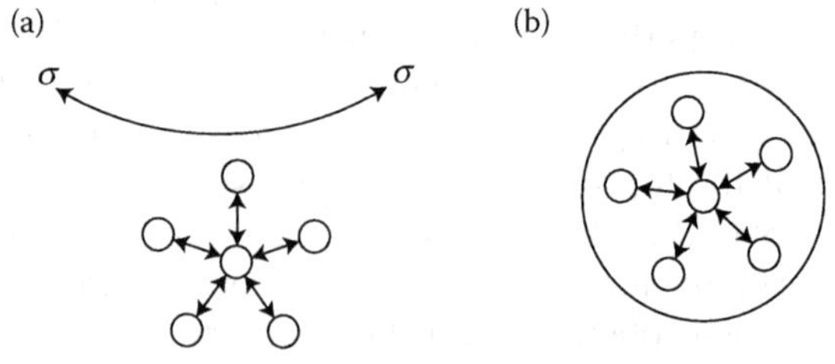

(a)   (b)

FIGURE 1.7
Formation of a liquid membrane illustrates the surface tension, σ. (a) Intermolecular forces acting at the top of and below a liquid surface. (b) Intermolecular forces acting on the interior molecule in a perfectly spherical shaped particle of liquid.

A surface tension, σ will always exist whenever there is a discontinuity (interface) in density between two fluids, such as between water and air or water and oil. As such, the magnitude of the surface tension, σ of the given liquid is a function of the nature of both substances (either a liquid and a gas or a liquid and a liquid), and in general, it is a function of temperature and only slightly a function of pressure. The surface tensions, σ for water in contact with air at standard sea-level atmospheric pressure are presented as a function of temperature in Table A.2 in Appendix A. And, although the surface tension, σ for some common liquids in contact with air will vary with temperature, for practical applications, their surface tension, σ at standard sea-level atmospheric pressure at room temperature (68°F or 20°C), in general, are given in Table A.4 in Appendix A. One may note that in general, when a second fluid is not specified at the interface, then one assumes that the liquid surface is in contact with air. Because the surface tension, σ is directly dependent on intermolecular cohesive forces, its magnitude will decrease as temperature increases. The magnitude of the surface tension for a liquid becomes zero at the critical temperature; thus, there is no distinct vapor–liquid interface at temperatures above the critical point; the vapor–liquid critical point specifies the temperature and pressure above which distinct gas and liquid phases no longer exist.

EXAMPLE PROBLEM 1.30

Using Table A.2 in Appendix A, (a) Graphically illustrate the variation of the surface tension of water with temperature assuming the BG system of units.

Mathcad Solution

(a) A graphical illustration of the variation of the surface tension of water with temperature assuming the BG system of units is given as follows:

$$
T := \begin{pmatrix} 32 \\ 40 \\ 50 \\ 60 \\ 70 \\ 80 \\ 90 \\ 100 \\ 110 \\ 120 \\ 130 \\ 140 \\ 150 \\ 160 \\ 170 \\ 180 \\ 190 \\ 200 \\ 212 \end{pmatrix} \qquad \sigma := \begin{pmatrix} 0.00518 \\ 0.00514 \\ 0.00509 \\ 0.00504 \\ 0.00498 \\ 0.00492 \\ 0.00486 \\ 0.00480 \\ 0.00473 \\ 0.00467 \\ 0.00460 \\ 0.00454 \\ 0.00447 \\ 0.00441 \\ 0.00434 \\ 0.00427 \\ 0.00420 \\ 0.00413 \\ 0.00404 \end{pmatrix}
$$

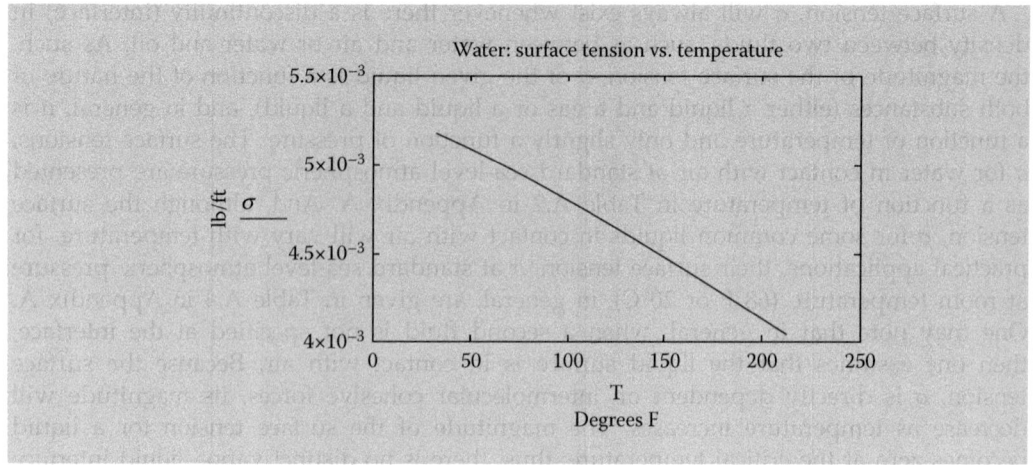

1.10.5.2 The Formation of a Droplet in the Atmosphere

The formation of a water droplet in the atmosphere is used to illustrate surface tension, σ involving a liquid–gas interface (such as for a liquid surface, bubble, or thin film), which deals with the molecular cohesive forces that resist the surface tension, σ. When gravity or other forces are negligible, an unconfined particle of liquid will assume a perfectly spherical shape, where molecular cohesion allows the liquid to resist tensile stress or surface tension, σ. This phenomenon is due to the attracting (cohesive) forces between the liquid molecules. Figure 1.7b illustrates that the molecule located at the interior of the perfectly spherical shape is surrounded by many other liquid molecules, where, on average, the attracting force is uniform in all directions. However, at the surface of the spherical shape, where there is an interface between the liquid and a gas, there is no outward attraction to balance the inward pull because there are no liquid molecules. Therefore, the surface liquid molecules are subjected to a net inward cohesive force/attraction, thus creating the perfectly spherical shape. Furthermore, the resulting compression effect causes the liquid to minimize the surface area of the sphere. As such, there is a tendency of a liquid droplet to attain a spherical shape, which is characterized by the minimum surface area for a given volume.

Consider half of a spherical droplet of fluid, as shown in Figure 1.8. Given that the droplet of fluid is in static equilibrium, the internal, p_i and external, p_e (atmospheric pressure when droplet is in the atmosphere) pressures acting on a perfectly spherical particle/droplet of

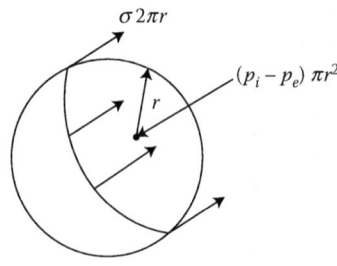

FIGURE 1.8
Half of a spherical droplet of fluid illustrates the surface tension force, $\sigma 2\pi r$ and the unbalanced net inward cohesive force, $(p_i - p_e)\, \pi r^2$.

liquid are balanced by the existence of surface tension, σ as follows:

$$\sum_s F = F_P - F_T = 0$$

$$\sum_s F = (p_i - p_e)A - \sigma P = 0$$

$$(p_i - p_e)A = \sigma P$$

$$(p_i - p_e)\pi r^2 = \sigma 2\pi r$$

$$(p_i - p_e) = \frac{2\sigma}{r} \tag{1.93}$$

where s = axis along the surface of the sphere, A = the cross-sectional area of the sphere, P = the perimeter of the sphere, and r = the radius of the sphere. The unbalanced net inward cohesive force, $(p_i - p_e)A$ acting on the liquid molecules along the surface of the spherical droplet causes the liquid surface to behave as a membrane/skin that is stretched over the liquid mass within the sphere and results in a tensile force that acts in the plane of the stretched surface along any line in the surface. The tensile force, $F_T = \sigma L = \sigma P$ represents the intensity of the molecular attraction/cohesion per unit length along any line in the stretched liquid surface, which is the liquid surface tension, σ.

Therefore, the fact that internal pressure, p_i exceeds the external pressure, p_e results in the occurrence of σ. Furthermore, when the surface tension force, $F_T = \sigma L$ is relatively significant in comparison to the inertia force, $F_I = \rho v^2 L^2$, then the droplet will increase in size. However, if a droplet is subjected to an air jet and there is a relative velocity between the droplet and the air/gas, inertia forces due to this relative velocity cause the droplet to deform. Thus, when the inertia force, F_I overcomes the surface tension force, F_T, the droplet bursts/breaks into smaller ones.

EXAMPLE PROBLEM 1.31

A spherical water droplet at a temperature of 10°C forms with a diameter of 0.5 cm during a rainfall at an elevation of 1 km above sea level. (a) Determine the internal pressure acting on the spherical water droplet (refer to Figure 1.8).

Mathcad Solution

(a) The surface tension for water at 10°C is read from Table A.2 in Appendix A, while the atmospheric pressure at an elevation of 1 km above sea level is read from Table A.1 in Appendix A. The internal pressure acting on the spherical water droplet is determined by applying Equation 1.93 as follows:

$$(p_i - p_e) = \frac{2\sigma}{r}$$

$$N := kg\frac{m}{sec^2} \qquad \sigma := 0.0742\frac{N}{m} \qquad P_{atm} := 89.976\,kPa = 8.998 \times 10^4\frac{N}{m^2}$$

$$P_e := P_{atm} = 8.998 \times 10^4\frac{N}{m^2}$$

$$D := 0.5\,cm = 5 \times 10^{-3}\,m \qquad\qquad r := \frac{D}{2} = 2.5 \times 10^{-3}\,m$$

Guess value: $\qquad\qquad\qquad\qquad p_i := 1\,\dfrac{N}{m^2}$

Given

$$(p_i - p_e) = \frac{2 \cdot \sigma}{r}$$

$$p_i := Find(p_i) = 9.004 \times 10^4\,\frac{N}{m^2}$$

1.10.5.3 The Formation of a Droplet on a Solid Surface and Capillarity

The formation of a water droplet or a droplet of mercury on a solid surface and the occurrence of a capillarity (rise of water or fall of mercury) are used to illustrate surface tension involving a liquid–gas–solid interface. As noted above, a liquid–gas–solid interface deals with both the molecular cohesive forces and the molecular adhesive forces. The molecular cohesion allows the liquid to resist tensile stress or surface tension, σ. Meanwhile, molecular adhesion allows the liquid to adhere to a solid surface as a result of the tensile stress or surface tension, σ. As a result, when the molecular adhesion is more significant than the molecular cohesion (such as for water), then the liquid will wet the solid surface that it comes in contact with. For instance, when a spherical water droplet comes in contact with a solid surface, the droplet will wet the solid surface and deform away from a spherical shape. And, when a thin tube is inserted in water, it will wet and rise in the tube, forming a concave meniscus (inward-curving liquid surface) (see Figure 1.9a). However, when molecular cohesion is more significant than the molecular adhesion (such as for mercury), then the liquid will not wet the solid surface that it comes in contact with. For instance, when a spherical mercury droplet comes in contact with a solid surface, the droplet will not wet the solid surface, will not deform away from a spherical shape, and will easily roll on the solid surface. And, when a thin tube is inserted in mercury, it will fall/depress in the tube, forming a convex meniscus (outward-bulging liquid surface) (see Figure 1.9b).

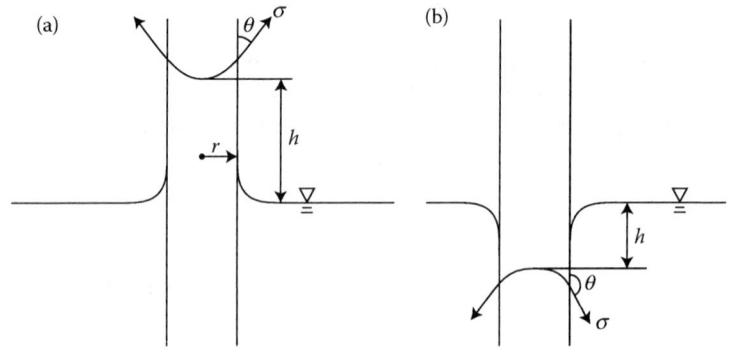

FIGURE 1.9
Capillary tubes. (a) Capillary rise of water, receding liquid, $\theta < 90°$. (b) Capillary depression of mercury, advancing liquid, $\theta > 90°$.

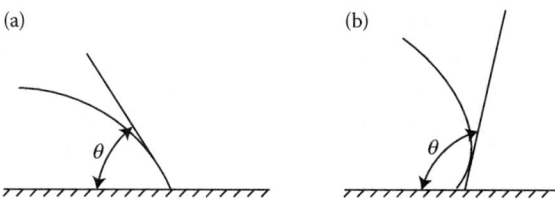

FIGURE 1.10
Surface tension contact angle, θ with the solid surface of a liquid in contact with a gas. (a) Receding liquid (e.g. water), $\theta < 90°$. (b) Advancing liquid (e.g., mercury), $\theta > 90°$.

1.10.5.4 The Formation of a Droplet on a Solid Surface

When a perfectly shaped spherical drop of liquid is placed on a solid surface, it will form a wetting angle or surface tension contact angle, θ with the solid surface, as illustrated in Figure 1.10. The magnitude of the contact angle, θ is a function of the attraction between the molecules of the specific liquid, gas, and solid surface material. Furthermore, the contact angle, θ is sensitive to slight contaminants such as detergents and the cleanliness of the solid surface. When the contact angle, θ is less than 90°, the liquid is described as "receding" over the solid surface and thus wets the solid surface (adhesive forces exceed cohesive forces) (see Figure 1.10a). On the other hand, when the contact angle, θ exceeds 90°, the liquid is described as "advancing" over the solid surface and thus does not wet the solid surface (cohesive forces exceed adhesive forces) (see Figure 1.10b). For instance, for water, air, and clean glass, the contact angle, θ is zero degrees, for which the water droplet will spread smoothly on the clean glass (wetting the glass surface); thus, the liquid is described as "receding" (adhesive forces exceed cohesive forces) over the solid. And similarly, for water, air, and ice, the contact angle, θ is 20°, so the liquid is described as "receding" over the solid, whereas, for water, air, paraffin wax, carnauba wax, or shellac, the contact angle, θ ranges between 105 and 110°, so the liquid is described as "advancing" over the solid. One may note that discrete water droplets will form when placed on a newly waxed surface. Furthermore, for mercury, air, and glass, the contact angle, θ is 129°, so the liquid is described as "advancing" (cohesive forces exceed adhesive forces) over the solid. One may note that mercury droplets form spherical balls that can be rolled around like a solid ball on a solid surface without wetting the solid surface.

1.10.5.5 Capillarity

The effects of surface tension, σ, coupled with the surface tension contact angle, θ, are important in problems involving the capillary rise or drop/depression of a liquid. The capillary rise of liquids occurs in narrow spaces, such as in the soil water zone or within the tissue of tall trees. Additionally, a capillary rise of a liquid such as water, but a capillary drop/ depression of a liquid such as mercury, occurs when a small tube is inserted into the liquid. As such, the capillary rise or fall of a liquid in a small tube inserted into a liquid is used to illustrate the surface tension, σ and the surface tension contact angle, θ that occur in capillarity. Figure 1.9a and b illustrate the capillary rise for water and the capillary drop/ depression for mercury, respectively. When the adhesive forces exceed the cohesive forces (receding liquids such as water), a net upward force causes water to wet and rise in the small tube, forming a concave meniscus (inward-curving liquid surface), and the contact angle, θ is less than 90°(see Figure 1.9a). However, when the cohesive forces exceed the adhesive forces (advancing liquids such as mercury), a net downward force causes the mercury not to wet the tube and to fall/depress in the small tube, forming a convex meniscus (outward-bulging liquid surface), and the contact angle, θ exceeds 90° (see Figure 1.9b). When the rise or depression, h is much larger than the radius, r of the tube, the interface separating the liquid from the adjacent gas, which is called the meniscus, may be assumed to be nearly spherical. Furthermore, the height, h of the capillary rise or drop is a function of the magnitude fluid surface tension, σ.

As such, the rise, h may be estimated by considering the cross section through a capillary rise in Figure 1.9a. Given that the capillary rise, h is in static equilibrium, the net upward force, which is due to molecular adhesion and is a result of the tensile stress or surface tension, σ, is balanced by the gravity force, F_G as follows:

$$\sum_s F = F_T - F_G = 0$$

$$\sum_s F = \sigma P \cos \theta - \gamma h A = 0$$

$$\gamma h A = \sigma P \cos \theta$$

$$\gamma h \pi r^2 = \sigma 2 \pi r \cos \theta$$

$$h = \frac{2\sigma \cos \theta}{\gamma r} \tag{1.94}$$

where s = axis along the capillary rise, A = the cross-sectional area of the small tube, P = the perimeter of the small tube, r = the radius of the small tube, and θ = surface tension contact angle. Therefore, the unbalanced net upward adhesive force, $\sigma P \cos \theta$ causes the water to wet and rise in the small tube, forming a concave meniscus. As such, the column of liquid, h supported by the fluid varies directly with the fluid surface tension, σ and the surface tension contact angle, θ, while it varies inversely with the fluid specific weight, γ and the radius, r of the small tube. Thus, lighter fluids experience greater capillary rise. And, thinner tubes allow a greater rise (and fall) of the liquid in the tube. As such, capillarity is typically negligible for tubes with a radius, r greater than 0.5 cm. Thus, when sizing the tubing for manometers and barometers, which are used to make pressure measurements (the rise of the fluid in the tube is a measure of pressure) (see Chapter 2), they should be designed with a sufficiently large radius, r in order to avoid capillary rise (and fall). Furthermore, when the surface tension contact angle, θ exceeds 90° ($-\cos \theta$) (such as for mercury), a depression (negative value for h) rather than a rise is obtained, as illustrated in Figure 1.9b for mercury. And, finally, Figure 1.9a and b illustrate a curved free surface when water and mercury surfaces contact a vertical glass surface, respectively, for which the surface tension on the free surface will support small loads; a small needle gently placed on the free surface will not sink, but will be supported by the tension in the liquid surface.

When the capillary rise has ceased in Figure 1.9a, the column of water is suspended from the meniscus, which is attached to the walls by hydrogen bonds. Therefore, the water is under "tension," which is defined as a negative pressure (i.e., it is less than atmospheric pressure). One may calculate this "tension" or negative pressure, p_m at the meniscus as follows. The negative pressure, p_m is defined as the weight, $W = F_G = \gamma h A$ of the water suspended below the plane that is tangent to the lowest point on the meniscus, acting over the area of the plane, $A = \pi\, r^2$, which yields:

$$p_m = \frac{-F_G}{A} = \frac{-\gamma h A}{A} = -\gamma h = \frac{-2\sigma \cos\theta}{r} \tag{1.95}$$

Thus, similar to the height of capillary rise, h, the pressure difference at the meniscus, p_m is inversely proportional to the radius of the small tube, r.

EXAMPLE PROBLEM 1.32

Water at a temperature of 40°C rises in a clean glass tube with a diameter of 0.75 cm, as illustrated in Figure EP 1.32. (a) Determine the capillary rise in the tube.

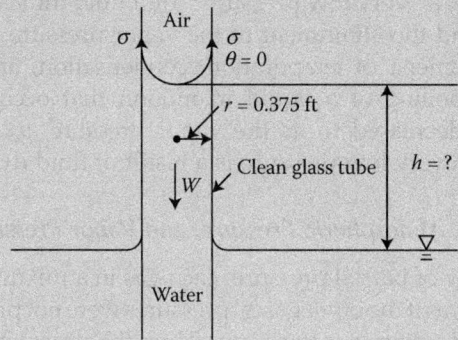

FIGURE EP 1.32
Capillary rise of water in clean glass tube.

Mathcad Solution

(a) The surface tension and the specific weight for water at 40°C are read from Table A.2 in Appendix A, and the capillary rise in the tube is determined by applying Equation 1.94 as follows:

$$h = \frac{2\sigma \cos\theta}{\gamma r}$$

Because the glass is clean, the contact angle, θ between the water, air, and glass is assumed to be zero degrees as follows:

$$N := kg\frac{m}{sec^2} \qquad \sigma := 0.0696\frac{N}{m} \qquad \gamma := 9.731\frac{kN}{m^3} \qquad \theta := 0\,deg$$

$$D := 0.75\,cm = 7.5 \times 10^{-3}m \qquad\qquad r := \frac{D}{2} = 3.75 \times 10^{-3}\,m$$

$$h := \frac{2 \cdot \sigma \cdot \cos(\theta)}{\gamma \cdot r} = 3.815 \times 10^{-3}\,m$$

1.10.6 Vapor Pressure

The pressure force, F_P plays an important role in both fluid statics and fluid flow/dynamics. When the fluid is either a gas or a liquid (static or dynamic), a significant change in pressure introduces the fluid property, compressibility, E_v (discussed in the following section), whereas when the fluid is a liquid (static or dynamic), a significant change in pressure introduces the liquid property, vapor pressure, p_v and an undesirable fluid phenomena, cavitation (only for a liquid in motion), which are discussed in the section herein. In particular, the vapor pressure, p_v of a liquid becomes important when very low pressures are involved in both fluid statics and fluid flow/dynamics. The vapor pressure, p_v of a liquid is measured in absolute pressure, where absolute pressure is the sum of the atmospheric pressure and the gage pressure (see Equation 1.24 and Chapter 2). As such, the vapor pressure, p_v of a liquid is measured relative to the atmospheric pressure. Furthermore, a "very low pressure" for a fluid is an indication of its absolute pressure relative to the atmospheric pressure (zero gage pressure). Therefore, a "very low pressure" for a fluid indicates a negative gage pressure. The definition of and the attainment of the vapor pressure, p_v of a liquid involves a description of the phenomena of evaporation, condensation, and boiling. Cavitation is defined as unintended boiling of a liquid in motion that occurs at room temperature when the pressure has decreased to its the vapor pressure, p_v. Furthermore, cavitation becomes important when very low pressure is a result of fluid dynamics.

1.10.6.1 Partial Pressure, Atmospheric Pressure, and Vapor Pressure

According to Dalton's law of partial pressure, each gas in a mixture of gases exerts its own pressure (partial pressure) as if the other gases' pressures were not present. For instance, atmospheric air is a mixture of dry air and water vapor. Thus, the atmospheric pressure is the sum of the partial pressure of the dry air (mostly nitrogen and oxygen) and the partial pressure of the water vapor (usually less than 3% of the atmospheric pressure). Furthermore, the sum of the partial pressures of the gases in the atmosphere equals the total atmospheric pressure, where the partial pressure of water vapor is called the vapor pressure, p_v of the liquid, water.

The vapor pressure, p_v of a liquid increases with an increase in temperature. The vapor pressure, p_v for water at standard sea-level atmospheric pressure is presented as a function of temperature in Table A.2 in Appendix A. And, although the vapor pressure, p_v for some common liquids will vary with temperature, for practical applications, their vapor pressures, p_v at standard sea-level atmospheric pressure at room temperature (68°F or 20°C), in general, are given in Table A.4 in Appendix A. One may note that at room temperature (68°F or 20°C), the vapor pressure, p_v of a liquid is much lower than the atmospheric pressure. For instance, one may note that at room temperature (68°F or 20°C), the vapor pressure, p_v of water is only 0.34 *psia* (*lb/in*^{2} absolute), while at the normal boiling point of water (212°F), the vapor pressure, p_v of water increases to the atmospheric pressure of 14.69 *psia*. Thus, at the standard/normal boiling point of a liquid, the vapor pressure, p_v of the liquid is equal to the atmospheric pressure. It is important to note that the vapor pressure, p_v of a liquid is measured in absolute pressure, where absolute pressure is the sum of the atmospheric pressure and the gage pressure (see Equation 1.24 and Chapter 2). Therefore, at room temperature, where the vapor pressure, p_v of a liquid is much lower than the atmospheric pressure, the gage pressure of the liquid is negative. Furthermore, at the normal boiling point of a liquid, the vapor pressure, p_v of the liquid increases to the atmospheric pressure; thus, the gage pressure of the liquid is zero. The dimensions and units for the vapor pressure, p_v in the English system are given as follows:

$$p_v = \frac{F_P}{A} = \frac{[F]}{[L^2]} = [lb/ft^2 \ absolute] = [psfa] \qquad (1.96)$$

or,

$$p_v = \frac{F_P}{A} = \frac{[F]}{[L^2]} = [lb/in^2 \ absolute] = [psia] \tag{1.97}$$

The dimensions and units for the vapor pressure, p_v in the metric system are given as follows:

$$p_v = \frac{F_P}{A} = \frac{[MLT^{-2}]}{[L^2]} = [N/m^2 \ absolute] \tag{1.98}$$

EXAMPLE PROBLEM 1.33

Using Table A.2 in Appendix A, (a) Graphically illustrate the variation of the vapor pressure of water with temperature assuming the SI system of units.

Mathcad Solution

(a) A graphical illustration of the variation of the vapor pressure of water with temperature assuming the SI system of units is given as follows:

$$T := \begin{pmatrix} 0 \\ 5 \\ 10 \\ 15 \\ 20 \\ 25 \\ 30 \\ 40 \\ 50 \\ 60 \\ 70 \\ 80 \\ 90 \\ 100 \end{pmatrix} \qquad Pv := \begin{pmatrix} 0.611 \\ 0.872 \\ 1.230 \\ 1.710 \\ 2.34 \\ 3.17 \\ 4.24 \\ 7.38 \\ 12.33 \\ 19.92 \\ 31.16 \\ 47.34 \\ 70.10 \\ 101.33 \end{pmatrix}$$

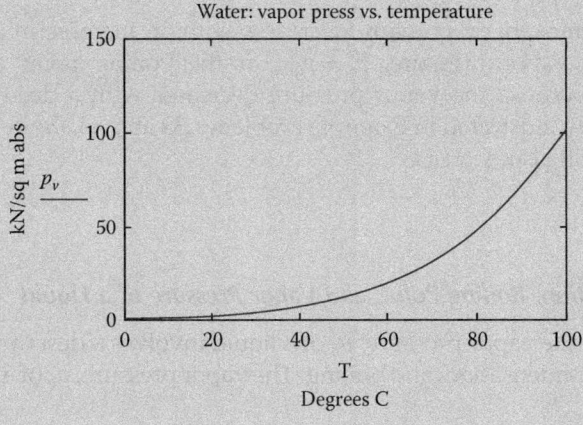

EXAMPLE PROBLEM 1.34

Water is placed in a container that is open to the atmosphere. (a) Determine the temperature and the vapor pressure at the boiling point at 0 ft above sea level. (b) Determine the temperature and the vapor pressure at the boiling point at 20,000 ft above sea level. (c) Discuss the results.

Solution

The boiling point of a liquid is defined as the temperature at which the vapor pressure of the liquid equals to the pressure above it. Thus, if the water is placed in a container that is open to the atmosphere, then the water will boil at the temperature at which the vapor pressure of the water is equal to the atmospheric pressure.

(a) The atmospheric pressure at 0 ft above sea level is read from Table A.1 in Appendix A as follows:

$p_{atm} = 14.6959 \, psia$

Thus, the vapor pressure at the boiling point is equal to the atmospheric pressure as follows:

$p_v = p_{atm} = 14.6959 \, psia$

Reading from Table A.2 in Appendix A, this magnitude of vapor pressure for water occurs at a temperature of 212°F. Thus, the temperature of water at the boiling point is 212°F.

(b) The atmospheric pressure at 20,000 ft above sea level is read from Table A.1 in Appendix A as follows:

$p_{atm} = 6.7588 \, psia$

Thus, the vapor pressure at the boiling point is equal to the atmospheric pressure as follows:

$p_v = p_{atm} = 6.7588 \, psia$

Reading from Table A.2 in Appendix A, this magnitude of vapor pressure for water occurs at a temperature of roughly 175°F. Thus, the temperature of water at the boiling point is roughly 175°F.

(c) Therefore, because the atmospheric pressure decreases with an increase in altitude, the corresponding vapor pressure, $p_v = p_{atm}$ at the boiling point also decreases. Furthermore, because the vapor pressure decreases with a decrease in temperature (graphically illustrated in Example Problem 1.33 above), the corresponding boiling point will also decrease.

1.10.6.2 Evaporation, Condensation, Boiling Point, and Vapor Pressure of a Liquid

The definition and attainment of the vapor pressure, p_v of a liquid involves a description of the phenomena of evaporation, condensation, and boiling. The vapor pressure, p_v of a liquid increases with an increase in temperature. Furthermore, one may note that at room temperature (68°F or 20°C), the vapor pressure, p_v of a liquid is much lower than the atmospheric pressure. In the case where a liquid at room temperature is open to the atmosphere, while evaporation will take place, the vapor pressure, p_v of the liquid will not be attained because the atmospheric pressure above the liquid is much higher than the vapor pressure, p_v of the liquid at room temperature. However, in the case where a liquid at room temperature is placed in an enclosed container and evacuated of air to form a vacuum (absolute pressure above liquid is at zero), evaporation and condensation will take place, and the vapor pressure, p_v of the liquid will be attained because the zero pressure above the liquid is lower than the vapor pressure, p_v of the liquid at room temperature. And, finally, in the case where a liquid that is open to the atmosphere is heated to its boiling point, evaporation, condensation, and boiling will take place. In this case, the vapor pressure, p_v of the liquid will be attained because the increase in temperature from room temperature to its boiling point increases the vapor pressure, p_v of the liquid such that it overcomes the atmospheric pressure.

In the surface phenomena of evaporation, the vapor pressure, p_v of a liquid (at room temperature and open to the atmosphere) is not attained. A liquid is placed in a container that is open to the atmosphere, as illustrated in Figure 1.11a. Some of the liquid molecules at the surface of the liquid have sufficient momentum (molecular kinetic energy) to overcome the intermolecular cohesive forces and thus escape into the atmosphere. This surface phenomena, where there is a change from the liquid phase to the gas phase, is called evaporation or vaporization, which is a potent cooling mechanism. The pressure above the liquid surface is atmospheric. And, because the vapor pressure, p_v of the liquid is low (much lower than the atmospheric pressure), it will not be attained during the ordinary evaporation process.

In the surface phenomena of evaporation and condensation, the vapor pressure, p_v of a liquid (at room temperature and subjected to zero pressure) is attained. A liquid is placed in a container that is enclosed and evacuated of air to form a vacuum, as illustrated in Figure 1.11b. The pressure above the liquid surface is initially at zero, and will increase and stabilize at the vapor pressure, p_v of the liquid when dynamic equilibrium is reached. A pressure will develop in the vacuum due to the vapor that is formed by the escaping molecules. Evaporation (liquid molecules leave the liquid surface) and condensation (gas molecules enter the liquid surface) will continue to take place until the liquid surface has reached a dynamic equilibrium or saturation. Specifically, the liquid surface will reach a dynamic equilibrium or saturation when the number of molecules leaving the liquid surface is equal to the number of molecules entering the liquid surface. As such, dynamic equilibrium or saturation of a liquid surface is reached when evaporation and condensation have balanced one another. At this point the vapor is said to be saturated, and the pressure of that vapor is called the saturated vapor pressure, or simply the vapor pressure, p_v of the liquid. Because

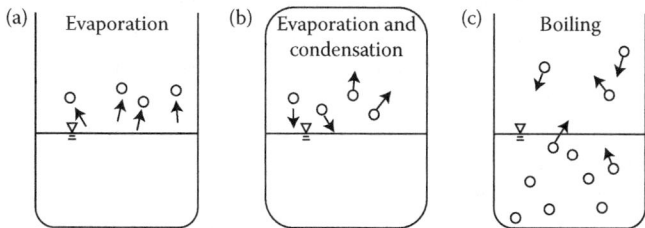

FIGURE 1.11
(a) Evaporation of liquid in open container. (b) Evaporation and condensation of liquid in enclosed and vacuumed container. (c) Boiling of liquid in open and heated container.

the molecular kinetic energy of a liquid is greater at a higher temperature, more molecules can escape from the liquid surface, so the corresponding vapor pressure, p_v of the liquid increases with an increase in temperature.

In the volume phenomena of evaporation, condensation, and boiling, the vapor pressure, p_v of a liquid (at boiling point and open to the atmosphere) is attained. A liquid placed in the container that is open to the atmosphere is heated to the boiling point of the liquid, as illustrated in Figure 1.11c. The vapor pressure, p_v of the liquid increases with an increase in temperature. Thus, at the standard/normal boiling point of a liquid, the vapor pressure, p_v of the liquid is equal to the atmospheric pressure. Bubbles will form and rise because the vapor pressure, p_v of the liquid can overcome the atmospheric pressure, and the evaporation or vaporization during the boiling process becomes a volume phenomena (as opposed to a surface phenomena in the evaporation cooling process). The boiling point is the temperature at which the vapor pressure, p_v of a liquid equals the external pressure above it. Therefore, one may reduce the boiling point of a liquid simply by reducing the pressure of the liquid to its vapor pressure, p_v at the corresponding temperature. Thus, the higher the vapor pressure, p_v of a liquid at a given temperature, the lower the normal boiling point (the boiling point at the atmospheric pressure) of the liquid. Furthermore, boiling of a liquid can be induced at a given vapor pressure, p_v of the liquid by raising the temperature, or at a given fluid temperature by lowering the atmospheric pressure. Thus, it is interesting to note that at higher altitudes (above sea level), the atmospheric pressure decreases. As a result, the corresponding vapor pressure, p_v and boiling point of a liquid also decrease. As such, it would take longer to cook food at a high altitude because the boiling point of water is lower. However, on the other hand, in a pressure cooker, where the pressure above the liquid is higher than atmospheric, the corresponding vapor pressure, p_v and boiling point of a liquid will increase. As such, food will cook more quickly in a pressure cooker.

1.10.6.3 Cavitation

Cavitation is defined as unintended boiling of a liquid in motion that occurs at room temperature when the pressure has decreased to its the vapor pressure, p_v. Cavitation becomes important when very low pressure is a result of fluid dynamics. In general fluid flow situations, if the differences in velocity (or elevation) are significant, then the differences in pressure can also be significant. Therefore, cavitation may occur in a flowing liquid if the local pressure falls to the vapor pressure, p_v of the liquid. Furthermore, it may be noted that cavitation does not occur in a gas flow because of its expansion property. A gas does not change phase at a low pressure, whereas a liquid does if the pressure is low enough. In the flow of a liquid, regions of high velocity (or high elevation) and thus low pressure, can cause cavitation of the liquid. Cavitation (unintended vaporization and thus boiling) results where the velocity of the flow (or elevation of the flow) is high and the liquid pressure is reduced to the vapor pressure, p_v at which the liquid boils at room temperature (or any temperature below the normal boiling point of the liquid). For instance, water flowing at a room temperature of 68°F will boil (cavitate) if the pressure is reduced to its vapor pressure, p_v of 0.34 *psia*. One may compare this to the standard that water at a temperature of 212°F (normal boiling point of water) will boil when it is at its corresponding standard vapor pressure, p_v of 14.69 *psia*. Therefore, one may reduce the boiling point of a liquid simply by reducing the pressure of the liquid to its vapor pressure, p_v at the corresponding temperature.

Cavitation, which occurs when the liquid pressure in the fluid flow is reduced to its vapor pressure, p_v, is modeled by the Bernoulli equation and is known as the "Bernoulli effect." As presented above, the vapor pressure, p_v of a liquid is measured in absolute pressure, where absolute pressure is the sum of the atmospheric pressure and the gage pressure (see Equation 1.24 and Chapter 2). Therefore, at room temperature, where the vapor pressure, p_v of a liquid is much lower than the atmospheric pressure, the gage pressure of the liquid is negative. The reduction in pressure to the vapor pressure, p_v can occur because of a significant increase in velocity (due to a reduction in cross-sectional area) or a significant increase in elevation (in a suction line of a pump or a turbine, or a siphon tube lifting water). For instance, when the diameter of a pipe is significantly reduced, as illustrated in Figure 1.12 (see Example Problem 4.5), the velocity increases (in accordance with the continuity equation presented in Chapter 3), while the corresponding pressure decreases in the pipe constriction (in accordance with the Bernoulli equation presented in Chapter 4). One may note that the decrease in pressure may be significant such that the liquid reaches its vapor pressure, p_v, which results in cavitation. Another practical example of cavitation as a result of a reduction in the cross-sectional area occurs when a garden hose is kinked, which results in a constriction in the flow area, and a hissing sound (which indicates the presence of cavitation). The resulting velocity at the kink will be significantly increased, while the corresponding pressure will be significantly decreased to the water's vapor pressure, p_v at the room temperature. An example of cavitation as a result of a significant increase in elevation is in a suction line of a pump or a siphon tube lifting water. Because the suction/siphon pressure is less than the atmospheric pressure, the pipe or tube can collapse, causing a restriction in the flow path, which can stop the pump/siphon from operating. Furthermore, if the pressure at the suction/siphon end of the pipe/tube is decreased to the water's vapor pressure, p_v at the room temperature, cavitation will occur (see Example Problem 4.17).

In general, flow situations that may result in cavitation include flow through complex passages in valves, flow through the suction side of a pump (see Example Problems 4.20 and 4.22), flow in a turbine (see Example Problems 4.25 and 4.27), flow through a siphon, flow in a pipeline, flow in the underflow structure of a high dam (if the overflow water on a high dam spillway breaks contact with the spillway surface, a vacuum will form at the point of separation and cavitation may occur), and in the high-speed motion of submarines and hydrofoils. Cavitation should be avoided or minimized in the flow system because it can cause reduction in performance and efficiency, annoying noise and vibration, and structural damage to the system. Furthermore, over a long period of time, the spikes in pressure, which result from the collapsing of the cavitation bubbles near a solid surface, may cause erosion, surface pitting, failure due to fatigue, and eventual destruction of the hydraulic system. Thus, in order to avoid cavitation, the design of the flow system must ensure that the local pressure is maintained above the vapor pressure, p_v. When the pressure in the flow decreases to the vapor pressure, p_v of the liquid, the resulting vapor bubbles are called cavitation bubbles because they form "cavities" in the liquid. As the vapor bubbles are swept

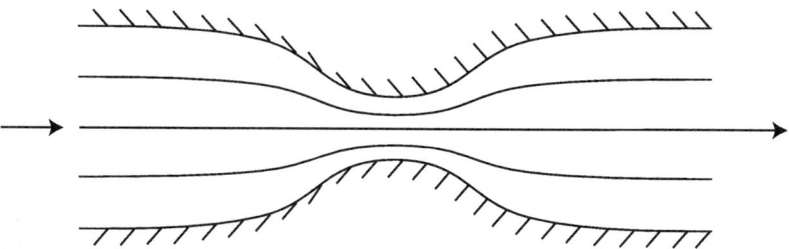

FIGURE 1.12
Cavitation of a liquid due to significant reduction in pipe diameter.

away from the low-pressure region and into a high-pressure region, they will suddenly collapse with sufficient intensity to actually cause structural damage. The formation of vapor bubbles and their subsequent collapse in a flowing liquid is called cavitation. Cavitation can produce dynamic effects called implosion (the opposite of explosion), which cause very large pressure transients in the vapor bubbles. When the bubbles collapse near a solid surface, they can, over time, cause damage to the surface in the cavitation area.

EXAMPLE PROBLEM 1.35

Water at 50°C is lifted from a tank to a higher elevation by the use of a siphon tube. (a) Determine the minimum pressure in the siphon tubing in order to avoid cavitation.

Solution

(a) Cavitation will occur if the pressure of the water in the siphon tubing system is allowed to drop below the vapor pressure of the water at 50°C, which is read from Table A.2 in Appendix A as follows:

$$p_v = 12.33 \, kN/m^2 \, abs$$

Thus, the minimum pressure in the siphon tubing in order to avoid cavitation is given as follows:

$$p_{min} = p_v = 12.33 \, kN/m^2 abs$$

1.10.7 Elasticity, Compressibility, or Bulk Modulus of Elasticity

The elastic force, F_E plays an important role in both fluid statics and fluid flow/dynamics when compressible fluids are involved. When the fluid is a liquid (static or dynamic), a significant change in pressure introduces the liquid property, vapor pressure, p_v and an undesirable fluid phenomena, cavitation (only for a liquid in motion), which were discussed in the preceding section. On the other hand, when the fluid is a either a gas or a liquid (static or dynamic), a significant change in pressure introduces the fluid property, elasticity, compressibility, or bulk modulus of elasticity, E_v, which is discussed in the section herein. Therefore, in both fluid statics and fluid flow/dynamics, the bulk modulus of elasticity, E_v becomes important when compressible fluids/flows are involved. For a given temperature, a change in density, ρ of a given fluid mass that is due to a change in pressure is a measure of the compressibility of the fluid. Although all fluids have a certain degree of compressibility, some fluids (liquids and gases) are assumed to be incompressible when the change in density due to a change in pressure is negligible. Therefore, for compressible fluids/flow, the fluid density is assumed to vary with pressure, while for incompressible fluids/flow, the fluid density is assumed to be a constant with respect to pressure. It is important to note that although liquids can be compressed (resulting in small changes in density) at very high pressures, they are typically assumed to be incompressible for most practical engineering applications. However, the fact that sound waves (pressure waves) can travel through liquids and cause water hammer in a pipe flow (discussed below) provides evidence of the compressibility or the elasticity of liquids. Furthermore, it is important to note that although gases are easily compressed (resulting in significant changes in density) at very high pressures, they are typically assumed to be incompressible (as indicated by the

Mach number, *M*) when the pressure variation is small in comparison to the absolute pressure (sum of atmospheric and gage pressure). Thus, while the bulk modulus of elasticity, E_v and the sonic velocity, c (discussed below) are physical fluid properties that measure the degree of compressibility of a fluid, the Mach number, *M* serves as a practical indicator of the degree of compressibility of fluids in general, and especially for gases.

Although the bulk modulus of elasticity, E_v can be determined for both liquids and gases, it is usually tabulated only for liquids. Determination of the bulk modulus of elasticity, E_v for gases involves the application of the ideal gas law and the first law of thermodynamics. Application of the ideal gas law results in the introduction of two additional gas properties, which include the gas constant, R, and the molecular weight (molar mass), *m*. Furthermore, applications of the first law of thermodynamics and the ideal gas law introduce three additional gas properties, which include the ideal specific heat at constant pressure, c_p; the ideal specific heat at constant volume, c_v; and the specific heat ratio, $k = \frac{c_p}{c_v}$, which are related to the gas constant, R. And, application of the first law of thermodynamics for incompressible fluids (liquids) introduces the definition for the specific heat, c for liquids. And, finally, application of the continuity and momentum equations introduces the sonic (acoustic) velocity, c.

1.10.7.1 Definition of Bulk Modulus of Elasticity

For a given temperature, a change in density ρ of a given fluid mass that is due to a change in pressure is a measure of the compressibility or elasticity of the fluid, as measured by the fluid property called the bulk modulus of elasticity, E_v. In general, the density, ρ of a fluid decreases with an increase in temperature, while it increases with an increase in pressure. Because the density of a gas is strongly dependent on both temperature and pressure, it is a variable; thus, gas flow is modeled as compressible flow (unless the pressure variation is small, for which it is modeled as incompressible flow). However, since the density of a liquid is more affected by the temperature than by the pressure, for practical changes in pressure, the density of a liquid is essentially constant and thus is accurately "approximated" as an incompressible flow. When pressure is applied to a given fluid mass, it will contract in volume (or increase in density), and when the pressure is released, it will expand in volume (or decrease in density). Furthermore, the amount of volume (or density) change is different for different fluids. Thus, the property that relates volume (or density) changes to the change in pressure is the elasticity or the compressibility of the fluid, which is called the bulk modulus of elasticity, and is denoted by E_v. The bulk modulus of elasticity, E_v is defined as the ratio of the change in unit pressure to the corresponding change per unit volume (or unit density) for a given temperature and is given as:

$$E_v = \frac{dp}{-\dfrac{dV}{V}} \tag{1.99}$$

where dp is the incremental change in pressure needed to create an incremental change in volume, dV of a volume of fluid, V (where dV/V is negative for a positive dp and thus a positive E_v). Since a decrease in volume of a given mass, $M = \rho V$ will result in an increase in density, ρ, E_v may also be expressed as follows:

$$E_v = \frac{dp}{\dfrac{d\rho}{\rho}} \tag{1.100}$$

where dp is the incremental change in density and ρ is the density of the fluid mass. The dimensions and units for the bulk modulus of elasticity, E_v in the English system are given as follows:

$$E_v = \frac{dp}{-\dfrac{dV}{V}} = \frac{[FL^{-2}]}{[L^3 L^{-3}]} = [lb/ft^2] = [psf] \tag{1.101}$$

or,

$$E_v = \frac{dp}{-\dfrac{dV}{V}} = \frac{[FL^{-2}]}{[L^3 L^{-3}]} = [lb/in^2] = [psi] \tag{1.102}$$

The dimensions and units for the bulk modulus of elasticity, E_v in the metric system are given as follows:

$$E_v = \frac{dp}{-\dfrac{dV}{V}} = \frac{[ML^{-1}T^{-2}]}{[L^3 L^{-3}]} = [N/m^2] \tag{1.103}$$

Thus, the dimensions and units for E_v are the same dimensions and units for pressure, p and are often expressed as absolute pressure (sum of atmospheric and gage pressure), in order to model a variable atmospheric pressure (especially in the determination of E_v for gases).

One may note that although the bulk modulus of elasticity, E_v can be determined for both liquids and gases, it is usually tabulated only for liquids. Determination of the bulk modulus of elasticity, E_v for gases involves the applications of the ideal gas law and the first law of thermodynamics, which are discussed below. The bulk modulus of elasticity, E_v for water at standard sea-level atmospheric pressure is presented as a function of temperature in Table A.2 in Appendix A (values for E_v are given in psi gage). And, although the bulk modulus of elasticity, E_v for some common liquids will vary with temperature, for practical applications, their bulk modulus of elasticity, E_v at standard sea-level atmospheric pressure at room temperature (68°F or 20°C), in general, are given in Table A.4 in Appendix A (values for E_v are given in psi gage). Therefore, although the bulk modulus of elasticity, E_v for liquids is a function of temperature and pressure (increases with pressure), the value for the bulk modulus of elasticity, E_v at or near the sea-level atmospheric pressure (and usually at room temperature) is the one of interest in most engineering applications. Furthermore, although liquids are typically viewed as incompressible for most practical engineering applications, they can be compressed at very high pressures; thus, the values for E_v for common liquids are large. Therefore, a large value for E_v for a liquid indicates that a large change in pressure is needed to cause a small fractional change in volume. Thus, in comparison to a compressible gas, a liquid with a large E_v is essentially incompressible and acts like an elastic solid with respect to pressure. For instance, for air (compressible gas) at room temperature (68°F), $E_v = 14.696 \, psia$ (assuming isothermal conditions, $E_v = p$, where p is the absolute pressure of the gas), while for water at room temperature, $E_v = 318,000 \, psi$ (see Table A.4 in Appendix A), and for steel (elastic solid) at room temperature (68°F), Young's modulus of elasticity is about $26,000,000 \, psi$. Thus, while the modulus of elasticity for water is about 80 times as compressible as steel, in comparison to air, it behaves as an incompressible fluid (elastic solid), where the modulus of elasticity for air is about 21,000 times as compressible as water. Furthermore, in comparison to mercury at room temperature (68°F), for which $E_v = 3,800,000\,psi$ (see Table A.4 in Appendix A), water is about 10 times as compressible as mercury. And, finally, if a fluid were truly incompressible, it would have a bulk modulus of elasticity, $E_v = \infty$.

EXAMPLE PROBLEM 1.36

A container with 190 gallons of water at 40°F and under an atmospheric pressure of 14.696 psia is compressed until the pressure is 3000 psia. (a) Determine the change in volume of the water. (b) Determine the change in density of the water.

Mathcad Solution

Table A.2 in Appendix A is used to read the bulk modulus of elasticity, E_v and the density, ρ.

(a) The change in volume of the water is determined by applying Equation 1.99 as follows:

$$E_v = \frac{dp}{-\dfrac{dV}{V}}$$

$1\,gal = 0.134\,ft^3 \quad E_v := 294000\,psi = 4.234 \times 10^7\,psf \quad V_1 := 190\,gal = 25.399\,ft^3$

$p_1 := 14.696\,psi \qquad p_2 := 3000\,psi \qquad \Delta p := p_2 - p_1 = 4.299 \times 10^5\,psf$

Guess value: $\qquad\qquad\qquad\qquad V_2 := 20\,ft^3$

Given

$$E_V = \frac{\Delta p}{-\dfrac{(V_2 - V_1)}{V_1}}$$

$V_2 := Find(V_2) = 188.071\,gal \qquad\qquad \Delta V := V_2 - V_1 = -1.929\,gal$

Thus, an increase in pressure causes a decrease in volume.

(b) The change in density of the water is determined by applying Equation 1.100 as follows:

$$E_v = \frac{dp}{\dfrac{d\rho}{\rho}}$$

$slug := 1\,lb\dfrac{sec^2}{ft} \qquad\qquad\qquad \rho_1 := 1.94\dfrac{slug}{ft^3}$

Guess value: $\qquad\qquad\qquad \rho_2 := 2\dfrac{slug}{ft^3}$

Given

$$E_V = \frac{\Delta p}{\dfrac{(\rho_2 - \rho_1)}{\rho_1}}$$

$$\rho_2 := \text{Find}(\rho_2) = 1.96\,\frac{\text{slug}}{\text{ft}^3} \qquad\qquad \Delta\rho := \rho_2 - \rho_1 = 0.02\,\frac{\text{slug}}{\text{ft}^3}$$

Thus, an increase in pressure causes an increase in density.

EXAMPLE PROBLEM 1.37

Using Table A.2 in Appendix A. (a) Graphically illustrate the variation of the bulk modulus of elasticity of water with temperature, assuming the SI system of units.

Mathcad Solution

(a) A graphical illustration of the variation of the bulk modulus of elasticity with temperature, assuming the SI system of units, is given as follows:

$$
T := \begin{pmatrix} 0 \\ 5 \\ 10 \\ 15 \\ 20 \\ 25 \\ 30 \\ 40 \\ 50 \\ 60 \\ 70 \\ 80 \\ 90 \\ 100 \end{pmatrix}
\qquad
E_V := \begin{pmatrix} 2.02 \\ 2.06 \\ 2.10 \\ 2.14 \\ 2.18 \\ 2.22 \\ 2.25 \\ 2.28 \\ 2.29 \\ 2.28 \\ 2.25 \\ 2.20 \\ 2.14 \\ 2.07 \end{pmatrix}
$$

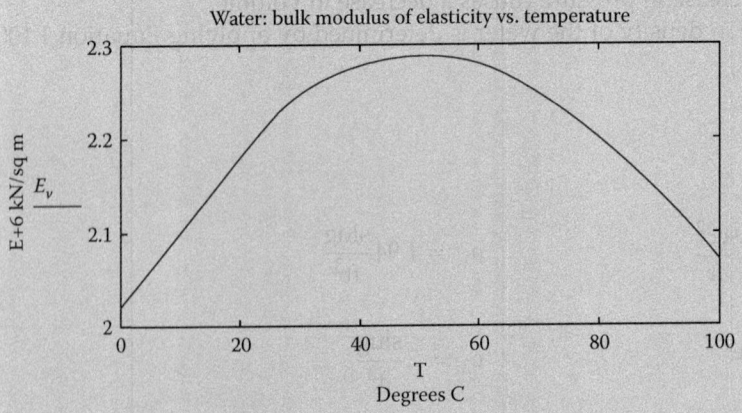

Although liquids are typically assumed to be incompressible for most practical engineering applications, it may be noted that small changes in the density of liquids due to large changes in pressure can have important practical consequences such as water hammer in a pipe flow. Water hammer is caused by the vibrations of the pipe generated by the reflection of sound waves (pressure waves) following the sudden closure (or opening) of a valve. The liquid in the pipe network encounters an abrupt flow restriction (due to the closing of the valve) and is thus locally compressed. The sound waves that are produced strike the pipe surfaces, bends, and valves as they propagate and reflect along the pipe, causing the pipe to vibrate and produce the sound of a water hammer. Therefore, the fact that sound waves (pressure waves) can travel through liquids, and cause water hammer in a pipe flow provides evidence of the compressibility or the elasticity of liquids.

1.10.7.2 Ideal Gas Law: Definition of Gas Constant and Molecular Weight

Determination of the bulk modulus of elasticity, E_v for gases involves the application of the ideal gas law (presented in this section) and application of the first law of thermodynamics (presented in Section 1.10.7.3). Application of the ideal gas law results in the introduction of two additional gas properties, which include the gas constant, R and the molecular weight (molar mass), m. Furthermore, applications of the first law of thermodynamics and the ideal gas law introduce three additional gas properties, which include the ideal specific heat at constant pressure, c_p; the ideal specific heat at constant volume, c_v; and the specific heat ratio, $k = (c_p/c_v)$, which are related to the gas constant, R.

For a given temperature, a change in density, ρ of a given fluid mass that is due to a change in pressure is a measure of the compressibility or elasticity of the fluid, as measured by the fluid property bulk modulus of elasticity, E_v. As noted above, in comparison to liquids, gases are highly compressible, resulting in significant changes in density, ρ at very high pressures for a given temperature. In general, the density, ρ of a fluid decreases with an increase in temperature, while it increases with an increase in pressure. Furthermore, any equation that relates the pressure, p; temperature, T; and density, ρ of a fluid mass is defined as an "equation of state." In particular, the simplest and most commonly applied equation of state for fluids in the gas phase is the ideal gas law. An ideal gas is defined as a hypothetical perfect gas that obeys the ideal gas law. Furthermore, a real gas at a low density closely approximates an ideal gas. In particular, at low pressures and high temperatures, the density of a real gas decreases and behaves like an ideal gas. Therefore, although the ideal gas law is defined for hypothetical perfect/ideal gases, it closely approximates the behavior of real gases when the conditions of the real gases are far removed from the liquid phase (state of saturation), where the density is low. As such, it is typical to assume that most real gases are perfect/ideal and are represented by the ideal gas law as follows:

$$p = \rho RT \tag{1.104}$$

or:

$$pv = RT \tag{1.105}$$

where p = absolute pressure (sum of atmospheric and gage pressure); ρ = the density of the fluid mass; $v = (1/\rho)$ = the specific volume of a fluid mass; R = a gas constant that is a function of the particular gas and is related to the molecular weight (molar mass), m of the gas;

and T = absolute temperature (measured above absolute zero) in degrees Rankine (English units) or degrees Kelvin (metric units). The gas constant, R is a function of the particular gas, and is related to the molecular weight (molar mass), m of the gas as follows:

$$R = \frac{R_u}{m} \tag{1.106}$$

where m = the molecular weight (molar mass) of a gas, and R_u = the universal gas constant. The definition for the molecular weight (molar mass), m of a gas is given as follows:

$$m = \frac{M}{N} \tag{1.107}$$

where M = mass and N = the number of moles of gas. The dimensions and units for the number of moles of gas, N in the English system are given as follows:

$$N = [slug - mol] \tag{1.108}$$

where a new dimension is defined herein as the "amount of matter," which has an English unit of a "mole" and is abbreviated by mol. The dimensions and units for the number of moles of gas, N in the metric system, are given as follows:

$$N = [kg - mol] \tag{1.109}$$

where the new dimension, "amount of matter" has a metric unit of a "mole" and is abbreviated by mol. Therefore, the dimensions and units for the molecular weight (molar mass), m in the English system are given as follows:

$$m = \frac{M}{N} = \frac{[FL^{-1}T^2]}{[FL^{-1}T^2N]} = \frac{[slug]}{[slug - mol]} \tag{1.110}$$

and the dimensions and units for the molecular weight (molar mass), m in the metric system are given as follows:

$$m = \frac{M}{N} = \frac{[M]}{[MN]} = \frac{[kg]}{[kg - mol]} \tag{1.111}$$

The dimensions and units for the gas constant, R in the English system are given as follows:

$$R = \frac{p}{\rho T} = \frac{[FL^{-2}]}{[FL^{-4}T^2{}^\circ R]} = \frac{[lb/ft^2]}{[slug/ft^3]} = [ft - lb/slug - {}^\circ R] \tag{1.112}$$

The dimensions and units for the gas constant, R in the metric system are given as follows:

$$R = \frac{p}{\rho T} = \frac{[ML^{-1}T^{-2}]}{[ML^{-3}{}^\circ K]} = \frac{[N/m^2]}{[kg/m^3{}^\circ K]} = [N - m/kg - {}^\circ K] = [J/kg - {}^\circ K] \tag{1.113}$$

The molar mass, m and the gas constant, R for some common gases at standard sea-level atmospheric pressure at room temperature (68°F or 20°C) are presented in Table A.5 in Appendix A. One may note that unlike some gas properties (such as density, ρ and absolute viscosity, μ) for the common gases illustrated in Table A.5 in Appendix A, the molar mass, m and the gas constant, R are independent of temperature and pressure. And, finally, the dimensions and units for the universal gas constant, R_u in the English system are given as follows:

$$R_u = Rm = [ft - lb/slug - °R]\frac{[slug]}{[slug - mol]} = [ft - lb/slug - mol - °R] \tag{1.114}$$

which has a value of 49,709 $(ft - lb)/(slug - mol - °R)$. Furthermore, the dimensions and units for the universal gas constant, R_u in the metric system are given as follows:

$$R_u = Rm = [N - m/kg - °K]\frac{[kg]}{[kg - mol]} = [N - m/kg - mol - °K] \tag{1.115}$$

which has a value of 8,314 $(N\text{-}m)/(kg - mol - °K)$. One final note: substituting the expressions for $\rho = (M/V)$, $R = (R_u/m)$, and $m = (M/N)$ into the ideal gas law $p = \rho RT$, the ideal gas law is commonly expressed as follows:

$$p = \frac{M}{V}\frac{R_u}{m}T = \frac{M}{V}R_u\frac{N}{M}T \tag{1.116}$$

$$pV = NR_uT \tag{1.117}$$

Finally, one may note that most of the common gases presented in Table A.5 in Appendix A, such as air, carbon dioxide, carbon monoxide, helium, hydrogen, methane, nitrogen, and oxygen, can be modeled by the ideal gas law. However, an exception among the common gases is water vapor at a high density (as a result of high pressures and low temperatures), where the real gas is near the liquid phase (state of saturation) and thus does not closely approximate an ideal gas.

EXAMPLE PROBLEM 1.38

A 300-ft^3 oxygen tank is compressed under a pressure of 500 psia and at a temperature of 40°F. (a) Determine the density of the oxygen in the tank. (b) Determine the weight of the oxygen in the tank.

Mathcad Solution

(a) To determine the density of the oxygen in the tank, one may apply either Equation 1.104 or Equation 1.117. Equation 1.104 is applied in this Example Problem as follows:

$$p = \rho RT$$

where the gas constant, R for oxygen is read from Table A.5 in Appendix A as follows:

$$slug := 1\,lb\frac{sec^2}{ft} \qquad\qquad p := 500\,psi = 7.2 \times 10^4\,psf$$

$$T := 40°F = 499.67°R \qquad\qquad R := 1554\,\frac{ft\,lb}{slug\,°R}$$

$$\text{Guess value:} \qquad\qquad \rho := 1\,\frac{slug}{ft^3}$$

Given

$$p = \rho R T$$

$$\rho := \text{Find}(\rho) = 0.093 \frac{\text{slug}}{\text{ft}^3}$$

(b) To determine the weight of the oxygen in the tank, Equation 1.44 is applied as follows:

$$\gamma = \frac{W}{V} = \frac{Mg}{V} = \rho g$$

$$V := 300 \, \text{ft}^3 \qquad\qquad\qquad g := 32.2 \frac{\text{ft}}{\text{sec}^2}$$

$$\gamma := \rho g = 2.991 \frac{\text{lb}}{\text{ft}^3} \qquad\qquad W := \gamma V = 897.385 \, \text{lb}$$

EXAMPLE PROBLEM 1.39

A 69-m^3 helium tank is compressed under a pressure of 800 kN/m^2 abs and at a temperature of 10°C. (a) Determine the density of the helium in the tank. (b) Determine the weight of the helium in the tank.

Mathcad Solution

(a) To determine the density of the helium in the tank, one may apply either Equation 1.104 or Equation 1.117. Equation 1.117 is applied in this Example Problem as follows:

$$pV = N R_u T$$

which is solved for N, the number of moles of helium, as follows:

$$N := \text{kg} \frac{\text{m}}{\text{sec}^2} \qquad\qquad p := 800 \frac{\text{kN}}{\text{m}^2} \qquad V := 69 \, \text{m}^3$$

$$T := 10°C = 283.15 \, \text{K} \qquad R_u := 8314 \frac{\text{Nm}}{\text{kg mol K}}$$

Guess value: $N_m := 1 \, \text{kg mol}$

Given

$$p \, V = N_m \, R_u \, T$$

$$N_m := \text{Find} \, (N_m) = 23.448 \, \text{kg mol}$$

Note that the number of moles, N is called N_m in Mathcad in order to distinguish between it and the force units of Newton, N. Then, the density, ρ of the helium is computed by applying Equation 1.33, as follows:

$$\rho = \frac{M}{V}$$

where the mass, M of the helium is computed by applying Equation 1.107 as follows:

$$m = \frac{M}{N}$$

where the molar mass, m for helium is read from Table A.5 in Appendix A as follows:

$$mm := 4.003 \, \frac{kg}{kg \, mol} \qquad\qquad M := mm \, N_m = 93.864 \, kg$$

$$\rho := \frac{M}{V} = 1.36 \, \frac{kg}{m^3}$$

Note that the molar mass, m is called *mm* in Mathcad in order to distinguish between it and the mass units of meters, m.

(b) To determine the weight of the helium in the tank, Equation 1.44 is applied as follows:

$$\gamma = \frac{W}{V} = \frac{Mg}{V} = \rho g$$

$$g := 9.81 \, \frac{m}{sec^2} \qquad \gamma := \rho g = 13.345 \, \frac{N}{m^3} \qquad W := \gamma \, V = 920.804 \, N$$

1.10.7.3 First Law of Thermodynamics: Definition of Specific Heat and Specific Heat Ratio

Determination of the bulk modulus of elasticity, E_v for gases involves the application of the ideal gas law (presented in Section 1.10.7.2) and application of the first law of thermodynamics (presented in this section). Application of the ideal gas law results in the introduction of two additional gas properties, which include the gas constant, R and the molecular weight (molar mass), m. Furthermore, applications of the first law of thermodynamics and the ideal gas law introduce three additional gas properties, which include the ideal specific heat at constant pressure, c_p; the ideal specific heat at constant volume, c_v; and the specific heat ratio, $k = (c_p/c_v)$, which are related to the gas constant, R. And, finally, application of the first law of thermodynamics for incompressible fluids (liquids) introduces the definition for the specific heat, c for liquids.

Application of the first law of thermodynamics introduces enthalpy, h which is a composite energy property of a fluid. Furthermore, application of the ideal gas law introduces the ideal specific heat at constant pressure, c_p; the ideal specific heat at constant volume, c_v; and the ideal specific heat ratio, $k = (c_p/c_v)$, which are related to the gas constant, R. The first law of thermodynamics states that "the energy of a system undergoing change can neither be created nor destroyed, it can only change forms." Specifically, the energy content, E of a fixed quantity of mass (a closed system) can be changed by two mechanisms: heat transfer, Q and work transfer, W, which is expressed as follows:

$$E_{sys} = Q_{sys} + W_{sys} \tag{1.118}$$

(see Chapter 4). In general, energy exists in numerous forms, including thermal, mechanical, kinetic, potential, magnetic, electrical, chemical, and nuclear, for which their sum is the total energy, E. One may note that a fluid property that is related to the total mass, M of a fluid system is known as an extensive fluid property and is usually represented by an uppercase letter. However, a fluid property that is independent of the mass, M of the fluid system is known as an intensive (or specific) fluid property, and is usually represented by a lowercase letter (see Chapter 3). The total energy per unit mass of a system is defined as $e = E/M$. There are two general types of energy: microscopic energy and macroscopic energy. Microscopic energy is related to the molecular structure and the degree of molecular activity of the system. The sum of all microscopic forms of energy is called the internal energy, U (where the internal energy per unit mass is defined as $u = U/M$) of the system, where the sensible and latent forms of internal energy are referred to as thermal energy (heat). The macroscopic energy is related to the motion and the influence of some external effects such as gravity, magnetism, electricity, and surface tension. In the absence of the effects of magnetism, electricity, and surface tension, a system is called a simple compressible system. Thus, the total energy, E of a simple compressible system consists of internal energy, U (microscopic energy); kinetic energy, KE (macroscopic energy); and potential energy, PE (macroscopic energy). The kinetic energy is a result of the motion, v of the system relative to some fixed frame of reference, where the kinetic energy per unit mass, $ke = KE/M = v^2/2$. And the potential energy is a result of the elevation, z of the system relative to some external reference point in a gravitational field (gravitational acceleration), g, where the potential energy per unit mass, $pe = PE/M = gz$. Therefore, the total energy per unit mass, e is defined as follows:

$$e = u + ke + pe \tag{1.119}$$

$$e = u + \frac{v^2}{2} + gz \tag{1.120}$$

However, one may note that a fluid entering or leaving a control volume possesses an additional form of energy, which is called the flow energy, or flow work, p/ρ, which is the energy per unit mass required to move/push the fluid and maintain the flow. Thus, the total energy of a flowing fluid (for a control volume), $e_{flowing}$ is defined as follows:

$$e_{flowing} = \frac{p}{\rho} + e \tag{1.121}$$

$$e_{flowing} = \underbrace{\frac{p}{\rho} + u}_{h = \text{enthalpy}} + \frac{v^2}{2} + gz \tag{1.122}$$

where the sum of the internal energy per unit mass, u and the flow energy (flow work), p/ρ is defined as fluid property enthalpy per unit mass, h as follows:

$$h = u + \frac{p}{\rho} \tag{1.123}$$

The dimensions and units for energy may be derived from the definitions of two mechanisms of transfer of energy: heat and work. The definition of energy in terms of work is force times distance. Thus, the dimensions and units for energy, E in the English system are given as follows:

$$E = (F)(d) = [FL] = [ft - lb] \tag{1.124}$$

where the British thermal unit (Btu) is defined as follows:

$$1 \, Btu = 777.649 \, ft - lb \tag{1.125}$$

which is defined as the energy required to raise the temperature of $1 \, lb$ of water at 68°F (room temperature) by 1°F. The dimensions and units for energy, E in the metric system are given as follows:

$$E = (F)(d) = [FL] = [N - m] = [J] \tag{1.126}$$

where

$$1 \, kJ = 1000 \, J \tag{1.127}$$

and where

$$1 \, Btu = 1.0543 \, kJ \tag{1.128}$$

Another metric unit of energy is called the calorie (cal), where:

$$1 \, cal = 4.1868 \, J \tag{1.129}$$

which is defined as the energy required to raise the temperature of $1 \, g$ of water at 14.5°C by 1°C.

Application of the ideal gas law in the definition flow energy (flow work), p/ρ in the expression for the enthalpy, h introduces three additional gas properties, which include the ideal specific heat at constant pressure, c_p; the ideal specific heat at constant volume, c_v; and the ideal specific heat ratio, $k = (c_p/c_v)$, which are related to the gas constant, R. Therefore, the flow energy, or flow work, p/ρ for an ideal gas is defined by the ideal gas law as follows:

$$\frac{p}{\rho} = RT \tag{1.130}$$

and thus, the enthalpy for an ideal gas may be expressed as follows:

$$h = u + RT \tag{1.131}$$

The internal energy per unit mass, u is due to the kinetic energy of molecular motion and the forces between molecules, and for an ideal gas is a function of temperature only. Furthermore, recall that R is independent of both temperature and pressure. Therefore, the enthalpy, h is considered to be a composite energy property of a fluid, where for an ideal gas, the enthalpy, h is a function of temperature only (and thus independent of pressure). Thus, differentiating the expression for the enthalpy, h with respect to temperature yields the following:

$$\frac{dh}{dT} = \frac{du}{dT} + \frac{d(RT)}{dT} \tag{1.132}$$

$$\frac{dh}{dT} = \frac{du}{dT} + R \tag{1.133}$$

where the increase in the enthalpy, h when the temperature of an ideal gas is increased one degree for a constant pressure is defined as the ideal specific heat at constant pressure, c_p as follows:

$$\frac{dh}{dT} = \left(\frac{\partial h}{\partial T}\right)_{p=constant} = c_p \tag{1.134}$$

where, although the value for c_p for a given gas varies with temperature, for moderate changes in temperature, it is reasonable to assume a constant value for the c_p for a given gas. And, the increase in internal energy, u when the temperature of an ideal gas is increased one degree with its volume, V (or specific volume, $v = (1/\rho)$ held constant is defined as the ideal specific heat at constant volume, c_v as follows:

$$\frac{du}{dT} = \left(\frac{\partial u}{\partial T}\right)_{v=constant} = c_v \tag{1.135}$$

where, although the value for c_v for a given gas varies with temperature, for moderate changes in temperature, it is reasonable to assume a constant value for the c_v for a given gas. Therefore, perfect/ideal gases are typically assumed to have constant specific heats, c_p and c_v. Furthermore, one may note the relationship between the ideal specific heats, c_p and c_v and the ideal gas constant, R, which is independent of both temperature and pressure, and is given as follows:

$$c_p - c_v = R \tag{1.136}$$

where $c_p > c_v$ thus, one may note that the difference between c_p and c_v is constant for an ideal gas; regardless of temperature. Additionally, the units for the ideal specific heats, c_p and c_v are the same as for the ideal gas constant, R. Specifically, the units for the ideal specific heats, c_p and c_v in the English system are given as follows:

$$[ft - lb/slug - {}^\circ R] \tag{1.137}$$

and the units for the ideal specific heats, c_p and c_v in the metric system are given as follows:

$$[N - m/kg - {}^\circ K] = [J/kg - {}^\circ K] \tag{1.138}$$

and, finally, the ideal specific heat ratio, k is defined as follows:

$$k = \frac{c_p}{c_v} \tag{1.139}$$

which is dimensionless. Combining the relationship between the ideal specific heats, c_p and c_v and R and the ideal specific heat ratio, k yields the following expressions for the ideal specific heats, c_p and c_v, respectively:

$$c_p = \frac{Rk}{k-1} \tag{1.140}$$

$$c_v = \frac{R}{k-1} \tag{1.141}$$

The ideal specific heats, c_p and c_v and the ideal specific heat ratio, k for some common gases at standard sea-level atmospheric pressure at room temperature (68°F or 20°C), are presented in Table A.5 in Appendix A. One may note that although c_p, c_v, and k for a given gas vary with temperature, for moderate changes in temperature, it is reasonable to assume a constant value for c_p, c_v, and k for a given gas.

EXAMPLE PROBLEM 1.40

Using Table A.5 in Appendix A, (a) What are the ideal specific heat at constant pressure, c_p; the ideal specific heat at constant volume, c_v; and the ideal specific heat ratio, $k = (c_p/c_v)$ for nitrogen at room temperature (68°F)? (b) What are the ideal specific heat at constant pressure, c_p; the ideal specific heat at constant volume, c_v; and the ideal specific heat ratio, $k = (c_p/c_v)$ for methane at room temperature (20°C)?

Solution

(a) The ideal specific heat at constant pressure, c_p; the ideal specific heat at constant volume, c_v; and the ideal specific heat ratio, $k = (c_p/c_v)$ for nitrogen at room temperature (68°F) are as follows:

$c_p = 6{,}210\, ft - lb/slug -°R$

$c_v = 4{,}437\, ft - lb/slug -°R$

$k = \frac{c_p}{c_v} = 1.40$

(b) The ideal specific heat at constant pressure, c_p; the ideal specific heat at constant volume, c_v; and the ideal specific heat ratio, $k = (c_p/c_v)$ for methane at room temperature (20°C) are as follows:

$c_p = 2{,}250\, N - m/kg -°K$

$c_v = 1{,}730\, N - m/kg -°K$

$k = \frac{c_p}{c_v} = 1.30$

And, finally, application of the first law of thermodynamics for incompressible fluids (liquids) introduces the definition for the specific heat, c for liquids. Differentiation of the expression for the enthalpy, h for incompressible fluids (liquids) with respect to temperature, T yields the definition for the specific heat, c for liquids. In the case of incompressible fluids (liquids), the expression for enthalpy, h is repeated as follows:

$$h = u + \frac{p}{\rho} \tag{1.142}$$

where the internal energy per unit mass, u is due to the kinetic energy of molecular motion and the forces between molecules and is a function of temperature, and the density of an incompressible fluid, ρ is a constant with respect to both temperature and pressure. Therefore, the enthalpy, h is considered to be a composite energy property of a fluid, where for an incompressible flow, the enthalpy, h is a function of both temperature and pressure. Thus, differentiating the expression for the enthalpy, h for an incompressible fluid with respect to temperature yields the following:

$$\frac{dh}{dT} = \frac{du}{dT} + \frac{d\left(\frac{p}{\rho}\right)}{dT} \tag{1.143}$$

$$\frac{dh}{dT} = \frac{du}{dT} + \frac{1}{\rho}\frac{dp}{dT} \tag{1.144}$$

where the increase in the enthalpy, h when the temperature of an incompressible fluid is increased one degree for a constant pressure is defined as the specific heat at constant pressure, c_p as follows:

$$\frac{dh}{dT} = \frac{du}{dT} + \frac{1}{\rho}\frac{dp}{dT} = \frac{du}{dT} + 0 \tag{1.145}$$

$$\frac{dh}{dT} = \frac{du}{dT} = \left(\frac{\partial h}{\partial T}\right)_{p=constant} = c_p \tag{1.146}$$

and the increase in internal energy, u when the temperature of an incompressible fluid is increased one degree with its volume, V [or specific volume, $v = (1/\rho)$] held constant is defined as the specific heat at constant volume, c_v as follows:

$$\frac{du}{dT} = \left(\frac{\partial u}{\partial T}\right)_{v=constant} = c_v \tag{1.147}$$

thus,

$$c_p = c_v \tag{1.148}$$

where the constant pressure-specific heat, c_p and the constant volume-specific heat, c_v for incompressible fluids (liquids) are equal; thus, the specific heat, c for liquids is defined as follows:

$$c_p = c_v = c \tag{1.149}$$

Therefore, for a constant-pressure process, a change in the enthalpy, h with respect to temperature, T is equal to a change in the internal energy, u with respect to temperature, T is represented by the specific heat, c for the liquid, and is given as follows:

$$\frac{dh}{dT} = \frac{du}{dT} = c \qquad (1.150)$$

$$dh = du = cdT \qquad (1.151)$$

However, for a constant-temperature process, a change in the enthalpy, h is equal to a change in the pressure, p and is given as follows

$$dh = du + \frac{dp}{\rho} \qquad (1.152)$$

$$dh = 0 + \frac{dp}{\rho} \qquad (1.153)$$

as the internal energy is a function of temperature. Although specific heat, c for some common liquids will vary with temperature, for practical applications, their specific heat, c at standard sea-level atmospheric pressure at room temperature (68°F or 20°C), in general, are given in Table A.4 in Appendix A.

EXAMPLE PROBLEM 1.41

Using Table A.4 in Appendix A, (a) What is the specific heat for mercury at room temperature (68°F)? (b) What is the specific heat for kerosene at room temperature (20°C)?

Solution

(a) The specific heat for mercury at room temperature (68°F) is as follows:

$c = 834 \, ft - lb/slug-°R$

(b) The specific heat for kerosene at room temperature (20°C) is as follows:

$c = 2000 \, N - m/kg-°K$

1.10.7.4 Determination of Bulk Modulus of Elasticity for Gases

Although the bulk modulus of elasticity E_v can be determined for both liquids and gases, it is usually tabulated only for liquids. Determination of the bulk modulus of elasticity E_v for gases involves applications of the ideal gas law and the first law of thermodynamics as presented above. Beginning with the definition of the bulk modulus of elasticity, E_v for a given temperature, it is summarized as follows:

$$E_v = -\frac{dp}{\dfrac{dV}{V}} = \frac{dp}{\dfrac{d\rho}{\rho}} \qquad (1.154)$$

where E_v and p are expressed as absolute pressure (sum of atmospheric and gage pressure). It is important to note that in accordance with the various laws of thermodynamics, when gases are compressed or expanded, the relationship between the pressure, p and the density, ρ will depend upon the nature of the compression or expansion process. Specifically, there are two types of compression or expansion processes: (1) isothermal (constant temperature), and (2) isentropic (frictionless and no heat exchange with the surroundings). Thus, the value of the bulk modulus of elasticity E_v for an ideal gas will depend upon the type of compression or expansion process involved (isothermal vs. isentropic). As such, if the compression or expansion of the gas takes place under isothermal conditions (constant temperature), then in accordance to ideal gas law in Equation 1.104:

$$\frac{p}{\rho} = C \tag{1.155}$$

where $C = $ a constant. However, if the compression or expansion of the gas takes place under isentropic (frictionless and no heat exchange with the surroundings), then:

$$\frac{p}{\rho^k} = C \tag{1.156}$$

where $C = $ a constant, and k is called an adiabatic exponent or the ideal specific heat ratio, $k = (c_p/c_v)$, which is the ratio of the ideal specific heat at constant pressure, c_p to the ideal specific heat at constant volume, c_v.

Therefore, assuming that the compression or expansion of the gas takes place under isothermal conditions (constant temperature), the bulk modulus of elasticity E_v for an ideal gas at a given temperature, T is determined as follows. Substituting the expression for the pressure, p Equation 1.155 in the definition for the bulk modulus of elasticity, E_v for a given temperature, T in Equation 1.154 yields the following:

$$E_v = \frac{dp}{-\dfrac{dV}{V}} = \frac{dp}{\dfrac{d\rho}{\rho}} = \rho\frac{dp}{d\rho} = \rho\frac{d(C\rho)}{d\rho} = C\rho\frac{d\rho}{d\rho} = C\rho = p \tag{1.157}$$

where under isothermal conditions (constant temperature), the bulk modulus of elasticity, E_v for an ideal gas at a given temperature, T is equal to the absolute pressure, p, and where E_v for the gas varies directly with the absolute pressure, p.

EXAMPLE PROBLEM 1.42

Refer to Example Problem 1.38, where for a 300-ft^3 oxygen tank compressed under a pressure of 500 psia and at a temperature of 40°F, the density of the oxygen in the tank was determined to be 0.093 slug/ft^3, and the weight of the oxygen in the tank was determined to be 897.385 lb. The oxygen in the tank is further compressed under isothermal conditions (constant temperature) to a volume of 100 ft^3. (a) Determine the resulting density of the oxygen in the tank. (b) Determine the resulting pressure of the oxygen in the tank. (c) Determine the resulting bulk modulus of elasticity of the oxygen in the tank.

Mathcad Solution

(a) To determine the resulting density of the oxygen in the tank, one must note that the mass of the oxygen in Equation 1.33, and thus, the weight of the oxygen Equation 1.44 remains a constant as follows:

$$slug := 1lb \frac{sec^2}{ft} \qquad p_1 := 500\,psi = 7.2 \times 10^4\,psf \qquad V_1 := 300\,ft^3$$

$$g := 32.2 \frac{ft}{sec^2} \qquad \rho_1 := 0.093 \frac{slug}{ft^3} \qquad W_1 := 897.385\,lb$$

$$V_2 := 100\,ft^3 \qquad W_2 := W_1 = 897.385\,lb \qquad \gamma_2 := \frac{W_2}{V_2} = 8.974 \frac{lb}{ft^3}$$

$$\rho_2 := \frac{\gamma_2}{g} = 0.279 \frac{slug}{ft^3}$$

(b) Given compression under isothermal conditions (constant temperature), to determine the resulting pressure of the oxygen in the tank, one may apply Equation 1.155, $\frac{p}{\rho} = C$ as follows:

$$\frac{p_1}{\rho_1} = 7.756 \times 10^5 \frac{ft^2}{s^2}$$

Guess value: $p_2 := 1\,psf$

Given

$$\frac{p_1}{\rho_1} = \frac{p_2}{\rho_2}$$

$$p_2 := Find(p_2) = 2.158 \times 10^5\,psf$$

(c) Given compression under isothermal conditions (constant temperature), the resulting bulk modulus of elasticity of the oxygen in the tank is equal to the absolute pressure of the oxygen in Equation 1.157. Thus:

$$E_v := p_2 = 2.158 \times 10^5\,psf$$

Alternatively, assuming that the compression or expansion of the gas takes place under isentropic (frictionless and no heat exchange with the surroundings), the bulk modulus of elasticity, E_v for an ideal gas at a given temperature, T is determined as follows. Substituting in the expression for the pressure, p Equation 1.156 in the definition for the bulk modulus of elasticity, E_v for a given temperature, T in Equation 1.154 yields the following:

$$E_v = \frac{dp}{-\frac{dV}{V}} = \frac{dp}{\frac{d\rho}{\rho}} = \rho \frac{dp}{d\rho} = \rho \frac{d(C\rho^k)}{d\rho} = \rho k C \rho^{k-1} \frac{d\rho}{d\rho} = kC\rho^k = kp \tag{1.158}$$

where under isentropic (frictionless and no heat exchange with the surroundings), the bulk modulus of elasticity, E_v for an ideal gas at a given temperature, T is equal to the product of the absolute pressure and the ideal specific heat ratio, kp.

EXAMPLE PROBLEM 1.43

Refer to Example Problem 1.39, where for a 69-m^3 helium tank compressed under a pressure of 800 kN/m^2 abs and at a temperature of 10°C, the density of the helium in the tank was determined to be 1.36 kg/m^3, and the weight of the helium in the tank was determined to be 920.804 N. The helium in the tank is further compressed under isentropic (frictionless and no heat exchange with the surroundings) to a volume of 35 m^3. (a) Determine the resulting density of the helium in the tank. (b) Determine the resulting pressure of the helium in the tank. (c) Determine the resulting bulk modulus of elasticity of the helium in the tank.

Mathcad Solution

(a) To determine the resulting density of the helium in the tank, one must note that the mass of the helium in Equation 1.33 and thus the weight of the helium in Equation 1.44 remains a constant as follows:

$$N := kg\frac{m}{sec^2} \qquad\qquad P_1 := 800\frac{kN}{m^2} \qquad\qquad V_1 := 69\,m^3$$

$$g := 9.81\frac{m}{sec^2} \qquad\qquad \rho_1 := 1.36\frac{kg}{m^3} \qquad\qquad W_1 := 920.804\,N^3$$

$$V_2 := 35\,m^3 \qquad\qquad W_2 = W_1 := 920.804\,N^3 \qquad \gamma_2 := \frac{W_2}{V_2} = 26.309\frac{N}{m^3}$$

$$\rho_2 := \frac{\gamma_2}{g} = 2.682\frac{kg}{m^3}$$

(b) Given compression under isentropic (frictionless and no heat exchange with the surroundings), to determine the resulting pressure of the helium in the tank, one may apply Equation 1.156, $\frac{p}{\rho^k} = C$ as follows:

$$k := 1.66 \qquad\qquad\qquad \frac{P_1}{(1.36)^k\frac{kg}{m^3}} = 4.802 \times 10^5\frac{m^2}{s^2}$$

Guess value: $\qquad\qquad\qquad\qquad P_2 := 1\frac{N}{m^2}$

Given

$$\frac{P_1}{\left[(1.36)^k\frac{kg}{m^3}\right]} = \frac{P_2}{\left[(2.682)^k\frac{kg}{m^3}\right]}$$

$$P_2 := Find(p_2) = 2.47 \times 10^6\frac{N}{m^2}$$

(c) Given compression under isentropic (frictionless and no heat exchange with the surroundings), the resulting bulk modulus of elasticity of the helium in the tank is equal to the product of the absolute pressure of the helium and the specific heat ratio, which is read from Table A.5 in Appendix A (Equation 1.158), as follows:

$$E_V := kp_2 = 4.1 \times 10^6\frac{N}{m^2}$$

1.10.7.5 Definition of the Sonic (Acoustic) Velocity for Fluids

For a compressible fluid (liquid or gas), the occurrence of a slight localized increase in pressure, p and density, ρ of the fluid results in a small wave called a wave of compression. As such, the velocity (celerity) of the small wave of compression through the compressible fluid may be determined by simultaneous application of the continuity and momentum equations. The resulting expression for the velocity of the small wave of compression is called the sonic or acoustic velocity, and is given as follows:

$$c = \sqrt{\frac{dp}{d\rho}} \tag{1.159}$$

Thus, the small pressure disturbance travels through the compressible fluid at a finite velocity, which is dependent upon the bulk modulus of elasticity, E_v of the fluid. Thus, substituting Equation 1.54 for E_v into Equation 1.159 for c yields the following expression for the sonic velocity:

$$c = \sqrt{\frac{dp}{d\rho}} = \sqrt{\frac{E_v}{\rho}} \tag{1.160}$$

where c is called the sonic velocity, because sound, which is a small pressure disturbance, travels at this velocity. As mentioned above, for practical engineering applications, although fluids (liquids and some gases) are typically modeled as incompressible, they can actually be compressed at very high pressures. Furthermore, if a fluid were truly incompressible, it would have a bulk modulus of elasticity, $E_v = \infty$, and thus, the sonic velocity, $c = \infty$. In the case of an ideal incompressible (inelastic) fluid, the sound (small pressure disturbance) would be transmitted instantaneously between two points in the fluid, which does not actually occur. The sonic velocity for the standard atmosphere at various elevations (and thus various temperatures and pressures) are given in Table A.1 in Appendix A. However, the sonic velocity, c for water and other common liquids may be computed by applying Equation 1.160. The bulk modulus of elasticity, E_v and the density, ρ for water at standard sea-level atmospheric pressure are given as a function of temperature in Table A.2 in Appendix A, and those for some common liquids at standard sea-level atmospheric pressure at room temperature (68°F or 20°C) are given in Table A.4 in Appendix A.

EXAMPLE PROBLEM 1.44

It is of interest to compare the sonic velocity for standard atmosphere and the sonic velocity for water, both at standard sea-level atmospheric pressure. (a) Determine the sonic velocity for standard atmosphere at 0 ft above sea level. (b) Determine the sonic velocity for water at standard sea-level atmospheric pressure.

Mathcad Solution

(a) Table A.1 in Appendix A is used to read the sonic velocity for standard atmosphere at 0 ft above sea level as follows:

$$c_{standardatm} := 1116.45 \frac{ft}{sec}$$

where the corresponding temperature at 0 ft above sea level is 59°F.

(b) The sonic velocity for standard atmosphere and the sonic velocity for water, both at standard sea-level atmospheric pressure, are compared at a common temperature of about 59°F or 60°F, respectively. Thus, the sonic velocity for water at 60°F at standard sea-level atmospheric pressure is determined by applying Equation 1.160. Table A.2 in Appendix A is used to read the bulk modulus of elasticity, E_v, and the density, ρ for water at standard sea-level atmospheric pressure, at 60°F, as follows:

$$\text{slug} := 1 \, \text{lb} \frac{\text{sec}^2}{\text{ft}} \qquad\qquad E_{vwater} := 311000 \frac{\text{lb}}{\text{in}^2} \qquad\qquad \rho_{water} := 1.938 \frac{\text{slug}}{\text{ft}^3}$$

$$c_{water} := \sqrt{\frac{E_{vwater}}{\rho_{water}}} = 4.807 \times 10^3 \frac{\text{ft}}{\text{s}} \qquad\qquad \frac{c_{water}}{c_{standardatm}} = 4.306$$

Thus, the sonic velocity for water is about four times the sonic velocity for standard atmosphere both at standard sea-level atmospheric pressure and at a temperature of about 59°F or 60°F.

And, finally, the sonic velocity, c for gases may be computed by applying Equation 1.160 above, with the appropriate equation for the bulk modulus of elasticity, E_v, either Equation 1.157 or Equation 1.158, depending upon if the ideal gas is assumed to be either isothermal (constant temperature) or isentropic (frictionless and no heat exchange with the surroundings), respectively. For the computation of the sonic velocity, c for gases, typically, the compression or expansion process of the ideal gas is assumed to be isentropic. The basis for the isentropic assumption is as follows. As stated above, a sound wave is a result of a slight localized increase in pressure, p and density, ρ of the fluid. As such, the disturbance caused by sound (small wave of compression, pressure disturbance, or sound wave) that moves through a fluid is very slight and fast, such that the heat exchange in the compression or expansion is negligible, so the process is assumed to be isentropic. Therefore, for an ideal gas, the sonic velocity is computed by substituting Equation 1.158 into Equation 1.160 as follows:

$$c = \sqrt{\frac{E_v}{\rho}} = \sqrt{\frac{kp}{\rho}} \tag{1.161}$$

Furthermore, for an ideal gas, one may substitute Equation 1.104 for the pressure into Equation 1.161 as follows:

$$c = \sqrt{\frac{E_v}{\rho}} = \sqrt{\frac{kp}{\rho}} = \sqrt{\frac{k\rho RT}{\rho}} = \sqrt{kRT} \tag{1.162}$$

where the sonic velocity, c for an ideal gas is independent of the pressure of the gas and thus dependent only on the temperature of the gas (c increases with an increase in temperature). Furthermore, recall that both k and R were each independent of both the pressure and the temperature of the gas. And, finally, k and R for some common gases at "standard sea-level atmospheric pressure at room temperature (68°F or 20°C)" (actually, at any pressure or temperature) are presented in Table A.5 in Appendix A. Recall that in water vapor at a high density (as a result of high pressures and low temperatures), the gas is near the liquid phase (state of saturation) and thus does not closely approximate an ideal gas.

EXAMPLE PROBLEM 1.45

It is of interest to compare the sonic velocity for air, carbon monoxide, and nitrogen at a specified temperature. (a) Determine the sonic velocity for air at 45°C. (b) Determine the sonic velocity for helium at 45°C. (c) Determine the sonic velocity for hydrogen at 45°C.

Mathcad Solution

The sonic velocities for the three common gases at 45°C are computed by applying Equation 1.162 as follows:

$$c = \sqrt{kRT}$$

where the values for k and R, which are each independent of both the pressure and the temperature of the gas, are read from Table A.5 in Appendix A.

(a) The sonic velocity for air at 45°C is determined as follows:

$$N := kg \frac{m}{sec^2} \qquad T := 45°C = 318.15 \, K$$

$$k_{air} := 1.4 \qquad R_{air} := 287 \frac{N\,m}{kg\,K} \qquad c_{air} := \sqrt{k_{air} R_{air} T} = 357.537 \frac{m}{s}$$

(b) The sonic velocity for helium at 45°C is determined as follows:

$$k_{He} := 1.66 \qquad R_{He} := 2077 \frac{N\,m}{kg\,K} \qquad c_{He} := \sqrt{k_{He} R_{He} T} = 1.047 \times 10^3 \frac{m}{s}$$

$$\frac{c_{He}}{c_{air}} = 2.929$$

   Thus, the sonic velocity of helium is about three times the sonic velocity of air, both at a temperature of 45°C.

(c) The sonic velocity for hydrogen at 45°C is determined as follows:

$$k_{H2} := 1.4 \qquad R_{H2} := 4120 \frac{N\,m}{kg\,K} \qquad c_{H2} := \sqrt{k_{H2} R_{H2} T} = 1.355 \times 10^3 \frac{m}{s}$$

$$\frac{c_{H2}}{c_{air}} = 3.789 \qquad \frac{c_{H2}}{c_{He}} = 1.293$$

   Thus, the sonic velocity of hydrogen is almost four times the sonic velocity of air, and about 1.3 times the sonic velocity of helium, all at a temperature of 45°C.

1.10.7.6 Definition of the Mach Number for Fluids: A Measure of Compressibility

Although fluids (liquids and some gases) are typically modeled as incompressible for practical engineering applications, they can actually be compressed at very high pressures. This was illustrated above by the fact that if a fluid were truly incompressible, it would have a bulk modulus of elasticity, $E_v = \infty$, and thus the sonic velocity, $c = \infty$, which does not occur. In particular, although gases are easily compressed (resulting in significant changes in density) at very high pressures, they are typically assumed to be incompressible when the pressure variation is small in comparison to the absolute pressure (sum of atmospheric and gage pressure). For instance, although the modulus of elasticity of gases such air ($E_v = 14.696\,psi$) is about 21,000 times as compressible as water ($E_v = 318,000\,psi$), gases are typically assumed to be incompressible when the pressure variation is small in comparison to the absolute pressure. As discussed above, while the bulk modulus of elasticity, E_v and the sonic velocity, c are physical fluid properties that measure the degree of compressibility of a fluid, the Mach number, M serves as a practical indicator of the degree of compressibility of fluid in general, especially for gases.

The level of variation in density in gas flows and thus the level of approximation made when modeling gas flows as incompressible depends on the Mach number, M, which is defined as the square root of the Cauchy number, C as follows:

$$C = \frac{F_I}{F_E} = \frac{\rho v^2 L^2}{E_v L^2} = \frac{v^2}{\dfrac{E_v}{\rho}} \tag{1.163}$$

$$\sqrt{C} = M = \frac{v}{\sqrt{\dfrac{E_v}{\rho}}} = \frac{v_a}{v_i} = \frac{v}{c} \tag{1.164}$$

where the Mach number, M is dimensionless and expresses the ratio of the actual velocity of the flow, v_a (or the actual velocity of a body moving in a stationary fluid) to the ideal velocity, v_i of the flow, which is equal to the sonic velocity of flow, c. Furthermore, the Mach number, M expresses the relative importance of the fluid's inertia force, F_I compared to its elastic force, F_E. The sonic velocity, c is the speed of sound, or the celerity, which is the speed of an infinitesimally small pressure (sonic) wave in a compressible or elastic fluid. One may note that in the case of sonic flow (ideal flow at sonic velocity), $M = 1$; however, in the case of subsonic flow (real flow at below the sonic velocity), $M < 1$, and in the case of supersonic flow (real flow at above the sonic velocity), $M > 1$. Furthermore, for extremely high values of M, where $M \gg 1$, the flow is called hypersonic. In practice, when $M \leq 0.5$ (subsonic flow), the inertia force caused by the fluid motion is not sufficiently large to cause a significant change in the fluid density, ρ; thus, the compressibility of the fluid is negligible (incompressible flow). This usually occurs when the pressure variation is small compared with the absolute pressure. However, when $M > 0.5$ and its velocity approaches the speed of sound, c, then one must treat the fluid as a compressible flow. For instance, Example Problem 1.44 above illustrated that the speed of sound (sonic velocity), c in water is much higher than in air. Specifically, at standard sea-level atmospheric pressure and a temperature of about 59°F or 60°F, the sonic velocity for water was 4,807 ft/sec, while the sonic velocity for standard atmosphere was only 1,116.45 ft/sec. Therefore, for water at a speed, v under 2,403.5 ft/sec (where $M \leq 0.5$), the compressibility effects of water can be neglected; however, for water at a speed, *v* over 2,403.5 *ft/sec* (where $M > 0.5$), the compressibility effects of water must be modeled, thus the assumption of compressible flow. However, for air at a speed, *v* under 558.225 *ft/sec* (where $M \leq 0.5$), the compressibility effects of air can be neglected; however, for air at a speed, *v* over 558.225 *ft/sec* (where $M > 0.5$), the compressibility effects of air must be modeled, thus the assumption of compressible flow. One final note is that while high speeds are not practically attained in liquids, they may be easier to attain in gases (see Chapter 11).

EXAMPLE PROBLEM 1.46

A plane travels at 700 mph at an altitude of 40,000 ft above sea level. (a) Determine the Mach number, assuming the BG system of units. (b) Determine the Mach number assuming the SI system of units. (c) Compare the results and characterize the speed of the plane (sonic, subsonic, supersonic, etc.) and type of flow (compressible or incompressible flow).

Mathcad Solution

(a) The sonic velocity for air at an altitude of 40,000 ft above sea level is read from Table A.1 in Appendix A. The Mach number, assuming the BG system of units, is determined by applying Equation 1.164 as follows:

$$c := 968.08 \frac{ft}{sec} \qquad v := 700 \, mph = 1.027 \times 10^3 \frac{ft}{s} \qquad M := \frac{v}{c} = 1.061$$

(b) The sonic velocity for air at an altitude of 40,000 ft (converted to meters) above sea level is read from Table A.1 in Appendix A. The Mach number, assuming the SI system of units, is determined by applying Equation 1.164 as follows:

$$40000 \, ft = 12.192 \, km \qquad c := 295.07 \frac{m}{sec} \qquad v := 700 \, mph = 312.928 \frac{m}{s}$$

$$M := \frac{v}{c} = 1.061$$

(c) First, it is important to note that because the Mach number is dimensionless, its magnitude for a given flow situation will be the same regardless of whether the BG or SI system of units is assumed. Second, because the Mach number, $M = 1.061 > 1$, the speed of the plane is supersonic. And, finally, because the Mach number, $M = 1.061 > 0.5$, the flow is compressible; thus, the compressibility effects of the air must be modeled.

End-of-Chapter Problems

Problems with a "C" are conceptual problems. Problems with a "BG" are in English units. Problems with an "SI" are in metric units. Problems with a "BG/SI" are in both English and metric units. All "BG" and "SI" problems that require computations are solved using Mathcad.

Introduction, Fundamental Principles of Fluid Mechanics, Types of Fluid Flow

1.1C List and define the three basic units of study in the fundamental principles of fluid mechanics. What important principle is each unit of study based on? What important equation(s) does each unit of study yield?

1.2C What are the two important principles that may be used to derive the energy equation?

1.3C The basic unit of study fluid dynamics includes what topic, which yields the resistance equations?

1.4C What is the difference between internal flow and external flow?

1.5C What are the three governing equations in fluid flow?

1.6C What is the difference between the Eulerian approach and the Lagrangian approach in the application of the governing equations?

1.7C The energy equation and the momentum equation play complementary roles in the analysis of a given flow situation. Explain.

1.8C What determines the type of fluid flow? List the seven classifications of fluid flow.

1.9C A real flow may be subdivided into laminar or turbulent flow depending upon the value of what dimensionless number?

1.10C What are seven important physical properties/characteristics of fluids?

Dimensions and Units

1.11C What are the four fundamental/primary dimensions used to express the physical quantities of fluid flow and the physical properties/characteristics of fluids?

1.12C What does it mean for an equation to be dimensionally homogenous? Give an example of a dimensionally homogenous equation.

1.13C What are two commonly used systems of units? What are the primary dimensions for each system of units? What are the units for the primary dimensions for each system of units?

1.14C The derivation of the M dimension for the BG system is done by applying what law? Give the derivation and the name of the derived M dimension for the BG system of units.

1.15C The derivation of the F dimension for the SI system is done by applying what law? Give the derivation and the name of the derived F dimension for the SI system of units.

1.16C For a given physical quantity or variable, how are secondary dimensions and units derived from the primary dimensions and units? Give an example.

1.17BG Using Table 1.1, based on their respective physical definitions, derive the secondary dimensions for the BG system of units (FLT) for the following variables: (a) frequency, ω. (b) discharge, $Q = vA$. (c) work, $W = Fd$. (d) density, $\rho = M/V$. (e) specific gravity, $s = \rho/\rho_{water}$.

1.18SI Using Table 1.1, based on their respective physical definitions, derive the secondary dimensions for the SI system of units (MLT) for the following variables: (a) frequency, ω. (b) discharge, $Q = vA$. (c) work, $W = Fd$. (d) density, $\rho = M/V$. (e) specific gravity, $s = \rho/\rho_{water}$.

1.19BG/SI Demonstrate that the head loss equation, $h_f = f(L/D)(v^2/2g)$ (see Chapter 8) is dimensionally homogenous, assuming (a) the BG system of units (FLT). (b) the SI system of units (MLT). Note that the Darcy–Weisbach friction factor, f is dimensionless.

1.20BG/SI Demonstrate that the minor head loss equation, $h_f = k(v^2/2g)$ (see Chapter 8) is dimensionally homogenous, assuming (a) the BG system of units (FLT). (b) The SI system of units (MLT). Note that the minor head loss coefficient, k is dimensionless.

1.21SI Manning's SI equation, $v = (1/n)R_h^{2/3}S_f^{1/2}$ is dimensionally homogenous (see Chapter 8 and Chapter 10) where the hydraulic radius, $R_h = (A/P)$, A = area, and P = wetted perimeter. (a) Derive the SI system of units (MLT) for Manning's roughness coefficient, n.

1.22BG/SI Demonstrate that the actual flow rate equation for orifice, nozzle, and venturi meters, $Q_a = C_d A_2 \sqrt{(2(p_1 - p_2))/(\rho[1 - (D_2/D_1)^4])}$ (see Chapter 8) is dimensionally homogenous, assuming (a) the BG system of units (FLT). (b) The SI system of units (MLT). Note that the discharge coefficient, C_d is dimensionless

1.23SI Using the front of this textbook, define: (a) a picometer, pm. (b) a petagram, Pg. (c) an attoNewton, aN.

1.24BG/SI Using the front of this textbook (some unit conversions are also available in Mathcad), define the variable and convert the units of the following quantities from the BG system of units to the SI system of units: (a) 6 quarts. (b) 88 *cfs*. (c) 54 *psi*. (d) 78 *hp*. (e) 55°F.

1.25BG/SI Using the front of this textbook (some unit conversions are also available in Mathcad), define the variable and convert the units of the following quantities from the SI system of units to the BG system of units: (a) 85 hectare. (b) 3 *Hz*. (c) 89 *N*. (d) 76 *cal*. (e) 33°K.

1.26BG (Refer to Problem 1.17) Based on their respective physical definitions, derive the BG units for the following variables: (a) frequency, ω. (b) Discharge, $Q = vA$. (c) Work, $W = Fd$. (d) Density, $\rho = M/V$. (e) Specific gravity, $s = \rho/\rho_{water}$.

1.27SI (Refer to Problem 1.18) Based on their respective physical definitions, derive the SI units for the following variables: (a) frequency, ω. (b) Discharge, $Q = vA$. (c) Work, $W = Fd$. (d) Density, $\rho = M/V$. (e) Specific gravity, $s = \rho/\rho_{water}$.

Pressure Scales

1.28C Most fluid properties vary with what two variables?

1.29C The acceleration due to gravity varies with what two variables?

1.30C What are the two basic types of pressure scales? What is the relationship between the two pressure scales?

1.31BG The atmospheric pressure at sea level is 14.6959 *psi* (abs). The gage pressure at a point in a tank of fluid is 567 *psi* (gage). (a) Determine the absolute pressure at the point in the tank of fluid.

1.32SI The atmospheric pressure at sea level is 101.325 *kPa* (abs). The gage pressure at a point in a tank of fluid is 43 *kPa* (gage). (a) Determine the absolute pressure at the point in the tank of fluid.

Standard Atmosphere

1.33C   Define the ICAO Standard Atmosphere.

1.34C   List seven important physical properties for the ICAO Standard Atmosphere presented as a function of elevation above sea level in Table A.1 in Appendix A.

1.35BG/SI   Using Table A.1 in Appendix A, (a) graphically illustrate the variation of the pressure of the standard atmosphere with altitude, assuming the BG system of units. (b) Graphically illustrate the variation of the temperature of the standard atmosphere with altitude, assuming the SI system of units. (c) Graphically illustrate the variation of the acceleration due to gravity of the standard atmosphere with altitude, assuming the SI system of units.

Standard Reference for Altitude, Pressure, Latitude, and Temperature

1.36C   The absolute pressure of a fluid and thus most of its fluid properties and the acceleration due to gravity, g vary with what variable?

1.37C   Define the standard atmospheric pressure and give its value in both the BG and the SI systems of units.

1.38C   Define the standard acceleration due to gravity, g and give its value in both the BG and the SI systems of units.

1.39C   Give the value of the temperature at the standard sea-level atmospheric pressure in both the BG and the SI systems of units.

1.40C   Most fluid properties vary the most with what single variable?

1.41BG/SI   Using Table A.2 in Appendix A and Table A.3 in Appendix A, respectively: (a) graphically illustrate the variation of the density of water with temperature, assuming the BG system of units. (b) Graphically illustrate the variation of the kinematic viscosity of air with temperature, assuming the SI system of units.

1.42C   Give the value of the room temperature assumed for common fluids in both the BG and the SI systems of units.

1.43BG/SI   Using Table A.4 in Appendix A and Table A.5 in Appendix A, respectively: (a) what are the fluid properties for SAE 30 oil at room temperature (68°F)? (b) What are the fluid properties for helium at room temperature (20°C)?

1.44C   What is the reference fluid for liquids? What is the reference fluid for gases?

1.45C   Give the value of the standard temperature assumed by physicists (and assumed in this textbook) for water at the standard sea-level atmospheric pressure in both the BG and the SI systems of units.

Newton's Second Law of Motion, Dynamic Forces Acting on a Fluid Element

1.46C   Based on Newton's second law of motion, give the six potential sources of dynamic forces acting on a fluid element.

1.47C   The actual forces acting on a fluid element will depend upon whether the fluid is in static equilibrium or in motion (fluid dynamics). (a) What are the potentially significant forces for fluid statics? (b) What are the potentially significant forces for fluids in motion?

1.48BG   A body has a mass of 78 slugs. (a) Determine the weight of the mass.

1.49SI   A body weighs 809 N. Determine the mass of the body.

1.50C   How are the dynamic forces acting on a fluid element (F_G, F_P, F_V, F_E, F_T, and F_I) and the fluid properties (mass density, ρ; specific gravity, s; specific weight, γ; viscosity, μ; surface tension, σ; vapor pressure, p_v; and bulk modulus, E_v) related?

Physical Properties of Fluids

1.51C   For a given fluid property, how are secondary dimensions and units derived from the primary dimensions and units? Give three examples from Table 1.1.

1.52C   What is the most important fluid property in fluid statics?

1.53C   What fluid property becomes important in both fluid statics and fluid dynamics when very low pressures are involved?

1.54C   What fluid property becomes important for compressible fluids?

1.55C   What fluid property becomes important in droplets, bubbles, thin films, capillarity, or sheets of fluid?

1.56C   What fluid property is significant once the fluid is in motion? What two fluid properties are most important in fluid flow?

1.57C   Most fluid properties (especially the density for a gas/compressible flow) vary with what two variables?

Density, Specific Gravity, and Specific Weight

1.58C   Why is the density considered an absolute fluid property?

1.59BG/SI   Using Table A.4 in Appendix A and Table A.5 in Appendix A, respectively, to read the fluid density: (a) determine the volume of 45 slug of glycerin at room temperature (68°F). (b) Determine the mass of 56 liter of nitrogen at room temperature (20°C).

1.60C   For a given temperature, define the compressibility of a fluid.

1.61C   What is the dimensionless number that indicates the compressibility of a gas?

1.62C   Give the value of the standard density (at standard temperature and standard pressure) for the reference fluid for liquids, water, in both the BG and the SI system of units.

1.63BG/SI   Refer to Problem 1.59 above: (a) determine the specific volume of glycerin at room temperature (68°F). (b) Determine the specific volume of nitrogen at room temperature (20°C).

1.64C   Why is the specific gravity considered an absolute fluid property?

1.65BG   Using Table A.4 in Appendix A to read the fluid specific gravity or fluid density, and Equation 1.40: (a) determine the specific gravity of seawater at room temperature (68°F). (b) Determine the specific gravity of SAE 10 oil at room temperature (68°F). (c) Determine the specific gravity of oxygen at room temperature (−320°F).

1.66SI   Using Table A.4 in Appendix A to read the fluid specific gravity or fluid density, and Equation 1.40: (a) determine the specific gravity of seawater at room temperature (20°C). (b) Determine the specific gravity of SAE 10 oil at room temperature (20°C). (c) Determine the specific gravity of oxygen at room temperature (-195°C). (d) Refer to Problem 1.65 and compare the results.

1.67C   Why does Table A.5 in Appendix A, which presents physical properties of some common gases at standard sea-level atmospheric pressure, not include the specific gravity, s for a given gas?

1.68C   Why is the specific weight not considered an absolute fluid property?

1.69BG/SI   Using Table A.2 in Appendix A and Table A.3 in Appendix A, respectively, to read the fluid specific weight: (a) determine the volume of 58 *lb* of water at 170°F. (b) Determine the weight of 444 m^3 of air at 60°C.

1.70BG/SI   Using Table A.4 in Appendix A and Table A.5 in Appendix A, respectively, to read the fluid density: (a) determine the volume of 99 *lb* of carbon tetrachloride at room temperature (68°F). (b) Determine the weight of 456 m^3 of carbon monoxide at room temperature (20°C).

1.71C   Give the value of the standard specific weight (at standard temperature, standard pressure, and standard gravity) for the reference fluid for liquids, water, in both the BG and the SI systems of units.

1.72BG   A 100-ft^3 volume of glycerin weighs about 7857 *lb*. (a) Determine the specific weight of the glycerin. (b) Determine the density of the glycerin. (c) Determine the specific gravity of the glycerin.

1.73BG   The specific gravity of a liquid is 0.91. (a) Determine the density of the liquid in BG units. (b) Determine the specific weight of the liquid in BG units.

1.74BG   The specific weight of a liquid is 90 lb/ft^3. (a) Determine the density of the liquid. (b) Determine the specific volume of the liquid. (c) Determine the specific gravity of the liquid relative to water at 80°F.

1.75SI   The specific weight of a gas is 19 N/m^3. (a) Determine the density of the gas. (b) Determine the specific volume of the gas. (c) Determine the specific gravity of the gas relative to air at 30°C.

1.76BG   For a weight of water of 90 *lbs* at standard sea-level atmospheric pressure, (a) determine the changes in its specific weight and in its corresponding volume due to an increase in temperature from 40°F to 200°F in increments of 10°F. (b) Plot the resulting specific weight vs. temperature. (c) Plot the resulting volume vs. temperature. (d) Explain each trend in (b) and (c).

1.77SI   For a weight of air of 800 N at standard sea-level atmospheric pressure, (a) determine the changes in its specific volume and in its corresponding volume due to an increase in temperature from 0 to 100°C in increments of 20°C. (b) Plot the resulting specific volume vs. temperature. (c) Plot the resulting volume vs. temperature. (d) Explain each trend in (b) and (c).

1.78BG   A 700-ft^3 tank that weighs 450 *lb* is filled with SAE 10 oil at a temperature of 68°F. (a) Determine the vertical force required to move the tank upward with an acceleration of 50 ft/sec^2.

Viscosity

1.79C What is the value assumed for the viscosity for an inviscid (ideal) fluid?

1.80C What is the most important fluid property in fluid dynamics?

1.81C The dimensions and units for the absolute or dynamic viscosity, μ are based on its physical definition, which is given by what law?

1.82C What is the relationship between the absolute or dynamic viscosity, μ and the kinematic viscosity, v?

1.83C Newton's law of viscosity provides a theoretical expression for the shear stress, τ (or friction/viscous force, $F_V = \tau A$, where $\tau = \mu(dv/dy)$) for laminar (deterministic) flow. What method is used to derive an empirical expression for the shear stress, τ (or friction/viscous force, $F_V = \tau A$, where $\tau_w = C_D \rho v^2$) for turbulent (stochastic/random) flow?

1.84C The theoretical modeling of the friction/viscous force, F_V by Newton's law of viscosity results in a theoretical (parabolic) velocity distribution for laminar flow. What does the empirical modeling of the friction/friction force, F_V for turbulent flow (empirical studies) yield for the empirical velocity distribution?

1.85C Define the term flow resistance. How is it modeled in Newton's second law of motion?

1.86C What are two molecular factors that produce viscosity?

1.87C How does temperature affect the viscosity of a liquid? How does temperature affect the viscosity of a gas?

1.88BG/SI Using Table A.2 in Appendix A and Table A.3 in Appendix A, respectively: (a) graphically illustrate the variation of both the density and the absolute viscosity of water with temperature, assuming the SI system of units. (b) Graphically illustrate the variation of both the density and the absolute viscosity of air with temperature, assuming the BG system of units. (c) Compare the results.

1.89BG/SI Using Table A.4 in Appendix A and Table A.5 in Appendix A, respectively: (a) read the absolute viscosity of benzene, crude oil, glycerin, mercury, SAE 30 oil, and fresh water, all at room temperature (20°C). (b) Read the absolute viscosity of air, carbon monoxide, helium, hydrogen, oxygen, and water vapor, all at room temperature (68°F).

1.90C A real flow may be subdivided into laminar (highly viscous) or turbulent flow depending upon the value of the Reynolds number, R, which is defined as the ratio of what two forces?

1.91C What range of values for the Reynolds number, R defines: (a) laminar flow? (b) Transitional flow? (c) Turbulent flow?

1.92BG/SI Glycerin at 68°F flows at $2\,ft/sec$ in a pipe with a diameter of $5\,ft$. (a) Determine the Reynolds number, assuming the BG system of units. (b) Determine the Reynolds number, assuming the SI system of units. (c) Compare the results.

1.93C Give the mathematical relationship between the shear stress, τ and the radial distance, r from the center of the pipe with a radius, r_o for laminar or turbulent flow. What law is this relationship based on?

1.94SI   As an illustration of Couette flow, the space between two very long parallel plates separated by distance 0.2 m is filled with SAE 30 oil at room temperature (20°C), as illustrated in Figure ECP 1.94. The lower plate is held stationary while a continuous tangential (shear) force, F_V is applied to the upper plate, thus moving the plate with a constant velocity of 0.15 m/sec. The dimensions of the upper plate are 3 m by 1.5 m. (a) Determine if the flow is laminar or turbulent. (b) Graphically illustrate the velocity profile at the cross section of flow. (c) Graphically illustrate the shear stress at the cross section of flow. (d) Determine the magnitude of the shear force required to keep the plate moving at the given constant velocity.

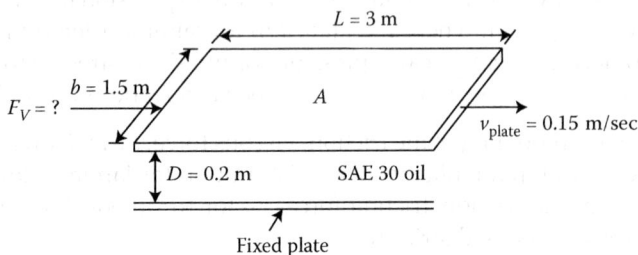

FIGURE ECP 1.94

1.95C   A plot of the shear stress, τ vs. the rate of shearing deformation, dv/dy (also called the velocity gradient) is given in Figure 1.4, where the viscosity, μ represents the slope of the linear or curvilinear function. Additionally, this plot indicates that according to Newton's law of viscosity, the shear stress, τ increases with an increase in the rate of shearing deformation, dv/dy, where the viscosity, μ predicts the behavior of the shear stress, τ with respect to the rate of shearing deformation, dv/dy and thus describes the nature of the fluid. Figure 1.4 illustrates five general types of fluids; describe the behavior of each general type of fluid, and give an example of each.

1.96C   While the dynamic viscosity, μ was assumed to be mostly independent of pressure, the kinematic pressure, $v=\mu/\rho$ varies inversely with pressure, especially for gases. Explain why.

1.97BG/SI   Using Table A.4 in Appendix A and Table A.5 in Appendix A, respectively: (a) read the absolute viscosity and the density of mercury at room temperature (68°F), and compute the corresponding kinematic viscosity. (b) Read the absolute viscosity and density of oxygen at room temperature (20°C), and compute the corresponding kinematic viscosity.

1.98C   When describing the parabolic velocity profile that is typical for laminar/viscous (nonturbulent) pipe flow (see Figure 1.2a), or laminar/viscous (nonturbulent) flow along a solid boundary (Figure 1.5a), what does the "no-slip" condition mean?

1.99C   For laminar pipe flow, the shear stress distribution in Figure 1.2b may be evaluated by applying one of two equations. Give those two equations, state what law each equation is based on, and state the limits of application for each equation.

1.100BG   SAE 10 oil at room temperature (68°F) flows at a flow rate of 5 cfs in a 4-ft-diameter pipe for a length of 20 ft, as illustrated in Figure ECP 1.100. (a) Determine if the flow is laminar or turbulent. (b) Graphically illustrate the velocity profile at the pipe cross section. (c) Evaluate and graphically illustrate the shear stress at the pipe cross section by applying Newton's second law of motion for either laminar or turbulent flow (Equation 1.53). (d) Evaluate and graphically illustrate the shear stress at the pipe cross section by applying Newton's law of viscosity for laminar flow, Equation 1.58. (e) Compare the results yielded by (c) and (d).

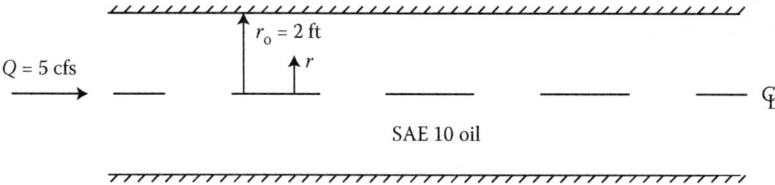

FIGURE ECP 1.100

1.101BG Glycerin at room temperature (68°F) flows at a depth of 0.35 ft and a maximum velocity of 0.55 ft/sec over a solid boundary, as illustrated in Figure ECP 1.101. (a) Determine if the flow is laminar or turbulent. (b) Graphically illustrate the velocity profile at the cross section of flow. (c) Graphically illustrate the shear stress at the cross section of flow.

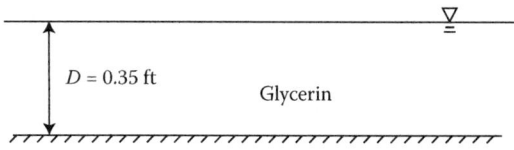

FIGURE ECP 1.101

1.102SI SAE 30 oil at room temperature (20°C) fills the annular space of a hydraulic lift as illustrated in Figure ECP 1.102. The inside moving cylinder is 350.00 mm in diameter, and the outside fixed cylinder is 350.25 mm in diameter. The velocity of the inside moving cylinder is 0.45 m/sec, and it moves a distance of 3.4 m along the fixed cylinder. (a) Determine the shear force along the inside moving cylinder. (b) Determine the shear force along the outside fixed cylinder.

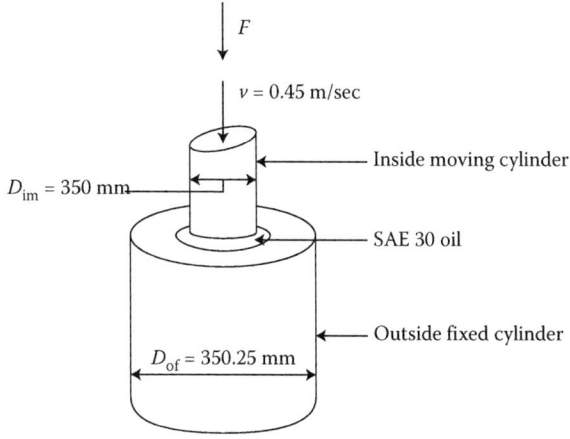

FIGURE ECP 1.102

1.103SI   A 40 *cm* cubical block that weighs 205 *N* is placed on an inclined plane ($\alpha = 15°$) as illustrated in Figure ECP 1.103. The static coefficient of friction between the block and the plane is 0.39. The inclined plane is held stationary, while a continuous force, F is applied to the block, thus moving the block with a constant velocity of 0.98 *m/sec*. (a) Determine the required force, F. (b) Assuming that a 0.7-*mm* film of crude at room temperature (20°C) is inserted between the block and the inclined plane, determine the required force, F. (c) Compare the results of (a) and (b) and discuss.

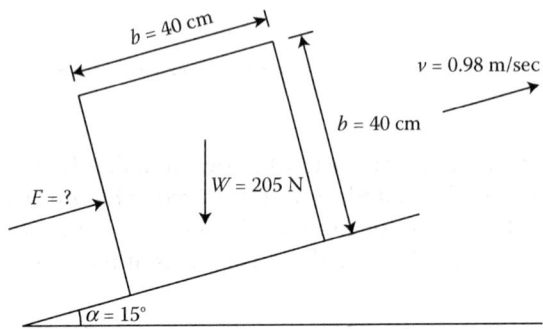

FIGURE ECP 1.103

1.104C   Provide a comparison between the theoretical expression for the shear stress for laminar flow [i.e., Newton's law of viscosity, $\tau = \mu(dv/dy)$] and the empirical expression for the shear stress for turbulent flow, $\tau_w = C_D \rho v^2$.

1.105C   The assumption made regarding the viscosity of the fluid will determine the resulting velocity profile for the flow. Discuss the different velocity profiles for: (a) ideal flow (see Figure 6.3a). (b) Real laminar flow (see Figure 6.3b) (give equation). (c) Real turbulent flow (see Figure 6.3c) (give equation).

1.106C   The wall shear stress for turbulent flow, $\tau_w = C_D \rho v^2$ is empirically derived using dimensional analysis, where C_D is a dimensionless drag coefficient that represents the flow resistance for turbulent flow. What are three more commonly used flow resistance factors/coefficients used to represent the drag coefficient, C_D?

Surface Tension

1.107C   Explain why the surface tension, σ is a liquid property and not a gas property.

1.108C   Give three examples for which the surface tension, σ is responsible for its formation or occurrence.

1.109C   Explain the difference between molecular cohesion and molecular adhesion. Give an example of when only molecular cohesion plays a role in its formation/occurrence. Give an example of when both molecular cohesion and molecular adhesion play a role in its formation/occurrence.

1.110C   As a result of the liquid property, σ, surface tension forces develop in the liquid surface, which cause the liquid surface to behave in what manner?

1.111C Explain the nature of the forces that cause the liquid surface to behave as a membrane/skin that is stretched over the air mass above it (see Figure 1.7a).

1.112C How does temperature affect the surface tension of a liquid? Why?

1.113SI Using Table A.2 in Appendix A, (a) graphically illustrate the variation of the surface tension of water with temperature assuming the SI system of units.

1.114C Explain the nature of the forces that create the perfectly spherical-shaped droplet in the atmosphere (see Figure 1.7b).

1.115C What happens to the formation of a droplet in the atmosphere when the inertia force, F_I overcomes the surface tension force, F_T?

1.116BG A spherical water droplet at a temperature of 40°F forms with a diameter of 0.01 in. during a rainfall at an elevation of 10,000 ft above sea level. (a) Determine the internal pressure acting on the spherical water droplet (refer to Figure 1.8).

1.117SI A spherical soap bubble forms in the atmosphere with a diameter of 6 mm. Assume the surface tension of the soap solution is 0.0510 N/m. (a) Derive the equation for the unbalanced net inward cohesive force, $(p_i - p_e)A$ acting on the liquid molecules along the surfaces of the spherical soap bubble. (b) Determine the internal pressure acting on the spherical soap bubble an elevation of 0 km above sea level. (Refer to Figure ECP 1.117).

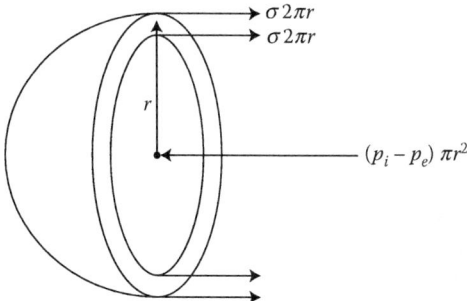

FIGURE ECP 1.117

1.118C Give two examples and explain what happens between a liquid–gas–solid interface when the molecular adhesion is more significant than the molecular cohesion (such as for water). And, give two examples and explain what happens between a liquid–gas–solid interface when molecular cohesion is more significant than the molecular adhesion (such as for mercury).

1.119C When a perfectly shaped spherical drop of liquid is placed on a solid surface, it will form a wetting angle or surface tension contact angle, θ with the solid surface, as illustrated in Figure 1.10. Explain the difference between a liquid that is described as "receding" over the solid surface vs. a liquid that is described as "advancing" over the solid surface. Give an example of each type of liquid–gas–solid interface.

1.120C The height, h of the capillary rise or drop is a function of the magnitude of what variables?

1.121C   What causes the water to wet and rise in the small tube (capillary), forming a con-cave meniscus?

1.122C   Can a free water surface support objects such as a small steel needle? Explain.

1.123C   Derive the expression for the "tension" or negative pressure, p_m at the meniscus of the capillary rise of water in Figure 1.9a. Discuss the relationship between the tension and the radius of the small tube.

1.124SI   Water at a temperature of 50°C rises in a glass tube with a diameter of 0.60 *cm*, as illustrated in Figure ECP 1.124. Assume a contact angle, θ of 10°. (a) Determine the capillary rise in the tube.

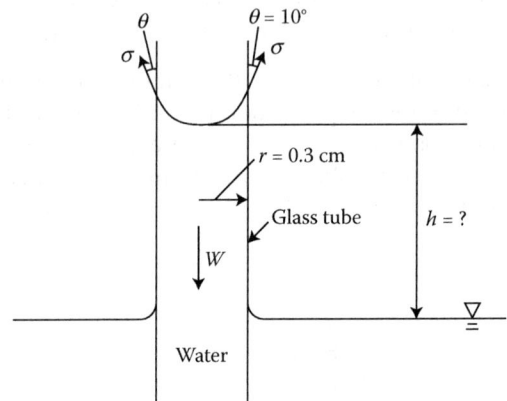

FIGURE ECP 1.124

1.125BG   Mercury at room temperature (68°F) falls in a glass tube with a diameter of 0.07 in, as illustrated in Figure ECP 1.125. Assume a contact angle, θ of 129°. (a) Determine the capillary fall/depression in the tube.

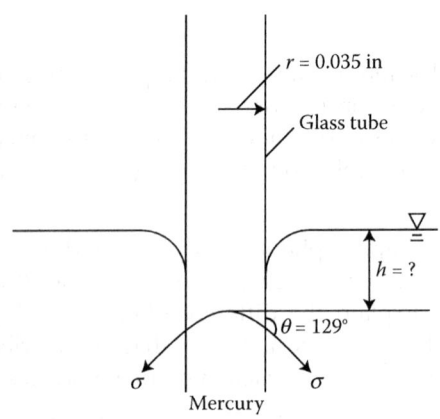

FIGURE ECP 1.125

1.126C A tube with a small diameter is inserted into a liquid with a contact angle, θ of 95°. Will the liquid rise or fall in the tube? Explain your answer.

1.127SI Kerosene at room temperature of 20°C rises in a glass tube with a diameter of 2.5 *mm*, as illustrated in Figure ECP 1.127. Assume a contact angle, θ of 25°. (a) Determine the capillary rise in the tube.

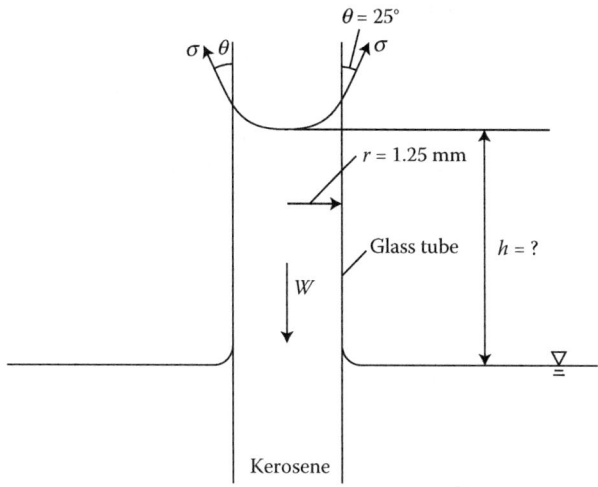

FIGURE ECP 1.127

1.128BG A small steel needle that is carefully placed on a fresh water (68°F) surface floats, as illustrated in Figure ECP 1.128. Assume the needle is 2 in. long and a contact angle, θ of 0°. (a) Determine the maximum weight of the steel needle that would float on the water.

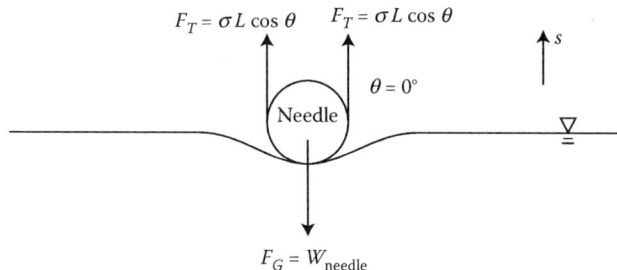

FIGURE ECP 1.128

Vapor Pressure

1.129C What are the typical conditions for which the vapor pressure, p_v of a liquid becomes important?

1.130C Define what cavitation is, and when it becomes important.

1.131C Is the vapor pressure, p_v of a liquid is measured in absolute pressure or gage pressure? Also, explain why, at room temperature, where the vapor pressure, p_v of a liquid is much lower than the atmospheric pressure, the gage pressure of the liquid is negative.

1.132C   Explain why the vapor pressure, p_v is a liquid property and not a gas property.

1.133C   What does Dalton's law of partial pressure state regarding each gas in a mixture of gases?

1.134C   How does the vapor pressure, p_v of a liquid vary with temperature?

1.135C   Illustrate that at room temperature (68°F or 20°C), the vapor pressure, p_v of a liquid is much lower than the atmospheric pressure.

1.136BG   Using Table A.2 in Appendix A, (a) graphically illustrate the variation of the vapor pressure of water with temperature, assuming the BG system of units.

1.137BG   Water is placed in a container that is open to the atmosphere. (a) Determine the temperature and the vapor pressure at the boiling point at 0 ft above sea level. (b) Determine the temperature and the vapor pressure at the boiling point at 80,000 ft above sea level. (c) Discuss the results.

1.138SI   Water is placed in an enclosed container and evacuated of air to form a vacuum. (a) Determine the absolute pressure at which the water at 60°C will boil. (b) Determine the absolute pressure at which the water at 90°C will boil. (c) Will water at a higher temperature boil at a higher pressure? (d) Discuss the magnitudes of the absolute pressures in (a) and (b) relative to the atmospheric pressure at sea level.

1.139BG   Water at a given elevation is placed in a container that is open to the atmosphere and boils at 150°F. (a) Determine the atmospheric pressure at this altitude.

1.140C   In the case where a liquid at room temperature is open to the atmosphere, will the vapor pressure, p_v of the liquid be attained (see Figure 1.11a)? Explain.

1.141C   In the case where a liquid at room temperature is placed in an enclosed container and evacuated of air to form a vacuum (absolute pressure above liquid is at zero), will the vapor pressure, p_v of the liquid be attained (see Figure 1.11b)? Explain.

1.142C   In the case where a liquid that is open to the atmosphere is heated to its boiling point, will the vapor pressure, p_v of the liquid be attained (see Figure 1.11c)? Explain.

1.143C   Explain why it would take longer to cook food at a high altitude (eg. 5,000 ft above sea level).

1.144C   Explain why food will cook more quickly in a pressure cooker.

1.145C   Explain why cavitation may occur in a liquid flow, but does not occur in a gas flow.

1.146C   Give an example of how one may reduce the boiling point of a flowing liquid simply by reducing the pressure of the liquid to its vapor pressure, p_v at the corresponding temperature.

1.147C   Give two flow situations where the reduction in pressure to the vapor pressure, p_v and thus potential cavitation can occur. Give an example for each flow situation.

1.148C   Why should cavitation be avoided? How should cavitation be avoided?

1.149SI   Water at 80°C is lifted from a tank to a higher elevation by the use of a siphon tube. (a) Determine the minimum pressure in the siphon tubing in order to avoid cavitation.

1.150BG   Water at 50°F flowing through the suction side of a pump drops to a pressure of 0.100 *psia*. (a) Determine if there is a potential for cavitation to occur for this pump.

1.151SI   Gasoline at 20°C flows through a complex passage in a valve. (a) Determine the minimum pressure in the valve in order to avoid cavitation.

Elasticity, Compressibility, or Bulk Modulus of Elasticity

1.152C   When the fluid is a liquid (static or dynamic), a significant change in pressure introduces the liquid property, vapor pressure, p_v and an undesirable fluid phenomena, cavitation (only for a liquid in motion). However, when the fluid is either a gas or a liquid (static or dynamic), a significant change in pressure introduces what fluid property?

1.153C   For a given temperature, define the compressibility of a fluid. Also, how is the compressibility of a fluid measured?

1.154C   Explain the difference between a compressible fluid and an incompressible fluid.

1.155C   Although liquids can be compressed (resulting in small changes in density) at very high pressures, they are typically assumed to be incompressible for most practical engineering applications. Give an example illustrating evidence of the compressibility or the elasticity of liquids.

1.156C   Although gases are easily compressed (resulting in significant changes in density) at very high pressures, they are typically assumed to be incompressible when the pressure variation is small in comparison to the absolute pressure (sum of atmospheric and gage pressure), as indicated by what dimensionless number?

1.157C   What is the relationship between the bulk modulus of elasticity, E_v and the Mach number, M? What does each number/quantity measure?

1.158C   Although the bulk modulus of elasticity, E_v can be determined for both liquids and gases, it is usually tabulated only for liquids. As such, the bulk modulus of elasticity, E_v for gases involves the applications of what two laws?

1.159C   In the determination of the bulk modulus of elasticity, E_v for gases, applications of the ideal gas law and the first law of thermodynamics introduce what additional five gas properties?

1.160C   The application of what law introduces the definition for the specific heat, c for liquids?

1.161C   The application of what equations introduces the sonic (acoustic) velocity, c for compressible fluids?

1.162C   In general, how do temperature and pressure affect the density, ρ of a fluid?

1.163C   When is it reasonable to model gas flow as an incompressible flow?

1.164C   Explain why it is reasonable to generally model liquid flow as an incompressible flow.

1.165C   The dimensions and units for E_v are the same dimensions and units for pressure, p. Why are the units for E_v often expressed as absolute pressure (sum of atmospheric and gage pressure)?

1.166C   Although the bulk modulus of elasticity, E_v for liquids is a function of temperature and pressure (increases with pressure), what is the value for the bulk modulus of elasticity, E_v that is of interest in most engineering applications?

1.167C As indicated by the definition of $E_v = (dp/(d\rho/\rho)) = (dp/-(dV/V))$, a large value for E_v for a fluid (or solid) indicates that a large change in pressure is needed to cause a small fractional change in volume. Demonstrate the definition of E_v by comparing the values of E_v for a gas, a liquid, and a solid.

1.168BG A container with $500\,ft^3$ of glycerin at 68°F and under an atmospheric pressure of 14.696 *psia* is compressed until the pressure is 4000 *psia*. (a) Determine the change in volume of the glycerin. (b) Determine the change in density of the glycerin.

1.169SI A container with $700\,m^3$ of benzene at 20°C and under an atmospheric pressure of 101.325 *kPa* abs is compressed until the volume decreases to $600\,m^3$. (a) Determine the change in pressure of the benzene. (b) Determine the change in density of the benzene.

1.170BG A tank contains $450\,ft^3$ of water at 40°F and under an atmospheric pressure of 14.696 *psia*. Determine the change in density and the change in volume of the water for the following situations. Also, state whether the situation illustrates the expandability or the compressibility of water and explain why. (a) The temperature of the water is raised to 100°F and the pressure is maintained at atmospheric pressure. (b) The temperature of the water is maintained at 40°F and the pressure is increased to 760 *psia*.

1.171SI Given that the pressure increases with depth, z as $dp = \gamma\,dz$ (see Chapter 2), determine the increase in density of the seawater at a depth of 3,590 *m* (see Figure ECP 1.171). Assume a temperature of 20°C and an atmospheric pressure of 101.325 *kPa* abs.

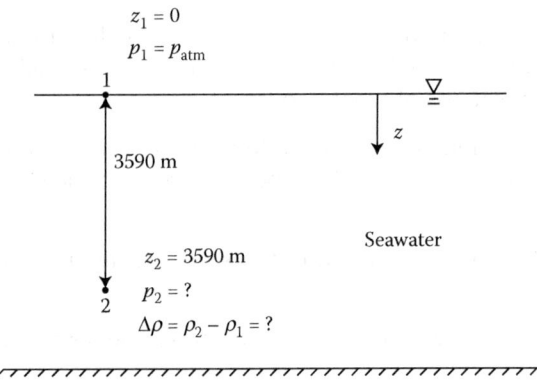

FIGURE ECP 1.171

1.172BG The volume of a liquid in a tank decreases from $10\,ft^3$ to $9\,ft^3$ when the pressure increases from 30 *psia* to 6,000 *psia*. (a) Determine the bulk modulus of elasticity of the liquid in the tank.

1.173BG Using Table A.2 in Appendix A, (a) graphically illustrate the variation of the bulk modulus of elasticity of water with temperature, assuming the BG system of units.

1.174C What causes water hammer? Give an example of how water hammer is generated in a pipe flowing under pressure. Explain what water hammer is. Also, state what water hammer is evidence of.

1.175C   Define an "equation of state." What is the simplest and most commonly applied equation state for fluids in the gas phase?

1.176C   Define an ideal gas.

1.177C   It is typical to assume that most real gases are perfect/ideal and are represented by the ideal gas law. When is this assumption correct?

1.178C   The dimensions and units for E_v are the same dimensions and units for pressure, p. The units for E_v are often expressed as absolute pressure (sum of atmospheric and gage pressure) in order to model a variable atmospheric pressure (especially in the determination of E_v for gases). However, tabulated values for E_v for liquids are typically given in psi gage (see Tables A.2 and A.4 in Appendix A). Explain why this may be so.

1.179C   The ideal gas law is given by Equation 1.104 as follows: $p = \rho RT$. What do the variables R and T represent in this equation, and what is R a function of?

1.180C   Equation 1.106 presents the expression for the gas constant $R = (R_u/m)$. What is the definition of the molecular weight (molar mass), m of a gas?

1.181C   Most gas properties vary with temperature and pressure. How do the molar mass, m and the gas constant, R vary with temperature and pressure?

1.182C   What is the value of the universal gas constant, R_u in both BG and SI units?

1.183C   Derive the common expression for the ideal gas law as given by Equation 1.117 as follows: $pV = NR_uT$

1.184C   Most of the common gases presented in Table A.5 in Appendix A, such as air, carbon dioxide, carbon monoxide, helium, hydrogen, methane, nitrogen, and oxygen, can be modeled by the ideal gas law. However, an exception among the common gases is water vapor at a high density (as a result of high pressures and low temperatures), where the real gas is near the liquid phase (state of saturation) and thus does not closely approximate an ideal gas; explain why this is so.

1.185BG   A 600-ft^3 helium tank is compressed under a pressure of 600 $psia$ and at a temperature of 30°F. Use the ideal gas law equation given by Equation 1.104. (a) Determine the density of the helium in the tank. (b) Determine the weight of the helium in the tank.

1.186SI   A 125-m^3 carbon dioxide tank is compressed under a pressure of 900 kN/m^2 abs and at a temperature of 15°C. Use the ideal gas law equation given by Equation 1.117. (a) Determine the density of the carbon dioxide in the tank. (b) Determine the weight of the carbon dioxide in the tank.

1.187BG   The pressure of a tire is a function of the temperature of the air in the tire. The pressure gage of a tire with a volume of 0.99 ft^3 reads 35 $psig$ when the air temperature is 75°F. Use the ideal gas law equation given by Equation 1.104, which assumes absolute pressure (sum of atmospheric and gage pressure). (a) Determine the density of the air in the tire. (b) If the air temperature increases to 90°F, what would the pressure gage read? (c) If a gage pressure of 35 psi is required to be maintained in the tire at a temperature of 90°F, determine the amount (mass) of air that must be released from the tire.

1.188SI   A 127-m^3 methane tank is compressed under a pressure of 750 kN/m^2 abs and at a temperature of 15°C. Use the ideal gas law equation given by Equation 1.104. (a) Determine the density of the methane in the tank. (b) Determine the specific volume of the methane in the tank. (c) Determine the specific weight of the methane in the tank. (d) Determine the weight of the methane in the tank.

1.189SI A 400-m^3 hydrogen tank is compressed under a pressure of 65 kN/m^2 abs and at a temperature of 40°C. Use the ideal gas law equation given by Equation 1.104. (a) Determine the density of the hydrogen in the tank. (b) If the hydrogen temperature is reduced to 20°C, what would the resulting pressure be?

1.190SI A spherical 8-m-diameter balloon is filled with air under a pressure of 700 kN/m^2 abs and at a temperature of 40°C. Use the ideal gas law equation given by Equation 1.117. (a) Determine the number of moles and the weight of the air in the balloon. (b) Graphically illustrate the effect of the diameter of the balloon on the weight of the air in the balloon by increasing the diameter to 10 m, then to 12 m, for a pressure of 800 kN/m^2 abs and 900 kN/m^2 abs, respectively.

1.191BG The pressure gage of a truck tire with a volume of 1.33 ft^3 reads 25 $psig$ when the air temperature is 80°F. Use the ideal gas law equation given by Equation 1.104, which assumes absolute pressure (sum of atmospheric and gage pressure). (a) Assuming that the temperature and the volume of the truck tire remain constant, determine the amount (mass) of the air that must be added to the tire in order to raise the pressure to 40 $psig$.

1.192BG The pressure gage of a tire with a volume of 0.99 ft^3 reads 35 $psig$ before a trip and 40 $psig$ after a trip. Assume the density of the air in the tire is 0.0075 $slug/ft^3$. Use the ideal gas law equation given by Equation 1.104, which assumes absolute pressure (sum of atmospheric and gage pressure). (a) Assuming that the volume of the tire remains constant, determine the change in temperature of the air in the tire.

1.193SI A 40-m^3 helium tank is compressed under a pressure of 655 kN/m^2 abs and at a temperature of 15°C. Some helium is released from the tank until the pressure drops to 555 kN/m^2 abs and the temperature drops to 10°C. Use the ideal gas law equation given by Equation 1.104. (a) Determine the amount (mass) of helium that was released from the tank.

1.194BG Nitrogen in a tank is compressed to a density of 0.004 $slug/ft^3$ under a pressure of 460 $psia$. Use the ideal gas law equation given by Equation 1.104. (a) Determine the temperature of the nitrogen in the tank.

1.195SI A 30-m^3 tank is filled with 196 N of gas and compressed under a pressure of 102 kN/m^2 abs and at a temperature of 21°C. Use the ideal gas law equation given by Equation 1.104. (a) Determine the type of gas in the tank.

1.196BG A tank is filled with 5 slugs of carbon dioxide and compressed under a pressure of 45 $psia$ and at a temperature of 95°F. Use the ideal gas law equation given by Equation 1.104. (a) Determine the volume of the carbon dioxide in the tank.

1.197C The application of what law introduces enthalpy, h? Also, define what enthalpy, h is.

1.198C What does the first law of thermodynamics state?

1.199C In general, energy exists in numerous forms. List eight forms of energy.

1.200C What is the difference between an extensive fluid property vs. an intensive fluid property? Give an example of each.

1.201C The total energy of a system is represented by E, while the total energy per unit mass of a system is defined as $e = E/M$. There are two general types of energy. What are they, and what is each type of energy related to?

1.202C Define a simple compressible system.

1.203C What type of energy does the total energy, E of a simple compressible system consist of?

1.204C The total energy per unit mass, e is defined by Equation 1.119 as follows: $e = u + ke + pe$ or by Equation 1.120 as follows: $e = u + (v^2/2) + gz$. Explain each term in the equations.

1.205C Define flow energy, or flow work.

1.206C Define the total energy of a flowing fluid (for a control volume).

1.207C Define is enthalpy.

1.208C Define a British thermal unit; and what is a calorie?

1.209BG/SI Use Table A.5 in Appendix A to illustrate (in both BG and SI units) the relationship between the ideal specific heats, c_p and c_v and the ideal gas constant, R, which is given by Equation 1.136 as follows: $c_p - c_v = R$, for air, carbon monoxide, hydrogen, and water vapor.

1.210BG/SI Use Table A.5 in Appendix A. (a) What are the ideal specific heat at constant pressure, c_p; the ideal specific heat at constant volume, c_v; and the ideal specific heat ratio, $k = (c_p/c_v)$ for air at room temperature (68°F)? (b) What are the ideal specific heat at constant pressure, c_p; the ideal specific heat at constant volume, c_v; and the ideal specific heat ratio, $k = (c_p/c_v)$ for helium at room temperature (20°C)?

1.211BG/SI Use Table A.4 in Appendix A. (a) What is the specific heat for gasoline at room temperature (68°F)? (b) What is the specific heat for benzene at room temperature (20°C)?

1.212C In accordance with the various laws of thermodynamics, when gases are compressed or expanded, the relationship between the pressure, p and the density, ρ will depend upon the nature of the compression or expansion process. Thus, the value of the bulk modulus of elasticity E_v for an ideal gas will depend upon the type of compression or expansion process involved. What are the two types of compression or expansion processes?

1.213BG Refer to Problem 1.185, where for a 600-ft^3 helium tank compressed under a pressure of 600 $psia$ and at a temperature of 30°F, the density of the helium in the tank was determined to be 0.0142 $slug/ft^3$, and the weight of the helium in the tank was determined to be 274.471 lb. The helium in the tank is further compressed under isothermal conditions (constant temperature) to a volume of 400 ft^3. (a) Determine the resulting density of the helium in the tank. (b) Determine the resulting pressure of the helium in the tank. (c) Determine the resulting bulk modulus of elasticity of the helium in the tank.

1.214BG What is the isothermal bulk modulus of elasticity of methane at 100°C and an absolute pressure of 7000 $psia$?

1.215BG What is the isentropic bulk modulus of elasticity of methane at 100°C and an absolute pressure of 7000 $psia$?

1.216SI   A 60-m^3 carbon monoxide tank under a pressure of 450 kN/m^2 abs and at a temperature of 20°C is expanded to a volume of 90 m^3. Assume a mass of 70 kg for the carbon monoxide. (a) Determine the resulting pressure and bulk modulus of elasticity of the carbon monoxide in the tank if it is expanded under isothermal conditions (constant temperature). (b) Determine the resulting pressure and bulk modulus of elasticity of the carbon monoxide in the tank if it is expanded under isentropic (frictionless and no heat exchange with the surroundings).

1.217BG   A 15-ft^3 oxygen tank under a pressure of 20 $psia$ and at a temperature of 60°F is compressed to a volume of 5 ft^3. Assume a mass of 0.04 slugs for the oxygen. (a) Determine the resulting pressure and the resulting bulk modulus of elasticity of the oxygen in the tank if it is compressed under isothermal conditions (constant temperature). (b) Determine the resulting pressure and the resulting bulk modulus of elasticity of the oxygen in the tank if it is compressed under isentropic (frictionless and no heat exchange with the surroundings).

1.218SI   Refer to Problem 1.186, where for a 125-m^3 carbon dioxide tank compressed under a pressure of 900 kN/m^2 abs and at a temperature of 15°C, the density of the carbon dioxide in the tank was determined to be 16.534 kg/m^3, and the weight of the carbon dioxide in the tank was determined to be $2.027 \times 10^4 \, N$. The carbon dioxide in the tank is further compressed under isentropic (frictionless and no heat exchange with the surroundings) to a volume of 45 m^3. (a) Determine the resulting density of the carbon dioxide in the tank. (b) Determine the resulting pressure of the carbon dioxide in the tank. (c) Determine the resulting bulk modulus of elasticity of the carbon dioxide in the tank.

1.219BG   Helium in a tank at 80°F and at atmospheric pressure is compressed under isentropic (frictionless and no heat exchange with the surroundings) to a pressure of 78 $psia$. (a) Determine the initial and resulting density of the helium in the tank. (b) Determine the resulting bulk modulus of elasticity of the helium in the tank. (c) Determine the resulting temperature of the helium in the tank.

1.220SI   Compare the isentropic bulk modulus of elasticity of nitrogen at 180 kN/m^2 abs to that of oxygen at the same pressure.

1.221C   For a compressible (elastic) fluid (liquid or gas), the occurrence of a slight localized increase in pressure, p and density, ρ of the fluid results in a small wave called a wave of compression. What is the velocity (celerity) of the small wave of compression called? How is the celerity related to the bulk modulus of elasticity, E_v of the fluid? What would be the value the bulk modulus of elasticity, E_v for an ideal incompressible (inelastic) fluid?

1.222BG   It is of interest to compare the sonic velocities for water, glycerin, and mercury, all at room temperature. (a) Determine the sonic velocity for water at room temperature. (b) Determine the sonic velocity for glycerin at room temperature. (c) Determine the sonic velocity for mercury at room temperature.

1.223C   The sonic velocity, c for gases may be computed by applying Equation 1.160, with the appropriate equation for the bulk modulus of elasticity, E_v, either Equation 1.157 or Equation 1.158, depending upon if the ideal gas is assumed to be either isothermal (constant temperature) or isentropic (frictionless and no heat exchange with the surroundings), respectively. What is the practical assumption made regarding the compression or expansion process of the ideal gas? Explain the basis for this assumption, and derive the sonic velocity, c for an ideal gas.

1.224SI It is of interest to compare the sonic velocity for carbon dioxide, methane, and oxygen at a specified temperature. (a) Determine the sonic velocity for carbon dioxide at 75°C. (b) Determine the sonic velocity for methane at 75°C. (c) Determine the sonic velocity for oxygen at 75°C.

1.225BG While the bulk modulus of elasticity, E_v and the sonic velocity, $c = \sqrt{E_v/\rho}$ are physical fluid properties that measure the degree of compressibility of a fluid, the Mach number, $M = \left(\sqrt{F_I}/\sqrt{F_E}\right) = \left(v/\sqrt{E_v/\rho}\right) = (v/c)$ serves as a practical indicator of the degree of compressibility of fluid in general, especially for gases. Referring to Problem 1.222 above, for each of the three fluids, water, glycerin, and mercury, use the Mach number, M to determine the limiting velocity at which the flow is usually assumed to be incompressible vs. compressible. State if high speeds are practically attainable in the three fluids.

1.226C The Mach number, M expresses the relative importance between what two forces in fluid flow? What is the value of the Mach number, M for sonic flow, subsonic flow, supersonic flow, and hypersonic flow? What is the value of the Mach number, M for incompressible flow and compressible flow?

1.227BG/SI A plane travels at 800 *mph* at an altitude of 50,000 *ft* above sea level. (a) Determine the Mach number, assuming the BG system of units. (b) Determine the Mach number, assuming the SI system of units. (c) Compare the results, and characterize the speed of the plane (sonic, subsonic, supersonic, etc.) and type of flow (compressible or incompressible flow).

1.228SI Planes usually fly at altitudes ranging between 0 and 12 *km*. Using Table A.1 in Appendix A, graphically illustrate the variation of the speed of sound with altitude, assuming the SI system of units.

2

Fluid Statics

2.1 Introduction

Fluid statics deals with problems associated with fluids at rest. The fluid can be either a liquid or a gas. In the study of fluids in static equilibrium, the significant forces acting on a fluid element include those due to gravity, $F_G = Mg = W = \gamma V$ and pressure, $F_p = \Delta pA$. Additionally, other potentially significant forces for fluid statics may include those due to elasticity, $F_E = E_v A$ (compressible fluids) and surface tension, $F_T = \sigma L$ (droplets, bubbles, thin films, and capillarity). However, the viscous force, $F_v = \tau A = \mu \, dv/dy$ does not exist in a fluid in static equilibrium, regardless of the fluid viscosity, μ, and the inertia force, $F_I = Ma = \rho v^2 L^2$ does not come into play in fluid statics. In order for a fluid element to be in static equilibrium (at rest), the sum of all of the external forces must be zero, as given by Newton's second law of motion, $\Sigma F = 0$. Fluid statics applies the principles of hydrostatics, which are based on Pascal's law and Newton's second law of motion. Furthermore, while hydrostatic fluid pressure exists in both fluid statics and in fluid flow (fluid dynamics) situations, Chapter 2 introduces the topic of fluid pressure for fluids in static equilibrium, while Chapter 4 and Chapter 5 introduce the topic of fluid pressure for fluids in motion.

There are a number of fluid properties that play an important role in fluid statics. The specific weight, $\gamma = \rho g$ is the most important fluid property in fluid statics. Vapor pressure, p_v becomes important when very low pressures are involved. The bulk modulus of elasticity, E_v becomes important when compressible fluids are involved. Furthermore, surface tension, σ is important in droplets, bubbles, thin films, and capillarity. Viscosity, μ only becomes important when the fluid is no longer in static equilibrium and thus there is a fluid flow. One may note that most fluid properties (especially the density for a gas/compressible flow) are a function of the temperature and pressure. Furthermore, while the density, ρ of a gas/compressible flow is assumed to vary with temperature and pressure, the density, ρ of a liquid/incompressible flow is assumed to be a constant with respect to temperature and pressure.

The topic of fluid statics, which is based on the principles of hydrostatics, addresses three topics: (1) the measurement of the hydrostatic pressure at a point (see Section 2.3), (2) the determination of the hydrostatic forces acting on a submerged surface (see Section 2.4), and (3) the role of the hydrostatic force in the buoyancy and stability of a floating or a neutrally buoyant body (see Section 2.5). The measurement of the hydrostatic pressure at a point is important in the determination of the atmospheric pressure (barometers), the pressure at a point in a tank of fluid at rest (manometers), and the pressure at a point in a pipe flowing under pressure (manometers). The determination of the hydrostatic force acting on a submerged surface is important in the design of storage tanks, dams, gates, boats, ships, bulkheads, hydrometers, buoys, pipes, and hydraulic structures. The role of the hydrostatic force in the buoyancy and stability of a floating or a neutrally buoyant body is important in the design of balloons, ships, boats, submarines, and other floating bodies or neutrally buoyant bodies.

2.2 The Principles of Hydrostatics

The principles of hydrostatics are based on Pascal's law and Newton's second law of motion for fluid statics, which yield the hydrostatic pressure equation that represents the scalar ordinary differential equation of fluid statics. While Pascal's law describes the hydrostatic pressure at a point, the hydrostatic pressure equation describes the variation of pressure from point to point. Application of the principles of hydrostatics requires the definition of the hydrostatic pressure, p; the hydrostatic pressure head, $h = p/\gamma$; and the hydrostatic pressure distribution for a fluid at rest. In order for a fluid element to be in static equilibrium (at rest), the sum of all of the external forces must be zero as given by Newton's second law of motion:

$$\sum F = 0 \qquad (2.1)$$

The inertia force, $F_I = Ma = \rho v^2 L^2$ does not come into play in fluid statics. The significant forces acting on a fluid element at rest include those due to gravity, $F_G = Mg = W = \gamma V$ and pressure, $F_p = \Delta p A$. Furthermore, the viscous/shear force, $F_v = \tau A = \mu \, dv/dy$ does not exist in a fluid in static equilibrium, regardless of the fluid viscosity, μ. Therefore, for fluids at rest, there are only normal forces due to gravity and hydrostatic pressure (see Figure 2.1), while there are no shear forces due to viscosity. The gravitational force, F_G acts normal

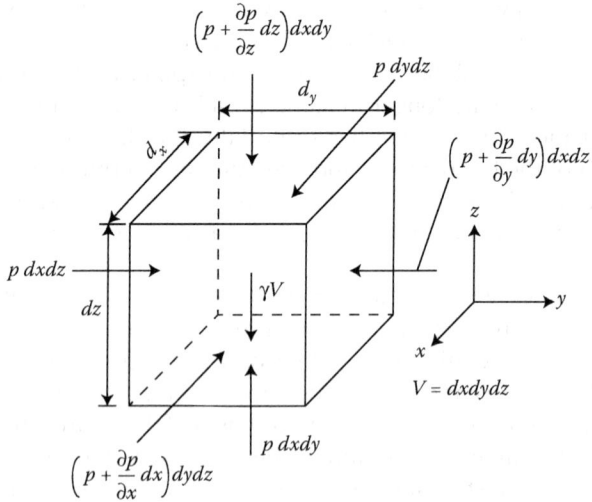

FIGURE 2.1
Significant forces acting on a fluid element in static equilibrium include the gravitational force, $F_G = \gamma V$ and the pressure force, $F_p = \Delta p A$.

through the center of gravity of the fluid element. The pressure force, F_p acts normal to the surface area of the fluid element, where the hydrostatic pressure, p is defined as the magnitude of this normal force per unit area of the surface as follows:

$$p = \frac{F_p}{A} \tag{2.2}$$

where $p =$ hydrostatic pressure, $F_p =$ pressure force, and $A =$ surface area. The dimensions and units for the pressure, p in the English system are given as follows:

$$p = \frac{F}{A} = \frac{[F]}{[L^2]} = [lb/ft^2] = [psf] \tag{2.3}$$

or

$$p = \frac{F}{A} = \frac{[F]}{[L^2]} = [lb/in^2] = [psi] \tag{2.4}$$

where *psf* is the abbreviation for lb/ft^2 and *psi* is the abbreviation for lb/in^2. The dimensions and units for the pressure, p in the metric system are given as follows:

$$p = \frac{F}{A} = \frac{[F]}{[L^2]} = [N/m^2] = [Pa] \tag{2.5}$$

where a Pascal (*Pa*) is the derived metric unit for pressure, p. Furthermore, a common unit for pressure, p in the metric system is the bar (*b*), which is defined as follows:

$$1\,N/m^2 = 1\,Pa = 10^{-5}\,bar \tag{2.6}$$

$$1\,kN/m^2 = 1\,kPa = 10\,mb \tag{2.7}$$

Definition of the hydrostatic pressure head, $h = p/\gamma$ and the hydrostatic pressure distribution for a fluid at rest are presented and discussed below.

2.2.1 Pascal's Law: The Hydrostatic Pressure at a Point

Pascal's law states that pressure acts equally in all directions at any point in a static body of fluid. Thus, for the fluid element given in Figure 2.1, as the size of the elemental cube becomes smaller and smaller, dx, dy, and dz each approach zero, the size of the element approaches a point; thus:

$$\frac{\partial p}{\partial x} = \frac{\partial p}{\partial y} = \frac{\partial p}{\partial z} = 0 \tag{2.8}$$

Because the magnitude of pressure at a point is the same in all directions and has no specific direction, the pressure has no vector sense and thus is a scalar quantity. However, the differential (unit) force, dF_p produced as a result of the action of pressure, p on a differential (unit) area, dA is a vector with both magnitude, which is proportional to the unit area, and a direction, which is normal to the unit area.

2.2.2 The Hydrostatic Pressure Equation: Variation of Pressure from Point to Point

While Pascal's law stated that the hydrostatic pressure acts equally in all directions at any point in a static body of fluid, it does not indicate how the hydrostatic pressure varies from one point to another within the body of fluid. However, the hydrostatic pressure equation, which is based on Newton's second law of motion for fluids in static equilibrium, describes the variation of pressure from point to point. The significant forces acting on a fluid element at rest include those due to gravity, $F_G = Mg = W = \gamma V$ and pressure, $F_p = \Delta pA$. Application of Newton's second law of motion for fluids in static equilibrium (Equation 2.1) to the fluid element given in Figure 2.1 in each of the three (x, y, z) directions yields the following:

$$\sum F_x = -\left(p + \frac{\partial p}{\partial x}dx\right)dy\,dz + p\,dy\,dz = 0 \tag{2.9}$$

$$\sum F_y = -\left(p + \frac{\partial p}{\partial y}dy\right)dx\,dz + p\,dx\,dz = 0 \tag{2.10}$$

$$\sum F_z = -\left(p + \frac{\partial p}{\partial z}dz\right)dx\,dy + p\,dx\,dy - \gamma\,dx\,dy\,dz = 0 \tag{2.11}$$

which results, respectively, in the following:

$$\frac{\partial p}{\partial x}dx\,dy\,dz = 0 \tag{2.12}$$

$$\frac{\partial p}{\partial y}dx\,dy\,dz = 0 \tag{2.13}$$

$$\frac{\partial p}{\partial z} = -\gamma \tag{2.14}$$

The elemental volume $dx\,dy\,dz$ does not equal zero, however; in order to satisfy Newton's second law of motion for fluids in static equilibrium, it must be that:

$$\frac{\partial p}{\partial x} = 0 \tag{2.15}$$

$$\frac{\partial p}{\partial y} = 0 \tag{2.16}$$

where the fluid is homogenous in γ and continuous in space (x, y, z). Furthermore, because the hydrostatic pressure is independent of x and y, the variation in the hydrostatic pressure in the z-direction can be written as a full derivative as follows:

$$\frac{dp}{dz} = -\gamma \tag{2.17}$$

Thus, Equations 2.15 and 2.16 indicate that there is no pressure variation in a horizontal plane for a given fluid, which is known as Pascal's law (stated above). Also, Equation 2.17 indicates that there is a pressure variation in the vertical direction, where the hydrostatic pressure increases linearly with depth, z, which is known as the hydrostatic pressure equation. The minus sign is a result of having defined the z-axis as negative in the downward direction in the application of Newton's second law of motion in Equation 2.11 above. One may alternatively define the z-axis as positive in the downward direction and state that the variation of hydrostatic pressure in the vertical direction is given as follows:

$$\frac{dp}{dz} = \gamma = \rho g \qquad (2.18)$$

where the variation in pressure, p in the vertical direction, z is due only to the weight, γ of the fluid. (Recall that the significant forces acting on a fluid element at rest include those due to gravity, $F_G = Mg = W = \gamma V$ and pressure, $F_p = \Delta pA$). Thus, the hydrostatic pressure equation given in Equation 2.18 represents the scalar ordinary differential equation of fluid statics. The solution of Equation 2.18 will depend upon how the density of the fluid, ρ is related to pressure, p and temperature, T and/or the height of the fluid, z. For compressible fluids (where the thermodynamics and heat transfer concepts are important), one may represent the density as a function of pressure and temperature using the ideal gas law, $\rho = p/RT$, where R is a gas constant, whereas for incompressible fluids (where the principles of mechanics are sufficient), one may assume that the density is a constant with respect to pressure and temperature.

The hydrostatic pressure equation illustrates that hydrostatic pressure increases linearly with depth, z and does not vary in the horizontal x–y plane. Furthermore application of the hydrostatic pressure equation commonly assumes an incompressible flow. Therefore, assuming that all liquids and some gases may be treated as incompressible, one may assume ρ to be a constant and integrate Equation 2.18 over the height of fluid, h in Figure 2.2 as follows:

$$\int_{p_1}^{p_2} dp = \int_{z_1}^{z_2} \gamma\, dz$$

$$(p_2 - p_1) = \gamma(z_2 - z_1)$$

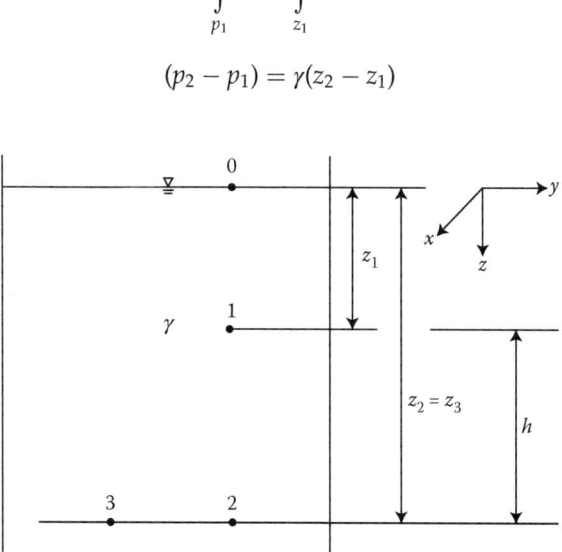

FIGURE 2.2
Hydrostatic pressure, p increases linearly with depth, z and does not vary in the horizontal x–y plane.

$$\Delta p = \gamma \Delta z = \gamma h \tag{2.19}$$

where Δp = the change in pressure between points 1 and 2, and $\Delta z = z_2 - z_1 = h$ = the height of the fluid. Thus, the change in pressure between any two points, Δp in the vertical direction equals the specific weight, γ times the change in depth, $\Delta z = h$, where the pressure increases linearly with depth. One may also express Equation 2.19 for Figure 2.2 as follows:

$$p_2 - p_1 = \gamma h \tag{2.20}$$

$$p_2 = p_1 + \gamma h \tag{2.21}$$

This form of the pressure equation illustrates that the pressure at point 2, p_2 equals the pressure at point 1, p_1 plus the increase in pressure, γh due to an increase in depth, h. Furthermore, it follows that the pressure at point 1, p_1 is given by:

$$p_1 = p_0 + \gamma z_1 \tag{2.22}$$

where p_0 = the pressure at the free surface in Figure 2.2, which would probably be equal to atmospheric pressure. The symbol ∇ on the free surface of the fluid is used to indicate that it is open to the atmosphere. Thus, alternatively, one may express the pressure at point 2, p_2 as follows:

$$p_2 = p_0 + \gamma z_1 + \gamma h = p_0 + \gamma z_1 + \gamma(z_2 - z_1) = p_0 + \gamma z_2 \tag{2.23}$$

Furthermore, according to Pascal's law, if there is no change in depth, then there is no change in pressure, such as between points 2 and 3 in Figure 2.2:

$$p_2 = p_3 \tag{2.24}$$

where $z_2 = z_3$. Finally, Figure 2.3a–c illustrates, various cases where the fluid is homogenous in γ and continuous in space (x, y, z). Once again, according to Pascal's law, because points 1 and 2 are at the same elevation, z, the pressures are equal: $p_1 = p_2$. Furthermore, according to Newton's second law of motion, because the fluid is in static equilibrium, the hydrostatic pressure increases linearly with depth, where:

$$p_3 = p_1 + \gamma h \tag{2.25}$$

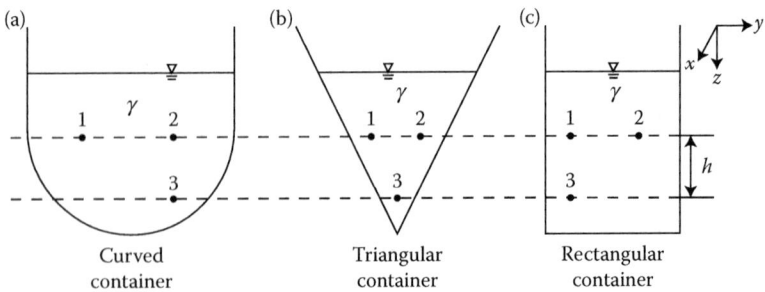

FIGURE 2.3
Hydrostatic pressure, p increases linearly with depth, z and does not vary in the horizontal x–y plane for a fluid that is homogenous in γ and continuous in space x, y, z.

or:

$$p_3 = p_2 + \gamma h \qquad (2.26)$$

which illustrates the hydrostatic pressure equation.

EXAMPLE PROBLEM 2.1

An open tank contains 357 ft of water, as illustrated in Figure EP 2.1. Assume that the atmospheric pressure at the free surface, p_0 is equal to zero. (a) Determine the pressure at points 1 and 2, located at 100 ft and 357 ft below the water surface, respectively.

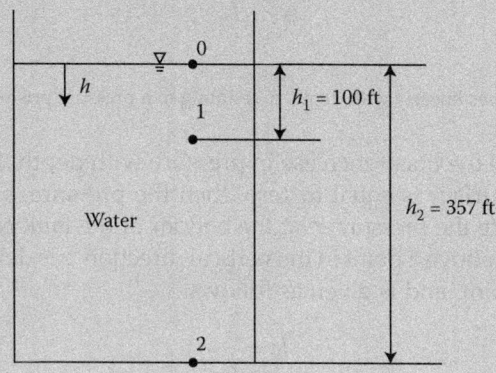

FIGURE EP 2.1
Hydrostatic pressure, p increases linearly with depth, h in a tank of water open to the atmosphere.

Mathcad Solution

(a) The pressures at points 1 and 2 are computed by applying the hydrostatic pressure equation as follows:

$$\gamma_w := 62.417 \frac{lb}{ft^3} \qquad p_0 := 0 \frac{lb}{ft^2} \qquad h_1 := 100 \text{ ft} \qquad h_2 := 357 \text{ ft}$$

$$p_1 := p_0 + \gamma_w \cdot h_1 = 6.242 \times 10^3 \frac{lb}{ft^2}$$

$$p_2 := p_1 + \gamma_w (h_2 - h_1) = 2.228 \times 10^4 \frac{lb}{ft^2}$$

Alternatively,

$$p_2 := p_0 + \gamma_w \cdot h_2 = 2.228 \times 10^4 \frac{lb}{ft^2}$$

2.2.3 The Hydrostatic Pressure Head

As stated above, the hydrostatic pressure in a body of fluid in static equilibrium increases linearly with depth, where the density (or the specific weight) is represented by the slope of the straight line, as illustrated in Figure 2.4. This figure illustrates an open tank filled with fluid to a depth H. The pressure prism on the side of the tank, which graphically depicts

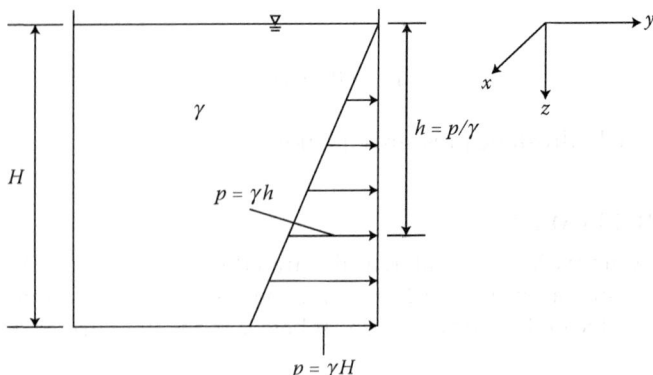

FIGURE 2.4
Hydrostatic pressure, p increases linearly with depth, z, resulting in a pressure prism on side of tank.

a straight line, illustrates the linear increase in pressure with depth. If one assumes that the pressure, p at the free surface is equal to zero, then the pressure, p at any depth, h along the z-axis equals γh, while the pressure, p at the bottom of the tank equals γH, where $h = H$. The height of the fluid, h above a point in the vertical direction, z is defined as the hydrostatic pressure head at that point, and is given as follows:

$$h = \frac{p}{\gamma} \tag{2.27}$$

where the dimension of the pressure head, h is length, L, with English units of ft and metric units of m. Thus, the pressure head, h is a vertical height of a column of fluid having a specific weight, γ that will cause a pressure, $p = \gamma h$ at the base of the column.

EXAMPLE PROBLEM 2.2

An open tank contains 200 ft of gasoline on top of 300 ft of water, as illustrated in Figure EP 2.2. Assume that the atmospheric pressure at the free surface, p_0 is equal

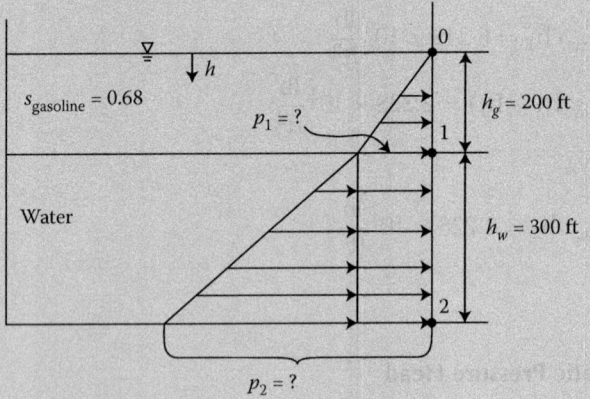

FIGURE EP 2.2
Hydrostatic pressure, p increases linearly with depth, h, resulting in a pressure prism in a tank of water layered with gasoline open to the atmosphere.

to zero, and the specific gravity of the gasoline is 0.68. (a) Determine the pressure at the interface of the two fluids at point 1 and at the bottom of the tank at point 2. (b) Illustrate the pressure prism on the side of the tank.

Mathcad Solution

(a) The pressures at points 1, 2 are computed by applying the hydrostatic pressure equation as follows:

$$\gamma_w := 62.417 \frac{lb}{ft^2} \qquad S_g := 0.68 \qquad p_0 := 0 \frac{lb}{ft^2} \qquad h_g := 200 \, ft \qquad h_w := 300 \, ft$$

$$\gamma_g := S_g \cdot \gamma_w = 42.444 \frac{lb}{ft^3} \qquad p_1 := p_0 + \gamma_g \cdot h_g = 8.489 \times 10^3 \frac{lb}{ft^2}$$

$$p_2 := p_1 + \gamma_w \cdot h_w = 2.721 \times 10^4 \frac{lb}{ft^2}$$

(b) The pressure prism on the side of the tank is illustrated in Figure EP 2.2, where the slope of the straight line in the gasoline layer represents its specific weight, γ_g and the slope of the straight line in the water layer represents its specific weight, γ_w.

2.2.4 The Hydrostatic Pressure Distribution

The hydrostatic pressure distribution in a fluid that is in static equilibrium remains a constant throughout the body of still water. This concept is illustrated by Figure 2.5. If one assumes that the pressure at the free surface, p is equal to zero, then the pressure at any depth h along the z-axis equals γh. Thus, while the pressure at point 1 is denoted by p_1, the pressure head at point 1 is denoted by $h_1 = p_1/\gamma$. The elevation of point 1 is denoted by z_1, which is called the elevation head at point 1. The sum of the pressure head, p_1/γ and the elevation head, z_1 is called the piezometric head at point 1, and is given as follows:

$$\frac{p_1}{\gamma} + z_1 \tag{2.28}$$

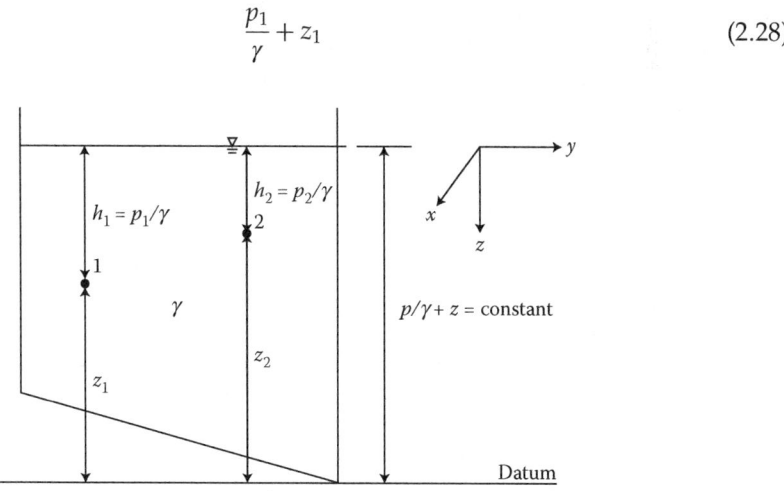

FIGURE 2.5
Piezometric head ($p/\gamma + z$) remains a constant throughout a body in static equilibrium.

Likewise, for point 2, the pressure head at point 2 is denoted by $h_2 = p_2/\gamma$. The elevation head of point 2 is denoted by z_2. The sum of the pressure head, p_2/γ and the elevation head, z_2 is called the piezometric head at point 2, and is given as follows:

$$\frac{p_2}{\gamma} + z_2 \qquad (2.29)$$

According to the principles of hydrostatics, the piezometric head $(p/\gamma + z)$ remains constant throughout a body of still water, so:

$$\frac{\partial\left(\frac{p}{\gamma} + z\right)}{\partial s} = 0 \qquad (2.30)$$

whatever the direction of s may be (see Figure 2.5). Alternatively, one may express the principles of hydrostatics by defining the piezometric pressure (static pressure) as follows:

$$p + \gamma z \qquad (2.31)$$

thus:

$$\frac{\partial(p + \gamma z)}{\partial s} = 0 \qquad (2.32)$$

2.2.5 Application of the Principles of Hydrostatics

In the application of the principles of hydrostatics, the goal is to measure the hydrostatic pressure at a point, the pressure difference between two points, or the hydrostatic pressure on a surface, which is defined as the hydrostatic force. While Section 2.3 focuses on measuring pressure at a point (barometers and manometers), Section 2.4 addresses determining the hydrostatic pressure force on a submerged surface, and Section 2.5 addresses the role of the hydrostatic force in the buoyancy and stability of a floating or a neutrally buoyant body. Practical application of the principles of hydrostatics to measure pressure differences include deep-sea submersibles and submarines, which are designed to withstand the extreme pressures experienced when diving to the bottom of the ocean. Other practical applications include the hydrostatic pressure of the water in a reservoir at the base of a dam, which must be considered in the design of the dam. An additional practical application of the principles of hydrostatics to measure the pressure difference between two points can be illustrated by a drinking straw and is given in the following paragraph.

A practical application of the principles of hydrostatics to measure the pressure difference between two points can be illustrated by a drinking straw. If a straw is submerged into a glass of water and then lifted out, the water will drop out of the straw. However, if the straw is submerged into the water and a finger is placed over the upper end of the straw (closing its top) and it is then lifted out, nearly all of the water will remain in the straw. It is of interest to examine what forces hold the water in the straw when it is closed at the top. As such, a free body diagram of the water in the straw when the top of the straw is open is illustrated in Figure 2.6a. Because the water drops out of the straw, the fluid mass is not in static equilibrium. The significant forces acting on a fluid element at rest include those due to gravity,

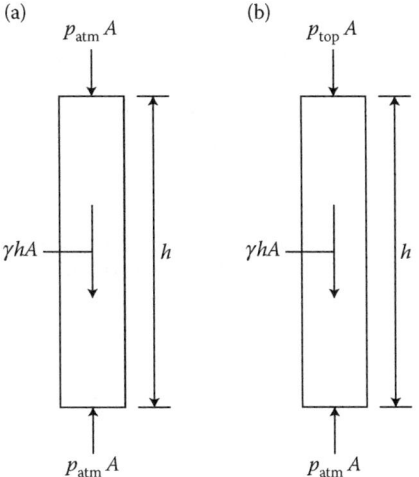

FIGURE 2.6
Free body diagram of a drinking straw submerged into a glass of water and lifted out. (a) Top of drinking straw is open, so water drops out of straw. (b) Top of drinking straw is closed, so water remains in straw.

$F_G = Mg = W = \gamma V$ and pressure, $F_p = \Delta pA$. The sum of the forces in the vertical direction, with a positive force assumed in the downward direction, yields the following expression of the lack of static equilibrium:

$$\sum F = p_{atm}A + W - p_{atm}A = W = \gamma hA \neq 0 \qquad (2.33)$$

where the unbalanced weight, W of the fluid mass causes the water to drop out of the straw, as the atmospheric pressure, p_{atm} acting on the bottom of the straw is canceled out by the atmospheric pressure, p_{atm} acting on the top of the straw. A free body diagram of the water remaining in the straw when the top of the straw is closed is illustrated in Figure 2.6b. The fluid mass of water remaining in the straw is in static equilibrium. The sum of the forces in the vertical direction, with a positive force assumed in the downward direction, yields the following expression of static equilibrium:

$$\sum F = p_{top}A + W - p_{atm}A = 0 \qquad (2.34)$$

where:

$$p_{top} = p_{atm} - \frac{W}{A} = p_{atm} - \gamma h \qquad (2.35)$$

which illustrates that the pressure applied at top of the straw, p_{top} is less than the atmospheric pressure, p_{atm}. Therefore, the pressure applied at the top of the straw, p_{top} is a negative gage pressure or a vacuum/suction (explained below). Furthermore, when the partially filled straw was lifted out of the water, water drained out of the straw, decreasing the height of the column of water, h until the pressure applied at top of the straw, p_{top} could balance the pressure difference, $\Delta p = p_{atm} - \gamma h$.

2.3 Measurment of Hydrostatic Pressure at a Point

The measurement of the hydrostatic pressure at a point is accomplished by the application of the principles of hydrostatics. The measurement of pressure at a point can be accomplished by using numerous methods with numerous apparati. The difference between the methods and apparati depend upon the type of pressure that is measured (atmospheric, tank, or pipe flow), the pressure scale assumed (absolute or gage), the range of pressures being measured, the type of fluids being used, and the desired accuracy in the measurement. Common types of apparatus include barometers, manometers, pressure gages, and pressure transducers. However, before discussing the various methods of pressure measurement, it is important to distinguish between the two types of pressure scales, which are absolute pressure and gage (relative to the atmospheric pressure) pressure.

2.3.1 Pressure Scales

When a pressure is being specified, there is a need to distinguish between the two types of pressure scales, which are absolute pressure and gage pressure. The absolute pressure of a fluid or the atmosphere is measured relative to an absolute zero pressure (perfect vacuum). The gage pressure of a fluid is measured relative to the atmospheric (barometric) pressure. Thus, the gage pressure (relative pressure) of a fluid at the atmospheric pressure registers as a zero pressure, while the gage pressure of a fluid above the atmospheric pressure registers as a positive pressure, and the gage pressure of a fluid below the atmospheric pressure registers as a negative pressure (suction or vacuum). The conditions in the atmosphere vary both temporally and spatially and thus are a function of the weather, the time of year (seasons), the altitude, and the latitude. Therefore, the atmospheric pressure will vary with the elevation of the particular surface on earth and the weather conditions; thus, it is a moveable datum/scale (for the gage pressure), as illustrated in Figure 2.7a. As such, it becomes important to establish a fixed datum relative to the moveable atmospheric pressure datum. This fixed datum is the absolute zero pressure scale, which occurs in a perfect vacuum such as in outer space. The moveable atmospheric pressure datum is measured relative to the fixed absolute zero pressure datum. Therefore, the variable atmospheric pressure above the fixed absolute zero is reported in absolute pressure units (psi [abs] or psia in the English system, and Pa [abs] in the metric system). Furthermore, the absolute pressure of a fluid above the fixed absolute zero is computed as the sum of the atmospheric pressure and the gage pressure of the fluid (measured relative to the atmospheric pressure) as follows:

$$p_{abs} = p_{atm} + p(gage) \tag{2.36}$$

where the absolute pressure, p_{abs} is the pressure that is measured relative to absolute zero pressure. Thus, at the moveable atmospheric pressure datum where the gage pressure, $p(gage)$ is equal to zero, the absolute pressure, p_{abs} is equal to the atmospheric pressure, p_{atm}, as illustrated in Figure 2.7a. However, when the gage pressure, $p(gage)$ is positive, the absolute pressure, p_{abs} is greater than the atmospheric pressure, p_{atm}, and when the gage pressure, $p(gage)$ is negative, the absolute pressure, p_{abs} is less than the atmospheric pressure, p_{atm}, as illustrated in Figure 2.7a. And, finally, the gage pressure is computed as follows:

$$p(gage) = p_{abs} - p_{atm} \tag{2.37}$$

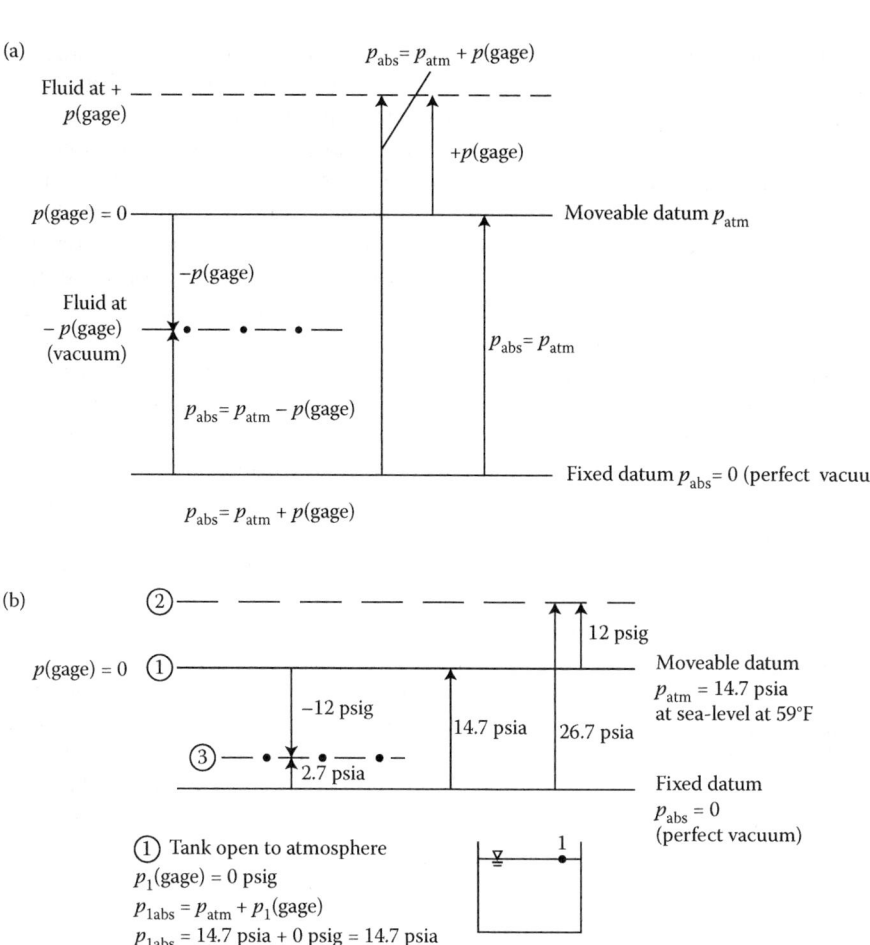

FIGURE 2.7
(a) Schematic relationship between the absolute pressure scale, p_{abs} and the gage pressure scale, $p(gage)$. The absolute pressure scale, p_{abs} is measured relative to a fixed datum of zero absolute pressure (perfect vacuum). The gage pressure scale, $p(gage)$ is measured relative to moveable datum of atmospheric (barometric) pressure, p_{atm}. (b) Illustration of the schematic relationship between the absolute pressure scale, p_{abs} and the gage pressure scale, $p(gage)$ for a tank of water. The absolute pressure scale, p_{abs} is measured relative to a fixed datum of zero absolute pressure (perfect vacuum). The gage pressure scale, $p(gage)$ is measured relative to moveable datum of atmospheric (barometric) pressure, p_{atm}.

Thus, when the absolute pressure is greater than the local atmospheric pressure, then the gage pressure registers as a positive pressure, whereas when the absolute pressure is less than the local atmospheric pressure, then the gage pressure registers as a negative (suction or vacuum) pressure. Furthermore, when the absolute pressure is equal to the local atmospheric pressure, the gage pressure registers as a zero pressure. Figure 2.7b illustrates this concept for a tank of water. Additionally, one may illustrate this concept using the following examples. The pressure gage for a completely inflated tire will read a positive gage pressure, $p(gage)$ above the atmospheric pressure, p_{atm} because $p_{abs} > p_{atm}$, while the pressure gage for a completely deflated tire will read a zero gage pressure, $p(gage)$ above the atmospheric pressure, p_{atm} because $p_{abs} = p_{atm}$. Furthermore, the pressure gage for a suction line will read a negative gage pressure, $p(gage)$ below the atmospheric pressure, p_{atm} because $p_{abs} < p_{atm}$. In order to distinguish between the absolute pressure and the gage pressure, the label "abs" (or "a") and "gage" (or "g") is added to the pressure reading, respectively. Therefore, in the English system of units, the absolute pressure is reported as psi (abs) or psia, while in the metric system of units, the absolute pressure is reported as Pa (abs). Furthermore, in the metric system of units, the gage pressure is reported as psi (gage) or psig, while in the metric system of units, the gage pressure is reported as Pa (gage).

In the application of the principles of hydrostatics in fluid mechanics problems, pressures are typically assumed to be gage pressures unless specifically indicated as absolute pressure, where a positive gage pressure is referred to simply as "gage pressure" or "pressure," while a negative gage pressure is referred to as a "vacuum." Thus, in the assumption of the gage pressure scale, the atmospheric pressure is at zero (gage). Working in the gage pressure scale is logical, especially when the magnitude of the atmospheric pressure is relatively small compared to the magnitude of the fluid pressure, where adding the atmospheric pressure to the gage pressure to yield the absolute pressure of the fluid would not make a significant difference. Therefore, in many applications, when the "atmospheric pressure is assumed to be zero," this implies that the gage pressure scale is assumed, where the atmospheric pressure is at zero (gage). Furthermore, working in the gage pressure scale allows the representation of a negative gage pressure or a "vacuum" when the top of a liquid or a gas in a tank (or at any point in a pipe flow system) is subjected to an absolute pressure that is lower than the atmospheric pressure, where the maximum possible negative gage pressure is equal to $-p_{atm}$, which occurs at $p_{abs} = 0$ (abs).

2.3.1.1 Standard Atmospheric Pressure

The local atmospheric (barometric) pressure varies with time and location on earth; thus, it will vary with the elevation of the particular surface on earth and the weather conditions that affect the density of the air. Thus, it may be noted that when the atmospheric pressure is reported without a reference to elevation, it is assumed that the elevation is zero ft or m above sea level, for which the location is at sea level (see Chapter 1). Thus, the standard atmospheric pressure at sea level at an atmospheric temperature of $59°$ Fahrenheit (English) or $15°$ Celsius (metric) is $14.696\,psi$ (abs) (English) or $101.325\,kPa$ (abs) (metric). Furthermore, at the standard sea-level atmospheric pressure, and a latitude of $45°$, the corresponding standard acceleration due to gravity, g is $32.174\,ft/sec^2$ in the English system or $9.807\,m/sec^2$ in the metric system. It may be noted that the standard acceleration due to gravity, g is used in the hydrostatic pressure equation, $(dp/dz) = \gamma = \rho g$.

2.3.2 Methods of Pressure Measurement

The measurement of pressure at a point can be accomplished by using numerous methods with numerous apparatus. The difference between the methods and apparatus depend upon the type of pressure that is measured (atmospheric, tank, or pipe flow), the pressure scale assumed (absolute or gage), the range of pressures being measured, the type of fluids being used, and the desired accuracy in the measurement. It is important to distinguish between the types of hydrostatic pressures that are measured, which include the atmospheric pressure, the hydrostatic pressure at a given point in a tank containing fluid at rest, and the hydrostatic pressure at a given cross section of a pipe containing fluid flowing under pressure. Common types of apparatus include barometers, manometers, pressure gages, and pressure transducers. While a brief description of the various types of pressure apparatus is given in the paragraphs below, details of the application of barometers and manometers in the measurement of fluid pressure are given in the following two sections below.

In general, a barometer is used to measure the absolute pressure, while a manometer is used to measure the gage pressure. Although barometers are typically used to measure the absolute pressure of the atmosphere (barometric pressure), they may also be used to directly measure the absolute pressure of any liquid. A manometer is typically used to directly measure the gage pressure in a fluid in a tank of fluid or in a pipe flow. Depending upon whether: (1) the fluid is a liquid or a gas, (2) the fluid is under a high or a low pressure, and (3) a pressure or a pressure differential is sought, there are various types of manometers. While manometers are both inexpensive and accurate, they cannot easily be used for measuring high pressures (above several atmospheres) or pressures that are changing rapidly with time. Additionally, because they require the measurement of the length(s) of a column of fluid(s), they cannot be read very quickly. Thus, for the measurement of high or rapidly changing pressures, pressure gages (discussed below) are used. It may be noted, however, that because of their mechanical limitations, pressure gages are not usually adequate for precise measurements of pressure. Therefore, when greater precision is required, manometers may be effectively used. Furthermore, a manometer is known as a primary standard and thus seldom needs to be calibrated.

Mechanical pressure-measuring devices known as pressure gages may be used to measure either absolute or gage pressure. Pressure gages are a necessary part of scuba diving gear and an obvious part of blood pressure measurement. The Bourdon pressure gage is typically used to measure gage pressure, while the aneroid pressure gage is typically used to measure absolute pressure.

Finally, there are more sophisticated electronic pressure measuring devices that may be used to measure pressure, and these are called pressure transducers. The advantages of pressure transducers over mechanical pressure gages is that they can transmit pressure readings to remote locations; they can display pressures in digital form; and they can record pressure as a function of time on magnetic tape, other storage medium, and hardcopy displays. Pressure transducers are used in automatic control of manufacturing systems. Two common types of pressure transducers are the strain gage pressure transducer and the linear variable differential transformer pressure transducer (LVDT).

2.3.3 Barometers

Although barometers are typically used to measure the absolute pressure of the atmosphere (barometric pressure), they may also be used to directly measure the absolute pressure of any liquid. While mercury (Hg) is usually used to measure the absolute pressure, other

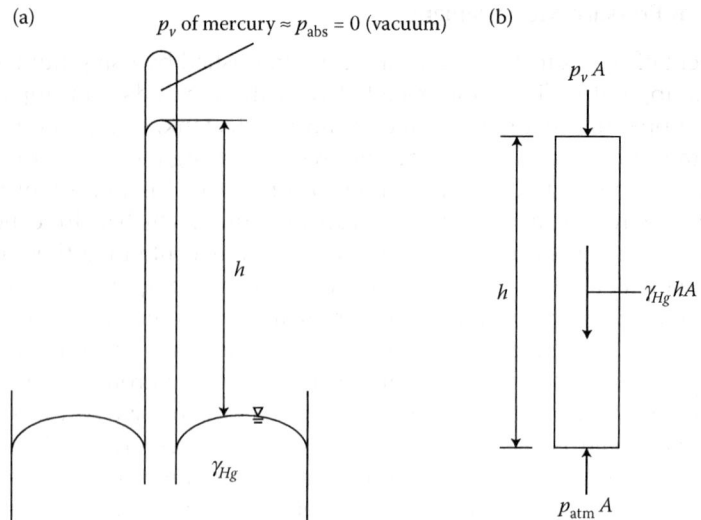

FIGURE 2.8
A mercury barometer may be used to measure the absolute pressure of the atmosphere (barometric pressure).
(a) Mercury barometer. (b) Free body diagram of mercury barometer.

liquids such as water may also be used. Figure 2.8a illustrates a typical mercury barometer used to measure the absolute pressure of the atmosphere. A glass tube with a cross-sectional area, A is filled with mercury and is inverted with its open end placed below the mercury surface in the container of mercury. Because there is a vacuum (absolute pressure is equal to zero for a perfect vacuum) at the top of the inverted tube, the vapor pressure, p_v of the mercury will be attained (see Chapter 1). Also, because the container is open to the atmosphere, the pressure on the free surface of the mercury in the container is atmospheric, p_{atm}. The symbol ∇ on the free surface of the mercury is used to indicate that it is open to the atmosphere. The free body diagram in Figure 2.8b illustrates a column of mercury with a cross-sectional area, A and a height, h. The significant forces acting on a fluid element at rest include those due to gravity, $F_G = Mg = W = \gamma V$ and pressure, $F_p = \Delta pA$. The sum of the forces in the vertical direction, with a positive force assumed in the downward direction, yields the following expression of static equilibrium:

$$\sum F = p_vA + W - p_{atm}A = p_vA + \gamma_{Hg}hA - p_{atm}A = 0 \tag{2.38}$$

where the absolute atmospheric pressure, p_{atm} is computed as follows:

$$p_{atm} = p_v + \gamma_{Hg}h \tag{2.39}$$

which illustrates that the application of Newton's second law of motion for fluid statics yields the hydrostatic pressure equation. However, the vapor pressure, p_v of the mercury is very small and is usually ignored; thus, the absolute atmospheric pressure, p_{atm} is given as follows:

$$p_{atm} = \gamma_{Hg}h \tag{2.40}$$

Therefore, the column of mercury, h will attain a position of static equilibrium when its weight, $W = F_G = \gamma_{Hg}\ hA$ is balanced by the force due to the atmospheric pressure,

p_{atm} A. As such, the atmospheric pressure, p_{atm} will support a column of mercury of height, h.

Because the height, h of the column of mercury is a direct measurement of the absolute pressure of the atmosphere, p_{atm}, it is conventional to specify the atmospheric pressure in terms of the height, h in inches or millimeters of mercury as follows:

$$h = \frac{p_{atm}}{\gamma_{Hg}} \tag{2.41}$$

Referring to Figure 2.8, the following two examples illustrate that the standard atmospheric pressure of 14.696 *psi* (abs) or 101.325 *kPa* (abs) supports a column of mercury (32°F or 0°C) of 29.92 *in* or 760 *mm* high, respectively.

EXAMPLE PROBLEM 2.3

Assuming English units, the standard atmospheric pressure, p_{atm} is 14.696 psi (abs). Assume the mercury is at 32°F, where the specific gravity of mercury is 13.595. (a) Determine the height, h that the column of mercury would rise to in a barometer.

Mathcad Solution

(a) The height, h that the column of mercury would rise to is computed from Equation 2.41 as follows:

$$P_{atm} := 14.696 \frac{lb}{in^2} \qquad s_{Hg} := 13.595 \qquad \gamma_w := 62.417 \frac{lb}{ft^3}$$

$$\gamma_{Hg} := s_{Hg} \cdot \gamma_w = 848.559 \frac{lb}{ft^3}$$

$$h := \frac{P_{atm}}{\gamma_{Hg}} = 29.927 \, in \qquad\qquad h := \frac{P_{atm}}{\gamma_{Hg}} = 2.494 \, ft$$

EXAMPLE PROBLEM 2.4

Assuming metric units, the standard atmospheric pressure, p_{atm} is 101.325 kPa (abs). Assume the mercury is at 0°C, where the specific gravity of mercury is 13.595. (a) Determine the height, h that the column of mercury would rise to in a barometer.

Mathcad Solution

(a) The height, h that the column of mercury would rise to is computed from Equation 2.41 as follows:

$$P_{atm} := 101.325 \, kPa \qquad s_{Hg} := 13.595 \qquad \gamma_w := 9810 \frac{N}{m^3}$$

$$\gamma_{Hg} := s_{Hg} \cdot \gamma_w = 1.334 \times 10^5 \frac{kg}{m^2 s^2}$$

$$h := \frac{P_{atm}}{\gamma_{Hg}} = 759.746 \, mm \qquad\qquad h := \frac{P_{atm}}{\gamma_{Hg}} = 0.76 \, m$$

One may note that in the metric units, the unit of *mm Hg* is called a *torr*. Thus:

$$1\,atm = 101.325\,kPa = 760\,torr = 760\,mm\,Hg$$

where:

$$1\,torr = 0.1333\,kPa = 133.3\,Pa$$

Thus, Example Problems 2.3 and 2.4 illustrate that the standard atmospheric pressure of 14.696 *psi* (abs) or 101.325 *kPa* (abs) supports a column of mercury of 29.92 *in* or 760 *mm* high, respectively. Furthermore, the level of mercury will rise and fall as the atmospheric pressure increases and decreases with the elevation and density of the air on earth, respectively. It may be noted that when an accurate measurement of the atmospheric pressure is required, one should make corrections to the barometer reading to account for both the vapor pressure, p_v of the mercury and capillarity (due to surface tension, σ of the mercury).

It is of interest to examine the case where water was used in the barometer instead of mercury. Once again, because the height, h of the column of water is a direct measurement of the absolute pressure of the atmosphere, p_{atm}, it is conventional to specify the atmospheric pressure in terms of the height, h in inches or millimeters of water as follows:

$$h = \frac{p_{atm}}{\gamma_w} \tag{2.42}$$

Referring to Figure 2.8, the next two examples illustrate that the standard atmospheric pressure of 14.696 *psi* (abs) or 101.325 *kPa* (abs) supports a column of water (39.2°F or 4°C) of 33.91 *ft* or 10.33 *m* high, respectively.

EXAMPLE PROBLEM 2.5

Assuming English units, the standard atmospheric pressure, p_{atm} is 14.696 psi (abs). Assume the water is at 39.2°F. (a) Determine the height, h that the column of water would rise to in a barometer.

Mathcad Solution

(a) The height, h that the column of water would rise to is computed from Equation 2.42 as follows:

$$p_{atm} := 14.696\,\frac{lb}{in^2} \qquad\qquad \gamma_w := 62.417\,\frac{lb}{ft^3}$$

$$h := \frac{p_{atm}}{\gamma_w} = 406.855\,in \qquad\qquad h := \frac{p_{atm}}{\gamma_w} = 33.905\,ft$$

EXAMPLE PROBLEM 2.6

Assuming metric units, the standard atmospheric pressure, p_{atm} is 101.325 kPa (abs). Assume the water is at 4°C. (a) Determine the height, h that the column of water would rise to in a barometer.

Mathcad Solution

(a) The height, h that the column of water would rise to is computed from Equation 2.42 as follows:

$$p_{atm} := 101.325 \, kPa \qquad\qquad \gamma_w := 9810 \frac{N}{m^3}$$

$$h := \frac{p_{atm}}{\gamma_w} = 1.033 \times 10^4 \, mm \qquad\qquad h := \frac{p_{atm}}{\gamma_w} = 10.329 \, m$$

Therefore, Example Problems 2.5 and 2.6 illustrate that the standard atmospheric pressure of 14.696 *psi* (abs) or 101.325 *kPa* (abs) supports a column of water of 33.91 *ft* or 10.33 *m* high, respectively. One may note that in the column of water (33.91 *ft* or 10.33 *m* high) in comparison to the corresponding column of mercury (29.92 *in* or 760 *mm* high), respectively, that the standard atmospheric pressure it supports is significantly larger. This is because the specific gravity, *s* of mercury is 13.595 times that of water. Furthermore, while the vapor pressure of mercury is negligibly small at ordinary temperatures, the same is not true of other fluids (including water). In fact, the height, *h* of the rise of the fluid (nonmercury) in the column would be less than the true atmospheric (barometric) height and thus one would have to make a correction to the reading (for both vapor pressure [especially] and capillarity due to surface tension, *σ* of the fluid). Thus, the high density (specific weight or specific gravity) of mercury (and thus the reasonably short rise in the barometric tube), coupled with its negligibly small vapor pressure, makes mercury the most convenient and accurate barometric fluid.

2.3.4 Manometers

A manometer is typically used to directly measure the gage pressure in a fluid in a tank of fluid or in a pipe flow. Depending upon whether: (1) the fluid is a liquid or a gas, (2) the fluid is under a high or a low pressure, and (3) a pressure or a pressure differential is sought, there are various types of manometers. The various types of manometers include the open piezometer column/tube (simplest manometer), the open U-tube manometer (simple manometer), and the differential manometer. An open piezometer column/tube, which is the simplest type of manometer, is used to measure moderate pressures of liquids. However, the open piezometer tube is too short for use with liquids under high pressure, will not work for fluids under negative pressures (because air would be sucked into the pipe system), and will not work for gases under any pressure (because the gas would escape). Therefore, the simple manometer (open U-tube manometer) may be used for liquids or gases under high pressures or under negative pressures. Furthermore, when the difference between two pressures in two different fluid systems or in a given fluid system is needed, a differential manometer may be used. Additionally, the type of differential manometer and the respective manometer fluid will vary depending upon whether the pressure difference is high (liquids or gases) or moderate (liquids only). And, finally, the range of pressure that a manometer can measure is dependent upon the height to which the manometer fluid can rise. Unlike the pressure gages (described above), manometers yield pressure measurements with high precision. Therefore, when greater precision is required, manometers may be effectively used. Furthermore, a manometer is known as a primary standard, so it seldom needs to be calibrated. The pressure in a manometer can be calculated directly if the specific gravity, s of the manometer fluid is known. It may be noted, however, that small errors may results from capillarity, unclean manometer glass tubing, or contaminants in the fluid.

2.3.4.1 Manometry

Devices based on the principles of hydrostatics are called manometers. Assuming a two-dimensional (y, z) model in fluid statics, application of the principles of hydrostatics includes the application of Pascal's law, $(\partial p/\partial y) = 0$ and the application of Newton's second law of motion for fluid statics, which yields the hydrostatic pressure equation, $(\partial p/\partial z) = \gamma$. The hydrostatic equation, which indicates that the pressure increases linearly with depth, z provides the means to measure pressure by the elevation of liquid levels in a tube. Furthermore, the concept of manometry is useful for understanding pressure measurement. The free body diagram in Figure 2.9a illustrates a column of fluid with a cross-sectional area, A and a height, h. The significant forces acting on a fluid element at rest include those due to gravity, $F_G = Mg = W = \gamma V$ and pressure, $F_p = \Delta pA$. Thus, a hydrostatic pressure force, p_1A acts down on the top of the fluid column, while another hydrostatic pressure force, p_2A acts upward on the bottom of the column. The gravitational force, F_G or the weight, W of the fluid acts downward. The sum of the forces in the vertical direction, with a positive force assumed in the downward direction, yields the following expression of static equilibrium:

$$\sum F = p_1A + W - p_2A = p_1A + \gamma hA - p_2A = 0 \tag{2.43}$$

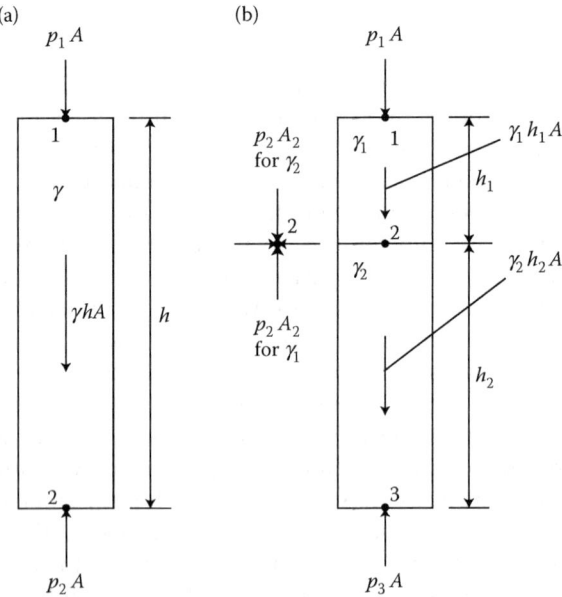

FIGURE 2.9
Significant forces acting on a column of fluid in static equilibrium. (a) Free body diagram of a column of a single fluid. (b) Free body diagram of a column of two fluids.

where the difference in pressure between the bottom and top of the fluid column is given as follows:

$$\Delta p = p_2 - p_1 = \gamma h \tag{2.44}$$

and solving for p_2 yields the following:

$$p_2 = p_1 + \gamma h \tag{2.45}$$

which illustrates that the application of Newton's second law of motion for fluid statics yields the hydrostatic pressure equation.

The free body diagram in Figure 2.9b illustrates a column that includes two fluids of different specific weights, γ_1 and γ_2 with a cross-sectional area, A and heights, h_1, and h_2, respectively. The pressure, p_2 at the bottom of the first fluid is given as follows:

$$p_2 = p_1 + \gamma_1 h_1 \tag{2.46}$$

The sum of the forces in the vertical direction, with a positive force assumed in the downward direction, yields the following expression of static equilibrium for the second fluid:

$$\sum F = p_2 A + \gamma_2 h_2 A - p_3 A = 0 \tag{2.47}$$

Solving for p_3 and substituting the value for p_2 (from Equation 2.46) yields the following:

$$p_3 = p_2 + \gamma_2 h_2 = p_1 + \gamma_1 h_1 + \gamma_2 h_2 \tag{2.48}$$

which illustrates that the application of Newton's second law of motion for fluid statics yields the hydrostatic pressure equation. One may note that both weight terms have positive signs, which indicates that the positive z-axis is downward, and that the pressure increases with depth. Solving for p_1 yields the following:

$$p_1 = p_3 - \gamma_1 h_1 - \gamma_2 h_2 \tag{2.49}$$

which indicates that as the point of interest rises, the pressure decreases linearly with the increase in height, h and the specific weight of the fluid, γ.

2.3.5 Piezometer Columns/Tubes: Simplest Manometers

An open piezometer column/tube, which is the simplest type of manometer, may be used to measure moderate pressures of liquids. A piezometer may be attached to either a container of fluid in static equilibrium or to a pipe flowing under pressure. The piezometer consists of a sufficiently long tube in which the liquid itself (system fluid) can rise freely without overflowing. As such, there is no additional manometer fluid used in a piezometer. The height of the liquid in the tube will give the value of the pressure head, p/γ for which p is the gage pressure at the base of the column. While this type of device is accurate and simple, in order to avoid capillary effects, a piezometer tube's diameter should be about 0.5 inches or 13 *mm* or greater. The fluid pressure to be measured by the open piezometer tube must be relatively small so that the required height of the column is reasonable. If the fluid was subjected to a high positive gage air pressure, p_g, the use of a piezometer would not be practical in this case, as the tube would have to be unreasonably tall to allow the liquid to rise to the high gage pressure (due to both the hydrostatic pressure and the gage air pressure, p_g) at a given point. The piezometer will not work for fluids under negative pressures, because air would be sucked into the pipe system. Furthermore, the piezometer will not work for gases under any pressure, because the gas would escape. For these situations, the simple manometer (open U-tube manometer) may be used and is described in the following section.

A piezometer may be attached to a tank of liquid in static equilibrium with a specific weight, γ that is at atmospheric pressure, as illustrated in Figure 2.10a. Because the pressure at the top of each open column/piezometer is at the atmospheric pressure, p_{atm}, the rise of the fluid in a given column is a measure of the gage pressure at the bottom of the column. Thus, the column of fluid in the first piezometer rises to a level $p_1/\gamma = 0$, for which $p_1 = 0$ is the gage pressure at point 1. The column of fluid in the second piezometer rises to a level of p_2/γ, for which p_2 is the gage pressure at point 2. And, the column of fluid in the third piezometer rises to a level of p_3/γ, for which p_3 is the gage pressure at point 3. If, however, the tank of fluid were subjected to a positive gage air pressure, p_g as illustrated in Figure 2.10b, then $p_1/\gamma = p_g/\gamma > 0$, for which the gage pressure at point 1, $p_1 = p_g > 0$ and the corresponding values for p_2/γ and p_3/γ would increase by the added pressure head, p/γ and are given as follows:

$$p_1 = p_g \tag{2.50}$$

$$p_2 = p_1 + \gamma(z_1 - z_2) = p_g + \gamma(z_1 - z_2) \tag{2.51}$$

$$p_3 = p_2 + \gamma(z_2 - z_3) = p_g + \gamma(z_1 - z_3) \tag{2.52}$$

which illustrates the application of the hydrostatic pressure equation, $(dp/dz) = \gamma$. Additionally, in accordance with the principles of hydrostatics, the piezometric head $(p/\gamma + z)$ remains a constant throughout the fluid body in Figure 2.10, thus:

$$\left(\frac{p_1}{\gamma} + z_1\right) = \left(\frac{p_2}{\gamma} + z_2\right) = \left(\frac{p_3}{\gamma} + z_3\right) \tag{2.53}$$

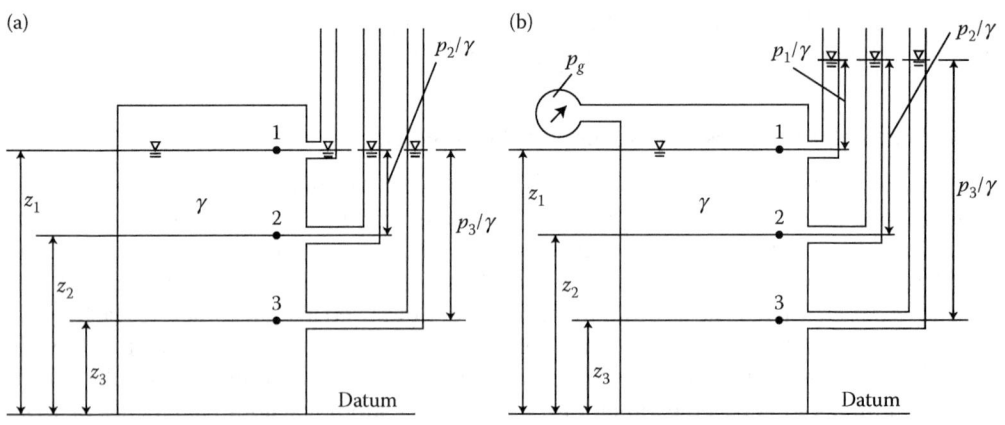

FIGURE 2.10
Piezometer. May be used to measure the gage pressure at any point in a tank of fluid. (a) Tank of fluid is at atmospheric pressure. (b) Tank of fluid is subjected to a positive gage air pressure, p_g.

EXAMPLE PROBLEM 2.7

Three piezometers are attached to a tank of water subjected to a positive gage air pressure, as illustrated in Figure EP 2.7. Points 1, 2, and 3 are located 5, 4, and 1 ft from the bottom of the tank. The pressure head (height of the water in the piezometers), p_1/γ at point 1 reads 11.535 ft. (a) Determine the pressure heads read at points 2 and 3. (b) Illustrate that the piezometric head $(p/\gamma + z)$ remains a constant throughout the tank of water.

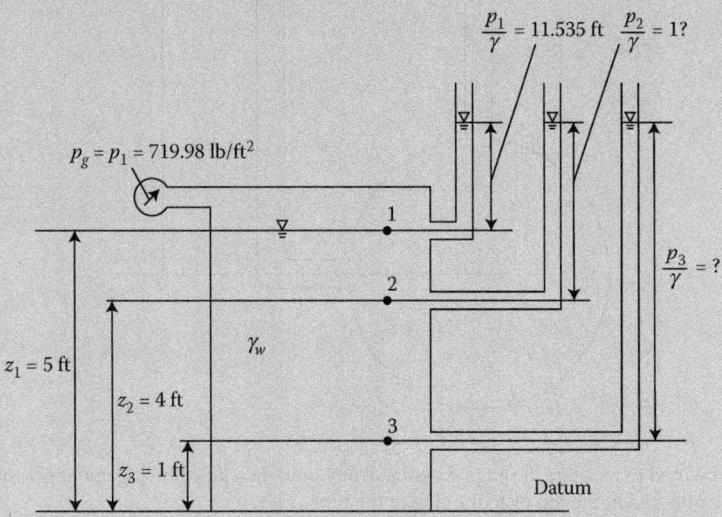

FIGURE EP 2.7
Piezometers are used to measure the gage pressure at several points in a tank of water subjected to a positive gage air pressure, p_g.

Mathcad Solution

(a) In order to determine the pressure at points 2 and 3, the principles of hydrostatics are applied as follows:

$$\gamma_w := 62.417 \frac{lb}{ft^3} \qquad h_1 := 11.535\ ft \qquad p_1 := \gamma_w \cdot h_1 = 719.98 \frac{lb}{ft^2}$$

$$z_1 := 5\ ft \qquad z_2 := 4\ ft \qquad z_3 := 1\ ft$$

$$p_2 := p_1 + \gamma_w\,(z_1 - z_2) = 782.397 \frac{lb}{ft^2} \qquad \frac{p_2}{\gamma_w} = 12.535\ ft$$

$$p_3 := p_2 + \gamma_w\,(z_2 - z_3) = 969.648 \frac{lb}{ft^2} \qquad \frac{p_3}{\gamma_w} = 15.535\ ft$$

(b) In accordance with the principle of hydrostatics, the piezometric head remains a constant throughout the tank of water as follows:

$$\frac{p_1}{\gamma_w} + z_1 = 16.535\ ft \qquad\qquad \frac{p_2}{\gamma_w} + z_2 = 16.535\ ft$$

$$\frac{p_3}{\gamma_w} + z_3 = 16.535\ ft$$

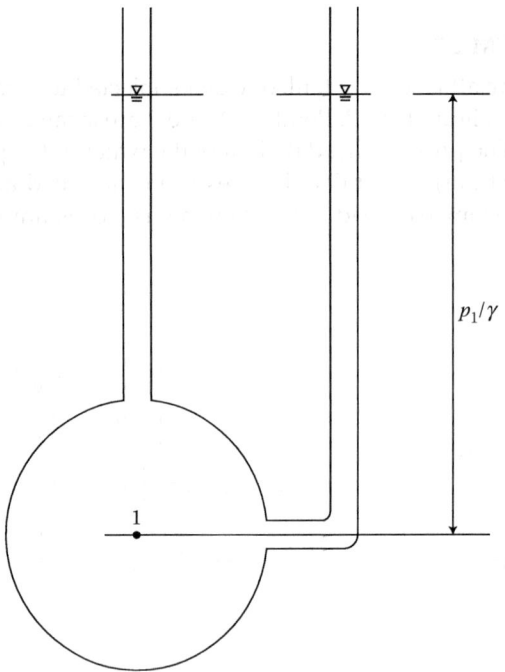

FIGURE 2.11
Piezometer. May be used to measure the gage pressure at any point in a pipe flowing under pressure. The piezometer may either be attached at the top or to the side of the pipe.

A piezometer may also be attached to a pipe, as illustrated in Figure 2.11. While the piezometer may either be attached at the top of the pipe or to the side of the pipe, it is used to measure the gage pressure, p_1 at the center of the pipe at point 1. The pipe of liquid with specific weight, γ is assumed to be flowing under pressure. Thus, while the fluid in the pipe is not assumed to be in static equilibrium, the pressure distribution at cross section 1 may assumed to be hydrostatic, so a piezometer (or a manometer in general) may be used to measure the gage pressure at the centerline of the pipe. Thus, the column of fluid in the piezometer rises to a level of p_1/γ, for which p_1 is the gage pressure at the center of the pipe, point 1.

EXAMPLE PROBLEM 2.8

A piezometer is attached to a pipe flowing under pressure carrying water, as illustrated in Figure EP 2.8. The height of the water in the piezometer at point 1 is 9 ft. (a) Determine the gage pressure at point 1.

Mathcad Solution

(a) In order to determine the gage pressure at point 1, the principles of hydrostatics are applied as follows:

$$h_1 := 9 \, \text{ft} \qquad \gamma_w := 62.417 \, \frac{\text{lb}}{\text{ft}^3} \qquad p_1 := \gamma_w \cdot h_1 = 561.753 \, \frac{\text{lb}}{\text{ft}^2}$$

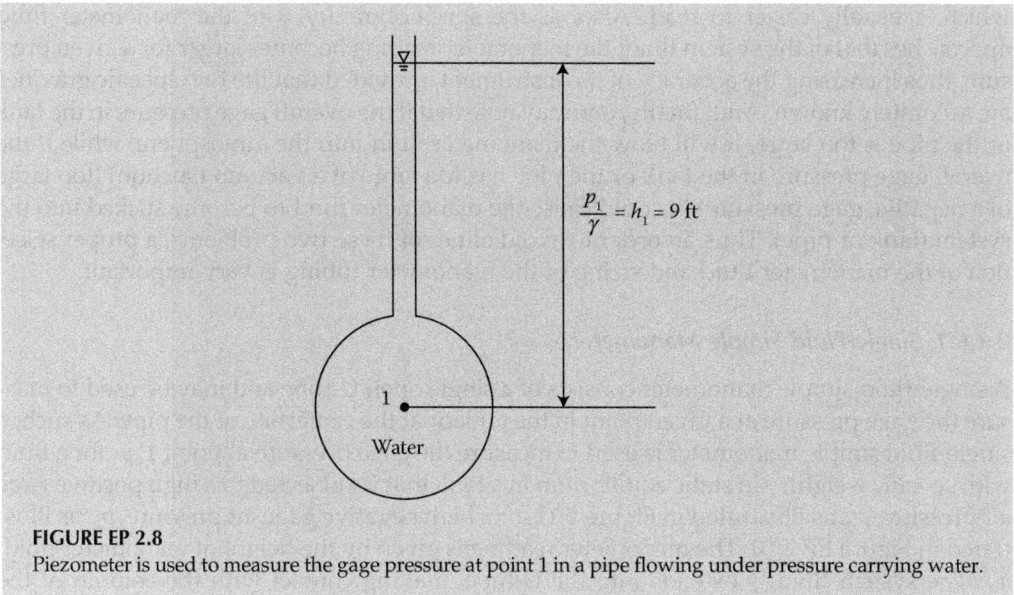

$$\frac{p_1}{\gamma} = h_1 = 9\ \text{ft}$$

1

Water

FIGURE EP 2.8
Piezometer is used to measure the gage pressure at point 1 in a pipe flowing under pressure carrying water.

2.3.6 Open U-Tube Manometers: Simple Manometers

When a tank of fluid in static equilibrium or a pipe of fluid flowing under pressure contains: (1) a liquid subjected to high positive gage air pressure or a negative air pressure or (2) a gas subjected to any air pressure, a piezometer can no longer be used to measure the gage pressure at a given point. Instead, a simple manometer (open U-tube) may be used to measure the gage pressure at a given point in the tank or at the centerline of the pipe. A simple manometer may be either a single-fluid or a multifluid manometer. A single-fluid simple manometer consists of a single glass or plastic bent tube (U-tube) with a single manometer fluid. A single-fluid simple manometer uses one manometer fluid (also known as the gage fluid) with a specific weight, γ_m to measure the gage pressure (due to both the hydrostatic pressure and the gage air pressure, p_g) at a given point in the fluid (system fluid) with a specific weight, γ. A multifluid simple manometer consists of more than one glass or plastic bent tube (U-tube) with more than one manometer fluid. A multifluid simple manometer uses more than one manometer fluid with specific weights, γ_{m1}, γ_{m2}, etc. to measure the gage pressure (due to both the hydrostatic pressure and the gage air pressure, p_g) at a given point in the fluid (system fluid) with a specific weight, γ. Similar to the piezometer, in order to avoid capillary effects, a manometer tube's diameter should be about 0.5 inches or 13 mm or greater. The manometer fluid may be colored water, alcohol, oil, carbon tetrachloride, mercury, or any other liquid with a known specific weight, γ_m that will not mix with the system fluid (γ). For a high accuracy in measurements, the change in temperature of both the system fluid and the manometer fluid(s) should be noted, as this will affect the specific gravity, s of the fluid.

Furthermore, if the overall pressure in the tank or the pipe is large, then a "heavy" (large specific gravity, s) manometer fluid such as mercury can be used while maintaining a reasonable manometer reading (column height), h_m. If, however, the overall pressure in the tank or the pipe is small, then a "lighter" (small specific gravity, s) manometer fluid such as water can be used, yielding a relatively large manometer reading (column height), h_m, which is usually easier to read. Also, as the specific gravity, s of the manometer fluid approaches that of the system fluid, the manometer reading becomes larger for a given pressure, thus increasing the accuracy of the instrument, provided that the two specific gravities are accurately known. And, finally, one may note that if the overall gage pressure in the tank or the pipe is too large, it will blow the manometer fluid into the atmosphere, while if the overall gage pressure in the tank or the pipe has too large of a vacuum (suction) (too large of a negative gage pressure), it could cause the manometer fluid to become sucked into the system (tank or pipe). Thus, in order to avoid either of these two problems, a proper selection of the manometer fluid and sizing of the manometer tubing is very important.

2.3.6.1 Single-Fluid Simple Manometer

A single-fluid simple manometer consists of a single open U-tube and may be used to measure the gage pressure at a given point in the tank or at the centerline of the pipe. As such, a single-fluid simple manometer is used to measure the gage pressure at point 1, p_1 for a fluid with specific weight, γ in static equilibrium in a tank that is subjected to a high positive gage air pressure, p_g, as illustrated in Figure 2.12, or a high negative gage air pressure, p_g, as illustrated in Figure EP 2.10. The manometer reading is given by the height of manometer fluid, h_m. The system fluid, γ extends into the U-tube, making contact with the column of the manometer fluid, γ_m. The fluids will attain an equilibrium configuration from which the gage pressure at a given point in the fluid system may be determined. Thus, the manometer measures the hydrostatic pressure and the gage air pressure, p_g in terms of the height of the manometer fluid, h_m. The pressure (due to both the hydrostatic pressure and the gage air pressure, p_g) to be measured is applied to the surface of the fluid, creating a force that must be balanced from the other side by the weight of the manometer fluid column. Thus, referring to Figure 2.12, in which the fluid is subjected to a positive gage air pressure, to find the gage pressure at point 1, p_1 in terms of the manometer reading, h_m, we start at one end of the system and work our way to the other end by applying the hydrostatic pressure equation, Equation 2.18, $(dp/dz) = \gamma$, and Pascal's law, Equation 2.16, $(\partial p/\partial y) = 0$ (for a fluid that is homogenous in γ and continuous in space $[x, y, z]$). Therefore, for the open U-tube

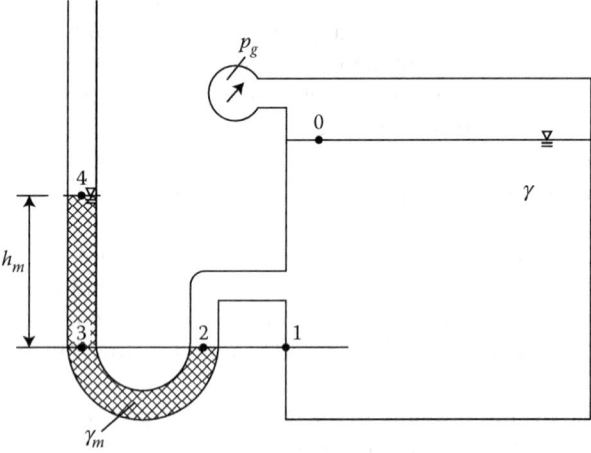

FIGURE 2.12
Open U-tube manometer. May be used to measure the gage pressure at a given point in a tank of fluid subjected to a high gage air pressure, p_g.

given in Figure 2.12, we will start at point 1 and work around to the open end at point 4. Application of $(\partial p/\partial y) = 0$ yields the following:

$$p_1 = p_2 \tag{2.54}$$

$$p_2 = p_3 \tag{2.55}$$

$$p_1 = p_2 = p_3 \tag{2.56}$$

Application of $(dp/dz) = \gamma$ yields the following:

$$p_3 = p_4 + \gamma_m h_m = p_{atm} + \gamma_m h_m \tag{2.57}$$

where $p_4 = p_{atm}$, as the open U-tube is "open" to the atmosphere. Thus,

$$p_1 = p_{atm} + \gamma_m h_m \tag{2.58}$$

where p_1 is the absolute pressure at point 1 and is indicated as follows:

$$p_1(abs) = p_{atm} + \gamma_m h_m \tag{2.59}$$

However, since we are seeking the gage pressure at point 1, it is computed by applying Equation 2.37, $p(gage) = p_{abs} - p_{atm}$ as follows:

$$p_1(gage) = p_1(abs) - p_{atm} = p_{atm} + \gamma_m h_m - p_{atm} = \gamma_m h_m \tag{2.60}$$

If the tank contains a liquid, then the variation of pressure in the tank will increase linearly with depth, $(dp/dz) = \gamma$. However, if the tank contains a gas, the contribution of the specific weight of the gas, γ is negligible compared to the specific weight of the manometer fluid, γ_m. Therefore, the pressure at any point within the tank containing a gas would be equal to the pressure at point 1, $p_1(abs) = p_{atm} + \gamma_m h_m$, or $p_1(gage) = \gamma_m h_m$.

EXAMPLE PROBLEM 2.9

Water in a tank is subjected to a positive gage air pressure, as illustrated in Figure EP 2.9. A single-fluid simple manometer is attached to the side of the tank. The manometer fluid is mercury and the manometer reading, h_m is 1.7 ft. Assume an atmospheric pressure of 14.7 psia. (a) Determine the gage pressure at point 1. (b) Given the depth to point 1, h_1 is 10 ft, determine the gage pressure at the water surface at point 0.

Mathcad Solution

(a) In order to determine the gage pressure at point 1, the principles of hydrostatics are applied, start at point 1 and work around to point 4. Furthermore, we are seeking the gage pressure at point 1, which is defined as follows:

$p(gage) = p_{abs} - p_{atm}$

$h_m := 1.7 \, ft$ $\qquad \gamma_w := 62.417 \dfrac{lb}{ft^3}$ $\qquad s_{Hg} := 13.56$

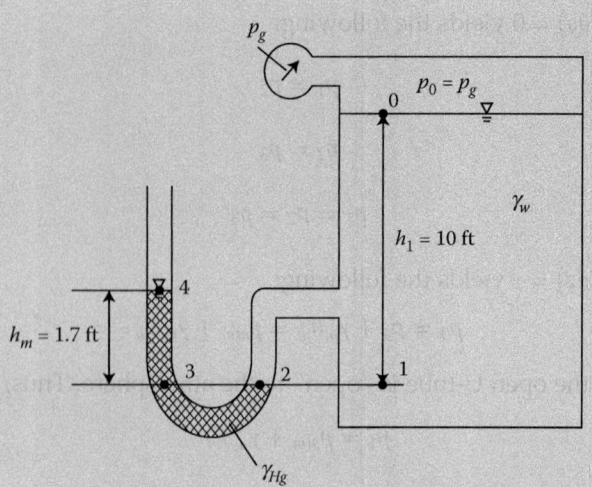

FIGURE EP 2.9
Open U-tube manometer is used to measure the gage pressure at points 1 and 0 in a tank of water subjected to a positive gage air pressure, p_g.

$$\gamma_{Hg} := s_{Hg} \cdot \gamma_w = 846.375 \frac{lb}{ft^3} \qquad P_{atm} := 14.7 \frac{lb}{in^2} \frac{(12\,in)^2}{(1\,ft)^2} = 2.117 \times 10^3 \frac{lb}{ft^2}$$

$$P_4 := P_{atm} - P_{atm} = 0 \frac{lb}{ft^2}$$

Guess value: $\qquad P_1 := 1 \frac{lb}{ft^2} \qquad P_2 := 1 \frac{lb}{ft^2} \qquad P_3 := 1 \frac{lb}{ft^2}$

Given

$$P_1 = P_2 \qquad\qquad P_2 = P_3 \qquad\qquad P_3 = P_4 + \gamma_{Hg} \cdot h_m$$

$$\begin{pmatrix} P_1 \\ P_2 \\ P_3 \end{pmatrix} := Find\ (P_1, P_2, P_3) = \begin{pmatrix} 1.439 \times 10^3 \\ 1.439 \times 10^3 \\ 1.439 \times 10^3 \end{pmatrix} \frac{lb}{ft^2}$$

(b) Given the depth to point 1, h_1 is 10 ft, in order to determine the gage pressure at the water surface at point 0, the principles of hydrostatics are applied as follows:

$$h_1 := 10\,ft$$

Guess value: $\qquad\qquad P_0 := 1 \frac{lb}{ft^2}$

Given

$$P_1 = P_0 + \gamma_w \cdot h_1$$

$$P_0 := Find\ (P_0) = 814.667 \frac{lb}{ft^2}$$

EXAMPLE PROBLEM 2.10

Water in a tank is subjected to a negative gage air pressure of −7 psi, as illustrated in Figure EP 2.10. A single-fluid simple manometer is attached to the side of the tank. The manometer fluid is mercury and the manometer reading, h_m is 0.9 ft. Assume an atmospheric pressure of 14.7 psia. (a) Determine the gage pressure at point 1. (b) Determine the depth of point 1 from the water surface.

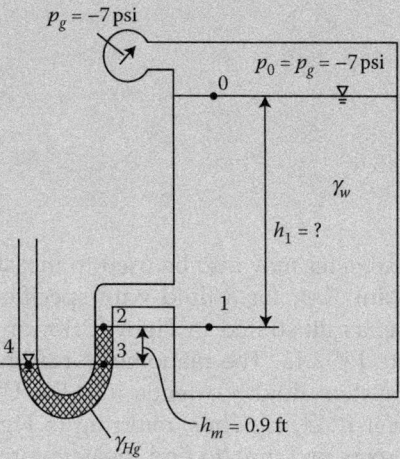

FIGURE EP 2.10

Open U-tube manometer is used to measure the gage pressure at point 1 in a tank of water subjected to a negative gage air pressure, p_g.

Mathcad Solution

(a) In order to determine the gage pressure at point 1, the principles of hydrostatics are applied, starting at point 1 and working around to point 4. Furthermore, we are seeking the gage pressure at point 1, which is defined as follows:

$$p(gage) = p_{abs} - p_{atm}$$

$$h_m := 0.9 \, \text{ft} \qquad \gamma_w := 62.417 \frac{\text{lb}}{\text{ft}^3} \qquad s_{Hg} := 13.56$$

$$\gamma_{Hg} := s_{Hg} \cdot \gamma_w = 846.375 \frac{\text{lb}}{\text{ft}^3} \qquad p_{atm} := 14.7 \frac{\text{lb}}{\text{in}^2} \frac{(12 \, \text{in})^2}{(1 \, \text{ft})^2} = 2.117 \times 10^3 \frac{\text{lb}}{\text{ft}^2}$$

$$p_4 := p_{atm} - p_{atm} = 0 \frac{\text{lb}}{\text{ft}^2}$$

Guess value: $\qquad p_1 := 1 \frac{\text{lb}}{\text{ft}^2} \qquad p_2 := 1 \frac{\text{lb}}{\text{ft}^2} \qquad p_3 := 1 \frac{\text{lb}}{\text{ft}^2}$

Given

$$p_1 = p_2 \qquad\qquad p_3 = p_2 + \gamma_{Hg} \cdot h_m \qquad\qquad p_3 = p_4$$

$$\begin{pmatrix} p_1 \\ p_2 \\ p_3 \end{pmatrix} := \text{Find} (p_1, p_2, p_3) = \begin{pmatrix} -761.737 \\ -761.737 \\ 0 \end{pmatrix} \frac{\text{lb}}{\text{ft}^2}$$

(b) In order to determine the depth of point 1 from the water surface, the principles of hydrostatics are applied as follows:

$$P_g := -7 \frac{lb}{in^2} \frac{(12 \, in)^2}{(1 \, ft)^2} = -1.008 \times 10^3 \frac{lb}{ft^2} \qquad\qquad P_0 := P_g = -1.008 \times 10^3 \frac{lb}{ft^2}$$

Guess value: $h_1 := 1 \, ft$

Given

$$P_1 = P_0 + \gamma_w \cdot h_1$$

$$h_1 := Find \, (h_1) = 3.945 \, ft$$

A single-fluid simple manometer may also be used to measure the gage pressure at the centerline of the pipe at point 1, p_1 for a fluid with specific weight, γ flowing under a high positive gage pressure, as illustrated in Figure 2.13, or a high negative gage pressure, as illustrated in Figure EP 2.12. The manometer reading is given by the height of manometer fluid, h_m. The system fluid γ extends into the U-tube, making contact with the column of the manometer fluid, γ_m. Thus, referring to Figure 2.13, in which the fluid is flowing under a positive gage pressure, to find the gage pressure at point 1, p_1 in terms of the manometer reading, h_m, we start at one end of the system and work our way to the other end by applying the hydrostatic pressure equation, Equation 2.18, $(dp/dz) = \gamma$, and Pascal's law, Equation 2.16, $(\partial p / \partial y) = 0$ (for a fluid that is homogenous in γ and continuous in space $[x, y, z]$). Therefore, for the pen U-tube given in Figure 2.13, we will start at

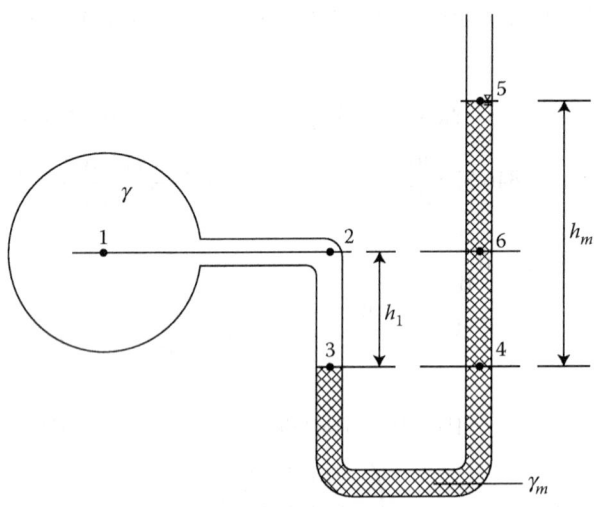

FIGURE 2.13
Open U-tube manometer. May be used to measure the gage pressure at any point in a pipe flowing under a high gage pressure.

point 1 and work around to the open end at point 5. Application of $(\partial p/\partial y) = 0$ yields the following:

$$p_1 = p_2 \tag{2.61}$$

Application of $(dp/dz) = \gamma$ yields the following:

$$p_3 = p_2 + \gamma h_1 \tag{2.62}$$

Application of $(\partial p/\partial y) = 0$ yields the following:

$$p_3 = p_4 \tag{2.63}$$

Application of $(dp/dz) = \gamma$ yields the following:

$$p_4 = p_5 + \gamma_m h_m = p_{atm} + \gamma_m h_m \tag{2.64}$$

where $p_5 = p_{atm}$, as the open U-tube is "open" to the atmosphere. Thus,

$$p_1 = p_2 = p_3 - \gamma h_1 = p_{atm} + \gamma_m h_m - \gamma h_1 \tag{2.65}$$

where p_1 is the absolute pressure at point 1 and is indicated as follows:

$$p_1(abs) = p_{atm} + \gamma_m h_m - \gamma h_1 \tag{2.66}$$

However, since we are seeking the gage pressure at point 1, it is computed by applying Equation 2.37, $p(gage) = p_{abs} - p_{atm}$ as follows:

$$p_1(gage) = p_1(abs) - p_{atm} = p_{atm} + \gamma_m h_m - \gamma h_1 - p_{atm} = \gamma_m h_m - \gamma h_1 \tag{2.67}$$

If the pipe contains a liquid, then the contribution of the specific weight of the liquid, γ is significant. However, if the pipe contains a gas, the contribution of the specific weight of the gas, γ is negligible compared to the specific weight of the manometer fluid, γ_m. There are two additional interesting points to note regarding Figure 2.13. First, it may be noted that the pressure at point 6 does not equal the pressure at point 1, even though they are at the same elevation. This is because these two points are not located in the same fluid. And second, if the gage pressure at point 1 were negative, then point 5 would be lower than point 3 and a new analysis would be required to yield the expression for the pressure at point 1 (see Example Problem 2.12 below).

EXAMPLE PROBLEM 2.11

Water flows in a pipe under a high positive gage pressure, as illustrated in Figure EP 2.11. A single-fluid simple manometer is attached to the pipe in order to measure the gage pressure at the centerline of the pipe at point 1. The distance between points 2 and 3 is 0.5 ft. The manometer fluid is mercury and the manometer reading, h_m is 2.98 ft. Assume an atmospheric pressure of 14.7 psia. (a) Determine the gage pressure at point 1.

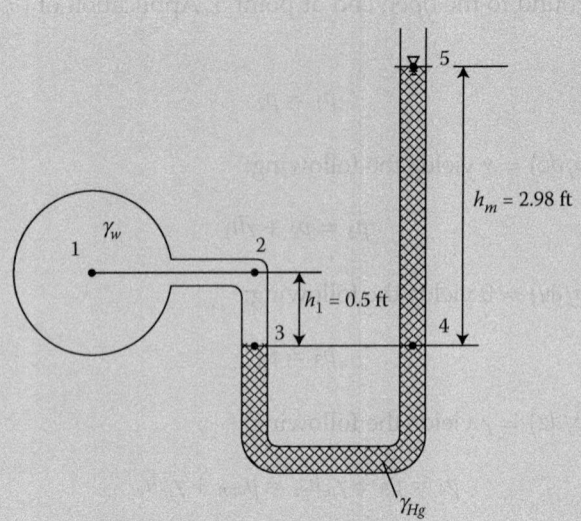

FIGURE EP 2.11
Open U-tube manometer is used to measure the gage pressure at point 1 in a pipe flowing under a high positive gage pressure carrying water.

Mathcad Solution

(a) In order to determine the gage pressure at point 1, the principles of hydrostatics are applied, starting at point 1 and working around to point 5. Furthermore, we are seeking the gage pressure at point 1, which is defined as follows:

$$p(gage) = p_{abs} - p_{atm}$$

$h_m := 2.98 \text{ ft}$ $h_1 := 0.5 \text{ ft}$ $\gamma_w := 62.417 \dfrac{\text{lb}}{\text{ft}^3}$

$s_{Hg} := 13.56$ $\gamma_{Hg} := s_{Hg} \cdot \gamma_w = 846.375 \dfrac{\text{lb}}{\text{ft}^3}$

$p_{atm} := 14.7 \dfrac{\text{lb}}{\text{in}^2} \dfrac{(12 \text{ in})^2}{(1 \text{ ft})^2} = 2.117 \times 10^3 \dfrac{\text{lb}}{\text{ft}^2}$ $p_5 := p_{atm} - p_{atm} = 0 \dfrac{\text{lb}}{\text{ft}^2}$

Guess value: $p_1 := 1 \dfrac{\text{lb}}{\text{ft}^2}$ $p_2 := 1 \dfrac{\text{lb}}{\text{ft}^2}$ $p_3 := 1 \dfrac{\text{lb}}{\text{ft}^2}$ $p_4 := 1 \dfrac{\text{lb}}{\text{ft}^2}$

Given

$p_1 = p_2$ $p_3 = p_2 + \gamma_w \cdot h_1$ $p_3 = p_4$ $p_4 = p_5 + \gamma_{Hg} \cdot h_m$

$$\begin{pmatrix} p_1 \\ p_2 \\ p_3 \\ p_4 \end{pmatrix} := \text{Find} (p_1, p_2, p_3, p_4) = \begin{pmatrix} 2.491 \times 10^3 \\ 2.491 \times 10^3 \\ 2.491 \times 10^3 \\ 2.491 \times 10^3 \end{pmatrix} \dfrac{\text{lb}}{\text{ft}^2}$$

EXAMPLE PROBLEM 2.12

Water flows in a pipe under a high negative gage pressure, as illustrated in Figure EP 2.12. A single-fluid simple manometer is attached to the pipe in order to measure the gage pressure at the centerline of the pipe at point 1. The distance between points 2 and 3 is 0.6 ft. The manometer fluid is mercury and the manometer reading, h_m is 2.0 ft. Assume an atmospheric pressure of 14.7 psia. (a) Determine the gage pressure at point 1.

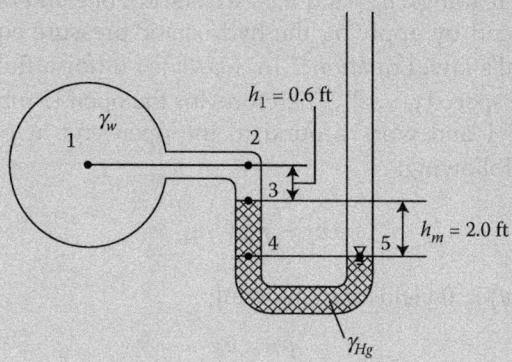

FIGURE EP 2.12
Open U-tube manometer is used to measure the gage pressure at point 1 in a pipe flowing under a high negative gage pressure carrying water.

Mathcad Solution

(a) In order to determine the gage pressure at point 1, the principles of hydrostatics are applied, starting at point 1 and working around to point 5. Furthermore, we are seeking the gage pressure at point 1, which is defined as follows:

$$p(gage) = p_{abs} - p_{atm}$$

$$h_m := 2.0 \, ft \qquad h_1 := 0.6 \, ft \qquad \gamma_w := 62.417 \frac{lb}{ft^3}$$

$$s_{Hg} := 13.56 \qquad\qquad \gamma_{Hg} := s_{Hg} \cdot \gamma_w = 846.375 \frac{lb}{ft^3}$$

$$p_{atm} := 14.7 \frac{lb}{in^2} \frac{(12 \, in)^2}{(1 \, ft)^2} = 2116.8 \frac{lb}{ft^2} \qquad p_5 := p_{atm} - p_{atm} = 0 \frac{lb}{ft^2}$$

Guess value: $\quad p_1 := 1 \frac{lb}{ft^2} \quad p_2 := 1 \frac{lb}{ft^2} \quad p_3 := 1 \frac{lb}{ft^2} \quad p_4 := 1 \frac{lb}{ft^2}$

Given

$$p_1 = p_2 \qquad p_3 = p_2 + \gamma_w \cdot h_1 \qquad p_4 = p_3 + \gamma_{Hg} \cdot h_m \qquad p_4 = p_5$$

$$\begin{pmatrix} p_1 \\ p_2 \\ p_3 \\ p_4 \end{pmatrix} := Find(p_1, p_2, p_3, p_4) = \begin{pmatrix} -1730.199 \\ -1730.199 \\ -1629.749 \\ 0 \end{pmatrix} \frac{lb}{ft^2}$$

2.3.6.2 Multifluid Simple Manometer

A multifluid simple manometer consists of more than one bent U-tube and may be used to measure the gage pressure at a given point in the tank or at the centerline of the pipe. As such, a multifluid simple manometer is used to measure the gage pressure at point 1, p_1 for a fluid with specific weight, γ in static equilibrium in a tank that is subjected to a high positive gage air pressure, p_g, as illustrated in Figure 2.14. The specific weight of the first manometer fluid is denoted by γ_{m1}. The specific weight of the second manometer fluid is denoted γ_{m2}. Thus, to find the gage air pressure at point 1, p_1 in terms of the two respective manometer readings, h_{m1} and h_{m2}, we start at one end of the system and work our way to the other end by applying the hydrostatic pressure equation, Equation 2.18, $(dp/dz) = \gamma$, and Pascal's law, Equation 2.16, $(\partial p/\partial y) = 0$ (for a fluid that is homogenous in γ and continuous in space $[x, y, z]$). Therefore, for the open U-tube given in Figure 2.14, we will start at point 1 and work around to the open end at point 7. Application of $(dp/dz) = \gamma$ yields the following:

$$p_2 = p_1 + \gamma h_1 \tag{2.68}$$

Application of $(\partial p/\partial y) = 0$ yields the following:

$$p_2 = p_3 \tag{2.69}$$

$$p_3 = p_4 \tag{2.70}$$

$$p_2 = p_3 = p_4 \tag{2.71}$$

Application of $(dp/dz) = \gamma$ yields the following:

$$p_5 = p_4 + \gamma_{m1} h_{m1} \tag{2.72}$$

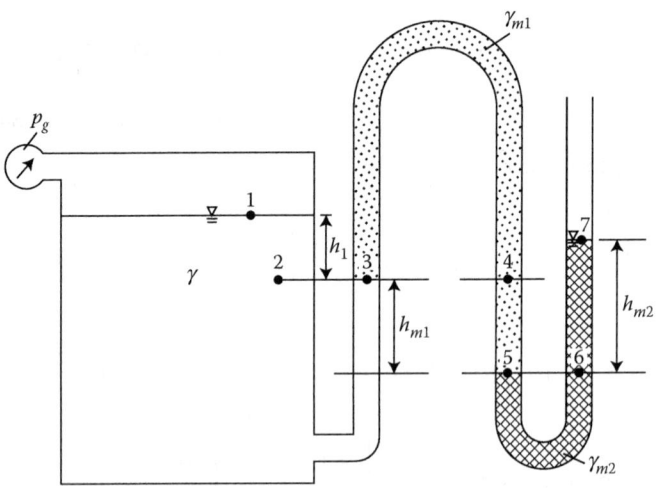

FIGURE 2.14
Multifluid manometer. May be used to measure the gage pressure at a given point in tank of fluid subjected to a high gage air pressure, p_g.

Application of $(\partial p/\partial y) = 0$ yields the following:

$$p_5 = p_6 \tag{2.73}$$

Application of $(dp/dz) = \gamma$ yields the following:

$$p_6 = p_7 + \gamma_{m2}h_{m2} = p_{atm} + \gamma_{m2}h_{m2} \tag{2.74}$$

where $p_7 = p_{atm}$, as the open U-tube is "open" to the atmosphere. Thus,

$$p_1 = p_2 - \gamma h_1 = p_4 - \gamma h_1 = p_6 - \gamma_{m1}h_{m1} - \gamma h_1 = p_{atm} + \gamma_{m2}h_{m2} - \gamma_{m1}h_{m1} - \gamma h_1 \tag{2.75}$$

where p_1 is the absolute pressure at point 1 and is indicated as follows:

$$p_1(abs) = p_{atm} + \gamma_{m2}h_{m2} - \gamma_{m1}h_{m1} - \gamma h_1 \tag{2.76}$$

However, since we are seeking the gage pressure at point 1, it is computed by applying Equation 2.37, $p(gage) = p_{abs} - p_{atm}$ as follows:

$$\begin{aligned} p_1(gage) &= p_1(abs) - p_{atm} = p_{atm} + \gamma_{m2}h_{m2} - \gamma_{m1}h_{m1} - \gamma h_1 - p_{atm} \\ &= \gamma_{m2}h_{m2} - \gamma_{m1}h_{m1} - \gamma h_1 \end{aligned} \tag{2.77}$$

EXAMPLE PROBLEM 2.13

Water in a tank is subjected to a positive gage air pressure, as illustrated in Figure EP 2.13. A multifluid simple manometer is attached to the side of the tank. The first manometer fluid is oil with a specific gravity of 0.78, and a manometer reading, h_{m1}

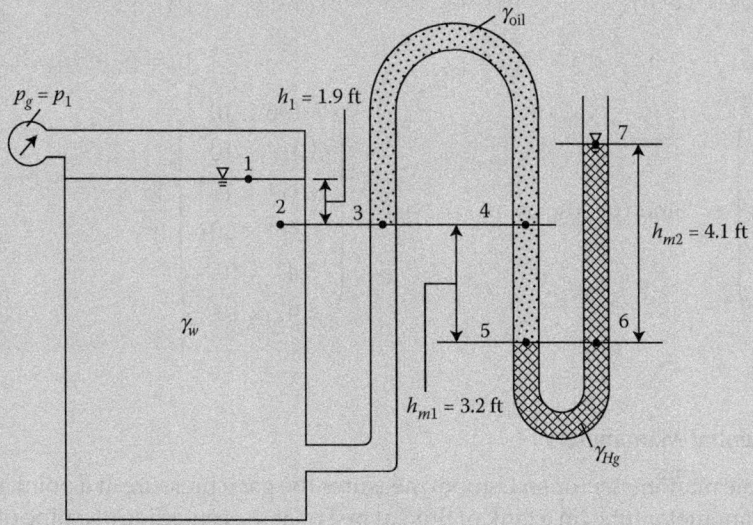

FIGURE EP 2.13
Multifluid manometer is used to measure the gage pressure at point 1 in a tank of water subjected to a positive gage air pressure.

of 3.2 ft. The second manometer fluid is mercury with a manometer reading, h_{m2} of 4.1 ft. The distance between points 1 and 2 is 1.9 ft. Assume an atmospheric pressure of 14.7 psia. (a) Determine the gage pressure at point 1.

Mathcad Solution

(a) In order to determine the gage pressure at point 1, the principles of hydrostatics are applied, starting at point 1 and working around to point 7. Furthermore, we are seeking the gage pressure at point 1, which is defined as follows:

$$p(gage) = p_{abs} - p_{atm}$$

$$h_1 := 1.9 \text{ ft} \qquad \gamma_w := 62.417 \frac{\text{lb}}{\text{ft}^3} \qquad h_{m1} := 3.2 \text{ ft} \qquad s_o := 0.78$$

$$\gamma_o := s_o \cdot \gamma_w = 48.685 \frac{\text{lb}}{\text{ft}^3} \qquad\qquad\qquad h_{m2} := 4.1 \text{ ft} \qquad s_{Hg} := 13.56$$

$$\gamma_{Hg} := s_{Hg} \cdot \gamma_w = 846.375 \frac{\text{lb}}{\text{ft}^3}$$

$$p_{atm} := 14.7 \frac{\text{lb}}{\text{in}^2} \frac{(12 \text{ in})^2}{(1 \text{ ft})^2} = 2.117 \times 10^3 \frac{\text{lb}}{\text{ft}^2} \qquad p_7 := p_{atm} - p_{atm} = 0 \frac{\text{lb}}{\text{ft}^2}$$

Guess value: $\quad p_1 := 1 \frac{\text{lb}}{\text{ft}^2} \qquad p_2 := 1 \frac{\text{lb}}{\text{ft}^2} \qquad p_3 := 1 \frac{\text{lb}}{\text{ft}^2} \qquad p_4 := 1 \frac{\text{lb}}{\text{ft}^2}$

$$p_5 := 1 \frac{\text{lb}}{\text{ft}^2} \qquad p_6 := 1 \frac{\text{lb}}{\text{ft}^2}$$

Given

$$p_2 = p_1 + \gamma_w \cdot h_1 \qquad\qquad p_3 = p_2 \qquad\qquad\qquad\qquad p_4 = p_3$$

$$p_5 = p_4 + \gamma_o \cdot h_{m1} \qquad\qquad p_5 = p_6 \qquad\qquad\qquad\qquad p_6 = p_7 + \gamma_{Hg} \cdot h_{m2}$$

$$\begin{pmatrix} p_1 \\ p_2 \\ p_3 \\ p_4 \\ p_5 \\ p_6 \end{pmatrix} := \text{Find } (p_1, p_2, p_3, p_4, p_5, p_6) = \begin{pmatrix} 3.196 \times 10^3 \\ 3.314 \times 10^3 \\ 3.314 \times 10^3 \\ 3.314 \times 10^3 \\ 3.47 \times 10^3 \\ 3.47 \times 10^3 \end{pmatrix} \frac{\text{lb}}{\text{ft}^2}$$

2.3.7 Differential Manometers

While a simple manometer (open U-tube) measures the gage pressure at a point at the closed end of the manometer tube (in a tank of fluid at rest or at the centerline of a pipe of fluid flowing under pressure), a differential manometer is used when the difference in pressure between two points at both ends of the manometer is needed. It may be noted that, because the differential manometer does not measure the pressure at a point relative to the atmosphere or to a perfect vacuum, it does not measure either gage or absolute pressure,

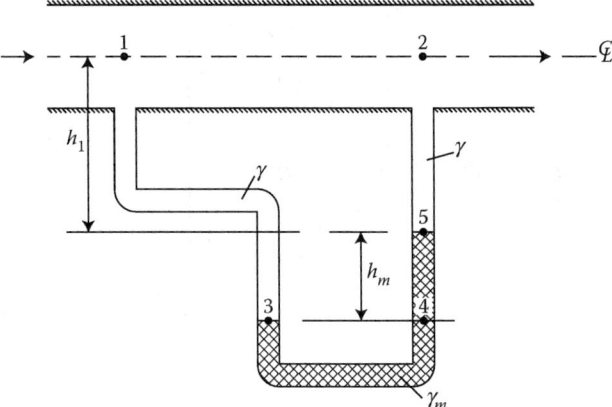

FIGURE 2.15
Differential manometer. May be used to measure the pressure difference between any two points within a single pipe flowing under pressure.

respectively. Rather, the differential manometer measures the difference in pressure between any two points. A differential manometer may be used to measure a pressure difference between two points within a single pipe flowing under pressure, as illustrated in Figure 2.15, or to measure a pressure difference between two pipes flowing under pressure (the more general application), as illustrated in Figure 2.16. Similar to a simple manometer, a differential manometer may be either a single-fluid or a multifluid differential manometer. When a single-fluid differential manometer is used to measure a small pressure difference, it consists of a single U-tube, as illustrated in Figures 2.15 and 2.16. However, when a single-fluid differential manometer is used to measure a large pressure difference, then it consists of a triple U-tube, as illustrated in Figure 2.17. Meanwhile, when a multifluid differential manometer is used to measure a small pressure difference, then it consists of a least a double U-tube, as illustrated in Figure 2.18. However, when a multifluid differential manometer is used to

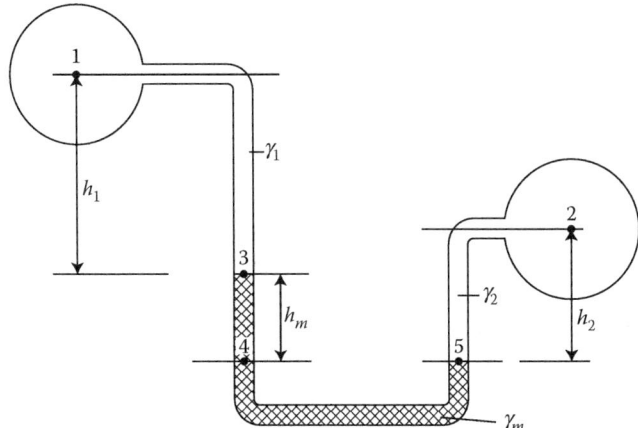

FIGURE 2.16
Differential manometer. May be used to measure the pressure difference between two pipes flowing under pressure.

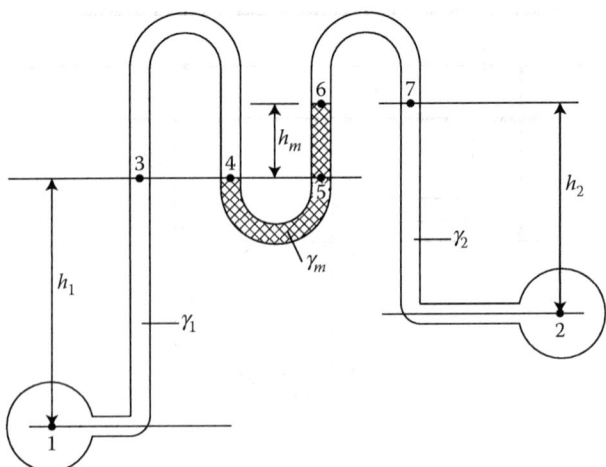

FIGURE 2.17
Differential manometer with a triple U-tube may be used to measure a large pressure difference between two pipes flowing under pressure.

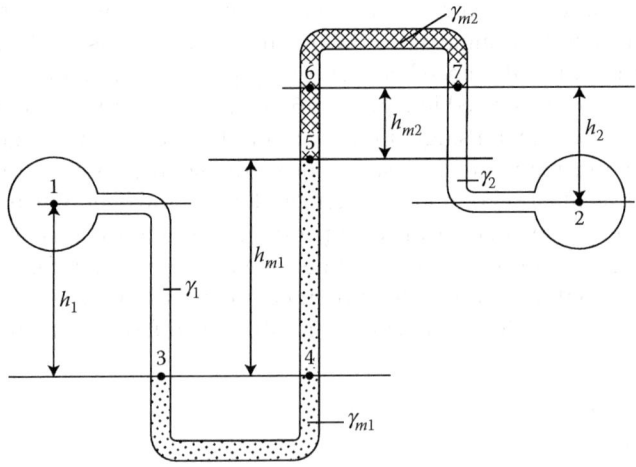

FIGURE 2.18
Multifluid differential manometer with a double U-tube may be used to measure a small pressure difference between two pipes flowing under pressure.

measure a large pressure difference, then it consists of more than three U-tubes. Furthermore, similar to the simple manometer, when the differential manometer is used to measure a large pressure difference, then a heavy manometer fluid such as mercury is used. However, for a small pressure difference, a light fluid such as oil or air may be used. And, finally, the manometer fluid must be one that will not mix with either of the two system fluids.

2.3.7.1 Single-Fluid Differential Manometer within a Single Pipe

A single-fluid differential manometer that is used to measure a small pressure difference within a single pipe consists of a single U-tube, as illustrated in Figure 2.15. There is a pressure drop from point 1 to point 2 due to the viscosity, μ (flow resistance) of the system fluid (γ). The manometer reading is given by the height of the manometer fluid (γ_m), h_m. Thus, to find the pressure drop between point 1 and point 2, $p_1 - p_2$ in terms of the manometer reading, h_m, we start at one end of the manometer system and work our way to the other end by applying the hydrostatic pressure equation, Equation 2.18, $(dp/dz) = \gamma$ and Pascal's law, Equation 2.16, $(\partial p/\partial y) = 0$ (for a fluid that is homogenous in γ and continuous in space $[x, y, z]$). Therefore, for the differential manometer given in Figure 2.15, we will start at point 1 and work around to point 2 in the pipe. Application of $(dp/dz) = \gamma$ yields the following:

$$p_3 = p_1 + \gamma(h_1 + h_m) \tag{2.78}$$

$$p_1 = p_3 - \gamma(h_1 + h_m) \tag{2.79}$$

Application of $(\partial p/\partial y) = 0$ yields the following:

$$p_3 = p_4 \tag{2.80}$$

Application of $(dp/dz) = \gamma$ yields the following:

$$p_4 = p_5 + \gamma_m h_m \tag{2.81}$$

$$p_5 = p_2 + \gamma h_1 \tag{2.82}$$

$$p_2 = p_5 - \gamma h_1 \tag{2.83}$$

Thus,

$$p_1 - p_2 = p_3 - \gamma(h_1 + h_m) - p_5 + \gamma h_1 = p_3 - \gamma h_m - p_5 = p_5 + \gamma_m h_m - \gamma h_m - p_5$$
$$= (\gamma_m - \gamma)h_m \tag{2.84}$$

Furthermore, if the fluid in the pipeline were a gas (γ), where the specific weight of the manometer fluid, γ_m was very large compared to the specific weight of the gas, γ, then we would get the following:

$$p_1 - p_2 = \gamma_m h_m \tag{2.85}$$

EXAMPLE PROBLEM 2.14

Water flows in a pipe, as illustrated in Figure EP 2.14. A single-fluid differential manometer is attached to the pipe at points 1 and 2 in order to measure the pressure drop between points 1 and 2. The distance between points 2 and 5 is 3 ft. The manometer fluid is mercury and the manometer reading, h_m is 1.5 ft. (a) Determine the pressure drop between points 1 and 2.

Mathcad Solution

(a) In order to determine the pressure drop between points 1 and 2, the principles of hydrostatics are applied, starting at point 1 and working around to point 2 as follows:

$$h_1 := 3 \text{ ft} \qquad \gamma_w := 62.417 \frac{\text{lb}}{\text{ft}^3} \qquad h_m := 1.5 \text{ ft} \qquad s_{Hg} := 13.56$$

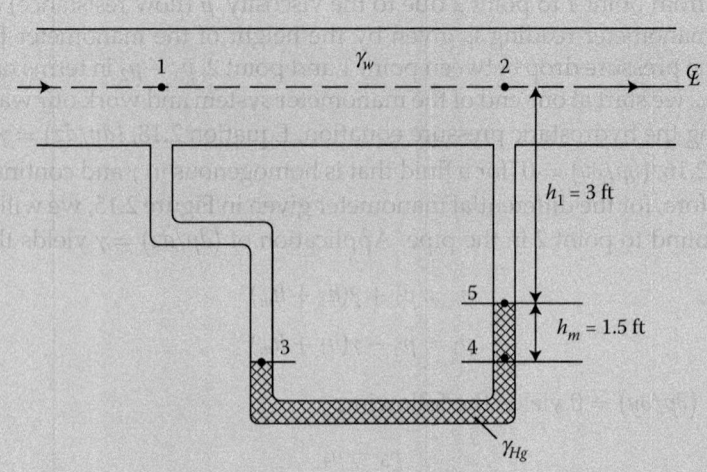

FIGURE EP 2.14
Differential manometer is used to measure the pressure difference between points 1 and 2 within a single pipe flowing under pressure carrying water.

$$\gamma_{Hg} := s_{Hg} \cdot \gamma_w = 846.375 \frac{lb}{ft^3}$$

Guess value: $p_1 := 1 \frac{lb}{ft^2}$ $p_2 := 1 \frac{lb}{ft^2}$ $p_3 := 1 \frac{lb}{ft^2}$ $p_4 := 1 \frac{lb}{ft^2}$

$$p_5 := 1 \frac{lb}{ft^2}$$

Given

$$p_3 = p_1 + \gamma_w(h_1 + h_m) \qquad\qquad p_3 = p_4 \qquad p_4 = p_5 + \gamma_{Hg} \cdot h_m$$

$$p_5 = p_2 + \gamma_w \cdot h_1$$

$$\begin{pmatrix} p_1 \\ p_2 \\ p_3 \\ p_4 \\ p_5 \end{pmatrix} := \text{Find}\,(p_1, p_2, p_3, p_4, p_5) = \begin{pmatrix} 254.725 \\ -921.211 \\ 535.601 \\ 535.601 \\ -733.96 \end{pmatrix} \frac{lb}{ft^2}$$

$$\Delta p := p_1 - p_2 = 1.176 \times 10^3 \frac{lb}{ft^2}$$

2.3.7.2 Single-Fluid Differential Manometer between Two Pipes

A single-fluid differential manometer that is used to measure a small pressure difference between two pipes consists of a single U-tube, as illustrated in Figure 2.16. There is a pressure differential between the two pipes 1 and 2. The first pipe contains a system fluid with a specific weight, γ_1, while the second pipe contains a system fluid with a specific weight, γ_2. The manometer reading is given by the height of the manometer fluid (γ_m), h_m. Thus, to find the pressure differential between point 1 and point 2, $p_1 - p_2$ in terms of the manometer reading, h_m, we start at one end of the manometer system and work our way to the other end by applying the hydrostatic pressure equation, Equation 2.18, $(dp/dz) = \gamma$, and Pascal's law, Equation 2.16, $(\partial p/\partial y) = 0$ (for a fluid that is homogenous in γ and continuous in space $[x, y, z]$). Therefore, for the differential manometer given in Figure 2.16, we will start at point 1 and work around to point 2 in the manometer system. Application of $(dp/dz) = \gamma$ yields the following:

$$p_3 = p_1 + \gamma_1 h_1 \tag{2.86}$$

$$p_1 = p_3 - \gamma_1 h_1 \tag{2.87}$$

$$p_4 = p_3 + \gamma_m h_m \tag{2.88}$$

Application of $(\partial p/\partial y) = 0$ yields the following:

$$p_4 = p_5 \tag{2.89}$$

Application of $(dp/dz) = \gamma$ yields the following:

$$p_5 = p_2 + \gamma_2 h_2 \tag{2.90}$$

$$p_2 = p_5 - \gamma_2 h_2 \tag{2.91}$$

Thus,

$$p_1 - p_2 = p_3 - \gamma_1 h_1 - p_5 + \gamma_2 h_2 = p_4 - \gamma_m h_m - \gamma_1 h_1 - p_4 + \gamma_2 h_2$$
$$= -\gamma_m h_m - \gamma_1 h_1 + \gamma_2 h_2 \tag{2.92}$$

Furthermore, if both pipe fluids (γ_1 and γ_2) were gases, where the specific weight of the manometer fluid, γ_m was very large compared to the specific weight of the gases, (γ_1 and γ_2), then we would get the following:

$$p_1 - p_2 = -\gamma_m h_m \tag{2.93}$$

EXAMPLE PROBLEM 2.15

Water flows in a pipe 1, and oil with a specific gravity of 0.7 flows in pipe 2, as illustrated in Figure EP 2.15. A single-fluid differential manometer is attached between pipe 1 and pipe 2 in order to measure the pressure difference between pipe 1 and pipe 2. The distance between points 1 and 3 is 2 ft, and the distance between points 2 and 5 is 1.5 ft. The manometer fluid is mercury and the manometer reading, h_m is 1.7 ft. (a) Determine the pressure difference between pipe 1 and pipe 2.

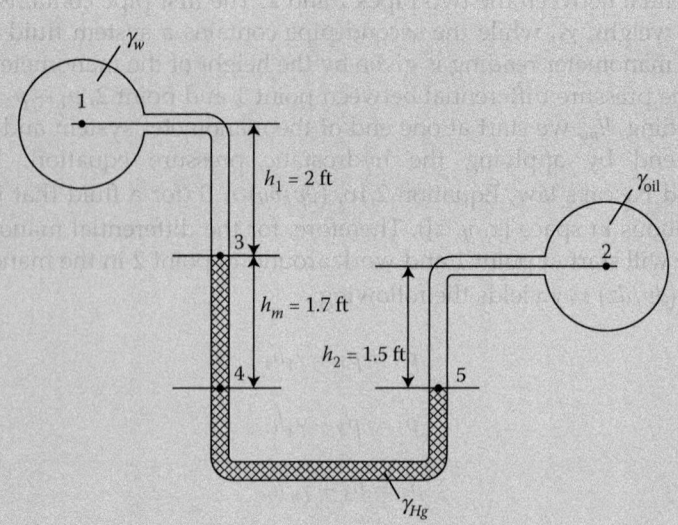

FIGURE EP 2.15
Differential manometer is used to measure the pressure difference between two pipes (pipe 1 carries water and pipe 2 carries oil) flowing under pressure.

Mathcad Solution

(a) In order to determine the pressure difference between pipe 1 and pipe 2, the principles of hydrostatics are applied, starting at point 1 and working around to point 2 as follows:

$h_1 := 2 \, ft$ $\qquad h_2 := 1.5 \, ft$ $\qquad \gamma_w := 62.417 \, \dfrac{lb}{ft^3}$ $\qquad h_m := 1.7 \, ft$

$s_{Hg} := 13.56$ $\qquad\qquad\qquad \gamma_{Hg} := s_{Hg} \cdot \gamma_w = 846.375 \, \dfrac{lb}{ft^3}$

$s_o: 0.7$ $\qquad\qquad\qquad\qquad \gamma_o := s_o \cdot \gamma_w = 43.692 \, \dfrac{lb}{ft^3}$

Guess value: $\qquad p_1 := 1 \, \dfrac{lb}{ft^2}$ $\qquad p_2 := 1 \, \dfrac{lb}{ft^2}$ $\qquad p_3 := 1 \, \dfrac{lb}{ft^2}$ $\qquad p_4 := 1 \, \dfrac{lb}{ft^2}$

$\qquad\qquad\qquad\qquad p_5 := 1 \, \dfrac{lb}{ft^2}$

Given

$p_3 = p_1 + \gamma_w \cdot h_1$ $\qquad\qquad\qquad p_4 = p_3 + \gamma_{Hg} \cdot h_m$ $\qquad\qquad\qquad p_4 = p_5$

$p_5 = p_2 + \gamma_o \cdot h_2$

$\begin{pmatrix} p_1 \\ p_2 \\ p_3 \\ p_4 \\ p_5 \end{pmatrix} := Find\,(p_1, p_2, p_3, p_4, p_5) = \begin{pmatrix} -976.511 \\ 521.622 \\ -851.677 \\ 587.159 \\ 587.159 \end{pmatrix} \dfrac{lb}{ft^2}$

$\Delta p := p_1 - p_2 = -1.498 \times 10^3 \, \dfrac{lb}{ft^2}$

A single-fluid differential manometer that is used to measure a large pressure difference between two pipes consists of a triple U-tube, as illustrated in Figure 2.17. There is a pressure differential between pipes 1 and 2. The first pipe contains a system fluid with a specific weight, γ_1, while the second pipe contains a system fluid with a specific weight, γ_2. The manometer reading is given by the height of the manometer fluid (γ_m), h_m. Thus, to find the pressure differential between point 1 and point 2, $p_1 - p_2$ in terms of the manometer reading, h_m, we start at one end of the manometer system and work our way to the other end by applying the hydrostatic pressure equation, Equation 2.18, $(dp/dz) = \gamma$, and Pascal's law, Equation 2.16, $(\partial p/\partial y) = 0$ (for a fluid that is homogenous in γ and continuous in space $[x, y, z]$). Therefore, for the differential manometer given in Figure 2.17, we will start at point 1 and work around to point 2 in the manometer system.

Application of $(dp/dz) = \gamma$ yields the following:

$$p_1 = p_3 + \gamma_1 h_1 \tag{2.94}$$

Application of $(\partial p/\partial y) = 0$ yields the following:

$$p_3 = p_4 \tag{2.95}$$

$$p_4 = p_5 \tag{2.96}$$

$$p_3 = p_4 = p_5 \tag{2.97}$$

Application of $(dp/dz) = \gamma$ yields the following:

$$p_5 = p_6 + \gamma_m h_m \tag{2.98}$$

Application of $(\partial p/\partial y) = 0$ yields the following:

$$p_6 = p_7 \tag{2.99}$$

Application of $(dp/dz) = \gamma$ yields the following:

$$p_2 = p_7 + \gamma_2 h_2 \tag{2.100}$$

Thus,

$$p_1 - p_2 = p_3 + \gamma_1 h_1 - p_7 - \gamma_2 h_2 = p_6 + \gamma_m h_m + \gamma_1 h_1 - p_6 - \gamma_2 h_2$$
$$= \gamma_m h_m + \gamma_1 h_1 - \gamma_2 h_2 \tag{2.101}$$

EXAMPLE PROBLEM 2.16

Water flows in pipe 1, and saltwater with a specific gravity of 1.023 flows in pipe 2, as illustrated in Figure EP 2.16. A single-fluid differential manometer is attached between pipe 1 and pipe 2 in order to measure the pressure difference between pipe 1 and pipe 2. The distance between points 1 and 3 is 5 ft, and the distance between points 2 and 7 is 3 ft. The manometer fluid is mercury and the manometer reading, h_m is 4 ft. (a) Determine the pressure difference between pipe 1 and pipe 2.

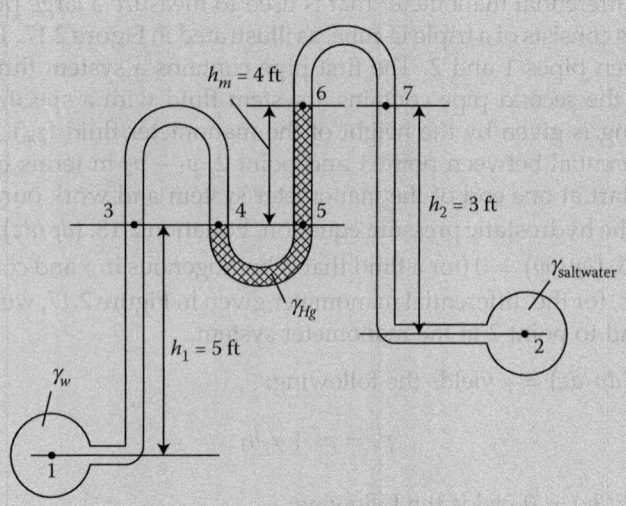

FIGURE EP 2.16
Differential manometer with a triple U-tube is used to measure a large pressure difference between two
pipes (pipe 1 carries water and pipe 2 carries saltwater) flowing under pressure.

Mathcad Solution

(a) In order to determine the pressure difference between pipe 1 and pipe 2, the principles of hydrostatics are applied, starting at point 1 and working around to point
2 as follows:

$h_1 := 5 \, ft$ $h_2 := 3 \, ft$ $\gamma_w := 62.417 \, \dfrac{lb}{ft^3}$ $h_m := 4 \, ft$

$s_{Hg} := 13.56$ $\gamma_{Hg} := s_{Hg} \cdot \gamma_w = 846.375 \, \dfrac{lb}{ft^3}$

$s_s := 1.023$ $\gamma_s := s_s \cdot \gamma_w = 63.853 \, \dfrac{lb}{ft^3}$

Guess value: $p_1 := 1 \, \dfrac{lb}{ft^2}$ $p_2 := 1 \, \dfrac{lb}{ft^2}$ $p_3 := 1 \, \dfrac{lb}{ft^2}$ $p_4 := 1 \, \dfrac{lb}{ft^2}$

$\quad\quad\quad\quad\quad p_5 := 1 \, \dfrac{lb}{ft^2}$ $p_6 := 1 \, \dfrac{lb}{ft^2}$ $p_7 := 1 \, \dfrac{lb}{ft^2}$

Given

$p_1 = p_3 + \gamma_w \cdot h_1$ $p_3 = p_4$ $p_4 = p_5$ $p_5 = p_6 + \gamma_{Hg} \cdot h_m$

$p_6 = p_7$ $p_2 = p_7 + \gamma_s \cdot h_2$

$$\begin{pmatrix} p_1 \\ p_2 \\ p_3 \\ p_4 \\ p_5 \\ p_6 \\ p_7 \end{pmatrix} := Find\,(p_1, p_2, p_3, p_4, p_5, p_6, p_7) = \begin{pmatrix} 1997.901 \\ -1508.124 \\ 1685.816 \\ 1685.816 \\ 1685.816 \\ -1699.682 \\ -1699.682 \end{pmatrix} \dfrac{lb}{ft^2}$$

$\Delta p := p_1 - p_2 = 3506.025 \, \dfrac{lb}{ft^2}$

2.3.7.3 *Multifluid Differential Manometer between Two Pipes*

A multifluid differential manometer that is used to measure a small pressure difference between two pipes consists of at least a double U-tube, as illustrated in Figure 2.18. There is a pressure differential between the two pipes, 1 and 2. The first pipe contains a system fluid with a specific weight, γ_1, while the second pipe contains a system fluid with a specific weight, γ_2. The specific weight of the first manometer fluid is denoted by γ_{m1}. The specific weight of the second manometer fluid is denoted γ_{m2}. Thus, to find the pressure differential between point 1 and point 2, $p_1 - p_2$ in terms of the two respective manometer readings, h_{m1} and h_{m2}, we start at one end of the system and work our way to the other end by applying the hydrostatic pressure equation, Equation 2.18, $(dp/dz) = \gamma$, and Pascal's law, Equation 2.16, $(\partial p/\partial y) = 0$ (for a fluid that is homogenous in γ and continuous in space [x, y, z]). Therefore, for the differential manometer given in Figure 2.18, we will start at point 1 and work around to point 2 in the manometer system.

Application of $(dp/dz) = \gamma$ yields the following:

$$p_3 = p_1 + \gamma_1 h_1 \tag{2.102}$$

$$p_1 = p_3 - \gamma_1 h_1 \tag{2.103}$$

Application of $(\partial p/\partial y) = 0$ yields the following:

$$p_3 = p_4 \tag{2.104}$$

Application of $(dp/dz) = \gamma$ yields the following:

$$p_4 = p_5 + \gamma_{m1} h_{m1} \tag{2.105}$$

$$p_5 = p_6 + \gamma_{m2} h_{m2} \tag{2.106}$$

Application of $(\partial p/\partial y) = 0$ yields the following:

$$p_6 = p_7 \tag{2.107}$$

Application of $(dp/dz) = \gamma$ yields the following:

$$p_2 = p_7 + \gamma_2 h_2 \tag{2.108}$$

Thus,

$$\begin{aligned} p_1 - p_2 &= p_3 - \gamma_1 h_1 - p_7 - \gamma_2 h_2 = p_6 + \gamma_{m2} h_{m2} + \gamma_{m1} h_{m1} - \gamma_1 h_1 - p_7 - \gamma_2 h_2 \\ &= \gamma_{m2} h_{m2} + \gamma_{m1} h_{m1} - \gamma_1 h_1 - \gamma_2 h_2 \end{aligned} \tag{2.109}$$

Furthermore, if the second manometer fluid were a gas, then it specific weight, γ_{m2} would be much smaller than the specific weights of the other fluids, then we would get the following:

$$p_1 - p_2 = \gamma_{m1} h_{m1} - \gamma_1 h_1 - \gamma_2 h_2 \tag{2.110}$$

EXAMPLE PROBLEM 2.17

Water flows in pipe 1 and pipe 2, as illustrated in Figure EP 2.17. A multifluid differential manometer is attached between pipe 1 and pipe 2 in order to measure the pressure difference between pipe 1 and pipe 2. The distance between points 1 and 3 is 2 ft, and the distance between points 2 and 7 is 1 ft. The first manometer fluid is mercury with manometer reading, h_{m1} of 2.2 ft. The second manometer fluid is oil with a specific gravity of 0.78, and a manometer reading, h_{m2} of 0.9 ft. (a) Determine the pressure difference between pipe 1 and pipe 2.

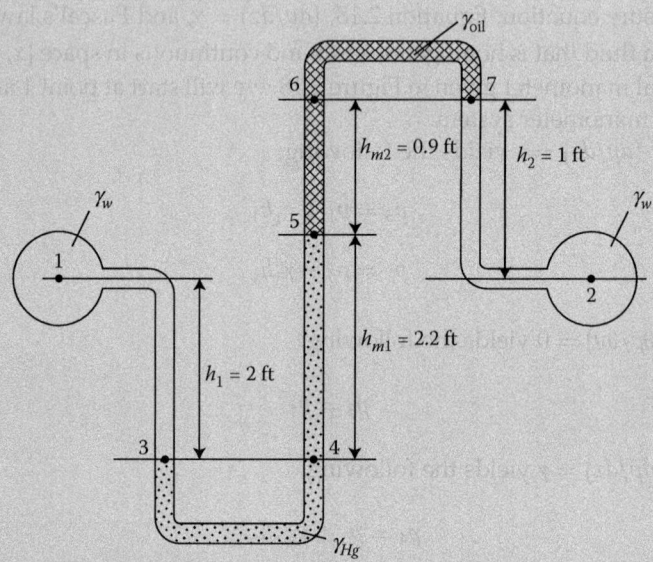

FIGURE EP 2.17
Multifluid differential manometer with a double U-tube is used to measure a small pressure difference between two pipes (pipe 1 carries water and pipe 2 carries water) flowing under pressure.

Mathcad Solution

(a) In order to determine the pressure difference between pipe 1 and pipe 2, the principles of hydrostatics are applied, starting at point 1 and working around to point 2 as follows:

$h_1 := 2 \text{ ft}$ $\qquad$ $h_2 := 1 \text{ ft}$ $\qquad$ $\gamma_w := 62.417 \dfrac{\text{lb}}{\text{ft}^3}$

$h_{m1} := 2.2 \text{ ft}$ $\qquad$ $s_{Hg} := 13.56$ $\qquad$ $\gamma_{Hg} := s_{Hg} \cdot \gamma_w = 846.375 \dfrac{\text{lb}}{\text{ft}^3}$

$h_{m2} := 0.9 \text{ ft}$ $\qquad$ $s_o := 0.78$ $\qquad$ $\gamma_o := s_o \cdot \gamma_w = 48.685 \dfrac{\text{lb}}{\text{ft}^3}$

Guess value: $\qquad$ $p_1 := 1\dfrac{\text{lb}}{\text{ft}^2}$ $\quad$ $p_2 := 1\dfrac{\text{lb}}{\text{ft}^2}$ $\quad$ $p_3 := 1\dfrac{\text{lb}}{\text{ft}^2}$ $\quad$ $p_4 := 1\dfrac{\text{lb}}{\text{ft}^2}$

$\qquad\qquad\qquad$ $p_5 := 1\dfrac{\text{lb}}{\text{ft}^2}$ $\quad$ $p_6 := 1\dfrac{\text{lb}}{\text{ft}^2}$ $\quad$ $p_7 := 1\dfrac{\text{lb}}{\text{ft}^2}$

Given

$p_3 = p_1 + \gamma_w \cdot h_1$ $\qquad p_3 = p_4$ $\qquad\qquad p_4 = p_5 + \gamma_{Hg} \cdot h_{m1}$

$p_5 = p_6 + \gamma_0 \cdot h_{m2}$ $\qquad p_6 = p_7$ $\qquad\qquad p_2 = p_7 + \gamma_w \cdot h_2$

$$
\begin{pmatrix} p_1 \\ p_2 \\ p_3 \\ p_4 \\ p_5 \\ p_6 \\ p_7 \end{pmatrix} := \text{Find } (p_1, p_2, p_3, p_4, p_5, p_6, p_7) = \begin{pmatrix} 557.82 \\ -1.161 \times 10^3 \\ 682.654 \\ 682.654 \\ -1.179 \times 10^3 \\ -1.223 \times 10^3 \\ -1.223 \times 10^3 \end{pmatrix} \frac{lb}{ft^2}
$$

$\Delta p := p_1 - p_2 = 1.719 \times 10^3 \dfrac{lb}{ft^2}$

2.4 Hydrostatic Forces on Submerged Surfaces

The determination of the hydrostatic pressure forces acting on a submerged surface is accomplished by the application of the principles of hydrostatics. The hydrostatic pressure distribution, p acting normal to the unit surface area, A that is submerged in a fluid is defined as the hydrostatic force, F_p or simply F, with units of force (lb or N). The calculation of the magnitude, direction, and location of the hydrostatic pressure force, F acting on a submerged surface area, A is important in the design of storage tanks, dams, gates, boats, ships, bulkheads, hydrometers, buoys, balloons, blimps, pipes, and other hydraulic structures. The magnitude of the hydrostatic force, F can be very large when the fluid is a liquid. The point of application (location) of the hydrostatic force on a submerged area is important when working with the moment resulting from this force. The hydrostatic pressure forces acting on the surfaces of a submerged body determine whether an anchored body will collapse or be upheld, while those acting on the surfaces of a floating or a neutrally buoyant body determine whether the body will sink or float (see Section 2.5). Furthermore, the ability of a floating or a neutrally buoyant body to return to its equilibrium position when the body is slightly disturbed depends on the pressure forces that act, in addition to characteristics of the body involved (see Section 2.5). And, finally, the related subject of buoyancy and flotation deals with the determination of the hydrostatic pressure forces acting on submerged bodies so that a stability analysis can be made (see Section 2.5).

Section 2.2 on the principles of hydrostatics presented that, according to Pascal's law, the hydrostatic pressure does not change in the horizontal plane, while according to Newton's second law of motion, the hydrostatic pressure increases linearly with depth, h and specific weight, γ, where $p = \gamma h$ and where the variation in pressure with depth, h is due only to the weight of the fluid. Furthermore, Section 2.3 on measurement of hydrostatic pressure at a point highlighted that when the fluid was a gas, the contribution of the specific weight of the gas was negligible compared to the specific weight of a liquid. Therefore, for a surface submerged in a gas, the hydrostatic pressure does not change in the horizontal plane (Pascal's law), and does not change significantly with depth, provided the depth is less than several hundred feet. Therefore, the hydrostatic pressure and thus the hydrostatic force throughout the gas can be considered to be a constant.

The submerged surface may be either a plane or a curved surface. While the submerged plane may be a horizontal, vertical, or a sloping surface, and may take on any geometric shape (rectangle, circle, triangle, etc.), the curved surface is typically assumed to be rectangular in geometric shape. Furthermore, while most of this section assumes that the surface is submerged in a single layer of fluid (a gas, a liquid open to the atmosphere, or a liquid enclosed and pressurized), surfaces submerged in a multilayered fluid is addressed in the last subsection (Section 2.4.8) of this section.

2.4.1 The Variation of the Hydrostatic Pressure along a Surface

Based on the principles of hydrostatics, the variation of the hydrostatic pressure, p acting normal to a submerged area, A may be graphically illustrated by a pressure prism. The shape of the pressure prism will vary depending upon the following variables: (1) the type of fluid: gas or liquid; (2) the type of surface: plane or curved; (3) the direction of slope of the plane: horizontal, vertical, or sloping; (4) a fluid open to the atmosphere versus a fluid enclosed and pressurized; and (5) the assumed pressure scale: gage or absolute. The decision to assume gage versus absolute pressure depends upon the following. If the other side of the submerged surface is open to the atmosphere (the dry side of the surface), and thus the atmospheric pressure, p_{atm} acts on both sides of the surface, this yields a zero resultant. Thus, in such cases, it is convenient to subtract the atmospheric pressure and work in the gage pressure scale (where p_{atm} [*gage*] $= 0$). Furthermore, pressures are assumed to be gage pressures unless specifically indicated as absolute pressure.

2.4.1.1 Variation of the Pressure Prism for a Surface Submerged in a Gas

Because the contribution of the specific weight of a gas is negligible compared to the specific weight of a liquid, the hydrostatic pressure does not change significantly with depth. Thus, because the hydrostatic pressure throughout a gas may considered to be a constant, the pressure prism for a surface submerged in a gas graphically depicts that the hydrostatic pressure, p is uniformly distributed along the submerged surface, A regardless of the type of surface (plane or curved) and the direction of the slope of the plane. This is illustrated in Figure 2.19a–d. Because the enclosed tank contains only a gas under pressure (no additional liquid in the tank), the pressure registered on the gage, p_g is the hydrostatic pressure measured in the gage scale, p (*gage*). Depending upon whether one assumes gage or absolute pressure, the hydrostatic pressure, p may either be $p = p(gage) = p_g$ or $p = p_{abs} = p(gage) + p_{atm}$, respectively, where p_g represents the hydrostatic pressure (gage) due to the pressurized gas. Because the contribution of the p_{atm} is a "gas" pressure, the shape of the pressure prism remains uniformly distributed along the submerged surface, A, while the magnitude of $p = p_{abs}$ will be greater than the magnitude of $p = p$ (*gage*) $= p_g$.

2.4.1.2 Variation of the Pressure Prism for a Surface Submerged in a Liquid Open to the Atmosphere

According to Pascal's law, the hydrostatic pressure does not change in the horizontal plane, while according to Newton's second law of motion, the hydrostatic pressure increases linearly with depth. Thus, because the hydrostatic pressure for a liquid does not change in the horizontal plane, the pressure prism for a horizontal plane submerged in a liquid graphically depicts that the pressure, p is uniformly distributed along the submerged surface, A, as

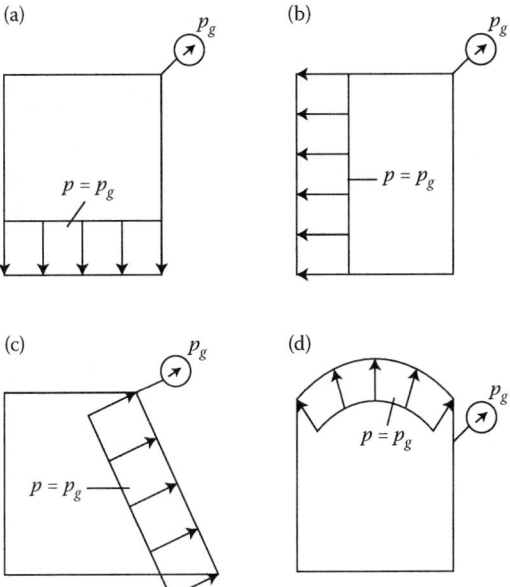

FIGURE 2.19
Uniform pressure prism for an enclosed surface submerged in a gas under gage pressure, p_g. (a) Horizontal plane. (b) Vertical plane. (c) Sloping plane. (d) Curved surface.

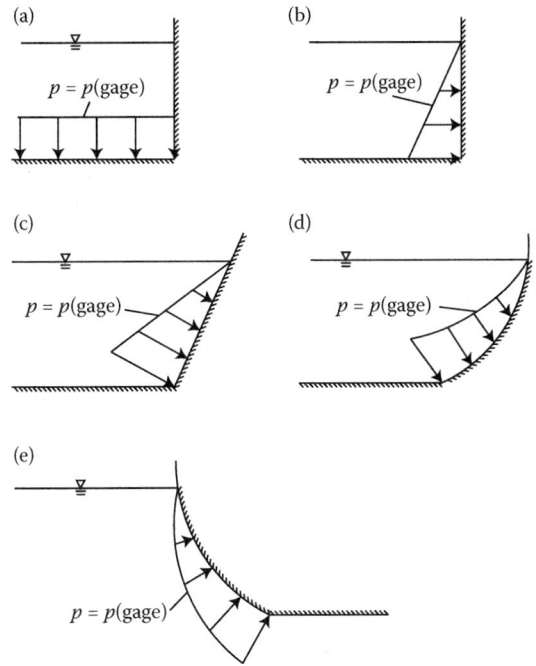

FIGURE 2.20
Variation of the pressure prism for surface submerged in a liquid open the atmosphere, assuming the gage pressure scale. (a) Horizontal plane. (b) Vertical plane. (c) Sloping plane. (d) Curved surface. (e) Curved surface.

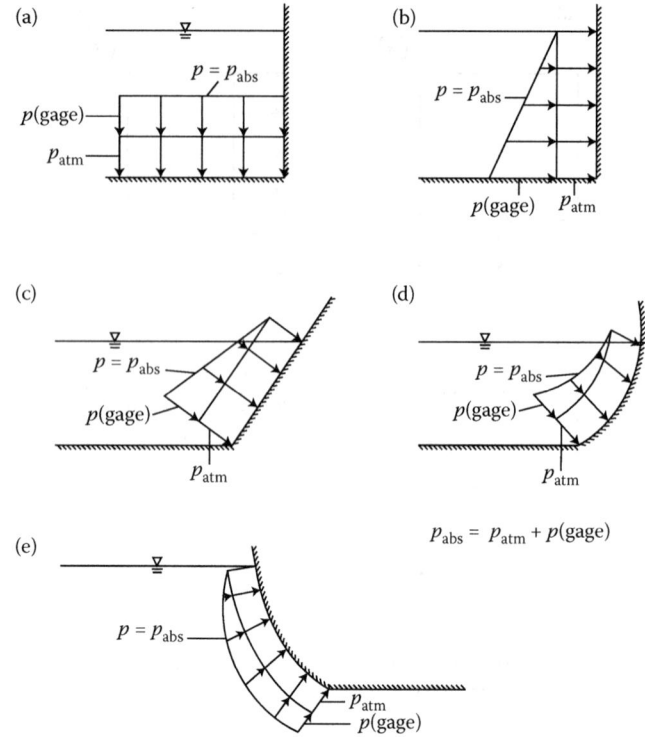

FIGURE 2.21
Variation of the pressure prism for surface submerged in a liquid open the atmosphere, assuming the absolute pressure scale. (a) Horizontal plane. (b) Vertical plane. (c) Sloping plane. (d) Curved surface. (e) Curved surface.

illustrated in Figure 2.20a. The hydrostatic pressure, p may either be $p = p(gage)$ or $p = p_{abs} = p(gage) + p_{atm}$. Furthermore, because the contribution of the p_{atm} is a "gas" pressure, the shape of the pressure prism remains uniformly distributed along the submerged surface, A, while the magnitude of $p = p_{abs}$ will be greater than the magnitude of $p = p(gage)$ as illustrated in Figure 2.21a.

Because the hydrostatic pressure increases linearly with depth, the pressure prism for a vertical or sloping plane surface or a curved surface submerged in a liquid graphically depicts that the pressure, p increases linearly with depth along the submerged surface, A, as illustrated in Figure 2.20b–e, respectively. The hydrostatic pressure, p depicted in Figure 2.20b–e is $p = p(gage)$, so the pressure prism begins with $p = p(gage) = p_{atm} = 0$ at the free surface and increases linearly with depth along the given surface. If, however, $p = p_{abs} = p(gage) + p_{atm}$, then the corresponding pressure prisms will be as illustrated in Figure 2.21b–e. One may note that because the contribution of the p_{atm} is a "gas" pressure, the corresponding pressure prisms in Figure 2.20a–e are now addended with a uniformly distributed pressure prism in Figure 2.21a–e, where the magnitude of $p = p_{abs}$ will be greater than the magnitude of $p = p(gage)$.

2.4.1.3 Variation of the Pressure Prism for a Surface Submerged in a Liquid, Enclosed, and Pressurized

If the liquid surfaces of Figure 2.20a–e were enclosed and subjected to a gage air pressure, p_g as illustrated in Figure 2.22a–e, then $p = p(gage) + p_g$ if one assumes gage pressure, where

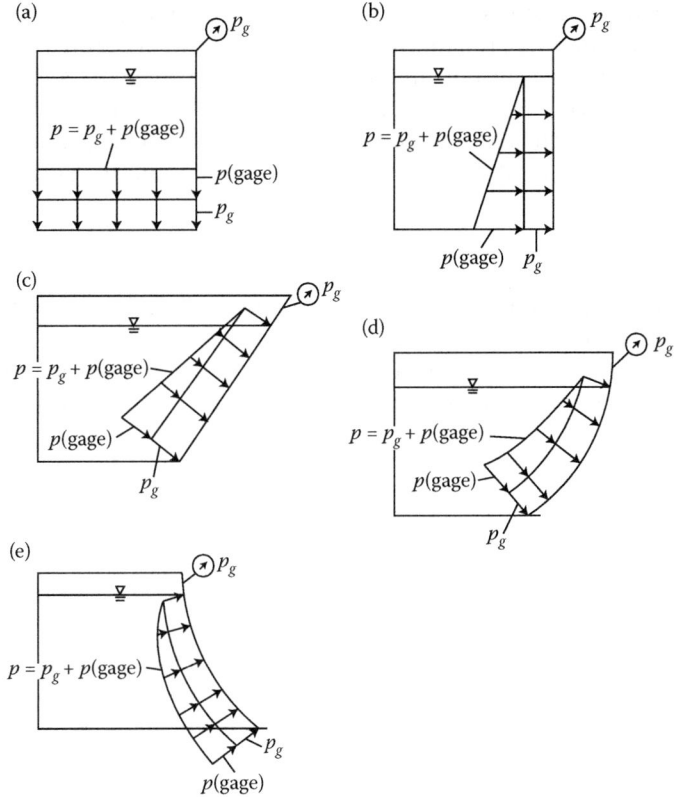

FIGURE 2.22
Variation of the pressure prism for surface submerged in a liquid enclosed in tank subjected to a gage air pressure, p_g, assuming the gage pressure scale. (a) Horizontal plane. (b) Vertical plane. (c) Sloping plane. (d) Curved surface. (e) Curved surface.

$p(gage)$ is due to the hydrostatic pressure of the liquid and p_g is due to the pressurized gas. However, if absolute pressure were assumed, then $p = p(gage) + p_g + p_{atm}$.

2.4.2 Magnitude and Location of the Hydrostatic Force for Plane Surfaces

Many submerged surfaces of interest are planar, such as a gate valve in a dam or the wall of a liquid storage tank. The distributed hydrostatic force of the pressure prism can be replaced by a point force, F in the analysis of static pressure forces. Furthermore, determination of both the magnitude and location of the point force that is equivalent to the distributed force is required. The shape of the pressure prism and thus the computation of the magnitude and the location of the resultant force will vary as a function of the type of fluid, the type of surface (in this case, plane), the direction of slope of the plane, an open or an enclosed fluid, and the assumed pressure scale. If the shape of the pressure prism is uniformly distributed over the submerged surface area, A of a plane (see Figure 2.19b, for instance), the resultant force, F is equal to the pressure, p times the area, A and is given as follows:

$$F = pA \qquad (2.111)$$

and the location of the resultant force, F is called the "center of pressure" and is located at the center of gravity of the pressure prism, which is at the centroid of the area, A. However, if the shape of the pressure prism in not uniformly distributed over the submerged area (see Figure 2.20b, for instance), in general, the magnitude of the resultant force, F acting on a plane surface area, A, which is submerged in a fluid, is equal to the product of the pressure acting at the centroid (center of area) of the surface area, p_{ca} and the surface area, A and is given as follows:

$$F = p_{ca}A \tag{2.112}$$

One may note that the pressures (p or p_{ca}) are assumed to be gage pressures unless specifically indicated as absolute pressure. Thus, in the case where the absolute pressure is assumed, or the plane is subjected to an air pressure, p_g (enclosing the plane in a pressurized tank), the magnitude of the resultant force, F acting on a plane surface area, A is given, respectively, as follows:

$$F = (p_{ca} + p_{atm})A \tag{2.113}$$

$$F = (p_{ca} + p_g)A \tag{2.114}$$

Furthermore, in general, the location of the resultant force, F is called the "center of pressure" and is located at the center of gravity of the pressure prism. While the details of the computation of the magnitude and location of the resultant force are given below for the three general categories of submerged planes described above (a plane submerged in a gas, a plane submerged in a liquid open to the atmosphere, and a plane submerged in an enclosed and pressurized liquid), the general approach is introduced in the following two sections below, assuming a vertical plane with a rectangular geometric shape submerged in liquid open to the atmosphere, and assuming gage pressure. Furthermore, nonrectangular planes are addressed in a separate section thereafter.

2.4.2.1 Magnitude of the Resultant Hydrostatic Force on Plane Surfaces

The distribution of the hydrostatic pressure, p acting normal to a submerged surface area, A is defined as the hydrostatic pressure force, F. The pressure prism illustrated the variation of the hydrostatic pressure along a horizontal, vertical, or sloping plane. Assuming that the h-axis is

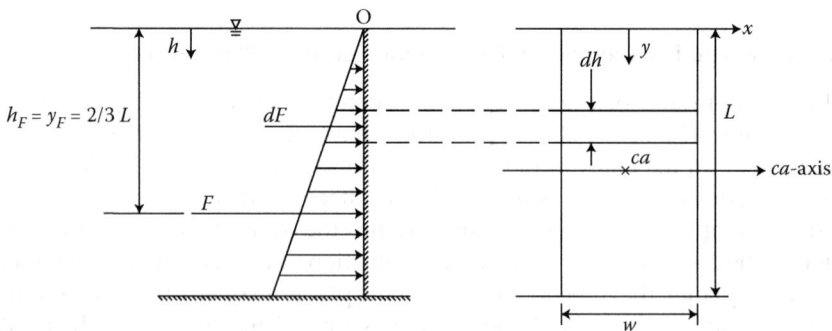

FIGURE 2.23
Incremental hydrostatic force distribution, dF acting on incremental areas, dA, and the resultant force, F and its location, h_F for a vertical plane with a rectangular geometric shape submerged at a depth, $h = 0$ in a liquid open to the atmosphere, assuming the gage pressure scale.

defined from the top of the free surface (of a liquid), where $p_{atm}(gage) = 0$, Figure 2.23 illustrates that the corresponding hydrostatic force distribution (distributed hydrostatic pressure force) consists of a series of incremental forces, dF acting on a series of incremental areas, dA with incremental heights, dh and width, w (illustration assumes a vertical plane with a rectangular geometric shape submerged [$h = 0$] in a liquid open to the atmosphere) where:

$$dF = p\,dA = (\gamma h)(w\,dh) \tag{2.115}$$

where $p = \gamma h$ and $dA = w\,dh$. Integration of the incremental forces, dF in Equation 2.115 over the entire height, L of the plane between the limits of integration 0 and L yields the magnitude of the resultant hydrostatic force, F as follows:

$$F = \int dF = \int p\,dA = \int_0^L (\gamma h)(w\,dh) \tag{2.116}$$

Using Mathcad to integrate yields:

$$\int_0^L (\gamma \cdot h)w\,dh \rightarrow \frac{\gamma \cdot L^2 \cdot w}{2}$$

And thus,

$$F = (\gamma)\left(\frac{L}{2}\right)(wL) = \gamma h_{ca} A \tag{2.117}$$

where the height to the center of area (rectangular planar area, A), h_{ca} is defined as follows:

$$h_{ca} = \frac{1}{A}\int_A h\,dA \tag{2.118}$$

and where the *first moment of the area* with respect to the *x-axis* (defined along the top of the free water surface) is defined as follows:

$$\int_A h\,dA \tag{2.119}$$

Thus, the resultant hydrostatic force, F is the product of pressure at the centroid (center of area) of the rectangular plane, $p_{ca} = \gamma h_{ca}$, and the surface area, $A = wL$, where $h_{ca} = L/2$ is the height to the centroid of the rectangular plane measured from the top of the free surface along the h-axis. Therefore, in general (for any type of fluid, and for vertical, horizontal, and sloping planes of any geometric shapes), the magnitude of the resultant hydrostatic force, F acting on a plane surface area, A, which is submerged in a fluid, is equal to the product of the pressure acting at the centroid of the surface area, p_{ca} and the surface area, A:

$$F = p_{ca} A = \gamma h_{ca} A \tag{2.120}$$

where the height to the center of area, h_{ca} is measured from the top of the free surface along the h-axis. Therefore, in order to compute the magnitude of the resultant hydrostatic force, F, one may either integrate Equation 2.116 or one may directly apply Equation 2.120. The direct application of Equation 2.120 to determine the hydrostatic force is straightforward and a logical choice when the definition of the h_{ca} is obvious, as it is for a submerged plane with any geometric shape. However, the application/ integration of Equation 2.116 would be more straightforward and a more logical choice when the definition of the h_{ca} is not so obvious, as is the case for a submerged plane with a nongeometric shape. It may be noted that in either case, care must be taken in the definition of the hydrostatic pressure variables, p and p_{ca}, in Equations 2.116 and 2.120, respectively, if either the absolute pressure scale is assumed, where $p_{atm}(abs) \neq 0$, or the plane is subjected to an air pressure, p_g (enclosing the plane in a pressurized tank). For such assumptions, p in the application of Equation 2.116 would be defined as $p = \gamma h + p_{atm}$ or $p = \gamma h + p_g$, respectively. Furthermore, in the application of Equation 2.120, in addition to p_{ca}, there is the additional p_{atm} or p_g, respectively, as follows:

$$F = (p_{ca} + p_{atm})A = (\gamma h_{ca} + p_{atm})A \tag{2.121}$$

$$F = (p_{ca} + p_g)A = (\gamma h_{ca} + p_g)A \tag{2.122}$$

EXAMPLE PROBLEM 2.18

A vertical plane with a width of 2 ft and a height of 6 ft is submerged in water as illustrated in Figure EP 2.18. The water is open to the atmosphere and assume gage pressure. (a) Determine the magnitude of the resultant hydrostatic force acting on the plane by applying Equation 2.116.

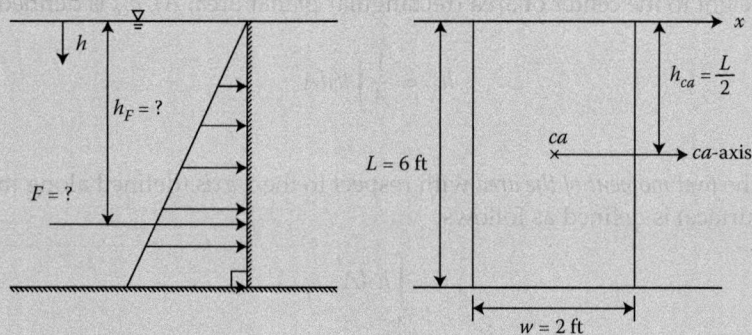

FIGURE EP 2.18
Triangular pressure prism, the resultant hydrostatic force, F, and its location, h_F for a vertical plane submerged at a depth, $h = 0$ in water open to the atmosphere.

Mathcad Solution

(a) In order to determine the magnitude of the hydrostatic resultant force acting on the plane, Equation 2.116 is applied as follows:

$$w := 2 \, ft \qquad L := 6 \, ft \qquad \gamma := 62.417 \, \frac{lb}{ft^3} \qquad F := \int_0^L (\gamma \cdot h) \, w \, dh = 2.247 \times 10^3 \, lb$$

EXAMPLE PROBLEM 2.19

Refer to Example Problem 2.18. A vertical plane with a width of 2 ft and a height of 6 ft is submerged in water as illustrated in Figure EP 2.18. The water is open to the atmosphere and assume gage pressure. (a) Determine the magnitude of the resultant hydrostatic force acting on the plane by applying Equation 2.120.

Mathcad Solution

(a) In order to determine the magnitude of the hydrostatic resultant force acting on the plane, Equation 2.120 is applied as follows:

$$w := 2 \text{ ft} \qquad L := 6 \text{ ft} \qquad A := w \cdot L = 12 \text{ ft}^2 \qquad h_{ca} := \frac{L}{2} = 3 \text{ ft}$$

$$\gamma := 62.417 \frac{\text{lb}}{\text{ft}^3} \qquad F = \gamma \cdot h_{ca} \cdot A = 2.247 \times 10^3 \text{ lb}$$

2.4.2.2 Location of the Resultant Hydrostatic Force on Plane Surfaces

The location of the resultant hydrostatic force, F is called the "center of pressure" or the moment arm and is located at the center of gravity of the pressure prism, regardless of the type of fluid, the direction of slope of the plane, an open or enclosed fluid, the assumed pressure scale, or the geometric shape of the plane. However, because the pressure prism is of uniform width about the h-axis for a rectangular plane, the center of gravity of the pressure prism may be represented by the center of area (centroid) of the pressure prism. Thus, for the vertical plane illustrated in Figure 2.23, the pressure prism is a triangle [because it was assumed that $p_{atm}(gage) = 0$] and thus the "center of pressure," h_F for the resultant hydrostatic force, F, measured from the top of the free surface along the h-axis, is located at $h_F = 2/3\,L$, which is the center of area of the triangular pressure prism measured from the h-axis.

Because it was assumed that $p_{atm}(gage) = 0$ at the top of the free surface, the resulting pressure prism was a triangle, for which the center of area relative to the top of the free surface along the h-axis is simply $2/3\,L$. If, however, either the absolute pressure scale is assumed or the plane is subjected to an air pressure, p_g (enclosing the plane in a pressurized tank), the resulting pressure prism would be a trapezoid as illustrated in Figures 2.21b and 2.22b, respectively. Once again, the "center of pressure" for the resultant hydrostatic force, F, measured from the top of the free surface along the h-axis, would be located at $h_F =$ "the center of area of the trapezoidal pressure prism measured from the h-axis," as illustrated in Figure 2.24. Because the computation of the center of area of a trapezoid is not as direct as that of a triangle, one may use a "computational" approach to determine the "center of pressure" for such resulting trapezoidal pressure prisms.

The "computational" approach to determine the "center of pressure," or the location of the resultant hydrostatic force, F is given as follows. Referring to Figure 2.23, (assuming the plane is submerged in a liquid open to the atmosphere, and assuming gage pressure), the location of the center of pressure, h_F may be computed by equating the moment, M about point O of the resultant hydrostatic force, F to the moment, about point O of the distributed

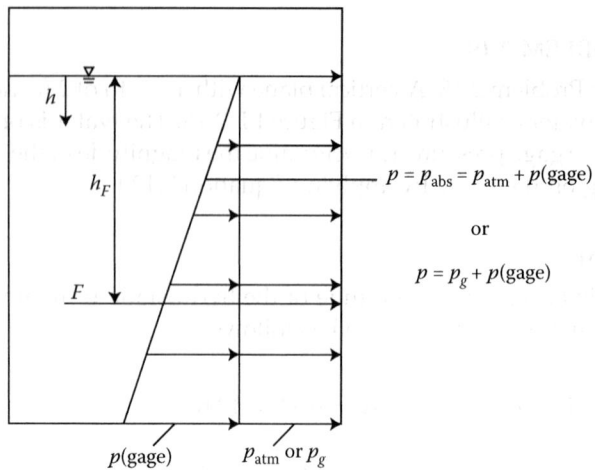

FIGURE 2.24
Trapezoidal pressure prism, the resultant hydrostatic force, F and its location, h_F for a vertical plane submerged at a depth, $h = 0$ in a liquid for either, assuming the absolute pressure scale or enclosing and subjecting to a gage air pressure, p_g.

hydrostatic pressure force, dF as follows:

$$M = h_F F = \int h\, dF = \int hp\, dA = \int_0^L h(\gamma h)(w\, dh) \tag{2.123}$$

Using Mathcad to integrate yields:

$$\int_0^L h\,(\gamma \cdot h)w\, dh \rightarrow \frac{\gamma \cdot L^3 \cdot w}{3}$$

And thus, the location of the center of pressure, h_F may be computed as follows:

$$h_F = \frac{M}{F} \tag{2.124}$$

Using Mathcad to solve for the center of pressure, h_F yields:

$$\frac{\dfrac{\gamma \cdot L^3 \cdot w}{3}}{\dfrac{\gamma \cdot L^2 \cdot w}{2}} \rightarrow \frac{2 \cdot L}{3}$$

where $F = \gamma h_{ca} A = \gamma w L^2 / 2$ acts at a distance h_F from the top of the free surface and has the same effect as the distributed force of the pressure prism. One may note that the h-axis is defined from the top of the free surface, the y-axis is defined from the top of the plane, and the x-axis is defined along the top of the free water surface. If the vertical plane is submerged at a vertical depth, $h = y_o$, then the relationship between the h-axis and the y-axis is $h = (y_o + y)$ (see section below on submerged vertical plane). However, in Figure 2.23, because the top of the free surface coincides with the top of the plane (where the height of submergence, $y_o = 0$), the h-axis and the y-axis coincide, where $h = y$. Thus, the location of the center of pressure with respect to the y-axis is $y_F = h_F$.

EXAMPLE PROBLEM 2.20

Refer to Example Problem 2.18 and Figure EP 2.18. (a) Determine the location of the resultant hydrostatic force acting on the vertical plane by applying Equation 2.124.

Mathcad Solution

(a) In order to determine the location of the resultant hydrostatic force acting on the vertical plane, Equation 2.124 is applied as follows:

$$w := 2\,\text{ft} \qquad L := 6\,\text{ft} \qquad \gamma := 62.417\,\frac{\text{lb}}{\text{ft}^3}$$

$$h_F := \frac{\displaystyle\int_0^L h\,(\gamma \cdot h)\,w\,dh}{\displaystyle\int_0^L (\gamma \cdot h)\,w\,dh} = 4\,\text{ft}$$

An alternative approach to compute the location of center of pressure, h_F is given as follows. From Figure 2.23 (assuming the plane is submerged in a liquid open to the atmosphere, and assuming gage pressure), one may realize that the moment, M about point O of the resultant hydrostatic force, F is given as follows:

$$M = \int h\,dF = \int_0^L h(\gamma h)(w\,dh) = \int_A \gamma h^2\,dA \qquad (2.125)$$

where the *second moment of the area (moment of inertia)* with respect to the x-axis, I_x is defined as follows:

$$I_x = \int_A h^2\,dA \qquad (2.126)$$

and thus, the moment, M is defined as follows:

$$M = \gamma I_x \qquad (2.127)$$

Furthermore, the location of the center of pressure, h_F may be computed as follows:

$$h_F = \frac{M}{F} = \frac{\displaystyle\int_A h(\gamma h)\,dA}{\gamma h_{ca} A} = \frac{\displaystyle\int_A \gamma h^2\,dA}{\gamma h_{ca} A} = \frac{\gamma I_x}{\gamma h_{ca} A} = \frac{I_x}{h_{ca} A} \qquad (2.128)$$

However, the application of Equation 2.128 to compute the location of the center of pressure, h_F typically proceeds by applying the parallel axis theorem to define the relationship between the *second moment of the area* with respect to the x-axis, I_x (*moment of inertia*) and the second moment of the area with respect to the axis passing the center of area, the *ca*-axis (center of area axis) and parallel to the *x*-axis, I_{x-ca} (*centroidal moment of inertia*) as follows:

$$I_x = I_{x-ca} + Ah_{ca}^2 \tag{2.129}$$

where $h_{ca} = y_{ca}$. The area, A; the *center of area (centroid)*, y_{ca}; and the *centroidal moment of inertia*, I_{x-ca} for various geometric plane shapes are given in Table B.1 in Appendix B. Thus, assuming the plane is submerged in a liquid open to the atmosphere and assuming gage pressure, the location of the center of pressure, h_F may be computed as follows:

$$h_F = \frac{I_x}{h_{ca}A} = \frac{I_{x-ca}}{h_{ca}A} + h_{ca} \tag{2.130}$$

Furthermore, in the case when the depth of submergence, $y_o = 0$, $h_{ca} = y_{ca}$, so $y_F = h_F$ as follows:

$$y_F = \frac{I_{x-ca}}{y_{ca}A} + y_{ca} \tag{2.131}$$

And, finally, one may note that Equation 2.131 illustrates that the location, y_F of the resultant force, F does not pass through the height to the center of area, y_{ca}, but is always *below* it, since the term $(I_{x-ca}/y_{ca}A) > 0$. From Table B.1 in Appendix B, for a rectangular plane, $I_{x-ca} = wL^3/12$, $y_{ca} = L/2$ and $A = wL$. Therefore, the center of pressure or the moment arm is computed as follows:

$$y_F = h_F = \frac{wL^3/12}{(L/2)(wL)} + \frac{L}{2} \tag{2.132}$$

Using Mathcad to solve for the center of pressure, $y_F = h_F$ yields:

$$\frac{\dfrac{w \cdot L^3}{12}}{\left(\dfrac{L}{2}\right)(w \cdot L)} + \frac{L}{2} \rightarrow \frac{2 \cdot L}{3}$$

which is the same result as derived from the integration of Equation 2.124. It is important to note that for the more general cases, where the plane is submerged in a liquid open to the atmosphere where absolute pressure is assumed, in an enclosed and pressurized liquid, or in a gas, the location of the center of pressure is determined in the respective sections below.

EXAMPLE PROBLEM 2.21

Refer to Example Problem 2.18 and Figure EP 2.18. (a) Determine the location of the resultant hydrostatic force acting on the vertical plane by applying Equation 2.130.

Mathcad Solution

(a) In order to determine the location of the resultant hydrostatic force acting on the vertical plane, Equation 2.130 is applied. Furthermore, the I_{x-ca} for a rectangular plane is given in Table B.1 in Appendix B.

$$w := 2 \text{ ft} \qquad L := 6 \text{ ft} \qquad A := w \cdot L = 12 \text{ ft}^2 \qquad h_{ca} := \frac{L}{2} = 3 \text{ ft}$$

$$I_{xca} := \frac{w \cdot L^3}{12} = 36 \text{ ft}^4 \qquad h_F := \frac{I_{xca}}{h_{ca} \cdot A} + h_{ca} = 4 \text{ ft}$$

2.4.3 Planes Submerged in a Gas

Planes submerged in a gas are assumed to be enclosed and subjected to a gas gage pressure, p_g. Plane surfaces (horizontal, vertical, or sloping) that are subjected to a gas pressure, p_g have essentially a uniform pressure prism over their entire surface area, as illustrated in Figure 2.19a–c. Therefore, the magnitude of the resultant force, $F = pA = p_g A$ and the location of the resultant force, F is called the "center of pressure" and is located at the center of gravity of the pressure prism, which is at the centroid of the area, A. Furthermore, pressures are assumed to be gage pressures unless specifically indicated as absolute pressure.

EXAMPLE PROBLEM 2.22

A sloping plane with a width of 1 ft and a length of 3 ft is submerged in a gas tank and subjected to a gas pressure of 8 psi, as illustrated in Figure EP 2.22. (a) Determine the magnitude of the resultant hydrostatic force acting on the sloping plane. (b) Determine the location of the resultant hydrostatic force acting on the sloping plane.

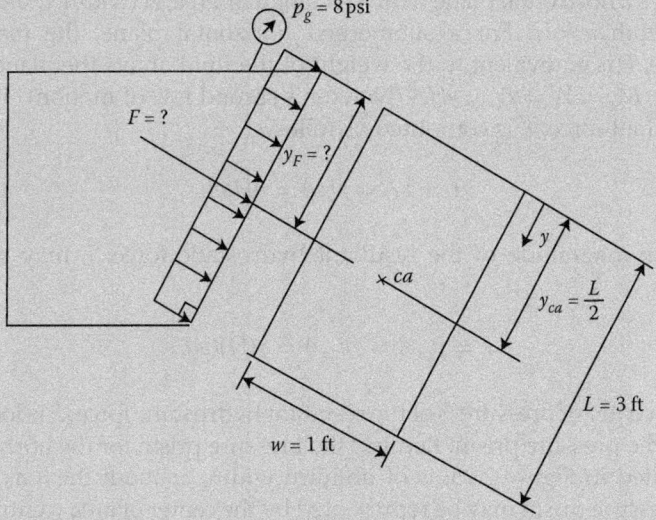

FIGURE EP 2.22
Uniform pressure prism, the resultant hydrostatic force, F, and its location, y_F for an enclosed sloping plane submerged at a depth, $y = 0$ in gas under a gage pressure, p_g.

Mathcad Solution

(a) The magnitude of the hydrostatic resultant force acting on the sloping plane is determined by applying Equation 2.111 as follows:

$$w := 1\,ft \qquad\qquad L := 3\,ft \qquad\qquad A := w \cdot L = 3\,ft^2$$

$$p_g := 8\frac{lb}{in^2}\frac{(12\,in)^2}{(1\,ft)^2} = 1.152 \times 10^3 \frac{lb}{ft^2} \qquad\qquad F := p_g \cdot A = 3.456 \times 10^3\,lb$$

(b) The location of the resultant hydrostatic force acting on the sloping plane is located at the center of area of the sloping rectangular plane and is computed as follows:

$$y_{ca} := \frac{L}{2} = 1.5\,ft \qquad\qquad y_F := y_{ca} = 1.5\,ft$$

2.4.4 Planes Submerged in a Liquid Open to the Atmosphere

If a plane (horizontal, vertical, or sloping) is submerged in a liquid that is open to the atmosphere, and if one assumes gage pressure, then $p = p(gage)$, where $p(gage)$ is due to the hydrostatic pressure of the liquid, as illustrated in Figure 2.20a–c. However, if absolute pressure were assumed, then $p = p(gage) + p_{atm}$, where $p(gage)$ is due to the hydrostatic pressure of the liquid, as illustrated in Figure 2.21a–c. Furthermore, pressures are assumed to be gage pressures unless specifically indicated as absolute pressure.

2.4.4.1 Submerged (h = H) Horizontal Plane (Open)

Given that the h-axis is defined from the top of the free surface, where $p_{atm}(gage) = 0$, Figure 2.25 illustrates a horizontal plane with a rectangular area, A (width, w and length, L) submerged at a depth $h = H$. For a submerged horizontal plane, the magnitude of the hydrostatic force, F is equivalent to the weight of the fluid above the plane or the gravitational force, $F_G = Mg = W = \gamma V = \gamma HA$ (Newton's second law of motion). Thus, the magnitude of the resultant force, F is computed as follows:

$$F = F_G = \gamma HA = \gamma HwL \tag{2.133}$$

Alternatively, the magnitude of the resultant hydrostatic force, F may be computed as follows:

$$F = p_{ca}A = \gamma h_{ca}A = \gamma(H)(wL) \tag{2.134}$$

The location ("center of pressure") of the resultant hydrostatic force, F is located at the center of gravity of the pressure prism. Because the pressure prism for the horizontal rectangular plane illustrated in Figure 2.25, is of uniform width, w about the h-axis, the center of gravity of the pressure prism may be represented by the center of area (centroid) of the pressure prism, which is a rectangle of length, L and width, w. Therefore, the "center of pressure" for the resultant hydrostatic force, F, measured from the top of the free surface along the h-axis, is located at $h_F = H$, and located at the center of area of the rectangular pressure prism at $L/2$ and $w/2$.

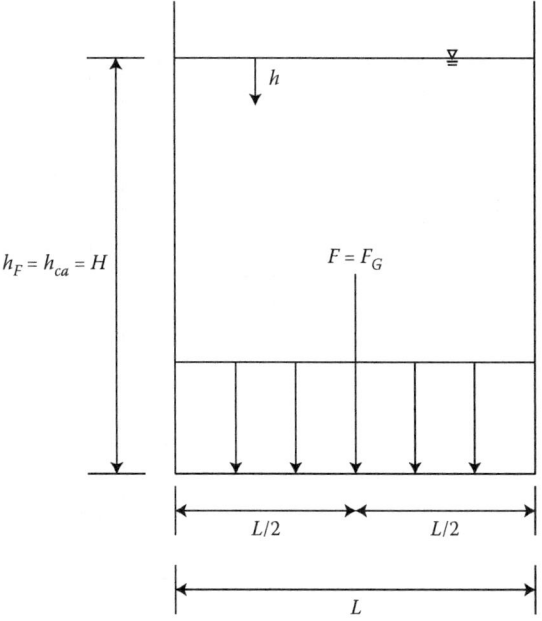

FIGURE 2.25
Resultant force, F and its location, h_F for a horizontal plane with a rectangular geometric shape submerged at a depth, $h = H$ in a liquid open to atmosphere, assuming the gage pressure scale.

EXAMPLE PROBLEM 2.23

A horizontal plane with a width of 2 ft and a length of 5 ft is submerged in water at a depth of 7 ft as illustrated in Figure EP 2.23. (a) Determine the magnitude of the resultant hydrostatic force acting on the horizontal plane. (b) Determine the location of the resultant hydrostatic force acting on the horizontal plane.

Mathcad Solution

(a) The magnitude of the hydrostatic resultant force acting on the horizontal plane is determined by applying Equation 2.133 as follows:

$$w := 2\,\text{ft} \qquad L := 5\,\text{ft} \quad A := w\,L = 10\,\text{ft}^2 \quad H := 7\,\text{ft}$$

$$\gamma := 62.417\,\frac{\text{lb}}{\text{ft}^3} \qquad F_G := \gamma\,H\,A = 4.369 \times 10^3\,\text{lb} \qquad F := F_G = 4.369 \times 10^3\,\text{lb}$$

Alternatively, the magnitude of the hydrostatic resultant force acting on the horizontal plane is determined by applying Equation 2.134 as follows:

$$h_{ca} := H = 7\,\text{ft} \qquad F := \gamma\,h_{ca}\,A = 4.369 \times 10^3\,\text{lb}$$

(b) The location of the resultant hydrostatic force acting on the horizontal plane is located at the center of area of the horizontal rectangular plane and is computed as follows:

$$h_F := h_{ca} = 7\,\text{ft} \qquad\qquad \frac{L}{2} = 2.5\,\text{ft} \qquad\qquad \frac{w}{2} = 1\,\text{ft}$$

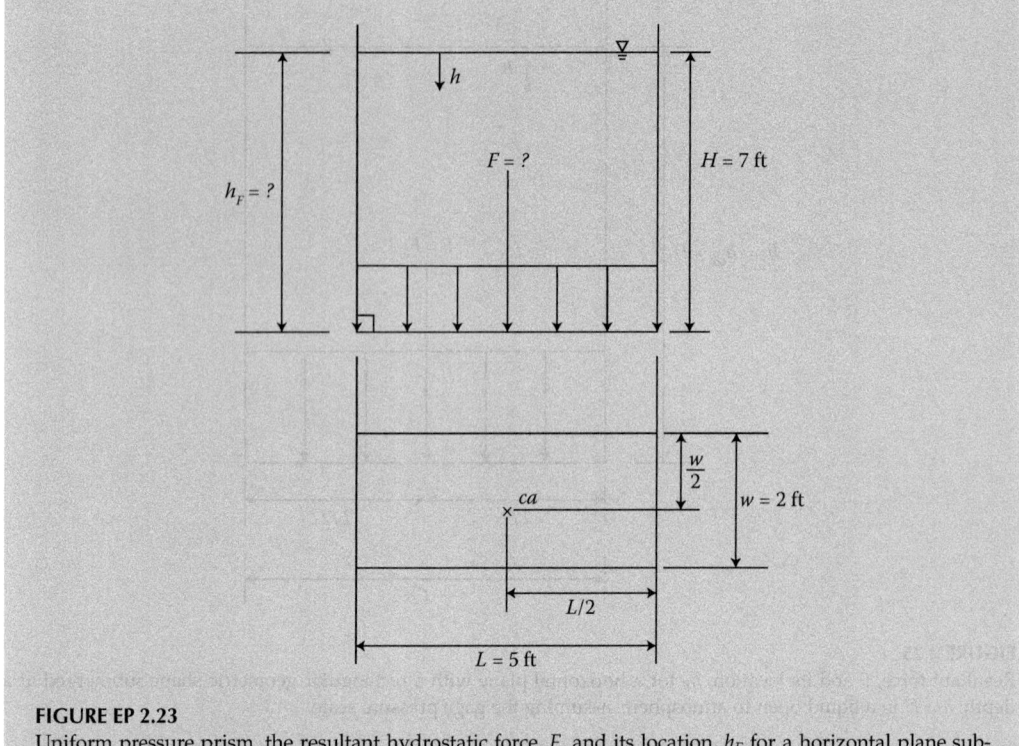

FIGURE EP 2.23
Uniform pressure prism, the resultant hydrostatic force, F, and its location, h_F for a horizontal plane submerged at a depth, $h = H$ in water open to the atmosphere.

2.4.4.2 Submerged ($h = y_o$) Vertical Plane (Open)

Given that the h-axis is defined from the top of the free surface, where $p_{atm}(gage) = 0$, Figure 2.26 illustrates a vertical plane with a rectangular area, A (width, w and length, L) submerged at a depth, $h = y_o$. Because the y-axis is defined from the top of the rectangular plane, the relationship between the h-axis and the y-axis is $h = (y_o + y)$. Furthermore, when $y_o = 0$,

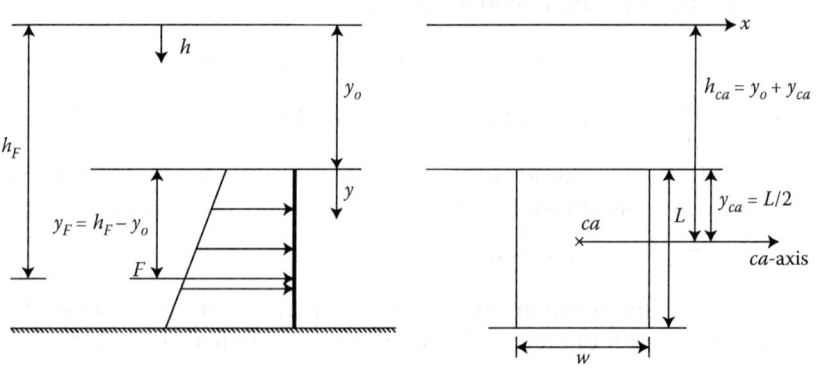

FIGURE 2.26
Resultant force, F and its location, h_F for a vertical plane with a rectangular geometric shape submerged at a depth, $h = y_o$ in a liquid open to atmosphere, assuming the gage pressure scale.

then the h-axis and the y-axis coincide, so $h = y$ (as for the case in Figure 2.23). For the case in Figure 2.26, the magnitude of the resultant hydrostatic force, F is computed by integrating over the length of the plane. The integration can be made using either the top of the free surface or the top of the plane as a reference point for integration. Integration over the length of the plane using the free surface as the reference is given as follows:

$$F = \int p \, dA = \int_{y_o}^{y_o+L} (\gamma h)(w \, dh) \tag{2.135}$$

Using Mathcad to integrate yields:

$$\int_{y_o}^{y_o+L} (\gamma \cdot h) w \, dh \rightarrow \frac{\gamma \cdot L \cdot w(L + 2 \cdot y_o)}{2}$$

Furthermore, integration over the length of the plane using the top of the plane as the reference yields the same results as follows:

$$F = \int p \, dA = \int_0^L (\gamma h(y))(w \, dy) = \int_0^L (\gamma(y_o + y))(w \, dy) \tag{2.136}$$

where $h(y) = y_o + y$, and where the use of the function $h(y)$ is required in the integration using Mathcad. Using Mathcad to integrate yields:

$$\int_0^L [\gamma \cdot (y_o + y)] w \cdot dy \rightarrow \frac{\gamma \cdot L \cdot w \, (L + 2 \cdot y_o)}{2}$$

Alternatively, one may compute the magnitude of the resultant force, F as follows:

$$F = p_{ca}A = \gamma h_{ca}A = \gamma(y_o + y_{ca})(wL) = \gamma\left(y_o + \frac{L}{2}\right)(wL) \tag{2.137}$$

where for a rectangular plane, $h_{ca} = (y_o + y_{ca})$, $y_{ca} = L/2$, and $A = wL$.

The location ("center of pressure") of the resultant hydrostatic force, F is located at the center of gravity of the pressure prism. Thus, for the vertical plane submerged at a depth, $h = y_o$ illustrated in Figure 2.26, the pressure prism is a trapezoid and thus the "center of pressure" or the moment arm, y_F for the resultant hydrostatic force, F, measured from the top of the plane along the y-axis, is computed as follows:

$$M = y_F F = \int_0^L y \, dF = \int_0^L yp \, dA = \int_0^L y(\gamma h(y))(w \, dy) = \int_0^L y(\gamma(y_o + y))(w \, dy) \tag{2.138}$$

where $h(y) = y_o + y$. Using Mathcad to integrate yields:

$$\int_0^L [y \cdot \gamma \cdot (y_o + y)]w\,dy \rightarrow \frac{\gamma \cdot L^2 \cdot w \cdot (2 \cdot L + 3 \cdot y_o)}{6}$$

Solving for the moment arm by using Equation 2.138 for the moment and Equation 2.136 for the force yields the following:

$$y_F = \frac{M}{F} \tag{2.139}$$

Using Mathcad to solve for the moment arm, y_F yields the following:

$$\frac{\dfrac{\gamma \cdot L^2 \cdot w \cdot (2 \cdot L + 3 \cdot y_o)}{6}}{\dfrac{\gamma \cdot L \cdot w \cdot (L + 2 \cdot y_o)}{2}} \rightarrow \frac{L \cdot (2 \cdot L + 3 \cdot y_o)}{3 \cdot (L + 2 \cdot y_o)}$$

Alternatively, the "center of pressure" or the moment arm, h_F for the resultant hydrostatic force, F, measured from the top of the free surface along the h-axis may be computed as follows:

$$h_F = \frac{M}{F} = \frac{\int_A h(\gamma h)\,dA}{\gamma h_{ca} A} = \frac{\int_A \gamma h^2\,dA}{\gamma h_{ca} A} = \frac{\gamma I_x}{\gamma h_{ca} A} = \frac{I_x}{h_{ca} A} \tag{2.140}$$

but:

$$I_x = I_{x-ca} + A h_{ca}^2 \tag{2.141}$$

and thus:

$$h_F = \frac{I_{x-ca}}{h_{ca} A} + h_{ca} = \frac{wL^3/12}{(y_o + L/2)(wL)} + (y_o + L/2) \tag{2.142}$$

where for a rectangular plane, $I_{x-ca} = wL^3/12$, $h_{ca} = (y_o + y_{ca})$, $y_{ca} = L/2$, and $A = wL$. Using Mathcad to simplify the expression for the center of pressure, h_F yields:

$$\frac{\dfrac{w \cdot L^3}{12}}{\left(y_o + \dfrac{L}{2}\right) \cdot (w \cdot L)} + \left(y_o + \frac{L}{2}\right) \rightarrow \frac{L}{2} + y_o + \frac{L^2}{12 \cdot \left(\dfrac{L}{2} + y_o\right)}$$

Furthermore, the "center of pressure" for the resultant force, F, measured from the top of the plane along the y-axis $y_F = h_F - y_o$ is given as follows:

$$y_F = h_F - y_o = \frac{L}{2} + \frac{L^2}{12\left(\dfrac{L}{2} + y_o\right)} \tag{2.143}$$

Using Mathcad to simplify the expression for y_F yields:

$$\frac{L}{2} + \frac{L^2}{12 \cdot \left(\frac{L}{2} + y_0\right)} \quad \text{simplify} \; \rightarrow \; \frac{L \cdot (2 \cdot L + 3 \cdot y_0)}{3 \cdot (L + 2 \cdot y_0)}$$

which is the same result yielded in Equation 2.139. The above Equations 2.139 and 2.143 for the moment arm, y_F may be used to demonstrate an important concept regarding the depth of the location of the moment arm, y_F as a function of the depth of submergence, y_o of the plane surface and the length, L of the plane.

EXAMPLE PROBLEM 2.24

A vertical planar gate with a width of 2 ft and a height of 6 ft is submerged at a depth of 4 ft in water open to the atmosphere, as illustrated in Figure EP 2.24. (a) Determine the magnitude of the resultant hydrostatic force acting on the vertical planar gate. (b) Determine the location of the resultant hydrostatic force acting on the vertical planar gate.

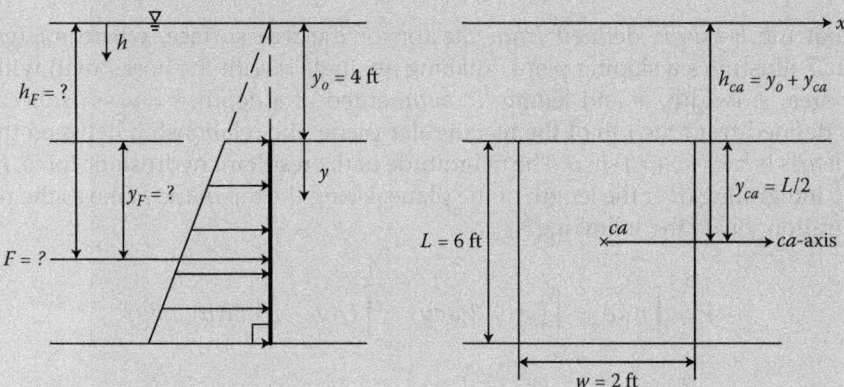

FIGURE EP 2.24
Trapezoidal pressure prism, the resultant hydrostatic force, F, and its location, h_F for a vertical plane submerged at a depth, $h = y_o$ in water open to the atmosphere.

Mathcad Solution

(a) The magnitude of the hydrostatic resultant force acting on the planar gate is determined by applying Equation 2.135, integrating over the length of the planar gate using the top of the free surface as a reference point for integration as follows:

$$w := 2 \, \text{ft} \qquad L := 6 \, \text{ft} \quad A := w \cdot L = 12 \, \text{ft}^2 \qquad\qquad y_o := 4 \, \text{ft}$$

$$\gamma := 62.417 \, \frac{\text{lb}}{\text{ft}^3} \qquad\qquad F := \int_{y_o}^{y_o + L} (\gamma \cdot h) \cdot w \, dh = 5.243 \times 10^3 \, \text{lb}$$

Alternatively, the magnitude of the hydrostatic force acting on the planar gate may be determined by applying Equation 2.137 as follows:

$$y_{ca} := \frac{L}{2} = 3 \, \text{ft} \qquad\qquad h_{ca} := y_o + y_{ca} = 7 \, \text{ft}$$

$$F := \gamma \cdot h_{ca} \cdot A = 5.243 \times 10^3 \, lb$$

(b) The location of the resultant hydrostatic force acting on the vertical planar gate is determined by applying Equation 2.139 as follows:

$$h(y) := y_o + y \qquad\qquad y_F = \frac{\int_0^L y \cdot (\gamma \cdot h(y)) \cdot w \, dy}{\int_0^L (\gamma \cdot h(y)) \cdot w \, dy} = 3.429 \, ft$$

$$h_F := y_0 + y_F = 7.429 \, ft$$

Alternatively, the location of the resultant force acting on the vertical planar gate may be determined by applying Equation 2.142 as follows:

$$I_{xca} := \frac{w \cdot L^3}{12} = 36 \, ft^4 \qquad\qquad h_F := \frac{I_{xca}}{h_{ca} \cdot A} + h_{ca} = 7.429 \, ft$$

2.4.4.3 Submerged ($h = y_o \sin \alpha$) Sloping Plane (Open)

Given that the h-axis is defined from the top of the free surface, where $p_{atm}(gage) = 0$, Figure 2.27 illustrates a sloping plane (making an angle α with the horizontal) with a rectangular area, A (width, w and length, L) submerged at a depth, $h = y_o \sin \alpha$. Because the y-axis is defined from the top of the rectangular plane, the relationship between the h-axis and the y-axis is $h = (y_o + y) \sin \alpha$. The magnitude of the resultant hydrostatic force, F is computed by integrating over the length of the plane. Using the top of the plane as the reference for integration yields the following:

$$F = \int p \, dA = \int_0^L (\gamma h(y))(w \, dy) = \int_0^L (\gamma(y_o + y) \sin \alpha)(w \, dy) \qquad (2.144)$$

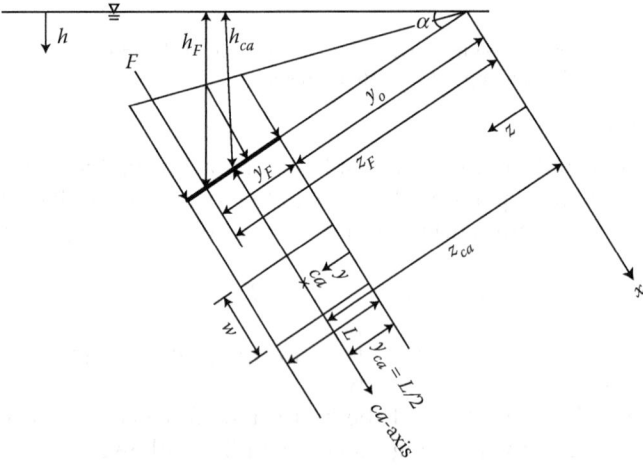

FIGURE 2.27
Resultant force, F and its location, h_F for a sloping plane with a rectangular geometric shape submerged at a depth, $h = y_o \sin \alpha$ in a liquid open to atmosphere, assuming the gage pressure scale.

where $h(y) = (y_o + y) \sin \alpha$. Using Mathcad to integrate yields:

$$\int_0^L [\gamma \cdot (y_o + y) \cdot \sin(\alpha)] w \, dy \rightarrow \frac{\gamma \cdot L \cdot w \cdot \sin(\alpha) \cdot (L + 2 \cdot y_o)}{2}$$

For the case where $y_o = 0$, which means that the top of the plane in Figure 2.27 is at the free surface, the integral that defines the force, F is defined as follows:

$$F = \int p \, dA = \int_0^L (\gamma h(y))(w \, dy) = \int_0^L (\gamma y \sin \alpha)(w \, dy) \tag{2.145}$$

where $h(y) = y \sin \alpha$. Using Mathcad to integrate yields:

$$\int_0^L (\gamma \cdot y \cdot \sin(\alpha)) w \, dy \rightarrow \frac{\gamma \cdot L^2 \cdot w \cdot \sin(\alpha)}{2}$$

where for the special case where $\alpha = 90°$, Equation 2.145 (above) simplifies to Equation 2.116, which is the submerged ($h = 0$) vertical plane. Alternatively, one may compute the magnitude of the resultant force, F as follows:

$$F = p_{ca}A = \gamma h_{ca}A = \gamma(y_o + y_{ca}) \sin \alpha(wL) = \gamma\left(y_o + \frac{L}{2}\right) \sin \alpha(wL) \tag{2.146}$$

where for a rectangular plane, $h_{ca} = (y_o + y_{ca}) \sin \alpha$, $y_{ca} = L/2$, and $A = wL$.

The location ("center of pressure") of the resultant hydrostatic force, F is located at the center of gravity of the pressure prism. Thus, for the sloping plane submerged at a depth, $h = y_o \sin \alpha$ illustrated in Figure 2.27, the pressure prism is a trapezoid. If the "center of pressure" for the resultant hydrostatic force, F is measured from the top of the plane along the y-axis, the moment, M and the moment arm, y_F are computed as follows:

$$M = y_F F = \int_0^L y \, dF = \int_0^L yp \, dA = \int_0^L y(\gamma h(y))(w \, dy) = \int_0^L y(\gamma(y_o + y) \sin \alpha)(w \, dy) \tag{2.147}$$

where $h(y) = (y_o + y) \sin \alpha$. Using Mathcad to integrate yields:

$$\int_0^L [y \cdot \gamma \cdot (y_o + y) \cdot \sin(\alpha)] w \, dy \rightarrow \frac{\gamma \cdot L^2 \cdot w \cdot \sin(\alpha) \cdot (2 \cdot L + 3 \cdot y_o)}{6}$$

Solving for the moment arm by using Equation 2.147 for the moment and Equation 2.144 for the force yields the following:

$$y_F = \frac{M}{F} \tag{2.148}$$

Using Mathcad to solve for the moment arm, y_F yields the following:

$$\frac{\dfrac{\gamma \cdot L^2 \cdot w \cdot \sin(\alpha) \cdot (2 \cdot L + 3 \cdot y_o)}{6}}{\dfrac{\gamma \cdot L \cdot w \cdot \sin(\alpha) \cdot (L + 2 \cdot y_o)}{2}} \rightarrow \frac{L \cdot (2 \cdot L + 3 \cdot y_o)}{3 \cdot (L + 2 \cdot y_o)}$$

where for the special case where $y_o = 0$, then $y_F = (2/3)L$, and for the special case where $y_o \gg L$, then $y_F = (L/2)$. Alternatively, the "center of pressure" may be measured from the x-axis along the z-axis, where $h = z \sin \alpha$, $z = (y_o + y)$, and $h = (y_o + y) \sin \alpha$, as follows:

$$z_F = \frac{M}{F} = \frac{\int_A z(\gamma h)\, dA}{\gamma h_{ca} A} = \frac{\int_A z(\gamma z \sin \alpha)\, dA}{\gamma(z_{ca} \sin \alpha) A} = \frac{\int_A (\gamma z^2 \sin \alpha)\, dA}{\gamma(z_{ca} \sin \alpha) A} = \frac{\int_A z^2\, dA}{z_{ca} A} = \frac{I_x}{z_{ca} A} \tag{2.149}$$

but:

$$I_x = I_{x-ca} + A z_{ca}^2 \tag{2.150}$$

and thus:

$$z_F = \frac{I_{x-ca} + A z_{ca}^2}{z_{ca} A} = \frac{I_{x-ca}}{z_{ca} A} + z_{ca} = \frac{wL^3/12}{(y_o + L/2)(wL)} + (y_o + L/2) \tag{2.151}$$

where for a rectangular plane, $I_{x-ca} = wL^3/12$, $z_{ca} = (y_o + y_{ca})$, $y_{ca} = L/2$, and $A = wL$. Using Mathcad to simplify the expression for the center of pressure, z_F yields:

$$\frac{\dfrac{w \cdot L^3}{12}}{\left(y_o + \dfrac{L}{2}\right) \cdot (w \cdot L)} + \left(y_o + \frac{L}{2}\right) \rightarrow \frac{L}{2} + y_o + \frac{L^2}{12 \cdot \left(\dfrac{L}{2} + y_o\right)}$$

Furthermore, the "center of pressure" for the resultant force, F, measured from the top of the plane along the y-axis, $y_F = z_F - y_o$ is given as follows:

$$y_F = z_F - y_o = \frac{L}{2} + \frac{L^2}{12\left(\dfrac{L}{2} + y_o\right)} \tag{2.152}$$

Using Mathcad to simplify the expression for y_F yields:

$$\frac{L}{2} + \frac{L^2}{12 \cdot \left(\dfrac{L}{2} + y_o\right)} \quad \text{simplify} \quad \rightarrow \frac{L \cdot (2 \cdot L + 3 \cdot y_o)}{3\,(L + 2 \cdot y_o)}$$

which is the same result yielded in Equation 2.148. Furthermore, the "center of pressure" for the resultant force, F, measured from the top of the free surface along the h-axis, h_F is computed as follows:

$$h_F = z_F \sin \alpha \tag{2.153}$$

EXAMPLE PROBLEM 2.25

A sloping planar gate with a width of 3 ft and a height of 5 ft is submerged at a depth of 6 ft in water open to the atmosphere, making an 30° angle with the horizontal, as illustrated in Figure EP 2.25. (a) Determine the magnitude of the resultant hydrostatic force acting on the sloping planar gate. (b) Determine the location of the resultant hydrostatic force acting on the sloping planar gate.

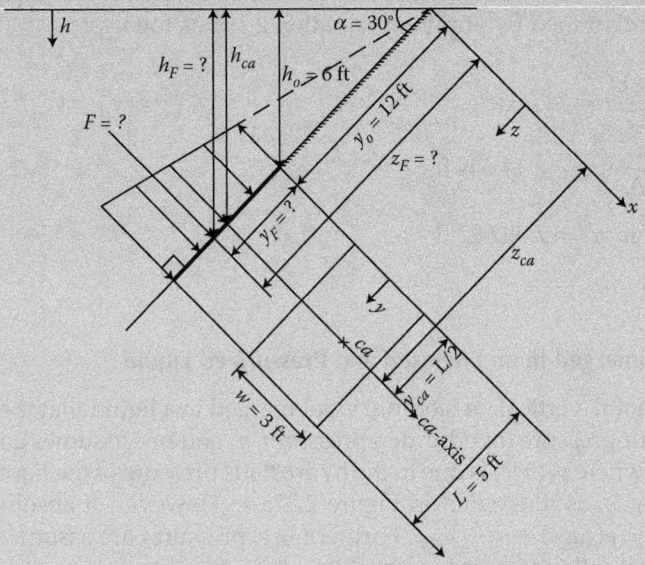

FIGURE EP 2.25

Trapezoidal pressure prism, the resultant hydrostatic force, F, and its location, h_F for a sloping plane submerged at a depth $h = y_o \sin \alpha$ in water open to the atmosphere.

Mathcad Solution

(a) The magnitude of the hydrostatic resultant force acting on the planar gate is determined by applying Equation 2.144, integrating over the length of the sloping planar gate using the top of the plane as a reference point for integration as follows:

$$w := 3\,\text{ft} \qquad L := 5\,\text{ft} \qquad A := w \cdot L = 15\,\text{ft}^2 \qquad h_o := 6\,\text{ft}$$

$$\alpha := 30\,\text{deg} \qquad y_o := \frac{h_o}{\sin(\alpha)} = 12\,\text{ft} \qquad h(y) := (y_o + y)\,\sin(\alpha)$$

$$\gamma := 62.417\,\frac{\text{lb}}{\text{ft}^3} \qquad\qquad F := \int_0^L (\gamma \cdot h(y)) \cdot w\,dy = 6787.849\,\text{lb}$$

Alternatively, the magnitude of the hydrostatic force acting on the sloping planar gate may be determined by applying Equation 2.146 as follows:

$$y_{ca} := \frac{L}{2} = 2.5\,\text{ft} \qquad\qquad h_{ca} := (y_o + y_{ca})\,\sin(\alpha) = 7.25\,\text{ft}$$

$$F := \gamma\,h_{ca}\,A = 6.788 \times 10^3\,\text{lb}$$

(b) The location of the resultant hydrostatic force acting on the sloping planar gate is determined by applying Equation 2.148 as follows:

$$y_F := \frac{\int_0^L y \cdot (\gamma \cdot h(y)) \cdot w \, dy}{\int_0^L (\gamma \cdot h(y)) \cdot w \, dy} = 2.644 \, ft \qquad h_F := (y_o + y_F) \cdot \sin(\alpha) = 7.322 \, ft$$

Alternatively, the location of the resultant force acting on the sloping planar gate may be determined by applying Equation 2.151 as follows:

$$I_{xca} := \frac{w \cdot L^3}{12} = 31.25 \, ft^4 \qquad\qquad z_{ca} := (y_o + y_{ca}) = 14.5 \, ft$$

$$z_F := \frac{I_{xca}}{z_{ca} \, A} + z_{ca} = 14.644 \, ft \qquad\qquad y_F := z_F - y_o = 2.644 \, ft$$

$$h_F := z_F \sin(\alpha) = 7.322 \, ft$$

2.4.5 Planes Submerged in an Enclosed and Pressurized Liquid

If a plane (horizontal, vertical, or sloping) is submerged in a liquid that is enclosed and subjected to a gage air pressure (overburden pressure), p_g and one assumes gage pressure, then $p = p(gage) + p_g$, where $p(gage)$ is due to the hydrostatic pressure of the liquid and p_g is due to the pressurized gas, as illustrated in Figure 2.22a–c. However, if absolute pressure were assumed, then $p = p(gage) + p_g + p_{atm}$. Furthermore, pressures are assumed to be gage pressures unless specifically indicated as absolute pressure.

2.4.5.1 Submerged (h = H) Horizontal Plane (Enclosed)

Given that the h-axis is defined from the top of the free surface, where $p = p_g$, Figure 2.28a illustrates a horizontal plane with a rectangular area, A (width, w and length, L) submerged at a depth, $h = H$ in an enclosed and pressurized liquid. For a submerged horizontal plane that is enclosed and pressurized, the magnitude of the hydrostatic force, F is equivalent to the weight of the fluid above the plane, or the gravitational force, $F_G = Mg = W = \gamma V = \gamma HA$ (Newton's second law of motion), plus the overburden pressure force, $p_g A$. Thus, the magnitude of the resultant force, F is computed as follows:

$$F = F_G + p_g A = \gamma HA + p_g A = \gamma HwL + p_g wL = (\gamma H + p_g)(wL) \qquad (2.154)$$

Alternatively, the magnitude of the resultant hydrostatic force, F may be computed as follows:

$$F = (p_{ca} + p_g)A = (\gamma h_{ca} + p_g)A = (\gamma H + p_g)(wL) \qquad (2.155)$$

The location ("center of pressure") of the resultant hydrostatic force, F is located at the center of gravity of the pressure prism, which is at the centroid of the area, A. Therefore, the "center of pressure" for the resultant hydrostatic force, F, measured from the top of the free surface along the h-axis, is located at $h_F = H$ and located at the center of area of the rectangular pressure prism at $L/2$ and $w/2$.

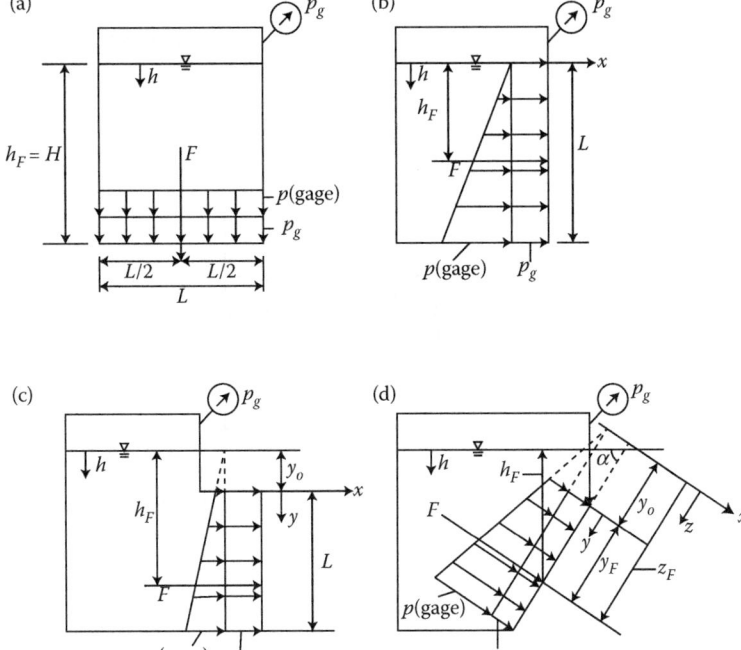

FIGURE 2.28
Resultant force, F and its location, h_F for rectangular plane submerged in a liquid enclosed in tank subjected to a gage air pressure, p_g, assuming the gage pressure scale. (a) Horizontal plane submerged at $h = H$. (b) Vertical plane submerged at $h = 0$. (c) Vertical plane submerged at $h = y_o$. (d) Sloping plane submerged at $h = y_o \sin \alpha$.

EXAMPLE PROBLEM 2.26

A horizontal plane with a width of 2 ft and a length of 5 ft is submerged in a tank of water at a depth of 7 ft and is subjected to gage air pressure of 10 psi, as illustrated in Figure EP 2.26. (a) Determine the magnitude of the resultant hydrostatic force acting on the horizontal plane. (b) Determine the location of the resultant hydrostatic force acting on the horizontal plane.

Mathcad Solution

(a) The magnitude of the hydrostatic resultant force acting on the horizontal plane is determined by applying Equation 2.154 as follows:

$$w := 2\,\text{ft} \qquad L := 5\,\text{ft} \qquad A := w \cdot L = 10\,\text{ft}^2 \qquad H := 7\,\text{ft} \qquad \gamma := 62.417\frac{\text{lb}}{\text{ft}^3}$$

$$F_G := \gamma \cdot H \cdot A = 4.369 \times 10^3\,\text{lb} \qquad p_g := 10\frac{\text{lb}}{\text{in}^2}\frac{(12\,\text{in})^2}{(1\,\text{ft})^2} = 1.44 \times 10^3\frac{\text{lb}}{\text{ft}^2}$$

$$F := F_G + p_g \cdot A = 1.877 \times 10^4\,\text{lb}$$

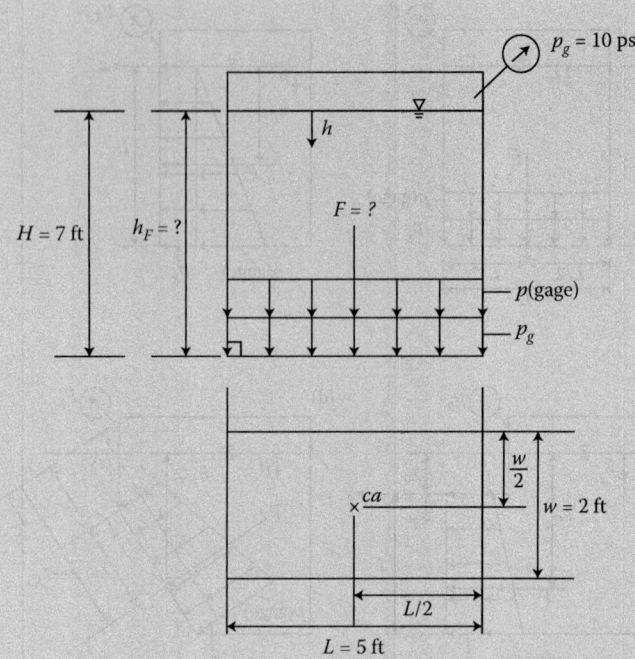

FIGURE EP 2.26
Uniform pressure prism; the resultant hydrostatic force, F; and its location, h_F for an enclosed horizontal plane submerged at a depth, $h = H$ in water subjected to a gage air pressure, p_g.

Alternatively, the magnitude of the hydrostatic resultant force acting on the horizontal plane is determined by applying Equation 2.155 as follows:

$$h_{ca} := H = 7\,\text{ft} \qquad\qquad F := (\gamma \cdot h_{ca} + p_g) \cdot A = 1.877 \times 10^4\,\text{lb}$$

(b) The location of the resultant hydrostatic force acting on the horizontal plane is located at the center of area of the horizontal rectangular plane and is computed as follows:

$$h_F := h_{ca} = 7\,\text{ft} \qquad\qquad \frac{L}{2} = 2.5\,\text{ft} \qquad\qquad \frac{w}{2} = 1\,\text{ft}$$

2.4.5.2 Submerged (h = 0) Vertical Plane (Enclosed)

Given that the h-axis is defined from the top of the free surface, where $p = p_g$, Figure 2.28b illustrates a vertical plane with a rectangular area, A (width, w and length, L) submerged at a depth, $h = 0$. The magnitude of the resultant force, F is computed by integrating over the length of the plane follows:

$$F = \int_0^L p\,dA = \int_0^L (\gamma h + p_g)(w\,dh) \qquad\qquad (2.156)$$

Using Mathcad to integrate yields:

$$\int_0^L (\gamma \cdot h + p_g) w \, dh \rightarrow \frac{L \cdot w \cdot (2 \cdot p_g + \gamma \cdot L)}{2}$$

where for $p_g = 0$, the above Equation 2.156 reduces to Equation 2.116. Alternatively, the magnitude of the resultant hydrostatic force, F may be computed as follows:

$$F = (p_{ca} + p_g)A = (\gamma h_{ca} + p_g)A = (\gamma L/2 + p_g)(wL) \tag{2.157}$$

The location ("center of pressure") of the resultant hydrostatic force, F is located at the center of gravity of the pressure prism. Thus, for the vertical plane submerged at a depth, $h = 0$ illustrated in Figure 2.28b, the pressure prism is a trapezoid and thus the "center of pressure," h_F for the resultant hydrostatic force, F, measured from the top of the free surface along the h-axis, is computed using integration as follows:

$$M = h_F F = \int_0^L h \, dF = \int_0^L hp \, dA = \int_0^L h(\gamma h + p_g)(w \, dh) \tag{2.158}$$

Using Mathcad to integrate yields:

$$\int_0^L h \cdot (\gamma \cdot h + p_g) w \, dh \rightarrow \frac{L^2 \cdot w \cdot (3 \cdot p_g + 2 \cdot \gamma \cdot L)}{6}$$

Solving for the moment arm, h_F yields the following:

$$h_F = \frac{M}{F} \tag{2.159}$$

Using Mathcad to solve for the moment arm, h_F yields the following:

$$\frac{\dfrac{L^2 \cdot w \cdot (3 \cdot p_g + 2 \cdot \gamma \cdot L)}{6}}{\dfrac{L \cdot w \cdot (2 \cdot p_g + \gamma \cdot L)}{2}} \rightarrow \frac{L \cdot (3 \cdot p_g + 2 \cdot \gamma \cdot L)}{3 \cdot (2 \cdot p_g + \gamma \cdot L)}$$

Alternatively, the center of pressure, h_F for the resultant hydrostatic force, F, measured from the top of the free surface along the h-axis, may be computed as follows:

$$h_F = \frac{M}{F} = \frac{\int_A h(\gamma h + p_g) \, dA}{(\gamma h_{ca} + p_g)A} = \frac{\int_A \gamma h^2 \, dA + \int_A hp_g \, dA}{(\gamma h_{ca} + p_g)A} = \frac{\gamma I_x + p_g h_{ca} A}{(\gamma h_{ca} + p_g)A} \tag{2.160}$$

but:

$$I_x = I_{x-ca} + Ah_{ca}^2 \tag{2.161}$$

and thus:

$$h_F = \frac{\gamma I_x + p_g h_{ca} A}{(\gamma h_{ca} + p_g)A} = \frac{\gamma(I_{x-ca} + Ah_{ca}^2) + p_g h_{ca} A}{(\gamma h_{ca} + p_g)A}$$

$$h_F = \frac{\gamma(I_{x-ca} + Ah_{ca}^2) + p_g h_{ca} A}{(\gamma h_{ca} + p_g)A} = \frac{\gamma(wL^3/12 + (wL)(L/2)^2) + p_g(L/2)(wL)}{(\gamma L/2 + p_g)wL} \tag{2.162}$$

where for a rectangular plane, $I_{x-ca} = wL^3/12$, $h_{ca} = L/2$, and $A = wL$. Using Mathcad to simplify the expression for the center of pressure, h_F yields:

$$\frac{\gamma \cdot \left[\left(w \cdot \dfrac{L^3}{12}\right) + (w \cdot L) \cdot \left(\dfrac{L}{2}\right)^2\right] + p_g \cdot \left(\dfrac{L}{2}\right) \cdot (w \cdot L)}{\left[\left(\dfrac{\gamma L}{2}\right) + p_g\right] \cdot (w \cdot L)} \quad \text{simplify} \;\rightarrow\; \frac{L \cdot (3 \cdot p_g + 2 \cdot \gamma \cdot L)}{3\,(2 \cdot p_g + \gamma L)}$$

which is the same result yielded in Equation 2.159.

EXAMPLE PROBLEM 2.27

A vertical plane with a width of 2 ft and a height of 6 ft is submerged in a tank of water and is subjected to gage air pressure of 13 psi, as illustrated in Figure EP 2.27.

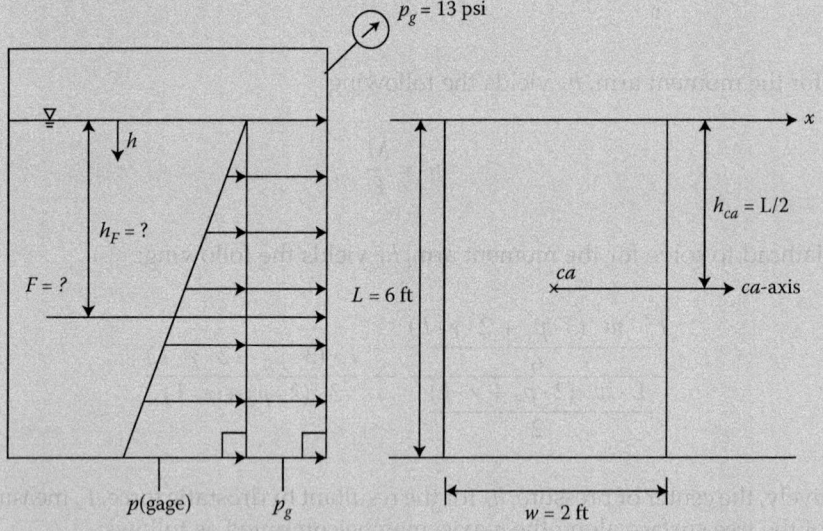

FIGURE EP 2.27

Trapezoidal pressure prism; the resultant hydrostatic force, F; and its location, h_F for an enclosed vertical plane submerged at a depth, $h = 0$ in water subjected to a gage air pressure, p_g.

(a) Determine the magnitude of the resultant hydrostatic force acting on the vertical plane. (b) Determine the location of the resultant hydrostatic force acting on the vertical plane.

Mathcad Solution

(a) The magnitude of the hydrostatic resultant force acting on the vertical plane is determined by applying Equation 2.156 as follows:

$$w := 2\,\text{ft} \qquad L := 6\,\text{ft} \qquad A := w \cdot L = 12\,\text{ft}^2 \qquad \gamma := 62.417\,\frac{\text{lb}}{\text{ft}^3}$$

$$p_g := 13\,\frac{\text{lb}}{\text{in}^2}\,\frac{(12\,\text{in})^2}{(1\,\text{ft})^2} = 1.872 \times 10^3\,\frac{\text{lb}}{\text{ft}^2}$$

$$F := \int_0^L (\gamma \cdot h + p_g) \cdot w\,dh = 2.471 \times 10^4\,\text{lb}$$

Alternatively, the magnitude of the hydrostatic resultant force acting on the vertical plane is determined by applying Equation 2.157 as follows:

$$h_{ca} := \frac{L}{2} = 3\,\text{ft} \qquad\qquad F := (\gamma \cdot h_{ca} + p_g) \cdot A = 2.471 \times 10^4\,\text{lb}$$

(b) The location of the resultant hydrostatic force acting on the vertical plane is determined by applying Equation 2.159 as follows:

$$h_F := \frac{\displaystyle\int_0^L h \cdot (\gamma \cdot h + p_g) \cdot w\,dh}{\displaystyle\int_0^L (\gamma \cdot h + p_g) \cdot w\,dh} = 3.091\,\text{ft}$$

Alternatively, the location of the resultant hydrostatic force acting on the vertical plane is determined by applying Equation 2.162 as follows:

$$I_{xca} := \frac{w \cdot L^3}{12} = 35\,\text{ft}^4 \qquad h_F := \frac{\gamma \cdot (I_{xca} + A \cdot h_{ca}^2) + p_g \cdot h_{ca} \cdot A}{(\gamma \cdot h_{ca} + p_g) \cdot A} = 3.091\,\text{ft}$$

2.4.5.3 Submerged (h = y_o) Vertical Plane (Enclosed)

Given that the h-axis is defined from the top of the free surface, where $p = p_g$, Figure 2.28c illustrates a vertical plane with a rectangular area, A (width, w and length, L) submerged at a depth, $h = y_o$. Because the y-axis is defined from the top of the rectangular plane, the relationship between the h-axis and the y-axis is $h = (y_o + y)$. The magnitude of the resultant hydrostatic force, F is computed by integrating over the length of the plane using the top of the plane as the reference as follows:

$$F = \int p\,dA = \int_0^L (\gamma h(y) + p_g)(w\,dy) = \int_0^L (\gamma(y_o + y) + p_g)(w\,dy) \qquad (2.163)$$

where $h(y) = y_o + y$. Using Mathcad to integrate yields:

$$\int_0^L [\gamma\,(y_o + y) + p_g]w\,dy \rightarrow \frac{L\,w\,(2\,p_g + \gamma\,L + 2\,\gamma\,y_o)}{2}$$

Alternatively, one may compute the magnitude of the resultant force as follows:

$$F = (p_{ca} + p_g)A = (\gamma h_{ca} + p_g)A = (\gamma(y_o + y_{ca}) + p_g)A = (\gamma(y_o + L/2) + p_g)(wL) \qquad (2.164)$$

The location ("center of pressure") of the resultant hydrostatic force, F is located at the center of gravity of the pressure prism. Thus, for the vertical plane submerged at a depth, $h = y_o$ illustrated in Figure 2.28c, the pressure prism is a trapezoid and thus the "center of pressure," y_F for the resultant hydrostatic force, F, measured from the top of the plane along the y-axis, is located at:

$$M = y_F F = \int_0^L y\,dF = \int_0^L yp\,dA = \int_0^L y(\gamma h(y) + p_g)(w\,dy) = \int_0^L y(\gamma(y_o + y) + p_g)(w\,dy) \qquad (2.165)$$

where $h(y) = y_o + y$. Using Mathcad to integrate yields:

$$\int_0^L [y \cdot [\gamma \cdot (y_o + y) + p_g]]w\,dy \rightarrow \frac{L^2 \cdot w \cdot (3 \cdot p_g + 2 \cdot \gamma \cdot L + 3 \cdot \gamma \cdot y_o)}{6}$$

where for $p_g = 0$, Equation 2.165 above becomes equal to Equation 2.138. The moment arm, y_F is computed as follows:

$$y_F = \frac{M}{F} \qquad (2.166)$$

Using Mathcad to solve for the moment arm, y_F yields the following:

$$\frac{\dfrac{L^2 \cdot w \cdot (3 \cdot p_g + 2 \cdot \gamma \cdot L + 3 \cdot \gamma \cdot y_o)}{6}}{\dfrac{L \cdot w \cdot (2 \cdot p_g + \gamma \cdot L + 2 \cdot \gamma \cdot y_o)}{2}} \rightarrow \frac{L \cdot (3 \cdot p_g + 2 \cdot \gamma \cdot L + 3 \cdot \gamma \cdot y_o)}{3 \cdot (2 \cdot p_g + \gamma \cdot L + 2 \cdot \gamma \cdot y_o)}$$

Alternatively, the "center of pressure," h_F for the resultant hydrostatic force, F, measured from the top of the free surface along the h-axis, is located at:

$$h_F = \frac{M}{F} = \frac{\int_A h(\gamma h + p_g)\,dA}{(\gamma h_{ca} + p_g)A} = \frac{\int_A \gamma h^2\,dA + \int_A h p_g\,dA}{(\gamma h_{ca} + p_g)A} = \frac{\gamma I_x + p_g h_{ca}A}{(\gamma h_{ca} + p_g)A} \qquad (2.167)$$

but:

$$I_x = I_{x-ca} + Ah_{ca}^2 \tag{2.168}$$

and thus:

$$h_F = \frac{\gamma I_x + p_g h_{ca} A}{(\gamma h_{ca} + p_g)A} = \frac{\gamma(I_{x-ca} + Ah_{ca}^2) + p_g h_{ca} A}{(\gamma h_{ca} + p_g)A}$$

$$h_F = \frac{\gamma(I_{x-ca} + Ah_{ca}^2) + p_g h_{ca} A}{(\gamma h_{ca} + p_g)A} = \frac{\gamma(wL^3/12 + (wL)(y_o + L/2)^2) + p_g(y_o + L/2)(wL)}{(\gamma(y_o + L/2) + p_g)wL} \tag{2.169}$$

where for a rectangular plane, $I_{x-ca} = wL^3/12$, $h_{ca} = y_o + y_{ca}$, $y_{ca} = L/2$, and $A = wL$. Using Mathcad to simplify the expression for the center of pressure, h_F yields:

$$\frac{\gamma \cdot \left[\left(w \cdot \frac{L^3}{12}\right) + (w \cdot L)\left(y_o + \frac{L}{2}\right)^2\right] + p_g \cdot \left(y_o + \frac{L}{2}\right) \cdot (w \cdot L)}{\left[\gamma \cdot \left(y_0 + \frac{L}{2}\right) + p_g\right] \cdot (w \cdot L)} \cdot \text{simplify}$$

$$\rightarrow \frac{L}{2} + y_o + \frac{\gamma \cdot L^2}{6\,(2 \cdot p_g + \gamma \cdot L + 2 \cdot \gamma \cdot y_o)}$$

Furthermore, the "center of pressure" for the resultant force, F, measured from the top of the plane along the y-axis, $y_F = h_F - y_o$ is given as follows:

$$y_F = h_F - y_o = \frac{L}{2} + \frac{\gamma L^2}{6(2p_g + \gamma L + 2\gamma y_o)} \tag{2.170}$$

Using Mathcad to simplify the expression for y_F yields:

$$\frac{L}{2} + \frac{\gamma \cdot L^2}{6 \cdot (2 \cdot p_g + \gamma \cdot L + 2 \cdot \gamma \cdot y_o)} \quad \text{simplify} \quad \rightarrow \quad \frac{L \cdot (3 \cdot p_g + 2 \cdot \gamma \cdot L + 3 \cdot \gamma \cdot y_o)}{3 \cdot (2 \cdot p_g + \gamma \cdot L + 2 \cdot \gamma \cdot y_o)}$$

which is the same result yielded in Equation 2.166.

EXAMPLE PROBLEM 2.28

A vertical plane with a width of 2.5 ft and a height of 3.5 ft is submerged at a depth of 4 ft in a tank of water and is subjected to a gage air pressure of 11 psi, as illustrated in Figure EP 2.28. (a) Determine the magnitude of the resultant hydrostatic force acting on the vertical plane. (b) Determine the location of the resultant hydrostatic force acting on the vertical plane.

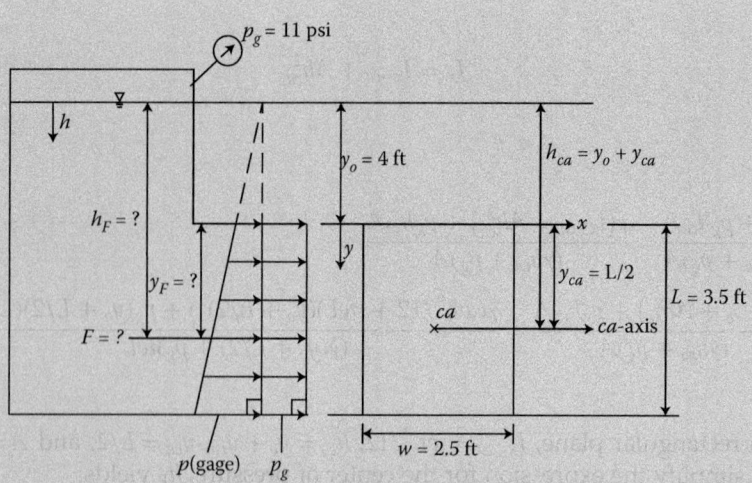

FIGURE EP 2.28
Trapezoidal pressure prism; the resultant hydrostatic force, F; and its location, h_F for an enclosed vertical plane submerged at a depth, $h = y_o$ in water subjected to a gage air pressure, p_g.

Mathcad Solution

(a) The magnitude of the hydrostatic resultant force acting on the vertical plane is determined by applying Equation 2.163, integrating over the length of the vertical plane using the top of the plane as a reference point for integration as follows:

$$w := 2.5\,\text{ft} \qquad L := 3.5\,\text{ft} \qquad A := w \cdot L = 8.75\,\text{ft}^2 \qquad y_o := 4\,\text{ft} \qquad \gamma := 62.417\frac{\text{lb}}{\text{ft}^3}$$

$$p_g := 11\frac{\text{lb}}{\text{in}^2}\frac{(12\,\text{in})^2}{(1\,\text{ft})^2} = 1.584 \times 10^3 \frac{\text{lb}}{\text{ft}^2} \qquad\qquad h(y) := y_o + y$$

$$F := \int_0^L (\gamma \cdot h(y) + p_g) \cdot w\,dy = 1.7 \times 10^4\,\text{lb}$$

Alternatively, the magnitude of the hydrostatic force acting on the vertical plane may be determined by applying Equation 2.164 as follows:

$$y_{ca} := \frac{L}{2} = 1.75\,\text{ft} \qquad\qquad\qquad\qquad h_{ca} := y_o + y_{ca} = 5.75\,\text{ft}$$

$$F := (\gamma \cdot h_{ca} + p_g) \cdot A = 1.7 \times 10^4\,\text{lb}$$

(b) The location of the resultant hydrostatic force acting on the vertical plane is determined by applying Equation 2.166 as follows:

$$y_F := \frac{\int_0^L y \cdot (\gamma \cdot h(y) + p_g) \cdot w\,dy}{\int_0^L (\gamma \cdot h(y) + p_g) \cdot w\,dy} = 1.783\,\text{ft} \qquad\qquad h_F := y_0 + y_F = 5.783\,\text{ft}$$

Alternatively, the location of the resultant force acting on the vertical plane may be determined by applying Equation 2.169 as follows:

$$I_{xca} := \frac{w \cdot L^3}{12} = 8.932\,\text{ft}^4 \qquad\qquad h_F := \frac{\gamma \cdot \left(I_{xca} + A \cdot h_{ca}^2\right) + p_g \cdot h_{ca} \cdot A}{(\gamma \cdot h_{ca} + p_g) \cdot A} = 5.783\,\text{ft}$$

2.4.5.4 Submerged ($h = y_o \sin \alpha$) Sloping Plane (Enclosed)

Given that the h-axis is defined from the top of the free surface, where $p = p_g$, Figure 2.28d illustrates a sloping plane (making an angle α with the horizontal) with a rectangular area, A (width, w and length, L) submerged at a depth, $h = y_o \sin \alpha$. Because the y-axis is defined from the top of the rectangular plane, the relationship between the h-axis and the y-axis is $h = (y_o + y) \sin \alpha$. The magnitude of the resultant hydrostatic force, F is computed by integrating over the length of the plane using the top of the plane as a reference as follows:

$$F = \int p\, dA = \int_0^L (\gamma h(y) + p_g)(w\, dy) = \int_0^L (\gamma(y_o + y)\sin \alpha + p_g)(w\, dy) \tag{2.171}$$

where $h(y) = (y_o + y)\sin \alpha$. Using Mathcad to integrate yields:

$$\int_0^L [\gamma \cdot (y_0 + y) \cdot \sin(\alpha) + p_g]w\, dy \to \frac{L \cdot w \cdot (2 \cdot p_g + 2 \cdot \gamma \cdot y_o \cdot \sin(\alpha) + \gamma \cdot L \cdot \sin(\alpha))}{2}$$

where for $p_g = 0$, Equation 2.171 above becomes equal to Equation 2.144. Alternatively, the magnitude of the resultant hydrostatic force, F may be computed as follows:

$$F = (p_{ca} + p_g)A = (\gamma h_{ca} + p_g)A = (\gamma(y_o + y_{ca})\sin \alpha + p_g)A$$
$$= (\gamma(y_o + L/2)\sin \alpha + p_g)(wL) \tag{2.172}$$

The location ("center of pressure") of the resultant hydrostatic force, F is located at the center of gravity of the pressure prism. Thus, for the sloping plane submerged at a depth, $h = y_o \sin \alpha$ illustrated in Figure 2.28d, the pressure prism is a trapezoid. If the "center of pressure" for the resultant hydrostatic force, F is measured from the top of the plane along the y-axis, the moment, M and the moment arm, y_F are computed as follows:

$$M = y_F F = \int_0^L y\, dF = \int_0^L yp\, dA = \int_0^L y(\gamma h(y) + p_g)(w\, dy) = \int_0^L y(\gamma(y_o + y)\sin \alpha + p_g)(w\, dy) \tag{2.173}$$

where $h(y) = (y_o + y)\sin \alpha$. Using Mathcad to integrate yields:

$$\int_0^L [y\, [\gamma \cdot (y_o + y) \cdot \sin(\alpha) + p_g]]w\, dy \to \frac{L^2 \cdot w \cdot (3 \cdot p_g + 3 \cdot \gamma \cdot y_o \cdot \sin(\alpha) + 2 \cdot \gamma \cdot L \cdot \sin(\alpha))}{6}$$

Solving for the moment arm by using Equation 2.173 for the moment and Equation 2.171 for the force yields the following:

$$y_F = \frac{M}{F} \tag{2.174}$$

Using Mathcad to solve for the moment arm, y_F yields the following:

$$\frac{\dfrac{L^2 \cdot w \cdot (3 \cdot p_g + 3 \cdot \gamma \cdot y_o \cdot \sin(\alpha) + 2 \cdot \gamma \cdot L \cdot \sin(\alpha))}{6}}{\dfrac{L \cdot w \cdot (2 \cdot p_g + 2 \cdot \gamma \cdot y_o \cdot \sin(\alpha) + \gamma \cdot L \cdot \sin(\alpha))}{2}} \rightarrow \frac{L \cdot (3 \cdot p_g + 3 \cdot \gamma \cdot y_o \cdot \sin(\alpha) + 2 \cdot \gamma \cdot L \cdot \sin(\alpha))}{3 \cdot (2 \cdot p_g + 2 \cdot \gamma \cdot y_o \cdot \sin(\alpha) + \gamma \cdot L \cdot \sin(\alpha))}$$

Alternatively, the "center of pressure" may be measured from the x-axis along the z-axis, where $h = z \sin \alpha$, $z = (y_o + y)$, and $h = (y_o + y) \sin \alpha$, as follows:

$$z_F = \frac{M}{F} = \frac{\int_A z(\gamma h + p_g)\,dA}{(\gamma h_{ca} + p_g)A} = \frac{\int_A z(\gamma z \sin\alpha + p_g)\,dA}{(\gamma(z_{ca}\sin\alpha) + p_g)A} = \frac{\int_A \gamma z^2 \sin\alpha\,dA + \int_A z p_g\,dA}{(\gamma(z_{ca}\sin\alpha) + p_g)A}$$
$$= \frac{\gamma I_x \sin\alpha + p_g z_{ca} A}{(\gamma(z_{ca}\sin\alpha) + p_g)A} \tag{2.175}$$

but:

$$I_x = I_{x-ca} + A z_{ca}^2 \tag{2.176}$$

and thus:

$$z_F = \frac{\gamma I_x \sin\alpha + p_g z_{ca} A}{(\gamma(z_{ca}\sin\alpha) + p_g)A} = \frac{\gamma(I_{x-ca} + A z_{ca}^2)\sin\alpha + p_g z_{ca} A}{(\gamma(z_{ca}\sin\alpha) + p_g)A}$$

$$z_F = \frac{\gamma(I_{x-ca} + A z_{ca}^2)\sin\alpha + p_g z_{ca} A}{(\gamma(z_{ca}\sin\alpha) + p_g)A} = \frac{\gamma(wL^3/12 + wL(y_o + L/2)^2)\sin\alpha + p_g(y_o + L/2)(wL)}{(\gamma(y_o + L/2)\sin\alpha + p_g)(wL)} \tag{2.177}$$

where for a rectangular plane, $I_{x-ca} = wL^3/12$, $z_{ca} = (y_o + y_{ca})$, $y_{ca} = L/2$, and $A = wL$. Using Mathcad to simplify the expression for the center of pressure, z_F yields:

$$\frac{\gamma \cdot \left[\left(w \cdot \dfrac{L^3}{12}\right) + (w \cdot L)\cdot \left(y_o + \dfrac{L}{2}\right)^2\right]\cdot \sin(\alpha) + p_g \cdot \left(y_o + \dfrac{L}{2}\right)\cdot (w \cdot L)}{\left[\gamma \cdot \left(y_o + \dfrac{L}{2}\right)\cdot \sin(\alpha) + p_g\right]\cdot (w \cdot L)} \quad \text{simplify}$$

$$\rightarrow \frac{L}{2} + y_o + \frac{\gamma \cdot L^2 \cdot \sin(\alpha)}{6 \cdot (2 \cdot p_g + 2 \cdot \gamma \cdot y_o \cdot \sin(\alpha) + \gamma \cdot L \cdot \sin(\alpha))}$$

Furthermore, the "center of pressure" for the resultant force, F, measured from the top of the plane along the y-axis, $y_F = z_F - y_o$ is given as follows:

$$y_F = z_F - y_o = \frac{L}{2} + \frac{\gamma L^2 \sin\alpha}{6(2 p_g + 2\gamma y_o \sin\alpha + \gamma L \sin\alpha)} \tag{2.178}$$

Using Mathcad to simplify the expression for y_F yields:

$$\frac{L}{2} + \frac{\gamma \cdot L^2 \cdot \sin(\alpha)}{6 \cdot (2 \cdot p_g + 2 \cdot \gamma \cdot y_o \cdot \sin(\alpha) + \gamma \cdot L \cdot \sin(\alpha))} \quad \text{factor} \rightarrow \frac{L \cdot (3 \cdot p_g + 3 \cdot \gamma \cdot y_o \cdot \sin(\alpha) + 2 \cdot \gamma \cdot L \cdot \sin(\alpha))}{3 \cdot (2 \cdot p_g + 2 \cdot \gamma \cdot y_o \cdot \sin(\alpha) + \gamma \cdot L \cdot \sin(\alpha))}$$

which is the same result yielded in Equation 2.174. Furthermore, the "center of pressure" for the resultant force, F, measured from the top of the free surface along the h-axis, h_F is computed as follows:

$$h_F = z_F \sin \alpha \tag{2.179}$$

EXAMPLE PROBLEM 2.29

A sloping plane with a width of 8 ft and a height of 10 ft is submerged at a depth of 6 ft in a tank of water and is subjected to a gage pressure of 14 psi, making an 40° angle with the horizontal, as illustrated in Figure EP 2.29. (a) Determine the magnitude of the resultant hydrostatic force acting on the sloping plane. (b) Determine the location of the resultant hydrostatic force acting on the sloping plane.

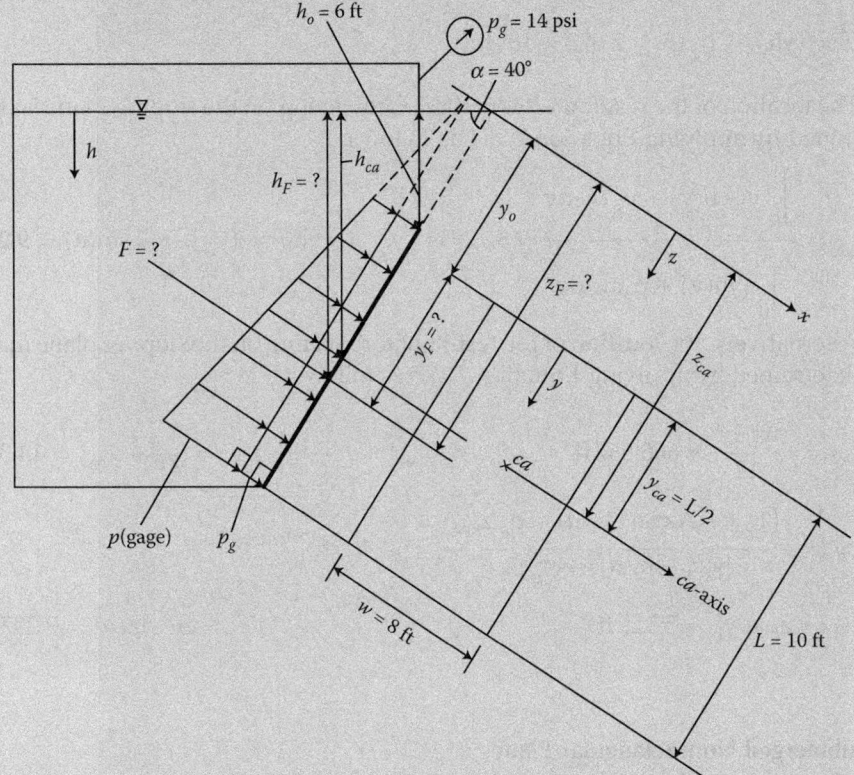

FIGURE EP 2.29
Trapezoidal pressure prism; the resultant hydrostatic force, F; and its location, h_F for an enclosed sloping plane submerged at a depth $h = y_o \sin \alpha$ in water subjected to a gage air pressure, p_g.

Mathcad Solution

(a) The magnitude of the hydrostatic resultant force acting on the sloping plane is determined by applying Equation 2.171, integrating over the length of the sloping planar gate using the top of the plane as a reference point for integration as follows:

$w := 8\,\text{ft}$ $L := 10\,\text{ft}$ $A := w\,L = 80\,\text{ft}^2$ $h_o := 6\,\text{ft}$

$\alpha := 40\,\text{deg}$ $y_o := \dfrac{h_o}{\sin(\alpha)} = 9.334\,\text{ft}$ $h(y) := (y_o + y)\,\sin(\alpha)$

$p_g := 14\dfrac{\text{lb}}{\text{in}^2}\dfrac{(12\,\text{in})^2}{(1\,\text{ft})^2} = 2.016 \times 10^3\,\dfrac{\text{lb}}{\text{ft}^2}$ $\gamma := 62.417\dfrac{\text{lb}}{\text{ft}^3}$

$F := \displaystyle\int_0^L (\gamma\,h(y) + p_g)w\,dy = 207288.51\,\text{lb}$

Alternatively, the magnitude of the hydrostatic force acting on the sloping plane may be determined by applying Equation 2.172 as follows:

$y_{ca} := \dfrac{L}{2} = 5\,\text{ft}$ $h_{ca} := (y_0 + y_{ca})\sin(\alpha) = 9.214\,\text{ft}$

$F := (\gamma h_{ca} + p_g)A = 2.073 \times 10^5\,\text{lb}$

(b) The location of the resultant hydrostatic force acting on the sloping plane is determined by applying Equation 2.174 as follows:

$y_F := \dfrac{\displaystyle\int_0^L y(\gamma\,h(y) + p_g)w\,dy}{\displaystyle\int_0^L (\gamma h(y) + p_g)w\,dy} = 5.129\,\text{ft}$ $h_F := (y_o + y_F)\sin(\alpha) = 9.297\,\text{ft}$

Alternatively, the location of the resultant force acting on the sloping plane may be determined by applying Equation 2.177 as follows:

$I_{xca} := \dfrac{w \cdot L^3}{12} = 666.667\,\text{ft}^4$ $z_{ca} := (y_0 + y_{ca}) = 14.334\,\text{ft}$

$z_F := \dfrac{\gamma\left(I_{xca} + Az_{ca}^2\right)\sin(\alpha) + p_g\,z_{ca}A}{[\gamma(z_{ca}\sin(\alpha)) + p_g]A} = 14.463\,\text{ft}$

$y_F := z_F - y_o = 5.129\,\text{ft}$ $h_F := z_F\,\sin(\alpha) = 9.297\,\text{ft}$

2.4.6 Submerged Nonrectangular Planes

A submerged horizontal, vertical, or sloping plane may take on any geometric shape such as a rectangle, a circle, a triangle, etc. While the sections above addressed the magnitude and location of the hydrostatic force on a rectangular plane, this section addresses the magnitude and location of the hydrostatic force on nonrectangular planes. Furthermore, while the general procedure to compute the magnitude and location of the hydrostatic force remains the same regardless of the geometric shape of the plane, the specific definition of the cross-sectional area, A; the height to the center of area, h_{ca}; and the center of gravity of the pressure prism vary with the geometric shape. And, finally, this section assumes that the nonrectangular plane is submerged in a liquid and open to the atmosphere. Furthermore, pressures are assumed to be gage pressures unless specifically indicated as absolute pressure.

2.4.6.1 Submerged (h = H) Horizontal Nonrectangular Plane

Given that the *h*-axis is defined from the top of the free surface, where $p_{atm}(gage) = 0$, Figure 2.29a and b illustrate two nonrectangular (circular and triangular, respectively) horizontal planes submerged at a depth, $h = H$. Computation of the magnitude of the resultant force remains the same as for a rectangular plane, except for the definition of the cross-sectional area, *A*. For a submerged horizontal plane, the magnitude of the hydrostatic force, *F* is equivalent to the weight of the fluid above the plane, or the gravitational force, $F_G = Mg = W = \gamma V = \gamma HA$ (Newton's second law of motion). Furthermore, the location ("center of pressure") of the resultant hydrostatic force, *F* is located at the center of gravity of the pressure prism.

2.4.6.2 Submerged (h = H) Horizontal Circular Plane

The magnitude of the resultant force, *F* for the for the circular plane in Figure 2.29a is computed as follows:

$$F = F_G = \gamma HA = \gamma H\left(\frac{\pi D^2}{4}\right) \tag{2.180}$$

Alternatively, the magnitude of the resultant hydrostatic force, *F* for the for the circular plane in Figure 2.29a may be computed as follows:

$$F = p_{ca}A = \gamma h_{ca}A = \gamma(H)\left(\frac{\pi D^2}{4}\right) \tag{2.181}$$

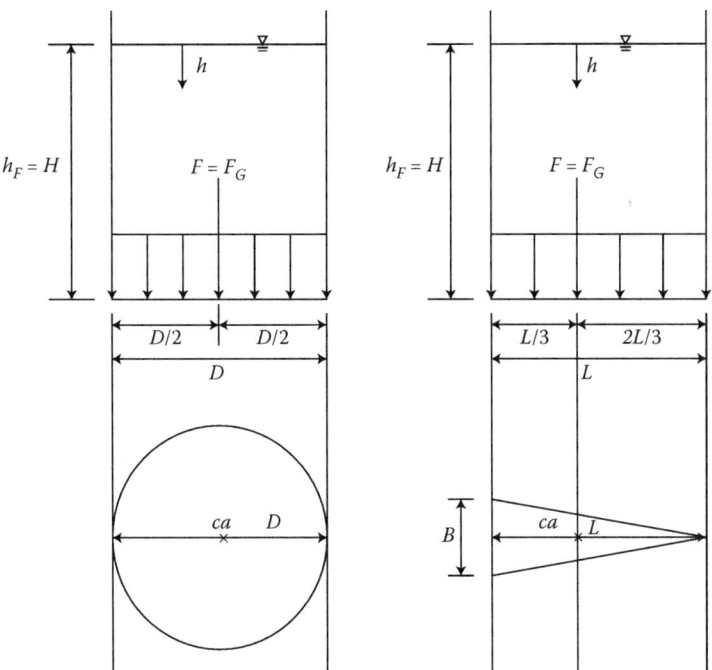

FIGURE 2.29
Resultant force, *F* and its location, h_F for a horizontal plane submerged at depth, $h = H$ in a liquid open to the atmosphere, assuming the gage pressure scale. (a) Circular plane. (b) Triangular plane.

where for the circular plane, $A = \pi D^2/4$. The location ("center of pressure") of the resultant hydrostatic force, F is located at the center of gravity of the pressure prism. For the submerged horizontal circular plane in Figure 2.29a, the center of gravity of the pressure prism is located at $h_F = H$, and located at the center of area of the circular plane at $D/2$.

EXAMPLE PROBLEM 2.30

A horizontal circular plane with a diameter of 5 ft is submerged in water open to the atmosphere at a depth of 9 ft as illustrated in Figure EP 2.30. (a) Determine the magnitude of the resultant hydrostatic force acting on the horizontal plane. (b) Determine the location of the resultant hydrostatic force acting on the horizontal plane.

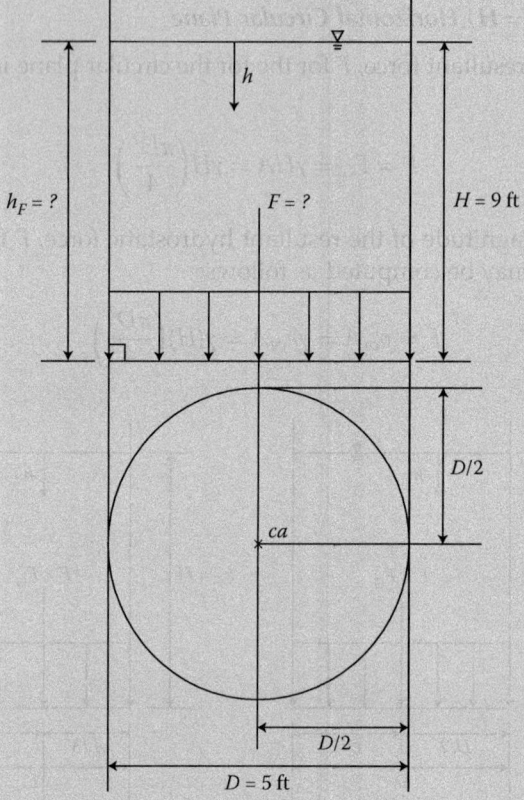

FIGURE EP 2.30
Uniform pressure prism; the resultant hydrostatic force, F; and its location, h_F for a horizontal circular plane submerged at a depth, $h = H$ in water open to the atmosphere.

Mathcad Solution

(a) The magnitude of the hydrostatic resultant force acting on the horizontal plane is determined by applying Equation 2.180 as follows:

$$D := 5 \text{ ft} \qquad\qquad A := \frac{\pi \cdot D^2}{4} = 19.635 \text{ ft}^2 \qquad H := 9 \text{ ft}$$

$$\gamma := 62.417 \frac{\text{lb}}{\text{ft}^3} \qquad\qquad F_G := \gamma H A = 1.103 \times 10^4 \text{ lb} \qquad F := F_G = 1.103 \times 10^4 \text{ lb}$$

Alternatively, the magnitude of the hydrostatic resultant force acting on the horizontal plane is determined by applying Equation 2.181 as follows:

$$h_{ca} := H = 9 \text{ ft} \qquad\qquad F := \gamma h_{ca} A = 1.103 \times 10^4 \text{ lb}$$

(b) The location of the resultant hydrostatic force acting on the horizontal plane is located at the center of area of the horizontal circular plane and is computed as follows:

$$h_F := h_{ca} = 9 \text{ ft} \qquad\qquad \frac{D}{2} = 2.5 \text{ ft}$$

2.4.6.3 Submerged (h = H) Horizontal Triangular Plane

The magnitude of the resultant force, F for the triangular plane in Figure 2.29b is computed as follows:

$$F = F_G = \gamma HA = \gamma H\left(\frac{BL}{2}\right) \tag{2.182}$$

Alternatively, the magnitude of the resultant hydrostatic force, F for the triangular plane in Figure 2.29b may be computed as follows:

$$F = p_{ca}A = \gamma h_{ca}A = \gamma(H)\left(\frac{BL}{2}\right) \tag{2.183}$$

where for the triangular plane, $A = BL/2$. The location ("center of pressure") of the resultant hydrostatic force, F is located at the center of gravity of the pressure prism. For the submerged horizontal triangular plane Figure 2.29b, the center of gravity of the pressure prism is located at $h_F = H$, and located at the center of area of the triangular plane at $L/3$ from the left, or $2L/3$ from the right.

EXAMPLE PROBLEM 2.31

A horizontal triangular plane with a base of 6 ft and a height of 8 ft is submerged in water open to the atmosphere at a depth of 11 ft as illustrated in Figure EP 2.31. (a) Determine the magnitude of the resultant hydrostatic force acting on the horizontal plane. (b) Determine the location of the resultant hydrostatic force acting on the horizontal plane.

Mathcad Solution

(a) The magnitude of the hydrostatic resultant force acting on the horizontal plane is determined by applying Equation 2.182 as follows:

$$B := 6 \text{ ft} \qquad\qquad L := 8 \text{ ft} \qquad A := \frac{BL}{2} = 24 \text{ ft}^2 \qquad H := 11 \text{ ft}$$

$$\gamma := 62.417 \frac{\text{lb}}{\text{ft}^3} \qquad F_G = \gamma HA = 1.648 \times 10^4 \text{ lb} \qquad F := F_G = 1.648 \times 10^4 \text{ lb}$$

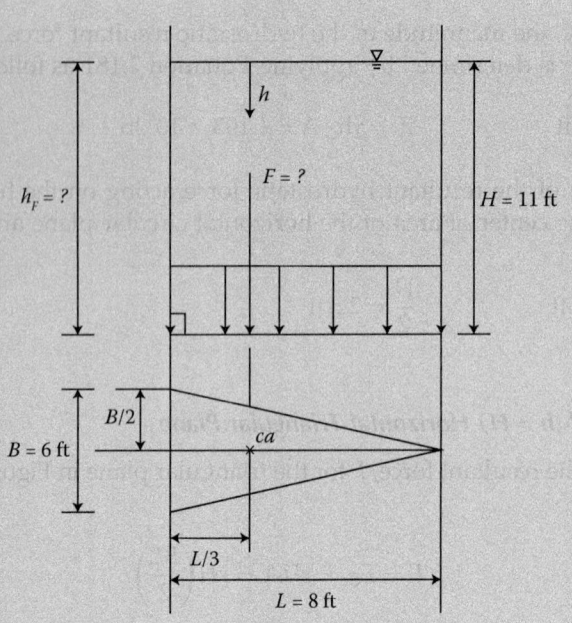

FIGURE EP 2.31
Uniform pressure prism; the resultant hydrostatic force, F; and its location, h_F for a horizontal triangular plane submerged at a depth, $h = H$ in water open to the atmosphere.

Alternatively, the magnitude of the hydrostatic resultant force acting on the horizontal plane is determined by applying Equation 2.183 as follows:

$$h_{ca} := H = 11 \text{ ft} \qquad\qquad F := \gamma h_{ca} A = 1.648 \times 10^4 \text{ lb}$$

(b) The location of the resultant hydrostatic force acting on the horizontal plane is located at the center of area of the horizontal triangular plane and is computed as follows:

$$h_F := h_{ca} = 11 \text{ ft} \qquad\qquad \frac{2L}{3} = 5.333 \text{ ft} \qquad\qquad \frac{L}{3} = 2.667 \text{ ft}$$

2.4.6.4 Submerged ($h = y_o$) Vertical Nonrectangular Plane

Given that the h-axis is defined from the top of the free surface, where $p_{atm}(gage) = 0$, Figure 2.30b and c illustrate two nonrectangular (circular and triangular, respectively) vertical planes submerged at a depth, $h = y_o$. The side view for either shape is illustrated in Figure 2.30a. Computation of the magnitude of the resultant force remains the same as for a rectangular plane, expect for the definition of the cross-sectional area, A and the height to the center of area, h_{ca}. In particular, for the nonrectangular planes, the cross-sectional area, $A = w(y)\,dy$, where $w(y)$ is specifically defined for a specific geometric shape (w is no longer a constant as in the case of a rectangular plane). The location ("center of pressure") of the resultant hydrostatic force, F is located at the center of gravity of the pressure prism. Thus, for the vertical nonrectangular planes (circular and triangular) submerged at a depth, $h = y_o$ illustrated in Figure 2.30b and c, respectively, the pressure prism is trapezoidal in side

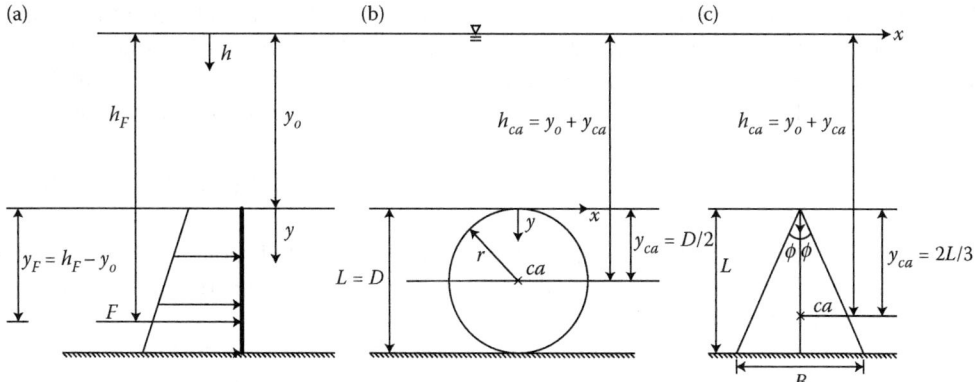

FIGURE 2.30

Resultant force, F and its location, h_F for a vertical plane submerged at depth $h = y_o$ in a liquid open to the atmosphere, assuming the gage pressure scale. (a) Side view for either plane. (b) Circular plane. (c) Triangular plane.

view only, as illustrated in Figure 2.30a. Therefore, depending upon the shape of the non-rectangular plane, the location of the center of gravity of the pressure prism will vary with the shape (circular or triangular) of the plane.

2.4.6.5 Submerged ($h = y_o$) Vertical Circular Plane

The magnitude of the resultant force, F for the circular plane in Figure 2.30b is computed by integrating over the length of the plane as follows:

$$F = \int p\,dA = \int_0^L (\gamma h(y))(w(y)\,dy) = \int_0^L (\gamma(y_o + y))(w(y)\,dy) \qquad (2.184)$$

where $h(y) = y_o + y$, and $w(y) = 2\sqrt{-y(y - 2r)}$, which is derived using Mathcad as follows:

> Given
>
> $r^2 = (x - 0)^2 + (y - r)^2$
>
> $w = 2x$

$$\text{Find } (x, w) \rightarrow \begin{bmatrix} \sqrt{-y(y - 2r)} & -\sqrt{-y(y - 2r)} \\ 2\sqrt{-y(y - 2r)} & -2\sqrt{-y(y - 2r)} \end{bmatrix}$$

where r is the radius of the circle, and for the definition of a circle, the x-axis is defined along the top of the circle, which has a center defined at ($x_c = 0$, $y_c = r$). One may note that while there are two solutions $(+, -)$ for x and w, it is the positive solution for w that we are seeking from the Mathcad solve block presented above. Therefore, the magnitude of the resultant force, F for the circular plane in Figure 2.30b is computed by using Mathcad to numerically integrate the following expression for the resultant force:

$$F = \int_0^L (\gamma(y_o + y))2\sqrt{-y(y - 2r)}\,dy \qquad (2.185)$$

It is important to note that because the above expression for the magnitude of the resultant force, F for the circular plane in Figure 2.30b cannot be analytically integrated, one must resort to a numerical integration for its solution. Alternatively, the magnitude of the resultant hydrostatic force, F for the for the circular plane in Figure 2.30b may be computed as follows:

$$F = p_{ca}A = \gamma h_{ca}A = \gamma(y_o + y_{ca})\left(\frac{\pi D^2}{4}\right) = \gamma(y_o + D/2)\left(\frac{\pi D^2}{4}\right) \tag{2.186}$$

where $y_{ca} = D/2$ and $A = \pi/D^2/4$ for the circular plane. Example Problem 2.32 below illustrates that either Equation 2.184 or Equation 2.186 for the force, F yields the same solution (numerically and not analytically).

The "center of pressure" or the moment arm, y_F for the resultant hydrostatic force, F measured from the top of the plane along the y-axis for the circular plane is computed as follows:

$$M = y_F F = \int_0^L y\,dF = \int_0^L yp\,dA = \int_0^L y(\gamma h(y))(w\,dy) = \int_0^L y(\gamma(y_o + y))(w\,dy) \tag{2.187}$$

where $h(y) = y_o + y$ and $w(y) = 2\sqrt{-y(y - 2r)}$. Therefore, the moment, M for the circular plane in Figure 2.30b is computed by using Mathcad to numerically integrate the following expression for the moment:

$$M = \int_0^L y(\gamma(y_o + y))\left(2\sqrt{-y(y - 2r)}\,dy\right) \tag{2.188}$$

It is important to note that because the above expression for the moment, M for the circular plane in Figure 2.30b cannot be analytically integrated, one must resort to a numerical integration for its solution. Solving for the moment arm by using Equation 2.188 for the moment and Equation 2.185 for the force yields the following:

$$y_F = \frac{M}{F} \tag{2.189}$$

Alternatively, the "center of pressure" or the moment arm, h_F for the resultant hydrostatic force, F, measured from the top of the free surface along the h-axis, may be computed as follows:

$$h_F = \frac{M}{F} = \frac{\int_A h(\gamma h)\,dA}{\gamma h_{ca}A} = \frac{\int_A \gamma h^2\,dA}{\gamma h_{ca}A} = \frac{\gamma I_x}{\gamma h_{ca}A} = \frac{I_x}{h_{ca}A} \tag{2.190}$$

but:

$$I_x = I_{x-ca} + Ah_{ca}^2 \tag{2.191}$$

and thus:

$$h_F = \frac{I_{x-ca}}{h_{ca}A} + h_{ca} = \frac{\pi D^4/64}{(y_o + D/2)(\pi D^2/4)} + (y_o + D/2) \tag{2.192}$$

where for a circular plane, $I_{x-ca} = \pi D^4/64$, $h_{ca} = (y_o + y_{ca})$, $y_{ca} = D/2$, and $A = \pi D^2/4$. Using Mathcad to simplify the expression for the center of pressure, h_F yields:

$$\frac{\frac{\pi \cdot D^4}{64}}{\left(y_o + \frac{D}{2}\right)\left(\frac{\pi \cdot D^2}{4}\right)} + \left(y_o + \frac{D}{2}\right) \rightarrow \frac{D}{2} + y_o + \frac{D^2}{16\left(\frac{D}{2} + y_o\right)}$$

Furthermore, the "center of pressure" for the resultant force, F measured from the top of the plane along the y-axis, $y_F = h_F - y_o$ is given as follows:

$$y_F = h_F - y_o = \frac{D}{2} + \frac{D^2}{16\left(\frac{D}{2} + y_o\right)} \tag{2.193}$$

Using Mathcad to simplify the expression for y_F yields:

$$\frac{D}{2} + \frac{D^2}{16\left(\frac{D}{2} + y_o\right)} \quad \text{simplify} \quad \rightarrow \quad \frac{D(5D + 8y_o)}{8(D + 2y_o)}$$

Example Problem 2.32 below illustrates that either Equation 2.189 or Equation 2.193 for the center of pressure, y_F yields the same solution (numerically, and not analytically).

EXAMPLE PROBLEM 2.32

A vertical circular plane gate with a diameter of 7 ft is submerged at a depth of 5 ft in water open to the atmosphere, as illustrated in Figure EP 2.32. (a) Determine the magnitude of the resultant hydrostatic force acting on the vertical circular plane gate. (b) Determine the location of the resultant hydrostatic force acting on the vertical circular plane gate.

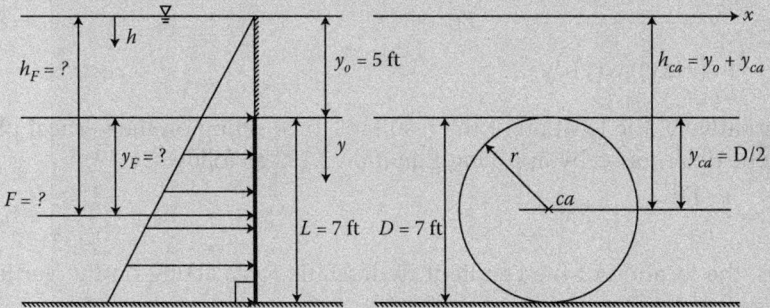

FIGURE EP 2.32
Trapezoidal pressure prism; the resultant hydrostatic force, F; and its location, h_F for a vertical circular plane submerged at a depth $h = y_o$ in water open to the atmosphere.

Mathcad Solution

(a) The magnitude of the hydrostatic resultant force acting on the plane gate is determined by applying Equation 2.184, integrating over the length of the plane gate using the top of the plane as a reference point for integration. Furthermore, because there is no analytical solution for the integral given in Equation 2.184 for the hydrostatic force, it must be numerically integrated. As such, in order to successfully numerically integrate Equation 2.184 using Mathcad, the units for all the variables have been temporarily eliminated throughout the entire example problem solution as follows:

$$D := 7 \qquad A := \frac{\pi \cdot D^2}{4} = 38.485 \qquad y_0 := 5 \qquad \gamma := 62.417$$

$$L := D = 7 \qquad r := \frac{D}{2} = 3.5 \qquad\qquad h(y) := y_0 + y \qquad w(y) := 2\sqrt{-y(y - 2r)}$$

$$F := \int_0^L \gamma h(y) w(y)\, dy \to 20417.745$$

Alternatively, the magnitude of the hydrostatic force acting on the plane gate may be determined by applying Equation 2.186 as follows:

$$y_{ca} := \frac{D}{2} = 3.5 \qquad\qquad\qquad h_{ca} := y_0 + y_{ca} = 8.5$$

$$F := \gamma h_{ca} A = 20417.745$$

Thus, the magnitude of the hydrostatic force acting on the plane gate is 20,417.745 lb.
(b) The location of the resultant hydrostatic force acting on the vertical plane gate is determined by applying Equation 2.189. Furthermore, because there is no analytical solution for the integral given in Equation 2.189 for the location of the hydrostatic force, it must be numerically integrated. As such, in order to successfully numerically integrate Equation 2.189 using Mathcad, the units for all the variables have been temporarily eliminated throughout the entire example problem solution as follows:

$$y_F := \frac{\int_0^L y(\gamma h(y))\, w(y)\, dy}{\int_0^L (\gamma\, h(y))\, w(y)\, dy} \to 3.86 \qquad\qquad h_F := y_0 + y_F = 8.86$$

Alternatively, the location of the resultant force acting on the vertical plane gate may be determined by applying Equation 2.192 as follows:

$$I_{xca} := \frac{\pi \cdot D^4}{64} = 117.859 \qquad\qquad h_F := \frac{I_{xca}}{h_{ca}\, A} + h_{ca} = 8.86$$

Thus, the location of the resultant hydrostatic force acting on the vertical plane gate is 8.86 ft below the water surface.

2.4.6.6 Submerged (h = y_o) Vertical Triangular Plane

The magnitude of the resultant force, F for the triangular plane in Figure 2.30c is computed by integrating over the length of the plane as follows:

$$F = \int p\, dA = \int_0^L (\gamma h(y))(w(y)\, dy) = \int_0^L (\gamma(y_o + y))(w(y)\, dy) \qquad (2.194)$$

where $h(y) = y_o + y$, and $w(y) = 2y \tan \phi = (yB/L)$, which is derived as follows:

$$\tan \phi = \frac{\frac{w(y)}{2}}{y} = \frac{\frac{B}{2}}{L}$$

$$\tan \phi = \frac{w(y)}{2y} = \frac{B}{2L}$$

(2.195)

where 2ϕ is the angle at the apex for the isosceles triangular plane in Figure 2.30c. Therefore, the magnitude of the resultant force, F for the triangular plane in Figure 2.30c is computed by integrating over the length of the plane as follows:

$$F = \int_0^L (\gamma(y_o + y)) \left(\left(\frac{yB}{L} \right) dy \right)$$

(2.196)

Using Mathcad to integrate yields:

$$\int_0^L [\gamma(y_o + y)] \left(\frac{yB}{L} \right) dy \rightarrow \frac{\gamma BL(2L + 3y_o)}{6}$$

Alternatively, the magnitude of the resultant hydrostatic force, F for the triangular plane in Figure 2.30c may be computed as follows:

$$F = p_{ca}A = \gamma h_{ca}A = \gamma(y_o + y_{ca}) \left(\frac{BL}{2} \right) = \gamma(y_o + 2L/3) \left(\frac{BL}{2} \right)$$

(2.197)

Using Mathcad to simplify yields the following:

$$\gamma \left(y_o + \frac{2L}{3} \right) \left(\frac{BL}{2} \right) \text{ simplify } \rightarrow \frac{\gamma BL \left(\frac{2L}{3} + y_o \right)}{2} \text{ factor } \rightarrow \frac{\gamma BL(2L + 3y_o)}{6}$$

which is the same result yielded in Equation 2.196, and where $y_{ca} = 2L/3$ and $A = BL/2$ for the triangular plane.

The "center of pressure" or the moment arm, y_F for the resultant hydrostatic force, F measured from the top of the plane along the y-axis for the triangular plane is computed as follows:

$$M = y_F F = \int_0^L y \, dF = \int_0^L yp \, dA = \int_0^L y(\gamma h(y))(w \, dy) = \int_0^L y(\gamma(y_o + y))(wdy)$$

(2.198)

where $h(y) = y_o + y$ and $w(y) = 2y \tan \phi = (yB/L)$. Therefore, the moment, M for the triangular plane in Figure 2.30c is computed as follows:

$$M = \int_0^L y(\gamma(y_o + y)) \left(\left(\frac{yB}{L} \right) dy \right)$$

(2.199)

Using Mathcad to integrate yields:

$$\int_0^L y[\gamma(y_o + y)]\left(\frac{yB}{L}\right) dy \rightarrow \frac{\gamma BL^2(3L + 4y_o)}{12}$$

Solving for the moment arm by using Equation 2.199 for the moment and Equation 2.196 for the force yields the following:

$$y_F = \frac{M}{F} \tag{2.200}$$

Using Mathcad to solve for the moment arm, y_F yields the following:

$$\frac{\dfrac{\gamma BL^2(3L + 4y_o)}{12}}{\dfrac{\gamma BL(2L + 3y_o)}{6}} \rightarrow \frac{L(3L + 4y_o)}{2(2L + 3y_o)}$$

Alternatively, the "center of pressure" or the moment arm, h_F for the resultant hydrostatic force, F, measured from the top of the free surface along the h-axis, may be computed as follows:

$$h_F = \frac{M}{F} = \frac{\displaystyle\int_A h(\gamma h)\, dA}{\gamma h_{ca}A} = \frac{\displaystyle\int_A \gamma h^2\, dA}{\gamma h_{ca}A} = \frac{\gamma I_x}{\gamma h_{ca}A} = \frac{I_x}{h_{ca}A} \tag{2.201}$$

but:

$$I_x = I_{x-ca} + Ah_{ca}^2 \tag{2.202}$$

and thus:

$$h_F = \frac{I_{x-ca}}{h_{ca}A} + h_{ca} = \frac{BL^3/36}{(y_o + 2L/3)(BL/2)} + (y_o + 2L/3) \tag{2.203}$$

where for a triangular plane, $I_{x-ca} = BL^3/36$, $h_{ca} = (y_o + y_{ca})$, $y_{ca} = 2L/3$, and $A = BL/2$.
   Using Mathcad to simplify the expression for the center of pressure, h_F yields:

$$\frac{\dfrac{BL^3}{36}}{\left(y_o + \dfrac{2L}{3}\right)\left(\dfrac{BL}{2}\right)} + \left(y_o + \frac{2L}{3}\right) \rightarrow \frac{2L}{3} + y_o + \frac{L^2}{18\left(\dfrac{2L}{3} + y_o\right)}$$

Furthermore, the "center of pressure" for the resultant force, F, measured from the top of the plane along the y-axis, $y_F = h_F - y_o$ is given as follows:

$$y_F = h_F - y_o = \frac{2L}{3} + \frac{L^2}{18\left(\dfrac{2L}{3} + y_o\right)} \tag{2.204}$$

Using Mathcad to simplify the expression for y_F yields:

$$\frac{2L}{3} + \frac{L^2}{18\left(\frac{2L}{3} + y_o\right)} \quad \text{simplify} \quad \rightarrow \quad \frac{L(3L + 4y_o)}{2(2L + 3y_o)}$$

which is the same result yielded in Equation 2.200.

EXAMPLE PROBLEM 2.33

A vertical triangular plane gate with a base of 6 ft and a height of 9 ft is submerged at a depth of 7 ft in water open to the atmosphere, as illustrated in Figure EP 2.33. (a) Determine the magnitude of the resultant hydrostatic force acting on the vertical triangular plane gate. (b) Determine the location of the resultant hydrostatic force acting on the vertical triangular plane gate.

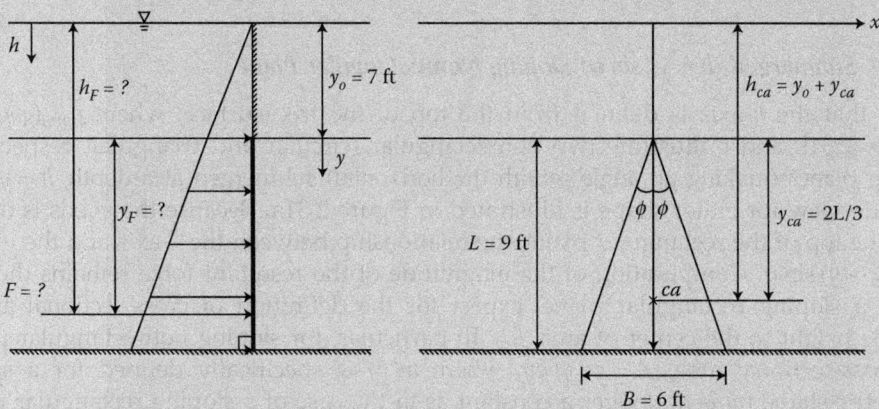

FIGURE EP 2.33
Trapezoidal pressure prism; the resultant hydrostatic force, F; and its location, h_F for a vertical triangular plane submerged at a depth, $h = y_o$ in water open to the atmosphere.

Mathcad Solution

(a) The magnitude of the hydrostatic resultant force acting on the plane gate is determined by applying Equation 2.194, integrating over the length of the plane gate using the top of the plane as a reference point for integration as follows:

$$B := 6 \text{ ft} \qquad L := 9 \text{ ft} \qquad A := \frac{BL}{2} = 27 \text{ ft}^2 \qquad y_o := 7 \text{ ft}$$

$$\phi := a\tan\left(\frac{\frac{B}{2}}{L}\right) = 18.435 \text{ deg} \qquad \gamma := 62.417 \frac{\text{lb}}{\text{ft}^3} \qquad h(y) := y_o + y$$

$$w(y) := 2y \tan(\phi) \qquad F := \int_0^L \gamma h(y) w(y)\, dy = 21908.367 \text{ lb}$$

Alternatively, the magnitude of the hydrostatic force acting on the plane gate may be determined by applying Equation 2.197 as follows:

$$y_{ca} := \frac{2L}{3} = 6 \text{ ft} \qquad\qquad h_{ca} := y_o + y_{ca} = 13 \text{ ft}$$

$$F := \gamma h_{ca} A = 21908.367 \text{ lb}$$

(b) The location of the resultant hydrostatic force acting on the vertical plane gate is determined by applying Equation 2.200 as follows:

$$y_F := \frac{\int_0^L y(\gamma h(y))w(y)\,dy}{\int_0^L (\gamma h(y))w(y)\,dy} = 6.346\,\text{ft} \qquad\qquad h_F := y_0 + y_F = 13.346\,\text{ft}$$

Alternatively, the location of the resultant force acting on the vertical plane gate may be determined by applying Equation 2.203 as follows:

$$I_{xca} := \frac{BL^3}{36} = 121.5\,\text{ft}^4 \qquad\qquad h_F := \frac{I_{xca}}{h_{ca}A} + h_{ca} = 13.346\,\text{ft}$$

2.4.6.7 Submerged ($h = y_o \sin \alpha$) Sloping Nonrectangular Plane

Given that the h-axis is defined from the top of the free surface, where $p_{atm}(gage) = 0$, Figures 2.31b and c illustrate two nonrectangular (circular and triangular, respectively) sloping planes (making an angle α with the horizontal) submerged at a depth, $h = y_o \sin \alpha$. The side view for either shape is illustrated in Figure 2.31a. Because the y-axis is defined from the top of the rectangular plane, the relationship between the h-axis and the y-axis is $h = (y_o + y)\sin \alpha$. Computation of the magnitude of the resultant force remains the same as for a sloping rectangular plane, expect for the definition of cross-sectional area, A and the height to the center of area, h_{ca}. In particular, for sloping nonrectangular planes, the cross-sectional area, $A = w(y)\,dy$, where $w(y)$ is specifically defined for a specific geometric shape (w is no longer a constant as in the case of a sloping rectangular plane). The location ("center of pressure") of the resultant hydrostatic force, F is located at the center of gravity of the pressure prism. Thus, for sloping nonrectangular planes (circular and triangular) submerged at a depth, $h = y_o \sin \alpha$ illustrated in Figures 2.31b and c, respectively, the pressure prism is a trapezoidal in side view only, as illustrated in Figure 2.31a. Therefore, depending upon the shape of the nonrectangular plane, the location of the center of gravity of the pressure prism will vary with the shape (circular or triangular) of the plane.

2.4.6.8 Submerged ($h = y_o \sin \alpha$) Sloping Circular Plane

The magnitude of the resultant force, F for the sloping circular plane in Figure 2.31b is computed by integrating over the length of the plane as follows:

$$F = \int p\,dA = \int_0^L (\gamma h(y))(w(y)\,dy) = \int_0^L (\gamma(y_o + y)\sin \alpha)(w(y)\,dy) \tag{2.205}$$

where $h(y) = (y_o + y)\sin \alpha$, and $w(y) = 2\sqrt{-y(y - 2r)}$, which was derived above using Mathcad (see Section 2.4.6.5), and where r is the radius of the circle, and for the definition of a circle, the x-axis is defined along the top of the circle, which has a center defined at ($x_c = 0$, $y_c = r$). Therefore, the magnitude of the resultant force, F for the sloping circular plane

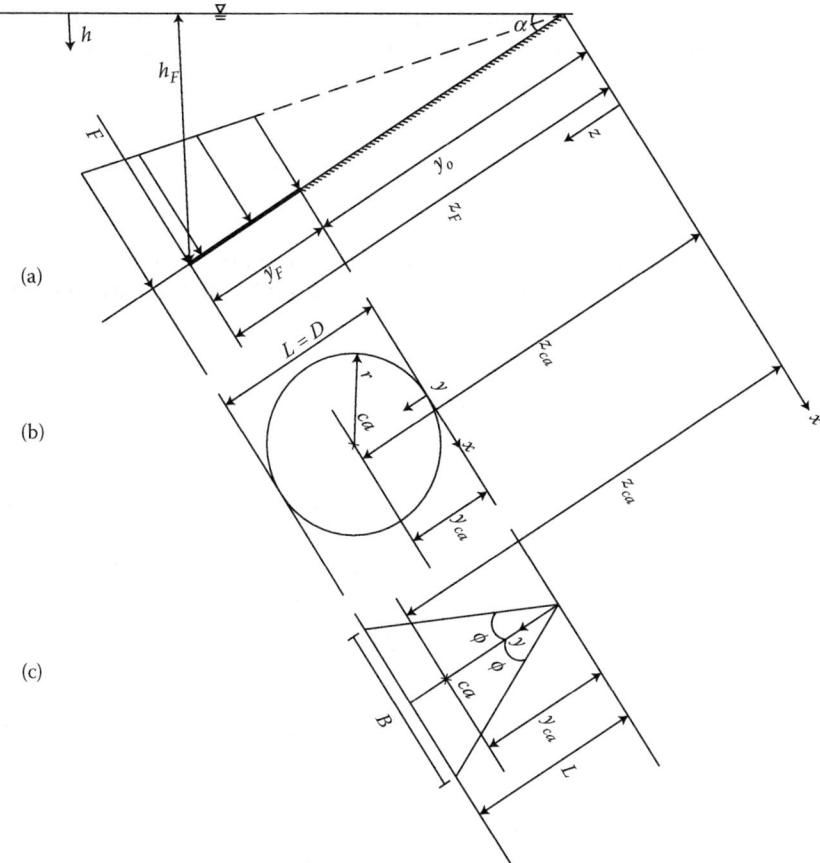

FIGURE 2.31
Resultant force, F and its location, h_F for a sloping plane submerged at depth, $h = y_o \sin \alpha$ in a liquid open to the atmosphere, assuming the gage pressure scale. (a) Side view for either plane. (b) Circular plane. (c) Triangular plane.

in Figure 2.31b is computed by using Mathcad to numerically integrate the following expression for the resultant force:

$$F = \int_0^L (\gamma(y_o + y) \sin \alpha) 2\sqrt{-y(y - 2r)} \, dy \qquad (2.206)$$

It is important to note that because the above expression for the magnitude of the resultant force, F for the sloping circular plane in Figure 2.31b cannot be analytically integrated, one must resort to a numerical integration for its solution. Alternatively, the magnitude of the resultant hydrostatic force, F for the for the sloping circular plane in Figure 2.31b may be computed as follows:

$$F = p_{ca}A = \gamma h_{ca}A = \gamma(y_o + y_{ca}) \sin \alpha \left(\frac{\pi D^2}{4}\right) = \gamma(y_o + D/2) \sin \alpha \left(\frac{\pi D^2}{4}\right) \qquad (2.207)$$

where $y_{ca} = D/2$ and $A = \pi/D^2/4$ for the circular plane. Example Problem 2.34 illustrates that either Equation 2.205 or Equation 2.207 for the force, F yields the same solution (numerically, and not analytically).

The "center of pressure" or the moment arm, y_F for the resultant hydrostatic force, F measured from the top of the plane along the y-axis for the sloping circular plane is computed as follows:

$$M = y_F F = \int_0^L y\, dF = \int_0^L yp\, dA = \int_0^L y(\gamma h(y))(w\, dy) = \int_0^L y(\gamma(y_o + y)\sin\alpha)(w\, dy) \tag{2.208}$$

where $h(y) = (y_o + y)\sin\alpha$ and $w(y) = 2\sqrt{-y(y - 2r)}$. Therefore, the moment, M for the circular plane in Figure 2.31b is computed by using Mathcad to numerically integrate the following expression for the moment:

$$M = \int_0^L y(\gamma(y_o + y)\sin\alpha)\left(2\sqrt{-y(y - 2r)}\, dy\right) \tag{2.209}$$

It is important to note that because the above expression for the moment, M for the sloping circular plane in Figure 2.31b cannot be analytically integrated, one must resort to a numerical integration for its solution. Solving for the moment arm by using Equation 2.209 for the moment and Equation 2.206 for the force yields the following:

$$y_F = \frac{M}{F} \tag{2.210}$$

Alternatively, the "center of pressure" may be measured from the x-axis along the z-axis, where $h = z\sin\alpha$, $z = (y_o + y)$, and $h = (y_o + y)\sin\alpha$, as follows:

$$z_F = \frac{M}{F} = \frac{\int_A z(\gamma h)\, dA}{\gamma h_{ca} A} = \frac{\int_A z(\gamma z\sin\alpha)\, dA}{\gamma(z_{ca}\sin\alpha)A} = \frac{\int_A (\gamma z^2\sin\alpha)\, dA}{\gamma(z_{ca}\sin\alpha)A} = \frac{\int_A z^2\, dA}{z_{ca}A} = \frac{I_x}{z_{ca}A} \tag{2.211}$$

but:

$$I_x = I_{x-ca} + Az_{ca}^2 \tag{2.212}$$

and thus:

$$z_F = \frac{I_{x-ca} + Az_{ca}^2}{z_{ca}A} = \frac{I_{x-ca}}{z_{ca}A} + z_{ca} = \frac{\pi D^4/64}{(y_o + D/2)(\pi D^2/4)} + (y_o + D/2) \tag{2.213}$$

where for a circular plane, $I_{x-ca} = \pi D^4/64$, $z_{ca} = (y_o + y_{ca})$, $y_{ca} = D/2$, and $A = \pi D^2/4$. Using Mathcad to simplify the expression for the center of pressure, z_F yields:

$$\frac{\dfrac{\pi \cdot D^4}{64}}{\left(y_o + \dfrac{D}{2}\right)\left(\dfrac{\pi \cdot D^2}{4}\right)} + \left(y_o + \dfrac{D}{2}\right) \rightarrow \frac{D}{2} + y_o + \frac{D^2}{16\left(\dfrac{D}{2} + y_o\right)}$$

Furthermore, the "center of pressure" for the resultant force, F, measured from the top of the plane along the y-axis, $y_F = z_F - y_o$ is given as follows:

$$y_F = z_F - y_o = \frac{D}{2} + \frac{D^2}{16\left(\dfrac{D}{2} + y_o\right)} \tag{2.214}$$

Using Mathcad to simplify the expression for y_F yields:

$$\frac{D}{2} + \frac{D^2}{16\left(\frac{D}{2} + y_o\right)} \quad \text{simplify} \; \rightarrow \; \frac{D(5D + 8y_o)}{8(D + 2y_o)}$$

Example Problem 2.34 illustrates that either Equation 2.210 or Equation 2.214 for the center of pressure, y_F yields the same solution (numerically, and not analytically).

EXAMPLE PROBLEM 2.34

A sloping circular gate with a diameter of 3 ft is submerged at a depth of 7 ft in water open to the atmosphere, making an 20° angle with the horizontal, as illustrated in Figure EP 2.34. (a) Determine the magnitude of the resultant hydrostatic force acting on the sloping circular gate. (b) Determine the location of the resultant hydrostatic force acting on the sloping circular gate.

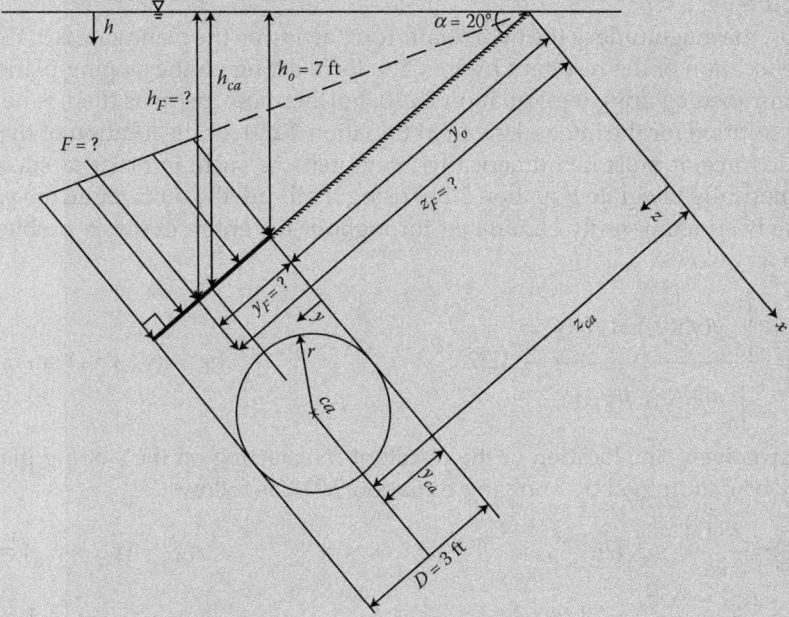

FIGURE EP 2.34
Trapezoidal pressure prism; the resultant hydrostatic force, F; and its location, h_F for a sloping circular plane submerged at a depth $h = y_o \sin \alpha$ in water open to the atmosphere.

Mathcad Solution

(a) The magnitude of the hydrostatic resultant force acting on the plane gate is determined by applying Equation 2.205, integrating over the length of the plane gate using the top of the plane as a reference point for integration. Furthermore, because there is no analytical solution for the integral given in Equation 2.205 for the hydrostatic force, it must be numerically integrated. As such, in order to successfully numerically integrate Equation 2.205 using Mathcad, the units for all the variables have been temporarily eliminated throughout the entire example problem solution as follows:

$$D := 3 \qquad A := \frac{\pi \cdot D^2}{4} = 7.069 \qquad L := D = 3 \qquad r := \frac{D}{2} = 1.5 \qquad h_o := 7$$

$$\gamma := 62.417 \qquad \alpha := 20 \text{ deg} \qquad \sin(\alpha) = 0.342 \qquad \sin \alpha := \sin(\alpha) = 0.342$$

$$y_o := \frac{h_o}{\sin \alpha} = 20.467 \qquad h(y) := (y_o + y)\sin \alpha \qquad w(y) := 2\sqrt{-y(y - 2r)}$$

$$F := \int_0^L \gamma h(y) w(y)\, dy \rightarrow 3314.747$$

Alternatively, the magnitude of the hydrostatic force acting on the plane gate may be determined by applying Equation 2.207 as follows:

$$y_{ca} := \frac{L}{2} = 1.5 \qquad\qquad\qquad h_{ca} := (y_o + y_{ca})\sin \alpha = 7.513$$

$$F := \gamma h_{ca} A = 3314.747$$

Thus, the magnitude of the hydrostatic force acting on the plane gate is 3,314.747 lb.

(b) The location of the resultant hydrostatic force acting on the sloping plane gate is determined by applying Equation 2.210. Furthermore, because there is no analytical solution for the integral given in Equation 2.210 for the location of the hydrostatic force, it must be numerically integrated. As such, in order to successfully numerically integrate Equation 2.210 using Mathcad, the units for all the variables have been temporarily eliminated throughout the entire example problem solution as follows:

$$y_F := \frac{\displaystyle\int_0^L y(\gamma h(y)) w(y)\, dy}{\displaystyle\int_0^L (\gamma h(y)) w(y)\, dy} \rightarrow 1.526 \qquad\qquad h_F := (y_o + y_F)\sin \alpha = 7.522$$

Alternatively, the location of the resultant force acting on the sloping plane gate may be determined by applying Equation 2.213 as follows:

$$I_{xca} := \frac{\pi \cdot D^4}{64} = 3.976 \qquad\qquad\qquad z_{ca} := (y_o + y_{ca}) = 21.967$$

$$z_F := \frac{I_{xca} + A z_{ca}^2}{z_{ca} A} = 21.992 \qquad\qquad\qquad y_F := z_F - y_o = 1.526$$

$$h_F := z_F \sin \alpha = 7.522$$

Thus, the location of the resultant hydrostatic force acting on the vertical plane gate is 7.522 ft below the water surface.

2.4.6.9 Submerged ($h = y_o \sin \alpha$) Sloping Triangular Plane

The magnitude of the resultant force, F for the sloping triangular plane in Figure 2.31c is computed by integrating over the length of the plane as follows:

$$F = \int p\, dA = \int_0^L (\gamma h(y))(w(y)\, dy) = \int_0^L (\gamma(y_o + y)\sin \alpha)(w(y)\, dy) \qquad (2.215)$$

where $h(y) = (y_o + y) \sin \alpha$, and $w(y) = 2y \tan \phi = yB/L$, which was derived above (see Equation 2.195), and where 2ϕ is the angle at the apex for the isosceles triangular plane in Figure 2.31c. Therefore, the magnitude of the resultant force, F for the sloping triangular plane in Figure 2.31c is computed by integrating over the length of the plane as follows:

$$F = \int_0^L (\gamma(y_o + y) \sin \alpha) \left(\left(\frac{yB}{L} \right) dy \right) \tag{2.216}$$

Using Mathcad to integrate yields:

$$\int_0^L [\gamma(y_o + y) \sin(\alpha)] \left(\frac{yB}{L} \right) dy \rightarrow \frac{\gamma BL \sin(\alpha)(2L + 3y_o)}{6}$$

Alternatively, the magnitude of the resultant hydrostatic force, F for the triangular plane in Figure 2.31c may be computed as follows:

$$F = p_{ca}A = \gamma h_{ca}A = \gamma(y_o + y_{ca}) \sin \alpha \left(\frac{BL}{2} \right) = \gamma(y_o + 2L/3) \sin \alpha \left(\frac{BL}{2} \right) \tag{2.217}$$

Using Mathcad to simplify yields the following:

$$\gamma \left(y_o + \frac{2L}{3} \right) \sin(\alpha) \left(\frac{BL}{2} \right) \text{ simplify } \rightarrow \frac{\gamma BL \sin(\alpha) \left(\frac{2L}{3} + y_o \right)}{2} \text{ factor } \rightarrow \frac{\gamma BL \sin(\alpha)(2L + 3y_o)}{6}$$

which is the same result yielded in Equation 2.216, and where $y_{ca} = 2L/3$ and $A = BL/2$ for the triangular plane.

The "center of pressure" or the moment arm, y_F for the resultant hydrostatic force, F measured from the top of the plane along the y-axis for the triangular plane is computed as follows:

$$M = y_F F = \int_0^L y \, dF = \int_0^L yp \, dA = \int_0^L y(\gamma h(y))(w \, dy) = \int_0^L y(\gamma(y_o + y) \sin \alpha)(w \, dy) \tag{2.218}$$

where $h(y) = (y_o + y) \sin \alpha$ and $w(y) = 2y \tan \phi = yB/L$. Therefore, the moment, M for the triangular plane in Figure 2.31c is computed as follows:

$$M = \int_0^L y(\gamma(y_o + y) \sin \alpha) \left(\left(\frac{yB}{L} \right) dy \right) \tag{2.219}$$

Using Mathcad to integrate yields:

$$\int_0^L y[\gamma(y_o + y)\sin(\alpha)]\left(\frac{yB}{L}\right) dy \rightarrow \frac{\gamma BL^2 \sin(\alpha)(3L + 4y_o)}{12}$$

Solving for the moment arm by using Equation 2.19 for the moment and Equation 2.16 for the force yields the following:

$$y_F = \frac{M}{F} \tag{2.220}$$

Using Mathcad to solve for the moment arm, y_F yields the following:

$$\frac{\dfrac{\gamma BL^2 \sin(\alpha)(3L + 4y_o)}{12}}{\dfrac{\gamma BL \sin(\alpha)(2L + 3y_o)}{6}} \rightarrow \frac{L(3L + 4y_o)}{2(2L + 3y_o)}$$

Alternatively, the "center of pressure" may be measured from the x-axis along the z-axis, where $h = z\sin\alpha$, $z = (y_o + y)$, and $h = (y_o + y)\sin\alpha$, as follows:

$$z_F = \frac{M}{F} = \frac{\int_A z(\gamma h)\, dA}{\gamma h_{ca} A} = \frac{\int_A z(\gamma z \sin\alpha)\, dA}{\gamma(z_{ca} \sin\alpha)A} = \frac{\int_A (\gamma z^2 \sin\alpha)\, dA}{\gamma(z_{ca}\sin\alpha)A} = \frac{\int_A z^2\, dA}{z_{ca}A} = \frac{I_x}{z_{ca}A} \tag{2.221}$$

but:

$$I_x = I_{x-ca} + Az_{ca}^2 \tag{2.222}$$

and thus:

$$z_F = \frac{I_{x-ca} + Az_{ca}^2}{z_{ca}A} = \frac{I_{x-ca}}{z_{ca}A} + z_{ca} = \frac{BL^3/36}{(y_o + 2L/3)(BL/2)} + (y_o + 2L/3) \tag{2.223}$$

where for a triangular plane, $I_{x-ca} = BL^3/36$, $z_{ca} = (y_o + y_{ca})$, $y_{ca} = 2L/3$, and $A = BL/2$.

Using Mathcad to simplify the expression for the center of pressure, h_F yields:

$$\frac{\dfrac{BL^3}{36}}{\left(y_o + \dfrac{2L}{3}\right)\left(\dfrac{BL}{2}\right)} + \left(y_o + \frac{2L}{3}\right) \rightarrow \frac{2L}{3} + y_o + \frac{L^2}{18\left(\dfrac{2L}{3} + y_o\right)}$$

Furthermore, the "center of pressure" for the resultant force, F, measured from the top of the plane along the y-axis, $y_F = z_F - y_o$ is given as follows:

$$y_F = z_F - y_o = \frac{2L}{3} + \frac{L^2}{18\left(\dfrac{2L}{3} + y_o\right)} \tag{2.224}$$

Using Mathcad to simplify the expression for y_F yields:

$$\frac{2L}{3} + \frac{L^2}{18\left(\dfrac{2L}{3} + y_o\right)} \quad \text{simplify} \; \rightarrow \; \frac{L(3L + 4y_o)}{2(2L + 3y_o)}$$

which is the same result yielded in Equation 2.220.

EXAMPLE PROBLEM 2.35

A sloping triangular gate with a base of 5 ft and a height of 8 ft is submerged at a depth of 6 ft in water open to the atmosphere, making an 30° angle with the horizontal, as illustrated in Figure EP 2.35. (a) Determine the magnitude of the resultant hydrostatic force acting on the sloping triangular gate. (b) Determine the location of the resultant hydrostatic force acting on the sloping triangular gate.

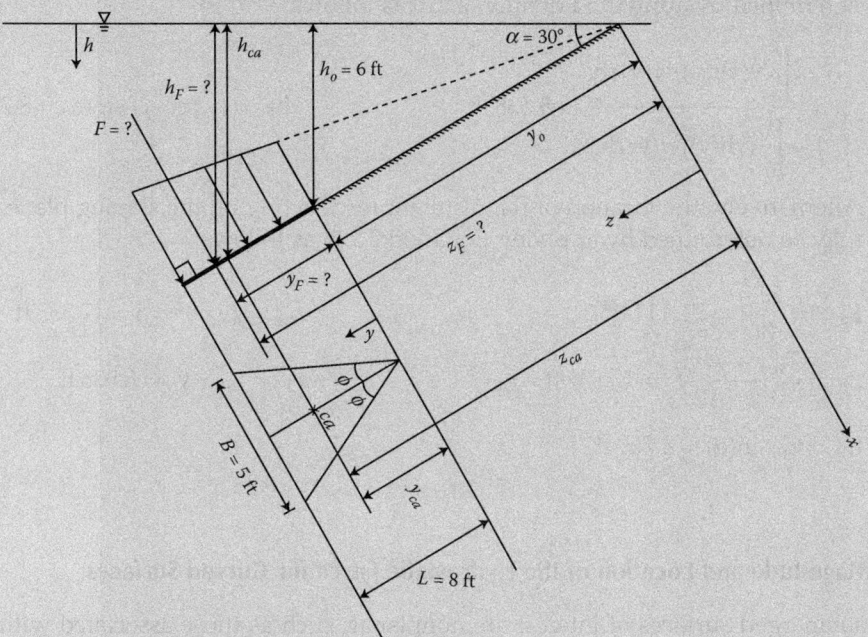

FIGURE EP 2.35
Trapezoidal pressure prism; the resultant hydrostatic force, F; and its location, h_F for a sloping triangular plane submerged at a depth $h = y_o \sin \alpha$ in water open to the atmosphere.

Mathcad Solution

(a) The magnitude of the hydrostatic resultant force acting on the plane gate is determined by applying Equation 2.215, integrating over the length of the plane gate using the top of the plane as a reference point for integration as follows:

$$B := 5 \, \text{ft} \qquad\qquad L := 8 \, \text{ft} \qquad A := \frac{BL}{2} = 20 \, \text{ft}^2 \qquad h_o := 6 \, \text{ft}$$

$$\phi := a\tan\left(\frac{\dfrac{B}{2}}{L}\right) = 17.354 \, \text{deg} \qquad \alpha := 30 \, \text{deg} \qquad y_o := \frac{h_o}{\sin(\alpha)} = 12 \, \text{ft}$$

$$\gamma := 62.417 \, \frac{\text{lb}}{\text{ft}^3} \qquad\qquad h(y) := (y_o + y) \sin(\alpha)$$

$$w(y) := 2y \tan(\phi) \qquad\qquad F := \int_0^L \gamma h(y) w(y) \, dy = 10818.947 \, \text{lb}$$

Alternatively, the magnitude of the hydrostatic force acting on the plane gate may be determined by applying Equation 2.217 as follows:

$$y_{ca} := \frac{2L}{3} = 5.333 \, \text{ft} \qquad\qquad h_{ca} := (y_o + y_{ca}) \sin(\alpha) = 8.667 \, \text{ft}$$

$$F := \gamma h_{ca} A = 10818.947 \, \text{lb}$$

(b) The location of the resultant hydrostatic force acting on the sloping plane gate is determined by applying Equation 2.220 as follows:

$$y_F := \frac{\displaystyle\int_0^L y(\gamma h(y)) w(y) \, dy}{\displaystyle\int_0^L (\gamma h(y)) w(y) \, dy} = 5.538 \, \text{ft} \qquad\qquad h_F := (y_o + y_F) \sin(\alpha) = 8.769 \, \text{ft}$$

Alternatively, the location of the resultant force acting on the sloping plane gate may be determined by applying Equation 2.223 as follows:

$$I_{xca} := \frac{BL^3}{36} = 71.111 \, \text{ft}^4 \qquad\qquad z_{ca} := (y_o + y_{ca}) = 17.333 \, \text{ft}$$

$$z_F := \frac{I_{xca} + A z_{ca}^2}{z_{ca} A} = 17.538 \, \text{ft} \qquad\qquad y_F := z_F - y_o = 5.538 \, \text{ft}$$

$$h_F := z_F \sin(\alpha) = 8.769 \, \text{ft}$$

2.4.7 Magnitude and Location of the Hydrostatic Force for Curved Surfaces

Many submerged surfaces of interest are nonplanar, such as those associated with pipes, storage tanks, ships, and dams. The distributed hydrostatic force of the pressure prism can be replaced by a point force in the analysis of static pressure forces. Furthermore, determination of both the magnitude and location of the point force that is equivalent to the distributed force is required. While the shape of the pressure prism and thus the computation of the magnitude and the location of the resultant force will vary as a function of the type of fluid, the type of surface (in this case, curved), an open or an enclosed fluid, and the assumed pressure scale, this section assumes that the curved surface is submerged in a liquid and open to the atmosphere. The same general approach can also be used for determining the force on curved surfaces that are enclosed in a tank and pressurized. Furthermore, pressures are assumed to be gage pressures unless specifically indicated as absolute pressure.

Given that the *h*-axis is defined from the top of the free surface, where $p_{atm}(gage) = 0$, Figure 2.32 illustrates a curved surface, *AB* with a width, *w* submerged at a depth, $h = y_o$. Although the calculation of the resultant hydrostatic force, F_R or *R* acting on a curved surface can be determined by integration over the curved surface, as was done for the plane

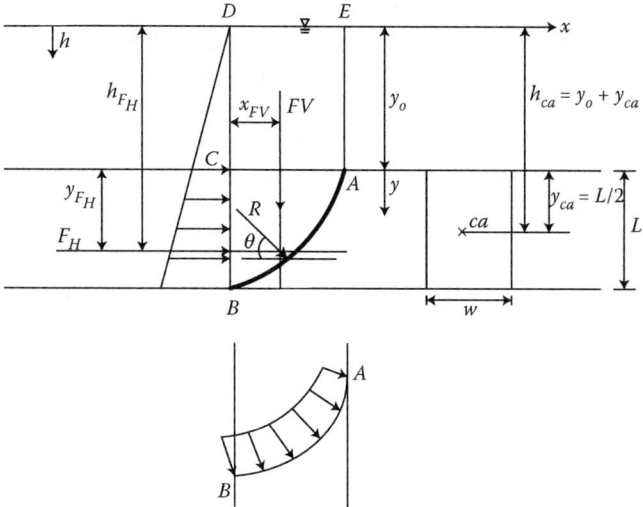

FIGURE 2.32
Resultant force, R and its location, θ for a curved surface submerged at depth, $h = y_o$ in a liquid open to the atmosphere, assuming the gage pressure scale.

surfaces above, the procedure is generally tedious and thus no general and simple formulas can be easily developed. As a result, an alternative approach to integration is to consider the static equilibrium of the fluid volume enclosed by the curved surface of interest, AB; the horizontal projection of the curved surface, DE (located along the free surface); and the vertical projection of the curved surface, CB (located below the free surface). Specifically, the magnitude of the resultant force, F_R or R is resolved into a horizontal component, F_H and a vertical component, F_V, where these two components are added vectorially to yield the magnitude of the total resultant hydrostatic force, R. The magnitude of the horizontal component, F_H is equal to the force acting on the vertical projection of the curved surface, CB. And, the magnitude of the vertical component, F_V is equal to the weight of the volume of liquid (real or imaginary [see End-of-Chapter Problems 2.154 through 2.156]) lying above the curved surface, AB to the horizontal projection of the curved surface, DE (located along the free surface). Furthermore, the location of the horizontal component, F_H, the "center of pressure" or the moment arm, is at the center of gravity of the trapezoidal pressure prism acting on the vertical projection of the curved surface, CB, while the location of the vertical component, F_V, the moment arm, is at the center of gravity of the volume of liquid (real or imaginary [see End-of-Chapter Problems 2.154 through 2.156]) lying above the curved surface, AB to the horizontal projection of the curved surface, DE. And, finally, the location of the resultant force, R is at the intersection of the lines of action of the horizontal, F_H and the vertical, F_V components of the force, R, making an angle, θ with the horizontal.

2.4.7.1 Magnitude and Location of the Horizontal Component of Force on Curved Surfaces

The magnitude of the horizontal component, F_H of the resultant force, R acting on the submerged curved surface, AB in Figure 2.32, is equal to the force acting on a vertical projection of the curved surface, CB. Based upon Pascal's law, the hydrostatic pressure does not change in the horizontal plane. Therefore, both, the magnitude and the location of the horizontal component, F_H of the resultant force, R acting on the submerged curved surface AB are equal to the respective magnitude and location of the hydrostatic force acting on the vertical projection, CB of the curved surface, AB. The vertical projection, CB is a vertical plane with a rectangular area, A (width, w and length, L), submerged at a depth, $h = y_o$. Because the y-axis is defined from the top of the rectangular plane (vertical projection, CB of the curved surface, AB), the relationship between the h-axis and the y-axis is $h = (y_o + y)$. The magnitude of the horizontal component, F_H is computed by integrating over the length of the rectangular plane (vertical projection, CB of the curved surface, AB) using the top of the plane as the reference as follows:

$$F_H = \int p\,dA = \int_0^L (\gamma h(y))(w\,dy) = \int_0^L (\gamma(y_o + y))(w\,dy) \tag{2.225}$$

where $h(y) = (y_o + y)$. Using Mathcad to integrate yields:

$$\int_0^L [\gamma(y_o + y)]w\,dy \rightarrow \frac{\gamma L w(L + 2y_o)}{2}$$

Alternatively, one may compute the magnitude of the horizontal component, F_H as follows:

$$F_H = p_{ca}A = \gamma h_{ca}A = \gamma(y_o + y_{ca})(wL) = \gamma\left(y_o + \frac{L}{2}\right)(wL) \tag{2.226}$$

where for a rectangular plane, $h_{ca} = (y_o + y_{ca})$, $y_{ca} = L/2$, and $A = wL$.

The location of the horizontal component, F_H, the "center of pressure" or the moment arm, is at the center of gravity of the trapezoidal pressure prism acting on the rectangular plane (vertical projection, CB of the curved surface, AB). Thus, the location of the horizontal component, F_H, measured from the top of the rectangular plane along the y-axis, is computed as follows:

$$M_H = y_{F_H}F_H = \int_0^L y\,dF = \int_0^L yp\,dA = \int_0^L y(\gamma h(y))(w\,dy) = \int_0^L y(\gamma(y_o + y))(w\,dy) \tag{2.227}$$

where $h(y) = (y_o + y)$. Using Mathcad to integrate yields:

$$\int_0^L [y\gamma(y_o + y)]w\,dy \rightarrow \frac{\gamma L^2 w(2L + 3y_o)}{6}$$

Solving for the moment arm by using Equation 2.227 for the moment and Equation 2.225 for the force yields the following:

$$y_{F_H} = \frac{M_H}{F_H} \tag{2.228}$$

Using Mathcad to solve for the moment arm, y_{F_H} yields the following:

$$\frac{\dfrac{\gamma L^2 w(2L + 3y_o)}{6}}{\dfrac{\gamma L w(L + 2y_o)}{2}} \rightarrow \frac{L(2L + 3y_o)}{3(L + 2y_o)}$$

Alternatively, the location of the horizontal component, F_H measured from the top of the free surface along the h-axis may be computed as follows:

$$h_{F_H} = \frac{I_{x-ca}}{h_{ca}A} + h_{ca} = \frac{wL^3/12}{(y_o + L/2)(wL)} + (y_o + L/2) \tag{2.229}$$

where for a rectangular plane, $I_{x-ca} = wL^3/12$, $h_{ca} = (y_o + y_{ca})$, $y_{ca} = L/2$, and $A = wL$. Using Mathcad to simplify the expression for the center of pressure, h_{F_H} yields:

$$\frac{\frac{wL^3}{12}}{\left(y_o + \frac{L}{2}\right)(wL)} + \left(y_o + \frac{L}{2}\right) \rightarrow \frac{L}{2} + y_o + \frac{L^2}{12\left(\frac{L}{2} + y_o\right)}$$

Furthermore, the "center of pressure" or the moment arm for the horizontal component, F_H measured from the top of the curved surface, AB along the y-axis is given as follows:

$$y_{F_H} = h_{F_H} - y_o = \frac{L}{2} + \frac{L^2}{12\left(\frac{L}{2} + y_o\right)} \tag{2.230}$$

Using Mathcad to simplify the expression for y_{F_H} yields:

$$\frac{L}{2} + \frac{L^2}{12\left(\frac{L}{2} + y_o\right)} \quad \text{simplify} \quad \rightarrow \quad \frac{L(2L + 3y_o)}{3(L + 2y_o)}$$

which is the same result yielded in Equation 2.228.

2.4.7.2 Magnitude and Location of the Vertical Component of Force on Curved Surfaces

The magnitude of the vertical component, F_V of the resultant force, R acting on the submerged curved surface, AB in Figure 2.32 is equal to the weight of the volume of liquid (real or imaginary [see End-of-Chapter Problems 2.154, 2.155, and 2.156]) lying above the curved surface, AB to the horizontal projection of the curved surface, DE (located along the free surface). Thus, based on Newton' second law of motion, the magnitude of the vertical component, F_V or the gravitational force, F_G is computed as follows:

$$F_V = F_G = \gamma V_{ABCDE} = \gamma w A_{ABCDE} \tag{2.231}$$

where the area of $ABCDE$, A_{ABCDE} binds the volume of liquid lying above the curved surface, AB. It is logical to break this area into two components, A_{ABC} and A_{ACDE}. Therefore the magnitude of the vertical component, F_V is computed as the sum of the weights of the two components of volumes as follows:

$$F_V = \gamma V_{ABC} + \gamma V_{ACDE} = \gamma w A_{ABC} + \gamma w A_{ACDE} \tag{2.232}$$

The location or the moment arm of the vertical component, F_V is located at the center of gravity of the volume of liquid (real or imaginary [see End-of-Chapter Problems 2.154, 2.155, and 2.156]), V_{ABCDE} lying above the curved surface, AB to the horizontal projection of the curved surface, DE. Because the width, w of the curved surface, AB is a constant, the center of gravity of V_{ABCDE} may be represented by the center of area of A_{ABCDE}. Thus, the location or the moment arm of the vertical component, F_V may be computed as the weighted average of the center of areas (moment arms) for A_{ABC} and A_{ACDE}. Measured from the origin of the x-axis (located at point D) along the x-axis, the moment arm of the vertical component, F_V is computed as follows:

$$x_{F_V} = \frac{(x_{ca})_{ABC}A_{ABC} + (x_{ca})_{ACDE}A_{ACDE}}{A_{ABCDE}} \tag{2.233}$$

where $(x_{ca})_{ABC}$ and $(x_{ca})_{ACDE} =$ center of area, measured from the origin of the x-axis along the x-axis, for the areas A_{ABC} and A_{ACDE}, respectively.

2.4.7.3 Magnitude and Location of the Resultant Force on Curved Surfaces

Given that the h-axis is defined from the top of the free surface, where $p_{atm}(gage) = 0$, Figure 2.32 illustrates a curved surface, AB of width, w submerged at a depth, $h = y_o$. Because the y-axis is defined from the top of the curved surface, AB, the relationship between the h-axis and the y-axis is $h = y_o + y$. The distributed force of the pressure prism can be replaced by resultant point force, R. In order to facilitate a simple procedure to determine the magnitude of the resultant force, R, it is resolved into a horizontal component, F_H and a vertical component, F_V, where these two components are added vectorially to yield the magnitude of the total resultant hydrostatic force, R. The resultant force, R is located at the intersection of the lines of action of the horizontal, F_H and the vertical, F_V components of the force, R, making an angle, θ with the horizontal.

Referring to Figure 2.32, the magnitude of the resultant force, R acting on the submerged curved surface, AB is equal to the vectorial sum of the horizontal, F_H and the vertical, F_V components, and is computed as follows:

$$R = \sqrt{F_H^2 + F_V^2} \tag{2.234}$$

The resultant force, R is located at the intersection of the lines of action of the horizontal, F_H and the vertical, F_V components, making an angle, θ with the horizontal and is computed as follows:

$$\theta = \tan^{-1}\left(\frac{F_V}{F_H}\right) \tag{2.235}$$

EXAMPLE PROBLEM 2.36

A curved surface, AB that has the shape of a quarter of a circle with a radius of 4 ft and a width of 5 ft is submerged at a depth of 8 ft in water open to the atmosphere, as illustrated in Figure EP 2.36. (a) Determine the magnitude and location of the horizontal component of the resultant hydrostatic force acting on the curved surface, AB. (b) Determine the magnitude and location of the vertical component of the resultant hydrostatic force acting on the curved surface, AB. (c) Determine the magnitude and the location of the resultant hydrostatic force acting on the curved surface, AB.

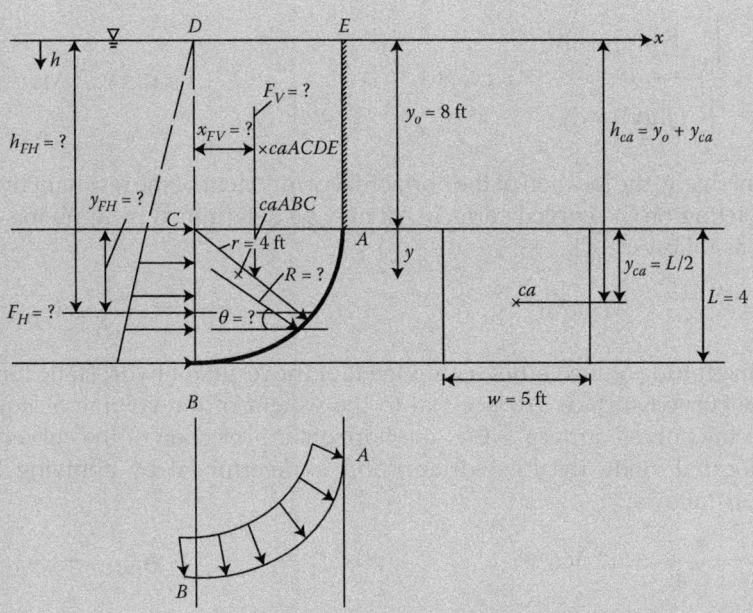

FIGURE EP 2.36
Curved pressure prism; the resultant hydrostatic force, R; and its location, θ for a curved surface submerged at a depth, $h = y_o$ in water open to the atmosphere.

Mathcad Solution

(a) The magnitude of the horizontal component of the resultant hydrostatic force acting on the curved surface, AB is determined by applying Equation 2.225, integrating over the length of the rectangular plane (vertical projection, CB of the curved surface, AB) using the top of the plane as a reference point for integration as follows:

$$w := 5\,\text{ft} \qquad r := 4\,\text{ft} \qquad L := r = 4\,\text{ft} \qquad A := w \cdot L = 20\,\text{ft}^2$$

$$y_o := 8\,\text{ft} \qquad \gamma := 62.417\,\frac{\text{lb}}{\text{ft}^3} \qquad h(y) := (y_o + y)$$

$$F_H := \int_0^L (\gamma h(y))w\,dy = 12483.4\,\text{lb}$$

Alternatively, the magnitude of the horizontal component of the resultant hydrostatic force acting on the curved surface, AB is determined by applying Equation 2.226 as follows:

$$y_{ca} := \frac{L}{2} = 2\,\text{ft} \qquad\qquad h_{ca} := y_0 + y_{ca} = 10\,\text{ft}$$

$$F_H := \gamma h_{ca} A = 12483.4\,\text{lb}$$

The location of the horizontal component of the resultant hydrostatic force acting on the curved surface, AB is determined by applying Equation 2.228 as follows:

$$y_{FH} := \frac{\int_0^L y(\gamma h(y))w \, dy}{\int_0^L (\gamma h(y))w \, dy} = 2.133 \, \text{ft} \qquad\qquad h_{FH} := y_o + y_{FH} = 10.133 \, \text{ft}$$

Alternatively, the location of the horizontal component of the resultant hydrostatic force acting on the curved surface, AB may be determined by applying Equation 2.229 as follows:

$$I_{xca} := \frac{w \cdot L^3}{12} = 26.667 \, \text{ft}^4 \qquad\qquad h_{FH} := \frac{I_{xca}}{h_{ca}A} + h_{ca} = 10.133 \, \text{ft}$$

(b) The magnitude of the vertical component of the resultant hydrostatic force acting on the curved surface, AB is equal to the weight of the volume of liquid lying above the curved surface, AB to the horizontal projection of the curved surface, DE (located along the free surface) and is determined by applying Equation 2.231 as follows:

$$A_{ABC} := \frac{\pi \cdot r^2}{4} = 12.566 \, \text{ft}^2 \qquad\qquad A_{ACDE} := ry_o = 32 \, \text{ft}^2$$

$$F_G := \gamma w(A_{ABC} + A_{ACDE}) = 13908.496 \, \text{lb} \qquad\qquad F_V := F_G = 13908.496 \, \text{lb}$$

The location (moment arm) of the vertical component of the resultant hydrostatic force acting on the curved surface, AB is located at the center of gravity of the liquid, V_{ABCDE} lying above the curved surface, AB to the horizontal projection of the curved surface, DE. However, because the width, w of the curved surface, AB is a constant, the center gravity of V_{ABCDE} may be represented by the center of area of A_{ABCDE}. Thus, the location of the vertical component of the resultant hydrostatic force is determined as the weighted average of the center of areas (moment arms) for A_{ABC} and A_{ACDE} and is computed by applying Equation 2.233. The center of area of a quarter of a circle and a rectangle is given in Table B.1 in Appendix B.

$$x_{caABC} := \frac{4 \cdot r}{3 \cdot \pi} = 1.698 \, \text{ft} \qquad\qquad x_{caACDE} := \frac{r}{2} = 2 \, \text{ft}$$

$$A_{ABCDE} := A_{ABC} + A_{ACDE} = 44.566 \, \text{ft}^2$$

$$x_{FV} := \frac{x_{caABC} \cdot A_{ABC} + x_{caACDE} \cdot A_{ACDE}}{A_{ABCDE}} = 1.915 \, \text{ft}$$

(c) The magnitude of the resultant hydrostatic force acting on the curved surface, AB is determined by applying Equation 2.234 as follows:

$$R := \sqrt{F_H^2 + F_V^2} = 18689.075 \, \text{lb}$$

The location of the resultant hydrostatic force acting on the curved surface, AB is determined by applying Equation 2.235 as follows:

$$\theta := a\tan\left(\frac{F_V}{F_H}\right) = 48.091 \, \text{deg}$$

2.4.8 Magnitude and Location of the Hydrostatic Force for Surfaces Submerged in a Multilayered Fluid

In some applications, planar and curved surfaces may be submerged in a multilayered fluid. The hydrostatic forces acting on a plane or curved surface submerged in a multilayered fluid of different specific weights, γ can be determined as follows. Considering the different parts of the surface in a different fluid as a different surface, determine the force acting on each part of the surface, then add them using vector addition. Given that the h-axis is defined from the top of the free surface, where $p_{atm}(gage) = 0$, Figure 2.33 illustrates a vertical plane with a rectangular area, A (width, w and length, L) submerged in two layers of fluid (γ_1 and γ_2). The plane of length, L is divided into two parts, L_1 submerged in the first fluid, γ_1 and L_2 submerged in the second fluid, γ_2. The resultant force acting on L_1 is denoted as F_1, while resultant force acting on L_2 is denoted as F_2. The location (center of pressure or moment arm), h_{F_1} of F_1 is at the center of gravity (center of area) of the pressure prism for γ_1, while the location, h_{F_2} of F_2 is at the center of gravity (center of area) of the pressure prism for γ_2. The resultant force, F is the vector sum of the forces F_1 and F_2. Furthermore, the location, h_F of the resultant force, F is at the center of gravity (center of area) of the entire pressure prism for both γ_1 and γ_2, and is computed as the weighted average of the center of pressures (moment arms) for the individual forces, F_1 and F_2. Furthermore, pressures are assumed to be gage pressures unless specifically indicated as absolute pressure.

2.4.8.1 Magnitude and Location of the Resultant Force Acting in First Fluid

Given that the h-axis is defined from the top of the free surface, where $p_{atm}(gage) = 0$, Figure 2.33 illustrates a vertical plane with a rectangular area, A (width, w and length, L) submerged in two layers of fluid (γ_1 and γ_2). The magnitude of the resultant force, F_1 acting on L_1, which is the first part of the plane that is submerged in the first fluid, γ_1 at a

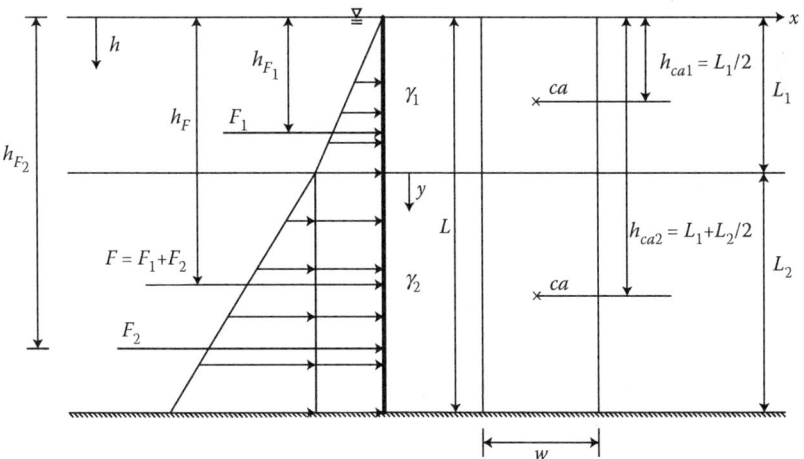

FIGURE 2.33

Resultant force, F and its location, h_F for a vertical plane with a rectangular geometric shape submerged at depth, $h = 0$ in a multilayered liquid open to the atmosphere, assuming the gage pressure scale.

depth, $h = 0$ is computed by integrating over the length, L_1 as follows:

$$F_1 = \int dF_1 = \int p_1 \, dA_1 = \int_0^{L_1} (\gamma_1 h)(w \, dh) \tag{2.236}$$

Using Mathcad to integrate yields:

$$\int_0^{L_1} (\gamma_1 h) w \, dh \rightarrow \frac{L_1^2 w \gamma_1}{2}$$

Alternatively, one may compute the magnitude of the resultant force, F_1 acting on L_1 as follows:

$$F_1 = p_{ca_1} A_1 = \gamma_1 h_{ca_1} A_1 = \gamma_1 \left(\frac{L_1}{2}\right)(w L_1) = \frac{\gamma_1 L_1^2 w}{2} \tag{2.237}$$

where for a rectangular plane, $h_{ca_1} = L_1/2$, and $A_1 = w L_1$.

The location (center of pressure or moment arm), h_{F_1} of F_1 is at the center of gravity (center of area) of the pressure prism for γ_1, which is triangular, as illustrated in Figure 2.33. The location, h_{F_1} of F_1, measured from the top of the free surface along the h-axis, is computed as follows:

$$M_1 = h_{F_1} F_1 = \int h \, dF_1 = \int h p_1 \, dA_1 = \int_0^{L_1} h(\gamma_1 h)(w \, dh) \tag{2.238}$$

Using Mathcad to integrate yields:

$$\int_0^{L_1} h(\gamma_1 h) w \, dh \rightarrow \frac{L_1^3 w \gamma_1}{3}$$

And thus, the location of the center of pressure, h_{F_1} may be computed as follows:

$$h_{F_1} = \frac{M_1}{F_1} \tag{2.239}$$

Using Mathcad to solve for the center of pressure, h_{F_1} yields:

$$\frac{\dfrac{L_1^3 w \gamma_1}{3}}{\dfrac{L_1^2 w \gamma_1}{2}} \rightarrow \frac{2 L_1}{3}$$

Alternatively, the location h_{F_1} of F_1, measured from the top of the free surface along the h-axis, may be computed as follows:

$$h_{F_1} = \frac{M_1}{F_1} = \frac{\displaystyle\int_{A_1} h(\gamma_1 h) \, dA_1}{\gamma_1 h_{ca_1} A_1} = \frac{\displaystyle\int_{A_1} \gamma_1 h^2 \, dA_1}{\gamma_1 h_{ca_1} A_1} = \frac{\gamma_1 I_x}{\gamma_1 h_{ca_1} A_1} = \frac{I_x}{h_{ca_1} A_1} \tag{2.240}$$

but:

$$I_x = I_{x-ca_1} + A_1 h_{ca_1}^2 \tag{2.241}$$

and thus:

$$h_{F_1} = \frac{I_{x-ca_1}}{h_{ca_1} A_1} + h_{ca_1} = \frac{wL_1^3/12}{(L_1/2)(wL_1)} + (L_1/2) \tag{2.242}$$

where for a rectangular plane, $I_{x-ca_1} = wL_1^3/12$, $h_{ca_1} = L_1/2$, and $A_1 = wL_1$. Using Mathcad to simplify the expression for the center of pressure, h_{F_1} yields:

$$\frac{\dfrac{wL_1^3}{12}}{\left(\dfrac{L_1}{2}\right)(wL_1)} + \frac{L_1}{2} \rightarrow \frac{2L_1}{3}$$

which is the same result as yielded by Equation 2.239.

2.4.8.2 Magnitude and Location of the Resultant Force Acting in Second Fluid

Given that the h-axis is defined from the top of the free surface, where $p_{atm}(gage) = 0$, Figure 2.33 illustrates a vertical plane with a rectangular area, A (width, w and length, L) submerged in two layers of fluid (γ_1 and γ_2). Because the y-axis is defined from the top of second part of the plane, L_2 that is submerged in the second fluid, γ_2, the relationship between the h-axis and the y-axis is $h = (L_1 + y)$. The magnitude of the resultant force, F_2 acting on L_2, which is the second part of the plane that is submerged in the second fluid, γ_2 at a depth $h = L_1$, is computed by integrating over the length, L_2 as follows:

$$F_2 = \int dF_2 = \int p_2 \, dA_2 = \int_0^{L_2} (\gamma_2 h(y) + \gamma_1 L_1)(w \, dy) = \int_0^{L_2} (\gamma_2(L_1 + y) + \gamma_1 L_1)(w \, dy) \tag{2.243}$$

where $h(y) = (L_1 + y)$. Using Mathcad to integrate yields:

$$\int_0^{L_2} [\gamma_2(L_1 + y) + \gamma_1 L_1] w \, dy \rightarrow \frac{L_2 w (2L_1 \gamma_1 + 2L_1 \gamma_2 + L_2 \gamma_2)}{2}$$

Alternatively, one may compute the magnitude of the resultant force, F_2 acting on L_2 as follows:

$$F_2 = (p_{ca_2} + \gamma_1 L_1)A_2 = (\gamma_2 h_{ca_2} + \gamma_1 L_1)A_2 = \left(\gamma_2\left(L_1 + \frac{L_2}{2}\right) + \gamma_1 L_1\right)(wL_2)$$
$$= \frac{wL_2(2\gamma_2 L_1 + \gamma_2 L_2 + 2\gamma_1 L_1)}{2} \tag{2.244}$$

where for a rectangular plane, $h_{ca_2} = (L_1 + L_2/2)$, and $A_2 = wL_2$.

The location (center of pressure or moment arm), h_{F_2} of F_2 is at the center of gravity (center of area) of the pressure prism for γ_2, which is trapezoidal, as illustrated in Figure 2.33.

The location, h_{F_2} of F_2, measured from the top of the free surface along the h-axis, is computed as follows:

$$M_2 = h_{F_2} F_2 = \int h \, dF_2 = \int h p_2 \, dA_2 = \int_{L_1}^{L_1+L_2} h(\gamma_2 h + \gamma_1 L_1)(w \, dh) \tag{2.245}$$

Using Mathcad to integrate yields:

$$\int_{L_1}^{L_1+L_2} \left[h(\gamma_2 h + \gamma_1 L_1) \right] w \, dh \text{ simplify} \rightarrow \frac{L_2 w(6L_1^2 \gamma_1 + 6L_1^2 \gamma_2 + 2L_2^2 \gamma_2 + 3L_1 L_2 \gamma_1 + 6L_1 L_2 \gamma_2)}{6}$$

And thus, the location of the center of pressure, h_{F_2} may be computed as follows:

$$h_{F_2} = \frac{M_2}{F_2} \tag{2.246}$$

Using Mathcad to solve for the center of pressure, h_{F_2} yields:

$$\frac{\dfrac{L_2 w(6L_1^2 \gamma_1 + 6L_1^2 \gamma_2 + 2L_2^2 \gamma_2 + 3L_1 L_2 \gamma_1 + 6L_1 L_2 \gamma_2)}{6}}{\dfrac{L_2 w(2L_1 \gamma_1 + 2L_1 \gamma_2 + L_2 \gamma_2)}{2}} \text{ simplify} \rightarrow L_1 + \frac{L_2}{2} + \frac{L_2^2 \gamma_2}{6(2L_1 \gamma_1 + 2L_1 \gamma_2 + L_2 \gamma_2)}$$

Alternatively, the location, h_{F_2} of F_2, measured from the top of the free surface along the h-axis, may be computed as follows:

$$h_{F_2} = \frac{M_2}{F_2} = \frac{\int_{A_2} h(\gamma_2 h + \gamma_1 L_1) \, dA_2}{(\gamma_2 h_{ca_2} + \gamma_1 L_1)A_2} = \frac{\int_{A_2} \gamma_2 h^2 \, dA_2 + \int_{A_2} h \gamma_1 L_1 \, dA_2}{(\gamma_2 h_{ca_2} + \gamma_1 L_1)A_2} = \frac{\gamma_2 I_x + \gamma_1 L_1 h_{ca_2} A_2}{(\gamma_2 h_{ca_2} + \gamma_1 L_1)A_2} \tag{2.247}$$

but:

$$I_x = I_{x-ca_2} + A_2 h_{ca_2}^2 \tag{2.248}$$

and thus:

$$h_{F_2} = \frac{\gamma_2 I_x + \gamma_1 L_1 h_{ca_2} A_2}{(\gamma_2 h_{ca_2} + \gamma_1 L_1)A_2} = \frac{\gamma_2 \left(I_{x-ca_2} + A_2 h_{ca_2}^2 \right) + \gamma_1 L_1 h_{ca_2} A_2}{(\gamma_2 h_{ca_2} + \gamma_1 L_1)A_2}$$

$$
\begin{aligned}
h_{F_2} &= \frac{\gamma_2 \left(I_{x-ca_2} + A_2 h_{ca_2}^2 \right) + \gamma_1 L_1 h_{ca_2} A_2}{(\gamma_2 h_{ca_2} + \gamma_1 L_1)A_2} \\
&= \frac{\gamma_2 \left(wL_2^3/12 + wL_2(L_1 + L_2/2)^2 \right) + \gamma_1 L_1 (L_1 + L_2/2)(wL_2)}{(\gamma_2 (L_1 + L_2/2) + \gamma_1 L_1)(wL_2)}
\end{aligned} \tag{2.249}
$$

where for a rectangular plane, $I_{x-ca_2} = wL_2^3/12$, $h_{ca_2} = (L_1 + L_2/2)$, and $A_2 = wL_2$. Using Mathcad to simplify the expression for the center of pressure, h_{F_2} yields:

$$\frac{\gamma_2\left[\left(\frac{wL_2^3}{12}\right) + wL_2\left(L_1 + \frac{L_2}{2}\right)^2\right] + \gamma_1 L_1\left(L_1\frac{L_2}{2}\right)(wL_2)}{\left[\gamma_2\left(L_1 + \frac{L_2}{2}\right) + \gamma L_1\right](wL_2)} \text{ simplify} \rightarrow L_1 + \frac{L_2}{2} + \frac{L_2^2\gamma_2}{6(2L_1\gamma_1 + 2L_1\gamma_2 + L_2\gamma_2)}$$

which is the same result as yielded by Equation 2.246.

2.4.8.3 Magnitude and Location of the Resultant Force Acting in Multilayer Fluid

The magnitude of the resultant force, F acting on L is computed as the vector sum of F_1 and F_2 as follows:

$$F = F_1 + F_2 \tag{2.250}$$

The location, h_F of the resultant force, F is at the center of gravity (center of area) of the entire pressure prism for both γ_1 and γ_2, and is computed as the weighted average of the center of pressures (moment arms) for the individual forces, F_1 and F_2, measured from the top of the free surface along the h-axis, as follows:

$$h_F = \frac{h_{F_1}F_1 + h_{F_2}F_2}{F} \tag{2.251}$$

Finally, while the above illustration was for a vertical plane with a rectangular area submerged in a multilayered liquid and open to the atmosphere, the same general approach can also be used for a horizontal or sloping plane with a nonrectangular geometric shape submerged in a multilayered liquid and enclosed in a tank and pressurized.

EXAMPLE PROBLEM 2.37

A vertical plane with a width of 4.5 ft and a height of 9 ft is submerged in two layers of fluid, as illustrated in Figure EP 2.37. The first fluid is oil with a specific gravity of 0.8 and a depth of 3 ft, and it is open to the atmosphere. The second fluid is water and has a depth of 6 ft. (a) Determine the magnitude and location of the resultant hydrostatic force acting on the vertical plane in the first fluid. (b) Determine the magnitude and location of the resultant hydrostatic force acting on the vertical plane in the second fluid. (c) Determine the magnitude and location of the resultant hydrostatic force acting on the vertical plane in the multilayer fluid.

Mathcad Solution

(a) The magnitude of the resultant hydrostatic force acting on the vertical plane in the first fluid is determined by applying Equation 2.236, integrating over the length of the first part of the plane using the top of the free surface as a reference point for integration as follows:

$w := 4.5 \text{ ft}$ $L := 9 \text{ ft}$ $\gamma_w := 62.417 \dfrac{\text{lb}}{\text{ft}^3}$

$L_1 := 3 \text{ ft}$ $s_1 := 0.8$ $\gamma_1 := s_1\gamma_w = 49.934 \dfrac{\text{lb}}{\text{ft}^3}$

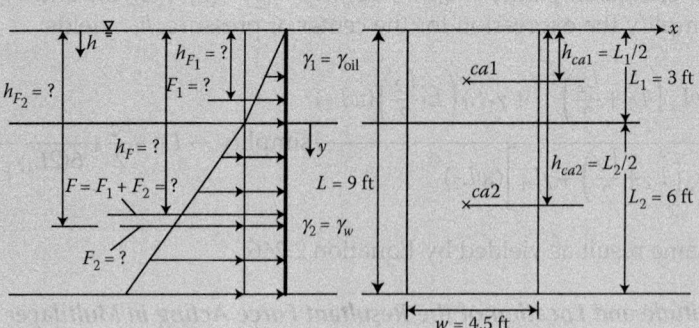

FIGURE EP 2.37
Pressure prism; the resultant hydrostatic force, F; and its location, h_F for a vertical plane submerged at a depth, $h = 0$ in water layered with oil open to the atmosphere.

$$F_1 := \int_0^{L_1} (\gamma_1 h) w \, dh = 1011.155 \, \text{lb}$$

Alternatively, the magnitude of the resultant hydrostatic force acting on the vertical plane in the first fluid may be determined by applying Equation 2.237 as follows:

$$A_1 := w L_1 = 13.5 \, \text{ft}^2 \qquad\qquad\qquad h_{cal} := \frac{L_1}{2} = 1.5 \, \text{ft}$$

$$F_1 := \gamma_1 \, h_{cal} \, A_1 = 1011.155 \, \text{lb}$$

The location of the resultant hydrostatic force acting on the vertical plane in the first fluid is determined by applying Equation 2.239 as follows:

$$h_{F1} := \frac{\int_0^{L_1} h(\gamma_1 h) w \, dh}{\int_0^{L_1} (\gamma_1 h) w \, dh} = 2 \, \text{ft}$$

Alternatively, the location of the resultant hydrostatic force acting on the vertical plane in the first fluid may be determined by applying Equation 2.242 as follows:

$$I_{xcal} := \frac{w L_1^3}{12} = 10.125 \, \text{ft}^4 \qquad\qquad h_{F1} := \frac{I_{xcal}}{h_{cal} A_1} + h_{cal} = 2 \, \text{ft}$$

(b) The magnitude of the resultant hydrostatic force acting on the vertical plane in the second fluid is determined by applying Equation 2.243, integrating over the length of the second part of the plane using the top of the second part of the plane as a reference point for integration as follows:

$$L_2 := 6 \, \text{ft} \qquad\qquad \gamma_2 := \gamma_w = 62.417 \, \frac{\text{lb}}{\text{ft}^3} \qquad\qquad h(y) := L_1 + y$$

$$F_2 := \int_0^{L_2} (\gamma_2 h(y) + \gamma_1 L_1) w \, dy = 14156.176 \, \text{lb}$$

Alternatively, the magnitude of the resultant hydrostatic force acting on the vertical plane in the second fluid may be determined by applying Equation 2.244 as follows:

$$A_2 := wL_2 = 27\,\text{ft}^2 \qquad\qquad h_{ca2} := L_1 + \frac{L_2}{2} = 6\,\text{ft}$$

$$F_2 := (\gamma_2 h_{ca2} + \gamma_1 L_1)A_2 = 14156.176\,\text{lb}$$

The location of the resultant hydrostatic force acting on the vertical plane in the second fluid is determined by applying Equation 2.246 as follows:

$$h_{F2} := \frac{\displaystyle\int_{L_1}^{L_1+L_2} h(\gamma_2 h + \gamma_1 L_1)w\,dh}{\displaystyle\int_0^{L_2} (\gamma_2 h(y) + \gamma_1 L_1)w\,dy} = 6.357\,\text{ft}$$

Alternatively, the location of the resultant hydrostatic force acting on the vertical plane in the second fluid may be determined by applying Equation 2.249 as follows:

$$I_{xca2} := \frac{wL_2^3}{12} = 81\,\text{ft}^4$$

$$h_{F2} := \frac{\gamma_2\left(I_{xca2} + A_2 h_{ca2}^2\right) + \gamma_1 L_1 h_{ca2} A_2}{(\gamma_2 h_{ca2} + \gamma_1 L_1)A_2} = 6.357\,\text{ft}$$

(c) The magnitude of the resultant hydrostatic force acting on the vertical plane in the multilayer fluid is computed as the vector sum of F_1 and F_2 and is determined by applying Equation 2.250 as follows:

$$F := F_1 + F_2 = 15167.331\,\text{lb}$$

The location of the resultant hydrostatic force acting on the vertical plane in the multilayer fluid is located at the center of gravity (center of area) of the entire pressure prism for both fluids and is computed as the weighted average of the center of pressures (moment arms) for the individual forces, F_1 and F_2 measured from the top of the free surface and is determined by applying Equation 2.251 as follows:

$$h_F := \frac{h_{F1}F_2 + h_{F2}F_2}{F} = 6.067\,\text{ft}$$

2.5 Buoyancy and Stability of a Floating or a Neutrally Buoyant Body

The role of the hydrostatic force in the buoyancy and stability of a floating or a neutrally buoyant body is important in the design of balloons, ships, boats, submarines, and other floating or neutrally buoyant bodies, and is based on the application of the principles of hydrostatics. The determination of the hydrostatic forces acting on a submerged surface (based on the principles of hydrostatics) was presented in Section 2.4. The magnitude of the hydrostatic force, F can be very large when the fluid is a liquid. The point of application (location) of the hydrostatic force on a submerged area is important when working with the moment resulting from this force. The hydrostatic pressure forces acting on the surfaces of a submerged body determine whether an anchored body will collapse or be upheld (see Section 2.4), while those acting on the surfaces of a floating or a neutrally buoyant body determine whether the body will sink or float (presented in Section 2.5, herein). Furthermore, the ability of a floating or a neutrally buoyant body to return to its equilibrium position when the body is slightly disturbed depends on the pressure forces that act in addition to characteristics of the body involved. As such, the subject of buoyancy and flotation deals with the determination of the hydrostatic pressure forces acting on submerged bodies so that a stability analysis can be made. The Archimedes principle provides the basis for the determination of the buoyant force. Furthermore, the vertical alignment of the line of action of the buoyant force and the line of action of the weight of the body (i.e., zero moment) provides the basis for determining the stability (stable equilibrium) of a floating or a neutrally buoyant body.

2.5.1 Buoyancy and Archimedes Principle

The linear increase of the hydrostatic pressure, p in a fluid with depth, z (hydrostatic pressure equation, $(dp/dz) = \gamma$) results in the existence of an upward vertical force called the buoyant force, F_B. One may observe the existence of the buoyant force, F_B when a body of any given density, ρ is submerged in the fluid. The buoyant force, F_B tends to lift the body upward. Thus, buoyancy or the buoyant force, F_B is defined as the hydrostatic force exerted by the fluid on the surface of a submerged body that is either floating or suspended (naturally or manually) in the fluid (as opposed to a submerged body that is anchored, as in Section 2.4). The weight, $F_G = Mg = W = \gamma V$ of a submerged floating or suspended body (object) appears to decrease due to the buoyant force, F_B. While the magnitude of the buoyant force, F_B can be significant when the fluid is a liquid, it is negligible in most applications when the fluid is a gas. The Archimedes principle provides the basis for the determination of the buoyant force, F_B and is stated as follows: the weight, W of a submerged body is reduced by an amount equal to the buoyant force, F_B, which has a magnitude equal to the weight of the fluid displaced by the body and acts vertically upward through the centroid of the displaced volume.

2.5.1.1 Magnitude and Location of the Buoyant Force

Depending upon the density of the body relative to the density of the fluid, one of three cases (occurrences) will take place; thus, the relationship between the buoyant force, F_B and the weight of the body, W is defined. The three cases include: (1) an immersed/completely submerged sinking body, (2) an immersed/completely submerged neutrally buoyant/suspended body, and (3) a partially submerged/floating body. The magnitude of the buoyant force, F_B is independent of the distance of the body from the free surface; the geometry of the body; and the density of the sinking, neutrally buoyant/suspended, or floating body. Regardless of whether the body is designed to remain floating or to remain neutrally buoyant/suspended in the fluid (or sink to the bottom), the weight, W of the body acts downward through the center of gravity, G of the body and is designed to be equal in magnitude to the weight of the fluid displaced by the body, the buoyant force, F_B, which acts through the "center of buoyancy," B. However, in the case of the buoyancy of a completely submerged sinking body (see Section 2.5.1.2), the weight, W of the body actually exceeds the buoyant force, F_B and thus is manually suspended in order to achieve static equilibrium.

The buoyant force, F_B acts vertically upward and passes through the centroid of the displaced volume, where the point through which the buoyant force acts is called the "center of buoyancy," B. The assumption is that the fluid has a constant density, ρ. If the body is either immersed or floats in a fluid in which the density, ρ varies with depth, the buoyant force still remains equal to the weight of the displaced fluid. However, in this case, the buoyant force does not pass through the centroid of the displaced volume, but rather passes through the center of gravity of the displaced volume. Furthermore, the buoyant force, F_B and the weight, W of the body are assumed to have the same line of action (i.e., B and G are in vertical alignment) in order to have a zero moment and thus be in a position of stable equilibrium.

2.5.1.2 Buoyancy of a Completely Submerged Sinking Body

For the first case, in which the density, ρ of the body is greater than the fluid density, the body will sink to the bottom (immersed/completely submerged sinking body). If this sinking body is externally attached to a scale (body is manually suspended in the fluid) and weighed in the fluid, its weight in the fluid, W_s will be less than its weight measured in air, W by an amount equal to the buoyant force, F_B, as illustrated in the free body diagram in Figure 2.34a. Application of Newton's second law of motion for fluids in static equilibrium to the manually suspended body in the vertical direction yields the following:

$$\sum F_z = W_s - W + F_B = 0$$

$$W_s = W - F_B$$

$$W_s = \gamma_b V_b - \gamma_f V_{dispfluid} = (\gamma_b - \gamma_f)V_b f$$

(2.252)

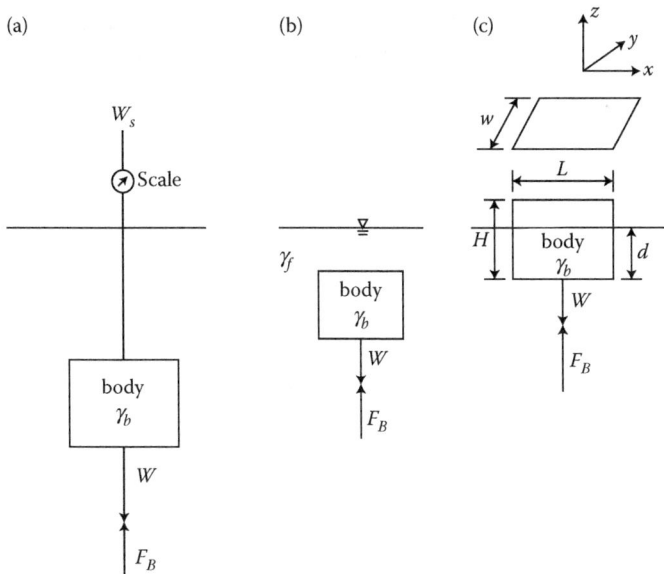

FIGURE 2.34
Free body diagram for buoyancy of a body. (a) Completely submerged sinking body. (b) Completely submerged neutrally buoyant body. (c) Partially submerged floating body.

where $\gamma_b = \rho_b g$ = the specific weight of the body, V_b = the volume of the body, $\gamma_f = \rho_f g$ = the specific weight of the fluid, and $V_{dispfluid}$ (or V_d) = the volume of the displaced fluid, which in the case of the sinking body equals to the volume of the body, V_b. Therefore, when the body is immersed in a fluid, it is buoyed up by a force equal to the weight of fluid displaced by the body (i.e., the buoyant force, $F_B = \gamma_f V_{dispfluid}$).

EXAMPLE PROBLEM 2.38

An object with a specific gravity of 1.78 is completely submerged in water. The object is attached to a scale and weighed in the water and registers 800 N. (a) Determine if the body of the object will sink, float, or become suspended (i.e., neutrally buoyant). (b) Draw the free body diagram for the manually suspended body in the water. (c) Determine the volume of the object (body).

Mathcad Solution

(a) In order to determine if the body of the object will sink, float, or become suspended (i.e., neutrally buoyant), the specific gravity (specific weight, or density) of the object is compared to the specific gravity (specific weight, or density) of the water as follows:

$$s_w := 1 \qquad\qquad\qquad\qquad s_b := 1.78$$

$$\gamma_w := 9810 \, \frac{N}{m^2} \qquad\qquad \gamma_b := s_b \, \gamma_w = 1.746 \times 10^4 \, \frac{N}{m^3}$$

$$\rho_w := 1000 \, \frac{kg}{m^3} \qquad\qquad \rho_b := s_b \, \rho_w = 1.78 \times 10^3 \, \frac{Kg}{m^3}$$

Therefore, because the specific gravity (specific weight and density) of the object is greater than the specific gravity (specific weight and density) of the water, the object will sink to the bottom unless it is manually suspended in the water.

(b) The free body diagram for the manually suspended body in the water is illustrated in Figure EP 2.38.

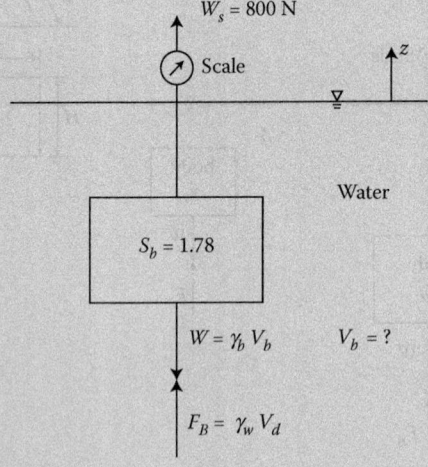

FIGURE EP 2.38
Free body diagram for buoyancy of a completely submerged sinking body suspended in water.

(c) The volume of the object (body) is determined by application of Newton's second law of motion for fluids in static equilibrium given in Equation 2.252 as follows:

$$\sum F_z = W_s - W + F_B = 0$$

where the volume of the displaced fluid, V_d is equal to the volume of the body, V_b because the body is completely submerged.

$W_s := 800 \, N$

Guess value: $\quad\quad V_b := 1 \, m^3 \quad\quad V_d := 1 \, m^3 \quad\quad W := 10 \, N \quad\quad F_B := 5 \, N$

Given

$$W_s - W + F_B = 0 \quad\quad\quad W = \gamma_b \cdot V_b \quad\quad F_B = \gamma_w \cdot V_d \quad\quad\quad\quad V_d = V_b$$

$$\begin{pmatrix} V_b \\ V_d \\ W \\ F_B \end{pmatrix} := Find \, (V_b, \, V_d, \, W, \, F_B)$$

$V_b = 0.105 \, m^3 \quad\quad\quad\quad\quad\quad\quad\quad\quad V_d = 0.105 \, m^3$

$W = 1.826 \times 10^3 \, N \quad\quad\quad\quad\quad\quad\quad F_B = 1.026 \times 10^3 \, N$

Furthermore, the buoyant force, F_B reduces the weight of the object by 1026 N as follows:

$F_B := W - W_s = 1.026 \times 10^3 \, N$

2.5.1.3 Buoyancy of a Completely Submerged Neutrally Buoyant Body

For the second case, in which the density of the body is equal to the density of the fluid (immersed/completely submerged neutrally buoyant/suspended body), the body will remain at rest at any/some point in the fluid (static equilibrium) (body is naturally suspended in the fluid), and displaces its own weight of the fluid in which it rests, as illustrated in the free body diagram in Figure 2.34b. Thus, its weight, W is equal to the buoyant force, F_B. Application of Newton's second law of motion for fluids in static equilibrium to the neutrally buoyant (naturally suspended) body in the vertical direction yields the following:

$$\sum F_z = -W + F_B = 0$$
$$W = F_B \quad\quad\quad\quad (2.253)$$
$$\gamma_b V_b = \gamma_f V_{dispfluid}$$

where in the case of the neutrally buoyant body: $\gamma_b = \rho_b g =$ the specific weight of the body is equal to $\gamma_f = \rho_f g =$ the specific weight of the fluid, and $V_b =$ the volume of the body is equal to $V_{dispfluid}$ = the volume of the displaced fluid. Therefore, when the body is immersed in a fluid, it is buoyed up by a force equal to the weight of fluid displaced by the body (i.e., the buoyant force, F_B).

One may note that neutral buoyancy occurs when the density of the body is equal to the density of the fluid. As such, the immersed/completely submerged neutrally buoyant/ naturally suspended body will not sink or float unless an external force is applied. Rather, it will remain at rest at any/some point in the fluid. Therefore, a sinking body can be made into a neutrally buoyant body by attaching floatation material in order to decrease the over-all density of the body–floatation material combination. An example of this would be plac-ing a life jacket on a person. Furthermore, a floating body can be made into a neutrally buoyant body by adding a weight in order to increase the overall density of the body–weight combination. An example of this would be attaching an anchor to a floating body.

EXAMPLE PROBLEM 2.39

An object with a specific gravity of 0.86 and a volume of 2 m^3 is completely submerged in crude oil with a specific gravity of 0.86. (a) Determine if the body of the object will sink, float, or become suspended (i.e., neutrally buoyant). (b) Draw the free body dia-gram for the body in the crude oil. (c) Determine the weight of the object (body) and the buoyant force.

Mathcad Solution

(a) In order to determine if the body of the object will sink, float, or become suspended (i.e., neutrally buoyant), the specific gravity (specific weight, or density) of the object is compared to the specific gravity (specific weight, or density) of the crude oil as follows:

$$\gamma_w := 9810 \ \frac{N}{m^3} \qquad\qquad\qquad \rho_w := 1000 \ \frac{kg}{m^3}$$

$$s_o := 0.86 \qquad\qquad\qquad\qquad s_b := 0.86$$

$$\gamma_o := s_o \cdot \gamma_w = 8.437 \times 10^3 \ \frac{N}{m^3} \qquad \gamma_b := s_b \cdot \gamma_w = 8.437 \times 10^3 \ \frac{N}{m^3}$$

$$\rho_o := s_o \cdot \rho_w = 860 \ \frac{kg}{m^3} \qquad\qquad \rho_b := s_b \cdot \rho_w = 860 \ \frac{kg}{m^3}$$

Therefore, because the specific gravity (specific weight and density) of the object is equal to the specific gravity (specific weight and density) of the crude oil, the object will become suspended or neutrally buoyant.

(b) The free body diagram for the neutrally buoyant body in the crude oil is illustrated in Figure EP 2.39.

(c) The weight of the object (body) is determined by application of Newton's second law of motion for fluids in static equilibrium given in Equation 2.253 as follows:

$$\sum F_z = -W + F_B = 0$$

where the volume of the displaced fluid, V_d is equal to the volume of the body, V_b because the body is completely submerged.

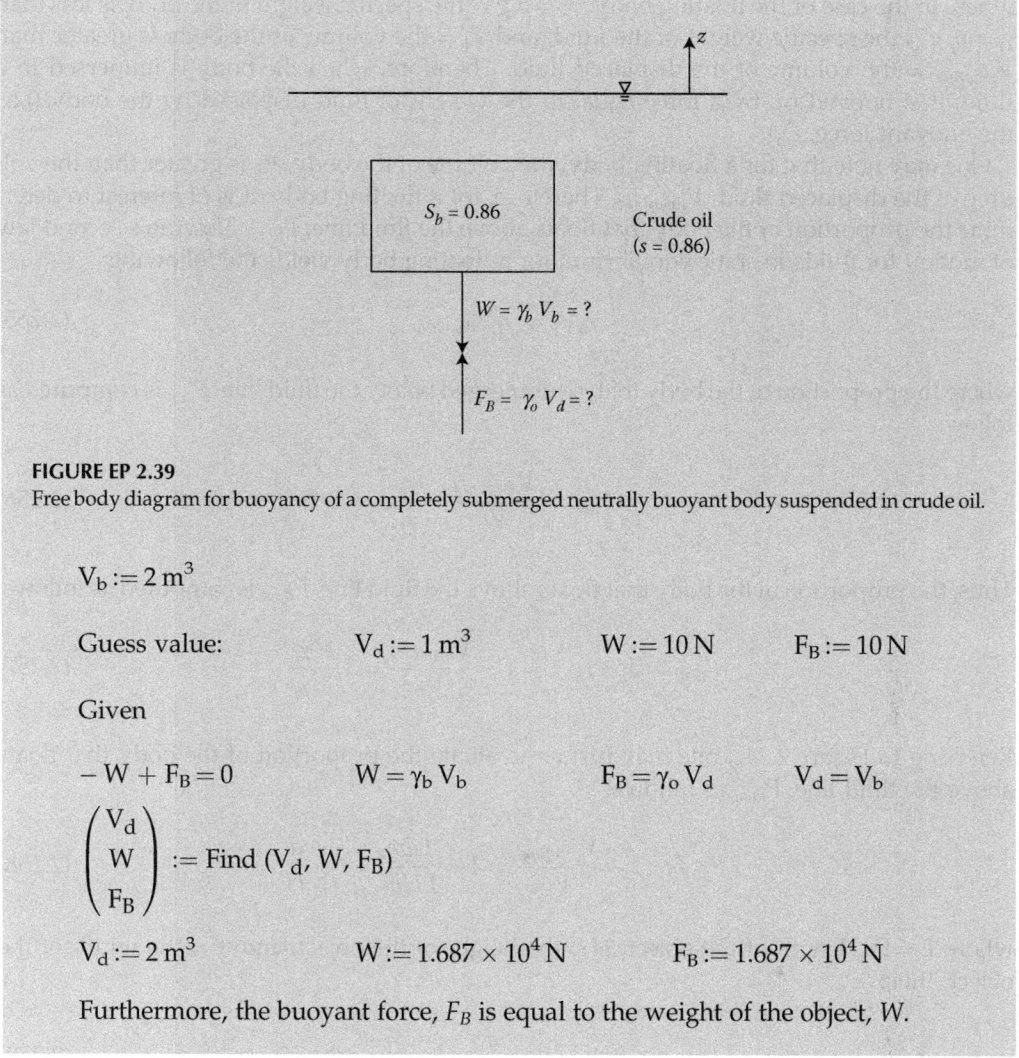

FIGURE EP 2.39
Free body diagram for buoyancy of a completely submerged neutrally buoyant body suspended in crude oil.

$$V_b := 2\,m^3$$

Guess value: $\qquad V_d := 1\,m^3 \qquad\qquad W := 10\,N \qquad F_B := 10\,N$

Given

$$-W + F_B = 0 \qquad W = \gamma_b\,V_b \qquad F_B = \gamma_o\,V_d \qquad V_d = V_b$$

$$\begin{pmatrix} V_d \\ W \\ F_B \end{pmatrix} := Find\ (V_d,\,W,\,F_B)$$

$$V_d := 2\,m^3 \qquad\qquad W := 1.687 \times 10^4\,N \qquad\qquad F_B := 1.687 \times 10^4\,N$$

Furthermore, the buoyant force, F_B is equal to the weight of the object, W.

2.5.1.4 Buoyancy of a Partially Submerged Floating Body

And, for the third case, in which the density of the body is less than the density of the fluid (partially submerged/floating body), the body will float at the fluid surface and displace its own weight of the fluid in which it floats, as illustrated in the free body diagram in Figure 2.34c. Thus, its weight, W is equal to the buoyant force, F_B. Application of Newton's second law of motion for fluids in static equilibrium to the floating body in the vertical direction yields the following:

$$\sum F_z = -W + F_B = 0$$
$$W = F_B \qquad\qquad (2.254)$$
$$\gamma_b V_b = \gamma_f V_{dispfluid}$$

where in the case of the floating body: $\gamma_b = \rho_b g$ = the specific weight of the body is less than $\gamma_f = \rho_f g$ = the specific weight of the fluid, and V_b = the volume of the body is greater than $V_{dispfluid}$ = the volume of the displaced fluid. Therefore, when the body is immersed in a fluid, it is buoyed up by a force equal to the weight of fluid displaced by the body (i.e., the buoyant force, F_B).

One may note that for a floating body, the volume of the body, V_b is greater than the volume of the displaced fluid, $V_{dispfluid}$. Therefore, for a floating body, it is of interest to determine the proportion of the body that floats above the fluid line, P_{float}. Newton's second law of motion for fluids in static equilibrium for a floating body yields the following:

$$\gamma_b V_b = \gamma_f V_{dispfluid} \tag{2.255}$$

where the proportion of the body that is submerged below the fluid line, P_{sub} is computed as follows:

$$P_{sub} = \frac{V_{dispfluid}}{V_b} = \frac{\gamma_b}{\gamma_f} \tag{2.256}$$

Thus, the proportion of the body that floats above the fluid line, P_{float} is computed as follows:

$$P_{float} = 1 - P_{sub} = 1 - \frac{V_{dispfluid}}{V_b} = 1 - \frac{\gamma_b}{\gamma_f} \tag{2.257}$$

Referring to Figure 2.34c, one may further evaluate the proportion of the body that floats above the fluid line, P_{float} as follows:

$$P_{float} = 1 - \frac{V_{dispfluid}}{V_b} = 1 - \frac{Ldw}{LHw} = 1 - \frac{d}{H} \tag{2.258}$$

where L = the length of the object, H = the height of the object, and w = the width of the object; thus:

$$P_{float} = 1 - \frac{d}{H} = 1 - \frac{\gamma_b}{\gamma_f} \tag{2.259}$$

EXAMPLE PROBLEM 2.40

A piece of plastic with a specific gravity of 0.65 and a volume of $5\,m^3$ ($L = 2.5\,m$, $H = 2\,m$, $w = 1\,m$) is partially submerged in water. (a) Determine if the plastic will sink, float, or become suspended (i.e., neutrally buoyant). (b) Draw the free body diagram for the plastic body in the water. (c) Determine the proportion of the plastic body that floats above the water line.

Mathcad Solution

(a) In order to determine if the plastic body will sink, float, or become suspended (i.e., neutrally buoyant), the specific gravity (specific weight, or density) of the plastic is compared to the specific gravity (specific weight, or density) of the water as follows:

$$s_w := 1 \qquad\qquad s_w := 0.65$$

$$\gamma_w := 9810 \, \frac{N}{m^3} \qquad\qquad \gamma_b := s_b \cdot \gamma_w = 6.377 \times 10^3 \, \frac{N}{m^3}$$

$$\rho_w := 1000 \, \frac{kg}{m^3} \qquad\qquad \rho_b := s_b \cdot \rho_w = 650 \, \frac{kg}{m^3}$$

Therefore, because the specific gravity (specific weight and density) of the plastic is less than the specific gravity (specific weight and density) of the water, the plastic will float.

(b) The free body diagram for floating plastic body in the water is illustrated in Figure EP 2.40.

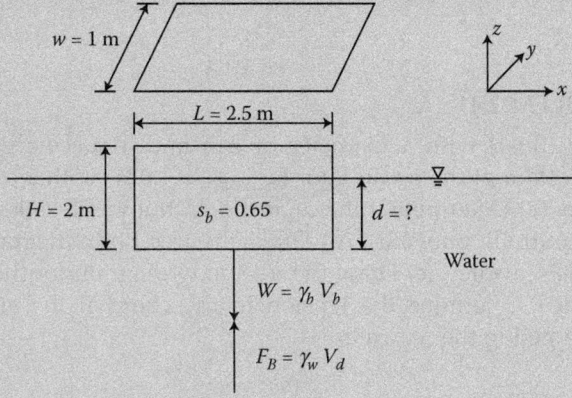

FIGURE EP 2.40
Free body diagram for buoyancy of a partially submerged floating body in water.

(c) The proportion of the plastic body that floats above the water line is determined by application of Newton's second law of motion for fluids in static equilibrium given in Equation 2.254 as follows:

$$\sum F_z = -W + F_B = 0$$

$$V_b := 5 \, m^3 \qquad L := 2.5 \, m \qquad H := 2 \, m \qquad w := 1 \, m$$

Guess value: $\qquad W: = 10 \, N \qquad F_B := 10 \, N \qquad V_d := 1 \, m^3$

$$d := 0.5 \, m \qquad P_{sub} := 0.5 \qquad P_{float} := 0.5$$

Given

$$-W + F_B = 0 \qquad W = \gamma_b \cdot V_b \qquad F_B = \gamma_w \cdot V_d \qquad P_{sub} = \frac{V_d}{V_b}$$

$$P_{float} = 1 - P_{sub} \qquad V_d = L \cdot d \cdot w \qquad V_b = L \cdot H \cdot w$$

$$\begin{pmatrix} W \\ F_B \\ V_d \\ d \\ P_{sub} \\ P_{float} \end{pmatrix} := \text{Find} (W, F_B, V_d, d, P_{sub}, P_{float})$$

$$W = 3.188 \times 10^4 \, N \qquad F_B = 3.188 \times 10^4 \, N \qquad V_d = 3.25 \, m^3$$

$$d = 1.3 \, m \qquad P_{sub} = 0.65 \qquad P_{float} = 0.35$$

Furthermore, the buoyant force, F_B is equal to the weight of the object, W.

EXAMPLE PROBLEM 2.41

A spherical buoy (float) with a diameter of 2 m and a specific gravity of 0.88 is anchored to the bottom of the ocean floor ($s_{seawater} = 1.024$) with a cord as illustrated in Figure EP 2.41b. (a) Determine if the unanchored buoy will sink, float, or become suspended (i.e., neutrally buoyant). (b) Draw the free body diagram for the unanchored buoy in the seawater. (c) Draw the free body diagram for the anchored buoy in the seawater. (d) Determine the tension in the chord if the anchored buoy is completely submerged in the seawater.

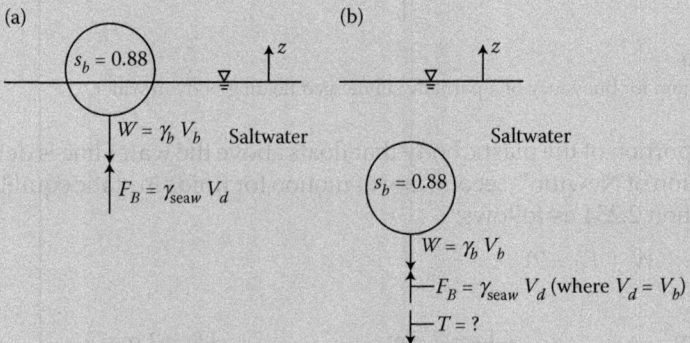

FIGURE EP 2.41
Free body diagram for buoyancy of a buoy in saltwater. (a) Partially submerged floating unanchored buoy. (b) Completely submerged anchored buoy.

Mathcad Solution

(a) In order to determine if the unanchored buoy will sink, float, or become suspended (i.e., neutrally buoyant), the specific gravity (specific weight, or density) of the buoy is compared to the specific gravity (specific weight, or density) of the seawater as follows:

$$s_{seaw} := 1.024 \qquad s_b := 0.88 \qquad \gamma_w := 9810\,\frac{N}{m^3} \qquad \rho_w := 1000\,\frac{kg}{m^3}$$

$$\gamma_b := s_b \cdot \gamma_w = 8.633 \times 10^3\,\frac{N}{m^3} \qquad\qquad \gamma_{seaw} := s_{seaw} \cdot \gamma_w = 1.005 \times 10^4\,\frac{N}{m^3}$$

$$\rho_b := s_b \cdot \rho_w = 880\,\frac{kg}{m^3} \qquad\qquad \rho_{seaw} := s_{seaw} \cdot \rho_w = 1.024 \times 10^3\,\frac{kg}{m^3}$$

Therefore, because the specific gravity (specific weight and density) of the buoy is less than the specific gravity (specific weight and density) of the seawater, the unanchored buoy will float.

(b) The free body diagram for the floating unanchored buoy in the seawater is illustrated in Figure EP 2.41a.

(c) The free body diagram for the anchored buoy in the seawater is illustrated in Figure EP 2.41b.

(d) The tension in the cord if the anchored buoy is completely submerged in the seawater is determined by application of Newton's second law of motion for fluids in static equilibrium as follows:

$$\sum F_z = -T - W + F_B = 0$$

$$D_b := 2\,m$$

Guess value: $\qquad W := 100\,N \quad F_B := 100\,N \quad T := 100\,N \quad V_b := 10\,m^3$

Given

$$-T - W + F_B = 0 \qquad\qquad V_b = \frac{\pi \cdot D_b^3}{6}$$

$$W = \gamma_b \cdot V_b \qquad\qquad F_B = \gamma_{seaw} \cdot V_b$$

$$\begin{pmatrix} W \\ F_B \\ T \\ V_b \end{pmatrix} := \text{Find } (W, F_B, T, V_b)$$

$$W = 3.616 \times 10^4\,N \qquad\qquad F_B = 4.208 \times 10^4\,N \qquad\qquad T = 5.917 \times 10^3\,N$$

$$V_b = 4.189\,m^3$$

2.5.1.5 The Role of the Buoyant Force in Practical Applications

The Archimedes principle and thus the realization of the existence of the buoyant force has numerous practical applications in engineering. Applications for floating bodies include the calculation of the draft of surface vessels, the increment in depth of flotation from the increment in weight of the ship's cargo, the lift of airships and balloons, swimmers, icebergs, and hydrometers (used to measure the specific gravity of a fluid). Applications for neutrally buoyant/suspended bodies include the design of submarines and satellites. Regardless of whether the body is designed to remain floating or to remain neutrally buoyant/suspended in the fluid (or sink to the bottom), the weight, W of the body acts downward through the center of gravity, G of the body, and is designed to be equal in magnitude to the weight of the fluid displaced by the body, the buoyant force, F_B, which acts through the "center of buoyancy," B.

In these practical applications, it is important to realize two basic points. First, the type of buoyancy of the body (sinking, neutrally buoyant/suspended, or floating body) is dependent upon the average density of the body in comparison to the density of the fluid. Thus, if the body is designed to remain floating, then its average density must remain less than the fluid density; otherwise, it will become neutrally buoyant/suspended, or it will sink to the bottom. However if the body is designed to remain neutrally buoyant/suspended, then its average density must remain equal to the density of the fluid at the design depth; otherwise, it will float to the fluid surface, or it will sink to the bottom. And, secondly, the stability of a floating or a neutrally buoyant/suspended body in a fluid is dependent upon the relative location of the buoyant force and the weight of the body. The buoyant force and the weight of the body must have the same line of action (i.e., B and G are in vertical alignment) in order to have a zero moment and thus be in a position of stable equilibrium (see Section 2.5.2).

2.5.2 Stability of a Floating or a Neutrally Buoyant Body

An important application of the buoyancy concept is the assessment of the stability of floating or a neutrally buoyant/suspended body. Stability considerations are important in the design of satellites, ships, submarines, bathyscaphes, and in the work of naval architecture. There are two types of stability to consider for both floating or neutrally buoyant/suspended bodies. The first is vertical stability and the second is rotational stability. For a floating or neutrally buoyant/suspended body that is in static equilibrium, because the weight, W of the body is equal to the buoyant force, F_B, they balance each other out, so these bodies are said to be inherently stable in the vertical direction. The rotational stability or instability of floating or neutrally buoyant/suspended bodies in a fluid will be determined by whether a restoring (stable) moment or an overturning (unstable) moment is developed when the center of gravity of the body (through which the weight of the body acts) and the center of buoyancy (through which the buoyant force acts) move out of vertical alignment due to a small rotation, θ. It is important to note that the stability analysis for a floating or neutrally buoyant body assumes a small angle of rotation, θ. As such, there is a maximum design angle of rotation, θ that the body can tilt, beyond which it will overturn (capsize) and either find a new position of equilibrium or sink (in the case of a floating body). Thus, given a small angle of rotation, θ, a rotational stability analysis for neutrally buoyant/suspended bodies depends upon the weight distribution of the body and thus the relative location of the center of gravity, G of the body and the center of buoyancy, B. And, given a small angle of rotation, θ, a rotational stability analysis for floating bodies depends upon the weight distribution, the particular geometry of the body, and the definition of the metacentric height (in the case of the stability of a top-heavy floating body). Furthermore, a stability analysis may be further complicated by the necessary inclusion of other types of external forces such as those induced by wind gusts or currents.

2.5.2.1 Vertical Stability of a Floating or Neutrally Buoyant Body

For a floating or neutrally buoyant/suspended body that is in static equilibrium, the weight, W of the body is equal to the buoyant force, F_B and thus they balance each other out.

Furthermore, these bodies are said to be inherently stable in the vertical direction. If a floating body is raised or lowered by an external vertical force, the body will return to its original position as soon as the external force is removed, so it is said to possess vertical stability. On the other hand, if a neutrally buoyant/suspended body is raised or lowered to a different depth, while the body will remain in static equilibrium at that new depth, it does not return to its original position after the disturbance and thus is said to possess neutral stability.

2.5.2.2 Rotational Stability of a Floating or Neutrally Buoyant Body

The rotational stability of a floating or neutrally buoyant/suspended body is determined by considering what happens when it is displaced from its position of stable equilibrium. Stability considerations are important for both floating or neutrally buoyant/suspended bodies because the centers of gravity and buoyancy do not necessarily coincide. Therefore, when a small rotation causes the two centers to move out of vertical alignment, it can result in either a restoring or overturning moment (couple). Furthermore, when a small rotation does not cause the two centers to move out of vertical alignment, it results in a zero moment (neutral stability). Thus, there are three positions/states of equilibrium with respect to rotational stability. A body is said to be in a position of stable equilibrium if, when displaced by a small rotation, it returns to its original position of stable equilibrium (due to a restoring moment). Conversely, a body is in a position of unstable equilibrium if, when displaced even slightly (i.e., a slight rotation), it moves to a new position of stable equilibrium (due to an overturning moment). Furthermore, a body is in a position of neutral stability if, when displaced by a small rotation, it will remain in static equilibrium in the new position of stable equilibrium (due to a zero moment).

2.5.2.3 Rotational Stability of Neutrally Buoyant/Suspended Bodies

Given a small angle of rotation, θ, the rotational stability of a neutrally buoyant/suspended body (such as a hot-air or helium balloon, a submarine, or a satellite) depends on the weight distribution of the body and thus the relative location of the center of gravity, G of the body and the center of buoyancy, B, which is the centroid of the displaced volume. For a small angle of rotation, θ, the rotational stability of a neutrally buoyant/suspended body will always be stable if G is either below B (bottom-heavy) or coincides with B (uniform density body; center-heavy). However, given a small angle of rotation, θ, the rotational stability of a top-heavy neutrally buoyant/suspended body will be unstable. One may note that for a neutrally buoyant/suspended body, while a slight rotation from its original position of equilibrium will cause the weight, W and the buoyant force, F_B to move out of vertical alignment (except in the case of a uniform density body), the position of both the center of gravity, G and the center of buoyancy, B will remain unchanged.

A neutrally buoyant/suspended body will be in position of stable equilibrium with respect to small rotations when its center of gravity, G is below the center of buoyancy, B (density of the body is bottom-heavy), as shown in Figure 2.35. If the body is given a slight rotation from its equilibrium position, W and F_B move out of vertical alignment by a horizontal distance, a. This will create a restoring moment formed by the weight, W and the buoyant force, F_B, which causes the body to rotate back to its original position. For instance, a stable design for a submarine requires for the engines and the cabins for the crew to be located at the lower half in order to shift the weight, W to the bottom as much as possible. Additionally, hot-air balloons are also stable because the cage that carries the load is located at the bottom (as illustrated in Figure 2.35c).

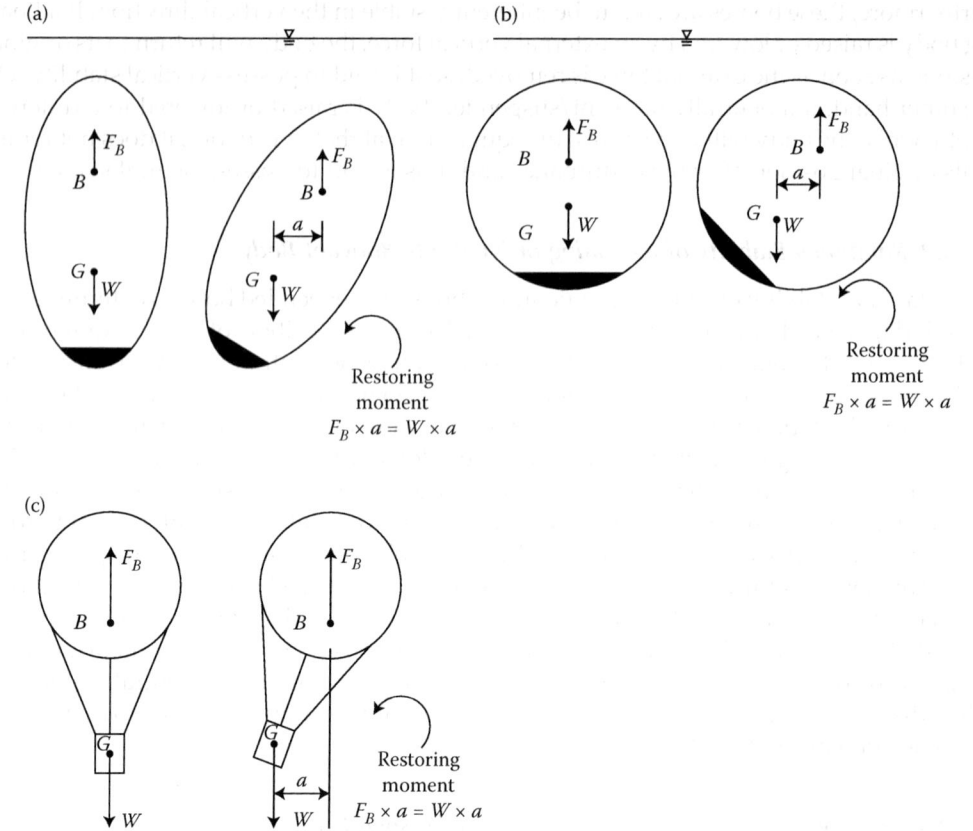

FIGURE 2.35
Neutrally buoyant/suspended body will be in position of stable equilibrium with respect to small rotations when its center of gravity, G is below the center of buoyancy, B (density of the body is bottom heavy), regardless of the geometry of the body. (a) Elliptical shaped body. (b) Spherical shaped body. (c) Hot-air balloon.

A neutrally buoyant/suspended body will also be in position of stable equilibrium when G and B coincide (constant density throughout the body; center-heavy) as shown in Figure 2.36; this is termed neutrally stable. Even with a slight rotation, the body will remain in an equilibrium potion, with G and B maintaining their coinciding positions, thus creating a zero moment; the body will remain in static equilibrium in the new position of stable equilibrium.

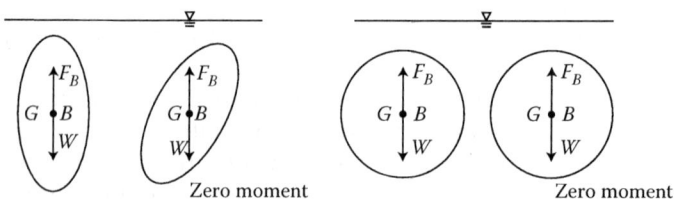

FIGURE 2.36
Neutrally buoyant/suspended body will be in position of stable equilibrium when G and B coincide (constant density throughout the body; center heavy); regardless of the geometry of the body. (a) Elliptical shaped body. (b) Spherical shaped body.

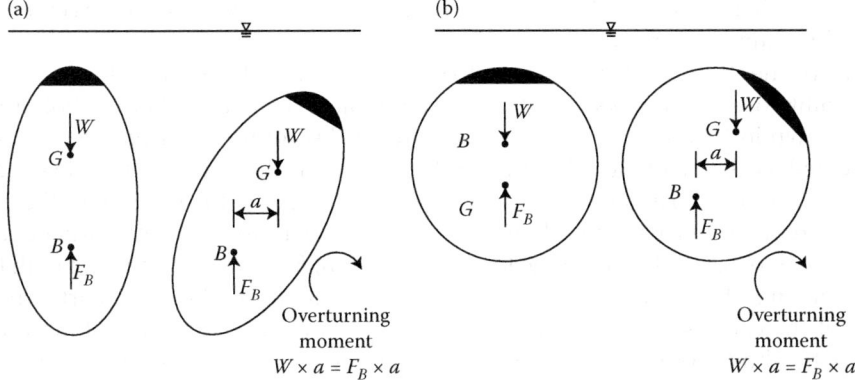

FIGURE 2.37
Neutrally buoyant/suspended body will be in a position of unstable equilibrium with respect to small rotations when G is above B (density of body is top heavy), regardless of the geometry of the body. (a) Elliptical shaped body. (b) Spherical shaped body.

However, a neutrally buoyant/suspended body will be in a position of unstable equilibrium with respect to small rotations when G is above B (density of body is top-heavy), such as shown in Figure 2.37. If the body is given a slight rotation from its equilibrium position, W and F_B move out of vertical alignment by a horizontal distance, a. This will create an overturning moment formed by the weight, W and the buoyant force, F_B, which causes the body to overturn and move to a new position of equilibrium.

2.5.2.4 Rotational Stability of Floating Bodies

Given a small angle of rotation, θ, the rotational stability of a floating body (such as a ship or a barge) depends upon the weight distribution, the particular geometry of the body, and the definition of the metacentric height (in the case of the stability of a top-heavy floating body). For a small angle of rotation, θ, the rotational stability of a floating body will always be stable if the floating body is bottom heavy, regardless of the particular geometry of the body (wide-based, relatively tall and slender, spherical-shaped, or streamlined/average height). However, given a small angle of rotation, θ, the rotational stability of a top-heavy floating body will depend upon the particular geometry (wide-based, relatively tall and slender, spherical-shaped, or streamlined/average height). For a top-heavy and wide-based body, a small rotational disturbance creates a restoring moment and returns the body to the original position of stable equilibrium. For a top-heavy and a relatively tall and slender-shaped body, a small rotational disturbance can create an overturning moment and bring the body to a new position of stable equilibrium. For a top-heavy and spherical-shaped body, a small rotational disturbance creates a zero moment; the body will remain in static equilibrium in the new position of stable equilibrium. For a top-heavy and streamlined/average-height body, the occurrence of the type of moment (restoring, overturning, or zero) due to a small rotational disturbance does not appear to be very obvious. Thus, in general, the rotational stability of a top-heavy floating body is typically measured/assessed through the definition of the metacentric height. Furthermore, one may note that for a floating body, a slight rotation from its original position of equilibrium will cause the weight, W and the buoyant force, F_B to move out of vertical alignment (except in the case of a spherical-shaped body; zero moment), and cause the center of buoyancy, B to change to a new position, B' (except in the case of a spherical-shaped body; zero moment), while the position of the center of gravity, G will remain unchanged.

Similar to a neutrally buoyant/suspended body, for a small angle of rotation, θ, the rotational stability of a floating body will always be stable if the floating body is bottom heavy, such as shown in Figure 2.38. This is the case regardless of the particular geometry of the body (wide-based, relatively tall and slender, spherical-shaped, or streamlined/average height). And, similar to a neutrally buoyant/suspended body, the position of the center of gravity, G remains unchanged. However, unlike a neutrally buoyant/suspended body, the center of buoyancy, B will change to a new position, B' that is a function of both the particular geometry of the body and the magnitude of the angle of rotation, θ. Furthermore, the location of the centers of buoyancy, B and B' do not determine the stability of a bottom-heavy floating body. When the body is given a slight rotation from its equilibrium position, W and F_B move out of vertical alignment by a horizontal distance, a. This will create a restoring moment formed by the weight, W and the buoyant force, F_B, which causes the body to rotate back to its original position.

Unlike a neutrally buoyant/suspended body, for a small angle of rotation, θ, a floating body may still be stable when G is directly above B (top-heavy) when the body has a wide-base geometry (such as a barge), as shown in Figure 2.39. This is because the center of buoyancy, B shifts to the side to a new position, B' during the small rotational disturbance (while the center of gravity, G remains unchanged), creating a restoring moment (due to the wide-based shape of the body) and returns the body to the original position of stable equilibrium.

If, however, the body is both top-heavy and relatively tall and slender (regardless of how high or low the body rides in the fluid) as shown in Figure 2.40, a small rotational disturbance can cause the center of buoyancy, B to shift to the side to a new position, B' (while the center of gravity, G remains unchanged), creating an overturning moment (due to the combination of both a top-heavy and a relatively tall and slender-shaped body) and bring the body to a new position of stable equilibrium.

Also, unlike a neutrally buoyant/suspended body, for a small angle of rotation, θ, a floating body may still be stable when G is directly above B (top-heavy) when the body is

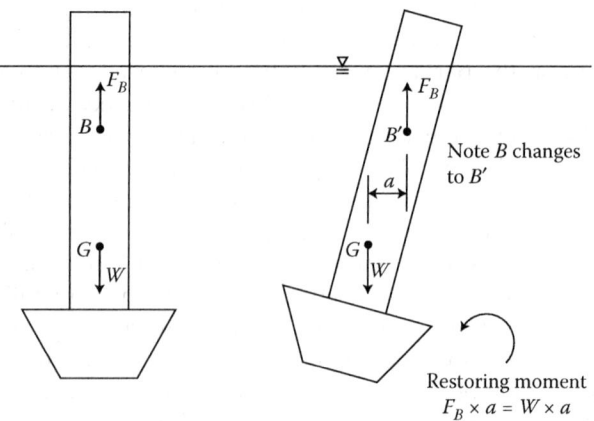

FIGURE 2.38
Floating body will be in a position of stable equilibrium with respect to small rotations when its center of gravity, G is below the center of buoyancy, B (density of body is bottom heavy), regardless of the geometry of the body.

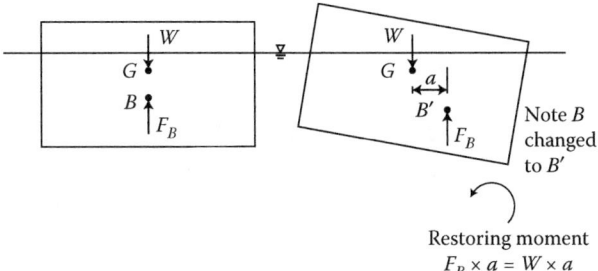

FIGURE 2.39
Floating body may still be in a position of stable equilibrium with respect to small rotations when its center of gravity, G is above the center of buoyancy, B (density of body is top heavy) when the body has a wide-base geometry (such as a barge).

spherically shaped, as shown in Figure 2.41. This is because the center of buoyancy, B does not shift to the side to a new position, B′ during the small rotational disturbance (and the center of gravity, G also remains unchanged), resulting in a zero moment (due to the spherical shape of the body); the body will remain in static equilibrium in the new position of stable equilibrium.

The three examples given above for a floating body with top-heavy weight distribution illustrated the occurrence of a restoring, an overturning, and a zero moment due to the three extreme/definitive geometries of a wide-based barge, a tall and slender body, and a spherical-shaped body, respectively. If however, the floating body is top-heavy, yet has a streamlined/average height geometry (not extreme/definitive in geometry), for such a ship or a boat, the occurrence of the type of moment (restoring, overturning, or zero) may not appear to be very obvious. Thus, in general, the rotational stability of a top-heavy

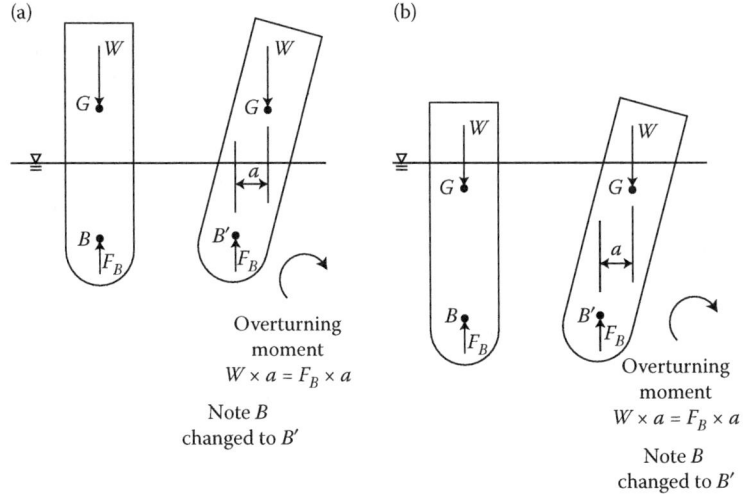

FIGURE 2.40
Floating body will be in a position of unstable equilibrium with respect to small rotations when its center of gravity, G is above the center of buoyancy, B (density of body is top heavy), and is relatively tall and slender in shape, regardless of how high or low the body rides in the fluid. (a) High-riding tall and slender body. (b) Low-riding tall and slender body.

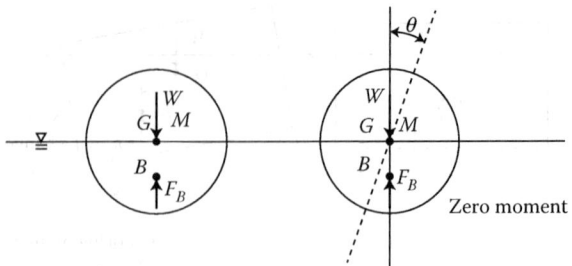

FIGURE 2.41
Floating body may still be in a position of stable equilibrium with respect to small rotations when G is above B (density of body is top heavy) when the body is spherically-shaped; a spherically-shaped floating body with a zero metacentric height, GM results in a zero moment (the center of buoyancy, B does not change).

floating body is typically measured/assessed through the definition of the metacentric height, as discussed below.

2.5.2.5 Maximum Design Angle of Rotation for a Top-Heavy Floating Body

For a given floating body that is top heavy, there is a maximum design rolling angle of rotation, θ by which the body can tilt, beyond which it will overturn (capsize) and either find a new position of equilibrium or, in the case of a boat or ship, sink. The maximum design small rolling angle of rotation by which the body can tilt before it will overturn decreases as the body becomes taller and more slender. Therefore, the stability of a top-heavy floating body is a function of the particular geometry of the body, the magnitude of the angle of rotation, θ, and the resulting location of the centers of buoyancy, B and B'. Thus, for a small (design) angle of rotation, θ: (1) a wide-based top-heavy body (as illustrated in Figure 2.39) will be in a position of stable equilibrium (restoring moment) due to its shape, (2) a tall and slender top-heavy body (as illustrated in Figure 2.40) will be in a position of unstable equilibrium (overturning moment) due to its shape, and (3) a spherical-shaped top-heavy body (as illustrated in Figure 2.41) will be in a position of neutral equilibrium (zero moment) due to its shape. However, for a streamlined/average-height top-heavy body (as illustrated in Figure 2.42), for a small (design) angle of rotation, θ, the type of moment (restoring, overturning, or zero) may not appear to be very obvious. Therefore, with the exception of a tall and slender shape, regardless of the shape (wide-based, spherical-shaped, or streamlined/average height) of the top-heavy floating body, there is a maximum design rolling angle of rotation, θ in order to maintain rotational stability. Furthermore, in particular, for the streamlined/average height top-heavy floating body, because the shape and thus the actual design angle of rotation, θ may not be good indicators of the stability of the floating body, instead, the metacentric height is a used as an indicator of the stability of the floating body. Therefore, in general, the rotational stability of a top-heavy floating body is typically measured/assessed through the definition of the metacentric height, GM, as discussed below. One may note that for most hull shapes, for a maximum design small rolling angle of rotation, θ of about 20° (with a typical range of 7–10° for ships), the metacenter, M and thus the metacentric height, GM may considered to be a fixed point.

2.5.2.6 Metacentric Height for a Top-Heavy Floating Body

The metacenter, M is the intersection point of the lines of action of the buoyant force, F_B through the body before and after rotation (B and B', respectively). The length of the

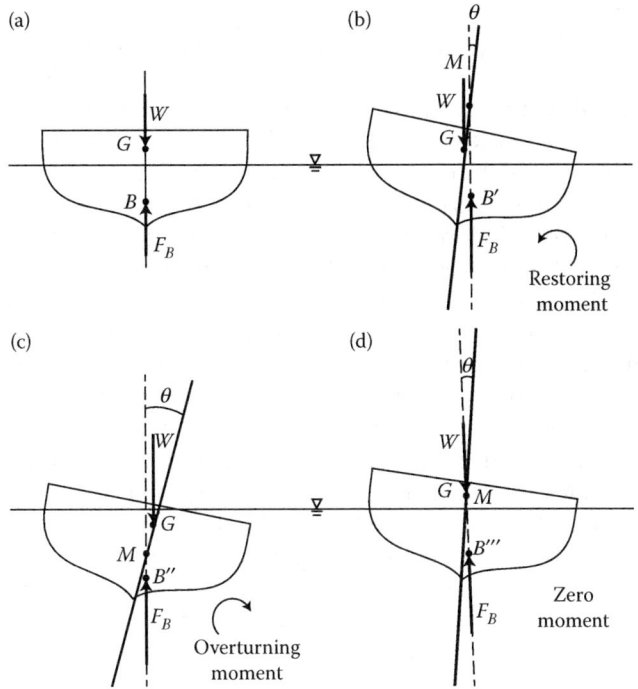

FIGURE 2.42

Stability of a streamlined/average height top-heavy (G is above B) floating body will be a function of the magnitude of the rotation, θ, and thus the magnitude/sign of the metacentric height, GM. (a) Original position of stable equilibrium. (b) Slight design rotation, θ and thus a positive GM results in a restoring moment (stable). (c) Greater than slight design rotation, θ and thus a negative GM results in an overturning moment (unstable). (d) Smaller than design rotation, θ and thus a zero GM results in zero moment (neutrally stable).

metacentric height, GM above G is a measure of rotational stability of a top-heavy floating body, where a positive, negative, or zero GM results in a stable, unstable, or neutral equilibrium, respectively. For a floating body, a rotation from its original position of equilibrium will cause the weight, W and the buoyant force, F_B to move out of vertical alignment (except in the case of a spherical-shaped body; zero moment). Furthermore, while the position of the center of gravity, G generally remains unchanged, the center of buoyancy, B will change to a new position B' (except in the case of a spherical-shaped body; zero moment) that is a function of both the particular geometry of the body and the magnitude of the angle of rotation, θ. Thus, the location of the centers of buoyancy, B and B' determines the stability of a top-heavy floating body, where B and B' are used to define the metacentric height, GM. Therefore, the rotational stability of a top-heavy floating body is a function of the metacentric height, GM, which is a function of: (1) the particular geometry of the body, (2) the magnitude of the angle of rotation, θ, and (3) the resulting location of the centers of buoyancy, B and B'. It may be noted that if liquid (liquid ballast or fuel oil) in the hull of a ship is not constrained and stored in tanks or bulkhead compartments, the center of gravity, G of the floating body may move toward the center of buoyancy, B when the ship rolls, thus decreasing the body's restoring moment and thus the stability of the floating body.

The length of the metacentric height, GM above G is a measure of rotational stability of a top-heavy floating body, where a positive, negative, or zero GM results in a stable, unstable, or neutral equilibrium, respectively. In most surface vessels (ships and boats), the floating body is usually top heavy, where the center of gravity, G is above the center of buoyancy, B. Figure 2.42 illustrates a floating hulled-shaped streamlined body, a boat, which is top heavy; thus, the center of gravity, G is above the center of buoyancy, B. Figure 2.42a illustrates the body in its original position of stable equilibrium. Figure 2.42b illustrates the body after it is given a slight/small (clockwise) rotational disturbance, whereby the center of buoyancy, B moves to the right of the center of gravity, G (which remains unchanged) to a new position, B'. The metacenter, M is the intersection point of the lines of action of the buoyant force, F_B through the body before and after rotation, as illustrated in Figure 2.42b. The metacentric height, GM is defined as the distance between the center of gravity, G and the metacenter, M. Because the resulting metacenter, M is above the center of gravity, G, the metacentric height, GM is positive, so the body is in a position of stable equilibrium. As a result, the moment formed by W and F_B is a restoring moment. One may note that a resulting positive GM (and thus a restoring moment), which is an indicator of rotational stability, is a result of the streamlined/average height geometric design of the body, the design angle of rotation, θ and thus the resulting location of the centers of buoyancy, B and B'.

Figure 2.42c illustrates the body after it is given a greater-than-designed slight/small (clockwise) rotational disturbance, whereby the center of buoyancy, B moves to the left of the center of gravity, G (which remains unchanged) to a new position, B''. Because the resulting metacenter, M is below the center of gravity, G, the metacentric height, GM is negative, and thus the body is in a position of unstable equilibrium. As a result, the moment formed by W and F_B is an overturning moment. In this case, one may note that a resulting negative GM (and thus an overturning moment), which is an indicator of rotational instability, is a result of exceeding the design angle of rotation, θ and thus the resulting location of the centers of buoyancy, B and B''; the rotational instability is not a function of the geometric design of the body, as it is streamlined/average height.

And, finally, Figure 2.42d illustrates the body after it is given a smaller-than-designed slight/small (clockwise) rotational disturbance, whereby the center of buoyancy, B moves only slightly to the right, aligning with the center of gravity, G (which remains unchanged) to a new position, B'''. The resulting metacenter, M is coincident with the center of gravity, G, and the metacentric height, GM is zero; thus, the body is in a position of neutral equilibrium. As a result, the moment formed by W and F_B is a zero moment. And, finally, in this case, one may note that a resulting zero GM (and thus a zero moment), which is an indicator of rotational neutral stability, is a result of the streamlined/average height geometric design of the body, a smaller-than-designed angle of rotation, θ and thus the resulting location of the centers of buoyancy, B and B'''.

2.5.2.7 Computation of the Metacentric Height and the Resulting Moment (Restoring, Overturning, or Zero) for a Top-Heavy Floating Body

When a top-heavy floating body is given a slight rotation from its equilibrium position, W and F_B move out of vertical alignment by a horizontal distance, a (except in the case of a spherical-shaped body; zero moment). Depending upon whether the metacentric height, GM is positive, negative, or zero, this will create a restoring, overturning, or zero (neutral) moment, respectively, formed by the weight, W and the buoyant force, F_B. Thus, the length of the metacentric height, GM above G is a measure of rotational stability of a top-heavy floating body. Furthermore, the larger the positive metacentric height, GM above G, the more stable the floating body. The length of the metacentric height, GM above G is a measure of stability of a top-heavy floating body because the magnitude (and direction: restoring, overturning, or zero) of the moment, M_{couple} of the couple is directly proportional to the

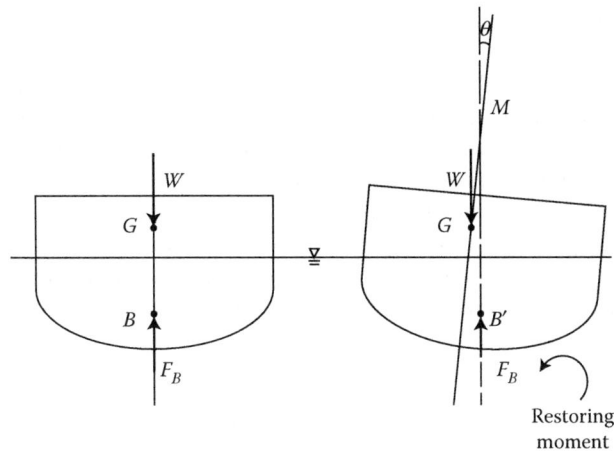

FIGURE 2.43
Top-heavy (G is above B) floating body with a positive metacentric height, GM results in a restoring moment (stable).

metacentric height GM. The moment, M_{couple} of the couple is computed as follows:

$$M_{couple} = (W)(GM)(\sin\theta) \tag{2.260}$$

where the horizontal distance between W and F_B is $a = (GM)(\sin\theta)$. A positive GM results in a restoring moment, a negative GM results in an overturning moment, and a zero GM results in a zero moment. Figure 2.43 illustrates a restoring moment, M_{couple} because GM is positive, while Figure 2.44 illustrates an overturning moment, M_{couple} because GM is negative. Furthermore, Figure 2.41 illustrates a zero moment, M_{couple} because GM is zero; when the

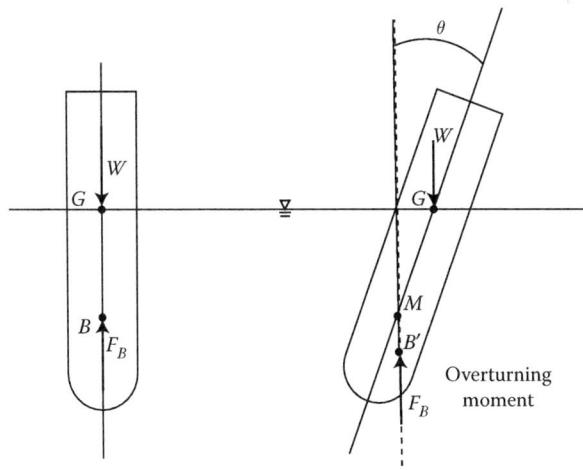

FIGURE 2.44
Top-heavy (G is above B) floating body with a negative metacentric height, GM results in an overturning moment (unstable).

body is rotated through an angle, θ, the center of buoyancy, B does not change due to the spherical shape of the floating body.

In order to apply Equation 2.260 to compute the moment, M_{couple} of the couple for a top-heavy floating body, one must determine the magnitude of the metacentric height, GM. The metacenter, M is the intersection point of the lines of action of the buoyant force, F_B through the body before and after rotation. The metacentric height, GM is the distance between the center of gravity, G and the metacenter, M. Figure 2.45b illustrates a top-heavy (G is above B) floating body in its original position of stable equilibrium (prior to a rotational disturbance). Figure 2.45c illustrates the top-heavy floating body after a small rotational disturbance. Assuming a small (design) angle of rotation, θ, the metacentric height, GM for a top-heavy floating body is computed as follows:

$$GM = BM - GB \tag{2.261}$$

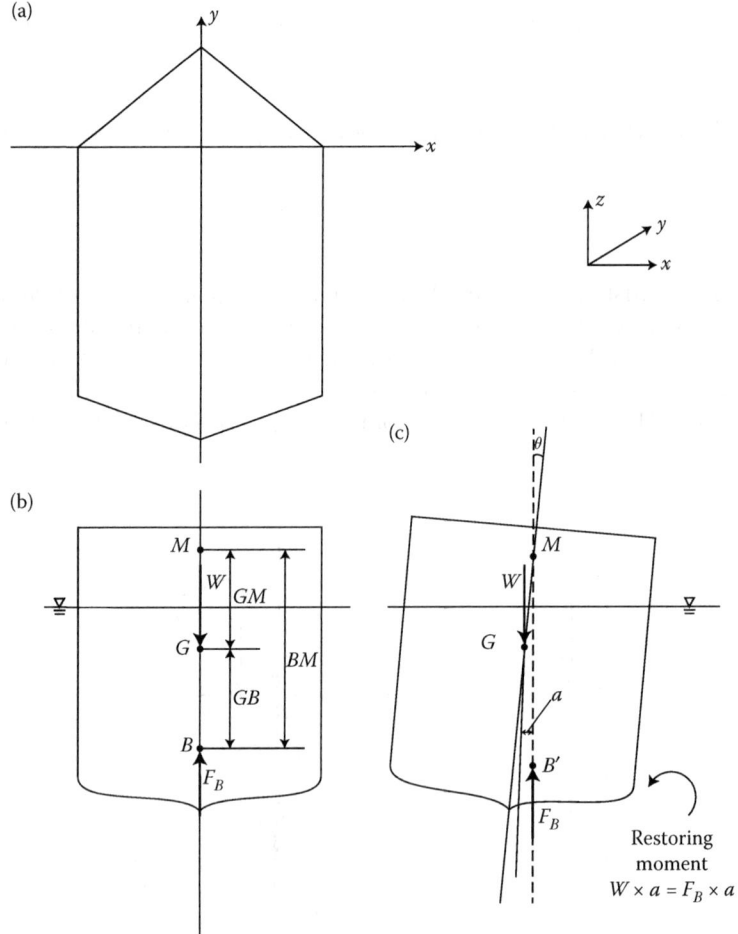

FIGURE 2.45
Computation of the metacentric height, GM for a top-heavy (G is above B) floating body subjected to a slight design rotation, θ. (a) Floating body's horizontal cross section; plan view of floating body at the waterline. (b) Original position of stable equilibrium. (c) Slight design rotation, θ, so a positive GM results in a restoring moment (stable).

where the distance GB (remains the same prior to and after the rotational disturbance) is the distance between the center of gravity, G and the center of buoyancy, B, and the distance, BM (computed below in Equation 2.263) is the distance between the center of buoyancy, B and the metacenter, M. One may note that Figure 2.45 illustrates the case where the meta-center, M is located above G and thus the metacentric height, GM is positive. In the case where the metacenter, M is located below G and thus the metacentric height, GM is negative (as illustrated in Figure 2.44), Equation 2.261 is used, yielding a negative value for GM. Furthermore, in the case where the metacenter, M is coincident with G and thus the metacentric height, GM is zero (as illustrated in Figure 2.41), Equation 2.261 is used, yielding a zero value for GM. However, one may note that in the situation where G is below B (bottom-heavy floating body), the floating body will always be stable and the metacentric height, GM is computed as follows:

$$GM = BM + GB \qquad (2.262)$$

Furthermore, regardless of whether the floating body is top-heavy or bottom-heavy, assuming a small (design) angle of rotation, θ, the distance BM is computed as follows:

$$BM = \frac{I_y}{V_{dispfluid}} \qquad (2.263)$$

where I_y is the moment of inertia about the y-axis of rotation of the body's horizontal cross section (plan view) at the waterline (prior the rotational disturbance), as illustrated in Figure 2.45a, and $V_{dispfluid}$ (or V_d) is the volume of the displaced fluid (which will remain the same prior to and after the rotational disturbance). The moments of inertia about the y-axis of rotation, I_y for various geometric shapes are given in Table B.1 in Appendix B. One may note that I_y represents the geometric design of the body, and the larger the value for I_y, the more rotationally stable the body is about the y-axis. Finally, it is important to note that the above equations for GM (Equation 2.261 and Equation 2.262) and the above equation for BM (Equation 2.263) are not valid for angles of rotation, θ larger or smaller than the design.

As noted above, the rotational stability of a top-heavy floating body is a function of the metacentric height, GM, which is a function of: (1) the particular geometry of the body, (2) the magnitude of the angle of rotation, θ, and (3) the resulting location of the centers of buoyancy, B and B' (and thus the resulting location of M). For the case of a top-heavy floating body that is relatively tall and slender as illustrated in Figure 2.44, for a small (design) rotational disturbance, θ the metacenter, M is located below G and the computed metacentric height, $GM = BM - GB$ is negative (overturning moment). Thus, the computed distance $BM = (I_y/V_{dispfluid})$ is less than the distance GB (remains the same prior to and after the rotational disturbance) due to the tall and slender geometric shape of the floating body.

For the case of a top-heavy floating body that is streamlined/average height body as illustrated in Figure 2.42b, for a small (design) rotational disturbance, θ the metacenter, M is located above G and the computed metacentric height, $GM = BM - GB$ is positive (restoring moment). Thus, the computed distance $BM = (I_y/V_{dispfluid})$ is greater than the distance GB (remains the same prior to and after the rotational disturbance) due to the streamlined/average height of the floating body.

For the case of a top-heavy floating body that is streamlined/average-height body, as illustrated in Figure 2.42c, for a larger-than-designed rotational disturbance, θ the resulting metacenter, M is located below G and thus the metacentric height, GM is negative (overturning moment). Thus, the resulting distance BM is less than the distance GB (remains the same prior to and after the rotational disturbance) due to the larger than design rotational disturbance, θ. One may note that because a larger than design rotational disturbance, θ has been applied, Equations 2.261 and 2.263 for GM and BM, respectively, are no longer valid; instead, an experimental/empirical approach may be used to compute GM and BM.

For the case of a top-heavy floating body that is streamlined/average height body, as illustrated in Figure 2.42d, for a smaller-than-designed rotational disturbance, θ the resulting metacenter, M is coincident with G and thus the metacentric height, GM is zero (zero moment). Thus, the resulting distance BM is equal to the distance GB (remains the same prior to and after the rotational disturbance) due to the smaller than design rotational disturbance, θ. One may note that because a smaller than design rotational disturbance, θ has been applied, Equations 2.261 and 2.263 for GM and BM, respectively, are no longer valid; instead, an experimental/empirical approach may be used to compute GM and BM.

And, finally, for the case of a top-heavy floating body that is spherical-shaped, as illustrated in Figure 2.41, for a small (design) rotational disturbance, θ the computed metacenter, M is coincident with G and thus, the metacentric height, $GM = BM - GB$ is zero (zero moment). Thus, the computed distance $BM = (I_y/V_{dispfluid})$ is equal to the distance GB (remains the same prior to and after the rotational disturbance) due to the spherical-shaped floating body.

EXAMPLE PROBLEM 2.42

A block of wood with uniform density weighs 400 N and floats in water. Its dimensions are $H = 40$ cm by $h = 40$ cm and $b = 80$ cm along the proposed axis of rotation, y as illustrated in Figure EP 2.42. (a) Determine if the floating block of wood is top heavy or bottom heavy. (b) Determine if the floating block will be in a position of rotational stable equilibrium about its longitudinal y-axis if the block is rotated by 10° (assumed to be a small rotational disturbance) clockwise about the y-axis. (c) Determine the magnitude and the type of moment that result from the rotational disturbance.

Mathcad Solution

(a) In order to determine if the block is top heavy or bottom heavy, one must first determine the location of the center of gravity, G and the location of the center of buoyancy, B. A free body diagram of the floating wood is illustrated in Figure EP 2.42b. Because the block of wood has a uniform density, the location of the center of gravity, G is at the centroid of the vertical rectangular cross section, which is located at $G = H/2 = 0.2$ m from the bottom of the block and at $h/2 = 0.2$ m from either side of the block as shown in Figure EP 2.42b. In order to determine the location of B, one must first determine the unknown submerged depth, d. Application of Newton's second law of motion for fluids in static equilibrium to the floating wood (body) in the vertical direction yields the following:

$$\sum F_z = -W + F_B = 0$$

The submerged depth, d and the buoyant force, $F_B = \gamma_w V_{dispfluid}$ are evaluated as follows:

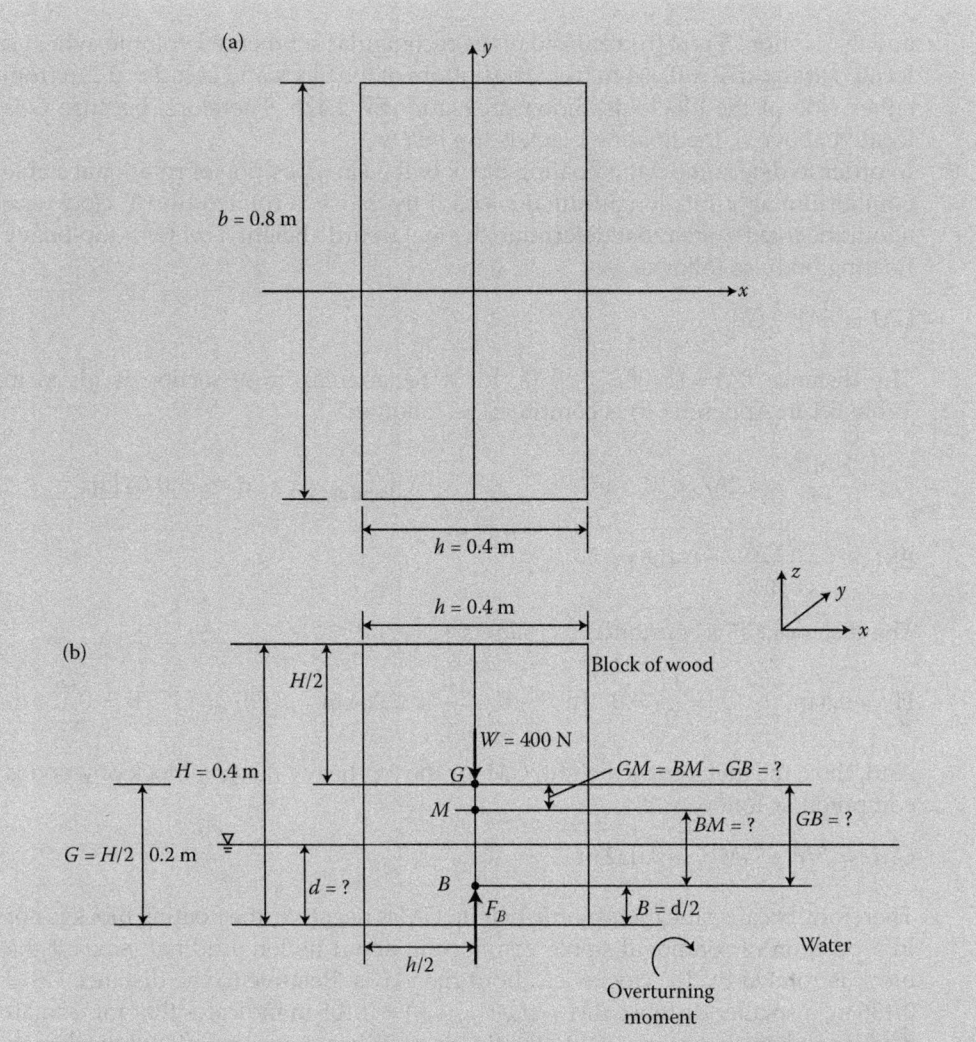

FIGURE EP 2.42
Buoyancy and stability of a floating body in water. (a) Floating body's horizontal cross section; plan view of floating body at the waterline. (b) Computation of the metacentric height, *GM* to determine the rotational stability of a top-heavy floating body for a small design rotation, θ.

$$W := 400 \, N \qquad\qquad \gamma_w := 9810 \, \frac{N}{m^3} \qquad\qquad h := 0.4 \, m \qquad\qquad b := 0.8 \, m$$

Guess value: $\qquad\qquad d := 0.1 \, m \qquad\qquad F_B := 100 \, N$

Given

$$-W + F_B = 0 \qquad\qquad\qquad\qquad F_B = \gamma_w \, h \, d \, b$$

$$\begin{pmatrix} d \\ F_B \end{pmatrix} := \text{Find } (d, F_B)$$

$$d = 0.127 \, m \qquad\qquad\qquad\qquad F_B = 400 \, N$$

and its location, B is at the centroid of the rectangular submerged volume, which is located at $B = d/2 = 0.064$ m from the bottom of the block and at $h/2 = 0.2$ m from either side of the block, as shown in Figure EP 2.42b. Therefore, because G is located above B, the floating block is top heavy.

(b) In order to determine if the floating block will be in a position of rotational stable equilibrium about its longitudinal y-axis if the block is rotated by $10°$ clockwise about the y-axis, one must determine the metacentric height, GM for a top-heavy floating body as follows:

$$GM = BM - GB$$

The distance $BM = (I_y/V_{dispfluid})$ (I_y for a rectangular cross section is given in Table B.1 in Appendix B) is computed as follows:

$$I_y := \frac{b \cdot h^3}{12} = 4.267 \times 10^{-3} \, m^4 \qquad\qquad V_{dispfluid} = h \cdot d \cdot b = 0.041 \, m^3$$

$$BM := \frac{I_y}{V_{dispfluid}} = 0.105 \, m$$

The distance GB is computed as follows:

$$H := 0.4 \, m \qquad G := \frac{H}{2} = 0.2 \, m \qquad B := \frac{d}{2} = 0.064 \, m \qquad GB := G - B = 0.136 \, m$$

And, thus, the metacentric height, GM for the top-heavy floating block of wood is computed as follows:

$$GM := BM - GM = -0.032 \, m$$

Therefore, because the metacentric height, GM is negative, the floating block is not in a position of rotational stable equilibrium about its longitudinal y-axis if the block is rotated by $10°$ clockwise about the y-axis. Relative to the distance $GB = 0.136$ m, a smaller distance $BM = (I_y/V_{dispfluid}) = 0.105$ m indicates that for a small rotational disturbance of $\theta = 10°$, the floating top-heavy unstreamlined block may be considered to be a "tall and slender body." One may recall that I_y represents the geometric design of the body, and the larger the value for I_y, the more rotationally stable the body is about the y-axis. Furthermore, because of the shape of the block of wood and thus the actual design angle of rotation, θ may not be a good indicator of the stability of the top-heavy floating block of wood; instead, the metacentric height, GM is used as an indicator of the stability of the floating block of wood.

(c) The magnitude of the moment, M_{couple} that results from the rotational disturbance is computed as follows:

$$\theta := 10° \qquad M_{couple} := W \, GM \, \sin(\theta) = -2.198 \, N \, m$$

And, finally, because the metacentric height, GM is negative, the floating block is not in a position of rotational stable equilibrium, so the resulting moment, M_{couple} is an overturning moment that is in the clockwise direction (as opposed to a restoring moment that would be in the counterclockwise direction).

EXAMPLE PROBLEM 2.43

A boat weighs 500,000 N and floats in water. Its horizontal cross section (plan view) at the water line and its vertical cross section, along with the dimensions, are illustrated in Figure EP 2.43. Assume that the boat is top heavy (i.e., G is located above B). (a) Determine the buoyant force, F_B and the volume of the displaced fluid, $V_{dispfluid}$. (b) Determine the maximum distance GB that the center of gravity, G may be above the center of buoyancy, B in order for the boat to be in a position of rotational stable equilibrium.

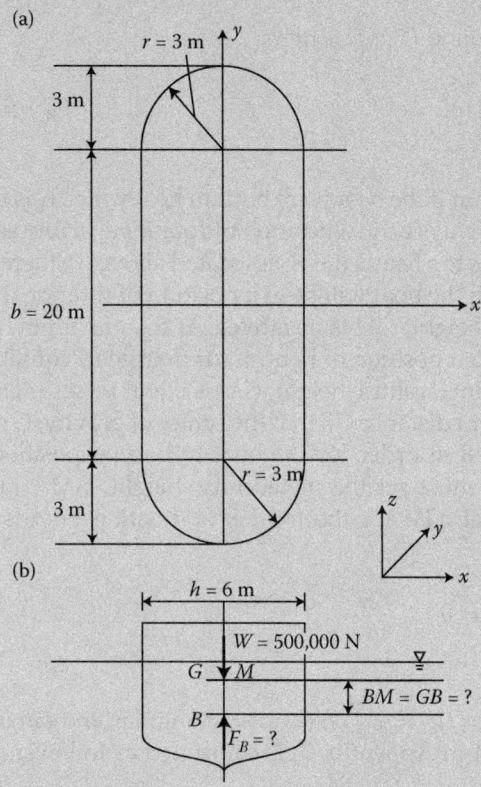

FIGURE EP 2.43
Buoyancy and stability of a floating body in water. (a) Floating body's horizontal cross section; plan view of floating body at the waterline. (b) Computation of the maximum distance GB for the rotational stability of a top-heavy floating body for a small design rotation, θ.

Mathcad Solution

(a) In order to determine the buoyant force, F_B and the volume of the displaced fluid, $V_{dispfluid}$, one may apply Newton's second law of motion for fluids in static equilibrium to the floating boat (body) in the vertical direction, which yields the following:

$$\sum F_z = -W + F_B = 0$$

The $V_{dispfluid}$ and the buoyant force, $F_B = \gamma_w V_{dispfluid}$ are determined as follows:

$$W := 500{,}000 \text{ N} \qquad\qquad \gamma_w := 9810 \frac{\text{N}}{\text{m}^3}$$

Guess value: $\qquad\qquad V_{dispfluid} = 10 \text{ m}^3 \qquad\qquad F_B := 1000000 \text{ N}$

Given

$$-W + F_B = 0 \qquad\qquad F_B = \gamma_w \cdot V_{dispfluid}$$

$$\begin{pmatrix} V_{dispfluid} \\ F_B \end{pmatrix} := \text{Find } (V_{dispfluid}, F_B)$$

$$V_{dispfluid} = 50.968 \text{ m}^3 \qquad\qquad\qquad\qquad F_B = 5 \times 10^5 \text{ N}$$

(b) It may be noted that if the boat were bottom heavy (i.e., G is located below B), then the boat would always be in a position of rotational stable equilibrium. However, because the boat is top heavy (i.e., G is located above B), there is a maximum height for G above which the boat will be in a position of rotational unstable equilibrium (i.e., metacentric height, GM is negative). At the maximum height for G above B, the boat will be in a position of neutral (or marginal) equilibrium. This condition occurs when the metacentric height, GM is equal to zero. Thus, in order to determine the maximum distance GB that the center of gravity, G may be above the center of buoyancy, B in order for the boat to be in a position of rotational stable equilibrium, one must set the metacentric height, GM equal to zero. Thus, the metacentric height, GM for the top-heavy floating boat is set equal to zero as follows:

$$GM = BM - GB = 0$$
$$BM = GB$$

The distance $BM = (I_y / V_{dispfluid})$ (I_y for rectangular and circular cross sections are given in Table B.1 in Appendix B) is computed as follows:

$$h := 6 \text{ m} \qquad\qquad b := 20 \text{ m} \qquad\qquad D := 6 \text{ m}$$

$$I_{yrectangle} := \frac{b \cdot h^3}{12} = 360 \text{ m}^4 \qquad\qquad I_{ycircle} := \frac{\pi \cdot D^4}{64} = 63.617 \text{ m}^4$$

$$I_y := I_{yrectangle} + I_{ycircle} = 423.617 \text{ m}^4$$

$$BM := \frac{I_y}{V_{dispfluid}} = 8.311 \text{ m}$$

Thus, the maximum distance GB that the center of gravity, G may be above the center of buoyancy, B in order for the boat to be in a position of rotational stable equilibrium is:

$$GB := BM = 8.311 \text{ m}$$

EXAMPLE PROBLEM 2.44

A rectangular barge floats in water. Its dimensions are $H = 3$ m by $h = 9$ m and $b = 15$ m along the proposed axis of rotation, y as illustrated in Figure EP 2.44. Assume that the barge is submerged at a depth, d (draft) of 2 m, and the center of gravity, G is located 2 m from the bottom of the barge and at $h/2 = 4.5$ m from either side of the barge, as shown in Figure EP 2.44b. (a) Determine if the floating barge is top heavy or bottom heavy. (b) Determine the weight of the barge, W; the buoyant force, F_B; and the volume of the displaced fluid, $V_{dispfluid}$. (c) Determine if the floating barge will be in a position of rotational stable equilibrium about its longitudinal y-axis if the barge is rotated by $11°$ (assume to be a small rotational disturbance) clockwise about the y-axis. (d) Determine the magnitude and the type of moment that results from the rotational disturbance.

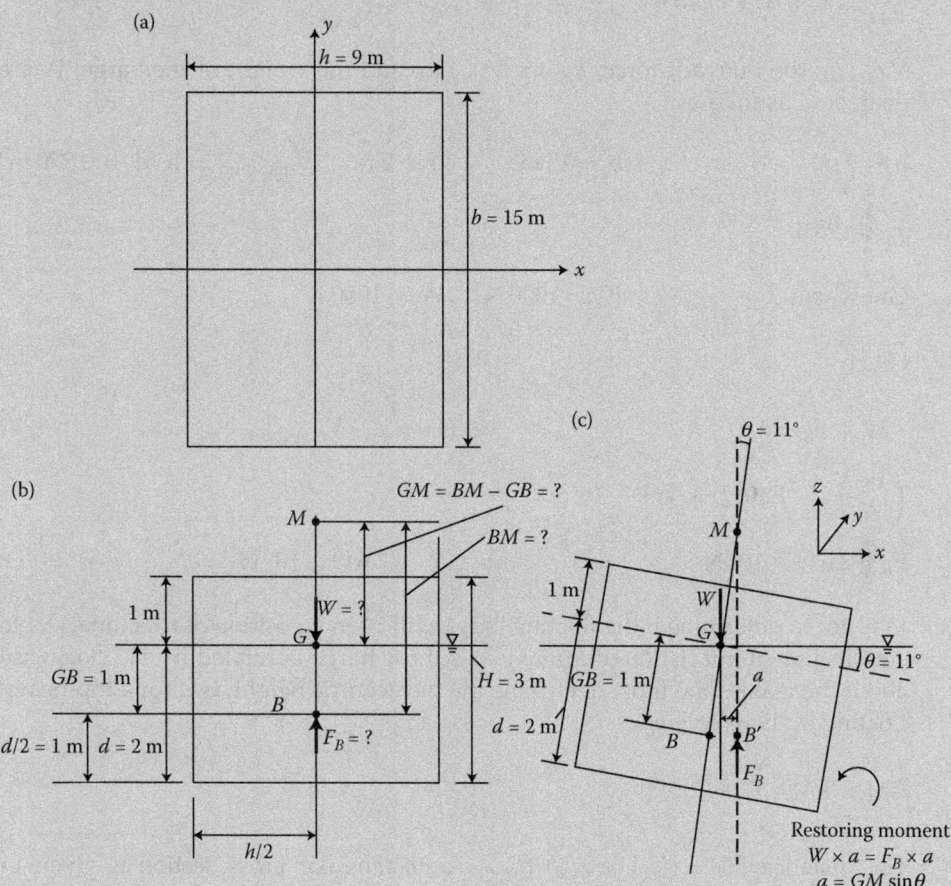

FIGURE EP 2.44
Computation of the metacentric height, GM to determine the rotational stability of a top-heavy floating body for a small design rotation, θ. (a) Floating body's horizontal cross section; plan view of floating body at the waterline. (b) Original position of stable equilibrium. (c) Small design rotation, θ.

Mathcad Solution

(a) In order to determine if the barge is top heavy or bottom heavy, one must first determine the location of the center of gravity, G and the location of the center of buoyancy, B. A free body diagram of the floating barge is illustrated in Figure EP 2.44b. The location of G was given at $G = 2$ m from the bottom of the barge. Given a draft of 2 m, the location of B is at the centroid of the rectangular submerged volume, which is located at $B = d/2 = 1$ m from the bottom of the barge and at $h/2 = 4.5$ m from either side of the barge, as shown in Figure EP 2.44b. Therefore, because G is located above B, the floating block is top heavy.

(b) In order to determine the weight of the barge, W; the buoyant force, F_B; and the volume of the displaced fluid, $V_{dispfluid}$, one may apply Newton's second law of motion for fluids in static equilibrium to the floating barge (body) in the vertical direction, which yields the following:

$$\sum F_z = -W + F_B = 0$$

$V_{dispfluid}$; the buoyant force, $F_B = \gamma_w V_{dispfluid}$; and the weight of the barge, W are evaluated as follows:

$h := 9\,\text{m}$ $b := 15\,\text{m}$ $d := 2\,\text{m}$ $V_{dispfluid} := h \cdot d \cdot b = 270\,\text{m}^3$

$\gamma_w := 9810\,\dfrac{\text{N}}{\text{m}^3}$

Guess value: $F_B := 1000\,\text{N}$ $W := 1000\,\text{N}$

Given

$-W + F_B = 0$ $F_B = \gamma_w \cdot V_{dispfluid}$

$\begin{pmatrix} F_B \\ W \end{pmatrix} := \text{Find}\,(F_B,\,W)$

$F_B = 2.649 \times 10^6\,\text{N}$ $W = 2.649 \times 10^6\,\text{N}$

(c) In order to determine if the floating barge will be in a position of rotational stable equilibrium about its longitudinal y-axis if the barge is rotated by $11°$ clockwise about the y-axis, one must determine the metacentric height, GM for a top-heavy floating body as follows:

$$GM = BM - GB$$

The distance $BM = (I_y/V_{dispfluid})$ (I_y for a rectangular cross section is given in Table B.1 in Appendix B) is computed as follows:

$I_y := \dfrac{b \cdot h^3}{12} = 911.25\,\text{m}^4$ $BM := \dfrac{I_y}{V_{dispfluid}} = 3.375\,\text{m}$

The distance GB is computed as follows:

$$G := 2\,\text{m} \qquad B := \frac{d}{2} = 1\,\text{m} \qquad GB := G - B = 1\,\text{m}$$

And, thus, the metacentric height, GM for the top-heavy floating barge is computed as follows:

$$GM := BM - GB = 2.375\,\text{m}$$

Therefore, because the metacentric height, GM is positive, the floating barge is in a position of rotational stable equilibrium about its longitudinal y-axis if the barge is rotated by $11°$ clockwise about the y-axis. Relative to the distance $GB = 1\,\text{m}$, a larger distance $BM = (I_y/V_{dispfluid}) = 3.375$ m indicates that for a small rotational disturbance of $\theta = 11°$, the top-heavy floating barge may be considered to be a "wide-based body." One may recall that I_y represents the geometric design of the body, and the larger the value for I_y, the more rotationally stable the body is about the y-axis. Furthermore, because the shape of the barge and thus the actual design angle of rotation, θ may not be good indicators of the stability of the top-heavy floating barge, instead, the metacentric height, GM is used as an indicator of the stability of the floating barge.

(d) The magnitude of the moment, M_{couple} that results from the rotational disturbance is computed as follows:

$$\theta := 11\ \text{deg} \qquad M_{couple} := W\,GM\,\sin(\theta) = 1.2 \times 10^6\,\text{N m}$$

And, finally, because the metacentric height, GM is positive, the floating barge is in a position of rotational stable equilibrium and thus the resulting moment, M_{couple} is a restoring moment that is in the counterclockwise direction (as opposed to an overturning moment that would be in the clockwise direction).

End-of-Chapter Problems

Problems with a "C" are conceptual problems. Problems with a "BG" are in English units. Problems with an "SI" are in metric units. Problems with a "BG/SI" are in both English and metric units. All "BG" and "SI" problems that require computations are solved using Mathcad.

Introduction

2.1C What type of problems does fluid statics deal with?

2.2C In the study of fluids in static equilibrium, what are the significant forces acting on a fluid element?

2.3C In the study of fluids in static equilibrium, describe the sum of all the external forces.

2.4C Fluid statics applies the principles of hydrostatics, which are based on what laws?

2.5C Hydrostatic fluid pressure exits in what types of flow situations?

2.6C What fluid properties play an important role in fluid statics?

2.7C Most fluid properties (especially the density for a gas/compressible flow) are a function of the temperature and pressure. What assumptions are made regarding the density of a gas/compressible flow vs. a liquid/incompressible flow?

2.8C The topic of fluid statics, which is based on the principles of hydrostatics, addresses what three topics?

2.9C When is the measurement of the hydrostatic pressure at a point important?

2.10C When is the determination of the hydrostatic force acting on a submerged surface important?

2.11C When is the role of the hydrostatic force in the buoyancy and stability of a floating or a neutrally buoyant body important?

The Principles of Hydrostatics

2.12C The principles of hydrostatics are based on Pascal's law and Newton's second law of motion for fluid statics and yield what equation?

2.13C While Pascal's law describes the hydrostatic pressure at a point, what does the hydrostatic pressure equation describe?

2.14C Application of the principles of hydrostatics requires the definition of what three basic terms?

2.15C Describe the application of Newton's second law of motion for a fluid element in static equilibrium (at rest).

2.16C What are two significant forces that act on a fluid element at rest? Also, describe if these forces are normal or tangential to the fluid element.

2.17C What are two forces that do not come into play in fluid statics?

2.18C Define the hydrostatic pressure, p.

2.19C What are the dimensions and units for the pressure, p in the English system?

2.20C What are the dimensions and units for the pressure, p in the metric system?

2.21C A common unit for pressure, p in the metric system is the bar (b). Define the metric pressure unit bar (b).

2.22C What does Pascal's law state? Illustrate Pascal's law.

2.23C Explain if the pressure at a point is a vector or a scalar.

2.24C Explain if the differential (unit) force, dF_p produced as a result of the action of pressure, p on a differential (unit) area, dA is a vector or a scalar.

2.25C While Pascal's law states that the hydrostatic pressure acts equally in all directions at any point in a static body of fluid, it does not indicate how the hydrostatic pressure varies from one point to another within the body of fluid. What equation and thus what law, describes how the hydrostatic pressure varies from one point to another within the body of fluid?

2.26C Illustrate how application of Newton's second law of motion for fluids in static equilibrium, $\sum F = 0$ (Equation 2.1) to the fluid element given in Figure 2.1 in each of the three (x, y, z) directions yields the derivation of the hydrostatic pressure equation, $(dp/dz) = \gamma = \rho g$ (Equation 2.18) that represents the scalar ordinary differential equation of fluid statics. Furthermore, highlight Pascal's law in the derivation process.

2.27C The solution of the hydrostatic pressure equation $(dp/dz) = \gamma = \rho g$ (Equation 2.18), which represents the scalar ordinary differential equation of fluid statics, will depend upon how density of the fluid, ρ is related to pressure, p and temperature, T and/or the height of the fluid, z. What assumptions are made regarding the density of a gas/compressible flow vs. a liquid/incompressible flow?

2.28C What does the hydrostatic pressure equation $(dp/dz) = \gamma = \rho g$ (Equation 2.18) illustrate/state regarding the variation of pressure in each of the three (x, y, z) directions? What specifically about the hydrostatic pressure equation supports your answer?

2.29C In the application of the hydrostatic pressure equation $(dp/dz) = \gamma = \rho g$ (Equation 2.18), what is a common assumption made regarding the compressibility of the flow? What is the implication of this common assumption?

2.30C Assuming that all liquids and some gases may be treated as incompressible, assume ρ to be a constant, and integrate the hydrostatic pressure equation $(dp/dz) = \gamma = \rho g$ (Equation 2.18) over the height of fluid, h in Figure 2.2, and explain the resulting equation.

2.31C What does the symbol $\triangledown$ on the free surface of the fluid indicate?

2.32C Referring to Figure 2.2, illustrate the application of Pascal's law and illustrate the application of the hydrostatic pressure equation.

2.33SI An open tank contains 400 m of water, as illustrated in Figure ECP 2.33. Assume that the atmospheric pressure at the free surface, p_0 is equal to zero. (a) Determine the pressure at points 1 and 2, located at 300 m and 400 m below the water surface, respectively.

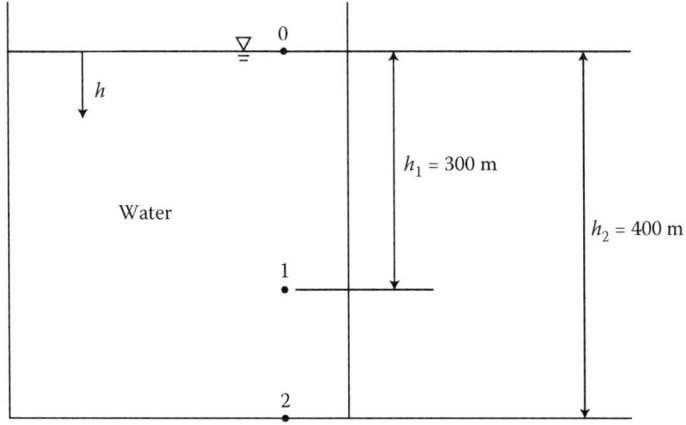

FIGURE ECP 2.33

2.34C The hydrostatic pressure in a body of fluid in static equilibrium increases linearly with depth, where the density (or the specific weight) is represented by the slope of the straight line. Graphically illustrate the pressure prism on the side of a tank of fluid that is open to the atmosphere, and evaluate the pressure at the free surface; the pressure at any depth, h along the z-axis; and the pressure at the bottom of the tank.

2.35C Referring to Figure 2.4, define the hydrostatic pressure head, and give its units in both the English and metric systems.

2.36BG For a given pressure, p, the hydrostatic pressure head, $h = (p/\gamma)$ will vary with the specific weight, γ of the fluid. (a) For a pressure of 600 psi, compare the magnitudes of the hydrostatic pressure head, $h = (p/\gamma)$ for the following fluids at 68°F: water, mercury, crude oil, and gasoline.

2.37BG A submarine is located at depth of 1200 ft under the sea. Determine the pressure the submarine will be subjected to at this depth.

2.38SI It is of interest to compare the hydrostatic pressure and the weight of the water at the bottom of two pools, each filled with 2.5 m of water. The first pool has a circular area of 40 m^2, and the second pool has a rectangular area of 60 m^2. (a) Compare the pressures at the bottom of the two pools. Explain. (b) Compare the weights of the water at the bottom of the two pools. Explain.

2.39BG A very small gold spherical charm is suspended by a very thin chain in a bottle of cleaning fluid ($s_{cf} = 1.25$) to a depth of 4 inches. (a) Determine the hydrostatic pressure on each side of the spherical charm. (b) State what law(s) govern(s) the determination of the hydrostatic pressure on each side of the spherical charm.

2.40SI An open tank contains 600 m of gasoline on top of 300 m of oil, as illustrated in Figure ECP 2.40. Assume that the atmospheric pressure at the free surface, p_0 is equal to zero, the specific gravity of the gasoline is 0.68, and the specific gravity of the oil is 0.89. (a) Determine the pressure at interface of the two fluids at point 1, and at the bottom of the tank at point 2. (b) Illustrate the pressure prism on the side of the tank.

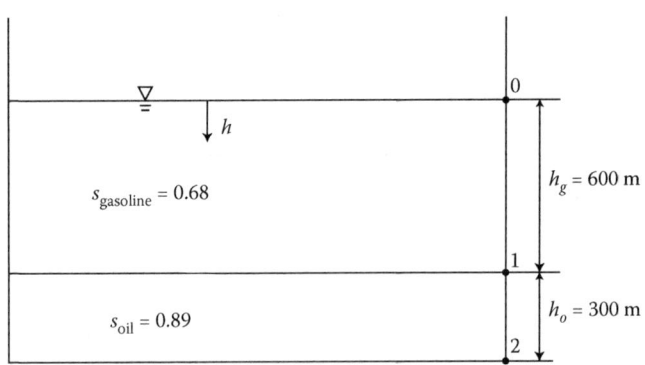

FIGURE ECP 2.40

2.41BG The pressure gage at an elevation of 300 ft at the side of a liquid storage tank reads 200 psi, while a second pressure gage at an elevation of 0 ft at the side of the tank reads 310 psi. (a) Determine the specific weight, density, and specific gravity of the liquid. (b) What type of liquid might be in the tank?

2.42C   Refer to Figure 2.5, and define the pressure at point 1, the pressure head at point 1, the elevation head at point 1, and the piezometric head at point 1.

2.43C   Refer to Figure 2.5, and illustrate that, in accordance to the principles of hydrostatics, the piezometric head, $(p/\gamma + z)$ remains constant throughout a body of still water, and thus: $(\partial((p/\gamma) + z)/\partial s) = 0$ (Equation 2.30) whatever the direction of s may be.

2.44C   List three basic goals in the application of the principles of hydrostatics.

2.45C   Give three practical applications of the principles of hydrostatics for each of the following three basic goals: to measure the hydrostatic pressure at a point, to measure the pressure difference between two points, or to measure the hydrostatic pressure on a surface, which is defined as the hydrostatic force.

Measurement of Hydrostatic Pressure at a Point

2.46C   The measurement of pressure at a point can be accomplished by using numerous methods with numerous apparati. Explain what the difference between the methods and apparati depend upon.

2.47C   Give some common types of apparati used to measure the pressure at a point.

2.48C   Distinguish the difference between the absolute pressure vs. the gage pressure.

2.49C   Describe the gage pressure (relative pressure) of a fluid: at the atmospheric pressure, above the atmospheric pressure, and below the atmospheric pressure.

2.50C   Refer to Figure 2.7, and explain why the datum for the atmospheric pressure is a moveable datum/scale (for the gage pressure).

2.51C   Refer to Figure 2.7, and explain the role and the definition of the fixed datum.

2.52C   Refer to Figure 2.7. The moveable atmospheric pressure datum is measured relative to the fixed absolute zero pressure datum. Therefore, the variable atmospheric pressure above the fixed absolute zero is reported in absolute pressure units (*psi* [abs] or psia in the English system, and *Pa* [abs] in the metric system). How is the absolute pressure of a fluid above the fixed absolute zero computed?

2.53C   Refer to Figure 2.7. Describe the absolute pressure of a fluid: at the atmospheric pressure (*p*[*gage*] is equal to zero), above the atmospheric pressure (gage pressure, *p*[*gage*] is positive), and below the atmospheric pressure (gage pressure, *p*[*gage*] is negative).

2.54C   Describe the reading of a pressure gage for a completely inflated tire vs. a completely deflated tire.

2.55SI   The gage pressure for a tire reads 350 kPa. Assume an atmospheric pressure of 101.325 *kPa* (abs). (a) Determine the absolute pressure in the tire.

2.56C   Describe the reading of a pressure gage for a suction line.

2.57BG   The pressure gage at the suction (inlet) side of a pump reads 5 *psi* vacuum. Assume an atmospheric pressure of 14.7 *psia*. (a) Determine the corresponding absolute pressure at the suction (inlet) side of a pump.

2.58C   In the application of the principles of hydrostatics in fluid mechanics problems, pressures are typically assumed to be gage pressures unless specifically indicated as absolute pressure, where a positive gage pressure is referred to simply as "gage pressure" or "pressure," while a negative gage pressure is referred to as a "vacuum." Explain why the assumption of the gage pressure scale is typically made.

2.59C Working in the gage pressure scale allows the representation of a negative gage pressure or a "vacuum" when the top of a liquid or a gas in a tank (or at any point in a pipe flow system) is subjected to an absolute pressure that is lower than the atmospheric pressure. What is the maximum possible negative pressure that the top of a liquid or a gas in a tank (or at any point in a pipe flow system) can be subjected to?

2.60C The local atmospheric (barometric) pressure varies with time and location on earth and thus will vary with the elevation of the particular surface on earth and the weather conditions that affect the density of the air. What is the standard atmospheric pressure at sea level at an atmospheric temperature of 59° Fahrenheit (English) or 15° Celsius (metric)?

2.61C What is the corresponding standard acceleration due to gravity at the standard sea-level atmospheric pressure and a latitude of 45°?

2.62SI A pressure gage attached to a tank reads 20 *kPa*. Assume an atmospheric pressure of 101.325 *kPa* (abs). (a) Determine the absolute pressure in the tank.

2.63SI A closed tank is partially filled with carbon tetrachloride to a depth of 4 *m*. The air pressure in the tank is 30 *kPa*. Assume an atmospheric pressure of 101.325 *kPa* (abs). (a) Determine the absolute pressure of the air pressure in the tank. (b) Determine the gage pressure, and the absolute pressure at the bottom of the tank.

2.64C In the measurement of the hydrostatic pressure at a point, it is important to distinguish between the types of hydrostatic pressures that are measured. What are three types of hydrostatic pressures that are measured at a point?

2.65C In general, what is the basic difference between the use of a barometer vs. a manometer?

2.66C The various types of manometers include the open piezometer column/tube (simplest manometer), the open U-tube manometer (simple manometer), and the differential manometer. How does one determine the appropriate type of manometer to use in order to measure the gage pressure in a fluid in a tank of fluid or in a pipe flow?

2.67C Compare the advantages and the disadvantages of using a manometer vs. a pressure gage (mechanical pressure-measuring devices) to directly measure the gage pressure in a fluid in a tank of fluid or in a pipe flow.

2.68C Mechanical pressure-measuring devices known as pressure gages may be used to measure either absolute or gage pressure. Give two types of pressure gages, and give two examples where they may be used to measure the pressure at a given point.

2.69C Compare pressure gages (mechanical pressure-measuring devices) vs. pressure transducers (more sophisticated electronic pressure-measuring devices) used to measure the pressure at a given point.

2.70C Give two types of pressure transducers, and give an example where they may be used to measure the pressure at a given point.

Barometers

2.71C   What are two typical fluids used in a barometer in order to measure the atmospheric (barometric) pressure?

2.72C   Assuming mercury for the barometric fluid, derive the atmospheric (barometric) pressure head, $h = (p_{atm}/\gamma_{Hg})$ (Equation 2.41).

2.73BG   Assuming the English units, the standard atmospheric pressure, p_{atm} is 14.696 *psi* (abs). Assume the barometer fluid is glycerin at room temperature (68°F). (a) Determine the height, h that the column of glycerin would rise to in a barometer.

2.74SI   Assume the barometric fluid is benzene at room temperature (20°C). (a) If the height, h that the column of benzene rises to in a barometer is 13.5 *m*, determine the atmospheric pressure.

2.75C   What is the definition of a torr?

2.76SI   A pressure gage attached to a tank reads 130 *mm Hg* vacuum. Assume an atmospheric pressure of 996 *mb* (abs). (a) Determine the absolute pressure in the tank.

2.77BG   A mercury barometer is used to measure the height of a mountain, where the average specific weight of the air is 0.025 lb/ft^3. The barometric reading at the foot of the mountain is 28 in *Hg*, and the barometric reading at the top of the mountain is 25 in *Hg*. (a) Determine the height of the mountain.

2.78C   In the derivation of the atmospheric (barometric) pressure head $h = (p_{atm}/\gamma_{Hg})$ (Equation 2.41), the significant forces assumed to be acting on a fluid element at rest included those due to gravity, $F_G = Mg = W = \gamma V$ and pressure, $F_p = \Delta pA$. Furthermore, the vapor pressure, p_v of the mercury was assumed to be very small and was ignored. State what modifications (or corrections) are made to these assumptions when an accurate measurement of the atmospheric pressure is required.

2.79BG   Assuming the English units, the standard atmospheric pressure, p_{atm} is 14.696 *psi* (abs). (a) Determine the height, h that the column of mercury (32°F, specific gravity is 13.595) would rise to in a barometer. (b) Determine the height, h that the column of water (39.2°F) would rise to in a barometer. (c) Compare the results of (a) and (b) and explain the difference in the magnitudes of h.

2.80C   Explain what makes mercury the most convenient and accurate barometric fluid.

Manometers

2.81C   What is an open piezometer column/tube, and when is it appropriate to use it?

2.82C   What is a simple manometer, and when is it appropriate to use it?

2.83C   What is a differential manometer, and when is it appropriate to use it?

2.84C   There are different types of differential manometers (single U-tube, triple U-tube, etc.; single fluid, multifluid). What determines the appropriate differential manometer to use?

2.85C   What type of manometer fluid used in a piezometer? Explain.

2.86BG   Three piezometers are attached to a tank of carbon tetrachloride subjected to a positive gage air pressure, as illustrated in Figure ECP 2.86. Points 1, 2, and 3 are located 6 *ft*, 5 *ft*, and $0.5\,ft$ from the bottom of the tank. The pressure head (height of the carbon tetrachloride in the piezometers), p_1/γ at point 1 reads $33\,ft$. (a) Determine the pressure heads read at points 2 and 3. (b) Illustrate that the piezometric head $(p/\gamma + z)$ remains a constant throughout the tank of carbon tetrachloride.

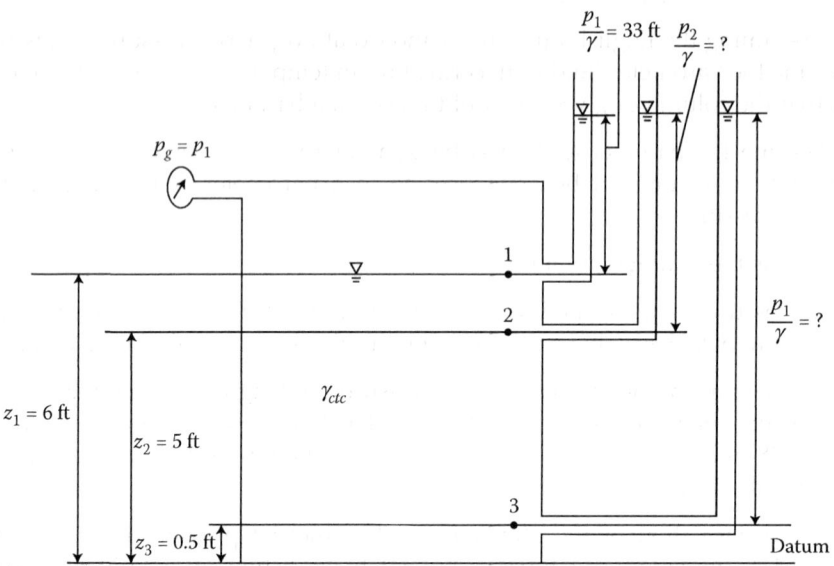

FIGURE ECP 2.86

2.87BG A piezometer is attached to a pipe flowing under pressure carrying glycerin, as illustrated in Figure ECP 2.87. The height of the glycerin in the piezometer at point 1 is $5\,ft$. (a) Determine the gage pressure at point 1.

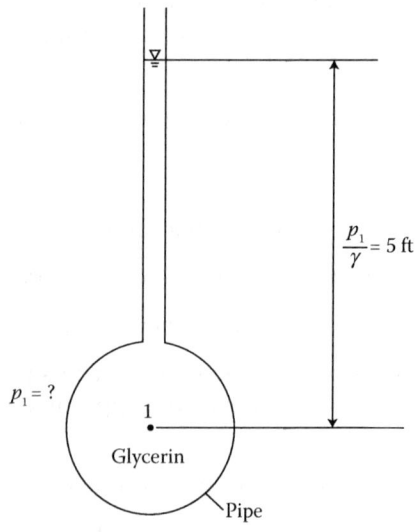

FIGURE ECP 2.87

2.88C A simple manometer may be either a single-fluid or a multifluid manometer. Discuss the difference between these two types of simple manometers.

2.89C Discuss what types of manometer fluids are typically used in a simple manometer.

2.90C What factors affect the choice of the manometer fluids for a simple manometer?

2.91BG A simple manometer contains glycerin and crude oil, both at 68°F, as illustrated in Figure ECP 2.91. The left portion of the tube contains glycerin only. The right portion of the tube contains 0.5 *ft* of glycerin and 4 *ft* of crude oil. (a) Determine the height of the glycerin in the left portion of the manometer.

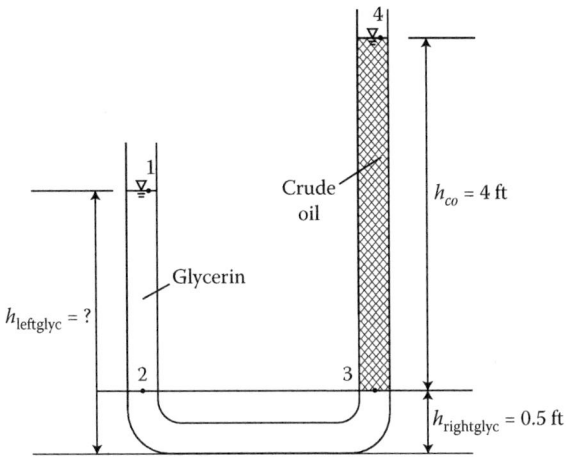

FIGURE ECP 2.91

2.92SI Seawater in tank is subjected to a positive gage air pressure, as illustrated in Figure ECP 2.92. A single-fluid simple manometer is attached to the side of the tank. The manometer fluid is mercury and the manometer reading, h_m is 0.8 *m*. Assume an atmospheric

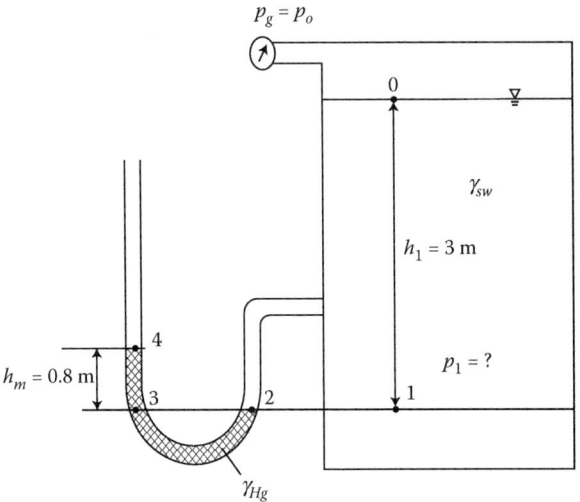

FIGURE ECP 2.92

pressure of 101.325 kPa (abs). (a) Determine the gage pressure at point 1. (b) Given the depth to point 1, h_1 is 3 m, determine the gage pressure at the seawater surface at point 0.

2.93BG   Water in tank is subjected to a negative gage air pressure of -9 psi, as illustrated in Figure ECP 2.93. A single-fluid simple manometer is attached to the side of the tank. The manometer fluid is mercury and the manometer reading, h_m is 0.45 ft. Assume an atmospheric pressure of 14.7 $psia$. (a) Determine the gage pressure at point 1. (b) Determine the depth of point 1 from the water surface.

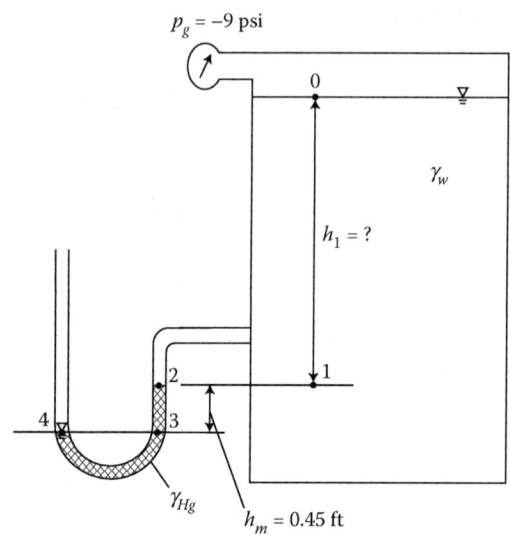

FIGURE ECP 2.93

2.94BG   Water flows in a pipe under a high positive gage pressure, as illustrated in Figure ECP 2.94. A single-fluid simple manometer is attached to the pipe in order to measure the gage pressure at the centerline of the pipe at point 1. The distance between points 2 and 3 is 0.8 ft. The manometer fluid is mercury and the manometer reading, h_m is 5.7 ft. Assume an atmospheric pressure of 14.7 $psia$. Determine the gage pressure at point 1.

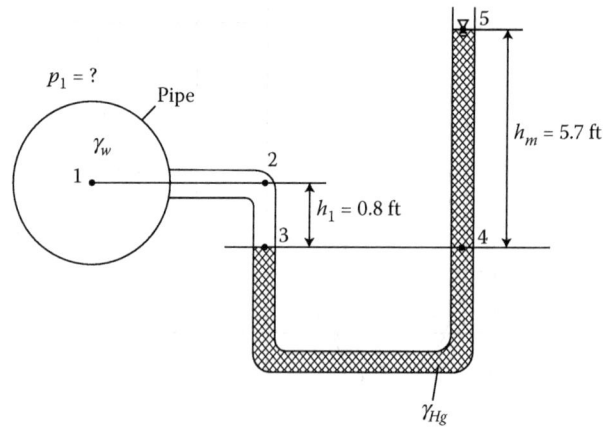

FIGURE ECP 2.94

2.95SI Water flows in a pipe under a high negative gage pressure, as illustrated in Figure ECP 2.95. A single-fluid simple manometer is attached to the pipe in order to measure the gage pressure at the centerline of the pipe at point 1. The distance between points 2 and 3 is 0.14 m. The manometer fluid is mercury and the manometer reading, h_m is 0.68 m. Assume an atmospheric pressure of 101.325 kPa (abs). Determine the gage pressure at point 1.

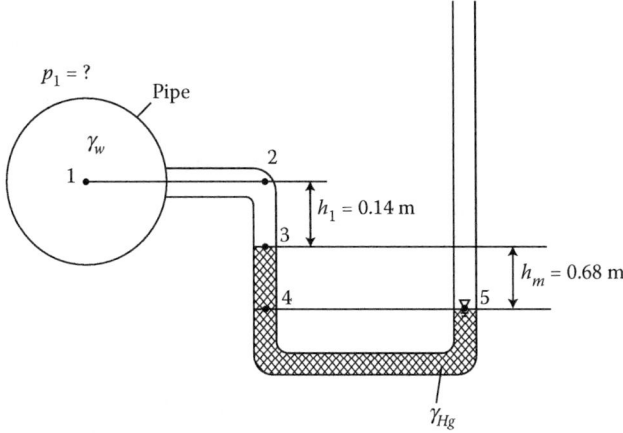

FIGURE ECP 2.95

2.96SI Water in tank is subjected to a positive gage air pressure, as illustrated in Figure ECP 2.96. A multifluid simple manometer is attached to the side of the tank. The first manometer fluid is glycerin with a specific gravity of 1.258, and a manometer reading, h_{m1} of 1.23 m. The second manometer fluid is mercury with a manometer reading, h_{m2} of 3.6 m. The distance between points 1 and 2 is 0.45 m. Assume an atmospheric pressure of 101.325 kPa (abs). Determine the gage pressure at point 1.

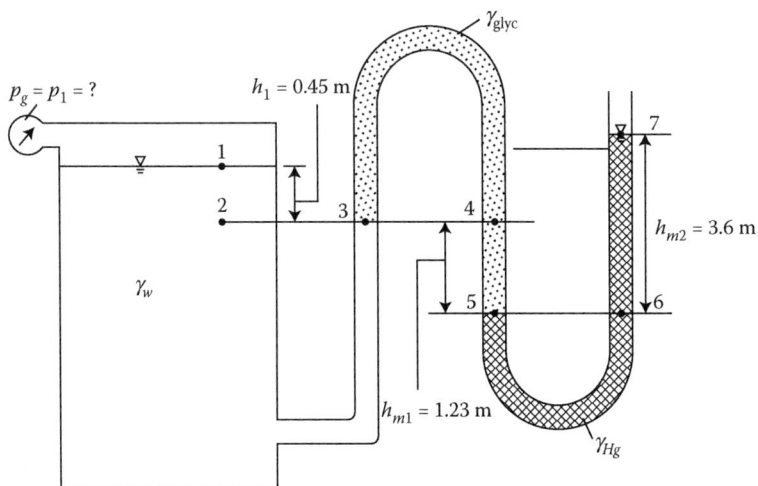

FIGURE ECP 2.96

2.97BG   Water flows in a pipe, as illustrated in Figure ECP 2.97. A single-fluid differential manometer is attached to the pipe at points 1 and 2 in order to measure the pressure drop between points 1 and 2. The distance between points 2 and 5 is 5 ft. The manometer fluid is mercury and the manometer reading, h_m is 2.5 ft. Determine the pressure drop between points 1 and 2.

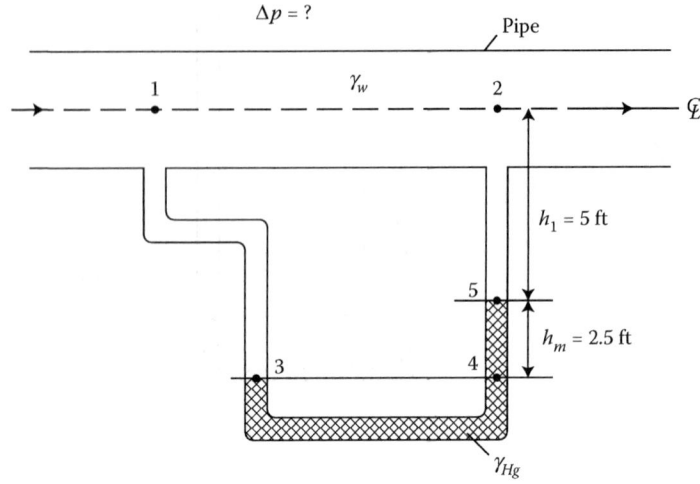

FIGURE ECP 2.97

2.98SI   Water flows in a pipe 1, and oil with a specific gravity of 0.89 flows in pipe 2, as illustrated in Figure ECP 2.98. A single-fluid differential manometer is attached between pipe 1 and pipe 2 in order to measure the pressure difference between pipe 1 and pipe 2. The distance between points 1 and 3 is 2.76 m, and the distance between points 2 and 5 is 1.4 m. The manometer fluid is mercury and the manometer reading, h_m is 0.57 m. Determine the pressure difference between pipe 1 and pipe 2.

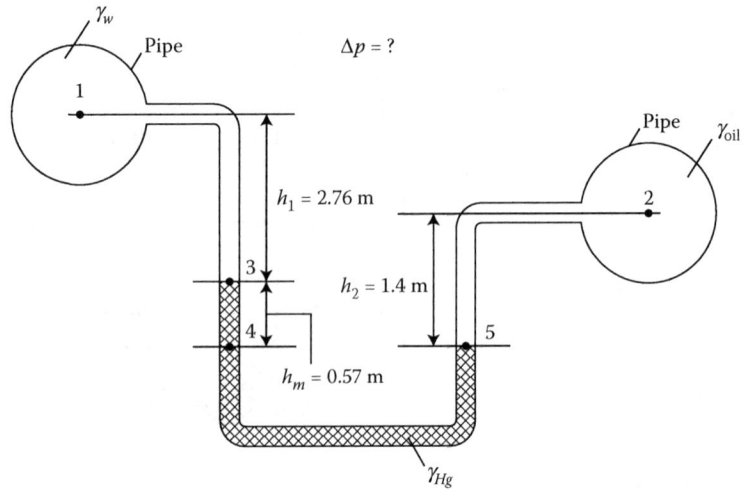

FIGURE ECP 2.98

2.99SI A tank containing carbon tetrachloride is connected to a tank containing kerosene by a single-fluid differential manometer, as illustrated in Figure ECP 2.99. The distance between points 1 and 3 is $0.8\,m$, and the distance between points 2 and 5 is $0.4\,m$. The manometer fluid is mercury and the manometer reading, h_m is $0.25\,m$. Determine the pressure difference between the air in tank 1 and the air in tank 2.

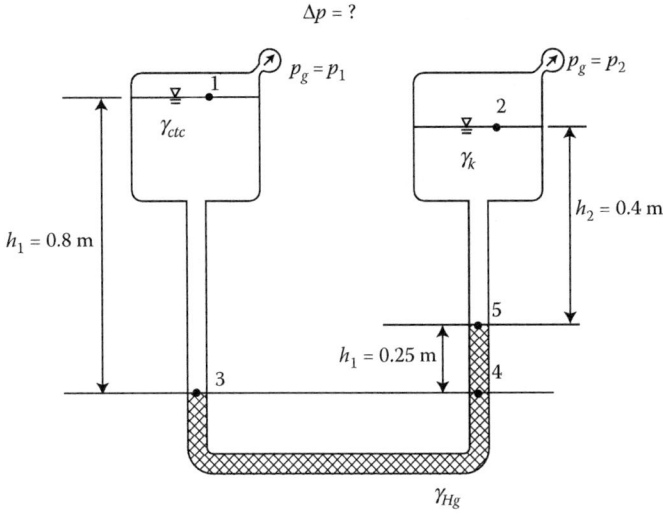

FIGURE ECP 2.99

2.100SI Tank A contains seawater, while tank B contains glycerin, as illustrated in Figure ECP 2.100. A single-fluid differential manometer is attached between tank A and tank B. The elevation in tank A is $3\,m$, and the elevation in tank B is $2.95\,m$. The distance between points 6 and 7 is $0.2\,m$. The manometer fluid is crude oil. Assume an atmospheric pressure of $101.325\,kPa$ (abs). Determine the manometer reading, h_m.

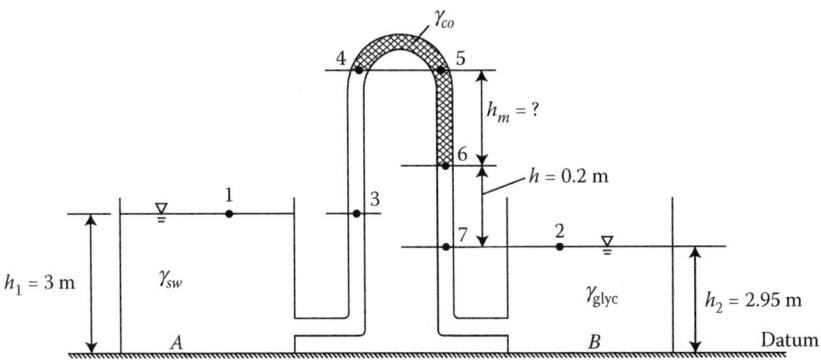

FIGURE ECP 2.100

2.101BG Water flows in a pipe 1, and saltwater with a specific gravity of 1.023 flows in pipe 2, as illustrated in Figure ECP 2.101. A single-fluid differential manometer is attached between pipe 1 and pipe 2 in order to measure the pressure difference between pipe 1 and pipe 2. The distance between points 1 and 3 is $6\,ft$, and the distance between points 2 and 7 is $4\,ft$. The manometer fluid is mercury and the manometer reading, h_m is $3\,ft$. Determine the pressure difference between pipe 1 and pipe 2.

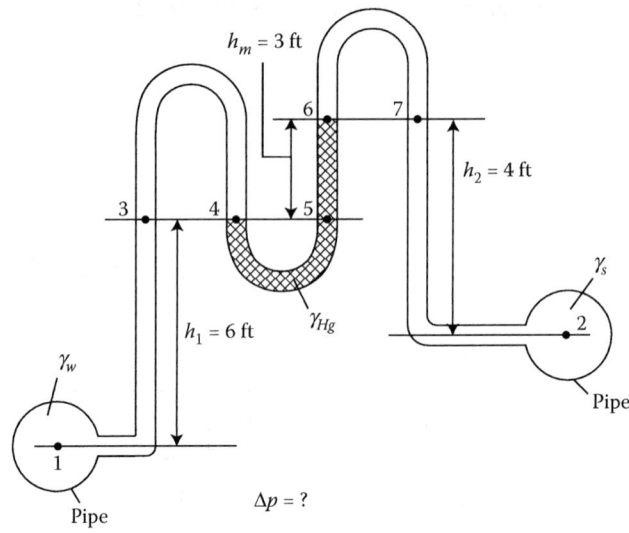

FIGURE ECP 2.101

2.102SI Water flows in pipe 1 and in pipe 2, as illustrated in Figure ECP 2.102. A multifluid differential manometer is attached between pipe 1 and pipe 2 in order to measure the pressure difference between pipe 1 and pipe 2. The distance between points 1 and 3 is 1.89 m, and the distance between points 2 and 7 is 0.33 m. The first manometer fluid is mercury with manometer reading, h_{m1} of 0.99 m. The second manometer fluid is oil with a specific gravity of 0.78, and a manometer reading, h_{m2} of 0.45 m. Determine the pressure difference between pipe 1 and pipe 2.

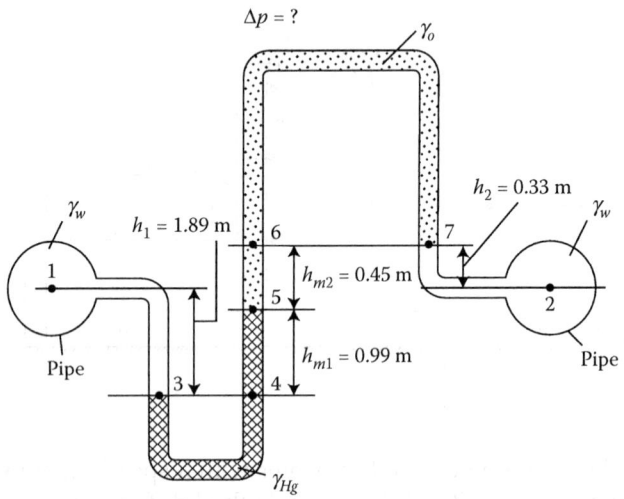

FIGURE ECP 2.102

Hydrostatic Forces on Submerged Surfaces

2.103C Define the hydrostatic pressure force.

2.104C When is the point of application (location) of the hydrostatic force on a submerged area is important?

2.105C Why is the hydrostatic pressure force important in the design of submerged surfaces/structures?

2.106C What are the two basic types of submerged surfaces of interest in the study of hydrostatic forces on submerged surfaces? What are two basic types of planes?

2.107C Based on the principles of hydrostatics, the variation of the hydrostatic pressure, p acting normal to a submerged area, A may be graphically illustrated by a pressure prism. List five variables that determine the shape of the pressure prism.

2.108C Explain why the pressure prism for a surface submerged in a gas, graphically illustrated in Figure 2.19a–d, depicts that the hydrostatic pressure, p is uniformly distributed along the submerged surface, A, regardless of the type of surface (plane or curved), and the direction of the slope of the plane.

Hydrostatic Force on Plane Surfaces

2.109C In the analysis of static pressure forces, the distributed hydrostatic force of the pressure prism can be replaced by a point force, F. What are two basic determinations that are required in the analysis of static pressure forces?

2.110C Define the "center of pressure."

2.111C Define the equation to compute the magnitude of the resultant hydrostatic force, F acting on a plane (rectangular or nonrectangular) surface area, A, which is submerged in a fluid.

2.112SI A vertical plane with a width of 3 m and a height of 9 m is submerged in water, as illustrated in Figure ECP 2.112. The water is open to the atmosphere and assume gage pressure. Determine the magnitude of the resultant hydrostatic force acting on the plane by applying Equation 2.116.

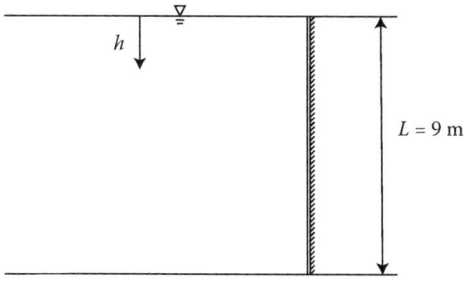

FIGURE ECP 2.112

2.113SI Refer to Problem 2.112. A vertical plane with a width of 3 m and a height of 9 m is submerged in water, as illustrated in Figure ECP 2.112. The water is open to the atmosphere and assume gage pressure. Determine the magnitude of the resultant hydrostatic force acting on the plane by applying Equation 2.120.

2.114C The "center of pressure" is the location of the resultant force, F (or the moment arm), and it is located at the center of gravity of the pressure prism. When can the center of gravity of the pressure prism be represented by the center of area (centroid) of the pressure prism?

2.115SI Refer to Problem 2.112 and Figure ECP 2.112. Determine the location of the resultant hydrostatic force acting on the vertical plane by applying Equation 2.124.

2.116SI Refer to Problem 2.112 and Figure ECP 2.112. Determine the location of the resultant hydrostatic force acting on the vertical plane by applying Equation 2.130.

2.117SI A sloping plane with a width of 0.67 *m* and a length of 2.9 *m* is submerged in a gas tank and subjected to a gas pressure of 10 *kPa*, as illustrated in Figure ECP 2.117. (a) Determine the magnitude of the resultant hydrostatic force acting on the sloping plane. (b) Determine the location of the resultant hydrostatic force acting on the sloping plane.

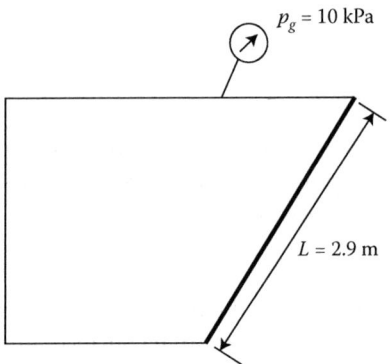

FIGURE ECP 2.117

2.118BG A horizontal plane with a width of 5 *ft* and a length of 9 *ft* is submerged in water at a depth of 9 *ft*, as illustrated in Figure ECP 2.118. (a) Determine the magnitude of the resultant hydrostatic force acting on the horizontal plane. (b) Determine the location of the resultant hydrostatic force acting on the horizontal plane.

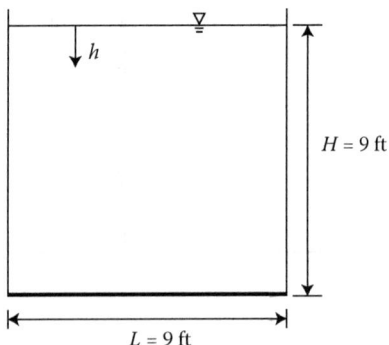

FIGURE ECP 2.118

2.119C   A vertical plane with a width, w and a length, L is submerged in water at a depth, y_o, as illustrated in Figure 2.26. (a) Derive the expression for the magnitude of the resultant hydrostatic force acting on the vertical plane. (b) Derive the expression for the location of the resultant hydrostatic force acting on the vertical plane.

2.120SI   A vertical planar gate with a width of $4\,m$ and a height of $5\,m$ is submerged at a depth $6\,m$ in water open to the atmosphere, as illustrated in Figure ECP 2.120. (a) Determine the magnitude of the resultant hydrostatic force acting on the vertical planar gate. (b) Determine the location of the resultant hydrostatic force acting on the vertical planar gate.

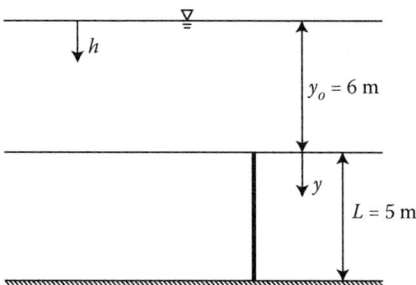

FIGURE ECP 2.120

2.121SI   An L-shaped gate with a width of $50\,m$ retains $6\,m$ water open to the atmosphere, is hinged a point O, and is controlled by a weight at point A, as illustrated in Figure ECP 2.121. (a) Determine the magnitude of the resultant hydrostatic force acting on the vertical portion of the gate. (b) Determine the location of the resultant hydrostatic force acting on the vertical portion of the gate. (c) Determine the magnitude of the weight at point A required to keep the gate closed.

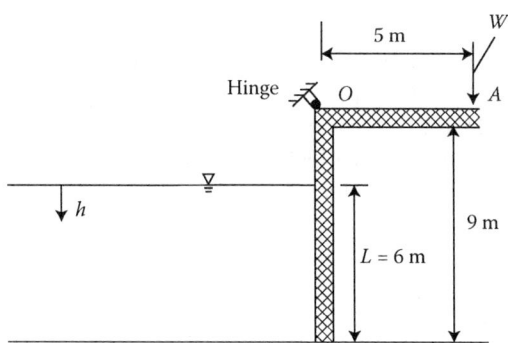

FIGURE ECP 2.121

2.122BG   A vertical planar gate with a width of $4\,ft$, a height of $10\,ft$, and a hinge at point A retains water open to the atmosphere, as illustrated in Figure ECP 2.122. Assume the height of water is $9\,ft$. (a) Determine the magnitude of the resultant hydrostatic force acting on the vertical gate. (b) Determine the location of the resultant hydrostatic force acting on the vertical gate. (c) Determine the magnitude of the torque required at point A to keep the vertical gate closed.

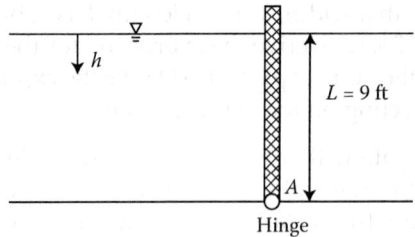

FIGURE ECP 2.122

2.123BG   A vertical planar floodgate with a width of 40 ft and a height of 7 ft retains 15 ft of water open to the atmosphere, as illustrated in Figure ECP 2.123. The floodgate is hinged at point A, and is held closed by a force, R applied at the bottom of the gate. (a) Determine the magnitude of the resultant hydrostatic force acting on the vertical planar floodgate. (b) Determine the location of the resultant hydrostatic force acting on the vertical planar floodgate. (c) Determine the magnitude of the force, R applied at the bottom of the gate required to keep the gate closed.

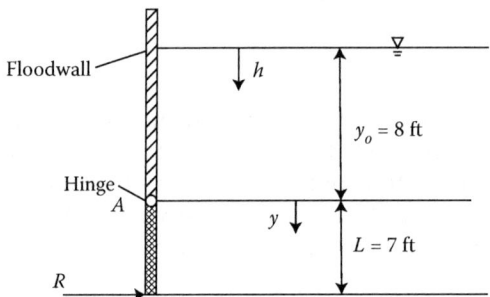

FIGURE ECP 2.123

2.124BG   A sloping planar gate with a width of 7 ft and a height of 9 ft is submerged at a depth 4 ft in water open to the atmosphere, making an 25° angle with the horizontal, as illustrated in Figure ECP 2.124. (a) Determine the magnitude of the resultant hydrostatic force acting on the sloping planar gate. (b) Determine the location of the resultant hydrostatic force acting on the sloping planar gate.

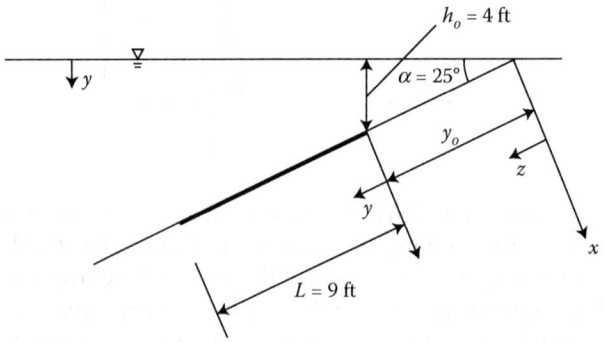

FIGURE ECP 2.124

2.125SI A sloping planar sliding gate to a pipe has a width of 2 *m* and a height of 4 *m* is submerged at a depth 3 *m* in a tank of water open to the atmosphere, at an 30° angle with the horizontal as illustrated in Figure ECP 2.125. The gate weighs 450 *N*, and the static coefficient of friction between the gate and the pipe opening is 0.54 (a) Determine the magnitude of the resultant hydrostatic force acting on the sloping planar gate. (b) Determine the location of the resultant hydrostatic force acting on the sloping planar gate. (c) Determine the magnitude of the force, *P* required to slide the gate open.

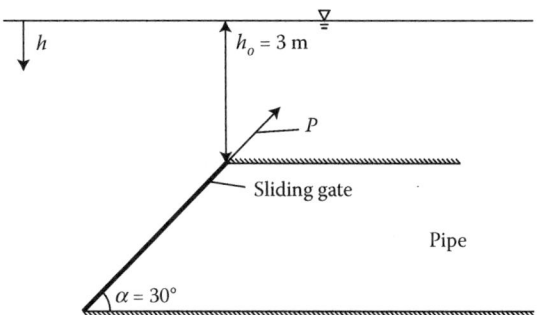

FIGURE ECP 2.125

2.126BG An earth dam with a width of 600 *ft* is built on a rock foundation and retains 12 *ft* water, as illustrated in Figure ECP 2.126. The sloping planar face makes an angle of 40° with the horizontal. Assume a static coefficient of friction between the earth dam and the rock foundation of 0.2. (a) Determine the magnitude of the resultant hydrostatic force acting on the sloping planar face. (b) Determine the location of the resultant hydrostatic force acting on the sloping planar face. (c) Determine the magnitude of the minimum required weight for the dam, *W* in order to keep the earth dam from sliding.

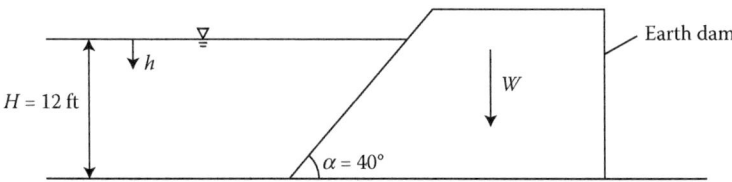

FIGURE ECP 2.126

2.127SI A sloping planar gate to a water tank has a width of 6 *m* and a height of 8 *m*, is hinged at point A, and makes an 40° angle with the horizontal. The sloping gate is controlled by a cable, as illustrated in Figure ECP 2.127. (a) Determine the magnitude of the resultant hydrostatic force acting on the sloping planar gate. (b) Determine the location of the resultant hydrostatic force acting on the sloping planar gate. (c) Determine the magnitude of the required tension, *T* in the cable in order to keep the sloping planar gate closed.

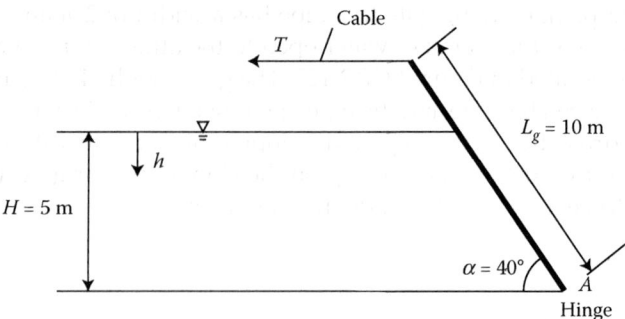

FIGURE ECP 2.127

2.128C   A sloping plane with a width, w and a length, L is submerged in water at a depth, y_o, as illustrated in Figure 2.27. Derive the expression for the magnitude of the resultant hydrostatic force acting on the sloping plane.

2.129SI   A horizontal plane with a width of 3 m and a length of 5 m is submerged in a tank of water at a depth of 8 m and is subjected to gage air pressure of 9 kPa, as illustrated in Figure ECP 2.129. (a) Determine the magnitude of the resultant hydrostatic force acting on the horizontal plane. (b) Determine the location of the resultant hydrostatic force acting on the horizontal plane.

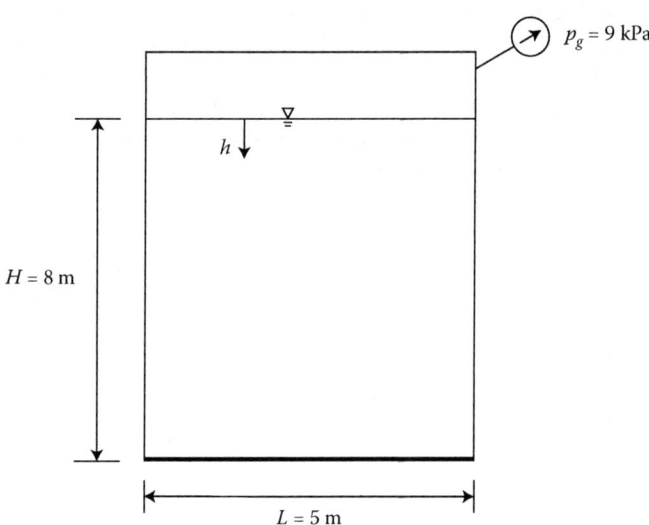

FIGURE ECP 2.129

2.130BG   A vertical plane with a width of 3.7 ft and a height of 7.8 ft is submerged in a tank of water and is subjected to gage air pressure of 8 psi, as illustrated in Figure ECP 2.130. (a) Determine the magnitude of the resultant hydrostatic force acting on the vertical plane. (b) Determine the location of the resultant hydrostatic force acting on the vertical plane.

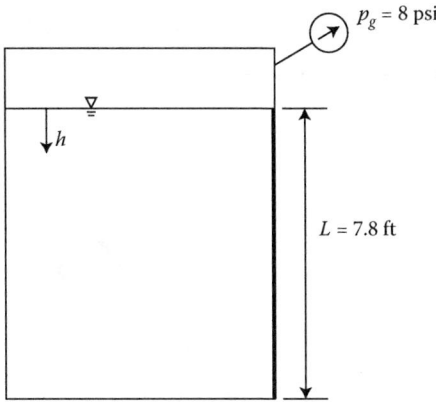

FIGURE ECP 2.130

2.131SI A vertical plane with a width of 4.3 *m* and a height of 6.9 *m* is submerged at a depth 7 *m* in a tank of water and is subjected to a gage air pressure of 11 *kPa*, as illustrated in Figure ECP 2.131. (a) Determine the magnitude of the resultant hydrostatic force acting on the vertical plane. (b) Determine the location of the resultant hydrostatic force acting on the vertical plane.

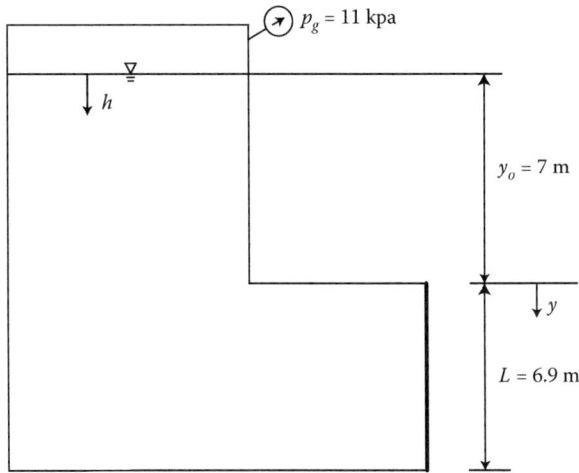

FIGURE ECP 2.131

2.132BG A sloping plane with a width of 5 *ft* and a height of 6 *ft* is submerged at a depth 5 *ft* in a tank of water and is subjected to a gage pressure of 13 *psi*, making an 30° angle with the horizontal, as illustrated in Figure ECP 2.132. (a) Determine the magnitude of the resultant hydrostatic force acting on the sloping plane. (b) Determine the location of the resultant hydrostatic force acting on the sloping plane.

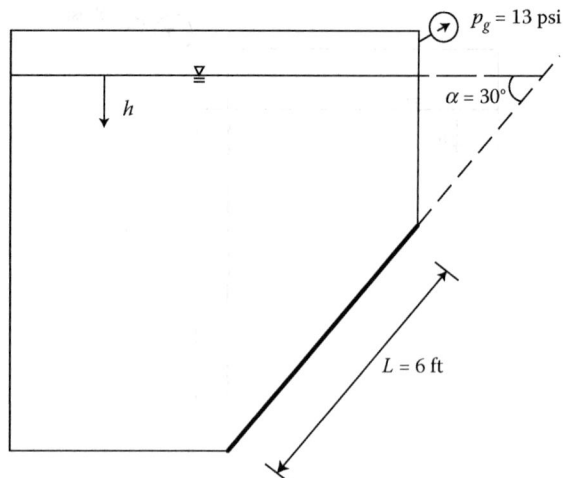

FIGURE ECP 2.132

2.133BG A horizontal circular plane with a diameter of 4.9 ft is submerged in water open to the atmosphere at a depth of 7.8 ft as illustrated in Figure ECP 2.133. (a) Determine the magnitude of the resultant hydrostatic force acting on the horizontal plane. (b) Determine the location of the resultant hydrostatic force acting on the horizontal plane.

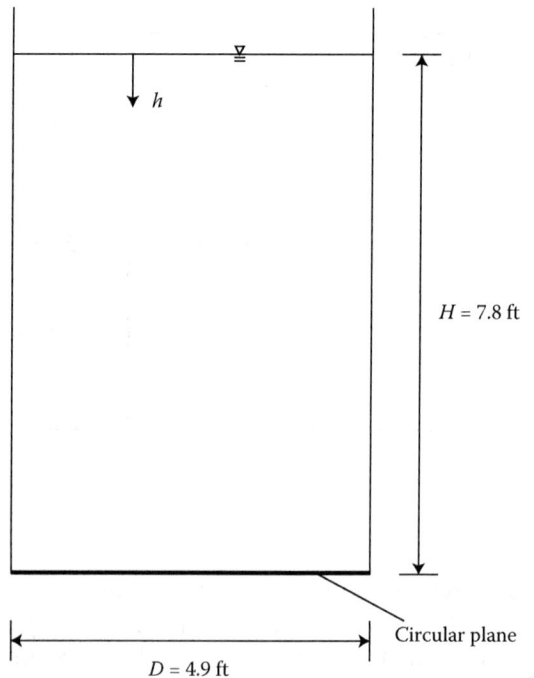

FIGURE ECP 2.133

2.134SI   A horizontal triangular plane with a base of 4.6 *m* and a height of 5 *m* is submerged in water open to the atmosphere at a depth of 6 *m* as illustrated in Figure ECP 2.134. (a) Determine the magnitude of the resultant hydrostatic force acting on the horizontal plane. (b) Determine the location of the resultant hydrostatic force acting on the horizontal plane.

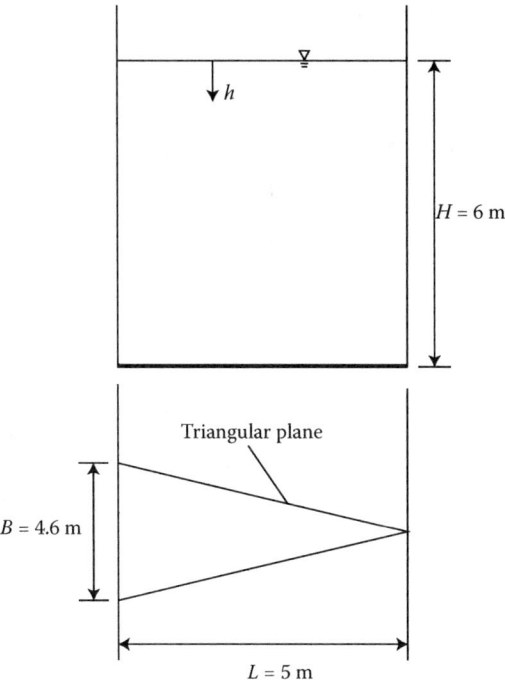

FIGURE ECP 2.134

2.135C   A vertical circular plane with a diameter, D is submerged in water at a depth, y_o, as illustrated in Figure 2.30b. (a) Derive the expression for the magnitude of the resultant hydrostatic force acting on the vertical plane. (b) Derive the expression for the location of the resultant hydrostatic force acting on the vertical plane.

2.136C/SI   A vertical semicircular plane with a diameter, D is submerged in water at a depth, y_o, as illustrated in Figure ECP 2.136. (a) Derive the expression for the magnitude of the resultant hydrostatic force acting on the vertical plane. (b) Derive the expression for the location of the resultant hydrostatic force acting on the vertical plane. (c) Assume a diameter, D of 6 *m* and a depth of submergence, y_o of 0 *m*, determine the magnitude and the location of the hydrostatic force acting on the vertical plane.

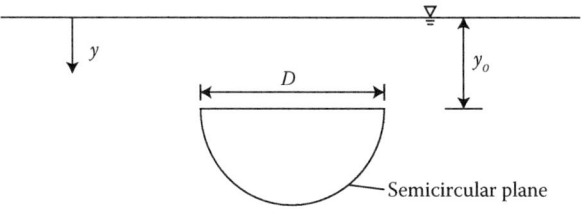

FIGURE ECP 2.136

2.137BG   A vertical circular plane gate with a diameter of 3 ft is submerged at a depth 0 ft in water open to the atmosphere, as illustrated in Figure ECP 2.137. (a) Determine the magnitude of the resultant hydrostatic force acting on the vertical circular plane gate. (b) Determine the location of the resultant hydrostatic force acting on the vertical circular plane gate. Discuss why the location of the resultant hydrostatic force is not at the center of area of the triangular pressure prism.

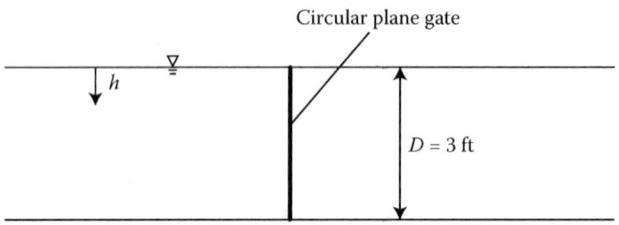

FIGURE ECP 2.137

2.138C   Three vertical triangular planes are submerged in water at a depth, y_o, as illustrated in Figure ECP 2.138. (a) Derive the expression for the location of the resultant hydrostatic force acting the vertical triangular plane illustrated in Figure ECP 2.138a. (b) Derive the expression for the location of the resultant hydrostatic force acting the vertical triangular plane illustrated in Figure ECP 2.138b. (c) Derive the expression for the location of the resultant hydrostatic force acting the vertical triangular plane illustrated in Figure ECP 2.138c. (d) Compare the results in (a), (b), and (c).

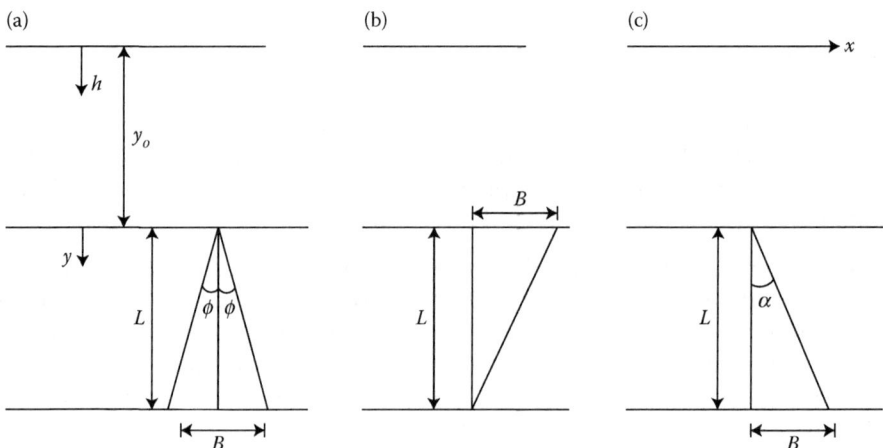

FIGURE ECP 2.138

2.139BG   A vertical triangular plane with a base of 7 ft and a height of 8 ft is submerged at a depth 3 ft in water open to the atmosphere, as illustrated in Figure ECP 2.139. (a) Determine the magnitude of the resultant hydrostatic force acting on the vertical triangular plane. (b) Determine the location of the resultant hydrostatic force acting on the vertical triangular plane.

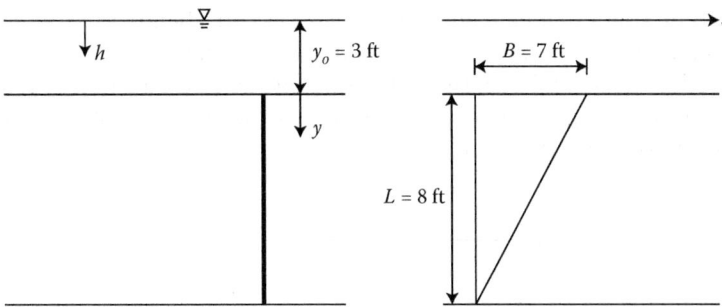

FIGURE ECP 2.139

2.140SI A vertical triangular plane gate with a base of 3 *m* and a height of 5 *m* is submerged at a depth 4 *m* in water open to the atmosphere, as illustrated in Figure ECP 2.140. (a) Determine the magnitude of the resultant hydrostatic force acting on the vertical triangular plane gate. (b) Determine the location of the resultant hydrostatic force acting on the vertical triangular plane gate.

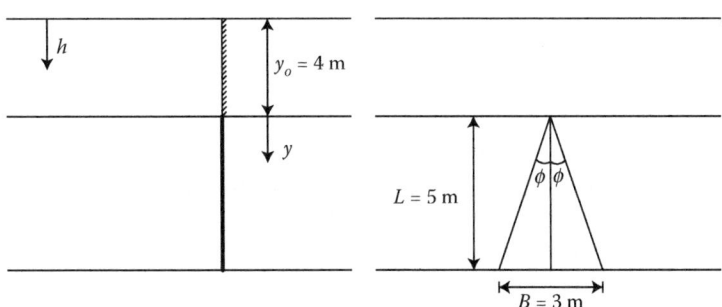

FIGURE ECP 2.140

2.141BG A sloping circular gate with a diameter of 3.75 *ft* is submerged at a depth 6.75 *ft* in water open to the atmosphere, making an 25° angle with the horizontal, as illustrated in Figure ECP 2.141. (a) Determine the magnitude of the resultant hydrostatic force acting on the sloping circular gate. (b) Determine the location of the resultant hydrostatic force acting on the sloping circular gate.

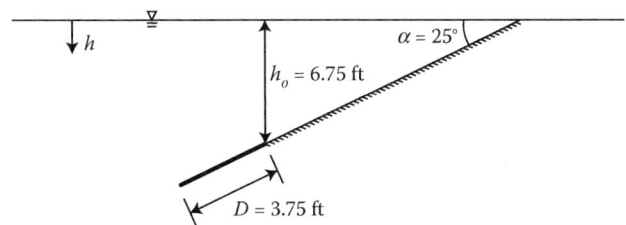

FIGURE ECP 2.141

2.142SI   A sloping triangular gate with a base of 3.3 m and a height of 6 m is submerged at a depth 4 m in water open to the atmosphere, making an 40° angle with the horizontal, as illustrated in Figure ECP 2.142. (a) Determine the magnitude of the resultant hydrostatic force acting on the sloping triangular gate. (b) Determine the location of the resultant hydrostatic force acting on the sloping triangular gate.

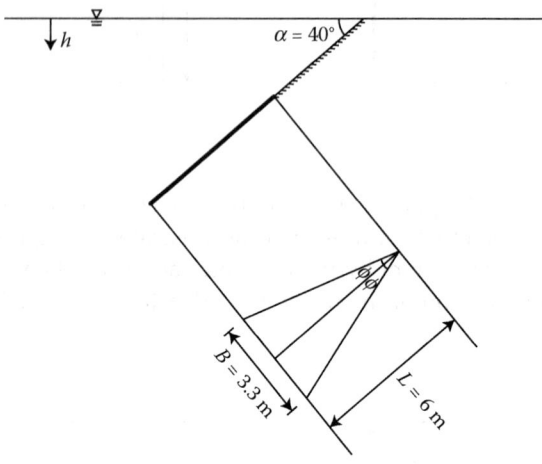

FIGURE ECP 2.142

2.143BG   An L-shaped vertical plane is submerged at a depth 4 ft in water open to the atmosphere, as illustrated in Figure ECP 2.143. (a) Determine the magnitude of the resultant hydrostatic force acting on the vertical plane. (b) Determine the location of the resultant hydrostatic force acting on the vertical plane.

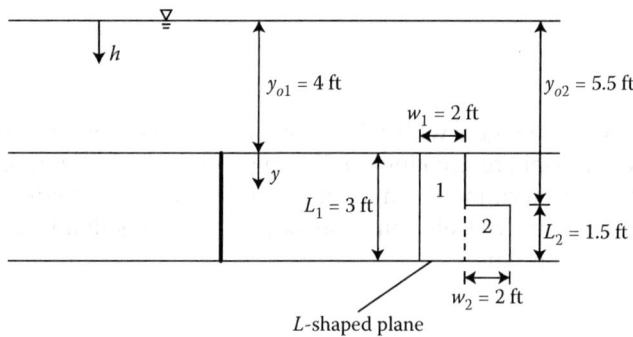

FIGURE ECP 2.143

2.144SI   A vertical semicircular gate with a diameter of 10 m is located at a depth of 7 m below the water surface on the side of a tank of water, as illustrated in Figure ECP 2.144. The vertical gate is hinged at point A. Refer to Problem 2.136. (a) Determine the magnitude of the resultant hydrostatic force acting on the vertical semicircular gate. (b) Determine the location of the resultant hydrostatic force acting on the vertical semicircular gate. (c) Determine the magnitude of the force, R at point B required to keep the gate closed.

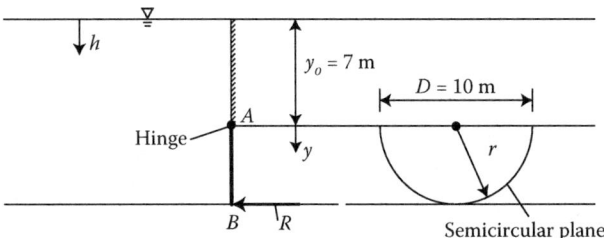

FIGURE ECP 2.144

2.145BG A sloping parabolic ($y = (L/B^2)x^2$) gate is located on the side of a tank of water, is hinged at point A, and makes an $25°$ angle with the horizontal, as illustrated in Figure ECP 2.145. (a) Determine the magnitude of the resultant hydrostatic force acting on the sloping parabolic gate. (b) Determine the location of the resultant hydrostatic force acting on the sloping parabolic gate. (c) Determine the magnitude of the force at point B required to keep the gate closed.

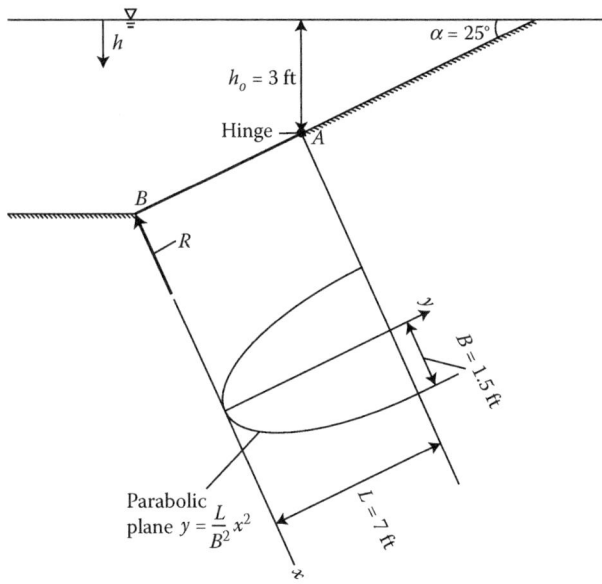

FIGURE ECP 2.145

2.146SI A sloping triangular gate is located on the side of a tank of water, is hinged along CD, and makes an $30°$ angle with the horizontal, as illustrated in Figure ECP 2.146. (a) Determine the magnitude of the resultant hydrostatic force acting on the sloping triangular gate. (b) Determine the location of the resultant hydrostatic force acting on the sloping triangular gate. (c) Determine the magnitude of the force, R at point A required to keep the gate closed.

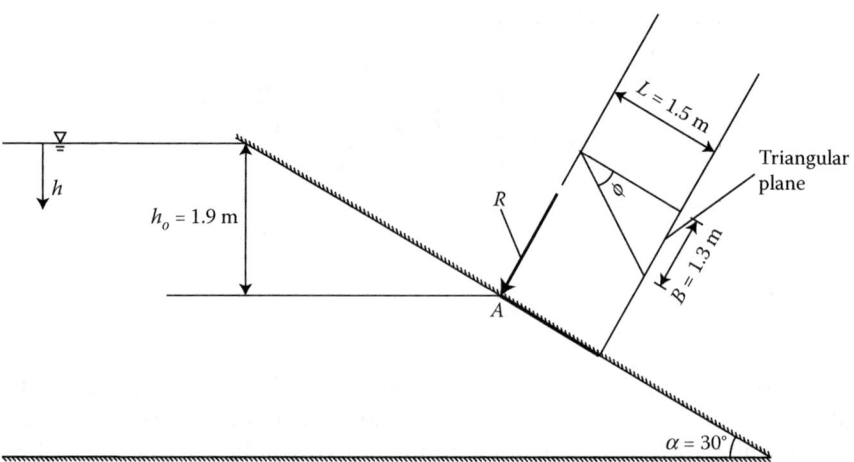

FIGURE ECP 2.146

2.147BG   A car is vertically submerged in water. The passenger's door is a quarter of a circle with a radius of 2.5 ft, as illustrated in Figure ECP 2.147. Assume the passenger is capable of pushing with a 350 lb force. (a) If the car contains air at atmospheric pressure, determine the maximum depth the car's roof can be submerged in order for the passenger to be able to open the car door, and determine the location of the resultant hydrostatic force. (b) If the car is filled with water, determine the maximum depth the car's roof can be submerged in order for the passenger to be able to open the car door, and determine the location of the resultant hydrostatic force.

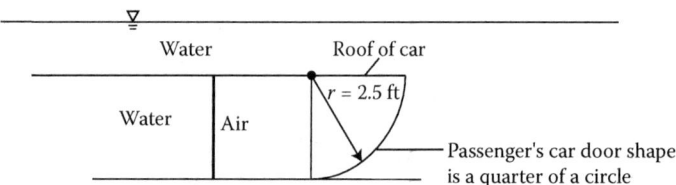

FIGURE ECP 2.147

Hydrostatic Force on Curved Surfaces

2.148BG   A curved surface, AB that has the shape of a parabola with B = 6 ft, L = 7 ft, and a width of 5 ft is submerged at a depth 9 ft in water open to the atmosphere, as illustrated in Figure ECP 2.148. (a) Determine the magnitude and location of the horizontal component of the resultant hydrostatic force acting on the curved surface, AB. (b) Determine the magnitude and location of the vertical component of the resultant hydrostatic force acting on the curved surface, AB. (c) Determine the magnitude and the location of the resultant hydrostatic force acting on the curved surface, AB.

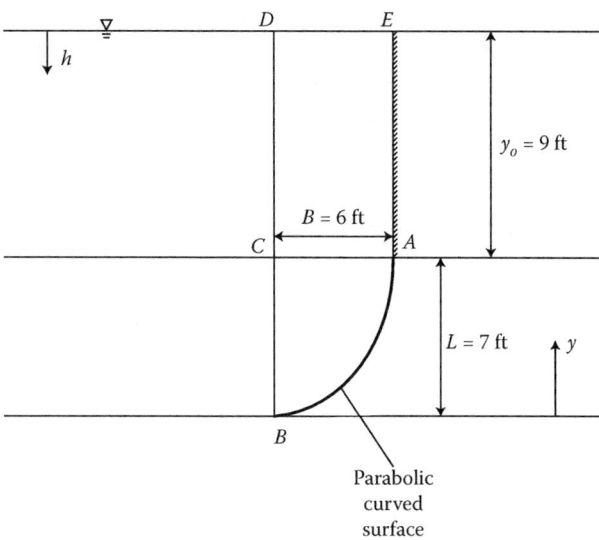

FIGURE ECP 2.148

2.149SI A tank containing gasoline at 20°C ($s = 0.68$) is $9\,m$ deep and $3\,m$ wide, has a curved surface, AB that has the shape of a quarter of a circle with a radius of $4\,m$, and is open to the atmosphere, as illustrated in Figure ECP 2.149. (a) Determine the magnitude and location of the horizontal component of the resultant hydrostatic force acting on the curved surface, AB. (b) Determine the magnitude and location of the vertical component of the resultant hydrostatic force acting on the curved surface, AB. (c) Determine the magnitude and the location of the resultant hydrostatic force acting on the curved surface, AB.

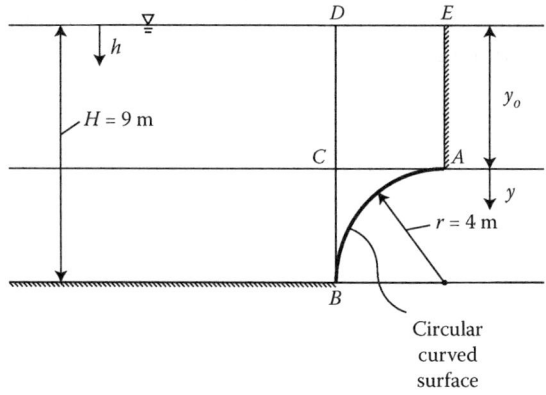

FIGURE ECP 2.149

2.150BG A tank containing glycerin at 68°F ($s = 1.258$) is $20\,ft$ deep and $5\,ft$ wide, has a curved surface, AB that has the shape of a quarter of an ellipse with B $= 8\,ft$, L $= 15\,ft$, and is open to the atmosphere, as illustrated in Figure ECP 2.150. (a) Determine the magnitude and location of the horizontal component of the resultant hydrostatic force acting on the curved surface, AB. (b) Determine the magnitude and location of the vertical component of the resultant hydrostatic force acting on the curved surface, AB. (c) Determine the magnitude and the location of the resultant hydrostatic force acting on the curved surface, AB.

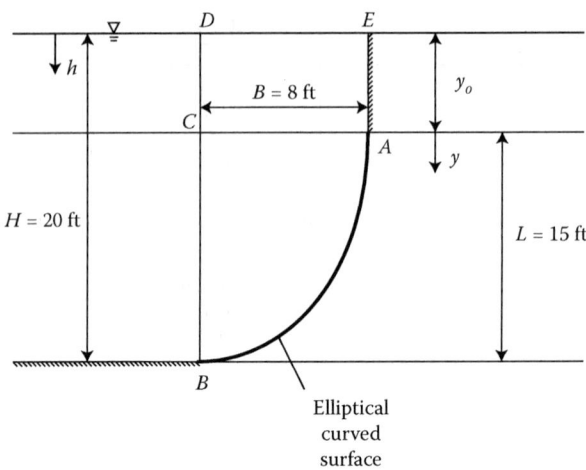

FIGURE ECP 2.150

2.151C/SI A tank containing seawater at 20°C ($s = 1.023$) is 5 m deep and 3 m wide, has a curved surface, AB that has the shape $y = (L/\sqrt{B})\sqrt{x}$, where $B = 4\,m$, and is open to the atmosphere, as illustrated in Figure ECP 2.151. (a) Derive the analytical expressions and evaluate the magnitude and location of the horizontal component of the resultant hydrostatic force acting on the curved surface, AB. (b) Derive the analytical expressions and evaluate the magnitude and location of the vertical component of the resultant hydrostatic force acting on the curved surface, AB. (c) Determine the magnitude and the location of the resultant hydrostatic force acting on the curved surface, AB.

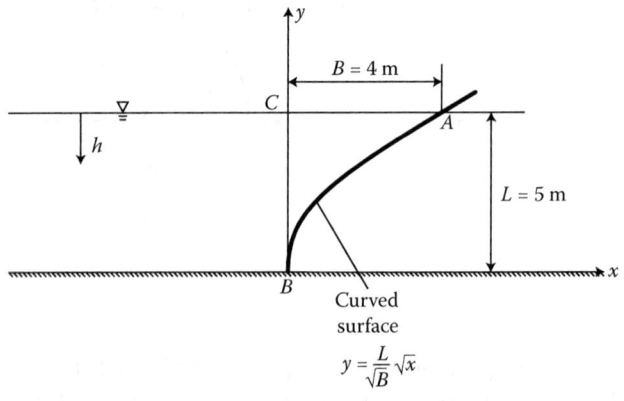

FIGURE ECP 2.151

2.152BG A tank containing crude oil at 68°F ($s = 0.856$) is 6 ft deep and 7 ft wide, has a curved gate, AB that has the shape of a parabola with B = 5 ft and is hinged at point B, and is open to the atmosphere, as illustrated in Figure ECP 2.152. (a) Determine the magnitude and location of the horizontal component of the resultant hydrostatic force acting on the curved surface, AB. (b) Determine the magnitude and location of the vertical component of the resultant hydrostatic force acting on the curved surface, AB. (c) Determine the magnitude and the location of the resultant hydrostatic force acting on the curved surface, AB. (d) Determine the magnitude of the force, F at point A required to keep the gate closed.

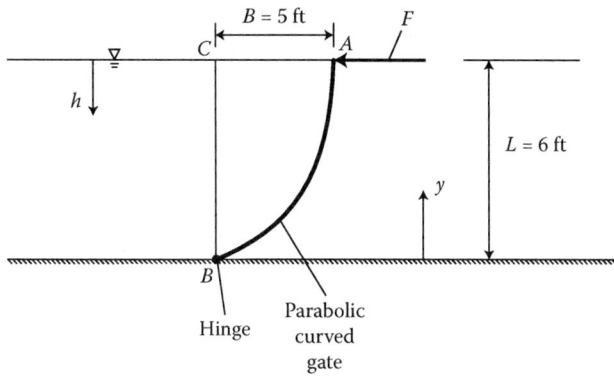

FIGURE ECP 2.152

2.153SI A dam retaining water is 60 m deep and 20 m wide, and has a curved surface, AB that has the shape of a quarter of a circle, as illustrated in Figure ECP 2.153. Assume a static coefficient of friction between the dam and its foundation is 0.33. (a) Determine the magnitude and location of the horizontal component of the resultant hydrostatic force acting on the curved surface, AB. (b) Determine the magnitude and location of the vertical component of the resultant hydrostatic force acting on the curved surface, AB. (c) Determine the magnitude and the location of the resultant hydrostatic force acting on the curved surface, AB. (d) Determine the magnitude the minimum required weight of the dam in order to keep the dam from sliding.

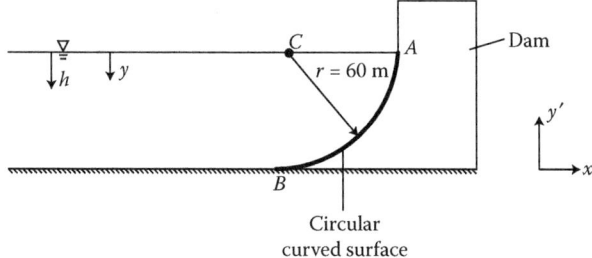

FIGURE ECP 2.153

2.154BG A tank of water 68 ft deep and 45 ft wide has a curved surface, AB that has the shape of a quarter of an ellipse with B = 5 ft and L = 7 ft, as illustrated in Figure ECP

2.154. (a) Determine the magnitude and location of the horizontal component of the resultant hydrostatic force acting on the curved surface, AB. (b) Determine the magnitude and location of the vertical component of the resultant hydrostatic force acting on the curved surface, AB. (c) Determine the magnitude and the location of the resultant hydrostatic force acting on the curved surface, AB.

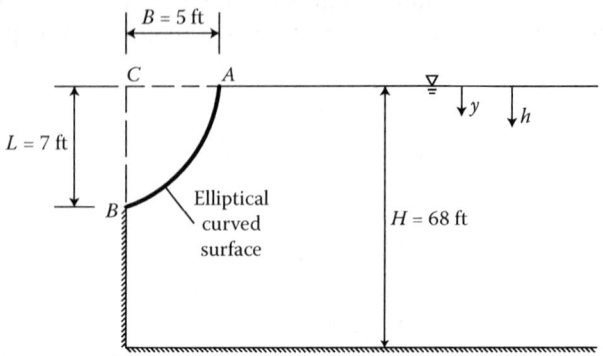

FIGURE ECP 2.154

2.155SI   A tank containing water is 8 *m* deep and 4 *m* wide, has a curved gate, AB that has the shape of a quarter of a circle with a radius of 5 *m* and hinged at point A, and is open to the atmosphere, as illustrated in Figure ECP 2.155. (a) Determine the magnitude and location of the horizontal component of the resultant hydrostatic force acting on the curved surface, AB. (b) Determine the magnitude and location of the vertical component of the resultant hydrostatic force acting on the curved surface, AB. (c) Determine the magnitude and the location of the resultant hydrostatic force acting on the curved surface, AB. (d) Determine the magnitude of the force, F at point B required to keep the gate closed.

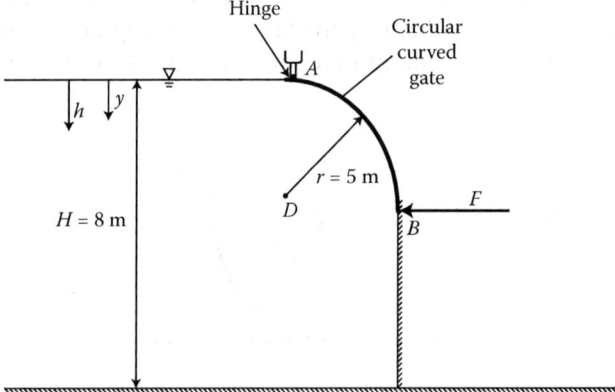

FIGURE ECP 2.155

2.156BG   A tank containing 10 *ft* of glycerin (*s* = 1.258) and 10 *ft* of water is 4 *ft* wide, has a curved gate, AB that has the shape of a quarter of circle with a radius of 10 *ft*, is hinged at point B, and is open to the atmosphere, as illustrated in Figure ECP 2.156. (a) Determine the magnitude and location of the horizontal and vertical components of the resultant hydrostatic force acting on the curved surface, AB due to the glycerin. (b) Determine the magnitude and location of the horizontal and vertical components of the resultant hydrostatic force acting on the curved surface, AB due to the water. (c) Determine the magnitude of the force, F at point A required to keep the gate closed.

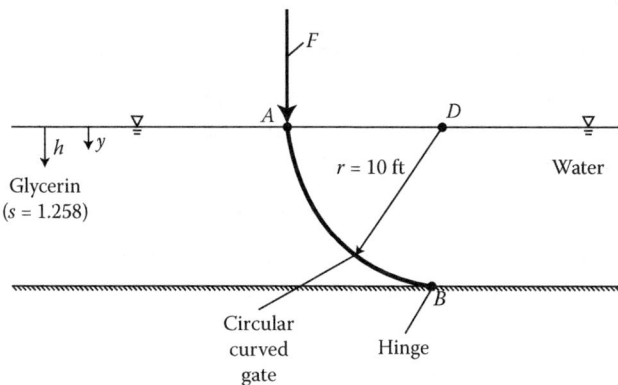

FIGURE ECP 2.156

Hydrostatic Force on Surfaces Submerged in a Multilayered Fluid

2.157SI A vertical plane with a width of $6\,m$ and a height of $11\,m$ is submerged in two layers of fluid, as illustrated in Figure ECP 2.157. The first fluid is gasoline with a specific gravity of 0.680, has a depth $6\,m$, and is open to the atmosphere The second fluid is glycerin with a specific gravity of 1.258 and has a depth of $5\,m$. (a) Determine the magnitude and location of the resultant hydrostatic force acting on the vertical plane in the first fluid. (b) Determine the magnitude and location of the resultant hydrostatic force acting on the vertical plane in the second fluid. (c) Determine the magnitude and location of the resultant hydrostatic force acting on the vertical plane in the multilayer fluid.

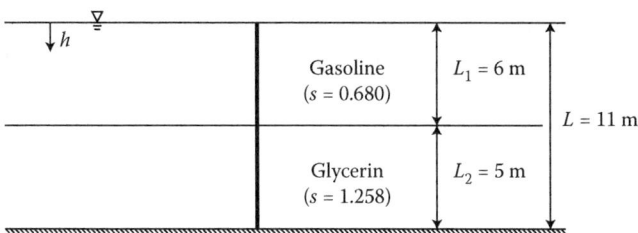

FIGURE ECP 2.157

Buoyancy and Stability of a Floating or a Neutrally Buoyant Body

2.158C Why is the hydrostatic pressure force important in the design of floating or a neutrally buoyant bodies?

2.159C What principle provides the basis for the determination of the buoyant force?

2.160C What provides the basis for determining the stability (stable equilibrium) of a floating or a neutrally buoyant body?

2.161C   What causes the existence of an upward vertical force called the buoyant force, F_B that acts on a floating or a neutrally buoyant body?

2.162C   What is buoyancy or the buoyant force, F_B?

2.163C   What causes the weight, $F_G = Mg = W = \gamma V$ of a submerged floating or suspended body (object) to appear to decrease?

2.164C   What does the Archimedes principle provide a basis for, and what does it state?

2.165C   Depending upon the density of the body relative to the density of the fluid, one of three cases (occurrences) will take place and thus the relationship between the buoyant force, F_B and the weight of the body, W is defined. What are these three cases?

2.166C   List three factors that the magnitude of the buoyant force, F_B is independent of.

2.167C   If two identical metal cubes are suspended in water at different depths, will the magnitude of the buoyant force acting on each cube be different or the same? Explain your answer.

2.168C   If two identical cubes, but one made of steel, and the other made of glass, are suspended in water, will the magnitude of the buoyant force acting on each cube be different or the same? Explain your answer.

2.169C   If two identical metal objects, but one is a cube, and the other is a sphere, are suspended in water, will the magnitude of the buoyant force acting on each object be different or the same? Explain your answer.

2.170C   Describe the design magnitude and the location of the weight, W in the design of a floating or neutrally buoyant body.

2.171SI   An object with a specific gravity of 1.56 is completely submerged in water. The object is attached to a scale and weighed in the water and registers 980 N. (a) Determine if the body of the object will sink, float, or become suspended (i.e., neutrally buoyant). (b) Draw the free body diagram for the manually suspended body in the water. (c) Determine the volume of the object (body).

2.172BG   A disk with a diameter of 2 ft and thickness of 0.5 ft, weighing 160 lb, sinks to the bottom of a tank containing 6 ft of crude oil ($s = 0.856$), as illustrated in Figure ECP 2.172.

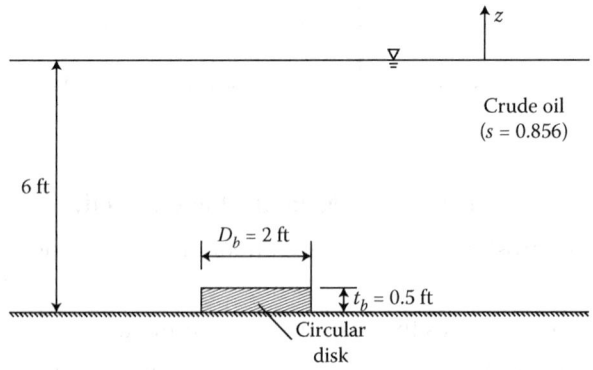

FIGURE ECP 2.172

(a) Draw the free body diagram for the manually suspended body/disk in the crude oil.
(b) Determine the required tension in the cable in order to lift the disk from the tank.

2.173SI An object weighs 700 N in gasoline ($s = 0.680$) and 500 N in crude oil ($s = 0.856$). (a) Draw the free body diagram for the manually suspended object in each fluid. (b) Determine the weight and the volume of the object.

2.174BG A 1.5-*ft* spherical ball weighing 250 *lb* is suspended by a cable that is connected to a 3 *ft* wide and 5 *ft* high rectangular gate hinged at point A, as illustrated in Figure ECP 2.174. (a) Draw the free body diagram for the gate and the manually suspended body in the water. (b) Determine the depth of water, *L* below which the gate will open.

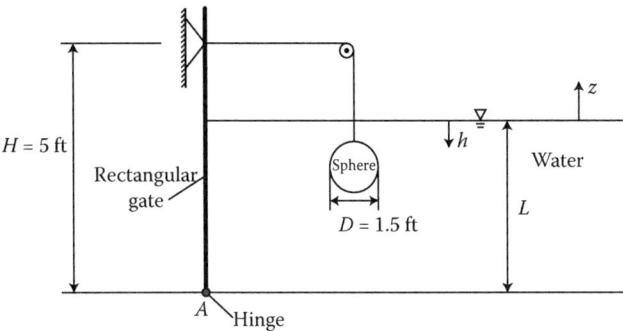

FIGURE ECP 2.174

2.175SI An iron disk ($s = 8$) weighing 1500 N is dropped into a swimming pool and drops to the bottom of the pool. A lifeguard dives into the pool to retrieve the weight. (a) Draw the free body diagram for the manually lifted body/iron disk in the water, and determine the force, *P* required to lift the iron disk from the bottom of the pool. (b) Draw the free body diagram for the manually lifted body/iron disk in the air, and determine the force, *P* required to lift the iron disk out of the pool and onto the pool deck.

2.176C Neutral buoyancy occurs when the density of the body is equal to the density of the fluid. As such, the immersed/completely submerged neutrally buoyant/naturally suspended body will not sink or float unless an external force is applied. (a) Explain how a sinking body can be made into a neutrally buoyant body, and give an example. (b) Explain how a floating body can be made into a neutrally buoyant body, and give an example.

2.177BG An object with a specific gravity of 1.258 and a volume of 5 ft^3 is completely submerged in glycerin with a specific gravity of 1.258. (a) Determine if the body of the object will sink, float, or become suspended (i.e., neutrally buoyant). (b) Draw the free body diagram for the body in the glycerin. (c) Determine the weight of the object (body) and the buoyant force.

2.178SI A piece of wood with a specific gravity of 0.36 and a volume of 8 m^3 ($L = 4\,m$, $H = 2\,m$, $w = 1\,m$) is partially submerged in water. (a) Determine if the wood will sink, float, or become suspended (i.e., neutrally buoyant). (b) Draw the free body diagram for the wood body in the water. (c) Determine the proportion of the wood body that floats above the water line.

2.179BG A piece of driftwood floats in the sea with one fourth of its volume above the seawater surface. (a) Draw the free body diagram for the wood body in the seawater.

(b) Determine the specific gravity of the driftwood relative to the seawater. (c) Determine the specific gravity of the driftwood relative to fresh water and the portion of its volume above a freshwater surface.

2.180SI   A hydrometer is used to measure the specific gravity of a fluid. A hydrometer consists of a 30-*mm*-diameter sphere weighing 0.1 N, and a 7-*mm*-diameter cylindrical stem with a length of 230 *mm* weighing 0.099 N, as illustrated in Figure ECP 2.180. (a) Draw the free body diagram for the hydrometer. (b) Determine the level of the fluid at the stem for fluid with a specific gravity of 0.9. (c) Determine the level of the fluid at the stem for fluid with a specific gravity of 1.3.

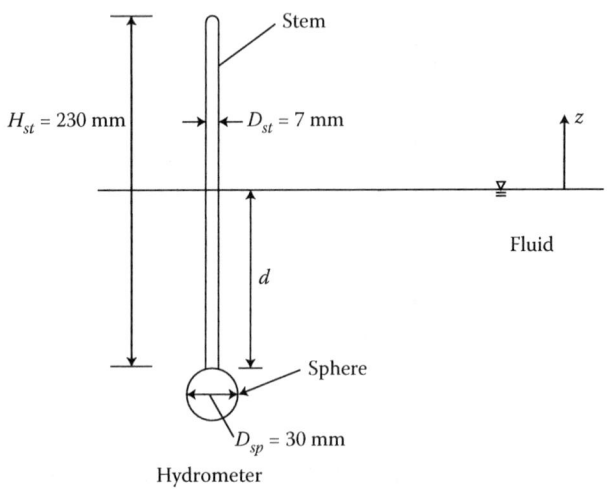

FIGURE ECP 2.180

2.181BG   A spherical buoy (float) with a diameter of 4 *ft* and a specific gravity of 0.75 is anchored to the bottom of the ocean floor ($s_{seawater} = 1.024$) with a cord, as illustrated in Figure ECP 2.181. (a) Determine if the unanchored buoy will sink, float, or become

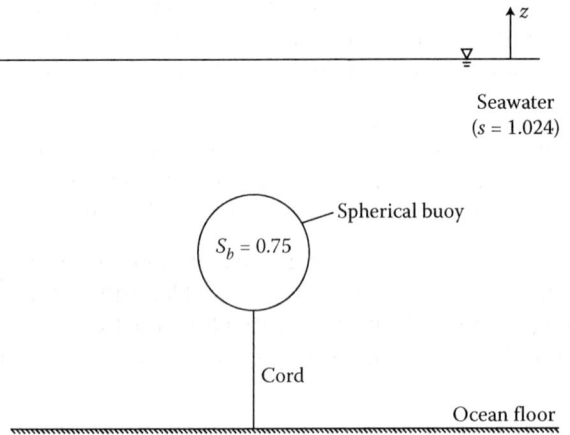

FIGURE ECP 2.181

suspended (i.e., neutrally buoyant). (b) Draw the free body diagram for the unanchored buoy in the seawater. (c) Draw the free body diagram for the anchored buoy in the seawater. (d) Determine the tension in the cord if the anchored buoy is completely submerged in the seawater.

2.182BG   A blimp weighing 400 *lbs* has a volume of 9000 ft^3, is filled with helium ($\gamma = 0.0115\ lb/ft^3$), and floats in the air ($\gamma = 0.09\ lb/ft^3$). (a) Draw the free body diagram for the blimp floating in the air and carrying a load. (b) Determine the maximum load that may be suspended from the blimp in order for it to remain in static equilibrium.

2.183SI   A 15-*m* cylindrical rod with a diameter of 0.33 *m* and a specific gravity of 0.68 is anchored to the bottom of a tank of fluid with a cord, making an 20° angle with the horizontal, as illustrated in Figure ECP 2.183. (a) Draw the free body diagram for the anchored rod in the tank of fluid. (b) Determine the specific weight of the fluid in the tank. (c) Determine the tension in the cord.

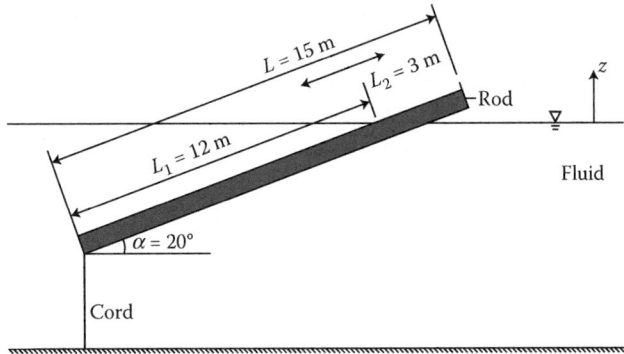

FIGURE ECP 2.183

2.184BG   An empty cargo ship weighs 2000 *lb* and has a hull with a volume of 5000 ft^3. (a) Draw the free body diagram for the ship in water and in seawater. (b) Determine the maximum load the cargo ship can carry without sinking in freshwater. (c) Determine the maximum load the cargo ship can carry without sinking in seawater.

2.185C   Give two examples of the application of the Archimedes principle and thus the realization of the existence of the buoyant force in the design of floating bodies.

2.186C   Give two examples of the application of the Archimedes principle and thus the realization of the existence of the buoyant force in the design of neutrally buoyant/suspended bodies.

2.187C   An important application of the buoyancy concept is the assessment of the stability of floating or a neutrally buoyant/suspended body. What are two types of stability to consider for both floating or neutrally buoyant/suspended bodies?

2.188C   Explain why there is an inherent vertical stability of floating or neutrally buoyant/suspended body that is in static equilibrium.

2.189C   Given a small angle of rotation, θ (displacement of the body from its position of stable equilibrium), what determines the rotational stability or instability of floating or neutrally buoyant/suspended bodies in a fluid?

2.190C The stability analysis for a floating or neutrally buoyant body assumes a small angle of rotation, θ. What will happen to the body if the maximum design angle of rotation, θ by which the body can tilt is exceeded?

2.191C Given a small angle of rotation, θ, what does the rotational stability analysis for a neutrally buoyant/suspended body depend upon?

2.192C Given a small angle of rotation, θ, what does the rotational stability analysis for a floating body depend upon?

2.193C Describe and explain what happens to a floating body if it is raised or lowered by an external vertical force.

2.194C Describe and explain what happens to a neutrally buoyant/suspended body if it is raised or lowered by an external vertical force.

2.195C Stability considerations are important for both floating or neutrally buoyant/suspended bodies because the centers of gravity and buoyancy do not necessarily coincide. Furthermore, the vertical alignment of the line of action of the buoyant force and the line of action of the weight of the body (i.e., zero moment) provides the basis for determining the stability (stable equilibrium) of a floating or a neutrally buoyant body. (a) Explain what happens when a small rotation causes the two centers to move out of vertical alignment. (b) Explain what happens when a small rotation does not cause the two centers to move out of vertical alignment.

2.196C What are the three positions/states of equilibrium with respect to rotational stability of a floating or neutrally buoyant/suspended body?

2.197C Given a small angle of rotation, θ, the rotational stability of a neutrally buoyant/suspended body (such as a hot-air or helium balloon, a submarine, or a satellite) depends on the weight distribution of the body and thus the relative location of the center of gravity, G of the body and the center of buoyancy, B, which is the centroid of the displaced volume. What weight distribution results in a rotationally stable vs. unstable neutrally buoyant/suspended body?

2.198C For a neutrally buoyant/suspended body, a slight rotation from its original position of equilibrium will cause the weight, W and the buoyant force, F_B to move out of vertical alignment (see Figures 2.35 and 2.37) (except in the case of a uniform density body [see Figure 2.36]). What happens to the position of the center of gravity, G and the position of the center of buoyancy, B?

2.199C Given a small angle of rotation, θ, the rotational stability of a floating body (such as a ship or a barge) depends upon the weight distribution, the particular geometry of the body, and the definition of the metacentric height (in the case of the stability of a top-heavy floating body). (a) What weight distribution will always result in a rotationally stable floating body, regardless of the particular geometry of the body (wide-based, relatively tall and slender, spherical-shaped, or streamlined/average height)? (b) Will a top-heavy floating body always be rotationally unstable? Explain.

2.200C For a top-heavy and streamlined/average height geometry (not extreme/definitive in geometry) floating body, such a ship or a boat, the occurrence of the type of moment (restoring, overturning, or zero) due to a small rotational disturbance does not appear to be very obvious. In such cases and thus in general, how is the rotational stability of a top-heavy floating body typically measured/assessed?

2.201C For a floating body, a slight rotation from its original position of equilibrium will cause the weight, W and the buoyant force, F_B to move out of vertical alignment (see Figures 2.38 through 2.40) (except in the case of a spherical-shaped body (see Figure 2.41); zero moment). What happens to the position of the center of gravity, G and the position of the center of buoyancy, B?

2.202C For a given floating body that is top heavy, there is a maximum design rolling angle of rotation, θ by which the body can tilt, beyond which it will overturn (capsize) and either find a new position of equilibrium, or, in the case of a boat or ship, it will sink. Describe how the maximum design small rolling angle of rotation, θ by which the body can tilt before it will overturn varies with the geometry of the floating body.

2.203C What does the rotation stability of a top-heavy floating body depend upon?

2.204C Explain what may occur if liquid (liquid ballast or fuel oil) in a hull of a ship is not constrained and stored in tanks or bulkhead compartments.

2.205C The length of the metacentric height, GM above G is a measure of rotational stability of a top-heavy floating body (see Figure 2.42). Explain the relationship between the magnitude ($+$, $-$, or 0) of the metacentric height, GM and the rotational stability of a top-heavy floating body.

2.206SI A block of wood with uniform density weighs $560\,N$ and floats in water. Its dimensions are $H = 60\,cm$ by $h = 60\,cm$, and $b = 90\,cm$ along the proposed axis of rotation, y as illustrated in Figure ECP 2.206. (a) Determine if the floating block of wood is top heavy

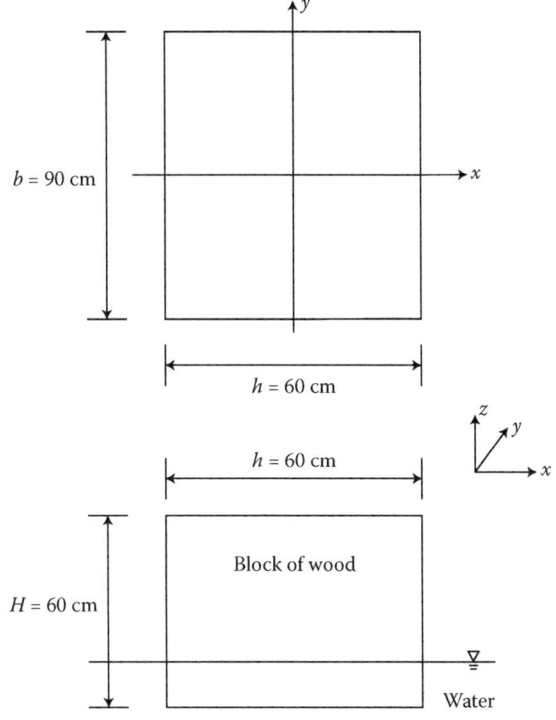

FIGURE ECP 2.206

or bottom heavy. (b) Determine if the floating block will be in a position of rotational stable equilibrium about its longitudinal y-axis if the block is rotated by $9°$ (assumed to be a small rotational disturbance) clockwise about the y-axis. (c) Determine the magnitude and the type of moment that results from the rotational disturbance.

2.207BG   A boat weighs $6{,}000$ *lb* and floats in water. Its horizontal cross section (plan view) at the water line and its vertical cross section, along with the dimensions, are illustrated in Figure ECP 2.207. Assume that the boat is top heavy (i.e., G is located above B). (a) Determine the buoyant force, F_B and the volume of the displaced fluid, $V_{dispfluid}$. (b) Determine the maximum distance GB that the center of gravity, G may be above the center of buoyancy, B in order for the boat to be in a position of rotational stable equilibrium.

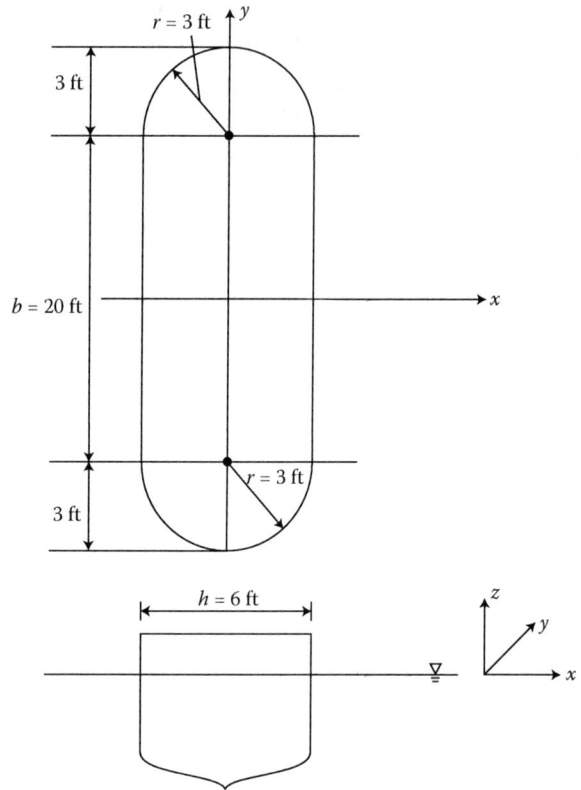

FIGURE ECP 2.207

2.208SI   A rectangular barge floats in water. Its dimensions are $H = 4\,m$ by $h = 10\,m$, and $b = 20\,m$ along the proposed axis of rotation, y, as illustrated in Figure ECP 2.208. Assume that the barge is submerged at a depth, d (draft) of $3\,m$, and the center of gravity, G is located $2\,m$ from the bottom of the barge and at $h/2 = 5\,m$ from either side of the barge, as shown in Figure ECP 2.208b. (a) Determine if the floating barge is top heavy or bottom heavy. (b) Determine the weight of the barge, W; the buoyant force, F_B; and the volume of the displaced fluid, $V_{dispfluid}$. (c) Determine if the floating barge will be in a position of rotational stable equilibrium about its longitudinal y-axis if the barge is rotated by $10°$ (assumed to be a small rotational disturbance) clockwise about the *y*-axis. (d) Determine the magnitude and the type of moment that results from the rotational disturbance.

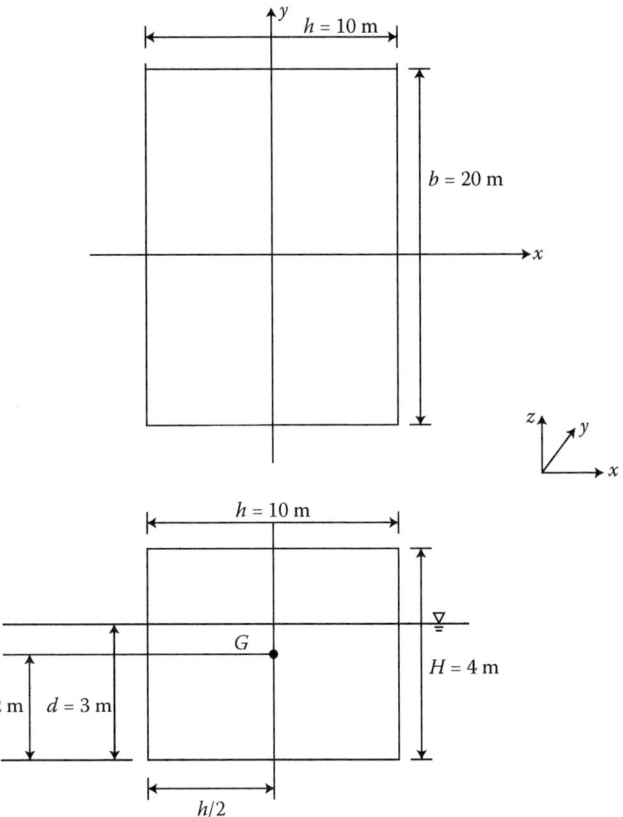

FIGURE ECP 2.208

3

Continuity Equation

3.1 Introduction

In the study of fluid mechanics, there are three basic units of study: fluid statics, fluid kinematics, and fluid dynamics. Fluid statics deals with fluids at rest, and is based upon the principles of hydrostatics (yielding the hydrostatic pressure equation), which was introduced and applied in Chapter 2. Fluid kinematics and fluid dynamics both deal with fluid in motion. Fluid kinematics is based upon the principle of conservation of mass, and yields the continuity equation, which is introduced in Chapter 3. Fluid dynamics is based upon the principle of conservation of momentum (Newton's second law of motion), and yields the equations of motion, known as the energy equation and the momentum equation. These equations are introduced in Chapters 4 and 5, respectively. Fluid dynamics also includes the topic of dimensional analysis, which yields the resistance equations, and is introduced in Chapter 6. The continuity equation, the equations of motion, and the results of dimensional analysis are applied to internal flows, which include pressure (pipe) flow in Chapter 8, and gravity (open channel) flow in Chapter 9. However, only the energy and momentum equations and the results of dimensional analysis are applied to external flows, which include the flow around objects in Chapter 10. And, finally, depending upon the internal flow problem, one may use either the Eulerian (integral) approach or the Lagrangian (differential) approach or both in the application of the three governing equations (continuity, energy, and momentum). It is important to note that while the continuity and momentum equations may be applied using either the integral or the differential approach, application of the energy equation is useful only using the integral approach, as energy is work, which is defined over a distance, L.

Therefore, depending on the unit of study, certain principles and thus certain governing equations will be applicable in the solution of problems. For fluids in static equilibrium, the principles of hydrostatics and thus the hydrostatic pressure equation would be applicable. For fluids in motion, the principles of the conservation of mass and the conservation of momentum and thus the continuity equation and the equations of motion (energy equation and momentum equation), respectively, would be applicable. Furthermore, it may be noted that for an internal fluid flow, while the application of the continuity equation will always be necessary, the energy equation and the momentum equation play complementary roles in the analysis of a given flow situation. Complementary roles mean that when one of the two equations of motion breaks down, the other may be used to find the additional unknown quantity.

3.2 Fluid Kinematics

The main focus of Chapter 3 is the topic of fluid kinematics, which is based on the principle of conservation of mass and yields the continuity equation. Specifically, fluid kinematics considers the type of fluid flow (Section 3.3), the flow visualization/geometry (Section 3.4), and the description and observation the fluid motion and thus the analysis of the velocity and acceleration of the fluid in motion (Section 3.5). Specifically, assuming internal flow, the fluid in motion can be described using either the Eulerian (integral) approach, the Lagrangian (differential) approach, both. The physical laws governing the fluid in motion (Section 3.6) yield the continuity equation (Section 3.7), the energy equation (Chapter 4), and the momentum equation (Chapter 5), which are the three governing equations in fluid flow. Thus, in Chapter 3 the various aspects of fluid motion (fluid kinematics) will be discussed without being concerned with the actual forces required to produce the motion. The dynamics of the motion (fluid dynamics), which is introduced in Chapters 4 and 5, considers the analysis of the specific forces necessary to produce the motion. Furthermore, the understanding of how to describe and observe the fluid motion (fluid kinematics) is a basic step to the thorough understanding of the cause of the fluid motion (fluid dynamics). And, finally, the measurement of the volume flowrate for internal flows (Section 3.8) is accomplished by using flow measurement devices (Chapters 8 and 9).

3.3 Types of Fluid Flow

In the study of both fluid kinematics and fluid dynamics, there are a number of different flow types to consider. The fluid properties, along with the fluid kinematics and the fluid dynamics will determine the type of flow depending upon the following: (1) the use of energy or work to move the fluid: internal versus external flow; (2) the cause of flow: pressure (pipe) versus gravity (open channel) flow; (3) the relative importance of the fluid viscosity: real versus ideal flow; (4) the spatial and temporal change in the fluid density: compressible versus incompressible flow; (5) the spatial variation of the fluid velocity: spatially varied (nonuniform) versus spatially uniform flow; (6) the temporal variation of the fluid velocity: unsteady versus steady flow; and (7) the assumption of spatial dimensionality: one-, two-, or three-dimensional flow.

3.3.1 Internal versus External Flow

In an internal flow, energy or work is used to move/force the fluid through a conduit, while in an external flow, energy or work is used to move a submerged body through the fluid, where the fluid is forced to flow over the surface of the body. For internal flows, one is interested in the determination of the energy or head losses, pressure drops, and cavitation where energy is dissipated. And, for external flows, one is interested in the determination of the flow pattern around the body, the lift, the drag (resistance to motion) on the body, and the patterns of viscous action in the fluid as it passes around the body. Internal flows include pipe (pressure) (see Chapter 8) and open channel (gravity) flow (see Chapter 9), while external flows include flow around objects (see Chapter 10). Both internal flows and external flows may be modeled as ideal or real flow; compressible (internal pipe flow and external flow only) or incompressible flow; nonuniform or uniform flow (internal flow only); unsteady or steady; and one-, two-, or three-dimensional flow.

3.3.2 Pressure versus Gravity Flow

Pressure (pipe) flow is a result of the existence of a hydraulic gradient (a pressure difference [pressure drop, Δp]) in a completely filled pipe, while gravity (open channel) flow is a result of invoking gravity in an open (to the atmosphere)/partially filled channel. The choice of the type of internal flow, pressure versus gravity flow, will depend upon the desired engineering application.

3.3.3 Real versus Ideal Flow

A real (viscous) flow assumes a nonzero value for fluid viscosity, while an ideal (inviscid or frictionless) flow assumes a zero value for fluid viscosity. The assumption of real flow models the existence of flow resistance, which is the shear stress, friction, or drag originating in the fluid viscosity. The validity of the assumption of a real versus an ideal flow will depend upon the physical flow situation. Furthermore, a real flow may be classified as a laminar or turbulent flow.

3.3.3.1 Laminar versus Turbulent Flow

A real flow may be further subdivided into laminar flow or turbulent flow depending on the value of the Reynolds number, $R = \rho v L / \mu$. Laminar flow is also called viscous or streamline flow. In laminar flow ($R < 2000$), the fluid particles move in definite and observable paths or streamlines, in which the flow is characteristic of a viscous fluid or of a fluid in which viscosity plays an important role. In turbulent flow ($R > 4000$), the fluid particles move in an irregular and erratic path, showing no observable pattern. It may be noted that ideal flows and real yet laminar flows tend to lend themselves to steady flows (described below), while real yet turbulent flows tend to lend themselves to unsteady flows (also described below).

3.3.4 Compressible versus Incompressible Flow

A compressible flow (gas or liquid) assumes a large spatial and/or temporal change in pressure and temperature (effects of thermodynamics) and thus fluid density, while an incompressible flow (gas or liquid) assumes a negligibly small spatial and/or temporal change in pressure and temperature and thus fluid density. This book illustrates that there are numerous practical engineering problems that involve gases and fluids whose densities may safely be considered constant and thus treated as incompressible flow. Furthermore, this book illustrates problems that involve gases whose densities are considered to vary spatially, without considering the effects of thermodynamics. While compressible flow may occur as either pressure (pipe) flow (Chapter 8) or external flow (Chapter 10), incompressible flow may occur as either pressure (pipe) flow or gravity (open channel) flow (Chapter 9).

3.3.5 Spatially Varied (Nonuniform) versus Spatially Uniform Flow

A spatial variation in the cross-sectional area of a pipe or open channel flow (or the existence of a control in an open channel) results in a spatial variation in the fluid velocity (convective acceleration) and thus spatially varied (nonuniform) flow. No spatial variation in the cross-sectional area of a pipe or open channel flow (or the nonexistence of a control in an open channel) results in no spatial variation in the fluid velocity and thus spatially uniform flow.

3.3.6 Unsteady versus Steady Flow

A temporal variation in the fluid velocity results in a fluid acceleration (local acceleration) and thus unsteady flow. No temporal variation in the fluid velocity results in no acceleration of the fluid and thus steady flow. Because most practical engineering problems involve steady flow, the focus of this textbook will be limited to steady flow problems.

3.3.7 One-, Two-, and Three-Dimensional Flows

A typical fluid flow involves a three-dimensional geometry; thus, the velocity may vary in all three dimensions (x, y, z) as: $v(x, y, z)$. Additionally, if the velocity of the flow varies with time, t then the velocity would be: $v(x, y, z, t)$. However, if the variation of velocity in certain directions is small relative to the variation in the other directions, it can be ignored with negligible error. Thus, in such a case, the flow can be modeled conveniently as a one- or two-dimensional flow, which is a lot easier to analyze. Therefore, the simplifying assumption of spatial dimensionality, from a three-dimensional flow to that of a one- or two-dimensional flow, is typically made in most engineering problems without sacrificing any required accuracy. It may be noted that if a one-dimensional flow is assumed, the flow may be alternatively defined along a streamline, s (defined in Section 3.4). Furthermore, the assumption regarding the spatial dimensionality of the flow is discussed in detail in a section below.

3.4 Flow Visualization/Geometry

While the quantitative study of fluid kinematics and fluid dynamics requires analytical/ computational mathematics, there are many types of flow patterns that can be visualized, both physically/experimentally and/or analytically/computationally. Thus, flow visualization enhances the study of physical experiments and the numerical solutions of computational fluid dynamics. For instance, in the study of steady nonuniform (spatially varied) open channel flow (Chapter 9), once a numerical solution of a surface water profile is completed, the flow is visualized by plotting the surface water profile (path lines or streamlines) in order to see the entire picture. Furthermore, a good understanding of fluid kinematics and flow visualization will facilitate the understanding of fluid dynamics.

Flow visualization/geometry is the visual examination of flow field features. Flow field features may be used to describe the geometry of the flow patterns of fluid in motion, and they include path lines, streamlines, and stream tubes. In the case of steady flow, a path line and a streamline are the same. However, in the case of unsteady flow, a path line and a streamline are different. Furthermore, for a differential cross-sectional area, a streamline and a stream tube are the same. It may be noted that while path lines are often used in experimental work, streamlines are often used in analytical work.

3.4.1 Path Lines

A "path line" is the spatial trace (actual path or trajectory) made by an individual fluid particle over a period of time and is illustrated in Figure 3.1. Because a path line is the plot of an individual fluid particle's spatial position over a finite time interval, it can be directly and visually observed. A path line is a Lagrangian concept (explained below) in that the path of an individual fluid particle is followed as it moves around the flow field. Furthermore, the path line of an individual particle shows the direction of velocity of the particle at successive instants of time. If the flow is steady, the path line of an individual fluid particle will be identical to the path lines formed by all other fluid particles that previously (in time) passed through the given spatial points.

3.4.2 Streamlines

A "streamline" is defined as the line drawn tangent to the velocity vectors of an individual fluid particle over a period of time. Alternatively stated, a tangent at any point on the streamline represents the instantaneous direction of the velocity of a fluid particle at that point in time. Furthermore, the velocity vector has a zero component normal to the streamline; thus, there can be no flow across the streamline at any point in time.

If the flow is steady (assuming either ideal or real and laminar flow) where there are no fluctuating velocity components, then the path line and the streamline will be identical (will coincide). If, however, the flow is unsteady (assuming real and turbulent flow), the (instantaneous) path line and the (instantaneous) streamline will not coincide. The path line, which is very irregular, simply represents the path followed by the individual fluid particle over a period of time, while the streamline represents a line whose tangent at any point in time is in the direction of the velocity at that point in time, which yields only an instantaneous flow representation. Figure 3.2 illustrates streamlines for either a steady flow or

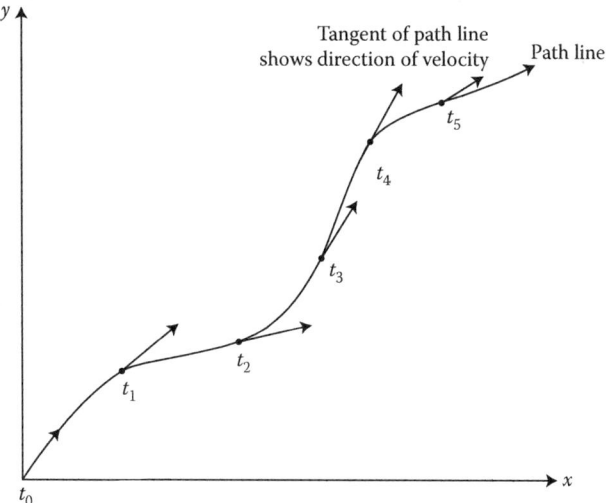

FIGURE 3.1
Path line; describes the trajectory of a fluid particle; a path line is the tracing of the spatial position (x, y) of a fluid particle as a function of time, t.

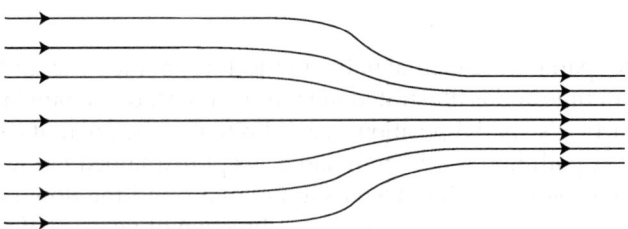

FIGURE 3.2
Streamline; line drawn tangent to the velocity vectors of an individual fluid particle over a period of time for either steady ideal or real and laminar flow or unsteady turbulent flow; streamlines coincide with path lines for steady flow but not for unsteady flow.

an unsteady flow, where they coincide with the path lines for the steady case but not for the unsteady case.

Assuming steady flow where the path line and the streamline coincide, an individual fluid particle will travel on a path whose tangent is always in the direction of the fluid velocity at any point in time. Plotting a series of streamlines yields a flow field (stream surface), as illustrated in Figure 3.3. Flow fields show the mean direction of velocity of a number of fluid particles at the same instant of time and serve both as qualitative and quantitative plots. Flow fields allow one to map the fluid flow through both experimental and mathematical determination of the streamlines and to locate regions of high and low velocity and thus zones of low and high pressure.

3.4.3 Stream Tubes

Assuming steady flow, if a series of streamlines are plotted contiguously along a closed curve, they will form a boundary across which fluid particles cannot pass because the velocity is always tangent to the boundary. Furthermore, the space between the streamlines yields a tube or a passage called a "stream tube." Thus, a stream tube may be viewed as a bundle of streamlines and is illustrated in Figure 3.4. It may be noted that while neither streamlines nor stream tubes have any physical significance, they are geometric illustrations that are drawn within the flowing field. The stream tube may be treated as if it were isolated

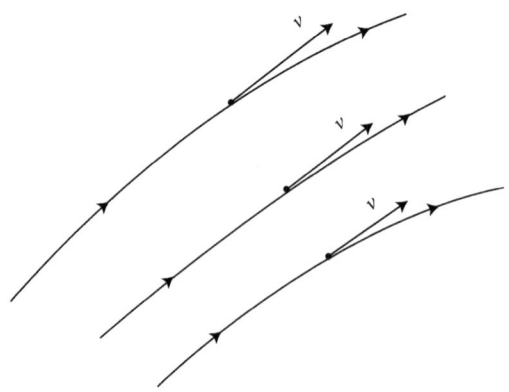

FIGURE 3.3
Flow field (stream surface); a series of streamlines.

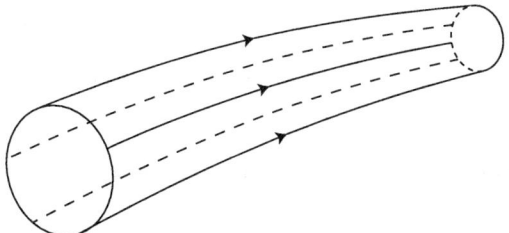

FIGURE 3.4
Stream tube; a bundle of streamlines.

from the adjacent fluid. Furthermore, because a stream tube of a differential size basically coincides with its axis, which is a streamline, many of the equations developed for a small stream tube will equally apply to a streamline.

A stream tube represents basic parts of a flowing field that are bounded by a group of streamlines that confine the flow. If the cross-sectional area of the stream tube is assumed to be sufficiently small (differential size), then the velocity at the center of any cross section may be assumed to represent the average velocity for the entire cross section. A stream tube with such a differential cross section may be used to derive the continuity equation assuming a one- or two-dimensional steady flow (given below).

3.5 Describing and Observing the Fluid Motion

Because the concept of a free-body diagram that was used in the study of fluid statics is inadequate for moving fluids, instead we use the concept of either a "fluid system" or a "control volume" to describe the motion of a fluid. Specifically, the fluid motion is described in terms of the velocity and the acceleration of the fluid particles. These two concepts are applicable for the study of both fluid kinematics and fluid dynamics. The difference between these two concepts is based on: (1) whether a constant mass (variable volume) or a constant volume (variable mass) of fluid is assumed and described, (2) how the motion of the fluid is observed and then further described and thus, (3) how the governing equations for fluid kinematics (continuity equation) and fluid dynamics (energy and momentum equations) are stated and applied. Application of the fluid system concept assumes a constant mass and is known as the Lagrangian point of view, while application of the control volume concept assumes a constant volume and is known as the Eulerian point of view.

A flow field is defined by field variables including velocity, density, pressure, and temperature. The assumption is that every fluid particle in a flow has an instantaneous value of velocity, density, pressure, temperature, and other fluid characteristics. As the fluid particle moves within the fluid, these above-listed fluid characteristics will change both spatially and temporally. If a fluid system (constant mass) is assumed, then the flow is described by the Lagrangian point of view by spatially and temporally following the detailed history of each individual fluid mass particle as it moves about. Thus, in the Lagrangian point of view, the spatial position, velocity, density, pressure, temperature, etc. are recorded as a function of time. If, however, a control volume (constant volume) is assumed, then the flow is described by the Eulerian point of view: the velocity, density, pressure, temperature, etc. at given fixed points in space are recorded as a function of time. A comparison of the Lagrangian and the Eulerian points of view, and both the convenience of and the need for their applications in fluid flow problems are described below.

3.5.1 Definition of a Fluid System: The Lagrangian Point of View

A fluid system assumes and describes a specific (predetermined) mass of fluid (such as a fluid particle or a collection of fluid particles) within a set of boundaries. The mass, shape, and boundaries of a fluid system are optionally selected. While the shape, boundaries, and volume of the fluid system may change with time (for instance, as the fluid flows through a constriction or an expansion), the fluid system cannot alter its mass. Additionally, energy in the form of heat or work can cross the boundaries. Therefore, for a fluid system, the mass remains constant (thus defining a closed system) and therefore, conservation of mass is automatically ensured.

In the Lagrangian point of view, a coordinate system is attached to the center of mass of the fluid system (constant mass of fluid), and the observer, who is located at the origin of the moving frame of reference, follows the motion of the fluid system. Each fluid particle (or a collection of fluid particles) is described by its initial spatial coordinates at some initial time. Then, as time passes, the path (spatial position), velocity, density, shape, and other fluid characteristics (such as volume, pressure, and temperature) of each individual particle(s) are traced. In particular, a plot of the spatial position of a fluid particle(s) as a function of time describes the trajectory of the particle and is called a "path line," and was illustrated above in Figure 3.1. Therefore, the Lagrangian point of view considers the dynamic behavior of a fluid particle(s), whereby the history of the particle(s) is analyzed: where it was, where it is, where it will be. The Lagrangian point of view is used to trace the gradual variation in depth of steady nonuniform open channel flow problems in Chapter 9. Furthermore, one may note that the Lagrangian point view is also used in the dynamic analyses of solid particles.

3.5.1.1 Definition of the Fluid Particle Velocity in the Lagrangian Point of View

While the various fluid characteristics of a flow field (such as velocity, density, volume, pressure, and temperature) may be spatially traced and defined as a function of time, the velocity is one of the most important fluid variables to define. One may note that because velocity is a vector, a change in either the magnitude or the direction of the velocity results in an acceleration of the fluid flow. In the Lagrangian point of view, the velocity of an individual fluid particle is obtained by differentiating the individual fluid particle's position vector with respect to time. The position vector, $l(t)$, which is used to track the position of a given fluid particle as time passes, is expressed as follows:

$$\vec{l}(t) = x\vec{i} + y\vec{j} + z\vec{k} \tag{3.1}$$

where i, j, and k are unit vectors in the x, y, and z directions, respectively. Then, differentiating Equation 3.1 with respect to time yields the total velocity vector of the fluid particle as follows:

$$\vec{v}(t) = \frac{d\vec{l}}{dt} = \frac{dx}{dt}\vec{i} + \frac{dy}{dt}\vec{j} + \frac{dz}{dt}\vec{k} \tag{3.2}$$

or:

$$\vec{v}(t) = u\vec{i} + v\vec{j} + w\vec{k} \tag{3.3}$$

where u, v, and w are the components of the fluid velocities in their respective coordinate directions. If a one-dimensional flow is assumed, one may alternatively express the position vector, $l(t)$ of an individual fluid particle as a function of the position, s along a streamline as follows:

$$\vec{l}(t) = \vec{s}(t) \tag{3.4}$$

And therefore, differentiating Equation 3.4 with respect to time yields the total velocity vector of the fluid particle as follows:

$$\vec{v}(t) = \frac{ds}{dt} \tag{3.5}$$

Because the motion of only one individual fluid particle would be inadequate to describe the entire flow field, the motion of all the fluid particles in the flow field must be considered simultaneously. Therefore, the motion in an entire flow field is defined by solving Newton's second law of motion, $F = Ma$ for each fluid particle in the flow field. An illustration of this is given when the Lagrangian point of view is applied to trace the gradual variation in depth (surface water profile) for the case of a steady nonuniform open channel flow problem in Chapter 9.

3.5.2 Definition of a Control Volume: The Eulerian Point of View

A control volume assumes and describes a specific (predetermined) volume of fluid within a set of boundaries in a fixed region in the fluid space that does not move. The volume and shape of the control volume are optionally selected, while the boundaries are usually selected to coincide with a solid or a natural flow boundary. While the mass of fluid flowing through the boundaries of a control volume may change with time (thus, defining an open system), with a fixed shape for rigid boundaries and a variable shape for flexible or no boundaries, the control volume cannot alter its volume. Additionally, energy in the form of heat or work can cross the boundaries.

Fluid flow problems that involve mass flow in and out of (or around) a system, and thus require the determination of forces put on (interaction with) a device or a structure, are modeled using a control volume as opposed to a fluid system. A control volume usually encloses a device/mechanism or structure that involves the flow of fluid mass, such as a pump, turbine, compressor, nozzle, water heater, car radiator, fan, airplane, car, spillway, weir, or any pipe or channel section.

While any arbitrary region in the fluid space can be selected as a control volume, a guided choice will make the analysis of the flow problem a lot easier. The control volume is usually selected as a region that fluid flows into and out of. The closed boundaries of a control volume are called the "control surface" and are illustrated in Figure 3.5. A control volume allows fluid mass to flow through its control surface, where the fluid flows into control face 1 and out of control face 2, as illustrated in Figure 3.5.

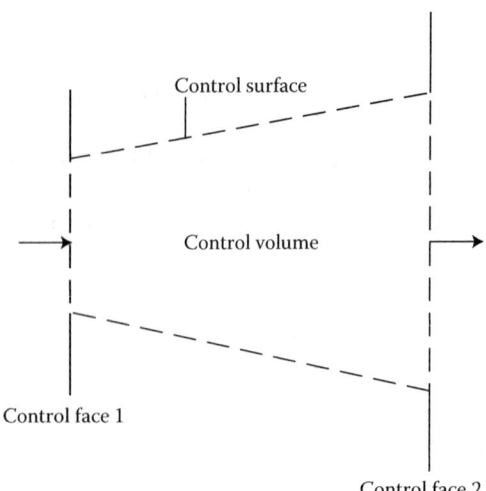

FIGURE 3.5
Definition of a control surface, control volume, and control faces.

In the Eulerian point of view, the motion of the individual fluid particle is not traced. Instead, the motion of a constant volume (variable mass) of fluid particles is observed from a fixed position in space, where a fixed position could be at the origin of an internal frame of reference. Therefore, the Eulerian point of view addresses the dynamic behavior of fluid motion at a fixed point in space and thus describes the motion of fluid as it moves through a given spatial region in the flow at a time of interest. The values and variations with time of the velocity, density, and other fluid characteristics (such as pressure and temperature) are determined at any desired spatial points in the flow, such as at control face 1 and control face 2 of the control volume in Figure 3.5 at a time of interest. Thus, the attention is focused on the particular spatial points in the space filled by the fluid at a particular time.

Because in the Eulerian point of view the observed and measured velocities are absolute velocities (relative to the earth or fixed ground), it is easy to determine if the fluid flow is steady or unsteady. In an unsteady flow state, the fluid velocity and other fluid characteristics will vary with time at the spatial points in the flow, while in a steady flow state, the fluid velocity and other fluid characteristics do not vary with time at a given spatial point in the flow; the fluid velocity and other fluid characteristics are only a function of the spatial position in the space filled by the fluid. Furthermore, in the Eulerian view, a steady flow state can still have convective accelerations that are caused by a spatial variation in the fluid velocity (spatially varied flow).

3.5.2.1 Definition of the Fluid Particle Velocity in the Eulerian Point of View

While the various fluid characteristics of a flow field (such as velocity, density, mass, pressure, and temperature) may be defined as a function of space and time, the velocity is one of the most important fluid variables to define. One may note that because velocity is a vector, a change in either the magnitude or the direction of the velocity results in an acceleration of the fluid flow. In the Eulerian point of view, where the motion of fluid particles passing a specific point of interest is observed, the fluid particle velocity vector depends on the point in space and time and is expressed as follows:

$$\vec{v} = \vec{v}(x, y, z, t) \tag{3.6}$$

which may be further expanded as follows:

$$\vec{v} = (u, v, w) = u(x, y, z, t)\vec{i} + v(x, y, z, t)\vec{j} + w(x, y, z, t)\vec{k} \tag{3.7}$$

where i, j, and k are unit vectors in the x, y, and z directions, respectively, and u, v, and w are the components of the fluid velocities in their respective coordinate directions. Thus, the velocity (and other flow field variables such as density, pressure, temperature, etc.) is defined at any location (x, y, z) in the control volume and at any instant of time t. If a one-dimensional flow is assumed, one may alternatively express the total velocity of a fluid particle as a function of the position, s along a streamline as follows:

$$\vec{v} = \vec{v}(s, t) \tag{3.8}$$

Furthermore, because we need to describe the volume of fluid within a set of boundaries, we must determine the fluid motion at certain specified points for the control volume.

3.5.2.2 Fixed versus Moving Control Volume in the Eulerian Point of View

As defined above, a control volume assumes and describes a specific (predetermined) volume of fluid within a set of boundaries in a fixed region in the fluid space that does not move. As such, in the Eulerian point of view, the velocity at a given control face, v_i is defined relative to the control volume. However, one may note that the absolute velocity at a given control face is defined relative to the earth (or fixed ground), $v_i(abs)$, and the absolute velocity of the control volume is also defined relative to the earth (or fixed ground), $v_{cv}(abs)$. Thus, assuming the same positive s-direction along a streamline, the velocity at a given control face relative to the control volume is defined as follows:

$$\vec{v}_i = \vec{v}_i(abs) - \vec{v}_{cv}(abs) \tag{3.9}$$

Furthermore, relative to the earth (or fixed ground), the control volume may be either a fixed or a moving control volume. A fixed control volume encloses a device/mechanism or structure that involves the flow of fluid mass through a *stationary* pump, turbine, compressor, nozzle, water heater, fan, spillway, weir, etc. However, a moving control volume encloses a device/mechanism that involves the flow of fluid mass through a *moving* pump, turbine, compressor, nozzle, water heater, fan, etc. The moving control volume is typically on board a car, ship, plane, etc. In the case where the control volume is fixed relative to the earth, its absolute velocity, $\vec{v}_{cv}(abs) = 0$. Thus, the velocity at the given control face relative to the control volume is given as follows:

$$\vec{v}_i = \vec{v}_i(abs) \tag{3.10}$$

However, in the case where the control volume is moving relative to the earth, its absolute velocity, $\vec{v}_{cv}(abs) \neq 0$. Thus, the velocity at the given control face relative to the control volume is given as follows:

$$\vec{v}_i = \vec{v}_i(abs) - \vec{v}_{cv}(abs) \tag{3.11}$$

It is important to note that the velocity is a vector with both magnitude and direction.

3.5.3 Deriving One View from the Other

It is important to note that if enough information in the Lagrangian form is available, then Eulerian information can be derived from the Lagrangian data, and vice versa. For instance, the velocity vector in the Lagrangian form was given above in Equation 3.2 as follows:

$$\vec{v}(t) = \frac{d\vec{l}}{dt} = \frac{dx}{dt}\vec{i} + \frac{dy}{dt}\vec{j} + \frac{dz}{dt}\vec{k} \tag{3.12}$$

Because the velocity vector of a fluid particle in the Lagrangian form is spatially traced in the flow field, the velocity vector as a function of position would not be known unless the location (x, y, z) of each fluid particle were known as a function of time. The velocity vector in the Eulerian form was given above in Equation 3.6 as follows:

$$\vec{v} = \vec{v}(x, y, z, t) \tag{3.13}$$

where the velocity of a fluid particle is the time rate of change of the position vector for that particle. Because the velocity vector of a fluid particle in the Eulerian form is given at fixed points in space, the velocity vector as function of time would not be known unless the location (x, y, z) of the fluid particle was known as a function of time. This means that whether the Lagrangian or the Eulerian form/point of view is used to observe and describe the motion of the fluid particle, the most basic data required in order to define the velocity vector (or any other fluid characteristic such as density, pressure, and temperature) is the definition of the position vector, $l(t)$ which describes the location (x, y, z) of each fluid particle as a function of time. The position vector was given above in Equation 3.1 as follows:

$$\vec{l}(t) = x\vec{i} + y\vec{j} + z\vec{k} \tag{3.14}$$

Therefore, in order to derive one form of information (Lagrangian or Eulerian) from the other form of data (Lagrangian or Eulerian), the position vector is required, at minimum.

3.5.4 Application of the Lagrangian View versus Application of the Eulerian View

Although the Eulerian approach is a more convenient and practical view for most engineering problems and is used in a great majority of experimental and fluid analysis problems, there are problems for which it is necessary to follow a fluid mass as it flows along and thus the Lagrangian approach is more convenient. While the Lagrangian point of view follows and observes the behavior of a fluid particle as it moves about, and tracks the position of a given fluid particle as a function of time, the Eulerian view remains stationary and simply observes the motion of fluid particles passing a specific point of interest as a function of time. In the flow of fluids, it is not easy to identify individual fluid particles from one another or the interactions between one another. Thus, keeping track of the positions of all particles in a flow field as a function of time, as done in the Lagrangian approach, is a large task (unlike for solid particles) because the relative positions of the fluid particles and their shape continuously change with time. Thus, when the engineering problem involves mass flow in and out of (or around) a system (as a large number of problems do) and thus one is interested in determining the forces put on (interaction with) a device or a structure, then they are modeled as a control volume (Eulerian view) as opposed to a fluid system (Lagrangian view). However, when the (numerical) solutions of certain fluid flow problems are based on determining the motion of individual fluid particles, which is based on the interactions among the fluid particles, it becomes necessary to describe the motion using the Lagrangian point of view. And, finally, it may be noted that the solutions of most problems presented in this textbook use the control volume approach, with the exception of a few cases. These cases include the steady nonuniform open channel flow problem (Chapter 9), where it is of interest to trace the gradual variation of the depth that takes place along a channel, or for any pipe or channel section (Chapters 8 and 9), where it is of interest to solve for an unknown quantity (velocity, pipe diameter or depth of channel flow, pipe or channel friction, or friction slope) at a given point in the fluid system. Additionally, several examples are presented in Chapter 3, herein. In such cases the fluid system approach is used.

3.6 The Physical Laws Governing Fluid in Motion

As mentioned above, the behavior of a fluid in motion is governed by a set of fundamental physical laws. These laws, which include the conservation of mass and the conservation of momentum (Newton's second law of motion), are adopted from solid mechanics and are stated in their basic form in terms of a "system" concept (Lagrangian point of view). These fundamental laws, as applied to a fluid flow, are approximated by an approximate set of governing equations including the continuity equation and the equations of motion (the energy equation and the momentum equation), which are also stated in their basic form in terms of a "fluid system" concept (Lagrangian point of view). For instance, definition of the conservation of mass includes a statement such as "the mass of a system remains constant," and definition of the conservation of momentum includes a statement such as "the time rate of change of momentum of a system is equal to the sum of the forces acting on the system." One may note that the word system as opposed to control volume is used in the statement of the laws. However, while the mass of a fluid system is conserved, the fluid system does not retain its position (because it's mobile) or its shape (because it's deformable). Furthermore, in most fluid flow problems, there is a need to define a more convenient object (concept) for analysis. This object is the control volume (Eulerian point of view), which is a volume fixed in space and through which matter, mass, momentum, energy, etc. flows across the boundaries. Thus, in order to apply the governing equations to a control volume, the laws must be appropriately rephrased. Therefore, in order to be able to shift from one concept (fluid system versus control volume) to another, one may use an analytical tool derived from the Reynolds transport theorem (discussed below). And finally, it may be noted that, while the equations of motion in the Lagrangian description (following the individual fluid particle) are well known (Newton's second law of motion), the equations of motion for fluid flow are not as readily apparent in the Eulerian description and must be carefully derived (see Chapters 4 and 5).

3.6.1 Statement of the Governing Physical Laws Assuming a Fluid System

The physical laws governing the behavior of a fluid in motion deal with the time rates of change of extensive properties and are expressed/stated for a fluid system. A fluid system assumes a constant mass, which may be as large as, for instance, the mass of the air in the earth's atmosphere, or may be as small as a single fluid particle. The fluid mass (fluid system) is tagged (either actually by a dye or visually in one's mind) so that it may be continually identified and tracked as it moves in the flow field. The fluid mass may interact with its surroundings through a transfer of energy (in the form of heat or work) across its boundaries. While it may continually change its size and shape, the fluid system maintains a constant mass. Therefore, the constant mass of a fluid system (closed system) automatically ensures the conservation of mass. Furthermore, the fluid mass in a system is isolated from its surroundings, whereby the surroundings are replaced by the equivalent actions they put on the fluid mass, and then Newton's second law of motion is applied (conservation of momentum).

3.6.1.1 Extensive and Intensive Fluid Properties

While the physical laws governing the behavior of a fluid in motion deal with time rates of change of extensive properties, the fluid system concept is related to both the extensive and intensive fluid properties. Furthermore, the distinction between extensive properties and intensive properties is very important in the derivation and the application of the control volume approach, in which a constant volume with a variable mass is assumed. Extensive properties are related to the total mass of the fluid system and are usually represented by uppercase letters, such as M for mass, W for weight ($W = Mg$), and V for volume ($V = \rho M$). And intensive properties are independent of the total mass of fluid and are usually represented by lowercase letters, such as ρ for mass density ($\rho = M/V$), p for pressure ($p = F/A$), and θ for temperature. In summary, the most basic extensive properties are mass, M; momentum, Mv; and energy, E. The corresponding intensive properties are mass per unit mass, which is unity; momentum per unit mass, which is velocity, v; and energy per unit of mass, which is defined as e, respectively.

3.6.2 The Reynolds Transport Theorem

Because the physical laws governing the behavior of a fluid in motion deal with the time rates of change of extensive properties and are expressed/stated for a fluid system, one may rephrase them for a control volume using an analytical tool derived from the Reynolds transport theorem. Specifically, the Reynolds transport theorem provides a relationship between the time rate of change of an extensive property for a fluid system and that for a control volume. Furthermore, this general relationship provides an important basis for rephrasing the continuity equation, the energy equation, and the momentum equation (which are stated in their basic form in terms of a system) for the control volume.

   In general, the physical laws governing the fluid in motion are stated in terms of various physical fluid parameters including velocity, acceleration, density, mass, energy, pressure, temperature, momentum, etc. But, more specifically, they are stated in terms of the time rates of change of extensive properties, which are related to the total mass, M of the fluid system. Let B represent any extensive property of the fluid, which is directly proportional to the mass, M of interest, and let $b = B/M$ represent the corresponding intensive property, which is that amount of parameter per unit mass, M and thus independent of the amount of mass, M. For instance if $B = M$, the mass, then $b = 1$ (where the mass per unit mass is unity).

If $B = Mv$, the momentum of the mass, then $b = v$ (where the momentum per unit mass is the velocity, v). Furthermore, if $B = Mv^2/2$, the kinetic energy of the mass, then $b = v^2/2$ (kinetic energy per unit mass). It may be noted that the extensive parameter, B and the intensive parameter, b may be either scalars (as in the first and third examples above) or vectors (as in the second example above). And, finally, it may be noted that because extensive properties are additive, the total amount of an extensive property (mass, momentum, kinetic energy, etc.) that a given fluid system possesses, B_{sys} at a given time, t can be determined by adding up the amount associated with each fluid particle in the fluid system.

The Reynolds transport theorem provides a link between the fluid system and the control volume approaches. Given below is a derivation of a simplified version of the Reynolds transport theorem, which is used to express the relationship between the time rates of change of an extensive property for a fluid system and for a control volume. Consider the one-dimensional uniform (at a given cross section) flow through a stream tube from left to right as illustrated in Figure 3.6. A control surface is defined by the upper and lower bounds of the stream tube and by control face 1 and control face 2. Furthermore, the control volume is fixed between sections 1 and 2 of the flow field. Thus, the control volume is represented by the given volume within the above-defined control volume, while the fluid system is represented by the given mass of fluid within the defined control volume. At an initial time, t, the fluid system between sections 1 and 2 coincides with the control volume between control face 1 and control face 2, and thus they are identical. Furthermore, given any extensive property, B, at the initial time, t, it will be equivalent for both the fluid system (sys) and the control volume (cv) as follows:

$$B_{sys}(t) = B_{cv}(t) \tag{3.15}$$

During a time interval, Δt, the fluid system between sections 1 and 2 *moves in time* to a new position bounded by sections 1′ and 2′, while the control volume *remains fixed in time* between control face 1 and control face 2. Because the extensive property of the fluid system, B is additive, B_{sys} at a time, $t + \Delta t$, $B_{sys}(t + \Delta t)$ (or B between sections 1′ and 2′, $B_{1'2'}$) can be determined by adding up the B_{cv} at time $t + \Delta t$, $B_{cv}(t + \Delta t)$ (or B_{12}) minus the amount B lost between 1 and 1′, $B_{11'}$ (inflow into cv during Δt), plus the amount B gained between 2

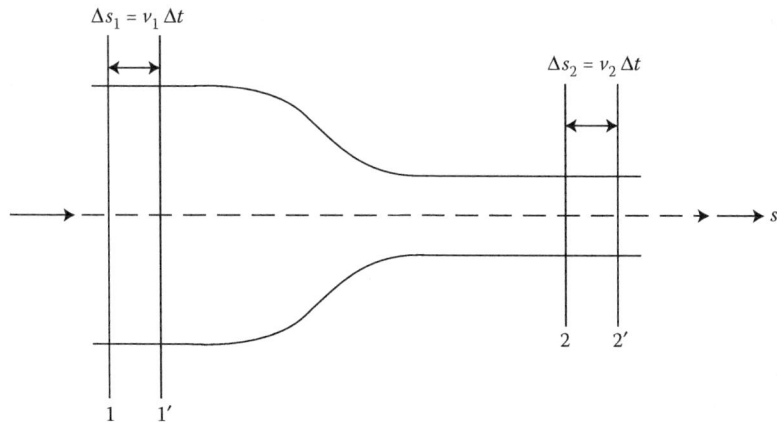

FIGURE 3.6
Definition of a control volume for flow through a stream tube.

and 2', $B_{22'}$ (outflow out of *cv* during Δt) as follows:

$$B_{sys}(t + \Delta t) = B_{cv}(t + \Delta t) - B_{11'} + B_{22'} \tag{3.16}$$

or:

$$B_{1'2'} = B_{12} - B_{11'} + B_{22'} \tag{3.17}$$

Thus, in order to state the general form of a physical law governing the behavior of a fluid in motion as the time rate of change of an extensive property, B expressed for a fluid system, we subtract the expression for B_{sys} at the initial time t (Equation 3.15) from the expression for B_{sys} at a time, $t + \Delta t$ (Equation 3.16) and divide by Δt and get the following:

$$\frac{B_{sys}(t + \Delta t) - B_{sys}(t)}{\Delta t} = \frac{B_{cv}(t + \Delta t) - B_{cv}(t)}{\Delta t} - \frac{B_{11'}}{\Delta t} + \frac{B_{22'}}{\Delta t} \tag{3.18}$$

Then, taking the limit as $\Delta t \to 0$, and using the definition of a derivative, we get the following:

$$\frac{dB_{sys}}{dt} = \frac{dB_{cv}}{dt} - \dot{B}_{in} + \dot{B}_{out} \tag{3.19}$$

which states that the "time rate of change of the extensive property, B of the fluid system is equal to the time rate of change of the extensive property, B of the control volume, plus the net flux of B out of the control volume (outflux minus the influx), which is caused by the fluid mass crossing the control surface." Alternatively stated: "the difference between the time rate of change of the extensive property, B within the fluid system and the control volume is equal to the net flux of B out of the control volume." Thus, Equation 3.19 is the sought-after simplified version of the Reynolds transport theorem, which provides a relationship between the time rate of change of an extensive property, B for a fluid system and that for a control volume and thus allows one to transfer from a fluid system approach (Lagrangian point of view) to a control volume approach (Eulerian point of view). Furthermore, Equation 3.19 is applicable at any instant in time, where it is assumed that the fluid system and the control volume occupy the same fluid space at that particular instant in time. And finally, the concept of a fluid system and a control volume occupying the same region of space at an instant of time (coincident condition) and the use of the Reynolds transport theorem (Equation 3.19) are the key elements in the derivation of the control volume governing equations (continuity, energy, and momentum equations).

It may be noted that the last two terms in Equation 3.18, which represent the net flux of B out of the control volume (outflux minus the influx), may be expressed in terms of the corresponding intensive property, $b = B/M$. Figure 3.6 illustrates that during a time interval, Δt, the fluid system between sections 1 and 2 *moves in time* to a new position bounded by sections 1' and 2', or by a distance, Δs_1 at section 1 and by a distance, Δs_2 at section 2. The velocities (measured relative to the fixed control volume) at the respective cross sections are $v_1 = \Delta s_1/\Delta t$ and $v_2 = \Delta s_2/\Delta t$, respectively, and the cross-sectional areas at the respective cross sections are A_1 and A_2, respectively, and in general, $b = B/M = B/\rho V = B/\rho \Delta s A = B/\rho v \Delta t A$. Thus, Equation 3.18 may be expressed in terms of the corresponding intensive property, b as follows:

$$\frac{B_{sys}(t + \Delta t) - B_{sys}(t)}{\Delta t} = \frac{B_{cv}(t + \Delta t) - B_{cv}(t)}{\Delta t} - b_1\rho_1 v_1 A_1 + b_2\rho_2 v_2 A_2 \tag{3.20}$$

Then, taking the limit as $\Delta t \rightarrow 0$, and using the definition of a derivative, we get the following:

$$\frac{dB_{sys}}{dt} = \frac{dB_{cv}}{dt} - b_1\rho_1 v_1 A_1 + b_2\rho_2 v_2 A_2 \tag{3.21}$$

This alternative statement of the Reynolds transport theorem not only expresses the time rate of change of the extensive property, B (for both a sys and a cv), but, additionally, it expresses the time rate of change of the intensive fluid property, b and the other intensive fluid properties (such as ρ, v, and A) for a control volume. Therefore, it may be noted that in the application of the control volume approach, the concept of a fluid system and both the extensive and intensive fluid properties related to the fluid system concept are used.

3.6.3 Statement of the Assumptions Made for the Type of Fluid Flow

In the statement of the governing equations for fluid kinematics (continuity equation) and for fluid dynamics (energy and momentum equations), there is a certain set of assumptions made regarding the type of fluid flow. As mentioned above, the flow may be (1) internal or external; (2) caused by pressure or gravity; (3) real or ideal; (4) compressible or incompressible; (5) nonuniform or uniform; (6) unsteady or steady; and (7) one-, two-, or three-dimensional. In reality, almost all flows (internal or external flows) are unsteady to a certain degree, characterized by behavior that is quite random and thus typically occurring as either two- or three-dimensional turbulent flows. Furthermore, the modeling and solution of compressible flow (gas or liquid) technically requires the effect of thermodynamics to be taken into account. However, simplifying assumptions regarding the type of flow are usually made in order to facilitate both the modeling and the solution of a flow situation without compromising the practical usefulness of the solution results. For instance, one may note that random behavior (described for turbulent flow) does not occur in laminar flow, where the flow is typically described as smooth and viscous and has a deterministic behavior. Thus, as mentioned above, while real and turbulent flows tend to lend themselves to unsteady flows, real and laminar flows (and ideal flows also) tend to lend themselves to steady flows. Furthermore, there are numerous practical engineering problems that involve either gases or liquids whose densities may safely be considered as constant and thus treated as incompressible flow, or involve gases whose densities are considered to vary spatially without considering the effects of thermodynamics.

In the derivation and the application of the three governing equations, the ultimate goal is to be able to make three basic simplifying assumptions: one-dimensional, incompressible, steady flow. The assumption of one-dimensional flow greatly simplifies both the modeling (mathematical equations) and the solution of the flow problem, while the assumption of steady and incompressible flow greatly simplifies the solution procedure. Therefore, derivations of the simplest yet most general form of the three governing equations are typically accomplished by assuming an idealized (ideal or real and laminar) one-dimensional flow in a stream tube or a pipe flowing under pressure. Because most practical engineering problems involve steady flow, the first assumption made is that the flow is steady, as the focus of this textbook will be limited to steady flow problems. A final assumption of incompressible flow is typically made to further simplify the solution procedure. And, finally, one may note that making the simplifying assumptions in a certain order yields logic in the simplifying assumptions that follow.

In the statement of the governing equations, one may assume an idealized flow (either real and laminar or ideal) as either a flow in a stream tube or a flow in a pipe under pressure (the pipe flow equations are then tailored to yield open channel flow equations). If one assumes a flow in a stream tube, beginning with the assumption of real and laminar flow makes the assumption of steady flow realistic, which in turn makes the assumption of a one-dimensional flow, defined along the center streamline, s also realistic. This is because the assumption of steady flow results in a uniform velocity profile across the entire cross section of the stream tube, and thus a one-dimensional flow. Furthermore, if one assumes a flow in a pipe under pressure, beginning with the assumption of ideal flow not only makes the assumption of steady flow realistic, it also makes the assumption of one-dimensional flow, defined along the centerline of the pipe, x realistic because the lack of viscosity in an ideal flow results in a lack of shear stress and thus no variation in velocity across the cross section. Finally, in either the assumption of a stream tube or a pipe under pressure, the flow may assumed to be either incompressible or compressible without modeling the effects of thermodynamics.

3.6.3.1 The Assumption of Spatial Dimensionality of the Fluid Flow

Although there are some flow situations for which it is necessary to analyze the flow in its complete three dimensions, such as all turbulent flow or the external flow of air around an airplane wing, for instance, there are numerous flow situations for which it is possible to make simplifying assumptions regarding the spatial dimensionality without loss of accuracy. In general, the spatial and temporal dimensionality of a fluid flow is a complex spatially three-dimensional, time-dependent phenomenon. As such, the velocity of the flow (and other fluid characteristics such as pressure, elevation, etc.) is expressed as follows:

$$\vec{v} = \vec{v}(x, y, z, t) = u\vec{i} + v\vec{j} + w\vec{k} \tag{3.22}$$

If the velocity component in one of the directions is small relative to the other two components, then it may be reasonable to neglect the smallest component and assume a two-dimensional flow; thus, the velocity of the flow would be expressed as follows:

$$\vec{v} = \vec{v}(x, y, t) = u\vec{i} + v\vec{j} \tag{3.23}$$

Furthermore, if two of the velocity components are negligible relative to the third one, then the velocity of the flow may be assumed to be a one-dimensional flow and would be expressed as follows:

$$\vec{v} = \vec{v}(x, t) = u\vec{i} \tag{3.24}$$

Alternatively, a one-dimensional flow may be defined along the streamline, s and the velocity of flow would be expressed as follows:

$$\vec{v} = \vec{v}(s, t) \tag{3.25}$$

It may be noted that in most engineering problems, one may approximate a three-dimensional real flow as a simple one- or two-dimensional flow. And, while there are very few flows that are actually one-dimensional, there are numerous flow situations for which the assumption of a one-dimensional flow is reasonable.

3.6.3.2 The Modeling of Two-Dimensional Flows

It may be noted that while the simplifying assumption of one-dimensional flow will facilitate the modeling and solution of a fluid flow problem, there are certain flow situations that would be most accurately modeled as a two-dimensional flow as opposed to a one-dimensional flow. When the streamlines are not essentially straight and parallel, spatial variations in velocity, pressure, temperature, and density are expected across a given cross section in a fluid flow situation. Specifically, if the flow is incompressible, an increase in pressure and a decrease in velocity occur with increasing distance from the center of curvature (which is basically along the centerline of the pipe or along the central streamline). And if the flow is compressible, such variations of velocity and pressure will produce variations of temperature and density as well. Figure 3.7 illustrates an example of a flow in a circular pipe under pressure with excessive variation in the cross section of the pipe (or excessive curvature in the streamlines of a stream tube) and thus significant changes in the velocity profile along the pipe (or the stream tube). Such a flow situation would be most accurately modeled as a two-dimensional flow as opposed to a one-dimensional flow. Two additional examples of two-dimensional flow are the flow past an airfoil of a constant cross section and the flow over a weir of uniform cross section, each infinitely long perpendicular to the plane of the paper, given in Figures 3.8 and 3.9, respectively. A final example is given in Figure 3.10, which illustrates a real (viscous) and laminar steady flow in a pipe flowing under pressure, and demonstrates that the actual velocity profile is typically parabolic, which is two-dimensional. The parabolic velocity profile illustrates that the velocity has two components, one in the x-direction and one in the y-direction. Furthermore, the parabolic velocity profile illustrates that the velocity is at a maximum at the centerline of the circular pipe, where the shear stress is at a minimum, while the velocity is at a minimum (zero) at either boundaries of the pipe, where the shear stress is at a maximum. One may note that the shear stress is due to the fluid viscosity. One may also note that the fluid at either boundaries of the pipe sticks to the boundaries and has a velocity of zero because of the "no-slip" condition; this implies that in a real and laminar steady flow situation, the velocity profile cannot be uniform (one-dimensional; same magnitude and same direction) across the cross section of the pipe. However, if the real (viscous) and laminar steady flow is at a well-rounded

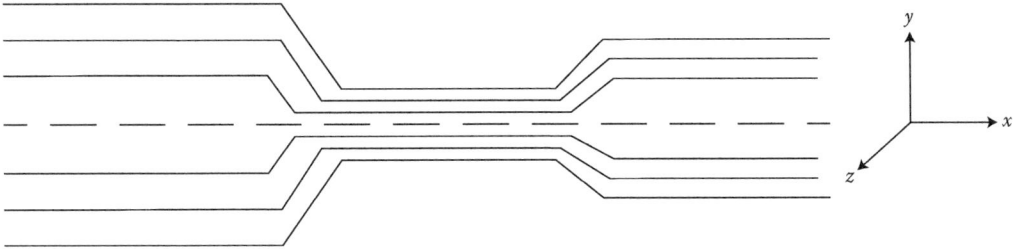

FIGURE 3.7
Two-dimensional nonuniform velocity profile occurs for a steady flow in either a circular pipe flowing under pressure or a stream tube of constant width (into the plane of the paper), with excessive variation in the cross section.

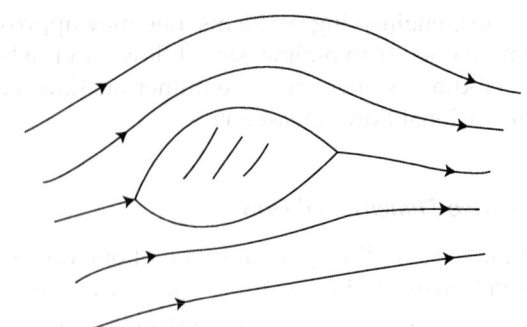

FIGURE 3.8
Two-dimensional flow past an airfoil of a constant cross section and infinitely long perpendicular to the plane of the paper.

entrance to a circular pipe flowing under pressure, as illustrated in Figure 3.11, then it may be modeled as a one-dimensional flow. The smooth curvature in the pipe results in a fairly uniform velocity profile across the entire cross section (especially close to the centerline of the pipe). Such a uniform velocity profile may be assumed to be (modeled as) a one-dimensional flow, where $v = v(x)$ and x is taken along the centerline of the pipe.

3.6.3.3 The Modeling of One-Dimensional Flows

The concept of one-dimensional flow is a very powerful and practical one that provides simplicity in flow analyses and accurate results in a wide range of engineering problems. In general, two- and three-dimensional flow engineering problems may be accurately modeled as one-dimensional problems when: (1) the variation of the cross section of the pipe or channel is not too excessive, (2) the velocity profile does not change significantly along the pipe or channel, and (3) the curvature of the streamlines is not too excessive (i.e., the streamlines are essentially straight and parallel). As such, the change in fluid properties (velocity, pressure, elevation, temperature, density, etc.) across a streamline is negligible compared to the change along the streamline. Average values of velocity (uniform velocity profile), pressure, elevation, etc. are considered to represent the flow as a whole at a given cross section, where

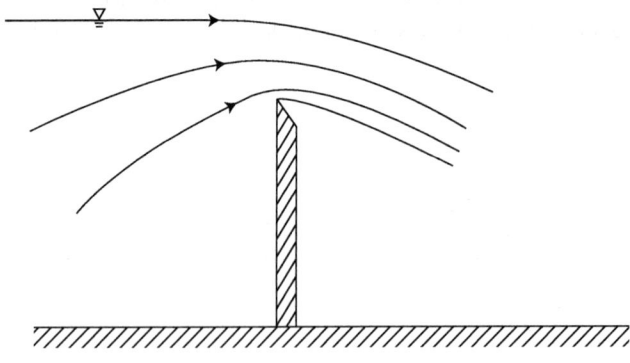

FIGURE 3.9
Two-dimensional flow over a weir of uniform cross section and infinitely long perpendicular to the plane of the paper.

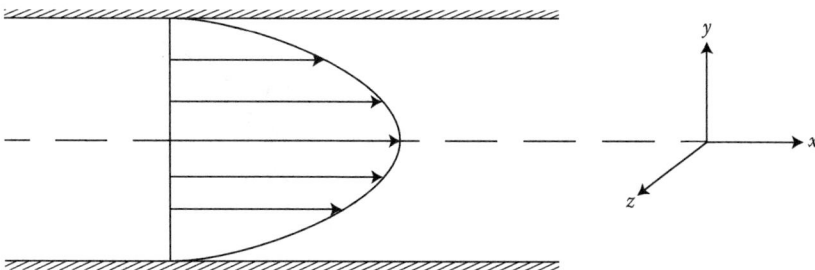

FIGURE 3.10
Two-dimensional parabolic velocity profile for a real (viscous) and laminar steady flow in a circular pipe flowing under pressure.

they are expressed as function of only one spatial coordinate. The streamline under consideration may be part of a stream tube, in a pipe flowing under pressure, or in an open channel flow. It may be noted/recalled that while a stream tube has no physical significance, it is simply a geometric illustration drawn within the flowing field (of either a pipe flowing under pressure, or an open channel flow). Furthermore, the assumption of one-dimensional flow is accurately made for idealized flow (either real and laminar, or ideal) as either a flow in a stream tube or a flow in a pipe under pressure as presented in the following two paragraphs.

Consider a real and laminar steady flow in a stream tube, as illustrated in Figure 3.12. The assumption of real and laminar flow yields to the assumption of steady flow, which results in a uniform velocity profile across the entire cross section of the stream tube. Such a uniform velocity profile may be modeled as a one-dimensional flow, where $v = v(s)$ and s is taken along the central streamline of the flow, and where the velocities and accelerations normal to the streamline are negligible. It may be noted that for a stream tube with a finite cross-sectional area, the fluid properties are assumed to be uniform across

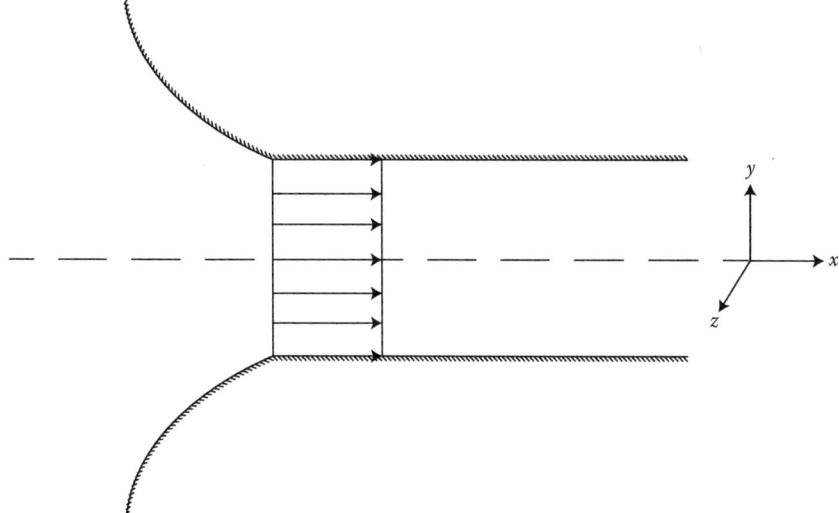

FIGURE 3.11
One-dimensional uniform velocity profile, $v(x)$ for a steady flow at a well-rounded entrance to a circular pipe flowing under pressure.

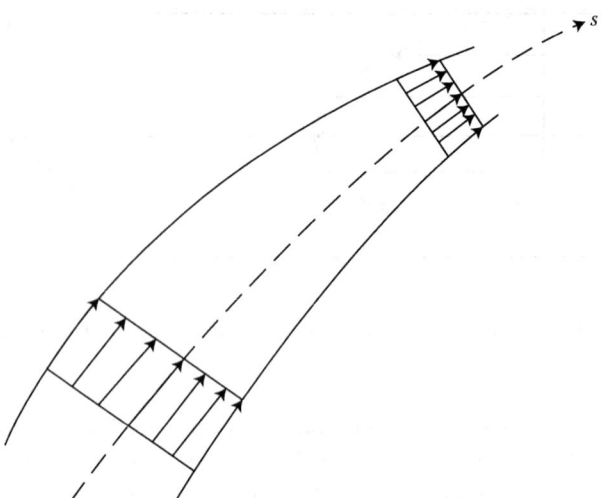

FIGURE 3.12
One-dimensional uniform velocity profile, $v(s)$ for a steady flow through a stream tube.

a given cross section. Therefore, flow along the individual streamlines (even if they are curved) of a stream tube is one-dimensional, measured along the center streamline, s. Furthermore, flow in a stream tube of a differential cross-sectional area is exactly one-dimensional because variations in the fluid properties diminish as the cross-sectional area approaches zero.

And, finally, consider an ideal (inviscid), incompressible steady flow in a pipe flowing under pressure, as illustrated in Figure 3.13. The assumption of ideal flow not only yields to the assumption of steady flow, but also yields to the assumption of one-dimensional flow measured along the centerline of the pipe, where the average fluid properties are used at each cross section. Specifically, the assumption of ideal flow results in no friction and thus no shear stress, and thus no variation in velocity across the entire cross section of the pipe, and thus a uniform velocity profile across the entire cross section. Furthermore, the additional assumption of incompressible flow results in no variation in fluid density across the cross section. Such a uniform velocity profile is considered to be a "true" one-dimensional flow, where $v = v(x)$ and x is taken along the centerline of the pipe.

It may be noted that the flow in pipes and channels is never truly one-dimensional because the velocity profile will vary across the cross section (usually parabolically, as

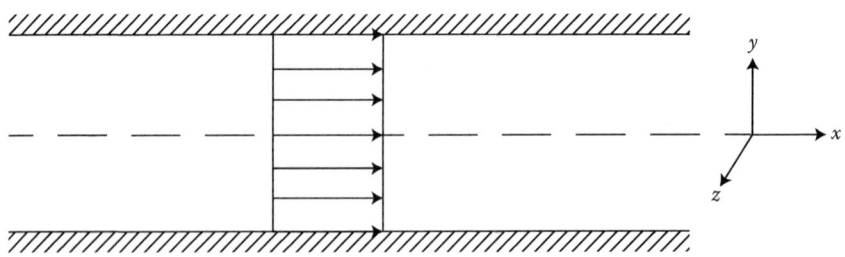

FIGURE 3.13
One-dimensional uniform velocity profile, $v(x)$ for an ideal (inviscid) incompressible steady flow in a circular pipe.

illustrated above in Figure 3.10). However, if the departure from one-dimensional is not too great, or if the changes in average fluid properties (velocity, pressure, elevation, temperature, density, etc.) along the length of the pipe or the channel, rather than at a given cross section, are of interest, then one-dimensional flow may be assumed to exist.

3.6.3.4 A Comparison of Velocity Profiles for Various Flow Types

The assumption made regarding the viscosity of the fluid will determine the resulting velocity profile for the flow. For an ideal flow through a pipe or a channel, the assumption of no viscosity results in a velocity profile with a rectangular distribution (see Figure 6.3a). However, for a real flow through a pipe or channel, modeling the viscosity results in the velocity of the fluid at the stationary solid boundary to be zero; the fluid at the boundary surface sticks to the boundary and has the velocity of the boundary (because of the "no-slip" condition). Thus, modeling of viscosity results in the assumption of real flow, which presents the possibility of two flow regimes. A real flow is subdivided into laminar flow or turbulent flow depending on the value of the Reynolds number, $R = (\rho v L / \mu)$. Additionally, the modeling of viscosity modifies the assumption of the velocity profile from the rectangular distribution assumed for ideal flow to a parabolic velocity distribution for laminar flow and a velocity distribution approximated by the seventh root velocity law for turbulent flow.

For a real laminar flow, the velocity of the fluid increases parabolically, reaching a maximum, v_{max} at the pipe centerline, as illustrated in Figure 6.3b. Figure 6.3b also illustrates that the fluid particles in laminar flow move in smooth, straight parallel (to the centerline of the pipe) observable paths (streamlines) with different velocities. In laminar flow, there are no mixing phenomena and eddies as are present in turbulent flow (described below), so it appears as a very smooth flow, such as in the flow of a viscous fluid such as honey. Furthermore, in laminar flow, the velocity is constant with time and thus is a steady flow. It may be noted that for real laminar flow, the average velocity (spatial average over the pipe cross section), $v_{ave} = (Q/A) = (v_{max}/2)$, for which the velocity profile is a rectangular distribution, as illustrated by the dotted line in Figure 6.3b. Thus, in Chapter 3, the application of the continuity equation assumes one-dimensional flow, where the velocity profile is a rectangular distribution, which further implies the assumption of ideal flow. Furthermore, it may be noted that the predictable (deterministic) nature of laminar flows results in a theoretical determination of the corresponding velocity profile (due to the theoretical relationship between the shear stress, τ and the velocity, v provided by Newton's law of viscosity) and thus a theoretical determination of the flow resistance coefficient in the equation for the major head loss, h_f due to flow resistance (see Chapter 6). The parabolic velocity distribution for laminar flow may be presented as a function of v_{max} as follows:

$$v = v_{max}\left[1 - \left(\frac{r}{r_0}\right)^2\right] \tag{3.26}$$

where $v_{max} = \left(\Delta p r_0^2 / 4 \mu L\right)$, $r_0 =$ pipe radius, and $v_{ave} = (Q/A) = (v_{max}/2) = \left(\Delta p r_0^2 / 8 \mu L\right)$.

For a real turbulent flow (for which the flowrate, Q is increased in order to reach a high enough velocity to qualify as turbulent flow), the empirically defined velocity profile of the fluid increases according to the seventh root velocity profile law, reaching a maximum, v_{max} at the pipe centerline, as illustrated in Figure 6.3c. The empirically defined velocity profile for turbulent flow may be presented as a function of v_{max} as follows:

$$v = v_{max}\left[1 - \frac{r}{r_o}\right]^{1/7} \tag{3.27}$$

where $v_{max} = 1.22\, v_{ave}$ and $v_{ave} = Q/A$. Because there is significantly more resistance in turbulent flow, there is a greater loss of energy than in laminar flow. Turbulent flow is characterized by mixing phenomena and eddies (flow turbulence) within the flow, such as is the case for open channel flow and atmospheric flow. The flow turbulence causes both (1) the resulting empirically defined velocity profile (spatial variation of the velocity, v across the pipe cross section), and (2) the unsteady flowrate (temporal variation of velocity, v). First, Figure 6.3c illustrates how the flow turbulence causes the movement of each fluid particle to deviate from a streamline and become random (fluid particles move in an irregular and erratic path, showing no observable pattern), fluctuating up and down in directions perpendicular and parallel to the centerline of the pipe, reaching fully turbulent flow with a steady increase in the flowrate. The flow turbulence transports the low-velocity fluid particles near the pipe wall to the pipe center, and the higher-velocity particles near the pipe center toward the pipe wall. The resulting velocity profile is flatter than the parabolic laminar velocity profile due to the "averaging" effect of the mixing of the fluid particles. Furthermore, it may be noted that the unpredictable (random) and complex nature of turbulent flows results in an experimental determination of the corresponding velocity profile (due to the empirical relationship between the shear stress, τ and the velocity, v provided by dimensional analysis) and thus an experimental determination of the flow resistance coefficients in the equation for the major head loss, h_f due to flow resistance (see Chapter 6). And, second, the unsteady flow caused by the flow turbulence is reflected in the jagged (noise) velocity profile illustrated in Figure 6.3c, where the solid (trend) velocity profile represents the time average velocity (temporal mean) over a long period of time for a given pipe cross section. Therefore, if the temporal mean velocity for a given pipe cross section is determined for a significant period of time, the resulting velocity may be considered to be constant with respect to time, thus the assumption of steady turbulent flow. Additionally, it may be noted that: (1) turbulent flow is more likely to occur in practice than laminar flow, and (2) the field of turbulent flow still remains the least understood topic of fluid mechanics. Furthermore, one may note that the use of the Bernoulli equation (see Chapter 4) assumes ideal flow, which has a rectangular velocity profile. And since most flows are turbulent, which has a nearly rectangular velocity profile, the usefulness of the application of the Bernoulli equation and the rectangular velocity profile is very practical, unless the need arises to account for flow properties that require the modeling of the viscous effects.

3.7 Conservation of Mass: The Continuity Equation

The conservation of mass principle (law of conservation of mass) is historically stated and applied for a fluid system, and then rephrased and extended to a control volume. The conservation of mass principle states that "the mass of a system undergoing a change can neither be created nor destroyed," which is expressed as follows:

$$M_{sys} = \text{constant} \tag{3.28}$$

Alternatively, the conservation of mass states that "the time rate of change of the extensive property, mass, M, of the fluid system is equal to zero" and is expressed as follows:

$$\frac{dM_{sys}}{dt} = 0 \tag{3.29}$$

The Reynolds transport theorem given in Equation 3.19 (for the one-dimensional uniform flow through a stream tube of Figure 3.6) may be applied to link the definition of the conservation of mass between a fluid system and a control volume as follows:

$$\frac{dM_{sys}}{dt} = \frac{dM_{cv}}{dt} - \dot{M}_{in} + \dot{M}_{out} = 0 \tag{3.30}$$

where the extensive property of the fluid, $B = M$, and thus the corresponding intensive property, $b = B/M = M/M = 1$ (unity). Thus, the conservation of mass for a control volume is expressed as follows:

$$\frac{dM_{cv}}{dt} = \dot{M}_{in} - \dot{M}_{out} \tag{3.31}$$

which states that the "time rate of change of mass within the control volume boundaries is equal to the total rate of mass flow (mass flowrate) into the control volume minus the total rate of mass flow (mass flowrate) out of the control volume" or, alternatively stated, the "time rate of change of mass within the control volume boundaries is equal to the net flux of mass into the control volume." Thus, it may be noted that for a closed system, the conservation of mass requires that the mass of a fluid system remains constant while undergoing a change, as expressed by Equation 3.29. On the other hand, for the open system of a control volume, because the mass can cross the boundaries, one must keep track of the amount of mass entering and leaving the control volume, as expressed by Equation 3.31.

3.7.1 The Continuity Equation

The conservation of mass principle in fluid mechanics is typically applied to either an infinitesimal (differential) control volume or a finite (integral) control volume, and is called the continuity equation, which deals with the mass flowrate of fluids. Application of the conservation of mass principle for a fluid system (Lagrangian approach) involves the application of the principle, stated in Equation 3.31 for a control volume, to an infinitesimal control volume, which yields the differential form of the continuity equation, while application of the conservation of mass for a control volume (Eulerian approach) involves application of the principle stated in Equation 3.31 for a control volume to a finite control volume, which yields the integral form of the continuity equation. Thus, one may note that, in general, the control volume is a convenient object (concept) of analysis because it is a volume fixed in space through which mass flows across its boundaries. As Figure 3.6 illustrated, the closed boundaries of a control volume are called the control surface, where the control surface is defined by the upper and lower bounds of the stream tube and by control face 1 and control face 2. Furthermore, the fixed control volume can be of any useful size (whether it is finite or infinitesimal in size) and shape, as long as the bounding control surface is a closed boundary (completely surrounding boundary). And, finally, neither the control volume nor the control surface changes shape or position with time.

As noted earlier in this chapter, the solution of most problems presented in this textbook use the control volume approach, with the exception of a few cases. These cases include the steady nonuniform open channel flow problem (Chapter 9), or for any pipe or channel section (Chapters 8 and 9). In such cases, the fluid system approach is used to solve for unknowns at a given point in the fluid system; thus, application of the Lagrangian (differential) form of the continuity equation becomes necessary in the solution of the unknowns.

3.7.1.1 The Continuity Equation for a Fluid System

Application of the conservation of mass principle, stated in Equation 3.31 for a control volume, to an infinitesimal (differential) control volume (differential analysis) yields a differential form of the continuity equation. Furthermore, a differential analysis involves application of the differential form of the continuity equation to any and every point in the flow field over a region called the flow domain. Thus, the differential approach may be viewed as the analysis of millions of infinitesimal control volumes, which make up the entire flow field. In the limit as the number of infinitesimal control volumes approaches infinity and thus the size of each control volume approaches zero (shrinks to a point), the continuity equation becomes a partial differential equation that is valid at any point in the flow field. Solution of this partial differential equation yields information for flow variables at every point throughout the entire flow domain, for which boundary conditions for the variables must be specified at all boundaries on the flow domain, including inlets, outlets, and stationary boundaries such as pipe walls, or channel boundaries.

In order to derive the differential form of the continuity equation, assume a control volume of an infinitesimal size represented by a differential section of a stream tube with a differential cross-sectional area, A and an differential length, ds, as illustrated in Figure 3.14, for which at the limit, the entire control volume shrinks to a point in the flow in the stream tube. We begin by applying the conservation of mass principle, stated in Equation 3.31 for a control volume, to the infinitesimal (differential) control volume illustrated in Figure 3.14 as follows:

$$\frac{dM_{cv}}{dt} = \dot{M}_{in} - \dot{M}_{out} \tag{3.32}$$

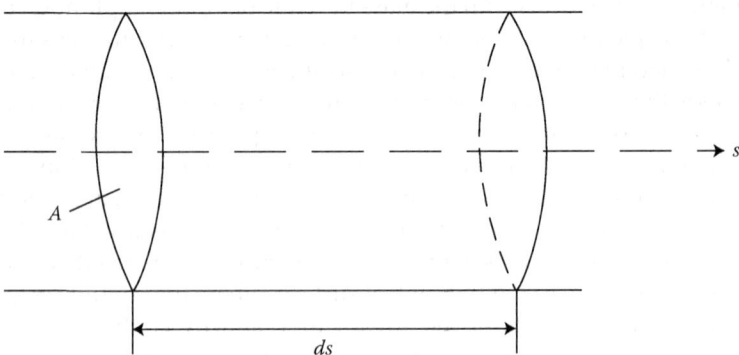

FIGURE 3.14
Infinitesimal control volume represented by a differential section, ds of a stream tube with a differential cross-sectional area, A is used to derive the differential form of the continuity equation for a fluid system.

which states that the "time rate of change of mass within the control volume is equal to the net mass flowrate into the control volume (mass flowrate into the control volume minus the mass flowrate out of the control volume)." One may note that because the mass flowrate may be changing with time, t as well as with the distance, s along the stream tube, it is necessary to express the equation in terms of partial derivatives. First, we examine the right-hand side of Equation 3.32:

$$\dot{M}_{in} - \dot{M}_{out} \tag{3.33}$$

By definition, the difference between the mass flowrate out of the cv and the mass flowrate into the cv defined across the differential distance, ds is given as follows:

$$\dot{M}_{out} - \dot{M}_{in} = \frac{\partial \dot{M}}{\partial s} ds \tag{3.34}$$

Therefore, Equation 3.32 expressing the conservation of mass may be rewritten as follows:

$$\frac{dM_{cv}}{dt} = -\frac{\partial \dot{M}}{\partial s} ds \tag{3.35}$$

Noting that the mass of the differential control volume, $M_{cv} = \rho dV = \rho A ds$ and the corresponding mass flowrate $= \rho A ds/dt = \rho v A = \rho Q$, where $Q = vA$ is defined as the volume flowrate, Equation 3.35 may be expressed as follows:

$$\frac{\partial(\rho A ds)}{\partial t} = -\frac{\partial(\rho Q)}{\partial s} ds \tag{3.36}$$

Canceling the ds term on both sides of the equation yields:

$$\frac{\partial(\rho A)}{\partial t} = -\frac{\partial(\rho Q)}{\partial s} \tag{3.37}$$

or:

$$\frac{\partial(\rho Q)}{\partial s} + \frac{\partial(\rho A)}{\partial t} = 0 \tag{3.38}$$

which indicates that the total change of mass consists of a convective change (first term in the equation) and a local change (second term in the equation), which is the general form of the partial differential continuity equation for a fluid system (Lagrangian approach) assuming a one-dimensional flow. The word general indicates that the continuity equation given by Equation 3.38 may be applied to the most general type of one-dimensional flow, which includes pressure or gravity flow, compressible (both spatial and temporal variation in the fluid density, ρ) or incompressible, nonuniform (spatial variation in the cross-sectional area, A and thus the velocity, v) or uniform, and unsteady (temporal variation in the velocity, v) or steady flow.

Beginning with the continuity equation given by Equation 3.38, which assumes a one-dimensional flow for either a pressure or a gravity flow, the ultimate goal is to make two additional simplifying assumptions, which are steady and incompressible flow.

First, Equation 3.38 may be further expanded to illustrate the potential for both the spatial and temporal variation in the fluid density, ρ as follows:

$$\left[\rho\frac{\partial Q}{\partial s} + Q\frac{\partial \rho}{\partial s}\right] + \left[\rho\frac{\partial A}{\partial t} + A\frac{\partial \rho}{\partial t}\right] = 0 \tag{3.39}$$

If one assumes steady flow, then the terms in the second brackets, which vary with time, are canceled and the one-dimensional continuity equation for compressible steady flow is given as follows:

$$\left[\rho\frac{\partial Q}{\partial s} + Q\frac{\partial \rho}{\partial s}\right] = 0 \tag{3.40}$$

or simply:

$$\frac{\partial(\rho Q)}{\partial s} = 0 \tag{3.41}$$

or finally:

$$\dot{M} = \rho Q = \rho v A = \text{constant} \tag{3.42}$$

Alternatively, one may define the weight flowrate, $G = \gamma Q = \rho g Q$ and thus express the one-dimensional continuity equation for compressible steady flow as follows:

$$G = \gamma Q = \gamma v A = \text{constant} \tag{3.43}$$

Therefore, the one-dimensional equation for compressible steady flow indicates that the mass flowrate (and the weight flowrate) is a constant over the differential distance, ds. If one further assumes incompressible flow, then:

$$\frac{\partial \rho}{\partial s} = 0 \tag{3.44}$$

Therefore, the one-dimensional continuity equation for incompressible steady flow is given as follows:

$$\left[\rho\frac{\partial Q}{\partial s}\right] = 0 \tag{3.45}$$

or simply:

$$\left[\frac{\partial Q}{\partial s}\right] = 0 \tag{3.46}$$

or finally:

$$Q = vA = \text{constant} \tag{3.47}$$

Therefore, the one-dimensional equation for incompressible steady flow indicates that the volume flowrate is a constant over the differential distance, ds.

The application of the continuity equation assumes one-dimensional flow, where the velocity profile is a rectangular distribution, which further implies the assumption of ideal flow. Thus, for any of the above one-dimensional continuity equations, the assumption of one-dimensional flow, measured along the centerline of the pipe, implies that the average fluid properties (velocity, pressure, elevation, temperature, etc.) are used at each cross section. Therefore, the velocity, v is the mean or average velocity, $v_{ave} = (Q/A) = (v_{max}/2)$ (mean over the pipe diameter, D or channel depth, y and mean over the pipe or channel differential distance, ds at a given cross section in the flow). Furthermore, one may note that for any of the above one-dimensional continuity equations, the distinction between the application for a pressure flow versus a gravity flow is in the definition of the cross-sectional area, A (normal to the velocity vector), where $A = \pi D^2/4$ for a pipe of diameter, D flowing under pressure, and $A = by$ for a rectangular open channel (for instance) of channel width, b and depth of flow, y. And finally, one may note that in this textbook, application of the one-dimensional continuity equation for a differential control volume is made to solve for an unknown (velocity, flowrate, or cross-sectional area) at a given point in the fluid system for the nonuniform, incompressible, steady open channel flow problem, or at a given point in the fluid system for any pipe or channel section.

EXAMPLE PROBLEM 3.1

Air at 10°C flows through a 40-cm pipe at a velocity of 5 m/s. (a) Determine the mass flowrate, $\dot{M}$; the weight flowrate, G; and the volume flowrate, Q.

Mathcad Solution

(a) The mass flowrate is computed from Equation 3.42, the weight flowrate is computed from Equation 3.43, and the volume flowrate is computed from Equation 3.47, where the density, ρ and the specific weight, γ for air at 10°C are given in Table A.3 in Appendix A as follows:

$$D := 40\,cm \qquad A := \frac{\pi D^2}{4} = 0.126\,m^2 \qquad v := 5\,\frac{m}{s}$$

$$\rho := 1.248\,\frac{kg}{m^3} \qquad \gamma := 12.24\,\frac{N}{m^3}$$

$$\text{Massflowrate} := \rho \cdot v \cdot A = 0.784\,\frac{kg}{s} \qquad G = \gamma \cdot v \cdot A = 7.691\,\frac{1}{s} \cdot N$$

$$Q := v \cdot A = 0.628\,\frac{m^3}{s}$$

EXAMPLE PROBLEM 3.2

Water flows at volume flowrate, Q of 50 ft³/s (cfs) through a pipe as illustrated in Figure EP 3.2. A conical diffuser is used to decelerate the fluid flow in order to achieve pressure recovery in the flow. The velocity of flow in the smaller pipe diameter is 30 ft/sec, and the velocity of flow in the larger pipe diameter is 10 ft/sec.

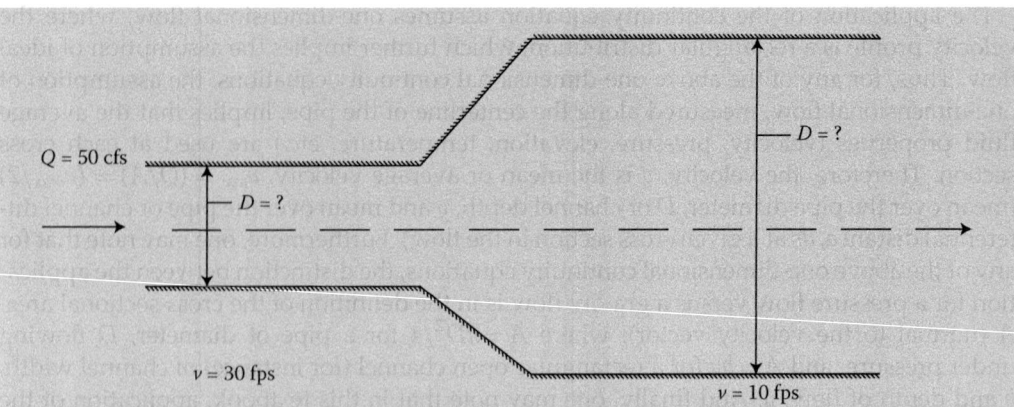

FIGURE EP 3.2
Conical diffuser used to decelerate the fluid flow in order to achieve pressure recovery in the flow of water.

(a) Determine the pipe size of the smaller pipe. (b) Determine the pipe size of the larger pipe.

Mathcad Solution

(a) The pipe size of the smaller pipe, where the velocity is 30 ft/sec is computed from Equation 3.47 as follows:

$$Q := 50 \; \frac{ft^3}{sec} \qquad A(D) := \frac{\pi \cdot D^2}{4} \qquad v := 30 \; \frac{ft}{sec}$$

Guess values: $D := 2 \; ft$

Given

$$Q = v \cdot A(D)$$

$$D := Find \; (D) = 1.457 \; ft$$

(b) The pipe size of the larger pipe, where the velocity is 10 ft/sec is computed from Equation 3.47 as follows:

$$v := 10 \frac{ft}{sec}$$

Guess values: $D := 4 \; ft$

Given

$$Q = v \cdot A(D)$$

$$D := Find \; (D) = 2.525 \; ft$$

EXAMPLE PROBLEM 3.3

Water flows in a rectangular open channel of width 5 ft at a volume flowrate (discharge) of 100 ft^3/s (cfs) and a depth of 2 ft. Determine the velocity of flow in channel.

Mathcad Solution

The velocity is computed from Equation 3.47 as follows:

$$y := 2 \, \text{ft} \qquad\qquad b := 5 \, \text{ft} \qquad\qquad A := b \cdot y = 10 \, \text{ft}^2$$

$$Q := 100 \, \frac{\text{ft}^3}{\text{s}} \qquad\qquad v := \frac{Q}{A} = 10 \, \frac{\text{ft}}{\text{s}}$$

EXAMPLE PROBLEM 3.4 (REFER TO EXAMPLE PROBLEM 9.15)

Water flows at a volume flowrate (discharge) of $30 \, \text{m}^3/\text{sec}$ in a trapezoidal channel as illustrated in Figure EP 3.4, where $b = 5$ m and $\theta = 30°$, with a channel bottom slope of 0.001 m/m, which terminates in a free overfall (brink), as illustrated in Figure EP 3.4. The uniform depth of flow at a distance 2500 m upstream of the brink is 2.216 m, while the critical depth of flow at the brink is 1.315 m, creating a nonuniform surface water profile. (a) Determine the velocity of the flow at a distance 2500 m upstream of the brink. (b) Determine the velocity of the flow at a distance 3.5 m upstream of the brink, where the depth of flow is 1.667 m. (c) Determine the velocity of the flow at the brink.

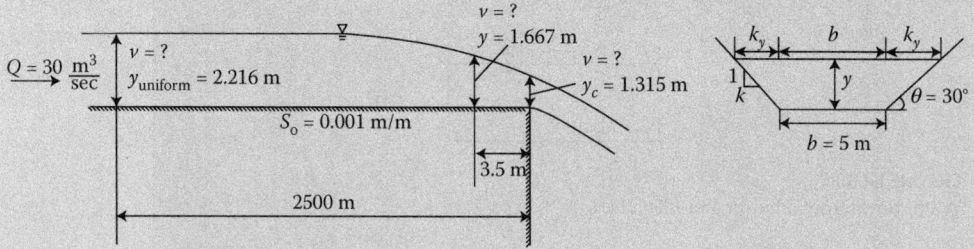

FIGURE EP 3.4
Water flows in a trapezoidal channel that terminates in a free overfall creating a nonuniform surface water profile.

Mathcad Solution

(a) The velocity of the flow at a distance 2500 m upstream of the brink, where the depth of flow is 2.216 m is computed from Equation 3.47 as follows:

$$Q := 30 \, \frac{\text{m}^3}{\text{sec}} \qquad\quad b := 5 \, \text{m} \qquad \theta := 30 \, \text{deg} \qquad k := \frac{1}{\tan(\theta)}$$

$$A(y) := y \cdot (k \cdot y + b) \qquad y := 2.216 \, \text{m} \qquad v := \frac{Q}{A(y)} = 1.523 \, \frac{\text{m}}{\text{s}}$$

(b) The velocity of the flow at a distance 3.5 m upstream of the brink, where the depth of flow is 1.667 m is computed from Equation 3.47 as follows:

$$y := 1.667 \, \text{m} \qquad v := \frac{Q}{A(y)} = 2.282 \, \frac{\text{m}}{\text{s}}$$

(c) The velocity of the flow at the brink, where the depth of flow is 1.315 m is computed from Equation 3.47 as follows:

$$y := 1.315 \, m \qquad v := \frac{Q}{A(y)} = 3.135 \, \frac{m}{s}$$

EXAMPLE PROBLEM 3.5

Water from a faucet with a 1 in diameter fills a 30 ft^3 tank in 2 minutes, as illustrated in Figure EP 3.5. (a) Determine the mass flowrate, $\dot{M}$, the weight flowrate, G and the volume flowrate, Q through the faucet. (b) Determine the velocity of flow through the faucet.

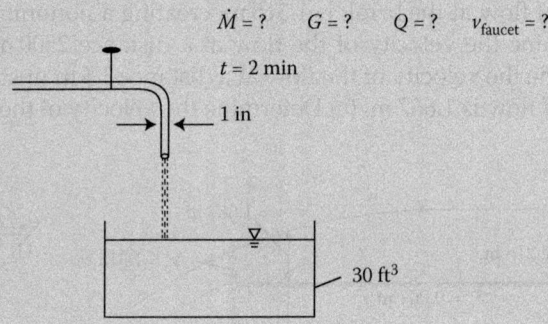

FIGURE EP 3.5
Water flows from a faucet and fills a tank.

Mathcad Solution

(a) The mass flowrate is computed from Equation 3.42, the weight flowrate is computed from Equation 3.43, and the volume flowrate is computed from Equation 3.47 as follows:

$$slug := lb \, \frac{sec^2}{ft} \qquad\qquad \rho := 1.94 \, \frac{slug}{ft^3} \qquad\qquad \gamma := 62.4 \, \frac{lb}{ft^3}$$

$$V := 30 \, ft^3 \qquad\qquad t := 2 \, min \qquad\qquad Q := \frac{V}{t} = 0.25 \, \frac{ft^3}{sec}$$

$$Massflowrate := \rho \cdot Q = 0.485 \, \frac{slug}{sec} \qquad\qquad G = \gamma \cdot Q = 15.6 \, \frac{lb}{s}$$

(b) The velocity of flow through the faucet is computed from Equation 3.47 as follows:

$$D := 1 \, in \qquad A := \frac{\pi \, D^2}{4} = 5.454 \times 10^{-3} \, ft^2 \qquad v := \frac{Q}{A} = 45.837 \, \frac{ft}{s}$$

EXAMPLE PROBLEM 3.6

Water from a faucet with a 5 cm diameter flows at a volume flowrate of $0.4 \, \text{m}^3/\text{sec}$ and fills a circular tank with a cross-sectional area of $3.5 \, \text{m}^2$ and 4 m deep, as illustrated in Figure EP 3.6. (a) Determine the velocity of flow through the faucet. (b) Determine the rate of change of the depth of water in the tank, $v_{tank} = (dh/dt)$ at any point in time.

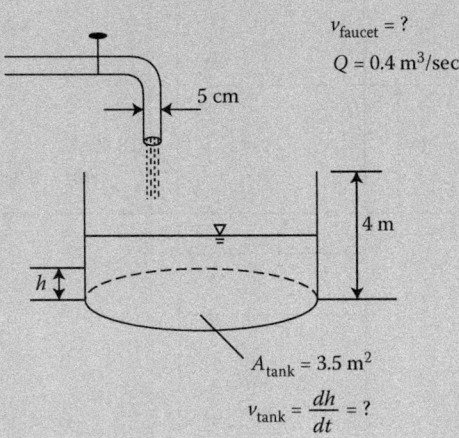

FIGURE EP 3.6
Water flows from a faucet and fills a circular tank.

Mathcad Solution

(a) The velocity of flow through the faucet is computed from Equation 3.47 as follows:

$$Q := 0.4 \, \frac{\text{m}^3}{\text{sec}} \qquad D_f := 5 \, \text{cm} \qquad A_f := \frac{\pi \cdot D_f^2}{4} = 1.936 \times 10^{-3} \, \text{m}^2$$

$$v_f := \frac{Q}{A_f} = 203.718 \, \text{m/s}$$

(b) The rate at of change of the depth of water in the tank, $v_{tank} = (dh/dt)$ at any point in time is computed from Equation 3.47 as follows:

$$A_{tank} := 3.5 \, \text{m}^2 \qquad V_{tank} := \frac{Q}{A_{tank}} = 0.114 \, \frac{\text{m}}{\text{s}}$$

EXAMPLE PROBLEM 3.7

A rectangular tank open to the atmosphere, with a cross-sectional area of $5 \, \text{ft}^2$, is filled with water to a depth of $h = h_1 = 7 \, \text{ft}$ above the centerline of the 0.8 in diameter plugged opening near the bottom of the tank, as illustrated in Figure EP 3.7. When the opening is unplugged, it will be open to the atmosphere, and a jet of water will flow at a variable velocity, $v(h) = \sqrt{2gh}$ (derived from applying the Bernoulli equation in Chapter 4). (a) Determine the variable volume flowrate of flow through the jet.

(b) Determine the time it takes for the water in the tank to reach a depth $h = h_2 = 4$ ft above the centerline of the jet. (c) Determine the rate of change of the depth of water in the tank, $v_{tank}(h) = (dh/dt)$ when the water in the tank is at a depth $h = h_2 = 4$ ft above the centerline of the jet.

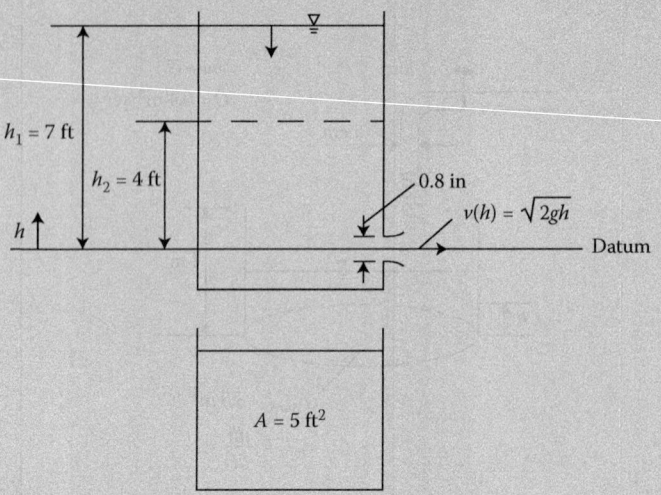

FIGURE EP 3.7
Jet of water flows out of a tank through an opening near bottom of tank.

Mathcad Solution

(a) The variable volume flowrate of flow through the jet is computed from Equation 3.47 as follows:

$$Q_j(h) = v_j(h)A_j = \sqrt{2gh}A_j$$

Thus, the volume flowrate through the jet is a function of the depth of the water in the tank, h.

(b) The time it takes for the water in the tank to reach a depth $h = h_2 = 4$ ft above the centerline of the jet is computed from Equation 3.47 as follows (note the negative sign for the rate of change of the depth of water in the tank, $v_{tank}(h) = (dh/dt)$ because it is decreasing with height, h):

$$v_{tank}(h) = -\frac{dh}{dt} = -\frac{Q_{tank}(h)}{A_{tank}} = -\frac{v_j(h)A_j}{A_{tank}} = -\frac{\sqrt{2gh}A_j}{A_{tank}}$$

$$\frac{dh}{dt} = -\frac{\sqrt{2gh}A_j}{A_{tank}}$$

$$\int_{h_1}^{h_2} -\left(\frac{A_{tank}}{A_j}\right)\frac{1}{\sqrt{2gh}}\,dh = \int_0^t dt$$

Using Mathcad to integrate yields the following:

$$\int_{h_1}^{h_2} -\left(\frac{A_{tank}}{A_j}\right) \cdot \frac{1}{\sqrt{2 \cdot g \cdot h}} dh \rightarrow \frac{\sqrt{2} \cdot A_{tank} \cdot \left(\sqrt{g \cdot h_1} - \sqrt{g \cdot h_2}\right)}{A_j \cdot g}$$

$$\int_0^t 1 \, dt \rightarrow t$$

Thus, the time it takes for the water in the tank to reach a depth $h = h_2 = 4$ ft above the centerline of the jet is evaluated as follows:

$h_1 := 7$ ft $\qquad h_2 := 4$ ft $\qquad g := 32.2 \dfrac{\text{ft}}{\text{sec}^2} \qquad D_j := 0.8$ in

$$A_j := \frac{\pi \cdot D_j^2}{4} = 3.491 \times 10^{-3} \text{ ft}^2 \qquad\qquad A_{tank} := 5 \text{ ft}^2$$

$$t := \frac{\sqrt{2} \cdot A_{tank} \cdot \left(\sqrt{g \cdot h_1} - \sqrt{g \cdot h_2}\right)}{A_j \cdot g} = 230.523 \text{ s}$$

(c) The rate of change of the depth of water in the tank, $v_{tank}(h) = (dh/dt)$ when the water in the tank is at a depth $h = h_2 = 4$ ft above the centerline of the jet is computed from Equation 3.47 as follows:

$v_j(h) := \sqrt{2 \cdot g \cdot h} \qquad\qquad Q_j(h) := v_j(h) \cdot A_j \qquad\qquad Q_{tank}(h) := Q_j(h)$

$$v_{tank}(h) := \frac{Q_{tank}(h)}{A_{tank}} \qquad\qquad v_{tank}(h_2) = 0.011 \tfrac{\text{ft}}{\text{s}}$$

3.7.1.2 The Continuity Equation for a Control Volume

Application of the conservation of mass principle, stated in Equation 3.31 for a control volume, to a finite control volume (integral analysis) yields an integral form of the continuity equation. Or more specifically/alternatively, we begin by integrating the differential form of the continuity equation given by Equation 3.38 between two points in a finite control volume such as between points 1 and 2 in Figure 3.6 as follows:

$$\frac{\partial(\rho Q)}{\partial s} + \frac{\partial(\rho A)}{\partial t} = 0 \tag{3.48}$$

or:

$$-\frac{\partial(\rho Q)}{\partial s} = \frac{\partial(\rho A)}{\partial t} \tag{3.49}$$

$$-\int_1^2 \frac{\partial(\rho Q)}{\partial s} ds = \int_1^2 \frac{\partial(\rho A)}{\partial t} ds \tag{3.50}$$

$$-\int_1^2 \partial(\rho Q) = \frac{d}{dt} \int_1^2 (\rho A) ds \tag{3.51}$$

where the mass flowrate $= \rho Q$ and the mass within the finite control volume, $M_{cv} = \rho A(s_2 - s_1)$; thus:

$$-\int_1^2 \partial \dot{M} = \frac{d}{dt} \int_1^2 (\rho A) ds \tag{3.52}$$

$$-(\dot{M}_2 - \dot{M}_1) = \frac{d[(\rho A)(s_2 - s_1)]}{dt} \tag{3.53}$$

$$-(\dot{M}_2 - \dot{M}_1) = \frac{dM_{cv}}{dt} \tag{3.54}$$

$$\dot{M}_1 - \dot{M}_2 = \frac{dM_{cv}}{dt} \tag{3.55}$$

which is the general form of the integral continuity equation for a control volume (Eulerian approach) assuming a one-dimensional flow. If one assumes steady flow, then the term on the right hand side of Equation 3.55 is canceled; thus, the one-dimensional continuity equation for compressible steady flow is given as follows:

$$\dot{M}_1 - \dot{M}_2 = 0 \tag{3.56}$$

$$\dot{M}_1 = \dot{M}_2 \tag{3.57}$$

$$\rho_1 v_1 A_1 = \rho_2 v_2 A_2 \tag{3.58}$$

Alternatively, one may use the definition of the weight flowrate, $G = \gamma Q = \rho g Q$ and thus, express the one-dimensional continuity equation for compressible steady flow as follows:

$$G_1 = G_2 \tag{3.59}$$

$$\gamma_1 v_1 A_1 = \gamma_2 v_2 A_2 \tag{3.60}$$

Therefore, the one-dimensional equation for compressible steady flow indicates that the mass flowrate (and the weight flowrate) are equal at sections 1 and 2 in the finite control volume. If one further assumes incompressible flow, then $\rho_1 = \rho_2$ and thus:

$$Q_1 = Q_2 \tag{3.61}$$

$$v_1 A_1 = v_2 A_2 \tag{3.62}$$

Therefore, the one-dimensional equation for incompressible steady flow indicates that the volume flowrate is equal at sections 1 and 2 in the finite control volume. One may note that while conservation of mass requires that $Q_1 = Q_2$, the corresponding velocities will be equal, $v_1 = v_2$ only if the corresponding cross-sectional areas (normal to the velocity vectors) are equal, $A_1 = A_2$. However, if the cross section of the stream tube increases (nonuniform flow), the velocity must decrease and vice versa in order to maintain continuity, $Q_1 = Q_2$. Furthermore, it may be noted that for the assumption of one-dimensional flow, the average velocity at a given cross section, $v_{ave} = (Q/A) = (v_{max}/2)$. And, finally, one may note that the streamlines of the stream tube are widely spaced (larger A) in regions of low velocity, while they are closely spaced (smaller A) in regions of high velocity.

The application of the continuity equation assumes one-dimensional flow, where the velocity profile is a rectangular distribution, which further implies the assumption of ideal flow. Thus, for any of the above one-dimensional continuity equations, the assumption of one-dimensional flow, measured along the centerline of the pipe, implies that the average fluid properties (velocity, pressure, elevation, temperature, etc.) are used at each cross section. Therefore, the velocity, v is the mean or average velocity, $v_{ave} = (Q/A) = (v_{max}/2)$ (mean over the pipe diameter, D or channel depth, y at a given cross section in the flow). Furthermore, one may note that for any of the above one-dimensional continuity equations, the distinction between the application for a pressure flow versus a gravity flow is in the definition of the cross-sectional area, A (normal to the velocity vector), where $A = \pi D^2/4$ for a pipe of diameter, D flowing under pressure, and $A = by$ for a rectangular open channel (for instance) of channel width, b and depth of flow, y. And, finally, one may note that in this textbook, the majority of flow problems will involve the application of the one-dimensional integral continuity equation for a finite control volume assuming either a compressible (either Equation 3.57 or Equation 3.59) or incompressible (Equation 3.61) steady flow.

It is important to note that a more general statement of the integral form of the continuity equation for a control volume (Eulerian approach) assuming a one-dimensional flow assumes there may be one or more control faces for the flow into and the flow out of the control volume, as illustrated in Figure 3.15. Thus, the more general statement of the one-dimensional continuity equation for compressible steady flow is given as follows:

$$\dot{M}_{in} - \dot{M}_{out} = 0 \tag{3.63}$$

$$\dot{M}_{in} = \dot{M}_{out} \tag{3.64}$$

Alternatively, one may use the definition of the weight flowrate, $G = \gamma Q = \rho g Q$ and thus express the one-dimensional continuity equation for compressible steady flow as follows:

$$G_{in} = G_{out} \tag{3.65}$$

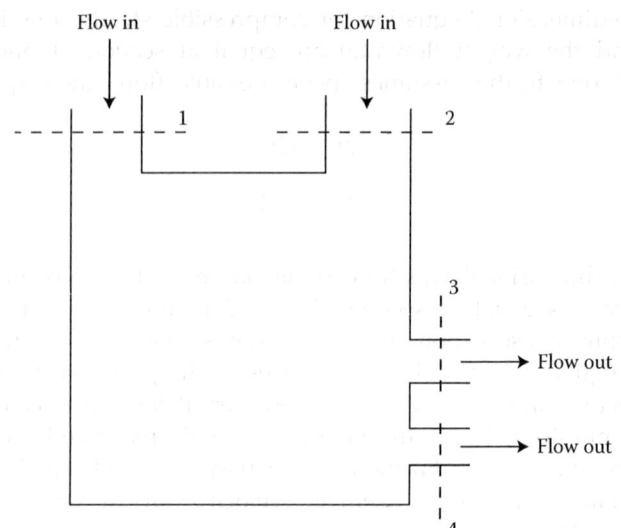

FIGURE 3.15
Control volume may have one or more control faces for the flow into and the flow out of the control volume.

Therefore, the more general statement of the one-dimensional equation for compressible steady flow indicates that the mass flowrate (and the weight flowrate) coming in is equal to the mass flowrate (and the weight flowrate) coming out of the finite control volume. If one further assumes incompressible flow, then $\rho_{in} = \rho_{out}$ and:

$$Q_{in} = Q_{out} \tag{3.66}$$

Therefore, the more general statement of the one-dimensional equation for incompressible steady flow indicates that the volume flowrate coming in is equal to the volume flowrate coming out of the finite control volume.

One final note regarding the application of the integral form of the continuity equation for a control volume (Eulerian approach) is the distinction between a fixed and a moving control volume (see Section 3.5.2.2 above). Specifically, in the case where the control volume is fixed relative to the earth, its absolute velocity, $\vec{v}_{cv}(abs) = 0$. Thus, the velocity at the given control face relative to the control volume is given as follows:

$$\vec{v}_i = \vec{v}_i(abs) \tag{3.67}$$

as illustrated in Example Problems 3.8 through 3.11. However, in the case where the control volume is moving relative to the earth, its absolute velocity, $\vec{v}_{cv}(abs) \neq 0$. Thus, the velocity at the given control face relative to the control volume is given as follows:

$$\vec{v}_i = \vec{v}_i(abs) - \vec{v}_{cv}(abs) \tag{3.68}$$

as illustrated in Example Problems 3.12 through 3.13. It is important to note that the velocity is a vector with both magnitude and direction and thus assume the same positive *s*-direction along the streamline.

EXAMPLE PROBLEM 3.8

Water flows through a pipe as illustrated in Figure EP 3.8. A conical contraction/nozzle is used to efficiently accelerate the fluid flow. At control face 1, the pipe diameter is 2 ft, and the velocity of flow is 20 ft/sec. At control face 2, the pipe diameter is 1 ft. (a) Determine the mass flowrate, $\dot{M}$; the weight flowrate, G; and the volume flowrate, Q through the pipe. (b) Determine the velocity at control face 2.

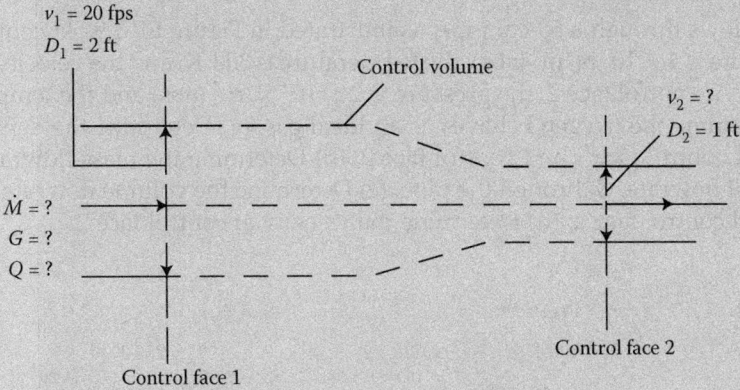

FIGURE EP 3.8
Water flows through a pipe with a conical contraction/nozzle used to efficiently accelerate the fluid flow.

Mathcad Solution

(a) The mass flowrate is computed from Equation 3.58, the weight flowrate is computed from Equation 3.60, and the volume flowrate is computed from Equation 3.62 as follows:

$$slug := lb\,\frac{sec^2}{ft} \qquad \rho := 1.94\,\frac{slug}{ft^3} \qquad \gamma := 62.4\,\frac{lb}{ft^3}$$

$$D_1 := 2\,ft \qquad A_1 := \frac{\pi\,D_1^2}{4} = 3.142\,ft^2 \qquad v_1 := 20\,\frac{ft}{s}$$

$$Massflowrate_1 := \rho \cdot v_1 \cdot A_1 = 121.894\,\frac{slug}{sec}$$

$$Massflowrate_2 := Massflowrate_1 = 121.894\,\frac{lb \cdot s}{ft}$$

$$G_1 = \gamma \cdot v_1 \cdot A_1 = 3.921 \times 10^3\,\frac{lb}{s} \qquad G_2 := G_1 = 3.921 \times 10^3\,\frac{lb}{s}$$

$$Q_1 := v_1 \cdot A_1 = 62.832\,\frac{ft^3}{sec} \qquad Q_2 := Q_1 = 62.832\,\frac{ft^3}{sec}$$

It is important to note that because water is assumed to be incompressible, continuity may be expressed using the volume flowrate, Q as well as the mass flowrate, $\dot{M}$ and the weight flowrate, G.

(b) The velocity at control face 2 is computed from Equation 3.62 as follows:

$$D_2 := 1\,\text{ft} \qquad\qquad A_2 := \frac{\pi D_2^2}{4} = 0.785\,\text{ft}^2 \qquad\qquad v_2 := \frac{Q_2}{A_2} = 80\,\frac{\text{ft}}{\text{s}}$$

EXAMPLE PROBLEM 3.9

Oxygen flows through a 50-cm pipe, as illustrated in Figure EP 3.9. At control face 1, the pressure is $4 \times 10^6\,\text{N/m}^2$ (abs), the temperature is 300°K, and the velocity of flow is 10 m/sec. At control face 2, the pressure is $2 \times 10^6\,\text{N/m}^2$ (abs) and the temperature is 250°K. Assume the oxygen behaves as an ideal gas. (a) Determine the density of the oxygen at control face 1 and control face 2. (b) Determine the mass flowrate, $\dot{M}$ and the weight flowrate, G through the pipe. (c) Determine the volume flowrate at control face 1 and control face 2. (d) Determine the velocity at control face 2.

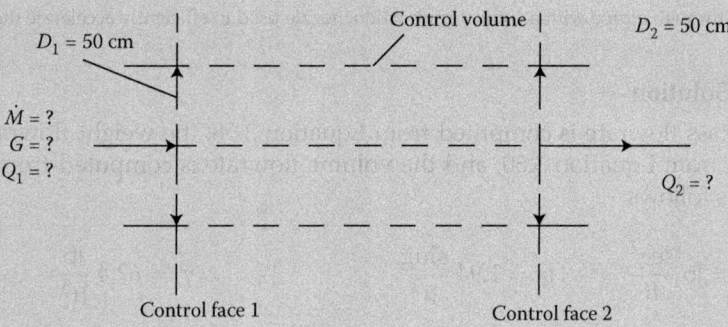

FIGURE EP 3.9
Oxygen flows through a pipe under pressure.

Mathcad Solution

(a) Assuming that the oxygen behaves as an ideal gas, the ideal gas law from Equation 1.104 is applied in order to determine the density of the oxygen at control face 1 and control face 2, where the gas constant, R for oxygen is given in Table A.5 in Appendix A as follows:

$$p_1 := 4 \times 10^6\,\frac{\text{N}}{\text{m}^2} \qquad\qquad T_1 := 300\,\text{K} \qquad\qquad v_1 := 10\,\frac{\text{m}}{\text{sec}}$$

$$p_2 := 2 \times 10^6\,\frac{\text{N}}{\text{m}^2} \qquad\qquad T_2 := 250\,\text{K} \qquad\qquad R := 260\,\text{N}\,\frac{\text{m}}{\text{kg} \cdot \text{K}}$$

$$\rho_1 := \frac{p_1}{R \cdot T_1} = 51.282\,\frac{\text{kg}}{\text{m}^3} \qquad\qquad \rho_2 := \frac{p_2}{R \cdot T_2} = 30.769\,\frac{\text{kg}}{\text{m}^3}$$

(b) The mass flowrate is computed from Equation 3.58, and the weight flowrate is computed from Equation 3.60 as follows:

$$D := 50\,cm \qquad\qquad A := \frac{\pi \cdot D^2}{4} = 0.196\,m^2 \qquad g := 9.81\,\frac{m}{sec^2}$$

$$\gamma_1 := \rho_1 \cdot g = 503.077\,\frac{kg}{m^2 \cdot s^2}$$

$$Massflowrate_1 := \rho_1 \cdot v_1 \cdot A = 100.692\,\frac{kg}{s}$$

$$Massflowrate_2 := Massflowrate_1 = 100.692\,\frac{kg}{s}$$

$$G_1 = \gamma_1 \cdot v_1 \cdot A = 987.789\,\frac{N}{s} \qquad\qquad G_2 := G_1 = 987.789\,\frac{N}{s}$$

It is important to note that because the oxygen flow is assumed to be compressible, continuity must be expressed using either the mass flowrate, $\dot{M}$ or the weight flowrate, G, but not the volume flowrate, Q (see part (c) below).

(c) The volume flowrate at each control face is computed from Equation 3.58 as follows:

$$Q_1 := \frac{Massflowrate_1}{\rho_1} = 1.963\,\frac{m^3}{s} \qquad\qquad Q_2 := \frac{Massflowrate_2}{\rho_2} = 3.272\,\frac{m^3}{s}$$

It is important to note that because the oxygen flow is assumed to be compressible, continuity must be expressed using either the mass flowrate, $\dot{M}$ or the weight flowrate, Q, but not the volume flowrate, Q. The volume flowrate, Q (and the corresponding velocity, v) will vary between control face 1 and control face 2 because the density, ρ varies between control face 1 and control face 2.

(d) The velocity at control face 2 is computed from Equation 3.58 as follows:

$$\text{Guess value:} \qquad v_2 := 20\,\frac{m}{sec}$$

Given

$$\rho_1 \cdot v_1 \cdot A = \rho_2 \cdot v_2 \cdot A$$

$$v_2 := Find\,(v_2) = 16.667\,\frac{m}{s}$$

Thus, the velocity, v will vary between control face 1 and control face 2 because the density, ρ varies between control face 1 and control face 2 as the pipe size diameter, D remains constant between the two control faces.

EXAMPLE PROBLEM 3.10

Water flows through a 0.8-m pipe that branches into a 0.6-m pipe and a 0.5-m pipe, as illustrated in Figure EP 3.10. The velocity of flow at control face 1 is 3 m/sec, and the velocity of flow at control face 2 is 2.5 m/sec. (a) Determine the volume flowrate at control face 1. (b) Determine the volume flowrate at control face 2. (c) Determine the volume flowrate at control face 3.

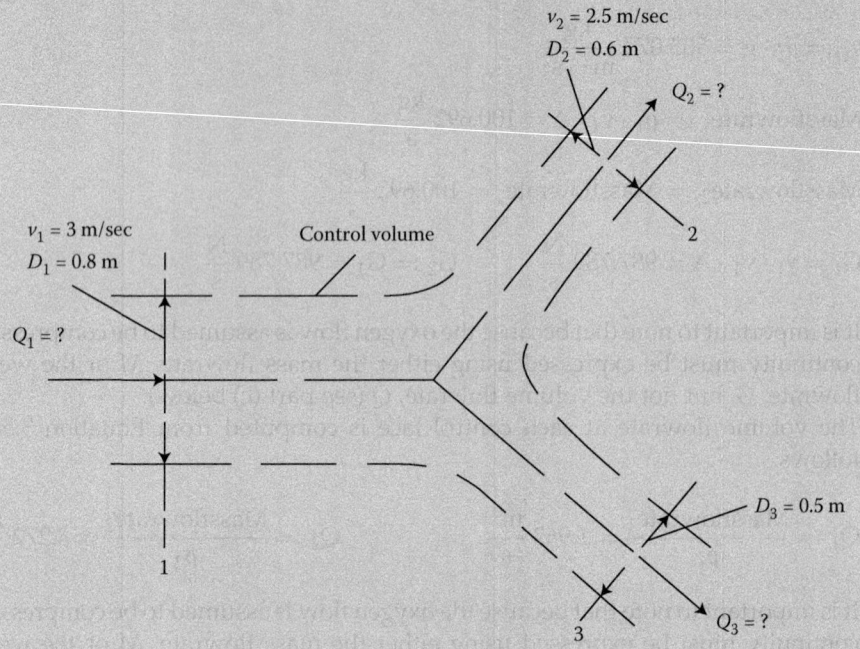

FIGURE EP 3.10
Water flows through a branching pipe.

Mathcad Solution

(a) The volume flowrate at control face 1 is computed from the definition of the volume flowrate, $Q = vA$ as follows:

$$D_1 := 0.8\,\text{m} \qquad\qquad A_1 := \frac{\pi \cdot D_1^2}{4} = 0.503\,\text{m}^2 \qquad\qquad v_1 := 3\,\frac{\text{m}}{\text{sec}}$$

$$Q_1 := v_1 \cdot A_1 = 1.508\,\frac{\text{m}^3}{\text{s}}$$

(b) The volume flowrate at control face 2 is computed from the definition of the volume flowrate, $Q = vA$ as follows:

$$D_2 := 0.6\,\text{m} \qquad\qquad A_2 := \frac{\pi \cdot D_2^2}{4} = 0.283\,\text{m}^2 \qquad\qquad v_2 := 2.5\,\frac{\text{m}}{\text{sec}}$$

$$Q_2 := v_2 \cdot A_2 = 0.707\,\frac{\text{m}^3}{\text{s}}$$

(c) The volume flowrate at control face 3 is computed from Equation 3.66 as follows:

$$Q_{in} = Q_{out}$$

$$Q_1 = Q_2 + Q_3$$

$$D_3 = 0.5\,m \qquad\qquad A_3 := \frac{\pi \cdot D_3^2}{4} = 0.196\,m^2$$

Guess values: $\qquad\qquad Q_3 := 1\,\dfrac{m^3}{sec}$

Given

$$Q_1 = Q_2 + Q_3$$

$$Q_3 := Find\,(Q_3) = 0.801\,\frac{m^3}{s}$$

where the continuity equation is satisfied (checks) as follows:

$$Q_2 + Q_3 = 1.508\,\frac{m^3}{s}$$

EXAMPLE PROBLEM 3.11

Water flows in a rectangular open channel of width 3 m at a velocity of 2.5 m/sec and depth of 1.8 m upstream (control face 1) of a spillway, as illustrated in Figure EP 3.11. The depth downstream (control face 2) of the spillway is 0.933 m. (a) Determine the volume flowrate in the channel. (b) Determine the velocity of flow at control face 2.

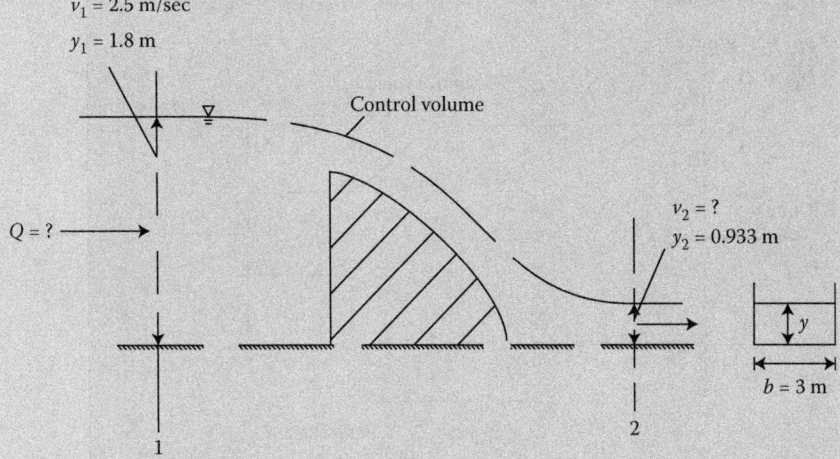

FIGURE EP 3.11
Water flows in a rectangular channel with a spillway.

Mathcad Solution

(a) The volume flowrate in the channel is computed from the definition of the volume flowrate, $Q = vA$ as follows:

$$y_1 := 1.8\,m \qquad\qquad b := 3\,m \qquad\qquad\qquad A_1 := b \cdot y_1 = 5.4\,m^2$$

$$v_1 := 2.5\,\frac{m}{sec} \qquad Q_1 := v_1 \cdot A_1 = 13.5\,\frac{m^3}{s} \qquad\qquad Q_2 := Q_1 = 13.5\,\frac{m^3}{s}$$

(b) The velocity of flow at control face 2 is computed from Equation 3.62 as follows:

$$y_2 := 0.933\,m$$

Guess values: $\qquad\qquad v_2 := 5\,\dfrac{m}{sec}$

Given

$$Q_2 = v_2 \cdot b \cdot y_2$$

$$v_2 := Find\,(v_2) = 4.823\,\frac{m}{s}$$

EXAMPLE PROBLEM 3.12

Air at a density of $0.9\,kg/m^3$ enters the intake (control face 1) of a car engine with a diameter of 0.2 m at an absolute velocity of 50 km/hr (relative to the earth), as illustrated in Figure EP 3.12. The exhaust at a density of $0.5\,kg/m^3$ leaves the tail of the car engine (control face 2) with a diameter of 0.1 m. The car (engine) (control volume)

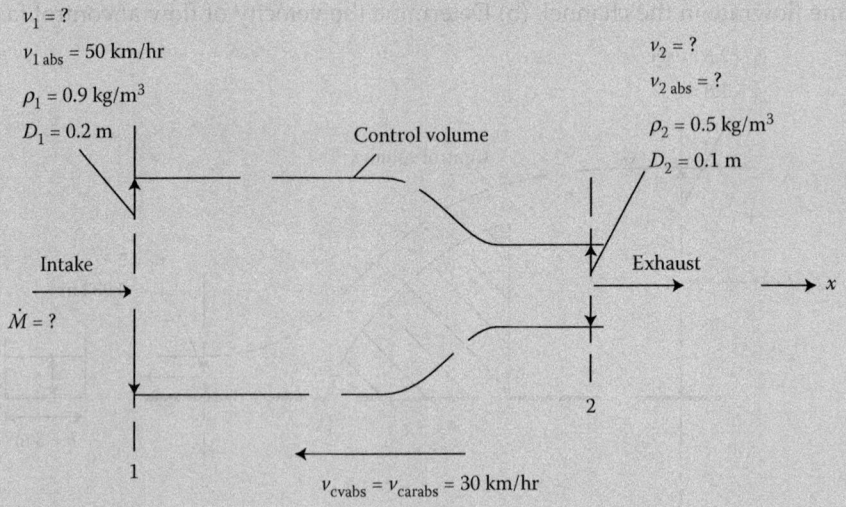

FIGURE EP 3.12
Air enters the intake of a car engine and exhaust leaves the tail of the car engine.

is traveling at an absolute velocity of 30 km/hr (relative to the earth). (a) Determine the velocity of the flow at control face 1 relative to the control volume, v_1. (b) Determine the mass flowrate through the car engine. (c) Determine the absolute velocity of the flow at control face 2, v_2 (abs). (d) Determine the velocity of the flow at control face 2 relative to the control volume, v_2.

Mathcad Solution

(a)–(b) The velocity of the flow at control face 1 relative to the control volume, v_1, and the mass flowrate through the car engine are computed from Equations 3.68 and 3.58, respectively, as follows (note that the positive x-direction is assumed to the right):

$$\rho_1 := 0.9 \frac{kg}{m^3} \qquad D_1 := 0.2\,m \qquad A_1 := \frac{\pi \cdot D_1^2}{4} = 0.031\,m^2$$

$$V_{labs} := 50 \frac{km}{hr} \qquad V_{carabs} := -30 \frac{km}{hr} \qquad V_{cvabs} := V_{carabs} = -30 \frac{km}{hr}$$

$$V_1 := V_{labs} - V_{cvabs} = 80 \frac{km}{hr}$$

$$Massflowrate_1 := \rho_1 \cdot V_1 \cdot A_1 = 0.628 \frac{kg}{s}$$

$$Massflowrate_2 := Massflowrate_1 = 0.628 \frac{kg}{s}$$

(c)–(d) The absolute velocity of the flow at control face 2, v_2 (abs), and the velocity at control face 2 relative to the control volume, v_2 are computed from Equations 3.68 and 3.58, respectively, as follows:

$$\rho_2 := 0.5 \frac{kg}{m^3} \qquad D_2 := 0.1\,m \qquad A_2 := \frac{\pi \cdot D_2^2}{4} = 7.854 \times 10^{-3}\,m^2$$

Guess values: $\qquad V_{2abs} := 100 \frac{km}{hr} \qquad V_2 := 200 \frac{km}{hr}$

Given

$$V_2 = V_{2abs} - V_{cvabs}$$

$$Massflowrate_2 = \rho_2 \cdot V_2 \cdot A_2$$

$$\begin{pmatrix} V_{2abs} \\ V_2 \end{pmatrix} := Find(V_{2abs}, V_2) = \begin{pmatrix} 546 \\ 576 \end{pmatrix} \frac{km}{hr}$$

EXAMPLE PROBLEM 3.13

Water flows through a hose at the base of a rotating sprinkler at a volume flowrate of 0.02 m³/sec, as illustrated in Figure EP 3.13. The radius of the sprinkler head, r_o is

0.25 m. There are two nozzles in the sprinkler head (control volume), each with a diameter of 9 mm. (a) Determine the volume flowrate in each of the nozzles (control faces 2 and 3). (b) Determine the velocity of the flow in each of the nozzles (control faces 2 and 3) relative to the control volume, v_2 and v_3, respectively. (c) Determine the absolute velocity (relative to the earth) of flow in each of the nozzles (control faces 2 and 3), v_2 (abs) and v_3 (abs), respectively, if the sprinkler head (control volume) is stationary. (d) Determine the absolute velocity (relative to the earth) of flow in each of the nozzles (control faces 2 and 3), v_2 (abs) and v_3 (abs), respectively, if the sprinkler head (control volume) is moving/rotating counterclockwise at an angular velocity, ω of 350 rpm.

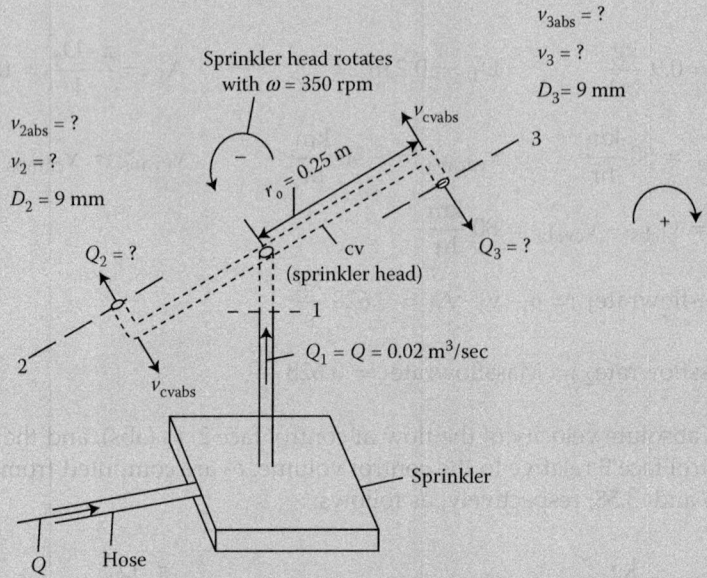

FIGURE EP 3.13
Water flows through a hose at the base of a rotating sprinkler head with two nozzles.

Mathcad Solution

(a) The volume flowrate in each of the nozzles (control faces 2 and 3) are computed from Equation 3.66 as follows:

$$Q_{in} = Q_{out}$$

$$Q_1 = Q_2 + Q_3$$

$Q_1 := 0.02\, \dfrac{m^3}{sec}$ $D_2 := 9\,mm$ $A_2 := \dfrac{\pi \cdot D_2^2}{4} = 6.362 \times 10^{-5}\, m^2$

$D_3 := 9\,mm$ $A_3 := \dfrac{\pi \cdot D_3^2}{4} = 6.362 \times 10^{-5} m^2$

Guess values: $Q_2 := 0.015\, \dfrac{m^3}{sec}$ $Q_3 := 0.005\, \dfrac{m^3}{sec}$

Given

$$Q_1 = Q_2 + Q_3$$

$$Q_2 = Q_3$$

$$\begin{pmatrix} Q_2 \\ Q_3 \end{pmatrix} := \text{Find}(Q_2, Q_3) = \begin{pmatrix} 0.01 \\ 0.01 \end{pmatrix} \frac{m^3}{s}$$

(b) The velocity of the flow in each of the nozzles (control faces 2 and 3) relative to the control volume, v_2 and v_3, respectively, are computed from the definition of the volume flowrate, $Q = vA$ as follows:

$$v_2 := \frac{Q_2}{A_2} = 157.19 \frac{m}{s} \qquad\qquad v_3 := \frac{Q_3}{A_3} = 157.19 \frac{m}{s}$$

(c) The absolute velocity (relative to the earth) of flow in each of the nozzles (control faces 2 and 3), v_2 (abs) and v_3 (abs), respectively, if the sprinkler head (control volume) is stationary are computed from Equation 3.68 as follows:

$$r_o := 0.25\,m \qquad \omega := 0\,rpm \qquad v_{cvabs} := \omega \cdot r_o = 0 \frac{m}{sec}$$

Guess value: $\qquad v_{2abs} := 0.1 \frac{m}{sec} \qquad v_{3abs} := 0.1 \frac{m}{sec}$

Given

$$v_2 = v_{2abs} - v_{cvabs}$$

$$v_3 = v_{3abs} - v_{cvabs}$$

$$\begin{pmatrix} v_{2abs} \\ v_{3abs} \end{pmatrix} := \text{Find}(v_{2abs}, v_{3abs}) = \begin{pmatrix} 157.19 \\ 157.19 \end{pmatrix} \frac{m}{s}$$

(d) The absolute velocity (relative to the earth) of flow in each of the nozzles (control faces 2 and 3), v_2 (abs) and v_3 (abs), respectively, if the sprinkler head (control volume) is moving/rotating counterclockwise at an angular velocity, ω of 350 rpm are computed from Equation 3.68 as follows:

$$r_o := 0.25\,m \qquad \omega := -350\,rpm \qquad v_{cvabs} := \omega \cdot r_o = -9.163 \frac{m}{sec}$$

Guess value: $\qquad v_{2abs} := 100 \frac{m}{sec} \qquad v_{3abs} := 100 \frac{m}{sec}$

Given

$$v_2 = v_{2abs} - v_{cvabs}$$

$$v_3 = v_{3abs} - v_{cvabs}$$

$$\begin{pmatrix} v_{2abs} \\ v_{3abs} \end{pmatrix} := \text{Find} (v_{2abs}, v_{3abs}) = \begin{pmatrix} 148.027 \\ 148.027 \end{pmatrix} \frac{m}{s}$$

3.8 Measurement of the Volume Flowrate

Assuming real flow, the volume flowrate, Q is measured for internal flows (pipe and open channel flows) using flow measurement devices, as presented in Chapter 8 on pipe flow and Chapter 9 on open channel flow. The assumption of real flow models the existence of flow resistance, which is the shear stress, friction, or drag originating in the fluid viscosity. Therefore, the existence of flow resistance causes the actual flow measured using a flow measurement device to be less than the ideal flowrate as explained in detail in Chapter 6 on flow resistance.

End-of-Chapter Problems

Problems with a "C" are conceptual problems. Problems with a "BG" are in English units. Problems with an "SI" are in metric units. Problems with a "BG/SI" are in both English and metric units. All "BG" and "SI" problems that require computations are solved using Mathcad.

Introduction

3.1C   In the study of fluid mechanics, what are the three basic units of study?

3.2C   What is the nature of the fluids that fluid statics deals with, and what principles is the study of fluid statics based upon?

3.3C   What is the nature of the fluids that fluid kinematics deals with, and what principle is the study of fluid kinematics based upon?

3.4C   What is the nature of the fluids that fluid dynamics deals with, and what principle is the study of fluid dynamics based upon?

3.5C   Fluid dynamics includes what additional topic, which yields the resistance equations, and is introduced in Chapter 6?

3.6C   What equations are applied to internal flows, which include pressure (pipe) flow in Chapter 8, and gravity (open channel) flow in Chapter 9?

3.7C   Is the continuity equation applied to external flows, which include the flow around objects in Chapter 10?

3.8C   What are the two approaches that may be used in the application of the continuity equation?

Types of Fluid Flow

3.9C In the study of both fluid kinematics and fluid dynamics, there are a number of different flow types to consider. The fluid properties, along with the fluid kinematics and the fluid dynamics, will determine the type of flow. What are the seven categories of flow types, and what determines the categories of flow types?

3.10C Describe the difference in the use of energy or work to move the fluid for internal vs. external flow.

3.11C Describe the variables sought in the study of internal vs. external flow.

3.12C Describe the differences and the similarities of internal vs. external flow.

3.13C Describe the difference between the occurrence of internal flow types, pressure (pipe) flow vs. gravity (open channel) flow. How does one choose between the types of internal flow, pressure vs. gravity flow?

3.14C Describe the difference between the assumptions made in real vs. ideal flow. What does the assumption of real flow model? What does the validity of the assumption of real vs. ideal flow depend upon?

3.15C A real flow may be classified as a laminar or turbulent flow. What is the dimensionless number and its range of values that further subdivides a real flow into laminar or turbulent flow?

3.16C Describe the nature of movement of the fluid particles in laminar vs. turbulent flow.

3.17C Describe the relationships between real vs. ideal flow, laminar vs. turbulent flow, and steady vs. unsteady flow.

3.18C Describe the difference in the assumptions made regarding the change in pressure and temperature (effects of thermodynamics) for compressible vs. incompressible flow.

3.19C Describe the difference in the occurrence of compressible vs. incompressible flow in internal flow.

3.20C Describe the difference in the occurrence of spatially varied (nonuniform) vs. spatially uniform flow.

3.21C Describe the difference between the fluid velocity in unsteady vs. steady flow.

3.22C Describe the general and the typical assumptions made in the spatial dimensionality of a typical fluid flow.

Flow Visualization/Geometry

3.23C Define flow field features and what they include.

3.24C Define and illustrate a path line. Explain why a path line is a Lagrangian concept. What does the path line of an individual particle visually illustrate?

3.25C Describe the characteristics of a path line for steady flow.

3.26C Define and illustrate a streamline.

3.27C In what case are a path line and a streamline the same? In what case are a path line and a streamline different?

3.28C   When does one typically use a path line vs. a streamline?

3.29C   Define and illustrate a flow field (stream surface). Explain what a flow field illustrates and what it allows one to map/locate.

3.30C   Define and illustrate a stream tube.

3.31C   When are a streamline and a stream tube the same?

3.32C   Explain the use/role of streamlines and stream tubes in analytical work.

Describing and Observing the Fluid Motion

3.33C   The concept of a free-body diagram that was used in the study of fluid statics is inadequate for moving fluids. Instead, what are the two concepts used to describe the motion of a fluid (fluid kinematics and fluid dynamics)? Describe how the fluid motion is described. Furthermore, explain the basic difference between these two concepts.

3.34C   A flow field (see Figure 3.3) is defined by field variables including velocity, density, pressure, and temperature. The assumption is that every fluid particle in a flow has an instantaneous value of velocity, density, pressure, temperature, and other fluid characteristics. As the fluid particle moves within the fluid, these above-listed fluid characteristics will change both spatially and temporally. Explain how the flow is described for a fluid system vs. a control volume.

3.35C   When does it become necessary to model a fluid flow problem using a control volume (Eulerian view) as opposed to a fluid system (Lagrangian view)? Give examples of such flow situations where a control volume (Eulerian view) is used.

3.36C   When does it become necessary to model the fluid flow problem using a fluid system (Lagrangian view) as opposed to a control volume (Eulerian view)? Give examples of such flow situations where a fluid system (Lagrangian view) is used.

The Physical Laws Governing Fluid in Motion

3.37C   The behavior of a fluid in motion is governed by a set of fundamental physical laws (yielding a set of governing equations). What are these laws (and these equations), and in what point of view (Eulerian vs. Lagrangian) is their basic form stated?

3.38C   Given that the governing equations (continuity, energy, and momentum) are stated in their basic form in terms of a system concept (Lagrangian point of view), explain why it becomes necessary to define a control volume (Eulerian point of view) for the analysis of most fluid flow problems. Also, explain what tool is used to shift from one concept (fluid system vs. control volume) to another.

3.39C   While the physical laws (conservation of mass and conservation of momentum) governing the behavior of a fluid in motion deal with time rates of change of extensive properties, the fluid system concept is related to both the extensive and intensive fluid properties. Furthermore, the distinction between extensive properties and intensive properties is very important in the derivation and the application of the control volume approach, in which a constant volume with a variable mass is assumed. Explain the difference between extensive fluid properties and intensive fluid properties, and give examples of each.

3.40C Explain how the Reynolds transport theorem is used to rephrase the physical laws governing the behavior of a fluid motion (dealing with the time rates of change of extensive properties, which are expressed/stated for a fluid system) for a control volume.

3.41C In reality, almost all flows (internal or external flows) are unsteady to a certain degree, characterized by behavior that is quite random and thus typically occur as either two- or three-dimensional turbulent flows. Furthermore, the modeling and solution of compressible flow (gas or liquid) technically requires the effect of thermodynamics to be taken into account. However, in the derivation and the application of the three governing equations, the ultimate goal is to be able to make three basic simplifying assumptions regarding the flow type, and they are: one-dimensional, incompressible, steady flow. Explain why these three simplifying assumptions are made.

3.42C While the simplifying assumption of one-dimensional flow will facilitate the modeling and solution of a fluid flow problem, there are certain flow situations that would be most accurately modeled as a two-dimensional flow as opposed to a one-dimensional flow. Explain what these certain flow situations are, and give an example.

3.43C The concept of one-dimensional flow is a very powerful and practical one that provides simplicity in flow analyses and accurate results in a wide range of engineering problems. Describe general flow situations when two- and three-dimensional flow engineering problems may be accurately modeled as one-dimensional problems, and explain why.

3.44C Give examples of a flow situation, which may be accurately modeled as a one-dimensional flow, and explain why.

3.45C The assumption made regarding the viscosity of the fluid will determine the resulting velocity profile for the flow. Explain and illustrate the difference between the velocity profile for an ideal flow vs. real flow (laminar flow vs. turbulent flow).

3.46C In Chapter 3, the application of the continuity equation assumes one-dimensional flow, where the velocity profile is a rectangular distribution, which further implies the assumption of ideal flow. Explain.

Conservation of Mass: The Continuity Equation

The Continuity Equation for a Fluid System

3.47C The application of the continuity equation (Equation 3.47) assumes one-dimensional flow, where the velocity profile is a rectangular distribution. Therefore, the velocity, v is the mean or average velocity, $v_{ave} = (Q/A) = (v_{max}/2)$ (mean over the pipe diameter, D or channel depth, y, and mean over the pipe or channel differential distance, ds at a given cross section in the flow). For the case of real laminar pipe flow for which the velocity profile is a parabolic distribution, prove that $v_{ave} = (Q/A) = (v_{max}/2)$.

3.48C The application of the continuity equation (Equation 3.47) assumes one-dimensional flow, where the velocity profile is a rectangular distribution. Therefore, the velocity, v is the mean or average velocity, $v_{ave} = (Q/A) = (v_{max}/2)$ (mean over the pipe diameter, D or channel depth, y, and mean over the pipe or channel differential distance, ds at a given cross section in the flow). For the case of real turbulent pipe flow for which the velocity profile is approximated by the seventh root velocity profile (see Equation 1.85 and Figure 6.3a), prove that $v_{max} = 1.22v_{ave} = 1.22(Q/A)$ or that $v_{ave} = (Q/A) = 0.8167v_{max}$.

3.49BG   The velocity at the centerline of a 1.8-*ft*-diameter pipe is measured to be 15 *ft*/*sec*. Assume real laminar pipe flow. (a) Plot the resulting velocity profile. (b) Determine the volume flowrate in the pipe. (c) Determine and illustrate the relationship between the mean or the average velocity and the maximum velocity. (d) Determine the velocities along the pipe radius at 0.3 *ft* increments.

3.50SI   The velocity at the centerline of a 1.2-*m*-diameter pipe is measured to be 0.4 *m*/*sec*. Assume real turbulent pipe flow. (a) Plot the resulting velocity profile. (b) Determine the volume flowrate in the pipe. (c) Determine and illustrate the relationship between the mean or the average velocity and the maximum velocity. (d) Determine the velocities along the pipe radius at 0.2 *m* increments.

3.51SI   Carbon dioxide at 20°C flows through an 80-*cm* pipe at a velocity of 7 *m*/*s*. Determine the mass flowrate, $\dot{M}$; the weight flowrate, G; and the volume flowrate, Q.

3.52BG   Water flows at volume flowrate, Q of 80 ft^3/*s* (cfs) through a pipe, as illustrated in Figure ECP 3.52. A conical diffuser is used to decelerate the fluid flow in order to achieve pressure recovery in the flow. The velocity of flow in the smaller pipe diameter is 40 *ft*/*sec*, and the velocity of flow in the larger pipe diameter is 12 *ft*/*sec*. (a) Determine the pipe size of the smaller pipe. (b) Determine the pipe size of the larger pipe.

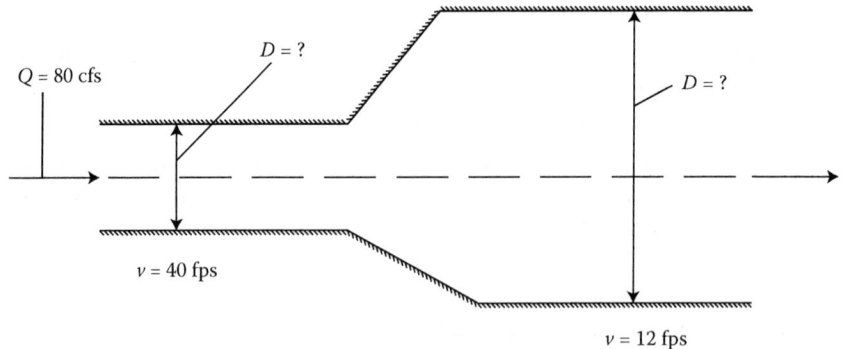

FIGURE ECP 3.52

3.53BG   Water flows in a rectangular open channel of width 7 *ft* at a volume flowrate (discharge) of 200 ft^3/*s* (cfs) and a depth of 3 *ft*. Determine the velocity of flow in channel.

3.54SI   (Refer to Example Problem 9.16.) Water flows at a discharge of 0.6 m^3/*sec* in a rectangular channel that is 1.6 m wide. There is a break in the channel slope, where the upstream channel section is characterized with a mild channel bottom slope of 0.0005 *m*/*m* and a length of 2000 *m*, and the downstream channel section is characterized with a steep channel bottom slope of 0.027 *m*/*m* and a length of 1500 *m*, as illustrated in Figure ECP 3.54. The uniform depth of flow at a distance 2000 *m* upstream of the break in channel slope is 0.485 *m*, and the critical depth of flow at the break in channel slope is 0.243 m, while the uniform depth of flow at a distance 1500 m downstream of the break is 0.129 m, creating a nonuniform surface water profile. (a) Determine the velocity of the flow at a distance 2000 m upstream of the break in channel slope. (b) Determine the velocity of the flow at the break in channel slope. (c) Determine the velocity of the flow at a distance 1500 m downstream of the break in channel slope.

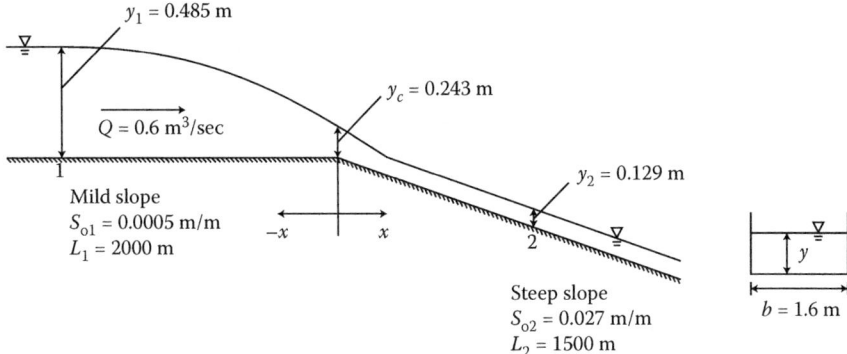

FIGURE ECP 3.54

3.55BG Water from a faucet with a 1.35-in diameter fills a 70 ft^3 tank in 4 minutes, as illustrated in Figure ECP 3.55. (a) Determine the mass flowrate, $\dot{M}$; the weight flowrate, G; and the volume flowrate, Q through the faucet. (b) Determine the velocity of flow through the faucet.

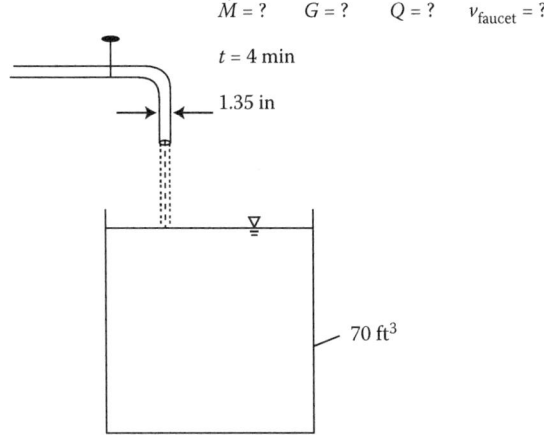

FIGURE ECP 3.55

3.56SI Water from a faucet with a 7-cm diameter flows at a volume flowrate of 0.8 m^3/sec and fills a circular tank with a cross-sectional area of 6 m^2 and 5 m deep, as illustrated in Figure ECP 3.56. (a) Determine the velocity of flow through the faucet. (b) Determine the rate of change of the depth of water in the tank, $v_{\tan k} = (dh/dt)$ at any point in time.

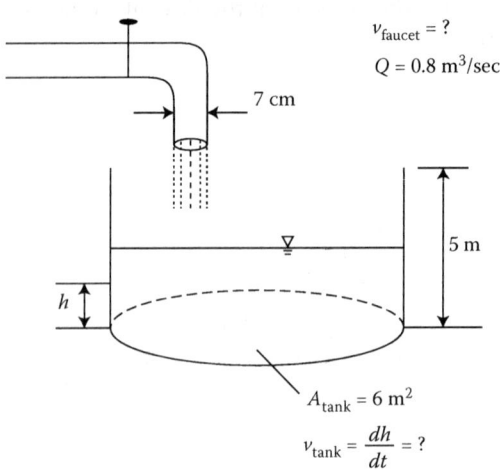

FIGURE ECP 3.56

3.57SI   A hose flowing at a rate of 9 *liters/sec* is used to fill a circular tank with a 10 *m* diameter. (a) Determine the time it takes for the water in the tank to reach a depth of 6 *m*.

3.58BG   A vehicle tire is filled with air. A small hole in the tire allows the air to leak at a rate of 0.08 *ft/sec*. (a) Plot the relationship between the volume flowrate and the size of the hole in the tire.

3.59SI   The water in pool with a volume of 50 m^3 needs to be replaced every 10 hrs without draining the pool. The water drains out of the pool through a 0.9-*m*-diameter pipe. (a) Determine the required average volume flowrate of the water into the pool. (b) Determine the average velocity of water drained out of the pool through the pipe.

3.60BG   Gasoline ($s = 0.68$) is siphoned from a car tank into a 5-gallon gas can using a 0.5-in-diameter flexible tube with a 0.35-in nozzle exit. The velocity of flow in the tube is 9 *ft/sec*. (a) Determine the mass flowrate, $\dot{M}$; the weight flowrate, G; and the volume flowrate, Q of the gasoline through the tube. (b) Determine the velocity of gasoline flow through the nozzle exit. (c) Determine how long it takes to fill the gas can with gasoline.

3.61BG   A circular tank open to the atmosphere, with a cross-sectional area of 6 ft^2, is filled with water to a depth of $h = h_1 = 9\,ft$ above the centerline of the 0.95-in-diameter plugged opening near the bottom of the tank, as illustrated in Figure ECP 3.61. When the opening is unplugged it will be open to the atmosphere, and a jet of water will flow at a variable velocity, $v(h) = \sqrt{2gh}$ (derived from applying the Bernoulli equation in Chapter 4). (a) Determine the variable volume flowrate of flow through the jet. (b) Determine the time it takes for the water in the tank to reach a depth $h = h_2 = 6\,ft$ above the centerline of the jet. (c) Determine the rate of change of the depth of water in the tank, $v_{\tan k}(h) = (dh/dt)$ when the water in the tank is at a depth $h = h_2 = 6\,ft$ above the centerline of the jet.

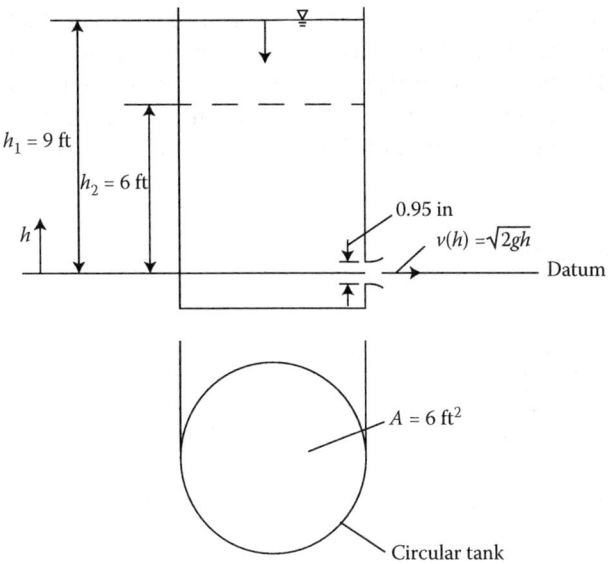

FIGURE ECP 3.61

3.62BG Water flows through a pipe at a volume flowrate of 8000 *cfs*. The velocity of flow in the pipe is not to exceed 50 *ft*/*sec*. (a) Determine the minimum allowable pipe size.

3.63BG Air with a density of 0.0020580 *slug*/*ft*^{3} enters a cooling system through a 1.3-*ft*-diameter pipe at 16 *ft*^{3}/*sec*. (a) Determine the mass flowrate, $\dot{M}$ and the weight flowrate, G through the pipe. (b) Determine the velocity through the pipe entrance.

The Continuity Equation for a Fixed Control Volume

3.64BG Water flows through a pipe as illustrated in Figure ECP 3.64. A conical contraction/nozzle is used to efficiently accelerate the fluid flow. At control face 1, the pipe diameter is 3 *ft*, and the velocity of flow is 17 *ft*/*sec*. At control face 2, the pipe diameter is 1.5 *ft*. (a) Determine the mass flowrate, $\dot{M}$; the weight flowrate, G; and the volume flowrate, Q through the pipe. (b) Determine the velocity at control face 2.

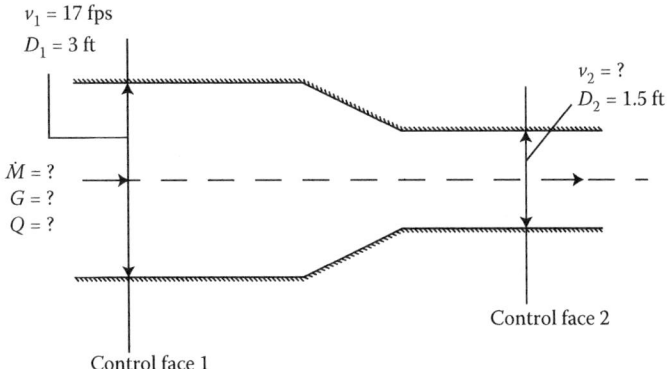

FIGURE ECP 3.64

3.65BG   Air at 0.002470 $slug/ft^3$ is forced into a 4-in blow-dryer duct by a mini fan at a velocity of $1\,ft/sec$ and heated by electric coils in the blow-dryer as illustrated in Figure ECP 3.65. The heated air at 0.002129 $slug/ft^3$ is forced out of a 2-in nozzle by the mini fan in the blow-dryer. (a) Determine the mass flowrate, $\dot{M}$; the weight flowrate, G; and the volume flowrate, Q through the blow-dryer. (b) Determine the velocity of flow as it exits the nozzle.

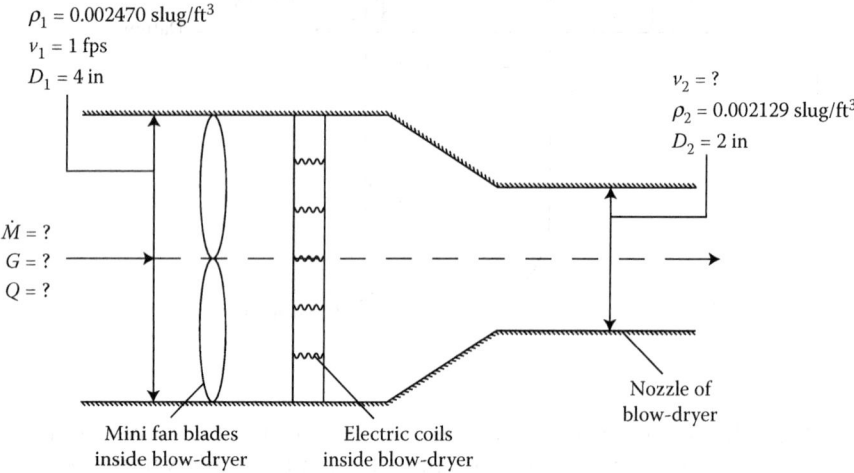

FIGURE ECP 3.65

3.66SI   Methane flows through a 40-*cm* pipe, as illustrated in Figure ECP 3.66. At control face 1, the pressure is $5 \times 10^6\,N/m^2$ (*abs*), the temperature is $295°K$, and the velocity of flow is $9\,m/sec$. At control face 2, the pressure is $3 \times 10^6\,N/m^2$ (*abs*) and the temperature is $260°K$. Assume the methane behaves as an ideal gas. (a) Determine the density of the methane at control face 1 and control face 2. (b) Determine the mass flowrate, $\dot{M}$ and the weight flowrate, G through the pipe. (c) Determine the volume flowrate at control face 1 and control face 2. (d) Determine the velocity at control face 2.

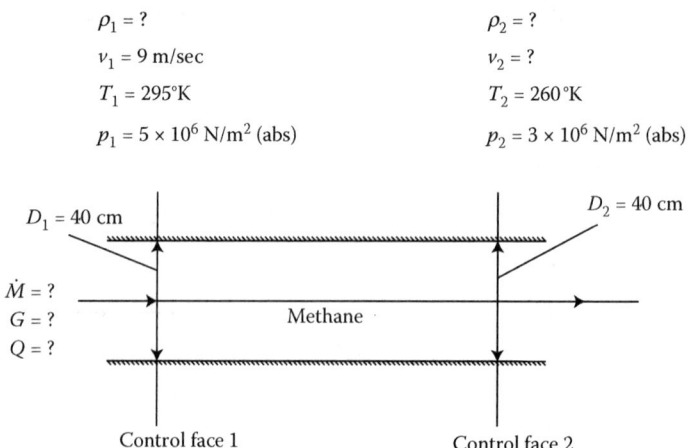

FIGURE ECP 3.66

3.67SI Helium at 20°C flows into a compressor through a 0.4-*m*-diameter pipe and at a velocity of 50 *m/sec* (control face 1), as illustrated in Figure ECP 3.67. Assume the helium remains at 20°C as it leaves the compressor through control face 2. The pressure at control face 1 is 3×10^6 N/m^2 (*abs*), and the pressure at control face 2 is 15 times the pressure at control face 1. Assume the helium behaves as an ideal gas and that the velocity of the helium is not to go below 60 *m/sec*. (a) Determine the maximum pipe size at control face 2.

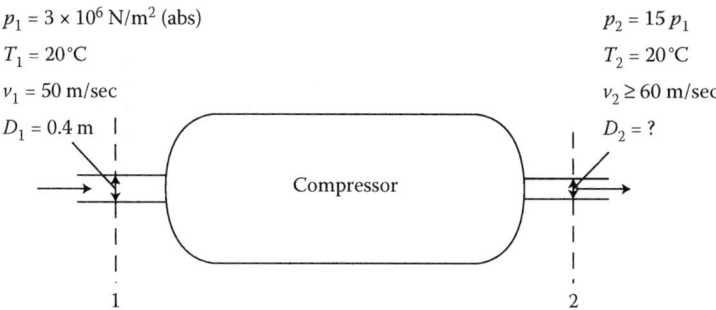

$p_1 = 3 \times 10^6$ N/m^2 (abs) $p_2 = 15\, p_1$

$T_1 = 20\,°C$ $T_2 = 20\,°C$

$v_1 = 50$ m/sec $v_2 \geq 60$ m/sec

$D_1 = 0.4$ m $D_2 = ?$

Compressor

1   2

FIGURE ECP 3.67

3.68BG Water flowing at a velocity of 5 *ft/sec* in rectangular open channel of width 5 *ft* and depth 4 *ft* (control face 1) diverges into two smaller rectangular open channels. The channel at control face 2 is of 3 *ft* width and depth 3.5 *ft*, and the channel at control face 3 is of 2 *ft* width and depth 1.5 *ft* and flows at a velocity of 8 *ft/sec*, as illustrated in Figure ECP 3.68. (a) Determine the volume flowrate at each of the three control faces and the velocity at control face 2.

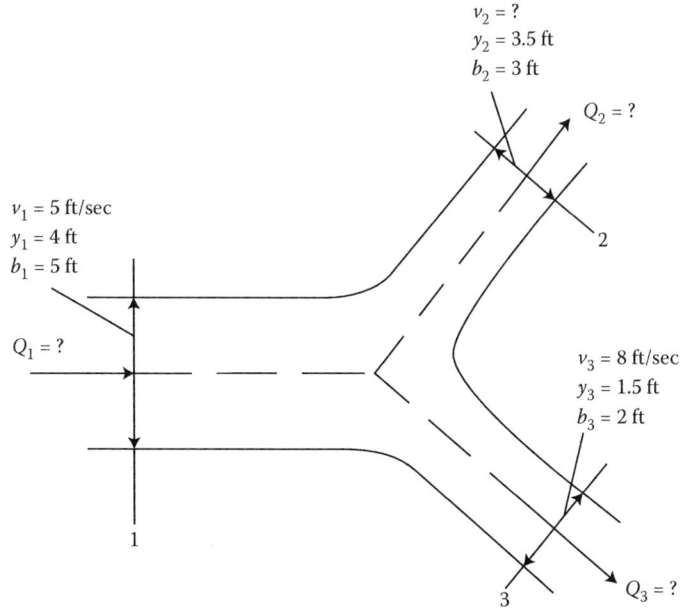

$v_2 = ?$
$y_2 = 3.5$ ft
$b_2 = 3$ ft

$Q_2 = ?$

$v_1 = 5$ ft/sec
$y_1 = 4$ ft
$b_1 = 5$ ft

2

$Q_1 = ?$

$v_3 = 8$ ft/sec
$y_3 = 1.5$ ft
$b_3 = 2$ ft

1

3       $Q_3 = ?$

FIGURE ECP 3.68

3.69SI A water and glycerin mixture ($\rho = 1128\,kg/m^3$) flows through a nozzle at $900\ m^3/sec$, as illustrated in Figure ECP 3.69. The pipe at control face 1 has a diameter of $1.8\ m$. (a) Determine the velocity at control face 1 and the minimum pipe size at control face 2 in order for the velocity not to exceed $700\ m/sec$. (b) Determine the mass flowrate and the weight flowrate through the nozzle.

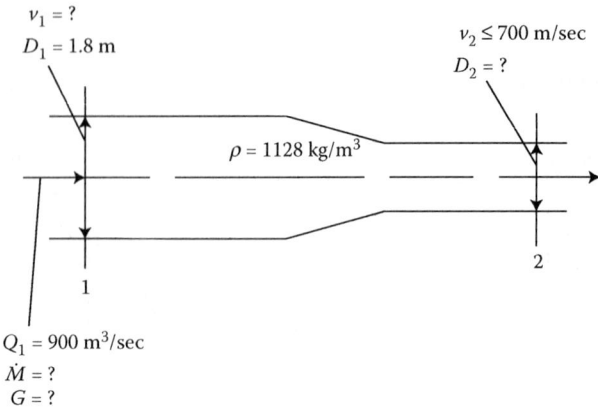

FIGURE ECP 3.69

3.70BG Water flows in a rectangular open channel of width $5\,ft$, as illustrated in Figure ECP 3.70. The upstream flow, as it is leaving the lake at control face 1, is $2\,ft$ deep and has a rectangular velocity profile. Downstream, there is a smooth step upward in the channel, which causes the flow to dip. Furthermore, the downstream flow, after the step at control face 2, is $1.4\,ft$ deep and has a parabolic velocity profile, with $v_{max} = 7\,ft/sec$. (a) Determine the relationship between v_{max} and v_{ave} at control face 2. (b) Determine the velocity at control face 1. (c) Determine v_{ave} at control face 2.

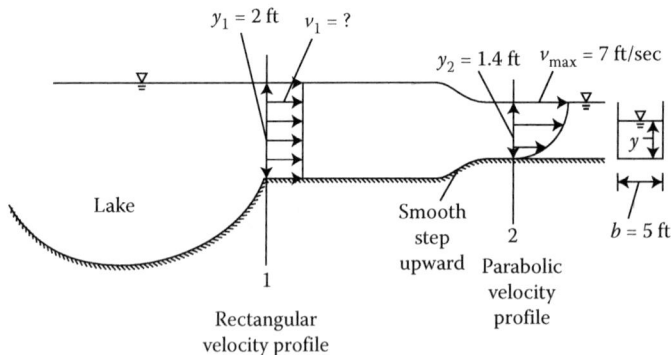

FIGURE ECP 3.70

3.71SI Water flows through a 0.95-m pipe that branches into a 0.7-m pipe and a 0.6-m pipe, as illustrated in Figure ECP 3.71. The velocity of flow at control face 1 is $2.9\ m/sec$, and the velocity of flow at control face 2 is $2.3\ m/sec$. (a) Determine the volume flowrate at control face 1. (b) Determine the volume flowrate at control face 2. (c) Determine the volume flowrate at control face 3.

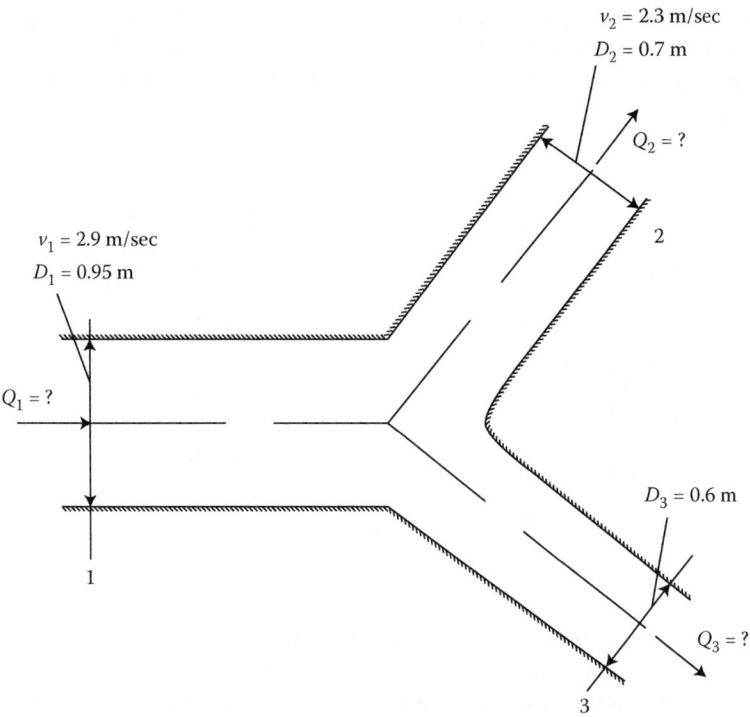

FIGURE ECP 3.71

3.72BG   Air flows into a living room through an 8 ft-by-11 ft door at a volume flowrate of 500 *cfs*, and flows out of the room through a 4 *ft*-by-6 *ft* window, as illustrated in Figure ECP 3.72. (a) Determine the velocity of the air as it flows out of the window.

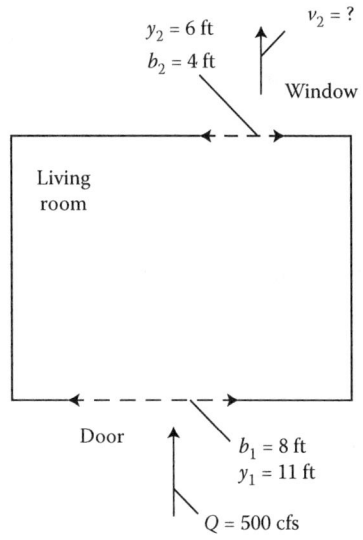

FIGURE ECP 3.72

3.73SI Water enters a square tank through a 0.6-m-diameter pipe at a velocity of 7 *m/sec*, and leaves the tank through two 0.4-*m*-diameter pipes as illustrated in Figure ECP 3.73. Assume that $v_2 = 3v_3$. (a) Determine the velocity of the water as it leaves the tank through the two pipes.

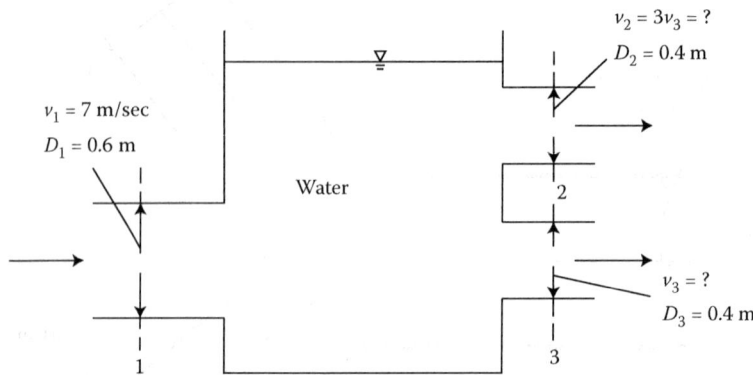

FIGURE ECP 3.73

3.74SI Water flows in a rectangular channel of width 6 *m* at a velocity of 1 *m/sec* and depth of 1.65 *m*. There is an abrupt upward step, Δz of 0.33 m in the channel bed, as illustrated in Figure ECP 3.74. The depth downstream (control face 2) of the abrupt step is 0.373 *m*. (a) Determine the volume flowrate in the channel. (b) Determine the velocity of flow at control face 2.

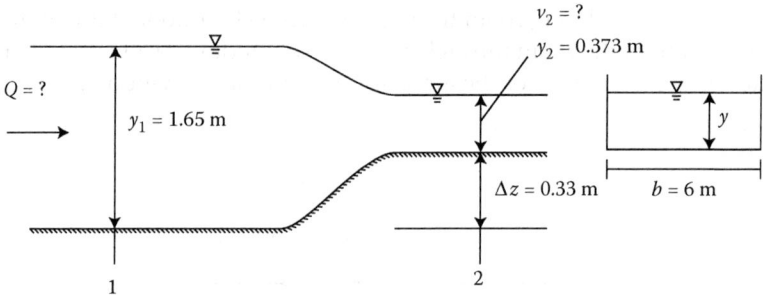

FIGURE ECP 3.74

The Continuity Equation for a Moving Control Volume

3.75SI Air at a density of 1.12 kg/m^3 enters the intake (control face 1) of a jet engine with a diameter of 1.5 *m* at an absolute velocity of 270 *km/hr* (relative to the earth), as illustrated in Figure ECP 3.75. The exhaust at a density of 0.95 kg/m^3 leaves the tail of the jet engine (control face 2) with a diameter of 1.1 *m*. The jet (engine) (control volume) is traveling at an absolute velocity of 500 *km/hr* (relative to the earth). (a) Determine the velocity of the flow at control face 1 relative to the control volume, v_1. (b) Determine the mass flowrate through the jet engine. (c) Determine the absolute velocity of flow at control face 2, v_2 (*abs*). (d) Determine the velocity of flow at control face 2 relative to the control volume, v_2.

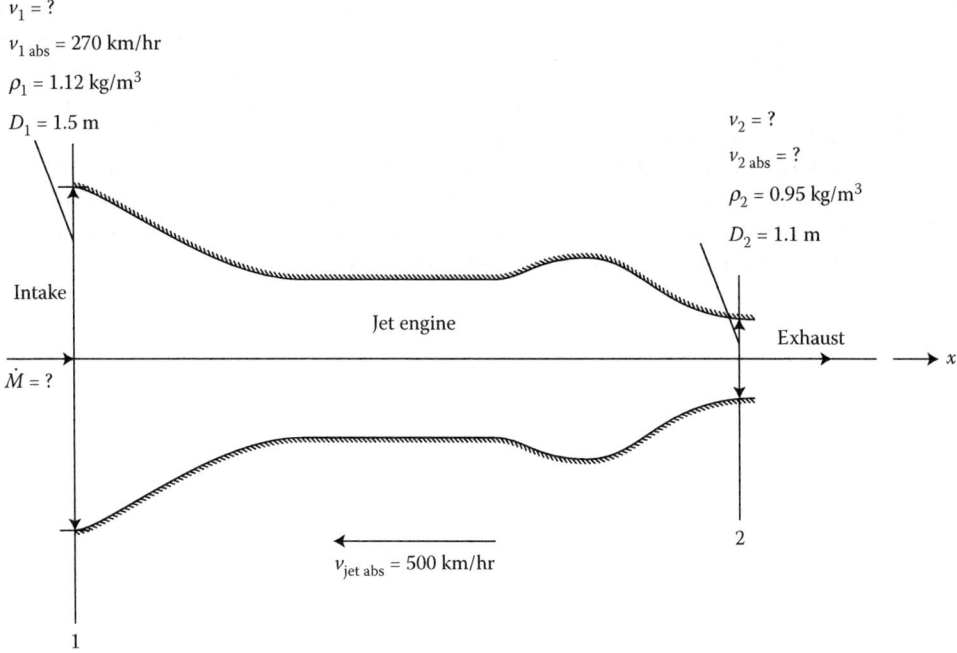

FIGURE ECP 3.75

3.76SI A pesticide solution flows through a garden hose at the base of a rotating sprinkler at a volume flowrate of $0.05\ m^3/sec$, as illustrated in Figure ECP 3.76. The radius of the

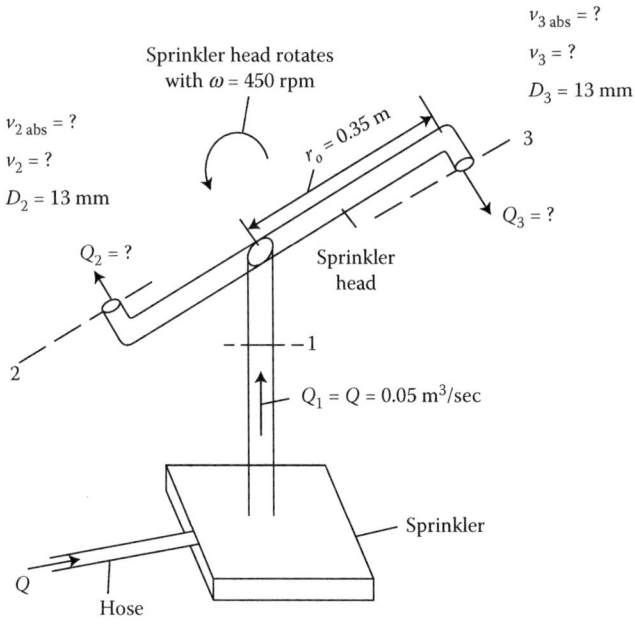

FIGURE ECP 3.76

sprinkler head, r_o is 0.35 m. There are two nozzles in the sprinkler head (control volume), each with a diameter of 13 mm. (a) Determine the volume flowrate in each of the nozzles (control faces 2 and 3). (b) Determine the velocity of the flow in each of the nozzles (control faces 2 and 3) relative to the control volume, v_2 and v_3, respectively. (c) Determine the absolute velocity (relative to the earth) of flow in each of the nozzles (control faces 2 and 3), v_2 (*abs*) and v_3 (*abs*), respectively, if the sprinkler head (control volume) is stationary. (d) Determine the absolute velocity (relative to the earth) of flow in each of the nozzles (control faces 2 and 3), v_2 (*abs*) and v_3 (*abs*), respectively, if the sprinkler head (control volume) is moving/rotating counterclockwise at an angular velocity, ω of 450 *rpm*.

4

Energy Equation

4.1 Introduction

For the study of fluids in motion, one must consider both the kinematics and the dynamics of the flow situation. Fluid kinematics is based on the principle of conservation of mass and yields the continuity equation, which was introduced in Chapter 3. Fluid dynamics is based upon the principle of conservation of momentum (Newton's second law of motion) and yields the equations of motion, known as the energy equation and the momentum equation. The energy equation is introduced and applied in Chapter 4, while the momentum equation will be introduced in Chapter 4 and applied in Chapter 5. Fluid dynamics also includes the topic of dimensional analysis, which is introduced in Chapter 7 and yields the resistance equations, which are introduced in Chapter 6. It may be noted that the derivation of the energy equation may alternatively be based on the principle of conservation of energy (the first law of thermodynamics). Therefore, in this chapter, the energy equation will be derived using both approaches: first by applying Newton's second law motion, and then by applying the first law of thermodynamics. As noted in Chapter 3, the continuity equation, the equations of motion, and the results of dimensional analysis are applied to internal flows, which include pressure (pipe) flow in Chapter 8 and gravity (open channel) flow in Chapter 9. However, only the energy and momentum equations and the results of dimensional analysis are applied to external flows, which include the flow around objects in Chapter 10. Depending upon the internal flow problem, one may use either the Eulerian (integral) approach or the Lagrangian (differential) approach, or both, in the application of the three governing equations (continuity, energy, and momentum). It is important to note that while the continuity and the momentum equations may be applied using either the integral or the differential approach, application of the energy equation is useful only using the integral approach, as energy is work, which is defined over a distance, L. Furthermore, as noted in Chapter 3, for an internal fluid flow, while the application of the continuity equation will always be necessary, the energy equation and the momentum equation play complementary roles in the analysis of a given flow situation; when one of the two equations of motion breaks down, the other may be used to find the additional unknown quantity.

4.2 Fluid Dynamics

Fluid dynamics is based upon the principle of conservation of momentum (Newton's second law of motion) (Section 4.4) and yields the equations of motion, known as the energy equation and the momentum equation. The energy equation is introduced and applied in Chapter 4 while the momentum equation will be introduced in Chapter 4 and applied in Chapter 5. Fluid dynamics also includes the topic of dimensional analysis (Chapter 7), which yields the resistance equations (Chapter 6). It may be noted that the derivation of the energy equation may alternatively be based on the principle of conservation of energy (the first law of thermodynamics) (Section 4.6). Therefore, in this chapter, the energy equation will be derived using both approaches (Section 4.3): first by applying Newton's second law of motion (Section 4.5), and then by applying the first law of thermodynamics (Section 4.7).

Fluid dynamics considers the analysis of the specific forces acting on a fluid element that produce the motion. In general, the forces that may act in a flow situation include those due to gravity, $F_G = Mg$; pressure, $F_P = \Delta p A$; viscosity, $F_V = \tau A$; elasticity, $F_E = E_v A$; and surface tension, $F_T = \sigma L$. Furthermore, according to Newton's second law of motion, if the summation of forces acting on a fluid element does not add up to zero, then the fluid element will accelerate. However, such an unbalanced force system can be transformed into a balanced system by adding an inertia force, $F_I = Ma$, that is equal and opposite to the resultant of the forces acting on the fluid element (along the streamline, s) as follows:

$$\sum_s F = F_G + F_P + F_V + F_E + F_T = -R_s \tag{4.1}$$

where

$$F_I = -R_s \tag{4.2}$$

and thus:

$$\sum_s F = F_G + F_P + F_V + F_E + F_T + F_I = 0 \tag{4.3}$$

Depending upon the fluid flow problem, some of the above-listed forces may not play a significant role in the derivation of the equations of motion (energy equation and momentum equation). Specifically, in the derivation of the equations of motion, the significant sources of forces acting on a fluid element include those due to gravity, $F_G = Mg = \gamma V$; pressure, $F_P = \Delta p A$; viscosity, $F_V = \tau A$; and inertia $F_I = Ma$. Thus, those due to elasticity, $F_E = E_v A$ and surface tension, $F_T = \sigma L$ are not considered in the derivation of the equations of motion. However, the consideration of the elastic and surface tension forces becomes important in the dimensional analysis (Chapter 7) and dynamic similitude and modeling (Chapter 11) of elastic/compressible fluids and thin sheets of fluid (and droplets, bubbles, and capillarity), respectively, yielding the Cauchy number ($C = \rho v^2 / E_v$) and the Weber number ($W = \rho v^2 L / \sigma$), respectively.

Assuming a real flow, the flow resistance is the shear stress, friction, or drag force originating in the fluid viscosity, which is modeled as the friction/viscous force, F_V in Equation 4.3. Regardless of whether the flow is internal or external, it is important to note that in the application of Newton's second law of motion (Equation 4.3), while the definition of the gravitational force, F_G and the pressure force, F_P can always be theoretically modeled, the friction/viscous force, F_V may not always be theoretically modeled. As such, the results of dimensional analysis (Chapter 7) are used to supplement the momentum theory (Chapter 5) in the definition of the friction/viscous force and thus the definition of the flow resistance (Chapter 6).

4.3 Derivation of the Energy Equation

As mentioned above, derivation of the energy equation may be based on either the principle of conservation of momentum (Newton's second law of motion), or the principle of conservation of energy (the first law of thermodynamics). Thus, the energy equation will be derived first by applying Newton's second law of motion and then by applying the first law of thermodynamics. Similar to the statement of the principle of conservation of mass in Chapter 3 (yielding the continuity equation), the principle of conservation of momentum (Newton's second law of motion) and the principle of conservation of energy (the first law of thermodynamics) are historically stated and applied for a fluid system, and then rephrased and extended to a control volume using the Reynolds transport theorem. Statement of the conservation of momentum principle deals with the time rate of change of the extensive property, momentum, Mv for a fluid system, whereas statement of the conservation of energy principle deals with the time rate of change of the extensive property, energy, E for a fluid system. Furthermore, one may note that while the energy equation that is derived based on the conservation of momentum principle allows for the modeling of unsteady flow, it does not directly consider the modeling of energy transfer by work (pumps, turbines, etc.). Finally, one may note that while the energy equation that is derived based on the conservation of energy principle assumes steady flow, it directly considers the modeling of energy transfer by work (pumps, turbines, etc.).

4.4 Conservation of Momentum Principle: Newton's Second Law of Motion

The conservation of momentum principle (Newton's second law of motion) states that "the time rate of change of the extensive property, momentum, Mv of a fluid system is equal to the sum of forces acting on the system" and is expressed as follows:

$$\frac{d(Mv)_{sys}}{dt} = \sum F_{sys} \tag{4.4}$$

The point of reference or the coordinate system for which Equation 4.4 is true is called "inertial." The next step is to express Newton's second law of motion for a control volume. To do this, consider the one-dimensional uniform (at a given cross section) flow through a stream tube from left to right, as illustrated in Figure 4.1 (the same as Figure 3.6). At an initial time t, the fluid system between sections 1 and 2 coincides with the control volume between control face 1 and control face 2, so they are identical. Thus, when a control volume is coincident with fluid system at an instant of time, the forces acting on the fluid system and the forces

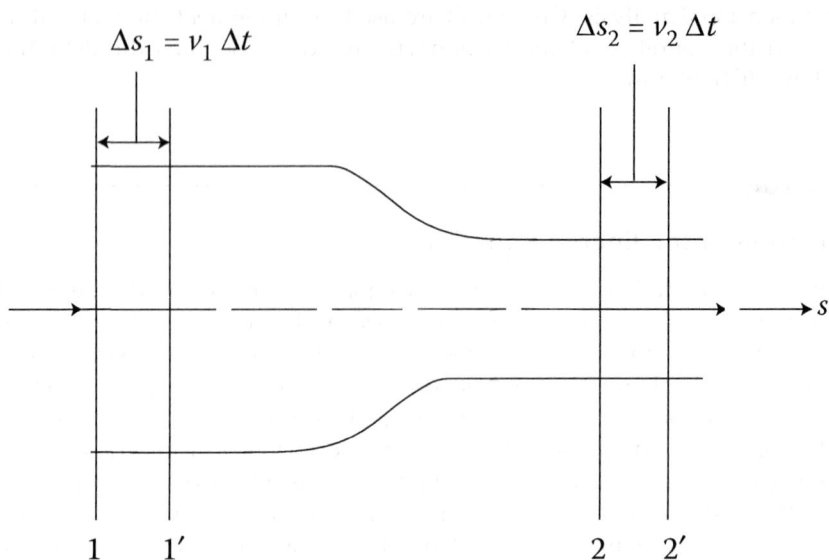

FIGURE 4.1
Finite control volume defined for a one-dimensional flow through a stream tube.

acting on the contents of the coincident control volume are instantaneously identical as follows:

$$\sum F_{sys} = \sum F_{cv} \tag{4.5}$$

Furthermore, the Reynolds transport theorem given in Equation 3.19 (for the one-dimensional uniform flow through a stream tube of Figure 4.1) may be applied to link the definition of the conservation of momentum between a fluid system and a control volume as follows:

$$\frac{d(Mv)_{sys}}{dt} = \frac{d(Mv)_{cv}}{dt} - (\dot{M}v)_{in} + (\dot{M}v)_{out} \tag{4.6}$$

where the extensive property of the fluid, $B = Mv$ and thus the corresponding intensive property, $b = B/M = Mv/M = v$, velocity. First substituting Equation 4.4 and then Equation 4.5 for the left-hand side of Equation 4.6 yields the conservation of momentum for a control volume as follows:

$$\sum F_{cv} = \frac{d(Mv)_{cv}}{dt} - (\dot{M}v)_{in} + (\dot{M}v)_{out} \tag{4.7}$$

which states that "the sum of forces acting on the control volume is equal to the time rate of change of momentum within the control volume boundaries, plus the net flux of momentum out of the control volume (outflux minus the influx)."

The conservation of momentum principle in fluid mechanics is typically applied to either a differential fluid system consisting of a differential mass, M or to a finite control volume. Application of the conservation of momentum principle for a fluid system (Lagrangian approach) involves the application of the principle stated in Equation 4.4 for a fluid system to a differential mass, M and yields the differential form of Newton's second law of motion. It may be noted that while application of the conservation of momentum principle stated in Equation 4.7 for a control volume to a infinitesimal control volume, ds would also yield the differential form of Newton's second law of motion, the fluid system approach is probably simpler. And, application of the conservation of momentum principle for a control volume (Eulerian approach) involves the application of the principle stated in Equation 4.7 for a control volume to a finite control volume and yields the integral form of Newton's second law of motion.

4.4.1 Newton's Second Law of Motion for a Fluid System

Application of the conservation of momentum principle stated in Equation 4.4 for a fluid system to a differential mass, M with a differential cross-sectional area, A and a differential section, ds defined along a stream tube, as illustrated in Figure 4.2, gives the differential form of Newton's second law of motion. Beginning with Equation 4.4 restated as follows:

$$\frac{d(Mv)_{sys}}{dt} = \sum F_{sys} \tag{4.8}$$

it may be expanded as follows:

$$\frac{M_{sys}d(v)_{sys}}{dt} + \frac{v_{sys}d(M_{sys})}{dt} = \sum F_{sys} \tag{4.9}$$

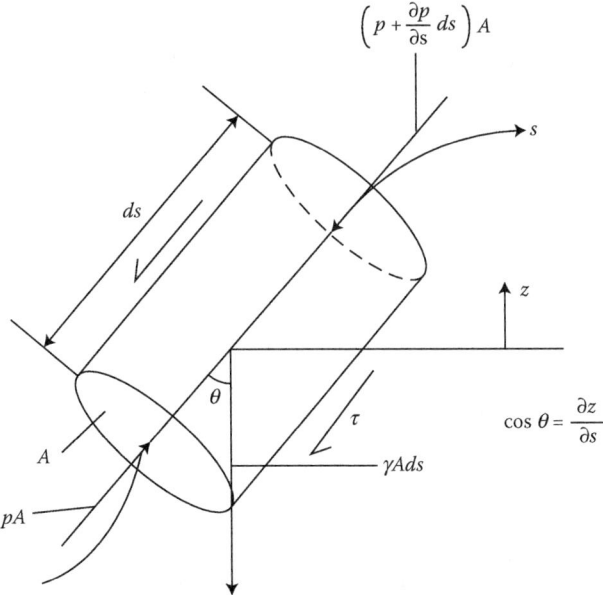

FIGURE 4.2
Differential mass, M with a differential cross-sectional area, A and a differential section, ds defined along a stream tube is used to derive the differential form of the conservation of momentum principle (Newton's second law of motion) for a fluid system.

One may note that from the conservation of mass principle for a fluid system, the second term on the left-hand side of Equation 4.9 equals zero because:

$$\frac{d(M_{sys})}{dt} = 0 \tag{4.10}$$

Therefore

$$\frac{M_{sys}d(v)_{sys}}{dt} = \sum F_{sys} \tag{4.11}$$

or

$$\sum F_{sys} = \frac{M_{sys}d(v)_{sys}}{dt} \tag{4.12}$$

Noting that the time rate of change of velocity for a system is defined as the acceleration of the system, a_{sys}, the differential form of Newton's second law of motion for a differential fluid element moving in the s-direction along the streamline is given as follows:

$$\sum F_s = Ma_s \tag{4.13}$$

Thus, Newton's second law of motion states that the sum of the forces, ΣF_s acting on the differential fluid element, ds of the stream tube will accelerate the differential fluid mass, M along the s-axis. First, we examine the left-hand side of Equation 4.13. The significant forces acting on the differential fluid element illustrated in Figure 4.2 include the gravitational force component along the s-axis, the pressure forces on the two ends of the fluid element, and the viscous force due to fluid friction as follows:

$$\sum F_s = F_G + F_P + F_V \tag{4.14}$$

where

$$F_G = -\gamma Ads\cos\theta = -\gamma Ads\frac{\partial z}{\partial s} \tag{4.15}$$

$$F_P = pA - \left(p + \frac{\partial p}{\partial s}ds\right)A = -\frac{\partial p}{\partial s}dsA \tag{4.16}$$

$$F_V = -\tau_w P_w ds \tag{4.17}$$

and P_w = the wet perimeter of the fluid element along which the shear stress, τ_w acts. Next, we examine the acceleration term in right hand side of Equation 4.13. The acceleration is defined is follows:

$$a_s = \frac{dv}{dt} = \frac{ds}{dt}\frac{\partial v}{\partial s} + \frac{\partial v}{\partial t} = v\frac{\partial v}{\partial s} + \frac{\partial v}{\partial t} \tag{4.18}$$

where the $\delta v/\delta s$ term represents convective acceleration, which represents the possibility of nonuniform flow, and the $\delta v/\delta t$ term represents local acceleration, which represents the possibility of unsteady flow. Finally, noting that the mass term on the right-hand side of Equation 4.13 is defined as $M = \rho A ds$, Equation 4.13 for the differential form of Newton's second law of motion for the most general flow type (compressible, nonuniform, unsteady flow for either pressure or gravity flow) may be written as follows:

$$-\gamma A ds \frac{\partial z}{\partial s} - \frac{\partial p}{\partial s} ds A - \tau_w P_w ds = (\rho A ds)\left(v\frac{\partial v}{\partial s} + \frac{\partial v}{\partial t}\right) \tag{4.19}$$

One may note that in the application of Equation 4.19, the distinction between pressure flow versus gravity flow is in the definition of the hydrostatic pressure, p; the cross-sectional area, A; and the wet perimeter, P_w (described in detail below). Because the definition of energy in terms of work is force times distance, multiplying the differential form of Newton's second law of motion (Equation 4.19) by the differential section, ds yields the differential form of the energy equation (see Section 4.5), which is not really useful (practical per se) until it is integrated and evaluated between two points to yield the integral form of the energy equation. Thus, integration of the differential form of the energy equation and evaluation between two points on the streamline, s yields the integral form of the energy equation (see Section 4.5), which may be solved for an unknown "energy head" term for one of the two points in a finite control volume. Finally, one may note that the differential form of Newton's second law of motion (Equation 4.19) (with some simple manipulation of terms) yields the differential form of the momentum equation (for a fluid system), which is used to solve for one of the following unknown quantities for a given point in the fluid system: friction slope, S_f; velocity, v; pipe or channel friction coefficient (Chezy coefficient, C; Darcy-Weisbach friction factor, f; Manning's roughness coefficient, n; or Hazen–Williams roughness, C_h); pipe diameter, D; or open channel flow depth, y (see Chapter 5 for complete derivation and application).

4.4.2 Newton's Second Law of Motion for a Control Volume

Application of the conservation of momentum principle stated in Equation 4.7 for a control volume to a finite control volume, as illustrated in Figure 4.1, assuming steady flow, gives the integral form of Newton's second law of motion, which is the integral form of the momentum equation (derived below and applied in Chapter 5). Begin with Equation 4.7 restated as follows:

$$\sum F_{cv} = \frac{d(Mv)_{cv}}{dt} - (\dot{M}v)_{in} + (\dot{M}v)_{out} \tag{4.20}$$

Because most practical engineering problems involve steady flow, and because the focus of this textbook will be limited to steady flow problems, the special case of steady flow will be assumed in the derivation of the integral form of Newton's second law of motion. As such, the time rate of change of momentum within the control volume boundaries (the first term on the right-hand side of Equation 4.20) is set equal to zero as follows:

$$\frac{d(Mv)_{cv}}{dt} = 0 \tag{4.21}$$

and thus Equation 4.20 becomes as follows:

$$\sum F_{cv} = -(\dot{M}v)_{in} + (\dot{M}v)_{out} \tag{4.22}$$

or

$$\sum F_{cv} = (\dot{M}v)_{out} - (\dot{M}v)_{in} \tag{4.23}$$

which yields Newton's second law of motion for a control volume assuming steady flow, which states that the sum of forces acting on the control volume is equal to the net flux of momentum out of the control volume (momentum flow rate [flux] out of the control volume minus the momentum flow rate [flux] into the control volume). Referring to Figure 4.1, one may expand the right-hand side of Equation 4.23 (net flux of momentum out of the control volume) as follows:

$$\sum F_{cv} = (\rho v^2 A)_{out} - (\rho v^2 A)_{in} \tag{4.24}$$

or

$$\sum F_{cv} = (\rho Q v)_{out} - (\rho Q v)_{in} \tag{4.25}$$

where $M = \rho V = \rho \Delta s A = \rho v \Delta t A = \rho Q \Delta t$, control face 1 represents the inflow cross section, and control face 2 represents the outflow cross section. Furthermore, assuming the forces acting on the control volume act along the center streamline, s, Equation 4.25 may be written as follows:

$$\sum F_s = (\rho Q v_s)_2 - (\rho Q v_s)_1 \tag{4.26}$$

where the summation of forces along the streamline, s (left-hand side of Equation 4.26) represents significant sources of forces as follows:

$$\sum F_s = (F_G + F_P + F_V + F_{other})_s \tag{4.27}$$

where F_{other} represents other forces acting on the control volume, such as reaction forces at bolts, cables, etc. Therefore, the integral form of Newton's second law of motion, assuming steady flow, states that the rate of change (increase) of fluid momentum across the control volume is accomplished by the action of a forward force, ΣF_s acting on the control volume as follows:

$$\sum F_s = (F_G + F_P + F_V + F_{other})_s = (\rho Q v_s)_2 - (\rho Q v_s)_1 \tag{4.28}$$

where the velocities are measured relative to the fixed control volume. It may be noted that this integral form of Newton's second law of motion (Equation 4.28) is the integral form of the momentum equation, assuming steady flow, and is used to solve for an unknown force acting on the finite control volume (applied in Chapter 5).

It is important to note that a more general statement of the integral form of the momentum equation for a control volume, assuming steady flow, assumes there may be one or more

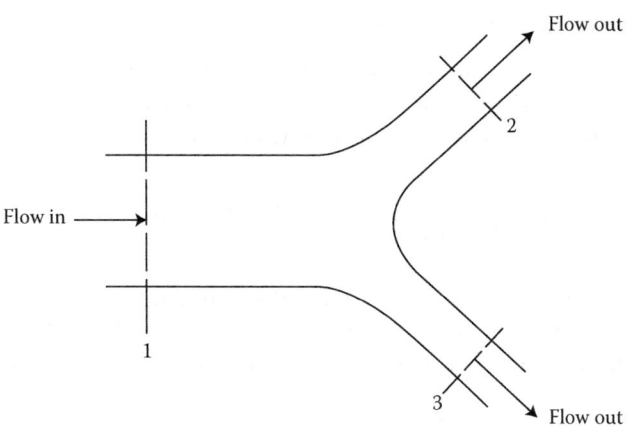

Flow out

2

Flow in

1

3

Flow out

FIGURE 4.3
Control volume: may have one or more control faces for the flow into and the flow out of the control volume.

control faces for the flow into and the flow out of the control volume, as illustrated in Figure 4.3. Thus, the more general statement of the momentum equation for a control volume, assuming steady flow along the streamline, s, is given as follows:

$$\sum F_s = (F_G + F_P + F_V + F_{other})_s = (\rho Q v_s)_{out} - (\rho Q v_s)_{in} \qquad (4.29)$$

4.5 The Energy Equation Based on Newton's Second Law of Motion

The definition of energy in terms of work is force times distance. Thus, when the differential form of Newton's second law of motion, given in Equation 4.19, is multiplied by the differential section, ds, this yields the differential form of the energy equation. Furthermore, when the differential form of the energy equation is integrated and evaluated between two points in a finite control volume such as between points 1 and 2 in Figure 4.1, the integral form of the energy equation for a control volume is derived. One may note that it is intriguing that the application of a Lagrangian approach (differential form of Newton's second law of motion) yields a Eulerian equation of motion, namely the integral form of the energy equation (one may recall that a similar approach was taken in the derivation of the integral form of the continuity equation in Chapter 3). Furthermore, unlike the continuity and momentum equations, application of the energy equation is useful only using the integral approach for a finite control volume. While application of the continuity equation will always be necessary for internal flow, the energy and momentum equations play complementary roles in the analysis of both internal and external flow. In the application of the governing equations, it is important to make the distinction between real flow, ideal flow, and a hydraulic jump. And for ideal flow, it is important to make the distinction between internal flow, external flow, and flow from a tank.

4.5.1 The Energy Equation for a Fluid System

Multiplying the differential form of Newton's second law of motion (Equation 4.19) by the differential section, ds yields the differential form of the energy equation for a fluid system. Beginning with Equation 4.19, restated as follows (each term with dimensions of F):

$$- \gamma A ds \frac{\partial z}{\partial s} - \frac{\partial p}{\partial s} ds A - \tau_w P_w ds = (\rho A ds)\left(v \frac{\partial v}{\partial s} + \frac{\partial v}{\partial t} \right) \tag{4.30}$$

Before multiplying Equation 4.30 by ds to yield the differential form of the energy equation, it will first be simplified. Begin by dividing both sides of Equation 4.30 by Ads (with dimensions of L^3), which yields the following (each term with dimensions of FL^{-3}):

$$- \gamma \frac{\partial z}{\partial s} - \frac{\partial p}{\partial s} - \frac{\tau_w P_w}{A} = \rho \left(v \frac{\partial v}{\partial s} + \frac{\partial v}{\partial t} \right) \tag{4.31}$$

where the hydraulic radius, $R_h = A/P_w$. Because we began with the most general form of Newton's second law of motion (Equation 4.30), this applies to compressible, nonuniform, unsteady flow. Thus, for compressible flow, the specific weight, $\gamma = \rho g$ of the fluid varies with respect to s and thus must be included with the partial derivative sign with respect to s as follows:

$$- \frac{\partial(p + \gamma z)}{\partial s} - \frac{\tau_w}{R_h} = \rho \left(v \frac{\partial v}{\partial s} + \frac{\partial v}{\partial t} \right) \tag{4.32}$$

One may recall from Chapter 2 that the term $(p + \gamma z)$ is called the piezometric pressure or the static pressure, and according to the principle of hydrostatics, it remains a constant throughout a body of still water (see Figure 2.5 in Chapter 2) and thus $\delta(p + \gamma z)/ds = 0$ whatever the direction of s may be. However, the presence of the friction term τ/R_h on the left-hand side of Equation 4.32 and the acceleration term on the right-hand side of Equation 4.32 indicates that if the water begins to move, the hydrostatic pressure distribution is disturbed and the piezometric pressure, $(p + \gamma z)$ no longer remains a constant throughout the body of water. One may note that the term $v\delta v = \delta(v^2)/2$ and thus Equation 4.32 may be expressed and rearranged as follows:

$$- \frac{\partial(p + \gamma z)}{\partial s} - \frac{\rho}{2} \partial \left(\frac{v^2}{\partial s} \right) = \frac{\tau_w}{R_h} + \rho \frac{\partial v}{\partial t} \tag{4.33}$$

And, recalling that for compressible flow, the density ρ of the fluid varies with respect to s, and thus must be included with the partial derivative sign with respect to s as follows:

$$- \frac{\partial \left(p + \gamma z + \frac{\rho v^2}{2} \right)}{\partial s} = \frac{\tau_w}{R_h} + \rho \frac{\partial v}{\partial t} \tag{4.34}$$

where the term $(\rho v^2/2)$ is called the dynamic pressure, and Equation 4.34 is known as the one-dimensional Navier–Stokes equation, which is basically the differential form of Newton's second law of motion (conservation of momentum principle). It may be noted that if the flow were assumed to be ideal (inviscid) then the first term on the right-hand side of

Equation 4.34, τ/R_h, which represents the friction due to the fluid viscosity, would be set to zero; in this case, Equation 4.34 without the friction term, τ_w/R_h is known as Euler's equation of motion. Furthermore, multiplying the differential form of Newton's second law of motion (the Navier–Stokes equation) (Equation 4.34) (each term with dimensions of FL^{-3}) by the differential section, ds (with dimensions of L) yields the differential form of the energy equation (each term with dimensions of FL^{-2}) as follows:

$$-\left[\frac{\partial\left(p + \gamma z + \frac{\rho v^2}{2}\right)}{\partial s}\right]ds = \left[\frac{\tau_w}{R_h} + \rho\frac{\partial v}{\partial t}\right]ds \tag{4.35}$$

This differential form of the energy equation is not really useful (practical per se) until it is integrated and evaluated between two points to yield the integral form of the energy equation (see Section 4.5.2). Thus, application of the energy equation is useful only using the integral approach (for a finite control volume), as energy is work, which is defined over distance, L.

4.5.2 The Energy Equation for a Control Volume

Integrating the differential form of the energy equation (Equation 4.35) and evaluating between two points in a finite control volume such as between points 1 and 2 in Figure 4.1, yields the integral form of the energy equation for a control volume. Beginning with the integration of Equation 4.35 yields the following:

$$-\int_1^2\left[\frac{\partial\left(p + \gamma z + \frac{\rho v^2}{2}\right)}{\partial s}\right]ds = \int_1^2\left[\frac{\tau_w}{R_h} + \rho\frac{\partial v}{\partial t}\right]ds \tag{4.36}$$

One may note that a compressible flow assumes a large spatial and/or temporal change in pressure, δp and temperature, T (effects of thermodynamics) and thus fluid density, ρ. Thus, if the effects of thermodynamics were modeled for compressible flow, then the ideal gas law equation ($p = \rho RT$, see Chapter 1 for details), which relates the fluid density, ρ to the pressure, p and the temperature, T, must be introduced before integration. However, for compressible flows in which δp is very small, the flow may be considered/treated as incompressible over s, and thus one may remove the specific weight, $\gamma = \rho g$ and the density, ρ out of the partial derivatives and integrate the equation as follows:

$$-\int_{p_1}^{p_2} \partial p - \int_{z_1}^{z_2}\gamma\partial z - \int_{v_1}^{v_2}\frac{\rho}{2}\partial(v^2) = \int_{s_1}^{s_2}\frac{\tau_w}{R_h}ds + \int_{s_1}^{s_2}\rho\frac{\partial v}{\partial t}ds \tag{4.37}$$

Liquids and gases at a Mach number, M less than approximately 0.5 may be assumed to be incompressible. Integration and evaluation between two points in the finite control volume of length, $L = (s_2 - s_1)$ yields the following:

$$(p_1 - p_2) + \gamma(z_1 - z_2) + \frac{\rho}{2}(v_1^2 - v_2^2) = \frac{\tau_w}{R_h}L + \rho L\frac{\partial v}{\partial t} \tag{4.38}$$

where each of the above terms is given in terms of pressure (or stress) (each term with dimensions of FL^{-2}). Thus, dividing the pressure terms in Equation 4.38 by the specific weight, γ yields "pressure head" terms, $h = p/\gamma$ with dimensions of length, L (m or ft) (recall from Chapter 2) and grouping all terms for point 1 and point 2 in the control volume yields the following:

$$\left(\frac{p}{\gamma} + z + \frac{v^2}{2g}\right)_1 - \left(\frac{p}{\gamma} + z + \frac{v^2}{2g}\right)_2 = \frac{\tau_w}{\gamma}\frac{L}{R_h} + \frac{L}{g}\frac{\partial v}{\partial t} \qquad (4.39)$$

Practical applications of the energy equation requires the definition of the terms in the equation, further simplifying assumptions, and the distinction between pipe and open channel flow in the definition of the terms in the equation.

4.5.2.1 Definition of the Terms in the Energy Equation

One may note that each term in Equation 4.39, which is the integral form of the energy equation for a control volume, is basically expressed in terms of energy per unit weight (each term with dimensions of $FL/F = L$). In the BG system of units, each term represents the ft-lb of energy per lb of fluid flowing. And, in the SI system of units, each term represents the N-m of energy per N of fluid flowing. Each of the two terms on the left-hand side of Equation 4.39 represents the total energy head, H at a given point (points 1 and 2) in the control volume, where the total energy head, H is defined as follows:

$$H = \left(\frac{p}{\gamma} + z + \frac{v^2}{2g}\right) \qquad (4.40)$$

and thus the integral form of the energy equation for a control volume (simply referred to the energy equation) may be written as follows:

$$H_1 - H_2 = \frac{\tau_w}{\gamma}\frac{L}{R_h} + \frac{L}{g}\frac{\partial v}{\partial t} \qquad (4.41)$$

The first term in the total energy head, H equation (Equation 4.40) is called the pressure head, p/γ and represents the energy per unit weight of fluid stored in the fluid by virtue of the pressure under which the fluid exists. The second term in the total energy head, H equation is called the elevation head, z and represents the potential energy per unit weight of fluid. One may note that the sum of the first two terms in the total energy head, H is called the piezometric head or the static head, $(p/\gamma + z)$. The third term in the total energy head, H equation is called the velocity head, $v^2/2g$ and represents the kinetic energy per unit weight of fluid, and is also called the dynamic head. Thus, the two terms on the left-hand side of Equation 4.41 represent the total energy heads, H_1 and H_2 at cross sections 1 and 2, respectively, where the difference is called the total energy head loss, $h_e = H_1 - H_2$. The first term on the right-hand side of Equation 4.41 is called the friction head loss, $h_f = \tau_w L/\gamma R_h$ and represents the energy per unit weight of fluid lost due to fluid friction (due to either pipe or channel resistance [Section 4.5.5], or due to turbulence [Section 4.5.8]; also discussed in detail in Chapters 6, 8, and 9). And, the second term on the right-hand side of Equation 4.41 is called the acceleration head loss, $h_a = (L/g)(\delta v/\delta t)$ and represents the energy per unit weight of fluid lost (if $\delta v/\delta t$ is positive) or gained (if $\delta v/\delta t$ is negative) due to fluid acceleration. Therefore,

Equation 4.39, which is the integral form of the energy equation for a control volume, may be written as follows:

$$h_e = h_f + h_a \tag{4.42}$$

This equation states that the energy difference (total head loss) between sections 1 and 2 in the control volume, h_e is equal to that energy required to overcome friction, h_f plus that energy required to produce acceleration, h_a, where h_e is the total energy lost due to h_f and h_a. It may be noted that the integral form of the energy equation given in Equation 4.39 (and thus Equation 4.42) already represents the assumption of incompressible flow (or compressible flows that may be treated as incompressible flows).

It is important to note that a more general statement of the integral form of the energy equation for a control volume, assuming steady flow, assumes there may be one or more control faces for the flow into and the flow out of the control volume as illustrated in Figure 4.3. Thus, the more general statement of the energy equation evaluated between two points in a control volume, assuming steady flow is given as follows:

$$\left(\frac{p}{\gamma} + z + \frac{v^2}{2g}\right)_{in} - \left(\frac{p}{\gamma} + z + \frac{v^2}{2g}\right)_{out} = \underbrace{\frac{\tau_w}{\gamma} \frac{L}{R_h}}_{h_f} \tag{4.43}$$

4.5.2.2 Practical Assumptions Made for the Energy Equation

It may be noted that because most practical engineering problems involve steady flow, the focus of this textbook will be limited to steady flow problems, and thus the next assumption made is that the flow is steady, and thus the term $\delta v / \delta t = 0$ in Equation 4.39. Therefore, the integral form of the energy equation (referred to from now on as simply the energy equation) for incompressible steady flow is given as follows:

$$\left(\frac{p}{\gamma} + z + \frac{v^2}{2g}\right)_1 - \left(\frac{p}{\gamma} + z + \frac{v^2}{2g}\right)_2 = \underbrace{\frac{\tau_w}{\gamma} \frac{L}{R_h}}_{h_f} \tag{4.44}$$

or expressed in terms of head losses:

$$h_e = h_f \tag{4.45}$$

where $h_a = 0$. Thus, the energy equation assumes incompressible and steady flow along a streamline and is applied between two points along a streamline. Furthermore, in addition to the assumptions of incompressible and steady flow for the energy equation given in Equation 4.44, if one makes a final assumption of ideal (inviscid) flow, then the friction term on the right-hand side of Equation 4.44 would be set to zero, and the equation would then be known as the Bernoulli's equation and is given as follows:

$$\left(\frac{p}{\gamma} + z + \frac{v^2}{2g}\right)_1 - \left(\frac{p}{\gamma} + z + \frac{v^2}{2g}\right)_2 = 0 \tag{4.46}$$

$$H_1 - H_2 = h_e = 0 \tag{4.47}$$

$$H_1 = H_2 \tag{4.48}$$

$$H = \frac{p}{\gamma} + z + \frac{v^2}{2g} = cons\tan t \tag{4.49}$$

Thus, the Bernoulli equation assumes incompressible, steady, and ideal flow along a streamline and is applied between two points along a streamline and where the total head, H along a streamline is constant.

4.5.2.3 Application of the Energy Equation for Pressure Flow versus Open Channel Flow

While the energy equation derived based on the conservation of momentum principle given by Equation 4.44 does not consider the modeling of energy transfer by work (pumps, turbines, etc.), the energy equation derived based on the conservation of energy principle given by Equation 4.181 does. One may note that the modeling of energy transfer by work (pumps, turbines, etc.) in the energy equation is typically applied for pipe flow as opposed to open channel flow. Therefore, a comparison between the application of the energy equation for pipe flow versus open channel flow will be made based on the energy equation given by Equation 4.44. Furthermore, a comparison between the effects of a pump and a turbine assuming pipe flow is made based on the energy equation given by Equation 4.181. The integral form of the energy equation for incompressible, real, steady flow given by Equation 4.44 may be used to solve for an unknown "energy head" term (p/γ, z, or $v^2/2g$) for one of the two points in a finite control volume, or an unknown head loss, h_f between two points in a finite control volume. In the application of Equation 4.44, the distinction between pressure (pipe) flow versus gravity (open channel) flow is in the definition of the hydrostatic pressure head, p/γ; the cross-sectional area, A; and the wet perimeter, P_w, for the hydraulic radius $R_h = A/P_w$ used in the h_f term. For pressure (pipe flowing full) flow, Equation 4.44 is applicable "as is" for the hydrostatic pressure head, p/γ, with the definition of the cross-sectional area, $A = \pi D^2/4$ for a pipe of diameter, D, and the definition of the wet perimeter $P_w = \pi D$, where points 1 and 2 are defined at the centerline of the pipe. However, for open channel flow the pressure head term, p/γ is replaced by the depth of flow in the channel, y, with the definition of the cross-sectional area, $A = by$ for a rectangular open channel (for instance) of channel width, b and depth of flow, y, and the definition of the wet perimeter $P_w = 2y + b$ for a rectangular open channel, where points 1 and 2 are defined at the bottom of the channel (see Chapter 9 for details and for nonrectangular cross sections).

4.5.3 The Bernoulli Equation

Application of the energy equation for incompressible, steady, and ideal flow results in the definition of the Bernoulli equation. The viscosity of real flow (a real fluid) introduces a great complexity in a fluid flow situation, namely friction due to the viscous shearing effects. The assumption of ideal flow (inviscid flow) assumes an ideal fluid that has no viscosity and thus experiences no shear stresses (friction) as it flows. Although this is an idealized flow situation that does not exist, such an assumption of ideal flow produces reasonably accurate results in flow systems where friction does not play a significant role. In such ideal flow situations, energy losses that are converted into heat due to friction represent a small percentage of the total energy of the fluid flow. Examples of situations where the assumption of ideal flow include fluids with a low viscosity, where the pipe or channel lengths are relatively short and the pipe diameters and the valve and fitting sizes are large enough to sufficiently handle the discharges. In an ideal flow situation, because there is no friction ($\mu = 0$ and thus $\tau = 0$), flow energy (flow work, or pressure energy) is converted to kinetic energy (or vice versa), with no energy lost to heat.

As noted above, while the energy equation derived based on the conservation of momentum principle given by Equation 4.44 does not consider the modeling of energy transfer by work (pumps, turbines, etc.), the energy equation derived based on the conservation of energy principle given by Equation 4.181 does and is given as follows:

$$\left(\frac{p_1}{\gamma} + \frac{v_1^2}{2g} + z_1\right) + h_{pump} = \left(\frac{p_2}{\gamma} + \frac{v_2^2}{2g} + z_2\right) + h_{turbine} + h_{f,total} \tag{4.50}$$

For the assumption of incompressible, steady, ideal (inviscid), and no shaft work (due to a pump or a turbine), the friction head loss term, and the pump and turbine head terms are set to zero and thus the energy equation applied between two points along a streamline results in the following equation:

$$\left(\frac{p}{\gamma} + z + \frac{v^2}{2g}\right)_1 - \left(\frac{p}{\gamma} + z + \frac{v^2}{2g}\right)_2 = 0 \tag{4.51}$$

which is known as the Bernoulli equation, and states the total head, H along a streamline is constant as follows:

$$H = \frac{p}{\gamma} + z + \frac{v^2}{2g} = cons \tan t \tag{4.52}$$

where the pressure head, p/γ represents the height of a column of fluid necessary to produce a pressure, p; the elevation head, z represents the potential energy of the fluid particle; and the velocity head, $v^2/2g$ represents the vertical distance needed for the fluid to fall freely in order to reach a velocity, v from rest, for instance, such as a flow from a tank. Thus, depending upon the ideal flow situation (internal, external, or flow from a tank), one or two forms of energy can be converted to another form(s) of energy. Furthermore, in addition to expressing the Bernoulli equation in terms of energy head terms, it may be expressed in terms of pressure terms, or energy terms, as discussed below.

The Bernoulli equation expressed in terms of pressure terms is given as follows:

$$\left(\underbrace{p}_{F_P} + \underbrace{\gamma z}_{F_G} + \underbrace{\frac{\rho v^2}{2}}_{F_I = Ma}\right)_1 - \left(\underbrace{p}_{static\,press} + \underbrace{\gamma z}_{hydrsta\,press} + \underbrace{\frac{\rho v^2}{2}}_{dyn\,pres}\right)_2 = \overbrace{\underbrace{\frac{\tau_w L}{R_h}}_{F_V}}^{flow\,res} = 0 \tag{4.53}$$

Because the viscous force is assumed to be negligible in comparison to the pressure, gravitational, and inertial forces, the Bernoulli equation provides an approximate relationship between the pressure, elevation, and velocity. The static pressure, p, which does not incorporate any dynamic effects, incorporates the actual thermodynamic pressure of the fluid. The hydrostatic pressure, γz accounts for the elevation effects (the effects of the fluid weight on the pressure). The dynamic pressure, $\rho v^2/2$ represents the pressure rise when the fluid in motion is brought to a stop, while it represents a pressure drop when the fluid is set in motion as it is released from a reservoir or a tank, for instance. Thus, the sum of the static pressure, the hydrostatic pressure, and the dynamic pressure is called the total pressure, which is constant along a streamline, as implied by the Bernoulli equation.

The Bernoulli equation expressed in terms of energy terms states that the sum of the flow energy (pressure energy stored in p), potential energy stored in γz, and kinetic energy of a fluid particle stored in $\rho v^2/2$ along a streamline is constant (no head loss for an ideal fluid) as follows:

$$\underbrace{\frac{p}{\rho}}_{Flow\ energy} + \underbrace{gz}_{Potential\ Energy} + \underbrace{\frac{v^2}{2}}_{Kinetic\ Energy} = cons\tan t \tag{4.54}$$

This implies that, for instance, the kinetic energy and the potential energy of the fluid can be converted to flow energy (pressure energy) and vice versa, thus causing the pressure to change. Assuming that $z_1 = z_2$, if the kinetic energy is converted to pressure energy, then there will be a pressure rise (as in the case of a pitot tube, the flow around a submerged body, or the flow into a reservoir or a pipe expansion where the fluid is brought to a stop). If, however, pressure energy is converted to kinetic energy, then there will be a pressure drop (as in the case of the flow from a reservoir or a tank, the flow through a pipe, and most pipe devices, where the fluid is set into motion).

While most applications of the Bernoulli equation illustrate the conversion of one form of energy to another, its simplest application illustrates no conversion of energy forms. Assume incompressible, steady, and ideal flow in a constant-diameter, horizontal pipe, as illustrated in Figure 4.4. Piezometric tubes are installed at points 1 and 2 in order to measure the pressure. The Bernoulli equation is applied to model the "change in pressure, Δp" between points 1 and 2 in the pipe as follows:

$$\frac{p_1}{\gamma} + z_1 + \frac{v_1^2}{2g} = \frac{p_2}{\gamma} + z_2 + \frac{v_2^2}{2g} \tag{4.55}$$

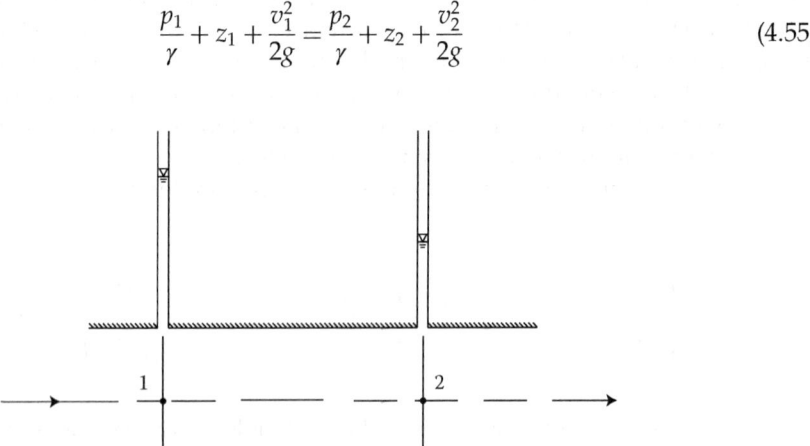

FIGURE 4.4
Horizontal pipe flowing under pressure with piezometric tubes to measure the pressure at points 1 and 2.

Because the pipe has a constant diameter, application of the continuity equation yields a constant velocity between points 1 and 2 as follows:

$$v_1 A_1 = v_2 A_2$$

$$v_1 = v_2 \tag{4.56}$$

Furthermore, because the pipe is horizontal, $z_1 = z_2$. Thus, application of the Bernoulli equation yields the following:

$$\frac{p_1}{\gamma} = \frac{p_2}{\gamma} \tag{4.57}$$

which implies that there is no "change in pressure, Δp" between points 1 and 2 in the pipe. In the assumption of real flow, the viscosity introduces friction due to the viscous shearing effects, which would result in a pressure drop, Δp between points 1 and 2 in the pipe, as would be illustrated by an actual difference in the piezometric heights at points 1 and 2. However, in the assumption of ideal flow, as modeled by the Bernoulli equation, there is no viscosity and thus there is no pressure drop, Δp between points 1 and 2 in the pipe. Therefore, if the actual difference in pressure as measured by the difference in piezometric heights at points 1 and 2 is small in comparison to the magnitude of pressure at points 1 and 2, then one can assume that the viscous force is negligible in comparison to the pressure, gravitational, and inertial forces. Finally, in its simplest application, the Bernoulli equation illustrates no conversion of energy forms.

4.5.3.1 Applications of the Bernoulli Equation

Application of the Bernoulli equation represents the assumption of ideal flow, which may be generally classified as internal flow (pipe or open channel), external flow around an object, or flow from a tank (or a water source). As such, there are numerous applications of the Bernoulli equation, which include: (1) flowrate measurement for pipe flow using ideal flow meters (ideal pipe flow); (2) flowrate measurement for open channel flow using ideal flow meters (ideal open channel flow); (3) gradual (vertical or horizontal) channel contractions/transitions (ideal open channel flow); 4) velocity (and pressure) measurement for both internal and external flow using a pitot-static tube (ideal pipe flow, ideal open channel flow, and ideal external flow); and (5) flow from a tank (or a water source) through a jet, siphon, or some other opening/connection to the tank (ideal flow from a tank). Application of the Bernoulli equation for velocity (and pressure) measurement for both internal and external flow using a pitot-static tube demonstrates how the dynamic pressure is modeled by a pressure rise, as presented in Section 4.5.3.2. The use of several pitot-static tubes along a pipe or a channel section allows schematic illustrations of the energy equation, which are known as the energy grade line, *EGL*, and the hydraulic grade line, *HGL*. Furthermore, application of the Bernoulli equation for ideal flow meters for internal flow, for ideal gradual channel contraction, and for ideal flow from a tank demonstrates how the dynamic pressure is modeled by a pressure drop, as presented in Section 4.5.3.3. Illustrations of the *EGL* and *HGL* are presented in Sections 4.5.3.4 and 4.5.3.5, while example problems illustrating the applications of the Bernoulli equation are presented in Sections 4.5.6 and 4.5.7.

4.5.3.2 The Pitot-Static Tube: Dynamic Pressure Is Modeled by a Pressure Rise

An application of the Bernoulli equation is the pitot-static tube (a combination of a pitot tube and a piezometer/static tube), which is a velocity-measuring device. The pitot-static tube may be used to measure the velocity (and pressure) for both internal and external flow. Assuming no head loss and that the fluid is momentarily brought to a stop at the downstream point (stagnation pressure), the dynamic pressure at the upstream point represents the ideal pressure rise (from static pressure to stagnation pressure). This application of Bernoulli's equation deals with the stagnation and the dynamic pressures and illustrates that these pressures arise from the conversion of kinetic energy (dynamic pressure) in a flowing fluid into an ideal pressure rise (stagnation pressure minus static pressure) as the fluid is momentarily brought to a rest. The downstream point at the pitot tube is the point of stagnation, which represents the conversion of all of the kinetic energy into an ideal pressure rise.

While the pitot-static tube may be used to measure the velocity (and pressure) for both internal and external flow (see Section 4.5.7.2), the following discussion assumes internal (pipe) flow. Figure 4.5 illustrates a horizontal pipe with a piezometer installed at point 1 and a pitot tube installed at point 2. The flow is from point 1 to point 2. One may note that both the piezometer and the pitot tube are pressure measuring devices. The piezometer is installed at the top of the pipe at point 1 and the rise of the fluid in the piezometer, p_1/γ measures the pressure at point 1. The piezometer does not interfere with the fluid flow and thus allows the fluid to travel without any change in the velocity, v_1. The pitot tube is installed at the top of the pipe at point 2 and continues inward towards the centerline

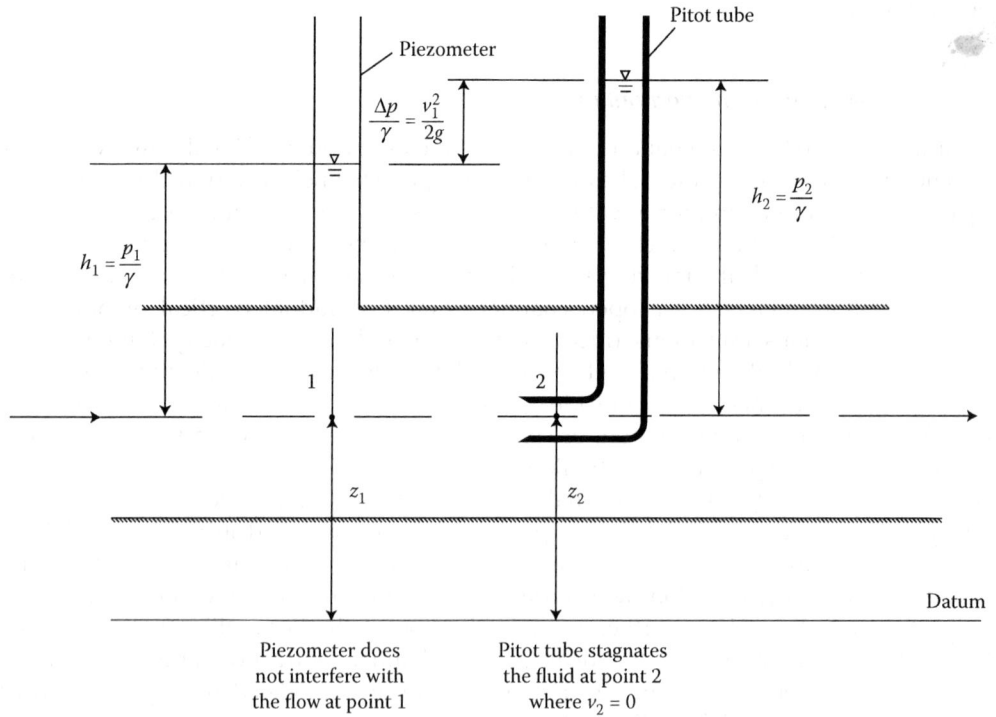

FIGURE 4.5
Pitot-static tube: used to measure the velocity (and pressure rise) for a pipe flowing under pressure.

of the pipe and then makes a 90° turn pointing upstream into the direction of flow. As a result, the fluid enters the pitot tube and momentarily stops (it stagnates), where $v_2 = 0$. The rise of the fluid in the pitot tube, p_2/γ measures the pressure at point 2. One may note that because the pitot tube stagnates the fluid at point 2, it registers a higher pressure than the piezometer does at point 1. As illustrated in Figure 4.5, the rise in pressure head, $\Delta p/\gamma$ is defined as follows:

$$\frac{\Delta p}{\gamma} = \frac{p_2}{\gamma} - \frac{p_1}{\gamma} \tag{4.58}$$

The Bernoulli equation between points 1 and 2 expressed in terms of pressure is given as follows:

$$p_1 + \gamma z_1 + \frac{\rho v_1^2}{2} = p_2 + \gamma z_2 + \frac{\rho v_2^2}{2} \tag{4.59}$$

Assuming that $z_1 = z_2$ and noting that $v_2 = 0$:

$$\underbrace{p_1}_{static\ press} + \underbrace{\frac{\rho v_1^2}{2}}_{dyn\ press} = \underbrace{p_2}_{stagn\ press} \tag{4.60}$$

where the sum of the static pressure (where the fluid is not slowed down) and the dynamic pressure equals the stagnation pressure (where the fluid is stagnated/momentarily slowed down), the static pressure, p_1 is measured by the piezometer; the stagnation pressure, p_2 is measured by the pitot tube; and the dynamic pressure, $\rho v_1^2/2$ is measured by the difference in the fluid levels in the pitot tube and the piezometer as follows:

$$\frac{v_1^2}{2g} = \frac{\Delta p}{\gamma} = \frac{p_2}{\gamma} - \frac{p_1}{\gamma} \tag{4.61}$$

thus, the dynamic pressure, $\rho v_1^2/2$ is given as follows:

$$\underbrace{\frac{\rho v_1^2}{2}}_{dyn\ press} = \underbrace{p_2}_{stagn\ press} - \underbrace{p_1}_{static\ press} = \underbrace{\Delta p}_{press\ rise} \tag{4.62}$$

Assuming no head loss and that the fluid is momentarily brought to a stop at the downstream point 2 where $v_2 = 0$, the velocity at the upstream point 1 is ideal, where $v_1 = v_i$, and the dynamic pressure at the upstream point, $\rho v_1^2/2$ represents the ideal pressure rise (from static pressure to stagnation pressure). Therefore, isolating the ideal upstream velocity, v_i defines the ideal velocity as follows:

$$v_i = \underbrace{\sqrt{\frac{2}{\rho}(\Delta p)}}_{ideal\ vel} = \underbrace{\sqrt{\frac{2}{\rho}\left(\underbrace{p_2}_{stag\ press} - \underbrace{p_1}_{static\ press}\right)}}_{ideal\ vel} \tag{4.63}$$

where the difference in the fluid levels, $\Delta p / \gamma = v_1^2 / 2g$ in the pitot tube and the piezometer is used to determine the unknown velocity, v_1 in the pipe, as illustrated in Figure 4.5. It is important to note that in the above illustration of the pitot-static tube, there was a rise in pressure, Δp, which was defined as follows:

$$\underbrace{\Delta p}_{press\ rise} = \underbrace{p_2}_{stag\ press} - \underbrace{p_1}_{static\ press} \tag{4.64}$$

Furthermore, the flowrate, Q is determined by applying the continuity equation, $Q_i = v_i A$. Note that in real flow, the actual velocity and the actual flowrate are less than the ideal velocity and the ideal flowrate, respectively. Thus, a velocity coefficient and a discharge coefficient (see Section 4.5.6), are applied, respectively.

4.5.3.3 Ideal Flow Meters, Ideal Gradual Channel Contractions, and Ideal Flow from a Tank: Dynamic Pressure Is Modeled by a Pressure Drop

While the conversion of the kinetic energy to pressure energy results in a pressure rise (as in the cases of a pitot tube, the flow around a submerged body, or the flow into a reservoir, a pipe or channel expansion, where the fluid is brought to a stop), the conversion of the pressure energy to kinetic energy results in a pressure drop (as in the cases of the flow from a reservoir or a tank, a pipe or channel contraction, the flow through a pipe or channel, and most pipe and open channel flow measuring devices, where the fluid is set in motion). Thus, another set of applications of the Bernoulli equation are ideal flow meters for internal flow, ideal gradual channel contractions, and ideal flow from a tank. Two piezometer tubes may be used to measure the velocity (and pressure) for both internal flow and flow from a tank. An illustration of a pressure drop is illustrated for the venturi tube (an ideal flow meter) presented in Figure 4.6. The Bernoulli equation between points 1 and 2 expressed in terms of pressure is given as follows:

$$p_1 + \gamma z_1 + \frac{\rho v_1^2}{2} = p_2 + \gamma z_2 + \frac{\rho v_2^2}{2} \tag{4.65}$$

Assuming that $z_1 = z_2$ and noting that $v_1 \cong 0$, while the downstream velocity, v_2 defines the ideal velocity, yields:

$$\underbrace{p_1}_{static\ press} = \underbrace{p_2}_{stagn\ press} + \underbrace{\frac{\rho v_2^2}{2}}_{dyn\ press} \tag{4.66}$$

The upstream stagnation pressure, p_1 is measured by a piezometer, and the downstream static pressure, p_2 is measured by a piezometer, where the pressure drop, Δp is defined as follows:

$$\underbrace{\frac{\rho v_2^2}{2}}_{dyn\ press} = \underbrace{p_1}_{stagn\ press} - \underbrace{p_2}_{static\ press} = \underbrace{\Delta p}_{press\ drop} \tag{4.67}$$

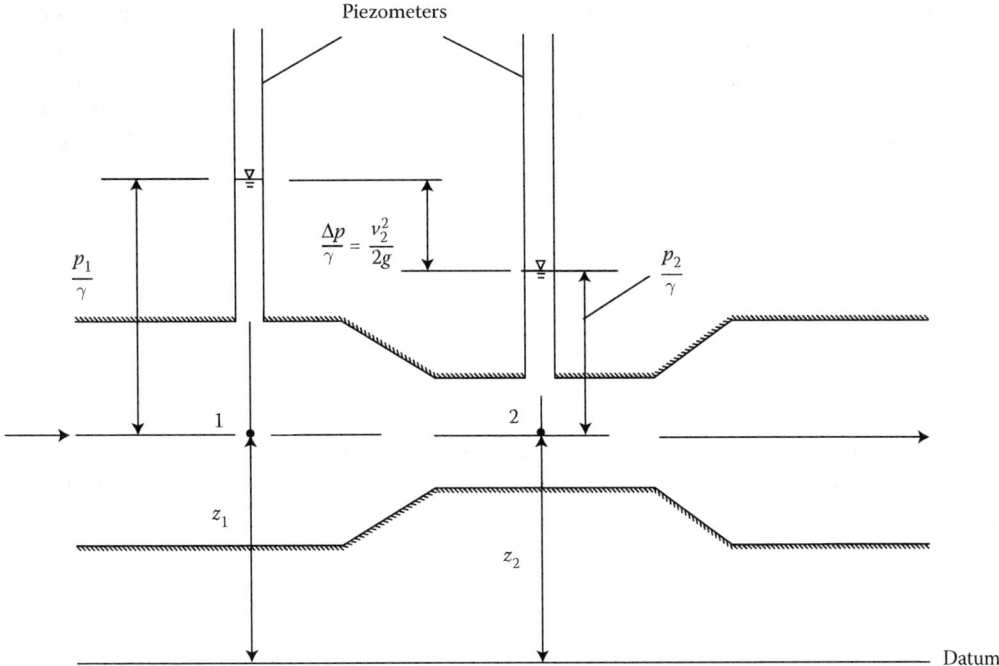

FIGURE 4.6
Venturi tube (ideal flow meter): causes a pressure drop in a pipe flow under pressure.

Thus, the pressure drop, Δp is exactly equal to the dynamic pressure, $\rho v^2 / 2$ (and not more), which is considered to be the ideal pressure drop. And the actual velocity, v_2 is exactly equal to the ideal velocity, v_i (and not less).

$$v_2 = \underbrace{\sqrt{\frac{2}{\rho}(\Delta p)}}_{ideal\ vel} \tag{4.68}$$

Furthermore, the flowrate, Q is determined by applying the continuity equation, $Q_i = v_i A$. Note that in real flow, the actual velocity and the actual flowrate are less than the ideal velocity and the ideal flowrate, respectively. Thus, a velocity coefficient and a discharge coefficient (see Section 4.5.6), are applied, respectively. Finally, one may note that depending upon the particular flow-measuring device or the flow from a particular tank, the expression for the ideal velocity that results from the application of the Bernoulli equation will slightly vary, as illustrated in example problems presented in Section 4.5.6 (and in Chapters 8 and 9).

4.5.3.4 The Energy Grade Line (EGL) and the Hydraulic Grade Line (HGL)

The energy grade line, *EGL* and hydraulic grade line, *HGL* are a schematic illustration of the energy equation and may be defined using several pitot-static tubes along a pipe or channel section. In general, the *EGL* and *HGL* will vary as a function of the following flow assumptions: (1) real or ideal flow (pipe or open channel flow), (2) pipe or open channel flow (real or ideal flow), and (3) a pump or turbine in real pipe flow. As illustrated above, the pitot-static tube (a combination of pitot tube and a piezometer/static tube), is a velocity-measuring device. Specifically, application of the Bernoulli equation at the centerline of the pipe (see Figure 4.5) between the piezometer at point 1 and the pitot tube at point 2 provided an expression for the ideal velocity at point 1 as given by Equation 4.63, and is repeated as follows:

$$\underbrace{v_1 = \sqrt{\frac{2}{\rho}(\Delta p)}}_{ideal\ vel} = \underbrace{\sqrt{\frac{2}{\rho}\left(\underbrace{p_2}_{stag\ press} - \underbrace{p_1}_{static\ press}\right)}}_{ideal\ vel} \tag{4.69}$$

The piezometer at point 1 measures the static pressure, p_1. The pitot tube at point 2 measures the stagnation pressure, p_2, which is equal to the static pressure, p_1 plus the dynamic pressure, $\rho v_1^2/2$ as follows:

$$\underbrace{p_2}_{stag\ press} = \underbrace{p_1}_{static\ press} + \underbrace{\frac{\rho v_1^2}{2}}_{dyn\ press} \tag{4.70}$$

Therefore, the height of the fluid in the piezometer at point 1 represents the pressure head at point 1, p_1/γ. And, the height of the fluid in the pitot tube at point 2 represents the pressure head at point 1, p_1/γ plus the velocity head at point 1, $v_1^2/2g$. Finally, the elevation head at point 1, z_1 is measured relative to an arbitrary fixed datum. Therefore, the total head at point 1, H_1 with respect to the fixed datum is given as follows:

$$H_1 = z_1 + \frac{p_1}{\gamma} + \frac{v_1^2}{2g} \tag{4.71}$$

where z_1 is measured from an arbitrary fixed datum to the centerline of the pipe, p_1/γ is measured by the height of the fluid in the piezometer at point 1, and $[(p_1/\gamma) + (v_1^2/2g)]$ is measured by the height of the fluid in the pitot tube at point 2. Furthermore, the static head (also known as the piezometric head) at point 1 with respect to the fixed datum is given as follows:

$$H_1 - \frac{v_1^2}{2g} = z_1 + \frac{p_1}{\gamma} \tag{4.72}$$

Thus, while the piezometer measures the static pressure at point 1, the pitot tube measures the static pressure plus the dynamic pressure at point 1.

The *EGL* and *HGL* will differ depending upon whether real or ideal flow (pipe or open channel flow) is assumed. Assuming real flow (fluid with viscosity), the total head, H and the piezometric head at several locations along the pipe may be determined using several pitot-static tubes, as illustrated in Figure 4.7. The line drawn between the total head, H (using the pitot tube) at the locations along the pipe is defined as the energy grade line, *EGL*. And, the line drawn between the piezometric head (using the

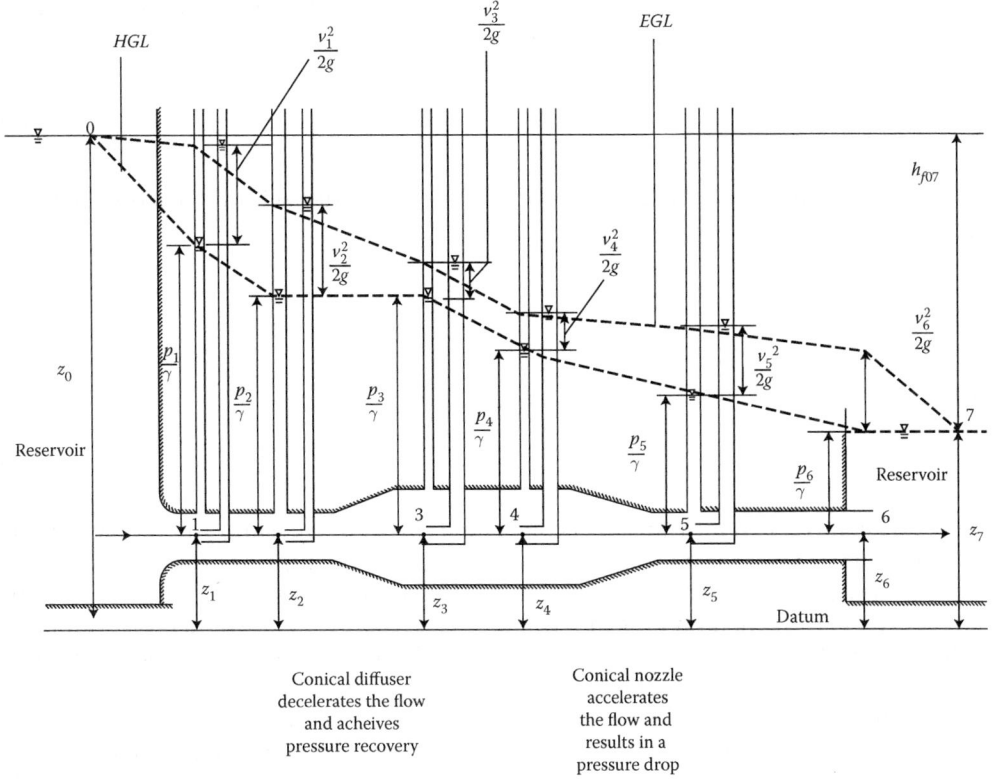

FIGURE 4.7
Energy grade line, *EGL*, and the hydraulic grade line, *HGL* are determined using several pitot-static tubes for a real flow in a pipe with components between two reservoirs.

piezometer) at the locations along the pipe is defined as the hydraulic grade line, *HGL*, where:

$$EGL - \frac{v^2}{2g} = HGL \qquad (4.73)$$

Although the *HGL* will increase and decrease along with the *EGL*, it cannot exceed the *EGL*. The *EGL* and the *HGL* converge as the velocity decreases and diverge as the velocity increases between two points; thus, the height of the *HGL* increases as the velocity decreases, while its height decreases as the velocity increases. In this case of real flow, the pressure drop measured by the piezometers and pitot tubes between the various locations along the pipe may be due to either an increase in velocity at a pipe constriction and/or flow resistance (pipe friction and or/ pipe components). However, the pressure rise (pressure recovery) measured by the piezometers or pitot tubes between the various locations along the pipe for real flow are due to a decrease in velocity at a pipe expansion. Furthermore, for real flow, the flow resistance (pipe friction and/or pipe components) causes the *EGL* and thus the *HGL* to decrease over the pipe or channel section in the direction of the flow, regardless of the existence of a pipe constriction. The decrease in the *EGL* between two points in the pipe or channel section for real flow is represented by the decreasing slope of the *EGL*, which is defined as follows:

$$S_e = \frac{h_e}{L} = \frac{H_1 - H_2}{L} \tag{4.74}$$

Furthermore, because $h_e = h_f$, the slope of the *EGL* coincides with the friction slope, S_f, where:

$$S_f = \frac{h_f}{L} = \frac{\tau_w}{\gamma R_h} \tag{4.75}$$

and thus

$$S_e = -S_f \tag{4.76}$$

where the negative sign indicates the slope of the *EGL* is decreasing, and the slope of the *EGL* (and thus the *HGL*) is a measure of the head loss in the pipe or channel section. The head loss may be due to either pipe friction or pipe components; pipe friction results in a gradual drop in the *EGL* (and thus the *HGL*) between two points, while a pipe component results in a sudden drop in the *EGL* (and thus the *HGL*) at the location of the component. Finally, it is important to note that the *EGL* (and thus the *HGL*) cannot increase along the pipe section in the direction of flow unless mechanical energy is added by a pump, h_{pump} (see Section 4.7.4). Figure 4.7 illustrates several important points regarding the variation of the *EGL* and the *HGL*. At the stationary free surface of the liquid in the tank at point 0, the *EGL* and the *HGL* coincide with the free surface of the liquid, where:

$$EGL = HGL = z \tag{4.77}$$

where the velocity head, $v_0^2/2g = 0$ and the pressure head, $p_0/\gamma = 0$ (open to the atmosphere). At the pipe entrance at point 1, although the sudden increase in velocity from zero in the tank to nonzero in the pipe causes the *HGL* to suddenly decrease, a well-rounded pipe entrance (pipe component) results in a very small minor head loss and thus a very gradual decrease in the *HGL*. At point 2, although there is no change (decrease) in pipe diameter, the *EGL* (and thus the *HGL*) drops due to pipe friction, where $v_2^2/2g = v_1^2/2g$. At point 3, although there is an initial recovery of pressure due to the gradual pipe expansion (conical diffuser), this is followed by a pressure drop due to pipe friction and the diffuser (pipe component), thus causing an overall drop in the *EGL* (and thus the *HGL*), where $v_3^2/2g < v_2^2/2g$ due to the pipe expansion. At point 4, although there is no change (decrease) in pipe diameter, the *EGL* (and thus the *HGL*) drops due to pipe friction, where $v_4^2/2g = v_3^2/2g$. At point 5, there is a pressure drop due to the gradual pipe contraction (conical nozzle) and pipe friction and an increase in velocity due to the nozzle. At the reentrant pipe exit at point 6, there is gradual decrease in the EGL due to pipe friction and the pipe exit. Finally, at point 7, there is a sudden decrease in velocity from nonzero in the pipe to zero in the tank, causing the *HGL* to suddenly decrease, where the *HGL* coincides with the free surface of the liquid in the tank.

Assuming ideal flow (inviscid) the total head, *H* and the piezometric head at several locations along the pipe may be determined using several pitot-static tubes, as illustrated

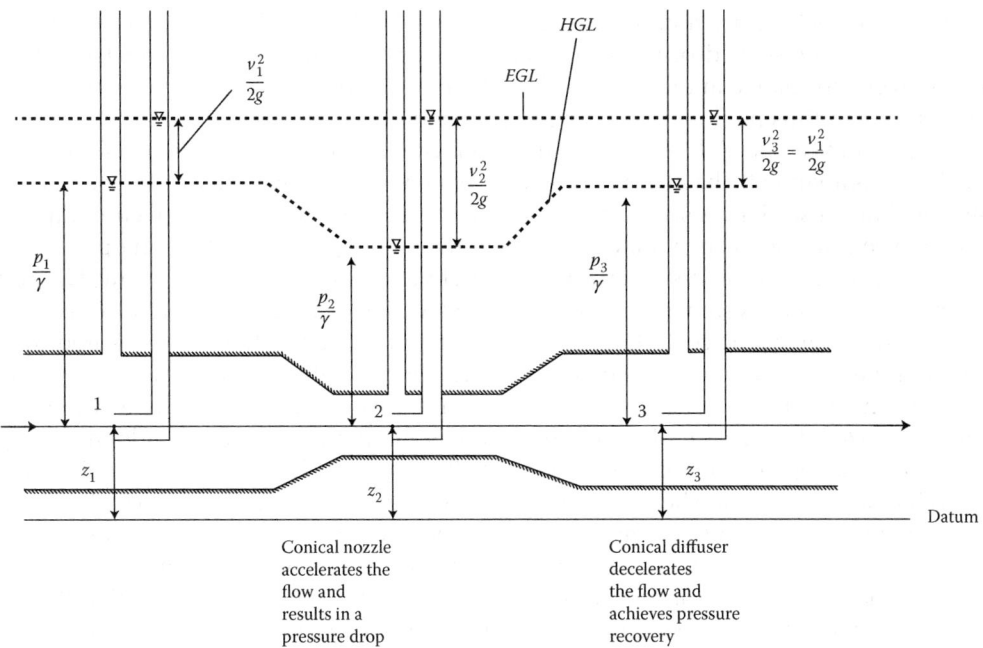

FIGURE 4.8
Energy grade line, *EGL* and the hydraulic grade line, *HGL* are determined using several pitot-static tubes for an ideal flow in a pipe with components.

in Figure 4.8. The line drawn between the total head, H (using the pitot tube) at the locations along the pipe is defined as the energy grade line, *EGL*. And, the line drawn between the piezometric head (using the piezometer) at the locations along the pipe is defined as the hydraulic grade line, *HGL*, where:

$$EGL - \frac{v^2}{2g} = HGL \qquad (4.78)$$

where the *EGL* and the *HGL* converge as the velocity decreases and diverge as the velocity increases between two points; thus, the height of the *HGL* increases as the velocity decreases, while its height decreases as the velocity increases. In this case of ideal flow, the pressure drop measured by piezometers and pitot tubes between the various locations along the pipe is due to an increase in velocity at a pipe constriction. And, the pressure rise measured by piezometers and pitot tubes between the various locations along the pipe for ideal flow is due to a decrease in velocity at a pipe expansion. Therefore, for ideal flow, because there is no flow resistance (pipe friction and or/pipe components) modeled, the *EGL* does not decrease over the pipe or channel section; rather, it remains a constant horizontal line parallel to the fixed datum. If the velocity is a constant, then the *HGL* will also remain a constant horizontal line, lower than the *EGL* and parallel to the fixed datum However, the *HGL* will decrease over the pipe or channel section if there is an increase in velocity at a pipe constriction. And, the *HGL* will increase over the pipe or channel section if there is a decrease in velocity at a pipe expansion. Practical instances of such ideal flow include ideal flow meters for both pipe and open channel flow or gradual channel constrictions; in such instances, a pipe or a channel constriction will cause an increase in the velocity and thus a decrease in pressure (i.e., a pressure drop), which will result in a flow in the pipe or channel (otherwise, if there were no change in the cross section, there would be no change in velocity, no pressure drop, and thus no flow in the pipe or channel).

The *EGL* and *HGL* will differ depending upon whether pipe or open channel flow (real or ideal flow) is assumed. The *EGL* and *HGL* are defined between two consecutive points in the pipe or channel section. A schematic illustration of the energy equation given in Equation 4.44 between two points in a finite control volume for a pipe flow and for an open channel flow are presented in Figures 4.9 and 4.10, respectively. Assuming real pipe flow, Figure 4.9 illustrates that for a given point in the pipe, the elevation head, z is measured from an arbitrary fixed datum to the centerline of the pipe; the pressure head, p/γ is measured by the height of the fluid in the piezometer at the given point; and the velocity head, $v^2/2g$ is measured by the difference in height between the piezometer and the pitot tube at the given point. The *HGL* in the case of a pipe flow represents the piezometric heads in the pipe. Assuming real open channel flow, Figure 4.10 illustrates that for a given point in the open channel, the elevation head, z is measured from an arbitrary fixed datum to the channel bottom; the pressure head, y is measured by the height of the fluid in the piezometer at the given point; and the velocity head, $v^2/2g$ is measured by the difference in height between the piezometer and the pitot tube at the given point. It is interesting to note that the *HGL* in the case of the open channel flow, which represents the piezometric heads in the channel, coincides with the free surface of the liquid.

The *EGL* and *HGL* will differ depending upon whether a pump or a turbine is assumed in real pipe flow. A schematic illustration of the energy equation given in Equation 4.181

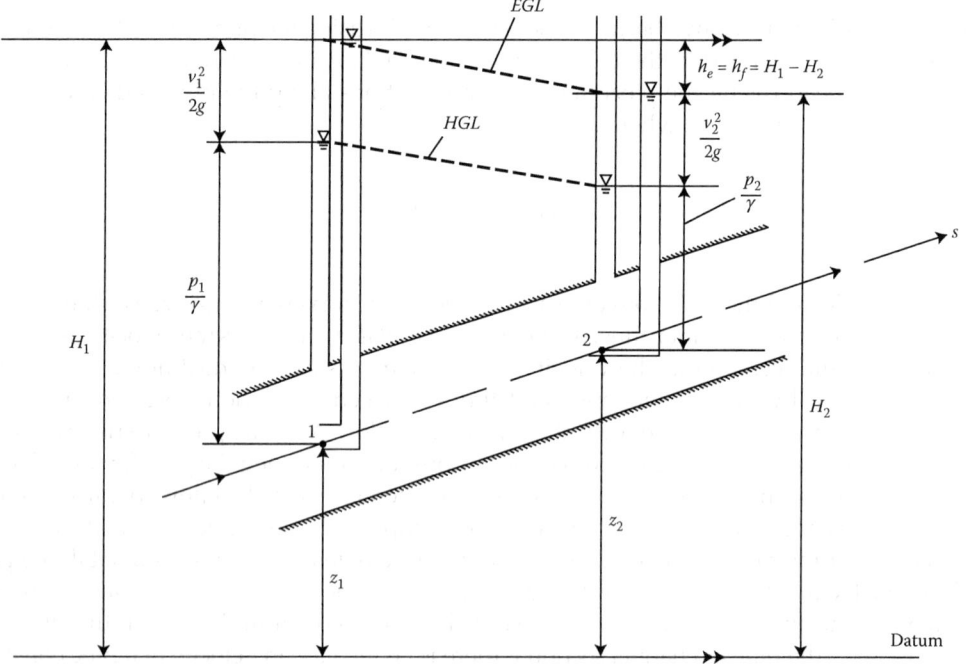

FIGURE 4.9
Schematic illustration of the energy equation, the energy grade line, *EGL* and the hydraulic grade line, *HGL* for an incompressible, real, steady pipe flow.

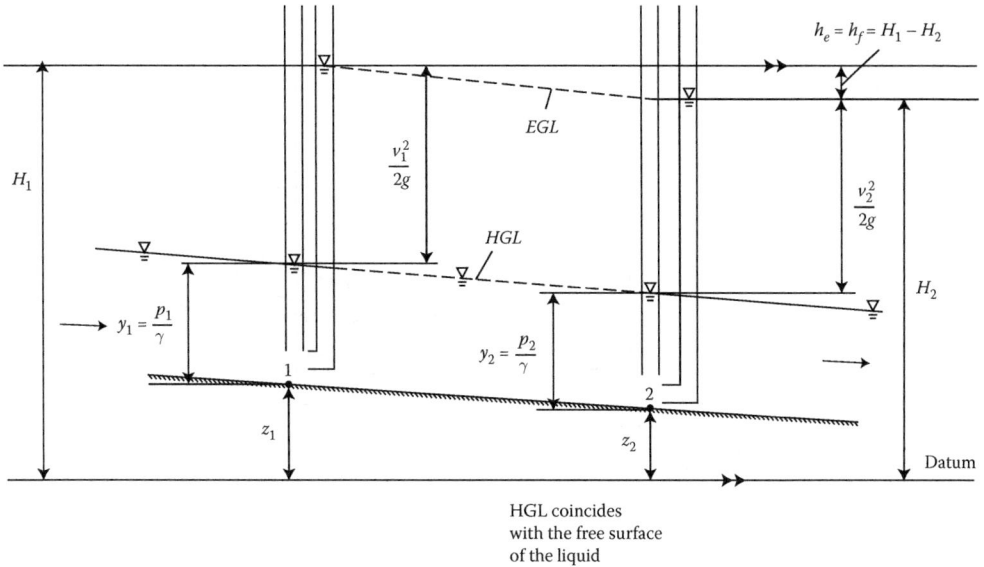

$$h_e = h_f = H_1 - H_2$$

HGL coincides
with the free surface
of the liquid

FIGURE 4.10
Schematic illustration of the energy equation, the energy grade line, *EGL* and the hydraulic grade line, *HGL* for an incompressible, real, steady open channel flow.

between two points in a finite control volume for pipe flow with both a pump and a turbine is presented in Figure 4.11 (see Section 4.6 for applications). The pump adds mechanical energy, h_{pump}, which causes a jump in both the *EGL* and the *HGL*. Additionally, the turbine removes mechanical energy, $h_{turbine}$, which causes a drop in both the *EGL* and the *HGL*. Furthermore, Figure 4.11 illustrates the gradual decrease in both the *EGL* and the *HGL* due to the friction head loss, h_f for the real pipe flow.

4.5.3.5 The Hydraulic Grade Line (HGL) and Cavitation

The *HGL* represents the piezometric heads (measured by the use of a piezometer). The *HGL* coincides with the free surface of the liquid, where the pressure head, $p/\gamma = 0$ (open to the atmosphere; thus, the gage pressure is equal to zero). Thus, the gage pressure of the fluid below the *HGL* is greater than zero (positive), while the gage pressure of the fluid above the *HGL* is less than zero (or negative). For open channel flow, the *HGL* coincides with the free surface of the liquid along the open channel flow. For pipe flow, the *HGL* coincides with the free surface of the liquid in the tank/reservoir just prior to the pipe entrance and with the free surface of the liquid at the pipe outlet. Furthermore, for pipe flow, the *HGL* coincides with the free surface of the liquid created by a suction line due to a siphon (see Figure 4.12) or a pump (see Figure 4.13), for instance. Therefore, drawing the *HGL* for a pipe system allows one to determine the regions of negative pressure and thus the regions of potential cavitation. Cavitation is defined as unintended boiling of a liquid in motion that occurs at room temperature when the pressure has decreased to its the vapor pressure, p_v. Cavitation becomes important when very low pressure is a result of fluid dynamics such as a siphon, the suction side of a pump or a turbine, or the flow through complex passages in valves.

A significant decrease in pressure for a liquid introduces the liquid property, vapor pressure, p_v, and an undesirable fluid phenomena called cavitation (see Chapter 1). In particular,

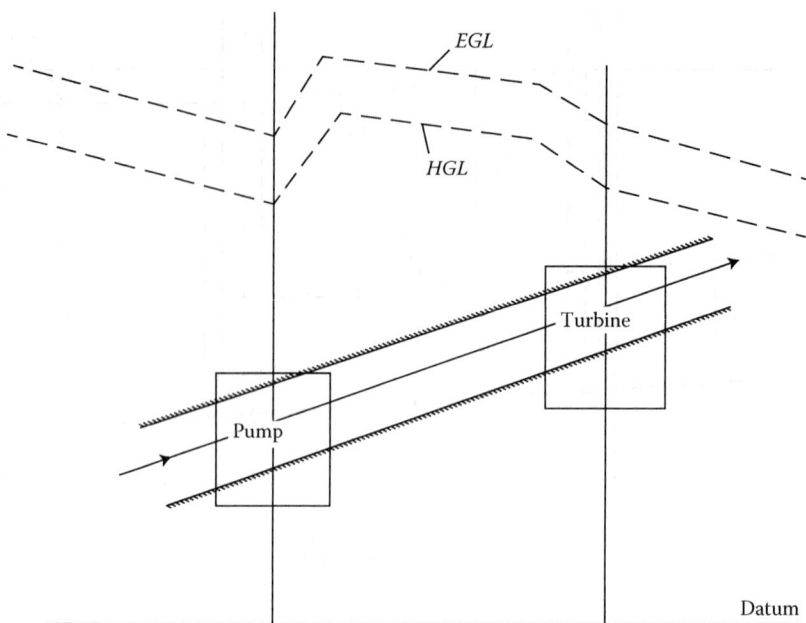

FIGURE 4.11
Schematic illustration of the energy equation, the energy grade line, *EGL* and the hydraulic grade line, *HGL* for an incompressible, real, steady pipe flow with a pump and a turbine.

the vapor pressure, p_v of a liquid becomes important when very low pressures are involved in both fluid statics and fluid flow/dynamics. The vapor pressure, p_v of a liquid is measured in absolute pressure, where absolute pressure is the sum of the atmospheric pressure and the gage pressure (see Chapter 2). As such, the vapor pressure, p_v of a liquid is measured relative to the atmospheric pressure. Furthermore, a "very low pressure" for a fluid is an indication

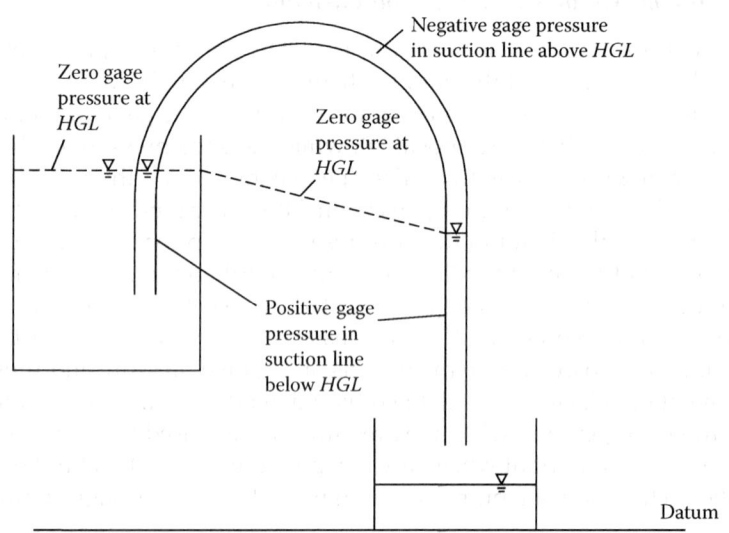

FIGURE 4.12
Hydraulic grade line, *HGL* coincides with the free surface of the liquid created by a suction line due to a siphon.

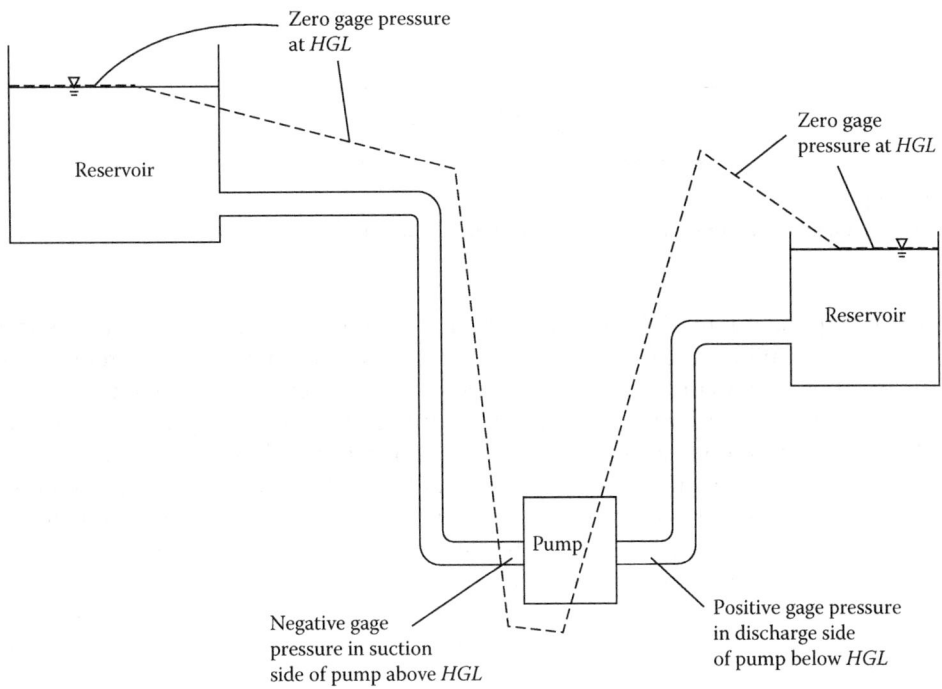

FIGURE 4.13
Hydraulic grade line, *HGL* coincides with the free surface of the liquid created by a suction line due to a pump.

of its absolute pressure relative to the atmospheric pressure (zero gage pressure). Therefore, a "very low pressure" for a fluid indicates a negative gage pressure. The definition and attainment of the vapor pressure, p_v of a liquid involves a description of the phenomena of evaporation, condensation, and boiling; the vapor pressure, p_v of a liquid increases with an increase in temperature, as illustrated in Table A.2 in Appendix A for water. Therefore, at room temperature, where the vapor pressure, p_v of a liquid is much lower than the atmospheric pressure, the gage pressure of the liquid is negative.

Cavitation is modeled by the Bernoulli equation, and is known as the "Bernoulli effect." In general fluid flow situations, if the differences in velocity (or elevation) are significant, then the differences in pressure can also be significant. In the flow of a liquid, regions of high velocity (or high elevation) and thus low pressure can cause cavitation of the liquid. Therefore, cavitation may occur in a flowing liquid if the local pressure is reduced to the vapor pressure, p_v of the liquid. The reduction in pressure to the vapor pressure, p_v can occur because of a significant increase in velocity (due to a reduction in cross-sectional area) or a significant increase in elevation (in a suction line of a pump or a turbine, or a siphon tube lifting water). Thus, for instance, when the diameter of a pipe is significantly reduced, as illustrated in Figure 4.14 (same as Figure 1.12) (see Example Problem 4.5), the velocity increases (in accordance with the continuity equation presented in Chapter 3), while the corresponding pressure decreases in the pipe constriction (in accordance with the Bernoulli equation). And, for instance, when there is a significant increase in elevation in a suction line of a pump or a siphon tube lifting water, the corresponding pressure decreases (in accordance with the Bernoulli equation). Because the suction/siphon pressure is less than the atmospheric pressure, the pipe or tube can collapse, causing a restriction in the flow path,

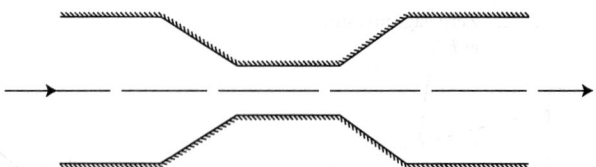

FIGURE 4.14
Cavitation of a liquid due to a significant reduction in pipe diameter.

which can stop the pump/siphon from operating. Therefore, if the pressure at the suction/siphon end of the pipe/tube is decreased to the water's vapor pressure, p_v at room temperature, cavitation will occur. As such, in the design of a siphon, consideration is given to the height of the tubing in order to avoid a pressure drop below the vapor pressure of the liquid and thus avoid cavitation (see Example Problem 4.17). In the design of a pump, consideration is given to making sure that the pressure at the suction side of a pump does not fall below the vapor pressure of the liquid to avoid cavitation (see Example Problems 4.20 and 4.22 in Section 4.7.4.2). Finally, in the design of a turbine, consideration is given to making sure that the pressure at the nozzle of an impulse turbine and the pressure in the turbine-draft tube system of a reaction turbine do not fall below the vapor pressure of the liquid in order to avoid cavitation (see Example Problems 4.25 and 4.27, respectively, in Section 4.7.4.3).

4.5.4 Application of the Energy Equation in Complement to the Momentum Equation

While the application of the continuity equation will always be necessary for internal flow, the energy equation and the momentum equations play complementary roles in the analysis of a given internal or external flow situation. The energy equation and the momentum equations are known as the equations of motion. They play complementary roles because when one of the two equations breaks down, the other may be used to find the additional unknown quantity. It is important to note that, similarly to the continuity equation, the momentum equation may be applied using either the integral (for a finite control volume) or the differential approach. However, application of the energy equation is useful only in the integral approach (for a finite control volume), as energy is work, which is defined over distance, L. In the application of the governing equations (continuity, energy, and momentum), it is important to make the distinction between real flow, ideal flow, and a hydraulic jump. And for ideal flow, it is important to make the distinction between internal (pipe or open channel flow), external flow, and the flow from a tank (jet and siphon flows) (see Table 4.1) (Section 4.5.4.1). Furthermore, in the application of the governing equations, although in some flow cases, one may ignore (or not need to model) the flow resistance, in most practical flow cases in general, one must account for the fluid viscosity and thus model the resulting shear stress and flow resistance. The modeling of flow resistance requires a "subset level" application of the governing equations (Section 4.5.4.2).

4.5.4.1 Application of the Governing Equations

In the application of the governing equations (continuity, energy, and momentum), it is important to make the distinction between real flow, ideal flow, and a hydraulic jump. And for ideal

TABLE 4.1

A Summary of Applications of the Governing Equations

	Real Flow	Ideal Flow	Hydraulic Jump
Applications of Governing Equations:	• Pipe flow resistance • Open channel flow resistance • Pipe components • Pipes with pumps or turbines	• Pipe flow-measuring devices (such as pitot-static tubes, orifice, nozzle, or venturi meters) • Open channel flow-measuring devices (such as pitot-static tubes, sluice gates, weirs, spillways, venturi flumes, and contracted openings) • Gradual channel transitions • Flow from a tank (jet and siphon flows) • External flow/velocity-measuring device	• Hydraulic jump
Integral Energy	$\left(\dfrac{p_1}{\gamma} + \dfrac{v_1^2}{2g} + z_1\right) - h_{f,maj} - h_{f,min} -$ $h_t + h_p = \left(\dfrac{p_2}{\gamma} + \dfrac{v_2^2}{2g} + z_2\right)$	$\left(\dfrac{p_1}{\gamma} + \dfrac{v_1^2}{2g} + z_1\right) = \left(\dfrac{p_2}{\gamma} + \dfrac{v_2^2}{2g} + z_2\right)$	$\left(\dfrac{p_1}{\gamma} + \dfrac{v_1^2}{2g} + z_1\right) - h_{f,maj} =$ $\left(\dfrac{p_2}{\gamma} + \dfrac{v_2^2}{2g} + z_2\right)$
Used to Solve for:	p/γ (or y), z, $v^2/2g$, $h_{f,min}$, $h_{f,maj}$, h_t or h_p	p/γ (or y), z, or $v^2/2g$	$h_{f,maj}$
Terms to Evaluate:	$h_p = \dfrac{(P_p)_{out}}{\gamma Q} \qquad h_t = \dfrac{(P_t)_m}{\gamma Q}$ • Pump and turbine heads (evaluated in Section 4.7.2.4)	$h_{f,min} = 0$ (do not model flow resistance) • Gradual channel transitions • Flow from a tank (jet and siphon flows)	
Flow Resistance Terms to Evaluate:	$h_{f,maj} = S_f L = \dfrac{\tau_w L}{\gamma R_h}$ • Pipe and open channel flow major head loss, $h_{f,maj}$		

(Continued)

TABLE 4.1 (*Continued*)

A Summary of Applications of the Governing Equations

	Real Flow	Ideal Flow	Hydraulic Jump
	$h_{f,min} = \dfrac{\tau_w L}{\gamma R_h}$ • Pipe component minor head loss, $h_{f,min}$	$h_{f,min} \neq 0$ Flow resistance due to minor head loss, $h_{f,min}$ is indirectly modeled as follows: • Actual discharge, $Q_a = (v_{ia}A_a/v_iA_1)Q_i$ for pipe and open channel flow measuring devices • Drag force, $F_D = (F_P + F_f)_s$ for external flow/velocity measuring device	
"Subset Level" Application of Governing Equations, with Focus Only on Element Causing Flow Resistance	• Pipe flow resistance • Open channel flow resistance • Pipe components	• Pipe flow measuring devices • Open channel flow measuring devices • External flow/velocity measuring device	
Integral Energy	$\left(\dfrac{p_1}{\gamma} + \dfrac{v_1^2}{2g} + z_1\right) - h_f = \left(\dfrac{p_2}{\gamma} + \dfrac{v_2^2}{2g} + z_2\right)$	$\left(\dfrac{p_1}{\gamma} + \dfrac{v_1^2}{2g} + z_1\right) = \left(\dfrac{p_2}{\gamma} + \dfrac{v_2^2}{2g} + z_2\right)$	
Used to Solve for:	$h_{f,maj} = S_f L = \dfrac{\tau_w L}{\gamma R_h} = \dfrac{\Delta p}{\gamma}$ (assume $z_1 = z_2$, $v_1 = v_2$) • Pipe and open channel flow major head loss, $h_{f,maj}$ $h_{f,min} = \dfrac{\tau_w L}{\gamma R_h} = \dfrac{\Delta p}{\gamma}$ (assume $z_1 = z_2$, $v_1 = v_2$) • Pipe component minor head loss, $h_{f,min}$	$v_i = \sqrt{\dfrac{2\Delta p}{\rho}}$ (assume $z_1 = z_2$, $v_1 \neq v_2$) • Pipe flow measuring device ideal velocity, v_i as a function of ideal pressure difference, Δp, which is directly measured $v_i = \sqrt{2g\Delta y}$ (assume $z_1 = z_2$, $v_1 \neq v_2$) • Open channel flow-measuring device ideal velocity, v_i as a function of ideal pressure difference, Δy, which is directly measured	

(Continued)

TABLE 4.1 (*Continued*)

A Summary of Applications of the Governing Equations

	Real Flow	Ideal Flow	Hydraulic Jump
		$v_i = \sqrt{\dfrac{2\Delta p}{\rho}}$ (assume $z_1 = z_2$, $v_1 \neq v_2$)	
		• External flow velocity-measuring device (e.g., pitot-static tube) ideal velocity, v_i as a function of ideal pressure difference, Δp, which is directly measured	
Differential Continuity		$Q = vA$	
Used to Solve for:		$Q_i = v_i A_i \qquad Q_a = \dfrac{v_a}{\sqrt{2\Delta p/\rho}}\dfrac{A_a}{A_i} Q_i$	
		• Pipe flow-measuring device ideal discharge, Q_i and actual discharge, Q_a	
		$Q_i = v_i A_i \qquad Q_a = \dfrac{v_a}{\sqrt{2g\Delta y}}\dfrac{A_a}{A_i} Q_i$	
		• Open channel flow-measuring device ideal discharge, Q_i and actual discharge, Q_a	
Integral Momentum	$\sum F_s = (F_G + F_P + F_V + F_{other})_s =$	$\sum F_s = (F_G + F_P + F_V + F_{other})_s =$	
	$(\rho Q v_s)_2 - (\rho Q v_s)_1$	$(\rho Q v_s)_2 - (\rho Q v_s)_1$	
Used to Solve for:	$\Delta p = \dfrac{\tau_w L}{R_h}$ (assume $z_1 = z_2$, $v_1 = v_2$)	$v_a = \sqrt{\dfrac{2}{\rho}\left(\Delta p - \dfrac{\tau_w L}{R_h}\right)}$ (assume $z_1 = z_2$, $v_1 \neq v_2$)	
	• Pipe and open channel flow pressure difference, Δp	• Pipe flow-measuring device actual velocity, v_a	

(*Continued*)

TABLE 4.1 (*Continued*)

A Summary of Applications of the Governing Equations

	Real Flow	Ideal Flow	Hydraulic Jump
	$\Delta p = \dfrac{\tau_w L}{R_h}$ (assume $z_1 = z_2$, $v_1 = v_2$) • Pipe component pressure difference, Δp	$v_a = \sqrt{2g\left(\Delta y - \dfrac{\tau_w L}{\gamma R_h}\right)}$ (assume $z_1 = z_2$, $v_1 \neq v_2$) • Open channel flow-measuring device actual velocity, v_a $v_a = \sqrt{\dfrac{2}{\rho}\left(\Delta p - \dfrac{\tau_w L}{R_h}\right)}$ (assume $z_1 = z_2$, $v_1 \neq v_2$) $F_D = (F_P + F_f)_s = (\Delta pA)_s = \left(\dfrac{\rho v^2}{2}A + \dfrac{\tau_w L}{R_h}A\right)_s$ (assume $z_1 = z_2$, $v_1 \neq v_2$) • External flow velocity-measuring device actual velocity, v_a and F_D	
Problem:	Cannot theoretically model the wall shear stress, τ_w in integral momentum for turbulent flow for pipe and open channel flow; thus, cannot analytically determine pressure drop, Δp Cannot theoretically model the wall shear stress, τ_w in integral momentum for turbulent flow for pipe components; thus, cannot analytically determine pressure drop, Δp	Cannot theoretically model the wall shear stress, τ_w in integral momentum for turbulent pipe flow-measuring device; thus, cannot analytically determine pressure drop, Δp and thus v_a and cannot measure actual area, A_a Cannot theoretically model the wall shear stress, τ_w in integral momentum for turbulent open channel flow-measuring device; thus, cannot analytically determine pressure drop, Δy and thus v_a and cannot measure actual area, A_a Cannot theoretically model the wall shear stress, τ_w in integral momentum for turbulent external flow; thus, cannot analytically determine pressure drop, Δp and thus v_a and cannot determine the exact component of the pressure and friction forces in the s-direction	
Solution: *Supplement Momentum*	Supplement the integral momentum (and the differential momentum for S_f) with dimensional analysis in order to empirically model the wall shear	Supplement the integral momentum with dimensional analysis in order to empirically model the wall shear stress, τ_w for turbulent pipe flow-measuring device and thus Δp and	

(Continued)

TABLE 4.1 (Continued)

A Summary of Applications of the Governing Equations

	Real Flow	Ideal Flow	Hydraulic Jump
Theory with Dimensional Analysis, in Order to Empirically Express Difficult-to-Measure τ_w in Terms of Easy-to-Compute v_i, by Measuring Δp in a Pitot-Static Tube or Other Velocity Measuring Device	stress, τ_w for turbulent flow for pipe and open channel flow and thus Δp by the use of a drag coefficient, C_D Differential momentum equation: $$\frac{\partial}{\partial s}\left(\frac{p}{\gamma}+z+\frac{v^2}{2g}\right) = -\frac{\tau_w}{\gamma R_h} - \frac{1}{g}\frac{\partial v}{\partial t}$$ $\underbrace{\quad}_{S_e}\quad\underbrace{\quad}_{S_f}\quad\underbrace{\quad}_{S_a}$ while S_e and S_a are expressed in terms of easy-to-measure velocity, v, S_f is expressed in terms of difficult-to-measure τ_w. Thus, use dimensional analysis to empirically interpret S_f (express τ_w) in terms of easy-to-measure velocity, v $$h_{f,maj} = S_f L = \frac{\tau_w L}{\gamma R_h} = \frac{\Delta p}{\gamma} = \frac{C_D v^2}{gR_h}L = \frac{v^2}{C^2 R_h}L = f\frac{L}{D}\frac{v^2}{2g}$$ • Pipe and open channel flow major head loss, $h_{f,maj}$ (can easily compute v_i by measuring Δp in a pitot-static tube) Supplement the integral momentum with dimensional analysis in order to empirically model the wall shear stress, τ_w for turbulent flow for pipe components and thus Δp by the use of a drag coefficient, C_D $$h_{f,min} = \frac{\tau_w L}{\gamma R_h} = \frac{\Delta p}{\gamma} = C_D\frac{v^2}{2g} = k\frac{v^2}{2g}$$	thus v_a and A_a by the use of a drag/discharge coefficient, C_D or C_d $$Q_a = \frac{\sqrt{(2/\rho)(\Delta p - (\tau_w L/R_h))}}{\sqrt{2\Delta p/\rho}}\frac{A_a}{A_i}\,Q_i = C_d Q_i$$ • Pipe flow measuring device actual discharge, Q_a (can easily compute v_i by measuring Δp in pipe flow-measuring device, and can easily measure A_i) Supplement the integral momentum with dimensional analysis in order to empirically model the wall shear stress, τ_w for turbulent open channel flow-measuring device and thus Δp and thus v_a and A_a by the use of a drag/discharge coefficient, C_D or C_d $$Q_a = \frac{\sqrt{2g(\Delta y - (\tau_w L/\gamma R_h))}}{\sqrt{2g\Delta y}}\frac{A_a}{A_i}\,Q_i = C_d Q_i$$	

(Continued)

TABLE 4.1 (Continued)

A Summary of Applications of the Governing Equations

	Real Flow	Ideal Flow	Hydraulic Jump
	• Pipe component minor head loss, $h_{f,min}$ (can easily compute v_i by measuring Δp in a pitot-static tube)	• Open channel flow-measuring device actual discharge, Q_a (can easily compute v_i by measuring Δy in open channel flow-measuring device, and can easily measure A_i) Supplement the integral momentum with dimensional analysis in order to empirically model the wall shear stress, τ_w for turbulent external flow and thus Δp and thus v_a and thus, F_D by the use of a drag coefficient, C_D $F_D = (F_P + F_f)_s = (\Delta p A)_s = \left(\dfrac{\rho v^2}{2} A + \dfrac{\tau_w L}{R_h} A\right)_s = C_D \dfrac{1}{2}\rho v^2 A$ • External flow drag force, F_D and actual velocity, v_a (can easily compute v_i by measuring Δp in a pitot-static tube)	
Integral Momentum	$\sum F_s = (F_G + F_P + F_V + F_{other})_s = (\rho Q v_s)_2 - (\rho Q v_s)_1$	$\sum F_s = (F_G + F_P + F_V + F_{other})_s = (\rho Q v_s)_2 - (\rho Q v_s)_1$	$\sum F_s = (F_G + F_P + F_V + F_{other})_s = (\rho Q v_s)_2 - (\rho Q v_s)_1$
Used to Solve for:	• Analytically (w/o use of dimensional analysis) solve for unknown force (typically a pressure force or a reaction force); unknown reaction force in the case of certain pipe components (such as contractions, nozzles, elbows, and bends) with a minor head loss; the viscous forces have already been accounted for indirectly through a minor head loss, $h_{f,min}$	• Analytically (w/o use of dimensional analysis) solve for unknown reaction force in the case of open channel flow-measuring devices (such as sluice gates, weirs, spillways, venturi flumes, and contracted openings); the viscous forces have already been accounted for indirectly through a discharge coefficient, C_d • Analytically (w/o use of dimensional analysis) solve for unknown reaction force in the case of a flow from a tank (jet and siphon flows) if there is an unknown reaction force; viscous force is directly computed as $F_f = \mu W$ • Analytically (w/o use of dimensional analysis) solve for unknown force in external flow; viscous and pressure forces are represented by the drag force, F_D	• Analytically (w/o use of dimensional analysis) solve for unknown depth of flow, y (hydraulic jump equation); there are no viscous forces to model

(Continued)

TABLE 4.1 (Continued)

A Summary of Applications of the Governing Equations

	Real Flow	Ideal Flow	Hydraulic Jump
Applications of Governing Equations:	• Nonuniform open channel flow (break in channel slope, hydraulic drop, free overfall)		
Differential Momentum	$\dfrac{dy}{ds} = \dfrac{S_o - S_f}{1 - F^2}$		
Used to Solve for:	y		
Flow Resistance Term to Evaluate:	$S_f = \dfrac{\tau_w}{\gamma R_h} = \dfrac{v^2}{C^2 R_h}$ (*evaluated above*)		
Integral Energy	$h_{f,maj} = S_f L = \dfrac{\tau_w L}{\gamma R_h} = \dfrac{v^2}{C^2 R_h} L$		
Used to Solve for:	$h_{f,maj}$		
Applications of Governing Equations:	• Straight uniform pipe section (single, series, parallel, branching, loop, network) • Straight uniform open channel flow section		
Differential Momentum	$S_f = \dfrac{\tau_w}{\gamma R_h} = \dfrac{v^2}{C^2 R_h} = f \dfrac{1}{D} \dfrac{v^2}{2g}$		
Used to Solve for:	$y, v, S_f,$ or C (or f)		
Integral Energy	$h_{f,maj} = S_f L = \dfrac{\tau_w L}{\gamma R_h} = \dfrac{v^2}{C^2 R_h} L = f \dfrac{L}{D} \dfrac{v^2}{2g}$		
Used to Solve for:	$h_{f,maj}$		

flow, it is important to make the distinction between internal flow, external flow, and flow from a tank (see Table 4.1). While the energy equation derived based on the conservation of momentum principle given by Equation 4.44 does not consider the modeling of energy transfer by work (pumps, turbines, etc.), the energy equation derived based on the conservation of energy principle given by Equation 4.181 does and is repeated as follows:

$$\left(\frac{p_1}{\rho_1 g} + \frac{v_1^2}{2g} + z_1\right) + h_{pump} = \left(\frac{p_2}{\rho_2 g} + \frac{v_2^2}{2g} + z_2\right) + h_{turbine} + h_{f,total} \tag{4.79}$$

Thus, the integral form of the energy equation for incompressible, real, steady flow (pipe or open channel flow) given by Equation 4.79 may be used to solve for an unknown "energy head" term (p/γ, z, or $v^2/2g$) for one of the two points in a finite control volume; an unknown friction head loss, $h_{f,total} = h_{f,maj} + h_{f,min}$ between two points in a finite control volume; an unknown head delivered to the fluid by a pump, h_{pump}; or an unknown head removed from the fluid by a turbine, $h_{turbine}$. Application of Equation 4.79 requires evaluation of the following terms: h_{min}, h_{maj}, h_{pump}, and $h_{turbine}$ (Section 4.5.5).

For the assumption of incompressible, ideal, steady flow (internal or external flow, or flow from a tank), the friction head loss terms and the pump and turbine head terms in Equation 4.79 would be set to zero, and the equation would then be known as the Bernoulli equation (Equation 4.46) and is repeated as follows:

$$\left(\frac{p}{\gamma} + z + \frac{v^2}{2g}\right)_1 - \left(\frac{p}{\gamma} + z + \frac{v^2}{2g}\right)_2 = 0 \tag{4.80}$$

which is applied for problems that affect the flow over a short pipe or channel section, external flow, or flow from a tank, where the major head loss due to flow resistance is assumed to be negligible. Thus, the Bernoulli equation may be solved for an unknown "energy head" term (p/γ, z, or $v^2/2g$) for one of the two points in a finite control volume (Sections 4.5.6 and 4.5.7).

Finally, for the assumption of incompressible, real, steady open channel flow that results in a hydraulic jump, while the friction head loss term, and the pump and turbine head terms in Equation 4.79 would be set to zero, there is a major head loss term to model. Specifically, while the major head loss due to a hydraulic jump is not due to frictional losses (flow resistance), it is due to flow turbulence due to the jump. Thus, the energy equation for a hydraulic jump is given as follows:

$$\left(\frac{p}{\gamma} + z + \frac{v^2}{2g}\right)_1 - h_{f,maj} = \left(\frac{p}{\gamma} + z + \frac{v^2}{2g}\right)_2 \tag{4.81}$$

which may be solved for the major head loss due to the jump (Section 4.5.8).

For the assumption of real flow, ideal flow, or a hydraulic jump, for the respective energy equation (Equation 4.79, Equation 4.80, or Equation 4.81), the integral form of the momentum equation (Equation 4.28) serves as the complementary equation of motion, and is repeated as follows:

$$\sum F_s = (F_G + F_P + F_V + F_{other})_s = (\rho Q v_s)_2 - (\rho Q v_s)_1 \tag{4.82}$$

which may be solved for an unknown force (typically, a pressure force [or depth of flow] or a reaction force); note that the viscous forces are typically accounted for on a "subset level" (indirectly) through the major and minor head loss flow resistance terms in the energy equation above (Equation 4.79) (see Sections 4.5.4.2 and 4.5.5), through an actual discharge (see Sections 4.5.4.2 and 4.5.6), or through a drag force (see Sections 4.5.4.2 and 4.5.7). As such, the integral form of the momentum equation is solved analytically, without the use of dimensional analysis (see Chapter 5 for its application). Thus, when one of these two equations of motion breaks down, the other may be used to find the additional unknown.

Furthermore, in the assumption of real flow, and for the specific case of a major head loss due to pipe or channel flow resistance, application of the differential form of the momentum equation may become necessary. Specifically, application of the differential form of the momentum equation will be necessary when it is of interest to trace the gradual variation of the steady nonuniform channel depth of flow (Equation 5.25) (see Chapter 9) and is given as follows:

$$\frac{dy}{ds} = \frac{S_o - S_f}{1 - F^2} \tag{4.83}$$

or to evaluate the flow (solve for an unknown quantity) at a given point in the fluid system for any uniform pipe or channel section (Equation 5.8) and is given as follows:

$$S_f = \frac{\tau_w}{\gamma R_h} \tag{4.84}$$

Application of Equations 4.83 and 4.84 requires evaluation of the friction slope term, S_f, which is modeled on a "subset level," where the major head loss due to flow resistance is determined as follows (see Sections 4.5.4.2 and 4.5.5):

$$h_{f,maj} = S_f L = \frac{\tau_w}{\gamma R_h} L \tag{4.85}$$

(see Chapters 8 and 9 for their applications).

4.5.4.2 Modeling Flow Resistance: A Subset Level Application of the Governing Equations

In the application of the governing equations (Section 4.5.4.1), although in some flow cases, one may ignore (or not need to model) the flow resistance, in most practical flow cases in general, one must account for the fluid viscosity and thus model the resulting shear stress and flow resistance. As such, in the application of the governing equations (continuity, energy, and momentum), it is important to make the distinction between real flow, ideal flow, and a hydraulic jump. And for ideal flow, it is important to make the distinction between internal (pipe or open channel flow), external flow, and flow from a tank. While the continuity, energy, and momentum equations are applied (as necessary) to internal flows (pipe and open channel flow) and to flow from a tank, etc., only the energy and momentum equations are applied to external flows. The application of the governing equations (continuity, energy, and momentum) for real flow (internal), and ideal flow (flow measuring devices for internal flow, and external flow around objects) requires the modeling of flow resistance (frictional losses). The modeling of flow resistance requires a "subset level" application of the governing equations. However, in the application of the governing equations (continuity, energy, and momentum) for ideal flow (flow from a tank, etc., and gradual channel transitions), and a hydraulic jump, there is no flow resistance to model. While in the case of ideal flow (flow from a tank, etc., and gradual channel transitions), the flow resistance is actually ignored, in the case of the hydraulic jump, the major head loss is due to flow turbulence due to the jump and not flow resistance (frictional losses).

Thus, the application of the governing equations (continuity, energy, and momentum) for real flow (internal) and ideal flow (flow measuring devices for internal flow, and external flow around objects) requires the modeling of flow resistance (frictional losses). Specifically, the application of the integral form of the energy equation for incompressible, real, steady flow (pipe or open channel flow) given by Equation 4.44 requires the evaluation of the major and minor head loss flow resistance terms, $h_{f,maj} = \tau_w L / \gamma R_h$ and $h_{f,min} = \tau_w L / \gamma R_h$, respectively. Furthermore, the application of the differential form of the momentum equation (pipe or open channel flow) given by Equation 5.15 requires the evaluation of the friction slope term, $S_f = \tau_w / \gamma R_h$. The application of the continuity equation for a flow-measuring device (pipe or open channel flow) given by Equation 3.47 requires the evaluation the actual discharge, $Q_a = v_a A_a$. Finally, the application of the integral form of the momentum equation for the case of external flow around an object given by Equation 4.28 (or Equation 5.27) requires the evaluation of the drag force, $F_D = (F_P + F_f)_s$.

The flow resistance in internal and external flow (major head loss, minor head loss, actual discharge, and drag force) is modeled by a "subset level" application of the appropriate governing equations, as presented in Chapter 6 (and summarized in Sections 4.5.5 through 4.5.7). A "subset level" application of the appropriate governing equations focuses only on the given element causing the flow resistance. Thus, assuming that flow resistance is to be accounted for depending upon the specific flow situation, the flow type is assumed to be either a "real" flow or an "ideal" flow. The distinction between ideal and real flow determines how the flow resistance is modeled in the "subset level" application of the appropriate governing equations. The assumption of "real" flow implies that the flow resistance is modeled in both the energy and the momentum equations. However, the assumption of "ideal" flow implies that the flow resistance is modeled only in the momentum equation (and thus the subsequent assumption of "real" flow). The flow resistance due to friction in pipe and open channel flow (modeled as a major head loss) and the flow resistance due to friction in most pipe devices (modeled as a minor head loss) are modeled by assuming "real" flow; application of the energy equation models a head loss that is accounted for by a drag coefficient (in the application of the momentum equation). However, the flow resistance due to friction in flow measuring devices for internal flow (pipe and open channel flow, modeled as an actual discharge), and the flow resistance due to friction in external flow around objects (modeled as a drag force) are modeled by assuming "ideal" flow; although application of the Bernoulli equation does not model a head loss in the determination of ideal velocity, the associated minor head loss with the velocity measurement is accounted for by a drag coefficient (in the application of the momentum equation, where a subsequent assumption of "real" flow is made). Furthermore, the use of dimensional analysis (Chapter 7) in supplement to a "subset level" application of the appropriate governing equations is needed to derive flow resistance equations when there is no theory to model the friction force (as in the case of turbulent internal and external flow). Finally, it is important to note that the use of dimensional analysis is not needed in the derivation of the major head loss term for laminar flow (friction force is theoretically modeled), and the minor head loss term for a sudden pipe expansion (friction force is ignored).

4.5.5 Application of the Energy and Momentum Equations for Real Internal Flow

Application of the integral form of the energy equation for incompressible, real, steady flow (pipe or open channel flow) given by Equation 4.79 requires the derivation/evaluation of the friction head loss term, $h_{f,total} = h_{f,maj} + h_{f,min}$; the head delivered to the fluid by a pump, h_{pump}; and the head removed from the fluid by a turbine, $h_{turbine}$. Evaluation of the major and minor head loss flow resistance terms requires a "subset level" application of the governing equations (Section 4.5.5.1). Then, application of the complementary integral momentum equation (Equation 4.82) is required only if there is an unknown reaction force, as in the case of certain pipe components (such as contractions, nozzles, elbows, and bends); note that the viscous forces have already been accounted for indirectly through the minor head loss flow resistance term in the energy equation. As such, the integral form of the momentum equation is solved analytically, without the use of dimensional analysis.

The evaluation of the total head loss, $h_{f,total} = h_{f,maj} + h_{f,min}$ may include both major and minor losses, which are due to flow resistance. The major head loss in pipe or open channel flow is due to the frictional losses/flow resistance caused by pipe friction or channel resistance, respectively. The minor head loss is due to flow resistance due various pipe components (valves, fittings, [tees, unions, elbows, and bends], entrances, exits, contractions, and expansions), pipe-flow measuring devices (pitot-static tube, orifice, nozzle, or venturi meters), or open channel flow-measuring devices (pitot-static tube, sluice gates, weirs, spillways, venturi flume, and contracted openings). One may note that in the case of a flow-measuring device for either pipe flow or open channel flow, although the associated minor head loss may be accounted for in the integral form of the energy equation (see Equation 4.79), it is typically accounted for by the use/calibration of a discharge coefficient to determine the actual discharge, where ideal flow (Bernoulli equation) is assumed (see Section 4.5.6). Applications of the governing equations for real internal flow are presented in Sections 4.5.5.3 through 4.5.5.5.

4.5.5.1 Evaluation of the Major and Minor Head Loss Terms

Evaluation of the major and minor head loss flow resistance terms requires a "subset level" of application of the governing equations. A "subset level" application of the governing equations focuses only on the given element causing the flow resistance. The assumption of real flow implies that the flow resistance is modeled in both the energy and momentum equations. The flow resistance equations for the major and minor head loss are derived in Chapters 6 and 7. In the derivation of one given source of head loss (major or minor), first, the integral form of the energy equation is applied to solve for the unknown head loss, h_f, which is given as follows:

$$\left(\frac{p}{\gamma} + z + \frac{v^2}{2g}\right)_1 - h_f = \left(\frac{p}{\gamma} + z + \frac{v^2}{2g}\right)_2 \tag{4.86}$$

One may note that in the derivation of the expression for one given source of head loss, h_f (major or minor head loss due to flow resistance), the control volume is defined between the two points that include only that one source of head loss. The derivation of the major head loss assumes $z_1 = z_2$, and $v_1 = v_2$, as follows:

$$\left(\frac{p}{\gamma} + z + \frac{v^2}{2g}\right)_1 - h_{f,maj} = \left(\frac{p}{\gamma} + z + \frac{v^2}{2g}\right)_2 \tag{4.87}$$

which yields $h_{f,maj} = S_f L = \tau_w L / \gamma R_h = \Delta p / \gamma$. And, the derivation of the minor head loss assumes $z_1 = z_2$, and $v_1 = v_2$, as follows:

$$\left(\frac{p}{\gamma} + z + \frac{v^2}{2g}\right)_1 - h_{f,min} = \left(\frac{p}{\gamma} + z + \frac{v^2}{2g}\right)_2 \tag{4.88}$$

which yields $h_{f,min} = \tau_w L / \gamma R_h = \Delta p / \gamma$. Then, second, the integral form of the momentum equation (supplemented by dimensional analysis) is applied to solve for an unknown pressure drop, Δp (in the case of major or minor head loss due to pipe flow resistance) or an unknown change in channel depth, Δy (in the case of major head loss due to open channel flow resistance), which was given in Equation 4.28 and is repeated as follows:

$$\sum F_s = (F_G + F_P + F_V + F_{other})_s = (\rho Q v_s)_2 - (\rho Q v_s)_1 \tag{4.89}$$

One may note that derivation of the major head loss involves both the integral and the differential momentum equations. Thus, the resulting flow resistance equation for the major head loss is given as follows:

$$h_{f,maj} = S_f L = \frac{\tau_w L}{\gamma R_h} = \frac{\Delta p}{\gamma} = \frac{v^2}{C^2 R_h} L = f \frac{L}{D} \frac{v^2}{2g} \tag{4.90}$$

where C is the Chezy coefficient and f is the Darcy–Weisbach friction factor. And, the resulting flow resistance equation for the minor head loss is given as follows:

$$h_{f,min} = \frac{\tau_w L}{\gamma R_h} = \frac{\Delta p}{\gamma} = k \frac{v^2}{2g} \tag{4.91}$$

where k is the minor head loss coefficient. Furthermore, each head loss flow resistance equation represents the simultaneous application of the two complementary equations of motion (on a "subset level"). It is important to note that the use of dimensional analysis is not needed in the derivation of the major head loss term for laminar flow and the minor head loss term for a sudden pipe expansion.

The flow resistance due to pipe or channel friction, which results in a major head loss term, $h_{f,major}$, is modeled as a resistance force (shear stress or drag force) in the integral form of the momentum equation (Equation 4.28), while it is modeled as a friction slope, S_f in the differential form of the momentum equation (Equation 4.84). As such, the energy equation may be used to solve for the unknown major head loss, h_f, while the integral momentum equation may be used to solve for the unknown pressure drop, Δp (or the unknown friction slope, S_f in the case of turbulent open channel flow). The major head loss, h_f is caused by both pressure and friction forces. Thus, when the friction/viscous forces can be theoretically modeled in the integral form of the momentum equation (application of Newton's law of viscosity in the laminar flow case), one can analytically determine the actual pressure drop, Δp from the integral form of the momentum equation. And, therefore, one may analytically derive an expression for the major head loss, h_f due to pipe or channel friction from the energy equation. However, when the friction/viscous forces cannot be theoretically modeled in the integral form of the momentum equation (as in the turbulent flow case), the friction/viscous forces, the actual pressure drop, Δp (or the unknown friction slope, S_f in the case of turbulent open channel flow) and thus the major head loss, h_f due to pipe or channel friction are determined empirically (see Chapter 7). Specifically, in the case of turbulent flow, the friction/viscous forces in the integral form of the momentum equation and thus the wall shear stress, τ_w cannot be theoretically modeled; thus, the integral form of the momentum equation cannot be directly applied to solve for the unknown pressure drop, Δp. Thus, an empirical interpretation (using dimensional analysis) of the wall shear stress, τ_w in the theoretical expression for the friction slope, $S_f = \tau_w / \gamma R_h$ in the differential form of the momentum equation is sought in terms of velocity, v and a flow resistance coefficient. This yields the Chezy equation $S_f = v^2 / C^2 R_h$ (which represents the differential form of the momentum equation, supplemented by dimensional analysis and guided by the integral momentum equation; the link between the differential and integral momentum equations), which is used to obtain an empirical evaluation for the pressure drop, Δp in Newton's second law of motion (integral form of the momentum equation). As such, one can then derive an empirical expression for the major head loss, h_f from the energy equation. Major losses are addressed in detail in Chapters 6, 8, and 9.

The flow resistance due to various pipe components, which results in a minor head loss term, $h_{f,minor}$, is modeled as a resistance force (shear stress or drag force) in the integral form of the momentum equation (Equation 4.28). As such, the energy equation may be used to solve for the unknown minor head loss, h_f, while the integral form of the momentum equation may be used to solve for the unknown pressure drop, Δp. The minor head loss, h_f is caused by both pressure and friction forces. Thus, when the friction/viscous forces are insignificant and can be ignored in the integral form of the momentum equation (as in the case of the sudden pipe expansion), one can analytically determine the actual pressure drop, Δp from the momentum equation. And, therefore, one may analytically derive an expression for the minor head loss, h_f due to pipe components from the energy equation. However, when the friction/viscous forces are significant and cannot be ignored in the integral form of the momentum equation (as in the case of the other pipe components), the friction/viscous forces; the actual pressure drop/rise, Δp; and thus the minor head loss, h_f are determined empirically (see Chapter 7). Specifically, in the case of turbulent pipe component flow, the friction/viscous forces in the integral form of the momentum equation and thus the wall shear stress, τ_w cannot be theoretically modeled; thus, the integral form of the momentum equation cannot be directly applied to solve for the unknown pressure drop, Δp. The evaluation of the minor head loss term, $h_{f,minor}$ involves supplementing the integral form of the momentum equation with dimensional analysis, as described in detail in Chapters 6 through 8.

4.5.5.2 Evaluation of the Pump and Turbine Head Terms

Evaluation of the head delivered to the fluid by a pump, h_{pump} and the head removed/extracted from the fluid by a turbine, $h_{turbine}$ require the application of the appropriate power equations, which are discussed in Section 4.7.2.4, and the resulting equations are summarized, respectively, as follows:

$$h_{pump} = \frac{(P_p)_{out}}{\gamma Q} \tag{4.92}$$

$$h_{turbine} = \frac{(P_t)_{in}}{\gamma Q} \tag{4.93}$$

4.5.5.3 Applications of the Governing Equations for Real Internal Flow

The governing equations (continuity, energy, and momentum) for real flow are applied in the analysis and design of both pipe systems and open channel flow systems. Applications of the governing equations for real pipe flow include the analysis and design of pipe systems that include a numerous arrangements of pipes, including: (1) single pipes, (2) pipes with components, (3) pipes with a pump or a turbine, (4) pipes in series, (5) pipes in parallel, (6) branching pipes, (7) pipes in a loop, and (8) pipe networks (see Chapter 8). And, applications of the governing equations for real open channel flow include the analysis and design of uniform and nonuniform flow (generated by controls such as a break in channel slope, a hydraulic drop, or a free overfall) (see Chapter 9). While the application of the continuity equation has already been presented in Chapter 3, application of the momentum equation is yet to be presented in Chapter 5. One may note from Section 4.5.5.1 that evaluation of the major and minor head loss terms in the energy equation requires a "subset level" application of the governing equations. Furthermore, the resulting definitions for the major and minor head loss (flow resistance) terms (see Equations 4.90 and 4.91, respectively) require details in the definition of variables (C, f, v, L, D, and k), which are yet to be presented in later chapters. Therefore, example problems illustrating the application of the energy equation for real pipe or open channel flow herein will assume that the minor and major head loss terms are either a given quantity (lumped value), computed from the given head loss equations, or directly solved for in the following energy equation:

$$\left(\frac{p_1}{\rho_1 g} + \frac{v_1^2}{2g} + z_1\right) + h_{pump} = \left(\frac{p_2}{\rho_2 g} + \frac{v_2^2}{2g} + z_2\right) + h_{turbine} + h_{f,maj} + h_{f,min} \qquad (4.94)$$

Furthermore, for the example problems involving certain pipe components, the illustration of the computation of the unknown reaction force, which requires the application of the complementary momentum equation, will be deferred to Chapter 5. Finally, for example problems involving a pump or a turbine, illustration of the evaluation of the head delivered by a pump and the head removed by a turbine, which require the application of the appropriate power equation, will be presented in Section 4.7.4.

4.5.5.4 Application of the Energy Equation for Real Pipe Flow

EXAMPLE PROBLEM 4.1

Water at 20°C flows in a 1.5-m-diameter inclined 18,000 m concrete pipe at a flowrate of 4 m³/sec, as illustrated in Figure EP 4.1. The pressure at point 1 is 27×10^5 N/m², and the pressure at point 2 is 15×10^5 N/m². Point 2 is 10 m higher than point 1. (a) Determine the major head loss due to the pipe friction between points 1 and 2. (b) Draw the energy grade line and the hydraulic grade line.

Mathcad Solution

(a) In order to determine the velocity of flow in the pipe, the continuity equation is applied as follows:

$$D := 1.5\,\mathrm{m} \qquad\qquad Q := 4\,\frac{\mathrm{m^3}}{\mathrm{sec}} \qquad\qquad A := \frac{\pi \cdot D^2}{4} = 1.767\,\mathrm{m^2}$$

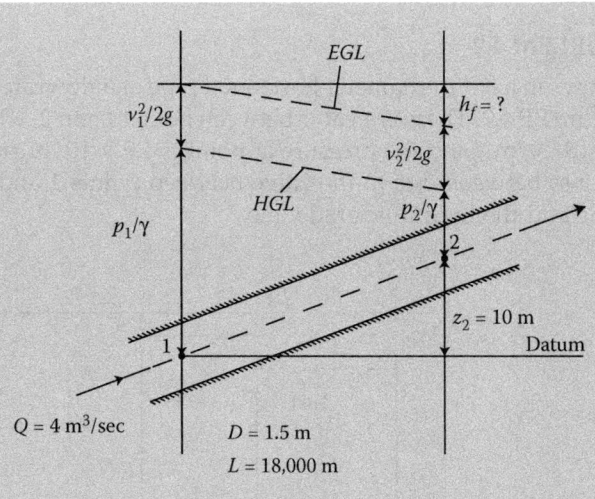

FIGURE EP 4.1
Water flows through an inclined pipe flowing under pressure (real flow).

$$v := \frac{Q}{A} = 2.264 \; \frac{m}{s} \qquad v_1 := v = 2.264 \; \frac{m}{s} \qquad v_2 := v = 2.264 \; \frac{m}{s}$$

In order to determine the major head loss between points 1 and 2, the energy equation is applied between points 1 and 2, where the datum is assumed to be at point 1, and where the density, ρ for water at $20°C$ is given in Table A.2 in Appendix A as follows:

$$p_1 := 27 \times 10^5 \; \frac{N}{m^2} \qquad p_2 := 15 \times 10^5 \; \frac{N}{m^2} \qquad z_1 := 0 \, m \qquad z_2 := 10 \, m$$

$$\rho := 998 \; \frac{kg}{m^3} \qquad g := 9.81 \; \frac{m}{sec^2} \qquad \gamma := \rho \cdot g = 9.79 \times 10^3 \; \frac{kg}{m^2 \, s^2}$$

Guess value: $\qquad h_f := 50 \, m$

Given

$$\frac{p_1}{\gamma} + z_1 + \frac{v_1^2}{2 \cdot g} - h_f = \frac{p_2}{\gamma} + z_2 + \frac{v_2^2}{2 \cdot g}$$

$$h_f := Find(h_f) = 112.569 \, m$$

Note that the pressure drop between points 1 and 2 is caused by the increase in elevation and the major head loss due to the pipe friction between points 1 and 2.
(b) The EGL and HGL are illustrated in Figure EP 4.1.

EXAMPLE PROBLEM 4.2

Water at 20°C flows in a 0.3-m-diameter horizontal pipe at a flowrate of 0.4 m³/sec, as illustrated in Figure EP 4.2. There is a valve between points 1 and 2, where the pressure at point 1 is 2×10^5 N/m², and the pressure at point is 1.9×10^5 N/m². (a) Determine the minor head loss between due to the valve between points 1 and 2. (b) Draw the energy grade line and the hydraulic grade line.

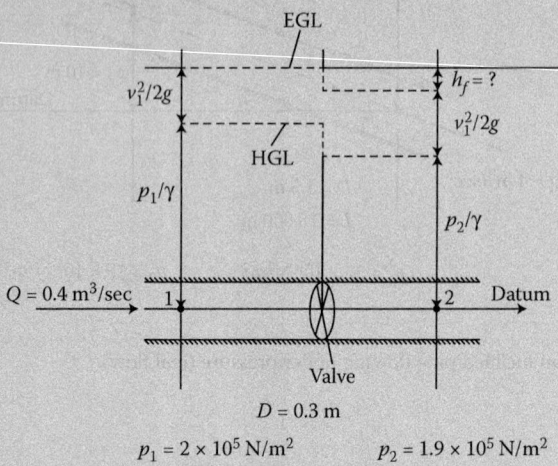

FIGURE EP 4.2
Water flows through a horizontal pipe with a valve flowing under pressure (real flow).

Mathcad Solution

(a) In order to determine the velocity of flow in the pipe, the continuity equation is applied as follows:

$$D := 0.3 \, m \qquad Q := 0.4 \, \frac{m^3}{sec} \qquad A := \frac{\pi \cdot D^2}{4} = 0.071 \, m^2$$

$$v := \frac{Q}{A} = 5.659 \, \frac{m}{s} \qquad v_1 := v = 5.659 \, \frac{m}{s} \qquad v_2 := v = 5.659 \, \frac{m}{s}$$

In order to determine the minor head loss between points 1 and 2, the energy equation is applied between points 1 and 2, where the datum is assumed to be at points 1 and 2, and where the density, ρ for water at 20°C is given in Table A.2 in Appendix A as follows:

$$p_1 := 2 \times 10^5 \, \frac{N}{m^2} \qquad p_2 := 1.9 \times 10^5 \, \frac{N}{m^2} \qquad z_1 := 0 \, m \qquad z_2 := 0 \, m$$

$$\rho := 998 \, \frac{kg}{m^3} \qquad g := 9.81 \, \frac{m}{sec^2} \qquad \gamma := \rho \cdot g = 9.79 \times 10^3 \, \frac{kg}{m^2 \, s^2}$$

Guess value: $\qquad h_f := 1 \, m$

Given

$$\frac{p_1}{\gamma} + z_1 + \frac{v_1^2}{2 \cdot g} - h_f = \frac{p_2}{\gamma} + z_2 + \frac{v_2^2}{2 \cdot g}$$

$h_f := \text{Find}(h_f) = 1.021\,\text{m}$

Note that the pressure drop between points 1 and 2 is caused by the minor head loss due to the valve between points 1 and 2.

(b) The EGL and HGL are illustrated in Figure EP 4.2.

EXAMPLE PROBLEM 4.3

Water at 20°C flows in a 0.85-m-diameter horizontal 11,000 m cast iron pipe at a flow-rate of 1.3 m³/sec. A 90° mitered bend is installed in between point 1 and 2 in order to change the direction of the flow by 90°, as illustrated in Figure EP 4.3. The pressure at point 1 is $15 \times 10^5\,\text{N/m}^2$. Assume a Darcy–Weisbach friction factor, f of 0.018, and assume a minor head loss coefficient due to the bend, k of 0.131. (a) Determine the major head loss due to the pipe friction between points 1 and 2. (b) Determine the minor head loss due to the bend between points 1 and 2. (c) Determine the pressure at point 2.

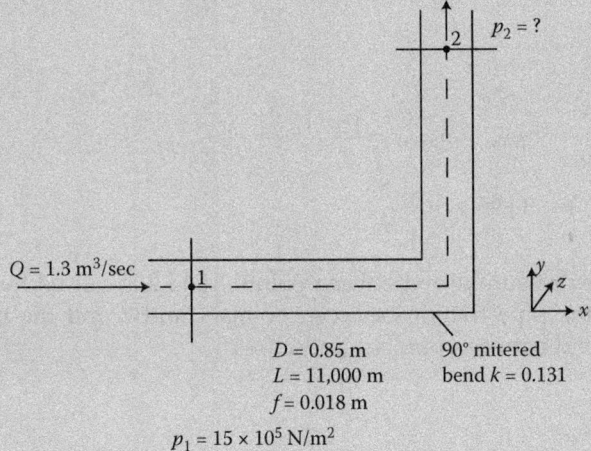

$D = 0.85\,\text{m}$ 90° mitered
$L = 11{,}000\,\text{m}$ bend $k = 0.131$
$f = 0.018\,\text{m}$

$p_1 = 15 \times 10^5\,\text{N/m}^2$

FIGURE EP 4.3
Water flows through a horizontal pipe with a 90° mitered bend flowing under pressure (real flow).

Mathcad Solution

(a) In order to determine the velocity of flow in the pipe, the continuity equation is applied as follows:

$$D := 0.85\,\text{m} \qquad Q := 1.3\,\frac{\text{m}^3}{\text{sec}} \qquad A := \frac{\pi \cdot D^2}{4} = 0.567\,\text{m}^2 \qquad v := \frac{Q}{A} = 2.291\,\frac{\text{m}}{\text{s}}$$

The major head loss due to the pipe friction between points 1 and 2 is determined by applying the Darcy–Weisbach head loss equation as follows:

$$g := 9.81 \, \frac{m}{sec^2} \qquad f := 0.018 \qquad L := 11{,}000 \, m \qquad h_{fmaj} := f \cdot \frac{L}{D} \cdot \frac{v^2}{2 \cdot g} = 62.313 \, m$$

(b) The minor head loss due to the bend between points 1 and 2 is determined by applying the minor head loss equation as follows:

$$k := 0.131 \qquad\qquad h_{fmin} := k \cdot \frac{v^2}{2 \cdot g} = 0.035 \, m$$

(c) In order to determine the pressure at point 2, the energy equation is applied between points 1 and 2, where the datum is assumed to be at points 1 and 2, and where the density, ρ for water at 20°C is given in Table A.2 in Appendix A as follows:

$$p_1 := 15 \times 10^5 \, \frac{N}{m^2} \qquad\qquad z_1 := 0 \, m \qquad z_2 := 0 \, m \qquad\qquad \rho := 998 \, \frac{kg}{m^3}$$

$$\gamma := \rho \cdot g = 9.79 \times 10^3 \, \frac{kg}{m^2 \, s^2} \qquad v_1 := v = 2.291 \, \frac{m}{s} \qquad v_2 := v = 2.291 \, \frac{m}{s}$$

Guess value: $\qquad\qquad p_2 := 15 \times 10^5 \, \frac{N}{m^2}$

Given

$$\frac{p_1}{\gamma} + z_1 + \frac{v_1^2}{2 \cdot g} - h_{fmaj} - h_{fmin} = \frac{p_2}{\gamma} + z_2 + \frac{v_2^2}{2 \cdot g}$$

$$p_2 := Find(p_2) = 8.896 \times 10^5 \, \frac{N}{m^2}$$

Note that the pressure drop between points 1 and 2 is caused by the major head loss due to the pipe friction between points 1 and 2 and the minor head loss due to the bend between points 1 and 2.

4.5.5.5 Application of the Energy Equation for Real Open Channel Flow

EXAMPLE PROBLEM 4.4

Water flows at a uniform flow at a discharge of 2.5 m³/sec in a rectangular channel that is 0.6 m wide, with a channel bottom slope of 0.025 m/m and a Chezy coefficient, C of 53 m$^{1/2}$ s^{-1}, as illustrated in Figure EP 4.4. For uniform open channel flow, the channel bottom slope, S_o is equal to the friction slope, S_f. (a) Determine the major head loss due to the channel resistance over a channel section length of 15 m between points 1 and 2. (b) Determine the uniform depth of flow in the channel. (c) Draw the energy grade line and the hydraulic grade line.

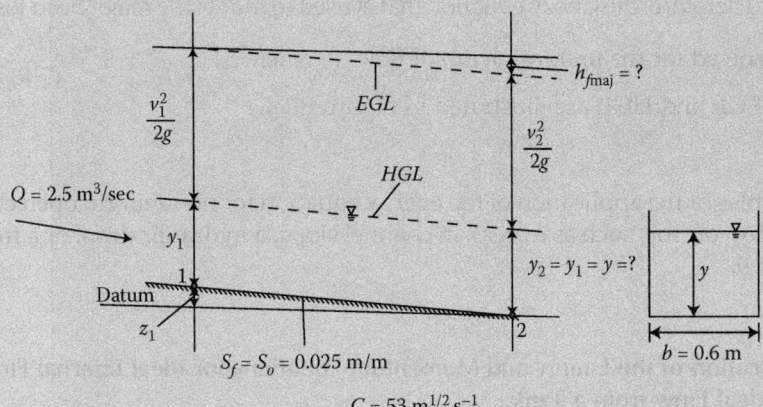

FIGURE EP 4.4
Water flows at a uniform flow in a rectangular channel (real flow).

Mathcad Solution

(a) The major head loss due to channel resistance between points 1 and 2 is determined by applying the head loss equation as follows:

$$S_o := 0.025 \frac{m}{m} \qquad S_f := S_o = 0.025 \frac{m}{m} \qquad L := 15\,m$$

$$h_{fmaj} := S_f \cdot L = 0.375\,m$$

(b) In order to determine the uniform depth of flow in the channel, the energy equation is applied between points 1 and 2, where the datum is assumed to be at point 2, and the major head loss is modeled by the Chezy head loss equation as follows:

$$Q := 2.5 \frac{m^3}{sec} \qquad b := 0.6\,m \qquad C := 53\,m^{\frac{1}{2}}\,sec^{-1}$$

$$z_2 := 0\,m \qquad z_1 := S_o \cdot L = 0.375\,m \qquad g := 9.81 \frac{m}{sec^2}$$

$$A(y) := b \cdot y \qquad v(y) := \frac{Q}{A(y)} \qquad v_1(y) := v(y) \qquad v_2(y) := v(y)$$

$$P_w(y) := 2 \cdot y + b \qquad R_h(y) := \frac{A(y)}{P_w(y)} \qquad \qquad S_f(y) := \frac{v(y)^2}{C^2\,R_h(y)}$$

Guess value: $\qquad y := 1\,m$

Given

$$y + z_1 + \frac{v_1(y)^2}{2 \cdot g} - S_f(y) \cdot L = y + z_2 + \frac{v_2(y)^2}{2 \cdot g}$$

$$y := Find(y) = 1.031\,m$$

It is important to note that for uniform flow, both the depth and the velocity head terms for points 1 and 2 are equal and thus cancel each other out in the energy equation. Therefore, the Chezy equation that is used to model the major head loss is actually solved for the uniform depth of flow, y, where $S_o = \dfrac{\Delta z}{L} = S_f = \dfrac{v^2}{C^2 R_h}$.

(c) The EGL and HGL are illustrated in Figure EP 4.4.

For examples of the application of the energy equation for nonuniform open channel flow (generated by controls such as a break in channel slope, a hydraulic drop, or a free overfall), see Chapter 9.

4.5.6 Application of the Energy and Momentum Equations for Ideal Internal Flow and Ideal Flow from a Tank

Application of the Bernoulli equation (Equation 4.80) assumes ideal flow, so there is no head loss term to evaluate in the energy equation. Therefore, for problems that affect the flow over a short pipe or channel section or flow from a tank, the major head loss due to flow resistance is assumed to be negligible, and one may apply the Bernoulli equation. Thus, the Bernoulli equation may be solved for an unknown "energy head" term (p/γ, z, or $v^2/2g$) for one of the two points in a finite control volume as follows:

$$\left(\frac{p}{\gamma} + z + \frac{v^2}{2g}\right)_1 - \left(\frac{p}{\gamma} + z + \frac{v^2}{2g}\right)_2 = 0 \qquad (4.95)$$

These types of flow problems include flow measuring devices for both pipe and open channel flow, which, however, have a minor loss associated with the flow measurement and gradual (vertical or horizontal) channel contractions/transitions, which do not have a minor loss. Also, the flow from a tank (jet flows and siphon flows) does not have a minor head loss. In the case of a flow-measuring device for either pipe flow or open channel flow, although the associated minor head loss may be accounted for in the integral form of the energy equation (see Equation 4.181 or Equation 4.44), it is typically accounted for by the use/calibration of a discharge coefficient to determine the actual discharge, where ideal flow (Bernoulli equation) is assumed. Thus, in the case of flow-measuring devices, one must indirectly model the minor head loss flow resistance term by the use of a discharge coefficient, which requires a "subset level" of application of the governing equations (Section 4.5.6.1). Then, application of the complementary integral momentum equation (Equation 4.82) is required only if there is an unknown reaction force, as in the case of open channel flow measuring devices; note that the viscous forces have already been accounted for indirectly through a minor head loss flow resistance term in the energy equation (or, actually, through the discharge coefficient). And in the case of a flow from a tank, the complementary integral momentum is applied if there is an unknown reaction force. As such, in either case, the integral form of the momentum equation is solved analytically, without the use of dimensional analysis. Applications of the governing equations for ideal internal flow and ideal flow from a tank are presented in Sections 4.5.6.2 through 4.5.6.5.

4.5.6.1 Evaluation of the Actual Discharge

In the case of a flow-measuring device for either pipe flow or open channel flow, the associated minor head is typically accounted for by the use/calibration of a discharge coefficient to determine the actual discharge. Evaluation of the actual discharge requires a "subset level" application of the governing equations. A "subset level" application of the governing equations focuses only on the given element causing the flow resistance. The assumption of "ideal" flow implies that the flow resistance is modeled only in the momentum equation (and thus the subsequent assumption of "real" flow). The flow resistance equation for the actual discharge is derived in Chapters 6 and 7. In the derivation of the actual discharge for a given flow-measuring device, first, ideal flow is assumed; thus, the Bernoulli equation is applied as follows:

$$\left(\frac{p}{\gamma} + z + \frac{v^2}{2g}\right)_1 - \left(\frac{p}{\gamma} + z + \frac{v^2}{2g}\right)_2 = 0 \tag{4.96}$$

Assuming $z_1 = z_2$, this equation has one unknown, which is the ideal velocity of the flow at the restriction in the flow-measuring device. Thus, this equation yields an expression for the ideal velocity, $v_i = \sqrt{2\Delta p/\rho}$ or $v_i = \sqrt{2g\Delta y}$, as a function of an ideal pressure difference, Δp or Δy, respectively, which is directly measured. Then, application of the continuity equation is used to determine an expression for the ideal discharge, $Q_i = v_i A_i$ and the actual discharge,

$$Q_a = \frac{v_a}{\sqrt{\dfrac{2\Delta p}{\rho}}} \frac{A_a}{A_i} Q_i \quad \text{or} \quad Q_a = \frac{v_a}{\sqrt{2g\Delta y}} \frac{A_a}{A_i} Q_i.$$

Finally, the integral form of the momentum equation (supplemented by dimensional analysis) is applied to solve for an unknown actual velocity, v_a and actual area, A_a which was given in Equation 4.28 and is repeated as follows:

$$\sum F_s = (F_G + F_P + F_V + F_{other})_s = (\rho Q v_s)_2 - (\rho Q v_s)_1 \tag{4.97}$$

Thus, the resulting flow resistance equations for the actual discharge for pipe flow-measuring devices, and open channel flow-measuring devices are given, respectively, as follows:

$$Q_a = \frac{\sqrt{\dfrac{2}{\rho}\left(\Delta p - \dfrac{\tau_w L}{R_h}\right)}}{\sqrt{\dfrac{2\Delta p}{\rho}}} \frac{A_a}{A_i} Q_i = C_d Q_i \tag{4.98a}$$

$$Q_a = \frac{\sqrt{2g\left(\Delta y - \dfrac{\tau_w L}{\gamma R_h}\right)}}{\sqrt{2g\Delta y}} \frac{A_a}{A_i} Q_i = C_d Q_i \tag{4.98b}$$

where the discharge coefficient, C_d accounts for (indirectly models) the minor head loss associated with the flow measurement.

The flow resistance (shear stress or drag force) due to a flow-measuring device is modeled as a reduced/actual discharge, Q_a in the differential form of the continuity equation. The flow resistance causes an unknown pressure drop, Δp in the case of pipe flow, and an unknown Δy in the case of open channel flow, which causes an unknown head loss, h_f, where the head loss is due to a conversion of kinetic energy to heat, which is modeled/displayed in the integral form of the energy equation. The head loss is caused by both pressure and friction forces. In the determination of the reduced/actual discharge, Q_a, which is less than the ideal discharge, Q_i, the exact reduction in the flowrate is unknown because the head loss, h_f causing it is unknown. Therefore, because one cannot model the exact reduction in the ideal flowrate, Q_i, one cannot derive an analytical expression for the reduced/actual discharge, Q_a from the continuity equation. Specifically, the complex nature of the flow does not allow a theoretical modeling of the existence of the viscous force (due to shear stress, τ_w) due to the flow measurement device in the momentum equation and thus does not allow an analytical derivation of the pressure drop, Δp or Δy and thus v_a and cannot measure A_a. Thus, an empirical expression (using dimensional analysis) for the actual discharge,

$$Q_a = \frac{\sqrt{\frac{2}{\rho}\left(\Delta p - \frac{\tau_w L}{R_h}\right)}}{\sqrt{\frac{2\Delta p}{\rho}}} \frac{A_a}{A_i} Q_i, \quad \text{or} \quad Q_a = \frac{\sqrt{2g\left(\Delta y - \frac{\tau_w L}{\gamma R_h}\right)}}{\sqrt{2g\Delta y}} \frac{A_a}{A_i} Q_i,$$

and thus Δp or Δy (and v_a and A_a) is derived as a function of the drag coefficient, C_D (discharge coefficient, C_d) that represents the flow resistance. This yields the actual discharge equation $Q_a = C_d Q_i$, which represents the integral momentum equation supplemented by dimensional analysis. Furthermore, although the differential form of the continuity equation is typically used to compute the reduced/actual discharge, Q_a for a flow measuring device, because the head loss, h_f causes the reduced/actual discharge, Q_a, the head loss may also be accounted for as a minor loss in the integral form of the energy equation, as illustrated in Chapter 8. Actual discharges are addressed in detail in Chapters 6, 8, and 9.

4.5.6.2 Applications of the Governing Equations for Ideal Internal Flow and Ideal Flow from a Tank

The governing equations (continuity, energy, and momentum) for ideal flow are applied in the analysis and design of pipe systems and open channel flow systems, and the flow from a tank. Applications of the governing equations for ideal pipe flow include the analysis and design of pipe systems that include a pipe flow measuring device such as pitot-static tubes, orifice, nozzle, or venturi meters, which are also known as ideal flow meters. Applications of the governing equations for ideal open channel flow include the analysis and design of an open channel flow-measuring device (such as pitot-static tubes, sluice gates, weirs, spillways, venturi flumes, and contracted openings), which are also known as ideal flow meters, and the analysis and design of a gradual (vertical or horizontal) channel contraction/ transition. And, applications of the governing equations for ideal flow include the flow from a tank (jet flows and siphon flows). While the application of the continuity equation has already been presented in Chapter 3, application of the momentum equation is yet to be presented in Chapter 5. One may note from Section 4.5.6.1 that the minor head loss term associated with a pipe or open channel flow-measuring device is indirectly modeled by the discharge coefficient, which requires a "subset level" application of the governing equations for its evaluation. Furthermore, the resulting definition for the actual discharge, $Q_a = C_d Q_i$ requires details in the definition of the discharge coefficient, C_d, which is yet to be presented in later chapters. Therefore, example problems illustrating the application of the Bernoulli equation for pipe or open channel flow-measuring devices herein will assume that the actual discharge is either a given quantity (lumped value) or computed from the actual discharge equation (see Chapters 8 and 9 for applications) as follows:

$$Q_a = v_a A_a \qquad (4.99)$$

Furthermore, for the example problems involving open channel flow-measuring devices, illustration of the computation of the unknown reaction force, which requires the application of the complementary momentum equation, will be deferred to Chapter 5. For problems involving the flow from a tank that have an unknown reaction force, illustration of the application of the complementary momentum equation will be deferred to Chapter 5. For the example problems involving a gradual (vertical or horizontal) channel contraction/transition, there is no minor head loss associated with the flow, and there are no reaction forces involved; thus, only the continuity and the Bernoulli equations are needed. Finally, for the example problems involving a flow-measuring device for pipe flow or for example problems involving a pitot-static tube for either pipe or open channel flow, there are no reaction forces involved. Example Problems 4.5 through 4.8 presented below focus on the application of the Bernoulli equation on a "subset level" and thus solve for the ideal velocity for pipe and open channel flow-measuring devices. However, Chapters 8 and 9 present example problems that apply the Bernoulli equation (on a "macro level," as opposed to a "subset level") and thus solve for any unknown "energy head" term (p/γ, z, or $v^2/2g$) for one of the two points in a finite control volume.

4.5.6.3 Application of the Bernoulli Equation for Ideal Pipe Flow

EXAMPLE PROBLEM 4.5

Water at 20°C flows in a 0.75-m-diameter horizontal pipe. A venturi meter with a 0.35-m throat diameter is inserted in the pipe at point 2 as illustrated in Figure EP 4.5

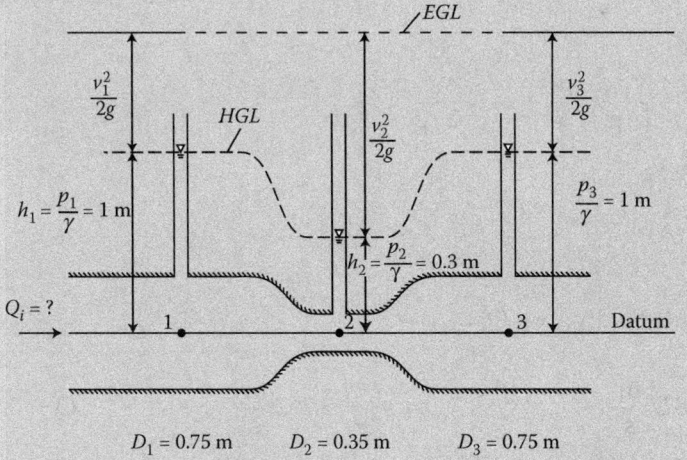

FIGURE EP 4.5
Water flows through a horizontal pipe with a venturi meter flowing under pressure (ideal flow).

in order to measure the flowrate in the pipe. Piezometric tubes are installed at point 1 just upstream of the venturi tube and at point 2, which is located at the throat of the venturi tube. The pressure head at point 1, p_1/γ is measured by the piezometer to be 1 m, and the pressure head at point 2, p_2/γ is measured by the piezometer to be 0.3 m. (a) Determine the ideal flowrate in the pipe. (b) Draw the energy grade line and the hydraulic grade line. (c) Given the same conditions at point 1 just upstream of the venturi tube, determine the minimum diameter of the venturi meter at point 2 without causing cavitation.

Mathcad Solution

(a) In order to determine the ideal velocity at either point 1 or point 2, the Bernoulli equation is applied between points 1 and 2. However, in order to determine the relationship between the ideal velocities at points 1 and 2, the continuity equation is applied between points 1 and 2; thus, we have two equations and two unknowns. Furthermore, in order to determine the ideal flowrate in the pipe, the continuity equation is applied at either points 1 or 2 as follows:

$D_1 := 0.75 \, m$ $D_2 := 0.35 \, m$

$A_1 := \dfrac{\pi \cdot D_1^2}{4} = 0.442 \, m^2$ $A_2 := \dfrac{\pi \cdot D_2^2}{4} = 0.096 \, m^2$

$h_1 := 1 \, m$ $h_2 := 0.3 \, m$ $z_1 := 0 \, m$ $z_2 := 0 \, m$

$\rho := 998 \, \dfrac{kg}{m^3}$ $g := 9.81 \, \dfrac{m}{sec^2}$ $\gamma := \rho \, g = 9.79 \times 10^3 \, \dfrac{kg}{m^2 \, s^2}$

$p_1 := \gamma \cdot h_1 = 9.79 \times 10^3 \, \dfrac{N}{m^2}$ $p_2 := \gamma \cdot h_2 = 2.937 \times 10^3 \, \dfrac{N}{m^2}$

Guess value: $v_1 := 1 \, \dfrac{m}{sec}$ $v_2 := 2 \, \dfrac{m}{sec}$ $Q := 1 \, \dfrac{m^3}{sec}$

Given

$$\dfrac{p_1}{\gamma} + z_1 + \dfrac{v_1^2}{2 \cdot g} = \dfrac{p_2}{\gamma} + z_2 + \dfrac{v_2^2}{2 \cdot g}$$

$$v_1 \cdot A_1 = v_2 \cdot A_2$$

$$Q = v_2 \cdot A_2$$

$$\begin{pmatrix} v_1 \\ v_2 \\ Q \end{pmatrix} := \text{Find}(v_1, v_2, Q)$$

$v_1 = 0.827 \, \dfrac{m}{s}$ $v_2 = 3.797 \, \dfrac{m}{s}$ $Q = 0.365 \, \dfrac{m^3}{s}$

Application of the Bernoulli equation for the venturi meter illustrates a conversion of pressure energy to kinetic energy.

(b) The EGL and HGL are illustrated in Figure EP 4.5.

(c) In order to determine the minimum diameter of the venturi meter at point 2 without causing cavitation, one must assume that the pressure at point 2 is the vapor pressure. Thus, from Table A.2 in Appendix A, for water at 20°C, the vapor pressure, p_v is $2.34 \times 10^3 \, N/m^2$ abs. However, because the vapor pressure is given in absolute pressure, the corresponding gage pressure is computed by subtracting the atmospheric pressure as follows: $p_{gage} = p_{abs} - p_{atm}$, where the standard atmospheric pressure is $101.325 \times 10^3 \, N/m^2$ abs.

$$p_v := 2.34 \times 10^3 \, \frac{N}{m^2} - 101.325 \times 10^3 \, \frac{N}{m^2} = -9.899 \times 10^4 \, \frac{N}{m^2}$$

$$p_2 := p_v = -9.899 \times 10^4 \, \frac{N}{m^2}$$

Guess value: $\qquad v_2 := 2 \, \dfrac{m}{sec} \qquad\qquad D_2 := 0.1 \, m \qquad Q := 1 \, \dfrac{m^3}{sec}$

Given

$$\frac{p_1}{\gamma} + z_1 + \frac{v_1^2}{2 \cdot g} = \frac{p_2}{\gamma} + z_2 + \frac{v_2^2}{2 \cdot g}$$

$$v_1 \cdot \frac{\pi \cdot D_1^2}{4} = v_2 \cdot \frac{\pi \cdot D_2^2}{4}$$

$$Q = v_2 \cdot \frac{\pi \cdot D_2^2}{4}$$

$$\begin{pmatrix} D_2 \\ v_2 \\ Q \end{pmatrix} := \text{Find}(D_2, v_2, Q)$$

$$D_2 = 0.177 \, m \qquad\qquad v_2 = 14.788 \, \frac{m}{s} \qquad\qquad Q = 0.365 \, \frac{m^3}{s}$$

Thus, the pipe diameter of the venturi meter at point 2 should be greater than 0.177 m in order to avoid cavitation.

EXAMPLE PROBLEM 4.6

Water at 20°C flows in a 0.95-m-diameter horizontal pipe. A piezometer is installed at point 1 and a pitot tube is installed at point 2 as illustrated in Figure EP 4.6 in order to measure the flowrate in the pipe. The static pressure head, p_1/γ measured by the piezometer is 1.2 m, and the stagnation pressure head, p_2/γ measured by the pitot tube is 1.8 m. (a) Determine the ideal flowrate in the pipe. (b) Draw the energy grade line and the hydraulic grade line.

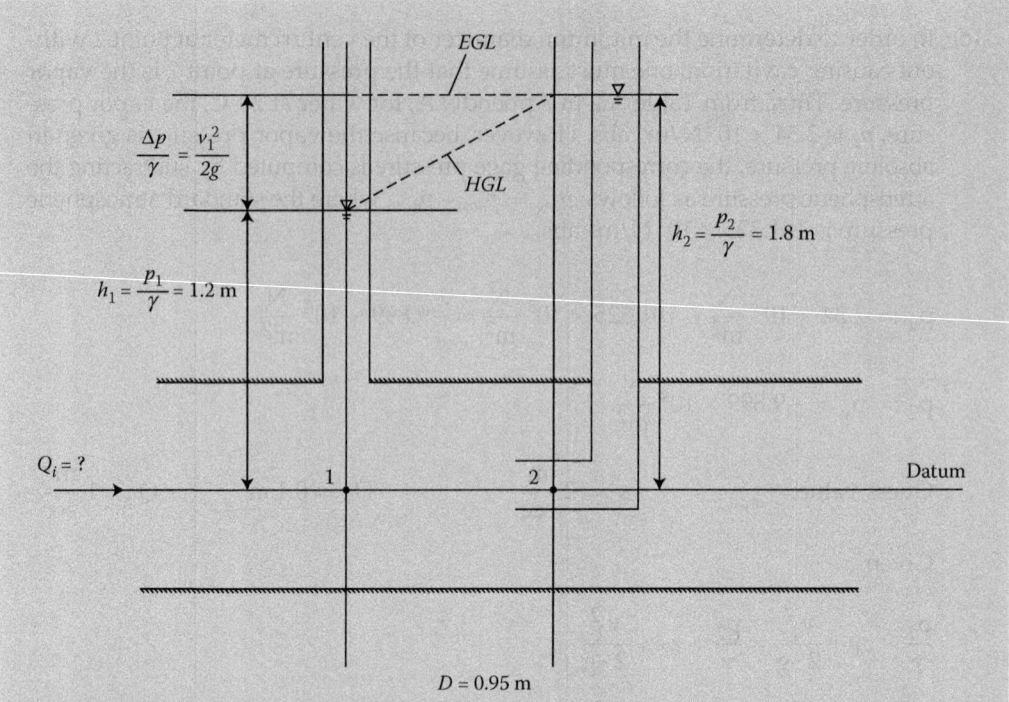

FIGURE EP 4.6
Water flows through a horizontal pipe with a pitot-static tube flowing under pressure (ideal flow).

Mathcad Solution

(a) Assume that the fluid is momentarily brought to a stop at the downstream point 2, where $v_2 = 0$. Then, in order to determine the ideal velocity at point 1, the Bernoulli equation is applied between points 1 and 2. Furthermore, in order to determine the ideal flowrate in the pipe, the continuity equation is applied at point 1 as follows:

$$D := 0.95\,m \qquad A := \frac{\pi \cdot D^2}{4} = 0.709\,m^2 \qquad v_2 := 0\,\frac{m}{sec}$$

$$h_1 := 1.2\,m \qquad h_2 := 1.8\,m \qquad z_1 := 0\,m \qquad z_2 := 0\,m$$

$$\rho := 998\,\frac{kg}{m^3} \qquad g := 9.81\,\frac{m}{sec^2} \qquad\qquad \gamma := \rho \cdot g = 9.79 \times 10^3\,\frac{kg}{m^2\,s^2}$$

$$p_1 := \gamma \cdot h_1 = 1.175 \times 10^4\,\frac{N}{m^2} \qquad\qquad p_2 := \gamma \cdot h_2 = 1.762 \times 10^4\,\frac{N}{m^2}$$

Guess value: $\qquad v_1 := 1\,\frac{m}{sec} \qquad Q := 1\,\frac{m^3}{sec}$

Given

$$\frac{p_1}{\gamma} + z_1 + \frac{v_1^2}{2 \cdot g} = \frac{p_2}{\gamma} + z_2 + \frac{v_2^2}{2 \cdot g}$$

$$Q = v_1 \cdot A$$

$$\begin{pmatrix} v_1 \\ Q \end{pmatrix} := \text{Find}(v_1, Q)$$

$$v_1 = 3.431 \, \frac{m}{s} \qquad\qquad Q = 2.432 \, \frac{m^3}{s}$$

Application of the Bernoulli equation for the pitot-static tube illustrates a conversion of kinetic energy to pressure energy.
(b) The EGL and HGL are illustrated in Figure EP 4.6.

4.5.6.4 Application of the Bernoulli Equation for Ideal Open Channel Flow

EXAMPLE PROBLEM 4.7

Water flows in a rectangular channel of width 3 m at a mild slope. A sluice gate is inserted in the channel as illustrated in Figure EP 4.7 in order to measure the flowrate in the channel. The depth of the flow upstream of the gate at point 1 is measured to be 2 m, and the depth of flow downstream of the gate at point 2 is measured to be 0.95 m. (a) Determine the ideal flowrate in the channel. (b) Draw the energy grade line and the hydraulic grade line.

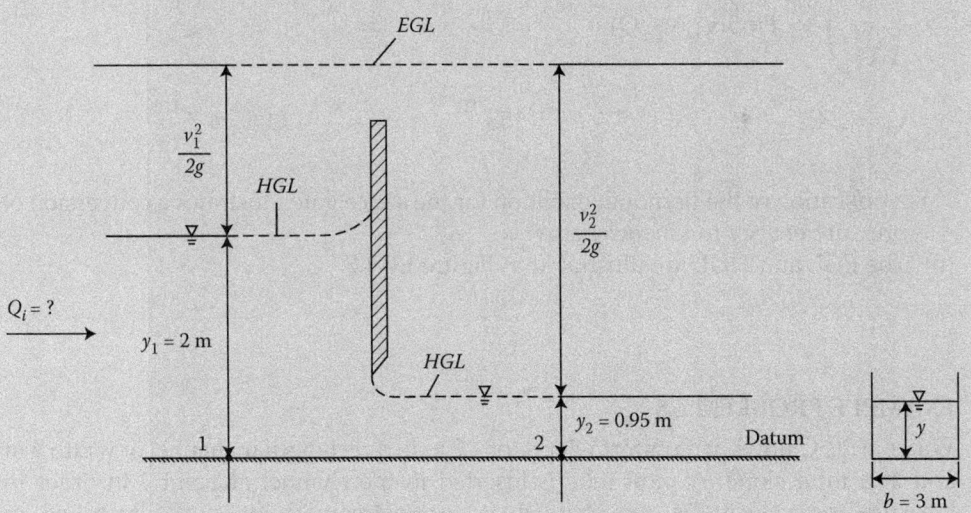

FIGURE EP 4.7
Water flows through a rectangular channel with a sluice gate (ideal flow).

Mathcad Solution

(a) In order to determine the ideal velocity at either point 1 or point 2, the Bernoulli equation is applied between points 1 and 2. However, in order to determine the relationship between the ideal velocities at points 1 and 2, the continuity equation is applied between points 1 and 2; thus, we have two equations and two unknowns. Furthermore, in order to determine the ideal flowrate in the pipe, the continuity equation is applied at either point 1 or 2 as follows:

$$b := 3\,m \qquad y_1 := 2\,m \qquad y_2 := 0.95\,m \qquad z_1 := 0\,m \qquad z_2 := 0\,m$$

$$g := 9.81\,\frac{m}{sec^2} \qquad A_1 := b \cdot y_1 = 6\,m^2 \qquad A_2 := b \cdot y_2 = 2.85\,m^2$$

$$\text{Guess value:} \qquad v_1 := 1\,\frac{m}{sec} \qquad v_2 := 2\,\frac{m}{sec} \qquad Q := 1\,\frac{m^3}{sec}$$

Given

$$y_1 + z_1 + \frac{v_1^2}{2 \cdot g} = y_2 + z_2 + \frac{v_2^2}{2 \cdot g}$$

$$v_1 \cdot A_1 = v_2 \cdot A_2$$

$$Q = v_2 \cdot A_2$$

$$\begin{pmatrix} v_1 \\ v_2 \\ Q \end{pmatrix} := \text{Find}(v_1, v_2, Q)$$

$$v_1 = 2.45\,\frac{m}{s} \qquad v_2 = 5.158\,\frac{m}{s} \qquad Q = 14.7\,\frac{m^3}{s}$$

Application of the Bernoulli equation for the sluice gate illustrates a conversion of pressure energy to kinetic energy.

(b) The EGL and HGL are illustrated in Figure EP 4.7.

EXAMPLE PROBLEM 4.8

Water at 20°C flows at a uniform depth of 1.7 m in a rectangular channel of width 4 m and at a mild slope. A pitot tube is inserted in the channel at point 2 in order to measure the velocity in the channel at an upstream point 1, as illustrated in Figure EP 4.8. The static pressure head at point 1, p_1/γ (the depth to point 1) is 1.5 m. The stagnation pressure head at point 2, p_2/γ measured by the pitot tube is 1.75 m. (a) Determine the ideal flowrate in the channel. (b) Draw the energy grade line and the hydraulic grade line.

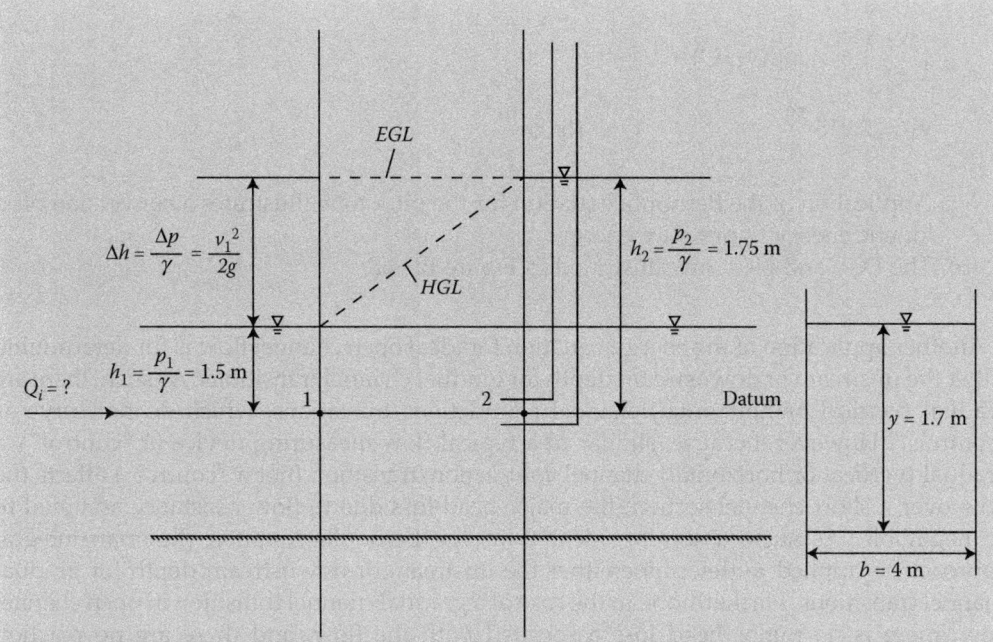

FIGURE EP 4.8
Water flows at a uniform flow through a rectangular channel with a pitot tube (ideal flow).

Mathcad Solution

(a) Assume that the fluid is momentarily brought to a stop at the downstream point 2, where $v_2 = 0$. Then, in order to determine the ideal velocity at point 1, the Bernoulli equation is applied between points 1 and 2. Furthermore, in order to determine the ideal flowrate in the channel, the continuity equation is applied at point 1 as follows:

$$b := 4\,m \qquad y := 1.7\,m \qquad A := b \cdot y = 6.8\,m^2 \qquad\qquad v_2 := 0\,\frac{m}{sec}$$

$$h_1 := 1.5\,m \qquad h_2 := 1.75\,m \qquad z_1 := 0\,m \qquad\qquad z_2 := 0\,m$$

$$\rho := 998\,\frac{kg}{m^3} \qquad g := 9.81\,\frac{m}{sec^2} \qquad\qquad \gamma := \rho \cdot g = 9.79 \times 10^3\,\frac{kg}{m^2\,s^2}$$

$$p_1 := \gamma \cdot h_1 = 1.469 \times 10^4\,\frac{N}{m^2} \qquad\qquad p_2 := \gamma \cdot h_2 = 1.713 \times 10^4\,\frac{N}{m^2}$$

Guess value: $\quad v_1 := 1\,\dfrac{m}{sec} \qquad Q := 1\,\dfrac{m^3}{sec}$

Given

$$\frac{p_1}{\gamma} + z_1 + \frac{v_1^2}{2 \cdot g} = \frac{p_2}{\gamma} + z_2 + \frac{v_2^2}{2 \cdot g}$$

$$Q = v_1 \cdot A$$

$$\begin{pmatrix} v_1 \\ Q \end{pmatrix} := \text{Find}(v_1, Q)$$

$$v_1 = 2.215 \, \frac{m}{s} \qquad\qquad Q = 15.06 \, \frac{m^3}{s}$$

Application of the Bernoulli equation for the pitot tube illustrates a conversion of kinetic energy to pressure energy.
(b) The EGL and HGL are illustrated in Figure EP 4.8.

Another application of the energy equation for ideal open channel flow is for determining either the upstream or downstream depth for gradual channel transitions. As such, there are gradual (vertical or horizontal) channel contractions/transitions, which do not serve as "controls." However, because, similar to a typical flow-measuring device (a "control"), a gradual (vertical or horizontal) channel contraction/transition (not a "control") affects the flow over a short channel section, the major head loss due to flow resistance assumed to be negligible. As such, assuming ideal flow, the Bernoulli equation (Eulerian/integral approach) is applied to determine either the upstream or downstream depth for gradual channel transitions. Furthermore, in the case of a gradual channel transition in open channel flow, there is no minor head loss associated with the flow, and there are no reaction forces involved.

EXAMPLE PROBLEM 4.9

Water flows in a rectangular channel of width 2 m at a velocity of 1 m/sec and depth of 1.75 m. There is a gradual upward step, Δz of 0.25 m in the channel bed as illustrated in Figure EP 4.9. (a) Determine the velocity and the depth of the water downstream of

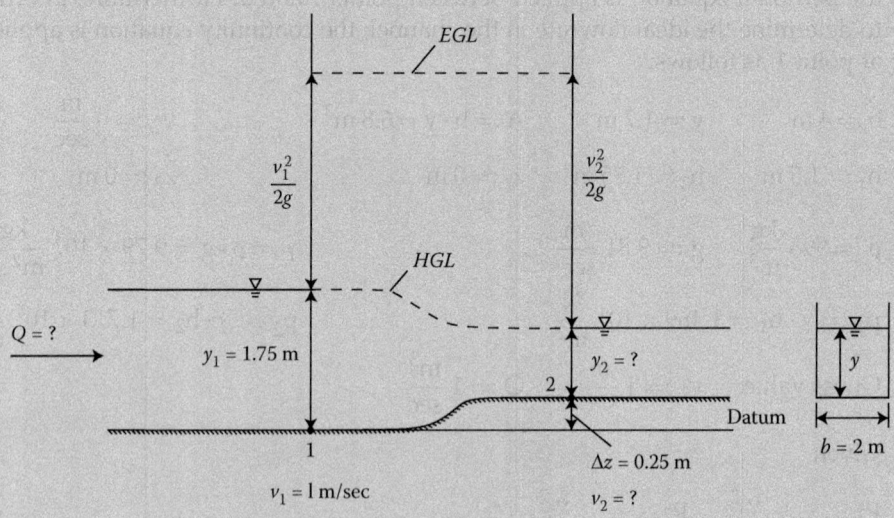

FIGURE EP 4.9
Water flows through a rectangular channel with a gradual step upward (ideal flow).

the step at point 2. (b) Determine the flowrate in the channel. (c) Draw the energy grade line and the hydraulic grade line.

Mathcad Solution

(a) In order to determine the velocity and the depth at point 2, the Bernoulli equation is applied between points 1 and 2. However, in order to determine the relationship between the velocities at points 1 and 2, the continuity equation is applied between points 1 and 2; thus, we have two equations and two unknowns as follows:

$$v_1 := 1 \frac{m}{sec} \qquad y_1 := 1.75\,m \qquad z_1 := 0\,m \qquad z_2 := 0.25\,m$$

$$b := 2\,m \qquad A_1 := b \cdot y_1 = 3.5\,m^2 \qquad g := 9.81 \frac{m}{sec^2}$$

$$\text{Guess value:} \qquad v_2 := 1.2 \frac{m}{sec} \qquad y_2 := 1.5\,m$$

Given

$$y_1 + z_1 + \frac{v_1^2}{2 \cdot g} = y_2 + z_2 + \frac{v_2^2}{2 \cdot g}$$

$$v_1 \cdot b \cdot y_1 = v_2 \cdot b \cdot y_2$$

$$\begin{pmatrix} v_2 \\ y_2 \end{pmatrix} := \text{Find}(v_2, y_2)$$

$$v_2 = 1.183 \frac{m}{s} \qquad y_2 = 1.48\,m$$

(b) In order to determine the flowrate in the channel, the continuity equation is applied at either points 1 or 2 as follows:

$$Q := v_1 \cdot A_1 = 3.5 \frac{m^3}{s}$$

Application of the Bernoulli equation for the gradual upward step illustrates a conversion of pressure energy to potential energy and kinetic energy.

(c) The EGL and HGL are illustrated in Figure EP 4.9.

4.5.6.5 Applications of the Bernoulli Equation for Ideal Flow from a Tank

Application of the energy equation for ideal flow from a tank (or water source) is given by the Bernoulli equation as follows:

$$\left(\frac{p}{\gamma} + z + \frac{v^2}{2g}\right)_1 - \left(\frac{p}{\gamma} + z + \frac{v^2}{2g}\right)_2 = 0 \tag{4.100}$$

There are a number of applications, including flow from a tank (or a water source) through a jet, a siphon, or some other opening/connection to the tank. Because the flow from a tank is assumed be ideal, the Bernoulli equation is used to solve for an unknown pressure, elevation, or velocity. In certain cases of a flow from a tank, the continuity equation is used to solve for the discharge as follows:

$$Q = vA \tag{4.101}$$

Furthermore, in certain cases of a flow from a tank, there may be an unknown reaction force, which may solved by applying the integral form of the momentum equation as follows:

$$\sum F_s = (F_G + F_P + F_V + F_{other})_s = (\rho Q v_s)_2 - (\rho Q v_s)_1 \tag{4.102}$$

which is solved analytically, without the use of dimensional analysis (see Chapter 5).

Applications of the Bernoulli equation for ideal flow from a tank (or a water source) may be classified as follows: (1) forced jets, (2) free jets (Torricelli's theorem), and (3) siphon flows. One may note that while forced jets are a result of pressure flow from a syringe or a pipe with a nozzle (or hose), free jets are a result of gravity flow from an open tank. Furthermore, siphon flows are a result of gravity flow from an open tank with a siphon.

EXAMPLE PROBLEM 4.10

Water at 20°C is forced from a 2-cm-diameter syringe through a 0.1 cm hollow needle as a result of a 0.75 N force applied to the piston, as illustrated in Figure EP 4.10. The final position of the top of the piston at point 1 is 3 cm below the tip of the needle at point 2. (a) Determine the ideal velocity of the forced jet as it leaves the tip of the needle at point 2. (b) Determine the ideal height of the forced jet trajectory at point 3.

Mathcad Solution

(a) In order to determine the ideal velocity of the forced jet as it leaves the tip of the needle at point 2, the Bernoulli equation is applied between points 1 and 2. However, in order to determine the relationship between the velocities at points 1 and 2, the continuity equation is applied between points 1 and 2; thus we have two equations and two unknowns as follows. Assuming the datum is at point 1 yields the following:

$$z_1 := 0 \, m \qquad z_2 := 3 \, cm \qquad F := 0.75 \, N \qquad D_1 := 2 \, cm \qquad D_2 := 0.1 \, cm$$

$$A_1 := \frac{\pi \cdot D_1^2}{4} = 3.142 \times 10^{-4} \, m^2 \qquad\qquad A_2 := \frac{\pi \cdot D_2^2}{4} = 7.854 \times 10^{-7} \, m^2$$

$$P_1 := \frac{F}{A_1} = 2.387 \times 10^3 \, \frac{N}{m^2} \qquad\qquad P_2 := 0 \, \frac{N}{m^2}$$

$$\rho := 998 \, \frac{kg}{m^3} \qquad\qquad g := 9.81 \, \frac{m}{sec^2} \qquad\qquad \gamma := \rho \cdot g = 9.79 \times 10^3 \, \frac{kg}{m^2 \, s^2}$$

$$\text{Guess value:} \quad v_1 := 1 \, \frac{m}{sec} \qquad\qquad\qquad v_2 := 3 \, \frac{m}{sec}$$

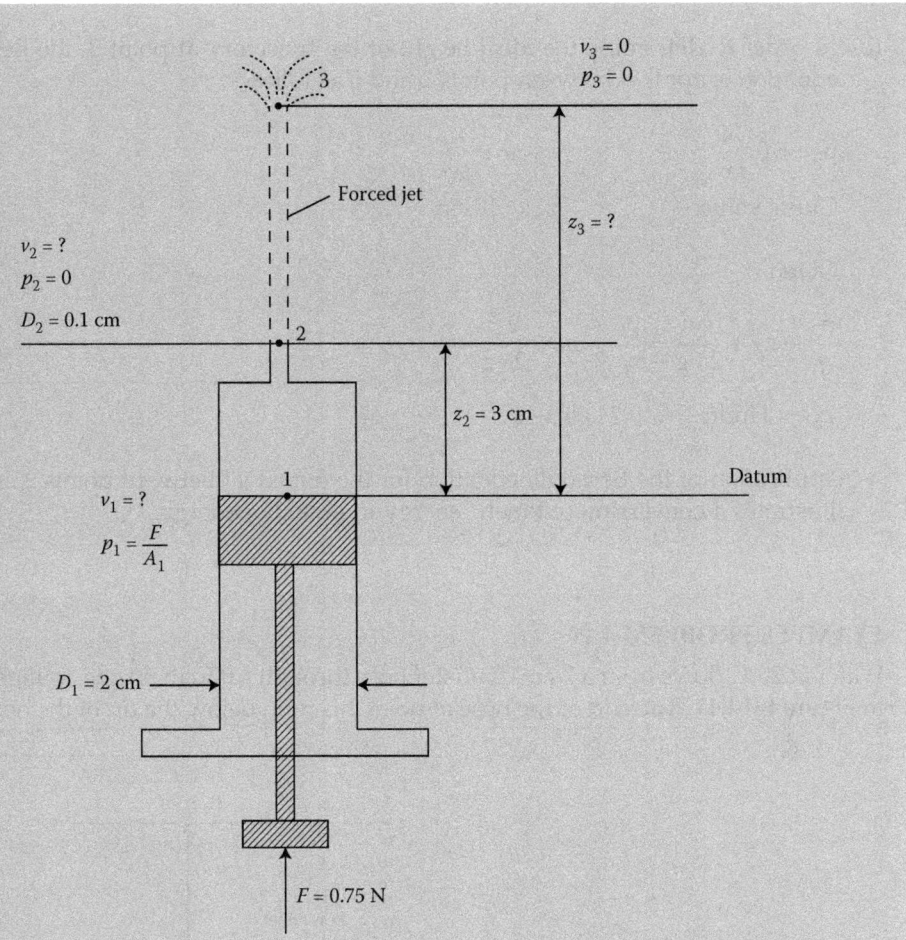

FIGURE EP 4.10
Water is forced from a syringe through a hollow needle by a force applied to the piston (ideal flow).

Given

$$\frac{p_1}{\gamma} + z_1 + \frac{v_1^2}{2 \cdot g} = \frac{p_2}{\gamma} + z_2 + \frac{v_2^2}{2 \cdot g}$$

$$v_1 \cdot A_1 = v_2 \cdot A_2$$

$$\begin{pmatrix} v_1 \\ v_2 \end{pmatrix} := Find(v_1, v_2)$$

$$v_1 = 5.121 \times 10^{-3} \, \frac{m}{s} \qquad\qquad v_2 = 2.048 \, \frac{m}{s}$$

Application of the Bernoulli equation for the forced jet between points 1 and 2 illustrates a conversion of pressure energy to kinetic energy.

(b) In order to determine the ideal height of the trajectory at point 3, the Bernoulli equation is applied between points 2 and 3 as follows:

$$p_3 := 0 \, \frac{N}{m^2} \qquad\qquad v_3 := 0 \, \frac{m}{sec}$$

Guess value: $\qquad\qquad z_3 := 10 \, cm$

Given

$$\frac{p_2}{\gamma} + z_2 + \frac{v_2^2}{2 \cdot g} = \frac{p_3}{\gamma} + z_3 + \frac{v_3^2}{2 \cdot g}$$

$$z_3 := Find(z_3) = 0.244 \, m$$

Application of the Bernoulli equation for the forced jet between points 2 and 3 illustrates a conversion of kinetic energy to potential energy.

EXAMPLE PROBLEM 4.11

Water at 20°C flows from a 3-cm-diameter hose through a 0.5 cm nozzle, as illustrated in Figure EP 4.11. The end of the hose at point 1 is 5 cm below the tip of the nozzle at

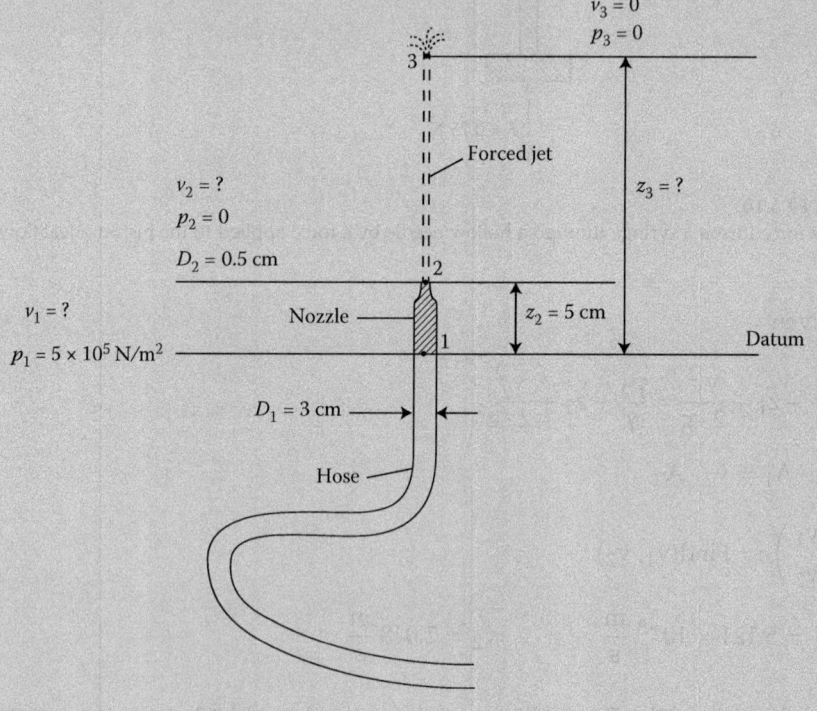

FIGURE EP 4.11
Water flows from a hose through a nozzle (ideal flow).

point 2. And the pressure at the end of the hose at point 1 is $5 \times 10^5 \, \text{N/m}^2$. (a) Determine the ideal velocity of the forced jet as it leaves the tip of the nozzle at point 2. (b) Determine the ideal height of the forced jet trajectory at point 3.

Mathcad Solution

(a) In order to determine the ideal velocity of the forced jet as it leaves the tip of the nozzle at point 2, the Bernoulli equation is applied between points 1 and 2. However, in order to determine the relationship between the velocities at points 1 and 2, the continuity equation is applied between points 1 and 2; thus, we have two equations and two unknowns as follows. Assuming the datum is at point 1 yields the following:

$$z_1 := 0 \, \text{m} \qquad z_2 := 5 \, \text{cm} \qquad D_1 := 3 \, \text{cm} \qquad D_2 := 0.5 \, \text{cm}$$

$$A_1 := \frac{\pi \cdot D_1^2}{4} = 7.069 \times 10^{-4} \, \text{m}^2 \qquad A_2 := \frac{\pi \cdot D_2^2}{4} = 1.963 \times 10^{-5} \, \text{m}^2$$

$$p_1 := 5 \times 10^5 \, \frac{\text{N}}{\text{m}^2} \qquad\qquad p_2 := 0 \, \frac{\text{N}}{\text{m}^2}$$

$$\rho := 998 \, \frac{\text{kg}}{\text{m}^3} \qquad g := 9.81 \, \frac{\text{m}}{\text{sec}^2} \qquad \gamma := \rho \cdot g = 9.79 \times 10^3 \, \frac{\text{kg}}{\text{m}^2 \, \text{s}^2}$$

Guess value: $\qquad v_1 := 1 \, \dfrac{\text{m}}{\text{sec}} \qquad v_2 := 10 \, \dfrac{\text{m}}{\text{sec}}$

Given

$$\frac{p_1}{\gamma} + z_1 + \frac{v_1^2}{2 \cdot g} = \frac{p_2}{\gamma} + z_2 + \frac{v_2^2}{2 \cdot g}$$

$$v_1 \cdot A_1 = v_2 \cdot A_2$$

$$\begin{pmatrix} v_1 \\ v_2 \end{pmatrix} := \text{Find}(v_1, v_2)$$

$$v_1 = 0.879 \, \frac{\text{m}}{\text{s}} \qquad\qquad v_2 = 31.651 \, \frac{\text{m}}{\text{s}}$$

Application of the Bernoulli equation for the forced jet between points 1 and 2 illustrates a conversion of pressure energy to kinetic energy.

(b) In order to determine the ideal height of the trajectory at point 3, the Bernoulli equation is applied between points 2 and 3 as follows:

$$p_3 := 0 \, \frac{\text{N}}{\text{m}^2} \qquad\qquad v_3 := 0 \, \frac{\text{m}}{\text{sec}}$$

Guess value: $\qquad z_3 := 20 \, \text{m}$

Given

$$\frac{p_2}{\gamma} + z_2 + \frac{v_2^2}{2 \cdot g} = \frac{p_3}{\gamma} + z_3 + \frac{v_3^2}{2 \cdot g}$$

$$z_3 := \text{Find}(z_3) = 51.11 \, \text{m}$$

Application of the Bernoulli equation for the forced jet between points 2 and 3 illustrates a conversion of kinetic energy to potential energy.

EXAMPLE PROBLEM 4.12

Water at 20°C flows from a tank through a 4-cm-diameter horizontal pipe with a 1 cm nozzle, as illustrated in Figure EP 4.12. The elevation of the water in the tank at point 1 is 8 m above the tip of the nozzle at point 3. (a) Determine the ideal velocity of the forced jet as it leaves the tip of the nozzle at point 3. (b) Determine the pressure in the pipe at point 2.

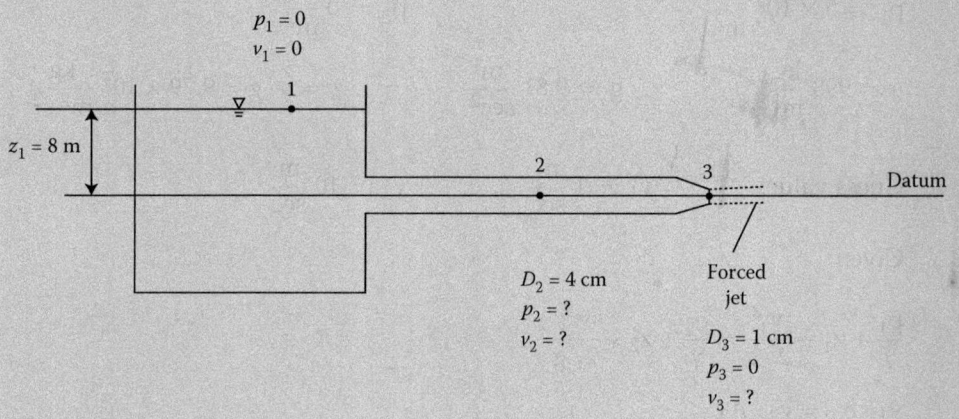

FIGURE EP 4.12
Water flows from a tank through a horizontal pipe with a nozzle (ideal flow).

Mathcad Solution

(a) In order to determine the ideal velocity of the forced jet as it leaves the tip of the nozzle at point 3, the Bernoulli equation is applied between points 1 and 3. Because the cross-sectional area of the tank at point 1 is much larger than the cross-sectional area of the pipe, the velocity at point 1 is assumed to be zero. Assuming the datum is through points 2 and 3 yields the following:

$$z_1 := 8 \, \text{m} \qquad z_3 := 0 \, \text{m} \qquad p_1 := 0 \, \frac{N}{m^2} \qquad p_3 := 0 \, \frac{N}{m^2} \qquad v_1 := 0 \, \frac{m}{\text{sec}}$$

$$\rho := 998 \, \frac{\text{kg}}{m^3} \qquad\qquad g := 9.81 \, \frac{m}{\text{sec}^2} \qquad\qquad \gamma := \rho \cdot g = 9.79 \times 10^3 \, \frac{\text{kg}}{m^2 \, s^2}$$

Guess value: $v_3 := 10 \, \dfrac{m}{sec}$

Given

$$\dfrac{p_1}{\gamma} + z_1 + \dfrac{v_1^2}{2 \cdot g} = \dfrac{p_3}{\gamma} + z_3 + \dfrac{v_3^2}{2 \cdot g}$$

$v_3 := \text{Find}(v_3) = 12.528 \, \dfrac{m}{s}$

Application of the Bernoulli equation for the forced jet between points 1 and 3 illustrates a conversion of potential energy to kinetic energy.

(b) In order to determine the pressure in the pipe at point 2, the Bernoulli equation is applied between points 2 and 3. However, in order to determine the relationship between the velocities at points 2 and 3, the continuity equation is applied between points 2 and 3; thus, we have two equations and two unknowns as follows:

$z_2 := 0 \, m$ $\qquad$ $D_2 := 4 \, cm$ $\qquad$ $D_3 := 1 \, cm$

$A_2 := \dfrac{\pi \cdot D_2^2}{4} = 1.257 \times 10^{-3} \, m^2$ $\qquad$ $A_3 := \dfrac{\pi \cdot D_3^2}{4} = 7.854 \times 10^{-5} \, m^2$

Guess value: $\qquad p_2 := 1 \times 10^3 \, \dfrac{N}{m^2}$ $\qquad$ $v_2 := 1 \, \dfrac{m}{sec}$

Given

$$\dfrac{p_2}{\gamma} + z_2 + \dfrac{v_2^2}{2 \cdot g} = \dfrac{p_3}{\gamma} + z_3 + \dfrac{v_3^2}{2 \cdot g}$$

$v_2 \cdot A_2 = v_3 \cdot A_3$

$\begin{pmatrix} p_2 \\ v_2 \end{pmatrix} := \text{Find}(p_2, v_2)$

$p_2 = 7.802 \times 10^4 \, \dfrac{N}{m^2}$ $\qquad\qquad\qquad$ $v_2 = 0.783 \, \dfrac{m}{s}$

Application of the Bernoulli equation for the forced jet between points 2 and 3 illustrates a conversion of pressure energy to kinetic energy. Application of the Bernoulli equation between points 1 and 2 would illustrate a conversion of potential energy to pressure energy (and kinetic energy).

EXAMPLE PROBLEM 4.13

Water at 20°C flows from a tank through a 5-cm-diameter vertical pipe with a 1.5-cm nozzle, as illustrated in Figure EP 4.13. The elevation of the water in the tank at point 1 is 25 m above the tip of the nozzle at point 5. The elevation of points 2, 3, and 4 are 15, 7, and 0.3 m above the tip of the nozzle at point 5, respectively. (a) Determine the ideal velocity of the forced jet as it leaves the tip of the nozzle at point 5. (b) Determine the pressure in the pipe at points 2, 3, and 4.

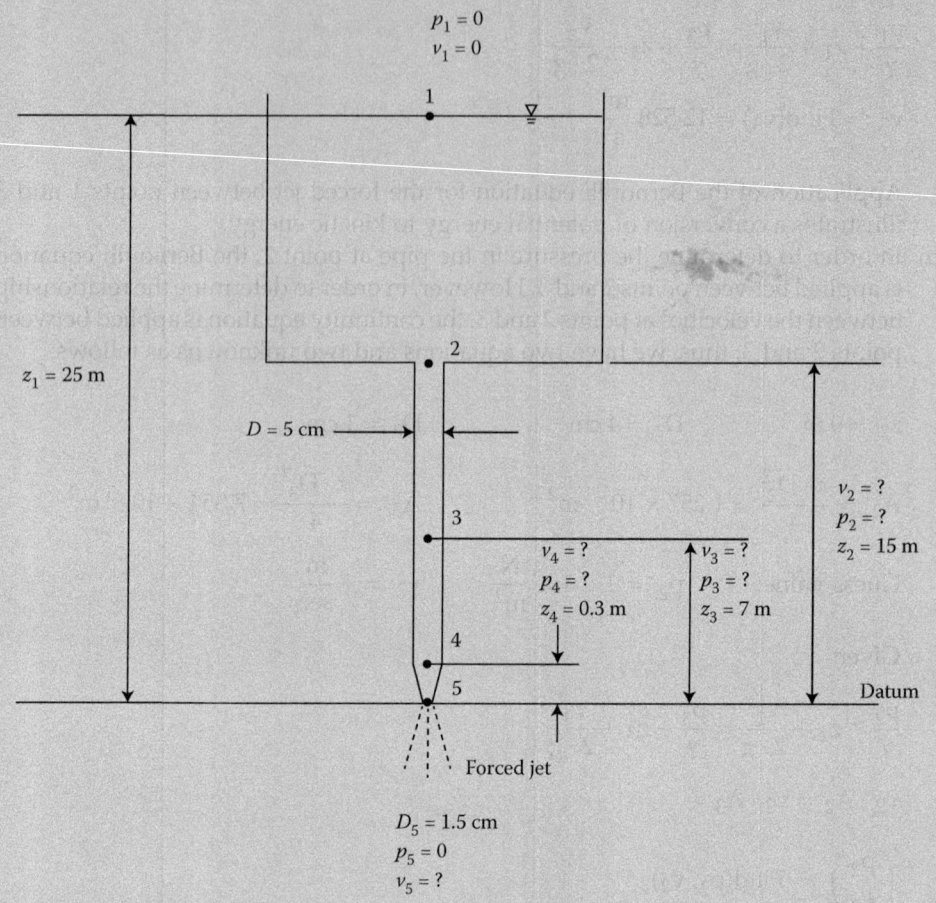

FIGURE EP 4.13
Water flows from a tank through a vertical pipe with a nozzle (ideal flow).

Mathcad Solution

(a) In order to determine the ideal velocity of the forced jet as it leaves the tip of the nozzle at point 5, the Bernoulli equation is applied between points 1 and 5. Because the cross-sectional area of the tank at point 1 is much larger than the cross-sectional area of the pipe, the velocity at point 1 is assumed to be zero. Assuming the datum is at point 5 yields the following:

$$z_1 := 25 \, m \qquad z_5 := 0 \, m \qquad p_1 := 0 \, \frac{N}{m^2} \qquad p_5 := 0 \, \frac{N}{m^2} \qquad v_1 := 0 \, \frac{m}{sec}$$

$$\rho := 998 \, \frac{kg}{m^3} \qquad\qquad g := 9.81 \, \frac{m}{sec^2} \qquad\qquad \gamma := \rho \cdot g = 9.79 \times 10^3 \, \frac{kg}{m^2 \, s^2}$$

$$\text{Guess value:} \qquad v_5 := 10 \, \frac{m}{sec}$$

Given

$$\frac{p_1}{\gamma} + z_1 + \frac{v_1^2}{2 \cdot g} = \frac{p_5}{\gamma} + z_5 + \frac{v_5^2}{2 \cdot g}$$

$$v_5 := Find(v_5) = 22.147 \frac{m}{s}$$

Application of the Bernoulli equation for the forced jet between points 1 and 5 illustrates a conversion of potential energy to kinetic energy.

(b) In order to determine the pressure in the pipe at point 2, the Bernoulli equation is applied between points 2 and 5. However, in order to determine the relationship between the velocities at points 2 and 5, the continuity equation is applied between points 2 and 5; thus, we have two equations and two unknowns as follows:

$$z_2 := 15 \, m \qquad D_2 := 5 \, cm \qquad\qquad D_5 := 1.5 \, cm$$

$$A_2 := \frac{\pi \cdot D_2^2}{4} = 1.963 \times 10^{-3} \, m^2 \qquad A_5 := \frac{\pi \cdot D_5^2}{4} = 1.767 \times 10^{-4} \, m^2$$

Guess value: $\qquad p_2 := 1 \times 10^3 \, \dfrac{N}{m^2} \qquad v_2 := 1 \, \dfrac{m}{sec}$

Given

$$\frac{p_2}{\gamma} + z_2 + \frac{v_2^2}{2 \cdot g} = \frac{p_5}{\gamma} + z_5 + \frac{v_5^2}{2 \cdot g}$$

$$v_2 \cdot A_2 = v_5 \cdot A_5$$

$$\binom{p_2}{v_2} := Find(p_2, v_2)$$

$$p_2 = 9.592 \times 10^4 \, \frac{N}{m^2} \qquad\qquad\qquad v_2 = 1.993 \, \frac{m}{s}$$

Application of the Bernoulli equation for the forced jet between points 2 and 5 illustrates a conversion of pressure energy (and potential energy) to kinetic energy. Application of the Bernoulli equation between points 1 and 2 would illustrate a conversion of potential energy to pressure energy (and kinetic energy). In order to determine the pressure in the pipe at point 3, the Bernoulli equation is applied between points 3 and 5. However, in order to determine the relationship between the velocities at points 3 and 5, the continuity equation is applied between points 3 and 5; thus, we have two equations and two unknowns as follows:

$$z_3 := 7 \, m \qquad D_3 := 5 \, cm \qquad\qquad D_5 := 1.5 \, cm$$

$$A_3 := \frac{\pi \cdot D_3^2}{4} = 1.963 \times 10^{-3} m^2 \qquad A_5 := \frac{\pi \cdot D_5^2}{4} = 1.767 \times 10^{-4} \, m^2$$

Guess value: $\qquad p_3 := 1 \times 10^3 \, \dfrac{N}{m^2} \qquad v_3 := 1 \, \dfrac{m}{sec}$

Given

$$\frac{p_3}{\gamma} + z_3 + \frac{v_3^2}{2 \cdot g} = \frac{p_5}{\gamma} + z_5 + \frac{v_5^2}{2 \cdot g}$$

$$v_3 \cdot A_3 = v_5 \cdot A_5$$

$$\begin{pmatrix} p_3 \\ v_3 \end{pmatrix} := Find(p_3, v_3)$$

$$p_3 = 1.742 \times 10^5 \, \frac{N}{m^2} \qquad\qquad v_3 = 1.993 \, \frac{m}{s}$$

Application of the Bernoulli equation for the forced jet between points 3 and 5 illustrates a conversion of pressure energy (and potential energy) to kinetic energy. In order to determine the pressure in the pipe at point 4, the Bernoulli equation is applied between points 4 and 5. However, in order to determine the relationship between the velocities at points 4 and 5, the continuity equation is applied between points 4 and 5; thus, we have two equations and two unknowns as follows:

$$z_4 := 0.3 \, m \qquad\qquad D_4 := 5 \, cm \qquad\qquad D_5 := 1.5 \, cm$$

$$A_4 := \frac{\pi \cdot D_4^2}{4} = 1.963 \times 10^{-3} \, m^2 \qquad\qquad A_5 := \frac{\pi \cdot D_5^2}{4} = 1.767 \times 10^{-4} \, m^2$$

Guess value: $\qquad p_4 := 1 \times 10^3 \, \frac{N}{m^2} \qquad v_4 := 1 \, \frac{m}{sec}$

Given

$$\frac{p_4}{\gamma} + z_4 + \frac{v_4^2}{2 \cdot g} = \frac{p_5}{\gamma} + z_5 + \frac{v_5^2}{2 \cdot g}$$

$$v_4 \cdot A_4 = v_5 \cdot A_5$$

$$\begin{pmatrix} p_4 \\ v_4 \end{pmatrix} := Find(p_4, v_4)$$

$$p_4 = 2.398 \times 10^5 \, \frac{N}{m^2} \qquad\qquad v_4 = 1.993 \, \frac{m}{s}$$

Application of the Bernoulli equation for the forced jet between points 4 and 5 illustrates a conversion of pressure energy (and potential energy) to kinetic energy.

Applications of the Bernoulli equation for ideal flow from a tank (or a water source) include free jets, which are a result of gravity flow from an open tank and are governed by Torricelli's theorem. Torricelli's theorem is a special case of the Bernoulli equation, which assumes incompressible, steady, and ideal flow. Specifically, Torricelli's theorem states that the velocity of a free jet is a result of gravity flow from an open tank, as illustrated in Figure 4.15. The open tank is large in comparison to the small opening in the side of the

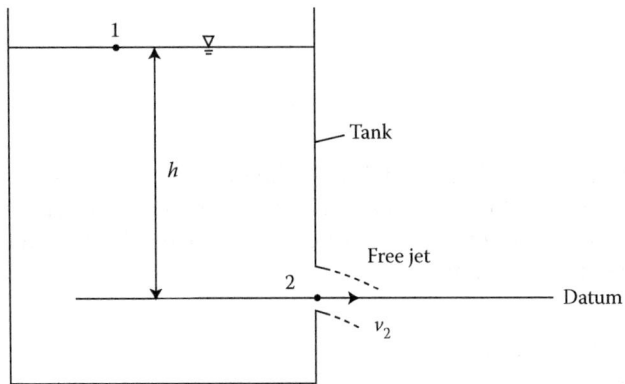

FIGURE 4.15
Free jet as a result of gravity flow from an open tank (Torricelli's theorem).

tank. The elevation of the fluid in the tank at point 1 is at a height, h above the centerline of the small opening in the side of the tank. The free jet of fluid flows into the atmosphere forming a trajectory. In order to derive Torricelli's theorem, the Bernoulli equation is applied between points 1 and 2. The pressure at points 1 and 2 are assumed to be atmospheric; thus:

$$p_1 = 0 \tag{4.103}$$

$$p_2 = 0 \tag{4.104}$$

The datum is assumed to be at point 2; thus:

$$z_1 = h \tag{4.105}$$

$$z_2 = 0 \tag{4.106}$$

The cross-sectional area of the tank at point 1 is assumed to be much larger in comparison to the cross-sectional area of the opening in the side of the tank at point 2; thus:

$$v_1 = 0 \tag{4.107}$$

Thus, application of the Bernoulli equation between points 1 and 2 yields the following:

$$\frac{p_1}{\gamma} + z_1 + \frac{v_1^2}{2g} = \frac{p_2}{\gamma} + z_2 + \frac{v_2^2}{2g} \tag{4.108}$$

Evaluation of the terms in the Bernoulli equation yields the following:

$$0 + h + 0 = 0 + 0 + \frac{v_2^2}{2g} \tag{4.109}$$

where

$$v_2 = \sqrt{2gh} \qquad\qquad (4.110)$$

which is Torricelli's theorem. Thus, Torricelli's theorem, which is a special case of the Bernoulli equation, specifically illustrates the conversion of potential energy stored in the height of the fluid in the tank, h to kinetic energy stored in the ideal velocity of the free jet of fluid at the opening in the side of the tank.

EXAMPLE PROBLEM 4.14

Water at 20°C flows from an open tank through a 5-cm opening on the side of the tank as illustrated in Figure EP 4.14. The elevation of the water in the tank at point 1 is 14 m above the centerline of the opening at point 2. (a) Determine the ideal velocity of the free jet as it leaves the opening at point 2. (b) Determine the ideal discharge from the open tank.

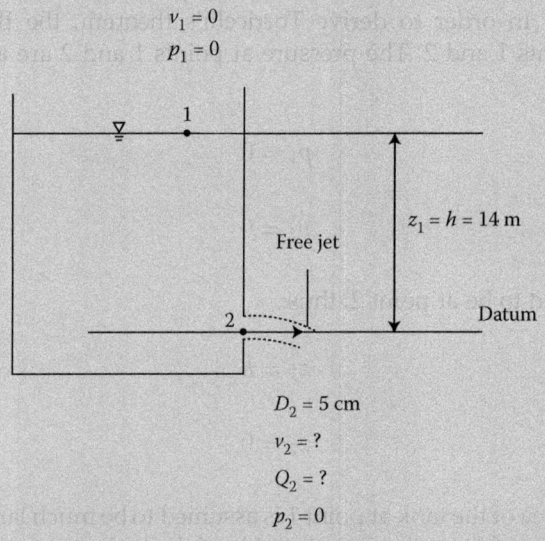

FIGURE EP 4.14
Water flows from an open tank through an opening on the side of the tank (ideal flow).

Mathcad Solution

(a) In order to determine the ideal velocity of the free jet at point 2, the Bernoulli equation is applied between points 1 and 2. Because the cross-sectional area of the tank at point 1 is much larger than the cross-sectional area of the opening, the velocity at point 1 is assumed to be zero. Assuming the datum is at point 2 yields the derivation and application of Torricelli's theorem as follows:

$$z_1 := 14\,\text{m} \qquad z_2 := 0\,\text{m} \qquad p_1 := 0\,\frac{\text{N}}{\text{m}^2} \qquad p_2 := 0\,\frac{\text{N}}{\text{m}^2} \qquad v_1 := 0\,\frac{\text{m}}{\text{sec}}$$

$$\rho := 998\,\frac{\text{kg}}{\text{m}^3} \qquad\qquad g := 9.81\,\frac{\text{m}}{\text{sec}^2} \qquad\qquad \gamma := \rho \cdot g = 9.79 \times 10^3\,\frac{\text{kg}}{\text{m}^2\,\text{s}^2}$$

Guess value: $v_2 := 5 \dfrac{m}{\sec}$

Given

$$\dfrac{P_1}{\gamma} + z_1 + \dfrac{v_1^2}{2 \cdot g} = \dfrac{P_2}{\gamma} + z_2 + \dfrac{v_2^2}{2 \cdot g}$$

$v_2 := \text{Find}(v_2) = 16.573 \dfrac{m}{s}$

Alternatively, one may directly apply Torricelli's theorem as follows:

$h := z_1 = 14\,m$

$v_2 := \sqrt{2 \cdot g \cdot h} = 16.573 \dfrac{m}{s}$

Thus, Torricelli's theorem, which is a special case of the Bernoulli equation, specifically illustrates the conversion of potential energy stored in the height of the fluid in the tank, *h* to kinetic energy stored in the ideal velocity of the free jet of fluid at the opening in the side of the tank.

(b) In order to determine the ideal discharge from the open tank, the continuity equation is applied at the free jet point 2 as follows:

$D_2 := 5\,cm$ $\qquad\qquad A_2 := \dfrac{\pi \cdot D_2^2}{4} = 1.963 \times 10^{-3}\,m^2$

$Q_2 := v_2 \cdot A_2 = 0.033 \dfrac{m^3}{s}$

EXAMPLE PROBLEM 4.15

Water flows from an open tank through a 3-cm opening on the side of the tank as illustrated in Figure EP 4.15. The elevation of the water in the tank at point 1 is 7 m above the centerline of the opening at point 2. The elevation of the centerline of the opening at point 2 is 5 m above the ground. (a) Determine the ideal velocity of the free jet as it leaves the opening at point 2. (b) Determine the ideal horizontal distance of the free jet trajectory as it hits the ground.

Mathcad Solution

(a) Because the cross-sectional area of the tank at point 1 is much larger than the cross-sectional area of the opening, the velocity at point 1 is assumed to be zero. In order to determine the ideal velocity of the free jet at point 2, Torricelli's theorem is applied as follows:

$z_1 := 7\,m$ $\qquad\qquad h := z_1 = 7\,m$ $\qquad\qquad g := 9.81 \dfrac{m}{\sec^2}$

$v_2 := \sqrt{2 \cdot g \cdot h} = 11.719 \dfrac{m}{s}$

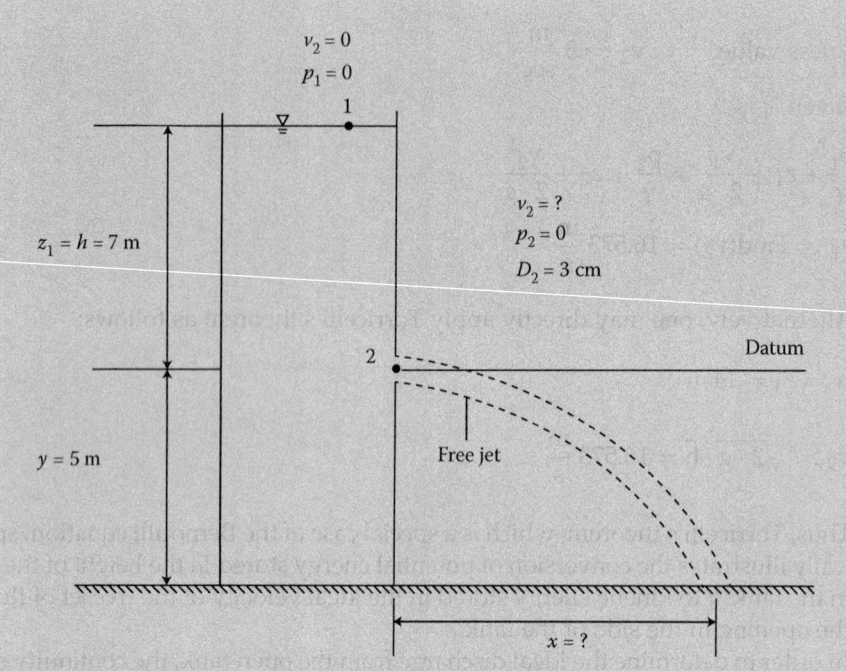

$v_2 = 0$
$p_1 = 0$

1

$z_1 = h = 7$ m

$v_2 = ?$
$p_2 = 0$
$D_2 = 3$ cm

2

Datum

$y = 5$ m

Free jet

$x = ?$

FIGURE EP 4.15
Water flows from an open tank through an opening on the side of the tank (ideal flow).

(b) In order to determine the ideal horizontal distance of the free jet trajectory, x as it hits the ground, one must apply Newton's equations of uniformly accelerated motion to the free jet trajectory. First, one must note that initially, as the flow leaves the opening in the tank, the ideal velocity of the free jet, v_2 is horizontal. Thus, the ideal horizontal distance, x of the free trajectory, which takes a time, t to hit the ground, is related to the constant ideal velocity, v_2 in the horizontal direction as follows:

$$x = v_2 t$$

Furthermore, one must note that the free jet trajectory is a body of fluid falling freely in the atmosphere to a vertical distance, y at a constant acceleration due to gravity, g; thus:

$$v_2 = \frac{dy}{dt} = gt$$

Thus, integrating and solving for the vertical distance, y yields the following relationship between y and t:

$$\int dy = \int gt\, dt$$

$$y = \frac{gt^2}{2}$$

Finally, substituting the expression $x = v_2 t$ into the expression $y = gt^2/2$ and eliminating the variable time, t yields the following relationship between the ideal horizontal distance of the free jet trajectory, x; the ideal velocity, v_2; and the vertical distance, y as follows:

$$x = v_2 \sqrt{\frac{2y}{g}}$$

Thus, for a vertical distance, y of 5 m, the ideal horizontal distance of the free jet trajectory, x is computed as follows:

$$y := 5\,m \qquad\qquad x := v_2 \sqrt{\frac{2 \cdot y}{g}} = 11.832\,m$$

EXAMPLE PROBLEM 4.16

Water at 20°C flows from an open tank through a 1.25-cm nozzle at the bottom of the tank as illustrated in Figure EP 4.16. The elevation of the water in the tank at point 1 is 13 m above the tip of the nozzle at point 2. The distance the water falls below the nozzle at point 3 is 0.75 m below the tip of the nozzle. (a) Determine the ideal velocity of the free jet as it leaves the nozzle at point 2. (b) Determine the ideal discharge from the open tank. (c) Determine the ideal velocity of the free jet just below the nozzle at point 3.

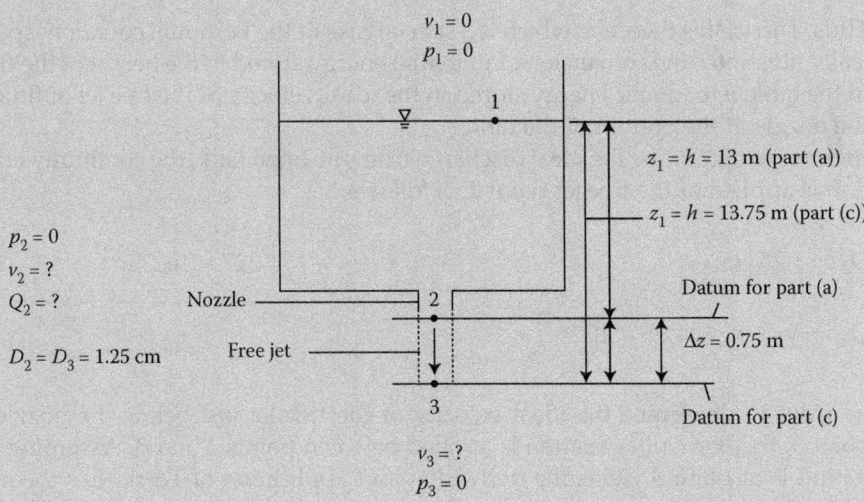

FIGURE EP 4.16
Water flows from an open tank through a nozzle at the bottom of the tank (ideal flow).

Mathcad Solution

(a) In order to determine the ideal velocity of the free jet at point 2, the Bernoulli equa-
    tion is applied between points 1 and 2. Because the cross-sectional area of the tank
    at point 1 is much larger than the cross-sectional area of the opening, the velocity at
    point 1 is assumed to be zero. Assuming the datum is at point 2 yields the deriva-
    tion and application of Torricelli's theorem as follows:

$$z_1 := 13\,m \qquad\qquad z_2 := 0\,m \qquad\qquad P_1 := 0\,\frac{N}{m^2} \qquad P_2 := 0\,\frac{N}{m^2} \qquad v_1 := 0\,\frac{m}{sec}$$

$$\rho := 998\,\frac{kg}{m^3} \qquad\qquad g := 9.81\,\frac{m}{sec^2} \qquad\qquad \gamma := \rho \cdot g = 9.79 \times 10^3\,\frac{kg}{m^2\,s^2}$$

Guess value: $\qquad v_2 := 5\,\dfrac{m}{sec}$

Given

$$\frac{P_1}{\gamma} + z_1 + \frac{v_1^2}{2 \cdot g} = \frac{P_2}{\gamma} + z_2 + \frac{v_2^2}{2 \cdot g}$$

$$v_2 := Find(v_2) = 15.971\,\frac{m}{s}$$

Alternatively, one may directly apply Torricelli's theorem as follows:

$$h := z_1 = 13\,m$$

$$v_2 := \sqrt{2 \cdot g \cdot h} = 15.971\,\frac{m}{s}$$

Thus, Torricelli's theorem, which is a special case of the Bernoulli equation, specifically illustrates the conversion of potential energy stored in the height of the fluid
in the tank, h to kinetic energy stored in the ideal velocity of the free jet of fluid at
the nozzle at the bottom of the tank.

(b) In order to determine the ideal discharge from the open tank, the continuity equa-
    tion is applied at the free jet point 2 as follows:

$$D_2 := 1.25\,cm \qquad\qquad\qquad\qquad A_2 := \frac{\pi \cdot D_2^2}{4} = 1.227 \times 10^{-4}\,m^2$$

$$Q_2 := v_2 \cdot A_2 = 1.96 \times 10^{-3}\,\frac{m^3}{s}$$

(c) In order to determine the ideal velocity of the free jet just below the nozzle at
    point 3, the Bernoulli equation is applied between points 1 and 3. Assuming the
    datum is at point 3 yields the derivation and application of Torricelli's theorem
    as follows:

$$z_1 := 13\,m + 0.75\,m = 13.75\,m \qquad z_3 := 0\,m \qquad\qquad P_1 := 0\,\frac{N}{m^2} \qquad P_3 := 0\,\frac{N}{m^2}$$

Guess value: $\qquad v_3 := 10 \dfrac{m}{sec}$

Given

$$\frac{P_1}{\gamma} + z_1 + \frac{v_1^2}{2 \cdot g} = \frac{P_3}{\gamma} + z_3 + \frac{v_3^2}{2 \cdot g}$$

$$v_3 := Find(v_3) = 16.425 \frac{m}{s}$$

Alternatively, one may directly apply Torricelli's theorem as follows:

$$h := z_1 = 13.75 \, m$$

$$v_3 := \sqrt{2 \cdot g \cdot h} = 16.425 \frac{m}{s}$$

Thus, just below the nozzle, the flow continues to fall as a free jet with an increase in velocity.

Applications of the Bernoulli equation for ideal flow from a tank (or a water source) include siphon flows, which are a result of gravity flow from an open tank with a siphon. A siphon is a small-diameter hose that is commonly used to transfer liquid from one tank to a second tank at a lower elevation without the use of a pump, as illustrated in Figure 4.16. In order for the siphon to work properly, there are several criteria that must be met. First, one end of the siphon is inserted into the tank full of fluid, and the free end of the siphon (point 3) must be at an elevation that is lower than the level of the fluid in the full tank (point 1) and may be inserted into a second tank at a lower elevation. Second, in order to siphon the fluid from the full tank into the second tank, a suction in the siphon must be established/initiated. One may initiate a suction in the siphon by temporarily suctioning (for instance, by sucking

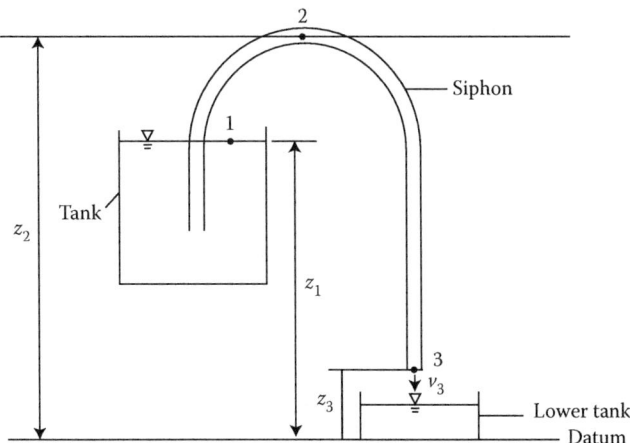

FIGURE 4.16
Siphon: used to transfer liquid from one tank to a second tank at a lower elevation.

on the free end of the tube) the free end of the siphon tube. As a result, the pressure difference created between the atmospheric pressure at point 1 and the negative pressure (vacuum) at point 2 causes the fluid to flow from the tank into the siphon tube, and then
finally out of the siphon tube into the second tank. And, third, in order to maintain the
flow in the siphon, the elevation of tubing at point 2 must be high enough above point 1
to maintain the established negative pressure (vacuum) at point 2. However, if the elevation
of point 2 is too high above point 1, the pressure at point 2 (suction end of the siphon) is
decreased to the liquid's vapor pressure, and cavitation will occur. Cavitation is unintended
boiling of a liquid in motion at room temperature. Furthermore, when evaporation of the
liquid occurs during cavitation, the vapor may cause a pocket to form in the suction line
of the siphon and thus stop the flow of the liquid. Therefore, in the design of a siphon, there
is a maximum allowable height for point 2 in the tubing in order to avoid a pressure drop
below the vapor pressure of the liquid and thus avoid cavitation and restriction of flow.
As such, determination of the maximum allowable height for point 2 assumes that the
pressure at point 2 is set equal to the vapor pressure of the liquid in the tank (see Example
Problem 4.17).

EXAMPLE PROBLEM 4.17

Water at 20°C is siphoned from an open tank through a 2-cm siphon tube to a second
tank at a lower elevation as illustrated in Figure EP 4.17. The elevation of the water in
the tank at point 1 is 7 m above the bottom of the second tank. The elevation of the
free end of the siphon tube at point 3 is 1.5 m above the bottom of the second tank.
(a) Determine the ideal velocity of the flow as it leaves the siphon tube at point 3.

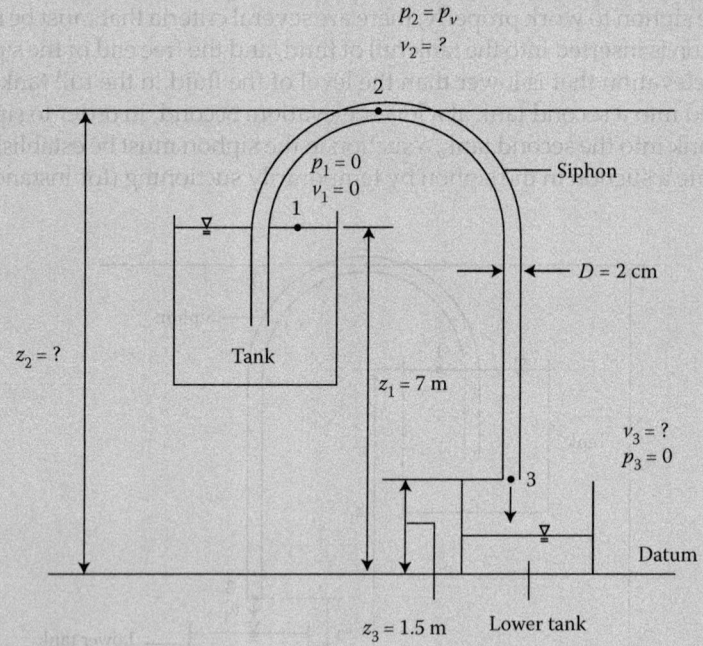

FIGURE EP 4.17
Water is siphoned from an open tank through a siphon tube to a second lower tank (ideal flow).

(b) Determine the ideal maximum allowable height of point 2 without causing cavitation.

Mathcad Solution

(a) In order to determine the ideal velocity of the flow as it leaves the siphon at point 3, the Bernoulli equation is applied between points 1 and 3. Because the cross-sectional area of the tank at point 1 is much larger than the cross-sectional area of the tubing, the velocity at point 1 is assumed to be zero. Assuming the datum is at the bottom of the second tank as follows:

$$z_1 := 7\,m \qquad z_3 := 1.5\,m \qquad p_1 := 0\,\frac{N}{m^2} \qquad p_3 := 0\,\frac{N}{m^2} \qquad v_1 := 0\,\frac{m}{sec}$$

$$\rho := 998\,\frac{kg}{m^3} \qquad g := 9.81\,\frac{m}{sec^2} \qquad \gamma := \rho \cdot g = 9.79 \times 10^3\,\frac{kg}{m^2\,s^2}$$

Guess value: $\quad v_3 := 10\,\dfrac{m}{sec}$

Given

$$\frac{p_1}{\gamma} + z_1 + \frac{v_1^2}{2 \cdot g} = \frac{p_3}{\gamma} + z_3 + \frac{v_3^2}{2 \cdot g}$$

$$v_3 := Find(v_3) = 10.388\,\frac{m}{s}$$

Application of the Bernoulli equation between points 1 and 3 illustrates the conversion of potential energy stored in the difference in elevation between points 1 and 3 to kinetic energy stored in the ideal velocity of the flow as it leaves the siphon at point 3.

(b) In order to determine the ideal maximum allowable height of point 2 without causing cavitation, the Bernoulli equation is applied between points 2 and 3, assuming that the pressure at point 2 is the vapor pressure of the water at 20°C. However, in order to determine the relationship between the ideal velocities at points 2 and 3, the continuity equation is applied between points 2 and 3. Thus, from Table A.2 in Appendix A, for water at 20°C, the vapor pressure, p_v is $2.34 \times 10^3\,N/m^2$ abs. However, because the vapor pressure is given in absolute pressure, the corresponding gage pressure is computed by subtracting the atmospheric pressure as follows: $p_{gage} = p_{abs} - p_{atm}$, where the standard atmospheric pressure is $101.325 \times 10^3\,N/m^2$ abs.

$$p_v := 2.34 \times 10^3\,\frac{N}{m^2} - 101.325 \times 10^3\,\frac{N}{m^2} = -9.899 \times 10^4\,\frac{N}{m^2}$$

$$p_2 := p_v = -9.899 \times 10^4\,\frac{N}{m^2} \qquad\qquad D_2 := 2\,cm \qquad\qquad D_3 := 2\,cm$$

Guess value: $\qquad\qquad v_2 := 10\,\dfrac{m}{sec} \qquad\qquad z_2 := 15\,m$

Given

$$\frac{p_2}{\gamma} + z_2 + \frac{v_2^2}{2 \cdot g} = \frac{p_3}{\gamma} + z_3 + \frac{v_3^2}{2 \cdot g}$$

$$v_2 \cdot \frac{\pi \cdot D_2^2}{4} = v_3 \cdot \frac{\pi \cdot D_3^2}{4}$$

$$\begin{pmatrix} z_2 \\ v_2 \end{pmatrix} := \text{Find}(z_2, v_2)$$

$$z_2 = 11.61 \, m \qquad\qquad v_2 = 10.388 \, \frac{m}{s}$$

Thus, the maximum allowable height of point 2 should be less than 11.61 m in order to avoid cavitation.

4.5.7 Application of the Energy and Momentum Equations for Ideal External Flow

Application of the Bernoulli equation (Equation 4.80) assumes ideal flow, so there is no head loss term to evaluate in the energy equation. Therefore, for problems that involve external flow around an object, the major head loss due to flow resistance is assumed to be negligible; thus, one may apply the Bernoulli equation. Thus, the Bernoulli equation may be solved for an unknown "energy head" term (p/γ, z, or $v^2/2g$) for one of the two points in a finite control volume. These types of flow problems include velocity measuring devices for external flow, which, however, have a minor loss associated with the velocity measurement. The associated minor head loss is typically accounted for by the use/calibration of a velocity/drag coefficient to determine the drag force, where ideal flow (Bernoulli equation) is assumed. Thus, in the case of velocity measuring devices (e.g., pitot-static tubes), one must indirectly model the minor head loss flow resistance term by the use of a velocity/drag coefficient, which requires a "subset level" of application of governing equations (Section 4.5.7.1). Then, application of the complementary integral momentum equation (Equation 4.82) is required only if there is an unknown reaction force acting on the object; note that the pressure and viscous forces are represented by the drag force, and the viscous forces have already been accounted for indirectly through the velocity/drag coefficient. As such, the integral form of the momentum equation is solved analytically, without the use of dimensional analysis. Applications of the governing equations for ideal external flow are presented in Sections 4.5.7.2 and 4.5.7.3.

4.5.7.1 Evaluation of the Drag Force

In the case of external flow around an object, the associated minor head loss is typically accounted for by the use/calibration of a drag coefficient to determine the drag force. Evaluation of the drag force requires a "subset level" application of the governing equations. A "subset level" application of the governing equations focuses only on the given element causing the flow resistance. The assumption of "ideal" flow implies that the flow resistance is modeled only in the momentum equation (and thus the subsequent assumption of "real" flow). The flow resistance equation for the drag force is derived in Chapters 6 and 7. In the derivation of the drag force for an external flow around an object, first, ideal flow is assumed and thus the Bernoulli equation is applied as follows:

$$\left(\frac{p}{\gamma} + z + \frac{v^2}{2g}\right)_1 - \left(\frac{p}{\gamma} + z + \frac{v^2}{2g}\right)_2 = 0 \tag{4.111}$$

Assuming $z_1 = z_2$, this equation has one unknown, which is the ideal velocity of the external flow around an object. Thus, this equation yields an expression for the ideal velocity, $v_i = \sqrt{2\Delta p/\rho}$ as a function of an ideal pressure difference, Δp, which is directly measured. Then, application of the momentum equation (supplemented by dimensional analysis) is used to determine an expression for the actual velocity, v_a and the drag force, F_D, which was given in Equation 4.28 and is repeated as follows:

$$\sum F_s = (F_G + F_P + F_V + F_{other})_s = (\rho Q v_s)_2 - (\rho Q v_s)_1 \tag{4.112}$$

Thus, the resulting flow resistance equation for the drag force is given as follows:

$$F_D = (F_P + F_f)_s = (\Delta p A)_s = \left(\frac{\rho v^2}{2} A + \frac{\tau_w L}{R_h} A\right)_s = C_D \frac{1}{2} \rho v^2 A \tag{4.113}$$

where F_D is equal to the sum of the pressure and friction force in the direction of the flow, s, and the drag coefficient, C_D accounts for (indirectly models) the minor head loss associated with the velocity measurement and the drag force.

The flow resistance (shear stress or drag force) in the external flow around an object is modeled as a resistance force/drag force, F_D in the integral form of the momentum equation. The flow resistance causes an unknown pressure drop, Δp, which causes an unknown head loss, h_f, where the head loss is due to a conversion of kinetic energy to heat, which is modeled/displayed in the integral form of the energy equation. However, although the head loss, h_f causes the drag force, F_D, the head loss is not actually determined in the design of external flow around an object. The assumption of ideal flow and thus applying the Bernoulli equation to measure the ideal velocity of flow, $v_i = \sqrt{2\Delta p/\rho}$ by the use of a pitot-static tube (see Section 4.5.7.3) yields an expression for the ideal velocity as a function of ideal pressure difference, Δp, which is directly measured. Therefore, the associated minor head loss with the velocity measurement is accounted for by the drag coefficient (in the application of the momentum equation). Thus, in the external flow around an object, the flow resistance is ultimately modeled as a drag force in the integral form of the momentum equation because the drag force is needed for the design of external flow around an object.

The head loss, h_f causing the actual velocity, $v_a = \sqrt{(2/\rho)(\Delta p - (\tau_w L/R_h))}$, and the drag force,

$$F_D = (F_P + F_f)_s = (\Delta p A)_s = \left(\frac{\rho v^2}{2} A + \frac{\tau_w L}{R_h} A\right)_s$$

is caused by both pressure and friction forces, where the drag force, F_D is equal to the sum of the pressure and friction force in the direction of the flow, s. However, because the friction/viscous forces (due to shear stress, τ_w) due to the external flow cannot be theoretically modeled in the integral momentum equation, the actual pressure drop, Δp

cannot be analytically determined; thus, the exact reduction in the actual velocity, v_a, which is less than the ideal velocity, v_i, cannot be theoretically determined. Furthermore, the exact component in the s-direction of the pressure and viscous forces cannot be theoretically determined. Therefore, one cannot derive an analytical expression for the drag force, F_D from the momentum equation. As a result, one must resort to dimensional analysis (which supplements the momentum theory) in order to derive an expression for the drag force, F_D that involves the definition of a drag coefficient, C_D that represents the flow resistance. Drag forces on external flow are addressed in detail in Chapters 6 and 10.

4.5.7.2 Applications of the Governing Equations for Ideal External Flow

The governing equations (energy and momentum) for ideal flow are applied in the analysis and design of external flow around an object. Applications of the governing equations for ideal external flow include the analysis and design of planes, for instance, that include a speed-measuring device such as a pitot-static tube, and the determination of a pressure difference in external flow from a velocity difference. One may note from Section 4.5.7.1 that the minor head loss associated with the velocity measurement in external flow is indirectly modeled by a drag coefficient, C_D, which requires a "subset level" application of the governing equations for its evaluation. Furthermore, the resulting definition of the drag force, $F_D = C_D(1/2)\rho v^2 A$ requires the definition of the drag coefficient, C_D, which is yet to be presented in later chapters. Therefore, example problems illustrating the application of the Bernoulli equation for external flow herein will assume that the drag force is either a given quantity or computed from the drag force equation (see Chapter 10 for applications). Furthermore, for the example problems involving external flow velocity measurements, illustration of the computation of an unknown force, which requires the application of the complementary momentum equation, will be deferred to Chapter 5.

4.5.7.3 Applications of the Bernoulli Equation for Ideal External Flow

While Section 4.5.3.2 illustrated a pitot-static tube to measure the velocity for internal (pipe) flow, this section illustrates a pitot-static tube to measure the velocity for external flow. Thus, Example Problem 4.18 presented below focuses on the application of the Bernoulli equation on a "subset level" and thus solving for the ideal velocity for external flow. However, Chapter 10 presents example problems that apply the Bernoulli equation (on a "macro level," as opposed to a "subset level"), thus solving for any unknown "energy head" term ($p/\gamma, z$, or $v^2/2g$) for one of the two points in a finite control volume. As such, Example Problem 4.18 presents an application of the Bernoulli equation for determining the speed of a plane using a pitot-static tube, which converts the velocity difference into a pressure difference, as follows:

$$\left(\frac{p}{\gamma} + z + \frac{v^2}{2g}\right)_1 - \left(\frac{p}{\gamma} + z + \frac{v^2}{2g}\right)_2 = 0 \tag{4.114}$$

The pitot-static tube is used in the measurement of the airplane's speed (or the velocity of external flow around an object), where the velocity-measuring device is mounted on the wing of the airplane. Because a pitot-static tube affects the airflow over a very short section (and the flow of air is at a high velocity), the major head loss due to flow resistance is assumed to be negligible. As such, assuming ideal flow, the Bernoulli's equation (Eulerian/

integral approach) is used to solve for the ideal velocity as a function of an ideal pressure difference, Δp, which is directly measured. The minor head loss associated with the velocity measurement is modeled by a drag coefficient, C_D.

Figure 4.17 illustrates a pitot-static tube mounted on the wing of a plane in order to measure the speed of the plane. Specifically, the velocity of the plane is measured relative to the surrounding air, where the pitot tube measures the local velocity by measuring the pressure difference, as modeled by the Bernoulli equation. The pitot-static tube, which is aligned with the flow, consists of an outer tube (static tube) and an inner tube (stagnation tube) that is placed inside the outer tube, both of which are connected to a differential pressure meter (manometer, pressure gage, or pressure transducer), which measures the differential gage pressure. The outer tube is sealed at the nose and contains a series of small holes around the circumference near the entrance (nose) of the tube, where the flow of air passes around the outside of the outer tube and thus is not stagnated. The point inside any of these small holes in the outer tube may be labeled as point 1. Finally, the outer tube is connected to a pressure port of the differential pressure meter, which registers the static pressure at point 1, p_1. It is important to note that because the fluid flow in this case of external flow represents the atmospheric air, the static pressure at point 1, p_1 is the standard atmospheric pressure at the given elevation above sea level. The inner tube is open to the flow at the nose, where the flow of air is stagnated, and the point inside the inner tube is labeled as point 2. Finally, the

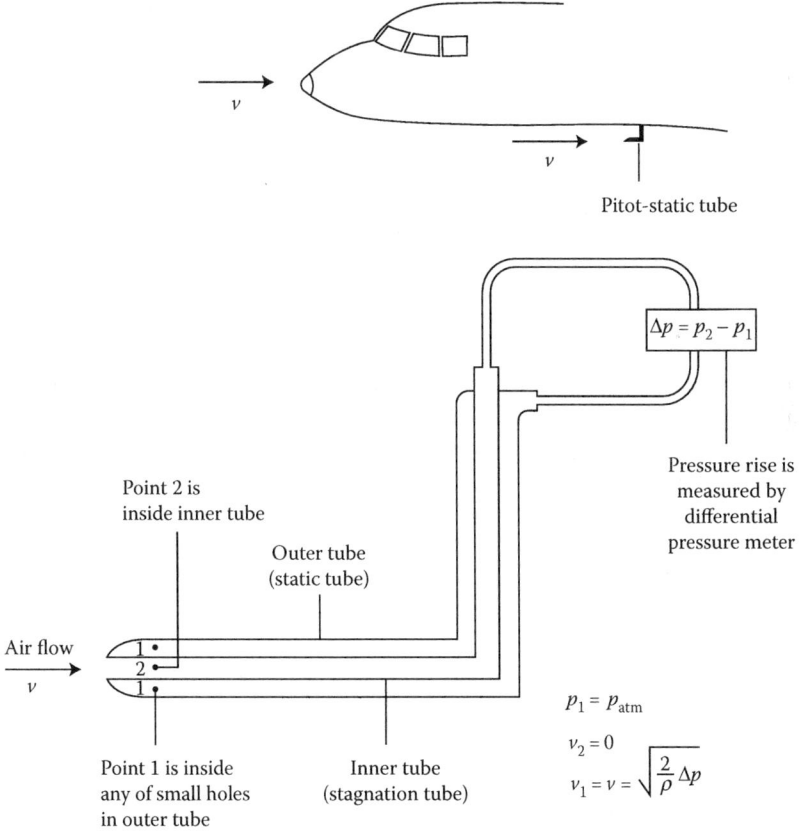

FIGURE 4.17
Pitot-static tube mounted on the wing of a plane to measure the speed of the plane.

inner tube is connected to a pressure port of the differential pressure meter, which registers the stagnation pressure at point 2, p_2. Thus, the differential pressure meter measures the pressure rise, $\Delta p = p_2 - p_1$, between points 1 and 2. The Bernoulli equation between points 1 and 2 expressed in terms of pressure is given as follows:

$$p_1 + \gamma z_1 + \frac{\rho v_1^2}{2} = p_2 + \gamma z_2 + \frac{\rho v_2^2}{2} \qquad (4.115)$$

which states that the total pressure is a constant along a streamline. Assuming that $z_1 = z_2$ and noting that $v_2 = 0$:

$$\underbrace{p_1}_{static\ press} + \underbrace{\frac{\rho v_1^2}{2}}_{dyn\ press} = \underbrace{p_2}_{stagn\ press} \qquad (4.116)$$

where the sum of the static pressure (where the fluid flow is not slowed down) and the dynamic pressure equals the stagnation pressure (where the fluid flow is stagnated/momentarily slowed down). The stagnation pressure is the largest pressure attainable along a given streamline. Furthermore, the stagnation pressure represents the conversion of kinetic energy into a pressure rise. The difference between the static pressure, p_1 and the stagnation pressure, p_2 is measured by the differential pressure meter, and the dynamic pressure, $\rho v_1^2/2$ is measured by the pressure rise as follows:

$$\underbrace{\frac{\rho v_1^2}{2}}_{dyn\ press} = \underbrace{p_2}_{stagn\ press} - \underbrace{p_1}_{static\ press} = \underbrace{\Delta p}_{press\ rise} \qquad (4.117)$$

One may note that the stagnation pressure, p_2 is greater than the static pressure, p_1 by the value of the dynamic pressure, $\rho v_1^2/2$; thus, the pressure rise. Additionally, one may note that the pressure rise will increase as the density of the atmospheric air, ρ or the velocity of the plane, v_1 increases. Thus, for a given density of the atmospheric air, ρ and a given static pressure (atmospheric pressure in this case of external flow), p_1, the stagnation pressure on the wing of the moving plane, p_2 will increase with an increase in the velocity of the plane, v_1. Assuming no head loss and that the fluid flow is momentarily brought to a stop at point 2, where $v_2 = 0$, the velocity at point 1 is ideal, and the dynamic pressure at point 1, $\rho v_1^2/2$ represents the ideal pressure rise (from static pressure to stagnation pressure). Therefore, isolating the velocity at point 1, v_1 defines the ideal velocity, v_i as follows:

$$v_1 = \underbrace{\sqrt{\frac{2}{\rho}(\Delta p)}}_{ideal\ vel} = \underbrace{\sqrt{\frac{2}{\rho}\left(\underbrace{p_2}_{stag\ press} - \underbrace{p_1}_{static\ press}\right)}}_{ideal\ vel} \qquad (4.118)$$

where the difference in the pressure rise, $\Delta p = p_2 - p_1$, between points 1 and 2 is used to determine the unknown velocity, v_1 of the plane as illustrated in Figure 4.17. Furthermore, calibration of the differential pressure meter for the density of the atmospheric air, ρ allows a direct reading of the velocity of the plane.

EXAMPLE PROBLEM 4.18

A plane flies at an altitude of 3000 m above sea level in a standard atmosphere. A pitot-static tube is mounted on the wing of a plane in order to measure the speed of the plane, as illustrated in Figure EP 4.18. The point inside any of the small holes in the outer tube of the pitot-static tube may be labeled as point 1, where the static pressure, p_1 is the standard atmospheric pressure at the given altitude above sea level. The inner tube is open to the flow at the nose of the pitot-static tube, where the flow of air is stagnated, and the point inside the inner tube is labeled as point 2. The inner and outer tubes are each connected to a pressure port of the differential pressure meter, which registers the differential gage pressure (rise), $\Delta p = p_2 - p_1 = 4.5 \times 10^3 \, \text{N/m}^2$ between points 1 and 2. (a) Determine the ideal speed of the plane, v_1. (b) Determine the stagnation pressure on the nose of the plane.

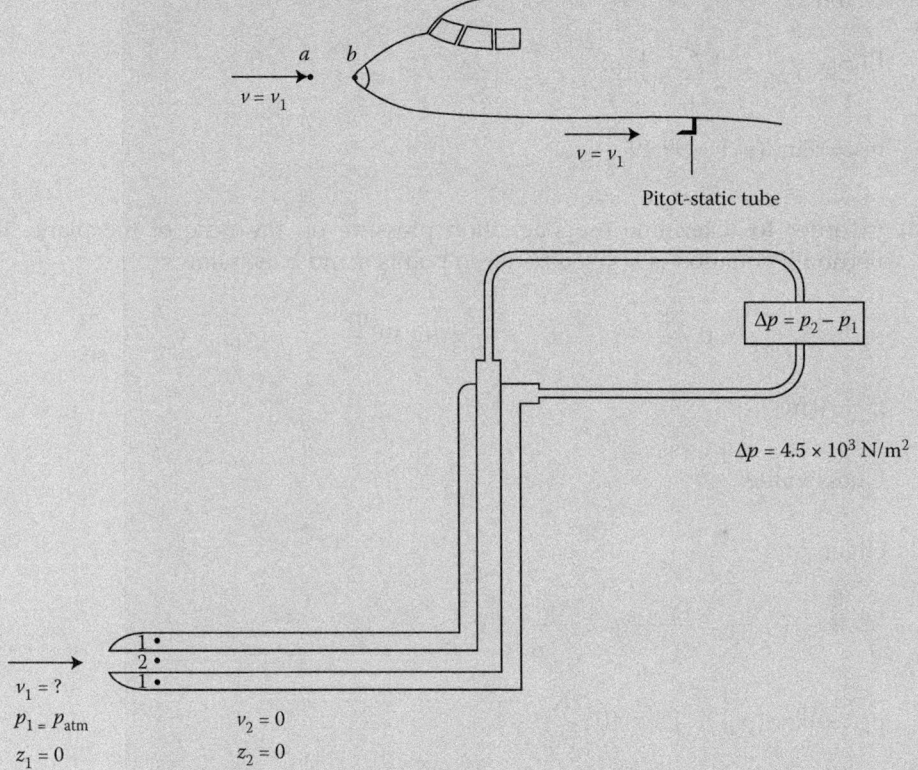

FIGURE EP 4.18
Pitot-static tube mounted on the wing on a plane to measure the speed of the plane (ideal flow).

Mathcad Solution

(a) Assume that the atmospheric airflow is momentarily brought to a stop at point 2, where $v_2 = 0$. Then, in order to determine the ideal velocity of the plane, v_1, the Bernoulli equation is applied between points 1 and 2. The standard atmospheric pressure (the static pressure at point 1, p_1) and the density of the standard atmospheric air, ρ at an altitude of 3000 m above sea level are given in Table A.1 in Appendix A.

$z_1 := 0 \, m$ $z_2 := 0 \, m$ $v_2 := 0 \, \dfrac{m}{sec}$ $\rho := 0.90925 \, \dfrac{kg}{m^3}$

$g := 9.81 \, \dfrac{m}{sec^2}$ $\gamma := \rho \cdot g = 8.92 \, \dfrac{kg}{m^2 \, s^2}$ $P_{atmabs} := 70.121 \times 10^3 \, \dfrac{N}{m^2}$

$P_{1abs} := P_{atmabs} = 7.012 \times 10^4 \, \dfrac{N}{m^2}$ $P_{1gage} := P_{1abs} - P_{atmabs} = 0 \, \dfrac{N}{m^2}$

$\Delta P_{gage} := 4.5 \times 10^3 \, \dfrac{N}{m^2}$ $P_{2gage} := \Delta P_{gage} + P_{1gage} = 4.5 \times 10^3 \, \dfrac{N}{m^2}$

Guess value: $v_1 := 50 \, \dfrac{m}{sec}$

Given

$$\dfrac{P_{1gage}}{\gamma} + z_1 + \dfrac{v_1^2}{2 \cdot g} = \dfrac{P_{2gage}}{\gamma} + z_2 + \dfrac{v_2^2}{2 \cdot g}$$

$v_1 := Find(v_1) = 99.49 \, \dfrac{m}{s}$

(b) In order to determine the stagnation pressure on the nose of the plane, the Bernoulli equation is applied between points a and b as follows:

$P_a := P_{1gage} = 0 \, \dfrac{N}{m^2}$ $v_a := v_1 = 99.49 \, \dfrac{m}{s}$ $v_b := v_2 = 0 \, \dfrac{m}{sec}$

$z_a := 0 \, m$ $z_b := 0 \, m$

Guess value: $P_b := 1 \, \dfrac{N}{m^2}$

Given

$$\dfrac{P_a}{\gamma} + z_a + \dfrac{v_a^2}{2 \cdot g} = \dfrac{P_b}{\gamma} + z_b + \dfrac{v_b^2}{2 \cdot g}$$

$P_b := Find(P_b) = 4.5 \times 10^3 \, \dfrac{N}{m^2}$

One may note that, similar to the pitot tube, there is a stagnation point on any stationary body that is placed into a flowing fluid (or alternatively, there is a stagnation point on any moving body in a stationary flow field). In the flow over a submerged body (see Figure 4.18), some of the fluid flows over the body and some flows under the body, while the fluid is brought to a stop at the nose of the body at the stagnation point, as illustrated in Figure 4.19. Furthermore, the dividing line is called the stagnation streamline and ends at the stagnation point on the submerged body, where the pressure is the stagnation pressure, as illustrated in Figure 4.20. The stagnation pressure is the largest pressure attainable along a given streamline. The Bernoulli equation between points 1 and 2 (along the stagnation streamline)

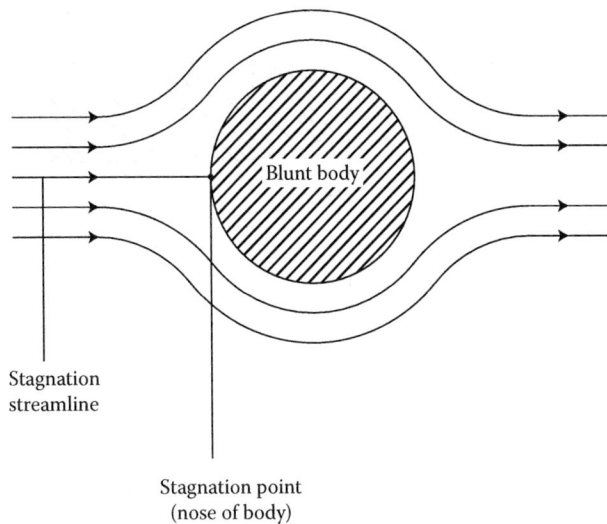

FIGURE 4.18
Stagnation point in the flow over a submerged blunt body.

expressed in terms of pressure is given as follows:

$$p_1 + \gamma z_1 + \frac{\rho v_1^2}{2} = p_2 + \gamma z_2 + \frac{\rho v_2^2}{2} \tag{4.119}$$

which states that the total pressure is a constant along a streamline. The point of stagnation (point 2) represents the conversion of all of the kinetic energy into an ideal pressure rise, where the fluid is momentarily brought to a stop. Therefore, at that stagnation point, the drag force, F_D is composed only of the resultant pressure force, F_P that acts in the direction of the flow (there is no shear force, F_f component in the direction of the flow).

4.5.8 Application of the Energy and Momentum Equations for a Hydraulic Jump

A hydraulic jump is a natural phenomenon that occurs when the open channel flow transitions from supercritical to subcritical flow (see Chapter 9). One may note that the head loss

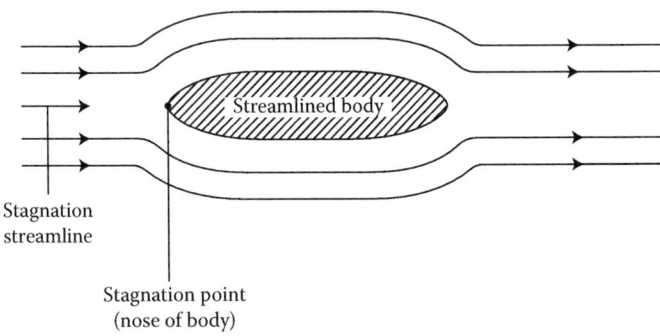

FIGURE 4.19
Stagnation point in the flow over a submerged streamlined body.

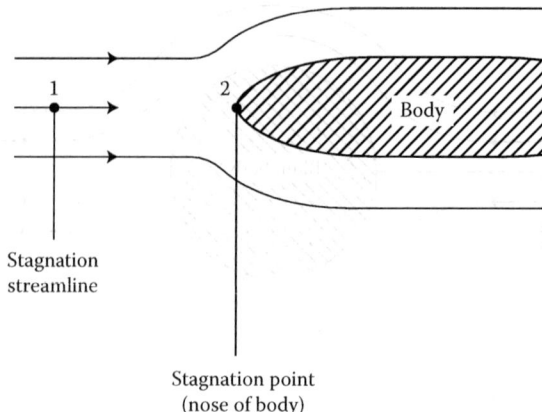

FIGURE 4.20
Stagnation streamline in the flow over a submerged body (Bernoulli's equation).

associated with a hydraulic jump is not due to frictional losses associated with the wall shear stress, τ_w, but is rather due to intense agitation and turbulence and thus results in a high energy loss. Therefore, there are no viscous forces to model in the integral form of the momentum equation. However, the unknown major head loss, $h_{f,major}$, which is due to a conversion of kinetic energy to heat (the fast-moving supercritical flow encounters the slow-moving subcritical flow and thus the flow at the jump becomes critical flow, with a minimum specific energy) is modeled in the integral form of the energy equation. Thus, for a hydraulic jump, the integral form of the momentum equation may be used to analytically solve for the unknown channel depth of flow (either upstream or downstream) as follows:

$$\sum F_s = (F_G + F_P + F_V + F_{other})_s = (\rho Q v_s)_2 - (\rho Q v_s)_1 \tag{4.120}$$

by deriving the hydraulic jump equations (see Chapter 9 for the derivations and applications). Then, the energy equation is applied to solve for the major head loss as follows:

$$\left(y + z + \frac{v^2}{2g}\right)_1 - h_{fmaj} = \left(y + z + \frac{v^2}{2g}\right)_2 \tag{4.121}$$

4.5.8.1 Applications of the Governing Equations for a Hydraulic Jump

The governing equations (continuity, energy, and momentum) are applied in the analysis and design of a hydraulic jump. While the application of the continuity equation has already been presented in Chapter 3, application of the momentum equation is yet to be presented in Chapter 5. One may note from the section above that evaluation of the unknown depth of flow requires the application of the integral form of the momentum equation in order to derive the hydraulic jump equations (see Equations 9.248 and 9.249). Therefore, example problems illustrating the application for the energy equation for a hydraulic jump will assume that the depths of flow are either a given quantity or assume that the hydraulic jump equation is given and is applied to solve for the unknown depth of flow, or directly solved for (assuming that the head loss due to the jump is a given quantity) in the following energy equation:

$$\left(y_1 + z_1 + \frac{v_1^2}{2g}\right) - h_{fmaj} = \left(y_2 + z_2 + \frac{v_2^2}{2g}\right) \tag{4.122}$$

4.5.8.2 Application of the Energy Equation for a Hydraulic Jump

EXAMPLE PROBLEM 4.19

Water flowing in a rectangular channel of width 3 m at a supercritical velocity of 4.823 m/sec and a supercritical depth of 0.933 m encounters a subcritical flow at a velocity of 2.666 m/sec and a depth of 1.688 m, as illustrated in Figure EP 4.19. As the flow transitions from supercritical to subcritical flow, a hydraulic jump occurs, where there is a major head loss due to intense agitation and turbulence, where the depth of flow is at a critical depth of 1.273 m. (a) Determine the head loss due to the flow turbulence at the jump. (b) Determine the flowrate in the channel. (c) Draw the energy grade line and the hydraulic grade line.

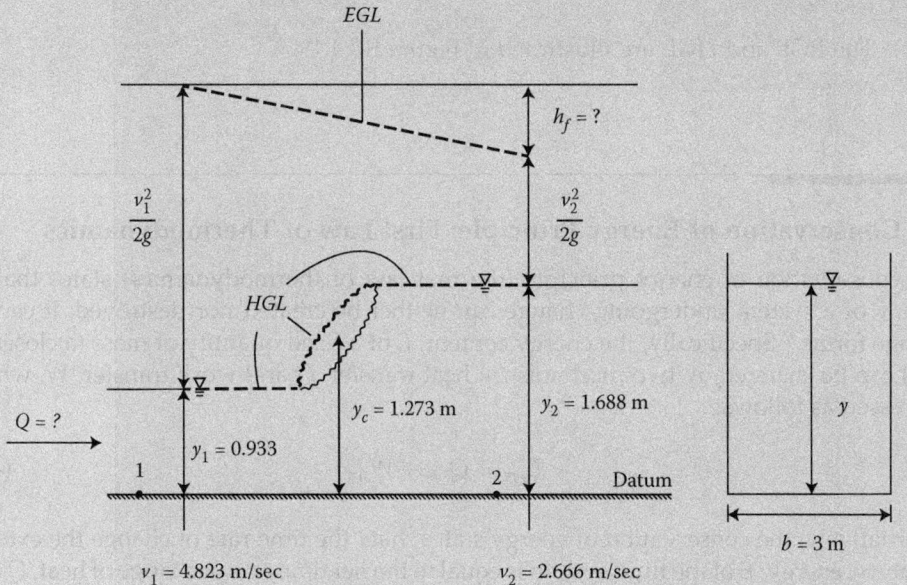

FIGURE EP 4.19
Water flows in a rectangular channel at a supercritical depth encounters a subcritical flow and thus forms a hydraulic jump.

Mathcad Solution

(a) In order to determine the major head loss due to the jump, the energy equation is applied between the upstream point 1 and the downstream point 2 as follows:

$$z_1 := 0 \text{ m} \qquad z_2 := 0 \text{ m} \qquad v_1 := 4.823 \, \frac{\text{m}}{\text{sec}} \qquad v_2 := 2.666 \, \frac{\text{m}}{\text{sec}}$$

$$y_1 := 0.933\,m \qquad y_2 := 1.688\,m \qquad g := 9.81\,\frac{m}{sec^2} \qquad b := 3\,m$$

Guess value: $h_f := 1\,m$

Given

$$y_1 + z_1 + \frac{v_1^2}{2 \cdot g} - h_f = y_2 + z_2 + \frac{v_2^2}{2 \cdot g}$$

$$h_f := Find(h_f) = 0.068\,m$$

(b) In order to determine the flowrate in the channel, the continuity equation is applied at either point 1 or point 2 as follows:

$$Q := v_1 \cdot b \cdot y_1 = 13.5\,\frac{m^3}{s}$$

(c) The EGL and HGL are illustrated in Figure EP 4.19.

4.6 Conservation of Energy Principle: First Law of Thermodynamics

The conservation of energy principle (the first law of thermodynamics) states that "the energy of a system undergoing change can neither be created nor destroyed, it can only change forms." Specifically, the energy content, E of a fixed quantity of mass (a closed system) can be changed by two mechanisms: heat transfer, Q and work transfer, W, which is expressed as follows:

$$E_{sys} = Q_{sys} + W_{sys} \tag{4.123}$$

Alternatively, the conservation of energy states that "the time rate of change the extensive property, energy, E of the fluid system is equal to the net time rate of change of heat, Q transfer to the system plus the net time rate of change of work, W transfer to the system" and is expressed as follows:

$$\frac{dE_{sys}}{dt} = (\dot{Q}_{in} - \dot{Q}_{out})_{sys} + (\dot{W}_{in} - \dot{W}_{out})_{sys} \tag{4.124}$$

$$\frac{dE_{sys}}{dt} = [(\dot{Q}_{net})_{in} + (\dot{W}_{net})_{in}]_{sys} \tag{4.125}$$

where it may be interpreted as "the time rate of increase of the total stored energy of the system equals the net time rate of energy addition by heat transfer into the system plus the net time rate of energy addition by work transfer into the system." Furthermore, if the net rate of heat transfer and the net rate of work transfer are "going into the system," then they are considered "+," while if they are "coming out of the system," then they are considered "−."

The next step is to express the first law of thermodynamics for a control volume. To do this, consider the one-dimensional uniform (at a given cross section) flow through a stream tube from left to right as illustrated in Figure 4.1. At an initial time t, the fluid system between sections 1 and 2 coincides with the control volume between control face 1 and control face 2, so they are identical. Thus, when a control volume is coincident with fluid system at an instant of time we get:

$$[(\dot{Q}_{net})_{in} + (\dot{W}_{net})_{in}]_{sys} = [(\dot{Q}_{net})_{in} + (\dot{W}_{net})_{in}]_{cv} \tag{4.126}$$

Furthermore, the Reynolds transport theorem given in Equation 3.19 (for the one-dimensional uniform flow through a stream tube of Figure 4.1) may be applied to link the definition of the conservation of energy between a fluid system and a control volume as follows:

$$\frac{dE_{sys}}{dt} = \frac{dE_{cv}}{dt} - (\dot{E})_{in} + (\dot{E})_{out} \tag{4.127}$$

where the extensive property of the fluid, $B = E$ and thus the corresponding intensive property, $b = B/M = E/M = e$. Substituting Equation 4.125 and then Equation 4.126 for the left-hand side of Equation 4.127 yields the conservation of energy for a control volume as follows:

$$[(\dot{Q}_{net})_{in} + (\dot{W}_{net})_{in}]_{cv} = \frac{dE_{cv}}{dt} - (\dot{E})_{in} + (\dot{E})_{out} \tag{4.128}$$

$$\frac{dE_{cv}}{dt} = [(\dot{Q}_{net})_{in} + (\dot{W}_{net})_{in}]_{cv} + (\dot{E})_{in} - (\dot{E})_{out} \tag{4.129}$$

which states that "the time rate of increase of the total stored energy of the control volume equals the net time rate of energy addition by heat transfer into the control volume plus the net time rate of energy addition by work transfer into the control volume, plus the net flux of energy into the control volume."

In order to apply the first law of thermodynamics to derive the energy equation, it is important to identify the various forms of energy (thermal, mechanical, kinetic, potential, magnetic, electrical, chemical, and nuclear) and the two mechanisms of transfer (heat and work) by which energy is transferred from one form to another.

4.6.1 Total Energy

In general, energy exists in numerous forms, including thermal, mechanical, kinetic, potential, magnetic, electrical, chemical, and nuclear, for which their sum is the total energy, E. The total energy per unit mass of a system is defined as $e = E/M$. There are two general types of energy: microscopic energy and macroscopic energy. Microscopic energy is related to the molecular structure and the degree of molecular activity of the system. The sum of all microscopic forms of energy is called the internal energy, U (where the internal energy per unit mass is defined as $u = U/M$) of the system, where the sensible and latent forms of internal energy are referred to as thermal energy (heat). Macroscopic energy is related to the motion and the influence of some external effects such as gravity, magnetism, electricity, and surface tension. In the absence of the effects of magnetism, electricity, and surface tension, a system is called a simple compressible system. Thus, the total energy, E of a simple compressible system consists of internal energy, U (microscopic energy); kinetic energy, KE (macroscopic energy); and potential energy, PE (macroscopic energy). The kinetic energy is a result of the motion, v of the system relative to some fixed frame of reference, where the kinetic energy per unit mass, $ke = KE/M = v^2/2$. And the potential energy is a result of the elevation, z of the system relative to some external reference point in a gravitational field (gravitational acceleration), g, where the potential energy per unit mass, $pe = PE/M = gz$. Therefore, the total energy per unit mass, e is defined as follows:

$$e = u + ke + pe \tag{4.130}$$

$$e = u + \frac{v^2}{2} + gz \tag{4.131}$$

However, one may note that a fluid entering or leaving a control volume possesses an additional form of energy, which is called the flow energy, or flow work, p/ρ, which is the energy per unit mass required to move/push the fluid and maintain the flow. Thus, the total energy of a flowing fluid (for a control volume), $e_{flowing}$ is defined as follows:

$$e_{flowing} = \frac{p}{\rho} + e \tag{4.132}$$

$$e_{flowing} = \underbrace{\frac{p}{\rho} + u}_{h=enthalpy} + \frac{v^2}{2} + gz \tag{4.133}$$

where the sum of the internal energy per unit mass, u and the flow energy (flow work), p/ρ is defined as fluid property enthalpy per unit mass, h as follows:

$$h = u + \frac{p}{\rho} \tag{4.134}$$

4.6.1.1 Total Energy for a Fluid Flow System

In the analysis and design of a fluid flow system, only mechanical forms of energy (kinetic, potential, flow/hydraulic, and shaft) and the frictional effects that cause mechanical energy to be lost are modeled in the energy equation. Specifically, in a flow system, using a pressure gradient or by invoking gravity, it is of interest to move the fluid from one location to another at a given flowrate, velocity, and elevation. Additionally, it may be of interest to add or subtract mechanical (shaft) energy by the use of a pump or turbine, respectively. One may note that a pump–motor system involves the conversion of electrical or chemical energy to mechanical energy, while a turbine–generator system involves the conversion of mechanical energy to electrical energy (see Section 4.7.2.2). However, a fluid flow system does not involve the conversion of electrical, nuclear, chemical, or thermal energy to mechanical energy. Furthermore, a flow system does not involve any significant heat transfer, as it operates at a constant temperature. Rather, a fluid flow system involves the conversion of various forms of mechanical energy (kinetic, potential, flow/hydraulic, and shaft) to other forms of mechanical energy, including the frictional effects that cause mechanical energy to be lost (see Section 4.7.2.5).

4.6.1.2 Dimensions and Units for Energy

The dimensions and units for energy may be derived from the definitions of the two mechanisms of transfer of energy: heat and work. The definition of energy in terms of work is force times distance. Thus, the dimensions and units for energy, E in the English system are given as follows:

$$E = (F)(d) = [FL] = [ft - lb] \tag{4.135}$$

where the British thermal unit (Btu) is defined as follows:

$$1\,Btu = 777.649\,ft - lb \tag{4.136}$$

which is defined as the energy required to raise the temperature of 1 slug of water at 68°F (room temperature) by 1°F. The dimensions and units for energy, E in the metric system are given as follows:

$$E = (F)(d) = [FL] = [N - m] = [J] \tag{4.137}$$

where

$$1\,kJ = 1000\,J \tag{4.138}$$

and where

$$1\,Btu = 1.0551\,kJ \tag{4.139}$$

Another metric unit of energy is called the calorie (cal), where:

$$1\,cal = 4.1868\,J \tag{4.140}$$

which is defined as the energy required to raise the temperature of $1\,g$ of water at 14.5°C by 1°C.

4.6.2 Energy Transfer by Heat

An energy transfer is accomplished by a heat transfer, Q if it is associated with the sensible and latent forms of internal energy, U, which are known as thermal energy (heat). Because a change in the thermal energy of a given mass results in a change in temperature, temperature is a good indicator of thermal energy. A temperature difference between systems causes the thermal energy to move in the direction of decreasing temperature; this transfer of thermal energy from one system to another is a called heat transfer, Q, while the time rate of heat transfer, $\dot{Q} = Q/\Delta t$ is called the heat transfer rate. Once temperature equality is established, the heat transfer process stops. Furthermore, the time rate of heat transfer (heat transfer rate), $\dot{Q}$ represents all of the ways in which thermal energy is exchanged between systems (or between the control volume and the surroundings) due to a temperature difference (from radiation, conduction, and/or convection), where heat transfer into the system (control volume) is considered positive, while heat transfer out of the system (control volume) is considered negative. It is interesting to note that in many engineering applications, the heat transfer rate, $\dot{Q}$ is zero, which is known as an adiabatic process. A process may be adiabatic due to either a well-insulated system (or control volume) or due to temperature equality. Furthermore, although there is no heat transfer during an adiabatic process, the energy content and thus the temperature of a system (or control volume) can still be changed by work transfer.

4.6.3 Energy Transfer by Work

An energy transfer is accomplished by a work transfer, W if it is associated with a force acting over a distance. The time rate of work transfer, $\dot{W} = W/\Delta t$, which is also called "power," $P = \dot{W} = W/\Delta t$, is considered positive when work is done on the system (or control volume) by the surroundings, while it is considered negative when work is done by the system (or control volume). For instance, a pump (fan, blower, or compressor in the case of gases) does work on the system (or control volume) (and thus transfers energy into the system or control volume), while a turbine does work by the system (or control volume) (and thus transfers energy out of the system or control volume). For a given system (or control volume), there are various forms of work, W, which include work transmitted by a rotating shaft, W_{shaft}; work done by pressure forces on the system (or control volume), $W_{pressure}$; work done by the normal and shear components of viscous forces on the control surface, $W_{viscous}$; and work done by other forces such as magnetic, electric, and surface tension (which are negligible for a simple compressible system or control volume), W_{other} and thus: $W_{total} = W_{shaft} + W_{pressure} + W_{viscous} + W_{other}$. Furthermore, it may be noted that although the $W_{viscous}$ component is usually negligible in comparison to the work done by a rotating shaft and by pressure forces, the shear component of the viscous force may be considered important in a detailed analysis of turbomachinery. Thus, $W_{total} = W_{shaft} + W_{pressure}$.

The work transmitted by a rotating shaft, W_{shaft} is considered important when the flow system (or control volume) involves devices/machinery such as a pump or a turbine. The power, $W_{shaft}/\Delta t$ is transferred to the flow system through a rotating shaft and is proportional to the shaft torque, T_{shaft}, where $W_{shaft}/\Delta t = \omega\, T_{shaft}$, and where $\omega =$ the angular speed of the shaft in radians/second.

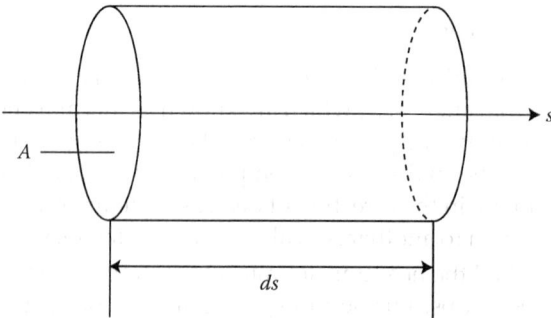

FIGURE 4.21
Infinitesimal control volume represented by a differential section, ds of a stream tube with a differential cross-sectional area, A is used to derive the differential form of the energy equation for a fluid system.

The work done by pressure forces on the system (or control volume), $W_{pressure}$ is considered important when the fluid is entering or leaving a control volume. Referring to Figure 4.21, which illustrates an infinitesimal control volume, the work done by the net pressure forces, $-dp\,A$ acting on the differential area, A acting over a differential distance, ds, is $W_{pressure} = -dp\,Ads$. And, the power, $W_{pressure}/\Delta t = -dp\,Ads/dt = -dp\,v\,A = -(dp/\rho)\rho\,vA$. Therefore, the net power transfer for a system (or control volume) is given as follows:

$$(\dot{W}_{net})_{in} = [(\dot{W}_{shaft})_{net}]_{in} + [(\dot{W}_{pressure})_{net}]_{in} \tag{4.141}$$

$$(\dot{W}_{net})_{in} = [(\dot{W}_{shaft})_{net}]_{in} - \frac{dp}{\rho}\rho vA \tag{4.142}$$

The SI unit of power is the watt, $W = \text{N-m/s} = \text{J/s}$, or kilowatt, kW ($1\,kW = 1000\,W$), while the BG unit is the horsepower, $hp = 550\,ft\text{-}lb/sec = 2546.12\,Btu/hr$.

4.7 The Energy Equation Based on the First Law of Thermodynamics

The conservation of energy principle in fluid mechanics is typically applied to either an infinitesimal (differential) control volume or a finite (integral) control volume, assuming steady flow, and is called the energy equation. Application of the conservation of energy principle for a fluid system (Lagrangian approach) involves the application of the principle, stated in Equation 4.129 for a control volume, to an infinitesimal control volume, which yields the differential form of the energy equation. And, application of the conservation of mass for a control volume (Eulerian approach) involves application of the principle, stated in Equation 4.129 for a control volume, to a finite control volume, which yields the integral form of the energy equation. Thus, one may note that, in general, the control volume is a convenient object (concept) of analysis because it is a volume fixed in space through which energy flows across its boundaries. In the application of the governing equations, it is important to make the distinction between real flow, ideal flow, and a hydraulic jump. And, for ideal flow, it is important to make the distinction between internal flow, external flow, and flow from a tank.

4.7.1 The Energy Equation for a Fluid System

Application of the conservation of energy principle, stated in Equation 4.129 for a control volume, assuming steady flow, to an infinitesimal (differential) control volume (differential analysis) (see Figure 4.21) yields the differential form of the energy equation. Beginning with Equation 4.129 and restated as follows:

$$\frac{dE_{cv}}{dt} = [(\dot{Q}_{net})_{in} + (\dot{W}_{net})_{in}]_{cv} + (\dot{E})_{in} - (\dot{E})_{out} \tag{4.143}$$

For the assumption of steady flow, the time rate of change of the total stored energy of the system term on the left-hand side of the equation is set equal to zero and thus:

$$[(\dot{Q}_{net})_{in} + (\dot{W}_{net})_{in}]_{cv} = (\dot{E})_{out} - (\dot{E})_{in} \tag{4.144}$$

By definition, the difference between the energy flow rate out of the control volume and the energy flow rate into the control volume defined across the differential distance, ds is given

as follows:

$$\dot{E}_{out} - \dot{E}_{in} = \frac{d\dot{E}}{ds}ds = d\dot{E} \tag{4.145}$$

and thus, Equation 4.144 may be rewritten as follows:

$$[(\dot{Q}_{net})_{in} + (\dot{W}_{net})_{in}]_{cv} = d\dot{E} \tag{4.146}$$

Substituting Equation 4.142 for the net power transfer for a system into Equation 4.146 yields the following:

$$(\dot{Q}_{net})_{in} + [(\dot{W}_{shaft})_{net}]_{in} = d\dot{E} + \frac{dp}{\rho}\rho v A \tag{4.147}$$

As noted above, in the analysis and design of a fluid flow system, only mechanical forms of energy (kinetic, potential, flow, and shaft), and the frictional effects that cause mechanical energy to be lost are modeled in the energy equation. Furthermore, one may note that the work done by the pressure forces on the control volume has now been combined with the energy of the fluid crossing the differential control. Substituting in the definition of total energy per unit mass, $e = u + v^2/2 + gz$ for the total energy, $E = eM = uM + M\ v^2/2 + gzM$, and $M = \rho\ A ds$ into Equation 4.147 yields the differential form of the energy equation (assuming steady flow) as follows:

$$(\dot{Q}_{net})_{in} + [(\dot{W}_{shaft})_{net}]_{in} = \frac{d\left(u + \dfrac{v^2}{2} + gz\right)\rho A\, ds}{dt} + \frac{dp}{\rho}\rho v A \tag{4.148}$$

$$(\dot{Q}_{net})_{in} + [(\dot{W}_{shaft})_{net}]_{in} = \frac{d\left(u + \dfrac{p}{\rho} + \dfrac{v^2}{2} + gz\right)\rho A ds}{dt} \tag{4.149}$$

This differential form of the energy equation is not really useful (practical per se) until it is evaluated between two points in a finite control volume to yield the integral form of the energy equation (see Section 4.7.2). Thus, application of the energy equation is useful only using the integral approach (for a finite control volume), as energy is work, which is defined over distance, L.

4.7.2 The Energy Equation for a Control Volume

Evaluating the differential form of the energy equation (Equation 4.149) between two points in a finite control volume such as between points 1 and 2 in Figure 4.1 yields the integral form of the energy equation (assuming steady flow) for a control volume as follows:

$$(\dot{Q}_{net})_{in} + [(\dot{W}_{shaft})_{net}]_{in} = \left[\left(u + \frac{p}{\rho} + \frac{v^2}{2} + gz\right)\dot{M}\right]_{out} - \left[\left(u + \frac{p}{\rho} + \frac{v^2}{2} + gz\right)\dot{M}\right]_{in} \tag{4.150}$$

This states that "the net rate of energy transfer to a control volume by heat, Q and work, W transfers for steady flow is equal to the difference between the rates of outgoing and incoming energy flows with the mass." Because there is only one inlet and one outlet in Figure 4.1, the equation may be written as follows:

$$(\dot{Q}_{net})_{in} + [(\dot{W}_{shaft})_{net}]_{in} = \left[\left(u + \frac{p}{\rho} + \frac{v^2}{2} + gz\right)\dot{M}\right]_2 - \left[\left(u + \frac{p}{\rho} + \frac{v^2}{2} + gz\right)\dot{M}\right]_1 \qquad (4.151)$$

where control face 1 represents the inflow cross section and control face 2 represents the outflow cross section. The energy equation given in Equation 4.151 is expressed in power terms. However, in order to easily define (group) the three basic components of mechanical energy, into, out of, and lost out of, a control volume, it is helpful to express the energy equation on a unit-mass basis. Thus, dividing both sides of Equation 4.151 by the mass flow rate, $\dot{M} = M/\Delta t$ yields each term defined in energy per unit mass as follows:

$$(q_{net})_{in} + [(w_{shaft})_{net}]_{in} = \left[u + \frac{p}{\rho} + \frac{v^2}{2} + gz\right]_2 - \left[u + \frac{p}{\rho} + \frac{v^2}{2} + gz\right]_1 \qquad (4.152)$$

Rearranging Equation 4.152, it may be expressed as follows:

$$\underbrace{[(w_{shaft})_{net}]_{in} + \frac{p_1}{\rho_1} + \frac{v_1^2}{2} + gz_1}_{mech\ energy\ into\ cv} = \underbrace{\frac{p_2}{\rho_2} + \frac{v_2^2}{2} + gz_2}_{mech\ energy\ out\ of\ cv} + \underbrace{[u_2 - u_1 - (q_{net})_{in}]}_{mech\ energy\ loss\ out\ of\ cv} \qquad (4.153)$$

which is valid for both compressible and incompressible flow. One may note that the left-hand side of Equation 4.153 represents the mechanical energy input to the control volume, while the first three terms on the right-hand side represent the mechanical energy output from the control volume. Furthermore, the last three terms on the right hand side of the equation represent the mechanical energy loss. It may be noted that if one assumes ideal flow (no friction), then the mechanical energy loss term equals zero, for which:

$$(q_{net})_{in} = u_2 - u_1 \qquad (4.154)$$

However, for real (viscous) flows, any increase in $u_2 - u_1$ above $q_{net\ in}$ is due to the irreversible conversion of mechanical energy to thermal energy, so the term $(u_2 - u_1 - q_{net\ in})$ represents the mechanical energy loss as follows:

$$(e_{mech})_{loss} = u_2 - u_1 - (q_{net})_{in} \qquad (4.155)$$

Therefore, the energy equation for steady flow, expressed in energy per unit mass terms, and given by Equation 4.153 can be conveniently expressed as mechanical energy balance as follows:

$$(e_{mech})_{in} = (e_{mech})_{out} + (e_{mech})_{loss} \qquad (4.156)$$

The first term in Equation 4.153 can be expanded as follows:

$$[(w_{shaft})_{net}]_{in} = (w_{shaft})_{in} - (w_{shaft})_{out} = w_{pump} - w_{turbine} \qquad (4.157)$$

where w_{pump} is the mechanical work input due to the presence of a pump and $w_{turbine}$ is the mechanical work output due to the presence of a turbine. Therefore, Equation 4.153 can be written as follows:

$$\underbrace{\frac{p_1}{\rho_1} + \frac{v_1^2}{2} + gz_1 + w_{pump}}_{\text{mech energy in}} = \underbrace{\frac{p_2}{\rho_2} + \frac{v_2^2}{2} + gz_2 + w_{turbine}}_{\text{mech energy out}} + \underbrace{(e_{mech})_{loss}}_{\text{mech energy loss}} \qquad (4.158)$$

4.7.2.1 The Energy Equation Expressed in Power Terms

The energy equation given in Equation 4.158 is expressed in energy per unit mass terms, which illustrates the three basic components (groups) of mechanical energy, into, out of, and lost out of, a control volume. However, in order to break down the total mechanical power loss into pump losses, turbine losses, and frictional losses in the piping network (major and minor losses), it is practical to re-express the energy equation in power terms. Specifically, this leads to the definition of the useful power supplied by the pump, the extracted power removed by the turbine, and the power loss due to friction losses in the piping network. Thus, multiplying the energy equation given in Equation 4.158 by the mass flow rate, $\dot{M} = M/\Delta t$ yields each term defined as a power term as follows:

$$\left(\frac{p_1}{\rho_1} + \frac{v_1^2}{2} + gz_1\right)\dot{M} + \dot{W}_{pump} = \left(\frac{p_2}{\rho_2} + \frac{v_2^2}{2} + gz_2\right)\dot{M} + \dot{W}_{turbine} + (\dot{E}_{mech})_{loss} \qquad (4.159)$$

where $\dot{W}_{pump} = \dot{W}_{shaft,in}$ is the shaft power input by the pump's shaft; $\dot{W}_{turbine} = \dot{W}_{shaft,out}$ is the shaft power output by the turbine's shaft; and $(\dot{E}_{mech})_{loss}$ is the total mechanical power loss due to pump losses, turbine losses, and frictional losses in the piping network (major losses and minor losses, which are discussed below), where:

$$(\dot{E}_{mech})_{loss} = [(\dot{E}_{mech})_{loss}]_{pump} + [(\dot{E}_{mech})_{loss}]_{turbine} + [(\dot{E}_{mech})_{loss}]_{piping} \qquad (4.160)$$

However, it may be noted that typically, the irreversible pump and turbine losses (the first and second terms in the right-hand side of Equation 4.160) are treated separately from the irreversible losses due to frictional losses in the piping network. Thus, substituting Equation 4.160 into Equation 4.159 yields the following:

$$\left(\frac{p_1}{\rho_1} + \frac{v_1^2}{2} + gz_1\right)\dot{M} + \underbrace{\dot{W}_{pump} - [(\dot{E}_{mech})_{loss}]_{pimp}}_{(\dot{W}_{pump})_u}$$

$$= \left(\frac{p_2}{\rho_2} + \frac{v_2^2}{2} + gz_2\right)\dot{M} + \underbrace{\dot{W}_{turbine} + [(\dot{E}_{mech})_{loss}]_{turbine}}_{(\dot{W}_{turbine})_e} + [(\dot{E}_{mech})_{loss}]_{piping} \qquad (4.161)$$

where $(\dot{W}_{pump})_u$ is the useful power supplied to the fluid by the pump, $(\dot{W}_{turbine})_e$ is the extracted power removed from the fluid by the turbine, and $[(\dot{E}_{mech})_{loss}]_{piping}$ is the power loss due to friction losses in the piping network. Thus, analogous to the energy equation, Equation 4.158, the energy equation, Equation 4.161 illustrates the three basic components of mechanical power, into, out of, and lost out of, a control volume for the fluid flow system.

One may note that the useful power supplied to the fluid by the pump, $(\dot{W}_{pump})_u$ and the extracted power removed from the fluid by the turbine, $(\dot{W}_{turbine})_e$ are directly modeled in the energy equation, Equation 4.161. However, the irreversible pump and turbine losses are indirectly modeled in the energy equation, Equation 4.161, by the pump and turbine efficiencies, respectively (see Section 4.7.2.2 for their definition). Furthermore, the irreversible loss due to friction losses in the piping network is directly modeled in the energy equation, Equation 4.161, by the major and minor head loss terms (see Sections 4.7.2.4 and 4.5.5.1 for their definition and evaluation).

4.7.2.2 Pump and Turbine Losses Are Modeled by Their Efficiencies

While the energy equation, Equation 4.161, represents the flow into, the flow out of, and loss of mechanical power for a control volume for a fluid flow system, the irreversible pump and turbine losses displayed in the energy equation, Equation 4.161, are indirectly modeled in the energy equation, Equation 4.161, by the pump and turbine efficiencies, respectively. However, in order to define the pump and turbine efficiencies, it is important to first discuss the role of pumps and turbines in a fluid flow and the shaft work regarding pumps and turbine. In a typical flow situation, the fluid is transported from one point to another at a specified flowrate, pressure, velocity, and elevation, while consuming mechanical work through the use of a pump or generating mechanical work through the use of a turbine. One may note that the pump transfers mechanical energy (hydraulic energy) to the fluid flow by increasing its pressure, velocity, and/or elevation, while the turbine extracts mechanical energy (hydraulic energy) from the fluid flow by dropping its pressure, velocity, and/or elevation. Furthermore, one may note that the transfer of mechanical energy is typically accomplished by a rotating shaft, and therefore, mechanical work is usually referred to as shaft work, W_{shaft} or shaft torque, T_{shaft}. The shaft power, $\dot{W}_{shaft}$ is transferred to the flow system through a rotating shaft, and is proportional to the shaft torque, T_{shaft}, and is defined as follows:

$$\dot{W}_{shaft} = \omega T_{shaft} \tag{4.162}$$

where ω = the angular speed of the shaft in radians/second, and T_{shaft} = the rotating shaft torque (shaft work) with dimensions of work (FL).

The irreversible pump and turbine losses displayed in the energy equation, Equation 4.161, are indirectly modeled in the energy equation, Equation 4.161, by the pump and turbine efficiencies, respectively. A pump–motor system consists of a motor and a pump and involves the conversion of electrical or chemical energy to mechanical energy (shaft and hydraulic energy), as illustrated in Figure 4.22. The electric or chemical motor drives the pump by a rotating shaft torque. The shaft rotates at an angular speed, ω and delivers a torque, T to the pump. Specifically, the pump receives shaft work (or shaft power) from a motor as follows:

$$\dot{W}_{pump} = \dot{W}_{shaft,in} = \omega T_{shaft,in} \tag{4.163}$$

and transfers it to the fluid as mechanical energy (useful pumping power or hydraulic power supplied to the fluid) as follows:

$$(\dot{W}_{pump})_u = \dot{W}_{pump} - [(\dot{E}_{mech})_{loss}]_{pump} = \eta_{pump}\dot{W}_{pump} = \eta_{pump}(\dot{W}_{shaft,in}) = \eta_{pump}(\omega T_{shaft,in})$$

$$\tag{4.164}$$

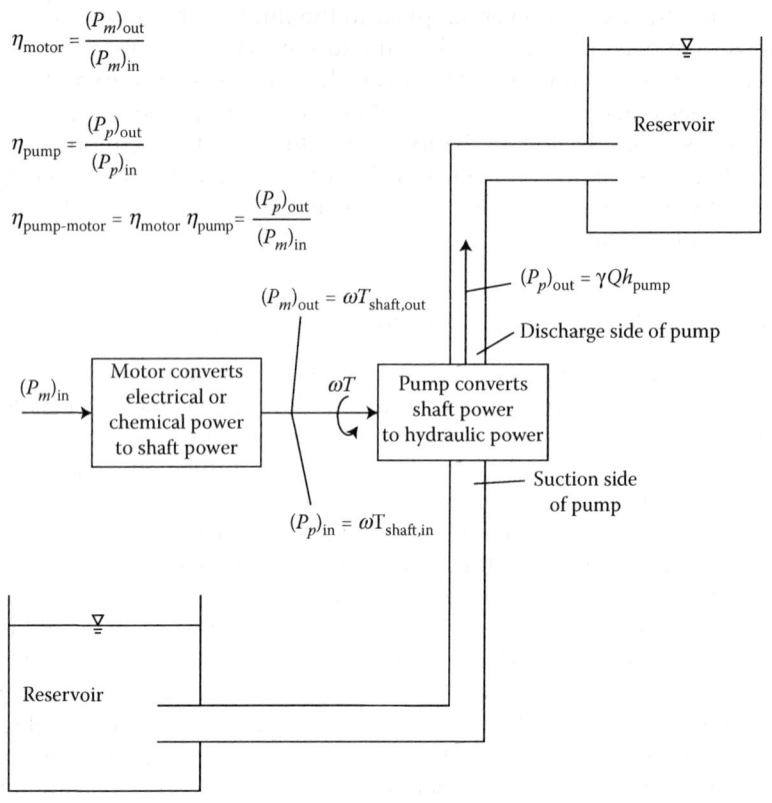

FIGURE 4.22
Pump–motor system: converts electrical or chemical power to mechanical power (shaft and hydraulic power).

where the pump efficiency, η_{pump} is defined as follows:

$$\eta_{pump} = \frac{hydraulic\ power}{shaft\ power} = \frac{(\dot{W}_{pump})_u}{\dot{W}_{pump}} = \frac{(\dot{W}_{pump})_u}{\dot{W}_{shaft,in}} = \frac{(\dot{W}_{pump})_u}{\omega T_{shaft,in}} = \frac{P_{out\ of\ pump}}{P_{in\ to\ pump}} = \frac{(P_p)_{out}}{(P_p)_{in}} \quad (4.165)$$

and represents the irreversible losses in the pump (see Section 4.7.2.3 for discussion of motor and system efficiencies).

A turbine–generator system consists of a turbine and a generator and involves the conversion of mechanical (hydraulic and shaft energy) energy to electrical energy, as illustrated in Figure 4.23. The turbine drives the generator by a rotating shaft torque. The shaft rotates at an angular speed, ω and delivers a torque, T to the generator. Specifically, a turbine converts the mechanical energy of the fluid (turbine extracted power or hydraulic power supplied by the fluid) to shaft work (or shaft power), which is used to drive a generator or a rotary device as follows:

$$(\dot{W}_{turbine})_e = \dot{W}_{turbine} + [(\dot{E}_{mech})_{loss}]_{turbine} = \frac{\dot{W}_{turbine}}{\eta_{turbine}} = \frac{(\dot{W}_{shaft,out})}{\eta_{turbine}} = \frac{(\omega T_{shaft,out})}{\eta_{turbine}} \quad (4.166)$$

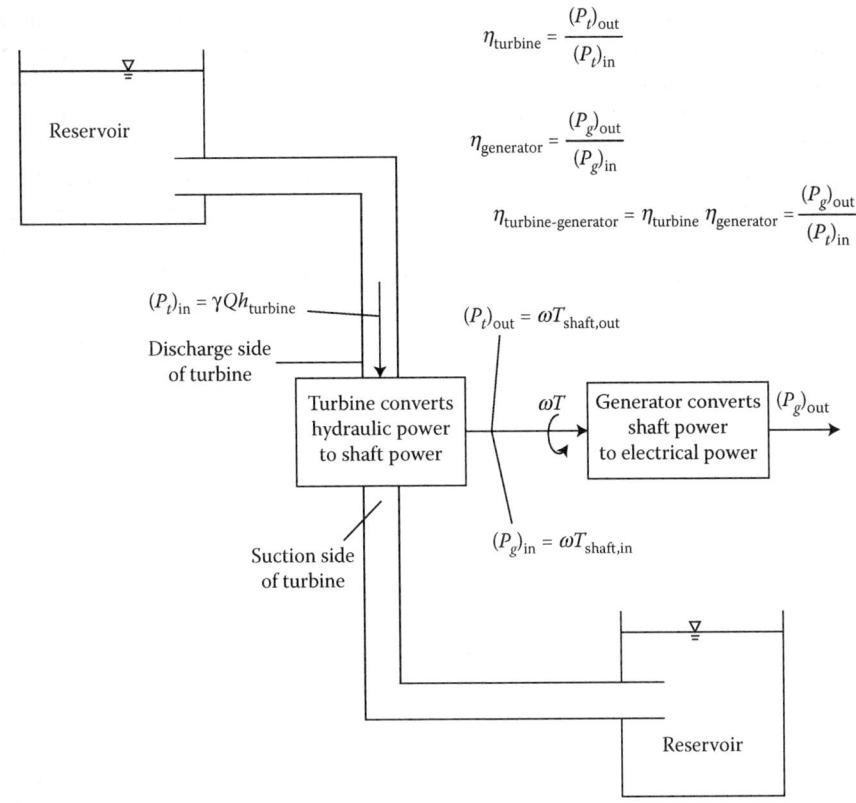

FIGURE 4.23
Turbine–generator system: converts mechanical power (hydraulic and shaft power) to electrical power.

where the turbine efficiency, $\eta_{turbine}$ is defined as follows:

$$\eta_{turbine} = \frac{shaft\ power}{hydraulic\ power} = \frac{\dot{W}_{turbine}}{(\dot{W}_{turbine})_e} = \frac{\dot{W}_{shaft,out}}{(\dot{W}_{turbine})_e} = \frac{\omega T_{shaft,out}}{(\dot{W}_{turbine})_e} = \frac{P_{out\ of\ turbine}}{P_{in\ to\ turbine}} = \frac{(P_t)_{out}}{(P_t)_{in}}$$

(4.167)

and represents the irreversible losses in a turbine (see Section 4.7.2.3 for discussion of generator and system efficiencies). Therefore, the energy equation given in Equation 4.161 expressed in power terms, may be rewritten as follows:

$$\underbrace{\left(\frac{p_1}{\rho_1} + \frac{v_1^2}{2} + gz_1\right)\dot{M} + (\dot{W}_{pump})_u}_{mech\ power\ in} = \underbrace{\left(\frac{p_2}{\rho_2} + \frac{v_2^2}{2} + gz_2\right)\dot{M} + (\dot{W}_{turbine})_e}_{mech\ power\ out} + \underbrace{[(\dot{E}_{mech})_{loss}]_{piping}}_{mech\ power\ loss}$$

(4.168)

4.7.2.3 Pump, Turbine, and System Efficiencies

The pump and turbine losses displayed in the energy equation, Equation 4.161 are indirectly modeled by the pump and turbine efficiencies, respectively, which were defined in Section

4.7.2.2. However, because the pump is part of a pump–motor system, it is necessary to evaluate the motor efficiency and the overall system efficiency in addition to the pump efficiency. And, because the turbine is part of a turbine–generator system, it is necessary to evaluate the generator efficiency and the overall system efficiency in addition to the turbine efficiency.

A pump–motor system consists of a motor and a pump. A schematic illustration of the power input, power output, and efficiency of a pump–motor system is presented in Figure 4.22. An electrical or a chemical motor usually supplies the mechanical power (shaft power) input to a pump. The motor efficiency, η_{motor} is computed as follows:

$$\eta_{motor} = \frac{shaft\ power\ output\ from\ motor}{electric\ power\ input\ from\ electical\ source} = \frac{\dot{W}_{shaft,out}}{\dot{W}_{elect,in}} = \frac{P_{out\ of\ motor}}{P_{in\ to\ motor}} = \frac{(P_m)_{out}}{(P_m)_{in}} \quad (4.169)$$

where:

$$(P_m)_{out} = \omega T_{shaft,out} = (P_p)_{in} = \omega T_{shaft,in} \quad (4.170)$$

The pump efficiency was given above in Equation 4.165 and is repeated as follows:

$$\eta_{pump} = \frac{hydraulic\ power}{shaft\ power} = \frac{(\dot{W}_{pump})_u}{\dot{W}_{pump}} = \frac{(\dot{W}_{pump})_u}{\dot{W}_{shaft,in}} = \frac{(\dot{W}_{pump})_u}{\omega T_{shaft,in}} = \frac{P_{out\ of\ pump}}{P_{in\ to\ pump}} = \frac{(P_p)_{out}}{(P_p)_{in}} \quad (4.171)$$

where, because of irreversible losses in the pump, the useful power supplied to the fluid by the pump, $(P_p)_{out}$ is less than $(P_p)_{in}$ by the factor η_{pump}. And, finally, the overall system efficiency of a pump–motor system is given as follows:

$$\eta_{pump-motor} = \frac{hydraulic\ power}{electric\ power} = \frac{(\dot{W}_{pump})_u}{\dot{W}_{elect,in}} = \frac{(P_p)_{out}}{(P_m)_{in}} = \eta_{pump}\eta_{motor} \quad (4.172)$$

A turbine–generator system consists of a turbine and a generator. A schematic illustration of the power input, power output, and efficiency of a turbine–generator system is presented in Figure 4.23. The flow supplies the shaft power input to a turbine. The turbine efficiency was given above in Equation 4.167 and is repeated as follows:

$$\eta_{turbine} = \frac{shaft\ power}{hydraulic\ power} = \frac{\dot{W}_{turbine}}{(\dot{W}_{turbine})_e} = \frac{\dot{W}_{shaft,out}}{(\dot{W}_{turbine})_e} = \frac{\omega T_{shaft,out}}{(\dot{W}_{turbine})_e} = \frac{P_{out\ of\ turbine}}{P_{in\ to\ turbine}} = \frac{(P_t)_{out}}{(P_t)_{in}}$$

$$(4.173)$$

where, because of irreversible losses in the turbine, $(P_t)_{out}$ is less than the extracted power removed from the fluid by the turbine, $(P_t)_{in}$ by the factor $\eta_{turbine}$. An electrical generator is driven by the shaft power output from a turbine. The generator efficiency, $\eta_{generator}$ is computed as follows:

$$\eta_{generator} = \frac{electrical\ power\ output\ from\ generator}{shaft\ power\ input\ from\ turbine} = \frac{\dot{W}_{elect,out}}{\dot{W}_{shaft,in}} = \frac{P_{out\ of\ gen}}{P_{in\ to\ gen}} = \frac{(P_g)_{out}}{(P_g)_{in}} \quad (4.174)$$

where

$$(P_t)_{out} = \omega T_{shaft,out} = (P_g)_{in} = \omega T_{shaft,in} \tag{4.175}$$

Finally, the overall system efficiency of a turbine–generator system is given as follows:

$$\eta_{turbine-generator} = \frac{electric\ power}{hydraulic\ power} = \frac{\dot{W}_{elect,out}}{(\dot{W}_{turbine})_e} = \frac{(P_g)_{out}}{(P_t)_{in}} = \eta_{turbine}\eta_{generator} \tag{4.176}$$

Therefore, the conversion of work (or power) by the pump–motor system (and its components), or the turbine–generator system (and its components) (i.e., the mechanical efficiency of the device or system) is less than 100% due to irreversible losses in the given device and overall system.

4.7.2.4 The Energy Equation Expressed in Energy Head Terms

The energy equation given in Equation 4.168 is expressed in power terms, which illustrates the three basic components of mechanical power, into, out of, and lost out of, a control volume. However, in the application of the energy equation to a control volume for a fluid flow system, it is typical/practical to express the energy equation in energy per unit weight terms or energy head terms. Thus, dividing the energy equation expressed in terms of power given in Equation 4.168 by mass flow rate, $\dot{M} = M/\Delta t$ would yield terms of energy per unit mass given in Equation 4.158. Then, dividing the energy per unit mass terms given in Equation 4.158 by g yields the energy equation expressed in terms of energy per unit weight or energy head (each term with dimensions of $FL/F = L$) as follows:

$$\left(\frac{p_1}{\rho_1 g} + \frac{v_1^2}{2g} + z_1\right) + \frac{(w_{pump})_u}{g} = \left(\frac{p_2}{\rho_2 g} + \frac{v_2^2}{2g} + z_2\right) + \frac{(w_{turbine})_e}{g} + \frac{[(e_{mech})_{loss}]_{piping}}{g} \tag{4.177}$$

The useful head delivered to the fluid by the pump, $(h_{pump})_u$ is defined as a function of the useful power supplied to the fluid by the pump, $(\dot{W}_{pump})_u$ and the weight flowrate, γQ as follows:

$$(h_{pump})_u = \frac{(w_{pump})_u}{g} = \frac{(\dot{W}_{pump})_u}{\dot{M}g} = \frac{(\dot{W}_{pump})_u}{(\rho V/\Delta t)g} = \frac{(\dot{W}_{pump})_u}{\gamma Q} = \frac{(P_p)_{out}}{\gamma Q} = \frac{(\eta_{pump})(P_p)_{in}}{\gamma Q}$$
$$= \frac{(\eta_{pump})(\omega T_{shaft,in})}{\gamma Q} \tag{4.178}$$

The extracted head removed from the fluid by the turbine, $(h_{turbine})_e$ is defined as a function of the extracted power removed from the fluid by the turbine, $(\dot{W}_{turbine})_e$ and the weight flowrate, γQ as follows:

$$(h_{turbine})_e = \frac{(w_{turbine})_e}{g} = \frac{(\dot{W}_{turbine})_e}{\dot{M}g} = \frac{(\dot{W}_{turbine})_e}{(\rho V/\Delta t)g} = \frac{(\dot{W}_{turbine})_e}{\gamma Q} = \frac{(P_t)_{in}}{\gamma Q} = \frac{(P_t)_{out}/\eta_{turbine}}{\gamma Q}$$
$$= \frac{\omega T_{shaft,out}/\eta_{turbine}}{\gamma Q} \tag{4.179}$$

And, the irreversible head loss between points 1 and 2 in the finite control volume due to the frictional losses in the piping network, h_f is defined as a function of the major head loss, $h_{f,maj}$ and the minor head loss, $h_{f,min}$ as follows:

$$h_f = h_{f,maj} + h_{f,min} = \frac{[(e_{mech})_{loss}]_{piping}}{g} \qquad (4.180)$$

where h_f represents the major losses and minor losses in all components in the piping system other than the pump or the turbine; the major and minor head loss terms are defined and evaluated in Section 4.5.5.1. Finally, substituting Equations 4.178 through 4.180 into Equation 4.177 yields the energy equation that is applied to a control volume for a fluid flow system, and is given as follows:

$$\underbrace{\left(\frac{p_1}{\rho_1 g} + \frac{v_1^2}{2g} + z_1\right) + (h_{pump})_u}_{\text{energy head in}} = \underbrace{\left(\frac{p_2}{\rho_2 g} + \frac{v_2^2}{2g} + z_2\right) + (h_{turbine})_e +}_{\text{energy head out}} \underbrace{h_f}_{\text{energy head loss}} \qquad (4.181)$$

Assuming incompressible flow, and removing the subscripts for the pump and turbine heads, the energy equation is given as follows:

$$\left(\frac{p_1}{\gamma} + \frac{v_1^2}{2g} + z_1\right) + h_{pump} = \left(\frac{p_2}{\gamma} + \frac{v_2^2}{2g} + z_2\right) + h_{turbine} + h_{f,total} \qquad (4.182)$$

where $h_{f,total} = h_{f,maj} + h_{f,min}$. Thus, the energy equation, Equation 4.181, represents the flow into, the flow out of, and the loss of mechanical energy head for a control volume for a fluid flow system. The energy equation given in Equation 4.181 models the useful head delivered to the fluid by the pump, the extracted head removed from the fluid by the turbine, and the major and minor frictional head losses in the piping network. As such, the irreversible pump and turbine losses displayed in the energy equation, Equation 4.161, are indirectly modeled by the pump and turbine efficiencies, respectively, as illustrated in Equations 4.178 and 4.179 for $(h_{pump})_u$ and $(h_{turbine})_e$, respectively.

One may note that while the energy equation that is derived based on the conservation of energy principle, Equation 4.181 assumes steady flow, it directly considers the modeling of energy transfer by work (pumps, turbines, etc.). Furthermore, it is important to note that a more general statement of the integral form of the energy equation for a control volume, assuming steady flow, assumes there may be one or more control faces for the flow into and the flow out of the control volume as illustrated in Figure 4.3. Thus, the more general statement of the energy equation evaluated between two points in a control volume, assuming steady flow, is given as follows:

$$\left(\frac{p}{\gamma} + \frac{v^2}{2g} + z\right)_{in} + h_{pump} = \left(\frac{p}{\gamma} + \frac{v^2}{2g} + z\right)_{out} + h_{turbine} + h_{f,total} \qquad (4.183)$$

4.7.2.5 Definition of the Terms in the Energy Equation

The energy equation given by Equation 4.181, which is applied between two points in a finite control volume, only considers the mechanical forms of energy, and the frictional effects that cause the mechanical energy to be lost. The energy equation given by Equation 4.181 is repeated as follows:

$$\underbrace{\left(\frac{p_1}{\gamma} + \frac{v_1^2}{2g} + z_1\right) + h_{pump}}_{\text{mech energy in}} = \underbrace{\left(\frac{p_2}{\gamma} + \frac{v_2^2}{2g} + z_2\right) + h_{turbine}}_{\text{mech energy out}} + \underbrace{h_{f,total}}_{\text{mech energy loss}} \qquad (4.184)$$

where the left-hand side represents the mechanical energy input to the control volume, while the first term on the right-hand side represents the mechanical energy output from the control volume, and the last term on the right-hand side represents the mechanical energy loss. The terms of Equation 4.184 are expressed in terms of energy per unit weight or energy head. Thus, the p/γ term represents the flow energy (pressure energy or flow work) per unit weight, the $v^2/2g$ term represents the kinetic energy per unit weight, and the z term represents the potential energy per unit weight. The $h_{f,total}$ term represents the mechanical energy per unit weight loss due to frictional losses in the piping network (major and minor losses). Furthermore, the h_{pump} term represents the useful mechanical energy (hydraulic energy) per unit weight (useful head) delivered to the fluid by a pump, while the $h_{turbine}$ term represents the extracted mechanical energy (hydraulic energy) per unit weight (extracted head) removed from the fluid by a turbine.

4.7.2.6 Practical Assumptions Made for the Energy Equation

It may be noted that when the piping losses, h_f are insignificant compared to the other terms in Equation 4.181 (i.e., the dissipation of mechanical energy into thermal energy is negligible), then it can be ignored and set equal to zero. Furthermore, if there are no mechanical work devices such as a pump, or a turbine, then the h_{pump} term and the $h_{turbine}$ term are each set equal to zero. As a result, Equation 4.181 reduces to the Bernoulli equation, which was also derived above using Newton's second law of motion, and is given as follows:

$$\left(\frac{p_1}{\gamma} + \frac{v_1^2}{2g} + z_1\right) = \left(\frac{p_2}{\gamma} + \frac{v_2^2}{2g} + z_2\right) \qquad (4.185)$$

Thus, the Bernoulli equation assumes incompressible, steady, and ideal flow along a streamline and is applied between two points along a streamline and where the total head, H along a streamline is constant (see Sections 4.5.3, 4.5.6 and 4.5.7 for applications of the Bernoulli equation).

4.7.3 Application of the Energy Equation in Complement to the Momentum Equation

As previously discussed in detail in Section 4.5.4, while the application of the continuity equation will always be necessary for internal flow, the energy equation and the momentum equations play complementary roles in the analysis of a given internal or external flow situation. The energy equation and the momentum equations are known as the equations of motion. They play complementary roles because when one of the two equations breaks down, the other may be used to find the additional unknown quantity. It is important to note that, similar to the continuity equation, the momentum equation may be applied using either the integral (for a finite control volume) or the differential approach. However, application of the energy equation is useful only in the integral approach (for a finite control volume), as energy is work, which is defined over distance, L. Furthermore, in the application of the governing equations (continuity, energy, and momentum), it is important to make the distinction between real flow, ideal flow, and a hydraulic jump. And, for ideal flow, it is important to make the distinction between internal (pipe or open channel flow), external flow, and the flow from a tank.

An outline of the application of the energy equation in complement to the momentum equation for the various flow types was presented in Section 4.5.4. Section 4.5.5.4 presented applications of the energy equation for real pipe flow. Section 4.5.5.5 presented applications of the energy equation for real open channel flow. Section 4.5.6.3 presented applications of the Bernoulli equation for ideal pipe flow. Section 4.5.6.4 presented applications of the Bernoulli equation for ideal open channel flow. Section 4.5.6.5 presented applications of the Bernoulli equation for ideal flow from a tank. Section 4.5.7.3 presented applications of the Bernoulli equation for ideal external flow. And, Section 4.5.8.2 presented applications of the energy equation for a hydraulic jump. However, it is important to note that the energy equation, Equation 4.39 derived in Section 4.5.2, based on the conservation of momentum does not consider the modeling of the energy transfer by work (pumps, turbines, etc.). Therefore, the applications of the energy equation for real pipe flow presented in Section 4.5.5.4 and the applications of the energy equation for real open channel flow presented in Section 4.5.5.5 did not illustrate example problems involving a pump or a turbine. As such, Section 4.7.4 presents the application of the energy equation, Equation 4.181, derived in Section 4.7.2.4, based on the conservation of energy for real internal flow (pipe or open channel flow) with a pump or a turbine.

4.7.4 Application of the Energy Equation for Real Internal Flow with a Pump or a Turbine

While Sections 4.5.5.4 and 4.5.5.5 presented applications of the energy equation, Equation 4.39, for real internal flow (pipe and open channel flow) without a pump or a turbine, this section presents applications of the energy equation, Equation 4.181, for real internal flow with a pump or a turbine. Application of the integral form of the energy equation for incompressible, real, steady flow (pipe or open channel flow) given by Equation 4.181 requires the evaluation of the friction head loss term, $h_{f,total} = h_{f,maj} + h_{f,min}$; the head delivered to the fluid by a pump, h_{pump}; and the head removed from the fluid by a turbine, $h_{turbine}$. Evaluation of the major and minor head loss flow resistance terms were presented in Section 4.5.5.1. As noted previously in Section 4.5.5.3, the resulting definitions for the major and minor head loss (flow resistance) terms (see Equations 4.90 and 4.91, respectively) require details in the definition of variables (C, f, v, L, D, and k), which are to be presented in later chapters. Therefore, example problems illustrating the application of the energy equation for real pipe or open channel flow will assume that the minor and major head loss terms are either a given quantity (lumped value) computed from the given head loss equations, or directly solved for in the energy equation, Equation 4.181. Furthermore, evaluation of the head delivered to the fluid by a pump, h_{pump} and the head removed from the fluid by a turbine, $h_{turbine}$ were presented in Section 4.7.2.4. However, unlike the major and minor head loss terms, the details of the derivation and the definition of the h_{pump}, and the $h_{turbine}$ terms were presented above in Section 4.7.2.4, and are summarized as follows:

$$h_{pump} = \frac{(P_p)_{out}}{\gamma Q} = \frac{(\eta_{pump})(P_p)_{in}}{\gamma Q} = \frac{(\eta_{pump})(\omega T_{shaft,in})}{\gamma Q} \tag{4.186}$$

where $(P_p)_{out} = (\eta_{pump})(\omega T_{shaft,in}) = \gamma Q h_{pump}$ is the hydraulic power added to the flow by the pump, and the discharge, Q is flowing in and out of the pump as illustrated in Figure 4.22. And

$$h_{turbine} = \frac{(P_t)_{in}}{\gamma Q} = \frac{(P_t)_{out}/\eta_{turbine}}{\gamma Q} = \frac{\omega T_{shaft,out}/\eta_{turbine}}{\gamma Q} \tag{4.187}$$

where

$$(P_t)_{in} = \frac{\omega T_{shaft,out}}{\eta_{turbine}} = \gamma Q h_{turbine}$$

is the hydraulic power removed from the flow by the turbine, and the discharge, Q is flowing in and out of the turbine as illustrated in Figure 4.23.

Therefore, substituting Equations 4.186 and 4.187 for the h_{pump} and $h_{turbine}$, respectively, into the energy equation, Equation 4.181, and rearranging, yields the following:

$$\left(\frac{p_1}{\gamma} + \frac{v_1^2}{2g} + z_1\right) + \underbrace{\left(\frac{(\eta_{pump})(\omega T_{shaft,in})}{\gamma Q}\right)}_{h_{pump}} - \underbrace{\left(\frac{\omega T_{shaft,out}/\eta_{turbine}}{\gamma Q}\right)}_{h_{turbine}} - h_{f,total} = \left(\frac{p_2}{\gamma} + \frac{v_2^2}{2g} + z_2\right)$$

$$\tag{4.188}$$

Thus, in a typical internal flow situation, the fluid is transported from point 1 to point 2 at a specified flowrate, pressure, velocity, and elevation, while consuming mechanical work through the use of a pump, or generating mechanical work through the use of a turbine, meanwhile sustaining a head loss due flow resistance. The head loss due to flow resistance, $h_{f,total}$ is a result of pipe or channel friction (major head loss), pipe components (minor head loss), and/or a flow measuring device. Specifically, the pump converts mechanical energy to hydraulic energy, which increases the fluid pressure (velocity, and/or elevation). The turbine converts hydraulic energy to mechanical energy, which drops the fluid pressure (elevation and/or velocity). The power losses of the pump and turbine are modeled by the pump and turbine efficiencies, respectively and thus are treated separately from the head loss due to flow resistance, $h_{f,total}$. However, it is important to note that the head loss due to flow resistance, $h_{f,total}$ contributes to the power losses of the fluid flow and thus the determination of the required efficiency of pumps and turbines. In particular, the total head loss, $h_{f,total}$ (major and minor) will increase the amount of power that a pump must deliver to the fluid. As such, the head loss, $h_{f,total}$ represents the additional height that the fluid needs to be elevated (or additional increase in pressure or velocity) by the pump in order to overcome the losses due to the flow resistance in the pipe or channel. Thus, some of the pumping power will be used to pump the fluid through the pipe or channel, while some will be used to compensate for the energy lost in the pressure drop, Δp caused by the pipe or channel friction, pipe component, or flow measuring device. A similar analogy may be deduced for evaluating the power of a turbine. In particular, the total head loss, $h_{f,total}$ (major and minor) will decrease the amount of power that a turbine can extract from the fluid. As such, the head loss, $h_{f,total}$ due to the flow resistance in the pipe or channel represents the lost pressure (and elevation and velocity) that the fluid can supply to the turbine. Thus, some of the fluid flow power will be extracted by the turbine, while some will be used to compensate for the energy lost in the pressure drop, Δp caused by the pipe or channel friction, pipe component, or flow measuring device.

The power losses of the pump and turbine are modeled by the pump and turbine efficiencies, respectively, and thus are treated separately from the head loss due to flow resistance, $h_{f,total}$. However, it is important to note that the head loss due to flow resistance, $h_{f,total}$ contributes to the power losses of the fluid flow and thus the determination of the required efficiency of pumps and turbines. In general, pumps have a smaller efficiency than turbines. This is attributed to a number of reasons. First, the pump converts mechanical (kinetic) energy to hydraulic (flow) energy, which increases the fluid pressure (velocity, and/or elevation). Because an increase in pressure is accompanied by significant head losses, it is considered to be an adverse pressure gradient (see Chapter 8). However, the turbine converts hydraulic (flow) energy to mechanical (kinetic) energy, which decreases the fluid pressure (elevation and/or velocity). Because a decrease in pressure is accompanied by relatively smaller head losses, it is considered to be a favorable pressure gradient (see Chapter 8). Thus, the significant head loss in a pump contributes to a more significant decrease in the pump efficiency. Second, pumps typically operate at higher angular velocities, ω than turbines do. The pump efficiency is inversely proportional to the angular velocity, ω (see Equation 4.165). However, the turbine efficiency is directly proportional to the angular velocity, ω (see Equation 4.167). Thus, the higher angular velocity of the pump, ω contributes to a direct decrease in the pump efficiency. Third, pumps are typically smaller in size than turbines, and thus the head losses in pumps play a more important role in pumps than in turbines. Finally, pumps typically operate over a wider range of flowrates, Q than turbines do. As such, the wider range of flowrates, Q for a pump result in a corresponding wider range of angular velocity, ω (see Equation 4.165). A wider operational range of angular velocity, ω for pumps leads to less efficient designs for pumps than for turbines. While pump efficiencies can range from 50% to 90%, highly efficient pumps can attain just over 90% efficiency. Furthermore, while impulse turbine efficiencies can approach 90%, and reaction turbines efficiencies can range from 90% to 95%, highly efficient turbines in general can attain just over 95% efficiency.

4.7.4.1 Dimensions and Units for the Pump and Turbine Head Terms

The energy equation, Equation 4.188, is expressed in energy per unit weight terms (with dimensions of FL/F), or energy head terms (with dimensions of L). While the dimensions and units for the pressure head, velocity head, and elevation head terms have been presented and discussed in Section 4.5.2, it is important to discuss the dimensions and units of the pump and turbine head and the variables that define them. The pump head given in Equation 4.186 is repeated as follows:

$$h_{pump} = \frac{(P_p)_{out}}{\gamma Q} = \frac{(\eta_{pump})(P_p)_{in}}{\gamma Q} = \frac{(\eta_{pump})(\omega T_{shaft,in})}{\gamma Q} \tag{4.189}$$

where the hydraulic power out of the pump and into the fluid, $(P_p)_{out}$ and the shaft power out of the motor and into pump, $(P_p)_{in}$ have dimensions of FLT^{-1}, SI units of N-m/s = W = J/s, and BG units of ft-lb/sec or hp = $550\,ft\text{-}lb/sec$; the pump efficiency, η_{pump} is expressed as a percent and is dimensionless; the angular speed of the rotating shaft of the motor, ω has dimensions of T^{-1} and units of radians/sec; the rotating shaft torque of the motor, $T_{shaft,in}$ has work dimensions of FL, SI units of N-m = J, and BG units of ft-lb; the fluid-specific weight, γ has dimensions of FL^{-3}, SI units of N/m^3, and BG units of lb/ft^3; and the discharge,

Q has dimensions of $L^3 T^{-1}$, SI units m^3/sec, and BG units of ft^3/sec. Thus, the pump head, h_{pump} has dimensions of L, SI units of m, and BG units of ft, and is expanded as follows:

$$h_{pump}[L] = \frac{\eta_{pump}[] \, \omega[T^{-1}] \, T_{shaft,in}[FL]}{\gamma[FL^{-3}] \, Q[L^3 T^{-1}]} \tag{4.190}$$

For instance, assuming SI units, the pump head, h_{pump} has units of m as follows:

$$h_{pump}[m] = \frac{\eta_{pump}[] \, \omega[rad/sec] \, T_{shaft,in}[N-m]}{\gamma[N/m^3] \, Q[m^3/sec]} \tag{4.191}$$

The turbine head given in Equation 4.187 is repeated as follows:

$$h_{turbine} = \frac{(P_t)_{in}}{\gamma Q} = \frac{(P_t)_{out}/\eta_{turbine}}{\gamma Q} = \frac{\omega T_{shaft,out}/\eta_{turbine}}{\gamma Q} \tag{4.192}$$

where the hydraulic power from the flow and into the turbine, $(P_t)_{in}$ and the shaft power out of the turbine and into the generator, $(P_t)_{out}$ have dimensions of FLT^{-1}, SI units of N-m/s = W = J/s, and BG units of ft-lb/sec or hp = 550 *ft-lb/sec*; the turbine efficiency, $\eta_{turbine}$ is expressed as a percent and is dimensionless; the angular speed of the rotating shaft of the turbine, ω has dimensions of T^{-1} and units of radians/sec; the rotating shaft torque of the turbine, $T_{shaft,out}$ has work dimensions of FL, SI units of N-m = J, and BG units of ft-lb; the fluid-specific weight, γ has dimensions of FL^{-3}, SI units of N/m^3, and BG units of lb/ft^3; and the discharge, Q has dimensions of $L^3 T^{-1}$, SI units m^3/sec, and BG units of ft^3/sec. Thus, the turbine head, $h_{turbine}$ has dimensions of L, SI units of m, and BG units of ft, and is expanded as follows:

$$h_{turbine}[L] = \frac{\dfrac{\omega[T^{-1}] \, T_{shaft,out}[FL]}{\eta_{turbine}[]}}{\gamma[FL^{-3}] \, Q[L^3 T^{-1}]} \tag{4.193}$$

For instance, assuming BG units, the turbine head, $h_{turbine}$ has units of ft as follows:

$$h_{turbine}[ft] = \frac{\dfrac{\omega[rad/sec] \, T_{shaft,out}[ft-lb]}{\eta_{turbine}[]}}{\gamma[lb/ft^3] \, Q[ft^3/sec]} \tag{4.194}$$

4.7.4.2 Applications of the Energy Equation for Real Internal Flow with a Pump

Application of the energy equation for real internal flow with a pump is given as follows:

$$\left(\frac{p_1}{\gamma} + \frac{v_1^2}{2g} + z_1\right) + \underbrace{\left(\frac{(\eta_{pump})(\omega T_{shaft,in})}{\gamma Q}\right)}_{h_{pump}} - h_{f,total} = \left(\frac{p_2}{\gamma} + \frac{v_2^2}{2g} + z_2\right) \tag{4.195}$$

A pump converts shaft power supplied by a motor to hydraulic power and supplies it to the fluid flow. Internal flow includes both pipe and open channel flow. However, in the utilization of a pump to transport the fluid from point 1 to point 2 at the specified flowrate,

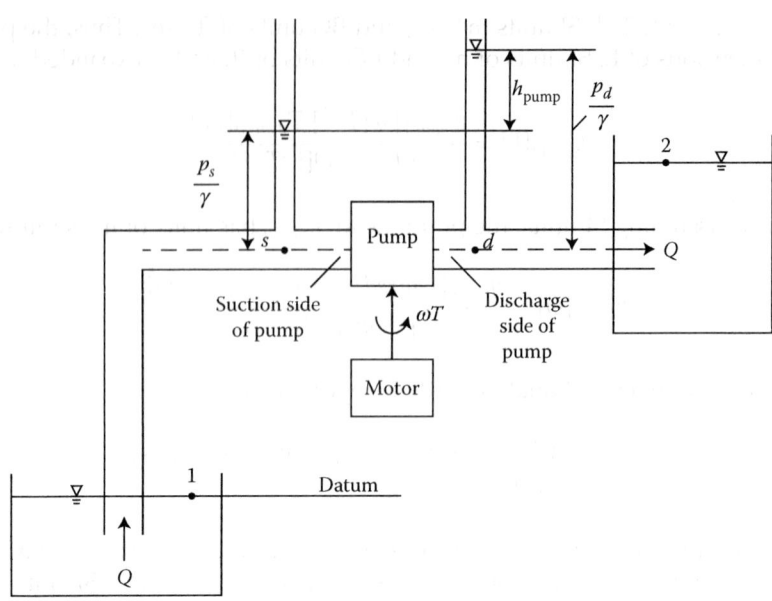

FIGURE 4.24
Application of the energy equation for pipe flow with a pump.

Q flowing in and out of the pump, the assumption is that there is some sort of piping between points 1 and 2. As such, the example problems presented in this section illustrate the application of the energy equation for real (or ideal) pipe flow with a pump, as illustrated in Figure 4.24. Furthermore, it is important to distinguish between the "analysis" problem versus the "design" problem of a pump–motor system. An analysis problem assumes that the magnitudes for the efficiency and power of the motor, pump, and overall system are given quantities, while the magnitudes for the discharge and head delivered by the pump are unknown quantities to be determined by their respective equations. And, a design problem assumes that the magnitudes for the required discharge and head to be delivered by a pump are given quantities, while the magnitudes for the efficiency and the required power of the motor, pump, and overall system are determined by their respective equations.

In the "analysis" problem of a pump–motor system, the magnitudes for the efficiency and power of the motor, pump, and overall system are given quantities, while the magnitudes for the discharge and head delivered by the pump are unknown quantities to be determined by their respective equations. The motor efficiency was defined in Equation 4.169 as the ratio of the shaft power output from the motor to the electric or chemical power input to the motor and is summarized as follows:

$$\eta_{motor} = \frac{shaft\ power}{electric\ power} = \frac{(P_m)_{out}}{(P_m)_{in}} = \frac{\omega T_{shaft,out}}{(P_m)_{in}} \tag{4.196}$$

The pump efficiency was defined in Equation 4.165 as the ratio of the hydraulic power output from the pump to the shaft power input to the pump and is summarized as follows:

$$\eta_{pump} = \frac{hydraulic\ power}{shaft\ power} = \frac{(P_p)_{out}}{(P_p)_{in}} = \frac{\gamma Q h_{pump}}{\omega T_{shaft,in}} \tag{4.197}$$

where $(P_m)_{out} = \omega T_{shaft,out} = (P_p)_{in} = \omega T_{shaft,in}$. And the overall pump–motor system efficiency was defined in Equation 4.172 as the ratio of the hydraulic power output from the pump to the electric or chemical power input to the motor and is summarized as follows:

$$\eta_{pump-motor} = \frac{hydraulic\ power}{electric\ power} = \frac{(P_p)_{out}}{(P_m)_{in}} = \frac{\gamma Q h_{pump}}{(P_m)_{in}} = \eta_{pump}\eta_{motor} \tag{4.198}$$

The magnitudes for the discharge, Q and head delivered by the pump, h_{pump} are related by Equation 4.186 as follows:

$$h_{pump} = \frac{(P_p)_{out}}{\gamma Q} = \frac{\eta_{pump}(P_p)_{in}}{\gamma Q} = \frac{\eta_{pump}\omega T_{shaft,in}}{\gamma Q} \tag{4.199}$$

Therefore, in the "analysis" problem, for the given magnitude of the hydraulic power output from the pump, $(P_p)_{out} = \eta_{pump}\omega T_{shaft,in} = \gamma Q h_{pump}$, there is a specific relationship between the magnitudes of the discharge, Q and the head delivered by the pump, h_{pump}. First, there is a maximum allowable velocity of the flow in the pipe, which is a function of the pressure at the suction side of the pump, as illustrated by application of the energy equation (see below). Thus, for a given pipe size, there is a corresponding maximum allowable discharge of flow in the pipe. Furthermore, for the given magnitude of the hydraulic power output from the pump, $(P_p)_{out} = \eta_{pump}\omega T_{shaft,in} = \gamma Q h_{pump}$ and the maximum allowable discharge, there is a corresponding maximum allowable head delivered by the pump. In the analysis of a pumping system, consideration is given to making sure that the pressure at the suction side of the pump (see Figure 4.24) does not fall below the vapor pressure of the liquid, thus avoiding cavitation (see Section 4.5.3.5 for further discussion). Cavitation is unintended boiling of a liquid in motion at room temperature. Furthermore, when evaporation of the liquid occurs during cavitation, the vapor may cause a pocket to form in the suction side of the pump, and thus stop the flow of the liquid. Thus, in the "analysis" problem, there is a maximum allowable velocity of the flow in the pipe in order to avoid a pressure drop below the vapor pressure of the liquid at the suction side of the pump and thus to avoid cavitation and restriction of flow. Determination of the maximum allowable velocity of the flow in the pipe assumes that the pressure at the suction side of the pump at point s is set equal to the vapor pressure, p_v of the liquid in the pipe. The maximum allowable velocity is determined by applying the energy equation between points 1 and s (see Figure 4.24), assuming the datum is a point 1 as follows:

$$\left(\frac{p_1}{\gamma} + \frac{v_1^2}{2g} + z_1\right) - h_{f,total} = \left(\frac{p_s}{\gamma} + \frac{v_s^2}{2g} + z_s\right) \tag{4.200}$$

where $p_1 = 0$, $v_1 = 0$, $z_1 = 0$, $p_s = p_v$ (negative gage pressure), and thus:

$$\frac{v_s^2}{2g} = -\frac{p_v}{\gamma} - h_{f,total} - z_s \tag{4.201}$$

There are two additional points to note. First, for the given vapor pressure, p_v of the liquid (and total head loss in the pipe, $h_{f,total}$), there is a maximum height, z_s to which the fluid may be pumped; otherwise, there will be no flow in the suction line at the suction side of the pump (this point is illustrated in Example Problem 4.22). And second, because the pump creates a suction in the pipe line, the pressure gradually decreases from atmospheric pressure at point 1 to the vapor pressure, p_v of the liquid at point s, as illustrated by the application of the energy equation between point 1 and any point inside the suction line (this point is illustrated in Example Problem 4.20). Thus, for a given pipe size, application of the continuity equation yields the maximum allowable discharge, as follows:

$$Q = v_s A \tag{4.202}$$

and the corresponding maximum allowable head delivered by the pump, h_{pump} is computed from Equation 4.186 for the given magnitude of the hydraulic power output from the pump, $(P_p)_{out} = \eta_{pump}\, \omega T_{shaft,in} = \gamma Q h_{pump}$. In the "analysis" problem, the actual discharge, Q delivered by the pump is measured by a pipe flow-measuring device (such as an orifice, nozzle, or venturi meter, as discussed in Chapter 8). Furthermore, the corresponding actual head delivered by the pump, h_{pump} is measured by installing a piezometer at the suction side of the pump at point s and at the discharge side of the pump at point d (see Figure 4.24). However, instead of installing a piezometer at the suction side of the pump, a manometer may be used in order to measure the vapor pressure (negative gage pressure) in the fluid flow, as illustrated in Example Problem 4.20. The actual head delivered by the pump, h_{pump} is measured by the application of the energy equation between point s and d (see Figure 4.24), assuming a horizontal pipe with a constant cross section, and ideal flow across the pump results in the following:

$$\left(\frac{p_s}{\gamma} + \frac{v_s^2}{2g} + z_s\right) + h_{pump} = \left(\frac{p_d}{\gamma} + \frac{v_d^2}{2g} + z_d\right) \tag{4.203}$$

where $v_s = v_d$, and $z_s = z_d$, and thus:

$$h_{pump} = \frac{p_d}{\gamma} - \frac{p_s}{\gamma} \tag{4.204}$$

where there is a pressure increase in the flow due to the pump. Furthermore, the maximum allowable head delivered by the pump, h_{pump} is defined by assuming $p_s = p_v$ (negative gage pressure); thus:

$$h_{pump} = \frac{p_d}{\gamma} - \frac{p_v}{\gamma} \tag{4.205}$$

EXAMPLE PROBLEM 4.20

Water at 20°C is pumped from reservoir 1 to reservoir 2 through a 0.95-m-diameter copper pipe 1300 m in length, fitted with three regular 90° flanged elbows in order to change the direction of the flow, as illustrated in Figure EP 4.20. The pump is located 10 m above the elevation of reservoir 1, at a distance of 100 m along the length of the pipe. The pipe entrance from reservoir 1 is reentrant, and the pipe exit to reservoir 2 is reentrant. Assume a Darcy–Weisbach friction factor, f of 0.15; a minor head loss coefficient due to an elbow, k of 0.3; a minor head loss coefficient due to a reentrant pipe entrance, k of 1.0; and a minor head loss coefficient due to a reentrant pipe exit, k of 1.0. The hydraulic power delivered to the flow by the pump is 500 kW. The pressure at the suction side of the pump is measured using a manometer (in order to measure

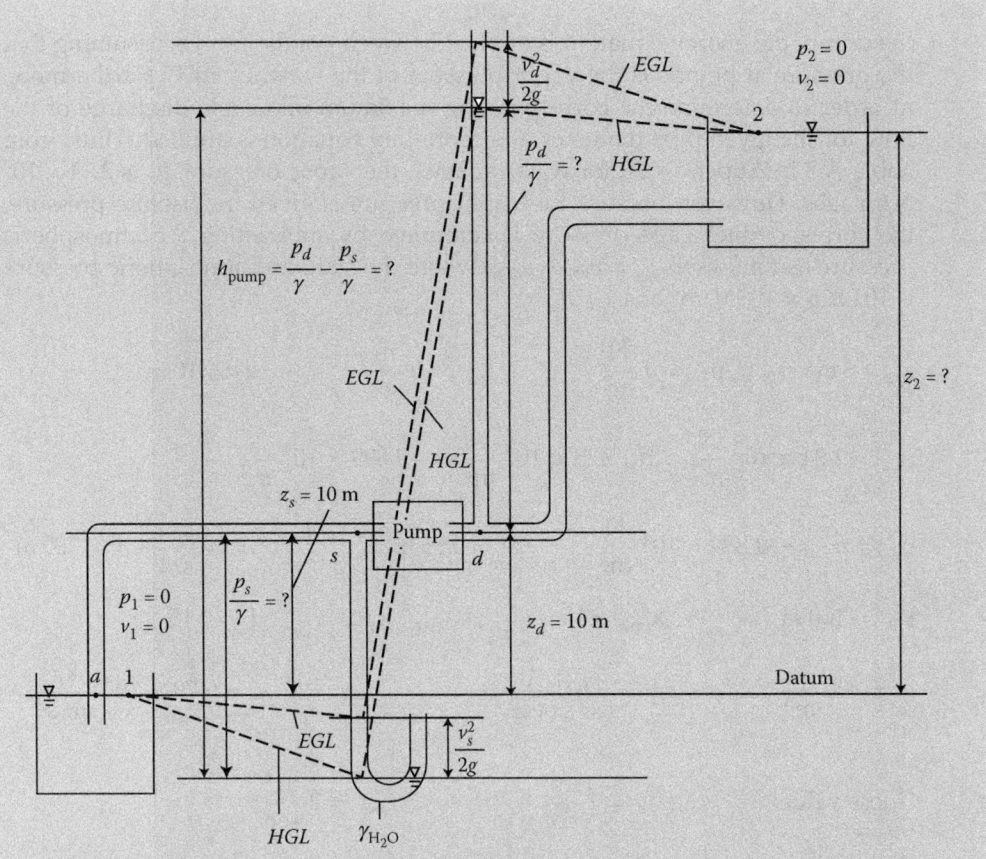

FIGURE EP 4.20
Water is pumped from reservoir 1 to reservoir 2 through a pipe with components.

a negative gage pressure in the fluid flow), while the pressure at the discharge side of the pump is measured using a piezometer. (a) Determine the maximum allowable velocity and discharge of the flow in the pipe without causing cavitation at the suction side, s of the pump. (b) Demonstrate that because the pump creates a suction in the pipe line, the pressure gradually decreases from atmospheric pressure at point 1 to the vapor pressure, p_v of the liquid at point s. (c) For the maximum allowable discharge, determine the corresponding maximum allowable head delivered to the flow by the pump. (d) Determine how the actual head delivered to the pump is measured. (e) If the maximum allowable head is delivered by the pump, determine the magnitude of the pressure head registered at the suction and discharge sides of the pump. (f) Determine the maximum allowable elevation to which the flow may be pumped to point 2. (h) Draw the energy grade line and the hydraulic grade line.

Mathcad Solution

(a) Assuming the datum is at point 1, in order to determine the maximum allowable velocity of the flow in the pipe without causing cavitation at the suction side, s of the pump, the energy equation is applied between points 1 and s, assuming that the pressure at point s is the vapor pressure of the water at 20°C. Furthermore, in order to determine the corresponding maximum allowable discharge of the flow for the given pipe diameter, the continuity equation is applied. Thus, from Table A.2 in Appendix A, for water at 20°C, the vapor pressure, p_v is 2.34×10^3 N/m² abs. However, because the vapor pressure is given in absolute pressure, the corresponding gage pressure is computed by subtracting the atmospheric pressure as follows: $p_{gage} = p_{abs} - p_{atm}$, where the standard atmospheric pressure is 101.325×10^3 N/m² abs.

$$z_1 := 0\,\text{m} \qquad p_1 := 0\,\frac{\text{N}}{\text{m}^2} \qquad v_1 := 0\,\frac{\text{m}}{\text{sec}} \qquad z_s := 10\,\text{m}$$

$$p_v := 2.34 \times 10^3\,\frac{\text{N}}{\text{m}^2} - 101.325 \times 10^3\,\frac{\text{N}}{\text{m}^2} = -9.899 \times 10^4\,\frac{\text{N}}{\text{m}^2}$$

$$p_s := p_v = -9.899 \times 10^4\,\frac{\text{N}}{\text{m}^2} \qquad D := 0.95\,\text{m} \qquad A := \frac{\pi \cdot D^2}{4} = 0.709\,\text{m}^2$$

$$L_{1s} := 100\,\text{m} \qquad k_{ent} := 1 \qquad k_{elbow} := 0.3 \qquad f := 0.15$$

$$\rho := 998\,\frac{\text{kg}}{\text{m}^3} \qquad g := 9.81\,\frac{\text{m}}{\text{sec}^2} \qquad \gamma := \rho \cdot g = 9.79 \times 10^3\,\frac{\text{kg}}{\text{m}^2\,\text{s}^2}$$

Guess value: $\qquad v_s := 1\,\dfrac{\text{m}}{\text{sec}} \qquad\qquad Q := 2\,\dfrac{\text{m}^3}{\text{sec}}$

$$h_{fmaj1s} := 10\,\text{m} \qquad\qquad h_{fent} := 1\,\text{m} \qquad h_{felbow} := 1\,\text{m}$$

Given

$$\frac{p_1}{\gamma} + z_1 + \frac{v_1^2}{2 \cdot g} - h_{fmaj1s} - h_{fent} - h_{felbow} = \frac{p_s}{\gamma} + z_s + \frac{v_s^2}{2 \cdot g}$$

$$h_{fmaj1s} = f \cdot \frac{L_{1s}}{D} \cdot \frac{v_s^2}{2 \cdot g} \qquad h_{fent} = k_{ent} \cdot \frac{v_s^2}{2 \cdot g} \qquad h_{felbow} = k_{elbow} \cdot \frac{v_s^2}{2 \cdot g}$$

$$Q = v_s \cdot A$$

$$\begin{pmatrix} v_s \\ Q \\ h_{fmaj1s} \\ h_{fent} \\ h_{felbow} \end{pmatrix} := \text{Find}(v_s, Q, h_{fmaj1s}, h_{fent}, h_{felbow})$$

$$v_s = 0.346\,\frac{\text{m}}{\text{s}} \qquad Q = 0.245\,\frac{\text{m}^3}{\text{s}}$$

$$h_{fmaj1s} = 0.096\,\text{m} \qquad h_{fent} = 6.105 \times 10^{-3}\,\text{m} \qquad h_{felbow} = 1.831 \times 10^{-3}\,\text{m}$$

Thus, the magnitude of the water vapor pressure limits the maximum velocity to 0.346 m/sec, and the given pipe size limits the corresponding maximum allowable discharge of flow in the pipe to 0.245 m³/sec.

(b) In order to demonstrate that the pump creates a suction in the pipe line where the pressure gradually decreases from atmospheric pressure at point 1 to the vapor pressure, p_v of the liquid at point s, the energy equation is applied between point 1 and any point inside the suction line. Assume that the point inside the suction line is at point a as illustrated in Figure EP 4.20.

$$z_a := 0 \, m \qquad\qquad v_a := v_s = 0.346 \, \frac{m}{s}$$

Guess value: $\qquad\qquad p_a := -1 \, \dfrac{N}{m^2}$

Given

$$\frac{p_1}{\gamma} + z_1 + \frac{v_1^2}{2 \cdot g} - h_{fent} = \frac{p_a}{\gamma} + z_a + \frac{v_a^2}{2 \cdot g}$$

$$p_a := Find(p_a) = -119.539 \, \frac{N}{m^2}$$

Thus, the gage pressure decreases from atmospheric pressure ($0 \, N/m^2$ gage) at point 1, to $-119.539 \, N/m^2$ at point a to the water vapor pressure, $p_v = -9.899 \times 10^4 \, N/m^2$ at point s.

(c) For the given magnitude of the hydraulic power output from the pump and the maximum allowable discharge, there is a corresponding maximum allowable head delivered by the pump, which is computed by applying Equation 4.186 as follows:

$$P_{pout} := 500 \, kW \qquad\qquad h_{pump} := \frac{P_{pout}}{\gamma \cdot Q} = 208.182 \, m$$

(d) The actual head delivered by the pump, h_{pump} is measured by reading the manometer at the suction side of the pump and the piezometer at the discharge side of the pump and applying the energy equation between point s and d, assuming a horizontal pipe with a constant cross section and ideal flow across the pump, which results in the following:

$$\left(\frac{p_s}{\gamma} + \frac{v_s^2}{2g} + z_s \right) + h_{pump} = \left(\frac{p_d}{\gamma} + \frac{v_d^2}{2g} + z_d \right)$$

where $v_1 = v_2$, and $z_1 = z_2$, and thus:

$$h_{pump} = \frac{p_d}{\gamma} - \frac{p_s}{\gamma}$$

(e) If the maximum allowable head is delivered by the pump, the magnitude of the pressure head registered on the manometer at the suction side of the pump would be the vapor pressure head of the water, and the magnitude of the pressure head registered on the piezometer at the discharge side of the pump may be calculated by applying the energy equation between the suction side, s and the discharge side, d of the pump as follows:

$$z_d := z_s = 10\,m \qquad v_d := v_s = 0.346\,\frac{m}{s} \qquad p_s := p_v = -9.899 \times 10^4\,\frac{N}{m^2}$$

$$\frac{p_s}{\gamma} = -10.11\,m \qquad h_{pump} = 208.182\,m$$

Guess value: $\qquad p_d := 20 \times 10^5\,\dfrac{N}{m^2}$

Given

$$\frac{p_s}{\gamma} + z_s + \frac{v_s^2}{2 \cdot g} + h_{pump} = \frac{p_d}{\gamma} + z_d + \frac{v_d^2}{2 \cdot g}$$

$$p_d := Find(p_d) = 1.939 \times 10^6\,\frac{N}{m^2}$$

$$\frac{p_d}{\gamma} = 198.072\,m$$

where the maximum allowable head developed by the pump, h_{pump} may be measured by reading the manometer and the piezometer and taking their difference as follows:

$$h_{pump} := \frac{p_d}{\gamma} - \frac{p_s}{\gamma} = 208.182\,m$$

which is a pressure increase in the flow due to the pump.

(f) In order to determine the maximum allowable elevation to which the flow may be pumped to point 2, the energy equation is applied between points 1 and 2 as follows:

$$p_2 := 0\,\frac{N}{m^2} \qquad v_2 := 0\,\frac{m}{sec} \qquad L_{d2} := 1200\,m \qquad k_{exit} := 1$$

Guess value: $\qquad z_2 := 300\,m \qquad h_{fmaj1s} = 10\,m \qquad h_{fmajd2} := 10\,m$

$$h_{fent} := 1\,m \qquad h_{felbow} := 1\,m \qquad h_{fexit} := 1\,m$$

Given

$$\frac{p_1}{\gamma} + z_1 + \frac{v_1^2}{2 \cdot g} - h_{fmaj1s} - h_{fmajd2} - h_{fent} - 3 \cdot h_{felbow} - h_{fexit} + h_{pump} = \frac{p_2}{\gamma} + z_2 + \frac{v_2^2}{2 \cdot g}$$

$$h_{fmaj1s} = f \cdot \frac{L_{1s}}{D} \cdot \frac{v_s^2}{2 \cdot g} \qquad h_{fmajd2} = f \cdot \frac{L_{d2}}{D} \cdot \frac{v_d^2}{2 \cdot g}$$

$$h_{fent} = k_{ent} \cdot \frac{v_s^2}{2 \cdot g} \qquad h_{felbow} = k_{elbow} \cdot \frac{v_s^2}{2 \cdot g} \qquad h_{fexit} = k_{exit} \cdot \frac{v_d^2}{2 \cdot g}$$

$$\begin{pmatrix} z_2 \\ h_{fmaj1s} \\ h_{fmajd2} \\ h_{fent} \\ h_{felbow} \\ h_{fexit} \\ h_{pump} \end{pmatrix} := \text{Find}(z_2, h_{fmaj1s}, h_{fmajd2}, h_{fent}, h_{felbow}, h_{fexit}, h_{pump})$$

$z_2 = 206.911\,\text{m}$ $\qquad\qquad h_{fmaj1s} = 0.096\,\text{m}$ $\qquad\qquad h_{fmajd2} = 1.157\,\text{m}$

$h_{fent} = 6.105 \times 10^{-3}\,\text{m}$ $\quad 3\,h_{felbow} = 5.494 \times 10^{-3}\,\text{m}$ $\quad h_{fexit} = 6.105 \times 10^{-3}\,\text{m}$

$h_{pump} = 208.182\,\text{m}$

$h_{ftotal} := h_{fmaj1s} + h_{fmajd2} + h_{fent} + 3\,h_{felbow} + h_{fexit} = 1.271\,\text{m}$

where the maximum allowable elevation to which the flow may be pumped to point 2 is $z_2 = 206.911$ m, and therefore, the maximum allowable head added by the pump, $h_{pump} = 208.182$ m is used to overcome the maximum difference in elevation between the two reservoirs, $\Delta z = 206.911$ m, and the total head loss due to pipe friction and pipe components between points 1 and 2, $h_{ftotal} = 1.271$ m.

(g) The EGL and HGL are illustrated in Figure EP 4.20.

EXAMPLE PROBLEM 4.21

The pump–motor system for Example Problem 4.20 is illustrated in Figure EP 4.21, where the hydraulic power delivered to the flow by the pump is 500 kW. The chemical power supplied to the motor is 700 kW, and the rotating shaft of the motor has an angular speed of 300 rad/sec and a shaft torque of 2000 N-m. (a) Determine the shaft power delivered by the motor to the pump. (b) Determine the motor efficiency. (c) Determine the pump efficiency. (d) Determine the pump–motor system efficiency.

Mathcad Solution

(a) The shaft power delivered by the motor to the pump, $(P_m)_{out} = \omega T_{shaft,out} = (P_p)_{in} = \omega T_{shaft,in}$ defined for Equation 4.169 is applied as follows:

$\omega := 300\,\dfrac{\text{rad}}{\text{sec}}$ $\qquad\qquad\qquad\qquad T_{shaftout} := 2000\,\text{Nm}$

$P_{mout} := \omega \cdot T_{shaftout} = 600\,\text{kW}$ $\qquad P_{pin} := P_{mout} = 600\,\text{kW}$

(b) The motor efficiency defined in Equation 4.169 is applied as follows:

$P_{min} := 700\,\text{kW}$ $\qquad\qquad\qquad\qquad \eta_{motor} := \dfrac{P_{mout}}{P_{min}} = 0.857$

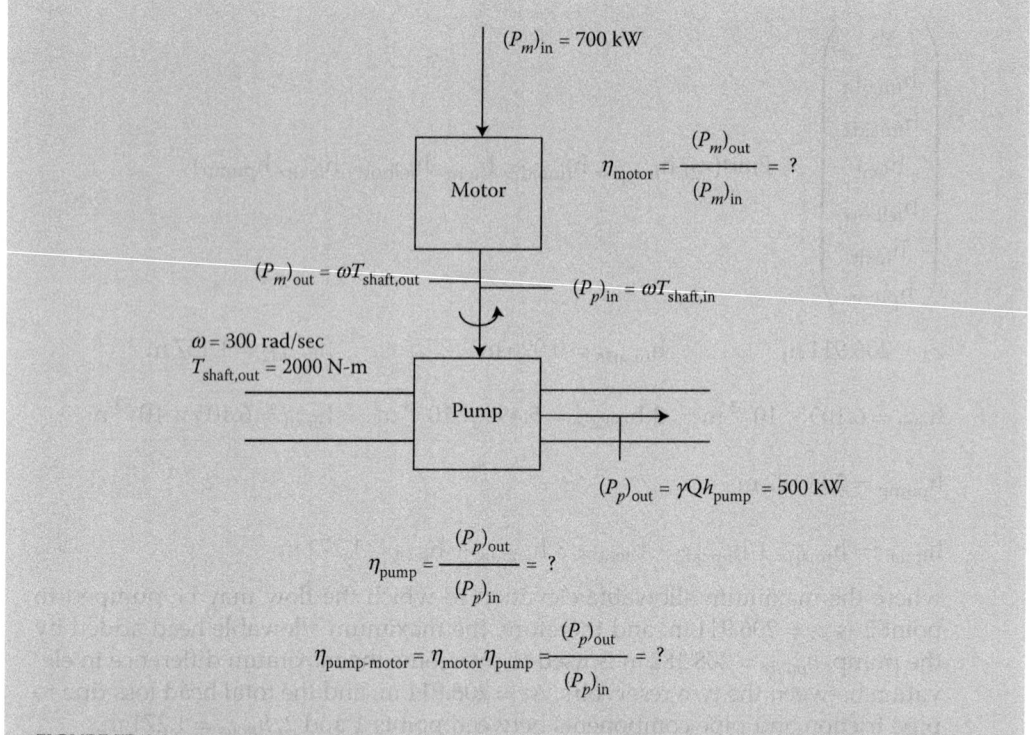

$(P_m)_{in} = 700\ kW$

$\eta_{motor} = \dfrac{(P_m)_{out}}{(P_m)_{in}} = ?$

Motor

$(P_m)_{out} = \omega T_{shaft,out}$

$(P_p)_{in} = \omega T_{shaft,in}$

$\omega = 300\ rad/sec$
$T_{shaft,out} = 2000\ N\text{-}m$

Pump

$(P_p)_{out} = \gamma Q h_{pump} = 500\ kW$

$\eta_{pump} = \dfrac{(P_p)_{out}}{(P_p)_{in}} = ?$

$\eta_{pump\text{-}motor} = \eta_{motor}\eta_{pump} = \dfrac{(P_p)_{out}}{(P_p)_{in}} = ?$

FIGURE EP 4.21
Pump–motor system: converts chemical power to mechanical power (shaft and hydraulic power).

Thus, the motor is 85.7% efficient.
(c) The pump efficiency defined in Equation 4.165 is applied as follows:

$P_{pout} := 500\ kW$ $\qquad\qquad$ $\eta_{pump} := \dfrac{P_{pout}}{P_{pin}} = 0.833$

Thus, the pump is 83.3% efficient.
(d) The pump–motor system efficiency defined in Equation 4.172 is applied as follows:

$\eta_{pumpmotor} := \dfrac{P_{pout}}{P_{min}} = 0.714$ $\qquad\qquad$ $\eta_{pumpmotor} := \eta_{pump}\cdot\eta_{motor} = 0.714$

Thus, the pump–motor system is 71.4% efficient.

In the "design" problem of a pump–motor system, the magnitudes for the required discharge, Q and head to be delivered by the pump, h_{pump} are given, while the magnitudes for the efficiency and the required power of the motor, pump, and overall system are determined from their respective equations. Although the determination of the magnitude for the required discharge is limited by the maximum allowable velocity (in order to avoid cavitation), it may be accommodated by adjusting the design pipe size. The maximum allowable velocity of the flow in the pipe assumes that the pressure at the suction side of the pump at point s is set equal to the vapor pressure, p_v of the liquid in the pipe. The maximum allowable velocity is determined by applying the energy equation between points 1 and s (see Figure 4.24), assuming the datum is a point 1 as follows:

$$\left(\frac{p_1}{\gamma} + \frac{v_1^2}{2g} + z_1\right) - h_{f,total} = \left(\frac{p_s}{\gamma} + \frac{v_s^2}{2g} + z_s\right) \tag{4.206}$$

where $p_1 = 0$, $v_1 = 0$, $z_1 = 0$, $p_s = p_v$ (negative gage pressure), and thus:

$$\frac{v_s^2}{2g} = -\frac{p_v}{\gamma} - h_{f,total} - z_s \tag{4.207}$$

where for the given vapor pressure, p_v of the liquid (and total head loss in the pipe, $h_{f,total}$), there is a maximum height, z_s to which the fluid may be pumped; otherwise, there will be no flow in the suction line at the suction side of the pump (this point is illustrated in Example Problem 4.22). The magnitude for the required discharge may be accommodated by adjusting the design pipe size diameter, and application of the continuity equation yields the magnitude of the required discharge to be delivered by the pump as follows:

$$Q = v_s A \tag{4.208}$$

The required head to be delivered by the pump, h_{pump} is determined by applying the energy equation between points 1 and 2 (see Figure 4.24), assuming the datum is at point 1 as follows:

$$\left(\frac{p_1}{\gamma} + \frac{v_1^2}{2g} + z_1\right) + h_{pump} - h_{f,total} = \left(\frac{p_2}{\gamma} + \frac{v_2^2}{2g} + z_2\right) \tag{4.209}$$

where $p_1 = p_2 = 0$, $v_1 = v_2 = 0$, $z_1 = 0$, and thus:

$$h_{pump} = z_2 + h_{f,total} \tag{4.210}$$

The magnitude of the required hydraulic power output from the pump is related to the magnitudes for the required discharge and head to be delivered by the pump by Equation 4.186 and thus is determined as follows:

$$(P_p)_{out} = \eta_{pump}(P_p)_{in} = \eta_{pump}\omega T_{shaaft,in} = \gamma Q h_{pump} \tag{4.211}$$

A pump converts shaft power supplied by a motor to hydraulic power and supplies it to the fluid flow. Meanwhile, the motor converts electrical or chemical power to shaft power and supplies it to the pump. Therefore, in the "design" problem, for the required magnitude of the hydraulic power output from the pump, $(P_p)_{out} = \gamma Q h_{pump}$, there is a specific relationship between the required magnitudes of the pump efficiency, η_{pump} and the shaft power input to the pump, $(P_p)_{in} = \omega T_{shaft,in}$, which is determined from Equation 4.165 as follows:

$$\eta_{pump} = \frac{hydraulic\ power}{shaft\ power} = \frac{(P_p)_{out}}{(P_p)_{in}} = \frac{\gamma Q h_{pump}}{\omega T_{shaft,in}} \tag{4.212}$$

where the electric or chemical motor drives the pump by a rotating shaft torque and the shaft rotates at an angular speed, ω and delivers a torque, $T_{shaft,in}$ to the pump. In particular, as the required pump efficiency decreases, the required magnitude of the shaft power input to the pump by the motor increases, and vice versa. Furthermore, for the required magnitude of the shaft power from the motor, $(P_m)_{out} = \omega T_{shaft,out} = (P_p)_{in} = \omega T_{shaft,in}$, there is a specific relationship between the required magnitudes of the motor efficiency, η_{motor} and the chemical or electric power, $(P_m)_{in}$, which is determined from Equation 4.169 as follows:

$$\eta_{motor} = \frac{shaft\ power}{electric\ power} = \frac{(P_m)_{out}}{(P_m)_{in}} = \frac{\omega T_{shaft,out}}{(P_m)_{in}} \tag{4.213}$$

In particular, as the required motor efficiency decreases, the required magnitude of the chemical or electric power input to the motor increases, and vice versa. Finally, for the required magnitude of the hydraulic power output from the pump, $(P_p)_{out} = \gamma Q h_{pump}$, there is a specific relationship between the required magnitudes of the overall pump–motor system efficiency, $\eta_{pump-motor}$, and the required electric power input to the motor, $(P_m)_{in}$, which is determined from Equation 4.172 as follows:

$$\eta_{pump-motor} = \frac{hydraulic\ power}{electric\ power} = \frac{(P_p)_{out}}{(P_m)_{in}} = \frac{\gamma Q h_{pump}}{(P_m)_{in}} = \eta_{pump}\eta_{motor} \tag{4.214}$$

In particular, as the required pump–motor system efficiency decreases, the required magnitude of the chemical or electric power input to the motor increases, and vice versa.

EXAMPLE PROBLEM 4.22

A pump is required to dewater a site as illustrated in Figure EP 4.22. It is required that the water (at 20°C) be pumped from reservoir 1 to reservoir 2 to an elevation of 150 m at a discharge of 0.45 m³/sec. Assume that the total head loss due to both major (pipe friction) and minor losses (pipe components) is 0.7 m between points 1 and the suction side of the pump at point s and 1.2 m between the discharge side of the pump at point d and point 2. (a) Determine the maximum height, z_s to which the fluid may be pumped; otherwise, there will be no flow in the suction line at the suction side of the pump. (b) Determine the maximum allowable velocity of the flow in the pipe without causing cavitation. (c) Determine the minimum required design pipe size diameter in order to accommodate the maximum allowable velocity and the required flowrate. (d) Determine the required head to be delivered by the pump. (e) Determine the required hydraulic power output from the pump. (f) Assuming a typical pump efficiency of 80%, determine the required shaft power input to the pump by the motor. (g) Assuming a typical motor efficiency of 85%, determine the required chemical or electric power input to the motor. (h) Determine the resulting overall pump–motor system efficiency.

Mathcad Solution

(a) Assuming the datum is at point 1, to determine the maximum height, z_s to which the fluid may be pumped (in order to establish a flow in the suction line), the

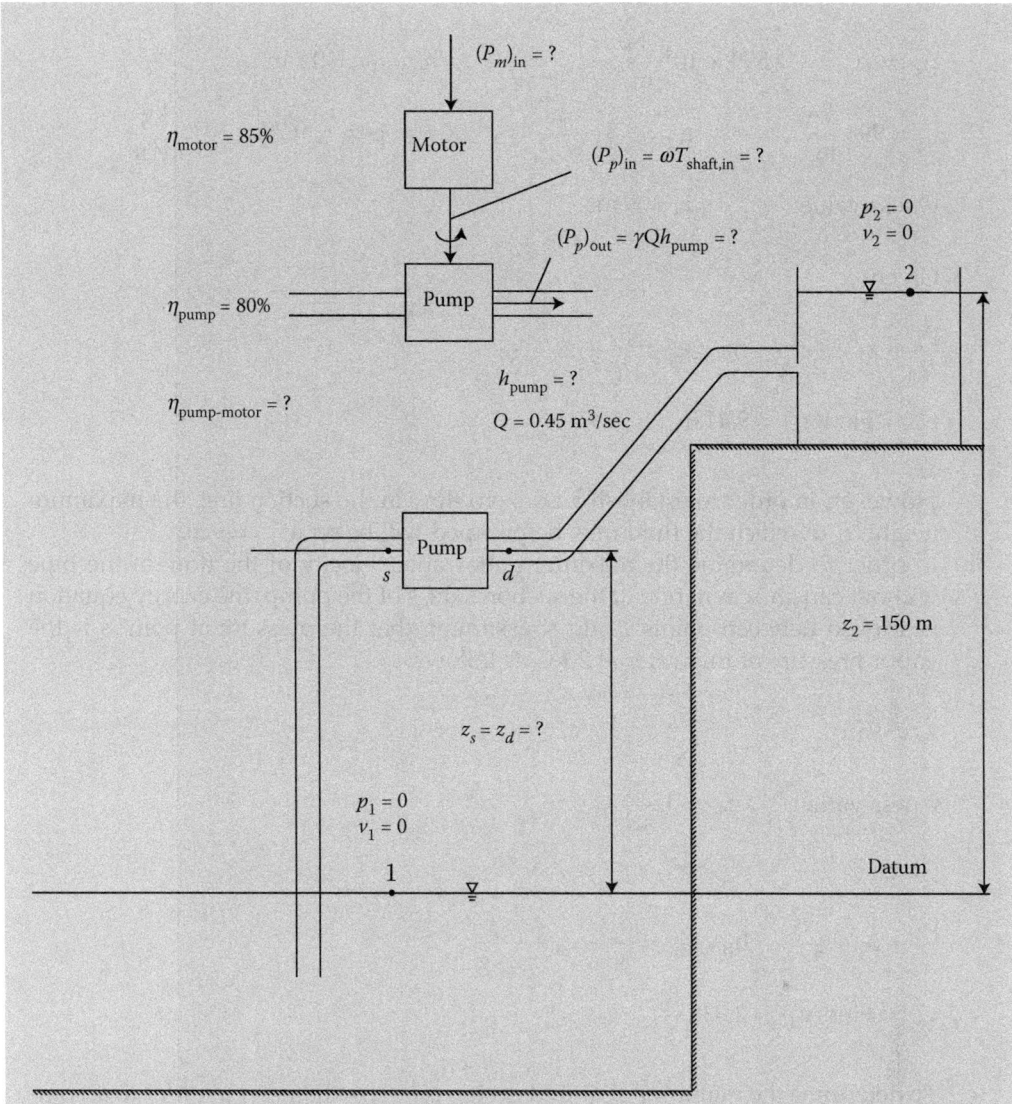

FIGURE EP 4.22
Pump used to dewater a site pumping water from reservoir 1 to reservoir 2 through a pipe with components.

energy equation is applied between points 1 and s, assuming a vapor pressure, p_v of the liquid at point s in order to avoid cavitation, and assuming no flow in the suction line as follows:

$$z_1 := 0 \text{ m} \qquad p_1 := 0 \, \frac{\text{N}}{\text{m}^2} \qquad v_1 := 0 \, \frac{\text{m}}{\text{sec}} \qquad v_s := 0 \, \frac{\text{m}}{\text{sec}}$$

$$p_v := 2.34 \times 10^3 \, \frac{\text{N}}{\text{m}^2} - 101.325 \times 10^3 \, \frac{\text{N}}{\text{m}^2} = -9.899 \times 10^4 \, \frac{\text{N}}{\text{m}^2}$$

$$p_s := p_v = -9.899 \times 10^4 \ \frac{N}{m^2} \qquad\qquad h_{ftotal1s} := 0.7 \ m$$

$$\rho := 998 \ \frac{kg}{m^3} \qquad g := 9.81 \ \frac{m}{sec^2} \qquad \gamma := \rho \cdot g = 9.79 \times 10^3 \ \frac{kg}{m^2 \, s^2}$$

Guess value: $z_s := 9 \ m$

Given

$$\frac{p_1}{\gamma} + z_1 + \frac{v_1^2}{2 \cdot g} - h_{flotal1s} = \frac{p_s}{\gamma} + z_s + \frac{v_s^2}{2 \cdot g}$$

$$z_s := Find(z_s) = 9.41 \ m$$

However, in order to establish a nonzero flow in the suction line, the maximum height, z_s to which the fluid may be pumped will be set at $z_s = 9 \ m$.

(b) In order to determine the maximum allowable velocity of the flow in the pipe without causing cavitation at the suction side, s of the pump, the energy equation is applied between points 1 and s, assuming that the pressure at point s is the vapor pressure of the water at 20°C as follows:

$$z_s := 9 \ m$$

Guess value: $v_s := 1 \ \frac{m}{sec}$

Given

$$\frac{p_1}{\gamma} + z_1 + \frac{v_1^2}{2 \cdot g} - h_{flotal1s} = \frac{p_s}{\gamma} + z_s + \frac{v_s^2}{2 \cdot g}$$

$$v_s := Find(v_s) = 2.838 \ \frac{m}{s}$$

(c) To determine the minimum required design pipe size diameter in order to accommodate the maximum allowable velocity and the required flowrate, the continuity equation is applied as follows:

$$Q := 0.45 \ \frac{m^3}{sec}$$

Guess value: $D := 1 \ m \qquad\qquad A := 1 \ m^2$

Given

$$Q = v_s \cdot A$$

$$A = \frac{\pi \cdot D^2}{4}$$

$$\begin{pmatrix} D \\ A \end{pmatrix} := \text{Find}(D, A)$$

$D = 0.449\,\text{m}$ $\qquad\qquad\qquad$ $A = 0.159\,\text{m}^2$

(d) The required head to be delivered by the pump is determined by applying the energy equation between points 1 and 2 as follows:

$p_2 := 0\,\dfrac{N}{m^2}$ $\qquad$ $v_2 := 0\,\dfrac{m}{sec}$ $\qquad$ $z_2 := 150\,\text{m}$

$h_{ftotal1s} := 0.7\,\text{m}$ $\qquad$ $h_{ftotald2} := 1.2\,\text{m}$

Guess value: $\qquad\qquad$ $h_{pump} := 200\,\text{m}$

Given

$$\dfrac{p_1}{\gamma} + z_1 + \dfrac{v_1^2}{2 \cdot g} - h_{ftotal1s} - h_{ftotald2} + h_{pump} = \dfrac{p_2}{\gamma} + z_2 + \dfrac{v_2^2}{2 \cdot g}$$

$h_{pump} := \text{Find}(h_{pump}) = 151.9\,\text{m}$

Thus, the required head to be delivered by the pump, $h_{pump} = 151.9\,\text{m}$ must overcome the difference in elevation between the two reservoirs, $\Delta z = 150\,\text{m}$ and the total head loss due to pipe friction and pipe components between points 1 and 2, $h_{f,total} = 0.7\,\text{m} + 1.2\,\text{m} = 1.9\,\text{m}$.

(e) The required hydraulic power output from the pump is computed by applying Equation 4.186 as follows:

$P_{pout} := \gamma \cdot Q \cdot h_{pump} = 669.221\,\text{kW}$

(f) Assuming a typical pump efficiency of 80%, the required shaft power input to the pump by the motor is computed by applying Equation 4.165 as follows:

$\eta_{pump} := 0.80$

Guess value: $\qquad\qquad$ $P_{pin} := 700\,\text{kW}$

Given

$$\eta_{pump} = \dfrac{P_{pout}}{P_{pin}}$$

$P_{pin} := \text{Find}(P_{pin}) = 836.527\,\text{kW}$

(g) Assuming a typical motor efficiency of 85%, the required chemical or electric power input to the motor is computed by applying Equation 4.169 as follows:

$$P_{mout} := P_{pin} = 836.527 \, kW \qquad\qquad \eta_{motor} := 0.85$$

Guess value: $\qquad\qquad\qquad\qquad\qquad P_{min} := 800 \, kW$

Given

$$\eta_{motor} = \frac{P_{mout}}{P_{min}}$$

$$P_{min} := Find(P_{min}) = 984.149 \, kW$$

(h) The resulting overall pump–motor system efficiency is computed by applying Equation 4.172 as follows:

$$\eta_{pumpmotor} := \frac{P_{pout}}{P_{min}} = 0.68 \qquad\qquad \eta_{pumpmotor} := \eta_{pump} \cdot \eta_{motor} = 0.68$$

4.7.4.3 Applications of the Energy Equation for Real Internal Flow with a Turbine

Application of the energy equation for real internal flow with a turbine is given as follows:

$$\left(\frac{p_1}{\gamma} + \frac{v_1^2}{2g} + z_1\right) - \underbrace{\left(\frac{\omega T_{shaft,out}/\eta_{turbine}}{\gamma Q}\right)}_{h_{turbine}} - h_{f,total} = \left(\frac{p_2}{\gamma} + \frac{v_2^2}{2g} + z_2\right) \qquad (4.215)$$

A turbine extracts freely available hydraulic power from the fluid flow and converts it to shaft power to drive a generator. Internal flow includes both pipe and open channel flow. Thus, in the utilization of the flow at the specified flowrate, Q flowing in and out of the turbine, to extract power from the flow, the assumption is that the flow between points 1 and the turbine may be either pipe flow (see Figure 4.25) or open channel flow (see Figure 4.26). However, first, it is important to distinguish between the two basic types of turbines, namely impulse turbines and reaction turbines. In an impulse turbine (also called the Pelton wheel), the fluid flows as a jet from a nozzle (pipe flow) and strikes the rotating blades (vanes) of the turbine, as illustrated in Figure 4.27. In a reaction turbine (such as the Francis turbine or the Kaplan turbine), the fluid flows (at a high pressure) from a pipe or open channel and rotates the vanes of the turbine, filling the turbine casing with fluid, and leaves the turbine through a draft tube, as illustrated in Figure 4.28. Furthermore, impulse turbines are most efficient for fluid flow systems with small flowrates, Q and large heads, $h_{turbine}$, while reaction turbines are more efficient for fluid flow systems with large flowrates, Q and small heads, $h_{turbine}$. In the case where the turbine extracts hydraulic power from a pipe flow, either an impulse turbine (see Figure 4.29) or a reaction turbine (see Figure 4.30) may be used. However, in the case where the turbine extracts hydraulic power from an open channel flow, a reaction

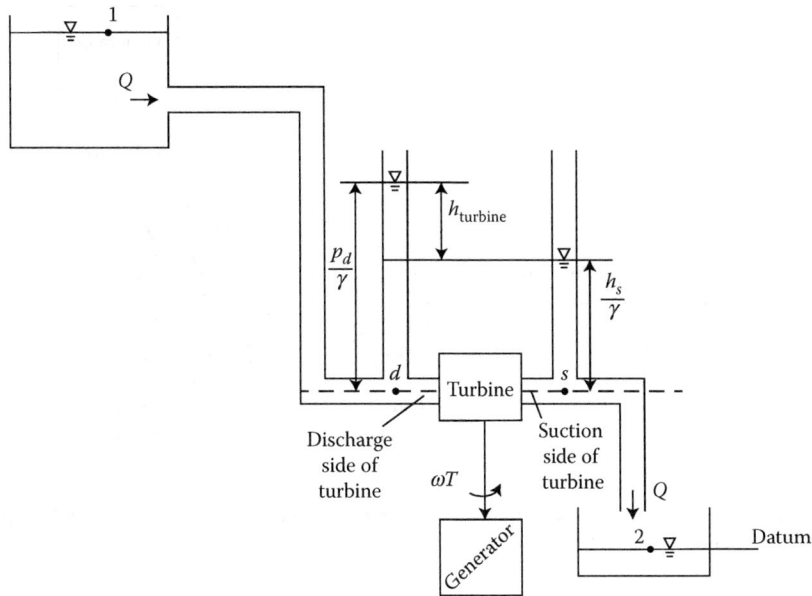

FIGURE 4.25
Application of the energy equation for pipe flow with a turbine.

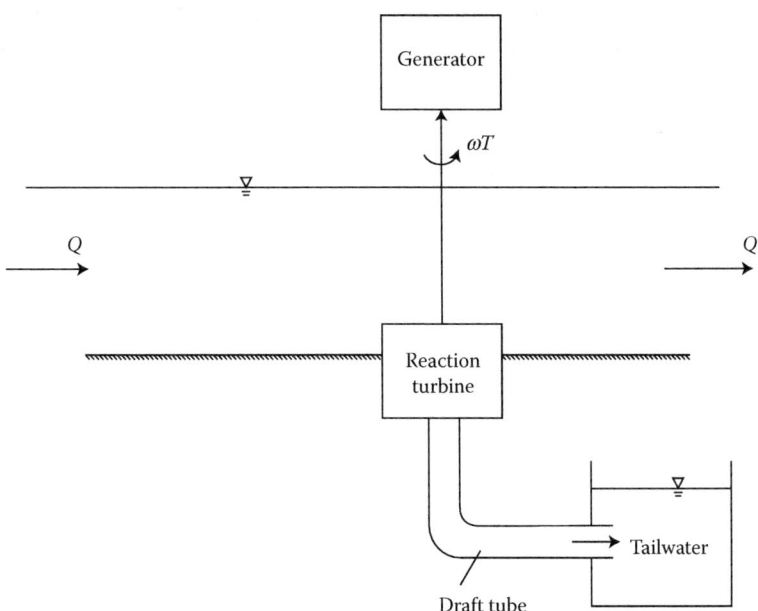

FIGURE 4.26
Reaction turbine: extracts hydraulic power from an open channel flow.

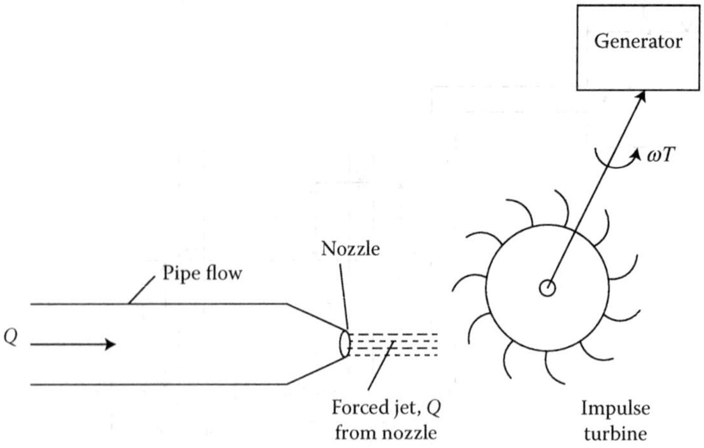

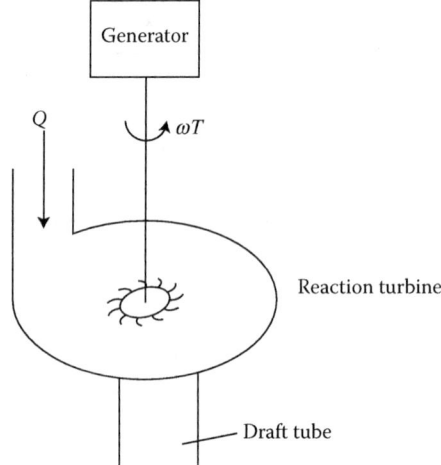

FIGURE 4.27
Impulse turbine (Pelton wheel): extracts hydraulic power from a forced jet flowing from a nozzle (pipe flow) that strikes the rotating blades (vanes) of the turbine.

turbine is used (see Figure 4.26). As such, the example problems presented in this section illustrate the application of the energy equation for real (or ideal) pipe or open channel flow with a turbine. Furthermore, it is important to distinguish between the "analysis" problem versus the "design" problem of a turbine–generator system. An analysis problem assumes that the magnitudes for the freely available discharge and head to be extracted by a turbine are given quantities, while the magnitudes for the efficiency and the freely generated power of the turbine, generator, and overall system are determined by their respective equations. And a design problem assumes that the magnitudes for the efficiency and the required power of the turbine, generator, and overall system are given quantities, while

FIGURE 4.28
Fluid at a high pressure from a pipe or open channel rotates the vanes of a reaction turbine, filling the turbine casing with fluid, and leaves the turbine through a draft tube.

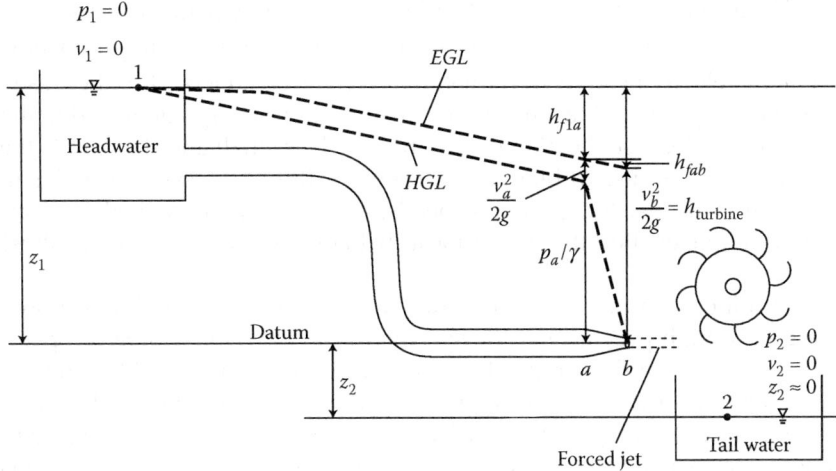

FIGURE 4.29
Application of the energy equation for an impulse turbine (Pelton wheel) extracting hydraulic power from a forced jet flowing from a nozzle (pipe flow) that strikes the rotating blades (vanes) of the turbine.

the magnitudes for the required freely available discharge and head to be extracted by the turbine are unknown quantities to be determined by their respective equations.

An impulse turbine, also called a Pelton wheel, extracts hydraulic power from a forced jet flowing at the specified flowrate, Q from a nozzle (pipe flow), which strikes the rotating blades (vanes) of the turbine and converts it to shaft power, as illustrated in

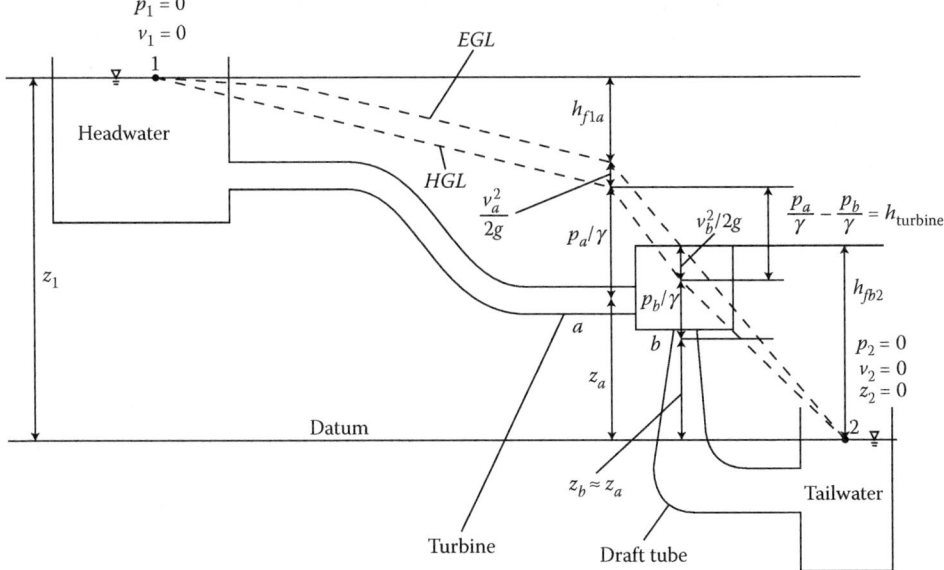

FIGURE 4.30
Application of the energy equation for a reaction turbine extracting hydraulic power from a high pressure pipe flow that rotates the vanes of the turbine, filling the turbine casing with fluid, and leaves the turbine through a draft tube.

Figure 4.27. The impulse of the jet of fluid striking the blades generates the torque. The total head, $H = (p/\gamma) + (v^2/2g) + z$ (minus any head losses) of the incoming fluid flow is converted to a large velocity head, $v^2/2g$ at the nozzle exit (forced jet). Thus, because the fluid in the forced jet is exposed to the atmosphere, the hydraulic energy delivered to the turbine is all kinetic energy. Because the diameter of the nozzle is small and thus limits the magnitude of the flowrate, impulse turbines are efficient for fluid flow systems with small flowrates, Q and large heads, $h_{turbine}$. Large heads, $h_{turbine}$ may be created by locating the turbine nozzle at an elevation that is significantly below the water source, as illustrated in Figure 4.29.

A reaction turbine, such as a Francis or a Kaplan turbine extracts hydraulic power from a fluid flowing (at a high pressure) at the specified flowrate, Q in a pipe or open channel, which rotates the vanes of the turbine, filling the casing with fluid, and leaving the turbine through a draft tube, converting it to shaft power, as illustrated in Figure 4.28. Because the fluid is flowing under a high pressure, the hydraulic energy delivered to the turbine is both pressure energy and kinetic energy. And, because the magnitude of the flowrate in the pipe or open channel is limited only by the selected design pipe or channel size, reaction turbines are efficient for fluid flow systems with large flowrates, Q and small heads, $h_{turbine}$. In order for the reaction turbine to operate properly, the flow leaving the turbine must be submerged as it leaves the turbine through a draft tube, which is an integral part of a reaction turbine, as illustrated in Figure 4.30. The draft tube has a gradual expansion in diameter (conical diffuser) and is designed to serve two main purposes. First, the conical diffuser of the draft tube serves to increase (recover) the pressure downstream of the turbine, which reduces the minor head loss in the submerged discharge and thus increases the available head to be extracted by the turbine. Second, the draft tube is used to elevate the turbine above the level of the tailwater, however, to a maximum allowable height without resulting in cavitation.

In the "analysis" problem of a turbine–generator system, the magnitudes for the freely available discharge, Q and head to be extracted by the turbine, $h_{turbine}$ are given quantities, while the magnitudes for the efficiency and the freely generated power of the turbine, generator, and overall system are determined by their respective equations. First, the determination of the magnitudes for the freely available discharge, Q and head to be extracted by the turbine, $h_{turbine}$ are a function of the type of turbine: impulse or reaction (presented below). And, finally, in the analysis of either an impulse turbine or a reaction turbine, the magnitude of the freely available hydraulic power that may be input to the turbine is related to the magnitudes of the freely available discharge and head to extracted by the turbine by Equation 4.187 and thus is determined as follows:

$$(P_t)_{in} = \frac{(P_t)_{out}}{\eta_{turbine}} = \frac{\omega T_{shaft,out}}{\eta_{turbine}} = \gamma Q h_{turbine} \qquad (4.216)$$

A turbine extracts freely available hydraulic power from the fluid flow and converts it to shaft power to drive a generator. Then, the generator converts the shaft power supplied by the turbine to electrical power. Therefore, in the "analysis" problem, for the freely available magnitude hydraulic power from the flow, $(P_t)_{in} = \gamma Q h_{turbine}$, there is a specific relationship between the magnitudes of the turbine efficiency, $\eta_{turbine}$ and the shaft power output from the turbine, $(P_t)_{out} = \omega T_{shaft,out}$, which is determined from Equation 4.167 as follows:

$$\eta_{turbine} = \frac{shaft\ power}{hydraulic\ power} = \frac{(P_t)_{out}}{(P_t)_{in}} = \frac{\omega T_{shaft,out}}{\gamma Q h_{turbine}} \qquad (4.217)$$

where the hydraulic power drives the turbine by a rotating shaft torque and the shaft rotates at an angular speed, ω and delivers a torque, $T_{shaft,out}$ to the generator. In particular, as the turbine efficiency decreases, the available magnitude of the shaft power output from the turbine to the generator also decreases. Furthermore, for the available magnitude of the shaft power from the turbine, $(P_t)_{out} = \omega T_{shaft,out} = (P_g)_{in} = \omega T_{shaft,in}$, there is a specific relationship between the magnitudes of the generator efficiency, $\eta_{generator}$ and the electric power, $(P_g)_{out}$, which is determined from Equation 4.174 as follows:

$$\eta_{generator} = \frac{electric\ power}{shaft\ power} = \frac{(P_g)_{out}}{(P_g)_{in}} = \frac{(P_g)_{out}}{\omega T_{shaft,in}} \tag{4.218}$$

In particular, as the generator efficiency decreases, the available magnitude of the electric power output from the generator also decreases. Finally, for the freely available magnitude of the hydraulic power input to the turbine, $(P_t)_{in} = \gamma Q h_{turbine}$, there is a specific relationship between the magnitudes of the overall turbine–generator system efficiency, $\eta_{turbine-generator}$ and the available electric power output from the generator, $(P_g)_{out}$, which is determined from Equation 4.176 as follows:

$$\eta_{turbine-generator} = \frac{electric\ power}{hydraulic\ power} = \frac{(P_g)_{out}}{(P_t)_{in}} = \frac{(P_g)_{out}}{\gamma Q h_{turbine}} = \eta_{turbine}\eta_{generator} \tag{4.219}$$

In particular, as the turbine–generator system efficiency decreases, the available magnitude of the electric power output from the generator also decreases.

In the analysis of an impulse turbine–generator system (see Figure 4.29), the magnitude of the freely available head to be extracted by the turbine, $h_{turbine}$ is determined by application of the energy equation. And, the magnitude of the freely available discharge is determined by application of the continuity equation. Application of the energy equation between points 1 and a, assuming the datum is through points a and b, yields the following:

$$\left(\frac{p_1}{\gamma} + \frac{v_1^2}{2g} + z_1\right) - h_{f,maj} - h_{f,min1a} = \left(\frac{p_a}{\gamma} + \frac{v_a^2}{2g} + z_a\right) \tag{4.220}$$

where $p_1 = 0$, $v_1 = 0$, and $z_a = 0$, and thus:

$$\frac{p_a}{\gamma} + \frac{v_a^2}{2g} = z_1 - h_{f,maj} - h_{f,min1a} \tag{4.221}$$

which illustrates a conversion of the potential energy at point 1 (minus the head losses) to flow and kinetic energy at point a. Next, application of the energy equation between points a and b yields the following:

$$\left(\frac{p_a}{\gamma} + \frac{v_a^2}{2g} + z_a\right) - h_{f,min\,ab} = \left(\frac{p_b}{\gamma} + \frac{v_b^2}{2g} + z_b\right) \tag{4.222}$$

where $z_a = z_b = 0$, and $p_b = 0$, and thus:

$$\frac{v_b^2}{2g} = \frac{p_a}{\gamma} + \frac{v_a^2}{2g} - h_{f,min\ ab} \qquad (4.223)$$

which illustrates a conversion of flow and kinetic energy at point a (minus the head loss) to kinetic energy at point b. And, finally, the freely available head to be extracted by the turbine, $h_{turbine}$ is determined by the application of the energy equation between points b and 2, and results in the following:

$$\left(\frac{p_b}{\gamma} + \frac{v_b^2}{2g} + z_b\right) - h_{turbine} = \left(\frac{p_2}{\gamma} + \frac{v_2^2}{2g} + z_2\right) \qquad (4.224)$$

where $p_b = p_2 = 0$, $v_2 = 0$, $z_b = 0$, z_2 is a small negative elevation (negligible), and thus:

$$h_{turbine} = \frac{v_b^2}{2g} \qquad (4.225)$$

which illustrates a conversion of kinetic energy at point b to hydraulic energy extracted by the turbine. Thus, for the given nozzle (integral part of the impulse turbine) diameter at point b, application of the continuity equation yields the freely available discharge as follows:

$$Q = v_b A_b \qquad (4.226)$$

In the "analysis" problem, the feely available discharge, Q extracted by the turbine is measured by a pipe flow measuring device (such as an orifice, nozzle, or venturi meter as discussed in Chapter 8), while the pressure head at point a is measured by installing a piezometer at point a, which is just upstream of the nozzle at point b in Figure 4.29.

EXAMPLE PROBLEM 4.23

An impulse turbine extracts hydraulic power from water at 20°C flowing from reservoir 1 though a 0.75-m riveted steel pipe, 200 m in length, fitted with two regular 90° threaded elbows in order to change the direction of the flow, and finally with a 0.3-m nozzle in order to create a forced jet, as illustrated in Figure EP 4.23a. The turbine is located 50 m below the elevation of the headwater at point 1 and just slightly above the elevation of the tailwater at point 2. The pipe entrance from reservoir 1 is square-edged, and the nozzle has a conical angle, $\theta = 30°$. Assume a Darcy–Weisbach friction factor, f of 0.13; a minor head loss coefficient due to a squared-edged pipe entrance, $k = 0.5$; a minor head loss coefficient due to an elbow, $k = 1.5$; and a minor head loss coefficient due to the nozzle, $k = 0.02$. The pressure at point a in the pipe is measured using a piezometer. (a) Determine the pressure and velocity at point a in the pipe and the velocity at point b at the nozzle. (b) Determine the magnitude of the freely available discharge to be extracted by the turbine. (c) Determine the magnitude of the freely available head to be extracted by the turbine. (d) Determine the

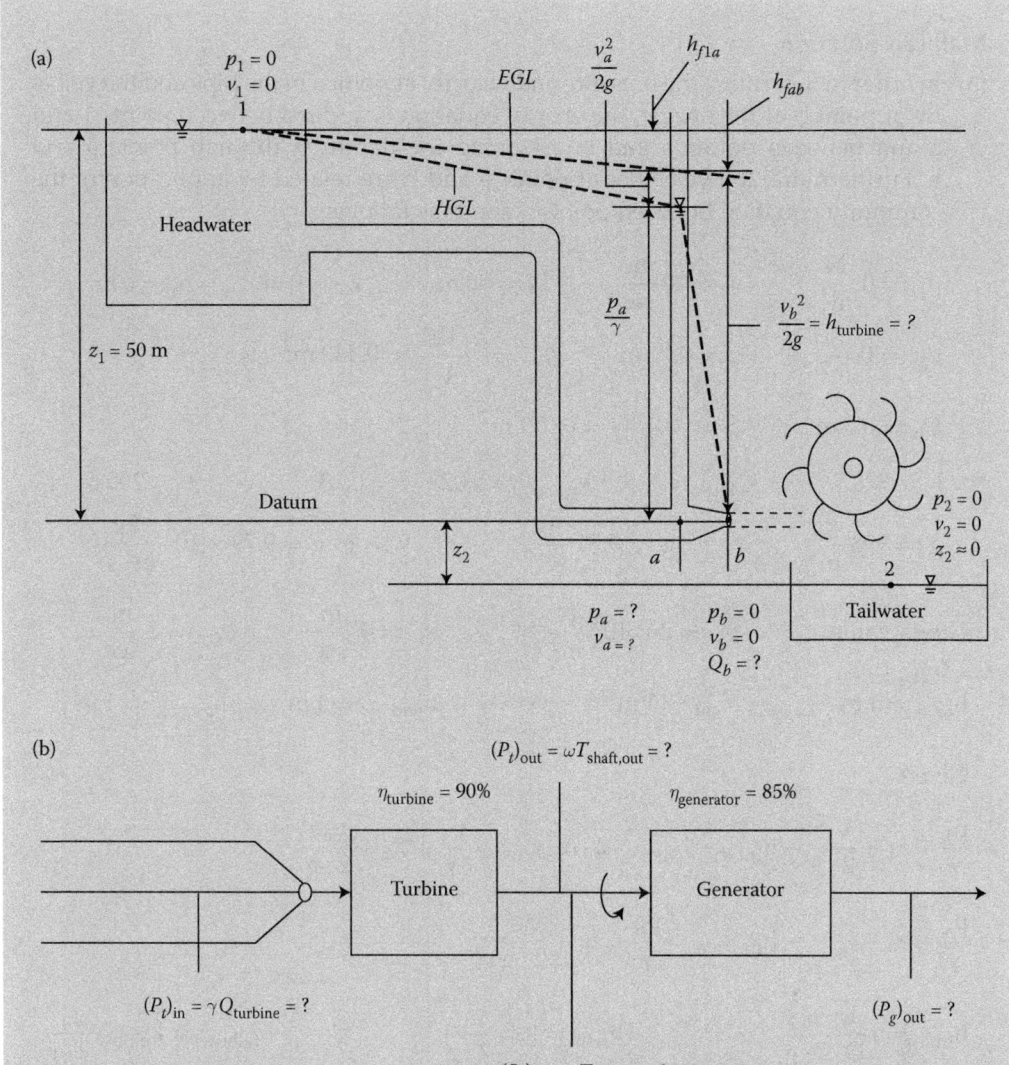

FIGURE EP 4.23
(a) Impulse turbine extracts hydraulic power from water flowing from reservoir 1 through a pipe with components ending in a nozzle that creates a forced jet. (b) Turbine–generator system converts mechanical power (hydraulic and shaft power) to electrical power.

magnitude of the freely available hydraulic power to be extracted by the turbine. (e) Draw the energy grade line and the hydraulic grade line. (f) Assuming a typical turbine efficiency of 90%, determine the freely generated shaft power output by the turbine. (g) Assuming a typical generator efficiency of 85%, determine the freely generated electric power output by the generator. (h) Determine the resulting overall turbine–generator system efficiency.

Mathcad Solution

(a) In order to determine the pressure and velocity at point a in the pipe and the velocity at point b at the nozzle, the energy equation is applied between points 1 and a and between points a and b, assuming the datum is through points a and b. Furthermore, the velocities at points a and b are related by application of the continuity equation between points a and b as follows:

$$p_1 := 0 \, \frac{N}{m^2} \qquad v_1 := 0 \, \frac{m}{sec} \qquad z_1 := 50 \, m \qquad z_a := 0 \, m \qquad z_b := 0 \, m$$

$$p_b := 0 \, \frac{N}{m^2} \qquad D_a := 0.75 \, m \qquad A_a := \frac{\pi \cdot D_a^2}{4} = 0.442 \, m^2$$

$$D_b := 0.3 \, m \qquad A_b := \frac{\pi \cdot D_b^2}{4} = 0.071 \, m^2$$

$$k_{ent} := 0.5 \qquad k_{elbow} := 1.5 \qquad k_{nozzle} := 0.02 \qquad f := 0.13 \qquad L := 200 \, m$$

$$\rho := 998 \, \frac{kg}{m^3} \qquad g := 9.81 \, \frac{m}{sec^2} \qquad \gamma := \rho \cdot g = 9.79 \times 10^3 \, \frac{kg}{m^2 \, s^2}$$

Guess value: $\qquad p_a := 2 \times 10^3 \, \frac{N}{m^2} \qquad v_a := 1 \, \frac{m}{sec} \qquad v_b := 2 \, \frac{m}{sec}$

$$h_{fmaj} := 1 \, m \qquad h_{fent} := 1 \, m \qquad h_{felbow} := 1 \, m \qquad h_{fnozzle} := 1 \, m$$

Given

$$\frac{p_1}{\gamma} + z_1 + \frac{v_1^2}{2 \cdot g} - h_{fmaj} - h_{fent} - 2 \cdot h_{felbow} = \frac{p_a}{\gamma} + z_a + \frac{v_a^2}{2 \cdot g}$$

$$\frac{p_a}{\gamma} + z_a + \frac{v_a^2}{2 \cdot g} - h_{fnozzle} = \frac{p_b}{\gamma} + z_b + \frac{v_b^2}{2 \cdot g}$$

$$h_{fmaj} = f \frac{L}{D_a} \frac{v_a^2}{2 \cdot g} \qquad\qquad h_{fent} = k_{ent} \frac{v_a^2}{2 \cdot g} \qquad\qquad h_{felbow} = k_{elbow} \frac{v_a^2}{2 \cdot g}$$

$$h_{fnozzle} = k_{nozzle} \frac{v_b^2}{2 \cdot g} \qquad v_a \cdot A_a = v_b \cdot A_b$$

$$\begin{pmatrix} p_a \\ v_a \\ v_b \\ h_{fmaj} \\ h_{fent} \\ h_{felbow} \\ h_{fnozzle} \end{pmatrix} := Find(p_a, v_a, v_b, h_{fmaj}, h_{fent}, h_{felbow}, h_{fnozzle})$$

$$p_a = 2.437 \times 10^5 \, \frac{N}{m^2} \qquad\qquad v_a = 3.546 \, \frac{m}{s} \qquad\qquad v_b = 22.163 \, \frac{m}{s}$$

$$h_{fmaj} = 22.219 \, m \qquad h_{fent} = 0.32 \, m \qquad 2 \cdot h_{felbow} = 1.923 \, m \qquad h_{fnozzle} = 0.501 \, m$$

(b) In order to determine the magnitude of the freely available discharge to be extracted by the turbine, the continuity equation is applied at point b at the nozzle as follows:

$$Q_b := v_b \cdot A_b = 1.567 \frac{m^3}{s}$$

(c) In order to determine the magnitude of the freely available head to be extracted by the turbine, the energy equation is applied between points b and 2 as follows:

$$p_2 := 0 \frac{N}{m^2} \qquad v_2 := 0 \frac{m}{sec} \qquad z_2 := 0 \, m$$

Guess value: $\quad h_{turbine} := 20 \, m$

Given

$$\frac{p_b}{\gamma} + z_b + \frac{v_b^2}{2 \cdot g} - h_{turbine} = \frac{p_2}{\gamma} + z_2 + \frac{v_2^2}{2 \cdot g}$$

$$h_{turbine} := Find(h_{turbine}) = 25.037 \, m$$

(d) In order to determine the magnitude of the freely available hydraulic power to be extracted by the turbine, Equation 4.187 is applied as follows:

$$Q := Q_b = 1.567 \frac{m^3}{s}$$

$$P_{tin} := \gamma \cdot Q \cdot h_{turbine} = 384.014 \, kW$$

(e) The EGL and HGL are illustrated in Figure EP 4.23a.
(f) Assuming a typical turbine efficiency of 90%, the freely generated shaft power output by the turbine is computed by applying Equation 4.167 as follows:

$$\eta_{turbine} := 0.90$$

Guess value: $\qquad P_{tout} := 350 \, kW$

Given

$$\eta_{turbine} = \frac{P_{tout}}{P_{tin}}$$

$$P_{tout} := Find(P_{tout}) = 345.613 \, kW$$

(g) Assuming a typical generator efficiency of 85%, the freely generated electric power output by the generator is computed by applying Equation 4.174 as follows:

$$P_{gin} := P_{tout} = 345.613 \, kW \qquad \eta_{generator} := 0.85$$

Guess value: $P_{gout} := 350\,kW$

Given

$$\eta_{generator} = \frac{P_{gout}}{P_{gin}}$$

$P_{gout} := Find(P_{gout}) = 293.771\,kW$

(h) The resulting overall turbine–generator system efficiency is computed by apply-ing Equation 4.176 as follows:

$$\eta_{turbinegenerator} := \frac{P_{gout}}{P_{tin}} = 0.765$$

$$\eta_{turbinegenerator} := \eta_{turbine} \cdot \eta_{generator} = 0.765$$

The turbine–generator system is illustrated in Figure EP 4.23b.

In the analysis of a reaction turbine–generator system, (see Figure 4.30), the magnitude of the free available head to be extracted by the turbine, $h_{turbine}$ is determined by application of the energy equation. And, the magnitude of the freely available discharge is determined by application of the continuity equation. Application of the energy equation between points 1 and a, assuming the datum is at point 2, yields the following:

$$\left(\frac{p_1}{\gamma} + \frac{v_1^2}{2g} + z_1\right) - h_{f,maj} - h_{f,min1a} = \left(\frac{p_a}{\gamma} + \frac{v_a^2}{2g} + z_a\right) \tag{4.227}$$

where $p_1 = 0$, and $v_1 = 0$, and thus:

$$\frac{p_a}{\gamma} + \frac{v_a^2}{2g} + z_a = z_1 - h_{f,maj} - h_{f,min1a} \tag{4.228}$$

which illustrates a conversion of the potential energy at point 1 (minus the head losses) to flow, kinetic, and potential energy at point a. Next, the freely available head to be extracted by the turbine, $h_{turbine}$ is determined by the application of the energy equation between points a and b and results in the following:

$$\left(\frac{p_a}{\gamma} + \frac{v_a^2}{2g} + z_a\right) - h_{turbine} = \left(\frac{p_b}{\gamma} + \frac{v_b^2}{2g} + z_b\right) \tag{4.229}$$

where $z_a = z_b$, and $v_a = v_b$, and thus:

$$h_{turbine} = \frac{p_a}{\gamma} - \frac{p_b}{\gamma} \tag{4.230}$$

which illustrates a conversion of flow energy at point a to hydraulic energy extracted by the turbine. Finally, the pressure at point b is determined by the application of the energy equation between points b and 2, as follows:

$$\left(\frac{p_b}{\gamma} + \frac{v_b^2}{2g} + z_b\right) - h_{f,minb2} = \left(\frac{p_2}{\gamma} + \frac{v_2^2}{2g} + z_2\right) \tag{4.231}$$

where $p_2 = 0$, $v_2 = 0$, $z_2 = 0$, and thus:

$$\frac{p_b}{\gamma} = h_{f,minb2} - \frac{v_b^2}{2g} - z_b \tag{4.232}$$

Furthermore, for the given pipe diameter at points a and b (assume $D_a = D_b$), application of the continuity equation between points a and b relates the velocities at points a and b as follows:

$$Q = v_a A_a = v_b A_b \tag{4.233}$$

which yields the velocities at points a and b and the freely available discharge. In the "analysis" problem, the freely available discharge, Q extracted by the turbine from a pipe flow is measured by a pipe flow-measuring device (such as an orifice, nozzle, or venturi meter as discussed in Chapter 8). And, the freely available discharge, Q extracted by the turbine from an open channel flow is measured by an open channel flow-measuring device (such as a pitot-static tube, sluice gate, weir, overflow spillway, the Parshall flume [venturi flume], and contracted opening, as discussed in Chapter 9). Furthermore, the pressure head at point a is measured by installing a piezometer just upstream of the turbine at point a (see Figure 4.30).

EXAMPLE PROBLEM 4.24

A reaction turbine extracts hydraulic power from water at 20°C flowing from reservoir 1 though a 0.9-m galvanized iron pipe 150 m in length, as illustrated in Figure EP 4.24. The turbine is located 30 m below the elevation of the headwater at point 1 and 2 m above the elevation of the tailwater at point 2. The pipe entrance from reservoir 1 is square-edged, and the draft tube has a conical angle, $\theta = 9°$. Assume a Darcy–Weisbach friction factor, f of 0.15; a minor head loss coefficient due to a squared-edged pipe entrance, $k = 0.5$; and a minor head loss coefficient due to the conical diffuser, $k = 0.5$. The pressure at point a in the pipe is measured using a piezometer. (a) Determine the pressure and velocity at points a and b. (b) Determine the magnitude of the freely available head to be extracted by the turbine. (c) Determine the magnitude of the freely available discharge to be extracted by the turbine. (d) Determine the magnitude of the freely available hydraulic power to be extracted by the turbine. (e) Draw the energy grade line and the hydraulic grade line. (f) Assuming a typical turbine efficiency of 90%, determine the freely generated shaft power output by the turbine. (g) Assuming a typical generator efficiency of 85%, determine the freely generated electric power output by the generator. (h) Determine the resulting overall turbine–generator system efficiency.

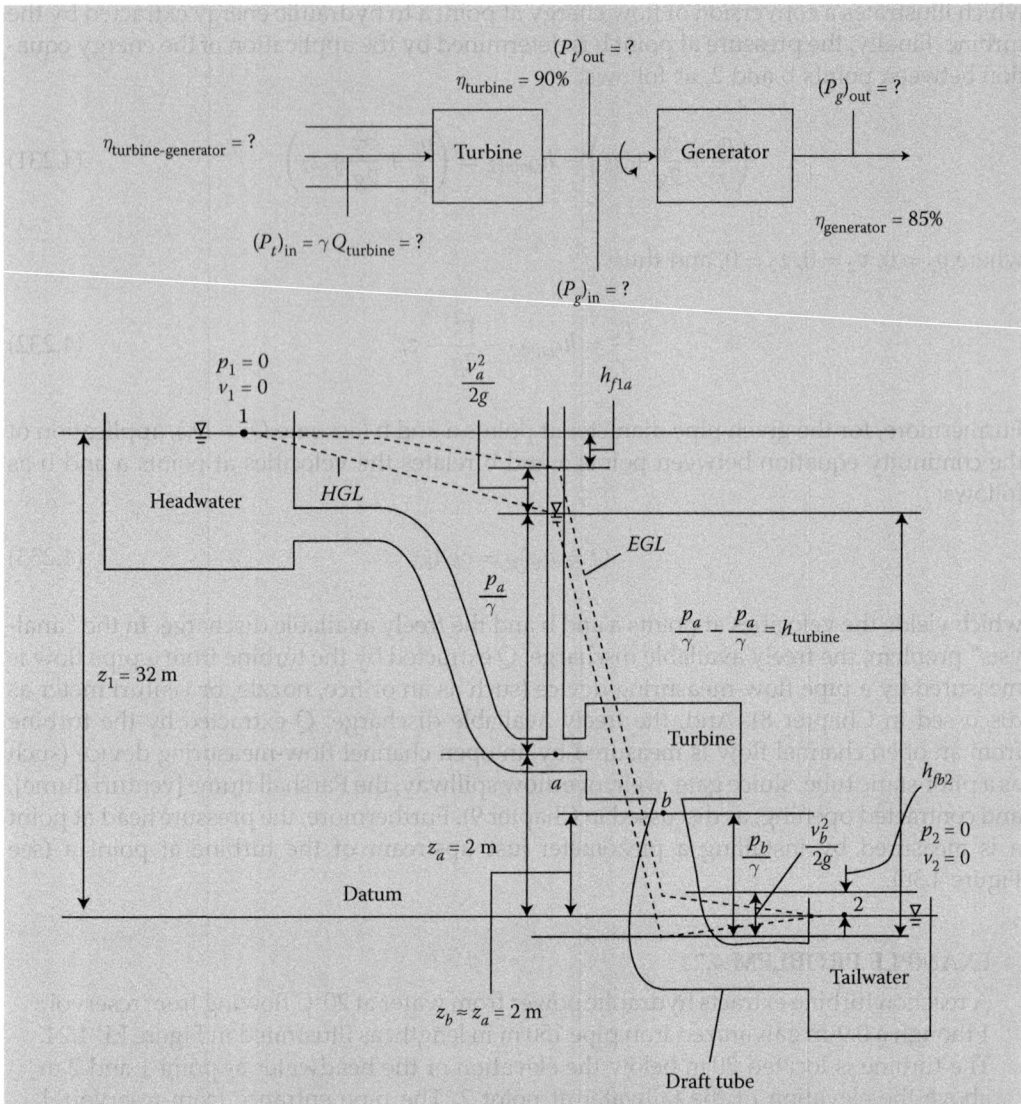

FIGURE EP 4.24
Reaction turbine extracts hydraulic power from water flowing from reservoir 1 through a pipe flowing under high pressure.

Mathcad Solution

(a)–(b) In order to determine the pressure and velocity at points a and b and the magnitude of the freely available head to be extracted by the turbine, the energy equation is applied between points 1 and a, between points a and b, and between points b and 2, assuming $z_a = z_b$ and the datum is at point 2. Furthermore, the velocities at points a and b are related by application of the continuity equation between points a and b as follows:

$$p_1 := 0 \ \frac{N}{m^2} \qquad v_1 := 0 \ \frac{m}{sec} \qquad z_1 := 32 \ m \qquad z_a := 2 \ m \qquad z_b := 2 \ m$$

$p_2 := 0 \, \dfrac{N}{m^2}$ $v_2 := 0 \, \dfrac{m}{sec}$ $z_2 := 0 \, m$

$D_a := 0.9 \, m$ $A_a := \dfrac{\pi \cdot D_a^2}{4} = 0.636 \, m^2$

$D_b := 0.9 \, m$ $A_b := \dfrac{\pi \cdot D_b^2}{4} = 0.636 \, m^2$

$k_{ent} := 0.5$ $k_{diffuser} := 0.5$ $f := 0.15$ $L := 150 \, m$

$\rho := 998 \, \dfrac{kg}{m^3}$ $g := 9.81 \, \dfrac{m}{sec^2}$ $\gamma := \rho \cdot g = 9.79 \times 10^3 \, \dfrac{kg}{m^2 \, s^2}$

Guess value: $p_a := 2 \times 10^3 \, \dfrac{N}{m^2}$ $p_b := 1 \times 10^3 \, \dfrac{N}{m^2}$ $v_a := 1 \, \dfrac{m}{sec}$

$v_b := 2 \, \dfrac{m}{sec}$ $h_{fmaj} := 1 \, m$ $h_{fent} := 1 \, m$ $h_{turbine} := 20 \, m$

$h_{fdiffuser} := 1 \, m$

Given

$$\frac{p_1}{\gamma} + z_1 + \frac{v_1^2}{2 \cdot g} - h_{fmaj} - h_{fent} = \frac{p_a}{\gamma} + z_a + \frac{v_a^2}{2 \cdot g}$$

$$\frac{p_a}{\gamma} + z_a + \frac{v_a^2}{2 \cdot g} - h_{turbine} = \frac{p_b}{\gamma} + z_b + \frac{v_b^2}{2 \cdot g}$$

$$\frac{p_b}{\gamma} + z_b + \frac{v_b^2}{2 \cdot g} - h_{fdiffuser} = \frac{p_2}{\gamma} + z_2 + \frac{v_2^2}{2 \cdot g}$$

$h_{fmaj} = f \dfrac{L}{D_a} \dfrac{v_a^2}{2 \cdot g}$ $h_{fent} = k_{ent} \dfrac{v_a^2}{2 \cdot g}$ $v_a \cdot A_a = v_b \cdot A_b$

$h_{fdiffuser} = k_{diffuser} \dfrac{v_b^2}{2 \cdot g}$

$$\begin{pmatrix} p_a \\ v_a \\ h_{turbine} \\ p_b \\ v_b \\ h_{fmaj} \\ h_{fent} \\ h_{fdiffuser} \end{pmatrix} := Find(p_a, v_a, h_{turbine}, p_b, v_b, h_{fmaj}, h_{fent}, h_{fdiffuser})$$

$p_a = 1.638 \times 10^5 \, \dfrac{N}{m^2}$ $v_a = 3.135 \, \dfrac{m}{s}$ $h_{turbine} = 18.977 \, m$

$p_b = -2.203 \times 10^4 \, \dfrac{N}{m^2}$ $v_b = 3.135 \, \dfrac{m}{s}$

$h_{fmaj} = 12.522 \, m$ $h_{fent} = 0.25 \, m$ $h_{fdiffuser} = 0.25 \, m$

where

$$\frac{P_a}{\gamma} = 16.727\,m \qquad \frac{P_b}{\gamma} = -2.25\,m \qquad h_{turbine} := \frac{P_a}{\gamma} - \frac{P_b}{\gamma} = 18.977\,m$$

(c) In order to determine the magnitude of the freely available discharge to be extracted by the turbine, the continuity equation is applied at point a as follows:

$$Q_a := v_a \cdot A_a = 1.994\,\frac{m^3}{s}$$

(d) In order to determine the magnitude of the freely available hydraulic power to be extracted by the turbine, Equation 4.187 is applied as follows:

$$Q := Q_a = 1.994\,\frac{m^3}{s}$$

$$P_{tin} := \gamma \cdot Q \cdot h_{turbine} = 370.529\,kW$$

(e) The EGL and HGL are illustrated in Figure EP 4.24.

(f) Assuming a typical turbine efficiency of 90%, the freely generated shaft power output by the turbine is computed by applying Equation 4.167 as follows:

$$\eta_{turbine} := 0.90$$

Guess value: $\qquad\qquad\qquad$ $P_{tout} := 350\,kW$

Given

$$\eta_{turbine} = \frac{P_{tout}}{P_{tin}}$$

$$P_{tout} := Find(P_{tout}) = 333.476\,kW$$

(g) Assuming a typical generator efficiency of 85%, the freely generated electric power output by the generator is computed by applying Equation 4.174 as follows:

$$P_{gin} := P_{tout} = 333.476\,kW \qquad\qquad \eta_{generator} := 0.85$$

Guess value: $\qquad\qquad\qquad$ $P_{gout} := 350\,kW$

Given

$$\eta_{generator} = \frac{P_{gout}}{P_{gin}}$$

$$P_{gout} := Find(P_{gout}) = 283.455\,kW$$

(h) The resulting overall turbine–generator system efficiency is computed by applying Equation 4.176 as follows:

$$\eta_{turbinegenerator} := \frac{P_{gout}}{P_{tin}} = 0.765$$

$$\eta_{turbinegenerator} := \eta_{turbine} \cdot \eta_{generator} = 0.765$$

In the "design" problem of a turbine–generator system, the magnitudes for the efficiency and required power of the turbine, generator, and overall system are given quantities, while the magnitudes for the required freely available discharge and head to be extracted by the turbine are unknown quantities to be determined by their respective equations. The turbine efficiency, $\eta_{turbine}$ was defined in Equation 4.167 as the ratio of the shaft power output from the turbine to the freely available hydraulic power extracted from the flow and is summarized as follows:

$$\eta_{turbine} = \frac{shaft\ power}{hydraulic\ power} = \frac{(P_t)_{out}}{(P_t)_{in}} = \frac{\omega T_{shaft,out}}{\gamma Q h_{turbine}} \qquad (4.234)$$

The generator efficiency, $\eta_{generator}$ was defined in Equation 4.174 as the ratio of electric power output from the generator to the shaft power input by the turbine and is summarized as follows:

$$\eta_{generator} = \frac{electric\ power}{shaft\ power} = \frac{(P_g)_{out}}{(P_g)_{in}} = \frac{(P_g)_{out}}{\omega T_{shaft,in}} \qquad (4.235)$$

where $(P_t)_{out} = \omega T_{shaft,out} = (P_g)_{in} = \omega T_{shaft,in}$. And the overall turbine–generator system efficiency was defined in Equation 4.176 as the ratio of the electric power output from the generator to the freely available hydraulic power extracted from the flow as follows:

$$\eta_{turbine-generator} = \frac{electric\ power}{hydraulic\ power} = \frac{(P_g)_{out}}{(P_t)_{in}} = \frac{(P_g)_{out}}{\gamma Q h_{turbine}} = \eta_{turbine}\eta_{generator} \qquad (4.236)$$

The magnitudes for the freely available discharge, Q and the head extracted by the turbine, $h_{turbine}$ are related by Equation 4.187 as follows:

$$h_{turbine} = \frac{(P_t)_{in}}{\gamma Q} = \frac{(P_t)_{out}/\eta_{turbine}}{\gamma Q} = \frac{\omega T_{shaft,out}/\eta_{turbine}}{\gamma Q} \qquad (4.237)$$

Therefore, in the "design" problem, for the given required magnitude of the hydraulic power input to the turbine,

$$(P_t)_{in} = \frac{\omega T_{shaft,out}}{\eta_{turbine}} = \gamma Q h_{turbine},$$

there is a specific relationship between the required magnitudes of the discharge, Q and the head extracted by the turbine, $h_{turbine}$, which is a function of the type of turbine: impulse or reaction (presented below).

In the "design" problem of an impulse turbine–generator system, for the required magnitude of the hydraulic power input to the turbine,

$$(P_t)_{in} = \frac{\omega T_{shaft,out}}{\eta_{turbine}} = \gamma Q h_{turbine},$$

there is a specific relationship between the required magnitudes of the discharge, Q and the head extracted by the turbine, $h_{turbine}$. Although the determination of the magnitude of the required freely available discharge is limited by the maximum allowable velocity of flow at the nozzle (in order to avoid cavitation), it may be accommodated by adjusting the design pipe and nozzle size. Furthermore, for the required magnitude of the hydraulic power input to the turbine,

$$(P_t)_{in} = \frac{\omega T_{shaft,out}}{\eta_{turbine}} = \gamma Q h_{turbine},$$

and the design magnitude of the freely available discharge, there is a corresponding design magnitude of the head to be extracted by the turbine, which is determined by applying Equation 4.187 as follows:

$$h_{turbine} = \frac{(P_t)_{in}}{\gamma Q} \tag{4.238}$$

In the design of an impulse turbine–generator system (see Figure 4.29), the design elevation of the headwater in reservoir 1 is determined by application of the energy equation. And, the design magnitude of the pipe and nozzle sizes are determined by application of the continuity equation. Application of the energy equation between points 1 and a, assuming the datum is through points a and b, yields the following:

$$\left(\frac{p_1}{\gamma} + \frac{v_1^2}{2g} + z_1\right) - h_{f,maj} - h_{f,min1a} = \left(\frac{p_a}{\gamma} + \frac{v_a^2}{2g} + z_a\right) \tag{4.239}$$

where $p_1 = 0$, $v_1 = 0$, and $z_a = 0$, and thus:

$$\frac{p_a}{\gamma} + \frac{v_a^2}{2g} = z_1 - h_{f,maj} - h_{f,min1a} \tag{4.240}$$

which illustrates a conversion of the potential energy at point 1 (minus the head losses) to flow and kinetic energy at point a. Next, application of the energy equation between points a and b yields the following:

$$\left(\frac{p_a}{\gamma} + \frac{v_a^2}{2g} + z_a\right) - h_{f,min\,ab} = \left(\frac{p_b}{\gamma} + \frac{v_b^2}{2g} + z_b\right) \tag{4.241}$$

where $z_a = z_b = 0$, and $p_b = 0$, and thus:

$$\frac{v_b^2}{2g} = \frac{p_a}{\gamma} + \frac{v_a^2}{2g} - h_{f,min\,ab} \qquad (4.242)$$

which illustrates a conversion of flow and kinetic energy at point a (minus the head loss) to kinetic energy at point b. However, in accordance with the Bernoulli equation, as the velocity increases, there is a corresponding decrease in pressure. Thus, in the design of a nozzle–turbine system, consideration is given to make sure that the pressure in the nozzle exit does not fall below the vapor pressure of the liquid in order to avoid cavitation. Cavitation causes pitting of the turbine vanes, mechanical vibration in the turbine–generator system, and a reduction in the turbine efficiency. As such, there is a maximum allowable velocity of flow at the nozzle in order to avoid a pressure drop below the vapor pressure of the liquid at the nozzle and thus avoid cavitation. Determination of the maximum allowable velocity of flow at the nozzle assumes that the pressure at the nozzle at point b in Figure 4.29 is set equal to the vapor pressure, p_v of the liquid in the pipe. The maximum allowable velocity at the nozzle at point b is determined by applying the energy equation between points a and b, assuming the datum is through points a and b as follows:

$$\left(\frac{p_a}{\gamma} + \frac{v_a^2}{2g} + z_a\right) - h_{f,min\,ab} = \left(\frac{p_b}{\gamma} + \frac{v_b^2}{2g} + z_b\right) \qquad (4.243)$$

where $z_a = z_b = 0$, and $p_b = p_v$ (negative gage pressure), and $h_{f,min\,ab}$ is due to the nozzle, and thus:

$$\frac{v_b^2}{2g} = \frac{p_a}{\gamma} + \frac{v_a^2}{2g} - h_{f,min\,ab} - \frac{p_v}{\gamma} \qquad (4.244)$$

Furthermore, the design magnitude of the freely available discharge may be accommodated by adjusting the design pipe and nozzle size, and application of the continuity equation yields the maximum allowable discharge as follows:

$$Q = v_b A_b \qquad (4.245)$$

And, finally, the design magnitude of the freely available head to be extracted by the turbine, $h_{turbine}$ is determined by the application of the energy equation between points b and 2, and results in the following:

$$\left(\frac{p_b}{\gamma} + \frac{v_b^2}{2g} + z_b\right) - h_{turbine} = \left(\frac{p_2}{\gamma} + \frac{v_2^2}{2g} + z_2\right) \qquad (4.246)$$

where $p_b = p_2 = 0$, $v_2 = 0$, $z_b = 0$, z_2 is a small negative elevation (negligible), and thus:

$$h_{turbine} = \frac{v_b^2}{2g} \qquad (4.247)$$

which illustrates a conversion of kinetic energy at point b to hydraulic energy extracted by the turbine. However, in order to avoid cavitation, the magnitude of the maximum allowable freely available head to be extracted by the turbine, $h_{turbine}$ is defined by assuming $p_b = p_v$ (negative gage pressure), and thus:

$$h_{turbine} = \frac{p_v}{\gamma} + \frac{v_b^2}{2g} \tag{4.248}$$

Finally, substituting Equations 4.240 and 4.244 into Equation 4.248 yields the magnitude of the maximum allowable freely available head to be extracted by the turbine, $h_{turbine}$ expressed in terms of the potential energy at point 1, z_1 as follows:

$$h_{turbine} = z_1 - h_{f,maj} - h_{f,min,1a} - h_{f,min,ab} \tag{4.249}$$

where $h_{f,min,1a}$ represents minor losses between points 1 and a, and $h_{f,min,ab}$ represents minor losses between points a and b (due to the nozzle). Thus, the design magnitude of the freely available head to be extracted by the turbine, $h_{turbine}$ may be accommodated by adjusting the design elevation of the headwater in reservoir 1 (i.e., adjusting the elevation of point 1, z_1).

EXAMPLE PROBLEM 4.25

The required hydraulic power to be extracted by an impulse turbine is determined to be 700 kW. The design magnitude of the freely available discharge is given to be $2 \, m^3/sec$. The hydraulic power is to be extracted from water at 20°C flowing from reservoir 1 though a riveted steel pipe, 200 m in length, fitted with two regular 90° threaded elbow, and finally with a nozzle in order to create a forced jet, as illustrated in Figure EP 4.23a. The pipe entrance from reservoir 1 is square-edged, and the nozzle has a conical angle, $\theta = 30°$. Assume a Darcy–Weisbach friction factor, f of 0.13; a minor head loss coefficient due to a squared-edged pipe entrance, $k = 0.5$; a minor head loss coefficient due to an elbow, $k = 1.5$; and a minor head loss coefficient due to the nozzle, $k = 0.02$. The pressure at point a in the pipe is measured using a piezometer. (a) Determine the design magnitude of the freely available head to be extracted by the turbine, $h_{turbine}$. (b) Determine the design level of headwater in reservoir 1 in order to accommodate the design magnitude of the freely available head to be extracted by the turbine, $h_{turbine}$. (c) Determine the pressure at point a. (d) Determine the maximum allowable velocity of flow at the nozzle without causing cavitation. (e) Determine the design pipe and nozzle size in order to accommodate the design magnitude of the required freely available discharge. (f) Draw the energy grade line and the hydraulic grade line.

Mathcad Solution

(a) In order to determine the design magnitude of the freely available head to be extracted by the turbine, $h_{turbine}$, Equation 4.187 is applied as follows:

$$\rho := 998 \, \frac{kg}{m^3} \qquad g := 9.81 \, \frac{m}{sec^2} \qquad \gamma := \rho \cdot g = 9.79 \times 10^3 \, \frac{kg}{m^2 \, s^2}$$

$$P_{tin} := 700 \, kW \qquad Q := 2 \, \frac{m^3}{sec}$$

Guess value: $h_{turbine} := 50 \, m$

Given

$$P_{tin} = \gamma \cdot Q \cdot h_{turbine}$$

$$h_{turbine} := Find(h_{turbine}) = 35.749\,m$$

(b)–(e) In order to determine the design level of headwater in reservoir 1 to accommodate the design magnitude of the freely available head to be extracted by the turbine, $h_{turbine}$, the energy equation is applied between points 1 and a and between
points b and 2. In order to determine the pressure at point a and the maximum
allowable velocity of flow at the nozzle at point b without causing cavitation, the
energy equation is applied between points a and b, assuming the pressure at point
b is the vapor pressure of water at 20°C. And, in order to determine the design pipe
and nozzle size in order to accommodate the design magnitude of the required
freely available discharge, the continuity equation is applied between points a
and b. Thus, from Table A.2 in Appendix A, for water at 20°C, the vapor pressure,
p_v is $2.34 \times 10^3\,N/m^2$ abs. However, because the vapor pressure is given in absolute
pressure, the corresponding gage pressure is computed by subtracting the atmospheric pressure as follows: $p_{gage} = p_{abs} - p_{atm}$, where the standard atmospheric
pressure is $101.325 \times 10^3\,N/m^2$ abs.

$$p_1 := 0\,\frac{N}{m^2} \quad v_1 := 0\,\frac{m}{sec} \qquad\qquad z_a := 0\,m \qquad\qquad z_b := 0\,m$$

$$p_2 := 0\,\frac{N}{m^2} \quad v_2 := 0\,\frac{m}{sec} \qquad\qquad z_2 := 0\,m$$

$$p_v := 2.34 \times 10^3\,\frac{N}{m^2} - 101.325 \times 10^3\,\frac{N}{m^2} = -9.899 \times 10^4\,\frac{N}{m^2}$$

$$p_b := p_v = -9.899 \times 10^4\,\frac{N}{m^2}$$

$$k_{ent} := 0.5 \qquad k_{elbow} := 1.5 \qquad k_{nozzle} := 0.02 \qquad f := 0.13 \qquad L := 200\,m$$

Guess value: $\quad z_1 := 100\,m \qquad p_a := 2 \times 10^3\,\frac{N}{m^2} \quad v_a := 1\,\frac{m}{sec} \quad v_b := 2\,\frac{m}{sec}$

$$D_a := 0.9\,m \qquad D_b := 0.4\,m$$

$$h_{fmaj} := 1\,m \qquad h_{fent} := 1\,m \qquad\qquad h_{felbow} := 1\,m \qquad h_{fnozzle} := 1\,m$$

Given

$$\frac{p_1}{\gamma} + z_1 + \frac{v_1^2}{2 \cdot g} - h_{fmaj} - h_{fent} - 2 \cdot h_{felbow} = \frac{p_a}{\gamma} + z_a + \frac{v_a^2}{2 \cdot g}$$

$$\frac{p_a}{\gamma} + z_a + \frac{v_a^2}{2 \cdot g} - h_{fnozzle} = \frac{p_b}{\gamma} + z_b + \frac{v_b^2}{2 \cdot g}$$

$$\frac{p_b}{\gamma} + z_b + \frac{v_b^2}{2 \cdot g} - h_{turbine} = \frac{p_2}{\gamma} + z_2 + \frac{v_2^2}{2 \cdot g}$$

$$h_{fmaj} = f \frac{L}{D_a} \frac{v_a^2}{2 \cdot g} \qquad h_{fent} = k_{ent} \frac{v_a^2}{2 \cdot g} \qquad h_{felbow} = k_{elbow} \frac{v_a^2}{2 \cdot g}$$

$$h_{fnozzle} = k_{nozzle} \frac{v_b^2}{2 \cdot g} \qquad Q = v_a \frac{\pi \cdot D_a^2}{4} \qquad Q = v_b \frac{\pi \cdot D_b^2}{4}$$

$$\begin{pmatrix} z_1 \\ p_a \\ v_a \\ v_b \\ D_a \\ D_b \\ h_{fmaj} \\ h_{fent} \\ h_{felbow} \\ h_{fnozzle} \end{pmatrix} := \text{Find}(z_1, p_a, v_a, v_b, D_a, D_b, h_{fmaj}, h_{fent}, h_{felbow}, h_{fnozzle})$$

$$z_1 = 57.004 \text{ m} \qquad p_a = 3.531 \times 10^5 \frac{N}{m^2} \qquad v_a = 3.44 \frac{m}{s} \qquad v_b = 29.996 \frac{m}{s}$$

$$D_a = 0.85 \text{ m} \qquad D_b = 0.291 \text{ m}$$

$$h_{fmaj} = 18.226 \text{ m} \qquad h_{fent} = 0.302 \text{ m} \qquad 2 \cdot h_{felbow} = 1.809 \text{ m} \qquad h_{fnozzle} = 0.917 \text{ m}$$

(f) The EGL and HGL are illustrated in Figure EP 4.23a.

EXAMPLE PROBLEM 4.26

The turbine–generator system for Example Problem 4.25 is illustrated in Figure EP 4.26, where the required hydraulic power to be extracted by the impulse turbine is 700 kW. The electric power output by the generator is 600 kW, and the rotating shaft of the turbine has an angular speed of 320 rad/sec and a shaft torque of 2100 N-m. (a) Determine the shaft power delivered to the generator by the turbine. (b) Determine the turbine efficiency. (c) Determine the generator efficiency. (d) Determine the turbine–generator system efficiency.

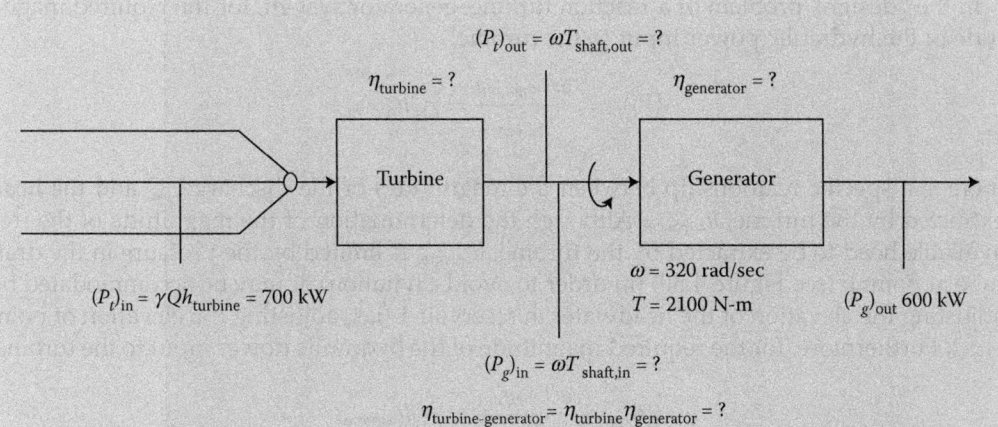

FIGURE EP 4.26
Turbine–generator system; converts mechanical power (hydraulic and shaft power) to electrical power.

Mathcad Solution

(a) The shaft power delivered to the generator by the turbine, $(P_t)_{out} = \omega T_{shaft,out} = (P_g)_{in} = \omega T_{shaft,in}$ defined for Equation 4.174 is applied as follows:

$$\omega := 320 \frac{rad}{sec} \qquad T_{shaftout} := 2100 \, N \cdot m$$

$$P_{tout} := \omega \cdot T_{shaftout} = 672 \, kW \qquad P_{gin} := P_{tout} = 672 \, kW$$

(b) The turbine efficiency defined in Equation 4.167 is applied as follows:

$$P_{tin} := 700 \, kW \qquad \eta_{turbine} := \frac{P_{tout}}{P_{tin}} = 0.96$$

Thus, the turbine is 96% efficient.

(c) The generator efficiency defined in Equation 4.174 is applied as follows:

$$P_{gout} := 600 \, kW \qquad \eta_{generator} := \frac{P_{gout}}{P_{gin}} = 0.893$$

Thus, the generator is 89.3% efficient.

(d) The turbine–generator system efficiency defined in Equation 4.176 is applied as follows:

$$\eta_{turbinegenerator} := \frac{P_{gout}}{P_{tin}} = 0.857$$

$$\eta_{turbinegenerator} := \eta_{turbine} \cdot \eta_{generator} = 0.857$$

Thus, the turbine–generator system is 85.7% efficient.

In the "design" problem of a reaction turbine–generator system, for the required magnitude of the hydraulic power input to the turbine,

$$(P_t)_{in} = \frac{\omega T_{shaft,out}}{\eta_{turbine}} = \gamma Q h_{turbine},$$

there is a specific relationship between the magnitudes of the discharge, Q and the head extracted by the turbine, $h_{turbine}$. Although the determination of the magnitude of the free available head to be extracted by the turbine, $h_{turbine}$ is limited by the pressure in the draft tube at point b (see Figure 4.30) (in order to avoid cavitation), it may be accommodated by adjusting the elevation of the headwater in reservoir 1 (i.e., adjusting the elevation of point 1, z_1). Furthermore, for the required magnitude of the hydraulic power input to the turbine,

$$(P_t)_{in} = \frac{\omega T_{shaft,out}}{\eta_{turbine}} = \gamma Q h_{turbine},$$

and the design magnitude of the freely available head, $h_{turbine}$ there is a corresponding design magnitude of the discharge to be extracted by the turbine, which is determined by applying Equation 4.187 as follows:

$$Q = \frac{(P_t)_{in}}{\gamma h_{turbine}} \tag{4.250}$$

In the design of a reaction turbine–generator system (see Figure 4.30), the design elevation of the headwater in reservoir 1 is determined by application of the energy equation. And, the design magnitude of the pipe diameter is determined by application of the continuity equation. Application of the energy equation between points 1 and a, assuming the datum is at point 2 yields the following:

$$\left(\frac{p_1}{\gamma} + \frac{v_1^2}{2g} + z_1\right) - h_{f,maj} - h_{f,min1a} = \left(\frac{p_a}{\gamma} + \frac{v_a^2}{2g} + z_a\right) \tag{4.251}$$

where $p_1 = 0$, and $v_1 = 0$, and thus:

$$\frac{p_a}{\gamma} + \frac{v_a^2}{2g} + z_a = z_1 - h_{f,maj} - h_{f,min1a} \tag{4.252}$$

which illustrates a conversion of the potential energy at point 1 (minus the head losses) to flow, kinetic, and potential energy at point a. Next, the freely available head to be extracted by the turbine, $h_{turbine}$ is determined by the application of the energy equation between points a and b and results in the following:

$$\left(\frac{p_a}{\gamma} + \frac{v_a^2}{2g} + z_a\right) - h_{turbine} = \left(\frac{p_b}{\gamma} + \frac{v_b^2}{2g} + z_b\right) \tag{4.253}$$

where $z_a = z_b$, and $v_a = v_b$, and thus:

$$h_{turbine} = \frac{p_a}{\gamma} - \frac{p_b}{\gamma} \tag{4.254}$$

which illustrates a conversion of flow energy at point a to hydraulic energy extracted by the turbine. In the design of a turbine–draft tube system, consideration is given to making sure that the pressure in the draft tube at point b (see Figure 4.30) does not fall below the vapor pressure of the liquid in order to avoid cavitation.

Therefore, the maximum allowable head to be extracted by the turbine is determined by setting the pressure at point b equal to the vapor pressure, p_v of the liquid in the flow as follows:

$$h_{turbine} = \frac{p_a}{\gamma} - \frac{p_v}{\gamma} \tag{4.255}$$

Additionally, substituting Equation 4.252 into Equation 4.255 yields the magnitude of the maximum allowable freely available head to be extracted by the turbine, $h_{turbine}$ expressed in terms of the potential energy at point 1, z_1 as follows:

$$h_{turbine} = z_1 - z_a - h_{f,maj} - h_{f,min1a} - \frac{v_a^2}{2g} - \frac{p_v}{\gamma} \tag{4.256}$$

where $h_{f,min,1a}$ represents minor losses between points 1 and a. Thus, the design magnitude of the freely available head to be extracted by the turbine, $h_{turbine}$ may be accommodated by adjusting the design elevation of the headwater in reservoir 1 (i.e., adjusting the elevation of point 1, z_1). Furthermore, in the design of a turbine–draft tube system, consideration is given to making sure that the pressure at point b (or anywhere in the turbine–draft tube system) does not fall below the vapor pressure of the liquid in order to avoid cavitation. As such, the draft tube is designed as follows: (1) with a gradual expansion in diameter (conical diffuser) in order for a pressure recovery to occur in the flow in the draft tube, to decrease the minor head loss, and thus to increase the available head to be extracted by the turbine; and (2) with a maximum allowable height of the draft tube, z_b in order to avoid a pressure drop below the vapor pressure of the liquid in the draft tube and thus to avoid cavitation. In order to prevent separation of the flow from the wall of the draft tube, the recommended conical angle, θ is typically less than 10°. Therefore, there is a maximum allowable height of the draft tube, z_b in order to avoid a pressure drop below the vapor pressure of the liquid in the draft tube at point b and thus to avoid cavitation. The maximum allowable height of the draft tube, z_b is determined by applying the energy equation between points b and 2, assuming the datum is at point 2, and the pressure at point b is set equal to the vapor pressure, p_v of the liquid in the flow as follows:

$$\left(\frac{p_b}{\gamma} + \frac{v_b^2}{2g} + z_b\right) - h_{f,min\,b2} = \left(\frac{p_2}{\gamma} + \frac{v_2^2}{2g} + z_2\right) \tag{4.257}$$

where $p_2 = 0$, $v_2 = 0$, $z_2 = 0$, $p_b = p_v$ (negative gage pressure), and $h_{f,minb2}$ is due to the conical diffuser, and thus:

$$z_b = h_{f,min\,b2} - \frac{p_v}{\gamma} - \frac{v_b^2}{2g} \tag{4.258}$$

Finally, the design magnitude of the freely available discharge may be accommodated by adjusting the design pipe, and application of the continuity equation yields the design discharge as follows:

$$Q = v_a A_a \tag{4.259}$$

EXAMPLE PROBLEM 4.27

The required hydraulic power to be extracted by a reaction turbine is determined to be 800 kW. The design freely available head to be extracted by the turbine is given to be 25 m. The hydraulic power is to be extracted from water at 20°C flowing from reservoir 1 though a galvanized iron pipe 150 m in length, as illustrated in Figure EP 4.24. The pipe entrance from reservoir 1 is square-edged, and the draft tube has a conical angle, $\theta = 9°$. Assume a Darcy–Weisbach friction factor, f of 0.15; a minor head loss coefficient due to a squared-edged pipe entrance, $k = 0.5$; and a minor head loss coefficient due to the conical diffuser, $k = 0.5$. The pressure at point a in the pipe is measured using a piezometer. (a) Determine the design magnitude of the freely available discharge. (b) Determine the design level of headwater in reservoir 1 in order to accommodate the design magnitude of the freely available head to be extracted by the turbine, $h_{turbine}$. (c) Determine the pressure and velocity at point a. (d) Determine the maximum allowable height of the draft tube, z_b without causing cavitation. (e) Determine the design pipe size in order to accommodate the design magnitude of the required freely available discharge. (f) Draw the energy grade line and the hydraulic grade line.

Mathcad Solution

(a) In order to determine the design magnitude of the freely available discharge, Equation 4.187 is applied as follows:

$$\rho := 998 \ \frac{kg}{m^3} \qquad g := 9.81 \ \frac{m}{sec^2} \qquad \gamma := \rho \cdot g = 9.79 \times 10^3 \ \frac{kg}{m^2 s^2}$$

$$P_{tin} := 800 \ kW \qquad h_{turbine} := 25 \ m$$

$$\text{Guess value:} \qquad Q := 2 \ \frac{m^3}{sec}$$

Given

$$P_{tin} = \gamma \cdot Q \cdot h_{turbine}$$

$$Q := Find(Q) = 3.269 \ \frac{m^3}{s}$$

(b)–(e) In order to determine the design level of headwater in reservoir 1 in order to accommodate the design magnitude of the freely available head to be extracted by the turbine, $h_{turbine}$ and the pressure and velocity at point a, the energy equation is applied between points 1 and a and between points a and b, assuming the pressure at point b is the vapor pressure of water at 20°C. In order to determine the maximum allowable height of the draft tube, z_b without causing cavitation, the energy equation is applied between points b and 2, assuming the pressure at point b is the vapor pressure of water at 20°C. And, in order to determine the design pipe size in order to accommodate the design magnitude of the required freely available discharge, the continuity equation is applied between points a and b. Thus, from Table A.2 in Appendix A, for water at 20°C, the vapor pressure, p_v is 2.34×10^3 N/m² abs. However, because the vapor pressure is given in absolute pressure, the corresponding gage pressure is computed by subtracting the atmospheric pressure as follows: $p_{gage} = p_{abs} - p_{atm}$, where the standard atmospheric pressure is 101.325×10^3 N/m² abs.

$$p_1 := 0 \, \frac{N}{m^2} \qquad v_1 := 0 \, \frac{m}{sec} \qquad p_2 := 0 \, \frac{N}{m^2} \qquad v_2 := 0 \, \frac{m}{sec} \qquad z_2 := 0 \, m$$

$$p_v := 2.34 \times 10^3 \, \frac{N}{m^2} - 101.325 \times 10^3 \, \frac{N}{m^2} = -9.899 \times 10^4 \, \frac{N}{m^2}$$

$$p_b := p_v = -9.899 \times 10^4 \, \frac{N}{m^2}$$

$$k_{ent} := 0.5 \qquad k_{diffuser} := 0.5 \qquad f := 0.15 \qquad L := 150 \, m$$

Guess value: $\quad z_1 := 100 \, m \quad p_a := 2 \times 10^3 \, \frac{N}{m^2} \quad v_a := 1 \, \frac{m}{sec} \quad v_b := 1 \, \frac{m}{sec}$

$$D_a := 0.5 \, m \qquad D_b := 0.5 \, m \qquad z_a := 2 \, m \qquad z_b := 2 \, m$$

$$h_{fmaj} := 1 \, m \qquad h_{fent} := 1 \, m \qquad h_{fdiffuser} := 1 \, m$$

Given

$$\frac{p_1}{\gamma} + z_1 + \frac{v_1^2}{2 \cdot g} - h_{fmaj} - h_{fent} = \frac{p_a}{\gamma} + z_a + \frac{v_a^2}{2 \cdot g}$$

$$\frac{p_a}{\gamma} + z_a + \frac{v_a^2}{2 \cdot g} - h_{turbine} = \frac{p_b}{\gamma} + z_b + \frac{v_b^2}{2 \cdot g}$$

$$\frac{p_b}{\gamma} + z_b + \frac{v_b^2}{2 \cdot g} - h_{fdiffuser} = \frac{p_2}{\gamma} + z_2 + \frac{v_2^2}{2 \cdot g}$$

$$h_{fmaj} = f \frac{L}{D_a} \frac{v_a^2}{2 \cdot g} \qquad h_{fent} = k_{ent} \frac{v_a^2}{2 \cdot g} \qquad h_{fdiffuser} = k_{diffuser} \frac{v_b^2}{2 \cdot g}$$

$$Q = v_a \frac{\pi \cdot D_a^2}{4} \qquad Q = v_b \frac{\pi \cdot D_b^2}{4} \qquad z_a = z_b \qquad D_a = D_b$$

$$
\begin{pmatrix}
z_1 \\
p_a \\
v_a \\
v_b \\
D_a \\
D_b \\
z_a \\
z_b \\
h_{fmaj} \\
h_{fent} \\
h_{fdiffuser}
\end{pmatrix}
:= Find(z_1, p_a, v_a, v_b, D_a, D_b, z_a, z_b, h_{fmaj}, h_{fent}, h_{fdiffuser})
$$

$z_1 = 57.405 \, m \qquad p_a = 1.458 \times 10^5 \, \dfrac{N}{m^2} \qquad v_a = 4.982 \, \dfrac{m}{s} \qquad v_b = 4.982 \, \dfrac{m}{s}$

$D_a = 0.914 \, m \qquad D_b = 0.914 \, m \qquad z_a = 9.478 \, m \qquad z_b = 9.478 \, m$

$h_{fmaj} = 31.14 \, m \qquad h_{fent} = 0.632 \, m \qquad h_{fdiffuser} = 0.632 \, m$

(f) The EGL and HGL are illustrated in Figure EP 4.24.

EXAMPLE PROBLEM 4.28

The turbine–generator system for Example Problem 4.27 is illustrated in Figure EP 4.24, where the required hydraulic power to be extracted by the reaction turbine is 800 kW. The electric power output by the generator is 700 kW, and the rotating shaft of the turbine has an angular speed of 350 rad/sec and a shaft torque of 2200 N-m. (a) Determine the shaft power delivered to the generator by the turbine. (b) Determine the turbine efficiency. (c) Determine the generator efficiency. (d) Determine the turbine–generator system efficiency.

Mathcad Solution

(a) The shaft power delivered to the generator by the turbine, $(P_t)_{out} = \omega T_{shaft,out} = (P_g)_{in} = \omega T_{shaft,in}$ defined for Equation 4.174 is applied as follows:

$\omega := 350 \, \dfrac{rad}{sec} \qquad\qquad T_{shaftout} := 2200 \, N \, m$

$P_{tout} := \omega \cdot T_{shaftout} = 770 \, kW \qquad P_{gin} := P_{tout} = 770 \, kW$

which is illustrated in Figure EP 4.28.

(b) The turbine efficiency defined in Equation 4.167 is applied as follows:

$P_{tin} := 800 \, kW \qquad\qquad\qquad \eta_{turbine} := \dfrac{P_{tout}}{P_{tin}} = 0.963$

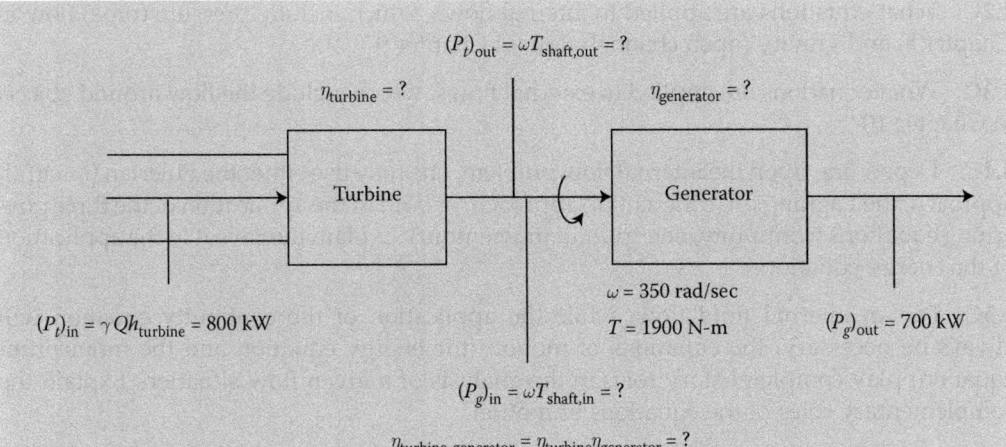

FIGURE EP 4.28
Turbine–generator system; converts mechanical power (hydraulic and shaft power) to electrical power.

Thus, the turbine is 96.3% efficient.

(c) The generator efficiency defined in Equation 4.174 is applied as follows:

$$P_{gout} := 700 \text{ kW} \qquad \eta_{generator} := \frac{P_{gout}}{P_{gin}} = 0.909$$

Thus, the generator is 90.9% efficient.

(d) The turbine–generator system efficiency defined in Equation 4.176 is applied as follows:

$$\eta_{turbinegenerator} := \frac{P_{gout}}{P_{tin}} = 0.875$$

$$\eta_{turbinegenerator} := \eta_{turbine} \cdot \eta_{generator} = 0.875$$

Thus, the turbine–generator system is 87.5% efficient.

End-of-Chapter Problems

Problems with a "C" are conceptual problems. Problems with a "BG" are in English units. Problems with an "SI" are in metric units. Problems with a "BG/SI" are in both English and metric units. All "BG" and "SI" problems that require computations are solved using Mathcad.

Introduction

4.1C The energy equation may be derived by two different approaches. What principles are these two different approaches based on?

4.2C What equations are applied to internal flows, which include pressure (pipe) flow in Chapter 8, and gravity (open channel) flow in Chapter 9?

4.3C What equations are applied to external flows, which include the flow around objects in Chapter 10?

4.4C Depending upon the internal flow problem, one may use either the Eulerian (integral) approach, the Lagrangian (differential) approach, or both in the application of the three governing equations (continuity, energy, and momentum). Explain the caveat in the application of the energy equation.

4.5C For an internal fluid flow, while the application of the continuity equation will always be necessary, the equations of motion (the energy equation and the momentum equation) play complementary roles in the analysis of a given flow situation. Explain the complementary roles of the equations of motion.

Fluid Dynamics

4.6C What principle is fluid dynamics based on, and what equations are derived from this principle? What additional topic does fluid dynamics include?

4.7C Fluid dynamics considers the analysis of the specific forces acting on a fluid element that produce the motion. In general, what are the forces that may act in a flow situation?

4.8C According to Newton's second law of motion, if the summation of forces acting on a fluid element does not add up to zero, then the fluid element will accelerate. Explain how such an unbalanced force system can be transformed into a balanced system.

4.9C Depending upon the fluid flow problem, some of the forces may not play a significant role in the derivation of the equations of motion (energy equation and momentum equation), which are based upon the principle of conservation of momentum (Newton's second law of motion). Explain.

4.10C Assuming a real flow, how is the flow resistance modeled in the application of the principle of conservation of momentum (Newton's second law of motion) (Equation 4.3)?

4.11C Regardless of whether the flow is internal or external, in the application of the principle of conservation of momentum (Newton's second law of motion) (Equation 4.3), while the definition of the gravitational force, F_G and the pressure force, F_P can always be theoretically modeled, the friction/viscous force, F_V may not always be theoretically modeled. Explain how the friction/viscous force, F_V is empirically modeled/defined.

Derivation of the Energy Equation

4.12C Derivation of the energy equation may be based on either the principle of conservation of momentum (Newton's second law of motion) or the principle of conservation of energy (the first law of thermodynamics). Similar to the statement of the principle of conservation of mass in Chapter 3 (yielding the continuity equation), the principle of conservation of momentum (Newton's second law of motion) and the principle of conservation of energy (the first law of thermodynamics) are historically stated and applied for a fluid system, and then rephrased and extended to a control volume using the Reynolds transport theorem. Describe the historically stated principles for a fluid system.

4.13C Discuss the basic difference between the energy equation that is derived based on the conservation of momentum principle vs. the energy equation that is derived based on the conservation of energy principle.

The Energy Equation Based on Newton's Second Law of Motion

4.14C In the application of the governing equations (continuity, energy, and momentum), it is important to make the distinction between types of flow. Assuming steady flow, explain what these flow types are.

The Energy Equation for a Fluid System

4.15C State the one-dimensional Navier–Stokes equation, and state what basic principle it represents.

4.16C State Euler's equation of motion, and state what equation it is a special case of.

4.17C State the differential form of the energy equation, and state how it is derived.

4.18C Explain why the differential form of the energy equation given by Equation 4.35 is not really useful (practical per se), and explain the steps necessary to make it practical, and why.

The Energy Equation for a Control Volume

4.19C Explain the simplifying assumption made regarding the compressibility of the flow in the derivation of the integral form of the energy equation for a control volume given by Equation 4.39 as follows:

$$\left(\frac{p}{\gamma} + z + \frac{v^2}{2g}\right)_1 - \left(\frac{p}{\gamma} + z + \frac{v^2}{2g}\right)_2 = \frac{\tau_w}{\gamma}\frac{L}{R_h} + \frac{L}{g}\frac{\partial v}{\partial t}$$

4.20C When is it appropriate to assume an incompressible flow, and thus apply the integral form of the energy equation given by Equation 4.39?

4.21C Explain how would one model compressible flow before the integration and evaluation of the differential form of the energy equation given by Equation 4.35 as follows:

$$-\left[\frac{\partial(p + \gamma z + (\rho v^2/2))}{\partial s}\right]ds = \left[\frac{\tau_w}{R_h} + \rho\frac{\partial v}{\partial t}\right]ds$$

4.22C Explain the units of each term in the integral form of the energy equation for a control volume given by Equation 4.39 as follows:

$$\left(\frac{p}{\gamma} + z + \frac{v^2}{2g}\right)_1 - \left(\frac{p}{\gamma} + z + \frac{v^2}{2g}\right)_2 = \frac{\tau_w}{\gamma}\frac{L}{R_h} + \frac{L}{g}\frac{\partial v}{\partial t}$$

4.23C Define the two terms on the left-hand side of the integral form of the energy equation for a control volume given by Equation 4.39 as follows:

$$\left(\frac{p}{\gamma}+z+\frac{v^2}{2g}\right)_1 - \left(\frac{p}{\gamma}+z+\frac{v^2}{2g}\right)_2 = \frac{\tau_w}{\gamma}\frac{L}{R_h}+\frac{L}{g}\frac{\partial v}{\partial t}$$

4.24C Define each term in equation for the total head, H given by Equation 4.40 as follows:

$$H = \left(\frac{p}{\gamma}+z+\frac{v^2}{2g}\right)$$

4.25C Define each term in the integral form of the energy equation for a control volume given by Equation 4.41 as follows:

$$H_1 - H_2 = \frac{\tau_w}{\gamma}\frac{L}{R_h}+\frac{L}{g}\frac{\partial v}{\partial t}$$

4.26C What is the head loss form of the integral form of the energy equation for a control volume given by Equation 4.39?

4.27C Assuming steady flow, state the more general form of the energy equation for a control volume given by Equation 4.39, which assumes there may be one or more control faces for the flow into and the flow out of the control volume as illustrated in Figure 4.3.

4.28C Begin with the energy equation for incompressible flow given by Equation 4.39 as follows:

$$\left(\frac{p}{\gamma}+z+\frac{v^2}{2g}\right)_1 - \left(\frac{p}{\gamma}+z+\frac{v^2}{2g}\right)_2 = \frac{\tau_w}{\gamma}\frac{L}{R_h}+\frac{L}{g}\frac{\partial v}{\partial t}$$

State the energy equation for incompressible steady state flow between two points along a streamline.

4.29C Explain how the energy equation for incompressible steady flow, which is given by Equation 4.44 as follows, is applied.

$$\left(\frac{p}{\gamma}+z+\frac{v^2}{2g}\right)_1 - \left(\frac{p}{\gamma}+z+\frac{v^2}{2g}\right)_2 = \underbrace{\frac{\tau_w}{\gamma}\frac{L}{R_h}}_{h_f}$$

4.30C Begin with the energy equation for incompressible steady flow given by Equation 4.44 as follows:

$$\left(\frac{p}{\gamma}+z+\frac{v^2}{2g}\right)_1 - \left(\frac{p}{\gamma}+z+\frac{v^2}{2g}\right)_2 = \underbrace{\frac{\tau_w}{\gamma}\frac{L}{R_h}}_{h_f}$$

State the energy equation for incompressible, steady state, ideal (inviscid) flow between two points along a streamline. What is the name of this equation? What do these assumptions for the energy equation imply?

4.31C The derivation of the energy equation for incompressible steady flow assumes a pipe flowing under pressure and is given by Equation 4.44 as follows:

$$\left(\frac{p}{\gamma} + z + \frac{v^2}{2g}\right)_1 - \left(\frac{p}{\gamma} + z + \frac{v^2}{2g}\right)_2 = \underbrace{\frac{\tau_w}{\gamma} \frac{L}{R_h}}_{h_f}$$

Compare the definition of terms in the above energy equation for pipe flow vs. open channel flow, and explain how one would modify the above energy equation in order to apply it to open channel (gravity) flow.

The Bernoulli Equation

4.32C What does the assumption of ideal flow (inviscid flow) imply, and when is it appropriate to assume ideal flow and thus apply the Bernoulli equation given by Equation 4.46?

4.33C The Bernoulli equation, which is given by Equation 4.46, implies that the total head, H along a streamline is constant as given by Equations 4.49 and 4.52 as follows:

$$H = \frac{p}{\gamma} + z + \frac{v^2}{2g} = constant$$

Explain each term in the equation for the total head, H along a streamline, and discuss the conversion of energy from one form to another.

4.34C Application of the Bernoulli equation represents the assumption of ideal flow, which may be generally classified as internal flow (pipe or open channel), external flow around an object, or flow from a tank (or a water source) (see Table 4.1). What are some of the applications of the Bernoulli equation for these three general flow classifications?

4.35C Application of the Bernoulli equation for velocity (and pressure) measurement for both internal and external flow using a pitot-static tube demonstrates how the dynamic pressure is modeled by a pressure rise (see Section 4.5.3.2). Explain how the use of several pitot-static tubes along a pipe or a channel section allows schematic illustrations of the energy equation.

4.36C Application of the Bernoulli equation for velocity (and pressure) measurement for both internal and external flow using a pitot-static tube demonstrates how the dynamic pressure is modeled by a pressure rise. Give an example of the application of the Bernoulli equation that demonstrates how the dynamic pressure is modeled by a pressure drop.

Application of the Energy Equation for Real Internal Flow

4.37C In the application of the energy equation (Equation 4.44) for real internal flow, the evaluation of the total head loss, $h_{f,total} = h_{f,maj} + h_{f,min}$ may include both major and minor losses, which are due to flow resistance. Explain what causes the major and minor head losses in real internal flow.

4.38SI   Water at 20°C flows in a 2.9-*m*-diameter inclined 22,000-*m* steel pipe at a flowrate of $6\,m^3/sec$, as illustrated in Figure ECP 4.38. The pressure at point 1 is $46 \times 10^5\,N/m^2$, and the pressure at point 2 is $30 \times 10^5\,N/m^2$. Point 2 is 8 *m* higher than point 1. (a) Determine the major head loss due to the pipe friction between points 1 and 2. (b) Draw the energy grade line and the hydraulic grade line.

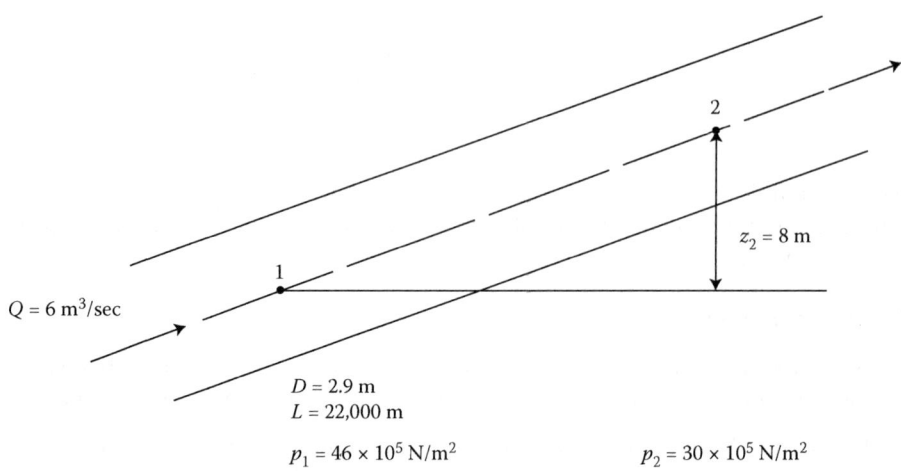

FIGURE ECP 4.38

4.39BG   Water at 50°F flows in a 5-*ft*-diameter inclined 1600-*ft* steel pipe at a flowrate of $78\,ft^3/sec$, as illustrated in Figure ECP 4.39. The pressure at point 1 is 20 *psi*, and the pressure at point 2 is 12 *psi*. Point 2 is 3 *ft* lower than point 1. (a) Determine the major head loss due to the pipe friction between points 1 and 2. (b) Determine the shear stress along the pipe wall. (c) Draw the energy grade line and the hydraulic grade line.

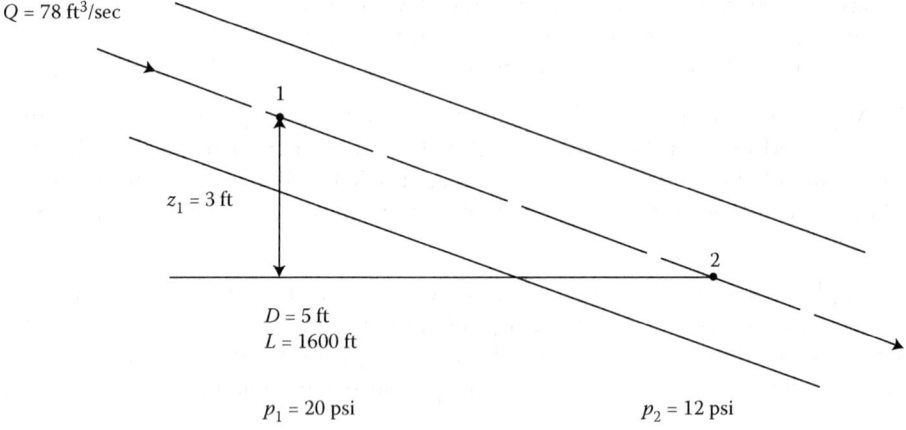

FIGURE ECP 4.39

4.40SI   Water at 20°C flows in a 0.7-*m*-diameter vertical pipe at a flowrate of $30\,m^3/sec$, as illustrated in Figure ECP 4.40. The head loss in the pipe is 0.1 *m/length* of pipe. The pressure at point 1 is $4 \times 10^5\,N/m^2$. (a) Determine the pressure at points 2 and 3.

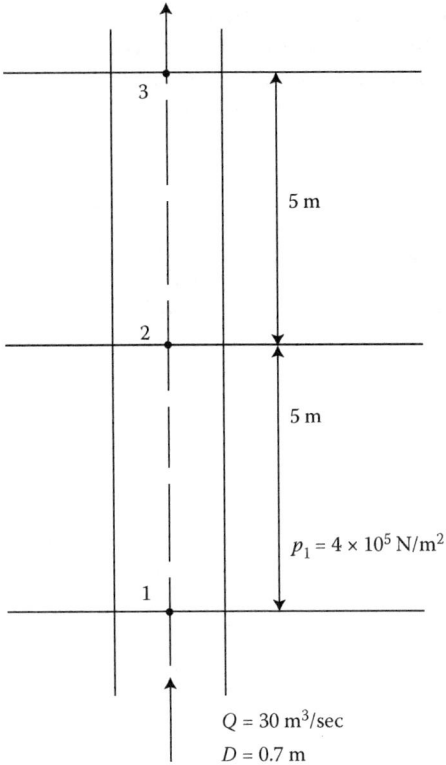

FIGURE ECP 4.40

4.41BG Water at $60°F$ flows at a discharge of $5\,ft^3/sec$ in a 2-*ft*-diameter horizontal pipe, and piezometers are used to measure the pressures in the pipe, as illustrated in Figure ECP 4.41. (a) Determine the major head loss due to the pipe friction between points 1 and 2. (b) Determine the direction of the flow.

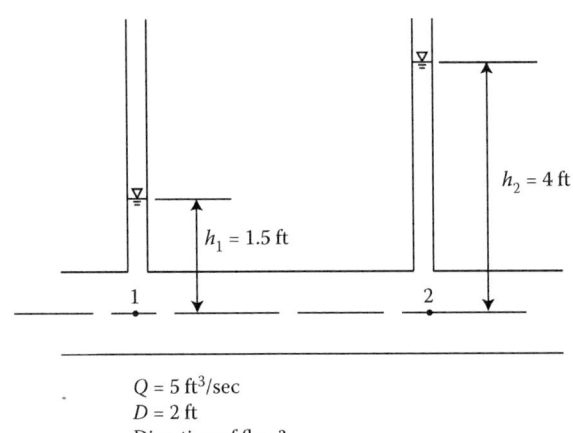

FIGURE ECP 4.41

4.42SI　Water at 20°C flows in a 0.4-*m*-diameter horizontal pipe at a flowrate of 0.55 m^3/sec, as illustrated in Figure ECP 4.42. There is a valve between points 1 and 2, where the pressure at point 1 is $3 \times 10^5 \, N/m^2$, and the pressure at point is $2.89 \times 10^5 \, N/m^2$. (a) Determine the minor head loss between due to the valve points 1 and 2. (b) Draw the energy grade line and the hydraulic grade line.

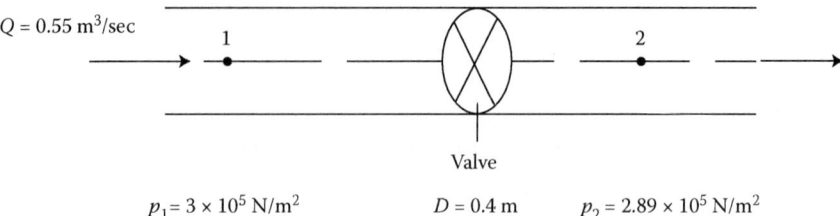

FIGURE ECP 4.42

4.43SI　Water at 20°C flows in a 1.95-*m*-diameter horizontal 13,000-*m* concrete pipe at a flowrate of 2.7 m^3/sec. A 90° mitered bend is installed in between point 1 and 2 in order to change the direction of the flow by 90°, as illustrated in Figure ECP 4.43. The pressure at point 1 is $19 \times 10^5 \, N/m^2$. Assume a Darcy–Weisbach friction factor, f of 0.019, and assume a minor head loss coefficient due to the bend, k of 0.131. (a) Determine the major head loss due to the pipe friction between points 1 and 2. (b) Determine the minor head loss due to the bend between points 1 and 2. (c) Determine the pressure at point 2.

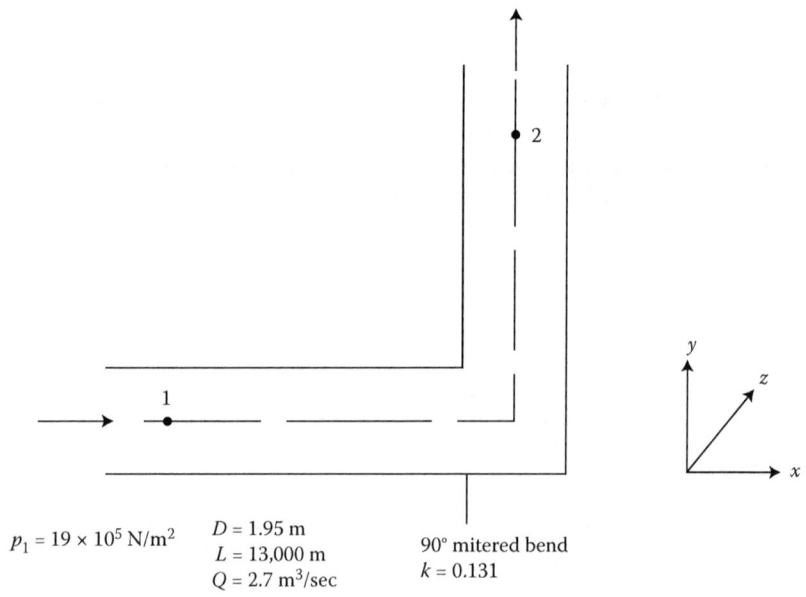

FIGURE ECP 4.43

4.44BG　Water at 90°F flows in a 3.75-*ft*-diameter vertical 40-*ft* concrete pipe at a flowrate of 80 ft^3/sec. A 90° mitered bend is installed in between point 1 and 2 in order to change the direction of the flow by 90°, as illustrated in Figure ECP 4.44. After the bend, the flow continues in a 3.75-*ft*-diameter horizontal 40-*ft* concrete pipe. The pressure at point 1 is 14,000 lb/ft^2. Assume a Darcy–Weisbach friction factor, f of 0.019, and assume a minor head loss coefficient due to the bend, k of 0.131. (a) Determine the major head loss due to the pipe friction between points 1 and 2. (b) Determine the minor head loss due to the bend between points 1 and 2. (c) Determine the pressure at point 2.

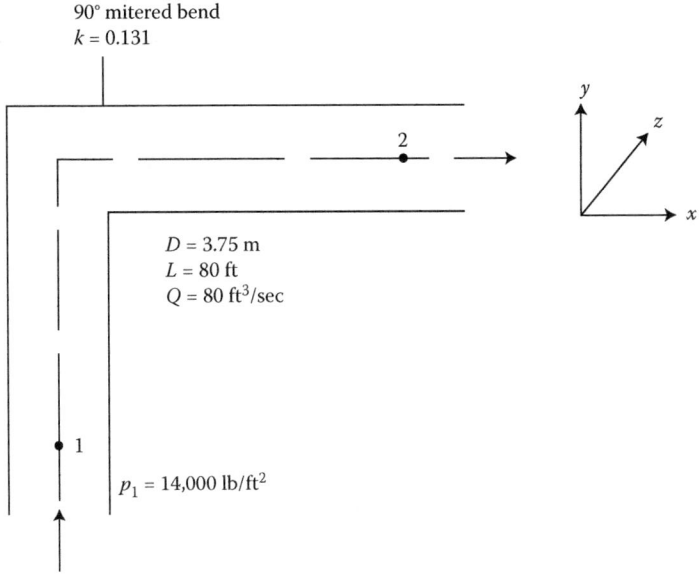

90° mitered bend
$k = 0.131$

2

$D = 3.75$ m
$L = 80$ ft
$Q = 80$ ft^3/sec

1

$p_1 = 14{,}000$ lb/ft^2

FIGURE ECP 4.44

4.45BG Water at 50°F flows in a 5-*ft*-diameter horizontal pipe at a flowrate of 150 *ft*^{3}/*sec*. A nozzle is installed between point 1 and 2 in order to accelerate the flow, as illustrated in Figure ECP 4.45. The pipe diameter after the nozzle is 2.8 *ft*. The pressure at point 2 is 1500 *lb/ft*^{2}. Assume a minor head loss coefficient due to the nozzle, k of 0.05. (a) Determine the minor head loss due to the nozzle between points 1 and 2. (b) Determine the pressure at point 1.

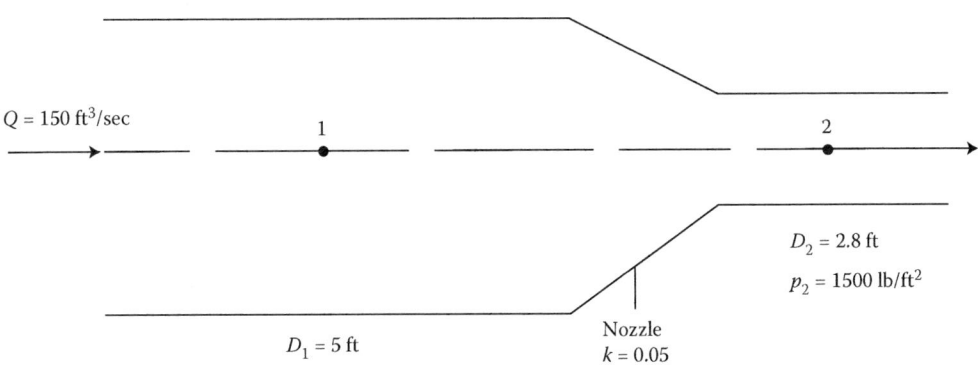

$Q = 150$ ft^3/sec

1

2

$D_2 = 2.8$ ft

$p_2 = 1500$ lb/ft^2

$D_1 = 5$ ft

Nozzle
$k = 0.05$

FIGURE ECP 4.45

4.46SI　Water at 25°C flows in a 1.56-*m*-diameter horizontal pipe at a flowrate of 4.5 m^3/sec. A conical diffuser is installed between point 1 and 2 in order to recover the pressure of the flow, as illustrated in Figure ECP 4.46. The pipe diameter after the conical diffuser is 3 *m*. The pressure at point 1 is $17 \times 10^5 \, N/m^2$. Assume a minor head loss coefficient due to the conical diffuser, *k* of 0.8. (a) Determine the minor head loss due to the conical diffuser between points 1 and 2. (b) Determine the pressure at point 2.

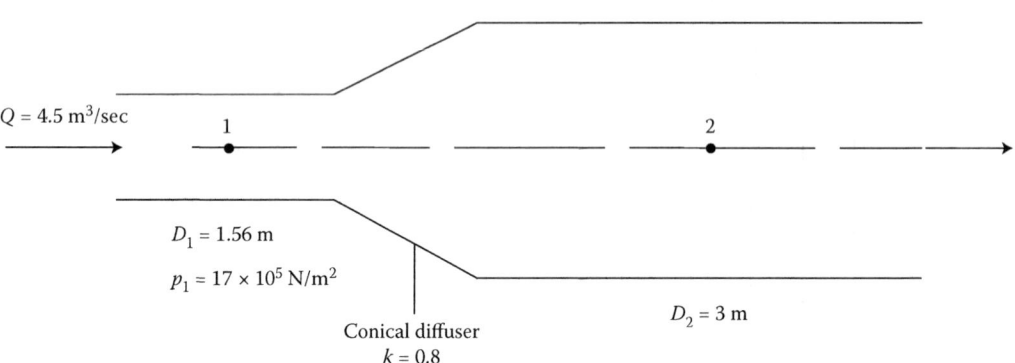

FIGURE ECP 4.46

4.47BG　Water at 68°F flows in a 2-*ft*-diameter horizontal pipe at a flowrate of 8 ft^3/sec. A valve is installed between point 1 and 2 in order to control the flow, as illustrated in Figure ECP 4.47. Assume a minor head loss coefficient due to the valve, *k* of 60. The pressure drop is measured by a mercury differential manometer. (a) Determine the pressure drop due to the valve. (b) Determine the manometer reading due to the valve between points 1 and 2. (c) Determine the pressure at points 1, 2, 3, 4, and 5.

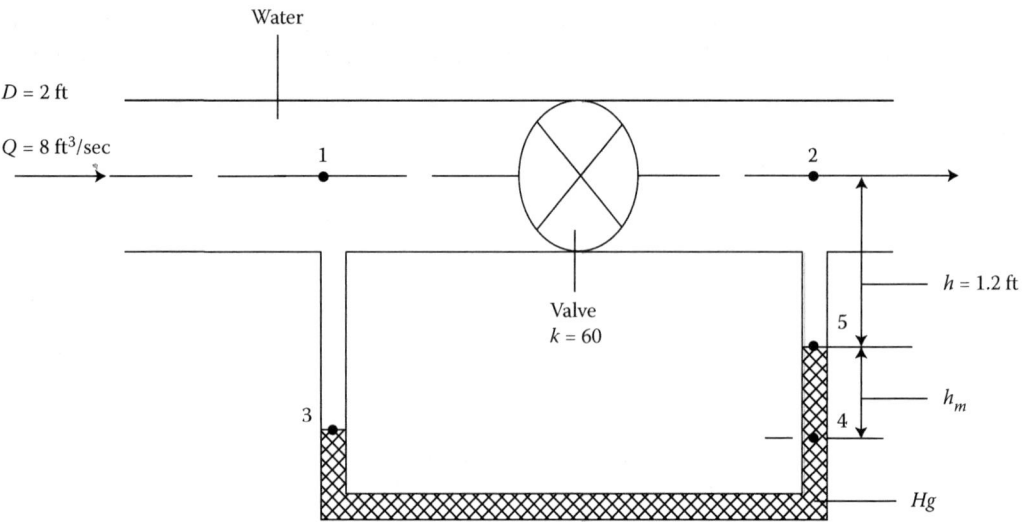

FIGURE ECP 4.47

4.48BG Water at 68°F flows in a 7-*ft*-diameter horizontal pipe at a flowrate of 9 ft^3/sec. A nozzle is installed between point 1 and 2 in order to accelerate the flow, as illustrated in Figure ECP 4.48. The pipe diameter after the nozzle is 3 *ft*, and the pressure drops from 19 *psi* to 16 *psi* due to the nozzle. Determine the minor head loss due to the nozzle.

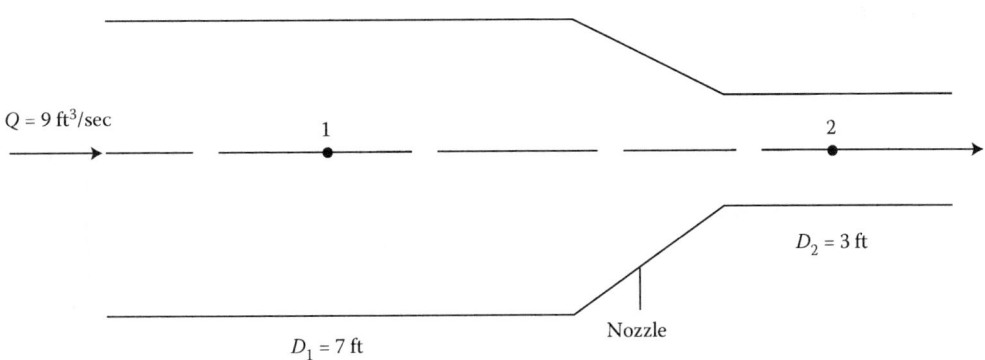

FIGURE ECP 4.48

4.49SI Water at 20°C flows in a 5-*m*-diameter horizontal pipe. A nozzle is installed between point 1 and 2 in order to accelerate the flow, as illustrated in Figure ECP 4.49. The pipe diameter after the nozzle is 1.9 *m*, and the pressure drop due to the nozzle is measured using a mercury differential manometer. Assume a minor head loss of 1 *m*. Determine the discharge in the pipe.

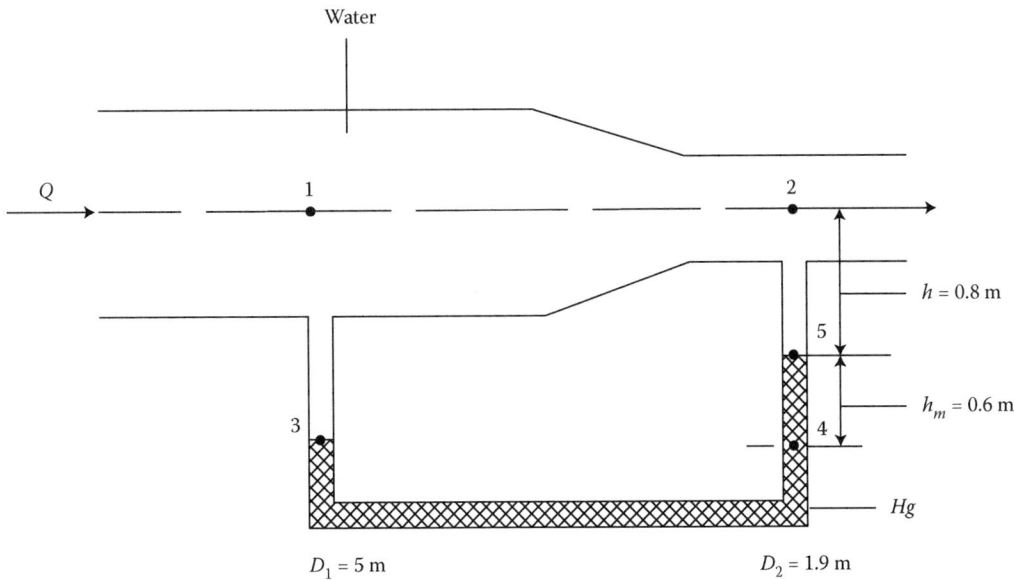

FIGURE ECP 4.49

4.50SI   Water flows at a uniform flow at a discharge of 3.9 m^3/sec in a rectangular channel that is 0.95 m wide, with a channel bottom slope of 0.03 m/m and a Chezy coefficient, C of 57 $m^{1/2} s^{-1}$, as illustrated in Figure ECP 4.50. For uniform open channel flow, the channel bottom slope, S_o is equal to the friction slope, S_f. (a) Determine the major head loss due to the channel resistance over a channel section length of 20 m between points 1 and 2. (b) Determine the uniform depth of flow in the channel. (c) Draw the energy grade line and the hydraulic grade line.

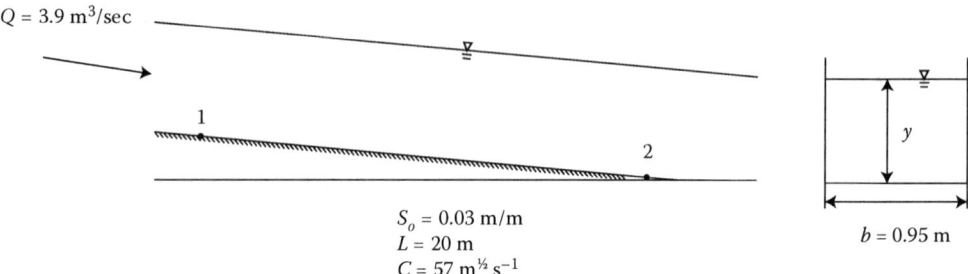

$$S_o = 0.03 \text{ m/m}$$
$$L = 20 \text{ m}$$
$$C = 57 \text{ m}^{\frac{1}{2}} \text{s}^{-1}$$

FIGURE ECP 4.50

Application of the Bernoulli Equation for Ideal Internal Flow and Ideal Flow from a Tank

4.51BG   Water at 70°F flows in a 4-*ft*-diameter horizontal pipe. A venturi meter with a 0.9-*ft* throat diameter is inserted in the pipe at point 2 as illustrated in Figure ECP 4.51 in order to measure the flowrate in the pipe. Piezometric tubes are installed at point 1 just upstream of the venturi tube, and at point 2, which is located at the throat of the venturi tube. The pressure head at point 1, p_1/γ is measured by the piezometer to be 3 *ft*, and the pressure head at

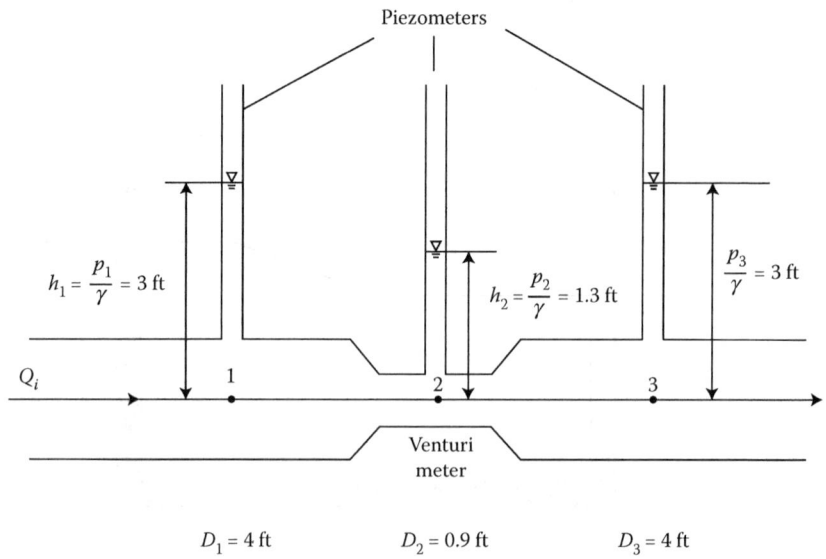

$$D_1 = 4 \text{ ft} \qquad D_2 = 0.9 \text{ ft} \qquad D_3 = 4 \text{ ft}$$

FIGURE ECP 4.51

point 2, p_2/γ is measured by the piezometer to be 1.3 ft. (a) Determine the ideal flowrate in the pipe. (b) Draw the energy grade line and the hydraulic grade line. (c) Given the same conditions at point 1 just upstream of the venturi tube, determine the minimum diameter of the venturi meter at point 2 without causing cavitation.

4.52C Oil flows in a horizontal pipe with a diameter D_1. A venturi meter with throat diameter D_2 is inserted in the pipe at point 2 as illustrated in Figure ECP 4.52 in order to measure the flowrate in the pipe. Piezometric tubes are installed at point 1 just upstream of the venturi tube, and at point 2, which is located at the throat of the venturi tube, in order to measure the pressure at points 1 and 2, respectively. Derive the expression for the ideal flowrate in the pipe.

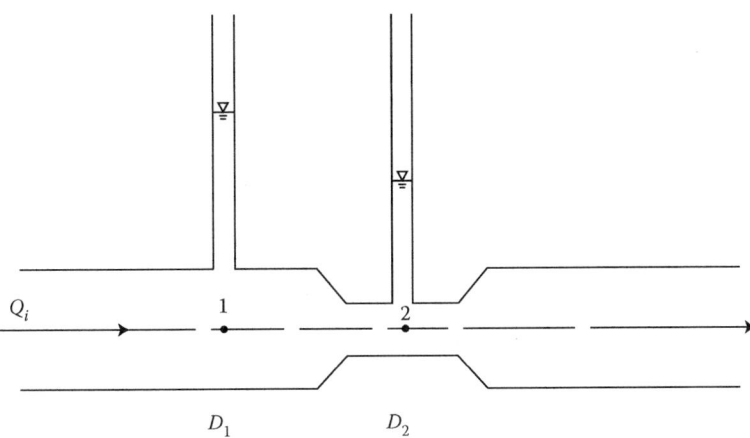

FIGURE ECP 4.52

4.53SI Water at 20°C flows in a 0.9-*m*-diameter horizontal pipe. A venturi meter with a 0.3-*m* throat diameter is inserted in the pipe at point 2 as illustrated in Figure ECP 4.53 in order to measure the flowrate in the pipe. Piezometric tubes are installed at point 1 just upstream of the venturi tube, and at point 2, which is located at the throat of the venturi

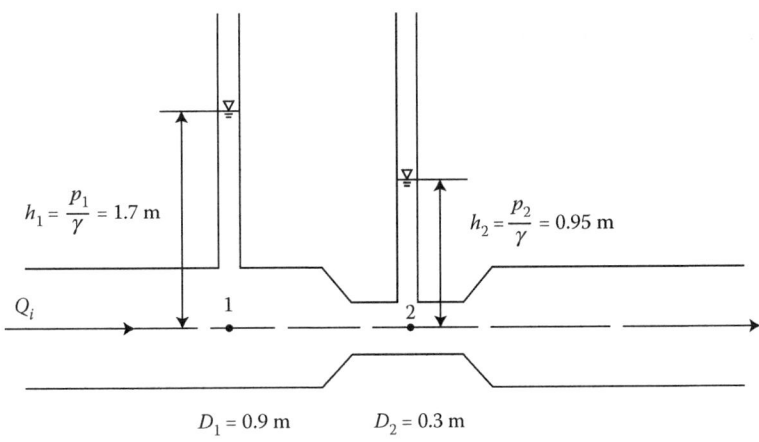

FIGURE ECP 4.53

tube. The pressure head at point 1, p_1/γ is measured by the piezometer to be 1.7 m, and the pressure head at point 2, p_2/γ is measured by the piezometer to be 0.95 m. Determine the ideal flowrate in the pipe. (b) Given the same conditions at point 1 just upstream of the venturi tube, determine the minimum diameter of the venturi meter at point 2 without causing cavitation.

4.54SI   Water at 20°C flows in a 1.33-m-diameter horizontal pipe. A piezometer is installed at point 1 and a pitot tube is installed at point 2 as illustrated in Figure ECP 4.54 in order measure the flowrate in the pipe. The static pressure head, p_1/γ measured by the piezometer is 1.5 m, and the stagnation pressure head, p_2/γ measured by the pitot tube is 2.4 m. (a) Determine the ideal flowrate in the pipe. (b) Draw the energy grade line and the hydraulic grade line.

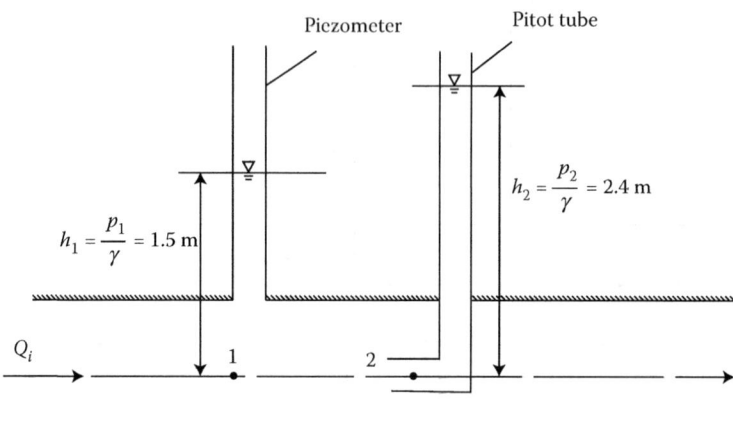

FIGURE ECP 4.54

4.55BG   Water flows in a rectangular channel of width 7 ft and at a mild slope. A sluice gate is inserted in the channel as illustrated in Figure ECP 4.55 in order to measure the flowrate in the channel. The depth of the flow upstream of the gate at point 1 is measured to be 6 ft, and the depth of flow downstream of the gate at point 2 is measured to be 3 ft. (a) Determine the ideal flowrate in the channel. (b) Draw the energy grade line and the hydraulic grade line.

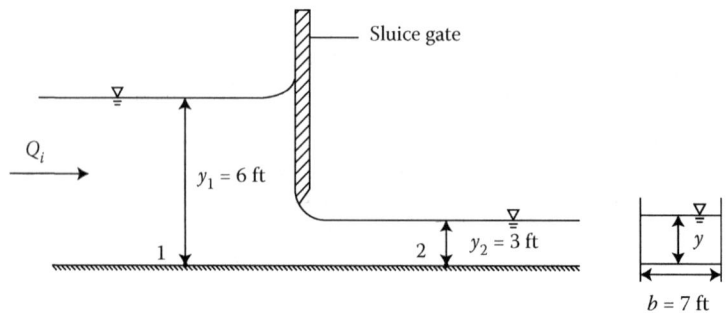

FIGURE ECP 4.55

4.56BG Water flows in a rectangular channel of width 4 *ft* and at a mild slope. A spillway is inserted in the channel as illustrated in Figure ECP 4.56 in order to measure the flowrate in the channel. The depth of the flow upstream of the spillway at point 1 is measured to be 7 *ft*, and the depth of flow downstream of the gate at point 2 is measured to be 4 *ft*. (a) Determine the ideal flowrate in the channel. (b) Draw the energy grade line and the hydraulic grade line.

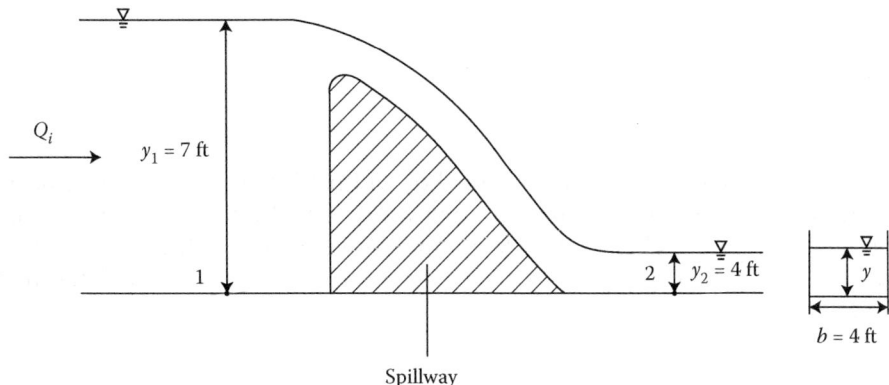

FIGURE ECP 4.56

4.57SI Water at 20°C flows at a uniform depth of 2.1 *m* in a rectangular channel of width 6 *m* and at a mild slope. A pitot tube is inserted in the channel at point 2 in order to measure the velocity in the channel at an upstream point 1, as illustrated in Figure ECP 4.57. The static pressure head at point 1, p_1/γ (the depth to point 1) is 1.9 *m*. The stagnation pressure head at point 2, p_2/γ measured by the pitot tube is 3 *m*. (a) Determine the ideal flowrate in the channel. (b) Draw the energy grade line and the hydraulic grade line.

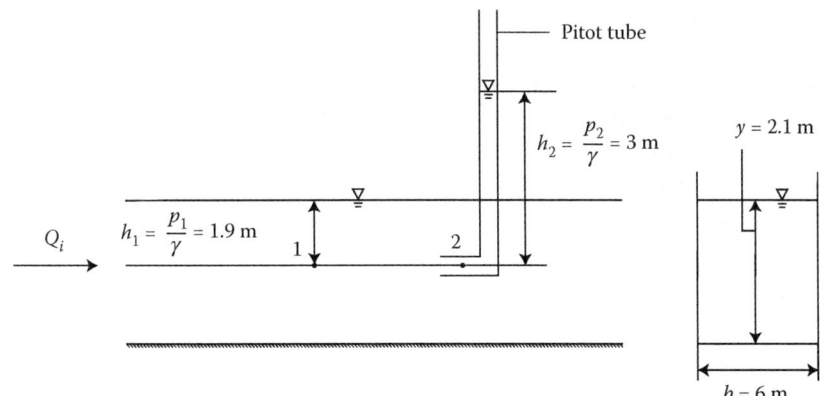

FIGURE ECP 4.57

4.58BG Water flows in a rectangular channel of width 3 *ft* at a velocity of 4 *ft/sec* and depth of 4 *ft*. There is a gradual upward step, Δz of 0.9 *ft* in the channel bed as illustrated in Figure ECP 4.58. (a) Determine the velocity and the depth of the water downstream of the step at point 2. (b) Determine the flowrate in the channel. (c) Draw the energy grade line and the hydraulic grade line.

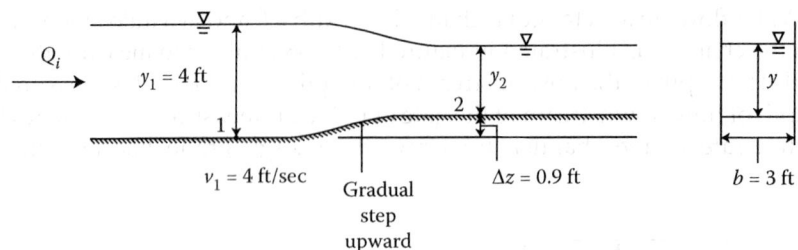

FIGURE ECP 4.58

4.59SI Water flows in a rectangular channel of width $1.3\,m$ at a discharge of $0.89\,m^3/sec$ and depth of $1.9\,m\,ft$. There is a gradual downward step, Δz of $0.3\,m$ in the channel bed as illustrated in Figure ECP 4.59. (a) Determine the velocity and the depth of the water downstream of the step at point 2. (b) Draw the energy grade line and the hydraulic grade line.

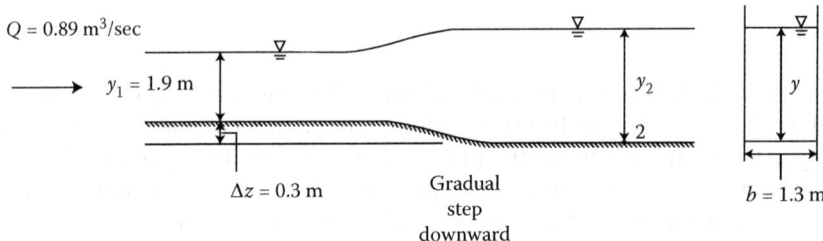

FIGURE ECP 4.59

4.60SI Water at 20°C is forced from a 2.5-*cm*-diameter syringe through a 0.15-*cm* hollow needle as a result of a $0.85\,N$ force applied to the piston, as illustrated in Figure ECP 4.60. The final position of the top of the piston at point 1 is 3.6 *cm* below the tip of the needle at point 2. (a) Determine the ideal velocity of the forced jet as it leaves the tip of the needle at point 2. (b) Determine the ideal height of the forced jet trajectory at point 3.

4.61BG Water at 68°F flows from a 1.5-in-diameter hose through a 0.25 in nozzle, as illustrated in Figure ECP 4.61. The end of the hose at point 1 is 2.5 in below the tip of the nozzle at point 2, and the pressure at the end of the hose at point 1 is 75 *psi*. (a) Determine the ideal velocity of the forced jet as it leaves the tip of the nozzle at point 2. (b) Determine the ideal height of the forced jet trajectory at point 3.

4.62BG Water at 90°F flows from a 2.8-in-diameter hose through a 0.33 in nozzle, as illustrated in Figure ECP 4.62. The end of the hose at point 2 is 3.4 in below the tip of the nozzle at point 3, and the pressure at the top of the tank at point 1 is 100 *psi*. (a) Determine the ideal velocity of the forced jet as it leaves the tip of the nozzle at point 3. (b) Determine the pressure and the velocity at point 2. (c) Determine the ideal height of the forced jet trajectory at point 4.

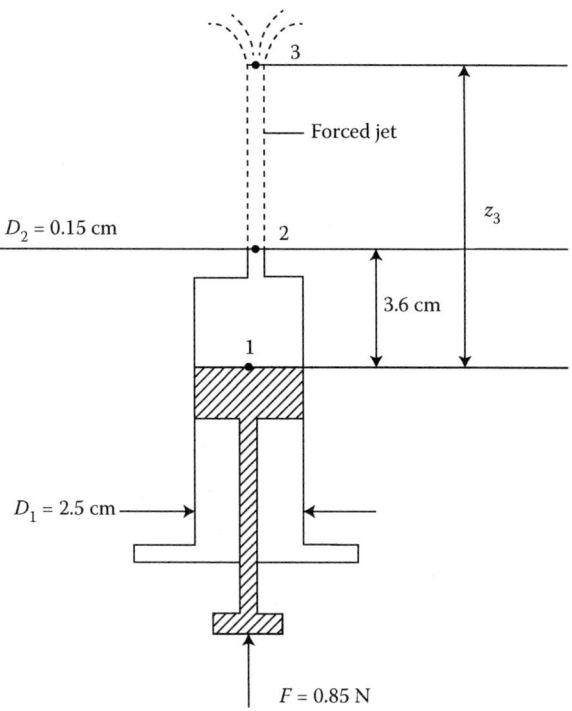

FIGURE ECP 4.60

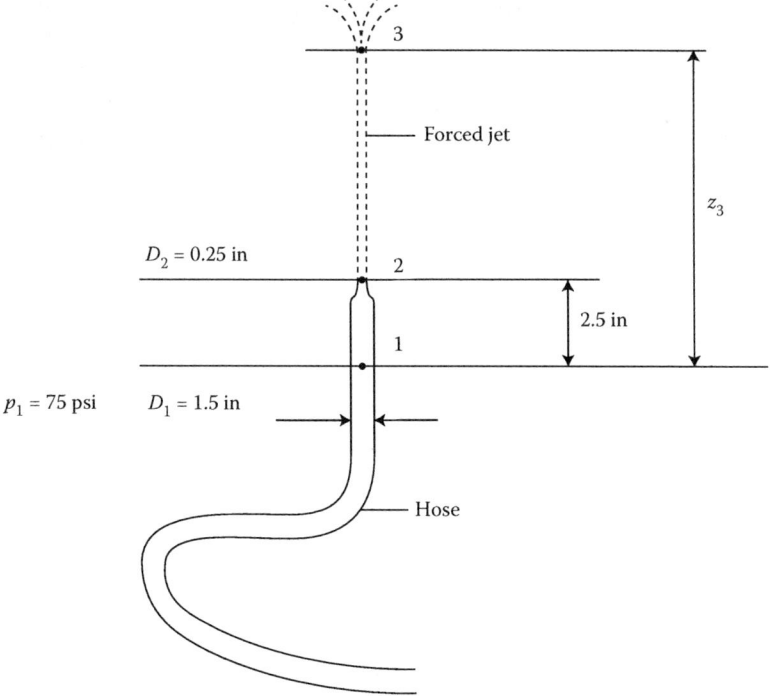

FIGURE ECP 4.61

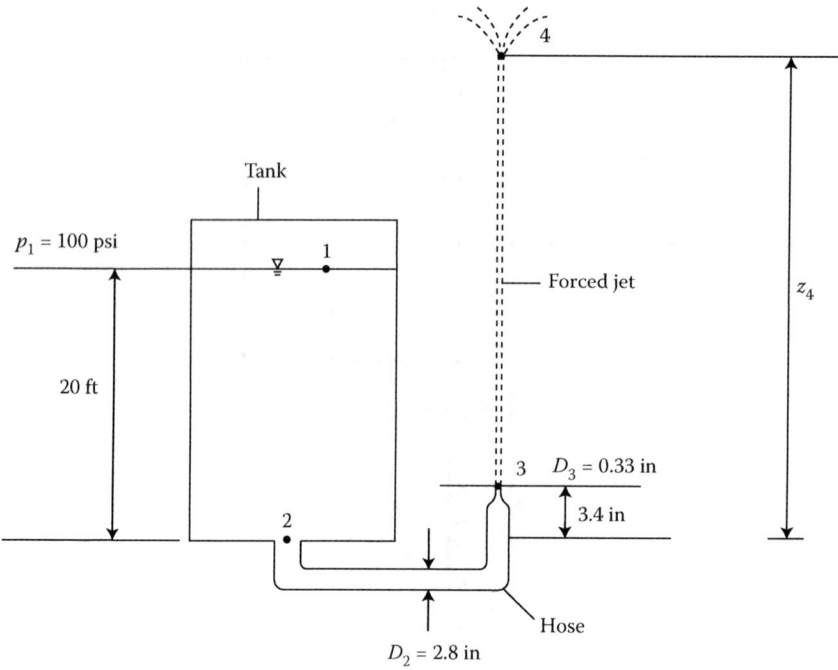

FIGURE ECP 4.62

4.63SI Water at 25°C flows through a 4-*cm*-diameter sprinkler hose at a velocity of 0.7 *m/sec*, as illustrated in Figure ECP 4.63. Water under a pressure of $2 \times 10^5 \, N/m^2$ shoots up from the holes in the sprinkler hose. (a) Determine the ideal velocity of the forced jets as they leave the holes of the sprinkler hose. (b) Determine the maximum number of holes in the sprinkler hose in order to maintain the steady flow assumption. (c) Determine the ideal height of the forced jet trajectories.

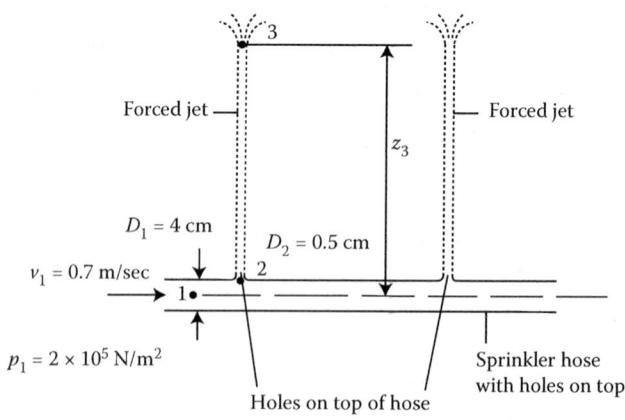

FIGURE ECP 4.63

4.64BG Water at 80°F flows from a hose through a nozzle, as illustrated in Figure ECP 4.64. The end of the hose at point 2 is below the tip of the nozzle at point 3, and the ideal height of the forced jet trajectory at point 4 is 550 ft. Determine the minimum required pressure at the top of the tank at point 1.

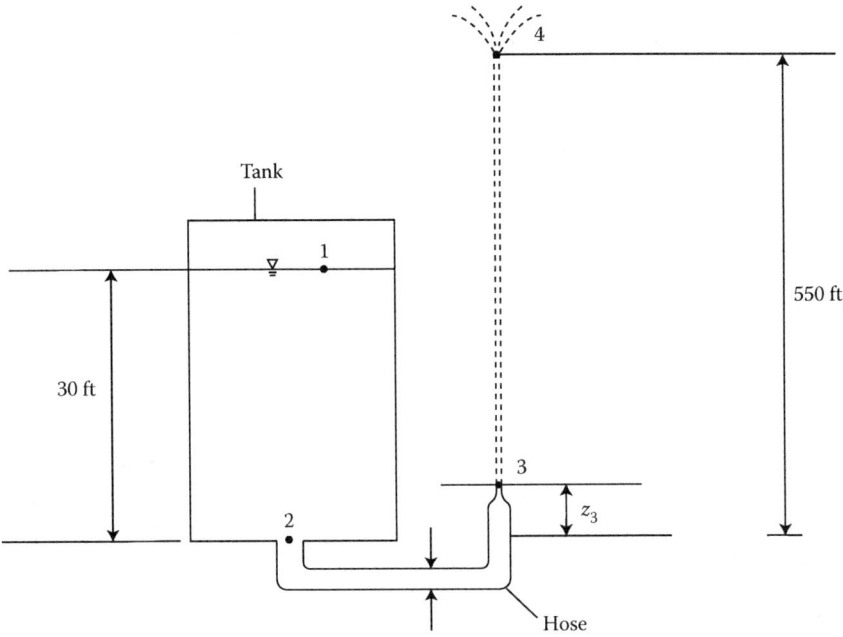

FIGURE ECP 4.64

4.65SI Water at 20°C flows from a tank through a 6-*cm*-diameter horizontal pipe with a 2-*cm* nozzle, as illustrated in Figure ECP 4.65. The elevation of the water in the tank at point 1 is 10 *m* above the tip of the nozzle at point 3. (a) Determine the ideal velocity of the forced jet as it leaves the tip of the nozzle at point 3. (b) Determine the pressure in the pipe at point 2.

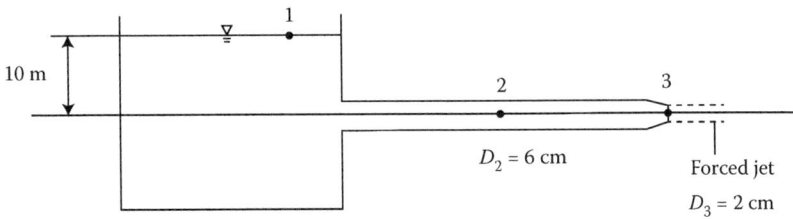

FIGURE ECP 4.65

4.66SI Water at 20°C flows from a pressurized tank through an 8-*cm*-diameter pipe in order to irrigate a field located at an elevation 24 *m* above the water level in the tank, as illustrated in Figure ECP 4.66. The pressure at the top of the water in the tank is $4 \times 10^5 \, N/m^2$. Determine the ideal velocity and the ideal discharge of the forced jet as it irrigates the field.

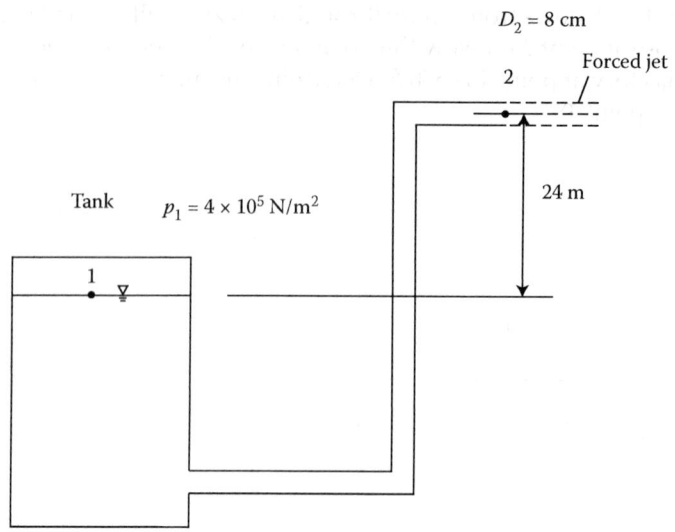

FIGURE ECP 4.66

4.67SI Water at 20°C flows from a pressurized tank through a pipe in order to create fountain, as illustrated in Figure ECP 4.67. The pressure at the top of the water in the tank is $6 \times 10^5 \, N/m^2$. (a) Determine the ideal velocity of the forced jet as it leaves the pipe at point 3. (b) Determine the ideal height of the forced jet trajectory at point 4. (c) Determine the pressure in the pipe at point 2.

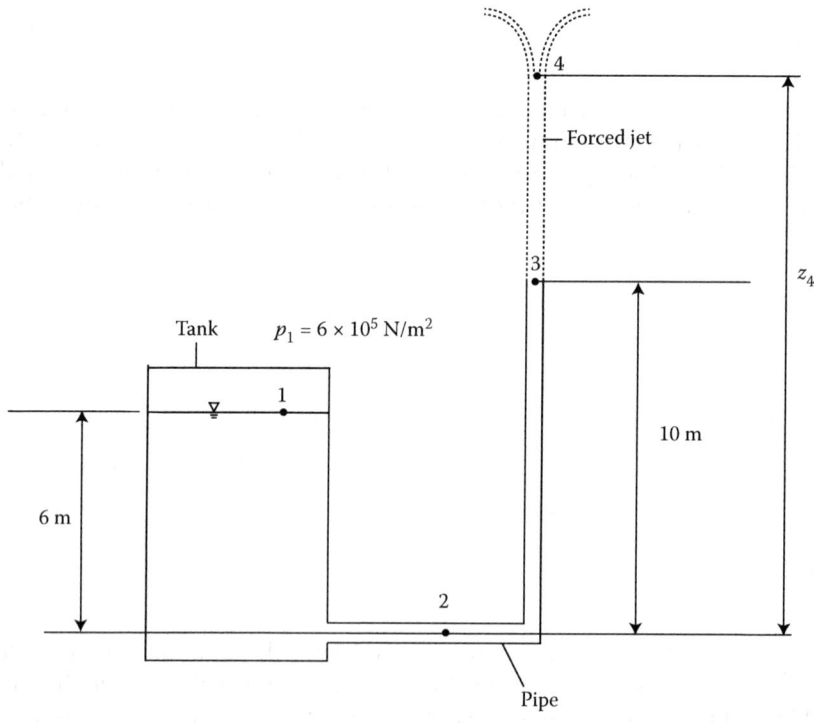

FIGURE ECP 4.67

4.68BG   Water at 68°F flows from a tank through a 3-in-diameter vertical pipe with a 0.9-in nozzle, as illustrated in Figure ECP 4.68. The elevation of the water in the tank at point 1 is 80 ft above the tip of the nozzle at point 5. The elevation of points 2, 3, and 4 are 40, 20, and 1 ft above the tip of the nozzle at point 5, respectively. (a) Determine the ideal velocity of the forced jet as it leaves the tip of the nozzle at point 5. (b) Determine the pressure in the pipe at points 2, 3, and 4.

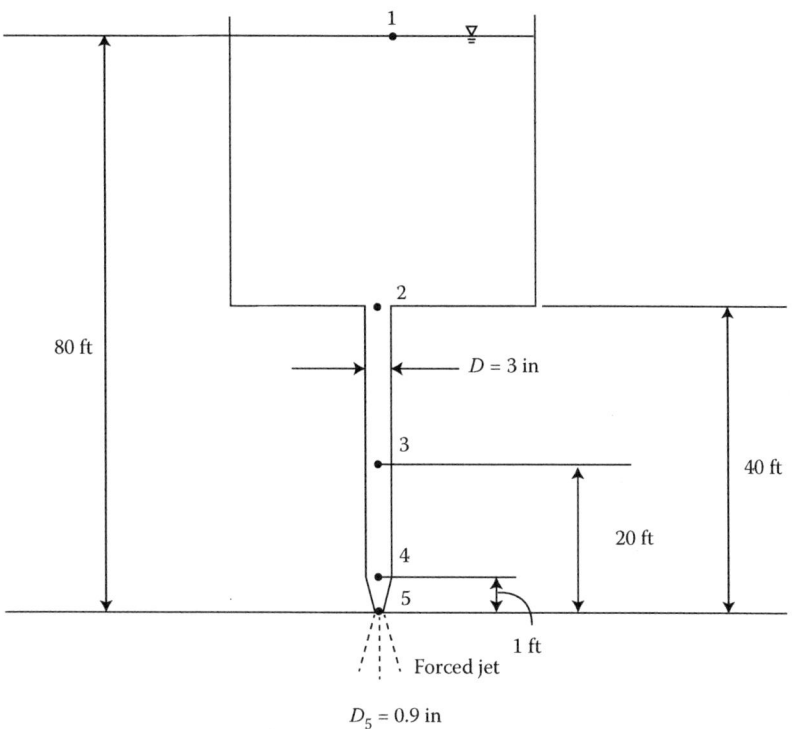

FIGURE ECP 4.68

4.69SI   Kerosene at 20°C is drained out of a pressurized tank through a 5-*cm*-diameter opening on the side of the tank as illustrated in Figure ECP 4.69. The initial elevation of the kerosene in the tank at point 1 is 17 *m* above the centerline of the opening at point 2. The pressure at the top of the kerosene is $4 \times 10^5 \, N/m^2$. (a) Determine the ideal velocity of the forced jet as it leaves the opening at point 2 when the tank is full. (b) Determine the ideal velocity of the forced jet as it leaves the opening at point 2 when the tank is half full.

4.70SI   Kerosene at 20°C is drained out of an open tank through a 5-*cm*-diameter opening on the side of the tank as illustrated in Figure ECP 4.70. The initial elevation of the kerosene in the tank at point 1 is 17 *m* above the centerline of the opening at point 2. (a) Determine the ideal velocity of the free jet as it leaves the opening at point 2 when the tank is full. (b) Determine the ideal velocity of the free jet as it leaves the opening at point 2 when the tank is half full. (c) Compare and discuss the results of this problem with the results of Problem 4.69SI.

$p_1 = 4 \times 10^5 \text{ N/m}^2$

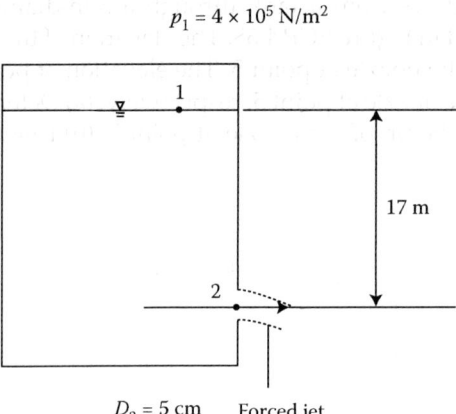

$D_2 = 5$ cm   Forced jet

FIGURE ECP 4.69

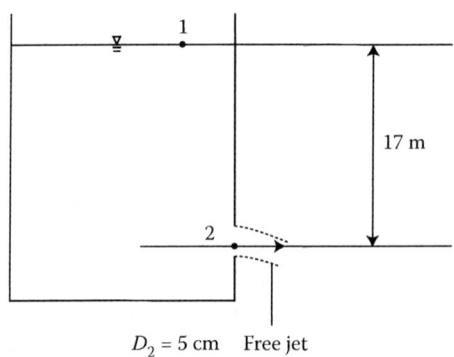

$D_2 = 5$ cm   Free jet

FIGURE ECP 4.70

4.71SI   Water at 20°C flows from an open tank through a 4.5-*cm* opening on the side of the tank as illustrated in Figure ECP 4.71. The elevation of the water in the tank at point 1 is 17 *m* above the centerline of the opening at point 2. (a) Determine the ideal velocity of the free jet as it leaves the opening at point 2. (b) Determine the ideal discharge from the open tank.

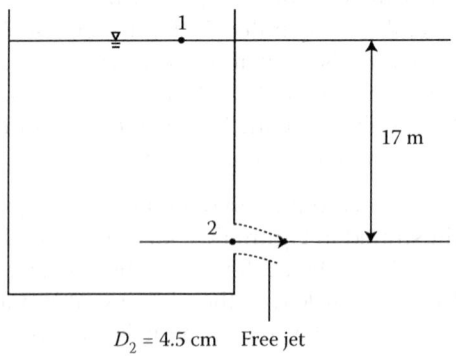

$D_2 = 4.5$ cm   Free jet

FIGURE ECP 4.71

4.72SI Laundry detergent ($\rho = 1100\,kg/m^3$) flows from an enclosed bottle through a 2.5-*cm* opening on the side of the tank as illustrated in Figure ECP 4.72. The elevation of the detergent in the bottle at point 1 is 0.3 *m* above the centerline of the opening at point 2. (a) Determine the ideal velocity of the free jet as it leaves the opening at point 2. (b) Determine the ideal discharge from the enclosed bottle. (c) Explain why the velocity of flow of the detergent at the opening at point 2 decreases with time. (d) Explain why loosening the cap of the enclosed bottle helps to increase the flow of the detergent.

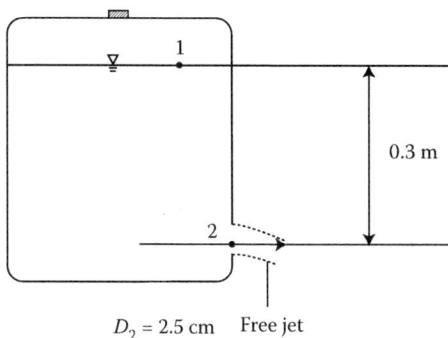

FIGURE ECP 4.72

4.73BG Water at 90°F is drained out of a 12-*ft*-diameter hot tub that is 4 *ft* high through a 2-*in*-diameter opening as illustrated in Figure ECP 4.73. The initial elevation of the water in the hot tub at point 1 is 3.9 *ft* above the centerline of the opening at point 2. (a) Determine the ideal velocity of the free jet as it leaves the opening at point 2 when the hot tub is full. (b) Determine the ideal velocity of the free jet as it leaves the opening at point 2 when the hot tub is half full. (c) Determine the ideal maximum discharge from the hot tub. (d) Determine the time it takes to empty the hot tub. (e) Determine the effect of the diameter of the hot tub on the time it takes to empty the hot tub. Assume hot tub diameters ranging from 5 *ft*, 6 *ft*, 7 *ft*, ...13 *ft*. (f) Determine the effect of the diameter of the opening on the time it takes to empty the hot tub. Assume opening diameters ranging from 1 *in*, 2 *in*, 3 *in*, ...5 *in*.

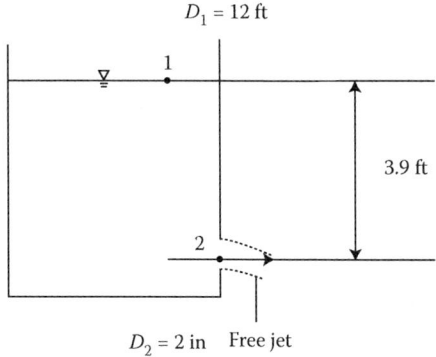

FIGURE ECP 4.73

4.74BG　Water flows from an open tank through a 1.5-*in* opening on the side of the tank as illustrated in Figure ECP 4.74. The elevation of the water in the tank at point 1 is 18 *ft* above the centerline of the opening at point 2. The elevation of the centerline of the opening at point 2 is 15 *ft* above the ground. (a) Determine the ideal velocity of the free jet as it leaves the opening at point 2. (b) Determine the ideal horizontal distance of the free jet trajectory as it hits the ground.

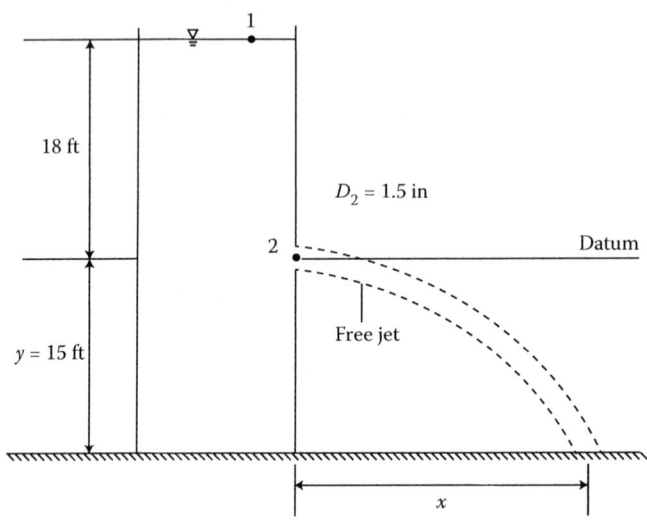

FIGURE ECP 4.74

4.75BG　Water flows from an open tank through an opening on the side of the tank as illustrated in Figure ECP 4.75. The elevation of the water in the tank at point 1 is *h* ft above the centerline of the opening at point 2. (a) Determine the ideal velocity of the free jet as it leaves the opening at point 2, and determine the corresponding height *h* for the following sets of (x, y) trajectories: $(2\,ft, 3\,ft)$, $(4\,ft, 5\,ft)$, and $(5\,ft, 6\,ft)$.

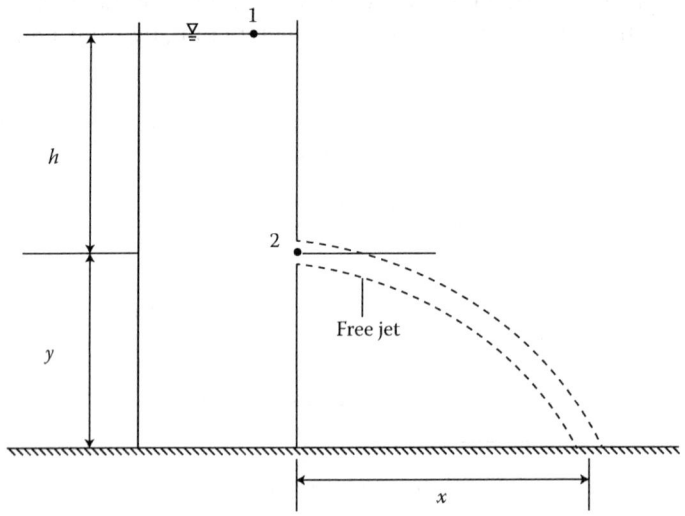

FIGURE ECP 4.75

4.76SI Water at 20°C flows from an open tank through a 1.45-*cm* nozzle at the bottom of the tank as illustrated in Figure ECP 4.76. The elevation of the water in the tank at point 1 is 15 *m* above the tip of the nozzle at point 2. The distance the water falls below the nozzle at point 3 is 0.95 *m* below the tip of the nozzle. (a) Determine the ideal velocity of the free jet as it leaves the nozzle at point 2. (b) Determine the ideal discharge from the open tank. (c) Determine the ideal velocity of the free jet just below the nozzle at point 3.

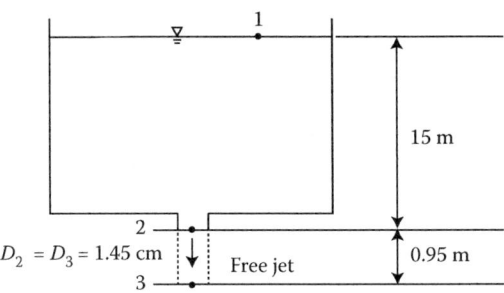

FIGURE ECP 4.76

4.77BG Water at 68°F flows from a 9-*ft*^{2} square open tank through a 1.3-*in* opening at the bottom of the tank as illustrated in Figure ECP 4.77. The elevation of the water in the tank at point 1 is 25 *ft* above the opening at point 2. (a) Determine the ideal velocity of the free jet as it leaves the opening at point 2. (b) Determine the ideal discharge from the open tank. (c) Derive the expression for the time it takes for the tank to empty. (d) Determine the time it takes for the tank to empty one third of the total water.

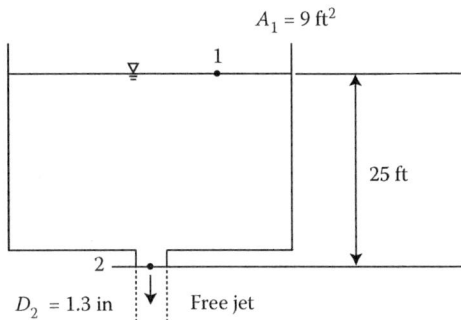

FIGURE ECP 4.77

4.78BG Water at 68°F from a faucet fills an empty unplugged tank with a diameter of 10 *ft* at a discharge of 1.8 *ft*^{3}/*sec* as illustrated in Figure ECP 4.78. The opening at the bottom of the tank has a diameter of 5 *in*. (a) Determine the ideal discharge from the open tank when the water reaches its final height in the tank at point 1. (b) Determine ideal velocity of the free jet as it leaves the opening at point 2, when the water reaches its final height in the tank at point 1. (c) Determine the final height of the water at point 1 in the tank. (d) Determine the relationship between the height, *h* of the water in the tank and the time, *t* as the tank is being filled with water.

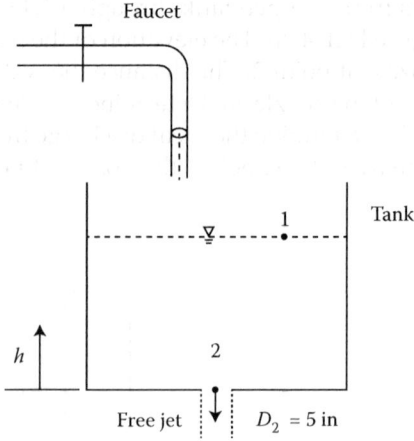

FIGURE ECP 4.78

4.79BG Water at 68°F fills a plugged tank with a diameter of 16 *ft* to a height of 78 *ft* as illustrated in Figure ECP 4.79. The opening at the bottom of the tank has a diameter of 7 *in*. Determine the ideal velocity of the free jet when the tank is completely full.

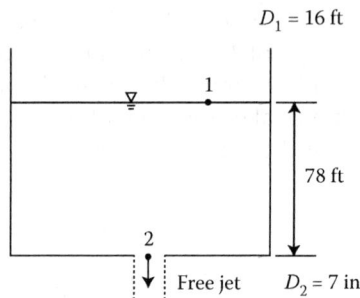

FIGURE ECP 4.79

4.80SI Water at 20°C is drained out of a pressurized 3-*m*-diameter tank through a 6-*cm*-diameter opening at the bottom of the tank, as illustrated in Figure ECP 4.80. The initial elevation of the water in the tank at point 1 is 30 *m* above the opening at point 2. The pressure at the top of the water is 3×10^5 N/m^2. (a) Determine the ideal velocity of the forced jet as it

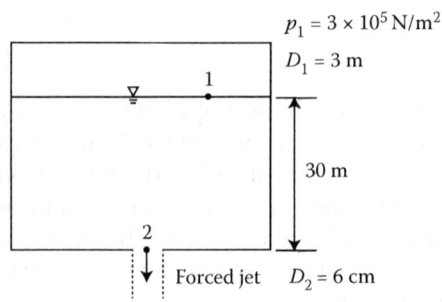

FIGURE ECP 4.80

leaves the opening at point 2. (b) Determine the ideal discharge from the open tank. (c) Determine how long it takes for the tank level to reach a level of 25 *m* above the opening at point 2. (d) Determine the level of the water after 600 seconds.

4.81BG Water at 60°F from a faucet fills a unplugged tank with a diameter of 33 *ft* as illustrated in Figure ECP 4.81. The elevation of the water in the tank at point 1 is 67 *ft* above the opening at point 2. The opening on the side of the tank is 3 *in*. (a) Determine the ideal velocity of the free jet as it leaves the opening at point 2. (b) Determine the ideal discharge from the open tank. (c) Determine the required discharge from the faucet in order to maintain the elevation of the water in the tank.

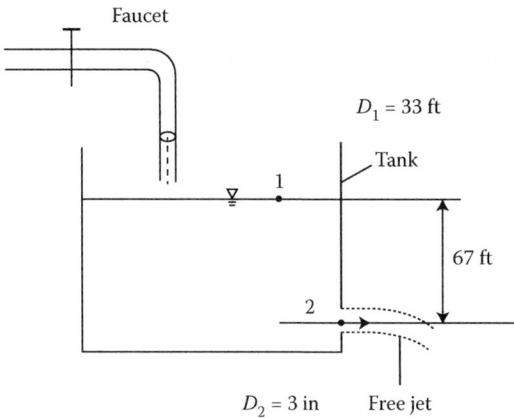

FIGURE ECP 4.81

4.82BG Water at 60°F is siphoned from an open tank through a 0.78-in siphon tube to a second tank at a lower elevation as illustrated in Figure ECP 4.82. The elevation of the water in the tank at point 1 is 25 *ft* above the bottom of the second tank. The elevation of the free end of the siphon tube at point 3 is 5 *ft* above the bottom of the second tank. (a) Determine the ideal velocity of the flow as it leaves the siphon tube at point 3. (b) Determine the ideal maximum allowable height of point 2 without causing cavitation.

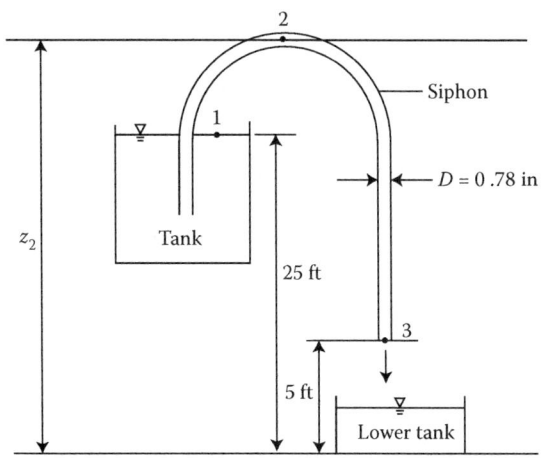

FIGURE ECP 4.82

4.83BG Gasoline at 68°F is siphoned from a full car tank through a 0.5-in siphon tube to a 2.5-ft^3 gas can at a lower elevation as illustrated in Figure ECP 4.83. The elevation of the gasoline in the car tank at point 1 is 2.8 ft above the bottom of the gas can. The height of the gasoline in the full car tank is 1 ft. The elevation of the free end of the siphon tube at point 3 is 1.3 ft above the bottom of the gas can. (a) Determine the ideal velocity of the flow as it leaves the siphon tube at point 3 when the car tank is full. (b) Determine the ideal maximum allowable height of point 2 without causing cavitation. (c) Determine how long it will take to fill the gas can when car tank is full. (d) Determine the ideal velocity of the flow as it leaves the siphon tube at point 3 when the car tank is half full. (e) Determine how long it will take to fill the gas can when car tank is half full.

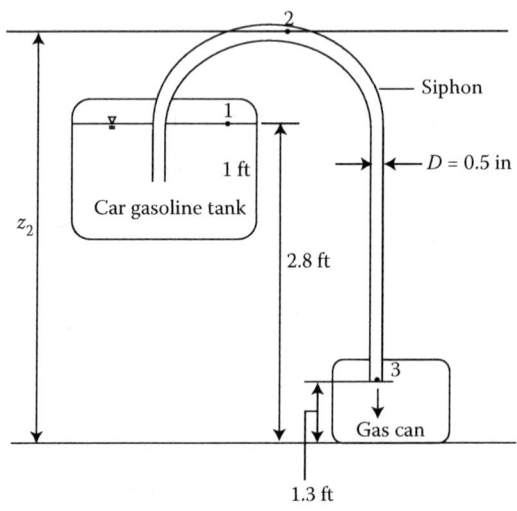

FIGURE ECP 4.83

4.84BG Kerosene at 68°F is siphoned from tank through a 1.3-in siphon tube to a second tank at a lower elevation as illustrated in Figure ECP 4.84. The initial level of kerosene in the tank is 40 ft above the bottom of the second tank. The elevation of point 2 is 44 ft above the bottom of the second tank. (a) Determine the relationship between ideal velocity of the flow

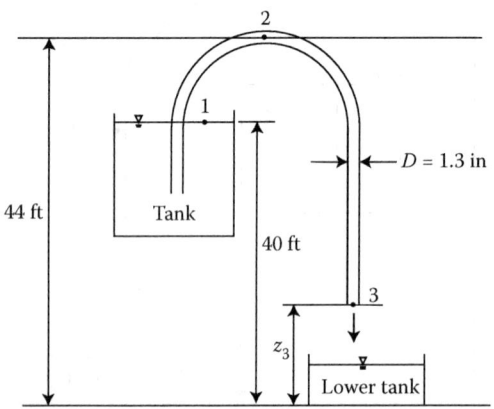

FIGURE ECP 4.84

as it leaves the siphon at point 3 and the difference in elevation between points 1 and 3. (b) Determine the relationship between the occurrence of cavitation and the difference in elevation between points 2 and 3. (c) Determine the maximum ideal velocity of the flow as it leaves the siphon at point 3 and thus the maximum ideal length of the siphon tube between points 2 and 3 without causing cavitation.

Application of the Bernoulli Equation for Ideal External Flow

4.85SI A plane flies at an altitude of 4000 m above sea level in a standard atmosphere. A pitot-static tube is mounted on the wing of a plane in order to measure the speed of the plane, as illustrated in Figure ECP 4.85. The point inside any of the small holes in the outer tube of the pitot-static tube may be labeled as point 1, where the static pressure, p_1 is the standard atmospheric pressure at the given altitude above sea level. The inner tube is open to the flow at the nose of the pitot-static tube, where the flow of air is stagnated, and the point inside the inner tube is labeled as point 2. The inner and outer tubes are each connected to a pressure port of the differential pressure meter, which registers the differential gage pressure (rise), $\Delta p = p_2 - p_1 = 6.3 \times 10^3 \, N/m^2$ between points 1 and 2. (a) Determine the ideal speed of the plane, v_1. (b) Determine the stagnation pressure on the nose of the plane.

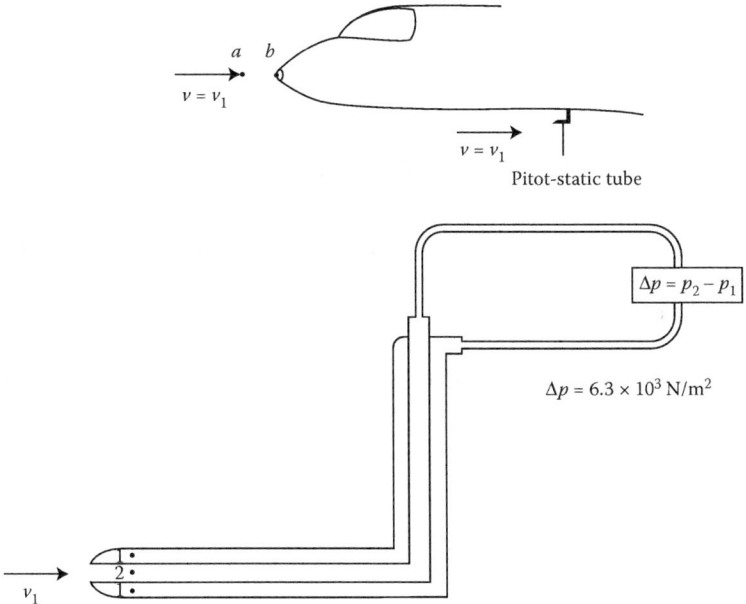

FIGURE ECP 4.85

4.86BG A shark at 40 ft below the surface swims at 2 ft/sec in seawater as illustrated in Figure ECP 4.86. Determine the stagnation pressure on the nose of the shark.

4.87BG A biker moves at 20 mph as illustrated in Figure ECP 4.87. Assume the biker is at sea level and the temperature of the air is 70°F. Determine the stagnation pressure on the nose of the biker.

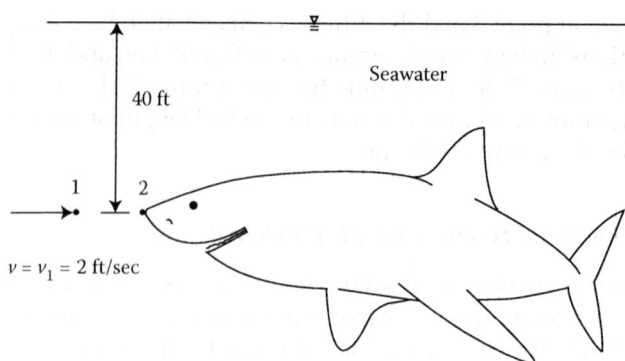

FIGURE ECP 4.86

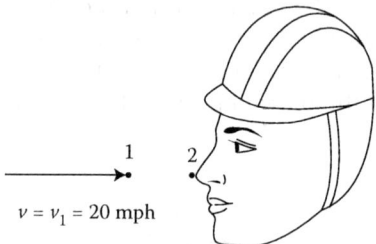

FIGURE ECP 4.87

Application of the Energy Equation for a Hydraulic Jump

4.88BG   Water flowing in a rectangular channel of width 9.842 *ft* at a supercritical velocity of 15.823 *ft/sec* and a supercritical depth of 3.061 *ft* encounters a subcritical flow at a velocity of 8.746 *ft/sec* and a depth of 5.538 *ft*, as illustrated in Figure ECP 4.88. As the flow transitions from supercritical to subcritical flow, a hydraulic jump occurs, where there is a major head loss due the intense agitation and turbulence and where the depth of flow is at a critical depth of 4.176 *ft*. (a) Determine the head loss due to the flow turbulence at the jump. (b) Determine the flowrate in the channel. (c) Draw the energy grade line and the hydraulic grade line.

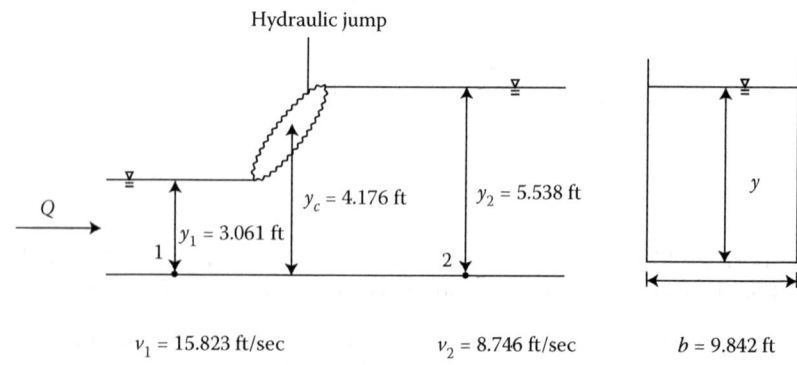

FIGURE ECP 4.88

The Energy Equation Based on the First Law of Thermodynamics

4.89C Define each term and explain the units of each term in the energy equation given by Equation 4.184 as follows:

$$\underbrace{\left(\frac{p_1}{\gamma} + \frac{v_1^2}{2g} + z_1\right) + h_{pump}}_{mech\ energy\ in} = \underbrace{\left(\frac{p_2}{\gamma} + \frac{v_2^2}{2g} + z_2\right) + h_{turbine}}_{mech\ energy\ out} + \underbrace{h_{f,total}}_{mech\ energy\ loss}$$

4.90C Give the expressions for the h_{pump} term and the $h_{turbine}$ term in the energy equation given by Equation 4.184.

Application of the Energy Equation for Real Internal Flow with a Pump

4.91C Explain the difference between the "analysis" problem vs. the "design" problem of a pump–motor system.

4.92SI Water at 20°C is pumped from reservoir 1 to reservoir 2 through a 1.2-*m*-diameter concrete pipe, 1600 *m* in length, fitted with three regular 90° flanged elbows in order to change the direction of the flow, as illustrated in Figure EP 4.20. The pump is located 9 *m* above the elevation of reservoir 1, at a distance of 150 *m* along the length of the pipe. The pipe entrance from reservoir 1 is reentrant, and the pipe exit to reservoir 2 is reentrant. Assume a Darcy–Weisbach friction factor, f of 0.17; a minor head loss coefficient due to an elbow, k of 0.3; a minor head loss coefficient due to a reentrant pipe entrance, k of 1.0; and a minor head loss coefficient due to a reentrant pipe exit, k of 1.0. The hydraulic power delivered to the flow by the pump is 600 *kW*. The pressure at the suction side of the pump is measured using a manometer (in order to measure a negative gage pressure in the fluid flow), while the pressure at the discharge side of the pump is measured using a piezometer. (a) Determine the maximum allowable velocity and discharge of the flow in the pipe; without causing cavitation at the suction side, s of the pump. (b) Demonstrate that because the pump creates a suction in the pipe line, the pressure gradually decreases from atmospheric pressure at point 1 to the vapor pressure, p_v of the liquid at point s. (c) For the maximum allowable discharge, determine the corresponding maximum allowable head delivered to the flow by the pump. (d) Determine how the actual head delivered to the pump is measured. (e) If the maximum allowable head is delivered by the pump, determine the magnitude of the pressure head registered at the suction and discharge sides of the pump. (f) Determine the maximum allowable elevation to which the flow may be pumped to point 2.

4.93SI The pump–motor system for Problem 4.92 is illustrated in Figure EP 4.21, where hydraulic power delivered to the flow by the pump is 600 *kW*. The chemical power supplied to the motor is 800 *kW*, and the rotating shaft of the motor has an angular speed of 350 *rad/sec* and a shaft torque of 1900 *N-m*. (a) Determine the shaft power delivered by the motor to the pump. (b) Determine the motor efficiency. (c) Determine the pump efficiency. (d) Determine the pump–motor system efficiency.

4.94SI Kerosene at room temperature is pumped at a flowrate of 0.98 m^3/sec from a tank as illustrated in Figure ECP 4.94. The pipe diameter at the suction side of the pump is 1.4 *m*, and the pipe diameter at the discharge side of the pump is 1.9 *m*. The pressure rises from 2×10^5 N/m^2 at the suction side to 6×10^5 N/m^2 at the discharge side due to the pump. Assume a total head loss of 2 *m*. The chemical power supplied to the motor is 538 *kW*, and the motor efficiency is 95%. (a) Determine the head delivered by the pump. (b) Determine the pump efficiency. (c) Determine the pump–motor system efficiency.

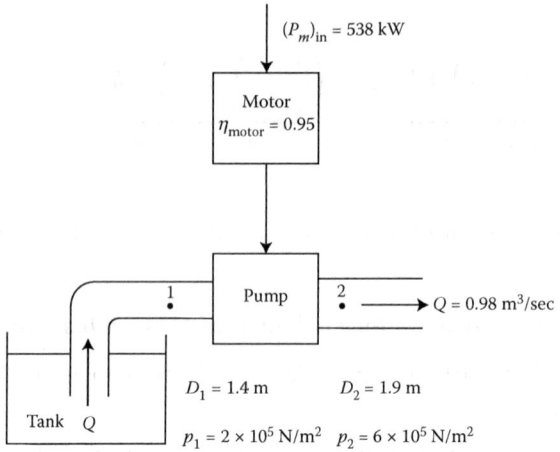

FIGURE ECP 4.94

4.95BG Seawater at 68°F is pumped at a flowrate of $6\,ft^3/sec$ from the sea to a tank at an elevation of $189\,ft$ as illustrated in Figure ECP 4.95. Assume a total head loss of $1.2\,ft$. The shaft power delivered by the motor to the pump is $170\,hp$. (a) Determine the head delivered by the pump. (b) Determine the pump efficiency. (c) Determine the effect of the magnitude of the total head loss on the magnitude of the pump efficiency for a head loss ranging from $0\,ft$ to $11\,ft$.

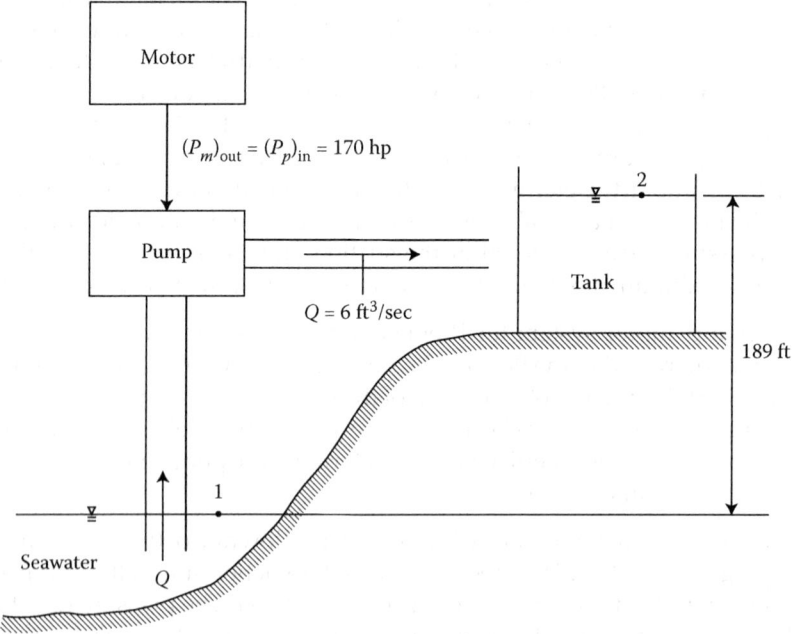

FIGURE ECP 4.95

4.96BG Water at 68°F is pumped from a lake to a tank at an elevation of 69 ft as illustrated in Figure ECP 4.96. Assume a total head loss of 1.66 ft. The shaft power delivered by the motor to the pump is 33 hp, and the pump efficiency is 86%. (a) Determine the head delivered by the pump. (b) Determine the hydraulic power supplied by the pump. (c) Determine the maximum allowable discharge.

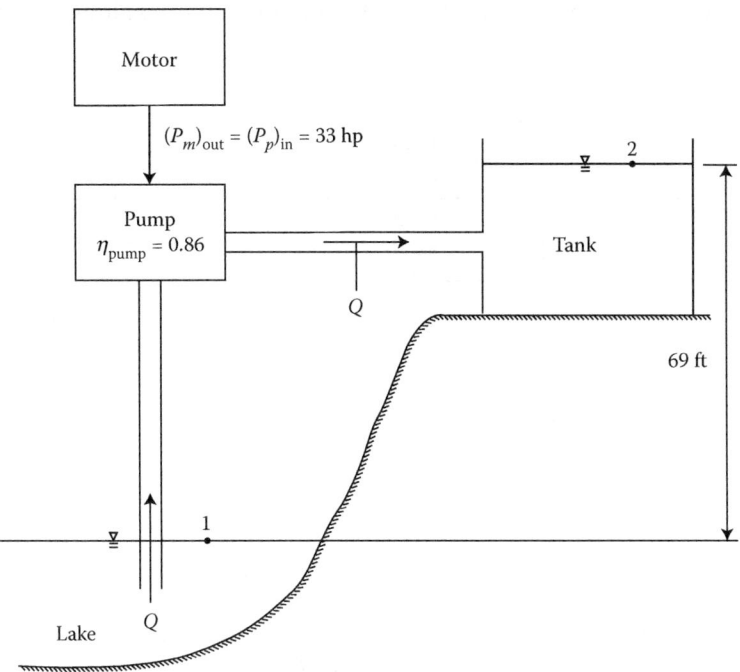

FIGURE ECP 4.96

4.97BG A pump is required to dewater a site as illustrated in Figure EP 4.22. It is required that the water (at 70°F) be pumped from reservoir 1 to reservoir 2 to an elevation of 450 ft at a discharge of 1.6 ft³/sec. Assume that the total head loss due to both major (pipe friction) and minor losses (pipe components) is 2 ft between point 1 and the suction side of the pump at point s and 3.9 ft between the discharge side of the pump at point d and point 2. (a) Determine the maximum height, z_s to which the fluid may be pumped; otherwise, there will be no flow in the suction line at the suction side of the pump. (b) Determine the maximum allowable velocity of the flow in the pipe without causing cavitation. (c) Determine the minimum required design pipe size diameter in order to accommodate the maximum allowable velocity and the required flowrate. (d) Determine the required head to be delivered by the pump. (e) Determine the required hydraulic power output from the pump. (f) Assuming a typical pump efficiency of 90%, determine the required shaft power input to the pump by the motor. (g) Assuming a typical motor efficiency of 95%, determine the required chemical or electric power input to the motor. (h) Determine the resulting overall pump–motor system efficiency.

4.98BG (Refer to Problem 3.65) A mini fan is required to force (pump) air into a 4-*in* blow dryer duct, which is heated by electric coils as illustrated in Figure ECP 4.98. The heated air is forced out of a 2-*in* nozzle by the mini fan in the blow dryer. It is required that the air be forced through the blow dryer at a discharge of $0.087\,ft^3/sec$. Assume the density of air to be $0.002470\ \text{slug/ft}^3$ and that no head loss between point 1 and point 2. (a) Determine the required head to be delivered by the fan. (b) Determine the required hydraulic power output from the fan. (c) Determine the pressure difference (increase) between point 3 and point 4. (d) Assuming a typical fan efficiency of 35%, determine the required shaft power input to the fan by the motor. (e) Assuming a typical motor efficiency of 45%, determine the required electric power input to the motor. (f) Determine the resulting overall fan–motor system efficiency.

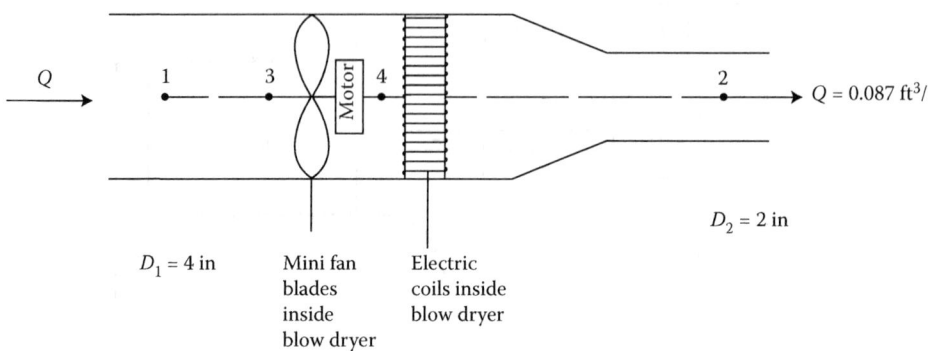

FIGURE ECP 4.98

4.99BG   Water at 68°F flows through a pump at a discharge of $4\,ft^3/sec$ as illustrated in Figure ECP 4.99. Assume that the required head to be delivered by the pump is $400\,ft$. (a) Determine the required hydraulic power output from the pump. (b) Assuming a typical pump efficiency of 85%, determine the required shaft power input to the pump by the motor. (c) Assuming a typical motor efficiency of 95%, determine the required chemical or electric power input to the motor. (d) Determine the resulting overall pump–motor system efficiency.

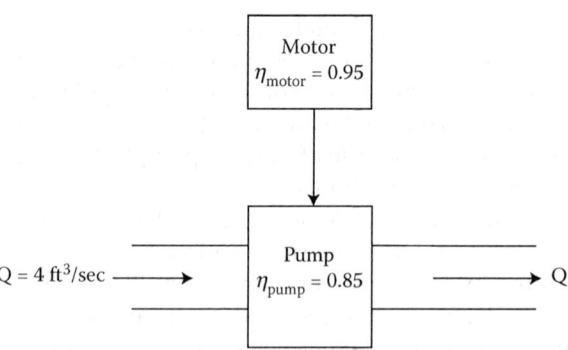

FIGURE ECP 4.99

4.100SI   Water at 20°C is pumped from a reservoir to a tank at an elevation of $30\,m$ at a discharge of $2\,m^3/sec$ through a 0.98-m-diameter pipe as illustrated in Figure ECP 4.100. Assume that the total head loss is $2\,m$, and the consumed chemical power is $900\,kW$.

(a) Determine the required hydraulic power output from the pump. (b) Determine the pressure difference between the suction side and the discharge side of the pump. (c) Assuming a typical pump efficiency of 83%, determine the required shaft power input to the pump by the motor. (d) Determine required the motor efficiency. (e) Determine the resulting overall pump–motor system efficiency.

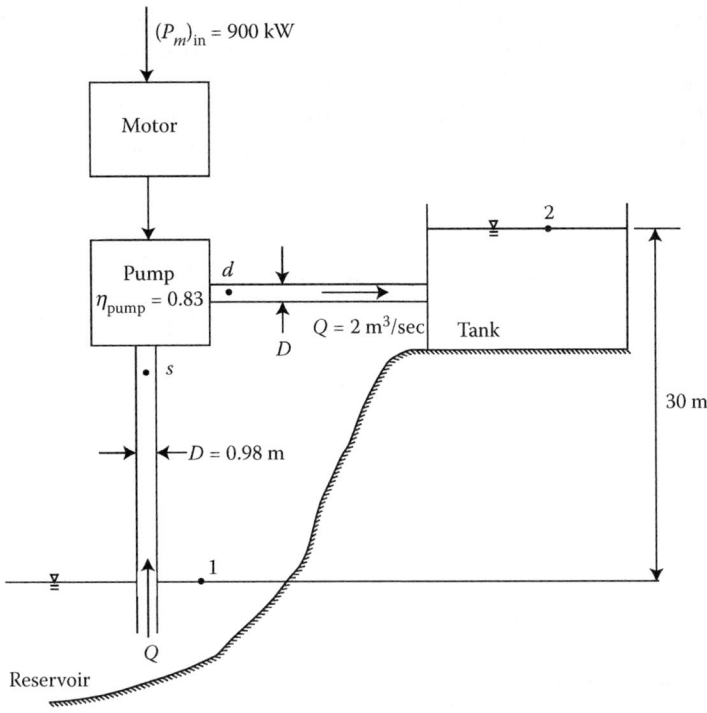

FIGURE ECP 4.100

4.101BG A kitchen that is $5000\, ft^3$ is ventilated by a fan every $5\, min$, as illustrated in Figure ECP 4.101. The diameter of the fan is $9\, in$. Assume the density of air to be 0.002470 $slug/ft^3$, and that there is no head loss between point 1 and point 2. (a) Determine the required head to be delivered by the fan. (b) Determine the required hydraulic power output

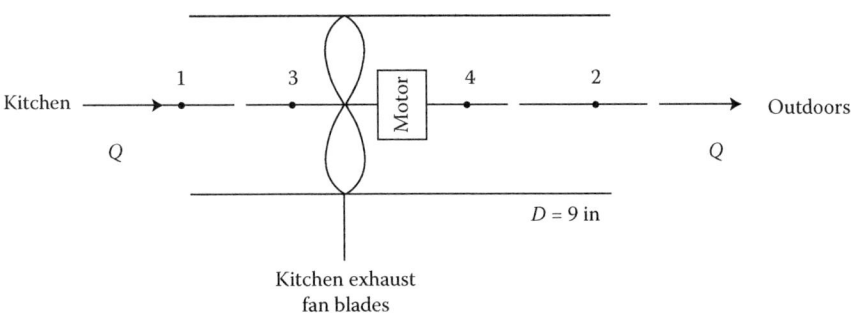

FIGURE ECP 4.101

from the fan. (c) Determine the pressure difference between point 3 and point 4. (d) Assuming a typical fan efficiency of 45%, determine the required shaft power input to the fan by the motor. (e) Assuming a typical motor efficiency of 55%, determine the required electric power input to the motor. (f) Determine the resulting overall fan–motor system efficiency.

4.102SI   Water at 20°C flows in a 0.6-*m*-diameter horizontal pipe at a flowrate of 0.66 m^3/ *sec*, as illustrated in Figure ECP 4.102. There is a valve between points 1 and 2, where the pressure at point 1 is $3.3 \times 10^5 \, N/m^2$ and the pressure at point is $2.3 \times 10^5 \, N/m^2$. (a) Determine the minor head loss due to the valve between points 1 and 2. (b) Determine the required head to be delivered by a pump between points 2 and 3 in order to overcome the pressure drop caused by the valve. (c) Determine the required hydraulic power output from the pump between points 2 and 3 in order to overcome the pressure drop caused by the valve.

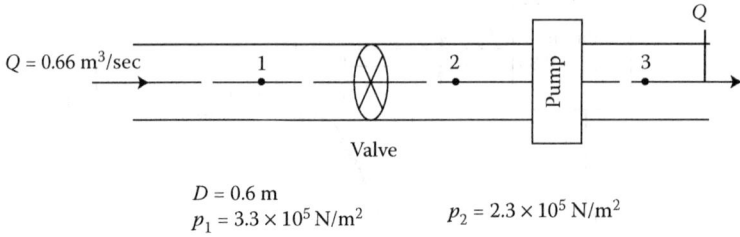

$$D = 0.6 \text{ m}$$
$$p_1 = 3.3 \times 10^5 \text{ N/m}^2 \qquad p_2 = 2.3 \times 10^5 \text{ N/m}^2$$

FIGURE ECP 4.102

4.103SI   A pump is used to dewater a site as illustrated in Figure ECP 4.103. Water (at 20°C) is pumped from reservoir 1 to reservoir 2 to an elevation of 29 *m* at a discharge of 0.77 m^3/*sec*. The hydraulic power output from the pump is 300 *kW*. (a) Determine the magnitude of the head delivered by the pump. (b) Determine the magnitude of the total head loss between points 1 and 2.

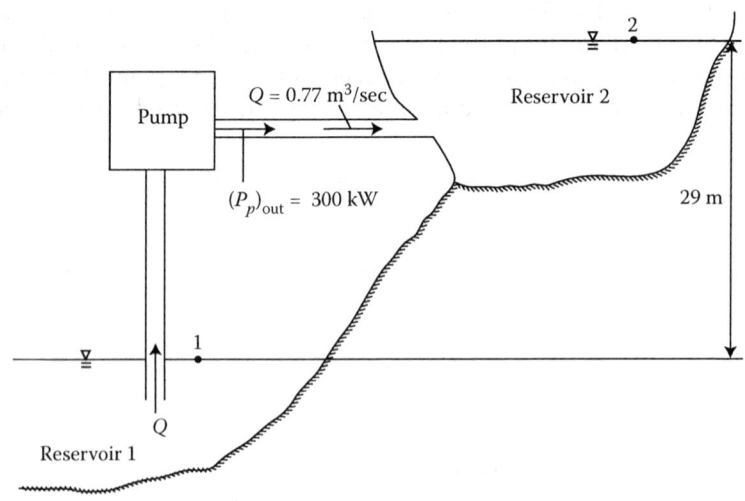

FIGURE ECP 4.103

4.104BG A pump is used to lift water from a lake to a water tower as illustrated in Figure ECP 4.104. Water (at 68°F) is pumped from the lake to the water tower to an elevation of 170 ft. Assume a total head loss of 2 ft. The shaft power input to the pump by the motor is 30 hp, and the pump efficiency is 79%. The pipe diameter at the suction side of the pump (point 3) is 5 in, and at the discharge side of the pump (point 4), it is 7 in. (a) Determine the required head to be delivered by the pump. (b) Determine the hydraulic power delivered by the pump. (c) Determine the maximum discharge in the pipe. (d) Determine the pressure difference (increase) across the pump between points 3 and 4.

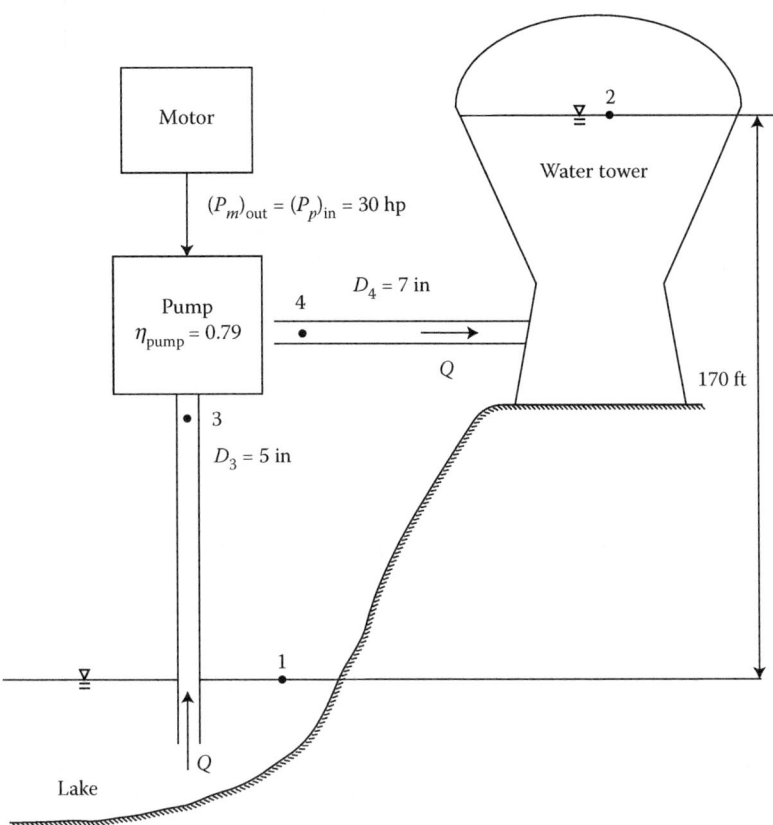

FIGURE ECP 4.104

4.105SI A pump is used to lift water from a tank to a reservoir as illustrated in Figure ECP 4.105. Water (at 20°C) is pumped from the tank to the reservoir to an elevation of 40 m at a discharge of 0.8 m^3/sec through a pipe with a constant diameter. The shaft power input to the pump by the motor is 600 kW, and the pump efficiency is 74%. (a) Determine the hydraulic power output from the pump. (b) Determine the magnitude of the head delivered by the pump. (c) Determine the magnitude of the total head loss between points 1 and 2. (d) Determine the hydraulic power output from the pump required to overcome the difference in elevation between points 1 and 2. (e) Determine the hydraulic power output from the pump required to overcome the head loss between points 1 and 2.

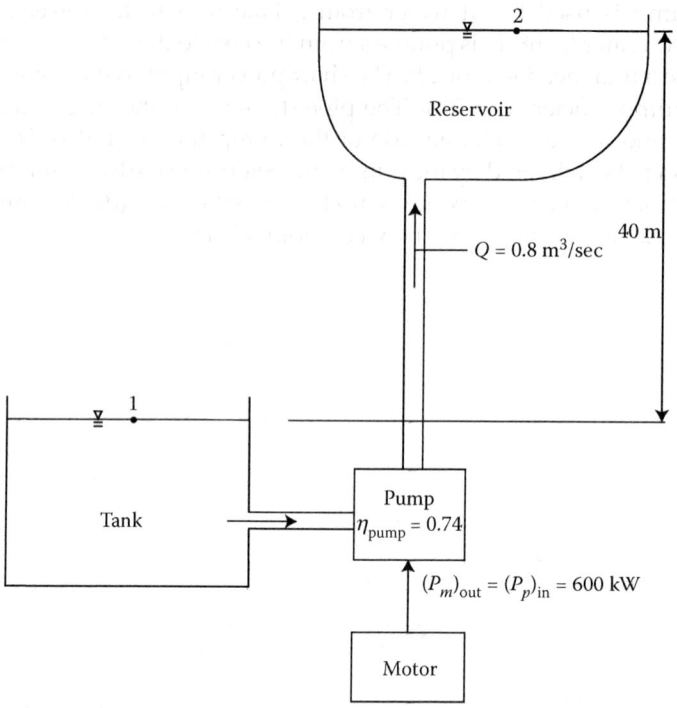

FIGURE ECP 4.105

Application of the Energy Equation for Real Internal Flow with a Turbine

4.106C   List the two basic types of turbines, and describe the fluid flow systems for which it is more efficient to use one over the over.

4.107C   Explain the difference between the "analysis" problem vs. the "design" problem of a turbine–generator system.

4.108SI   An impulse turbine extracts hydraulic power from water at 20°C flowing from reservoir 1 though a 0.85-*m* riveted steel pipe, 200 *m* in length, fitted with two regular 90° threaded elbows in order to change the direction of the flow, and finally with a 0.35-*m* nozzle in order to create a forced jet, as illustrated in Figure EP 4.23a. The turbine is located 55 *m* below the elevation of the headwater at point 1 and just slightly above the elevation of the tailwater at point 2. The pipe entrance from reservoir 1 is square-edged, and the nozzle has a conical angle, $\theta = 30°$. Assume a Darcy–Weisbach friction factor, f of 0.13; a minor head loss coefficient due to a squared-edged pipe entrance, $k = 0.5$; a minor head loss coefficient due to an elbow, $k = 1.5$; and a minor head loss coefficient due to the nozzle, $k = 0.02$. The pressure at point a in the pipe is measured using a piezometer. (a) Determine the pressure and velocity at point a in the pipe and the velocity at point b at the nozzle. (b) Determine the magnitude of the freely available discharge to be extracted by the turbine. (c) Determine the magnitude of the freely available head to be extracted by the turbine. (d) Determine the magnitude of the freely available hydraulic power to be extracted by the turbine. (e) Assuming a typical turbine efficiency of 92%, determine the freely generated shaft power output by the turbine. (f) Assuming a typical generator efficiency of 87%, determine the freely generated electric power output by the generator. (g) Determine the resulting overall turbine–generator system efficiency.

4.109BG   A reaction turbine extracts hydraulic power from water at 60°F flowing from reservoir 1 though a 3-*ft* concrete pipe 500 *ft* in length, as illustrated in Figure EP 4.24. The turbine is located 100 *ft* below the elevation of the headwater at point 1 and 6 *ft* above the elevation of the tailwater at point 2. The pipe entrance from reservoir 1 is square-edged, and the draft tube has a conical angle, $\theta = 9°$. Assume a Darcy-Weisbach friction factor, f of 0.18; a minor head loss coefficient due to a squared-edged pipe entrance, $k = 0.5$; and a minor head loss coefficient due to the conical diffuser, $k = 0.5$. The pressure at point a in the pipe is measured using a piezometer. (a) Determine the pressure and velocity at points a and b. (b) Determine the magnitude of the freely available head to be extracted by the turbine. (c) Determine the magnitude of the freely available discharge to be extracted by the turbine. (d) Determine the magnitude of the freely available hydraulic power to be extracted by the turbine. (e) Assuming a typical turbine efficiency of 88%, determine the freely generated shaft power output by the turbine. (f) Assuming a typical generator efficiency of 83%, determine the freely generated electric power output by the generator. (g) Determine the resulting overall turbine–generator system efficiency.

4.110BG   A wind turbine with 150-*ft* blades extracts hydraulic power from wind blowing at 25 *ft/sec*, as illustrated in Figure ECP 4.110. Assume the density of the air to be 0.00233 slug/ft^3. (a) Determine the magnitude of the freely available discharge to be extracted by the turbine. (b) Determine the magnitude of the freely available head to be extracted by the turbine. (c) Determine the magnitude of the freely available hydraulic power to be extracted by the turbine. (d) Assuming a typical turbine efficiency of 85%, determine the freely generated shaft power output by the turbine. (e) Assuming a typical generator efficiency of 80%, determine the freely generated electric power output by the generator. (f) Determine the resulting overall turbine–generator system efficiency. (g) Determine and plot the effect of the wind speed and the turbine blade diameter on the performance of the wind turbine (freely generated electric power output by the generator), for a range of wind speeds up to 100 *ft/sec* and a turbine blade diameter ranging from 50 *ft* to 300 *ft*.

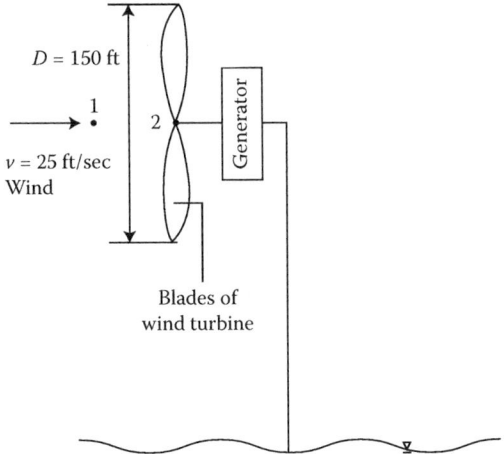

FIGURE ECP 4.110

4.111SI A reaction turbine extracts hydraulic power from water at 20°C flowing at a discharge of 4 m^3/sec from a reservoir at an elevation of 250 m above the turbine, as illustrated in Figure EP 4.24. Assume the turbine generates a shaft power of 8000 kW, the electric generated power by the generator is 7600 kW, and ignore any head loss. (a) Determine the magnitude of the freely available head to be extracted by the turbine. (b) Determine the magnitude of the freely available hydraulic power to be extracted by the turbine. (c) Determine the turbine efficiency. (d) Determine the generator efficiency. (e) Determine the resulting overall turbine–generator system efficiency.

4.112BG A reaction turbine extracts hydraulic power from water flowing at a discharge of 26 ft^3/sec in an open channel at an elevation of 66 ft above the pool level, as illustrated in Figure ECP 4.112. Ignore any head loss. (a) Determine the magnitude of the freely available head to be extracted by the turbine. (b) Determine the magnitude of the freely available hydraulic power to be extracted by the turbine.

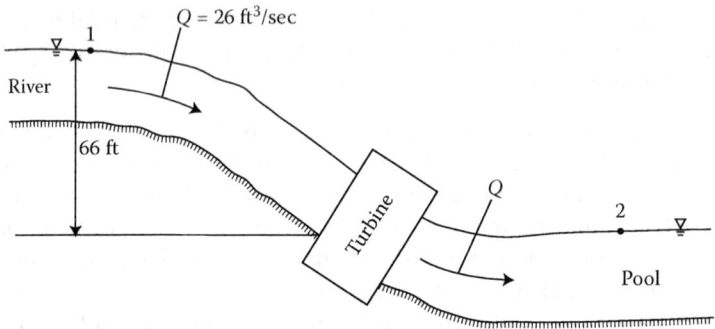

FIGURE ECP 4.112

4.113BG A reaction turbine extracts hydraulic power from seawater flowing at a discharge of 35 ft^3/sec flowing in a 2-ft-diameter pipe, as illustrated in Figure ECP 4.113. The seawater leaves the turbine through a 1.5-ft pipe. The pressure drop across the turbine is 150 psi. Ignore any head loss. (a) Determine the magnitude of the freely available head to be

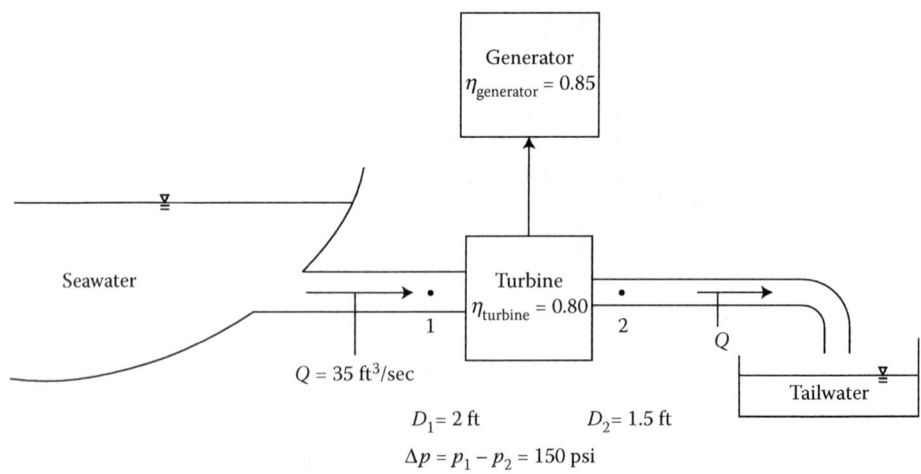

FIGURE ECP 4.113

extracted by the turbine. (b) Determine the magnitude of the freely available hydraulic power to be extracted by the turbine. (c) Assuming a typical turbine efficiency of 80%, determine the freely generated shaft power output by the turbine. (d) Assuming a typical generator efficiency of 85%, determine the freely generated electric power output by the generator. (e) Determine the resulting overall turbine–generator system efficiency.

4.114SI The required hydraulic power to be extracted by an impulse turbine is determined to be 750 kW. The design magnitude of the freely available discharge is given to be 3 m^3/sec. The hydraulic power is to be extracted from water at 20°C flowing from reservoir 1 though a cast iron pipe, 250 m in length, fitted with two regular 90° threaded elbow, and finally with a nozzle in order to create a forced jet, as illustrated in Figure EP 4.23a. The pipe entrance from reservoir 1 is square-edged, and the nozzle has a conical angle, $\theta = 30°$. Assume a Darcy–Weisbach friction factor, f of 0.15; a minor head loss coefficient due to a squared-edged pipe entrance, $k = 0.5$; a minor head loss coefficient due to an elbow, $k = 1.5$; and a minor head loss coefficient due to the nozzle, $k = 0.02$. The pressure at point a in the pipe is measured using a piezometer. (a) Determine the design magnitude of the freely available head to be extracted by the turbine, $h_{turbine}$. (b) Determine the design level of headwater in reservoir 1 in order to accommodate the design magnitude of the freely available head to be extracted by the turbine, $h_{turbine}$. (c) Determine the pressure at point a. (d) Determine the maximum allowable velocity of flow at the nozzle without causing cavitation. (e) Determine the design pipe and nozzle size in order to accommodate the design magnitude of the required freely available discharge.

4.115BG The required electric power output by a generator from a reaction turbine is determined to be 80 hp. Water flows from a tank at an elevation of 169 ft, as illustrated in Figure ECP 4.115. Assume frictional losses add up to 3 ft and a turbine–generator system efficiency of 87%. (a) Determine the magnitude of the design freely available head to be extracted by the turbine, $h_{turbine}$. (b) Determine the magnitude of the design-required hydraulic power to be extracted by the turbine. (c) Determine the magnitude of the design-required freely available discharge to be extracted by the turbine, Q.

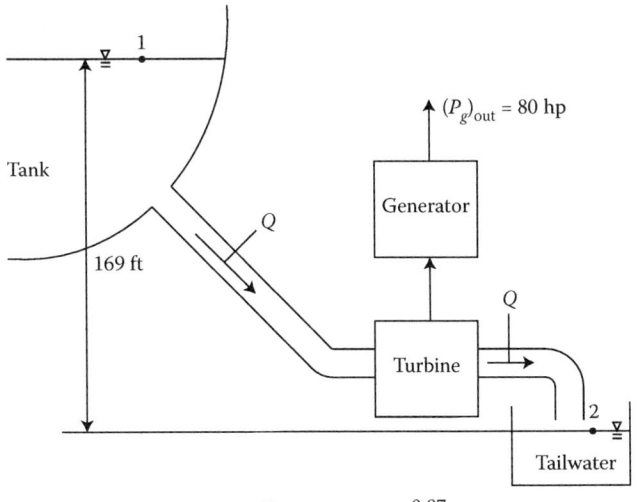

FIGURE ECP 4.115

4.116BG   The freely available head to be extracted by a turbine is $170\,ft$, and the freely available discharge to be extracted by the turbine is $1.5\,ft^3/sec$, as illustrated in Figure 4.28. Assume a turbine–generator system efficiency of 77%. (a) Determine the magnitude of the freely available hydraulic power to be extracted by the turbine. (b) Determine the magnitude of the electric power output by the generator from the turbine.

4.117SI   The turbine–generator system for Problem 4.114 is illustrated in Figure EP 4.26, where the required hydraulic power to be extracted by the impulse turbine is $750\,kW$. The electric power output by the generator is $650\,kW$, and the rotating shaft of the turbine has an angular speed of $350\,rad/sec$ and a shaft torque of $1900\,N\text{-}m$. (a) Determine the shaft power delivered to the generator by the turbine. (b) Determine the turbine efficiency. (c) Determine the generator efficiency. (d) Determine the turbine–generator system efficiency.

4.118BG   The required hydraulic power to be extracted by a reaction turbine is determined to be $1180\,hp$. The designed freely available head to be extracted by the turbine is given to be $100\,ft$. The hydraulic power is to be extracted from water at $60°F$ flowing from reservoir 1 though a concrete pipe $500\,ft$ in length, as illustrated in Figure EP 4.24. The pipe entrance from reservoir 1 is square-edged, and the draft tube has a conical angle, $\theta = 9°$. Assume a Darcy–Weisbach friction factor, f of 0.18; a minor head loss coefficient due to a squared-edged pipe entrance, $k = 0.5$; and a minor head loss coefficient due to the conical diffuser, $k = 0.5$. The pressure at point a in the pipe is measured using a piezometer. (a) Determine the design magnitude of the freely available discharge. (b) Determine the design level of headwater in reservoir 1 in order to accommodate the design magnitude of the freely available head to be extracted by the turbine, $h_{turbine}$. (c) Determine the pressure and velocity at point a. (d) Determine the maximum allowable height of the draft tube, z_b without causing cavitation. (e) Determine the design pipe size in order to accommodate the design magnitude of the required freely available discharge.

4.119BG   The turbine–generator system for Problem 4.118 is illustrated in Figure EP 4.24, where the required hydraulic power to be extracted by the reaction turbine is $1180\,hp$. The electric power output by the generator is $1000\,hp$, and the rotating shaft of the turbine has an angular speed of $340\,rad/sec$ and a shaft torque of $1800\,ft\text{-}lb$. (a) Determine the shaft power delivered to the generator by the turbine. (b) Determine the turbine efficiency. (c) Determine the generator efficiency. (d) Determine the turbine–generator system efficiency.

5

Momentum Equation

5.1 Introduction

In the study of fluids in motion, consideration of the fluid kinematics and the fluid dynamics yields three governing equations that are applied in the solution of a fluid flow problem. The study of fluid kinematics is based on the principle of conservation of mass and yields the continuity equation, which was introduced in Chapter 3. The study of fluid dynamics is based on the principle of conservation of momentum (Newton's second law of motion) and yields the equations of motion, known as the energy equation and the momentum equation. Although both of the equations of motion were introduced in Chapter 4, only the energy equation was applied to practical flow problems. In Chapter 5, the momentum equation will be formally introduced and applied to practical flow problems in complement to the energy equation. The study of fluid dynamics also includes the topic of dimensional analysis, which is introduced in Chapter 7 and yields the resistance equations, which are introduced in Chapter 6. As noted in Chapter 3, the continuity equation, the equations of motion, and the results of dimensional analysis are applied to internal flows, which include pressure (pipe) flow in Chapter 8 and gravity (open channel) flow in Chapter 9. However, only the energy and momentum equations and the results of dimensional analysis are applied to external flows, which include the flow around objects in Chapter 10. Depending upon the internal flow problem, one may use either the Eulerian (integral) approach, the Lagrangian (differential) approach, or both, in the application of the three governing equations (continuity, energy, and momentum). It is important to note that while the continuity and the momentum equations may be applied using either the integral or the differential approach, application of the energy equation is useful only using the integral approach, as energy is work, which is defined over a distance, L. Furthermore, as noted in Chapter 3 and Chapter 4, for an internal fluid flow, while the application of the continuity equation will always be necessary, the energy equation and the momentum equation play complementary roles in the analysis of a given flow situation; when one of the two equations of motion breaks down, the other may be used to find the additional unknown quantity. In the application of the governing equations (continuity, energy, and momentum), it is important to make the distinction between real flow, ideal flow, and a hydraulic jump. And for ideal flow, it is important to make the distinction between internal flow (pipe or open channel flow), external flow, and flow from a tank (jet and siphon flows) (see Table 4.1, Section 4.5.4.1, and Section 5.3). Furthermore, in the application of the governing equations, although in some flow cases, one may ignore (or not need model) the flow resistance, in most practical flow cases in general, one must account for the fluid viscosity and thus model the resulting shear stress and flow resistance. The modeling of flow resistance requires a "subset level" application of the governing equations (Section 5.3.2). Section 5.2 presents the derivation of the momentum equation (and example problems applying the two forms of the momentum equation), Section 5.3.1 presents application of the governing equations, Section 5.3.2 presents modeling of flow resistance (a "subset level" application of the governing equations), and Sections 5.3.3 to 5.3.6 present applications of the governing equations (example problems in the analysis and design of internal and external flow).

5.2 Derivation of the Momentum Equation

Derivation of the momentum equation is based on the principle of conservation of momentum (Newton's second law of motion). As presented in Chapter 4, the principle of conservation of momentum is historically stated and applied for a fluid system (Section 5.2.1), and then rephrased and extended to a control volume system (Section 5.2.2) using the Reynolds transport theorem. Statement of the conservation of momentum principle deals with the time rate of change of the extensive property, momentum, Mv for a fluid system.

The momentum equation is applied in complement to the energy equation, which is useful only for a control volume. However, depending upon the fluid flow problem, one may apply the differential form of the momentum equation (for a fluid system) or the integral form of the momentum equation (for a control volume). For the assumption of real flow, ideal flow, or a hydraulic jump, for the respective energy equation (Equations 4.79, 4.80, or 4.81), the integral form of the momentum equation (Equation 4.28 or Equation 5.27) serves as the complementary governing equation, which may be solved for an unknown force (typically, a pressure force [or depth of flow] or a reaction force); note that the viscous forces are typically accounted for on a "subset level" (indirectly) through the major and minor head loss flow resistance terms in the energy equation above (see Sections 5.3.2 and 5.3.3 below), through an actual discharge (see Sections 5.3.2 and 5.3.4 below), or through a drag force (see Sections 5.3.2 and 5.3.5 below). As such, the integral form of the momentum equation is solved analytically, without the use of dimensional analysis. Thus, when one of these two equations of motion breaks down, the other may be used to find the additional unknown. Furthermore, in the assumption of real flow, and for the specific case of a major head loss due to pipe or channel flow resistance, application of the differential form of the momentum equation (Equation 5.18) may become necessary. Specifically, application of the differential form of the momentum equation will be necessary when it is of interest to trace the gradual variation of the steady nonuniform channel depth of flow (Equation 5.25) (see Chapter 9) or evaluate the flow (solve for an unknown quantity) at a given point in the fluid system for any uniform pipe or channel section (Equation 5.8). Application of Equation 5.25 and Equation 5.8 requires evaluation of the friction slope term, S_f, which is modeled by a "subset level" application of the governing equations, where the major head loss due to flow resistance is determined by application of Equation 4.85 (see Sections 5.3.2 and 5.3.3 below).

5.2.1 The Momentum Equation for a Fluid System: Differential Form of the Momentum Equation

As presented in Chapter 4, application of the conservation of momentum principle stated in Equation 4.4 for a fluid system to a differential mass, M with a differential cross-sectional area, A and a differential section, ds defined along a stream tube, as illustrated in Figure 4.2, yielded the differential form of Newton's second law of motion given by Equation 4.19 (each term with dimensions of F) and repeated as follows:

$$-\gamma A ds \frac{\partial z}{\partial s} - \frac{\partial p}{\partial s} ds A - \tau_w P_w ds = (\rho A ds)\left(v\frac{\partial v}{\partial s} + \frac{\partial v}{\partial t}\right) \quad (5.1)$$

After some simple manipulation of the terms in Equation 5.1, one may easily yield the differential form of the momentum equation. First we begin by dividing both sides of Equation 5.1 by $P_w ds$ (with dimensions of L^2), which yields the following (each term with dimensions of FL^{-2}):

$$-\gamma \frac{A}{P_w}\frac{\partial z}{\partial s} - \frac{\partial p}{\partial s}\frac{A}{P_w} - \tau_w = \left(\frac{\rho A}{P_w}\right)\left(v\frac{\partial v}{\partial s} + \frac{\partial v}{\partial t}\right) \quad (5.2)$$

where the hydraulic radius, $R_h = A/P_w$ and thus:

$$-\gamma R_h \frac{\partial z}{\partial s} - R_h \frac{\partial p}{\partial s} - \tau_w = (\rho R_h)\left(v\frac{\partial v}{\partial s} + \frac{\partial v}{\partial t}\right) \quad (5.3)$$

Isolating the shear stress term, τ_w, factoring out the γR_h term, and realizing that $\gamma = \rho g$, we get:

$$\tau_w = \gamma R_h \left[-\frac{\partial z}{\partial s} - \frac{1}{\gamma}\frac{\partial p}{\partial s} - \left(\frac{1}{g}\right)\left(v\frac{\partial v}{\partial s} + \frac{\partial v}{\partial t}\right)\right] \quad (5.4)$$

recalling that $v\delta v = \delta(v^2)/2$, Equation 5.4 may thus be expressed and rearranged as follows:

$$\tau_w = \gamma R_h \left[-\frac{1}{\gamma}\frac{\partial p}{\partial s} - \frac{\partial z}{\partial s} - \frac{1}{2g}\frac{\partial(v^2)}{\partial s} - \frac{1}{g}\frac{\partial v}{\partial t}\right] \quad (5.5)$$

Noting that for compressible flow, the specific weight, $\gamma = \rho g$ of the fluid varies with respect to s and thus must be included with the partial derivative sign with respect to s and regrouping the terms to define the total energy head, $H = (p/\gamma + z + v^2/2g)$ as follows:

$$\tau_w = \gamma R_h \left[-\frac{\partial}{\partial s}\left(\frac{p}{\gamma} + z + \frac{v^2}{2g}\right) - \frac{1}{g}\frac{\partial v}{\partial t}\right] \quad (5.6)$$

Dividing both sides by the γR_h term (with dimensions of FL^{-2}) yields the following dimensionless slope terms:

$$\frac{\tau_w}{\gamma R_h} = \left[-\frac{\partial H}{\partial s} - \frac{1}{g}\frac{\partial v}{\partial t}\right] \quad (5.7)$$

where the term on the left-hand side of Equation 5.7 is the friction slope, S_f:

$$S_f = \frac{\tau_w}{\gamma R_h} \quad (5.8)$$

the first term on the right-hand side of Equation 5.7 is the slope of the total energy (H or EGL), S_e:

$$S_e = \frac{\partial H}{\partial s}$$ (5.9)

and the second term on the right-hand side of Equation 5.7 is the acceleration slope, S_a:

$$S_a = \frac{1}{g}\frac{\partial v}{\partial t}$$ (5.10)

Therefore, Equation 5.7 may be restated as follows:

$$S_f = -S_e - S_a$$ (5.11)

or:

$$S_e = -(S_f + S_a)$$

which states that the slope of the total energy (H or EGL) is equal to the sum of the friction slope and the acceleration slope, where the negative sign indicates that the slope of the EGL is decreasing (see Figures 4.9 and 4.10).

5.2.1.1 Differential Form of the Momentum Equation

The differential form of the momentum equation is basically the derivative of the integral form of the energy equation (as it should be because the integral form of the energy equation is the integral of the momentum equation; see Chapter 4). Begin with Equation 5.11, which is restated as follows:

$$S_e = -(S_f + S_a)$$ (5.12)

Substituting the respective expression for each slope term, Equation 5.9 for S_e, Equation 5.8 for S_f, and Equation 5.10 for S_a yield the following:

$$\frac{\partial H}{\partial s} = -\frac{\tau_w}{\gamma R_h} - \frac{1}{g}\frac{\partial v}{\partial t}$$ (5.13)

Expanding the expression for the total energy head, H, we get:

$$\frac{\partial}{\partial s}\left(\frac{p}{\gamma}\right) + \frac{\partial z}{\partial s} + \frac{v}{g}\frac{\partial v}{\partial s} = -\frac{\tau_w}{\gamma R_h} - \frac{1}{g}\frac{\partial v}{\partial t}$$ (5.14)

Isolating the term S_f and strategically rearranging the terms, we get:

$$S_f = -\frac{\partial z}{\partial s} - \frac{\partial}{\partial s}\left(\frac{p}{\gamma}\right) - \frac{v}{g}\frac{\partial v}{\partial s} - \frac{1}{g}\frac{\partial v}{\partial t}$$ (5.15)

which is the differential form of the momentum equation for the most general flow type (compressible, nonuniform, unsteady flow for either pressure or gravity flow). One may note that the individual terms in Equation 5.15 relate directly to the differential form of Newton's second law of motion for a differential fluid element moving in the *s*-direction along the streamline, as given in Equation 4.13 and repeated and expanded as follows:

$$\sum F_s = F_G + F_P + F_V = Ma_s \tag{5.16}$$

Thus, interpretations of the individual terms in Equation 5.15 are given as follows. The friction slope, S_f on the left-hand side of Equation 5.15 represents the frictional force, F_V; the first term on the right-hand side of Equation 5.15, $\delta z/\delta s$ represents the gravitational force, F_G; the second term on the right-hand side of Equation 5.15, $\delta(p/\gamma)/\delta s$ represents the pressure force, F_P; the third term on the right-hand side of Equation 5.15, $v\delta v/g\delta s$ represents the convective acceleration; and the fourth and last term on the right-hand side of Equation 5.15, $\delta v/g\delta t$ represents the local acceleration. Therefore, the friction slope, S_f on the left-hand side of Equation 5.15 represents the frictional force, F_V (flow resistance, which results in the right hand side of Equation 5.15), which causes an unknown pressure drop, Δp, which causes an unknown head loss, $h_f = S_f L$, where the head loss is due to a conversion of kinetic energy to heat and is modeled/displayed in the integral form of the energy equation given by Equation 4.44.

The differential form of the momentum equation for the most general flow type (compressible, nonuniform, unsteady flow for either pressure or gravity flow) was given by Equation 5.15 and is repeated as follows:

$$\underbrace{\frac{\tau_w}{\gamma R_h}}_{S_f} = \underbrace{-\frac{\partial z}{\partial s} - \frac{\partial}{\partial s}\left(\frac{p}{\gamma}\right) - \frac{v}{g}\frac{\partial v}{\partial s}}_{-S_e} \underbrace{- \frac{1}{g}\frac{\partial v}{\partial t}}_{-S_a} \tag{5.17}$$

If the assumption of steady flow is made, the $\delta v/g\delta t$ term on the right-hand side of Equation 5.17 would be set equal to zero, and the resulting equation would represent compressible, nonuniform, *steady* flow, where $S_f = -S_e$ as follows:

$$\underbrace{\frac{\tau_w}{\gamma R_h}}_{S_f} = \underbrace{-\frac{\partial z}{\partial s} - \frac{\partial}{\partial s}\left(\frac{p}{\gamma}\right) - \frac{v}{g}\frac{\partial v}{\partial s}}_{-S_e} \tag{5.18}$$

5.2.1.2 Application of the Differential Form of the Momentum Equation

In the assumption of real flow, and for the specific case of a major head loss due to pipe or channel flow resistance, application of the differential form of the momentum equation (Equation 5.18) may become necessary. Specifically, application of the differential form of the momentum equation will be necessary when it is of interest to trace the gradual variation of the steady nonuniform channel depth of flow (Equation 5.25) (see Chapter 9) or to evaluate the flow (solve for an unknown quantity) at a given point in the fluid system for any uniform pipe or channel section (Equation 5.8). Application of Equations 5.25 and 5.8 requires evaluation of the friction slope term, S_f, which is modeled by a "subset level" application of the governing equations and yields the Chezy equation (Equation 5.20), where the major head loss due to flow resistance is determined by application of Equation 4.85

(see Sections 5.3.2 and 5.3.3 below). In the application of the differential momentum equation, the distinction between pressure (pipe) flow versus gravity (open channel) flow is in the definition of the hydrostatic pressure head, p/γ and thus the $\delta(p/\gamma)/\delta s$ term in the $\delta z/\delta s$ term and in the definition of the cross-sectional area, A and the wet perimeter, P_w for the hydraulic radius $R_h = A/P_w$ used in the S_f term. Furthermore, additional assumptions (presented below) are made regarding the application of the differential momentum equation (Equation 5.18) for pressure versus open channel flow.

In the application of the differential momentum equation (Equation 5.18) for pressure flow (pipe flowing full), the hydrostatic pressure head, p/γ remains "as is," with the definition of the cross-sectional area, $A = \pi D^2/4$ for a pipe of diameter, D and the definition of the wet perimeter $P_w = \pi D$. Because application of the differential form of the momentum equation (for a fluid system) is used to solve for an unknown quantity for a given point in the fluid system (pipe flow), additional assumptions of a straight pipe section of length, L, with a constant diameter, D and thus a constant velocity, v (and thus, $\delta v/\delta s = 0$) with a specific pipe friction coefficient and a friction slope, S_f are made. Therefore, the resulting differential form of the momentum equation would represent compressible, *uniform*, steady flow, where $S_f = -S_e$ as follows:

$$\underbrace{\frac{\tau_w}{\gamma R_h}}_{S_f} = \underbrace{-\frac{\partial z}{\partial s} - \frac{\partial}{\partial s}\left(\frac{p}{\gamma}\right)}_{-S_e} \tag{5.19}$$

where evaluation of the friction slope term, S_f (Equation 5.8) is modeled by a "subset level" application of the governing equations (see Sections 5.3.2 and 5.3.3 below, Chapters 6 and 7). Specifically, the expression for S_f is interpreted by supplementing the differential form of the momentum equation with dimensional analysis to yield the Chezy equation as follows:

$$S_f = \frac{v^2}{C^2 R_h} \tag{5.20}$$

where C is the Chezy coefficient, which is derived from dimensional analysis and evaluated empirically. Thus, the Chezy equation may be used to solve for one of the following unknown quantities for a given point in the fluid system (pipe system): friction slope, S_f; velocity, v; pipe friction coefficient (Chezy coefficient, C; Darcy–Weisbach friction factor, f; Manning's roughness coefficient, n; or Hazen–Williams roughness, C_h); or pipe diameter, D. Example problems applying the Chezy equation for pipe flow are presented in Chapter 8, Section 5.3.3.4 below, and as follows.

EXAMPLE PROBLEM 5.1

Water at 20°C flows in a 3.5-m-diameter inclined straight pipe section at a flowrate of 2.6 m³/sec, as illustrated in Figure EP 5.1. The pipe length between points 1 and 2 is 900 m, and the Chezy coefficient, C for the pipe is 60 m$^{1/2}$/s^{-1}. The elevations of points 1 and 2 are 1.3 m and 1.8 m, respectively, and the pressure at point 1 is 3×10^4 N/m². (a) Determine the friction slope at any point in the straight pipe section. (b) Determine the head loss due to pipe friction between points 1 and 2. (c) Determine the pressure drop at point 2 due to the pipe friction.

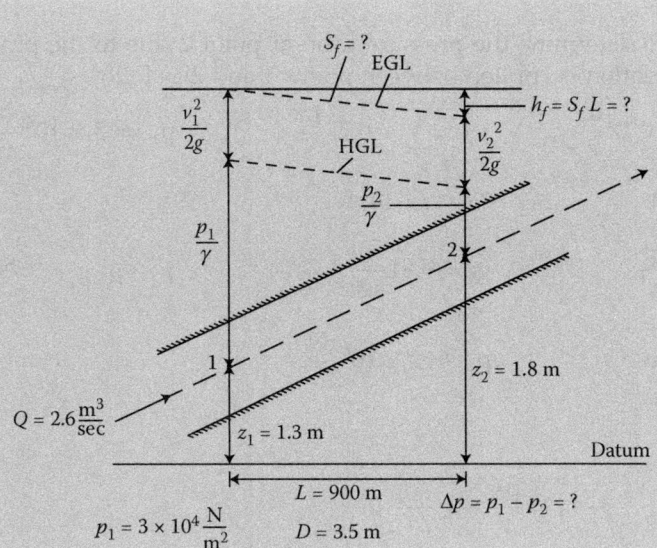

FIGURE EP 5.1
Water flows through an inclined pipe flowing under pressure.

Mathcad Solution

(a) In order to determine the friction slope at any point in the straight pipe section, the Chezy equation is applied. Furthermore, in order to determine the velocity of flow in the pipe, the continuity equation is applied as follows:

$D := 3.5 \, m$ $\qquad A := \dfrac{\pi \cdot D^2}{4} = 9.621 \, m^2$ $\qquad P_w := \pi \cdot D = 10.996 \, m$

$R_h := \dfrac{A}{P_w} = 0.875 \, m$ $\qquad Q := 2.6 \dfrac{m^3}{sec}$ $\qquad C := 60 \, m^{\frac{1}{2}} sec^{-1}$

Guess values: $\qquad\qquad S_f := 0.5 \dfrac{m}{m}$ $\qquad\qquad v := 1 \dfrac{m}{sec}$

Given

$$S_f = \dfrac{v^2}{C^2 \cdot R_h}$$

$$v = \dfrac{Q}{A}$$

$$\begin{pmatrix} v \\ S_f \end{pmatrix} := Find \, (v, S_f)$$

$v = 0.27 \dfrac{m}{s}$ $\qquad\qquad S_f = 2.318 \times 10^{-5} \dfrac{m}{m}$

(b) In order to determine the head loss due to pipe friction between points 1 and 2, the head loss due to pipe friction equation is applied as follows:

$L := 900 \, m$ $\qquad\qquad h_f := S_f \cdot L = 0.021 \, m$

(c) In order to determine the pressure drop at point 2 due to the pipe friction, the
     energy equation is applied between points 1 and 2 as follows:

$$v_1 := v = 0.27 \frac{m}{s} \qquad\qquad v_2 := v = 0.27 \frac{m}{s} \qquad\qquad p_1 := 3 \times 10^4 \frac{N}{m^2}$$

$$z_1 := 1.3 \, m \qquad\qquad\qquad z_2 := 1.8 \, m$$

$$\rho := 998 \frac{kg}{m^3} \qquad\qquad g := 9.81 \frac{m}{sec^2} \qquad\qquad \gamma := \rho \cdot g = 9.79 \times 10^3 \frac{kg}{m^2 \cdot s^2}$$

Guess value: $\qquad\qquad p_2 := 2 \times 10^4 \dfrac{N}{m^2}$

Given

$$\frac{p_1}{\gamma} + z_1 + \frac{v_1^2}{2 \cdot g} - h_f = \frac{p_2}{\gamma} + z_2 + \frac{v_2^2}{2 \cdot g}$$

$$p_2 := \text{Find}(p_2) = 2.49 \times 10^4 \frac{N}{m^2}$$

$$\Delta p := p_1 - p_2 = 5.099 \times 10^3 \frac{N}{m^2}$$

In the application of the differential momentum equation (Equation 5.18) for open channel
flow, the pressure head term, p/γ is replaced by the depth of flow in the channel, y; thus, the
$\delta(p/\gamma)/\delta s$ term becomes $\delta y/\delta s$, and $-\delta z/\delta s = S_o$, where S_o = the channel bottom slope, which
must be decreasing in order to invoke gravity. The definition of the cross-sectional area, $A =$
by for a rectangular open channel (for instance) of channel width, b and depth of flow, y, for
which the definition of the wet perimeter, $P_w = 2y + b$ (see Chapter 9 for details and for nonrectangular cross sections). Thus, the differential momentum equation (Equation 5.18) for
open channel flow becomes as follows:

$$\underbrace{\frac{\tau_w}{\gamma R_h}}_{S_f} = \underbrace{S_o - \frac{\partial}{\partial s}(y) - \frac{v}{g}\frac{\partial v}{\partial s}}_{-S_e} \tag{5.21}$$

Furthermore, if the assumption of uniform flow were made, then both the $v\delta v/g\delta s$ term
and the $\delta y/\delta s$ terms on the right-hand side of the equation would be set equal to zero,
and the resulting equation would represent incompressible, *uniform*, steady flow, where
$S_f = -S_e = S_o$ and is given as follows:

$$\underbrace{\frac{\tau_w}{\gamma R_h}}_{S_f} = \underbrace{S_o}_{-S_e} \tag{5.22}$$

where, similar to pipe flow, the expression for S_f (Equation 5.8) is interpreted by the Chezy
equation (Equation 5.20) as follows:

$$S_f = S_o = \frac{v^2}{C^2 R_h} \tag{5.23}$$

Thus, assuming uniform flow, the Chezy equation may be used to solve for one of the following unknown quantities for a given point in the fluid system (open channel system): friction slope, $S_f =$ channel slope, S_o; velocity, v; channel friction coefficient (Chezy coefficient, C; Darcy–Weisbach friction factor, f; Manning's roughness coefficient, n; or Hazen–Williams roughness, C_h); or open channel flow depth, y. Example problems applying the Chezy equation for uniform open channel flow are presented in Chapter 9, Section 5.3.3.4 below, and as follows.

EXAMPLE PROBLEM 5.2

Water at 20°C flows at a discharge of 5.4 m³/sec in a rectangular channel that is 0.8 m wide, with a channel bottom slope of 0.035 m/m, as illustrated in Figure EP 5.2. The channel length between points 1 and 2 is 1200 m, and the Chezy coefficient is $80\,\mathrm{m}^{1/2}/\mathrm{s}^{-1}$. (a) Determine the uniform depth of flow in the channel. (b) Determine the head loss due to channel friction between points 1 and 2.

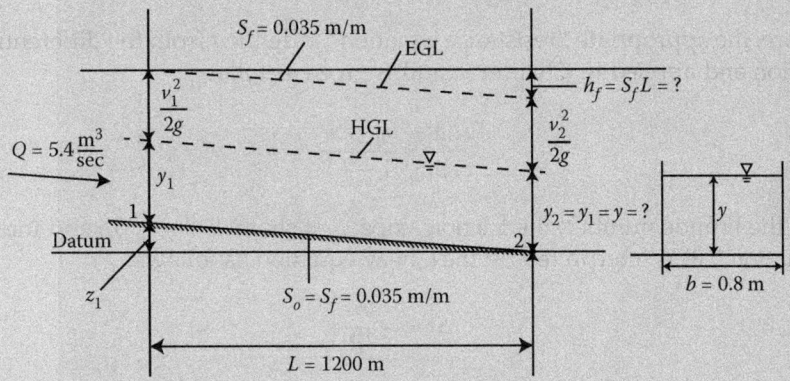

FIGURE EP 5.2
Water flows at a uniform flow in a rectangular channel.

Mathcad Solution

(a) In order to determine the uniform depth of flow in the channel, the Chezy equation is applied. Furthermore, in order to determine the velocity of flow in the pipe, the continuity equation is applied as follows:

$$Q := 5.4\,\frac{\mathrm{m}^3}{\mathrm{sec}} \qquad b := 0.8\,\mathrm{m} \qquad S_o := 0.035\,\frac{\mathrm{m}}{\mathrm{m}} \qquad C := 80\,\mathrm{m}^{\frac{1}{2}}\,\mathrm{sec}^{-1}$$

$$A(y) := b \cdot y \qquad P_w(y) := (2 \cdot y + b) \qquad R_h(y) := \frac{A(y)}{P_w(y)} \qquad v(y) := \frac{Q}{A(y)}$$

Guess values: $\qquad y := 1\,\mathrm{m}$

Given

$$S_o = \frac{v(y)^2}{C^2 \cdot R_h(y)}$$

$$y := \text{Find } (y) = 0.863 \, \text{m} \qquad\qquad v(y) = 7.824 \, \frac{\text{m}}{\text{s}}$$

(b) In order to determine the head loss due to channel friction between points 1 and 2, the head loss due to channel friction equation is applied as follows:

$$L := 1200 \, \text{m} \qquad\qquad h_f := S_o \cdot L = 42 \, \text{m}$$

If however, the assumption of nonuniform flow is made, then the differential form of the momentum equation (Equation 5.21) remains as follows:

$$\underbrace{\frac{\tau_w}{\gamma R_h}}_{S_f} = \underbrace{S_o - \frac{\partial}{\partial s}(y) - \frac{v}{g}\frac{\partial v}{\partial s}}_{-S_e} \tag{5.24}$$

Furthermore, the appropriate "resistance equation" is defined from the differential momentum equation and applied in Chapter 9, and is given as follows:

$$\frac{dy}{ds} = \frac{S_o - S_f}{1 - F^2} \tag{5.25}$$

where $F =$ the Froude number, the friction slope, $S_f \neq$ channel slope, S_o, and the expression for S_f (Equation 5.8) is interpreted by the Chezy equation as follows:

$$S_f = \frac{v^2}{C^2 R_h} \tag{5.26}$$

Equation 5.25 is numerically solved for the unknown depth, y for a given point in the channel, whereby solution of this equation for a number of points along the channel yields the surface water profile. It is interesting to note that while the application of the Chezy equation (Equation 5.26) for incompressible, uniform, steady open channel flow may "technically" be applied for very short "finite control volume" channel sections, this becomes too tedious, and instead the "resistance equation" $dy/ds = (S_o - S_f)/(1 - F^2)$ is numerically solved for the unknown depth, y for a given point in the channel for a series of points along the channel. And, finally, one may note that the "resistance equation" always involves the friction slope term, S_f. Example problems applying the resistance equation for nonuniform open channel flow are presented in Chapter 9.

5.2.2 The Momentum Equation for a Control Volume: Integral Form of the Momentum Equation

As presented in Chapter 4, application of the conservation of momentum principle stated in Equation 4.7 for a control volume to a finite control volume, as illustrated in Figure 4.1, assuming steady flow, yielded the integral form of Newton's second law of motion, which is the integral form of the momentum equation given by Equation 4.28 and repeated as follows:

$$\sum F_s = (F_G + F_P + F_V + F_{other})_s = (\rho Q v_s)_2 - (\rho Q v_s)_1 \tag{5.27}$$

which states that the rate of change (increase) of fluid momentum across the control volume along the s-axis is accomplished by the action of a forward force, ΣF_s acting on the control volume along the s-axis. The significant forces acting on the control volume illustrated in Figure 4.1 include the component along the s-axis of the gravitational force, $F_G = Mg$; the pressure forces, $F_P = \Delta pA$ on the two ends of the control volume; the viscous force, $F_V = \tau_w A$ due to fluid friction; and other forces, F_{other} acting on the control volume, such as reaction forces at bolts, cables, etc. Furthermore, the velocities are measured relative to the fixed control volume (see Section 5.2.2.1 below). The integral momentum equation (Equation 5.27) is used to solve for an unknown force acting on a finite control volume.

It is important to note that a more general statement of the integral form of the momentum equation for a control volume, assuming steady flow, assumes there may be one or more control faces for the flow into and the flow out of the control volume, as illustrated in Figure 4.3. Thus, the more general statement of the momentum equation for a control volume, assuming steady flow along the streamline, s is given as follows:

$$\sum F_s = (F_G + F_P + F_V + F_{other})_s = (\rho Q v_s)_{out} - (\rho Q v_s)_{in} \tag{5.28}$$

An example problem applying the integral form of the momentum equation modeling more than one control face for the flow out of the control volume is presented in Example Problem 5.5 below.

5.2.2.1 Fixed versus Moving Control Volume in the Eulerian (Integral) Point of View

As defined in Chapter 3, a control volume assumes and describes a specific (predetermined) volume of fluid within a set of boundaries in a fixed region in the fluid space that does not move. As such, in the Eulerian point of view, the velocity at a given control face, v_i is defined relative to the control volume. However, one may note that the absolute velocity at a given control face is defined relative to the earth (or fixed ground), v_i (abs), and the absolute velocity of the control volume is also defined relative to the earth (or fixed ground), v_{cv} (abs). Thus, assuming the same positive s-direction along a streamline, the velocity at a given control face (see Figure 4.1) relative to the control volume is defined as follows:

$$\vec{v_i} = \vec{v_i}(abs) - \vec{v_{cv}}(abs) \tag{5.29}$$

Furthermore, relative to the earth (or fixed ground), the control volume may be either a fixed or a moving control volume. A fixed control volume encloses a device/mechanism or structure that involves the flow of fluid mass through a *stationary* pump, turbine, compressor, nozzle, water heater, fan, spillway, weir, etc. However, a moving control volume encloses a device/mechanism that involves the flow of fluid mass through a *moving* pump, turbine, compressor, nozzle, water heater, fan, etc. The moving control volume is typically on board a car, ship, plane, etc. In the case where the control volume is fixed relative to the earth, its absolute velocity, $\vec{v_{cv}}(abs) = 0$. Thus, the velocity at the given control face relative to the control volume is given as follows:

$$\vec{v_i} = \vec{v_i}(abs) \tag{5.30}$$

However, in the case where the control volume is moving relative to the earth, its absolute velocity, $\vec{v_{cv}}(abs) \neq 0$. Thus, the velocity at the given control face relative to the control volume is given as follows:

$$\vec{v}_i = \vec{v}_i(abs) - \vec{v}_{cv}(abs) \tag{5.31}$$

It is important to note that the velocity is a vector with both magnitude and direction.

Example problems applying the integral form of the momentum equation for a fixed control volume are presented in Example Problems 5.3 through 5.6 below and example problems in Section 5.3 below. An example problem applying the integral form of the momentum equation for a moving control volume is presented in Example Problem 5.7 below. A forced jet hitting a vane will be used to illustrate the difference between a stationary control volume in Example Problem 5.6 and a moving control volume in Example Problem 5.7 below. In the case of a jet hitting a vane, there are two points to consider regarding the flow. The first point is that because the jet is open to the atmosphere, the pressure at control faces 1 and 2 (see Figures EP 5.6 and EP 5.7 below) is atmospheric and thus the gage pressure is zero. The second point is that the friction (flow resistance) between the fluid jet and the vane is typically assumed to be negligible and thus the flow is assumed to be ideal. However, in cases where the velocity of the jet is high, ignoring the minor head loss due to friction may introduce a significant error in the problem analysis or design. Therefore, in such cases, it is typical to model the minor head loss by a reduced velocity at the control face(s) for the flow out of the control volume.

5.2.2.2 Application of the Integral Form of the Momentum Equation

For the assumption of real flow, ideal flow, or a hydraulic jump for the respective energy equation (Equations 4.79, Equation 4.80, or Equation 4.81), the integral form of the momentum equation (Equation 4.28 or Equation 5.27) serves as the complementary governing equation, which may be solved for an unknown force (typically, a pressure force [or depth of flow] or a reaction force); note that the viscous forces are typically accounted for on a "subset level" (indirectly) through the major and minor head loss flow resistance terms in the energy equation above (see Sections 5.3.2 and 5.3.3 below), through an actual discharge (see Sections 5.3.2 and 5.3.4 below), or through a drag force (see Sections 5.3.2 and 5.3.5 below). As such, the integral form of the momentum equation is solved analytically, without the use of dimensional analysis. Thus, when one of these two equations of motion breaks down, the other may be used to find the additional unknown. Example problems applying the integral form of the momentum equation for pipe flow are presented in Chapter 8, Sections 5.3.3.5, and 5.3.4.3, and this section (Example Problems 5.3 through 5.5). Example problems applying the integral form of the momentum equation for open channel flow are presented in Chapter 9 and Section 5.3.4.4. Example problems applying the integral form of the momentum equation for flow from a tank (jets) are presented in Section 5.3.4.5 and this section (Example Problems 5.6 and 5.7). Example problems applying the integral form of the momentum equation for external flow around a body are presented in Chapter 10 and Section 5.3.5.2. And, finally, example problems applying the integral form of the momentum equation for a hydraulic jump are presented in Chapter 9 and Section 5.3.6.1.

EXAMPLE PROBLEM 5.3

Water at 20°C flows in a 0.35-m-diameter pipe that is fitted with a 180° threaded bend in order to reverse the direction of flow, as illustrated in Figure EP 5.3. Assume that the pipe and the bend lie in a horizontal plane, (x, y), and the flow discharges into the

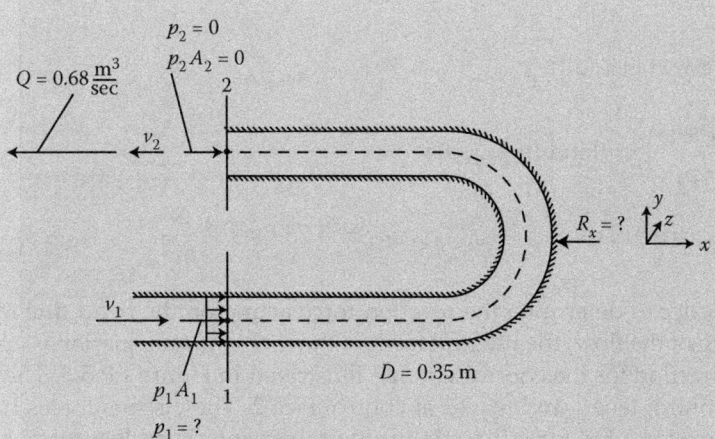

FIGURE EP 5.3
Water flows in a pipe fitted with a 180° threaded bend.

atmosphere at a rate of 0.68 m³/sec. Also, assume a minor head loss coefficient due to the 180° threaded bend, k of 1.5. (a) Determine the minor head loss due to the bend. (b) Determine the pressure at point 1. (c) Determine the reaction force acting on the bend due to the thrust force of the flow (anchoring force required to keep the bend from moving).

Mathcad Solution

(a)–(b) The minor head loss due to the bend is evaluated by applying the minor head loss equation, and the pressure at point 1 is determined by applying the energy equation between points 1 and 2, and as follows:

$$D := 0.35\,m \qquad A := \frac{\pi \cdot D^2}{4} = 0.096\,m^2 \qquad k_{tbend} := 1.5 \qquad Q := 0.68\,\frac{m}{sec}$$

$$v := \frac{Q}{A} = 7.068\,\frac{m}{s} \qquad v_1 := v = 7.068\,\frac{m}{s} \qquad v_2 := v = 7.068\,\frac{m}{s}$$

$$p_2 := 0\,\frac{N}{m^2} \qquad z_1 := 0\,m \qquad z_2 := 0\,m$$

$$\rho := 998\,\frac{kg}{m^3} \qquad g := 9.81\,\frac{m}{sec^2} \qquad \gamma := \rho \cdot g = 9.79 \times 10^3\,\frac{kg}{m^2 \cdot s^2}$$

Guess value: $\qquad h_{fmin} := 1\,m \qquad\qquad p_1 := 2 \times 10^4\,\frac{N}{m^2}$

Given

$$\frac{p_1}{\gamma} + z_1 + \frac{v_1^2}{2g} - h_{fmin} = \frac{p_2}{\gamma} + z_2 + \frac{v_2^2}{2g}$$

$$h_{fmin} = k_{tbend} \frac{v^2}{2 \cdot g}$$

$$\begin{pmatrix} h_{fmin} \\ p_1 \end{pmatrix} := \text{Find } (h_{fmin}, p_1)$$

$$h_{fmin} = 3.819 \, m \qquad\qquad p_1 = 3.739 \times 10^4 \frac{N}{m^2}$$

(c) In order to determine the reaction force acting on the bend due to the thrust force of the flow, the integral form of the momentum equation is applied in the *x* direction for the control volume illustrated in Figure EP 5.3. The flow enters at control face 1 and leaves at control face 2. The viscous forces have already been accounted for indirectly through the minor head loss term, $h_{f,minor}$ in the energy equation, and there is no gravitational force component along the *x-axis* because the pipe is horizontal ($z_1 = z_2 = 0$).

$$\sum F_x = (p_1 A_1)_x + (p_2 A_2)_x + R_x = \rho Q (v_{2x} - v_{1x})$$

$$A_1 := A = 0.096 \, m^2 \qquad\qquad A_2 := A = 0.096 \, m^2$$

$$p_{1x} := p_1 = 3.739 \times 10^4 \frac{N}{m^2} \qquad\qquad p_{2x} := p_2 = 0 \frac{N}{m^2}$$

$$v_{1x} := v_1 = 7.068 \frac{m}{s} \qquad\qquad v_{2x} := -v_2 = -7.068 \frac{m}{s}$$

Guess value: $\qquad\qquad\qquad\qquad\qquad R_x := 1000 \, N$

Given

$$p_{1x} \cdot A_1 + p_{2x} \cdot A_2 - R_x = \rho \cdot Q (v_{2x} - v_{1x})$$

$$R_x := \text{Find } (R_x) = 1.319 \times 10^4 N$$

Thus, the reaction force acting on the bend, R_x due to the thrust force of the flow acts to the left as assumed and illustrated in Figure EP 5.3.

EXAMPLE PROBLEM 5.4

Water at 20°C flows in a 0.65-m-diameter horizontal pipe at a flowrate of 1.5 m³/sec. A 90° mitered bend is installed in between point 1 and 2 in order to change the direction of the flow by 90°, as illustrated in Figure EP 5.4. Assume that the pipe and the bend lie in a horizontal plane, (*x, y*). The pressure at point 1 is 5×10^5 N/m², and assume a minor head loss coefficient due to the bend, *k* of 0.131. (a) Determine the minor head loss due to the bend between points 1 and 2. (b) Determine the pressure at point 2. (c) Determine the reaction force acting on the bend due to the thrust force of the flow (anchoring force required to keep the bend from moving).

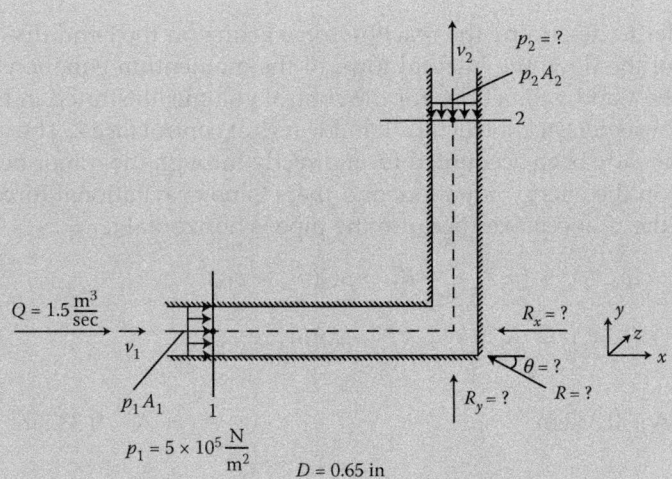

FIGURE EP 5.4
Water flows in a horizontal pipe with a 90° mitered bend.

Mathcad Solution

(a)–(b) The minor head loss due to the bend is evaluated by applying the minor head loss equation, and the pressure at point 2 is determined by applying the energy equation between points 1 and 2, and as follows:

$$D := 0.65\,m \qquad A := \frac{\pi \cdot D^2}{4} = 0.332\,m^2 \qquad k_{bend} := 0.131 \qquad Q := 1.5\frac{m^3}{sec}$$

$$v := \frac{Q}{A} = 4.52\frac{m}{s} \qquad v_1 := v = 4.52\frac{m}{s} \qquad v_2 := v = 4.52\frac{m}{s}$$

$$p_1 := 5 \times 10^5\frac{N}{m^2} \qquad z_1 := 0\,m \qquad z_2 := 0\,m$$

$$\rho := 998\frac{kg}{m^3} \qquad g := 9.81\frac{m}{sec^2} \qquad \gamma := \rho \cdot g = 9.79 \times 10^3\frac{kg}{m^2 \cdot s^2}$$

Guess value: $\qquad h_{fmin} := 1\,m \qquad\qquad p_2 := 1 \times 10^4\frac{N}{m^2}$

Given

$$\frac{p_1}{\gamma} + z_1 + \frac{v_1^2}{2 \cdot g} - h_{fmin} = \frac{p_2}{\gamma} + z_2 + \frac{v_2^2}{2 \cdot g}$$

$$h_{fmin} = k_{bend}\frac{v^2}{2 \cdot g}$$

$$\begin{pmatrix} h_{fmin} \\ p_2 \end{pmatrix} := \text{Find}\,(h_{fmin}, p_2)$$

$$h_{fmin} = 0.136\,m \qquad\qquad p_2 := 4.987 \times 10^5\frac{N}{m^2}$$

(c) In order to determine the reaction force acting on the bend due to the thrust force of the flow, the integral form of the momentum equation is applied in both the x and y directions for the control volume illustrated in Figure EP 5.4. The flow enters at control face 1 and leaves at control face 2. The viscous forces have already been accounted for indirectly through the minor head loss term, $h_{f,minor}$ in the energy equation, and there is no gravitational force component along the x- and y-axes because the pipe is horizontal ($z_1 = z_2 = 0$).

$$\sum F_x = (p_1 A_1)_x + (p_2 A_2)_x + R_x = \rho Q(v_{2x} - v_{1x})$$

$$\sum F_y = (p_1 A_1)_y + (p_2 A_2)_y + R_y = \rho Q(v_{2y} - v_{1y})$$

$$A_1 := A = 0.332 \, \text{m}^2 \qquad\qquad\qquad A_2 := A = 0.332 \, \text{m}^2$$

$$p_{1x} := p_1 = 5 \times 10^5 \, \frac{\text{N}}{\text{m}^2} \qquad\qquad\qquad p_{2x} := 0 \, \frac{\text{N}}{\text{m}^2}$$

$$v_{1x} := v_1 = 4.52 \, \frac{\text{m}}{\text{s}} \qquad\qquad\qquad v_{2x} := 0 \, \frac{\text{m}}{\text{sec}}$$

$$p_{1y} := 0 \, \frac{\text{N}}{\text{m}^2} \qquad\qquad\qquad p_{2y} := -p_2 = -4.987 \times 10^5 \, \text{m} \frac{\text{N}}{\text{m}}$$

$$v_{1y} := 0 \, \frac{\text{m}}{\text{sec}} \qquad\qquad\qquad v_{2y} := v_2 = 4.52 \, \frac{\text{m}}{\text{s}}$$

Guess value: $\qquad\qquad\qquad\qquad R_x := 1000 \, \text{N} \qquad R_y := 1000 \, \text{N}$

Given

$$p_{1x} \cdot A_1 + p_{2x} \cdot A_2 - R_x = \rho \cdot Q(v_{2x} - v_{1x})$$

$$p_{1y} \cdot A_1 + p_{2y} \cdot A_2 + R_y = \rho \cdot Q(v_{2y} - v_{1y})$$

$$\begin{pmatrix} R_x \\ R_y \end{pmatrix} := \text{Find} \, (R_x, R_y) = \begin{pmatrix} 1.727 \times 10^5 \\ 1.722 \times 10^5 \end{pmatrix} \text{N}$$

$$R := \sqrt{R_x{}^2 + R_y{}^2} = 2.439 \times 10^5 \, \text{N} \qquad\qquad \theta := \text{atan} \left(\frac{|R_y|}{|R_x|} \right) = 44.926 \, \text{deg}$$

Thus, the reaction force acting on the bend, R_x acts to the left, and the reaction force acting on the bend, R_y acts upwards, both as assumed and illustrated in Figure EP 5.4.

EXAMPLE PROBLEM 5.5

Water at 20°C flows in a 0.8-m-diameter pipe at a flowrate of 1.6 m³/sec. A double nozzle is installed downstream of point 1 in order to divide the flow in two, as illustrated in Figure EP 5.5. Assume that the pipe and the nozzle lie in a horizontal plane, (x, y). The diameter of the nozzle at point 2 is 0.45 m and is at an angle of

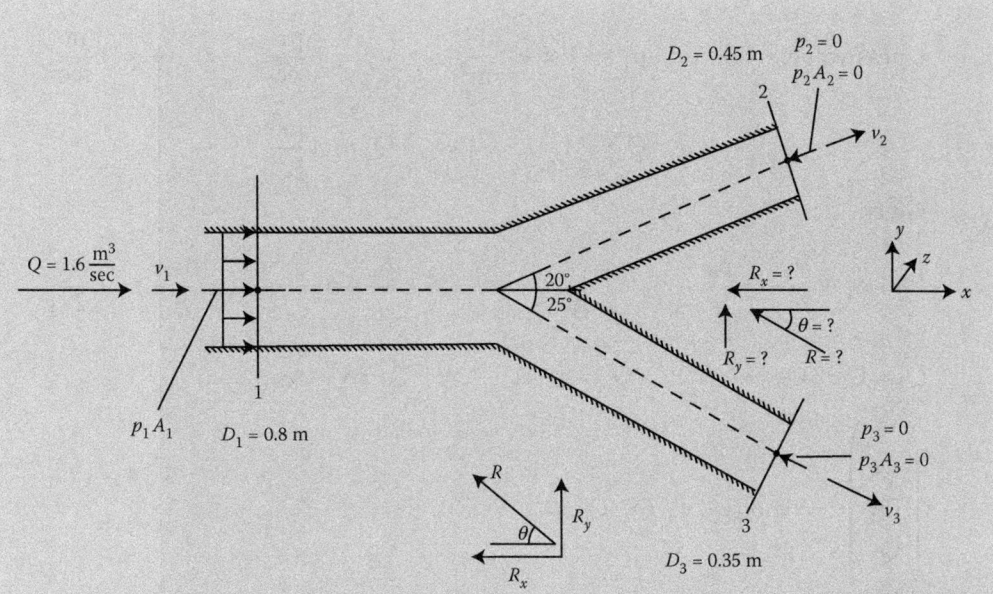

FIGURE EP 5.5
Water flows in a horizontal pipe with a double nozzle.

$20°$ with the x-axis, and the diameter of the nozzle at point 3 is $0.35\,\text{m}$ and is at an angle of $25°$ with the x-axis. The flows at points 2 and 3 are open to the atmosphere. Assume that the flow is ideal and thus ignore any minor head loss due to the nozzle. (a) Determine the pressure at point 1. (b) Determine the reaction force acting on the nozzle due to the thrust force of the flow (anchoring force required to keep nozzle from moving).

Mathcad Solution

(a) In order to determine the pressure at point 1, the energy equation is applied between points 1 and 2 and between points 1 and 3. Furthermore, in order to determine the velocity of flow in the pipe and the double nozzle, the continuity equation is applied as follows:

$$D_1 := 0.8\,\text{m} \qquad A_1 := \frac{\pi \cdot D_1^2}{4} = 0.503\,\text{m}^2 \qquad Q_1 := 1.6\,\frac{\text{m}^3}{\text{sec}}$$

$$v_1 := \frac{Q_1}{A_1} = 3.183\,\frac{\text{m}}{\text{s}} \qquad D_2 := 0.45\,\text{m} \qquad A_2 := \frac{\pi D_2^2}{4} = 0.159\,\text{m}^2$$

$$D_3 := 0.35\,\text{m} \qquad A_3 := \frac{\pi D_3^2}{4} = 0.096\,\text{m}^2 \qquad p_2 := 0\,\frac{\text{N}}{\text{m}^2}$$

$$p_3 := 0\,\frac{\text{N}}{\text{m}^2} \qquad z_1 := 0\,\text{m} \qquad z_2 := 0\,\text{m} \qquad z_3 := 0\,\text{m}$$

$$\rho := 998\,\frac{\text{kg}}{\text{m}^3} \qquad g := 9.81\,\frac{\text{m}}{\text{sec}^2} \qquad \gamma := \rho \cdot g = 9.79 \times 10^3\,\frac{\text{kg}}{\text{m}^2 \cdot \text{s}^2}$$

Guess value: $\quad\quad p_1 := 1 \times 10^4 \dfrac{N}{m^2} \quad\quad v_2 := 1\dfrac{m}{sec} \quad\quad v_3 := 1\dfrac{m}{sec}$

$$Q_2 := 1\dfrac{m^3}{sec} \quad\quad Q_3 := 1\dfrac{m^3}{sec}$$

Given

$$\dfrac{p_1}{\gamma} + z_1 + \dfrac{v_1^2}{2 \cdot g} = \dfrac{p_2}{\gamma} + z_2 + \dfrac{v_2^2}{2 \cdot g} \quad\quad\quad \dfrac{p_1}{\gamma} + z_1 + \dfrac{v_1^2}{2 \cdot g} = \dfrac{p_3}{\gamma} + z_3 + \dfrac{v_3^2}{2 \cdot g}$$

$$Q_1 = Q_2 + Q_3 \quad\quad\quad Q_2 = v_2 \cdot A_2 \quad\quad\quad Q_3 = v_3 \cdot A_3$$

$$\begin{pmatrix} v_2 \\ v_3 \\ Q_2 \\ Q_3 \\ p_1 \end{pmatrix} := \text{Find } (v_2, v_3, Q_2, Q_3, p_1)$$

$$v_2 = 6.268\dfrac{m}{s} \quad\quad\quad v_3 = 6.268\dfrac{m}{s} \quad\quad Q_2 = 0.997\dfrac{m^3}{s} \quad\quad Q_3 = 0.603\dfrac{m^3}{s}$$

$$p_1 = 1.455 \times 10^4 \dfrac{N}{m^2}$$

(b) In order to determine the reaction force acting on the nozzle due to the thrust force of the flow, the integral form of the momentum equation is applied in both the x and y directions for the control volume illustrated in Figure EP 5.5. The flow enters at control face 1 and leaves at control face 2 and control face 3. The viscous forces are ignored because of the assumption of ideal flow, and there is no gravitational force component along the x- and y-axes because the pipe is horizontal ($z_1 = z_2 = 0$).

$$\sum F_x = (p_1 A_1)_x + (p_2 A_2)_x + (p_3 A_3)_x + R_x = (\rho Q_2 v_{2x} + \rho Q_3 v_{3x}) - (\rho Q_1 v_{1x})$$

$$\sum F_y = (p_1 A_1)_y + (p_2 A_2)_y + (p_3 A_3)_y + R_y = (\rho Q v_{2y} + \rho Q_3 v_{3y}) - (\rho Q_1 v_{1y})$$

$p_{1x} := p_1 = 1.455 \times 10^4 \dfrac{N}{m^2} \quad\quad\quad\quad p_{2x} := -p_2 \cdot \cos(20\ deg) = 0\dfrac{N}{m^2}$

$p_{3x} := -p_3 \cdot \cos(25\ deg) = 0\dfrac{N}{m^2} \quad\quad\quad v_{1x} := v_1 = 3.183\dfrac{m}{s}$

$v_{2x} := v_2 \cdot \cos(20\ deg) = 5.89\dfrac{m}{s} \quad\quad\quad v_{3x} := v_3 \cdot \cos(25\ deg) = 5.681\dfrac{m}{s}$

$p_{1y} := 0\dfrac{N}{m^2} \quad\quad\quad\quad\quad\quad\quad\quad\quad p_{2y} := -p_2 \cdot \sin(20\ deg) = 0\dfrac{N}{m^2}$

$p_{3y} := p_3 \cdot \cos(25\ deg) = 0\dfrac{N}{m^2} \quad\quad\quad v_{1y} := 0\dfrac{m}{sec}$

$$v_{2y} := v_2 \cdot \sin(20 \text{ deg}) = 2.144 \frac{m}{s} \qquad\qquad v_{3y} := -v_3 \cdot \sin(25 \text{ deg}) = -2.649 \frac{m}{s}$$

Guess value: $\qquad\qquad R_x := 1000\,N \qquad R_y := 1000\,N$

Given

$$p_{1x} \cdot A_1 + p_{2x} \cdot A_2 + p_{3x} \cdot A_3 - R_x = (\rho \cdot Q_2 \cdot v_{2x} + \rho \cdot Q_3 \cdot v_{3x}) - (\rho \cdot Q_1 \cdot v_{1x})$$

$$p_{1y} \cdot A_1 + p_{2y} \cdot A_2 + p_{3y} \cdot A_3 + R_y = (\rho \cdot Q_2 \cdot v_{2y} + \rho \cdot Q_3 \cdot v_{3y}) - (\rho \cdot Q_1 \cdot v_{1y})$$

$$\binom{R_x}{R_y} := \text{Find}\,(R_x, R_y) = \binom{3.117 \times 10^3}{538.595}\,N$$

$$R := \sqrt{R_x{}^2 + R_y{}^2} = 3.163 \times 10^3\,N$$

$$\theta := \text{atan}\left(\frac{|R_y|}{|R_x|}\right) = 9.804 \text{ deg}$$

Thus, the reaction force acting on the nozzle, R_x acts to the left as assumed, and the reaction force, R_y acts upwards as assumed, as illustrated in Figure EP 5.5.

EXAMPLE PROBLEM 5.6

A forced jet of water at 20°C hits a stationary vane at a velocity of 6 m/sec and a flow-rate of 0.09 m³/sec, as illustrated in Figure EP 5.6. The vane angle α is 55°, and assume

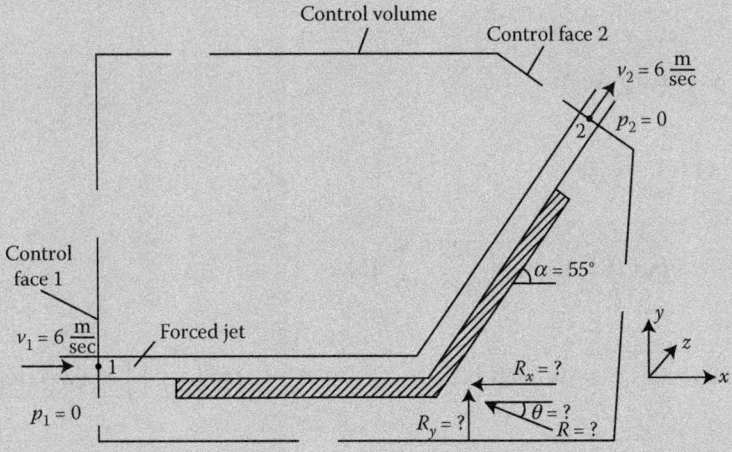

FIGURE EP 5.6
Forced jet of water hits a stationary horizontal vane angled at 55°.

that the jet and the vane lie in a horizontal plane (x, y), and that flow is ideal. (a) Determine the reaction force acting on the vane due to the thrust force of the jet (anchoring force required to keep vane from moving).

Mathcad Solution

(a) In order to determine the reaction force acting on the vane due to the thrust force of the jet, the integral form of the momentum equation is applied in both the x and y directions for the control volume illustrated in Figure EP 5.6. The flow enters at control face 1 and leaves at control face 2. The viscous forces are ignored because of the assumption of ideal flow, there is no pressure force component because the jet flow is open to the atmosphere $(p_1 = p_2 = 0\,\text{N/m}^2)$, and there is no gravitational force component along the x- and y-axes because the pipe is horizontal $(z_1 = z_2 = 0)$.

$$\sum F_x = R_x = \rho Q(v_{2x} - v_{1x})$$

$$\sum F_y = R_y = \rho Q(v_{2y} - v_{1y})$$

$$v_1 := 6\,\frac{m}{sec} \qquad v_2 := 6\,\frac{m}{sec} \qquad Q := 0.09\,\frac{m^3}{sec} \qquad \alpha := 55\,deg$$

$$\rho := 998\,\frac{kg}{m^3} \qquad v_{1x} := v_1 = 6\,\frac{m}{s} \qquad v_{1y} := 0\,\frac{m}{sec}$$

$$v_{2x} := v_2 \cos(\alpha) = 3.441\,\frac{m}{s} \qquad\qquad v_{2y} := v_2 \sin(\alpha) = 4.915\,\frac{m}{s}$$

Guess value: $R_x := 1000\,N \qquad R_y := 1000\,N$

Given

$$-R_x = \rho \cdot Q\,(v_{2x} - v_{1x})$$

$$R_y = \rho \cdot Q\,(v_{2y} - v_{1y})$$

$$\begin{pmatrix} R_x \\ R_y \end{pmatrix} := \text{Find}\,(R_x, R_y) = \begin{pmatrix} 229.808 \\ 441.457 \end{pmatrix} N$$

$$R := \sqrt{R_x^2 + R_y^2} = 497.691\,N \qquad\qquad \theta := \text{atan}\left(\frac{|R_y|}{|R_x|}\right) = 62.5\,deg$$

Thus, the reaction force acting on the vane, R_x acts to the left, and the reaction force acting on the vane, R_y acts upwards, both as assumed and illustrated in Figure EP 5.6.

EXAMPLE PROBLEM 5.7

A forced jet of water at 20°C hits a moving vane, which is moving at a velocity of 4 m/sec (abs), at a velocity of 7 m/sec (abs), and a flowrate of $0.075 \text{ m}^3/\text{sec}$, as illustrated in Figure EP 5.7. The vane angle α is 45°, and assume that the jet and the vane lie in a horizontal plane (x, y) and that flow is ideal. (a) Determine the reaction force acting on the moving vane due to the thrust force of the jet (anchoring force required to keep the moving vane from accelerating).

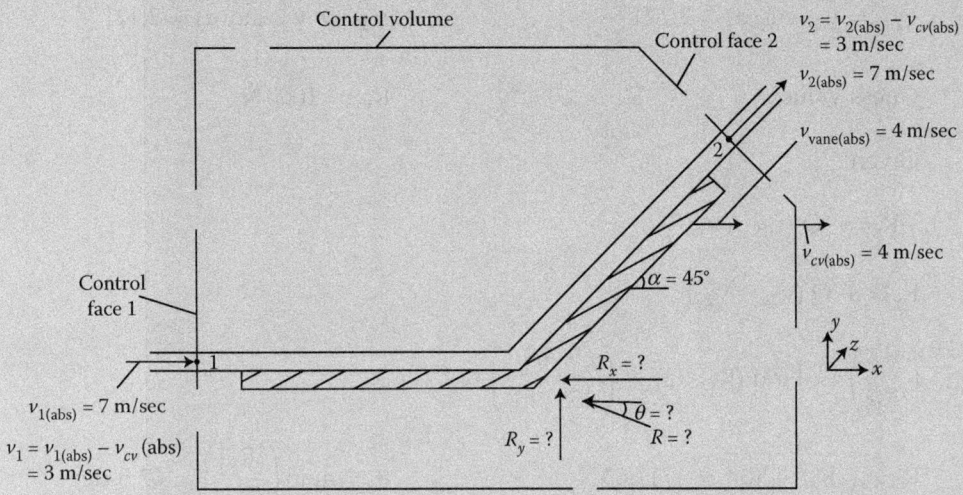

FIGURE EP 5.7
Forced jet of water hits a moving horizontal vane angled at 45°.

Mathcad Solution

(a) In order to determine the reaction force acting on the moving vane due to the thrust force of the jet, the integral form of the momentum equation is applied in both the x and y directions for the control volume illustrated in Figure EP 5.7. The flow enters at control face 1 and leaves at control face 2. The viscous forces are ignored because of the assumption of ideal flow, there is no pressure force component because the jet flow is open to the atmosphere $(p_1 = p_2 = 0 \text{ N/m}^2)$, and there is no gravitational force component along the x- and y-axes because the pipe is horizontal $(z_1 = z_2 = 0)$. Furthermore, the velocity of the jet flow at control faces 1 and 2 relative to the control volume, v_1 and v_2, respectively, are computed by applying Equation 5.29.

$$\sum F_x = R_x = \rho Q(v_{2x} - v_{1x})$$

$$\sum F_y = R_y = \rho Q(v_{2y} - v_{1y})$$

$$v_{1abs} := 7\frac{m}{sec} \qquad v_{2abs} := 7\frac{m}{sec} \qquad Q := 0.075\frac{m^3}{sec} \qquad \alpha := 45 \deg$$

$$v_{vaneabs} := 4 \frac{m}{sec} \qquad\qquad v_{cvabs} := v_{vaneabs} := 4 \frac{m}{s}$$

$$v_1 := v_{1abs} - v_{cvabs} = 3 \frac{m}{s} \qquad\qquad v_2 := v_{2abs} - v_{cvabs} = 3 \frac{m}{s}$$

$$\rho := 998 \frac{kg}{m^3} \qquad v_{1x} := v_1 = 3 \frac{m}{s} \qquad\qquad v_{1y} := 0 \frac{m}{sec}$$

$$v_{2x} := v_2 \cdot \cos(\alpha) = 2.121 \frac{m}{s} \qquad\qquad v_{2y} := v_2 \cdot \sin(\alpha) = 2.121 \frac{m}{s}$$

Guess value: $\qquad\qquad R_x := 1000 \, N \qquad R_y := 1000 \, N$

Given

$$-R_x = \rho \cdot Q \, (v_{2x} - v_{1x})$$

$$R_y = \rho \cdot Q \, (v_{2y} - v_{1y})$$

$$\begin{pmatrix} R_x \\ R_y \end{pmatrix} := Find \, (R_x, R_y) = \begin{pmatrix} 65.769 \\ 158.781 \end{pmatrix} N$$

$$R := \sqrt{R_x^2 + R_y^2} = 171.863 \, N \qquad\qquad \theta := atan\left(\frac{|R_y|}{|R_x|}\right) = 67.5 \, deg$$

Thus, the reaction force acting on the moving vane, R_x acts to the left, and the reaction force acting on the moving vane, R_y acts upwards, both as assumed and illustrated in Figure EP 5.7.

5.3 Application of the Energy Equation in Complement to the Momentum Equation

While the application of the continuity equation will always be necessary for internal flow, the energy equation and the momentum equations play complementary roles in the analysis of a given internal or external flow situation. The energy equation and the momentum equations are known as the equations of motion. They play complementary roles because when one of the two equations breaks down, the other may be used to find the additional unknown quantity. It is important to note that, similar to the continuity equation, the momentum equation may be applied using either the integral (for a finite control volume) or the differential approach. However, application of the energy equation is useful only in the integral approach (for a finite control volume), as energy is work, which is defined over distance, L. In the application of the governing equations (continuity, energy, and momentum), it is important to make the distinction between real flow, ideal flow, and a hydraulic jump. And for ideal flow, it is important to make the distinction between internal flow (pipe or open channel flow), external flow, and the flow from a tank (jet and siphon flows) (see Table 4.1) (Section 5.3.1). Furthermore, in the application of the governing equations, although in some flow cases, one may ignore (or not need to model) the flow resistance, in most practical flow cases in general, one must account for the fluid viscosity and thus model the resulting shear stress and flow resistance. The modeling of flow resistance requires a "subset level" application of the governing equations (Section 5.3.2).

5.3.1 Application of the Governing Equations

In the application of the governing equations (continuity, energy, and momentum), it is important to make the distinction between real flow, ideal flow, and a hydraulic jump. And for ideal flow, it is important to make the distinction between internal flow, external flow, and flow from a tank (see Table 4.1). While the energy equation derived based on the conservation of momentum principle given by Equation 4.44 does not consider the modeling of energy transfer by work (pumps, turbines, etc.), the energy equation derived based on the conservation of energy principle given by Equation 4.181 does and is repeated as follows:

$$\left(\frac{p_1}{\rho_1 g} + \frac{v_1^2}{2g} + z_1\right) + h_{pump} = \left(\frac{p_2}{\rho_2 g} + \frac{v_2^2}{2g} + z_2\right) + h_{turbine} + h_{f,total} \tag{5.32}$$

Thus, the integral form of the energy equation for incompressible, real, steady flow (pipe or open channel flow) given by Equation 5.32 may be used to solve for an unknown "energy head" term (p/γ, z, or $v^2/2g$) for one of the two points in a finite control volume; an unknown friction head loss, $h_{f,total} = h_{f,maj} + h_{f,min}$ between two points in a finite control volume; an unknown head delivered to the fluid by a pump, h_{pump}; or an unknown head removed from the fluid by a turbine, $h_{turbine}$. Application of Equation 5.32 requires evaluation of the following terms: h_{min}, h_{maj}, h_{pump}, and $h_{turbine}$ (Section 5.3.3).

For the assumption of incompressible, ideal, steady flow (internal or external flow, or flow from a tank), the friction head loss terms, and the pump and turbine head terms in Equation 5.32 would be set to zero, and the equation would then be known as the Bernoulli equation (Equation 4.46) and is repeated as follows:

$$\left(\frac{p}{\gamma} + z + \frac{v^2}{2g}\right)_1 - \left(\frac{p}{\gamma} + z + \frac{v^2}{2g}\right)_2 = 0 \tag{5.33}$$

which is applied for problems that affect the flow over a short pipe or channel section, external flow, or flow from a tank, where the major head loss due to flow resistance is assumed to be negligible. Thus, the Bernoulli equation may be solved for an unknown "energy head" term (p/γ, z, or $v^2/2g$) for one of the two points in a finite control volume (Sections 5.3.4 and 5.3.5).

And finally, for the assumption of incompressible, real, steady open channel flow that results in a hydraulic jump, while the friction head loss term and the pump and turbine head terms in Equation 5.32 would be set to zero, there is a major head loss term to model. Specifically, while the major head loss due to a hydraulic jump is not due to frictional losses (flow resistance), it is due to flow turbulence due to the jump. Thus, the energy equation for a hydraulic jump is given as follows:

$$\left(\frac{p}{\gamma}+z+\frac{v^2}{2g}\right)_1 -h_{f,maj} = \left(\frac{p}{\gamma}+z+\frac{v^2}{2g}\right)_2 \tag{5.34}$$

which may be solved for the major head loss due to the jump (Section 5.3.6).

For the assumption of real flow, ideal flow, or a hydraulic jump, for the respective energy equation (Equations 5.32, 5.33, or Equation 5.34), the integral form of the momentum equation (Equation 5.27) serves as the complementary equation of motion, and is repeated as follows:

$$\sum F_s = (F_G + F_P + F_V + F_{other})_s = (\rho Q v_s)_2 - (\rho Q v_s)_1 \tag{5.35}$$

which may be solved for an unknown force (typically, a pressure force (or depth of flow) or a reaction force); note that the viscous forces are typically accounted for on a "subset level" (indirectly) through the major and minor head loss flow resistance terms in the energy equation above (Equation 5.32) (see Sections 5.3.2 and 5.3.3 below), through an actual discharge (see Sections 5.3.2 and 5.3.4 below), or through a drag force (see Sections 5.3.2 and 5.3.5 below). As such, the integral form of the momentum equation is solved analytically, without the use of dimensional analysis. Thus, when one of these two equations of motion breaks down, the other may be used to find the additional unknown.

Furthermore, in the assumption of real flow, and for the specific case of a major head loss due to pipe or channel flow resistance, application of the differential form of the momentum equation may become necessary. Specifically, application of the differential form of the momentum equation will be necessary when it is of interest to trace the gradual variation of the steady nonuniform channel depth of flow (Equation 5.25) (see Chapter 9), and it is given as follows:

$$\frac{dy}{ds} = \frac{S_o - S_f}{1 - F^2} \tag{5.36}$$

or to evaluate the flow (solve for an unknown quantity) at a given point in the fluid system for any uniform pipe or channel section (Equation 5.8), which is given as follows:

$$S_f = \frac{\tau_w}{\gamma R_h} \tag{5.37}$$

Application of Equations 5.36 and 5.37 requires evaluation of the friction slope term, S_f, which is modeled on a "subset level," where the major head loss due to flow resistance is determined by application of Equation 4.85, which is repeated as follows (see Sections 5.3.2 and 5.3.3 below):

$$h_{f,maj} = S_f L = \frac{\tau_w}{\gamma R_h} L \tag{5.38}$$

(see Chapters 8 and 9 for their applications).

5.3.2 Modeling Flow Resistance: A Subset Level Application of the Governing Equations

In the application of the governing equations (Section 5.3.1), although in some flow cases, one may ignore (or not need to model) the flow resistance, in most practical flow cases in general, one must account for the fluid viscosity and thus model the resulting shear stress and flow resistance. As such, in the application of the governing equations (continuity, energy, and momentum), it is important to make the distinction between real flow, ideal flow, and a hydraulic jump. And for ideal flow, it is important to make the distinction between internal (pipe or open channel flow), external flow, and flow from a tank. While the continuity, energy, and momentum equations are applied (as necessary) to internal flows (pipe and open channel flow) and to flow from a tank, etc., only the energy and momentum equations are applied to external flows. The application of the governing equations (continuity, energy, and momentum) for real flow (internal), and ideal flow (flow-measuring devices for internal flow, and external flow around objects) requires the modeling of flow resistance (frictional losses). The modeling of flow resistance requires a "subset level" application of the governing equations. However, in the application of the governing equations (continuity, energy, and momentum) for ideal flow (flow from a tank, etc., and gradual channel transitions), and a hydraulic jump, there is no flow resistance to model. While in the case of ideal flow (flow from a tank, etc., and gradual channel transitions) the flow resistance is actually ignored, in the case of the hydraulic jump, the major head loss is due to flow turbulence due to the jump and not flow resistance (frictional losses).

Thus, the application of the governing equations (continuity, energy, and momentum) for real flow (internal) and ideal flow (flow-measuring devices for internal flow and external flow around objects) requires the modeling of flow resistance (frictional losses). Specifically, the application of the integral form of the energy equation for incompressible, real, steady flow (pipe or open channel flow) given by Equation 4.44 requires the evaluation of the major and minor head loss flow resistance terms, $h_{f,maj} = \tau_w L/\gamma R_h$ and $h_{f,min} = \tau_w L/\gamma R_h$, respectively. Furthermore, the application of the differential form of the momentum equation (pipe or open channel flow) given by Equation 5.15 requires the evaluation of the friction slope term, $S_f = \tau_w/\gamma R_h$. The application of the continuity equation for a flow-measuring device (pipe or open channel flow) given by Equation 3.47 requires the evaluation the actual discharge, $Q_a = v_a A_a$. And, finally, the application of the integral form of the momentum equation for the case of external flow around an object given by Equation 5.27 requires the evaluation of the drag force, $F_D = (F_P + F_f)_s$.

The flow resistance in internal and external flow (major head loss, minor head loss, actual discharge, and drag force) is modeled by a "subset level" application of the appropriate governing equations, as presented in Chapter 6 (and summarized in Sections 5.3.3, 5.3.4, and 5.3.5 below). A "subset level" application of the appropriate governing equations focuses only on the given element causing the flow resistance. Thus, assuming that flow resistance is to be accounted for, depending upon the specific flow situation, the flow type is assumed to be either a "real" flow or an "ideal" flow. The distinction between ideal and real flow determines how the flow resistance is modeled in the "subset level" application of the appropriate governing equations. The assumption of "real" flow implies that the flow resistance is modeled in both the energy and the momentum equations. However, the assumption of "ideal" flow implies that the flow resistance is modeled only in the momentum equation (and thus the subsequent assumption of "real" flow). The flow resistance due to friction in pipe and open channel flow (modeled as a major head loss) and the flow resistance due to friction in most pipe devices (modeled as a minor head loss) are modeled by assuming "real" flow; application of the energy equation models a head loss that is accounted for by a drag coefficient (in the application of the momentum equation). However, the flow resistance due to friction in flow-measuring devices for internal flow (pipe and open channel flow, modeled as an actual discharge) and the flow resistance due to friction in external flow around objects (modeled as a drag force) are modeled by assuming "ideal" flow; although application of the Bernoulli equation does not model a head loss in the determination of ideal velocity, the associated minor head loss with the velocity measurement is accounted for by a drag coefficient (in the application of the momentum equation, where a subsequent assumption of "real" flow is made). Furthermore, the use of dimensional analysis (Chapter 7) in supplement to a "subset level" application of the appropriate governing equations is needed to derive flow resistance equations when there is no theory to model the friction force (as in the case of turbulent internal and external flow). And, finally, it is important to note that the use of dimensional analysis is not needed in the derivation of the major head loss term for laminar flow (friction force is theoretically modeled), and the minor head loss term for a sudden pipe expansion (friction force is ignored).

5.3.3 Application of the Energy and Momentum Equations for Real Internal Flow

Application of the integral form of the energy equation for incompressible, real, steady flow (pipe or open channel flow) given by Equation 5.32 requires the derivation/evaluation of the friction head loss term, $h_{f,total} = h_{f,maj} + h_{f,min}$; the head delivered to the fluid by a pump, h_{pump}; and the head removed from the fluid by a turbine, $h_{turbine}$. Evaluation of the major and minor head loss flow resistance terms requires a "subset level" application of the governing equations (Section 5.3.3.1). Then, application of the complementary integral momentum equation (Equation 5.35) is required only if there is an unknown reaction force, as in the case of certain pipe components (such as contractions, nozzles, elbows, and bends); note that the viscous forces have already been accounted for indirectly through the minor head loss flow resistance term in the energy equation. As such, the integral form of the momentum equation is solved analytically, without the use of dimensional analysis.

The evaluation of the total head loss, $h_{f,total} = h_{f,maj} + h_{f,min}$ may include both major and minor losses, which are due to flow resistance. The major head loss in pipe or open channel flow is due to the frictional losses/flow resistance caused by pipe friction or channel resistance, respectively. The minor head loss is due to flow resistance due various pipe components (valves, fittings [tees, unions, elbows, and bends], entrances, exits, contractions, and expansions), pipe flow-measuring devices (pitot-static tube, orifice, nozzle, or venturi meters), or open channel flow-measuring devices (pitot-static tube, sluice gates, weirs, spillways, venturi flume, and contracted openings). One may note that in the case of a flow-measuring device for either pipe flow or open channel flow, although the associated minor head loss may be accounted for in the integral form of the energy equation (see Equation 5.32), it is typically accounted for by the use/calibration of a discharge coefficient to determine the actual discharge, where ideal flow (Bernoulli equation) is assumed (see Section 5.3.4 below). Applications of the governing equations for real internal flow are presented in Sections 5.3.3.3, 5.3.3.4, and 5.3.3.5 below.

5.3.3.1 Evaluation of the Major and Minor Head Loss Terms

Evaluation of the major and minor head loss flow resistance terms requires a "subset level" of application of the governing equations. A "subset level" application of the governing equations focuses only on the given element causing the flow resistance. The assumption of real flow implies that the flow resistance is modeled in both the energy and momentum equations. The flow resistance equations for the major and minor head loss are derived in Chapters 6 and 7. In the derivation of one given source of head loss (major or minor), first, the integral form of the energy equation is applied to solve for the unknown head loss, h_f, which is given as follows:

$$\left(\frac{p}{\gamma} + z + \frac{v^2}{2g}\right)_1 - h_f = \left(\frac{p}{\gamma} + z + \frac{v^2}{2g}\right)_2 \tag{5.39}$$

One may note that in the derivation of the expression for one given source of head loss, h_f (major or minor head loss due to flow resistance), the control volume is defined between the two points that include only that one source of head loss. The derivation of the major head loss assumes $z_1 = z_2$, and $v_1 = v_2$, as follows:

$$\left(\frac{p}{\gamma} + z + \frac{v^2}{2g}\right)_1 - h_{f,maj} = \left(\frac{p}{\gamma} + z + \frac{v^2}{2g}\right)_2 \tag{5.40}$$

which yields $h_{f,maj} = S_f L = (\tau_w L / \gamma R_h) = (\Delta p / \gamma)$. And, the derivation of the minor head loss assumes $z_1 = z_2$, and $v_1 = v_2$, as follows:

$$\left(\frac{p}{\gamma} + z + \frac{v^2}{2g}\right)_1 - h_{f,min} = \left(\frac{p}{\gamma} + z + \frac{v^2}{2g}\right)_2 \tag{5.41}$$

which yields $h_{f,min} = (\tau_w L / \gamma R_h) = (\Delta p / \gamma)$. Then, second, the integral form of the momentum equation (supplemented by dimensional analysis) is applied to solve for an unknown pressure drop, Δp (in the case of major or minor head loss due to pipe flow resistance) or an unknown change in channel depth, Δy (in the case of major head loss due to open channel flow resistance), which was given in Equation 5.27 and is repeated as follows:

$$\sum F_s = (F_G + F_P + F_V + F_{other})_s = (\rho Q v_s)_2 - (\rho Q v_s)_1 \tag{5.42}$$

One may note that derivation of the major head loss involves both the integral and the differential momentum equations. Thus, the resulting flow resistance equation for the major head loss is given as follows:

$$h_{f,maj} = S_f L = \frac{\tau_w L}{\gamma R_h} = \frac{\Delta p}{\gamma} = \frac{v^2}{C^2 R_h} L = f \frac{L}{D} \frac{v^2}{2g} = \left(\frac{vn}{R_h^{2/3}}\right)^2 L = \left[\frac{v}{0.849 C_h R_h^{0.63}}\right]^{1/0.54} L \tag{5.43}$$

where C is the Chezy coefficient, f is the Darcy–Weisbach friction factor, n is the Manning's roughness coefficient, and C_h is the Hazen–Williams roughness coefficient. And, the resulting flow resistance equation for the minor head loss is given as follows:

$$h_{f,min} = \frac{\tau_w L}{\gamma R_h} = \frac{\Delta p}{\gamma} = k \frac{v^2}{2g} \tag{5.44}$$

where k is the minor head loss coefficient. Furthermore, each head loss flow resistance equation represents the simultaneous application of the two complementary equations of motion (on a "subset level"). It is important to note that the use of dimensional analysis is not needed either in the derivation of the major head loss term for laminar flow or for the minor head loss term for a sudden pipe expansion.

The flow resistance due to pipe or channel friction, which results in a major head loss term, $h_{f,major}$, is modeled as a resistance force (shear stress or drag force) in the integral form of the momentum equation (Equation 5.27), while it is modeled as a friction slope, S_f in the differential form of the momentum equation (Equation 5.37). As such, the energy equation may be used to solve for the unknown major head loss, h_f, while the integral momentum equation may be used to solve for the unknown pressure drop, Δp (or the unknown friction slope, S_f in the case of turbulent open channel flow). The major head loss, h_f is caused by both pressure and friction forces. Thus, when the friction/viscous forces can be theoretically modeled in the integral form of the momentum equation (application of Newton's law of viscosity in the laminar flow case), one can analytically determine the actual pressure drop, Δp from the integral form of the momentum equation. And, therefore, one may analytically derive an expression for the major head loss, h_f due to pipe or channel friction from the energy equation. However, when the friction/viscous forces cannot be theoretically modeled in the integral form of the momentum equation (as in the turbulent flow case), the friction/viscous forces, the actual pressure drop, Δp (or the unknown friction slope, S_f in the case of turbulent open channel flow) and thus the major head loss, h_f due to pipe or channel friction are determined empirically (see Chapter 7). Specifically, in the case of turbulent flow, the friction/viscous forces in the integral form of the momentum equation; thus, the wall shear stress, τ_w cannot be theoretically modeled, so the integral form of the momentum equation cannot be directly applied to solve for the unknown pressure drop, Δp. Thus, an empirical interpretation (using dimensional analysis) of the wall shear stress, τ_w in the theoretical expression for the friction slope, $S_f = \tau_w/\gamma R_h$ in the differential form of the momentum equation is sought in terms of velocity, v and a flow resistance coefficient. This yields the Chezy equation ($S_f = v^2/C^2 R_h$, which represents the differential form of the momentum equation, supplemented by dimensional analysis and guided by the integral momentum equation; link between the differential and integral momentum equations), which is used to obtain an empirical evaluation for the pressure drop, Δp in Newton's second law of motion (integral form of the momentum equation). As such, one can then derive an empirical expression for the major head loss, h_f from the energy equation. Major losses are addressed in detail in Chapters 6, 8, and 9.

The flow resistance due to various pipe components, which results in a minor head loss term, $h_{f,minor}$, is modeled as a resistance force (shear stress or drag force) in the integral form of the momentum equation (Equation 5.27). As such, the energy equation may be used to solve for the unknown minor head loss, h_f, while the integral form of the momentum equation may be used to solve for the unknown pressure drop, Δp. The minor head loss, h_f is caused by both pressure and friction forces. Thus, when the friction/viscous forces are insignificant and can be ignored in the integral form of the momentum equation (as in the case of the sudden pipe expansion), one can analytically determine the actual pressure drop, Δp from the momentum equation. Therefore, one may analytically derive an expression for the minor head loss, h_f due to pipe components from the energy equation. However, when the friction/viscous forces are significant and cannot be ignored in the integral form of the momentum equation (as in the case of the other pipe components), the friction/viscous forces; the actual pressure drop/rise, Δp; and thus the minor head loss, h_f are determined empirically (see Chapter 7). Specifically, in the case of turbulent pipe component flow, the friction/viscous forces in the integral form of the momentum equation and thus the wall shear stress, τ_w cannot be theoretically modeled; thus, the integral form of the momentum equation cannot be directly applied to solve for the unknown pressure drop, Δp. The evaluation of the minor head loss term, $h_{f,minor}$ involves supplementing the integral form of the momentum equation with dimensional analysis, as described in detail in Chapters 6, 7, and 8.

5.3.3.2 Evaluation of the Pump and Turbine Head Terms

Evaluation of the head delivered to the fluid by a pump, h_{pump} and the head removed/ extracted from the fluid by a turbine, $h_{turbine}$ require the application of the appropriate power equation, which are discussed in Section 4.7.2.4, and the resulting equations are summarized, respectively, as follows:

$$h_{pump} = \frac{(P_p)_{out}}{\gamma Q} \tag{5.45}$$

$$h_{turbine} = \frac{(P_t)_{in}}{\gamma Q} \tag{5.46}$$

5.3.3.3 Applications of the Governing Equations for Real Internal Flow

Real internal flow includes pipe or open channel flow, and may be treated as either a differential fluid element or a finite control volume. The governing equations (continuity, energy, and momentum) for real flow are applied in the analysis and design of both pipe systems and open channel flow systems. Applications of the governing equations for real pipe flow include the analysis and design of pipe systems that include a numerous arrangements of pipes including: (1) single pipes, (2) pipes with components, (3) pipes with a pump or a turbine, (4) pipes in series, (5) pipes in parallel, (6) branching pipes, (7) pipes in a loop, and (8) pipe networks (see Chapter 8). And, applications of the governing equations for real open channel flow include the analysis and design of uniform and nonuniform flow (generated by controls such as a break in channel slope, a hydraulic drop, or a free overfall) (see Chapter 9). One may note from Section 5.3.2 above that evaluation of the major and minor head loss terms in the energy equation requires a "subset level" application of the governing equations. Furthermore, the resulting definitions for the major and minor head loss (flow resistance) terms (see Equations 5.43 and 5.44, respectively) require details in the definition of variables (C, f, v, L, D, and k), which are to be presented in later chapters. Therefore, example problems illustrating the application of the energy equation for real pipe or open channel flow in Chapter 5 will assume that the minor and major head loss terms are either a given quantity (lumped value) computed from the given head loss equations or directly solved for in the following energy equation:

$$\left(\frac{p_1}{\rho_1 g} + \frac{v_1^2}{2g} + z_1 \right) + h_{pump} = \left(\frac{p_2}{\rho_2 g} + \frac{v_2^2}{2g} + z_2 \right) + h_{turbine} + h_{f,maj} + h_{f,min} \tag{5.47}$$

For the example problems involving a pump or a turbine, illustration of the evaluation of the head delivered by a pump and the head removed by a turbine, which require the application of the appropriate power equation, were presented in Section 4.7.4. Furthermore, for the example problems involving certain pipe components, illustration of the computation of the unknown reaction force, which requires the application of the complementary integral momentum equation, are presented below. And, finally, example problems illustrating the application of the differential momentum equation are presented below. Thus, illustrated in

Chapter 5 are applications of the governing equations for the simplest example problems of real pipe and open channel flow.

5.3.3.4 Applications of the Governing Equations for Real Internal Flow: Differential Fluid Element

The governing equations (continuity, energy, and momentum) for real internal flow, for a differential fluid element, are applied in the analysis and design of both pipe systems and open channel flow systems. Application of the differential form of the momentum equation for real internal flow will be necessary when it is of interest to trace the gradual variation of the steady nonuniform channel depth of flow or to evaluate the flow (solve for an unknown quantity) at a given point in the fluid system for any uniform pipe or channel section. Evaluation of the friction slope, S_f requires a "subset level" application of the governing equations, where the major head loss due to flow resistance is determined by application of Equation 5.38. Example problems involving nonuniform flow are presented in Chapter 9, while example problems involving uniform pipe or channel sections were presented in Example Problems 5.1 and 5.2 above and are presented in Example Problems 5.8 to 5.12 below and in Chapters 8 and 9.

The governing equations are applied in the analysis and design of real pipe flow problems. The continuity equation is applied to solve for an unknown velocity, discharge, or pipe size. The differential form of the momentum equation (the Chezy equation, the Darcy–Weisbach equation, the Manning's equation, or the Hazen–Williams equation) is applied to solve for an unknown quantity in a straight pipe section. The energy equation is applied to solve for an unknown major head loss in the straight pipe section.

EXAMPLE PROBLEM 5.8

Water at 20°C flows in a 4-m-diameter horizontal straight pipe section, as illustrated in Figure EP 5.8. The pipe length between points 1 and 2 is 1300 m; the friction slope between points 1 and 2 is 0.0005 m/m; and the Chezy coefficient, C for the

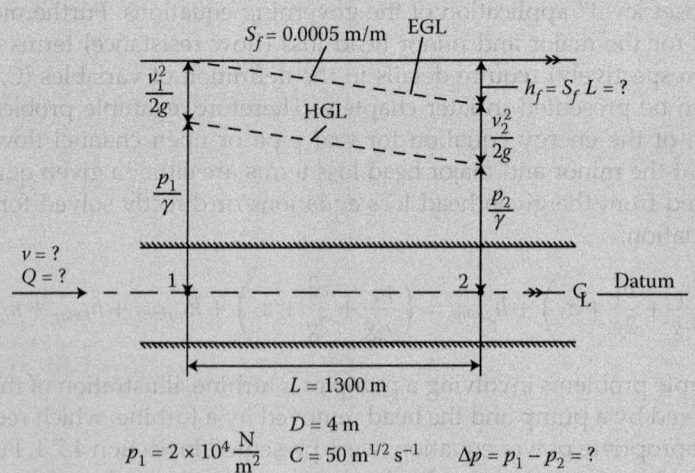

FIGURE EP 5.8
Water flows through a horizontal pipe flowing under pressure.

pipe is $50\,\mathrm{m}^{1/2}\,\mathrm{s}^{-1}$. The pressure at point 1 is $2 \times 10^4\,\mathrm{N/m}^2$. (a) Determine the velocity of flow and discharge in the straight pipe section. (b) Determine the head loss due to pipe friction between points 1 and 2. (c) Determine the pressure drop at point 2 due to the pipe friction.

Mathcad Solution

(a) In order to determine the velocity at any point in the straight pipe section, the Chezy equation is applied. Furthermore, in order to determine the discharge in the pipe, the continuity equation is applied as follows:

$$D := 4\,\mathrm{m} \qquad\qquad A := \frac{\pi \cdot D^2}{4} = 12.566\,\mathrm{m}^2 \qquad\qquad P_w := \pi \cdot D = 12.566\,\mathrm{m}$$

$$R_h := \frac{A}{P_w} = 1\,\mathrm{m} \qquad\qquad S_f := 0.0005\,\frac{\mathrm{m}}{\mathrm{m}} \qquad\qquad C := 50\,\mathrm{m}^{\frac{1}{2}}\,\mathrm{sec}^{-1}$$

$$\text{Guess values:} \qquad\qquad v := 1\,\frac{\mathrm{m}}{\mathrm{sec}} \qquad\qquad Q := 1\,\frac{\mathrm{m}^3}{\mathrm{sec}}$$

Given

$$S_f = \frac{v^2}{C^2 \cdot R_h}$$

$$Q = v \cdot A$$

$$\begin{pmatrix} v \\ Q \end{pmatrix} := \text{Find}\,(v, Q)$$

$$v = 1.118\,\frac{\mathrm{m}}{\mathrm{s}} \qquad\qquad Q = 14.05\,\frac{\mathrm{m}^3}{\mathrm{s}}$$

(b) In order to determine the head loss due to pipe friction between points 1 and 2, the head loss due to pipe friction equation is applied as follows:

$$L := 1300\,\mathrm{m} \qquad\qquad h_f := S_f \cdot L = 0.65\,\mathrm{m}$$

(c) In order to determined the pressure drop at point 2 due to the pipe friction, the energy equation is applied between points 1 and 2 as follows:

$$v_1 := v = 1.118\,\frac{\mathrm{m}}{\mathrm{s}} \qquad v_2 := v = 1.118\,\frac{\mathrm{m}}{\mathrm{s}} \qquad p_1 := 2 \times 10^4\,\frac{\mathrm{N}}{\mathrm{m}^2}$$

$$z_1 := 0\,\mathrm{m} \qquad\qquad z_2 := 0\,\mathrm{m}$$

$$\rho := 998\,\frac{\mathrm{kg}}{\mathrm{m}^3} \qquad g := 9.81\,\frac{\mathrm{m}}{\mathrm{sec}^2} \qquad \gamma := \rho \cdot g = 9.79 \times 10^3\,\frac{\mathrm{kg}}{\mathrm{m}^2 \cdot \mathrm{s}^2}$$

$$\text{Guess value:} \qquad\qquad p_2 := 1 \times 10^4\,\frac{\mathrm{N}}{\mathrm{m}^2}$$

Given

$$\frac{p_1}{\gamma} + z_1 + \frac{v_1{}^2}{2 \cdot g} - h_f = \frac{p_2}{\gamma} + z_2 + \frac{v_2{}^2}{2 \cdot g}$$

$$p_2 := \text{Find } (p_2) = 1.364 \times 10^4 \frac{N}{m^2}$$

$$\Delta p := p_1 - p_2 = 6.364 \times 10^3 \frac{N}{m^2}$$

EXAMPLE PROBLEM 5.9

Water at 20°C flows in a 6-m-diameter horizontal straight pipe section, as illustrated in Figure EP 5.9. The pipe length between points 1 and 2 is 800 m; the friction slope between points 1 and 2 is 0.0009 m/m; and the friction factor, f for the pipe is 0.018. The pressure at point 1 is 4×10^4 N/m². (a) Determine the velocity of flow and discharge in the straight pipe section. (b) Determine the head loss due to pipe friction between points 1 and 2. (c) Determine the pressure drop at point 2 due to the pipe friction.

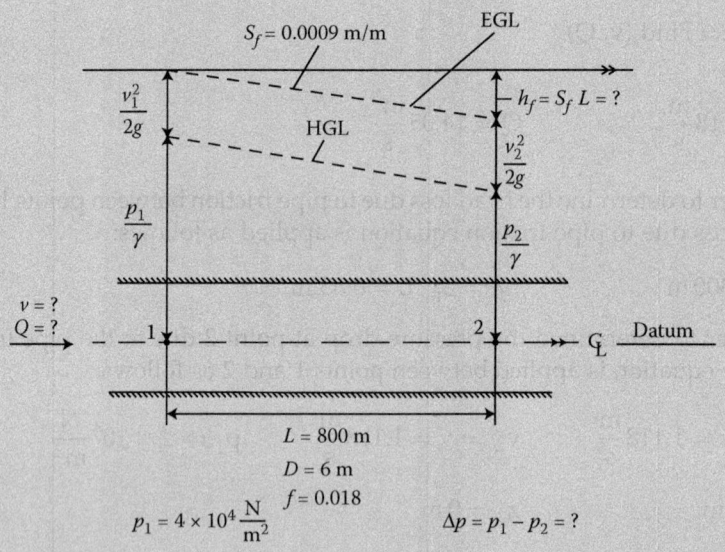

FIGURE EP 5.9
Water flows through a horizontal pipe flowing under pressure.

Mathcad Solution

(a)–(b) In order to determine the velocity at any point in the straight pipe section, the Darcy–Weisbach equation is applied. And in order to determine the discharge in the pipe, the continuity equation is applied. Furthermore, in order to determine the head loss due to pipe friction between points 1 and 2, the head loss due to pipe friction equation is applied as follows:

$$D := 6\,m \qquad A := \frac{\pi \cdot D^2}{4} = 28.274\,m^2 \qquad P_w := \pi \cdot D = 18.85\,m$$

$$R_h := \frac{A}{P_w} = 1.5\,m \qquad S_f := 0.0009\,\frac{m}{m} \qquad f := 0.018 \qquad L := 800\,m$$

Guess values:
$$v := 1\,\frac{m}{sec} \qquad Q := 1\,\frac{m^3}{sec} \qquad h_f := 1\,m$$

Given

$$h_f = f\frac{L}{D}\frac{v^2}{2 \cdot g} \qquad h_f = S_f \cdot L \qquad Q = v \cdot A$$

$$\begin{pmatrix} v \\ Q \\ h_f \end{pmatrix} := \text{Find }(v, Q, h_f)$$

$$v = 2.426\,\frac{m}{s} \qquad Q = 68.597\,\frac{m^3}{s} \qquad h_f = 0.72\,m$$

(c) In order to determine the pressure drop at point 2 due to the pipe friction, the energy equation is applied between points 1 and 2 as follows:

$$v_1 := v = 2.426\,\frac{m}{s} \qquad v_2 := v = 2.426\,\frac{m}{s} \qquad p_1 := 4 \times 10^4\,\frac{N}{m^2}$$

$$z_1 := 0\,m \qquad z_2 := 0\,m$$

$$\rho := 998\,\frac{kg}{m^3} \qquad g := 9.81\,\frac{m}{sec^2} \qquad \gamma := \rho \cdot g = 9.79 \times 10^3\,\frac{kg}{m^2 \cdot s^2}$$

Guess value:
$$p_2 := 1 \times 10^4\,\frac{N}{m^2}$$

Given

$$\frac{p_1}{\gamma} + z_1 + \frac{v_1{}^2}{2 \cdot g} - h_f = \frac{p_2}{\gamma} + z_2 + \frac{v_2{}^2}{2 \cdot g}$$

$$p_2 := \text{Find }(p_2) = 3.295 \times 10^4\,\frac{N}{m^2}$$

$$\Delta p := p_1 - p_2 = 7.049 \times 10^3\,\frac{N}{m^2}$$

EXAMPLE PROBLEM 5.10

Water at 20°C flows at a velocity of 2.6 m/sec in a horizontal straight pipe section, as illustrated in Figure EP 5.10. The pipe length between points 1 and 2 is 600 m; the friction slope between points 1 and 2 is 0.0075 m/m; and the Hazen–Williams roughness coefficient, C_h for the pipe is 120. The pressure at point 1 is $5.4 \times 10^4 \, \text{N/m}^2$. (a) Determine the pipe size diameter and discharge in the straight pipe section. (b) Determine the head loss due to pipe friction between points 1 and 2. (c) Determine the pressure drop at point 2 due to the pipe friction.

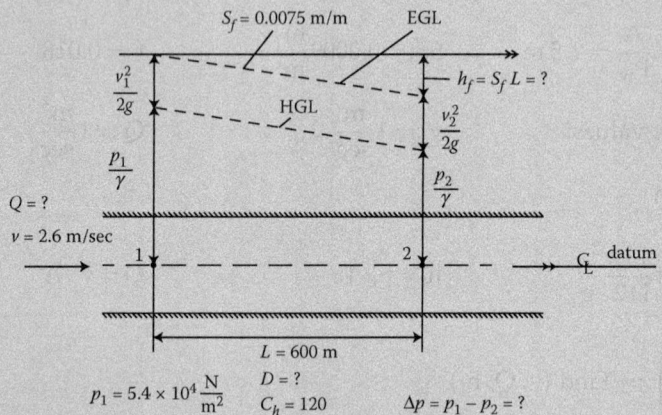

FIGURE EP 5.10
Water flows through a horizontal pipe flowing under pressure.

Mathcad Solution

(a)–(b) In order to determine the pipe size diameter of the straight pipe section, the Hazen–Williams equation is applied. And in order to determine the discharge in the pipe, the continuity equation is applied. Furthermore, in order to determine the head loss due to pipe friction between points 1 and 2, the head loss due to pipe friction equation is applied. It is important to note that the empirically calibrated coefficients (0.849, 0.63, and 0.54) in the Hazen–Williams equation represent the assumed metric units for the variables (v [m], R_h [m], and S_f [m/m]) and the specific ranges assumed for the individual variables (v, D, and water temperature). Furthermore, although the empirically calibrated coefficients are not dimensionless, their units are embedded in the equation, and thus, unknown. As such, in order to successfully apply the Mathcad numerical solve block to solve for the unknown pipe diameter, D, the units for all the variables (except for C_h, which is already dimensionless) have been temporarily eliminated as follows:

$$v := 2.6 \qquad\qquad S_f := 0.0075 \qquad\qquad C_h := 120 \qquad\qquad L := 600$$

$$A(D) := \frac{\pi \cdot D^2}{4} \qquad\qquad P_w(D) := \pi \cdot D \qquad R_h(D) := \frac{A(D)}{P_w(D)}$$

Guess values: $D := 1$ $Q := 1$ $h_f := 1$

Given

$$h_f = \left(\frac{v}{0.849 \cdot C_h \cdot R_h(D)^{0.63}}\right)^{\frac{1}{0.54}} \cdot L \qquad h_f = S_f \cdot L \qquad Q = v \cdot A(D)$$

$$\begin{pmatrix} D \\ Q \\ h_f \end{pmatrix} := \text{Find} \, (D, Q, h_f)$$

$$D = 0.785 \qquad\qquad Q = 1.257 \qquad\qquad h_f = 4.5$$

Thus, the design pipe size diameter is 0.785 m, the discharge is 1.257 m^3/sec, and the head loss is 4.5 m.

(c) In order to determined the pressure drop at point 2 due to the pipe friction, the energy equation is applied between points 1 and 2 as follows:

$$v_1 := 2.6 \frac{m}{\sec} \qquad v_2 := 2.6 \frac{m}{\sec} \qquad p_1 := 5.4 \times 10^4 \frac{N}{m^2}$$

$$z_1 := 0 \, m \qquad\qquad z_2 := 0 \, m \qquad\qquad h_f := 4.5 \, m$$

$$\rho := 998 \frac{kg}{m^3} \qquad g := 9.81 \frac{m}{\sec^2} \qquad \gamma := \rho \cdot g = 9.79 \times 10^3 \frac{kg}{m^3 \cdot s^2}$$

Guess value: $\qquad p_2 := 1 \times 10^4 \frac{N}{m^2}$

Given

$$\frac{p_1}{\gamma} + z_1 + \frac{v_1^2}{2 \cdot g} - h_f = \frac{p_2}{\gamma} + z_2 + \frac{v_2^2}{2 \cdot g}$$

$$p_2 := \text{Find}(p_2) = 9.943 \times 10^3 \frac{N}{m^2}$$

$$\Delta p := p_1 - p2 = 4.406 \times 10^4 \frac{N}{m^2}$$

The governing equations are applied in the analysis and design of real open channel flow problems. The continuity equation is applied to solve for an unknown velocity, discharge, or channel size. The differential form of the momentum equation (the Chezy equation, or the Manning's equation) is applied to solve for an unknown quantity in a uniform open channel flow section. The energy equation is applied to solve for an unknown head loss in the uniform flow channel section.

EXAMPLE PROBLEM 5.11

Water at 20°C flows at a depth of 1.3 m and a velocity of 3 m/sec in a rectangular channel that is 0.95 m wide, as illustrated in Figure EP 5.11. The channel length between points 1 and 2 is 2400 m, and the Chezy coefficient, C is $90 \, m^{1/2}/s^{-1}$. (a) Determine the channel bottom slope. (b) Determine the head loss due to channel friction between points 1 and 2.

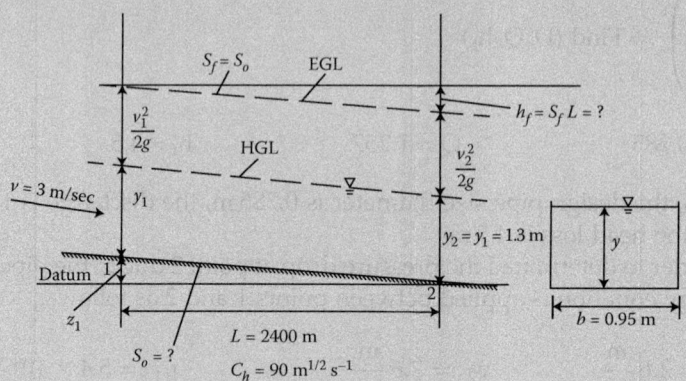

FIGURE EP 5.11
Water flows at uniform flow in a rectangular channel.

Mathcad Solution

(a) In order to determine the channel bottom slope, the Chezy equation is applied as follows:

$$y := 1.3 \, m \qquad\qquad v := 3 \frac{m}{sec} \qquad b := 0.95 \, m \qquad C := 90 \, m^{\frac{1}{2}} sec^{-1}$$

$$A := b \cdot y = 1.235 \, m^2 \quad P_w := 2 \cdot y + b = 3.55 \, m \qquad\qquad R_h := \frac{A}{P_w} = 0.348 \, m$$

Guess values: $\qquad\qquad S_o := 0.015 \dfrac{m}{m}$

Given

$$S_o = \frac{v^2}{C^2 \cdot R_h}$$

$$S_o := Find \, (S_o) = 3.194 \times 10^{-3}$$

(b) In order to determine the head loss due to channel friction between points 1 and 2, the head loss due to channel friction equation is applied as follows:

$$L := 2400 \, m \qquad\qquad\qquad h_f := S_o \, L = 7.665 \, m$$

EXAMPLE PROBLEM 5.12

Water at 20°C flows at a depth of 4.9 m in a rectangular channel that is 1.3 m wide, as illustrated in Figure EP 5.12. The channel length between points 1 and 2 is 1600 m, and the Manning's roughness coefficient, n is 0.018 $m^{-1/3}$s. The channel bottom slope is 0.025 m/m. (a) Determine the velocity of the flow. (b) Determine the head loss due to channel friction between points 1 and 2.

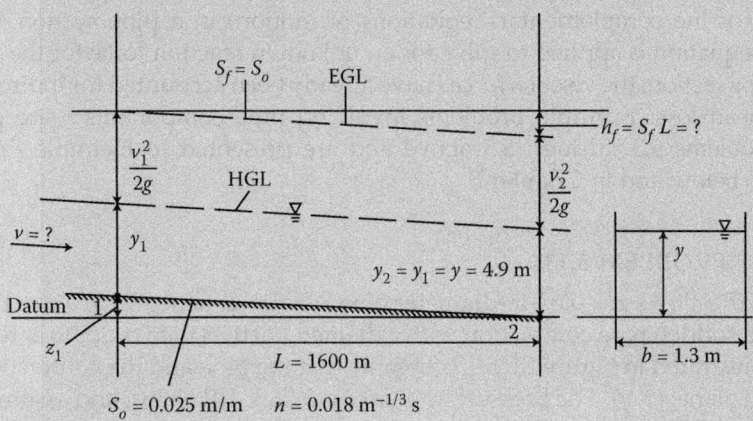

FIGURE EP 5.12
Water flows at uniform flow in a rectangular channel.

Mathcad Solution

(a) In order to determine the velocity of the flow, the Manning's equation is applied as follows:

$$y := 4.9\,m \qquad\qquad b := 1.3\,m \quad n := 0.018\,m^{\frac{-1}{3}}\,sec \qquad S_o := 0.025\,\frac{m}{m}$$

$$A := b \cdot y = 6.37\,m^2 \qquad P_w := 2 \cdot y + b = 11.1\,m \qquad R_h := \frac{A}{P_w} = 0.574\,m$$

Guess values: $\qquad\qquad v := 1\,\dfrac{m}{sec}$

Given

$$v = \frac{1}{n} \cdot R_h^{\frac{2}{3}} \cdot S_o^{\frac{1}{2}}$$

$$v := Find\,(v) = 6.066\,\frac{m}{s}$$

(b) In order to determine the head loss due to channel friction between points 1 and 2, the head loss due to channel friction equation is applied as follows:

$$L := 1600\,m \qquad h_f := S_o \cdot L = 40\,m$$

5.3.3.5 Applications of the Governing Equations for Real
Internal Flow: Finite Control Volume

The integral form of the governing equations (continuity, energy, and momentum) for real internal flow are applied to a finite control volume in the analysis and design of certain pipe components (such as contractions, nozzles, tees, elbows, and bends). The continuity equation is applied to solve for an unknown velocity, discharge, or pipe size. The energy equation is applied to solve for an unknown head or minor head loss (requires a "subset level" applications of the complementary equations of motion) in a pipe section. The integral momentum equation is applied to solve for an unknown reaction force for the pipe components in a pipe section; the viscous forces have already been accounted for indirectly through the minor head loss. Example problems involving pipe components were presented in Example Problems 5.3 through 5.5 above and are presented in Example Problems 5.13 through 5.15 below and in Chapter 8.

EXAMPLE PROBLEM 5.13

Water at 20°C flows in a 0.65-m-diameter pipe at a flowrate of 0.98 m³/sec. The pipe is fitted with sudden pipe contraction with a diameter of 0.35 m in order to accelerate the flow, as illustrated in Figure EP 5.13. Assume that the pipe and the contraction lie in a horizontal plane, (x, y). The pressure at point 2 is 3.56×10^4 N/m², and assume a minor head loss coefficient due to the sudden pipe contraction, k of 0.25. (a) Determine the minor head loss due to the contraction. (b) Determine the pressure at point 1. (c) Determine the reaction force acting on the pipe contraction due to the thrust force of the flow (anchoring force required to keep the pipe contraction from moving).

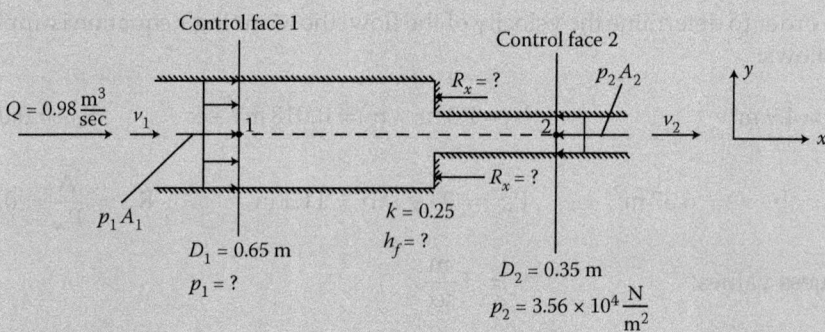

FIGURE EP 5.13
Water flows through a horizontal pipe with a sudden pipe contraction to accelerate the flow.

Mathcad Solution

(a)–(b) The minor head loss due to the pipe contraction is evaluated by applying the minor head loss equation, and the pressure at point 1 is determined by applying the energy equation between points 1 and 2, and as follows:

$$k_{contrc} := 0.25 \qquad Q := 0.98\,\frac{m^3}{sec} \qquad D_1 := 0.65\,m \qquad A_1 := \frac{\pi \cdot D_1{}^2}{4} = 0.332\,m^2$$

$$D_2 := 0.35\,m \qquad A_2 := \frac{\pi \cdot D_2^2}{4} = 0.096\,m^2 \qquad v_1 := \frac{Q}{A_1} = 2.953\,\frac{m}{s}$$

$$v_2 := \frac{Q}{A_2} = 10.186\,\frac{m}{s} \qquad p_2 := 3.56 \times 10^4\,\frac{N}{m^2} \quad z_1 := 0\,m \qquad z_2 := 0\,m$$

$$\rho := 998\,\frac{kg}{m^3} \qquad g := 9.81\,\frac{m}{sec^2} \qquad \gamma := \rho \cdot g = 9.79 \times 10^3\,\frac{kg}{m^2\,s^2}$$

Guess value: $\qquad h_{fmin} := 1\,m \qquad p_1 := 6 \times 10^4\,\dfrac{N}{m^2}$

Given

$$\frac{p_1}{\gamma} + z_1 + \frac{v_1^2}{2 \cdot g} - h_{fmin} = \frac{p_2}{\gamma} + z_2 + \frac{v_2^2}{2 \cdot g}$$

$$h_{fmin} = k_{contrc}\frac{v_2^2}{2 \cdot g}$$

$$\binom{h_{fmin}}{p_1} := Find\,(h_{fmin}, p_1)$$

$$h_{fmin} = 1.322\,m \qquad\qquad p_1 = 9.596 \times 10^4\,\frac{N}{m^2}$$

(c) In order to determine the reaction force acting on the pipe contraction due to the thrust force of the flow, the integral form of the momentum equation is applied in the x direction for the control volume illustrated in Figure EP 5.13. The flow enters at control face 1 and leaves at control face 2. The viscous forces have already been accounted for indirectly through the minor head loss term, $h_{f,minor}$ in the energy equation, and there is no gravitational force component along the x-axis because the pipe is horizontal ($z_1 = z_2 = 0$).

$$\sum F_x = (p_1 A_1)_x + (p_2 A_2)_x + R_x = \rho Q(v_{2x} - v_{1x})$$

$$p_{1x} := p_1 = 9.596 \times 10^4\,\frac{N}{m^2} \qquad\qquad p_{2x} := -p_2 = -3.56 \times 10^4\,\frac{N}{m^2}$$

$$v_{1x} := v_1 = 2.953\,\frac{m}{s} \qquad\qquad v_{2x} := v_2 = 10.186\,\frac{m}{s}$$

Guess value: $\qquad\qquad\qquad\qquad R_x := 1000\,N$

Given

$$p_{1x} \cdot A_1 + p_{2x} \cdot A_2 - R_x = \rho \cdot Q\,(v_{2x} - v_{1x})$$

$$R_x := Find\,(R_x) = 2.134 \times 10^4\,N$$

Thus, the reaction force acting on the pipe contraction, R_x due to the thrust force of the flow acts to the left as assumed and illustrated in Figure EP 5.13.

EXAMPLE PROBLEM 5.14

Water at 20°C flows in a 0.45-m-diameter pipe at a flowrate of 0.34 m³/sec. The pipe is fitted with a gradual pipe contraction (nozzle) with a diameter of 0.15 m in order to accelerate the flow, as illustrated in Figure EP 5.14. Assume that the pipe and the nozzle lie in a horizontal plane, (x, y). The pressure at point 1 is 3×10^5 N/m², and assume a minor head loss coefficient due to the nozzle, k of 0.07. (a) Determine the minor head loss due to the nozzle. (b) Determine the pressure at point 2. (c) Determine the reaction force acting on the nozzle due to the thrust force of the flow (anchoring force required to keep the nozzle from moving).

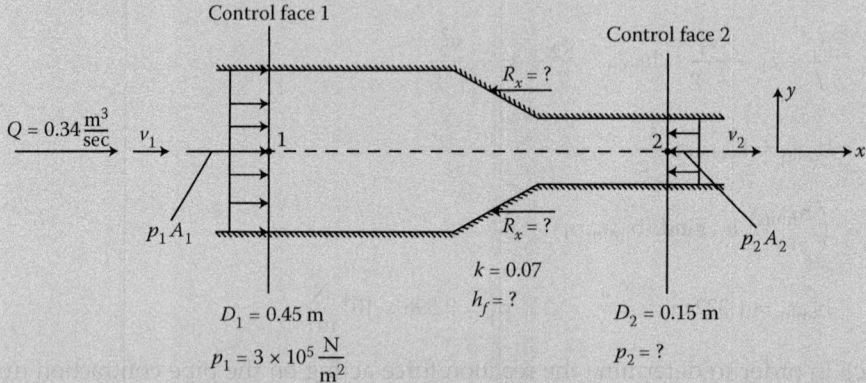

FIGURE EP 5.14
Water flows through a horizontal pipe with a gradual pipe contraction (nozzle) to accelerate the flow.

Mathcad Solution

(a)–(b) The minor head loss due to the nozzle is evaluated by applying the minor head loss equation, and the pressure at point 2 is determined by applying the energy equation between points 1 and 2, and as follows:

$$k_{nozzle} := 0.07 \qquad Q := 0.34 \frac{m^3}{sec} \qquad D_1 := 0.45\,m \qquad A_1 := \frac{\pi \cdot D_1^2}{4} = 0.159\,m^2$$

$$D_2 := 0.15\,m \qquad A_2 := \frac{\pi \cdot D_2^2}{4} = 0.018\,m^2 \qquad v_1 := \frac{Q}{A_1} = 2.138 \frac{m}{s}$$

$$v_2 := \frac{Q}{A_2} = 19.24 \frac{m}{s} \qquad P_1 := 3 \times 10^5 \frac{N}{m^2} \qquad z_1 := 0\,m \qquad z_2 := 0\,m$$

$$\rho := 998 \frac{kg}{m^3} \qquad g := 9.81 \frac{m}{sec^2} \qquad \gamma := \rho \cdot g = 9.79 \times 10^3 \frac{kg}{m^2 \cdot s^2}$$

Guess value: $\qquad h_{fmin} := 1\,m \qquad P_2 := 1 \times 10^4 \frac{N}{m^2}$

Given

$$\frac{P_1}{\gamma} + z_1 + \frac{v_1^2}{2 \cdot g} - h_{fmin} = \frac{P_2}{\gamma} + z_2 + \frac{v_2^2}{2 \cdot g}$$

$$h_{fmin} = k_{nozzle} \frac{v_2^2}{2 \cdot g}$$

$$\begin{pmatrix} h_{fmin} \\ p_2 \end{pmatrix} := \text{Find } (h_{fmin}, p_2)$$

$$h_{fmin} = 1.321 \text{ m} \qquad p_2 = 1.046 \times 10^5 \frac{N}{m^2}$$

(c) In order to determine the reaction force acting on the nozzle due to the thrust force of the flow, the integral form of the momentum equation is applied in the x direction for the control volume illustrated in Figure EP 5.14. The flow enters at control face 1 and leaves at control face 2. The viscous forces have already been accounted for indirectly through the minor head loss term, $h_{f,minor}$ in the energy equation, and there is no gravitational force component along the *x-axis* because the pipe is horizontal ($z_1 = z_2 = 0$).

$$\sum F_x = (p_1 A_1)_x + (p_2 A_2)_x + R_x = \rho Q (v_{2x} - v_{1x})$$

$$p_{1x} := p_1 = 3 \times 10^5 \frac{N}{m^2} \qquad\qquad p_{2x} := -p_2 = -1.046 \times 10^5 \frac{N}{m^2}$$

$$v_{1x} := v_1 = 2.138 \frac{m}{s} \qquad\qquad v_{2x} := v_2 = 19.24 \frac{m}{s}$$

Guess value: $\qquad\qquad\qquad\qquad\qquad R_x := 1000 \text{ N}$

Given

$$p_{1x} \cdot A_1 + p_{2x} \cdot A_2 - R_x = \rho \cdot Q \, (v_{2x} - v_{1x})$$

$$R_x := \text{Find } (R_x) = 4.006 \times 10^4 \text{ N}$$

Thus, the reaction force acting on the nozzle, R_x due to the thrust force of the flow acts to the left, as assumed and illustrated in Figure EP 5.14.

EXAMPLE PROBLEM 5.15

Water at 20°C flows in a 0.85-m-diameter horizontal pipe at a flowrate of 2.9 m³/sec. A regular 45° threaded elbow is installed in between point 1 and 2 in order to change the direction of the flow by 45°, as illustrated in Figure EP 5.15. Assume that the pipe and the elbow lie in a horizontal plane, (x, y). The pressure at point 1 is 3.9×10^5 N/m², and assume a minor head loss coefficient due to the elbow, k of 0.4. (a) Determine the minor head loss due to the elbow between points 1 and 2. (b) Determine the pressure at point 2. (c) Determine the reaction force acting on the elbow due to the thrust force of the flow (anchoring force required to keep the elbow from moving).

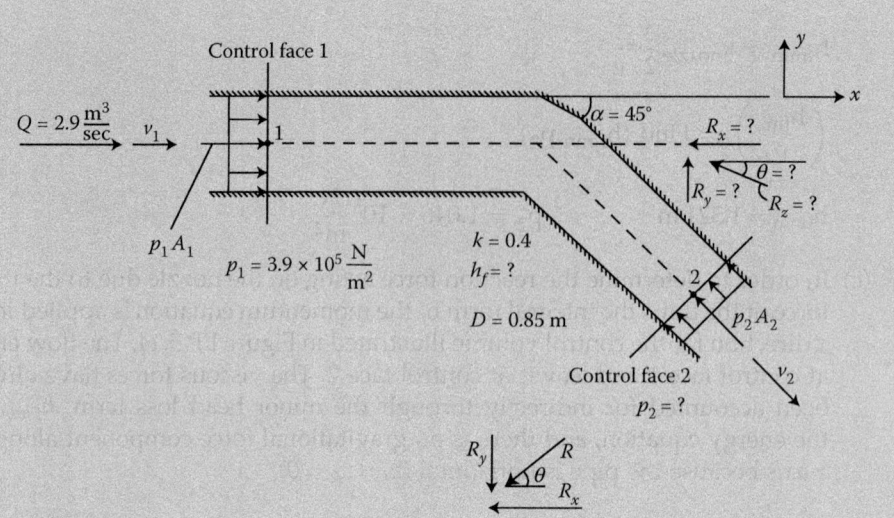

FIGURE EP 5.15
Water flows through a horizontal pipe with a 45° threaded elbow to change the direction of flow.

Mathcad Solution

(a)–(b) The minor head loss due to the elbow is evaluated by applying the minor head loss equation, and the pressure at point 2 is determined by applying the energy equation between points 1 and 2, and as follows:

$$D := 0.85\,m \qquad A := \frac{\pi \cdot D^2}{4} = 0.567\,m^2 \quad k_{elbow} := 0.4 \qquad Q := 2.9\,\frac{m^3}{sec}$$

$$v := \frac{Q}{A} = 5.111\,\frac{m}{s} \qquad v_1 := v = 5.111\,\frac{m}{s} \qquad v_2 := v = 5.111\,\frac{m}{s}$$

$$p_1 := 3.9 \times 10^5\,\frac{N}{m^2} \qquad z_1 := 0\,m \qquad\qquad z_2 := 0\,m$$

$$\rho := 998\,\frac{kg}{m^3} \qquad g := 9.81\,\frac{m}{sec^2} \qquad \gamma := \rho \cdot g = 9.79 \times 10^3\,\frac{kg}{m^2 \cdot s^2}$$

$$\text{Guess value:} \qquad h_{fmin} := 1\,m \qquad\qquad p_2 := 1 \times 10^4\,\frac{N}{m^2}$$

Given

$$\frac{p_1}{\gamma} + z_1 + \frac{v_1^2}{2 \cdot g} - h_{fmin} = \frac{p_2}{\gamma} + z_2 + \frac{v_2^2}{2 \cdot g}$$

$$h_{fmin} = k_{elbow}\,\frac{v^2}{2 \cdot g}$$

$$\begin{pmatrix} h_{fmin} \\ p_2 \end{pmatrix} := \text{Find}\,(h_{fmin}, p_2)$$

$$h_{fmin} = 0.532\,m \qquad p_2 := 3.848 \times 10^5\,\frac{N}{m^2}$$

(c) In order to determine the reaction force acting on the elbow due to the thrust force of the flow, the integral form of the momentum equation is applied in both the x and y directions for the control volume illustrated in Figure EP 5.15. The flow enters at control face 1 and leaves at control face 2. The viscous forces have already been accounted for indirectly through the minor head loss term, $h_{f,minor}$ in the energy equation, and there is no gravitational force component along the x- and y-axes because the pipe is horizontal ($z_1 = z_2 = 0$).

$$\sum F_x = (p_1 A_1)_x + (p_2 A_2)_x + R_x = \rho Q(v_{2x} - v_{1x})$$

$$\sum F_y = (p_1 A_1)_y + (p_2 A_2)_y + R_y = \rho Q(v_{2y} - v_{1y})$$

$$A_1 := A = 0.567\,m^2 \qquad\qquad A_2 := A = 0.567\,m^2 \quad \alpha := 45\,deg$$

$$p_{1x} := p_1 = 3.9 \times 10^5\,\frac{N}{m^2} \qquad p_{2x} := -p_2 \cdot \cos(\alpha) = -2.721 \times 10^5\,\frac{N}{m^2}$$

$$v_{1x} := v_1 = 5.111\,\frac{m}{s} \qquad\qquad v_{2x} := v_2 \cdot \cos(\alpha) = 3.614\,\frac{m}{s}$$

$$p_{1y} := 0\,\frac{N}{m^2} \qquad\qquad\qquad p_{2y} := p_2 \cdot \sin(\alpha) = 2.721 \times 10^5 m\,\frac{N}{m^3}$$

$$v_{1y} := 0\,\frac{m}{sec} \qquad\qquad\qquad v_{2y} := -v_2 \cdot \sin(\alpha) = -3.614\,\frac{m}{s}$$

Guess value: $\qquad R_x := 1000\,N \qquad R_y := 1000\,N$

Given

$$p_{1x} \cdot A_1 + p_{2x} \cdot A_2 - R_x = \rho \cdot Q\,(v_{2x} - v_{1x})$$

$$p_{1y} \cdot A_1 + p_{2y} \cdot A_2 + R_y = \rho \cdot Q\,(v_{2y} - v_{1y})$$

$$\begin{pmatrix} R_x \\ R_y \end{pmatrix} := \text{Find}\,(R_x, R_y) = \begin{pmatrix} 7.124 \times 10^4 \\ -1.649 \times 10^5 \end{pmatrix}\,N$$

$$R := \sqrt{R_x^2 + R_y^2} = 1.796 \times 10^5\,N \qquad \theta := \text{atan}\left(\frac{|R_y|}{|R_x|}\right) = 66.628\,deg$$

Thus, the reaction force acting on the elbow, R_x acts to the left as assumed, and the reaction force acting on the elbow, R_y acts downward (not upward as assumed), as illustrated in Figure EP 5.15.

5.3.4 Application of the Energy and Momentum Equations for Ideal Internal Flow and Ideal Flow from a Tank

Application of the Bernoulli equation (Equation 5.33) assumes ideal flow; thus there is no head loss term to evaluate in the energy equation. Therefore, for problems that affect the flow over a short pipe or channel section or flow from a tank, the major head loss due to flow resistance is assumed to be negligible and thus one may apply the Bernoulli equation. Thus, the Bernoulli equation may be solved for an unknown "energy head" term (p/γ, z, or $v^2/2g$) for one of the two points in a finite control volume as follows:

$$\left(\frac{p}{\gamma} + z + \frac{v^2}{2g}\right)_1 - \left(\frac{p}{\gamma} + z + \frac{v^2}{2g}\right)_2 = 0 \tag{5.48}$$

These types of flow problems include flow-measuring devices for both pipe and open channel flow, which, however, have a minor loss associated with the flow measurement and gradual (vertical or horizontal) channel contractions/transitions, which do not have a minor loss. Also, the flow from a tank (jet flows and siphon flows) does not have a minor head loss. In the case of a flow-measuring device for either pipe flow or open channel flow, although the associated minor head loss may be accounted for in the integral form of the energy equation (see Equation 5.32), it is typically accounted for by the use/calibration of a discharge coefficient to determine the actual discharge, where ideal flow (Bernoulli equation) is assumed. Thus, in the case of flow-measuring devices, one must indirectly model the minor head loss flow resistance term by the use of a discharge coefficient, which requires a "subset level" of application of the governing equations (Section 5.3.4.1). Then, application of the complementary integral momentum equation (Equation 5.35) is required only if there is an unknown reaction force, as in the case of open channel flow-measuring devices; note that the viscous forces have already been accounted for indirectly through a minor head loss flow resistance term in the energy equation (or, actually, through the discharge coefficient). And in the case of a flow from a tank, the complementary integral momentum is applied if there is an unknown reaction force. As such, in either case, the integral form of the momentum equation is solved analytically, without the use of dimensional analysis. Applications of the governing equations for ideal internal flow and ideal flow from a tank are presented in Sections 5.3.4.2–5.3.4.5 below.

5.3.4.1 Evaluation of the Actual Discharge

In the case of a flow-measuring device for either pipe flow or open channel flow, the associated minor head is typically accounted for by the use/calibration of a discharge coefficient to determine the actual discharge. Evaluation of the actual discharge requires a "subset level" application of the governing equations. A "subset level" application of the governing equations focuses only on the given element causing the flow resistance. The assumption of "ideal" flow implies that the flow resistance is modeled only in the momentum equation (and thus, the subsequent assumption of "real" flow). The flow resistance equation for the actual discharge is derived in Chapters 6 and 7. In the derivation of the actual discharge for a given flow-measuring device, first, ideal flow is assumed, so the Bernoulli equation is applied as follows:

$$\left(\frac{p}{\gamma} + z + \frac{v^2}{2g}\right)_1 - \left(\frac{p}{\gamma} + z + \frac{v^2}{2g}\right)_2 = 0 \tag{5.49}$$

Assuming $z_1 = z_2$, this equation has one unknown, which is the ideal velocity of the flow at the restriction in the flow-measuring device. Thus, this equation yields an expression for the ideal velocity, $v_i = \sqrt{2\Delta p/\rho}$ or $v_i = \sqrt{2g\Delta y}$, as a function of an ideal pressure difference, Δp or Δy, respectively, which is directly measured. Then, application of the continuity equation is used to determine an expression for the ideal discharge, $Q_i = v_i A_i$ and the actual discharge, $Q_a = (v_a/\sqrt{(2\Delta p/\rho)})(A_a/A_i)(Q_i)$ or $Q_a = (v_a/\sqrt{(2g\Delta y)})(A_a/A_i)Q_i$. And, finally, the integral form of the momentum equation (supplemented by dimensional analysis) is applied to solve for an unknown actual velocity, v_a, and actual area, A_a, which was given in Equation 5.27 and is repeated as follows:

$$\sum F_s = (F_G + F_P + F_V + F_{other})_s = (\rho Q v_s)_2 - (\rho Q v_s)_1 \tag{5.50}$$

Thus, the resulting flow resistance equations for the actual discharge for pipe flow-measuring devices, and open channel flow-measuring devices are given, respectively, as follows:

$$Q_a = \frac{\sqrt{\dfrac{2}{\rho}\left(\Delta p - \dfrac{\tau_w L}{R_h}\right)}}{\sqrt{\dfrac{2\Delta p}{\rho}}} \frac{A_a}{A_i} Q_i = C_d Q_i \tag{5.51a}$$

$$Q_a = \frac{\sqrt{2g\left(\Delta y - \dfrac{\tau_w L}{\gamma R_h}\right)}}{\sqrt{2g\Delta y}} \frac{A_a}{A_i} Q_i = C_d Q_i \tag{5.51b}$$

where the discharge coefficient, C_d accounts for (indirectly models) the minor head loss associated with the flow measurement.

The flow resistance (shear stress or drag force) due to a flow-measuring device is modeled as a reduced/actual discharge, Q_a in the differential form of the continuity equation. The flow resistance causes an unknown pressure drop, Δp in the case of pipe flow, and an unknown Δy in the case of open channel flow, which causes an unknown head loss, h_f, where the head loss is due to a conversion of kinetic energy to heat, which is modeled/displayed in the integral form of the energy equation. The head loss is caused by both pressure and friction forces. In the determination of the reduced/actual discharge, Q_a, which is less than the ideal discharge, Q_i, the exact reduction in the flowrate is unknown because the head loss, h_f causing it is unknown. Therefore, because one cannot model the exact reduction in the ideal flowrate, Q_i, one cannot derive an analytical expression for the reduced/actual discharge, Q_a from the continuity equation. Specifically, the complex nature of the flow does not allow a theoretical modeling of the existence of the viscous force (due to shear stress, τ_w) due to the flow-measurement device in the momentum equation and thus does not allow an analytical derivation of the pressure drop, Δp or Δy and thus, v_a, and cannot measure A_a. Thus, an empirical expression (using dimensional analysis) for the actual discharge,

$$Q_a = \frac{\sqrt{\frac{2}{\rho}\left(\Delta p - \frac{\tau_w L}{R_h}\right)}}{\sqrt{\frac{2\Delta p}{\rho}}} \frac{A_a}{A_i} Q_i, \quad \text{or} \quad Q_a = \frac{\sqrt{2g\left(\Delta y - \frac{\tau_w L}{\gamma R_h}\right)}}{\sqrt{2g\Delta y}} \frac{A_a}{A_i} Q_i;$$

thus, Δp or Δy (and v_a and A_a) is derived as a function of the drag coefficient, C_D (discharge coefficient, C_d) that represents the flow resistance. This yields the actual discharge equation $Q_a = C_d Q_i$, which represents the integral momentum equation supplemented by dimensional analysis. Furthermore, although the differential form of the continuity equation is typically used to compute the reduced/actual discharge, Q_a for a flow-measuring device, because the head loss, h_f causes the reduced/actual discharge, Q_a, the head loss may also be accounted for as a minor loss in the integral form of the energy equation, as illustrated in Chapter 8. Actual discharges are addressed in detail in Chapters 6, 8, and 9.

5.3.4.2 Applications of the Governing Equations for Ideal Internal Flow and Ideal Flow from a Tank

The integral form of the governing equations (continuity, energy, and momentum) for ideal flow are applied to a finite control volume in the analysis and design of pipe systems, open channel flow systems, and the flow from a tank. Applications of the governing equations for ideal pipe flow include the analysis and design of pipe systems that include a pipe flow-measuring device such as pitot-static tubes, orifice, nozzle, or venturi meters, which are also known as ideal flow meters (Section 5.3.4.3). Applications of the governing equations for ideal open channel flow include the analysis and design of open channel flow-measuring devices (such as pitot-static tubes, sluice gates, weirs, spillways, venturi flumes, and contracted openings), which are also known as ideal flow meters (Section 5.3.4.4), and the analysis and design of a gradual (vertical or horizontal) channel contraction/transition (see Chapter 9 for applications). And, applications of the governing equations for ideal flow include the flow from a tank (jet flows and siphon flows) (Section 5.3.4.5). One may note from Section 5.3.4.1 above that the minor head loss term associated with a pipe or open channel flow-measuring device is indirectly modeled by the discharge coefficient, which requires a "subset level" application of the governing equations for its evaluation. Furthermore, the resulting definition for the actual discharge, $Q_a = C_d Q_i$ requires details in the definition of the discharge coefficient, C_d, which is to be presented in later chapters. Therefore, example problems illustrating the application of the Bernoulli equation for pipe or open channel flow-measuring devices in Chapter 5 will assume that the actual discharge is either a given quantity (lumped value) or computed from the actual discharge equation (see Chapters 8 and 9 for applications) as follows:

$$Q_a = v_a A_a \tag{5.52}$$

Furthermore, for the example problems involving open channel flow-measuring devices, illustrations of the computation of the unknown reaction force, which require the application of the complementary integral momentum equation, are presented below. And, for ideal flow from a tank, example problems involve forced jets or free jets. While application of the integral momentum equation is needed for an unknown force for problems involving forced and free jets, it is not needed for problems involving siphon flows. Specifically, in the case of a forced jet, computation of the unknown reaction force due to the impact of the jet hitting a vane (stationary or moving) or a stationary flat plate requires the application of the complementary integral momentum equation. And, in the case of a free jet, computation of the unknown reaction force (required to keep the tank from moving) requires the application of the complementary integral momentum equation. For the example problems involving a gradual (vertical or horizontal) channel contraction/transition, there is no minor head loss associated with the flow, and there are no reaction forces involved; thus, only the continuity and the Bernoulli equations are needed. And, finally, for the example problems involving a flow-measuring device for pipe flow or for example problems involving a pitot-static tube for either pipe or open channel flow, there are no reaction forces involved.

5.3.4.3 Applications of the Governing Equations for Ideal Pipe Flow

The governing equations are applied in the analysis and design of ideal pipe flow problems involving flow-measuring devices, such as orifice, nozzle, or venturi meters, which are also known as ideal flow meters. The Bernoulli equation is applied to solve for the ideal velocity, which is expressed as a function of the ideal pressure difference, which is directly measured. Then, the continuity equation is applied to solve for the ideal discharge, which is corrected to reflect the actual discharge by the application of discharge coefficient (requires a "subset level" application of the governing equations). The integral momentum equation is not needed because there are no reaction forces to compute in a pipe flow-measuring device. Example problems involving pipe flow-measuring devices are presented in Example Problem 5.16 below and in Chapter 8.

EXAMPLE PROBLEM 5.16

Water at 20°C flows in a 0.7-m-diameter horizontal pipe. An orifice meter with a 0.3-m diameter is mounted in the flanged pipe just upstream of point 2 as illustrated in Figure EP 5.16 in order to measure the flowrate in the pipe. A pressure tap, p_1 is installed at point 1, where the full pipe flow occurs just upstream of the orifice, and a pressure tap, p_2 is installed at point 2, where the vena contracta occurs just downstream of the orifice. The pressure at point 1, p_1 is measured to be $3 \times 10^5 \, \text{N/m}^2$, and the pressure at point 2, p_2 is measured to be $2.5 \times 10^5 \, \text{N/m}^2$. The orifice discharge coefficient, C_o is 0.6. (a) Determine the ideal flowrate. (b) Determine the actual flowrate.

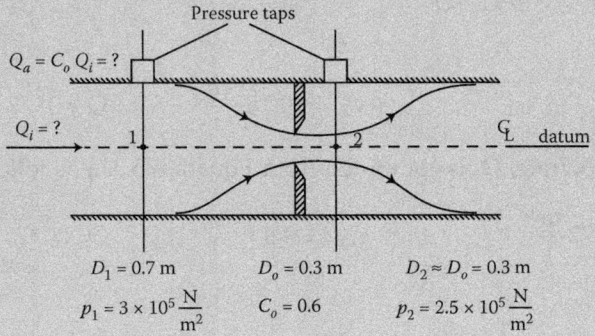

Pressure taps

$Q_a = C_o \, Q_i = ?$

$Q_i = ?$

1

2

datum

$D_1 = 0.7 \, \text{m}$ $D_o = 0.3 \, \text{m}$ $D_2 \approx D_o = 0.3 \, \text{m}$

$p_1 = 3 \times 10^5 \, \dfrac{\text{N}}{\text{m}^2}$ $C_o = 0.6$ $p_2 = 2.5 \times 10^5 \, \dfrac{\text{N}}{\text{m}^2}$

FIGURE EP 5.16
Water flows through a horizontal pipe with an orifice meter to measure the flowrate.

Mathcad Solution

(a) In order to determine the ideal velocity at either point 1 or point 2, the Bernoulli equation is applied between points 1 and 2. However, in order to determine the relationship between the ideal velocities at points 1 and 2, the continuity equation is applied between points 1 and 2; thus, we have two equations and two unknowns. Furthermore, in order to determine the ideal flowrate in the pipe, the continuity equation is applied at either points 1 or 2. And, finally, the diameter of flow at point 2 is assumed to be equal to the diameter of the orifice meter, because the vena contracta occurs a point 2.

$$D_1 := 0.7\,m \qquad\qquad D_o := 0.3\,m \qquad\qquad D_2 := D_o = 0.3\,m$$

$$A_1 := \frac{\pi \cdot D_1^2}{4} = 0.385\,m^2 \qquad\qquad\qquad A_2 := \frac{\pi \cdot D_2^2}{4} = 0.071\,m^2$$

$$P_1 := 3 \times 10^5\,\frac{N}{m^3} \qquad P_2 := 2.5 \times 10^5\,\frac{N}{m^2} \quad z_1 := 0\,m \qquad\qquad z_2 := 0\,m$$

$$\rho := 998\,\frac{kg}{m^3} \qquad\qquad g := 9.81\,\frac{m}{sec^2} \qquad \gamma := \rho \cdot g = 9.79 \times 10^3\,\frac{kg}{m^2 \cdot s^2}$$

Guess value: $\qquad\qquad v_1 := 1\,\dfrac{m}{sec} \qquad\quad v_2 := 2\,\dfrac{m}{sec} \qquad Q := 1\,\dfrac{m^3}{sec}$

Given

$$\frac{P_1}{\gamma} + z_1 + \frac{v_1^2}{2 \cdot g} = \frac{P_2}{\gamma} + z_2 + \frac{v_2^2}{2 \cdot g}$$

$$v_1 \cdot A_1 = v_2 \cdot A_2$$

$$Q = v_2 \cdot A_2$$

$$\begin{pmatrix} v_1 \\ v_2 \\ Q \end{pmatrix} := Find\,(v_1, v_2, Q)$$

$$v_1 = 1.87\,\frac{m}{s} \qquad\qquad\qquad v_2 = 10.183\,\frac{m}{s} \qquad Q = 0.72\,\frac{m^3}{s}$$

(b) The actual flowrate, Q_a is computed from Equation 5.51a as follows:

$$Q_i := Q = 0.72\,\frac{m^3}{s} \qquad\qquad C_o := 0.6 \qquad\qquad Q_a := C_o \cdot Q_i = 0.432\,\frac{m^3}{s}$$

5.3.4.4 Applications of the Governing Equations for Ideal Open Channel Flow

The governing equations are applied in the analysis and design of ideal open channel flow problems involving flow-measuring devices (such as sluice gates, weirs, spillways, venturi flumes, and contracted openings), which are also known as ideal flow meters. The Bernoulli equation is applied to solve for the ideal velocity, which is expressed as a function of the ideal pressure difference, which is directly measured. Then, the continuity equation is applied to solve for the ideal discharge, which is corrected to reflect the actual discharge by the application of discharge coefficient (requires a "subset level" application of the governing equations). The integral momentum equation is applied to solve for an unknown reaction force for the open channel flow-measuring device. Example problems involving open channel flow-measuring devices are presented in Example Problem 5.17 below and in Chapter 9.

EXAMPLE PROBLEM 5.17

Water at 20°C flows at a depth of 3.7 m in a rectangular channel of width 4 m at a mild slope. A weir is inserted in the channel as illustrated in Figure EP 5.17 in order to measure the flowrate. The depth of flow downstream of the weir is 1.5 m. The weir discharge coefficient, C_d is 0.66. (a) Determine the ideal flow rate. (b) Determine the actual flowrate. (c) Determine the reaction force acting on the weir due to the thrust force of flow (anchoring force required to keep the weir from moving).

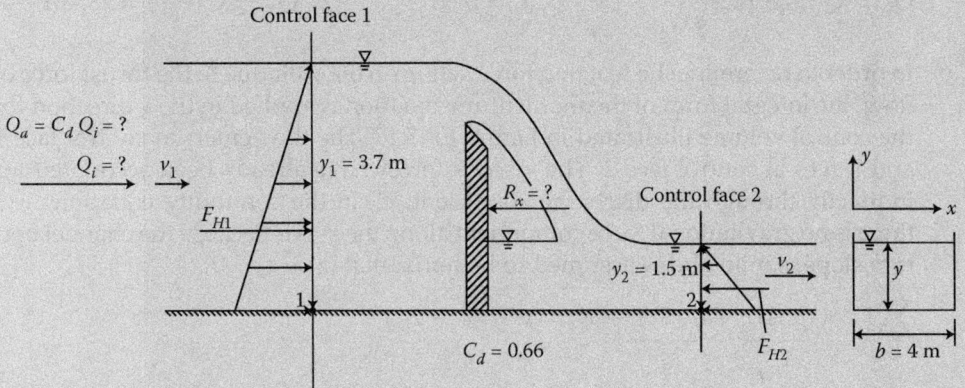

FIGURE EP 5.17
Water flows in a rectangular channel with a weir to measure the flowrate.

Mathcad Solution

(a) In order to determine the ideal velocity at either point 1 or point 2, the Bernoulli equation is applied between points 1 and 2. However, in order to determine the relationship between the ideal velocities at points 1 and 2, the continuity equation is applied between points 1 and 2; thus, we have two equations and two unknowns. Furthermore, in order to determine the ideal flowrate in the pipe, the continuity equation is applied at either points 1 or 2.

$$y_1 := 3.7\,m \qquad y_2 := 1.5\,m \qquad b := 4\,m \qquad g := 9.81\,\frac{m}{sec^2}$$

$$A_1 := b \cdot y_1 = 14.8\,m^2 \qquad A_2 := b \cdot y_2 = 6\,m^2 \qquad z_1 := 0\,m \qquad z_2 := 0\,m$$

$$\text{Guess value:} \qquad v_1 := 1\,\frac{m}{sec} \qquad v_2 := 2\,\frac{m}{sec} \qquad Q := 1\,\frac{m^3}{sec}$$

Given

$$y_1 + z_1 + \frac{v_1^2}{2 \cdot g} = y_2 + z_2 + \frac{v_2^2}{2 \cdot g}$$

$$v_1 \cdot b \cdot y_1 = v_2 \cdot b \cdot y_2$$

$$Q = v_1 \cdot A_1$$

$$\begin{pmatrix} v_1 \\ v_2 \\ Q \end{pmatrix} := \text{Find}\,(v_1, v_2, Q)$$

$$v_1 = 2.914\,\frac{m}{s} \qquad\qquad v_2 = 7.187\,\frac{m}{s} \qquad\qquad Q = 43.122\,\frac{m^3}{s}$$

(b) The actual flowrate, Q_a is computed from Equation 5.51b as follows:

$$Q_i := Q = 43.122\,\frac{m^3}{s} \qquad\qquad C_d := 0.66 \qquad\qquad Q_a := C_d \cdot Q_i = 28.461\,\frac{m^3}{s}$$

(c) In order to determine the reaction force acting on the weir due to the thrust force of flow, the integral form of the momentum equation is applied in the x direction for the control volume illustrated in Figure EP 5.17. The flow enters at control face 1 and leaves at control face 2. The viscous forces have already been accounted for indirectly through the discharge coefficient, C_d in the continuity equation, and there is no gravitational force component along the x-axis because the channel bottom slope is mild and is assumed to be horizontal ($z_1 = z_2 = 0$).

$$\sum F_x = \underbrace{(p_1 A_1)_x}_{F_{H1}} + \underbrace{(p_2 A_2)_x}_{F_{H2}} + R_x = \rho Q (v_{2x} - v_{1x})$$

$$\rho := 998\,\frac{kg}{m^3} \qquad\qquad\qquad \gamma := \rho \cdot g = 9.79 \times 10^3\,\frac{kg}{m^2 \cdot s^2}$$

$$h_{ca1} := \frac{y_1}{2} = 1.85\,m \qquad\qquad\qquad h_{ca2} := \frac{y_2}{2} = 0.75\,m$$

$$p_1 := \gamma \cdot h_{ca1} = 1.811 \times 10^4\,\frac{N}{m^2} \qquad\qquad p_2 := \gamma \cdot h_{ca2} = 7.343 \times 10^3\,\frac{N}{m^2}$$

$$p_{1x} := p_1 = 1.811 \times 10^4\,\frac{N}{m^2} \qquad\qquad p_{2x} := -p_2 = -7.343 \times 10^3\,\frac{N}{m^2}$$

$$F_{H1} := p_{1x} \cdot A_1 = 2.681 \times 10^5\,N \qquad\qquad F_{H2} := p_{2x} \cdot A_2 = -4.406 \times 10^4\,N$$

$$v_{1x} := v_1 = 2.914\,\frac{m}{s} \qquad\qquad\qquad v_{2x} := v_2 = 7.187\,\frac{m}{s}$$

Guess value: $R_x := 1000\,N$

Given

$$F_{H1} + F_{H2} - R_x = \rho \cdot Q \, (v_{2x} - v_{1x})$$

$$R_x := \text{Find } (R_x) = 4.01 \times 10^4 \, \text{N}$$

Thus, the reaction force acting on the weir, R_x acts to the left as assumed and is illustrated in Figure EP 5.17.

5.3.4.5 Applications of the Governing Equations for Ideal Flow from a Tank

The governing equations are applied in the analysis and design of ideal flow from a tank (or a water source). The ideal flow from a tank (or a water source) includes: (1) forced jets and (2) free jets (Torricelli's theorem). As noted in Chapter 4, forced jets are a result of pressure flow from a syringe or a pipe with a nozzle (hose), while free jets are a result of gravity flow from an open tank. Furthermore, siphon flows are a result of gravity flow from an open tank with a siphon. The Bernoulli equation is applied to solve for the unknown pressure, elevation, or velocity. The continuity equation is applied to solve for the discharge. While application of the momentum equation is needed to solve for an unknown force for problems involving forced and free jets, it is not needed for problems involving siphon flows.

In the case of a forced jet, the integral momentum equation is applied to solve for an unknown reaction force due to the impact of the jet hitting a vane (stationary or moving) (see Figures EP 5.6 and EP 5.7) or a stationary flat plate (see Figures EP 5.18 and EP 5.19 below). In the case of a forced jet hitting a vane or plate, there are two points to consider regarding the flow. The first point is that because the jet is open to the atmosphere, the pressure at the control faces for the flow into and out of the control volume are atmospheric, and thus the gage pressure is zero. The second point is that the friction (flow resistance) between the fluid jet and the vane or plate is typically assumed to be negligible, so the flow is assumed to be ideal. However, in cases where the velocity of the jet hitting a vane or plate is high, ignoring the minor head loss due to friction may introduce a significant error in the problem analysis or design. Therefore, in such cases, it is typical to indirectly model the minor head loss by a reduced velocity at the control face(s) for the flow out of the control volume. Example problems involving forced jets were presented in Example Problems 5.6 (stationary vane) and 5.7 (moving vane) above, and are presented in Example Problems 5.18 (stationary sloping flat plate) and 5.19 (stationary vertical flat plate) below.

EXAMPLE PROBLEM 5.18

A forced jet of water at 20°C with a diameter of 0.02 m and a flowrate of 0.076 m^3/sec hits a stationary sloping flat plate, where the jet splits into two paths, as illustrated in Figure EP 5.18. The plate angle α is 60°, assume that the jet and the plate lie in a horizontal plane, (x, y), and assume ideal flow. (a) Determine the velocity of the flow at points 2 and 3. (b) Determine the flowrate at control faces 2 and 3. (c) Determine the reaction force acting on the plate due to the thrust force of the flow (anchoring force required to keep plate from moving).

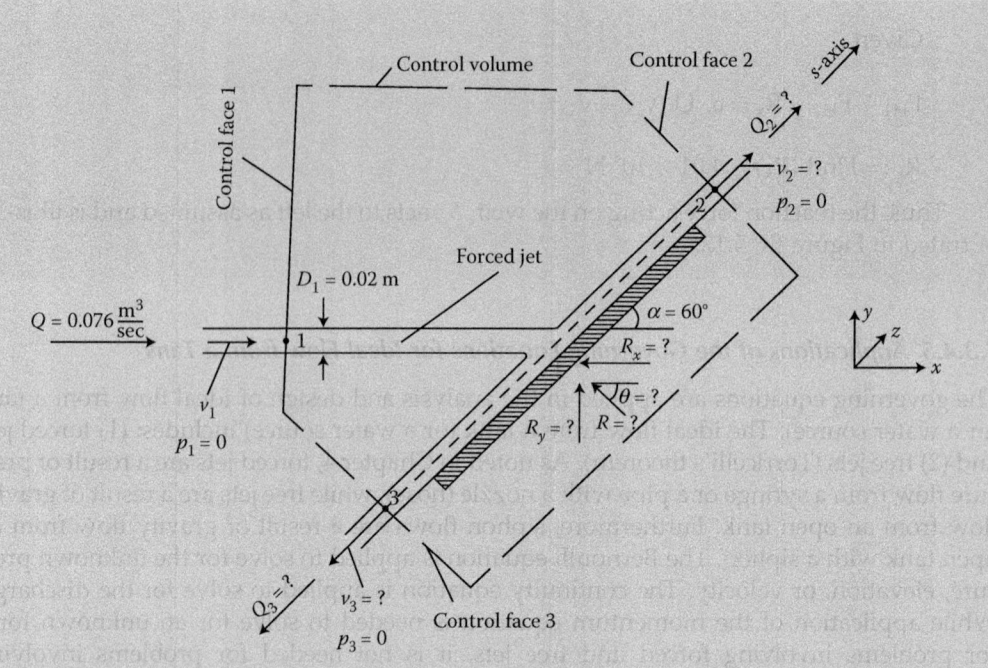

FIGURE EP 5.18
Forced jet of water hits a stationary sloping flat plate angled at 60° and splits into two paths.

Mathcad Solution

(a) In order to determine the velocity of flow at points 2 and 3, the Bernoulli equation is applied between points 1 and 2, and between points 1 and 3 as follows:

$$Q_1 := 0.076 \frac{m^3}{sec} \qquad\qquad D_1 := 0.02\,m \qquad\qquad A_1 := \frac{\pi \cdot D_1^2}{4} = 3.142 \times 10^{-4}\,m^2$$

$$v_1 := \frac{Q_1}{A_1} = 241.916\,\frac{m}{s} \qquad\qquad p_1 := 0\,\frac{N}{m^2} \qquad\qquad p_2 := 0\,\frac{N}{m^2}$$

$$p_3 := 0\,\frac{N}{m^2} \qquad\qquad z_1 := 0\,m \qquad\qquad z_2 := 0\,m \qquad\qquad z_3 := 0\,m$$

$$\rho := 998\,\frac{kg}{m^3} \qquad\qquad g := 9.81\,\frac{m}{sec^2} \qquad\qquad \gamma := \rho \cdot g = 9.79 \times 10^3\,\frac{kg}{m^2 \cdot s^2}$$

Guess value: $\qquad\qquad v_2 := 1\,\frac{m}{sec} \qquad\qquad v_3 := 1\,\frac{m}{sec}$

Given

$$\frac{p_1}{\gamma} + z_1 + \frac{v_1^2}{2 \cdot g} = \frac{p_2}{\gamma} + z_2 + \frac{v_2^2}{2 \cdot g}$$

$$\frac{p_1}{\gamma} + z_1 + \frac{v_1^2}{2 \cdot g} = \frac{p_3}{\gamma} + z_3 + \frac{v_3^2}{2 \cdot g}$$

$$\begin{pmatrix} v_2 \\ v_3 \end{pmatrix} := \text{Find } (v_2, v_3) = \begin{pmatrix} 241.916 \\ 241.916 \end{pmatrix} \frac{m}{s}$$

Thus, $v_1 = v_2 = v_3$. However, because the velocity of the jet hitting the plate is high, a minor head loss is indirectly modeled by a reduced velocity at control faces 2 and 3 for the flow out of the control volume. Assume a 10% reduction in the velocity at control faces 2 and 3 as follows:

$$v_2 := 0.90 \cdot v_2 = 217.724 \frac{m}{s} \qquad v_3 := 0.90 \cdot v_3 = 217.724 \frac{m}{s}$$

(b) In order to determine the flowrate at control faces 2 and 3, in addition to the continuity equation, $Q_1 = Q_2 + Q_3$, a second equation is required to solve for the two unknowns (Q_2 and Q_3). That second equation is the integral form of the momentum equation applied along the s-axis of the sloping plate; because the friction (flow resistance) between the fluid jet and the sloping plate is assumed to be negligible (ideal flow), there is no reaction force acting on the plate along the s-axis of the sloping plate. The flow enters at control face 1 and leaves at control face 2 and control face 3. Furthermore, the viscous forces are ignored because of the assumption of ideal flow, there is no pressure force component because the jet flow is open to the atmosphere ($p_1 = p_2 = p_3 = 0 \, N/m^2$), and there is no gravitational force component along the s-axis because the pipe is horizontal ($z_1 = z_2 = z_3 = 0$).

$$\sum F_s = R_s = 0 = (\rho Q_2 v_{2s} + \rho Q_3 v_{3s}) - (\rho Q_1 v_{1s})$$

$\alpha := 60 \deg$ $\qquad\qquad\qquad v_{1s} := v_1 \cdot \cos(\alpha) = 120.958 \frac{m}{s}$

$v_{2s} := v_2 = 217.724 \frac{m}{s}$ $\qquad\qquad v_{3s} := -v_3 = -217.724 \frac{m}{s}$

Guess value: $\quad Q_2 := 1 \frac{m^3}{sec} \qquad Q_3 := 1 \frac{m^3}{sec} \qquad A_2 := 1 \, m^2 \qquad A_3 := 1 \, m^2$

Given

$$Q_1 = Q_2 + Q_3$$

$$0 = (\rho \cdot Q_2 \cdot v_{2s} + \rho \cdot Q_3 \cdot v_{3s}) - (\rho \cdot Q_1 \cdot v_{1s})$$

$$A_2 = \frac{Q_2}{v_2} \qquad\qquad\qquad A_3 = \frac{Q_3}{v_3}$$

$$\begin{pmatrix} Q_2 \\ Q_3 \\ A_2 \\ A_3 \end{pmatrix} := \text{Find } (Q_2, Q_3, A_2, A_3)$$

$Q_2 = 0.059 \frac{m^3}{s}$ $\qquad\qquad Q_3 = 0.017 \frac{m^3}{s}$ $\qquad\qquad Q_2 + Q_3 = 0.076 \frac{m^3}{s}$

$A_2 = 2.715 \times 10^{-4} \, \mathrm{m}^2$ $\qquad\qquad$ $A_3 = 7.757 \times 10^{-5} \, \mathrm{m}^2$

(c) In order to determine the reaction force acting on the plate due to the thrust force of the flow, the integral form of the momentum equation is applied in both the x and y directions for the control volume illustrated in Figure EP 5.18. The flow enters at control face 1 and leaves at control face 2 and control face 3. The viscous forces are ignored because of the assumption of ideal flow, there is no pressure force component because the jet flow is open to the atmosphere ($p_1 = p_2 = p_3 = 0 \, \mathrm{N/m}^2$), and there is no gravitational force component along the x- and y-axes because the pipe is horizontal ($z_1 = z_2 = z_3 = 0$).

$$\sum F_x = R_x = (\rho Q_2 v_{2x} + \rho Q_3 v_{3x}) - (\rho Q_1 v_{1x})$$

$$\sum F_y = R_y = (\rho Q v_{2y} + \rho Q_3 v_{3y}) - (\rho Q_1 v_{1y})$$

$v_{1x} := v_1 = 241.916 \dfrac{\mathrm{m}}{\mathrm{s}}$ $\qquad\qquad$ $v_{1y} := 0 \dfrac{\mathrm{m}}{\mathrm{sec}}$

$v_{2x} := v_2 \cdot \cos(\alpha) = 108.862 \dfrac{\mathrm{m}}{\mathrm{s}}$ $\qquad$ $v_{2y} := v_2 \cdot \sin(\alpha) = 188.554 \dfrac{\mathrm{m}}{\mathrm{s}}$

$v_{3x} := -v_3 \cdot \cos(\alpha) = -108.862 \dfrac{\mathrm{m}}{\mathrm{s}}$ $\qquad$ $v_{3y} := -v_3 \cdot \sin(\alpha) = -188.554 \dfrac{\mathrm{m}}{\mathrm{s}}$

Guess value: $\qquad\qquad$ $R_x := 1000 \, \mathrm{N}$ $\qquad$ $R_y := 1000 \, \mathrm{N}$

Given

$$-R_x = (\rho \cdot Q_2 \cdot v_{2x} + \rho \cdot Q_3 \cdot v_{3x}) - (\rho \cdot Q_1 \cdot v_{1x})$$

$$R_y = (\rho \cdot Q_2 \cdot v_{2y} + \rho \cdot Q_3 \cdot v_{3y}) - (\rho \cdot Q_1 \cdot v_{1y})$$

$$\begin{pmatrix} R_x \\ R_y \end{pmatrix} := \mathrm{Find}\,(R_x, R_y) = \begin{pmatrix} 1.376 \times 10^4 \\ 7.945 \times 10^3 \end{pmatrix} \mathrm{N}$$

$R := \sqrt{R_x{}^2 + R_y{}^2} = 1.589 \times 10^4 \, \mathrm{N}$ $\qquad$ $\theta := \mathrm{atan}\left(\dfrac{|R_y|}{|R_x|}\right) = 30 \, \mathrm{deg}$

Thus, the reaction force acting on the plate, R_x acts to the left, and the reaction force, R_y acts upward, both as assumed, and they are illustrated in Figure EP 5.18. Note that the resultant force, R acts at an angle, $\theta = 30°$ with the horizontal, which is normal (at an angle of $90°$) to the plate. Such would be expected because there is no reaction force acting parallel to the plate (along the s-axis of the sloping plate), as illustrated in (b) above.

EXAMPLE PROBLEM 5.19

A forced jet of water at 20°C with a diameter of 0.06 m and a flowrate of 0.095 m³/sec hits a stationary horizontal flat plate, where the jet splits into two paths, as illustrated in Figure EP 5.19. Assume that the jet and the plate lie in a horizontal plane, (x, y), and assume ideal flow. (a) Determine the velocity of the flow at points 2 and 3. (b) Determine the flowrate at control faces 2 and 3. (c) Determine the reaction force acting on the plate due to the thrust force of the flow (anchoring force required to keep plate from moving).

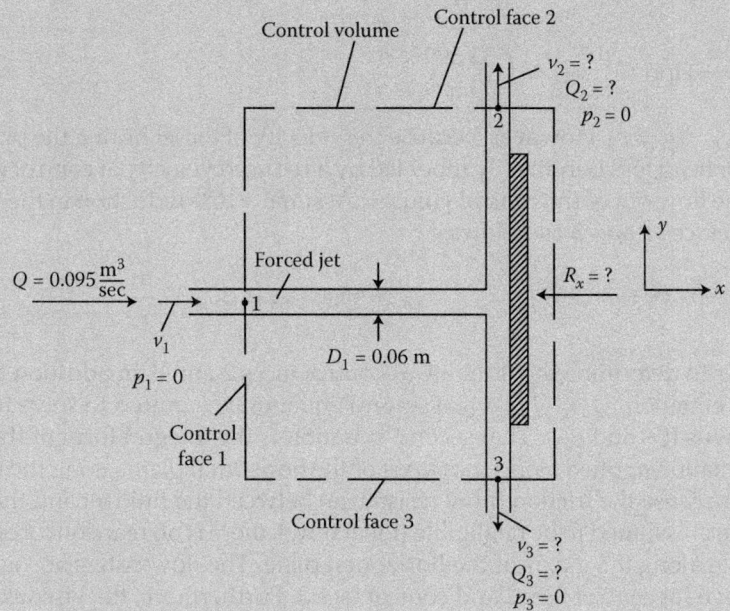

FIGURE EP 5.19
Forced jet of water hits a stationary horizontal plate and splits into two paths.

Mathcad Solution

(a) In order to determine the velocity of flow at points 2 and 3, the Bernoulli equation is applied between points 1 and 2, and between points 1 and 3 as follows:

$$Q_1 := 0.095 \frac{m^3}{sec} \qquad D_1 := 0.06 \, m \qquad A_1 := \frac{\pi \cdot D_1^2}{4} = 2.827 \times 10^{-3} \, m^2$$

$$v_1 := \frac{Q_1}{A_1} = 33.599 \frac{m}{s} \qquad\qquad P_1 := 0 \frac{N}{m^2} \qquad\qquad P_2 := 0 \frac{N}{m^2}$$

$$P_3 := 0 \frac{N}{m^2} \qquad\qquad z_1 := 0 \, m \qquad z_2 := 0 \, m \qquad z_3 := 0 \, m$$

$$\rho := 998 \frac{kg}{m^3} \qquad g := 9.81 \frac{m}{sec^2} \qquad \gamma := \rho \cdot g = 9.79 \times 10^3 \frac{kg}{m^2 \cdot s^2}$$

Guess value: $\qquad$ $v_2 := 1 \dfrac{\text{m}}{\text{sec}}$ $\qquad$ $v_3 := 1 \dfrac{\text{m}}{\text{sec}}$

Given

$$\frac{p_1}{\gamma} + z_1 + \frac{v_1^2}{2g} = \frac{p_2}{\gamma} + z_2 + \frac{v_2^2}{2g}$$

$$\frac{p_1}{\gamma} + z_1 + \frac{v_1^2}{2 \cdot g} = \frac{p_3}{\gamma} + z_3 + \frac{v_3^2}{2 \cdot g}$$

$$\begin{pmatrix} v_2 \\ v_3 \end{pmatrix} := \text{Find } (v_2, v_3) = \begin{pmatrix} 33.599 \\ 33.599 \end{pmatrix} \frac{\text{m}}{\text{s}}$$

Thus, $v_1 = v_2 = v_3$. However, because the velocity of the jet hitting the plate is high, a minor head loss is indirectly modeled by a reduced velocity at control faces 2 and 3 for the flow out of the control volume. Assume a 10% reduction in the velocity at control faces 2 and 3 as follows:

$$v_2 := 0.90 \cdot v_2 = 30.239 \, \frac{\text{m}}{\text{s}} \qquad\qquad v_3 := 0.90 \cdot v_3 = 30.239 \, \frac{\text{m}}{\text{s}}$$

(b) In order to determine the flowrate at control faces 2 and 3, in addition to the continuity equation, $Q_1 = Q_2 + Q_3$, a second equation is required to solve for the two unknowns (Q_2 and Q_3). That second equation is the integral form of the momentum equation applied along the y-axis of the horizontal plate (along the horizontal plate); because the friction (flow resistance) between the fluid jet and the horizontal plate is assumed to be negligible (ideal flow), there is no reaction force acting on the plate along the y-axis of the horizontal plate. The flow enters at control face 1 and leaves at control face 2 and control face 3. Furthermore, the viscous forces are ignored because of the assumption of ideal flow, there is no pressure force component because the jet flow is open to the atmosphere ($p_1 = p_2 = p_3 = 0 \, \text{N/m}^2$), and there is no gravitational force component along the y-axis because the pipe is horizontal ($z_1 = z_2 = z_3 = 0$).

$$\sum F_y = R_y = 0 = (\rho Q_2 v_{2y} + \rho Q_3 v_{3y}) - (\rho Q_1 v_{1y})$$

$v_{1y} := 0 \dfrac{\text{m}}{\text{sec}}$ $\qquad$ $v_{2y} := v_2 = 30.239 \dfrac{\text{m}}{\text{s}}$ $\qquad$ $v_{3y} := -v_3 = -30.239 \dfrac{\text{m}}{\text{s}}$

Guess value: $\quad Q_2 := 1 \dfrac{\text{m}^3}{\text{sec}}$ $\qquad$ $Q_3 := 1 \dfrac{\text{m}^3}{\text{sec}}$ $\qquad$ $A_2 := 1 \, \text{m}^2$ $\qquad$ $A_3 := 1 \, \text{m}^2$

Given

$$Q_1 = Q_2 + Q_3$$

$$0 = (\rho \cdot Q_2 \cdot v_{2y} + \rho \cdot Q_3 \cdot v_{3y}) - (\rho \cdot Q_1 \cdot v_{1y})$$

$$A_2 = \frac{Q_2}{v_2} \qquad\qquad A_3 = \frac{Q_3}{v_3}$$

$$\begin{pmatrix} Q_2 \\ Q_3 \\ A_2 \\ A_3 \end{pmatrix} := \text{Find } (Q_2, Q_3, A_2, A_3)$$

$$Q_2 = 0.048\,\frac{m^3}{s} \qquad\qquad Q_3 = 0.048\,\frac{m^3}{s} \qquad\qquad Q_2 + Q_3 = 0.095\,\frac{m^3}{s}$$

$$A_2 = 1.571 \times 10^{-3}\,m^2 \qquad\quad A_3 = 1.571 \times 10^{-3}\,m^2$$

(c) In order to determine the reaction force acting on the plate due to the thrust force of the flow, the integral form of the momentum equation is applied in the x direction for the control volume illustrated in Figure EP 5.19. The flow enters at control face 1 and leaves at control face 2 and control face 3. The viscous forces are ignored because of the assumption of ideal flow, there is no pressure force component because the jet flow is open to the atmosphere ($p_1 = p_2 = p_3 = 0\,N/m^2$), and there is no gravitational force component along the *x-axis* because the pipe is horizontal ($z_1 = z_2 = z_3 = 0$).

$$\sum F_x = R_x = (\rho Q_2 v_{2x} + \rho Q_3 v_{3x}) - (\rho Q_1 v_{1x})$$

$$v_{1x} := v_1 = 33.599\,\frac{m}{s} \qquad\qquad v_{2x} := 0\,\frac{m}{sec} \qquad\qquad v_{3x} := 0\,\frac{m}{sec}$$

Guess value: $\qquad\qquad\qquad\qquad\qquad R_x := 1000\,N$

Given

$$-R_x = (\rho \cdot Q_2 \cdot v_{2x} + \rho \cdot Q_3 \cdot v_{1x}) - (\rho \cdot Q_1 \cdot v_{1x})$$

$$R_x := \text{Find } (R_x) = 3.186 \times 10^{-3}\,N$$

Thus, the reaction force acting on the plate, R_x acts to the left, as assumed, and is illustrated in Figure EP 5.19.

And, finally, in the case of a free jet (gravity flow from an open tank), the integral momentum equation is applied to solve for an unknown reaction force (required to keep the tank from moving). It is important to note that a free jet acting to the right (see Figure EP 5.20) will cause the tank to move to the left (reaction of the tank to the free jet). Therefore, the force required to keep the tank from moving will act in the direction of the free jet. An example problem is presented in Example Problem 5.20 below.

EXAMPLE PROBLEM 5.20

Water at 20°C flows from an 2-m-diameter open tank through a 3-cm opening on the side of the tank (close to the bottom of the tank) as illustrated in Figure EP 5.20. The elevation of the water in the tank at point 1 is 20 m above the centerline of the opening at point 2. The empty tank weights 3 N, and assume a static coefficient of friction between the tank and the floor, μ of 0.44. (a) Determine the ideal velocity and flowrate of the free jet as it leaves the opening at point 2. (b) Determine the anchoring force required to keep the tank from moving to the left due to the free jet acting to the right. (c) Determine the friction force, F_f acting between the tank and the floor due to the weight of the tank and the water. (d) State if the friction force acting between the tank and the floor, F_f is sufficient to keep tank from moving to the left.

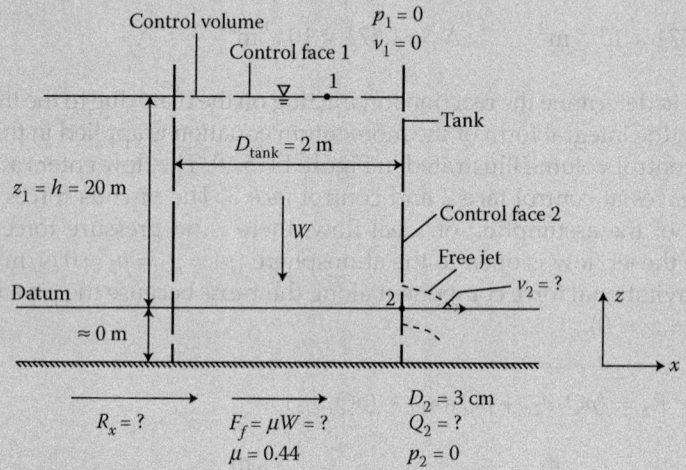

FIGURE EP 5.20
Water flows from an open tank through an opening on the side of the tank.

Mathcad Solution

(a) In order to determine the ideal velocity of the free jet at point 2, the Bernoulli equation is applied between points 1 and 2. Because the cross-sectional area of the tank at point 1 is much larger than the cross-sectional area of the opening, the velocity at point 1 is assumed to be zero. Assuming the datum is at point 2 yields the derivation and application of Torricelli's theorem as follows:

$$z_1 := 20 \text{ m} \qquad z_2 := 0 \text{ m} \qquad p_1 := 0 \frac{N}{m^2} \qquad p_2 := 0 \frac{N}{m^2} \qquad v_1 := 0 \frac{m}{\sec}$$

$$\rho := 998 \frac{kg}{m^3} \qquad\qquad g := 9.81 \frac{m}{\sec^2} \qquad \gamma := \rho \cdot g = 9.79 \times 10^3 \frac{kg}{m^2 \cdot s^2}$$

Guess value: $\qquad\qquad v_2 := 1 \frac{m}{\sec}$

Given

$$\frac{p_1}{\gamma} + z_1 + \frac{v_1^2}{2 \cdot g} = \frac{p_2}{\gamma} + z_2 + \frac{v_2^2}{2 \cdot g}$$

$v_2 :=$ Find $(v_2) = 19.809 \dfrac{m}{s}$

Alternatively, one may directly apply Torricelli's theorem as follows:

$h := z_1 = 20 \, m$ $\qquad\qquad v_2 := \sqrt{2 \cdot g \cdot h} = 19.809 \dfrac{m}{s}$

Thus, Torricelli's theorem, which is a special case of the Bernoulli equation, specifically illustrates the conversion of potential energy stored in the height of the fluid in the tank, h to kinetic energy stored in the ideal velocity of the free jet of fluid at the opening in the side of the tank. Furthermore, in order to determine the flowrate, the continuity equation is applied at point 2 as follows:

$D_2 := 3 \, cm$ $\qquad\qquad A_2 := \dfrac{\pi \cdot D_2^2}{4} = 7.069 \times 10^{-4} \, m^2$

$Q_2 := v_2 \cdot A_2 = 0.014 \dfrac{m^3}{s}$

(b) In order to determine the anchoring force required to keep the tank from moving to the left due to the free jet acting to the right, the integral form of the momentum equation is applied in the x direction for the control volume illustrated in Figure EP 5.20. The flow enters at control face 1 and leaves at control face 2. The viscous forces are ignored because of the assumption of ideal flow, there is no pressure force component because the open tank flow is open to the atmosphere ($p_1 = p_2 = 0 \, N/m^2$), and there is no gravitational force component along the *x-axis* because the flow at point 2 is assumed to be horizontal.

$$\sum F_x = R_x = (\rho Q_2 v_{2x}) - (\rho Q_1 v_{1x})$$

$v_{1x} := v_1 = 0$ $\qquad\qquad v_{2x} := v_2 = 19.809 \dfrac{m}{s}$ $\qquad Q := Q_2 = 0.014 \dfrac{m^3}{s}$

Guess value: $\qquad\qquad\qquad\qquad R_x := 1000 \, N$

Given

$R_x = \rho \cdot Q \, (v_{2x} - v_{1x})$

$R_x :=$ Find $(R_x) = 276.816 \, N$

Thus, the anchoring force required to keep the tank from moving to the left, R_x acts to the right as assumed and is illustrated in Figure EP 5.20.

(c) The friction force, $F_f = \mu W$ acting between the tank and the floor due to the weight of the tank and the water is computed as follows:

$\mu := 0.44$ $\qquad\qquad W_{tank} := 3 \, N$ $\qquad\qquad D_{tank} := 2 \, m$

$$A_{tank} := \frac{\pi \cdot D_{tank}^2}{4} = 3.142 \, m^2 \qquad\qquad V_{water} := A_{tank} \cdot h = 62.832 \, m^3$$

$$W_{water} := \gamma \cdot V_{water} = 6.151 \times 10^5 \, N \qquad\qquad W := W_{tank} + W_{water} = 6.152 \times 10^5 \, N$$

$$F_f := \mu \cdot W = 2.707 \times 10^5 \, N$$

(d) Thus, the friction force acting between the tank and the floor, F_f is (more than) sufficient to keep tank from moving to the left.

5.3.5 Application of the Energy and Momentum Equations for Ideal External Flow

Application of the Bernoulli equation (Equation 5.33) assumes ideal flow; thus, there is no head loss term to evaluate in the energy equation. Therefore, for problems that involve external flow around an object, the major head loss due to flow resistance is assumed to be negligible, so one may apply the Bernoulli equation. Thus, the Bernoulli equation may be solved for an unknown "energy head" term (p/γ, z, or $v^2/2g$) for one of the two points in a finite control volume as follows:

$$\left(\frac{p}{\gamma} + z + \frac{v^2}{2g}\right)_1 - \left(\frac{p}{\gamma} + z + \frac{v^2}{2g}\right)_2 = 0 \tag{5.53}$$

These types of flow problems include velocity measuring devices for external flow, which however, have a minor loss associated with the velocity measurement. The associated minor head loss is typically accounted for by the use/calibration of a velocity/drag coefficient to determine the drag force, where ideal flow (Bernoulli equation) is assumed. Thus, in the case of velocity measuring devices (e.g., pitot-static tubes), one must indirectly model the minor head loss flow resistance term by the use of a velocity/drag coefficient, which requires a "subset level" of application of the governing equations (Section 5.3.5.1). Then, application of the complementary integral momentum equation (Equation 5.35) is required only if there is an unknown reaction force acting on the object; note that the pressure and viscous forces are represented by the drag force, and the viscous forces have already been accounted for indirectly through the velocity/drag coefficient. As such, the integral form of the momentum equation is solved analytically, without the use of dimensional analysis. Applications of the governing equations for ideal external flow are presented in Section 5.3.5.2 below.

5.3.5.1 Evaluation of the Drag Force

In the case of external flow around an object, the associated minor head loss is typically accounted for by the use/calibration of a drag coefficient to determine the drag force. Evaluation of the drag force requires a "subset level" application of the governing equations. A "subset level" application of the governing equations focuses only on the given element causing the flow resistance. The assumption of "ideal" flow implies that the flow resistance is modeled only in the momentum equation (and thus, the subsequent assumption of "real" flow). The flow resistance equation for the drag force is derived in Chapters 6 and 7. In the derivation of the drag force for an external flow around an object, first, ideal flow is assumed and thus the Bernoulli equation is applied as follows:

$$\left(\frac{p}{\gamma} + z + \frac{v^2}{2g}\right)_1 - \left(\frac{p}{\gamma} + z + \frac{v^2}{2g}\right)_2 = 0 \tag{5.54}$$

Assuming $z_1 = z_2$, this equation has one unknown, which is the ideal velocity of the external flow around an object. Thus, this equation yields an expression for the ideal velocity, $v_i = \sqrt{2\Delta p/\rho}$ as a function of an ideal pressure difference, Δp, which is directly measured. Then, application of the momentum equation (supplemented by dimensional analysis) is used to determine an expression for the actual velocity, v_a and the drag force, F_D, which was given in Equation 5.27 and is repeated as follows:

$$\sum F_s = (F_G + F_P + F_V + F_{other})_s = (\rho Q v_s)_2 - (\rho Q v_s)_1 \tag{5.55}$$

Thus, the resulting flow resistance equation for the drag force is given as follows:

$$F_D = (F_P + F_f)_s = (\Delta p A)_s = \left(\frac{\rho v^2}{2} A + \frac{\tau_w L}{R_h} A\right)_s = C_D \frac{1}{2} \rho v^2 A \tag{5.56}$$

where F_D is equal to the sum of the pressure and friction force in the direction of the flow, s, and the drag coefficient, C_D accounts for (indirectly models) the minor head loss associated with the velocity measurement, and the drag force.

The flow resistance (shear stress or drag force) in the external flow around an object is modeled as a resistance force/drag force, F_D in the integral form of the momentum equation. The flow resistance causes an unknown pressure drop, Δp, which causes an unknown head loss, h_f, where the head loss is due to a conversion of kinetic energy to heat, which is modeled/displayed in the integral form of the energy equation. However, although the head loss, h_f causes the drag force, F_D, the head loss is not actually determined in the design of external flow around an object. The assumption of ideal flow and thus applying the Bernoulli equation to measure the ideal velocity of flow, $v_i = \sqrt{2\Delta p/\rho}$ by the use of a pitot-static tube (see Section 4.5.7.3) yields an expression for the ideal velocity as a function of ideal pressure difference, Δp, which is directly measured. Therefore, the associated minor head loss with the velocity measurement is accounted for by the drag coefficient (in the application of the momentum equation). Thus, in the external flow around an object, the flow resistance is ultimately modeled as a drag force in the integral form of the momentum equation because the drag force is needed for the design of external flow around an object. The head loss, h_f causing the actual velocity, $v_a = \sqrt{(2/\rho)(\Delta p - \tau_w L/R_h)}$ and the drag force, $F_D = (F_p + F_f)_s = (\Delta p A)_s = \left(((\rho v^2/2)A) + ((\tau_w L/R_h)A)\right)_s$ is caused by both pressure and friction forces, where the drag force, F_D is equal to the sum of the pressure and friction force in the direction of the flow, s. However, because the friction/viscous forces (due to shear stress, τ_w) due to the external flow cannot be theoretically modeled in the integral momentum equation, the actual pressure drop, Δp cannot be analytically determined, and thus the exact reduction in the velocity actual velocity, v_a, which is less than the ideal velocity, v_i, cannot be theoretically determined. Furthermore, the exact component in the s-direction of the pressure and viscous forces cannot be theoretically determined. Therefore, one cannot derive an analytical expression for the drag force, F_D from the momentum equation. As a result, one must resort to dimensional analysis (which supplements the momentum theory) in order to derive an expression for the drag force, F_D, which involves the definition of a drag coefficient, C_D that represents the flow resistance. Drag forces on external flow are addressed in detail in Chapters 6 and 10.

5.3.5.2 Applications of the Governing Equations for Ideal External Flow

The integral form of the governing equations (energy and momentum) for ideal flow are applied to a finite control volume in the analysis and design of external flows around a body. Applications of the governing equations for ideal external flow include the analysis and design of planes or street signs, for instance, which include a speed measuring device such as a pitot-static tube, and the determination of a pressure difference in external flow from a velocity difference. The continuity equation is not applied in external flow. The Bernoulli equation is applied to measure the velocity of the flow by the use of a pitot-static tube, which yields an expression for the ideal velocity as a function of ideal pressure difference, which is directly measured. The minor head loss associated with the velocity measurement is accounted for by the drag force. Then, application of the complementary momentum equation is required only if there is an unknown force acting on the object; note that the pressure and viscous forces are represented by the drag force. One may note from Section 5.3.5.1 above that the minor head loss associated with the velocity measurement in external flow is indirectly modeled by a drag coefficient, C_D, which requires a "subset level" application of governing equations for its evaluation. Furthermore, the resulting definition of the drag force, $F_D = C_D(1/2)\rho v^2 A$ requires the definition of the drag coefficient, C_D, which is to be presented in later chapters. Therefore, example problems illustrating the application of the governing equations for ideal external flow in Chapter 5 will assume that the drag force is either a given quantity, computed from the given drag force equation, or directly solved for in the integral momentum equation. Furthermore, for the example problems involving external flow velocity measurements, illustration of the computation of an unknown force, which requires the application of the complementary integral momentum equation are presented in Example Problem 5.21 below and in Chapter 10.

EXAMPLE PROBLEM 5.21

A rectangular street sign that is 2 m wide and 1 m high is secured to a post that is secured into the ground, as illustrated in Figure EP 5.21. The sign is at sea level, and is designed to withstand a maximum wind speed of 35 m/sec. Assume a drag coefficient for the sign, C_d of 1.2. (a) Determine the drag force on the street sign. (b) Determine the reaction force acting on the sign due to the thrust force of the maximum wind flow (anchoring force required to keep sign from detaching from the post).

Mathcad Solution

(a) In order to determine the drag force on the street sign, the drag force equation is applied. The density of the standard atmospheric air, ρ at sea level is given in Table A.1 in Appendix A.

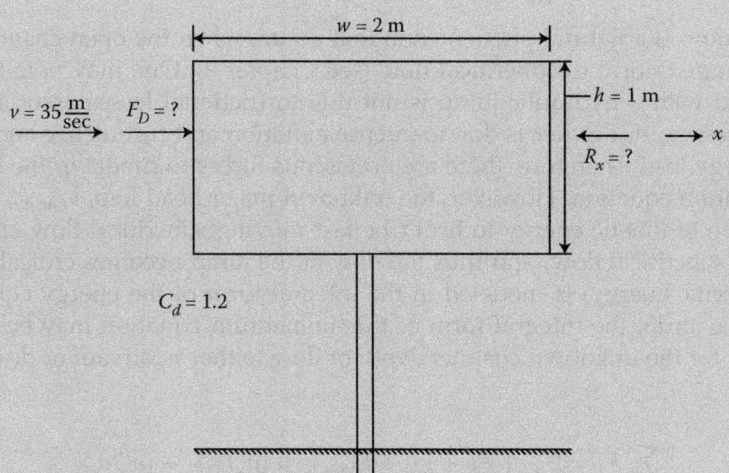

FIGURE EP 5.21
Wind flows past a rectangular street sign secured to a post secured into ground.

$$\rho := 1.225\,\frac{\text{kg}}{\text{m}^3} \qquad C_d := 1.2 \qquad w := 2\,\text{m} \qquad h := 1\,\text{m}$$

$$A := w \cdot h = 2\,\text{m}^2 \qquad v := 35\,\frac{\text{m}}{\text{sec}} \qquad F_D := C_d\frac{1}{2}\rho \cdot v^2 \cdot A = 1.801 \times 10^3\,\text{N}$$

(b) In order to determine the reaction force acting on the sign due to the thrust force of the maximum wind flow (anchoring force required to keep sign from detaching from the post), the integral form of the momentum equation is applied in the x direction (direction of air flow) for the sign illustrated in Figure EP 5.21. The pressure and viscous forces in the direction of wind flow are represented by the drag force. And there is no gravitational force component along the x-axis because the wind flow (also, negligible weight) is assumed to be horizontal.

$$\sum F_x = F_D + R_x = 0$$

Guess value: $\qquad\qquad\qquad\qquad R_x := 1000\,\text{N}$

Given

$$F_D - R_x = 0$$

$$R_x := \text{Find}\,(R_x) = 1.801 \times 10^3\,\text{N}$$

Thus, the anchoring force required to keep the sign from detaching from the post, R_x acts to the left, as assumed and illustrated in Figure EP 5.21.

5.3.6 Application of the Energy and Momentum Equations for a Hydraulic Jump

A hydraulic jump is a natural phenomenon that occurs when the open channel flow transitions from supercritical to subcritical flow (see Chapter 9). One may note that the head loss associated with a hydraulic jump is not due to frictional losses associated with the wall shear stress, τ_w but rather is due to intense agitation and turbulence and thus results in a high-energy loss. Therefore, there are no viscous forces to model in the integral form of the momentum equation. However, the unknown major head loss, $h_{f,major}$, which is due to a conversion of kinetic energy to heat (the fast-moving subcritical flow encounters the slow-moving subcritical flow, and thus the flow at the jump becomes critical flow, with a minimum specific energy) is modeled in the integral form of the energy equation. Thus, for a hydraulic jump, the integral form of the momentum equation may be used to analytically solve for the unknown channel depth of flow (either upstream or downstream) as follows:

$$\sum F_s = (F_G + F_P + F_V + F_{other})_s = (\rho Q v_s)_2 - (\rho Q v_s)_1 \qquad (5.57)$$

by deriving the hydraulic jump equations (see Chapter 9 for the derivations and applications). Then, the energy equation is applied to solve for the major head loss as follows:

$$\left(y + z + \frac{v^2}{2g} \right)_1 - h_{f,maj} = \left(y + z + \frac{v^2}{2g} \right)_2 \qquad (5.58)$$

5.3.6.1 Applications of the Governing Equations for a Hydraulic Jump

The integral form of the governing equations (continuity, energy, and momentum) are applied to a finite control volume in the analysis and design of a hydraulic jump. The integral form of the momentum equation is applied in order to derive the hydraulic jump equations (see Equations 9.248 and 9.249), which evaluate the unknown depth of flow. Therefore, example problems illustrating the application for the momentum equation for a hydraulic jump in Chapter 5 will assume that the hydraulic jump equation is given and is applied to solve for the unknown depth of flow. Then, the complementary energy equation is applied to solve for the unknown major head loss due to the jump. Example problems involving a hydraulic jump are presented in Example Problem 5.22 below and in Chapter 9.

EXAMPLE PROBLEM 5.22

Water flowing in a rectangular channel of width 3 m at a supercritical velocity of 4.823 m/sec and a supercritical depth of 0.933 m encounters a subcritical flow, as illustrated in Figure EP 5.22. As the flow transitions from supercritical to subcritical flow, a hydraulic jump occurs, where there is a major head loss due the intense agitation and turbulence, where the depth of flow is at critical depth of 1.273 m. (a) Determine the subcritical depth of flow and the subcritical velocity of flow after the jump. (b) Determine the head loss due to the flow turbulence at the jump.

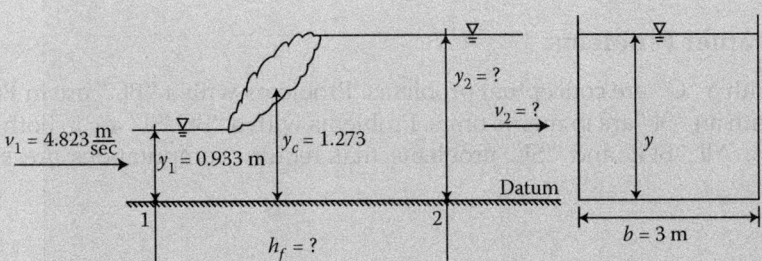

FIGURE EP 5.22
Water flows in a rectangular channel at a supercritical depth encounters a subcritical flow and thus forms a hydraulic jump.

Mathcad Solution

(a) In order to determine the subcritical depth of flow after the jump, the hydraulic jump equation (see Equation 9.248), which is derived by applying the integral form of the momentum equation, is applied. Given that the upstream conditions (depth and velocity) are known, the hydraulic jump equation (Equation 9.248), $y_2 = \dfrac{y_1}{2}\left(-1 + \sqrt{1 + 8F_1^2}\right)$ requires the definition of the Froude number for the supercritical depth, y_1 upstream of the jump, $F_1 = \dfrac{v_1}{\sqrt{g y_1}}$. Furthermore, the continuity equation is applied to determine the subcritical velocity of flow.

$$v_1 := 4.823\,\frac{m}{sec} \qquad y_1 := 0.933\,m \qquad b := 3\,m \qquad g := 9.81\,\frac{m}{sec^2}$$

$$F_1 := \frac{v_1}{\sqrt{g \cdot y_1}} = 1.594 \qquad\qquad y_2 := \frac{y_1}{2}(-1 + \sqrt{1 + 8 \cdot F_1^2} = 1.688\,m$$

$$Q := v_1 \cdot b \cdot y_1 = 13.5\,\frac{m^3}{s} \qquad\qquad v_2 := \frac{Q}{b \cdot y_2} = 2.666\,\frac{m}{s}$$

(b) In order to determine the major head loss due to the jump, the energy equation is applied between the upstream point 1 and the downstream point 2 as follows:

$$z_1 := 0\,m \qquad\qquad\qquad z_2 := 0\,m$$

Guess value: $\qquad\qquad\qquad\qquad h_f := 1\,m$

Given

$$y_1 + z_1 + \frac{v_1^2}{2 \cdot g} - h_f = y_2 + z_2 + \frac{v_2^2}{2 \cdot g}$$

$$h_f := \text{Find}\,(h_f) = 0.068\,m$$

End-of-Chapter Problems

Problems with a "C" are conceptual problems. Problems with a "BG" are in English units. Problems with an "SI" are in metric units. Problems with a "BG/SI" are in both English and metric units. All "BG" and "SI" problems that require computations are solved using Mathcad.

Introduction

5.1C In the study of fluids in motion, consideration of the fluid kinematics and the fluid dynamics yields three governing equations that are applied in the solution of a fluid flow problem. Discuss this point in further detail.

5.2C What equations are applied to internal flows, which include pressure (pipe) flow in Chapter 8 and gravity (open channel) flow in Chapter 9?

5.3C What equations are applied to external flows, which include the flow around objects in Chapter 10?

5.4C Depending upon the internal flow problem, one may use either the Eulerian (integral) approach, the Lagrangian (differential) approach, or both in the application of the three governing equations (continuity, energy, and momentum). Explain the caveat in the application of the energy equation.

5.5C For an internal fluid flow, while the application of the continuity equation will always be necessary, the equations of motion (the energy equation and the momentum equation) play complementary roles in the analysis of a given flow situation. Explain the complementary roles of the equations of motion.

5.6C In the application of the governing equations (continuity, energy, and momentum), it is important to make the distinction between types of flow. Assuming steady flow, explain what these flow types are.

Derivation of The Momentum Equation

5.7C Derivation of the momentum equation is based on the principle of conservation of momentum (Newton's second law of motion). As presented in Chapter 4, the principle of conservation of momentum is historically stated and applied for a fluid system (Section 5.2.1), and then rephrased and extended to a control volume system (Section 5.2.2) using the Reynolds transport theorem. Describe the historically stated principle for a fluid system.

5.8C The momentum equation is applied in complement to the energy equation, which is useful only for a control volume. However, depending upon the fluid flow problem, one may apply the differential form of the momentum equation (for a fluid system) or the integral form of the momentum equation (for a control volume). Explain this in detail.

The Momentum Equation for a Fluid System: Differential Form of the Momentum Equation

5.9C Explain why the differential form of the momentum equation given by Equation 5.14 as follows: $\dfrac{\partial}{\partial s}\left(\dfrac{p}{\gamma}\right) + \dfrac{\partial z}{\partial s} + \dfrac{v}{g}\dfrac{\partial v}{\partial s} = -\dfrac{\tau_w}{\gamma R_h} - \dfrac{1}{g}\dfrac{\partial v}{\partial t}$ is basically the derivative of the integral form of the energy equation given by Equation 4.39 as follows:

$$\left(\frac{p}{\gamma} + z + \frac{v^2}{2g}\right)_1 - \left(\frac{p}{\gamma} + z + \frac{v^2}{2g}\right)_2 = \frac{\tau_w}{\gamma}\frac{L}{R_h} + \frac{L}{g}\frac{\partial v}{\partial t}$$

5.10C Explain how the individual terms in the differential form of the momentum equation for the most general flow type (compressible, nonuniform, unsteady flow for either pressure or gravity flow) given by Equation 5.15 as follows:

$$S_f = -\frac{\partial z}{\partial s} - \frac{\partial}{\partial s}\left(\frac{p}{\gamma}\right) - \frac{v}{g}\frac{\partial v}{\partial s} - \frac{1}{g}\frac{\partial v}{\partial t}$$

directly relate to the differential form of Newton's second law of motion for a differential fluid element moving in the *s*-direction along the streamline given by Equation 5.16 as follows:

$$\sum F_s = F_G + F_P + F_V = Ma_s$$

5.11C Beginning with the differential form of the momentum equation for the most general flow type (compressible, nonuniform, unsteady flow for either pressure or gravity flow) given by Equation 5.17 as follows:

$$\underbrace{\frac{\tau_w}{\gamma R_h}}_{S_f} = \underbrace{-\frac{\partial z}{\partial s} - \frac{\partial}{\partial s}\left(\frac{p}{\gamma}\right)}_{-S_e} \underbrace{-\frac{v}{g}\frac{\partial v}{\partial s} - \frac{1}{g}\frac{\partial v}{\partial t}}_{-S_a}$$

state the differential form of the momentum equation for compressible, nonuniform, steady flow.

5.12C The derivation of the differential form of the momentum equation for compressible, nonuniform, *steady* flow assumes a pipe flowing under pressure, and is given by Equation 5.18 as follows:

$$\underbrace{\frac{\tau_w}{\gamma R_h}}_{S_f} = \underbrace{-\frac{\partial z}{\partial s} - \frac{\partial}{\partial s}\left(\frac{p}{\gamma}\right)}_{-S_e} - \frac{v}{g}\frac{\partial v}{\partial s}$$

state the assumptions made regarding the application of the differential momentum equation (Equation 5.18) for pressure flow.

5.13C Explain how the friction slope term, S_f (Equation 5.8) is evaluated in the application of the differential form of the momentum equation (Equation 5.19) for pipe flow.

5.14C Explain how the Chezy (Equation 5.20) equation may be applied in pipe flow.

5.15BG Water at 68° F flows in a 10-ft-diameter inclined straight pipe section at a flowrate of 7 ft^3/sec, as illustrated in Figure ECP 5.15. The pipe length between points 1 and 2 is 2500 ft, and the Chezy coefficient, C for the pipe is 120 $ft^{1/2} s^{-1}$. The elevations of points 1 and 2 are 5 ft and 9 ft, respectively, and the pressure at point 1 is 45 psi. (a) Determine the friction slope at any point in the straight pipe section. (b) Determine the head loss due to pipe friction between points 1 and 2. (c) Determine the pressure drop at point 2 due to the pipe friction.

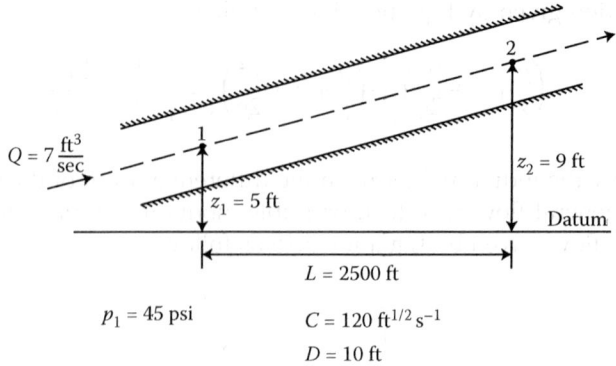

FIGURE ECP 5.15

5.16C The derivation of the differential form of the momentum equation for compressible, nonuniform, steady flow assumes a pipe flowing under pressure, and is given by Equation 5.18 as follows:

$$\underbrace{\frac{\tau_w}{\gamma R_h}}_{S_f} = \underbrace{-\frac{\partial z}{\partial s} - \frac{\partial}{\partial s}\left(\frac{p}{\gamma}\right) - \frac{v}{g}\frac{\partial v}{\partial s}}_{-S_e}$$

State the assumptions made regarding the application of the differential momentum equation (Equation 5.18) for open channel flow.

5.17C Begin with the differential momentum equation for incompressible, nonuniform, steady flow open channel flow given by Equation 5.21 as follows:

$$\underbrace{\frac{\tau_w}{\gamma R_h}}_{S_f} = S_o - \underbrace{\frac{\partial}{\partial s}(y) - \frac{v}{g}\frac{\partial v}{\partial s}}_{-S_e}$$

and state the differential momentum equation for incompressible, uniform, steady flow open channel flow.

5.18C Explain how the friction slope term, S_f (Equation 5.8) is evaluated in the application of the differential form of the momentum equation (Equation 5.22) for uniform open channel flow.

5.19C Explain how the Chezy Equation (Equation 5.23) may be applied in uniform open channel flow.

5.20C Explain how the differential momentum equation for incompressible, nonuniform, steady flow open channel flow given by Equation 5.25 (the "resistance equation") as follows is applied in open channel flow:

$$\frac{dy}{ds} = \frac{S_o - S_f}{1 - F^2}$$

5.21C Explain if the Chezy equation (Equation 5.26) for incompressible, uniform, steady open channel flow may be applied to incompressible, nonuniform, steady open channel flow.

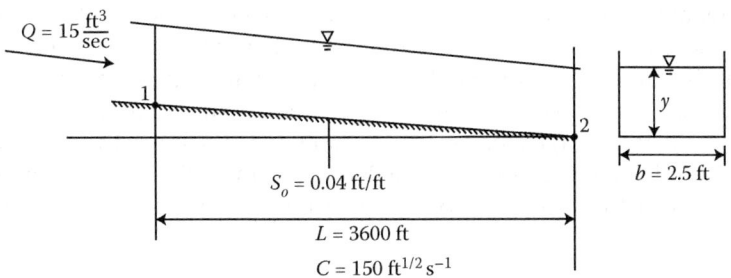

FIGURE ECP 5.22

5.22 BG Water at 68° F flows at a discharge of $15\,ft^3/sec$ in a rectangular channel that is 2.5 *ft* wide, with a channel bottom slope of $0.04\,ft/ft$, as illustrated in Figure ECP 5.22. The channel length between points 1 and 2 is $3600\,ft$, and the Chezy coefficient of $150\,ft^{1/2}\,s^{-1}$. (a) Determine the uniform depth of flow in the channel. (b) Determine the head loss due to channel friction between points 1 and 2.

The Momentum Equation for a Control Volume: Integral Form of the Momentum Equation

5.23C Explain what the integral form of the momentum equation, which is given by Equation 5.27 as follows, states and how it is applied:

$$\sum F_s = (F_G + F_P + F_V + F_{other})_s = (\rho Q v_s)_2 - (\rho Q v_s)_1$$

5.24C Assuming steady flow, state the more general statement of the integral form of the momentum equation for a control volume given by Equation 5.27, which assumes there may be one or more control faces for the flow into and the flow out of the control volume as illustrated in Figure 4.3.

5.25 SI Water at 20°C flows in a 0.48-*m*-diameter pipe that is fitted with a 180° threaded bend in order to reverse the direction of flow, as illustrated in Figure ECP 5.25. Assume that the pipe and the bend lie in a horizontal plane, (x, y), and the flow discharges into the atmosphere at a rate of $0.99\,m^3/sec$. Also, assume a minor head loss coefficient due to the 180° threaded bend, k of 1.5. (a) Determine the minor head loss due to the bend. (b) Determine the pressure at point 1. (c) Determine the reaction force acting on the bend due to the thrust force of the flow (anchoring force required to keep the bend from moving).

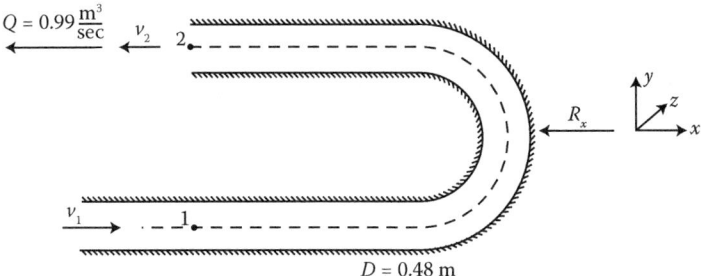

FIGURE ECP 5.25

5.26BG   Water at 68°F flows in a 1.6-*ft*-diameter horizontal pipe at a flowrate of 5 *ft*³/*sec*. A 90° mitered bend is installed in between point 1 and 2 in order to change the direction of the flow by 90°, as illustrated in Figure ECP 5.26. Assume that the pipe and the bend lie in a horizontal plane, (x, y). The pressure at point 1 is 60 *psi*, and assume a minor head loss coefficient due to the bend, k of 0.131. (a) Determine the minor head loss due to the bend between points 1 and 2. (b) Determine the pressure at point 2. (c) Determine the reaction force acting on the bend due to the thrust force of the flow (anchoring force required to keep the bend from moving).

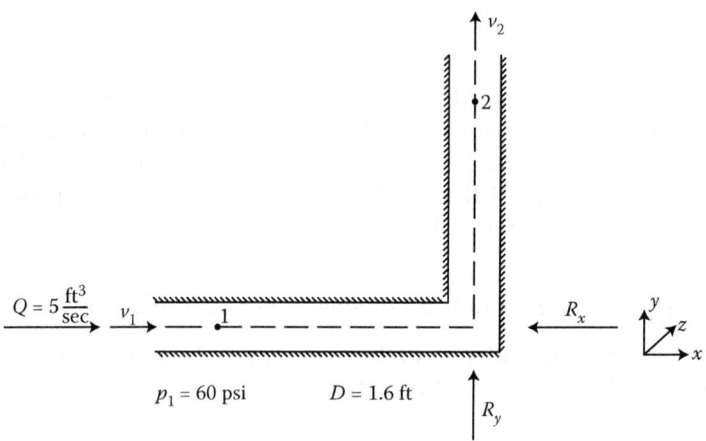

FIGURE ECP 5.26

5.27SI   Water at 20°C flows in a 1.4-*m*-diameter pipe at a flowrate of 2.7 *m*³/*sec*. A double nozzle is installed downstream of point 1 in order to divide the flow in two, as illustrated in Figure ECP 5.27. Assume that the pipe and the nozzle lie in a horizontal plane, (x, y). The diameter of the nozzle at point 2 is 0.66 *m* and is at an angle of 30° with the x-axis, and the diameter of the nozzle at point 3 is 0.77 *m* and is at an angle of 20° with the x-axis. The flow at points 2 and 3 are open to the atmosphere. Assume that the flow is ideal and

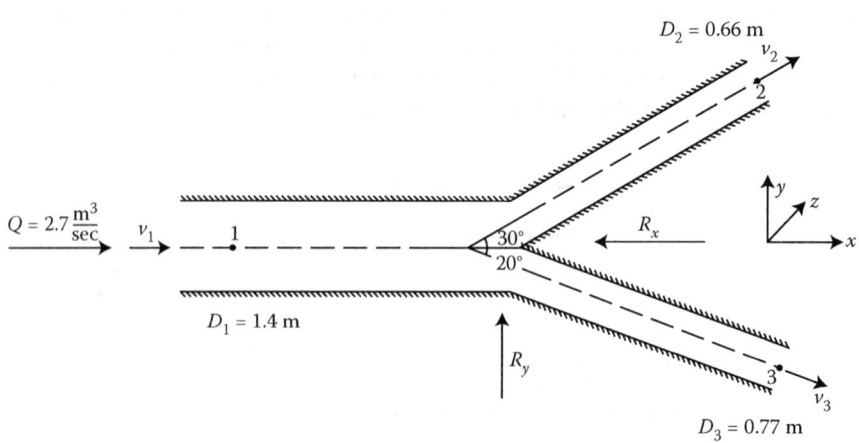

FIGURE ECP 5.27

thus ignore any minor head loss due to the nozzle. (a) Determine the pressure at point 1. (b) Determine the reaction force acting on the nozzle due to the thrust force of the flow (anchoring force required to keep nozzle from moving).

5.28BG Water at 68 °F flows at a depth of 5 ft in a rectangular channel of width 4 ft at a mild slope. A weir is inserted in the channel as illustrated in Figure ECP 5.28 in order to measure the flowrate, which is 0.78 ft³/sec. The depth of flow downstream of the weir is 2 ft. Determine the reaction force acting on the weir due to the thrust force of flow (anchoring force required to keep the weir from moving).

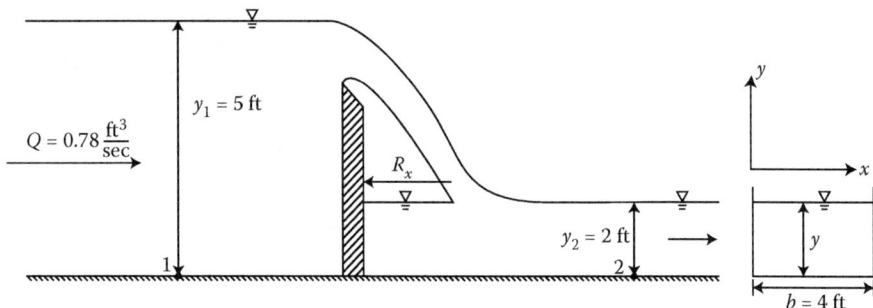

FIGURE ECP 5.28

5.29BG A forced jet of water at 80° F hits a stationary vane at a velocity of 16 ft/sec and a flowrate of 1.3 ft³/sec, as illustrated in Figure ECP 5.29. The vane angle α is 65°, and assume that the jet and the vane lie in a horizontal plane (x, y) and that flow is ideal. Determine the reaction force acting on the vane due to the thrust force of the jet (anchoring force required to keep vane from moving).

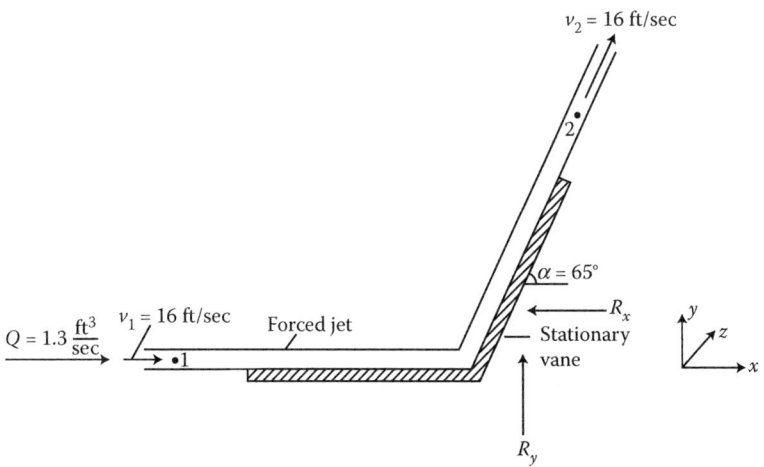

FIGURE ECP 5.29

5.30SI   A forced jet of water at 20°C hits a moving vane, which is moving at a velocity of 3 m/sec (abs), at a velocity of 8 m/sec (abs) and a flowrate of 1.5 m³/sec, as illustrated in Figure ECP 5.30. The vane angle α is 55°, and assume that the jet and the vane lie in a horizontal plane (x, y) and that flow is ideal. Determine the reaction force acting on the moving vane due to the thrust force of the jet (anchoring force required to keep the moving vane from accelerating).

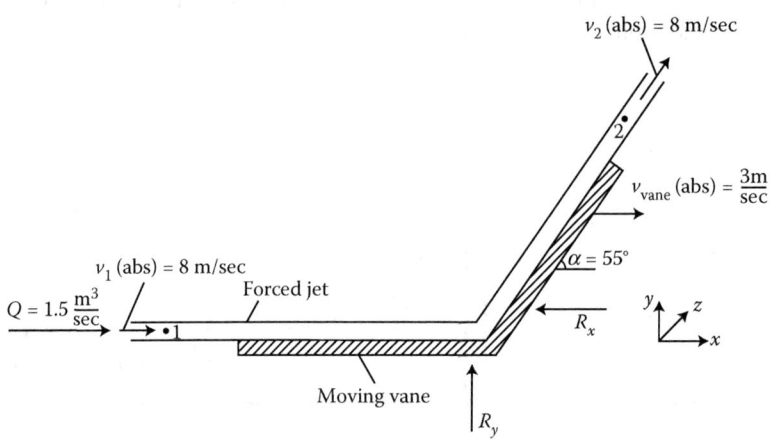

FIGURE ECP 5.30

5.31BG   Air at a density of 0.00237 slug/ft³ and at a pressure of 10 psi enters the intake (control face 1) of a car engine with a diameter of 5 in at an absolute velocity of 20 m/hr (relative to the earth), as illustrated in Figure ECP 5.31. The exhaust at a pressure of 0 psi leaves the tail of the car engine (control face 2) with a diameter of 3 in. The car (engine) (control volume) is parked in the driveway. Determine the reaction force acting on the car engine due to the thrust force of the air intake (anchoring force required to keep the car engine from accelerating).

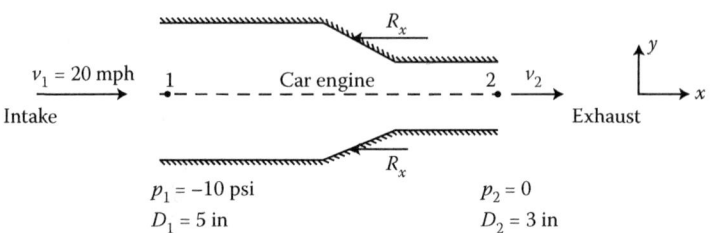

FIGURE ECP 5.31

5.32SI   Air at a density of 1.12 kg/m³ and at a pressure of -2.5×10^4 N/m² enters the intake (control face 1) of a jet engine with a diameter of 1.5 m at an absolute velocity of 270 km/hr (relative to the earth), as illustrated in Figure ECP 5.32. The exhaust at a pressure of 0 N/m² leaves the tail of the jet engine (control face 2) with a diameter of 1.1 m. The jet (engine) (control volume) is traveling at an absolute velocity of 500 km/hr (relative to the earth). Determine the reaction force acting on the moving jet engine due to the thrust force of the air intake (anchoring force required to keep the moving jet engine from accelerating).

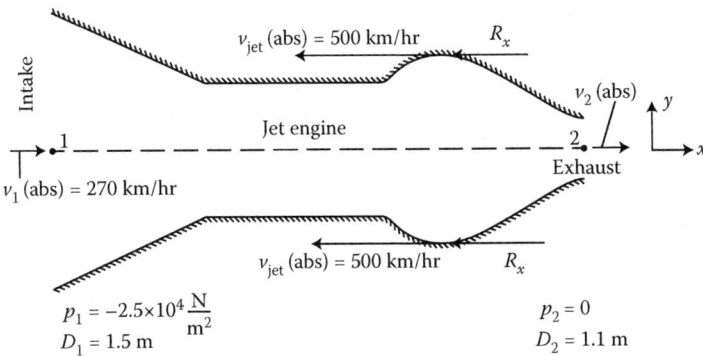

FIGURE ECP 5.32

Application of the Energy Equation in Complement to the Momentum Equation

Modeling Flow Resistance: A Subset Level Application of the Governing Equations

5.33C   In the application of the governing equations (continuity, energy, and momentum), it is important to make the distinction between real flow, ideal flow, and a hydraulic jump. And, for ideal flow, it is important to make the distinction between internal flow (pipe or open channel flow), external flow, and the flow from a tank (jet and siphon flows) (see Table 4.1) (Section 5.3.1). Furthermore, in the application of the governing equations, although in some flow cases, one may ignore (or not need to model) the flow resistance (viscous force), in most practical flow cases in general, one must account for the fluid viscosity and thus model the resulting shear stress and flow resistance. Explain when it is necessary to model flow resistance, and explain how the flow resistance (viscous force) is modeled. Also, explain when it is not necessary to model flow resistance.

5.34C   The application of the governing equations (continuity, energy, and momentum) for real flow (internal), and ideal flow (flow-measuring devices for internal flow and external flow around objects) requires the modeling of flow resistance (frictional losses). The modeling of flow resistance requires a "subset level" application of the governing equations (Section 5.3.2). Identity the flow resistance (frictional losses) terms that require to be modeled for a given flow type.

5.35C   The application of the governing equations (continuity, energy, and momentum) for real flow (internal), and ideal flow (flow-measuring devices for internal flow and external flow around objects) requires the modeling of flow resistance (frictional losses). The modeling of flow resistance requires a "subset level" application of the governing equations (Section 5.3.2). Furthermore, the use of dimensional analysis (Chapter 7) in supplement to a "subset level" application of the appropriate governing equations is needed to derive flow resistance equations when there is no theory to model the friction force. Explain the flow types for which it is necessary, and the flow types for which it is unnecessary, to use dimensional analysis (Chapter 7) in supplement to a "subset level" application of the appropriate governing equations.

Application of the Energy and Momentum Equations for Real Internal Flow

5.36C   The application of the integral form of the energy equation for incompressible, real, steady flow (pipe or open channel flow) given by Equation 4.44 requires the evaluation of the major and minor head loss flow resistance terms, $h_{f,maj} = \tau_w L/\gamma R_h$ and $h_{f,min} = \tau_w L/\gamma R_h$, respectively. Evaluation of the major and minor head loss flow resistance terms requires a "subset level" of application of the governing equations, which are supplemented by dimensional analysis (Chapter 7) when there is no theory to model the friction force (as in the case of turbulent internal and external flow). Give the expressions for the evaluated major and minor head loss flow resistance terms.

5.37C The application of the differential form of the momentum equation (pipe or open channel flow) given by Equation 5.15 requires the evaluation of the friction slope term, $S_f = \tau_w/\gamma R_h$. Give the expression for the evaluated friction slope flow resistance term.

Applications of the Governing Equations for Real Internal Flow: Differential Fluid Element

5.38BG Water at 68°F flows in a 11-*ft*-diameter horizontal straight pipe section, as illustrated in Figure ECP 5.38. The pipe length between points 1 and 2 is 1700 *ft*, the friction slope between points 1 and 2 is 0.0006 *ft*/ft, and the Chezy coefficient, C for the pipe is $80 \, ft^{1/2} \, s^{-1}$. The pressure at point 1 is 69 *psi*. (a) Determine the velocity of flow and discharge in the straight pipe section. (b) Determine the head loss due to pipe friction between points 1 and 2. (c) Determine the pressure drop at point 2 due to the pipe friction. (d) Draw the energy grade line and the hydraulic grade line.

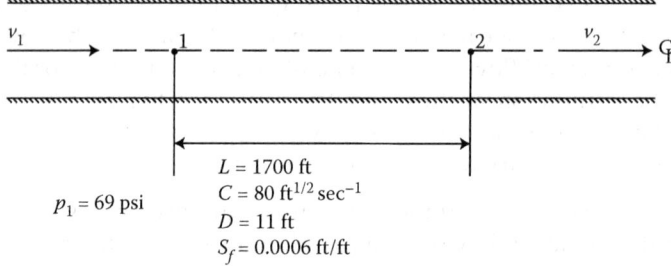

$$p_1 = 69 \text{ psi}$$

$$L = 1700 \text{ ft}$$
$$C = 80 \text{ ft}^{1/2} \text{ sec}^{-1}$$
$$D = 11 \text{ ft}$$
$$S_f = 0.0006 \text{ ft/ft}$$

FIGURE ECP 5.38

5.39SI Water at 20°C flows in a 7-*m*-diameter horizontal straight pipe section, as illustrated in Figure ECP 5.39. The pipe length between points 1 and 2 is 900 *m*, the friction slope between points 1 and 2 is 0.0008 *m/m*, and the friction factor, f for the pipe is 0.019. The

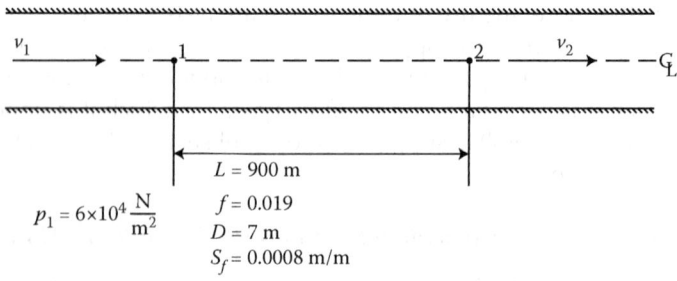

$$p_1 = 6 \times 10^4 \, \frac{N}{m^2}$$

$$L = 900 \text{ m}$$
$$f = 0.019$$
$$D = 7 \text{ m}$$
$$S_f = 0.0008 \text{ m/m}$$

FIGURE ECP 5.39

pressure at point 1 is $6 \times 10^4 \, N/m^2$. (a) Determine the velocity of flow and discharge in the straight pipe section. (b) Determine the head loss due to pipe friction between points 1 and 2. (c) Determine the pressure drop at point 2 due to the pipe friction. (d) Draw the energy grade line and the hydraulic grade line.

5.40SI Water at 20°C flows at a velocity of 3 m/sec in a horizontal straight pipe section, as illustrated in Figure ECP 5.40. The pipe length between points 1 and 2 is 600 m; the friction slope between points 1 and 2 is 0.0085 m/m; and the Hazen–Williams roughness coefficient, C_h for the pipe is 130. The pressure at point 1 is $7 \times 10^4 \, N/m^2$. (a) Determine the pipe size diameter and discharge in the straight pipe section. (b) Determine the head loss due to pipe friction between points 1 and 2. (c) Determine the pressure drop at point 2 due to the pipe friction. (d) Draw the energy grade line and the hydraulic grade line.

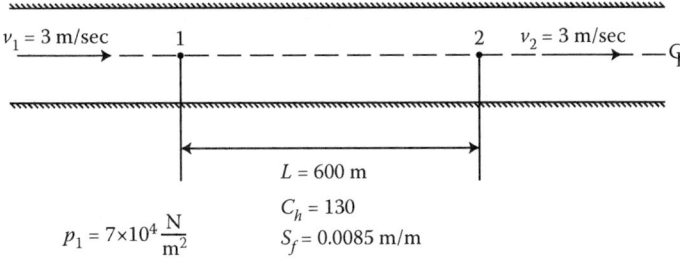

FIGURE ECP 5.40

5.41BG Water at 68°F flows at a depth of 5 ft and a velocity of 8 ft/sec in a rectangular channel that is 4 ft wide, as illustrated in Figure ECP 5.41. The channel length between points 1 and 2 is 3000 ft, and the Chezy coefficient, C is 90 $ft^{1/2} \, s^{-1}$. (a) Determine the channel bottom slope. (b) Determine the head loss due to channel friction between points 1 and 2. (c) Draw the energy grade line and the hydraulic grade line.

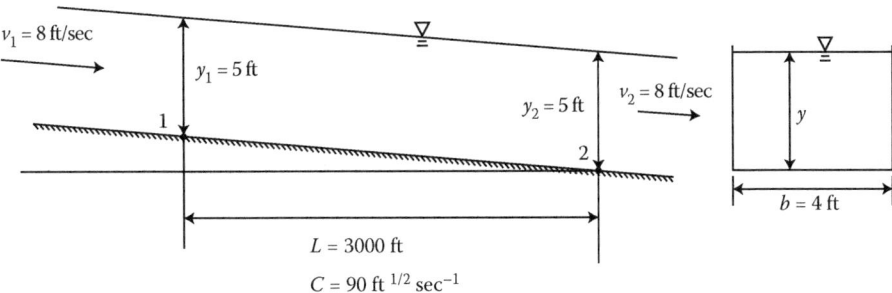

FIGURE ECP 5.41

5.42SI Water at 20°C flows at a depth of 6.5 m in a rectangular channel that is 2.7 m wide, as illustrated in Figure ECP 5.42. The channel length between points 1 and 2 is 1800 m, and the Manning's roughness coefficient, n is 0.019 $m^{-1/3} \, s$. The channel bottom slope is 0.027 m/m. (a) Determine the velocity of the flow. (b) Determine the head loss due to channel friction between points 1 and 2. (c) Draw the energy grade line and the hydraulic grade line.

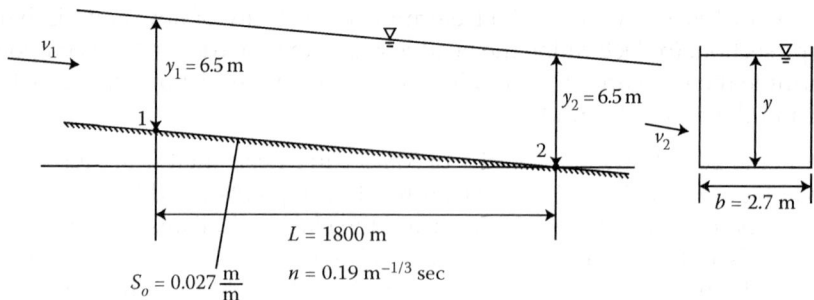

FIGURE ECP 5.42

Applications of the Governing Equations for Real Internal Flow: Finite Control Volume

5.43BG   Water at 68°F flows in a 3-*ft*-diameter pipe at a flowrate of 2 ft^3/sec. The pipe is fitted with sudden pipe contraction with a diameter of 1 *ft* in order to accelerate the flow, as illustrated in Figure ECP 5.43. Assume that the pipe and the contraction lie in a horizontal plane, (x, y). The pressure, at point 2 is 56 *psi*, and assume a minor head loss coefficient due to the sudden pipe contraction, k of 0.25. (a) Determine the minor head loss due to the contraction. (b) Determine the pressure at point 1. (c) Determine the reaction force acting on the pipe contraction due to the thrust force of the flow (anchoring force required to keep the pipe contraction from moving).

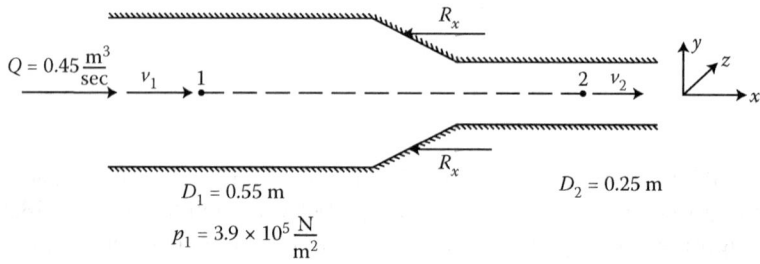

FIGURE ECP 5.43

5.44SI   Water at 20°C flows in a 0.55-*m*-diameter pipe at a flowrate of 0.45 m^3/sec. The pipe is fitted with gradual pipe contraction (nozzle) with a diameter of 0.25 *m* in order to accelerate the flow, as illustrated in Figure ECP 5.44. Assume that the pipe and the nozzle lie in a

FIGURE ECP 5.44

horizontal plane, (x, y). The pressure at point 1 is $3.9 \times 10^5 \, N/m^2$, and assume a minor head loss coefficient due to the nozzle, k of 0.07. (a) Determine the minor head loss due to the nozzle. (b) Determine the pressure at point 2. (c) Determine the reaction force acting on the nozzle due to the thrust force of the flow (anchoring force required to keep the nozzle from moving).

5.45BG A fan is required to force (pump) air into a 5-*ft* duct, as illustrated in Figure ECP 5.45. The air is forced out of a 3-*ft* duct by the fan. It is required that the air be forced through the duct at a discharge of $25 \, ft^3/sec$. Assume the density of air to be $0.0025 \, slug/ft^3$, and that there is no head loss between point 1 and point 2. Assume that the fan lies in a horizontal plane, (x, y). (a) Determine the reaction force acting on the fan due to the thrust force of the flow (anchoring force required to keep the fan from moving).

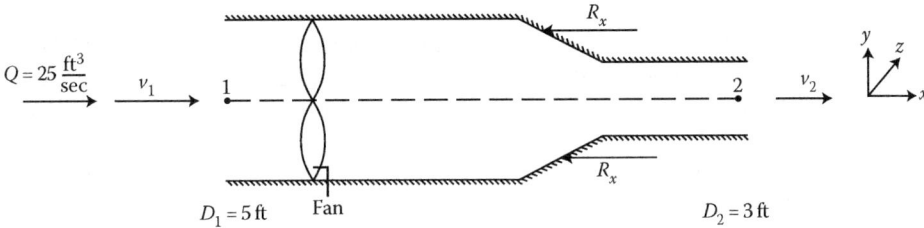

FIGURE ECP 5.45

5.46BG (Refer to Problem 3.65) A mini fan is required to force (pump) air into a 4-in blow dryer duct, which is heated by electric coils as illustrated in Figure ECP 5.46. The heated air is forced out of a 2-in nozzle by the mini fan in the blow dryer. It is required that the air be forced through the blow dryer at a discharge of $0.087 \, ft^3/sec$. Assume the density of air to be 0.002470 $slug/ft^3$, and there is no head loss between point 1 and point 2. Assume that the nozzle lies in a horizontal plane, (x, y). (a) Determine the reaction force acting on the nozzle due to the thrust force of the flow (anchoring force required to keep the blow dryer from moving).

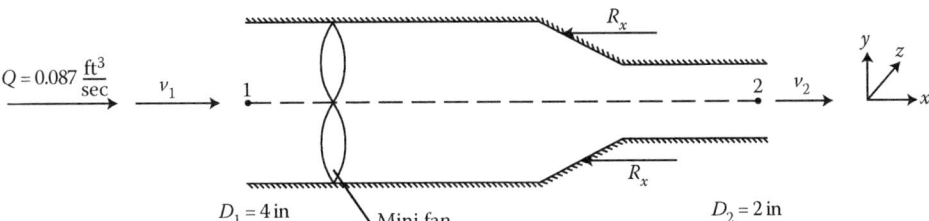

FIGURE ECP 5.46

5.47SI Water at 20°C flows in a 0.35-*m*-diameter pipe at a flowrate of 0.44 m^3/sec. The pipe is fitted with gradual pipe expansion (conical diffuser) with a diameter of 0.75 m in order to decelerate the flow, as illustrated in Figure ECP 5.47. Assume that the pipe and the conical diffuser lie in a horizontal plane, (x, y). The pressure, at point 1 is $4 \times 10^5 \, N/m^2$, and assume a minor head loss coefficient due to the conical diffuser, k of 3. (a) Determine the minor head loss due to the conical diffuser. (b) Determine the pressure at point 2. (c) Determine the reaction force acting on the conical diffuser due to the thrust force of the flow (anchoring force required to keep the conical diffuser from moving).

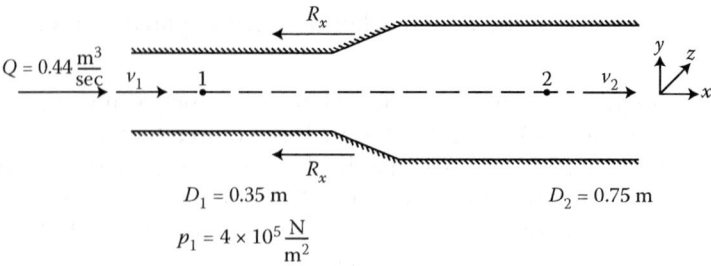

FIGURE ECP 5.47

5.48BG Water at 68°F flows in a 1.2-*ft*-diameter horizontal pipe at a flowrate of 13 ft^3/sec. A regular 45° threaded elbow is installed between point 1 and 2 in order to change the direction of the flow by 45°, as illustrated in Figure ECP 5.48. Assume that the pipe and the elbow lie in a horizontal plane, (x, y). The pressure at point 1 is 88 *psi*, and assume a minor head loss coefficient due to the elbow, k of 0.4. (a) Determine the minor head loss due to the elbow between points 1 and 2. (b) Determine the pressure at point 2. (c) Determine the reaction force acting on the elbow due to the thrust force of the flow (anchoring force required to keep the elbow from moving).

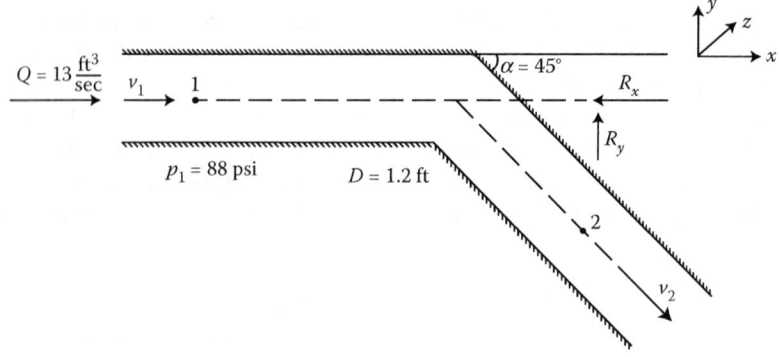

FIGURE ECP 5.48

5.49BG Water at 68°F flows in a 0.99-in-diameter flanged faucet with an fully open globe valve at a flowrate of 0.02 ft^3/sec, as illustrated in Figure ECP 5.49. Assume that the faucet and the water weigh 15 *lb*. The pressure at point 1 is 16 *psi*. (a) Determine the reaction force acting on the flange due to the thrust force of the flow (anchoring force required to keep the flange from moving).

5.50SI Water at 20°C flows in a 3-*m*-diameter horizontal pipe at a flowrate of 9 m^3/sec. A threaded branch flow tee is installed downstream of point 1 in order to divide the flow in two, as illustrated in Figure ECP 5.50. Assume that the pipe and the tee lie in a horizontal plane, (x, y). The diameter of the pipe at point 2 is 1.9 *m*, and the diameter of the pipe at point 3 is 1.5 *m*. The flow at points 2 and 3 are open to the atmosphere. Assume that the flow is ideal and thus ignore any minor head loss due to the tee. (a) Determine the pressure at

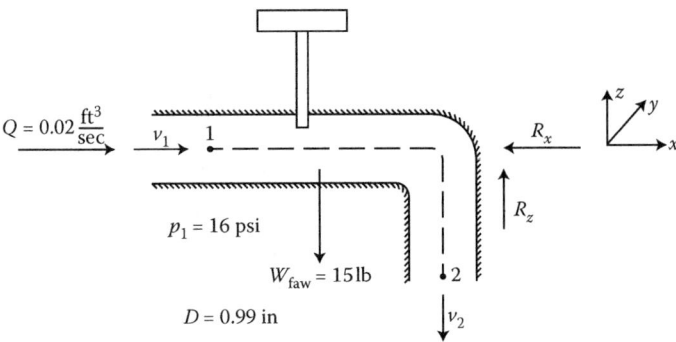

FIGURE ECP 5.49

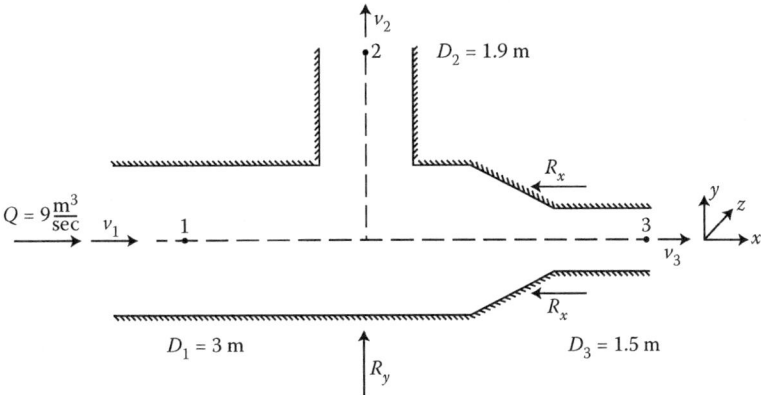

FIGURE ECP 5.50

point 1. (b) Determine the reaction force acting on the tee due to the thrust force of the flow (anchoring force required to keep tee from moving).

5.51BG Water at 68°F flows in a 3.6-*ft*-diameter horizontal pipe at a flowrate of $3.2 ft^3/sec$. A regular 90° threaded elbow is installed in between point 1 and 2 in order to change the direction of the flow by 90° as illustrated in Figure ECP 5.51. The pipe at point 2 is fitted with gradual pipe contraction (nozzle) with a diameter of $0.9 ft$ in order to accelerate the flow, and point 2 is $0.8 ft$ above point 1. Assume that the pipe, elbow, nozzle, and the water weigh 20 *lb*. The pressure at point 1 is 270 *psi*, and assume a minor head loss coefficient due to the elbow and nozzle, *k* of 1.5. (a) Determine the minor head loss due to the elbow and nozzle between points 1 and 2. (b) Determine the pressure at point 2. (c) Determine the reaction force acting on the elbow and nozzle due to the thrust force of the flow (anchoring force required to keep the elbow and nozzle from moving).

5.52SI Water at 20°C flows in a 2.3-*m*-diameter pipe at a flowrate of $14 m^3/sec$. A symmetrical double nozzle is installed downstream of point 1 in order to divide the flow in two, as illustrated in Figure ECP 5.52. Assume that the pipe and the nozzle lie in a horizontal plane, (x, y). The diameter of the nozzle at point 2 is $1.5 m$ and is at an angle, α of 30° with the *x*-axis, and the diameter of the nozzle at point 3 is $1.5 m$ and is at an angle, α of 30° with the *x*-axis. The flow at points 2 and 3 are open to the atmosphere. Assume that the flow is ideal and thus

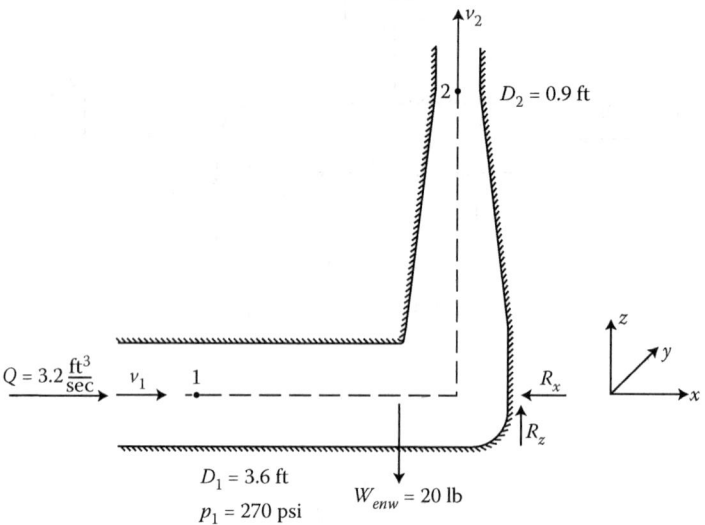

FIGURE ECP 5.51

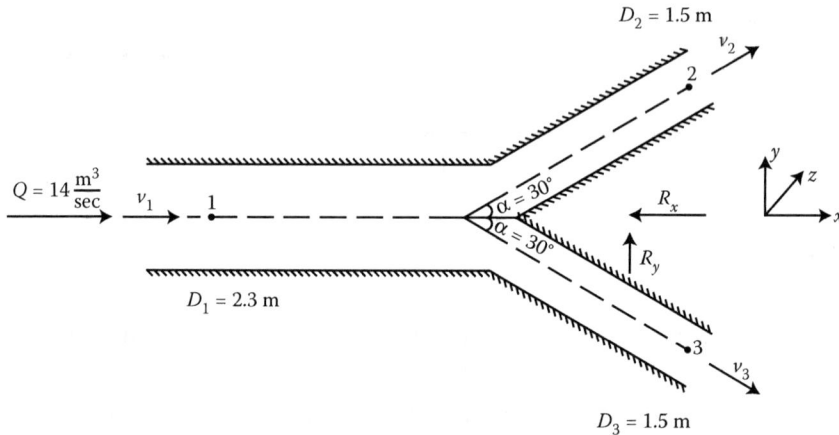

FIGURE ECP 5.52

ignore any minor head loss due to the nozzle. (a) Determine the pressure at point 1. (b) Determine the reaction force acting on the nozzle due to the thrust force of the flow (anchoring force required to keep nozzle from moving). (c) Assume that the angle, α varies from $10°$ to $90°$, and determine the resulting reaction force acting on the nozzle due to the thrust force of the flow (anchoring force required to keep nozzle from moving). Plot and discuss the results.

Application of the Energy and Momentum Equations for Ideal Internal Flow and Ideal Flow from a Tank

5.53C   The application of the continuity equation for a flow-measuring device (pipe or open channel flow) given by Equation 3.47 requires the evaluation the actual discharge, $Q_a = v_a A a$. Give the expression for the evaluated actual discharge flow resistance term.

Applications of the Governing Equations for Ideal Pipe Flow

5.54SI Water at 20°C flows in a 0.98-*m*-diameter horizontal pipe. An orifice meter with a 0.45 *m* diameter is mounted in the flanged pipe just upstream of point 2 as illustrated in Figure ECP 5.54 in order to measure the flowrate in the pipe. A pressure tap, p_1 is installed at point 1, where the full pipe flow occurs just upstream of the orifice, and a pressure tap, p_2 is installed at point 2, where the vena contracta occurs just downstream of the orifice. The pressure at point 1, p_1 is measured to be $3.3 \times 10^5 \, N/m^2$, and the pressure at point 2, p_2 is measured to be $2.9 \times 10^5 \, N/m^2$. The orifice discharge coefficient, C_o is 0.6. (a) Determine the ideal flowrate. (b) Determine the actual flowrate.

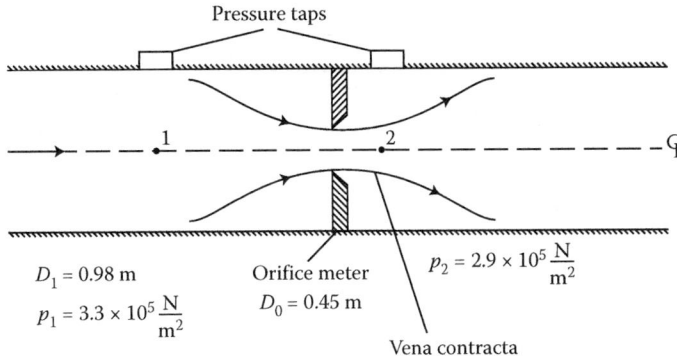

FIGURE ECP 5.54

5.55BG Water at 68°F flows in a 3-*ft*-diameter horizontal pipe. A nozzle meter with a 0.9-*ft* throat is mounted in the flanged pipe just upstream of point 2 as illustrated in Figure ECP 5.55 in order to measure the flowrate in the pipe. A pressure tap, p_1 is installed at point 1 just upstream of the nozzle flange, and a pressure tap, p_2 is installed at point 2, which is located just downstream of the nozzle flange. The pressure at point 1, p_1 is measured to be 50 *psi*, and the pressure at point 2, p_2 is measured to be 45 *psi*. The nozzle discharge coefficient, C_n is 0.99. (a) Determine the ideal flowrate. (b) Determine the actual flowrate.

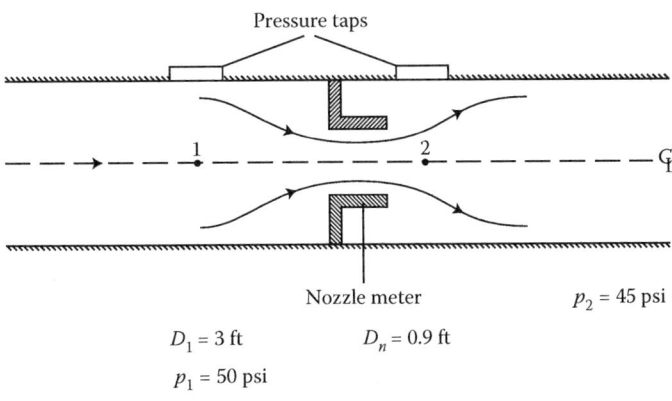

FIGURE ECP 5.55

5.56SI Water at 20°C flows in a 0.8-*m*-diameter horizontal pipe. A venturi meter with a 0.23-*m* throat is mounted in the pipe at point 2 as illustrated in Figure ECP 5.56 in order to measure the flowrate in the pipe. A pressure tap, p_1 is installed at point 1 just upstream of the venturi tube, and a pressure tap, p_2 is installed at point 2, which is located at the throat of the venturi tube. The pressure at point 1, p_1 is measured to be $2.3 \times 10^5 \, N/m^2$, and the pressure at point 2, p_2 is measured to be $1.7 \times 10^5 \, N/m^2$. The venturi discharge coefficient, C_v is 0.98. (a) Determine the ideal flowrate. (b) Determine the actual flowrate.

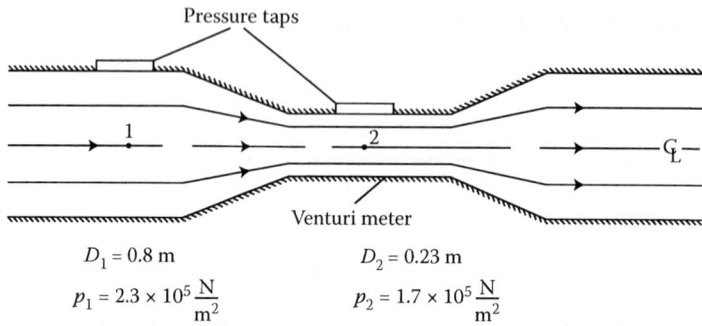

FIGURE ECP 5.56

Applications of the Governing Equations for Ideal Open Channel Flow

5.57BG Water at 68°F flows at a depth of 8 *ft* in a rectangular channel of width 7 *ft* and at a mild slope. A weir is inserted in the channel as illustrated in Figure ECP 5.57 in order to measure the flowrate. The depth of flow downstream of the weir is 4 *ft*. The weir discharge coefficient, C_d is 0.66. (a) Determine the ideal flow rate. (b) Determine the actual flowrate. (c) Determine the reaction force acting on the weir due to the thrust force of flow (anchoring force required to keep the weir from moving).

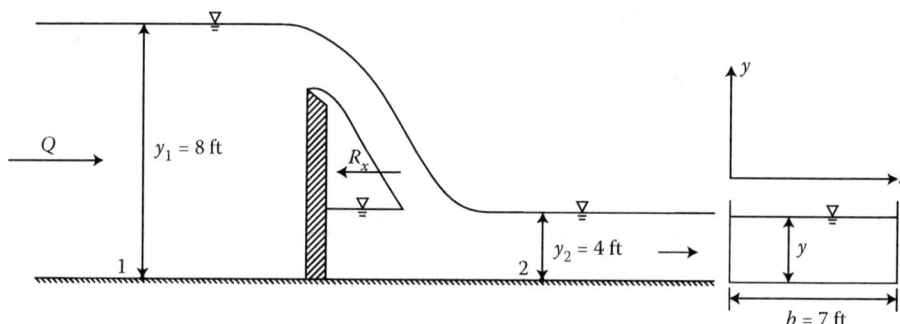

FIGURE ECP 5.57

5.58SI Water at 20°C flows at a depth of 5 *m* in a rectangular channel of width 5 *m* and at a mild slope. A sluice gate is inserted in the channel as illustrated in Figure ECP 5.58 in order to measure the flowrate. The depth of flow downstream of the gate is 1.7 *m*. The sluice gate discharge coefficient, C_d is 0.55. (a) Determine the ideal flow rate. (b) Determine the actual flowrate. (c) Determine the reaction force acting on the gate due to the thrust force of flow (anchoring force required to keep the gate from moving).

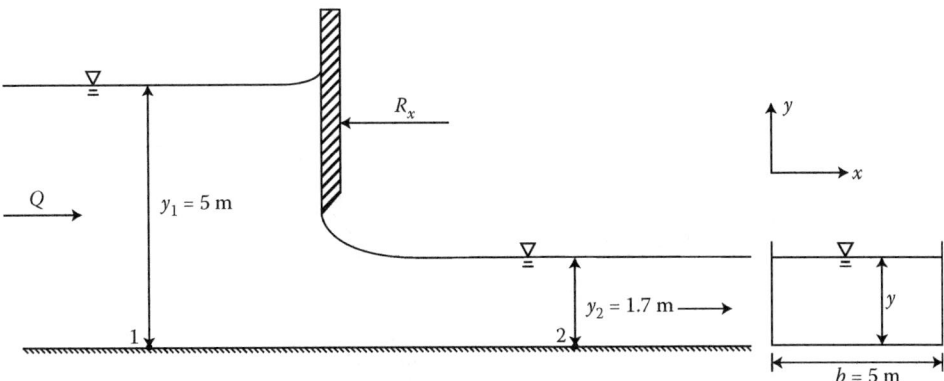

FIGURE ECP 5.58

5.59SI Water at 20°C flows at a depth of 6 m in a rectangular channel of width 4.9 m and at a mild slope. A spillway is inserted in the channel as illustrated in Figure ECP 5.59 in order to measure the flowrate. The depth of flow downstream of the spillway is 2.8 m. The spillway discharge coefficient, C_d is 0.67. (a) Determine the ideal flow rate. (b) Determine the actual flowrate. (c) Determine the reaction force acting on the spillway due to the thrust force of flow (anchoring force required to keep the spillway from moving).

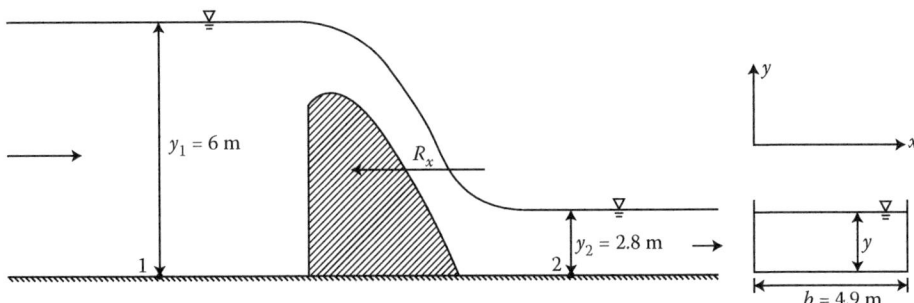

FIGURE ECP 5.59

Applications of the Governing Equations for Ideal Flow from a Tank (or a Water Source)

5.60SI A forced jet of water at 20°C with a diameter of 0.03 m and a flowrate of 0.099 m^3/sec hits a stationary sloping flat plate, where the jet splits into two paths, as illustrated in Figure ECP 5.60. The plate angle α is 65°, assume that the jet and the plate lie in a horizontal plane, (x, y), and assume ideal flow. (a) Determine the velocity of the flow at points 2 and 3. (b) Determine the flowrate at control faces 2 and 3. (c) Determine the reaction force acting on the plate due to the thrust force of the flow (anchoring force required to keep plate from moving).

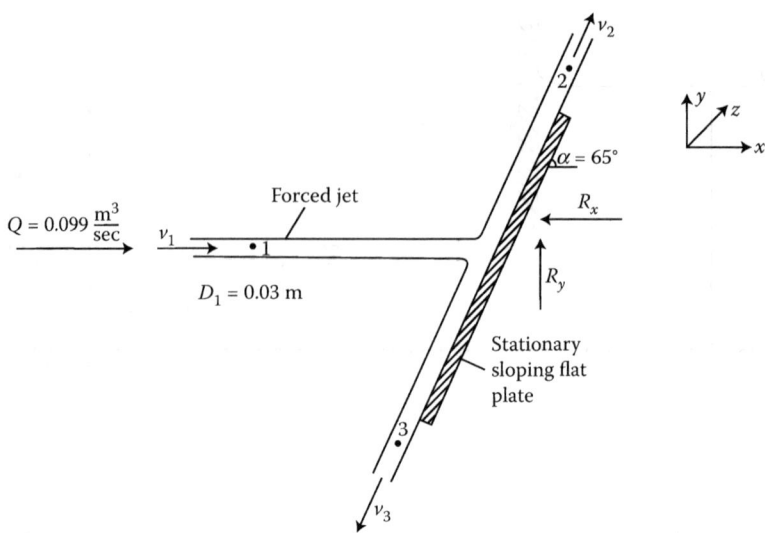

FIGURE ECP 5.60

5.61BG   A forced jet of water at 68°F with a diameter of 0.9 ft and a flowrate of 3 ft³/sec hits a stationary horizontal flat plate, where the jet splits into two paths, as illustrated in Figure ECP 5.61. Assume that the jet and the plate lie in a horizontal plane, (x, y), and assume ideal flow. (a) Determine the velocity of the flow at points 2 and 3. (b) Determine the flowrate at control faces 2 and 3. (c) Determine the reaction force acting on the plate due to the thrust force of the flow (anchoring force required to keep plate from moving).

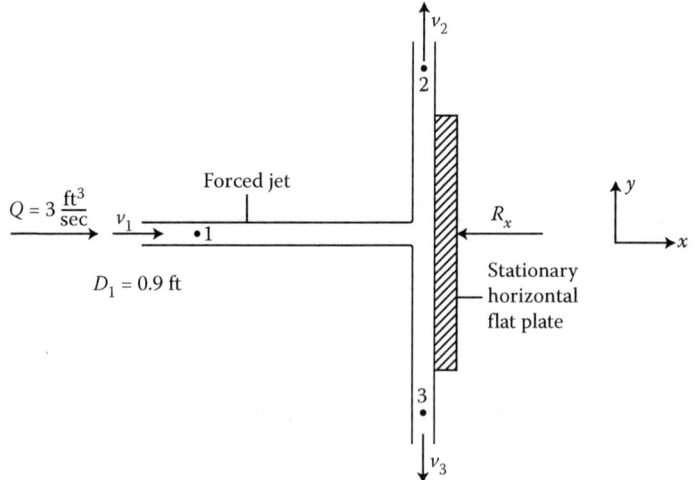

FIGURE ECP 5.61

5.62BG   A forced jet of water at 68°F with a diameter of 0.55 ft and a flowrate of 7 ft³/sec hits a stationary wedge, where the jet splits into two equal paths, as illustrated in Figure ECP 5.62. Assume that the jet and the wedge lie in a horizontal plane, (x, y), and assume ideal flow. (a) Determine the velocity of the flow at points 2 and 3. (b) Determine the flowrate at control faces 2 and 3. (c) Determine the reaction force acting on the wedge due to the thrust force of the flow (anchoring force required to keep the wedge from moving).

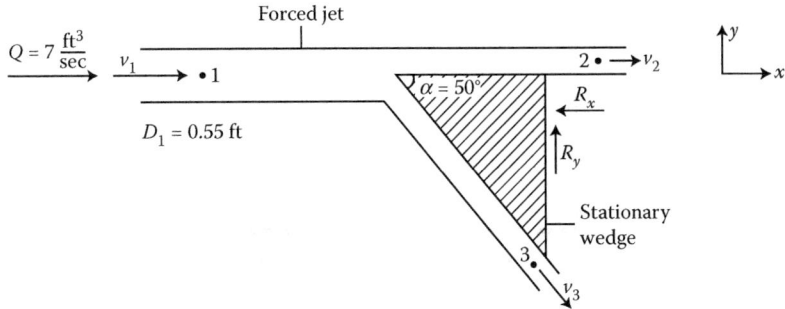

FIGURE ECP 5.62

5.63SI A forced jet of water at 20°C with a diameter of 0.25 m and a flowrate of 0.3 m^3/sec hits a stationary conical wedge, where the jet splits into two equal paths, as illustrated in Figure ECP 5.63. Assume that the jet and the wedge lie in a horizontal plane, (x, y), and assume ideal flow. (a) Determine the velocity of the flow at points 2 and 3. (b) Determine the flowrate at control faces 2 and 3. (c) Determine the reaction force acting on the wedge due to the thrust force of the flow (anchoring force required to keep wedge from moving).

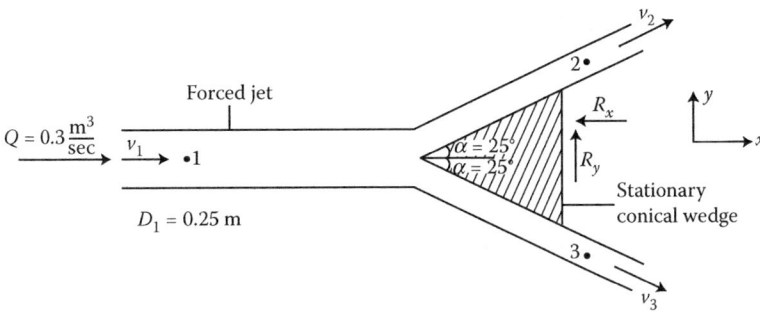

FIGURE ECP 5.63

5.64SI Water at 20°C flows from an 2.6-m-diameter open tank through a 3.1-cm opening on the side of the tank (close to the bottom of the tank) as illustrated in Figure ECP 5.64. The elevation of the water in the tank at point 1 is 25 m above the centerline of the opening at point 2. The empty tank weights 4 N, and assume a static coefficient of friction between the tank and the floor, μ of 0.44. (a) Determine the ideal velocity and flowrate of the free jet as it leaves the opening at point 2. (b) Determine the anchoring force required to keep the tank from moving to the left due to the free jet acting to the right. (c) Determine the friction force, F_f acting between the tank and the floor due to the weight of the tank and the water. (d) State if the friction force acting between the tank and the floor, F_f is sufficient to keep tank from moving to the left.

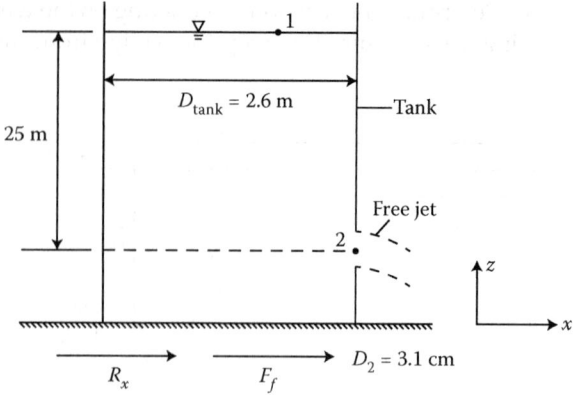

FIGURE ECP 5.64

5.65BG   Water at 60°F from a faucet fills a unplugged tank with a diameter of 30 *ft* as illustrated in Figure ECP 5.65. The elevation of the water in the tank at point 1 is maintained at 65 *ft* above the opening at point 2. The opening on the side of the tank is 2.8 in. The empty tank weights 1000 *lb*, and assume a static coefficient of friction between the tank and the floor, μ of 0.45. (a) Determine the ideal velocity and flowrate of the free jet as it leaves the opening at point 2. (b) Determine the anchoring force required to keep the tank from moving to the left, due to the free jet acting to the right. (c) Determine the friction force, F_f acting between the tank and the floor due to the weight of the tank and the water. (d) State if the friction force acting between the tank and the floor, F_f is sufficient to keep tank from moving to the left.

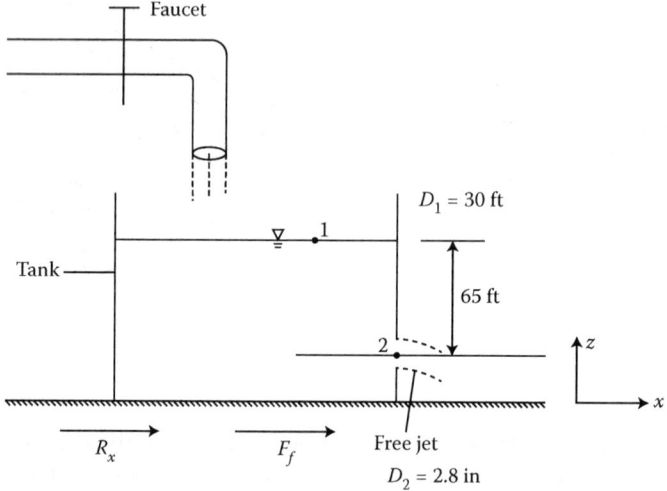

FIGURE ECP 5.65

Application of the Energy and Momentum Equations for Ideal External Flow

5.66C   The application of the integral form of the momentum equation for the case of external flow around an object given by Equation 5.27 requires the evaluation of the drag force, $F_D = (F_P + F_f)_s$. Give the expression for the evaluated drag force flow resistance term.

Applications of the Governing Equations for Ideal External Flow

5.67BG A rectangular street sign that is 6 *ft* wide and 3 *ft* high is secured to a post that is secured into the ground, as illustrated in Figure ECP 5.67. The sign is at sea level, and is designed to withstand a maximum wind speed of 100 *ft/sec*. Assume a drag coefficient for the sign, C_d of 1.2. (a) Determine the drag force on the street sign. (b) Determine the reaction force acting on the sign due to the thrust force of the maximum wind flow (anchoring force required to keep sign from detaching from the post).

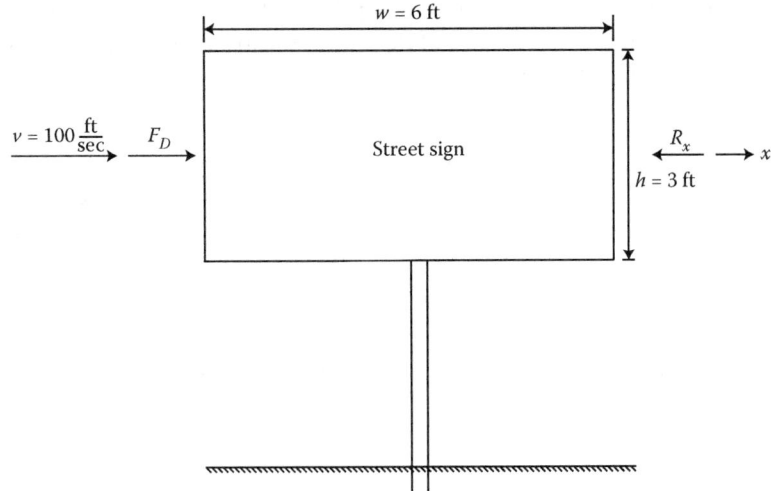

FIGURE ECP 5.67

5.68SI A decorative spherical dome structure with a diameter of 1.8 *m* is secured into the ground, as illustrated in Figure ECP 5.68. The dome is at sea level, and is designed to withstand a maximum wind speed of 50 *km/hr*). Assume a drag coefficient for the sign, C_d of 0.15. (a) Determine the drag force on the dome. (b) Determine the reaction force acting on the dome due to the thrust force of the maximum wind flow (anchoring force required to keep dome from detaching from the ground).

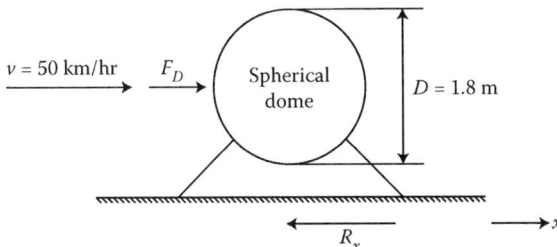

FIGURE ECP 5.68

Application of the Energy and Momentum Equations for a Hydraulic Jump

5.69C A hydraulic jump is a natural phenomenon that occurs when the open channel flow transitions from supercritical to subcritical flow (see Chapter 9). The head loss associated with a hydraulic jump is not due to frictional losses associated with the wall shear stress, τ_w. Explain this point, and explain how the head loss is modeled.

Applications of the Governing Equations for a Hydraulic Jump

5.70BG   Water flowing in a rectangular channel of width 9.842 *ft* at a supercritical velocity of 15.823 *ft/sec* and a supercritical depth of 3.061 *ft* encounters a subcritical flow, as illustrated in Figure ECP 5.70. As the flow transitions from supercritical to subcritical flow, a hydraulic jump occurs, where there is a major head loss due the intense agitation and turbulence, where the depth of flow is at critical depth of 4.176 *ft*. (a) Determine the subcritical depth of flow and the subcritical velocity of flow after the jump. (b) Determine the head loss due to the flow turbulence at the jump.

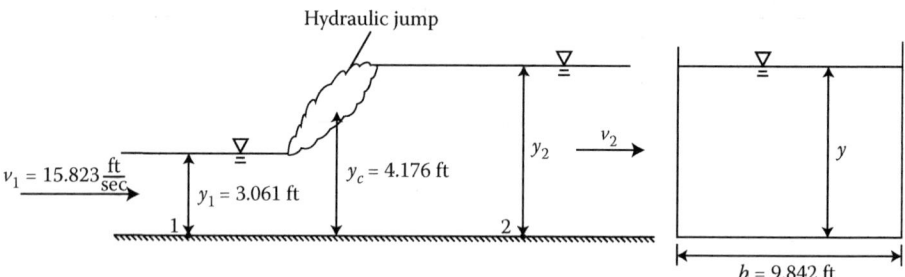

FIGURE ECP 5.70

5.71SI   Water flowing in a rectangular channel of width 3 *m* at a supercritical velocity of 5 *m/sec* and a supercritical depth of 1 *m* encounters a subcritical flow, as illustrated in Figure ECP 5.71. As the flow transitions from supercritical to subcritical flow, a hydraulic jump occurs, where there is a major head loss due the intense agitation and turbulence, where the depth of flow is at critical depth of 1.25 *m*. (a) Determine the subcritical depth of flow and the subcritical velocity of flow after the jump. (b) Determine the head loss due to the flow turbulence at the jump.

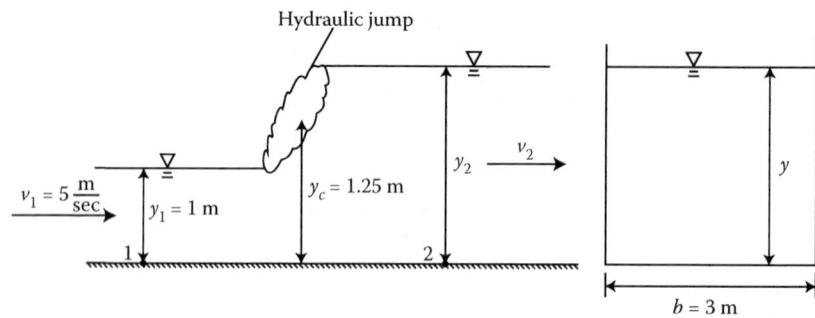

FIGURE ECP 5.71

6

Flow Resistance Equations

6.1 Introduction

In the application of the governing equations, although in some flow cases, one may ignore (or not need to model) the flow resistance, in most practical flow cases in general, one must account for the fluid viscosity and thus model the resulting shear stress and flow resistance. As such, in the application of the governing equations (continuity, energy, and momentum), it is important to make the distinction between real flow, ideal flow, and a hydraulic jump. And, for ideal flow, it is important to make the distinction between internal flow (pipe or open channel flow), external flow, and flow from a tank (see Table 4.1). While the continuity, energy, and momentum equations are applied (as necessary) to internal flows (pipe and open channel flow) and to flow from a tank, etc., only the energy and momentum equations are applied to external flows. The application of the governing equations (continuity, energy, and momentum) for real flow (internal) and ideal flow (flow-measuring devices for internal flow, and external flow around objects) requires the modeling of flow resistance (frictional losses). The modeling of flow resistance requires a "subset level" application of the governing equations. However, in the application of the governing equations (continuity, energy, and momentum) for ideal flow (flow from a tank, etc., and gradual channel transitions) and a hydraulic jump, there is no flow resistance to model. While in the case of ideal flow (flow from a tank, etc., and gradual channel transitions), the flow resistance is actually ignored, in the case of the hydraulic jump, the major head loss is due to flow turbulence due to the jump and not due to flow resistance (frictional losses).

Thus, the application of the governing equations (continuity, energy, and momentum) for real flow (internal), and ideal flow (flow-measuring devices for internal flow, and external flow around objects) requires the modeling of flow resistance (frictional losses). Specifically, the application of the integral form of the energy equation for incompressible, real, steady flow (pipe or open channel flow) given by Equation 4.44 requires the evaluation of the major and minor head loss flow resistance terms, $h_{fmaj} = \tau_w L / \gamma R_h$ and $h_{fmin} = \tau_w L / \gamma R_h$, respectively. Furthermore, the application of the differential form of the momentum equation (pipe or open channel flow) given by Equation 5.15 requires the evaluation of the friction slope term, $S_f = \tau_w / \gamma R_h$. The application of the continuity equation for a flow-measuring device (pipe or open channel flow) given by Equation 3.47 requires the evaluation the actual discharge, $Q_a = v_a A_a$. And, finally, the application of the integral form of the momentum equation for the case of external flow around an object given by Equation 4.28 (or Equation 5.27) requires the evaluation of the drag force, $F_D = (F_P + F_f)_s$.

The flow resistance in internal and external flow (major head loss, minor head loss, actual discharge, and drag force) is modeled by a "subset level" application of the appropriate governing equations, as presented in this chapter. A "subset level" application of the appropriate governing equations focuses only on the given element causing the flow resistance. Thus, assuming that flow resistance is to be accounted for, depending upon the specific flow situation, the flow type is assumed to be either a "real" flow or an "ideal" flow. The distinction between ideal and real flow determines how the flow resistance is modeled in the "subset level" application of the appropriate governing equations. The assumption of "real" flow implies that the flow resistance is modeled in both the energy and the momentum equations. However, the assumption of "ideal" flow implies that the flow resistance is modeled only in the momentum equation (and thus the subsequent assumption of "real" flow). The flow resistance due to friction in pipe and open channel flow (modeled as a major head loss) and the flow resistance due to friction in most pipe devices (modeled as a minor head loss) are modeled by assuming "real" flow; application of the energy equation models a head loss that is accounted for by a drag coefficient (in the application of the momentum equation). However, the flow resistance due to friction in flow-measuring devices for internal flow (pipe and open channel flow, modeled as an actual discharge) and the flow resistance due to friction in external flow around objects (modeled as a drag force) are modeled by assuming "ideal" flow; although application of the Bernoulli equation does not model a head loss in the determination of ideal velocity, the associated minor head loss with the velocity measurement is accounted for by a drag coefficient (in the application of the momentum equation, where a subsequent assumption of "real" flow is made). Furthermore, the use of dimensional analysis (Chapter 7) in supplement to a "subset level" application of the appropriate governing equations is needed to derive flow resistance equations when there is no theory to model the friction force (as in the case of turbulent internal and external flow). And, finally, it is important to note that the use of dimensional analysis is not needed in the derivation of the major head loss term for laminar flow (friction force is theoretically modeled) and the minor head loss term for a sudden pipe expansion (friction force is ignored).

Thus, regardless of whether the flow type is external or internal, the distinction between ideal and real flow determines how flow resistance is modeled in the "subset level" application of the appropriate governing equations. Furthermore, assuming real flow, the distinction between laminar and turbulent flow determines the need to supplement the momentum equation with dimensional analysis in the derivation of the flow resistance equation (major head loss equation). As such, Section 6.2 presents the types of flow in the modeling of flow resistance.

6.1.1 Modeling Flow Resistance: A Subset Level Application of the Governing Equations

Derivation of a given flow resistance equation involves a "subset level" application of the appropriate governing equations and in some cases supplementing the momentum equation with dimensional analysis. A "subset level" application of the appropriate governing equations focuses only on the given element causing the flow resistance. Assuming a real flow, the flow resistance is the shear stress, friction, or drag force originating in the fluid viscosity, which is modeled as the friction/viscous force, F_V in the integral form of the momentum equation (Equation 5.27). Regardless of whether the flow is internal or external, it is important to note that the application of the integral form of the momentum equation (Equation 5.27), while the definition of the gravitational force, F_G and the pressure force, F_P can always be theoretically modeled, the friction/viscous force, F_V may not always be theoretically modeled. As such, the results of dimensional analysis (Chapter 7) are used to supplement the momentum theory (Chapter 5) in the definition of the friction/viscous force, and thus the definition of the flow resistance, which is discussed in this chapter.

In the flow of any real fluid, the fluid has to do work against flow resistance, which results in a continual dissipation of energy in the momentum transfer. The flow resistance causes an unknown pressure drop, Δp (or S_f), which causes an unknown head loss, h_f, where the head loss is due to a conversion of kinetic energy to heat, which is modeled/ displayed in the integral form of the energy equation (Equation 4.44). The flow resistance is modeled as a drag force in external flows, while it is modeled as a head loss in internal flows. In the external flow around an object, the flow resistance is ultimately modeled as a drag force in the integral form of the momentum equation (see Section 6.3) because the drag force is needed for the design of external flow around an object. In the internal flow in a pipe, the flow resistance is ultimately modeled as a major head loss due to pipe friction (see Sections 6.4 through 6.7) and as a minor head loss due to pipe transitions (see Section 6.8) in the integral form of the energy equation because the major and minor head losses are needed for the design of internal flow in a pipe. In the internal flow in an open channel, the flow resistance is ultimately modeled as a major head loss due to channel friction in the integral form of the energy equation (see Sections 6.4 through 6.7) because the major head loss is needed for the design of internal flow in an open channel. And, in the internal flow in both a pipe and an open channel, the flow resistance is ultimately modeled as a reduced/actual flowrate due to a flow-measurement device in the differential form of the continuity equation (see Section 6.9) because the actual flowrate is needed for the design of a flow-measuring device. Furthermore, when the friction/viscous forces can either be theoretically modeled (as in the case of the major head loss in laminar flow) or ignored (as in the case of the minor head loss due to a sudden pipe expansion) in the momentum equation, the respective flow resistance equations can be analytically derived (a theoretical expression for the pressure drop, Δp is derived). However, when the friction/ viscous forces cannot either be theoretically modeled or ignored in the momentum equation, the momentum theory will be exhausted due to an unknown head loss, h_f. As a result, the momentum theory is supplemented with dimensional analysis (Chapter 7) in order to derive an empirical expression for the pressure drop, Δp and thus for the drag force, the major head loss, the minor head loss, and the actual discharge. The flow resistance (unknown head loss, h_f), represented by each of the above-listed flow resistance equations, is modeled by the definition of a "drag coefficient, C_D," which is defined by dimensional analysis, and evaluated empirically.

6.1.2 Derivation of the Flow Resistance Equations and the Drag Coefficients

When the friction/viscous forces can either be theoretically modeled (as in the case of the major head loss in laminar flow) or ignored (as in the case of the minor head loss due to a sudden pipe expansion) in the integral form of the momentum equation, the respective flow resistance equations can be analytically derived. However, when the friction/viscous forces cannot either be theoretically modeled or ignored in the integral form of the momentum equation, the momentum theory will be exhausted due to an unknown head loss, h_f. As a result, the momentum theory is supplemented with dimensional analysis in order to derive an expression for the drag force, the major head loss, the minor head loss, and the actual discharge. The flow resistance (unknown head loss, h_f) represented by each of the above-listed resistance equations is modeled by the definition of a "drag coefficient, C_D," which is defined by dimensional analysis and evaluated empirically. As such, the results of dimensional analysis introduce a drag coefficient, C_D, which expresses the ratio of a real flow variable (such as velocity, pressure, force, etc.) to its corresponding ideal flow variable.

The appropriate flow resistance equations and their corresponding "drag coefficient, C_D" for both external and internal flow are derived in Chapter 7 and presented in this chapter. Specifically, the drag force equation along with the coefficient of drag, C_D (and lift coefficient, C_L) is derived for the external flow around an object (see Section 6.3). The major head loss equation along with the major head loss friction coefficients (Chezy, C; Darcy–Weisbach, f; and Manning's, n) for internal flows (pipe and open channel flow) are derived for both laminar and turbulent flow (see Sections 6.4 through 6.7). The minor head loss equation, along with the minor loss coefficient, k, is derived for pipe flow (see Section 6.8). And, the actual discharge equation, along with the coefficient of discharge, C_d, is derived for both pipe and open channel flow (see Section 6.9). The determination of the efficiency of pumps and turbines are presented in Chapter 4, and in Sections 6.1.3 and 6.10.

6.1.2.1 Empirical Evaluation of the Drag Coefficients and Application of the Flow Resistance Equations

Although the flow resistance equations are presented in this chapter, the empirical evaluation of the respective drag coefficients and application of the respective flow resistance equations are presented in the following chapters. Empirical evaluation of the major head loss friction coefficients (Chezy, C; Darcy–Weisbach, f; and Manning's, n) and the application of the corresponding major head loss equation are presented in Chapter 8 and in Chapter 9. Empirical evaluation of the minor head loss coefficient, k and the application of the corresponding minor head loss equation are presented in Chapter 8. Empirical evaluation of the coefficient of discharge, C_d and the application of the corresponding actual discharge equation are presented in Chapters 8 and 9. Furthermore, empirical evaluation of the coefficient of drag, C_D (and lift coefficient, C_L) and application of the corresponding drag equation (lift equation) are presented in Chapter 10. And, finally, application of similitude and modeling for the flow resistance prediction equations are presented in Chapter 11.

6.1.3 Modeling the Flow Resistance as a Loss in Pump and Turbine Efficiency in Internal Flow

And, finally, in cases where the flow resistance is due to pipe or channel friction (major head loss), pipe component (minor head loss), and/or a flow-measuring device (minor head loss), the flow resistance contributes to the power losses and thus the determination of the efficiency of pumps and turbines (see Chapter 4). In particular, the total head loss, h_f (major and minor) will increase the amount of power that a pump must deliver to the fluid; the head loss, h_f represents the additional height that the fluid needs to be elevated by the pump in order to overcome the losses due to the flow resistance in the pipe or channel. Some of the pumping power will be used to pump the fluid through the pipe or channel, while some will be used to compensate for the energy lost in the pressure drop, Δp caused by the pipe or channel friction, pipe component, or flow-measuring device. A similar analogy may be deduced for evaluating the efficiency of a turbine.

The efficiency, η of a pump or a turbine is represented by the "drag coefficient, C_D" that is derived using dimensional analysis. The derivation of the efficiency, η of a pump or a turbine is presented in Chapter 7 and in this chapter (Section 6.10). The empirical evaluation of efficiency, η of a pump or a turbine and its application are presented in Chapter 4. And, finally, the application of the similitude and modeling in the application of the efficiency, η of a pump or a turbine are presented in Chapter 11.

6.2 Types of Flow

Regardless of whether the flow type is external or internal (pipe or open channel flow), the distinction between ideal and real flow determines how flow resistance is modeled. Furthermore, in order to fully understand real flow, it is important to study the behavior of ideal flow. Assuming real flow (both external and internal flow), depending upon the value for the Reynolds number, R, the flow may be described as either laminar or turbulent flow. The distinction between laminar and turbulent flow determines the need to supplement the momentum equation with dimensional analysis in the derivation of the flow resistance equation (major head loss equation). Regardless of whether the flow is laminar or turbulent, the flow may be described as developing or developed flow. Furthermore, in the flow of any real fluid (internal or external flow, pressure or gravity flow, laminar or turbulent flow, and developing or developed flow), the fluid has to do work against flow resistance (shear stress or drag force) originating in the fluid viscosity. As a result, energy is continually being dissipated in the momentum transfer, where the magnitude of the shear stress depends on the friction caused by the fluid viscosity.

6.2.1 Internal Flow versus External Flow

The results sought in a given flow analysis will depend upon whether the flow is internal or external. In an internal flow, energy or work is used to move/force the fluid through a conduit. Thus, one is interested in the determination of the forces (reaction forces, pressure drops, flow depths), energy or head losses, and cavitation where energy is dissipated. In pipe flow, the conduit is completely filled with fluid and the flow is mainly a result of invoking a pressure difference (pressure drop, Δp). However, in open channel flow, the conduit is only partially filled with fluid and the flow is a result of invoking only gravity. In an external flow, energy or work is used to move a submerged body through the fluid, where the fluid is forced to flow over the surface of the body. Thus, one is interested in the determination of the flow pattern around the body, the lift, the drag (resistance to motion) on the body, and the patterns of viscous action in the fluid as it passes around the body.

6.2.2 Pipe Flow versus Open Channel Flow

For the derivation of the major head loss equation, the laws of flow resistance are essentially the same in pipe flow and open channel flow. Specifically, application of Newton's second law of motion (integral form of the momentum equation) represents an equilibrium between the propulsive pressure and/or gravity forces against the retarding shear forces needed to move each fluid particle along the pipe or open channel with a constant velocity. However, it is important to note that in pipe flow, the propulsive force is provided by a pressure gradient $\Delta p/L$ (which results in an unknown Δp or S_f), while in open channel flow, the propulsive force is provided by invoking gravity, and thus by the weight of the fluid along a channel slope (channel gradient, $\Delta z/L = S_o$) (which results in an unknown S_f). Furthermore, the boundary conditions are different for pipe flow and open channel flow. In pipe flow, the shear stress restraining the fluid motion is distributed around the entire boundary of the cross section. However, in open channel flow, there is a free surface, on which the shear stress is insignificant; thus, the flow cross section is assumed to be equivalent to the lower half of a pipe flow. One may note that this equivalence is not exact because while the maximum velocity for pipe flow is at the centerline of the pipe, the maximum velocity for open

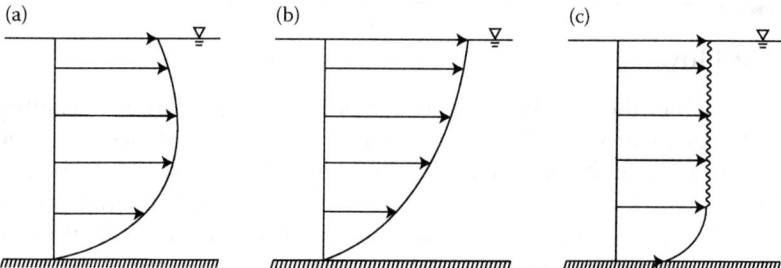

FIGURE 6.1
Velocity distribution for open channel flow. (a) Maximum velocity is actually below the free surface. (b) Parabolic velocity distribution assumed for laminar flow. (c) Empirical velocity distribution assumed for turbulent flow.

channel flow is at a point somewhat below the free surface as illustrated in Figure 6.1a. This is due to the existence of circulating currents in the plane of the channel cross section caused by floating objects. However, in practice, the velocity profile for open channel flow is assumed to be equivalent to the lower half of a pipe flow as illustrated in Figures 6.1b and c for laminar and turbulent flow, respectively.

One may note that although in this chapter, the laws of flow resistance are developed assuming pipe flow (internal flow under pressure), the general results are also applicable for open channel flow (internal flow under gravity). Furthermore, while major head losses are applicable to both pipe flow and open channel flow, minor losses are modeled and thus determined differently for pipe flow versus open channel flow. The present chapter (and Chapter 8) addresses the minor head loss for pipe flow due to pipe transitions and due to pipe flow-measuring devices, while it (and Chapter 9) also addresses the minor head losses for open channel flow due to flow-measuring devices. And, finally, in open channel flow, there is a difference between a smooth transition (no head loss) and an abrupt transition/control/flow-measurement device, for which there is a minor head loss and thus the occurrence of critical depth, y_c and the minimum specific energy, E_{min} at the control.

6.2.3 Real Flow versus Ideal Flow

A real (viscous) flow assumes a nonzero value for fluid viscosity, while an ideal (inviscid or frictionless) flow assumes a zero value for fluid viscosity. The validity of the assumption of a real versus an ideal flow will depend upon the physical flow situation. Although in some flow cases, one may ignore flow resistance, in most practical flow cases in general, one must account for the fluid viscosity and thus model the resulting shear stress and flow resistance. Thus, assuming that flow resistance is to be accounted for, depending upon the specific flow situation, the flow type is assumed to be either a "real" flow or an "ideal" flow. The distinction between ideal and real flow determines how the flow resistance is modeled in the "subset level" application of the appropriate governing equations. The assumption of "real" flow implies that the flow resistance is modeled in both the energy and the momentum equations. However, the assumption of "ideal" flow implies that the flow resistance is modeled only in the momentum equation (and thus the subsequent assumption of "real" flow). The flow resistance for pipe and open channel flow, and the flow through most pipe devices, is modeled by assuming "real" flow; application of the energy equation models a head loss that is accounted for by a drag coefficient (in the application of the momentum equation). However, the flow resistance for flow-measuring devices for internal flow (pipe and open channel flow) and external flow around objects is modeled by assuming "ideal" flow; although application of the Bernoulli equation does not model a head loss in the determination of ideal velocity, the associated minor head loss with the velocity measurement is accounted for by a drag coefficient (in the application of the momentum equation, where a subsequent assumption of "real" flow is made).

Regardless of whether the flow is assumed to be real or ideal, the conversion of energy from one type to another will result in either a pressure drop or a pressure rise. In the cases of the flow through a pipe or channel, the flow from a reservoir or a tank, the flow through most pipe devices, and the flow around a submerged body, where the fluid is set into motion, the conversion of the pressure energy to kinetic energy results in a pressure drop. While the flow through a pipe or channel results in a major head loss, the flow from a reservoir and the flow through most pipe devices and flow-measurement devices result in a minor head loss. Finally, the flow around a submerged body results in a drag force. And, in the cases of a pitot-static tube or the flow into a reservoir or a pipe expansion, where the fluid is brought to a stop, the conversion of the kinetic energy to pressure energy results in a pressure rise. The flow in a pitot-static tube and the flow into a reservoir or a pipe expansion result in a minor head loss.

6.2.4 Ideal Flow

The viscosity of real flow (a real fluid) introduces a great complexity in a fluid flow situation, namely friction due to the viscous shearing effects. Thus, in order to fully understand real flow, it is important to study the behavior of ideal flow. The assumption of ideal flow (inviscid flow) assumes an ideal fluid that has no viscosity and thus experiences no shear stresses (friction) as it flows. Although this is an idealized flow situation, such an assumption of ideal flow produces reasonably accurate results in flow systems where friction does not play a significant role. In such ideal flow situations, energy losses that are converted into heat due to friction represent a small percentage of the total energy of the fluid flow. Finally, the assumption of ideal flow may either ignore flow resistance or subsequently model the flow resistance (by the subsequent assumption of real flow). Ignoring the flow resistance implies application of the Bernoulli equation and assuming no viscous force ($F_V = 0$) in the application of the momentum equation. However, subsequently modeling the flow resistance implies application of the Bernoulli equation and modeling the viscous force ($F_V \neq 0$) in the application of the momentum equation.

6.2.4.1 Ignoring the Flow Resistance in Ideal Flow

Ignoring the flow resistance implies application of the Bernoulli equation and assuming no viscous force ($F_V = 0$) in the application of the momentum equation. Examples of situations where such an assumption of ideal flow is made include fluids with a low viscosity, when the pipe or channel lengths are relatively short, and when the pipe diameters and the valve and fitting sizes are large enough to sufficiently handle the discharges. Such an assumption of ideal flow is made for flow from a tank (jet and siphon flows) and gradual channel transitions (see Table 4.1). In such an ideal flow situation, because there is no friction, $F_V = 0$, there is no flow resistance. As a result, there is no drag force in external flows, and there is no head loss in internal flows. As such, the drag force, $F_D = 0$ because there is no flow resistance due to $F_V = 0$. And, according to d'Alembert's paradox, without viscosity, there could be no drag forces at all. For instance, the flow in a frictionless flow (ideal flow) around a body produces opposing stagnation points at the nose and the tail of the body (see Figure 4.19). Thus, the pressure distribution, as computed from the Bernoulli equation and integrated over the entire body, always adds up to zero in the direction of the flow. Additionally, for internal flow, the major head loss, $h_{f,\ major} = 0$ because there is no flow resistance due to $F_V = 0$; there is no minor head loss, $h_{f,\ minor} = 0$ because there is no flow resistance due to $F_V = 0$; and the actual flow rate equals the ideal flow rate, $Q_a = Q_i$ because there is no flow resistance due to $F_V = 0$.

6.2.4.2 Subsequently Modeling the Flow Resistance in Ideal Flow

Subsequently modeling the flow resistance in ideal flow implies application of the Bernoulli equation, then subsequently modeling the viscous force ($F_V \neq 0$) in the application of the momentum equation (subsequent assumption of real flow). Such an assumption of ideal flow is made in the subsequent modeling of flow resistance for flow-measuring devices for internal flow (pipe and open channel flow) (see Section 6.9) and external flow around objects (see Section 6.3); the ideal velocity is determined from the Bernoulli equation, where its associated minor head loss is accounted for by a drag coefficient (in the application of the momentum equation [see Table 4.1]). The conversion of energy from one type to another will result in either a pressure drop or a pressure rise. Flow-measuring devices for pipe flow include pitot-static tubes, orifice meters, nozzle meters, and venturi meters. Flow-measuring devices for open channel flow include pitot-static tubes, sluice gates, weirs, spillways, venturi flumes, and contracted openings. And, the external flow velocity around objects is determined by the use of pitot-static tubes. In the cases of flow-measuring devices (except for pitot-static tubes) for internal flow (pipe and open channel flow) or the flow around a submerged body, where the fluid is set into motion, assuming that $z_1 = z_2$, the conversion of the pressure energy to kinetic energy results in a pressure drop. While the flow-measuring devices result in a minor head loss, the flow around a submerged body results in a drag force. And, in the case of a pitot-static tube (flow-measuring device for both internal and external flow), where the fluid is brought to a stop, assuming that $z_1 = z_2$, the conversion of the kinetic energy to pressure energy results in a pressure rise. The flow in a pitot-static tube results in a minor head loss.

In an ideal flow situation, because there is no friction ($\mu = 0$ and thus $\tau = 0$), flow energy (flow work, or pressure energy) and potential energy are converted to kinetic energy (or vice versa), with no energy lost to heat. Thus, the energy equation assuming no head loss due to friction is called the Bernoulli equation (see Equation 4.46) and is given in terms of pressure terms as follows:

$$\left(\underbrace{p}_{F_P} + \underbrace{\gamma z}_{F_G} + \underbrace{\frac{\rho v^2}{2}}_{Ma} \right)_1 - \left(\underbrace{p}_{staticpres\ s} + \underbrace{\gamma z}_{hydrstapre\ ss} + \underbrace{\frac{\rho v^2}{2}}_{dynpres} \right)_2 = \underbrace{\overbrace{\frac{\tau_w L}{R_h}}^{flowres}}_{F_V} = 0 \qquad (6.1)$$

The static pressure, p, which does not incorporate any dynamic effects, incorporates the actual thermodynamic pressure of the fluid. The hydrostatic pressure, γz accounts for the elevation effects (the effects of the fluid weight on the pressure). The dynamic pressure, $\rho v^2/2$ represents a pressure drop when the fluid is set in motion, while it represents a pressure rise when the fluid in motion is brought to a stop. Thus, the sum of the static pressure, the hydrostatic pressure, and the dynamic pressure is called the total pressure. And, for ideal flow, the Bernoulli equation states that the total pressure along a streamline is constant. The Bernoulli equation states that the sum of the flow energy (pressure energy) per unit weight, p/γ; potential energy per unit weight, z; and kinetic energy per unit weight, $v^2/2g$ of a fluid particle along a streamline is constant (no head loss for an ideal fluid). This implies that the flow energy (pressure energy) and the potential energies of the fluid can be converted to kinetic energy and vice versa, thus causing the pressure to change. Presented below are two applications of the Bernoulli equation. The first application is the pitot-static tube, which illustrates that the dynamic pressure is modeled by a pressure rise. The second application is the venturi meter, which illustrates that the dynamic pressure is modeled by a pressure drop.

6.2.4.3 The Pitot-Static Tube: Dynamic Pressure Is Modeled by a Pressure Rise

One application of the Bernoulli equation is the pitot-static tube (a combination of pitot-tube and a piezometer/static tube), which is a velocity-measuring device that illustrates that the dynamic pressure is modeled by a pressure rise. Assuming no head loss and that the fluid is momentarily brought to a stop at the downstream point (stagnation pressure), the dynamic pressure at the upstream point represents the ideal pressure rise (from static pressure to stagnation pressure). This application of Bernoulli's equation deals with the stagnation and the dynamic pressures and illustrates that these pressures arise from the conversion of kinetic energy (dynamic pressure) in a flowing fluid into an ideal pressure rise (stagnation pressure minus static pressure) as the fluid is momentarily brought to a rest. The downstream point at the pitot tube is the point of stagnation, which represents the conversion of all of the kinetic energy into an ideal pressure rise.

While the pitot-static tube may be used to measure the velocity (and pressure) for both internal and external flow (see Sections 4.5.6 and 4.5.7, respectively), the following discussion assumes internal (pipe) flow. Figure 6.2 (the same as Figure 4.5) illustrates a horizontal pipe with a piezometer installed at point 1 and a pitot tube installed at point 2. The flow is from point 1 to point 2. One may note that both the piezometer and the pitot tube are pressure measuring devices. The piezometer is installed at the top of the pipe at point 1, and the rise of the fluid in the piezometer, p_1/γ measures the pressure at point 1. The piezometer does not interfere with the fluid flow and thus allows the fluid to travel without any change in the velocity, v_1. The pitot-tube is installed at the top of the pipe at point 2 continues inward towards the centerline of the pipe and then makes a $90°$ turn pointing upstream into the direction of flow. As a result, the fluid enters the pitot tube and momentarily stops (it stagnates), where $v_2 = 0$. The rise of the fluid in the pitot tube, p_2/γ measures the pressure at point 2. One may note that because the pitot tube stagnates the fluid at point 2, it registers a higher pressure than the piezometer does at point 1. As illustrated in Figure 6.2, the rise in pressure head, $\Delta p/\gamma$ is defined as follows:

$$\frac{\Delta p}{\gamma} = \frac{p_2}{\gamma} - \frac{p_1}{\gamma} \tag{6.2}$$

The Bernoulli equation between points 1 and 2 expressed in terms of pressure is given as follows:

$$p_1 + \gamma z_1 + \frac{\rho v_1^2}{2} = p_2 + \gamma z_2 + \frac{\rho v_2^2}{2} \tag{6.3}$$

Assuming that $z_1 = z_2$ and noting that $v_2 = 0$:

$$\underbrace{p_1}_{staticpres} + \underbrace{\frac{\rho v_1^2}{2}}_{dynpress} = \underbrace{p_2}_{stagnpress} \tag{6.4}$$

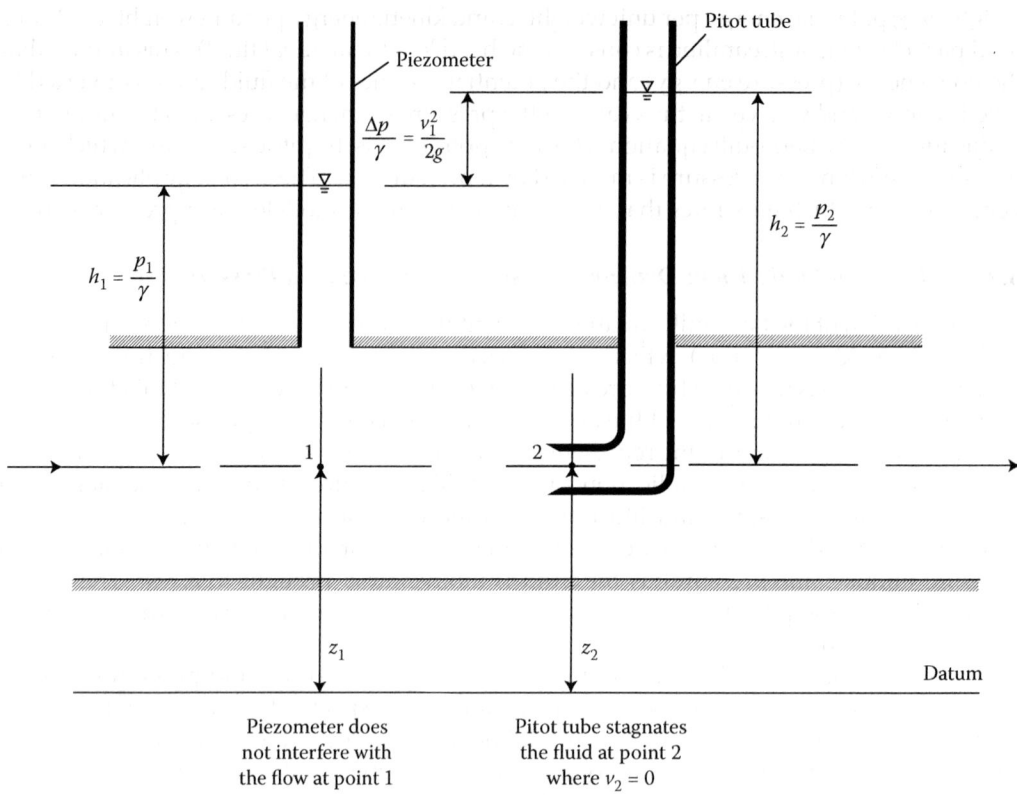

FIGURE 6.2
Pitot-static tube used to measure the velocity (and pressure rise) for a pipe flowing under pressure (same as Figure 4.5).

where the sum of the static pressure (where the fluid is not slowed down) and the dynamic pressure equals the stagnation pressure (where the fluid is stagnated/momentarily slowed down), the static pressure, p_1 is measured by the piezometer; the stagnation pressure, p_2 is measured by the pitot tube; and the dynamic pressure, $\rho v_1^2/2$ is measured by the difference in the fluid levels in the pitot tube and the piezometer as follows:

$$\frac{v_1^2}{2g} = \frac{\Delta p}{\gamma} = \frac{p_2}{\gamma} - \frac{p_1}{\gamma} \tag{6.5}$$

thus, the dynamic pressure, $\rho v_1^2/2$ is given as follows:

$$\underbrace{\frac{\rho v_1^2}{2}}_{dynpress} = \underbrace{p_2}_{stagnpress} - \underbrace{p_1}_{staticpres} = \underbrace{\Delta p}_{pressrise} \tag{6.6}$$

Assuming no head loss and that the fluid is momentarily brought to a stop at the downstream point 2 where $v_2 = 0$, the velocity at the upstream point 1 is ideal, where $v_1 = v_i$, and the dynamic pressure at the upstream point, $\rho v_1^2/2$ represents the ideal pressure rise (from static pressure to stagnation pressure). Therefore, isolating the ideal upstream velocity, v_i defines the ideal velocity as follows:

$$v_i = \underbrace{\sqrt{\frac{2}{\rho}(\Delta p)}}_{idealvel} = \underbrace{\sqrt{\frac{2}{\rho}\left(\underbrace{p_2}_{stagpress} - \underbrace{p_1}_{staticpress}\right)}}_{idealvel} \tag{6.7}$$

where the difference in the fluid levels, $\Delta p/\gamma = v_1^2/2g$ in the pitot tube and the piezometer is used to determine the unknown velocity, v_1 in the pipe as illustrated in Figure 6.2. One may note that the pitot-static tube is also used in the measurement of the airplane's speed, where the velocity measuring device is mounted on the wing of the airplane.

It is important to note that, similar to the pitot tube, there is a stagnation point on any stationary body that is placed into a flowing fluid. In the flow over a submerged body, some of the fluid flows over the body and some flows under the body, while the fluid is brought to a stop at the nose of the body at the stagnation point, as illustrated in Figures 4.18 and 4.19. Furthermore, the dividing line is called the stagnation streamline and ends at the stagnation point on the submerged body, where the pressure is the stagnation pressure. The point of stagnation represents the conversion of all of the kinetic energy into an ideal pressure rise. It is important to note that in the above illustration of the pitot-static tube, there was a rise in pressure, Δp, which was defined as follows:

$$\underbrace{\Delta p}_{pressrise} = \underbrace{p_2}_{stagpress} - \underbrace{p_1}_{staticpress} \tag{6.8}$$

6.2.4.4 The Venturi Meter: Dynamic Pressure Is Modeled by a Pressure Drop

A second application of the Bernoulli equation is the venturi meter (see Figure 4.6), which is a flow-measuring device that illustrates that the dynamic pressure is modeled by a pressure drop. The Bernoulli equation between points 1 and 2 expressed in terms of pressure is given as follows:

$$p_1 + \gamma z_1 + \frac{\rho v_1^2}{2} = p_2 + \gamma z_2 + \frac{\rho v_2^2}{2} \tag{6.9}$$

Assuming that $z_1 = z_2$ and noting that $v_1 \cong 0$, while the downstream velocity, v_2 defines the ideal velocity, yields:

$$\underbrace{p_1}_{staticpres} = \underbrace{p_2}_{stagnpress} + \underbrace{\frac{\rho v_2^2}{2}}_{dynpress} \tag{6.10}$$

The upstream stagnation pressure, p_1 is measured by a piezometer, and the downstream static pressure, p_2 is measured by a piezometer, where the pressure drop, Δp is defined as follows:

$$\underbrace{\frac{\rho v_2^2}{2}}_{dynpress} = \underbrace{p_1}_{stagnpress} - \underbrace{p_2}_{staticpres} = \underbrace{\Delta p}_{pressddrop} \tag{6.11}$$

Thus, the pressure drop, Δp is exactly equal to the dynamic pressure, $\rho v^2/2$ (and not more), which is considered to be the ideal pressure drop. And the actual velocity, v_2 is exactly equal to the ideal velocity, v_i (and not less).

$$v_2 = \underbrace{\sqrt{\frac{2}{\rho}(\Delta p)}}_{idealvel} \tag{6.12}$$

6.2.5 Real Flow

The viscosity of real flow (a real fluid) introduces a great complexity in a fluid flow situation, namely friction due to the viscous shearing effects. A real (viscous) flow assumes a nonzero value for the fluid viscosity. Modeling of viscosity results in the assumption of real flow, which presents the possibility of two flow regimes. A real flow is subdivided into laminar flow or turbulent flow depending on the value of the Reynolds number, $R = \rho v L/\mu$. Additionally, the modeling of viscosity modifies the assumption of the velocity profile from the rectangular distribution assumed for ideal flow to a parabolic velocity distribution for laminar flow and a velocity distribution approximated by the seventh root velocity law for turbulent flow. Regardless of whether the flow is laminar or turbulent, the flow may be described as developing or developed flow. Finally, the assumption of real flow may either directly or subsequently model the existence of flow resistance. Directly modeling the flow resistance implies modeling the flow resistance in both the energy and momentum equations. Subsequently modeling the flow resistance implies application of the Bernoulli equation, then modeling the viscous forces (flow resistance) in the application of the momentum equation.

6.2.5.1 Directly and Subsequently Modeling the Flow Resistance in Real Flow

The assumption of real flow may either directly or subsequently model the existence of flow resistance. Directly modeling the flow resistance implies modeling the flow resistance in both the energy and momentum equations. Subsequently modeling the flow resistance implies application of the Bernoulli equation, then modeling the viscous forces (flow resistance) in the application of the momentum equation. The direct assumption of real flow is made in the direct modeling of flow resistance in pipe and open channel flow (see Sections 6.4 through 6.7) and the flow through most pipe devices (see Section 6.8) (see Table 4.1). And, the subsequent assumption of real flow is made in the subsequent modeling of flow resistance for flow-measuring devices for internal flow (pipe and open channel flow) (see Section 6.9) and external flow around objects (see Section 6.3) (see Table 4.1). In a real flow situation, tangential or shearing forces always develop whenever there is motion relative to a body; this creates fluid friction, because these shearing forces oppose the motion of one fluid particle past another. It is these frictional forces that result in the fluid viscosity. Furthermore, in a real flow situation, because there is friction, $F_V \neq 0$, there is flow resistance. As such, in the external flow around an object, the drag force, $F_D \neq 0$ because there is flow resistance due to the friction, $F_V \neq 0$. Additionally, for internal flow, the major head loss, $h_{f,major} \neq 0$ because there is flow resistance due to $F_V \neq 0$, the minor head loss, $h_{f,minor} \neq 0$ because there is flow resistance due to $F_V \neq 0$, and the actual flowrate is less than the ideal flowrate, $Q_a < Q_i$ because there is flow resistance due to $F_V \neq 0$.

The flow of a real fluid is a lot more complex than the flow of an ideal fluid because of the existence of viscosity. The fluid viscosity causes the resistance to the fluid in motion by causing shear or friction forces both between the fluid particles and between the fluid and solid boundary. In order for flow to occur, work is done against these resistance forces whereby energy is converted to heat. In the flow of any real fluid, energy is continually being dissipated $(H_1 - h_f = H_2)$ in the momentum transfer $[\sum F_s = (F_G + F_P + F_V + F_{other})_s = (\rho Q v_s)_2 - (\rho Q v_s)_1]$ because the fluid has to do work against resistance originating in the fluid viscosity. Thus, the magnitude of the shear stress, τ depends on the friction caused by the fluid viscosity, μ. Regardless of whether the flow is laminar or turbulent, the basic resistance mechanism is the shear stress, τ by which a slow-moving layer of fluid exerts a retarding force on adjacent layer of faster-moving fluid. The retardation/resistance mechanism is in the form of a shear force because it acts parallel to the interface between the two layers. In laminar flow, the momentum transfer occurs on a microscopic level through the random movement of molecules. However, in turbulent flow, the momentum transfer occurs on a macroscopic level through random fluctuations in velocity of the fluid particles between the adjacent layers. Additionally, whether the flow is laminar or turbulent, the momentum is dependent on a velocity difference. Furthermore, when a fluid flow is in contact with a solid surface, neither the viscous nor turbulent shear can be said to originate a force-resisting motion; rather, they just transmit such a shear force from its origin (the solid surface). Thus, the flow resistance depends as much on the presence of a solid surface as on the strength of the viscosity or turbulence of the flow.

6.2.5.2 Directly Modeling the Flow Resistance in Real Flow

Directly modeling the flow resistance in real flow implies modeling the flow resistance in both the energy and momentum equations. Such an assumption of real flow is made in the direct modeling of flow resistance for pipe and open channel flow and the flow through most pipe devices. Application of the energy equation assumes a head loss that is accounted for by a drag coefficient (in the application of the momentum equation). The conversion of energy from one type to another will result in either a pressure drop or a pressure rise. Pipe and channel sections are assumed to be of significant length. Pipe devices/components include valves, fittings (tees, unions, elbows, and bends), entrances, exits, contractions, and expansions. In the case of pipe and open channel flow, and the flow through most pipe devices (except the flow into a reservoir [pipe exit] or a pipe expansion), where the fluid is set into motion, assuming that $z_1 = z_2$, the conversion of the pressure energy to kinetic energy results in a pressure drop and a head loss. Specifically, while the flow through a pipe or channel results in a major head loss, the flow from a reservoir (pipe entrance) and the flow through most pipe devices result in a minor head loss. And, in the case of the flow into a reservoir (pipe exit) or a pipe expansion, where the fluid is brought to a stop, assuming that $z_1 = z_2$, the conversion of the kinetic energy to pressure energy results in a pressure rise. The flow into a reservoir or a pipe expansion results in a minor head loss.

In a real flow situation, because there is friction ($\mu \neq 0$ and thus $\tau \neq 0$), flow energy (flow work, or pressure energy) and potential energy are converted to kinetic energy (or vice versa) plus energy lost to heat (due to flow resistance). Thus, the energy equation assuming a head loss due to friction (see Equation 4.44) is given in terms of pressure terms as follows:

$$\left(\underbrace{p}_{F_P} + \underbrace{\gamma z}_{F_G} + \underbrace{\frac{\rho v^2}{2}}_{Ma} \right)_1 - \left(\underbrace{p}_{statispress} + \underbrace{\gamma z}_{hydrstapress} + \underbrace{\frac{\rho v^2}{2}}_{dynpres} \right)_2 = \overbrace{\underbrace{\frac{\tau_w L}{R_h}}_{F_V}}^{flowres} \qquad (6.13)$$

In most pipe and channel flow situations (except in the case of a pipe expansion or a pipe exit), the fluid is set in motion, and there is usually a pressure drop, Δp, where the pressure drop, Δp is defined as follows:

$$\underbrace{\Delta p}_{presdrop} = p_1 - p_2 \tag{6.14}$$

The energy equation between points 1 and 2 expressed in terms of pressure is given as follows:

$$p_1 + \gamma z_1 + \frac{\rho v_1^2}{2} - \frac{\tau_w L}{R_h} = p_2 + \gamma z_2 + \frac{\rho v_2^2}{2} \tag{6.15}$$

Furthermore, assuming that $z_1 = z_2$, the upstream velocity, $v_1 = 0$, while the downstream velocity $v_2 = v$, yields:

$$p_1 - \frac{\tau_w L}{R_h} = p_2 + \frac{\rho v_2^2}{2} \tag{6.16}$$

$$\underbrace{\frac{\rho v_2^2}{2}}_{dynpress} + \underbrace{\frac{\tau_w L}{R_h}}_{flowres} = p_1 - p_2 = \underbrace{\Delta p}_{pressddrop} \tag{6.17}$$

Thus, the pressure drop, Δp is equal to the dynamic pressure, $\rho v^2/2$ plus an additional pressure drop, $\tau_w L / R_h$ due to the flow resistance, which makes the final pressure drop more than the ideal pressure drop. And, the actual velocity, v is equal to the ideal velocity, $\sqrt{(2/\rho)(\Delta p)}$ minus an amount due to the flow resistance, which makes the actual velocity less than the ideal velocity as follows:

$$v = \sqrt{\frac{2}{\rho}\left(\Delta p - \frac{\tau_w L}{R_h}\right)} \tag{6.18}$$

6.2.5.3 Laminar Flow versus Turbulent Flow

The basic difference between laminar and turbulent flow is that laminar flow is deterministic in nature, while turbulent flow is mostly stochastic in nature. However, although the movement of laminar flow fluid particles is deterministic in nature, there is stochastic/random movement at a molecular level. The deterministic nature of laminar flow allows a two-dimensional spatial description of the velocity, the pressure, the shear stress, the temperature, and any other variable that has a field description. As a result, real laminar flows (and ideal flows) tend to lend themselves to steady flows. Additionally, a theoretical approach may be used to model laminar flow. The movement of turbulent flow fluid particles is stochastic/random in nature. The stochastic nature of turbulent flow requires a three-dimensional spatial and a temporal description for the corresponding variables that have a field description. As a result, real turbulent flows tend to lend themselves to unsteady flows. Additionally, empirical approaches are needed to model turbulent flow, even when the flow is fully developed. The field of turbulent flow remains the least understood areas of fluid mechanics; turbulence is the most complex type of motion. And, although turbulent flow seems to be chaotic/stochastic, turbulent motion does possess a reasonable degree of order/ deterministic behavior in order to achieve a certain degree of comprehension/modeling (partly stochastic and partly deterministic). Furthermore, although fluid viscosity dampens out turbulence and is one of the factors that make a flow well behaved, turbulence is the result of both fluid viscosity and surface friction; surface friction is a local phenomenon, unlike laminar friction, which is spread out. And, finally, laminar flow is a special case of turbulent flow, with turbulent flow more likely to occur in practice than laminar flow; laminar flows typically occur for fluids with a high viscosity in pipe flow with small cross sections or open channel flow with small cross sections.

Whether a real flow is an external or internal flow, its behavior may be described and classified as laminar or turbulent depending upon the range of values for the Reynolds number, $R = \rho v L / \mu$. Furthermore, the particular range of values for the Reynolds number, R depends upon whether the flow is an internal or an external flow (see Chapter 10 for external flow classifications). In a laminar internal flow ($R < 2000$), the fluid particles move in definite and observable paths or streamlines, in which the flow is characteristic of a viscous fluid or of a fluid in which viscosity plays an important role. The flow of fluids with a high viscosity, μ moving at low velocities is typically characterized as laminar flow. In turbulent internal flow ($R > 4000$), the fluid particles move in an irregular and erratic path, showing no observable pattern. The flow of fluids with a low viscosity, μ moving at high velocities and high velocity fluctuations is typically characterized as turbulent flow. In the critical zone ($2000 < R < 4000$), the flow can be either laminar or turbulent, where any flow disturbance can cause the flow to change from laminar to turbulent. And because the energy loss in turbulent flow is greater than in laminar flow, the flow in the critical zone is assumed to be turbulent. Fully turbulent flow, where the flow everywhere is turbulent, is highly dependent on the pipe inlet condition, the inherent level of turbulence on the fluid, and on the pipe roughness. As such, for very rough pipes, turbulent flow may exist at $R = 10^4$, while for very smooth pipes, turbulent flow may exist at $R = 10^8$. Thus, determination of the flow type (laminar or turbulent) is important in the design of a pump, as the head loss, h_f will increase the amount of power that a pump must deliver to the fluid.

6.2.5.4 The Velocity Profiles for Laminar and Turbulent Internal Flows

The assumption made regarding the viscosity of the fluid will determine the resulting velocity profile for the flow. For an ideal flow through a pipe or a channel, the assumption of no viscosity results in a velocity profile with a rectangular distribution (see Figure 6.3a). However, for a real flow through a pipe or channel, modeling the viscosity results in the velocity of the fluid at the stationary solid boundary to be zero; the fluid at the boundary surface sticks to the boundary and has the velocity of the boundary (because of the "no-slip" condition).

For a real laminar flow, the velocity of the fluid increases parabolically, reaching a maximum, v_{max} at the pipe centerline as illustrated in Figure 6.3b. Figure 6.3b also illustrates that the fluid particles in laminar flow move in smooth straight parallel (to the centerline of the pipe) observable paths (streamlines) with different velocities. In laminar flow, there is no mixing phenomena and eddies as are present in turbulent flow (described below); thus, it appears as a very smooth flow, such as the flow of a viscous fluid such as honey. Furthermore, in laminar flow, the velocity is constant with time, and thus is a steady flow. It may be noted that for real laminar flow, the average velocity (spatial average over the pipe cross section), $v_{ave} = Q/A = v_{max}/2$, for which the velocity profile is a rectangular distribution as illustrated by the dotted line in Figure 6.3b. One may recall from Chapter 3

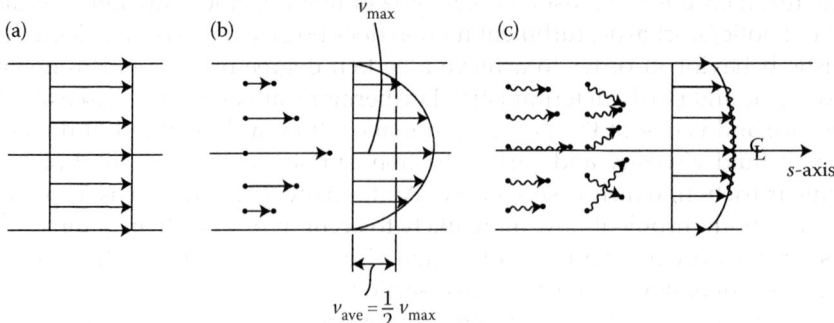

FIGURE 6.3
Velocity distribution for pipe flow. (a) Rectangular velocity distribution for ideal flow. (b) Parabolic velocity distribution for laminar flow. (c) Empirical (seventh root profile law) velocity distribution for turbulent flow.

that the application of the continuity equation assumes one-dimensional flow, where the velocity profile is a rectangular distribution, which further implies the assumption of ideal flow. Furthermore, it may be noted that the predictable (deterministic) nature of laminar flows results in a theoretical determination of the corresponding velocity profile (due to the theoretical relationship between the shear stress, τ and the velocity, v provided by Newton's law of viscosity) and thus a theoretical determination of the flow resistance coefficient in the equation for the major head loss, h_f due to flow resistance.

For a real turbulent flow (for which the flow rate, Q is increased in order to reach a high enough velocity to qualify as turbulent flow), the velocity profile of the fluid increases according to the seventh root velocity profile law (given below), reaching a maximum, v_{max} at the pipe centerline as illustrated in Figure 6.3c. Because there is significantly more resistance in turbulent flow, there is a greater loss of energy than in laminar flow. Turbulent flow is characterized by mixing phenomena and eddies (flow turbulence) within the flow, such as is the case for open channel flow and atmospheric flow. The flow turbulence causes both: (1) the resulting empirically defined velocity profile (spatial variation of the velocity, v across the pipe cross section) and (2) the unsteady flow rate (temporal variation of velocity, v). First, Figure 6.3c illustrates how the flow turbulence causes the movement of each fluid particle to deviate from a streamline and become random (fluid particles move in an irregular and erratic path, showing no observable pattern), fluctuating up and down in a directions perpendicular and parallel to the centerline of the pipe, reaching fully turbulent flow with a steady increase in the flowrate. The flow turbulence transports the low-velocity fluid particles near the pipe wall to the pipe center, and the higher-velocity particles near the pipe center toward the pipe wall. The resulting velocity profile is flatter than the parabolic laminar velocity profile due to the "averaging" effect of the mixing of the fluid particles. Furthermore, it may be noted that the unpredictable (random) and complex nature of turbulent flows results in an experimental determination of the corresponding velocity profile (due to the empirical relationship between the shear stress, τ_w and the velocity, v provided by dimensional analysis) and thus an experimental determination of the flow resistance coefficients in the equation for the major head loss, h_f due to flow resistance. And, second, the unsteady flow caused by the flow turbulence is reflected in the jagged (noise) velocity profile illustrated in Figure 6.3c, where the solid (trend) velocity profile represents the time average velocity (temporal mean) over a long period of time for a given pipe cross section. Therefore, if the temporal mean velocity for a given pipe cross section is determined for a significant period of time, the resulting velocity may be considered to be constant with respect to time, thus the assumption of steady turbulent flow. Additionally, it may be noted that: (1) turbulent flow is more likely to occur in practice than laminar flow, and (2) the field of turbulent flow still remains the least understood topic of fluid mechanics. Furthermore, one may note that the use of the Bernoulli equation assumes ideal flow, which has a rectangular velocity profile. And since most flows are turbulent and have a nearly rectangular velocity profile, the usefulness of the application of the Bernoulli equation and the rectangular velocity profile is very practical, unless the need arises to account for flow properties that require the modeling of the viscous effects.

6.2.5.5 Developing Flow versus Developed Flow

As explained in detail in Section 6.5.3 below, regardless of whether the flow is laminar or turbulent, at the pipe entrance, the velocity profile, the pressure drop, and the wall shear stress will vary with the length of the pipe until the flow changes from developing flow to fully developed flow. Given that the pipe is of sufficient length, over the length of the pipe, the flow becomes fully developed and thus there is no variation in the velocity profile, the pressure drop, and the wall shear stress with the length of the pipe, until a change occurs (a change in diameter, a bend, a tee, a valve, or another pipe component).

6.3 Modeling the Flow Resistance as a Drag Force in External Flow

In the external flow around an object, energy or work is used to move the submerged body through the fluid, where the fluid is forced to flow over the surface of the body. As such, in general, one is interested in the determination of the flow pattern around the body, the lift, the drag (flow resistance) on the body, and the patterns of viscous action in the fluid as it passes around the body. However, only if the body is moving at a high speed (and depending upon the shape of the body and the angle it makes with the direction of flow) does the lift force, F_L become dominant. Thus, the main focus of discussion in this section will be the drag force, F_D. The derivation of the drag force equation along with the coefficient of drag, C_D (and lift coefficient, C_L) for the external flow around an object is presented below. Furthermore, empirical evaluation of the coefficient of drag, C_D (and lift coefficient, C_L) and application of the corresponding drag equation (lift equation) are presented in Chapter 10.

In external flow, the flow resistance causes an unknown pressure drop, Δp, which causes an unknown head loss, h_f. However, although the head loss, h_f causes the drag force, F_D, the head loss is not actually determined in the design of external of around an object. Rather, the flow resistance is ultimately modeled as a drag force $F_D = (F_P + F_f)_s = (\Delta pA)_s$ in the integral form of the momentum equation. Because the friction/viscous forces (due to shear stress, τ_w) due to the external turbulent flow cannot be theoretically modeled in the integral momentum equation, the actual pressure drop, Δp cannot be analytically determined. Furthermore, the exact component in the s-direction of the pressure and viscous forces cannot be theoretically determined. Therefore, one cannot derive an analytical expression for the drag force, F_D (or the actual velocity, v_a), from the momentum equation. As a result, one must resort to dimensional analysis (supplements the momentum theory) in order to empirically model the wall shear stress, τ_w for turbulent external flow and thus Δp, v_a, and F_D by the use of a drag coefficient, C_D that represents the flow resistance. Thus, the associated minor head loss is typically accounted for by the use/calibration of a drag coefficient in the determination of the drag force.

6.3.1 Evaluation of the Drag Force

In the case of external flow around an object, the flow resistance is ultimately modeled as a drag force in the integral form of the momentum equation because the drag force is needed for the design of external flow around an object. Thus, the associated minor head loss is typically accounted for by the use/calibration of a drag coefficient to determine the drag force. Evaluation of the drag force requires a "subset level" application of the governing equations. A "subset level" application of the governing equations focuses only on the given element causing the flow resistance. The assumption of "ideal" flow implies that the flow resistance is modeled only in the momentum equation (and thus the subsequent assumption of "real" flow) (see Table 4.1). Thus, in the derivation of the drag force for an external flow around an object, first, ideal flow is assumed in order to determine an expression for the ideal velocity by applying the Bernoulli equation. Then, a subsequent assumption of real flow is made in order to determine an expression for the actual velocity, v_a and the drag force, F_D by application of the momentum equation (supplemented by dimensional analysis). Thus, the flow resistance (shear stress or drag force) in the external flow around an object is modeled as a resistance force/drag force, F_D in the integral form of the momentum equation. The flow resistance causes an unknown pressure drop, Δp which causes an unknown head loss, h_f, where the head loss is due to a conversion of kinetic energy to heat, which is modeled/ displayed in the integral form of the energy equation. However, although the head loss, h_f causes the drag force, F_D, the head loss is not actually determined in the design of external flow around an object. The assumption of ideal flow and thus applying the Bernoulli equation to measure the ideal velocity of flow, $v_i = \sqrt{2\Delta p/\rho}$ by the use of a pitot-static tube (see Section 4.5.7.3) yields an expression for the ideal velocity as a function of ideal pressure difference, Δp, which is directly measured. Therefore, the associated minor head loss with the velocity measurement is accounted for by the drag coefficient (in the application of the momentum equation). The head loss, h_f causing the actual velocity,

$$v_a = \sqrt{\frac{2}{\rho}\left(\Delta p - \frac{\tau_w L}{R_h}\right)},$$

and the drag force,

$$F_D = (F_P + F_f)_s = (\Delta p A)_s = \left(\frac{\rho v^2}{2}A + \frac{\tau_w L}{R_h}A\right)_s$$

is caused by both pressure and friction forces, where the drag force, F_D is equal to the sum of the pressure and friction force in the direction of the flow, s. However, because the friction/ viscous forces (due to shear stress, τ_w) due to the external flow cannot be theoretically modeled in the integral momentum equation, the actual pressure drop, Δp cannot be analytically determined, so the exact reduction in the velocity, actual velocity, v_a which is less than the ideal velocity, v_i, cannot be theoretically determined. Furthermore, the exact component in the s-direction of the pressure and viscous forces cannot be theoretically determined. Therefore, one cannot derive an analytical expression for the drag force, F_D from the momentum equation. As a result, one must resort to dimensional analysis (which supplements the momentum theory) in order to derive an expression for the drag force, F_D, which involves the definition of a drag coefficient, C_D that represents the flow resistance. The flow resistance equation for the drag force is derived below and in Chapter 7.

6.3.2 Application of the Bernoulli Equation: Derivation of the Ideal Velocity

In the derivation of the drag force for an external flow around an object, first, ideal flow is assumed in order to determine an expression for the ideal velocity by applying the Bernoulli equation as follows:

$$\left(\frac{p}{\gamma} + z + \frac{v^2}{2g}\right)_1 - \left(\frac{p}{\gamma} + z + \frac{v^2}{2g}\right)_2 = 0 \qquad (6.19)$$

Assuming $z_1 = z_2$, this equation has one unknown, which is the ideal velocity of the external flow around an object. The Bernoulli equation yields an expression for the ideal velocity, $v_i = \sqrt{2\Delta p/\rho}$ as a function of an ideal pressure difference, Δp, which is directly measured by the use of a pitot-static tube (see Section 4.5.7.3). Furthermore, the ideal velocity is used in the definition of the drag force below.

6.3.3 Application of the Momentum Equation and Dimensional Analysis: Derivation of the Actual Velocity, Drag Force Equation, and the Drag Coefficient

In the derivation of the drag force for an external flow around an object, the subsequent assumption of real flow is made in order to determine an expression for the actual velocity, v_a and the drag force, F_D by applying the integral momentum equation (supplemented by dimensional analysis) as follows:

$$\sum F_s = (F_G + F_P + F_V + F_{other})_s = (\rho Q v_s)_2 - (\rho Q v_s)_1 \qquad (6.20)$$

Assuming $z_1 = z_2$, and that only the pressure and viscous forces are important, the momentum equation is used to solve for the external flow actual velocity as follows:

$$v_a = \sqrt{\frac{2}{\rho}\left(\Delta p - \frac{\tau_w L}{R_h}\right)} \qquad (6.21)$$

and the drag force as follows:

$$F_D = (F_P + F_f)_s = (\Delta p A)_s = \left(\frac{\rho v^2}{2} A + \frac{\tau_w L}{R_h} A\right)_s \qquad (6.22)$$

However, the problem is that one cannot theoretically model the wall shear stress, τ_w in the integral momentum equation for turbulent external flow. Thus, one cannot analytically determine pressure drop, Δp and thus v_a in Equation 6.21. Furthermore, one cannot theoretically determine the exact component of the pressure and friction forces in the s-direction (direction of flow) in Equation 6.22. Thus, the integral momentum equation is supplemented with dimensional analysis in order to empirically model the wall shear stress, τ_w for turbulent external flow and thus Δp, v_a, and F_D by the use of a drag coefficient, C_D. The resulting flow resistance equation for the drag force is given as follows:

$$F_D = (F_P + F_f)_s = (\Delta p A)_s = \left(\frac{\rho v^2}{2} A + \frac{\tau_w L}{R_h} A\right)_s = C_D \frac{1}{2}\rho v^2 A \qquad (6.23)$$

where F_D is equal to the sum of the pressure and friction force in the direction of the flow, s and the drag coefficient, C_D accounts for (indirectly models) the minor head loss associated with the ideal velocity measurement and the drag force. Furthermore, one can easily compute v_i used in Equation 6.23 by measuring Δp using a pitot-static tube.

6.3.4 Modeling the Drag Force in the Momentum Equation

In a real flow situation, because there is friction, $F_V \neq 0$, there is flow resistance. As such, in the external flow around an object, the drag force, $F_D \neq 0$ because there is flow resistance due to the friction, $F_V \neq 0$. If a submerged body were moving in a stationary homogenous real fluid, then, due to the relative motion between the body and fluid, the fluid exerts pressure forces normal to the surface and shear forces parallel to the surface along the outer surface of the body. The resultant of the combined pressure and shear (viscous) forces acting on the surface of the body is due to the pressure drop, Δp and has two components. Specifically, the component of the resultant pressure, F_p and shear, F_f (also referred to as F_V) forces that acts in the direction of the flow, s is called the drag force, F_D. And, the component of the resultant pressure and shear forces that acts normal to the direction of the flow is called the lift force, F_L, which tends to move the submerged body in that direction. As noted above, only if the body is moving at a high speed (and depending upon the shape of the body and the angle it makes with the direction of flow) does the lift force, F_L become dominant.

Thus, the drag force, F_D on a submerged body has two components, a pressure drag (form drag), F_p and a friction drag (surface drag), F_f, where $F_D = (F_p + F_f)_s$. The pressure drag, F_p is due to the dynamic pressure (and is highly dependent on the form or the shape of the body) and acts normal to the surface of the body, while the friction drag, F_f is due to wall shear stress, τ_w and acts along the surface of the body. One may note that if the body is blunt in shape (spherical, for instance), then the pressure drag, F_p predominates the friction drag, F_f. Furthermore, if the body is more streamlined, then the friction drag, F_f predominates the pressure drag, F_p. And, finally, it is important to note that in order for the drag force, F_D to exist, either the submerged body or the fluid must be moving, and the assumed fluid must be a real fluid. As such, it is of interest to examine the nonexistence of the drag force, F_D in the case of a submerged stationary body in a stationary real or ideal fluid, and the case of a submerged moving body in a stationary ideal fluid, both of which illustrate d'Alembert's paradox.

6.3.4.1 Submerged Stationary Body in a Stationary Real or Ideal Fluid

In the case of a submerged stationary body in a stationary fluid, because there are no viscous forces, there will be no drag force, F_D. If a submerged body were not moving in a stationary homogenous fluid (either ideal or real), then only pressure forces, F_p would be exerted on the surface of the body, as illustrated by the momentum equation as follows:

$$\sum F = \Delta pA = F_P \tag{6.24}$$

where the pressure drop is given as follows:

$$\underbrace{\Delta p}_{presdrop} = \underbrace{p + \gamma z}_{staticpres} \tag{6.25}$$

and thus the pressure force is defined as:

$$F_P = (p + \gamma z)A \tag{6.26}$$

and the friction (viscous) force is defined as follows:

$$F_f = F_V = 0 \tag{6.27}$$

where there is no viscous force, because there is no movement of flow; additionally, in one case, the fluid is ideal. Therefore, there is no drag force, because there is no flow resistance due to $F_V = 0$ (d'Alembert's paradox), and thus:

$$F_D = \sum F_s = (\Delta p A)_s = (F_P)_s = 0 \tag{6.28}$$

As noted above, according to d'Alembert's paradox, without viscosity, there could be no drag forces at all. The flow in a frictionless flow (ideal flow) around a body produces opposing stagnation points at the nose and the tail of the body. Thus, the pressure distribution, as computed from the Bernoulli equation and integrated over the entire body, always adds up to zero in the direction of the flow.

6.3.4.2 Submerged Moving Body in a Stationary Ideal Fluid

In the case of a submerged moving body in a stationary ideal fluid, because there are no viscous forces, there will be no drag force, F_D. If a submerged body were moving in a stationary homogenous ideal fluid, then only pressure forces, F_p would be exerted on the surface of the body, as illustrated by the momentum equation as follows:

$$\sum F = \Delta p A = F_P \tag{6.29}$$

where the pressure drop is given as follows:

$$\underbrace{\Delta p}_{presdrop} = \underbrace{\frac{\rho v^2}{2}}_{dynpres} \tag{6.30}$$

Thus, the pressure drop, Δp is exactly equal to the dynamic pressure, $\rho v^2/2$ (and not more), which is considered to be the ideal pressure drop. Thus, the pressure force is defined as:

$$F_P = \frac{\rho v^2}{2} A \tag{6.31}$$

and the friction (viscous) force is defined as follows:

$$F_f = F_V = 0 \tag{6.32}$$

where there is no viscous force, because the fluid is ideal. Therefore, there is no drag force, because there is no flow resistance due to $F_V = 0$ (d'Alembert's paradox), and thus:

$$F_D = \sum F_s = (\Delta p A)_s = (F_P)_s = 0 \tag{6.33}$$

Once again, as noted above, according to d'Alembert's paradox, without viscosity, there could be no drag forces at all. The flow in a frictionless flow (ideal flow) around a body produces opposing stagnation points at the nose and the tail of the body. Therefore, the pressure distribution, as computed from the Bernoulli equation and integrated over the entire body, always adds up to zero in the direction of the flow.

6.3.4.3 Submerged Moving Body in a Stationary Real Fluid

In the case of a submerged moving body in a stationary real fluid, because there are viscous forces, there will be a drag force, F_D for which one will resort to dimensional analysis to derive. If a submerged body were moving in a stationary homogenous real fluid, then both pressure forces, F_p and viscous forces, F_V (shear forces, F_f) would be exerted on the surface of the body as illustrated by the momentum equation as follows:

$$\sum F = \Delta p A = F_p + F_f \tag{6.34}$$

where the pressure drop is given as follows:

$$\underbrace{\Delta p}_{presdrop} = \underbrace{\frac{\rho v^2}{2}}_{dynpres} + \underbrace{\frac{\tau_w L}{R_h}}_{flowres} \tag{6.35}$$

Thus, the pressure drop, Δp is equal to the dynamic pressure, $\rho v^2/2$ plus an additional pressure drop, $\tau_w L/R_h$ due to the flow resistance, which makes the final pressure drop more than the ideal pressure drop. And thus, the pressure force is defined as:

$$F_P = \frac{\rho v^2}{2} A \tag{6.36}$$

and the friction (viscous) force is defined as follows:

$$F_f = F_V = \frac{\tau_w L}{R_h} A \tag{6.37}$$

Furthermore, the total force due to the pressure drop, Δp acting in the direction of the flow, s is called the drag force, F_D and is given from theory as follows:

$$F_D = \sum F_s = (\Delta p A)_s = \left(\underbrace{\frac{\rho v^2}{2} A}_{F_P} \right)_s + \left(\underbrace{\frac{\tau_w L}{R_h} A}_{F_f} \right)_s$$

$$F_D = (F_P)_s + (F_f)_s \tag{6.38}$$

Furthermore, from Equation 6.38, isolating and working with the definition for

$$\Delta p = \left(\frac{\rho v^2}{2} \right) + \left(\frac{\tau_w L}{R_h} \right),$$

or working directly from Equation 6.35, one may isolate the expression given from theory for the external flow actual velocity, v_a as follows:

$$v_a = \sqrt{\frac{2}{\rho}\left(\Delta p - \frac{\tau_w L}{R_h}\right)} \tag{6.39}$$

6.3.5 Supplementing the Momentum Equation with Dimensional Analysis: Derivation of the Drag Force and the Drag Coefficient

The flow resistance in the external flow around an object is considered to be turbulent flow, for which one cannot analytically determine the actual pressure drop, Δp (the momentum theory has been exhausted due to an unknown head loss, h_f). Specifically, one cannot theoretically model the wall shear stress, τ_w in the integral momentum equation; thus, one cannot analytically determine the pressure drop, Δp and thus v_a in Equation 6.21. Furthermore, one cannot theoretically determine the exact component of the pressure and friction forces in the s-direction (direction of flow) in Equation 6.22 (or Equation 6.38). Thus, the integral momentum equation is supplemented with dimensional analysis in Chapter 7 in order to empirically model the wall shear stress, τ_w for turbulent external flow and thus Δp, v_a, and F_D by the use of a drag coefficient, C_D. Thus, the flow resistance (unknown head loss, h_f) represented by the drag force equation is modeled by the definition of a "drag coefficient, C_D," which is defined by dimensional analysis and evaluated empirically. The result of the derivation of the drag force equation along with the coefficient of drag, C_D for the external flow around an object is presented as follows:

$$\frac{F_D}{\frac{\rho v^2}{2}A} = \frac{F_D}{F_{dynpres}} = \frac{F_P}{F_{dynpres}} = \frac{\Delta p A}{\frac{\rho v^2}{2}A} = \frac{\Delta p}{\frac{\rho v^2}{2}} = \frac{2}{E} = C_D = \phi\left(R, C, F, \frac{\varepsilon}{L}, \frac{L_i}{L}\right) \tag{6.40}$$

where E = the Euler number, R = the Reynolds number, C = the Cauchy number, F = the Froude number, ε/L is the relative roughness, and L_i/L is the geometry (see Chapter 7 for applicability of the above-listed dimensionless numbers/pi terms). Thus, the resulting drag force, F_D derived from dimensional analysis is given as follows:

$$F_D = C_D \frac{1}{2}\rho v^2 A \tag{6.41}$$

And, following a similar approach as done for the drag force, one may derive a similar expression for the lift force as follows:

$$F_L = C_L \frac{1}{2}\rho v^2 A \tag{6.42}$$

where C_L is the lift coefficient (see Chapter 7).

Furthermore, examining the standard form of the drag coefficient, C_D yields the following:

$$\frac{F_D}{F_{dynpres}} = \frac{\Delta p A}{\frac{\rho v^2}{2}A} = \frac{\Delta p}{\frac{\rho v^2}{2}} = C_D = \frac{Actual\ Drag\ Force}{Ideal\ Force} = \frac{Actual\ Presdrop}{Ideal\ Presdrop} = \frac{staticpres}{dynpres} \tag{6.43}$$

where the results of dimensional analysis introduce a drag coefficient, C_D, which expresses the ratio of a real flow variable (in this case: force or pressure) to its corresponding ideal flow variable. Depending upon the particular external flow situation, the pressure drop, Δp may: (1) be equal to the dynamic pressure (ideal pressure), $\rho v^2/2$; (2) be greater than the dynamic pressure; or (3) be less than the dynamic pressure. In the case when the actual pressure drop, Δp is exactly equal to the dynamic pressure, $\rho v^2/2$ (when there is no friction drag), then the drag force, F_D is equal to the dynamic (pressure) force, $F_{dynpres}$, so:

$$\frac{F_D}{F_{dynpres}} = \frac{F_{dynpres}}{F_{dynpres}} = \frac{\Delta p A}{\frac{\rho v^2}{2} A} = \frac{\Delta p}{\frac{\rho v^2}{2}} = \frac{\frac{\rho v^2}{2}}{\frac{\rho v^2}{2}} = C_D = 1 \qquad (6.44)$$

Specifically, this is the case of an ideal flow, where the pressure drop, Δp is exactly equal to the dynamic pressure, $\rho v^2/2$ and there is no drag force, because there is no flow resistance due to $F_V = 0$; thus, d'Alembert's paradox applies here. However, in most cases, the actual pressure drop, Δp is either larger than or smaller than the dynamic pressure, $\rho v^2/2$; thus, the drag force, F_D is correspondingly either larger than or smaller than the dynamic (pressure) force, $F_{dynpres}$, for which:

$$\underbrace{\Delta p}_{presdrop} = \underbrace{\overbrace{\frac{\rho v^2}{2}}^{F_P/A}}_{dynpres} + \underbrace{\overbrace{\frac{\tau_w L}{R_h}}^{F_f/A}}_{flowres} \qquad (6.45)$$

where $C_D > 1$ when the actual pressure drop is greater than the dynamic pressure drop, and $C_D < 1$ when the actual pressure drop is smaller than the dynamic pressure drop. One may recall that if the body is blunt in shape (spherical, for instance), then the pressure drag, F_p predominates the friction drag, F_f and thus, in general, $C_D > 1$, whereas, if the body is more streamlined, then the friction drag, F_f predominates the pressure drag, F_p, and thus, in general, $C_D < 1$. And finally, because the actual pressure drop, Δp is unknown, the drag coefficient, C_D is determined empirically from experimentation. Empirical evaluation of the drag coefficient, C_D (and the lift coefficient, C_L) and application of the drag force equation (and lift force equation) in external flow are presented in Chapter 10.

6.4 Modeling the Flow Resistance as a Major Head Loss in Internal Flow

In the internal flow through a pipe or a channel, energy or work is used to move/force the fluid through a conduit. As such, one is interested in the determination of the energy or head losses, pressure drops, and cavitation where energy is dissipated. The flow of a real fluid (liquid or a gas) through a pipe under pressure is used in heating and cooling systems and in distribution systems of fluids. In the case of a gas, the fluid is typically forced through the system by the use of a fan, whereas in the case of a liquid, the fluid is typically forced through the system by the use of a pump. Furthermore, in either case, the energy dissipated by the viscous forces within the fluid is supplied by the excess work done by the pressure and gravity forces. The determination of the corresponding head loss, h_f is one of the most important problems in fluid mechanics, where the head loss, h_f will increase the amount of power that a pump must deliver to the fluid; the head loss, h_f represents the additional height that the fluid needs to be elevated by the pump in order to overcome the losses due to friction in the pipe. Some of the pumping power will be used to pump the fluid through the pipe, while some will be used to compensate for the energy lost in the pressure drop caused by the friction of the fluid flowing past the pipe surface (pipe friction). The pressure drop, Δp and the head loss, h_f are dependent on the wall shear stress, τ_w between the fluid and the pipe. Furthermore, the wall shear stress, τ_w and the pressure drop, Δp and thus the head loss, h_f and the power requirements for pumping are considerably lower for laminar flow than for turbulent flow.

6.4.1 Evaluation of the Major Head Loss

In the case where the flow resistance is due to pipe or channel friction in internal flow, the flow resistance is ultimately modeled as a major head loss due to pipe friction in the integral form of the energy equation because the major loss is needed for the design of internal flow in a pipe or open channel. The flow resistance (shear stress or drag force) is modeled as a resistance force (shear stress or drag force) in the integral form of the momentum equation (see Section 6.5), while it is modeled as a friction slope, S_f in the differential form of the momentum equation (see Section 6.7). One may note that the propulsive force for a pipe flow is provided by a pressure gradient, $\Delta p/L$ (which results in an unknown Δp or S_f), while the propulsive force for an open channel flow is provided by invoking gravity and thus by the weight of the fluid along a channel slope (channel gradient, $\Delta z/L = S_o$) (which results in an unknown S_f). The flow resistance causes an unknown pressure drop, Δp (or S_f), which causes an unknown major head loss, h_f, where the head loss is due to a conversion of kinetic energy to heat, which is modeled/displayed in the integral form of the energy equation. Evaluation of the major head loss flow resistance term requires a "subset level" of application of the governing equations. A "subset level" application of the governing equations focuses only on the given element causing the flow resistance. The assumption of real flow implies that the flow resistance is modeled in both the energy and momentum equations (see Table 4.1). Thus, in the derivation of the major head loss, first, the integral form of the energy equation is applied to determine an expression for the unknown head loss, h_f. Then, the integral momentum and the differential momentum equations (supplemented by dimensional analysis) are applied in order to determine an expression for the pressure drop, Δp (or S_f) and the shear stress, τ_w. The flow resistance due to pipe or channel friction, which results in a major head loss term, $h_{f,major}$, is modeled as a resistance force (shear stress or drag force) in the integral form of the momentum equation (Equation 4.28), while it is modeled as a friction slope, S_f in the differential form of the momentum equation (Equation 4.84). As such, the energy equation may be used to solve for the unknown major head loss,

$$h_{f,maj} = S_f L = \frac{\tau_w L}{\gamma R_h} = \frac{\Delta p}{\gamma},$$

while the integral momentum equation may be used to solve for the unknown pressure drop, $\Delta p = \tau_w L/R_h$ (or the unknown friction slope, $S_f = \tau_w/\gamma R_h$ in the case of turbulent open channel flow).

The major head loss, h_f is caused by both pressure and friction forces. Thus, when the friction/viscous forces can be theoretically modeled in the integral form of the momentum equation (application of Newton's law of viscosity in the laminar flow case), one can analytically determine the actual pressure drop, Δp (or S_f) from the integral form of the momentum equation. And, therefore, one may analytically derive an expression for the major head loss, h_f due to pipe or channel friction from the energy equation (see Section 6.6). However, when the friction/viscous forces cannot be theoretically modeled in the integral form of the momentum equation (as in the turbulent flow case), the friction/viscous forces, the actual pressure drop, Δp (or the unknown friction slope, S_f in the case of turbulent open channel flow) and thus the major head loss, h_f due to pipe or channel friction are determined empirically (see Section 6.7). Specifically, in the case of turbulent flow, the friction/viscous forces in the integral form of the momentum equation and thus the wall shear stress, τ_w cannot be theoretically modeled; thus, the integral form of the momentum equation cannot be directly applied to solve for the unknown pressure drop, Δp. Thus, an empirical interpretation (using dimensional analysis) of the wall shear stress, τ_w in the theoretical expression for the friction slope, $S_f = \tau_w/\gamma R_h$ in the differential form of the momentum equation is sought in terms of velocity, v and a flow resistance coefficient. This yields the Chezy equation ($S_f = v^2/C^2 R_h$, which represents the differential form of the momentum equation, supplemented by dimensional analysis and guided by the integral momentum equation; link between the differential and integral momentum equations), which is used to obtain an empirical evaluation for the pressure drop, Δp in Newton's second law of motion (integral form of the momentum equation). As such, one can then derive an empirical expression for the major head loss, h_f from the energy equation. The flow resistance equation for the major head loss is derived below and in Chapter 7.

6.4.2 Application of the Energy Equation: Derivation of the Head Loss

In the derivation of the major head loss for pipe or open channel flow, real flow is assumed in order to determine an expression for the unknown head loss, h_f by applying the integral form of the energy equation as follows:

$$\left(\frac{p}{\gamma} + z + \frac{v^2}{2g}\right)_1 - h_{f,\,maj} = \left(\frac{p}{\gamma} + z + \frac{v^2}{2g}\right)_2 \tag{6.46}$$

Assuming $z_1 = z_2$, and $v_1 = v_2$, the energy equation has one unknown, which is the major head loss,

$$h_{f,maj} = S_f L = \frac{\tau_w L}{\gamma R_h} = \frac{\Delta p}{\gamma},$$

which is expressed as a function of the pressure drop, Δp.

6.4.3 Application of the Momentum Equations and Dimensional Analysis: Derivation of the Pressure Drop, Shear Stress, and the Drag Coefficient

In the derivation of the major head loss for pipe or open channel flow, real flow is assumed in order to determine an expression for the pressure drop, Δp (in the case of major head loss due to pipe flow resistance) or the change in channel depth, Δy (in the case of major head loss due to open channel flow resistance) and the shear stress, τ_w by applying the integral form of the momentum equation (supplemented by dimensional analysis) as follows:

$$\sum F_s = (F_G + F_P + F_V + F_{other})_s = (\rho Q v_s)_2 - (\rho Q v_s)_1 \tag{6.47}$$

Assuming $z_1 = z_2$, and $v_1 = v_2$, the integral momentum equation is used to solve for the pressure drop as a function of the shear stress as follows (derived in Section 6.5):

$$\Delta p = \frac{\tau_w L}{R_h} \tag{6.48}$$

However, the problem is that one cannot theoretically model the wall shear stress, τ_w in the integral momentum for turbulent flow for pipe and open channel flow; thus, one cannot analytically determine the pressure drop, Δp in Equation 6.48. Furthermore, in the differential momentum equation, which is given as follows:

$$\underbrace{\frac{\partial}{\partial s}\left(\frac{p}{\gamma} + z + \frac{v^2}{2g}\right)}_{S_e} = \underbrace{-\frac{\tau_w}{\gamma R_h}}_{S_f} \underbrace{-\frac{1}{g}\frac{\partial v}{\partial t}}_{S_a} \tag{6.49}$$

while S_e and S_a are expressed in terms of easy to measure velocity, v, S_f is expressed in terms of difficult-to-measure τ_w. Thus, the integral momentum equation (and the differential momentum equation for S_f) is supplemented with dimensional analysis in order to empirically model the wall shear stress, τ_w for turbulent pipe and open channel flow and thus Δp (and Δy), by the use of a drag coefficient, C_D. Specifically, dimensional analysis is used to empirically interpret S_f (express τ_w) in terms of easy to measure velocity, v. Thus, the resulting flow resistance equation for the major head loss is given as follows:

$$h_{f,maj} = S_f L = \frac{\tau_w L}{\gamma R_h} = \frac{\Delta p}{\gamma} = \frac{C_D v^2}{g R_h} L = \frac{v^2}{C^2 R_h} L = f \frac{L}{D}\frac{v^2}{2g} \tag{6.50}$$

where the drag coefficient, C_D is expressed in terms of the more commonly defined Chezy coefficient, C and Darcy–Weisbach friction factor, f. Furthermore, one can easily compute v_i used in Equation 6.50 by measuring Δp using a pitot-static tube. And, finally, it is important to note that the use of dimensional analysis is not needed in the derivation of the major head loss term for laminar flow (see Section 6.6).

6.4.4 Application of the Governing Equations to Derive the Major Head Loss Equations for Laminar and Turbulent Flow

Application of the three governing equations yields an expression for the major head loss for both laminar and turbulent flow. First, the integral form of the energy equation (Equation 6.46) yields the following relationship between the major head loss, h_f; the wall shear stress, τ_w; and the pressure drop, Δp:

$$h_{f,maj} = S_f L = \frac{\tau_w L}{\gamma R_h} = \frac{\Delta p}{\gamma} \tag{6.51}$$

Then, regardless of whether the flow is laminar or turbulent, application of Newton's second law of motion (integral form of the momentum equation) (Equation 6.47) yields a linear relationship between the shear stress, τ and the radial distance, r, which also describes the theoretical relationship between the pressure drop, Δp (or S_f) and the shear stress, τ; this relationship plays a key role in the derivation of the major head loss equations for laminar and turbulent flow. Specifically, the integral form of the momentum equation yields the following relationship between the pressure drop, Δp and the wall shear stress, τ_w (derived in Section 6.5):

$$\tau = \frac{r}{2}\frac{\Delta p}{L} \qquad (6.52)$$

Evaluated for the wall shear stress, τ_w at $r = r_o = D/2$, where $R_h = D/4$, and expressed as a function of the pressure drop, Δp and yields:

$$\Delta p = \frac{\tau_w L}{R_h} \qquad (6.53)$$

Furthermore, the differential form of the momentum equation (Equation 6.49) yields the following relationship between the friction slope, S_f and the wall shear stress, τ_w:

$$S_f = \frac{\tau_w}{\gamma R_h} \qquad (6.54)$$

However, because all three terms, the pressure drop, Δp (Equation 6.53); the friction slope, S_f (Equation 6.54); and the major head loss, h_f (Equation 6.51) are expressed in terms of the wall shear stress, τ_w, in order to evaluate the respective expressions, one must first obtain an expression for the wall shear stress, τ_w. Depending upon whether the flow is laminar or turbulent, this will determine whether the friction/viscous forces and thus the wall shear stress, τ_w can be theoretically modeled and whether application of the three governing equations will yield an analytical or an empirical expression for the major head loss, h_f.

In the case of laminar flows, first, application of the integral form of the energy equation yields an expression for the major head loss (Equation 6.51). Then, because the friction/viscous forces in the integral form of the momentum equation and thus the wall shear stress, τ_w can be theoretically modeled (by Newton's law of viscosity, where $\tau = \mu\, dv/dy$) in the case of laminar flow, one can analytically solve for the unknown pressure drop, Δp from the momentum equation (Equation 6.52); thus, one can analytically derive an expression for the major head loss, h_f from the energy equation. Specifically, application of the integral form of the momentum equation, where the wall shear stress, τ_w is theoretically modeled by Newton's law of viscosity, yields a theoretical (parabolic) expression for the velocity distribution for laminar flow, expressed as a function of the pressure drop, Δp. Next, application of the continuity equation by integrating the resulting parabolic velocity distribution over the pipe diameter yields an analytical expression for the actual pressure drop, Δp (Poiseuille's law), which is substituted into the integral form of the energy equation (Equation 6.51), yielding an analytical expression for the head loss, h_f (Poiseuille's law expressed in terms of the head loss) (see Section 6.6).

In the case of turbulent flows, first, application of the integral form of the energy equation yields an expression for the major head loss (Equation 6.51). Then, however, because the friction/viscous forces in the integral form of the momentum equation and thus the wall shear stress, τ_w cannot be theoretically modeled in the case of turbulent flow, one cannot analytically solve for the unknown pressure drop, Δp from the integral form of the momentum equation (Equation 6.52). Therefore, the unknown pressure drop, Δp (or the unknown friction slope, S_f in the case of turbulent open channel flow) and thus the major head loss, h_f due to pipe or channel friction are determined empirically. Specifically, an empirical interpretation of the wall shear stress, τ_w in the theoretical expression for the friction slope (Equation 6.54) in the differential form of the momentum equation is sought in terms of velocity, v and a flow resistance coefficient. This is accomplished by using dimensional analysis in order to derive an empirical expression for the wall shear stress, τ_w as a function of the velocity, v which involves the definition of a drag coefficient, C_D that represents the flow resistance (Chezy coefficient, C; Darcy–Weisbach friction factor, f; and Manning's roughness coefficient, n) and is empirically evaluated; this is analogous to Newton's law of viscosity, where the shear stress is expressed as a function of velocity, v and flow resistance, μ. One may note that the theoretical relationship between the pressure drop, Δp and the shear stress, τ provided by Newton's second law of motion (integral form of the momentum equation) (Equation 6.53) is used in the dimensional analysis procedure, in which the friction forces are modeled empirically. Dimensional analysis yields an empirical expression for the friction slope, $S_f = v^2/C^2 R_h$ (the Chezy equation) in the differential form of the momentum equation. The Chezy equation represents the differential form of the momentum equation, supplemented by dimensional analysis and guided by the integral momentum equation, a link between the differential and integral momentum equations. The Chezy equation is used to obtain an empirical evaluation for the pressure drop, Δp in Newton's second law of motion (integral form of the momentum equation). And, finally, the Chezy equation is substituted into the integral form of the energy equation (Equation 6.51), yielding an empirical expression for the head loss, h_f (the Chezy head loss equation) (see Section 6.7). It is interesting to note that in the case of turbulent flow, the lack of a theoretical expression for the friction/viscous forces leads to the inability to derive a theoretical expression for the velocity distribution for turbulent flow. Instead, empirical studies have shown that the velocity varies according to the seventh root profile law for turbulent flow. As a result, the empirical expression for the friction forces yields an empirical expression for the head loss, with the need to define an empirical flow resistance.

The derivation of the major head loss equation will depend upon the type of internal flow. The assumption made in the derivation of the major head loss equation is that the flow is under pressure in a circular pipe. Although the laws of flow resistance are essentially the same for pipe and open channel flow (adjustments for open channel flow are presented in Section 6.7.11), the variation of the shear stress, τ and the variation of the velocity, v across an open channel flow section may be modeled by the lower half of the pipe flow section. Furthermore, noncircular pipe flow is addressed in Section 6.7.9.1. Additionally, regardless of whether the flow is laminar or turbulent, the pipe flow at the entrance is modeled as ideal flow, followed by real flow; the ideal developing flow at the pipe entrance accelerates (both local and convective), and the real developing flow near the pipe entrance and along the pipe length does not accelerate (steady and uniform developed flow—the rate of velocity and the velocity profile does not change with pipe length). Furthermore, while Section 6.5 presents the flow characteristics of laminar and turbulent flow, Sections 6.6 and 6.7 present the derivation of the major head loss for laminar flow and turbulent flow, respectively.

6.5 Laminar and Turbulent Internal Flow Characteristics

Regardless of whether the flow is laminar or turbulent, the two types of flow share some similar flow characteristics. As stated above, the application of Newton's second law of motion (integral form of the momentum equation) plays a role in the determination of the major head loss, h_f for both laminar and turbulent flow. Specifically, the application of Newton's second law of motion yields a linear relationship between the shear stress, τ and the radial distance, r, which also describes the theoretical relationship between the pressure drop, Δp and the shear stress, τ for both laminar and turbulent flow; this relationship plays a key role in the derivation of the major head loss equations for laminar and turbulent flow (Section 6.5.1). Additionally, this section illustrates the variation of the velocity with time for laminar and turbulent flow (Section 6.5.2). And, finally, this section illustrates the variation of the velocity, pressure drop, and the wall shear stress with length of pipe (developing vs. developed flow) for laminar and turbulent flow (Section 6.5.3).

The development of the equations in the section herein for laminar and turbulent flow assumes the flow is steady (no local acceleration, $\partial v / \partial t = 0$) and fully developed/uniform (no convective acceleration, $\partial v / \partial s = 0$). Further simplifying assumptions regarding the flow are made below. And, as explained in detail below (Section 6.5.3), regardless of whether the flow is laminar or turbulent, at the pipe entrance, the velocity profile, the pressure drop, and the wall shear stress will vary with the length of the pipe until the flow changes from developing flow to fully developed flow. Given that the pipe is of sufficient length, over the length of the pipe, the flow becomes fully developed and thus there is no variation in the velocity profile, the pressure drop, and the wall shear stress with the length of the pipe until a change occurs (a change in diameter, a bend, a tee, a valve, or another pipe component). Thus, this section assumes that the pipe is sufficiently long relative to the entry length that the entrance effects can be ignored.

6.5.1 Determining the Variation of Shear Stress with Radial Distance for Laminar and Turbulent Flow

Application of Newton's second law of motion (integral form of the momentum equation) yields a linear relationship between the shear stress, τ and the radial distance, r, which also describes the theoretical relationship between the pressure drop, Δp and the shear stress, τ for both laminar and turbulent flow. As illustrated in the respective sections below on laminar (Section 6.6) and turbulent flow (Section 6.7), this relationship plays a key role in the derivation of the major head loss equations for laminar and turbulent flow, respectively. Figure 6.4 illustrates the forces acting on a cylindrical element of fluid, ds that is oriented coaxially with the pipe, assuming steady (no local acceleration) and fully developed/uniform (no convective acceleration) flow. The assumption of uniform flow implies that the streamlines are straight and parallel; therefore, the pressure across any section of the pipe will be hydrostatically distributed; as a result, the pressure force acting on an end face of the fluid element will be the product of the pressure at the center of the element (also the center of the pipe) and the area of the face of the element. And, the assumption of uniform, steady flow further implies equilibrium between the pressure, gravity, and shear forces as illustrated below. Beginning with the differential form of Newton's second law of motion for the most general flow type (compressible, nonuniform, unsteady flow for either pressure or gravity flow) for the cylindrical element of fluid, ds moving in the s-direction along the streamline illustrated in Figure 6.4, it is given as follows:

$$\sum F_s = Ma_s \tag{6.55}$$

The significant forces acting on the cylindrical element of fluid, ds illustrated in Figure 6.4 include the gravitational force component along the s-axis, the pressure forces on the two

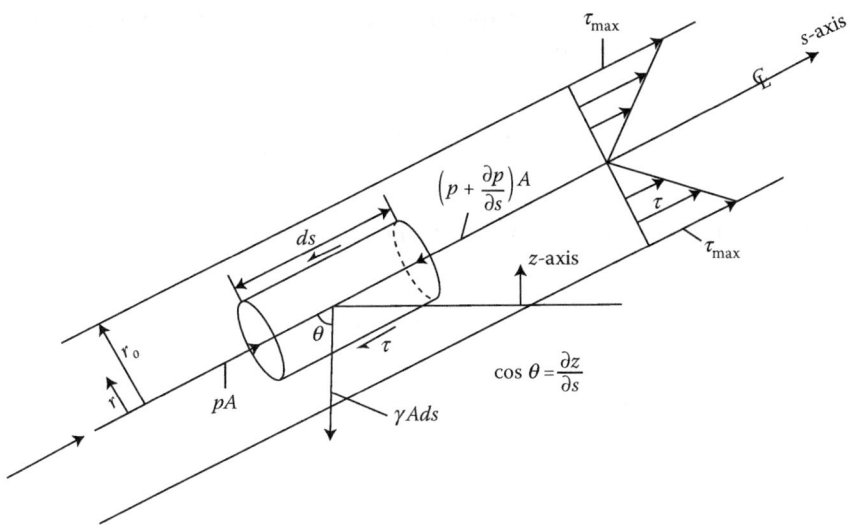

FIGURE 6.4
Significant forces acting on a cylindrical fluid element in a pipe flow, and resulting linear shear stress distribution for both laminar and turbulent flow.

ends of the fluid element, and the viscous force due to fluid friction as follows:

$$\sum F_s = F_G + F_P + F_V \tag{6.56}$$

where

$$F_G = -\gamma A \, ds \cos \theta = -\gamma A \, ds \frac{\partial z}{\partial s} \tag{6.57}$$

$$F_P = pA - \left(p + \frac{\partial p}{\partial s} ds \right) A = -\frac{\partial p}{\partial s} ds \, A \tag{6.58}$$

$$F_V = -\tau P_w \, ds \tag{6.59}$$

The acceleration is defined is follows:

$$a_s = \frac{dv}{dt} = \frac{ds}{dt} \frac{\partial v}{\partial s} + \frac{\partial v}{\partial t} = v \frac{\partial v}{\partial s} + \frac{\partial v}{\partial t} \tag{6.60}$$

And finally, noting that the mass term on the right hand side of Newton's second law of motion is defined as $M = \rho A \, ds$ yields:

$$-\gamma A \, ds \frac{\partial z}{\partial s} - \frac{\partial p}{\partial s} ds \, A - \tau P_w \, ds = (\rho A \, ds) \left(v \frac{\partial v}{\partial s} + \frac{\partial v}{\partial t} \right) \tag{6.61}$$

Furthermore, noting that $P_w = 2\pi r$ is the wet perimeter of the cylindrical element of fluid, ds along which the shear stress, τ acts, as illustrated in Figure 6.4, yields the differential form of Newton's second law of motion for the most general flow type (compressible, nonuniform, unsteady flow for either pressure or gravity flow) as follows:

$$-\gamma A\,ds\frac{\partial z}{\partial s} - \frac{\partial p}{\partial s}ds\,A - \tau(2\pi r)ds = (\rho A\,ds)\left(v\frac{\partial v}{\partial s} + \frac{\partial v}{\partial t}\right) \tag{6.62}$$

Next, assuming uniform, steady flow yields the following:

$$-\gamma A\,ds\frac{dz}{ds} - \frac{dp}{ds}ds\,A - \tau(2\pi r)ds = 0 \tag{6.63}$$

Then, dividing by $A\,ds$, where $A = \pi r^2$, yields the following:

$$-\gamma\frac{dz}{ds} - \frac{dp}{ds} - \frac{\tau(2\pi r)}{\pi r^2} = 0 \tag{6.64}$$

And finally, isolating the shear stress, τ yields the following:

$$\tau = \frac{r}{2}\left[-\frac{d}{ds}(p + \gamma z)\right] \tag{6.65}$$

First, one may note that the following term is the slope of the total energy, H, which is negative (decreasing slope):

$$-\frac{d(p + \gamma z)}{ds} = -\frac{dH}{ds} \tag{6.66}$$

Second, assuming that $z_1 = z_2$, and $ds = L$ yields:

$$\tau = \frac{r}{2}\left[-\frac{d}{ds}(p)\right] = \frac{r}{2}\frac{\Delta p}{L} \tag{6.67}$$

which yields an expression for the variation of the shear stress, τ across the pipe diameter (or actually as a function of the radial distance, r from the center of the pipe) that also describes the theoretical relationship between the pressure drop, Δp and the shear stress, τ, and for both laminar and turbulent flow; it represents an equilibrium among the pressure, gravity, and shear forces needed to move each fluid particle along the pipe with a constant velocity.

Figure 6.4 illustrates how the shear stress, τ varies across the pipe diameter for both laminar and turbulent flow (linear distribution) as follows:

$$\tau = \frac{r}{2}\frac{\Delta p}{L} \tag{6.68}$$

Specifically, Figure 6.4 illustrates that the shear stress, τ increases linearly with the radial distance, r to a maximum shear stress, τ_{max} at the pipe wall/boundary (or the wall shear stress, τ_w) where $r = r_o$, and is given as follows:

$$\tau_{max} = \tau_w = \frac{r_o}{2}\frac{\Delta p}{L} \qquad (6.69)$$

and the shear stress, τ at the centerline of the pipe $r = 0$ is equal to zero.

The linear relationship between the shear stress, τ and the radial distance, r may be illustrated by isolating the pressure drop, Δp over the length, L term from the expressions for the variation of the shear stress, τ with the radial distance, r from the center of the pipe and the expression for the maximum shear stress, τ_{max} as follows (for both laminar and turbulent flow):

$$\frac{\Delta p}{L} = \frac{2\tau}{r} = \frac{2\tau_{max}}{r_o} \qquad (6.70)$$

Thus:

$$\frac{\tau}{r} = \frac{\tau_{max}}{r_o} \qquad (6.71)$$

$$\tau = \tau_{max}\frac{r}{r_o} \qquad (6.72)$$

where the shear stress is a maximum at the pipe wall ($\tau_w = \tau_{max}$) and there is no shear stress at the centerline of the pipe. One may note that the linear dependence of the shear stress, τ on the radial distance, r is a direct result of the fact that the shear force, $F_V = -\tau P_w\, ds = -\tau 2\pi r\, ds$, which acts on the surface area of the cylindrical element of fluid, ds, is linearly dependent on the radial distance, r. Furthermore, the relationship $\Delta p/L = 2\tau_{max}/r_o$ indicates that if the viscosity were zero, then there would be no shear stress and thus the pressure drop would be zero (constant pressure in the pipe). Thus, a small shear stress can produce a significant pressure drop for a relatively long pipe ($L \gg D$).

Additionally, the following expression for the maximum shear stress, τ_{max} indicates that it is a function of the shear stress throughout the fluid, τ, which is a function of the fluid viscosity, μ:

$$\tau_{max} = \tau\frac{r_o}{r} \qquad (6.73)$$

Furthermore, it is the shear stress at the wall, τ_w that is responsible for the head loss, h_f.

6.5.2 The Variation of the Velocity with Time for Laminar and Turbulent Flow

Figure 6.5 illustrates the three temporal phases of flow in the development of steady flow (laminar and turbulent) in a pipe flow with a valve to control the flow. The first phase is prior to any flow, $v = 0$ (for which the valve is closed). The second phase is the beginning of actual flow, $dv/dt = a \neq 0$ (for which the valve is just opened), which is unsteady. And the third and final phase is the developed flow, which becomes steady, $dv/dt = 0$ (valve has been open for some time).

6.5.3 The Variation of the Velocity, Pressure Drop, and the Wall Shear Stress with Length of Pipe for Laminar and Turbulent Flow: Developing versus Developed Flow

Regardless of whether the flow is laminar or turbulent, at the pipe entrance, the velocity profile, the pressure drop, and the wall shear stress will vary with the length of the pipe until the

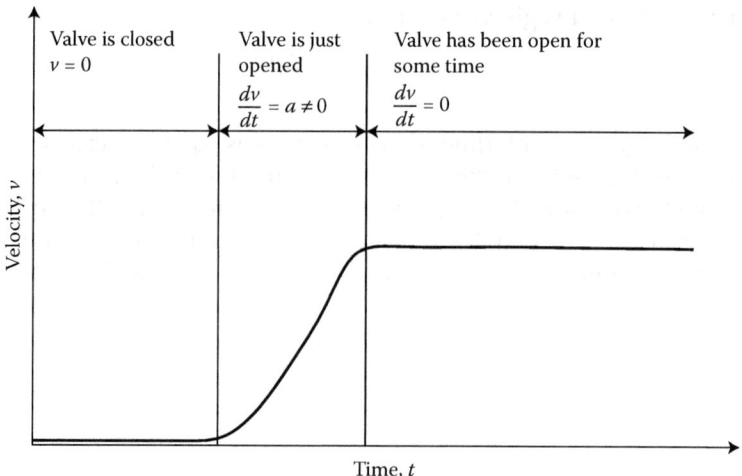

FIGURE 6.5
Variation of velocity with time for laminar and turbulent pipe flow with a valve to control the flow. (Adapted from Esposito, A., 1998. *Fluid Mechanics with Applications*. New Jersey: Prentice Hall, 188.)

flow changes from developing flow to fully developed flow. Given that the pipe is of sufficient length, over the length of the pipe, the flow becomes fully developed and thus there is no variation in the velocity profile, the pressure drop, and the wall shear stress with the length of the pipe, until a change occurs (a change in diameter, a bend, a tee, a valve, or another pipe component). It may be noted that although the following discussion assumes a pipe flow, it is also applicable to open channel flow.

6.5.3.1 The Extent of Developing Flow in the Pipe Length

Figures 6.6 (laminar flow) and 6.7 (turbulent flow) illustrate a large tank of fluid with a rounded entrance connected to a long pipe of constant diameter. Pressure is applied to the flow so that a steady flow results. The type of flow, laminar versus turbulent, in a pipe is highly a function of pipe inlet (entrance) conditions, the inherent turbulence (or lack of it) of the flow at the pipe inlet, and the roughness of the pipe wall. Regardless of whether the flow is laminar or turbulent, Figures 6.6 (laminar flow) and 6.7 (turbulent flow) illustrate that the flow at the pipe entrance is characterized as "developing flow" (as the velocity profile is developing), while the flow much further downstream in the pipe is characterized as "developed flow" (as the velocity profile is fully developed). For laminar flow (Figure 6.6), the length of the developing region, known as the entrance length, L_e is defined as a function of the pipe diameter, D and the Reynolds number, R as follows:

$$L_e = 0.058DR \qquad (6.74)$$

On the other hand, for turbulent flow (Figure 6.7), the length of the developing region, L_e is independent of the Reynolds number, R and is defined as follows:

$$L_e = 50D \qquad (6.75)$$

One may note that the length of the developing region, L_e is shorter for laminar flow than for turbulent flow. Additionally, turbulent flow is more likely to occur in practice

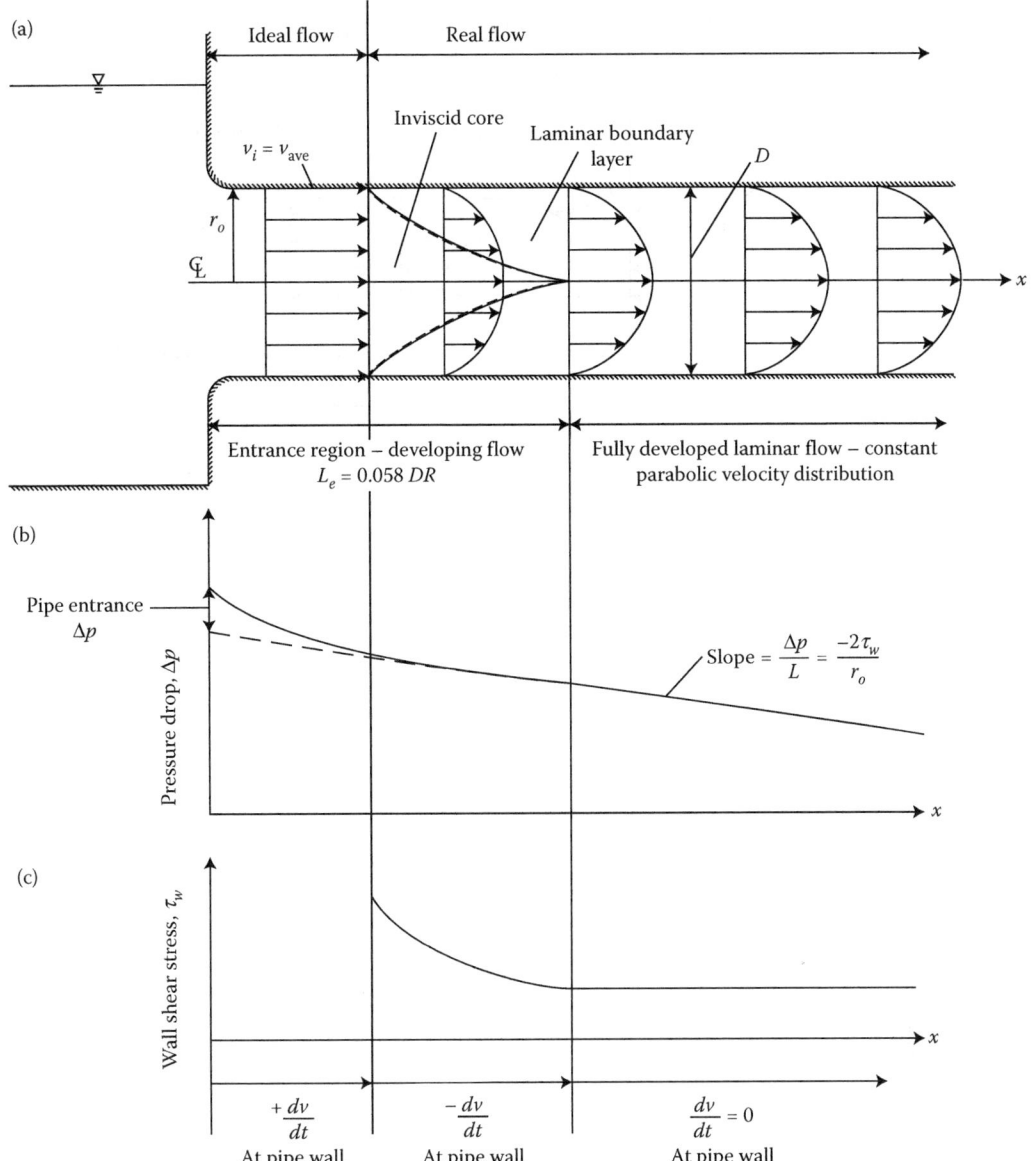

FIGURE 6.6
Developing versus developed laminar flow. (a) The variation of the velocity distribution with pipe length. (b) The variation of the pressure drop with pipe length. (c) The variation of the wall shear stress with pipe length.

than laminar flow. However, in the calculation of the frictional head losses in the entrance region, the following simplifications are made. Because pipes used in practice are typically several times the length of the developing region, L_e, the flow through the pipe is often assumed to be fully developed for the entire length of the pipe. While such a simplistic approach yields reasonable results for long pipes, it may yield poor results for short pipes because it underpredicts the wall shear stress and thus the friction factor.

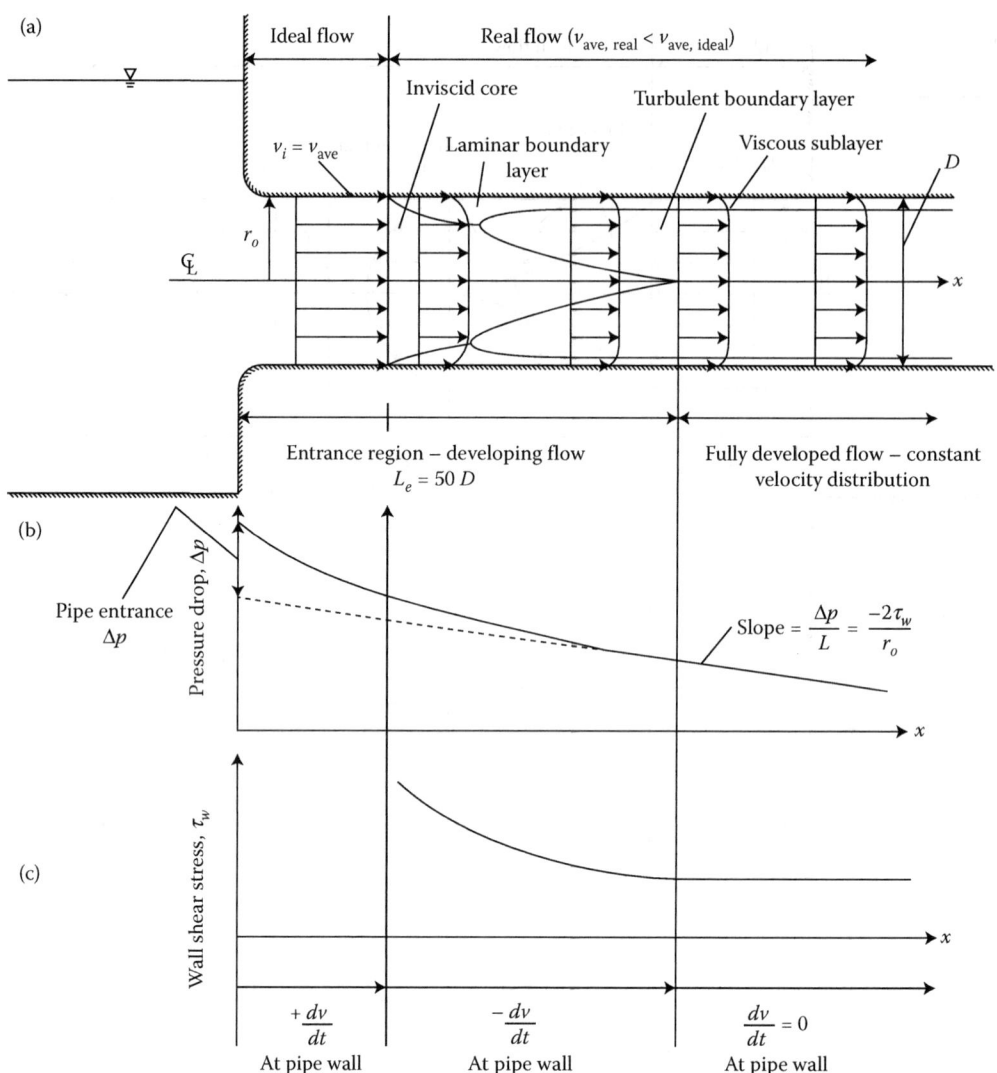

FIGURE 6.7
Developing versus developed turbulent flow. (a) The variation of the velocity distribution with pipe length. (b) The variation of the pressure drop with pipe length. (c) The variation of the wall shear stress with pipe length.

6.5.3.2 The Variation of the Velocity Profile with Pipe Length

Figures 6.6 (laminar flow) and 6.7 (turbulent flow) illustrate that at the pipe entrance (the beginning of the developing flow region), the flow behaves as an inviscid (ideal) flow, with a rectangular velocity distribution, $v_{ave} = Q/A$. Because there are no viscous effects in this inviscid section at the pipe entrance, there is an acceleration of flow, where $v_1 = 0$ and $v_2 = v$ over a pipe section between x_1 (at the pipe entrance) and x_2 (downstream of the pipe entrance). As the fluid moves through the pipe near the beginning of the developing flow region, viscous effects (caused by the fluid viscosity) cause the fluid to stick to the pipe wall (the no-slip boundary condition) (this occurs whether the fluid is relatively inviscid or highly viscous). As a result, a boundary layer (Prandtl boundary layer theory) in which viscous effects are important is produced along the pipe wall, which causes the velocity profile to gradually deviate from the initial rectangular shape as the flow moves through the pipe. Downstream of the entrance of the pipe, the flow has not been subjected to the influence of viscosity for long, so the velocity profile is still somewhat constant across the pipe cross section, except for the region near the pipe wall, where the velocity will have been decreased due to the wall shear stress. The boundary layer continues to grow in thickness in the direction of the flow until it completely fills the pipe—note the difference in the boundary layer for laminar versus turbulent flow illustrated in Figures 6.6 (laminar flow) and 6.7 (turbulent flow). One may note that while the viscous effects are very important within the boundary layer, the viscous effects are negligible outside of the boundary layer, which is called the "inviscid core." The inviscid core surrounds the centerline of the pipe until the boundary layer fills the pipe, which marks the end of the developing region and thus the beginning of the fully developed region. One may note that the section of the velocity profile within the inviscid core is uniform while decreasing to zero within the boundary layer moving towards the pipe wall.

As the flow proceeds downstream, the effect of viscosity will increase. This causes the affected region near the wall to increase and thus causes the velocity to decrease more near the pipe wall, deviating from a somewhat uniform velocity profile to: (1) a parabolic velocity profile in the case of laminar flow or (2) a velocity profile that varies according to the seventh root profile law in the case of turbulent flow. It is important to note that the decrease in the velocity near the pipe wall is balanced by an increase in velocity near the pipe center (i.e., the development of a velocity gradient along the pipe length) in order to maintain a steady flow in the pipe. The end of the inviscid core, where the boundary layer has filled the pipe, marks the beginning of the developed flow region, in which the velocity profile will no longer change with distance. Thus, while the velocity profile undergoes changes with distance moving downstream, eventually, however, the velocity distribution will no longer change with distance. Therefore, an equilibrium velocity profile will be established (the velocity profile will be independent of distance/pipe length). The established velocity profile will either be laminar or turbulent depending upon the Reynolds number, R characterizing the flow in the pipe. Thus, for long pipes, the velocity profile becomes fully established or uniform along the pipe length. The established flow continues in the pipe until a change occurs. Such a change may include a change in diameter, a bend, a tee, a valve, or another pipe component. When a change occurs, the established velocity profile will change and will thus require more pipe flow length to return to established/developed flow. Thus once beyond the interruption, the flow returns to the fully developed flow region until the next pipe component (change) occurs. It may be noted that in most cases, the pipe length following a change in the pipe component is long enough that fully developed flow is achieved. However, in some cases where the distances between one pipe component of the pipe system and the next component are too short, fully developed flow is never quite achieved.

6.5.3.3 *The Variation of the Pressure Drop and the Wall Shear Stress with Pipe Length*

Figures 6.6 (laminar flow) and 6.7 (turbulent flow) illustrate that at the pipe entrance (the beginning of the developing flow region), where the flow behaves as an inviscid (ideal) flow, the wall shear stress, $\tau_w = 0$ and thus the pressure drop, Δp over the pipe length, x is exactly equal to the dynamic pressure, $\rho v^2/2$, which is considered to be the ideal pressure drop. For a horizontal pipe, it is this pressure difference, Δp across a pipe section that forces the fluid through the pipe in order to overcome pipe friction. Because there are no viscous effects in this inviscid section at the pipe entrance, there is an acceleration of flow, where $v_1 = 0$ and $v_2 = v$ over a pipe section between x_1 (at the pipe entrance) and x_2 (downstream of the pipe entrance). The magnitude of the pressure gradient, $\Delta p/L$ is a maximum at the pipe entrance, decreasing in the developing flow region and finally becoming a constant in the developed flow region. In the developed flow region, for a given wall shear stress, τ_w and pipe radius, r_o, the pressure drop, Δp over the length, L is linear with a slope of -2 τ_w/r_o. Furthermore, in the developed flow region, there is no acceleration of flow. It is interesting to note that the effect of the maximum pressure drop at the pipe entrance is always to increase the average friction factor for the entire pipe length, where such an increase may be significant for short pipes, but negligible for longer pipes. Furthermore, the decrease in pressure in the developing flow region is accompanied by a corresponding increase in the kinetic energy.

Figures 6.6 (laminar flow) and 6.7 (turbulent flow) illustrate that downstream of the inviscid pipe flow section, the boundary layer in which viscous effects are important is produced along the pipe wall. At the point in the pipe section where the boundary layer begins, the wall shear stress, τ_w is a maximum, decreasing in the developing flow region and finally reaching a constant value in the developed flow region. One may note that the wall shear stress, τ_w is a function of the slope of the velocity profile, $-dv/dr$. Thus, in accordance with Newton's law of viscosity, $\tau = -\mu \, dv/dr$, the wall shear stress, τ_w decreases with a decrease in the slope of the velocity profile, while the wall shear stress, τ_w remains a constant with a constant slope of the velocity profile. It may be noted that Newton's law of viscosity is applicable in the viscous sublayer in the turbulent flow case, as well as of course the laminar flow case. Figures 6.6 (laminar flow) and 6.7 (turbulent flow) illustrate that (beginning with a maximum slope of the velocity profile at the beginning of the boundary layer and thus the occurrence of the maximum wall shear stress, τ_w) in the developing flow region, the slope of the velocity profile decreases as the velocity profile deviates from a somewhat uniform velocity profile to: (1) a parabolic velocity profile in the case of laminar flow or (2) a velocity profile that varies according to the seventh root profile law in the case of turbulent flow. Furthermore, Figures 6.6 (laminar flow) and 6.7 (turbulent flow) illustrate that in the developed flow region, the slope of the velocity profile remains a constant, as the velocity profile no longer changes with distance. And, finally, for a given flow rate, Q, turbulent flow has a higher velocity gradient, $-dv/dr$ at the pipe wall than laminar flow; thus, a higher wall shear stress, τ_w and a high friction loss, h_f may be expected as the Reynolds number, R increases. As the flow becomes more turbulent, the internal mixing activities intensify, which indicates an increasing rate of viscous dissipation in the flow. Thus, for turbulent flow, the rate of energy loss varies with both the Reynolds number, R and the pipe roughness, ε/D.

6.5.3.4 Laminar versus Turbulent Flow

Because the pressure drop and the shear stress differ for laminar and turbulent flow, the maximum shear stress also differs for laminar and turbulent flow. It is important to note that the nature of the pipe flow is highly dependent on whether the flow is laminar or turbulent. The difference in nature of the flow (laminar vs. turbulent) is a direct result of the differences in nature of the shear stress in the two flow types. The shear stress in laminar flow is a result of a microscopic phenomenon in which there is a momentum transfer among the randomly moving molecules. And the shear stress in turbulent flow is a result of a macroscopic phenomenon in which there is a momentum transfer among the randomly moving fluid particles; the turbulent mixing during random fluctuations usually overshadows the effects of the molecular activity. As a result, the physical properties of the shear stress are different for the two types of flow. Specifically, the intense mixing of the fluid in turbulent flow increases the friction force on the boundary (and thus the wall shear stress, τ_w), which increases the friction factor and corresponding head loss, h_f, and thus will increase the amount of power that a pump must deliver to the fluid.

6.6 Derivation of the Major Head Loss Equation for Laminar Flow

In the case of laminar flows, first, application of the integral form of the energy equation yields an expression for the major head loss (Equation 6.51). Then, because the friction/ viscous forces in the integral form of the momentum equation and thus the wall shear stress, τ_w can be theoretically modeled (by Newton's law of viscosity, where $\tau = \mu \, dv/dy$) in the case of laminar flow, one can analytically solve for the unknown pressure drop, Δp from the momentum equation (Equation 6.52) and analytically derive an expression for the major head loss, h_f from the energy equation. Specifically, application of the integral form of the momentum equation, where the wall shear stress, τ_w is theoretically modeled by Newton's law of viscosity, yields a theoretical (parabolic) expression for the velocity distribution for laminar flow, expressed as a function of the pressure drop, Δp. Next, application of the continuity equation by integrating the resulting parabolic velocity distribution over the pipe diameter yields an analytical expression for the actual pressure drop, Δp (Poiseuille's law), which is substituted into the integral form of the energy equation (Equation 6.51), yielding an analytical expression for the head loss, h_f (Poiseuille's law expressed in terms of the head loss). The development of the equations for this section on laminar flow assumes the flow is steady and fully developed. Further simplifying assumptions regarding the flow are made below. It is interesting to note that in the case of laminar flow, the theoretical expression for the friction/viscous forces allows a theoretical derivation of the velocity profile, which directly yields an analytical expression for the major head loss, h_f (expressed as a function of the flow resistance, μ), without the need to define an empirical flow resistance coefficient. Furthermore, the derivation of the major head loss equation for laminar pipe flow is presented below, and application of the corresponding major head loss equation is presented in Chapter 8.

6.6.1 Application of the Integral Form of the Momentum Equation: Deriving the Velocity Profile for Laminar Flow

Once the integral form of the energy equation has been applied to yield an expression for the major head loss (Equation 6.51), the integral form of the momentum equation is then applied in order to solve for the unknown pressure drop, Δp. The theoretical expression for the friction/viscous forces (Newton's law of viscosity) in the integral momentum equation allows a theoretical derivation of the parabolic velocity profile for laminar flow, expressed as a function of the pressure drop, Δp. As such, application of Newton's second law of motion (integral form of the momentum equation) in Section 6.5.1 above yielded the following expression for the unknown pressure drop, Δp (Equation 6.52):

$$\tau = \frac{r}{2} \frac{\Delta p}{L} \tag{6.76}$$

which yields a relationship between the pressure drop, Δp and the wall shear stress, τ_w. In laminar flow, the friction/viscous forces in the integral form of the momentum equation can be theoretically modeled by Newton's law of viscosity and is given as follows:

$$\tau = \mu \frac{dv}{dy} \tag{6.77}$$

which provides a relationship between the shear stress, τ and the velocity, v and a flow resistance coefficient for laminar flow, where the fluid viscosity, μ represents the flow resistance for laminar flow. However, in order to apply Newton's law of viscosity for laminar pipe flow, one may note that:

$$\frac{dv}{dy} = -\frac{dv}{dr} \tag{6.78}$$

Thus, substituting the expression for the shear stress, τ as given by Newton's law of viscosity (Equation 6.77) into the momentum equation (Equation 6.76) yields the following:

$$\tau = \frac{r}{2}\frac{\Delta p}{L} = -\mu \frac{dv}{dr} \tag{6.79}$$

Next, in order to determine how the velocity varies across the pipe radius, r one may integrate over the radial distance, r from the center of the pipe and evaluate the constant of integration, C at the boundary condition, where the velocity of the fluid at the stationary solid boundary is zero; $v = 0$ at $r = r_o$, as the fluid at the boundary surface sticks to the boundary and has the velocity of the boundary (because of the "no-slip" condition) as follows:

$$\frac{r}{2}\frac{\Delta p}{L} = -\mu \frac{dv}{dr} \tag{6.80}$$

$$\frac{\Delta p}{2\mu L} r\, dr = -dv \tag{6.81}$$

$$\int dv = -\frac{\Delta p}{2\mu L} \int r\, dr \tag{6.82}$$

$$v = -\frac{\Delta p}{2\mu L}\left(\frac{r^2}{2}\right) + C \tag{6.83}$$

Evaluating the constant of integration, C yields:

$$C = \frac{\Delta p}{2\mu L}\left(\frac{r_o^2}{2}\right) \tag{6.84}$$

Thus, the velocity varies across the radial distance, r from the center of the pipe as follows:

$$v = -\frac{\Delta p}{2\mu L}\left(\frac{r^2}{2}\right) + \frac{\Delta p}{2\mu L}\left(\frac{r_o^2}{2}\right) \tag{6.85}$$

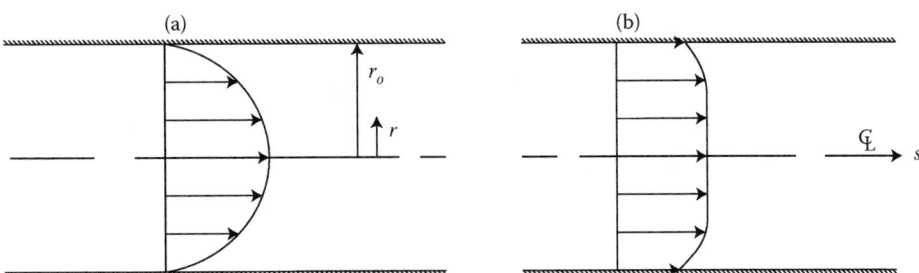

FIGURE 6.8
Velocity distribution for pipe flow. (a) Parabolic velocity distribution for laminar flow. (b) Empirical (seventh root profile law) velocity distribution for turbulent flow.

$$v = \frac{\Delta p}{2\mu L}\left(\frac{r_o^2 - r^2}{2}\right) = \frac{\Delta p}{4\mu L}(r_o^2 - r^2) \tag{6.86}$$

which indicates that the velocity distribution for laminar flow in a pipe is parabolic across the pipe section and is expressed as a function of the pressure drop, Δp, with the maximum velocity, v_{max} at the center of the pipe, where $r = 0$, and $v = 0$ at $r = r_o$, as illustrated in Figure 6.8a. Thus, the expression for v_{max} for laminar flow is:

$$v_{max} = \frac{\Delta p r_o^2}{4\mu L} \tag{6.87}$$

Alternately, the parabolic velocity distribution for laminar flow may be presented as a function of v_{max} as follows:

$$v = v_{max}\left[1 - \left(\frac{r}{r_o}\right)^2\right] \tag{6.88}$$

6.6.2 Application of the Differential form of the Continuity Equation: Deriving Poiseuille's Law

Next, the continuity equation is applied by integrating the resulting parabolic velocity distribution (Equation 6.86) over the pipe diameter in order to derive an analytical expression for the actual pressure drop, Δp (Poiseuille's law). As such, the expression for the mean velocity, v_{ave}, which is assumed in the application of the continuity equation (and the head loss equation), may be evaluated as follows:

$$Q = \int v\,d\Lambda = \int_{r=0}^{r=r_o} \frac{\Delta p}{4\mu L}(r_o^2 - r^2)(2\pi r\,dr) \tag{6.89}$$

Using Mathcad to symbolically integrate yields:

$$\int_0^{r_o} \frac{\Delta p}{4\mu L} \cdot (r_o^2 - r^2)2\pi \cdot r\,dr \rightarrow \frac{\pi \cdot r_o^4 \cdot \Delta p}{8 \cdot \mu L}$$

Thus:

$$Q = \frac{\pi r_0{}^4 \, \Delta p}{8\mu L} = \frac{\pi D^4 \, \Delta p}{128\mu L} \tag{6.90}$$

or expressed in terms of the a pressure drop, Δp:

$$\Delta p = \frac{128\mu QL}{\pi D^4} \tag{6.91}$$

which is known as Poiseuille's law, or the Hagan–Poiseuille law, for a pressure drop, Δp in a pipe due to viscosity, μ. One may note that (although independently experimentally discovered by Hagan and Poiseuille) Poiseuille's law represents the theoretical resistance law for laminar flow through circular pipes.

Poiseuille's law indicates the following properties of steady and uniform laminar flow in a straight and horizontal pipe: (1) the pressure drop, Δp is directly proportional to the absolute viscosity, μ, the discharge, Q, and the pipe length, L; and (2) the pressure drop, Δp is indirectly proportional to the pipe diameter to the fourth power D^4. The first property indicates the pressure drop, Δp is completely due to pipe friction (flow resistance), which causes the major head loss, h_f, where the corresponding loss in pressure energy (flow energy or flow work) is converted to kinetic energy plus energy lost due to heat (due to flow resistance). Furthermore, the pressure drop, Δp (and thus the shear stress at the wall, τ_w) is not at all a function of the fluid density, ρ (as in the case of turbulent flow, see Section 6.7), thus leaving the viscosity, μ as the only important fluid property in laminar flow. And, the second property indicates that a small change in the pipe diameter, D results in a large change in pressure drop, Δp.

It may be noted that Poiseuille's law assumes the average velocity, v_{ave}, where the average velocity, v_{ave} may be computed by dividing the discharge, Q by the pipe cross-sectional area, A as follows:

$$v_{ave} = \frac{Q}{A} = \frac{\pi r_0{}^4 \, \Delta p}{8\mu L} \frac{1}{\pi r_0{}^2} = \frac{r_0{}^2 \, \Delta p}{8\mu L} \tag{6.92}$$

where

$$v_{ave} = \frac{v_{max}}{2} = \frac{\Delta p r_0{}^2}{8\mu L} \tag{6.93}$$

Thus, alternatively, the parabolic velocity distribution for laminar flow may be presented as a function of v_{ave} as follows:

$$v = 2v_{ave}\left[1 - \left(\frac{r}{r_0}\right)^2\right] \tag{6.94}$$

6.6.3 Substituting Poiseuille's Law into the Integral Form of the Energy Equation: Deriving the Major Head Loss Equation for Laminar Flow (Poiseuille's Law)

And, finally, the analytical expression for the actual pressure drop, Δp (Poiseuille's law) (Equation 6.91) is substituted into the integral form of the energy equation (Equation 6.51), yielding an analytical expression for the head loss, h_f (Poiseuille's law expressed in terms of the head loss) (assuming the average velocity, v_{ave}), without the need to define an empirical flow resistance coefficient as follows:

$$h_f = S_f L = \frac{\tau_w L}{\gamma R_h} = \frac{\Delta p}{\gamma} = \frac{128\mu QL}{\pi D^4 \gamma} = \frac{128\mu v_{ave}\pi D^2 L}{\pi D^4 \rho g 4} = \underbrace{\left[\frac{32\mu v_{ave} L}{D^2}\right]}_{\Delta p}\frac{1}{\gamma} = \underbrace{\left[\frac{32\mu v_{ave}}{D^2 \gamma}\right]}_{S_f} L \quad (6.95)$$

where the major head loss equation for laminar is an alternate expression of Poiseuille's law, which represents the theoretical resistance law for laminar flow through circular pipes. Application of the major head loss equation for laminar flow is presented in Chapter 8.

6.6.4 Interpretation of Poiseuille's Law Expressed in Terms of the Pressure Drop

Examination of Poiseuille's law expressed in terms of the pressure drop, Δp, indicates a fundamental difference between laminar and turbulent flow as follows:

$$\Delta p = \frac{\tau_w L}{R_h} = \frac{32\mu v L}{D^2} \quad (6.96)$$

Poiseuille's law indicates that the pressure drop, Δp (and thus τ_w) in laminar flow is mainly a function of the fluid viscosity, μ and not at all a function of the fluid density, ρ (as in the case of turbulent flow, see Section 6.7). Furthermore, for laminar flow the pressure drop, Δp is independent of the relative pipe roughness, ε/D. However, for turbulent flow (see Section 6.7), the pressure drop, Δp is a function of the friction factor, f, which is a function of the Reynolds number, R and the relative pipe roughness, ε/D. Because laminar flow is a special case of turbulent flow, the friction factor, f evaluated for laminar flow (see Section 6.7) is only a function of the Reynolds number, R. Laminar flow occurring over either smooth or rough boundaries is characterized by the same properties: the velocity is zero at the boundary surface, and Newton's law of viscosity models the variation of shear stress with velocity. Therefore, in laminar flow, the boundary roughness, ε/D has no effect on the flow (as long as the absolute roughness, ε is small compared to the pipe size diameter, D). A comparison of the velocity profiles for the laminar flow and the turbulent flow given in Figure 6.8 illustrates the velocity of the fluid particles near the pipe wall is very small for laminar flow compared to turbulent flow. Therefore, for turbulent flow, the physical interaction of the fluid particles with the pipe wall roughness plays an important role in the definition of the empirical flow resistance coefficient. And, finally, in laminar flow, the pressure drop, Δp (and thus the head loss, h_f) is linear in velocity and nonlinear in pipe diameter as $1/D^2$ (whereas, in turbulent flow, the pressure drop, Δp [and thus the head loss, h_f] is nonlinear in velocity as v^2; see Section 6.7).

Because the pressure drop, Δp differs for laminar and turbulent flow, the shear stress, τ and thus the maximum shear stress, τ_{max} also differ for laminar and turbulent flow. As such, the wall shear stress, τ_w (i.e., τ_{max}) for laminar flow may be expressed by substituting Equation 6.96 for Δp into Equation 6.69 for τ_{max} as follows:

$$\tau_{max} = \tau_w = \frac{D}{4}\frac{\Delta p}{L} = \frac{D}{4}\frac{32\mu v L}{D^2}\frac{1}{L} = \frac{8\mu v}{D} \quad (6.97)$$

where one may compute the maximum shear stress, τ_{max} for laminar flow in a circular pipe for a given value of viscosity, μ (see Section 6.7 on turbulent flow for a comparison).

6.7 Derivation of the Major Head Loss Equation for Turbulent Flow

In the case of turbulent flows, first, application of the integral form of the energy equation yields an expression for the major head loss (Equation 6.51). Then, however, because the friction/viscous forces in the integral form of the momentum equation and thus the wall shear stress, τ_w cannot be theoretically modeled in the case of turbulent flow, one cannot analytically solve for the unknown pressure drop, Δp from the integral form of the momentum equation (Equation 6.52). Therefore, the unknown pressure drop, Δp (or the unknown friction slope, S_f in the case of turbulent open channel flow) and thus the major head loss, h_f due to pipe or channel friction are determined empirically. Specifically, an empirical interpretation of the wall shear stress, τ_w in the theoretical expression for the friction slope (Equation 6.54) in the differential form of the momentum equation is sought in terms of velocity, v and a flow resistance coefficient. This is accomplished by using dimensional analysis to derive an empirical expression for the wall shear stress, τ_w as a function of the velocity, v which involves the definition of a drag coefficient, C_D that represents the flow resistance (Chezy coefficient, C; Darcy–Weisbach friction factor, f; and Manning's roughness coefficient, n) and is empirically evaluated; this is analogous to Newton's law of viscosity, where the shear stress is expressed as a function of velocity, v and flow resistance, μ. One may note that the theoretical relationship between the pressure drop, Δp and the shear stress, τ provided by Newton's second law of motion (integral form of the momentum equation) (Equation 6.53) is used in the dimensional analysis procedure, in which the friction forces are modeled empirically. Dimensional analysis yields an empirical expression for the friction slope, $S_f = v^2/C^2 R_h$ (the Chezy equation) in the differential form of the momentum equation. The Chezy equation represents the differential form of the momentum equation, supplemented by dimensional analysis and guided by the integral momentum equation, a link between the differential and integral momentum equations. The Chezy equation is used to obtain an empirical evaluation for the pressure drop, Δp in Newton's second law of motion (integral form of the momentum equation). And, finally, the Chezy equation is substituted into the integral form of the energy equation (Equation 6.51), yielding an empirical expression for the head loss, h_f (the Chezy head loss equation). The development of the equations for this section on turbulent flow assumes the flow is steady and fully developed. Further simplifying assumptions regarding the flow are made below. It is interesting to note that in the case of turbulent flow, the lack of a theoretical expression for the friction/viscous forces leads to the inability to derive a theoretical expression for velocity distribution for turbulent flow. Instead, empirical studies have shown that the velocity varies according to the seventh root profile law for turbulent flow. As a result, the empirical expression for the friction forces yields an empirical expression for the head loss, with the need to define an empirical flow resistance.

The drag coefficient, C_D representing the flow resistance that is derived by dimensional analysis, is a nonstandard form of the drag coefficient. More standard forms of the drag coefficient include empirical flow resistance coefficients including the Chezy coefficient, C; the Darcy–Weisbach friction factor, f; and the Manning's roughness coefficient, n. In addition to these three standard empirical flow resistance coefficients, there is another commonly used one, which is the Hazen–Williams roughness coefficient, C_h.

The derivation of the major head loss equation for turbulent flow along with the coefficient of drag, C_D (Chezy coefficient, C; Darcy–Weisbach friction factor, f; Manning's roughness coefficient, n; and the Hazen–Williams roughness coefficient, C_h) for pipe and open channel flow is presented below. Empirical evaluation of the Chezy coefficient, C; the

Darcy–Weisbach friction factor, f; the Manning's roughness coefficient, n; and the Hazen–Williams roughness coefficient, C_h is also presented below. A comparison between laminar flow and turbulent flow is conducted (see Section 6.7.9). Additionally, determination of the velocity profile for turbulent flow is presented below (see Section 6.7.10). And, finally, the application of the major head loss equation for open channel flow is addressed in detail (see Section 6.7.11). Application of the corresponding major head loss equation is presented in Chapters 8 and 9.

6.7.1 Application of the Integral Momentum Equation: Deriving an Expression for the Pressure Drop

Once the integral form of the energy equation has been applied to yield an expression for the major head loss (Equation 6.51), the integral form of the momentum equation is then applied in order to solve for the unknown pressure drop, Δp. Thus, application of Newton's second law of motion (integral form of the momentum equation) (Equation 6.47) yields a linear relationship between the shear stress, τ and the radial distance, r, which also describes the theoretical relationship between the pressure drop, Δp (or S_f) and the shear stress, τ; this relationship plays a key role in the derivation of the major head loss equations for (laminar and) turbulent flow. Specifically, the integral form of the momentum equation yields the following relationship between the pressure drop, Δp and the wall shear stress, τ_w (derived in Section 6.5):

$$\tau = \frac{r}{2}\frac{\Delta p}{L} \qquad (6.98)$$

Evaluated for the wall shear stress, τ_w at $r = r_o = D/2$, where $R_h = D/4$, and expressed as a function of the pressure drop, Δp and yields:

$$\Delta p = \frac{\tau_w L}{R_h} \qquad (6.99)$$

However, the problem is that (unlike laminar flow) one cannot theoretically model the wall shear stress, τ_w in the integral momentum for turbulent flow for pipe and open channel flow and thus cannot analytically determine pressure drop, Δp in Equation 6.99. Therefore, the unknown pressure drop, Δp (or the unknown friction slope, S_f in the case of turbulent open channel flow) and thus the major head loss, h_f due to pipe or channel friction are determined empirically (see sections below).

6.7.2 Application of the Differential Momentum Equation: Interpreting the Friction Slope in Turbulent Flow

Next, the differential momentum equation is applied in order to interpret the friction slope in turbulent flow. As such, the differential momentum equation given in Equation 6.49 models the friction slope, S_f due to flow resistance, and is repeated as follows:

$$\underbrace{\frac{\partial}{\partial s}\left(\frac{p}{\gamma} + z + \frac{v^2}{2g}\right)}_{S_e} = -\underbrace{\frac{\tau_w}{\gamma R_h}}_{S_f} - \underbrace{\frac{1}{g}\frac{\partial v}{\partial t}}_{S_a} \qquad (6.100)$$

$$S_e = -(S_f + S_a) \qquad (6.101)$$

where S_e and S_a are expressed in terms of easy-to-measure velocity, v, and S_f is expressed in terms of difficult-to-measure τ_w as follows:

$$S_f = \frac{\tau_w}{\gamma R_h} \tag{6.102}$$

However, because all three terms, the pressure drop, Δp (Equation 6.99); the friction slope, S_f (Equation 6.102); and the major head loss, h_f (Equation 6.51) are expressed in terms of the wall shear stress, τ_w, in order to evaluate the respective expressions, one must first obtain an expression for the wall shear stress, τ_w. One may note that the velocity distribution in a pipe/channel is linked to the shear stress distribution; specifically, the velocity distribution combined with the fluid viscosity, μ produces the shear stress. Therefore, because the wall shear stress, τ_w, which causes the flow resistance, is difficult to measure, instead, one may model it by measuring the velocity, v and defining a flow resistance coefficient, which relates τ_w to v. Thus, in order to interpret the expression (Equation 6.102) in terms of velocity, v and a flow resistance coefficient, information about the relationship between the magnitude of the shear stress, τ and the velocity, v and a flow resistance coefficient is needed. Because there is no similar theory (to Newton's law of viscosity) for the more complex turbulent flow (and the momentum theory has been exhausted), one may resort to dimensional analysis in order to derive a relationship between the shear stress, τ and the velocity, v and a flow resistance coefficient. Thus, dimensional analysis (see section below) is used to empirically interpret S_f (express τ_w) in terms of easy-to-measure velocity, v. Specifically, an empirical interpretation of the wall shear stress, τ_w in the theoretical expression for the friction slope (Equation 6.102) in the differential form of the momentum equation is sought in terms of velocity, v and a flow resistance coefficient.

6.7.3 Application of Dimensional Analysis: Empirically Interpreting Friction Slope (Empirically Deriving an Expression for Wall Shear Stress) in Turbulent Flow

Dimensional analysis is applied to empirically interpret S_f (express τ_w) in terms of easy-to-measure velocity, v. Specifically, an empirical interpretation of the wall shear stress, τ_w in the theoretical expression for the friction slope (Equation 6.102) in the differential form of the momentum equation is sought in terms of velocity, v and a flow resistance coefficient. This is accomplished by using dimensional analysis in order to derive an empirical expression for the wall shear stress, τ_w as a function of the velocity, v (specifically: $\tau_w = C_D \rho v^2$), which involves the definition of a drag coefficient, C_D that represents the flow resistance (Chezy coefficient, C; Darcy–Weisbach friction factor, f; and Manning's roughness coefficient, n) and is empirically evaluated; this is analogous to Newton's law of viscosity, where the shear stress is expressed as a function of velocity, v and flow resistance, μ. One may note that the theoretical relationship between the pressure drop, Δp and the shear stress, τ provided by Newton's second law of motion (integral form of the momentum equation) (Equations 6.99 and 6.53) is used in the dimensional analysis procedure, in which the friction forces are modeled empirically. Finally, dimensional analysis yields an empirical expression for the friction slope, $S_f = v^2/C^2 R_h$ (the Chezy equation) in the differential form of the momentum equation. The Chezy equation represents the differential form of the momentum equation, supplemented by dimensional analysis and guided by the integral momentum equation, a link between the differential and integral momentum equations. The Chezy equation is used to obtain an empirical evaluation for the pressure drop, Δp in Newton's second law of motion (integral form of the momentum equation).

6.7.3.1 Derivation of an Empirical Expression for the Wall Shear Stress as a Function of the Velocity

Dimensional analysis, guided by the theoretical relationship between the pressure drop, Δp and the shear stress, τ provided by Newton's second law of motion (integral form of the momentum equation) (Equations 6.99 and 6.53), yields an empirical expression for the wall shear stress, $\tau_w = C_D\, \rho v^2$, expressed as a function of the average velocity, v_{ave} and a drag coefficient, C_D for turbulent flow. One may note that this is an expression for only the wall shear stress, τ_w and thus is not a general relationship between shear stress, τ and the velocity, v. It is important to note that more standard forms of the drag coefficient, C_D for the major head loss due to pipe friction include the empirical flow resistance coefficients including the Chezy coefficient, C; the Darcy–Weisbach friction coefficient, f; and the Manning's roughness coefficient, n, and are derived in the sections below. Thus, the momentum theory is supplemented with dimensional analysis in Chapter 7 in order to derive an expression for the wall shear stress for turbulent pipe and open channel flow, for which the result is presented as follows:

$$\frac{\tau_w}{\rho v^2} = \frac{F_f}{F_I} = \phi\left(R, \frac{\varepsilon}{D}\right) \tag{6.103}$$

where F_f is the wall friction force and F_I is the inertia force. Also, R = the Reynolds number and ε/D is the relative pipe roughness. Note that the theoretical relationship between the pressure drop, Δp and the shear stress, τ, evaluated for the wall shear stress, τ_w at $r = r_o = D/2$, where $R_h = D/4$ (derived from Newton's second law of motion for both laminar and turbulent flow), $\Delta p = \tau_w L/R_h$ for the wall shear stress, τ_w yields an expression for the wall shear stress, τ_w expressed as a function of the pressure drop, Δp as follows:

$$\tau_w = \frac{\Delta p R_h}{L} \tag{6.104}$$

One may note that, typically, the pressure drop, Δp is equal to the dynamic pressure, $\rho v^2/2$ plus an additional pressure drop, $\tau_w L/R_h$, due to the flow resistance. But since $v_1 = v_2$ (assumption of uniform flow made in the application of Newton's second law of motion), the pressure drop, Δp is equal to the pressure drop, $\tau_w L/R_h$, due to the flow resistance, which causes the major head loss, h_f. Thus, the pressure drop, Δp in pipe flow is mostly due to the wall shear stress, τ_w, unlike in the computation of the drag force, F_D in the external flow around an object, for which the pressure drop, Δp is due to both the dynamic pressure, $\rho v^2/2$ and the wall shear stress, τ_w. Thus, the drag coefficient, C_D is defined as follows:

$$\frac{\tau_w}{\rho v^2} = \frac{F_f}{F_I} = \frac{F_P}{F_I} = \frac{\Delta p R_h}{L}\frac{1}{\rho v^2} = \frac{R_h}{L}\frac{\Delta p}{\rho v^2} = \frac{R_h}{L}\frac{1}{E} = C_D = \phi\left(R, \frac{\varepsilon}{D}\right) \tag{6.105}$$

which is a nonstandard form of the drag coefficient, C_D, and where E = the Euler number. One may note that the drag coefficient, C_D is dimensionless and is a function of boundary roughness, ε/D; the Reynolds number, R; and the cross-sectional shape or the hydraulic radius, R_h. More standard forms of the drag coefficient, C_D include empirical flow resistance coefficients including the Chezy coefficient, C; the Darcy–Weisbach friction coefficient, f; and the Manning's roughness coefficient, n and are derived below. Furthermore, the resulting empirical relationship between the wall shear stress, τ_w and the velocity, v is related by a flow resistance coefficient from dimensional analysis is given as follows:

$$\tau_w = C_D \rho v^2 \tag{6.106}$$

6.7.3.2 Derivation of the Chezy Equation and Evaluation of the Chezy Coefficient, C

The Chezy equation is derived, and the Chezy coefficient, C is evaluated as follows. Substituting the empirically derived expression for the wall shear stress (Equation 6.106) into the theoretical expression for the friction slope (Equation 6.102) in the differential form of the momentum equation yields an empirical expression for the friction slope as follows:

$$S_f = \frac{\tau_w}{\gamma R_h} = \frac{C_D \rho v^2}{\gamma R_h} = \frac{C_D v^2}{g R_h} \tag{6.107}$$

The Chezy coefficient, C, which is a more standard form of the drag coefficient, C_D, is defined as follows:

$$C = \sqrt{\frac{g}{C_D}} \tag{6.108}$$

Thus, we finally get an expression for the friction slope, S_f expressed as a function of velocity and a flow resistance coefficient (a result of the application of both the momentum theory and dimensional analysis) as follows:

$$S_f = \frac{v^2}{C^2 R_h} \tag{6.109}$$

which is known as the Chezy equation. The friction slope, S_f is now expressed as a function of velocity, v, and the Chezy flow resistance coefficient, C, which models the flow resistance. Therefore, because the wall shear stress, τ_w, which causes the flow resistance, is difficult to measure, instead, now one may model it by measuring the velocity, v and an empirically evaluated drag coefficient, C_D (or a more standard form of the drag coefficient, such as the Chezy coefficient, C; Darcy–Weisbach friction factor, f; or the Manning's roughness coefficient, n). The Chezy coefficient, C has dimensions of $L^{1/2} T^{-1}$ and is a function of the boundary roughness, ε/D; the Reynolds number, R; and the cross-sectional shape or the hydraulic radius, R_h. Empirical findings illustrate that the Chezy coefficient, C ranges between $30 \, m^{1/2} \, s^{-1}$ (or $60 \, ft^{1/2} \, sec^{-1}$) for small channels with rough boundary surfaces to $90 \, m^{1/2} \, s^{-1}$ (or $160 \, ft^{1/2} \, sec^{-1}$) for large channels with smooth boundary surfaces. One may note that although the derivation/formulation of the Chezy equation assumes no specific units (SI or BG), the Chezy coefficient, C has dimensions of $L^{1/2} T^{-1}$. Furthermore, because the values for the Chezy coefficient, C have been provided/calibrated in both units (see above), one may apply the Chezy equation assuming either SI or BG units. Although the Chezy equation is used for both open channel flow and pipe flow, it has a drawback in that the Chezy coefficient, C is not dimensionless like the Darcy–Weisbach friction factor, f (defined below). Therefore, one must attach the appropriate metric or English units to the Chezy coefficient, C prior to using it in the Chezy equation.

6.7.3.3 Application of the Chezy Equation

One may recall that because all three terms, the pressure drop, Δp (Equation 6.99); the friction slope, S_f (Equation 6.102); and the major head loss, h_f (Equation 6.51), are expressed in terms of the wall shear stress, τ_w, in order to evaluate the respective expressions, an expression for the wall shear stress, τ_w was sought. Specifically, because the wall shear stress, τ_w, which causes flow resistance, is difficult to measure, instead, one may model it by measuring the velocity, v and defining a flow resistance coefficient, which relates τ_w to v

(accomplished by using dimensional analysis). As such, application of the differential form of the momentum equation, supplemented by dimensional analysis and guided by the integral momentum equation, resulted in derivation of the Chezy equation, which represents an empirical evaluation of the friction slope, S_f, and is repeated as follows:

$$S_f = \frac{\tau_w}{\gamma R_h} = \frac{v^2}{C^2 R_h} \tag{6.110}$$

which may be solved for one of the following unknown quantities: S_f, v, C, or $D = 4\,R_h$, where the Chezy equation for pipe flow assumes a straight pipe section of length, L, with a constant diameter, D and thus a constant velocity of flow, v, with a specific Chezy coefficient, C and a friction slope, S_f. Furthermore, the Chezy equation may be used to obtain an empirical evaluation for the pressure drop, Δp in Newton's second law of motion (integral form of the momentum equation) (Equation 6.99), as follows:

$$\Delta p = \frac{\tau_w L}{R_h} = \gamma \underbrace{\left[\frac{v^2}{C^2 R_h} \right]}_{S_f} L = \gamma S_f L \tag{6.111}$$

Additionally, an empirical evaluation of the major head loss, h_f (Equation 6.51) using the Chezy equation to define the friction slope, S_f is presented in the following section below. And, finally, the Chezy equation yields an empirical expression for the wall shear stress, τ_w is expressed as a function of velocity, v as follows:

$$\tau_w = \gamma R_h S_f = \gamma R_h \underbrace{\left[\frac{v^2}{C^2 R_h} \right]}_{S_f} = \frac{\gamma v^2}{C^2} = \frac{g}{C^2} \rho v^2 \tag{6.112}$$

6.7.4 Substituting the Chezy Equation into the Integral Form of the Energy Equation: Deriving the Major Head Loss Equation for Turbulent Flow

And, finally, assuming that the Chezy coefficient, C is chosen to model the empirical flow resistance, the Chezy equation for the friction slope (Equation 6.109) is substituted into the integral form of the energy equation (Equation 6.51), yielding an empirical expression for the head loss, h_f (the Chezy head loss equation) as follows:

$$h_f = S_f L = \frac{\tau_w L}{\gamma R_h} = \frac{\Delta p}{\gamma} = \underbrace{\left[\frac{v^2}{C^2 R_h} \right]}_{S_f} L \tag{6.113}$$

where the expression for the head loss, h_f is an expression of the mechanical energy per unit weight, expressed in terms of the kinetic energy. One may refer to this equation as the Chezy head loss equation. Furthermore, although values for the Chezy coefficient, C have been empirically evaluated and tabulated (limited tables available) for various combinations of boundary roughness, ε/D; Reynolds number, R; and cross-sectional shape (or the hydraulic radius, R_h), it has been found that it is easier to model the effects of the Chezy coefficient, C (i.e., the drag coefficient, C_D) by using the Darcy–Weisbach friction coefficient, f or the Manning's roughness coefficient, n. As such, the Darcy–Weisbach friction coefficient, f; the Manning's roughness coefficient, n; and the Hazen–Williams roughness coefficient,

C_h, along with the corresponding head loss equations, are derived and discussed in the following three sections.

6.7.5 The Darcy–Weisbach Equation

The Darcy–Weisbach friction coefficient (factor), f, which represents a dimensionless flow resistance coefficient, is derived using dimensional analysis, and expressed as a function of the Chezy coefficient, C as: $C = \sqrt{8g/f}$. The empirical evaluation of the friction factor, f is presented in both graphical (Moody diagram) and mathematical forms. And, finally, the Chezy equation, expressed as a function of the friction factor, f, is substituted into the integral form of the energy equation (Equation 6.51), yielding an empirical expression for the head loss, $h_f = f(L/D)(v^2/2g)$ (the Darcy–Weisbach head loss equation).

6.7.5.1 Derivation of the Darcy–Weisbach Friction Coefficient, f

From the definition of pi terms (see Chapter 7) one may introduce a new pi term, the Darcy friction factor, f, which, like the Chezy coefficient, C, is a more standard form of the drag coefficient, C_D; however, unlike the Chezy coefficient, C, a is dimensionless. Thus, defining a new pi term, the Darcy friction factor, f expressed as a function of the drag coefficient, C_D (a pi term) is given as follows:

$$\frac{F_f}{F_I} = f = 8C_D \tag{6.114}$$

The Darcy–Weisbach friction coefficient, f is typically defined for pipe flow; thus, if one assumes a circular pipe flow, then:

$$R_h = \frac{A}{P} = \frac{\pi D^2}{4}\frac{1}{\pi D} = \frac{D}{4} \tag{6.115}$$

Thus, the drag coefficient, C_D is defined as follows:

$$\frac{\tau_w}{\rho v^2} = \frac{R_h}{L}\frac{\Delta p}{\rho v^2} = \frac{D}{4L}\frac{\Delta p}{\rho v^2} = \frac{D}{4L}\frac{1}{E} = C_D = \phi\left(R, \frac{\varepsilon}{D}\right) \tag{6.116}$$

and the Darcy–Weisbach friction coefficient, f is defined as follows:

$$\frac{F_f}{F_I} = f = 8C_D = 8\left[\frac{\tau_w}{\rho v^2}\right] = 8\left[\frac{D}{4L}\frac{\Delta p}{\rho v^2}\right] = \frac{D}{L}\frac{\Delta p}{\rho v^2/2} = \frac{D}{L}\frac{2}{E} = \phi\left(R, \frac{\varepsilon}{D}\right) \tag{6.117}$$

Thus, examining the Darcy–Weisbach friction coefficient, f yields the following:

$$\frac{F_f}{F_I} = f = 8C_D = \frac{D}{L}\frac{\Delta p}{\rho v^2/2} = \frac{Actual\ Pres\ Drop}{Ideal\ Pres\ Drop} = \frac{staticpres}{dynpres} \tag{6.118}$$

where the results of dimensional analysis introduce a drag coefficient, C_D that expresses the ratio of a real flow variable (in this case: force or pressure) to its corresponding ideal flow variable. Thus, using the Darcy–Weisbach friction coefficient, f, which is a more standard definition of the drag coefficient, C_D, the resulting relationship between the wall shear stress,

τ_w and the velocity, v from dimensional analysis is given as follows:

$$\tau_w = \frac{f}{8}\rho v^2 \qquad (6.119)$$

where the Darcy–Weisbach friction coefficient, f has no dimensions and is a function of the boundary roughness, ε/D; the Reynolds number, R; and the cross-sectional shape or the hydraulic radius, R_h. One may note that ε is the absolute pipe roughness, while ε/D is the relative pipe roughness. Furthermore, Table 8.3 presents typical values for the absolute pipe roughness, ε for various pipe material, assuming new and clean pipes. It is important to note that over the lifetime of a pipe design, the absolute pipe roughness, ε values will change (increase) due to the buildup of deposits on the pipe wall and thus cause an increase in the frictional head loss. Thus, in order to minimize potential problems associated with head loss, the design of a pipe system should consider the temporal change of the absolute pipe roughness, ε for a given pipe material.

Based on the definition of the newly defined pi term, f (the Darcy–Weisbach friction factor) and the definition of the Chezy coefficient, C one may define the relationship between the Chezy coefficient, C and the Darcy–Weisbach friction factor, f as follows:

Noting that:

$$C = \sqrt{\frac{g}{C_D}} \qquad (6.120)$$

$$f = 8C_D \qquad (6.121)$$

yields:

$$C = \sqrt{\frac{8g}{f}} \qquad (6.122)$$

Furthermore, while the behavior of the Darcy–Weisbach friction factor, f in circular pipe flow has been thoroughly explored, a similarly complete investigation of the behavior of the Chezy coefficient, C has never been made. This is because in addition to the extra variables involved in the open channel flow case, there is an extremely wide range of surface roughness sizes and types found in practice, and it is difficult to attain steady, uniform, fully developed flow outside of the controlled laboratory environment. Additionally, empirical studies have illustrated that the behavior of the Chezy coefficient, C and the Darcy–Weisbach friction factor, f are least affected by the cross-sectional shape (or the effects are at least within the limits of accuracy that is normally accepted in practice).

6.7.5.2 Evaluation of the Darcy–Weisbach Friction Coefficient, f

As presented above, the results of dimensional analysis, assuming turbulent flow, (see Chapter 7) indicated that the Darcy–Weisbach friction coefficient, f has no dimensions and is a function of the boundary roughness, ε/D; the Reynolds number, R; and the cross-sectional shape or the hydraulic radius, R_h. However, at very high velocities (and thus very high values for R), the f for completely turbulent flow (rough pipes) is a function only of the relative pipe roughness, ε/D and thus independent of R. Furthermore, because laminar flow is a special case of turbulent flow, a theoretical evaluation of the Darcy–Weisbach friction coefficient, f for laminar "circular" pipe flow (given in a section below) illustrates that f for laminar flow is only a function of the Reynolds number, R (where: $f = 64/R$).

Unlike laminar flow, the friction factor, f for turbulent flow cannot be evaluated theoretically. Instead, however, it is evaluated using experiments. Specifically, J. Nikuradse (Moody 1944) computed the friction factor, f from measurements of the flow rate, Q and the pressure drop, Δp. Using pipes with simulated absolute pipe roughness, ε, the pressure drop, Δp needed to produce a specified flow rate, Q was measured. The data were then converted into the friction factor, f for the corresponding Reynolds number, R and the relative pipe roughness, ε/D. Numerous tests were conducted for a wide range of R and ε/D in order to determine an empirical relationship between the friction factor, f and both the Reynolds number, R and the relative pipe roughness, ε/D. Further experimental studies conducted by L. F. Moody resulted in the Moody diagram illustrated in Figure 8.1 presented in Chapter 8, which presents the Darcy–Weisbach friction coefficient, f as a function of the Reynolds number, R and the relative pipe roughness, ε/D (in both graphical and mathematical forms).

6.7.5.3 The Darcy–Weisbach Head Loss Equation

Assuming that the Darcy–Weisbach friction factor, f is chosen to model the empirical flow resistance, the Chezy equation, expressed as a function of the friction factor, f, is substituted into the integral form of the energy equation (Equation 6.51), yielding an empirical expression for the head loss, $h_f = f(L/D)(v^2/2g)$ (the Darcy–Weisbach head loss equation). Choosing the Darcy–Weisbach friction factor, f to model the empirical flow resistance is a typical choice for a pipe flow (where $R_h = D/4$). As such, one would substitute the relationship between the Chezy coefficient, C and the Darcy–Weisbach friction factor, f (Equation 6.122) into the Chezy head loss equation (Equation 6.113), which yields the Darcy–Weisbach head loss equation as follows:

$$C = \sqrt{\frac{8g}{f}} \tag{6.123}$$

$$h_f = S_f L = \frac{\tau_w L}{\gamma R_h} = \frac{\Delta p}{\gamma} = \frac{v^2 L}{C^2 R_h} = \frac{v^2 L}{\dfrac{8g}{f}\dfrac{D}{4}} = f\frac{L}{D}\frac{v^2}{2g} \tag{6.124}$$

which is expressed in terms of the velocity (kinetic) head, $v^2/2g$. One may note that the derivation/formulation of the Darcy–Weisbach equation assumes no specific units (SI or BG), and furthermore, the Darcy–Weisbach friction factor, f is dimensionless. Therefore, one may apply the Darcy–Weisbach head loss equation assuming either SI or BG units.

6.7.6 Manning's Equation

The Manning's roughness coefficient, n is empirically derived and expressed as a function of the Chezy coefficient, C as: $C = R_h^{1/6}/n$. The empirical evaluation of the Manning's roughness coefficient, n is presented in tabular form. Substituting the expression for the Chezy coefficient, C as a function of the Manning's roughness coefficient, n into the Chezy equation yields the Manning's equation. And, finally, the Chezy equation, expressed as a function of the Manning's roughness coefficient, n is substituted into the integral form of the energy equation (Equation 6.51), yielding an empirical expression for the head loss (the Manning's head loss equation).

6.7.6.1 Derivation and Evaluation of the Manning's Roughness Coefficient, n

Empirical studies have indicated that a more accurate dependence of the Chezy coefficient, C on the hydraulic radius, R_h is defined by the definition of the Manning's roughness coefficient, n. One may recall that the dimensions of the Chezy coefficient, C are $L^{1/2}\,T^{-1}$, and it is empirically evaluated as a function of the boundary roughness, ε/D; the Reynolds number, R; and the cross-sectional shape or the hydraulic radius, R_h. However, further empirical studies (assuming the metric system) by Manning indicated that a more accurate dependence of the Chezy coefficient, C on the hydraulic radius, R_h is given as follows:

$$C = \frac{R_h^{1/6}}{n} \tag{6.125}$$

where n is the Manning's roughness coefficient with dimensions of $L^{-1/3}\,T$ and is empirically evaluated only as a function of the boundary roughness, ε/D. One may note that the dependency of the Manning's roughness coefficient, n on the hydraulic radius, R_h has been removed by its relationship with the Chezy coefficient, C. Furthermore, one may note that the dependency of the Manning's roughness coefficient, n on the Reynolds number, R has been removed by assuming turbulent flow. One may note that the most common occurrence of flow resistance in open channel and pipe flow is in long channels or long pipe sections. In the open channel flow case, the boundary fills the whole channel section, while in the pipe flow case, the boundary fills the whole pipe section; in such cases, the Reynolds number, R is usually so large that it has virtually no influence on the resistance coefficient, n. Empirical values for the Manning roughness coefficient, n as a function of the boundary roughness, ε/D are typically given assuming metric units for n of $m^{-1/3}$ s (see Table 8.6).

6.7.6.2 Manning's Equation

Substituting the expression for the Chezy coefficient, C as a function of the Manning's roughness coefficient, n (Equation 6.125) into the Chezy equation (Equation 6.109) yields the Manning's equation as follows:

$$v = \frac{R_h^{1/6}}{n}\sqrt{R_h S_f} = \frac{1}{n} R_h^{2/3} S_f^{1/2} \tag{6.126}$$

One may note that although the derivation/formulation of the Manning's equation assumes no specific units (SI or BG), the Manning's roughness coefficient, n has dimensions of $L^{-1/3}\,T$. Furthermore, because the values for the Manning's roughness coefficient, n have been provided/calibrated in the SI units $m^{-1/3}$ s (see Table 8.6), the Manning's equation as given above assumes the metric (SI) system. The corresponding Manning's equation for the English (BG) system is given as follows:

$$v = \frac{1.486}{n} R_h^{2/3} S_f^{1/2} \tag{6.127}$$

which also uses the Manning roughness coefficient, n with units $m^{-1/3}$ s (see Table 8.6). One may note the value of 1.486 simply converts the Manning roughness coefficient, n with units $m^{-1/3}$ s (from Table 8.6) to units of $ft^{-1/3}$ sec, because $3.281\,ft = 1\,m$ and thus:

$$\frac{\left[\dfrac{3.281ft}{1m}\right]^{1/3}}{n\,[m^{-1/3}s]} = \frac{1.486}{n\,[ft^{-1/3}\,sec]} \tag{6.128}$$

As illustrated in Table 8.6, the rougher the boundary (wetted perimeter), the larger the value for Manning roughness coefficient, n.

6.7.6.3 Manning's Head Loss Equation

Assuming that the Manning roughness coefficient, n is chosen to model the empirical flow resistance, the Chezy equation, expressed as a function of the Manning's roughness coefficient, n, is substituted into the integral form of the energy equation (Equation 6.51), yielding an empirical expression for the head loss (the Manning's head loss equation). Substituting the relationship between the Chezy coefficient, C and the Manning's roughness coefficient, n (Equation 6.125) into the Chezy head loss equation (Equation 6.113), yields the Manning's head loss equation (assuming the metric system) as follows:

$$C = \frac{R_h^{1/6}}{n} \tag{6.129}$$

$$h_f = S_f L = \frac{\tau_w L}{\gamma R_h} = \frac{\Delta p}{\gamma} = \frac{v^2 L}{C^2 R_h} = \frac{v^2}{\left(\frac{R_h^{1/6}}{n}\right)^2 R_h} \frac{L}{R_h} = \left(\frac{vn}{R_h^{2/3}}\right)^2 L \tag{6.130}$$

The corresponding Manning's head loss equation for the English (BG) system is given as follows:

$$h_f = S_f L = \frac{\tau_w L}{\gamma R_h} = \frac{\Delta p}{\gamma} = \frac{v^2 L}{C^2 R_h} = \frac{v^2}{\left(\frac{1.486 R_h^{1/6}}{n}\right)^2 R_h} \frac{L}{R_h} = \left(\frac{vn}{1.486 R_h^{2/3}}\right)^2 L \tag{6.131}$$

6.7.7 The Hazen–Williams Equation

In addition to the standard empirical flow resistance coefficients including the Chezy coefficient, C; the Darcy–Weisbach friction factor, f; and the Manning's roughness coefficient, n, there is another commonly used flow resistance coefficient, namely the Hazen–Williams roughness coefficient, C_h. The empirically derived Hazen–Williams equation is presented. The empirical evaluation of the Hazen–Williams roughness coefficient, C_h is presented in tabular form. And, finally, the Hazen–Williams equation for the friction slope, S_f is substituted into the integral form of the energy equation (Equation 6.51), yielding the Hazen–Williams head loss equation.

6.7.7.1 The Hazen–Williams Equation

The Hazen–Williams equation was empirically derived assuming the English (BG) system and is given as follows:

$$v = 1.318 C_h R_h^{0.63} S_f^{0.54} \tag{6.132}$$

It was specifically developed for the analysis and design water flow in larger pipes ($2 \text{ in} \leq D \leq 6 \text{ ft}$) within a moderate range of velocity ($v \leq 10 \text{ ft/sec}$) for water at 60°F. The corresponding Hazen–Williams equation for the SI (metric) system is given as follows:

$$v = 0.849 C_h R_h^{0.63} S_f^{0.54} \tag{6.133}$$

with the corresponding units: velocity (m/s), R_h (m), and S_f (m/m). It is important to note that, because the Chezy equation, $v = C\sqrt{R_h S_f}$ (and the corresponding Darcy–Weisbach and Manning's equations) was derived (formulated without the need for calibration) using dimensional analysis, there are no limitations for the ranges assumed for the individual variables (v, D, and water temperature). However, although the Hazen–Williams equation was formulated based on the results of dimensional analysis (the predictor variables, C_h, R_h, and S_f are based on the Chezy equation), it was "calibrated" for a specific range of variables (v, D, and water temperature) as listed above, and it assumes only pipe flow (not open channel flow), thus the limitation of its application.

6.7.7.2 Evaluation of the Hazen–Williams Equation Roughness Coefficient, C_h

The Hazen–Williams roughness coefficient, C_h values typically range from 140 for very smooth and straight pipes to 80 or 90 for old unlined pipes, with a value of 100 assumed for average conditions. Similar to the Darcy–Weisbach friction factor, f, the Hazen–Williams roughness coefficient, C_h is dimensionless. Furthermore, similar to the Manning's roughness coefficient, n the Hazen–Williams roughness coefficient, C_h is not a function of the flow conditions (i.e., the Reynolds number, R) and is empirically evaluated only as of the boundary roughness, ε/D. One may note that the formulation of the Hazen–Williams equation assumes no specific units (SI or BG), and the Hazen–Williams roughness coefficient, C_h is dimensionless. However, the Hazen–Williams equation was initially calibrated in BG units and then calibrated in SI units (see Table 8.8).

6.7.7.3 The Hazen–Williams Head Loss Equation

Assuming that the Hazen–Williams roughness coefficient, C_h is chosen to model the empirical flow resistance, then one would substitute the Hazen–Williams expression for the friction slope, S_f (derived from Equation 6.132) into the integral form of the energy equation (Equation 6.51), yielding the Hazen–Williams head loss equation (assuming the English system) as follows:

$$h_f = S_f L = \frac{\tau_w L}{\gamma R_h} = \frac{\Delta p}{\gamma} = \left[\frac{v}{1.318 C_h R_h^{0.63}}\right]^{1/0.54} L \tag{6.134}$$

noting that:

$$S_f = \left[\frac{v}{1.318 C_h R_h^{0.63}}\right]^{1/0.54} \tag{6.135}$$

The corresponding Hazen–Williams head loss equation for the SI (metric) system is given as follows:

$$h_f = S_f L = \frac{\tau_w L}{\gamma R_h} = \frac{\Delta p}{\gamma} = \left[\frac{v}{0.849 C_h R_h^{0.63}}\right]^{1/0.54} L \tag{6.136}$$

noting that:

$$S_f = \left[\frac{v}{0.849 C_h R_h^{0.63}}\right]^{1/0.54} \tag{6.137}$$

6.7.8 The Relationship between the Drag Coefficient, C_D; the Chezy Coefficient, C; the Darcy–Weisbach Friction Factor, f; and Manning's Roughness Coefficient, n

The sections above illustrated that the drag coefficient, C_D is more commonly modeled by the more standard forms including the Chezy Coefficient, C; the Darcy–Weisbach friction factor, f; or the Manning's roughness coefficient, n. These are basically flow resistance coefficients that empirically relate the wall shear stress, τ_w (or flow resistance, which causes the pressure drop, Δp and thus a head loss, h_f) to the velocity, v (average velocity, v_{ave}) as follows:

$$\frac{\tau_w}{\rho v^2} = \frac{Wall\ Friction\ Force}{Inertia\ Force} = \frac{R_h}{L}\frac{\Delta p}{\rho v^2} = \frac{Pressure\ Force}{Inertia\ Force} = \frac{R_h}{L}\frac{1}{E} = C_D = \phi\left(R, \frac{\varepsilon}{D}\right) \quad (6.138)$$

where the flow resistance coefficients are related as follows:

$$C = \sqrt{\frac{g}{C_D}} \quad (6.139)$$

$$f = 8C_D \quad (6.140)$$

$$C = \sqrt{\frac{8g}{f}} \quad (6.141)$$

$$C = \frac{R_h^{1/6}}{n} \quad (6.142)$$

and these flow resistance coefficients are a function of the boundary roughness, ε/D; the Reynolds number, R; and the cross-sectional shape or the hydraulic radius, R_h (note: the Manning's roughness coefficient, n is a function only of the boundary roughness and assumes turbulent flow).

Furthermore, although the three standard flow resistance coefficients are related as illustrated above, for laminar or turbulent pipe flow, the use of the Darcy–Weisbach friction factor, f is preferred over the Chezy coefficient, C because f is dimensionless, while C has dimensions of $L^{1/2}\,T^{-1}$ and thus one must specify either SI or BG units for C. Finally for open channel flow (usually turbulent flow) and turbulent pipe flow, the use of the Manning's roughness coefficient, n (which has dimensions of $L^{-1/3}\,T$) is preferred over the Chezy coefficient, C because it is independent of both the Reynolds number, R (by assuming turbulent flow), and the cross-sectional shape or the hydraulic radius, R_h (by its relationship with the Chezy coefficient, C).

6.7.9 A Comparison between Laminar and Turbulent Flow Using the Darcy–Weisbach Head Loss Equation

It is of interest to conduct a comparison between laminar and turbulent flow. Because the Darcy–Weisbach head loss equation is commonly used in the analysis and design of both laminar and turbulent pipe flow, it is used to compare the two flow types. In the case of laminar flow, Poiseuille's law provides an analytical expression for the major head loss, h_f (expressed as a function of the flow resistance, μ), without the need to define an empirical flow resistance coefficient. However, in the case of turbulent flow, the Chezy equation provided an empirical expression for the friction slope, which yielded an empirical expression for the major head loss, h_f, with the need to define an empirical flow resistance. Because laminar flow is a special case of turbulent flow, it is of interest to evaluate the Darcy–Weisbach friction factor, f for laminar flow. Furthermore, it is of interest to compare and interpret the corresponding expressions for the major head loss, h_f; the pressure drop, Δp; and the wall shear stress, τ_{max} for laminar and turbulent flow. However, prior to conducting the comparisons between the two flow types, it is important to note that because the Darcy–Weisbach head loss equation and the expression for the Reynolds number, R assume a circular pipe, it is important to first address their respective definitions for noncircular pipes.

6.7.9.1 The Darcy–Weisbach Head Loss Equation and the Reynolds Number, R for Noncircular Pipes

Circular pipes are used for the transport of liquids under high pressure, because a circular pipe cross section is more capable of withstanding a large pressure difference between the inside and outside of the pipe without the risk of a pipe failure (such as a pipe distortion). However, noncircular pipes may be used for the transport of gases under low pressure, where a typical pipe section may be rectangular for a heating and cooling system using pressurized air. As mentioned above, the derivation of both, the Darcy–Weisbach equation and the definition of the Reynolds number, R for pipe flow has assumed a circular pipe with a diameter, D. As such, the hydraulic radius, R_h, for a circular pipe with diameter, D was defined as:

$$R_h = \frac{A}{P} = \frac{\pi D^2}{4} \frac{1}{\pi D} = \frac{D}{4} \tag{6.143}$$

And, thus, the Darcy–Weisbach equation and the definition of the Reynolds number, R for a circular pipe with a diameter, D were given respectively, as follows:

$$h_f = f \frac{L}{D} \frac{v^2}{2g} \tag{6.144}$$

and

$$R = \frac{\rho v D}{\mu} \tag{6.145}$$

However, when noncircular pipes, such as rectangular and other cross-sectional shapes, are used in practice, one may define the "equivalent diameter" or the "hydraulic diameter," D_h as follows:

$$D_h = 4R_h \tag{6.146}$$

Therefore, the Darcy–Weisbach equation and the definition of the Reynolds number, R for a noncircular pipe are given, respectively, as follows:

$$h_f = f \frac{L}{(4R_h)} \frac{v^2}{2g} \tag{6.147}$$

and

$$R = \frac{\rho v (4R_h)}{\mu} \tag{6.148}$$

Additionally, one may use the "equivalent diameter" or the "hydraulic diameter," D_h in the case of noncircular pipes in the definition of the relative pipe roughness, ε/D_h and with any other equation or chart (such as the Moody diagram), that assumes a circular pipe with a diameter, D.

6.7.9.2 Evaluating the Darcy–Weisbach Friction Coefficient, f for Laminar Pipe Flow

It may be illustrated that the "theoretical flow resistance coefficient" for laminar flow is a special case of the empirical flow resistance coefficient for turbulent flow ($f = 64/R$). Noting that laminar flow is a special case of turbulent flow, one may evaluate the Darcy–Weisbach friction coefficient, f for laminar pipe flow as follows. Equating Poiseuille's law expressed in terms of the head loss (Equation 6.95) (derived for laminar flow) to the Darcy–Weisbach head loss equation (Equation 6.124) (derived for turbulent flow), and isolating the friction coefficient, f yields the Darcy–Weisbach friction coefficient, f for laminar "circular" pipe flow as follows:

$$h_f = S_f L = \frac{\tau_w L}{\gamma R_h} = \frac{\Delta p}{\gamma} = \frac{32 \mu v L}{D^2 \rho g} = f \frac{L}{D} \frac{v^2}{2g} \tag{6.149}$$

$$f = \frac{64 \mu}{\rho v D} = \frac{64}{R} \tag{6.150}$$

where the Reynolds number, $R = \rho v D / \mu$. In the case where the pipe is noncircular, then one may use the appropriate relationship given in Table 8.4 for various noncircular pipe sections. One may note that the relationships between the friction factor, f and the Reynolds number, $R = \rho v(4R_h)/\mu$ for the various noncircular pipe sections are derived from theory and/or empirical studies. Thus, f for laminar flow is only a function of the Reynolds number, R, while f for turbulent flow is a function of both the Reynolds number, R and the relative pipe roughness, ε/D. However, at very high velocities (and thus very high values for R), the f for completely turbulent flow (rough pipes) is a function only of the relative pipe roughness, ε/D and thus independent of R.

6.7.9.3 A Comparison between Laminar and Turbulent Flow Using the Darcy–Weisbach Head Loss Equation

The Darcy–Weisbach head loss equation may used to conduct a comparison between laminar and turbulent flow. However, first, one may recall that regardless of whether the flow is laminar or turbulent, the flow resistance (shear stress, τ) causes an unknown pressure drop, Δp which causes an unknown head loss, h_f, where the head loss is due to a conversion of kinetic energy to heat. Furthermore, one may recall that application of Newton's second law of motion yielded a linear relationship between the shear stress, τ and the radial distance, r which also describes the theoretical relationship between the pressure drop, Δp and the shear stress, τ for both laminar and turbulent flow, and is repeated here as follows:

$$\tau = \frac{r}{2} \frac{\Delta p}{L} \tag{6.151}$$

where the maximum shear stress, τ_{max} at the pipe wall/boundary, where $R_h = D/4$ is given as follows:

$$\tau_w = \frac{D}{4} \frac{\Delta p}{L} = R_h \frac{\Delta p}{L} \tag{6.152}$$

See Figure 6.4 for the variation of shear stress for both laminar and turbulent pipe flow. However, because the dissipation of energy (i.e., h_f) is considerably less for laminar flow than for turbulent flow, the resulting magnitude of the pressure drop, Δp and thus the maximum shear stress, τ_{max} is also considerably less for laminar flow in comparison to turbulent flow. Therefore, it is of interest to compare and interpret the corresponding expressions for the major head loss, h_f, the pressure drop, Δp and the wall shear stress, τ_{max} for laminar and turbulent flow.

First, it is interesting to compare the corresponding expressions for the major head loss, h_f for laminar flow and turbulent flow as follows. Poiseuille's law expressed in terms of the major head loss, h_f was analytically derived for laminar flow and is repeated here as follows:

$$h_f = \frac{32\mu v L}{D^2 \rho g} \tag{6.153}$$

The Darcy–Weisbach head loss equation for turbulent flow is given as follows:

$$h_f = f\frac{L}{D}\frac{v^2}{2g} \tag{6.154}$$

A comparison between the head loss equation for laminar and turbulent flow indicates that the head loss, h_f in laminar flow is linear in velocity as v, and nonlinear in the pipe diameter as $1/D^2$, whereas in turbulent flow, it is nonlinear in velocity as v^2, and linear in the pipe diameter as $1/D$.

Next, it is interesting to compare the corresponding expressions for the pressure drop, Δp for laminar and turbulent flow as follows. Poiseuille's law expressed in terms of the pressure drop, Δp was analytically derived for laminar flow and is repeated here as follows:

$$\Delta p = \frac{\tau_w L}{R_h} = h_f \gamma = \frac{32\mu v L}{D^2} \tag{6.155}$$

In contrast, for turbulent flow substituting the theoretical relationship between the pressure drop, Δp and the shear stress, τ_w (derived from Newton's second law of motion for both laminar and turbulent flow) for a pipe flow (where $R_h = D/4$) into the Darcy–Weisbach head loss equation yields the following expression for the pressure drop, Δp for turbulent flow as follows:

$$\tau_w = \frac{D}{4}\frac{\Delta p}{L} = R_h \frac{\Delta p}{L} \tag{6.156}$$

$$h_f = \frac{\tau_w L}{\gamma R_h} = R_h \frac{\Delta p}{L}\frac{L}{\gamma R_h} = \frac{\Delta p}{\gamma} = f\frac{L}{D}\frac{v^2}{2g} \tag{6.157}$$

$$\Delta p = f\frac{L}{D}\frac{v^2}{2g}\gamma = f\frac{L}{D}\frac{v^2}{2}\rho \tag{6.158}$$

A comparison between the pressure drop equation for laminar flow (Equation 6.155) and for turbulent flow (Equation 6.158) indicates that the pressure drop, Δp (and thus the wall shear stress, τ_w) in laminar flow is a function of the fluid viscosity, μ, whereas in turbulent flow it is a function of the fluid density, ρ and the friction factor, f. Furthermore, because the friction factor, f in turbulent flow is a function of the fluid viscosity (as the Reynolds number, R is a function of viscosity) and the relative pipe roughness, ε/D, so is the pressure drop in turbulent flow. Thus, although the pressure drop in turbulent flow is a function of both the fluid density and the fluid viscosity, it is a stronger function of fluid density than it is of fluid viscosity. Furthermore, while the fluid density, ρ and the relative pipe roughness, ε/D are important in the determination of the pressure drop, Δp in turbulent flow, only the fluid viscosity, μ is needed in the case of laminar flow.

And, finally, it is interesting to compare the corresponding expressions for the wall shear stress, τ_w (i.e., τ_{max}) for laminar and turbulent flow as follows. The wall shear stress, τ_w for laminar flow was theoretically derived (Equation 6.97) as follows:

$$\tau_{max} = \tau_w = \frac{D}{4}\frac{\Delta p}{L} = \frac{D}{4}\frac{32\mu v L}{D^2}\frac{1}{L} = \frac{8\mu v}{D} \tag{6.159}$$

The wall shear stress, τ_w for turbulent flow was empirically derived from dimensional analysis (Equation 6.119) as follows:

$$\tau_{max} = \tau_w = \frac{D}{4}\frac{\Delta p}{L} = \frac{D}{4}f\frac{L}{D}\frac{v^2}{2}\rho\frac{1}{L} = \frac{f\rho v^2}{8} \tag{6.160}$$

A comparison between the wall shear stress equation for laminar flow and for turbulent flow indicates that for laminar flow, one may compute the maximum shear stress, τ_{max} in a circular pipe for a given value of viscosity, μ, whereas for turbulent flow, one may compute the maximum shear stress, τ_{max} in a circular pipe for a given value of the Darcy–Weisbach friction factor, f.

6.7.10 Determining the Velocity Profile for Turbulent Flow

While the deterministic nature of laminar flow analytically yielded a parabolic velocity profile for laminar flow, as illustrated in Figure 6.8a, the more stochastic nature of turbulent flow yields an empirically defined velocity profile for turbulent flow, as illustrated in Figure 6.8b (and Figure 6.3c). A comparison between these two velocity profiles illustrates that the velocity profile for turbulent flow is flatter than the parabolic laminar velocity profile due to the "averaging" effect of the mixing of the fluid particles (a transfer of momentum among the fluid particles). One may note that the front of the turbulent velocity profile is serrated; this is due to the axial velocity fluctuations, while near the pipe wall, it is smooth and laminar (viscous sublayer). Thus, the unpredictable nature of turbulent flow results in the need to use dimensional analysis, experimentation, and semiempirical theoretical procedures in the determination of the corresponding velocity profile. Empirical studies have shown that the velocity profile for turbulent flow may be approximated by the seventh root velocity profile law as follows:

$$v = v_{max}\left[1 - \frac{r}{r_o}\right]^{1/7} \tag{6.161}$$

where $v_{max} = 1.22\,v_{ave}$, and $v_{ave} = Q/A$. Thus, alternatively, the velocity profile for turbulent flow may be approximated by the seventh root velocity profile law as a function of v_{ave} as follows:

$$v = 1.22v_{ave}\left[1 - \frac{r}{r_o}\right]^{1/7} \tag{6.162}$$

Figure 6.9, which was generated using Mathcad, illustrates a comparison between the parabolic velocity profile for laminar flow (see Equations 6.87 and 6.88) and the velocity profile for turbulent flow, which is approximated by the seventh root velocity profile law (see Equation 6.162), for a given average velocity, $v_{ave} = 5$, and pipe radius, $r_o = 10$.

$$v_{ave} := 5 \qquad\qquad r_o := 10$$

$$vl_{max} := 2\,v_{ave} \qquad\qquad vl(r) := vl_{max}\left[1 - \left(\frac{|r|}{r_o}\right)^2\right]$$

$$vt_{max} := 1.22\,v_{ave} \qquad vt(r) := vt_{max}\left(1 - \frac{|r|}{r_o}\right)^{\frac{1}{7}}$$

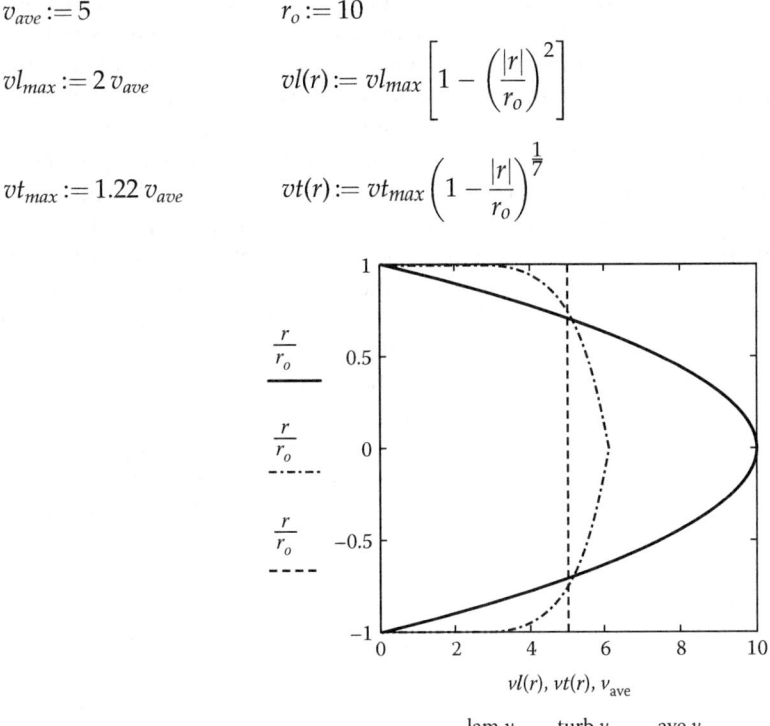

$$vl(r),\ vt(r),\ v_{ave}$$

___ lam v _._ turb v ___ ave v

FIGURE 6.9
Comparison between the parabolic velocity profile for laminar flow and the seventh root velocity profile for turbulent flow.

One may note: (1) for a given average velocity, v_{ave}, the maximum velocity for laminar flow, $v_{max} = 2\,v_{ave}$, while the maximum velocity for turbulent flow, $v_{max} = 1.22\,v_{ave}$; (2) thus, the velocity profile for the turbulent flow is significantly flatter than the laminar velocity profile; (3) while the parabolic velocity profile for laminar flow can be used to calculate the shear stress across the entire pipe section (using Newton's law of viscosity, $\tau = -\mu\,dv/dr$), the seventh root velocity profile law for turbulent flow cannot be used to calculate the wall shear stress, τ_w (which is contained within the thin viscous sublayer, where Newton's law of viscosity is applicable) because it yields a velocity gradient, dv/dr of infinity at $r = r_o$ (instead, $\tau_w = f\rho v^2/8$ from dimensional analysis is used); and (4) because the seventh root velocity profile law for turbulent flow does not yield a zero slope at the centerline of the pipe (at $r = 0$), the mathematically generated turbulent velocity profile is not actually flat at the pipe centerline (see Figures 6.3c and 6.8b); however, schematically the turbulent velocity profile is drawn as flat at the pipe centerline. Furthermore, in spite of the last two points regarding the seventh root velocity profile law for turbulent flow, it provides highly accurate results for turbulent flow.

6.7.10.1 A Comparison between the Velocity Profiles for Laminar Flow and Turbulent Flow

Because the velocity profile for laminar flow is determined analytically, it provides a basis of comparison for the velocity profile for turbulent flow. In laminar flow, the viscous effects are important across the entire pipe section, resulting in a zero velocity at the boundary (see Figure 6.8a) and the shear stress throughout the flow being described by Newton's law of viscosity, $\tau = -\mu \, dv/dr$. Furthermore, Newton's law of viscosity coupled with Newton's second law of motion and the continuity equation yielded Poiseuille's law for laminar flow, which indicates that the flow resistance; thus, the head loss is mainly a function of the fluid viscosity, μ and not a function of the relative pipe roughness, ε/D (as in the case of turbulent flow). In laminar flow, very little of the flow comes in contact with the projection at the pipe wall; the affected flow and eddies resulting from the projections in the pipe wall are quickly damped out by significantly large viscous dissipation forces. It may be noted that a relatively small wall roughness element, ε has negligible effects on laminar flow as long as the roughness projections are small relative to the pipe size, as illustrated in Figure 6.10; the entire boundary layer consists entirely of a laminar boundary layer and not just a thin viscous sublayer as in the case of turbulent flow. However, one may note that a relatively large wall roughness element, ε (for which the relative pipe roughness, $\varepsilon/D \geq 0.1$) may affect the structure of laminar flow and its properties (the pressure drop, Δp; the wall shear stress, τ_w; the flow resistance; and the head loss). Typical pipes in practice are characterized by a relative pipe roughness in the range $0 \leq \varepsilon/D \leq 0.05$, for which the roughness projections, ε are small relative to the pipe diameter, D and thus the relative pipe roughness, ε/D has no effect on the flow properties.

In turbulent flow, in addition to the turbulent boundary layer, a thin viscous sublayer has been experimentally observed near the pipe wall, for which the shear stress is described by Newton's law of viscosity, $\tau = -\mu \, dv/dr$. However, because the seventh root velocity profile law for turbulent flow yields a velocity gradient, dv/dr of infinity at $r = r_o$, it cannot be used to calculate the wall shear stress, τ_w (which is contained within the thin viscous sublayer, where Newton's law of viscosity is applicable); instead, $\tau_w = f\rho v^2/8$ from dimensional analysis is used. Furthermore, for turbulent flow, if a typical wall roughness element, ε protrudes sufficiently far into or through the thin viscous sublayer, the structure and properties (the pressure drop, Δp; the wall shear stress, τ_w; the flow resistance; and the head loss) will be different than if the wall were smooth, as illustrated by a comparison of Figure 6.11a and b. As a result, for turbulent flow, the pressure drop, Δp; the wall shear stress, τ_w; the head loss, h_f; and thus the friction factor, f will be a function of the wall roughness element, ε.

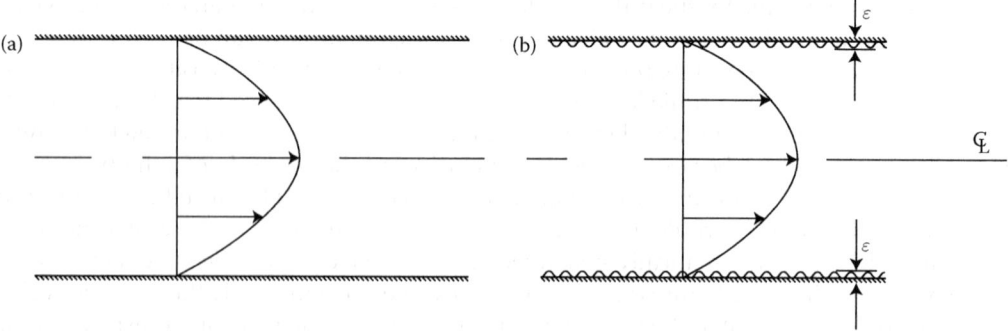

FIGURE 6.10
Flow resistance in laminar pipe flow is mostly independent of the pipe roughness, ε. (a) Smooth pipe wall. (b) Rough pipe wall.

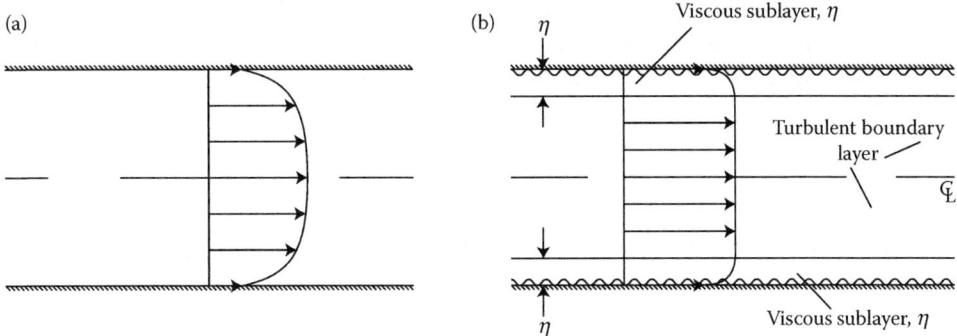

FIGURE 6.11
Flow resistance in turbulent pipe flow is dependent on the pipe roughness, ε. (a) Smooth pipe wall. (b) Rough pipe wall. (Adapted from Vennard, J. K., and R. L. Street. 1975. *Elementary Fluid Mechanics*, 5th ed. New York: John Wiley and Sons, 309.)

Therefore, for turbulent flow, the physical interaction of the fluid particles with the pipe wall roughness, ε plays an important role in the definition of the empirical flow resistance coefficient (as empirically derived from dimensional analysis) and thus the head loss equation (unlike laminar flow, where the flow resistance is mainly a function of the fluid viscosity, μ as theoretically derived by Poiseuille's law). In turbulent flow, the higher velocities cause the fluid particles to come into contact with the projections of the pipe wall. As a result, the projections in the pipe wall affect the flow, causing a greater loss of energy than for laminar flow.

6.7.10.2 The Role of the Boundary Roughness in Laminar Flow and Turbulent Flow

The effect of the roughness of the boundary, ε on the physical flow properties is dependent upon the relative size of the wall roughness element, ε and the viscous sublayer, η. Because the thickness of the boundary layer, η depends upon certain properties of the flow, it is possible for the same boundary surface to behave as a smooth or a rough boundary, depending upon the magnitude of the Reynolds number, R and the thickness of the boundary layer, η. Furthermore, because the thickness of the boundary layer, η decreases with an increase in the Reynolds number, R the change from a smooth surface to a rough one will result from an increase of the Reynolds number, R which is caused by an increase in velocity of the flow. For cases where the Reynolds number, R is very large, the boundary layer, η becomes so thin that the boundary roughness, ε dominates the character of the flow near the pipe wall. As a result, the friction factor, f becomes independent of the Reynolds number, R and only dependent of the relative pipe roughness, ε/D; the pressure drop, Δp is a result of an inertia-dominated turbulent shear stress, as opposed to a viscosity-dominated laminar shear stress typically found in the thin viscous sublayer. The boundary surface may be described as smooth if its projections are completely submerged in the viscous sublayer (see Figure 6.11b); thus, they have no effect on the structure of the turbulence. However, experiments have shown that a boundary roughness heights, $\varepsilon > \eta/3$ will increase the turbulence and have an effect on the flow. Therefore, the thickness of the boundary layer, η is the criterion of effective roughness.

The surface roughness, ε increases the turbulence of the flow and decreases the viscous effects. Therefore, in turbulent flow over rough surfaces, energy is dissipated by the work done in the generation of turbulence caused by the roughness projections. The energy involved in the turbulence consists of the kinetic energy of fluid masses, which is proportional to the velocity squared ($KE = Mv^2/2$). Thus, the energy dissipation (head loss, h_f) and flow resistance caused by rough surfaces should also vary with the square of the velocities for turbulent flow

$$\left(h_f = \frac{\tau_w L}{\gamma R_h} = \frac{\Delta p}{\gamma} = f \frac{L}{D} \frac{v^2}{2g} \right),$$

whereas for laminar flow over boundary surfaces, energy dissipation and resistance will be directly proportional to the velocities

$$\left(h_f = \frac{\tau_w L}{\gamma R_h} = \frac{\Delta p}{\gamma} = \frac{32 \mu v L}{D^2 \rho g} \right).$$

Evaluation of the Darcy–Weisbach friction coefficient, $f = 64/R$ for laminar pipe flow (a special case of turbulent flow) illustrates that f for laminar flow is only a function of the Reynolds number, R (given below), while f for turbulent flow is a function of both the Reynolds number, R and the relative pipe roughness, ε/D (given above from dimensional analysis). However, at very high velocities (and thus very high values for R), the f for completely turbulent flow (rough pipes) is a function only of the relative pipe roughness, ε/D and thus independent of R.

6.7.11 Application of the Major Head Loss Equation for Open Channel Flow

As noted above, for the derivation of the major head loss equation, although the laws of flow resistance are essentially the same in pipe flow and open channel flow, there is a slight modification needed for the application in open channel flow. First, application of the integral form of the energy equation yielded the following relationship between the major head loss, h_f and the wall shear stress, τ_w (Equation 6.51):

$$h_f = \frac{\tau_w L}{\gamma R_h} \tag{6.163}$$

Then, application of Newton's second law of motion (integral form of the momentum equation) represents an equilibrium between the propulsive pressure and/or gravity forces against the retarding shear forces needed to move each fluid particle along the pipe or open channel with a constant velocity (Equation 6.65) and is repeated here as follows:

$$\tau = \frac{r}{2} \left[-\frac{d}{ds} (p + \gamma z) \right] \tag{6.164}$$

Because the propulsive force for a pipe flow is provided by a pressure gradient $\Delta p/L$, it was assumed that $z_1 = z_2$, and $ds = L$ and thus yielded the following:

$$\tau = \frac{r}{2} \left[-\frac{d}{ds} (p) \right] = \frac{r}{2} \frac{\Delta p}{L} \tag{6.165}$$

which yields an expression for the variation of the shear stress, τ across the pipe diameter (or actually as a function of the radial distance, r from the center of the pipe) that also describes the theoretical relationship between the pressure drop, Δp and the shear stress, τ for both laminar and turbulent flow. Evaluated for the wall shear stress, τ_w at $r = r_o = D/2$, where $R_h = D/4$ yielded the following:

$$\tau_w = R_h \frac{\Delta p}{L} \tag{6.166}$$

However, because the propulsive force for an open channel flow is provided by invoking gravity and thus by the weight of the fluid along a channel slope (channel gradient, $\Delta z/L = S_o$), assuming a constant velocity implies that $p_1 = p_2$, which yields the following:

$$\tau = \frac{r}{2}\left[-\frac{d}{ds}(\gamma z)\right] = \frac{r}{2}\gamma\frac{\Delta z}{L} = \frac{r}{2}\gamma S_o \tag{6.167}$$

which yields an expression for the variation of the shear stress, τ across the "pipe diameter" (or actually as a function of the radial distance, r from the center of the pipe, where the flow cross section in an open channel is assumed to be equivalent to the lower half of a pipe flow), which also describes the theoretical relationship between the elevation drop, Δz and the shear stress, τ for both laminar and turbulent flow. Evaluated for the wall shear stress, τ_w at $r = r_o = D/2$, where $R_h = D/4$ yields the following:

$$\tau_w = \gamma R_h S_o \tag{6.168}$$

or

$$S_o = \frac{\tau_w}{\gamma R_h} \tag{6.169}$$

Furthermore, the differential form of the momentum equation yielded the following relationship between the friction slope, S_f and the wall shear stress, τ_w (Equation 6.54):

$$S_f = \frac{\tau_w}{\gamma R_h} \tag{6.170}$$

Thus, for open channel flow, one may note that substitution of the two above forms of the momentum equation (Equations 6.169 and 6.170) into the energy equation (Equation 6.163) yields the following expression for the major head loss, h_f:

$$h_f = \frac{\tau_w L}{\gamma R_h} = \frac{\gamma R_h S_o L}{\gamma R_h} = S_o L \tag{6.171}$$

$$h_f = \frac{\tau_w L}{\gamma R_h} = \frac{\gamma R_h S_f L}{\gamma R_h} = S_f L \tag{6.172}$$

where the channel slope (channel gradient, $\Delta z/L = S_o$) is equal to the friction slope, S_f for the assumption of uniform open channel flow as follows:

$$S_o = S_f = \frac{\tau_w}{\gamma R_h} \tag{6.173}$$

6.7.11.1 Application of the Chezy Equation for Open Channel Flow

Because all three terms, the channel slope, S_o (Equation 6.169); the friction slope, S_f (Equation 6.170); and the major head loss, h_f (Equation 6.163) are expressed in terms of the wall shear stress, τ_w, in order to evaluate the respective expressions, the Chezy equation (Equation 6.109), which provides an empirical evaluation of the friction slope, S_f, is applied as follows:

$$S_f = \frac{v^2}{C^2 R_h} \tag{6.174}$$

$$S_o = S_f = \frac{\tau_w}{\gamma R_h} = \frac{v^2}{C^2 R_h} \tag{6.175}$$

where

$$\tau_w = \gamma R_h S_o = \gamma R_h S_f = \gamma R_h \underbrace{\left[\frac{v^2}{C^2 R_h} \right]}_{S_o = S_f} = \frac{\gamma v^2}{C^2} = \frac{g}{C^2} \rho v^2 \tag{6.176}$$

The Chezy equation for open channel flow assumes an open channel section of length, L, with a channel slope, S_o with a uniform flow depth, y and thus a constant velocity of flow, v and a friction slope, $S_f = S_o$, with a specific Chezy coefficient, C. The Chezy equation (representing the momentum equation supplemented by dimensional analysis) may be solved for one of the following unknown quantities: $S_f = S_o$, v, C, or y (which is defined as a function of R_h).

Furthermore, assuming that the Chezy coefficient, C is chosen to model the empirical flow resistance, the Chezy equation for the channel and friction slopes (Equation 6.175) is substituted into the integral form of the energy equation (Equation 6.163), yielding an empirical expression for the head loss, h_f (the Chezy head loss equation for open channel flow) as follows:

$$h_f = \frac{\tau_w L}{\gamma R_h} = S_o L = S_f L = \underbrace{\left[\frac{v^2}{C^2 R_h} \right]}_{S_o = S_f} L \tag{6.177}$$

Alternatively, modeling the flow resistance using the Darcy–Weisbach friction factor, f (although typically used only for pipe flow) and thus substituting the relationship between the Chezy coefficient, C and the Darcy–Weisbach friction factor, f into the Chezy head loss equation for open channel flow (using the "equivalent diameter" or the "hydraulic diameter," $D_h = 4 R_h$) yields the Darcy–Weisbach head loss equation for open channel flow as follows:

$$C = \sqrt{\frac{8g}{f}} \tag{6.178}$$

$$h_f = \frac{\tau_w L}{\gamma R_h} = S_o L = S_f L = \frac{v^2 L}{C^2 R_h} = \frac{v^2 L}{\dfrac{8g}{f} (4R_h)} = f \frac{L}{(4R_h)} \frac{v^2}{2g} \tag{6.179}$$

Or, alternatively, modeling the flow resistance using the Manning's roughness coefficient, n and thus substituting the relationship between the Chezy coefficient, C and the Manning's roughness coefficient, n into the Chezy head loss equation, yields the Manning's head loss equation for open channel flow as follows:

$$C = \frac{R_h^{1/6}}{n} \tag{6.180}$$

$$h_f = \frac{\tau_w L}{\gamma R_h} = S_o L = S_f L = \frac{v^2 L}{C^2 R_h} = \frac{v^2}{\left(\dfrac{R_h^{1/6}}{n}\right)^2 R_h} \frac{L}{R_h} = \left(\frac{vn}{R_h^{2/3}}\right)^2 L \tag{6.181}$$

and the corresponding Manning's head loss equation for the English (BG) system for open channel flow is given as follows:

$$h_f = \frac{\tau_w L}{\gamma R_h} = S_o L = S_f L = \frac{v^2 L}{C^2 R_h} = \frac{v^2}{\left(\dfrac{1.486 R_h^{1/6}}{n}\right)^2 R_h} \frac{L}{R_h} = \left(\frac{vn}{1.486 R_h^{2/3}}\right)^2 L \tag{6.182}$$

And, finally, the Hazen–Williams equation is intended for pipe flow only and not open channel flow.

6.8 Modeling the Flow Resistance as a Minor Head Loss in Pipe Flow

While the major head loss term, $h_{f,major}$ in the energy equation models the head loss due to pipe friction in the straight section of a pipe, the minor head loss term, $h_{f,minor}$ in the energy equation models the head loss due to various pipe components in the pipe system. The minor head loss term, $h_{f,minor}$ in the energy equation is significant in very short pipes and is due to different pipe appurtenances (components, devices, or transitions), which include valves, fittings (tees, unions, elbows, and bends), entrances, exits, contractions, and expansions. The pipe components interrupt the smooth fluid flow and result in minor head losses that are caused by the flow separation and mixing induced by the various pipe components. As such, one is interested in the determination of the minor head loss and pressure drops where energy is dissipated.

The pipe length usually determines the role/significance of major and minor losses. For very short pipes, minor losses are usually significant, while major losses due to pipe friction are usually ignored. For very long pipes, minor losses are usually ignored, while the major losses due to pipe friction are significant. For instance, the minor losses may be insignificant when pipe devices are separated by approximately 1000 diameters. Furthermore, the minor head loss associated with a completely open valve may be negligible. However, in some cases, both minor and major head losses are significant and thus both are accounted for. Furthermore, although in general, major losses due to pipe friction are greater than minor losses due to pipe components, in some cases, such as in pipe systems where there are valves and numerous turns in a short pipe distance, minor losses can add up and be greater than major losses. Additionally, a partially closed valve, which results in a drop in the flowrate, may cause the most significant head loss in the pipe system.

6.8.1 Evaluation of the Minor Head Loss

In the case where the flow resistance is due to various pipe components, transitions, or devices, the flow resistance is ultimately modeled as a minor head loss in the integral form of the energy equation because the minor head loss is needed for the design of internal flow in a pipe transition. The flow resistance (shear stress or drag force) is modeled as a resistance force in the integral form of the momentum equation. The flow resistance causes an unknown pressure drop, Δp, which causes an unknown minor head loss, h_f, where the head loss is due to a conversion of kinetic energy to heat, which is modeled/displayed in the integral form of the energy equation. Evaluation of the minor head loss flow resistance term requires a "subset level" of application of the governing equations. A "subset level" application of the governing equations focuses only on the given element causing the flow resistance. The assumption of real flow implies that the flow resistance is modeled in both the energy and momentum equations (see Table 4.1). Thus, in the derivation of the minor head loss, first the integral form of the energy equation is applied to determine an expression for the unknown head loss, h_f. Then, the integral momentum equation (supplemented by dimensional analysis) is applied in order to determine an expression for the pressure drop, Δp and the shear stress, τ_w. The flow resistance due to pipe components, which results in a minor head loss term, $h_{f,minor}$, is modeled as a resistance force (shear stress or drag force) in the integral form of the momentum equation (Equation 4.28). As such, the energy equation may be used to solve for the unknown minor head loss,

$$ h_{f,min} = \frac{\tau_w L}{\gamma R_h} = \frac{\Delta p}{\gamma}, $$

while the integral momentum equation may be used to solve for the unknown pressure drop, $\Delta p = \tau_w L / R_h$.

The minor head loss, h_f is caused by both pressure and friction forces. Thus, when the friction/viscous forces are insignificant and can be ignored in the momentum equation (as in the case of a sudden pipe expansion), one can analytically determine the actual pressure drop, Δp from the momentum equation. Therefore, one may analytically derive an expression for the minor head loss, h_f due to pipe components from the energy equation (see Section 6.8.4). However, when the friction/viscous forces are significant and cannot be ignored in the momentum equation (as in the case of pipe components in general), the friction/viscous forces; the actual pressure drop, Δp, and thus, the minor head loss, h_f are determined empirically using dimensional analysis (see Section 6.8.5). Specifically, in the case of pipe components in general, the friction/viscous forces in the integral form of the momentum equation and thus the wall shear stress, τ_w cannot be ignored; thus, the integral form of the momentum equation cannot be directly applied to solve for the unknown pressure drop, Δp. Thus, an empirical interpretation (using dimensional analysis) of the wall shear stress, τ_w in the theoretical expression for the pressure drop, $\Delta p = \tau_w L / R_h$ in the integral form of the momentum equation is sought in terms of velocity, v and a flow resistance coefficient. This yields $\Delta p = k(\rho v^2 / 2)$, which represents the integral form of the momentum equation supplemented by dimensional analysis and is used to obtain an empirical evaluation for the pressure drop, Δp in Newton's second law of motion (integral form of the momentum equation). As such, one can then derive an empirical expression for the minor head loss, h_f from the energy equation. The flow resistance equation for the minor head loss is derived below and in Chapter 7. The minor losses computed in this section are for pipe transitions in pipe flow. Minor losses for open channel flow are addressed in Chapter 9.

6.8.2 Application of the Energy Equation: Derivation of the Head Loss

The objective here is to derive an expression for the minor head loss, $h_{f,minor}$, noting that, in general, both pressure forces, F_p and viscous forces, F_V (shear forces, F_f) may be exerted on the pipe wall as a result of a pipe component. As noted above, in general, the flow resistance causes an unknown pressure drop, Δp, which causes an unknown minor head loss, $h_{f,minor}$, where the head loss is due to a conversion of kinetic energy to heat, which is modeled/displayed in the integral form of the energy equation. As such, one may recall from a section above that in the cases of the flow through a pipe (major head loss), the flow from a reservoir or a tank or a pipe contraction (minor head loss), the flow through most pipe devices (minor head loss), and the flow around a submerged body (drag force), where the fluid is set into motion, the conversion of the pressure energy to kinetic energy results in a pressure drop. However, in the cases of a pitot tube (no head loss) or the flow into a reservoir (minor head loss) or a pipe expansion (minor head loss) where the fluid is brought to a stop (even momentarily), the conversion of the kinetic energy to pressure energy results in a pressure rise. Furthermore, one may note that, in particular, while the pressure rise and the resulting minor head loss for the case of a sudden pipe expansion may be analytically derived, the pressure drop/rise and the resulting minor head loss in the cases of the other pipe components/devices/transitions, in general, are empirically derived.

Regardless of whether one may analytically or empirically determine the actual pressure drop/rise and the resulting minor head loss due to a pipe component, it is of interest to determine the expression for the pressure drop/rise and the corresponding minor head loss and to compare them to the actual pressure drop and the resulting major head loss due to pipe friction. Expressing the energy equation in terms of pressures yields the following:

$$\left(\underbrace{p}_{F_P} + \underbrace{\gamma z}_{F_G} + \underbrace{\frac{\rho v^2}{2}}_{ma} \right)_1 - \left(\underbrace{p}_{staticpress} + \underbrace{\gamma z}_{hydrostapress} + \underbrace{\frac{\rho v^2}{2}}_{dynpres} \right)_2 = \underbrace{\overbrace{\frac{\tau_w L}{R_h}}^{flowres}}_{F_V} \tag{6.183}$$

Assuming a horizontal pipe ($z_1 = z_1$) and thus that F_G is not significant, the energy equation yields the following:

$$p_1 - p_2 = \Delta p = \underbrace{\left(\frac{\rho v_2{}^2}{2} - \frac{\rho v_1{}^2}{2} \right)}_{dynpres} + \underbrace{\frac{\tau_w L}{R_h}}_{flowres} \tag{6.184}$$

In the cases of the flow from a reservoir or a tank or a pipe contraction and the flow through most pipe devices, where the fluid is set into motion ($v_1 = 0$ and $v_2 = v$), the conversion of the pressure energy to kinetic energy results in a pressure drop (and a minor head loss). Thus, if one assumes that $v_1 = 0$ and $v_2 = v$, the energy equation yields a pressure drop as follows:

$$p_1 - p_2 = \underbrace{\Delta p}_{presdrop} = \underbrace{\overbrace{\frac{\rho v^2}{2}}^{F_P}}_{dynpres} + \underbrace{\overbrace{\frac{\tau_w L}{R_h}}^{F_V}}_{flowres} \tag{6.185}$$

where the pressure drop, Δp is equal to the dynamic pressure term plus the flow resistance term. However, in the cases of the flow into a reservoir or a pipe expansion, where the fluid is momentarily brought to a stop ($v_1 = v$ and $v_2 = 0$), the conversion of the kinetic energy to pressure energy results in a pressure rise (and a minor head loss). Thus, if one assumes that $v_1 = v$ and $v_2 = 0$, the energy equation yields a pressure rise as follows:

$$p_2 - p_1 = \underbrace{\Delta p}_{presrise} = \underbrace{\overbrace{\frac{\rho v^2}{2}}^{F_P}}_{dynpres} - \underbrace{\overbrace{\frac{\tau_w L}{R_h}}^{F_V}}_{flowres} \tag{6.186}$$

where the pressure rise, Δp is equal to the dynamic pressure term minus the flow resistance term. Furthermore, in the case of the steady uniform flow through a pipe ($v_1 = v_2$), the conversion of the pressure energy to kinetic energy results in a pressure drop (and a major head loss). Thus, if one assumes that $v_1 = v_2$, the energy equation yields a pressure drop as follows:

$$p_1 - p_2 = \underbrace{\Delta p}_{presdrop} = \underbrace{\frac{\tau_w L}{R_h}}_{flowres} \tag{6.187}$$

where the pressure drop, Δp is equal to the flow resistance term. One may note that this is the same result yielded by applying Newton's second law of motion for a steady and uniform flow though a horizontal pipe, which describes the theoretical relationship between the pressure drop, Δp and the shear stress, τ, which played a key role in the derivation of the major head loss equation for laminar and turbulent flow and plays a key role in the derivation of the minor head loss equation for pipe devices in general. Therefore, in comparison, the pressure drop, Δp due to the flow through a pipe is mostly due to the wall shear stress, τ_w, while the pressure drop/rise, Δp due to a pipe component is due to both the dynamic pressure, $\rho v^2/2$ and the wall shear stress, τ_w as illustrated above.

Next, it is of interest to compare the resulting minor head loss due to a pipe component to the major head loss due to pipe friction. Expressing the energy equation in terms of energy heads yields the following:

$$\left(\frac{p_1}{\gamma} + z_1 + \frac{v_1^2}{2g}\right) - \left(\frac{p_2}{\gamma} + z_2 + \frac{v_2^2}{2g}\right) = \frac{\tau_w L}{\gamma R_h} = h_f \tag{6.188}$$

Assuming a horizontal pipe ($z_1 = z_2$) and thus that F_G is not significant, the energy equation yields the following:

$$\left(\frac{p_1}{\gamma} + \frac{v_1^2}{2g}\right) - \left(\frac{p_2}{\gamma} + \frac{v_2^2}{2g}\right) = \frac{\tau_w L}{\gamma R_h} = h_f \tag{6.189}$$

In the cases of the flow from a reservoir or a tank or a pipe contraction and the flow through most pipe devices, where the fluid is set into motion ($v_1 = 0$ and $v_2 = v$), the conversion of the pressure energy to kinetic energy results in a pressure drop, where the pressure drop, Δp is equal to the dynamic pressure term plus the flow resistance term, the energy equation yields the minor head loss as follows:

$$h_{f,min} = \underbrace{\frac{\tau_w L}{\gamma R_h}}_{flowres} = \underbrace{\frac{\Delta p}{\gamma}}_{presdrop} - \underbrace{\frac{v^2}{2g}}_{dynpres} \tag{6.190}$$

However, in the cases of the flow into a reservoir or a pipe expansion where the fluid is momentarily brought to a stop ($v_2 = 0$ and $v_1 = v$), the conversion of the kinetic energy to pressure energy results in a pressure rise, where the pressure rise, Δp is equal to the dynamic pressure term minus the flow resistance term, the energy equation yields the minor head loss as follows:

$$h_{f,min} = \underbrace{\frac{\tau_w L}{\gamma R_h}}_{flowres} = \underbrace{\frac{v^2}{2g}}_{dynpres} - \underbrace{\frac{\Delta p}{\gamma}}_{presrise} \tag{6.191}$$

Furthermore, in the case of the steady uniform flow through a pipe ($v_1 = v_2$), the conversion of the pressure energy to kinetic energy results in a pressure drop, where the pressure drop, Δp is equal to the flow resistance term, the energy equation yields the major head loss as follows:

$$h_{f,maj} = \underbrace{\frac{\tau_w L}{\gamma R_h}}_{flowres} = \underbrace{\frac{\Delta p}{\gamma}}_{presdrop} \tag{6.192}$$

Therefore, in comparison, the major head loss, $h_{f,major}$ due to the flow through a pipe is mostly due to the pressure head drop, $\Delta p/\gamma$, while the minor head loss, $h_{f,minor}$ due to a pipe component is due to both the pressure head drop/rise, $\Delta p/\gamma$ and the dynamic pressure (velocity) head, $v^2/2g$ as illustrated above.

6.8.3 Application of the Momentum Equation and Dimensional Analysis: Derivation of the Pressure Drop, Shear Stress, and the Drag Coefficient

Once the integral form of the energy equation has been applied to yield an expression for the minor head loss, $h_{f,minor}$, the integral form of the momentum equation is then applied in order to solve for the unknown pressure drop, Δp. However, because the flow though the various devices, in general, is very complex, it is not possible to theoretically model the existence of the viscous force due to the wall shear stress in the momentum equation, and thus one cannot theoretically define the momentum equation. The flow through these pipe devices is complex, as the fluid flows through variable areas and thus nonlinear flow torturous flow paths. In the special case of a sudden pipe expansion, the friction/viscous forces are considered to be insignificant and thus can be ignored in the momentum equation. As a result, one can analytically determine the actual pressure drop, Δp from the momentum equation; therefore, one may analytically derive an expression for the minor head loss, h_f due to a sudden pipe expansion from the energy equation. However, in the case of the remaining pipe devices, in general, the friction/viscous forces are significant and cannot be ignored in the momentum equation. As a result, the friction/viscous forces, the actual pressure drop, Δp and thus the minor head loss, h_f are determined empirically.

Thus, the particular type of pipe component will determine whether the friction/viscous forces are insignificant and can be ignored and thus determine whether application of the three governing equations will yield an analytical or an empirical expression for the minor head loss, h_f. In particular, in the case of a sudden pipe expansion, the friction/viscous forces in the integral form of the momentum equation can be ignored. Then, application of the continuity equation yields an analytical expression for the actual pressure drop, Δp. Finally, the analytical expression for the actual pressure drop, Δp is substituted into the integral form of the energy equation, yielding an analytical expression for the minor head loss, h_f. However, in the case of the other pipe components/devices/transitions, in general, the friction/viscous forces in the integral form of the momentum equation cannot be ignored. Instead, the friction forces are modeled empirically. Specifically, because one cannot ignore the viscous force in the integral form of the momentum equation, one cannot derive an analytical expression for the actual pressure drop, Δp from application of the momentum equation and the continuity equation, so one cannot derive an analytical expression for the minor head loss, h_f from application of the integral form of the energy equation. As a result, one must resort to dimensional analysis in order to derive an expression for actual pressure drop, Δp and thus the minor head loss, h_f, which involves the definition of a drag coefficient, C_D (minor loss coefficient, k) that represents the flow resistance and which is empirically evaluated. And, finally, it is important to note that the use of dimensional analysis is not needed in the derivation of the minor head loss for a sudden pipe expansion (see Section 6.8.4).

6.8.4 Analytical Derivation of the Minor Head Loss Equation due to a Sudden Pipe Expansion

For the particular case of a sudden pipe expansion, once the integral form of the energy equation has been applied to yield an expression for the minor head loss, $h_{f,minor}$, the integral form of the momentum equation is then applied in order to analytically solve for the unknown pressure drop, Δp. For the case of a sudden pipe expansion, the application of the three governing equations yields a theoretical expression for the actual pressure drop, Δp and for the minor head loss, $h_{f,minor}$. The analytical derivation of the minor head loss is given in Chapter 8 and summarized as follows. Specifically, assuming a horizontal pipe ($z_1 = z_1$) and thus that F_G is not significant, and that the viscous force, F_V due to the wall shear stress is not significant, the integral form of the momentum equation is used to derive an expression for the actual pressure drop, Δp. Applying the integral form of the continuity equation and substituting $v_2 = v_1 A_1 / A_2$, the expression for the actual pressure drop, Δp from the momentum equation is substituted into the energy equation in order to solve for $h_{f,minor}$. The minor head loss, $h_{f,minor}$ is analytically evaluated using the upstream velocity, v_1 and is given as follows:

$$h_f = \left[1 - \frac{A_1}{A_2}\right]^2 \left(\frac{v_1{}^2}{2g}\right) = \left[1 - \left(\frac{D_1}{D_2}\right)^2\right]^2 \left(\frac{v_1{}^2}{2g}\right) \tag{6.193}$$

$$h_f = k\left(\frac{v_1{}^2}{2g}\right) \tag{6.194}$$

where the minor loss coefficient, k due to a sudden expansion is evaluated analytically as follows:

$$k = \left[1 - \frac{A_1}{A_2}\right]^2 = \left[1 - \left(\frac{D_1}{D_2}\right)^2\right]^2 \tag{6.195}$$

One may note that although the application of the three governing equations yielded a general expression for the minor head loss, $h_f = k(v^2/2g)$, an analytical evaluation of the minor loss coefficient, k due to all other pipe devices (other than a sudden pipe expansion) is not possible because the flow through the different devices is very complex. Instead, the minor loss coefficient, k is evaluated empirically as illustrated in Section 6.8.5.

6.8.5 Empirical Derivation of the Minor Head Loss Equation due to Pipe Components in General

For pipe components in general, once the integral form of the energy equation has been applied to yield an expression for the minor head loss, $h_{f,minor}$, the integral form of the momentum equation and dimensional analysis are then applied in order to empirically solve for the unknown pressure drop, Δp. Finally, the empirical expression for the actual pressure drop, Δp is substituted into the integral form of the energy equation, yielding an empirical expression for the minor head loss, h_f. Thus, for pipe components in general, one may not make the simplifying assumption that that the viscous force, F_V due to the wall shear stress is not significant (as done for the case of the sudden pipe expansion, see Section 6.8.4). As a result, the momentum equation may not be solved for a theoretical expression for the actual pressure drop/rise, Δp. Thus, the energy equation may not be solved for a theoretical expression for the minor head loss, $h_{f,minor}$. Instead, dimensional analysis guided by the theoretical relationship between the pressure drop/rise, Δp and the shear stress, τ provided by Newton's second law of motion (for a horizontal pipe for steady uniform flow), and application of the energy equation; thus,

$$h_{f,min} = \frac{\tau_w L}{\gamma R_h} = \frac{\Delta p}{\gamma},$$

yields an empirical expression for the pressure drop/rise, $\Delta p = k(\rho v^2/2)$ and for the minor head loss, $h_{f,min} = k(v^2/2g)$, where the actual pressure drop/rise, Δp is indirectly evaluated, through an empirical determination of a minor head loss coefficient, k from experimentation.

6.8.5.1 Application of the Energy Equation: Derivation of the Head Loss

In the derivation of the minor head loss for pipe flow, real flow is assumed in order to determine an expression for the unknown head loss, h_f by applying the integral form of the energy equation as follows:

$$\left(\frac{p}{\gamma} + z + \frac{v^2}{2g}\right)_1 - h_{f,min} = \left(\frac{p}{\gamma} + z + \frac{v^2}{2g}\right)_2 \tag{6.196}$$

For pipe components in general, assuming $z_1 = z_2$, and $v_1 = v_2$, the energy equation has one unknown, which is the minor head loss as follows:

$$h_{f,min} = \frac{\tau_w L}{\gamma R_h} = \frac{\Delta p}{\gamma} \tag{6.197}$$

which is expressed as a function of the wall shear stress, τ_w and the pressure drop, Δp.

6.8.5.2 Application of the Momentum Equation and Dimensional Analysis: Derivation of the Pressure Drop, Shear Stress, and the Drag Coefficient

In the derivation of the minor head loss for pipe flow, real flow is assumed in order to determine an expression for the pressure drop, Δp and the shear stress, τ_w by applying the integral form of the momentum equation (supplemented by dimensional analysis) as follows:

$$\sum F_s = (F_G + F_P + F_V + F_{other})_s = (\rho Q v_s)_2 - (\rho Q v_s)_1 \qquad (6.198)$$

For pipe components in general, assuming $z_1 = z_2$, and $v_1 = v_2$, the integral momentum equation is used to solve for the pressure drop as a function of the shear stress as follows (derived in Section 6.5):

$$\Delta p = \frac{\tau_w L}{R_h} \qquad (6.199)$$

However, the problem is that one cannot theoretically model (or ignore) the wall shear stress, τ_w in the integral momentum for pipe flow with components and thus cannot analytically determine the pressure drop, Δp in Equation 6.199. Furthermore, because the head loss (Equation 6.197) and the pressure drop (Equation 6.199) are expressed in terms of the wall shear stress, τ_w, in order to evaluate the respective expressions, one must first empirically interpret Δp (express τ_w) in terms of easy-to-measure velocity, v. Thus, the integral momentum equation is supplemented with dimensional analysis in order to empirically model the wall shear stress, τ_w for pipe flow with components, and thus Δp by the use of a drag coefficient, C_D. Dimensional analysis is used to empirically interpret Δp (express τ_w) in terms of easy-to-measure velocity, v. Specifically, the momentum theory is supplemented with dimensional analysis in Chapter 7 in order to derive an expression for the actual pressure drop/rise, Δp for the pipe components in general, for which the result is presented as follows:

$$\frac{\Delta p}{\dfrac{\rho v^2}{2}} = \frac{F_P}{F_{dynpres}} = \frac{2}{E} = C_D = k = \phi\left(R, \frac{\varepsilon}{D}, \frac{L_i}{L}\right) \qquad (6.200)$$

where E = the Euler number, R = the Reynolds number, ε/D is the relative roughness, and L_i/L is the geometry (see Chapter 7 for applicability of the above-listed dimensionless numbers/pi terms). Finally, the empirical expression for the actual pressure drop (Equation 6.200) is substituted into the integral form of the energy equation (Equation 6.197), yielding an empirical expression for the minor head loss, h_f as follows:

$$h_{f,min} = \frac{\tau_w L}{\gamma R_h} = \frac{\Delta p}{\gamma} = C_D \frac{v^2}{2g} = k \frac{v^2}{2g} \qquad (6.201)$$

where the drag coefficient, C_D is expressed in terms of the more commonly defined minor head loss coefficient, k. Furthermore, one can easily compute v_i used in Equation 6.201 by measuring Δp using a pitot-static tube.

6.8.5.3 Evaluation of the Minor Head Loss Coefficient, k

One may recall that, in general, both pressure forces, F_p and viscous forces, F_V (shear forces, F_f) may be exerted on the pipe wall as a result of a pipe component. Therefore, regardless of whether the particular flow device results in a pressure drop or a pressure rise, the actual pressure drop/rise, Δp may: (1) be equal to the dynamic pressure (ideal pressure), $\rho v^2/2$; (2) be greater than the dynamic pressure; or (3) be less than the dynamic pressure. As noted above, because, in general, the actual pressure drop/rise, Δp is unknown, k is determined empirically from experimentation. Examining the standard form of the minor head loss coefficient, k yields the following:

$$\frac{F_P}{F_{dynpres}} = \frac{\Delta p}{\frac{\rho v^2}{2}} = C_D = k = \frac{Actual\ Pressure\ Force}{Ideal\ Force} = \frac{Actual\ Presdrop}{Ideal\ Presdrop} = \frac{staticpres}{dynpress} \quad (6.202)$$

where the results of dimensional analysis introduce a drag coefficient, C_D which expresses the ratio of a real flow variable (in this case force or pressure) to its corresponding ideal flow variable. In the cases where the pipe component results in a pressure drop, Δp, the energy equation yields the following:

$$p_1 - p_2 = \underbrace{\Delta p}_{presdrop} = \underbrace{\left(\frac{\rho v_2{}^2}{2} - \frac{\rho v_1{}^2}{2}\right)}_{dynpress} + \underbrace{\frac{\tau_w L}{R_h}}_{flowres} \quad (6.203)$$

Thus, the minor head loss coefficient, k is derived as follows:

$$\frac{F_P}{F_{dynpres}} = \frac{\Delta p}{\frac{\rho v^2}{2}} = \frac{\left(\frac{\rho v_2{}^2}{2} - \frac{\rho v_1{}^2}{2}\right) + \frac{\tau_w L}{R_h}}{\frac{\rho v^2}{2}} = k \quad (6.204)$$

where $k > 1$ in the case when the actual pressure drop, Δp is greater than the dynamic pressure, and $k < 1$ when the actual pressure drop, Δp is less than the dynamic pressure. In the cases where the pipe component results in a pressure rise, Δp the energy equation yields the following:

$$p_2 - p_1 = \underbrace{\Delta p}_{presrise} = \underbrace{\left(\frac{\rho v_1{}^2}{2} - \frac{\rho v_2{}^2}{2}\right)}_{dynpress} - \underbrace{\frac{\tau_w L}{R_h}}_{flowres} \quad (6.205)$$

Thus, the minor head loss coefficient, k is derived as follows:

$$\frac{F_P}{F_{dynpres}} = \frac{\Delta p}{\frac{\rho v^2}{2}} = \frac{\left(\frac{\rho v_1{}^2}{2} - \frac{\rho v_2{}^2}{2}\right) - \frac{\tau_w L}{R_h}}{\frac{\rho v^2}{2}} = k \quad (6.206)$$

where $k > 1$ in the case when the actual pressure rise, Δp is greater than the dynamic pressure, and $k < 1$ when the actual pressure rise, Δp is less than the dynamic pressure. In particular, for the case of a pipe exit into a reservoir, $v_1 = v$ and $v_2 = 0$, where there is a dissipation of kinetic energy into internal energy and heat transfer, and the wall shear stress, τ_w is not significant, and the energy equation yields the actual pressure rise, Δp as follows:

$$p_2 - p_1 = \underbrace{\Delta p}_{presrise} = \frac{\rho v^2}{2} \quad (6.207)$$

where the actual pressure rise, Δp is exactly equal to the dynamic pressure, $\rho v^2/2$. Thus, the minor head loss coefficient, k is derived as follows:

$$\frac{F_P}{F_{dynpres}} = \frac{F_{dynpres}}{F_{dynpres}} = \frac{\Delta p}{\frac{\rho v^2}{2}} = \frac{\frac{\rho v^2}{2}}{\frac{\rho v^2}{2}} = k = 1 \tag{6.208}$$

It is interesting to examine the case of the sudden pipe expansion for which the minor head loss coefficient, k was analytically derived. Assuming a horizontal pipe ($z_1 = z_1$) and thus that F_G is not significant, the viscous force, F_V due to the wall shear stress is not significant, and that the minor head loss due to a sudden expansion is evaluated using the upstream velocity, v_1, the analytical expression for the actual pressure rise, Δp derived by applying the three governing equations yielded:

$$\Delta p = \gamma h_{f,min} = \gamma \left[1 - \frac{A_1}{A_2}\right]^2 \left(\frac{v_1^2}{2g}\right) = \gamma \left[1 - \left(\frac{D_1}{D_2}\right)^2\right]^2 \left(\frac{v_1^2}{2g}\right) \tag{6.209}$$

Thus, the minor head loss coefficient, k is derived as follows:

$$\frac{F_P}{F_{dynpres}} = \frac{\Delta p}{\frac{\rho v^2}{2}} = \frac{\frac{\rho v_1^2}{2}\left[1 - \frac{A_1}{A_2}\right]^2}{\frac{\rho v^2}{2}} = \left[1 - \frac{A_1}{A_2}\right]^2 = k \tag{6.210}$$

The value for k changes from the limiting/extreme case of a square-edged exit with an area ratio of $A_1/A_2 = 0$ and $v_1 = v$ and $v_2 = 0$, where $k = 1$ (a pipe exit into a reservoir), to the other extreme case of a constant diameter pipe with an area ratio of $A_1/A_2 = 1$, and $v_1 = v_2 = v$ where $k = 0$ (flow through a pipe). Evaluation of the velocity and the minor head loss coefficient, k and determination of the minor head loss due to various pipe components in a pipe system are presented in Chapter 8.

6.9 Modeling the Flow Resistance as a Loss in Flowrate in Internal Flow

One of the major applications of the study of fluid mechanics is the determination of the rate of flow of fluids. As a result, numerous flow rate measuring devices/flow meters have been developed, which differ depending upon whether the flow is through a pipe or through an open channel. Additionally, the flow rate measuring devices will vary depending on various factors including cost, space, size, accuracy, precision, sophistication, versatility, capacity, pressure drop, and the governing principle of operation. Furthermore, most of the flow measurement devices can also serve as flow control devices. The most common types of flow measurement devices used in pipe flow include pitot-static tubes, orifice meters, nozzle meters, and venturi meters, whereas, the most common types of flow measurement devices used in open channel flow include pitot-static tubes, sluice gates, weirs, overflow spillways, the Parshall flume (venturi flume), and contracted openings.

Most flow measurement devices (flow meters) are basically velocimeters that directly measure an ideal pressure difference, which is related to the velocity by the Bernoulli equation; thus, the velocity is indirectly measured. Then, application of the continuity equation is used to determine the flowrate. Thus, many flow measurement devices used to measure the velocity and thus the flowrate for pipe and open channel flow apply the Bernoulli equation (assumes ideal flow) and are considered to be ideal flow meters. However, in real flow, the effects of viscosity, compressibility, and other real flow effects cause the actual velocity, v_a and thus the actual flowrate, Q_a to be less than the ideal velocity and the ideal flowrate, respectively. As a result, corrections for these real flow effects are made by applying a velocity or a discharge coefficient, respectively. The discharge coefficient, C_d represents both a velocity coefficient, which accounts for head loss due to friction, and a contraction coefficient, which accounts for the contraction of the streamlines of the fluid flow (and thus the actual cross-sectional area, A_a).

6.9.1 Evaluation of the Actual Discharge

In the case where the flow resistance is due to a flow measurement device for either pipe flow or open channel flow, the flow resistance (shear stress or drag force) is ultimately modeled as a reduced/actual discharge, Q_a in the differential form of the continuity equation because the actual flowrate is needed for the design of a flow-measuring device. Thus, the associated minor head loss is typically accounted for by the use/calibration of a discharge coefficient to determine the actual discharge. Evaluation of the actual discharge requires a "subset level" application of the governing equations. A "subset level" application of the governing equations focuses only on the given element causing the flow resistance. The assumption of "ideal" flow implies that the flow resistance is modeled only in the momentum equation (and thus the subsequent assumption of "real" flow) (see Table 4.1). Thus, in the derivation of the actual discharge, first ideal is assumed in order to determine and expression for the ideal velocity by applying the Bernoulli equation. Then, application of the continuity equation is used to determine an expression for the ideal discharge and the actual discharge. And, finally, the integral form of the momentum equation (supplemented by dimensional analysis) is applied to solve for an unknown actual velocity, v_a, and actual area, A_a. Thus, the flow resistance (shear stress or drag force) is ultimately modeled as a reduced/actual discharge, Q_a in the differential form of the continuity equation. The flow resistance causes an unknown pressure drop, Δp in the case of pipe flow and an unknown Δy in the case of open channel flow, which causes an unknown head loss, h_f where the head loss is due to a conversion of kinetic energy to heat, which is modeled/displayed in the integral form of the energy equation. In the determination of the reduced/actual discharge, Q_a, which is less than the ideal discharge, Q_i, the exact reduction in the flowrate is unknown because the head loss, h_f causing it is unknown. Furthermore, although the differential form of the continuity equation is typically used to compute the reduced/actual discharge, Q_a for a flow-measuring device, because the head loss, h_f causes the reduced/actual discharge, Q_a, the head loss may also be accounted for as a minor loss in the integral form of the energy equation, as illustrated in Chapter 8; a minor head loss is not typically modeled in the energy equation for open channel flow. The assumption of ideal flow and thus applying the Bernoulli equation to measure the ideal velocity of flow, $v_i = \sqrt{2\Delta p/\rho}$ or $v_i = \sqrt{2g\Delta y}$ by using a flow-measuring device (see Chapters 8 and 9) yields an expression for ideal velocity as a function of an ideal pressure difference, Δp or Δy, respectively, which is directly measured. Therefore, the associated minor head loss with the velocity measurement is accounted for by the discharge coefficient (in the application of the continuity equation). The head loss, h_f causing the actual velocity,

$$v_a = \sqrt{\frac{2}{\rho}\left(\Delta p - \frac{\tau_w L}{R_h}\right)} \quad \text{or} \quad v_a = \sqrt{2g\left(\Delta y - \frac{\tau_w L}{\gamma R_h}\right)},$$

and the actual area, A_a and thus the actual discharge, Q_a is caused by both pressure and friction forces, where the reduced/actual discharge, Q_a is less than the ideal discharge, Q_i. However, because the friction/viscous forces (due to shear stress, τ_w) due to the internal flow measurement cannot be theoretically modeled in the integral momentum equation, the actual pressure drop, Δp or Δy, cannot be analytically determined, thus the reduction in the discharge, actual discharge, Q_a, which is less than the ideal discharge, Q_i, cannot be theoretically determined. Therefore, one cannot derive an analytical expression for the actual discharge, Q_a from the continuity equation. Specifically, the complex nature of the flow does not allow a theoretical modeling of the existence of the viscous force due to the flow measurement device in the momentum equation and thus does not allow a theoretical derivation of the pressure drop. Thus, the friction/viscous forces; the actual pressure drop, Δp or Δy; the minor head loss, h_f; and the actual the reduced/actual discharge, Q_a are determined empirically using dimensional analysis. As such, one must resort to dimensional analysis (which supplements the momentum theory) in order to derive an expression for the actual discharge, Q_a, which involves the definition of a discharge coefficient, C_d that represents the flow resistance. The flow resistance equation for the actual discharge is derived below and in Chapter 7. Actual discharges and the evaluation of the discharge coefficients are addressed in detail in Chapters 8 and 9.

6.9.2 Application of the Bernoulli Equation: Derivation of the Ideal Velocity

In the derivation of the actual discharge for a given flow-measuring device, first, ideal flow is assumed in order to determine an expression for the ideal velocity by applying the Bernoulli equation as follows:

$$\left(\frac{p}{\gamma} + z + \frac{v^2}{2g}\right)_1 - \left(\frac{p}{\gamma} + z + \frac{v^2}{2g}\right)_2 = 0 \tag{6.211}$$

Assuming $z_1 = z_2$, this equation has one unknown, which is the ideal velocity of the flow at the restriction in the flow-measuring device. The Bernoulli equation yields an expression for the ideal velocity, $v_i = \sqrt{2\Delta p/\rho}$ or $v_i = \sqrt{2g\Delta y}$, as a function of an ideal pressure difference, Δp or Δy, respectively, which is directly measured by the use of the flow-measuring device. Furthermore, the ideal velocity is used in the definition of the actual discharge below. And, finally, the exact expression for the ideal velocity will depend upon the particular flow measurement device (see Chapters 8 and 9 for details).

6.9.3 Application of the Continuity Equation: Derivation of the Ideal Discharge and the Actual Discharge

In the derivation of the actual discharge for a given flow-measuring device, next, application of the differential continuity equation is used to determine an expression for the ideal discharge, and the actual discharge as follows:

$$Q = vA \tag{6.212}$$

Thus, application of the continuity equation yields an expression for the ideal discharge, and the actual discharge for pipe flow-measuring devices and open channel flow-measuring devices, respectively, as follows:

$$Q_i = v_i A_i \tag{6.213}$$

$$Q_a = \frac{v_a}{\sqrt{\dfrac{2\Delta p}{\rho}}} \frac{A_a}{A_i} Q_i \tag{6.214}$$

$$Q_a = \frac{v_a}{\sqrt{2g\Delta y}} \frac{A_a}{A_i} Q_i \tag{6.215}$$

6.9.4 Application of the Momentum Equation and Dimensional Analysis: Derivation of the Actual Velocity, Actual Area, Actual Discharge, and the Discharge Coefficient

In the derivation of the actual discharge for a given flow-measuring device, the subsequent assumption of real flow is made in order to determine an expression for the actual velocity, v_a by applying the integral momentum (supplemented by dimensional analysis) as follows:

$$\sum F_s = (F_G + F_P + F_V + F_{other})_s = (\rho Q v_s)_2 - (\rho Q v_s)_1 \tag{6.216}$$

Assuming $z_1 = z_2$, and that only the pressure and viscous forces are important, the momentum equation is used to solve for the actual velocity for pipe flow-measuring devices and open channel flow-measuring devices, respectively, as follows:

$$v_a = \sqrt{\frac{2}{\rho}\left(\Delta p - \frac{\tau_w L}{R_h}\right)} \tag{6.217}$$

$$v_a = \sqrt{2g\left(\Delta y - \frac{\tau_w L}{\gamma R_h}\right)} \tag{6.218}$$

Substituting Equation 6.217 into Equation 6.214 and Equation 6.218 into Equation 6.215 for the actual velocity, respectively, yields the following flow resistance equations for the actual discharge for pipe flow-measuring devices, and open channel flow-measuring devices, respectively, as follows:

$$Q_a = \frac{\sqrt{\dfrac{2}{\rho}\left(\Delta p - \dfrac{\tau_w L}{R_h}\right)}}{\sqrt{\dfrac{2\Delta p}{\rho}}} \frac{A_a}{A_i} Q_i \tag{6.219}$$

$$Q_a = \frac{\sqrt{2g\left(\Delta y - \dfrac{\tau_w L}{\gamma R_h}\right)}}{\sqrt{2g\Delta y}} \frac{A_a}{A_i} Q_i \tag{6.220}$$

However, the problem is that one cannot theoretically model the wall shear stress, τ_w in the integral momentum equation for flow-measuring devices. Thus, one cannot analytically determine pressure drop, Δp or Δy and thus v_a, and cannot measure actual area, A_a in Equations 6.219 and 6.220. Thus, the integral momentum equation is supplemented with dimensional analysis in order to empirically model the wall shear stress, τ_w for the flow-measuring devices, and thus Δp or Δy and thus v_a and Q_a, by the use of a discharge coefficient, C_d. The drag coefficient, C_D

(discharge coefficient, C_d) represents the flow resistance and is empirically evaluated. The resulting flow resistance equations for the actual discharge for pipe flow-measuring devices, and open channel flow-measuring devices are given, respectively, as follows:

$$Q_a = \frac{\sqrt{\dfrac{2}{\rho}\left(\Delta p - \dfrac{\tau_w L}{R_h}\right)}}{\sqrt{\dfrac{2\Delta p}{\rho}}} \frac{A_a}{A_i} Q_i = C_d Q_i \tag{6.221}$$

$$Q_a = \frac{\sqrt{2g\left(\Delta y - \dfrac{\tau_w L}{\gamma R_h}\right)}}{\sqrt{2g\Delta y}} \frac{A_a}{A_i} Q_i = C_d Q_i \tag{6.222}$$

where the discharge coefficient, C_d accounts for (indirectly models) the minor head loss associated with the ideal flow measurement, and the actual discharge. Furthermore, one can easily compute v_i used in Equation 6.221 by measuring Δp in a pipe flow-measuring device and can easily measure A_i. As well, one can easily compute v_i used in Equation 6.222 by measuring Δy in an open channel flow-measuring device and can easily measure A_i.

Finally, one may note that because the dominant forces and the important variables to consider in the dimensional analysis procedure to derive an empirical expression for Q_a differ for pipe flow versus open channel, these two derivations are done separately in Chapter 7 and the results presented below.

6.9.4.1 Supplementing the Momentum Equation with Dimensional Analysis: Derivation of the Reduced/Actual Discharge Equation for Pipe Flow

The integral momentum equation is supplemented with dimensional analysis in order to empirically model the wall shear stress, τ_w for the pipe flow-measuring devices and thus Δp, v_a, and Q_a by the use of a drag coefficient, C_D (discharge coefficient, C_d). Therefore, the flow resistance (unknown head loss, h_f) represented by the actual discharge equation is modeled by the definition of a "drag coefficient, C_D," which is defined by dimensional analysis, and evaluated empirically. The result of the derivation of the actual discharge equation along with the a drag coefficient, C_D (discharge coefficient, C_d) for pipe flow-measuring devices is presented as follows:

$$\frac{Q_a}{Q_i} = \frac{F_{dynpres}}{F_P} = \frac{v_a A_a}{v_i A_i} = \frac{v_a}{\sqrt{\dfrac{2\Delta p}{\rho}}} \frac{A_a}{A_i} = \sqrt{\frac{E}{2}} \frac{A_a}{A_i} = C_v C_c = C_d = \phi\left(R, \frac{L_i}{L}\right) \tag{6.223}$$

where E = the Euler number, R = the Reynolds number, and L_i/L is the geometry (see Chapter 7 for applicability of the above listed dimensionless numbers/pi terms). Thus, the resulting actual discharge, Q_a derived from dimensional analysis is given as follows:

$$Q_a = \left[\underbrace{\left(\frac{v_a}{\sqrt{2\Delta p/\rho}}\right)}_{C_v}\underbrace{\left(\frac{A_a}{A_i}\right)}_{C_c}\right] Q_i \tag{6.224}$$

$$\underbrace{\phantom{\left[\left(\frac{v_a}{\sqrt{2\Delta p/\rho}}\right)\left(\frac{A_a}{A_i}\right)\right]}}_{C_d}$$

Furthermore, the actual area, A_a is determined as a function of the contraction coefficient, C_c and the ideal area, A_i as follows:

$$A_a = C_c A_i \tag{6.225}$$

And thus, from continuity, the actual discharge, Q_a is computed as a function of the discharge coefficient, C_d and the ideal discharge, Q_i as follows:

$$Q_a = v_a A_a = (C_v v_i)(C_c A_i) = C_v C_c v_i A_i = C_v C_c Q_i = C_d Q_i \tag{6.226}$$

where the discharge coefficient, C_d is defined as the product of the velocity coefficient, C_v and the contraction coefficient, C_c as follows:

$$C_d = C_v C_c \tag{6.227}$$

Evaluation of the discharge coefficient, C_d and determination of the actual discharge in flow measurement devices in pipe flow are presented in Chapter 8.

6.9.4.2 Supplementing the Momentum Equation with Dimensional Analysis: Derivation of the Reduced/Actual Discharge Equation for Open Channel Flow

The integral momentum equation is supplemented with dimensional analysis in order to empirically model the wall shear stress, τ_w for the open channel flow-measuring devices, and thus Δy, v_a, and Q_a by the use of a drag coefficient, C_D (discharge coefficient, C_d). Thus, the flow resistance (unknown head loss, h_f) represented by the actual discharge equation is modeled by the definition of a "drag coefficient, C_D," which is defined by dimensional analysis and evaluated empirically. The result of the derivation of the actual discharge equation along with the a drag coefficient, C_D (discharge coefficient, C_d) for open channel flow-measuring devices is presented as follows:

$$\frac{Q_a}{Q_i} = \frac{F_{dynpres}}{F_P} = \frac{v_a A_a}{v_i A_i} = \frac{v_a}{\sqrt{2g\Delta y}} \frac{A_a}{A_i} = \sqrt{\frac{E}{2}} \frac{A_a}{A_i} = \sqrt{\frac{F}{2}} \frac{A_a}{A_i} = C_v C_c = C_d = \phi\left(R, W, \frac{L_i}{L}\right) \tag{6.228}$$

where E = the Euler number, F = the Froude number, R = the Reynolds number, W = the Weber number, and L_i/L is the geometry (see Chapter 7 for applicability of the above listed dimensionless numbers/pi terms). Therefore, the resulting actual discharge, Q_a derived from dimensional analysis is given as follows:

$$Q_a = \left[\underbrace{\underbrace{\left(\frac{v_a}{\sqrt{2g\Delta y}}\right)}_{C_v} \underbrace{\left(\frac{A_a}{A_i}\right)}_{C_c}}_{C_d} \right] Q_i \tag{6.229}$$

Furthermore, the actual area, A_a is determined as a function of the contraction coefficient, C_c and the ideal area, A_i as follows:

$$A_a = C_c A_i \tag{6.230}$$

And thus, from continuity, the actual discharge, Q_a is computed as a function of the discharge coefficient, C_d and the ideal discharge, Q_i as follows:

$$Q_a = v_a A_a = (C_v v_i)(C_c A_i) = C_v C_c v_i A_i = C_v C_c Q_i = C_d Q_i \qquad (6.231)$$

where the discharge coefficient, C_d is defined as the product of the velocity coefficient, C_v and the contraction coefficient, C_c as follows:

$$C_d = C_v C_c \qquad (6.232)$$

Evaluation of the discharge coefficient, C_d and determination of the actual discharge in flow measurement devices in open flow is presented in Chapter 9.

6.9.5 A Comparison of the Velocity Profiles for Ideal and Real Flows

The actual discharge, Q_a (measured) in real flow is less than the ideal discharge, Q_i assumed in ideal flow due to the effects of the flow resistance in real flow. As a result, the average velocity for real flow, $v_{ave,a} = Q_a/A$ will be less than the average velocity for ideal flow, $v_{ave,i} = Q_i/A$. The velocity profile for an ideal flow is a rectangular distribution as illustrated in Figures 6.3a and 6.12a. This rectangular distribution is a direct result of assuming an ideal

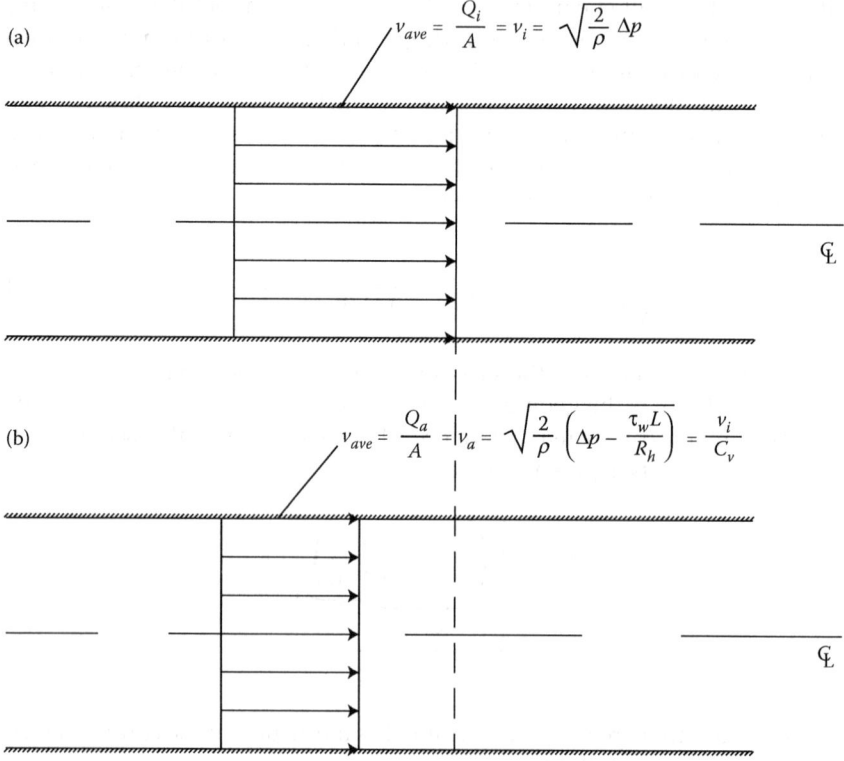

FIGURE 6.12
Comparison between rectangular velocity distributions. (a) Rectangular velocity distribution for ideal flow. (b) Rectangular velocity distribution for the average velocity for real (laminar or turbulent) flow.

flow with no viscosity. The rectangular distribution represents ideal velocity, v_i, which is modeled by the average velocity, $v_{ave,i} = Q_i/A$ and is given as follows:

$$v_{ave,i} = \frac{Q_i}{A} = v_i = \underbrace{\sqrt{\frac{2}{\rho}(\Delta p)}}_{idealvel} \tag{6.233}$$

For a real flow, regardless of whether the flow is laminar or turbulent, the actual velocity, v_a, which is modeled by the average velocity, $v_{ave,a} = Q_a/A$, is represented by a rectangular distribution as illustrated in Figure 6.12b and is given as follows:

$$v_{ave,a} = \frac{Q_a}{A} = v_a = \sqrt{\frac{2}{\rho}\left(\Delta p - \frac{\tau_w L}{R_h}\right)} \tag{6.234}$$

A comparison of the rectangular velocity profile for the average velocity for real flow, $v_a = v_{ave,a} = Q_a/A = C_v v_i$ illustrated in Figure 6.12b to the rectangular velocity profile for ideal flow, $v_i = v_{ave,i} = Q_i/A$ illustrated in Figure 6.12a indicates that the average velocity, $v_a = v_{ave,a} = Q_a/A = C_v v_i$ in the real flow case is less than the average velocity, $v_i = v_{ave,i} = Q_i/A$ in the ideal flow case. This is due to the effects of the flow resistance, which is modeled by the coefficient of velocity term, C_v, where the actual discharge, Q_a is less than the ideal discharge, Q_i. Furthermore, the velocity profile for an ideal flow is a rectangular distribution, which is a direct result of assuming an ideal flow with no viscosity. However, in the case of a real flow (either laminar or turbulent), the resulting corresponding velocity profiles deviate from the ideal rectangular distribution, which is a direct result of modeling the flow resistance in real flow.

6.10 Modeling the Flow Resistance as a Loss in Pump and Turbine Efficiency in Internal Flow

In the cases where the flow resistance is due to pipe or channel friction (major head loss), pipe component (minor head loss), and/or a flow-measuring device (minor head loss), the flow resistance contributes to the power losses and thus the determination of the efficiency of pumps and turbines (see Chapter 4). In particular, the total head loss, h_f (major and minor) will increase the amount of power that a pump must deliver to the fluid; the head loss, h_f represents the additional height that the fluid needs to be elevated by the pump in order to overcome the losses due to the flow resistance in the pipe or channel. Some of the pumping power will be used to pump the fluid through the pipe or channel, while some will be used to compensate for the energy lost in the pressure drop, Δp caused by the pipe or channel friction, pipe component, or flow-measuring device. A similar analogy may be deduced for evaluating the efficiency of a turbine. Although the derivations of the expressions for the efficiency of pumps, η_{pump} and turbines, $\eta_{turbine}$ are determined analytically in Chapter 4, dimensional analysis is applied in order to derive empirical expressions for the efficiency of pumps and turbines. Empirical expressions for the efficiency of pumps and turbines will allow their respective empirical evaluations. Because a turbine is the exact opposite of a pump, an empirical derivation of the efficiency of a pump is actually conducted, and the empirical derivation of the efficiency of a turbine is interpreted from the results.

6.10.1 Evaluation of the Efficiency of Pumps and Turbines

The efficiency of pumps and turbines is discussed in detail in Chapter 4. The efficiency of a pump (Equation 4.197) expresses the ratio of a real/actual flow variable (hydraulic power output from the pump) to its corresponding ideal flow variable (shaft power input to the pump). And, the efficiency of a turbine (Equation 4.217) expresses the ratio of a real/actual flow variable (shaft power output from the turbine) to its corresponding ideal flow variable (hydraulic power input to the turbine). Dimensional analysis is applied in order to derive empirical expressions for the efficiency of pumps and turbines, which will allow their respective empirical evaluations.

6.10.1.1 Supplementing the Momentum Equation with Dimensional Analysis: Derivation of the Efficiency of Pumps and Turbines

An empirical expression for the efficiency of a pump, η_{pump} in the theoretical expression for the efficiency of a pump (Equation 4.197) is sought in order to allow its empirical evaluation; the empirical expression of the efficiency of a turbine (Equation 4.217) is interpreted from the results. This is accomplished by using dimensional analysis in order to derive an expression for the efficiency of a pump, η_{pump}, which involves the definition of a drag coefficient C_D (pump efficiency, η_{pump}) that represents flow resistance, and which is empirically evaluated. Thus, the momentum theory is supplemented with dimensional analysis in order to derive an expression for the efficiency of a pump, η_{pump}. The results of the derivation of the empirical expression for the efficiency of a pump, η_{pump} is presented as follows:

$$\eta_{pump} = \frac{(P_p)_{out}}{(P_p)_{in}} = \frac{\gamma Q h_{pump}}{\omega T_{shaft,in}} = \frac{\gamma Q \frac{\Delta p}{\gamma}}{\omega T_{shaft,in}} = \frac{Q \Delta p}{\omega T_{shaft,in}} = \frac{v L^2 \Delta p}{\frac{vFL}{L}} = \frac{\Delta p L^2}{F} = \frac{F_P}{F_I}$$

$$= \frac{\Delta p L^2}{\rho v^2 L^2} = \frac{\Delta p}{\rho v^2} = \frac{1}{E} = C_D = \phi \left(C_H, C_Q, C_P, R, C, \frac{\varepsilon}{D} \right) \qquad (6.235)$$

where E = the Euler number, C_H = the head coefficient, C_Q = the capacity coefficient, C_P = the power coefficient, R = the Reynolds number, C = the Cauchy number, and ε/D = the relative blade surface roughness (see Chapter 7 for applicability of the above listed dimensionless numbers/pi terms). And, finally, the empirical expression of the efficiency of a turbine is interpreted by its definition (exact opposite of a pump) as follows:

$$\eta_{turbine} = \frac{(P_t)_{out}}{(P_t)_{in}} = \frac{\omega T_{shaft,out}}{\gamma Q h_{turbine}} = \frac{F_I}{F_P} = \frac{\rho v^2 L^2}{\Delta p L^2} = \frac{\rho v^2}{\Delta p} = E = C_D$$

$$= \phi \left(C_H, C_Q, C_P, R, C, \frac{\varepsilon}{D} \right) \qquad (6.236)$$

End-of-Chapter Problems

Problems with a "C" are conceptual problems.

Introduction

6.1C In the application of the governing equations, although in some flow cases, one may ignore (or not need to model) the flow resistance, in most practical flow cases in general, one must account for the fluid viscosity and thus model the resulting shear stress and flow resistance. Explain the various flow cases, explain when it is necessary to model flow resistance, and explain how the flow resistance (viscous force) is modeled. Also, explain when it is not necessary to model flow resistance.

6.2C The application of the governing equations (continuity, energy, and momentum) for real flow (internal) and ideal flow (flow-measuring devices for internal flow, and external flow around objects) requires the modeling of flow resistance (frictional losses). The modeling of flow resistance requires a "subset level" application of the governing equations. Identify the flow resistance (frictional losses) terms that need to be modeled for a given flow type.

6.3C The flow resistance in internal and external flow (major head loss, minor head loss, actual discharge, and drag force) is modeled by a "subset level" application of the appropriate governing equations, as presented in this chapter. Explain what a "subset level" application of the appropriate governing equations implies.

6.4C Assuming that flow resistance is to be accounted for, depending upon the specific flow situation, the flow type is assumed to be either a "real" flow or an "ideal" flow. The distinction between ideal and real flow determines how the flow resistance is modeled in the "subset level" application of the appropriate governing equations. Explain how the flow resistance is modeled for the assumption of "real" flow vs. "ideal" flow.

6.5C The application of the governing equations (continuity, energy, and momentum) for real flow (internal) and ideal flow (flow-measuring devices for internal flow, and external flow around objects) requires the modeling of flow resistance (frictional losses). The modeling of flow resistance requires a "subset level" application of the governing equations. Furthermore, the use of dimensional analysis (Chapter 7) in supplement to a "subset level" application of the appropriate governing equations is needed to derive flow resistance equations when there is no theory to model the friction force. Explain the flow types for which it is necessary, and the flow types for which it is unnecessary, to use dimensional analysis (Chapter 7) in supplement to a "subset level" application of the appropriate governing equations.

Modeling Flow Resistance: A Subset Level Application of the Governing Equations

6.6C The derivation of a given flow resistance equation (major head loss, minor head loss, actual discharge, and drag force) involves a "subset level" application of the appropriate governing equations and in some cases supplementing the momentum equation with dimensional analysis. A "subset level" application of the appropriate governing equations focuses only on the given element causing the flow resistance. Explain how the momentum equation is supplemented with dimensional analysis.

6.7C The flow resistance is modeled as a drag force in external flows, while it is modeled as a head loss (major head loss, minor head loss, or actual discharge) in internal flows. Furthermore, in the flow of any real fluid, the fluid has to do work against flow resistance, which results in a continual dissipation of energy in the momentum transfer. The flow resistance causes an unknown pressure drop, Δp (or S_f), which causes an unknown head loss, h_f where the head loss is due to a conversion of kinetic energy to heat, which is modeled/displayed in the integral form of the energy equation (Equation 4.44). Explain how the expression for the unknown pressure drop, Δp (and thus for the drag force, the major head loss, the minor head loss, and the actual discharge) is derived. Also, explain how the flow resistance (unknown head loss, h_f) represented by each of the above listed flow resistance equations is modeled.

6.8C   Explain why the flow resistance is modeled as a drag force in external flows, while it is modeled as a head loss (major head loss, minor head loss, or actual discharge) in internal (pipe and open channel) flows.

Derivation and Application of the Flow Resistance Equations, and Derivation and Evaluation of the Drag Coefficients

6.9C   When the friction/viscous forces cannot either be theoretically modeled or ignored in the integral form of the momentum equation, the momentum theory will be exhausted due to an unknown head loss, h_f. As a result, in order to derive an expression for the drag force, the major head loss, the minor head loss, and the actual discharge, the momentum theory is supplemented with dimensional analysis. The flow resistance (unknown head loss, h_f) represented by each of the above listed resistance equations is modeled by the definition of a "drag coefficient, C_D," which is defined by dimensional analysis and evaluated empirically. Explain what the drag coefficient, C_D represents.

6.10C   The appropriate flow resistance equations (drag force, the major head loss, the minor head loss, and the actual discharge) and their corresponding "drag coefficient, C_D" for both external and internal flow are derived in Chapter 7 and presented in this chapter. List each of the flow resistance equations and their specific corresponding "drag coefficients, C_D" for both external and internal flow.

6.11C   The flow resistance equations (drag force, the major head loss, the minor head loss, and the actual discharge) are presented in this chapter. List which chapters present the empirical evaluation of the respective drag coefficients and application of the respective flow resistance equations.

Modeling the Flow Resistance as a Loss in Pump and Turbine Efficiency in Internal Flow

6.12C   In the cases where the flow resistance is due to pipe or channel friction (major head loss), pipe component (minor head loss), and/or a flow-measuring device (minor head loss), explain how the flow resistance (total head loss) contributes to the power losses and thus the determination of the efficiency of pumps and turbines (Chapter 4). Also, explain how the efficiency, η of a pump or a turbine is represented in dimensional analysis.

Types of Flow

6.13C   Given the following types of flow: ideal or real flow, internal or external flow, pressure or gravity flow, laminar or turbulent flow, and developing or developed flow, highlight the importance of the distinction between ideal and real flow, and the distinction between laminar and turbulent flow, in the modeling of flow resistance.

Internal Flow versus External Flow

6.14C   The results sought in a given flow analysis will depend upon whether the flow is internal (pipe or open channel flow) or external. Explain what results are sought in the two flow types.

Pipe Flow versus Open Channel Flow

6.15C   For the derivation of the major head loss equation, list the similarities and the differences between pipe and open channel flow.

6.16C   In the derivation of the major head loss equation, the boundary conditions are different for pipe flow and open channel flow. Specifically, in pipe flow, the shear stress restraining the fluid motion is distributed around the entire boundary of the cross section. However, in open channel flow, there is a free surface on which the shear stress is insignificant, and thus the flow cross section is assumed to be equivalent to the lower half of a pipe flow. Furthermore, one may note that this equivalence is not exact because while the maximum velocity for pipe flow is at the centerline of the pipe, the maximum velocity for open channel flow is at a point somewhat below the free surface as illustrated in Figure 6.1a. Explain why the maximum velocity for open channel flow is at a point somewhat below the free surface, and state what is assumed in practice.

6.17C   Although this chapter discusses the laws of flow resistance assuming pipe flow (internal flow under pressure), the general results are also applicable for open channel flow (internal flow under gravity). Furthermore, while major head losses are applicable to both pipe flow and open channel flow, minor losses are modeled and thus determined differently for pipe flow vs. open channel flow. Explain the difference in modeling minor head losses in pipe flow vs. open channel flow.

Real Flow versus Ideal Flow

6.18C   State the basic difference between the assumption made for real flow vs. ideal flow, and discuss the validity of the assumed type of flow.

6.19C   Regardless of whether the flow is assumed to be real or ideal, the conversion of energy from one type to another will result in either a pressure drop or a pressure rise. Assuming that flow resistance is to be accounted for, list the flow cases for which there will be a pressure drop vs. a pressure rise, and state what type of flow resistance results in each flow case.

Ideal Flow

6.20C   Although in some flow cases, one may ignore (or not need to model) flow resistance, in most practical flow cases in general, one must account for the fluid viscosity and thus model the resulting shear stress and flow resistance. List the flow cases for which it is necessary to model flow resistance, and the flow cases for which it is not necessary to model flow resistance.

6.21C   The viscosity of real flow (a real fluid) introduces a great complexity in a fluid flow situation, namely friction due to the viscous shearing effects. The assumption of ideal flow (inviscid flow) assumes an ideal fluid that has no viscosity and thus experiences no shear stresses (friction) as it flows. Although this is an idealized flow situation that does not exist, such an assumption of ideal flow produces reasonably accurate results in flow systems where friction does not play a significant role. Explain the role of the assumption of ideal flow in the study of real flow.

6.22C   In the assumption of ideal flow, the case of ignoring the flow resistance implies application of the Bernoulli equation and assuming no viscous force ($F_V = 0$) in the application of the momentum equation. Such an assumption of ideal flow is made for flow from a tank (jet and siphon flows) and gradual channel transitions (see Table 4.1). Explain the theoretical consequences when flow resistance is ignored for both internal and external flow.

6.23C   In the assumption of ideal flow, the case of subsequently modeling the flow resistance implies application of the Bernoulli equation, then subsequently modeling the viscous force ($F_V \neq 0$) in the application of the momentum equation (subsequent assumption of real flow). The conversion of energy from one type to another will result in either a pressure drop or a pressure rise. Such an assumption of ideal flow is made in the subsequent modeling of flow resistance for flow-measuring devices for internal flow (pipe and open channel flow) (see Section 6.9) and external flow around objects (see Section 6.3); the ideal velocity is determined from the Bernoulli equation, where its associated minor head loss is accounted for by a drag coefficient—in the application of the momentum equation (see Table 4.1). List and explain the flow cases for which there will be a pressure drop vs. a pressure rise in the conversion of energy from one type to another, and state what type of flow resistance results in each flow case.

6.24C   In an ideal flow situation, because there is no friction ($\mu = 0$ and thus $\tau = 0$), flow energy (flow work, or pressure energy) and potential energy are converted to kinetic energy (or vice versa), with no energy lost to heat. Thus, the energy equation assuming no head loss due to friction is called the Bernoulli equation (see Equation 4.46) and is given in terms of pressure terms by Equation 6.1 as follows:

$$\left(\underbrace{p}_{F_P} + \underbrace{\gamma z}_{F_G} + \underbrace{\frac{\rho v^2}{2}}_{Ma} \right)_1 - \left(\underbrace{p}_{staticpress} + \underbrace{\gamma z}_{hydrstapress} + \underbrace{\frac{\rho v^2}{2}}_{dynpres} \right)_2 = \underbrace{\overbrace{\frac{\tau_w L}{R_h}}^{flowres}}_{F_V} = 0$$

For the Bernoulli equation given above, explain what each pressure term represents; furthermore, explain the implication of fact that the sum of the pressure terms along a streamline is constant.

Real Flow

6.25C   The viscosity of real flow (a real fluid) introduces a great complexity in a fluid flow situation, namely friction due to the viscous shearing effects. The modeling of viscosity results in the assumption of real flow, which presents the possibility of two flow regimes. List the two possible flow regimes for real flow.

6.26C   Explain how the modeling of viscosity modifies the assumption of the velocity profile from the rectangular distribution assumed for ideal flow.

6.27C   The assumption of real flow may either directly or subsequently model the existence of flow resistance. Explain the difference in how the flow resistance is modeled in the application of the governing equations.

6.28C   In the assumption of real flow, directly modeling the flow resistance implies modeling the flow resistance in both the energy and momentum equations. List the flow cases for which the direction assumption of real flow is made in the modeling of flow resistance, and state what type of flow resistance is modeled.

6.29C In the assumption of real flow, subsequently modeling the flow resistance implies application of the Bernoulli equation, then modeling the viscous forces (flow resistance) in the application of the momentum equation. List the flow cases for which the subsequent assumption of real flow is made in the subsequent modeling of flow resistance, and state what type of flow resistance is modeled.

6.30C In a real flow situation (assumption of real flow), tangential or shearing forces always develop whenever there is motion relative to a body; this creates fluid friction, because these shearing forces oppose the motion of one fluid particle past another. It is these frictional forces that result in the fluid viscosity. Furthermore, in a real flow situation, because there is friction, $F_V \neq 0$, there is flow resistance. Explain the practical consequences when flow resistance is modeled for both internal and external flow.

6.31C Directly modeling the flow resistance in real flow implies modeling the flow resistance in both the energy and momentum equations. The conversion of energy from one type to another will result in either a pressure drop or a pressure rise. Such an assumption of real flow is made in the direct modeling of flow resistance for pipe and open channel flow and the flow through most pipe devices. Application of the energy equation assumes a head loss that is accounted for by a drag coefficient (in the application of the momentum equation). List and explain the flow cases for which there will be a pressure drop vs. a pressure rise in the conversion of energy from one type to another, and state the type of flow resistance results in each flow case.

6.32C In a real flow situation, because there is friction ($\mu \neq 0$ and thus $\tau \neq 0$), flow energy (flow work, or pressure energy) and potential energy are converted to kinetic energy (or vice versa) plus energy lost to heat (due to flow resistance). Thus, the energy equation assuming a head loss due to friction (see Equation 4.44) is given in terms of pressure terms by Equation 6.13 as follows:

$$\left(\underbrace{p}_{F_P} + \underbrace{\gamma z}_{F_G} + \underbrace{\frac{\rho v^2}{2}}_{Ma} \right)_1 - \left(\underbrace{p}_{statispress} + \underbrace{\gamma z}_{hydrstapress} + \underbrace{\frac{\rho v^2}{2}}_{dynpres} \right)_2 = \underbrace{\overbrace{\frac{\tau_w L}{R_h}}^{flowres}}_{F_V}$$

Beginning with the above energy equation, derive the expression for the actual velocity, and compare it to the expression for the ideal velocity.

6.33C What is the basic difference between the nature of laminar and turbulent flow?

6.34C Describe how the difference in the nature of laminar flow (deterministic) and turbulent flow (stochastic) affects the spatial and temporal description of field variables (velocity, pressure, shear stress, temperature, etc.).

6.35C Describe how the difference in the nature of laminar flow (deterministic) and turbulent flow (stochastic) affects the modeling approach of the flow.

6.36C Discuss the basic differences and the similarities between laminar flow and turbulent flow, and state which type of flow is more likely to occur in practice.

6.37C A real flow is subdivided into laminar flow or turbulent flow depending upon the range of values for the Reynolds number, $R = \rho v L/\mu$. Furthermore, the particular range of values for the Reynolds number, R depends upon whether the flow is an internal or an external flow (see Chapter 10 for external flow classifications). Assuming internal flow: (a) Discuss the range of values for the Reynolds number, $R = \rho v L / \mu$; the nature of movement of the fluid particles; the viscosity of the fluid; and the velocity of the flow; for both laminar and turbulent flow. (b) Discuss the range of values for the Reynolds number, $R = \rho v L / \mu$ for flow in the critical zone, and state the classification of flow type (laminar or turbulent).

6.38C Assuming internal flow, fully turbulent flow, where the flow everywhere is turbulent, is highly dependent on the pipe inlet condition, the inherent level of turbulence on the fluid, and the pipe roughness. Discuss the role of the pipe roughness in the classification of the flow type (laminar or turbulent) and the implication of the flow type in the design of a pump.

6.39C The assumption made regarding the viscosity of the fluid will determine the resulting velocity profile for the flow (internal flow). Discuss and illustrate the velocity profile for ideal flow vs. real laminar flow and real turbulent flow.

6.40C For a real laminar flow, the velocity of the fluid increases parabolically, reaching a maximum, v_{max} at the pipe centerline as illustrated in Figure 6.3b. (a) Discuss the nature of movement of the fluid particles and the spatial and temporal distribution of the laminar flow parabolic velocity profile. (b) What is the average velocity (spatial average over the pipe cross-section) for real laminar flow?

6.41C Discuss the implications of the predictable (deterministic) nature of laminar flow on the determination of the resulting velocity profile and the resulting flow resistance coefficient in the equation for the major head loss, h_f due to flow resistance.

6.42C For a real turbulent flow, the velocity profile of the fluid increases according to the seventh root velocity profile law, reaching a maximum, v_{max} at the pipe centerline as illustrated in Figure 6.3c. Furthermore, turbulent flow is characterized by mixing phenomena and eddies (flow turbulence) within the flow, such as is the case for open channel flow and atmospheric flow. Discuss the effects of the flow turbulence on the resulting velocity profile and the resulting temporal variation of the velocity.

6.43C Discuss the implications of the unpredictable (stochastic) nature of turbulent flow on determination of the resulting velocity profile and the resulting flow resistance coefficients in the equation for the major head loss, h_f due to flow resistance.

6.44C Turbulent flow is more likely to occur in practice than laminar flow. Furthermore, the flow turbulence causes an unsteady flow rate (temporal variation of velocity, v). Explain how turbulent flow can be modeled as a steady state flow.

6.45C For a real flow (laminar or turbulent), explain the difference between developing and developed flow.

Modeling the Flow Resistance as a Drag Force in External Flow

6.46C In the external flow around an object, energy or work is used to move the submerged body through the fluid, where the fluid is forced to flow over the surface of the body. Explain what results are sought in external flow.

6.47C Explain when the lift force, F_L becomes dominant in external flow.

6.48C The derivation of the drag force equation (and the lift force equation) along with the coefficient of drag, C_D (and lift coefficient, C_L) for the external flow around an object is presented in this chapter. List what chapter presents the empirical evaluation of the respective flow resistance coefficients and application of the respective flow resistance equations.

Evaluation of the Drag Force

6.49C In external flow, the flow resistance causes an unknown pressure drop, Δp, which causes an unknown head loss, h_f, where the head loss is due to a conversion of kinetic energy to heat, which is modeled/displayed in the integral form of the energy equation. However, although the head loss, h_f causes the drag force, F_D, the head loss is not actually determined in the design of external flow around an object. Explain why the flow resistance is modeled as a drag force (as opposed to a minor head loss) in external flows.

6.50C Evaluation of the drag force requires a "subset level" application of the governing equations. A "subset level" application of the governing equations focuses only on the given element causing the flow resistance. The assumption of "ideal" flow implies that the flow resistance is modeled only in the momentum equation (and thus, the subsequent assumption of "real" flow). Explain the assumption of ideal flow, and the subsequent modeling of the flow resistance.

6.51C In the derivation of the drag force for an external flow around an object, first, ideal flow is assumed in order to determine an expression for the ideal velocity $v_i = \sqrt{2\Delta p/\rho}$ by applying the Bernoulli equation, where the ideal pressure difference, Δp is directly measured by the use of a pitot-static tube (see Section 4.5.7.3). Then, a subsequent assumption of real flow is made in order to determine the expressions given from theory for the actual velocity,

$$v_a = \sqrt{\frac{2}{\rho}\left(\Delta p - \frac{\tau_w L}{R_h}\right)},$$

thus, for the drag force,

$$F_D = (F_P + F_f)_s = (\Delta p A)_s = \left(\frac{\rho v^2}{2}A + \frac{\tau_w L}{R_h}A\right)_s;$$

this is done by application of the momentum equation

$$\sum F_s = (F_G + F_P + F_V + F_{other})_s = (\rho Q v_s)_2 - (\rho Q v_s)_1,$$

assuming $z_1 = z_2$ and that only the pressure and viscous forces are important. Finally, the momentum theory is supplemented by dimensional analysis in order to derive an expression for the drag force, $F_D = C_D(1/2)\rho v^2 A$, which involves the definition of a drag coefficient, C_D that represents the flow resistance and uses the ideal velocity, v_i in its equation. Explain why it is necessary to supplement the momentum theory with dimensional analysis.

6.52C In external flow, the flow resistance causes an unknown pressure drop, Δp, which causes an unknown head loss, h_f. However, although the head loss, h_f causes the drag force, F_D, the head loss is not actually determined in the design of external flow around an object. Rather, the flow resistance is ultimately modeled as a drag force $F_D = (F_P + F_f)_s = (\Delta p A)_s$ in the integral form of the momentum equation. Explain the relationship between the drag force, F_D and the lift force, F_L and explain when it is important to model the lift force, F_L.

6.53C The drag force, F_D on a submerged body has two components, a pressure drag (form drag), F_p and a friction drag (surface drag), F_f, where $F_D = (F_p + F_f)_s$. Explain the roles and the relative importance of the pressure drag and the friction drag in the definition of the drag force. Also, explain the flow conditions/type for the existence of the drag force.

6.54C Explain d'Alembert's paradox, and list two flow situations that illustrate d'Alembert's paradox.

6.55C The flow resistance (unknown head loss, h_f) represented by the drag force equation is modeled by the definition of a "drag coefficient, C_D," which is defined by dimensional analysis (see Chapter 7), and evaluated empirically (Chapter 10). Give the result of the derivation of the drag force equation along with the definition of the coefficient of drag, C_D for the external flow around an object.

6.56C Following a similar approach as done for the drag force (supplementing the momentum equation with dimensional analysis), one may derive a similar expression for the lift force. Give the resulting equation for the lift force.

6.57C Examining the standard form of the drag coefficient, C_D given by Equation 6.40 yields Equation 6.43 and is given as follows:

$$\frac{F_D}{F_{dynpres}} = \frac{\Delta p A}{(\rho v^2/2)A} = \frac{\Delta p}{\rho v^2/2} = C_D = \frac{Actual\ Drag\ Force}{Ideal\ Force} = \frac{Actual\ Presdrop}{Ideal\ Presdrop} = \frac{staticpres}{dynpres}$$

where the results of dimensional analysis introduce a drag coefficient, C_D which expresses the ratio of a real flow variable (in this case: force or pressure) to its corresponding ideal flow variable. Depending upon the particular external flow situation, the pressure drop, Δp may: (1) be equal to the dynamic pressure (ideal pressure), $\rho v^2/2$; (2) be greater than the dynamic pressure; or (3) be less than the dynamic pressure. Discuss these three possible flow situations and the range of the value(s) of the drag coefficient, C_D. Also, explain why the drag coefficient, C_D is determined empirically from experimentation.

Modeling the Flow Resistance as a Major Head Loss in Internal Flow

6.58C In the internal flow through a pipe or a channel, energy or work is used to move/force the fluid through a conduit. Explain what results are sought in internal flow.

6.59C The flow of a real fluid (liquid or a gas) through a pipe under pressure is used in heating and cooling systems and in distribution system of fluids. Explain how energy or work is used to move/force the fluid through a conduit.

6.60C In internal flow, the flow resistance causes an unknown pressure drop, Δp (or S_f), which causes an unknown major head loss, h_f, where the head loss is due to a conversion of kinetic energy to heat, which is modeled/displayed in the integral form of the energy equation. The energy dissipated by the viscous forces within the fluid is supplied by the excess work done by the pressure and gravity forces, which are modeled in the integral form of the momentum equation. Furthermore, the pressure drop, Δp and the head loss, h_f are dependent on the wall shear stress, τ_w between the fluid and the pipe. Explain the practical importance in the determination of the head loss, h_f in fluid flow (and pumping of fluid flow).

6.61C Discuss the difference in the power requirements for pumping laminar vs. turbulent flow.

Evaluation of the Major Head Loss

6.62C In internal flow, the flow resistance causes an unknown pressure drop, Δp (or S_f), which causes an unknown major head loss, h_f, where the head loss is due to a conversion of kinetic energy to heat, which is modeled/displayed in the integral form of the energy equation. Explain how (and why) the flow resistance (due to pipe or channel friction) is ultimately modeled in internal flows.

6.63C Evaluation of the major head loss flow resistance term requires a "subset level" of application of the governing equations. A "subset level" application of the governing equations focuses only on the given element causing the flow resistance. The assumption of real flow implies that the flow resistance is modeled in both the energy and momentum equations. Explain the assumption of real flow in the modeling of the flow resistance.

6.64C In the derivation of the major head loss, first, the integral form of the energy equation is applied to determine an expression for the unknown head loss, h_f. Then, the integral momentum and the differential momentum equations (supplemented by dimensional analysis) are applied in order to determine an expression for the pressure drop, Δp (or S_f) and the shear stress, τ_w. The flow resistance due to pipe or channel friction, which results in a major head loss term, $h_{f,major}$, is modeled as a resistance force (shear stress or drag force) in the integral form of the momentum equation (Equation 4.28), while it is modeled as a friction slope, S_f in the differential form of the momentum equation (Equation 4.84). As such, the energy equation may be used to solve for the unknown major head loss,

$$h_{f,maj} = S_f L = \frac{\tau_w L}{\gamma R_h} = \frac{\Delta p}{\gamma},$$

while the integral momentum equation may be used to solve for the unknown pressure drop, $\Delta p = \tau_w L / R_h$ (or the unknown friction slope, $S_f = \tau_w / \gamma R_h$ in the case of turbulent open channel flow). Explain when it is necessary to supplement the momentum theory with dimensional analysis.

6.65C The derivation of the major head loss for pipe or open channel flow yields the Chezy equation, $S_f = v^2 / C^2 R_h$, which represents the differential form of the momentum equation, supplemented by dimensional analysis and guided by the integral momentum equation; a link between the differential and integral momentum equations, which is used to obtain an empirical evaluation for the pressure drop, Δp in Newton's second law of motion (integral form of the momentum equation). Thus, the integral momentum equation (and the differential momentum equation for S_f) is supplemented with dimensional analysis in order to empirically model the wall shear stress, τ_w for turbulent pipe and open channel flow and thus Δp (and Δy) by the use of a drag coefficient, C_D. Specifically explain how dimensional analysis is used to model the wall shear stress, τ_w for turbulent pipe and open channel flow; give the final practical equation for the head loss; explain how the velocity is measured.

6.66C Application of the three governing equations yields an expression for the major head loss equation for both laminar and turbulent flow. First, the integral form of the energy equation (Equation 6.46) yields the following relationship between the major head loss, h_f;

the wall shear stress, τ_w; and the pressure drop, Δp given by Equation 6.51, which is given as follows:

$$h_{f,maj} = S_f L = \frac{\tau_w L}{\gamma R_h} = \frac{\Delta p}{\gamma}.$$

Then, regardless of whether the flow is laminar or turbulent, application of Newton's second law of motion (integral form of the momentum equation) (Equation 6.47) yields a linear relationship between the shear stress, τ and the radial distance, r, which also describes the theoretical relationship between the pressure drop, Δp (or S_f) and the shear stress, τ; this relationship plays a key role in the derivation of the major head loss equations for laminar and turbulent flow. Specifically, the integral form of the momentum equation yields the relationship between the pressure drop, Δp and the wall shear stress, τ_w (derived in Section 6.5) given by Equation 6.52, which is given as follows: $\tau = (r/2)(\Delta p/L)$. Equation 6.52 is evaluated for the wall shear stress, τ_w at $r = r_o = D/2$, where $R_h = D/4$, and expressed as a function of the pressure drop, Δp, it yields Equation 6.53, which is given as follows: $\Delta p = \tau_w L/R_h$. Furthermore, the differential form of the momentum equation (Equation 6.49) yields the relationship between the friction slope, S_f and the wall shear stress, τ_w given by Equation 6.54, which is given as follows: $S_f = \tau_w/\gamma R_h$. However, because all three terms, the pressure drop, Δp (Equation 6.53); the friction slope, S_f (Equation 6.54); and the major head loss, h_f (Equation 6.51), are expressed in terms of the wall shear stress, τ_w, in order to evaluate the respective expressions, one must first obtain an expression for the wall shear stress, τ_w. Explain the procedures used to derive the expression for the major head loss for (a) laminar flow vs. (b) turbulent flow.

6.67C The assumptions made in the derivation of the major head loss equation will depend upon the type of internal flow (pipe flow vs. open channel flow and laminar flow vs. turbulent flow). Explain the differences and the similarities in the assumptions made in the derivation of the major head loss equation, depending upon the type of internal flow (pipe flow vs. open channel flow and laminar flow vs. turbulent flow).

Laminar and Turbulent Internal Flow Characteristics

6.68C Application of the three governing equations yields an expression for the major head loss equation for both laminar and turbulent flow. In particular, the application of Newton's second law of motion (integral form of the momentum equation) yields a linear relationship between the shear stress, τ and the radial distance, r, which also describes the theoretical relationship between the pressure drop, Δp and the shear stress, τ for both laminar and turbulent flow; this relationship plays a key role in the derivation of the major head loss equations for laminar and turbulent flow (see Section 6.5.1). Specifically, the integral form of the momentum equation yields the relationship between the pressure drop, Δp and the wall shear stress, τ_w (derived in Section 6.5) given by Equation 6.52, which is given as follows: $\tau = (r/2)(\Delta p/L)$. Evaluated for the wall shear stress, τ_w at $r = r_o = D/2$, where $R_h = D/4$, and expressed as a function of the pressure drop, Δp, this yields Equation 6.53, which is given as follows: $\Delta p = \tau_w L/R_h$. Therefore, regardless of whether the flow is laminar or turbulent, the two types of flow share some similar flow characteristics. However, explain the main difference between the derivation of the major head loss for laminar vs. turbulent flow.

6.69C Regardless of whether the flow is laminar or turbulent, the pipe flow at the pipe entrance is modeled as ideal flow, followed by real flow. Explain these assumptions and their practical implications on the flow characteristics (variation of the velocity, pressure drop, and the wall shear stress with length of pipe).

6.70C Although turbulent flow is more likely to occur in practice than laminar flow, the type of flow (laminar vs. turbulent) in a pipe is very much a function of pipe inlet (entrance) conditions and the inherent turbulence (or lack of it) of the flow at the pipe inlet and on the roughness of the pipe wall. Additionally, because the pressure drop and the shear stress differ for laminar and turbulent flow, the maximum shear stress also differs for laminar and turbulent flow. Thus, the nature of the pipe flow is highly dependent on whether the flow is laminar or turbulent. Explain the main cause of the difference in nature of laminar vs. turbulent flow, and practical implications for pumping of the flow.

Derivation of the Major Head Loss Equation for Laminar Flow

The Hagan–Poiseuille Law

6.71C In the case of laminar flows, first, application of the integral form of the energy equation yields an expression for the major head loss (Equation 6.51). Then, because the friction/viscous forces in the integral form of the momentum equation and the wall shear stress, τ_w can be theoretically modeled (by Newton's law of viscosity, where $\tau = \mu \, dv/dy$) in the case of laminar flow, one can analytically solve for the unknown pressure drop, Δp from the momentum equation (Equation 6.52); thus, one can analytically derive an expression for the major head loss, h_f from the energy equation. Specifically, application of the integral form of the momentum equation, where the wall shear stress, τ_w is theoretically modeled by Newton's law of viscosity, yields a theoretical (parabolic) expression for velocity distribution for laminar flow, expressed as a function of the pressure drop, Δp. Next, application of the continuity equation by integrating the resulting parabolic velocity distribution over the pipe diameter yields an analytical expression for the actual pressure drop, Δp (Poiseuille's law), which is substituted into the integral form of the energy equation (Equation 6.51), yielding an analytical expression for the head loss, h_f (Poiseuille's law expressed in terms of the head loss). (a) Derive the theoretical (parabolic) expression for the velocity distribution for laminar flow, expressed as a function of the pressure drop, Δp. (b) Derive the analytical expression for the actual pressure drop, Δp (Poiseuille's law). (c) Derive the analytical expression for the head loss, h_f (Poiseuille's law expressed in terms of the head loss), where the head loss, h_f is expressed as a function of the flow resistance, μ, without the need to define an empirical flow resistance coefficient.

6.72C The analytical expression for the actual pressure drop, Δp (Poiseuille's law) is given by Equation 6.91 as follows: $\Delta p = 128 \mu QL/\pi D^4$. Poiseuille's law indicates the following properties of steady and uniform laminar flow in a straight and horizontal pipe: (1) the pressure drop, Δp is directly proportional to the absolute viscosity, μ; the discharge, Q; and the pipe length, L; and (2) the pressure drop, Δp is indirectly proportional to the pipe diameter to the fourth power D^4. Explain and discuss these two properties of steady and uniform laminar flow in a straight and horizontal pipe.

6.73C Derive the parabolic velocity distribution for laminar flow presented as a function of v_{ave} as given by Equation 6.94 as follows:

$$v = 2v_{ave}\left[1 - \left(\frac{r}{r_o}\right)^2\right].$$

Derivation of the Major Head Loss Equation for Turbulent Flow

6.74C   In the case of turbulent flows, first, application of the integral form of the energy equation yields an expression for the major head loss (Equation 6.51). Derive the expression for the unknown pressure drop, Δp, explain why one cannot analytically determine the pressure drop, Δp (or the unknown friction slope, S_f in the case of turbulent open channel flow), and state how it is actually determined.

6.75C   Explain, specifically, why the unknown pressure drop, Δp (or the unknown friction slope, S_f in the case of turbulent open channel flow) and thus the major head loss, h_f due to pipe or channel friction are determined empirically for turbulent flow.

6.76C   Explain how dimensional analysis is used to empirically interpret the friction slope (Equation 6.102) in the differential form of the momentum equation (i.e. empirically derive an expression for the wall shear stress, τ_w) in terms of easy to measure velocity, v, and thus obtain an empirical evaluation for the pressure drop, Δp in Newton's second law of motion (integral form of the momentum equation).

6.77C   In the case of turbulent flow, what are two major implications of the lack of a theoretical expression for the friction/viscous forces?

6.78C   In the case of turbulent flow, dimensional analysis is used to derive an empirical expression for the friction forces, which yields an empirical expression for the head loss, with the need to define an empirical flow resistance. The drag coefficient, C_D representing the flow resistance that is derived by dimensional analysis is a nonstandard form of the drag coefficient. What are more standard forms of the drag coefficient?

6.79C   Briefly outline the procedure used to derive the empirical expression for the wall shear stress, $\tau_w = C_D \rho v^2$, expressed as a function of the average velocity, v_{ave}, and a drag coefficient, C_D for turbulent flow.

The Chezy Equation

6.80C   Derive the Chezy equation and the Chezy coefficient, C. Furthermore, discuss the practical implications of the Chezy equation in modeling the flow resistance in turbulent pipe and open channel flow.

6.81C   The Chezy coefficient, C models the flow resistance. What are its dimensions, what is it a function of, and what are the typical ranges for its empirical values?

6.82C   State how the Chezy equation (Equation 6.109) may be practically applied in pipe flow.

6.83C   Because all three terms, the pressure drop, Δp (Equation 6.99); the friction slope, S_f (Equation 6.102); and the major head loss, h_f (Equation 6.51), are expressed in terms of the wall shear stress, τ_w, in order to evaluate the respective expressions, an expression for the wall shear stress, τ_w was sought. Specifically, because the wall shear stress, τ_w, which causes the flow resistance, is difficult to measure, instead, one may model it by measuring the velocity, v and defining a flow resistance coefficient, which relates τ_w to v (accomplished by using dimensional analysis). As such, application of the differential form of the momentum equation, supplemented by dimensional analysis and guided by the integral momentum equation, resulted in derivation of the Chezy equation, which represents an empirical evaluation of the friction slope, S_f and is given by

Equation 6.110 as follows:

$$S_f = \frac{\tau_w}{\gamma R_h} = \frac{v^2}{C^2 R_h}.$$

Illustrate how the Chezy equation is used to obtain empirical expressions for: (a) the pressure drop, Δp in Newton's second law of motion (integral form of the momentum equation) (Equation 6.99); (b) the wall shear stress, τ_w; and (c) the major head loss, h_f in the integral form of the energy equation (Equation 6.51).

6.84C Although values for the Chezy coefficient, C have been empirically evaluated and tabulated (limited tables available) for various combinations of boundary roughness, ε/D; Reynolds number, R; and cross-sectional shape (or the hydraulic radius, R_h), it has been found that it is easier to model the effects of the Chezy coefficient, C (i.e., the drag coefficient, C_D) by using what two other friction coefficients?

The Darcy–Weisbach Equation

6.85C The Darcy–Weisbach friction coefficient (factor), f, which represents a dimensionless flow resistance coefficient, is derived using dimensional analysis, and expressed as a function of the Chezy coefficient, C as: $C = \sqrt{8g/f}$. Briefly outline the procedure used to derive the Darcy–Weisbach friction coefficient (factor), f.

6.86C Using the Darcy–Weisbach friction coefficient, f (which is a more standard definition of the drag coefficient, C_D), derive the resulting relationship between the wall shear stress, τ_w and the velocity, v from dimensional analysis.

6.87C The Darcy–Weisbach friction coefficient, f models the flow resistance. What are its dimensions and what is it a function of?

6.88C Define the relationship between the Chezy coefficient, C and the Darcy–Weisbach friction factor, f.

6.89C Briefly state the practical advantages of using the Darcy–Weisbach friction factor, f as opposed to using the Chezy coefficient, C to model the effects of the flow resistance (i.e., the drag coefficient, C_D).

6.90C The Darcy–Weisbach friction coefficient, f has no dimensions and (in general) is a function of the boundary roughness, ε/D; the Reynolds number, R; and the cross-sectional shape or the hydraulic radius, R_h, where ε is the absolute pipe and ε/D is the relative pipe roughness. When is the Darcy–Weisbach friction coefficient, f independent of R, and when is the Darcy–Weisbach friction coefficient, f independent of the boundary roughness, ε/D?

6.91C Briefly explain how the friction factor, f for turbulent flow is evaluated.

6.92C Derive the Darcy–Weisbach head loss equation.

Manning's Equation

6.93C The dimensions of the Chezy coefficient, C are $L^{1/2} T^{-1}$, and it is empirically evaluated as a function of the boundary roughness, ε/D; the Reynolds number, R; and the cross-sectional shape or the hydraulic radius, R_h. However, further empirical studies (assuming the metric system) by Manning have indicated that a more accurate dependence of the Chezy coefficient, C on the hydraulic radius, R_h is defined by the definition of the Manning's roughness coefficient, n. Define the empirical relationship between the Chezy coefficient, C and the Manning's roughness coefficient, n. Also, what are the dimensions and units of the Manning's roughness coefficient, n, and what is it a function of?

6.94C Modeling the flow resistance by the Manning's roughness coefficient, n assumes that the pipe or open channel flow is turbulent and thus the Reynolds number, R is so large that it has virtually no influence on the resistance coefficient, n. Discuss the flow characteristics for which the assumption of turbulent pipe and open channel flow is accurate.

6.95C Derive the Manning's equation.

6.96C Although the derivation/formulation of the Manning's equation assumes no specific units (SI or BG), the Manning's roughness coefficient, n has dimensions of $L^{-1/3}$ T. Furthermore, because the values for the Manning's roughness coefficient, n have been provided/calibrated in the SI units $m^{-1/3}$ s (see Table 8.6), the Manning's equation,

$$v = \frac{R_h^{1/6}}{n}\sqrt{R_h S_f} = \frac{1}{n}R_h^{2/3}S_f^{1/2} \qquad (6.126)$$

assumes the metric (SI) system. Give the corresponding Manning's equation for the English (BG) system.

6.97C Derive the Manning's head loss equation, for the metric (SI) system and the English (BG) system.

The Hazen–Williams Equation

6.98C In addition to the standard empirical flow resistance coefficients including the Chezy coefficient, C; the Darcy–Weisbach friction factor, f; and the Manning's roughness coefficient, n, there is another commonly used flow resistance coefficient, namely the Hazen–Williams roughness coefficient, C_h. (a) Give the derived Hazen–Williams equation. (b) State any limitations in modeling the flow resistance using the Hazen–Williams roughness coefficient, C_h vs. using the standard empirical flow resistance coefficients including the Chezy coefficient, C; the Darcy–Weisbach friction factor, f; and the Manning's roughness coefficient, n.

6.99C The Hazen–Williams roughness coefficient, C_h models the flow resistance for pipe flow. What are its dimensions, what is it a function of, and what are the typical ranges for its empirical values?

6.100C Derive the Hazen–Williams head loss equation, for the metric (SI) system, and the English (BG) system.

6.101C The drag coefficient, C_D is more commonly modeled by the more standard forms including the Chezy coefficient, C; the Darcy–Weisbach friction factor, f; or the Manning's roughness coefficient, n. These are basically flow resistance coefficients that empirically relate the wall shear stress, τ_w (or flow resistance, which causes the pressure drop, Δp; and thus; a head loss, h_f) to the velocity, v (average velocity, v_{ave}) as given by Equation 6.138 as follows:

$$\frac{\tau_w}{\rho v^2} = \frac{\text{Wall Friction Force}}{\text{Inertia Force}} = \frac{R_h}{L}\frac{\Delta p}{\rho v^2} = \frac{\text{Pressure Force}}{\text{Inertia Force}} = \frac{R_h}{L}\frac{1}{E} = C_D = \phi\left(R, \frac{\varepsilon}{D}\right).$$

How are the four above-listed flow resistance coefficients related?

6.102C The three standard flow resistance coefficients (the Chezy coefficient, C; the Darcy–Weisbach friction factor, f; or the Manning's roughness coefficient, n) are related as illustrated by Equations 6.139 through 6.142. (a) What is the preferred flow resistance coefficient for laminar or turbulent pipe flow? (b) What is the preferred flow resistance coefficient for open channel flow (usually turbulent flow) and turbulent pipe flow?

A Comparison between Laminar and Turbulent Flow Using the Darcy–Weisbach Head Loss Equation

6.103C Because laminar flow is a special case of turbulent flow, the "theoretical flow resistance coefficient" for laminar flow is a special case of the empirical flow resistance coefficient for turbulent flow ($f = 64/R$). Evaluate the Darcy–Weisbach friction coefficient, f for laminar pipe flow.

6.104C Regardless of whether the flow is laminar or turbulent, the flow resistance (shear stress, τ) causes an unknown pressure drop, Δp, which causes an unknown head loss, h_f, where the head loss is due to a conversion of kinetic energy to heat. Furthermore, application of Newton's second law of motion yielded a linear relationship between the shear stress, τ and the radial distance, r, which also describes the theoretical relationship between the pressure drop, Δp and the shear stress, τ for both laminar and turbulent flow (Equation 6.151), where the maximum shear stress, τ_{max} at the pipe wall/boundary, where $R_h = D/4$ is given as shown in Equation 6.152 (see Figure 6.4). However, because the dissipation of energy (i.e., h_f) is considerably less for laminar flow than for turbulent flow, the resulting magnitude of the pressure drop, Δp and thus the maximum shear stress, τ_{max} is also considerably less for laminar flow in comparison to turbulent flow. Compare and interpret (for laminar vs. turbulent flow) the corresponding expressions for: (a) the major head loss, h_f; (b) the pressure drop, Δp; and (c) the wall shear stress, τ_{max}.

A Comparison between the Velocity Profiles for Laminar Flow and Turbulent Flow

6.105C While the deterministic nature of laminar flow analytically yielded a parabolic velocity profile for laminar flow, the more stochastic nature of turbulent flow yields an empirically defined velocity profile for turbulent flow. (a) Mathematically and visually illustrate the difference between the velocity profiles for laminar vs. turbulent flow. (b) Compare and discuss the visual differences between the velocity profiles for laminar vs. turbulent flow.

6.106C Figure 6.9, which was generated using Mathcad, illustrates a comparison between the parabolic velocity profile for laminar flow (see Equation 6.88 [and Equations 6.92 and 6.93]) and the velocity profile for turbulent flow, which is approximated by the seventh root velocity profile law (see Equations 6.161 and 6.162) for a given average velocity, $v_{ave} = 5$, and pipe radius, $r_o = 10$. (a) Compare the shape of the velocity profiles (laminar vs. turbulent) for a given average velocity, v_{ave}. (b) Compare the approach used to calculate the shear stress across the entire pipe section (laminar vs. turbulent).

6.107C Because the velocity profile for laminar flow is determined analytically, it provides a basis of comparison for the velocity profile for turbulent flow. In laminar flow, the viscous effects are important across the entire pipe section, resulting in a zero velocity at the boundary (see Figure 6.8a) and the shear stress throughout the flow being described by Newton's law of viscosity, $\tau = -\mu \, dv/dr$. Furthermore, Newton's law of viscosity coupled with Newton's second law of motion and the continuity equation yielded Poiseuille's law for laminar flow, which indicates that the flow resistance and thus the head loss is mainly a function of the fluid viscosity, μ and not a function of the relative pipe roughness, ε/D (as in the case of turbulent flow). (a) Give the head loss equation for laminar flow (special case of turbulent flow). (b) In laminar flow, very little of the flow comes in contact with the projection at the pipe wall; the affected flow and eddies resulting from the projections in the pipe wall are quickly damped out by significantly large viscous dissipation forces. Explain the effect/ role of the wall roughness element, ε for laminar flow.

6.108C   In turbulent flow, in addition to the turbulent boundary layer, a thin viscous sub-layer has been experimentally observed near the pipe wall, for which the shear stress is described by Newton's law of viscosity, $\tau = -\mu\, dv/dr$. However, because the seventh root velocity profile law for turbulent flow yields a velocity gradient, dv/dr of infinity at $r = r_o$, it cannot be used to calculate the wall shear stress, τ_w (which is contained within the thin vis-cous sublayer, where Newton's law of viscosity is applicable); instead, $\tau_w = f\rho v^2/8$ from dimensional analysis is used. Furthermore, for turbulent flow, if a typical wall roughness element, ε protrudes sufficiently far into or through the thin viscous sublayer, the structure and properties (the pressure drop, Δp; the wall shear stress, τ_w; the flow resistance; and the head loss) will be different than if the wall were smooth, as illustrated by a comparison of Figures 6.11a and b. As a result, for turbulent flow, the pressure drop, Δp; the wall shear stress, τ_w; the head loss, h_f; and thus the friction factor, f will be a function of the wall rough-ness element, ε. (a) Give the head loss equation for turbulent flow. (b) Explain the effect/role of the wall roughness element, ε for turbulent flow.

6.109C   Evaluation of the Darcy–Weisbach friction coefficient, $f = 64/R$ for laminar pipe flow (a special case of turbulent flow) illustrates that f for laminar flow is only a function of the Reynolds number, R, while f for turbulent flow, it is a function of both the Rey-nolds number, R and the relative pipe roughness, ε/D. However, at very high velocities (and thus very high values for R), the f for completely turbulent flow (rough pipes) is a function only of the relative pipe roughness, ε/D and thus independent of R. However, a relatively large wall roughness element, ε (for which the relative pipe roughness, $\varepsilon/D \geq 0.1$) may affect the structure of laminar flow and its properties (the pressure drop, Δp; the wall shear stress, τ_w; the flow resistance; and the head loss). Furthermore, in turbu-lent flow, in addition to the turbulent boundary layer, a thin viscous sublayer has been experimentally observed near the pipe wall. However, in laminar flow, the entire boun-dary layer consists entirely of a laminar boundary layer and not just thin viscous sub-layer as in the case of turbulent flow. (a) Explain how the thickness of the boundary layer, η serves as the criterion of effective roughness. (b) For the flow over boundary surfaces, compare the energy dissipation (head loss, h_f) and flow resistance for turbulent vs. laminar flow.

Application of the Major Head Loss Equation for Open Channel Flow

6.110C   For the derivation of the major head loss equation, although the laws of flow resis-tance are essentially the same in pipe flow and open channel flow, there is a slight modifi-cation needed for the application in open channel flow. Highlight and explain the slight modification needed in the derivation of the major head loss equation for the application in open channel flow.

6.111C   All three terms, the channel slope, S_o (Equation 6.169); the friction slope, S_f (Equa-tion 6.170); and the major head loss, h_f (Equation 6.163), are expressed in terms of the wall shear stress, τ_w. Illustrate how the Chezy equation (Equations 6.109 and 6.174), $S_f = v^2/C^2 R_h$, which provides an empirical evaluation of the friction slope, S_f, may be applied to obtain empirical expressions for: (a) the channel slope, S_o; (b) the wall shear stress, τ_w; and (c) the major head loss, h_f.

6.112C State how the Chezy equation (Equations 6.109 and 6.174), $S_f = v^2/C^2 R_h$ may be practically applied in open channel flow.

6.113C Derive the Darcy–Weisbach head loss equation for open channel flow.

6.114C Derive the Manning's head loss equation for open channel flow, for the metric (SI) system, and the English (BG) system.

Modeling the Flow Resistance as a Minor Head Loss in Pipe Flow

6.115C Explain the difference between the major head loss term, $h_{f,major}$ in the energy equation vs. the minor head loss term, $h_{f,minor}$ in the energy equation.

6.116C Explain when the minor head loss term, $h_{f,minor}$ in the energy equation is significant, and explain in detail what causes it.

6.117C The pipe length usually determines the role/significance of major and minor losses. Explain this point.

6.118C Compare the significance between the minor head loss vs. the major head loss.

Evaluation of the Minor Head Loss

6.119C In internal pipe flow with pipe appurtenances (components, devices, or transitions), which include valves, fittings (tees, unions, elbows, and bends), entrances, exits, contractions, and expansions, the flow resistance causes an unknown pressure drop, Δp, which causes an unknown minor head loss, h_f, where the head loss is due to a conversion of kinetic energy to heat, which is modeled/displayed in the integral form of the energy equation. Explain how (and why) the flow resistance due to the pipe components is ultimately modeled in internal pipe flow.

6.120C Evaluation of the minor head loss flow resistance term requires a "subset level" of application of the governing equations. A "subset level" application of the governing equations focuses only on the given element causing the flow resistance. The assumption of real flow implies that the flow resistance is modeled in both the energy and momentum equations. Explain the assumption of real flow in the modeling of the flow resistance.

6.121C In the derivation of the minor head loss, first the integral form of the energy equation is applied to determine an expression for the unknown head loss, h_f. Then, the integral momentum equation (supplemented by dimensional analysis) is applied in order to determine an expression for the pressure drop, Δp and the shear stress, τ_w. The flow resistance due to pipe components, which results in a minor head loss term, $h_{f,minor}$, is modeled as a resistance force (shear stress or drag force) in the integral form of the momentum equation (Equation 4.28). As such, the energy equation may be used to solve for the unknown minor head loss, $h_{f, min} = \tau_w L/\gamma R_h = \Delta p/\gamma$, while the integral momentum equation may be used to solve for the unknown pressure drop, $\Delta p = \tau_w L/R_h$. Explain when is it necessary to supplement the momentum theory with dimensional analysis.

6.122C In general, the flow resistance due to various pipe components, transitions, or devices causes an unknown pressure drop, Δp, which causes an unknown minor head loss, $h_{f,minor}$, where the head loss is due to a conversion of kinetic energy to heat, which is modeled/displayed in the integral form of the energy equation. Explain when the flow results in a pressure drop vs. a pressure rise.

6.123C While the pressure rise and the resulting minor head loss for the case of a sudden pipe expansion may be analytically derived, the pressure drop/rise and the resulting minor head loss in the cases of the other pipe components/devices/transitions, in general, are empirically derived. (a) Derive the expression for the pressure drop/rise that results in a minor head loss due to pipe components, regardless of whether one may analytically or empirically determine the actual pressure drop/rise. (b) Derive the expression for the pressure drop that results in a major head loss due to pipe friction. (c) Compare the pressure drop/rise and the resulting minor head loss in (a) to the pressure drop and the resulting major head loss in (b).

Analytical Derivation of the Minor Head Loss due to a Sudden Pipe Expansion

6.124C Application of the three governing equations yields a general expression for the minor head loss, $h_f = k(v^2/2g)$, where an analytical evaluation of the minor loss coefficient, k due to a sudden pipe expansion is possible (details of the analytical derivation of the minor head loss is given in Chapter 8). Briefly summarize the derivation of the analytical expression for the minor head loss due to a sudden pipe expansion.

Empirical Derivation of the Minor Head Loss due to Pipe Components in General

6.125C Application of the three governing equations yields a general expression for the minor head loss, $h_f = k(v^2/2g)$, where an analytical evaluation of the minor loss coefficient, k due to all other pipe devices (other than a sudden pipe expansion) is not possible because the flow through the different devices is very complex. Instead, the minor loss coefficient, k is evaluated empirically. Briefly summarize the derivation of the empirical expression for the minor head loss due to pipe components/devices in general.

6.126C Explain, specifically, why the unknown pressure drop/rise, Δp and thus the minor head loss, h_f due to pipe components in general is determined empirically.

6.127C In the derivation of the empirical expression for the minor head loss, h_f due to pipe components in general, the integral momentum equation is supplemented with dimensional analysis in order to empirically model the wall shear stress, τ_w for pipe flow with components and thus Δp by the use of a drag coefficient, C_D. Briefly outline the procedure used to empirically interpret Δp (express τ_w) in terms of easy-to-measure velocity, v and give the resulting empirical expression for the minor head loss, h_f.

6.128C Examining the standard form of the minor head loss coefficient, k given by Equation 6.200 yields Equation 6.202 and is given as follows:

$$\frac{F_P}{F_{dynpres}} = \frac{\Delta p}{\rho v^2 / 2} = C_D = k = \frac{Actual\ Pressure\ Force}{Ideal\ Force} = \frac{Actual\ Presdrop}{Ideal\ Presdrop} = \frac{staticpres}{dynpress}$$

where the results of dimensional analysis introduce a drag coefficient, C_D that expresses the ratio of a real flow variable (in this case: force or pressure) to its corresponding ideal flow variable. In general, both pressure forces, F_p and viscous forces, F_V (shear forces, F_f) may be exerted on the pipe wall as a result of a pipe component. Therefore, regardless of whether the particular flow device results in a pressure drop or a pressure rise, the actual pressure drop/rise, Δp may: (1) be equal to the dynamic pressure (ideal pressure), $\rho v^2/2$; (2) be greater than the dynamic pressure; or (3) be less than the dynamic pressure. Discuss these three possible flow situations and the range of the value(s) of the minor head loss coefficient, k. Also explain why the minor head loss coefficient, k is determined empirically from experimentation.

6.129C Examining the standard form of the minor head loss coefficient, k given by Equation 6.200 yields Equation 6.202 and is given as follows:

$$\frac{F_P}{F_{dynpres}} = \frac{\Delta p}{\dfrac{\rho v^2}{2}} = C_D = k = \frac{Actual\ Pressure\ Force}{Ideal\ Force} = \frac{Actual\ Presdrop}{Ideal\ Presdrop} = \frac{staticpres}{dynpress}$$

where the results of dimensional analysis introduce a drag coefficient, C_D that expresses the ratio of a real flow variable (in this case: force or pressure) to its corresponding ideal flow variable. However, in the case of the sudden pipe expansion, the minor head loss coefficient, k was analytically derived. Briefly outline the analytical derivation of the minor head loss coefficient, k for a sudden pipe expansion, and discuss the range of the value (s) of the minor head loss coefficient, k for a sudden pipe expansion.

Modeling the Flow Resistance as a Loss in Flowrate in Internal Flow

6.130C One of the major applications of the study of fluid mechanics is the determination of the rate of flow of fluids. As a result, numerous flow rate measuring devices/flow meters have been developed, which differ depending upon whether the flow is through a pipe or through an open channel. List the most common types of flow measurement devices used in pipe flow vs. open channel flow.

6.131C Most flow measurement devices (flow meters) are basically velocimeters that directly measure an ideal pressure difference, which is related to the velocity by the Bernoulli equation; thus, the velocity is indirectly measured. Then, application of the continuity equation is used to determine the flowrate. Thus, many flow-measurement devices used to measure the velocity and thus the flowrate for pipe and open channel flow apply the Bernoulli equation (assumes ideal flow) and are considered to be ideal flow meters. Explain how the flow resistance (real flow) is accounted for in the flow measurement.

Evaluation of the Actual Discharge

6.132C In the case where the flow resistance is due to a flow-measurement device for either pipe flow or open channel flow, the flow resistance causes an unknown pressure drop, Δp in the case of pipe flow and an unknown Δy in the case of open channel flow, which causes an unknown head loss, h_f, where the head loss is due to a conversion of kinetic energy to heat, which is modeled/displayed in the integral form of the energy equation. Explain how (and why) the flow resistance due to a flow-measurement device for either pipe flow or open channel flow is ultimately modeled.

6.133C Evaluation of the actual discharge requires a "subset level" application of the governing equations. A "subset level" application of the governing equations focuses only on the given element causing the flow resistance. The assumption of "ideal" flow implies that the flow resistance is modeled only in the momentum equation (and thus the subsequent assumption of "real" flow). Explain the assumption of ideal flow and the subsequent modeling of the flow resistance.

6.134C In the derivation of the actual discharge for a given flow-measuring device, first, ideal flow is assumed in order to determine an expression for the ideal velocity $v_i = \sqrt{2\Delta p/\rho}$ or $v_i = \sqrt{2g\Delta y}$ by applying the Bernoulli equation, where the ideal pressure difference, Δp or Δy, is directly measured by the use of the flow-measuring device; the exact expression for the ideal velocity will depend upon the particular flow measurement device (see Chapters 8 and 9 for details). Then, application of the continuity equation yields an expression for the ideal discharge, $Q_i = v_i A_i$ and the actual discharge for pipe flow-measuring devices,

$$Q_a = \frac{v_a}{\sqrt{\dfrac{2\Delta p}{\rho}}} \frac{A_a}{A_i} Q_i$$

and open channel flow-measuring devices,

$$Q_a = \frac{v_a}{\sqrt{2g\Delta y}} \frac{A_a}{A_i} Q_i.$$

Next, a subsequent assumption of real flow is made in order to determine an expression for the actual velocity for pipe flow-measuring devices,

$$v_a = \sqrt{\frac{2}{\rho}\left(\Delta p - \frac{\tau_w L}{R_h}\right)}$$

and open channel flow-measuring devices,

$$v_a = \sqrt{2g\left(\Delta y - \frac{\tau_w L}{\gamma R_h}\right)},$$

and thus, for the actual discharge for pipe flow-measuring devices,

$$Q_a = \frac{\sqrt{\dfrac{2}{\rho}\left(\Delta p - \dfrac{\tau_w L}{R_h}\right)}}{\sqrt{\dfrac{2\Delta p}{\rho}}} \frac{A_a}{A_i} Q_i$$

and open channel flow-measuring devices,

$$Q_a = \frac{\sqrt{2g\left(\Delta y - \dfrac{\tau_w L}{\gamma R_h}\right)}}{\sqrt{2g\Delta y}} \frac{A_a}{A_i} Q_i,$$

this is done by application of the integral momentum

$$\sum F_s = (F_G + F_P + F_V + F_{other})_s = (\rho Q v_s)_2 - (\rho Q v_s)_1$$

assuming $z_1 = z_2$ and that only the pressure and viscous forces are important. And, finally, the momentum theory is supplemented by dimensional analysis in order to derive an expression for the actual discharge for pipe flow-measuring devices,

$$Q_a = \frac{\sqrt{\frac{2}{\rho}\left(\Delta p - \frac{\tau_w L}{R_h}\right)}}{\sqrt{\frac{2\Delta p}{\rho}}} \frac{A_a}{A_i} Q_i = C_d Q_i$$

and the actual discharge for open channel flow-measuring devices,

$$Q_a = \frac{\sqrt{2g\left(\Delta y - \frac{\tau_w L}{\gamma R_h}\right)}}{\sqrt{2g\Delta y}} \frac{A_a}{A_i} Q_i = C_d Q_i,$$

which involve the definition of a drag coefficient, C_D (discharge coefficient, C_d) that accounts for flow resistance. Explain why it is necessary to supplement the momentum theory with dimensional analysis.

6.135C   In the case where the flow resistance is due to a flow measurement device for either pipe flow or open channel flow, the flow resistance causes an unknown pressure drop, Δp in the case of pipe flow and an unknown Δy in the case of open channel flow, which causes an unknown head loss, h_f. The flow resistance (shear stress or drag force) due to a flow measurement device for either pipe flow or open channel flow is ultimately modeled as a reduced/actual discharge, Q_a in the differential form of the continuity equation because the actual flowrate is needed for the design of a flow-measuring device. Thus, the associated minor head loss is typically accounted for by the use/calibration of a discharge coefficient to determine the actual discharge. Explain if there is an alternative approach to modeling the flow resistance due to a flow measurement device for either pipe flow or open channel flow.

6.136C   The flow resistance (unknown head loss, h_f) represented by the actual discharge equation for pipe flow-measuring devices is modeled by the definition of a "drag coefficient, C_D," which is defined by dimensional analysis (see Chapter 7), and evaluated empirically (Chapter 8). Give the result of the derivation of the actual discharge equation along with the drag coefficient, C_D (discharge coefficient, C_d) for pipe flow-measuring devices.

6.137C   The flow resistance (unknown head loss, h_f) represented by the actual discharge equation for open channel flow-measuring devices is modeled by the definition of a "drag coefficient, C_D," which is defined by dimensional analysis (see Chapter 7) and evaluated empirically (Chapter 9). Give the result of the derivation of the actual discharge equation along with the drag coefficient, C_D (discharge coefficient, C_d) for open channel flow-measuring devices.

6.138C   The actual discharge, Q_a (measured) in real flow is less than the ideal discharge, Q_i assumed in ideal flow due to the effects of the flow resistance in real flow. As a result, the average velocity for real flow, $v_{ave,a} = Q_a/A$ will be less than the average velocity for ideal flow, $v_{ave,i} = Q_i/A$. Illustrate this point both graphically and mathematically.

Modeling the Flow Resistance as a Loss in Pump and Turbine Efficiency in Internal Flow

6.139C   The derivations of the expressions for the efficiency of pumps, η_{pump} and turbines, $\eta_{turbine}$ are determined analytically in Chapter 4. The efficiency of a pump,

$$\eta_{pump} = \frac{(P_p)_{out}}{(P_p)_{in}} = \frac{\gamma Q h_{pump}}{\omega T_{shaft,in}} \tag{4.197}$$

expresses the ratio of a real/actual flow variable (hydraulic power output from the pump) to its corresponding ideal flow variable (shaft power input to the pump). And, the efficiency of a turbine,

$$\eta_{turbine} = \frac{(P_t)_{out}}{(P_t)_{in}} = \frac{\omega T_{shaft,out}}{\gamma Q h_{turbine}} \tag{4.217}$$

expresses the ratio of a real/actual flow variable (shaft power output from the turbine) to its corresponding ideal flow variable (hydraulic power input to the turbine). Explain why dimensional analysis is applied.

6.140C   The empirical expressions for the efficiency of pumps and turbines are respectively modeled by the definition of respective drag coefficients C_D, which are derived using dimensional analysis (see Chapter 7) and evaluated empirically (Chapter 4). Give the results of the derivation of the empirical expressions for the efficiency of pumps and turbines along with their respective drag coefficients C_D.

7

Dimensional Analysis

7.1 Introduction

Dimensional analysis is an essential and invaluable tool in the deterministic (as opposed to stochastic) mathematical modeling of fluid flow problems (or any problems other than fluid flow). Deterministic mathematical models are used in order to predict the performance of physical fluid flow situations. As such, in the deterministic mathematical modeling of physical fluid flow situations (both internal and external flow), while some fluid flow situations may be modeled by the application of pure theory (e.g., creeping flow, laminar flow, sonic flow, or critical flow), some (actually, most, due to the assumption of real turbulent flow, which is stochastic in nature) fluid flow situations are too complex for theoretical modeling—for instance, turbulent flow, subsonic (or supersonic or hypersonic) flow, subcritical (or supercritical) flow. However, regardless of the complexity of the fluid flow situation, the analysis phase of mathematical modeling involves the formulation, calibration, and verification of the mathematical model, while the subsequent synthesis or design phase of mathematical modeling involves application of the mathematical model in order to predict the performance of the fluid flow situation. Analysis, by definition, is "to break apart" or "to separate into its fundamental constituents," where synthesis, by definition, is "to put together" or "to combine separate elements to form a whole." As such, dimensional analysis is an essential and invaluable tool used in both the analysis and synthesis phases of the deterministic mathematical modeling of fluid flow problems.

Regardless of the complexity of the fluid flow situation, dimensional analysis plays an important role in both the analysis and the synthesis phases of the mathematical modeling of a fluid flow problem (or any problem other than fluid flow). In the deterministic mathematical modeling of a fluid flow (or any) process, the model formulation step begins by applying the theories/principles (conservation of momentum, conservation of mass, conservation of energy, etc.) (see Chapters 3 through 5) that govern the fluid flow situation. Then, the initial form of the model is determined by identifying and defining the physical forces and physical quantities that play an important role in the given flow situation. In the case where each physical force and physical quantity can be theoretically modeled, one may easily analytically/theoretically derive/formulate the mathematical model and apply the model without the need to calibrate or verify the model (e.g., see the definitions of Stokes Law, Poiseuille's law, sonic velocity, critical velocity, etc.). Additionally, if one wishes to experimentally formulate, calibrate, and verify the theoretical mathematical model, one may use the dimensional analysis procedure itself (Buckingham π theorem) and the results of dimensional analysis (laws of dynamic similarity/similitude) to do so. However, in the case where there is at least one physical force or one physical quantity that cannot be theoretically modeled, one may not analytically/theoretically derive/formulate the mathematical model (as in the case of empirically modeling the flow resistance in real turbulent flow: major head loss, minor head loss, actual discharge, drag force, and the efficiency of a pump or turbine), and, furthermore, one needs to actually calibrate and verify the model before applying the model. In such a case, the dimensional analysis procedure itself (Buckingham π theorem) (see Chapter 7) is actually necessary in order to formulate (derive) the mathematical model while reducing/minimizing the number of variables required to be calibrated. Furthermore, the results of dimensional analysis (laws of dynamic similarity/similitude) (see Chapter 11) are necessary in order to design a geometrically scaled model (of convenient size and fluids and of economic and time proportions) and allow the design of a systematic and well-guided approach to conducting and interpreting the experimental investigations (greatly reducing the number of experiments) required to calibrate the mathematical model.

7.1.1 The Role of Dimensional Analysis in the Modeling of Fluid Flow

The important role that dimensional analysis plays in the deterministic mathematical modeling of fluid flow problems is highlighted in the following five steps conducted in the modeling process. First, the dimensional analysis procedure itself (Buckingham π theorem) is used to formulate a mathematical model of the fluid flow situation by arranging/reducing the two or more significant variables (identified from the governing equations) into one or more dimensionless coefficients/π terms ($\pi_1 = \phi(nothing) = constant$ for a theoretical model, or $\pi_1 = \phi(\pi_2, \pi_3, \ldots)$ for an empirical model), which represents the uncalibrated mathematical model. It is important to note that the dimensional analysis procedure is capable (as a very powerful tool) of reducing a large number of significant variables into one or more dimensionless coefficients/π terms (see Chapter 7), thus reducing the number of variables required to be calibrated (experimentally/empirically evaluated). Second, this uncalibrated mathematical model is ideally calibrated (i.e., the π terms are evaluated) by conducting experiments using the full-scale physical prototype of the fluid flow situation. However, using the full-scale physical prototype to conduct experiments may present challenges such as inconvenient size (too large or too small), inconvenient fluids, economic constraints, time constraints, etc. Therefore, in order to solve this problem, one may instead design a geometrically scaled (either smaller or larger) physical model of the fluid flow situation. Fortunately, the uncalibrated mathematical model resulting from the dimensional analysis procedure ($\pi_1 = \phi(nothing) = constant$ for a theoretical model, or $\pi_1 = \phi(\pi_2, \pi_3, \ldots)$) for an empirical model) may be used to develop criteria/laws that govern dynamic similarity/similitude between two flow situations that are geometrically similar but different in size; the uncalibrated mathematical model is ready for the initial model application (synthesis) phase. These developed laws of dynamic similarity (similitude) may be applied in order to design a geometrically scaled model and thus allow the design of experiments using a model of convenient size and fluids and of economic and time proportions, which can predict the performance of the full-scale prototype (see Chapter 11). Therefore, the uncalibrated mathematical model is actually/practically calibrated (i.e., the π terms are evaluated) by conducting experiments using the geometrically scaled (either smaller or larger) physical model of the fluid flow situation. It is important to note that although the dimensional analysis procedure itself cannot directly calibrate the mathematical model, the results of the dimensional analysis procedure are used in the development of laws of dynamic similarity, which allow the design of a geometrically scaled physical model that is used to directly calibrate the mathematical model by conducting experiments. Third, the uncalibrated mathematical model is calibrated (i.e., the π terms are evaluated) by conducting experiments using the scaled physical model of the flow situation. The use of the dimensional analysis procedure to reduce a large number of significant variables into one or more dimensionless coefficients/π terms (conducted in the model formulation) drastically reduces the number of variables required to be calibrated (experimentally evaluated). For instance, assume one is conducting an experiment with one dependent variable and six independent variables and wishes to test the response of the dependent variable to changes made in the independent variables at three different levels. In order to conduct a complete set of experiments, one must test all possible combinations of the three test levels for each of the six independent variables; that would be $3^6 = 729$ experiments. However, if one uses dimensional analysis to reduce the seven significant variables to, let's say, two π terms, that reduces the number of independent variables from six to only one. Therefore, in order to conduct a complete set of experiments, one must test all possible combinations of the three test levels for now only one independent variable; that would be $3^1 = 3$ experiments. As such, by using dimensional analysis to reduce six independent variables to only one independent variable, the total number of required experiments would be reduced from 729 to only 3 experiments. Furthermore, the resulting one or more dimensionless coefficients/π terms may be used to design a systematic and well-guided approach to conducting and interpreting the experimental investigations, thus greatly reducing the number of experiments required to calibrate the resulting one or more dimensionless coefficients/π terms (see historically documented calibrated π terms in Chapters 4, 5, and 8 through 10). Fourth (ideally), the calibrated mathematical model is verified using test data. Fifth and finally, the calibrated mathematical model is ready for the final model application (synthesis) phase; one can now use the calibrated mathematical model to predict the performance of either the geometrically scaled model or the full-scale prototype of the fluid flow situation (see applications of theoretical and empirical equations throughout the textbook).

Therefore, the dimensional analysis procedure itself, and the results of the procedure, including the laws of similitude, play an important role in the deterministic mathematical modeling of fluid flow problems. Specifically, the dimensional analysis procedure itself may be used to formulate both theoretical and empirical mathematical models. The results of the procedure, including the laws of similitude, may be used to design both the geometrically scaled models and the actual experiments using the scaled models that are used to calibrate the mathematical model. Thus, the procedure of dimensional analysis and its results that are presented in Chapter 7 are used as follows: (1) use dimensional analysis to empirically formulate the flow resistance equations in internal and external flow (Sections 7.3 through 7.6), (2) use dimensional analysis to empirically formulate the efficiency equations of pumps and turbines (Section 7.7), (3) use dimensional analysis to experimentally formulate theoretical mathematical models (Section 7.8), (4) use the results of dimensional analysis to develop criteria/laws that govern dynamic similarity/similitude between two flow situations (internal and external) in order to design a geometrically scaled model (of convenient size and fluids and of economic and time proportions) that is used to calibrate the mathematical model (Chapter 11), and (5) use the results of dimensional analysis to design systematic and well-guided experiments in order to calibrate the mathematical model.

The relative importance of the role of the dimensional analysis procedure itself and the results of the procedure (including the laws of similitude) in the modeling of fluid flow may be summarized as follows. The dimensional analysis procedure and its results (including the laws of similitude) may be used as an academic illustrative exercise if one wishes to experimentally formulate, calibrate, and verify a theoretical mathematical model. However, the dimensional analysis procedure and its results (including the laws of similitude) are actually necessary in order to formulate, calibrate, and verify the empirical modeling of flow resistance in internal and external flow and the efficiency of pumps and turbines. As such, the historically documented calibration results and applications of the respective flow resistance equations and the efficiency of pumps and turbines are presented in Chapters 4, 5, and 8 through 10.

7.1.2 The Role of Dimensional Analysis in the Empirical Modeling of Flow Resistance in Real Fluid Flow

A major application of dimensional analysis in fluid flow is for problems that are too complex to be modeled by the application of pure theory. In particular, the modeling of flow resistance (due to turbulent flow) in internal and external real fluid flow is too complex for theoretical solution. Therefore, a very important application of dimensional analysis in fluid flow is in the modeling of flow resistance in internal and external (real) flow. Specifically, dimensional analysis is applied in the derivation of a given flow resistance equation in internal and external flow (major head loss, minor head loss, actual discharge, and drag force). Furthermore, in the cases where the flow resistance is due to pipe or channel friction (major head loss), pipe component (minor head loss), and/or a flow-measuring device (minor head loss), the flow resistance contributes to the power losses and thus the determination of the efficiency of pumps and turbines. As such, dimensional analysis is also applied in the derivation of the efficiency of pumps and turbines. The respective results of dimensional analysis (laws of dynamic similarity/similitude) are necessary in order to design a geometrically scaled model (of convenient size and fluids and of economic and time proportions) and allow the design of a systematic and well-guided approach to conducting and interpreting the experimental investigations (greatly reducing the number of experiments) required to calibrate the mathematical model. Thus, once the flow resistance equations (major head loss, minor head loss, actual discharge, and drag force) and the efficiency of pumps and turbines are derived, calibrated, and verified, they may be directly applied in the appropriate flow situations.

7.1.2.1 Modeling of Flow Resistance in Real Fluid Flow

In the application of the governing equations, although in some flow cases, one may ignore the flow resistance (flow from a tank, etc., and gradual channel transitions) or not need to model the flow resistance (a hydraulic jump), in most practical flow cases in general, one must account for the fluid viscosity and thus model the resulting shear stress and flow resistance. As such, the application of the governing equations (continuity, energy, and momentum) for real flow (internal) and ideal flow (flow-measuring devices for internal flow and external flow around objects) requires the modeling of flow resistance (frictional losses). One may note that the assumption of "real" flow implies that the flow resistance is modeled in both the energy and the momentum equations, while the assumption of "ideal" flow implies that the flow resistance is modeled only in the momentum equation (and thus the subsequent assumption of "real" flow). Specifically, the application of the integral form of the energy equation for incompressible, real, steady flow (pipe or open channel flow) given by Equation 4.44 requires the evaluation of the major and minor head loss flow resistance terms, $h_{fmaj} = \tau_w L / \gamma R_h$ and $h_{fmin} = \tau_w L / \gamma R_h$, respectively. Furthermore, the application of the differential form of the momentum equation (pipe or open channel flow) given by Equation 5.15 requires the evaluation of the friction slope term, $S_f = \tau_w / \gamma R_h$. The application of the continuity equation for a flow-measuring device (pipe or open channel flow) given by

Equation 3.47 requires the evaluation the actual discharge, $Q_a = v_a A_a$. Finally, the application of the integral form of the momentum equation for the case of external flow around an object given by Equation 4.28 (or Equation 5.27) requires the evaluation of the drag force, $F_D = (F_P + F_f)_s$.

7.1.2.2 Using Dimensional Analysis in the Empirical Modeling of Flow Resistance in Real Fluid Flow

Dimensional analysis is used in the empirical modeling of flow resistance in real fluid flow. Specifically, the results of the dimensional analysis procedure are used to supplement the momentum theory in the definition of the friction/viscous force and thus the definition of the flow resistance equations. In the application of the governing equations, it is assumed that the important sources of forces acting on the fluid element have been identified; however, not all forces may be theoretically modeled. In particular, assuming a real flow, the flow resistance is the shear stress, friction, or drag force originating in the fluid viscosity, which is modeled as the friction/viscous force, F_V in the integral form of the momentum equation (Equation 5.27). Regardless of whether the flow is internal or external, it is important to note that in the application of the integral form of the momentum equation (Equation 5.27), while the definition of the gravitational force, F_G and the pressure force, F_P can always be theoretically modeled, the friction/viscous force, F_V may not always be theoretically modeled (in particular, the modeling of flow resistance due to turbulent internal and external flow). Thus, the results of dimensional analysis (Chapter 7) are used to supplement the momentum theory (Chapter 5) in the definition of the friction/viscous force and thus the definition of the flow resistance equations, which were presented in Chapter 6. As such, dimensional analysis is actually necessary in order to arrange/reduce the resulting two or more significant variables (identified from the governing equations) into two or more dimensionless coefficients/π terms, which represents the formulated but uncalibrated mathematical model for the flow resistance (i.e., the friction/viscous force, F_V). It is important to note that the use of dimensional analysis is not needed in the derivation of the major head loss term for laminar flow (friction force is theoretically modeled) or the minor head loss term for a sudden pipe expansion (friction force is ignored).

The important role that dimensional analysis plays in the empirical modeling of flow resistance in real fluid flow is highlighted in the following five steps conducted in the modeling process. First, the dimensional analysis procedure itself (Buckingham π theorem) is used to formulate a mathematical model of the fluid flow situation by arranging/reducing the two or more significant variables (identified from the governing equations) into two or more dimensionless coefficients/π terms [$\pi_1 = \phi(\pi_2, \pi_3, \ldots)$], which represents the uncalibrated mathematical model; the dependent π term, π_1 is a function of two or more independent π terms, $\pi_2, \pi_3, \ldots$. It is important to note that the dimensional analysis procedure is capable of reducing a large number of significant variables into one or more dimensionless coefficients/π terms (see Chapter 7), thus reducing the number of variables required to be calibrated (experimentally evaluated). Second, this uncalibrated mathematical model is ideally calibrated (i.e., the π terms are evaluated) by conducting experiments using the full-scale physical prototype of the fluid flow situation. However, using the full-scale physical prototype to conduct experiments may present challenges such as inconvenient size (too large or too small), inconvenient fluids, economic constraints, time constraints, etc. Therefore, in order to solve this problem, one may instead design a geometrically scaled (either smaller or larger) physical model of the fluid flow situation. Fortunately, the uncalibrated mathematical model resulting from the dimensional analysis procedure ($\pi_1 = \phi(\pi_2, \pi_3, \ldots)$) may be used to develop criteria/laws that govern dynamic similarity/similitude between two flow situations that are geometrically similar but different in size; the uncalibrated mathematical model is ready for the initial model application (synthesis) phase. These developed laws of dynamic similarity (similitude) may be applied in order to design a geometrically scaled model and thus allow the design of experiments using a model of convenient size and fluids and of economic and time proportions, which can predict the performance of the full-scale prototype (see Chapter 11). Therefore, the uncalibrated mathematical model is actually calibrated (i.e., the π terms are evaluated) by conducting experiments using the geometrically scaled (either smaller or larger) physical model of the fluid flow situation. It is important to note that although the dimensional analysis procedure itself cannot directly calibrate the mathematical model, the results of the dimensional analysis procedure are used in the development of laws of dynamic similarity, which allow the design of a geometrically scaled physical model that is used to directly calibrate the mathematical model by conducting experiments. Third, the uncalibrated mathematical model is calibrated (i.e., the π terms are evaluated) by conducting experiments using the scaled physical model of the flow situation. The use of the dimensional analysis procedure to reduce a large number of significant variables into one or more dimensionless coefficients/π terms (conducted in the model formulation) drastically reduces the number of variables required to be calibrated (experimentally evaluated). Furthermore, the resulting two or more dimensionless coefficients/π terms may be used to design a systematic and well-guided approach to conducting and interpreting the experimental investigations, thus greatly reducing the number of experiments required to calibrate the resulting two or more dimensionless coefficients/π terms—note that while experimental investigations to calibrate the π terms were not conducted in the textbook, the historically documented calibrated π terms (drag coefficients, C_D) are presented and applied in Chapters 4, 5, and 8 through 10. It is important to note that in this case, where the fluid flow situation is too complex to be modeled by the application of pure theory (e.g., turbulent flow, subsonic [or supersonic or hypersonic] flow, or subcritical [or supercritical] flow), the resulting dependent π term is a function of two or more independent π terms [i.e., $\pi_1 = \phi(\pi_2, \pi_3, \ldots)$; this represents an uncalibrated mathematical model]. As such, calibration of the two or more π terms will require a well-guided limited number of experiments. The calibration results of the experiments may be graphically illustrated as follows. If $\pi_1 = \phi(\pi_2)$, then a plot of the dependent π term, π_1 versus the only one independent π term, π_2 will yield either a single linear or a single nonlinear plot. For instance, a single linear plot is illustrated on the Moody diagram in Figure 8.1 for laminar flow where the friction factor, $f = 64/R$, where $\pi_1 = f$ and $\pi_2 = R$. A single nonlinear plot is illustrated in Figure 10.9 for the drag coefficient for external flow, C_D, where $\pi_1 = C_D$ and $\pi_2 = L/D$. Another single nonlinear plot is illustrated in Figure 10.16 for the drag coefficient for external flow, $C_D = \phi(R)$, where $\pi_1 = C_D$ and $\pi_2 = R$. However, if $\pi_1 = \phi(\pi_2, \pi_3)$, then a plot of the dependent π term, π_1 versus two independent π terms, π_2 and π_3 will yield a family of lines or a family of curves. For instance, a family of curves is illustrated in the Moody diagram in Figure 8.1 for transitional and turbulent flow where the friction factor, $f = \phi(R, (\varepsilon/D))$ for various values of ε/D, where $\pi_1 = f$, $\pi_2 = R$, and $\pi_3 = \varepsilon/D$. Another family of curves is illustrated in Figure 10.18 for the drag coefficient for external flow, $C_D = \phi(R, (\varepsilon/L))$ for various values of ε/L, where $\pi_1 = C_D$, $\pi_2 = R$, and $\pi_3 = \varepsilon/L$. Fourth (ideally), the calibrated mathematical model is verified using test data. Fifth and finally, the calibrated mathematical model is ready for the final model application (synthesis) phase; one can now use the calibrated empirical mathematical model to predict the performance of either the geometrically scaled model or the full-scale prototype of the fluid flow situation (see Chapters 4, 5, and 8 through 11).

7.1.3 Flow Types and Dimensional Analysis

In the application of dimensional analysis to model the fluid flow, it is important to consider the type of flow. First, the flow may be defined as internal flow, which includes pipe and open channel flow, or it may be external flow, which includes the flow around objects. Second, as already mentioned, internal flow may be pipe or open channel flow. Third, the flow may be defined as ideal or real flow. Finally, real flow may be defined as laminar or turbulent flow. The assumption of real flow addresses the existence of flow resistance, which models the friction/viscous, or drag force and thus the potential need for the use of dimensional analysis in its definition. As such, the results of dimensional analysis introduce a drag coefficient, C_D, which expresses the ratio of a real flow variable (such as velocity, pressure, force, etc.) to its corresponding ideal flow variable (see Chapter 6 for definitions and details).

7.1.4 Internal Flow versus External Flow

The results sought in a given flow analysis will depend upon whether the flow is internal or external. While the flow of fluid in both pipes (pressure/closed conduit flow) and open channels (gravity/open conduit flow) are contained in a conduit and thus are referred to as internal flows, the flow of fluid over a body that is immersed in the fluid is referred to as external flow. In an internal flow, energy or work is used to move/force the fluid through a conduit; thus, one is interested in the determination of the forces (reaction forces, pressure drops, flow depths), energy or head losses, and cavitation where energy is dissipated (see Chapters 3, 4, 5, 8, and 9). However, in an external flow (Chapter 10), energy or work is used to move a submerged body through the fluid, where the fluid is forced to flow over the surface of the body. Thus, one is interested in the determination of the flow pattern around the body; the lift force, F_L; the drag force; F_D (resistance to motion) on the body; and the patterns of viscous action in the fluid as it passes around the body. If a submerged body were moving in a stationary homogenous real fluid, then, due to the relative motion between the body and fluid, the fluid exerts pressure forces normal to the surface and shear/friction forces parallel to the surface along the outer surface of the body—one may note that if the submerged body were not moving in the fluid, then only pressure forces would be exerted on the surface of the body. The resultant of the combined pressure and shear (viscous) forces acting on the surface of the body is due to the pressure drop, Δp and has two components. Specifically, the component of the resultant pressure, F_p and shear, F_f (also referred to as F_V) forces that acts in the direction of the flow, s is called the drag force, F_D. The component of the resultant pressure and shear forces that acts normal to the direction of the flow is called the lift force, F_L, which tends to move the submerged body in that direction. However, only if the body is moving at a high speed (and depending upon the shape of the body and the angle it makes with the direction of flow) does the lift force, F_L become dominant.

7.1.5 Modeling Flow Resistance: A Subset Level Application of the Governing Equations

Derivation of a given flow resistance equation involves a "subset level" application of the appropriate governing equations and in some cases supplementing the momentum equation with dimensional analysis. A "subset level" application of the appropriate governing equations focuses only on the given element causing the flow resistance. The friction/

viscous force, F_V, which is modeled in the integral form of the momentum equation (Equation 5.27), represents the flow resistance, which is the shear stress, friction, or drag force originating in the fluid viscosity. Thus, in the flow of any real fluid, the fluid has to do work against flow resistance, which results in a continual dissipation of energy in the momentum transfer. The flow resistance causes an unknown pressure drop, Δp, which causes an unknown head loss, h_f, where the head loss is due to a conversion of kinetic energy to heat which is modeled/displayed in the integral form of the energy equation (Equation 4.44). The flow resistance is modeled as a drag force in external flows, while it is modeled as a head loss in internal flows. Specifically, in the external flow around an object, the flow resistance is ultimately modeled as a drag force in the integral form of the momentum equation because the drag force is needed for the design of external flow around an object. In the internal flow in a pipe, the flow resistance is ultimately modeled as a major head loss due to pipe friction and as a minor head loss due to pipe transitions in the integral form of the energy equation because the major and minor head losses are needed for the design of internal flow in a pipe. In the internal flow in an open channel, the flow resistance is ultimately modeled as a major head loss due to channel friction in the integral form of the energy equation because the major head loss is needed for the design of internal flow in an open channel. And, in the internal flow in both a pipe and an open channel, the flow resistance is ultimately modeled as a reduced/actual flow rate due to a flow measurement device in the differential form of the continuity equation because the actual flow rate is needed for the design of a flow-measuring device.

7.1.6 Supplementing the Momentum Theory with Dimensional Analysis

When the friction/viscous forces can either be theoretically modeled (as in the case of the major head loss in laminar flow) or ignored (as in the case of the minor head loss due to a sudden pipe expansion) in the momentum equation, the respective flow resistance equation can be analytically derived. However, when the friction/viscous forces cannot either be theoretically modeled or ignored in the momentum equation, the momentum theory will be exhausted due to an unknown head loss, h_f. As a result, the momentum theory is supplemented with dimensional analysis (Section 7.2) in order to derive an expression for the drag force, the major head loss, the minor head loss, and the actual discharge. The flow resistance (unknown head loss, h_f) represented by each of the above listed resistance equations is modeled by the definition of a "drag coefficient, C_D," which is defined by dimensional analysis, and evaluated empirically.

7.1.7 Derivation of the Flow Resistance Equations and the Drag Coefficients

The derivation of the appropriate flow resistance equations and their corresponding "drag coefficients, C_D" for both external and internal flow are presented in Chapter 7 (also see Chapter 6). Specifically, the drag force equation along with the coefficient of drag, C_D (and the lift force along with the lift coefficient, C_L) is derived for the external flow around an object (Section 7.3). The major head loss equation along with the major head loss friction coefficients (Chezy, C; Darcy–Weisbach, f; and Manning's, n) for internal flows (pipe and open channel flow) is derived for both laminar and turbulent flow (Section 7.4). The minor head loss equation along with the minor loss coefficient, k is derived for pipe flow (Section 7.5). And, the actual discharge equation along with the coefficient of discharge, C_d is derived for both pipe and open channel flow (Section 7.6).

7.1.7.1 Empirical Evaluation of the Drag Coefficients and Application of the Flow Resistance Equations

Although the derivation of the flow resistance equations and their corresponding "drag coefficient, C_D" for both external and internal flow are presented in Chapter 7 (also see Chapter 6), the empirical evaluation (calibration) of the respective drag coefficients and application of the respective flow resistance equations are presented in the following chapters. Empirical evaluation of the major head loss friction coefficients (Chezy, C; Darcy–Weisbach, f; and Manning's, n), and the application of the corresponding major head loss equation are presented in Chapter 8 and 9. Empirical evaluation of the minor head loss coefficient, k and the application of the corresponding minor head loss equation are presented in Chapter 8. Empirical evaluation of the coefficient of discharge, C_d and the application of the corresponding actual discharge equation are presented in Chapters 8 and 9. Furthermore, empirical evaluation of the coefficient of drag, C_D (and lift coefficient, C_L) and application of the corresponding drag equation (lift equation) are presented in Chapter 10. Finally, application of the flow resistance prediction equations for similitude and modeling are presented in Chapter 11.

7.1.8 Derivation of the Efficiency of Pumps and Turbines

In the cases where the flow resistance is due to pipe or channel friction (major head loss), pipe component (minor head loss), and/or a flow-measuring device (minor head loss), the flow resistance contributes to the power losses and thus the determination of the efficiency of pumps and turbines (see Chapter 4). As such, dimensional analysis is also applied in the derivation of the efficiency of pumps and turbines. In particular, the total head loss, h_f (major and minor) will increase the amount of power that a pump must deliver to the fluid; the head loss, h_f represents the additional height that the fluid needs to be elevated by the pump in order to overcome the losses due to the flow resistance in the pipe or channel. Some of the pumping power will be used to pump the fluid through the pipe or channel, while some will be used to compensate for the energy lost in the pressure drop, Δp caused by the pipe or channel friction, pipe component, or flow-measuring device. A similar analogy may be deduced for evaluating the efficiency of a turbine.

The efficiency, η of a pump or a turbine is represented by the "drag coefficient, C_D" that is derived using dimensional analysis. The derivation of the efficiency, η of a pump or a turbine is presented in Chapter 7 (Section 7.7) (also see Chapter 6). The empirical evaluation of efficiency, η of a pump or a turbine and its application are presented in Chapter 4. Finally, the application of the efficiency, η of a pump or a turbine for similitude and modeling are presented in Chapter 11.

7.2 Dimensional Analysis of Fluid Flow

The basis of the dimensional analysis theory is the Buckingham π theorem (a formalization of the Rayleigh method). Briefly, it states that if a functional relationship exists between n physical quantities (variables), all of which can be expressed in terms of m fundamental dimensions (mass or force, length, and time), then $n\text{-}m$ dimensionless numbers (or π terms) can be formed from the n original physical quantities. Specifically, a functional relationship means that if values can be assigned to all but one of the physical variables in the relationship, the unknown value of the remaining variable may "theoretically" be determined.

However, in practice, a functional relationship means that when all but one of the physical variables have values assigned to them, the unknown value for the remaining variable will be fixed by natural physical laws. Furthermore, because the natural physical laws are either unexplained by existing theory or too complex to be explained, the unknown value for the remaining variable is not calculated, but determined empirically through experimentation (calibration) (as represented by a "drag coefficient," C_D in fluid flow).

Based on a general understanding of (and experience with and judgment of) the fluid flow phenomena, one may identify the physical forces and physical quantities that play an important role (dominant forces) in a given flow situation and thus identify the n important physical quantities (variables) in the fluid flow situation (use Table 11.7 for external flow and Table 11.8 for internal flow as a guide). As such, one may then derive a functional relationship (which includes the n-m dimensionless numbers, or π terms) between the n physical quantities in the fluid flow situation. However, before describing the procedure to derive the functional relationship and thus the n-m dimensionless numbers, it is important to first identify all of the possible sources of forces that may act in any given flow situation and second present the two possible dimensional systems that may be used to express the m fundamental dimensions (mass or force, length, and time) for a given physical quantity. Furthermore, while there is sufficient theory to analytically model most of the forces in fluid flow (pressure, gravity, elastic, surface tension, and inertia), there is insufficient theory on the friction/viscous forces (flow resistance) in most types of internal and external flow. As such, dimensional analysis is applied to empirically model the respective flow resistance equations resulting from both internal and external flow. Finally, guidelines in the derivation of the flow resistance equations, and their corresponding drag coefficients are presented below.

7.2.1 Dynamic Forces Acting on a Fluid Element

The dynamic forces that may act in a flow situation include those due to pressure, F_P; gravity, F_G; viscosity, F_V; elasticity, F_E; and surface tension, F_T. Furthermore, according to Newton's second law of motion, if the summation of the forces that act on a fluid element does not add up to zero, the fluid element will accelerate. However, such an unbalanced force system can be transformed into a balanced system by adding an inertial force, F_I that is equal and opposite to the resultant, R_s of the acting forces as follows:

$$\sum_s F = F_P + F_G + F_V + F_E + F_T = R_s \tag{7.1}$$

and

$$F_I = -R_s \tag{7.2}$$

and thus:

$$\sum_s F = F_P + F_G + F_V + F_E + F_T + F_I = 0 \tag{7.3}$$

The simplest mathematical expressions for these forces are given as follows:

Pressure Force: $F_P = \Delta p A = \Delta p L^2$ $\tag{7.4}$

Gravity Force: $F_G = Mg = \rho V g = \rho L^3 g$ $\tag{7.5}$

Viscous Force: $F_V = \tau A = \mu \dfrac{dv}{dy} A = \mu \dfrac{v}{L} L^2 = \mu v L$ (7.6)

Elastic Force: $F_E = E_v A = E_v L^2$ (7.7)

Surface Tension Force: $F_T = \sigma L$ (7.8)

Inertia Force: $F_I = Ma = \rho V \dfrac{L}{T^2} = \rho L^3 \dfrac{L}{T^2} = \rho \dfrac{L^4}{T^2} = \rho v^2 L^2$ (7.9)

where F = force, M = mass, L = length, T = time, A = cross-sectional area, V = volume, v = velocity, a = acceleration, g = acceleration due to gravity, Δp = fluid pressure difference, τ = shear stress, ρ = fluid density, γ = fluid specific weight, μ = dynamic or absolute fluid viscosity, E_v = fluid bulk modulus of elasticity, and σ = fluid surface tension. Furthermore, one may note that depending upon the fluid flow problem, while the inertia force will always be important (assuming steady state flow; no acceleration), some of the above listed forces may not play a significant role, as discussed in a detail in sections below.

7.2.2 Two-Dimensional Systems

The dimensions of a given physical quantity may be expressed in either the force-length-time (FLT) system or in the mass-length-time (MLT) system. These two systems are related through Newton's second law, which is given as follows:

$$F = Ma \qquad (7.10)$$

where F = force, M = mass, and a = acceleration, or expressed in dimensions as follows:

$$F = M \dfrac{L}{T^2} \qquad (7.11)$$

Thus, using this above relationship, one may convert from one dimensional system to the other, where the dimensions used in either system may be English units or metric units. Given below in Table 7.1 are the dimensions (in both systems) and units (for both units) of some common physical quantities in fluid flow problems.

7.2.3 Deriving a Functional Relationship/Dimensionless Numbers

Based on the Buckingham π theorem, the procedure to derive a functional relationship between the n physical variables in the fluid flow situation and thus the n-m dimensionless numbers (π terms) for a given flow situation is given in the following five steps:

Step 1: Based on a general understanding of (and experience with and judgment of) the fluid flow phenomena, identify the dominant/significant physical forces and dominant physical quantities in the flow situation (use Table 11.7 for external flow and Table 11.8 for internal flow as a guide). It is important to note that the inertia force, F_I will always be a dominant force (assumption of steady state flow; no acceleration). Furthermore, there will be at least one more additional dominant force or dominant physical quantity.

TABLE 7.1

Dimensions and Units of Some Common Physical Quantities in Fluid Flow

Physical Quantity/Property	FLT System: Primary Dimensions for BG Units	MLT System: Primary Dimensions for SI Units	BG (English) Units	SI (metric) Units
Geometrics				
Length, L	L	L	ft	m
Area, A	L^2	L^2	ft^2	m^2
Volume, V	L^3	L^3	ft^3	m^3
Kinematics				
Time, T	T	T	sec	sec
Frequency, ω	T^{-1}	T^{-1}	cycle/sec = sec^{-1}	cycle/sec = sec^{-1} = Hz
Velocity, v	LT^{-1}	LT^{-1}	ft/sec (fps)	m/sec
Acceleration, a	LT^{-2}	LT^{-2}	ft/sec^2	m/sec^2
Discharge, Q	$L^3 T^{-1}$	$L^3 T^{-1}$	ft^3/sec (cfs)	m^3/sec
Dynamics				
Mass, M	$FL^{-1}T^2$	M	slug = lb-sec^2/ft	kg
Force, F	F	MLT^{-2}	lb	N = kg-m/sec^2
Pressure, p	FL^{-2}	$ML^{-1}T^{-2}$	lb/ft^2 (psf)	Pa = N/m^2
Momentum, Mv or Impulse, FT	FT	MLT^{-1}	lb-sec	N-sec
Energy, E or Work, W	FL	ML^2T^{-2}	ft-lb	J = N-m
Power, P	FLT^{-1}	ML^2T^{-3}	ft-lb/sec	W = N-m/sec
Fluid Properties				
Density, ρ	$FL^{-4}T^2$	ML^{-3}	slug/ft^3	kg/m^3
Specific gravity, s	dimensionless	dimensionless	no units	no units
Specific weight, γ	FL^{-3}	$ML^{-2}T^{-2}$	lb/ft^3	N/m^3
Absolute (dynamic) viscosity, μ	$FL^{-2}T$	$ML^{-1}T^{-1}$	lb-sec/ft^2	N-sec/m^2
Kinematic viscosity, $\upsilon = \mu/\rho$	L^2T^{-1}	L^2T^{-1}	ft^2/sec	m^2/sec
Surface tension, σ	FL^{-1}	MT^{-2}	lb/ft	N/m
Vapor pressure, p_v	FL^{-2}	$ML^{-1}T^{-2}$	lb/ft^2 (psf)	Pa = N/m^2
Bulk modulus of elasticity, E_v	FL^{-2}	$ML^{-1}T^{-2}$	lb/ft^2	N/m^2
Temperature, θ	θ	θ	°F	°C

Note the following abbreviations: Hertz (Hz), Newtons (N), Pascal (Pa), Joules (J), Watts (W), Fahrenheit (F), and Celsius (C).

Step 2: Use the simplest mathematical expressions for the identified dominant physical forces and dominant physical quantities and identify the n physical quantities, all of which can be expressed in terms of m fundamental dimensions (M or F, L, and T). In order to identify the n physical quantities, choose variables that relate to force, mass, kinematics, and geometry of the flow situation (see Table 7.1). If unimportant variables are identified in this step, the Buckingham π theorem will automatically remove the redundant variables. Furthermore, unimportant variables may also be eliminated during the laboratory/experimental analyses (to calibrate the resulting equation/relationship) that often follow the application of the Buckingham π theorem. However, if an important variable is omitted, the results will be incomplete and may lead to inaccurate conclusions. Therefore, it is better to include a variable than not to include it. It is important to note that the inertia force, $F_I = \rho v^2 L^2$, which will always be a dominant force represents the variables that relate to mass (ρ), kinematics (v), and geometry (L). Furthermore, the first/main additional dominant force or dominant physical quantity represents the variable that relates to force. Additional dominant forces or dominant physical quantities may be added.

Step 3: Determine the number of dimensionless numbers, which is n-m.

Step 4: Apply the Buckingham π theorem to derive the n-m dimensionless numbers (π terms). In order to derive each dimensionless number (π term), choose primary variables that relate to mass (ρ), kinematics (v), and geometry (L) of the flow situation, from among the n physical quantities; these primary variables will be the same for each of the π terms and represent the inertia force, $F_I = \rho v^2 L^2$. Then, for each of the π terms, choose a secondary variable that represents one specific dominant dynamic force or dominant physical quantity for that dimensionless number (π term), from among the n physical quantities. Additional dominant forces or physical quantities can be represented by their key variables from their simplest mathematical expression. The final form of the π terms are achieved by conducting a dimensional analysis of each π term through the application of the Buckingham π theorem, which is best illustrated by examples, which are presented below.

Step 5: Interpret the resulting functional relationship between the n physical variables and thus, the n-m dimensionless numbers (π terms).

7.2.4 Main Pi Terms

The five main dimensionless numbers (π terms), as derived from dimensional analysis, are defined as follows:

Euler Number: $E = \dfrac{F_I}{F_P} = \dfrac{\rho v^2}{\Delta p}$ \hfill (7.12)

Froude Number: $F = \dfrac{F_I}{F_G} = \dfrac{v^2}{gL}$ \hfill (7.13)

Reynolds Number: $R = \dfrac{F_I}{F_V} = \dfrac{\rho v L}{\mu}$ \hfill (7.14)

Cauchy Number: $C = \dfrac{F_I}{F_E} = \dfrac{\rho v^2}{E_v}$ \hfill (7.15)

Weber Number: $W = \dfrac{F_I}{F_T} = \dfrac{\rho v^2 L}{\sigma}$ \hfill (7.16)

One may note that the Froude number, F may alternatively be expressed as follows:

$$\sqrt{F} = F = \frac{F_I}{F_G} = \frac{v}{\sqrt{gL}} \qquad (7.17)$$

Furthermore, the Mach number, M is defined as a function of the Cauchy number, C as follows:

$$\sqrt{C} = M = \frac{F_I}{F_E} = \frac{v}{\sqrt{\dfrac{E_v}{\rho}}} \qquad (7.18)$$

7.2.5 The Definition of New Pi Terms

Because dimensional analysis may not yield the final expression for the drag coefficient, C_D, one may need to interpret the resulting dimensionless numbers (π terms) through manipulation and empirical studies. Furthermore, one may note that the rules for alternately redefining a given dimensionless number (π term) to yield a new π term are given as follows:

1. Reciprocal rule: $\pi_2 = a/\pi_1$
2. Power rule: $\pi_2 = \pi_1^a$
3. Multiplication rule: $\pi_3 = \pi_1 \pi_2$
4. Factor rule: $\pi_2 = a\pi_1$

7.2.6 Guidelines in the Derivation of the Flow Resistance Equations and the Drag Coefficients

Dimensional analysis is used to derive the flow resistance equations and their corresponding drag coefficient, C_D, which expresses the ratio of a real flow variable (such as velocity, pressure, force, etc.) to its corresponding ideal flow variable. In the dimensional analysis of fluid flow, there are always at least four parameters involved: ρ, v, L (or D or y; characteristic length dimension), and Δp (or Δy, Δz, or $\Delta H = h_f = \tau_w L/\gamma R_h = \Delta p/\gamma = S_f L$; a representation of energy loss or some form of flow resistance).

> *Step 1:* Based on a general understanding of (and experience with and judgment of) the fluid flow phenomena, identify the dominant/significant physical forces and dominant physical quantities in the flow situation (use Table 11.7 for external flow and Table 11.8 for internal flow as a guide). The inertia force, $F_I = Ma = \rho v^2 L^2$, which will always be a dominant force, represents the primary variables that relate to mass (ρ), kinematics (v), and geometry, (L [or D or y; characteristic length dimension]) of the flow. Furthermore, there will be at least one more additional dominant force or dominant physical quantity. One may note that the momentum theorem is supplemented with dimensional analysis because the friction/viscous forces in the case of turbulent flow, F_f or $F_V = \tau A$ cannot be either theoretically modeled (as in the case of laminar flow) or ignored (as in the case of a sudden pipe expansion). Therefore, the first/main additional dominant force is the friction/viscous force for turbulent flow, F_f or F_V, which can be represented by a dominant physical quantity (the first/main secondary variable, which represents an energy loss or some form of

flow resistance), such as the drag force in external flow, F_D (or Δp); the major head loss in pipe or open channel flow, h_f (or τ_w); the minor head loss in pipe flow, h_f (or Δp); or the actual discharge in pipe or open channel flow, Q_a (or h_f).

Step 2: Use the simplest mathematical expressions for the identified dominant physical forces and dominant physical quantities and identify the n physical quantities, all of which can be expressed in terms of m fundamental dimensions (M or F, L, and T). In order to identify the n physical quantities, choose variables that relate to force, mass, kinematics, and geometry of the flow situation (see Table 7.1). The inertia force, $F_I = Ma = \rho v^2 L^2$, represents the primary variables relate to mass (ρ), kinematics (v), and geometry (L [or D or y; characteristic length dimension]) of the flow. Furthermore, the first/main additional dominant force or dominant physical quantity represents the variable that relates to force and is the friction/viscous force for turbulent flow, F_f or $F_V = \tau A$, which can be represented by a dominant physical quantity (the first/main secondary variable, which represents an energy loss or some form of flow resistance), such as one of the following:

- The drag force in external flow, F_D (or Δp), where $F_D = \Delta p A$ is a result of Δp, and the secondary variable is specifically $F_D = \Delta p A$ (see Table 7.2).

- The major head loss in pipe flow, h_f (or τ_w), where $h_f = \tau_w L/\gamma R_h = \Delta p/\gamma = S_f L$ is a result of Δp, and the secondary variable is specifically $\tau_w = \Delta p R_h/L$ (see Table 7.3).

TABLE 7.2

Guidelines in the Derivation of the Drag Force Equation

Drag Force, $F_D = \Delta p A$ Is a Result of Δp	Dominant Force or Dominant Physical Quantity	Dominant Force or Dominant Physical Quantity	Dimensional Analysis and Empirical Studies Result
	$F_I = \rho v^2 L^2$ is always dominant and results in the three primary variables	$F_V = \tau A$ is dominant and results in main secondary variable	
Mass Primary Variable	ρ		
Kinematics Primary Variable	v		
Geometric Primary Variable	L		
Main Secondary Variable		$F_D = \Delta p A$	
Additional Secondary Variables			$\mu, E_v, g, \varepsilon, L_i$
π Term or Drag Coefficient			$\dfrac{F_D}{\rho v^2 L^2} = \dfrac{F_D}{F_I} = \dfrac{F_P}{F_I} = \dfrac{\Delta p}{\rho v^2} = \dfrac{1}{E}$ $= C_D = \phi\left(R, C, F, \dfrac{\varepsilon}{L}, \dfrac{L_i}{L}\right)$
Conclusion			F_D (or Δp) represents F_P and F_f (or F_V) in the direction of the flow, s and is a result of Δp; thus, C_D is mainly a function of E

TABLE 7.3

Guidelines in the Derivation of the Major Head Loss in Pipe Flow Equation

Major Head Loss in Pipe Flow, $h_f = \tau_w L/\gamma R_h = \Delta p/\gamma = S_f L$ Is a Result of Δp	Dominant Force or Dominant Physical Quantity	Dominant Force or Dominant Physical Quantity	Dimensional Analysis and Empirical Studies Result
	$F_I = \rho v^2 L^2$ is always dominant and results in the three primary variables	$F_V = \tau A$ is dominant and results in main secondary variable	
Mass Primary Variable	ρ		
Kinematics Primary Variable	v		
Geometric Primary Variable	L		
Main Secondary Variable		$\tau_w = \dfrac{\Delta p R_h}{L}$	
Additional Secondary Variables			μ, ε
π Term or Drag Coefficient			$\dfrac{\tau_w}{\rho v^2} = \dfrac{F_f}{F_I} = \dfrac{F_P}{F_I} = \dfrac{R_h}{L}\dfrac{\Delta p}{\rho v^2} = \dfrac{R_h}{L}\dfrac{1}{E}$ $= C_D = \phi\left(R, \dfrac{\varepsilon}{D}\right)$
Conclusion			h_f (or τ_w) represents F_V and is a result of Δp; thus, C_D is mainly a function of E

- The major head loss in open channel flow, h_f (or τ_w), where $h_f = \tau_w L/\gamma R_h = S_o L = S_f L$ is a result of gravity, g, and the secondary variable is specifically $\tau_w = \gamma R_h S_f = \rho g R_h S_f$ (see Table 7.4).
- The minor head loss in pipe flow, h_f (or Δp), where $h_f = \tau_w L/\gamma R_h = \Delta p/\gamma$ is a result of Δp, and the secondary variable is specifically Δp (see Table 7.5).
- The actual discharge in pipe flow, Q_a (or h_f), where:

$$Q_a = \frac{v_a}{\sqrt{\dfrac{2\Delta p}{\rho}}}\frac{A_a}{A_i}Q_i$$

is a result of Δp, and the secondary variable is specifically Δp (see Table 7.6).

- The actual discharge in open channel flow, Q_a (or h_f), where:

$$Q_a = \frac{v_a}{\sqrt{2g\Delta y}}\frac{A_a}{A_i}Q_i$$

is a result of Δy and gravity, g, and the secondary variable is specifically gravity, g (see Table 7.7).

Additional dominant forces or dominant physical quantities may be added.

Step 3: Determine the number of dimensionless numbers, which is $n\text{-}m$. In the dimensional analysis of fluid flow, there are always at least four parameters ($n \geq 4$) involved: ρ, v, L (or D or y; characteristic length dimension), and Δp (or Δy, Δz, or

TABLE 7.4

Guidelines in the Derivation of the Major Head Loss in Open Channel Flow Equation

Major Head Loss in Open Channel Flow, $h_f = \tau_w L/\gamma R_h = S_o L = S_f L$ Is a Result of g	Dominant Force or Dominant Physical Quantity	Dominant Force or Dominant Physical Quantity	Dimensional Analysis and Empirical Studies Result
	$F_I = \rho v^2 L^2$ is always dominant and results in the three primary variables	$F_V = \tau A$ is dominant and results in main secondary variable	
Mass Primary Variable	ρ		
Kinematics Primary Variable	v		
Geometric Primary Variable	L		
Main Secondary Variable		$\tau_w = \gamma R_h S_f$ or $\tau_w = \rho g R_h S_f$	
Additional Secondary Variables			μ, ε
π Term or Drag Coefficient			$\dfrac{\tau_w}{\rho v^2} = \dfrac{F_f}{F_I} = \dfrac{\gamma R_h S_f}{\rho v^2} = \dfrac{\rho g y S_f}{2\rho v^2} = \dfrac{S_f}{2}\dfrac{g y}{v^2}$ $= \dfrac{S_f}{2}\dfrac{1}{F} = C_D = \phi\left(R, \dfrac{\varepsilon}{y}\right)$
Conclusion			h_f (or τ_w) represents F_V and is a result of g; thus, C_D is mainly a function of F

TABLE 7.5

Guidelines in the Derivation of the Minor Head Loss in Pipe Flow Equation

Minor Head Loss in Pipe Flow, $h_f = \tau_w L/\gamma R_h = \Delta p/\gamma$ Is a Result of Δp	Dominant Force or Dominant Physical Quantity	Dominant Force or Dominant Physical Quantity	Dimensional Analysis and Empirical Studies Result
	$F_I = \rho v^2 L^2$ is always dominant and results in the three primary variables	$F_V = \tau A$ is dominant and results in main secondary variable	
Mass Primary Variable	ρ		
Kinematics Primary Variable	v		
Geometric Primary Variable	L		
Main Secondary Variable		Δp	
Additional Secondary Variables			μ, ε, L_i
π Term or Drag Coefficient			$\dfrac{\Delta p}{\rho v^2} = \dfrac{F_P}{F_I} = \dfrac{1}{E} = C_D = k = \phi\left(R, \dfrac{\varepsilon}{D}, \dfrac{L_i}{L}\right)$
Conclusion			h_f (or Δp) represents F_V and is a result of Δp; thus, C_D is mainly a function of E

TABLE 7.6

Guidelines in the Derivation of the Actual Discharge in Pipe Flow Equation

Actual Discharge in Pipe Flow, $Q_a = \dfrac{v_a}{\sqrt{2\Delta p/\rho}}\dfrac{A_a}{A_i}Q_i$ Is a Result of Δp	Dominant Force or Dominant Physical Quantity	Dominant Force or Dominant Physical Quantity	Dimensional Analysis and Empirical Studies Result
	$F_I = \rho v^2 L^2$ is always dominant and results in the three primary variables	$F_V = \tau A$ is dominant and results in main secondary variable	
Mass Primary Variable	ρ		
Kinematics Primary Variable	$Q_a = v_a A_a$		
Geometric Primary Variable	L		
Main Secondary Variable		Δp	
Additional Secondary Variables		μ, L_i	
π Term or Drag Coefficient			$\dfrac{Q_a}{Q_i} = \dfrac{F_I}{F_P} = \dfrac{v_a A_a}{v_i A_i} = \dfrac{v_a}{\sqrt{\dfrac{\Delta p}{\rho}}}\dfrac{A_a}{A_i} = \sqrt{E}\dfrac{A_a}{A_i}$ $= C_v C_c = C_d = \phi\left(R, \dfrac{L_i}{L}\right)$
Conclusion			Q_a (or h_f) represents F_V and is a result of Δp; thus, C_D is mainly a function of E

$\Delta H = h_f = \tau_w L/\gamma R_h = \Delta p/\gamma = S_f L$; a representation of energy loss or some form of flow resistance), and three fundamental dimensions ($m = 3$). Thus, there will be at least one dimensionless number ($n\text{-}m \geq 1$), which is the drag coefficient, C_D, which expresses the ratio of a real flow variable (such as velocity, pressure, force, etc.) to its corresponding ideal flow variable.

Step 4: Apply the Buckingham π theorem to derive the $n\text{-}m$ dimensionless numbers (π terms). The dimensionless numbers (π terms) for the resistance equations are derived as follows:

- For the drag force in external flow, F_D, the primary variables are: ρ, v, and L, and the secondary variable is $F_D = \Delta p A$ (see Table 7.2).
- For the major head loss in pipe flow, h_f, the primary variables are: ρ, v, and L, and the secondary variable is $\tau_w = \Delta p R_h/L$ (see Table 7.3).
- For the major head loss in open channel flow, h_f, the primary variables are: ρ, v, and L, and the secondary variable is $\tau_w = \gamma R_h S_f = \rho g R_h S_f$ (see Table 7.4).
- For the minor head loss in pipe flow, h_f, the primary variables are: ρ, v, and L, and the secondary variable is Δp (see Table 7.5).
- For the actual discharge in pipe flow, Q_a, the primary variables are: ρ, $Q_a = v_a A_a$, and L, and the secondary variable is Δp (see Table 7.6).
- For the actual discharge in open channel flow, Q_a, the primary variables are: ρ, $Q_a = v_a A_a$, and L, and the secondary variable is gravity, g (see Table 7.7).

TABLE 7.7

Guidelines in the Derivation of the Actual Discharge in Open Channel Flow Equation

Actual Discharge in Open Channel Flow, $Q_a = \dfrac{v_a}{\sqrt{2g\Delta y}}\dfrac{A_a}{A_i}Q_i$ Is a Result of Δy and g	Dominant Force or Dominant Physical Quantity	Dominant Force or Dominant Physical Quantity	Dimensional Analysis and Empirical Studies Result
	$F_I = \rho v^2 L^2$ is always dominant and results in the three primary variables	$F_V = \tau A$ is dominant and results in main secondary variable	
Mass Primary Variable	ρ		
Kinematics Primary Variable	$Q_a = v_a A_a$		
Geometric Primary Variable	L		
Main Secondary Variable		g	
Additional Secondary Variables		$\mu,\ \sigma,\ L_i$	
π Term or Drag Coefficient			$\dfrac{Q_a}{Q_i} = \dfrac{F_I}{F_P} = \dfrac{v_a A_a}{v_i A_i} = \underbrace{\dfrac{v_a}{\sqrt{g\Delta y}}}_{v_i = v_c}\dfrac{A_a}{A_i} = \sqrt{E}\dfrac{A_a}{A_i}$ $= \sqrt{F}\dfrac{A_a}{A_i} = C_v C_c = C_d = \phi\left(R, W, \dfrac{L}{L_i}\right)$
Conclusion			Q_a (or h_f) represents F_V and is a result of Δy and g; thus, C_D is mainly a function of $E = F$

Additional secondary variables may be added as represented by their simplest mathematical expressions for the dominant forces or physical quantities

Step 5: Interpret the resulting functional relationship between the n physical variables and thus the n-m dimensionless numbers (π terms). In fluid flow, there will be at least one π term, which represents a drag coefficient, C_D. Depending upon the dominant physical quantity that represents the first/main additional dominant force (i.e., the friction/viscous force for turbulent flow, F_f or F_V), the π term will be interpreted accordingly as follows:

- In external flow, the propulsive force for the fluid flow is provided by a pressure drop, Δp, which results in an unknown Δp. Thus, in the derivation of the drag force, $F_D = \Delta p A$, the following four variables, ρ, v, L, and $F_D = \Delta p A$, result in one dimensionless number, namely: $\Delta p / \rho v^2 = 1/E = C_D$, where $E = F_I/F_P =$ the Euler number and $C_D =$ the drag coefficient that represents the flow resistance. Therefore, the drag force, F_D (or Δp), which represents F_P and F_V in the direction of the flow, s is a result of Δp; thus, C_D is mainly a function of E (see Table 7.2).

- In pipe flow, the propulsive force for the fluid flow is provided by a pressure gradient, $\Delta p/L$, which results in an unknown Δp or S_f. Thus, in the derivation of the major head loss in pipe flow, $h_f = \tau_w L/\gamma R_h = \Delta p/\gamma = S_f L$, the following four variables, ρ, v, L, and $\tau_w = \Delta p R_h/L$, result in one dimensionless number, namely: $(R_h/L)(\Delta p/\rho v^2) = (R_h/L)(1/E) = C_D$, where $E = F_I/F_P =$ the Euler number and $C_D =$ the drag coefficient that represents the flow resistance. Therefore, the major head loss in pipe flow, h_f (or τ_w), which represents F_V, is a result of Δp and thus C_D is mainly a function of E (see Table 7.3).

- In open channel flow, the propulsive force is provided by invoking gravity, g and thus by the weight of the fluid along the channel slope (channel gradient, $S_o = \Delta z/L$), which results in an unknown S_f. Thus, in the derivation of the major head loss in open channel flow, $h_f = \tau_w L/\gamma R_h = S_o L = S_f L$, the following four variables, ρ, v, L, and $\tau_w = \gamma R_h S_f = \rho g R_h S_f$, result in one dimensionless number, namely: $(S_f/2)(gy/v^2) = (S_f/2)(1/F) = C_D$, where $F = F_I/F_G =$ the Froude number, and $C_D =$ the drag coefficient that represents the flow resistance. Therefore, the major head loss in open channel flow, h_f (or τ_w), which represents F_V, is a result of g; thus, C_D is mainly a function of F (see Table 7.4).

- In pipe flow, the propulsive force for the fluid flow is provided by a pressure gradient, $\Delta p/L$, which results in an unknown Δp or S_f. Thus, in the derivation of the minor head loss in pipe flow, $h_f = \tau_w L/\gamma R_h = \Delta p/\gamma$, the following four variables, ρ, v, L, and Δp, result in one dimensionless number, namely $\Delta p/\rho v^2 = 1/E = C_D = C_p$ where $E = F_I/F_P =$ the Euler number, $C_D =$ the drag coefficient that represents the flow resistance, and $C_p =$ the pressure coefficient. Therefore, the minor head loss in pipe flow, h_f (or Δp), which represents F_V, is a result of Δp and thus C_D is mainly a function of E (see Table 7.5).

- In pipe flow, the propulsive force for the fluid flow is provided by a pressure gradient, $\Delta p/L$, which results in an unknown Δp or S_f. Thus, in the derivation of the actual discharge in pipe flow,

$$Q_a = \frac{v_a}{\sqrt{\dfrac{2\Delta p}{\rho}}} \frac{A_a}{A_i} Q_i,$$

the following four variables, ρ, $Q_a = v_a A_a$, L, and Δp, result in one dimensionless number, namely:

$$\frac{v_a}{\sqrt{\dfrac{\Delta p}{\rho}}} \frac{A_a}{A_i} = \sqrt{E}\frac{A_a}{A_i} = C_v C_c = C_D$$

where $E = F_I/F_P =$ the Euler number, $C_v =$ the velocity coefficient, $C_c =$ the contraction coefficient, and $C_D =$ the drag coefficient that represents the flow resistance. Therefore, the actual discharge in pipe flow, Q_a (or h_f), which represents F_V, is a result of Δp and thus C_D is mainly a function of E (see Table 7.6).

- In open channel flow, the propulsive force is provided by invoking gravity, g and thus by the weight of the fluid along the channel slope (channel gradient, $S_o = \Delta z/L$), which results in an unknown Δy or S_f. Thus in the derivation of the actual discharge in open channel flow,

$$Q_a = \frac{v_a}{\sqrt{2g\Delta y}} \frac{A_a}{A_i} Q_i,$$

the following four variables, ρ, $Q_a = v_a A_a$, L, and g, result in one dimensionless number, namely:

$$\frac{v_a}{\sqrt{g\Delta y}}\frac{A_a}{A_i} = \sqrt{E}\frac{A_a}{A_i} = \sqrt{F}\frac{A_a}{A_i} = C_v C_c = C_D$$

where $E = F_I/F_P =$ the Euler number, $F = F_I/F_G =$ the Froude number, $C_v =$ the velocity coefficient, $C_c =$ the contraction coefficient, $C_D =$ the drag coefficient that represents the flow resistance, and $E = F$. Therefore, the actual discharge in open channel flow, Q_a (or h_f), which represents F_V, is a result of Δy and g, and thus C_D is mainly a function of $E = F$. It is important to note that in the case of a flow-measuring device in open channel flow, the ideal velocity, $v_i = \sqrt{g\Delta y}$ equals the critical velocity, $v_c = \sqrt{gy_c}$, which results in $E = F$. As such, the flow through a flow-measuring device in open channel flow (and the resulting friction/viscous force, F_V) is a result of both the pressure force, $F_P = \Delta p A = \gamma \Delta y A$ and the gravitational force, $F_G = Mg$ (see Table 7.7).

7.2.7 The Definition of the Drag Coefficient

The definition of the drag coefficient, C_D involves at least two dominant forces. The dimensional analysis of fluid flow assumes that the inertia force, $F_I = Ma = \rho v^2 L^2$ will always be a dominant force and results in the three primary variables: ρ, v, and L. Furthermore, the friction/viscous force for turbulent flow, F_f or $F_V = \tau A$ will always be the first/main additional dominant force, which can be represented by a dominant physical quantity (F_D or h_f) that represents an energy loss or some form of flow resistance and results in an appropriate secondary variable (F_D, τ_w, Δp, or g). One may note that the modeling of only one additional dominant force (i.e., the friction/viscous force for turbulent flow) results in only one dimensionless number (π term), which is the drag coefficient, C_D. Furthermore, modeling additional dominant forces or dominant physical quantities results in additional secondary variables and will introduce an additional dimensionless number (π term) for each new secondary variable added.

In summary, one may note that in the dimensional analysis of fluid flow, although the first/main additional dominant force is the friction/viscous force for turbulent flow, $F_V = \tau A$, the type of flow will determine the cause of the friction/viscous force and thus the definition of the drag coefficient, C_D. As such, in external flow and in pipe flow, the first/main additional dominant friction/viscous force for turbulent flow, $F_V = \tau A$ is a result of the pressure force, $F_P = \Delta p A$. Therefore, in external and pipe flow, the drag coefficient, C_D will mainly be a function of the Euler number, $E = F_I/F_P = \rho v^2/\Delta p$. However, in uniform open channel flow, the first/main additional dominant friction/viscous force for turbulent flow, $F_V = \tau A$ is a result of the gravity force, $F_G = Mg$. Therefore, in uniform open channel flow, the drag coefficient, C_D will mainly be a function of the Froude number, $F = F_I/F_G = v^2/gL$. Finally, for flow-measuring devices in open channel flow, the first/main additional dominant friction/viscous force for turbulent flow, $F_V = \tau A$ is a result of both the pressure force, $F_P = \Delta p A = \gamma \Delta y A$ and the gravitational force, $F_G = Mg$. Therefore, for flow-measuring devices in open channel flow, the drag coefficient, C_D will mainly be a function of the Euler number, $E = F_I/F_P = \rho v^2/\Delta p$, which in this case is equal to the Froude number, $F = F_I/F_G = v^2/gL$.

The modeling of additional dominant forces or dominant physical quantities results in additional secondary variables and will introduce an additional dimensionless number for each new secondary variable added. It is important to note that the viscous force for turbulent flow, $F_V = \tau A$, which is introduced as the first/main additional dominant force, represents turbulent flow, thus the need for dimensional analysis for its empirical derivation (as represented by the resistance equations). Thus, in addition to the pressure force, F_P and the gravity force, F_G (both which may cause the viscous force for turbulent flow, $F_V = \tau A$), additional dominant forces include the viscous force for laminar flow, $F_V = \tau A = \mu vL$; the elastic force, F_E; and the surface tension force, F_T. The pressure force, F_P and the gravity force, F_G result in the Euler number, E and the Froude number, F, respectively. The viscous force for laminar flow, $F_V = \tau A = \mu vL$ results in the Reynolds number, R. Furthermore, the elastic force, F_E and the surface tension force, F_T result in the Cauchy number, C and the Weber number, W, respectively. Finally, in general, additional independent π terms may also include relative surface roughness, ε/L; cross-sectional shape; hydraulic radius, R_h; and geometry, L_i/L (design detail) of the flow device or flow-measuring device.

Therefore, for instance, in the derivation of the drag force, F_D in external flow, if, in addition to the inertia force, F_I and the pressure force, F_P (which causes the viscous force for turbulent flow, $F_V = \tau A$), other forces, including the viscous force for laminar flow, $F_V = \tau A = \mu vL$; the elastic force, F_E; the gravity force, F_G; and other physical quantities including the absolute surface roughness, ε and the geometry, L_i, also play an important role in the external fluid flow, then the drag coefficient, C_D will also be a function of the Reynolds number, R; the Cauchy number, C; the Froude number, F; the absolute surface roughness, ε; and the geometry, L_i, respectively, as follows:

$$\frac{F_D}{\rho v^2 L^2} = \frac{\Delta p A}{\rho v^2 L^2} = \frac{\Delta p}{\rho v^2} = \frac{1}{E} = C_D = \phi\left(R, C, F, \frac{\varepsilon}{L}, \frac{L_i}{L}\right) \tag{7.19}$$

where the drag coefficient, C_D as modeled by the Euler number, E is the dependent π term, as it models the dependent variable, Δp that cannot be derived analytically from theory, while the Reynolds number, R; the Cauchy number, C; the Froude number, F; the relative surface roughness, ε/L; and the geometry, L_i/L are the independent π terms. One may note that the symbol, ϕ means that the drag coefficient, C_D is an unknown function of R, C, F, ε/L, and L_i/L and thus the drag coefficient, C_D is determined empirically from experimentation. Similar illustrations in the derivation of the other flow resistance equations are presented in the sections below, and the guidelines are summarized in Tables 7.2 through 7.7.

7.2.8 Guidelines in the Derivation of the Efficiency of Pumps and Turbines

Dimensional analysis is used to derive the efficiency, η of a pump or a turbine, which is represented by the "drag coefficient, C_D." In general, the drag coefficient, C_D expresses the ratio of a real flow variable (such as velocity, pressure, force, power, etc.) to its corresponding ideal flow variable. In particular, the efficiency of a pump, $\eta_{pump} = (P_p)_{out}/(P_p)_{in} = \gamma Q h_{pump}/\omega T_{shaft,in}$ (Equation 4.197) expresses the ratio of a real/actual flow variable (hydraulic power output from the pump) to its corresponding ideal flow variable (shaft power input to the pump). And, the efficiency of a turbine, $\eta_{turbine} = (P_t)_{out}/(P_t)_{in} = \omega T_{shaft,out}/\gamma Q h_{turbine}$ (Equation 4.217) expresses the ratio of a real/actual flow variable (shaft power output from the turbine) to its corresponding ideal flow variable (hydraulic power input to the turbine). In the dimensional analysis of fluid flow, there are always at least four parameters involved: ρ, v, L (or D or y; characteristic length dimension), and Δp (or Δy, Δz, or $\Delta H = h_f = \tau_w L/\gamma R_h = \Delta p/\gamma = S_f L$; a representation of energy loss or some form of flow resistance). For pumps, the characteristic length dimension is the diameter of the impeller blades, D. For turbines, the characteristic length dimension is the diameter of the runner blades, D. For pumps and turbines, the representation of energy loss or some form of flow resistance is a head loss, h_f or η.

Step 1: Based on a general understanding of (and experience with and judgment of) the fluid flow phenomena, identify the dominant/significant physical forces and dominant physical quantities in the flow situation (use Table 11.8 for internal flow as a guide). The inertia force, $F_I = Ma = \rho v^2 L^2$, which will always be a dominant force, represents the primary variables that relate to mass (ρ), kinematics (v), and geometry (L [or D or y; characteristic length dimension]) of the flow. Furthermore, there will be at least one more additional dominant force or dominant physical quantity. One may note that the momentum theorem is supplemented with dimensional analysis because the friction/viscous forces in the case of turbulent flow, F_f or $F_V = \tau A$ cannot be either theoretically modeled (as in the case of laminar flow) or ignored (as in the case of a sudden pipe expansion). Therefore, the first/main additional dominant force is the friction/viscous force for turbulent flow, F_f or F_V, which can be represented by a dominant physical quantity (the first/main secondary variable, which represents an energy loss or some form of flow resistance), such as: $\eta_{pump} = (P_p)_{out}/(P_p)_{in} = \gamma Q h_{pump}/\omega T_{shaft,in}$ for a pump, or $\eta_{turbine} = (P_t)_{out}/(P_t)_{in} = \omega T_{shaft,out}/\gamma Q h_{turbine}$ for a turbine.

Step 2: Use the simplest mathematical expressions for the identified dominant physical forces and dominant physical quantities and identify the n physical quantities, all of which can be expressed in terms of m fundamental dimensions (M or F, L, and T). In order to identify the n physical quantities, choose variables that relate to force, mass, kinematics, and geometry of the flow situation (see Table 7.1). The inertia force, $F_I = Ma = \rho v^2 L^2$ represents the primary variables relate to mass (ρ), kinematics ($v = \omega D$), and geometry (L [or D or y; characteristic length dimension]) of the flow. Furthermore, the first/main additional dominant force or dominant physical quantity represents the variable that relates to force and is the friction/viscous force for turbulent flow, F_f or $F_V = \tau A$, which can be represented by a dominant physical quantity (the first/main secondary variable, which represents an energy loss or some form of flow resistance), such as the following for a pump (a turbine is the exact opposite of a pump, as illustrated in Section 7.7):

- The efficiency of a pump, $\eta_{pump} = (P_p)_{out}/(P_p)_{in} = \gamma Q h_{pump}/\omega T_{shaft,in}$ represents a head loss, h_f, which is a result of Δp, and the first secondary variable is specifically, η_{pump}.

  Additional dominant forces or dominant physical quantities (including γ, Q, h_{pump}, and T) may be added (see Table 7.8).

Step 3: Determine the number of dimensionless numbers, which is n-m. In the dimensional analysis of fluid flow, there are always at least four parameters ($n \geq 4$) involved: ρ, v, L (or D or y; characteristic length dimension), and Δp (or Δy, Δz, or $\Delta H = h_f = \tau_w L/\gamma R_h = \Delta p/\gamma = S_f L$; a representation of energy loss or some form of flow resistance) and three fundamental dimensions ($m = 3$). Thus, there will be at least one dimensionless number (n-$m \geq 1$), which is the drag coefficient, C_D, which

TABLE 7.8

Guidelines in the Derivation of the Efficiency of a Pump

Efficiency of a Pump, $\eta_{pump} =$ $(P_p)_{out}/(P_p)_{in} =$ $\gamma Q h_{pump}/\omega T_{shaft,in}$ Is a Result of Δp	Dominant Force or Dominant Physical Quantity	Dominant Force or Dominant Physical Quantity	Dimensional Analysis and Empirical Studies Result
	$F_I = \rho v^2 L^2$ is always dominant and results in the three primary variables	$F_V = \tau A$ is dominant and results in main secondary variable	
Mass Primary Variable	ρ		
Kinematics Primary Variable	ω		
Geometric Primary Variable	D		
Main Secondary Variables		η	
Additional Secondary Variables			$\gamma, Q, h_{pump}, T, \mu, E_v, \varepsilon$
π Term or Drag Coefficient			$\eta_{pump} = \dfrac{(P_p)_{out}}{(P_p)_{in}} = \dfrac{\gamma Q h_{pump}}{\omega T_{shaft,in}} = \dfrac{F_P}{F_I} = \dfrac{\Delta p}{\rho v^2}$ $= \dfrac{1}{E} = C_D = \phi\left(C_H, C_Q, C_P, R, C, \dfrac{\varepsilon}{D}\right)$
Conclusion			The efficiency of a pump, $C_D = \eta$ represents F_V and is a result of Δp; thus, $C_D = \eta$ is mainly a function of E

expresses the ratio of a real flow variable (such as velocity, pressure, force, power, etc.) to its corresponding ideal flow variable.

Step 4: Apply the Buckingham π theorem to derive the *n-m* dimensionless numbers (π terms). The dimensionless numbers (π terms) for the efficiency of a pump, η are derived as follows:

- For the efficiency of a pump, $\eta_{pump} = (P_p)_{out}/(P_p)_{in} = \gamma Q h_{pump}/\omega T_{shaft,in}$, the primary variables are: ρ, $v = \omega D$, which is the characteristic velocity—where ω is the pump impeller rotational speed (or the turbine runner rotational speed), and D, which is the characteristic length, L, and is the diameter of the pump impeller blades (or the diameter of the turbine runner blades); the first secondary variable is η, and additional secondary variables include $\gamma, Q, h_{pump}, T, \mu, E_v$, and ε.

  Additional secondary variables may be added as represented by their simplest mathematical expression for the dominant forces or physical quantities (see Table 7.8).

Step 5: Interpret the resulting functional relationship between the *n* physical variables and thus, the *n-m* dimensionless numbers (π terms). In fluid flow, there will be at least one π term, which represents a drag coefficient, C_D. Depending upon the dominant physical quantity that represents the first/main additional dominant force (i.e., the friction/viscous force for turbulent flow, F_f or F_V), the π term will be interpreted accordingly as follows:

- In the flow through a pump, the propulsive force for the fluid flow is provided by a pressure drop, Δp, which results in an unknown Δp. Thus, in the derivation of the efficiency of a pump, $\eta_{pump} = (P_p)_{out}/(P_p)_{in} = \gamma Q h_{pump}/\omega T_{shaft,in}$, the following four variables, ρ, ω, D, and η, result in one dimensionless number, namely $\Delta p/\rho v^2 = 1/E = C_D = \eta$, where $E = F_I/F_P =$ the Euler number and $C_D = \eta =$ the drag coefficient that represents the flow resistance. Therefore, the efficiency of a pump, η, which represents a head loss, h_f is a result of Δp and thus $C_D = \eta$ is mainly a function of E (see Table 7.8).

7.2.9 The Definition of the Pump (or Turbine) Efficiency

The definition of the pump efficiency, η (drag coefficient, C_D) (a turbine is the exact opposite of a pump, as illustrated in Section 7.7) involves at least two dominant forces. The dimensional analysis of fluid flow assumes that the inertia force, $F_I = Ma = \rho v^2 L^2$ will always be a dominant force and results in the three primary variables: ρ, ω, and D. Furthermore, the friction/viscous force for turbulent flow, F_f or $F_V = \tau A$ will always be the first/main additional dominant force, which can be represented by a dominant physical quantity (η or h_f) that represents an energy loss or some form of flow resistance and results in the appropriate secondary variable η. One may note that the modeling of only one additional dominant force (i.e., the friction/viscous force for turbulent flow) results in only one dimensionless number (π term), which is the drag coefficient, C_D (pump efficiency, η). Furthermore, modeling additional dominant forces or dominant physical quantities results in additional secondary variables, and will introduce an additional dimensionless number (π term) for each new secondary variable added.

In summary, one may note that in the dimensional analysis of fluid flow, although the first/main additional dominant force is the friction/viscous force for turbulent flow, $F_V = \tau A$, the type of flow will determine the cause of the friction/viscous force and thus the definition of the drag coefficient, C_D. As such, in the flow through a pump, the first/main additional dominant friction/viscous force for turbulent flow, $F_V = \tau A$ is a result of the pressure force, $F_P = \Delta p A$. Therefore, in the flow through a pump, the drag coefficient, C_D (pump efficiency, η) will mainly be a function of the Euler number, $E = F_I/F_P = \rho v^2/\Delta p$.

The modeling of additional dominant forces or dominant physical quantities results in additional secondary variables and will introduce an additional dimensionless number for each new secondary variable added. It is important to note that the viscous force for turbulent flow, $F_V = \tau A$, which is introduced as the first/main additional dominant force, represents turbulent flow, thus the need for dimensional analysis for its empirical derivation (as represented by the pump efficiency, η). Thus, in addition to the pressure force, F_P, which causes the viscous force for turbulent flow, $F_V = \tau A$, additional dominant forces include the viscous force for laminar flow, $F_V = \tau A = \mu v L$ and the elastic force, F_E. The pressure force, F_P results in the Euler number, E. The viscous force for laminar flow, $F_V = \tau A = \mu v L$ results in the Reynolds number, R. Furthermore, the elastic force, F_E results in the Cauchy number, C. Additional dominant physical quantities include γ, Q, h_{pump}, T, and ε. Thus, additional independent π terms will also include the power coefficient, C_P; the capacity coefficient, C_Q; the head coefficient, C_H; and the relative surface roughness, ε/D.

Therefore, in the derivation of the pump efficiency, η if in addition to the inertia force, F_I and the pressure force, F_P (which causes the viscous force for turbulent flow, $F_V = \tau A$), other forces including the viscous force for laminar flow, $F_V = \tau A = \mu v L$, and the elastic force, F_E, and other physical quantities including γ, Q, h_{pump}, T and absolute blade surface roughness, ε, also play an important role in the flow through a pump, then the drag coefficient, C_D- (pump efficiency, η) will also be a function of the Reynolds number, R; the Cauchy number, C; the power coefficient, C_P; the capacity coefficient, C_Q; the head coefficient, C_H; and the relative blade surface roughness, ε/D, as follows:

$$\eta_{pump} = \frac{(P_p)_{out}}{(P_p)_{in}} = \frac{\gamma Q h_{pump}}{\omega T_{shaft,in}} = \frac{F_P}{F_I} = \frac{\Delta p}{\rho v^2} = \frac{1}{E} = C_D = \phi\left(C_H, C_Q, C_P, R, C, \frac{\varepsilon}{D}\right)$$

where the drag coefficient, C_D (pump efficiency, η) as modeled by the Euler number, E is the dependent π term, as it models the dependent variable, Δp that cannot be derived analytically from theory, while the head coefficient, C_H; the capacity coefficient, C_Q; the power coefficient, C_P; the Reynolds number, R; the Cauchy number, C; and the relative blade surface roughness, ε/D are the independent π terms. One may note that the symbol ϕ means that the drag coefficient, C_D (pump efficiency, η) is an unknown function of C_H, C_Q, C_P, R, C, and ε/D and thus the drag coefficient, C_D (pump efficiency, η) is determined empirically from experimentation. The derivation of the pump efficiency, η is presented in Section 7.7, and the guidelines are summarized in Table 7.8.

7.2.10 Specific Guidelines and Summary in the Application of Dimensional Analysis for Example Problems and End-of-Chapter Problems

Although some overall general assumptions have been made in the application of dimensional analysis in Chapter 7, more specific guidelines are given for the application of dimensional analysis for the example problems and the end-of-chapter problems. Furthermore, in order to facilitate the application of the dimensional analysis procedure in the example problems and the end-of-chapter problems, a summary of the expected results of the procedure has been provided. The application of dimensional analysis in the empirical modeling (derivation) of the flow resistance equations in internal and external flow (Sections 7.3 through 7.6) and in the empirical modeling (derivation) of the efficiency of pumps and turbines (Section 7.7) makes two important and overall general assumptions, as follows. First, in Step 1 of dimensional analysis, it is assumed that the reader has a general understanding of (and experience with and judgment of) the fluid flow phenomena and thus may easily identify the dominant/significant physical forces and dominant physical quantities in the flow situation. Second, for the most thorough application of dimensional analysis, the most general flow situation is assumed in the identification of the dominant/significant physical forces and dominant physical quantities in the flow situation (use Table 11.7 for external flow and Table 11.8 for internal flow as a guide). As such, the empirically modeled (derived) flow resistance equations in internal and external flow and the empirically modeled (derived) efficiency of pumps and turbines will represent the most general flow situation. However, in the application of dimensional analysis for the example problems and the end-of-chapter problems, more specific guidelines are given as follows. In Step 1 of dimensional analysis, it is assumed that the reader may not have a general understanding of (and experience with and judgment of) the fluid flow phenomena and thus may not be able to easily identify the dominant/significant physical forces and dominant physical quantities in the flow situation. Thus, in order to guide the reader in the identification of the dominant/ significant physical forces and dominant physical quantities in the given specific flow situation, the reader is referred to Table 11.7 for external flow and Table 11.8 for internal flow. Furthermore, the dominant/significant physical forces and dominant physical quantities in the given flow situation are actually specified. Finally, a summary of the expected results of the dimensional analysis procedure is given in the following paragraph.

In order to facilitate the application of the dimensional analysis procedure in the example problems and the end-of-chapter problems, a summary of the expected results of the procedure is given as follows: First, in Step 1 of dimensional analysis, the identified dominant/significant physical forces in the flow situation will always include the inertia force, $F_I = Ma = \rho v^2 L^2$, and the friction/viscous force for turbulent flow, F_f or $F_V = \tau A$. The inertia force represents the assumption of steady state flow (no acceleration) and represents the primary variables that relate to mass (ρ), kinematics (v), and geometry (L [or D or y; characteristic length dimension]) of the flow. The friction/viscous forces represent an energy loss or some form of flow resistance (which cannot be theoretically modeled; thus, the need for dimensional analysis) and represent the first/main secondary variables that relate to one of the following: the drag force in external flow, F_D (or Δp); the major head loss in pipe or open channel flow, h_f (or τ_w or Δp); the minor head loss in pipe flow, h_f (or Δp); the actual discharge in pipe or open channel flow, Q_a (or h_f); the efficiency of a pump or turbine, η; or the head loss in a pump or turbine (actual values for: discharge, head, pressure rise/drop, hydraulic power, and shaft power). The inertia force and the friction/viscous force will always result in the first π term, π_1, which is the dependent π term, and will either be the Euler number, E or the Froude number, F. It is important to note that if the first/main secondary variable is theoretically related to another first/main secondary variable, then it will always yield the dependent π term. For instance, when using dimensional analysis to derive the major head loss in pipe or open channel flow, because Newton's second law of motion yields $\tau_w = \Delta p R_h / L$, using either τ_w or Δp as the first/main secondary variable will always yield the dependent π term as follows: $(\tau_w / \rho v^2) = (R_h / L)(\Delta p / \rho v^2) = C_D$. Another example is when using dimensional analysis to derive the efficiency of a pump or turbine, η, because, theoretically, for instance $\eta_{pump} = (P_p)_{out} / (P_p)_{in} = \gamma Q h_{pump} / \omega T_{shaft,in}$, using the following variables: γ, Q, h_{pump}, T, and ω as secondary variables will yield the dependent π term as follows (see Equation 7.147)

$$\eta_{pump} = \frac{(P_p)_{out}}{(P_p)_{in}} = \frac{\gamma Q h_{pump}}{\omega T_{shaft,in}} = \frac{C_Q C_H}{C_P} = C_D$$

Second, in Step 1 of dimensional analysis, additional identified dominant/significant physical forces or dominant physical quantities in the flow situation will depend upon what additional forces/physical quantities govern the given flow situation. Each additional force/physical quantity results in an additional secondary variable, which yields an independent π term as follows: if the viscous force for laminar flow $F_V = \tau A = \mu v L$ is important, then it results in a secondary variable, μ, which yields the Reynolds number, R. If the elastic force for compressible flow $F_V = E_v A = E_v L^2$ is important, then it results in a secondary variable, E_v, which yields the Cauchy number, C. If the gravity force for gravity flow $F_V = Mg = \rho L^3 g$ is important, then it results in a secondary variable, g, which yields the Froude number, F. If the surface tension force $F_V = \sigma L$ is important, then it results in a secondary variable, σ, which yields the Weber number, W. If the absolute roughness is important, then it results in a secondary variable, ε, which yields the relative roughness, ε/L. Finally, if the geometry, L_i ($i = 1, 2, \ldots$) is important, then it results in a secondary variable, L_i, which yields the geometry L_i/L. Finally, understanding the possible expected results of the dimensional procedure allows an accurate interpretation of the resulting π terms in Step 5 of dimensional analysis.

7.3 Modeling the Flow Reistance as a Drag Force in External Flow

In the external flow around an object, energy or work is used to move the submerged body through the fluid, where the fluid is forced to flow over the surface of the body. As such, in general, one is interested in the determination of the flow pattern around the body, the lift, the drag (flow resistance) on the body, and the patterns of viscous action in the fluid as it passes around the body. However, only if the body is moving at a high speed (and depending upon the shape of the body and the angle it makes with the direction of flow) does the lift force, F_L become dominant. Thus, the main focus of discussion in this section will be the drag force, F_D. The derivation of the drag force equation along with the coefficient of drag, C_D (and lift coefficient, C_L) for the external flow around an object is presented below. Furthermore, empirical evaluation of the coefficient of drag, C_D (and lift coefficient, C_L) and application of the corresponding drag equation (lift equation) are presented in Chapter 10.

7.3.1 Evaluation of the Drag Force

In the case of external flow around an object, the flow resistance is ultimately modeled as a drag force in the integral form of the momentum equation because the drag force is needed for the design of external flow around an object. Thus, the associated minor head loss is typically accounted for by the use/calibration of a drag coefficient to determine the drag force. Evaluation of the drag force requires a "subset level" application of the governing equations. A "subset level" application of the governing equations focuses only on the given element causing the flow resistance. The assumption of "ideal" flow implies that the flow resistance is modeled only in the momentum equation (and thus the subsequent assumption of "real" flow) (see Table 4.1). Thus, in the derivation of the drag force for an external flow around an object, first, ideal flow is assumed in order to determine an expression for the ideal velocity by applying the Bernoulli equation. Then, a subsequent assumption of real flow is made in order to determine an expression for the actual velocity, v_a and the drag force, F_D by application of the momentum equation (supplemented by dimensional analysis). Thus, the flow resistance (shear stress or drag force) in the external flow around an object is modeled as a resistance force/drag force, F_D in the integral form of the momentum equation. The flow resistance causes an unknown pressure drop, Δp, which causes an unknown head loss, h_f, where the head loss is due to a conversion of kinetic energy to heat, which is modeled/displayed in the integral form of the energy equation. However, although the head loss, h_f causes the drag force, F_D, the head loss is not actually determined in the design of external of around an object. The assumption of ideal flow and thus applying the Bernoulli equation to measure the ideal velocity of flow, $v_i = \sqrt{2\Delta p/\rho}$ by the use of a pitot-static tube (see Section 4.5.7.3) yields an expression for the ideal velocity as a function of ideal pressure difference, Δp, which is directly measured. Therefore, the associated minor head loss with the velocity measurement is accounted for by the drag coefficient (in the application of the momentum equation).

The head loss, h_f causing the actual velocity, $v_a = \sqrt{(2/\rho)(\Delta p - (\tau_w L/R_h))}$ and the drag force,

$$F_D = (F_P + F_f)_s = (\Delta p A)_s = \left(\frac{\rho v^2}{2} A + \frac{\tau_w L}{R_h} A \right)_s$$

is caused by both pressure and friction forces, where the drag force, F_D is equal to the sum of the pressure and friction force in the direction of the flow, s. However, because the friction/

viscous forces (due to shear stress, τ_w) due to the external flow cannot be theoretically modeled in the integral momentum equation, the actual pressure drop, Δp cannot be analytically determined and thus the exact reduction in the velocity, actual velocity, v_a, which is less than the ideal velocity, v_i, cannot be theoretically determined. Furthermore, the exact component in the s-direction of the pressure and viscous forces cannot be theoretically determined. Therefore, one cannot derive an analytical expression for the drag force, F_D from the momentum equation. As a result, one must resort to dimensional analysis (supplements the momentum theory) in order to derive an expression for the drag force, F_D, which involves the definition of a drag coefficient, C_D that represents the flow resistance. The flow resistance equation for the drag force is derived below and in Chapter 6.

7.3.2 Application of the Bernoulli Equation: Derivation of the Ideal Velocity

In the derivation of the drag force for an external flow around an object, first, ideal flow is assumed in order to determine an expression for the ideal velocity by applying the Bernoulli equation as follows:

$$\left(\frac{p}{\gamma} + z + \frac{v^2}{2g}\right)_1 - \left(\frac{p}{\gamma} + z + \frac{v^2}{2g}\right)_2 = 0 \tag{7.20}$$

Assuming $z_1 = z_2$, this equation has one unknown, which is the ideal velocity of the external flow around an object. The Bernoulli equation yields an expression for the ideal velocity, $v_i = \sqrt{2\Delta p/\rho}$ as a function of an ideal pressure difference, Δp, which is directly measured by the use of a pitot-static tube (see Section 4.5.7.3). Furthermore, the ideal velocity is used in the definition of the drag force below.

7.3.3 Application of the Momentum Equation and Dimensional Analysis: Derivation of the Actual Velocity, Drag Force Equation, and the Drag Coefficient

In the derivation of the drag force for an external flow around an object, the subsequent assumption of real flow is made in order to determine an expression for the actual velocity, v_a and the drag force, F_D by applying the integral momentum equation (supplemented by dimensional analysis) as follows:

$$\sum F_s = (F_G + F_P + F_V + F_{other})_s = (\rho Q v_s)_2 - (\rho Q v_s)_1 \tag{7.21}$$

Assuming $z_1 = z_2$, and that only the pressure and viscous forces are important, the momentum equation is used to solve for the external flow actual velocity as follows:

$$v_a = \sqrt{\frac{2}{\rho}\left(\Delta p - \frac{\tau_w L}{R_h}\right)} \tag{7.22}$$

and the drag force as follows:

$$F_D = (F_P + F_f)_s = (\Delta p A)_s = \left(\frac{\rho v^2}{2}A + \frac{\tau_w L}{R_h}A\right)_s \tag{7.23}$$

However, the problem is that one cannot theoretically model the wall shear stress, τ_w in the integral momentum equation for turbulent external flow. Thus, one cannot analytically determine pressure drop, Δp and thus v_a in Equation 7.22. Furthermore, one cannot theoretically determine the exact component of the pressure and friction forces in the s-direction (direction of flow) in Equation 7.23. Thus, the integral momentum equation is supplemented with dimensional analysis in order to empirically model the wall shear stress, τ_w for turbulent external flow and thus Δp, v_a, and F_D by the use of a drag coefficient, C_D. The resulting flow resistance equation for the drag force is given as follows:

$$F_D = (F_P + F_f)_s = (\Delta pA)_s = \left(\frac{\rho v^2}{2}A + \frac{\tau_w L}{R_h}A\right)_s = C_D \frac{1}{2}\rho v^2 A \qquad (7.24)$$

where F_D is equal to the sum of the pressure and friction force in the direction of the flow, s and the drag coefficient, C_D accounts for (indirectly models) the minor head loss associated with the ideal velocity measurement, and the drag force. Furthermore, one can easily compute v_i used in Equation 7.24 by measuring Δp using a pitot-static tube.

7.3.3.1 Application of Dimensional Analysis: Derivation of the Drag Force and the Drag Coefficient

One may note that while dimensional analysis is used to derive an empirical expression for the drag force, F_D, (and the lift force, F_L), the practical application of the expression for the drag force, F_D (and the lift force, F_L) is accomplished by the further application of the principle of conservation of momentum and the definition of the dynamic pressure term, $\rho v^2/2$. Presented below is the derivation of the drag force, F_D for the most general assumptions of external flow. Assume that the submerged body illustrated in Figure 10.2 is moving at a low speed in a stationary, homogeneous, viscous, compressible fluid, with possible wave action at the free surface. As a result, the fluid would exert a drag force, F_D on the body, while the effect of the lift force, F_L would be negligible. As such, an empirical expression for the drag force, F_D in the theoretical expression for the drag force,

$$F_D = (F_P + F_f)_s = (\Delta pA)_s = \left(\frac{\rho v^2}{2}A + \frac{\tau_w L}{R_h}A\right)_s$$

in the integral form of the momentum equation is sought in terms of velocity, v and a flow resistance coefficient. This is accomplished by using dimensional analysis in order to derive an empirical expression for the drag force, F_D as a function of the velocity, v, which involves the definition of a drag coefficient, C_D that represents the flow resistance and is empirically evaluated. Thus, the momentum theory is supplemented with dimensional analysis in order to derive an expression for the drag force, F_D for external flow. Thus, for the derivation of the drag force, F_D, in addition to the inertia force, F_I and the pressure force, F_P (causes the viscous force for turbulent flow, $F_V = \tau A$, which is represented by the drag force, F_D), other forces including the viscous force for laminar flow, $F_V = \tau A = \mu v L$; the elastic force, F_E; and the gravity force, F_G also play an important role in the external fluid flow. Additionally, other physical quantities include the absolute surface roughness, ε and the geometry, L_i.

Dimensional analysis is applied in order to derive an empirical expression for the drag force, $F_D = \Delta pA$ (or Δp), which involves the definition of a drag coefficient, C_D as follows (see Table 7.2):

Step 1: Based on a general understanding of (and experience with and judgment of) the fluid flow phenomena, identify the dominant/significant physical forces and dominant physical quantities in the flow situation (use Table 11.7 for external flow as a guide).

a. The inertia force, $F_I = Ma = \rho v^2 L^2$ will always be a dominant force and results in the three primary variables: ρ, v, and L.

b. The friction/viscous force for turbulent flow, F_f or $F_V = \tau A$ will always be the first/main additional dominant force, which in this case is represented by a dominant physical quantity: the drag force, F_D. The drag force is caused by the pressure force and represents an energy loss or some form of flow resistance and results in a secondary variable: $F_D = \Delta p A$.

c. In the external flow around an object, the propulsive force for the fluid flow is provided by a pressure drop, Δp, which results in an unknown Δp. Thus, in the derivation of the drag force, $F_D = \Delta p A$, the following four variables, ρ, v, L, and $F_D = \Delta p A$, result in one dimensionless number, namely $\Delta p / \rho v^2 = 1/E = C_D$, where $E = F_I/F_P$ = the Euler number and C_D = the drag coefficient that represents the flow resistance (dependent π term). Therefore, the drag force, F_D (or Δp), which represents F_P and F_V in the direction of the flow, s is a result of Δp, thus C_D is mainly a function of E.

d. Modeling additional dominant forces or dominant physical quantities results in additional secondary variables and will introduce an additional dimensionless number (π term) for each new secondary variable added.

  1. The viscous force for laminar flow, $F_V = \tau A = \mu v L$ results in a secondary variable, μ, which introduces an independent π term, R.

  2. The elastic force, F_E results in a secondary variable, E_v, which introduces an independent π term, C.

  3. The gravity force, F_G results in a secondary variable, g, which introduces an independent π term, F.

  4. The absolute roughness, ε results in a secondary variable, ε, which introduces an independent π term, ε/L.

  5. The geometry, L_i ($i = 1, 2, \ldots$) results in one or more secondary variables, L_i, which introduces one or more independent π terms, L_i/L.

Step 2: Use the simplest mathematical expressions for the identified dominant physical forces and dominant physical quantities and identify the n physical quantities, all of which can be expressed in terms of m fundamental dimensions (F, L, and T).

Dominant Physical Force or Quantity	Primary Variable and Dimensions	Primary Variable and Dimensions	Primary Variable and Dimensions	Secondary Variable and Dimensions
$F_I = Ma = \rho v^2 L^2$	$\rho = [FL^{-4}T^2]$	$v = [LT^{-1}]$	$L = [L]$	
$F_D = \Delta p A$				$F_D = [F]$
$F_V = \tau A = \mu v L$				$\mu = [FL^{-2}T]$
$F_E = E_v A = E_v L^2$				$E_v = [FL^{-2}]$
$F_G = Mg = \rho V g = \rho L^3 g$				$g = [LT^{-2}]$
ε				$\varepsilon = [L]$
L_i				$L_i = [L]$

where $n = 9$ and $m = 3$.

Step 3: The number of dimensionless numbers (π terms) is $n-m = 9-3 = 6$.

Step 4: The Buckingham π theorem is applied in order to derive the $n-m = 6$ π terms as follows:

$$\pi_1 = \rho^{a_1} v^{b_1} L^{c_1} F_D^{d_1}$$

$$\pi_2 = \rho^{a_2} v^{b_2} L^{c_2} \mu^{d_2}$$

$$\pi_3 = \rho^{a_3} v^{b_3} L^{c_3} E_v^{d_3}$$

$$\pi_4 = \rho^{a_4} v^{b_4} L^{c_4} g^{d_4}$$

$$\pi_5 = \rho^{a_5} v^{b_5} L^{c_5} \varepsilon^{d_5}$$

$$\pi_6 = \rho^{a_6} v^{b_6} L^{c_6} L_i^{d_6}$$

where the exponents are unknown quantities to be determined by applying the Buckingham π theorem.

Working with the π_1 term, the dimensional equations for the *FLT* system are set up as follows:

$$\pi_1 = \rho^{a_1} v^{b_1} L^{c_1} F_D^{d_1}$$

$$F^0 L^0 T^0 = (FL^{-4}T^2)^{a_1}(LT^{-1})^{b_1}(L)^{c_1}(F)^{d_1}$$

Equating the exponents for the dimensions and solving for three of the unknown exponents in terms of the fourth unknown exponent using the Mathcad symbolic solve block yields the following:

Given

F: $0 = a_1 + d_1$

L: $0 = -4a_1 + b_1 + c_1$

T: $0 = 2a_1 - b_1$

$$\text{Find}(a_1, b_1, c_1) \rightarrow \begin{pmatrix} -d_1 \\ -2d_1 \\ -2d_1 \end{pmatrix}$$

Thus:

$$\pi_1 = \rho^{-d_1} v^{-2d_1} L^{-2d_1} F_D^{d_1}$$

$$\pi_1 = \frac{F_D^{d_1}}{\rho^{d_1} v^{2d_1} L^{2d_1}}$$

From experimentation $d_1 = 1$; therefore:

$$\pi_1 = \frac{F_D}{\rho v^2 L^2}$$

which yields the empirical expression for the drag force, F_D, where L^2 is represented by A, the frontal area of the body.

Working with the π_2 term, the dimensional equations for the *FLT* system are set up as follows:

$$\pi_2 = \rho^{a_2} v^{b_2} L^{c_2} \mu^{d_2}$$

$$F^0 L^0 T^0 = (FL^{-4}T^2)^{a_2} (LT^{-1})^{b_2} (L)^{c_2} (FL^{-2}T)^{d_2}$$

Equating the exponents for the dimensions and solving for three of the unknown exponents in terms of the fourth unknown exponent using the Mathcad symbolic solve block yields the following:

Given

F: $\quad 0 = a_2 + d_2$

L: $\quad 0 = -4a_2 + b_2 + c_2 - 2d_2$

T: $\quad 0 = 2a_2 - b_2 + d_2$

Find $(a_2, b_2, c_2) \rightarrow \begin{pmatrix} -d_2 \\ -d_2 \\ -d_2 \end{pmatrix}$

Thus:

$$\pi_2 = \rho^{-d_2} v^{-d_2} L^{-d_2} \mu^{d_2}$$

$$\pi_2 = \frac{\mu^{d_2}}{\rho^{d_2} v^{d_2} L^{d_2}}$$

From experimentation $d_2 = -1$; therefore:

$$\pi_2 = \frac{\rho v L}{\mu}$$

which is the Reynolds number, **R**.

Working with the π_3 term, the dimensional equations for the *FLT* system are set up as follows:

$$\pi_3 = \rho^{a_3} v^{b_3} L^{c_3} E_v^{d_3}$$

$$F^0 L^0 T^0 = (FL^{-4}T^2)^{a_3} (LT^{-1})^{b_3} (L)^{c_3} (FL^{-2})^{d_3}$$

Equating the exponents for the dimensions and solving for three of the unknown exponents in terms of the fourth unknown exponent using the Mathcad symbolic solve block yields the following:

Given

F: $\quad 0 = a_3 + d_3$

L: $\quad 0 = -4a_3 + b_3 + c_3 - 2d_3$

T: $\quad 0 = 2a_3 - b_3$

Find $(a_3, b_3, c_3) \rightarrow \begin{pmatrix} -d_3 \\ -2 \cdot d_3 \\ 0 \end{pmatrix}$

Thus:

$$\pi_3 = \rho^{-d_3} v^{-2d_3} L^0 E_v^{d_3}$$

$$\pi_3 = \frac{E_v^{d_3}}{\rho^{d_3} v^{2d_3}}$$

From experimentation $d_3 = -1$; therefore:

$$\pi_3 = \frac{\rho v^2}{E_v}$$

which is the Cauchy number, C.

Working with the π_4 term, the dimensional equations for the *FLT* system are set up as follows:

$$\pi_4 = \rho^{a_4} v^{b_4} L^{c_4} g^{d_4}$$

$$F^0 L^0 T^0 = (FL^{-4}T^2)^{a_4} (LT^{-1})^{b_4} (L)^{c_4} (LT^{-2})^{d_4}$$

Equating the exponents for the dimensions and solving for three of the unknown exponents in terms of the fourth unknown exponent using the Mathcad symbolic solve block yields the following:

Given

F: $0 = a_4$

L: $0 = -4a_4 + b_4 + c_4 + d_4$

T: $0 = 2a_4 - b_4 - 2d_4$

Find $(a_4, b_4, c_4) \rightarrow \begin{pmatrix} 0 \\ -2 \cdot d_4 \\ d_4 \end{pmatrix}$

Thus:

$$\pi_4 = \rho^0 v^{-2d_4} L^{d_4} g^{d_4}$$

$$\pi_4 = \frac{L^{d_4} g^{d_4}}{v^{2d_4}}$$

From experimentation $d_4 = -1$; therefore:

$$\pi_4 = \frac{v^2}{gL}$$

which is the Froude number, F.

Working with the π_5 term, the dimensional equations for the *FLT* system are set up as follows:

$$\pi_5 = \rho^{a_5} v^{b_5} L^{c_5} \varepsilon^{d_5}$$

$$F^0 L^0 T^0 = (FL^{-4}T^2)^{a_5} (LT^{-1})^{b_5} (L)^{c_5} (L)^{d_5}$$

Equating the exponents for the dimensions and solving for three of the unknown exponents in terms of the fourth unknown exponent using the Mathcad symbolic solve block yields the following:

Given

F: $0 = a_5$

L: $0 = -4a_5 + b_5 + c_5 + d_5$

T: $0 = 2a_5 - b_5$

$$\text{Find}(a_5, b_5, c_5) \rightarrow \begin{pmatrix} 0 \\ 0 \\ -d_5 \end{pmatrix}$$

Thus:

$$\pi_5 = \rho^0 v^0 L^{-d_5} \varepsilon^{d_5}$$

$$\pi_5 = \frac{\varepsilon^{d_5}}{L^{d_5}}$$

From experimentation $d_5 = 1$; therefore:

$$\pi_5 = \frac{\varepsilon}{L}$$

which is the relative roughness.

Working with the π_6 term, the dimensional equations for the *FLT* system are set up as follows:

$$\pi_6 = \rho^{a_6} v^{b_6} L^{c_6} L_i^{d_6}$$

$$F^0 L^0 T^0 = (FL^{-4}T^2)^{a_6} (LT^{-1})^{b_6} (L)^{c_6} (L)^{d_6}$$

Equating the exponents for the dimensions and solving for three of the unknown exponents in terms of the fourth unknown exponent using the Mathcad symbolic solve block yields the following:

Given

F: $0 = a_6$

L: $0 = -4a_6 + b_6 + c_6 + d_6$

T: $0 = 2a_6 - b_6$

$$\text{Find}(a_6, b_6, c_6) \rightarrow \begin{pmatrix} 0 \\ 0 \\ -d_6 \end{pmatrix}$$

Thus:

$$\pi_6 = \rho^0 v^0 L^{-d_6} L_i^{d_6}$$

$$\pi_6 = \frac{L_i^{d_6}}{L^{d_6}}$$

From experimentation $d_6 = 1$; therefore:

$$\pi_6 = \frac{L_i}{L}$$

which is the geometry.

The resulting functional relationship between the $n\text{-}m = 6\ \pi$ terms is expressed as follows:

$$\phi(\pi_1, \pi_2, \pi_3, \pi_4, \pi_5, \pi_6) = 0$$

$$\phi\left(\frac{F_D}{\rho v^2 L^2}, \frac{\rho v L}{\mu}, \frac{\rho v^2}{E_v}, \frac{v^2}{gL}, \frac{\varepsilon}{L}, \frac{L_i}{L}\right) = 0$$

$$\phi\left(\frac{F_D}{\rho v^2 L^2}, R, C, F, \frac{\varepsilon}{L}, \frac{L_i}{L}\right) = 0$$

or:

$$\frac{F_D}{\rho v^2 L^2} = \phi\left(R, C, F, \frac{\varepsilon}{L}, \frac{L_i}{L}\right)$$

Step 5: Interpret the resulting functional relationship between the $n\text{-}m = 6\ \pi$ terms. Recall the definition of the drag force, F_D (from theory in Chapter 6):

$$F_D = \sum F_s = (\Delta p A)_s = \underbrace{\left(\frac{\rho v^2}{2} A\right)}_{F_P}{}_s + \underbrace{\left(\frac{\tau_w L}{R_h} A\right)}_{F_f}{}_s \tag{7.25}$$

Define the drag coefficient, C_D and note the definition of the inertia force, F_I as follows:

$$\frac{F_D}{\rho v^2 L^2} = \frac{F_D}{F_I} = \frac{F_P}{F_I} = \frac{\Delta p A}{\rho v^2 L^2} = \frac{\Delta p}{\rho v^2} = \frac{1}{E} = C_D = \phi\left(R, C, F, \frac{\varepsilon}{L}, \frac{L_i}{L}\right) \tag{7.26}$$

where the drag coefficient, C_D as modeled by the Euler number, E is the dependent π term, as it models the dependent variable, Δp that cannot be derived analytically from theory, while the Reynolds number, R; the Cauchy number, C; the Froude number, F; the relative surface roughness, ε/L; and the geometry, L_i/L are the independent π terms. One may note that because the drag coefficient, C_D represents the flow resistance, which causes an unknown pressure drop, Δp, the pressure force will always play an important role and thus the definition of the drag coefficient, C_D is mainly a function of E. The viscous force will play an important role for streamlined bodies and in the case of low and intermediate velocities (laminar flow); thus, the definition of the drag coefficient, C_D will be dependent on R. However, for blunt bodies (unstreamlined bodies), the pressure force predominates the viscous force; thus, the definition of the drag coefficient, C_D will be mostly independent of R and mainly a function of E. Furthermore, for low and intermediate velocities (laminar flow), the drag coefficient, C_D is independent of the relative surface roughness, ε/L. In the case of high velocities (turbulent flow), the drag coefficient, C_D will be dependent on the relative surface roughness, ε/L and less dependent on R, but dependent on the Mach number,

$$M = \sqrt{C} = \frac{v}{\sqrt{\dfrac{E_v}{\rho}}} = \frac{v_a}{c},$$

where c is the sonic velocity; thus, the elastic force (compressibility effects) will play an important role. The gravity force will play an important role if there is a wave action at the free surface; thus, the definition of the drag coefficient, C_D will be dependent on F; such examples include ships and open channel flow-measuring devices/structures such as weirs and spillways.

Finally, one may note that dimensional analysis yields a nonstandard form of the drag coefficient, C_D, using inertia force, F_I in its definition, which yields the corresponding expression for the drag force, F_D as follows:

$$F_D = C_D \rho v^2 L^2 = C_D \rho v^2 A \qquad (7.27)$$

If however, one considers the fact that inertia force, $F_I = \rho v^2 A$ may be further defined as the dynamic (pressure) force, $F_{dynpres} = \rho v^2 A/2$, then the drag coefficient, C_D would be defined as follows. The practical application of the expression for the drag force, F_D (given by Equation 7.26) is accomplished by the application of the principle of conservation of momentum and the definition of the dynamic pressure term, $\rho v^2/2$. The inertia force, F_I term in Equation 7.26 may be further defined by the application of the principle of conservation of momentum (Newton's second law of motion) as follows:

$$F_I = Ma = M\frac{dv}{dt} \qquad (7.28)$$

Work or energy is defined as force times distance, so multiplying both sides of Equation 7.28 by a differential distance, ds and integrating yields:

$$\int_{s_1}^{s_2} F_I ds = \int_{s_1}^{s_2} M\frac{dv}{dt}ds$$

$$\int_{s_1}^{s_2} F_I ds = \int_{v_1}^{v_2} M\frac{dv}{dt}v dt$$

$$\int_{s_1}^{s_2} F_I ds = \int_{v_1}^{v_2} Mv\,dv$$

$$\underbrace{F_I \Delta s}_{PE} = \underbrace{\frac{1}{2} M \Delta v^2}_{KE} \tag{7.29}$$

which indicates that the work done on a body as it moves from s_1 to s_2 (the potential energy, PE) is equal to the kinetic energy, KE acquired by the body. Furthermore, the inertia force may be further defined as the dynamic (pressure) force as follows:

$$Pressure = \frac{Energy}{Volume} = \frac{Energy}{\dfrac{M}{\rho}} \tag{7.30}$$

$$Dynamic\ Pressure = \frac{Kinetic\ Energy}{Volume} = \frac{\dfrac{1}{2} M v^2}{\dfrac{M}{\rho}} = \frac{\rho v^2}{2} \tag{7.31}$$

$$Dynamic\ Pressure\ Force = \frac{\rho v^2 A}{2} \tag{7.32}$$

Therefore, substituting the dynamic pressure force given by Equation 7.32 for the inertia force in Equation 7.26 yields the drag coefficient, C_D as follows:

$$\frac{F_D}{\dfrac{\rho v^2}{2} A} = \frac{F_D}{F_{dynpres}} = \frac{F_P}{F_{dynpres}} = \frac{\Delta p A}{\dfrac{\rho v^2}{2} A} = \frac{\Delta p}{\dfrac{\rho v^2}{2}} = \frac{2}{E} = C_D = \phi\left(R, C, F, \frac{\varepsilon}{L}, \frac{L_i}{L}\right) \tag{7.33}$$

which is the standard form of the drag coefficient, C_D resulting from replacing the inertia force, $F_I = \rho v^2 A$ used in the dimensional analysis with the dynamic (pressure) force, $F_{dynpres} = \rho v^2 A/2$. Thus, the resulting drag force, F_D derived from dimensional analysis is given as follows:

$$F_D = C_D \tfrac{1}{2} \rho v^2 A \tag{7.34}$$

Furthermore, examining the standard form of the drag coefficient, C_D yields the following:

$$\frac{F_D}{F_{dynpres}} = \frac{\Delta p A}{\dfrac{\rho v^2}{2} A} = \frac{\Delta p}{\dfrac{\rho v^2}{2}} = C_D = \frac{Actual\ Drag\ Force}{Ideal\ Force} = \frac{Actual\ Presdrop}{Ideal\ Presdrop}$$

$$= \frac{staticpres}{dynpres} \tag{7.35}$$

where the results of dimensional analysis introduce a drag coefficient, C_D, which expresses the ratio of a real flow variable (in this case: force or pressure) to its corresponding ideal flow variable. Finally, empirical evaluation of the drag coefficient, C_D and application of the drag force equation, F_D for external flow are presented in Chapter 10.

7.3.3.2 Application of Dimensional Analysis to Derive the Drag Force and Drag Coefficient for More Specific Assumptions of External Flow

The application above of dimensional analysis to derive an empirical equation for the drag force for external flow (Equation 7.26) is for the most general assumptions of external flow; it was assumed that the submerged body illustrated in Figure 10.2 is moving at a low speed in a stationary, homogeneous, viscous, compressible fluid, with a possible wave action at the free surface. As a result, while the drag coefficient, C_D, in general, is a mainly a function of E, it will selectively vary with R, L_i/L, ε/L, M or C, and F, depending mainly on the velocity of flow, the shape of the body, and whether there is a wave action at the free surface (see Tables 11.7 [the same as Table 10.1] and 11.9; see Section 11.7.2), and is generally summarized as follows. The viscous force will play an important role for streamlined bodies and in the case of low and intermediate velocities (laminar flow); thus, the definition of the drag coefficient, C_D will be dependent on R. However, for blunt bodies (unstreamlined bodies), the pressure force predominates the viscous force; thus, the definition of the drag coefficient, C_D will be mostly independent of R and mainly a function of E. Furthermore, for low and intermediate velocities (laminar flow), the drag coefficient, C_D is independent of the relative surface roughness, ε/L. In the case of high velocities (turbulent flow), the drag coefficient, C_D will be dependent on the relative surface roughness, ε/L and less dependent on R, but dependent on the Mach number,

$$M = \sqrt{C} = \frac{v}{\sqrt{\dfrac{E_v}{\rho}}} = \frac{v_a}{c},$$

where c is the sonic velocity; thus, the elastic force (compressibility effects) will play an important role. The gravity force will play an important role if there is a wave action at the free surface; thus, the definition of the drag coefficient, C_D will be dependent on F; such examples include ships and open channel flow-measuring devices/structures such as weirs and spillways.

EXAMPLE PROBLEM 7.1

Derive an empirical expression for the drag force for creeping flow over a body of any shape. The inertia force ($F_I = Ma = \rho v^2 L^2$) and the drag force, F_D will always be important. Using Table 11.7 as a guide, the additional dominant/significant physical force in the flow situation is the viscous force for laminar flow, $F_V = \tau A = \mu v L$. Thus, the important physical quantities include ρ, v, L, F_D, and μ. What are the important π terms used to study and describe creeping flow?

Mathcad Solution

In order to derive an empirical expression for the drag force for creeping flow over a body of any shape, the five steps of dimensional analysis are applied as follows:

Step 1: Using the given information and Table 11.7 as a guide, the dominant/significant physical forces and dominant physical quantities in the flow situation are identified as follows:

The inertia force, $F_I = Ma = \rho v^2 L^2$ will always be a dominant force and results in the three primary variables: ρ, v, and L. The friction/viscous force for turbulent flow,

F_f or $F_V = \tau A$ will always be the first/main additional dominant force, which in this case is represented by a dominant physical quantity: the drag force, F_D. The drag force is caused by the pressure force and represents an energy loss or some form of flow resistance and results in a secondary variable: $F_D = \Delta p A$. In the external flow around an object, the propulsive force for the fluid flow is provided by a pressure drop, Δp, which results in an unknown Δp. Thus, in the derivation of the drag force, $F_D = \Delta p A$, the following four variables, ρ, v, L, and $F_D = \Delta p A$, result in one dimensionless number, namely $\Delta p / \rho v^2 = 1/E = C_D$, where $E = F_I/F_P =$ the Euler number, and $C_D =$ the drag coefficient that represents the flow resistance (dependent π term). Therefore, the drag force, F_D (or Δp), which represents F_P and F_V in the direction of the flow, s is a result of Δp; thus, C_D is mainly a function of E. Modeling additional dominant forces or dominant physical quantities results in additional secondary variables and will introduce an additional dimensionless number (π term) for each new secondary variable added. Using Table 11.7 as a guide, the viscous force for laminar flow, $F_V = \tau A = \mu v L$ results in a secondary variable, μ, which introduces an independent π term, R.

Step 2: Use the simplest mathematical expressions for the identified dominant physical forces and dominant physical quantities and identify the n physical quantities, all of which can be expressed in terms of m fundamental dimensions (F, L, and T).

Dominant Physical Force or Quantity	Primary Variable and Dimensions	Primary Variable and Dimensions	Primary Variable and Dimensions	Secondary Variable and Dimensions
$F_I = Ma = \rho v^2 L^2$	$\rho = [FL^{-4}T^2]$	$v = [LT^{-1}]$	$L = [L]$	
$F_D = \Delta p A$				$F_D = [F]$
$F_V = \tau A = \mu v L$				$\mu = [FL^{-2}T]$

where $n = 5$ and $m = 3$.

Step 3: The number of dimensionless numbers (π terms) is $n-m = 5-3 = 2$.

Step 4: The Buckingham π theorem is applied in order to derive the $n-m = 2$ π terms as follows:

$$\pi_1 = \rho^{a_1} v^{b_1} L^{c_1} F_D^{d_1}$$

$$\pi_2 = \rho^{a_2} v^{b_2} L^{c_2} \mu^{d_2}$$

where the exponents are unknown quantities to be determined by applying the Buckingham π theorem.

Working with the π_1 term, the dimensional equations for the FLT system are set up as follows:

$$\pi_1 = \rho^{a_1} v^{b_1} L^{c_1} F_D^{d_1}$$

$$F^0 L^0 T^0 = (FL^{-4}T^2)^{a_1} (LT^{-1})^{b_1} (L)^{c_1} (F)^{d_1}$$

Equating the exponents for the dimensions and solving for three of the unknown exponents in terms of the fourth unknown exponent using the Mathcad symbolic solve block yields the following:

Given

F: $\qquad 0 = a_1 + d_1$

L: $\qquad 0 = -4a_1 + b_1 + c_1$

T: $\qquad 0 = 2a_1 - b_1$

Find $(a_1, b_1, c_1) \rightarrow \begin{pmatrix} -d_1 \\ -2d_2 \\ -2d_2 \end{pmatrix}$

Thus:

$$\pi_1 = \rho^{-d_1} v^{-2d_1} L^{-2d_1} F_D^{d_1}$$

$$\pi_1 = \frac{F_D^{d_1}}{\rho^{d_1} v^{2d_1} L^{2d_1}}$$

From experimentation $d_1 = 1$; therefore:

$$\pi_1 = \frac{F_D}{\rho v^2 L^2}$$

which yields the empirical expression for the drag force, F_D, and where L^2 is represented by A, the frontal area of the body.

Working with the π_2 term, the dimensional equations for the *FLT* system are set up as follows:

$$\pi_2 = \rho^{a_2} v^{b_2} L^{c_2} \mu^{d_2}$$

$$F^0 L^0 T^0 = (FL^{-4}T^2)^{a_2} (LT^{-1})^{b_2} (L)^{c_2} (FL^{-2}T)^{d_2}$$

Equating the exponents for the dimensions and solving for three of the unknown exponents in terms of the fourth unknown exponent using the Mathcad symbolic solve block yields the following:

Given

F: $\qquad 0 = a_2 + d_2$

L: $\qquad 0 = -4a_2 + b_2 + c_2 - 2d_2$

T: $\qquad 0 = 2a_2 - b_2 + d_2$

Find $(a_2, b_2, c_2) \rightarrow \begin{pmatrix} -d_2 \\ -d_2 \\ -d_2 \end{pmatrix}$

Thus:

$$\pi_2 = \rho^{-d_2} v^{-d_2} L^{-d_2} \mu^{d_2}$$

$$\pi_2 = \frac{\mu^{d_2}}{\rho^{d_2} v^{d_2} L^{d_2}}$$

From experimentation $d_2 = -1$; therefore:

$$\pi_2 = \frac{\rho v L}{\mu}$$

which is the Reynolds number, R.

The resulting functional relationship between the $n\text{-}m = 2$ π terms is expressed as follows:

$$\phi(\pi_1, \pi_2) = 0$$

$$\phi\left(\frac{F_D}{\rho v^2 L^2}, \frac{\rho v L}{\mu}\right) = 0$$

$$\phi\left(\frac{F_D}{\rho v^2 L^2}, R\right) = 0$$

or:

$$\frac{F_D}{\rho v^2 L^2} = \phi(R)$$

Step 5: Interpret the resulting functional relationship between the $n\text{-}m = 2$ π terms. Recall the definition of the drag force, F_D (from theory in Chapter 6):

$$F_D = \sum F_s = (\Delta p A)_s = \underbrace{\left(\frac{\rho v^2}{2} A\right)}_{F_P}_s + \underbrace{\left(\frac{\tau_w L}{R_h} A\right)}_{F_f}_s$$

Define the drag coefficient, C_D and note the definition of the inertia force, F_I as follows:

$$\frac{F_D}{\rho v^2 L^2} = \frac{F_D}{F_I} = \frac{F_P}{F_I} = \frac{\Delta p A}{\rho v^2 L^2} = \frac{\Delta p}{\rho v^2} = \frac{1}{E} = C_D = \phi(R)$$

where the drag coefficient, C_D as modeled by the Euler number, E is the dependent π term, as it models the dependent variable, Δp that cannot be derived analytically from theory, while the Reynolds number, R is the independent π term. One may note that because the drag coefficient, C_D represents the flow resistance, which causes an unknown pressure drop, Δp, the pressure force will always play an important role and thus the definition of the drag coefficient, C_D is mainly a function of E. The viscous force will play an important role for creeping flow and thus the definition of the drag coefficient, C_D will be dependent on R. Finally, dimensional analysis yields a nonstandard form of the drag coefficient, C_D using inertia force, F_I in its definition, which yields the corresponding expression for the drag force, F_D as follows: $F_D = C_D \rho v^2 L^2 = C_D \rho v^2 A$.

7.3.4 Derivation of the Lift Force and the Lift Coefficient

The component of the resultant pressure, F_p and shear/friction forces, F_f that acts normal to the direction of the flow is called the lift force, F_L, which tends to move the submerged body in that direction. However, because the exact component of the pressure and viscous forces acting normal to the direction of flow is unknown and thus cannot be theoretically modeled, one cannot analytically determine the actual pressure drop, Δp from the integral form of the momentum equation. Therefore, one cannot derive an analytical expression for the lift force, F_L from the momentum equation. As a result, one must resort to dimensional analysis (follow a similar approach as done for the drag force, F_D) in order to derive an expression for the lift force, F_L that involves the definition of a lift coefficient, C_L that represents the lift and is empirically evaluated. The resulting lift force, F_L derived from dimensional analysis is given as follows:

$$F_L = C_L \frac{1}{2}\rho v^2 A \tag{7.36}$$

Empirical evaluation of the lift coefficient, C_L and application of the lift force equation, F_L for external flow for external flow are presented in Chapter 10.

EXAMPLE PROBLEM 7.2

Derive an empirical expression for the lift force acting on an airfoil moving at a high speed (turbulent flow). The inertia force ($F_I = Ma = \rho v^2 L^2$) and the lift force, F_L will always be important. Using Section 10.5.5 as a guide, the additional dominant/significant physical quantities in the flow situation are the geometry (shape of the body), L_i and the angle of attack, α. Thus, the important physical quantities include $\rho, v, L, F_L, L_i,$ and α. What are the important π terms used to study and describe this flow situation?

Mathcad Solution

In order to derive an empirical expression for the lift force for an airfoil moving at a high speed, the five steps of dimensional analysis are applied as follows:

Step 1: Using the given information and Section 10.5.5 as a guide, the dominant/significant physical forces and dominant physical quantities in the flow situation are identified as follows:

The inertia force, $F_I = Ma = \rho v^2 L^2$ will always be a dominant force and results in the three primary variables: $\rho, v,$ and L. The friction/viscous force for turbulent flow, F_f or $F_V = \tau A$ will always be the first/main additional dominant force, which in this case is represented by a dominant physical quantity: the lift force, F_L. The lift force is caused by the pressure force and represents an energy loss or some form of flow resistance and results in a secondary variable: $F_L = \Delta pA$. In the external flow around an object, the propulsive force for the fluid flow is provided by a pressure drop, Δp, which results in an unknown Δp. Thus, in the derivation of the lift force, $F_D = \Delta pA$ the following four variables, $\rho, v, L,$ and $F_L = \Delta pA$, result in one dimensionless number, namely $\Delta p/\rho v^2 = 1/E = C_L$, where $E = F_I/F_P =$ the Euler number and $C_L =$ the lift coefficient that represents the flow resistance (dependent π term). Therefore, the lift force, F_L (or Δp), which represents F_P

and F_V normal to the direction of the flow, n, is a result of Δp; thus, C_L is mainly a function of E. Modeling additional dominant forces or dominant physical quantities results in additional secondary variables and will introduce an additional dimensionless number (π term) for each new secondary variable added. Using Section 10.5.5 as a guide, the geometry (shape of the body), L_i and the angle of attack, α are two additional secondary variables, which introduce two independent π terms, L_i/L and α.

Step 2: Use the simplest mathematical expressions for the identified dominant physical forces and dominant physical quantities and identify the n physical quantities, all of which can be expressed in terms of m fundamental dimensions (M, L, and T).

Dominant Physical Force or Quantity	Primary Variable and Dimensions	Primary Variable and Dimensions	Primary Variable and Dimensions	Secondary Variable and Dimensions
$F_I = Ma = \rho v^2 L^2$	$\rho = [ML^{-3}]$	$v = [LT^{-1}]$	$L = [L]$	
$F_L = \Delta p A$				$F_L = [MLT^{-2}]$
L_i				$L_i = [L]$
α				$\alpha = [LL^{-1}]$

where $n = 6$ and $m = 3$.

Step 3: The number of dimensionless numbers (π terms) is $n-m = 6-3 = 3$.

Step 4: The Buckingham π theorem is applied in order to derive the $n-m = 3$ π terms as follows:

$$\pi_1 = \rho^{a_1} v^{b_1} L^{c_1} F_L^{d_1}$$

$$\pi_2 = \rho^{a_2} v^{b_2} L^{c_2} L_i^{d_2}$$

$$\pi_3 = \rho^{a_3} v^{b_3} L^{c_3} \alpha^{d_3}$$

where the exponents are unknown quantities to be determined by applying the Buckingham π theorem.

Working with the π_1 term, the dimensional equations for the MLT system are set up as follows:

$$\pi_1 = \rho^{a_1} v^{b_1} L^{c_1} F_L^{d_1}$$

$$M^0 L^0 T^0 = (ML^{-3})^{a_1} (LT^{-1})^{b_1} (L)^{c_1} (MLT^{-2})^{d_1}$$

Equating the exponents for the dimensions and solving for three of the unknown exponents in terms of the fourth unknown exponent using the Mathcad symbolic solve block yields the following:

Given

M: $0 = a_1 + d_1$

L: $0 = -3a_1 + b_1 + c_1 + d_1$

T: $0 = -b_1 - 2 \cdot d_1$

Find $(a_1, b_1, c_1) \rightarrow \begin{pmatrix} -d_1 \\ -2 \cdot d_1 \\ -2 \cdot d_1 \end{pmatrix}$

Thus:

$$\pi_1 = \rho^{-d_1} v^{-2d_1} L^{-2d_1} F_L^{d_1}$$

$$\pi_1 = \frac{F_L^{d_1}}{\rho^{d_1} v^{2d_1} L^{2d_1}}$$

From experimentation $d_1 = 1$; therefore:

$$\pi_1 = \frac{F_L}{\rho v^2 L^2}$$

which yields the empirical expression for the lift force, F_L, and where L^2 is represented by A, the planform area of the body.

Working with the π_2 term, the dimensional equations for the MLT system are set up as follows:

$$\pi_2 = \rho^{a_2} v^{b_2} L^{c_2} L_i^{d_2}$$

$$M^0 L^0 T^0 = (ML^{-3})^{a_2} (LT^{-1})^{b_2} (L)^{c_2} (L)^{d_2}$$

Equating the exponents for the dimensions and solving for three of the unknown exponents in terms of the fourth unknown exponent using the Mathcad symbolic solve block yields the following:

Given

M: $0 = a_2$

L: $0 = -3a_2 + b_2 + c_2 + d_2$

T: $0 = -b_1$

Find $(a_2, b_2, c_2) \rightarrow \begin{pmatrix} 0 \\ 0 \\ -d_2 \end{pmatrix}$

Thus:

$$\pi_2 = \rho^0 v^0 L^{-d_2} L_i^{d_2}$$

$$\pi_2 = \frac{L_i^{d_2}}{L^{d_2}}$$

From experimentation $d_2 = 1$; therefore:

$$\pi_2 = \frac{L_i}{L}$$

which is the geometry.

Working with the π_3 term, the dimensional equations for the MLT system are set up as follows:

$$\pi_3 = \rho^{a_3} v^{b_3} L^{c_3} \alpha^{d_3}$$

$$M^0 L^0 T^0 = (ML^{-3})^{a_3} (LT^{-1})^{b_3} (L)^{c_3} (LL^{-1})^{d_3}$$

Equating the exponents for the dimensions and solving for three of the unknown exponents in terms of the fourth unknown exponent using the Mathcad symbolic solve block yields the following:

Given

M: $0 = a_3$

L: $0 = -3a_3 + b_3 + c_3 + d_3 - d_3$

T: $0 = -b_3$

$$\text{Find } (a_3, b_3, c_3) \rightarrow \begin{pmatrix} 0 \\ 0 \\ 0 \end{pmatrix}$$

Thus:

$$\pi_3 = \rho^0 v^0 L^0 \alpha^{d_3}$$

$$\pi_2 = \alpha^{d_3}$$

From experimentation $d_3 = 1$; therefore:

$$\pi_2 = \alpha$$

which is the angle of attack.

The resulting functional relationship between the $n\text{-}m = 3$ π terms is expressed as follows:

$$\phi(\pi_1, \pi_2, \pi_3) = 0$$

$$\phi\left(\frac{F_L}{\rho v^2 L^2}, \frac{L_i}{L}, \alpha \right) = 0$$

or:

$$\frac{F_L}{\rho v^2 L^2} = \phi\left(\frac{L_i}{L}, \alpha\right)$$

Step 5: Interpret the resulting functional relationship between the $n-m = 3$ π terms. Recall the definition of the lift force, F_L (from theory in Chapter 6):

$$F_L = \sum F_n = (\Delta p A)_n = \underbrace{\left(\frac{\rho v^2}{2} A\right)}_{F_P}{}_n + \underbrace{\left(\frac{\tau_w L}{R_h} A\right)}_{F_f}{}_n$$

Define the lift coefficient, C_L and note the definition of the inertia force, F_I as follows:

$$\frac{F_L}{\rho v^2 L^2} = \frac{F_L}{F_I} = \frac{F_P}{F_I} = \frac{\Delta p A}{\rho v^2 L^2} = \frac{\Delta p}{\rho v^2} = \frac{1}{E} = C_L = \phi\left(\frac{L_i}{L}, \alpha\right)$$

where the lift coefficient, C_L as modeled by the Euler number, E is the dependent π term, as it models the dependent variable, Δp that cannot be derived analytically from theory, while the geometry, L_i/L and the angle of attack, α are the independent π terms. One may note that because the lift coefficient, C_L represents the flow resistance, which causes an unknown pressure drop, Δp, the pressure force will always play an important role and thus the definition of the lift coefficient, C_L is mainly a function of E. The geometry, L_i/L and the angle of attack, α will play an important role and thus the definition of the lift coefficient, C_L will be dependent on L_i/L and α. Finally, dimensional analysis yields a nonstandard form of the lift coefficient, C_L using inertia force, F_I in its definition, which yields the corresponding expression for the lift force, F_L as follows: $F_L = C_L \rho v^2 L^2 = C_L \rho v^2 A$.

7.4 Modeling the Flow Resistance as a Major Head Loss in Internal Flow

In the internal flow through a pipe or a channel, energy or work is used to move/force the fluid through a conduit. As such, one is interested in the determination of the energy or head losses, pressure drops, and cavitation where energy is dissipated. Thus, the determination of the corresponding head loss, h_f is one of the most important problems in fluid mechanics, where the head loss, h_f will increase the amount of power that a pump must deliver to the fluid; the head loss, h_f represents the additional height that the fluid needs to be elevated by the pump in order to overcome the losses due to friction in the pipe. The derivation of the major head loss equation along with the drag coefficient, C_D for pipe flow is presented below. Furthermore, one may note that while the derivation of the major head loss equation assumes a pipe flow, it is applied for open channel flow with a slight modification as presented in Section 7.4.8 for open channel flow.

7.4.1 Evaluation of the Major Head Loss

In the case where the flow resistance is due to pipe or channel friction in internal flow, the flow resistance is ultimately modeled as a major head loss due to pipe friction in the integral form of the energy equation because the major loss is needed for the design of internal flow in a pipe or open channel. The flow resistance (shear stress or drag force) is modeled as a resistance force (shear stress or drag force) in the integral form of the momentum equation (see Section 6.5), while it is modeled as a friction slope, S_f in the differential form of the momentum equation (see Section 6.7). One may note that the propulsive force for a pipe flow is provided by a pressure gradient, $\Delta p/L$ (which results in an unknown Δp or S_f), while the propulsive force for an open channel flow is provided by invoking gravity and thus by the weight of the fluid along a channel slope (channel gradient, $\Delta z/L = S_o$) (which results in an unknown S_f). The flow resistance causes an unknown pressure drop, Δp (or S_f), which causes an unknown major head loss, h_f, where the head loss is due to a conversion of kinetic energy to heat, which is modeled/displayed in the integral form of the energy equation. Evaluation of the major head loss flow resistance term requires a "subset level" of application of the governing equations. A "subset level" application of the governing equations focuses only on the given element causing the flow resistance. The assumption of real flow implies that the flow resistance is modeled in both the energy and momentum equations (see Table 4.1). Thus, in the derivation of the major head loss, first the integral form of the energy equation is applied to determine an expression for the unknown head loss, h_f. Then, the integral momentum and the differential momentum equations (supplemented by dimensional analysis) are applied in order to determine an expression for the pressure drop, Δp (or S_f) and the shear stress, τ_w. The flow resistance due to pipe or channel friction, which results in a major head loss term, $h_{f,major}$, is modeled as a resistance force (shear stress or drag force) in the integral form of the momentum equation (Equation 4.28), while it is modeled as a friction slope, S_f in the differential form of the momentum equation (Equation 4.84). As such, the energy equation may be used to solve for the unknown major head loss, $h_{f,maj} = S_f L = \tau_w L/\gamma R_h = \Delta p/\gamma$, while the integral momentum equation may be used to solve for the unknown pressure drop, $\Delta p = \tau_w L/R_h$ (or the unknown friction slope, $S_f = \tau_w/\gamma R_h$ in the case of turbulent open channel flow).

The major head loss, h_f is caused by both pressure and friction forces. Thus, when the friction/viscous forces can be theoretically modeled in the integral form of the momentum equation (application of Newton's law of viscosity in the laminar flow case), one can analytically determine the actual pressure drop, Δp (or S_f) from the integral form of the momentum equation. Therefore, one may analytically derive an expression for the major head loss, h_f due to pipe or channel friction from the energy equation (see Sections 6.6 and 7.4.4). However, when the friction/viscous forces cannot be theoretically modeled in the integral form of the momentum equation (as in the turbulent flow case), the friction/viscous forces, the actual pressure drop, Δp (or the unknown friction slope, S_f in the case of turbulent open channel flow) and thus the major head loss, h_f due to pipe or channel friction are determined empirically (see Sections 6.7 and 7.4.5). Specifically, in the case of turbulent flow, the friction/viscous forces in the integral form of the momentum equation and thus the wall shear stress, τ_w cannot be theoretically modeled; therefore, the integral form of the momentum equation cannot be directly applied to solve for the unknown pressure drop, Δp. Thus, an empirical interpretation (using dimensional analysis) of the wall shear stress, τ_w in the theoretical expression for the friction slope, $S_f = \tau_w/\gamma R_h$ in the differential form of the momentum equation is sought in terms of velocity, v and a flow resistance coefficient. This yields the Chezy equation ($S_f = v^2/C^2 R_h$, which represents the differential form of the momentum equation, supplemented by dimensional analysis and guided by the integral momentum equation; a link between the differential and integral momentum equations), which is used to obtain an empirical evaluation for the pressure drop, Δp in Newton's second law of motion (integral form of the momentum equation). As such, one can then derive an empirical expression for the major head loss, h_f from the energy equation. The flow resistance equation for the major head loss is derived below and in Chapter 6.

7.4.2 Application of the Energy Equation: Derivation of the Head Loss

In the derivation of the major head loss for pipe or open channel flow, real flow is assumed in order to determine an expression for the unknown head loss, h_f by applying the integral form of the energy equation as follows:

$$\left(\frac{p}{\gamma} + z + \frac{v^2}{2g}\right)_1 - h_{f,maj} = \left(\frac{p}{\gamma} + z + \frac{v^2}{2g}\right)_2 \tag{7.37}$$

Assuming $z_1 = z_2$, and $v_1 = v_2$, the energy equation has one unknown, which is the major head loss and is given as follows:

$$h_{f,maj} = S_f L = \frac{\tau_w L}{\gamma R_h} = \frac{\Delta p}{\gamma} \tag{7.38}$$

which is expressed as a function of the pressure drop, Δp.

7.4.3 Application of the Momentum Equations and Dimensional Analysis: Derivation of the Pressure Drop, Shear Stress, and the Drag Coefficient

In the derivation of the major head loss for pipe or open channel flow, real flow is assumed in order to determine an expression for the pressure drop, Δp (in the case of major head loss due to pipe flow resistance) or the change in channel depth, Δy (in the case of major head loss due to open channel flow resistance) and the shear stress, τ_w by applying the integral form of the momentum equation (supplemented by dimensional analysis) as follows:

$$\sum F_s = (F_G + F_P + F_V + F_{other})_s = (\rho Q v_s)_2 - (\rho Q v_s)_1 \tag{7.39}$$

Assuming $z_1 = z_2$, and $v_1 = v_2$, the integral momentum equation is used to solve for the pressure drop as a function of the shear stress as follows. As such, regardless of whether the flow is laminar or turbulent, application of Newton's second law of motion (integral form of the momentum equation) (Equation 7.39) yields a linear relationship between the shear stress, τ and the radial distance, r, which also describes the theoretical relationship between the pressure drop, Δp (or S_f) and the shear stress, τ; this relationship plays a key role in the derivation of the major head loss equations for laminar and turbulent flow. Specifically, the integral form of the momentum equation yields the following relationship between the pressure drop, Δp and the wall shear stress, τ_w (derived in Section 6.5):

$$\tau = \frac{r}{2} \frac{\Delta p}{L} \tag{7.40}$$

Evaluated for the wall shear stress, τ_w at $r = r_o = D/2$, where $R_h = D/4$, and expressed as a function of the pressure drop, Δp yields:

$$\Delta p = \frac{\tau_w L}{R_h} \qquad (7.41)$$

However, the problem is that one cannot theoretically model the wall shear stress, τ_w in the integral momentum for turbulent flow for pipe and open channel flow and thus cannot analytically determine the pressure drop, Δp in Equation 7.41. Furthermore, in the differential momentum equation, which is given as follows:

$$\underbrace{\frac{\partial}{\partial s}\left(\frac{p}{\gamma} + z + \frac{v^2}{2g}\right)}_{S_e} = -\underbrace{\frac{\tau_w}{\gamma R_h}}_{S_f} - \underbrace{\frac{1}{g}\frac{\partial v}{\partial t}}_{S_a} \qquad (7.42)$$

while S_e and S_a are expressed in terms of easy-to-measure velocity, v, S_f is expressed in terms of difficult-to-measure τ_w. Specifically, the differential form of the momentum equation (Equation 7.42) yields the following relationship between the friction slope, S_f and the wall shear stress, τ_w:

$$S_f = \frac{\tau_w}{\gamma R_h} \qquad (7.43)$$

However, because all three terms, the pressure drop, Δp (Equation 7.41); the friction slope, S_f (Equation 7.43); and the major head loss, h_f (Equation 7.38) are expressed in terms of the wall shear stress, τ_w, in order to evaluate the respective expressions, one must first obtain an expression for the wall shear stress, τ_w. Whether the flow is laminar (see Section 7.4.4) or turbulent (see Section 7.4.5) will determine whether the friction/viscous forces and thus the wall shear stress, τ_w can be theoretically modeled and thus determine whether application of the three governing equations will yield an analytical or an empirical expression for the major head loss, h_f. Specifically, in the case of turbulent flow (see Section 7.4.5), the integral momentum equation (and the differential momentum equation for S_f) is supplemented with dimensional analysis in order to empirically model the wall shear stress, τ_w for turbulent pipe and open channel flow and thus Δp (and Δy) by the use of a drag coefficient, C_D. Specifically, dimensional analysis is used to empirically interpret S_f (express τ_w) in terms of easy-to-measure velocity, v. Thus, the resulting flow resistance equation for the major head loss is given as follows:

$$h_{f,maj} = S_f L = \frac{\tau_w L}{\gamma R_h} = \frac{\Delta p}{\gamma} = \frac{C_D v^2}{g R_h} L = \frac{v^2}{C^2 R_h} L = f \frac{L}{D}\frac{v^2}{2g} \qquad (7.44)$$

where the drag coefficient, C_D is expressed in terms of the more commonly defined Chezy coefficient, C and Darcy–Weisbach friction factor, f. Furthermore, one can easily compute v_i used in Equation 7.44 by measuring Δp using a pitot-static tube. Finally, it is important to note that the use of dimensional analysis is not needed in the derivation of the major head loss term for laminar flow (see Section 7.4.4).

7.4.4 Derivation of the Major Head Loss Equation for Laminar Flow

In the case of laminar flows, first, application of the integral form of the energy equation yields an expression for the major head loss (Equation 7.38). Then, because the friction/viscous forces in the integral form of the momentum equation and thus the wall shear stress, τ_w can be theoretically modeled (by Newton's law of viscosity, where $\tau = \mu dv/dy$) in the case of laminar flow, one can analytically solve for the unknown pressure drop, Δp from the momentum equation (Equation 7.40) and thus analytically derive an expression for the major head loss, h_f from the energy equation. Specifically, application of the integral form of the momentum equation, where the wall shear stress, τ_w is theoretically modeled by Newton's law of viscosity, yields a theoretical (parabolic) expression for velocity distribution for laminar flow, expressed as a function of the pressure drop, Δp. Next, application of the continuity equation by integrating the resulting parabolic velocity distribution over the pipe diameter yields an analytical expression for the actual pressure drop, Δp (Poiseuille's law), which is substituted into the integral form of the energy equation (Equation 7.38), yielding an analytical expression for the head loss, h_f (Poiseuille's law expressed in terms of the head loss) (see Section 6.6 for details) as follows:

$$h_f = \frac{\tau_w L}{\gamma R_h} = \frac{\Delta p}{\gamma} = S_f L = \underbrace{\left[\frac{32 \mu v_{ave} L}{D^2}\right]}_{\Delta p} \frac{1}{\gamma} = \underbrace{\left[\frac{32 \mu v_{ave}}{D^2 \gamma}\right]}_{S_f} L \tag{7.45}$$

7.4.5 Derivation of the Major Head Loss Equation for Turbulent Flow

In the case of turbulent flows, first, application of the integral form of the energy equation yields an expression for the major head loss (Equation 7.38). Then, however, because the friction/viscous forces in the integral form of the momentum equation and thus the wall shear stress, τ_w cannot be theoretically modeled in the case of turbulent flow, one cannot analytically solve for the unknown pressure drop, Δp from the integral form of the momentum equation (Equation 7.40). Therefore, the unknown pressure drop, Δp (or the unknown friction slope, S_f in the case of turbulent open channel flow) and thus the major head loss, h_f due to pipe or channel friction are determined empirically. Specifically, an empirical interpretation of the wall shear stress, τ_w in the theoretical expression for the friction slope (Equation 7.43) in the differential form of the momentum equation is sought in terms of velocity, v and a flow resistance coefficient. This is accomplished by using dimensional analysis to derive an empirical expression for the wall shear stress, τ_w as a function of the velocity, v, which involves the definition of a drag coefficient, C_D that represents the flow resistance (more standard forms of the drag coefficient, C_D include: Chezy coefficient, C; Darcy–Weisbach friction factor, f; and Manning's roughness coefficient, n) and is empirically evaluated; this is analogous to Newton's law of viscosity, where the shear stress is expressed as a function of velocity, v and flow resistance, μ. One may note that the theoretical relationship between the pressure drop, Δp and the shear stress, τ provided by Newton's second law of motion (integral form of the momentum equation) (Equation 7.41) is used in the dimensional analysis procedure, in which the friction forces are modeled empirically. Dimensional analysis yields an empirical expression for the friction slope, $S_f = v^2/C^2 R_h$ (the Chezy equation) in the differential form of the momentum equation. The Chezy equation represents the differential form of the momentum equation, supplemented by dimensional analysis and guided by the integral momentum equation; a link between the differential and integral momentum equations. The Chezy equation is used to obtain an empirical evaluation for the pressure drop, Δp in Newton's second law of motion (integral form of the momentum equation). Finally, the Chezy equation is substituted into the integral form of the energy equation (Equation 7.38), yielding an empirical expression for the head loss, h_f (the Chezy head loss equation) (see Sections 6.7 and 7.4.5.3 below). It is interesting to note that in the case of turbulent flow, the lack of a theoretical expression for the friction/viscous forces leads to the inability to derive a theoretical expression for velocity distribution for turbulent flow. Instead, empirical studies have shown that the velocity varies according to the seventh root profile law for turbulent flow. As a result, the empirical expression for the friction forces yields an empirical expression for the head loss, with the need to define an empirical flow resistance (using dimensional analysis).

7.4.5.1 Application of Dimensional Analysis: Derivation of the Wall Shear Stress and Drag Coefficient in Turbulent Flow

An empirical expression for the wall shear stress, τ_w in the theoretical expression for the friction slope (Equation 7.43) in the differential form of the momentum equation is sought in terms of velocity, v and a flow resistance coefficient. This is accomplished by using dimensional analysis in order to derive an empirical expression for the wall shear stress, τ_w as a function of the velocity, v, which involves the definition of a drag coefficient, C_D that represents the flow resistance (more standard forms of the drag coefficient, C_D include: Chezy coefficient, C; Darcy–Weisbach friction factor, f; and Manning's roughness coefficient, n) and is empirically evaluated; this is analogous to Newton's law of viscosity, where the shear stress is expressed as a function of velocity, v and flow resistance, μ. Thus, the momentum theory is supplemented with dimensional analysis in order to derive an expression for the wall shear stress, τ_w for turbulent pipe and open channel flow. Therefore, for the derivation of the major head loss in turbulent pipe flow, h_f, in addition to the inertia force, F_I and the pressure force, F_P (causes the viscous force for turbulent flow, $F_V = \tau A$, which is represented by the major head loss, h_f), other forces including the viscous force for laminar flow, $F_V = \tau A = \mu v L$, and other physical quantities, including the absolute surface roughness, ε and the cross-sectional shape, R_h, also play an important role in the pipe fluid flow.

Dimensional analysis is applied in order to derive an empirical expression for the wall shear stress, $\tau_w = \Delta p R_h / L$ (or Δp) in turbulent pipe flow, which involves the definition of a drag coefficient, C_D as follows (see Table 7.3):

> *Step 1:* Based on a general understanding of (and experience with and judgment of) the fluid flow phenomena identify, the dominant/significant physical forces and dominant physical quantities in the flow situation (use Table 11.8 for internal flow as a guide).
>
> a. The inertia force, $F_I = Ma = \rho v^2 L^2$ will always be a dominant force and results in the three primary variables: ρ, v, and L.
>
> b. The friction/viscous force for turbulent flow, F_f or $F_V = \tau A$ will always be the first/main additional dominant force, which in this case is represented by a dominant physical quantity: the major head loss, h_f. The major head loss is caused by the pressure force and represents an energy loss or some form of flow resistance and results in a secondary variable: $\tau_w = \Delta p R_h / L$.
>
> c. In pipe flow, the propulsive force for the fluid flow is provided by a pressure gradient, $\Delta p / L$, which results in an unknown Δp or S_f. Thus, in the derivation of the major head loss in pipe flow, $h_f = \tau_w L / \gamma R_h = \Delta p / \gamma = S_f L$, the following

four variables, ρ, v, L, and $\tau_w = \Delta p R_h/L$, result in one dimensionless number, namely $(R_h/L)(\Delta p/\rho v^2) = (R_h/L)(1/E) = C_D$, where $E = F_I/F_P =$ the Euler number and $C_D =$ the drag coefficient that represents the flow resistance (dependent π term). Therefore, the major head loss in pipe flow, h_f (or τ_w), which represents F_V or F_f, is a result of Δp; thus, C_D is mainly a function of E.

d. Modeling additional dominant forces or dominant physical quantities results in additional secondary variables and will introduce an additional dimensionless number (π term) for each new secondary variable added.

1. The viscous force for laminar flow, $F_V = \tau A = \mu v L$ results in a secondary variable, μ, which introduces an independent π term, R.

2. The absolute roughness, ε results in a secondary variable, ε, which introduces an independent π term, ε/D.

Step 2: Use the simplest mathematical expressions for the identified dominant physical forces and dominant physical quantities and identify the n physical quantities, all of which can be expressed in terms of m fundamental dimensions (F, L, and T).

Dominant Physical Force or Quantity	Primary Variable and Dimensions	Primary Variable and Dimensions	Primary Variable and Dimensions	Secondary Variable and Dimensions
$F_I = Ma = \rho v^2 L^2$	$\rho = [FL^{-4}T^2]$	$v = [LT^{-1}]$	$L = [L]$	
$h_f = \tau_w L/\gamma R_h =$ $\Delta p/\gamma = S_f L$				$\tau_w = [FL^{-2}]$
$F_V = \tau A = \mu v L$				$\mu = [FL^{-2}T]$
ε				$\varepsilon = [L]$

where $n = 6$ and $m = 3$.

Step 3: The number of dimensionless numbers (π terms) is $n\text{-}m = 6\text{-}3 = 3$.

Step 4: The Buckingham π theorem is applied in order to derive the $n\text{-}m = 3$ π terms as follows:

$$\pi_1 = \rho^{a_1} v^{b_1} L^{c_1} \tau_w^{d_1}$$

$$\pi_2 = \rho^{a_2} v^{b_2} L^{c_2} \mu^{d_2}$$

$$\pi_3 = \rho^{a_3} v^{b_3} L^{c_3} \varepsilon^{d_3}$$

where the exponents are unknown quantities to be determined by applying the Buckingham π theorem.

Working with the π_1 term, the dimensional equations for the *FLT* system are set up as follows:

$$\pi_1 = \rho^{a_1} v^{b_1} L^{c_1} \tau_w^{d_1}$$

$$F^0 L^0 T^0 = (FL^{-4}T^2)^{a_1}(LT^{-1})^{b_1}(L)^{c_1}(FL^{-2})^{d_1}$$

Equating the exponents for the dimensions and solving for three of the unknown exponents in terms of the fourth unknown exponent using the Mathcad symbolic solve block yields the following:

Given

F: $0 = a_1 + d_1$

L: $0 = -4a_1 + b_1 + c_1 - 2d_1$

T: $0 = 2a_1 - b_1$

Find $(a_1, b_1, c_1) \rightarrow \begin{pmatrix} -d_1 \\ -2 \cdot d_1 \\ 0 \end{pmatrix}$

Thus:

$$\pi_1 = \rho^{-d_1} v^{-2d_1} L^0 \tau_w^{d_1}$$

$$\pi_1 = \frac{\tau_w^{d_1}}{\rho^{d_1} v^{2d_1}}$$

From experimentation $d_1 = 1$; therefore:

$$\pi_1 = \frac{\tau_w}{\rho v^2}$$

which yields the empirical expression for the shear stress, τ_w.

Working with the π_2 term, the dimensional equations for the *FLT* system are set up as follows:

$$\pi_2 = \rho^{a_2} v^{b_2} L^{c_2} \mu^{d_2}$$

$$F^0 L^0 T^0 = (FL^{-4}T^2)^{a_2} (LT^{-1})^{b_2} (L)^{c_2} (FL^{-2}T)^{d_2}$$

Equating the exponents for the dimensions and solving for three of the unknown exponents in terms of the fourth unknown exponent using the Mathcad symbolic solve block yields the following:

Given

F: $0 = a_2 + d_2$

L: $0 = -4a_2 + b_2 + c_2 - 2d_2$

T: $0 = 2a_2 - b_2 + d_2$

Find $(a_2, b_2, c_2) \rightarrow \begin{pmatrix} -d_2 \\ -d_2 \\ -d_2 \end{pmatrix}$

Thus:

$$\pi_2 = \rho^{-d_2} v^{-d_2} L^{-d_2} \mu^{d_2}$$

$$\pi_2 = \frac{\mu^{d_2}}{\rho^{d_2} v^{d_2} L^{d_2}}$$

From experimentation $d_2 = -1$; therefore:

$$\pi_2 = \frac{\rho v L}{\mu}$$

which is the Reynolds number, **R** and where L represents D, the pipe diameter.

Working with the π_3 term, the dimensional equations for the *FLT* system are set up as follows:

$$\pi_3 = \rho^{a_3} v^{b_3} L^{c_3} \varepsilon^{d_3}$$

$$F^0 L^0 T^0 = (F L^{-4} T^2)^{a_3} (L T^{-1})^{b_3} (L)^{c_3} (L)^{d_3}$$

Equating the exponents for the dimensions and solving for three of the unknown exponents in terms of the fourth unknown exponent using the Mathcad symbolic solve block yields the following:

Given

F: $\qquad 0 = a_3$

L: $\qquad 0 = -4a_3 + b_3 + c_3 + d_3$

T: $\qquad 0 = 2a_3 - d_3$

Find $(a_3, b_3, c_3) \rightarrow \begin{pmatrix} 0 \\ 0 \\ -d_3 \end{pmatrix}$

Thus:

$$\pi_3 = \rho^0 v^0 L^{-d_3} \varepsilon^{d_3}$$

$$\pi_3 = \frac{\varepsilon^{d_3}}{L^{d_3}}$$

From experimentation $d_3 = 1$; therefore:

$$\pi_3 = \frac{\varepsilon}{L}$$

which is the relative roughness, and L is represented by D.

The resulting functional relationship between the $n\text{-}m = 3$ π terms is expressed as follows:

$$\phi(\pi_1, \pi_2, \pi_3) = 0$$

$$\phi\left(\frac{\tau_w}{\rho v^2}, \frac{\rho v D}{\mu}, \frac{\varepsilon}{D}\right) = 0$$

$$\phi\left(\frac{\tau_w}{\rho v^2}, R, \frac{\varepsilon}{D}\right) = 0$$

or:

$$\frac{\tau_w}{\rho v^2} = \phi\left(R, \frac{\varepsilon}{D}\right)$$

Step 5: Interpret the resulting functional relationship between the $n-m = 3$ π terms. Note that the theoretical relationship between the pressure drop, Δp and the shear stress, τ evaluated for the wall shear stress, τ_w at $r = r_o = D/2$, where $R_h = D/4$ (derived from Newton's second law of motion for both laminar and turbulent flow) (Equation 7.41) for the wall shear stress, τ_w yields an expression for the wall shear stress, τ_w expressed as a function of the pressure drop, Δp as follows:

$$\tau_w = \frac{\Delta p R_h}{L} \tag{7.46}$$

One may note that, typically, the pressure drop, Δp is equal to the dynamic pressure, $\rho v^2/2$ plus an additional pressure drop, $\tau_w L/R_h$ due to the flow resistance. But since $v_1 = v_2$ (assumption of uniform flow made in the application of Newton's second law of motion), the pressure drop, Δp is equal to the pressure drop, $\tau_w L/R_h$ due to the flow resistance which causes the major head loss, h_f. Thus, the pressure drop, Δp in pipe flow is mostly due to the wall shear stress, τ_w, unlike in the computation of the drag force, F_D in the external flow around an object, for which the pressure drop, Δp is due to both the dynamic pressure, $\rho v^2/2$ and the wall shear stress, τ_w. Thus, the drag coefficient, C_D is defined as follows:

$$\frac{\tau_w}{\rho v^2} = \frac{F_f}{F_I} = \frac{F_P}{F_I} = \frac{\Delta p R_h}{L}\frac{1}{\rho v^2} = \frac{R_h}{L}\frac{\Delta p}{\rho v^2} = \frac{R_h}{L}\frac{1}{E} = C_D = \phi\left(R, \frac{\varepsilon}{D}\right) \tag{7.47}$$

where the drag coefficient, C_D as modeled by the Euler number, E is the dependent π term, as it models the dependent variable, Δp that cannot be derived analytically from theory, while the Reynolds number, R and the relative surface roughness, ε/D are the independent π terms. One may note that because the drag coefficient, C_D represents the flow resistance, which causes an unknown pressure drop, Δp, the pressure force will always play an important role and thus the definition of the drag coefficient, C_D is mainly a function of E. Furthermore, one may note that the drag coefficient, C_D, which is a nonstandard form of the flow resistance, is dimensionless and is a function of boundary roughness, ε/D; the Reynolds number, R; and the cross-sectional shape or the hydraulic radius, R_h. More standard forms of the drag coefficient, C_D include empirical flow resistance coefficients including the Chezy coefficient, C; the Darcy–Weisbach friction coefficient, f; and the Manning's roughness coefficient, n and are derived below. Furthermore, the resulting empirical relationship between the wall shear stress, τ_w and the velocity, v is related by a flow resistance coefficient from dimensional analysis, which is given as follows:

$$\tau_w = C_D \rho v^2 \tag{7.48}$$

7.4.5.2 Derivation of the Chezy Equation and Evaluation of the Chezy Coefficient, C

The Chezy equation is derived and the Chezy coefficient, C is evaluated as follows. Substituting the empirically derived expression for the wall shear stress (Equation 7.48) into the theoretical expression for the friction slope (Equation 7.43) in the differential form of the momentum equation yields an empirical expression for the friction slope as follows:

$$S_f = \frac{\tau_w}{\gamma R_h} = \frac{C_D \rho v^2}{\gamma R_h} = \frac{C_D v^2}{g R_h} \tag{7.49}$$

The Chezy coefficient, C, which is a more standard form of the drag coefficient, C_D, is defined as follows:

$$C = \sqrt{\frac{g}{C_D}} \tag{7.50}$$

Thus, we finally get an expression for the friction slope, S_f expressed as a function of velocity and a flow resistance coefficient (a result of the application of both the momentum theory and dimensional analysis) as follows:

$$S_f = \frac{v^2}{C^2 R_h} \tag{7.51}$$

which is known as the Chezy equation. The friction slope, S_f is now expressed as a function of velocity, v and the Chezy flow resistance coefficient, C, which models the flow resistance. Therefore, because the wall shear stress, τ_w, which causes the flow resistance, is difficult to measure, instead, now one may model it by measuring the velocity, v and an empirically evaluated drag coefficient, C_D (or a more standard form of the drag coefficient, such as the Chezy coefficient, C; Darcy–Weisbach friction factor, f; or the Manning's roughness coefficient, n). The Chezy coefficient, C has dimensions of $L^{1/2} T^{-1}$ and is a function of the boundary roughness, ε/D; the Reynolds number, R; and the cross-sectional shape or the hydraulic radius, R_h. Empirical findings illustrate that the Chezy coefficient, C ranges between $30 \, \text{m}^{1/2} \, \text{s}^{-1}$ (or $60 \, \text{ft}^{1/2} \, \text{sec}^{-1}$) for small channels with rough boundary surfaces to $90 \, \text{m}^{1/2} \, \text{s}^{-1}$ (or $160 \, \text{ft}^{1/2} \, \text{sec}^{-1}$) for large channels with smooth boundary surfaces. One may note that although the derivation/formulation of the Chezy equation assumes no specific units (SI or BG), the Chezy coefficient, C has dimensions of $L^{1/2} T^{-1}$. Furthermore, because the values for the Chezy coefficient, C have been provided/calibrated in both units (see above), one may apply the Chezy equation assuming either SI or BG units. Although the Chezy equation is used for both open channel flow and pipe flow, it has a drawback in that the Chezy coefficient, C is not dimensionless like the Darcy–Weisbach friction factor, f (defined below). Therefore, one must attach the appropriate metric or English units to the Chezy coefficient, C prior to using it in the Chezy equation.

7.4.5.3 Substituting the Chezy Equation into the Energy Equation: Deriving the Major Head Loss Equation for Turbulent Flow

Finally, assuming that the Chezy coefficient, C is chosen to model the empirical flow resistance, the Chezy equation for the friction slope (Equation 7.51) is substituted into the integral form of the energy equation (Equation 7.38), yielding an empirical expression for the head loss, h_f (the Chezy head loss equation) as follows:

$$h_f = S_f L = \frac{\tau_w L}{\gamma R_h} = \frac{\Delta p}{\gamma} = \underbrace{\left[\frac{v^2}{C^2 R_h}\right]}_{S_f} L \tag{7.52}$$

where the expression for the head loss, h_f is an expression of the mechanical energy per unit weight, expressed in terms of the kinetic energy. One may refer to this equation as the Chezy head loss equation. Furthermore, although values for the Chezy coefficient, C have been empirically evaluated and tabulated (limited tables available) for various combinations of boundary roughness, ε/D; Reynolds number, R; and cross-sectional shape (or the hydraulic radius, R_h), it has been found that it is easier to model the effects of the Chezy coefficient, C (i.e., the drag coefficient, C_D) by using the Darcy–Weisbach friction coefficient, f or the Manning's roughness coefficient, n. As such, the Darcy–Weisbach friction coefficient, f; the Manning's roughness coefficient, n; and the Hazen–Williams roughness coefficient, C_h, along with the corresponding head loss equations, are derived and discussed in the following three sections below.

7.4.6 The Darcy–Weisbach Equation

The Darcy–Weisbach friction coefficient (factor), f, which represents a dimensionless flow resistance coefficient, is derived using dimensional analysis and expressed as a function of the Chezy coefficient, C as: $C = \sqrt{8g/f}$. The empirical evaluation of the friction factor, f is presented in both graphical (Moody diagram) and mathematical forms. Finally, the Chezy coefficient, C expressed as a function of the friction factor, f is substituted into the integral form of the energy equation (Equation 7.38), yielding an empirical expression for the head loss, $h_f = f(L/D)(v^2/2g)$ (the Darcy–Weisbach head loss equation).

7.4.6.1 Derivation of the Darcy–Weisbach Friction Coefficient, f

From the definition of π terms (see Section 7.2.5) one may introduce a new π term, the Darcy friction factor, f, which, like the Chezy coefficient, C, is a more standard form of the drag coefficient, C_D; however, unlike the Chezy coefficient, C, it is dimensionless. Thus, defining a new π term, the Darcy friction factor, f expressed as a function of the drag coefficient, C_D (a π term) is given as follows:

$$\frac{F_f}{F_I} = f = 8C_D \tag{7.53}$$

The Darcy–Weisbach friction coefficient, f is typically defined for pipe flow; thus, if one assumes a circular pipe flow, then:

$$R_h = \frac{A}{P} = \frac{\pi D^2}{4} \frac{1}{\pi D} = \frac{D}{4} \tag{7.54}$$

Thus, the drag coefficient, C_D is defined as follows:

$$\frac{\tau_w}{\rho v^2} = \frac{R_h}{L} \frac{\Delta p}{\rho v^2} = \frac{D}{4L} \frac{\Delta p}{\rho v^2} = \frac{D}{4L} \frac{1}{E} = C_D = \phi\left(R, \frac{\varepsilon}{D}\right) \tag{7.55}$$

And, the Darcy–Weisbach friction coefficient, f is defined as follows:

$$\frac{F_f}{F_I} = f = 8C_D = 8\left[\frac{\tau_w}{\rho v^2}\right] = 8\left[\frac{D}{4L}\frac{\Delta p}{\rho v^2}\right] = \frac{D}{L}\frac{\Delta p}{\frac{\rho v^2}{2}} = \frac{D}{L}\frac{2}{E} = \phi\left(R, \frac{\varepsilon}{D}\right) \qquad (7.56)$$

Thus, examining the Darcy–Weisbach friction coefficient, f yields the following:

$$\frac{F_f}{F_I} = f = 8C_D = \frac{D}{L}\frac{\Delta p}{\frac{\rho v^2}{2}} = \frac{Actual\ PresDrop}{Ideal\ PresDrop} = \frac{staticpres}{dynpres} \qquad (7.57)$$

where the results of dimensional analysis introduce a drag coefficient, C_D, which expresses the ratio of a real flow variable (in this case: force or pressure) to its corresponding ideal flow variable. Thus, using the Darcy–Weisbach friction coefficient, f, which is a more standard definition of the drag coefficient, C_D, the resulting relationship between the wall shear stress, τ_w and the velocity, v from dimensional analysis is given as follows:

$$\tau_w = \frac{f}{8}\rho v^2 \qquad (7.58)$$

where the Darcy–Weisbach friction coefficient, f has no dimensions, and is a function of the boundary roughness, ε/D; the Reynolds number, R; and the cross-sectional shape or the hydraulic radius, R_h. One may note that ε is the absolute pipe roughness, while ε/D is the relative pipe roughness. Furthermore, Table 8.3 presents typical values for the absolute pipe roughness, ε for various pipe material, assuming new and clean pipes. It is important to note that over the lifetime of a pipe design, the absolute pipe roughness, ε values will change (increase) due to the buildup of deposits on the pipe wall and thus cause an increase in the frictional head loss. Thus, in order to minimize potential problems associated with head loss, the design of a pipe system should consider the temporal change of the absolute pipe roughness, ε for a given pipe material.

Based on the definition of the newly defined π term, f (the Darcy–Weisbach friction factor) and the definition of the Chezy coefficient, C, one may define the relationship between the Chezy coefficient, C and the Darcy–Weisbach friction factor, f as follows:

Noting that:

$$C = \sqrt{\frac{g}{C_D}} \qquad (7.59)$$

$$f = 8C_D \qquad (7.60)$$

yields:

$$C = \sqrt{\frac{8g}{f}} \qquad (7.61)$$

Furthermore, while the behavior of the Darcy–Weisbach friction factor, f in circular pipe flow has been thoroughly explored, a similarly complete investigation of the behavior of the Chezy coefficient, C has never been made. This is because in addition to the extra variables involved in the open channel flow case, there is an extremely wide range of surface roughness sizes and types found in practice, and it is difficult to attain steady, uniform, fully developed flow outside of the controlled laboratory environment. Additionally, empirical studies have illustrated that the behavior of the Chezy coefficient, C and the Darcy–Weisbach friction factor, f are least affected by the cross-sectional shape (or the effects are at least within the limits of accuracy that is normally accepted in practice).

7.4.6.2 Evaluation of the Darcy–Weisbach Friction Coefficient, f

As presented above, the results of dimensional analysis, assuming turbulent flow, indicated that the Darcy–Weisbach friction coefficient, f has no dimensions and is a function of the boundary roughness, ε/D; the Reynolds number, R; and the cross-sectional shape or the hydraulic radius, R_h. However, at very high velocities (and thus very high values for R) the f for completely turbulent flow (rough pipes) is a function only of the relative pipe roughness, ε/D and thus independent of R. Furthermore, because laminar flow is a special case of turbulent flow, a theoretical evaluation of the Darcy–Weisbach friction coefficient, f for laminar "circular" pipe flow (given in a section below) illustrates that f for laminar flow is only a function of the Reynolds number, R (where: $f = 64/R$).

Unlike laminar flow, the friction factor, f for turbulent flow cannot be evaluated theoretically. Instead, it is evaluated using experiments. Specifically, J. Nikuradse (Moody 1944) computed the friction factor, f from measurements of the flowrate, Q and the pressure drop, Δp. Using pipes with simulated absolute pipe roughness, ε, the pressure drop, Δp needed to produce a specified flowrate, Q was measured. The data were then converted into the friction factor, f for the corresponding Reynolds number, R and the relative pipe roughness, ε/D. Numerous tests were conducted for a wide range of R and ε/D in order to determine an empirical relationship between the friction factor, f and both the Reynolds number, R and the relative pipe roughness, ε/D. Further experimental studies conducted by L. F. Moody resulted in the Moody diagram illustrated in Figure 8.1 presented in Chapter 8, which presents the Darcy–Weisbach friction coefficient, f as a function of the Reynolds number, R and the relative pipe roughness, ε/D (in both graphical and mathematical forms).

7.4.6.3 The Darcy–Weisbach Head Loss Equation

Assuming that the Darcy–Weisbach friction factor, f is chosen to model the empirical flow resistance, the Chezy coefficient, C, expressed as a function of the friction factor, f, is substituted into the integral form of the energy equation (Equation 7.38), yielding an empirical expression for the head loss, $h_f = f(L/D)(v^2/2g)$ (the Darcy–Weisbach head loss equation). Choosing the Darcy–Weisbach friction factor, f to model the empirical flow resistance is a typical choice for a pipe flow (where $R_h = D/4$). As such, one would substitute the relationship between the Chezy coefficient, C and the Darcy–Weisbach friction factor, f (Equation 7.61) into the Chezy head loss Equation 7.52, which yields the Darcy–Weisbach head loss equation as follows:

$$C = \sqrt{\frac{8g}{f}} \tag{7.62}$$

$$h_f = S_f L = \frac{\tau_w L}{\gamma R_h} = \frac{\Delta p}{\gamma} = \frac{v^2 L}{C^2 R_h} = \frac{v^2 L}{\dfrac{8g}{f}\dfrac{D}{4}} = f\frac{L}{D}\frac{v^2}{2g} \tag{7.63}$$

which is expressed in terms of the velocity (kinetic) head, $v^2/2g$. One may note that the derivation/formulation of the Darcy–Weisbach equation assumes no specific units (SI or BG), and, furthermore, the Darcy–Weisbach friction factor, f is dimensionless. Therefore, one may apply the Darcy–Weisbach head loss equation assuming either the SI or BG units.

7.4.6.4 Evaluating the Darcy–Weisbach Friction Coefficient, f for Laminar Pipe Flow

It may be illustrated that the "theoretical flow resistance coefficient" for laminar flow is a special case of the empirical flow resistance coefficient for turbulent flow ($f = 64/R$). Noting that laminar flow is a special case of turbulent flow, one may evaluate the Darcy–Weisbach friction coefficient, f for laminar pipe flow as follows. Equating Poiseuille's law expressed in terms of the head loss (Equation 7.45) (derived for laminar flow) to the Darcy–Weisbach head loss equation (Equation 7.63) (derived for turbulent flow) and isolating the friction coefficient, f yields the Darcy–Weisbach friction coefficient, f for laminar "circular" pipe flow as follows:

$$h_f = S_f L = \frac{\tau_w L}{\gamma R_h} = \frac{\Delta p}{\gamma} = \frac{32 \mu v L}{D^2 \rho g} = f \frac{L}{D} \frac{v^2}{2g} \tag{7.64}$$

$$f = \frac{64 \mu}{\rho v D} = \frac{64}{R} \tag{7.65}$$

where the Reynolds number, $R = \rho v D / \mu$. In the case where the pipe is noncircular, then one may use the appropriate relationship given in Table 8.4 for various noncircular pipe sections. One may note that the relationships between the friction factor, f and the Reynolds number, $R = \rho v (4R_h)/\mu$ for the various noncircular pipe sections are derived from theory and/or empirical studies. Thus, f for laminar flow is only a function of the Reynolds number, R, while f for turbulent flow is a function of both the Reynolds number, R and the relative pipe roughness, ε/D. However, at very high velocities (and thus very high values for R), the f for completely turbulent flow (rough pipes) is a function only of the relative pipe roughness, ε/D and thus independent of R.

7.4.7 Manning's Equation

The Manning's roughness coefficient, n is empirically derived and expressed as a function of the Chezy coefficient, C as: $C = R_h^{1/6}/n$. The empirical evaluation of the Manning's roughness coefficient, n is presented in tabular form. Substituting the expression for the Chezy coefficient, C as a function of the Manning's roughness coefficient, n into the Chezy equation yields the Manning's equation. Finally, the Chezy equation, expressed as a function of the Manning's roughness coefficient, n, is substituted into the integral form of the energy equation (Equation 7.38), yielding an empirical expression for the head loss (the Manning's head loss equation).

7.4.7.1 Derivation and Evaluation of the Manning's Roughness Coefficient, n

Empirical studies have indicated that a more accurate dependence of the Chezy coefficient, C on the hydraulic radius, R_h is defined by the definition of the Manning's roughness coefficient, n. One may recall that the dimensions of the Chezy coefficient, C are $L^{1/2} T^{-1}$, and it is empirically evaluated as a function of the boundary roughness, ε/D; the Reynolds number,

R; and the cross-sectional shape or the hydraulic radius, R_h. However, further empirical studies (assuming the metric system) by Manning indicated that a more accurate dependence of the Chezy coefficient, C on the hydraulic radius, R_h is given as follows:

$$C = \frac{R_h^{1/6}}{n} \tag{7.66}$$

where n is the Manning's roughness coefficient with dimensions of $L^{-1/3}T$ and is empirically evaluated only as a function of the boundary roughness, ε/D. One may note that the dependency of the Manning's roughness coefficient, n on the hydraulic radius, R_h has been removed by its relationship with the Chezy coefficient, C. Furthermore, one may note that the dependency of the Manning's roughness coefficient, n on the Reynolds number, R has been removed by assuming turbulent flow. One may note that the most common occurrence of flow resistance in open channel and pipe flow is in long channels or long pipe sections. In the open channel flow case, the boundary fills the whole channel section, while in the pipe flow case, the boundary fills the whole pipe section; in such cases, the Reynolds number, R is usually so large that it has virtually has no influence on the resistance coefficient, n. Empirical values for the Manning roughness coefficient, n as a function of the boundary roughness, ε/D are typically given assuming metric units for n of $m^{-1/3}$ s (see Table 8.6).

7.4.7.2 Manning's Equation

Substituting the expression for the Chezy coefficient, C as a function of the Manning's roughness coefficient, n (Equation 7.66) into the Chezy equation (Equation 7.51) yields the Manning's equation as follows:

$$v = \frac{R_h^{1/6}}{n}\sqrt{R_h S_f} = \frac{1}{n}R_h^{2/3}S_f^{1/2} \tag{7.67}$$

One may note that although the derivation/formulation of the Manning's equation assumes no specific units (SI or BG), the Manning's roughness coefficient, n has dimensions of $L^{-1/3}$ T. Furthermore, because the values for the Manning's roughness coefficient, n have been provided/calibrated in the SI units $m^{-1/3}$ s (see Table 8.6), the Manning's equation as given above assumes the metric (SI) system. The corresponding Manning's equation for the English (BG) system is given as follows:

$$v = \frac{1.486}{n}R_h^{2/3}S_f^{1/2} \tag{7.68}$$

which also uses the Manning roughness coefficient, n with units $m^{-1/3}$ s (see Table 8.6). One may note the value of 1.486 simply converts the Manning roughness coefficient, n with units $m^{-1/3}$ s (from Table 8.6) to units of $ft^{-1/3}$ sec, because 3.281 ft = 1 m and thus:

$$\frac{\left[\dfrac{3.281 \text{ ft}}{1\text{ m}}\right]^{1/3}}{n[m^{-1/3}\text{ s}]} = \frac{1.486}{n[ft^{-1/3}\text{ sec}]} \tag{7.69}$$

As illustrated in Table 8.6, the rougher the boundary (wetted perimeter), the larger the value for Manning roughness coefficient, n.

7.4.7.3 Manning's Head Loss Equation

Assuming that the Manning roughness coefficient, n is chosen to model the empirical flow resistance, the Chezy equation, expressed as a function of the Manning's roughness coefficient, n is substituted into the integral form of the energy equation (Equation 7.38), yielding an empirical expression for the head loss (the Manning's head loss equation). Substituting the relationship between the Chezy coefficient, C and the Manning's roughness coefficient, n (Equation 7.66) into the Chezy head loss equation (Equation 7.52) yields the Manning's head loss equation (assuming the metric system) as follows:

$$C = \frac{R_h^{1/6}}{n} \tag{7.70}$$

$$h_f = S_f L = \frac{\tau_w L}{\gamma R_h} = \frac{\Delta p}{\gamma} = \frac{v^2 L}{C^2 R_h} = \frac{v^2}{\left(\dfrac{R_h^{1/6}}{n}\right)^2} \frac{L}{R_h} = \left(\frac{vn}{R_h^{2/3}}\right)^2 L \tag{7.71}$$

The corresponding Manning's head loss equation for the English (BG) system is given as follows:

$$h_f = S_f L = \frac{\tau_w L}{\gamma R_h} = \frac{\Delta p}{\gamma} = \frac{v^2 L}{C^2 R_h} = \frac{v^2}{\left(\dfrac{1.486 R_h^{1/6}}{n}\right)^2} \frac{L}{R_h} = \left(\frac{vn}{1.486 R_h^{2/3}}\right)^2 L \tag{7.72}$$

Finally, one may note that because modeling the flow resistance using the empirically derived and calibrated Hazen–Williams roughness, C_h does not involve dimensional analysis, it is not addressed Chapter 7. Rather, it was presented in Chapter 6.

7.4.8 Application of the Major Head Loss Equation for Open Channel Flow

As noted above, for the derivation of the major head loss equation, although the laws of flow resistance are essentially the same in pipe flow and open channel flow, there is a slight modification needed for the application in open channel flow. First, application of the integral form of the energy equation yielded the following relationship between the major head loss, h_f and the wall shear stress, τ_w (Equation 7.38):

$$h_f = \frac{\tau_w L}{\gamma R_h} \tag{7.73}$$

Then, application of Newton's second law of motion (integral form of the momentum equation) represents an equilibrium between the propulsive pressure, and/or gravity forces against the retarding shear forces needed to move each fluid particle along the pipe or open channel with a constant velocity (Equation 6.65) and is repeated here as follows:

$$\tau = \frac{r}{2} \left[-\frac{d}{ds}(p + \gamma z) \right] \tag{7.74}$$

Because the propulsive force for a pipe flow is provided by a pressure gradient $\Delta p/L$, it was assumed that $z_1 = z_2$, and $ds = L$ and thus yielded the following:

$$\tau = \frac{r}{2}\left[-\frac{d}{ds}(p)\right] = \frac{r}{2}\frac{\Delta p}{L} \tag{7.75}$$

which yields an expression for the variation of the shear stress, τ across the pipe diameter (or actually as a function of the radial distance, r from the center of the pipe) that also describes the theoretical relationship between the pressure drop, Δp and the shear stress, τ for both laminar and turbulent flow. Evaluated for the wall shear stress, τ_w at $r = r_o = D/2$, where $R_h = D/4$ yielded the following:

$$\tau_w = R_h \frac{\Delta p}{L} \tag{7.76}$$

However, because the propulsive force for an open channel flow is provided by invoking gravity and thus by the weight of the fluid along a channel slope (channel gradient, $\Delta z/L = S_o$), assuming a constant velocity implies that $p_1 = p_2$, which yields the following:

$$\tau = \frac{r}{2}\left[-\frac{d}{ds}(\gamma z)\right] = \frac{r}{2}\gamma\frac{\Delta z}{L} = \frac{r}{2}\gamma S_o \tag{7.77}$$

which yields an expression for the variation of the shear stress, τ across the "pipe diameter" (or actually as a function of the radial distance, r from the center of the pipe, where the flow cross section in an open channel is assumed to be equivalent to the lower half of a pipe flow) that also describes the theoretical relationship between the elevation drop, Δz and the shear stress, τ for both laminar and turbulent flow. Evaluated for the wall shear stress, τ_w at $r = r_o = D/2$, where $R_h = D/4$ yields the following:

$$\tau_w = \gamma R_h S_o \tag{7.78}$$

or:

$$S_o = \frac{\tau_w}{\gamma R_h} \tag{7.79}$$

Furthermore, the differential form of the momentum equation yielded the following relationship between the friction slope, S_f and the wall shear stress, τ_w (Equation 7.43):

$$S_f = \frac{\tau_w}{\gamma R_h} \tag{7.80}$$

Thus, for open channel flow, one may note that substitution of the two above forms of the momentum equation (Equations 7.79 and 7.80) into the energy equation (Equation 7.73) yields the following expression for the major head loss, h_f:

$$h_f = \frac{\tau_w L}{\gamma R_h} = \frac{\gamma R_h S_o L}{\gamma R_h} = S_o L \tag{7.81}$$

$$h_f = \frac{\tau_w L}{\gamma R_h} = \frac{\gamma R_h S_f L}{\gamma R_h} = S_f L \tag{7.82}$$

where the channel slope (channel gradient, $\Delta z/L = S_o$) is equal to the friction slope, S_f for the assumption of uniform open channel flow as follows:

$$S_o = S_f = \frac{\tau_w}{\gamma R_h} \tag{7.83}$$

However, because all three terms, the channel slope, S_o (Equation 7.79); the friction slope, S_f (Equation 7.80); and the major head loss, h_f (Equation 7.73) are expressed in terms of the wall shear stress, τ_w, in order to evaluate the respective expressions, one must first obtain an expression for the wall shear stress, τ_w. Application of dimensional analysis in Section 7.4.5.1 above yielded a functional relationship, $(\tau_w/\rho v^2) = \phi(R, (\varepsilon/D))$, which was interpreted (in Step 5 of the dimensional analysis procedure) for pipe flow (see Equation 7.47). However, the functional relationship, $(\tau_w/\rho v^2) = \phi(R, (\varepsilon/D))$ requires a slightly different interpretation in the application for open channel flow.

7.4.8.1 Interpretation of the Results of Dimensional Analysis for Open Channel Flow

The functional relationship, $(\tau_w/\rho v^2) = \phi(R, (\varepsilon/D))$ derived from dimensional analysis in Section 7.4.5.1 above requires a slightly different interpretation in the application for open channel flow. As such, in addition to the inertia force, F_I and the gravity force, F_G (causes the viscous force for turbulent flow, $F_V = \tau A$ which is represented by the major head loss, h_f), other forces including the viscous force for laminar flow, $F_V = \tau A = \mu v L$ and other physical quantities including the absolute surface roughness, ε also play an important role in the open channel fluid flow (see Table 7.4). Noting that the definition of R_h assumes a pipe where an open channel is considered to be the lower half of a pipe where $y = D/2$, so that $R_h = D/4 = y/2$, and that $S_f = \tau_w/\gamma R_h$ (from the differential form of the momentum equation), one may now interpret the expression for the wall shear stress, τ_w derived above from dimensional analysis, $(\tau_w/\rho v^2) = \phi(R, (\varepsilon/D))$, for open channel flow as follows:

$$\frac{\tau_w}{\rho v^2} = \frac{F_f}{F_I} = \frac{\gamma R_h S_f}{\rho v^2} = \frac{\rho g y S_f}{2\rho v^2} = \frac{S_f}{2} \frac{g y}{v^2} = \frac{S_f}{2} \frac{1}{F} = C_D = \phi\left(R, \frac{\varepsilon}{y}\right) \tag{7.84}$$

where the drag coefficient, C_D as modeled by the Froude number, F is the dependent π term, as it models the dependent variable, S_f that cannot be derived analytically from theory, while the Reynolds number, R and the relative surface roughness, ε/y are the independent π terms. One may note that because the drag coefficient, C_D represents the flow resistance which causes an unknown friction slope, S_f the gravity force (invoked by the channel gradient, $S_o = \Delta z/L$) will always play an important role and thus the definition of the drag coefficient, C_D is mainly a function of F. One may recall that in the case of pipe flow, the definition of the drag coefficient, C_D was mainly a function of E. As in the case of pipe flow, the drag coefficient, C_D is dimensionless and is a function of boundary roughness, ε/y; the Reynolds number, R; and the cross-sectional shape or the hydraulic radius, R_h. However, because the flow is mostly turbulent in open channels, the Reynolds number, R is usually so large that it has virtually no influence on the drag coefficient, C_D. And, as in the case of pipe flow, the more standard forms of the drag coefficient, C_D which include the Chezy coefficient, C; the Darcy–Weisbach friction coefficient, f; and the Manning's roughness coefficient, n are interpreted/applied for open channel flow below. Finally, the resulting empirical relationship between the wall shear stress, τ_w and the velocity, v is related by a flow resistance coefficient from dimensional analysis remains the same as derived for pipe flow (Equation 7.48) and is given once again as follows:

$$\tau_w = C_D \rho v^2 \tag{7.85}$$

Furthermore, derivation of the Chezy equation (see Section 7.4.5.2) remains the same as derived for pipe flow (Equation 7.51).

7.4.8.2 Application of the Chezy Equation for Open Channel Flow

Because all three terms, the channel slope, S_o (Equation 7.79); the friction slope, S_f (Equation 7.80); and the major head loss, h_f (Equation 7.73), are expressed in terms of the wall shear stress, τ_w, in order to evaluate the respective expressions, the Chezy equation (Equation 7.51) which provides an empirical evaluation of the friction slope, S_f is applied as follows:

$$S_f = \frac{v^2}{C^2 R_h} \tag{7.86}$$

$$S_o = S_f = \frac{\tau_w}{\gamma R_h} = \frac{v^2}{C^2 R_h} \tag{7.87}$$

where:

$$\tau_w = \gamma R_h S_o = \gamma R_h S_f = \gamma R_h \underbrace{\left[\frac{v^2}{C^2 R_h}\right]}_{S_o = S_f} = \frac{\gamma v^2}{C^2} = \frac{g}{C^2} \rho v^2 \tag{7.88}$$

The Chezy equation for open channel flow assumes an open channel section of length, L with a channel slope, S_o with a uniform flow depth, y and thus a constant velocity of flow, v and a friction slope, $S_f = S_o$ with a specific Chezy coefficient, C. The Chezy equation (which represents the momentum equation, supplemented by dimensional analysis) may be solved for one of the following unknown quantities: $S_f = S_o$, v, C, or y (which is defined as a function of R_h).

Furthermore, assuming that the Chezy coefficient, C is chosen to model the empirical flow resistance, the Chezy equation for the channel and friction slopes (Equation 7.87) is substituted into the integral form of the energy equation (Equation 7.73), yielding an empirical expression for the head loss, h_f (the Chezy head loss equation for open channel flow) as follows:

$$h_f = \frac{\tau_w L}{\gamma R_h} = S_o L = S_f L = \underbrace{\left[\frac{v^2}{C^2 R_h}\right]}_{S_o = S_f} L \tag{7.89}$$

7.4.8.3 Application of the Darcy–Weisbach Equation for Open Channel Flow

Alternatively, modeling the flow resistance using the Darcy–Weisbach friction factor, f, a new π term was defined by Equation 7.53 and is repeated as follows:

$$\frac{F_f}{F_I} = f = 8C_D \tag{7.90}$$

Thus, substituting Equation 7.90 for f into the expression for the wall shear stress, τ_w derived above from dimensional analysis, $(\tau_w/\rho v^2) = \phi(R, (\varepsilon/D))$, which was interpreted for open channel flow in Equation 7.84, yields the following:

$$\frac{F_f}{F_I} = f = 8C_D = 8\left[\frac{\tau_w}{\rho v^2}\right] = 8\left[\frac{S_f}{2}\frac{gy}{v^2}\right] = 4S_f\frac{gy}{v^2} = 4S_f\frac{1}{F} = \phi\left(R, \frac{\varepsilon}{y}\right) \tag{7.91}$$

However, because the flow is mostly turbulent in open channels, the Reynolds number, R is usually so large that it has virtually no influence on the drag coefficient, C_D (and thus, f). Thus, modeling the flow resistance using the Darcy–Weisbach friction factor, f (although typically used only for pipe flow) and thus substituting the relationship between the Chezy coefficient, C and the Darcy–Weisbach friction factor, f (Equation 7.61) into the Chezy head loss equation for open channel flow (Equation 7.89) (using the "equivalent diameter" or the "hydraulic diameter," $D_h = 4\,R_h$) yields the Darcy–Weisbach head loss equation for open channel flow as follows:

$$C = \sqrt{\frac{8g}{f}} \tag{7.92}$$

$$h_f = \frac{\tau_w L}{\gamma R_h} = S_o L = S_f L = \frac{v^2 L}{C^2 R_h} = \frac{v^2 L}{\dfrac{8g\,(4R_h)}{f}} = f\frac{L}{(4R_h)}\frac{v^2}{2g} \tag{7.93}$$

One may note that the Darcy–Weisbach friction factor, f is typically applied for pipe flow only.

7.4.8.4 Application of Manning's Equation for Open Channel Flow

Or, alternatively, modeling the flow resistance using the Manning's roughness coefficient, n and thus substituting the relationship between the Chezy coefficient, C and the Manning's roughness coefficient, n (Equation 7.70) into the Chezy head loss equation (Equation 7.89) yields the Manning's head loss equation for open channel flow (assuming the metric system) as follows:

$$C = \frac{R_h^{1/6}}{n} \tag{7.94}$$

$$h_f = \frac{\tau_w L}{\gamma R_h} = S_o L = S_f L = \frac{v^2 L}{C^2 R_h} = \frac{v^2}{\left(\dfrac{R_h^{1/6}}{n}\right)^2 R_h}L = \left(\frac{vn}{R_h^{2/3}}\right)^2 L \tag{7.95}$$

and the corresponding Manning's head loss equation for the English (BG) system for open channel flow is given as follows:

$$h_f = \frac{\tau_w L}{\gamma R_h} = S_o L = S_f L = \frac{v^2 L}{C^2 R_h} = \frac{v^2}{\left(\dfrac{1.486 R_h^{1/6}}{n}\right)^2} \frac{L}{R_h} = \left(\frac{vn}{1.486 R_h^{2/3}}\right)^2 L \qquad (7.96)$$

Finally, the Hazen–Williams equation is intended for pipe flow only and not open channel flow.

7.4.9 Application of Dimensional Analysis to Derive the Wall Shear Stress and Drag Coefficient for More Specific Assumptions of Internal Flow

The above application of dimensional analysis to derive an empirical equation for the major head loss (in terms of the wall shear stress, τ_w) for internal flow (Equation 7.47 for pipe flow and Equation 7.84 for open channel flow) was for the most general assumptions of internal flow; it was assumed that the flow may be either turbulent or laminar (special case of turbulent) flow. As a result, while the drag coefficient, C_D, in general, is a mainly a function of the hydraulic radius, R_h and E (for pipe flow) or F (for open channel flow), it will selectively vary with R and/or ε/L depending mainly on the velocity of flow (see Tables 11.8 and 11.9; see Sections 11.7.3 and 11.7.4), and is generally summarized as follows. In the case of low velocities (laminar flow), the viscous force will play a dominant role; thus, the definition of the drag coefficient, C_D will be totally dependent on R and thus independent of the relative surface roughness, ε/L. In the case of intermediate velocities (transitional flow), the viscous force and the relative surface roughness will both play dominant roles and thus the definition of the drag coefficient, C_D will be dependent on both R, and the relative surface roughness, ε/L. However, in the case of very high velocities (and thus very high values for R; completely turbulent flow), the relative pipe roughness plays a dominant role and thus the definition of the drag coefficient, C_D for completely turbulent flow (rough pipes or channels) is a function only of the relative pipe roughness, ε/L and thus independent of R.

EXAMPLE PROBLEM 7.3

Derive an empirical expression for the wall shear stress, τ_w for turbulent open channel flow. The inertia force ($F_I = Ma = \rho v^2 L^2$), and the wall shear stress, τ_w will always be important. Using Table 11.8 as a guide, the additional dominant/significant physical quantity in the flow situation includes ε. Thus, the important physical quantities include: ρ, v, L, τ_w, and ε. What are the important π terms used to study and describe turbulent open channel flow? What are the corresponding expressions for the friction slope, S_f and the major head loss, $h_{f,major}$?

Mathcad Solution

In order to derive an empirical expression for the wall shear stress, τ_w for turbulent open channel flow, the five steps of dimensional analysis are applied as follows:

Step 1: Using the given information, and Table 11.8 as a guide, the dominant/ significant physical forces and dominant physical quantities in the flow situation are identified as follows:

The inertia force, $F_I = Ma = \rho v^2 L^2$, will always be a dominant force and results in the three primary variables: ρ, v, and L. The friction/viscous force for turbulent flow, F_f or $F_V = \tau A$ will always be the first/main additional dominant force, which in this case is represented by a dominant physical quantity: the major head loss, h_f. The major head loss is caused by the gravity force and represents an energy loss or some form of flow resistance and results in a secondary variable: $\tau_w = \gamma R_h S_f$. In open channel flow, the propulsive force for the fluid flow is provided by invoking gravity and thus by the weight of the fluid along a channel slope (channel gradient, $\Delta z/L = S_o$), which results in an unknown S_f. Thus, in the derivation of the major head loss in open channel flow, $h_f = \tau_w L/\gamma R_h = \gamma R_h S_o/\gamma R_h = S_o L = S_f L$ (where $S_o = S_f$ for uniform open channel flow), the following four variables, ρ, v, L, and $\tau_w = \gamma R_h S_f$, result in one dimensionless number, namely:

$$\frac{\gamma R_h S_f}{\rho v^2} = \frac{\rho gy S_f}{2\rho v^2} = \frac{S_f}{2}\frac{gy}{v^2} = \frac{S_f}{2}\frac{1}{F} = C_D$$

where $F = F_I/F_G =$ the Froude number and $C_D =$ the drag coefficient that represents the flow resistance (dependent π term). Therefore, the major head loss in open channel flow, h_f (or τ_w), which represents F_V or F_f, is a result of S_f; thus, C_D is mainly a function of F. Modeling additional dominant forces or dominant physical quantities results in additional secondary variables; and will introduce an additional dimensionless number (π term) for each new secondary variable added. Using Table 11.8 as a guide, the absolute roughness, ε results in a secondary variable, ε, which introduces an independent π term, ε/L.

Step 2: Use the simplest mathematical expressions for the identified dominant physical forces and dominant physical quantities and identify the n physical quantities, all of which can be expressed in terms of m fundamental dimensions (F, L, and T).

Dominant Physical Force or Quantity	Primary Variable and Dimensions	Primary Variable and Dimensions	Primary Variable and Dimensions	Secondary Variable and Dimensions
$F_I = Ma = \rho v^2 L^2$	$\rho = [FL^{-4}T^2]$	$v = [LT^{-1}]$	$L = [L]$	
$h_f = \tau_w L/\gamma R_h = S_o L = S_f L$				$\tau_w = [FL^{-2}]$
ε				$\varepsilon = [L]$

where $n = 5$ and $m = 3$

Step 3: The number of dimensionless numbers (π terms) is $n-m = 5-3 = 2$.

Step 4: The Buckingham π theorem is applied in order to derive the $n-m = 2$ π terms as follows:

$$\pi_1 = \rho^{a_1} v^{b_1} L^{c_1} \tau_w^{d_1}$$

$$\pi_2 = \rho^{a_2} v^{b_2} L^{c_2} \varepsilon^{d_2}$$

where the exponents are unknown quantities to be determined by applying the Buckingham π theorem.

Working with the π_1 term, the dimensional equations for the *FLT* system are set up as follows:

$$\pi_1 = \rho^{a_1} v^{b_1} L^{c_1} \tau_w^{d_1}$$

$$F^0 L^0 T^0 = (FL^{-4}T^2)^{a_1} (LT^{-1})^{b_1} (L)^{c_1} (FL^{-2})^{d_1}$$

Equating the exponents for the dimensions and solving for three of the unknown exponents in terms of the fourth unknown exponent using the Mathcad symbolic solve block yields the following:

Given

F: $0 = a_1 + d_1$

L: $0 = -4a_1 + b_1 + c_1 - 2d_1$

T: $0 = 2a_1 - b_1$

Find $(a_1, b_1, c_1) \rightarrow \begin{pmatrix} -d_1 \\ -2 \cdot d_1 \\ 0 \end{pmatrix}$

Thus:

$$\pi_1 = \rho^{-d_1} v^{-2d_1} L^0 \tau_w^{d_1}$$

$$\pi_1 = \frac{\tau_w^{d_1}}{\rho^{d_1} v^{2d_1}}$$

From experimentation $d_1 = 1$; therefore:

$$\pi_1 = \frac{\tau_w}{\rho v^2}$$

which yields the empirical expression for the shear stress, τ_w.

Working with the π_2 term, the dimensional equations for the *FLT* system are set up as follows:

$$\pi_2 = \rho^{a_2} v^{b_2} L^{c_2} \varepsilon^{d_2}$$

$$F^0 L^0 T^0 = (FL^{-4}T^2)^{a_2} (LT^{-1})^{b_2} (L)^{c_2} (L)^{d_2}$$

Equating the exponents for the dimensions and solving for three of the unknown exponents in terms of the fourth unknown exponent using the Mathcad symbolic solve block yields the following:

Given

F: $\qquad 0 = a_2$

L: $\qquad 0 = -4a_2 + b_2 + c_2 + d_2$

T: $\qquad 0 = 2a_2 - b_2$

Find $(a_2, b_2, c_2) \rightarrow \begin{pmatrix} 0 \\ 0 \\ -d_2 \end{pmatrix}$

Thus:

$$\pi_2 = \rho^0 v^0 L^{-d_2} \varepsilon^{d_2}$$

$$\pi_2 = \frac{\varepsilon^{d_2}}{L^{d_2}}$$

From experimentation $d_2 = 1$; therefore:

$$\pi_2 = \frac{\varepsilon}{L}$$

which is the relative roughness, and L is represented by y.

The resulting functional relationship between the $n\text{-}m = 2$ π terms is expressed as follows:

$$\phi(\pi_1, \pi_2) = 0$$

$$\phi\left(\frac{\tau_w}{\rho v^2}, \frac{\varepsilon}{y}\right) = 0$$

or:

$$\frac{\tau_w}{\rho v^2} = \phi\left(\frac{\varepsilon}{y}\right)$$

Step 5: Interpret the resulting functional relationship between the $n\text{-}m = 2$ π terms. Thus, the drag coefficient, C_D is defined as follows:

$$\frac{\tau_w}{\rho v^2} = \frac{F_f}{F_I} = \frac{\gamma R_h S_f}{\rho v^2} = \frac{\rho g y S_f}{2\rho v^2} = \frac{S_f}{2}\frac{gy}{v^2} = \frac{S_f}{2}\frac{1}{F} = C_D = \phi\left(\frac{\varepsilon}{y}\right)$$

where the drag coefficient, C_D as modeled by the Froude number, F is the dependent π term, as it models the dependent variable, S_f that cannot be derived analytically from theory, while the relative surface roughness, ε/y is the independent π term. One may note that because the drag coefficient, C_D represents the flow resistance, which causes an unknown friction slope, S_f, the gravity force (invoked by the channel gradient, $S_o = \Delta z/L$) will always play an important role and thus the definition of the drag coefficient, C_D is mainly a function of F. The drag coefficient, C_D is dimensionless and is a function of boundary roughness, ε/y and the cross-sectional shape or the hydraulic radius, $R_h = y/2$. Finally, dimensional analysis yields a nonstandard form of the drag coefficient, C_D, which yields the resulting expression for the wall shear stress, τ_w as follows: $\tau_w = C_D \rho v^2$, where the more standard forms of the drag coefficient, C_D include the Chezy coefficient, C; the Darcy–Weisbach friction coefficient, f; and the Manning's roughness coefficient, n. Furthermore, the corresponding expressions for the friction slope is $S_f = \tau_w/\gamma R_h = C_D \rho v^2/\gamma R_h$ and for the major head loss is $h_{f,maj} = S_f L = \tau_w L/\gamma R_h = C_D \rho v^2 L/\gamma R_h$.

7.5 Modeling the Flow Resistance as a Minor Head Loss in Pipe Flow

While the major head loss term, $h_{f,major}$ in the energy equation models the head loss due to pipe friction in the straight section of a pipe, the minor head loss term, $h_{f,minor}$ in the energy equation models the head loss due to various pipe components in the pipe system. The minor head loss term, $h_{f,minor}$ in the energy equation is significant in very short pipes and is due to different pipe appurtenances (components, devices, or transitions), which include valves, fittings (tees, unions, elbows, and bends), entrances, exits, contractions, and expansions. The pipe components interrupt the smooth fluid flow and result in minor head losses that are caused by the flow separation and mixing induced by the various pipe components. As such, one is interested in the determination of the minor head loss and pressure drops where energy is dissipated. The derivation of the minor head loss equation along with the drag coefficient, C_D for pipe flow is presented below.

7.5.1 Evaluation of the Minor Head Loss

In the case where the flow resistance is due to various pipe components, transitions, or devices, the flow resistance is ultimately modeled as a minor head loss in the integral form of the energy equation because the minor head loss is needed for the design of internal flow in a pipe transition. The flow resistance (shear stress or drag force) is modeled as a resistance force in the integral form of the momentum equation. The flow resistance causes an unknown pressure drop, Δp, which causes an unknown minor head loss, h_f, where the head loss is due to a conversion of kinetic energy to heat, which is modeled/displayed in the integral form of the energy equation. Evaluation of the minor head loss flow resistance term requires a "subset level" of application of the governing equations. A "subset level" application of the governing equations focuses only on the given element causing the flow resistance. The assumption of real flow implies that the flow resistance is modeled in both the energy and momentum equations (see Table 4.1). Thus, in the derivation of the minor head loss, first the integral form of the energy equation is applied to determine an expression for the unknown head loss, h_f. Then, the integral momentum equation (supplemented by dimensional analysis) is applied in order to determine an expression for the pressure drop, Δp and the shear stress, τ_w. The flow resistance due to pipe components, which results in a minor head loss term, $h_{f,minor}$ is modeled as a resistance force (shear stress or drag force) in the integral form of the momentum equation (Equation 4.28). As such, the energy equation may be used to solve for the unknown minor head loss, $h_{f,min} = \tau_w L/\gamma R_h = \Delta p/\gamma$, while the integral momentum equation may be used to solve for the unknown pressure drop, $\Delta p = \tau_w L/R_h$.

The minor head loss, h_f is caused by both pressure and friction forces. Thus, when the friction/viscous forces are insignificant and can be ignored in the momentum equation (as in the case of a sudden pipe expansion), one can analytically determine the actual pressure drop, Δp from the momentum equation. Therefore, one may analytically derive an expression for the minor head loss, h_f due to pipe components from the energy equation (see Section 6.8.4 and Section 7.5.4). However, when the friction/viscous forces are significant and cannot be ignored in the momentum equation (as in the case of pipe components in general), the friction/viscous forces, the actual pressure drop, Δp and thus the minor head loss, h_f are determined empirically using dimensional analysis (see Section 6.8.5 and Section 7.5.5). Specifically, in the case of pipe components in general, the friction/viscous forces in the integral form of the momentum equation and thus the wall shear stress, τ_w cannot be ignored; thus, the integral form of the momentum equation cannot be directly applied to solve for the unknown pressure drop, Δp. Thus, an empirical interpretation (using dimensional analysis) of the wall shear stress, τ_w in the theoretical expression for the pressure drop, $\Delta p = \tau_w L/R_h$ in the integral form of the momentum equation is sought in terms of velocity, v and a flow resistance coefficient. This yields $\Delta p = k(\rho v^2/2)$, which represents the integral form of the momentum equation, supplemented by dimensional analysis, and is used to obtain an empirical evaluation for the pressure drop, Δp in Newton's second law of motion (integral form of the momentum equation). As such, one can then derive an empirical expression for the minor head loss, h_f from the energy equation. The flow resistance equation for the minor head loss is derived below and in Chapter 6. The minor losses computed in this section are for pipe transitions in pipe flow. Minor losses for open channel flow are addressed in Chapter 9.

7.5.2 Application of the Energy Equation: Derivation of the Head Loss

The objective here is to derive an expression for the minor head loss, $h_{f,minor}$, noting that, in general, both pressure forces, F_p and viscous forces, F_V (shear forces, F_f) may be exerted on the pipe wall as a result of a pipe component. Thus, regardless of whether one may analytically or empirically determine the actual pressure drop/rise and the resulting minor head loss due to a pipe component, it is of interest to determine the expression for the pressure drop/rise and the corresponding minor head loss. Thus, in the derivation of the minor head loss for pipe flow, real flow is assumed in order to determine an expression for the unknown head loss, h_f by applying the integral form of the energy equation as follows:

$$\left(\frac{p_1}{\gamma} + z_1 + \frac{v_1^2}{2g}\right) - h_{f,min} = \left(\frac{p_2}{\gamma} + z_2 + \frac{v_2^2}{2g}\right) \tag{7.97}$$

Assuming a horizontal pipe ($z_1 = z_2$) and thus that F_G is not significant, the energy equation yields the following:

$$h_{f,min} = \frac{\tau_w L}{\gamma R_h} = \left(\frac{p_1}{\gamma} + \frac{v_1^2}{2g}\right) - \left(\frac{p_2}{\gamma} + \frac{v_2^2}{2g}\right) \tag{7.98}$$

In the cases of the flow from a reservoir or a tank or a pipe contraction and the flow through most pipe devices, where the fluid is set into motion ($v_1 = 0$ and $v_2 = v$), the conversion of the pressure energy to kinetic energy results in a pressure drop, where the pressure drop, Δp is equal to the dynamic pressure term plus the flow resistance term, the energy equation yields the minor head loss as follows:

$$h_{f,min} = \underbrace{\frac{\tau_w L}{\gamma R_h}}_{flowres} = \underbrace{\frac{\Delta p}{\gamma}}_{presdrop} - \underbrace{\frac{v^2}{2g}}_{dynpres} \tag{7.99}$$

However, in the cases of: the flow into a reservoir or a pipe expansion where the fluid is momentarily brought to a stop ($v_2 = 0$ and $v_1 = v$), the conversion of the kinetic energy to pressure energy results in a pressure rise, where the pressure rise, Δp is equal to the dynamic pressure term minus the flow resistance term, and the energy equation yields the minor head loss as follows:

$$h_{f,min} = \underbrace{\frac{\tau_w L}{\gamma R_h}}_{flowres} = \underbrace{\frac{v^2}{2g}}_{dynpres} - \underbrace{\frac{\Delta p}{\gamma}}_{presrise} \tag{7.100}$$

Therefore, in comparison, the major head loss, $h_{f,major}$ due to the flow through a pipe is mostly due to the pressure head drop, $\Delta p/\gamma$ (see Equation 7.38), while the minor head loss, $h_{f,minor}$ due to a pipe component is due to both the pressure head drop/rise, $\Delta p/\gamma$ and the dynamic pressure (velocity) head, $v^2/2g$ as illustrated by Equations 7.99 and 7.100.

7.5.3 Application of the Momentum Equation and Dimensional Analysis: Derivation of the Pressure Drop, Shear Stress, and the Drag Coefficient

In the derivation of the minor head loss for pipe flow, real flow is assumed in order to determine an expression for the pressure drop, Δp by applying the integral form of the momentum equation (supplemented by dimensional analysis). However, because the flow through the various devices, in general, is very complex, it is not possible to theoretically model the existence of the viscous force due to the wall shear stress in the momentum equation and thus one cannot theoretically define the momentum equation. The flow through these pipe devices is complex, as the fluid flows through variable areas and thus nonlinear flow torturous flow paths. In the special case of a sudden pipe expansion, the friction/ viscous forces are considered to be insignificant and thus can be ignored in the momentum equation. As a result, one can analytically determine the actual pressure drop, Δp from the momentum equation, and therefore one may analytically derive an expression for the minor head loss, h_f due to a sudden pipe expansion from the energy equation. However, in the case of the remaining pipe devices, in general, the friction/viscous forces are significant and cannot be ignored in the momentum equation. As a result, the friction/viscous forces; the actual pressure drop, Δp; and thus the minor head loss, h_f are determined empirically.

Thus, the particular type of pipe component will determine whether the friction/viscous forces are insignificant and can be ignored and thus determine whether application of the three governing equations will yield an analytical or an empirical expression for the minor head loss, h_f. In particular, in the case of a sudden pipe expansion, the friction/viscous forces in the integral form of the momentum equation can be ignored. Then, application of the continuity equation yields an analytical expression for the actual pressure drop, Δp. Finally, the analytical expression for the actual pressure drop, Δp is substituted into the integral form of the energy equation, yielding an analytical expression for the minor head loss, h_f. However, in the case of the other pipe components/devices/transitions, in general, the friction/viscous forces in the integral form of the momentum equation cannot be ignored. Instead, the friction forces are modeled empirically. Specifically, because one cannot ignore the viscous force in the integral form of the momentum equation, one cannot derive an analytical expression for the actual pressure drop, Δp from application of the momentum equation and the continuity equation; thus, one cannot derive an analytical expression for the minor head loss, h_f from application of the integral form of the energy equation. As a result, one must resort to dimensional analysis in order to derive an expression for actual pressure drop, Δp and thus the minor head loss, h_f, which involves the definition of a drag coefficient, C_D (minor loss coefficient, k) that represents the flow resistance and is empirically evaluated. Finally, it is important to note that the use of dimensional analysis is not needed in the derivation of the minor head loss for a sudden pipe expansion (see Section 7.5.4).

7.5.4 Analytical Derivation of the Minor Head Loss Equation due to a Sudden Pipe Expansion

For the case of a sudden pipe expansion, the application of the three governing equations yields a theoretical expression for the actual pressure drop, Δp and for the minor head loss, $h_{f,minor}$. The analytical derivation of the minor head loss is given in Chapter 8 and summarized as follows. For the case of a sudden pipe expansion, first, application of the integral energy equation yields an expression for the minor head loss (Equation 7.100). Then, the integral form of the momentum equation is applied in order to analytically solve for the unknown pressure drop, Δp. Specifically, assuming a horizontal pipe ($z_1 = z_1$) and thus F_G is not significant, and that the viscous force, F_V due to the wall shear stress is not significant, the integral form of the momentum equation is used to derive an expression for the actual pressure drop, Δp. Applying the integral form of the continuity equation and substituting $v_2 = v_1(A_1/A_2)$, the expression for the actual pressure drop, Δp from the momentum equation is substituted into the energy equation, in order to solve for the $h_{f,minor}$. The minor head loss, $h_{f,minor}$ is analytically evaluated using the upstream velocity, v_1 and is given as follows:

$$h_f = \underbrace{\left[1 - \frac{A_1}{A_2}\right]^2}_{k} \left(\frac{v_1^2}{2g}\right) = \underbrace{\left[1 - \left(\frac{D_1}{D_2}\right)^2\right]^2}_{k} \left(\frac{v_1^2}{2g}\right) \tag{7.101}$$

$$h_f = k\left(\frac{v_1^2}{2g}\right) \tag{7.102}$$

where the minor loss coefficient, k due to a sudden expansion is evaluated analytically as follows:

$$k = \left[1 - \frac{A_1}{A_2} \right]^2 = \left[1 - \left(\frac{D_1}{D_2} \right)^2 \right]^2 \tag{7.103}$$

One may note that although the application of the three governing equations yielded a general expression for the minor head loss, $h_f = k(v^2/2g)$, an analytical evaluation of the minor loss coefficient, k due to all other pipe devices (other than a sudden pipe expansion) is not possible because the flow through the different devices is very complex. Instead, the minor loss coefficient, k is evaluated empirically as illustrated in Section 7.5.5.

7.5.5 Empirical Derivation of the Minor Head Loss Equation due to Pipe Components in General

For pipe components in general, first, application of the integral form of the energy equation yields an expression for the minor head loss (Equation 7.98). However, a further simplifying assumption of a horizontal pipe with steady uniform flow yields an expression for the minor head loss, $h_{f,min} = \tau_w L / \gamma R_h = \Delta p / \gamma$ (see Section 6.8.5 for details). Then, however, because the friction/viscous forces in the integral form of the momentum equation and thus the wall shear stress, τ_w cannot be theoretically modeled or ignored (as done for the case of the sudden pipe expansion, see Section 7.5.4 above), one cannot analytically solve for the unknown pressure drop/rise, Δp from the integral momentum equation. However, the further simplifying assumption of a horizontal pipe with steady uniform flow yields an expression for the pressure drop, Δp as a function of the shear stress: $\Delta p = \tau_w L / R_h$ (see Section 6.8.5 for details). Therefore, the unknown pressure drop, Δp and thus the minor head loss, h_f are determined empirically. Specifically, an empirical interpretation of the wall shear stress, τ_w in the theoretical expression for the pressure drop, $\Delta p = \tau_w L / R_h$ in the integral momentum equation is sought in terms of velocity, v and a flow resistant coefficient. This is accomplished by using dimensional analysis in order to derive an empirical expression for the pressure drop, Δp as a function of velocity, v, which involves the definition of a drag coefficient, C_D (minor head loss coefficient, k) that represents the flow resistance and is empirically evaluated. Dimensional analysis yields an empirical expression for the pressure drop/rise, $\Delta p = k(\rho v^2/2)$ in the integral momentum equation, where the actual pressure drop/rise, Δp is indirectly evaluated through an empirical determination of a minor head loss coefficient, k from experimentation. Finally, the empirical expression for the actual pressure drop, Δp is substituted into the integral form of the energy equation, $h_{f,min} = \tau_w L / \gamma R_h = \Delta p / \gamma$ yielding an empirical expression for the minor head loss, $h_{f,min} = k(v^2/2g)$.

7.5.5.1 Application of Dimensional Analysis: Derivation of the Pressure Drop and Drag Coefficient for Pipe Components in General

An empirical expression for the pressure drop, Δp in the theoretical expression for the pressure drop, $\Delta p = \tau_w L / R_h$ in the integral momentum equation is sought in terms of velocity, v and a flow resistant coefficient. This is accomplished by using dimensional analysis in order to derive an empirical expression for the pressure drop, Δp as a function of the velocity, v, which involves the definition of a drag coefficient, C_D (minor head loss coefficient, k) that represents the flow resistance and is empirically evaluated. Thus, the momentum theory is supplemented with dimensional analysis in order to derive an expression for the actual pressure drop/rise, Δp for the pipe components in general. Thus, for the derivation of the minor head loss in pipe flow, h_f, in addition to the inertia force, F_I and the pressure force, F_P (causes the viscous force for turbulent flow, $F_V = \tau A$, which is represented by the minor head loss, h_f), other forces, including the viscous force for laminar flow, $F_V = \tau A = \mu v L$ and other physical quantities including the geometry of the pipe device, L_i and the absolute surface roughness, ε, also play an important role in the pipe fluid flow.

Dimensional analysis is applied in order to derive an empirical expression for the pressure drop/rise, Δp for the pipe components in general, which involves the definition of a drag coefficient, C_D as follows (see Table 7.5):

Step 1: Based on a general understanding of (and experience with and judgment of) the fluid flow phenomena, identify the dominant/significant physical forces and dominant physical quantities in the flow situation (use Table 11.8 for internal flow as a guide).

a. The inertia force, $F_I = Ma = \rho v^2 L^2$ will always be a dominant force and results in the three primary variables: ρ, v, and L.

b. The friction/viscous force for turbulent flow, F_f or $F_V = \tau A$ will always be the first/main additional dominant force, which in this case is represented by a dominant physical quantity: the minor head loss, h_f. The minor head loss is caused by the pressure force and represents an energy loss or some form of flow resistance and results in a secondary variable: Δp.

c. In pipe flow, the propulsive force for the fluid flow is provided by a pressure gradient, $\Delta p/L$, which results in an unknown Δp or S_f. Thus, in the derivation of the minor head loss in pipe flow, $h_f = \tau_w L/\gamma R_h = \Delta p/\gamma$, the following four variables, ρ, v, L, and Δp, result in one dimensionless number, namely $\Delta p/\rho v^2 = 1/E = C_D = C_p$ where $E = F_I/F_P =$ the Euler number, $C_D =$ the drag coefficient that represents the flow resistance, and $C_p =$ the pressure coefficient (dependent π term). Therefore, the minor head loss in pipe flow, h_f (or Δp), which represents F_V, is a result of Δp; thus, C_D is mainly a function of E.

d. Modeling additional dominant forces or dominant physical quantities results in additional secondary variables and will introduce an additional dimensionless number (π term) for each new secondary variable added.

1. The viscous force for laminar flow, $F_V = \tau A = \mu v L$ results in a secondary variable, μ, which introduces an independent π term, R.

2. The absolute roughness, ε results in a secondary variable, ε, which introduces an independent π term, ε/L.

3. The geometry, L_i ($i = 1, 2, \ldots$) results in one or more secondary variables, L_i, which introduces one or more independent π terms, L_i/L.

Step 2: Use the simplest mathematical expressions for the identified dominant physical forces and dominant physical quantities and identify the n physical quantities, all of which can be expressed in terms of m fundamental dimensions (F, L, and T).

Dominant Physical Force or Quantity	Primary Variable and Dimensions	Primary Variable and Dimensions	Primary Variable and Dimensions	Secondary Variable and Dimensions
$F_I = Ma = \rho v^2 L^2$	$\rho = [FL^{-4}T^2]$	$v = [LT^{-1}]$	$L = [L]$	
$h_f = \tau_w L / \gamma R_h = \Delta p / \gamma$				$\Delta p = [FL^{-2}]$
$F_V = \tau A = \mu v L$				$\mu = [FL^{-2}T]$
ε				$\varepsilon = [L]$
L_i				$L_i = [L]$

where $n = 7$ and $m = 3$.

Step 3: The number of dimensionless numbers (π terms) is $n-m = 7-3 = 4$.

Step 4: The Buckingham π theorem is applied in order to derive the $n-m = 4$ π terms as follows:

$$\pi_1 = \rho^{a_1} v^{b_1} L^{c_1} \Delta p^{d_1}$$

$$\pi_2 = \rho^{a_2} v^{b_2} L^{c_2} \mu^{d_2}$$

$$\pi_3 = \rho^{a_3} v^{b_3} L^{c_3} \varepsilon^{d_3}$$

$$\pi_4 = \rho^{a_4} v^{b_4} L^{c_4} L_i^{d_4}$$

where the exponents are unknown quantities to be determined by applying the Buckingham π theorem.

Working with the π_1 term, the dimensional equations for the *FLT* system are set up as follows:

$$\pi_1 = \rho^{a_1} v^{b_1} L^{c_1} \Delta p^{d_1}$$

$$F^0 L^0 T^0 = (FL^{-4}T^2)^{a_1} (LT^{-1})^{b_1} (L)^{c_1} (FL^{-2})^{d_1}$$

Equating the exponents for the dimensions and solving for three of the unknown exponents in terms of the fourth unknown exponent using the Mathcad symbolic solve block yields the following:

Given

F: $0 = a_1 + d_1$

L: $0 = -4a_1 + b_1 + c_1 - 2d_1$

T: $0 = 2a_1 - b_1$

$$\text{Find}(a_1, b_1, c_1) \rightarrow \begin{pmatrix} -d_1 \\ -2 \cdot d_1 \\ 0 \end{pmatrix}$$

Thus:

$$\pi_1 = \rho^{-d_1} v^{-2d_1} L^0 \Delta p^{d_1}$$

$$\pi_1 = \frac{\Delta p^{d_1}}{\rho^{d_1} v^{2d_1}}$$

From experimentation $d_1 = 1$; therefore:

$$\pi_1 = \frac{\Delta p}{\rho v^2}$$

which yields the empirical expression for the pressure drop/rise, Δp.

Working with the π_2 term, the dimensional equations for the *FLT* system are set up as follows:

$$\pi_2 = \rho^{a_2} v^{b_2} L^{c_2} \mu^{d_2}$$

$$F^0 L^0 T^0 = (FL^{-4}T^2)^{a_2} (LT^{-1})^{b_2} (L)^{c_2} (FL^{-2}T)^{d_2}$$

Equating the exponents for the dimensions and solving for three of the unknown exponents in terms of the fourth unknown exponent using the Mathcad symbolic solve block yields the following:

Given

F: $\qquad 0 = a_2 + d_2$

L: $\qquad 0 = -4a_2 + b_2 + c_2 - 2d_2$

T: $\qquad 0 = 2a_2 - b_2 + d_2$

Find $(a_2, b_2, c_2) \rightarrow \begin{pmatrix} -d_2 \\ -d_2 \\ -d_2 \end{pmatrix}$

Thus:

$$\pi_2 = \rho^{-d_2} v^{-d_2} L^{-d_2} \mu^{d_2}$$

$$\pi_2 = \frac{\mu^{d_2}}{\rho^{d_2} v^{d_2} L^{d_2}}$$

From experimentation $d_2 = -1$; therefore:

$$\pi_2 = \frac{\rho v L}{\mu}$$

which is the Reynolds number, *R* and where L represents D, the pipe diameter.

Working with the π_3 term, the dimensional equations for the FLT system are set up as follows:

$$\pi_3 = \rho^{a_3} v^{b_3} L^{c_3} \varepsilon^{d_3}$$

$$F^0 L^0 T^0 = (FL^{-4}T^2)^{a_3} (LT^{-1})^{b_3} (L)^{c_3} (L)^{d_3}$$

Equating the exponents for the dimensions and solving for three of the unknown exponents in terms of the fourth unknown exponent using the Mathcad symbolic solve block yields the following:

Given

F: $0 = a_3$

L: $0 = -4a_3 + b_3 + c_3 + d_3$

T: $0 = 2a_3 - d_3$

Find $(a_3, b_3, c_3) \rightarrow \begin{pmatrix} 0 \\ 0 \\ -d_3 \end{pmatrix}$

Thus:

$$\pi_3 = \rho^0 v^0 L^{-d_3} \varepsilon^{d_3}$$

$$\pi_3 = \frac{\varepsilon^{d_3}}{L^{d_3}}$$

From experimentation $d_3 = 1$; therefore:

$$\pi_3 = \frac{\varepsilon}{L}$$

which is the relative roughness, and L is represented by D.

Working with the π_4 term, the dimensional equations for the FLT system are set up as follows:

$$\pi_4 = \rho^{a_4} v^{b_4} L^{c_4} L_i^{d_4}$$

$$F^0 L^0 T^0 = (FL^{-4}T^2)^{a_4} (LT^{-1})^{b_4} (L)^{c_4} (L)^{d_4}$$

Equating the exponents for the dimensions and solving for three of the unknown exponents in terms of the fourth unknown exponent using the Mathcad symbolic solve block yields the following:

Given

F: $0 = a_4$

L: $0 = -4a_4 + b_4 + c_4 + d_4$

T: $0 = 2a_4 - d_4$

Find $(a_4, b_4, c_4) \rightarrow \begin{pmatrix} 0 \\ 0 \\ -d_4 \end{pmatrix}$

Thus:

$$\pi_4 = \rho^0 v^0 L^{-d_4} L_i^{d_4}$$

$$\pi_4 = \frac{L_i^{d_4}}{L^{d_4}}$$

From experimentation $d_4 = 1$; therefore:

$$\pi_4 = \frac{L_i}{L}$$

which is the geometry.

The resulting functional relationship between the $n\text{-}m = 4$ π terms is expressed as follows:

$$\phi(\pi_1, \pi_2, \pi_3, \pi_4) = 0$$

$$\phi\left(\frac{\Delta p}{\rho v^2}, \frac{\rho v D}{\mu}, \frac{\varepsilon}{D}, \frac{L_i}{L}\right) = 0$$

$$\phi\left(\frac{\Delta p}{\rho v^2}, R, \frac{\varepsilon}{D}, \frac{L_i}{L}\right) = 0$$

or:

$$\frac{\Delta p}{\rho v^2} = \phi\left(R, \frac{\varepsilon}{D}, \frac{L_i}{L}\right)$$

Step 5: Interpret the resulting functional relationship between the $n\text{-}m = 4$ π terms. Define the drag coefficient, C_D/minor head loss coefficient, k, and note the definition of the inertia force, F_I as follows:

$$\frac{\Delta p}{\rho v^2} = \frac{F_P}{F_I} = \frac{1}{E} = C_D = k = \phi\left(R, \frac{\varepsilon}{D}, \frac{L_i}{L}\right) \qquad (7.104)$$

where the drag coefficient, C_D as modeled by the Euler number, E is the dependent π term, as it models the dependent variable, Δp that cannot be derived analytically from theory, while the Reynolds number, R; the relative surface roughness, ε/D; and the geometry of the pipe device, L_i/L are the independent π terms. One may note that because the drag coefficient, C_D/minor head loss coefficient, k represents the flow resistance, which causes an unknown pressure drop, Δp, the pressure force will always play an important role and thus the definition of the drag coefficient, C_D is mainly a function of E. The minor head loss coefficient, k is almost always evaluated using experimental results, where no distinction is made between laminar and turbulent flow. While the loss coefficient, k in general depends on both the geometry (design details) of the flow device and the Reynolds number, R (and in some cases on the relative pipe roughness, ε/D), it is usually assumed to be independent of the Reynolds number, as most pipe flows in practice usually have large Reynolds numbers and the loss coefficients tend to be independent of R when its values are large. Although of minor importance in turbulent flow, one may note that, similar to the pipe friction factor, f, the head loss coefficient, k tends to increase with an increase in the relative pipe roughness, ε/D and a decrease in the Reynolds number, R. Therefore, the magnitude of the head loss coefficient, k is mainly a function of the flow geometry produced by the pipe device. Additionally, it is important to note that because the minor loss coefficients, k for the various pipe devices are dependent upon the particular manufacturer of the device, the final design of the pipe system should rely on the manufacturer's information rather than on typical experimental values in handbooks.

Finally, one may note that dimensional analysis yields a nonstandard form of the drag coefficient/minor head coefficient, k, using inertia force, F_I in its definition, which yields the corresponding expression for the pressure drop/rise, Δp as follows:

$$\Delta p = k\rho v^2 \qquad (7.105)$$

If, however, one considers the fact that inertia force, $F_I = \rho v^2 A$ may be further defined as the dynamic (pressure) force, $F_{dynpres} = \rho v^2 A/2$ (see Section 7.3.3.1 above on the drag force for details), then the drag coefficient/minor head loss coefficient, k would be defined as follows:

$$\frac{\Delta p}{\dfrac{\rho v^2}{2}} = \frac{F_P}{F_{dynpres}} = \frac{2}{E} = C_D = k = \phi\left(R, \frac{\varepsilon}{D}, \frac{L_i}{L}\right) \qquad (7.106)$$

which is the standard form of the drag coefficient/minor head loss coefficient, k, resulting from replacing the inertia force, $F_I = \rho v^2 A$ used in the dimensional analysis, with the dynamic (pressure) force, $F_{dynpres} = \rho v^2 A/2$. Thus, the standard empirical expression for the pressure drop/rise is given as follows:

$$\Delta p = k\frac{\rho v^2}{2} \qquad (7.107)$$

Furthermore, examining the standard form of the minor head loss coefficient, k yields the following:

$$\frac{F_P}{F_{dynpres}} = \frac{\Delta p}{\dfrac{\rho v^2}{2}} = C_D = k = \frac{Actual\ PresForce}{Ideal\ Force} = \frac{Actual\ Presdrop}{Ideal\ Presdrop} = \frac{staticpres}{dynpres} \qquad (7.108)$$

where the results of dimensional analysis introduce a drag coefficient, C_D, which expresses the ratio of a real flow variable (in this case: force or pressure) to its corresponding ideal flow variable. Finally, evaluation of the velocity and empirical evaluation of the minor head loss coefficient, k due to various pipe components in a pipe system (along with the application of the minor head loss equation for pipe flow, h_f) are presented in Chapter 8.

One may note that k is sometimes referred to as the pressure (recovery) coefficient, C_p (especially in the case of a gradual pipe expansion) and is defined as follows:

$$\frac{\Delta p}{\dfrac{\rho v^2}{2}} = \frac{2}{E} = k = C_p = \frac{\Delta p_{actual}}{\Delta p_{ideal}} \qquad (7.109)$$

For ideal flow, the flow through a contraction is the conversion of flow energy (pressure energy) to kinetic energy, while the flow through an expansion is the conversion of kinetic energy to flow energy (pressure energy). Furthermore, there is the possibility of cavitation (high velocity and thus low pressure) in the pipe flow, such as in the complex passages that may exist in valves. As such, Δp is defined as $(p_r - p_v)$, where p_r is some reference pressure and p_v is the vapor pressure. Therefore, the pressure coefficient, C_p becomes the dimensionless quantity known as the cavitation number, C_a as follows:

$$\frac{\Delta p}{\frac{\rho v^2}{2}} = \frac{2}{E} = k = C_p = C_a = \frac{p_r - p_v}{\frac{\rho v^2}{2}} \tag{7.110}$$

where both pressures must be absolute.

7.5.5.2 Substituting the Pressure Drop into the Energy Equation: Deriving the Minor Head Loss Equation for Pipe Components in General

And, finally, the empirical expression for the actual pressure drop (Equation 7.107) is substituted into the integral form of the energy equation $h_{f,min} = \tau_w L/\gamma R_h = \Delta p/\gamma$, yielding an empirical expression for the minor head loss, h_f as follows:

$$h_{f,min} = \frac{\tau_w L}{\gamma R_h} = \frac{\Delta p}{\gamma} = C_D \frac{v^2}{2g} = k \frac{v^2}{2g} \tag{7.111}$$

where the drag coefficient, C_D is expressed in terms of the more commonly defined minor head loss coefficient, k, where the actual pressure drop/rise, Δp is indirectly evaluated through an empirical determination of a minor head loss coefficient, k (evaluated from experimentation). Furthermore, one can easily compute v_i used in Equation 7.111 by measuring Δp using a pitot-static tube.

7.5.5.3 Evaluation of the Minor Head Loss Coefficient, k

One may recall that, in general, both pressure forces, F_p and viscous forces, F_V (shear forces, F_f) may be exerted on the pipe wall as a result of a pipe component. Therefore, regardless of whether the particular flow device results in a pressure drop or a pressure rise, the actual pressure drop/rise, Δp may: (1) be equal to the dynamic pressure (ideal pressure), $\rho v^2/2$; (2) be greater than the dynamic pressure; or (3) be less than the dynamic pressure. As noted above, because, in general, the actual pressure drop/rise, Δp is unknown, k is determined empirically from experimentation. Examining the standard form of the minor head loss coefficient, k yields the following:

$$\frac{F_P}{F_{dynpres}} = \frac{\Delta p}{\frac{\rho v^2}{2}} = C_D = k = \frac{Actual\ Pressure\ Force}{Ideal\ Force} = \frac{Actual\ Presdrop}{Ideal\ Presdrop} = \frac{staticpres}{dynpress} \tag{7.112}$$

where the results of dimensional analysis introduce a drag coefficient, C_D that expresses the ratio of a real flow variable (in this case: force or pressure) to its corresponding ideal flow variable. In the cases where the pipe component results in a pressure drop, Δp, the energy equation yields the following:

$$p_1 - p_2 = \underbrace{\Delta p}_{presdrop} = \underbrace{\left(\frac{\rho v_2^2}{2} - \frac{\rho v_1^2}{2}\right)}_{dynpress} + \underbrace{\frac{\tau_w L}{R_h}}_{flowres} \tag{7.113}$$

Thus, the minor head loss coefficient, k is derived as follows:

$$\frac{F_P}{F_{dynpres}} = \frac{\Delta p}{\frac{\rho v^2}{2}} = \frac{\left(\dfrac{\rho v_2^2}{2} - \dfrac{\rho v_1^2}{2}\right) + \dfrac{\tau_w L}{R_h}}{\dfrac{\rho v^2}{2}} = k \tag{7.114}$$

where $k > 1$ in the case when the actual pressure drop, Δp is greater than the dynamic pressure, and $k < 1$ when the actual pressure drop, Δp is less than the dynamic pressure. In the cases where the pipe component results in a pressure rise, Δp the energy equation yields the following:

$$p_2 - p_1 = \underbrace{\Delta p}_{presrise} = \underbrace{\left(\frac{\rho v_1^2}{2} - \frac{\rho v_2^2}{2}\right)}_{dynpress} - \underbrace{\frac{\tau_w L}{R_h}}_{flowres} \tag{7.115}$$

Thus, the minor head loss coefficient, k is derived as follows:

$$\frac{F_P}{F_{dynpres}} = \frac{\Delta p}{\frac{\rho v^2}{2}} = \frac{\left(\dfrac{\rho v_1^2}{2} - \dfrac{\rho v_2^2}{2}\right) - \dfrac{\tau_w L}{R_h}}{\dfrac{\rho v^2}{2}} = k \tag{7.116}$$

where $k > 1$ in the case when the actual pressure rise, Δp is greater than the dynamic pressure, and $k < 1$ when the actual pressure rise, Δp is less than the dynamic pressure. In particular, for the case of a pipe exit into a reservoir, $v_1 = v$ and $v_2 = 0$, where there is a dissipation of kinetic energy into internal energy and heat transfer, and the wall shear stress, τ_w is not significant, and the energy equation yields the actual pressure rise, Δp as follows:

$$p_2 - p_1 = \underbrace{\Delta p}_{presrise} = \frac{\rho v^2}{2} \tag{7.117}$$

where the actual pressure rise, Δp is exactly equal to the dynamic pressure, $\rho v^2/2$. Thus, the minor head loss coefficient, k is derived as follows:

$$\frac{F_P}{F_{dynpres}} = \frac{F_{dynpres}}{F_{dynpres}} = \frac{\Delta p}{\frac{\rho v^2}{2}} = \frac{\frac{\rho v^2}{2}}{\frac{\rho v^2}{2}} = k = 1 \tag{7.118}$$

It is interesting to examine the case of the sudden pipe expansion, for which the minor head loss coefficient, k was analytically derived. Assuming a horizontal pipe ($z_1 = z_1$) and thus that F_G is not significant, the viscous force, F_V due to the wall shear stress is not significant, and that the minor head loss due to a sudden expansion is evaluated using the upstream velocity, v_1, the analytical expression for the actual pressure rise, Δp derived by applying the three governing equations yielded:

$$\Delta p = \gamma h_{f,min} = \gamma \left[1 - \frac{A_1}{A_2} \right]^2 \left(\frac{v_1^2}{2g} \right) = \gamma \left[1 - \left(\frac{D_1}{D_2} \right)^2 \right]^2 \left(\frac{v_1^2}{2g} \right) \tag{7.119}$$

Thus, the minor head loss coefficient, k is derived as follows:

$$\frac{F_P}{F_{dynpres}} = \frac{\Delta p}{\frac{\rho v^2}{2}} = \frac{\frac{\rho v_1^2}{2} \left[1 - \frac{A_1}{A_2} \right]^2}{\frac{\rho v^2}{2}} = \left[1 - \frac{A_1}{A_2} \right]^2 = k \tag{7.120}$$

The value for k changes from the limiting/extreme case of square-edged exit with an area ratio of $A_1/A_2 = 0$ and $v_1 = v$ and $v_2 = 0$, where $k = 1$ (a pipe exit into a reservoir) to the other extreme case of a constant diameter pipe with an area ratio of $A_1/A_2 = 1$, and $v_1 = v_2 = v$ where $k = 0$ (flow through a pipe). Evaluation of the velocity, and the minor head loss coefficient, k and determination of the minor head loss due to various pipe components in a pipe system are presented in Chapter 8.

7.5.5.4 Application of Dimensional Analysis to Derive the Pressure Drop and Drag Coefficient for More Specific Assumptions for Pipe Flow Components

The above application of dimensional analysis to derive an empirical equation for the minor head loss (in terms of the pressure drop, Δp) for pipe flow components (Equation 7.104) was for the most general assumptions of pipe flow components; it was assumed that the flow may be either turbulent or laminar (special case of turbulent) flow. As a result, while the drag coefficient, C_D, in general, is a mainly a function of the Euler number, E and the geometry of the pipe device, L_i/L, it will selectively vary with the Reynolds number, R and/or the relative surface roughness, ε/D depending mainly on the particular type of pipe component/device (see Tables 11.8 and 11.9; see Section 11.7.5; see Section 8.6.4 for valves and fittings [tees, unions, elbows, and bends], and see Section 8.6.5 for entrances, exits, contractions, and expansions). One may note that because the drag coefficient, C_D/minor head loss coefficient, k represents the flow resistance, which causes an unknown pressure drop, Δp, the pressure force will always play an important role and thus the definition of the drag coefficient, C_D is mainly a function of E. The minor head loss coefficient, k is almost always evaluated using experimental results, where no distinction is made between laminar and turbulent flow. While the loss coefficient, k, in general, depends on both the geometry (design details of the pipe device, L_i/L) of the flow device and the Reynolds number, R (and in some cases on the relative pipe roughness, ε/D), it is usually assumed to be independent of the Reynolds number, as most pipe flows in practice usually have large Reynolds numbers and the loss coefficients tend to be independent of R when its values are large. Although of minor importance in turbulent flow, one may note that, similar to the pipe friction factor, f, the head loss coefficient, k tends to increase with an increase in the relative pipe roughness,

ε/D and a decrease in the Reynolds number, R. Therefore, the magnitude of the head loss coefficient, k is mainly a function of the flow geometry, which is produced by the pipe device.

EXAMPLE PROBLEM 7.4

Derive an empirical expression for the pressure drop, Δp across a globe valve, assuming turbulent flow. The inertia force ($F_I = Ma = \rho v^2 L^2$), and the pressure drop, Δp will always be important. Using Table 11.8 and Section 8.6.4 as a guide, the additional dominant/significant physical quantities in the flow situation include ε and L_i. Thus, the important physical quantities include: ρ, v, L, Δp, ε, and L_i. What are the important π terms used to study and describe turbulent flow across a globe valve? What are the corresponding expressions for the wall shear stress, τ_w and the minor head loss, $h_{f,min}$?

Mathcad Solution

In order to derive an empirical expression for the pressure drop, Δp across a globe valve, assuming turbulent flow, the five steps of dimensional analysis are applied as follows:

Step 1: Using the given information, and Table 11.8 and Section 8.6.4 as a guide, the dominant/significant physical forces and dominant physical quantities in the flow situation are identified as follows:

The inertia force, $F_I = Ma = \rho v^2 L^2$ will always be a dominant force and results in the three primary variables: ρ, v, and L. The friction/viscous force for turbulent flow, F_f or $F_V = \tau A$ will always be the first/main additional dominant force, which in this case is represented by a dominant physical quantity: the minor head loss, h_f. The minor head loss is caused by the pressure force and represents an energy loss or some form of flow resistance and results in a secondary variable: Δp. In pipe flow, the propulsive force for the fluid flow is provided by a pressure gradient, $\Delta p/L$ which results in an unknown Δp or S_f. Thus, in the derivation of the minor head loss in pipe flow, $h_f = \tau_w L/\gamma R_h = \Delta p/\gamma$, the following four variables, ρ, v, L, and Δp, result in one dimensionless number, namely $\Delta p/\rho v^2 = 1/E = C_D = C_p$ where $E = F_I/F_P =$ the Euler number; $C_D =$ the drag coefficient that represents the flow resistance; and $C_p =$ the pressure coefficient (dependent π term). Therefore, the minor head loss in pipe flow, h_f (or Δp), which represents F_V, is a result of Δp and thus C_D is mainly a function of E. Modeling additional dominant forces or dominant physical quantities results in additional secondary variables and will introduce an additional dimensionless number (π term) for each new secondary variable added. Using Table 11.8 and Section 8.6.4 as a guide, the absolute roughness, ε results in a secondary variable, ε, which introduces an independent π term, ε/L. And, the geometry, L_i ($i = 1, 2, ...$) results in an additional one or more secondary variables, L_i, which introduces an additional one or more independent π terms, L_i/L.

Step 2: Use the simplest mathematical expressions for the identified dominant physical forces and dominant physical quantities and identify the n physical quantities, all of which can be expressed in terms of m fundamental dimensions (F, L, and T).

Dominant Physical Force or Quantity	Primary Variable and Dimensions	Primary Variable and Dimensions	Primary Variable and Dimensions	Secondary Variable and Dimensions
$F_I = Ma = \rho v^2 L^2$	$\rho = [FL^{-4}T^2]$	$v = [LT^{-1}]$	$L = [L]$	
$h_f = \tau_w L / \gamma R_h = \Delta p / \gamma$				$\Delta p = [FL^{-2}]$
ε				$\varepsilon = [L]$
L_i				$L_i = [L]$

where $n = 6$ and $m = 3$.

Step 3: The number of dimensionless numbers (π terms) is $n - m = 6 - 3 = 3$.

Step 4: The Buckingham π theorem is applied in order to derive the $n - m = 3$ π terms as follows:

$$\pi_1 = \rho^{a_1} v^{b_1} L^{c_1} \Delta p^{d_1}$$

$$\pi_2 = \rho^{a_2} v^{b_2} L^{c_2} \varepsilon^{d_2}$$

$$\pi_3 = \rho^{a_3} v^{b_3} L^{c_3} L_i^{d_3}$$

where the exponents are unknown quantities to be determined by applying the Buckingham π theorem.

Working with the π_1 term, the dimensional equations for the *FLT* system are set up as follows:

$$\pi_1 = \rho^{a_1} v^{b_1} L^{c_1} \Delta p^{d_1}$$

$$F^0 L^0 T^0 = (FL^{-4}T^2)^{a_1} (LT^{-1})^{b_1} (L)^{c_1} (FL^{-2})^{d_1}$$

Equating the exponents for the dimensions and solving for three of the unknown exponents in terms of the fourth unknown exponent using the Mathcad symbolic solve block yields the following:

Given

F: $\quad 0 = a_1 + d_1$

L: $\quad 0 = -4a_1 + b_1 + c_1 - 2d_1$

T: $\quad 0 = 2a_1 - b_1$

Find $(a_1, b_1, c_1) \rightarrow \begin{pmatrix} -d_1 \\ -2 \cdot d_1 \\ 0 \end{pmatrix}$

Thus:

$$\pi_1 = \rho^{-d_1} v^{-2d_1} L^0 \Delta p^{d_1}$$

$$\pi_1 = \frac{\Delta p^{d_1}}{\rho^{d_1} v^{2d_1}}$$

From experimentation $d_1 = 1$; therefore:

$$\pi_1 = \frac{\Delta p}{\rho v^2}$$

which yields the empirical expression for the pressure drop, Δp.

Working with the π_2 term, the dimensional equations for the FLT system are set up as follows:

$$\pi_2 = \rho^{a_2} v^{b_2} L^{c_2} \varepsilon^{d_2}$$

$$F^0 L^0 T^0 = (FL^{-4}T^2)^{a_2} (LT^{-1})^{b_2} (L)^{c_2} (L)^{d_2}$$

Equating the exponents for the dimensions and solving for three of the unknown exponents in terms of the fourth unknown exponent using the Mathcad symbolic solve block yields the following:

Given

F: $0 = a_2$

L: $0 = -4a_2 + b_2 + c_2 + d_2$

T: $0 = 2a_2 - b_2$

$$\text{Find}(a_2, b_2, c_2) \rightarrow \begin{pmatrix} 0 \\ 0 \\ -d_2 \end{pmatrix}$$

Thus:

$$\pi_3 = \rho^0 v^0 L^{-d_2} \varepsilon^{d_2}$$

$$\pi_2 = \frac{\varepsilon^{d_2}}{L^{d_2}}$$

From experimentation $d_2 = 1$; therefore:

$$\pi_2 = \frac{\varepsilon}{L}$$

which is the relative roughness, and L is represented by D.

Working with the π_3 term, the dimensional equations for the FLT system are set up as follows:

$$\pi_3 = \rho^{a_3} v^{b_3} L^{c_3} L_i^{d_3}$$

$$F^0 L^0 T^0 = (FL^{-4}T^2)^{a_3}(LT^{-1})^{b_3}(L)^{c_3}(L)^{d_3}$$

Equating the exponents for the dimensions and solving for three of the unknown exponents in terms of the fourth unknown exponent using the Mathcad symbolic solve block yields the following:

Given

F: $\quad 0 = a_3$

L: $\quad 0 = -4a_3 + b_3 + c_3 + d_3$

T: $\quad 0 = 2a_3 - b_3$

Find $(a_3, b_3, c_3) \rightarrow \begin{pmatrix} 0 \\ 0 \\ -d_3 \end{pmatrix}$

Thus:

$$\pi_4 = \rho^0 v^0 L^{-d_3} L_i^{d_3}$$

$$\pi_3 = \frac{L_i^{d_3}}{L^{d_3}}$$

From experimentation $d_3 = 1$; therefore:

$$\pi_3 = \frac{L_i}{L}$$

which is the geometry.

The resulting functional relationship between the $n\text{-}m = 3$ π terms is expressed as follows:

$$\phi(\pi_1, \pi_2, \pi_3) = 0$$

$$\phi\left(\frac{\Delta p}{\rho v^2}, \frac{\varepsilon}{D}, \frac{L_i}{L}\right) = 0$$

or:

$$\frac{\Delta p}{\rho v^2} = \phi \left(\frac{\varepsilon}{D}, \frac{L_i}{L} \right)$$

Step 5: Interpret the resulting functional relationship between the $n-m=3$ π terms. Define the drag coefficient, C_D/minor head loss coefficient, k, and note the definition of the inertia force, F_I as follows:

$$\frac{\Delta p}{\rho v^2} = \frac{F_P}{F_I} = \frac{1}{E} = C_D = k = \phi \left(\frac{\varepsilon}{D}, \frac{L_i}{L} \right)$$

where the drag coefficient, C_D as modeled by the Euler number, E is the dependent π term, as it models the dependent variable, Δp that cannot be derived analytically from theory, while the relative surface roughness, ε/D and the geometry of the pipe device, L_i/L are the independent π terms. One may note that because the drag coefficient, C_D/minor head loss coefficient, k represents the flow resistance, which causes an unknown pressure drop, Δp, the pressure force will always play an important role and thus the definition of the drag coefficient, C_D is mainly a function of E. Thus, the loss coefficient, k in turbulent flow depends on the geometry (design details) of the flow device and in some cases on the relative pipe roughness, ε/D. Finally, dimensional analysis yields a nonstandard form of the drag coefficient/minor head coefficient, k, using inertia force, F_I in its definition, which yields the corresponding expression for the pressure drop, Δp as follows: $\Delta p = k\rho v^2$. Furthermore, the corresponding expressions for the wall shear stress is $\tau_w = \Delta p R_h/L = k\rho v^2 R_h/L$, and the minor head loss is $h_f = \tau_w L/\gamma R_h = \Delta p/\gamma = k\rho v^2/\gamma = kv^2/g$.

7.6 Modeling the Flow Resistance as a Loss in Flowrate in Internal Flow

One of the major applications of the study of fluid mechanics is the determination of the rate of flow of fluids. As a result, numerous flowrate measuring devices/flow meters have been developed, which differ depending upon whether the flow is through a pipe or through an open channel. The most common types of flow measurement devices used in pipe flow include pitot-static tubes, orifice meters, nozzle meters, and venturi meters, whereas, the most common types of flow measurement devices used in open channel flow include pitot-static tubes, sluice gates, weirs, overflow spillways, the Parshall flume (venturi flume), and contracted openings. Most flow measurement devices (flow meters) are basically velocimeters that directly measure an ideal pressure difference, which is related to the velocity by the Bernoulli equation; thus, the velocity is indirectly measured. Then, application of the continuity equation is used to determine the flowrate. Thus, many flow measurement devices used to measure the velocity and thus the flowrate for pipe and open channel flow apply the Bernoulli equation (assumes ideal flow) and are considered to be ideal flow meters. However, in real flow, the effects of viscosity, compressibility, and other real flow effects cause the actual velocity, v_a and thus the actual flowrate, Q_a to be less than the ideal velocity and the ideal flowrate, respectively. As a result, corrections for these real flow effects are made by applying a velocity or a discharge coefficient. The discharge coefficient, C_d represents both a velocity coefficient, which accounts for head loss due to friction, and a contraction coefficient, which accounts for the contraction of the streamlines of the fluid flow (and thus the actual cross-sectional area, A_a).

7.6.1 Evaluation of the Actual Discharge

In the case where the flow resistance is due to a flow measurement device for either pipe flow or open channel flow, the flow resistance (shear stress or drag force) is ultimately modeled as a reduced/actual discharge, Q_a in the differential form of the continuity equation because the actual flowrate is needed for the design of a flow-measuring device. Thus, the associated minor head loss is typically accounted for by the use/calibration of a discharge coefficient to determine the actual discharge. Evaluation of the actual discharge requires a "subset level" application of the governing equations. A "subset level" application of the governing equations focuses only on the given element causing the flow resistance. The assumption of "ideal" flow implies that the flow resistance is modeled only in the momentum equation (and thus, the subsequent assumption of "real" flow) (see Table 4.1). Thus, in the derivation of the actual discharge, first, ideal is assumed in order to determine an expression for the ideal velocity by applying the Bernoulli equation. Then, application of the continuity equation is used to determine an expression for the ideal discharge and the actual discharge. Finally, the integral form of the momentum equation (supplemented by dimensional analysis) is applied to solve for an unknown actual velocity, v_a and actual area, A_a. Thus, the flow resistance (shear stress or drag force) is ultimately modeled as a reduced/actual discharge, Q_a in the differential form of the continuity equation. The flow resistance causes an unknown pressure drop, Δp in the case of pipe flow and an unknown Δy in the case of open channel flow, which causes an unknown head loss, h_f, where the head loss is due to a conversion of kinetic energy to heat, which is modeled/displayed in the integral form of the energy equation. In the determination of the reduced/actual discharge, Q_a, which is less than the ideal discharge, Q_i, the exact reduction in the flowrate is unknown because the head loss, h_f causing it is unknown. Furthermore, although the differential form of the continuity equation is typically used to compute the reduced/actual discharge, Q_a for a flow-measuring device, because the head loss, h_f causes the reduced/actual discharge, Q_a, the head loss may also be accounted for as a minor loss in the integral form of the energy equation, as illustrated for pipe flow in Chapter 8; a minor head loss is not typically modeled in the energy equation for open channel flow. The assumption of ideal flow and thus applying the Bernoulli equation to measure the ideal velocity of flow, $v_i = \sqrt{2\Delta p/\rho}$ or $v_i = \sqrt{2g\Delta y}$, by using a flow-measuring device (see Chapters 8 and 9) yields an expression for ideal velocity as a function of an ideal pressure difference, Δp or Δy, for pipe and open channel flow, respectively, which is directly measured. Therefore, the associated minor head loss with the velocity measurement is accounted for by the discharge coefficient (in the application of the continuity equation).

The head loss, h_f causing the actual velocity, $v_a = \sqrt{(2/\rho)(\Delta p - (\tau_w L/R_h))}$ or $v_a = \sqrt{2g(\Delta y - (\tau_w L/\gamma R_h))}$, and the actual area, A_a and thus the actual discharge, Q_a is caused by both pressure and friction forces, where the reduced/actual discharge, Q_a is less than the ideal discharge, Q_i. However, because the friction/viscous forces (due to shear stress, τ_w) due to the internal flow measurement cannot be theoretically modeled in the integral momentum equation, the actual pressure drop, Δp or Δy cannot be analytically determined; thus, the reduction in the discharge, actual discharge, Q_a, which is less than the ideal discharge, Q_i, cannot be theoretically determined. Therefore, one cannot derive an analytical expression for the actual discharge, Q_a from the continuity equation. Specifically, the complex nature of the flow does not allow a theoretical modeling of the existence of the viscous force due to the flow measurement device in the momentum equation and thus does not allow a theoretical derivation of the pressure drop. Thus, the friction/viscous forces; the actual pressure drop, Δp or Δy; the minor head loss, h_f; and the actual reduced/actual discharge, Q_a are determined empirically using dimensional analysis. As such, one must resort to dimensional analysis (supplements the momentum theory) in order to derive an expression for the actual discharge, Q_a, which involves the definition of a discharge coefficient, C_d that represents the flow resistance. The flow resistance equation for the actual discharge is derived below and in Chapter 6. Furthermore, one may note that because the dominant forces and the important variables to consider in the dimensional analysis procedure to derive an empirical expression for Q_a differ for pipe flow versus open channel, these two derivations are done separately and are presented below in Sections 7.6.4.1 and 7.6.4.2, respectively. Actual discharges and the evaluation of the discharge coefficients are addressed in detail in Chapters 8 and 9.

7.6.2 Application of the Bernoulli Equation: Derivation of the Ideal Velocity

In the derivation of the actual discharge for a given flow-measuring device, first, ideal flow is assumed in order to determine an expression for the ideal velocity by applying the Bernoulli equation as follows:

$$\left(\frac{p}{\gamma} + z + \frac{v^2}{2g}\right)_1 - \left(\frac{p}{\gamma} + z + \frac{v^2}{2g}\right)_2 = 0 \tag{7.121}$$

Assuming $z_1 = z_2$, this equation has one unknown, which is the ideal velocity of the flow at the restriction in the flow-measuring device. The Bernoulli equation yields an expression for the ideal velocity, $v_i = \sqrt{2\Delta p/\rho}$ or $v_i = \sqrt{2g\Delta y}$, as a function of an ideal pressure difference, Δp or Δy, respectively, which is directly measured by the use of the flow-measuring device. Furthermore, the ideal velocity is used in the definition of the actual discharge below. Finally, the exact expression for the ideal velocity will depend upon the particular flow measurement device (see Chapters 8 and 9 for details).

7.6.3 Application of the Continuity Equation: Derivation of the Ideal Discharge and the Actual Discharge

In the derivation of the actual discharge for a given flow-measuring device, next, application of the differential continuity equation is used to determine an expression for the ideal discharge and the actual discharge as follows:

$$Q = vA \tag{7.122}$$

Thus, application of the continuity equation yields an expression for the ideal discharge and the actual discharge for pipe flow-measuring devices and open channel flow-measuring devices, respectively, as follows:

$$Q_i = v_i A_i \tag{7.123}$$

$$Q_a = \frac{v_a}{\sqrt{\dfrac{2\Delta p}{\rho}}} \frac{A_a}{A_i} Q_i \tag{7.124}$$

$$Q_a = \frac{v_a}{\sqrt{2g\Delta y}} \frac{A_a}{A_i} Q_i \tag{7.125}$$

7.6.4 Application of the Momentum Equation and Dimensional Analysis: Derivation of the Actual Velocity, Actual Area, Actual Discharge, and the Discharge Coefficient

In the derivation of the actual discharge for a given flow-measuring device, the subsequent assumption of real flow is made in order to determine an expression for the actual velocity, v_a by applying the integral momentum (supplemented by dimensional analysis) as follows:

$$\sum F_s = (F_G + F_P + F_V + F_{other})_s = (\rho Q v_s)_2 - (\rho Q v_s)_1 \tag{7.126}$$

Assuming $z_1 = z_2$, and that only the pressure and viscous forces are important, the momentum equation is used to solve for the actual velocity for pipe flow-measuring devices, and open channel flow-measuring devices, respectively, as follows:

$$v_a = \sqrt{\frac{2}{\rho}\left(\Delta p - \frac{\tau_w L}{R_h}\right)} \tag{7.127}$$

$$v_a = \sqrt{2g\left(\Delta y - \frac{\tau_w L}{\gamma R_h}\right)} \tag{7.128}$$

Substituting Equation 7.127 into Equation 7.124, and Equation 7.128 into Equation 7.125 for the actual velocity, respectively, yields the following flow resistance equations for the actual discharge for pipe flow-measuring devices and open channel flow-measuring devices, respectively, as follows:

$$Q_a = \frac{\sqrt{\dfrac{2}{\rho}\left(\Delta p - \dfrac{\tau_w L}{R_h}\right)}}{\sqrt{\dfrac{2\Delta p}{\rho}}} \frac{A_a}{A_i} Q_i \tag{7.129}$$

$$Q_a = \frac{\sqrt{2g\left(\Delta y - \dfrac{\tau_w L}{\gamma R_h}\right)}}{\sqrt{2g\Delta y}} \frac{A_a}{A_i} Q_i \tag{7.130}$$

However, the problem is that one cannot theoretically model the wall shear stress, τ_w in the integral momentum equation for flow-measuring devices. Thus, one cannot analytically determine pressure drop, Δp, or Δy and thus v_a, and cannot measure actual area, A_a in Equations 7.129 and 7.130. Thus, the integral momentum equation is supplemented with dimensional analysis in order to empirically model the wall shear stress, τ_w for the flow-measuring devices and thus Δp or Δy, v_a, and Q_a by the use of a discharge coefficient, C_d. The drag coefficient, C_D (discharge coefficient, C_d) represents the flow resistance and is empirically evaluated. The resulting flow resistance equations for the actual discharge for pipe flow-measuring devices and open channel flow-measuring devices are given, respectively, as follows:

$$Q_a = \frac{\sqrt{\frac{2}{\rho}\left(\Delta p - \frac{\tau_w L}{R_h}\right)}}{\sqrt{\frac{2\Delta p}{\rho}}} \frac{A_a}{A_i} Q_i = C_d Q_i \tag{7.131}$$

$$Q_a = \frac{\sqrt{2g\left(\Delta y - \frac{\tau_w L}{\gamma R_h}\right)}}{\sqrt{2g\Delta y}} \frac{A_a}{A_i} Q_i = C_d Q_i \tag{7.132}$$

where the discharge coefficient, C_d accounts for (indirectly models) the minor head loss associated with the ideal flow measurement and the actual discharge. Furthermore, one can easily compute v_i used in Equation 7.131 by measuring Δp in pipe flow-measuring device and can easily measure A_i. And, one can easily compute v_i used in Equation 7.132 by measuring Δy in open channel flow-measuring device and can easily measure A_i.

Finally, one may note that because the dominant forces and the important variables to consider in the dimensional analysis procedure to derive an empirical expression for Q_a differ for pipe flow versus open channel, these two derivations are done separately and are presented below in Sections 7.6.4.1 and 7.6.4.2, respectively.

7.6.4.1 Application of Dimensional Analysis: Derivation of the Reduced/Actual Discharge and Drag Coefficient for Pipe Flow Measuring Devices

An empirical expression for the actual discharge, Q_a for a pipe flow-measuring device in the theoretical expression for the actual discharge (Equation 7.124) in the differential form of the continuity equation is sought in terms of ideal discharge, Q_i and a flow resistance coefficient. This is accomplished by using dimensional analysis in order to derive an expression for the actual discharge, Q_a as a function of the ideal discharge, Q_i, which involves the definition of a drag coefficient C_D (discharge coefficient, C_d) that represents flow resistance and is empirically evaluated. Thus, the momentum theory is supplemented with dimensional analysis in order to derive an expression for the actual discharge, Q_a for pipe flow. Thus, assuming the most general flow conditions for the derivation of the actual discharge in pipe flow-measuring devices, Q_a, in addition to the inertia force, F_I and the pressure force, F_P (causes the viscous force for turbulent flow, $F_V = \tau A$, which is represented by the actual discharge, Q_a), other forces including the viscous force for laminar flow, $F_V = \tau A = \mu v L$ and other physical quantities including the geometry (design detail) of the pipe device also play an important role in the pipe fluid flow.

Dimensional analysis is applied in order to derive an empirical expression for the actual discharge, $Q_a = (v_a/\sqrt{2\Delta p/\rho})(A_a/A_i)Q_i$—and thus Δp (and A_a and v_a), which involves the definition of a drag coefficient, C_D/discharge coefficient, C_d as follows (see Table 7.6):

Step 1: Based on a general understanding of (and experience with and judgment of) the fluid flow phenomena identify, the dominant/significant physical forces and dominant physical quantities in the flow situation (use Table 11.8 for internal flow as a guide).

a. The inertia force, $F_I = Ma = \rho v^2 L^2$ will always be a dominant force and results in the three primary variables: ρ, v, and L.

b. The friction/viscous force for turbulent flow, F_f or $F_V = \tau A$ will always be the first/main additional dominant force, which in this case is represented by a dominant physical quantity: actual discharge, Q_a. The actual discharge is caused by the pressure force and represents an energy loss or some form of flow resistance and results in a secondary variable: Δp.

c. In pipe flow, the propulsive force for the fluid flow is provided by a pressure gradient, $\Delta p/L$, which results in an unknown Δp or S_f. Thus, in the derivation of the actual discharge in pipe flow, $Q_a = (v_a/\sqrt{2\Delta p/\rho})(A_a/A_i)Q_i$, the following four variables, ρ, $Q_a = v_a A_a$, L, and Δp, result in one dimensionless number, namely:

$$\frac{v_a}{\sqrt{\dfrac{\Delta p}{\rho}}} \frac{A_a}{A_i} = \sqrt{E}\frac{A_a}{A_i} = C_v C_c = C_D$$

where $E = F_I/F_P =$ the Euler number, $C_v =$ the velocity coefficient, $C_c =$ the contraction coefficient, and $C_D =$ the drag coefficient that represents the flow resistance (dependent π term).

d. Modeling additional dominant forces or dominant physical quantities results in additional secondary variables and will introduce an additional dimensionless number (π term) for each new secondary variable added.

   1. The viscous force for laminar flow, $F_V = \tau A = \mu v L$ results in a secondary variable, μ, which introduces an independent π term, R.

   2. The geometry, L_i ($i = 1, 2, \ldots$) results in one or more secondary variables, L_i, which introduces one or more independent π terms, L_i/L.

Step 2: Use the simplest mathematical expressions for the identified dominant physical forces and dominant physical quantities and identify the n physical quantities, all of which can be expressed in terms of m fundamental dimensions (F, L, and T).

Dominant Physical Force or Quantity	Primary Variable and Dimensions	Primary Variable and Dimensions	Primary Variable and Dimensions	Secondary Variable and Dimensions
$F_I = Ma = \rho v^2 L^2$	$\rho = [FL^{-4}T^2]$	$Q_a = [L^3 T^{-1}]$	$L = [L]$	
$Q_a = \dfrac{v_a}{\sqrt{2\Delta p/\rho}}\dfrac{A_a}{A_i}Q_i$				$\Delta p = [FL^{-2}]$
$F_V = \tau A = \mu v L$				$\mu = [FL^{-2}T]$
L_i				$L_i = [L]$

where $n = 6$ and $m = 3$.

Step 3: The number of dimensionless numbers (π terms) is $n-m = 6-3 = 3$.

Step 4: The Buckingham π theorem is applied in order to derive the $n-m = 3$ π terms as follows:

$$\pi_1 = \rho^{a_1} Q_a^{b_1} L^{c_1} \Delta p^{d_1}$$

$$\pi_2 = \rho^{a_2} Q_a^{b_2} L^{c_2} \mu^{d_2}$$

$$\pi_3 = \rho^{a_3} Q_a^{b_3} L^{c_3} L_i^{d_3}$$

where the exponents are unknown quantities to be determined by applying the Buckingham π theorem.

Working with the π_1 term, the dimensional equations for the *FLT* system are set up as follows:

$$\pi_1 = \rho^{a_1} Q_a^{b_1} L^{c_1} \Delta p^{d_1}$$

$$F^0 L^0 T^0 = (FL^{-4}T^2)^{a_1} (L^3 T^{-1})^{b_1} (L)^{c_1} (FL^{-2})^{d_1}$$

Equating the exponents for the dimensions and solving for three of the unknown exponents in terms of the fourth unknown exponent using the Mathcad symbolic solve block yields the following:

Given

F: $0 = a_1 + d_1$

L: $0 = -4a_1 + 3b_1 + c_1 - 2d_1$

T: $0 = 2a_1 - b_1$

$$\text{Find}(a_1, b_1, c_1) \rightarrow \begin{pmatrix} -d_1 \\ -2 \cdot d_1 \\ 4 \cdot d_1 \end{pmatrix}$$

Thus:

$$\pi_1 = \rho^{-d_1} Q_a^{-2d} L^{4d_1} \Delta p^{d_1}$$

$$\pi_1 = \frac{\Delta p^{d_1} L^{4d_1}}{\rho^{d_1} Q_a^{2d_1}}$$

From experimentation $d_1 = -1/2$; therefore:

$$\pi_1 = \frac{\sqrt{\rho} Q_a}{\sqrt{\Delta p L^2}}$$

$$\pi_1 = \frac{Q_a}{\sqrt{\frac{\Delta p}{\rho}} L^2}$$

which yields the empirical expression for the actual discharge in pipe flow, Q_a and where L^2 is represented by A_i, the ideal cross-sectional area.

Working with the π_2 term, the dimensional equations for the *FLT* system are set up as follows:

$$\pi_2 = \rho^{a_2} Q_a^{b_2} L^{c_2} \mu^{d_2}$$

$$F^0 L^0 T^0 = (FL^{-4}T^2)^{a_2} (L^3 T^{-1})^{b_2} (L)^{c_2} (FL^{-2}T)^{d_2}$$

Equating the exponents for the dimensions and solving for three of the unknown exponents in terms of the fourth unknown exponent using the Mathcad symbolic solve block yields the following:

Given

F: $\quad 0 = a_2 + d_2$

L: $\quad 0 = -4a_2 + 3b_2 + c_2 - 2d_2$

T: $\quad 0 = 2a_2 - b_2 + d_2$

Find $(a_2, b_2, c_2) \rightarrow \begin{pmatrix} -d_2 \\ -d_2 \\ d_2 \end{pmatrix}$

Thus:

$$\pi_2 = \rho^{-d_2} Q_a^{-d_2} L^{d_2} \mu^{d_2}$$

$$\pi_2 = \frac{L^{d_2} \mu^{d_2}}{\rho^{d_2} Q_a^{d_2}} = \frac{L^{d_2} \mu^{d_2}}{\rho^{d_2} v^{d_2} L^{2d_2}} = \frac{\mu^{d_2}}{\rho^{d_2} v^{d_2} L^{d_2}}$$

From experimentation $d_2 = -1$; therefore:

$$\pi_2 = \frac{\rho v L}{\mu}$$

which is the Reynolds number, $\mathbf{R}$ and where L is represented by D, the pipe diameter.

Working with the π_3 term, the dimensional equations for the FLT system are set up as follows:

$$\pi_3 = \rho^{a_3} Q_a^{b_3} L^{c_3} L_i^{d_3}$$

$$F^0 L^0 T^0 = (FL^{-4}T^2)^{a_3} (L^3 T^{-1})^{b_3} (L)^{c_3} (L)^{d_3}$$

Equating the exponents for the dimensions and solving for three of the unknown exponents in terms of the fourth unknown exponent using the Mathcad symbolic solve block yields the following:

Given

F: $0 = a_3$

L: $0 = -4a_3 + 3b_3 + c_3 + d_3$

T: $0 = 2a_3 - b_3$

Find $(a_3, b_3, c_3) \rightarrow \begin{pmatrix} 0 \\ 0 \\ -d_3 \end{pmatrix}$

Thus:

$$\pi_3 = \rho^0 Q_a^0 L^{-d_3} L_i^{d_3}$$

$$\pi_3 = \frac{L_i^{d_3}}{L^{d_3}}$$

From experimentation $d_3 = 1$; therefore:

$$\pi_3 = \frac{L_i}{L}$$

which is the geometry.

The resulting functional relationship between the $n-m = 3$ π terms is expressed as follows:

$$\phi(\pi_1, \pi_2, \pi_3) = 0$$

$$\phi\left(\frac{Q_a}{\sqrt{\frac{\Delta p}{\rho}} A_i}, \frac{\rho v D}{\mu}, \frac{L_i}{L}\right) = 0$$

$$\phi\left(\frac{Q_a}{Q_i}, R, \frac{L_i}{L}\right) = 0$$

or:

$$\frac{Q_a}{Q_i} = \phi\left(R, \frac{L_i}{L}\right)$$

Step 5: Interpret the resulting functional relationship between the $n-m=3$ π terms.
Define the discharge coefficient, C_d and note the definition of the inertia force, F_I as follows:

$$\frac{Q_a}{Q_i} = \frac{F_I}{F_P} = \frac{v_a A_a}{v_i A_i} = \frac{v_a}{\sqrt{\frac{\Delta p}{\rho}}} \frac{A_a}{A_i} = \sqrt{E}\frac{A_a}{A_i} = C_v C_c = C_d = \phi\left(R, \frac{L_i}{L}\right) \qquad (7.133)$$

where the velocity coefficient, $C_v = v_a/v_i = \sqrt{E}$. For ideal flow, the flow through the flow-measuring device (a constriction) is the conversion of flow energy (pressure energy) to kinetic energy. In real flow, the effects of viscosity, compressibility, and other real flow effects cause the actual velocity, v_a and thus the actual flowrate to be less than the ideal velocity, v_i and the ideal flowrate, respectively. The discharge coefficient, C_d is defined as the product of the velocity coefficient, C_v and the contraction coefficient, C_c as follows: $C_d = C_v C_c$, where $C_c = A_a/A_i$ (models the geometry of the flow-measuring device). As a result, corrections for these real flow effects are made by applying a velocity or a discharge coefficient, respectively. The discharge coefficient, C_d represents both a velocity coefficient, C_v, which accounts for head loss due to friction, and a contraction coefficient, C_c, which accounts for the contraction of the streamlines of the fluid flow (due to frictional pressure losses). Although the contraction coefficient, C_c and the velocity coefficient, C_v which model the two nonideal flow effects, may be individually experimentally determined, their combined effect is usually modeled by the discharge coefficient, C_d, which is determined experimentally and is given as follows: $C_d = C_v C_c$. The drag coefficient, C_D (discharge coefficient, C_d), as modeled by the Euler number, E, is the dependent π term, as it models the dependent variable, Δp that cannot be derived analytically from theory, while the Reynolds number, R and the geometry of the flow-measuring pipe device, L_i/L are the independent π terms. One may note that because the drag coefficient, C_D (discharge coefficient, C_d) represents the flow resistance, which causes an unknown pressure drop, Δp, the pressure force will always play an important role and thus the definition of the drag coefficient, C_D (discharge coefficient, C_d) is mainly a function of E. The discharge coefficient, C_d is a function of geometry of the flow-measuring device, L_i/L; the Reynolds number, R; the location of the pressure taps; and the conditions of the upstream and downstream flow in the pipe. For low to moderate values of the Reynolds number, R where the viscous effects are significant, the experimentally determined discharge coefficient, C_d models both the frictional effects (C_v,) and the contraction effects (C_c), whereas, for relatively high values of the Reynolds number, R, where the viscous effects are insignificant, the experimentally determined discharge coefficient, C_d models mostly the contraction effects (C_c). Thus, on-site calibration is usually recommended in addition to the manufacturer's data. One may note that for real flow, the pressure drop, $\Delta p = p_1 - p_2$ is a function of the flowrate, Q; the density,

ρ; the areas, A_1 and A_2; and the fluid viscosity, μ, which is modeled by the discharge coefficient, C_d.

Finally, one may note that dimensional analysis yields a nonstandard from of the velocity coefficient, C_v (and thus the discharge coefficient, C_d), using the inertia force, F_I in its definition, which yields the corresponding expression for the actual discharge, Q_a as follows:

$$Q_a = \underbrace{\left[\underbrace{\left(\frac{v_a}{\sqrt{\frac{\Delta p}{\rho}}} \right)}_{C_v} \underbrace{\left(\frac{A_a}{A_i} \right)}_{C_c} \right]}_{C_d} Q_i \qquad (7.134)$$

If, however, one considers the fact that inertia force, $F_I = \rho v^2 A$ may be further defined as the dynamic (pressure) force, $F_{dynpres} = \rho v^2 A/2$ (see Section 7.3.3.1 above on the drag force for details), then the velocity coefficient, C_v (and thus the discharge coefficient, C_d) would be defined as follows:

$$\frac{Q_a}{Q_i} = \frac{F_{dynpres}}{F_P} = \frac{v_a A_a}{v_i A_i} = \frac{v_a}{\sqrt{\frac{2\Delta p}{\rho}}} \frac{A_a}{A_i} = \sqrt{\frac{E}{2}} \frac{A_a}{A_i} = C_v C_c = C_d = \phi\left(R, \frac{L_i}{L} \right) \qquad (7.135)$$

which is the standard form of the velocity coefficient, C_v (and thus the discharge coefficient, C_d), resulting from replacing the inertia force, $F_I = \rho v^2 A$ used in the dimensional analysis, with the dynamic (pressure) force, $F_{dynpres} = \rho v^2 A/2$. Thus, the resulting actual discharge, Q_a derived from dimensional analysis is given as follows:

$$Q_a = \underbrace{\left[\underbrace{\left(\frac{v_a}{\sqrt{\frac{2\Delta p}{\rho}}} \right)}_{C_v} \underbrace{\left(\frac{A_a}{A_i} \right)}_{C_c} \right]}_{C_d} Q_i \qquad (7.136)$$

Furthermore, examining the discharge coefficient, C_d yields the following:

$$\frac{v}{\sqrt{\frac{2\Delta p}{\rho}}} \frac{A_a}{A_i} = \sqrt{\frac{E}{2}} \frac{A_a}{A_i} = C_v C_a = C_d = \frac{actual Q}{ideal Q} \qquad (7.137)$$

where the results of dimensional analysis introduce a drag coefficient, C_D/discharge coefficient, C_d, which expresses the ratio of a real variable (in this case: velocity or discharge) to its corresponding ideal flow variable. Finally, empirical evaluation of the discharge coefficient, C_d in flow measurement devices in pipe flow and application of the actual discharge equation for pipe flow, Q_a are presented in Chapter 8.

The application of dimensional analysis to derive the reduced/actual discharge and drag coefficient for a pipe flow-measuring device above was for the most general assumptions for flow through a pipe flow-measuring device (see Tables 11.8 and 11.9; see Sections 11.7.6 and 8.8.3), and resulted in Equation 7.133. Specifically, Equation 7.133 models the possibility of either laminar or turbulent flow through the device. The following example below will demonstrate the application of dimensional analysis to derive the reduced/actual discharge and drag coefficient for a pipe flow-measuring device for more specific assumptions for a pipe flow-measuring device.

EXAMPLE PROBLEM 7.5

Derive an empirical expression for the reduced/actual discharge, Q_a and drag coefficient, C_D for an orifice meter, assuming turbulent flow. The inertia force ($F_I = Ma = \rho v^2 L^2$), and the reduced/actual discharge, Q_a will always be important. Using Table 11.8, Section 11.7.6, and Section 8.8.3 as a guide, the additional dominant/significant physical quantities in the flow situation include Δp and L_i. Thus, the important physical quantities include: ρ, Q_a, L, Δp, and L_i. What are the important π terms used to study and describe the turbulent flow through an orifice meter? What are the corresponding expressions for the actual velocity, v_a and the actual pressure drop?

Mathcad Solution

In order to derive an empirical expression for the reduced/actual discharge, Q_a and drag coefficient, C_D for an orifice meter, assuming turbulent flow, the five steps of dimensional analysis are applied as follows:

Step 1: Using the given information, and Table 11.8 and Sections 11.7.6 and 8.8.3 as a guide, the dominant/significant physical forces and dominant physical quantities in the flow situation are identified as follows:

The inertia force, $F_I = Ma = \rho v^2 L^2$ will always be a dominant force and results in the three primary variables: ρ, v, and L. The friction/viscous force for turbulent flow, F_f or $F_V = \tau A$ will always be the first/main additional dominant force, which in this case is represented by a dominant physical quantity: actual discharge, Q_a. The actual discharge is caused by the pressure force and represents an energy loss or some form of flow resistance and results in a secondary variable: Δp. In pipe flow, the propulsive force for the fluid flow is provided by a pressure gradient, $\Delta p/L$, which results in an unknown Δp or S_f. Thus, in the derivation of the actual discharge in pipe flow, $Q_a = \dfrac{v_a}{\sqrt{2\Delta p/\rho}}\dfrac{A_a}{A_i}Q_i$, the following four variables,

ρ, $Q_a = v_a A_a$, L, and Δp, result in one dimensionless number, namely $\dfrac{v_a}{\sqrt{\Delta p/\rho}}\dfrac{A_a}{A_i} = \sqrt{E}\dfrac{A_a}{A_i} = C_v C_c = C_D$, where $E = F_I/F_P =$ the Euler number, $C_v =$ the velocity coefficient, $C_c =$ the contraction coefficient, and $C_D =$ the drag coefficient that represents the flow resistance (dependent π term). Modeling additional dominant forces or dominant physical quantities results in additional secondary variables and will introduce an additional dimensionless number (π term) for each new secondary variable added. Using Table 11.8 and Sections 11.7.6 and 8.8.3 as a guide, the geometry, L_i ($i = 1, 2, ...$) results in one or more secondary variables, L_i, which introduces one or more independent π terms, L_i/L.

Step 2: Use the simplest mathematical expressions for the identified dominant physical forces and dominant physical quantities and identify the n physical quantities, all of which can be expressed in terms of m fundamental dimensions (F, L, and T).

Dominant Physical Force or Quantity	Primary Variable and Dimensions	Primary Variable and Dimensions	Primary Variable and Dimensions	Secondary Variable and Dimensions
$F_I = Ma = \rho v^2 L^2$	$\rho = [FL^{-4}T^2]$	$Q_a = [L^3T^{-1}]$	$L = [L]$	
$Q_a = \dfrac{v_a}{\sqrt{\dfrac{2\Delta p}{\rho}}}\dfrac{A_a}{A_i}Q_i$				$\Delta p = [FL^{-2}]$
L_i				$L_i = [L]$

where $n = 5$ and $m = 3$.

Step 3: The number of dimensionless numbers (π terms) is $n-m = 5-3 = 2$.

Step 4: The Buckingham π theorem is applied in order to derive the $n-m = 2$ π terms as follows:

$$\pi_1 = \rho^{a_1} Q_a^{b_1} L^{c_1} \Delta p^{d_1}$$

$$\pi_2 = \rho^{a_2} Q_a^{b_2} L^{c_2} L_i^{d_2}$$

where the exponents are unknown quantities to be determined by applying the Buckingham π theorem.

Working with the π_1 term, the dimensional equations for the *FLT* system are set up as follows:

$$\pi_1 = \rho^{a_1} Q_a^{b_1} L^{c_1} \Delta p^{d_1}$$

$$F^0 L^0 T^0 = (FL^{-4}T^2)^{a_1}(L^3T^{-1})^{b_1}(L)^{c_1}(FL^{-2})^{d_1}$$

Equating the exponents for the dimensions and solving for three of the unknown exponents in terms of the fourth unknown exponent using the Mathcad symbolic solve block yields the following:

Given

F: $\quad 0 = a_1 + d_1$

L: $\quad 0 = -4a_1 + 3b_1 + c_1 - 2d_1$

T: $\quad 0 = 2a_1 - b_1$

Find $(a_1, b_1, c_1) \rightarrow \begin{pmatrix} -d_1 \\ -2 \cdot d_1 \\ 4 \cdot d_1 \end{pmatrix}$

Thus:

$$\pi_1 = \rho^{-d_1} Q_a^{-2d} L^{4d_1} \Delta p^{d_1}$$

$$\pi_1 = \frac{\Delta p^{d_1} L^{4d_1}}{\rho^{d_1} Q_a^{2d_1}}$$

From experimentation $d_1 = -1/2$; therefore:

$$\pi_1 = \frac{\sqrt{\rho} Q_a}{\sqrt{\Delta p L^2}}$$

$$\pi_1 = \frac{Q_a}{\sqrt{\frac{\Delta p}{\rho}} L^2}$$

which yields the empirical expression for the actual discharge in pipe flow, Q_a, where L^2 is represented by A_i, the ideal cross-sectional area.

Working with the π_2 term, the dimensional equations for the FLT system are set up as follows:

$$\pi_2 = \rho^{a_2} Q_a^{b_2} L^{c_2} L_i^{d_2}$$

$$F^0 L^0 T^0 = (FL^{-4}T^2)^{a_2}(L^3 T^{-1})^{b_2}(L)^{c_2}(L)^{d_2}$$

Equating the exponents for the dimensions and solving for three of the unknown exponents in terms of the fourth unknown exponent using the Mathcad symbolic solve block yields the following:

Given

F: $\quad 0 = a_2$

L: $\quad 0 = -4a_2 + 3b_2 + c_2 + d_2$

T: $\quad 0 = 2a_2 - b_2$

Find $(a_2, b_2, c_2) \rightarrow \begin{pmatrix} 0 \\ 0 \\ -d_2 \end{pmatrix}$

Thus:

$$\pi_2 = \rho^0 Q_a^0 L^{-d_2} L_i^{d_2}$$

$$\pi_2 = \frac{L_i^{d_2}}{L^{d_2}}$$

From experimentation $d_2 = 1$; therefore:

$$\pi_2 = \frac{L_i}{L}$$

which is the geometry.

The resulting functional relationship between the $n\text{-}m = 2\ \pi$ terms is expressed as follows:

$$\phi(\pi_1, \pi_2) = 0$$

$$\phi\left(\frac{Q_a}{\sqrt{\frac{\Delta p}{\rho}} A_i}, \frac{L_i}{L}\right) = 0$$

$$\phi\left(\frac{Q_a}{Q_i}, \frac{L_i}{L}\right) = 0$$

or:

$$\frac{Q_a}{Q_i} = \phi\left(\frac{L_i}{L}\right)$$

Step 5: Interpret the resulting functional relationship between the $n\text{-}m = 2\ \pi$ terms.

Define the discharge coefficient, C_d and note the definition of the inertia force, F_I as follows:

$$\frac{Q_a}{Q_i} = \frac{F_I}{F_P} = \frac{v_a A_a}{v_i A_i} = \frac{v_a}{\sqrt{\frac{\Delta p}{\rho}}} \frac{A_a}{A_i} = \sqrt{E} \frac{A_a}{A_i} = C_v C_c = C_d = \phi\left(\frac{L_i}{L}\right)$$

where the velocity coefficient, $C_v = v_a/v_i = \sqrt{E}$. For ideal flow, the flow through the flow-measuring device (a constriction) is the conversion of flow energy (pressure energy) to kinetic energy. In real flow, the effects of viscosity, compressibility, and other real flow effects cause the actual velocity, v_a and thus the actual flowrate to be less than the ideal velocity, v_i and the ideal flowrate, respectively. The discharge coefficient, C_d is defined as the product of the velocity coefficient, C_v and the contraction coefficient, C_c as follows: $C_d = C_v C_c$, where $C_c = A_a/A_i$ (models the geometry of the flow-measuring device). As a result, corrections for these real flow effects are made by applying a velocity or a discharge coefficient. The discharge coefficient, C_d represents both a velocity coefficient, C_v, which accounts for head loss due to friction, and a contraction coefficient, C_c, which accounts for the contraction of the streamlines of the fluid flow (due to frictional pressure losses). Although the contraction coefficient, C_c and the velocity coefficient, C_v, which model the two nonideal flow effects, may be individually experimentally determined, their combined effect is usually modeled by the discharge coefficient, C_d, which is determined experimentally and is given as follows: $C_d = C_v C_c$. The drag coefficient, C_D (discharge coefficient, C_d) as modeled by the Euler number, E is the dependent π term, as it models the dependent variable, Δp that cannot be derived analytically from theory, while the geometry of the flow-measuring pipe device, L_i/L is the independent π term. One may note that because the drag coefficient, C_D (discharge coefficient, C_d) represents the flow resistance that causes an unknown pressure drop, Δp, the pressure force will always play an important role and thus the definition of the drag coefficient, C_D (discharge coefficient, C_d) is mainly a function of E. The discharge coefficient, C_d is a function of geometry of the flow-measuring device, L_i/L, the location of the pressure taps, and the conditions of the upstream and downstream flow in the pipe. For relatively high values of the Reynolds number, R (assumption of turbulent flow), where the viscous effects are insignificant, the experimentally determined discharge coefficient, C_d models mostly the contraction effects (C_c). Finally, dimensional analysis yields a nonstandard from of the velocity coefficient, C_v (and thus the discharge coefficient, C_d), using the inertia force, F_I in its definition, which yields the corresponding expression for the actual discharge, Q_a as follows:

$$Q_a = \underbrace{\left[\underbrace{\left(\frac{v_a}{\sqrt{\frac{\Delta p}{\rho}}} \right)}_{C_v} \underbrace{\left(\frac{A_a}{A_i} \right)}_{C_c} \right]}_{C_d} Q_i.$$

An orifice meter is a velocimeter that directly measures an ideal pressure drop, Δp, and the ideal velocity, $v_i = \sqrt{\Delta p/\rho}$ is computed by applying the Bernoulli equation. Furthermore, the corresponding expressions for the actual velocity is $v_a = C_v\sqrt{\Delta p/\rho}$, and the actual pressure drop is $\Delta p = \rho v_a^2/C_v^2$.

7.6.4.2 Application of Dimensional Analysis: Derivation of the Reduced/Actual Discharge and Drag Coefficient for Open Channel Flow-Measuring Devices

An empirical expression for the actual discharge, Q_a for an open channel flow-measuring device in the theoretical expression for actual discharge (Equation 7.125) in the differential form of the continuity equation is sought in terms of ideal discharge, Q_i and a flow resistance coefficient. This is accomplished by using dimensional analysis in order to derive an expression for the actual discharge, Q_a as a function of the ideal discharge, Q_i, which involves the definition of a drag coefficient C_D (discharge coefficient, C_d) that represents flow resistance and is empirically evaluated. Thus, the momentum theory is supplemented with dimensional analysis in order to derive an expression for the actual discharge, Q_a for open channel flow. Thus, assuming the most general flow conditions for the derivation of the actual discharge in open channel flow-measuring devices, Q_a, in addition to the inertia force, F_I and the gravity force, $F_G = Mg$ (and the pressure force, $F_P = \Delta pA = \gamma \Delta yA$) (both of which cause the viscous force for turbulent flow, $F_V = \tau A$, which is represented by the actual discharge, Q_a), other forces including the viscous force for laminar flow, $F_V = \tau A = \mu vL$; the surface tension force, F_T; and other physical quantities including the geometry (design details) of the flow-measuring device also play an important role in the open channel fluid flow.

Dimensional analysis is applied in order to derive an empirical expression for the actual discharge, $Q_a = \dfrac{v_a}{\sqrt{2g\Delta y}}\dfrac{A_a}{A_i}Q_i$—and thus Δy (and A_a and v_a), which involves the definition of a drag coefficient, C_D/discharge coefficient, C_d as follows (see Table 7.7):

Step 1: Based on a general understanding of (and experience with and judgment of) the fluid flow phenomena, identify the dominant/significant physical forces and dominant physical quantities in the flow situation (use Table 11.8 for internal flow as a guide).

a. The inertia force, $F_I = Ma = \rho v^2 L^2$ will always be a dominant force and results in the three primary variables: ρ, v, and L (represented by the head of the weir or spillway, H).

b. The friction/viscous force for turbulent flow, F_f or $F_V = \tau A$ will always be the first/main additional dominant force, which in this case is represented by a dominant physical quantity: actual discharge, Q_a. The actual discharge is a result of both the pressure force and the gravity force and represents an energy loss or some form of flow resistance and results in a secondary variable: g.

c. In open channel flow, the propulsive force is provided by invoking gravity, g and thus by the weight of the fluid along the channel slope (channel gradient, $S_o = \Delta z/L$), which results in an unknown Δy or S_f. Thus, in the derivation of the actual discharge in open channel flow, $Q_a = \dfrac{v_a}{\sqrt{2g\Delta y}}\dfrac{A_a}{A_i}Q_i$, the following four variables, ρ, $Q_a = v_aA_a$, L, and g, result in one dimensionless number, namely:

$$\frac{v_a}{\sqrt{g\Delta y}}\frac{A_a}{A_i} = \sqrt{E}\frac{A_a}{A_i} = \sqrt{F}\frac{A_a}{A_i} = C_vC_c = C_D$$

where $E = F_I/F_P =$ the Euler number, $F = F_I/F_G =$ the Froude number, $C_v =$ the velocity coefficient, $C_c =$ the contraction coefficient, $C_D =$ the drag coefficient that represents the flow resistance, and $E = F$. Therefore, the actual discharge in open channel flow, Q_a (or h_f), which represents F_V, is a result of Δy and g; thus, C_D is mainly a function of $E = F$ (dependent π term). It is important to note that in the case of a flow-measuring device in open channel flow, the ideal velocity, $v_i = \sqrt{g\Delta y}$ equals the critical velocity, $v_c = \sqrt{gy_c}$ which results in $E = F$. As such, the flow through a flow-measuring device in open channel flow (and the resulting friction/viscous force, F_V) is a result of both the pressure force, $F_P = \Delta pA = \gamma\Delta yA$ and the gravitational force, $F_G = Mg$.

d. Modeling additional dominant forces or dominant physical quantities results in additional secondary variables and will introduce an additional dimensionless number (π term) for each new secondary variable added.

1. The viscous force for laminar flow, $F_V = \tau A = \mu vL$ results in a secondary variable, μ, which introduces an independent π term, R.

2. The surface tension force, F_T results in a secondary variable, σ, which introduces an independent π term, W.

3. The geometry, L_i ($i = 1, 2, \ldots$) results in one or more secondary variables, L_i, which introduces one or more independent π terms L/L_i.

Step 2: Use the simplest mathematical expressions for the identified dominant physical forces and dominant physical quantities and identify the n physical quantities, all of which can be expressed in terms of m fundamental dimensions (F, L, and T).

Dominant Physical Force or Quantity	Primary Variable and Dimensions	Primary Variable and Dimensions	Primary Variable and Dimensions	Secondary Variable and Dimensions
$F_I = Ma = \rho v^2 L^2$	$\rho = [FL^{-4}T^2]$	$Q_a = [L^3T^{-1}]$	$L = [L]$	
$Q_a = \dfrac{v_a}{\sqrt{2g\Delta y}} \dfrac{A_a}{A_i} Q_i$				$g = [LT^{-2}]$
$F_V = \tau A = \mu vL$				$\mu = [FL^{-2}T]$
$F_T = \sigma L$				$\sigma = [FL^{-1}]$
L_i				$L_i = [L]$

where $n = 7$ and $m = 3$.

Step 3: The number of dimensionless numbers (π terms) is $n\text{-}m = 7\text{-}3 = 4$.

Step 4: The Buckingham π theorem is applied in order to derive the $n\text{-}m = 3$ π terms as follows:

$$\pi_1 = \rho^{a_1} Q_a^{b_1} L^{c_1} g^{d_1}$$

$$\pi_2 = \rho^{a_2} Q_a^{b_2} L^{c_2} \mu^{d_2}$$

$$\pi_3 = \rho^{a_3} Q_a^{b_3} L^{c_3} \sigma^{d_3}$$

$$\pi_4 = \rho^{a_4} Q_a^{b_4} L^{c_4} L_i^{d_4}$$

where the exponents are unknown quantities to be determined by applying the Buckingham π theorem.

Working with the π_1 term, the dimensional equations for the *FLT* system are set up as follows:

$$\pi_1 = \rho^{a_1} Q_a^{b_1} L^{c_1} g^{d_1}$$

$$F^0 L^0 T^0 = (FL^{-4}T^2)^{a_1} (L^3 T^{-1})^{b_1} (L)^{c_1} (LT^{-2})^{d_1}$$

Equating the exponents for the dimensions and solving for three of the unknown exponents in terms of the fourth unknown exponent using the Mathcad symbolic solve block yields the following:

Given

F: $0 = a_1$

L: $0 = -4a_1 + 3b_1 + c_1 + d_1$

T: $0 = 2a_1 - b_1 - 2d_1$

Find $(a_1, b_1, c_1) \rightarrow \begin{pmatrix} 0 \\ -2 \cdot d_1 \\ 5 \cdot d_1 \end{pmatrix}$

Thus:

$$\pi_1 = \rho^0 Q_a^{-2d_1} L^{5d_1} g^{d_1}$$

$$\pi_1 = \frac{L^{5d_1} g^{d_1}}{Q_a^{2d_1}}$$

From experimentation $d_1 = -1/2$; therefore:

$$\pi_1 = \frac{Q_a}{\sqrt{g} L^{\frac{5}{2}}}$$

$$\pi_1 = \frac{Q_a}{\sqrt{g L L^2}}$$

which yields the empirical expression for the actual discharge in open channel flow, Q_a, where L^2 is represented by A_i, the ideal cross-sectional area, and where L is represented by Δy, the change in flow depth, or y_c, the critical depth of flow.

Working with the π_2 term, the dimensional equations for the *FLT* system are set up as follows:

$$\pi_2 = \rho^{a_2} Q_a^{b_2} L^{c_2} \mu^{d_2}$$

$$F^0 L^0 T^0 = (FL^{-4}T^2)^{a_2} (L^3 T^{-1})^{b_2} (L)^{c_2} (FL^{-2}T)^{d_2}$$

Equating the exponents for the dimensions and solving for three of the unknown exponents in terms of the fourth unknown exponent using the Mathcad symbolic solve block yields the following:

Given

F: $0 = a_2 + d_2$

L: $0 = -4a_2 + 3b_2 + c_2 - 2d_2$

T: $0 = 2a_2 - b_2 + d_2$

Find $(a_2, b_2, c_2) \rightarrow \begin{pmatrix} -d_2 \\ -d_2 \\ d_2 \end{pmatrix}$

Thus:

$$\pi_2 = \rho^{-d_2} Q_a^{-d_2} L^{d_2} \mu^{d_2}$$

$$\pi_2 = \frac{L^{d_2} \mu^{d_2}}{\rho^{d_2} Q_a^{d_2}} = \frac{L^{d_2} \mu^{d_2}}{\rho^{d_2} v^{d_2} L^{2d_2}} = \frac{\mu^{d_2}}{\rho^{d_2} v^{d_2} L^{d_2}}$$

From experimentation $d_2 = -1$; therefore:

$$\pi_2 = \frac{\rho v L}{\mu}$$

which is the Reynolds number, R and where L in open channel flow is typically represented by R_h, the hydraulic radius. However, one may note that in the case of a weir or a spillway, L is represented by H—the head of the weir or spillway. Furthermore, in the case of a sluice gate, L is represented by y, the upstream or downstream depth.

Working with the π_3 term, the dimensional equations for the FLT system are set up as follows:

$$\pi_3 = \rho^{a_3} Q_a^{b_3} L^{c_3} \sigma^{d_3}$$

$$F^0 L^0 T^0 = (FL^{-4}T^2)^{a_3} (L^3 T^{-1})^{b_3} (L)^{c_3} (FL^{-1})^{d_3}$$

Equating the exponents for the dimensions and solving for three of the unknown exponents in terms of the fourth unknown exponent using the Mathcad symbolic solve block yields the following:

Given

F: $0 = a_3 + d_3$

L: $0 = -4a_3 + 3b_3 + c_3 - 2d_3$

T: $0 = 2a_3 - b_3$

Find $(a_3, b_3, c_3) \rightarrow \begin{pmatrix} -d_3 \\ -2 \cdot d_3 \\ 3 \cdot d_3 \end{pmatrix}$

Thus:

$$\pi_3 = \rho^{-d_3} Q_a^{-2d_3} L^{3d_3} \sigma^{d_3}$$

$$\pi_3 = \frac{L^{3d_3} \sigma^{d_3}}{\rho^{d_3} Q_a^{2d_3}} = \frac{L^{3d_3} \sigma^{d_3}}{\rho^{d_3} v^{2d_3} L^{4d_3}} = \frac{\sigma^{d_3}}{\rho^{d_3} v^{2d_3} L^{d_3}}$$

From experimentation $d_3 = -1$; therefore:

$$\pi_3 = \frac{\rho v^2 L}{\sigma}$$

which is the Weber number, **W**. One may note that in the case of a weir or a spillway, L represents H, the head of the weir or spillway.

Working with the π_4 term, the dimensional equations for the *FLT* system are set up as follows:

$$\pi_4 = \rho^{a_4} Q_a^{b_4} L^{c_4} L_i^{d_4}$$

$$F^0 L^0 T^0 = (FL^{-4}T^2)^{a_4} (L^3 T^{-1})^{b_4} (L)^{c_4} (L)^{d_4}$$

Equating the exponents for the dimensions and solving for three of the unknown exponents in terms of the fourth unknown exponent using the Mathcad symbolic solve block yields the following:

Given

F: $0 = a_4$

L: $0 = -4a_4 + 3b_4 + c_4 + d_4$

T: $0 = 2a_4 - b_4$

Find $(a_4, b_4, c_4) \rightarrow \begin{pmatrix} 0 \\ 0 \\ -d_4 \end{pmatrix}$

Thus:

$$\pi_4 = \rho^0 Q_a^0 L^{-d_4} L_i^{d_4}$$

$$\pi_4 = \frac{L_i^{d_4}}{L^{d_4}}$$

From experimentation $d_4 = 1$; therefore:

$$\pi_4 = \frac{L_i}{L}$$

which is the geometry. One may note that in the case of a weir or a spillway, L is represented by P, the height of the weir or spillway, and L_i is represented by $H \approx y_c$, the head of the weir or spillway. Thus, $\pi_4 = L_i/L = H/P$ in the case of a weir or spillway. Furthermore, in the case of a sluice gate, L is represented by the opening of the sluice gate, a and L_i is represented by the upstream depth, y_1 and the downstream depth, y_2. Thus, $\pi_4 = L_1/L = y_1/a$ and $\pi_4 = L_2/L = y_2/a$ in the case of a sluice gate.

The resulting functional relationship between the $n-m = 4$ π terms is expressed as follows:

$$\phi(\pi_1, \pi_2, \pi_3, \pi_4) = 0$$

$$\phi\left(\frac{Q_a}{\sqrt{gL}A_i}, \frac{\rho v L}{\mu}, \frac{\rho v^2 L}{\sigma}, \frac{L_i}{L}\right) = 0$$

$$\phi\left(\frac{Q_a}{Q_i}, R, W, \frac{L_i}{L}\right) = 0$$

or:

$$\frac{Q_a}{Q_i} = \phi\left(R, W, \frac{L_i}{L}\right)$$

Step 5: Interpret the resulting functional relationship between the $n-m = 4$ π terms. One may note that in the empirical expression for the actual discharge in open channel flow, Q_a (i.e., the π_1 term), L is represented by Δy, the change in flow depth, or y_c, the critical depth of flow. Define the discharge coefficient, C_d and note the definition of the inertia force, F_I as follows:

$$\frac{Q_a}{Q_i} = \frac{F_I}{F_P} = \frac{v_a A_a}{v_i A_i} = \underbrace{\frac{v_a}{\sqrt{g\Delta y}}}_{v_i = v_c} \frac{A_a}{A_i} = \sqrt{E}\frac{A_a}{A_i} = C_v C_c = C_d = \phi\left(R, W, \frac{L_i}{L}\right) \qquad (7.138)$$

where:

$$\sqrt{F} = \frac{v}{\sqrt{gL}} = \frac{v}{\sqrt{gy_c}} = \frac{v_a}{v_c} = \frac{v_a}{v_i} = C_v = \sqrt{E} \qquad (7.139)$$

thus:

$$\frac{Q_a}{Q_i} = \frac{v_a A_a}{v_i A_i} = \frac{v_a}{\sqrt{g\Delta y}}\frac{A_a}{A_i} = \sqrt{E}\frac{A_a}{A_i} = \sqrt{F}\frac{A_a}{A_c} = C_v C_c = C_d = \phi\left(R, W, \frac{L_i}{L}\right) \qquad (7.140)$$

where the ideal velocity, v_i equals the critical velocity, v_c, the ideal area, A_i equals the critical area, A_c, and Q_i will vary for the various flow-measuring devices.

The velocity coefficient, $C_v = \dfrac{v_a}{v_i} = \sqrt{E} = \dfrac{v_a}{v_c} = \sqrt{F}$. For ideal flow, the flow through the flow-measuring device (a constriction) is the conversion of flow energy (pressure energy) to kinetic energy. In real flow, the effects of viscosity, compressibility, and other real flow effects cause the actual velocity, v_a and thus the actual flowrate to be less than the ideal velocity, v_i and the ideal flowrate, respectively. The discharge coefficient, C_d is defined as the product of the velocity coefficient, C_v and the contraction coefficient, C_c as follows: $C_d = C_v C_c$ where $C_c = A_a/A_i = A_a/A_c$ (models the geometry of the flow-measuring device). As a result, corrections for these real flow effects are made by applying a velocity or a discharge coefficient, respectively. The discharge coefficient, C_d represents both a velocity coefficient, C_v, which accounts for head loss due to friction, and a contraction coefficient, C_c, which accounts for the contraction of the streamlines of the fluid flow (due to frictional pressure losses). Although the contraction coefficient, C_c and the velocity coefficient, C_v, which model the two nonideal flow effects, may be individually experimentally determined, their combined effect is usually modeled by the discharge coefficient, C_d, which is determined experimentally and is given as follows: $C_d = C_v C_c$. The drag coefficient, C_D (discharge coefficient, C_d) as modeled by the Euler number, $E =$ the Froude number, F is the dependent π term, as it models the dependent variable, Δy or S_f that cannot be derived analytically from theory, while the Reynolds number, R; the Weber number, W; and the geometry of the flow-measuring device, L_i/L are the independent π terms. One may note that the drag coefficient, C_D (discharge coefficient, C_d) represents the flow resistance which causes an unknown pressure drop, Δy and an unknown S_f. As such, both the pressure force and the gravitational force will always play an important role and thus the definition of the drag coefficient, C_D (discharge coefficient, C_d) is mainly a function of E and F, where $E = F$.

Finally, one may note that dimensional analysis yields a nonstandard from of the velocity coefficient, C_v (and thus the discharge coefficient, C_d), using the inertia force, F_I in its definition, which yields the corresponding expression for the actual discharge, Q_a as follows:

$$Q_a = \left[\underbrace{\left(\frac{v_a}{\sqrt{g\Delta y}} \right)}_{C_v} \underbrace{\left(\frac{A_a}{A_i} \right)}_{C_c} \right] Q_i \qquad (7.141)$$

$$\underbrace{\phantom{\left[\left(\frac{v_a}{\sqrt{g\Delta y}} \right) \left(\frac{A_a}{A_i} \right) \right]}}_{C_d}$$

If, however, one considers the fact that inertia force, $F_I = \rho v^2 A$ may be further defined as the dynamic (pressure) force, $F_{dynpres} = \rho v^2 A/2$ (see Section 7.3.3.1 above on drag force for details), then the velocity coefficient, C_v (and thus the discharge coefficient, C_d) would be defined as follows:

$$\frac{Q_a}{Q_i} = \frac{F_{dynpres}}{F_P} = \frac{v_a A_a}{v_i A_i} = \frac{v_a}{\sqrt{2g\Delta y}} \frac{A_a}{A_i} = \sqrt{\frac{E}{2}} \frac{A_a}{A_i} = \sqrt{\frac{F}{2}} \frac{A_a}{A_i} = C_v C_c = C_d = \phi \left(R, W, \frac{L_i}{L} \right)$$

$$(7.142)$$

which is the standard form of the velocity coefficient, C_v (and thus the discharge coefficient, C_d), resulting from replacing the inertia force, $F_I = \rho v^2 A$ used in the dimensional analysis with the dynamic (pressure) force, $F_{dynpres} = \rho v^2 A / 2$. Thus, the resulting actual discharge, Q_a derived from dimensional analysis is given as follows:

$$Q_a = \left[\underbrace{\left(\frac{v_a}{\sqrt{2g\Delta y}} \right)}_{C_v} \underbrace{\left(\frac{A_a}{A_i} \right)}_{C_c} \right] Q_i \qquad (7.143)$$

$$\underbrace{\phantom{\left[\left(\frac{v_a}{\sqrt{2g\Delta y}} \right) \left(\frac{A_a}{A_i} \right) \right]}}_{C_d}$$

Furthermore, examining the discharge coefficient, C_d yields the following:

$$\frac{v}{\sqrt{2g\Delta y}} \frac{A_a}{A_i} = \sqrt{\frac{E}{2}} \frac{A_a}{A_i} = \sqrt{\frac{F}{2}} \frac{A_a}{A_i} = C_v C_a = C_d = \frac{actualQ}{idealQ} \qquad (7.144)$$

where the results of dimensional analysis introduce a drag coefficient, C_D/discharge coefficient, C_d which expresses the ratio of a real variable (in this case: velocity or discharge) to its corresponding ideal flow variable. Finally, empirical evaluation of the discharge coefficient, C_d in flow measurement devices in open channel flow and application of the actual discharge equation for open channel flow, Q_a are presented in Chapter 9.

7.6.4.3 Application of Dimensional Analysis to Derive the Reduced/Actual Discharge and Drag Coefficient for More Specific Assumptions for Open Channel Flow-Measuring Devices

The above application of dimensional analysis to derive an empirical expression for the actual discharge, Q_a for an open channel flow-measuring device (Equation 7.140) was for the most general flow conditions for open channel flow-measuring devices; it was assumed that the head on the weir or spillway may be either high or low. As a result, while the drag coefficient, C_D (discharge coefficient, C_d), in general, is a mainly a function of the Euler number, E = the Froude number, F and the geometry of the flow-measuring device, L_i/L, it will selectively vary with the Reynolds number, R and the Weber number, W depending mainly on the magnitude of the head on the weir or spillway (see Tables 11.8 and 11.9; see Sections 9.17 and 11.7.7).

EXAMPLE PROBLEM 7.6

Derive an empirical expression for the reduced/actual discharge, Q_a and drag coefficient, C_D for a sluice gate. The inertia force ($F_I = Ma = \rho v^2 L^2$), and the reduced/actual discharge, Q_a will always be important. Using Table 11.8 and Sections 11.7.7 and 9.17 as a guide, the additional dominant/significant physical quantities in the flow situation include g and L_i. Thus, the important physical quantities include: ρ, Q_a, L, g, and L_i. What are the important π terms used to study and describe this flow process for flow through a sluice gate?

Mathcad Solution

In order to derive an empirical expression for the reduced/actual discharge, Q_a and drag coefficient, C_D for a sluice, the five steps of dimensional analysis are applied as follows:

Step 1: Using the given information, and Table 11.8 and Section 11.7.7 and 9.17 as a guide, the dominant/significant physical forces and dominant physical quantities in the flow situation are identified as follows:

The inertia force, $F_I = Ma = \rho v^2 L^2$ will always be a dominant force and results in the three primary variables: ρ, v, and L (represented by the opening of the sluice gate, a). The friction/viscous force for turbulent flow, F_f or $F_V = \tau A$ will always be the first/main additional dominant force, which in this case is represented by a dominant physical quantity: actual discharge, Q_a. The actual discharge is a result of both the pressure force and the gravity force and represents an energy loss or some form of flow resistance and results in a secondary variable: g. In open channel flow, the propulsive force is provided by invoking gravity, g and thus by the weight of the fluid along the channel slope (channel gradient, $S_o = \Delta z/L$), which results in an unknown Δy or S_f. Thus in the derivation of the actual discharge in open channel flow, $Q_a = \dfrac{v_a}{\sqrt{2g\Delta y}}\dfrac{A_a}{A_i} Q_i$, the following four variables, ρ, $Q_a = v_a A_a$, L, and g, result in one dimensionless number, namely:

$$\frac{v_a}{\sqrt{g\Delta y}}\frac{A_a}{A_i} = \sqrt{E}\frac{A_a}{A_i} = \sqrt{F}\frac{A_a}{A_i} = C_v C_c = C_D$$

where $E = F_I/F_P =$ the Euler number, $F = F_I/F_G =$ the Froude number, $C_v =$ the velocity coefficient, $C_c =$ the contraction coefficient, $C_D =$ the drag coefficient that represents the flow resistance, and $E = F$. Therefore, the actual discharge in open channel flow, Q_a (or h_f), which represents F_V, is a result of Δy and g and thus C_D is mainly a function of $E = F$ (dependent π term). It is important to note that in the case of a flow-measuring device in open channel flow, the ideal velocity, $v_i = \sqrt{g\Delta y}$ equals the critical velocity, $v_c = \sqrt{gy_c}$, which results in $E = F$. As such, the flow through a flow-measuring device in open channel flow (and the resulting friction/viscous force, F_V) is a result of both the pressure force, $F_P = \Delta p A = \gamma \Delta y A$ and the gravitational force, $F_G = Mg$. Modeling additional dominant forces or dominant physical quantities results in additional secondary variables and will introduce an additional dimensionless number (π term) for each new secondary variable added. Using Table 11.8 and Sections 11.7.7 and 9.17 as a guide, the geometry, L_i ($i = 1, 2, \ldots$) results in one or more secondary variables, L_i, which introduces one or more independent π terms L/L_i.

Step 2: Use the simplest mathematical expressions for the identified dominant physical forces and dominant physical quantities and identify the n physical quantities, all of which can be expressed in terms of m fundamental dimensions (F, L, and T).

Dominant Physical Force or Quantity	Primary Variable and Dimensions	Primary Variable and Dimensions	Primary Variable and Dimensions	Secondary Variable and Dimensions
$F_I = Ma = \rho v^2 L^2$	$\rho = [FL^{-4}T^2]$	$Q_a = [L^3T^{-1}]$	$L = [L]$	
$Q_a = \dfrac{v_a}{\sqrt{2g\Delta y}}\dfrac{A_a}{A_i}Q_i$				$g = [LT^{-2}]$
L_i				$L_i = [L]$

where $n = 5$ and $m = 3$.

Step 3: The number of dimensionless numbers (π terms) is $n\text{-}m = 5\text{-}3 = 2$.

Step 4: The Buckingham π theorem is applied in order to derive the $n\text{-}m = 2$ π terms as follows:

$$\pi_1 = \rho^{a_1} Q_a^{b_1} L^{c_1} g^{d_1}$$

$$\pi_2 = \rho^{a_2} Q_a^{b_2} L^{c_2} L_i^{d_2}$$

where the exponents are unknown quantities to be determined by applying the Buckingham π theorem.

Working with the π_1 term, the dimensional equations for the *FLT* system are set up as follows:

$$\pi_1 = \rho^{a_1} Q_a^{b_1} L^{c_1} g^{d_1}$$

$$F^0 L^0 T^0 = (FL^{-4}T^2)^{a_1}(L^3T^{-1})^{b_1}(L)^{c_1}(LT^{-2})^{d_1}$$

Equating the exponents for the dimensions and solving for three of the unknown exponents in terms of the fourth unknown exponent using the Mathcad symbolic solve block yields the following:

Given

F: $0 = a_1$

L: $0 = -4a_1 + 3b_1 + c_1 + d_1$

T: $0 = 2a_1 - b_1 - 2d_1$

Find $(a_1, b_1, c_1) \rightarrow \begin{pmatrix} 0 \\ -2 \cdot d_1 \\ 5d_1 \end{pmatrix}$

Thus:

$$\pi_1 = \rho^0 Q_a^{-2d_1} L^{5d_1} g^{d_1}$$

$$\pi_1 = \frac{L^{5d_1} g^{d_1}}{Q_a^{2d_1}}$$

From experimentation $d_1 = -1/2$; therefore:

$$\pi_1 = \frac{Q_a}{\sqrt{g}L^{\frac{5}{2}}}$$

$$\pi_1 = \frac{Q_a}{\sqrt{gL}L^2}$$

which yields the empirical expression for the actual discharge in open channel flow, Q_a, where L^2 is represented by A_i, the ideal cross-sectional area, and where L is represented by Δy, the change in flow depth, or y_c, the critical depth of flow.

Working with the π_2 term, the dimensional equations for the *FLT* system are set up as follows:

$$\pi_2 = \rho^{a_2} Q_a^{b_2} L^{c_2} L_i^{d_2}$$

$$F^0 L^0 T^0 = (FL^{-4}T^2)^{a_2}(L^3 T^{-1})^{b_2}(L)^{c_2}(L)^{d_2}$$

Equating the exponents for the dimensions and solving for three of the unknown exponents in terms of the fourth unknown exponent using the Mathcad symbolic solve block yields the following:

Given

F: $0 = a_2$

L: $0 = -4a_2 + 3b_2 + c_2 + d_2$

T: $0 = 2a_2 - b_2$

$$\text{Find } (a_2, b_2, c_2) \rightarrow \begin{pmatrix} 0 \\ 0 \\ -d_2 \end{pmatrix}$$

Thus:

$$\pi_2 = \rho^0 Q_a^0 L^{-d_2} L_i^{d_2}$$

$$\pi_2 = \frac{L_i^{d_2}}{L^{d_2}}$$

From experimentation $d_2 = 1$; therefore:

$$\pi_2 = \frac{L_i}{L}$$

which is the geometry. One may note that in the case of a sluice gate, L is represented by the opening of the sluice gate, a and L_i is represented by the upstream depth, y_1 and the downstream depth, y_2. Thus, $\pi_2 = L_1/L = y_1/a$ and $\pi_2 = L_2/L = y_2/a$ in the case of a sluice gate.

The resulting functional relationship between the n-m = 2 π terms is expressed as follows:

$$\phi(\pi_1, \pi_2) = 0$$

$$\phi\left(\frac{Q_a}{\sqrt{gLA_i}}, \frac{L_i}{L}\right) = 0$$

$$\phi\left(\frac{Q_a}{Q_i}, \frac{L_i}{L}\right) = 0$$

or:

$$\frac{Q_a}{Q_i} = \phi\left(\frac{L_i}{L}\right)$$

Step 5: Interpret the resulting functional relationship between the n-m = 2 π terms. One may note that in the empirical expression for the actual discharge in open channel flow, Q_a (i.e., the π_1 term), L is represented by Δy, the change in flow depth or y_c, the critical depth of flow. Define the discharge coefficient, C_d and note the definition of the inertia force, F_I as follows:

$$\frac{Q_a}{Q_i} = \frac{F_I}{F_P} = \frac{v_a A_a}{v_i A_i} = \underbrace{\frac{v_a}{\sqrt{g\Delta y}}}_{v_i = v_c} \frac{A_a}{A_i} = \sqrt{E}\frac{A_a}{A_i} = C_v C_c = C_d = \phi\left(\frac{L_i}{L}\right)$$

where:

$$\sqrt{F} = \frac{v}{\sqrt{gL}} = \frac{v}{\sqrt{gy_c}} = \frac{v_a}{v_c} = \frac{v_a}{v_i} = C_v = \sqrt{E},$$

and thus:

$$\frac{Q_a}{Q_i} = \frac{v_a A_a}{v_i A_i} = \frac{v_a}{\sqrt{g\Delta y}}\frac{A_a}{A_i} = \sqrt{E}\frac{A_a}{A_i} = \sqrt{F}\frac{A_a}{A_c} = C_v C_c = C_d = \phi\left(\frac{L_i}{L}\right)$$

where the ideal velocity, v_i equals the critical velocity, v_c, and the ideal area, A_i equals the critical area, A_c, and Q_i will vary for the particular flow-measuring device (sluice gate). The velocity coefficient, $C_v = \frac{v_a}{v_i} = \sqrt{E} = \frac{v_a}{v_c} = \sqrt{F}$. For ideal flow, the flow through the flow-measuring device (sluice gate) (a constriction) is the conversion of flow energy (pressure energy) to kinetic energy. In real flow, the effects of viscosity, compressibility, and other real flow effects cause the actual velocity, v_a and thus the actual flowrate to be less than the ideal velocity, v_i and the ideal flowrate, respectively. The discharge coefficient, C_d is defined as the product of the velocity coefficient, C_v and the contraction coefficient, C_c as follows: $C_d = C_v C_c$ where $C_c = A_a/A_i = A_a/A_c$ (models the geometry of the flow-measuring device). As a result, corrections for these real flow effects are made by applying a velocity or a discharge coefficient, respectively. The discharge coefficient, C_d represents both a velocity coefficient, C_v, which accounts for head loss due to friction, and a contraction coefficient, C_c, which accounts for the contraction of the streamlines of the fluid flow (due to frictional pressure losses). Although the contraction coefficient, C_c and the velocity coefficient, C_v, which model the two nonideal flow effects may be individually experimentally determined, their combined effect is usually modeled by the discharge coefficient, C_d, which is determined experimentally and is given as follows: $C_d = C_v C_c$. The drag coefficient, C_D (discharge coefficient, C_d) as modeled by the Euler number, E = the Froude number, F is the dependent π term, as it models the dependent variable, Δy or S_f that cannot be derived analytically from theory, while the geometry of the flow-measuring device, L_i/L is the independent π term. One may note that the drag coefficient, C_D (discharge coefficient, C_d) represents the flow resistance which causes an unknown pressure drop, Δy and an unknown S_f. As such, both the pressure force and the gravitational force will always play an important role and thus the definition of the drag coefficient, C_D (discharge coefficient, C_d) is mainly a function of E and F, where $E = F$. Finally, dimensional analysis yields a nonstandard from of the velocity coefficient, C_v (and thus, the discharge coefficient, C_d), using the inertia force, F_I in its definition, which yields the corresponding expression for the actual discharge, Q_a as follows:

$$Q_a = \left[\underbrace{\left(\frac{v_a}{\sqrt{g \Delta y}} \right)}_{C_v} \underbrace{\left(\frac{A_a}{A_i} \right)}_{C_c} \right] Q_i.$$
$$\underbrace{}_{C_d}$$

A sluice gate is a velocimeter that directly measures an ideal pressure drop, Δy and the ideal velocity, $v_i = \sqrt{g \Delta y}$ is computed by applying the Bernoulli equation. Furthermore, the corresponding expressions for the actual velocity is $v_a = C_v \sqrt{g \Delta y}$ and the actual pressure drop is $\Delta y = v_a^2/g C_v^2$.

7.7 Modeling the Flow Resistance as a Loss in Pump and Turbine Efficiency in Internal Flow

In the cases where the flow resistance is due to pipe or channel friction (major head loss), pipe component (minor head loss), and/or a flow-measuring device (minor head loss), the flow resistance contributes to the power losses and thus the determination of the efficiency of pumps and turbines (see Chapter 4). In particular, the total head loss, h_f (major and minor) will increase the amount of power that a pump must deliver to the fluid; the head loss, h_f represents the additional height that the fluid needs to be elevated by the pump in order to overcome the losses due to the flow resistance in the pipe or channel. Some of the pumping power will be used to pump the fluid through the pipe or channel, while some will be used to compensate for the energy lost in the pressure drop, Δp caused by the pipe or channel friction, pipe component, or flow-measuring device. A similar analogy may be deduced for evaluating the efficiency of a turbine. Although the derivations of the expressions for the efficiency of pumps, η_{pump} and turbines, $\eta_{turbine}$ are determined analytically in Chapter 4, dimensional analysis is applied in order to derive empirical expressions for the efficiency of pumps and turbines. Empirical expressions for the efficiency of pumps and turbines will allow their respective empirical evaluations. Because a turbine is the exact opposite of a pump, an empirical derivation of the efficiency of a pump is actually conducted, and the empirical derivation of the efficiency of a turbine is interpreted from the results.

7.7.1 Evaluation of the Efficiency of Pumps and Turbines

The efficiency of pumps and turbines is discussed in detail in Chapter 4. The efficiency of a pump, $\eta_{pump} = (P_p)_{out}/(P_p)_{in} = \gamma Q h_{pump}/\omega T_{shaft,in}$ (Equation 4.197) expresses the ratio of a real/actual flow variable (hydraulic power output from the pump) to its corresponding ideal flow variable (shaft power input to the pump). And, the efficiency of a turbine, $\eta_{turbine} = (P_t)_{out}/(P_t)_{in} = \omega T_{shaft,out}/\gamma Q h_{turbine}$ (Equation 4.217) expresses the ratio of a real/actual flow variable (shaft power output from the turbine) to its corresponding ideal flow variable (hydraulic power input to the turbine). Dimensional analysis is applied in order to derive empirical expressions for the efficiency of pumps and turbines, which will allow their respective empirical evaluations.

7.7.1.1 Application of Dimensional Analysis: Derivation of the Efficiency of Pumps and Turbines

An empirical expression for the efficiency of a pump, η_{pump} in the theoretical expression for the efficiency of a pump, $\eta_{pump} = (P_p)_{out}/(P_p)_{in} = \gamma Q h_{pump}/\omega T_{shaft,in}$ (Equation 4.197) is sought in order to allow its empirical evaluation; the empirical expression of the efficiency of a turbine, $\eta_{turbine} = (P_t)_{out}/(P_t)_{in} = \omega T_{shaft,out}/\gamma Q h_{turbine}$ (Equation 4.217) is interpreted from the results. This is accomplished by using dimensional analysis in order to derive an expression for the efficiency of a pump, η_{pump} which involves the definition of a drag coefficient $C_D (\eta_{pump})$ that represents flow resistance and is empirically evaluated. Thus, the momentum theory is supplemented with dimensional analysis in order to derive an expression for the efficiency of a pump, η_{pump}. Thus, assuming the most general flow conditions, for the derivation of the efficiency of a pump, η_{pump}, in addition to the inertia force, F_I and the pressure force, F_P (which causes the viscous force for turbulent flow, $F_V = \tau A$), other forces including the viscous force for laminar flow, $F_V = \tau A = \mu v L$, the elastic force, F_E, and other physical quantities, including γ, Q, h_{pump}, T, and absolute blade surface roughness, ε, also play an important role in the flow through a pump.

Dimensional analysis is applied in order to derive an empirical expression for the efficiency of a pump, $\eta_{pump} = (P_p)_{out}/(P_p)_{in} = \gamma Q h_{pump}/\omega T_{shaft,in}$ (and thus Δp), which involves the definition of a drag coefficient, $C_D (\eta_{pump})$ as follows (see Table 7.8):

> *Step 1:* Based on a general understanding of (and experience with and judgment of) the fluid flow phenomena identify, the dominant/significant physical forces and

dominant physical quantities in the flow situation (use Table 11.8 for internal flow as a guide).

a. The inertia force, $F_I = Ma = \rho v^2 L^2$ will always be a dominant force and results in the three primary variables: ρ, ω (where $v = \omega D$), which is the characteristic velocity (where ω is the pump impeller rotational speed—or the turbine runner rotational speed), and D, which is the characteristic length, L, and is the diameter of the pump impeller blades (or the diameter of the turbine runner blades).

b. The friction/viscous force for turbulent flow, F_f or $F_V = \tau A$ will always be the first/main additional dominant force, which in this case is represented by a dominant physical quantity: efficiency of a pump, η_{pump}. The efficiency of a pump, η_{pump} is a result of the pressure force and represents an energy loss or some form of flow resistance and results in a secondary variable: η_{pump}.

c. In the flow through a pump, the propulsive force is provided by a pressure drop, Δp, which results in an unknown pressure drop, Δp. Thus, in the derivation of the efficiency of a pump, $\eta_{pump} = (P_p)_{out}/(P_p)_{in} = \gamma Q h_{pump}/\omega T_{shaft,in}$, the following four variables, ρ, ω, D, and η, result in one dimensionless number, namely $\Delta p / \rho v^2 = 1/E = C_D = \eta$, where $E = F_I/F_P$ = the Euler number, and $C_D = \eta$ = the drag coefficient that represents the flow resistance. Therefore, the efficiency of a pump, η, which represents a head loss, h_f is a result of Δp and thus $C_D = \eta$ is mainly a function of E.

d. Modeling additional dominant forces or dominant physical quantities results in additional secondary variables and will introduce an additional dimensionless number (π term) for each new secondary variable added.

1. The hydraulic power output from the pump, $(P_p)_{out} = \gamma Q h_{pump}$ results in three secondary variables: the specific weight of the fluid, γ; the flowrate (discharge), Q; and the head added by the pump, h_{pump}. The γ and the h_{pump} each introduce an independent π term, which are combined to define a new π term, C_H. The flowrate (discharge), Q introduces an independent π term, C_Q.

2. The shaft power input to the pump, $(P_p)_{in} = \omega T_{shaft,in}$ results in two secondary variables: the pump impeller rotational speed, ω and the rotating shaft torque (shaft work), T. Because the pump impeller rotational speed, ω is a primary variable, it is not repeated. However, the T introduces an independent π term, C_P.

3. The viscous force for laminar flow, $F_V = \tau A = \mu v L$ results in a secondary variable, μ, which introduces an independent π term, R.

4. The elastic force, F_E results in a secondary variable, E_v, which introduces an independent π term, C.

5. The absolute pump impeller blade (or turbine runner blade) surface roughness height, ε results in a secondary variable, ε, which introduces an independent π term ε/D.

6. The geometry, L_i ($i = 1, 2, \ldots$) may result in one or more secondary variables, L_i, which may introduce one or more independent π terms, L_i/L. Specifically, the geometry may represent the gap thickness between the pump impeller blade tips (or turbine runner blade tips) and the pump (or turbine) housing, and the pump impeller blade thickness (or turbine runner blade thickness). However, because these geometry variables are usually of insignificant importance, they are not usually modeled.

Step 2: Use the simplest mathematical expressions for the identified dominant physical forces and dominant physical quantities and identify the n physical quantities, all of which can be expressed in terms of m fundamental dimensions (F, L, and T).

Dominant Physical Force or Quantity	Primary Variable and Dimensions	Primary Variable and Dimensions	Primary Variable and Dimensions	Secondary Variable and Dimensions
$F_I = Ma = \rho v^2 L^2$	$\rho = [FL^{-4}T^2]$	$\omega = [T^{-1}]$	$D = [L]$	
$\eta_{pump} = \dfrac{(P_p)_{out}}{(P_p)_{in}} = \dfrac{\gamma Q h_{pump}}{\omega T_{shaft,in}}$				$\eta_{pump} = []$
$(P_p)_{out} = \gamma Q h_{pump}$				$\gamma = [FL^{-3}]$ $Q = [L^3 T^{-1}]$ $h_{pump} = [L]$
$(P_p)_{in} = \omega T_{shaft,in}$				$\omega = [T^{-1}]$ $T = [FL]$
$F_V = \tau A = \mu v L$				$\mu = [FL^{-2}T]$
$F_E = E_v A = E_v L^2$				$E_v = [FL^{-2}]$
ε				$\varepsilon = [L]$

where $n = 11$ and $m = 3$.

Step 3: The number of dimensionless numbers (π terms) is $n\text{-}m = 11\text{-}3 = 8$.

Step 4: The Buckingham π theorem is applied in order to derive the $n\text{-}m = 8$ π terms as follows:

$$\pi_1 = \rho^{a_1} \omega^{b_1} D^{c_1} \eta_{pump}^{d_1}$$

$$\pi_2 = \rho^{a_2} \omega^{b_2} D^{c_2} \gamma^{d_2}$$

$$\pi_3 = \rho^{a_3} \omega^{b_3} D^{c_3} Q^{d_3}$$

$$\pi_4 = \rho^{a_4} \omega^{b_4} D^{c_4} h_{pump}^{d_4}$$

$$\pi_5 = \rho^{a_5} \omega^{b_5} D^{c_5} T^{d_5}$$

$$\pi_6 = \rho^{a_6} \omega^{b_6} D^{c_6} \mu^{d_6}$$

$$\pi_7 = \rho^{a_7} \omega^{b_7} D^{c_7} E_v^{d_7}$$

$$\pi_8 = \rho^{a_8} \omega^{b_8} D^{c_8} \varepsilon^{d_8}$$

where the exponents are unknown quantities to be determined by applying the Buckingham π theorem.

Working with the π_1 term, the dimensional equations for the *FLT* system are set up as follows:

$$\pi_1 = \rho^{a_1}\omega^{b_1}D^{c_1}\eta_{pump}^{d_1}$$

$$F^0L^0T^0 = (FL^{-4}T^2)^{a_1}(T^{-1})^{b_1}(L)^{c_1}()^{d_1}$$

Equating the exponents for the dimensions and solving for three of the unknown exponents in terms of the fourth unknown exponent using the Mathcad symbolic solve block yields the following:

Given

F: $0 = a_1$

L: $0 = -4a_1 + c_1$

T: $0 = 2a_1 - b_1$

Find $(a_1, b_1, c_1) \rightarrow \begin{pmatrix} 0 \\ 0 \\ 0 \end{pmatrix}$

Thus:

$$\pi_1 = \rho^0\omega^0D^0\eta_{pump}^{d_1}$$

$$\pi_1 = \eta_{pump}^{d_1}$$

From experimentation $d_1 = 1$; therefore:

$$\pi_1 = \eta_{pump}$$

which yields the empirical expression for the efficiency of a pump.

Working with the π_2 term, the dimensional equations for the *FLT* system are set up as follows:

$$\pi_2 = \rho^{a_2}\omega^{b_2}D^{c_2}\gamma^{d_2}$$

$$F^0L^0T^0 = (FL^{-4}T^2)^{a_2}(T^{-1})^{b_2}(L)^{c_2}(FL^{-3})^{d_2}$$

Equating the exponents for the dimensions and solving for three of the unknown exponents in terms of the fourth unknown exponent using the Mathcad symbolic solve block yields the following:

Given

F: $0 = a_2 + d_2$

L: $0 = -4a_2 + c_2 - 3d_2$

T: $0 = 2a_2 - b_2$

Find $(a_2, b_2, c_2) \rightarrow \begin{pmatrix} -d_2 \\ -2 \cdot d_2 \\ -d_2 \end{pmatrix}$

Thus:

$$\pi_2 = \rho^{-d_2}\omega^{-2d_2}D^{-d_2}\gamma^{d_2}$$

$$\pi_2 = \frac{\gamma^{d_2}}{\rho^{d_2}\omega^{2d_2}D^{d_2}} = \frac{g^{d_2}}{\omega^{2d_2}D^{d_2}}$$

From experimentation $d_2 = 1$; therefore:

$$\pi_2 = \frac{g}{\omega^2 D}$$

Working with the π_3 term, the dimensional equations for the FLT system are set up as follows:

$$\pi_3 = \rho^{a_3}\omega^{b_3}D^{c_3}Q^{d_3}$$

$$F^0L^0T^0 = (FL^{-4}T^2)^{a_3}(T^{-1})^{b_3}(L)^{c_3}(L^3T^{-1})^{d_3}$$

Equating the exponents for the dimensions and solving for three of the unknown exponents in terms of the fourth unknown exponent using the Mathcad symbolic solve block yields the following:

Given

F: $0 = a_3$

L: $0 = -4a_3 + c_3 + 3d_3$

T: $0 = 2a_3 - b_3 - d_3$

Find $(a_3, b_3, c_3) \rightarrow \begin{pmatrix} 0 \\ -d_3 \\ -3 \cdot d_3 \end{pmatrix}$

Thus:

$$\pi_3 = \rho^0\omega^{-d_3}D^{-3d_3}Q^{d_3}$$

$$\pi_3 = \frac{Q^{d_3}}{\omega^{d_3}D^{3d_3}}$$

From experimentation $d_3 = 1$; therefore:

$$\pi_3 = \frac{Q}{\omega D^3}$$

Working with the π_4 term, the dimensional equations for the *FLT* system are set up as follows:

$$\pi_4 = \rho^{a_4}\omega^{b_4}D^{c_4}h_{pump}^{d_4}$$

$$F^0L^0T^0 = (FL^{-4}T^2)^{a_4}(T^{-1})^{b_4}(L)^{c_4}(L)^{d_4}$$

Equating the exponents for the dimensions and solving for three of the unknown exponents in terms of the fourth unknown exponent using the Mathcad symbolic solve block yields the following:

Given

F: $0 = a_4$

L: $0 = -4a_4 + c_4 + d_4$

T: $0 = 2a_4 - b_4$

Find $(a_4, b_4, c_4) \rightarrow \begin{pmatrix} 0 \\ 0 \\ -d_4 \end{pmatrix}$

Thus:

$$\pi_4 = \rho^0 \omega^0 D^{-d_4} h_{pump}^{d_4}$$

$$\pi_4 = \frac{h_{pump}^{d_4}}{D^{d_4}}$$

From experimentation $d_4 = 1$; therefore:

$$\pi_4 = \frac{h_{pump}}{D}$$

Working with the π_5 term, the dimensional equations for the *FLT* system are set up as follows:

$$\pi_5 = \rho^{a_5} \omega^{b_5} D^{c_5} T^{d_5}$$

$$F^0 L^0 T^0 = (FL^{-4} T^2)^{a_5} (T^{-1})^{b_5} (L)^{c_5} (FL)^{d_5}$$

Equating the exponents for the dimensions and solving for three of the unknown exponents in terms of the fourth unknown exponent using the Mathcad symbolic solve block yields the following:

Given

F: $0 = a_5 + d_5$

L: $0 = -4a_5 + c_5 + d_5$

T: $0 = 2a_5 - b_5$

Find $(a_5, b_5, c_5) \rightarrow \begin{pmatrix} -d_5 \\ -2 \cdot d_5 \\ -5 \cdot d_5 \end{pmatrix}$

Thus:

$$\pi_5 = \rho^{-d_5} \omega^{-2d_5} D^{-5d_5} T^{d_5}$$

$$\pi_5 = \frac{T^{d_5}}{\rho^{d_5} \omega^{2d_5} D^{5d_5}}$$

From experimentation $d_5 = 1$; therefore:

$$\pi_5 = \frac{T}{\rho \omega^2 D^5}$$

Working with the π_6 term, the dimensional equations for the *FLT* system are set up as follows:

$$\pi_6 = \rho^{a_6} \omega^{b_6} D^{c_6} \mu^{d_6}$$

$$F^0 L^0 T^0 = (FL^{-4}T^2)^{a_6}(T^{-1})^{b_6}(L)^{c_6}(FL^{-2}T)^{d_6}$$

Equating the exponents for the dimensions and solving for three of the unknown exponents in terms of the fourth unknown exponent using the Mathcad symbolic solve block yields the following:

Given

F: $\qquad 0 = a_6 + d_6$

L: $\qquad 0 = -4a_6 + c_6 - 2d_6$

T: $\qquad 0 = 2a_6 - b_6 + d_6$

Find $(a_6, b_6, c_6) \rightarrow \begin{pmatrix} -d_6 \\ -d_6 \\ -2 \cdot d_6 \end{pmatrix}$

Thus:

$$\pi_6 = \rho^{-d_6} \omega^{-d_6} D^{-2d_6} \mu^{d_6}$$

$$\pi_6 = \frac{\mu^{d_6}}{\rho^{d_6} \omega^{d_6} D^{2d_6}}$$

From experimentation $d_6 = -1$; therefore:

$$\pi_6 = \frac{\rho \omega D^2}{\mu}$$

which is the Reynolds number, **R**.

Working with the π_7 term, the dimensional equations for the *FLT* system are set up as follows:

$$\pi_7 = \rho^{a_7} \omega^{b_7} D^{c_7} E_v^{d_7}$$

$$F^0 L^0 T^0 = (FL^{-4}T^2)^{a_7}(T^{-1})^{b_7}(L)^{c_7}(FL^{-2})^{d_7}$$

Equating the exponents for the dimensions and solving for three of the unknown exponents in terms of the fourth unknown exponent using the Mathcad symbolic solve block yields the following:

Given

F: $0 = a_7 + d_7$

L: $0 = -4a_7 + c_7 - 2d_7$

T: $0 = 2a_7 - b_7$

Find $(a_7, b_7, c_7) \rightarrow \begin{pmatrix} -d_7 \\ -2 \cdot d_7 \\ -2 \cdot d_7 \end{pmatrix}$

Thus:

$$\pi_7 = \rho^{-d_7} \omega^{-2d_7} D^{-2d_7} E_v^{d_7}$$

$$\pi_7 = \frac{E_v^{d_7}}{\rho^{d_7} \omega^{2d_7} D^{2d_7}}$$

From experimentation $d_7 = -1$; therefore:

$$\pi_7 = \frac{\rho \omega^2 D^2}{E_v}$$

which is the Cauchy number, C.

Working with the π_8 term, the dimensional equations for the *FLT* system are set up as follows:

$$\pi_8 = \rho^{a_8} \omega^{b_8} D^{c_8} \varepsilon^{d_8}$$

$$F^0 L^0 T^0 = (FL^{-4} T^2)^{a_8} (T^{-1})^{b_8} (L)^{c_8} (L)^{d_8}$$

Equating the exponents for the dimensions and solving for three of the unknown exponents in terms of the fourth unknown exponent using the Mathcad symbolic solve block yields the following:

Given

F: $0 = a_8$

L: $0 = -4a_8 + c_8 + d_8$

T: $0 = 2a_8 - b_8$

Find $(a_8, b_8, c_8) \rightarrow \begin{pmatrix} 0 \\ 0 \\ -d_8 \end{pmatrix}$

Thus:

$$\pi_8 = \rho^0 \omega^0 D^{-d_8} \varepsilon^{d_8}$$

$$\pi_8 = \frac{\varepsilon^{d_8}}{D^{d_8}}$$

From experimentation $d_8 = 1$; therefore:

$$\pi_8 = \frac{\varepsilon}{D}$$

which is the relative blade surface roughness.

The resulting functional relationship between the $n\text{-}m = 8$ π terms is expressed as follows:

$$\phi(\pi_1, \pi_2, \pi_3, \pi_4, \pi_5, \pi_6, \pi_7, \pi_8) = 0$$

$$\phi\left(\eta_{pump}, \frac{g}{\omega^3 D}, \frac{Q}{\omega D^3}, \frac{h_{pump}}{D}, \frac{T}{\rho\omega^2 D^5}, \frac{\rho\omega D^3}{\mu}, \frac{\rho\omega^2 D^2}{E_v}, \frac{\varepsilon}{D}\right) = 0$$

$$\phi\left(\eta_{pump}, \frac{g}{\omega^3 D}, \frac{Q}{\omega D^3}, \frac{h_{pump}}{D}, \frac{T}{\rho\omega^2 D^5}, R, C, \frac{\varepsilon}{D}\right) = 0$$

where laboratory results (experimentation) yield the following pump design coefficients:

$$\pi_9 = \pi_2\pi_4 = \frac{g}{\omega^2 D}\frac{h_{pump}}{D} = \frac{gh_{pump}}{\omega^2 D^2} = C_H = \text{ the head coefficient}$$

$$\pi_3 = \frac{Q}{\omega D^3} = C_Q = \text{the capacity coefficient}$$

$$\pi_5 = \frac{T}{\rho\omega^2 D^5} = \frac{\omega T}{\rho\omega^3 D^5} = C_P = \text{the power coefficient} = \text{Power number} = \frac{Power}{Rotational\ inertia}$$

thus:

$$\phi\left(\eta_{pump}, C_H, C_Q, C_P, R, C, \frac{\varepsilon}{D}\right) = 0$$

or:

$$\eta_{pump} = \phi\left(C_H, C_Q, C_P, R, C, \frac{\varepsilon}{D}\right)$$

Step 5: Interpret the resulting functional relationship between the $n\text{-}m = 8$ π terms (now 7 π terms). Define the drag coefficient, C_D (i.e., η_{pump}) and note the definition of the inertia force, F_I:

$$\eta_{pump} = \frac{(P_p)_{out}}{(P_p)_{in}} = \frac{\gamma Q h_{pump}}{\omega T_{shaft,in}} = \frac{\gamma Q \frac{\Delta p}{\gamma}}{\omega T_{shaft,in}} = \frac{Q\Delta p}{\omega T_{shaft,in}} = \frac{vL^2\Delta p}{\frac{vFL}{L}} = \frac{\Delta p L^2}{F} = \frac{F_P}{F_I}$$

$$= \frac{\Delta p L^2}{\rho v^2 L^2} = \frac{\Delta p}{\rho v^2} = \frac{1}{E} = C_D = \phi\left(C_H, C_Q, C_P, R, C, \frac{\varepsilon}{D}\right) \tag{7.145}$$

where the head added by the pump, $h_{pump} = \Delta p/\gamma$, and Δp is the actual pressure rise due to the pump. The drag coefficient, C_D (pump efficiency, η_{pump}) as modeled by the Euler number, E is the dependent π term, as it models the dependent variable, Δp that cannot be derived analytically from theory, while the head coefficient, C_H; the capacity coefficient, C_Q; the power coefficient, C_P; the Reynolds number, R; the Cauchy number, C; and the relative blade surface roughness, ε/D are the independent π terms. One may note that the symbol ϕ means that the drag coefficient, C_D (pump efficiency, η) is an unknown function of C_H, C_Q, C_P, R, C, and ε/D; thus, the drag coefficient, C_D (pump efficiency, η) is determined empirically from experimentation. Also, because the efficiency of a pump, η represents a head loss, h_f due to a pressure drop, Δp, the pressure force will always play an important role and thus the definition of the drag coefficient, $C_D = \eta$ is mainly a function of E.

Furthermore, examining the drag coefficient, C_D (pump efficiency, η_{pump}) yields the following:

$$\eta_{pump} = \frac{(P_p)_{out}}{(P_p)_{in}} = \frac{\gamma Q h_{pump}}{\omega T_{shaft,in}} = \frac{F_P}{F_I} = \frac{\Delta p}{\rho v^2} = \frac{1}{E} = C_D = \frac{actual\ Power}{ideal\ Power} \qquad (7.146)$$

where the results of dimensional analysis introduce a drag coefficient, C_D (pump efficiency, η_{pump}), which expresses the ratio of a real/actual flow variable (hydraulic power output from the pump) to its corresponding ideal flow variable (shaft power input to the pump). Finally, empirical evaluation of the drag coefficient, C_D (pump efficiency, η_{pump}) and application of pump efficiency is presented in Chapter 4.

The results from dimensional analysis are used in the development of criteria that govern dynamic similarity/similitude between two flow situations that are geometrically similar but different in size. Specifically, the application of the efficiency, η of a pump for similitude and modeling are presented in Chapter 11. Based on the theoretical expression for the efficiency of a pump, $\eta_{pump} = (P_p)_{out}/(P_p)_{in} = \gamma Q h_{pump}/ \omega T_{shaft,in}$ (Equation 4.197), the empirical expression for the efficiency of a pump derived from dimensional analysis (Equation 7.145) can be expressed as a function of the pump design coefficients (C_H, C_Q, and C_P) as follows:

$$\eta_{pump} = \frac{(P_p)_{out}}{(P_p)_{in}} = \frac{\gamma Q h_{pump}}{\omega T_{shaft,in}} = \frac{(\rho Q)(g h_{pump})}{(\omega T_{shaaft,in})} = \frac{(\rho \omega D^3 C_Q)(\omega^2 D^2 C_H)}{(\rho \omega^3 D^5 C_P)} = \frac{C_Q C_H}{C_P}$$
$$= C_D = \phi \left(R, C, \frac{\varepsilon}{D} \right) \qquad (7.147)$$

where the pump design coefficients (C_H, C_Q, and C_P) each express the ratio of a real/ actual flow variable (head, discharge, and power) to its corresponding ideal flow variable (head, discharge, and power), respectively. The efficiency of a pump, η_{pump} represents a head loss, h_f due to a pressure drop, Δp. Furthermore, because one cannot analytically derive the expression for the pressure drop, Δp, one must resort to an empirical derivation of the resulting actual variables (head, pressure rise due to pump, discharge, hydraulic power, and shaft power) for a pump, which are given, respectively, as follows:

$$h_{pump} = \frac{C_H \omega^2 D^2}{g}$$

$$\Delta p = \gamma h_{pump} = \frac{\rho g C_H \omega^2 D^2}{g} = C_H \rho \omega^2 D^2$$

$$Q = C_Q \omega D^3$$

$$(P_p)_{out} = \gamma Q h_{pump} = C_Q C_H \rho \omega^3 D^5$$

$$(P_p)_{in} = \omega T_{shaft,in} = C_P \rho \omega^3 D^5$$

Finally, the empirical expression of the efficiency of a turbine is interpreted by its definition (exact opposite of a pump), and where:

$$\eta_{turbine} = \frac{(P_t)_{out}}{(P_t)_{in}} = \frac{\omega T_{shaft,out}}{\gamma Q h_{turbine}} = \frac{F_I}{F_P} = \frac{\rho v^2 L^2}{\Delta p L^2} = \frac{\rho v^2}{\Delta p} = E = C_D = \phi\left(C_H, C_Q, C_P, R, C, \frac{\varepsilon}{D}\right)$$

(7.148)

where the head removed by the turbine, $h_{turbine} = \Delta p / \gamma$, and Δp is the actual pressure drop due to the turbine and where:

$$\pi_9 = \pi_2 \pi_4 = \frac{g}{\omega^2 D} \frac{h_{turbine}}{D} = \frac{g h_{turbine}}{\omega^2 D^2} = C_H = \text{ the head coefficient}$$

Thus,

$$\eta_{turbine} = \frac{(P_t)_{out}}{(P_t)_{in}} = \frac{\omega T_{shaft,out}}{\gamma Q h_{turbine}} = \frac{F_I}{F_P} = \frac{\rho v^2}{\Delta p} = E = C_D = \frac{actual\ Power}{ideal\ Power}$$

(7.149)

where the results of dimensional analysis introduce a drag coefficient, C_D (turbine efficiency, $\eta_{turbine}$), which expresses the ratio of a real/actual flow variable (shaft power output from the turbine) to its corresponding ideal flow variable (hydraulic power input to the turbine). Finally, empirical evaluation of the drag coefficient, C_D (turbine efficiency, $\eta_{turbine}$) and application of turbine efficiency is presented in Chapter 4.

The application of the efficiency, η of a turbine for similitude and modeling are presented in Chapter 11. Based on the theoretical expression for the efficiency of a turbine, $\eta_{turbine} = (P_t)_{out}/(P_t)_{in} = \omega T_{shaft,out}/\gamma Q h_{turbine}$ (Equation 4.217), the empirical expression for the efficiency of a turbine derived from dimensional analysis (Equation 7.148) can be expressed as a function of the turbine design coefficients (C_H, C_Q, and C_P) as follows:

$$\eta_{turbine} = \frac{(P_t)_{out}}{(P_t)_{in}} = \frac{\omega T_{shaft,out}}{\gamma Q h_{turbine}} = \frac{(\omega T_{shaft,out})}{(\rho Q)(g h_{turbine})} = \frac{(\rho \omega^3 D^5 C_P)}{(\rho \omega D^3 C_Q)(\omega^2 D^2 C_H)}$$

$$= \frac{C_P}{C_Q C_H} = C_D = \phi\left(R, C, \frac{\varepsilon}{D}\right)$$

(7.150)

where the turbine design coefficients (C_H, C_Q, and C_P) each express the ratio of a real/actual flow variable (head, discharge, and power) to its corresponding ideal flow variable (head, discharge, and power), respectively. The efficiency of a turbine, $\eta_{turbine}$ represents a head loss, h_f due to a pressure drop, Δp. Furthermore, because one cannot analytically derive the expression for the pressure drop, Δp, one must resort to an empirical derivation of the resulting actual variables (head, pressure drop due to turbine, discharge, shaft power, and hydraulic power) for a turbine which are given, respectively, as follows:

$$h_{turbine} = \frac{C_H \omega^2 D^2}{g}$$

$$\Delta p = \gamma h_{turbine} = \frac{\rho g C_H \omega^2 D^2}{g} = C_H \rho \omega^2 D^2$$

$$Q = C_Q \omega D^3$$

$$(P_t)_{out} = \omega T_{shaft,out} = C_P \rho \omega^3 D^5$$

$$(P_t)_{in} = \gamma Q h_{turbine} = C_Q C_H \rho \omega^3 D^5$$

7.7.1.2 Application of Dimensional Analysis to Derive the Efficiency of Pumps and Turbines for More Specific Flow Assumptions

The above application of dimensional analysis to derive an empirical expression for the efficiency of a pump, η_{pump} (Equation 7.145)—and for the efficiency of a turbine, $\eta_{turbine}$ (Equation 7.148) was for the most general flow conditions; it was assumed that the flow may be either incompressible or compressible and either laminar or turbulent. As a result, while the drag coefficient, $C_D = \eta$, in general, is mainly a function of E, it will (theoretically) selectively vary with R, ε/L, and C depending mostly on the velocity of flow and the type of fluid through the pump or turbine (see Tables 11.8, 11.9, and 7.8; see Section 11.7.8). However, in practice, the drag coefficient, $C_D = \eta$ is assumed to be independent of C (assumption of incompressible flow), and will vary with R for laminar flow, while it will vary with ε/D for turbulent flow in the case where the prototype pump or turbine is significantly larger than the model pump or turbine (or vice versa). Furthermore, it is important to note the definitions of the pump design coefficients C_H, C_Q, and C_P and the head added by the pump, $h_{pump} = \Delta p/\gamma$. Finally, it is important to note the definitions of the turbine design coefficients C_H, C_Q, and C_P, and the head removed by the turbine, $h_{turbine} = \Delta p/\gamma$.

EXAMPLE PROBLEM 7.7

Derive an empirical expression for the for the efficiency of a pump, $\eta_{pump} = (P_p)_{out}/(P_p)_{in} = \gamma Q h_{pump}/\omega T_{shaft,in}$ (the drag coefficient, C_D), assuming incompressible, turbulent flow. The inertia force ($F_I = Ma = \rho v^2 L^2$), and the efficiency of a pump, η_{pump} will always be important. Note that $v = \omega D$, which is the characteristic velocity (where ω is the pump impeller rotational speed), and D, which is the characteristic length, L and is the diameter of the pump impeller blades. Using Tables 11.8, 11.9, and 7.8, as well as Section 11.7.8, as a guide, the additional dominant/significant forces and physical quantities in the flow situation include the hydraulic power output from the pump, $(P_p)_{out} = \gamma Q h_{pump}$; the shaft power into the pump, $(P_p)_{in} = \omega T_{shaft,in}$; and the absolute pump impeller blade surface roughness height, ε. Thus, the important physical quantities include: ρ, ω, D, η_{pump}, γ, Q, h_{pump}, T, and ε. What are the important π terms used to study and describe this flow process for incompressible, turbulent flow through a pump?

Mathcad Solution

In order to derive an empirical expression for the for the efficiency of a pump, η_{pump} (the drag coefficient, C_D), assuming incompressible, turbulent flow, the five steps of dimensional analysis are applied as follows:

Step 1: Using the given information, and Tables 11.8, 11.9, and 7.8, as well as Section 11.7.8, as a guide, the dominant/significant physical forces and dominant physical quantities in the flow situation are identified as follows:

The inertia force, $F_I = Ma = \rho v^2 L^2$ will always be a dominant force and results in the three primary variables: ρ, ω, and D (where $v = \omega D$). The friction/viscous force for turbulent flow, F_f or $F_V = \tau A$ will always be the first/main additional dominant force, which in this case is represented by a dominant physical quantity: efficiency of a pump, η_{pump}. The efficiency of a pump, η_{pump} is a result of the pressure force and represents an energy loss or some form of flow resistance and results in a secondary variable: η_{pump}. In the flow through a pump, the propulsive force is provided by a pressure drop, Δp, which results in an unknown pressure drop, Δp. Thus, in the derivation of the efficiency of a pump, $\eta_{pump} = (P_p)_{out}/(P_p)_{in} = \gamma Q h_{pump}/\omega T_{shaft,in}$, the following four variables, ρ, ω, D, and η, result in one dimensionless number, namely $\Delta p/\rho v^2 = 1/E = C_D = \eta$, where $E = F_I/F_P =$ the Euler number and $C_D = \eta =$ the drag coefficient that represents the flow resistance. Therefore, the efficiency of a pump, η, which represents a head loss, h_f is a result of Δp; thus, $C_D = \eta$ is mainly a function of E. Modeling additional dominant forces or dominant physical quantities results in additional secondary variables and will introduce an additional dimensionless number (π term) for each new secondary variable added. Using Tables 11.8, 11.9, and 7.8, as well as Section 11.7.8, as a guide, the hydraulic power output from the pump, $(P_p)_{out} = \gamma Q h_{pump}$ results in three secondary variables: the specific weight of the fluid, γ; the flowrate (discharge), Q; and the head added by the pump, h_{pump}. The γ and the h_{pump} each introduce an independent π term, which are combined to define a new π term, C_H. The flowrate (discharge), Q introduces an independent π term, C_Q. Furthermore, the shaft power input to the pump, $(P_p)_{in} = \omega T_{shaft,in}$ results in two secondary variables: the pump impeller rotational speed, ω and the rotating shaft torque (shaft work), T. Because the pump impeller rotational speed, ω is a primary variable, it is not repeated. However, the T introduces an independent π term, C_P. And, the absolute pump impeller blade surface roughness height, ε results in a secondary variable, ε, which introduces an independent π term ε/D.

Step 2: Use the simplest mathematical expressions for the identified dominant physical forces and dominant physical quantities and identify the n physical quantities, all of which can be expressed in terms of m fundamental dimensions (F, L, and T).

Dominant Physical Force or Quantity	Primary Variable and Dimensions	Primary Variable and Dimensions	Primary Variable and Dimensions	Secondary Variable and Dimensions
$F_I = Ma = \rho v^2 L^2$	$\rho = [FL^{-4}T^2]$	$\omega = [T^{-1}]$	$D = [L]$	
$\eta_{pump} = \dfrac{(P_p)_{out}}{(P_p)_{in}} = \dfrac{\gamma Q h_{pump}}{\omega T_{shaft,in}}$				$\eta_{pump} = []$
$(P_p)_{out} = \gamma Q h_{pump}$				$\gamma = [FL^{-3}]$
				$Q = [L^3 T^{-1}]$
				$h_{pump} = [L]$
$(P_p)_{in} = \omega T_{shaft,in}$				$\omega = [T^{-1}]$
				$T = [FL]$
ε				$\varepsilon = [L]$

where $n = 9$ and $m = 3$.

Step 3: The number of dimensionless numbers (π terms) is $n\text{-}m = 9\text{-}3 = 6$.

Step 4: The Buckingham π theorem is applied in order to derive the $n\text{-}m = 6$ π terms as follows:

$$\pi_1 = \rho^{a_1} \omega^{b_1} D^{c_1} \eta_{pump}^{d_1}$$

$$\pi_2 = \rho^{a_2} \omega^{b_2} D^{c_2} \gamma^{d_2}$$

$$\pi_3 = \rho^{a_3} \omega^{b_3} D^{c_3} Q^{d_3}$$

$$\pi_4 = \rho^{a_4} \omega^{b_4} D^{c_4} h_{pump}^{d_4}$$

$$\pi_5 = \rho^{a_5} \omega^{b_5} D^{c_5} T^{d_5}$$

$$\pi_6 = \rho^{a_6} \omega^{b_6} D^{c_6} \varepsilon^{d_6}$$

where the exponents are unknown quantities to be determined by applying the Buckingham π theorem.

Working with the π_1 term, the dimensional equations for the *FLT* system are set up as follows:

$$\pi_1 = \rho^{a_1} \omega^{b_1} D^{c_1} \eta_{pump}^{d_1}$$

$$F^0 L^0 T^0 = (FL^{-4}T^2)^{a_1} (T^{-1})^{b_1} (L)^{c_1} ()^{d_1}$$

Equating the exponents for the dimensions and solving for three of the unknown exponents in terms of the fourth unknown exponent using the Mathcad symbolic solve block yields the following:

Given

F: $\qquad 0 = a_1$

L: $\qquad 0 = -4a_1 + c_1$

T: $\qquad 0 = 2a_1 - b_1$

Find $(a_1, b_1, c_1) \rightarrow \begin{pmatrix} 0 \\ 0 \\ 0 \end{pmatrix}$

Thus:

$$\pi_1 = \rho^0 \omega^0 D^0 \eta_{pump}^{d_1}$$

$$\pi_1 = \eta_{pump}^{d_1}$$

From experimentation $d_1 = 1$; therefore:

$$\pi_1 = \eta_{pump}$$

which yields the empirical expression for the efficiency of a pump.

Working with the π_2 term, the dimensional equations for the *FLT* system are set up as follows:

$$\pi_2 = \rho^{a_2} \omega^{b_2} D^{c_2} \gamma^{d_2}$$

$$F^0 L^0 T^0 = (FL^{-4}T^2)^{a_2} (T^{-1})^{b_2} (L)^{c_2} (FL^{-3})^{d_2}$$

Equating the exponents for the dimensions and solving for three of the unknown exponents in terms of the fourth unknown exponent using the Mathcad symbolic solve block yields the following:

Given

F: $\qquad 0 = a_2 + d_2$

L: $\qquad 0 = -4a_2 + c_2 - 3d_2$

T: $\qquad 0 = 2a_2 - b_2$

Find $(a_2, b_2, c_2) \rightarrow \begin{pmatrix} -d_2 \\ 2 \cdot d_2 \\ -d_2 \end{pmatrix}$

Thus:

$$\pi_2 = \rho^{-d_2} \omega^{-2d_2} D^{-d_2} \gamma^{d_2}$$

$$\pi_2 = \frac{\gamma^{d_2}}{\rho^{d_2} \omega^{2d_2} D^{d_2}} = \frac{g^{d_2}}{\omega^{2d_2} D^{d_2}}$$

From experimentation $d_2 = 1$; therefore:

$$\pi_2 = \frac{g}{\omega^2 D}$$

Working with the π_3 term, the dimensional equations for the FLT system are set up as follows:

$$\pi_3 = \rho^{a_3} \omega^{b_3} D^{c_3} Q^{d_3}$$

$$F^0 L^0 T^0 = (FL^{-4}T^2)^{a_3} (T^{-1})^{b_3} (L)^{c_3} (L^3 T^{-1})^{d_3}$$

Equating the exponents for the dimensions and solving for three of the unknown exponents in terms of the fourth unknown exponent using the Mathcad symbolic solve block yields the following:

Given

F: $0 = a_3$

L: $0 = -4a_3 + c_3 + 3d_3$

T: $0 = 2a_3 - b_3 - d_3$

Find $(a_3, b_3, c_3) \rightarrow \begin{pmatrix} 0 \\ -d_3 \\ -3 \cdot d_3 \end{pmatrix}$

Thus:

$$\pi_3 = \rho^0 \omega^{-d_3} D^{-3d_3} Q^{d_3}$$

$$\pi_3 = \frac{Q^{d_3}}{\omega^{d_3} D^{3d_3}}$$

From experimentation $d_3 = 1$; therefore:

$$\pi_3 = \frac{Q}{\omega D^3}$$

Working with the π_4 term, the dimensional equations for the *FLT* system are set up as follows:

$$\pi_4 = \rho^{a_4}\omega^{b_4}D^{c_4}h_{pump}^{d_4}$$

$$F^0L^0T^0 = (FL^{-4}T^2)^{a_4}(T^{-1})^{b_4}(L)^{c_4}(L)^{d_4}$$

Equating the exponents for the dimensions and solving for three of the unknown exponents in terms of the fourth unknown exponent using the Mathcad symbolic solve block yields the following:

Given

F:	$0 = a_4$
L:	$0 = -4a_4 + c_4 + d_4$
T:	$0 = 2a_4 - b_4$

$$\text{Find } (a_4, b_4, c_4) \rightarrow \begin{pmatrix} 0 \\ 0 \\ -d_4 \end{pmatrix}$$

Thus:

$$\pi_4 = \rho^0\omega^0 D^{-d_4}h_{pump}^{d_4}$$

$$\pi_4 = \frac{h_{pump}^{d_4}}{D^{d_4}}$$

From experimentation $d_4 = 1$; therefore:

$$\pi_4 = \frac{h_{pump}}{D}$$

Working with the π_5 term, the dimensional equations for the *FLT* system are set up as follows:

$$\pi_5 = \rho^{a_5}\omega^{b_5}D^{c_5}T^{d_5}$$

$$F^0L^0T^0 = (FL^{-4}T^2)^{a_5}(T^{-1})^{b_5}(L)^{c_5}(FL)^{d_5}$$

Equating the exponents for the dimensions and solving for three of the unknown exponents in terms of the fourth unknown exponent using the Mathcad symbolic solve block yields the following:

Given

F:	$0 = a_5 + d_5$
L:	$0 = -4a_5 + c_5 + d_5$

T: $0 = 2a_5 - b_5$

Find $(a_5, b_5, c_5) \rightarrow \begin{pmatrix} -d_5 \\ -2 \cdot d_5 \\ -5 \cdot d_5 \end{pmatrix}$

Thus:

$$\pi_5 = \rho^{-d_5} \omega^{-2d_5} D^{-5d_5} T^{d_5}$$

$$\pi_5 = \frac{T^{d_5}}{\rho^{d_5} \omega^{2d_5} D^{5d_5}}$$

From experimentation $d_5 = 1$; therefore:

$$\pi_5 = \frac{T}{\rho \omega^2 D^5}$$

Working with the π_6 term, the dimensional equations for the *FLT* system are set up as follows:

$$\pi_6 = \rho^{a_6} \omega^{b_6} D^{c_6} \varepsilon^{d_6}$$

$$F^0 L^0 T^0 = (FL^{-4}T^2)^{a_6}(T^{-1})^{b_6}(L)^{c_6}(L)^{d_6}$$

Equating the exponents for the dimensions and solving for three of the unknown exponents in terms of the fourth unknown exponent using the Mathcad symbolic solve block yields the following:

Given

F: $0 = a_6$

L: $0 = -4a_6 + c_6 + d_6$

T: $0 = 2a_6 - b_6$

Find $(a_6, b_6, c_6) \rightarrow \begin{pmatrix} 0 \\ 0 \\ -d_6 \end{pmatrix}$

Thus:

$$\pi_6 = \rho^0 \omega^0 D^{-d_6} \varepsilon^{d_6}$$

$$\pi_6 = \frac{\varepsilon^{d_6}}{D^{d_6}}$$

From experimentation $d_6 = 1$; therefore:

$$\pi_6 = \frac{\varepsilon}{D}$$

which is the relative blade surface roughness.

The resulting functional relationship between the $n\text{-}m = 6$ π terms is expressed as follows:

$$\phi(\pi_1, \pi_2, \pi_3, \pi_4, \pi_5, \pi_6) = 0$$

$$\phi\left(\eta_{pump}, \frac{g}{\omega^3 D}, \frac{Q}{\omega D^3}, \frac{h_{pump}}{D}, \frac{T}{\rho\omega^2 D^5}, \frac{\varepsilon}{D}\right) = 0$$

where laboratory results (experimentation) yield the following pump design coefficients:

$$\pi_7 = \pi_2\pi_4 = \frac{g}{\omega^2 D}\frac{h_{pump}}{D} = \frac{gh_{pump}}{\omega^2 D^2} = C_H = \text{ the head coefficient}$$

$$\pi_3 = \frac{Q}{\omega D^3} = C_Q = \text{ the capacity coefficient}$$

$$\pi_5 = \frac{T}{\rho\omega^2 D^5} = \frac{\omega T}{\rho\omega^3 D^5} = C_P = \text{ the power coefficient} = \text{Power number} = \frac{Power}{Rotational\ inertia}$$

thus:

$$\phi\left(\eta_{pump}, C_H, C_Q, C_P, \frac{\varepsilon}{D}\right) = 0$$

or:

$$\eta_{pump} = \phi\left(C_H, C_Q, C_P, \frac{\varepsilon}{D}\right)$$

Step 5: Interpret the resulting functional relationship between the $n\text{-}m = 6$ π terms (now 5 π terms). Define the drag coefficient, C_D (i.e., η_{pump}) and note the definition of the inertia force, F_I:

$$\eta_{pump} = \frac{(P_p)_{out}}{(P_p)_{in}} = \frac{\gamma Q h_{pump}}{\omega T_{shaft,in}} = \frac{\gamma Q \frac{\Delta p}{\gamma}}{\omega T_{shaft,in}} = \frac{Q\Delta p}{\omega T_{shaft,in}} = \frac{vL^2\Delta p}{\frac{vFL}{L}} = \frac{\Delta p L^2}{F} = \frac{F_P}{F_I} = \frac{\Delta p L^2}{\rho v^2 L^2} = \frac{\Delta p}{\rho v^2}$$

$$= \frac{1}{E} = C_D = \phi\left(C_H, C_Q, C_P, \frac{\varepsilon}{D}\right)$$

where the head added by the pump, $h_{pump} = \Delta p/\gamma$, and Δp is the actual pressure rise due to the pump. The drag coefficient, C_D (pump efficiency, η_{pump}) as modeled by the Euler number, E is the dependent π term, as it models the dependent variable, Δp that cannot be derived analytically from theory, while the head coefficient,

C_H; the capacity coefficient, C_Q; the power coefficient, C_P; and the relative blade surface roughness, ε/D are the independent π terms. One may note that the symbol, ϕ means that the drag coefficient, C_D (pump efficiency, η) is an unknown function of C_H, C_Q, C_P, and ε/D; thus, the drag coefficient, C_D (pump efficiency, η) is determined empirically from experimentation. Also, because the efficiency of a pump, η represents a head loss, h_f due to a pressure drop, Δp, the pressure force will always play an important role and thus the definition of the drag coefficient, $C_D = \eta$ is mainly a function of E.

EXAMPLE PROBLEM 7.8

Derive an empirical expression for the for the efficiency of a turbine, $\eta_{turbine} = \dfrac{(P_t)_{out}}{(P_t)_{in}} = \dfrac{\omega T_{shaft,out}}{\gamma Q h_{turbine}}$ (the drag coefficient, C_D), assuming incompressible, turbulent flow. The inertia force ($F_I = Ma = \rho v^2 L^2$) and the efficiency of a turbine, $\eta_{turbine}$ will always be important. Note that $v = \omega D$, which is the characteristic velocity (where ω is the turbine runner rotational speed), and D, which is the characteristic length, L and is the diameter of the turbine runner blades. Using Tables 11.8, 11.9, and Table 7.8, as well as Section 11.7.8, as a guide, the additional dominant/significant forces and physical quantities in the flow situation include the shaft power output from the turbine, $(P_t)_{out} = \omega T_{shaft,out}$; the hydraulic power input to the turbine, $(P_t)_{in} = \gamma Q h_{turbine}$; and the absolute turbine runner blade surface roughness height, ε. Thus, the important physical quantities include: ρ, ω, D, $\eta_{turbine}$, γ, Q, $h_{turbine}$, T, and ε. What are the important π terms used to study and describe this flow process for incompressible, turbulent flow through a turbine?

Mathcad Solution

In order to derive an empirical expression for the for the efficiency of a turbine, $\eta_{turbine}$ (the drag coefficient, C_D), assuming incompressible, turbulent flow, the five steps of dimensional analysis are applied as follows:

Step 1: Using the given information, and Tables 11.8, 11.9, and 7.8, as well as Section 11.7.8, as a guide, the dominant/significant physical forces and dominant physical quantities in the flow situation are identified as follows:

The inertia force, $F_I = Ma = \rho v^2 L^2$ will always be a dominant force and results in the three primary variables: ρ, ω, and D (where $v = \omega D$). The friction/viscous force for turbulent flow, F_f or $F_V = \tau A$ will always be the first/main additional dominant force, which in this case is represented by a dominant physical quantity: efficiency of a turbine, $\eta_{turbine}$. The efficiency of a turbine, $\eta_{turbine}$ is a result of the pressure force and represents an energy loss or some form of flow resistance and results in a secondary variable: $\eta_{turbine}$. In the flow through a turbine, the propulsive force is provided by a pressure drop, Δp, which results in an unknown pressure drop, Δp. Thus, in the derivation of the efficiency of a turbine, $\eta_{turbine} = \dfrac{(P_t)_{out}}{(P_t)_{in}} = \dfrac{\omega T_{shaft,out}}{\gamma Q h_{turbine}}$, the following four variables, ρ, ω, D, and η, result in one dimensionless number, namely

$\frac{\rho v^2}{\Delta p} = E = C_D = \eta$, where $E = F_I/F_P =$ the Euler number and $C_D = \eta =$ the drag coefficient that represents the flow resistance. Therefore, the efficiency of a turbine, η, which represents a head loss, h_f is a result of Δp; thus, $C_D = \eta$ is mainly a function of E. Modeling additional dominant forces or dominant physical quantities results in additional secondary variables and will introduce an additional dimensionless number (π term) for each new secondary variable added. Using Tables 11.8, 11.9, and 7.8, as well as Section 11.7.8, as a guide, the hydraulic power input to the turbine, $(P_t)_{in} = \gamma Q h_{turbine}$ results in three secondary variables: the specific weight of the fluid, γ; the flowrate (discharge), Q; and the head removed by the turbine, $h_{turbine}$. The γ and the $h_{turbine}$ each introduce an independent π term, which are combined to define a new π term, C_H. The flowrate (discharge), Q introduces an independent π term, C_Q. Furthermore, the shaft power output from the turbine, $(P_t)_{out} = \omega T_{shaft,out}$ results in two secondary variables: the turbine runner rotational speed, ω and the rotating shaft torque (shaft work), T. Because the turbine runner rotational speed, ω is a primary variable, it is not repeated. However, the T introduces an independent π term, C_P. And, the absolute turbine runner blade surface roughness height, ε results in a secondary variable, ε, which introduces an independent π term ε/D.

Step 2: Use the simplest mathematical expressions for the identified dominant physical forces and dominant physical quantities and identify the n physical quantities, all of which can be expressed in terms of m fundamental dimensions (F, L, and T).

Dominant Physical Force or Quantity	Primary Variable and Dimensions	Primary Variable and Dimensions	Primary Variable and Dimensions	Secondary Variable and Dimensions
$F_I = Ma = \rho v^2 L^2$	$\rho = [FL^{-4}T^2]$	$\omega = [T^{-1}]$	$D = [L]$	
$\eta_{turbine} = \dfrac{(P_t)_{out}}{(P_t)_{in}}$ $= \dfrac{\omega T_{shaft,out}}{\gamma Q h_{turbine}}$				$\eta_{turbine} = []$
$(P_t)_{in} = \gamma Q h_{turbine}$				$\gamma = [FL^{-3}]$ $Q = [L^3 T^{-1}]$ $h_{turbine} = [L]$
$(P_t)_{out} = \omega T_{shaft,out}$				$\omega = [T^{-1}]$ $T = [FL]$
ε				$\varepsilon = [L]$

where $n = 9$ and $m = 3$.

Step 3: The number of dimensionless numbers (π terms) is $n-m = 9-3 = 6$.

Step 4: The Buckingham π theorem is applied in order to derive the $n-m = 6$ π terms as follows:

$$\pi_1 = \rho^{a_1} \omega^{b_1} D^{c_1} \eta_{turbine}^{d_1}$$

$$\pi_2 = \rho^{a_2} \omega^{b_2} D^{c_2} \gamma^{d_2}$$

$$\pi_3 = \rho^{a_3} \omega^{b_3} D^{c_3} Q^{d_3}$$

$$\pi_4 = \rho^{a_4} \omega^{b_4} D^{c_4} h_{turbine}^{d_4}$$

$$\pi_5 = \rho^{a_5} \omega^{b_5} D^{c_5} T^{d_5}$$

$$\pi_6 = \rho^{a_6} \omega^{b_6} D^{c_6} \varepsilon^{d_6}$$

where the exponents are unknown quantities to be determined by applying the Buckingham π theorem.

Working with the π_1 term, the dimensional equations for the *FLT* system are set up as follows:

$$\pi_1 = \rho^{a_1} \omega^{b_1} D^{c_1} \eta_{turbine}^{d_1}$$

$$F^0 L^0 T^0 = (FL^{-4}T^2)^{a_1} (T^{-1})^{b_1} (L)^{c_1} ()^{d_1}$$

Equating the exponents for the dimensions and solving for three of the unknown exponents in terms of the fourth unknown exponent using the Mathcad symbolic solve block yields the following:

Given

F: $0 = a_1$

L: $0 = -4a_1 + c_1$

T: $0 = 2a_1 - b_1$

Find $(a_1, b_1, c_1) \rightarrow \begin{pmatrix} 0 \\ 0 \\ 0 \end{pmatrix}$

Thus:

$$\pi_1 = \rho^0 \omega^0 D^0 \eta_{turbine}^{d_1}$$

$$\pi_1 = \eta_{turbine}^{d_1}$$

From experimentation $d_1 = 1$; therefore:

$$\pi_1 = \eta_{turbine}$$

which yields the empirical expression for the efficiency of a turbine.

Working with the π_2 term, the dimensional equations for the *FLT* system are set up as follows:

$$\pi_2 = \rho^{a_2}\omega^{b_2}D^{c_2}\gamma^{d_2}$$

$$F^0L^0T^0 = (FL^{-4}T^2)^{a_2}(T^{-1})^{b_2}(L)^{c_2}(FL^{-3})^{d_2}$$

Equating the exponents for the dimensions and solving for three of the unknown exponents in terms of the fourth unknown exponent using the Mathcad symbolic solve block yields the following:

Given

F: $\qquad 0 = a_2 + d_2$

L: $\qquad 0 = -4a_2 + c_2 - 3d_2$

T: $\qquad 0 = 2a_2 - b_2$

$$\text{Find } (a_2, b_2, c_2) \rightarrow \begin{pmatrix} -d_2 \\ 2 \cdot d_2 \\ -d_2 \end{pmatrix}$$

Thus:

$$\pi_2 = \rho^{-d_2}\omega^{-2d_2}D^{-d_2}\gamma^{d_2}$$

$$\pi_2 = \frac{\gamma^{d_2}}{\rho^{d_2}\omega^{2d_2}D^{d_2}} = \frac{g^{d_2}}{\omega^{2d_2}D^{d_2}}$$

From experimentation $d_2 = 1$; therefore:

$$\pi_2 = \frac{g}{\omega^2 D}$$

Working with the π_3 term, the dimensional equations for the FLT system are set up as follows:

$$\pi_3 = \rho^{a_3}\omega^{b_3}D^{c_3}Q^{d_3}$$

$$F^0L^0T^0 = (FL^{-4}T^2)^{a_3}(T^{-1})^{b_3}(L)^{c_3}(L^3T^{-1})^{d_3}$$

Equating the exponents for the dimensions and solving for three of the unknown exponents in terms of the fourth unknown exponent using the Mathcad symbolic solve block yields the following:

Given

F: $\qquad 0 = a_3$

L: $\qquad 0 = -4a_3 + c_3 + 3d_3$

T: $\qquad 0 = 2a_3 - b_3 - d_3$

$$\text{Find } (a_3, b_3, c_3) \rightarrow \begin{pmatrix} 0 \\ -d_3 \\ -3 \cdot d_3 \end{pmatrix}$$

Thus:

$$\pi_3 = \rho^0 \omega^{-d_3} D^{-3d_3} Q^{d_3}$$

$$\pi_3 = \frac{Q^{d_3}}{\omega^{d_3} D^{3d_3}}$$

From experimentation $d_3 = 1$; therefore:

$$\pi_3 = \frac{Q}{\omega D^3}$$

Working with the π_4 term, the dimensional equations for the *FLT* system are set up as follows:

$$\pi_4 = \rho^{a_4} \omega^{b_4} D^{c_4} h_{turbine}^{d_4}$$

$$F^0 L^0 T^0 = (FL^{-4}T^2)^{a_4} (T^{-1})^{b_4} (L)^{c_4} (L)^{d_4}$$

Equating the exponents for the dimensions and solving for three of the unknown exponents in terms of the fourth unknown exponent using the Mathcad symbolic solve block yields the following:

Given

F: $0 = a_4$

L: $0 = -4a_4 + c_4 + d_4$

T: $0 = 2a_4 - b_4$

Find $(a_4, b_4, c_4) \rightarrow \begin{pmatrix} 0 \\ 0 \\ -d_4 \end{pmatrix}$

Thus:

$$\pi_4 = \rho^0 \omega^0 D^{-d_4} h_{turbine}^{d_4}$$

$$\pi_4 = \frac{h_{turbine}^{d_4}}{D^{d_4}}$$

From experimentation $d_4 = 1$; therefore:

$$\pi_4 = \frac{h_{turbine}}{D}$$

Working with the π_5 term, the dimensional equations for the *FLT* system are set up as follows:

$$\pi_5 = \rho^{a_5} \omega^{b_5} D^{c_5} T^{d_5}$$

$$F^0 L^0 T^0 = (FL^{-4}T^2)^{a_5} (T^{-1})^{b_5} (L)^{c_5} (FL)^{d_5}$$

Equating the exponents for the dimensions and solving for three of the unknown exponents in terms of the fourth unknown exponent using the Mathcad symbolic solve block yields the following:

Given

F: $\qquad 0 = a_5 + d_5$

L: $\qquad 0 = -4a_5 + c_5 + d_5$

T: $\qquad 0 = 2a_5 - b_5$

Find $(a_5, b_5, c_5) \rightarrow \begin{pmatrix} -d_5 \\ -2 \cdot d_5 \\ -5 \cdot d_5 \end{pmatrix}$

Thus:

$$\pi_5 = \rho^{-d_5} \omega^{-2d_5} D^{-5d_5} T^{d_5}$$

$$\pi_5 = \frac{T^{d_5}}{\rho^{d_5} \omega^{2d_5} D^{5d_5}}$$

From experimentation $d_5 = 1$; therefore:

$$\pi_5 = \frac{T}{\rho \omega^2 D^5}$$

Working with the π_6 term, the dimensional equations for the *FLT* system are set up as follows:

$$\pi_6 = \rho^{a_6} \omega^{b_6} D^{c_6} \varepsilon^{d_6}$$

$$F^0 L^0 T^0 = (FL^{-4}T^2)^{a_6} (T^{-1})^{b_6} (L)^{c_6} (L)^{d_6}$$

Equating the exponents for the dimensions and solving for three of the unknown exponents in terms of the fourth unknown exponent using the Mathcad symbolic solve block yields the following:

Given

F: $\qquad 0 = a_6$

L: $\qquad 0 = -4a_6 + c_6 + d_6$

T: $\qquad 0 = 2a_6 - b_6$

Find $(a_6, b_6, c_6) \rightarrow \begin{pmatrix} 0 \\ 0 \\ -d_6 \end{pmatrix}$

Thus:

$$\pi_6 = \rho^0 \omega^0 D^{-d_6} \varepsilon^{d_6}$$

$$\pi_6 = \frac{\varepsilon^{d_6}}{D^{d_6}}$$

From experimentation $d_6 = 1$; therefore:

$$\pi_6 = \frac{\varepsilon}{D}$$

which is the relative blade surface roughness.

The resulting functional relationship between the $n-m = 6$ π terms is expressed as follows:

$$\phi(\pi_1, \pi_2, \pi_3, \pi_4, \pi_5, \pi_6) = 0$$

$$\phi\left(\eta_{turbine}, \frac{g}{\omega^3 D}, \frac{Q}{\omega D^3}, \frac{h_{turbine}}{D}, \frac{T}{\rho \omega^2 D^5}, \frac{\varepsilon}{D}\right) = 0$$

where laboratory results (experimentation) yield the following turbine design coefficients:

$$\pi_7 = \pi_2 \pi_4 = \frac{g}{\omega^2 D} \frac{h_{turbine}}{D} = \frac{g h_{turbine}}{\omega^2 D^2} = C_H = \text{ the head coefficient}$$

$$\pi_3 = \frac{Q}{\omega D^3} = C_Q = \text{ the capacity coefficient}$$

$$\pi_5 = \frac{T}{\rho \omega^2 D^5} = \frac{\omega T}{\rho \omega^3 D^5} = C_P = \text{ the power coefficient} = \text{ Power number} = \frac{Power}{Rotational\ inertia}$$

thus:

$$\phi\left(\eta_{turbine}, C_H, C_Q, C_P, \frac{\varepsilon}{D}\right) = 0$$

or:

$$\eta_{turbine} = \phi\left(C_H, C_Q, C_P, \frac{\varepsilon}{D}\right)$$

Step 5: Interpret the resulting functional relationship between the $n-m = 6$ π terms (now 5 π terms). Define the drag coefficient, C_D (i.e., $\eta_{turbine}$) and note the definition of the inertia force, F_I:

$$\eta_{turbine} = \frac{(P_t)_{out}}{(P_t)_{in}} = \frac{\omega T_{shaft,out}}{\gamma Q h_{turbine}} = \frac{F_I}{F_P} = \frac{\rho v^2 L^2}{\Delta p L^2} = \frac{\rho v^2}{\Delta p} = E = C_D = \phi\left(C_H, C_Q, C_P, \frac{\varepsilon}{D}\right)$$

where the head removed by the turbine, $h_{turbine} = \Delta p/\gamma$, and Δp is the actual pressure drop due to the turbine. The drag coefficient, C_D (turbine efficiency, $\eta_{turbine}$) as modeled by the Euler number, E is the dependent π term, as it models the dependent variable, Δp that cannot be derived analytically from theory, while the head coefficient, C_H; the capacity coefficient, C_Q; the power coefficient, C_P; and the relative blade surface roughness, ε/D are the independent π terms. One may note that the symbol ϕ means that the drag coefficient, C_D (turbine efficiency, η) is an unknown function of C_H, C_Q, C_P, and ε/D, and thus the drag coefficient, C_D (turbine efficiency, η) is determined empirically from experimentation. Also, because the efficiency of a turbine, η represents a head loss, h_f due to a pressure drop, Δp, the pressure force will always play an important role and thus the definition of the drag coefficient, $C_D = \eta$ is mainly a function of E.

7.8 Experimental Formulation of Theoretical Equations

The dimensional analysis procedure itself, and the results of the procedure, including the laws of similitude, play an important role in the deterministic mathematical modeling of fluid flow problems. Thus far in Chapter 7, application of the procedure of dimensional analysis has been used as follows: (1) to empirically formulate the flow resistance equations in internal and external flow (Sections 7.3 through 7.6), and (2) to empirically formulate the efficiency equations of pumps and turbines (Section 7.7). Furthermore, the results of dimensional analysis are used as follows: (1) to develop criteria/laws that govern dynamic similarity/similitude between two flow situations (internal and external) in order to design a geometrically scaled model (of convenient size and fluids and of economic and time proportions) that is used to calibrate the mathematical model (Chapter 11), and (2) to design systematic and well-guided experiments in order to calibrate the mathematical model. Finally, in Section 7.8, application of the procedure of dimensional analysis will be used to experimentally formulate theoretical mathematical models. While mathematically formulating equations from pure theory involves applying the five steps used in dimensional analysis to derive a functional relationship/dimensionless numbers (i.e., applying the Buckingham π theorem), experimentally calibrating and verifying equations derived from pure theory involves collecting data from an experiment.

In the case where the fluid flow situation may be modeled by the application of pure theory (for e.g., creeping flow, or laminar flow, or sonic flow, or critical flow), each physical force and physical quantity can be theoretically modeled. Thus, one may easily analytically/theoretically derive/formulate the mathematical model and apply the model without the need to calibrate or verify the model. Additionally, if one wishes to experimentally formulate, calibrate, and verify the theoretical mathematical model, one may use the dimensional analysis procedure itself (Buckingham π theorem) and the results of dimensional analysis (laws of dynamic similarity/similitude) to do so. The important role that dimensional analysis plays in the theoretical deterministic mathematical modeling is highlighted in the following five steps conducted in the modeling process. First, the dimensional analysis procedure itself (Buckingham π theorem) is used to formulate a mathematical model of the fluid flow situation by arranging/reducing the two or more significant variables (identified from the governing equations) into one dimensionless coefficient/π term ($\pi_1 = \phi(nothing) = constant$), which represents the uncalibrated mathematical model. The resulting dependent π term is a function of no other independent π terms and thus is equal to a constant. It is important to note that the dimensional analysis procedure is capable of reducing a large number of significant variables into one dimensionless coefficient/π term (see Section 7.2.3), thus reducing the number of variables required to be calibrated (experimentally evaluated). Second, this uncalibrated mathematical model is ideally calibrated (i.e., the π term is evaluated) by conducting experiments using the full-scale physical prototype of the fluid flow situation. However, using the full-scale physical prototype to conduct experiments may present challenges such as inconvenient size (too large or too small), inconvenient fluids, economic constraints, time constraints, etc. Therefore, in order to solve this problem, one may instead design a geometrically scaled (either smaller or larger) physical model of the fluid flow situation. Fortunately, the uncalibrated mathematical model resulting from the dimensional analysis procedure ($\pi_1 = \phi(nothing) = constant$) may be used to develop criteria/laws that govern dynamic similarity/similitude between two flow situations that are geometrically similar but different in size; the uncalibrated mathematical model is ready for the initial model application (synthesis) phase. These developed laws of dynamic similarity (similitude) may be applied in order to design a geometrically scaled model and thus allow the design of experiments using a model of convenient size and fluids and of economic and time proportions, which can predict the performance of the full-scale prototype (see Chapter 11). Therefore, the uncalibrated mathematical model is actually calibrated (i.e., the π term is evaluated) by conducting experiments using the geometrically scaled (either smaller or larger) physical model of the fluid flow situation. It is important to note that, although the dimensional analysis procedure itself cannot directly calibrate the mathematical model, the results of the dimensional analysis procedure are used in the development of laws of dynamic similarity, which allow the design of a geometrically scaled physical model that is used to directly calibrate the mathematical model by conducting experiments. Third, the uncalibrated mathematical model is calibrated (i.e., the π term is evaluated) by conducting experiments using the scaled physical model of the flow situation. The use of the dimensional analysis procedure to reduce a large number of significant variables into one or more dimensionless coefficients/π terms (conducted in the model formulation) drastically reduces the number of variables required to be calibrated (experimentally evaluated). Furthermore, the resulting one dimensionless coefficient/π term may be used to design a systematic and well-guided approach to conducting and interpreting the experimental investigations, thus greatly reducing the number of experiments required to calibrate the resulting one dimensionless coefficient/π term. It is important to note that in this case, where the fluid flow situation may be modeled by the application of pure theory (e.g., for creeping flow, laminar flow, sonic flow, or critical flow), the resulting dependent π term is a function of no other independent π terms and thus is equal to a constant (i.e., $\pi_1 = \phi(nothing) = constant$) (e.g., see the definitions of Stokes Law, Poiseuille's law, sonic velocity, critical velocity, etc.). As such, calibration of the π term, which is a constant, will actually require only one experiment. Fourth (ideally), the calibrated mathematical model is verified using test data. Fifth and finally, the calibrated mathematical model is ready for the final model application (synthesis) phase; one can now use the calibrated mathematical model to predict the performance of either the geometrically scaled model or the full-scale prototype of the fluid flow situation (see applications of theoretical equations throughout the textbook).

EXAMPLE PROBLEM 7.9

Use dimensional analysis to derive an expression for the shear stress, τ for laminar Couette flow. In order to determine the dominant/significant physical quantities in the fluid situation, one may examine Figure 1.3 (laminar Couette flow). Using Section 1.10.4.3 and Equation 1.59, which is $\tau = \mu \frac{dv}{dy} = \mu \frac{v_{plate}}{D} = constant$ (Newton's law of viscosity for laminar flow), as a guide, the only dominant/significant physical force is the viscous or frictional force for laminar flow, $F_V = \tau A = \mu v L$. Although the fluid is in motion, because the focus of this problem is the frictional force for laminar flow, the inertia force is not relevant. Thus, the important physical quantities include: τ, μ, v_{plate}, and D.

Mathcad Solution

In order to derive an expression for the shear stress, τ for laminar Couette flow, the five steps of dimensional analysis are applied as follows:

Step 1: Using Section 1.10.4.3 and Equation 1.59, which is $\tau = \mu \frac{dv}{dy} = \mu \frac{v_{plate}}{D} = constant$

(Newton's law of viscosity for laminar flow), as a guide, the only dominant/ significant physical force is the viscous or frictional force for laminar flow, $F_V = \tau A = \mu v L$.

Thus, in the derivation of the shear stress, τ for laminar Couette flow, the following four variables, τ, μ, v_{plate}, and D, result in one dimensionless number, namely:

$$\frac{\tau}{\frac{\mu v_{plate}}{D}} = \pi_1 = \phi(nothing) = constant.$$

Step 2: Use the simplest mathematical expressions for the identified dominant physical force and dominant physical quantities and identify the n physical quantities, all of which can be expressed in terms of m fundamental dimensions (F, L, and T).

Dominant Physical Force or Quantity	Variable and Dimensions	Variable and Dimensions	Variable and Dimensions	Variable and Dimensions
$F_V = \tau A = \mu v L$	$\mu = [FL^{-2}T]$	$v_{plate} = [LT^{-1}]$	$D = [L]$	$\tau = [FL^{-2}]$

where $n = 4$ and $m = 3$.

Step 3: The number of dimensionless numbers (π terms) is $n-m = 4-3 = 1$.

Step 4: The Buckingham π theorem is applied in order to derive the $n-m = 1$ π term as follows:

$$\pi_1 = \mu^{a_1} v_{plate}^{b_1} D^{c_1} \tau^{d_1}$$

where the exponents are unknown quantities to be determined by applying the Buckingham π theorem.

Working with the π_1 term, the dimensional equations for the *FLT* system are set up as follows:

$$\pi_1 = \mu^{a_1} v_{plate}^{b_1} D^{c_1} \tau^{d_1}$$

$$F^0 L^0 T^0 = (FL^{-2}T)^{a_1}(LT^{-1})^{b_1}(L)^{c_1}(FL^{-2})^{d_1}$$

Equating the exponents for the dimensions and solving for three of the unknown exponents in terms of the fourth unknown exponent using the Mathcad symbolic solve block yields the following:

Given

F:　　　$0 = a_1 + d_1$

L:　　　$0 = -2a_1 + b_1 + c_1 - 2 \cdot d_1$

T:　　　$0 = a_1 - b_1$

$$\text{Find } (a_1, b_1, c_1) \rightarrow \begin{pmatrix} -d_1 \\ -d_1 \\ d_1 \end{pmatrix}$$

Thus:

$$\pi_1 = \mu^{a_1} v_{plate}^{b_1} D^{c_1} \tau^{d_1}$$

$$\pi_1 = \mu^{-d_1 1} v_{plate}^{-d_1} D^{d_1} \tau^{d_1}$$

$$\pi_1 = \frac{\tau^{d_1}}{\dfrac{\mu^{d_1} v_{plate}^{d_1}}{D^{d_1}}}$$

From experimentation $d_1 = 1$; therefore:

$$\pi_1 = \frac{\tau}{\dfrac{\mu \, v_{plate}}{D}}$$

which yields the empirical expression for the shear stress, τ for laminar Couette flow.

The resulting functional relationship between the $n\text{-}m = 1$ π terms is expressed as follows:

$$\phi(\pi_1) = 0$$

$$\phi\left(\frac{\tau}{\dfrac{\mu \, v_{plate}}{D}}\right) = 0$$

or:

$$\frac{\tau}{\frac{\mu\, v_{plate}}{D}} = \phi(nothing) = constant$$

Step 5: Interpret the resulting functional relationship between the n-$m = 1$ π term. The constant may be evaluated (i.e., π_1 is calibrated) by conducting one experiment to be equal to 1, and thus $\tau = \mu\frac{dv}{dy} = \mu\frac{v_{plate}}{D} = constant$ (Newton's law of viscosity for laminar flow) as given in Equation 1.59.

EXAMPLE PROBLEM 7.10

Use dimensional analysis to derive an expression for the unbalanced net inward cohesive pressure, $\Delta p = (p_i - p_e)$ acting on the liquid molecules along the surfaces of the spherical soap bubble. In order to determine the dominant/significant physical quantities in the fluid situation, consider half of a spherical soap bubble of fluid as shown in Figure ECP 1.117. Assume that the soap bubble is neutrally buoyant in the air and thus the internal, p_i and external, p_e (atmospheric pressure when bubble is in the atmosphere) pressures acting on a perfectly spherical bubble are balanced by the existence of surface tension, σ, which acts on both the inside and outside surfaces. Using Section 1.10.5 and Figure ECP 1.117 as a guide, the dominant/significant forces acting on the neutrally buoyant (i.e., the gravitational force, $F_G =$ the buoyant force, F_B and thus, they cancel each other out) soap bubble are due to the unbalanced net inward cohesive pressure, $\Delta p = (p_i - p_e)$, (i.e., $F_P = \Delta p A = \Delta p L^2$) and the surface tension force $F_T = \sigma L$. Because the bubble is not in motion, the inertia force and the viscous/frictional force are not relevant. Finally, because the soap is assumed to be incompressible, the elastic force is also not relevant. Thus, the important physical quantities include: Δp; the fluid surface tension, σ; and the radius of the soap bubble, r. Hint: from the solution to ECP 1.117, $\Delta p = (p_i - p_e) = \frac{4\sigma}{r}$.

Mathcad Solution

In order to derive an expression for the unbalanced net inward cohesive pressure, $\Delta p = (p_i - p_e)$ acting on the liquid molecules along the surfaces of the spherical soap bubble, the five steps of dimensional analysis are applied as follows:

Step 1: Using Section 1.10.5 and Figure ECP 1.117 as a guide, the important physical quantities include: Δp; the fluid surface tension, σ; and the radius of the soap bubble, r. Thus, in the derivation of the Δp, the following three variables: $\Delta p, \sigma$, and r result in one dimensionless number, namely:

$$\frac{\Delta p}{\frac{\sigma}{r}} = \pi_1 = \phi(nothing) = constant.$$

Step 2: Use the simplest mathematical expressions for the identified dominant physical force and dominant physical quantities and identify the n physical quantities, all of which can be expressed in terms of m fundamental dimensions $(F, L, \text{ and } T)$.

Dominant Physical Force or Quantity	Variable and Dimensions	Variable and Dimensions	Variable and Dimensions
$F_P = \Delta p A = \Delta p L^2$	$\Delta p = [FL^{-2}]$	$r = [L]$	
$F_T = \sigma L$			$\sigma = [FL^{-1}]$

where $n = 3$ and $m = 2$.

Step 3: The number of dimensionless numbers (π terms) is $n\text{-}m = 3\text{-}2 = 1$.

Step 4: The Buckingham π theorem is applied in order to derive the $n\text{-}m = 1$ π term as follows:

$$\pi_1 = \Delta p^{a_1} r^{b_1} \sigma^{c_1}$$

where the exponents are unknown quantities to be determined by applying the Buckingham π theorem.

Working with the π_1 term, the dimensional equations for the FLT system are set up as follows:

$$\pi_1 = \Delta p^{a_1} r^{b_1} \sigma^{c_1}$$

$$F^0 L^0 T^0 = (FL^{-2})^{a_1} (L)^{b_1} (FL^{-1})^{c_1}$$

Equating the exponents for the dimensions and solving for two of the unknown exponents in terms of the third unknown exponent using the Mathcad symbolic solve block yields the following:

Given

F: $0 = a_1 + c_1$

L: $0 = -2a_1 + b_1 - c_1$

T: $0 = 0 + 0 + 0$

Find $(a_1, b_1) \rightarrow \begin{pmatrix} -c_1 \\ -c_1 \end{pmatrix}$

Thus:

$$\pi_1 = \Delta p^{a_1} r^{b_1} \sigma^{c_1}$$

$$\pi_1 = \Delta p^{-c_1} r^{-c_1} \sigma^{c_1}$$

$$\pi_1 = \frac{\sigma^{c_1}}{\Delta p^{c_1} r^{c_1}}$$

From experimentation $c_1 = -1$; therefore:

$$\pi_1 = \frac{\Delta p}{\dfrac{\sigma}{r}}$$

which yields the empirical expression for the unbalanced net inward cohesive pressure, $\Delta p = (p_i - p_e)$ acting on a soap bubble.

The resulting functional relationship between the $n\text{-}m = 1$ π terms is expressed as follows:

$$\phi(\pi_1) = 0$$

$$\phi\left(\frac{\Delta p}{\dfrac{\sigma}{r}}\right) = 0$$

or:

$$\frac{\Delta p}{\dfrac{\sigma}{r}} = \phi(nothing) = constant$$

Step 5: Interpret the resulting functional relationship between the $n\text{-}m = 1$ π term. The constant may be evaluated (i.e., π_1 is calibrated) by conducting one experiment to be equal to 4, and thus $\Delta p = (p_i - p_e) = \frac{4\sigma}{r}$.

EXAMPLE PROBLEM 7.11

Use dimensional analysis to derive an expression for the terminal (free-fall) velocity for a sphere falling very slowly (creeping flow at $R \leq 1$) in a fluid. In order to determine the dominant/significant physical quantities in the fluid flow situation, one may examine Figure EP 10.8b (falling sphere viscometer). Using Section 10.5.3.7 and Figure EP 10.8b as a guide, the dominant/significant forces acting on the falling sphere are the inertia force (assumption of steady state flow), $F_I = Ma = \rho_s v^2 L^2$; the gravitational force, $F_G = Mg = \rho_s L^3 g$; the pressure (buoyant) force, $F_B = \Delta p A = \gamma_f V = \gamma_f L^3 = \rho_f g L^3$; and the flow resistance (drag) force, $F_D = \tau A = \mu v L$ (as modeled by Stokes law, $F_D = 3\pi\mu v D$). However, because there is a vast amount of fluid, the surface tension force is not relevant. Finally, because the fluid is assumed to be incompressible, the elastic force is also not relevant. Thus, the important physical quantities include: $\Delta \gamma = (\rho_s - \rho_f) g, v, D$, and μ. Hint: from the solution to Example Problem 10.8 $v = \dfrac{D^2(\gamma_s - \gamma_f)}{18\mu}$.

Mathcad Solution

In order to derive an expression for the terminal (free-fall) velocity for a sphere falling very slowly (creeping flow at $R \leq 1$) in a fluid, the five steps of dimensional analysis are applied as follows:

Step 1: Using Section 10.5.3.7 and Figure EP 10.8b as a guide, the important physical quantities include: $\Delta\gamma = (\rho_s - \rho_f)g$, v, D, and μ. Thus, in the derivation of the v, the following four variables, $\Delta\gamma = (\rho_s - \rho_f)g$, v, D, and μ, result in one dimensionless number, namely:

$$\frac{v}{\dfrac{D^2 \Delta\gamma}{\mu}} = \pi_1 = \phi(nothing) = constant.$$

Step 2: Use the simplest mathematical expressions for the identified dominant physical force and dominant physical quantities and identify the n physical quantities, all of which can be expressed in terms of m fundamental dimensions (F, L, and T).

Dominant Physical Force or Quantity	Variable and Dimensions	Variable and Dimensions	Variable and Dimensions	Variable and Dimensions
$F_I = Ma = \rho_s v^2 L^2$		$v = [LT^{-1}]$	$D = [L]$	
$F_G = Mg = \rho_s L^3 g$				
$F_B = \Delta p A = \gamma_f V = \rho_f g L^3$	$\Delta\gamma = [FL^{-3}]$			
$F_D = \tau A = \mu v L$				$\mu = [FL^{-2}T]$

where $n = 4$ and $m = 3$.

Step 3: The number of dimensionless numbers (π terms) is $n-m = 4-3 = 1$.

Step 4: The Buckingham π theorem is applied in order to derive the $n-m = 1$ π term as follows:

$$\pi_1 = v^{a_1} D^{b_1} \Delta\gamma^{c_1} \mu^{d_1}$$

where the exponents are unknown quantities to be determined by applying the Buckingham π theorem.

Working with the π_1 term, the dimensional equations for the *FLT* system are set up as follows:

$$\pi_1 = v^{a_1} D^{b_1} \Delta\gamma^{c_1} \mu^{d_1}$$

$$F^0 L^0 T^0 = (LT^{-1})^{a_1} (L)^{b_1} (FL^{-3})^{c_1} (FL^{-2}T)^{d_1}$$

Equating the exponents for the dimensions and solving for three of the unknown exponents in terms of the fourth unknown exponent using the Mathcad symbolic solve block yields the following:

Given

F: $0 = c_1 + d_1$

L: $0 = a_1 + b_1 - 3c_1 - 2{\cdot}d_1$

T: $0 = -a_1 + d_1$

$$\text{Find } (a_1, b_1, c_1) \rightarrow \begin{pmatrix} d_1 \\ -2 \cdot d_1 \\ -d_1 \end{pmatrix}$$

Thus:

$$\pi_1 = v^{a_1} D^{b_1} \Delta\gamma^{c_1} \mu^{d_1}$$

$$\pi_1 = v^{d_1} D^{-2d_1} \Delta\gamma^{-d_1} \mu^{d_1}$$

$$\pi_1 = \frac{v^{d_1}}{\dfrac{D^{2d_1} \Delta\gamma^{d_1}}{\mu^{d_1}}}$$

From experimentation $d_1 = 1$; therefore:

$$\pi_1 = \frac{v}{\dfrac{D^2 \Delta\gamma}{\mu}}$$

which yields the empirical expression for the terminal (free-fall) velocity for a sphere falling very slowing (creeping flow at $R \leq 1$) in a fluid.

The resulting functional relationship between the $n\text{-}m = 1$ π terms is expressed as follows:

$$\phi(\pi_1) = 0$$

$$\phi\left(\frac{v}{\dfrac{D^2 \Delta\gamma}{\mu}}\right) = 0$$

or:

$$\frac{v}{\dfrac{D^2 \Delta\gamma}{\mu}} = \phi(nothing) = constant$$

Step 5: Interpret the resulting functional relationship between the $n\text{-}m = 1$ π term. The constant may be evaluated (i.e., π_1 is calibrated) by conducting one experiment to be equal to 1/18, and thus $v = \dfrac{D^2 \Delta\gamma}{18\mu} = \dfrac{D^2(\gamma_s - \gamma_f)}{18\mu}$.

End-of-Chapter Problems

Problems with a "C" are conceptual problems. All problems that require computations are solved using Mathcad.

Introduction

7.1C   The dimensional analysis procedure itself, and the results of the procedure, including the laws of similitude, play an important role in the deterministic mathematical modeling of fluid flow problems. Explain in detail the role of dimensional in the deterministic modeling of fluid flow problems.

7.2C   A major application of dimensional analysis in fluid flow is for problems that are too complex to be modeled by the application of pure theory. Explain and give examples.

7.3C   In the application of the governing equations, although in some flow cases, one may ignore the flow resistance (flow from a tank, etc., and gradual channel transitions) or not need to model the flow resistance (a hydraulic jump), in most practical flow cases in general, one must account for the fluid viscosity and thus model the resulting shear stress and flow resistance. List five basic applications (practical flow cases) of the governing equations for which one must account for flow resistance.

7.4C   The application of the governing equations (continuity, energy, and momentum) for real flow (internal), and ideal flow (flow-measuring devices for internal flow and external flow around objects) requires the modeling of flow resistance (frictional losses). Explain the implication of the assumption of real flow vs. ideal flow in the modeling of flow resistance.

7.5C   Dimensional analysis is used in the empirical modeling of flow resistance in real fluid flow. Specifically, the results of the dimensional analysis procedure are used to supplement the momentum theory in the definition of the friction/viscous force and thus the definition of the flow resistance equations. Explain why dimensional analysis is used to supplement the momentum theory in the definition of the friction/viscous force.

7.6C   Derivation of a given flow resistance equation in internal and external flow (major head loss, minor head loss, actual discharge, and drag force) involves a "subset level" application of the appropriate governing equations, and in some cases supplementing the momentum equation with dimensional analysis. Explain what a "subset level" application of the appropriate governing equations implies.

Supplementing the Momentum Theory with Dimensional Analysis: Derivation of the Flow Resistance Equations and the Drag Coefficients

7.7C   The flow resistance causes an unknown pressure drop, Δp, which causes an unknown head loss, h_f, where the head loss is due to a conversion of kinetic energy to heat, which is modeled/displayed in the integral form of the energy equation (Equation 4.44). The flow resistance is modeled as a drag force in external flows, while it is modeled as a head loss in internal flows. When the friction/viscous forces can either be theoretically modeled (as in the case of the major head loss in laminar flow) or ignored (as in the case of the minor head loss due to a sudden pipe expansion) in the momentum equation, the respective flow resistance equation can be analytically derived. However, when the friction/viscous forces cannot either be theoretically modeled or ignored in the momentum equation, the momentum theory will be exhausted due to an unknown head loss, h_f. Explain, specifically, how the results of dimensional analysis are used to model the unknown pressure drop, Δp in the derivation of the flow resistance equation in internal and external flow (major head loss, minor head loss, actual discharge, and drag force).

7.8C   The derivation of the appropriate flow resistance equations in internal and external flow (major head loss, minor head loss, actual discharge, and drag force) and their corresponding "drag coefficient, C_D" is presented in Chapter 7, herein (also see Chapter 6). Briefly describe the resistance equations and their corresponding "drag coefficient, C_D." Also, state which chapters in this textbook present the empirical evaluation of the respective drag coefficients and application of the respective flow resistance equations.

Derivation of the Efficiency of Pumps and Turbines

7.9C The derivation of the efficiency, η of a pump or a turbine is represented by the "drag coefficient, C_D" and is presented in Chapter 7 (also see Chapter 6). State which chapters in this textbook present the empirical evaluation of the efficiency, η of a pump or a turbine and application of the efficiency, η of a pump or a turbine.

Dimensional Analysis of Fluid Flow

7.10C Explain the basis of the dimensional analysis theory, and briefly give what it states.

7.11C Define what is meant by a functional relationship in dimensional analysis.

7.12C What is the first step required in the procedure to derive a functional relationship between n physical variables in a given flow situation and thus the n-m dimensionless numbers (or π terms) for the given flow situation?

Dynamic Forces Acting on a Fluid Element

7.13C List the dynamic forces that may act in a flow situation, and give the simplest mathematical expressions for these forces.

7.14C Assuming steady state flow (no acceleration), state and mathematically illustrate which force will always be important.

Two-Dimensional Systems

7.15C What are the two basic dimensional systems (for a given physical quantity), and how are they related? What are the two basic systems of units?

Deriving a Functional Relationship/Dimensionless Numbers (Pi Terms), Main Pi Terms, and Definition of New Pi Terms

7.16C List and describe the five steps in the procedure to derive a functional relationship between the n physical variables in the fluid flow situation and thus the n-m dimensionless numbers (π terms) for a given flow situation.

7.17C List the five main dimensionless numbers (π terms) derived from dimensional analysis, and give the expressions for each π term.

7.18C Because dimensional analysis may not yield the final expression for the drag coefficient, C_D, one may need to interpret the resulting dimensionless numbers (π terms) through manipulation and empirical studies. Give the four basic rules for alternatively redefining a given dimensionless number (π term) to yield a new π term.

Guidelines in the Derivation of the Flow Resistance Equations and the Drag Coefficients

7.19C Dimensional analysis is used to derive the flow resistance equations in internal and external flow (major head loss, minor head loss, actual discharge, and drag force) and their corresponding drag coefficient, C_D, which expresses the ratio of a real flow variable (such as velocity, pressure, force, etc.) to its corresponding ideal flow variable. In the dimensional analysis of fluid flow, there are always at least four parameters involved, ρ, v, L (or D or y; characteristic length dimension), and Δp (or Δy, Δz, or $\Delta H = h_f = \tau_w L / \gamma R_h = \Delta p / \gamma = S_f L$; a representation of energy loss or some form of flow resistance). Explain Step 1 in the five-step procedure to derive a functional relationship between the n physical variables in the fluid flow situation and thus the n-m dimensionless numbers (π terms) for a given internal or external flow situation.

7.20C   Dimensional analysis is used to derive the flow resistance equations in internal and external flow (major head loss, minor head loss, actual discharge, and drag force) and their corresponding drag coefficient, C_D, which expresses the ratio of a real flow variable (such as velocity, pressure, force, etc.) to its corresponding ideal flow variable. Explain Step 2 in the five-step procedure to derive a functional relationship between the n physical variables in the fluid flow situation and thus the n-m dimensionless numbers (π terms) for a given internal or external flow situation.

7.21C   Dimensional analysis is used to derive the flow resistance equations in internal and external flow (major head loss, minor head loss, actual discharge, and drag force) and their corresponding drag coefficient, C_D, which expresses the ratio of a real flow variable (such as velocity, pressure, force, etc.) to its corresponding ideal flow variable. Explain Step 3 in the five-step procedure to derive a functional relationship between the n physical variables in the fluid flow situation and thus the n-m dimensionless numbers (π terms) for a given internal or external flow situation.

7.22C   Dimensional analysis is used to derive the flow resistance equations in internal and external flow (major head loss, minor head loss, actual discharge, and drag force) and their corresponding drag coefficient, C_D, which expresses the ratio of a real flow variable (such as velocity, pressure, force, etc.) to its corresponding ideal flow variable. Explain Step 4 in the five-step procedure to derive a functional relationship between the n physical variables in the fluid flow situation and thus the n-m dimensionless numbers (π terms) for a given internal or external flow situation.

7.23C   Dimensional analysis is used to derive the flow resistance equations in internal and external flow (major head loss, minor head loss, actual discharge, and drag force) and their corresponding drag coefficient, C_D, which expresses the ratio of a real flow variable (such as velocity, pressure, force, etc.) to its corresponding ideal flow variable. Explain Step 5 in the five-step procedure to derive a functional relationship between the n physical variables in the fluid flow situation and thus the n-m dimensionless numbers (π terms) for a given internal or external flow situation.

Definition of the Drag Coefficient

7.24C   Conceptually explain how the drag coefficient, C_D is defined in the dimensional analysis of fluid flow.

7.25C   The drag coefficient, C_D is defined in the dimensional analysis of fluid flow. The dimensional analysis of fluid flow assumes that the inertia force, $F_I = Ma = \rho v^2 L^2$ will always be a dominant force and results in the three primary variables: ρ, v, and L. Furthermore, the friction/viscous force for turbulent flow, F_f or $F_V = \tau A$ will always be the first/main additional dominant force, which can be represented by a dominant physical quantity (F_D or h_f) that represents an energy loss or some form of flow resistance and results in an appropriate secondary variable (F_D, τ_w, Δp, or g). Although the first/main additional dominant force is the friction/viscous force for turbulent flow, $F_V = \tau A$, the type of flow will determine the cause of the friction/viscous force, and thus the definition of the drag coefficient, C_D. Explain the cause of the friction/viscous force and its implication in the definition of the drag coefficient, C_D for external and internal fluid flow.

7.26C The definition of the drag coefficient, C_D involves at least two dominant forces. The dimensional analysis of fluid flow assumes that the inertia force, $F_I = Ma = \rho v^2 L^2$ will always be a dominant force and results in the three primary variables: ρ, v, and L. Furthermore, the friction/viscous force for turbulent flow, F_f or $F_V = \tau A$ will always be the first/main additional dominant force, which can be represented by a dominant physical quantity (F_D or h_f) that represents an energy loss or some form of flow resistance and results in an appropriate secondary variable (F_D, τ_w, Δp, or g). The modeling of only one additional dominant force (i.e., the friction/viscous force for turbulent flow) results in only one dimensionless number (π term), which is the drag coefficient, C_D. Furthermore, modeling additional dominant forces or dominant physical quantities results in additional secondary variables, and will introduce an additional dimensionless number (π term) for each new secondary variable added. Explain the results and implications of the modeling of additional dominant forces or dominant physical quantities in the definition of the drag coefficient, C_D in fluid flow.

7.27C In the derivation of the drag force, F_D in external flow, assume that in addition to the inertia force, F_I and the pressure force, F_P (which causes the viscous force for turbulent flow, $F_V = \tau A$), other forces including the viscous force for laminar flow, $F_V = \tau A = \mu v L$, the elastic force, F_E and the gravity force, F_G and other physical quantities including the absolute surface roughness, ε and the geometry, L_i also play an important role in the external fluid flow. Explain what dimensionless quantities (π terms) the drag coefficient, C_D will be a function of.

Guidelines in the Derivation of the Efficiency of Pumps and Turbines

7.28C Dimensional analysis is used to derive the efficiency, η of a pump or a turbine, which is represented by the "drag coefficient, C_D." In general, the drag coefficient, C_D expresses the ratio of a real flow variable (such as velocity, pressure, force, power, etc.) to its corresponding ideal flow variable. In particular, the efficiency of a pump, $\eta_{pump} = (P_p)_{out}/(P_p)_{in} = \gamma Q h_{pump}/\omega T_{shaft,in}$ (Equation 4.197) expresses the ratio of a real/actual flow variable (hydraulic power output from the pump) to its corresponding ideal flow variable (shaft power input to the pump). And, the efficiency of a turbine, $\eta_{turbine} = (P_t)_{out}/(P_t)_{in} = \omega T_{shaft,out}/\gamma Q h_{turbine}$ (Equation 4.217) expresses the ratio of a real/actual flow variable (shaft power output from the turbine) to its corresponding ideal flow variable (hydraulic power input to the turbine). In the dimensional analysis of fluid flow, there are always at least four parameters involved, ρ, v, L (or D or y; characteristic length dimension), and Δp (or Δy, Δz, or $\Delta H = h_f = \tau_w L/\gamma R_h = \Delta p/\gamma = S_f L$; a representation of energy loss or some form of flow resistance). For pumps, the characteristic length dimension is the diameter of the impeller blades, D. For turbines, the characteristic length dimension is the diameter of the runner blades, D. For pumps and turbines, the representation of energy loss or some form of flow resistance is a head loss, h_f or η. Explain Step 1 in the five-step procedure to derive a functional relationship between the n physical variables in the fluid flow situation and thus the n-m dimensionless numbers (π terms) for a given flow situation.

7.29C Dimensional analysis is used to derive the efficiency, η of a pump or a turbine, which is represented by the "drag coefficient, C_D." The efficiency of a pump, $\eta_{pump} = (P_p)_{out}/(P_p)_{in} = \gamma Q h_{pump}/\omega T_{shaft,in}$ (Equation 4.197) expresses the ratio of a real/actual flow variable (hydraulic power output from the pump) to its corresponding ideal flow variable (shaft power input to the pump). And, the efficiency of a turbine, $\eta_{turbine} = (P_t)_{out}/(P_t)_{in} = \omega T_{shaft,out}/\gamma Q h_{turbine}$ (Equation 4.217) expresses the ratio of a real/actual flow variable (shaft power output from the turbine) to its corresponding ideal flow variable (hydraulic power input to the turbine). Explain Step 2 in the five-step procedure to derive a functional relationship between the n physical variables in the fluid flow situation and thus the n-m dimensionless numbers (π terms) for a given flow situation.

7.30C Dimensional analysis is used to derive the efficiency, η of a pump or a turbine, which is represented by the "drag coefficient, C_D." The efficiency of a pump, $\eta_{pump} = (P_p)_{out}/(P_p)_{in} = \gamma Q h_{pump}/\omega T_{shaft,in}$ (Equation 4.197) expresses the ratio of a real/actual flow variable (hydraulic power output from the pump) to its corresponding ideal flow variable (shaft power input to the pump). And, the efficiency of a turbine, $\eta_{turbine} = (P_t)_{out}/(P_t)_{in} = \omega T_{shaft,out}/\gamma Q h_{turbine}$ (Equation 4.217) expresses the ratio of a real/actual flow variable (shaft power output from the turbine) to its corresponding ideal flow variable (hydraulic power input to the turbine). Explain Step 3 in the five-step procedure to derive a functional relationship between the n physical variables in the fluid flow situation and thus the n-m dimensionless numbers (π terms) for a given flow situation.

7.31C Dimensional analysis is used to derive the efficiency, η of a pump or a turbine, which is represented by the "drag coefficient, C_D." The efficiency of a pump, $\eta_{pump} = (P_p)_{out}/(P_p)_{in} = \gamma Q h_{pump}/\omega T_{shaft,in}$ (Equation 4.197) expresses the ratio of a real/actual flow variable (hydraulic power output from the pump) to its corresponding ideal flow variable (shaft power input to the pump). And, the efficiency of a turbine, $\eta_{turbine} = (P_t)_{out}/(P_t)_{in} = \omega T_{shaft,out}/\gamma Q h_{turbine}$ (Equation 4.217) expresses the ratio of a real/actual flow variable (shaft power output from the turbine) to its corresponding ideal flow variable (hydraulic power input to the turbine). Explain Step 4 in the five-step procedure to derive a functional relationship between the n physical variables in the fluid flow situation and thus the n-m dimensionless numbers (π terms) for a given flow situation.

7.32C Dimensional analysis is used to derive the efficiency, η of a pump or a turbine, which is represented by the "drag coefficient, C_D." The efficiency of a pump, $\eta_{pump} = (P_p)_{out}/(P_p)_{in} = \gamma Q h_{pump}/\omega T_{shaft,in}$ (Equation 4.197) expresses the ratio of a real/actual flow variable (hydraulic power output from the pump) to its corresponding ideal flow variable (shaft power input to the pump). And, the efficiency of a turbine, $\eta_{turbine} = (P_t)_{out}/(P_t)_{in} = \omega T_{shaft,out}/\gamma Q h_{turbine}$ (Equation 4.217) expresses the ratio of a real/actual flow variable (shaft power output from the turbine) to its corresponding ideal flow variable (hydraulic power input to the turbine). Explain Step 5 in the five-step procedure to derive a functional relationship between the n physical variables in the fluid flow situation and thus the n-m dimensionless numbers (π terms) for a given flow situation.

The Definition of the Pump (or Turbine) Efficiency

7.33C Conceptually explain how the pump efficiency, η (drag coefficient, C_D) is defined in the dimensional analysis of fluid flow.

7.34C The pump efficiency, η (drag coefficient, C_D) is defined in the dimensional analysis of fluid flow. The dimensional analysis of fluid flow assumes that the inertia force, $F_I = Ma = \rho v^2 L^2$ will always be a dominant force and results in the three primary variables: ρ, ω, and D. Furthermore, the friction/viscous force for turbulent flow, F_f or $F_V = \tau A$ will always be the first/main additional dominant force, which can be represented by a dominant physical quantity (η or h_f) that represents an energy loss or some form of flow resistance and results in the appropriate secondary variable η. Although the first/main additional dominant force is the friction/viscous force for turbulent flow, $F_V = \tau A$, the type of flow will determine the cause of the friction/viscous force; thus, the definition of the drag coefficient, C_D. Explain the cause of the friction/viscous force and its implication in the definition of the drag coefficient, C_D (pump efficiency, η).

7.35C The definition of the pump efficiency, η (drag coefficient, C_D) involves at least two dominant forces. The dimensional analysis of fluid flow assumes that the inertia force, $F_I = Ma = \rho v^2 L^2$ will always be a dominant force and results in the three primary variables: ρ, ω, and D. Furthermore, the friction/viscous force for turbulent flow, F_f or $F_V = \tau A$ will always be the first/main additional dominant force, which can be represented by a dominant physical quantity (η or h_f) that represents an energy loss or some form of flow resistance and results in the appropriate secondary variable η. The modeling of only one additional dominant force (i.e., the friction/viscous force for turbulent flow) results in only one dimensionless number (π term), which is the drag coefficient, C_D (pump efficiency, η). Furthermore, modeling additional dominant forces or dominant physical quantities results in additional secondary variables and will introduce an additional dimensionless number (π term) for each new secondary variable added. Explain the results and implications of the modeling of additional dominant forces or dominant physical quantities in the definition of the pump efficiency, η (drag coefficient, C_D) in fluid flow.

7.36C In the derivation of the pump efficiency, η, assume that in addition to the inertia force, F_I and the pressure force, F_P (which causes the viscous force for turbulent flow, $F_V = \tau A$), other forces, including the viscous force for laminar flow, $F_V = \tau A = \mu v L$; the elastic force, F_E; and other physical quantities, including γ, Q, h_{pump}, T, as well as absolute blade surface roughness, ε, also play an important role in the flow through a pump. Explain what dimensionless quantities (π terms) the drag coefficient, C_D (pump efficiency, η) will also be a function of.

Modeling the Flow Reistance as a Drag Force in External Flow

7.37C In the external flow around an object, energy or work is used to move the submerged body through the fluid, where the fluid is forced to flow over the surface of the body. Thus, in general, what dominant forces/physical quantities are typically sought in the dimensional analysis of external flow around an object?

Application of Dimensional Analysis: Derivation of the Drag Force and the Drag Coefficient

7.38 Use dimensional analysis to derive an empirical expression for the drag force, F_D for the most general assumptions as follows: assume that the submerged body illustrated in Figure 10.2 is moving at a low speed in a stationary, homogeneous, viscous, compressible fluid, with a possible wave action at the free surface. As a result, the fluid would exert a drag force, F_D on the body, while the effect of the lift force, F_L would be negligible (low speed). The inertia force ($F_I = Ma = \rho v^2 L^2$) and the drag force, F_D will always be important. Using Table 11.7 as a guide, the additional dominant/significant physical forces and dominant physical quantities in the flow situation are the viscous force for laminar flow, $F_V = \tau A = \mu v L$; the elastic force, $F_E = E_v A = E_v L^2$; the gravitational force, $F_G = Mg = \rho L^3 g$; the absolute surface roughness, ε; and the geometry, L_i. Thus, the important physical quantities include: $\rho, v, L, F_D, \mu, E_v, g, \varepsilon$, and L_i.

7.39C   While dimensional analysis is used to derive an expression for the drag force, F_D, as given by Equation 7.27 as follows: $F_D = C_D \rho v^2 L^2 = C_D \rho v^2 A$, the practical application of the expression for the drag force, F_D is given by Equation 7.34 as follows: $F_D = C_D \frac{1}{2} \rho v^2 A$. Explain, and provide the mathematical derivation of Equation 7-34.

7.40C   Dimensional analysis is used to derive the drag force (flow resistance equation in external flow) $F_D = C_D \frac{1}{2} \rho v^2 A$ (Equation 7.34), and its corresponding drag coefficient, C_D, which expresses the ratio of a real flow variable (such as velocity, pressure, force, etc.) to its corresponding ideal flow variable. Mathematically illustrate the expression for the drag coefficient, C_D.

7.41   Derive an empirical expression for the drag force for laminar flow over a long elliptical (streamlined) rod. The inertia force $(F_I = Ma = \rho v^2 L^2)$, and the drag force, F_D will always be important. Using Table 11.7 as a guide, the additional dominant/significant physical forces and dominant physical quantities in the flow situation are the viscous force for laminar flow, $F_V = \tau A = \mu v L$, and the geometry, L_i. Thus, the important physical quantities include: ρ, v, L, F_D, μ, and L_i. Experiments are performed using glycerin to determine the drag force acting on the long elliptical rod. What are the important π terms used to organize the experimental data?

7.42   (Refer to Example Problem 7.1, and use Table 11.7 as a guide.) Infer the empirical expression for the drag force for laminar flow over a short (blunt) cylinder. Experiments are performed using oil to determine the drag force acting on the short cylinder. What are the important π terms used to organize the experimental data?

7.43   (Refer to Example Problem 7.1, and use Table 11.7 as a guide.) Infer the empirical expression for the drag force for turbulent flow over a long square shaped (blunt) cylinder. Experiments are performed using water to determine the drag force acting on the long square shaped cylinder. What are the important π terms used to organize the experimental data?

7.44   Derive an empirical expression for the drag force for an airplane (turbulent flow over a streamlined body) flying at subsonic speed ($M = v/c < 1$, where c is the sonic velocity; see Chapter 10). The inertia force ($F_I = Ma = \rho v^2 L^2$), and the drag force, F_D will always be important. Using Table 11.7 as a guide, the additional dominant/significant physical forces and dominant physical quantities in the flow situation are the viscous force for laminar flow, $F_V = \tau A = \mu v L$; the elastic force, $F_E = E_v A = E_v L^2$; the absolute surface roughness, ε; and the geometry, L_i. Thus, the important physical quantities include: ρ, v, L, F_D, μ, E_v, ε, and L_i. Refer to Section 10.5.3.12, and explain the role of each derived π term, depending on the speed of the airplane.

7.45   (Refer to Example Problem 7.1, and use Table 11.7 as a guide.) Infer the empirical expression for the drag force for laminar flow over a circular cylinder. Experiments are performed using oil to determine the drag force acting on the circular cylinder. What are the important π terms used to organize the experimental data?

7.46   (Refer to Example Problem 7.1, and use Table 11.7 as a guide.) Infer the empirical expression for the drag force for turbulent flow over a sphere. Experiments are performed using various fluids to determine the drag force acting on the sphere. What are the important π terms used to study and describe this flow process for compressible vs. incompressible flow?

7.47   (Refer to Example Problem 7.1, and use Table 11.7 as a guide.) Infer the empirical expression for the drag force for flow over a long streamlined hull. Experiments are performed using water to determine the drag force acting on the long streamlined hull. What are the important π terms used to study and describe this flow process for laminar vs. turbulent flow?

Application of Dimensional Analysis: Derivation of the Lift Force and the Lift Coefficient

7.48C In the external flow around an object, if the body is moving at a high speed (and depending upon the shape of the body and the angle it makes with the direction of flow), in addition to the drag force, F_D, the lift force, F_L also becomes dominant. Define the lift force, F_L and explain why dimensional analysis is used to derive its expression.

7.49 (Refer to Example Problem 7.2.) Infer the empirical expression for the lift force acting on an airfoil. Experiments are performed using a wind tunnel to determine the lift force acting on the airfoil. Refer to Section 10.5.5, and explain the role of each derived π term, depending on the speed of the airfoil.

Modeling the Flow Resistance as a Major Head Loss in Internal Flow

7.50C In the internal flow through a pipe or a channel, energy or work is used to move/force the fluid through a conduit. Thus, what dominant forces/physical quantities are typically sought in the dimensional analysis of internal flow?

Application of Dimensional Analysis: Derivation of the Wall Shear Stress and Drag Coefficient in Turbulent Flow

7.51 Use dimensional analysis to derive an empirical expression for the wall shear stress, τ_w for use in the definition of the expression for the major head loss (Equation 7.38). Assume either laminar or turbulent flow. The inertia force ($F_I = Ma = \rho v^2 L^2$) and the wall shear stress, τ_w will always be important. Using Table 11.8 as a guide, the additional dominant/significant forces and physical quantities in the flow situation include the viscous force for laminar flow, $F_V = \tau A = \mu v L$ and the absolute surface roughens, ε. Thus, the important physical quantities include: ρ, v, L, τ_w, μ, and ε.

The Chezy Equation

7.52C Derive the Chezy equation and the Chezy coefficient, C. Furthermore, discuss the practical implications of the Chezy equation in modeling the flow resistance in turbulent pipe and open channel flow.

7.53C The Chezy coefficient, C models the flow resistance. What are its dimensions, what is it a function of, and what are typical ranges for its empirical values?

7.54C Assuming that the Chezy coefficient, C is chosen to model the empirical flow resistance, derive the empirical expression for the major head loss, h_f equation for turbulent flow (the Chezy head loss equation).

The Darcy–Weisbach Equation

7.55C The Darcy–Weisbach friction coefficient (factor), f, which represents a dimensionless flow resistance coefficient, is derived using dimensional analysis, and expressed as a function of the Chezy coefficient, C as: $C = \sqrt{8g/f}$. Briefly outline the procedure used to derive the Darcy–Weisbach friction coefficient (factor), f.

7.56C   Using the Darcy–Weisbach friction coefficient, f (which is a more standard definition of the drag coefficient, C_D), derive the resulting relationship between the wall shear stress, τ_w and the velocity, v from dimensional analysis.

7.57C   The Darcy–Weisbach friction coefficient, f models the flow resistance. What are its dimensions and what is it a function of?

7.58C   Define the relationship between the Chezy coefficient, C and the Darcy–Weisbach friction factor, f.

7.59C   Briefly state the practical advantages of using the Darcy–Weisbach friction factor, f as opposed to using the Chezy coefficient, C to model the model the effects of the flow resistance (i.e., the drag coefficient, C_D).

7.60C   The Darcy–Weisbach friction coefficient, f has no dimensions, and (in general) is a function of the boundary roughness, ε/D; the Reynolds number, R; and the cross-sectional shape or the hydraulic radius, R_h where ε is the absolute pipe and ε/D is the relative pipe roughness. When is the Darcy–Weisbach friction coefficient, f independent of R and when is the Darcy–Weisbach friction coefficient, f independent of the boundary roughness, ε/D?

7.61C   Briefly explain how the friction factor, f for turbulent flow is evaluated.

7.62C   Assuming that the Darcy–Weisbach friction factor, f is chosen to model the empirical flow resistance, derive the empirical expression for the head loss, h_f equation for turbulent flow (the Darcy–Weisbach head loss equation).

7.63C   Because laminar flow is a special case of turbulent flow, the "theoretical flow resistance coefficient" for laminar flow is a special case of the empirical flow resistance coefficient for turbulent flow ($f = 64/R$). Evaluate the Darcy–Weisbach friction coefficient, f for laminar pipe flow.

Manning's Equation

7.64C   The dimensions of the Chezy coefficient, C are $L^{1/2} T^{-1}$, and it is empirically evaluated as a function of the boundary roughness, ε/D; the Reynolds number, R; and the cross-sectional shape or the hydraulic radius, R_h. However, further empirical studies (assuming the metric system) by Manning indicated that a more accurate dependence of the Chezy coefficient, C on the hydraulic radius, R_h is defined by the definition of the Manning's roughness coefficient, n. Define the empirical relationship between the Chezy coefficient, C and the Manning's roughness coefficient, n. Also, what are the dimensions and units of the Manning's roughness coefficient, n and what is it a function of?

7.65C   Begin with the Chezy equation (Equation 7.51), derive the Manning's equation.

7.66C   Although the derivation/formulation of the Manning's equation assumes no specific units (SI or BG), the Manning's roughness coefficient, n has dimensions of $L^{-1/3}$ T. Furthermore, because the values for the Manning's roughness coefficient, n have been provided/calibrated in the SI units $m^{-1/3}$ s (see Table 8.6), the Manning's equation (Equation 7.67) above assumes the metric (SI) system. Give the corresponding Manning's equation for the English (BG) system.

7.67C   Assuming that the Manning roughness coefficient, n is chosen to model the empirical flow resistance, derive the Manning's head loss equation for the metric (SI) system and the English (BG) system.

Application of the Major Head Loss Equation for Open Channel Flow

7.68C   The functional relationship, $(\tau_w/\rho v^2) = \phi(R, (\varepsilon/D))$ derived from dimensional analysis in Section 7.4.5.1 requires a slightly different interpretation in the application for open channel flow. Thus, provide the interpretation of the results of dimensional analysis (Step 5) for open channel flow.

7.69C   Assuming that the Chezy coefficient, C is chosen to model the empirical flow resistance, derive the Chezy head loss equation for open channel flow.

7.70C   The functional relationship, $(\tau_w/\rho v^2) = \phi(R, (\varepsilon/D))$ derived from dimensional analysis in Section 7.4.5.1 requires a slightly different interpretation in the application for open channel flow. Assuming that the Darcy–Weisbach friction factor, f is chosen to model the flow resistance, provide the interpretation of the results of dimensional analysis (Step 5) for open channel flow.

7.71C   Assuming that the Darcy–Weisbach friction factor, f is chosen to model the empirical flow resistance, derive the Darcy–Weisbach head loss equation for open channel flow.

7.72C   Assuming that the Manning's roughness coefficient, n is chosen to model the empirical flow resistance, derive the Manning's head loss equation for open channel flow.

7.73   (Refer to Example Problem 7.3 that used the F, L, T fundamental dimensions.) Using the M, L, T fundamental dimensions, derive an empirical expression for the wall shear stress, τ_w for turbulent open channel flow. The inertia force $(F_I = Ma = \rho v^2 L^2)$ and the wall shear stress, τ_w will always be important. Using Table 11.8 as a guide, the additional dominant/significant physical quantity in the flow situation includes ε. Thus, the important physical quantities include: $\rho, v, L, \tau_w,$ and ε. What are the important π terms used to study and describe turbulent open channel flow? What are the corresponding expressions for the friction slope, S_f and the major head loss, $h_{f,major}$? Compare your results with Example Problem 7.3, which used the F, L, T fundamental dimensions; they should be identical.

Modeling the Flow Resistance as a Minor Head Loss in Pipe Flow

7.74C   In the internal flow through a pipe, energy or work is used to move/force the fluid through a conduit. Furthermore, in the dimensional analysis of internal flow through a pipe, one is interested in the determination of the energy or head losses, pressure drops, and cavitation where energy is dissipated. While the major head loss term, $h_{f,major}$ in the energy equation models the head loss due to pipe friction in the straight section of a pipe, the minor head loss term, $h_{f,minor}$ in the energy equation models the head loss due to various pipe components in the pipe system. Explain the importance of modeling the minor head loss, $h_{f,minor}$ in the dimensional analysis of pipe flow.

Application of Dimensional Analysis: Derivation of the Pressure Drop and Drag Coefficient for Pipe Components in General

7.75   Use dimensional analysis to derive an empirical expression for the pressure drop, Δp for use in the definition of the expression for the minor head loss, $h_{f,min} = \tau_w L/\gamma R_h = \Delta p/\gamma$ for pipe components in general. The inertia force $(F_I = Ma = \rho v^2 L^2)$ and the pressure drop, Δp will always be important. Using Table 11.8 and Section 8.6.4 as a guide, the additional dominant/significant physical forces and quantities in the flow situation include the viscous force for laminar flow, $F_V = \tau A = \mu v L$; the absolute surface roughness, ε; and the geometry, L_i. Thus, the important physical quantities include: $\rho, v, L, \Delta p, \mu, \varepsilon,$ and L_i.

7.76C Dimensional analysis is used the derive the pressure drop/rise, Δp and its corresponding drag coefficient, C_D, which expresses the ratio of a real flow variable (in this case: force or pressure) to its corresponding ideal flow variable. Mathematically illustrate the expression for the drag coefficient, C_D (minor head loss coefficient, k).

7.77C The minor head loss coefficient, k is sometimes referred to as the pressure [recovery] coefficient, C_p (especially in the case of a gradual pipe expansion) and is defined by Equation 7.109 as follows:

$$\frac{\Delta p}{\frac{\rho v^2}{2}} = \frac{2}{E} = k = C_p = \frac{\Delta p_{actual}}{\Delta p_{ideal}}.$$

Define the cavitation number, C_a.

7.78C Given the standard empirical expression for the pressure drop/rise (Equation 7.107), derive the empirical expression for the minor head loss, h_f.

7.79 (Refer to Example Problem 7.4, which used the F, L, T fundamental dimensions.) Using the M, L, T fundamental dimensions, derive an empirical expression for the pressure drop, Δp across a globe valve, assuming turbulent flow. The inertia force ($F_I = Ma = \rho v^2 L^2$) and the pressure drop, Δp will always be important. Using Table 11.8 and Section 8.6.4 as a guide, the additional dominant/significant physical quantities in the flow situation include ε and L_i. Thus, the important physical quantities include: $\rho, v, L, \Delta p, \varepsilon,$ and L_i. What are the important π terms used to study and describe turbulent flow across a globe valve? What are the corresponding expressions for the wall shear stress, τ_w and the minor head loss, $h_{f,min}$? Compare your results with Example Problem 7.4, which used the F, L, T fundamental dimensions; they should be identical.

Modeling the Flow Resistance as a Loss in Flowrate in Internal Flow

7.80C One of the major applications of the study of fluid mechanics is the determination of the rate of flow of fluids. As a result, numerous flowrate measuring devices/flow meters have been developed, which differ depending upon whether the flow is through a pipe or through an open channel. Most flow measurement devices (flow meters) are basically velocimeters that directly measure an ideal pressure difference, which is related to the velocity by the Bernoulli equation; thus, the velocity is indirectly measured. Then, application of the continuity equation is used to determine the flowrate. Explain the role of dimensional analysis in flowrate measuring devices/flow meters.

Application of Dimensional Analysis: Derivation of the Reduced/Actual Discharge and Drag Coefficient for Pipe Flow

7.81 Use dimensional analysis to derive an empirical expression for the actual discharge, Q_a for pipe flow-measuring devices (pitot-static tubes, orifice meters, nozzle meters, and venturi meters). As such, an empirical expression for the actual discharge, Q_a for pipe flow, in the theoretical expression for the actual discharge (Equation 7.124) in the differential form of the continuity equation is sought in terms of ideal discharge, Q_i and a flow resistance coefficient. The inertia force ($F_I = Ma = \rho v^2 L^2$) and the reduced/actual discharge, Q_a will always be important. Using Table 11.8 as well as Sections 11.7.6 and 8.8.3 as a guide, the additional dominant/significant forces and physical quantities in the flow situation include

Δp; the viscous force for laminar flow, $F_V = \tau A = \mu v L$; and the geometry, L_i. Thus, the important physical quantities include: ρ, Q_a, L, Δp, μ, and L_i.

7.82C Dimensional analysis is used to derive the actual discharge equation for pipe flow, Q_a and its corresponding drag coefficient, C_D/discharge coefficient, C_d, which expresses the ratio of a real variable (in this case: velocity or discharge) to its corresponding ideal flow variable. Mathematically illustrate the expression for the drag coefficient, C_D/discharge coefficient, C_d.

7.83 (Refer to Example Problem 7.5, which used the F, L, T fundamental dimensions.) Using the M, L, T fundamental dimensions, derive an empirical expression for the reduced/actual discharge, Q_a and drag coefficient, C_D for an orifice meter, assuming turbulent flow. The inertia force $(F_I = Ma = \rho v^2 L^2)$ and the reduced/actual discharge, Q_a will always be important. Using Table 11.8 as well as Sections 11.7.6 and 8.8.3 as a guide, the additional dominant/significant physical quantities in the flow situation include Δp and L_i. Thus, the important physical quantities include: ρ, Q_a, L, Δp, and L_i. What are the important π terms used to study and describe the turbulent flow through an orifice meter? What are the corresponding expressions for the actual velocity, v_a and the actual pressure drop? Compare your results with Example Problem 7.5 which used the F, L, T fundamental dimensions; they should be identical.

Application of Dimensional Analysis: Derivation of the Reduced/Actual Discharge and Drag Coefficient for Open Channel Flow

7.84 Use dimensional analysis to derive an empirical expression for the actual discharge, Q_a for open channel flow-measuring devices (pitot-static tubes, sluice gates, weirs, overflow spillways, the Parshall flume [venturi flume], and contracted openings). As such, an empirical expression for the actual discharge, Q_a for open channel flow in the theoretical expression for actual discharge (Equation 7.125) in the differential form of the continuity equation is sought in terms of ideal discharge, Q_i and a flow resistance coefficient. The inertia force $(F_I = Ma = \rho v^2 L^2)$ and the reduced/actual discharge, Q_a will always be important. Using Table 11.8 as well as Sections 11.7.7 and 9.17 as a guide, the additional dominant/significant physical forces and quantities in the flow situation include the gravity force, $F_G = Mg$; the viscous force for laminar flow, $F_V = \tau A = \mu v L$; the surface tension force, $F_T = \sigma L$; and the geometry (design details) of the flow-measuring device, L_i. Thus, the important physical quantities include: ρ, Q_a, L, g, μ σ, and L_i.

7.85C Dimensional analysis is used to derive the actual discharge equation for open channel flow, Q_a, and its corresponding drag coefficient, C_D/ discharge coefficient, C_d, which expresses the ratio of a real variable (in this case: velocity or discharge) to its corresponding ideal flow variable. Mathematically illustrate the expression for the drag coefficient, C_D/discharge coefficient, C_d.

7.86 (Refer to Example Problem 7.6, which used the F, L, T fundamental dimensions.) Using the M, L, T fundamental dimensions, derive an empirical expression for the reduced/actual discharge, Q_a and drag coefficient, C_D for a sluice gate. The inertia force $(F_I = Ma = \rho v^2 L^2)$ and the reduced/actual discharge, Q_a will always be important. Using Table 11.8 as well as Sections 11.7.7 and 9.17 as a guide, the additional dominant/significant physical quantities in the flow situation include g and L_i. Thus, the important physical quantities include: ρ, Q_a, L, g, and L_i. What are the important π terms used to study and describe this flow process for flow through a sluice gate? Compare your results with Example Problem 7.6, which used the F, L, T fundamental dimensions; they should be identical.

Modeling the Flow Resistance as a Loss in Pump and Turbine Efficiency in Internal Flow

7.87C In the cases where the flow resistance is due to pipe or channel friction (major head loss), pipe component (minor head loss), and/or a flow-measuring device (minor head loss), the flow resistance contributes to the power losses and thus the determination of the efficiency of pumps and turbines. Explain this point further.

7.88C The derivations of the expressions for the efficiency of pumps, η_{pump} and turbines, $\eta_{turbine}$ are determined analytically in Chapter 4. Explain why dimensional analysis is used to derive empirical expressions for the efficiency of pumps and turbines.

Application of Dimensional Analysis: Derivation of the Efficiency of Pumps and Turbines

7.89 Use dimensional analysis to derive an empirical expression for the efficiency of a pump, η_{pump} in the theoretical expression for the efficiency of a pump, $\eta_{pump} = (P_p)_{out}/(P_p)_{in} = \gamma Q h_{pump}/\omega T_{shaft,in}$ (Equation 4.197). The inertia force ($F_I = Ma = \rho v^2 L^2$) and the efficiency of a pump, η_{pump} will always be important. Note that $v = \omega D$, which is the characteristic velocity (where ω is the pump impeller rotational speed), and D, which is the characteristic length, L and is the diameter of the pump impeller blades. Using Tables 11.8, 11.9, and 7.8 and Section 11.7.8 as a guide, the additional dominant/significant forces and physical quantities in the flow situation include the hydraulic power output from the pump, $(P_p)_{out} = \gamma Q h_{pump}$; the shaft power into the pump, $(P_p)_{in} = \omega T_{shaft,in}$; the laminar viscous force, $F_V = \tau A = \mu v L$; the elastic force, $F_E = E_v A = R_v L^2$; and the absolute pump impeller blade surface roughness height, ε. Thus, the important physical quantities include: ρ, ω, D, η_{pump}, γ, Q, h_{pump}, T, μ, E_v, and ε.

7.90C Dimensional analysis is used to derive an empirical expression for the efficiency of a pump, η_{pump} in the theoretical expression for the efficiency of a pump, $\eta_{pump} = (P_p)_{out}/(P_p)_{in} = \gamma Q h_{pump}/\omega T_{shaft,in}$ (Equation 4.197), which expresses the ratio of a real/actual flow variable (hydraulic power output from the pump) to its corresponding ideal flow variable (shaft power input to the pump). Mathematically illustrate the expression for the efficiency of a pump, η_{pump}.

7.91C Give the mathematical expression for the efficiency of a pump as a function of the pump design coefficients (C_H, C_Q, and C_P).

7.92 Derive an empirical expression for the actual pressure rise, $\Delta p = \gamma h_{pump}$ added by a pump, assuming incompressible, turbulent flow. The inertia force ($F_I = Ma = \rho v^2 L^2$) and the actual pressure rise, Δp will always be important. Note that $v = \omega D$, which is the characteristic velocity (where ω is the pump impeller rotational speed), and D, which is the characteristic length, L and is the diameter of the pump impeller blades. Using Tables 11.8, 11.9 and 7.8 and Section 11.7.8 as a guide, the additional dominant/significant forces and physical quantities in the flow situation include the definition of the actual pressure rise, $\Delta p = \gamma h_{pump}$ and the absolute pump impeller blade surface roughness height, ε. Thus, the important physical quantities include: ρ, ω, D, Δp, γ, h_{pump}, and ε. What are the important π terms used to study and describe this flow process for incompressible, turbulent flow through a pump?

7.93 Derive an empirical expression for the actual discharge, Q from a pump, assuming incompressible, laminar flow. The inertia force $(F_I = Ma = \rho v^2 L^2)$ and the actual discharge, Q will always be important. Note that $v = \omega D$, which is the characteristic velocity (where ω is the pump impeller rotational speed) and D, which is the characteristic length, L and is the diameter of the pump impeller blades. Using Tables 11.8, 11.9, and 7.8 and Section 11.7.8 as a guide, the additional dominant/significant forces and physical quantities in the flow situation include the laminar viscous force, $F_V = \tau A = \mu v L$. Thus, the important physical quantities include: ρ, ω, D, Q, and μ. What are the important π terms used to study and describe this flow process for incompressible, laminar flow through a pump?

7.94 Derive an empirical expression for the actual hydraulic power output from a pump, $(P_p)_{out}$ assuming incompressible, laminar flow. The inertia force $(F_I = Ma = \rho v^2 L^2)$ and the actual hydraulic power, $(P_p)_{out}$ will always be important. Note that $v = \omega D$, which is the characteristic velocity (where ω is the pump impeller rotational speed), and D, which is the characteristic length, L and is the diameter of the pump impeller blades. Using Tables 11.8, 11.9, and 7.8 and Section 11.7.8 as a guide, the additional dominant/significant forces and physical quantities in the flow situation include the definition of the actual hydraulic power output from a pump, $(P_p)_{out} = \gamma Q h_{pump}$, and the laminar viscous force, $F_V = \tau A = \mu v L$. Thus, the important physical quantities include: ρ, ω, D, $(P_p)_{out}$, γ, Q, h_{pump}, and μ. What are the important π terms used to study and describe this flow process for incompressible, laminar flow through a pump?

7.95 Derive an empirical expression for the for the actual shaft power, $(P_p)_{in}$ input to a pump, assuming incompressible, turbulent flow. The inertia force $(F_I = Ma = \rho v^2 L^2)$ and the actual shaft power, $(P_p)_{in}$ will always be important. Note that $v = \omega D$, which is the characteristic velocity (where ω is the pump impeller rotational speed), and D, which is the characteristic length, L, and is the diameter of the pump impeller blades. Using Tables 11.8, 11.9, and 7.8 and Section 11.7.8 as a guide, the additional dominant/significant forces and physical quantities in the flow situation include the definition of the actual shaft power input to a pump, $(P_p)_{in} = \omega T_{shaft,in}$ and the absolute pump impeller blade surface roughness height, ε. Thus, the important physical quantities include: ρ, ω, D, $(P_p)_{in}$, T, and ε. What are the important π terms used to study and describe this flow process for incompressible, turbulent flow through a pump?

7.96C Because a turbine is the exact opposite of a pump, an empirical derivation of the efficiency of a pump is actually conducted, and the empirical derivation of the efficiency of a turbine is interpreted from the results. Give the empirical expression of the efficiency of a turbine as interpreted by its definition (exact opposite of a pump).

7.97C Dimensional analysis is used (actually, because a turbine is the exact opposite of a pump, an empirical derivation of the efficiency of a pump is actually conducted, and the empirical derivation of the efficiency of a turbine is interpreted from the results) to derive an empirical expression for the efficiency of a turbine, $\eta_{turbine}$ in the theoretical expression for the efficiency of a turbine, $\eta_{turbine} = (P_t)_{out}/(P_t)_{in} = \omega T_{shaft,out}/\gamma Q h_{turbine}$ (Equation 4.217), which expresses the ratio of a real/actual flow variable (shaft power output from the turbine) to its corresponding ideal flow variable (hydraulic power input to the turbine). Mathematically illustrate the expression for the efficiency of a turbine, $\eta_{turbine}$.

7.98C Give the mathematical expression for the efficiency of a turbine as a function of the turbine design coefficients (C_H, C_Q, and C_P).

7.99 Derive an empirical expression for the for the actual pressure drop, $\Delta p = \gamma h_{turbine}$ removed by a turbine, assuming incompressible, turbulent flow. The inertia force ($F_I = Ma = \rho v^2 L^2$) and the actual pressure drop, Δp will always be important. Note that $v = \omega D$, which is the characteristic velocity (where ω is the turbine runner rotational speed), and D, which is the characteristic length, L and is the diameter of the turbine runner blades. Using Tables 11.8, 11.9, and 7.8 and Section 11.7.8 as a guide, the additional dominant/significant forces and physical quantities in the flow situation include the definition of the actual pressure drop, $\Delta p = \gamma h_{turbine}$ and the absolute turbine runner blade surface roughness height, ε. Thus, the important physical quantities include: ρ, ω, D, Δp, γ, $h_{turbine}$, and ε. What are the important π terms used to study and describe this flow process for incompressible, turbulent flow through a turbine?

Experimental Formulation of Theoretical Equations

7.100C Explain how dimensional analysis may be used to experimentally formulate theoretical mathematical models.

7.101 Use dimensional analysis to derive an expression for the capillary rise, h of water in a small tube. In order to determine the dominant/significant physical quantities in the fluid situation, one may examine Figure 1.9a. Using Section 1.10.5.5 as a guide, the dominant/significant forces acting on the capillary rise of water are due to the surface tension force $F_T = \sigma L$, and the gravitational force, $F_G = Mg = \rho L^3 g^3$. However, because the capillary rise is due to the unbalanced net upward adhesive force due to the surface tension, σ, the pressure force has already been indirectly modeled. And, because the capillary rise is not in motion, the inertia force and the viscous/frictional force are not relevant. Finally, because the water is assumed to be incompressible, the elastic force is also not relevant. Assume that the capillary rise, h is in static equilibrium, and thus, the net upward force which is due to the molecular adhesion and is a result of the tensile stress or surface tension σ, is balanced by the gravitational force. Thus, the important physical quantities include: σ, the height of the capillary rise, h, ρ, g, the radius of the small tube, r and the surface tension contact angle, θ. Hint: from Equation 1.94 $h = 2\sigma\cos\theta/\gamma r$.

7.102 Use dimensional analysis to derive an expression for the drag force acting on a spherical particle falling very slowly (creeping flow at $R \leq 1$) in a fluid (i.e., Stokes law). In order to determine the dominant/significant physical quantities in the fluid flow situation, one may examine Figure EP 10.8b (falling sphere viscometer). Using Section 10.5.3.7 and Equation 10.17, which is $F_D = 3\pi\mu v D$ (Stokes law) as a guide, the only dominant/significant physical force is the viscous or frictional force for laminar flow, $F_V = \tau A = \mu v L$. Although the fluid is in motion, because the focus of this problem is the frictional force for laminar flow, the inertia force is not relevant. Thus, the important physical quantities include: F_D, μ, v, and D.

7.103 Use dimensional analysis to derive an expression for the pressure drop, Δp for laminar incompressible flow in a horizontal pipe (Poiseuille's law; momentum and continuity equations). In order to determine the dominant/significant physical quantities in the fluid flow situation, one may examine Figure 6.4 (significant forces acting on a cylindrical fluid element in a pipe flow). Using Section 6.6.2 and Figure 6.4 as a guide, the dominant/significant forces acting on the laminar incompressible flow in a horizontal pipe are the inertia force (assumption of steady state flow), $F_I = Ma = \rho v^2 L^2$; the pressure force, $F_P = \Delta p A = \Delta p L^2$; and the viscous force, $F_V = \tau A = \mu v L$ (as modeled by Newton's law of viscosity). However, because the pipe is assumed to be horizontal, the gravity force is not relevant. And, because there is a vast amount of fluid, the surface tension force is not relevant. Finally, because the fluid is assumed to be incompressible, the elastic force is also not relevant. Thus, the important physical quantities include: ρ, v, D, Δp, L, and μ. Hint: from Equation 6.96 $\Delta p = 32\mu vL/D^2$ (Poiseuille's law; momentum and continuity equations).

7.104 Use dimensional analysis to derive an expression for the velocity of flow of water from an open tank through an opening in the side of the tank (Torricelli's theorem). In order to determine the dominant/significant physical quantities in the fluid flow situation, one may examine Figure 4.15 (free jet as a result of gravity flow from an open tank). Using Section 4.5.6.5 and Figure 4.15 as a guide, the dominant/significant forces acting on the flow of water from an open tank through an opening in the side of the tank are the inertia force (assumption of steady state flow), $F_I = Ma = \rho_s v^2 L^2$ and the gravitational force, $F_G = Mg = \rho L^3 g$. Because the tank and the opening in the side of the tank are open to the atmosphere, the pressures are assumed to be zero and thus the pressure force is not modeled. Because ideal flow is assumed, the viscous force is not modeled. Because there is a vast amount of fluid, the surface tension force is not relevant. Finally, because the fluid is assumed to be incompressible, the elastic force is also not relevant. Thus, the important physical quantities include: ρ, v, h, and g. Hint: from Equation 4.110 $v_2 = \sqrt{2gh}$ (Torricelli's theorem).

7.105 Use dimensional analysis to derive an expression for the speed of sound (sonic velocity), c for a compressible fluid (liquid or gas). For a compressible fluid (liquid or gas), the occurrence of a slight localized increase in pressure, p and density, ρ of the fluid results in a small wave called a wave of compression. As such, the velocity (celerity) of the small wave of compression through the compressible fluid may be determined by simultaneous application of the continuity and momentum equations. The resulting expression for the velocity of the small wave of compression is called the sonic or acoustic velocity, c. Using Section 1.10.7.5 as a guide, the dominant/significant forces acting on a wave of compression are the inertia force (assumption of steady state flow), $F_I = Ma = \rho v^2 L^2$; the pressure force, $F_P = dpA = dpL^2$; and the elastic force, $F_E = E_v A = E_v L^2$. However, because there is no change in elevation, the gravitational force is not relevant. And, due to a small wave, the viscous force is not relevant. Furthermore, because there is a vast amount of fluid, the surface tension force is not relevant. Thus, the important physical quantities include: $d\rho$, c, L, dp, and E_v. Hint: from Equation 1.160, $c = \sqrt{dp/d\rho} = \sqrt{E_v/\rho}$ (sonic velocity for fluids; momentum and continuity equations).

7.106 Use dimensional analysis to derive an expression for the critical velocity for the critical depth of flow occurring at the spillway. In order to determine the dominant/significant physical quantities in the fluid flow situation, one may examine Figure EP 9.8 (flow over a spillway). Using Section 9.9.4.2 and Figure EP 9.8 as a guide, the dominant/significant forces acting on the flow of water from an open tank through an opening in the side of the tank are the inertia force (assumption of steady state flow), $F_I = Ma = \rho_s v^2 L^2$, and the gravitational force, $F_G = Mg = \rho L^3 g$. Because the open channel flow is open to the atmosphere, the pressures are assumed to be zero, and thus, the pressure force is not modeled. Because ideal flow is assumed, the viscous force is not modeled. Because there is a significant head on the spillway, the surface tension force is not relevant. Finally, because the fluid is assumed to be incompressible, the elastic force is also not relevant. Thus, the important physical quantities include: ρ, v_c, y_c, and g. Hint: from Equation 9.43 $v_c = \sqrt{gy_c}$ (critical velocity).

8

Pipe Flow

8.1 Introduction

The flow of fluid in both pipes (pressure/closed conduit flow) and open channels (gravity/open conduit flow) is contained in a conduit and thus is referred to as internal flow, whereas the flow of fluid over a body that is immersed in the fluid is referred to as external flow (see Chapter 10). In pipe flow, the conduit is completely filled with fluid and the flow is mainly a result of invoking a pressure difference (pressure drop, Δp) and is addressed in Chapter 8. Pipe flow occurs in fresh water supply pipes, natural gas supply pipes, oil supply pipes, and any other fluid flowing under pressure. However, in open channel flow the conduit is only partially filled with fluid and the flow is a result of invoking only gravity, and is addressed in Chapter 9. Furthermore, the results sought in a given flow analysis will depend upon whether the flow is internal or external. In an internal flow, energy or work is used to move/force the fluid through a conduit. Thus, one is interested in the determination of the forces (reaction forces, pressure drops, flow depths), energy or head losses, and cavitation where energy is dissipated. The governing equations (continuity, energy, and momentum) and the results of dimensional analysis are applied to internal flows. Depending upon the flow problem, one may use either the Eulerian (integral) approach or the Lagrangian (differential) approach, or both, in the application of the three governing equations (Section 8.2). It is important to note that while the continuity and the momentum equations may be applied using either the integral or the differential approach, application of the energy equation is useful only using the integral approach, as energy is work, which is defined over a distance, L. Furthermore, as noted in previous chapters, for an internal fluid flow, while the application of the continuity equation will always be necessary, the energy equation and the momentum equation play complementary roles in the analysis of a given flow situation; when one of the two equations of motion breaks down, the other may be used to find the additional unknown quantity.

In the application of the governing equations (continuity, energy, and momentum) for pipe flow, it is important to make the distinction between real pipe flow and ideal pipe flow (see Table 4.1, and Section 8.4). The application of the governing equations for real pipe flow and ideal pipe flow (flow-measuring devices) requires the modeling of flow resistance (frictional losses). The modeling of flow resistance requires a "subset level" application of the governing equations (Section 8.3). The distinction between ideal and real flow determines how the flow resistance is modeled in the "subset level" application of the governing equations. The assumption of "real" flow implies that the flow resistance is modeled in both the energy and the momentum equations. However, the assumption of "ideal" flow implies that the flow resistance is modeled only in the momentum equation (and thus the subsequent assumption of "real" flow). The flow resistance due to pipe friction (modeled as a major head loss; see Section 8.5), and the flow resistance due to most pipe devices (modeled as a minor head loss; see Section 8.6) are modeled by assuming "real" flow; application of the energy equation models a major head loss and a minor head loss, respectively, that are accounted for by a drag coefficient (in the application of the momentum equation). However, the flow resistance due to friction for flow-measuring devices for pipe flow (modeled as an actual discharge, Section 8.8) is modeled by assuming "ideal" flow; although application of the Bernoulli equation does not model a head loss in the determination of ideal velocity, the associated minor head loss with the velocity measurement is accounted for by a drag coefficient (in the application of the momentum equation, where a subsequent assumption of "real" flow is made). Applications of the governing equations for real pipe flow include the analysis and design of pipe systems that include a numerous arrangements of pipes including: (1) single pipes (Section 8.5), (2) pipes with components (Section 8.6), (3) pipes with a pump or a turbine (Section 8.7), (4) pipes in series (Section 8.7), (5) pipes in parallel (Section 8.7), (6) branching pipes (Section 8.7), (7) pipes in a loop (Section 8.7), and (8) pipe networks (Section 8.7). And, applications of the governing equations for ideal pipe flow include the analysis and design of pipe systems that include a pipe flow-measuring device such as orifice, nozzle, or venturi meters, which are also known as ideal flow meters (Section 8.8). One last note is that the scope of this textbook will be limited to steady pipe flow problems.

8.2 Application of the Eulerian (Integral) versus Lagrangian (Differential) Forms of the Governing Equations

Depending upon the flow problem, one may use either the Eulerian (integral) approach or the Lagrangian (differential) approach, or both, in the application of the three governing equations. It is important to note that while the continuity and the momentum equations may be applied using either the integral (for a finite control volume) or the differential approach, application of the energy equation is useful only in the integral approach (for a finite control volume), as energy is work, which is defined over distance, L. While the application of the continuity equation will always be necessary, the energy equation and the momentum equations (equations of motion) play complementary roles in the analysis of a given flow situation. The momentum equation is applied in complement to the energy equation, which is useful only for a control volume. However, depending upon the fluid flow problem, one may apply the differential form of the momentum equation (for a fluid system) or the integral form of the momentum equation (for a control volume). Thus, when one of the two equations of motion breaks down, the other may be used to find the additional unknown quantity.

8.2.1 Eulerian (Integral) Approach for Pipe Flow Problems

In the case where the problem involves mass flow in and out of a pipe system, then the problem is modeled as a finite control volume. The integral form of the three governing equations are applied for the assumption of both real and ideal pipe flow problems. In the assumption of real pipe flow, if one is interested in determining an unknown "energy head" term (p/γ, z, or $v^2/2g$) between two points, an unknown head loss (major and/or minor) due to friction between two points, an unknown head delivered to the fluid by a pump, or an unknown head removed from the fluid by a turbine, then the integral form of the energy equation is applied. For a real flow, if one is interested in determining the forces put on (interaction with) a device, then the integral form of the momentum equation is applied (Section 8.6.6). For a real flow, if one is interested in determining an unknown velocity or discharge between two points, then the integral form of the continuity equation is applied. Furthermore, for a real flow, evaluation of the major and minor head loss flow resistance terms in the energy equation requires a "subset level" application of the governing equations (Section 8.3); the momentum equation is supplemented by dimensional analysis in the derivation of the major head loss for turbulent flow (Section 8.4.1.2) and the minor head loss for the pipe devices in general (Section 8.4.1.2). However, the use of dimensional analysis is not needed in the derivation of the major head loss for laminar flow and the minor head loss for a sudden pipe expansion.

In the assumption of ideal flow, if one is interested in determining an unknown "energy head" term (p/γ, z, or $v^2/2g$) between two points, then the integral form of the energy equation (Bernoulli equation) is applied. A typical assumption of ideal pipe flow is made for flow-measuring devices, which are considered to be ideal flow meters. Furthermore, because there are no reaction forces for pipe flow-measuring devices, there is no need to apply the integral form of the momentum equation. For the ideal flow, if one is interested in determining an unknown velocity or discharge between two points, then the integral form of the continuity equation is applied. Furthermore, for an ideal flow meter, the minor head loss (flow resistance) associated with the flow measurement is indirectly modeled by a discharge coefficient, which requires a "subset level" application of the governing equations (Section 8.3); the momentum equation is supplemented by dimensional analysis in the derivation of the actual discharge (Section 8.4.2.1).

8.2.2 Lagrangian (Differential) Approach for Pipe Flow Problems

In the assumption of real flow, and for the specific case of a major head loss due to pipe flow resistance, application of the differential form of the momentum equation may become necessary. Specifically, application of the differential form of the momentum equation will be necessary when it is of interest to evaluate the flow (solve for an unknown quantity) at a given point in the fluid system for any pipe section. Evaluation of the friction slope term (flow resistance terms) in the differential momentum equation requires a "subset level" application of the governing equations (Section 8.3); the momentum equation is supplemented by dimensional analysis in the derivation of the major head loss (or friction slope) for turbulent flow (Section 8.4.1.2).

8.3 Modeling Flow Resistance in Pipe Flow: A Subset Level Application of the Governing Equations

The application of the governing equations for real pipe flow and ideal pipe flow (flow-measuring devices) (Section 8.4) requires the modeling of flow resistance (frictional losses). The modeling of flow resistance requires a "subset level" application of the governing equations. The distinction between ideal and real flow determines how the flow resistance is modeled in the "subset level" application of the governing equations. Specifically, the application of the integral form of the energy equation for incompressible, real, steady flow pipe flow given by Equation 4.44 requires the evaluation of the major and minor head loss flow resistance terms, $h_{f,maj} = (\tau_w L / \gamma R_h)$ and $h_{f,min} = (\tau_w L / \gamma R_h)$, respectively. Furthermore, the application of the differential form of the momentum equation for pipe flow given by Equation 5.15 requires the evaluation of the friction slope term, $S_f = (\tau_w / \gamma R_h)$. Finally, the application of the continuity equation for a pipe flow-measuring device given by Equation 3.47 requires the evaluation the actual discharge, $Q_a = v_a A_a$.

The flow resistance in pipe flow (major head loss, minor head loss, and actual discharge) is modeled by a "subset level" application of the appropriate governing equations, as presented in Chapter 6 (and summarized in Sections 8.4.1.2 and 8.4.2.1). A "subset level" application of the appropriate governing equations focuses only on the given element causing the flow resistance. Thus, assuming that flow resistance is to be accounted for, depending upon the specific flow situation, the flow type is assumed to be either a "real" flow or an "ideal" flow. The distinction between ideal and real flow determines how the flow resistance is modeled in the "subset level" application of the appropriate governing equations. The assumption of "real" flow implies that the flow resistance is modeled in both the energy and the momentum equations. However, the assumption of "ideal" flow implies that the flow resistance is modeled only in the momentum equation (and thus the subsequent assumption of "real" flow). The flow resistance due to friction in pipe flow (modeled as a major head loss) and the flow resistance due to friction in most pipe devices (modeled as a minor head loss) are modeled by assuming "real" flow; application of the energy equation models a head loss that is accounted for by a drag coefficient (in the application of the momentum equation). However, the flow resistance due to friction in flow-measuring devices for pipe flow (modeled as an actual discharge) is modeled by assuming "ideal" flow; although application of the Bernoulli equation does not model a head loss in the determination of ideal velocity, the associated minor head loss with the velocity measurement is accounted for by a drag coefficient (in the application of the momentum equation, where a subsequent assumption of "real" flow is made). Furthermore, the use of dimensional analysis (Chapter 7) in supplement to a "subset level" application of the appropriate governing equations is needed to derive flow resistance equations when there is no theory to model the friction force (as in the case of turbulent pipe flow). Finally, it is important to note that the use of dimensional analysis is not needed in the derivation of the major head loss term for laminar pipe flow (friction force is theoretically modeled), and the minor head loss term for a sudden pipe expansion (friction force is ignored).

8.4 Application of the Governing Equations in Pipe Flow

The governing equations applied in the analysis and design of pipe flow include the continuity, energy, and momentum (in some cases supplemented with dimensional analysis) equations. While the application of the continuity equation will always be necessary, the energy equation and the momentum equations (equations of motion) play complementary roles in the analysis of a given flow situation. Thus, when one of the two equations of motion breaks down, the other may be used to find the additional unknown quantity. Furthermore, in the application of the governing equations for pipe flow, it is important to make the distinction between real pipe flow and ideal pipe flow (see Table 4.1). The application of the governing equations for real pipe flow and ideal pipe flow (flow-measuring devices) requires the modeling of flow resistance (frictional losses). The modeling of flow resistance requires a "subset level" application of the governing equations (Sections 8.3, 8.4.1.2, and 8.4.2.1). The distinction between ideal and real flow determines how the flow resistance is modeled in the "subset level" application of the governing equations.

8.4.1 Application of the Governing Equations for Real Pipe Flow

The momentum equation is applied in complement to the energy equation, which is useful only for a control volume. However, depending upon the (real) fluid flow problem, one may apply the differential form of the momentum equation (for a fluid system) or the integral form of the momentum equation (for a control volume). Thus, when one of the two equations of motion breaks down, the other may be used to find the additional unknown quantity.

8.4.1.1 Integral Approach for Real Pipe Flow Problems

Assuming a finite control volume for the assumption of incompressible, real, steady pipe flow, first, the integral energy equation is applied, which was given by Equation 4.181, and is repeated as follows:

$$\left(\frac{p_1}{\rho_1 g} + \frac{v_1^2}{2g} + z_1\right) + h_{pump} = \left(\frac{p_2}{\rho_2 g} + \frac{v_2^2}{2g} + z_2\right) + h_{turbine} + h_{f,total} \tag{8.1}$$

which may be used to solve for an unknown "energy head" term (p/γ, z, or $v^2/2g$) for one of the two points in a finite control volume; an unknown friction head loss, $h_{ftotal} = h_{f,maj} + h_{f,min}$ between two points in a finite control volume; an unknown head delivered to the fluid by a pump, h_{pump}; or an unknown head removed from the fluid by a turbine, $h_{turbine}$. Application of Equation 8.1 requires evaluation of the following terms: h_{min}, h_{maj}, h_{pump}, and $h_{turbine}$. Evaluation of the major and minor head loss flow resistance terms requires a "subset level" application of the governing equations (Section 8.4.1.2). Then, second, application of the complementary integral momentum equation is required only if there is an unknown reaction force, as in the case of certain pipe components (such as contractions, nozzles, elbows, and bends) (see Section 8.6.6), which was given by Equation 5.27 and is repeated as follows:

$$\sum F_s = (F_G + F_P + F_V + F_{other})_s = (\rho Q v_s)_2 - (\rho Q v_s)_1 \tag{8.2}$$

which may be solved for an unknown force (typically a reaction force). Note that the viscous forces have already been accounted for indirectly through the minor head loss flow resistance term in the energy equation, Equation 8.1. As such, the integral form of the momentum equation is solved analytically, without the use of dimensional analysis.

8.4.1.2 Evaluation of the Major and Minor Head Loss Terms

The evaluation of the total head loss, $h_{ftotal} = h_{f,maj} + h_{f,min}$ may include both major and minor losses, which are due to flow resistance. The major head loss in pipe flow is due to the frictional losses/flow resistance caused by pipe friction. The minor head loss is due to flow resistance due to various pipe components—valves, fittings (tees, unions, elbows, and bends), entrances, exits, contractions, and expansions, or pipe flow-measuring devices (orifice, nozzle, or venturi meters). One may note that in the case of a flow-measuring device, although the associated minor head loss may be accounted for in the integral form of the energy equation, it is typically accounted for by the use/calibration of a discharge coefficient to determine the actual discharge, where ideal flow (Bernoulli equation) is assumed (see Section 8.4.2).

Evaluation of the major and minor head loss flow resistance terms requires a "subset level" application of the governing equations (see Chapters 6 and 7). A "subset level" application of the governing equations focuses only on the given element causing the flow resistance. The assumption of real flow implies that the flow resistance is modeled in both the energy and momentum equations. The flow resistance equations for the major and minor head loss are derived in Chapters 6 and 7. In the derivation of one given source of head loss (major or minor), first, the integral form of the energy equation is applied to solve for the unknown head loss, h_f, which is given as follows:

$$\left(\frac{p}{\gamma} + z + \frac{v^2}{2g}\right)_1 - h_f = \left(\frac{p}{\gamma} + z + \frac{v^2}{2g}\right)_2 \tag{8.3}$$

One may note that in the derivation of the expression for one given source of head loss, h_f (major or minor head loss due to flow resistance), the control volume is defined between the two points that include only that one source of head loss. The derivation of the major head loss assumes $z_1 = z_2$, and $v_1 = v_2$, as follows:

$$\left(\frac{p}{\gamma} + z + \frac{v^2}{2g}\right)_1 - h_{f,maj} = \left(\frac{p}{\gamma} + z + \frac{v^2}{2g}\right)_2 \tag{8.4}$$

which yields $h_{f,maj} = S_f L = (\tau_w L/\gamma R_h) = (\Delta p/\gamma)$. And, the derivation of the minor head loss assumes $z_1 = z_2$, and $v_1 = v_2$, as follows:

$$\left(\frac{p}{\gamma} + z + \frac{v^2}{2g}\right)_1 - h_{f,min} = \left(\frac{p}{\gamma} + z + \frac{v^2}{2g}\right)_2 \tag{8.5}$$

which yields $h_{f,min} = (\tau_w L/\gamma R_h) = (\Delta p/\gamma)$. And then, second, the integral form of the momentum equation (supplemented by dimensional analysis) is applied to solve for an unknown pressure drop, Δp (in the case of major or minor head loss due to pipe flow resistance), which was given in Equation 5.27 and is repeated as follows:

$$\sum F_s = (F_G + F_P + F_V + F_{other})_s = (\rho Q v_s)_2 - (\rho Q v_s)_1 \tag{8.6}$$

One may note that derivation of the major head loss involves both the integral and the differential momentum equations. Thus, the resulting flow resistance equation for the major head loss is given as follows:

$$h_{f,maj} = S_f L = \frac{\tau_w L}{\gamma R_h} = \frac{\Delta p}{\gamma} = \frac{v^2}{C^2 R_h} L = f\frac{L}{D}\frac{v^2}{2g} = \left(\frac{vn}{R_h^{2/3}}\right)^2 L = \left[\frac{v}{0.849 C_h R_h^{0.63}}\right]^{1/0.54} L \tag{8.7}$$

where C is the Chezy coefficient, f is the Darcy–Weisbach friction factor, n is the Manning's roughness coefficient, and C_h is the Hazen–Williams roughness coefficient. In the case of laminar flow, in lieu of the Chezy equation (turbulent flow), one would apply Poiseuille's law as follows:

$$h_{f,maj} = S_f L = \frac{\Delta p}{\gamma} = \underbrace{\left[\frac{32\mu v}{\gamma D^2}\right]}_{S_f} L \tag{8.8}$$

and the resulting flow resistance equation for the minor head loss is given as follows:

$$h_{f,\min} = \frac{\tau_w L}{\gamma R_h} = \frac{\Delta p}{\gamma} = k\frac{v^2}{2g} \tag{8.9}$$

where k is the minor head loss coefficient. Furthermore, each head loss flow resistance equation represents the simultaneous application of the two complementary equations of motion (on a "subset level"). It is important to note that the use of dimensional analysis is not needed in the derivation of the major head loss term for laminar flow and the minor head loss term for a sudden pipe expansion.

The flow resistance due to pipe friction, which results in a major head loss term, $h_{f,major}$, is modeled as a resistance force (shear stress or drag force) in the integral form of the momentum equation, Equation 5.27, while it is modeled as a friction slope, S_f in the differential form of the momentum equation, Equation 5.37. As such, the energy equation may be used to solve for the unknown major head loss, h_f, while the integral momentum equation may be used to solve for the unknown pressure drop, Δp (or the unknown friction slope, S_f). The major head loss, h_f is caused by both pressure and friction forces. Thus, when the friction/viscous forces can be theoretically modeled in the integral form of the momentum equation (application of Newton's law of viscosity in the laminar flow case), one can analytically determine the actual pressure drop, Δp from the integral form of the momentum equation. And, therefore, one may analytically derive an expression for the major head loss, h_f due to pipe or channel friction from the energy equation. However, when the friction/viscous forces cannot be theoretically modeled in the integral form of the momentum equation (as in the turbulent flow case), the friction/viscous forces, the actual pressure drop, Δp (or the unknown friction slope, S_f) and thus the major head loss, h_f due to pipe or channel friction are determined empirically (see Chapter 7). Specifically, in the case of turbulent flow, the friction/viscous forces in the integral form of the momentum equation and thus the wall shear stress, τ_w cannot be theoretically modeled; thus, the integral form of the momentum equation cannot be directly applied to solve for the unknown pressure drop, Δp. Thus, an empirical interpretation (using dimensional analysis) of the wall shear stress, τ_w in the theoretical expression for the friction slope, $S_f = (\tau_w/\gamma R_h)$ in the differential form of the momentum equation is sought in terms of velocity, v and a flow resistance coefficient. This yields the Chezy equation $S_f = (v^2/C^2 R_h)$, which represents the differential form of the momentum equation, supplemented by dimensional analysis and guided by the integral momentum equation (link between the differential and integral momentum equations), which is used to obtain an empirical evaluation for the pressure drop, Δp in Newton's second law of motion (integral form of the momentum equation). As such, one can then derive an empirical expression for the major head loss, h_f from the energy equation. Major losses in pipe flow are addressed in detail in Chapters 6 through 8.

The flow resistance due to various pipe components, which results in a minor head loss term, $h_{f,minor}$, is modeled as a resistance force (shear stress or drag force) in the integral form of the momentum equation, Equation 5.27. As such, the energy equation may be used to solve for the unknown minor head loss, h_f, while the integral form of the momentum equation may be used to solve for the unknown pressure drop, Δp. The minor head loss, h_f is caused by both pressure and friction forces. Thus, when the friction/viscous forces are insignificant and can be ignored in the integral form of the momentum equation (as in the case of the sudden pipe expansion), one can analytically determine the actual pressure drop, Δp from the momentum equation. And, therefore, one may analytically derive an expression for the minor head loss, h_f due to pipe components from the energy equation. However, when the friction/viscous forces are significant and cannot be ignored in the integral form of the momentum equation (as in the case of the other pipe components), the friction/viscous forces; the actual pressure drop/rise, Δp; and thus the minor head loss, h_f are determined empirically (see Chapter 7). Specifically, in the case of turbulent pipe component flow, the friction/viscous forces in the integral form of the momentum equation and thus the wall shear stress, τ_w cannot be theoretically modeled; thus, the integral form of the momentum equation cannot be directly applied to solve for the unknown pressure drop, Δp. The evaluation of the minor head loss term, $h_{f,minor}$ involves supplementing the integral form of the momentum equation with dimensional analysis, as described in detail in Chapters 6 through 8.

8.4.1.3 Evaluation of the Pump and Turbine Head Terms

Evaluation of the head delivered to the fluid by a pump, h_{pump} and the head removed/extracted from the fluid by a turbine, $h_{turbine}$ require the application of the appropriate power equation, which is discussed in Section 4.7.2.4, and the resulting equations are summarized, respectively, as follows:

$$h_{pump} = \frac{(P_p)_{out}}{\gamma Q} \tag{8.10}$$

$$h_{turbine} = \frac{(P_t)_{in}}{\gamma Q} \tag{8.11}$$

8.4.1.4 Differential Approach for Real Pipe Flow Problems

Furthermore, in the assumption of real flow, and for the specific case of a major head loss due to pipe flow resistance, application of the differential form of the momentum equation may become necessary. Specifically, application of the differential form of the momentum equation will be necessary when it is of interest to evaluate the flow (solve for an unknown quantity) at a given point in the fluid system for any uniform pipe or channel section (Equation 5.8) and is given as follows:

$$S_f = \frac{\tau_w}{\gamma R_h} \tag{8.12}$$

Application of Equation 8.12 requires evaluation of the friction slope term, S_f, which is modeled on a "subset level" (see Section 8.4.1.2 above) that yielded the Chezy equation as follows:

$$S_f = \frac{\tau_w}{\gamma R_h} = \frac{v^2}{C^2 R_h} \tag{8.13}$$

where the major head loss due to flow resistance is determined by application of Equation 8.7, which is repeated as follows:

$$h_{f,maj} = S_f L = \frac{\tau_w L}{\gamma R_h} = \frac{\Delta p}{\gamma} = \frac{v^2}{C^2 R_h} L = f \frac{L}{D} \frac{v^2}{2g} = \left(\frac{vn}{R_h^{2/3}}\right)^2 L = \left[\frac{v}{0.849 C_h R_h^{0.63}}\right]^{1/0.54} L \quad (8.14)$$

It is important to note that in the case of laminar flow, in lieu of the Chezy equation (turbulent flow), one would apply Poiseuille's law as follows:

$$S_f = \frac{\Delta p}{\gamma L} = \frac{32 \mu v}{\gamma D^2} \quad (8.15)$$

which represents the differential form of the momentum equation.

8.4.1.5 Applications of the Governing Equations for Real Pipe Flow Problems

Applications of the governing equations for real pipe flow include the analysis and design of pipe systems that include a numerous arrangements of pipes, including: (1) single pipes (Section 8.5), (2) pipes with components (Section 8.6), (3) pipes with a pump or a turbine (Section 8.7), (4) pipes in series (Section 8.7), (5) pipes in parallel (Section 8.7), (6) branching pipes (Section 8.7), (7) pipes in a loop (Section 8.7), and (8) pipe networks (Section 8.7). A single pipe forms the basic component for any pipe system. Furthermore, in the analysis and design of a pipe system, there may be pipe friction, pipe components, transitions, or devices in the pipe flow that will determine the significance of a major head loss, a minor head loss, an added or removed head, and a reaction force. Finally, the momentum equation is applied in complement to the energy equation, which is useful only for a control volume. However, depending upon the (real) fluid flow problem, one may apply the differential form of the momentum equation (for a fluid system) or the integral form of the momentum equation (for a control volume).

8.4.2 Application of the Governing Equations for Ideal Pipe Flow

When one of the two equations of motion breaks down, the other may be used to find the additional unknown. Assuming a finite control volume, for the assumption of incompressible, ideal, and steady pipe flow, first, the integral energy equation (Bernoulli equation) is applied, which was given by Equation 4.46 and is repeated as follows:

$$\left(\frac{p_1}{\rho_1 g} + \frac{v_1^2}{2g} + z_1\right) = \left(\frac{p_2}{\rho_2 g} + \frac{v_2^2}{2g} + z_2\right) \quad (8.16)$$

which may be used to solve for an unknown "energy head" term (p/γ, z, or $v^2/2g$) for one of the two points in a finite control volume. Application of the Bernoulli equation assumes ideal flow and thus that there is no head loss term to evaluate in the energy equation. Therefore, for problems that affect the flow over a short pipe section, the major head loss due to flow resistance is assumed to be negligible and thus one may apply the Bernoulli equation. A typical assumption of ideal pipe flow is made for flow-measuring devices, which are considered to be ideal flow meters. In the case of pipe flow-measuring devices, although the associated minor head loss may be accounted for in the integral form of the energy equation, Equation 8.1, it is typically accounted for by the use/calibration of a discharge coefficient to determine the actual discharge, where ideal flow (Bernoulli equation) is assumed. Thus, in the case of flow-measuring devices, one must indirectly model the minor head loss flow resistance term by the use of a discharge coefficient to determine the actual discharge, which requires a "subset level" of application of the governing equations (Section 8.4.2.1). Then, because there are no reaction forces for pipe flow-measuring devices, there is no need to apply the integral form of the momentum equation.

8.4.2.1 Evaluation of the Actual Discharge

In the case of a flow-measuring device for pipe flow, the associated minor head is typically accounted for by the use/calibration of a discharge coefficient to determine the actual discharge. Evaluation of the actual discharge requires a "subset level" application of the governing equations. A "subset level" application of the governing equations focuses only on the given element causing the flow resistance. The assumption of "ideal" flow implies that the flow resistance is modeled only in the momentum equation (and thus the subsequent assumption of "real" flow). The flow resistance equation for the actual discharge is derived in Chapters 6 and 7. In the derivation of the actual discharge for a given flow-measuring device, first, ideal flow is assumed and thus the Bernoulli equation is applied as follows:

$$\left(\frac{p}{\gamma}+z+\frac{v^2}{2g}\right)_1 - \left(\frac{p}{\gamma}+z+\frac{v^2}{2g}\right)_2 = 0 \tag{8.17}$$

Assuming $z_1 = z_2$, this equation has one unknown, which is the ideal velocity of the flow at the restriction in the flow-measuring device. Thus, this equation yields an expression for the ideal velocity, $v_i = \sqrt{(2\Delta p/\rho)}$, as a function of an ideal pressure difference, Δp, which is directly measured. Then, application of the continuity equation is used to determine an expression for the ideal discharge, $Q_i = v_i A_i$ and the actual discharge, $Q_a = ((v_a)/\sqrt{(2\Delta p/\rho)})(A_a/A_i)Q_i$. Finally, the integral form of the momentum equation (supplemented by dimensional analysis) is applied to solve for an unknown actual velocity, v_a and actual area, A_a, which was given in Equation 5.27 and is repeated as follows:

$$\sum F_s = (F_G + F_P + F_V + F_{other})_s = (\rho Q v_s)_2 - (\rho Q v_s)_1 \tag{8.18}$$

Thus, the resulting flow resistance equation for the actual discharge for pipe flow-measuring devices is given as follows:

$$Q_a = \frac{\sqrt{\frac{2}{\rho}\left(\Delta p - \frac{\tau_w L}{R_h}\right)}}{\sqrt{\frac{2\Delta p}{\rho}}}\frac{A_a}{A_i}Q_i = C_d Q_i \tag{8.19}$$

where the discharge coefficient, C_d accounts for (indirectly models) the minor head loss associated with the flow measurement.

The flow resistance (shear stress or drag force) due to a flow-measuring device is modeled as a reduced/actual discharge, Q_a in the differential form of the continuity equation. The flow resistance causes an unknown pressure drop, Δp in the case of pipe flow, which causes an unknown head loss, h_f, where the head loss is due to a conversion of kinetic energy to heat, which is modeled/displayed in the integral form of the energy equation. The head loss is caused by both pressure and friction forces. In the determination of the reduced/actual discharge, Q_a, which is less than the ideal discharge, Q_i, the exact reduction in the flowrate is unknown because the head loss, h_f causing it is unknown. Therefore, because one cannot model the exact reduction in the ideal flowrate, Q_i, one cannot derive an analytical expression for the reduced/actual discharge, Q_a from the continuity equation. Specifically, the complex nature of the flow does not allow a theoretical modeling of the existence of the viscous force (due to shear stress, τ_w) due to the flow measurement device in the momentum equation and thus does not allow an analytical derivation of the pressure drop, Δp and thus v_a and cannot measure A_a. Thus, an empirical expression (using dimensional analysis) for the actual discharge,

$$Q_a = \frac{\sqrt{\dfrac{2}{\rho}\left(\Delta p - \dfrac{\tau_w L}{R_h}\right)}}{\sqrt{\dfrac{2\Delta p}{\rho}}} \frac{A_a}{A_i} Q_i,$$

and thus Δp (and v_a and A_a) is derived as a function of the drag coefficient, C_D (discharge coefficient, C_d) that represents the flow resistance. This yields the actual discharge equation $Q_a = C_d Q_i$, which represents the integral momentum equation supplemented by dimensional analysis. Furthermore, although the differential form of the continuity equation is typically used to compute the reduced/actual discharge, Q_a for a flow-measuring device, because the head loss, h_f causes the reduced/actual discharge, Q_a, the head loss may also be accounted for as a minor loss in the integral form of the energy equation, as illustrated in Section 8.8.3.5. Actual discharges are addressed in detail in Chapters 6 through 8.

8.4.2.2 Applications of the Governing Equations for Ideal Pipe Flow

Applications of the governing equations for ideal pipe flow include the analysis and design of pipe systems that include a pipe flow-measuring device such as orifice, nozzle, or venturi meters, which are also known as ideal flow meters (see Section 8.8). The governing equations are applied assuming a finite control volume.

8.5 Single Pipes: Major Head Loss in Real Pipe Flow

Applications of the governing equations for real pipe flow include the analysis and design of a single pipe, which forms the basic component for any pipe system (see Section 8.7). While the application of the continuity equation will always be necessary, the energy equation and the momentum equation play complementary roles; when one of the two equations of motion breaks down, the other may be used to find the additional unknown quantity. Furthermore, while the energy equation is useful only for a control volume, depending upon the general fluid flow problem, one may apply the continuity and momentum equations using either the integral or differential approach. However, for the specific case of a single straight pipe section without pipe components (addressed in Section 8.5), only the differential continuity and momentum equations for a fluid system will be required; the integral continuity and momentum equations for a control volume are required for pipes with components (see Section 8.6). As such, for a single straight pipe section, the energy equation, Equation 8.1, models the major head loss, $h_{f,maj}$ due to pipe friction for a given pipe length, while the differential momentum equation, Equation 8.12, models the friction slope, S_f due to pipe friction for a given point in the pipe section.

While the major head loss may be ignored for short pipes, it is significant for very long pipes. The major head loss term, $h_{f,maj}$ in the energy equation, Equation 8.1, models the head loss due to pipe friction for the flow of a real fluid (liquid or gas) in the straight section of a single pipe. When a real fluid flows through a pipe, there is a pressure drop, Δp over a pipe distance, L, which is caused by the friction of the fluid against the pipe wall, where the pressure drop, Δp is directly related to the wall shear stress, τ_w. The major head loss term, $h_{f,maj}$ accounts for any energy loss associated with the flow, and is a direct result of the viscous dissipation that occurs throughout the fluid in the pipe.

8.5.1 Evaluation of the Major Head Loss Term in the Energy Equation

Evaluation of the major head loss flow resistance term requires a "subset level" application of the governing equations (see Chapters 6 and 7 and Section 8.4.1.2). Depending upon the value for the Reynolds number, R, real flow may be described as either laminar or turbulent. As such, the major head loss due to pipe friction modeled in the energy equation, Equation 8.1, is evaluated by the Chezy head loss equation, Equation 8.7, for turbulent pipe flow, and by the Poiseuille head loss equation, Equation 8.8, for laminar flow. And, the friction slope modeled in the differential momentum equation, Equation 8.12, is evaluated by the Chezy equation, Equation 8.13, for turbulent pipe flow, and by the Poiseuille's law Equation 8.15 for laminar flow. The Poiseuille head loss equation and Poiseuille's law are applied for single pipes in Section 8.5.1.1, while the Chezy head loss equation and the Chezy equation are applied for single pipes in Section 8.5.1.2. Furthermore, in addition to the Chezy equation, additional turbulent pipe flow resistance equations (Section 8.5.2) include the Darcy–Weisbach equation (Section 8.5.3), the Manning's equation (Section 8.5.4), and the Hazen–Williams equation (Section 8.5.5). Therefore, presented in the sections below are applications of the governing equations for a single pipe for both laminar and turbulent flow, where the major head loss term, $h_{f,maj}$ in the energy equation is evaluated and schematically illustrated by the energy grade line (*EGL*) and the hydraulic grade line (*HGL*).

8.5.1.1 Laminar Pipe Flow: Poiseuille's Law

Evaluation of the major head loss flow resistance term requires a "subset level" application of the governing equations (see Chapters 6 and 7 and Section 8.4.1.2). When the friction/viscous forces can be theoretically modeled in the momentum equation, as in the laminar flow case, the friction/viscous forces (Newton's law of viscosity), the actual pressure drop, Δp and thus the major head loss, h_f due to pipe friction are determined analytically. One can analytically determine the actual pressure drop, Δp between two points in a finite control volume from the integral form of the momentum equation (Newton's second law of motion) as follows (see Chapter 6 for its derivation and Chapter 8 herein, for its application):

$$\Delta p = \frac{32\mu v L}{D^2} \tag{8.20}$$

which is known as Poiseuille's law. Furthermore, in order to transition from the integral point of view to the differential point of view, one may define the pressure gradient as follows:

$$\frac{\Delta p}{L} = \gamma S_f \tag{8.21}$$

for which

$$S_f = \frac{\tau_w}{\gamma R_h} = \frac{\Delta p}{\gamma L} = \frac{32\mu v}{\gamma D^2} \tag{8.22}$$

This differential form of Poiseuille's law (differential momentum) expressed in terms of a friction slope may be used to solve for one of the unknown quantities (S_f, v, μ, D) in the case of laminar flow. One may note that because Poiseuille's law assumes a straight pipe section of length, L, with a constant diameter, D; a fluid with a specific value for μ (and γ); and a friction slope, S_f, it can be applied to solve for one of the unknown quantities at a given point in the fluid system. It is important to note that equating the analytical expression for the friction slope for laminar flow (Poiseuille's law, $S_f = [32\mu v/\gamma D^2]$) to the empirical expression for the friction slope for turbulent flow (the Chezy equation, $S_f = [v^2/C^2 R_h]$) and noting that $C = \sqrt{(8g/f)}$ yields an a expression for the Darcy–Weisbach friction factor, f for laminar flow, $f = (64\mu/\rho v D) = (64/R)$, which is a special case of turbulent flow; this expression links the empirically defined friction factor to the analytically defined friction factor for the special case of laminar flow.

Finally, substituting the analytical expression for the actual pressure drop, Δp Equation 8.20 into the integral form of the energy equation, Equation 8.4, yields an analytical expression for the major head loss, h_f due to pipe friction as follows (see Chapters 6 and 8 for its application):

$$h_f = \frac{\tau_w L}{\gamma R_h} = \frac{\Delta p}{\gamma} = S_f L = \underbrace{\left[\frac{32\mu v_{ave}L}{D^2}\right]}_{\Delta p}\frac{1}{\gamma} = \underbrace{\left[\frac{32\mu v_{ave}}{D^2 \gamma}\right]}_{S_f} L \tag{8.23}$$

which is Poiseuille's law expressed in terms of the major head loss and provides an analytical expression for the major head loss, h_f (expressed as a function of the flow resistance, μ), without the need to define an empirical flow resistance coefficient. One may note that although the theoretical relationship between the pressure drop, Δp and the shear stress, τ_w was derived assuming a horizontal pipe—derived from Newton's second law of motion (integral form of momentum equation) for both laminar and turbulent flow, Poiseuille's law expressed in terms of the major head loss, h_f is valid for any fully developed, steady, incompressible pipe flow, whether the pipe is horizontal or inclined; this is because the governing equations are applied on a "subset level" (focusing only on the given element causing the flow resistance) in the derivation of Poiseuille's law. Furthermore, one may note that the velocity, v

in Poiseuille's law is the mean velocity, v_{ave} (mean over both the pipe diameter, D and the pipe section length, L). Thus, Poiseuille's "head loss equation" is used to evaluate the major head loss in the application of the integral form of the energy equation, Equation 8.1, as follows:

$$\underbrace{\left(\frac{p}{\gamma}+z+\frac{v^2}{2g}\right)_1 - \left(\frac{p}{\gamma}+z+\frac{v^2}{2g}\right)_2}_{h_e} = \underbrace{\frac{\tau_w}{\gamma}\frac{L}{R_h} = \left[\frac{32\mu v_{ave}}{D^2\gamma}\right]L}_{h_{f,major}} \qquad (8.24)$$

where, in general, the pressure head drop, $\Delta p/\gamma$ is due to the change in elevation, the change in the velocity head (dynamic pressure), and the head loss due to the flow resistance, μ.

Poiseuille's law is applied in the analysis and design of laminar pipe flow. Specifically, assuming that the fluid viscosity, μ and fluid specific weight, γ are specified, there are three types of problems in the application of Poiseuille's law, are outlined in Table 8.1 below. Poiseuille's law, expressed as a head loss equation, Equation 8.23, may be applied to solve for an unknown head loss, h_f, which is a Type 1 problem (analysis). The differential form of Poiseuille's law Equation 8.22, may be applied to solve for an unknown velocity of flow, v which is a Type 2 problem (analysis). Furthermore, the differential form of Poiseuille's law Equation 8.22, may be applied to solve for an unknown pipe diameter, D, which is a Type 3 problem (design).

TABLE 8.1

Summary of Three Types of Problems for Laminar Pipe Flow (Poiseuille's Law)

Problem Type	Given	Find	Resistance Equation	Governing Equation
Type 1	μ, γ, L, v, D	h_f	$h_f = \dfrac{\Delta p}{\gamma} = S_f L = \underbrace{\left[\dfrac{32\mu v_{ave}L}{D^2}\right]}_{\Delta p}\dfrac{1}{\gamma}$	**Poiseuille's Law** expressed as a head loss equation
Type 2	μ, γ, L, S_f, D	v	$S_f = \dfrac{h_f}{L} = \dfrac{\Delta p}{\gamma L} = \dfrac{32\mu v}{\gamma D^2}$	Differential momentum form of **Poiseuille's Law**
Type 3	μ, γ, L, v, S_f	D	$S_f = \dfrac{h_f}{L} = \dfrac{\Delta p}{\gamma L} = \dfrac{32\mu v}{\gamma D^2}$	Differential momentum form of **Poiseuille's Law**

EXAMPLE PROBLEM 8.1

Crude oil at 20°C flows in a 2-m-diameter inclined pipe at a flowrate of 0.02 m³/sec, as illustrated in Figure EP 8.1. The pressure at point 1 is 2×10^5 N/m². The increase in elevation between points 1 and 2 is 0.03 m, and the pipe length between points 1 and 2 is 8000 m. (a) Determine the flow regime. (b) Determine the major head loss due to the pipe friction (flow resistance) between points 1 and 2. (c) Determine the pressure drop between points 1 and 2 due to the increase in elevation and the pipe friction (flow resistance) between points 1 and 2.

Mathcad Solution

(a) In order to determine the flow regime, one must compute the Reynolds number, R where the density, ρ and the viscosity, μ for crude oil at 20°C are given in Table A.4 in Appendix A as follows:

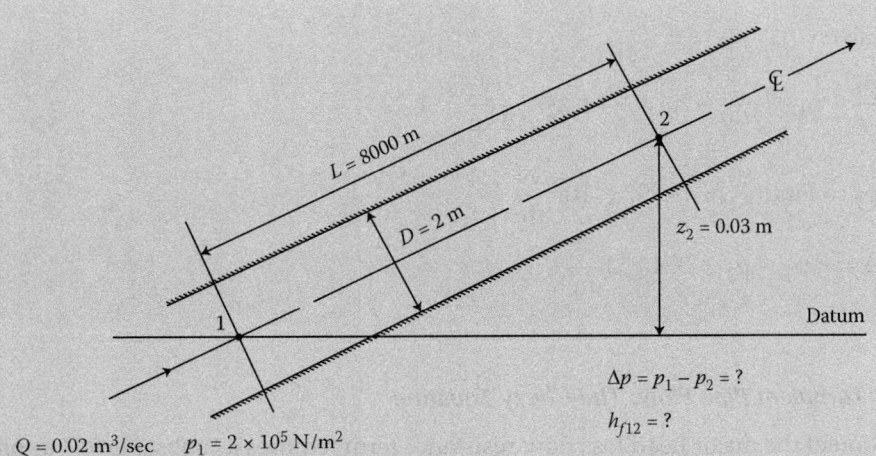

FIGURE EP 8.1
Crude oil flows in an inclined pipe flowing under pressure.

$$\rho := 856 \frac{kg}{m^3} \qquad \mu := 7.2 \times 10^{-3} \, N \frac{sec}{m^2} \qquad D := 2\,m$$

$$Q := 0.02 \frac{m^3}{sec} \qquad A := \frac{\pi \cdot D^2}{4} = 3.142 \, m^2 \qquad v := \frac{Q}{A} = 6.366 \times 10^{-3} \frac{m}{s}$$

$$R := \frac{\rho \cdot v \cdot D}{\mu} = 1.514 \times 10^3$$

Thus, because the Reynolds number, R is less than 2000, the flow is laminar.

(b) Assuming laminar flow, the major head loss due to pipe friction is computed from Equation 8.23 (Poiseuille's law expressed in terms of the major head loss) as a Type 1 problem as follows:

$$L := 8000\,m \qquad g := 9.81 \frac{m}{sec^2} \qquad \gamma := \rho \cdot g = 8.397 \times 10^3 \frac{kg}{m^2 \cdot s^2}$$

$$h_f := \left(\frac{32\mu \cdot v \cdot L}{D^2} \right) \cdot \frac{1}{\gamma} = 3.493 \times 10^{-4}\,m$$

(c) Assuming the datum is at point 1, the pressure drop between points 1 and 2 is computed by applying the energy equation between points 1 and 2 as follows:

$$p_1 := 2 \times 10^5 \frac{N}{m^2} \qquad z_1 := 0\,m \qquad z_2 := 0.03\,m$$

$$v_1 := v = 6.366 \times 10^{-3} \frac{m}{s} \qquad v_2 := v = 6.366 \times 10^{-3} \frac{m}{s}$$

Guess value: $\qquad p_2 := 1 \times 10^5 \frac{N}{m^2}$

Given

$$\frac{p_1}{\gamma} + z_1 + \frac{v_1{}^2}{2 \cdot g} - h_f = \frac{p_2}{\gamma} + z_2 + \frac{v_2{}^2}{2 \cdot g}$$

$$p_2 := \text{Find}(p_2) = 1.997 \times 10^5 \ \frac{N}{m^2}$$

$$\Delta p := p_1 - p_2 = 254.854 \ \frac{N}{m^2}$$

8.5.1.2 Turbulent Pipe Flow: The Chezy Equation

Evaluation of the major head loss flow resistance term requires a "subset level" application of the governing equations (see Chapters 6 and 7 and Section 8.4.1.2). When the friction/viscous forces cannot be theoretically modeled in the momentum equation, as in the turbulent flow case, the friction/viscous forces; the actual pressure drop, Δp; and thus the major head loss, h_f due to pipe friction are determined empirically. Application of the differential form of the momentum equation (supplemented with dimensional analysis and guided by the integral form of the momentum equation; a link between the two forms of the momentum equation), yields an expression for the friction slope, S_f and is given as follows (see Chapters 6 and 7 for its derivation and Chapter 8 for its application):

$$S_f = \frac{\tau_w}{\gamma R_h} = \frac{\Delta p}{\gamma L} = \frac{v^2}{C^2 R_h} \tag{8.25}$$

which is known as the Chezy equation (also known as the "resistance equation"), and C is the Chezy coefficient, which is derived from dimensional analysis and evaluated empirically. The differential form of the Chezy equation expressed in terms of a friction slope may be used to solve for one of the unknown quantities (S_f, v, C, $D = 4 R_h$) in the case of turbulent flow. One may note that because the Chezy equation assumes a straight pipe section of length, L, with a constant diameter, D, with a specific Chezy coefficient, C and a friction slope, S_f, it can be applied to solve for one of the unknown quantities at a given point in the fluid system. The Chezy coefficient, C has dimensions of $L^{1/2} T^{-1}$ and is a function of the boundary roughness, ε/D; the Reynolds number, R; and the cross-sectional shape or the hydraulic radius, R_h. Empirical findings illustrate that the Chezy coefficient, C ranges between $30 \ m^{1/2} \ s^{-1}$ (or $60 \ ft^{1/2} \ sec^{-1}$) for small channels with rough boundary surfaces to $90 \ m^{1/2} \ s^{-1}$ (or $160 \ ft^{1/2} \ sec^{-1}$) for large channels with smooth boundary surfaces. One may note that although the derivation/formulation of the Chezy equation assumes no specific units (SI or BG), the Chezy coefficient, C has dimensions of $L^{1/2} T^{-1}$. Furthermore, because the values for the Chezy coefficient, C have been provided/calibrated in both units (see above), one may apply the Chezy equation assuming either the SI or BG units. Although the Chezy equation is used for both open channel flow and pipe flow, it has a drawback in that the Chezy coefficient, C is not dimensionless like the Darcy–Weisbach friction factor, f (presented in Chapter 6 and also below). Therefore, one must attach the appropriate metric or English units to the Chezy coefficient, C prior to using it in the Chezy equation.

Finally, substituting the empirical expression for the friction slope (Chezy equation, Equation 8.25) into the integral form of the energy equation, Equation 8.4, yields an empirical expression for the major head loss, h_f due to pipe friction as follows (see Chapters 6 and 7 for its derivation and Chapter 8 for its application):

$$h_f = \frac{\tau_w L}{\gamma R_h} = \frac{\Delta p}{\gamma} = S_f L = \underbrace{\left[\frac{v^2}{C^2 R_h}\right]}_{S_f} L \tag{8.26}$$

which is the Chezy equation expressed in terms of the major head loss and provides an empirical expression for the major head loss, h_f (expressed as a function of an empirical flow resistance coefficient, C). One may note that although the theoretical relationship between the pressure drop, Δp and the shear stress, τ_w was derived assuming a horizontal pipe—derived from Newton's second law of motion (integral form of momentum equation) for both laminar and turbulent flow, the Chezy head loss equation (and the other empirical head loss equations for turbulent flow that are presented in the sections below) is valid for any fully developed, steady, incompressible pipe flow, whether the pipe is horizontal or inclined; this is because the governing equations are applied on a "subset level" (focusing only on the given element causing the flow resistance) in the derivation of the Chezy head loss equation. Furthermore, one may note that the velocity, v in the Chezy equation is the mean velocity, v_{ave} (mean over both the pipe diameter, D and the pipe section length, L). Thus, the Chezy "head loss equation" is used to evaluate the major head loss in the application of the integral form of the energy equation, Equation. 8.1, as follows:

$$\underbrace{\left(\frac{p}{\gamma} + z + \frac{v^2}{2g}\right)_1 - \left(\frac{p}{\gamma} + z + \frac{v^2}{2g}\right)_2}_{h_e} = \underbrace{\frac{\tau_w}{\gamma} \frac{L}{R_h} = \left[\frac{v^2}{C^2 R_h}\right] L}_{h_{f,major}} \tag{8.27}$$

where, in general, the pressure head drop, $\Delta p / \gamma$ is due to the change in elevation, the change in the velocity head (dynamic pressure), and the head loss due to pipe friction, which is represented by the Chezy coefficient, C.

The Chezy equation is applied in the analysis and design of turbulent pipe flow. Specifically, assuming that the Chezy coefficient, C and fluid specific weight, γ are specified, there are three types of problems in the application of the Chezy equation and are outlined in Table 8.2 below. The Chezy equation, expressed as a head loss equation, Equation 8.26, may be applied to solve for an unknown head loss, h_f, which is a Type 1 problem (analysis). The differential form of the Chezy equation, Equation 8.25, may be applied to solve for an unknown velocity of flow, v, which is a Type 2 problem (analysis). Furthermore, the differential form of the Chezy equation, Equation 8.25, may be applied to solve for an unknown pipe diameter, D, which is a Type 3 problem (design).

TABLE 8.2

Summary of Three Types of Problems for Turbulent Pipe Flow (Chezy Equation)

Problem Type	Given	Find	Resistance Equation	Governing Equation
Type 1	C, γ, L, v, D	h_f	$h_f = \frac{\Delta p}{\gamma} = S_f L = \underbrace{\left[\frac{v^2}{C^2 R_h}\right]}_{S_f} L$	**Chezy Equation** expressed as a head loss equation
Type 2	C, γ, L, S_f, D	v	$S_f = \frac{h_f}{L} = \frac{\Delta p}{\gamma L} = \frac{v^2}{C^2 R_h}$	**Chezy Equation**
Type 3	C, γ, L, v, S_f	D	$S_f = \frac{h_f}{L} = \frac{\Delta p}{\gamma L} = \frac{v^2}{C^2 R_h}$	**Chezy Equation**

EXAMPLE PROBLEM 8.2

Crude oil at 20°C flows in a 1-m-diameter inclined pipe at a flowrate of 2 m³/sec, as illustrated in Figure EP 8.2. The pressure at point 1 is 10×10^5 N/m². The increase in elevation between points 1 and 2 is 0.05 m, and the pipe length between points 1 and 2 is 10,000 m. The Chezy coefficient, C for the pipe is 50 m$^{1/2}$ s^{-1}. (a) Determine the flow regime. (b) Determine the major head loss due to the pipe friction (flow resistance) between points 1 and 2. (c) Determine the pressure drop between points 1 and 2 due to the increase in elevation and the pipe friction (flow resistance) between points 1 and 2. (d) Determine the friction slope at either point 1 or 2.

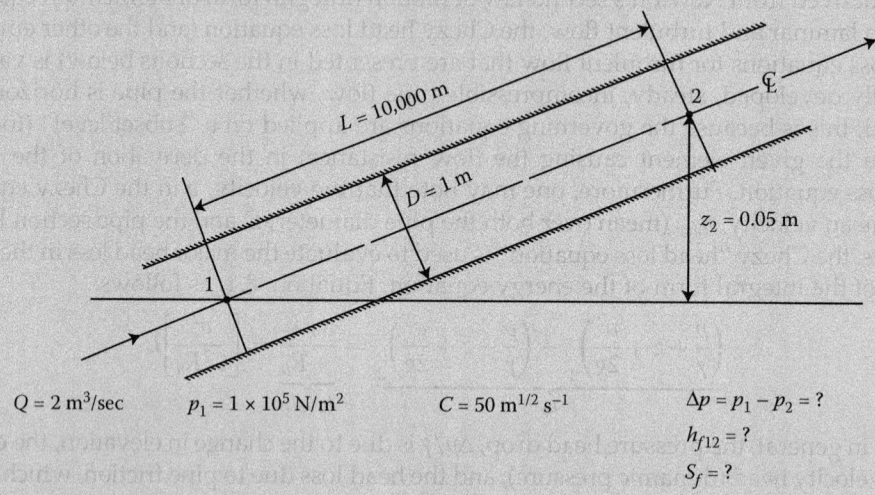

$Q = 2\ \text{m}^3/\text{sec}$ $p_1 = 1 \times 10^5\ \text{N/m}^2$ $C = 50\ \text{m}^{1/2}\ \text{s}^{-1}$ $\Delta p = p_1 - p_2 = ?$
$h_{f12} = ?$
$S_f = ?$

FIGURE EP 8.2
Crude oil flows in an inclined pipe flowing under pressure.

Mathcad Solution

(a) In order to determine the flow regime, one must compute the Reynolds number, **R** where the density, ρ and the viscosity, μ for crude oil at 20°C are given in Table A.4 in Appendix A as follows:

$$\rho := 856\ \frac{\text{kg}}{\text{m}^3} \qquad\qquad \mu := 7.2 \times 10^{-3}\ \text{N}\ \frac{\text{sec}}{\text{m}^2} \qquad\qquad D := 1\ \text{m}$$

$$Q := 2\ \frac{\text{m}^3}{\text{sec}} \qquad\qquad A := \frac{\pi \cdot D^2}{4} = 0.785\ \text{m}^2 \qquad\qquad v := \frac{Q}{A} = 2.546\ \frac{\text{m}}{\text{s}}$$

$$R := \frac{\rho \cdot v \cdot D}{\mu} = 3.027 \times 10^5$$

Thus, because the Reynolds number, **R** is greater than 4000, the flow is turbulent.

(b) Assuming turbulent flow, the major head loss due to pipe friction is computed from Equation 8.26 (Chezy equation expressed in terms of the major head loss) as a Type 1 problem as follows:

$$L := 10000 \text{ m} \qquad\qquad C := 50 \text{ m}^{\frac{1}{2}} \text{ sec}^{-1} \qquad\qquad R_h := \frac{D}{4} = 0.25 \text{ m}$$

$$h_f := \left(\frac{v^2}{C^2 \cdot R_h}\right) \cdot L = 103.753 \text{ m}$$

(c) Assuming the datum is at point 1, the pressure drop between points 1 and 2 is computed by applying the energy equation between points 1 and 2 as follows:

$$g := 9.81 \frac{\text{m}}{\text{sec}^2} \qquad\qquad \gamma := \rho \cdot g = 8.397 \times 10^3 \frac{\text{kg}}{\text{m}^2 \cdot \text{s}^2}$$

$$p_1 := 10 \times 10^5 \frac{\text{N}}{\text{m}^2} \qquad\qquad z_1 := 0 \text{ m} \qquad\qquad z_2 := 0.05 \text{ m}$$

$$v_1 := v = 2.546 \frac{\text{m}}{\text{s}} \qquad\qquad\qquad\qquad v_2 := v = 2.546 \frac{\text{m}}{\text{s}}$$

Guess value: $\qquad\qquad p_2 := 1 \times 10^5 \dfrac{\text{N}}{\text{m}^2}$

Given

$$\frac{p_1}{\gamma} + z_1 + \frac{v_1{}^2}{2 \cdot g} - h_f = \frac{p_2}{\gamma} + z_2 + \frac{v_2{}^2}{2 \cdot g}$$

$$p_2 := \text{Find}(p_2) = 1.283 \times 10^5 \frac{\text{N}}{\text{m}^2}$$

$$\Delta p := p_1 - p_2 = 8.717 \times 10^5 \frac{\text{N}}{\text{m}^2}$$

(d) The friction slope at either point 1 or 2 is computed from Equation 8.25, (Chezy equation) as follows:

$$S_f := \frac{v^2}{C^2 R_h} = 0.01 \frac{\text{m}}{\text{m}}$$

8.5.2 Turbulent Pipe Flow Resistance Equations and Their Roughness Coefficients

In addition to the Chezy equation, additional turbulent pipe flow resistance equations include the Darcy–Weisbach equation, the Manning's equation, and the Hazen–Williams equation. One may note that the drag coefficient, C_D representing the flow resistance that is derived by dimensional analysis in Chapter 7 is a nonstandard form of the drag coefficient and is summarized as follows:

$$\frac{\tau_w}{\rho v^2} = \frac{F_f}{F_I} = \frac{F_P}{F_I} = \frac{\Delta p R_h}{L} \frac{1}{\rho v^2} = \frac{R_h}{L} \frac{\Delta p}{\rho v^2} = \frac{R_h}{L} \frac{1}{E} = C_D = \phi\left(R, \frac{\varepsilon}{D}\right) \tag{8.28}$$

One may note that because the drag coefficient, C_D represents the flow resistance, which causes an unknown pressure drop, Δp, the pressure force will always play an important role and thus the definition of the drag coefficient, C_D is mainly a function of E. Furthermore, one may note that the drag coefficient, C_D, which is a nonstandard form of the flow resistance, is dimensionless and is a function of boundary roughness, ε/D; the Reynolds number,

R; and the cross-sectional shape or the hydraulic radius, R_h. More standard forms of the drag coefficient include empirical flow resistance coefficients including the Chezy coefficient, C; the Darcy–Weisbach friction factor, f; and the Manning's roughness coefficient, n. In addition to these three standard empirical flow resistance coefficients, there is another commonly used one, which is the Hazen–Williams roughness coefficient, C_h. As presented and discussed in detail in Chapters 6 and 7, from dimensional analysis, the drag coefficient drag coefficient, C_D is related to the Chezy coefficient, C and the Darcy–Weisbach friction factor, f, respectively, as follows:

$$C = \sqrt{\frac{g}{C_D}} \tag{8.29}$$

$$f = 8C_D \tag{8.30}$$

where the drag coefficient, C_D; the Chezy coefficient, C; and the Darcy–Weisbach friction factor, f are a function of the boundary roughness, ε/D; the Reynolds number, R; and the cross-sectional shape or the hydraulic radius, R_h. Thus, the Chezy coefficient, C and the Darcy–Weisbach friction factor, f are related as follows:

$$C = \sqrt{\frac{8g}{f}} \tag{8.31}$$

Furthermore, empirical studies (assuming the metric system) have indicated that a more accurate dependence of the Chezy coefficient, C on the hydraulic radius, R_h is defined by the definition of the Manning's roughness coefficient, n as follows:

$$C = \frac{R_h^{1/6}}{n} \tag{8.32}$$

where the Manning's roughness coefficient, n is independent of both the Reynolds number, R (flow conditions) (because it assumes turbulent flow) and the cross-sectional shape or the hydraulic radius, R_h (flow geometry) (by its relationship with the Chezy coefficient, C). As such, the Manning's roughness coefficient, n is a function only of the boundary roughness, ε/D. And, similarly to the Manning's roughness coefficient, n, the Hazen–Williams roughness coefficient, C_h is a function only of the boundary roughness, ε/D. Therefore, while the Chezy coefficient, C and the Darcy–Weisbach friction factor, f represent both laminar and turbulent flow, the Manning's roughness coefficient, n and the Hazen–Williams roughness coefficient, C_h represent turbulent flow only.

8.5.2.1 A Comparison of the Three Standard Empirical Flow Resistance Coefficients

It is important to compare the three standard empirical flow resistance coefficients. Although values for the Chezy coefficient, C have been empirically evaluated and tabulated (limited tables available) for various combinations of boundary roughness, ε/D; Reynolds number, R; and cross-sectional shape (or the hydraulic radius, R_h), it has been found that it is easier to model the effects of the Chezy coefficient, C (i.e., the drag coefficient, C_D) by using the Darcy–Weisbach friction coefficient, f, or the Manning's roughness coefficient, n. While the behavior of the Darcy–Weisbach friction factor, f in circular pipe flow has been thoroughly explored, a similarly complete investigation of the behavior of the Chezy coefficient, C has never been made. Therefore, in the cases of laminar or turbulent flow, the use of the Darcy–Weisbach friction factor, f is preferred over the Chezy coefficient, C. Additionally, f is dimensionless, while C has dimensions of $L^{1/2} T^{-1}$ and thus one must specify either SI or BG units for C. In the case of turbulent flow, the use of the Manning's roughness coefficient, n (which has dimensions of $L^{-1/3} T$) is preferred over the Chezy coefficient, C because it is independent of both the Reynolds number, R (by assuming turbulent flow) and the cross-sectional shape or the hydraulic radius, R_h (by its relationship with the Chezy coefficient, C). And, finally, although f and n are independent of R for the case of turbulent flow, the use of the Darcy–Weisbach friction factor, f is preferred over the use of the Manning's roughness coefficient, because f is dimensionless, while n has dimensions of $L^{-1/3} T$, and thus one must specify either SI or BG units for n.

8.5.2.2 A Comparison between Manning's and Hazen–Williams Roughness Coefficients

For the case of turbulent flow, although n and C_h are a function only of the boundary roughness, ε/D, the Manning's roughness coefficient, n has dimensions of $L^{-1/3} T$, while the Hazen–Williams roughness coefficient, C_h is dimensionless. Because the Manning's equation was derived (formulated without the need for calibration) using dimensional analysis, there are no limitations for the ranges assumed for the individual variables (v, D, and water temperature). However, although the Hazen–Williams equation was formulated based on the results of dimensional analysis (the predictor variables, C_h, R_h, and S_f are based on the Chezy equation), it was "calibrated" for a specific range of variables (v, D, and water temperature), and thus there is a limitation of its practical application.

8.5.2.3 Turbulent Pipe Flow Resistance Equations

The turbulent pipe flow resistance equations, along with the corresponding resistance coefficients and head loss equations, are derived and discussed in Chapter 6 and summarized and applied in the sections below. The Chezy equation was presented in Section 8.5.1.2, the Darcy–Weisbach equation is presented in Section 8.5.3, the Manning's equation is presented in Section 8.5.4, and the Hazen–Williams equation is presented in Section 8.5.5. In summary, the Darcy–Weisbach friction factor, f represents the most accurate and the most commonly used pipe flow resistance coefficient for both laminar and turbulent flow. The accuracy and general representation of the flow type of the Darcy–Weisbach friction factor, f is a result of its dependence on the boundary roughness, ε/D; the Reynolds number, R; and the cross-sectional shape or the hydraulic radius, R_h. The resulting tediousness in evaluation of the Darcy–Weisbach friction factor, f is overcome by the use of mathematical software such as Mathcad to solve the mathematical model for f, which includes the Colebrook equation and the Haaland equation (see Equations 8.35, and 8.38, respectively), and is graphically illustrated by the Moody diagram (see Figure 8.1). Additionally, because f is dimensionless, there is no need to specify either SI or BG units for f. Finally, because the Darcy–Weisbach equation was derived (formulated without the need for calibration) using dimensional analysis, there are no limitations for the ranges assumed for the individual variables (v, D, and water temperature). For laminar flow, which is a special case of turbulent flow, the Darcy–Weisbach friction factor, $f = 64/R$ and thus the Darcy–Weisbach equation becomes Poiseuille's law expressed in terms of the major head loss. And, for turbulent flow, one may choose to apply the Darcy–Weisbach equation, the Manning's equation, or the Hazen–Williams equation, depending upon the desired accuracy, simplicity, and practical constraints in the range of variables (v, D, and water temperature). It is important to note that the Hazen–Williams equation is calibrated (and the coefficient, C_h is evaluated) specifically for the fluid water.

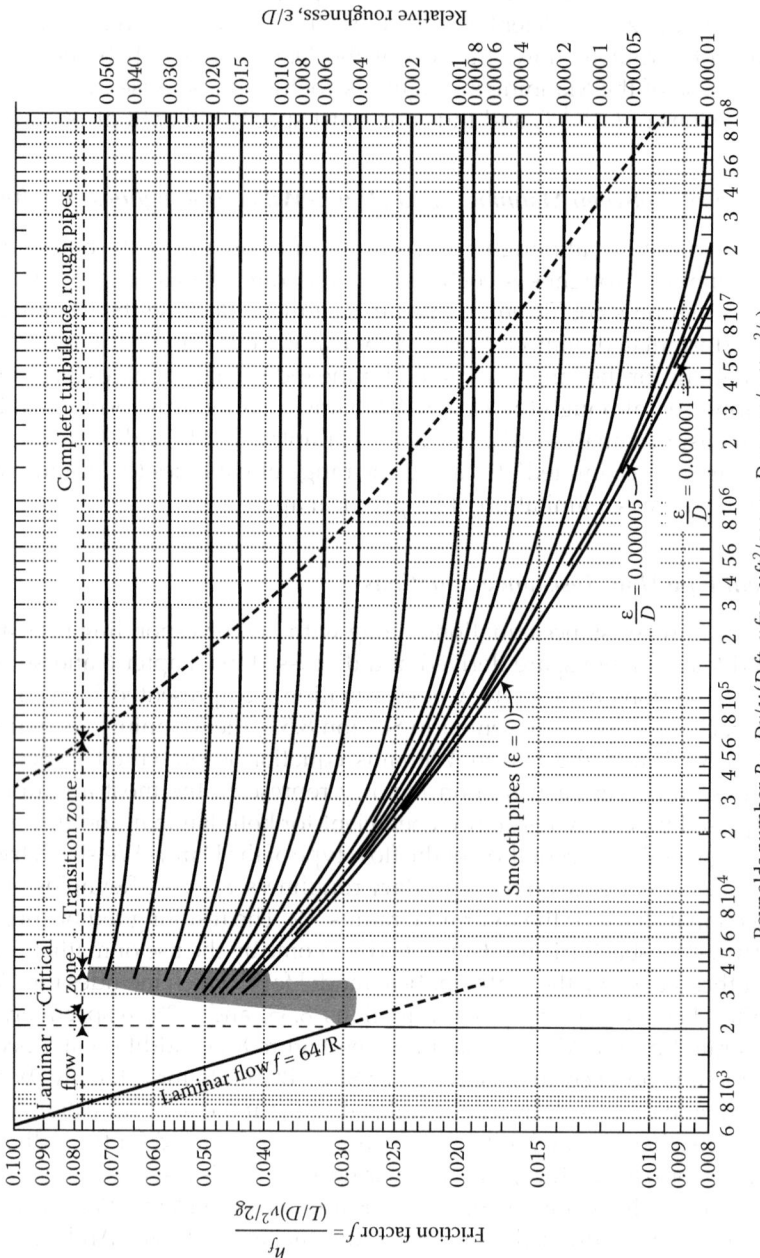

FIGURE 8.1
Moody diagram. (Adapted from Moody, L. F., *Transactions of the American Society of Mechanical Engineers*, 66, 672, 1944.)

8.5.3 The Darcy–Weisbach Friction Coefficient, *f* and the Darcy–Weisbach Equation

The empirical derivation of the Darcy–Weisbach friction coefficient/factor, *f*, which represents a dimensionless flow resistance coefficient, is derived using dimensional analysis and is expressed as a function of the Chezy coefficient, *C*, as presented in Chapters 6 and 7, and is summarized as follows:

$$C = \sqrt{\frac{8g}{f}} \tag{8.33}$$

where the Darcy friction factor, *f*, which, like the Chezy coefficient, *C* is a more standard form of the drag coefficient, C_D. However, unlike the Chezy coefficient, *C*, *f* is dimensionless and is a function of the boundary roughness, ε/D; the Reynolds number, *R*; and the cross-sectional shape or the hydraulic radius, R_h. One may note that ε is the absolute pipe roughness, while ε/D is the relative pipe roughness. Furthermore, Table 8.3 presents typical values for the absolute pipe roughness, ε for various pipe material, assuming new and clean pipes. It is important to note that over the lifetime of a pipe design, the absolute pipe roughness, ε values will change (increase) due to the buildup of deposits on the pipe wall and thus cause an increase in the frictional head loss. Thus, in order to minimize potential problems associated with head loss, the design of a pipe system should consider the temporal change of the absolute pipe roughness, ε for a given pipe material.

Furthermore, while the behavior of the Darcy–Weisbach friction factor, *f* in circular pipe flow has been thoroughly explored, a similarly complete investigation of the behavior of the Chezy coefficient, *C* has never been made. This is because in addition to the extra variables involved in the open channel flow case, there is an extremely wide range of surface roughness sizes and types found in practice, and it is difficult to attain steady, uniform, fully developed flow outside of the controlled laboratory environment. Additionally, empirical studies have illustrated that the behavior of the Chezy coefficient, *C* and the Darcy–Weisbach

TABLE 8.3

Typical Values for the Absolute Pipe Roughness, ε for Various Pipe Material for New Pipes

Pipe Material	Absolute Pipe Roughness, ε (Feet)	Absolute Pipe Roughness, ε (Millimeters)
Glass, plastic (smooth)	0.0	0.0
Transite, brass, drawn tubing, bituminous lining, lead, copper, centrifugally spun cement	0.000 005	0.0015
Rubber (smoothed)	0.000 033	
Commercial steel, wrought iron, welded-steel pipe	0.000 15	0.046
Asphalted cast iron	0.000 4	0.12
Galvanized iron	0.000 5	0.15
Wood stave	0.000 6–0.003	0.18–0.9
Cast iron	0.000 85	0.25
Concrete	0.001–0.01	0.3–3
Riveted steel	0.003–0.03	0.9–9
Rusted steel	0.007	2.0

Source: Moody, L. F., *Transactions of the American Society of Mechanical Engineers*, 66, 681, 1944; Finnemore, E. J., and J. B., Franzini, *Fluid Mechanics with Engineering Applications*, 10th ed, McGraw Hill, New York, 2002, p. 286; White, F. M., *Fluid Mechanics*, 7th ed, McGraw Hill, New York, 2011, p. 371.

friction factor, f are least affected by the cross-sectional shape (or the effects are at least within the limits of accuracy that is normally accepted in practice).

Substituting the relationship between the Chezy coefficient, C and the Darcy–Weisbach friction factor, f Equation 8.33 into the Chezy head loss equation, Equation 8.26, yields the Darcy–Weisbach head loss equation, as presented in Chapter 6 and repeated as follows:

$$h_f = \frac{\tau_w L}{\gamma R_h} = \frac{\Delta p}{\gamma} = S_f L = \frac{v^2 L}{C^2 R_h} = \frac{v^2 L}{\left(\frac{8g}{f}\right)\left(\frac{D}{4}\right)} = f \frac{L}{D} \frac{v^2}{2g} \tag{8.34}$$

One may note that the derivation/formulation of the Darcy–Weisbach equation assumes no specific units (SI or BG), and, furthermore, the Darcy–Weisbach friction factor, f is dimensionless. Therefore, one may apply the Darcy–Weisbach equation assuming either the SI or BG units.

8.5.3.1 Evaluation of the Darcy–Weisbach Friction Coefficient, f

The empirical evaluation of the friction factor, f is presented in both graphical (Moody diagram) and mathematical forms. As mentioned above, the results of dimensional analysis, assuming turbulent flow (see Chapter 7), indicated that the Darcy–Weisbach friction coefficient, f has no dimensions, and is a function of the boundary roughness, ε/D; the Reynolds number, R; and the cross-sectional shape or the hydraulic radius, R_h. However, at very high velocities (and thus very high values for R) the f for completely turbulent flow (rough pipes) is a function only of the relative pipe roughness, ε/D and thus independent of R. Furthermore, because laminar flow is a special case of turbulent flow, a theoretical evaluation of the Darcy–Weisbach friction coefficient, f for laminar "circular" pipe flow (given in Section 8.5.3.2 below) illustrates that f for laminar flow is only a function of the Reynolds number, R (where $f = 64/R$).

Unlike laminar flow, the friction factor, f for turbulent flow cannot be evaluated theoretically. Instead, however, it is evaluated using experiments. Specifically, Johann Nikuradse computed the friction factor, f from measurements of the flowrate, Q and the pressure drop, Δp. Using pipes with simulated absolute pipe roughness, ε, the pressure drop, Δp needed to produce a specified flowrate, Q was measured. The data was then converted into the friction factor, f for the corresponding Reynolds number, R and the relative pipe roughness, ε/D. Numerous tests were conducted for a wide range of R and ε/D in order to determine an empirical relationship between the friction factor, f and both the Reynolds number, R and the relative pipe roughness, ε/D. Further experimental studies conducted by L. F. Moody resulted in the Moody diagram illustrated in Figure 8.1 that presents the Darcy–Weisbach friction coefficient, f as a function of the Reynolds number, R and the relative pipe roughness, ε/D.

In order to use the Moody diagram to estimate the Darcy–Weisbach friction coefficient, f, one may enter the diagram with the value of the Reynolds number, R and the relative pipe roughness, ε/D and read off the value for f. The Moody diagram assumes that the flow is steady, fully developed, incompressible flow. There are a number of important points that one should note about the Moody diagram, given as follows: (1) a log-log plot is used in order to accommodate the large range in values for f and R; (2) for the laminar flow region (a special case of turbulent flow) ($R < 2000$), $f = 64/R$ plots as a straight line (f decreases linearly with an increase in R and is independent of the relative pipe roughness, ε/D); (3) for the critical zone ($2000 < R < 4000$), the flow can be either laminar or turbulent, and thus there is no curve; (4) for turbulent flow ($R > 4000$), f is generally a function of both the Reynolds number, R and the relative pipe roughness, ε/D (f initially decreases nonlinearly with an increase in R and a decrease in ε/D, then f becomes independent of R); (5) the right vertical axis represents a given value for ε/D associated with a given nonlinear curve for f; (6) the lowest value of $\varepsilon/D = 0$ is modeled by the last nonlinear curve at the bottom of the collection of curves, which represents a smooth pipe, for which the $f \neq 0$ (because the fluid at the pipe wall sticks to the wall regardless of whether the pipe is smooth or rough due to the "no-slip" condition); and (7) the turbulent flow region ($R > 4000$) modeled by the nonlinear curves is further subdivided into two zones: (a) the first is the transition zone, which starts at the beginning of turbulence ($R = 4000$) and ends at the dashed line and is characterized by nonlinear curves; thus, f is a function of both the Reynolds number, R and the relative pipe roughness, ε/D; and (b) the second is the complete turbulence (rough pipe) zone, which starts at the dashed line and ends at the largest values for R and is characterized by horizontal lines; thus, f is a function only of the relative pipe roughness, ε/D and thus independent of the Reynolds number, R.

The Moody diagram, which presents a graphical approach to estimating the Darcy–Weisbach friction coefficient, f may be presented by two mathematical formulas in order to expedite calculations in the design of pipes. First, the friction factor, f for laminar flow region ($R < 2000$) of the Moody diagram was already given by the theoretical formula, $f = 64/R$, which is derived in Section 8.5.3.2 below. And second, the friction factor, f for the turbulent flow region ($R > 4000$) of the Moody diagram has been mathematically modeled by numerous empirical formulas for certain ranges of the Reynolds number, R. The most common formula for turbulent flow ($R > 4000$) (which includes both the transition zone and the complete turbulence/rough pipe zone for both rough and smooth pipes) is the Colebrook equation and is given as follows:

$$\frac{1}{\sqrt{f}} = -2.0 \log_{10} \left(\frac{\frac{\varepsilon}{D}}{3.7} + \frac{2.51}{R\sqrt{f}} \right) \tag{8.35}$$

Furthermore, because the Colebrook equation fits the turbulent ($R > 4000$) pipe flow experimental data in the Moody diagram so well, it is commonly used to draw the Moody diagram in the turbulent flow region. One may note that because the friction factor, f appears on both sides of the Colebrook equation, it is not explicit in f and thus must be solved by either by a conventional trial-and-error approach (Esposito, 1998) or by a mathematical numerical solution such as by Mathcad, which is the approach taken in this textbook.

Beginning with the above Colebrook formula for turbulent flow, one may make specific assumptions and further tailor the Colebrook formula for: (1) a smooth pipe ($\varepsilon/D = 0$) (Prandtl equation), and (2) complete turbulence/rough pipe zone ($R \to \infty$)(von Karman equation), respectively, as follows:

$$\frac{1}{\sqrt{f}} = 2.0 \log_{10}\left(R\sqrt{f}\right) - 0.8 \quad \text{smooth pipe, } \varepsilon/D = 0 \,(\text{Prandtl equation}) \tag{8.36}$$

$$\frac{1}{\sqrt{f}} = -2.0 \log_{10}\left(\frac{\frac{\varepsilon}{D}}{3.7} \right) \quad \text{completely rough zone, } R \to \infty \,(\text{von Karman equation}) \tag{8.37}$$

Furthermore, while the Colebrook equation given above for turbulent flow ($R > 4000$) Equation 8.35 is not explicit in the friction factor, f, the Haaland equation for turbulent flow ($R > 4000$) is explicit in the friction factor, f and is given as follows:

$$\frac{1}{\sqrt{f}} \cong -1.8 \log_{10} \left[\left(\frac{\frac{\varepsilon}{D}}{3.7} \right)^{1.11} + \frac{6.9}{R} \right] \tag{8.38}$$

Application of the Haaland equation, which is explicit in the friction factor, f would not require the use of mathematical software such as Mathcad, as the application of the Colebrook equation would.

8.5.3.2 Application of the Darcy–Weisbach Equation

The Darcy–Weisbach equation for pipe flow assumes a straight pipe section of length, L, with a constant diameter, D and thus a constant velocity of flow, v with a specific Darcy–Weisbach friction factor, f and a friction slope, S_f, which may be solved for one of the following unknown quantities (integral approach for the head loss, while differential approach for the other quantities): h_f, S_f, v, f, or $D = 4\,R_h$. Choosing the Darcy–Weisbach friction factor, f to model the empirical flow resistance is a typical choice for a pipe flow (where $R_h = D/4$).

It is important to note that because the Darcy–Weisbach head loss equation and the expression for the Reynolds number, $R = \rho v D/\mu$ assume a circular pipe, it is important to address their respective definitions for noncircular pipes as presented in Chapter 6 and repeated as follows. When noncircular pipes, such rectangular and other cross-sectional shapes are used in practice, one may define the "equivalent diameter" or the "hydraulic diameter," D_h as follows:

$$D_h = 4R_h \tag{8.39}$$

Therefore, the Darcy–Weisbach equation and the definition of the Reynolds number, R for a noncircular pipe are given, respectively, as follows:

$$h_f = f \frac{L}{(4R_h)} \frac{v^2}{2g} \tag{8.40}$$

$$R = \frac{\rho v (4R_h)}{\mu} \tag{8.41}$$

Additionally, one may use the "equivalent diameter" or the "hydraulic diameter," D_h in the case of noncircular pipes in the definition of the relative pipe roughness, ε/D_h and with any other equation or chart (such as the Moody diagram) that assumes a circular pipe with a diameter, D.

Finally, the Darcy–Weisbach head loss equation is commonly used in the analysis and design of both laminar and turbulent pipe flow. In the case of laminar flow, Poiseuille's law provided an analytical expression for the major head loss, h_f (expressed as a function of the flow resistance, μ), without the need to define an empirical flow resistance coefficient. However, because laminar flow is a special case of turbulent flow, it is of interest to evaluate the Darcy–Weisbach friction factor, f for laminar flow as presented in Chapter 6 and repeated as follows. Equating Poiseuille's law expressed in terms of the head loss Equation 8.23 (derived for laminar flow) to the Darcy–Weisbach head loss equation (Equation 8.34) (derived for turbulent flow) and isolating the friction coefficient, f yields the Darcy–Weisbach friction coefficient, f for laminar "circular" pipe flow as follows:

$$f = \frac{64\mu}{\rho v D} = \frac{64}{R} \tag{8.42}$$

where the Reynolds number, $R = \rho v D/\mu$. However, in the case where the pipe is noncircular, then one may use the appropriate relationship between the friction factor, f and the Reynolds number, R for various noncircular pipe cross sections given in Table 8.4. One

TABLE 8.4

Relationship between Darcy–Weisbach Friction Factor, f and Reynolds Number, $R = \rho v (4R_h)/\mu$ for Fully Developed Laminar Flow for Various Pipe Cross Sections

Pipe Cross Section	a/b or θ	Darcy–Weisbach Friction Factor, f
Circle	—	$64.00/R$
Ellipse	**a/b** 1	$64.00/R$
	2	$67.28/R$
	4	$72.96/R$
	8	$76.60/R$
	16	$78.16/R$
Rectangle	**a/b** 1	$56.91/R$
	1.33	$57.89/R$
	2	$62.19/R$
	2.5	$65.47/R$
	3	$68.36/R$
	4	$72.93/R$
	6	$78.81/R$
	8	$82.34/R$
	10	$84.68/R$
	20	$89.91/R$
	∞	$96.00/R$
Isosceles triangle	**θ** 0°	$48.00/R$
	10°	$50.80/R$
	20°	$51.60/R$
	30°	$52.28/R$
	40°	$52.90/R$
	60°	$53.30/R$
	80°	$52.90/R$
	90°	$52.60/R$
	100°	$52.00/R$
	120°	$51.10/R$

Source: Adapted from Cengel, Y. A., and J. M., Cimbala, *Fluid Mechanics Fundamentals and Applications*, 3rd ed, McGraw Hill, New York, 2014, p. 358; White, F. M., *Fluid Mechanics*, 7th ed, McGraw Hill, New York, 2011, p. 387.

TABLE 8.5

Summary of Three Types of Problems for Turbulent Pipe Flow (Darcy–Weisbach Equation)

Problem Type	Given	Find	Resistance Equation	Governing Equation
Type 1	f, γ, L, v, D	h_f	$h_f = \dfrac{\Delta p}{\gamma} = S_f L = f \dfrac{L}{D} \dfrac{v^2}{2g}$	Darcy–Weisbach Equation
Type 2	f, γ, L, S_f, D	v	$S_f = \dfrac{h_f}{L} = \dfrac{\Delta p}{\gamma L} = f \dfrac{1}{D} \dfrac{v^2}{2g}$	Darcy–Weisbach Equation
Type 3	f, γ, L, v, S_f	D	$S_f = \dfrac{h_f}{L} = \dfrac{\Delta p}{\gamma L} = f \dfrac{1}{D} \dfrac{v^2}{2g}$	Darcy–Weisbach Equation

may note that the relationships between the friction factor, f and the Reynolds number, $R = \rho v (4R_h)/\mu$ for the various noncircular pipe sections are derived from theory and/or empirical studies. Thus, f for laminar flow is only a function of the Reynolds number, R, while f for turbulent flow is a function of both the Reynolds number, R and the relative pipe roughness, ε/D. However, at very high velocities (and thus very high values for R), the f for completely turbulent flow (rough pipes) is a function only of the relative pipe roughness, ε/D and thus independent of R.

The Darcy–Weisbach equation is applied in the analysis and design of turbulent pipe flow. Specifically, assuming that the Darcy–Weisbach friction factor, f and fluid specific weight, γ are specified, there are three types of problems in the application of the Darcy–Weisbach equation and are outlined in Table 8.5. The Darcy–Weisbach equation, Equation 8.34, may be applied to solve for an unknown head loss, h_f, which is a Type 1 problem (analysis). The Darcy–Weisbach equation, Equation 8.34, may be applied to solve for an unknown velocity of flow, v which is a Type 2 problem (analysis). Furthermore, the Darcy–Weisbach equation, Equation. 8.34, may be applied to solve for an unknown pipe diameter, D which is a Type 3 problem (design).

It is important to note that when solving numerical (as opposed to analytical) problems in Mathcad, especially those problems that have many unknowns to solve for, the solution procedure is highly sensitive to the guessed values. Thus, one may have to continue to adjust the guess values until the numerical procedure converges to a solution. This is in contrast to solving an analytical problem such as the specific energy equation for open channel flow, which is cubic in y (see Chapter 9). The Mathcad numerical solution procedure is especially sensitive to the assumed guess values for f and R (and h_f and v). Thus, "good" guess values to assume are: $f = 0.01$, $R = 400,000$ (and $h_f = 10$ m and $v = 5$ m/sec), as they seem to work for most assumed values for L, D, ε, Q, etc.

EXAMPLE PROBLEM 8.3

Water at 20°C flows in a 1.5-m-diameter horizontal concrete pipe at a flowrate of $4 \, \text{m}^3/\text{sec}$, as illustrated in Figure EP 8.3. The pressure at point 1 is $25 \times 10^5 \, \text{N/m}^2$, and the pipe length between points 1 and 2 is 20,000 m. Assume the Darcy–Weisbach friction factor is chosen to model the flow resistance. (a) Determine the flow regime. (b) Determine the major head loss due to the pipe friction (flow resistance) between points 1 and 2. (c) Determine the pressure drop between points 1 and 2 due to the pipe friction (flow resistance) between points 1 and 2. (d) Determine the friction slope at either point 1 or 2.

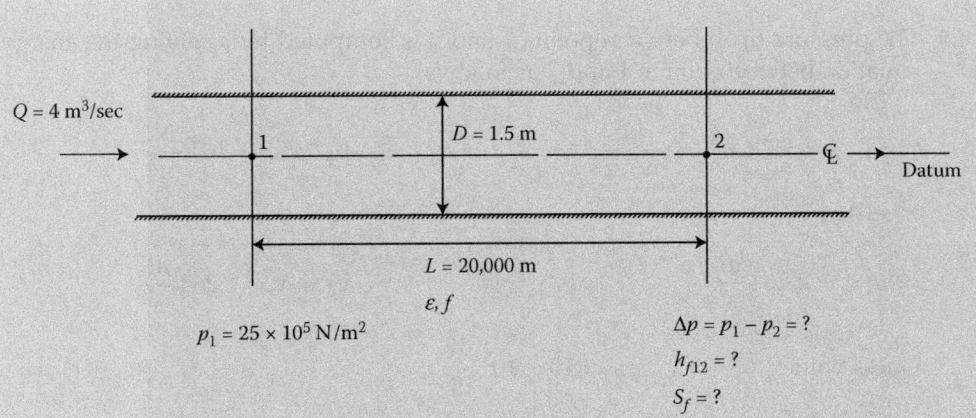

FIGURE EP 8.3
Water flows in a horizontal concrete pipe flowing under pressure.

Mathcad Solution

(a) In order to determine the flow regime, one must compute the Reynolds number, R where the density, ρ and the viscosity, μ for water at 20°C are given in Table A.4 in Appendix A as follows:

$$\rho := 998 \frac{kg}{m^3} \qquad \mu := 1 \times 10^3 \, N \, \frac{sec}{m^2} \qquad D := 1.5 \, m$$

$$Q := 4 \frac{m^3}{sec} \qquad A := \frac{\pi \cdot D^2}{4} = 1.767 \, m^2 \qquad v := \frac{Q}{A} = 2.264 \frac{m}{s}$$

$$R := \frac{\rho \cdot v \cdot D}{\mu} = 3.389 \times 10^6$$

Thus, because the Reynolds number, R is greater than 4000, the flow is turbulent.

(b) Assuming turbulent flow, the major head loss due to pipe friction is computed from Equation 8.34 (Darcy–Weisbach head loss equation) as a Type 1 problem. The absolute pipe roughness for concrete is given in Table 8.3, and the Colebrook equation for the friction factor is assumed as follows:

$$L := 20,000 \, m \qquad \varepsilon := 0.3 \, mm \qquad \frac{\varepsilon}{D} = 2 \times 10^{-4} \qquad g := 9.81 \frac{m}{sec^2}$$

Guess values: $\quad f := 0.01 \qquad \qquad h_f := 100 \, m$

Given

$$\frac{1}{\sqrt{f}} = -2 \cdot \log\left(\frac{\frac{\varepsilon}{D}}{3.7} + \frac{2.51}{R \cdot \sqrt{f}}\right)$$

$$h_f = f \cdot \frac{L}{D} \frac{v^2}{2 \cdot g}$$

$$\begin{pmatrix} f \\ h_f \end{pmatrix} := Find\,(f, h_f) \qquad f = 0.014 \qquad h_f = 48.888 \, m$$

(c) The pressure drop between points 1 and 2 is computed by applying the energy equation between points 1 and 2 as follows:

$$\gamma := \rho \cdot g = 9.79 \times 10^3 \; \frac{kg}{m^2 \cdot s^2} \qquad\qquad p_1 := 25 \times 10^5 \; \frac{N}{m^2}$$

$$z_1 := 0 \, m \qquad\qquad z_2 := 0 \, m$$

$$v_1 := v = 2.264 \; \frac{m}{s} \qquad\qquad v_2 := v = 2.264 \; \frac{m}{s}$$

Guess value: $\qquad\qquad p_2 := 10 \times 10^5 \; \dfrac{N}{m^2}$

Given

$$\frac{p_1}{\gamma} + z_1 + \frac{v_1{}^2}{2 \cdot g} - h_f = \frac{p_2}{\gamma} + z_2 + \frac{v_2{}^2}{2 \cdot g}$$

$$p_2 := Find(p_2) = 2.021 \times 10^6 \; \frac{N}{m^2}$$

$$\Delta p := p_1 - p_2 = 4.786 \times 10^5 \; \frac{N}{m^2}$$

(d) The friction slope at either point 1 or 2 is computed from Equation 8.34 (Darcy–Weisbach equation) as follows:

$$S_f := f \cdot \frac{1}{D} \cdot \frac{v^2}{2 \cdot g} = 2.444 \times 10^{-3} \; \frac{m}{m}$$

EXAMPLE PROBLEM 8.4

Water at 80°C flows in a 2.5-m-diameter horizontal riveted steel pipe, as illustrated in Figure EP 8.4. The pressure at point 1 is $29 \times 10^5 \, N/m^2$, the major head loss due to the pipe friction is 75 m, and the pipe length between points 1 and 2 is 3,000 m. Assume the Darcy–Weisbach friction factor is chosen to model the flow resistance. (a) Determine velocity at either point 1 or 2. (b) Determine the flow regime. (c) Determine the pressure drop between points 1 and 2 due to the pipe friction (flow resistance) between points 1 and 2.

Mathcad Solution

(a) The velocity at either point 1 or 2 is computed from Equation 8.34 (Darcy–Weisbach equation) as a Type 2 problem. The absolute pipe roughness for riveted steel is given in Table 8.3, and the Colebrook equation for the friction factor is assumed. The density, ρ and the viscosity, μ for water at 80°C are given in Table A.4 in Appendix A as follows:

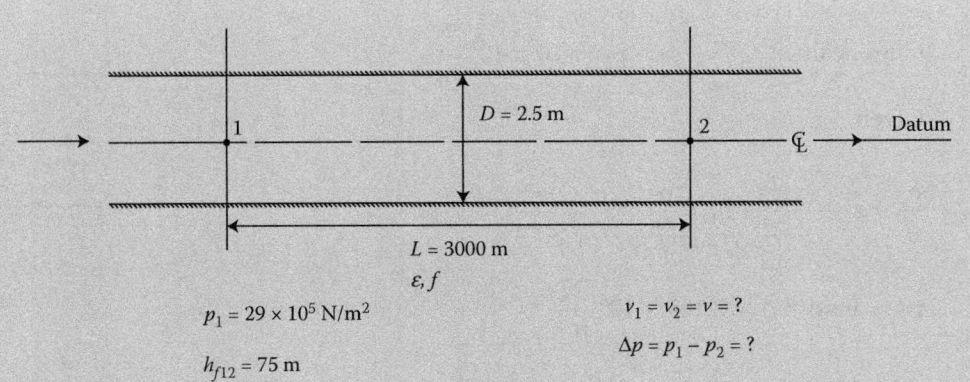

FIGURE EP 8.4
Water flows in a horizontal riveted steel pipe flowing under pressure.

$D := 2.5 \, m$ $\qquad$ $L := 3000 \, m$ $\quad$ $\varepsilon := 0.9 \, mm$ $\qquad$ $\dfrac{\varepsilon}{D} = 3.6 \times 10^{-4}$

$h_f := 75 \, m$ $\qquad$ $S_f := \dfrac{h_f}{L} = 0.025 \, \dfrac{m}{m}$ $\qquad$ $g := 9.81 \, \dfrac{m}{sec^2}$

$\rho := 971.8 \, \dfrac{kg}{m^3}$ $\qquad$ $\mu := 0.354 \times 10^{-3} \, N \, \dfrac{sec}{m^2}$

Guess values: $\qquad$ $f := 0.01$ $\qquad$ $R := 400{,}000$ $\qquad$ $v := 2 \, \dfrac{m}{sec}$

Given

$$\frac{1}{\sqrt{f}} = -2 \cdot \log\left(\frac{\frac{\varepsilon}{D}}{3.7} + \frac{2.51}{R \cdot \sqrt{f}}\right)$$

$$R = \frac{\rho \cdot v \cdot D}{\mu}$$

$$S_f = f \cdot \frac{1}{D} \frac{v^2}{2 \cdot g}$$

$$\begin{pmatrix} f \\ R \\ v \end{pmatrix} := Find(f, R, v) \quad f = 0.016 \quad R = 6.096 \times 10^7 \quad v = 8.882 \, \frac{m}{s}$$

(b) Thus, because the Reynolds number, *R* is greater than 4000, the flow is turbulent.
(c) The pressure drop between points 1 and 2 is computed by applying the energy equation between points 1 and 2 as follows:

$\gamma := \rho \cdot g = 9.533 \times 10^3 \, \dfrac{kg}{m^2 \cdot s^2}$ $\qquad\qquad$ $p_1 := 29 \times 10^5 \, \dfrac{N}{m^2}$

$z_1 := 0 \, m$ $\qquad\qquad\qquad\qquad\qquad\qquad$ $z_2 := 0 \, m$

$v_1 := v = 8.882 \, \dfrac{m}{s}$ $\qquad\qquad\qquad\qquad$ $v_2 := v = 8.882 \, \dfrac{m}{s}$

Guess value: $\qquad$ $p_2 := 10 \times 10^5 \; \dfrac{N}{m^2}$

Given

$$\dfrac{p_1}{\gamma} + z_1 + \dfrac{v_1{}^2}{2 \cdot g} - h_f = \dfrac{p_2}{\gamma} + z_2 + \dfrac{v_2{}^2}{2 \cdot g}$$

$$p_2 := \text{Find}(p_2) = 2.185 \times 10^6 \; \dfrac{N}{m^2}$$

$$\Delta p := p_1 - p_2 = 7.15 \times 10^5 \; \dfrac{N}{m^2}$$

EXAMPLE PROBLEM 8.5

Water at 15°C flows in an inclined galvanized iron pipe at a velocity of 3 m/sec, as illustrated in Figure EP 8.5. The pressure at point 1 is $15 \times 10^5 \; N/m^2$, and the major head loss due to the pipe friction is 80 m. The increase in elevation between points 1 and 2 is 0.07 m, and the pipe length between points 1 and 2 is 10,000 m. Assume the Darcy–Weisbach friction factor is chosen to model the flow resistance. (a) Determine pipe diameter at either point 1 or 2. (b) Determine the flow regime. (c) Determine the pressure drop between points 1 and 2 due to the increase in elevation and the pipe friction (flow resistance) between points 1 and 2.

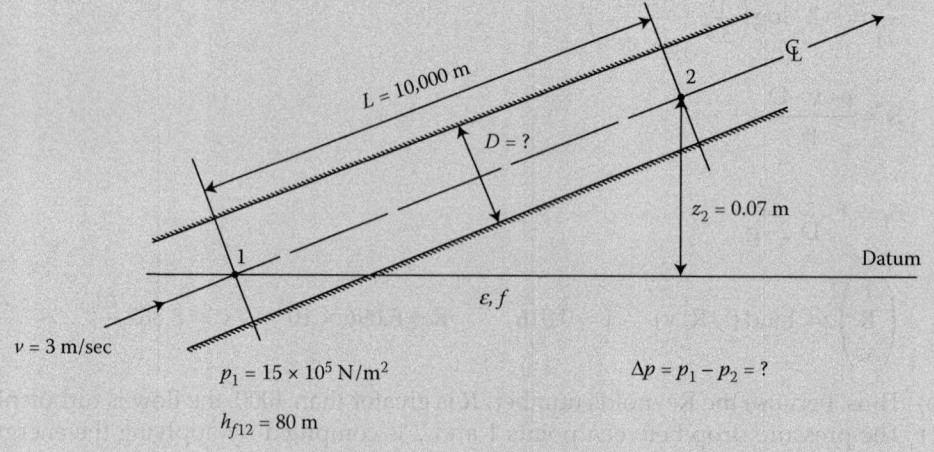

FIGURE EP 8.5
Water flows in an inclined galvanized iron pipe flowing under pressure.

Mathcad Solution

(a) The pipe diameter at either point 1 or 2 is computed from Equation 8.34 (Darcy–Weisbach equation) as a Type 3 problem. The absolute pipe roughness for galvanized iron is given in Table 8.3, and the Colebrook equation for the friction factor is assumed. The density, ρ and the viscosity, μ for water at 15°C are given in Table A.4 in Appendix A as follows:

$L := 10{,}000\,\text{m}$ $\varepsilon := 0.15\,\text{mm}$ $v := 3\,\dfrac{\text{m}}{\text{sec}}$

$h_f := 80\,\text{m}$ $S_f := \dfrac{h_f}{L} = 8 \times 10^{-3}\,\dfrac{\text{m}}{\text{m}}$ $g := 9.81\,\dfrac{\text{m}}{\text{sec}^2}$

$\rho := 999\,\dfrac{\text{kg}}{\text{m}^3}$ $\mu := 1 \times 10^{-3}\,\text{N}\,\dfrac{\text{sec}}{\text{m}^2}$

Guess values: $f := 0.01$ $R := 400{,}000$ $D := 2\,\text{m}$

Given

$$\frac{1}{\sqrt{f}} = -2 \cdot \log\left(\frac{\frac{\varepsilon}{D}}{3.7} + \frac{2.51}{R \cdot \sqrt{f}}\right)$$

$$R = \frac{\rho \cdot v \cdot D}{\mu}$$

$$S_f = f \cdot \frac{1}{D}\frac{v^2}{2 \cdot g}$$

$$\begin{pmatrix} f \\ R \\ D \end{pmatrix} := \text{Find}\,(f, R, D) \quad f = 0.014 \quad R = 2.405 \times 10^6 \quad D = 0.802\,\text{m}$$

(b) Thus, because the Reynolds number, **R** is greater than 4000, the flow is turbulent.
(c) Assuming the datum is at point 1, the pressure drop between points 1 and 2 is computed by applying the energy equation between points 1 and 2 as follows:

$\gamma := \rho \cdot g = 9.8 \times 10^3\,\dfrac{\text{kg}}{\text{m}^2 \cdot \text{s}^2}$ $p_1 := 15 \times 10^5\,\dfrac{\text{N}}{\text{m}^2}$

$z_1 := 0\,\text{m}$ $z_2 := 0.07\,\text{m}$

$v_1 := v = 3\,\dfrac{\text{m}}{\text{s}}$ $v_2 := v = 3\,\dfrac{\text{m}}{\text{s}}$

Guess value: $p_2 := 10 \times 10^5\,\dfrac{\text{N}}{\text{m}^2}$

Given

$$\frac{p_1}{\gamma} + z_1 + \frac{v_1^2}{2 \cdot g} - h_f = \frac{p_2}{\gamma} + z_2 + \frac{v_2^2}{2 \cdot g}$$

$$p_2 := \text{Find}\,(p_2) = 7.153 \times 10^5\,\frac{\text{N}}{\text{m}^2}$$

$$\Delta p := p_1 - p_2 = 7.847 \times 10^5\,\frac{\text{N}}{\text{m}^2}$$

8.5.4 Manning's Roughness Coefficient, *n* and Manning's Equation

Empirical studies have indicated that a more accurate dependence of the Chezy coefficient, C on the hydraulic radius, R_h is defined by the definition of the Manning's roughness coefficient, n. One may recall that the dimensions of the Chezy coefficient; C are $L^{1/2} T^{-1}$, and it is empirically evaluated as a function of the boundary roughness, ε/D; the Reynolds number, R; and the cross-sectional shape or the hydraulic radius, R_h. However, further empirical studies (assuming the metric system) by Manning indicated that a more accurate dependence of the Chezy coefficient, C on the hydraulic radius, R_h was presented in Chapter 6 and repeated here as follows:

$$C = \frac{R_h^{1/6}}{n} \tag{8.43}$$

where n is the Manning's roughness coefficient with dimensions of $L^{-1/3} T$, and is empirically evaluated only as a function of the boundary roughness, ε/D. One may note that the dependency of the Manning's roughness coefficient, n on the hydraulic radius, R_h has been removed by its relationship with the Chezy coefficient, C. Furthermore, one may note that the dependency of the Manning's roughness coefficient, n on the Reynolds number, R has been removed by assuming turbulent flow. One may note that the most common occurrence of flow resistance in open channel and pipe flow is in long channels or long pipe sections. In the open channel flow case, the boundary fills the whole channel section, while in the pipe flow case, the boundary fills the whole pipe section; in such cases, the Reynolds number, R is usually so large that it has virtually no influence on the resistance coefficient, n. Empirical values for the Manning roughness coefficient, n as a function of the boundary roughness, ε/D are typically given assuming metric units for n of $m^{-1/3}$ s (see Table 8.6).

Substituting the expression for the Chezy coefficient, C as a function of the Manning's roughness coefficient, n, Equation 8.43, into the Chezy equation, Equation 8.25, and isolating the velocity, v yields the Manning's equation as presented in Chapter 6 and repeated here as follows:

$$v = \frac{R_h^{1/6}}{n} \sqrt{R_h S_f} = \frac{1}{n} R_h^{2/3} S_f^{1/2} \tag{8.44}$$

One may note that although the derivation/formulation of the Manning's equation assumes no specific units (SI or BG), the Manning's roughness coefficient, n has dimensions of $L^{-1/3} T$. Furthermore, because the values for the Manning's roughness coefficient, n have been provided/calibrated in the SI units $m^{-1/3}$ s (see Table 8.6), the Manning's equation as given above assumes the metric (SI) system. The corresponding Manning's equation for the English (BG) system is given as follows:

$$v = \frac{1.486}{n} R_h^{2/3} S_f^{1/2} \tag{8.45}$$

which also uses the Manning roughness coefficient, n with units $m^{-1/3}$ s (see Table 8.6). One may note the value of 1.486 simply converts the Manning roughness coefficient, n with units $m^{-1/3}$ s (from Table 8.6) to units of $ft^{-1/3}$ sec, because 3.281 ft $= 1$ m and thus:

$$\frac{\left[\dfrac{3.281 \text{ ft}}{1 \text{ m}}\right]^{1/3}}{n[m^{-1/3} \text{ s}]} = \frac{1.486}{n[ft^{-1/3} \text{ sec}]} \tag{8.46}$$

TABLE 8.6

Empirical Values for the Manning Roughness Coefficient, n ($m^{-1/3}$ s) for Various Boundary Surfaces

Boundary Surface	Manning Roughness Coefficient, n (Min)	Manning Roughness Coefficient, n (Max)
Lucite	0.008	0.010
Brass	0.009	0.013
Glass	0.009	0.013
Wood stave pipe	0.010	0.013
Neat cement surface	0.010	0.013
Plank flumes, planed	0.010	0.014
Vitrified sewer pipe	0.010	0.017
Concrete, precast	0.011	0.013
Cement mortar surfaces	0.011	0.015
Metal flumes, smooth	0.011	0.015
Plank flumes, unplaned	0.011	0.015
Common-clay drainage tile	0.011	0.017
Concrete, monolithic	0.012	0.016
Brick with cement mortar	0.012	0.017
Cast iron, new	0.013	0.017
Wood laminated	0.015	0.020
Asphalt, rough	0.016	0.016
Riveted steel	0.017	0.020
Cement rubble surfaces	0.017	0.030
Canals and ditches, smooth earth	0.017	0.025
Corrugated metal pipe	0.021	0.030
Metal flumes, corrugated	0.022	0.030
Gravel bottom with rubble	0.023	0.036
Canals (excavated or dredged)		
Dredged in earth, smooth	0.025	0.033
In rock cuts, smooth	0.025	0.035
Rough beds and weeds on sides	0.025	0.040
Rock cuts, jagged and irregular	0.035	0.045
Dense weeds	0.050	0.120
Dense brush	0.080	0.140
Natural streams		
Smoothest (clean, straight)	0.025	0.033
Roughest	0.045	0.060
Very weedy	0.075	0.150
Floodplains (brush)	0.035	0.070
streams	0.025	0.060

Source: Chow, V. T., *Open Channel Hydraulics*, McGraw Hill, New York, 1959, pp. 110–113; Finnemore, E. J., and J. B., Franzini, *Fluid Mechanics with Engineering Applications*, 10th ed, McGraw Hill, New York, 2002, p. 286.

As illustrated in Table 8.6, the rougher the boundary (wetted perimeter), the larger the value for Manning roughness coefficient, n.

Finally, substituting the relationship between the Chezy coefficient, C and the Manning's roughness coefficient, n, Equation 8.43, into the Chezy head loss equation, Equation 8.26, yields the Manning's head loss equation (assuming the metric system) as presented in Chapter 6 and repeated as follows:

$$h_f = \frac{\tau_w L}{\gamma R_h} = \frac{\Delta p}{\gamma} = S_f L = \frac{v^2 L}{C^2 R_h} = \frac{v^2}{\left(\dfrac{R_h^{1/6}}{n}\right)^2 R_h} \frac{L}{} = \left(\frac{vn}{R_h^{2/3}}\right)^2 L \qquad (8.47)$$

The corresponding Manning's head loss equation for the English (BG) system as presented in Chapter 6 and repeated as follows:

$$h_f = \frac{\tau_w L}{\gamma R_h} = \frac{\Delta p}{\gamma} = S_f L = \frac{v^2 L}{C^2 R_h} = \frac{v^2}{\left(\dfrac{1.486 R_h^{1/6}}{n}\right)^2 R_h} \frac{L}{} = \left(\frac{vn}{1.486 R_h^{2/3}}\right)^2 L \qquad (8.48)$$

8.5.4.1 Application of Manning's Equation

The Manning's equation for pipe flow assumes turbulent flow in a straight pipe section of length, L, with a constant diameter, D and thus a constant velocity of flow, v with a specific Manning's roughness coefficient, n and a friction slope, S_f, which may be solved for one of the following unknown quantities (differential approach): S_f, v, n, or $D = 4\,R_h$. The Manning's equation is applied in the analysis and design of turbulent pipe flow. Specifically, assuming that the Manning's roughness coefficient, n and fluid specific weight, γ are specified, there are three types of problems in the application of the Manning's equation and are outlined in Table 8.7. The Manning's equation, expressed as a head loss equation, Equation 8.47, may be applied to solve for an unknown head loss, h_f which is a Type 1 problem (analysis). The Manning's equation, Equation 8.44, may be applied to solve for an unknown velocity of flow, v, which is a Type 2 problem (analysis). Furthermore, the Manning's equation, Equation 8.44, may be applied to solve for an unknown pipe diameter, D, which is a Type 3 problem (design).

TABLE 8.7

Summary of Three Types of Problems for Turbulent Pipe Flow (Manning's Equation)

Problem Type	Given	Find	Resistance Equation	Governing Equation
Type 1	n, γ, L, v, D	h_f	$h_f = \dfrac{\Delta p}{\gamma} = S_f L = \left(\dfrac{vn}{R_h^{2/3}}\right)^2 L$	Manning's Equation expressed as a head loss equation
Type 2	n, γ, L, S_f, D	v	$S_f = \dfrac{h_f}{L} = \dfrac{\Delta p}{\gamma L} = \left(\dfrac{vn}{R_h^{2/3}}\right)^2$	Manning's Equation
Type 3	n, γ, L, v, S_f	D	$S_f = \dfrac{h_f}{L} = \dfrac{\Delta p}{\gamma L} = \left(\dfrac{vn}{R_h^{2/3}}\right)^2$	Manning's Equation

EXAMPLE PROBLEM 8.6

Water at 80°C flows in a 2.5-m-diameter horizontal riveted steel pipe, as illustrated in Figure EP 8.6. The pressure at point 1 is $29 \times 10^5 \, \text{N/m}^2$, the major head loss due to the pipe friction is 75 m, and the pipe length between points 1 and 2 is 3000 m. Assume the Manning's roughness coefficient is chosen to model the flow resistance. (a) Determine the velocity at either point 1 or 2. (b) Determine the flow regime. (c) Determine the pressure drop between points 1 and 2 due to the pipe friction (flow resistance) between points 1 and 2. (d) Compare the results between assuming the Manning's coefficient (this example problem) versus assuming the Darcy–Weisbach coefficient in Example Problem 8.4 above.

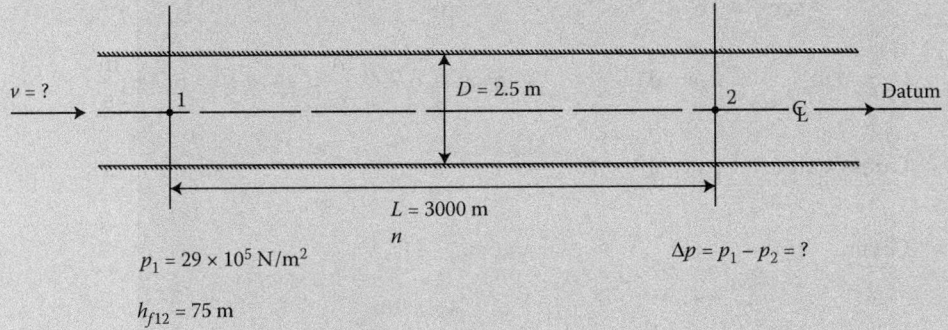

$p_1 = 29 \times 10^5 \, \text{N/m}^2$

$h_{f12} = 75 \, \text{m}$

$\Delta p = p_1 - p_2 = ?$

FIGURE EP 8.6
Water flows in a horizontal riveted steel pipe flowing under pressure.

Mathcad Solution

(a) The velocity at either point 1 or 2 is computed from Equation 8.44 (Manning's equation), as a Type 2 problem, where the Manning's roughness coefficient for riveted steel is given in Table 8.6 as follows:

$$D := 2.5 \, \text{m} \qquad L := 3000 \, \text{m} \qquad n := 0.017 \, \text{m}^{\frac{-1}{3}} \, \text{sec} \qquad h_f := 75 \, \text{m}$$

$$S_f := \frac{h_f}{L} = 0.025 \, \frac{\text{m}}{\text{m}} \qquad A := \frac{\pi \cdot D^2}{4} = 4.909 \, \text{m}^2 \qquad R_h := \frac{D}{4} = 0.625 \, \text{m}$$

Guess values: $\quad v := 2 \, \dfrac{\text{m}}{\text{sec}}$

Given

$$S_f = \left(\frac{v \cdot n}{R_h^{\frac{2}{3}}} \right)$$

$$v := \text{Find}(v) = 6.799 \, \frac{\text{m}}{\text{s}}$$

(b) In order to determine the flow regime, one must compute the Reynolds number, R where the density, ρ and the viscosity, μ for water at 80°C are given in Table A.4 in Appendix A as follows:

$$\rho := 971.8 \ \frac{kg}{m^3} \qquad \mu := 0.354 \times 10^{-3} \, N \ \frac{sec}{m^2}$$

$$R := \frac{\rho \cdot v \cdot D}{\mu} = 4.666 \times 10^7$$

Thus, because the Reynolds number, R is greater than 4000, the flow is turbulent.

(c) The pressure drop between points 1 and 2 is computed by applying the energy equation between points 1 and 2 as follows:

$$g := 9.81 \ \frac{m}{sec^2} \qquad \gamma := \rho \cdot g = 9.533 \times 10^3 \ \frac{kg}{m^2 \cdot s^2} \qquad p_1 := 29 \times 10^5 \ \frac{N}{m^2}$$

$$z_1 := 0 \, m \qquad\qquad z_2 := 0 \, m \qquad\qquad v_1 := v = 6.799 \ \frac{m}{s} \quad v_2 := v = 6.799 \ \frac{m}{s}$$

Guess value: $p_2 := 10 \times 10^5 \ \dfrac{N}{m^2}$

Given

$$\frac{p_1}{\gamma} + z_1 + \frac{v_1{}^2}{2 \cdot g} - h_f = \frac{p_2}{\gamma} + z_2 + \frac{v_2{}^2}{2 \cdot g}$$

$$p_2 := \text{Find}(p_2) = 2.185 \times 10^6 \ \frac{N}{m^2}$$

$$\Delta p := p_1 - p_2 = 7.15 \times 10^5 \ \frac{N}{m^2}$$

(d) The computed value for the velocity, $v = 6.799$ m/sec and thus the computed value for the Reynolds number, $R = 4.666 \times 10^7$ assuming the Manning's roughness coefficient, n in this example are slightly lower than the velocity, $v = 8.882$ m/sec and thus the computed value for the Reynolds number, $R = 6.096 \times 10^7$ assuming the Darcy–Weisbach friction factor, f in Example Problem 8.4. However, for a given head loss, $h_f = 75$ m, because the velocities at points 1 and 2 cancel out in the energy equation, the pressure drop between points 1 and 2, $\Delta p = p_1 - p_2 = 7.15 \times 10^5 \, N/m^2$, regardless of whether the Manning's roughness coefficient, n, or the Darcy–Weisbach friction factor, f is chosen to model the flow resistance. The advantages of applying the Darcy–Weisbach friction factor, f is that it is applicable to both laminar and turbulent flow (whereas the Manning's roughness coefficient, n assumes turbulent flow only; thus, its independence from the Reynolds number, R), and the friction factor, f is dimensionless. Given that the flow is turbulent, the advantages of the Manning's roughness coefficient, n is that it is empirically evaluated only as a function of the boundary roughness, ε/D (whereas the Darcy–Weisbach friction factor, f is evaluated empirically as a function of the boundary roughness, ε/D and the Reynolds number, R).

8.5.5 The Hazen–Williams Roughness Coefficient, C_h and the Hazen–Williams Equation

In addition to the standard empirical flow resistance coefficients including the Chezy coefficient, C, the Darcy–Weisbach friction factor, f and the Manning's roughness coefficient, n there is another commonly used flow resistance coefficient; namely the Hazen–Williams roughness coefficient, C_h (see Table 8.8). The Hazen–Williams roughness coefficient, C_h

TABLE 8.8

Empirical Values for the Hazen–Williams Roughness Coefficient, C_h for Various Pipe Materials

Pipe Material	Hazen–Williams Roughness Coefficient, C_h
Glass	140
Plastic	140–150
Wood stave	120
Fiber	140
Fiberglass	150
Cement-asbestos	140
Brass	130–140
Brick sewer	100
Tin	130
Vitrified clay (good condition)	110–140
Cast iron	
New, unlined	130
10 yr old	107–113
20 yr old	89–100
30 yr old	75–90
40 yr old	64–83
Tar (asphalt) coated	100
Cement lined	140
Concrete or concrete lined	
Steel forms	140
Wooded forms	120
Centrifugally spun	135
Copper	130–140
Galvanized iron	120
Lead	130–140
Steel	
Coal-tar enamel lined	145–150
New unlined	140–150
Riveted	110
Corrugated	60

Source: Williams, G. S., and A., Hazen, *Hydraulic Tables*, 2nd ed, John Wiley and Sons, New York, 1914; Hwang, N. H. C., and R. J., Houghtalen, *Fundamentals of Hydraulic Engineering Systems*, 3rd ed, Prentice Hall, New Jersey, 1996, p. 67.

values typically range from 140 for very smooth and straight pipes to 80 or 90 for old unlined pipes, with a value of 100 assumed for average conditions. Similar to the Darcy–Weisbach friction factor, f, the Hazen–Williams roughness coefficient, C_h is dimensionless. Furthermore, similar to the Manning's roughness coefficient, n, the Hazen–Williams roughness coefficient, C_h is not a function of the flow conditions (i.e., the Reynolds number, R) and thus assumes turbulent flow and is empirically evaluated only as a function of the boundary roughness, ε/D. One may note that the formulation of the Hazen–Williams equation assumes no specific units (SI or BG), and the Hazen–Williams roughness coefficient, C_h is dimensionless. However, the Hazen–Williams equation was initially calibrated in the BG units and then calibrated in the SI units.

The Hazen–Williams equation was empirically derived assuming the English (BG) system and is given as follows:

$$v = 1.318 C_h R_h^{0.63} S_f^{0.54} \tag{8.49}$$

with the corresponding units: velocity (ft/s), R_h (ft), and S_f (ft/ft). It was specifically developed for the analysis and design water flow in larger pipes (2 in $\leq D \leq$ 6 ft or 50 mm $\leq D \leq$ 2 m) within a moderate range of velocity ($v \leq$ 10 ft/sec or $v \leq$ 3 m/sec) for water at 60°F (or 15°C). The corresponding Hazen–Williams equation for the SI (metric) system is given as follows:

$$v = 0.849 C_h R_h^{0.63} S_f^{0.54} \tag{8.50}$$

with the corresponding units: velocity (m/s), R_h (m), and S_f (m/m). It is important to note that because the Chezy equation, $v = C\sqrt{R_h S_f}$ (and the corresponding Darcy–Weisbach and Manning's equations) was derived (formulated without the need for calibration) using dimensional analysis, there are no limitations for the ranges assumed for the individual variables (v, D, and water temperature). However, although the Hazen–Williams equation was formulated based on the results of dimensional analysis (the predictor variables, C_h, R_h, and S_f, are based on the Chezy equation), it was "calibrated" for a specific range of variables (v, D, and water temperature) as listed above, thus the limitation of its application.

Assuming that the Hazen–Williams roughness coefficient, C_h is chosen to model the empirical flow resistance, then one would substitute the Hazen–Williams equation, Equation 8.49, into the Chezy head loss equation, Equation 8.26, which yields the Hazen–Williams head loss equation (assuming the English system) as follows:

$$h_f = \frac{\tau_w L}{\gamma R_h} = \frac{\Delta p}{\gamma} = S_f L = \frac{v^2 L}{C^2 R_h} = \left[\frac{v}{1.318 C_h R_h^{0.63}} \right]^{1/0.54} L \tag{8.51}$$

The corresponding Hazen–Williams head loss equation for the SI (metric) system is given as follows:

$$h_f = \frac{\tau_w L}{\gamma R_h} = \frac{\Delta p}{\gamma} = S_f L = \frac{v^2 L}{C^2 R_h} = \left[\frac{v}{0.849 C_h R_h^{0.63}} \right]^{1/0.54} L \tag{8.52}$$

8.5.5.1 Application of the Hazen–Williams Equation

The Hazen–Williams equation for pipe flow assumes a straight pipe section of length, L with a constant diameter, D and thus a constant velocity of flow, v with a specific Hazen Williams roughness coefficient, C_h and a friction slope, S_f, which may be solved for one of the following unknown quantities (differential approach): S_f, v, C_h, or $D = 4\ R_h$. The Hazen–Williams equation is applied in the analysis and design of turbulent pipe flow. Specifically, assuming that the Hazen–Williams roughness coefficient, C_h and fluid specific weight, γ are specified, there are three types of problems in the application of the Hazen–Williams equation, which are outlined in Table 8.9. The Hazen–Williams equation, expressed as a head loss equation, Equation 8.52, may be applied to solve for an unknown head loss, h_f, which is a Type 1 problem (analysis). The Hazen–Williams equation, Equation 8.50 may be applied to solve for an unknown velocity of flow, v, which is a Type 2 problem (analysis). Furthermore, the Hazen–Williams equation, Equation 8.50 may be applied to solve for an unknown pipe diameter, D, which is a Type 3 problem (design).

TABLE 8.9

Summary of Three Types of Problems for Turbulent Pipe Flow (Hazen–Williams Equation)

Problem Type	Given	Find	Resistance Equation	Governing Equation
Type 1	C_h, γ, L, v, D	h_f	$h_f = \dfrac{\Delta p}{\gamma} = S_f L = \left(\dfrac{v}{0.849 C_h R_h^{0.63}}\right)^{1/0.54} L$	**Hazen–Williams Equation** expressed as a head loss equation
Type 2	$C_h, \gamma, L, S_f,$ D	v	$S_f = \dfrac{h_f}{L} = \dfrac{\Delta p}{\gamma L} = \left(\dfrac{v}{0.849 C_h R_h^{0.63}}\right)^{1/0.54}$	**Hazen–Williams Equation**
Type 3	C_h, γ, L, v, S_f	D	$S_f = \dfrac{h_f}{L} = \dfrac{\Delta p}{\gamma L} = \left(\dfrac{v}{0.849 C_h R_h^{0.63}}\right)^{1/0.54}$	**Hazen–Williams Equation**

EXAMPLE PROBLEM 8.7

Water at 15°C flows in an inclined galvanized iron pipe at a velocity of 3 m/sec, as illustrated in Figure EP 8.7. The pressure at point 1 is $15 \times 10^5\,\text{N/m}^2$, and the major head loss due to the pipe friction is 80 m. The increase in elevation between points 1 and 2 is 0.07 m, and the pipe length between points 1 and 2 is 10,000 m. Assume the Hazen–Williams roughness coefficient is chosen to model the flow resistance. (a) Determine pipe diameter at either point 1 or 2. (b) Determine the flow regime. (c) Determine the pressure drop between points 1 and 2 due to the increase in elevation and the pipe friction (flow resistance) between points 1 and 2. (d) Compare the results between assuming the Hazen–Williams roughness coefficient (this example problem) versus assuming the Darcy–Weisbach coefficient in Example Problem 8.5 above.

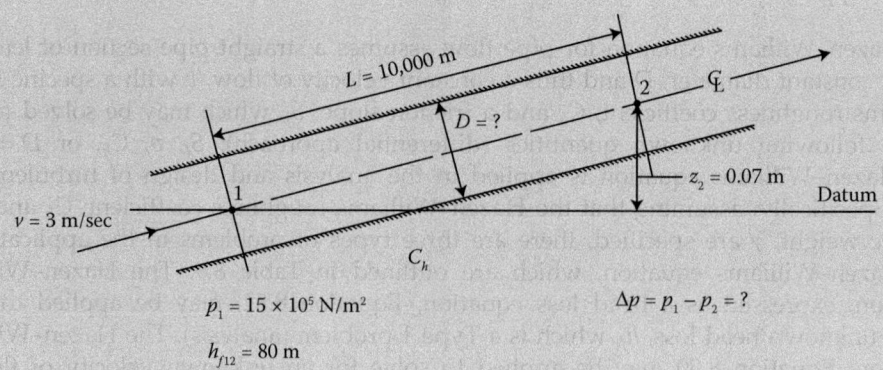

FIGURE EP 8.7
Water flows in an inclined galvanized iron pipe flowing under pressure.

Mathcad Solution

(a) The pipe diameter at either point 1 or 2 is computed from Equation 8.50, (Hazen–Williams equation), as a Type 3 problem, where the Hazen–Williams roughness coefficient for galvanized iron is given in Table 8.8. It is important to note that the empirically calibrated coefficients (0.849, 0.63, and 0.54) in the Hazen–Williams equation represent the assumed metric units for the variables v (m), R_h (m), and S_f (m/m), and the specific ranges assumed for the individual variables (v, D, and water temperature). Furthermore, although the empirically calibrated coefficients are not dimensionless, their units are embedded in the equation and thus unknown. As such, in order to successfully apply the Mathcad numerical solve block to solve for the unknown pipe diameter, D, the units for all the variables (except for C_h, which is already dimensionless) have been temporarily eliminated as follows:

$L := 10,000$ $C_h := 120$ $v := 3$

$h_f := 80$ $S_f := \dfrac{h_f}{L} = 8 \times 10^{-3}$ $R_h(D) := \dfrac{D}{4}$

Guess values: $D := 2$

Given

$$S_f = \left(\frac{v}{0.849 \cdot C_h \cdot R_h(D)^{0.63}} \right)^{\frac{1}{0.54}}$$

$D := \text{Find} (D) = 0.932$

(b) In order to determine the flow regime, one must compute the Reynolds number, R where the density, ρ and the viscosity, μ for water at 15°C are given in Table A.4 in Appendix A as follows:

$$\rho := 999 \, \frac{kg}{m^3} \qquad\qquad \mu := 1 \times 10^{-3} \, N \, \frac{sec}{m^2} \qquad v := 3 \, \frac{m}{sec} \qquad D := 0.932$$

$$R := \frac{\rho \cdot v \cdot D}{\mu} = 2.793 \times 10^6$$

Thus, because the Reynolds number, R is greater than 4000, the flow is turbulent.

(c) Assuming the datum is at point 1, the pressure drop between points 1 and 2 is computed by applying the energy equation between points 1 and 2 as follows:

$$g := 9.81 \, \frac{m}{sec^2} \qquad\qquad \gamma := \rho \cdot g = 9.8 \times 10^3 \, \frac{kg}{m^2 \cdot s^2} \qquad\qquad p_1 := 15 \times 10^5 \, \frac{N}{m^2}$$

$$z_1 := 0 \, m \qquad\qquad\qquad z_2 := 0.07 \, m$$

$$v_1 := v = 3 \, \frac{m}{s} \qquad\qquad v_2 := v = 3 \, \frac{m}{s} \qquad\qquad\qquad h_f := 80 \, m$$

Guess value: $\qquad\qquad p_2 := 10 \times 10^5 \, \frac{N}{m^2}$

Given

$$\frac{p_1}{\gamma} + z_1 + \frac{v_1{}^2}{2 \cdot g} - h_f = \frac{p_2}{\gamma} + z_2 + \frac{v_2{}^2}{2 \cdot g}$$

$$p_2 := Find(p_2) = 7.153 \times 10^5 \, \frac{N}{m^2}$$

$$\Delta p := p_1 - p_2 = 7.847 \times 10^5 \, \frac{N}{m^2}$$

(d) The computed value for the velocity, $D = 0.932 \, m$ and thus the computed value for the Reynolds number, $R = 2.793 \times 10^6$ assuming the Hazen–Williams roughness coefficient, C_h in this example are slightly higher than the velocity, $D = 0.802 \, m$ and thus the computed value for the Reynolds number, $R = 2.405 \times 10^6$ assuming the Darcy–Weisbach friction factor, f in Example Problem 8.5. However, for a given head loss, $h_f = 80 \, m$, because the velocity at points 1 and 2 cancel out in the energy equation, the pressure drop between points 1 and 2, $\Delta p = p_1 - p_2 = 7.847 \times 10^5 \, N/m^2$, regardless of whether the Hazen–Williams roughness coefficient, C_h or the Darcy–Weisbach friction factor, f is chosen to model the flow resistance. The advantages of applying the Darcy–Weisbach friction factor, f is that it is applicable to both laminar and turbulent flow (whereas the Hazen–Williams roughness coefficient, C_h assumes turbulent flow only, thus its independence from the Reynolds number, R). The friction factor, f is dimensionless. Furthermore, because the Darcy–Weisbach friction equation was derived (formulated without the need for calibration) using dimensional analysis, there are no limitations for the ranges assumed for the individual variables (v, D, and water temperature). However, although the Hazen–Williams equation was formulated based on the results of dimensional analysis (the predictor variables, C_h, R_h, and S_f, are based on the Chezy equation), it was "calibrated" for a specific range of variables (v, D, and water temperature) as listed above, thus the limitation of its application. Given that the flow is turbulent, the advantages of the Hazen–Williams roughness coefficient, C_h is that it is empirically evaluated only as a function of the boundary roughness, ε/D (whereas the Darcy–Weisbach friction factor, f is evaluated empirically as a function of the boundary roughness, ε/D and the Reynolds number, R). The Hazen–Williams roughness coefficient, C_h is also dimensionless.

8.6 Pipes with Components: Minor Head Losses and Reaction Forces in Real Pipe Flow

Applications of the governing equations for real pipe flow include the analysis and design of pipes with components, which form the basic component for any pipe system (see Section 8.7). While the application of the continuity equation will always be necessary, the energy equation and the momentum equation play complementary roles; when one of the two equations of motion breaks down, the other may be used to find the additional unknown quantity. Furthermore, while the energy equation is useful only for a control volume, depending upon the general fluid flow problem, one may apply the continuity and momentum equations using either the integral or differential approach. However, for the specific case of pipes with components (addressed in Section 8.6, herein), only the integral continuity and momentum equations for a control volume are required; the differential continuity and momentum equations for a fluid system are required for single straight pipe sections (see Section 8.5). As such, for pipes with components, the energy equation, Equation 8.1 models the minor head loss, $h_{f,min}$ due to pipe components (Sections 8.6.1 through 8.6.5), while the integral momentum equation, Equation 8.2 models the reaction force due to certain pipe components (Section 8.6.6).

The minor head loss term, $h_{f,minor}$ in the energy equation models the head loss due to various pipe appurtenances (components, devices, or transitions) in the pipe system. Pipe components may include valves, fittings (tees, unions, elbows, and bends), entrances, exits, contractions, and expansions. The pipe components induce flow separation and mixing and thus, interrupt the smooth fluid flow and result in minor head losses and pressure drops/rises where energy is dissipated. The minor head loss term, $h_{f,minor}$ in the energy equation is significant in very short pipes, whereas the major head loss term, $h_{f,major}$ in the energy equation (models the head loss due to pipe friction in the straight section of a pipe) is significant for very long pipes. Therefore, for very short pipes, minor losses are usually significant, while major losses due to pipe friction are usually ignored. And, for very long pipes, minor losses are usually ignored, while the major losses due to pipe friction are significant. For instance, the minor losses may be insignificant when pipe devices are separated by approximately 1000 diameters. Furthermore, the minor head loss associated with a completely open valve may be negligible. However, in some cases, both minor and major head losses are significant and thus both are accounted for. Additionally, although in general, major losses due to pipe friction are greater than minor losses due to pipe components, in some cases such as in pipe systems where there are valves and numerous turns in a short pipe distance, minor losses can add up and be greater than major losses. Furthermore, a partially closed valve, which results in a drop in the flowrate, may cause the most significant head loss in the pipe system. Finally, the minor head loss caused by pipe components where the pipe diameter does not change (valves and fittings) may be modeled by an equivalent major head caused by a section of the pipe of length (Section 8.6.2).

It is of interest to examine the role of the various pipe components and their corresponding effect on the change in pressure (drop or rise). The pipe components may be subdivided into two categories: (1) valves and fittings (tees, unions, elbows, and bends) and (2) entrances, exits, contractions, and expansions. The valves control the flowrate, while the fittings (tees, unions, elbows, and bends) cause a change in the direction of the flow both, without a change in the pipe diameter. The entrances, exits, contractions, and expansions cause a change in the cross section of the flow and thus cause a change in the velocity of the flow. In the cases of the flow from a reservoir or a tank (pipe entrance) or a pipe contraction and the flow through valves and fittings, where the fluid is set into motion, the conversion of the pressure energy to kinetic energy results in a pressure drop (and a minor head loss). However, in the cases of the flow into a reservoir, a tank (pipe exit), a pipe expansion and the flow through valves and fittings where the fluid is brought to a stop (even momentarily or just a reduction in velocity), the conversion of the kinetic energy to pressure energy results in a pressure rise (and a minor head loss).

In the case of certain pipe components (such as pipe contractions, elbow, and bends; including jets, vanes, or nozzles) there are reaction forces to compute, which may be solved by applying the integral form of the momentum equation (solved analytically, without the use of dimensional analysis) (see Section 8.6.6 and Chapter 5); note that the viscous forces have already been accounted for indirectly through the minor head loss term, $h_{f,minor}$ in the energy equation.

8.6.1 Evaluation of the Minor Head Loss Term in the Energy Equation

Evaluation of the minor head loss flow resistance term requires a "subset level" application of the governing equations (see Chapters 6 and 7 and Section 8.4.1.2). The minor head loss due to a pipe component modeled in the energy equation, Equation 8.1 is evaluated by the minor head loss equation, Equation 8.9 for turbulent pipe flow. If the friction/viscous forces are insignificant and can be ignored (as in the case of a sudden pipe expansion; see Section 8.6.5.3), the minor head loss coefficient, k and thus the minor head loss are evaluated analytically. However, if the friction/viscous forces are significant and cannot be ignored (as in the case of the other pipe components, see Sections 8.6.4 and 8.6.5 in general), the minor head loss coefficient, k and thus the minor head loss are evaluated empirically. Therefore, presented in the sections below, are applications of the governing equations for a pipes with components, where the minor head loss term, $h_{f,min}$ in the energy equation is evaluated and schematically illustrated by the energy grade line (*EGL*) and the hydraulic grade line (*HGL*).

8.6.1.1 Evaluation of the Minor Head Loss

Evaluation of the minor head loss flow resistance term requires a "subset level" application of the governing equations (see Chapters 6 and 7 and Section 8.4.1.2). Regardless of whether the minor head loss, h_f is determined analytically (Section 8.6.5.3) or empirically (Chapter 7), the expression for the minor head loss term, $h_{f,minor}$ in the energy equation due to the different pipe devices, which include valves, fittings (tees, unions, elbows, and bends), entrances, exits, contractions, and expansions, which was given in

Equation 8.9, is repeated as follows:

$$h_{f,min} = \frac{\tau_w L}{\gamma R_h} = \frac{\Delta p}{\gamma} = k \frac{v^2}{2g} \tag{8.53}$$

where the minor head loss, $h_{f,minor}$ is proportional to the square of the average pipe velocity, v^2 and thus proportional to the velocity head of the pipe, and where the constant of proportionality is the head loss coefficient, k due to the specific pipe device. It is interesting to note while the majority of the minor head loss occurs near the pipe component, some of the minor head loss occurs for several pipe diameters downstream of the component, before the flow and thus the velocity profile, returns to fully developed flow. For instance, a valve produces swirling turbulent eddies in the pipe flow that continue downstream (approximately 10–20 pipe diameters), eventually dissipate into heat, and result in head loss before the flow resumes to fully developed flow. Another example is the case of an elbow, where the effect of the resulting swirling turbulent eddies and thus the effect of the associated minor head loss do not cease until considerably further downstream (approximately 10–20 pipe diameters) of the elbow. As Figure 8.2 illustrates, the primary effect of a pipe device is to cause the EGL to drop an amount equal to the head loss produced by that device. In general, this drop occurs over a distance of several diameters downstream of the transition. As explained and illustrated below, in the case of a square-edged entrance, the HGL also drops significantly immediately downstream of the entrance because of the high-velocity flow in the vena contracta. Furthermore, although many pipe devices produce EGLs and HGLs with local detail (such as the case of a square-edged entrance), for the sake of simplicity, it is common to show only the abrupt changes in the EGL and thus ignore local departures between the EGL and the HGL due to local changes in the velocity. Figure 8.2 illustrates a simplified HGL and EGL plot for a pipe system with a pipe entrance, a partially closed valve, and a pipe exit. One may note that the EGL includes both the major head losses due to pipe friction and the minor head losses due to the various pipe devices.

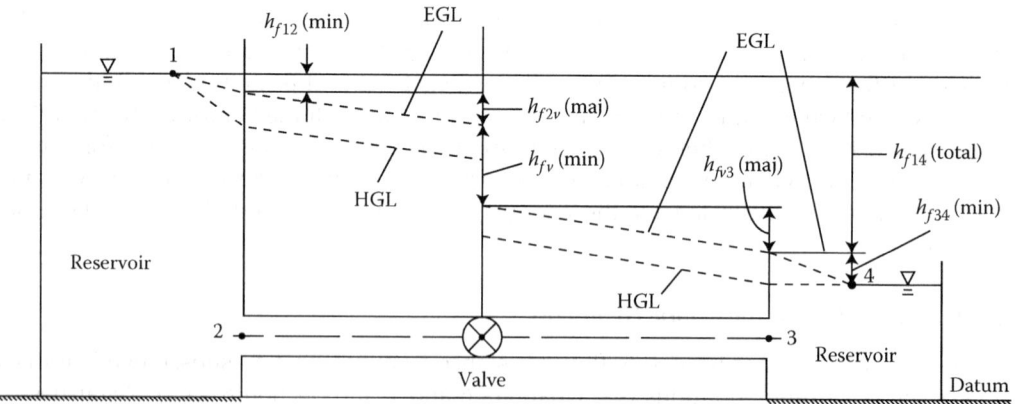

FIGURE 8.2
Simplified EGL and HGL for a pipe system with a pipe entrance, a partially closed valve, and a pipe exit. (Adapted from Roberson, J. A., and C. T., Crowe, *Engineering Fluid Mechanics*, 6th ed, John Wiley and Sons, New York, 1997, p. 379.)

8.6.1.2 *Derivation of the Minor Head Loss Coefficient,* k

Derivation of the minor head loss coefficient, k for the special case of a sudden pipe expansion is analytically achieved (see Section 8.6.5.3 below on sudden pipe expansions), whereas for the remaining pipe devices, the minor head loss coefficient, k is empirically derived. The empirical derivation of the minor head loss coefficient, k for the case of the other pipe devices in general results from the corresponding empirical derivation of the minor head loss equation, Equation 8.53, as presented in Chapter 7, and is summarized as follows:

$$\frac{\Delta p}{\rho v^2/2} = \frac{F_P}{F_{dynpres}} = \frac{2}{E} = C_D = k = \frac{staticpres}{dynpres} = \phi\left(R, \frac{\varepsilon}{D}, \frac{L_i}{L}\right) \tag{8.54}$$

One may note that because the drag coefficient, C_D/minor head loss coefficient, k represents the flow resistance, which causes an unknown pressure drop, Δp, the pressure force will always play an important role and thus the definition of the drag coefficient, C_D is mainly a function of the Euler number, E. Furthermore, R = the Reynolds number, ε/D = the relative pipe roughness, and L_i/L = the geometry of the pipe device. Therefore, the empirical expression for the minor head loss is given by Equation 8.53. Because, in general, the actual pressure drop/rise, Δp is unknown, it is indirectly evaluated through an empirical determination of a minor head loss coefficient, k (evaluated from experimentation). One may note that the static pressure is the pressure drop/rise, Δp across the pipe device, while the dynamic pressure is at the larger of the two velocities (either the entrance velocity or the exit velocity) of the device. Thus, the minor head loss, $h_{f,min}$ is caused by a pressure drop/rise, Δp, where the minor head loss coefficient, k represents the fraction of the kinetic energy, $\rho v^2/2$ lost due to the pipe device. It may be noted that the minor head loss coefficient, k is sometimes referred to as the pressure recovery coefficient, C_p; especially in the case of a gradual pipe expansion.

8.6.1.3 *Evaluation of the Minor Head Loss Coefficient,* k

Evaluation of the minor head loss coefficient, k for the special case of a sudden pipe expansion is analytically evaluated using Equation 8.74 (analytically derived in Section 8.6.5.3 below on sudden pipe expansions), whereas for the remaining pipe devices, the minor head loss coefficient, k is evaluated using experimental results, as presented in sections below on a particular pipe component (Sections 8.6.4 and 8.6.5), and discussed in general in the section herein. Thus, the minor head loss coefficient, k is almost always evaluated using experimental results, where no distinction is made between laminar and turbulent flow. One may note that while the loss coefficient, k in general depends on both the geometry, L_i/L (design details) of the flow device and the Reynolds number, R (and in some cases on the relative pipe roughness, ε/D), it is usually assumed to be independent of the Reynolds number, as most pipe flows in practice usually have large Reynolds numbers and the loss coefficients tend to be independent of R at large values of R. Although of minor importance in turbulent flow, one may note that, similar to the pipe friction factor, f, the head loss coefficient, k tends to increase with an increase in the relative pipe roughness, ε/D and a decrease in the Reynolds number, R. Therefore, the magnitude of the head loss coefficient, k is mainly a function of the flow geometry, L_i/L, which is produced by the pipe device. Additionally, it is important to note that because the minor loss coefficient, k for the various pipe devices are dependent upon the particular manufacturer of the device, the final design of the pipe system should rely on the manufacturer's information rather than on typical experimental values in handbooks.

Noting that the Reynolds number, $R = F_I/F_V$, for many practical applications, the Reynolds number is large enough that the flow through the pipe device is dominated by inertia (dynamic) effects and thus the viscous effects become secondary. Such is a result of the relatively large accelerations and decelerations that occur in the flow through pipe devices; the flow through these pipe devices is complex as the fluid flows through variable areas and thus nonlinear flow torturous flow paths. Furthermore, while the flow through the device may be accelerated efficiently and thus with a minimal head loss, it is very difficult to decelerate it efficiently; thus, a minor head loss occurs though a pipe device; in the case of ideal flow, the fluid would accelerate and decelerate efficiently and thus there would be no head loss. Thus, for the real flow case, the actual pressure drop/rise, Δp is mostly due to the dynamic pressure term and less due to the flow resistance term (see Equations 6.203 and 6.205). Therefore, in most pipe devices, it is not the viscous effects due to the wall shear stress that causes most of the minor head loss. Rather it is the dissipation of kinetic energy (dynamic pressure term in Equations 6.203 and 6.205), which is another type of viscous effect as the fluid decelerates inefficiently and is eventually dissipated by the shear stresses within the fluid. Furthermore, the minor losses are caused by the flow separation and mixing (generation of additional turbulence) induced by the various pipe components. One may also note that whenever the average velocity of turbulent flow is altered in either direction or magnitude, such as in the case of most pipe devices, large eddies or eddy currents form, which cause a loss of energy in addition to the head loss due to the pipe friction. Also, while the pipe device itself is usually confined to a very short length of the fluid flow, the resulting effects continue for a considerable distance downstream of the device.

8.6.2 Alternative Modeling of the Minor Head Loss Term in the Energy Equation

The minor head loss caused by pipe components where the pipe diameter does not change (valves and fittings) may be modeled by an equivalent major head caused by a section of the pipe of length, L_{eq}, where the contribution of the pipe component to the overall head loss of the pipe system may be modeled by adding L_{eq} to the total pipe length, L. One may recall that the empirical derivation of the Darcy–Weisbach friction factor, f presented in Chapter 7 on dimensional analysis yielded the following:

$$\frac{F_f}{F_I} = \frac{D}{L} \frac{\Delta p}{\frac{\rho v^2}{2}} = f = \frac{staticpres}{dynpres} \tag{8.55}$$

and the empirical derivation of the minor head loss coefficient, k for pipe devices in general presented in Chapter 7 on dimensional analysis yielded the following:

$$\frac{F_P}{F_{dynpres}} = \frac{\Delta p}{\frac{\rho v^2}{2}} = k = \frac{staticpres}{dynpress} \tag{8.56}$$

A comparison of the equations for the friction factor, f and for the minor head loss coefficient, k indicates the following relationship:

$$\frac{\Delta p}{\frac{\rho v^2}{2}} = k = f \frac{L}{D} \tag{8.57}$$

Therefore, for pipe components where the pipe diameter does not change (valves and fittings), the minor head loss is sometimes computed in terms of an equivalent length of pipe, L_{eq} as follows:

$$h_{f,min} = \frac{\Delta p}{\gamma} = f \frac{L_{eq}}{D} \frac{v^2}{2g} \tag{8.58}$$

where

$$L_{eq} = \frac{kD}{f} \tag{8.59}$$

and D and f are for the pipe length containing the pipe device. Thus, the minor head loss caused by the pipe component is equivalent to the major head loss caused by a section of the pipe of length, L_{eq}, where the contribution of the pipe component to the overall head loss of the pipe system may be modeled by adding L_{eq} to the total pipe length, L. It may be noted that while both approaches may be used to model the minor head loss, the direct modeling of the minor head loss using the minor head loss coefficient, k is the most common approach.

Furthermore, it is important to recall from Chapter 4 (and above) that the total head loss of the pipe system consists of a major head loss and a minor head loss, where the total head loss, $h_{f,total} = h_{f,major} + h_{f,minor}$. Assuming that there are n_i pipe sections with different pipe diameters, and n_j pipe components in a pipe system, the total head loss in the pipe system is computed as follows:

$$h_{f,total} = h_{f,maj} + h_{f,min} = \sum_{i=1}^{n_i} f_i \frac{L_i}{D_i} \frac{v_i^2}{2g} + \sum_{j=1}^{n_j} k_j \frac{v_j^2}{2g} \tag{8.60}$$

In the cases where there is no change in the pipe diameter, D and thus no change in the average velocity, v as in the case of valves and pipe fittings, and assuming that there are n_i pipe sections with potentially different friction factors, f_i and thus pipe lengths, L_i, with n_j pipe components in a pipe system, the total head loss in the pipe system is computed as follows:

$$h_{f,total} = h_{f,maj} + h_{f,min} = \left[\sum_{i=1}^{n_i} f_i \frac{L_i}{D} + \sum_{j=1}^{n_j} k_j \right] \frac{v^2}{2g} \tag{8.61}$$

8.6.3 Evaluation of the Minor Head Loss due to Pipe Components

Presented in the sections below is the detailed evaluation of the minor head loss due to the various pipe flow components, including evaluation of the velocity term, v and evaluation of the minor head loss coefficient, k. The particular category of pipe components: (1) valves and fittings (tees, unions, elbows, and bends) (Section 8.6.4), and (2) entrances, exits, contractions, and expansions (Section 8.6.5) will determine how the velocity term in the minor head loss equation is evaluated and the complexity in the experimental determination of the minor loss coefficient, k. In the case of valves and pipe fittings, because there is no change in pipe diameter, there is no change in the corresponding velocity, which leads to a simpler determination of the minor loss coefficient, k; recall that the minor head loss caused by this category of pipe components may be modeled by an equivalent major head caused by a section of the pipe of length, L_{eq}. However, in the case of entrances, exits, contractions, and expansions, because there is a change in the cross section of the flow, there is a change in the velocity of the flow, which leads to a more complicated determination of the minor loss coefficient, k.

8.6.4 Minor Losses in Valves and Fittings (Tees, Unions, Elbows, and Bends)

Evaluation of the minor head loss for valves and fittings (tees, unions, elbows, and bends) involves evaluating the velocity term, v and the minor head loss coefficient, k for the particular pipe component. The installation of valves in the pipe system control the flowrate, while the installation of fittings (tees, unions, elbows, and bends) in the pipe system causes a change in the direction of flow, both without a change in the pipe diameter. Thus, the average velocity, v_{ave} of the either the upstream or downstream developed flow is used to compute the associated minor losses, $h_{f,minor}$ for valves and fittings as follows:

$$h_f = k \left(\frac{v_{ave}^2}{2g} \right) \tag{8.62}$$

The pipe component introduces turbulence in the flow section. Because the nature of turbulence is somewhat random, experimental results are generally used to determine the minor loss coefficient, k and thus the associated minor losses. Furthermore, one may note that many pipe fittings cause a reduction in the pressure of the pipe system, thus requiring additional energy from a pump. Tables 8.10 and 8.11 present head loss coefficients, k for some typical valves and fittings, respectively, which are illustrated in Figures 8.3 and 8.4,

TABLE 8.10

Minor Head Loss Coefficients, k for Various Types of Valves

Valves	Minor Head Loss Coefficient, k
Ball valve, fully open	0.05
Ball valve, 1/3 closed	5.5
Ball valve, 2/3 closed	210
Globe, fully open	10
Angle, fully open	2–5
Gate, fully open	0.15–0.02
Gate, 1/4 closed	0.26
Gate, 1/2 closed	2.1–5.6
Gate, 3/4 closed	17
Swing check, forward flow	2
Swing check, backward flow	∞

Source: Adapted from Munson, B. R., Young, D. F., and T. H., Okiishi, *Fundamentals of Fluid Mechanics*, 3rd ed, John Wiley and Sons, New York, 1998, p. 505; Streeter, V. L., Ed., *Handbook of Fluid Dynamics*, McGraw Hill, New York, 1961, pp. 3–23.

TABLE 8.11

Minor Head Loss Coefficients, k for Various Types of Fittings (Tees, Unions, Elbows, and Bends; See Figures 8.4 and 8.8 for Illustrations)

Pipe Fitting	Minor Head Loss Coefficient, k
Tees	
Line flow, flanged	0.2
Line flow, threaded	0.9
Branch flow, flanged	1.0
Branch flow, threaded	2.0
Union, threaded	0.08
Elbows	
Regular 90°, flanged	0.3
Regular 90°, threaded	1.5
Short radius 90°	0.9
Long radius 90°	0.6
Long radius 90°, flanged	0.2
Long radius 90°, threaded	0.7
Long radius 45°, flanged	0.2
Regular 45°, threaded	0.4
Bends	
90° mitered without guide vanes	1.1–1.3
90° mitered with guide vanes	0.2
180° return bend, flanged	0.2
180° return bend, threaded	1.5

Source: Adapted from Munson, B. R., Young, D. F., and T. H., Okiishi, *Fundamentals of Fluid Mechanics*, 3rd ed, John Wiley and Sons, New York, 1998, p. 505; ASHRAE, *ASHRAE Handbook of Fundamentals*, American Society of Heating, Refrigeration and Air-Conditioning Engineers, Inc., Atlanta, 1981, 4.11.

respectively. It is important to note that the minor head loss coefficient, k for the various valves and fittings are designed to optimize both the cost and the ease of their manufacturing, rather than for the reduction in the associated minor head loss. One may note that the smaller the head loss coefficient, k, the less flow resistance the pipe component presents and thus the smaller the minor head loss, $h_{f,minor}$ (and generally the more expensive the cost). Furthermore, one may note that values for the minor loss coefficient, k depend highly on the shape of the pipe device and much less on the Reynolds number, R for typical large values of R. As explained in more detail below, the value for the minor loss coefficient, k for a given type of valve depends on both the type of valve and whether it is fully open, partially open, or closed, while the value for the minor loss coefficient, k for a given type of fitting (tees, unions, elbows, and bends) depends on whether it is flanged or threaded.

8.6.4.1 *Valves*

A valve is typically installed in a pipe system in order to control both, the magnitude and the direction of the flow, where regulating the flowrate is achieved by opening or closing the

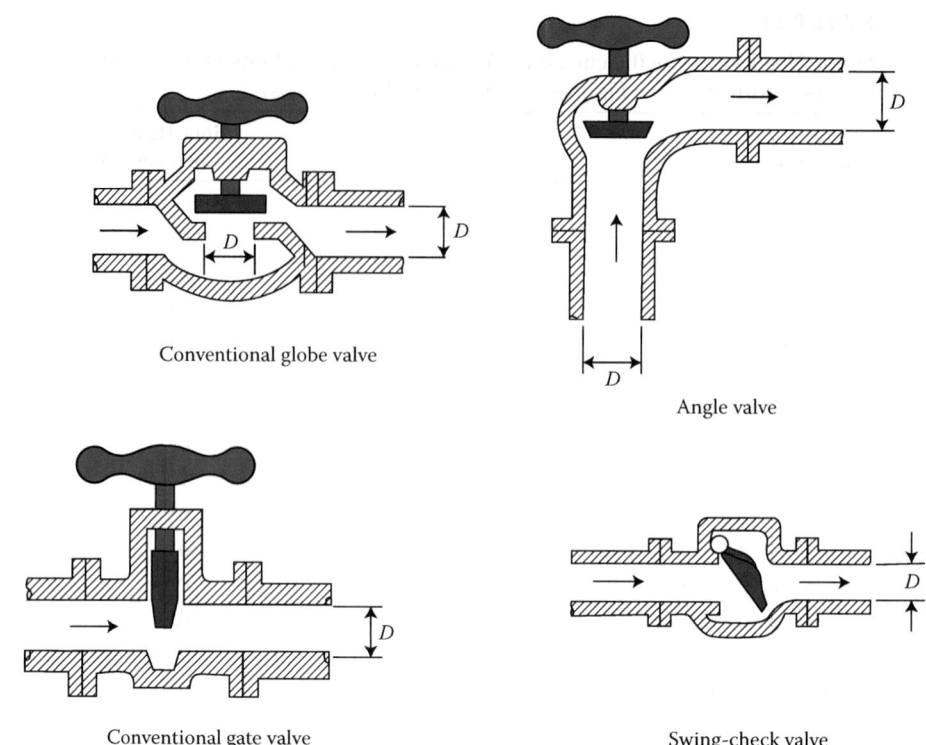

Conventional globe valve

Angle valve

Conventional gate valve

Swing-check valve

FIGURE 8.3
Typical cross sections of various types of valves. (Adapted from Blevins, R. D., *Applied Fluid Dynamics Handbook*, Von Nostrand Reinhold, New York, 1984, pp. 85–86; White, F. M., *Fluid Mechanics*, 7th ed, McGraw Hill, New York, 2011, p. 390.)

valve. Thus, a valve is used to control the fluid flow by throttling the flow, directing the flow, preventing back flow, and turning the flow on or off. The flow resistance due to a valve has the potential to contribute significantly to the flow resistance of the pipe system. A valve controls the flowrate by directly adjusting the minor loss coefficient, k of the entire pipe system. The flow resistance and associated minor losses vary depending upon whether the valve is open or closed and the type of valve used. Thus, regardless of the type of valve used, closing a valve significantly increases the flow resistance, where a closed valve results in an infinite flow resistance and thus no flow. And, opening the valve to a specified degree (fully open or partially open) controls the flow resistance, which achieves the desired flowrate. There are various types of valves, depending upon the desired outcome. A gate, plug, or a ball valve may be used if the objective is to turn a flow on or off. A globe, needle, or butterfly valve may be used if the objective is to be able to carefully regulate or throttle the flow. And, a lift-, swing-, or ball-check valve is used if the objective is to guarantee there is no back flow. Figure 8.3 illustrates typical cross sections of various types of valves. Table 8.10 presents head loss coefficients, k for the various types of valves. One may note that the deviation in the minor loss coefficients for valves for different manufactures (and thus different authors) is greater for valves than most other pipe devices because of their complex geometries. It is important to note that the minor head loss due to valves is mainly a result of the dissipation of kinetic energy of a high-speed portion of the flow, which is at the valve itself (see Figure 8.5).

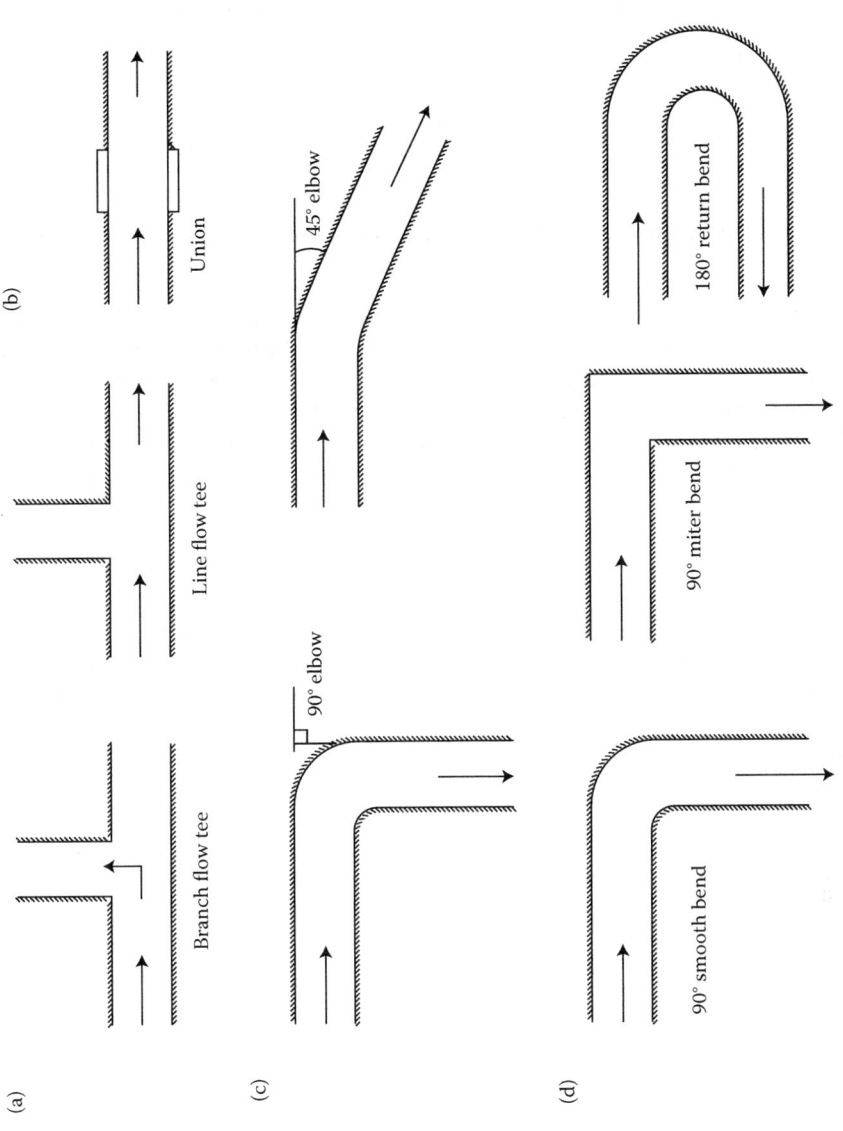

FIGURE 8.4

Four basic types of pipe fittings: (a) Tees, (b) Union, (c) Elbows, (d) Bends.

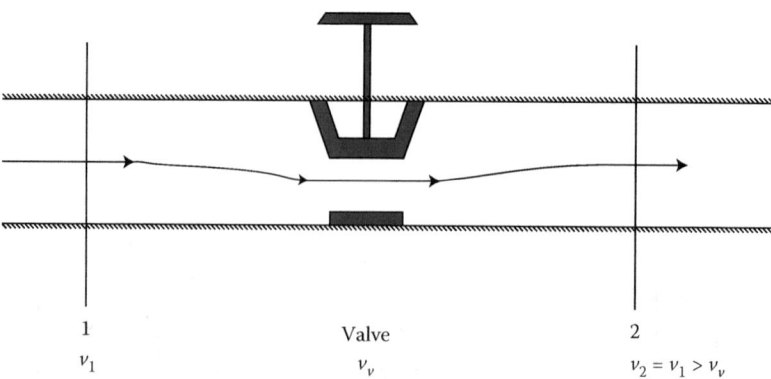

FIGURE 8.5
Minor head loss due to valves is mainly a result of dissipation of kinetic energy of high-speed flow at the valve itself.

While a gate, plug, or a ball valve may be used if the objective is to turn a flow on or off, the gate valve is one of the best types of valves for limiting the head loss. A gate valve slides up and down like a gate and is used to control the flowrate, Q where less frequent valve operation is required, and to allow the flow to move in straight paths. As a result, the gate valve results in less flow resistance (lower head loss coefficient, k) and thus, less pressure loss than the globe valve. Table 8.10 presents head loss coefficients, k for a fully open or a partially open gate valve.

EXAMPLE PROBLEM 8.8

Water flows in a 0.5-m-diameter horizontal pipe at a flowrate of $0.3 \, \text{m}^3/\text{sec}$. A gate valve is installed in between points 1 and 2 as illustrated in Figure EP 8.8 in order to turn the flowrate on or off in the pipe. The pressure at point 1 is $2 \times 10^5 \, \text{N/m}^2$. (a) Determine the minor head loss due to the gate valve. (b) Determine the pressure at point 2.

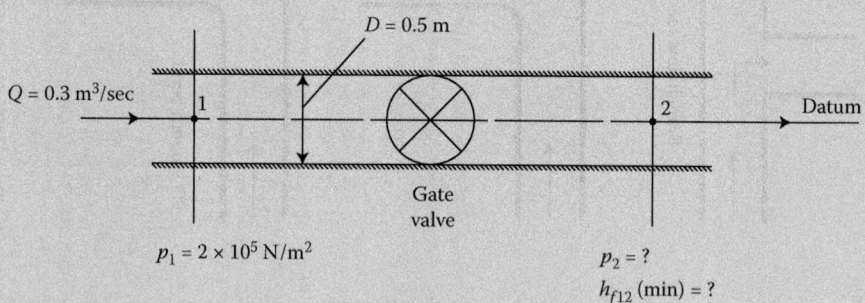

FIGURE EP 8.8
Water flows in a horizontal pipe with a gate valve to turn flow on and off.

Mathcad Solution

(a) The minor head loss due to the gate valve is computed from Equation 8.62, where the minor head loss coefficient, k is determined from Table 8.10. However, because the gate valve may be fully open, 1/4 closed, 1/2 closed, or 3/4 closed, there will be a range of values for the minor head loss coefficient, k and a corresponding range of values for the minor head loss due to the gate valve, h_f as follows:

$$D := 0.5 \, m \qquad Q := 0.3 \, \frac{m^3}{sec} \qquad g := 9.81 \, \frac{m}{sec^2}$$

$$A := \frac{\pi \cdot D^2}{4} = 0.196 \, m^2 \qquad v := \frac{Q}{A} = 1.528 \, \frac{m}{s}$$

$$k_{fopen} := 0.15 \qquad h_{ffopen} := k_{fopen} \cdot \left(\frac{v^2}{2 \cdot g}\right) = 0.018 \, m$$

$$k_{qclosed} := 0.26 \qquad h_{fqclosed} := k_{qclosed} \cdot \left(\frac{v^2}{2 \cdot g}\right) = 0.031 \, m$$

$$k_{hclosed} := 2.1 \qquad h_{fhclosed} := k_{hclosed} \cdot \left(\frac{v^2}{2 \cdot g}\right) = 0.25 \, m$$

$$k_{tqclosed} := 17 \qquad h_{ftqclosed} := k_{tqclosed} \cdot \left(\frac{v^2}{2 \cdot g}\right) = 2.023 \, m$$

One may note that as the gate valve incrementally closes from open to 3/4 closed, the minor head loss coefficient, k increases and thus the corresponding values for the minor head loss due to the gate valve, h_f also increase.

(b) The corresponding range of values for the pressure at point 2 is computed by applying the energy equation between points 1 and 2 as follows:

$$p_1 = 2 \times 10^5 \, \frac{N}{m^2} \qquad \gamma := 9810 \, \frac{N}{m^3} \qquad z_1 := 0 \, m \qquad z_2 := 0 \, m$$

$$v_1 := v = 1.528 \, \frac{m}{s} \qquad v_2 := v = 1.528 \, \frac{m}{s}$$

Guess values:

$$P_{2fopen} := 1 \times 10^5 \, \frac{N}{m^2} \qquad P_{2qclosed} := 1 \times 10^5 \, \frac{N}{m^2}$$

$$P_{2hclosed} := 1 \times 10^5 \, \frac{N}{m^2} \qquad P_{2tqclosed} := 1 \times 10^5 \, \frac{N}{m^2}$$

Given

$$\frac{p_1}{\gamma} + z_1 + \frac{v_1^2}{2 \cdot g} - h_{ffopen} = \frac{P_{2fopen}}{\gamma} + z_2 + \frac{v_2^2}{2 \cdot g}$$

$$\frac{p_1}{\gamma} + z_1 + \frac{v_1^2}{2 \cdot g} - h_{fqclosed} = \frac{P_{2qclosed}}{\gamma} + z_2 + \frac{v_2^2}{2 \cdot g}$$

$$\frac{p_1}{\gamma} + z_1 + \frac{v_1^2}{2 \cdot g} - h_{fhclosed} = \frac{P_{2hclosed}}{\gamma} + z_2 + \frac{v_2^2}{2 \cdot g}$$

$$\frac{P_1}{\gamma} + z_1 + \frac{v_1{}^2}{2 \cdot g} - h_{ftqclosed} = \frac{P_{2tqclosed}}{\gamma} + z_2 + \frac{v_2{}^2}{2 \cdot g}$$

$$\begin{pmatrix} P_{2fopen} \\ P_{2qclosed} \\ P_{2hclosed} \\ P_{2tqclosed} \end{pmatrix} := Find \left(P_{2fopen}, P_{2qclosed}, P_{2hclosed}, P_{2tqclosed} \right) = \begin{pmatrix} 1.998 \times 10^5 \\ 1.997 \times 10^5 \\ 1.975 \times 10^5 \\ 1.802 \times 10^5 \end{pmatrix} \frac{N}{m^2}$$

Therefore, as the gate valve incrementally closes from open to 3/4 closed, the minor head loss coefficient, k increases, the corresponding values for the minor head loss due to the gate valve, h_f increase, causing an increase in the pressure drop between points 1 and 2.

While a globe, needle, or butterfly valve may be used if the objective is to be able to carefully regulate or throttle the flow, the globe valve is one of the most common valves. Although it is relatively inexpensive, it yields a relatively large head loss. A globe valve closes a hole placed in the valve and is used to control the flowrate, Q where frequent valve operation is required and to change the direction of flow. Although the globe valve results in a larger flow resistance (higher head loss coefficient, k) than the gate valve, it also allows a closer regulation (throttling) of the flow, as it provides convenient control between the extremes of fully closed and fully open. Table 8.10 presents head loss coefficients, k for a fully open or a partially open globe valve. Furthermore, the needle valve is designed to provide a very fine control on the flowrate.

A lift-, swing-, or ball-check valve is used if the objective is to guarantee there is no back flow. As such, a check valve is used to control the flowrate, Q, allowing the flow to pass through in only one direction (providing a diode type operation) and thus is used to prevent backflow in the pipe system. A swing-check valve provides a modest restriction of flow, while a ball-check valve causes more restriction because the must flow completely around the ball. Table 8.10 presents head loss coefficients, k for a forward flow and a backward flow swing-check valve and for a ball-check valve.

8.6.4.2 Fittings (Tees, Unions, Elbows, and Bends)

A pipe fitting is typically installed in the pipe system in order to cause a change in the direction of flow. There are four basic types of pipe fitting depending upon the desired function. Pipe fittings may include tees, unions, elbows, and bends as illustrated in Figure 8.4. The information on tees and unions is brief and thus will be included as follows, while a more lengthy discussion on elbows and bends is presented below. A tee is used to change the direction of flow. There are two types of tees: (1) a branch flow tee, and (2) a line flow tee. A tee may be either flanged or threaded. Table 8.11 presents the minor head loss coefficients, k for both types of tees, for both the flanged and threaded types. There is a pipe device called a union, which connects two pipes of the same diameter without changing the direction of flow. Table 8.11 presents the minor head loss coefficients, k for a threaded union.

When the flow leaves the elbow or bend, the eddies produced by the flow separation create considerable head loss. The resistance to flow in a bend is dependent on the round

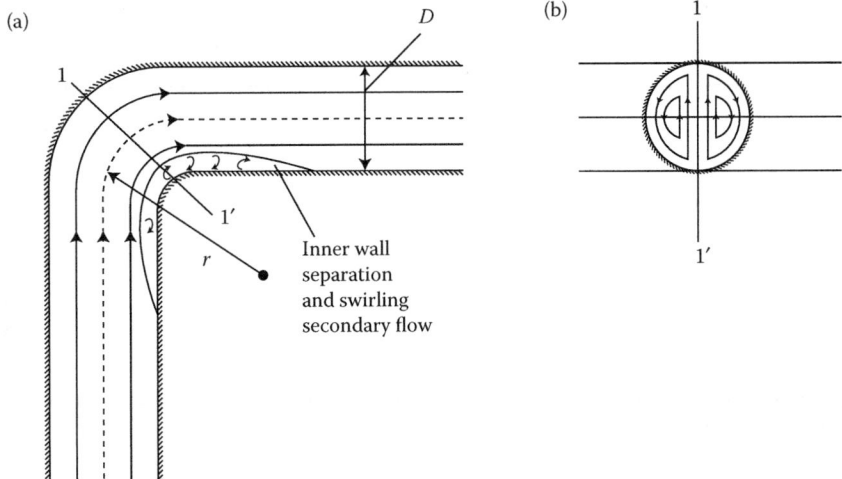

FIGURE 8.6
Minor head loss due to elbow or bend is a result of eddies produced by flow separation on inner wall, wall friction, and swirling (twin-eddy) secondary flow. (a) Flow separation on inner wall, (b) Swirling (twin-eddy) secondary flow.

radius, r to pipe diameter, D ratio, r/D, as illustrated in Figure 8.6a. The losses in elbows and bends are a due to flow separation on the inner wall (see Figure 8.6a), wall friction, and the swirling (twin-eddy) secondary flows (see Figure 8.6b) caused by different path lengths. Because of centrifugal effects, pipe flow around a bend experiences an increase in pressure along the outer wall, while experiencing a decrease of pressure along the inner wall, where the flow separates, as illustrated in Figure 8.6a. Furthermore, the velocity near the inner wall is lower than that at the outer wall. The velocity and pressure will resume their normal distributions a certain distance downstream from the bend, where the inner wall pressure and the velocity both increase back to their normal values when the flow reattaches to the pipe wall. One may note that the unbalanced pressure at the bend causes a secondary current, which, coupled with the axial velocity, forms a pair of spiral flows that persist approximately 100 diameters downstream of the bend. Therefore, the head loss at the bend is combined with the distorted flow conditions downstream from the bend until the spiral flows are dissipated by viscous friction. One may note that for bends of large radius of curvature, r the wall friction and the swirling secondary flows will predominate, while for bends with a small radius of curvature, r the separation and swirling secondary flow will play a more important role. Therefore, when the flow leaves the elbow or bend, the eddies produced by the flow separation create considerable head loss. The losses during changes of direction may be minimized by making the turn easier for the fluid flow to handle by using a 90-degree elbow as opposed to a miter bend, as illustrated in Figure 8.7a and b, respectively. However, the use of the sharp miter bends instead of a 90-degree elbow or a smooth bend is necessary in situations where space in the pipe system is limited; the result is a higher minor head loss coefficient, k. In such cases, using guide vanes in order to help the flow turn in an orderly manner, without being thrown off the pipe centerline, may minimize the minor head loss by impeding the formation of secondary flow. Furthermore, in addition to the minor head loss due to elbows and bend, the major head loss to the pipe friction in the axial length of the bend must also be added in the final calculation of the minor head loss due to the pipe device.

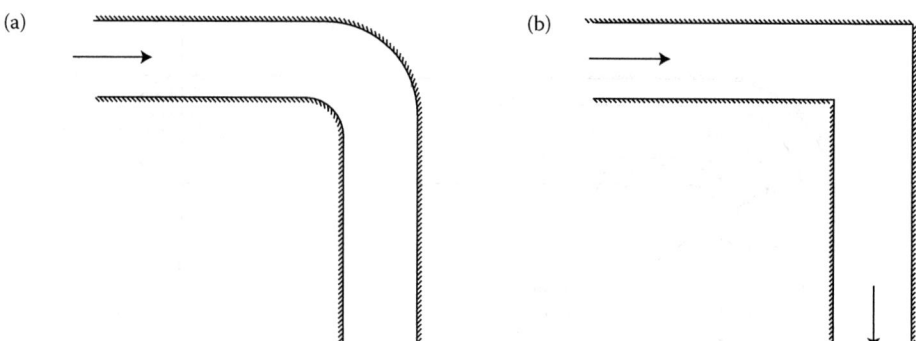

FIGURE 8.7
Minor losses due to change in direction of flow may be minimized by using a 90° elbow instead of a 90° miter bend.
(a) 90° elbow, (b) 90° miter bend.

While elbows and bends are used in a pipe system in order to change the direction of flow without a change in pipe diameter, it is usually more convenient to bend a pipe than to install an elbow. There are two types of elbows: (1) a 90-degree elbow and (2) a 45-degree elbow. An elbow may be either a regular elbow or a long radius elbow. Additionally, an elbow be either flanged or threaded. Table 8.11 presents the minor head loss coefficients, k for both the 90- and 45-degress elbows (regular and long radius) for both the flanged and threaded types. There are two general types of bends: (1) a 90-degree bend, and (2) a 180-degree return bend. One may note that minor head loss due to a bend is greater than the major head loss if the pipe were straight. There are two types of 90-degree bends. The first type is a 90-degree smooth bend (elbow) as illustrated in Figure 8.7a. The second type is a 90-degree mitered bend as illustrated in Figure 8.7b. Furthermore, a 90-degree mitered bend may either be without or with guide vanes, as illustrated in Figure 8.8a and b, respectively. The minor head loss in a bend is due to the separated region of the flow near the inside of the bend and the swirling secondary flow that occurs as a result of the

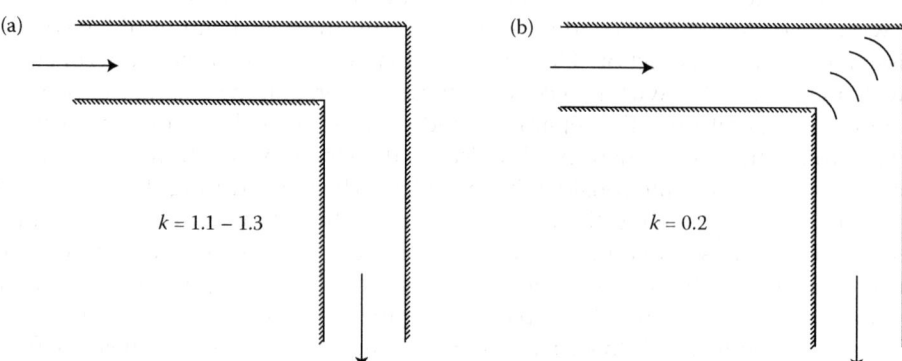

FIGURE 8.8
A 90° miter bend may either be without or with guide vanes. (a) 90° miter bend without guide vanes, (b) 90° miter bend with guide vanes. (Adapted from Streeter, V. L., Ed., *Handbook of Fluid Dynamics*, McGraw Hill, New York, 1961, p. 3–23; ASHRAE, *ASHRAE Handbook of Fundamentals*, American Society of Heating, Refrigeration and Air-Conditioning Engineers, Inc., Atlanta, 1981, 4.11.)

imbalance of the centripetal forces caused by the curvature of the pipe centerline. The flow separation is especially marked if the bend is sharp (a mitered bend without a guide vane). The minor head loss coefficient, k for a 90-degree smooth bend is a function of the relative pipe roughness, ε/D and the round radius, r to pipe diameter, D ratio, r/D for large values of the Reynolds number, R as illustrated in Figure 8.9. One may note that the minor head loss coefficient, k increases with an increase in the relative pipe roughness, ε/D. Furthermore, if the radius, r of the bend is very short, then the minor head loss coefficient, k may be quite high. However, for a larger radius, the minor head loss coefficient, k decrease until a minimum value is found at about an r/D ratio of 6. It is interesting to note that for a ratio, r/D larger than 6, there is an increase in the minor head loss coefficient, k because the bend becomes longer than the bends with small values for the ratio, r/D, where the greater length creates an additional head loss. The miter bend is used instead of a smooth bend in situations where space in the pipe system is limited, yet a change in flow direction is required. The minor head loss coefficients, k for a 90-degree miter bend without and with a guide are illustrated in Figures 8.8a and b and Table 8.11, respectively. It is important to note that the use of miter bend with a vane significantly reduces the minor head loss coefficient, k, as the vanes help direct the flow with a minimum amount of swirling secondary flow and flow disturbances. The 180-degree return bend may be either flanged or threaded. Table 8.11 presents the minor head loss coefficients, k for 180-degree bend for both the flanged and threaded types.

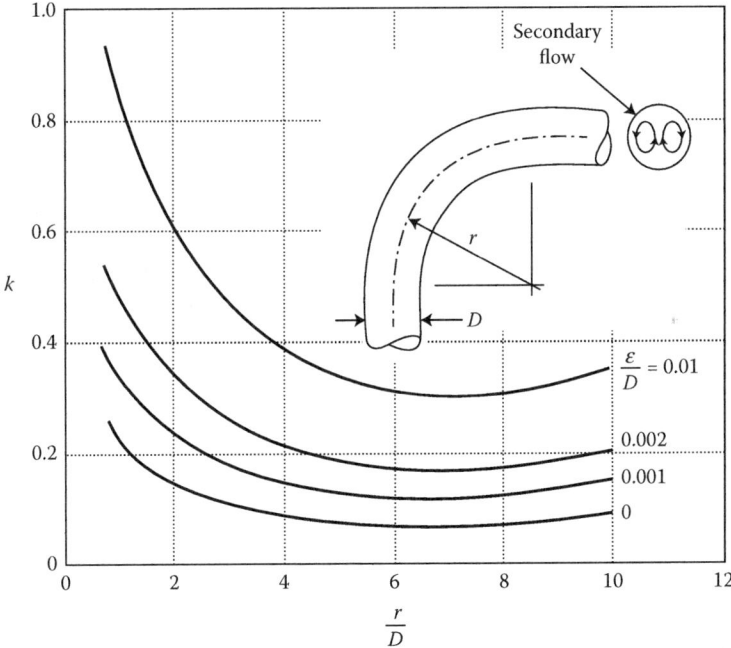

FIGURE 8.9

Minor head loss coefficient, k for a 90° smooth bend is a function relative pipe roughness, ε/D and round radius, r to pipe diameter, D ratio, r/D for large values of the Reynolds number, R. (Adapted from Munson, B. R., Young, D. F., and T. H., Okiishi, *Fundamentals of Fluid Mechanics*, 3rd ed, John Wiley and Sons, New York, 1998, p. 503; White, F. M., *Fluid Mechanics*, McGraw Hill, New York, 1979, p. 355.)

EXAMPLE PROBLEM 8.9

Water flows in a 0.5-m-diameter horizontal pipe at a flowrate of 0.3 m³/sec. A 90°
mitered bend is installed in between points 1 and 2 as illustrated in Figure EP 8.9
in order to change the direction of the flowrate by 90°. The pressure at point 1 is 2 ×
10^5 N/m². (a) Determine the minor head loss due to the 90° mitered bend. (b) Determine the pressure at point 2.

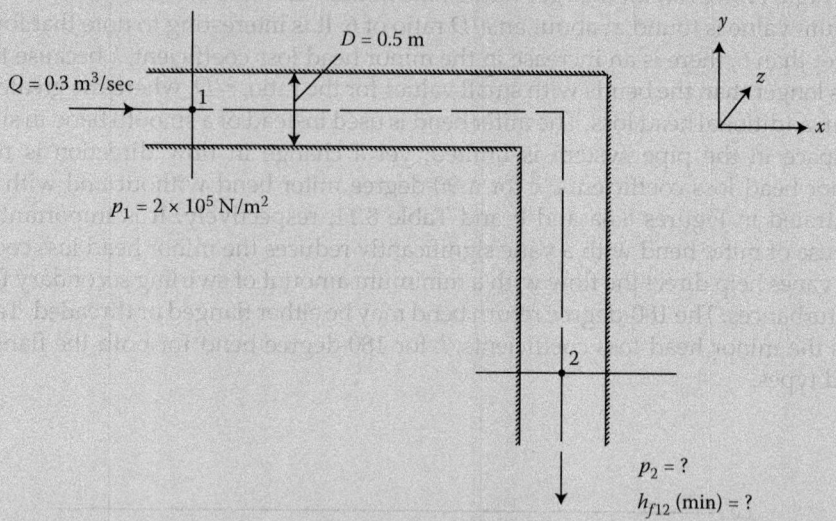

FIGURE EP 8.9
Water flows in a horizontal pipe with a 90° miter bend to change direction of flow.

Mathcad Solution

(a) The minor head loss due to the 90° mitered bend is computed from Equation 8.62
where the minor head loss coefficient, k is determined from Table 8.11. However,
because the 90° mitered bend may be without vanes or with vanes, there will be
two values for the minor head loss coefficient, k and a corresponding two values
for the minor head loss due to the gate valve, h_f as follows:

$$D := 0.5\,m \qquad\qquad Q := 0.3\,\frac{m^3}{sec} \qquad\qquad g := 9.81\,\frac{m}{sec^2}$$

$$A := \frac{\pi \cdot D^2}{4} = 0.196\,m^2 \qquad\qquad v := \frac{Q}{A} = 1.528\,\frac{m}{s}$$

$$k_{wov} := 1.1 \qquad\qquad h_{fwov} := k_{wov} \cdot \left(\frac{v^2}{2 \cdot g}\right) = 0.131\,m$$

$$k_{wv} := 0.26 \qquad\qquad h_{fwv} := k_{wv} \cdot \left(\frac{v^2}{2 \cdot g}\right) = 0.031\,m$$

One may note that for the as the 90° mitered bend changes from without vanes to
with vanes, the minor head loss coefficient, k decreases and thus the corresponding value for the minor head loss due to the 90° mitered bend, h_f also decreases.

(b) The corresponding values for the pressure at point 2 are computed by applying the energy equation between points 1 and 2 as follows:

$$p_1 := 2 \times 10^5 \frac{N}{m^2} \qquad \gamma := 9810 \frac{N}{m^3} \qquad\qquad z_1 := 0\,m \qquad\qquad z_2 := 0\,m$$

$$v_1 := v = 1.528 \frac{m}{s} \qquad v_2 := v = 1.528 \frac{m}{s}$$

Guess values: $\qquad\qquad p_{2wov} := 1 \times 10^2 \frac{N}{m^2} \qquad p_{2wv} := 1 \times 10^2 \frac{N}{m^2}$

Given

$$\frac{p_1}{\gamma} + z_1 + \frac{v_1{}^2}{2 \cdot g} - h_{fwov} = \frac{p_{2wov}}{\gamma} + z_2 + \frac{v_2{}^2}{2 \cdot g}$$

$$\frac{p_1}{\gamma} + z_1 + \frac{v_1{}^2}{2 \cdot g} - h_{fwv} = \frac{p_{2wv}}{\gamma} + z_2 + \frac{v_2{}^2}{2 \cdot g}$$

$$\binom{p_{2wov}}{p_{2wv}} := Find\,(p_{2wov},\,p_{2wv}) = \binom{1.987 \times 10^5}{1.997 \times 10^5} \frac{N}{m^2}$$

Therefore, as the 90° mitered bend changes from without vanes to with vanes, the minor head loss coefficient, k decreases, the corresponding value for the minor head loss due to the 90° mitered bend, h_f decreases, causing a decrease in the pressure drop between points 1 and 2.

8.6.5 Minor Losses in Entrances, Exits, Contractions, and Expansions

Evaluation of the minor head loss for entrances, exits, contractions, and expansions involves evaluating the velocity term, v and the minor head loss coefficient, k for the particular pipe transition section. The installation of entrances, exits, contractions, and expansions causes a change in the cross section of the flow and thus a change in the velocity of the flow. Such changes in pipe cross sections may occur suddenly or gradually in the pipe system. The flow through a square-edged pipe entrance from a reservoir and the flow through a square-edged pipe exit to a reservoir are considered to be the extreme cases of a sudden pipe contraction and a sudden pipe expansion, respectively. Furthermore, because there is a change in the pipe diameter and thus a change in the velocity of the flow, the larger of the two values of the upstream average developed flow velocity, v_{ave} or the downstream average developed flow velocity, v_{ave} (to or from the pipe component/device) is used to compute the associated minor losses, $h_{f,minor}$, as follows:

$$h_f = k\left(\frac{v_{ave}^2}{2g}\right) \tag{8.63}$$

Similar to the installation of valves and fittings, the pipe transition sections introduce turbulence in the flow section, for which experimental results are generally used to determine the minor loss coefficient, k and thus the associated minor losses, $h_{f,minor}$. However, unlike the valves and fittings, the pipe diameter downstream of the component changes in the case of entrances, exits, contractions, and expansions. As a result, determination of the minor head loss coefficient, k is more complicated due to the change in pipe diameter and thus the corresponding velocity. Furthermore, one may note that the minor loss coefficient, k (and thus the associated minor losses, $h_{f,minor}$) is evaluated analytically for the case of a sudden expansion.

8.6.5.1 Pipe Entrances

A pipe flow system is typically supplied by a reservoir as illustrated in Figure 8.10, where the flow enters the pipe at the pipe entrance. There are various types of pipe entrances, including a reentrant entrance, a square-edged entrance, and a well-rounded entrance. The various geometrical shapes of the different entrances result in different values of minor loss coefficient, k and thus the associated minor losses, $h_{f,minor}$ as illustrated in Figure 8.10. The fluid accelerates from a negligible velocity to the velocity of the flow in the pipe. Thus, the ease with which the acceleration takes place will determine the amount of head loss. Therefore, the value of the minor loss coefficient, k for a pipe entrance is highly a function of the geometry of the device (entrance flow geometry) and the Reynolds number, R. As presented below, the values of minor loss coefficient, k assumes a large Reynolds number, R and is reduced as the geometry of the entrance changes from a reentrant, to a square-edged, and then to a well-rounded entrance; $k = 1, 0.5, 0.04$, respectively (see Figure 8.10). Therefore, rounding the edges of a pipe entrance can result in a significant reduction in the minor head loss coefficient, k. Furthermore, the average velocity, v_{ave} of the downstream developed flow is used to compute the associated minor losses, $h_{f,minor}$, as follows:

$$h_f = k\left(\frac{v_{ave}^2}{2g}\right) \tag{8.64}$$

A reentrant entrance from the reservoir has the pipe protruding inside the reservoir and thus presents an abrupt change in flow area inside the reservoir. Although more turbulence (which results in a minor head loss) is introduced than in the square-edged pipe entrance, it is contained mostly inside the reservoir. As a result, although there is no flow separation form the pipe wall just downstream of the pipe entrance, the minor head loss coefficient is rather high because some of the fluid near the entrance edge is forced to make a 180-degree turn, as illustrated in Figure 8.10a. Beginning at point 1, the velocity of the flow in the reservoir is assumed to be zero ($v_1 = 0$). At point 2, where the flow remains attached to the pipe wall, the flow accelerates to an average pipe flow velocity ($v_2 = v_{ave} = Q/A$). Then, downstream of point 2, at point 3, the flow maintains the average pipe flow velocity ($v_3 = v_{ave} = Q/A$). Experimental results have yielded a minor loss coefficient, $k = 1.0$ (see Figure 8.10), where the average velocity, v_{ave} is used to compute the associated minor losses, $h_{f,minor}$.

A square-edged entrance from the reservoir presents an abrupt change in flow area, which causes the flow to separate from the pipe wall just downstream of the pipe entrance, as illustrated in Figure 8.10b, introducing turbulence, which results in a minor head loss. Beginning at point 1, the velocity of the flow in the reservoir is assumed to be zero ($v_1 = 0$). Because the fluid cannot turn a sharp-angled corner of the pipe entrance, the streamlines converge and the flow separates due to the sharp corner. At point 2, where the flow separates from the pipe wall, a vena contracta is created (somewhat similar to that of an orifice), where the flow efficiently accelerates to a maximum pipe flow velocity ($v_2 = v_{max}$), where the pressure

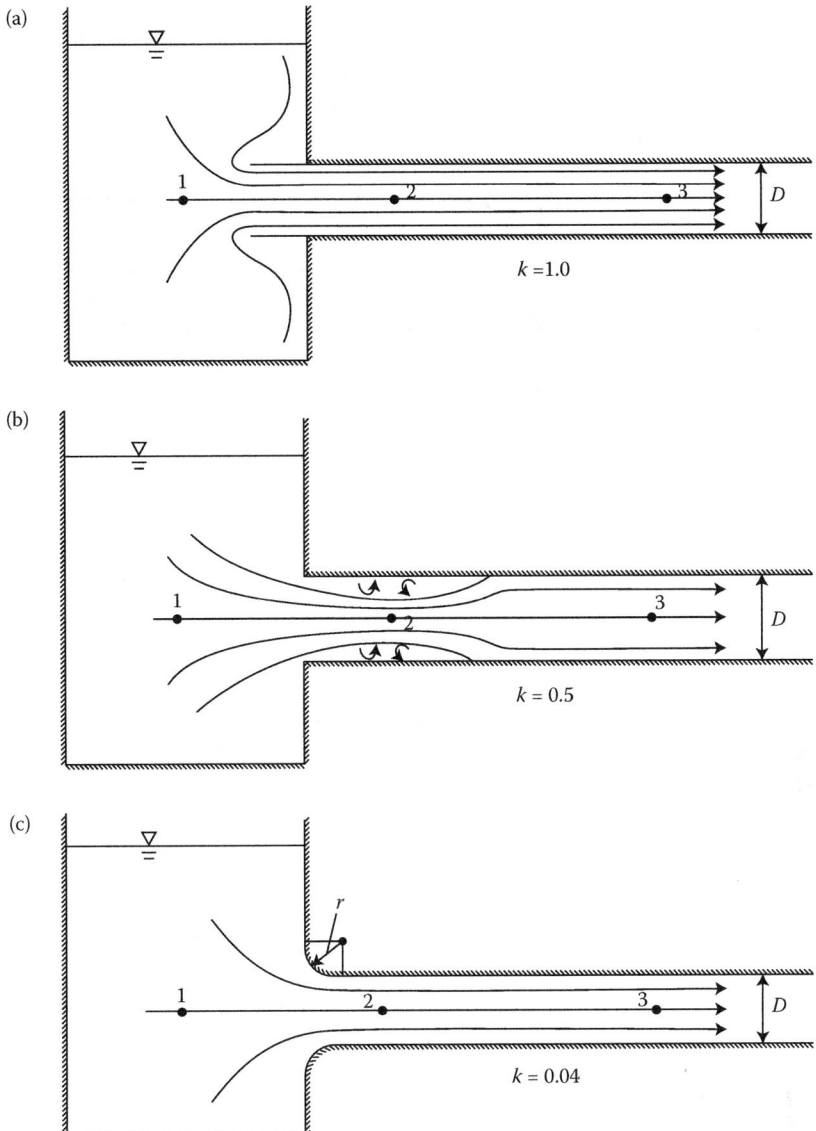

FIGURE 8.10
Three types of pipe entrances from a reservoir and typical minor head loss coefficients, k. (a) Reentrant entrance, (b) Square-edged entrance, (c) Well-rounded entrance. (Adapted from Esposito, A., *Fluid Mechanics with Applications*, Prentice Hall, New Jersey, 1998, p. 339.)

drops to a minimum; the efficient acceleration implies that the pressure head (pressure energy or flow energy) at point 1 is converted to a velocity head (kinetic energy) at point 2, without a head loss as illustrated in Figure 8.11. Then, downstream of the vena contracta at point 3, the streamlines diverge and the flow reattaches to the pipe wall (completely filling the pipe cross section), and the flow inefficiently decelerates to the average pipe flow velocity ($v_3 = v_{ave} = Q/A$). The inefficient deceleration implies that the velocity head (kinetic energy) at point 2 is converted to a pressure head (pressure energy or flow energy) at point 3, with a minor head loss as illustrated in Figure 8.11. However, if the flow were to efficiently

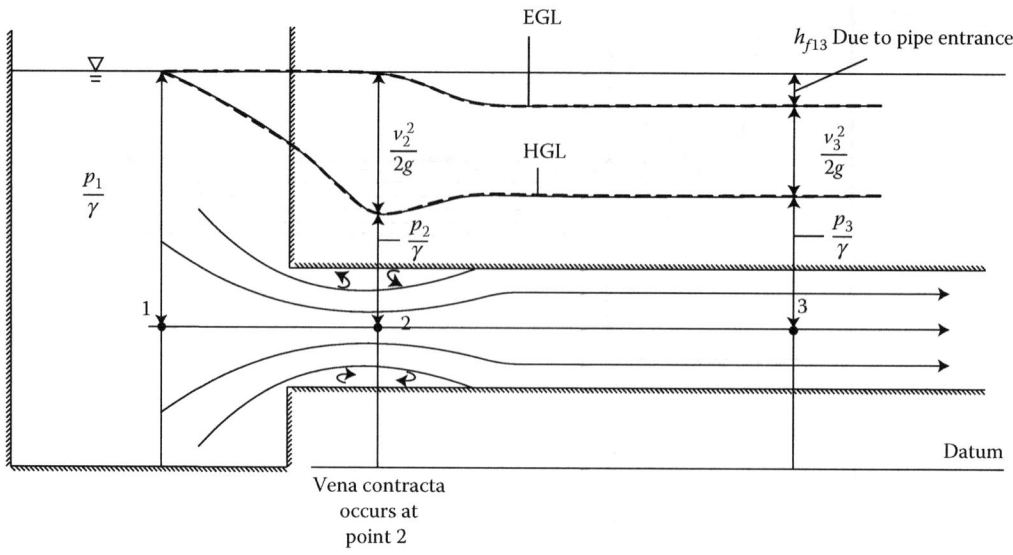

FIGURE 8.11
EGL and HGL for a square-edged pipe entrance from a reservoir. (Adapted from Cengel, Y. A., and J. M., Cimbala, *Fluid Mechanics Fundamentals and Applications*, 3rd ed, McGraw Hill, New York, 2014, 378.)

decelerate, then the velocity head (kinetic energy) at point 2 would be converted to a larger pressure head (pressure energy or flow energy) at point 3, without a minor head loss, and thus an ideal pressure would result. Instead, the additional kinetic energy at point 2 is mostly lost as frictional heating at point 3 due to viscous dissipation caused by intense mixing and turbulent eddies, and, as a result, the pressure at point 3 does not return to its ideal value; there is a drop in velocity without much pressure recovery, and the minor head loss is a measure of this irreversible pressure drop, Δp. Furthermore, one may note that resulting minor head loss is due mostly to inertia effects that are eventually dissipated by the shear stresses within the fluid. And, therefore, only a small part of the head loss is due to the wall shear stress near the entrance region. One may note that Figure 8.11 illustrates the EGL drops by an amount equal to the head loss produced by the pipe entrance, where the drop occurs over a distance of several diameters downstream of the entrance. The HGL also drops significantly immediately downstream of the entrance because of the high-velocity flow in the vena contracta at point 3. Furthermore, as the turbulent mixing occurs even farther downstream, energy is lost due to viscous action occurring in the mixing process. Therefore, the EGL at the entrance is steeper than it is farther downstream, where the flow becomes uniform, fully developed flow. Finally, as the additional head loss due to the pipe entrance dissipates, the EGL assumes a slope (head loss) generated by the pipe friction. Experimental results have yielded a minor loss coefficient, $k = 0.5$ (see Figure 8.10), where the average velocity, v_{ave} is used to compute the associated minor losses, $h_{f,minor}$.

A well-rounded entrance from the reservoir presents a gradual change in flow area, which does not cause the flow to separate from the pipe wall just downstream of the pipe entrance, as illustrated in Figure 8.10c, yet introducing a small amount of turbulence, which results in a minor head loss. Thus, rounding the entrance will prevent the formation of a vena contracta and thus, eliminating a head loss due to deceleration of flow. Beginning at point 1, the velocity of the flow in the reservoir is assumed to be zero ($v_1 = 0$). At point 2, where the flow remains attached to the pipe wall, the flow accelerates to an average pipe flow

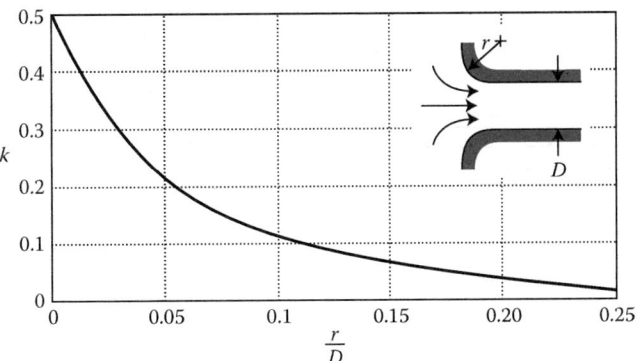

FIGURE 8.12
Minor head loss coefficient, k for a rounded pipe entrance is a function of the round radius, r to pipe diameter, D ratio, r/D. (From ASHRAE, *ASHRAE Handbook of Fundamentals*, American Society of Heating, Refrigeration and Air-Conditioning Engineers, Inc., Atlanta, 1981, 33.28.)

velocity ($v_2 = v_{ave} = Q/A$). Then downstream of point 2, at point 3, the flow maintains the average pipe flow velocity ($v_3 = v_{ave} = Q/A$), where the average velocity, v_{ave} is used to compute the associated minor losses, $h_{f,minor}$. Experimental results indicate the minor loss coefficient, k varies depending upon the round radius, r to pipe diameter, D ratio, r/D, as illustrated in Figure 8.12. Specifically, for a well-rounded pipe entrance, $r/D \geq 0.2$, $k = 0.04$. For a slightly rounded pipe entrance, $r/D = 0.05$, $k = 0.2$. And, for the limiting/extreme case of a rounded pipe entrance, which is the case of a square-edged pipe entrance, $r/D = 0$, $k = 0.5$ (see above). Therefore, one may note that an effective way to reduce the entrance loss is to round the entrance region.

8.6.5.2 Pipe Exists

A pipe flow system may supply fluid to a reservoir as illustrated in Figure 8.13, where the flow leaves the pipe at the pipe exit. Similar to pipe entrances, there are various types of pipe exits, including a reentrant exit, a square-edged exit, and a rounded exit. However, unlike the turbulence introduced at a pipe entrance, which causes the minor head loss, the minor head loss at the pipe exit is caused by a dissipation of all of the kinetic energy (into internal energy and heat transfer, and thus there is a pressure rise) of the pipe flow due to the viscous effects of the stationary fluid in the reservoir (as a result of an abrupt change/increase in flow area); the wall shear stress, τ_w is not significant. The EGL and HGL for a reentrant pipe exit are illustrated in Figure 8.14. Furthermore, unlike the various pipe entrances, the various geometrical shapes of the different exits result in the same value of minor loss coefficient, $k = 1.0$ and thus the associated minor losses, $h_{f,minor}$ is only a function of the average velocity, v_{ave} ($v_1 = v_{ave} = Q/A$) in the pipe section upstream of the pipe exit and reservoir (see Figure 8.13, where the average velocity, v_{ave} is used to compute the associated minor losses, $h_{f,minor}$, as follows:

$$h_f = k\left(\frac{v_{ave}^2}{2g}\right) \tag{8.65}$$

Therefore, unlike a pipe entrance, rounding the edges of a pipe exit does not result in any change in the minor head loss coefficient, k.

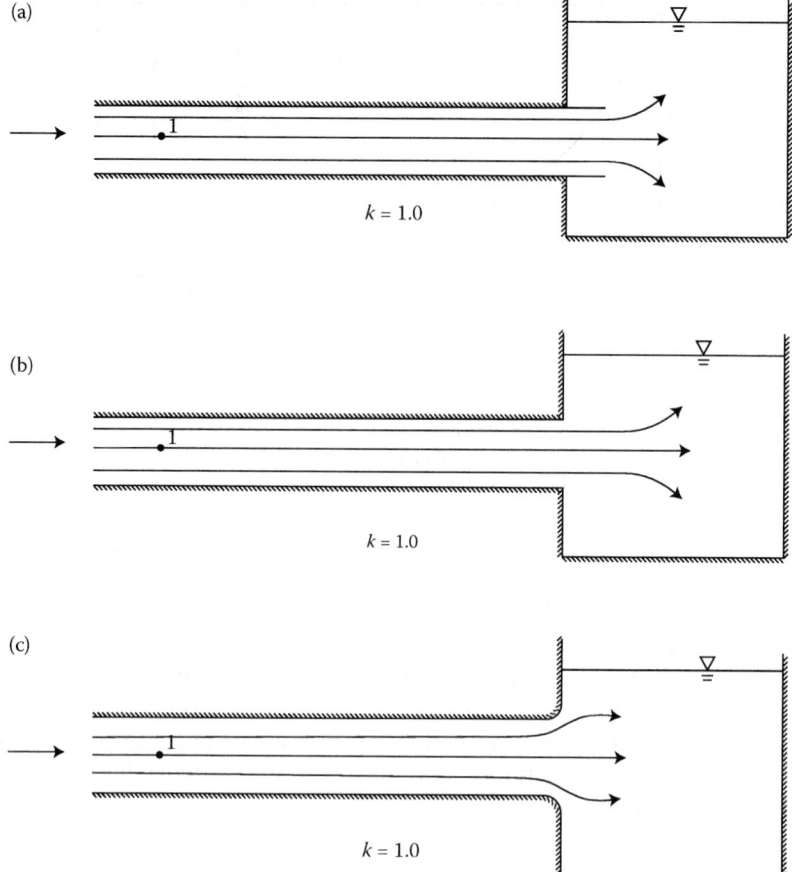

FIGURE 8.13

Three types of pipe exits to a reservoir and typical minor head loss coefficients. (a) Reentrant exit, (b) Square-edged exit, (c) Rounded exit. (Adapted from Esposito, A., *Fluid Mechanics with Applications*, Prentice Hall, New Jersey, 1998, p. 341.)

8.6.5.3 Sudden Pipe Expansions

The flow through a square-edged pipe exit to a reservoir is considered to be the extreme cases of a sudden pipe expansion. In general, the flow separation due to a sudden expansion or a sudden contraction results in a relatively greater head loss than for a gradual expansion or a gradual contraction. And, the head loss due to a sudden expansion is greater than the head loss due to a sudden contraction because of flow separation; it is the diverging flow (flow reattaches to the pipe wall and thus, inefficiently decelerates) that causes the formation of eddies within the flow and leads to a head loss. Furthermore, depending upon the design/choice of the conical angle, θ, the minor losses may be significantly reduced by the use of a gradual expansion or a gradual contraction as opposed to a sudden expansion and a sudden contraction, respectively.

The flow pattern and geometry of a sudden pipe expansion is similar to a square-edged exit (described above), which is the limiting/extreme case of a sudden pipe expansion with an area ratio $A_1/A_2 = 0$, where $k = 1$ (see details below). A change in the cross section of flow may occur as a sudden expansion as illustrated in Figure 8.15a, where there is a

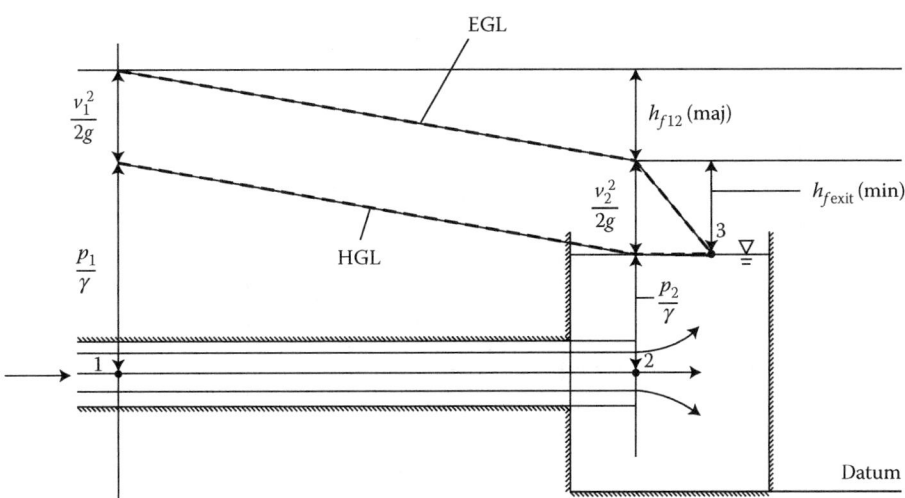

FIGURE 8.14
EGL and HGL for a reentrant pipe exit.

sudden increase in pipe size diameter. At point 1, the expansion occurs, where the pipe section increases from D_1 to D_2 causing the velocity to decrease from v_1 to eventually v_2 at point 2. As the flow leaves point 1 in the pipe, the flow separates from the pipe wall introducing turbulence (viscous effects due to wall shear stress), which contributes insignificantly to the minor head loss. Then as the flow approaches point 2, the flow reattaches to the pipe wall, inefficiently decelerates to v_2, and becomes fully developed at point 2, where there is a pressure recovery as a result of the deceleration of the flow. The EGL and the HGL for a pipe with a sudden expansion are illustrated in Figure 8.15a. The rapid deceleration from v_1 to v_2 is accompanied by large-scale turbulence, which may continue in the larger pipe for a distance of approximately 50 diameters before the normal turbulence pattern of fully established flow is restored. However, it may be noted that similar to the square-edged exit, the minor head loss in a sudden expansion is mostly due to the dissipation of kinetic energy (into internal energy and heat transfer) due to the viscous effects of the slower moving fluid as a result of an increase in flow area from A_1 to A_2. Thus, because the separation of flow (and thus flow turbulence) between the upstream and downstream cross sections (A_1 and A_2, respectively) in a sudden expansion contributes insignificantly to the minor head loss, the minor loss coefficient, k may be evaluated analytically as follows.

In the case of a sudden pipe expansion, because the friction/viscous forces are insignificant and can be ignored in the integral form of the momentum equation, one can analytically determine the actual pressure rise, Δp from the momentum equation. As a result, one may analytically derive an expression for the minor head loss, h_f (and thus an analytical expression for the corresponding minor head loss coefficient, k) due to the sudden pipe expansion from the energy equation. One may assume a control volume between points 1 and 2 (along the outer most separation path of flow from the pipe wall), as illustrated in Figure 8.15a, and thus one may apply the three governing equations. Assuming a horizontal pipe, there is no component of the gravitational force, F_G in the direction of flow. Furthermore, assuming that the turbulence due to the flow separation takes place outside of the control volume, there is no component of the viscous force due the wall shear stress, F_V in the direction of the flow. Noting that the pressure at control face 1 applies across the entire expanded cross-sectional

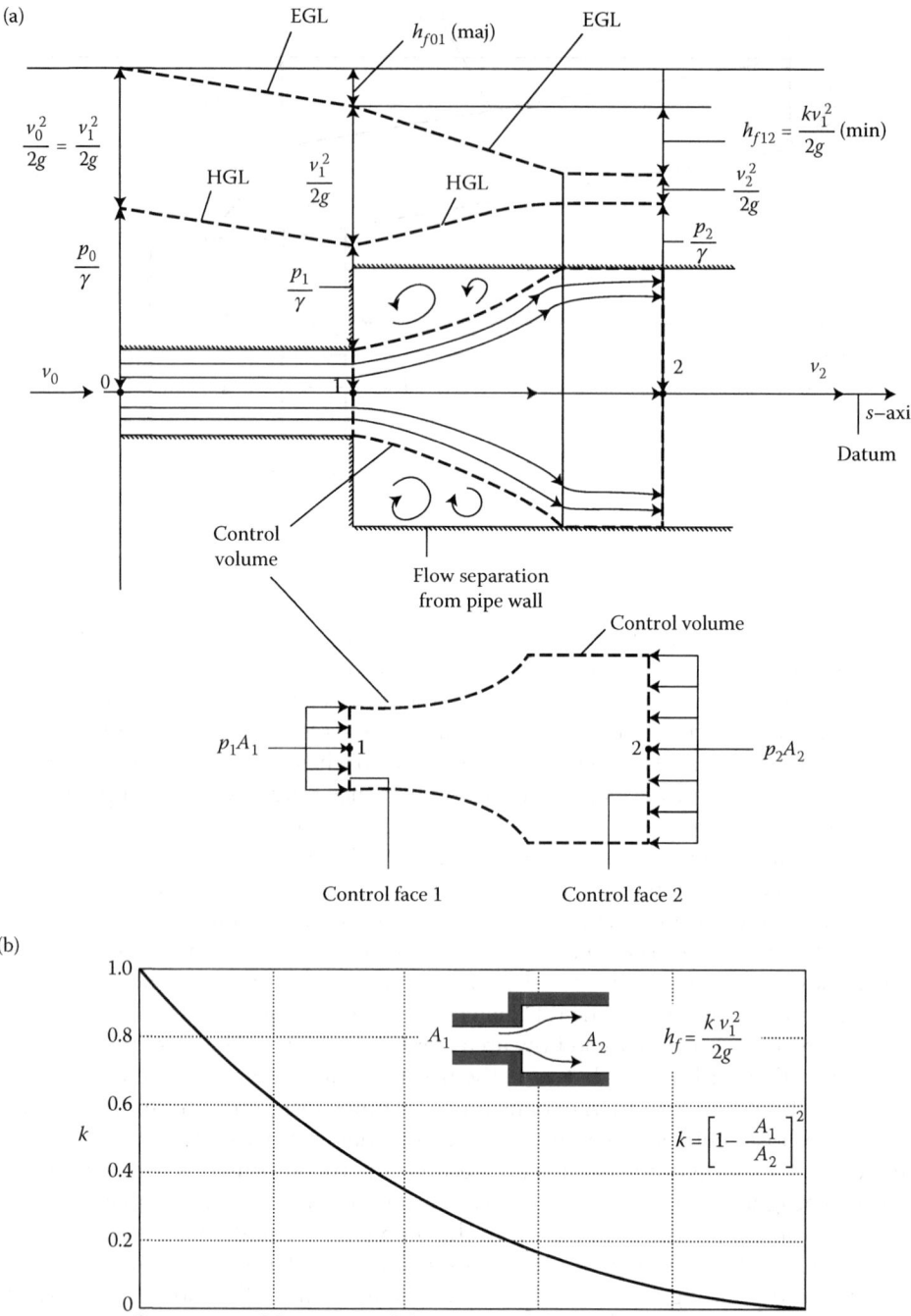

FIGURE 8.15
(a) Control volume and EGL and HGL for sudden pipe expansion, (b) Minor head loss coefficient, k for a sudden pipe expansion is a function of area ratio, A_1/A_2. (See analytical expression for k given in Equation 8.74.)

area, A_2, application of the momentum equation for the control volume between control face 1 and control face 2 yields the following:

$$\sum F_s = (F_G + F_P + F_V + F_{other})_s = (\rho Q v_s)_2 - (\rho Q v_s)_1 \tag{8.66}$$

$$p_1 A_2 - p_2 A_2 = \rho Q (v_2 - v_1) \tag{8.67}$$

Application of the integral form of the continuity equation for steady flow between points 1 and 2 yields the following:

$$Q = v_1 A_1 = v_2 A_2 \tag{8.68}$$

Substituting the continuity equation, Equation 8.68 (assuming that the minor head loss due to a sudden expansion is evaluated using the upstream velocity, v_1) into the momentum equation, Equation 8.67, dividing by g, and rearranging yields an analytical expression of the actual pressure rise, Δp as follows:

$$\left(\frac{p_1 - p_2}{\rho}\right) A_2 = v_2 A_2 (v_2 - v_1)$$

$$\left(\frac{p_1 - p_2}{\rho g}\right) = \left(\frac{v_2^2}{g} - \frac{v_1 v_2}{g}\right)$$

$$\left(\frac{p_1 - p_2}{\rho g}\right) = \left(\frac{v_1^2 \left(\dfrac{A_1}{A_2}\right)^2}{g} - v_1^2 \frac{A_1}{A_2}{g}\right)$$

$$\left(\frac{p_1 - p_2}{\rho g}\right) = \frac{v_1^2}{g}\left(\frac{A_1}{A_2}\right)\left(\frac{A_1}{A_2} - 1\right) \tag{8.69}$$

Application of the energy equation between points 1 and 2 yields the following:

$$\left(\frac{p_1}{\gamma} + z_1 + \frac{v_1^2}{2g}\right) - \left(\frac{p_2}{\gamma} + z_2 + \frac{v_2^2}{2g}\right) = h_{f,min} \tag{8.70}$$

Assuming a horizontal pipe, $z_1 = z_2$ and thus:

$$\left(\frac{p_1}{\gamma} + \frac{v_1^2}{2g}\right) - \left(\frac{p_2}{\gamma} + \frac{v_2^2}{2g}\right) = h_{f,min}$$

$$\left(\frac{p_1}{\gamma} - \frac{p_2}{\gamma}\right) + \left(\frac{v_1^2}{2g} - \frac{v_2^2}{2g}\right) = h_{f,min} \tag{8.71}$$

Substituting the expression for the analytical expression of the actual pressure rise, Δp from the momentum equation above Equation 8.69, and substituting $v_2 = v_1 A_1/A_2$ from the continuity equation, Equation 8.68, into the energy equation, Equation 8.71, yields an analytical expression for the minor head loss, h_f (and thus an analytical expression for the corresponding minor head loss coefficient, k) as follows:

$$\frac{v_1^2}{g}\left(\frac{A_1}{A_2}\right)\left(\frac{A_1}{A_2}-1\right)+\left(\frac{v_1^2}{2g}-\frac{v_2^2}{2g}\right)=h_{f,min}$$

$$\frac{v_1^2}{g}\left(\frac{A_1}{A_2}\right)\left(\frac{A_1}{A_2}-1\right)+\left(\frac{v_1^2}{2g}-\frac{v_1^2\left(\frac{A_1}{A_2}\right)^2}{2g}\right)=h_{f,min}$$

$$\frac{v_1^2}{g}\left(\frac{A_1}{A_2}\right)^2-\frac{v_1^2}{g}\left(\frac{A_1}{A_2}\right)+\frac{v_1^2}{2g}-\frac{v_1^2}{2g}\left(\frac{A_1}{A_2}\right)^2=h_{f,min}$$

$$\frac{v_1^2}{2g}\left(\frac{A_1}{A_2}\right)^2-\frac{v_1^2}{g}\left(\frac{A_1}{A_2}\right)+\frac{v_1^2}{2g}=h_{f,min}$$

$$\frac{v_1^2}{2g}\left[\left(\frac{A_1}{A_2}\right)^2-2\left(\frac{A_1}{A_2}\right)+1\right]=h_{f,min}$$

$$\frac{v_1^2}{2g}\left[1-2\left(\frac{A_1}{A_2}\right)+\left(\frac{A_1}{A_2}\right)^2\right]=h_{f,min}$$

$$\frac{v_1^2}{2g}\left[1-\left(\frac{A_1}{A_2}\right)\right]^2=\frac{v_1^2}{2g}\left[1-\left(\frac{D_1}{D_2}\right)^2\right]^2=h_{f,min} \tag{8.72}$$

Therefore, assuming that the minor head loss due to a sudden expansion is evaluated using the upstream velocity, v_1, the analytical expression is given as follows:

$$h_f=k\left(\frac{v_1^2}{2g}\right) \tag{8.73}$$

where the minor loss coefficient, k due to a sudden expansion is evaluated analytically as follows:

$$k=\left[1-\frac{A_1}{A_2}\right]^2=\left[1-\left(\frac{D_1}{D_2}\right)^2\right]^2 \tag{8.74}$$

One may note that the flow pattern and geometry of a sudden pipe expansion is similar to a square-edged exit (described above), which is the limiting/extreme case of a sudden pipe expansion with an area ratio $A_1/A_2=0$, where the minor loss coefficient due to a sudden pipe expansion, k is a function of the area ratio, A_1/A_2 as given by the analytical Equation 8.74 and graphically illustrated in Figure 8.15b. The value for k changes from the limiting/

extreme case of square-edged exit with an area ratio of $A_1/A_2 = 0$, where $k = 1$, to the other extreme case of a constant diameter pipe with an area ratio of $A_1/A_2 = 1$, where $k = 0$.

8.6.5.4 Sudden Pipe Contractions

The flow through a square-edged a pipe entrance from a reservoir is considered to be the extreme case of a sudden pipe contraction. A change in the cross section of flow may occur as a sudden contraction as illustrated in Figure 8.16a, where there is a sudden decrease in

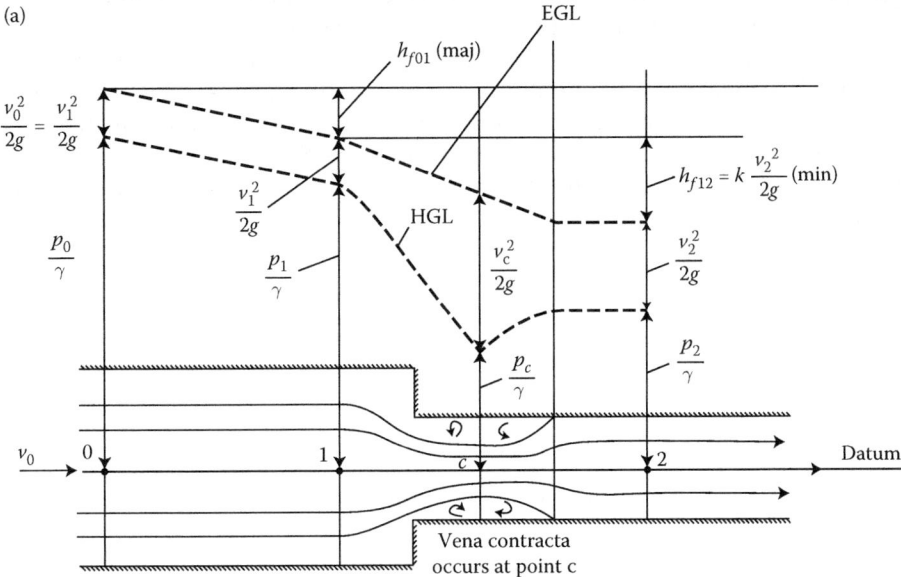

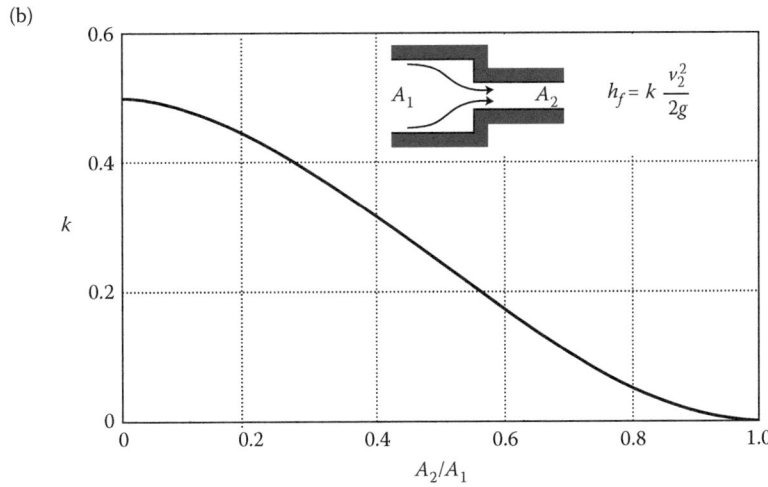

FIGURE 8.16
(a) EGL and HGL for a sudden pipe contraction, (b) Minor head loss coefficient, k for a sudden pipe contraction is a function of area ratio, A_2/A_1. (From Streeter, V. L., Ed., *Handbook of Fluid Dynamics*, McGraw Hill, New York, 1961, pp. 3–21.)

pipe size diameter. Similar to a square-edged entrance, a sudden contraction results in the formation of a vena contracta due to flow separation, where the flow accelerates, then subsequently decelerates where the flow reattaches to the pipe wall. A sudden contraction causes a significant drop in pressure, Δp (at the vena contracta), which is due both an increase in velocity and the loss of energy due to turbulence. Once the flow reattaches to the wall, there is a pressure recovery accompanied by the deceleration of the flow. Figure 8.16a illustrates the EGL and the HGL resulting from a sudden pipe contraction. One may note that the flow pattern and geometry of a sudden pipe contraction is similar to a square-edged entrance (described above), which is the limiting/extreme case of a sudden pipe contraction with an area ratio $A_2/A_1 = 0$, where the minor loss coefficient due to a sudden pipe contraction, k is a function of the area ratio, A_2/A_1 as illustrated in Figure 8.16b. The value for k changes from the limiting/extreme case of square-edged entrance with an area ratio of $A_2/A_1 = 0$ where $k = 0.5$ to the other extreme of a constant diameter pipe with an area ratio of $A_2/A_1 = 1$ where $k = 0$. Furthermore, the minor head loss due to a sudden contraction is evaluated using the, the larger velocity, which is the downstream velocity, v_2 as follows:

$$h_f = k\left(\frac{v_2^2}{2g}\right)$$ (8.75)

8.6.5.5 Gradual Pipe Expansions

The head loss due to a pipe expansion may be significantly reduced by the use of a gradual pipe expansion. A change in the cross section of flow may occur as a gradual expansion as illustrated in Figure 8.17a, where there is a gradual increase in pipe size diameter through a conical diffuser. One may note that a conical diffuser is used to decelerate the fluid flow in order to achieve pressure recovery (a pressure rise, as opposed to the typical pressure drop, Δp in the direction of flow), where a decrease in velocity is accompanied by an increase in static pressure (recall: $k = $ static pressure/dynamic pressure); the function of a conical diffuser is to convert kinetic energy to pressure energy by decelerating the flow as it moves from the smaller to the larger pipe. Thus, the pressure rise, Δp is defined as follows:

$$p_2 - p_1 = \Delta p = \left(\frac{\rho v_1^2}{2} - \frac{\rho v_2^2}{2}\right) - \frac{\tau_w L}{R_h}$$ (8.76)

One may note that the pressure will continue to rise for a few pipe diameters downstream from point 2; this is due to the recovery of the velocity distribution from a decelerated developing flow at the diffuser to developed turbulent downstream of point 2. The EGL and the HGL for a gradual pipe expansion is illustrated in Figure 8.17a. For such a gradual pipe transition, the minor loss coefficient, k due to a gradual pipe expansion (sometimes referred to as the pressure recovery coefficient, C_p) is a function of the conical diffuser area ratio, A_2/A_1 and the conical angle, θ as illustrated in Figure 8.17b. It is important to note that because of the large surface of the conical diffuser that comes into contact with the fluid flow, the minor loss coefficient, k is a result of the pipe wall friction as well as the large-scale turbulence. One may note that Figure 8.17b presents the variation of the minor loss coefficient, k with the conical angle, θ for a given constant value of the area ratio, A_2/A_1. It is interesting to note, that a very small conical angle, θ (less than 5°) results in a very long gradual pipe expansion, which results in large pressure losses, where most of the head loss is due to

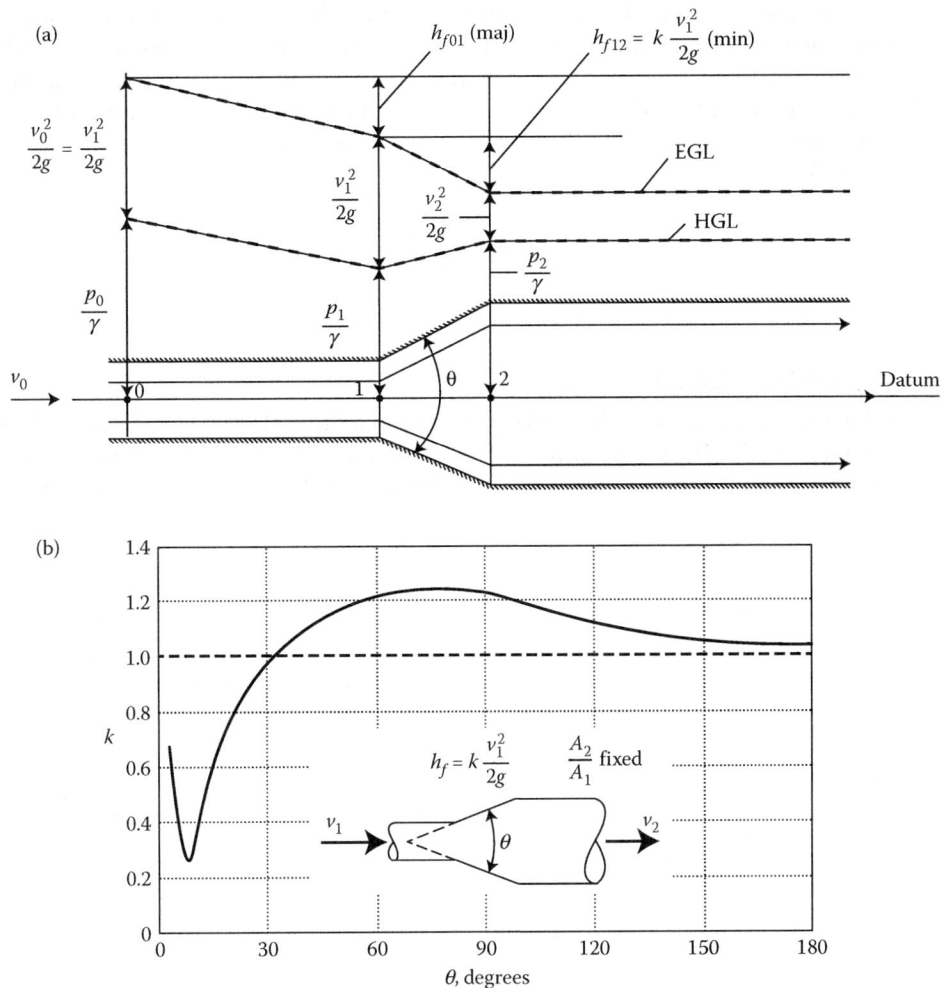

FIGURE 8.17
(a) EGL and HGL for a gradual pipe expansion, (b) Minor head loss coefficient, k for a gradual pipe expansion is a function of conical angle, θ for a given constant conical diffuser area ratio, A_2/A_1. (Adapted from White, F. M., *Fluid Mechanics*, McGraw Hill, New York, 1979, p. 358.)

the wall shear stress (similar to the major head loss due to pipe friction in fully developed flow). As a result, a very small conical angle, θ does not yield the minimum value for k, as one may have expected. Instead, the minimum value for k occurs at an optimum conical angle, θ of approximately 8°, which results in a long conical diffuser for which the combination of the effect of surface friction and eddying turbulence is at a minimum. At moderate to large values for the conical angle, θ, the expansion becomes more sudden, thus reducing the surface area. As a result, flow separation occurs, which produces large eddies; thus, the minor head loss is mainly due to a dissipation of the kinetic energy of the flow leaving the upstream smaller diameter pipe. Furthermore, a conical angle, θ of greater than approximately 35–40° yields values for $k > 1$ (see Figure 8.17b), which exceeds the $k = 1$ value for the limiting/extreme case of a sudden pipe expansion (square-edged pipe exit). Thus, the experimental results for minor loss coefficient, k for a gradual pipe expansion illustrate the difficulty of efficiently decelerating the fluid flow. Furthermore, in addition to dependence of k on the conical diffuser area ratio, A_2/A_1 and the conical angle, θ, it is highly a function of both the specific geometry of the device and the Reynolds number, $\boldsymbol{R}$. The minor head loss due to a gradual expansion is evaluated using the larger velocity, which is the upstream velocity, v_1 as follows:

$$h_f = k\left(\frac{v_1^2}{2g}\right) \tag{8.77}$$

EXAMPLE PROBLEM 8.10

Water flows in a 0.04-m-diameter horizontal pipe at a flowrate of 0.02 m^3/sec, as illustrated in Figure EP 8.10. The pressure at point 1 is 1.8×10^5 N/m^2. An expansion to a 0.09 m diameter pipe at point 2 is sought to decelerate the fluid flow in order to achieve

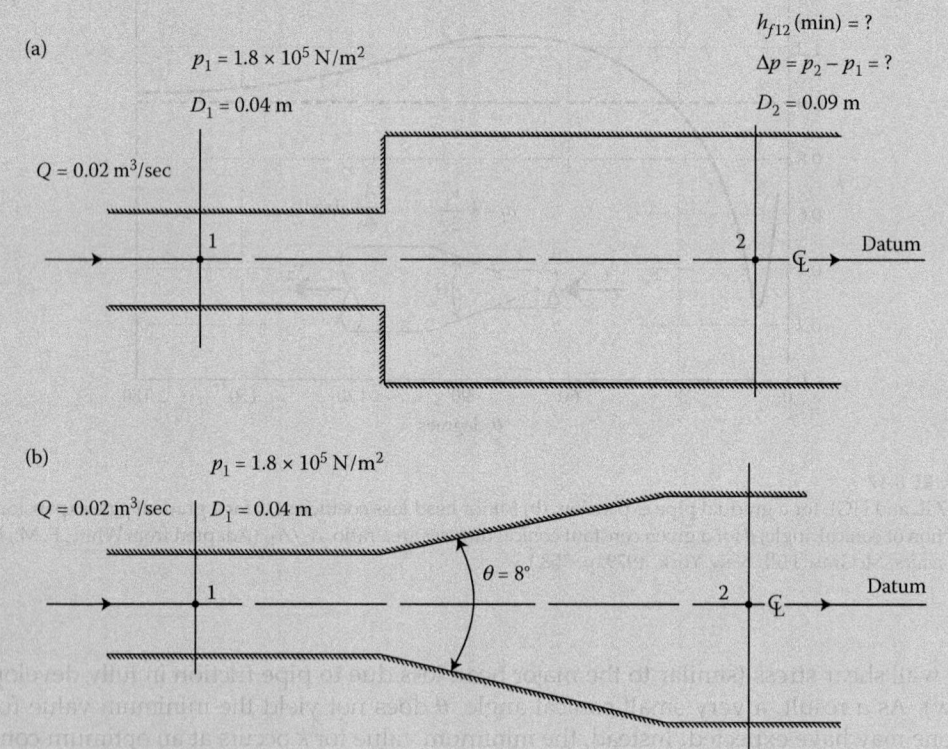

FIGURE EP 8.10
Water flows in a horizontal pipe with pipe expansion to decelerate flow. (a) Sudden pipe expansion, (b) Gradual pipe expansion.

pressure recovery. Two potential solutions include a sudden pipe expansion, as illustrated in Figure EP 8.10a, or a gradual pipe expansion/conical diffuser, as illustrated in Figure EP 8.10b, to be installed in between points 1 and 2. (a) Determine the minor head loss and the pressure rise between points 1 and 2 due to the sudden pipe expansion. (b) Determine the minor head loss and the pressure rise between points 1 and 2 due to the gradual pipe expansion. (c) Conduct a comparison between the two types of pipe expansions.

Mathcad Solution

(a) The minor head loss due to the sudden pipe expansion is computed from Equation 8.73, where the minor head loss coefficient, k is determined from Equation 8.74 (and graphically illustrated in Figure 8.15b, as follows:

$$D_1 := 0.04 \, m \qquad D_2 := 0.09 \, m \qquad Q := 0.02 \, \frac{m^3}{sec} \qquad g := 9.81 \, \frac{m}{sec^2}$$

$$A_1 := \frac{\pi \cdot D_1^2}{4} = 1.257 \times 10^{-3} \, m^2 \qquad v_1 := \frac{Q}{A_1} = 15.915 \, \frac{m}{s}$$

$$A_2 := \frac{\pi \cdot D_2^2}{4} = 6.362 \times 10^{-3} \, m^2 \qquad v_2 = \frac{Q}{A_2} = 3.144 \, \frac{m}{s}$$

$$k_{spe} := \left(1 - \frac{A_1}{A_2}\right)^2 = 0.644 \qquad h_{fspe} := k_{spe}\left(\frac{v_1^2}{2 \cdot g}\right) = 8.314 \, m$$

The pressure rise between point 1 and point 2 for the sudden pipe expansion is computed by applying the energy equation between points 1 and 2 as follows:

$$p_1 := 1.8 \times 10^5 \, \frac{N}{m^2} \qquad \gamma := 9810 \, \frac{N}{m^3} \qquad z_1 := 0 \, m \qquad z_2 := 0 \, m$$

$$v_1 = 15.915 \, \frac{m}{s} \qquad v_2 = 3.144 \, \frac{m}{s}$$

Guess values: $\qquad p_{2spe} := 2 \times 10^5 \, \frac{N}{m^2}$

Given

$$\frac{p_1}{\gamma} + z_1 + \frac{v_1^2}{2 \cdot g} - h_{fspe} = \frac{p_{2spe}}{\gamma} + z_2 + \frac{v_2^2}{2 \cdot g}$$

$$p_{2spe} := Find(p_{2spe}) = 2.202 \times 10^5 \, \frac{N}{m^2}$$

$$\Delta p_{spe} := p_{2spe} - p_1 = 4.015 \times 10^4 \, \frac{N}{m^2}$$

(b) The minor head loss due to the gradual pipe expansion is computed from Equation 8.77, where the minor head loss coefficient, k is determined from Figure 8.17b, which presents the variation of the minor loss coefficient, k with the conical angle, θ for a given constant value of the area ratio, A_2/A_1. The minimum value for k occurs at an optimum conical angle, θ of approximately 8° that results in a long conical diffuser for which the combination of the effect of surface friction and eddying turbulence is at a minimum.

$$\theta := 8 \text{ deg} \qquad\qquad\qquad \frac{A_2}{A_1} = 5.063$$

$$k_{gpe} := 0.28 \qquad\qquad h_{fgpe} := k_{gpe} \cdot \left(\frac{v_1{}^2}{2 \cdot g}\right) = 3.615 \text{ m}$$

The pressure rise between point 1 and point 2 for the gradual pipe expansion is computed by applying the energy equation between points 1 and 2 as follows:

$$p_1 := 1.8 \times 10^5 \ \frac{N}{m^2} \qquad \gamma := 9810 \ \frac{N}{m^3} \qquad\qquad z_1 := 0 \text{ m} \qquad z_2 := 0 \text{ m}$$

$$v_1 = 15.915 \ \frac{m}{s} \qquad v_2 = 3.144 \ \frac{m}{s}$$

Guess values: $\qquad\qquad p_{2gpe} := 2 \times 10^5 \ \dfrac{N}{m^2}$

Given

$$\frac{p_1}{\gamma} + z_1 + \frac{v_1{}^2}{2 \cdot g} - h_{fgpe} = \frac{p_{2gpe}}{\gamma} + z_2 + \frac{v_2{}^2}{2 \cdot g}$$

$$p_{2gpe} := \text{Find}(p_{2gpe}) = 2.662 \times 10^5 \ \frac{N}{m^2}$$

$$\Delta p_{gpe} := p_{2gpe} - p_1 = 8.625 \times 10^4 \ \frac{N}{m^2}$$

(c) The choice of a sudden pipe expansion results in a value for $k = 0.644$, a minor head loss, $h_f = 8.314$ m, and a pressure rise, $\Delta p = 4.015 \times 10^4$ N/m². In comparison, the choice of a gradual pipe expansion, assuming an optimum conical angle, $\theta = 8°$, results in the minimum value for $k = 0.28$ (for the given constant value of $A_2/A_1 = 5.063$); a minor head loss, $h_f = 3.615$ m; and a pressure rise, $\Delta p = 8.625 \times 10^4$ N/m². Therefore, the use of a gradual pipe expansion with an optimum conical angle, $\theta = 8°$ instead of a sudden pipe expansion significantly reduces the minor head loss, which significantly increases the pressure recovery for a given increase in pipe size diameter to $D_2 = 0.09$ m and a given decrease in velocity to $v_2 = 3.144$ m/sec.

8.6.5.6 Gradual Pipe Contractions

The head loss due to a pipe contraction may be significantly reduced by the use of a gradual pipe contraction. A change in the cross section of flow may occur as a gradual contraction as illustrated in Figure 8.18, where there is a gradual decrease in pipe size diameter through a conical contraction/nozzle. Unlike the conical diffuser, the nozzle is less complex and is used to efficiently accelerate the fluid flow. The head loss is due to both the local turbulence, which is caused by flow separation, and pipe friction. As a result, the head loss is so small that it is sometimes neglected in practical engineering problems. Experimental results for the minor head loss coefficient, k due to a nozzle range from $k = 0.02$ for a conical angle, $\theta = 30°$, to $k = 0.07$ for a conical angle, $\theta = 60°$ (Munson et al., 1998). The minor head loss due to a gradual contraction is evaluated using the larger velocity, which is the downstream velocity, v_2 as follows:

$$h_f = k\left(\frac{v_2^2}{2g}\right) \tag{8.78}$$

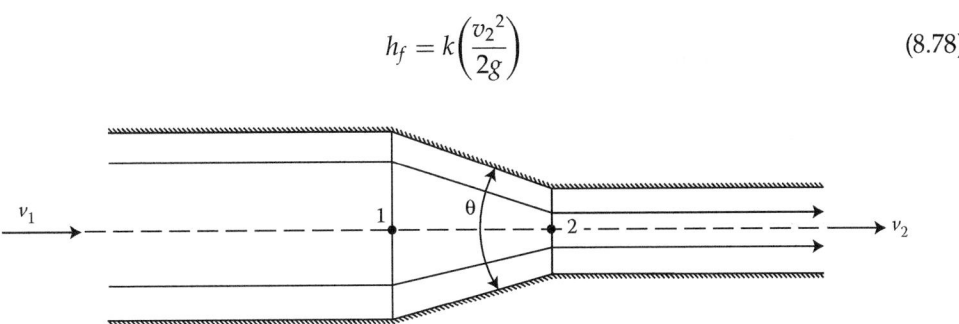

FIGURE 8.18
Gradual pipe contraction/nozzle.

EXAMPLE PROBLEM 8.11

Water flows in a 0.08-m-diameter horizontal pipe at a flowrate of 0.01 m³/sec, as illustrated in Figure EP 8.11. The pressure at point 1 is 2.5×10^5 N/m². A contraction to a 0.03 m diameter pipe at point 2 is sought in order to accelerate the fluid flow (which will result in a pressure drop). Two potential solutions include a sudden pipe contraction, as illustrated in Figure EP 8.11a, or a gradual pipe contraction/conical contraction (nozzle), as illustrated in Figure EP 8.11b, to be installed in between points 1 and 2. (a) Determine the minor head loss and the pressure drop between points 1 and 2 due to the sudden pipe contraction. (b) Determine the minor head loss and the pressure drop between points 1 and 2 due to the gradual pipe contraction. (c) Conduct a comparison between the two types of pipe contractions.

Mathcad Solution

(a) The minor head loss due to the sudden pipe contraction is computed from Equation 8.75, where the minor head loss coefficient, k is a function of the area ratio, A_2/A_1 as illustrated in Figure 8.16b, as follows:

$$D_1 := 0.08\,\text{m} \qquad D_2 := 0.03\,\text{m} \qquad Q := 0.01\,\frac{\text{m}^3}{\text{sec}} \qquad g := 9.81\,\frac{\text{m}}{\text{sec}^2}$$

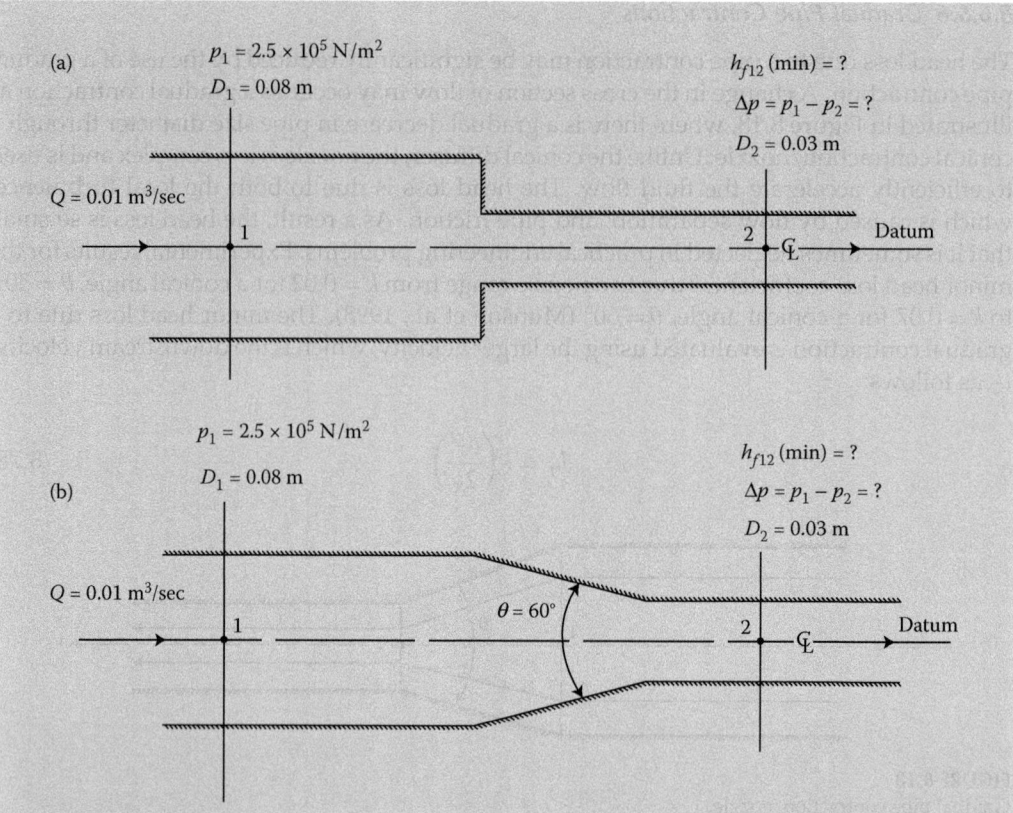

FIGURE EP 8.11
Water flows in a horizontal pipe with pipe contraction to accelerate flow. (a) Sudden pipe contraction, (b) Gradual pipe contraction.

$$A_1 := \frac{\pi \cdot D_1^2}{4} = 5.027 \times 10^{-3}\,\text{m}^2 \qquad\qquad v_1 := \frac{Q}{A_1} = 1.989\,\frac{\text{m}}{\text{s}}$$

$$A_2 := \frac{\pi \cdot D_2^2}{4} = 7.069 \times 10^{-4}\,\text{m}^2 \qquad\qquad v_2 := \frac{Q}{A_2} = 14.147\,\frac{\text{m}}{\text{s}}$$

$$\frac{A_2}{A_1} = 0.141 \qquad k_{spc} := 0.46 \qquad h_{fspc} := k_{spc} \cdot \left(\frac{v_2^2}{2 \cdot g}\right) = 4.692\,\text{m}$$

The pressure drop between point 1 and point 2 for the sudden pipe contraction is computed by applying the energy equation between points 1 and 2 as follows:

$$p_1 := 2.5 \times 10^5\,\frac{\text{N}}{\text{m}^2} \qquad \gamma := 9810\,\frac{\text{N}}{\text{m}^3} \qquad z_1 := 0\,\text{m} \qquad z_2 := 0\,\text{m}$$

$$v_1 = 1.989\,\frac{\text{m}}{\text{s}} \qquad\qquad v_2 = 14.147\,\frac{\text{m}}{\text{s}}$$

Guess values: $\qquad\qquad p_{2spc} := 1 \times 10^5\,\frac{\text{N}}{\text{m}^2}$

Given

$$\frac{p_1}{\gamma} + z_1 + \frac{v_1{}^2}{2 \cdot g} - h_{fspc} = \frac{p_{2spc}}{\gamma} + z_2 + \frac{v_2{}^2}{2 \cdot g}$$

$$p_{2spc} := \text{Find}(p_{2spc}) = 1.059 \times 10^5 \, \frac{N}{m^2}$$

$$\Delta p_{spc} := p_1 - p_{2spc} = 1.441 \times 10^5 \, \frac{N}{m^2}$$

(b) The minor head loss due to the gradual pipe contraction/nozzle is computed from Equation 8.78, where the minor head loss coefficient, k ranges from $k = 0.02$ for a conical angle, $\theta = 30°$, to $k = 0.07$ for a conical angle, $\theta = 60°$. Assuming a conical angle, $\theta = 60°$:

$$\theta := 60 \, \text{deg} \qquad k_{gpc} := 0.07 \qquad h_{fgpc} := k_{gpc} \cdot \left(\frac{v_2{}^2}{2 \cdot g}\right) = 0.714 \, m$$

The pressure drop between point 1 and point 2 for the gradual pipe contraction/nozzle is computed by applying the energy equation between points 1 and 2 as follows:

$$p_1 := 2.5 \times 10^5 \, \frac{N}{m^2} \qquad \gamma := 9810 \, \frac{N}{m^3} \qquad z_1 := 0 \, m \qquad z_2 := 0 \, m$$

$$v_1 = 1.989 \, \frac{m}{s} \qquad v_2 = 14.147 \, \frac{m}{s}$$

Guess values: $\qquad p_{2gpc} := 1 \times 10^5 \, \frac{N}{m^2}$

Given

$$\frac{p_1}{\gamma} + z_1 + \frac{v_1{}^2}{2 \cdot g} - h_{fgpc} = \frac{p_{2gpc}}{\gamma} + z_2 + \frac{v_2{}^2}{2 \cdot g}$$

$$p_{2gpc} := \text{Find}(p_{2gpc}) = 1.449 \times 10^5 \, \frac{N}{m^2}$$

$$\Delta p_{gpc} := p_1 - p_{2gpc} = 1.051 \times 10^5 \, \frac{N}{m^2}$$

(c) The choice of a sudden pipe contraction results in a value for $k = 0.46$; a minor head loss, $h_f = 4.692 \, m$; and a pressure drop, $\Delta p = 1.441 \times 10^5 \, N/m^2$. In comparison, the choice of a gradual pipe contraction/nozzle, assuming a conical angle, $\theta = 60°$, results in a value for $k = 0.07$; a minor head loss, $h_f = 0.714 \, m$; and a pressure drop, $\Delta p = 1.051 \times 10^5 \, N/m^2$. Therefore, the use of a gradual pipe contraction/nozzle with a conical angle, $\theta = 60°$ instead of a sudden pipe contraction significantly reduces the minor head loss, which significantly decreases the pressure drop for a given decrease in pipe size diameter to $D_2 = 0.03$ m and a given increase in velocity to $v_2 = 14.147$ m/sec.

8.6.6 Reaction Forces on Pipe Components

In the case of certain pipe components (such as pipe contractions, elbow, and bends; including jets, vanes, or nozzles) there are reaction forces to compute, which may be solved by applying the integral form of the momentum Equation 8.2 (solved analytically, without the use of dimensional analysis) (see Chapter 5); note that the viscous forces have already been accounted for indirectly through the minor head loss term, $h_{f,minor}$ in the energy equation.

EXAMPLE PROBLEM 8.12

Referring to Example Problem 8.9 above, water flows in a 0.5-m-diameter horizontal pipe at a flowrate of 0.3 m³/sec. A 90° mitered bend is installed in between points 1 and 2 as illustrated in Figure EP 8.12 in order to change the direction of the flowrate by 90°. The pressure at point 1 is 2×10^5 N/m². From the solution of Example Problem 8.9 above, the minor head loss due to the 90° mitered bend without vanes is 0.131 m and with vanes is 0.031 m, and the respective values for the pressure at point 2 are 1.987×10^5 N/m² and 1.997×10^5 N/m². Assume the 90° mitered bend without vanes, and a very short pipe section between points 1 and 2. (a) Determine the reaction force acting on the bend due to the thrust of the flow.

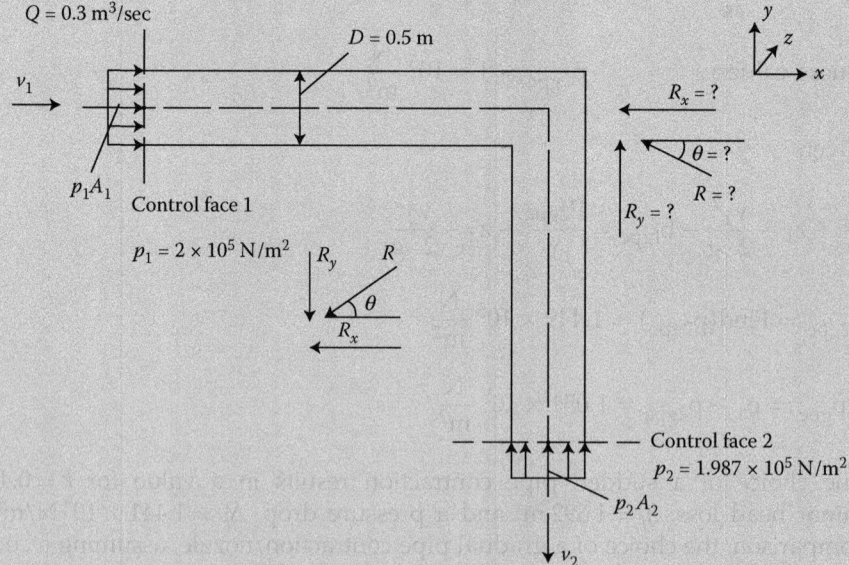

FIGURE EP 8.12
Water flows in a horizontal pipe with a 90° miter bend to change direction of flow.

Mathcad Solution

(a) The reaction force acting on the bend due to the thrust of the flow is solved by applying the integral form of the momentum equation (this time solved analytically, without the use of dimensional analysis), Equation 8.6, noting that the viscous forces have already been accounted for indirectly through the minor head loss term, $h_{f,minor} = 0.131$ m in the energy equation, and that there is no gravitational force component along the x- and y-axes because the pipe is horizontal ($z_1 = z_2 = 0$), as follows:

$$\sum F_x = (p_1 A_1)_x + (p_2 A_2)_x + R_x = \rho Q(v_{2x} - v_{1x})$$

$$\sum F_y = (p_1 A_1)_y + (p_2 A_2)_y + R_y = \rho Q(v_{2y} - v_{1y})$$

$h_{fmin} := 0.131 \text{ m}$ $\qquad$ $D := 0.5 \text{ m}$ $\qquad$ $Q := 0.3 \dfrac{m^3}{\sec}$ $\quad$ $\rho := 1000 \dfrac{kg}{m^3}$

$A := \dfrac{\pi \cdot D^2}{4} = 0.196 \text{ m}^2$ $\quad$ $v := \dfrac{Q}{A} = 1.528 \dfrac{m}{s}$ $\quad$ $z_1 := 0 \text{ m}$ $\qquad$ $z_2 := 0 \text{ m}$

$v_1 := v = 1.528 \dfrac{m}{s}$ $\qquad$ $v_2 := v = 1.528 \dfrac{m}{s}$ $\qquad$ $p_1 := 2 \times 10^5 \dfrac{N}{m^2}$

$p_2 := 1.987 \times 10^5 \dfrac{N}{m^2}$ $\qquad$ $A_1 := A = 0.196 \text{ m}^2$ $\qquad$ $A_2 := A = 0.196 \text{ m}^2$

$p_{1x} := p_1 = 2 \times 10^5 \dfrac{N}{m^2}$ $\quad$ $p_{2x} := 0 \dfrac{N}{m^2}$ $\qquad$ $v_{1x} := v_1 = 1.528 \dfrac{m}{s}$

$v_{2x} := 0 \dfrac{m}{\sec}$ $\qquad$ $p_{1y} := 0 \dfrac{N}{m^2}$ $\qquad$ $p_{2y} := p_2 = 1.987 \times 10^5 \dfrac{N}{m^2}$

$v_{1y} := 0 \dfrac{m}{\sec}$ $\qquad$ $v_{2y} := -v_2 = -1.528 \dfrac{m}{s}$

Guess values: $\qquad$ $R_x := 1000 \text{ N}$ $\qquad$ $R_y := 1000 \text{ N}$

Given

$$p_{1x} \cdot A_1 + p_{2x} \cdot A_2 - R_x = \rho \cdot Q \cdot (v_{2x} - v_{1x})$$

$$p_{2x} \cdot A_2 + p_{2y} \cdot A_2 - R_y = \rho \cdot Q \cdot (v_{2y} - v_{1y})$$

$$\begin{pmatrix} R_x \\ R_y \end{pmatrix} := \text{Find}(R_x, R_y) = \begin{pmatrix} 3.973 \times 10^4 \\ -3.947 \times 10^4 \end{pmatrix} \text{N}$$

$R := \sqrt{R_x^2 + R_y^2} = 5.6 \times 10^4 \text{ N}$ $\qquad\qquad$ $\theta := \text{atan}\left(\dfrac{|R_y|}{|R_x|}\right) = 44.815 \text{ deg}$

Thus, the reaction force, R_x acts to the left as assumed, and the reaction force, R_y acts downward (the opposite of what was assumed and thus the negative sign), as illustrated in Figure EP 8.12.

8.7 Pipe Systems: Major and Minor Head Losses in Real Pipe Flow

Applications of the governing equations for real pipe flow include the analysis and design of pipe systems that include a numerous arrangements of pipes, including: (1) single pipes (Sections 8.7.1 and 8.5), (2) pipes with components (Sections 8.7.2 and 8.6), (3) pipes with a pump or a turbine (Section 8.7.3), (4) pipes in series (Section 8.7.4), (5) pipes in parallel (Section 8.7.5), (6) branching pipes (Section 8.7.6), (7) pipes in a loop (Section 8.7.7), and (8) pipe networks (Section 8.7.8). A single pipe forms the basic component for any pipe system. Furthermore, in the analysis and design of a pipe system, there may be pipe friction, pipe components, transitions, or devices in the pipe flow that will determine the significance of a major head loss, a minor head loss, an added or removed head, and a reaction force.

While the application of the continuity equation will always be necessary, the energy equation and the momentum equation play complementary roles; when one of the two equations of motion breaks down, the other may be used to find the additional unknown quantity. Furthermore, while the energy equation is useful only for a control volume, depending upon the general fluid flow problem, one may apply the continuity and momentum equations using either the integral or differential approach. For the specific case of a single straight pipe section without pipe components (addressed in Section 8.5), only the differential continuity and momentum equations for a fluid system will be required. However, for the specific case of pipes with components, only the integral continuity and momentum equations for a control volume are required (see Section 8.6). As such, for a single straight pipe section, the energy equation, Equation 8.1, models the major head loss, $h_{f,maj}$ due to pipe friction for a given pipe length, while the differential momentum equation, Equation 8.12, models the friction slope, S_f due to pipe friction for a given point in the pipe section. And, for pipes with components, the energy equation, Equation 8.1, models the minor head loss, $h_{f,min}$ due to pipe components, while the integral momentum equation, Equation 8.2, models the reaction force due to certain pipe components (Section 8.6.6). Finally, for pipes with pumps or turbines, the energy equation, Equation 8.1, models the head delivered by a pump, h_{pump} or the head removed by a turbine, $h_{turbine}$.

Thus, the energy equation, Equation 8.1, models the total head loss, $h_{f,total} = h_{f,maj} + h_{f,min}$, which may include both major and minor losses, and are due to flow resistance, and models the pump and turbine heads. The major head loss in pipe flow is due to the frictional losses/flow resistance caused by pipe friction. And, the minor head loss is due to flow resistance due various pipe components—valves, fittings (tees, unions, elbows, and bends), entrances, exits, contractions, and expansions), or pipe flow-measuring devices (orifice, nozzle, or venturi meters). Evaluation of the major and minor head loss flow resistance terms requires a "subset level" application of the governing equations (see Chapters 6 and 7 and Section 8.4.1.2) (see Equations 8.7 and 8.8 for the major head loss equations and Equation 8.9 for the minor head loss equation). And, evaluation of the head delivered to the fluid by a pump, h_{pump}, and the head removed/extracted from the fluid by a turbine, $h_{turbine}$ requires the application of the appropriate power equation (see Section 4.7.2.4) (see Equations 8.10, and 8.11 for pump head, and the turbine head equations, respectively). Finally, the differential momentum equation, Equation 8.12, models the friction slope, S_f due to pipe friction for a given point in the pipe section. Evaluation of the friction slope requires a "subset level" application of the governing equations (see Chapters 6 and 7 and Section 8.4.1.4) (see Equation 8.13 for the Chezy equation).

8.7.1 Single Pipes

A single pipe forms the basic component for any pipe system. Examples of the analysis and design of a single pipe (Problem Types 1, 2, and 3), for which the major head loss term, $h_{f,major}$ due to pipe friction is significant were presented above in Section 8.5 and below for the remaining pipe systems.

8.7.2 Pipes with Components

In the analysis and design of a pipe system consisting of more than a single pipe, including very long pipes with pipe components—valves, fittings (tees, unions, elbows, and bends), entrances, exits, contractions, and expansions, both major losses (Problem Types 1, 2, and 3) and minor losses may be significant, as illustrated by the example below. Examples where the minor losses are computed were presented in Section 8.6. Furthermore, examples where the reaction forces are computed for certain pipe components were presented in Section 8.6.6 and Chapter 5.

EXAMPLE PROBLEM 8.13

Water at 20°C flows from reservoir 1 to reservoir 2 through a 0.5-m-diameter commercial steel pipe at a flowrate of 2 m³/sec, as illustrated in Figure EP 8.13. The pipe system

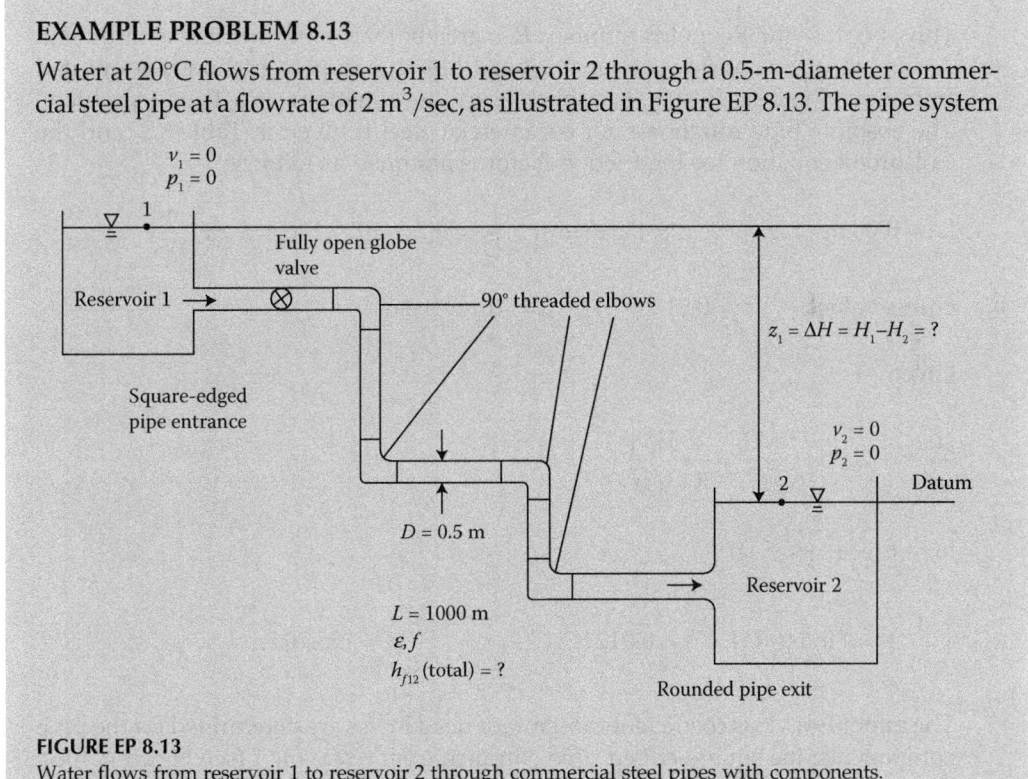

FIGURE EP 8.13
Water flows from reservoir 1 to reservoir 2 through commercial steel pipes with components.

contains two reservoirs, five single pipes with a total length of 1000 m, one fully open globe valve to regulate the flow, and four regular 90° threaded elbows. The pipe entrance from reservoir 1 is square-edged, and the pipe exit to reservoir 2 is rounded. Assume the Darcy–Weisbach friction factor is chosen to model the flow resistance. (a) Determine the flow regime. (b) Determine the total head loss due to pipe friction and pipe components between points 1 and 2. (c) Determine the difference in the total head, ΔH between the two reservoirs in order to overcome the head loss due to pipe friction and pipe components between points 1 and 2.

Mathcad Solution

(a) In order to determine the flow regime, one must compute the Reynolds number, R where the density, ρ and the viscosity, μ for water at 20°C are given in Table A.4 in Appendix A as follows:

$$\rho := 998 \ \frac{kg}{m^3} \qquad\qquad \mu := 1 \times 10^{-3} \times N \ \frac{sec}{m^2} \qquad D := 0.5 \ m$$

$$Q := 2 \ \frac{m^3}{sec} \qquad\qquad A := \frac{\pi \cdot D^2}{4} = 0.196 \ m^2 \qquad v := \frac{Q}{A} = 10.186 \ \frac{m}{s}$$

$$R := \frac{\rho \cdot v \cdot D}{\mu} = 5.083 \times 10^6$$

Thus, because the Reynolds number, R is greater than 4000, the flow is turbulent.

(b) Assuming turbulent flow, the major head loss due to pipe friction is computed from Equation. 8.34 (Darcy–Weisbach head loss equation), as a Type 1 problem. The absolute pipe roughness for commercial steel is given in Table 8.3, and the Colebrook equation for the friction factor is assumed as follows:

$$L := 1000 \ m \qquad\qquad \varepsilon := 0.046 \ mm \qquad \frac{\varepsilon}{D} = 9.2 \times 10^{-5} \qquad g := 9.81 \ \frac{m}{sec^2}$$

Guess values: $f := 0.01$ $h_f := 100 \ m$

Given

$$\frac{1}{\sqrt{f}} = -2 \cdot \log\left(\frac{\varepsilon/D}{3.7} + \frac{2.51}{R \cdot \sqrt{f}}\right)$$

$$h_f = f \cdot \frac{L}{D} \frac{v^2}{2 \cdot g}$$

$$\binom{f}{h_f} := \text{Find}(f, h_f) \qquad f = 0.012 \qquad\qquad\qquad h_f = 128.704 \ m$$

The minor head loss coefficients and minor head losses are determined for the pipe components: the square-edged pipe entrance from reservoir 1 (see Figure 8.10b), the fully open globe valve (see Table 8.10), the four regular 90° threaded elbows (see Table 8.11), and the rounded pipe exit to reservoir 2 (see Figure 8.13c) as follows:

$$k_{ent} := 0.5 \qquad\qquad h_{fent} := k_{ent}\left(\frac{v^2}{2 \cdot g}\right) = 2.644 \text{ m}$$

$$k_{valve} := 10 \qquad\qquad h_{fvalve} := k_{valve}\left(\frac{v^2}{2 \cdot g}\right) = 52.881 \text{ m}$$

$$k_{elbow} := 1.5 \qquad\qquad h_{felbow} := k_{elbow}\left(\frac{v^2}{2 \cdot g}\right) = 7.932 \text{ m}$$

$$k_{exit} := 1 \qquad\qquad h_{fexit} := k_{exit}\left(\frac{v^2}{2 \cdot g}\right) = 5.288 \text{ m}$$

Thus, the total head loss due to pipe friction and pipe components between points 1 and 2 is computed as follows:

$$h_{ftotal} := h_f + h_{fent} + h_{fvalve} + 4 \cdot h_{felbow} + h_{fexit} = 221.246 \text{ m}$$

(c) Assuming the datum is at point 1, the difference in the total head, ΔH between the two reservoirs in order to overcome the head loss due to pipe friction and pipe components between points 1 and 2 is computed by applying the energy equation between points 1 and 2 as follows:

$$\gamma := \rho \cdot g = 9.79 \times 10^3 \frac{kg}{m^2 \cdot s^2} \qquad p_1 := 0 \frac{N}{m^2} \qquad p_2 := 0 \frac{N}{m^2}$$

$$z_2 := 0 \text{ m} \qquad\qquad v_1 := 0 \frac{m}{sec} \qquad v_2 := 0 \frac{m}{sec}$$

Guess values: $\qquad z_1 := 200 \text{ m}$

Given

$$\frac{p_1}{\gamma} + z_1 + \frac{v_1^2}{2 \cdot g} - h_{ftotal} = \frac{p_2}{\gamma} + z_2 + \frac{v_2^2}{2 \cdot g}$$

$$z_1 := \text{Find}(z_1) = 221.246 \text{ m}$$

$$\Delta H := z_1 - z_2 = 221.246 \text{ m}$$

Thus, the elevation of reservoir 1, $z_1 = 221.246$ m must be high enough to overcome the total head loss due to pipe friction and pipe components between points 1 and 2, $h_{ftotal} = 221.246$ m.

8.7.3 Pipes with a Pump or a Turbine

In the analysis and design of a pipe system with a pump or a turbine consisting of more than a single pipe, including very long pipes with pipe components, in addition to the added or removed head, major losses (Problem Types 1, 2, and 3) and minor losses may be significant. Examples of pipes with a pump or a turbine for which there is a head added due to the pump, h_{pump} or a head removed due to the turbine, $h_{turbine}$ are presented below and in Chapter 4.

EXAMPLE PROBLEM 8.14

Gasoline at 20°C is pumped to an elevation of 500 m from reservoir 1 to reservoir 2 through a 0.25-m-diameter copper pipe, as illustrated in Figure EP 8.14. The pipe system contains two reservoirs, two single pipes with a total length of 1500 m, a pump, and one regular 90° threaded elbow. The pipe entrance from reservoir 1 is reentrant, and the pipe exit to reservoir 2 is square-edged. The hydraulic power added to the flow by the pump is 350 kW. Assume the Darcy–Weisbach friction factor is chosen to model the flow resistance. (a) Determine the velocity and the discharge of the flow in the pipe. (b) Determine the flow regime. (c) Determine the total head loss due to pipe friction and pipe components between points 1 and 2. (d) Determine the head added by the pump in order to overcome the difference in elevation between the two reservoirs, and the total head loss due to pipe friction and pipe components between points 1 and 2.

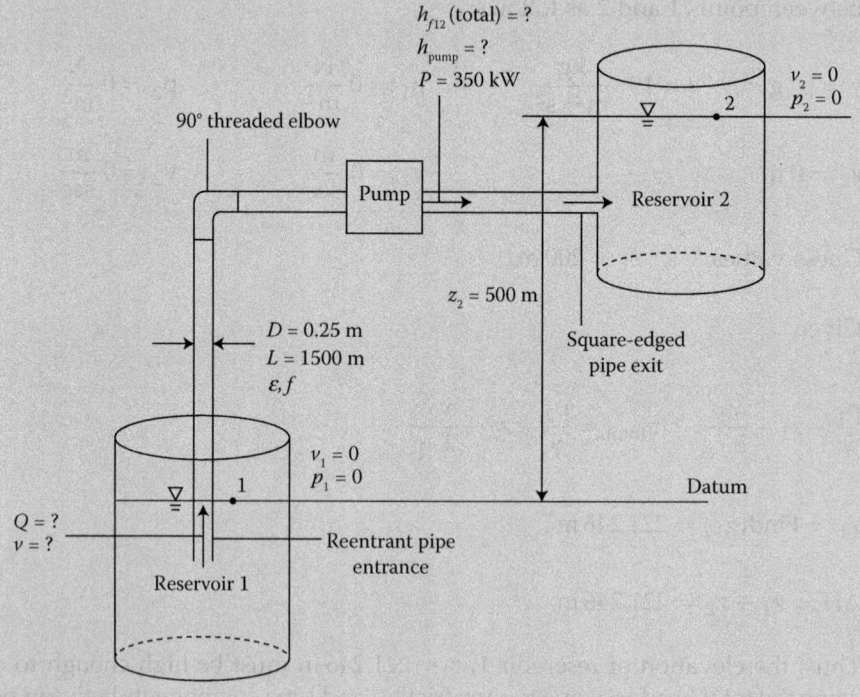

FIGURE EP 8.14
Gasoline is pumped from reservoir 1 to reservoir 2 through copper pipes with components.

Mathcad Solution

(a) Assuming the datum is at point 1, the velocity of flow in the pipe is computed by applying the energy equation between points 1 and 2 as follows:

$$\frac{p_1}{\gamma} + z_1 + \frac{v_1^2}{2g} - h_{fmaj} - h_{fent} - h_{felbow} - h_{fexit} + h_{pump} = \frac{p_2}{\gamma} + z_2 + \frac{v_2^2}{2g}$$

The discharge is computed from the continuity equation. The major head loss due to pipe friction is computed from Equation 8.34 (Darcy–Weisbach head loss equation) as a Type 1 problem. The absolute pipe roughness for copper is given in Table 8.3, and the Colebrook equation for the friction factor is assumed. The minor head loss coefficients and minor head losses are determined for the following pipe components: the reentrant entrance from reservoir 1 (see Figure 8.10a), the regular 90° threaded elbows (see Table 8.11), and the square-edged pipe exit to reservoir 2 (see Figure 8.13b). The head added by the pump is determined from Equation 8.10 (power equation). The density, ρ and the viscosity, μ for gasoline at 20°C are given in Table A.4 in Appendix A. Thus, in total, nine equations are solved simultaneously for nine unknowns as follows:

$$D := 0.25\,m \qquad A := \frac{\pi \cdot D^2}{4} = 0.049\,m^2 \qquad L := 1500\,m$$

$$\varepsilon := 0.0015\,mm \qquad \frac{\varepsilon}{D} = 6 \times 10^{-6} \qquad P := 350\,kW$$

$$\rho := 680\,\frac{kg}{m^3} \qquad \mu := 0.29 \times 10^{-3}\,N\,\frac{sec}{m^2} \qquad g := 9.81\,\frac{m}{sec^2}$$

$$\gamma := \rho \cdot g = 6.671 \times 10^3\,\frac{kg}{m^2 \cdot s^2} \qquad p_1 := 0\,\frac{N}{m^2} \qquad p_2 := 0\,\frac{N}{m^2}$$

$$z_1 := 0\,m \qquad z_2 := 500\,m \qquad v_1 := 0\,\frac{m}{sec} \qquad v_2 := 0\,\frac{m}{sec}$$

$$k_{ent} := 1 \qquad k_{elbow} := 1.5 \qquad k_{exit} := 1$$

Guess values:
$$f := 0.01 \qquad R := 400{,}000 \qquad v := 2\,\frac{m}{sec}$$

$$Q := 1\,\frac{m^3}{sec} \qquad h_f := 100\,m \qquad h_{fent} := 1\,m$$

$$h_{felbow} := 1\,m \qquad h_{fexit} := 1\,m \qquad h_{pump} := 800\,m$$

Given

$$\frac{p_1}{\gamma} + z_1 + \frac{v_1^2}{2 \cdot g} - h_f - h_{fent} - h_{felbow} - h_{fexit} + h_{pump} = \frac{p_2}{\gamma} + z_2 + \frac{v_2^2}{2 \cdot g}$$

$$\frac{1}{\sqrt{f}} = -2 \cdot \log\left(\frac{\varepsilon/D}{3.7} + \frac{2.51}{R \cdot \sqrt{f}}\right) \qquad\qquad R = \frac{\rho \cdot v \cdot D}{\mu}$$

$$h_f = f \cdot \frac{L}{D} \frac{v^2}{2 \cdot g} \qquad h_{fent} = h_{ent} \frac{v^2}{2 \cdot g} \qquad h_{felbow} = k_{elbow} \frac{v^2}{2 \cdot g}$$

$$h_{fexit} = k_{exit} \frac{v^2}{2 \cdot g} \qquad h_{pump} = \frac{P}{\gamma \cdot Q} \qquad\qquad Q = v \cdot A$$

$$\begin{pmatrix} f \\ R \\ v \\ Q \\ h_f \\ h_{fent} \\ h_{felbow} \\ h_{fexit} \\ h_{pump} \end{pmatrix} := \text{Find}\,(f,\, R,\, v,\, Q,\, h_f,\, h_{fent},\, h_{felbow},\, h_{fexit},\, h_{pump})$$

$$f = 0.011 \qquad\qquad R = 1.215 \times 10^6 \qquad\qquad v = 2.072\,\frac{m}{s}$$

$$Q = 0.102\,\frac{m^3}{s} \qquad\qquad h_f = 14.999\,m \qquad\qquad h_{fent} = 0.219\,m$$

$$h_{felbow} = 0.328\,m \qquad h_{fexit} = 0.219\,m \qquad\qquad h_{pump} = 515 \cdot 766\,m$$

(b) Thus, because the Reynolds number, R is greater than 4000, the flow is turbulent.
(c) The total head loss due to pipe friction and pipe components between points 1 and 2 is computed as follows:

$$h_{fftotal} := h_f + h_{fent} + h_{felbow} + h_{fexit} = 15.766\,m$$

(d) Thus, in order to overcome the difference in elevation between the two reservoirs, $\Delta z = 500\,m$, and the total head loss due to pipe friction and pipe components between points 1 and 2, $h_{fftotal} = 15.766\,m$, the head added by the pump, $h_{pump} = \Delta z + h_{fftotal} = 515.766\,m$ (while the hydraulic power output by the pump is 350 kW, the power input to the pump by the motor is a function of the pump efficiency (see Chapter 4)).

EXAMPLE PROBLEM 8.15

Water at 20°C flowing from a reservoir at an elevation of 300 m through a galvanized iron pipe at a velocity of 3 m/sec is used to supply power to a turbine, as illustrated in Figure EP 8.15. The pipe system contains one reservoir, two single pipes with a total length of 1000 m, and a turbine. The pipe entrance from the reservoir is square-edged.

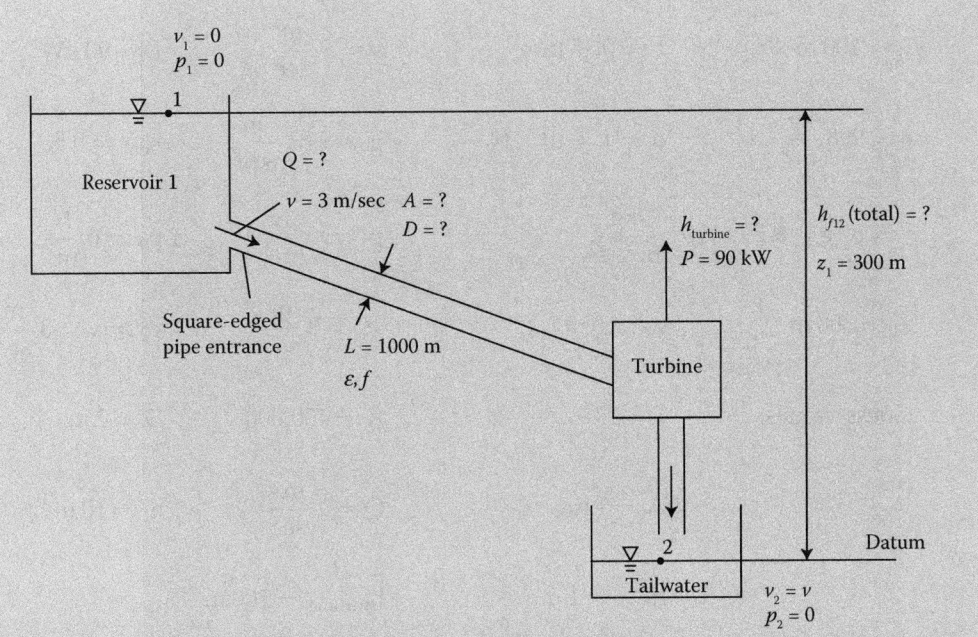

FIGURE EP 8.15
Water flows from reservoir 1 through galvanized iron pipes to supply power to a turbine.

The hydraulic power removed from the flow by the turbine is 90 kW. Assume the Darcy–Weisbach friction factor is chosen to model the flow resistance. (a) Determine the pipe diameter, cross-sectional area, and the discharge. (b) Determine the flow regime. (c) Determine the total head loss due to pipe friction and pipe component between points 1 and 2. (d) Determine the head removed by the turbine in order to supply it with 90 kW of hydraulic power.

Mathcad Solution

(a) Assuming the datum is at point 2, the diameter of the pipe is computed by applying the energy equation between points 1 and 2 as follows:

$$\frac{p_1}{\gamma} + z_1 + \frac{v_1^2}{2g} - h_{fmaj} - h_{fent} - h_{turbine} = \frac{p_2}{\gamma} + z_2 + \frac{v_2^2}{2g}$$

The discharge is computed from the continuity equation. The major head loss due to pipe friction is computed from Equation 8.34 (Darcy–Weisbach head loss equation), as a Type 1 problem. The absolute pipe roughness for galvanized iron is given in Table 8.3, and the Haaland equation for the friction factor is assumed. The minor head loss coefficient and minor head loss is determined for the square-edged entrance from the reservoir (see Figure 8.10b). The head removed by the turbine is determined from Equation 8.11 (power equation). The density, ρ and the viscosity, μ for water at 20°C are given in Table A.4 in Appendix A. Thus, in total, eight equations are solved simultaneously for eight unknowns as follows:

$$L := 100 \, m \qquad \varepsilon := 0.15 \, mm \qquad v := 3 \, \frac{m}{sec} \qquad P := 90 \, kW$$

$$\rho := 998 \, \frac{kg}{m^3} \qquad \mu := 1 \times 10^{-3} \, N \, \frac{sec}{m^2} \qquad g := 9.81 \, \frac{m}{sec^2} \qquad k_{ent} := 0.5$$

$$\gamma := \rho \cdot g = 9.79 \times 10^3 \, \frac{kg}{m^2 \cdot s^2} \qquad\qquad P_1 := 0 \, \frac{N}{m^2} \qquad p_2 := 0 \, \frac{N}{m^2}$$

$$z_1 := 300 \, m \qquad z_2 := 0 \, m \qquad v_1 := 0 \, \frac{m}{sec} \qquad v_2 := v = 3 \, \frac{m}{sec}$$

Guess values: $\qquad f := 0.01 \qquad\qquad R := 400{,}000 \qquad D := 2 \, m$

$$A := 1 \, m^2 \qquad\qquad Q := 1 \, \frac{m^3}{sec} \qquad h_f := 10 \, m$$

$$h_{fent} := 1 \, m \qquad\qquad h_{turbine} := 100 \, m$$

Given

$$\frac{p_1}{\gamma} + z_1 + \frac{v_1^2}{2 \cdot g} - h_f - h_{fent} - h_{turbine} = \frac{p_2}{\gamma} + z_2 + \frac{v^2}{2 \cdot g}$$

$$\frac{1}{\sqrt{f}} = -1.8 \cdot \log\left[\left(\frac{\varepsilon/D}{3.7}\right)^{1.11} + \frac{6.9}{R}\right] \qquad R = \frac{\rho \cdot v \cdot D}{\mu} \qquad A = \frac{\pi \cdot D^2}{4}$$

$$Q = v \cdot A \qquad h_f = f \cdot \frac{L}{D} \frac{v^2}{2 \cdot g} \qquad h_{fent} = k_{ent} \frac{v^2}{2 \cdot g} \qquad h_{turbine} = \frac{P}{\gamma \cdot Q}$$

$$\begin{pmatrix} f \\ R \\ D \\ A \\ Q \\ h_f \\ h_{fent} \\ h_{turbine} \end{pmatrix} := \text{Find} \, (f, R, D, A, Q, h_f, h_{fent}, h_{turbine})$$

$$f = 0.021 \qquad R = 3.932 \times 10^5 \quad D = 0.131 \, m \qquad A = 0.014 \, m^2$$

$$Q = 0.041 \, \frac{m^3}{s} \qquad h_f = 73.097 \, m \qquad h_{fent} = 0.229 \, m \qquad h_{turbine} = 226.215 \, m$$

(b) Thus, because the Reynolds number, **R** is greater than 4000, the flow is turbulent.

(c) The total head loss due to pipe friction and pipe component between points 1 and 2 is computed as follows:

$$h_{ftotal} := h_f + h_{fent} = 73.326\,\text{m}$$

(d) In order to supply the turbine with 90 kW of hydraulic power, the head removed by the turbine, $h_{turbine} = 226.215$ m. Furthermore, one may note that the potential energy provided by the elevation of the reservoir, $z_1 = 300$ m is converted to the head loss due to pipe friction, $h_f = 73.097$ m, the head loss due to the pipe component, $h_{fent} = 0.229$ m, the head supplied to the turbine, $h_{turbine} = 226.215$ m (potential hydroelectric power, which is a function of the turbine efficiency [see Chapter 4]), and the velocity head at point 2, $v^2/2g = 0.459$ m (kinetic energy), where $z_1 = h_f + h_{fent} + h_{turbine} + v^2/2g = 300$ m.

8.7.4 Pipes in Series

Pipes in series include two or more pipes with different lengths and diameters connected to each other in series as illustrated in Figure 8.19. In the analysis and design of a pipe system containing pipes in series, including very long pipes with pipe components, both major losses (Problem Types 1, 2, and 3) and minor losses may be significant. There is only one

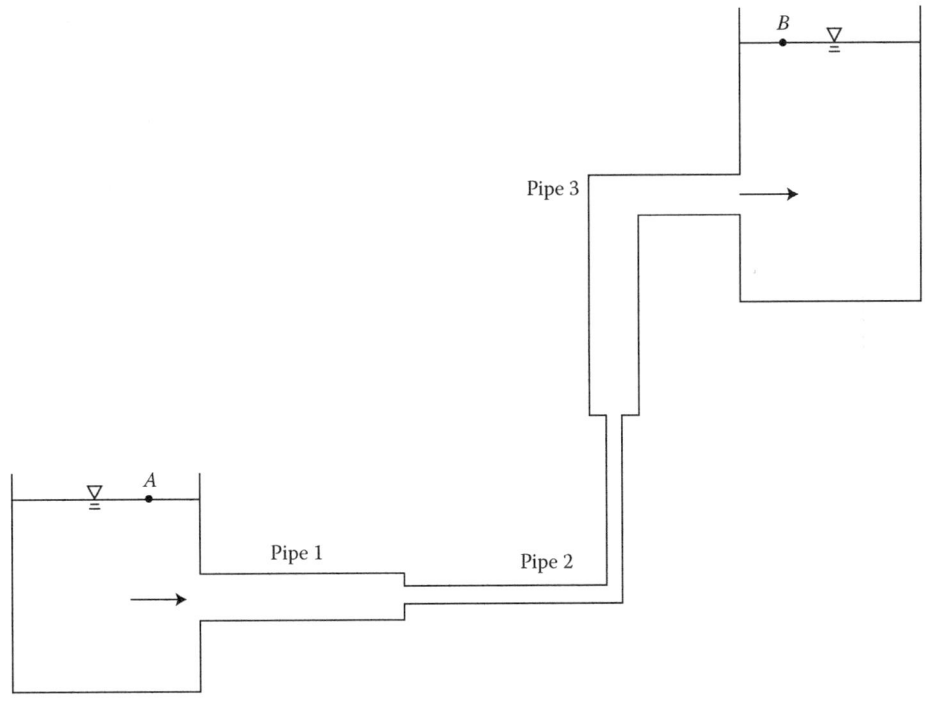

FIGURE 8.19
Pipes in series.

path for the fluid to take in the pipe system. As such, from the conservation of mass principle (integral continuity equation), the flowrate through the entire pipe system remains constant regardless of the diameters of the individual pipes in series as follows:

$$Q = Q_1 = Q_2 = Q_3 = \cdots \tag{8.79}$$

which is the first governing principle. One may note that the above governing flowrate equation, which requires that the flowrate be the same in each pipe in series, establishes the relative velocities and thus the relative head losses in the pipes in series. This governing flowrate equation can be applied to any number of pipes in series. From the conservation of energy principle, all the significant head losses are modeled in the energy equation applied between points A and B as follows:

$$\frac{p_A}{\gamma} + z_A + \frac{v_A^2}{2g} - \sum h_{f,maj} - \sum h_{f,min} = \frac{p_B}{\gamma} + z_B + \frac{v_B^2}{2g} \tag{8.80}$$

which is the second governing principle, where the total head loss in the pipe system between points A and B is equal to the sum of the head losses (major and minor) in the individual pipes and pipe components in series as follows:

$$h_{f total} = \sum h_{f,maj} + \sum h_{f,min} \tag{8.81}$$

The pressure head drop, $\Delta p/\gamma$ between points A and B is due to the change in elevation, Δz; the change in velocity head, $\Delta v^2/2g$; and the total head loss (major and minor) due to the flow resistance (pipe friction and pipe components), $h_{f total}$. Furthermore, because there is only one path for the fluid to take for pipes in series, the total pressure drop across the entire pipe system, Δp is equal to the sum of the pressure drops across each pipe component. Evaluation of the major and minor head loss flow resistance terms requires a "subset level" application of the governing equations (see Chapters 6 and 7 and Section 8.4.1.2). Depending upon whether the flow is laminar or turbulent—and depending upon the desired accuracy, simplicity, and practical constraints in the range of the variables (v, D, and water temperature)—the major head loss due to pipe friction, $h_{f,major}$ is evaluated by the appropriate resistance equation (see Section 8.5). And, depending upon the particular type of pipe component, the minor head loss due to the pipe components, $h_{f,min} = k(v_{ave}^2/2g)$ is evaluated by applying the appropriate head loss coefficient, k (see Section 8.6). Furthermore, because there is a change in the pipe diameter and thus a change in the velocity of the flow, the larger of the two values of the upstream average developed flow velocity, v_{ave} or the downstream average developed flow velocity, v_{ave} (to or from the pipe component/device) is used to compute the associated minor losses, $h_{f,minor}$. Finally, from the conservation of momentum principle, the friction slope, S_f due to pipe friction at a given point in a straight pipe section may be modeled in the differential momentum equation, Equation 8.12, which is the third governing principle. Evaluation of the friction slope requires a "subset level" application of the governing equations (see Chapters 6 and 7 and Section 8.4.1.4) (see Equation 8.13 for the Chezy equation).

EXAMPLE PROBLEM 8.16

Water at 20°C flows from point A to point B through three cast iron pipes in series at a flowrate of 0.5 m³/sec, as illustrated in Figure EP 8.16. The pipe system contains three single pipes, one sudden pipe contraction, and one sudden pipe expansion. The first pipe is 0.7 m in diameter and 100 m long. The second pipe is 0.3 m in diameter and

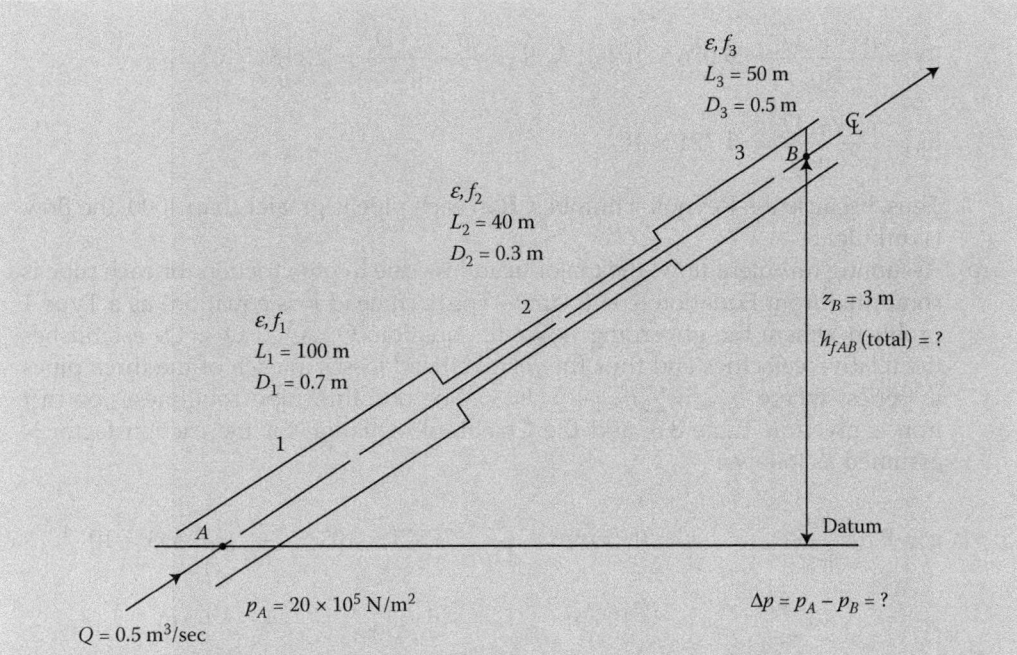

FIGURE EP 8.16
Water flows through cast iron pipes in series with components.

40 m long. The third pipe is 0.5 m in diameter and 50 m long. The pressure at point A is $20 \times 10^5 \, N/m^2$, and the difference in elevation between points A and B is 3 m. Assume the Darcy–Weisbach friction factor is chosen to model the flow resistance. (a) Determine the flow regime. (b) Determine the total head loss due to pipe friction and pipe components between points A and B. (c) Determine the pressure drop, Δp due to the pipe friction, pipe components, difference in elevation, and change in velocity between points A and B.

Mathcad Solution

(a) In order to determine the flow regime, one must compute the Reynolds number, R for each pipe where the density, ρ and the viscosity, μ for water at 20°C are given in Table A.4 in Appendix A as follows:

$$\rho := 998 \, \frac{kg}{m^3} \qquad \mu := 1 \times 10^3 \, N \, \frac{sec}{m^2} \qquad Q := 0.5 \, \frac{m^3}{sec}$$

$$D_1 := 0.7 \, m \qquad A_1 := \frac{\pi \cdot D_1{}^2}{4} = 0.385 \, m^2 \qquad v_1 := \frac{Q}{A_1} = 1.299 \, \frac{m}{s}$$

$$D_2 := 0.3 \, m \qquad A_2 := \frac{\pi \cdot D_2{}^2}{4} = 0.071 \, m^2 \qquad v_2 := \frac{Q}{A_2} = 7.074 \, \frac{m}{s}$$

$$D_3 := 0.5 \, m \qquad A_3 := \frac{\pi \cdot D_3^2}{4} = 0.196 \, m^2 \qquad v_3 := \frac{Q}{A_3} = 2.546 \, \frac{m}{s}$$

$$R_1 := \frac{\rho \cdot v_1 \cdot D_1}{\mu} = 9.076 \times 10^5 \qquad R_2 := \frac{\rho \cdot v_2 \cdot D_2}{\mu} = 2.118 \times 10^6$$

$$R_3 := \frac{\rho \cdot v_3 \cdot D_3}{\mu} = 1.271 \times 10^6$$

Thus, because the Reynolds number, **R** in each pipe is greater than 4000, the flow is turbulent.

(b) Assuming turbulent flow, the major head loss due to pipe friction for each pipe is computed from Equation 8.34 (Darcy–Weisbach head loss equation) as a Type 1 problem, where the governing flowrate equation, $Q = Q_1 = Q_2 = Q_3$ establishes the relative velocities and thus the relative head losses in each of the three pipes in series, where $h_{ftotal} = \sum h_{fmaj} + \sum h_{fmin}$. The absolute pipe roughness for cast iron is given in Table 8.3, and the Colebrook equation for the friction factor is assumed as follows:

$$g := 9.81 \frac{m}{sec^2} \qquad \varepsilon := 0.25 \, mm \qquad \frac{\varepsilon}{D_1} = 3.571 \times 10^{-4} \qquad \frac{\varepsilon}{D_2} = 8.333 \times 10^{-4}$$

$$\frac{\varepsilon}{D_3} = 5 \times 10^{-4} \qquad L_1 := 100 \, m \qquad L_2 := 40 \, m \qquad L_3 := 50 \, m$$

Guess values: $\qquad f_1 := 0.01 \qquad f_2 := 0.01 \qquad f_3 := 0.01$

$$h_{f1} := 10 \, m \qquad h_{f2} := 10 \, m \qquad h_{f3} := 10 \, m$$

Given

$$\frac{1}{\sqrt{f_1}} = -2 \cdot \log\left(\frac{\varepsilon/D_1}{3.7} + \frac{2.51}{R_1 \cdot \sqrt{f_1}}\right) \qquad h_{f_1} = f_1 \cdot \frac{L_1}{D_1} \frac{v_1^2}{2 \cdot g}$$

$$\frac{1}{\sqrt{f_2}} = -2 \cdot \log\left(\frac{\varepsilon/D_2}{3.7} + \frac{2.51}{R_2 \cdot \sqrt{f_2}}\right) \qquad h_{f_2} = f_2 \cdot \frac{L_2}{D_2} \frac{v_2^2}{2 \cdot g}$$

$$\frac{1}{\sqrt{f_3}} = -2 \cdot \log\left(\frac{\varepsilon/D_3}{3.7} + \frac{2.51}{R_3 \cdot \sqrt{f_3}}\right) \qquad h_{f_3} = f_3 \cdot \frac{L_3}{D_3} \frac{v_3^2}{2 \cdot g}$$

$$\begin{pmatrix} f_1 \\ f_2 \\ f_3 \\ h_{f_1} \\ h_{f_2} \\ h_{f_3} \end{pmatrix} := \text{Find}\,(f_1, f_2, f_3, h_{f_1}, h_{f_2}, h_{f_3})$$

$$f_1 = 0.016 \qquad f_2 = 0.019 \qquad f_3 = 0.017$$

$h_{f1} = 0.199 \, \text{m}$ $h_{f2} = 6.447 \, \text{m}$ $h_{f3} = 0.565 \, \text{m}$

Thus, the total major head loss due to pipe friction is computed as follows:

$$h_{fmaj} := h_{f_1} + h_{f_2} + h_{f_3} = 7.212 \, \text{m}$$

The minor head loss coefficients and minor head losses are determined for the sudden pipe contraction (see Figure 8.16b, where k is a function of the area ratio A_2/A_1) and the sudden pipe expansion (see Equation 8.74 or Figure 8.15b, where k is a function of the area ratio A_1/A_2) as follows:

$$\frac{A_2}{A_1} = 0.184 \qquad k_{spc} := 0.44 \qquad h_{fspc} := k_{spc} \cdot \left(\frac{v_2^2}{2 \cdot g} \right) = 1.122 \, \text{m}$$

$$\frac{A_2}{A_3} = 0.36 \qquad k_{spe} := \left(1 - \frac{A_2}{A_3} \right)^2 = 0.41 \quad h_{fspe} := k_{spe} \cdot \left(\frac{v_2^2}{2 \cdot g} \right) = 1.045 \, \text{m}$$

Thus, the total minor head loss due to pipe components is computed as follows:

$$h_{fmin} := h_{fspc} + h_{fspe} = 2.167 \, \text{m}$$

And, the total head loss due to pipe friction and pipe components between points A and B is computed as follows:

$$h_{ftotal} := h_{fmaj} + h_{fmin} = 9.378 \, \text{m}$$

(c) Assuming the datum is at point A, the pressure drop, Δp due to the pipe friction, pipe components, difference in elevation, and change in velocity between points A and B is computed by applying the energy equation between points A and B as follows:

$$\gamma := \rho \cdot g = 9.79 \times 10^3 \, \frac{\text{kg}}{\text{m}^2 \cdot \text{s}^2} \qquad p_A := 20 \times 10^5 \, \frac{\text{N}}{\text{m}^2} \qquad z_A := 0 \, \text{m}$$

$$z_B := 3 \, \text{m} \qquad v_A := v_1 = 1.299 \, \frac{\text{m}}{\text{s}} \qquad v_B := v_3 = 2.546 \, \frac{\text{m}}{\text{s}}$$

Guess value: $\qquad p_B := 10 \times 10^5 \, \frac{\text{N}}{\text{m}^2}$

Given

$$\frac{p_A}{\gamma} + z_A + \frac{v_A^2}{2 \cdot g} - h_{ftotal} = \frac{p_B}{\gamma} + z_B + \frac{v_B^2}{2 \cdot g}$$

$$p_B := \text{Find} \, (p_B) = 1.876 \times 10^6 \, \text{Pa}$$

$$\Delta p := p_A - p_B = 1.236 \times 10^5 \, \text{Pa}$$

EXAMPLE PROBLEM 8.17

Kerosene at 20°C flows from reservoir A to reservoir B through two galvanized iron pipes in series, as illustrated in Figure EP 8.17. The pipe system contains two reservoirs, two single pipes, and one gradual pipe contraction/nozzle with a conical angle, $\theta = 30°$ and thus $k = 0.02$. The pipe entrance from reservoir A is reentrant, and the pipe exit to reservoir B is square-edged. The first pipe is 0.2 m in diameter and 900 m long, and the second pipe is 0.1 m in diameter and 1200 m long. The difference in elevation between points A and B is 50 m. Assume the Darcy–Weisbach friction factor is chosen to model the flow resistance. (a) Determine the velocity in each pipe, and the discharge of the flow in the pipes (b) Determine the flow regime. (c) Determine the total head loss due to pipe friction and pipe components between points A and B.

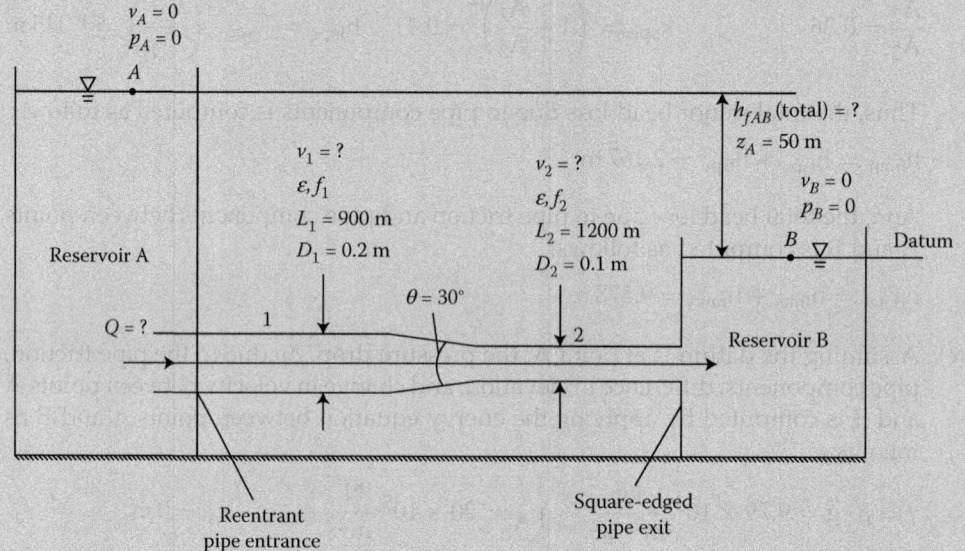

FIGURE EP 8.17
Kerosene flows from reservoir A to reservoir B through galvanized iron pipes with components.

Mathcad Solution

(a) Assuming the datum is at point B, the velocity of flow in each pipe is computed by applying the energy equation between points A and B as follows:

$$\frac{p_A}{\gamma} + z_A + \frac{v_A^2}{2g} - h_{f1maj} - h_{f2maj} - h_{fent} - h_{fgpc} - h_{fexit} = \frac{p_B}{\gamma} + z_B + \frac{v_B^2}{2g}$$

The discharge is computed from the continuity equation. The major head loss due to pipe friction for each pipe is computed from Equation 8.34 (Darcy–Weisbach head loss equation) as a Type 1 problem, where the governing flowrate equation, $Q = Q_1 = Q_2$ establishes the relative velocities and thus the relative head losses in each of the two pipes in series, where $h_{ftotal} = \sum h_{fmaj} + \sum h_{f\,min}$. The absolute pipe roughness for galvanized iron is given in Table 8.3, and the Colebrook equation for the friction factor is assumed. The minor head loss coefficients and minor head losses are determined for the following pipe components: the reentrant entrance from reservoir A (see Figure 8.10a), the gradual pipe contraction/nozzle with a conical angle, $\theta = 30°$ and thus, $k = 0.02$, and the square-edged pipe exit to reservoir B (see Figure 8.13b). The density, ρ and the viscosity, μ for kerosene at 20°C are given in Table A.4 in Appendix A. Thus, in total, 12 equations are solved simultaneously for 12 unknowns as follows:

$$D_1 := 0.2\,m \qquad A_1 := \frac{\pi \cdot D_1^2}{4} = 0.031\,m^2 \qquad L_1 := 900\,m$$

$$D_2 := 0.1\,m \qquad A_2 := \frac{\pi \cdot D_2^2}{4} = 7.854 \times 10^{-3}\,m^2 \qquad L_2 := 1200\,m$$

$$\varepsilon := 0.15\,mm \qquad \frac{\varepsilon}{D_1} = 7.5 \times 10^{-4} \qquad \frac{\varepsilon}{D_2} = 1.5 \times 10^{-3}$$

$$\rho := 808\,\frac{kg}{m^3} \qquad \mu := 1.92 \times 10^{-3}\,N\frac{sec}{m^2} \qquad g := 9.81\,\frac{m}{sec^2}$$

$$\gamma := \rho \cdot g = 7.926 \times 10^3\,\frac{kg}{m^2 \cdot s^2} \qquad P_A := 0\,\frac{N}{m^2} \qquad P_B := 0\,\frac{N}{m^2}$$

$$z_A := 50\,m \qquad z_B := 0\,m \quad v_A := 0\,\frac{m}{sec} \qquad v_B := 0\,\frac{m}{sec}$$

$$k_{ent} := 1 \qquad k_{gpc} := 0.02 \quad k_{exit} := 1$$

Guess values: $\qquad f_1 := 0.01 \qquad R_1 := 400{,}000 \qquad v_1 := 1\,\frac{m}{sec}$

$$f_2 := 0.01 \qquad R_2 := 400{,}000 \qquad v_2 := 2\,\frac{m}{sec}$$

$$Q := 1\,\frac{m^3}{sec} \quad h_{f1} := 10\,m \qquad h_{f2} := 10\,m$$

$$h_{fent} := 1\,m \quad h_{fgpc} := 1\,m \qquad h_{fexit} := 1\,m$$

Given

$$\frac{P_A}{\gamma} + z_A + \frac{v_A^2}{2 \cdot g} - h_{f_1} - h_{f_2} - h_{fent} - h_{fgpc} - h_{fexit} = \frac{P_B}{\gamma} + z_B + \frac{v_B^2}{2 \cdot g}$$

$$\frac{1}{\sqrt{f_1}} = -2 \cdot \log\left(\frac{\varepsilon/D_1}{3.7} + \frac{2.51}{R_1 \cdot \sqrt{f_1}}\right) \qquad\qquad R_1 = \frac{\rho \cdot v_1 \cdot D_1}{\mu}$$

$$\frac{1}{\sqrt{f_2}} = -2 \cdot \log\left(\frac{\varepsilon/D_2}{3.7} + \frac{2.51}{R_2 \cdot \sqrt{f_2}}\right) \qquad\qquad R_2 = \frac{\rho \cdot v_2 \cdot D_2}{\mu}$$

$$h_{f_1} = f_1 \cdot \frac{L_1}{D_1}\frac{v_1^2}{2 \cdot g} \qquad\qquad h_{f_2} = f_2 \cdot \frac{L_2}{D_2}\frac{v_2^2}{2 \cdot g} \qquad\qquad h_{fent} = k_{ent}\frac{v_1^2}{2 \cdot g}$$

$$h_{fgpc} = k_{gpc}\frac{v_2^2}{2 \cdot g} \qquad\qquad h_{fexit} = k_{exit}\frac{v_2^2}{2 \cdot g} \qquad v_1 = \frac{Q}{A_1} \qquad v_2 = \frac{Q}{A_2}$$

$$\begin{pmatrix} f_1 \\ f_2 \\ R_1 \\ R_2 \\ v_1 \\ v_2 \\ Q \\ h_{f_1} \\ h_{f_2} \\ h_{fent} \\ h_{fgpc} \\ h_{fexit} \end{pmatrix} := \text{Find}\,(f_1, f_2, R_1, R_2, v_1, v_2, Q, h_{f_1}, h_{f_2}, h_{fent}, h_{fgpc}, h_{fexit})$$

$$f_1 = 0.024 \qquad\qquad f_2 = 0.024 \qquad\qquad R_1 = 3.811 \times 10^4 \qquad\qquad R_2 = 7.621 \times 10^4$$

$$v_1 = 0.453\,\frac{m}{s} \qquad\qquad\qquad v_2 = 1.811\,\frac{m}{s} \qquad\qquad Q = 0.014\,\frac{m^3}{s}$$

$$h_{f1} = 1.145\,m \qquad\qquad\qquad h_{f2} = 48.674\,m$$

$$h_{fent} = 0.01\,m \qquad\qquad\qquad h_{fgpc} = 3.343 \times 10^{-3}\,m \quad h_{fexit} = 0.167\,m$$

(b) Thus, because the Reynolds number, R is greater than 4000, the flow is turbulent.
(c) The total head loss due to pipe friction and pipe components between points A and B is computed as follows:

$$h_{ftotal} := h_{f_1} + h_{f_2} + h_{fent} + h_{fgpc} + h_{fexit} = 50\,m$$

where the potential energy, $z_A = 50$ m is converted into the total head loss, $h_{ftotal} = 50$ m.

8.7.5 Pipes in Parallel

Pipes in parallel include two or more pipes with different lengths and diameters connected to each other in parallel as illustrated in Figure 8.20. In the analysis and design of a pipe system containing pipes in parallel, including very long pipes with pipe components, both major losses (Problem Types 1, 2, and 3) and minor losses may be significant. There is more than one path for the fluid to take in the pipe system. The fluid flows from a single path (pipe) and enters an upstream junction at point A (inlet). As the fluid leaves the junction at point A, it flows in multiple parallel paths/branches (pipes), where the paths rejoin as they enter a second (downstream) junction at point B (outlet). The fluid flows to a single path (pipe) as it leaves the junction at point B. As such, from the conservation of mass principle (integral continuity equation), the total flowrate entering a junction is equal to the total flowrate leaving the junction for all junctions in the system regardless of the diameters of the individual pipes in parallel as follows:

$$Q = Q_A = Q_1 + Q_2 + Q_3 + \cdots$$
$$Q_1 + Q_2 + Q_3 + \cdots = Q_B = Q \tag{8.82}$$

which is the first governing principle. From the conservation of energy principle, all the significant head losses are modeled in the energy equation applied between points A and B as follows:

$$\frac{p_A}{\gamma} + z_A + \frac{v_A^2}{2g} - \sum h_{f,maj} - \sum h_{f,min} = \frac{p_B}{\gamma} + z_B + \frac{v_B^2}{2g} \tag{8.83}$$

which is the second governing principle, where the total head loss in the pipe system between points A and B is equal to the sum of the head losses (major and minor) in the individual pipes and pipe components in series, for a given path (pipe), as follows:

$$h_{ftotal} = \sum h_{f,maj} + \sum h_{f,min} \tag{8.84}$$

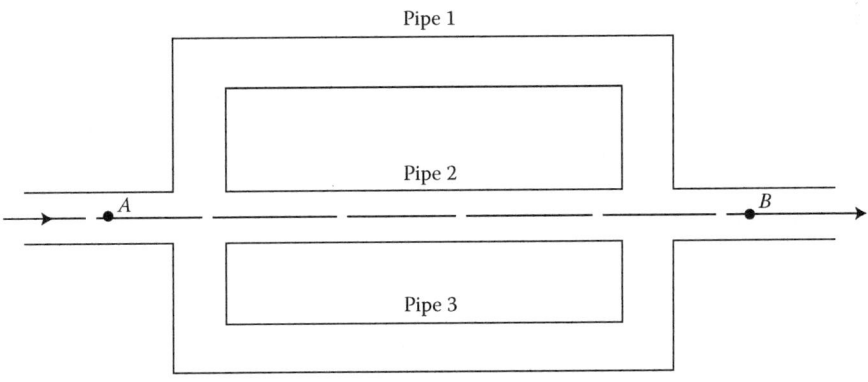

FIGURE 8.20
Pipes in parallel.

The pressure head drop, $\Delta p/\gamma$ between points A and B is due to the change in elevation, Δz; the change in velocity head, $\Delta v^2/2g$; and the total head loss (major and minor) due to the flow resistance (pipe friction and pipe components), h_{ftotal}. Furthermore, although there is more than one path for the fluid to take for pipes in parallel, each path has the same inlet point A and the same outlet point B. Therefore, the total pressure drop across the entire pipe system, Δp_{AB} is equal to the pressure drop across each pipe connecting the two junctions A and B as follows:

$$\Delta p_{AB} = \Delta p_{pipe1} = \Delta p_{pipe2} = \Delta p_{pipe3} = \cdots \qquad (8.85)$$

As a result, the total head loss (major and minor) due to the flow resistance (pipe friction and pipe components) across the entire pipe system $h_{ftotalAB}$ is equal to the total head loss across each pipe connecting the two junction A and B as follows:

$$h_{ftotalAB} = h_{ftotalpipe1} = h_{ftotalpipe2} = h_{ftotalpipe3} = \cdots \qquad (8.86)$$

which is equivalent to Equation 8.85 above in the case where the change in elevation, $\Delta z = 0$ and the change in velocity head, $\Delta v^2/2g = 0$. One may note that the above governing head loss equation, which requires that the head loss be the same in each parallel pipe, establishes the relative velocities and thus the relative flowrates in the parallel pipes. This governing head loss equation can be applied to any number of pipes in parallel. Furthermore, one may note that the application of the above governing head loss equation is accomplished by requiring that the algebraic sum of head losses for a given loop be equal to zero, where a head loss is assumed positive for flow in the clockwise direction and negative for flow in the counterclockwise direction (see Section 8.7.8). Evaluation of the major and minor head loss flow resistance terms requires a "subset level" application of the governing equations (see Chapters 6 and 7 and Section 8.4.1.2). Depending upon whether the flow is laminar or turbulent—and depending upon the desired accuracy, simplicity, and practical constraints in the range of the variables (v, D, and water temperature), the major head loss due to pipe friction, $h_{f,major}$ is evaluated by the appropriate resistance equation (see Section 8.5). And, depending upon the particular type of pipe component, the minor head loss due to the pipe components, $h_{f,min} = k(v_{ave}^2/2g)$ is evaluated by applying the appropriate head loss coefficient, k (see Section 8.6). Furthermore, because there is a change in the pipe diameter and thus a change in the velocity of the flow, the larger of the two values of the upstream average developed flow velocity, v_{ave} or the downstream average developed flow velocity, v_{ave} (to or from the pipe component/device) is used to compute the associated minor losses, $h_{f,minor}$. Finally, from the conservation of momentum principle, the friction slope, S_f due to pipe friction at a given point in a straight pipe section may be modeled in the differential momentum equation, Equation 8.12, which is the third governing principle. Evaluation of the friction slope requires a "subset level" application of the governing equations (see Chapters 6 and 7 and Section 8.4.1.4) (see Equation 8.13 for the Chezy equation).

EXAMPLE PROBLEM 8.18

Water at 20°C flows from point A to point B through three cast iron pipes in parallel at a total flowrate of 2 m³/sec, as illustrated in Figure EP 8.18. The pipes at points A and B are also cast iron and are 1.2 m in diameter. The pipe system contains five single pipes, two threaded branch flow tees, and four regular 90° threaded elbows. The first pipe is 0.8 m in diameter and 1000 m long. The second pipe is 0.7 m in diameter and 600 m long. The third pipe is 0.5 m in diameter and 1000 m long. The pressure at point A is $15 \times 10^5 \, \text{N/m}^2$, and the difference in elevation between points A and B is 0 m. Assume the Darcy–Weisbach friction factor is chosen to model the flow resistance. (a) Determine the pressure drop, Δp due to the pipe friction and pipe components, between points A and B. (b) Determine the total head loss due to pipe friction and pipe components between points A and B. (c) Determine the flow regime. (d) Determine the flow-rate in each of the three pipes.

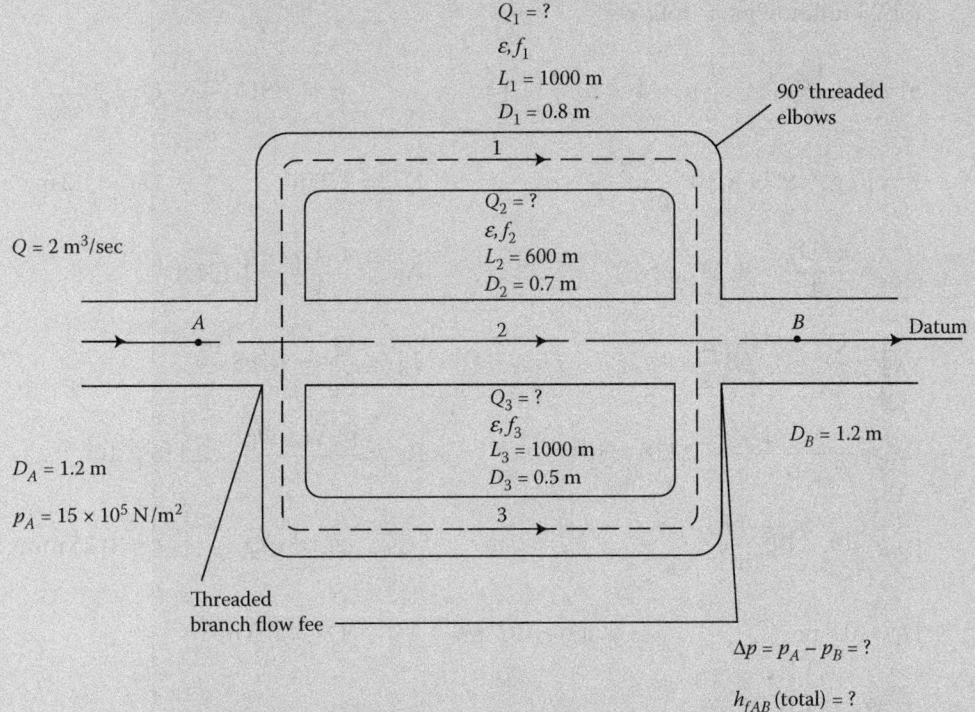

FIGURE EP 8.18
Water flows through cast iron pipes in parallel with components.

Mathcad Solution

(a) Assuming the datum is at points A and B, the pressure drop, Δp due to the pipe friction and pipe components, between points A and B is computed by applying the energy equation between points A and B as follows:

$$\frac{p_A}{\gamma} + z_A + \frac{v_A^2}{2g} - \sum h_{f,maj} - \sum h_{f,min} = \frac{p_B}{\gamma} + z_B + \frac{v_B^2}{2g}$$

The velocity of flow is computed from the continuity equation. The major head loss due to pipe friction for each pipe is computed from Equation 8.34 (Darcy–Weisbach head loss equation) as a Type 1 problem. The absolute pipe roughness for cast iron is given in Table 8.3, and the Colebrook equation for the friction factor is assumed. The minor head loss coefficients and minor head losses are determined for the two threaded branch flow tees, and four regular 90° threaded elbows (see Table 8.11). The density, ρ and the viscosity, μ for water at 20°C are given in Table A.2 in Appendix A. The governing head loss equation, $h_{ftotalAB} = h_{ftotalpipe1} = h_{ftotalpipe2} = h_{ftotalpipe3}$ establishes the relative velocities and thus the relative flowrates in each of the three pipes in parallel, solved as three independent Type 2 problems, where $Q = Q_1 + Q_2 + Q_3$. Thus, in total, 23 equations are solved for 23 unknowns as follows:

$$\rho := 998 \,\frac{kg}{m^3} \qquad \mu := 1 \times 10^{-3}\, N\,\frac{sec}{m^2} \qquad\qquad g := 9.81\,\frac{m}{sec^2} \quad Q := 2\,\frac{m^3}{sec}$$

$$\gamma := \rho \cdot g = 9.79 \times 10^3\,\frac{kg}{m^2 \cdot s^2} \qquad\qquad D_A := 1.2\,m \qquad\qquad D_B := 1.2\,m$$

$$A_A := \frac{\pi \cdot D_A^2}{4} = 1.131\,m^2 \qquad\qquad A_B := \frac{\pi \cdot D_B^2}{4} = 1.131\,m^2$$

$$v_A := \frac{Q}{A_A} = 1.768\,\frac{m}{s} \qquad\qquad v_B := \frac{Q}{A_B} = 1.768\,\frac{m}{s}$$

$$R_A := \frac{\rho \cdot v_A \cdot D_A}{\mu} = 2.118 \times 10^6 \qquad\qquad R_B := \frac{\rho \cdot v_B \cdot D_B}{\mu} = 2.118 \times 10^6$$

$$p_A := 15 \times 10^5\,\frac{N}{m^2} \qquad\qquad z_A := 0\,m \qquad\qquad z_B := 0\,m \qquad \varepsilon := 0.25\,mm$$

$$D_1 := 0.8\,m \qquad\qquad D_2 := 0.7\,m \qquad\qquad D_3 := 0.5\,m$$

$$L_1 := 1000\,m \qquad\qquad L_2 := 600\,m \qquad\qquad L_3 := 1000\,m$$

$$\frac{\varepsilon}{D_1} = 3.125 \times 10^{-4} \qquad\qquad \frac{\varepsilon}{D_2} = 3.571 \times 10^{-4} \qquad \frac{\varepsilon}{D_3} = 5 \times 10^{-4}$$

$$A_1 := \frac{\pi \cdot D_1^2}{4} = 0.503\,m^2 \qquad\qquad\qquad A_2 := \frac{\pi \cdot D_2^2}{4} = 0.385\,m^2$$

$$A_3 := \frac{\pi \cdot D_3^2}{4} = 0.196\,m^2 \qquad\qquad k_{felbow} := 1.5 \qquad k_{tee} := 2$$

Guess values: $f_1 := 0.01$ $f_2 := 0.01$ $f_3 := 0.01$

$$R_1 := 400{,}000 \qquad R_2 := 400{,}000 \qquad R_3 := 400{,}000$$

$$v_1 := 1 \, \frac{m}{sec} \qquad v_2 := 1 \, \frac{m}{sec} \qquad v_3 := 1 \, \frac{m}{sec}$$

$$h_{f1} := 10 \, m \qquad h_{f2} := 10 \, m \qquad h_{f3} := 10 \, m$$

$$h_{fmin1} := 1 \, m \qquad h_{fmin2} := 1 \, m \qquad h_{fmin3} := 1 \, m$$

$$h_{ftotal} := 20 \, m \qquad h_{ftotal1} := 20 \, m \qquad h_{ftotal2} := 20 \, m$$

$$h_{ftotal3} := 20 \, m \qquad p_B := 10 \times 10^5 \, N/m^2$$

$$Q_1 := 1 \, \frac{m^3}{sec} \qquad Q_2 := 1 \, \frac{m^3}{sec} \qquad Q_3 := 1 \, \frac{m^3}{sec}$$

Given

$$\frac{p_A}{\gamma} + z_A + \frac{v_A^2}{2 \cdot g} - h_{ftotal} = \frac{p_B}{\gamma} + z_B + \frac{v_B^2}{2 \cdot g}$$

$$\frac{1}{\sqrt{f_1}} = -2 \cdot \log\left(\frac{\varepsilon/D_1}{3.7} + \frac{2.51}{R_1 \cdot \sqrt{f_1}}\right) \qquad R_1 = \frac{\rho \cdot v_1 \cdot D_1}{\mu} \qquad h_{f1} = f_1 \cdot \frac{L_1}{D_1} \frac{v_1^2}{2 \cdot g}$$

$$\frac{1}{\sqrt{f_2}} = -2 \cdot \log\left(\frac{\varepsilon/D_2}{3.7} + \frac{2.51}{R_2 \cdot \sqrt{f_2}}\right) \qquad R_2 = \frac{\rho \cdot v_2 \cdot D_2}{\mu} \qquad h_{f2} = f_2 \cdot \frac{L_2}{D_2} \frac{v_2^2}{2 \cdot g}$$

$$\frac{1}{\sqrt{f_3}} = -2 \cdot \log\left(\frac{\varepsilon/D_3}{3.7} + \frac{2.51}{R_3 \cdot \sqrt{f_3}}\right) \qquad R_3 = \frac{\rho \cdot v_3 \cdot D_3}{\mu} \qquad h_{f3} = f_3 \cdot \frac{L_3}{D_3} \frac{v_3^2}{2 \cdot g}$$

$$h_{fmin1} = 2 \cdot k_{tee} \left(\frac{v_1^2}{2 \cdot g}\right) + 2 \cdot k_{elbow} \left(\frac{v_1^2}{2 \cdot g}\right) \qquad h_{fmin2} = 2 \cdot k_{tee} \cdot \frac{v_2^2}{2 \cdot g}$$

$$h_{fmin3} = 2 \cdot k_{tee} \left(\frac{v_3^2}{2 \cdot g}\right) + 2 \cdot k_{elbow} \left(\frac{v_3^2}{2 \cdot g}\right)$$

$$h_{ftotal1} = h_{f1} + h_{fmin1} \qquad h_{ftotal2} = h_{f2} + h_{fmin2} \qquad h_{ftotal3} = h_{f3} + h_{fmin3}$$

$$h_{ftotal} = h_{ftotal1} \qquad h_{ftotal} = h_{ftotal2} \qquad h_{ftotal} = h_{ftotal3}$$

$$v_1 = \frac{Q_1}{A_1} \qquad v_2 = \frac{Q_2}{A_2} \qquad v_3 = \frac{Q_3}{A_3} \qquad Q = Q_1 + Q_2 + Q_3$$

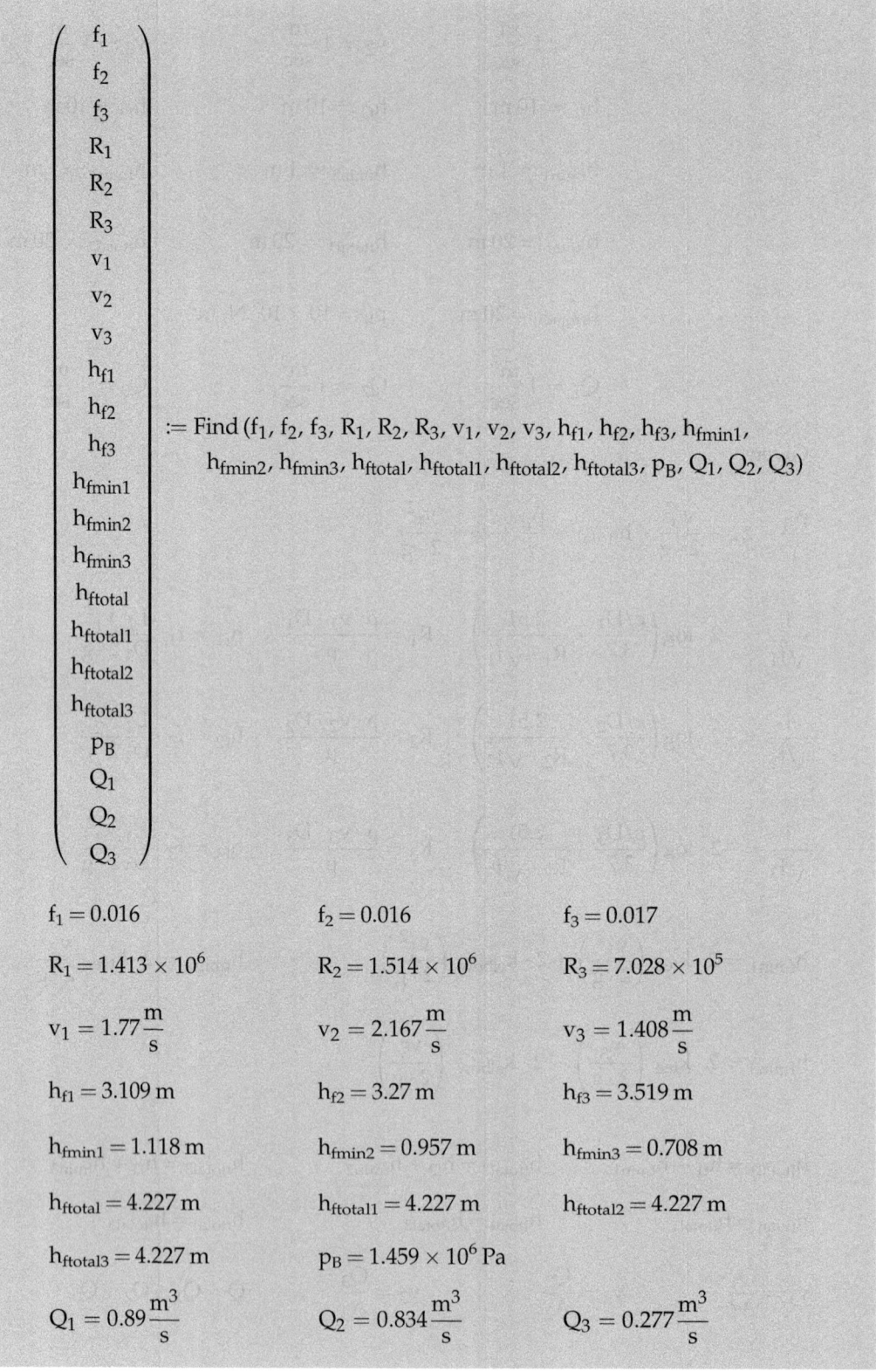

$$
\begin{pmatrix}
f_1 \\
f_2 \\
f_3 \\
R_1 \\
R_2 \\
R_3 \\
v_1 \\
v_2 \\
v_3 \\
h_{f1} \\
h_{f2} \\
h_{f3} \\
h_{fmin1} \\
h_{fmin2} \\
h_{fmin3} \\
h_{ftotal} \\
h_{ftotal1} \\
h_{ftotal2} \\
h_{ftotal3} \\
p_B \\
Q_1 \\
Q_2 \\
Q_3
\end{pmatrix}
:= \text{Find} \, (f_1, f_2, f_3, R_1, R_2, R_3, v_1, v_2, v_3, h_{f1}, h_{f2}, h_{f3}, h_{fmin1},
$$
$$
h_{fmin2}, h_{fmin3}, h_{ftotal}, h_{ftotal1}, h_{ftotal2}, h_{ftotal3}, p_B, Q_1, Q_2, Q_3)
$$

$f_1 = 0.016$

$f_2 = 0.016$

$f_3 = 0.017$

$R_1 = 1.413 \times 10^6$

$R_2 = 1.514 \times 10^6$

$R_3 = 7.028 \times 10^5$

$v_1 = 1.77 \dfrac{m}{s}$

$v_2 = 2.167 \dfrac{m}{s}$

$v_3 = 1.408 \dfrac{m}{s}$

$h_{f1} = 3.109 \, m$

$h_{f2} = 3.27 \, m$

$h_{f3} = 3.519 \, m$

$h_{fmin1} = 1.118 \, m$

$h_{fmin2} = 0.957 \, m$

$h_{fmin3} = 0.708 \, m$

$h_{ftotal} = 4.227 \, m$

$h_{ftotal1} = 4.227 \, m$

$h_{ftotal2} = 4.227 \, m$

$h_{ftotal3} = 4.227 \, m$

$p_B = 1.459 \times 10^6 \, Pa$

$Q_1 = 0.89 \dfrac{m^3}{s}$

$Q_2 = 0.834 \dfrac{m^3}{s}$

$Q_3 = 0.277 \dfrac{m^3}{s}$

where the pressure drop, Δp due to the pipe friction and pipe components, between points A and B is computed as follows:

$$\Delta p := p_A - p_B = 4.138 \times 10^4 \, Pa$$

(b) Thus, the total head loss due to pipe friction and pipe components between points A and B is computed as follows:

$$h_{ftotal} := \frac{\Delta p}{\gamma} = 4.227 \, m$$

(c) Thus, because the Reynolds number, R is greater than 4000, the flow is turbulent.

(d) Thus, the flowrate in each of the three pipes are given as follows:

$$Q_1 = 0.89 \frac{m^3}{s} \qquad Q_2 = 0.834 \frac{m^3}{s} \qquad Q_3 = 0.277 \frac{m^3}{s}$$

where the continuity governing equation is satisfied (checks) as follows:

$$Q_1 + Q_2 + Q_3 = 2 \frac{m^3}{s}$$

EXAMPLE PROBLEM 8.19

Crude oil at 20°C flows from point A to point B through three riveted steel pipes in parallel, as illustrated in Figure EP 8.19. The pipe system contains two reservoirs

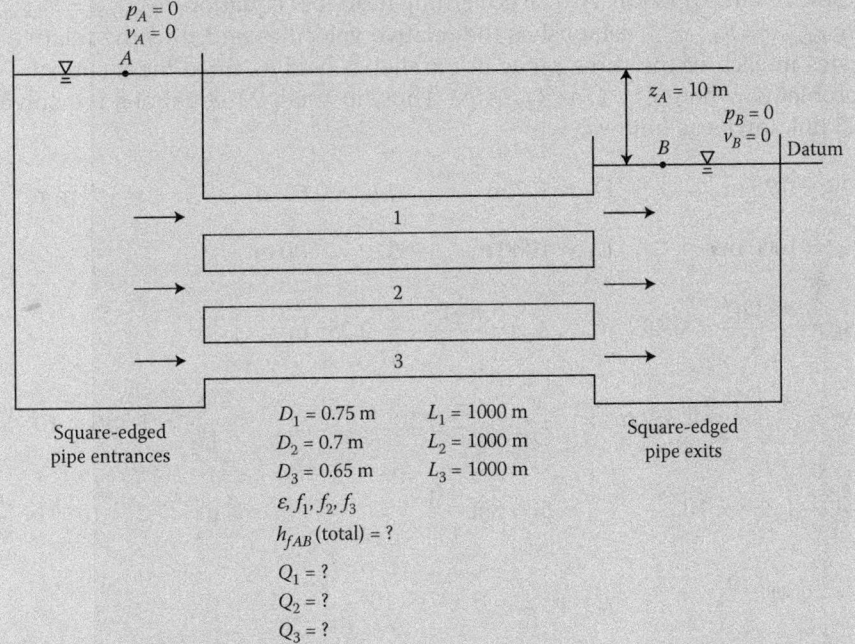

$p_A = 0$
$v_A = 0$
A

$z_A = 10 \, m$
$p_B = 0$
$v_B = 0$
B
Datum

1

2

3

Square-edged pipe entrances

$D_1 = 0.75 \, m \qquad L_1 = 1000 \, m$
$D_2 = 0.7 \, m \qquad L_2 = 1000 \, m$
$D_3 = 0.65 \, m \qquad L_3 = 1000 \, m$
$\varepsilon, f_1, f_2, f_3$
$h_{fAB} (total) = ?$
$Q_1 = ?$
$Q_2 = ?$
$Q_3 = ?$

Square-edged pipe exits

FIGURE EP 8.19
Crude oil from reservoir A to reservoir B through riveted steel pipes in parallel with components.

and three single pipes. The three pipe entrances from reservoir A are square-edged, and the three pipe exits to reservoir B are square-edged. The first pipe is 0.75 m in diameter and 1000 m long. The second pipe is 0.7 m in diameter and 1000 m long, and the third pipe is 0.65 m in diameter and 1000 m long. The difference in elevation between points A and B is 10 m. Assume the Darcy–Weisbach friction factor is chosen to model the flow resistance. (a) Determine the total flowrate and the flowrate in each of the three pipes. (b) Determine the flow regime. (c) Determine the total head loss due to pipe friction and pipe components between points A and B.

Mathcad Solution

(a) Assuming the datum is at point B, the velocity in each pipe is computed by applying the energy equation between points A and B as follows:

$$\frac{p_A}{\gamma} + z_A + \frac{v_A^2}{2g} - \sum h_{f,maj} - \sum h_{f,min} = \frac{p_B}{\gamma} + z_B + \frac{v_B^2}{2g}$$

The discharge is computed from the continuity equation. The major head loss due to pipe friction for each pipe is computed from Equation 8.34 (Darcy–Weisbach head loss equation) as a Type 1 problem. The absolute pipe roughness for riveted steel is given in Table 8.3, and the Colebrook equation for the friction factor is assumed. The minor head loss coefficients and minor head losses are determined for the following pipe components: the three square-edged entrances from reservoir A (see Figure 8.10b) and the three square-edged pipe exits to reservoir B (see Figure 8.13b). The density, ρ and the viscosity, μ for crude oil at 20°C are given in Table A.4 in Appendix A. The governing head loss equation, $h_{ftotalAB} = h_{ftotalpipe1} = h_{ftotalpipe2} = h_{ftotalpipe3}$ establishes the relative velocities and thus the relative flowrates in each of the three pipes in parallel, solved as three independent Type 2 problems, where $Q = Q_1 + Q_2 + Q_3$. Thus, in total, 23 equations are solved for 23 unknowns as follows:

$D_1 := 0.75 \text{ m}$ $D_2 := 0.7 \text{ m}$ $D_3 := 0.65 \text{ m}$ $\varepsilon := 0.9 \text{ mm}$

$L_1 := 1000 \text{ m}$ $L_2 := 1000 \text{ m}$ $L_3 := 1000 \text{ m}$

$A_1 := \dfrac{\pi \cdot D_1^2}{4} = 0.442 \text{ m}^2$ $A_2 := \dfrac{\pi \cdot D_2^2}{4} = 0.385 \text{ m}^2$

$A_3 := \dfrac{\pi \cdot D_3^2}{4} = 0.332 \text{ m}^2$ $\dfrac{\varepsilon}{D_1} = 1.2 \times 10^{-3}$ $\dfrac{\varepsilon}{D_2} = 1.286 \times 10^{-3}$

$\dfrac{\varepsilon}{D_3} = 1.385 \times 10^{-3}$ $\rho := 856 \dfrac{\text{kg}}{\text{m}^3}$ $\mu := 7.2 \times 10^{-3} \text{ N} \dfrac{\text{sec}}{\text{m}^2}$

$g := 9.81 \dfrac{\text{m}}{\text{sec}^2}$ $\gamma := \rho \cdot g = 8.397 \times 10^3 \dfrac{\text{kg}}{\text{m}^2 \cdot \text{s}^2}$ $p_A := 0 \dfrac{\text{N}}{\text{m}^2}$

$p_B := 0 \dfrac{\text{N}}{\text{m}^2}$ $z_A := 10 \text{ m}$ $z_B := 0 \text{ m}$

$$v_A := 0\,\frac{m}{sec} \qquad v_B := 0\,\frac{m}{sec} \qquad k_{ent} := 0.5 \qquad k_{exit} := 1$$

Guess values:
$$f_1 := 0.01 \qquad f_2 := 0.01 \qquad f_3 := 0.01$$

$$R_1 := 400{,}000 \qquad R_2 := 400{,}000 \qquad R_3 := 400{,}000$$

$$v_1 := 1\,\frac{m}{sec} \qquad v_2 := 1\,\frac{m}{sec} \qquad v_3 := 1\,\frac{m}{sec}$$

$$h_{f1} := 10\,m \qquad h_{f2} := 10\,m \qquad h_{f3} := 10\,m$$

$$h_{fmin1} := 1\,m \qquad h_{fmin2} := 1\,m \qquad h_{fmin3} := 1\,m$$

$$h_{ftotal} := 20\,m \qquad h_{ftotal1} := 20\,m \qquad h_{ftotal2} := 20\,m$$

$$h_{ftotal3} := 60\,m \qquad Q := 3\,\frac{m^3}{sec}$$

$$Q_1 := 1\,\frac{m^3}{sec} \qquad Q_2 := 1\,\frac{m^3}{sec} \qquad Q_3 := 1\,\frac{m^3}{sec}$$

Given

$$\frac{p_A}{\gamma} + z_A + \frac{v_A^2}{2 \cdot g} - h_{ftotal} = \frac{p_B}{\gamma} + z_B + \frac{v_B^2}{2 \cdot g}$$

$$\frac{1}{\sqrt{f_1}} = -2 \cdot \log\left(\frac{\varepsilon/D_1}{3.7} + \frac{2.51}{R_1 \cdot \sqrt{f_1}}\right) \qquad R_1 = \frac{\rho \cdot v_1 \cdot D_1}{\mu} \qquad h_{f1} = f_1 \cdot \frac{L_1}{D_1}\frac{v_1^2}{2 \cdot g}$$

$$\frac{1}{\sqrt{f_2}} = -2 \cdot \log\left(\frac{\varepsilon/D_2}{3.7} + \frac{2.51}{R_2 \cdot \sqrt{f_2}}\right) \qquad R_2 = \frac{\rho \cdot v_2 \cdot D_2}{\mu} \qquad h_{f2} = f_2 \cdot \frac{L_2}{D_2}\frac{v_2^2}{2 \cdot g}$$

$$\frac{1}{\sqrt{f_3}} = -2 \cdot \log\left(\frac{\varepsilon/D_3}{3.7} + \frac{2.51}{R_3 \cdot \sqrt{f_3}}\right) \qquad R_3 = \frac{\rho \cdot v_3 \cdot D_3}{\mu} \qquad h_{f3} = f_3 \cdot \frac{L_3}{D_3}\frac{v_3^2}{2 \cdot g}$$

$$h_{fmin1} = k_{ent}\left(\frac{v_1^2}{2 \cdot g}\right) + k_{exit}\left(\frac{v_1^2}{2 \cdot g}\right) \qquad h_{fmin2} = k_{ent}\left(\frac{v_2^2}{2 \cdot g}\right) + k_{exit}\left(\frac{v_2^2}{2 \cdot g}\right)$$

$$h_{fmin3} = k_{ent}\left(\frac{v_3^2}{2 \cdot g}\right) + k_{exit}\left(\frac{v_3^2}{2 \cdot g}\right)$$

$$h_{ftotal1} = h_{f1} + h_{fmin1} \qquad h_{ftotal2} = h_{f2} + h_{fmin2} \qquad h_{ftotal3} = h_{f3} + h_{fmin3}$$

$$h_{ftotal} = h_{ftotal1} \qquad h_{ftotal} = h_{ftotal2} \qquad h_{ftotal} = h_{ftotal3}$$

$$v_1 = \frac{Q_1}{A_1} \qquad v_2 = \frac{Q_2}{A_2} \qquad v_3 = \frac{Q_3}{A_3} \qquad Q = Q_1 + Q_2 + Q_3$$

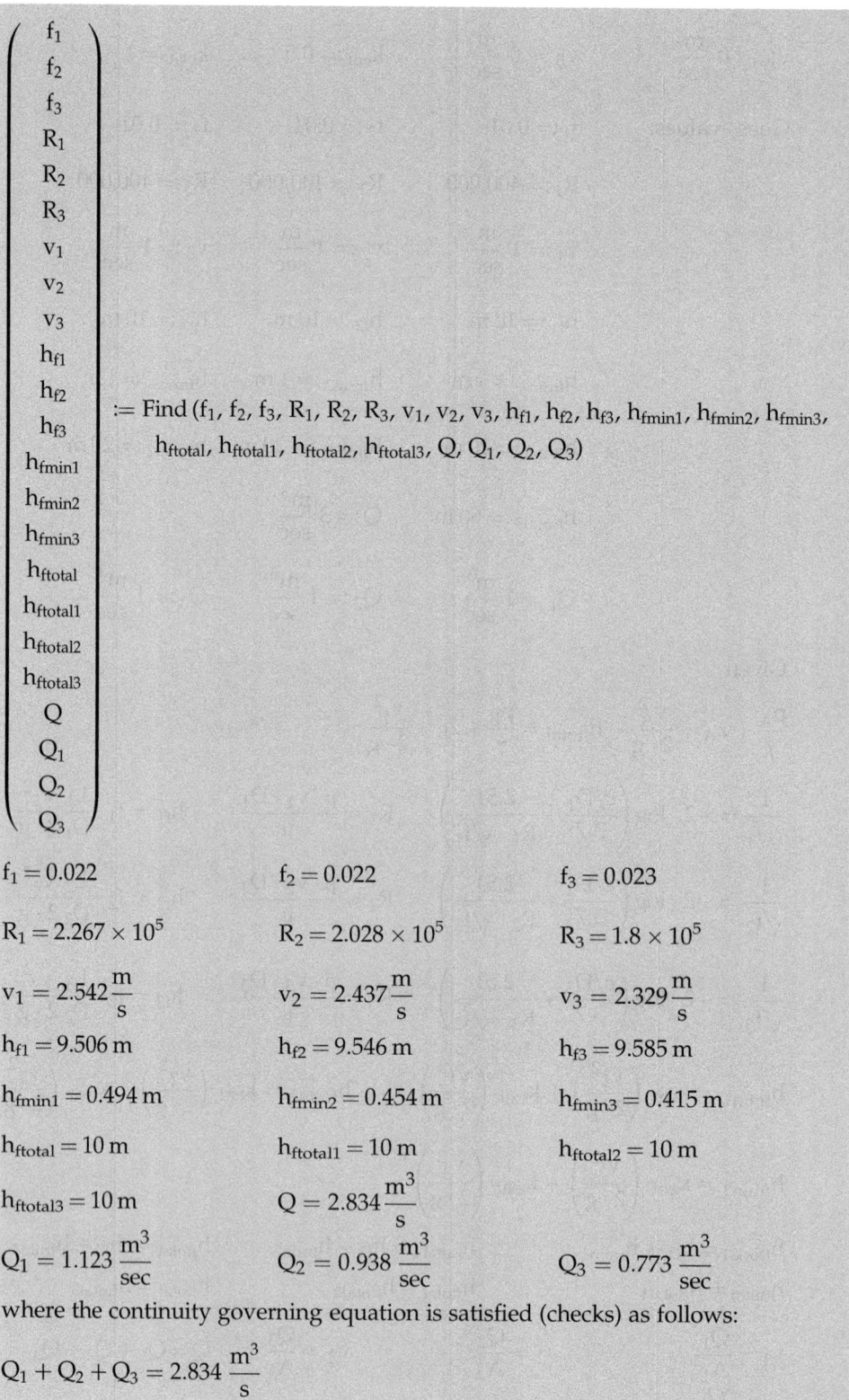

$$\begin{pmatrix} f_1 \\ f_2 \\ f_3 \\ R_1 \\ R_2 \\ R_3 \\ v_1 \\ v_2 \\ v_3 \\ h_{f1} \\ h_{f2} \\ h_{f3} \\ h_{fmin1} \\ h_{fmin2} \\ h_{fmin3} \\ h_{ftotal} \\ h_{ftotal1} \\ h_{ftotal2} \\ h_{ftotal3} \\ Q \\ Q_1 \\ Q_2 \\ Q_3 \end{pmatrix} := \text{Find} \, (f_1, f_2, f_3, R_1, R_2, R_3, v_1, v_2, v_3, h_{f1}, h_{f2}, h_{f3}, h_{fmin1}, h_{fmin2}, h_{fmin3},$$
$$h_{ftotal}, h_{ftotal1}, h_{ftotal2}, h_{ftotal3}, Q, Q_1, Q_2, Q_3)$$

$f_1 = 0.022$ $\qquad\qquad$ $f_2 = 0.022$ $\qquad\qquad$ $f_3 = 0.023$

$R_1 = 2.267 \times 10^5$ $\qquad$ $R_2 = 2.028 \times 10^5$ $\qquad$ $R_3 = 1.8 \times 10^5$

$v_1 = 2.542 \dfrac{m}{s}$ $\qquad\qquad$ $v_2 = 2.437 \dfrac{m}{s}$ $\qquad\qquad$ $v_3 = 2.329 \dfrac{m}{s}$

$h_{f1} = 9.506 \, m$ $\qquad\qquad$ $h_{f2} = 9.546 \, m$ $\qquad\qquad$ $h_{f3} = 9.585 \, m$

$h_{fmin1} = 0.494 \, m$ $\qquad\quad$ $h_{fmin2} = 0.454 \, m$ $\qquad\quad$ $h_{fmin3} = 0.415 \, m$

$h_{ftotal} = 10 \, m$ $\qquad\qquad$ $h_{ftotal1} = 10 \, m$ $\qquad\qquad$ $h_{ftotal2} = 10 \, m$

$h_{ftotal3} = 10 \, m$ $\qquad\qquad$ $Q = 2.834 \dfrac{m^3}{s}$

$Q_1 = 1.123 \dfrac{m^3}{sec}$ $\qquad\quad$ $Q_2 = 0.938 \dfrac{m^3}{sec}$ $\qquad\quad$ $Q_3 = 0.773 \dfrac{m^3}{sec}$

where the continuity governing equation is satisfied (checks) as follows:

$$Q_1 + Q_2 + Q_3 = 2.834 \dfrac{m^3}{s}$$

(b) Thus, because the Reynolds number, R is greater than 4000, the flow is turbulent.

(c) Thus, the total head loss due to pipe friction and pipe components between points A and B is given as follows:

$$h_{ftotal3} = 10\,m$$

where the potential energy, $z_A = 10\,m$ is converted into the total head loss, $h_{ftotal} = 10\,m$.

EXAMPLE PROBLEM 8.20

Water at 20°C is pumped from reservoir A to reservoir B through two cast iron pipes in parallel at a total flowrate of $1.2\,m^3/sec$, as illustrated in Figure EP 8.20. The pipe system contains two reservoirs, two single pipes, a pump, two threaded branch flow tees, and four regular 90° threaded elbows. The pipe entrance from reservoir A is square-edged and the pipe exit to reservoir B is square-edged. The first pipe is 0.8 m in diameter and 600 m long. The second pipe is 0.7 m in diameter and 600 m long. The hydraulic power added to the flow by the pump is 400 kW. Assume the Darcy–Weisbach friction factor is chosen to model the flow resistance. (a) Determine the elevation of reservoir B. (b) Determine the total head loss due to pipe friction and pipe components between points A and B. (c) Determine the head added by the pump in order to

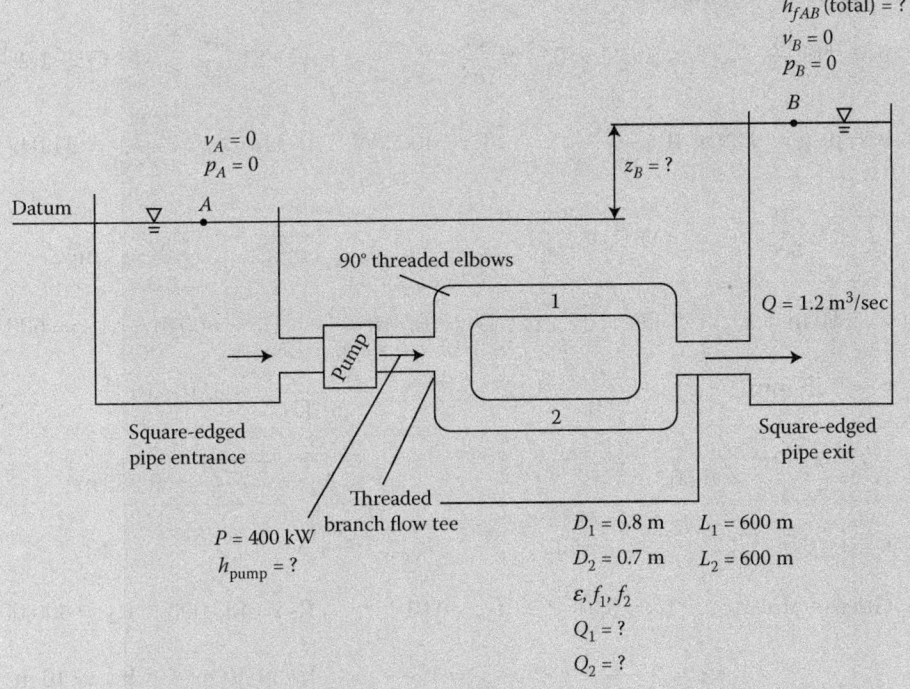

FIGURE EP 8.20

Water is pumped from reservoir A to reservoir B through cast iron pipes in parallel with components.

overcome the difference in elevation between the two reservoirs, and the total head loss due to the pipe friction and pipe components, between points A and B. (d) Determine the flow regime. (e) Determine the flowrate in each of the two pipes.

Mathcad Solution

(a) Assuming the datum is at point A, the elevation of reservoir B is computed by applying the energy equation between points A and B as follows:

$$\frac{p_A}{\gamma} + z_A + \frac{v_A^2}{2g} - \sum h_{f,maj} - \sum h_{f,min} + h_{pump} = \frac{p_B}{\gamma} + z_B + \frac{v_B^2}{2g}$$

The velocity of flow is computed from the continuity equation. The major head loss due to pipe friction for each pipe is computed from Equation 8.34 (Darcy–Weisbach head loss equation) as a Type 1 problem. The absolute pipe roughness for cast iron is given in Table 8.3, and the Colebrook equation for the friction factor is assumed. The minor head loss coefficients and minor head losses are determined for the two threaded branch flow tees and four regular 90° threaded elbows (see Table 8.11). The density, ρ and the viscosity, μ for water at 20°C are given in Table A.4 in Appendix A. The head added by the pump is determined from Equation 8.10 (power equation). The governing head loss equation, $h_{ftotalAB} = h_{ftotalpipe1} = h_{ftotalpipe2}$ establishes the relative velocities and thus the relative flowrates in each of the two pipes in parallel, solved as two independent Type 2 problems, where $Q = Q_1 + Q_2$. Thus, in total, 16 equations are solved for 16 unknowns as follows:

$$\rho := 998 \, \frac{kg}{m^3} \qquad \mu := 1 \times 0^{-3} \, N \frac{sec}{m^2} \qquad g := 9.81 \, \frac{sec}{m^2} \qquad Q := 1.2 \, \frac{m^3}{sec}$$

$$\gamma := \rho \cdot g = 9.79 \times 10^3 \, \frac{kg}{m^2 \cdot s^2} \qquad P := 400 \, kW \qquad h_{pump} := \frac{P}{\gamma \cdot Q} = 34.047 \, m$$

$$v_A := 0 \, \frac{m}{sec} \qquad v_B := 0 \, \frac{m}{sec} \qquad p_A := 0 \, \frac{N}{m^2} \qquad p_B := 0 \, \frac{N}{m^2}$$

$$z_A := 0 \, m \qquad D_1 := 0.8 \, m \qquad D_2 := 0.7 \, m \qquad L_1 := 600 \, m \qquad L_2 := 600 \, m$$

$$\varepsilon := 0.25 \, mm \qquad \frac{\varepsilon}{D_1} = 3.125 \times 10^{-4} \qquad \frac{\varepsilon}{D_2} = 3.571 \times 10^{-4}$$

$$A_1 := \frac{\pi \cdot D_1^2}{4} = 0.503 \, m^2 \qquad\qquad A_2 := \frac{\pi \cdot D_2^2}{4} = 0.385 \, m^2$$

$$k_{ent} := 0.5 \qquad k_{exit} := 1 \qquad k_{elbow} := 1.5 \qquad k_{tee} := 2$$

Guess values: $\qquad f_1 := 0.01 \qquad f_2 := 0.01 \qquad R_1 := 400,000 \qquad R_2 := 400,000$

$$v_1 := 1 \, \frac{m}{sec} \qquad v_2 := 1 \, \frac{m}{sec} \qquad h_{f1} := 10 \, m \qquad h_{f2} := 10 \, m$$

$$h_{fmin1} := 1 \, m \qquad h_{fmin2} := 1 \, m \qquad h_{ftotal} := 20 \, m \qquad h_{ftotal1} := 20 \, m$$

$$h_{ftotal2} := 20\,m \qquad z_B := 200\,m \qquad Q_1 := 1\,\frac{m^3}{sec} \qquad Q_2 := 1\,\frac{m^3}{sec}$$

Given

$$\frac{p_A}{\gamma} + z_A + \frac{v_A^2}{2 \cdot g} - h_{ftotal} + h_{pump} = \frac{p_B}{\gamma} + z_B + \frac{v_B^2}{2 \cdot g}$$

$$\frac{1}{\sqrt{f_1}} = -2 \cdot \log\left(\frac{\varepsilon/D_1}{3.7} + \frac{2.51}{R_1 \cdot \sqrt{f_1}}\right) \qquad R_1 = \frac{\rho \cdot v_1 \cdot D_1}{\mu} \qquad h_{f1} = f_1 \cdot \frac{L_1}{D_1} \frac{v_1^2}{2 \cdot g}$$

$$\frac{1}{\sqrt{f_2}} = -2 \cdot \log\left(\frac{\varepsilon/D_2}{3.7} + \frac{2.51}{R_2 \cdot \sqrt{f_2}}\right) \qquad R_2 = \frac{\rho \cdot v_2 \cdot D_2}{\mu} \qquad h_{f2} = f_2 \cdot \frac{L_2}{D_2} \frac{v_2^2}{2 \cdot g}$$

$$h_{fmin1} = k_{ent}\left(\frac{v_1^2}{2 \cdot g}\right) + 2 \cdot k_{tee}\left(\frac{v_1^2}{2 \cdot g}\right) + 2 \cdot k_{elbow}\left(\frac{v_1^2}{2 \cdot g}\right) + k_{exit}\left(\frac{v_1^2}{2 \cdot g}\right)$$

$$h_{fmin2} = k_{ent}\left(\frac{v_2^2}{2 \cdot g}\right) + 2 \cdot k_{tee}\left(\frac{v_2^2}{2 \cdot g}\right) + 2 \cdot k_{elbow}\left(\frac{v_2^2}{2 \cdot g}\right) + k_{exit}\left(\frac{v_2^2}{2 \cdot g}\right)$$

$$h_{ftotal1} = h_{f1} + h_{fmin1} \qquad\qquad h_{ftotal2} = h_{f2} + h_{fmin2}$$
$$h_{ftotal} = h_{ftotal1} \qquad\qquad h_{ftotal} = h_{ftotal2}$$

$$v_1 = \frac{Q_1}{A_1} \qquad\qquad v_2 = \frac{Q_2}{A_2} \qquad\qquad Q = Q_1 + Q_2$$

$$\begin{pmatrix} f_1 \\ f_2 \\ R_1 \\ R_1 \\ v_1 \\ v_2 \\ h_{f1} \\ h_{f2} \\ h_{fmin1} \\ h_{fmin2} \\ h_{ftotal} \\ h_{ftotal1} \\ h_{ftotal2} \\ z_B \\ Q_1 \\ Q_2 \end{pmatrix} := Find\,(f_1, f_2, R_2, v_1, v_2, h_{f1}, h_{f2}, h_{fmin1}, h_{fmin2}, h_{ftotal}, \\ h_{ftotal1}, h_{ftotal2}, z_B, Q_1, Q_2)$$

$$f_1 = 0.016 \qquad f_2 = 0.016 \qquad R_1 = 1.103 \times 10^6 \quad R_2 = 9.183 \times 10^5$$

$$v_1 = 1.381 \frac{m}{s} \qquad v_2 = 1.314 \frac{m}{s} \qquad h_{f1} = 1.145\,m \qquad h_{f2} = 1.223\,m$$

$$h_{fmin1} = 0.826\,m \quad h_{fmin2} = 0.749\,m \quad h_{ftotal} = 1.972\,m \quad h_{ftotal1} = 1.972\,m$$

$$h_{ftotal2} = 1.972\,m \quad z_B = 32.075\,m \quad Q_1 = 0.694\frac{m^3}{s} \quad Q_2 = 0.506\frac{m^3}{s}$$

where the elevation of reservoir B, z_B is computed as follows:

$$z_B = h_{pump} - h_{ftotal} = 32.075\,m$$

(b) The total head loss due to pipe friction and pipe components between points A and B is given as follows:

$$h_{ftotal} = 1.972\,m$$

(c) The head added by the pump in order to overcome the difference in elevation between the two reservoirs, and the total head loss due to the pipe friction and pipe components, between points A and B is computed as follows:

$$h_{pump} := \frac{P}{\gamma \cdot Q} = 34.047\,m$$

(d) Thus, because the Reynolds number, R is greater than 4000, the flow is turbulent.
(e) Thus, the flowrate in each of the two pipes is given as follows:

$$Q_1 = 0.694\frac{m^3}{s} \qquad\qquad Q_2 = 0.506\frac{m^3}{s}$$

where the continuity governing equation is satisfied (checks) as follows:

$$Q_1 + Q_2 = 1.2\frac{m^3}{s}$$

8.7.6 Branching Pipes

Branching pipes typically include three (or more) pipes with different lengths and diameters connected to each other through one common junction as illustrated in Figure 8.21. One may note that there are two general kinds of branching pipe systems. The first kind of branching pipe system is where the three branching pipes are connected to three reservoirs (three-reservoir problem) as illustrated in Figure 8.21a. The elevation of reservoir A is higher than the elevation of reservoirs B and C, and the elevation of reservoir B is higher than the elevation of reservoir C. Thus, the flow is always from reservoir A to reservoir C. However, the total head at junction J, relative to the total head at point B, is unknown. Therefore, the direction of flow in pipe 2 is also unknown. Thus, depending upon the magnitude of the total head at junction J relative to the magnitude of the total head of reservoir B, the direction of flow in pipe 2 may be: (1) from J to B, (2) no flow, or (3) from B to J. As such, the computations to solve for the unknowns are a bit involved. The second kind of branching pipe system is where the three branching pipes are connected to a water supply source under pressure at point A, as illustrated in Figure 8.21b. The total head at point A (source) is greater than the

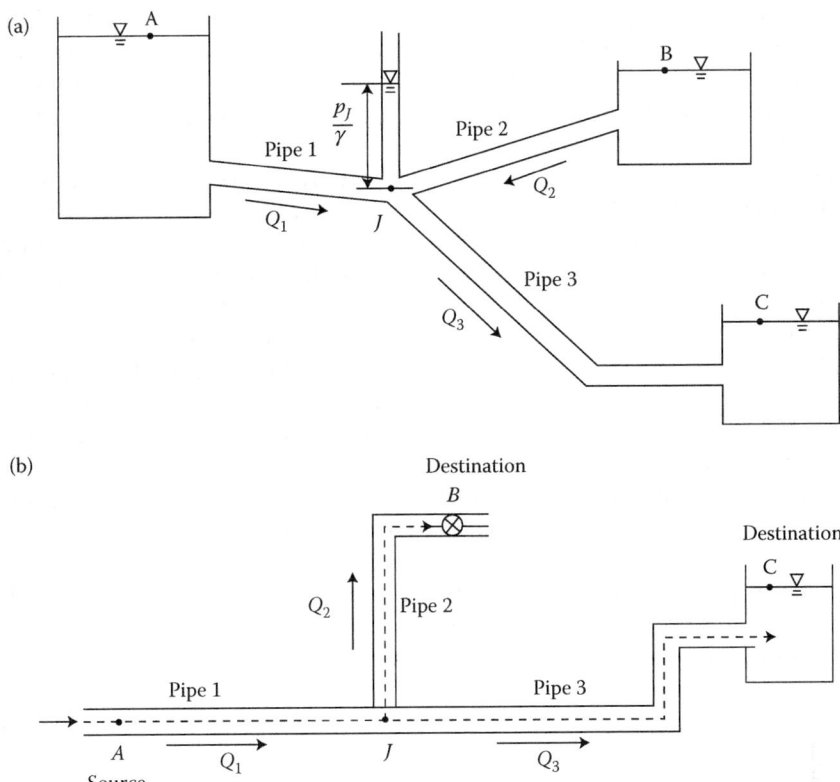

FIGURE 8.21
Branching pipes. (a) Branching pipes connected to three reservoirs (three-reservoir problem), (b) Branching pipes connected to a water supply source under pressure at point A.

total head at junction J. Furthermore, the total head at junction J is larger than the total head at point B (destination), and larger than the total head at point C (destination). Thus, the flow is always from point A to point B and from point A to point C. As such, the computations for the unknowns are a bit more straightforward. Thus, these two kinds of branching pipe systems are addressed separately in the following two sections below (Sections 8.7.6.1 and 8.7.6.2). In general, in the analysis and design of a pipe system containing branching pipes, including very long pipes with pipe components, both major losses (Problem Types 1 and 2) and minor losses may be significant. However, minor losses may be insignificant when pipe components are separated by approximately 1000 diameters and/or when there is a large pipe length-to-pipe diameter ratio. As such, to simplify calculations, minor losses and velocity heads will be assumed negligible in most branching pipe problems.

8.7.6.1 Branching Pipes Connected to Three Reservoirs

The first kind of branching pipe system is where the three branching pipes are connected to three reservoirs (three-reservoir problem) as illustrated in Figure 8.21a. The elevation of reservoir A is higher than the elevation of reservoirs B and C, and the elevation of reservoir B is higher than the elevation of reservoir C. Thus, the flow is always from reservoir A to reservoir C. However, because the total head at junction J, relative to the total head at point B, is unknown, the direction of flow in pipe 2 is also unknown. Thus, the magnitude of the total head at junction J relative to the magnitude of the total head of reservoir B will determine the direction of flow in pipe 2. The governing equations are outlined as follows. The datum is assumed to be below point C. Because points A, B, and C are open to the atmosphere, $p_A = p_B = p_C = 0$, and because the reservoirs are large compared to the pipe diameters, $v_A = v_B = v_C = 0$. Thus, the total head, H for points A, B, and C are equal to their respective elevations: $H_A = z_A$, $H_B = z_B$, and $H_C = z_C$, respectively. Furthermore, the assumption is that the velocity head at junction J is negligible; thus, $v_J^2/2g = 0$. And, the assumption is that the total head at junction J, $H_J = (p_J/\gamma + z_J)$ is lower than the total head of reservoir A, $H_A = z_A$ (thus, flow is always from reservoir A to junction J in pipe 1), and higher than the total head of reservoir C, $H_C = z_C$ (thus, flow is always from junction J to reservoir C in pipe 3). Finally, the assumption is that the total head at junction J, $H_J = (p_J/\gamma + z_J)$ may be: (1) higher than the total head of reservoir B, $H_B = z_B$ (then flow would be from junction J to reservoir B in pipe 2); (2) equal to the total head of reservoir B, $H_B = z_B$ (then there would be no flow in pipe 2); or (3) lower than the total head of reservoir B, $H_B = z_B$ (then flow would be from reservoir B to junction J in pipe 2). Therefore, from the conservation of mass principle (integral continuity equation; first governing principle), the total flowrate entering junction J is equal to the total flowrate leaving junction J regardless of the diameters of the individual branching pipes, which is the first governing equation and is given in Table 8.12 below. From the conservation of energy principle (second governing principle), all the significant head losses are modeled in the energy equation applied between two points, which is the second governing equation(s) and is given in Table 8.12 below. One may note that the appropriate governing continuity, and governing energy equations depend upon the magnitude of total head at junction J, H_J relative to the magnitude of the total head at reservoir B, H_B as presented in Table 8.12 below.

Evaluation of the major (and minor) head loss flow resistance terms requires a "subset level" application of the governing equations (see Chapters 6 and 7 and Section 8.4.1.2). Depending upon whether the flow is laminar or turbulent—and depending upon the desired accuracy, simplicity, and practical constraints in the range of the variables (v, D, and water temperature), the major head loss due to pipe friction, $h_{f,major}$ is evaluated by

TABLE 8.12

Governing Equations for Branching Pipes Connected to Three Reservoirs

Constant Assumptions for H_A, H_J, and H_C	Constant Direction of Flow in Pipe	Governing Continuity Equation at Junction J	Governing Energy Equations
$H_A > H_J$	Flow is always from A to J in pipe 1		$\dfrac{p_A}{\gamma} + z_A + \dfrac{v_A^2}{2g} - h_{f1} = \dfrac{p_J}{\gamma} + z_J + \dfrac{v_J^2}{2g}$
$H_J > H_C$	Flow is always from J to C in pipe 3		$\dfrac{p_J}{\gamma} + z_J + \dfrac{v_J^2}{2g} - h_{f3} = \dfrac{p_C}{\gamma} + z_C + \dfrac{v_C^2}{2g}$
Variable Assumptions for H_J Relative to H_B	**Variable Direction of Flow in Pipe 2**		
$H_J > H_B$	Flow is from J to B in pipe 2	$Q_1 = Q_2 + Q_3$	$\dfrac{p_J}{\gamma} + z_J + \dfrac{v_J^2}{2g} - h_{f2} = \dfrac{p_B}{\gamma} + z_B + \dfrac{v_B^2}{2g}$
$H_J = H_B$	No flow in pipe 2	$Q_1 = Q_3$	
$H_J < H_B$	Flow is from B to J in pipe 2	$Q_1 = Q_2 + Q_3$	$\dfrac{p_B}{\gamma} + z_B + \dfrac{v_B^2}{2g} - h_{f2} = \dfrac{p_J}{\gamma} + z_J + \dfrac{v_J^2}{2g}$

the appropriate resistance equation (see Section 8.5). Finally, from the conservation of momentum principle, the friction slope, S_f due to pipe friction at a given point in a straight pipe section may be modeled in the differential momentum equation, Equation 8.12, which is the third governing principle. Evaluation of the friction slope requires a "subset level" application of the governing equations (see Chapters 6 and 7 and Section 8.4.1.4) (see Equation 8.13 for the Chezy equation).

In the analysis and design of a pipe system containing branching pipes connected to three reservoirs (three-reservoir problem), assuming that the pipe material, pipe lengths, pipe diameters, and the fluid type and temperature are specified, there are three cases that are outlined in Table 8.13.

It is important to note that for each of the three cases outlined above, the four governing equations, that is, the continuity equation at junction J, energy equation between A and J, energy equation between J and C, and energy equation between J and B (or energy equation between B and J), are solved for the four unknowns. A case 1 problem for branching pipes connected to three reservoirs is illustrated in Example 8.21, a case 2 problem for branching pipes connected to three reservoirs is illustrated in Example 8.22, and a case 3 problem for branching pipes connected to three reservoirs is illustrated in Example 8.23.

TABLE 8.13

Summary of Three Cases for Branching Pipes Connected to Three Reservoirs

Case	Given	Find	One Governing Continuity Equation at Junction J	Three Governing Energy Equations
Case 1	$\gamma, \varepsilon, L, D, z$ for two reservoirs, Q for one of these reservoirs	Q for two reservoirs, z for one of these reservoirs, H_J	See Table 8.12 above	See Table 8.12 above
Case 2	$\gamma, \varepsilon, L, D, z$ for two reservoirs, Q for third reservoir	Q for two reservoirs, z for third reservoir, H_J	See Table 8.12 above	See Table 8.12 above
Case 3	$\gamma, \varepsilon, L, D, z$ for all three reservoirs	Q for all three reservoirs, H_J ("classic three-reservoir problem")	See Table 8.12 above	See Table 8.12 above

EXAMPLE PROBLEM 8.21

Water at 20°C flows between reservoirs A, B, and C through three cast iron branching pipes with a common junction J as illustrated in Figure EP 8.21. The pipe system contains three reservoirs and three single pipes. Pipe 1 is 0.45 m in diameter and 1800 m long. Pipe 2 is 0.5 m in diameter and 2000 m long. Pipe 3 is 0.65 m in diameter and 1030 m long. The elevation of reservoir A is 320 m, the elevation of reservoir B is 280 m, and the discharge in pipe 1 is 0.9 m³/sec. (a) Determine the discharge in pipe 2. (b) Determine the discharge in pipe 3. (c) Determine the elevation of reservoir C. (d) Determine the total head for junction J.

Mathcad Solution

(a)–(d) This example demonstrates a case 1 problem. The datum is assumed to be below point C. A constant assumption for branching pipes is that the

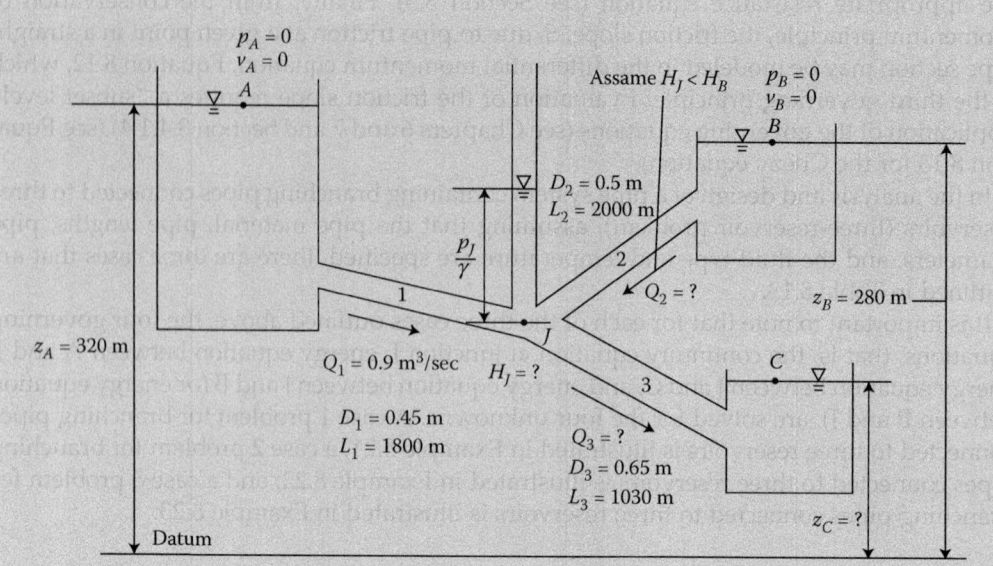

FIGURE EP 8.21
Water flows between reservoirs A, B, and C through three cast iron branching pipes with common junction J (case 1 problem).

total head at junction J, $H_J = (p_J/\gamma + z_J)$ is lower than the total head of reservoir A, $H_A = z_A$; thus, the flow is always from reservoir A to junction J in pipe 1, and thus, the energy equation is applied between points A and J as follows:

$$\frac{p_A}{\gamma} + z_A + \frac{v_A^2}{2g} - h_{f1} = \frac{p_J}{\gamma} + z_J + \frac{v_J^2}{2g}$$

Additionally, another constant assumption for branching pipes is that that the total head at junction J, $H_J = (p_J/\gamma + z_J)$ is higher than the total head of reservoir C, $H_C = z_C$; thus, the flow is always from junction J to reservoir C in pipe 3, and thus, the energy equation is applied between points J and C as follows:

$$\frac{p_J}{\gamma} + z_J + \frac{v_J^2}{2g} - h_{f3} = \frac{p_C}{\gamma} + z_C + \frac{v_C^2}{2g}$$

However, because the total head at junction J, $H_J = (p_J/\gamma + z_J)$, relative to the total head of reservoir B, $H_B = z_B$, is typically unknown for branching pipes, one must make an "assumption." For this problem, it will be "assumed" that the total head at junction J, $H_J = (p_J/\gamma + z_J)$ is lower than the total head of reservoir B, $H_B = z_B$, and thus, the flow is "assumed" to be from reservoir B to junction J in pipe 2; thus, the energy equation is applied between points B and J as follows:

$$\frac{p_B}{\gamma} + z_B + \frac{v_B^2}{2g} - h_{f2} = \frac{p_J}{\gamma} + z_J + \frac{v_J^2}{2g}$$

As such, the corresponding appropriate governing continuity equation is given as follows:

$$Q_1 + Q_2 = Q_3$$

It is important to note that if the correct "assumption" regarding the magnitude of H_J relative to H_B was made, then the flow computed for pipe 2, Q_2 will have a positive sign. However, if an incorrect "assumption" regarding the magnitude of H_J relative to H_B was made, then the flow computed for pipe 2, Q_2 will have a negative sign. Finally, the above four governing equations are solved for the four unknowns: Q_2, Q_3, z_C, and H_J. The velocity of flow is computed from the continuity equation. The major head loss due to pipe friction for each pipe is computed from Equation 8.34 (Darcy–Weisbach head loss equation), as a Type 1 problem. The absolute pipe roughness for cast iron is given in Table 8.3, and the Colebrook equation for the friction factor is assumed. The density, ρ and the viscosity, μ for water at 20°C are given in Table A.2 in Appendix A. The governing continuity $Q_1 + Q_2 = Q_3$ establishes the relative flowrates in each of the three branching pipes, solved as three (actually there are only two unknown discharges) independent Type 2 problems. Thus, in total, 15 equations are solved for 15 unknowns as follows:

$$\rho := 998\,\frac{\text{kg}}{\text{m}^3} \qquad \mu := 1 \times 10^{-3}\,\text{N}\frac{\text{sec}}{\text{m}^2} \qquad g := 9.81\,\frac{\text{m}}{\text{sec}^2} \quad \varepsilon := 0.25\,\text{mm}$$

$$\gamma := \rho \cdot g = 9.79 \times 10^3\,\frac{\text{kg}}{\text{m}^2 \cdot \text{s}^2} \quad D_1 := 0.45\,\text{m} \quad D_2 := 0.5\,\text{m} \quad D_3 := 0.65\,\text{m}$$

$$\frac{\varepsilon}{D_1} = 5.556 \times 10^{-4} \qquad\qquad \frac{\varepsilon}{D_2} = 5 \times 10^{-4} \qquad\qquad \frac{\varepsilon}{D_3} = 3.845 \times 10^{-4}$$

$$L_1 := 1800\,\text{m} \qquad L_2 := 2000\,\text{m} \quad L_3 := 1030\,\text{m} \quad A_1 := \frac{\pi \cdot D_1^2}{4} = 0.159\,\text{m}^2$$

$$A_2 := \frac{\pi \cdot D_2^2}{4} = 0.196\,\text{m}^2 \qquad A_3 := \frac{\pi \cdot D_3^2}{4} = 0.332\,\text{m}^2$$

$$v_A := 0\,\frac{\text{m}}{\text{sec}} \qquad p_A := 0\,\frac{\text{N}}{\text{m}^2} \qquad z_A := 320\,\text{m}$$

$$p_B := 0\,\frac{\text{N}}{\text{m}^2} \qquad v_B := 0\,\frac{\text{m}}{\text{sec}} \qquad z_B := 280\,\text{m}$$

$$p_C := 0\,\frac{\text{N}}{\text{m}^2} \qquad v_C := 0\,\frac{\text{m}}{\text{sec}} \qquad Q_1 := 0.9\,\frac{\text{m}^3}{\text{sec}} \qquad v_1 := \frac{Q_1}{A_1} = 5.659\,\frac{\text{m}}{\text{s}}$$

Guess values: $\quad f_1 := 0.01 \qquad\qquad f_2 := 0.01 \qquad\qquad f_3 := 0.01 \qquad R_1 := 400{,}000$

$$R_2 := 400{,}000 \qquad R_3 := 400{,}000 \qquad h_{f1} := 10\,\text{m} \qquad h_{f2} := 10\,\text{m}$$

$$h_{f3} := 10\,\text{m} \qquad v_2 := 5\,\frac{\text{m}}{\text{sec}} \qquad v_3 := 5\,\frac{\text{m}}{\text{sec}} \qquad Q_2 := 1\,\frac{\text{m}^3}{\text{sec}}$$

$$Q_3 := 1\,\frac{\text{m}^3}{\text{sec}} \qquad z_C := 200\,\text{m} \qquad H_J := 270\,\text{m}$$

Given

$$\frac{p_A}{\gamma} + z_A + \frac{v_A^2}{2 \cdot g} - h_{f1} = H_J \qquad\qquad H_J - h_{f3} = \frac{p_C}{\gamma} + z_C + \frac{v_C^2}{2 \cdot g}$$

$$\frac{p_B}{\gamma} + z_B + \frac{v_B^2}{2 \cdot g} - h_{f2} = H_J \qquad\qquad Q_1 + Q_2 = Q_3 \qquad v_2 = \frac{Q_2}{A_2} \qquad v_3 = \frac{Q_3}{A_3}$$

$$\frac{1}{\sqrt{f_1}} = -2 \cdot \log\left(\frac{\varepsilon/D_1}{3.7} + \frac{2.51}{R_1 \cdot \sqrt{f_1}}\right) \qquad R_1 = \frac{\rho \cdot v_1 \cdot D_1}{\mu} \qquad h_{f1} = f_1 \frac{L_1}{D_1} \frac{v_1^2}{2 \cdot g}$$

$$\frac{1}{\sqrt{f_2}} = -2 \cdot \log\left(\frac{\varepsilon/D_2}{3.7} + \frac{2.51}{R_2 \cdot \sqrt{f_2}}\right) \qquad R_2 = \frac{\rho \cdot v_2 \cdot D_2}{\mu} \qquad h_{f2} = f_2 \frac{L_2}{D_2} \frac{v_2^2}{2 \cdot g}$$

$$\frac{1}{\sqrt{f_3}} = -2 \cdot \log\left(\frac{\varepsilon/D_3}{3.7} + \frac{2.51}{R_3 \cdot \sqrt{f_3}}\right) \qquad R_3 = \frac{\rho \cdot v_3 \cdot D_3}{\mu} \qquad h_{f3} = f_3 \frac{L_3}{D_3} \frac{v_3^2}{2 \cdot g}$$

$$\begin{pmatrix} f_1 \\ f_2 \\ f_3 \\ R_1 \\ R_2 \\ R_3 \\ h_{f1} \\ h_{f2} \\ h_{f3} \\ v_2 \\ v_3 \\ Q_2 \\ Q_3 \\ z_C \\ H_J \end{pmatrix} := \text{Find}\,(f_1, f_2, f_3, R_1, R_2, R_3, h_{f1}, h_{f2}, h_{f3}, v_2, v_3, Q_2, Q_3, z_C, H_J)$$

$f_1 = 0.017 \qquad\qquad f_2 = 0.017 \qquad\qquad f_3 = 0.016 \qquad\qquad R_1 = 2.541 \times 10^6$

$R_2 = 2.293 \times 10^6 \qquad R_3 = 3.524 \times 10^6 \qquad h_{f1} = 112.893\,\text{m} \qquad h_{f2} = 72.893\,\text{m}$

$h_{f3} = 37.983\,\text{m} \qquad v_2 = 4.596\,\dfrac{\text{m}}{\text{s}} \qquad v_3 = 5.432\,\dfrac{\text{m}}{\text{s}} \qquad Q_2 = 0.902\,\dfrac{\text{m}^3}{\text{s}}$

$Q_3 = 1.802\,\dfrac{\text{m}^3}{\text{s}} \qquad z_C = 169.123\,\text{m} \qquad H_J = 207.107\,\text{m}$

Thus, because the flow computed for pipe 2, Q_2 has a positive sign, the correct "assumption" regarding the magnitude of H_J relative to H_B was made. Furthermore, the governing continuity equation is satisfied (checks) as follows:

$$Q_1 + Q_2 = 1.802 \, \frac{m^3}{s}$$

EXAMPLE PROBLEM 8.22

Water at 20°C flows between reservoirs A, B, and C through three concrete branching pipes with a common junction J as illustrated in Figure EP 8.22. The pipe system contains three reservoirs and three single pipes. Pipe 1 is 0.5 m in diameter and 1200 m long. Pipe 2 is 0.65 m in diameter and 900 m long. Pipe 3 is 0.8 m in diameter and 750 m long. The elevation of reservoir A is 330 m, the elevation of reservoir B is 270 m, and the discharge in pipe 3 is 0.85 m³/sec. (a) Determine the discharge in pipe 1. (b) Determine the discharge in pipe 2. (c) Determine the elevation of reservoir C. (d) Determine the total head for junction J.

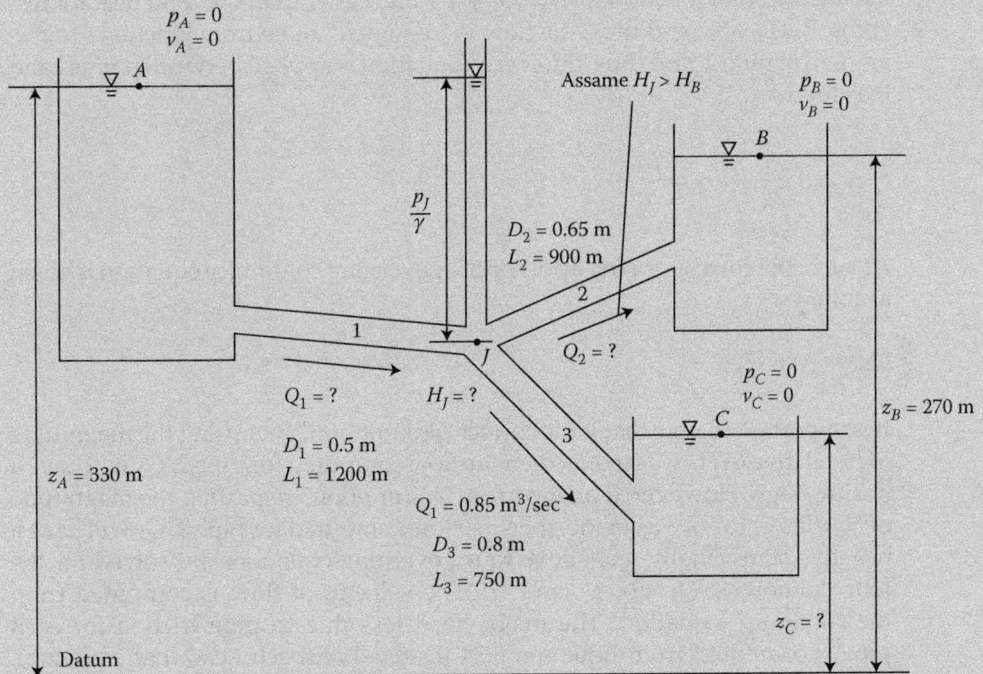

FIGURE EP 8.22
Water flows between reservoirs A, B, and C through three concrete branching pipes with common junction J (case 2 problem).

Mathcad Solution

(a)–(d) This example demonstrates a case 2 problem. The datum is assumed to be below point C. A constant assumption for branching pipes is that the total head at junction J, $H_J = (p_J/\gamma + z_J)$ is lower than the total head of reservoir A, $H_A = z_A$, thus, the flow is always from reservoir A to junction J in pipe 1, and thus, the energy equation is applied between points A and J as follows:

$$\frac{p_A}{\gamma} + z_A + \frac{v_A^2}{2g} - h_{f1} = \frac{p_J}{\gamma} + z_J + \frac{v_J^2}{2g}$$

Additionally, another constant assumption for branching pipes is that that the total head at junction J, $H_J = (p_J/\gamma + z_J)$ is higher than the total head of reservoir C, $H_C = z_C$; thus, the flow is always from junction J to reservoir C in pipe 3, and thus, the energy equation is applied between points J and C as follows:

$$\frac{p_J}{\gamma} + z_J + \frac{v_J^2}{2g} - h_{f3} = \frac{p_C}{\gamma} + z_C + \frac{v_C^2}{2g}$$

However, because the total head at junction J, $H_J = (p_J/\gamma + z_J)$, relative to the total head of reservoir B, $H_B = z_B$, is typically unknown for branching pipes, one must make an "assumption." For this problem, it will be "assumed" that the total head at junction J, $H_J = (p_J/\gamma + z_J)$ is higher than the total head of reservoir B, $H_B = z_B$, and thus, the flow is "assumed" to be from junction J to reservoir B in pipe 2, and thus, the energy equation is applied between points J and B as follows:

$$\frac{p_J}{\gamma} + z_J + \frac{v_J^2}{2g} - h_{f2} = \frac{p_B}{\gamma} + z_B + \frac{v_B^2}{2g}$$

As such, the corresponding appropriate governing continuity equation is given as follows:

$$Q_1 = Q_2 + Q_3$$

It is important to note that if the correct "assumption" regarding the magnitude of H_J relative to H_B was made, then the flow computed for pipe 2, Q_2 will have a positive sign. However, if an incorrect "assumption" regarding the magnitude of H_J relative to H_B was made, then the flow computed for pipe 2, Q_2 will have a negative sign. Finally, the above four governing equations are solved for the four unknowns: Q_1, Q_2, z_C, and H_J. The velocity of flow is computed from the continuity equation. The major head loss due to pipe friction for each pipe is computed from Equation 8.34 (Darcy–Weisbach head loss equation), as a Type 1 problem. The absolute pipe roughness for concrete is given in Table 8.3, and the Colebrook equation for the friction factor is assumed. The density, ρ and the viscosity, μ for water at 20°C are given in Table A.2 in Appendix A. The governing continuity $Q_1 = Q_2 + Q_3$ establishes the relative flowrates in each of the three branching pipes, solved as three (actually there are only two unknown discharges) independent Type 2 problems. Thus, in total, 15 equations are solved for 15 unknowns as follows:

$\rho := 998 \dfrac{kg}{m^3}$ $\qquad$ $\mu := 1 \times 10^{-3} \, N \, \dfrac{sec}{m^2}$ $\qquad$ $g := 9.81 \dfrac{m}{sec^2}$ $\quad \varepsilon := 0.3 \, mm$

$\gamma := \rho \cdot g = 9.79 \times 10^3 \dfrac{kg}{m^2 \cdot s^2}$ $\qquad$ $D_1 := 0.5 \, m$ $\qquad$ $D_2 := 0.65 \, m$ $\qquad$ $D_3 := 0.8 \, m$

$\dfrac{\varepsilon}{D_1} = 6 \times 10^{-4}$ $\qquad\qquad\qquad$ $\dfrac{\varepsilon}{D_2} = 4.615 \times 10^{-4}$ $\quad$ $\dfrac{\varepsilon}{D_3} = 3.75 \times 10^{-4}$

$L_1 := 1200 \, m$ $\qquad$ $L_2 := 900 \, m$ $\qquad$ $L_3 := 750 \, m$ $\qquad$ $A_1 := \dfrac{\pi \cdot D_1^2}{4} = 0.196 \, m^2$

$A_2 := \dfrac{\pi \cdot D_2^2}{4} = 0.332 \, m^2$ $\qquad\qquad$ $A_3 := \dfrac{\pi \cdot D_3^2}{4} = 0.503 \, m^2$

$v_A := 0 \dfrac{m}{sec}$ $\qquad$ $p_A := 0 \dfrac{N}{m^2}$ $\qquad$ $z_A := 330 \, m$

$p_B := 0 \dfrac{N}{m^2}$ $\qquad$ $v_B := 0 \dfrac{m}{sec}$ $\qquad$ $z_B := 270 \, m$

$p_C := 0 \dfrac{N}{m^2}$ $\qquad$ $v_C := 0 \dfrac{m}{sec}$ $\qquad$ $Q_3 := 0.85 \dfrac{m^3}{sec}$ $\qquad$ $v_3 := \dfrac{Q_3}{A_3} = 1.691 \dfrac{m}{s}$

Guess values: $\quad f_1 := 0.01$ $\qquad$ $f_2 := 0.01$ $\qquad$ $f_3 := 0.01$ $\qquad$ $R_1 := 400{,}000$

$R_2 := 400{,}000$ $\quad$ $R_3 := 400{,}000$ $\quad$ $h_{f1} := 10 \, m$ $\quad$ $h_{f2} := 10 \, m$

$h_{f3} := 10 \, m$ $\quad$ $v_1 := 5 \dfrac{m}{sec}$ $\quad$ $v_2 := 5 \dfrac{m}{sec}$ $\quad$ $Q_1 := 1 \dfrac{m^3}{sec}$

$Q_2 := 1 \dfrac{m^3}{sec}$ $\quad$ $z_C := 200 \, m$ $\quad$ $H_J := 300 \, m$

Given

$\dfrac{p_A}{\gamma} + z_A + \dfrac{v_A^2}{2 \cdot g} - h_{f1} = H_J$ $\qquad\qquad$ $H_J - h_{f3} = \dfrac{p_C}{\gamma} + z_C + \dfrac{v_C^2}{2 \cdot g}$

$H_J - h_{f2} = \dfrac{p_B}{\gamma} + z_B + \dfrac{v_B^2}{2 \cdot g}$ $\qquad\qquad$ $Q_1 = Q_2 + Q_3$ $\qquad$ $v_1 = \dfrac{Q_1}{A_1}$ $\qquad$ $v_2 = \dfrac{Q_2}{A_2}$

$\dfrac{1}{\sqrt{f_1}} = -2 \cdot \log\left(\dfrac{\varepsilon/D_1}{3.7} + \dfrac{2.51}{R_1 \cdot \sqrt{f_1}}\right)$ $\qquad$ $R_1 = \dfrac{\rho \cdot v_1 \cdot D_1}{\mu}$ $\qquad$ $h_{f1} = f_1 \dfrac{L_1}{D_1} \dfrac{v_1^2}{2 \cdot g}$

$\dfrac{1}{\sqrt{f_2}} = -2 \cdot \log\left(\dfrac{\varepsilon/D_2}{3.7} + \dfrac{2.51}{R_2 \cdot \sqrt{f_2}}\right)$ $\qquad$ $R_2 = \dfrac{\rho \cdot v_2 \cdot D_2}{\mu}$ $\qquad$ $h_{f2} = f_2 \dfrac{L_2}{D_2} \dfrac{v_2^2}{2 \cdot g}$

$$\frac{1}{\sqrt{f_3}} = -2 \cdot \log\left(\frac{\varepsilon/D_3}{3.7} + \frac{2.51}{R_3 \cdot \sqrt{f_3}}\right) \qquad R_3 = \frac{\rho \cdot v_3 \cdot D_3}{\mu} \qquad h_{f3} = f_3 \frac{L_3}{D_3} \frac{v_3^2}{2 \cdot g}$$

$$\begin{pmatrix} f_1 \\ f_2 \\ f_3 \\ R_1 \\ R_2 \\ R_3 \\ h_{f1} \\ h_{f2} \\ h_{f3} \\ v_1 \\ v_2 \\ Q_1 \\ Q_2 \\ z_C \\ H_J \end{pmatrix} := \text{Find}(f_1, f_2, f_3, R_1, R_2, R_3, h_{f1}, h_{f2}, h_{f3}, v_1, v_2, Q_1, Q_2, z_C, H_J)$$

$f_1 = 0.018$ $\qquad$ $f_2 = 0.018$ $\qquad$ $f_3 = 0.016$ $\qquad$ $R_1 = 2.627 \times 10^6$

$R_2 = 3.594 \times 10^5$ $\quad$ $R_3 = 1.35 \times 10^6$ $\quad$ $h_{f1} = 59.615\,\text{m}$ $\quad$ $h_{f2} = 0.385\,\text{m}$

$h_{f3} = 2.205\,\text{m}$ $\quad$ $v_1 = 5.265\,\dfrac{\text{m}}{\text{s}}$ $\quad$ $v_2 = 0.554\,\dfrac{\text{m}}{\text{s}}$ $\quad$ $Q_1 = 1.034\,\dfrac{\text{m}^3}{\text{s}}$

$Q_2 = 0.184\,\dfrac{\text{m}^3}{\text{s}}$ $\quad$ $z_C = 268.179\,\text{m}$ $\quad$ $H_J = 270.385\,\text{m}$

Thus, because the flow computed for pipe 2, Q_2 has a positive sign, the correct "assumption" regarding the magnitude of H_J relative to H_B was made. Furthermore, the governing continuity equation is satisfied (checks) as follows:

$$Q_2 + Q_3 = 1.034\,\frac{\text{m}^3}{\text{s}}$$

EXAMPLE PROBLEM 8.23

Water at 20°C flows between reservoirs A, B, and C through three riveted steel branching pipes with a common junction J as illustrated in Figure EP 8.23. The pipe system contains three reservoirs and three single pipes. Pipe 1 is 0.8 m in diameter and 1100

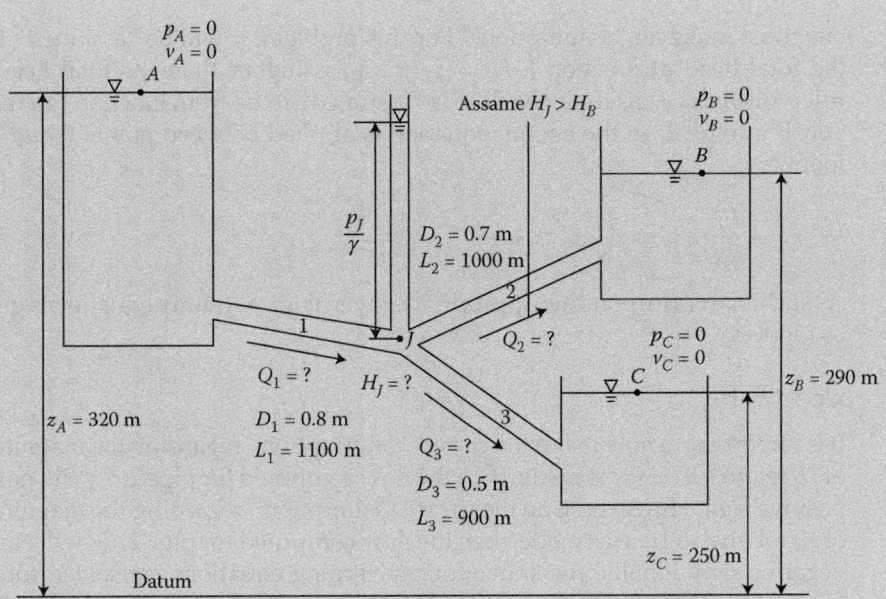

FIGURE EP 8.23
Water flows between reservoirs A, B, and C through three riveted steel branching pipes with common junction J (case 3 "classic three-reservoir" problem).

m long. Pipe 2 is 0.7 m in diameter and 1000 m long. Pipe 3 is 0.5 m in diameter and 900 m long. The elevation of reservoir A is 320 m, the elevation of reservoir B is 290 m, and elevation of reservoir C is 250 m (a) Determine the discharge in pipe 1. (b) Determine the discharge in pipe 2. (c) Determine the discharge in pipe 3. (d) Determine the total head for junction J.

Mathcad Solution

(a)–(d) This example demonstrates a case 3 ("classic three-reservoir) problem. The datum is assumed to be below point C. A constant assumption for branching pipes is that the total head at junction J, $H_J = (p_J/\gamma + z_J)$ is lower than the total head of reservoir A, $H_A = z_A$; thus, the flow is always from reservoir A to junction J in pipe 1; thus, the energy equation is applied between points A and J as follows:

$$\frac{p_A}{\gamma} + z_A + \frac{v_A^2}{2g} - h_{f1} = \frac{p_J}{\gamma} + z_J + \frac{v_J^2}{2g}$$

Additionally, another constant assumption for branching pipes is that that the total head at junction J, $H_J = (p_J/\gamma + z_J)$ is higher than the total head of reservoir C, $H_C = z_C$; thus, the flow is always from junction J to reservoir C in pipe 3, and thus, the energy equation is applied between points J and C as follows:

$$\frac{p_J}{\gamma} + z_J + \frac{v_J^2}{2g} - h_{f3} = \frac{p_C}{\gamma} + z_C + \frac{v_C^2}{2g}$$

However, because the total head at junction J, $H_J = (p_J/\gamma + z_J)$, relative to the total head of reservoir B, $H_B = z_B$, is typically unknown for branching pipes, one must make an "assumption." For this problem, it will be "assumed" that the total head at junction J, $H_J = (p_J/\gamma + z_J)$ is higher than the total head of reservoir B, $H_B = z_B$; thus, the flow is "assumed" to be from junction J to reservoir B in pipe 2, so the energy equation is applied between points B and J as follows:

$$\frac{p_J}{\gamma} + z_J + \frac{v_J^2}{2g} - h_{f2} = \frac{p_B}{\gamma} + z_B + \frac{v_B^2}{2g}$$

As such, the corresponding appropriate governing continuity equation is given as follows:

$$Q_1 = Q_2 + Q_3$$

It is important to note that if the correct "assumption" regarding the magnitude of H_J relative to H_B was made, then the flow computed for pipe 2, Q_2 will have a positive sign. However, if an incorrect "assumption" regarding the magnitude of H_J relative to H_B was made, then the flow computed for pipe 2, Q_2 will have a negative sign. Finally, the above four governing equations are solved for the four unknowns: Q_1, Q_2, Q_3, and H_J. The velocity of flow is computed from the continuity equation. The major head loss due to pipe friction for each pipe is computed from Equation 8.34 (Darcy–Weisbach head loss equation), as a Type 1 problem. The absolute pipe roughness for riveted steel is given in Table 8.3, and the Colebrook equation for the friction factor is assumed. The density, ρ and the viscosity, μ for water at 20°C are given in Table A.4 in Appendix A. The governing continuity $Q_1 = Q_2 + Q_3$ establishes the relative flowrates in each of the three branching pipes, solved as three independent Type 2 problems. Thus, in total, 16 equations are solved for 16 unknowns as follows:

$$\rho := 998 \, \frac{kg}{m^3} \qquad \mu := 1 \times 10^{-3} \, N \, \frac{sec}{m^2} \qquad g := 9.81 \, \frac{m}{sec^2} \qquad \varepsilon := 0.3 \, mm$$

$$\gamma := \rho \cdot g = 9.79 \times 10^3 \, \frac{kg}{m^2 \cdot s^2} \qquad D_1 := 0.8 \, m \qquad D_2 := 0.7 \, m \qquad D_3 := 0.5 \, m$$

$$\frac{\varepsilon}{D_1} = 1.125 \times 10^{-3} \qquad \frac{\varepsilon}{D_2} = 1.286 \times 10^{-3} \qquad \frac{\varepsilon}{D_3} = 1.8 \times 10^{-3}$$

$$L_1 := 1100 \, m \qquad L_2 := 1000 \, m \qquad L_3 := 900 \, m \qquad A_1 := \frac{\pi \cdot D_1^{\,2}}{4} = 0.503 \, m^2$$

$$A_2 := \frac{\pi \cdot D_2^{\,2}}{4} = 0.385 \, m^2 \qquad\qquad A_3 := \frac{\pi \cdot D_3^2}{4} = 0.196 \, m^2$$

$$v_A := 0 \, \frac{m}{sec} \qquad p_A := 0 \, \frac{N}{m^2} \qquad z_A := 320 \, m$$

$$p_B := 0 \, \frac{N}{m^2} \qquad v_B := 0 \, \frac{m}{sec} \qquad z_B := 290 \, m$$

$$p_C := 0 \, \frac{N}{m^2} \qquad v_C := 0 \, \frac{m}{sec} \qquad z_C := 250 \, m$$

Guess values: $f_1 := 0.01$ $f_2 := 0.01$ $f_3 := 0.01$ $R_1 := 400,000$

$R_2 := 400,000$ $R_3 := 400,000$ $h_{f1} := 10\,m$ $h_{f2} := 10\,m$

$h_{f3} := 10\,m$ $v_1 := 5\,\dfrac{m}{sec}$ $v_2 := 5\,\dfrac{m}{sec}$ $v_3 := 5\,\dfrac{m}{sec}$

$Q_1 := 1\,\dfrac{m^3}{sec}$ $Q_2 := 1\,\dfrac{m^3}{sec}$ $Q_3 := 1\,\dfrac{m^3}{sec}$ $H_J := 295\,m$

Given

$$\frac{P_A}{\gamma} + z_A + \frac{v_A^2}{2\cdot g} - h_{f1} = H_J \qquad H_J - h_{f3} = \frac{P_C}{\gamma} + z_C + \frac{v_C^2}{2\cdot g} \qquad v_1 = \frac{Q_1}{A_1}$$

$$H_J - h_{f2} = \frac{P_B}{\gamma} + z_B + \frac{v_B^2}{2\cdot g} \qquad Q_1 = Q_2 + Q_3 \qquad v_2 = \frac{Q_2}{A_2} \qquad v_3 = \frac{Q_3}{A_3}$$

$$\frac{1}{\sqrt{f_1}} = -2\cdot \log\left(\frac{\varepsilon/D_1}{3.7} + \frac{2.51}{R_1\cdot \sqrt{f_1}}\right) \quad R_1 = \frac{\rho\cdot v_1\cdot D_1}{\mu} \quad h_{f1} = f_1\cdot \frac{L_1}{D_1}\frac{v_1^2}{2\cdot g}$$

$$\frac{1}{\sqrt{f_2}} = -2\cdot \log\left(\frac{\varepsilon/D_2}{3.7} + \frac{2.51}{R_2\cdot \sqrt{f_2}}\right) \quad R_2 = \frac{\rho\cdot v_2\cdot D_2}{\mu} \quad h_{f2} = f_2\cdot \frac{L_2}{D_2}\frac{v_2^2}{2\cdot g}$$

$$\frac{1}{\sqrt{f_3}} = -2\cdot \log\left(\frac{\varepsilon/D_3}{3.7} + \frac{2.51}{R_3\cdot \sqrt{f_3}}\right) \quad R_3 = \frac{\rho\cdot v_3\cdot D_3}{\mu} \quad h_{f3} = f_3\cdot \frac{L_3}{D_3}\frac{v_3^2}{2\cdot g}$$

$$\begin{pmatrix} f_1 \\ f_2 \\ f_3 \\ R_1 \\ R_2 \\ R_3 \\ h_{f1} \\ h_{f2} \\ h_{f3} \\ v_1 \\ v_2 \\ v_3 \\ Q_1 \\ Q_2 \\ Q_3 \\ H_J \end{pmatrix} := Find\,(f_1, f_2, f_3, R_1, R_2, R_3, h_{f1}, h_{f2}, h_{f3}, v_1, v_2, v_3, Q_1, Q_2, Q_3, H_J)$$

$f_1 = 0.02$ $f_2 = 0.021$ $f_3 = 0.023$ $R_1 = 3.033 \times 10^6$

$R_2 = 1.736 \times 10^6$ $R_3 = 2.423 \times 10^6$ $h_{f1} = 20.541 \text{ m}$ $h_{f2} = 9.459 \text{ m}$

$h_{f3} = 49.459 \text{ m}$ $v_1 = 3.799 \dfrac{\text{m}}{\text{s}}$ $v_2 := 2.485 \dfrac{\text{m}}{\text{s}}$ $v_3 := 4.856 \dfrac{\text{m}}{\text{s}}$

$Q_1 := 1.91 \dfrac{\text{m}^3}{\text{s}}$ $Q_2 := 0.956 \dfrac{\text{m}^3}{\text{s}}$ $Q_3 := 0.953 \dfrac{\text{m}^3}{\text{s}}$ $H_J = 299.459 \text{ m}$

Thus, because the flow computed for pipe 2, Q_2 has a positive sign, the correct "assumption" regarding the magnitude of H_J relative to H_B was made. Furthermore, the governing continuity equation is satisfied (checks) as follows:

$$Q_2 + Q_3 = 1.802 \dfrac{\text{m}^3}{\text{s}}$$

8.7.6.2 Branching Pipes Connected to a Water Supply Source under Pressure

The second kind of branching pipe system is where the three branching pipes are connected to a water supply source under pressure at point A, as illustrated in Figure 8.21b. The total head at point A (supply/source) is greater than the total head at junction J. Furthermore, the total head at junction J is larger than the total head at point B (demand/destination/point of discharge) and larger than the total head at point C (demand/destination/point of discharge). Thus, the flow is always from point A to point B, and from point A to point C. The governing equations are outlined as follows. The assumption is that the velocity heads, $v^2/2g$ are negligible; thus, the total heads are $H = (p/\gamma + z_J)$. Therefore, from the conservation of mass principle (integral continuity equation; first governing principle), the total flowrate entering junction J is equal to the total flowrate leaving junction J regardless of the diameters of the individual branching pipes, which is the first governing equation and is given as follows:

$$Q_1 = Q_2 + Q_3 \tag{8.87}$$

From the conservation of energy principle (second governing principle), all the significant head losses are modeled in the energy equation applied between two points, which is the second set of governing equations and are given as follows:

$$\frac{p_A}{\gamma} + z_A + \frac{v_A^2}{2g} - h_{f1} - h_{f2} = \frac{p_B}{\gamma} + z_B + \frac{v_B^2}{2g} \tag{8.88}$$

$$\frac{p_A}{\gamma} + z_A + \frac{v_A^2}{2g} - h_{f1} - h_{f3} = \frac{p_C}{\gamma} + z_C + \frac{v_C^2}{2g} \tag{8.89}$$

Evaluation of the major (and minor) head loss flow resistance terms requires a "subset level" application of the governing equations (see Chapters 6 and 7 and Section 8.4.1.2). Depending upon whether the flow is laminar or turbulent, and depending upon the desired accuracy, simplicity, and practical constraints in the range of the variables (v, D, and water temperature), the major head loss due to pipe friction, $h_{f,major}$ is evaluated by the appropriate

TABLE 8.14

Summary of Three Cases for Branching Pipes Connected to a Water Supply System under Pressure

Case	Given	Find	One Governing Continuity Equation at Junction J	Two Governing Energy Equations
Case 1	$\gamma, \varepsilon, L, D$, H for two points, Q for one of these points' pipes	Q for two pipes, H for one of these points	$Q_1 = Q_2 + Q_3$	$\dfrac{p_A}{\gamma} + z_A + \dfrac{v_A^2}{2g} - h_{f1} - h_{f2} = \dfrac{p_B}{\gamma} + z_B + \dfrac{v_B^2}{2g}$ $\dfrac{p_A}{\gamma} + z_A + \dfrac{v_A^2}{2g} - h_{f1} - h_{f3} = \dfrac{p_C}{\gamma} + z_C + \dfrac{v_C^2}{2g}$
Case 2	$\gamma, \varepsilon, L, D$, H for two points, Q for third point's pipe	Q for two points' pipes, H for third point	$Q_1 = Q_2 + Q_3$	$\dfrac{p_A}{\gamma} + z_A + \dfrac{v_A^2}{2g} - h_{f1} - h_{f2} = \dfrac{p_B}{\gamma} + z_B + \dfrac{v_B^2}{2g}$ $\dfrac{p_A}{\gamma} + z_A + \dfrac{v_A^2}{2g} - h_{f1} - h_{f3} = \dfrac{p_C}{\gamma} + z_C + \dfrac{v_C^2}{2g}$
Case 3	$\gamma, \varepsilon, L, D$, H for all three points	Q for all three pipes	$Q_1 = Q_2 + Q_3$	$\dfrac{p_A}{\gamma} + z_A + \dfrac{v_A^2}{2g} - h_{f1} - h_{f2} = \dfrac{p_B}{\gamma} + z_B + \dfrac{v_B^2}{2g}$ $\dfrac{p_A}{\gamma} + z_A + \dfrac{v_A^2}{2g} - h_{f1} - h_{f3} = \dfrac{p_C}{\gamma} + z_C + \dfrac{v_C^2}{2g}$

resistance equation (see Section 8.5). Finally, from the conservation of momentum principle, the friction slope, S_f due to pipe friction at a given point in a straight pipe section may be modeled in the differential momentum equation, Equation 8.12, which is the third governing principle. Evaluation of the friction slope requires a "subset level" application of the governing equations (see Chapters 6 and 7 and Section 8.4.1.4) (see Equation 8.13 for the Chezy equation).

In the analysis and design of a pipe system containing branching pipes connected to a water supply source under pressure at point A, assuming that the pipe material, pipe lengths, pipe diameters, and the fluid type and temperature are specified, there are three cases, outlined in Table 8.14.

It is important to note that for each of the three cases outlined above, the three governing equations (continuity equation at junction J, energy equation between A and B, and energy equation between A and C) are solved for the three unknowns. A case 1 problem for branching pipes connected to a water supply source under pressure is illustrated in Example 8.24, a case 2 problem for branching pipes connected to a water supply source under pressure is illustrated in Example 8.25, and a case 3 problem for branching pipes connected to a water supply source under pressure is illustrated in Example 8.26.

EXAMPLE PROBLEM 8.24

Water at 20°C flows from point A (water supply source under pressure) to point B (destination is kitchen faucet), and from point A to point C (destination is laundry room faucet) through three copper branching pipes with a common junction J as illustrated in Figure EP 8.24. The pipe system contains a source, two destinations, three single pipes, three regular 90° threaded elbows, and two fully open globe valves. Pipe 1 is 0.08 m in diameter and 100 m long. Pipe 2 is 0.05 m in diameter and 90 m long. Pipe 3 is 0.065 m in diameter and 150 m long. At point A, the pressure is $2.8 \times 10^5 \, \text{N/m}^2$, and the elevation is 0 m. At point B, the pressure is $0 \, \text{N/m}^2$, and the elevation is 20 m. The discharge in pipe 1 is $0.009 \, \text{m}^3/\text{sec}$. (a) Determine the discharge in pipe 2. (b) Determine the discharge in pipe 3. (c) Given that the pressure at point C is $0 \, \text{N/m}^2$ determine the maximum elevation at point C.

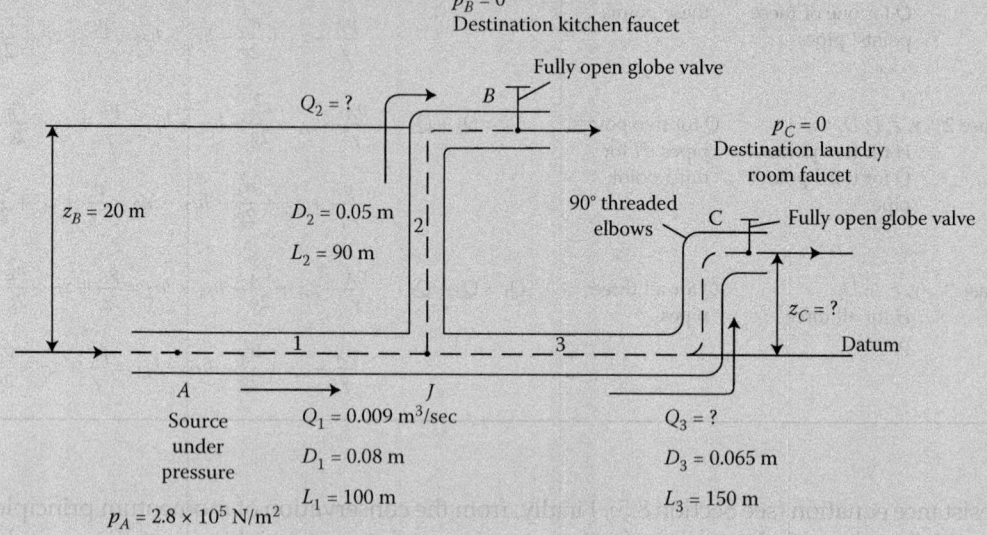

FIGURE EP 8.24
Water flows from source A to destinations B and C through three copper branching pipes with common junction J (case 1 problem).

Mathcad Solution

(a)–(c) This example demonstrates a case 1 problem. The datum is assumed to be through point A and junction J. The flow is always from point A to point B; thus, the energy equation is applied between points A and B as follows:

$$\frac{p_A}{\gamma} + z_A + \frac{v_A^2}{2g} - h_{f1} - h_{f2} = \frac{p_B}{\gamma} + z_B + \frac{v_B^2}{2g}$$

Additionally, the flow is always from point A to point C; thus, the energy equation is applied between points A and C as follows:

$$\frac{p_A}{\gamma} + z_A + \frac{v_A^2}{2g} - h_{f1} - h_{f3} = \frac{p_C}{\gamma} + z_C + \frac{v_C^2}{2g}$$

As such, the corresponding appropriate governing continuity equation is given as follows:

$$Q_1 = Q_2 + Q_3$$

Finally, the above three governing equations are solved for the three unknowns: Q_2, Q_3, and H_C. The velocity of flow is computed from the continuity equation. The major head loss due to pipe friction for each pipe is computed from Equation 8.34 (Darcy–Weisbach head loss equation) as a Type 1 problem. The absolute pipe roughness for copper is given in Table 8.3, and the Colebrook equation for the friction factor is assumed. The density, ρ and the viscosity, μ for water at 20°C are given in Table A.4 in Appendix A. The governing continuity $Q_1 = Q_2 + Q_3$ establishes the relative flowrates in each of the three branching pipes, solved as three (actually there are only two unknown discharges) independent Type 2 problems. Thus, in total, 14 equations are solved for 14 unknowns as follows:

$$\rho := 998 \, \frac{kg}{m^3} \qquad \mu := 1 \times 10^{-3} \, N\frac{sec}{m^2} \qquad g := 9.81 \, \frac{m}{sec^2} \qquad \varepsilon := 0.0015 \, mm$$

$$\gamma := \rho \cdot g = 9.79 \times 10^3 \, \frac{kg}{m^2 \cdot s^2} \qquad D_1 := 0.08 \, m \qquad D_2 := 0.05 \, m \qquad D_3 := 0.065 \, m$$

$$\frac{\varepsilon}{D_1} = 1.875 \times 10^{-5} \qquad \frac{\varepsilon}{D_2} = 3 \times 10^{-5} \qquad \frac{\varepsilon}{D_3} = 2.308 \times 10^{-5}$$

$$L_1 := 100 \, m \qquad L_2 := 90 \, m \qquad L_3 := 150 \, m \qquad A_1 := \frac{\pi \cdot D_1{}^2}{4} = 5.027 \times 10^{-3} \, m^2$$

$$A_2 := \frac{\pi \cdot D_2{}^2}{4} = 1.963 \times 10^{-3} \, m^2 \qquad A_3 := \frac{\pi \cdot D_3^2}{4} = 3.318 \times 10^{-3} \, m^2$$

$$v_A := 0 \, \frac{m}{sec} \qquad P_A := 2.8 \times 10^5 \, \frac{N}{m^2} \qquad z_A := 0 \, m$$

$$v_B := 0 \, \frac{m}{sec} \qquad P_B := 0 \, \frac{N}{m^2} \qquad z_B := 20 \, m$$

$$P_C := 0 \, \frac{N}{m^2} \qquad v_C := 0 \, \frac{m}{sec} \qquad Q_1 := 0.009 \, \frac{m^3}{sec} \qquad v_1 := \frac{Q_1}{A_1} = 1.79 \, \frac{m}{s}$$

Guess values: $\qquad f_1 := 0.01 \qquad f_2 := 0.01 \qquad f_3 := 0.01 \qquad R_1 := 400,000$

$$R_2 := 400,000 \qquad R_3 := 400,000 \qquad h_{f1} := 10 \, m \qquad h_{f2} := 10 \, m$$

$$h_{f3} := 10 \, m \qquad v_2 := 5 \, \frac{m}{sec} \qquad v_3 := 5 \, \frac{m}{sec} \qquad Q_2 := 1 \, \frac{m^3}{sec}$$

$$Q_3 := 1 \, \frac{m^3}{sec} \qquad z_C := 15 \, m$$

Given

$$\frac{P_A}{\gamma} + z_A + \frac{v_A^2}{2 \cdot g} - h_{f1} - h_{f2} = \frac{P_B}{\gamma} + z_B + \frac{v_B^2}{2 \cdot g}$$

$$\frac{p_A}{\gamma} + z_A + \frac{v_A^2}{2 \cdot g} - h_{f1} - h_{f3} = \frac{p_C}{\gamma} + z_C + \frac{v_C^2}{2 \cdot g}$$

$$Q_1 = Q_2 + Q_3 \qquad\qquad v_2 = \frac{Q_2}{A_2} \qquad v_3 = \frac{Q_3}{A_3}$$

$$\frac{1}{\sqrt{f_1}} = -2 \cdot \log\left(\frac{\varepsilon/D_1}{3.7} + \frac{2.51}{R_1 \cdot \sqrt{f_1}}\right) \quad R_1 = \frac{\rho \cdot v_1 \cdot D_1}{\mu} \quad h_{f1} = f_1 \cdot \frac{L_1}{D_1}\frac{v_1^2}{2 \cdot g}$$

$$\frac{1}{\sqrt{f_2}} = -2 \cdot \log\left(\frac{\varepsilon/D_2}{3.7} + \frac{2.51}{R_2 \cdot \sqrt{f_2}}\right) \quad R_2 = \frac{\rho \cdot v_2 \cdot D_2}{\mu} \quad h_{f2} = f_2 \cdot \frac{L_2}{D_2}\frac{v_2^2}{2 \cdot g}$$

$$\frac{1}{\sqrt{f_3}} = -2 \cdot \log\left(\frac{\varepsilon/D_3}{3.7} + \frac{2.51}{R_3 \cdot \sqrt{f_3}}\right) \quad R_3 = \frac{\rho \cdot v_3 \cdot D_3}{\mu} \quad h_{f3} = f_3 \cdot \frac{L_3}{D_3}\frac{v_3^2}{2 \cdot g}$$

$$\begin{pmatrix} f_1 \\ f_2 \\ f_3 \\ R_1 \\ R_2 \\ R_3 \\ h_{f1} \\ h_{f2} \\ h_{f3} \\ v_2 \\ v_3 \\ Q_2 \\ Q_3 \\ z_C \end{pmatrix} := \text{Find}\,(f_1, f_2, f_3, R_1, R_2, R_3, h_{f1}, h_{f2}, h_{f3}, v_2, v_3, Q_2, Q_3, z_C)$$

$$f_1 = 0.017 \qquad\qquad f_2 = 0.019 \qquad\qquad f_3 = 0.018 \qquad R_1 = 1.43 \times 10^5$$

$$R_2 = 8.655 \times 10^4 \qquad\qquad R_3 = 1.094 \times 10^5 \quad h_{f1} = 3.44\,\text{m} \quad h_{f2} = 5.16\,\text{m}$$

$$h_{f3} = 5.947\,\text{m} \qquad\qquad v_2 = 1.735\,\frac{\text{m}}{\text{s}} \qquad v_3 = 1.686\,\frac{\text{m}}{\text{s}}$$

$$Q_2 = 3.406 \times 10^{-3}\,\frac{\text{m}^3}{\text{s}} \quad Q_2 = 5.594 \times 10^{-3}\,\frac{\text{m}^3}{\text{s}} \qquad z_C = 19.213\,\text{m}$$

Furthermore, the governing continuity equation is satisfied (checks) as follows:

$$Q_2 + Q_3 = 9 \times 10^{-3} \, \frac{m^3}{s}$$

EXAMPLE PROBLEM 8.25

Water at 20°C flows from point A (water supply source under pressure) to point B (destination is kitchen faucet) and from point A to point C (destination is laundry room faucet) through three riveted steel branching pipes with a common junction J as illustrated in Figure EP 8.25. The pipe system contains a source, two destinations, three single pipes, three regular 90° threaded elbows, and two fully open globe valves. Pipe 1 is 0.08 m in diameter and 100 m long. Pipe 2 is 0.05 m in diameter and 90 m long. Pipe 3 is 0.065 m in diameter and 150 m long. At point A, the elevation is 0 m. At point B, the pressure is $0 \, N/m^2$, and the elevation is 25 m. At point C, the pressure is $0 \, N/m^2$, and the elevation is 21 m. The discharge in pipe 1 is $0.02 \, m^3/sec$. (a) Determine the minimum required pressure at point A in order to overcome the elevations at points B and C, and the head loss due to pipe friction in pipes 1, 2, and 3. (b) Determine the discharge in pipe 2. (c) Determine the discharge in pipe 3.

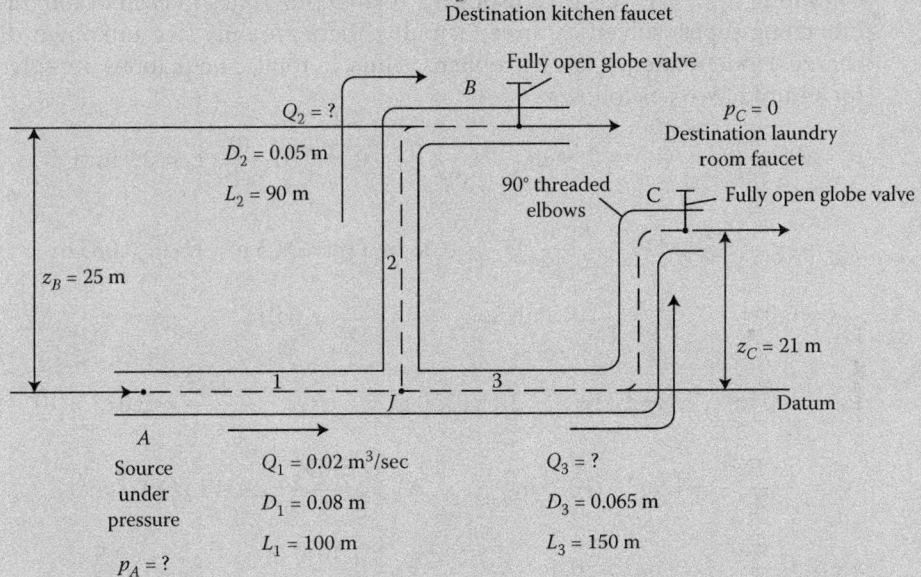

FIGURE EP 8.25
Water flows from source A to destinations B and C through three riveted steel branching pipes with common junction J (case 2 problem).

Mathcad Solution

(a)–(c) This example demonstrates a case 2 problem. The datum is assumed to be through point A and junction J. The flow is always from point A to point B;

thus, the energy equation is applied between points A and B as follows:

$$\frac{p_A}{\gamma} + z_A + \frac{v_A^2}{2g} - h_{f1} - h_{f2} = \frac{p_B}{\gamma} + z_B + \frac{v_B^2}{2g}$$

Additionally, the flow is always from point A to point C; thus, the energy equation is applied between points A and C as follows:

$$\frac{p_A}{\gamma} + z_A + \frac{v_A^2}{2g} - h_{f1} - h_{f3} = \frac{p_C}{\gamma} + z_C + \frac{v_C^2}{2g}$$

As such, the corresponding appropriate governing continuity equation is given as follows:

$$Q_1 = Q_2 + Q_3$$

Finally, the above three governing equations are solved for the three unknowns: Q_2, Q_3, and p_A. The velocity of flow is computed from the continuity equation. The major head loss due to pipe friction for each pipe is computed from Equation 8.34 (Darcy–Weisbach head loss equation) as a Type 1 problem. The absolute pipe roughness for riveted steel is given in Table 8.3, and the Colebrook equation for the friction factor is assumed. The density, ρ and the viscosity, μ for water at 20°C are given in Table A.4 in Appendix A. The governing continuity $Q_1 = Q_2 + Q_3$ establishes the relative flowrates in each of the three branching pipes, solved as three (actually, there are only two unknown discharges) independent Type 2 problems. Thus, in total, 14 equations are solved for 14 unknowns as follows:

$$\rho := 998\,\frac{kg}{m^3} \qquad \mu := 1 \times 10^{-3}\,N\frac{sec}{m^2} \qquad g := 9.81\,\frac{m}{sec^2} \quad \varepsilon := 0.9\,mm$$

$$\gamma := \rho \cdot g = 9.79 \times 10^3\,\frac{kg}{m^2 \cdot s^2} \quad D_1 := 0.08\,m \quad D_2 := 0.05\,m \quad D_3 := 0.065\,m$$

$$\frac{\varepsilon}{D_1} = 0.011 \qquad\qquad \frac{\varepsilon}{D_2} = 0.018 \qquad\qquad \frac{\varepsilon}{D_3} = 0.014$$

$$L_1 := 100\,m \quad L_2 := 90\,m \quad L_3 := 150\,m \quad A_1 := \frac{\pi \cdot D_1^2}{4} = 5.027 \times 10^{-3}\,m^2$$

$$A_2 := \frac{\pi \cdot D_2^2}{4} = 1.963 \times 10^{-3}\,m^2 \qquad A_3 := \frac{\pi \cdot D_3^2}{4} = 3.318 \times 10^{-3}\,m^2$$

$$v_A := 0\,\frac{m}{sec} \qquad z_A := 0\,m \qquad v_B := 0\,\frac{m}{sec} \qquad p_B := 0\,\frac{N}{m^2}$$

$$z_B := 25\,m \qquad p_C := 0\,\frac{N}{m^2} \qquad v_C := 0\,\frac{m}{sec} \qquad z_C := 21\,m$$

$$Q_1 := 0.02\,\frac{m^3}{sec} \qquad\qquad v_1 := \frac{Q_1}{A_1} = 3.979\,\frac{m}{s}$$

Guess values:　　$f_1 := 0.01$　　　　$f_2 := 0.01$　　　　$f_3 := 0.01$　　　$R_1 := 400{,}000$

$$R_2 := 400{,}000 \quad R_3 := 400{,}000 \quad h_{f1} := 10\,\text{m} \quad h_{f2} := 10\,\text{m}$$

$$h_{f3} := 10\,\text{m} \quad v_2 := 5\,\frac{\text{m}}{\text{sec}} \quad v_3 := 5\,\frac{\text{m}}{\text{sec}} \quad Q_2 := 1\,\frac{\text{m}^3}{\text{sec}}$$

$$Q_3 := 1\,\frac{\text{m}^3}{\text{sec}} \quad p_A := 3 \times 10^5\,\frac{\text{N}}{\text{m}^2}$$

Given

$$\frac{p_A}{\gamma} + z_A + \frac{v_A^2}{2 \cdot g} - h_{f1} - h_{f2} = \frac{p_B}{\gamma} + z_B + \frac{v_B^2}{2 \cdot g}$$

$$\frac{p_A}{\gamma} + z_A + \frac{v_A^2}{2 \cdot g} - h_{f1} - h_{f3} = \frac{p_C}{\gamma} + z_C + \frac{v_C^2}{2 \cdot g}$$

$$Q_1 = Q_2 + Q_3 \qquad\qquad v_2 = \frac{Q_2}{A_2} \qquad v_3 = \frac{Q_3}{A_3}$$

$$\frac{1}{\sqrt{f_1}} = -2 \cdot \log\left(\frac{\varepsilon/D_1}{3.7} + \frac{2.51}{R_1 \cdot \sqrt{f_1}}\right) \qquad R_1 = \frac{\rho \cdot v_1 \cdot D_1}{\mu} \qquad h_{f1} = f_1 \cdot \frac{L_1}{D_1}\frac{v_1^2}{2 \cdot g}$$

$$\frac{1}{\sqrt{f_2}} = -2 \cdot \log\left(\frac{\varepsilon/D_2}{3.7} + \frac{2.51}{R_2 \cdot \sqrt{f_2}}\right) \qquad R_2 = \frac{\rho \cdot v_2 \cdot D_2}{\mu} \qquad h_{f2} = f_2 \cdot \frac{L_2}{D_2}\frac{v_2^2}{2 \cdot g}$$

$$\frac{1}{\sqrt{f_3}} = -2 \cdot \log\left(\frac{\varepsilon/D_3}{3.7} + \frac{2.51}{R_3 \cdot \sqrt{f_3}}\right) \qquad R_3 = \frac{\rho \cdot v_3 \cdot D_3}{\mu} \qquad h_{f3} = f_3 \cdot \frac{L_3}{D_3}\frac{v_3^2}{2 \cdot g}$$

$$\begin{pmatrix} f_1 \\ f_2 \\ f_3 \\ R_1 \\ R_2 \\ R_3 \\ h_{f1} \\ h_{f2} \\ h_{f3} \\ v_2 \\ v_3 \\ Q_2 \\ Q_3 \\ p_A \end{pmatrix} := \text{Find}\,(f_1, f_2, f_3, R_1, R_2, R_3, h_{f1}, h_{f2}, h_{f3}, v_2, v_3, Q_2, Q_3, p_A)$$

$$f_1 = 0.04 \qquad\qquad f_2 = 0.047 \qquad\qquad f_3 = 0.043 \qquad R_1 = 3.177 \times 10^5$$

$$R_2 = 1.945 \times 10^5 \qquad\qquad R_3 = 2.414 \times 10^5 \quad h_{f1} = 39.979 \text{ m} \quad h_{f2} = 65.444 \text{ m}$$

$$h_{f3} = 69.444 \text{ m} \qquad\qquad v_2 = 3.898\,\frac{\text{m}}{\text{s}} \qquad v_3 = 3.721\,\frac{\text{m}}{\text{s}}$$

$$Q_2 = 7.654 \times 10^{-3}\,\frac{\text{m}^3}{\text{s}} \quad Q_3 = 0.012\,\frac{\text{m}^3}{\text{s}} \qquad\qquad\qquad p_A = 1.277 \times 10^6\,\frac{\text{N}}{\text{m}^2}$$

Furthermore, the governing continuity equation is satisfied (checks) as follows:

$$Q_2 + Q_3 = 0.02\,\frac{\text{m}^3}{\text{s}}$$

EXAMPLE PROBLEM 8.26

Water at 20°C flows from point A (water supply source under pressure) to point B (destination is kitchen faucet) and from point A to point C (destination is laundry room faucet) through three galvanized iron branching pipes with a common junction J as illustrated in Figure EP 8.26. The pipe system contains a source, two destinations,

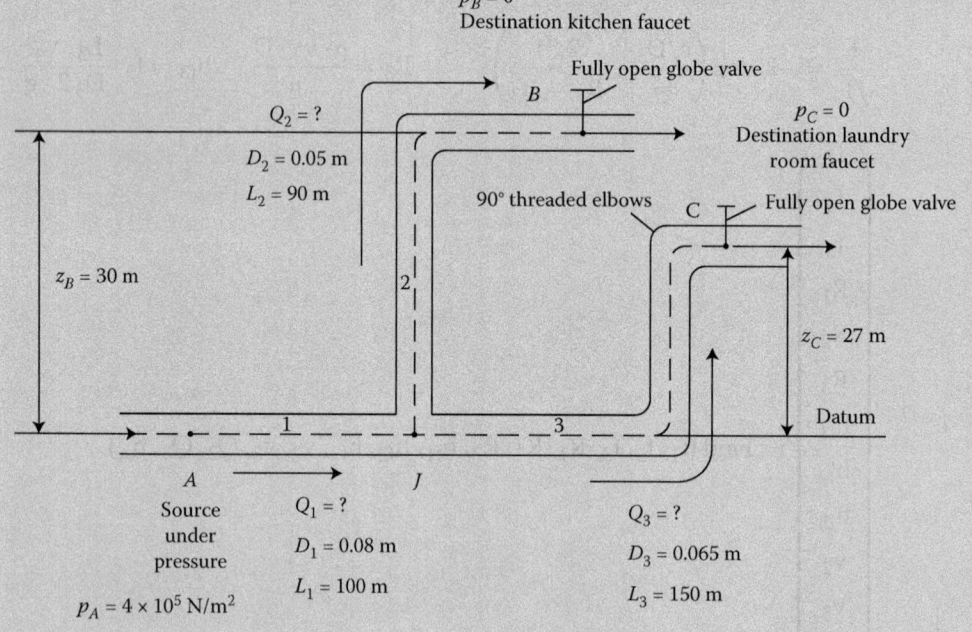

FIGURE EP 8.26
Water flows from source A to destinations B and C through three galvanized iron branching pipes with common junction J (case 3 problem).

three single pipes, three regular 90° threaded elbows, and two fully open globe valves. Pipe 1 is 0.08 m in diameter and 100 m long. Pipe 2 is 0.05 m in diameter and 90 m long. Pipe 3 is 0.065 m in diameter and 150 m long. At point A, the pressure is $4 \times 10^5 \, \text{N/m}^2$, and the elevation is 0 m. At point B, the pressure is $0 \, \text{N/m}^2$, and the elevation is 30 m. At point C, the pressure is $0 \, \text{N/m}^2$, and the elevation is 27 m. (a) Determine the discharge in pipe 1. (b) Determine the discharge in pipe 2. (c) Determine the discharge in pipe 3.

Mathcad Solution

(a)–(c) This example demonstrates a case 3 problem. The datum is assumed to be through point A and junction J. The flow is always from point A to point B; thus, the energy equation is applied between points A and B as follows:

$$\frac{p_A}{\gamma} + z_A + \frac{v_A^2}{2g} - h_{f1} - h_{f2} = \frac{p_B}{\gamma} + z_B + \frac{v_B^2}{2g}$$

Additionally, the flow is always from point A to point C; thus, the energy equation is applied between points A and C as follows:

$$\frac{p_A}{\gamma} + z_A + \frac{v_A^2}{2g} - h_{f1} - h_{f3} = \frac{p_C}{\gamma} + z_C + \frac{v_C^2}{2g}$$

As such, the corresponding appropriate governing continuity equation is given as follows:

$$Q_1 = Q_2 + Q_3$$

Finally, the above three governing equations are solved for the three unknowns: Q_1, Q_2, and Q_3. The velocity of flow is computed from the continuity equation. The major head loss due to pipe friction for each pipe is computed from Equation 8.34 (Darcy–Weisbach head loss equation) as a Type 1 problem. The absolute pipe roughness for galvanized iron is given in Table 8.3, and the Colebrook equation for the friction factor is assumed. The density, ρ and the viscosity, μ for water at 20°C are given in Table A.4 in Appendix A. The governing continuity $Q_1 = Q_2 + Q_3$ establishes the relative flowrates in each of the three branching pipes, solved as three independent Type 2 problems. Thus, in total, 15 equations are solved for 15 unknowns as follows:

$$\rho := 998 \, \frac{\text{kg}}{\text{m}^3} \qquad \mu := 1 \times 10^{-3} \, \text{N} \frac{\text{sec}}{\text{m}^2} \qquad g := 9.81 \, \frac{\text{m}}{\text{sec}^2} \qquad \varepsilon := 0.15 \, \text{mm}$$

$$\gamma := \rho \cdot g = 9.79 \times 10^3 \, \frac{\text{kg}}{\text{m}^2 \cdot \text{s}^2} \qquad D_1 := 0.08 \, \text{m} \qquad D_2 := 0.05 \, \text{m} \qquad D_3 := 0.065 \, \text{m}$$

$$\frac{\varepsilon}{D_1} = 1.875 \times 10^{-3} \qquad \frac{\varepsilon}{D_2} = 3 \times 10^{-3} \qquad \frac{\varepsilon}{D_3} = 2.308 \times 10^{-3}$$

$$L_1 := 100 \, \text{m} \qquad L_2 := 90 \, \text{m} \qquad L_3 := 150 \, \text{m} \qquad A_1 := \frac{\pi \cdot D_1^2}{4} = 5.027 \times 10^{-3} \, \text{m}^2$$

$$A_2 := \frac{\pi \cdot D_2^{\,2}}{4} = 1.963 \times 10^{-3} \, m^2 \qquad\qquad A_3 := \frac{\pi \cdot D_3^{\,2}}{4} = 3.318 \times 10^{-3} \, m^2$$

$$v_A := 0 \, \frac{m}{sec} \qquad\qquad z_A := 0 \, m \qquad\qquad p_A := 4 \times 10^5 \, \frac{N}{m^2}$$

$$v_B := 0 \, \frac{m}{sec} \qquad\qquad z_B := 30 \, m \qquad\qquad p_B := 0 \, \frac{N}{m^2}$$

$$v_C := 0 \, \frac{m}{sec} \qquad\qquad p_C := 0 \, \frac{N}{m^2} \qquad\qquad z_C := 27 \, m$$

Guess values: $f_1 := 0.01$ $f_2 := 0.01$ $f_3 := 0.01$ $R_1 := 400{,}000$

$$R_2 := 400{,}000 \quad R_3 := 400{,}000 \quad h_{f1} := 10 \, m \quad h_{f2} := 10 \, m$$

$$h_{f3} := 10 \, m \quad v_1 := 5 \, \frac{m}{sec} \quad v_2 := 5 \, \frac{m}{sec} \quad v_3 := 5 \, \frac{m}{sec}$$

$$Q_1 := 1 \, \frac{m^3}{sec} \quad Q_2 := 1 \, \frac{m^3}{sec} \quad Q_3 := 1 \, \frac{m^3}{sec}$$

Given

$$\frac{p_A}{\gamma} + z_A + \frac{v_A^2}{2 \cdot g} - h_{f1} - h_{f2} = \frac{p_B}{\gamma} + z_B + \frac{v_B^2}{2 \cdot g}$$

$$\frac{p_A}{\gamma} + z_A + \frac{v_A^2}{2 \cdot g} - h_{f1} - h_{f3} = \frac{p_C}{\gamma} + z_C + \frac{v_C^2}{2 \cdot g}$$

$$Q_1 = Q_2 + Q_3 \qquad\qquad v_1 = \frac{Q_1}{A_1} \qquad v_2 = \frac{Q_2}{A_2} \qquad v_3 = \frac{Q_3}{A_3}$$

$$\frac{1}{\sqrt{f_1}} = -2 \cdot \log\left(\frac{\varepsilon/D_1}{3.7} + \frac{2.51}{R_1 \cdot \sqrt{f_1}}\right) \quad R_1 = \frac{\rho \cdot v_1 \cdot D_1}{\mu} \qquad h_{f1} = f_1 \cdot \frac{L_1}{D_1} \frac{v_1^{\,2}}{2 \cdot g}$$

$$\frac{1}{\sqrt{f_2}} = -2 \cdot \log\left(\frac{\varepsilon/D_2}{3.7} + \frac{2.51}{R_2 \cdot \sqrt{f_2}}\right) \quad R_2 = \frac{\rho \cdot v_2 \cdot D_2}{\mu} \qquad h_{f2} = f_2 \cdot \frac{L_2}{D_2} \frac{v_2^{\,2}}{2 \cdot g}$$

$$\frac{1}{\sqrt{f_3}} = -2 \cdot \log\left(\frac{\varepsilon/D_3}{3.7} + \frac{2.51}{R_3 \cdot \sqrt{f_3}}\right) \quad R_3 = \frac{\rho \cdot v_3 \cdot D_3}{\mu} \qquad h_{f3} = f_3 \cdot \frac{L_3}{D_3} \frac{v_3^{\,2}}{2 \cdot g}$$

$$\begin{pmatrix} f_1 \\ f_2 \\ f_3 \\ R_1 \\ R_2 \\ R_3 \\ h_{f1} \\ h_{f2} \\ h_{f3} \\ v_1 \\ v_2 \\ v_3 \\ Q_1 \\ Q_2 \\ Q_3 \end{pmatrix} := \text{Find}\,(f_1, f_2, f_3, R_1, R_2, R_3, h_{f1}, h_{f2}, h_{f3}, v_1, v_2, v_3, Q_1, Q_2, Q_3)$$

$f_1 = 0.024$ $\qquad f_2 = 0.028$ $\qquad f_3 = 0.026$ $\qquad R_1 = 1.397 \times 10^5$

$R_2 = 7.721 \times 10^4$ $\qquad R_3 = 1.126 \times 10^5$ $\qquad h_{f1} = 4.745 \text{ m}$ $\qquad h_{f2} = 6.111 \text{ m}$

$h_{f3} = 9.111 \text{ m}$ $\qquad v_1 = 1.75\,\dfrac{\text{m}}{\text{s}}$ $\qquad v_2 = 1.547\,\dfrac{\text{m}}{\text{s}}$ $\qquad v_3 = 1.735\,\dfrac{\text{m}}{\text{s}}$

$Q_1 = 8.797 \times 10^{-3}\,\dfrac{\text{m}^3}{\text{s}}$ $\qquad Q_2 = 3.038 \times 10^{-3}\,\dfrac{\text{m}^3}{\text{s}}$ $\qquad Q_3 = 5.759 \times 10^{-3}\,\dfrac{\text{m}^3}{\text{s}}$

Furthermore, the governing continuity equation is satisfied (checks) as follows:

$$Q_2 + Q_3 = 8.797 \times 10^{-3}\,\dfrac{\text{m}^3}{\text{s}}$$

8.7.7 Pipes in a Loop

Pipes in a loop include three or more pipes with different lengths and diameters connected to each other in a loop created by the multiple pipes and the two reservoirs, as illustrated in Figure 8.22. In the analysis and design of a pipe system containing pipes in a loop, including very long pipes with pipe components, both major losses (Problem Types 1 and 2) and minor losses may be significant. However, minor losses may be insignificant when pipe components are separated by approximately 1000 diameters, and/or when there is a large pipe length-to-pipe diameter ratio. As such, to simplify calculations, minor losses and velocity

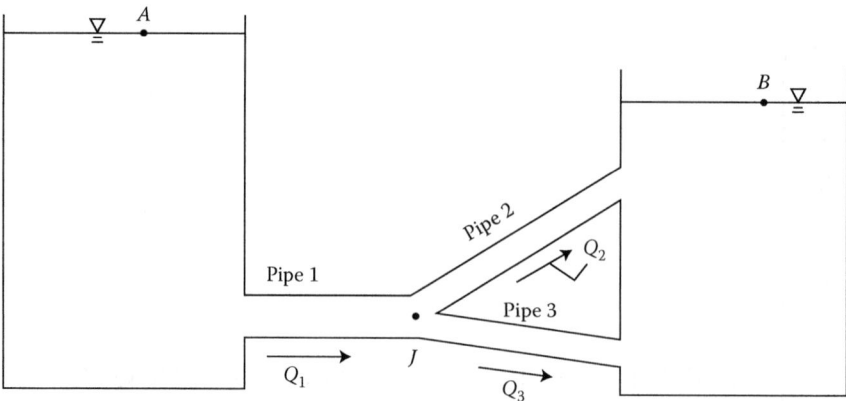

FIGURE 8.22
Pipes in a loop.

heads will be assumed negligible in most pipes in a loop problems. Once the fluid leaves reservoir A, there is more than one path for the fluid to take in the pipe system to reach reservoir B. The fluid flows from reservoir A in a single path in pipe 1, and enters a downstream junction J. As the fluid leaves junction J, it flows in multiple paths in pipe 2 and pipe 3, into reservoir B, forming a loop. The governing equations are outlined as follows. The datum is assumed to be at point B. Because points A and B are open to the atmosphere, $p_A = p_B = 0$, and because the reservoirs are large compared to the pipe diameters, $v_A = v_B = 0$. Thus, the total head, H for points A and B is equal to their respective elevations: $H_A = z_A$, and $H_B = z_B$ respectively. Furthermore, the assumption is that the velocity head at junction J is negligible; thus, $v_J^2/2g = 0$. Therefore, from the conservation of mass principle (integral continuity equation; first governing principle), the total flowrate entering junction J is equal to the total flowrate leaving junction J regardless of the diameters of the individual pipes in the loop, which is the first governing equation and is given as follows:

$$Q_1 = Q_2 + Q_3 \tag{8.90}$$

From the conservation of energy principle (second governing principle), all the significant head losses are modeled in the energy equation applied between two points, which is the second set of governing equations and are given as follows:

$$\frac{p_A}{\gamma} + z_A + \frac{v_A^2}{2g} - h_{f1} - h_{f2} = \frac{p_B}{\gamma} + z_B + \frac{v_B^2}{2g} \tag{8.91}$$

$$\frac{p_A}{\gamma} + z_A + \frac{v_A^2}{2g} - h_{f1} - h_{f3} = \frac{p_B}{\gamma} + z_B + \frac{v_B^2}{2g} \tag{8.92}$$

which requires that the head loss for pipe 2 must equal to the head loss for pipe 3 regardless of the diameters of the individual pipes in the loop as follows:

$$h_{f2} = h_{f3} \tag{8.93}$$

which is the governing head loss equation that establishes the relative velocities and thus, the relative flowrates in the pipes in a loop. Evaluation of the major (and minor) head loss flow resistance terms requires a "subset level" application of the governing equations (see Chapters 6 and 7 and Section 8.4.1.2). Depending upon whether the flow is laminar or turbulent (and depending upon the desired accuracy, simplicity, and practical constraints in the range of the variables (v, D, and water temperature)), the major head loss due to pipe friction, $h_{f,major}$ is evaluated by the appropriate resistance equation (see Section 8.5). Finally, from the conservation of momentum principle, the friction slope, S_f due to pipe friction at a given point in a straight pipe section may be modeled in the differential momentum equation, Equation 8.12, which is the third governing principle. Evaluation of the friction slope requires a "subset level" application of the governing equations (see Chapters 6 and 7 and Section 8.4.1.4) (see Equation 8.13 for the Chezy equation).

EXAMPLE PROBLEM 8.27

Water at 20°C flows from point A to point B through three concrete pipes in a loop at a total flowrate of 1.8 m³/sec, as illustrated in Figure EP 8.27. The pipe system contains three single pipes and two reservoirs. The first pipe is 0.9 m in diameter and 700 m long. The second pipe is 0.75 m in diameter and 600 m long. And the third pipe is 0.65 m in diameter and 600 m long. (a) Determine the difference in elevation between points A and B. (b) Determine the total head loss due to pipe friction between points A and B. (c) Determine the flowrate in pipe 2 and pipe 3.

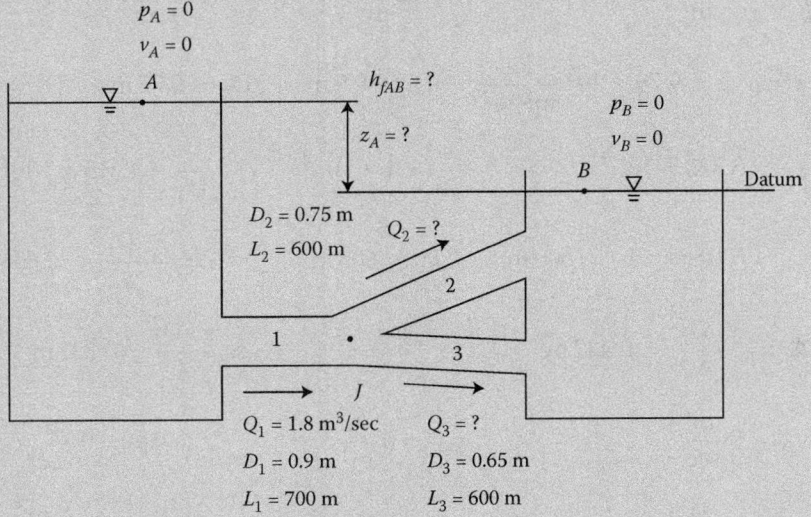

FIGURE EP 8.27
Water flows from reservoir A to reservoir B through three concrete pipes in a loop with common junction J.

Mathcad Solution

(a)–(c) Assuming the datum is at point B, the difference in elevation between points A and B is computed by applying the two governing energy equations between points A and B as follows:

$$\frac{p_A}{\gamma} + z_A + \frac{v_A^2}{2g} - h_{f1} - h_{f2} = \frac{p_B}{\gamma} + z_B + \frac{v_B^2}{2g}$$

$$\frac{p_A}{\gamma} + z_A + \frac{v_A^2}{2g} - h_{f1} - h_{f3} = \frac{p_B}{\gamma} + z_B + \frac{v_B^2}{2g}$$

which requires that the head loss for pipe 2 must equal the head loss for pipe 3 regardless of the diameters of the individual pipes in the loop as follows:

$$h_{f2} = h_{f3}$$

The velocity of flow is computed from the continuity equation. The major head loss due to pipe friction for each pipe is computed from Equation 8.34 (Darcy–Weisbach head loss equation) as a Type 1 problem. The absolute pipe roughness for concrete is given in Table 8.3, and the Colebrook equation for the friction factor is assumed. The density, ρ and the viscosity, μ for water at 20°C are given in Table A.4 in Appendix A. The governing head loss equation, $h_{f2} = h_{f3}$ establishes the relative velocities and thus the relative flowrates in pipe 2 and pipe 3, solved as two independent Type 2 problems, where $Q_1 = Q_2 + Q_3$. Thus, in total, 14 equations are solved for 14 unknowns as follows:

$$\rho := 998 \, \frac{kg}{m^3} \qquad \mu := 1 \times 10^{-3} \, N \, \frac{sec}{m^2} \qquad g := 9.81 \, \frac{m}{sec^2} \qquad \varepsilon := 0.3 \, mm$$

$$\gamma := \rho \cdot g = 9.79 \times 10^3 \, \frac{kg}{m^2 \cdot s^2} \qquad D_1 := 0.9 \, m \qquad D_2 := 0.75 \, m \qquad D_3 := 0.65 \, m$$

$$\frac{\varepsilon}{D_1} = 3.333 \times 10^{-4} \qquad\qquad \frac{\varepsilon}{D_2} = 4 \times 10^{-4} \qquad\qquad \frac{\varepsilon}{D_3} = 4.615 \times 10^{-4}$$

$$L_1 := 700 \, m \qquad L_2 := 600 \, m \qquad L_3 := 600 \, m \qquad A_1 := \frac{\pi \cdot D_1^2}{4} = 0.636 \, m^2$$

$$A_2 := \frac{\pi \cdot D_2^2}{4} = 0.442 \, m^2 \qquad\qquad\qquad A_3 := \frac{\pi \cdot D_3^2}{4} = 0.332 \, m^2$$

$$v_A := 0 \, \frac{m}{sec} \qquad\qquad p_A := 0 \, \frac{N}{m^2} \qquad\qquad v_B := 0 \, \frac{m}{sec}$$

$$z_B := 0 \, m \qquad\qquad p_B := 0 \, \frac{N}{m^2} \qquad\qquad Q_1 := 1.8 \, \frac{m^3}{sec}$$

$$v_1 := \frac{Q_1}{A_1} = 2.829 \, \frac{m}{s}$$

Guess values: $f_1 := 0.01$ $f_2 := 0.01$ $f_3 := 0.01$ $R_1 := 400{,}000$

$$R_2 := 400{,}000 \quad R_3 := 400{,}000 \quad h_{f1} := 10 \, m \quad h_{f2} := 10 \, m$$

$$h_{f3} := 10\,m \qquad v_2 = 5\,\frac{m}{sec} \qquad v_3 = 5\,\frac{m}{sec} \quad Q_2 := 1\,\frac{m^3}{sec}$$

$$Q_3 := 1\,\frac{m^3}{sec} \qquad z_A := 20\,m$$

Given

$$\frac{p_A}{\gamma} + z_A + \frac{v_A^2}{2 \cdot g} - h_{f1} - h_{f2} = \frac{p_B}{\gamma} + z_B + \frac{v_B^2}{2 \cdot g}$$

$$\frac{p_A}{\gamma} + z_A + \frac{v_A^2}{2 \cdot g} - h_{f1} - h_{f3} = \frac{p_B}{\gamma} + z_B + \frac{v_B^2}{2 \cdot g}$$

$$Q_1 = Q_2 + Q_3 \qquad\qquad v_2 = \frac{Q_2}{A_2} \qquad\qquad v_3 = \frac{Q_3}{A_3}$$

$$\frac{1}{\sqrt{f_1}} = -2 \cdot \log\left(\frac{\varepsilon/D_1}{3.7} + \frac{2.51}{R_1 \cdot \sqrt{f_1}}\right) \qquad R_1 = \frac{\rho \cdot v_1 \cdot D_1}{\mu} \qquad h_{f1} = f_1 \cdot \frac{L_1}{D_1}\frac{v_1^2}{2 \cdot g}$$

$$\frac{1}{\sqrt{f_2}} = -2 \cdot \log\left(\frac{\varepsilon/D_2}{3.7} + \frac{2.51}{R_2 \cdot \sqrt{f_2}}\right) \qquad R_2 = \frac{\rho \cdot v_2 \cdot D_2}{\mu} \qquad h_{f2} = f_2 \cdot \frac{L_2}{D_2}\frac{v_2^2}{2 \cdot g}$$

$$\frac{1}{\sqrt{f_3}} = -2 \cdot \log\left(\frac{\varepsilon/D_3}{3.7} + \frac{2.51}{R_3 \cdot \sqrt{f_3}}\right) \qquad R_3 = \frac{\rho \cdot v_3 \cdot D_3}{\mu} \qquad h_{f3} = f_3 \cdot \frac{L_3}{D_3}\frac{v_3^2}{2 \cdot g}$$

$$\begin{pmatrix} f_1 \\ f_2 \\ f_3 \\ R_1 \\ R_2 \\ R_3 \\ h_{f1} \\ h_{f2} \\ h_{f3} \\ v_2 \\ v_3 \\ Q_2 \\ Q_3 \\ z_A \end{pmatrix} := Find\,(f_1, f_2, f_3, R_1, R_2, R_3, h_{f1}, h_{f2}, h_{f3}, v_2, v_3, Q_2, Q_3, z_A)$$

$f_1 = 0.016$ $\qquad\qquad$ $f_2 = 0.016$ $\qquad\qquad$ $f_3 = 0.017$ $\qquad$ $R_1 = 2.541 \times 10^6$

$R_2 = 1.807 \times 10^6$ $\qquad$ $R_3 = 1.434 \times 10^6$ $\quad$ $h_{f1} = 4.937\,m$ $\quad$ $h_{f2} = 3.858\,m$

$h_{f3} = 3.858\,m$ $\qquad\qquad$ $v_2 = 2.414\,\dfrac{m}{s}$ $\qquad$ $v_3 = 2.21\,\dfrac{m}{s}$ $\qquad$ $Q_2 = 1.067\,\dfrac{m^3}{s}$

$Q_3 = 0.733\,\dfrac{m^3}{s}$ $\qquad$ $z_A = 8.795\,m$

where the total head loss due to pipe friction is computed as follows:

$h_{ftotal} := h_{f1} + h_{f2} = 8.795\,m$ $\qquad\qquad$ $h_{ftotal} := h_{f1} + h_{f3} = 8.795\,m$

Furthermore, the governing continuity equation is satisfied (checks) as follows:

$$Q_2 + Q_3 = 1.8\,\dfrac{m^3}{s}$$

EXAMPLE PROBLEM 8.28

Water at 20°C flows from point A to point B through three galvanized iron pipes in a loop, as illustrated in Figure EP 8.28. The pipe system contains three single pipes and two reservoirs. The first pipe is 0.9 m in diameter and 700 m long. The second pipe is 0.75 m in diameter and 600 m long, and the third pipe is 0.65 m in diameter and 600 m long. The difference in elevation between points A and B is 15 m. (a) Determine the flowrate in each of the three pipes. (b) Determine the total head loss due to pipe friction between points A and B.

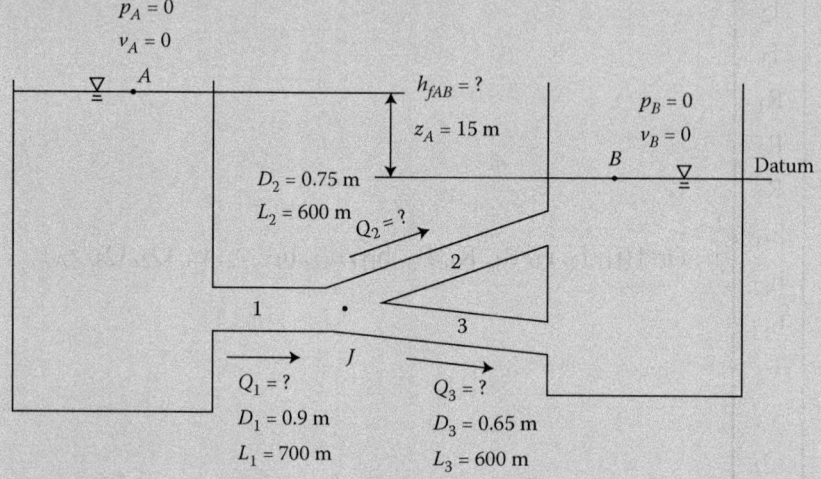

FIGURE EP 8.28
Water flows from reservoir A to reservoir B through three galvanized iron pipes in a loop with common junction J.

Mathcad Solution

(a)–(c) Assuming the datum is at point B, the velocity in each pipe is computed by applying the two governing energy equations between points A and B as follows:

$$\frac{p_A}{\gamma} + z_A + \frac{v_A^2}{2g} - h_{f1} - h_{f2} = \frac{p_B}{\gamma} + z_B + \frac{v_B^2}{2g}$$

$$\frac{p_A}{\gamma} + z_A + \frac{v_A^2}{2g} - h_{f1} - h_{f3} = \frac{p_B}{\gamma} + z_B + \frac{v_B^2}{2g}$$

which requires that the head loss for pipe 2 must equal to the head loss for pipe 3 regardless of the diameters of the individual pipes in the loop as follows:

$$h_{f2} = h_{f3}$$

The discharge is computed from the continuity equation. The major head loss due to pipe friction for each pipe is computed from Equation 8.34 (Darcy–Weisbach head loss equation) as a Type 1 problem. The absolute pipe roughness for galvanized iron is given in Table 8.3, and the Colebrook equation for the friction factor is assumed. The density, ρ and the viscosity, μ for water at 20°C are given in Table A.4 in Appendix A. The governing head loss equation, $h_{f2} = h_{f3}$ establishes the relative velocities and thus the relative flowrates in pipe 1, pipe 2, and pipe 3, solved as three independent Type 2 problems, where $Q_1 = Q_2 + Q_3$. Thus, in total, 15 equations are solved for 15 unknowns as follows:

$$\rho := 998 \, \frac{kg}{m^3} \qquad \mu := 1 \times 10^{-3} \, N \, \frac{sec}{m^2} \qquad g := 9.81 \, \frac{m}{sec^2} \qquad \varepsilon := 0.15 \, mm$$

$$\gamma := \rho \cdot g = 9.79 \times 10^3 \, \frac{kg}{m^2 \cdot s^2} \qquad D_1 := 0.9 \, m \qquad D_2 := 0.75 \, m \qquad D_3 := 0.65 \, m$$

$$\frac{\varepsilon}{D_1} = 1.667 \times 10^{-4} \qquad \frac{\varepsilon}{D_2} = 2 \times 10^{-4} \qquad \frac{\varepsilon}{D_3} = 2.308 \times 10^{-4}$$

$$L_1 := 700 \, m \qquad L_2 := 600 \, m \qquad L_3 := 600 \, m \qquad A_1 := \frac{\pi \cdot D_1^2}{4} = 0.636 \, m^2$$

$$A_2 := \frac{\pi \cdot D_2^2}{4} = 0.442 \, m^2 \qquad\qquad A_3 := \frac{\pi \cdot D_3^2}{4} = 0.332 \, m^2$$

$$v_A := 0 \, \frac{m}{sec} \qquad p_A := 0 \, \frac{N}{m^2} \qquad z_A := 15 \, m$$

$$v_B := 0 \, \frac{m}{sec} \qquad p_B := 0 \, \frac{N}{m^2} \qquad z_B := 0 \, m$$

Guess values: $\quad f_1 := 0.01 \qquad f_2 := 0.01 \qquad f_3 := 0.01 \qquad R_1 := 400,000$

$$R_2 := 400,000 \qquad R_3 := 400,000 \qquad h_{f1} := 10 \, m \qquad h_{f2} := 10 \, m$$

$$h_{f3} := 10 \text{ m} \qquad v_1 := 5 \frac{m}{sec} \qquad v_2 := 5 \frac{m}{sec} \qquad v_3 := 5 \frac{m}{sec}$$

$$Q_1 := 1 \frac{m^3}{sec} \qquad Q_2 := 1 \frac{m^3}{sec} \qquad Q_3 := 1 \frac{m^3}{sec}$$

Given

$$\frac{p_A}{\gamma} + z_A + \frac{v_A^2}{2 \cdot g} - h_{f1} - h_{f2} = \frac{p_B}{\gamma} + z_B + \frac{v_B^2}{2 \cdot g}$$

$$\frac{p_A}{\gamma} + z_A + \frac{v_A^2}{2 \cdot g} - h_{f1} - h_{f3} = \frac{p_B}{\gamma} + z_B + \frac{v_B^2}{2 \cdot g}$$

$$Q_1 = Q_2 + Q_3 \qquad\qquad v_1 = \frac{Q_1}{A_1} \qquad\qquad v_2 = \frac{Q_2}{A_2} \qquad\qquad v_3 = \frac{Q_3}{A_3}$$

$$\frac{1}{\sqrt{f_1}} = -2 \cdot \log\left(\frac{\varepsilon/D_1}{3.7} + \frac{2.51}{R_1 \cdot \sqrt{f_1}}\right) \quad R_1 = \frac{\rho \cdot v_1 \cdot D_1}{\mu} \quad h_{f1} = f_1 \cdot \frac{L_1}{D_1} \frac{v_1^2}{2 \cdot g}$$

$$\frac{1}{\sqrt{f_2}} = -2 \cdot \log\left(\frac{\varepsilon/D_2}{3.7} + \frac{2.51}{R_2 \cdot \sqrt{f_2}}\right) \quad R_2 = \frac{\rho \cdot v_2 \cdot D_2}{\mu} \quad h_{f2} = f_2 \cdot \frac{L_2}{D_2} \frac{v_2^2}{2 \cdot g}$$

$$\frac{1}{\sqrt{f_3}} = -2 \cdot \log\left(\frac{\varepsilon/D_3}{3.7} + \frac{2.51}{R_3 \cdot \sqrt{f_3}}\right) \quad R_3 = \frac{\rho \cdot v_3 \cdot D_3}{\mu} \quad h_{f3} = f_3 \cdot \frac{L_3}{D_3} \frac{v_3^2}{2 \cdot g}$$

$$\begin{pmatrix} f_1 \\ f_2 \\ f_3 \\ R_1 \\ R_2 \\ R_3 \\ h_{f1} \\ h_{f2} \\ h_{f3} \\ v_1 \\ v_2 \\ v_3 \\ Q_1 \\ Q_2 \\ Q_3 \end{pmatrix} := \text{Find}\,(f_1, f_2, f_3, R_1, R_2, R_3, h_{f1}, h_{f2}, h_{f3}, v_1, v_2, v_3, Q_1, Q_2, Q_3)$$

$f_1 = 0.014$ $\qquad$ $f_2 = 0.014$ $\qquad$ $f_3 = 0.015$ $\qquad$ $R_1 = 3.555 \times 10^6$

$R_2 = 2.527 \times 10^6$ $\qquad$ $R_3 = 2.007 \times 10^6$ $\qquad$ $h_{f1} = 8.429\,m$ $\qquad$ $h_{f2} = 6.571\,m$

$h_{f3} = 6.571\,m$ $\qquad$ $v_1 = 3.958\,\dfrac{m}{s}$ $\qquad$ $v_2 = 3.376\,\dfrac{m}{s}$ $\qquad$ $v_2 = 3.093\,\dfrac{m}{s}$

$Q_1 = 2.518\,\dfrac{m^3}{s}$ $\qquad$ $Q_2 = 1.419\,\dfrac{m^3}{s}$ $\qquad$ $Q_3 = 1.026\,\dfrac{m^3}{s}$

where the total head loss due to pipe friction is computed as follows:

$h_{ftotal} := h_{f1} + h_{f2} = 15\,m$ $\qquad\qquad$ $h_{ftotal} := h_{f1} + h_{f3} = 15\,m$

Furthermore, the governing continuity equation is satisfied (checks) as follows:

$$Q_2 + Q_3 = 2.518\,\dfrac{m^3}{s}$$

8.7.8 Pipe Networks

Pipe networks such as water distribution systems include one or more sources (inlets) of water supply, numerous arrangements of pipes in series and pipes in parallel, junctions or nodes where two or more pipes meet, and one or more points of discharge (destination/demand/outlet), as illustrated in Figure 8.23. In the analysis and design of a pipe network, there may be pipe friction (Problem Types 1, 2, and 3), pipe components, transitions, or devices in the pipe flow that will determine the significance of a major head loss, a minor head loss, an added or removed head, and a reaction force. Because a pipe network consists of numerous arrangements of pipes in series and pipes in parallel, their respective governing equations are applied as discussed and summarized in the sections below (Sections 8.7.8.1 through 8.7.8.4). Furthermore, one may note that in order to simplify calculations, minor losses and velocity heads will be assumed negligible in most pipe network problems.

8.7.8.1 Continuity Principle

From the conservation of mass principle (integral continuity equation; first governing principle), the total flowrate entering a junction/node is equal to the total flowrate leaving the junction for all junctions in the system regardless of the diameters of the individual pipes in parallel as follows:

$$Q = Q_1 + Q_2 + Q_3 + \cdots \tag{8.94}$$

Furthermore, from the conservation mass principle (continuity equation; first governing principle), the flowrate through the entire pipe system remains constant regardless of the diameters of the individual pipes in series as follows:

$$Q = Q_1 = Q_2 = Q_3 = \cdots \tag{8.95}$$

One may note that the above governing flowrate equation, which requires that the flowrate be the same in each pipe in series, establishes the relative velocities and thus the relative

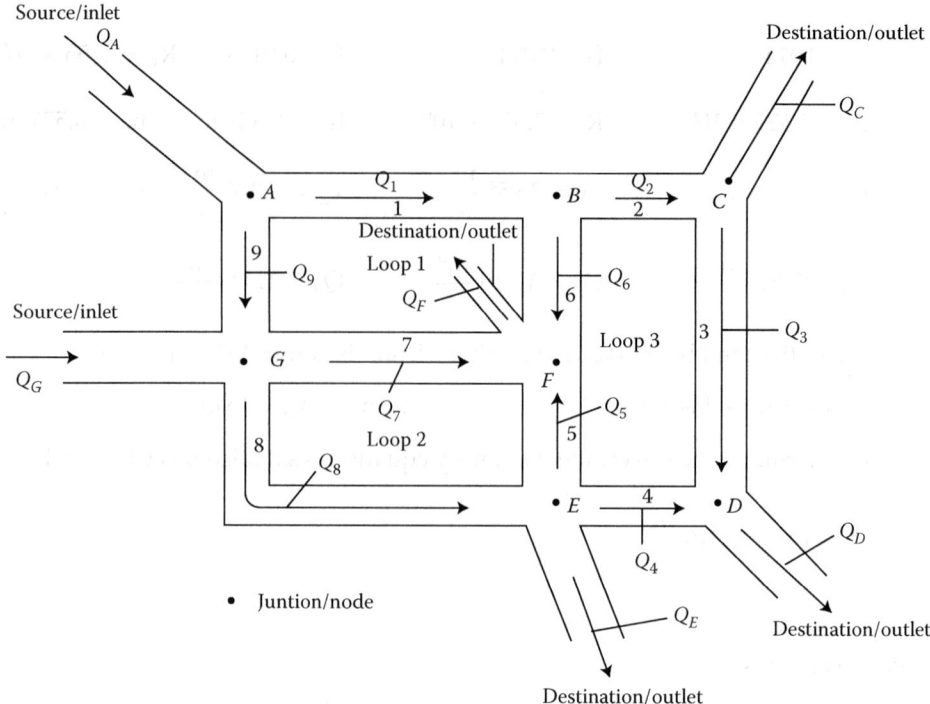

FIGURE 8.23
Pipe network.

head losses in the pipes in series. This governing flowrate equation can be applied to any number of pipes in series. Therefore, these two continuity equations, Equation 8.94 for pipes in parallel and Equation 8.95 for pipes in series, respectively, are the first set of governing equations.

8.7.8.2 Energy Principle

From the conservation of energy principle, all the significant head losses are modeled in the energy equation applied between points A and B as follows:

$$\frac{p_A}{\gamma} + z_A + \frac{v_A^2}{2g} - \sum h_{f,maj} - \sum h_{f,min} = \frac{p_B}{\gamma} + z_B + \frac{v_B^2}{2g} \tag{8.96}$$

which is the second governing principle, where the total head loss in between points A and B (which may be two junctions/nodes) is equal to the sum of the head losses (major and minor) in the individual pipes and pipe components in series for a given path (pipe), as follows:

$$h_{ftotal} = \sum h_{f,maj} + \sum h_{f,min} \tag{8.97}$$

The pressure head drop, $\Delta p/\gamma$ between points A and B is due to the change in elevation, Δz; the change in velocity head, $\Delta v^2/2g$; and the total head loss (major and minor) due to the flow resistance (pipe friction and pipe components), h_{ftotal}. Furthermore, although there is more than one path for the fluid to take for pipes in parallel, each path has the same (sub) inlet point A and the same (sub) outlet point B. Therefore, the total pressure drop between junction A and junction B in the pipe system, Δp_{AB} is equal to the pressure drop across each path connecting the two junctions A and B as follows:

$$\Delta p_{AB} = \Delta p_{path1} = \Delta p_{path2} = \Delta p_{path3} = \cdots \tag{8.98}$$

As a result, the total head loss (major and minor) due to the flow resistance (pipe friction and pipe components) between junctions A and B, $h_{ftotalAB}$ is equal to the total head loss across each path connecting the two junctions A and B as follows:

$$h_{ftotalAB} = h_{ftotalpath1} = h_{ftotalpath2} = h_{ftotalpath3} = \cdots \tag{8.99}$$

which is equivalent to Equation 8.98 above in the case where the change in elevation, $\Delta z = 0$ and the change in velocity head, $\Delta v^2/2g = 0$. One may note that the above governing head loss equation, which requires that the head loss be the same in each parallel pipe/path, establishes the relative velocities and thus the relative flowrates in the parallel pipes/paths. This governing head loss equation can be applied to any number of pipes/paths in parallel. Therefore, these two energy head loss equations (Equations 8.97 for pipes in series and Equation 8.99 for pipes in parallel, respectively) are the second set of governing equations. Furthermore, one may note that the simultaneous application of the above two governing head loss equations, Equations 8.97 and 8.99, is accomplished by requiring that the algebraic sum of head losses for a given loop, for all loops, be equal to zero as follows:

$$\sum_{loop} h_{ftotal} = 0 \tag{8.100}$$

where a head loss is assumed positive for flow in the clockwise direction and negative for flow in the counterclockwise direction.

Evaluation of the major (and minor) head loss flow resistance terms (in the energy equation, Equation 8.96) requires a "subset level" application of the governing equations (see Chapters 6 and 7 and Section 8.4.1.2). Depending upon whether the flow is laminar or turbulent, and depending upon the desired accuracy, simplicity, and practical constraints in the range of the variables (v, D, and water temperature), the major head loss due to pipe friction, $h_{f,major}$ is evaluated by the appropriate resistance equation (see Section 8.5). One may note because the flowrate in the pipes significantly vary throughout a given day, the value for the Darcy–Weisbach friction factor, f, (and the C coefficient for the Chezy equation) which is function of R, which is a function of v, will also vary throughout the day. However, because the water supply capacity/demand requirements for most pipe networks are difficult to predict, it is common in practice to assume a constant value for the Darcy–Weisbach friction factor, f for a given pipe in the network.

In the case where a constant value for the Darcy–Weisbach friction factor, f for a given pipe in the network is assumed, the head loss equation is simplified as follows:

$$h_f = f\frac{L}{D}\frac{v^2}{2g} = f\frac{L}{D}\frac{Q^2}{A^2 2g} = f\frac{L}{D}\frac{Q^2}{(\pi D^2/4)^2 2g} = \underbrace{\left[\frac{8fL}{\pi^2 g D^5}\right]}_{K} Q^2 = KQ^a \tag{8.101}$$

where a represents a constant exponent of 2 for the Darcy–Weisbach equation, and K represents a constant value for a given pipe as follows:

$$K = \frac{8fL}{\pi^2 g D^5} \tag{8.102}$$

Unlike the friction factor, f for the Darcy–Weisbach equation (and the C coefficient for the Chezy equation), the flow resistance (μ, n, and C_h) represented in the other flow resistance equations in Section 8.5 (for the Poiseuille, Manning, and Hazen–Williams equations, respectively) are not a function of the variable flowrates in the pipes and thus do not vary throughout the day. However, similar to the simplified representation for the Darcy–Weisbach equation, the other flow resistance equations may also be simplified as follows. The Poiseuille equation can be simplified as follows:

$$h_f = \frac{32\mu L v}{\gamma D^2} = \frac{32\mu L Q}{\gamma D^2 \frac{\pi D^2}{4}} = \underbrace{\left[\frac{128\mu L}{\pi\gamma D^4}\right]}_{K} Q = KQ^a \tag{8.103}$$

where a represents a constant exponent of 1 for the Poiseuille equation, and K represents a constant value for a given pipe as follows:

$$K = \frac{128\mu L}{\pi\gamma D^4} \tag{8.104}$$

The Chezy equation can be simplified as follows:

$$h_f = \frac{Lv^2}{C^2 R_h} = \frac{LQ^2}{C^2 R_h \left(\frac{\pi D^2}{4}\right)^2} = \frac{LQ^2}{C^2(D/4)\left(\frac{\pi^2 D^4}{16}\right)} = \underbrace{\left[\frac{64L}{C^2\pi^2 D^5}\right]}_{K} Q^2 = KQ^a \tag{8.105}$$

where a represents a constant exponent of 2 for the Chezy equation, and K represents a constant value for a given pipe as follows:

$$K = \frac{64L}{C^2\pi^2 D^5} \tag{8.106}$$

The Manning equation in SI units can be simplified as follows:

$$h_f = \left(\frac{vn}{R_h^{2/3}}\right)^2 L = \frac{Ln^2 v^2}{\left(\frac{D}{4}\right)^{4/3}} = \frac{Ln^2 Q^2}{\left(\frac{D^{4/3}}{6.35}\right)\left(\frac{\pi D^2}{4}\right)^2} = \frac{Ln^2 Q^2}{\left(\frac{D^{4/3}}{6.35}\right)\left(\frac{\pi^2 D^4}{16}\right)} = \underbrace{\left[\frac{10.30Ln^2}{D^{16/3}}\right]}_{K} Q^2 = KQ^a \tag{8.107}$$

where a represents a constant exponent of 2 for the Manning equation, and K represents a constant value for a given pipe as follows:

$$K = \frac{10.30Ln^2}{D^{16/3}} \tag{8.108}$$

The Manning equation in BG units can be simplified as follows:

$$h_f = \left(\frac{vn}{1.486R_h^{2/3}}\right)^2 L = \frac{Ln^2 v^2}{2.208\left(\frac{D}{4}\right)^{4/3}} = \frac{Ln^2 Q^2}{2.208\left(\frac{D^{4/3}}{6.35}\right)\left(\frac{\pi D^2}{4}\right)^2}$$

$$= \frac{Ln^2 Q^2}{2.208\left(\frac{D^{4/3}}{6.35}\right)\left(\frac{\pi^2 D^4}{16}\right)} = \underbrace{\left[\frac{4.66Ln^2}{D^{16/3}}\right]}_{K} Q^2 = KQ^a \tag{8.109}$$

where *a* represents a constant exponent of 2 for the Manning equation, and *K* represents a constant value for a given pipe as follows:

$$K = \frac{4.66Ln^2}{D^{16/3}} \tag{8.110}$$

The Hazen–Williams equation in SI units can be simplified as follows:

$$h_f = \left(\frac{v}{0.849C_hR_h^{0.63}}\right)^{1/0.54} \quad L = \frac{Lv^{1.852}}{0.738C_h^{1.852}\left(\frac{D}{4}\right)^{1.167}} = \frac{LQ^{1.852}}{0.738C_h^{1.852}\left(\frac{D^{1.167}}{5.042}\right)\left(\frac{\pi D^2}{4}\right)^{1.852}}$$

$$= \frac{LQ^{1.852}}{0.738C_h^{1.852}\left(\frac{D^{1.167}}{5.042}\right)\left(\frac{\pi^{1.852}D^{3.704}}{13.032}\right)} = \underbrace{\left[\frac{10.687L}{C_h^{1.852}D^{4.871}}\right]}_{K}Q^{1.852} = KQ^a \tag{8.111}$$

where *a* represents a constant exponent of 1.852 for the Hazen–Williams equation, and *K* represents a constant value for a given pipe as follows:

$$K = \frac{10.687L}{C_h^{1.852}D^{4.871}} \tag{8.112}$$

The Hazen–Williams equation in BG units can be simplified as follows:

$$h_f = \left(\frac{v}{1.318C_hR_h^{0.63}}\right)^{1/0.54} \quad L = \frac{Lv^{1.852}}{1.668C_h^{1.852}\left(\frac{D}{4}\right)^{1.167}} = \frac{LQ^{1.852}}{1.668C_h^{1.852}\left(\frac{D^{1.167}}{5.042}\right)(\pi D^2/4)^{1.852}}$$

$$= \frac{LQ^{1.852}}{1.668C_h^{1.852}\left(\frac{D^{1.167}}{5.042}\right)\left(\frac{\pi^{1.852}D^{3.704}}{13.032}\right)} = \underbrace{\left[\frac{4.728L}{C_h^{1.852}D^{4.871}}\right]}_{K}Q^{1.852} = KQ^a \tag{8.113}$$

where *a* represents a constant exponent of 1.852 for the Hazen–Williams equation, and *K* represents a constant value for a given pipe as follows:

$$K = \frac{4.728L}{C_h^{1.852}D^{4.871}} \tag{8.114}$$

8.7.8.3 Momentum Principle

And, finally, from the conservation of momentum principle, the friction slope, S_f due to pipe friction at a given point in a straight pipe section may be modeled in the differential momentum equation, Equation 8.12, which is the third governing principle. Evaluation of the friction slope requires a "subset level" application of the governing equations (see Chapters 6 and 7 and Section 8.4.1.4) (see Equation 8.13 for the Chezy equation).

8.7.8.4 Summary of Governing Equations for Pipe Networks

Thus, the governing equations along with their respective applied criteria in the analysis and design of a pipe network are summarized in Table 8.15 as follows:

TABLE 8.15

Summary of Governing Equations for Pipe Networks

Governing Principle	Governing Equation	Applied Equation/Criteria
Continuity for Pipes in Parallel	$Q = Q_1 + Q_2 + Q_3 + \cdots$	The flowrate entering a given *junction/node* must equal to the flowrate leaving the *junction/node*
Continuity for Pipes in Series	$Q = Q_1 = Q_2 = Q_3 = \cdots$ Establishes velocities and thus head losses in pipes in series	The flowrate through the *inlets* must be equal to the flowrate through the *outlets*
Energy for Pipes in Series	$h_{ftotal} = \sum h_{f,maj} + \sum h_{f,min}$	
Energy for Pipes in Parallel	$h_{ftotalAB} = h_{ftotalpath1} = h_{ftotalpath2} = h_{ftotalpath3} = \cdots$ Establishes velocities and thus flowrates in pipes/paths in parallel	
Energy for Pipes in Series and Pipes in Parallel	$\sum_{loop} h_{ftotal} = 0$	The algebraic sum of head losses for a given *loop* must be equal to zero
"Subset Level" Application of the Governing Equations	$h_f = f \dfrac{L}{D} \dfrac{v^2}{2g}$	The head loss in a given *pipe* is modeled by the appropriate flow resistance equation

Based upon the total number of junctions, j and the total number of pipes, i in the pipe network, one may determine the resulting total number of loops, l as follows:

$$l = i - j + 1 \tag{8.115}$$

Furthermore, based upon the number of pipes, i in the pipe network, and the whether a constant or a variable Darcy–Weisbach friction factor, f is assumed, one may determine the total number of unknowns and thus the total number of required equations, e accordingly. In the case where a variable value for the Darcy–Weisbach friction factor, f for a given pipe in the network is assumed, the total number of required equations, e is computed as follows:

$$e = 5i \tag{8.116}$$

Thus, for instance, the pipe network illustrated in Figure EP 8.29 for Example Problem 8.29 below contains a total of four junctions and five pipes. Therefore, there are $l = 5 - 4 + 1 = 2$ loops. And, $e = 5 \times 5 = 25$ required equations in order to solve for 25 unknowns. The breakdown of the required 25 equations is given in Table 8.16.

However, in the case where a constant value for the Darcy–Weisbach friction factor, f for a given pipe in the network is assumed, the total number of required equations, e is computed as follows:

$$e = 2i \tag{8.117}$$

Thus, for instance, the pipe network illustrated in Figure EP 8.29 for Example Problem 8.29 below contains a total of four junctions and five pipes. Therefore, there are $l = 5 - 4 + 1 = 2$ loops. And, $e = 5 \times 2 = 10$ required equations in order to solve for 10 unknowns. The breakdown of the required 10 equations (and corresponding unknowns solved for) is given in Table 8.16, except for the last three table entries for v, f, and $\boldsymbol{R}$.

TABLE 8.16

Required Governing Equations for Pipe Network in Example Problem 8.29

Governing Principle	Governing Equation	Applied Principle	Required Equations for Example Problem 8.29	Unknowns Solved for Example Problem 8.29
Continuity for Pipes in Series	$\sum Q_{inlets} = \sum Q_{outlets}$	The flowrate through the *inlets* must be equal to the flowrate through the *outlets*	Continuity is implied by continuity for pipes in parallel	
Continuity for Pipes in Parallel	$Q = Q_1 + Q_2 + Q_3 + \cdots$	Required for all but one $(j-1)$ *junction*; continuity is implied at the last junction	$j - 1 = 3$ required equations for 3 junctions	Q for 3 pipes
Energy for Pipes in Series and Pipes in Parallel	$\sum\limits_{loop} h_{ftotal} = 0$	The algebraic sum of head losses for each of the *l* loops must be equal to zero	$l = 2$ required equations for 2 loops	Q for 2 pipes
"Subset Level" Application of the Governing Equations	$h_f = f\dfrac{L}{D}\dfrac{v^2}{2g}$	The head loss in each *pipe* is modeled by a variable Darcy–Weisbach friction factor, f	$i = 5$ required equations for 5 pipes	h_f for 5 pipes
	$v = \dfrac{Q}{A}$		$i = 5$ required equations for 5 pipes	v for 5 pipes
	$\dfrac{1}{\sqrt{f}} = -2.0\log_{10}\left(\dfrac{\varepsilon/D}{3.7} + \dfrac{2.51}{R\sqrt{f}}\right)$		$i = 5$ required equations for 5 pipes	f for 5 pipes
	$R = \dfrac{\rho v D}{\mu}$		$i = 5$ required equations for 5 pipes	R for 5 pipes

EXAMPLE PROBLEM 8.29

Water at 20°C flows in a galvanized iron pipe network system, as illustrated in Figure EP 8.29. The pipe system contains one inlet, five pipes, four junctions, two loops, and two outlets. The first pipe is 0.9 m in diameter and 100 m long. The second pipe is 0.8 m in diameter and 95 m long. The third pipe is 0.95 m in diameter and 185 m long. The fourth pipe is 0.85 m in diameter and 100 m long. And the fifth pipe is 0.9 m in diameter and 190 m long. Assume that the change in elevation between all points in the pipe system, $\Delta z = 0$. The inflow at the inlet at junction A is 3 m³/sec, the outflow at the outlet at junction C is 1.7 m³/sec, and the outflow at the outlet at junction D is 1.3 m³/sec. Assume that the Darcy–Weisbach friction factor, f varies with the flowrate. (a) Determine the flowrate in each of the five pipes. (b) Determine the head loss in each of the five pipes.

Mathcad Solution

(a)–(b) Application of the continuity equation for pipes in series requires that the flowrate through the *inlets* must be equal to the flowrate through the

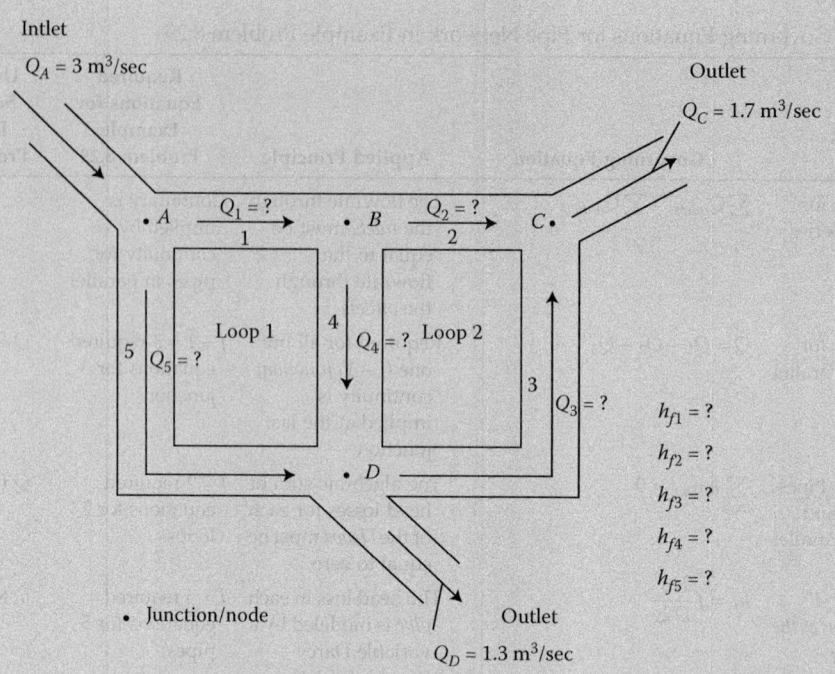

FIGURE EP 8.29
Water flows through a galvanized iron pipe network.

outlets as follows:

$$\sum Q_{inlets} = \sum Q_{outlets}$$

where the flow through inlet A is equal to the sum of the flows through outlets C and D as follows:

$$Q_A = Q_C + Q_D$$

which establishes the relative velocities and thus the relative head losses in the pipes in series. Application of the continuity equation for pipes in parallel requires that the flowrate entering a given *junction/node* must equal the flowrate leaving the *junction/node* for all junctions. One may note that application of this principle is accomplished by writing the appropriate continuity equation for all but one junction, as they imply continuity at the last junction. "Assuming the direction of flow" in the pipes yields the following continuity equations for three of the four junctions as follows:

Continuity at Junction A: $Q_A = Q_1 + Q_5$

Continuity at Junction B: $Q_1 = Q_2 + Q_4$

Continuity at Junction C: $Q_C = Q_2 + Q_3$

where the continuity at junction D, $Q_D = Q_5 + Q_4 - Q_3$ is implied by the above continuity equations at junctions A, B, and C. One may note that if the "assumed direction of flow" is incorrect, there will be a negative sign for the computed flowrate for the given pipe. Simultaneous application of the energy equations for pipes in series and pipes in parallel is accomplished by requiring that the algebraic sum of head losses for a given *loop*, for each of the two loops, be equal to zero as follows:

$$\sum_{loop1} h_{ftotal} = 0$$

$$\sum_{loop2} h_{ftotal} = 0$$

which establishes the relative velocities and thus the relative flowrates in the pipes/paths in parallel as five independent Type 2 problems, as guided by the continuity equations listed above. Finally, the major head loss due to pipe friction for a given *pipe* is computed from Equation 8.34 (Darcy–Weisbach head loss equation), as a Type 1 problem as follows:

$$h_f = f \frac{L}{D} \frac{v^2}{2g}$$

The absolute pipe roughness for galvanized iron is given in Table 8.3, and the Colebrook equation for the friction factor is assumed. The density, ρ and the viscosity, μ for water at 20°C are given in Table A.4 in Appendix A. Thus, in total, 25 equations are solved for 25 unknowns as follows:

$$\rho := 998 \frac{kg}{m^3} \quad \mu := 1 \times 10^{-3} N \frac{sec}{m^2} \quad g := 9.81 \frac{m}{sec^2} \quad \varepsilon := 0.15\,mm$$

$$D_1 := 0.9\,m \qquad D_2 := 0.8\,m \qquad D_3 := 0.95\,m$$

$$D_4 := 0.85\,m \quad D_5 := 0.9\,m \quad \frac{\varepsilon}{D_1} = 1.667 \times 10^{-4} \quad \frac{\varepsilon}{D_2} = 1.875 \times 10^{-4}$$

$$\frac{\varepsilon}{D_3} = 1.579 \times 10^{-4} \qquad \frac{\varepsilon}{D_4} = 1.765 \times 10^{-4} \qquad \frac{\varepsilon}{D_5} = 1.667 \times 10^{-4}$$

$$L_1 := 100\,m \qquad L_2 := 95\,m \qquad L_3 := 185\,m \qquad L_4 := 100\,m \qquad L_5 := 190\,m$$

$$A_1 := \frac{\pi \cdot D_1^2}{4} = 0.636\,m^2 \qquad A_2 := \frac{\pi \cdot D_2^2}{4} = 0.503\,m^2$$

$$A_3 := \frac{\pi \cdot D_3^2}{4} = 0.709\,m^2 \qquad A_4 := \frac{\pi \cdot D_4^2}{4} = 0.567\,m^2$$

$$A_5 := \frac{\pi \cdot D_5^2}{4} = 0.636\,m^2 \qquad Q_A := 3 \frac{m^3}{sec}$$

$$Q_C := 1.7 \, \frac{m^3}{sec} \qquad\qquad Q_D := 1.3 \, \frac{m^3}{sec}$$

Guess values: $f_1 := 0.01 \qquad f_2 := 0.01 \qquad f_3 := 0.01 \qquad f_4 := 0.01$

$$f_5 := 0.01 \qquad R_1 = 400{,}000 \qquad R_2 := 400{,}000$$

$$R_3 := 400{,}000 \qquad R_4 := 400{,}000 \qquad R_5 := 400{,}000$$

$$h_{f1} := 10 \, m \qquad h_{f2} := 10 \, m \qquad h_{f3} := 10 \, m \qquad h_{f4} := 10 \, m$$

$$h_{f5} := 10 \, m \qquad v_1 := 5 \, \frac{m}{sec} \qquad v_2 := 5 \, \frac{m}{sec} \qquad v_3 := 5 \, \frac{m}{sec}$$

$$v_4 := 5 \, \frac{m}{sec} \qquad v_5 := 5 \, \frac{m}{sec} \qquad Q_1 := 1.5 \, \frac{m^3}{sec} \quad Q_5 := 1.5 \, \frac{m^3}{sec}$$

$$Q_2 := 0.75 \, \frac{m^3}{sec} \quad Q_4 := 0.75 \, \frac{m^3}{sec} \quad Q_3 := 0.95 \, \frac{m^3}{sec}$$

Given

$$Q_A = Q_1 + Q_5 \qquad\qquad Q_1 = Q_2 + Q_4 \qquad\qquad Q_C = Q_2 + Q_3$$

$$v_1 = \frac{Q_1}{A_1} \qquad v_2 = \frac{Q_2}{A_2} \qquad v_3 = \frac{Q_3}{A_3} \qquad v_4 = \frac{Q_4}{A_4} \qquad v_5 = \frac{Q_5}{A_5}$$

$$\frac{1}{\sqrt{f_1}} = -2 \cdot \log\left(\frac{\varepsilon/D_1}{3.7} + \frac{2.51}{R_1 \cdot \sqrt{f_1}}\right) \quad R_1 = \frac{\rho \cdot v_1 \cdot D_1}{\mu} \quad h_{f1} = f_1 \cdot \frac{L_1}{D_1} \frac{v_1^2}{2 \cdot g}$$

$$\frac{1}{\sqrt{f_2}} = -2 \cdot \log\left(\frac{\varepsilon/D_2}{3.7} + \frac{2.51}{R_2 \cdot \sqrt{f_2}}\right) \quad R_2 = \frac{\rho \cdot v_2 \cdot D_2}{\mu} \quad h_{f2} = f_2 \cdot \frac{L_2}{D_2} \frac{v_2^2}{2 \cdot g}$$

$$\frac{1}{\sqrt{f_3}} = -2 \cdot \log\left(\frac{\varepsilon/D_3}{3.7} + \frac{2.51}{R_3 \cdot \sqrt{f_3}}\right) \quad R_3 = \frac{\rho \cdot v_3 \cdot D_3}{\mu} \quad h_{f3} = f_3 \cdot \frac{L_3}{D_3} \frac{v_3^2}{2 \cdot g}$$

$$\frac{1}{\sqrt{f_4}} = -2 \cdot \log\left(\frac{\varepsilon/D_4}{3.7} + \frac{2.51}{R_4 \cdot \sqrt{f_4}}\right) \quad R_4 = \frac{\rho \cdot v_4 \cdot D_4}{\mu} \quad h_{f4} = f_4 \cdot \frac{L_4}{D_4} \frac{v_4^2}{2 \cdot g}$$

$$\frac{1}{\sqrt{f_5}} = -2 \cdot \log\left(\frac{\varepsilon/D_5}{3.7} + \frac{2.51}{R_5 \cdot \sqrt{f_5}}\right) \quad R_5 = \frac{\rho \cdot v_5 \cdot D_5}{\mu} \quad h_{f5} = f_5 \cdot \frac{L_5}{D_5} \frac{v_5^2}{2 \cdot g}$$

$$h_{f1} + h_{f4} - h_{f5} = 0 \qquad\qquad h_{f2} - h_{f3} - h_{f4} = 0$$

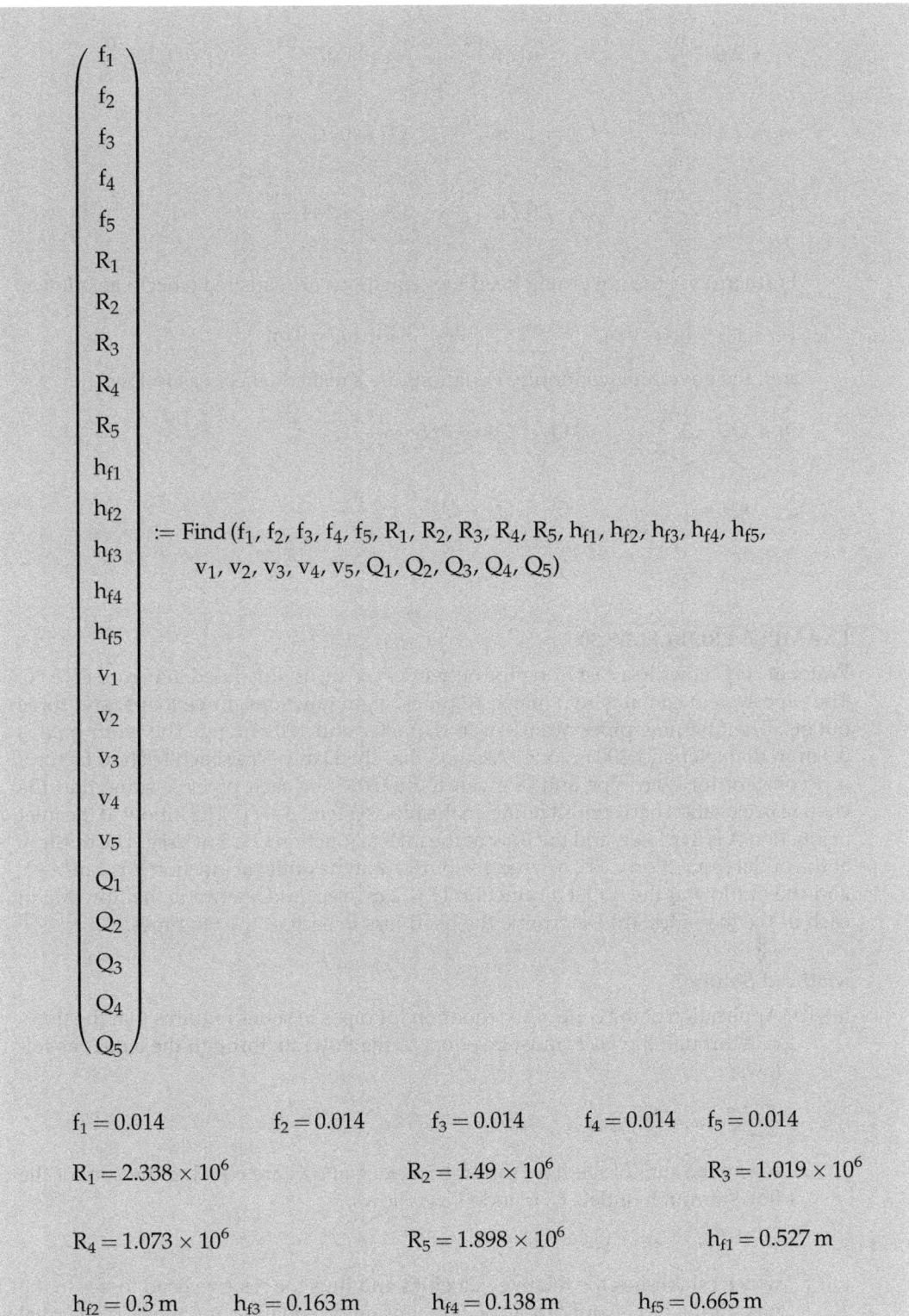

$$\begin{pmatrix} f_1 \\ f_2 \\ f_3 \\ f_4 \\ f_5 \\ R_1 \\ R_2 \\ R_3 \\ R_4 \\ R_5 \\ h_{f1} \\ h_{f2} \\ h_{f3} \\ h_{f4} \\ h_{f5} \\ v_1 \\ v_2 \\ v_3 \\ v_4 \\ v_5 \\ Q_1 \\ Q_2 \\ Q_3 \\ Q_4 \\ Q_5 \end{pmatrix} := \text{Find}\,(f_1,\, f_2,\, f_3,\, f_4,\, f_5,\, R_1,\, R_2,\, R_3,\, R_4,\, R_5,\, h_{f1},\, h_{f2},\, h_{f3},\, h_{f4},\, h_{f5}, \\ v_1,\, v_2,\, v_3,\, v_4,\, v_5,\, Q_1,\, Q_2,\, Q_3,\, Q_4,\, Q_5)$$

$f_1 = 0.014 \qquad f_2 = 0.014 \qquad f_3 = 0.014 \qquad f_4 = 0.014 \qquad f_5 = 0.014$

$R_1 = 2.338 \times 10^6 \qquad R_2 = 1.49 \times 10^6 \qquad R_3 = 1.019 \times 10^6$

$R_4 = 1.073 \times 10^6 \qquad R_5 = 1.898 \times 10^6 \qquad h_{f1} = 0.527\,\text{m}$

$h_{f2} = 0.3\,\text{m} \qquad h_{f3} = 0.163\,\text{m} \qquad h_{f4} = 0.138\,\text{m} \qquad h_{f5} = 0.665\,\text{m}$

$$v_1 = 2.603 \; \frac{m}{s} \qquad v_2 = 1.866 \; \frac{m}{s} \qquad v_3 = 1.075 \; \frac{m}{s} \qquad v_4 = 1.265 \; \frac{m}{s}$$

$$v_5 = 2.113 \; \frac{m}{s} \qquad Q_1 = 1.656 \; \frac{m^3}{s} \qquad Q_2 = 0.938 \; \frac{m^3}{s}$$

$$Q_3 = 0.762 \; \frac{m^3}{s} \qquad Q_4 = 0.718 \; \frac{m^3}{s} \qquad Q_5 = 1.344 \; \frac{m^3}{s}$$

Furthermore, the governing head loss equations are satisfied (check) as follows:

$$h_{f1} + h_{f4} - h_{f5} = 0 \; m \qquad\qquad h_{f2} - h_{f3} - h_{f4} = 0 \; m$$

and, the governing continuity equations are satisfied (check) as follows:

$$Q_1 + Q_5 = 3 \; \frac{m^3}{s} \qquad Q_2 + Q_4 = 1.656 \; \frac{m^3}{s}$$

$$Q_2 + Q_3 = 1.7 \; \frac{m^3}{s} \qquad Q_5 + Q_4 - Q_3 = 1.3 \; \frac{m^3}{s}$$

EXAMPLE PROBLEM 8.30

Water at 20°C flows in a cast iron pipe network system, as illustrated in Figure EP 8.30. The pipe system contains two inlets, 10 pipes, eight junctions, three loops, and three outlets. The first nine pipes are 0.9 m in diameter and 100 m long. The tenth pipe is 0.9 m in diameter and 200 m long. Assume that the Darcy–Weisbach friction factor, f is a constant for each pipe and is given to be 0.015 for each pipe. Assume that the change in elevation between all points in the pipe system, $\Delta z = 0$. The inflow at the inlet at junction A is 3 m³/sec, and the flow at the inlet at junction G is 5 m³/sec. The outflow at the outlet at junction C is 2 m³/sec, the outflow at the outlet at junction E is 4 m³/sec, and the outflow at the outlet at junction H is 2 m³/sec. (a) Determine the flowrate in each of the ten pipes. (b) Determine the head loss in each of the ten pipes.

Mathcad Solution

(a)–(b) Application of the continuity equation for pipes in series requires that the flow-rate through the *inlets* must be equal to the flowrate through the *outlets* as follows:

$$\sum Q_{inlets} = \sum Q_{outlets}$$

where the sum of the flow through inlets A and G are equal to the sum of the flows through outlets C, E and H as follows:

$$Q_A + Q_G = Q_C + Q_E + Q_H$$

which establishes the relative velocities and thus the relative head losses in the pipes in series. Application of the continuity equation for pipes in parallel

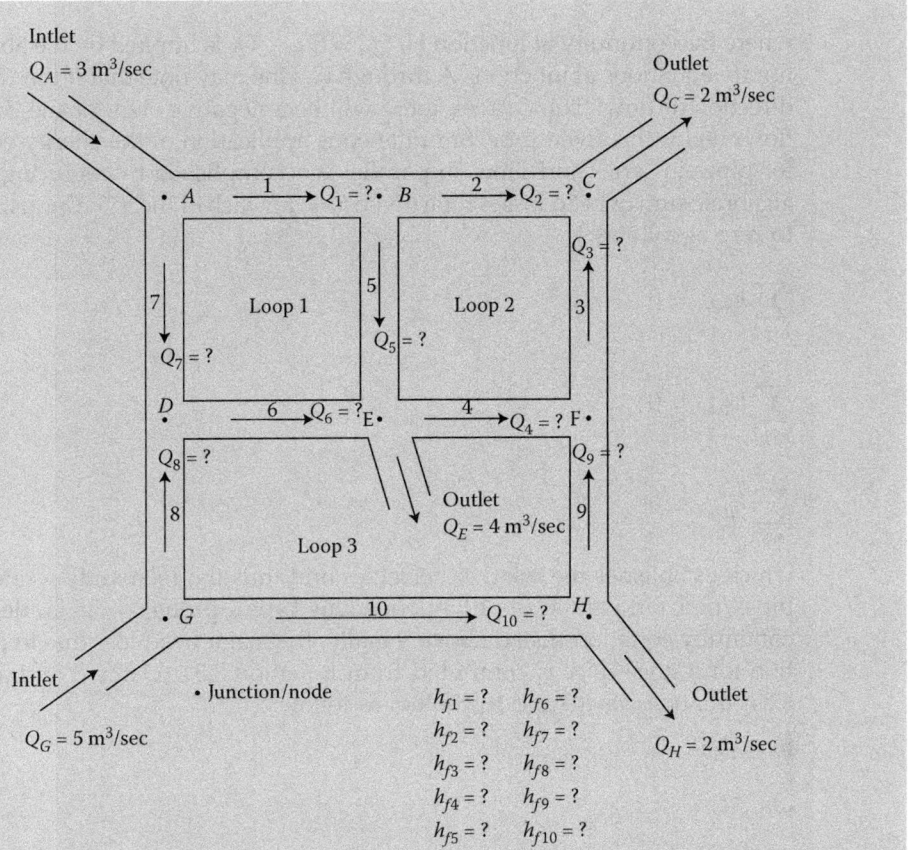

FIGURE EP 8.30
Water flows through a cast iron pipe network.

requires that the flowrate entering a given *junction/node* must equal to the flowrate leaving the *junction/node*, for all junctions. One may note that application of
this principle is accomplished by writing the appropriate continuity equation
for all but one junction, as they imply continuity at the last junction. "Assuming
the direction of flow" in the pipes yields the following continuity equations for
seven of the eight junctions as follows:

Continuity at Junction A: $Q_A = Q_1 + Q_7$

Continuity at Junction B: $Q_1 = Q_2 + Q_5$

Continuity at Junction C: $Q_C = Q_2 + Q_3$

Continuity at Junction D: $Q_6 = Q_7 + Q_8$

Continuity at Junction E: $Q_E = Q_6 + Q_5 - Q_4$

Continuity at Junction F: $Q_3 = Q_4 + Q_9$

Continuity at Junction G: $Q_G = Q_8 + Q_{10}$

where the continuity at junction H, $Q_H = Q_{10} - Q_9$ is implied by the above continuity equations at junctions A through G. One may note that if the "assumed direction of flow" is incorrect, there will be a negative sign for the computed flowrate for the given pipe. Simultaneous application of the energy equations for pipes in series and pipes in parallel is accomplished by requiring that the algebraic sum of head losses for a given *loop*, for each of the three loops, be equal to zero as follows:

$$\sum_{loop1} h_{f total} = 0$$

$$\sum_{loop2} h_{f total} = 0$$

$$\sum_{loop3} h_{f total} = 0$$

which establishes the relative velocities and thus the relative flowrates in the pipes/paths in parallel, as 10 independent Type 2 problems, as guided by the continuity equations listed above. Finally, the major head loss due to pipe friction for a given *pipe* is computed from Equation 8.34 (Darcy–Weisbach head loss equation), as a Type 1 problem as follows:

$$h_f = KQ^2$$

where

$$K = \frac{8fL}{\pi^2 g D^5}$$

Thus, in total, $e = 2 \times 10 = 20$ equations are solved for 20 unknowns as follows:

$$g := 9.81 \frac{m}{\sec^2} \qquad D := 0.9\,m \qquad L := 100\,m \qquad L_{10} := 200\,m$$

$$f := .015 \qquad K := \frac{8 \cdot f \cdot L}{\pi^2 \cdot g \cdot D^5} = 0.21 \frac{s^2}{m^5} \qquad K_{10} := \frac{8 \cdot f \cdot L_{10}}{\pi^2 \cdot g \cdot D^5} = 0.42 \frac{s^2}{m^5}$$

$$Q_A := 3 \frac{m^3}{\sec} \qquad Q_G := 5 \frac{m^3}{\sec} \qquad Q_C := 2 \frac{m^3}{\sec} \qquad Q_E := 4 \frac{m^3}{\sec} \qquad Q_H := 2 \frac{m^3}{\sec}$$

Guess values: $\qquad h_{f1} := 10\,m \qquad h_{f2} := 10\,m \qquad h_{f3} := 10\,m \qquad h_{f4} := 10\,m$

$$h_{f5} := 10\,m \qquad h_{f6} := 10\,m \qquad h_{f7} := 10\,m \qquad h_{f8} := 10\,m$$

$$h_{f9} := 10\,m \qquad h_{f10} := 10\,m \qquad Q_1 := 1.5 \frac{m^3}{\sec} \qquad Q_7 := 1.5 \frac{m^3}{\sec}$$

$$Q_2 := 0.75 \frac{m^3}{\sec} \qquad Q_5 := 0.75 \frac{m^3}{\sec} \qquad Q_3 := 1.25 \frac{m^3}{\sec}$$

$$Q_4 := 0.75 \, \frac{m^3}{sec} \qquad Q_6 := 4 \, \frac{m^3}{sec} \qquad Q_8 := 2.5 \, \frac{m^3}{sec}$$

$$Q_9 := 0.5 \, \frac{m^3}{sec} \qquad\qquad Q_{10} := 2.5 \, \frac{m^3}{sec}$$

Given

$$Q_A = Q_1 + Q_7 \qquad Q_1 = Q_2 + Q_5 \qquad Q_C = Q_2 + Q_3 \qquad Q_6 = Q_7 + Q_8$$

$$Q_E = Q_6 + Q_5 - Q_4 \qquad Q_3 = Q_4 + Q_9 \qquad Q_G = Q_8 + Q_{10} \qquad h_{f1} = K \cdot Q_1^{\,2}$$

$$h_{f2} = K \cdot Q_2^{\,2} \qquad\qquad h_{f3} = K \cdot Q_3^2 \qquad h_{f4} = K \cdot Q_4^2 \qquad h_{f5} = K \cdot Q_5^2$$

$$h_{f6} = K \cdot Q_6^2 \qquad\qquad h_{f7} = K \cdot Q_7^2 \qquad h_{f8} = K \cdot Q_8^2 \qquad h_{f9} = K \cdot Q_9^2$$

$$h_{f10} = K_{10} \cdot Q_{10}^2 \qquad\qquad h_{f1} + h_{f5} - h_{f6} - h_{f7} = 0$$

$$h_{f2} - h_{f3} - h_{f4} - h_{f5} = 0 \qquad h_{f6} + h_{f4} - h_{f9} - h_{f10} - h_{f8} = 0$$

$$\begin{pmatrix} h_{f1} \\ h_{f2} \\ h_{f3} \\ h_{f4} \\ h_{f5} \\ h_{f6} \\ h_{f7} \\ h_{f8} \\ h_{f9} \\ h_{f10} \\ Q_1 \\ Q_2 \\ Q_3 \\ Q_4 \\ Q_5 \\ Q_6 \\ Q_7 \\ Q_8 \\ Q_9 \\ Q_{10} \end{pmatrix} := \text{Find} \, (h_{f1}, h_{f2}, h_{f3}, h_{f4}, h_{f5}, h_{f6}, h_{f7}, h_{f8}, h_{f9}, h_{f10}, Q_1, Q_2, \\ Q_3, Q_4, Q_5, Q_6, Q_7, Q_8, Q_9, Q_{10})$$

$$h_{f1} = 1.401 \text{ m} \qquad h_{f2} = 0.394 \text{ m} \qquad h_{f3} = 0.084 \text{ m} \qquad h_{f4} = 3.242 \times 10^{-4} \text{ m}$$

$$h_{f5} = 0.31 \text{ m} \qquad h_{f6} = 1.675 \text{ m} \qquad h_{f7} = 0.036 \text{ m} \qquad h_{f8} = 1.218 \text{ m}$$

$$h_{f9} = 0.073 \text{ m} \qquad h_{f10} = 2.819 \text{ m} \qquad Q_1 = 2.584 \, \frac{\text{m}^3}{\text{s}} \qquad Q_2 = 1.369 \, \frac{\text{m}^3}{\text{s}}$$

$$Q_3 = 0.631 \, \frac{\text{m}^3}{\text{s}} \qquad Q_4 = 0.039 \, \frac{\text{m}^3}{\text{s}} \qquad Q_5 = 1.215 \, \frac{\text{m}^3}{\text{s}} \qquad Q_6 = 2.825 \, \frac{\text{m}^3}{\text{s}}$$

$$Q_7 = 0.416 \, \frac{\text{m}^3}{\text{s}} \qquad Q_8 = 2.409 \, \frac{\text{m}^3}{\text{s}} \qquad Q_9 = 0.591 \, \frac{\text{m}^3}{\text{s}} \qquad Q_{10} = 2.591 \, \frac{\text{m}^3}{\text{s}}$$

Furthermore, the governing head loss equations are satisfied (check) as follows:

$$h_{f1} + h_{f5} - h_{f6} - h_{f7} = 0 \text{ m} \qquad\qquad h_{f2} - h_{f3} - h_{f4} - h_{f5} = 0 \text{ m}$$

$$h_{f6} + h_{f4} - h_{f9} - h_{f10} + h_{f8} = 0 \text{ m}$$

and, the governing continuity equations are satisfied (check) as follows;

$$Q_1 + Q_7 = 3 \, \frac{\text{m}^3}{\text{s}} \qquad Q_2 + Q_5 = 2.584 \, \frac{\text{m}^3}{\text{s}} \qquad Q_2 + Q_3 = 2 \, \frac{\text{m}^3}{\text{s}}$$

$$Q_7 + Q_8 = 2.825 \, \frac{\text{m}^3}{\text{s}} \qquad Q_6 + Q_5 - Q_4 = 4 \, \frac{\text{m}^3}{\text{s}} \qquad Q_4 + Q_9 = 0.631 \, \frac{\text{m}^3}{\text{s}}$$

$$Q_8 + Q_{10} = 5 \, \frac{\text{m}^3}{\text{s}} \qquad Q_{10} - Q_9 = 2 \, \frac{\text{m}^3}{\text{s}}$$

8.8 Pipe Flow Measurement and Control Devices: Actual Flowrate in Ideal Flow Meters

Applications of the governing equations for ideal pipe flow include the analysis and design of pipe systems that include a pipe flow-measuring device such as pitot-static tubes, orifice, nozzle, or venturi meters. As such, the flow measurement devices in pipe flow are considered to be ideal flow meters. While the application of the continuity equation will always be necessary, the energy equation and the momentum equation play complementary roles; when one of the two equations of motion breaks down, the other may be used to find the additional unknown quantity. Furthermore, while the energy equation is useful only for a control volume, depending upon the general fluid flow problem, one may apply the continuity and momentum equations using either the integral or differential approach. However, for the specific case of ideal flow meters (addressed in Section 8.8), while both the differential and integral continuity equations are required, neither the differential nor the integral momentum equation for a control volume are required. As such, for ideal flow meters, the Bernoulli equation, Equation 8.16 assumes ideal flow; thus, there is no head loss to evaluate in the energy equation. Although the associated minor head loss may be modeled in the energy equation (Equation 8.1) (see Section 8.8.3.5; integral continuity equation is applied), it is typically accounted for by the use/calibration of a discharge coefficient to determine the actual discharge, where ideal flow is assumed. Thus, in the case of flow-measuring devices, one must indirectly model the minor head loss flow resistance term by the use of a discharge coefficient to determine the actual discharge, which requires a "subset level" of application of the governing equations (Sections 8.4.2.1, and 8.8.1).

The actual discharge is evaluated for pipe flow-measuring devices. There are numerous flow measurement devices for pipe flow, most of which can also serve as flow control devices. The most common types of flow measurement devices used in pipe flow include pitot-static tubes (also used in open channel flow), orifice meters, nozzle meters, and venturi meters. The flowrate measuring devices will vary depending on various factors including cost, space, size, accuracy, precision, sophistication, versatility, capacity, pressure drop, and the governing principle of operation. The flow measurement devices used in pipe flow may be subdivided into two categories: (1) pitot-static tubes (not a control) (also used in open channel flow) and (2) pipe flow controls: orifice meters, nozzle meters, and venturi meters. While a control in the pipe flow can serve as a flow-measuring device, a flow-measuring device is not necessarily considered a control. For instance, although a pitot-static tube is a common flow-measuring device, it is not a control. Because the pitot-static tube and the pipe flow controls (orifice meters, nozzle meters, and venturi meters) affect the flow over a short pipe section, the major loss due to the pipe flow resistance is assumed to be negligible. Thus, the flow measurement devices used in pipe flow are considered to be "ideal flow" meters (or basically velocimeters).

8.8.1 Evaluation of the Actual Flowrate for Ideal Flow Meters in Pipe Flow

Evaluation of the actual discharge flow resistance term requires a "subset level" application of the governing equations (see Chapters 6 and 7 and Section 8.4.2.1). Regardless of the type of pipe flow-measuring device (pitot-static tubes, orifice, nozzle, or venturi meters), the expression for the actual discharge in the differential continuity equation, which was given in Equation 8.19 is repeated as follows:

$$Q_a = \sqrt{\frac{(2/\rho)\left(\frac{\Delta p - \tau_w L}{R_h}\right)}{\sqrt{2\left(\frac{\Delta p}{\rho}\right)}}} \frac{A_a}{A_i} Q_i = C_d Q_i \tag{8.118}$$

where the discharge coefficient, C_d accounts for the minor head loss due to the flow measurement, which is due to the flow resistance, and flow contraction for one of the controls (an orifice meter). The assumption of ideal flow and thus the application of the Bernoulli equation between any two points yields an expression for the ideal velocity as a function of an ideal pressure difference, Δp, which is directly measured. Then, application of the differential continuity equation yields the ideal flowrate, Q_i. However, in real flow, the effects of viscosity, compressibility, and other real flow effects cause the actual velocity, v_a and actual area, A_a and thus the actual flowrate, Q_a to be less than the ideal velocity, v_i and ideal area, A_i and thus the ideal flowrate, Q_i, respectively. As a result, corrections for these real flow effects are made by applying a velocity or a discharge coefficient, respectively. The discharge coefficient represents both a velocity coefficient, which accounts for head loss due to friction, and a contraction coefficient, which accounts for the contraction of the streamlines of the fluid flow. As a result, the assumption of ideal flow is corrected by the use of a discharge coefficient, $C_d = C_c C_v$ (derived by application of the integral momentum supplemented by dimensional analysis), where application of the differential continuity equation yields the ideal flowrate. It is important to note that in the case of an orifice meter, because there is a contraction in the flow at the control, there is a need to apply a contraction coefficient, C_c to the computed discharge, and, because ideal flow was assumed, there is a need to apply a velocity coefficient, C_v to the computed discharge (there is a minor head loss associated with the flow measurement) and thus the discharge coefficient, $C_d = C_c C_v$. However, in the case of a pitot-static tube, a nozzle meter, and a venturi meter, because there is no contraction in the flow in the device, there is no need to apply a contraction coefficient, C_c to the computed discharge. Thus, in these cases, only a velocity coefficient, C_v or the discharge coefficient, $C_d = C_v$ is applied to the computed discharge in order to account for the minor head loss associated with the flow measurement.

In the case of a pitot tube, the fluid is momentarily brought to a stop, where the conversion of kinetic energy (dynamic pressure) in the flowing fluid into an ideal pressure rise (stagnation pressure minus static pressure) results in the ideal pressure rise. However, in the cases of an orifice meter, nozzle meter, and venturi meter, the fluid is in motion through some type of restriction/obstruction, where the conversion of the pressure energy to kinetic energy results in an ideal pressure drop. Thus, the determination of the ideal velocity and thus the ideal discharge will vary depending upon whether there is an ideal pressure rise or pressure drop. As such, details of the evaluation of the actual flowrate for a pitot-static tube is presented in Section 8.8.2, while details of the evaluation of the actual flowrate for orifice, nozzle, and venturi meters are presented in Section 8.8.3.

8.8.1.1 Derivation of the Discharge Coefficient, C_d

The empirical derivation of the discharge coefficient, C_d, which is defined as the product of the velocity coefficient, C_v and the contraction coefficient, C_c, was derived in Chapter 7 and is summarized as follows:

$$C_d = C_v C_c \tag{8.119}$$

where the discharge coefficient, C_d is defined from dimensional analysis as follows:

$$\frac{Q_a}{Q_i} = \frac{F_{dynpres}}{F_P} = \frac{v_a A_a}{v_i A_i} = \frac{v_a}{\sqrt{2\left(\frac{\Delta p}{\rho}\right)}} \frac{A_a}{A_i} = \sqrt{\frac{E}{2}} \frac{A_a}{A_i} = C_v C_c = C_d = \phi\left(R, \frac{L_i}{L}\right) \tag{8.120}$$

where:

$$\frac{v_a}{v_i} = C_v = \sqrt{\frac{E}{2}} \tag{8.121}$$

$$\frac{A_a}{A_i} = C_c \tag{8.122}$$

where v_a is the actual velocity, v_i is the ideal velocity, A_a is the actual flow area, A_i is the ideal area, and Q_i will vary for the various flow-measuring devices. The discharge coefficient, C_d represents both a velocity coefficient, C_v and a contraction coefficient, C_c; thus the discharge coefficient, C_d is defined as the product of the velocity coefficient, C_v and the contraction coefficient, C_c as follows: $C_d = C_v C_c$. The velocity coefficient, $C_v = (v_a/v_i) = \sqrt{E/2}$ accounts for minor head loss due to friction of the fluid flow.

For ideal flow, the flow through the flow-measuring device (a constriction) is the conversion of flow energy (pressure energy) to kinetic energy. In real flow, the effects of viscosity, compressibility, and other real flow effects cause the actual velocity, v_a and thus the actual flowrate to be less than the ideal velocity, v_i and the ideal flowrate, respectively. The contraction coefficient, $C_c = A_a/A_i$ accounts for the contraction of the streamlines of the fluid flow (due to frictional pressure losses). As a result, corrections for these real flow effects are made by applying a velocity or a discharge coefficient, respectively. Although the contraction coefficient, C_c and the velocity coefficient, C_v, which model the two nonideal flow effects, may be individually experimentally determined, their combined effect is usually modeled by the discharge coefficient, C_d, which is determined experimentally and is given as follows: $C_d = C_v C_c$.

The drag coefficient, C_D (discharge coefficient, C_d), as modeled by the Euler number, E, is the dependent π term, as it models the dependent variable, Δp that cannot be derived analytically from theory, while the Reynolds number, R and the geometry of the flow-measuring pipe device, L_i/L (L represents the characteristic length of the flow meter and L_i represents the geometry of the flow meter, where $i = 1, 2, 3, \ldots$) are the independent π terms. One may note that because the drag coefficient, C_D (discharge coefficient, C_d) represents the flow resistance, which causes an unknown pressure drop, Δp, the pressure force will always play an important role; thus the definition of the drag coefficient, C_D (discharge coefficient, C_d) is mainly a function of E.

8.8.1.2 *Evaluation of the Discharge Coefficient, C_d*

Evaluation of the discharge coefficient, C_d for flow measurement devices in pipe flow is determined experimentally and presented in the sections below on a particular flow measurement device. The discharge coefficient, C_d is a function of the geometry of the flow-measuring device, L_i/L; the Reynolds number, R; the location of the pressure taps; and the conditions of the upstream and downstream flow in the pipe. For low to moderate values of the Reynolds number, R where the viscous effects are significant, the experimentally determined discharge coefficient, C_d models both the frictional effects (C_v) and the contraction effects (C_c), whereas for relatively high values of the Reynolds number, R, where the viscous effects are insignificant, the experimentally determined discharge coefficient, C_d mostly models the contraction effects (C_c). Thus, on-site calibration is usually recommended in addition to the manufacturer's data. One may note that for real flow, the pressure drop, $\Delta p = p_1 - p_2$ is a function of the flowrate, Q; the density, ρ; the areas, A_1 and A_2; and the fluid viscosity, μ, which is modeled by the discharge coefficient, C_d.

8.8.2 Evaluation of the Actual Flowrate for a Pitot-Static Tube

Evaluation of the actual flowrate and the discharge coefficient for a pitot-static tube is presented in the current section (also see Section 8.4.2.1). An effective method to measure the velocity in a pipe flow (and open channel flow) is to attach a pitot-static tube device to the pipe as illustrated in Figure 8.24 (same as Figure 6.2 in Chapter 6). A pitot-static tube is a combination of pitot tube and a piezometer/static tube. Assuming no head loss and that the fluid is momentarily brought to a stop at the downstream point (stagnation pressure), the dynamic pressure at the upstream point represents the ideal pressure rise (from static pressure to stagnation pressure). This application of Bernoulli's equation deals with the stagnation and the dynamic pressures and illustrates that these pressures arise from the conversion of kinetic energy (dynamic pressure) in a flowing fluid into an ideal pressure

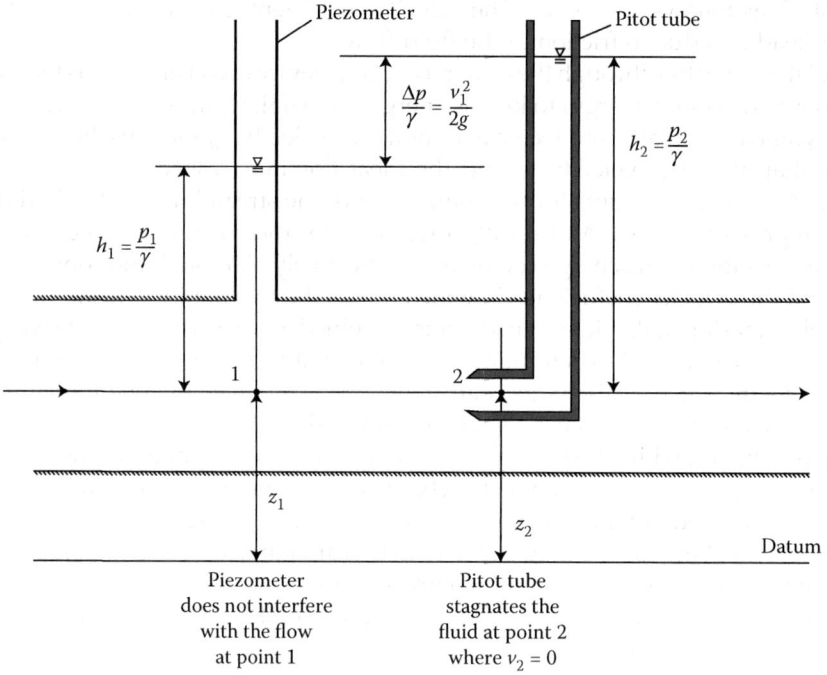

FIGURE 8.24
Pitot-static tube used to measure the velocity (and pressure rise) for a pipe flowing under pressure (same as Figures 6.2 and 4.5).

rise (stagnation pressure minus static pressure) as the fluid is momentarily brought to a rest. The downstream point at the pitot tube is the point of stagnation, which represents the conversion of all of the kinetic energy into an ideal pressure rise.

Figure 8.24 illustrates a horizontal pipe with a piezometer installed at point 1 and a pitot tube installed at point 2. The flow is from point 1 to point 2. One may note that both the piezometer and the pitot tube are pressure measuring devices. The piezometer is installed at the top of the pipe at point 1 and the rise of the fluid in the piezometer, p_1/γ measures the pressure at point 1. The piezometer does not interfere with the fluid flow and thus allows the fluid to travel without any change in the velocity, v_1. The pitot tube is installed at the top of the pipe at point 2 continues inward towards the centerline of the pipe and then makes a 90 degree turn pointing upstream into the direction of flow. As a result, the fluid enters the pitot tube and momentarily stops (it stagnates), where $v_2 = 0$. The rise of the fluid in the pitot tube, p_2/γ measures the pressure at point 2. One may note that because the pitot tube stagnates the fluid at point 2, it registers a higher pressure than the piezometer does at point 1. As illustrated in Figure 8.24, the rise in pressure head, $\Delta p/\gamma$ is defined as follows:

$$\frac{\Delta p}{\gamma} = \frac{p_2}{\gamma} - \frac{p_1}{\gamma} \tag{8.123}$$

The Bernoulli equation between points 1 and 2 expressed in terms of pressure is given as follows:

$$p_1 + \gamma z_1 + \frac{\rho v_1^2}{2} = p_2 + \gamma z_2 + \frac{\rho v_2^2}{2} \tag{8.124}$$

Assuming that $z_1 = z_2$ and noting that $v_2 = 0$:

$$\underbrace{p_1}_{staticpres} + \underbrace{\frac{\rho v_1{}^2}{2}}_{dynpress} = \underbrace{p_2}_{stagnpress} \qquad (8.125)$$

where the sum of the static pressure (where the fluid is not slowed down) and the dynamic pressure equals the stagnation pressure (where the fluid is stagnated/momentarily slowed down), the static pressure, p_1 is measured by the piezometer; the stagnation pressure, p_2 is measured by the pitot tube; and the dynamic pressure, $\rho v_1^2/2$ is measured by the difference in the fluid levels in the pitot tube and the piezometer as follows:

$$\frac{v_1{}^2}{2g} = \frac{\Delta p}{\gamma} = \frac{p_2}{\gamma} - \frac{p_1}{\gamma} \qquad (8.126)$$

thus, the dynamic pressure, $\rho v_1^2/2$ is given as follows:

$$\underbrace{\frac{\rho v_1{}^2}{2}}_{dynpress} = \underbrace{p_2}_{stagnpress} - \underbrace{p_1}_{staticpres} = \underbrace{\Delta p}_{pressrise} \qquad (8.127)$$

Assuming no head loss and that the fluid is momentarily brought to a stop at the downstream point 2 where $v_2 = 0$, the velocity at the upstream point 1 is ideal, where $v_1 = v_i$ and the dynamic pressure at the upstream point, $\rho v_1^2/2$ represents the ideal pressure rise (from static pressure to stagnation pressure). Therefore, isolating the ideal upstream velocity, v_i defines the ideal velocity as follows:

$$v_i = \underbrace{\sqrt{\frac{2}{\rho}(\Delta p)}}_{idealvel} = \underbrace{\sqrt{\frac{2}{\rho}\left(\underbrace{p_2}_{stagpress} - \underbrace{p_1}_{staticpress} \right)}}_{idealvel} \qquad (8.128)$$

Figure 8.25 illustrates the use of pitot tubes to determine the unknown velocities in a pipe, where, for instance, the difference in the fluid levels, $\Delta p/\gamma = v_1^2/2g$ in the pitot tube and the piezometer is used to determine the unknown velocity in the pipe at point 1, v_1. Next, application of the differential continuity equation yields an expression for the ideal flowrate. Assuming that the velocity profile at points 1 and 2 is uniform (ideal velocity profile), and that $A_1 = A_2 = A$, the ideal flowrate is computed as follows:

$$Q_i = v_i A = A\sqrt{\frac{2}{\rho}(p_2 - p_1)} \qquad (8.129)$$

and, finally, the reduced/actual discharge, Q_a is determined by application the differential continuity equation, and the momentum equation, which is supplemented by dimensional analysis in order to define the discharge coefficient; the discharge coefficient, C_d is applied to the ideal discharge, Q_i in order to define the actual discharge, Q_a. In real flow however, the actual velocity at point 1, $v_1 = v$ will be less than the ideal velocity, v_i, due to the frictional losses due to the fluid viscosity by an unknown amount as modeled by a velocity coefficient,

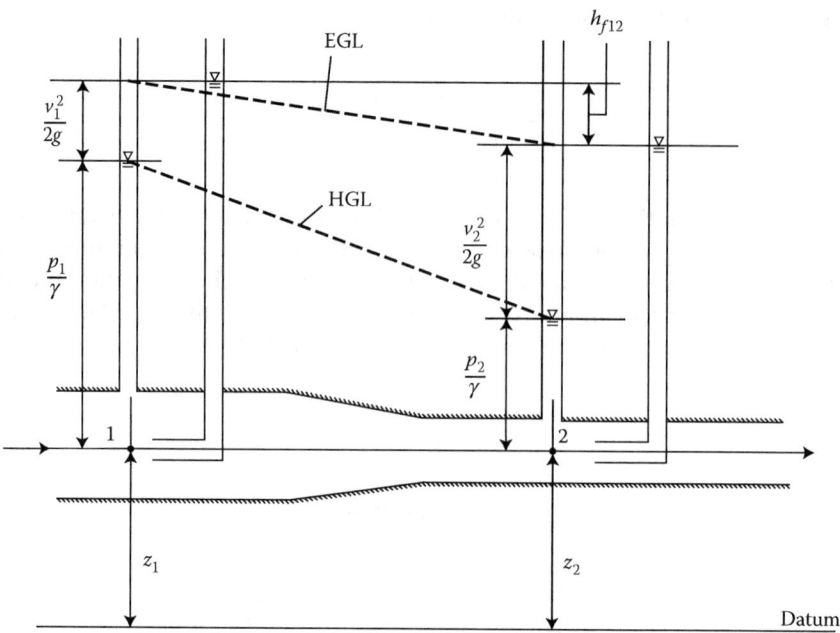

FIGURE 8.25
Pitot-static tubes are used to measure the velocities in a pipe flowing under pressure.

$C_v = v_a/v_i$. Furthermore, because there is no contraction in flow, there is no need to apply a contraction coefficient, C_c to the computed discharge. Thus, in this case, only a velocity coefficient, C_v or the discharge coefficient, $C_d = C_v$ is applied to the computed discharge in order to account for the minor head loss associated with the flow measurement. Therefore, the actual discharge, Q_a will be less than the ideal discharge, Q_i and is experimentally determined by the use of a discharge coefficient. As a result, an experimentally determined discharge coefficient, $C_d = C_v$ is applied to the ideal discharge, Q_i in order to determine the actual discharge, Q_a as follows:

$$Q_a = C_d Q_i = C_v A \sqrt{\frac{2}{\rho}(p_2 - p_1)} \qquad (8.130)$$

where the discharge coefficient, $C_d = C_v$ models only one effect of real (nonideal) flow, which is flow resistance, as there is no contraction in flow. The value of $C_d = C_v = v_a/v_i$ is determined from experimentation to be in the range of 0.98–0.995.

EXAMPLE PROBLEM 8.31

Water flows in a 1-m-diameter horizontal pipe. A piezometer is installed at point 1 and a pitot tube is installed at point 2 as illustrated in Figure EP 8.31 in order to measure the flowrate in the pipe. The static pressure head, p_1/γ measured by the piezometer is 1.3 m, and the stagnation pressure head, p_2/γ measured by the pitot tube is 1.6 m. Assume a discharge coefficient, C_d of 0.985. (a) Determine the ideal flowrate. (b) Determine the actual flowrate.

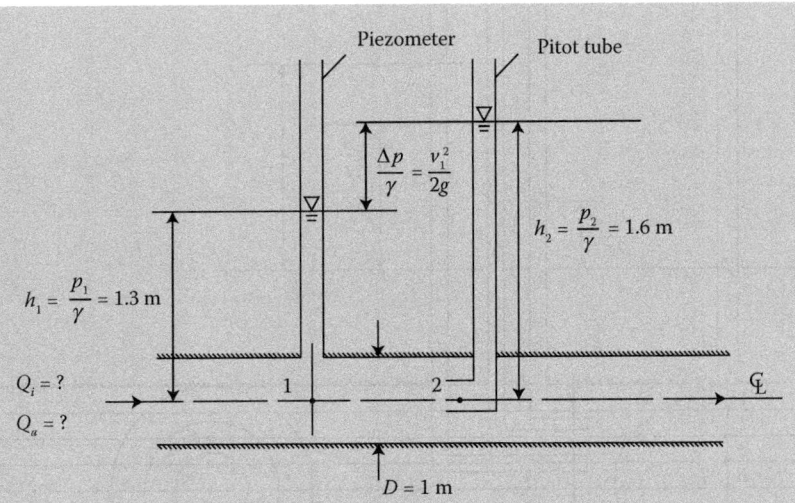

FIGURE EP 8.31
Pitot-static tube used to measure the velocity (and pressure rise) for water flowing in a horizontal pipe under pressure.

Mathcad Solution

(a) The ideal flowrate, Q_i is computed from Equation 8.129 as follows:

$$D := 1\,m \qquad\qquad \gamma := 9810\,\frac{N}{m^3} \qquad \rho := 1000\,\frac{kg}{m^3}$$

$$h_1 := 1.3\,m^3 \qquad\qquad h_2 := 1.6\,m \qquad A := \frac{\pi \cdot D^2}{4} = 0.785\,m^2$$

$$p_1 := \gamma\,h_1 = 1.275 \times 10^4\,\frac{N}{m^2} \qquad\qquad p_2 := \gamma\,h_2 = 1.57 \times 10^4\,\frac{N}{m^2}$$

$$Q_i := A \cdot \sqrt{\frac{2}{\rho} \cdot (p_2 - p_1)} = 1.905\,\frac{m^2}{sec}$$

(b) The actual flowrate, Q_a is computed from Equation 8.130 as follows:

$$C_d := 0.985 \qquad\qquad Q_a := C_d \cdot A \cdot \sqrt{\frac{2}{\rho} \cdot (p_2 - p_1)} = 1.877\,\frac{m^3}{sec}$$

The assumption of ideal flow assumes a uniform velocity profile, for which the ideal velocity, v_i is expressed as a function of an ideal pressure difference, Δp, which is directly measured. In real flow, however, (1) the actual velocity, v_a will be less than the ideal velocity, v_i, and (2) the velocity profile is characterized by a nonuniform (typically parabolic) velocity profile, for which the ideal velocity, v_i will vary across the pipe diameter. As a result, it is necessary to take a number of pressure readings across the pipe diameter at cross section B (see Figure 8.26) in order to model the actual nonuniform velocity profile. Integration of

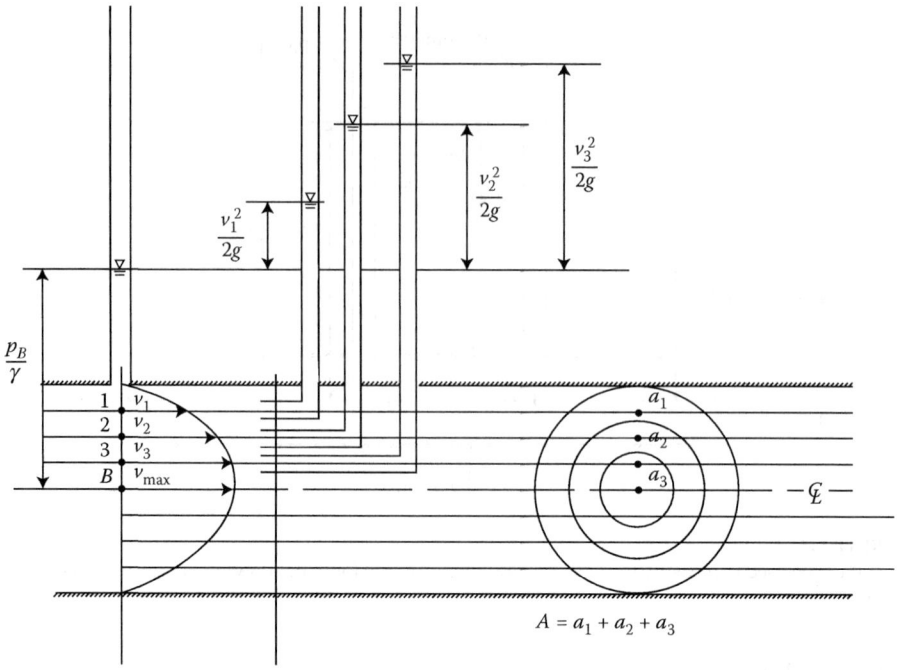

FIGURE 8.26
Pitot-static tubes are used to measure the parabolic velocity distribution (profile) in a pipe flowing under pressure.

the numerous ideal velocities across the pipe diameter at cross section B is done in order to obtain the total ideal flowrate, Q_i as follows:

$$Q_i = v_i A = \sum_j v_{ji} a_j = v_{1i} a_1 + v_{2i} a_2 + v_{3i} a_3 + \cdots \cdots \tag{8.131}$$

and thus, the actual discharge, Q_a is computed as follows:

$$Q_a = C_d Q_i = C_v v_i A \tag{8.132}$$

where

$$A = a_1 + a_2 + a_3 + \cdots \cdots \tag{8.133}$$

$$v_{ji} = \sqrt{\frac{2}{\rho}(p_{j2} - p_1)} \tag{8.134}$$

8.8.3 Evaluation of the Actual Flowrate for Orifice, Nozzle, and Venturi Meters

Evaluation of the actual flowrate, the discharge coefficient, and the minor head loss for an orifice, a nozzle, and a venturi meter is presented in the current section (also see Section 8.4.2.1). An effective method to measure the flowrate (of gases and liquids) in a pipe flow is to install some type of restriction/obstruction (an orifice, a nozzle, or a venturi), known as obstruction flow meters, within the pipe, as illustrated in Figure 8.27, and to measure the pressure difference between the low velocity, high pressure at the upstream point 1 and the high velocity, low pressure at the downstream point 2, which is at the point of restriction, using a differential manometer (or by mounting pressure gages at points 1 and 2). These three flow-measuring devices are based on the Bernoulli principle that an increase in velocity causes an ideal drop in pressure, where the measurement of the ideal pressure drop is used to determine the average velocity of ideal flow. It may be noted that the actual velocity is less than the ideal velocity due to frictional losses in real flow. Then, application of the continuity equation is used to

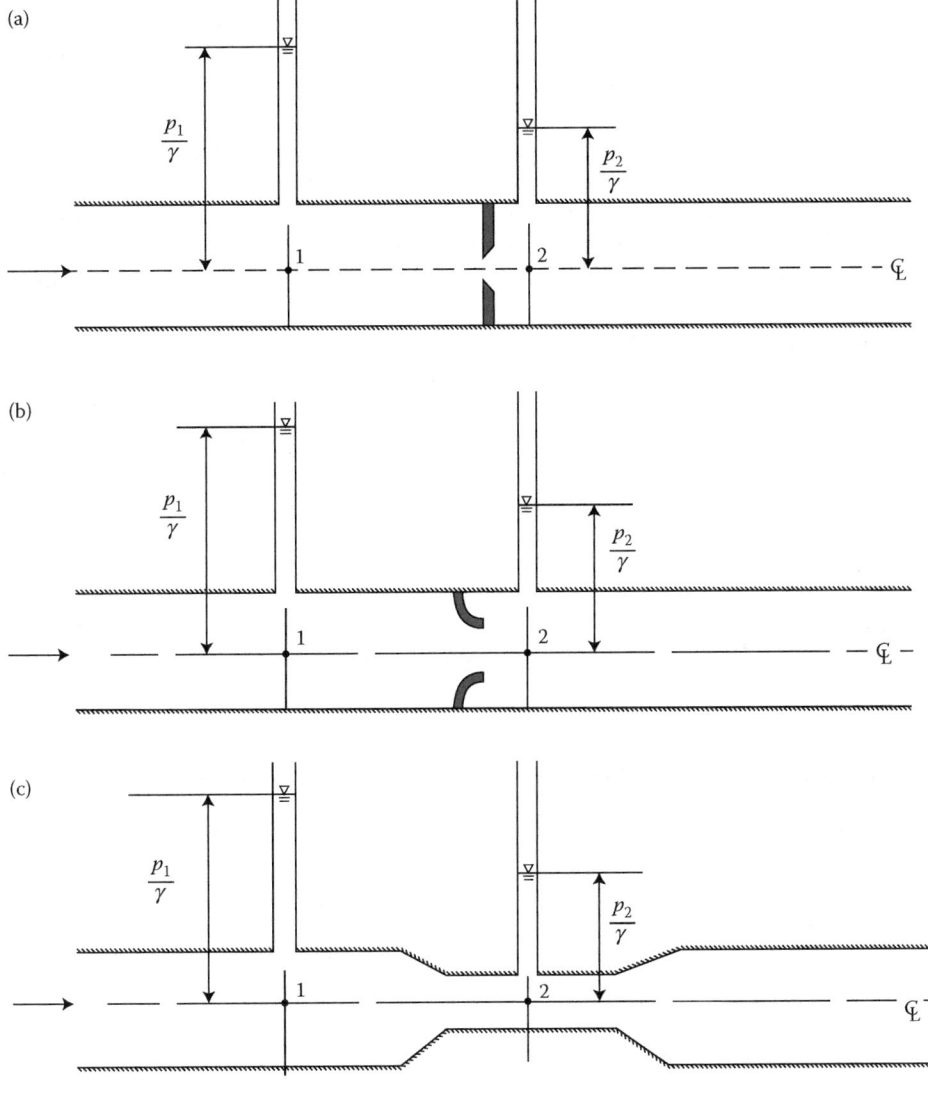

FIGURE 8.27
Pipe flow-measuring devices. (a) Orifice meter, (b) Nozzle meter, (c) Venturi meter.

determine the flowrate. Because this flowrate assumes ideal flow, it is the ideal flowrate, Q_i. Furthermore, the actual flowrate, Q_a, which accounts for frictional losses due to viscous effects, is determined by applying a correction factor to the ideal flowrate, Q_i. Finally, this correction factor is the discharge coefficient, C_d (empirically evaluated), which may account for both a velocity coefficient, C_v, which accounts for head loss due to friction, and a contraction coefficient, C_c, which accounts for the contraction of the streamlines of the fluid flow.

   While the difference between these three devices (an orifice, a nozzle, or a venturi) is a function of cost, space, accuracy, and precision (how closely their actual operation obeys the idealized flow assumptions made by the Bernoulli equation), application of the Bernoulli equation to illustrate that an increase in velocity causes an ideal drop in pressure involves the same calculation. Thus, while the determination of the ideal velocity and thus the ideal discharge is the same for the three devices (an orifice, a nozzle, or a venturi), determination of the discharge coefficient is unique for a given device. The orifice meter is the least expensive, requires the least space, and yields the least accuracy of these three meters. On the other hand the venturi meter is the most expensive, requires the most space, and yields the most accuracy (closest to ideal flow) of these three meters. Unlike the orifice and nozzle meters, the streamlining of the venturi meter eliminates any jet contraction (vena contracta) downstream of the smallest flow section, and thus, produces the least amount of head loss. One may note that the orifice and nozzle meters produce a significant amount of head loss because most of the pressure energy that is converted into kinetic energy at the vena contracta cannot be recovered. Furthermore, the pressure distribution in the vena contracta is not hydrostatic (it is dynamic). In the venturi meter, the gradual contraction is an efficient device for converting pressure energy to kinetic energy, while the gradual enlargement converts kinetic energy to pressure energy with a slight head loss due to friction. Furthermore, in order for the flow meter/flow-measuring device to operate properly, it should be installed in a section of the pipe where there is fully developed/established flow for about 30 diameters upstream of the flow-measuring device. Finally, it is interesting to note that the sluice gate and the weir for open channel flow are analogous to the orifice meter for pipe flow, where the head loss is due to both flow resistance (as modeled by the velocity coefficient, C_v) and flow contraction (as modeled by the contraction coefficient, C_c), where $C_d = C_c C_v$. Additionally, the venturi flume for open channel flow is analogous to a venturi meter for pipe flow, where the head loss is due only to flow resistance (as modeled by the velocity coefficient, C_v), (there is no contraction in the flow) where $C_d = C_v$.

8.8.3.1 Evaluation of the Actual Flowrate for an Orifice, a Nozzle, or a Venturi Meter

The empirical derivation of the reduced/actual discharge, Q_a for an orifice, a nozzle, or a venturi involves the application of the three governing equations (where the momentum equation is supplemented by dimensional analysis). Application of the Bernoulli equation yields an expression for the ideal velocity, v_i as a function of an ideal pressure difference, Δp, which is directly measured. Referring to Figure 8.27, assuming ideal flow, the Bernoulli equation between points 1 and 2 expressed in terms of pressure is given as follows:

$$p_1 + \gamma z_1 + \frac{\rho v_1^2}{2} = p_2 + \gamma z_2 + \frac{\rho v_2^2}{2} \qquad (8.135)$$

Although the effects of a nonhorizontal pipe may be easily included in the Bernoulli equation, assuming that $z_1 = z_2$ yields:

$$p_1 + \frac{\rho v_1^2}{2} = p_2 + \frac{\rho v_2^2}{2} \tag{8.136}$$

$$p_1 - p_2 = \underbrace{\Delta p}_{pressdrop} = \frac{\rho v_2^2}{2} - \frac{\rho v_1^2}{2} = \frac{\rho}{2}\left(v_2^2 - v_1^2\right) \tag{8.137}$$

Assuming that the velocity profiles at the upstream point 1 and the downstream point 2 are uniform (ideal flow velocity profile), then the continuity equation between points 1 and 2 is given as:

$$Q = v_1 A_1 = v_2 A_2 \tag{8.138}$$

Applying the continuity equation and solving for the ideal velocity at the restriction at point 2, v_{2i} yields the following:

$$p_1 - p_2 = \frac{\rho}{2}\left[v_2^2 - v_2^2\left(\frac{A_2}{A_1}\right)^2\right] = \frac{\rho v_2^2}{2}\left[1 - \left(\frac{A_2}{A_1}\right)^2\right] \tag{8.139}$$

$$v_{2i} = \sqrt{\frac{2(p_1 - p_2)}{\rho\left[1 - \left(\dfrac{A_2}{A_1}\right)^2\right]}} \tag{8.140}$$

Next, application of the continuity equation yields an expression for the ideal (theoretical) flowrate, Q_i is as follows:

$$Q_i = A_2\sqrt{\frac{2(p_1 - p_2)}{\rho\left[1 - \left(\dfrac{A_2}{A_1}\right)^2\right]}} \tag{8.141}$$

One may note that for ideal flow, the pressure drop, $\Delta p = p_1 - p_2$ is a function of the flowrate, Q; the density, ρ; and the areas, A_1 and A_2.

And, finally, the reduced/actual discharge, Q_a is determined by application of the differential continuity equation and the momentum equation, which is supplemented by dimensional analysis in order to define the discharge coefficient; the discharge coefficient, C_d is applied to the ideal discharge, Q_i in order to define the actual discharge, Q_a. In real flow, the actual velocity, v_a will be less than the ideal velocity, v_i due to the frictional losses due to the fluid viscosity. Therefore, the actual discharge, Q_a will be less than the ideal discharge, Q_i, and is experimentally determined by the use of a discharge coefficient. As a result, an experimentally determined discharge coefficient, C_d is applied to the ideal discharge, Q_i in order to determine the actual discharge, Q_a as follows:

$$Q_a = C_d Q_i = C_d A_2\sqrt{\frac{2(p_1 - p_2)}{\rho\left[1 - \left(\dfrac{A_2}{A_1}\right)^2\right]}} = C_d A_2\sqrt{\frac{2(p_1 - p_2)}{\rho\left[1 - \left(\dfrac{D_2}{D_1}\right)^4\right]}} \tag{8.142}$$

where D_1 and D_2 are the pipe diameters at points 1 and 2, respectively, and the discharge coefficient, C_d which is <1, is experimentally determined (tailored) depending upon which of the three devices (an orifice, a nozzle, or a venturi) is used to measure the flowrate, as presented below. The discharge coefficient, C_d may significantly vary from one flow-measuring device to another. Furthermore, the discharge coefficient, C_d is a function of geometry of the device, L_i/L; the Reynolds number, R; the location of the pressure taps; and the conditions of the upstream and downstream flow in the pipe. Thus, on-site calibration is usually recommended in addition to the manufacturer's data. One may note that for real flow, the pressure drop, $\Delta p = p_1 - p_2$ is a function of the flowrate, Q; the density, ρ; the areas, A_1 and A_2; and the fluid viscosity, μ, which is modeled by the discharge coefficient, C_d. Furthermore, unlike the pitot-static tube, a correction (discharge coefficient, C_d) for real flow is easily applied to the ideal discharge, Q_i without the need to take a number of pressure readings across the pipe diameter in order to model the actual nonuniform velocity profile, where integration of the numerous velocities across the pipe diameter is done in order to obtain the total ideal flowrate, Q_i.

8.8.3.2 Actual Flowrate for an Orifice Meter

Although the orifice flow meter is easier and less expensive to manufacture and requires less space than the nozzle and venture meter, it causes higher frictional losses and yields the least accuracy of these three meters. One may note that the sudden change in the flow area in an orifice meter causes a significant amount of swirling motion and thus a significant pressure drop (head loss). The orifice meter, which measures the flowrate in the pipe, consists of a circular plate with a sharp-edged or rounded opening of any shape (circular, square, rectangular, etc.) with a closed perimeter through which a fluid flows. The orifice meter is usually mounted in a flanged pipe as illustrated in Figure 8.28. Upstream of the orifice at point 1 the flow is still attached to the pipe wall. Upstream of the orifice opening, the flow contracts and separates from the pipe wall as it approaches the orifice (of diameter, D_o), forming a vena contracta (minimum flow area due to the contraction of the streamlines) just downstream of the orifice at point 2. The pressure difference may be measured using either a differential manometer or by mounting pressure gages at points 1 and 2. A pressure tap, p_1 is installed at point 1, where the full pipe flow occurs just upstream of the orifice, and a pressure tap, p_2 is installed at point 2, where the vena contracta occurs just downstream of the orifice. The

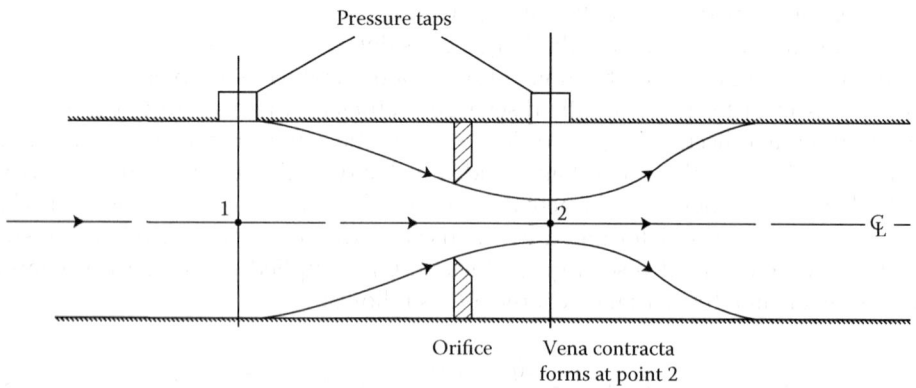

FIGURE 8.28
Orifice meter.

pressure at point 2 is less than the pressure at point 1. The actual discharge, Q_a measured through the orifice meter is given as follows:

$$Q_a = C_o Q_i = C_o A_o \sqrt{\frac{2(p_1 - p_2)}{\rho \left[1 - \left(\dfrac{D_o}{D_1} \right)^4 \right]}} \qquad (8.143)$$

where D_1 is the inlet pipe diameter at point 1, D_o is the orifice diameter, A_o is area of the orifice opening, and C_o is the orifice discharge coefficient, which is experimentally determined. One may note that due to the abrupt change in flow area, the value of the experimentally determined orifice discharge coefficient, C_o is significantly smaller than that for the nozzle and venturi meters. Furthermore, unlike the experimentally determined discharge coefficients for the nozzle and the venturi meters (models only the effect of frictional pressure losses), the orifice discharge coefficient, C_o models two effects of real (nonideal) flow. The first effect is the fact that the area of the vena contracta, A_2 is less than the area of the orifice, A_o by an unknown amount (because one cannot easily measure the diameter of the vena contracta, D_2) where:

$$C_c = \frac{A_2}{A_o} = \frac{D_2{}^2}{D_o^2} \qquad (8.144)$$

where C_c is the contraction coefficient, which is <1. The second effect is due to frictional pressure losses, which occur as a result of the turbulence and swirling motion near the downstream face of the orifice, yielding a head loss, which cannot be determined analytically. Instead, the frictional effects may be expressed as a velocity coefficient, C_v as follows:

$$C_v = \frac{v_{oa}}{v_{oi}} \qquad (8.145)$$

where v_{oa} is the actual velocity at the orifice, v_{oi} is the ideal velocity at the orifice, and C_v is the velocity coefficient. It is important to note that although the contraction coefficient, C_c and the velocity coefficient, C_v, which model the two nonideal flow effects, may be individually experimentally determined, their combined effect is usually modeled by the orifice discharge coefficient, C_o, which is determined experimentally and is given as follows:

$$C_o = C_c C_v \qquad (8.146)$$

where the orifice discharge coefficient, C_o is a function of the diameter ratio, D_o/D_1 and the Reynolds number, R (for which the upstream velocity, $v_1 = Q_i/v_1$ is used to determine $R = \rho v_1 D_1/\mu$) at low values of R. However, for large values of R, C_o is only a function of the diameter ratio, D_o/D_1. Experimental values for the orifice discharge coefficient, C_o are presented in Figure 8.29 where typical values range from about 0.6–0.65. It is important to note that the value of the orifice discharge coefficient, C_o is a function of the specific construction of the orifice meter, including the installation of the pressure taps and the shape and the type of edge of the orifice. Furthermore, it is important to note that for low to moderate values of the Reynolds number, R, where the viscous effects are significant, the experimentally determined orifice discharge coefficient, C_o models both the frictional effects and the contraction effects, whereas for relatively high values of the Reynolds number, R, where the viscous effects are insignificant, the experimentally determined orifice discharge coefficient, C_o models mostly the contraction effects.

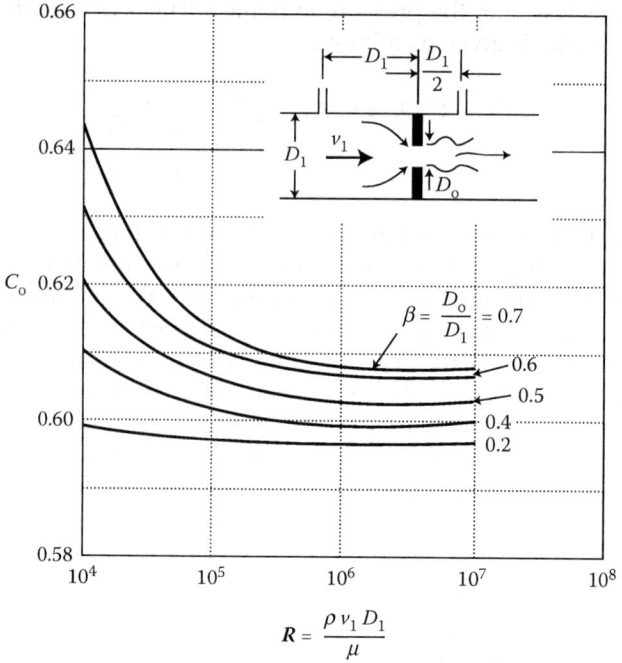

FIGURE 8.29

Orifice meter discharge coefficient, C_o is a function of location of pressure taps, shape and type of edge of orifice, and Reynolds number, $\mathbf{R}$. (From International Organization for Standardization (ISO), *Measurement of Fluid Flow by Means of Orifice Plates, Nozzles, and Venturi Tubes Inserted in Circular Cross Section Conduits Running Full*, 1st ed, ISO 5167-1980 (E), International Organization for Standardization, Geneva, 1980, 18.19, p.38.)

EXAMPLE PROBLEM 8.32

Water at 5°C flows in a 0.5-m-diameter horizontal pipe. An orifice meter with a 0.2-m diameter is mounted in the flanged pipe just upstream of point 2 as illustrated in Figure EP 8.32 in order to measure the flowrate in the pipe. A pressure tap, p_1 is installed at point 1, where the full pipe flow occurs just upstream of the orifice, and a pressure tap, p_2 is installed at point 2, where the vena contracta occurs just downstream of the orifice. The pressure at point 1, p_1 is measured to be $2 \times 10^5 \, \text{N/m}^2$, and the pressure at point 2, p_2 is measured to be $1.5 \times 10^5 \, \text{N/m}^2$. The absolute or dynamic viscosity for water at 5°C is $0.001518 \, \text{N-sec/m}^2$ (see Table A.4 in Appendix A). (a) Determine the ideal flowrate. (b) Determine the actual flowrate.

Mathcad Solution

(a) The ideal flowrate, Q_i is computed from Equation 8.141 as follows:

$$D_1 := 0.5 \, \text{m} \qquad\qquad D_o := 0.2 \, \text{m} \qquad\qquad \rho := 1000 \, \frac{\text{kg}}{\text{m}^3}$$

$$p_1 := 2 \times 10^5 \, \frac{\text{N}}{\text{m}^2} \qquad\qquad p_2 := 1.5 \times 10^5 \, \frac{\text{N}}{\text{m}^2} \qquad\qquad A_o := \frac{\pi \cdot D_o^2}{4} = 0.031 \, \text{m}^2$$

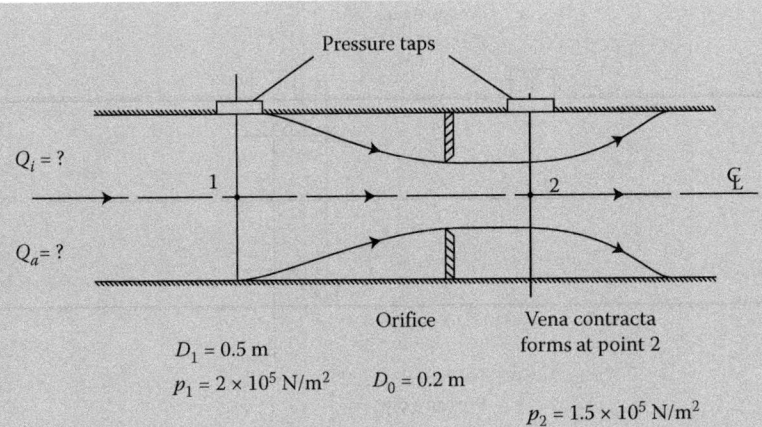

FIGURE EP 8.32
Orifice meter used to measure the flowrate of water flowing in a horizontal pipe under pressure.

$$Q_i := A_o \cdot \sqrt{\frac{2 \cdot (p_1 - p_2)}{\rho \cdot \left[1 - \left(\frac{D_o}{D_1}\right)^4\right]}} = 0.318 \frac{m^3}{sec}$$

(b) The actual flowrate, Q_a is computed from Equation 8.143 where the orifice discharge coefficient, C_o is a function of the diameter ratio, D_o/D_1 and the Reynolds number, R and is determined from Figure 8.29 as follows:

$$\mu := 0.001518 \, N \frac{sec}{m^2} \qquad A_1 := \frac{\pi \cdot D_1^2}{4} = 0.196 \, m^2 \qquad v_1 := \frac{Q_i}{A_1} = 1.621 \frac{m}{s}$$

$$R := \frac{\rho \cdot v_1 \cdot D_1}{\mu} = 5.339 \times 10^5 \qquad\qquad \frac{D_o}{D_1} = 0.4$$

$$C_o := 0.6 \qquad\qquad Q_a := C_o \cdot A_o \cdot \sqrt{\frac{2 \cdot (p_1 - p_2)}{\rho \cdot \left[1 - \left(\frac{D_o}{D_1}\right)^4\right]}} = 0.191 \frac{m^3}{sec}$$

8.8.3.3 Actual Flowrate for a Nozzle Meter

An orifice with prolonged sides in the direction of flow, which may have a constant or a variable cross section, serves as a tube or a nozzle flow meter in the pipeline as illustrated in Figure 8.30. One may note that a nozzle is used for the creation of jets and streams as well as a flow-measuring device. As illustrated in Figure 8.30, there are three typical variations of a nozzle meter construction. A nozzle flow meter, which is used to measure the flowrate in the pipe, contains a contoured nozzle that has a flange on its upstream face, where the nozzle flange is installed between pipe flanges in order to hold the nozzle concentric with the centerline of the pipe. One may note that although the nozzle occupies more space

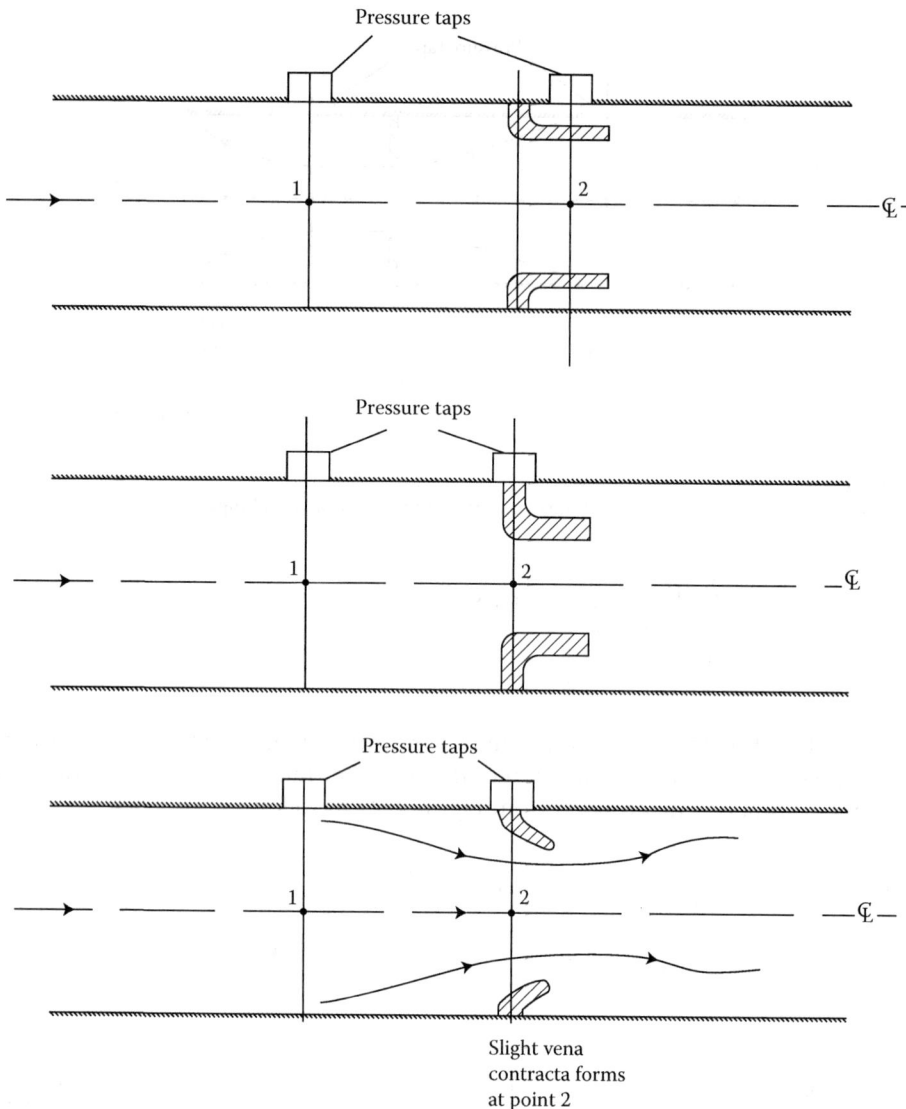

FIGURE 8.30
Three typical variations of nozzle meter construction.

than the orifice meter presented above, it occupies less space than the venturi meter presented below. Furthermore, the nozzle has a gradual converging section (streamlined), which results in a flow pattern that is closer to ideal flow than the orifice meter, but not as close as the venturi meter. In comparison to the orifice meter, there is only a slight vena contracta in the nozzle meter, where the secondary flow separation is less severe. However, the frictional losses due the viscous effects still occur, which are higher than those for the venturi meter. The pressure difference may be measured using either a differential manometer or by mounting pressure gages at points 1 and 2. A pressure tap, p_1 is installed at point 1 just upstream of the nozzle, and a throat pressure tap, p_2 is installed at point 2, which can either be located at the flange or just downstream of the flange as illustrated in

Figure 8.30. The actual discharge, Q_a measured through the nozzle meter is given as follows:

$$Q_a = C_n Q_i = C_n A_2 \sqrt{\dfrac{2(p_1 - p_2)}{\rho \left[1 - \left(\dfrac{D_2}{D_1} \right)^4 \right]}} \tag{8.147}$$

where D_1 is the inlet pipe diameter at point 1, D_2 is the throat diameter of the nozzle (minimum diameter at the exit of the nozzle), A_2 is the minimum area at the exit of the nozzle, and C_n is the nozzle discharge coefficient, which is experimentally determined. The nozzle discharge coefficient, C_n mostly models the effect of frictional pressure losses, where:

$$C_n = C_v = \dfrac{v_{2a}}{v_{2i}} \tag{8.148}$$

where v_{2a} is the actual velocity at throat of the nozzle, v_{2i} is the ideal velocity at the throat of the nozzle, and C_v is the velocity coefficient. The nozzle discharge coefficient, C_n is a function of the nozzle diameter ratio, D_2/D_1 (exit throat diameter to inlet diameter) and the Reynolds number, R (for which the upstream velocity, $v_1 = Q_i/v_1$ is used to determine $R = \rho v_1 D_1/\mu$) at low values of R. However, for large values of R, C_n is only a function of the diameter ratio, D_2/D_1. Experimental values for the nozzle discharge coefficient, C_n are presented in Figure 8.31. It is important to note that the value of the nozzle discharge coefficient, C_n is a function of the specific construction of the nozzle meter, including the installation of the pressure taps, and the shape and the type of edge of the nozzle. Furthermore, it is important to note that because the nozzle meter is closer to ideal flow than the orifice meter (less frictional effects), the experimental values for the nozzle discharge coefficient, C_n are larger (closer to 1) than the orifice discharge coefficient, C_o.

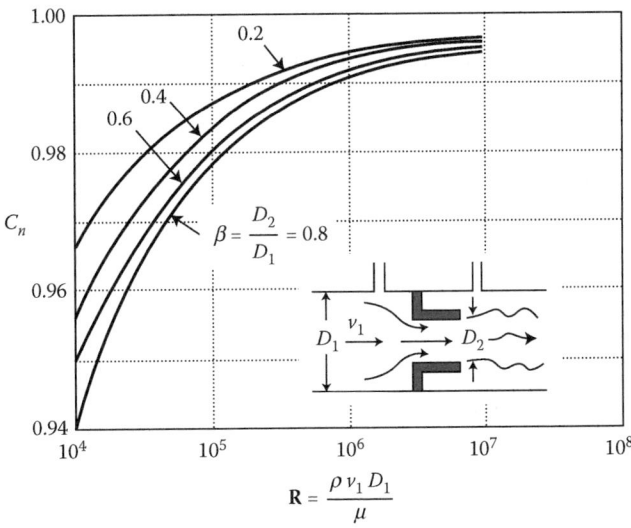

FIGURE 8.31
Nozzle meter discharge coefficient, C_n is a function of location of pressure taps, shape and type of edge of nozzle, and Reynolds number, R. (From International Organization for Standardization (ISO), *Measurement of Fluid Flow by Means of Orifice Plates, Nozzles, and Venturi Tubes Inserted in Circular Cross Section Conduits Running Full*, 1st ed. ISO 5167-1980 (E), International Organization for Standardization, Geneva, 1980, pp. 23, 56.)

EXAMPLE PROBLEM 8.33

Water at 5°C flows in a 0.5-m-diameter horizontal pipe. A nozzle meter with a 0.2-m throat diameter is mounted in the flanged pipe just upstream of point 2 as illustrated in Figure EP 8.33 in order to measure the flowrate in the pipe. A pressure tap, p_1 is installed at point 1 just upstream of the nozzle flange, and a pressure tap, p_2 is installed at point 2, which is located just downstream of the nozzle flange. The pressure at point 1, p_1 is measured to be 2×10^5 N/m², and the pressure at point 2, p_2 is measured to be 1.5×10^5 N/m². The absolute or dynamic viscosity for water at 5°C is 0.001518 N-sec/m² (Table A.4 in Appendix A). (a) Determine the ideal flowrate. (b) Determine the actual flowrate.

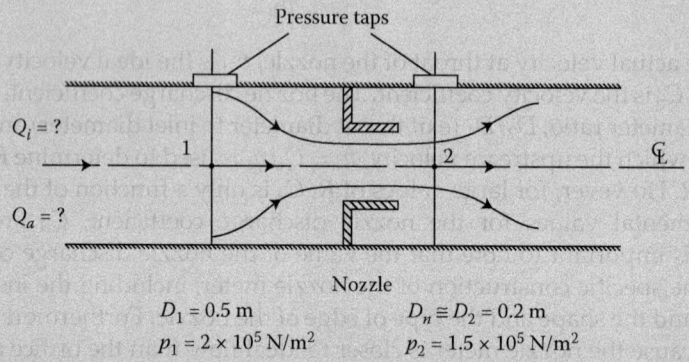

Pressure taps

$Q_i = ?$

$Q_a = ?$

Nozzle

$D_1 = 0.5$ m $D_n \cong D_2 = 0.2$ m
$p_1 = 2 \times 10^5$ N/m² $p_2 = 1.5 \times 10^5$ N/m²

FIGURE EP 8.33
Nozzle meter used to measure the flowrate of water flowing in a horizontal pipe under pressure.

Mathcad Solution

(a) The ideal flowrate, Q_i is computed from Equation 8.141 as follows:

$$D_1 := 0.5\,m \qquad\qquad D_2 := 0.2\,m \qquad\qquad \rho := 1000\,\frac{kg}{m^3}$$

$$p_1 := 2 \times 10^5\,\frac{N}{m^2} \qquad p_2 := 1.5 \times 10^5\,\frac{N}{m^2} \qquad A_2 := \frac{\pi \cdot D_2^2}{4} = 0.031\,m^2$$

$$Q_i := A_2 \cdot \sqrt{\frac{2 \cdot (p_1 - p_2)}{\rho \cdot \left[1 - \left(\frac{D_2}{D_1}\right)^4\right]}} = 0.318\,\frac{m^3}{sec}$$

(b) The actual flowrate, Q_a is computed from Equation. 8.147 where the nozzle discharge coefficient, C_n is a function of the diameter ratio, D_2/D_1 and the Reynolds number, R and is determined from Figure 8.31 as follows:

$$\mu := 0.001518\,N\,\frac{sec}{m^2} \qquad A_1 := \frac{\pi \cdot D_1^2}{4} = 0.196\,m^2 \qquad v_1 := \frac{Q_i}{A_1} = 1.621\,\frac{m}{s}$$

$$R := \frac{\rho \cdot v_1 \cdot D_1}{\mu} = 5.339 \times 10^5 \qquad \frac{D_2}{D_1} = 0.4$$

$$C_n := 0.99 \qquad Q_a := C_n \cdot A_2 \cdot \sqrt{\frac{2 \cdot (p_1 - p_2)}{\rho \cdot \left[1 - \left(\frac{D_2}{D_1}\right)^4\right]}} = 0.315 \, \frac{m^3}{sec}$$

8.8.3.4 Actual Flowrate for a Venturi Meter

A venturi flow meter is a tube that is installed into a pipeline, which is used to measure the flowrate as illustrated in Figure 8.32. The venturi tube consists of a gradual/conical contraction followed by a constant diameter section (the throat), followed by a gradual/conical diffuser, which is a gradual enlargement to the original size of the pipe. Although typical values for the ratio of the inlet pipe diameter to the throat diameter, D_2/D_1 varies from ¼ to ¾, a common ratio is ½. And while a small ratio yields an increase in accuracy, it produces a higher head loss due to friction accompanied by an undesirable low pressure at the throat of the venturi tube. One may recall that at a sufficiently low pressure (vapor pressure), a phenomenon known as cavitation occurs, where vaporization of the liquid may occur. The venturi meter is the most expensive and most precise of the three restriction type flow meters. As such, it is designed to reduce the head loss to a minimum. In order to keep the frictional pressure losses to a minimum, the maximum recommended angle of the gradual contraction is about 21–22°, while the maximum recommended angle of the gradual diffuser is about 5–7°. One may note that if these angles are made much smaller then the maximum recommended values, then the venturi tube would become too long and expensive to serve as an efficient flowrate meter. The head loss is reduced to a minimum in the venturi meter by providing a relatively streamlined contraction throat, followed by a streamline expansion, which eliminates flow separation and swirling in the decelerating gradual/conical diffuser portion of the venturi meter. And, unlike the orifice and nozzle meters, the streamlining of the venturi meter eliminates any jet contraction (vena contracta) downstream of the smallest flow section. Thus, most of the head loss that occurs in an optimally designed venturi meter is a result of frictional losses along the pipe walls as opposed to those associated with flow separation and the swirling motion. Therefore, venturi meters are used when large pressure drops in

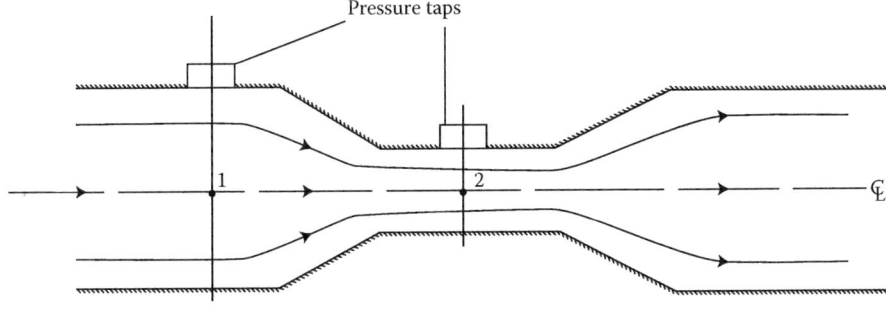

FIGURE 8.32
Venturi meter.

the pipe cannot be tolerated. The pressure difference may be measured using either a differential manometer or by mounting pressure gages at points 1 and 2. A pressure tap, p_1 is installed at point 1 just upstream of the venturi tube and a throat pressure tap, p_2 is installed at point 2, which is located at the throat of the venturi tube as illustrated in Figure 8.32. The actual discharge, Q_a measured through the venturi meter is given as follows:

$$Q_a = C_v Q_i = C_v A_2 \sqrt{\dfrac{2(p_1 - p_2)}{\rho \left[1 - \left(\dfrac{D_2}{D_1} \right)^4 \right]}} \qquad (8.149)$$

where D_1 is the inlet pipe diameter at point 1, D_2 is the throat diameter of the venturi tube, A_2 is the area at the throat of the venturi tube, and C_v is the venturi discharge coefficient, which is experimentally determined. The venturi tube discharge coefficient, C_v models only the effect of frictional pressure losses (as there is no contraction of flow), where:

$$C_v = C_v = \dfrac{v_{2a}}{v_{2i}} \qquad (8.150)$$

where v_{2a} is the actual velocity at throat of the venturi tube, v_{2i} is the ideal velocity at the throat of the venturi tube, and C_v is the velocity coefficient. The venturi discharge coefficient, C_v is a function of the geometry of the venturi tube (including the converging and diverging angles, and the inlet, throat, and outlet pipe diameters) and the Reynolds number, R (for which the upstream velocity, $v_1 = Q_i/v_1$ is used to determine $R = \rho v_1 D_1/\mu$) at low values of R. However, for large values of R, C_v is only a function of the geometry of the venturi tube. One may note that because of the smooth and gradual flow transition through a venturi flow meter, the value of the venturi discharge coefficient, C_v is $\cong 1$. Experimental values for

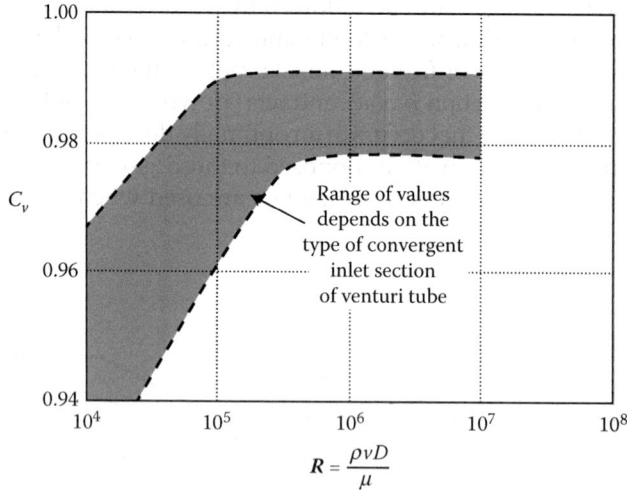

FIGURE 8.33
Venturi meter discharge coefficient, C_v is a function of geometry of venturi tube, and Reynolds number, R. (Adapted from Munson, B. R., Young, D. F., and T. H. Okiishi, *Fundamentals of Fluid Mechanics*, 3rd ed, John Wiley and Sons, New York, 1998, p. 532; Bean, H. S., *Report of ASME Research Committee on Fluid Meters*, 6th ed, American Society of Mechanical Engineers, New York, 1971, 64, 232.)

the venturi discharge coefficient, C_v are presented in Figure 8.33, where typical values range form about 0.94–0.99. Additionally, one may note that for the venturi meter, one may directly determine the actual discharge, Q_a by using a pressure-measuring device that is calibrated to account for the discharge coefficient, C_v, the fluid density, ρ and the inlet and throat pipe diameters, A_1 and A_2, respectively, to measure the pressure drop, Δp.

EXAMPLE PROBLEM 8.34

Water at 5°C flows in a 0.5-m-diameter horizontal pipe. A venturi meter with a 0.2-m throat diameter is inserted in the pipe at point 2 as illustrated in Figure EP 8.34 in order to measure the flowrate in the pipe. A pressure tap, p_1 is installed at point 1 just upstream of the venturi tube, and a pressure tap, p_2 is installed at point 2, which is located at the throat of the venturi tube. The pressure at point 1, p_1 is measured to be $2 \times 10^5 \, \text{N/m}^2$, and the pressure at point 2, p_2 is measured to be $1.5 \times 10^5 \, \text{N/m}^2$. The absolute or dynamic viscosity for water at 5°C is $0.001518 \, \text{N-sec/m}^2$ (see Table A.4 in Appendix A). (a) Determine the ideal flowrate. (b) Determine the actual flowrate.

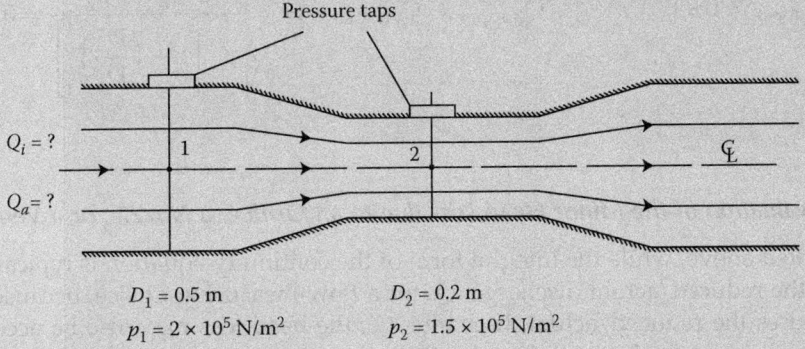

FIGURE EP 8.34
Venturi meter used to measure the flowrate of water flowing in a horizontal pipe under pressure.

Mathcad Solution

(a) The ideal flowrate, Q_i is computed from Equation 8.141 as follows:

$$D_1 := 0.5 \, \text{m} \qquad D_2 := 0.2 \, \text{m} \qquad \rho := 1000 \, \frac{\text{kg}}{\text{m}^3}$$

$$p_1 := 2 \times 10^5 \, \frac{\text{N}}{\text{m}^2} \qquad p_2 := 1.5 \times 10^5 \, \frac{\text{N}}{\text{m}^2} \qquad A_2 := \frac{\pi \cdot D_2{}^2}{4} = 0.031 \, \text{m}^2$$

$$Q_i := A_2 \cdot \sqrt{\frac{2 \cdot (p_1 - p_2)}{\rho \cdot \left[1 - \left(\dfrac{D_2}{D_1}\right)^4\right]}} = 0.318 \, \frac{\text{m}^3}{\text{sec}}$$

(b) The actual flowrate, Q_a is computed from Equation 8.149 where the venturi discharge coefficient, C_v is a function of the geometry of the venturi tube, and the Reynolds number, R and is determined from Figure 8.33. For a given value for the Reynolds number, R, the venturi discharge coefficient, C_v and thus the actual discharge, Q_a will have a range of values, depending upon the converging and diverging angles, and the inlet, throat, and outlet pipe diameters as follows:

$$\mu := 0.001518 \, N \frac{sec}{m^2}$$

$$A_1 := \frac{\pi \cdot D_1^2}{4} = 0.196 \, m^2$$

$$v_1 := \frac{Q_i}{A_1} = 1.621 \, \frac{m}{s}$$

$$R := \frac{\rho \cdot v_1 \cdot D_1}{\mu} = 5.339 \times 10^5$$

$$C_{vlower} := 0.977$$

$$Q_{alower} := C_{vlower} \cdot A_2 \cdot \sqrt{\frac{2 \cdot (p_1 - p_2)}{\rho \cdot \left[1 - \left(\frac{D_2}{D_1}\right)^4\right]}} = 0.311 \, \frac{m^3}{sec}$$

$$C_{vupper} := 0.99$$

$$Q_{aupper} := C_{vupper} \cdot A_2 \cdot \sqrt{\frac{2 \cdot (p_1 - p_2)}{\rho \cdot \left[1 - \left(\frac{D_2}{D_1}\right)^4\right]}} = 0.315 \, \frac{m^3}{sec}$$

8.8.3.5 Evaluation of the Minor Head Loss due to an Orifice, a Nozzle, or a Venturi Meter

As illustrated above, while the integral form of the continuity equation is typically used to compute the reduced/actual discharge, Q_a for a flow-measuring device, because the head loss, h_f causes the reduced/actual discharge, Q_a, the head loss may also be accounted for as a minor loss in the integral form of the energy equation, Equation 8.1. Specifically, although the frictional losses due to the fluid viscosity are typically accounted for through the use of a discharge coefficient, C_d in the continuity equation, additionally, one may compute the head loss, h_f due the flow-measuring device by noting that the energy equation for real flow between points 1 and 2 is given as follows:

$$\left(\frac{p_1}{\gamma} + z_1 + \frac{v_1^2}{2g}\right) - \left(\frac{p_2}{\gamma} + z_2 + \frac{v_2^2}{2g}\right) = \frac{\tau_w L}{\gamma R_h} = h_f \tag{8.151}$$

Beginning with the expression for the actual discharge, Q_a Equation 8.142, substituting $Q_a = v_2 \, A_2$, rearranging it to express the head loss, h_f, and substituting $v_1 = v_2 A_2/A_1$ yields an exact expression for the head loss, h_f between points 1 and 2 as follows:

$$Q_a = C_d A_2 \sqrt{\frac{2(p_1 - p_2)}{\rho \left[1 - \left(\frac{A_2}{A_1}\right)^2\right]}} \tag{8.152}$$

$$v_2 A_2 = C_d A_2 \sqrt{\frac{2(p_1 - p_2)}{\rho \left[1 - \left(\frac{A_2}{A_1}\right)^2\right]}} \tag{8.153}$$

$$v_2{}^2 = C_d^2 \frac{2(p_1 - p_2)}{\rho\left[1 - \left(\frac{A_2}{A_1}\right)^2\right]} \tag{8.154}$$

$$\frac{v_2{}^2}{2g} = C_d^2 \frac{2(p_1 - p_2)}{2\rho g\left[1 - \left(\frac{A_2}{A_1}\right)^2\right]} \tag{8.155}$$

$$\frac{v_2{}^2}{2g} = \frac{\left(\frac{p_1}{\gamma} - \frac{p_2}{\gamma}\right)}{\left(\frac{1}{C_d^2}\right)\left[1 - \left(\frac{A_2}{A_1}\right)^2\right]} \tag{8.156}$$

Although the effects of a nonhorizontal pipe may be easily included in the energy equation, assuming that $z_1 = z_2$ yields:

$$\frac{p_1}{\gamma} + \frac{v_1{}^2}{2g} - \frac{p_2}{\gamma} - \frac{v_2{}^2}{2g} = h_f = \frac{1}{C_d^2}\left[1 - \left(\frac{A_2}{A_1}\right)^2\right]\frac{v_2{}^2}{2g} + \frac{v_1{}^2}{2g} - \frac{v_2{}^2}{2g} \tag{8.157}$$

$$h_f = \frac{1}{C_d^2}\left[1 - \left(\frac{A_2}{A_1}\right)^2\right]\frac{v_2{}^2}{2g} + \frac{v_2{}^2}{2g}\left(\frac{A_2}{A_1}\right)^2 - \frac{v_2{}^2}{2g} \tag{8.158}$$

$$h_f = \left[\frac{1}{C_d^2}\left[1 - \left(\frac{A_2}{A_1}\right)^2\right] + \left(\frac{A_2}{A_1}\right)^2 - 1\right]\frac{v_2{}^2}{2g} \tag{8.159}$$

$$h_f = \left[\frac{1}{C_d^2} - \frac{1}{C_d^2}\left(\frac{A_2}{A_1}\right)^2 + \left(\frac{A_2}{A_1}\right)^2 - 1\right]\frac{v_2{}^2}{2g} \tag{8.160}$$

$$h_f = \underbrace{\left[\left(\frac{1}{C_d^2} - 1\right)\left(1 - \left(\frac{A_2}{A_1}\right)^2\right)\right]}_{k}\frac{v_2{}^2}{2g} = k\frac{v_2{}^2}{2g} \tag{8.161}$$

where v_2 is the actual velocity and k is the "minor head loss coefficient" due to the flow-measuring device, which is a function of the discharge coefficient, C_d which is a function of geometry of the flow-measuring device and the Reynolds number, R.

End-of-Chapter Problems

Problems with a "C" are conceptual problems. Problems with a "BG" are in English units. Problems with an "SI" are in metric units. Problems with a "BG/SI" are in both English and metric units. All "BG" and "SI" problems that require computations are solved using Mathcad.

Introduction

8.1C   Explain the basic difference between pipe flow and open channel flow.

8.2C   In the application of the governing equations (continuity, energy, and momentum) for pipe flow, it is important to make the distinction between real pipe flow and ideal pipe flow (see Table 4.1 and Section 8.4). Furthermore, the application of the governing equations for real pipe flow and ideal pipe flow (flow-measuring devices) requires the modeling of flow resistance (frictional losses). The modeling of flow resistance requires a "subset level" application of the governing equations (Section 8.3). Explain how the flow resistance is modeled for real flow vs. ideal flow in the application of the governing equations.

8.3C   Explain how the flow resistance flow is specifically modeled for real pipe flow vs. ideal pipe flow.

8.4C   List specific applications of the governing equations for real pipe flow vs. ideal pipe flow.

Application of the Eulerian (Integral) versus Lagrangian Differential) Forms of the Governing Equations

8.5C   Depending upon the flow problem, one may use either the Eulerian (integral) approach or the Lagrangian (differential) approach, or both, in the application of the three governing equations (continuity, energy, and momentum). Explain further.

8.6C   In the case where the problem involves mass flow in and out of a pipe system, then the problem is modeled as a finite control volume. The integral (Eulerian) form of the three governing equations are applied for the assumption of both real and ideal pipe flow problems. Explain why and how one would apply the integral form of the three governing equations for the assumption of: (a) real pipe flow and (b) ideal pipe flow.

8.7C   In the assumption of real flow, and for the specific case of a major head loss due to pipe flow resistance, application of the differential (Lagrangian) form of the momentum equation may become necessary. List specific applications of the differential form of the momentum equation for real pipe flow.

Modeling Flow Resistance in Pipe Flow: A Subset Level Application of the Governing Equations

8.8C   The application of the governing equations for real pipe flow and ideal pipe flow (flow-measuring devices) (Section 8.4) requires the modeling of flow resistance (frictional losses). The modeling of flow resistance requires a "subset level" application of the governing equations. The distinction between ideal and real flow determines how the flow resistance is modeled in the "subset level" application of the governing equations. Explain how the flow resistance is evaluated in the application of the governing equations (continuity, energy, and momentum).

8.9C   The flow resistance in pipe flow (major head loss, minor head loss, and actual discharge) is modeled by a "subset level" application of the appropriate governing equations, as presented in Chapter 6 (and summarized in Sections 8.4.1.2 and 8.4.2.1). Explain what a "subset level" application of the appropriate governing equations implies, and explain when the use of dimensional analysis (Chapter 7) is needed in the derivation of the flow resistance equations.

Single Pipes: Major Head Loss in Real Pipe Flow

8.10C Applications of the governing equations for real pipe flow include the analysis and design of a single pipe, which forms the basic component for any pipe system (see Section 8.7). Explain the application of the governing equations (integral vs. differential) for the case of a single straight pipe section without pipe components.

8.11C While the major head loss may be ignored for short pipes, it is significant for very long pipes. Explain the importance of modeling the major head loss in pipe flow.

Evaluation of the Major Head Loss: Poiseuille's Law for Laminar Flow and the Chezy Equation for Turbulent Flow

8.12C Evaluation of the major head loss flow resistance term in the energy equation requires a "subset level" application of the governing equations (see Chapters 6 and 7 and Section 8.4.1.2). Depending upon the value for the Reynolds number, R, real flow may be described as either laminar or turbulent. How is the major head loss due to pipe friction modeled in the energy equation, Equation 8.1 evaluated for laminar vs. turbulent flow?

8.13C Evaluation of the major head loss flow resistance term requires a "subset level" application of the governing equations (see Chapters 6 and 7 and Section 8.4.1.2). Explain why the major head loss for laminar flow may be determined analytically (i.e., without the need for dimensional analysis).

8.14C Begin with the integral form of Poiseuille's law given by Equation 8.20 as follows: $\Delta p = 32 \mu v L / D^2$. Illustrate how would one transition from the integral point of view to the differential point of view. Explain how the differential form of Poiseuille's law is applied in laminar flow.

8.15C Derive the expression for the Darcy–Weisbach friction factor, f for laminar flow.

8.16C Derive the analytical expression for the major head loss, h_f due to pipe friction for laminar flow.

8.17C Explain how Poiseuille's law is applied in the analysis and design of laminar pipe flow.

8.18SI Glycerin at $20°$ C flows in an inclined pipe at a flowrate of $0.01\ m^3/sec$, as illustrated in Figure EP 8.1. The pressure at point 1 is $1.6 \times 10^5\ N/m^2$. The increase in elevation between points 1 and 2 is $0.05\ m$, the pipe length between points 1 and 2 is $7500\ m$, and the friction slope is $0.0003\ m/m$. Assume laminar flow. (a) Determine the major head loss due to the pipe friction (flow resistance) between points 1 and 2. (b) Determine the pipe diameter. (c) Confirm that the flow is laminar. (d) Determine the pressure drop between points 1 and 2 due to the increase in elevation and the pipe friction (flow resistance) between points 1 and 2.

8.19C Evaluation of the major head loss flow resistance term requires a "subset level" application of the governing equations (see Chapters 6 and 7 and Section 8.4.1.2). Explain why the major head loss for turbulent flow must be determined empirically (i.e., with the need for dimensional analysis).

8.20C Explain how the differential form of the Chezy equation, $S_f = v^2/C^2 R_h$, Equation 8.25, is applied in turbulent flow.

8.21C Discuss the Chezy coefficient, C: what is it a function of, its range of values, its units and dimensions, its application in fluid flow, and compare it to the Darcy–Weisbach friction factor, f.

8.22C Derive the empirical expression for the major head loss, h_f due to pipe friction for turbulent flow; assume the Chezy coefficient, C is used to model the flow resistance.

8.23C Explain how the Chezy equation is applied in the analysis and design of turbulent pipe flow.

8.24SI Seawater at $20°C$ flows in a $6\,m$ diameter inclined pipe, as illustrated in Figure EP 8.2. The pressure at point 1 is $5 \times 10^5 \, N/m^2$. The increase in elevation between points 1 and 2 is $0.088\,m$, the pipe length between points 1 and 2 is $16000\,m$, and the head loss due to pipe friction is $2\,m$. The Chezy coefficient, C for the pipe is $70\,m^{1/2}\,s^{-1}$. Assume turbulent flow. (a) Determine the friction slope at either points 1 or 2. (b) Determine the velocity and discharge in the pipe. (c) Confirm that the flow is turbulent. (d) Determine the pressure drop between points 1 and 2 due to the increase in elevation and the pipe friction (flow resistance) between points 1 and 2.

Turbulent Pipe Flow Resistance Equations and Their Roughness Coefficients

8.25C In addition to the Chezy equation, list three additional turbulent pipe flow resistance equations.

8.26C The drag coefficient, C_D representing the flow resistance that is derived by dimensional analysis in Chapter 7 is a nonstandard form of the drag coefficient and is summarized in Equation 8.28 as follows:

$$\frac{\tau_w}{\rho v^2} = \frac{F_f}{F_I} = \frac{F_P}{F_I} = \frac{\Delta p R_h}{L} \frac{1}{\rho v^2} = \frac{R_h}{L} \frac{\Delta p}{\rho v^2} = \frac{R_h}{L} \frac{1}{E} = C_D = \phi \left(R, \frac{\varepsilon}{D} \right).$$

Because the drag coefficient, C_D represents the flow resistance, which causes an unknown pressure drop, Δp, the pressure force will always play an important role; thus, the definition of the drag coefficient, C_D is mainly a function of E. Furthermore, one may note that the drag coefficient, C_D, which is a nonstandard form of the flow resistance, is dimensionless and is a function of boundary roughness, ε/D the Reynolds number, R and the cross-sectional shape or the hydraulic radius, R_h. List four more standard forms of the drag coefficient.

8.27C How is the drag coefficient drag coefficient, C_D related to the Chezy coefficient, C, and the Darcy–Weisbach friction factor, f, and what factors are these coefficients a function of?

8.28C How are the Chezy coefficient, C and the Darcy–Weisbach friction factor, f related?

8.29C How are the Chezy coefficient, C and the Manning's roughness coefficient, n related?

8.30C Explain how the Manning's roughness coefficient, n and the Hazen–Williams roughness coefficient, C_h are similar to one another. Furthermore, explain how the Manning's roughness coefficient, n and the Hazen–Williams roughness coefficient, C_h are different from the Chezy coefficient, C and the Darcy–Weisbach friction factor, f.

The Darcy–Weisbach Equation

8.31C Beginning with the relationship between the Chezy coefficient, C and the Darcy–Weisbach friction factor, f, $C = \sqrt{8g/f}$, Equation 8.33, derive the Darcy–Weisbach head loss equation. Also, is the derived equation in SI or BG units, or both?

8.32C Choosing the Darcy–Weisbach friction factor, f to model the empirical flow resistance is a typical choice for a pipe flow (where $R_h = D/4$). Explain how the Darcy–Weisbach equation, $h_f = f(L/D)(v^2/2g)$, Equation 8.34, is applied in pipe flow.

8.33C Explain how the Darcy–Weisbach head loss equation is applied in the analysis and design of turbulent pipe flow.

8.34SI Water at 20°C flows in a 3.9-m square horizontal riveted steel pipe at a flowrate of 6 m^3/sec, as illustrated in Figure EP 8.3. The pressure at point 1 is $37 \times 10^5 \, N/m^2$, and the pipe length between points 1 and 2 is 18,000 m. Assume the Darcy–Weisbach friction factor is chosen to model the flow resistance. (a) Determine the flow regime. (b) Determine the major head loss due to the pipe friction (flow resistance) between points 1 and 2. (c) Determine the pressure drop between points 1 and 2 due to the pipe friction (flow resistance) between points 1 and 2. (d) Determine the friction slope at either point 1 or 2.

8.35SI Gasoline at 20°C flows in a 3.75-m-diameter horizontal plastic pipe, as illustrated in Figure EP 8.4. The pressure at point 1 is $33 \times 10^5 \, N/m^2$, the major head loss due to the pipe friction is 34 m, and the pipe length between points 1 and 2 is 2200 m. Assume the Darcy–Weisbach friction factor is chosen to model the flow resistance. (a) Determine velocity at either point 1 or 2. (b) Determine the flow regime. (c) Determine the pressure drop between points 1 and 2 due to the pipe friction (flow resistance) between points 1 and 2.

8.36SI Kerosene at 20°C flows in an inclined asphalt-dipped cast iron pipe at a velocity of 8 m/sec, as illustrated in Figure EP 8.5. The pressure at point 1 is $6 \times 10^5 \, N/m^2$, and the major head loss due to the pipe friction is 33 m. The increase in elevation between points 1 and 2 is 0.08 m, and the pipe length between points 1 and 2 is 11,000 m. Assume the Darcy–Weisbach friction factor is chosen to model the flow resistance. (a) Determine pipe diameter at either point 1 or 2. (b) Determine the flow regime. (c) Determine the pressure drop between points 1 and 2 due to the increase in elevation and the pipe friction (flow resistance) between points 1 and 2.

Manning's Equation

8.37C Beginning with the Chezy equation, $S_f = v^2/C^2 R_h$, Equation 8.25, derive the Manning's equation (expressed as a function of the velocity, v). Furthermore, derive the Manning's head loss equation.

8.38C Explain how the Manning's equation is applied in turbulent pipe flow.

8.39C Explain how the Manning's Equation is applied in the analysis and design of turbulent pipe flow.

8.40SI Seawater at 20°C flows at a flowrate of 1.2 m^3/sec in a horizontal pipe with an isosceles triangular cross section (each side, s is 2.33 m), as illustrated in Figure EP 8.6. The pressure at point 1 is $17 \times 10^5 \, N/m^2$, the major head loss due to the pipe friction is 22 m, and the pipe length between points 1 and 2 is 7000 m. Assume the Manning's roughness coefficient is chosen to model the flow resistance. (a) Determine the Manning's roughness coefficient for the pipe, and estimate the pipe material. (b) Determine the flow regime. (c) Determine the pressure drop between points 1 and 2 due to the pipe friction (flow resistance) between points 1 and 2.

The Hazen–Williams Equation

8.41C In addition to the standard empirical flow resistance coefficients including the Chezy coefficient, C; the Darcy–Weisbach friction factor, f; and the Manning's roughness coefficient, n, there is another commonly used flow resistance coefficient; namely the Hazen–Williams roughness coefficient, C_h (see Table 8.8). Give the empirically derived Hazen–Williams equation derived assuming the BG system.

8.42C Beginning with the Chezy head loss equation,

$$h_f = \frac{\tau_w L}{\gamma R_h} = \frac{\Delta p}{\gamma} = S_f L = \underbrace{\left[\frac{v^2}{C^2 R_h}\right]}_{S_f} L$$

Equation 8.26, derive the Hazen–Williams head loss equation (assuming the English system).

8.43C Explain how the Hazen–Williams equation is applied in turbulent pipe flow.

8.44C Explain how the Hazen–Williams equation is applied in the analysis and design of turbulent pipe flow.

8.45SI Crude oil at 20°C flows in a 6-m-diameter inclined asbestos cement pipe at a velocity of $2\,m/sec$, as illustrated in Figure EP 8.7. The pressure at point 1 is $5 \times 10^5\,N/m^2$. The increase in elevation between points 1 and 2 is $0.04\,m$, and the pipe length between points 1 and 2 is $14{,}000\,m$. Assume the Hazen–Williams roughness coefficient is chosen to model the flow resistance. (a) Determine the major head loss due to the pipe friction (flow resistance) between points 1 and 2. (b) Determine the flow regime. (c) Determine the pressure drop between points 1 and 2 due to the increase in elevation and the pipe friction (flow resistance) between points 1 and 2.

Pipes with Components: Minor Head Losses and Reaction Forces in Real Pipe Flow

8.46C Applications of the governing equations for real pipe flow include the analysis and design of pipes with components, which form the basic component for any pipe system (see Section 8.7). Explain the application of the governing equations (integral vs. differential) for the case of pipes with components.

8.47C Explain the importance of modeling the minor head loss in pipe flow.

Evaluation of the Minor Head Loss

8.48C Evaluation of the minor head loss flow resistance term requires a "subset level" application of the governing equations (see Chapters 6 and 7 and Section 8.4.1.2). Explain when the minor head loss due to a pipe component modeled in the energy equation, Equation 8.1, may be evaluated analytically vs. empirically.

8.49C Evaluation of the minor head loss flow resistance term requires a "subset level" application of the governing equations (see Chapters 6 and 7 and Section 8.4.1.2). What is the expression for the minor head loss term, $h_{f,minor}$ in the energy equation due to the different pipe devices?

8.50C   Explain how the minor head loss coefficient, k, and thus the minor head loss for the case of pipe devices in general is empirically derived. Give and explain the resulting empirical expression.

8.51C   Given the empirical expression for the minor head loss in Equation 8.53 as follows:

$$h_{f,min} = \frac{\tau_w L}{\gamma R_h} = \frac{\Delta p}{\gamma} = k\frac{v^2}{2g}.$$

What is the difference between the static pressure and the dynamic pressure, and what is another term used for the minor head loss coefficient, k?

Minor Losses in Valves and Fittings (Tees, Unions, Elbows, and Bends)

8.52C   Evaluation of the minor head loss for valves and fittings (tees, unions, elbows, and bends) involves evaluating the velocity term, v and the minor head loss coefficient, k for the particular pipe component in the minor head loss, $h_{f,min} = k(v^2/2g)$. Explain the basic difference and similarity between valves and fittings.

8.53SI   Water flows in a 0.9-*m*-diameter horizontal pipe at a flowrate of 0.7 m^3/sec. A globe valve is installed in between points 1 and 2 as illustrated in Figure EP 8.8 in order to regulate or throttle the flowrate in the pipe. The pressure at point 1 is $4 \times 10^5 \, N/m^2$. (a) Determine the minor head loss due to the globe valve. (b) Determine the pressure at point 2.

8.54SI   Water flows in a 0.35-*m*-diameter horizontal pipe at a flowrate of 0.79 m^3/sec. A flanged line flow tee is installed in between points 1 and 2 as illustrated in Figure ECP 8.54 in order to change the direction of the flowrate. The pressure at point 1 is 6×10^5 N/m^2. (a) Determine the minor head loss due to the flanged line flow tee. (b) Determine the pressure at point 2.

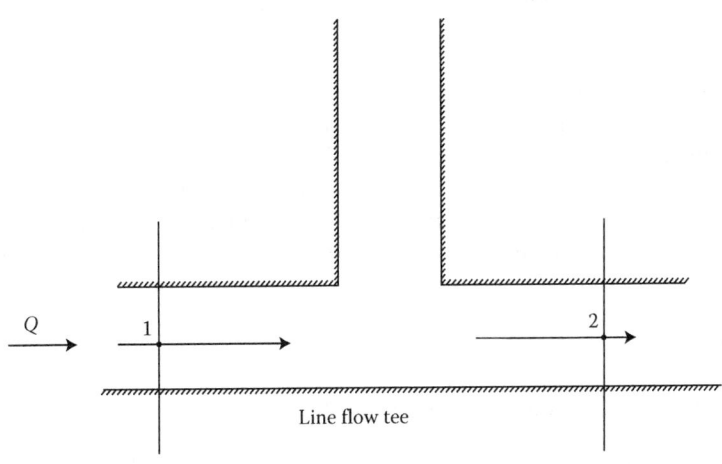

Line flow tee

FIGURE ECP 8.54

Minor Losses in Entrances, Exits, Contractions, and Expansions

8.55C   Evaluation of the minor head loss for entrances, exits, contractions, and expansions involves evaluating the velocity term, v and the minor head loss coefficient, k for the particular pipe transition section in the minor head loss, $h_{f,min} = k(v^2/2g)$. Explain the roles and the relationship between (1) entrances and exits and (2) contractions and expansions.

8.56SI   Water flows in a 0.66-m-diameter horizontal pipe at a flowrate of 0.45 m^3/sec and exits into a reservoir through a reentrant pipe exit, as illustrated in Figure ECP 8.56. (a) Determine the minor head loss due to the reentrant pipe exit. (b) Would the magnitude of the minor head loss change if the water exits into the reservoir through a square-edged exit, or a rounded exit?

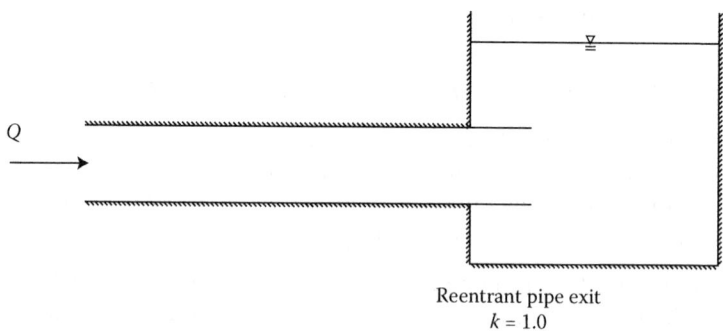

Reentrant pipe exit
$k = 1.0$

FIGURE ECP 8.56

8.57SI   Water flow enters a 0.06-m-diameter horizontal pipe at a flowrate of 0.07 m^3/sec from a reservoir, as illustrated in Figure 8.10. (a) Compare the magnitudes of the minor head loss assuming a reentrant entrance, a square-edged entrance, and a well-rounded entrance.

Reaction Forces on Pipe Components

8.58C   When is it necessary to model the reaction forces on pipe components?

8.59SI   Referring to Example Problem 8.11, water flows in a 0.08-m-diameter horizontal pipe at a flowrate of 0.01 m^3/sec, as illustrated in Figure EP 8.11b. A gradual pipe contraction/conical contraction (nozzle) to a 0.03-m diameter pipe at point 2 is installed in order to accelerate the fluid flow (which will result in a pressure drop). The pressure at point 1 is $2.5 \times 10^5 \, N/m^2$. From the solution of Example Problem 8.11b, the minor head loss due to gradual pipe contraction/conical contraction (nozzle) is 0.714 m, and the pressure at point 2 is $1.051 \times 10^5 \, N/m^2$. Assume a very short pipe section between points 1 and 2. Determine the reaction force acting on the gradual pipe contraction/conical contraction (nozzle) due to the thrust of the flow.

Pipe Systems: Major and Minor Head Losses in Real Pipe Flow

8.60C   List specific applications of the governing equations for real pipe flow.

Single Pipes

8.61C Explain the important role of single pipes in a pipe system.

Pipes with Components

8.62SI Water at 20°C flows from reservoir 1 to reservoir 2 through a 0.8-*m*-diameter commercial steel pipe, as illustrated in Figure EP 8.13. The difference in the total head, ΔH between the two reservoirs is 150 *m*. The pipe system contains two reservoirs, five single pipes with a total length of 1200 *m*, one fully open globe valve to regulate the flow, and four regular 90° threaded elbows. The pipe entrance from reservoir 1 is square-edged, and the pipe exit to reservoir 2 is rounded. Assume the Darcy–Weisbach friction factor is chosen to model the flow resistance. (a) Determine the total head loss due to pipe friction and pipe components between points 1 and 2. (b) Determine the velocity in the pipe.

Pipes with a Pump or a Turbine

8.63SI Gasoline at 20°C is pumped to an elevation of 700 *m* from reservoir 1 to reservoir 2 through a 0.44-*m*-diameter copper pipe at a discharge of 0.3 m^3/sec, as illustrated in Figure EP 8.14. The pipe system contains two reservoirs, two single pipes with a total length of 1600 *m*, a pump, and one regular 90° threaded elbow. The pipe entrance from reservoir 1 is reentrant, and the pipe exit to reservoir 2 is square-edged. Assume the Darcy–Weisbach friction factor is chosen to model the flow resistance. (a) Determine the head added by the pump in order to overcome the difference in elevation between the two reservoirs, and the total head loss due to pipe friction and pipe components between points 1 and 2. (b) Determine the hydraulic power added to the flow by the pump.

8.64SI Water at 20°C flowing from a reservoir at an elevation z_1 through a galvanized iron 0.25-*m*-diameter pipe at a velocity of 4 *m/sec* is used to supply power to a turbine, as illustrated in Figure EP 8.15. The pipe system contains one reservoir, two single pipes with a total length of 1300 *m*, and a turbine. The pipe entrance from the reservoir is square-edged. The hydraulic power required to be removed from the flow by the turbine is 110 *kW*. Assume the Darcy–Weisbach friction factor is chosen to model the flow resistance. (a) Determine the head removed by the turbine in order to supply it with 110 *kW* of hydraulic power. (b) Determine the elevation of the reservoir, z_1 that is required to supply the necessary hydraulic power to the turbine.

Pipes in Series

8.65SI Water at 20°C flows from point A to point B through three cast iron pipes in series at a flowrate of 0.9 m^3/sec, as illustrated in Figure EP 8.16. The pipe system contains three single pipes, one sudden pipe contraction, and one sudden pipe expansion. The first pipe is 0.8 *m* in diameter and 200 *m* long. The second pipe is 0.4 *m* in diameter and 50 *m* long. The third pipe is 0.6 *m* in diameter and 60 *m* long. The pressure at point A is $17 \times 10^5\ N/m^2$, and the pressure at point B is $15 \times 10^5\ N/m^2$; the pressure drop, Δp is due to the pipe friction, pipe components, difference in elevation, and change in velocity between points A and B. Assume the Darcy–Weisbach friction factor is chosen to model the flow resistance. (a) Determine the flow regime. (b) Determine the total head loss due to pipe friction and pipe components between points A and B. (c) Determine the difference in elevation between points A and B.

8.66SI Kerosene at 20°C flows from reservoir A to reservoir B through two galvanized iron pipes in series, as illustrated in Figure EP 8.17. The pipe system contains two reservoirs, two single pipes, and one gradual pipe contraction/nozzle with a conical angle, $\theta = 30°$; thus, $k = 0.02$. The pipe entrance from reservoir A is reentrant, and the pipe exit to reservoir B is square-edged. The first pipe is 0.2 m in diameter and 900 m long, and the second pipe is 0.1 m in diameter and 1200 m long. The maximum allowable total head loss due to pipe friction and pipe components between points A and B is 70 m. Assume the Darcy–Weisbach friction factor is chosen to model the flow resistance. (a) Determine the required difference in elevation between points A and B in order to maintain the maximum allowable total head loss. (b) Determine the velocity in each pipe and the discharge of the flow in the pipes.

Pipes in Parallel

8.67SI Water at 20°C flows from point A to point B through three cast iron pipes in parallel at a total flowrate of 4 m^3/sec, as illustrated in Figure EP 8.18. The pipes at points A and B are also cast iron and are 1.2 m in diameter. The pipe system contains five single pipes, two threaded branch flow tees, and four regular 90° threaded elbows. The first pipe is 0.8 m in diameter and 1000 m long. The second pipe is 0.7 m in diameter and 600 m long. The third pipe is 0.5 m in diameter and 1000 m long. The pressure at point B is $11 \times 10^5 N/m^2$, and the difference in elevation between points A and B is 0 m. Assume the Darcy–Weisbach friction factor is chosen to model the flow resistance. (a) Determine the pressure at point A, and the pressure drop, Δp due to the pipe friction and pipe components, between points A and B. (b) Determine the total head loss due to pipe friction and pipe components between points A and B. (c) Determine the flow regime. (d) Determine the flowrate in each of the three pipes.

8.68SI Glycerin at 20°C flows from point A to point B through three riveted steel pipes in parallel, as illustrated in Figure EP 8.19. The pipe system contains two reservoirs and three single pipes. The three pipe entrances from reservoir A are square-edged, and the three pipe exits to reservoir B are square-edged. The first pipe is 0.75 m in diameter and 1000 m long. The second pipe is 0.7 m in diameter and 1000 m long. The third pipe is 0.65 m in diameter and 1000 m long. The difference in elevation between points A and B is 22 m. Assume the Darcy–Weisbach friction factor is chosen to model the flow resistance. (a) Determine the total flowrate and the flowrate in each of the three pipes. (b) Determine the flow regime. (c) Determine the total head loss due to pipe friction and pipe components between points A and B.

8.69SI Water at 20°C is pumped from reservoir A to reservoir B through two cast iron pipes in parallel at a total flowrate of 3.33 m^3/sec, as illustrated in Figure EP 8.20. The pipe system contains two reservoirs, two single pipes, a pump, two threaded branch flow tees, and four regular 90° threaded elbows. The pipe entrance from reservoir A is square-edged and the pipe exit to reservoir B is square-edged. The first pipe is 0.8 m in diameter and 600 m long. The second pipe is 0.7 m in diameter and 600 m long. The elevation of reservoir B is 55 m. Assume the Darcy–Weisbach friction factor is chosen to model the flow resistance. (a) Determine the head added by the pump in order to overcome the difference in elevation between the two reservoirs, and the total head loss due to the pipe friction and pipe components, between points A and B. (b) Determine the hydraulic power added to the flow by the pump. (c) Determine the total head loss due to pipe friction and pipe components between points A and B. (d) Determine the flow regime. (e) Determine the flowrate in each of the two pipes.

Branching Pipes Connected to Three Reservoirs

8.70SI Benzene at 20°C flows between reservoirs A, B, and C through three cast iron branching pipes with a common junction J as illustrated in Figure EP 8.21. The pipe system contains three reservoirs and three single pipes. Pipe 1 is 0.45 m in diameter and 1800 m long. Pipe 2 is 0.5 m in diameter and 2000 m long. Pipe 3 is 0.65 m in diameter and 1030 m long. The elevation of reservoir A is 430 m, the elevation of reservoir B is 320 m, and the discharge in pipe 1 is 0.9 m^3/sec. (a) Determine the discharge in pipe 2. (b) Determine the discharge in pipe 3. (c) Determine the elevation of reservoir C. (d) Determine the total head for junction J.

8.71SI Water at 20°C flows between reservoirs A, B, and C through three wood stave branching pipes with a common junction J as illustrated in Figure EP 8.22. The pipe system contains three reservoirs and three single pipes. Pipe 1 is 0.77 m in diameter and 1200 m long. Pipe 2 is 0.88 m in diameter and 900 m long. Pipe 3 is 0.99 m in diameter and 750 m long. The elevation of reservoir A is 330 m, the elevation of reservoir B is 270 m, and the discharge in pipe 3 is 0.85 m^3/sec. (a) Determine the discharge in pipe 1. (b) Determine the discharge in pipe 2. (c) Determine the elevation of reservoir C. (d) Determine the total head for junction J.

8.72SI Water at 20°C flows between reservoirs A, B, and C through three riveted steel branching pipes with a common junction J as illustrated in Figure EP 8.23. The pipe system contains three reservoirs and three single pipes. Pipe 1 is 0.8 m in diameter and 1100 m long. Pipe 2 is 0.7 m in diameter and 1000 m long. Pipe 3 is 0.5 m in diameter and 900 m long. The elevation of reservoir A is 530 m, the elevation of reservoir B is 330 m, and elevation of reservoir C is 230 m (a) Determine the discharge in pipe 1. (b) Determine the discharge in pipe 2. (c) Determine the discharge in pipe 3. (d) Determine the total head for junction J.

Branching Pipes Connected to a Water Supply Source under Pressure

8.73SI Water at 20°C flows from point A (water supply source under pressure) to point B (destination is kitchen faucet), and from point A to point C (destination is laundry room faucet) through three copper branching pipes with a common junction J as illustrated in Figure EP 8.24. The pipe system contains a source, two destinations, three single pipes, three regular 90° threaded elbows, and two fully open globe valves. Pipe 1 is 0.07 m in diameter and 85 m long. Pipe 2 is 0.06 m in diameter and 80 m long. Pipe 3 is 0.05 m in diameter and 70 m long. At point A, the pressure is $2.8 \times 10^5 \, N/m^2$, and the elevation is 0 m. At point B, the pressure is $0 \, N/m^2$, and the elevation is 20 m. The discharge in pipe 1 is 0.008 m^3/sec. (a) Determine the discharge in pipe 2. (b) Determine the discharge in pipe 3. (c) Given that the pressure at point C is $0 \, N/m^2$ determine the maximum elevation at point C.

8.74SI Water at 20°C flows from point A (water supply source under pressure) to point B (destination is kitchen faucet), and from point A to point C (destination is laundry room faucet) through three riveted steel branching pipes with a common junction J as illustrated in Figure EP 8.25. The pipe system contains a source, two destinations, three single pipes, three regular 90° threaded elbows, and two fully open globe valves. Pipe 1 is 0.08 m in diameter and 100 m long. Pipe 2 is 0.05 m in diameter and 90 m long. Pipe 3 is 0.065 m in diameter and 150 m long. At point A, the elevation is 0 m. At point B, the pressure is $0 \, N/m^2$, and the elevation is 25 m. At point C, the pressure is $0 \, N/m^2$, and the elevation is 21 m. The discharge in pipe 1 is 0.035 m^3/sec. (a) Determine the minimum required pressure at point A in order to overcome the elevations at points B and C, and the head loss due to pipe friction in pipes 1, 2, and 3. (b) Determine the discharge in pipe 2. (c) Determine the discharge in pipe 3.

8.75SI Water at 20°C flows from point A (water supply source under pressure) to point B (destination is kitchen faucet) and from point A to point C (destination is laundry room faucet) through three galvanized iron branching pipes with a common junction J as illustrated in Figure EP 8.26. The pipe system contains a source, two destinations, three single pipes, three regular 90° threaded elbows, and two fully open globe valves. Pipe 1 is 0.08 m in diameter and 100 m long. Pipe 2 is 0.05 m in diameter and 90 m long. Pipe 3 is 0.065 m in diameter and 150 m long. At point A the pressure is $4 \times 10^5\, N/m^2$, and the elevation is 0 m. At point B, the pressure is $0\, N/m^2$, and the elevation is 25 m. At point C, the pressure is $0\, N/m^2$, and the elevation is 16 m. (a) Determine the discharge in pipe 1. (b) Determine the discharge in pipe 2. (c) Determine the discharge in pipe 3.

Pipes in a Loop

8.76SI Water at 20°C flows from point A to point B three wood stave pipes in a loop at a total flowrate of 3.5 m^3/sec, as illustrated in Figure EP 8.27. The pipe system contains three single pipes and two reservoirs. The first pipe is 0.9 m in diameter and 700 m long. The second pipe is 0.75 m in diameter and 600 m long. The third pipe is 0.65 m in diameter and 600 m long. (a) Determine the difference in elevation between points A and B. (b) Determine the total head loss due to pipe friction between points A and B. (c) Determine the flowrate in pipe 2 and pipe 3.

8.77SI Water at 20°C flows from point A to point B three galvanized iron pipes in a loop, as illustrated in Figure EP 8.28. The pipe system contains three single pipes and two reservoirs. The first pipe is 0.9 m in diameter and 700 m long. The second pipe is 0.75 m in diameter and 600 m long. The third pipe is 0.65 m in diameter and 600 m long. The difference in elevation between points A and B is 35 m. (a) Determine the flowrate in each of the three pipes. (b) Determine the total head loss due to pipe friction between points A and B.

Pipe Networks

8.78SI Water at 20°C flows in a galvanized iron pipe network system, as illustrated in Figure EP 8.29. The pipe system contains one inlet, five pipes, four junctions, two loops, and two outlets. The first pipe is 0.9 m in diameter and 100 m long. The second pipe is 0.8 m in diameter and 95 m long. The third pipe is 0.95 m in diameter and 185 m long. The fourth pipe is 0.85 m in diameter and 100 m long. The fifth pipe is 0.9 m in diameter and 190 m long. Assume that the change in elevation between all points in the pipe system, $\Delta z = 0$. The inflow at the inlet at junction A is 6 m^3/sec, the outflow at the outlet at junction C is 3.4 m^3/sec, and the outflow at the outlet at junction D is 2.6 m^3/sec. Assume that the Darcy–Weisbach friction factor, f varies with the flowrate. (a) Determine the flowrate in each of the five pipes. (b) Determine the head loss in each of the five pipes.

8.79SI Water at 20°C flows in a cast iron pipe network system, as illustrated in Figure EP 8.30. The pipe system contains two inlets, ten pipes, eight junctions, three loops, and three outlets. The first nine pipes are 1.3 m in diameter and 100 m long. The tenth pipe is 1.3 m in diameter and 200 m long. Assume that the Darcy–Weisbach friction factor, f is a constant for each pipe and is given to be 0.018 for each pipe. Assume that the change in elevation between all points in the pipe system, $\Delta z = 0$. The inflow at the inlet at junction A is 3 m^3/sec, and the flow at the inlet at junction G is 5 m^3/sec. The outflow at the outlet at junction C is 2 m^3/sec, the outflow at the outlet at junction E is 4 m^3/sec, and the outflow at the outlet at junction H is 2 m^3/sec. (a) Determine the flowrate in each of the 10 pipes. (b) Determine the head loss in each of the 10 pipes.

Pipe Flow Measurement and Control Devices: Actual Flowrate in Ideal Flow Meters

8.80C   List specific applications of the governing equations for ideal pipe flow.

Evaluation of the Actual Flowrate for Ideal Flow Meters in Pipe Flow

8.81C   Evaluation of the actual discharge flow resistance term requires a "subset level" application of the governing equations (see Chapters 6 and 7 and Section 8.4.2.1). Give the expression for the actual discharge for pipe flow-measuring devices.

8.82C   Explain how the discharge coefficient, C_d and thus the actual discharge for pipe flow-measuring devices is empirically derived. Give, and explain the resulting expression.

Evaluation of the Actual Flowrate for a Pitot-Static Tube

8.83SI   Water flows in a 1-m-diameter horizontal pipe. A piezometer is installed at point 1 and a pitot tube is installed at point 2 as illustrated in Figure EP 8.31 in order to measure the flowrate in the pipe. The static pressure head, p_1/γ measured by the piezometer is 2.4 m, and the stagnation pressure head, p_2/γ measured by the pitot tube is 3.7 m. Assume a discharge coefficient, C_d of 0.995. (a) Determine the ideal flowrate. (b) Determine the actual flowrate.

Evaluation of the Actual Flowrate for Orifice, Nozzle, and Venturi Meters

8.84SI   Water at 5°C flows in a 0.88-m-diameter horizontal pipe. An orifice meter with a 0.5-m diameter is mounted in the flanged pipe just upstream of point 2 as illustrated in Figure EP 8.32 in order to measure the flowrate in the pipe. A pressure tap, p_1 is installed at point 1, where the full pipe flow occurs just upstream of the orifice, and a pressure tap, p_2 is installed at point 2, where the vena contracta occurs just downstream of the orifice. The pressure at point 1, p_1 is measured to be $3 \times 10^5 \, N/m^2$, and the pressure at point 2, p_2 is measured to be $2.4 \times 10^5 \, N/m^2$. The absolute or dynamic viscosity for water at 5°C is $0.001518 \, N\text{-}sec/m^2$ (see Table A.4 in Appendix A). (a) Determine the ideal flowrate. (b) Determine the actual flowrate.

8.85SI   Water at 10°C flows in a 0.77-m-diameter horizontal pipe. A nozzle meter with a 0.44-m throat diameter is mounted in the flanged pipe just upstream of point 2 as illustrated in Figure EP 8.33 in order to measure the flowrate in the pipe. A pressure tap, p_1 is installed at point 1 just upstream of the nozzle flange, and a pressure tap, p_2 is installed at point 2, which is located just downstream of the nozzle flange. The pressure at point 1, p_1 is measured to be $2.3 \times 10^5 \, N/m^2$, and the pressure at point 2, p_2 is measured to be $1.7 \times 10^5 \, N/m^2$. The absolute or dynamic viscosity for water at 5°C is $0.001307 \, N\text{-}sec/m^2$ (Table A.4 in Appendix A). (a) Determine the ideal flowrate. (b) Determine the actual flowrate.

8.86SI   Water at 25°C flows in a 0.96-m-diameter horizontal pipe. A venturi meter with a 0.66-m throat diameter is inserted in the pipe at point 2 as illustrated in Figure EP 8.34 in order to measure the flowrate in the pipe. A pressure tap, p_1 is installed at point 1 just upstream of the venturi tube, and a pressure tap, p_2 is installed at point 2, which is located at the throat of the venturi tube. The pressure at point 1, p_1 is measured to be $4 \times 10^5 \, N/m^2$, and the pressure at point 2, p_2 is measured to be $3.6 \times 10^5 \, N/m^2$. The absolute or dynamic viscosity for water at 5°C is $0.00089 \, N\text{-}sec/m^2$ (see Table A.4 in Appendix A). (a) Determine the ideal flowrate. (b) Determine the actual flowrate.

9

Open Channel Flow

9.1 Introduction

The flow of fluid in both pipes (pressure/closed conduit flow) and open channels (gravity/ open conduit flow) are contained in a conduit and thus are referred to as internal flows, whereas the flow of fluid over a body that is immersed in the fluid is referred to as external flow (see Chapter 10). In pipe flow, the conduit is completely filled with fluid and the flow is mainly a result of invoking a pressure difference (pressure drop, Δp), which was addressed in Chapter 8, "Pipe Flow." However, open channel flow—where the conduit is only partially filled with fluid and the top of the surface is open to the atmosphere—is addressed in Chapter 9. Open channel flow is a result of invoking gravity and occurs in natural rivers, streams, channels and in artificial channels, canals, and waterways. Furthermore, the results sought in a given flow analysis will depend upon whether the flow is internal or external. In an internal flow, energy or work is used to move/force the fluid through a conduit. Thus, one is interested in determination of the forces (reaction forces, pressure drops, flow depths), energy or head losses, and cavitation where energy is dissipated. The governing equations (continuity, energy, and momentum) and the results of dimensional analysis are applied to internal flows. Depending upon the flow problem, one may use either the Eulerian (integral) approach or the Lagrangian (differential) approach or both in the application of the three governing equations (Section 9.2). It is important to note that while the continuity and the momentum equations may be applied using either the integral or the differential approach, application of the energy equation is useful only using the integral approach, as energy is work, which is defined over a distance, L. Furthermore, as noted in previous chapters, for an internal fluid flow, while the application of the continuity equation will always be necessary, the energy equation and the momentum equation play complementary roles in the analysis of a given flow situation; when one of the two equations of motion breaks down, the other may be used to find the additional unknown quantity.

In the application of the governing equations (continuity, energy, and momentum) for open channel flow, it is important to make the distinction between real flow, ideal flow, and a hydraulic jump (see Table 4.1 and Section 9.5). Furthermore, in the application of the governing equations, although in some flow cases one may ignore (or not need to model) the flow resistance, in most practical flow cases, one must account for the fluid viscosity and thus model the resulting shear stress and flow resistance. The application of the governing equations for real open channel flow and ideal open channel flow (flow-measuring devices) requires the modeling of flow resistance (frictional losses). The modeling of flow resistance requires a "subset level" application of the governing equations (Section 9.4). However, in the application of the governing equations (continuity, energy, and momentum) for ideal flow (gradual channel transitions) and a hydraulic jump, there is no flow resistance to model. While in the case of ideal flow (gradual channel transitions) the flow resistance is actually ignored, in the case of the hydraulic jump, the major head loss is because of flow turbulence due to the jump and not the flow resistance (frictional losses). The distinction between ideal and real flow determines how the flow resistance is modeled in the "subset level" application of the governing equations. The assumption of "real" flow implies that the flow resistance is modeled in both the energy and the momentum equations. However, the assumption of "ideal" flow implies that the flow resistance is modeled only in the momentum equation (and thus the subsequent assumption of "real" flow). The flow resistance due to channel friction (modeled as a major head loss; see Section 9.6) is modeled by assuming "real" flow; application of the energy equation models a major head loss that is accounted for by a drag coefficient (in the application of the momentum equation). However, the flow resistance due to friction for flow-measuring devices for open channel flow (modeled as an actual discharge; see Sections 9.16 and 9.17) is modeled by assuming "ideal" flow; although application of the Bernoulli equation does not model a head loss in the determination of ideal velocity, the associated minor head loss with the velocity measurement is accounted for by a drag coefficient (in the application of the momentum equation, where a subsequent assumption of "real" flow is made).

It is important to note that there are specific topics of interest in the application of the governing equations in open channel flow. Furthermore, it is assumed that the flow in open channel flow is typically turbulent as opposed to laminar flow. The occurrence or absence of transitions and controls in the open channel (Section 9.8) will define the flow type (uniform vs. nonuniform) (Section 9.7); define the flow regime (critical, subcritical, or supercritical) (Section 9.7); and determine the significance of a major head loss (Sections 9.3 and 9.6), a minor head loss (Sections 9.16 and 9.17), and a reaction force (Section 9.12). The energy (Section 9.9) and momentum (Section 9.10) concepts are tailored specifically for open channel flow and define the specific energy and the momentum function, respectively. Both rectangular and nonrectangular channel sections (Section 9.11) are modeled.

Applications of the governing equations for real open channel flow include the analysis and design of uniform (Section 9.14) and nonuniform flow (generated by controls such as a break in the channel slope, a hydraulic drop, or a free overfall) (Section 9.15). Applications of the governing equations for ideal open channel flow include the analysis and design of an open channel flow-measuring device (such as sluice gates, weirs, spillways, venturi flumes, and contracted openings) (Sections 9.12, 9.16, and 9.17), which are also known as ideal flow meters, or the analysis and design of a gradual (vertical or horizontal) channel contractions/transition (Section 9.12). And, applications of the governing equations for open channel flow are made for the analysis and design of a hydraulic jump (Section 9.13). One last note is that the scope of this textbook will be limited to steady open channel flow problems.

9.2 Application of the Eulerian (Integral) versus Lagrangian (Differential) Forms of the Governing Equations

Depending upon the flow problem, one may use either the Eulerian (integral) approach, the Lagrangian (differential) approach, or both in the application of the three governing equations. It is important to note that while the continuity and the momentum equations may be applied using either the integral (for a finite control volume) or the differential approach, application of the energy equation is useful only in the integral approach (for a finite control volume), as energy is work, which is defined over distance, *L*. While the application of the continuity equation will always be necessary, the energy equation and the momentum equations (equations of motion) play complementary roles in the analysis of a given flow situation. The momentum equation is applied in complement to the energy equation, which is useful only for a control volume. However, depending upon the fluid flow problem, one may apply the differential form of the momentum equation (for a fluid system) or the integral form of the momentum equation (for a control volume). Thus, when one of the two equations of motion breaks down, the other may be used to find the additional unknown quantity.

9.2.1 Eulerian (Integral) Approach for Open Channel Flow Problems

In the case where a problem involves mass flow in and out of an open channel flow system, then this problem is modeled as a finite control volume. The integral forms of the three governing equations are applied for the assumption of real flow, ideal flow, and hydraulic jump problems. In the assumption of real open channel flow, if one is interested in determining an unknown "energy head" term (y, z, or $v^2/2g$) between two points or an unknown head loss (major) due to friction between two points, as well as an unknown head delivered to the fluid by a pump or an unknown head removed from the fluid by a turbine, then the integral form of the energy equation is applied. Furthermore, because there are no reaction forces for real open channel flow, there is no need to apply the integral form of the momentum equation. For a real flow, if one is interested in determining an unknown velocity or discharge between two points, then the integral form of the continuity equation is applied. Furthermore, for a real flow, evaluation of the major head loss flow resistance term in the energy equation requires a "subset level" application of the governing equations (Section 9.4); the momentum equation is supplemented by dimensional analysis in the derivation of the major head loss for turbulent flow (Section 9.5.1.2). However, although the use of dimensional analysis is not needed in the derivation of the major head loss for laminar flow, the flow in open channel flow is typically assumed to be turbulent.

In the assumption of ideal flow, if one is interested in determining an unknown "energy head" term (y, z, or $v^2/2g$) between two points, then the integral form of the energy equation (Bernoulli equation) is applied. A typical assumption of ideal open channel flow is made for flow-measuring devices, which are considered to be ideal flow meters, or gradual (vertical or horizontal) channel contractions/transitions. For the ideal flow, if one is interested in determining the forces put on (interaction with) a device, then the integral form of the momentum equation is applied. For the ideal flow, if one is interested in determining an unknown velocity or discharge between two points, then the integral form of the continuity equation is applied. Furthermore, for an ideal flow meter, the minor head loss (flow resistance) associated with the flow measurement is indirectly modeled by a discharge coefficient, which requires a "subset level" application of the governing equations (Section 9.4); the momentum equation is supplemented by dimensional analysis in the derivation of the actual discharge (Section 9.5.2.1). In the assumption of a hydraulic jump problem, if one is interested in determining an unknown channel depth, y in the energy equation, then the integral form of the momentum equation is applied (hydraulic jump equations). In the assumption of a hydraulic jump problem, if one is interested in determining the unknown head loss (major) due to flow turbulence between two points, then the integral form of the energy equation is applied.

9.2.2 Lagrangian (Differential) Approach for Open Channel Flow Problems

In the assumption of real flow, and for the specific case of a major head loss due to channel flow resistance, application of the differential form of the momentum equation may become necessary. Specifically, application of the differential form of the momentum equation will be necessary when it is of interest to trace the gradual variation of the steady nonuniform channel depth of flow or to evaluate the flow (solve for an unknown quantity) at a given point in the fluid system for any channel section. Evaluation of the friction slope term (flow resistance terms) in the differential momentum equation requires a "subset level" application of the governing equations (Section 9.4); the momentum equation is supplemented by dimensional analysis in the derivation of the major head loss (or friction slope) for turbulent flow (Section 9.5.1.2).

9.3 The Occurrence of a Major Head Loss in Open Channel Flow

It is important to note that in addition to a major head loss due to flow resistance (Section 9.6), there is an additional source of major head loss in open channel flow that is due to the flow turbulence in the occurrence of a hydraulic jump (Section 9.13). Given that uniform flow will occur in the open channel in the absence of a channel or flow transition, whereas nonuniform flow will occur in the open channel as a result of a channel or flow transition, it is important to note the following points regarding the occurrence/modeling of a major head loss. The major head loss due to flow resistance is significant for the occurrence of uniform flow (a "control") and nonuniform flow resulting from certain "controls" (such as a break in channel bottom slope, a hydraulic drop, or a free overfall); this is because their occurrence generates a surface water profile and thus, affects the flow over a long channel section. However, the major head loss due to flow resistance is assumed to be negligible for (1) a gradual (vertical or horizontal) channel contraction/transition (not a "control"), (2) "controls" such as an abrupt vertical or horizontal contraction (abrupt step upward or abrupt decrease in channel width), or (3) a flow-measuring device (a "control"); this is because these affect the flow over a short channel section. Furthermore, although the occurrence of a hydraulic jump (not a "control," but rather an energy dissipator) affects the flow over a short channel section, the major head loss due to the flow turbulence in the jump is significant.

9.4 Modeling Flow Resistance in Open Channel Flow: A Subset Level Application of the Governing Equations

The application of the governing equations for real open channel flow and ideal open channel flow (flow-measuring devices) (Section 9.5) requires the modeling of flow resistance (frictional losses). The modeling of flow resistance requires a "subset level" application of the governing equations. The distinction between ideal and real flow determines how the flow resistance is modeled in the "subset level" application of the governing equations. Specifically, the application of the integral form of the energy equation for incompressible, real, steady flow open channel flow given by Equation 4.44 requires the evaluation of the major and minor head loss flow resistance term, $h_{fmaj} = \tau_w L / \gamma R_h$. Furthermore, the application of the differential form of the momentum equation for open channel flow given by

Equation 5.15 requires the evaluation of the friction slope term, $S_f = \tau_w / \gamma R_h$. Finally, the application of the continuity equation for an open channel flow-measuring device given by Equation 3.47 requires the evaluation the actual discharge, $Q_a = v_a A_a$.

The flow resistance in open channel flow (major head loss, and actual discharge) is modeled by a "subset level" application of the appropriate governing equations, as presented in Chapter 6 (and summarized in Sections 9.5.1.2 and 9.5.2.1 below). A "subset level" application of the appropriate governing equations focuses only on the given element causing the flow resistance. Thus, assuming that flow resistance is to be accounted for, depending upon the specific flow situation, the flow type is assumed to be either a "real" flow or an "ideal" flow. The distinction between ideal and real flow determines how the flow resistance is modeled in the "subset level" application of the appropriate governing equations. The assumption of "real" flow implies that the flow resistance is modeled in both the energy and the momentum equations. However, the assumption of "ideal" flow implies that the flow resistance is modeled only in the momentum equation (and thus the subsequent assumption of "real" flow). The flow resistance due to friction in open channel flow (modeled as a major head loss) is modeled by assuming "real" flow; application of the energy equation models a head loss that is accounted for by a drag coefficient (in the application of the momentum equation). However, the flow resistance due to friction in flow-measuring devices for open channel flow (modeled as an actual discharge) is modeled by assuming "ideal" flow; although application of the Bernoulli equation does not model a head loss in the determination of ideal velocity, the associated minor head loss with the velocity measurement is accounted for by a drag coefficient (in the application of the momentum equation, where a subsequent assumption of "real" flow is made). Furthermore, the use of dimensional analysis (Chapter 7) in supplement to a "subset level" application of the appropriate governing equations is needed to derive flow resistance equations when there is no theory to model the friction force (as in the case of turbulent open channel flow). Finally, it is important to note that the use of dimensional analysis is not needed in the derivation of the major head loss term for laminar flow (friction force is theoretically modeled).

9.5 Application of the Governing Equations in Open Channel Flow

The governing equations applied in the analysis and design of open channel flow include the continuity, energy, and momentum (in some cases supplemented with dimensional analysis) equations. While the application of the continuity equation will always be necessary, the energy equation and the momentum equations (equations of motion) play complementary roles in the analysis of a given flow situation. Thus, when one of the two equations of motion breaks down, the other may be used to find the additional unknown quantity. In the application of the governing equations for open channel flow, it is important to make the distinction between real flow, ideal flow, and a hydraulic jump (see Table 4.1). Furthermore, in the application of the governing equations, although in some flow cases, one may ignore (or not need to model) the flow resistance, in most practical flow cases, one must account for the fluid viscosity and thus model the resulting shear stress and flow resistance. The application of the governing equations for real open channel flow and ideal open channel flow (flow-measuring devices) requires the modeling of flow resistance (frictional losses). The modeling of flow resistance requires a "subset level" application of the governing equations (Sections 9.4, 9.5.1.2, and 9.5.2.1). However, in the application of the governing equations (continuity, energy, and momentum) for ideal flow (gradual channel transitions) and a hydraulic jump, there is no flow resistance to model. While in the case of ideal flow (gradual channel transitions), the flow resistance is actually ignored, in the case of the hydraulic jump, the major head loss is because of flow turbulence due to the jump and not the flow resistance (frictional losses). The distinction between ideal and real flow determines how the flow resistance is modeled in the "subset level" application of the governing equations.

9.5.1 Application of the Governing Equations for Real Open Channel Flow

The momentum equation is applied in complement to the energy equation, which is useful only for a control volume. However, depending upon the (real) fluid flow problem, one may apply the differential form of the momentum equation (for a fluid system) or the integral form of the momentum equation (for a control volume). Thus, when one of the two equations of motion breaks down, the other may be used to find the additional unknown quantity.

9.5.1.1 Integral Approach for Real Open Channel Flow Problems

Assuming a finite control volume, for the assumption of incompressible, real, steady open channel flow, first, the integral energy equation is applied, which was given by Equation 4.181, and is repeated as follows:

$$\left(y_1 + \frac{v_1^2}{2g} + z_1\right) + h_{pump} = \left(y_2 + \frac{v_2^2}{2g} + z_2\right) + h_{turbine} + h_{f,total} \tag{9.1}$$

which may be used to solve for an unknown "energy head" term (y, z, or $v^2/2g$) for one of the two points in a finite control volume; an unknown friction head loss, $h_{ftotal} = h_{f,maj}$ between two points in a finite control volume; an unknown head delivered to the fluid by a pump, h_{pump}; or an unknown head removed from the fluid by a turbine, $h_{turbine}$. Application of Equation 9.1 requires evaluation of the following terms: h_{maj}, h_{pump}, and $h_{turbine}$. Evaluation of the major head loss flow resistance term requires a "subset level" application of the governing equations (Section 9.5.1.2). Then, second, application of the complementary integral momentum equation (Equation 5.27) is required only if there is an unknown force reaction force. However, because there are no reaction forces in the case of real open channel flow problems (uniform and nonuniform open channel flow), there is no need to apply the integral momentum equation.

9.5.1.2 Evaluation of the Major Head Loss Term

The major head loss in open channel flow is due to the frictional losses/flow resistance caused by channel friction. One may note that in the case of a flow-measuring device, although the associated minor head loss may be accounted for in the integral form of the energy equation, it is typically accounted for by the use/calibration of a discharge coefficient to determine the actual discharge, where ideal flow (Bernoulli equation) is assumed (see Section 9.5.2). Evaluation of the major head loss flow resistance term requires a "subset level" application of the governing equations (see Chapters 6 and 7). A "subset level" application of the governing equations focuses only on the given element causing the flow resistance. The assumption of real flow implies that the flow resistance is modeled in both the energy and momentum equations. The flow resistance equations for the major head loss is derived in Chapters 6 and 7. In the derivation of one given source of head loss (major), first, the integral form of the energy equation is applied to solve for the unknown head loss, h_f, which is given as follows:

$$\left(y + z + \frac{v^2}{2g}\right)_1 - h_{f,maj} = \left(y + z + \frac{v^2}{2g}\right)_2 \tag{9.2}$$

One may note that in the derivation of the expression for one given source of head loss, h_f (major head loss due to flow resistance), the control volume is defined between the two points that include only that one source of head loss. One may also note that the propulsive force for a pipe flow is provided by a pressure gradient, $\Delta p/L$, while the propulsive force for an open channel flow is provided by invoking gravity and thus by the weight of the fluid along a channel slope (channel gradient, $\Delta z/L = S_o$). Assuming uniform flow, the derivation of the major head loss assumes $y_1 = y_2$, and $v_1 = v_2$ and the channel gradient, $\Delta z/L = S_o$ which yields $h_{f,\ maj} = \tau_w L/\gamma R_h = S_o L = S_f L$. And then, second, the integral form of the momentum equation (supplemented by dimensional analysis) is applied to solve for an unknown friction slop, S_f (in the case of major head loss due to channel flow resistance), which was given in Equation 5.27 and is repeated as follows:

$$\sum F_s = (F_G + F_P + F_V + F_{other})_s = (\rho Q v_s)_2 - (\rho Q v_s)_1 \tag{9.3}$$

One may note that derivation of the major head loss involves both the integral and the differential momentum equations. Thus, the resulting flow resistance equation for the major head loss is given as follows:

$$h_{f,maj} = \frac{\tau_w L}{\gamma R_h} = S_o L = S_f L = \frac{v^2}{C^2 R_h} L = f \frac{L}{D} \frac{v^2}{2g} = \left(\frac{vn}{R_h^{2/3}}\right)^2 L \tag{9.4}$$

where C is the Chezy coefficient, f is the Darcy–Weisbach friction factor, and n is the Manning's roughness. In the case of laminar flow, in lieu of the Chezy equation (turbulent flow), one would apply Poiseuille's law as follows:

$$h_{f,maj} = S_f L = \frac{\Delta p}{\gamma} = \underbrace{\left[\frac{32\mu v}{\gamma D^2}\right]}_{S_f} L \tag{9.5}$$

Furthermore, each head loss flow resistance equation represents the simultaneous application of the two complementary equations of motion (on a "subset level"). It is important to note that although the use of dimensional analysis is not needed in the derivation of the major head loss term for laminar flow (Poiseuille's law), the flow in open channel flow is typically assumed to be turbulent.

The flow resistance due to channel friction, which results in a major head loss term, $h_{f,major}$, is modeled as a resistance force (shear stress or drag force) in the integral form of the momentum equation (Equation 5.27), while it is modeled as a friction slope, S_f in the differential form of the momentum equation (Equation 5.37). As such, the energy equation may be used to solve for the unknown major head loss, h_f, while the integral momentum equation may be used to solve for the unknown pressure drop, Δp (or the unknown friction slope, S_f). The major head loss, h_f is caused by both pressure and friction forces. Thus, when the friction/viscous forces can be theoretically modeled in the integral form of the momentum equation (application of Newton's law of viscosity in the laminar flow case), one can analytically determine the actual pressure drop, Δp from the integral form of the momentum equation. And, therefore, one may analytically derive an expression for the major head loss, h_f due to channel friction from the energy equation. However, when the friction/viscous forces cannot be theoretically modeled in the integral form of the momentum equation (as in the turbulent flow case), the friction/viscous forces, the actual pressure drop, Δp (or the unknown friction slope, S_f), and thus the major head loss, h_f due to channel friction are determined empirically (see Chapter 7). Specifically, in the case of turbulent flow, the friction/viscous forces in the integral form of the momentum equation and thus the wall shear stress, τ_w cannot be theoretically modeled; therefore, the integral form of the momentum equation cannot be directly applied to solve for the unknown pressure drop, Δp (or the unknown friction slope, S_f). Thus, an empirical interpretation (using dimensional analysis) of the wall shear stress, τ_w in the theoretical expression for the friction slope, $S_f = \tau_w/\gamma R_h$ in the differential form of the momentum equation is sought in terms of velocity, v and a flow resistance coefficient. This yields the Chezy equation ($S_f = v^2/C^2 R_h$, which represents the differential form of the momentum equation, supplemented by dimensional analysis, and guided by the integral momentum equation; a link between the differential and integral momentum equations) that is used to obtain an empirical evaluation for the pressure drop, Δp in Newton's second law of motion (integral form of the momentum equation). As such, one can then derive an empirical expression for the major head loss, h_f from the energy equation. Major losses in open channel flow are addressed in detail in Chapters 6 and 9.

9.5.1.3 Evaluation of the Pump and Turbine Head Terms

Evaluation of the head delivered to the fluid by a pump, h_{pump} and the head removed from the fluid by a turbine, $h_{turbine}$ requires the application of the appropriate power equation, which are discussed in Section 4.7.2.4, and the resulting equations are summarized, respectively, as follows:

$$h_{pump} = \frac{(P_p)_{out}}{\gamma Q} \tag{9.6a}$$

$$h_{turbine} = \frac{(P_t)_{in}}{\gamma Q} \tag{9.6b}$$

It may be noted that examples involving pumps and turbines are presented for pipe flow in Chapters 4 and 8.

9.5.1.4 Differential Approach for Real Open Channel Flow Problems

Furthermore, in the assumption of real flow, and for the specific case of a major head loss due to open channel flow resistance, application of the differential form of the momentum equation may become necessary. Specifically, application of the differential form of the momentum equation will be necessary when it is of interest to trace the gradual variation of the steady nonuniform channel depth of flow (Equation 5.25 or 9.101) and is repeated as follows:

$$\frac{dy}{ds} = \frac{S_o - S_f}{1 - F^2} \tag{9.7}$$

or to evaluate the flow (solve for an unknown quantity) at a given point in the fluid system for any uniform flow channel section (Equation 5.8) and is repeated as follows:

$$S_f = \frac{\tau_w}{\gamma R_h} \tag{9.8}$$

Application of Equations 9.7 and 9.8 require evaluation of the friction slope term, S_f, which is modeled on a "subset level" (see Section 9.5.1.2 above), and yielded the Chezy equation as follows:

$$S_f = \frac{\tau_w}{\gamma R_h} = \frac{v^2}{C^2 R_h} \tag{9.9}$$

where the major head loss due to flow resistance is determined by application of Equation 9.4, which is repeated as follows:

$$h_{f,maj} = \frac{\tau_w L}{\gamma R_h} = S_o L = S_f L = \frac{v^2}{C^2 R_h} L = f \frac{L}{D} \frac{v^2}{2g} = \left(\frac{vn}{R_h^{2/3}}\right)^2 L \tag{9.10}$$

however, unlike the uniform flow case, the friction slope, $S_f \neq$ channel slope, S_o in the nonuniform flow case. Although the flow in open channel flow is typically assumed to be turbulent, it is important to note that in the case of laminar flow, in lieu of the Chezy equation (turbulent flow), one would apply Poiseuille's law as follows:

$$S_f = \frac{\Delta p}{\gamma L} = \frac{32 \mu v}{\gamma D^2} \tag{9.11}$$

which represents the differential form of the momentum equation.

9.5.1.5 Applications of the Governing Equations for Real Open Channel Flow

Applications of the governing equations for real open channel flow include the analysis and design of uniform (see Section 9.14; also see Example Problems 5.2, 5.11, and 5.12) and nonuniform (see Section 9.15) flow (generated by controls such as a break in the channel slope, a hydraulic drop, or a free overfall). Furthermore, in the analysis and design of an open channel flow system, there may be channel friction (resistance) or controls (break in the channel slope, hydraulic drop, or free overfall) in the open channel flow that will determine the significance of a major head loss and a nonuniform surface water profile.

9.5.2 Application of the Governing Equations for Ideal Open Channel Flow

When one of the two equations of motion breaks down, the other may be used to find the additional unknown. Assuming a finite control volume, for the assumption of incompressible, ideal, steady open channel flow, first, the integral energy equation (Bernoulli equation) is applied and is given as follows:

$$\left(y_1 + \frac{v_1^2}{2g} + z_1\right) = \left(y_2 + \frac{v_2^2}{2g} + z_2\right) \tag{9.12}$$

which may be used to solve for an unknown "energy head" term (y, z, or $v^2/2g$) for one of the two points in a finite control volume. Application of the Bernoulli equation assumes ideal flow; thus, there is no head loss term to evaluate in the energy equation. Therefore, for problems that affect the flow over a short channel section, the major head loss due to flow resistance is assumed to be negligible, and one may apply the Bernoulli equation. These types of flow problems include flow-measuring devices for open channel flow, which, however, have a minor loss associated with the flow measurement, and gradual (vertical or horizontal) channel contractions/transition, which do not have a minor loss. Thus, a typical assumption of ideal open channel flow is made for flow-measuring devices, which are considered to be ideal flow meters, and another typical assumption of ideal open channel flow is made for gradual (vertical or horizontal) channel contractions/transition. In the case of flow-measuring devices, although the associated minor head loss may be accounted for in the integral form of the energy equation (Equation 9.1), it is typically accounted for by the use/calibration of a discharge coefficient to determine the actual discharge (Section 9.5.2.1), where ideal flow (Bernoulli equation) is assumed. Thus, in the case of flow-measuring devices, one must indirectly model the minor head loss flow resistance term by the use of a discharge coefficient to determine the actual discharge, which requires a "subset level" of application of the governing equations (Section 9.5.2.1). Then, application of the complementary integral momentum equation is required only if there is an unknown reaction force, as in the case of open channel flow-measuring devices (such as sluice gates, weirs, spillways, venturi flumes, and contracted openings) (see Section 9.12), which was given by Equation 5.27 and is repeated as follows:

$$\sum F_s = (F_G + F_P + F_V + F_{other})_s = (\rho Q v_s)_2 - (\rho Q v_s)_1 \tag{9.13}$$

which may be solved for an unknown force (typically a reaction force). Note that the viscous forces have already been accounted for indirectly through a minor head loss flow resistance term in the energy equation (or, actually, through the discharge coefficient, Section 9.5.2.1).

9.5.2.1 Evaluation of the Actual Discharge

In the case of a flow-measuring device for open channel flow, the associated minor head is typically accounted for by the use/calibration of a discharge coefficient to determine the actual discharge. Evaluation of the actual discharge requires a "subset level" application of the governing equations. A "subset level" application of the governing equations focuses only on the given element causing the flow resistance. The assumption of "ideal" flow implies that the flow resistance is modeled only in the momentum equation (and thus the subsequent assumption of "real" flow). The flow resistance equation for the actual discharge is derived in Chapters 6 and 7. In the derivation of the actual discharge for a given flow-measuring device, first, ideal flow is assumed and thus the Bernoulli equation is applied

as follows:

$$\left(y + z + \frac{v^2}{2g}\right)_1 - \left(y + z + \frac{v^2}{2g}\right)_2 = 0 \tag{9.14}$$

Assuming $z_1 = z_2$, this equation has one unknown, which is the ideal velocity of the flow at the restriction in the flow-measuring device. Thus, this equation yields an expression for the ideal velocity, $v_i = \sqrt{2g\Delta y}$, as a function of an ideal pressure difference, Δy, which is directly measured. Then, application of the continuity equation is used to determine an expression for the ideal discharge, $Q_i = v_i A_i$ and the actual discharge, $Q_a = (v_a/\sqrt{2g\Delta y})(A_a/A_i)Q_i$. Finally, the integral form of the momentum equation (supplemented by dimensional analysis) is applied to solve for an unknown actual velocity, v_a and actual area, A_a which was given in Equation 5.27 and is repeated as follows:

$$\sum F_s = (F_G + F_P + F_V + F_{other})_s = (\rho Q v_s)_2 - (\rho Q v_s)_1 \tag{9.15}$$

Thus, the resulting flow resistance equations for the actual discharge for open channel flow-measuring devices is given as follows:

$$Q_a = \frac{\sqrt{2g\left(\Delta y - \frac{\tau_w L}{\gamma R_h}\right)}}{\sqrt{2g\Delta y}} \frac{A_a}{A_i} Q_i = C_d Q_i \tag{9.16}$$

where the discharge coefficient, C_d accounts for (indirectly models) the minor head loss associated with the flow measurement.

The flow resistance (shear stress or drag force) due to a flow-measuring device is modeled as a reduced/actual discharge, Q_a in the differential form of the continuity equation. The flow resistance causes an unknown Δy in the case of open channel flow, which causes an unknown head loss, h_f, where the head loss is due to a conversion of kinetic energy to heat, that is modeled/displayed in the integral form of the energy equation. The head loss is caused by both pressure and friction forces. In the determination of the reduced/actual discharge, Q_a which is less than the ideal discharge, Q_i, the exact reduction in the flowrate is unknown because the head loss, h_f causing it is unknown. Therefore, because one cannot model the exact reduction in the ideal flowrate, Q_i, one cannot derive an analytical expression for the reduced/actual discharge, Q_a from the continuity equation. Specifically, the complex nature of the flow does not allow a theoretical modeling of the existence of the viscous force (due to shear stress, τ_w) due to the flow measurement device in the momentum equation and thus does not allow an analytical derivation of the pressure drop, Δy and thus v_a and cannot measure A_a. Thus, an empirical expression (using dimensional analysis) for the actual discharge, $Q_a = (\sqrt{2g(\Delta y - \tau_w L/\gamma R_h)}/\sqrt{2g\Delta y})(A_a/A_i)Q_i$ and thus Δy (and v_a and A_a) is derived as a function of the drag coefficient, C_D (discharge coefficient, C_d) that represents the flow resistance. This yields the actual discharge equation $Q_a = C_d Q_i$, which represents the integral momentum equation supplemented by dimensional analysis. Actual discharges are addressed in detail in Chapters 6 and 9.

9.5.2.2 Applications of the Governing Equations for Ideal Open Channel Flow

Applications of the governing equations for ideal open flow include the analysis and design of open channel flow systems that include an open channel flow-measuring device (sluice gates, weirs, spillways, venturi flumes, and contracted openings) (see Sections 9.12, 9.16, and 9.17), or a gradual (vertical or horizontal) channel contractions/transitions (see Section 9.12). Furthermore, while the flow-measuring devices have a minor loss associated with the flow measurement, and a reaction force, the gradual (vertical or horizontal) channel contractions/transitions do not.

9.5.3 Application of the Governing Equations for a Hydraulic Jump

A hydraulic jump is a natural phenomenon that occurs when the open channel flow transitions from supercritical to subcritical flow (see Section 9.13). One may note that the head loss associated with a hydraulic jump is not due to frictional losses associated with the wall shear stress, τ_w but rather is due to intense agitation and turbulence and thus results in a high energy loss. Therefore, there are no viscous forces to model in the integral form of the momentum equation. However, the unknown major head loss, $h_{f,major}$, which is due to a conversion of kinetic energy to heat (the fast-moving supercritical flow encounters the slow-moving subcritical flow, and thus the flow at the jump becomes critical flow, with a minimum specific energy) is modeled in the integral form of the energy equation. Thus, for a hydraulic jump, assuming a finite control volume, the integral form of the momentum equation may be used to analytically solve for the unknown channel depth of flow (either upstream or downstream) as follows:

$$\sum F_s = (F_G + F_P + F_V + F_{other})_s = (\rho Q v_s)_2 - (\rho Q v_s)_1 \tag{9.17}$$

by deriving the hydraulic jump equations (Equations 9.248 and 9.249; see Section 9.13 for the derivations and applications). Then, the energy equation is applied to solve for the major head loss as follows:

$$\left(\frac{p}{\gamma} + z + \frac{v^2}{2g}\right)_1 - h_{f,maj} = \left(\frac{p}{\gamma} + z + \frac{v^2}{2g}\right)_2 \tag{9.18}$$

Applications of the integral governing equations (continuity, energy, and momentum) for a finite control volume are made in the analysis and design of a hydraulic jump (see Section 9.13).

9.6 Major Head Loss due to Flow Resistance in Real Open Channel Flow

Applications of the governing equations for real open channel flow include the analysis and design of uniform (see Section 9.14; also see Example Problems 5.2, 5.11, and 5.12) and nonuniform (see Section 9.15) flow (generated by controls such as a break in the channel slope, a hydraulic drop, or a free overfall). Furthermore, in the analysis and design of an open channel flow system, there may be channel friction (resistance) or controls (break in channel slope, hydraulic drop, or free overfall) in the open channel flow that will determine the significance of a major head loss and a nonuniform surface water profile. While the application of the continuity equation will always be necessary, the energy equation and the momentum equation play complementary roles; when one of the two equations of motion breaks down, the other may be used to find the additional unknown quantity. Furthermore, while the energy equation is useful only for a control volume, depending upon the general fluid flow problem, one may apply the continuity and momentum equations using either the integral or differential approach. However, for the specific case of a straight channel section without flow-measuring devices (addressed in Section 9.6), only the differential continuity and momentum equations for a fluid system will be required; the integral continuity and momentum equations for a control volume are required for channel sections with flow-measuring devices (see Section 9.12). As such, for a straight channel section, the energy equation (Equation 9.1) models the major head loss, $h_{f,maj}$ due to channel friction for a given channel length, while the differential momentum equations (Equation 9.7 and Equation 9.8) model the friction slope, S_f due to channel friction for a given point in the channel section for nonuniform and uniform flow, respectively.

When a real fluid flows through a channel, there is a friction slope, S_f over a channel distance, L, which is caused by the friction of the fluid against the channel wall, where the friction slope, S_f is directly related to the wall shear stress, τ_w. The head loss, h_f accounts for any energy loss associated with the flow and is a direct result of the viscous dissipation that occurs throughout the fluid in the channel. The flow in open channel flow is typically turbulent.

9.6.1 Uniform versus Nonuniform Open Channel Flow

As previously discussed in Chapter 3, the spatial variation in the cross-sectional area of the flow (or the existence of a channel or flow transition in an open channel, explained in Section 9.8 below) results in a spatial variation in the fluid velocity (convective acceleration) and thus spatially varied (nonuniform) flow. However, no spatial variation in the cross-sectional area of the flow (or the nonexistence of a channel or flow transition in an open channel) results in no spatial variation in the fluid velocity and thus spatially uniform flow. As such, it is important to highlight the differences and similarities between the solutions of uniform versus nonuniform flow problems. In the case of uniform flow, if the depth, y is unknown, either the Chezy equation (Equation 9.9) or the Manning's equation ("resistance equations" for turbulent uniform flow) is used to solve for an unknown uniform depth, y (differential approach). However, in the case of nonuniform flow, if the depth, y is unknown, the "resistance equation" for turbulent nonuniform open channel flow $dy/ds = (S_o - S_f)/(1 - F^2)$ (Equation 9.7) is numerically solved for an unknown nonuniform depth, y (differential approach), where $S_f = v^2/C^2R_h$ (Chezy equation, Equation 9.9). Furthermore, in the case of either uniform or nonuniform flow, if the friction slope, S_f; the velocity, v; or the channel friction coefficient (Chezy coefficient, C or Manning's roughness coefficient, n) is unknown, then the Chezy equation (Equation 9.9) (or the Manning's equation) is used to solve for the unknown quantity (differential approach). Finally, the determination of the associated major head loss for either uniform or nonuniform flow is determined using the major head loss equation, $h_{f,maj} = (\tau_w L/\gamma R_h) = S_o L = S_f L = (v^2/C^2R_h)L = f(L/D)(v^2/2g) = (vn/R_h^{2/3})^2L$ (Equation 9.10) (either the Chezy head loss equation or the Manning's head loss equation), however unlike the uniform flow case, the friction slope, $S_f \neq$ channel slope, S_o in the nonuniform flow case.

9.6.2 Evaluation of the Major Head Loss Term in the Energy Equation

Evaluation of the major head loss flow resistance term requires a "subset level" application of the governing equations (see Chapters 6 and 7 and Section 9.5.1.2). Depending upon the value for the Reynolds number, R, real flow may be described as either laminar or turbulent. As such, the major head loss due to channel friction modeled in the energy equation (Equation 9.1) is evaluated by the Chezy head loss equation (Equation 9.4) for turbulent pipe flow and by the Poiseuille head loss equation (Equation 9.5) for laminar flow. And, the friction slope modeled in the differential momentum equation (Equation 9.8) is evaluated by the Chezy equation (Equation 9.9) for turbulent pipe flow and by Poiseuille's law (Equation 9.11) for laminar flow. However, because it is assumed that the flow in open channel flow is typically turbulent flow as opposed to laminar flow, only the Chezy equations are applied in Chapter 9. Furthermore, in addition to the Chezy equation (Section 9.6.3), additional turbulent pipe flow resistance equations (Section 9.6.4) include the Darcy–Weisbach equation (Chapter 8), and the Manning's equation, (Section 9.6.5). Therefore, applications of the governing equations for uniform flow are presented in Section 9.14 (also see Example Problems 5.2, 5.11, and 5.12), while they are applied for nonuniform flow in Section 9.15, where the major head loss term, $h_{f,maj}$ in the energy equation is evaluated and schematically illustrated by the energy grade line (EGL) and the hydraulic grade line (HGL) (see Sections 9.14 and 9.15).

9.6.3 The Chezy Equation and Evaluation of the Chezy Coefficient, C

Evaluation of the major head loss flow resistance term requires a "subset level" application of the governing equations (see Chapters 6 and 7 and Section 9.5.1.2). When the friction/viscous forces cannot be theoretically modeled in the momentum equation, as in the turbulent flow case, the friction/viscous forces, the actual pressure drop, Δp (or friction slope, S_f) and thus the major head loss, h_f due to pipe friction are determined empirically. Application of the differential form of the momentum equation (supplemented with dimensional analysis and guided by the integral form of the momentum equation; a link between the two forms of the momentum equation) yields an expression for the friction slope, S_f and is given as follows (see Chapters 6 and 7 for its derivation and Chapter 9 for its application):

$$S_f = \frac{\tau_w}{\gamma R_h} = \frac{v^2}{C^2 R_h} \tag{9.19}$$

which is known as the Chezy equation (also known as the "resistance equation"), and C is the Chezy coefficient, which is derived from dimensional analysis and evaluated empirically. The Chezy coefficient, C has dimensions of $L^{1/2} T^{-1}$ and is a function of the boundary roughness, ε/y; the Reynolds number, R; and the cross-sectional shape or the hydraulic radius, R_h. Empirical findings illustrate that the Chezy coefficient, C ranges between 30 $m^{1/2} s^{-1}$ (or 60 $ft^{1/2} sec^{-1}$) for small channels with rough boundary surfaces to 90 $m^{1/2} s^{-1}$ (or 160 $ft^{1/2} sec^{-1}$) for large channels with smooth boundary surfaces. One may note that although the derivation/formulation of the Chezy equation assumes no specific units (SI or BG), the Chezy coefficient, C has dimensions of $L^{1/2} T^{-1}$. Furthermore, because the values for the Chezy coefficient, C have been provided/calibrated in both units (see above), one may apply the Chezy equation assuming either SI or BG units. Although the Chezy equation is used for both open channel flow and pipe flow, it has a drawback in that the Chezy coefficient, C is not dimensionless like the Darcy–Weisbach friction factor, f (presented in

Chapters 6 and 8, and discussed below). Therefore, one must attach the appropriate metric or English units to the Chezy coefficient, C prior to using it in the Chezy equation. The Chezy equation for open channel flow assumes a straight channel section of length, L, with a constant depth, y (uniform flow) and thus a constant velocity of flow, v, with a specific Chezy coefficient, C and a friction slope, $S_f = S_o$ (channel slope). Thus, the Chezy equation may be solved for one of the following unknown quantities (differential approach): S_f (friction slope) $= S_o$ (channel slope), v, C, or y at a given point in the fluid system.

Finally, substituting the empirical expression for the friction slope (Chezy equation, Equation 9.19) into the integral form of the energy equation (Equation 9.2) yields an empirical expression for the major head loss, h_f due to pipe friction as follows (see Chapters 6 and 7 for its derivation and Chapter 9 for its application):

$$h_{f,maj} = \frac{\tau_w L}{\gamma R_h} = S_o L = S_f L = \frac{v^2}{C^2 R_h} L \qquad (9.20)$$

which is the Chezy equation expressed in terms of the major head loss, and provides an empirical expression for the major head loss, h_f (expressed as a function of an empirical flow resistance coefficient, C).

9.6.4 Turbulent Channel Flow Resistance Equations and Their Roughness Coefficients

One may note that in addition to Chezy coefficient, C alternative standard forms of the drag coefficient, C_D include empirical flow resistance coefficients including the Darcy–Weisbach friction factor, f (typically for pipe flow only), and the Manning's roughness coefficient, n (presented in detail in Chapter 6 and summarized in Chapter 8). Although the three standard flow resistance coefficients are related, the use of the Darcy–Weisbach friction factor, f is typically preferred over the Chezy Coefficient, C for both laminar and turbulent pipe flow because f is dimensionless, while C has dimensions of $L^{1/2} T^{-1}$ and thus one must specify either SI or BG units for C. Furthermore, the use of the Manning's roughness coefficient, n, (which has dimensions of $L^{-1/3} T$) is preferred over the Chezy Coefficient, C for open channel flow (usually turbulent flow) and turbulent pipe flow because it is independent of both the Reynolds number, R (by assuming turbulent flow) and the cross-sectional shape or the hydraulic radius, R_h (by its relationship with the Chezy coefficient, C).

Furthermore, although values for the Chezy coefficient, C have been empirically evaluated and tabulated (limited tables available) for various combinations of boundary roughness, ε/y; Reynolds number, R; and cross-sectional shape (or the hydraulic radius, R_h), it has been found that it is easier to model the effects of the Chezy coefficient, C (i.e., the drag coefficient, C_D) by using the Darcy–Weisbach friction coefficient, f (typically for pipe flow only) or the Manning's roughness coefficient, n (for turbulent pipe and open channel flow). As such, while the Darcy–Weisbach friction coefficient, f (typically for pipe flow only) was already presented in Chapter 8, the Manning's roughness coefficient, n is presented for turbulent open channel flow below.

9.6.5 Manning's Equation and Evaluation of Manning's Roughness Coefficient, *n*

Empirical studies have indicated that a more accurate dependence of the Chezy coefficient, C on the hydraulic radius, R_h is defined by the definition of the Manning's roughness coefficient, n. One may recall that the dimensions of the Chezy coefficient, C are $L^{1/2} T^{-1}$, and it is empirically evaluated as a function of the boundary roughness, ε/y; the Reynolds number, R; and the cross-sectional shape or the hydraulic radius, R_h. However, further empirical studies (assuming the metric system) by Manning indicated that a more accurate dependence of the Chezy coefficient, C on the hydraulic radius, R_h was presented in Chapter 6 and repeated here as follows:

$$C = \frac{R_h^{1/6}}{n} \tag{9.21}$$

where n is the Manning's roughness coefficient with dimensions of $L^{-1/3} T$, and is empirically evaluated only as a function of the boundary roughness, ε/y. One may note that the dependency of the Manning's roughness coefficient, n on the hydraulic radius, R_h has been removed by its relationship with the Chezy coefficient, C. Furthermore, one may note that the dependency of the Manning's roughness coefficient, n on the Reynolds number, R has been removed by assuming turbulent flow. One may note that the most common occurrence of flow resistance in open channel and pipe flow is in long channels or long pipe sections. In the open channel flow case, the boundary fills the whole channel section, while in the pipe flow case, the boundary fills the whole pipe section; in such cases, the Reynolds number, R is usually so large that it has virtually no influence on the resistance coefficient, n. Empirical values for the Manning roughness coefficient, n as a function of the boundary roughness, ε/y are typically given assuming metric units for n of $m^{-1/3}$ s (see Table 8.6).

Substituting the expression for the Chezy coefficient, C as a function of the Manning's roughness coefficient, n (Equation 9.21) into the Chezy equation (Equation 9.19), and isolating the velocity, v yields the Manning's equation as presented in Chapter 6 and repeated here as follows:

$$v = \frac{R_h^{1/6}}{n}\sqrt{R_h S_f} = \frac{1}{n} R_h^{2/3} S_f^{1/2} \tag{9.22a}$$

One may note that although the derivation/formulation of the Manning's equation assumes no specific units (SI or BG), the Manning's roughness coefficient, n has dimensions of $L^{-1/3} T$. Furthermore, because the values for the Manning's roughness coefficient, n have been provided/calibrated in the SI units $m^{-1/3}$ s (see Table 8.6), the Manning's equation as given above assumes the metric (SI) system. The corresponding Manning's equation for the English (BG) system is given as follows:

$$v = \frac{1.486}{n} R_h^{2/3} S_f^{1/2} \tag{9.22b}$$

which also uses the Manning roughness coefficient, n with units $m^{-1/3}$ s (see Table 8.6). One may note the value of 1.486 simply converts the Manning roughness coefficient, n with units $m^{-1/3}$ s (from Table 8.6) to units of $ft^{-1/3}$ sec, because 3.281 ft $= 1$ m and thus:

$$\frac{\left[\dfrac{3.281ft}{1m}\right]^{1/3}}{n[m^{-1/3}s]} = \frac{1.486}{n[ft^{-1/3}\sec]} \tag{9.23}$$

As illustrated in Table 8.6, the rougher the boundary (wetted perimeter), the larger the value for Manning roughness coefficient, n. The Manning's equation for open channel flow assumes a straight channel section of length, L with a constant depth, y (uniform flow) and thus a constant velocity of flow, v with a specific Manning's roughness coefficient, n and a friction slope, $S_f = S_o$ (channel slope). Thus, the Manning's equation may be solved for one of the following unknown quantities (differential approach): S_f (friction slope) = S_o (channel slope), v, n, or y at a given point in the fluid system.

Finally, substituting the relationship between the Chezy coefficient, C and the Manning's roughness coefficient, n (Equation 9.21) into the Chezy head loss equation (Equation 9.20), yields the Manning's head loss equation (assuming the metric system) as presented in Chapter 6 and repeated as follows:

$$h_f = \frac{\tau_w L}{\gamma R_h} = S_o L = S_f L = \frac{v^2 L}{C^2 R_h} = \frac{v^2}{\left(\frac{R_h^{1/6}}{n}\right)^2 R_h} \frac{L}{} = \left(\frac{vn}{R_h^{2/3}}\right)^2 L \qquad (9.24a)$$

The corresponding Manning's head loss equation for the English (BG) system as presented in Chapter 6 and repeated as follows:

$$h_f = \frac{\tau_w L}{\gamma R_h} = S_o L = S_f L = \frac{v^2 L}{C^2 R_h} = \frac{v^2}{\left(\frac{1.486 R_h^{1/6}}{n}\right)^2 R_h} \frac{L}{} = \left(\frac{vn}{1.486 R_h^{2/3}}\right)^2 L \qquad (9.24b)$$

9.7 Flow Type (State) and Flow Regime

The flow type (state) in an open channel may be described as either uniform or nonuniform, while the flow regime in an open channel flow may be described as subcritical, critical, or supercritical. Uniform flow is the state the flow will tend to assume in the open channel in the absence of a channel (particular localized features) or flow transition. However, if there are any channel or flow transitions in the channel, they will tend to pull the flow away from the uniform state of flow, transitioning (gradually or abruptly) into a nonuniform state of flow. The Froude number, F is used to indicate the flow regime, as it is the ratio of the inertia force, F_I to the gravity force, F_G which is invoked in open channel flow. The Froude number is defined as $F = F_I/F_G = v/v_c$, where v is the velocity of the flow and v_c is the critical velocity.

9.7.1 The Role and Significance of Uniform Flow

Although a uniform state of flow rarely occurs in an open channel flow, uniform flow is considered in the design of all channel problems. Uniform flow is defined by the resistance (Chezy or Manning's) equation, which provides a relationship between the depth and discharge. However, regardless of whether the channel section is in a natural or artificial condition, the existence of abrupt channel transitions (controls) will result in a relationship between the depth and discharge, as dictated by the critical flow, that is different than that provided by the resistance equation (unless the uniform flow is at a critical depth). Furthermore, in the design of such controls, it becomes necessary to compare the two respective depth–discharge relationships (resistance equation and critical flow equation). Specifically, one may note that the design of a particular channel is based on the uniform state of flow, which provides the criterion for the minimum cross-sectional area required (based upon the channel friction, the channel slope, and the discharge). Then, the channel controls and transitions are added as needed by the design.

9.7.2 The Definition of Flow Regimes for Uniform Flow

Given that the flow in the open channel is uniform (there are no channel or flow transitions), the occurrence of a particular flow regime is dictated by the classification of the channel bottom slope, S_o (mild, steep, or critical). The classification of the slope will depend on the channel roughness, the actual magnitude of the slope, and the discharge. Thus, if the channel bottom slope is mild, then the uniform open channel flow is a deep slow moving flow, known as subcritical flow (most common flow regime in open channel flow), and $F = v/v_c < 1$. If the channel bottom slope is steep, then the uniform open channel flow is a shallow fast-moving flow, known as supercritical flow, and $F = v/v_c > 1$. However, if the channel bottom slope is in between mild and steep, and at a critical slope, then the uniform open channel flow is at a critical depth and at a critical velocity, known as critical flow, and $F = v/v_c = 1$.

9.7.3 The Occurrence of Nonuniform Flow and Changes in the Flow Regime

Assuming a subcritical, critical, or supercritical flow in a uniform open channel flow, if there is the occurrence of either a channel or flow transition in the open channel, then the flow will change from uniform to nonuniform. If the channel transition is a gradual vertical or horizontal contraction (gradual step upward or gradual decrease in channel width), although there will be a change (slight decrease) in the depth of flow, the flow regime will remain at subcritical. However, if the channel transition is an abrupt vertical or horizontal contraction (abrupt step upward or abrupt decrease in channel width, a typical flow-measuring device [such as a sluice gate, weir, overflow spillway, venturi flume, and contracted opening], a break in channel bottom slope, a hydraulic drop, or a free overfall; each transition is considered to be a "control"), there will be a change in the depth of flow, and a change in flow regime. Specifically, the flow will change from subcritical upstream of the control to critical at the control to supercritical downstream of the control. Furthermore, if there is an abrupt flow transition in the open channel as posed by supercritical flow encountering subcritical flow (for instance, as a result of the insertion of two sluice gates in the open channel), a natural phenomena known as a hydraulic jump occurs, occurring at the critical depth of flow.

9.8 Transitions and Controls in Open Channel Flow

Uniform flow will occur in the open channel in the absence of a channel or flow transition, whereas nonuniform flow will occur in the open channel as a result of a channel or flow transition. A channel transition may be defined as a change in the direction, slope, or cross section of the channel that results in a change in the state of flow from uniform flow to nonuniform flow. And, a flow transition may be defined as a change in the flow regime (such as from supercritical to subcritical; thus, the occurrence of a hydraulic jump).

Although channel bends produce only transient changes in the flow, most channel transitions produce a permanent change in the state of flow (from uniform to nonuniform). Furthermore, although most practical channel transitions are relatively short channel features (a gradual or abrupt vertical or horizontal contraction, a typical flow-measuring device [such as a sluice gate, weir, overflow spillway, venturi flume, and contracted opening], a break in channel bottom slope, a hydraulic drop, or a free overfall), they may affect the flow for a great distance upstream or downstream of the transition. However, as noted above, because a gradual or abrupt vertical or horizontal contraction or a flow-measuring device affects the flow over a short channel section, the major head loss due to flow resistance is assumed to be negligible; the Bernoulli equation is applied to determine either the upstream or downstream depth. Additionally, as noted above, because the nonuniform flow resulting from channel transitions such as a break in channel bottom slope, a hydraulic drop, or a free overfall, as well as the occurrence of uniform flow, generate a surface water profile and thus affect the flow over a long channel section, the major head loss due to flow resistance is significant; the appropriate resistance equation is applied to determine the water surface profile (depth of flow). Finally, although the occurrence of a hydraulic jump affects the flow over a short channel section, the major head loss due to the flow turbulence in the jump is significant; the momentum equation is applied to determine either the upstream or downstream depth.

9.8.1 The Definition of a Control in Open Channel Flow

A control in an open channel is a mechanism or feature (not necessarily a channel or flow transition) that determines or dictates the relationship between the flow depth and discharge in the channel; the discharge can be calculated once the depth is known. It is important to note that a critical depth of flow occurs at the more obvious "controls," while a subcritical, supercritical, or critical depth of flow occurs at the less obvious "control." The most obvious controls are associated with particular localized features in the channel such as an abrupt vertical or horizontal contraction (abrupt step upward or abrupt decrease in channel width), a typical flow-measuring device (such as a sluice gate, weir, overflow spillway, venturi flume, and contracted opening), a break in the channel bottom slope, a hydraulic drop, or a free overfall. Critical flow, which is defined by an analytical relationship between the flow depth and the discharge in the channel, occurs at these more obvious types of "controls"; these are considered to be controls of a special kind. A less obvious control is due the flow resistance due to the channel roughness that occurs in a uniform open channel flow, where the flow may be subcritical, supercritical, or critical. As such, because the resistance equation for uniform flow (Chezy or Manning's equation, expressed as the continuity equation) provides a relationship between the depth and discharge, uniform flow is considered a "control" in open channel flow. Furthermore, it is important to note that a gradual vertical or horizontal channel transition does not result in a control in the channel and does not produce critical flow. Finally, it is important to note that although critical flow occurs in a hydraulic jump, it is not considered a control, but rather an energy dissipator (see section below on hydraulic jumps).

Additionally, it is interesting to highlight the differences and similarities between the roles of uniform versus certain nonuniform flows as controls in open channel flow. In the case of the control mechanism or feature provided by the channel roughness that occurs in a uniform surface water profile, the relationship between the flow depth and channel discharge is expressed by the resistance equation for uniform flow (Chezy or Manning's equation). However, in the case of controls provided by a break in the channel slope, a hydraulic drop, or a free overfall, which all generate a nonuniform surface water profile, the corresponding relationship between the flow depth and the channel discharge is expressed by the analytical expression for critical depth for a given discharge (similar to the remaining controls that generate nonuniform flow, but affect the flow only over a short channel section). Furthermore, in the case of either uniform or nonuniform flow, because a water surface profile is generated and thus affects the flow over the long channel section, the major head loss due to flow resistance is significant. As such, the appropriate resistance equation is applied to determine the water surface profile (depth of flow).

Given that there is a control mechanism in the open channel, the flow in the channel will be influenced by the control, as it determines the depth–discharge relationship. In the case of the obvious types of controls/controls of a special kind, the flow is critical at the control. Therefore, the definite relationship between the depth and the discharge is the analytical expression for the critical depth of flow, y_c as a function of discharge, Q (critical equations are derived below). As such, the setting of the particular control (obstruction, constriction, break in slope, etc.) is guided by the analytical expression for y_c as a function of discharge, Q. Thus, for a certain setting of the control, there is a certain relationship between the upstream flow depth and the discharge, and, likewise, there is a certain relationship between the downstream depth and the discharge. However, in the case of the less obvious type of control/channel roughness in uniform flow, the flow is not necessarily critical at the control. Depending upon the classification of the slope of the channel: mild, steep, or critical, the uniform flow will be subcritical, supercritical, or critical, respectively. The classification of the slope will depend on the channel roughness, the actual magnitude of the slope, and the discharge. The definite relationship between the depth and the discharge is determined by the resistance equation for uniform flow (Chezy or Manning's equation, expressed as the continuity equation). One may note that the critical equations (derived below) are a special case of the resistance equation for uniform flow. As such, the setting of the slope is guided by the resistance equation. Thus, for a certain setting of the slope (mild, steep, or critical), there is a certain relationship between the uniform flow depth and the discharge, for a given flow regime: subcritical, supercritical, or critical, respectively.

9.8.1.1 The Occurrence of Critical Flow at Controls

It is of interest to focus on the occurrence of critical flow at controls (associated with particular localized features in the channel; excluding the flow resistance due to channel roughness in uniform flow as a control). Prior to the occurrence of critical flow at a control, it is assumed that the flow has previously been restrained/forced (by a control in general, by the banks of a lake, or by the roughness of the channel bottom [classifying it as a subcritical slope]) into the subcritical flow regime. Once the flow is subsequently released into a region where there is no further restraint and is no longer forced into the subcritical condition, critical flow occurs at a point of release, which is defined as a control. Thus, the control provides a mechanism of release from subcritical flow to critical flow, then to supercritical flow. Furthermore, it is important to note that regardless of the type of control causing the occurrence of critical flow (including uniform flow at a critical slope), the analytical critical flow equations (derived below) will apply. Finally, although the hydraulic jump is not considered a control, the occurrence of critical flow at the jump is also described by the analytical critical flow equations as well.

Finally, it is important to note that while some types of controls (break in the channel slope, a hydraulic drop, a free overfall, uniform flow) generate a surface water profile over the long channel section, other types of controls, such as an abrupt vertical or horizontal contraction (abrupt step upward or abrupt decrease in channel width), or a typical flow-measuring device (such as a sluice gate, weir, overflow spillway, venturi flume, and contracted opening) do not. As such, while the appropriate resistance equation is applied to determine the water surface profile (depth of flow) for the former type of controls, the Bernoulli equation is applied to determine either the upstream or the downstream depth for the latter types of controls.

9.8.2 The Definition of a Flow-Measuring Device in Open Channel Flow

While a "control" in the open channel can serve as a flow-measuring device, a flow-measuring device is not necessarily considered a "control." For instance, although a pitot-static tube is a typical flow-measuring device, it is not a "control." Furthermore, while the flow is critical at most controls (the flow is subcritical, supercritical, or critical for uniform flow as a control), the occurrence of critical flow does not necessarily provide a control. For instance, although the flow is critical at a hydraulic jump, it is not a control. Thus, in addition to the typical flow-measuring devices (such as sluice gates, weirs, overflow spillways, venturi flumes, and contracted openings), controls including, a hydraulic drop, a break in channel bottom slope, a free overfall, and uniform flow may also serve as "flow-measuring devices." Furthermore, any of the "controls" listed above can serve as a flow-measuring device, as each provides a relationship between the flow depth and discharge in the channel. As such, there are four types of controls. The first is an abrupt vertical channel contraction, which includes typical flow-measuring devices, such as sluice gates, weirs, and overflow spillways, and it includes a hydraulic drop. The second is an abrupt horizontal channel contraction, which includes typical flow-measuring devices, such as venturi flumes and contracted openings. The third is a change in channel bottom slope, and includes a break in the channel slope and a free overfall. And the fourth is uniform flow. One may note that a free overfall is a special case of the sharp-crested weir, where the height of the weir, $P = 0$. Furthermore, one may note that: (1) the typical flow-measuring devices have a minor head loss associated with the flow measurement; (2) a break in channel bottom slope, a hydraulic drop, or a free overfall have a small minor head loss associated with the flow measurement; and (3) the uniform flow actually has a major head loss associated with the flow measurement. Additionally, a gradual vertical or horizontal channel transition does not result in a "control" and thus does not serve as a flow-measuring device. And, finally, the hydraulic jump, which is not considered a "control," does not serve as a "flow-measuring device," but rather, it serves an "energy dissipator," resulting in major head loss.

9.9 Energy Concepts in Open Channel Flow

Although application of the energy equation is useful only using an integral approach, the differential form of the energy equation may be used to illustrate the occurrence of hydrostatic pressure, or the deviation from hydrostatic pressure. One may note that the application of the energy equation, using an integral approach, in open channel flow assumes a hydrostatic pressure distribution at the given channel cross section, and requires the definition of the specific energy equation.

The differential form of the energy equation for a fluid system, assuming compressible, nonuniform, unsteady flow was given in Equation 4.32 and is repeated here as follows:

$$-\frac{\partial(p + \gamma z)}{\partial s} - \frac{\tau_w}{R_h} = \rho \left(\underbrace{v \frac{\partial v}{\partial s}}_{a_s} + \underbrace{\frac{\partial v}{\partial t}}_{a_t} \right) \tag{9.25}$$

Assuming steady flow (local acceleration, $a_t = \delta v / \delta t = 0$) yields:

$$-\frac{\partial(p + \gamma z)}{\partial s} - \frac{\tau_w}{R_h} = \rho a_s \tag{9.26}$$

where the $(p + \gamma z)$ term is called the piezometric pressure or the static pressure, the τ_w / R_h term represents flow resistance, the term a_s represents the convective acceleration ($\delta v / \delta s$), and s is measured along any straight line in the field of flow. According to the principle of hydrostatics, the piezometric pressure, $(p + \gamma z)$ remains a constant throughout a body of still water; thus, $\delta(p + \gamma z)/ds = 0$, whatever the direction of s may be (see Figure 2.5 in Chapter 2). It is important to note that the presence of the friction term, τ_w / R_h on the left-hand side of Equation 9.26 and the acceleration term, ρa_s on the right-hand side of Equation 9.26 indicates that if the water begins to move, the hydrostatic pressure distribution is disturbed and the piezometric pressure, $(p + \gamma z)$ no longer remains a constant throughout the body of water. However, one may note that while there are numerous special cases in open channel flow that may be modeled assuming a hydrostatic pressure distribution, there are a few cases that may not (pressure distribution strongly deviates from a hydrostatic pressure distribution).

9.9.1 Hydrostatic Pressure Distribution

In order to model a hydrostatic pressure distribution in open channel flow, one may assume short channel section with a nearly horizontal channel bottom slope. Thus, assuming a short channel section, one may ignore the effects of the flow resistance, and thus the τ_w / R_h term, which yields the Euler's equation of motion as follows:

$$\frac{\partial(p + \gamma z)}{\partial z} + \rho a_z = 0 \tag{9.27}$$

where the direction of the z-axis is in the vertical direction (see Figure 9.1). Furthermore, assuming a nearly horizontal channel bottom results in small vertical flow components, negligible curvature of the streamlines, and therefore, negligible acceleration in the z-direction, $a_z = 0$ and thus, yielding a hydrostatic pressure distribution as follows:

$$\frac{\partial(p + \gamma z)}{\partial z} = 0 \tag{9.28}$$

Therefore, when the channel conditions (small slope and negligible vertical curvature and accelerations) result in a hydrostatic pressure distribution (the case in most open channel

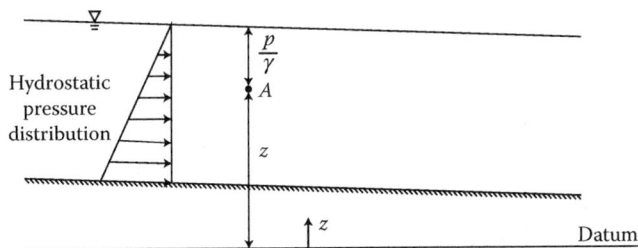

FIGURE 9.1
Hydrostatic pressure distribution implies a constant piezometric head, $(p/\gamma + z)$ for any point A along the vertical z-axis for a short channel section with a mild slope.

flow problems), the piezometric pressure, $(p + \gamma z)$ or the piezometric head, $(p/\gamma + z)$ remains a constant as follows:

$$\frac{p}{\gamma} + z = constant \tag{9.29}$$

for which at any point A in Figure 9.1, the pressure head, p/γ equals the depth of point A below the water surface (as defined in principles of hydrostatics), the elevation head, z equals the height of point A above the datum, and the piezometric head, $(p/\gamma + z)$ equals the height of the water surface above the datum. As such, the free water surface in open channel flow represents the hydraulic grade line—i.e., the piezometric head, $(p/\gamma + z)$ for any point in a given cross section and the flow in general. Furthermore, it may be noted that in open channel flow, the elevation head, z is always defined as the height of the channel bottom above the datum. As such, the pressure head, p/γ is always defined as the vertical distance from the channel bottom to water surface, which equals to the depth of flow, y. Therefore, the expression for the total energy head, H for an open channel flow assuming a hydrostatic pressure distribution is given as follows:

$$H = y + \frac{v^2}{2g} + z \tag{9.30}$$

9.9.2 Deviation from a Hydrostatic Pressure Distribution

There are a few cases where the pressure distribution strongly deviates from a hydrostatic pressure distribution. Assuming a short channel section and thus that one may ignore the effects of the flow resistance, these include: (1) cases where the channel bottom slope is very steep (see Figure 9.2), and (2) cases which include a sharp-crested weir, and a free overfall (a special case of the sharp-crested weir, where the height of the weir, $P = 0$) (see Figures 9.3 and 9.4, respectively). In the cases where the channel bottom slope is very steep, there are vertical flow components and therefore significant acceleration in the vertical direction, $a_z \neq 0$ and thus, the pressure distribution is not hydrostatic. Furthermore, in the cases that include a sharp-crested weir and a free overfall, due to the sharp-edged crest and the brink, respectively, gravity creates a pronounced curvature of the streamlines in the vertical plane, forming a falling jet. As such, there are vertical flow components and therefore, significant acceleration in the vertical direction, $a_z \neq 0$ and thus, the pressure distribution is not hydrostatic at the sharp-edged crest and the brink.

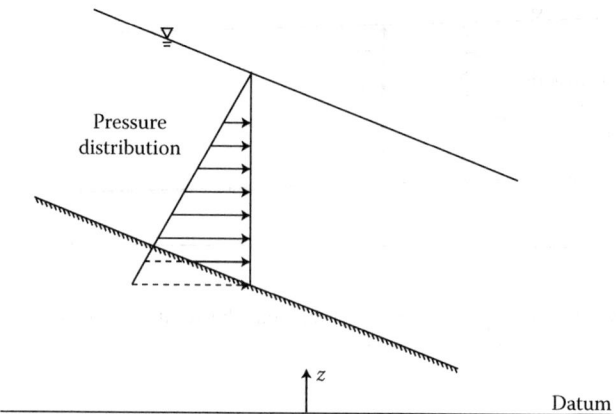

FIGURE 9.2
Deviation from a hydrostatic pressure distribution occurs for a short channel section with a very steep slope. (Adapted from Henderson, F. M., *Open Channel Flow*, Macmillan, New York, 1966, p. 28.)

9.9.2.1 Sharp-Crested Weir and Free Overfall

It is important to note that critical depth is derived based on the assumptions that (1) there is no contraction in the depth of flow, and (2) the flow pressure distribution is hydrostatic (see derivation below). Due to the sharp-edged crest of the weir and the brink of the free overall, the fluid viscosity causes a contraction in the depth of flow, while gravity creates a pronounced curvature of the streamlines (forming a falling jet) in the vertical plane and thus a deviation from a hydrostatic pressure distribution at the sharp-edged crest and the brink (see Figures 9.3 and 9.4, respectively). Specifically, at the sharp-edged crest of the weir and at the brink of the overfall, $\partial(p + \gamma z)/\partial z + \rho a_z = 0$, which is a strong deviation from a hydrostatic pressure distribution, with a mean pressure that is significantly less than hydrostatic (see Figures 9.3 and 9.4). Thus, while the critical flow condition does not disappear, it simply retreats upstream into the region where (1) there is no contraction in the depth of flow and (2) the pressure distribution is hydrostatic. As such, the occurrence of critical depth of flow,

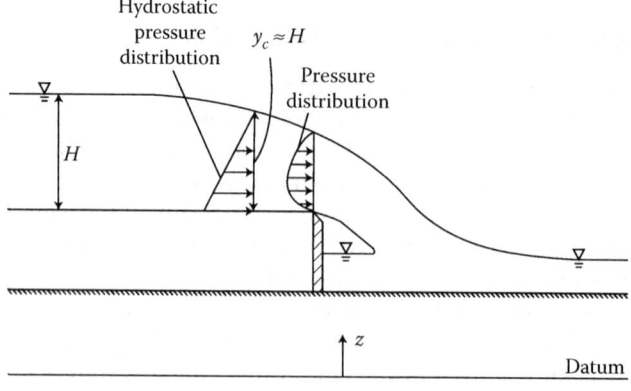

FIGURE 9.3
Deviation from a hydrostatic pressure distribution occurs for a short channel section with a sharp-crested weir. (Adapted from Henderson, F. M., *Open Channel Flow*, Macmillan, New York, 1966, p. 175.)

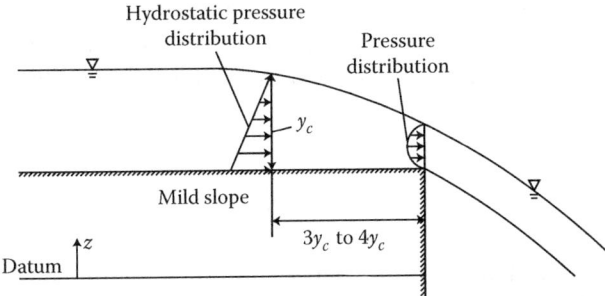

FIGURE 9.4
Deviation from a hydrostatic pressure distribution occurs for a short channel section with a brink of a free overfall. (Adapted from Henderson, F. M., *Open Channel Flow*, Macmillan, New York, 1966, p. 28, p. 192.)

y_c occurs just upstream of the sharp-edged crest of the weir, and just upstream of the brink of the overfall. Experimental evidence has shown that the depth at the sharp-edged crest of the weir and at the brink of the overfall is approximately $5y_c/7$. Furthermore, in the case of a sharp-crested weir, $y_c \approx H$ (the head of the weir), which occurs just upstream of the sharp-edged crest of the weir. And, in the case of the free overfall, the critical depth of flow, y_c occurs at a distance of approximately $3y_c$ to $4y_c$ upstream of the brink, depending on the value of the discharge.

Furthermore, it is important to note some differences and similarities between the various controls, regarding (1) contraction in the depth of flow, and (2) the occurrence of hydrostatic pressure distribution at the control. Unlike the sharp-crested weir and the free overfall, for the remaining controls (a sluice gate, a spillway, a broad-crested weir, a venturi flume, a contracted opening, a break in channel bottom slope, a hydraulic drop, and critical flow resulting from the flow resistance due to channel roughness in uniform flow), there is no deviation from a hydrostatic pressure distribution at the control. Furthermore, unlike a sharp-crested weir and a free overfall, in the cases of a broad-crested weir, a venturi flume, a contracted opening, a hydraulic drop, a break in channel bottom slope, and critical flow due to flow resistance in uniform flow, there is no contraction in the depth of flow at the control, thus allowing the critical depth of flow, y_c to occur at the control; these are known as "critical depth" meters. However, similar to a sharp-crested weir and a free overfall, in the cases of a sluice gate and a spillway, the contraction in the depth of flow at the control causes the critical depth of flow, y_c to occur just upstream of the control.

9.9.3 Specific Energy in Open Channel Flow

The application of the energy equation in open channel flow assumes a hydrostatic pressure distribution at the given channel cross section and requires the definition of the specific energy equation. Assuming steady open channel flow with a hydrostatic pressure distribution at a given channel cross section, the energy equation defined between two point in a finite control volume is given as follows:

$$\underbrace{\left(y_1 + \frac{v_1^2}{2g} + z_1\right)}_{H_1} = \underbrace{\left(y_2 + \frac{v_2^2}{2g} + z_2\right)}_{H_2} + h_f \qquad (9.31)$$

where y is the depth of flow in the channel, $v^2/2g$ is the velocity head, z is the elevation of the channel bottom/bed above a given datum, and h_f is the head loss, as illustrated in Figure 4.10. The total energy head, H at a given cross section in the control volume is defined as follows:

$$H = y + \frac{v^2}{2g} + z \tag{9.32}$$

The specific energy, E is defined as the energy with respect to the channel bed as the datum and is defined as follows:

$$E = y + \frac{v^2}{2g} \tag{9.33}$$

which represents a simple concept that is key to even the most complex of open channel flow phenomena. The shape of the channel cross section may be irregular, which typically occurs in natural rivers, or it may be regular or prismatic (rectangular, trapezoidal, or circular, for instance), which typically occurs in artificial channels. However, because the rectangular channel is the simplest type of channel cross section, the specific energy concept and its applications will be first developed assuming a rectangular cross section; then it will be extended to apply to other prismatic channel cross sections.

9.9.4 Specific Energy for Rectangular Channel Cross Sections

The assumption of a rectangular cross section allows the definition of a unit discharge, q; the definition of the specific energy curve, E versus y for a given q; and the definition of the discharge–depth curve, q versus y for a given E. Additionally, the critical flow equations, specifically for a rectangular cross section, are derived. Assuming a rectangular channel cross section with a width, b, one may express the specific energy E as a function of a unit discharge, q which is the discharge, Q per unit width, b. The unit discharge, q is given as follows:

$$q = \frac{Q}{b} = \frac{vA}{b} = \frac{vby}{b} = vy \tag{9.34}$$

Thus, the specific energy, E is expressed as follows:

$$E = y + \frac{q^2}{2gy^2} \tag{9.35}$$

9.9.4.1 The Specific Energy Curve for Rectangular Channel Sections

It is of interest to examine how the specific energy, E varies with the depth of flow, y for a given constant value of unit discharge, q. Thus, one may define the specific energy curve, E versus y for a given constant value of q as follows. The specific energy equation is rearranged in order to isolate the constant term as follows:

$$(E - y)y^2 = \frac{q^2}{2g} = constant \tag{9.36}$$

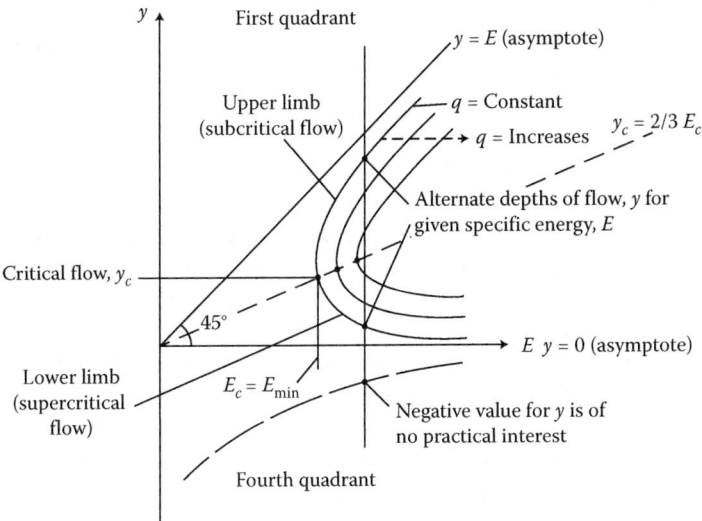

FIGURE 9.5
Specific energy curve, E-y plot for a rectangular channel section.

which has asymptotes of $(E-y)=0$ (making a $45°$ angle) and $y=0$. As illustrated in Figure 9.5, one section of the E-y plot is in the first quadrant (solid curve between the two asymptotes), while another section of the E-y plot is in the fourth quadrant (broken curve), which is of no practical interest because it yields negative values for y.

The E-y plot for a given constant value of q may be used to graphically illustrate the three solutions for y in the cubic specific energy equation, which is solve numerically using Mathcad. Specifically, there are three values for y for a given E and q. Thus, drawing a vertical line corresponding to a given value for E on the E-y plot illustrates that there are three solutions for y in the specific energy equation. While the two positive solutions for y in the first quadrant are physically real and are of practical interest, the third negative solution for y in the fourth quadrant is not (and will not be discussed any further). Thus, focusing only on the solid curve in the first quadrant, the two positive real solutions for y for a given E and q are known as alternate depths of flow; one flow depth is on the upper limb (subcritical flow) and the other is on the lower limb (supercritical flow). One may note that E-y plot graphically illustrates three possible flow regimes. The deep slow moving flow, known as subcritical flow, is illustrated on the upper limb, the shallow fast moving flow, known as supercritical flow, is illustrated on the lower limb, and the critical flow is illustrated at the crest, where the upper and lower limbs meet.

And, finally, as illustrated in Figure 9.5, one may plot additional curves on the E-y plot for other given constant values of q. It is important to note that because for a given value of y, E increases with q (see Equation 9.36 above), curves representing higher values of q occur inside and to the right of those curves having a lower value of q; creating a family of E-y curves corresponding to a number of different values for a constant unit discharge, q.

9.9.4.2 Derivation of Critical Flow for Rectangular Channel Sections

While the occurrence of critical flow in open channel flow was presented and discussed above, the derivation of the analytical critical flow equations assuming a rectangular channel section is presented in this section. It is important to note that the derived analytical critical flow equations will be applicable regardless of the nature of the channel or flow transition causing the occurrence of critical flow. As discussed above, the occurrence of critical flow may be due to any type of channel control, as well as due to the occurrence of a hydraulic jump, which is not considered a control.

As illustrated in Figure 9.5, the subcritical flow on the upper limb and the supercritical flow on the lower limb are separated by the crest, where the flow is critical. Furthermore, for a given E and q there are two possible depths of flow, y, known as alternate depths of flow (one is subcritical and the other is supercritical). One may note that the transition from one flow regime (subcritical or supercritical) to the other flow regime (supercritical or subcritical) can only occur under certain special conditions. These special conditions are such that there is a channel or flow transition in the channel (such as a control or a hydraulic jump) that results in the occurrence of critical flow. It is interesting to note that for a control, the flow regime changes from subcritical, to critical at the control, to supercritical. And, for a hydraulic jump, the flow regime changes from supercritical, to critical at the jump, to subcritical. In either special condition, the critical flow is poised between two alternative flow regimes (subcritical and supercritical), where the term critical indicates that the specific energy, E is at a minimum for a given constant value of unit discharge, q as illustrated in Figure 9.5.

The derivation of the analytical critical flow equations (expressions for y_c, v_c, and E_c) for a rectangular channel section considers the variation of E with y for a given constant value of unit discharge, q, at which the specific energy, E is a minimum. One begins with the specific energy equation for a rectangular channel section, which is given as follows:

$$E = y + \frac{q^2}{2gy^2} \tag{9.37}$$

The specific energy, E is a minimum for a given unit discharge, q at critical flow, and is defined by differentiating the specific energy equation with respect to y and setting it equal to zero as follows:

$$\frac{dE}{dy} = 1 - \frac{q^2}{gy^3} = 0 \tag{9.38}$$

and thus:

$$q^2 = gy_c^{\ 3} \tag{9.39}$$

or

$$y_c = \sqrt[3]{\frac{q^2}{g}} \tag{9.40}$$

which indicates that for a given unit discharge, q, the minimum specific energy, E occurs at the critical depth of flow, y_c, where y_c increases with q. This equation also indicates that for a given unit discharge, q, there is a corresponding/unique critical depth of flow, y_c (see Figure 9.5). One may note that this analytical relationship between the critical depth of flow, y_c and the unit discharge, q provides: (1) one of the analytical critical flow equations, which is applicable to a channel control as well as a hydraulic jump, and (2) the unique relationship between y and q sought in the case of a control, which can serve as a flow-measuring device. Next, beginning with Equation 9.39, one may derive an analytical relationship between the critical velocity, v_c and the critical depth of flow, y_c as follows:

$$q^2 = gy_c^3 \tag{9.41}$$

$$v_c^2 y_c^2 = gy_c^3 \tag{9.42}$$

$$v_c = \sqrt{gy_c} \tag{9.43}$$

Finally, beginning with Equation 9.42, one may derive an analytical relationship between the minimum specific energy, E_c and the critical depth of flow, y_c as follows:

$$v_c^2 y_c^2 = gy_c^3 \tag{9.44}$$

$$v_c^2 = gy_c \tag{9.45}$$

$$\frac{v_c^2}{2g} = \frac{y_c}{2} \tag{9.46}$$

Substituting this expression for the velocity head into the specific energy equation yields the following:

$$E_c = y_c + \frac{v_c^2}{2g} = y_c + \frac{y_c}{2} = \frac{3}{2}y_c \tag{9.47}$$

or:

$$y_c = \frac{2}{3}E_c \tag{9.48}$$

which indicates that for a given unit discharge, q, the critical depth of flow, y_c is linearly related to the minimum specific energy, E_c with a slope of 2/3 and a y-intercept of 0 (line goes through the origin). As such, it is important to note that plotting additional curves on the E-y plot for other given constant values of q, and connecting the crests of the E-y curves for all values of q, graphically illustrates this linear relationship, $y_c = 2/3E_c$; this is illustrated by the broken line in Figure 9.5. Therefore, the derivation of the analytical critical flow equations considered the variation of E with y for a given constant value of unit discharge, q, at which the specific energy, E is at a minimum. Furthermore, as presented in the following section, it is of interest to examine how the unit discharge, q varies with the depth of flow, y for a given specific energy, E.

9.9.4.3 The Depth–Discharge Curve for Rectangular Cross Sections

In order to examine how the unit discharge, q varies with the depth of flow, y for a given specific energy, E, one may begin with the family of E-y curves corresponding to a number of different values for a constant unit discharge, q, as illustrated in Figure 9.5. A general observation was: the family of E-y curves indicates that because for a given value of y, E increases with q, curves representing higher values of q occur inside and to the right of those curves having a lower value of q (and thus, y_c increases with q as indicated by Equation 9.40 $y_c = \sqrt[3]{q^2/g}$). However, for a given specific energy, $E = E_o$, more specific observations are: (1) when $y \to E_o$, $q \to 0$, (2) when $y \to 0$, $q \to 0$, and (3) when $y = 2/3E_o$, $y = y_c$, and $q = q_{max}$; the last observation may be analytically derived and is given below. Thus, with these three specific observations, one may plot the depth–discharge curve, q versus y for a given specific energy, $E = E_o$ as illustrated in Figure 9.6. Furthermore, the observation that when $y = 2/3E_o$, $y = y_c$, and $q = q_{max}$ may be analytically derived as follows. One begins with the specific energy equation for a rectangular channel section, which is given as follows:

$$E = y + \frac{q^2}{2gy^2} \tag{9.49}$$

The derivation of the maximum unit discharge, q_{max} for a rectangular channel section considers the variation of q with y for a given specific energy, $E = E_o$. As such, the specific energy equation is rearranged in order to express the unit discharge, q as a function of the depth of flow, y as follows ("unit discharge equation"):

$$q^2 = 2gy^2(E_o - y) \tag{9.50}$$

The maximum unit discharge, q_{max} for a given specific energy, $E = E_o$ is defined by differentiating the "unit discharge equation" with respect to y and setting it equal to zero as follows:

$$\frac{d(q^2)}{dy} = 2q\frac{dq}{dy} = 4gyE_o - 6gy^2 = 0 \tag{9.51}$$

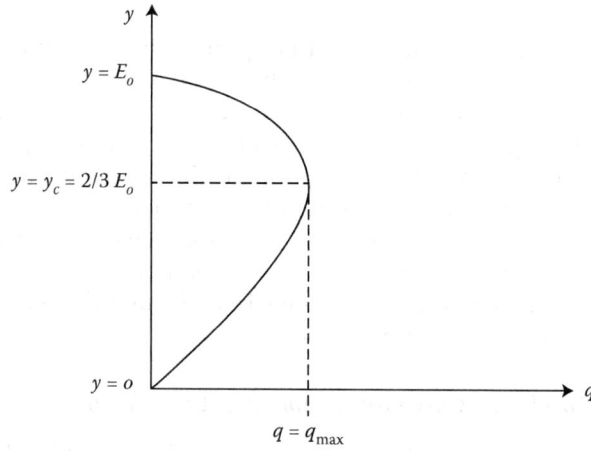

FIGURE 9.6
Depth–discharge curve, q versus y for a given specific energy, $E = E_o$.

and thus:

$$4gyE_o = 6gy^2 \tag{9.52}$$

or

$$y = \frac{2}{3}E_o \tag{9.53}$$

which represents critical flow as given by Equation 9.48 above. Therefore, the maximum unit discharge, q_{max} occurs at the critical depth of flow, y_c, for given specific energy, $E = E_o$. As such, one may highlight an important property of critical flow: critical flow $(y = y_c)$ not only connotes minimum specific energy, E_c for a given unit discharge, q, but also maximum unit discharge, q_{max} for a given specific energy, $E = E_o$.

9.9.5 Specific Energy for Nonrectangular Channel Cross Sections

It is important to note that while the use of a rectangular channel section in artificial channels is not typically common in practical applications, the existence of an irregular channel section in natural channels and the use of a trapezoidal channel section in artificial channels are more common. A trapezoidal section is usually preferred over a rectangular section because of its bank stability and economic benefits. As such, the assumption of a nonrectangular cross section does not allow the definition of a unit discharge, q; the definition of a specific energy curve, E versus y for a given q; or the definition of a discharge–depth curve, q versus y for a given E. However, the critical flow equations for a nonrectangular cross section (see Figure 9.7) may be derived. Furthermore, the occurrence of critical flow (discussed above) are equally true regardless of the shape of the channel cross section.

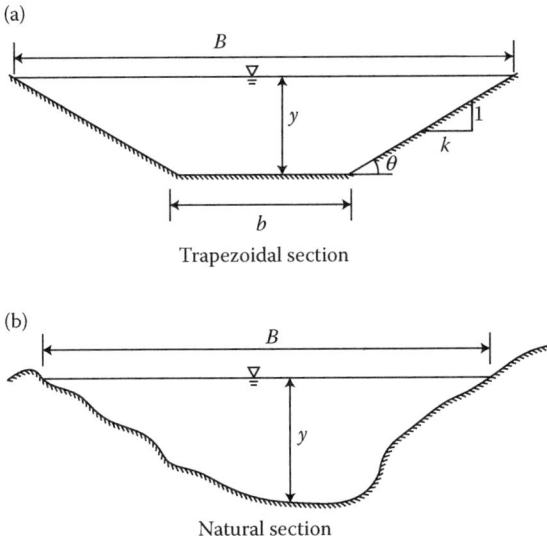

FIGURE 9.7
Nonrectangular channel sections. (a) Trapezoidal section, (b) Natural section.

9.9.5.1 Derivation of Critical Flow for Nonrectangular Channel Sections

While the occurrence of critical flow in open channel flow was presented and discussed above, the derivation of the analytical critical flow equations assuming a nonrectangular channel section is presented in this section. It is important to note that the derived analytical critical flow equations will be applicable regardless of the nature of the channel or flow transition causing the occurrence of critical flow. As discussed above, the occurrence of critical flow may be due to any type of channel control, as well as due to the occurrence of a hydraulic jump, which is not considered a control. The derivation of the analytical critical flow equations (expressions for y_c, v_c, and E_c) for a nonrectangular channel section considers the variation of E with y for a given constant value of discharge, Q, at which the specific energy, E is at a minimum. One begins with the specific energy equation for a nonrectangular channel section, which is given as follows:

$$E = y + \frac{v^2}{2g} \tag{9.54}$$

where $v = Q/A$, $Q =$ the total discharge, $A =$ the whole area of the cross section, and thus:

$$E = y + \frac{Q^2}{2gA^2} \tag{9.55}$$

which is quintic in y and where y is solved for numerically using Mathcad. The specific energy, E is a minimum for a given discharge, Q at critical flow and is defined by differentiating the specific energy equation with respect to y and setting it equal to zero as follows:

$$\frac{dE}{dy} = 1 - \frac{Q^2}{gA^3}\frac{dA}{dy} = 0 \tag{9.56}$$

where $dA = Bdy$, $B =$ surface width of the channel (as illustrated in Figure 9.7); thus:

$$\frac{dE}{dy} = 1 - \frac{Q^2 B}{gA^3} = 0 \tag{9.57}$$

and therefore:

$$Q^2 B_c = gA_c{}^3 \tag{9.58}$$

or

$$\frac{A_c}{B_c} = \frac{Q^2}{A_c{}^2}\frac{1}{g} \tag{9.59}$$

One may note that although the term A_c/B_c is the generalization of the depth, y_c (as for a rectangular channel), the actual value for y_c for a nonrectangular channel is solved numerically using Mathcad. Equation 9.59 indicates that for a given discharge, Q, the minimum specific energy, E occurs at the critical depth of flow, y_c, which increases with Q. This equation also indicates that for a given discharge, Q, there is a corresponding/unique value for y_c. Furthermore, it is important to note that the numerically solved value for y_c does not equal to A_c/B_c. One may note that this analytical relationship between A_c/B_c and the discharge, Q provides (1) one of the analytical critical flow equations, which is applicable to a channel control as well as a hydraulic jump, and (2) the unique relationship between A_c/B_c and Q sought in the case of a control, which can serve as a flow-measuring device. Next, beginning with Equation 9.59, one may derive an analytical relationship between the critical velocity, v_c and A_c/B_c for a nonrectangular channel as follows:

$$\frac{A_c}{B_c} = \frac{Q^2}{A_c^2}\frac{1}{g} = \frac{v_c^2}{g} \tag{9.60}$$

$$\frac{A_c}{B_c} = \frac{v_c^2}{g} \tag{9.61}$$

$$v_c = \sqrt{g\frac{A_c}{B_c}} \tag{9.62}$$

And, finally, beginning with the specific energy equation for a nonrectangular channel section (Equation 9.54), one may derive an analytical relationship between the minimum specific energy, E_c and A_c/B_c for a nonrectangular channel as follows:

$$E = y + \frac{v^2}{2g} \tag{9.63}$$

Substituting the expression for A_c/B_c (generalization of the depth, y_c) for critical flow and the expression for the velocity head (Equation 9.61) for critical flow into the specific energy equation yields the following:

$$E_c = \frac{A_c}{B_c} + \frac{1}{2}\frac{A_c}{B_c} = \frac{3}{2}\frac{A_c}{B_c} \tag{9.64}$$

or:

$$\frac{A_c}{B_c} = \frac{2}{3}E_c \tag{9.65}$$

9.9.6 Subcritical and Supercritical Flow

As mentioned above, subcritical flow is deep slow-moving flow, while supercritical flow is shallow fast-moving flow. Furthermore, critical flow occurs at a depth in between subcritical and supercritical, where the specific energy is a minimum for a given discharge and where the discharge is a maximum for a given specific energy. It is important to point out that the subcritical and supercritical flow regimes are defined in relation to the critical flow regime. As such, the critical flow equations were derived above for both rectangular and nonrectangular channel sections. Also, as mentioned above, the Froude number, F is used to indicate the flow regime, as it is the ratio of the inertia force, F_I to the gravity force, F_G, which is invoked in open channel flow. The Froude number is defined as $F = F_I/F_G = v/v_c$, where v is the velocity of the flow and v_c is the critical velocity. Thus, if the flow regime is subcritical, $F = v/v_c < 1$, if the flow regime is supercritical flow, $F = v/v_c > 1$, and if the flow regime is critical, $F = v/v_c = 1$. Thus, for a rectangular channel section, the Froude number, F is defined as follows:

$$F = \frac{v}{v_c} = \frac{v}{\sqrt{gy_c}} = \frac{q}{y}\frac{1}{\sqrt{gy_c}} \tag{9.66}$$

And, for a nonrectangular channel section, the Froude number, F is defined as follows:

$$F = \frac{v}{v_c} = \frac{v}{\sqrt{g\dfrac{A_c}{B_c}}} = \frac{Q}{A}\frac{1}{\sqrt{g\dfrac{A_c}{B_c}}} \tag{9.67}$$

It is important to note that the Froude number, F presents a very convenient dimensionless parameter in the manipulation of equations used to describe and solve problems in open channel flow. As such, one may note that the derivative of the specific energy equation may be expressed as a function of the Froude number, F for a rectangular channel section as follows:

$$\frac{dE}{dy} = 1 - \frac{q^2}{gy^3} = 1 - F^2 \tag{9.68}$$

and for a nonrectangular channel section as follows:

$$\frac{dE}{dy} = 1 - \frac{Q^2 B}{gA^3} = 1 - F^2 \tag{9.69}$$

Furthermore, as presented in several sections below, many other equations describing open channel flow problems can be expressed in terms of the Froude number, F. As such, because the Froude number, F indicates the flow regime, once it is evaluated, one may accordingly describe a particular flow situation.

9.9.7 Analysis of the Occurrence of Critical Flow at Controls

It is of interest to analyze the occurrence of critical flow at controls (associated with particular localized features in the channel; excluding the flow resistance due to channel roughness in uniform flow as a control). The controls associated with particular localized features in the channel, where the flow is critical, include an abrupt vertical or horizontal contraction (abrupt step upward or abrupt decrease in channel width), a typical flow-measuring device (such as a sluice gate, weir, overflow spillway, venturi flume, and contracted opening), a break in channel bottom slope, a hydraulic drop, or a free overfall. The flow in the channel will be influenced by the control, as the control determines the depth–discharge relationship. Therefore, the definite relationship between the depth and the discharge is the analytical expression for the critical depth of flow, y_c as a function of discharge, Q. As such, the setting of the particular control (obstruction, constriction, break in slope, etc.) is guided by the analytical expression for y_c as a function of discharge, Q. Thus, for a certain setting of the control, there is a certain relationship between the upstream flow depth and the discharge, and, likewise, there is a certain relationship between the downstream depth and the discharge (this is illustrated in Figure 9.5 for a rectangular channel section). Thus, the control mechanism dictates what the flow depth must be for a given discharge and what the discharge must be for a given flow depth along the channel. Furthermore, it is important to note that because a control determines the relationship between the flow depth and the discharge, a control in the open channel can also serve as a flow-measuring device; the discharge can be calculated once the depth is known. Finally, the typical flow-measuring devices have a minor head loss associated with the flow measurement, while a break in channel bottom slope, a hydraulic drop, or a free overfall has a small minor head loss associated with the flow measurement.

In order to analyze the occurrence of critical flow at the various types of controls, it important to subdivide the controls into four categories: (1) abrupt/maximum vertical constriction (hydraulic drop, sluice gate, weir, and overflow spillway), (2) abrupt/maximum horizontal constriction (venturi flume and contracted opening), (3) change in channel bottom slope (break in channel bottom slope and free overfall), and (4) uniform flow. Prior to the occurrence of critical flow at a control, it is assumed that the flow has previously been restrained/forced into the subcritical flow regime by a control in general, by the banks of a lake, or by the roughness of the channel bottom (classifying it as a subcritical slope). Once the flow is subsequently released into a region where there is no further restraint, and it is no longer forced into the subcritical condition, critical flow occurs at a point of release, which is defined as a control; then supercritical flow occurs downstream of the control. Thus, the control provides a mechanism of release from subcritical flow to critical flow, then to supercritical flow. Specifically, the flow prior to the hydraulic drop was forced into the subcritical flow regime by the banks of the lake. In the case of a subcritical channel bottom slope, the flow prior to the control was already forced into subcritical flow by the channel bottom roughness. And, in the case of either a critical or supercritical channel bottom slope, the flow prior to the control was forced into subcritical flow by the control itself.

Assuming a rectangular channel section, there are numerous practical open channel flow problems that involve the occurrence of critical flow at a control, in which both the unit discharge, q and the specific energy, E are initially given. However, in the analysis of the occurrence of critical flow at controls, it is of interest to (depending upon the category of the control): (1) given a value for the unit discharge, q, define the factors that determine the specific energy, E and thus the depth of flow, y, or (2) given a value for the specific energy, E, define the factors that determine the unit discharge, q and thus the depth of flow, y. Using the four above-listed categories of control mechanisms as a guide, one may appropriately apply the energy equation in order to analytically define these factors; this specifically involves analytically defining the physical description/nature of the control features. One may note that these analytical results and their implications may be applied to a nonrectangular channel provided the corresponding variables are defined for a nonrectangular channel section.

9.9.7.1 Critical Flow at Controls due to an Abrupt/Maximum Vertical Constriction

In order to analytically define the physical description/nature of the control features due to an abrupt/maximum vertical constriction (hydraulic drop, sluice gate, weir, and overflow spillway), one may apply the energy equation assuming no head loss due to channel roughness as follows:

$$H = \underbrace{y + \frac{v^2}{2g}}_{E} + z = E + z = constant \qquad (9.70)$$

Given a value for the unit discharge, q, in order to determine how the specific energy, E and thus the depth of flow, y varies along the channel bottom, one may differentiate the total energy, H with respect to x-axis, which is defined along the channel bottom, yielding the momentum equation as follows:

$$\frac{dE}{dx} + \frac{dz}{dx} = 0 \tag{9.71}$$

However, because the specific energy, E varies with the depth of flow, y, one may introduce its variation with y as follows:

$$\frac{dE}{dy}\frac{dy}{dx} + \frac{dz}{dx} = 0 \tag{9.72}$$

One may recall from above that:

$$\frac{dE}{dy} = 1 - F^2 \tag{9.73}$$

So:

$$\frac{dy}{dx}(1 - F^2) + \frac{dz}{dx} = 0 \tag{9.74}$$

where the Froude number, F plays a significant role in defining the physical description/nature of the control features due to an abrupt/maximum vertical constriction. It is important to note that Equation 9.74 mathematically presents the same result that is graphically illustrated on the E-y plot for a given unit discharge, q (see Figure 9.5) for a rectangular channel section. Furthermore, one may note that Equation 9.74 may be applied to a nonrectangular channel provided the corresponding variables are defined for a nonrectangular channel section.

First, in the case where dz/dx is positive, there would be a gradual upward step in the channel bed, so the term $(dy/dx)(1 - F^2)$ must be negative. If the upstream depth were subcritical, then $F = v/v_c < 1$; thus, dy/dx must be negative, and therefore, the flow depth would decrease over the upward step as illustrated in Figure 9.8. However, if the upstream depth were supercritical, then $F = v/v_c > 1$; thus, dy/dx must be positive, and therefore, the flow depth would increase over the upward step as illustrated in Figure 9.9.

Second, in the case where dz/dx is negative, there would be a gradual downward step in the channel bed, and thus the term $(dy/dx)(1 - F^2)$ must be positive. If the upstream depth were subcritical, then $F = v/v_c < 1$ and thus dy/dx must be positive; therefore, the flow depth

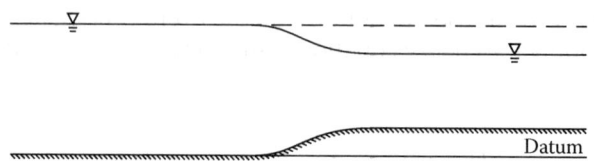

FIGURE 9.8
Decrease in flow depth over gradual step upward in channel bed for subcritical flow.

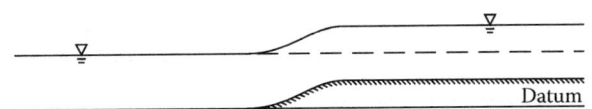

FIGURE 9.9
Increase in flow depth over gradual step upward in channel bed for supercritical flow.

would increase over the downward step as illustrated in Figure 9.10. However, if the upstream depth were supercritical, then $F = v/v_c > 1$; thus, dy/dx must be negative; therefore, the flow depth would decrease over the downward step as illustrated in Figure 9.11.

Finally, in the case where dz/dx is zero, there would no change in elevation of the lake or channel bed; thus, the term $(dy/dx)(1 - F^2)$ must be zero as follows:

$$\frac{dy}{dx}(1 - F^2) = 0 \tag{9.75}$$

This implies that either:

$$\frac{dy}{dx} = 0 \quad \text{(uniform flow at any flow regime: subcritical, supercritical, or critical flow)} \tag{9.76}$$

or:

$$F = 1 \quad \text{(critical flow at either flow type: uniform or nonuniform flow)} \tag{9.77}$$

or both:

$$\frac{dy}{dx} = 0 \quad \text{and} \quad F = 1 \text{(uniform flow type at the critical flow regime)} \tag{9.78}$$

One may note that the objective of this section was to analytically define the physical description/nature of the control features due to an abrupt/maximum vertical constriction (hydraulic drop, sluice gate, weir, and overflow spillway), as opposed to uniform flow as a control. Therefore, the appropriate choice would be that: $F = 1$ (critical flow at either flow type: uniform or nonuniform flow). Additionally, one may note that, as illustrated by Figure 9.12, the flow at a given control is accelerating (convective acceleration); thus, $dy/dx \neq 0$; thus, the $F = 1$ (critical flow at a nonuniform flow) would be the appropriate specific choice. So, in summary, $dz/dx = 0$, $dy/dx \neq 0$, and $F = 1$ would analytically define the

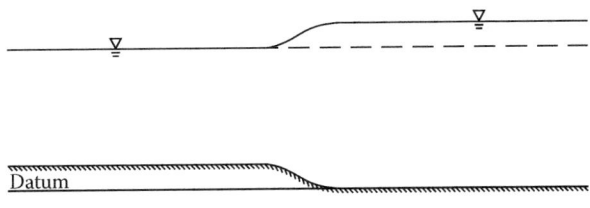

FIGURE 9.10
Increase in flow depth over gradual step downward in channel bed for subcritical flow.

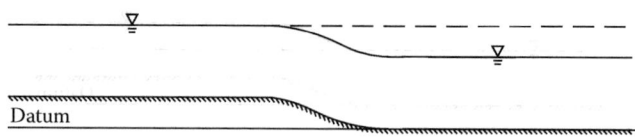

FIGURE 9.11
Decrease in flow depth over gradual step downward in channel bed for supercritical flow.

physical nature of the control features due to an abrupt/maximum vertical constriction (hydraulic drop, sluice gate, weir, and overflow spillway). One may recall that critical depth is derived based on the assumptions that (1) there is no contraction in the depth of flow, and (2) the flow pressure distribution is hydrostatic. Because in the cases of the hydraulic drop and the broad-crested weir, there is no contraction of the flow and the flow pressure distribution is hydrostatic at the control, critical flow occurs at the control as illustrated in Figure 9.12a and d, respectively. One may note that although in practice the critical depth of flow occurs in the close vicinity of the control, the hydraulic drop and the broad-crested weir are considered to be "critical depth" meters. In the case of the hydraulic drop, the flow

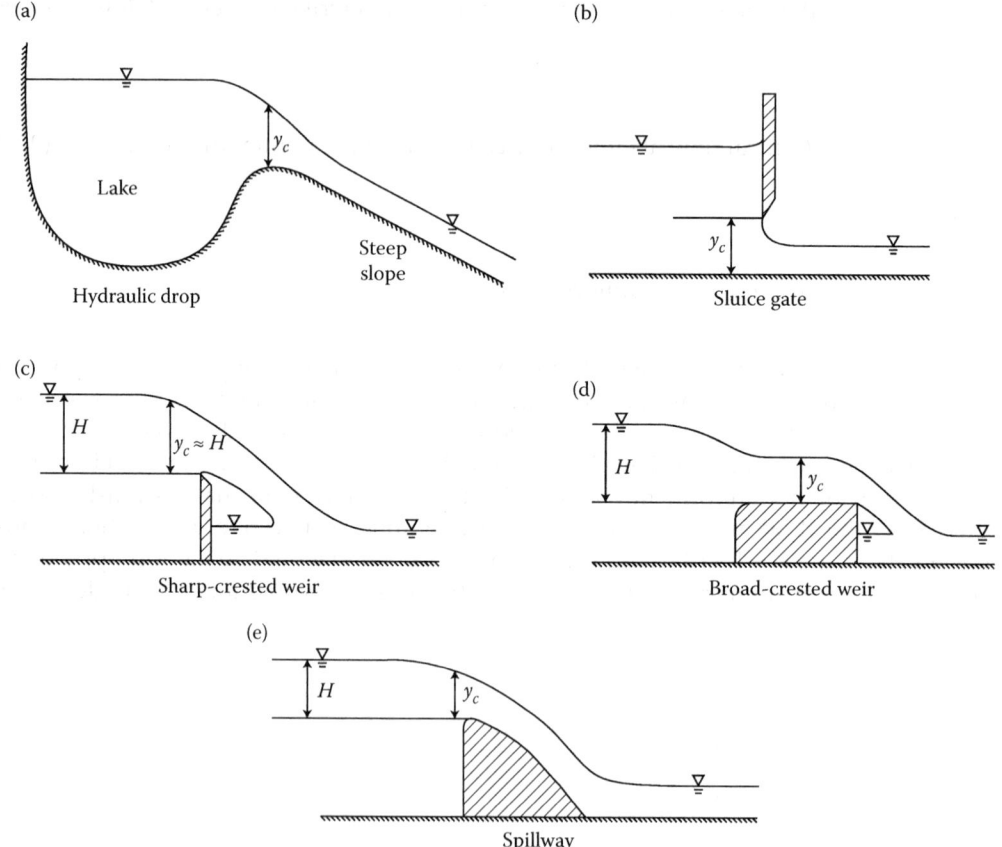

FIGURE 9.12
Critical flow at controls due to an abrupt/maximum vertical constriction. (a) Hydraulic drop, (b) Sluice gate, (c) Sharp-crested weir, (d) Broad-crested weir, (e) Overflow spillway.

is forced into subcritical flow by the banks of a lake, which is followed by a short crest (where $dz/dx = 0$, $dy/dx \neq 0$, and $F = 1$), then a channel with a steep slope where the flow is supercritical; the control provides a mechanism of release from subcritical to critical flow, followed by a steep channel bottom slope such that the channel resistance/roughness does not impose an effective restraint on the flow. In the case of the broad-crested weir, the flow is forced into subcritical flow (upstream depth of flow) by the channel roughness, then at the broad-crested weir the flow becomes critical (where $dz/dx = 0$, $dy/dx \neq 0$, and $F = 1$), followed by supercritical flow (downstream depth is alternate to the upstream depth) in the mildly sloped channel; the control provides a mechanism of release from subcritical to critical flow, followed by supercritical flow downstream of the control. In the cases of the sluice gate and the spillway (see Figure 9.12b and e, respectively), although the flow pressure distribution is hydrostatic at the control, there is a contraction in the depth of flow at the control, which causes the critical depth of flow to occur just upstream of the control. And, finally, in the case of the sharp-crested weir (see Figure 9.12c), there is a contraction in the depth of flow and the flow pressure distribution is not hydrostatic at the control. As a result, the critical flow retreats just upstream of the sharp-edged crest, where there is no contraction in the depth of flow and where the pressure distribution is hydrostatic. Therefore, in the cases of the sluice gate, the spillway, and the sharp-crested weir, the flow is forced into subcritical flow (upstream depth of flow) by the channel roughness, then just upstream of the control the flow becomes critical (where $dz/dx = 0$, $dy/dx \neq 0$, and $F = 1$), followed by supercritical flow (downstream depth is alternate to the upstream depth) in the mildly sloped channel; the control provides a mechanism of release from subcritical to critical flow, followed by supercritical flow downstream of the control. Furthermore, in the cases of the sluice gate, the spillway, and the sharp-crested weir, because of the typical practical assumption of ideal flow made for flow-measuring devices in general (ideal flow meters), it is assumed that the critical flow occurs at the control.

9.9.7.2 Critical Flow at Controls due to an Abrupt/Maximum Horizontal Constriction

In order to analytically define the physical description/nature of the control features due to an abrupt/maximum horizontal constriction (venturi flume and contracted opening), one may apply the energy equation assuming no head loss due to channel roughness as follows:

$$H = y + \underbrace{\frac{v^2}{2g}}_{E} + z = E + z = constant \tag{9.79}$$

$$H = y + \frac{q^2}{2gy^2} + z = constant \tag{9.80}$$

Given a value for specific energy, E, in order to determine how the unit discharge, $q = vy$ and thus the depth of flow, y varies along the channel bottom, one may differentiate the total energy, H with respect to x-axis, which is defined along the channel bottom, yielding the momentum equation as follows:

$$\frac{dy}{dx} - \frac{2q^2}{2gy^3}\frac{dy}{dx} + \frac{2q}{2gy^2}\frac{dq}{dx} + \frac{dz}{dx} = 0 \tag{9.81}$$

$$\frac{dy}{dx} - \frac{q^2}{gy^3}\frac{dy}{dx} + \frac{q}{gy^2}\frac{dq}{dx} + \frac{dz}{dx} = 0 \tag{9.82}$$

Note that from continuity:

$$qb = Q = constant \tag{9.83}$$

for which:

$$b\frac{dq}{dx} + q\frac{db}{dx} = 0 \tag{9.84}$$

Noting that the variation of the unit discharge, q with respect to the x-axis, dq/dx may be expressed as a function of the variation of the channel width, b with respect to x-axis, db/dx as follows:

$$\frac{dq}{dx} = -\frac{q}{b}\frac{db}{dx} \tag{9.85}$$

the expression for the total energy, H differentiated with respect to the x-axis becomes:

$$\frac{dy}{dx} - \frac{q^2}{gy^3}\frac{dy}{dx} - \frac{q}{gy^2}\frac{q}{b}\frac{db}{dx} + \frac{dz}{dx} = 0 \tag{9.86}$$

One may recall from above that:

$$\frac{dE}{dy} = 1 - \frac{q^2}{gy^3} = 1 - F^2 \tag{9.87}$$

where

$$\frac{q^2}{gy^3} = F^2 \tag{9.88}$$

Thus, the expression for the total energy, H differentiated with respect to the x-axis becomes:

$$\frac{dy}{dx}(1 - F^2) - F^2\frac{y}{b}\frac{db}{dx} + \frac{dz}{dx} = 0 \tag{9.89}$$

where the Froude number, F plays a significant role in defining the physical description/nature of the control features due to an abrupt/maximum horizontal constriction. Assuming that there is no change in elevation of the channel bottom and thus $dz/dx = 0$, the expression for the total energy, H differentiated with respect to the x-axis becomes as follows:

$$\frac{dy}{dx}(1 - F^2) - F^2\frac{y}{b}\frac{db}{dx} = 0 \tag{9.90}$$

It is important to note that Equation 9.90 mathematically presents the same result that is graphically illustrated on the E-y plot for a given specific energy, E (see Figure 9.13c) for a rectangular channel section. One may note that objective of this section was to analytically define the physical description/nature of the control features due to an abrupt/maximum horizontal constriction (venturi flume and contracted opening). Thus, at a cross section of maximum horizontal constriction, $db/dx = 0$, and thus:

$$\frac{dy}{dx}(1 - F^2) = 0 \qquad (9.91)$$

This implies that either:

$$\frac{dy}{dx} = 0 \quad \text{(uniform flow at any flow regime: subcritical, supercritical, or critical flow)}$$

$$(9.92)$$

or:

$$F = 1 \quad \text{(critical flow at either flow type: uniform or nonuniform flow)} \qquad (9.93)$$

or both:

$$\frac{dy}{dx} = 0 \quad \text{and} \quad F = 1 \text{(uniform flow type at the critical flow regime)} \qquad (9.94)$$

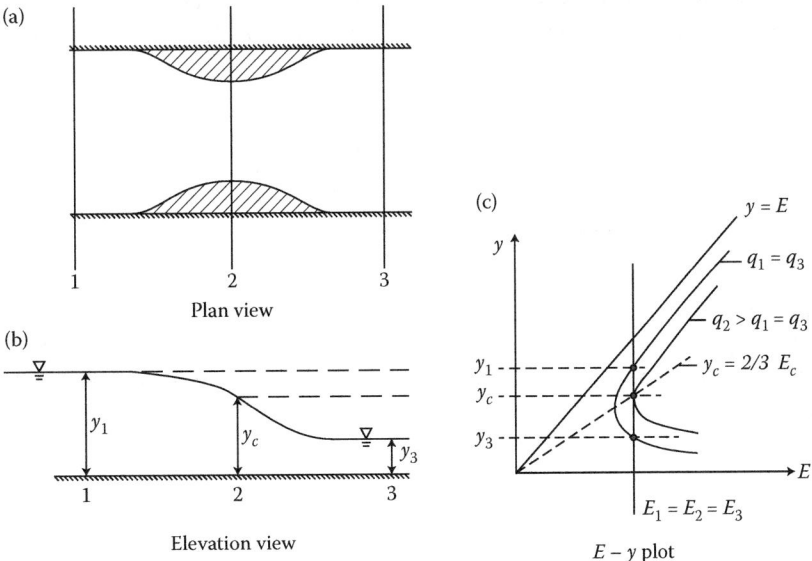

FIGURE 9.13
Critical flow at controls due to an abrupt/maximum horizontal constriction (venturi flume or contracted opening). (a) Plan view, (b) Elevation view, (c) E-y plot. (Adapted from Henderson, F. M., *Open Channel Flow*, Macmillan, New York, 1966, p. 33.)

However, since critical flow occurs at a section of maximum horizontal constriction, the appropriate choice would be that: $F = 1$ (critical flow at either flow type: uniform or non-uniform flow). Additionally, one may note that, as illustrated by Figure 9.13b, the flow at a given control is accelerating (convective acceleration); thus, $dy/dx \neq 0$, and the $F = 1$ (critical flow at a nonuniform flow) would be the appropriate specific choice. So, in summary, $dz/dx = 0$, $db/dx = 0$, $dy/dx \neq 0$, and $F = 1$ would analytically define the physical nature of the control features due to an abrupt/maximum horizontal constriction (venturi flume and contracted opening). One may recall that critical depth is derived based on the assumptions that (1) there is no contraction in the depth of flow, and (2) the flow pressure distribution is hydrostatic. Because in the cases of the venturi flume and the contracted opening there is no contraction of the flow and the flow pressure distribution is hydrostatic at the control, critical flow occurs at the control as illustrated in Figure 9.13b. One may note that although in practice, the critical depth of flow occurs in the close vicinity of the control, the venturi flume and the contracted opening are considered to be "critical depth" meters. In the case of the venturi flume and the contracted opening, the flow is forced into subcritical flow (upstream depth of flow) by the channel roughness, then at the venturi flume or the contracted opening, the flow becomes critical (where $dz/dx = 0$, $db/dx = 0$, $dy/dx \neq 0$, and $F = 1$), followed by supercritical flow (downstream depth is alternate to the upstream depth) in the mildly sloped channel; the control provides a mechanism of release from subcritical to critical flow, followed by supercritical flow downstream of the control. Furthermore, one may note that Equation 9.91 may be applied to a nonrectangular channel provided the corresponding variables are defined for a nonrectangular channel section.

9.9.7.3 Critical Flow at Controls due to a Change in Channel Bottom Slope

In order to analytically define the physical description/nature of the control features due to a change in the channel bottom slope (break in channel bottom slope and free overfall), one may apply the energy equation assuming a head loss, h_f due to channel roughness as follows:

$$\underbrace{\left(\underbrace{y_1 + \frac{v_1^2}{2g}}_{E_1} + z_1\right)}_{H_1} = \underbrace{\left(\underbrace{y_2 + \frac{v_2^2}{2g}}_{E_2} + z_2\right)}_{H_2} + \underbrace{h_f}_{S_f L} \tag{9.95}$$

Given a value for the unit discharge, q, in order to determine how the specific energy, E and thus the depth of flow, y varies along the channel bottom, one may differentiate the total energy, H with respect to the x-axis, which is defined along the channel bottom, yielding the momentum equation as follows:

$$\underbrace{\frac{dH}{dx}}_{S_e} = \frac{d}{dx}\left(y + \frac{v^2}{2g} + z\right) = -S_f = -\frac{v^2}{CR_h} \tag{9.96}$$

$$\frac{d}{dx}\left(y + \frac{v^2}{2g}\right) = -\underbrace{\frac{dz}{dx}}_{S_o} - S_f \tag{9.97}$$

$$\frac{dE}{dx} = S_o - S_f \tag{9.98}$$

However, because the specific energy, E varies with the depth of flow, y, one may introduce its variation with y as follows:

$$\frac{dE}{dy}\frac{dy}{dx} = S_o - S_f \tag{9.99}$$

One may recall from above that:

$$\frac{dE}{dy} = 1 - F^2 \tag{9.100}$$

So:

$$\frac{dy}{dx}(1 - F^2) = S_o - S_f \tag{9.101}$$

where the Froude number, F plays a significant role in defining the physical description/ nature of the control features due to a change in the channel bottom slope. One may note that Equations 9.98 and 9.101 are two forms of the resistance equation (momentum equation) for nonuniform flow (see section below on nonuniform flow). It is important to note that Equation 9.101 is a generalization of Equation 9.74, for which $S_f = 0$ (i.e., $h_f = 0$). Furthermore, one may note that Equations 9.98 and 9.101 may be applied to a nonrectangular channel provided the corresponding variables are defined for a nonrectangular channel section.

The objective of this section was to analytically define the physical description/nature of the control features due to a change in the channel bottom slope (break in channel bottom slope and free overfall). Thus, consider the special case in Equation 9.101 where $S_o = S_f$, which yields:

$$\frac{dy}{dx}(1 - F^2) = 0 \tag{9.102}$$

This implies that either:

$$\frac{dy}{dx} = 0 \quad \text{(uniform flow at any flow regime: subcritical, supercritical, or critical flow)}$$
$$\tag{9.103}$$

or:

$$F = 1 \quad \text{(critical flow at either flow type: uniform or nonuniform flow)} \tag{9.104}$$

or both:

$$\frac{dy}{dx} = 0 \quad \text{and} \quad F = 1 \text{(uniform flow type at the critical flow regime)} \tag{9.105}$$

However, since critical flow occurs at a section where there is a change in the channel bottom slope (the control changes the classification of the slope from subcritical to supercritical), the appropriate choice would be that: $F = 1$ (critical flow at either flow type: uniform or nonuniform flow). Additionally, one may note that, as illustrated by Figure 9.14a and b, the flow at a given control is accelerating (convective acceleration) and thus $dy/dx \neq 0$ and thus the $F = 1$ (critical flow at a nonuniform flow) would be the appropriate specific choice. In summary, $S_o = S_f$, $dy/dx \neq 0$, and $F = 1$ would analytically define the physical nature of the control features due to a change in the channel bottom slope (break in the channel bottom slope and free overfall). One may recall that critical depth is derived based on the assumptions that (1) there is no contraction in the depth of flow, and (2) the flow pressure distribution is hydrostatic. Because in the case of the break in the channel slope, there is no contraction of the flow and the flow pressure distribution is hydrostatic at the control, critical flow occurs at the control as illustrated in Figure 9.14a. One may note that a break in the channel bottom slope constitutes an ideal critical depth meter, for which the depth of flow at the break is definitively critical. In the case of the break in the channel slope, the flow is forced into subcritical flow (upstream depth of flow) by the channel roughness, then at the break in the channel slope the flow becomes critical (where $S_o = S_f$, $dy/dx \neq 0$, and $F = 1$), followed by steep channel, where the flow is supercritical (downstream depth of flow); the control provides a mechanism of release from subcritical to critical flow, followed by supercritical flow downstream of the control. However, in the case of the free overfall, there is a contraction in the depth of flow (to a depth of approximately $5y_c/7$) and the flow pressure distribution is not hydrostatic at the control. As a result, the critical flow retreats just upstream of the brink (to a distance of approximately $3y_c$ to $4 y_c$), where there is no contraction in the depth of flow and where the pressure distribution is hydrostatic as illustrated in Figure 9.14b. Therefore, in the case of the free overall, the flow is forced into subcritical flow (upstream depth of flow) by the channel roughness, then just upstream of

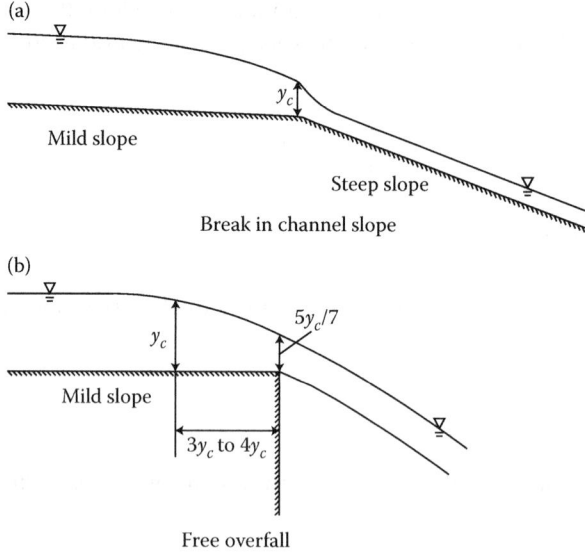

FIGURE 9.14
Critical flow at controls due to a change in channel bottom slope. (a) Break in channel slope, (b) Free overfall. (Adapted from Henderson, F. M., *Open Channel Flow*, Macmillan, New York, 1966, p. 107, p. 192, p. 211.)

the control the flow becomes critical (where $S_o = S_f$, $dy/dx \neq 0$, and $F = 1$), followed by a falling jet; the control provides a mechanism of release from subcritical to critical flow, followed by a falling jet at the brink. It is interesting to note that although a break in the channel slope constitutes an ideal critical depth meter, in practical open channel flow situations, a long downstream supercritical slope is typically not available. As a result, a very steep and shorter downstream subcritical slope may be used. Thus, the case of a break in the channel slope approaches the case of a free overfall as described above. Furthermore, in the case of a free overfall, because of the typical practical assumption of ideal flow made for flow-measuring devices in general (ideal flow meters), it is assumed that the critical flow occurs at the control/brink.

It is interesting to note that a steep channel slope (for which the uniform flow is supercritical) ending in a free overfall does not produce a draw-down curve (as in the case of a typical channel control) and thus does not serve as a channel control as illustrated in Figure 9.15.

Furthermore, one may note that unlike the typical flow-measuring devices, a break in the channel slope, a free overfall, and a hydraulic drop may serve as flow-measuring devices with a small minor head loss associated with the flow measurement. Finally, unlike most of the typical flow-measuring devices, a break in the channel slope, a free overfall, a hydraulic drop, a venturi flume, and a contracted opening may serve as flow-measuring devices without the existence of a dead water region upstream of the control where silt and debris can accumulate and significantly affect the depth–discharge relationship.

9.9.8 Analysis of the Occurrence of Uniform Flow as a Control

It is of interest to analyze the occurrence of uniform flow as a control, which is due to the channel roughness, and where the flow may be subcritical, supercritical, or critical. The flow in the channel will be influenced by the control, as the control determines the depth–discharge relationship. Therefore, the definite relationship between the depth and the discharge is given by the resistance equation for uniform flow (Chezy or Manning's equation, expressed as the continuity equation), where the critical equations are a special case of the resistance equation for uniform flow. The uniform flow will be subcritical, supercritical, or critical depending upon the classification of the slope of the channel: mild, steep, or critical, respectively. The classification of the slope will depend on the channel roughness, the actual magnitude of the slope, and the discharge. As such, the setting of the slope is guided by the resistance equation. Thus, for a certain setting of the slope (mild, steep, or critical), there is a certain relationship between the uniform flow depth and the discharge, for a given flow regime: subcritical, supercritical, or critical, respectively. Because uniform flow as a control

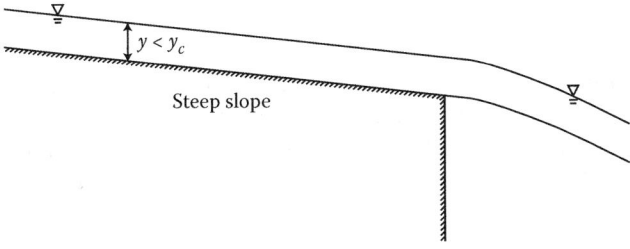

FIGURE 9.15
A steep channel slope ending in a free overfall does not produce a draw down curve and thus does not serve as a channel control.

is directly associated with the roughness of the channel, the analysis of uniform flow as a control assumes a head loss due to the channel roughness. Finally, the uniform flow may be used as a flow-measuring device, for which there is a major head loss associated with the flow measurement.

In the analysis of the occurrence of uniform flow as a control, given a value for the unit discharge, q, it is of interest to define the factors that determine the specific energy, E and thus the depth of flow, y. One may appropriately apply the energy equation in order to analytically define these factors; this specifically involves analytically defining the physical description/nature of the control features. One may note that these analytical results and their implications may be applied to a nonrectangular channel provided the corresponding variables are defined for a nonrectangular channel section. In order to analytically define the physical description/nature of the control features due to the channel roughness, one may apply the energy equation assuming a head loss, h_f due to channel roughness as follows:

$$\underbrace{\underbrace{\left(y_1 + \frac{v_1^2}{2g} + z_1\right)}_{E_1}}_{H_1} = \underbrace{\underbrace{\left(y_2 + \frac{v_2^2}{2g} + z_2\right)}_{E_2}}_{H_2} + \underbrace{h_f}_{S_f L} \tag{9.106}$$

Given a value for the unit discharge, q, in order to determine how the specific energy, E and thus the depth of flow, y varies along the channel bottom, one may differentiate the total energy, H with respect to the x-axis, which is defined along the channel bottom, yielding the momentum equation as follows:

$$\underbrace{\frac{dH}{dx}}_{S_e} = \frac{d}{dx}\left(y + \frac{v^2}{2g} + z\right) = -S_f = -\frac{v^2}{CR_h} \tag{9.107}$$

However, because the theory presented above to analytically define the physical description/nature of the control features due to a change in the channel bottom slope (break in channel bottom slope and free overfall) is the same as the theory for analysis of the occurrence of uniform flow as a control, the final results are presented as follows:

$$\frac{dy}{dx}(1 - F^2) = S_o - S_f \tag{9.108}$$

where the Froude number, F plays a significant role in defining the physical description/ nature of the control features due to a the channel roughness.

The objective of this section was to analytically define the physical description/nature of the control features due to the channel. Thus, consider the special case in Equation 9.108 where $S_o = S_f$, which yields:

$$\frac{dy}{dx}(1 - F^2) = 0 \tag{9.109}$$

This implies that either:

$$\frac{dy}{dx} = 0 \quad \text{(uniform flow at any flow regime: subcritical, supercritical, or critical flow)}$$
(9.110)

or:

$$F = 1 \quad \text{(critical flow at either flow type: uniform or nonuniform flow)} \qquad (9.111)$$

or both:

$$\frac{dy}{dx} = 0 \quad \text{and} \quad F = 1 \text{(uniform flow type at the critical flow regime)} \qquad (9.112)$$

However, since the focus of this section was to analyze uniform flow as a control, the appropriate choice would be uniform flow at any flow regime: subcritical, supercritical, or critical flow; thus, $dy/dx = 0$, as illustrated by Figure 9.16. In summary, $S_o = S_f$, $dy/dx = 0$, and either: $F < 1$, $F > 1$, or $F = 1$ for subcritical, supercritical, or critical flow, respectively, would analytically define the physical nature of the control features for

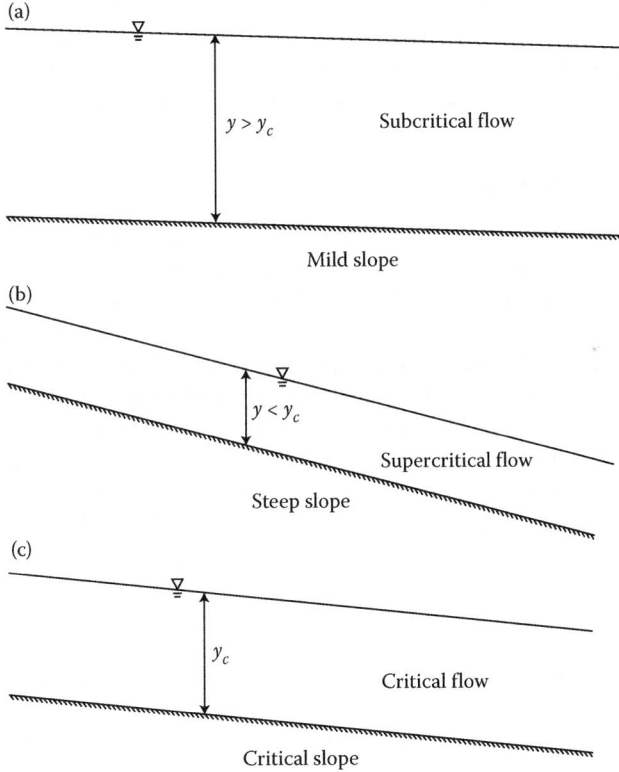

FIGURE 9.16
Uniform flow as a control. (a) Mild slope, (b) Steep slope, (c) Critical slope.

uniform flow as a control due to channel roughness. Furthermore, in the case of uniform flow, there is no contraction of the flow and the flow pressure distribution is hydrostatic at the control; thus, depending upon the classification of the channel bottom slope, subcritical, supercritical, or critical flow occurs at the control (defined along the entire channel bottom) as illustrated in Figure 9.16. Thus, critical uniform flow is considered to be a "critical depth meter."

9.10 Momentum Concepts in Open Channel Flow

Application of the momentum equation is useful using either an integral approach or a differential approach. Application of the momentum equation using the differential approach (for a given point in the fluid system) is appropriate for uniform or nonuniform flow that affects the flow over a long channel section and thus the major head loss due to flow resistance is significant. As such, the flow resistance is modeled by the appropriate resistance equation, depending on whether the flow is uniform or nonuniform. Furthermore, application of the momentum equation using the integral approach (between two points in a finite control volume) is appropriate for flow over a short channel section and, consequently, the major head loss due to flow resistance is insignificant. One may note that application of the appropriate resistance equations is discussed in detail in the sections below on uniform and nonuniform flow. As such, the focus of this section is to discuss the application of integral form of the momentum equation, which assumes that the flow resistance is negligible (note that the viscous forces have already been accounted for indirectly through the minor head loss term or the discharge coefficient term), and to define the momentum function in open channel flow.

9.10.1 The Momentum Function

Application of the momentum equation using the integral approach (between two points in a finite control volume) is appropriate for any channel or flow transition that affects the flow over a short channel section and thus assumes the flow resistance to be negligible. However, the momentum function in open channel flow is useful to define and apply over a short channel section when (1) the major head loss due to flow resistance is negligible, and (2) one flow depth is supercritical and one flow depth is subcritical, and they are thus conjugate to one another. Such cases include certain types of controls (abrupt channel transitions) and a flow transition such as the hydraulic jump. Specifically, the certain types of controls include an abrupt vertical or horizontal contraction (abrupt step upward or abrupt decrease in channel width), and a typical flow-measuring device (such as a sluice gate, weir, overflow spillway, venturi flume, and contracted opening). However, controls such as a break in the channel bottom slope, a hydraulic drop, or a free overfall are not included because they affect the flow over a long channel section. Furthermore, the momentum function is not useful to define and apply for gradual channel transitions, which are not considered "controls" (and thus the upstream and downstream depths are not conjugate to one another), or for uniform or nonuniform flow that affect the flow over a long channel section (thus, the major head loss due to flow resistance is not negligible). While abrupt channel transitions have been introduced above, the formation of a hydraulic jump is introduced below. Finally, the momentum function will be defined first for rectangular cross sections and then for nonrectangular cross sections.

9.10.1.1 The Use of Controls in the Formation of a Hydraulic Jump

A hydraulic jump is a natural phenomenon that occurs when the open channel flow transitions from supercritical to subcritical flow. Numerous flow situations may result in the formation of a hydraulic jump. For instance, the insertion of two controls in series in an open channel typically results in the occurrence of a hydraulic jump, as illustrated in Figure 9.17; to generate a hydraulic jump, one may insert the downstream control first, then the upstream control second. As discussed above, a control determines or dictates a certain depth–discharge relationship at the control and in the vicinity upstream and downstream of the control. As such, while the control produces critical flow at the control, it produces subcritical flow upstream of the control, and supercritical flow downstream of the control. For instance, the sluice gate presented in Figure 9.17a, illustrates how a control influences both the upstream (subcritical) and downstream (supercritical) depths of flow. One may note that in the design of a control structure, there is often a need to provide an energy dissipator in order to dissipate the excess kinetic energy possessed by the downstream supercritical flow. As such, one may insert an additional control structure downstream of the supercritical flow as illustrated by the second sluice gate in Figure 9.17a. The second sluice gate (a control) produces subcritical flow upstream of the control and supercritical flow downstream of the control (unless it is placed near the end of the channel, so as to not significantly affect the flow downstream of it). The supercritical flow produced by the first control encounters subcritical flow produced by the second control. In order for the flow to

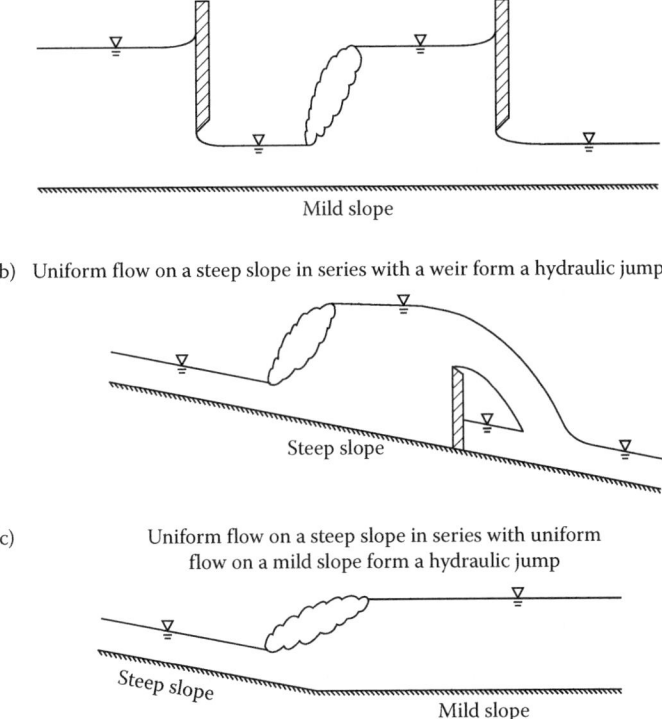

(a)   Two sluice gates in series form a hydraulic jump

Mild slope

(b)   Uniform flow on a steep slope in series with a weir form a hydraulic jump

Steep slope

(c)   Uniform flow on a steep slope in series with uniform flow on a mild slope form a hydraulic jump

Steep slope

Mild slope

FIGURE 9.17
The use of controls in series in the formation of a hydraulic jump.

transition from supercritical flow to subcritical flow, it must go through critical flow. As such, in order for this to occur, a natural phenomena called a hydraulic jump, which is a natural energy dissipator, occurs. The flow at the jump is critical and at a minimum specific energy, where a maximum dissipation of energy (major head loss) occurs due to the intense agitation and turbulence in the flow at the jump (and not due to flow resistance).

9.10.1.2 Definition of the Momentum Function for Rectangular Channel Sections

In order to define the momentum function, one begins with the application of the momentum equation using the integral approach (between two points in a finite control volume), which is appropriate for any channel section or flow transition that affects the channel over a short channel section and thus assumes that the flow resistance (viscous force, F_V) is negligible. Furthermore, assuming a short channel section as illustrated in Figure 9.18 also allows one to also ignore the effects of the gravitational force, F_G component. Thus, beginning with integral form of the momentum equation as follows:

$$\sum F_x = (F_G + F_P + F_V + F_{other})_x = (\rho Q v_x)_2 - (\rho Q v_x)_1 \tag{9.113}$$

and applying the assumptions regarding F_V and F_G:

$$\sum F_x = (F_P + F_{other})_x = (\rho Q v_x)_2 - (\rho Q v_x)_1 \tag{9.114}$$

where the x-axis is defined along the channel bottom, F_P is resultant hydrostatic thrust (pressure) force acting on the control volume between points 1 and 2 (specifically, F_{H1} = hydrostatic thrust force acting on control face 1, and F_{H2} = hydrostatic thrust force acting on control face 2), and F_{other} is specifically the reaction force, R_x acting on the control volume due to the thrust of the flow on an open channel flow structure (such as a sluice gate, weir, overflow spillway, venturi flume, contracted opening, or a sill that is sometimes used in assisting the formation of the hydraulic jump), and $v_x = v$; thus,

$$F_{H1} - F_{H2} - R_x = (\rho Q v)_2 - (\rho Q v)_1 \tag{9.115}$$

where the left-hand side of the momentum equation is the sum of forces acting on the control volume, and the right-hand side is the rate of change of momentum. Evaluating the hydrostatic forces:

$$\gamma h_{ca1} A_1 - \gamma h_{ca2} A_2 - R_x = (\rho Q v)_2 - (\rho Q v)_1 \tag{9.116}$$

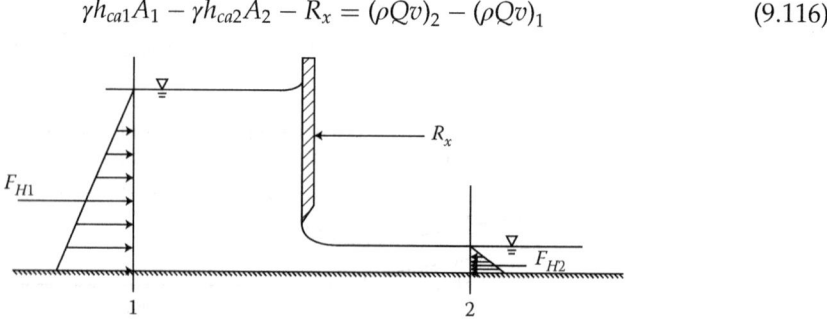

FIGURE 9.18
Application of the integral momentum equation for a finite control volume.

Assuming a rectangular channel section of width, b and thus defining each term per unit width of channel as follows:

$$\frac{\gamma h_{ca1} A_1}{b} - \frac{\gamma h_{ca2} A_2}{b} - \frac{R_x}{b} = \frac{(\rho Q v)_2}{b} - \frac{(\rho Q v)_1}{b} \tag{9.117}$$

Noting that $v = Q/b = vA/b = vby/b = q/y$ yields:

$$\frac{\gamma y_1 b y_1}{2b} - \frac{\gamma y_2 b y_2}{2b} - \frac{R_x}{b} = \frac{\rho v_2^2 b y_2}{b} - \frac{\rho v_1^2 b y_1}{b} \tag{9.118}$$

$$\frac{\gamma y_1^2}{2} - \frac{\gamma y_2^2}{2} - \frac{R_x}{b} = \frac{\rho q^2}{y_2} - \frac{\rho q^2}{y_1} \tag{9.119}$$

Rearranging the momentum equation and grouping the terms for each control face (1 and 2) yields:

$$\frac{R_x}{b} = \left(\frac{\rho q^2}{y_1} + \frac{\gamma y_1^2}{2} \right) - \left(\frac{\rho q^2}{y_2} + \frac{\gamma y_2^2}{2} \right) \tag{9.120}$$

Finally, dividing each term by the specific weight of the fluid, γ yields:

$$\frac{R_x}{\gamma b} = \left(\frac{\rho q^2}{\gamma y_1} + \frac{\gamma y_1^2}{\gamma 2} \right) - \left(\frac{\rho q^2}{\gamma y_2} + \frac{\gamma y_2^2}{\gamma 2} \right) \tag{9.121}$$

$$\frac{R_x}{\gamma b} = \underbrace{\left(\frac{q^2}{g y_1} + \frac{y_1^2}{2} \right)}_{M_1} - \underbrace{\left(\frac{q^2}{g y_2} + \frac{y_2^2}{2} \right)}_{M_2} \tag{9.122}$$

where the first term in the momentum function, M is defined as the of the rate of change of momentum divided by γ and the second term in the momentum function, M is defined as the hydrostatic thrust force divided by the γ as follows:

$$M = \left(\frac{q^2}{g y} + \frac{y^2}{2} \right) \tag{9.123}$$

Therefore, Equation 9.122 (the integral form of the momentum equation) may be used to solve for the unknown reaction force, R_x acting on the control volume due to the thrust of the flow on an open channel flow structure, such as a sluice gate, weir, overflow spillway, venturi flume, contracted opening (or sill sometimes used in the formation of the hydraulic jump), as presented in a section below; the unknown depth of flow (either upstream or downstream) is determined from the Bernoulli equation, also presented in a section below. Furthermore, Equation 9.122 (the integral form of the momentum equation) is used to solve for the unknown channel depth of flow, either upstream or downstream of a simple hydraulic jump ($R_x = 0$) (hydraulic jump equations are derived in a section below). Then, the integral form of the energy equation is used to solve for the unknown major head loss due to the flow turbulence in the hydraulic jump (see section below on hydraulic jumps).

9.10.1.3 The Momentum Function Curve for Rectangular Channel Sections

Similar to the specific energy curve, E versus y, it is of interest to examine how the momentum function, M varies with the depth of flow, y for a given constant value of unit discharge, q. Thus, one may define the momentum function curve, M versus y for a given constant value of q as follows. The momentum function equation is rearranged in order to isolate the constant term as follows:

$$\left(M - \frac{y^2}{2}\right)y = \frac{q^2}{g} = constant \tag{9.124}$$

which has only one asymptote of $y = 0$. As illustrated in Figure 9.19, the M-y plot is presented for a given value of q. Focusing only on the two positive solutions for the cubic momentum function equation, they are illustrated in the first quadrant. Similar to the E-y plot, for a given value of q, the M-y plot has an upper limb where the flow is subcritical, a lower limb where the flow is supercritical flow, and a crest where the momentum function, M is a minimum at the critical depth of flow. This may be proven analytically (as done for the specific energy, E) by differentiating the momentum function, M given by Equation 9.123 with respect to y and setting it equal to zero as follows:

$$M = \left(\frac{q^2}{gy} + \frac{y^2}{2}\right) \tag{9.125}$$

$$\frac{dM}{dy} = -\frac{q^2}{gy^2} + \frac{2y}{2} = 0 \tag{9.126}$$

$$\frac{q^2}{g} = y^3 \tag{9.127}$$

$$y = \sqrt[3]{\frac{q^2}{g}} \tag{9.128}$$

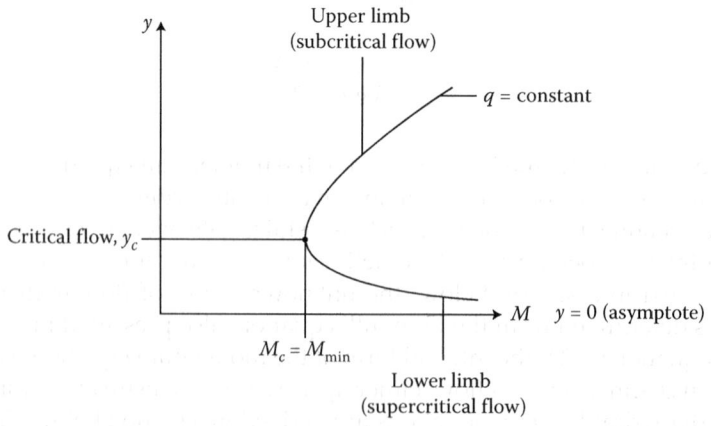

FIGURE 9.19
Momentum function curve, M-y plot for a rectangular channel section.

which is identical to the expression for the critical depth of flow, y_c derived in Equation 9.40 for minimum specific energy. Thus, the minimum momentum function, M_c is defined as follows:

$$M_c = \left(\frac{q^2}{gy_c} + \frac{y_c^2}{2}\right) \tag{9.129}$$

In particular, Figure 9.20 illustrates the M-y plot and the corresponding E-y plot for a sluice gate that is followed by a hydraulic jump. The total energy equation for the sluice gate between points 1 and 2 is given as follows:

$$\underbrace{y_1 + \frac{v_1^2}{2g} + z_1}_{E_1} - h_f = \underbrace{y_2 + \frac{v_2^2}{2g} + z_2}_{E_2} \tag{9.130}$$

Because $z_1 = z_2$, and the head loss due to flow resistance, h_f acting within the control volume is negligible, the specific energy equation for the sluice gate between points 1 and 2 is given as follows:

$$E_1 = E_2 \tag{9.131}$$

thus, y_1 (subcritical) and y_2 (supercritical) are alternate depths of flow as illustrated on the E-y plot. The momentum equation for the sluice gate between points 1 and 2 is given as follows:

$$\frac{R_x}{\gamma b} = \underbrace{\left(\frac{q^2}{gy_1} + \frac{y_1^2}{2}\right)}_{M_1} - \underbrace{\left(\frac{q^2}{gy_2} + \frac{y_2^2}{2}\right)}_{M_2} \tag{9.132}$$

Because the reaction force, R_x acting on the control volume between points 1 and 2 due to the thrust of flow on the sluice gate is not zero, the momentum function, M decreases

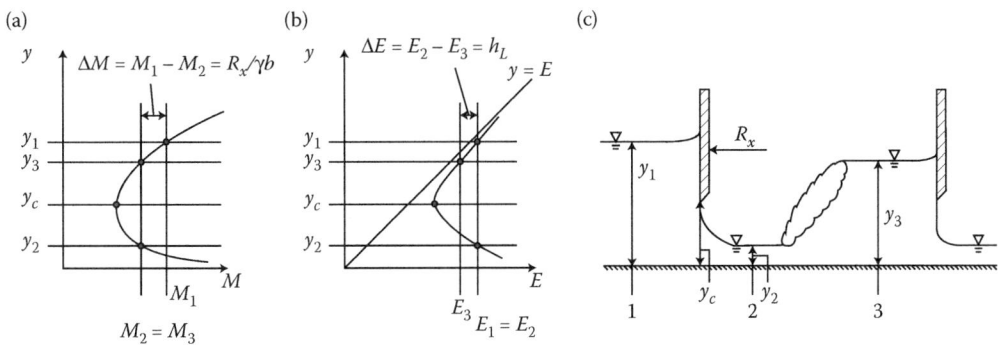

FIGURE 9.20
The M-y plot and the E-y plot for a sluice gate followed by a hydraulic jump. (a) M-y plot, (b) E-y plot, (c) Sluice gate followed by a hydraulic jump. (Adapted from Henderson, F. M., *Open Channel Flow*, Macmillan, New York, 1966, p. 71.)

between points 1 and 2 is given as follows:

$$M_1 \neq M_2 \tag{9.133}$$

where the decrease in the momentum function, ΔM is given as follows:

$$\Delta M = \frac{R_x}{\gamma b} \tag{9.134}$$

as illustrated on the M-y plot. The total energy equation for the hydraulic jump between points 2 and 3 is given as follows:

$$\underbrace{y_2 + \frac{v_2^2}{2g} + z_2}_{E_2} - h_L = \underbrace{y_3 + \frac{v_3^2}{2g} + z_3}_{E_3} \tag{9.135}$$

Because $z_2 = z_3$, the specific energy equation for the hydraulic between points 2 and 3 is given as follows:

$$E_2 - h_L = E_3 \tag{9.136}$$

Because the head loss, h_L acting within the control volume between points 2 and 3 due to the flow turbulence in the hydraulic jump is not zero, the specific energy, E decreases between points 2 and 3 and is given as follows:

$$E_2 \neq E_3 \tag{9.137}$$

where the decrease in the specific energy, ΔE is given as follows:

$$\Delta E = h_L \tag{9.138}$$

thus, y_2 (supercritical) and y_3 (subcritical) are not alternate depths of flow as illustrated on the E-y plot. The momentum equation for the hydraulic jump between points 2 and 3 is given as follows:

$$\frac{R_x}{\gamma b} = \underbrace{\left(\frac{q^2}{g y_2} + \frac{y_2^2}{2} \right)}_{M_2} - \underbrace{\left(\frac{q^2}{g y_3} + \frac{y_3^2}{2} \right)}_{M_3} \tag{9.139}$$

Because, for a simple hydraulic jump, there is no reaction force, R_x acting on the control volume between points 2 and 3, the momentum function, M equation for the hydraulic jump between points 2 and 3 is given as follows:

$$M_2 = M_3 \tag{9.140}$$

One may note that y_2 (supercritical) and y_3 (subcritical) are known as conjugate depths of flow and are illustrated on the M-y plot. Conjugate depths of flow represent possible combinations of depths that could occur before and after the hydraulic jump. Furthermore, one may note (see Figure 9.20a) that for a given unit discharge, q, if the upstream depth, y_2 (supercritical) is increased, then its conjugate downstream depth, y_3 (subcritical) will decrease, and vice versa.

It is important to note that conjugate depths of flow also occur in the case of an assisted hydraulic jump and may also occur in the case of the sluice gate (and the weir, overflow spillway, venturi flume, and contracted openings). In the particular case of a simple hydraulic jump (illustrated in Figure 9.20), because there was no sill assisting the formation of the jump, $R_x = 0$ and thus $M_2 = M_3$, where y_2 (supercritical) and y_3 (subcritical) are conjugate depths of flow. However, in the case of an assisted hydraulic jump, the reaction force, R_x acting on the control volume between points 2 and 3 due to the thrust of flow on the sill is not zero (see Figure 9.21). The momentum function, M decreases between points 2 and 3 and thus $M_2 \neq M_3$, where there is a decrease in the momentum function, $\Delta M = R_x/\gamma b$ and where still, however, y_2 (supercritical) and y_3 (subcritical) are conjugate depths of flow. Furthermore, similar to the case of a simple hydraulic jump, in the case of an assisted hydraulic jump, for a given unit discharge, q if the upstream depth, y_2 (supercritical) is increased, then its conjugate downstream depth, y_3 (subcritical) will decrease, and vice versa. Thus, one may conclude that a hydraulic jump will be produced only if there is a certain relationship between the upstream depth, y_2 (supercritical) and its conjugate downstream depth, y_3 (subcritical). Furthermore, one may conclude that in the case of the sluice gate (illustrated in Figure 9.20c) (and the weir, overflow spillway, venturi flume, and contracted opening), similar to the assisted hydraulic jump, $M_1 \neq M_2$, where the alternate depths of flow, y_1 (subcritical) and y_2 (supercritical) may also qualify as conjugate depths of flow, as they represent possible combinations of depths that could occur before and after the hydraulic jump. And, analogous to the simple hydraulic jump and the assisted hydraulic jump, in the case of the sluice gate (and the weir, overflow spillway, venturi flume, and contracted opening), for a given unit discharge, q, if the upstream depth, y_1 (subcritical) is increased, then its conjugate downstream depth, y_2 (supercritical) will decrease, and vice versa. Finally, one may note (as explained above), that the formation of a hydraulic jump

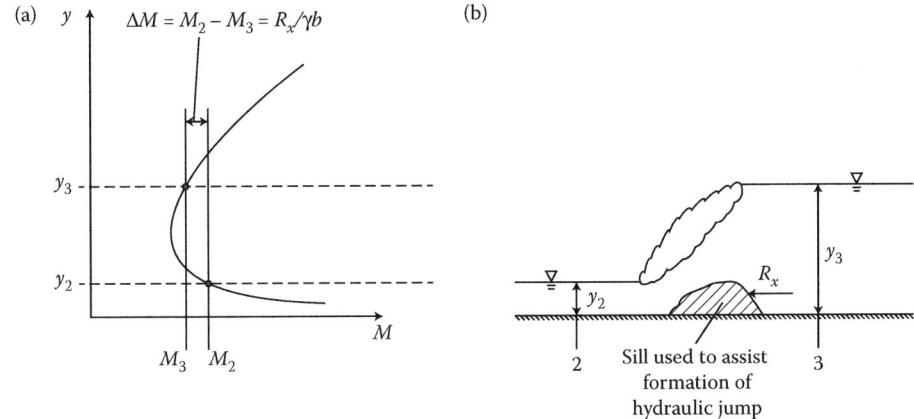

FIGURE 9.21
The M-y plot for an assisted hydraulic jump. (a) M-y plot, (b) Assisted hydraulic jump.

may result from the insertion of two controls, such as a sluice gate (or a weir, overflow spillway, venturi flume, or contracted opening), in series in an open channel.

9.10.1.4 Definition of the Momentum Function for Nonrectangular Channel Sections

The basic difference between the definitions of the momentum function for a rectangular channel section versus a nonrectangular channel section is that the assumption of a nonrectangular channel section does not allow the definition of a unit discharge, q and thus the definition of a momentum function curve, M versus y for a given q. Therefore, similar to the definition of the specific energy, E in the definition of the momentum function, M, the nonrectangular channel section must be treated as a whole because a unit width, b does not represent the entire channel cross section. However, regardless of the shape of the channel section, the first term in the momentum function, M is defined as the of the rate of change of momentum divided by γ, and the second term in the momentum function, M is defined as the hydrostatic thrust force divided by the γ. Thus, the momentum function, M for a nonrectangular channel section is given as follows:

$$M = \left(\frac{\rho Q v}{\gamma} + \frac{\gamma h_{ca} A}{\gamma} \right) \tag{9.141}$$

$$M = \left(\frac{Q^2}{gA} + h_{ca} A \right) \tag{9.142}$$

where the resulting second term, $h_{ca}A$ is the moment of area of the nonrectangular channel section about the fluid surface, which may, in general, be determined using numerical integration. However, for the assumption of a typical geometric trapezoidal channel section, the moment of area may be determined by adding the components that represent the two triangular components (A_1 and A_3) and the rectangular component (A_2) as illustrated in Figure 9.22 as follows:

$$h_{ca}A = h_{ca1}A_1 + h_{ca2}A_2 + h_{ca3}A_3 \tag{9.143}$$

$$h_{ca}A = \frac{y}{3}\frac{ky^2}{2} + \frac{y}{2}by + \frac{y}{3}\frac{ky^2}{2} \tag{9.144}$$

$$h_{ca}A = \frac{y}{3}ky^2 + \frac{y}{2}by \tag{9.145}$$

$$h_{ca}A = \frac{y^2}{6}(2ky + 3b) \tag{9.146}$$

which is the moment of area for the whole channel section, and where $k = 1/\tan\theta$ (θ is expressed in degrees); thus,

$$h_{ca}A = \frac{y^2}{6}\left(\frac{2y}{\tan\theta} + 3b \right) \tag{9.147}$$

One may note that because it is of interest to determine the moment of area of the entire cross section, $h_{ca}A$, it is unnecessary to determine h_{ca} separately from A. Finally, one may

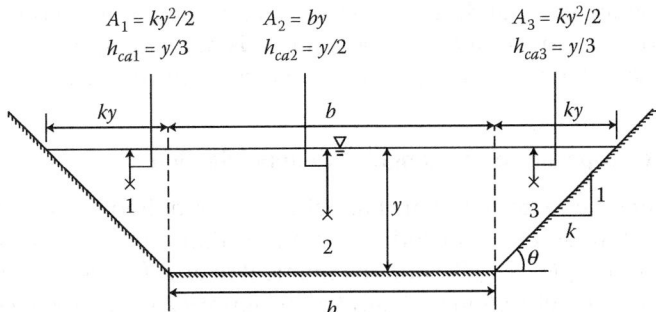

FIGURE 9.22
Computation of the moment of area, $h_{ca}A$ for a trapezoidal channel section.

note that similar to the specific energy, E, the momentum function, M is a minimum at critical flow as follows:

$$\frac{A_c}{B_c} = \frac{Q^2}{A_c^2}\frac{1}{g} \tag{9.148}$$

$$v_c = \sqrt{g\frac{A_c}{B_c}} \tag{9.149}$$

This may be proven analytically (as done for the specific energy, E) by differentiating the momentum function, M given by Equation 9.142 with respect to y and setting it equal to zero, assuming that $d(h_{ca}A)/dy = A$.

9.11 Geometric Properties of Some Common Channel Sections

Although the simplest channel section in artificial channels is the rectangular section, the existence of irregular channel sections in natural channels, the use of trapezoidal channel sections in artificial canals, and the use of a partially filled circular channel section or a triangular channel section in artificial drainage systems requires the need to define the geometric properties of nonrectangular sections as well. Geometric properties of interest include the cross-sectional area, A; the wetted perimeter, P_w; the hydraulic radius, $R_h = A/P_w$; and the moment of area for the whole channel section about the fluid surface, $h_{ca}A$. One may note that while numerical integration may be used to determine the geometric properties of irregular channel sections, the definition of the geometric properties for some common geometric sections (rectangular, trapezoidal, triangular, and partially filled circular channel sections) are presented as follows. One may note that for a nonrectangular channel section, the moment of area for the whole channel section, $h_{ca}A$ is required in the determination of (1) the reaction force, R_x acting on the control volume, which may include a typical flow-measuring device and thus an open channel flow structure, and (2) the unknown depth of flow, either upstream or downstream of a hydraulic jump (see the definition of the momentum function, M above). As such, because a partially filled circular channel section is typically used in an artificial drainage system as opposed to in an open channel with a flow-measuring device, and/or a hydraulic jump, it is not necessary to define the moment of area for the whole channel section, $h_{ca}A$ for a partially filled circular channel section.

9.11.1 Geometric Properties of Rectangular Channel Sections

The geometric properties of rectangular channel sections required in the computation of typical open channel flow problems include the cross-sectional area, A; the wetted perimeter, P_w; the hydraulic radius, $R_h = A/P_w$; and the moment of area for the whole channel section, $h_{ca}A$. A rectangular channel section is illustrated in Figure 9.23. The cross-sectional area is by; the wetted perimeter, $P_w = 2y + b$; and the hydraulic radius, $R_h = A/P_w = by/(2y + b)$. Furthermore, the moment of area for the whole channel section is $h_{ca}A = by^2/2$.

9.11.2 Geometric Properties of Trapezoidal Channel Sections

The geometric properties of trapezoidal channel sections required in the computation of typical open channel flow problems include the cross-sectional area, A; the wetted perimeter, P_w; the hydraulic radius, $R_h = A/P_w$; and the moment of area for the whole channel section, $h_{ca}A$. A trapezoidal channel section is illustrated in Figure 9.24. The cross-sectional area is given as follows:

$$A = A_1 + A_2 + A_3 \tag{9.150}$$

$$A = \frac{ky^2}{2} + by + \frac{ky^2}{2} \tag{9.151}$$

$$A = y(ky + b) \tag{9.152}$$

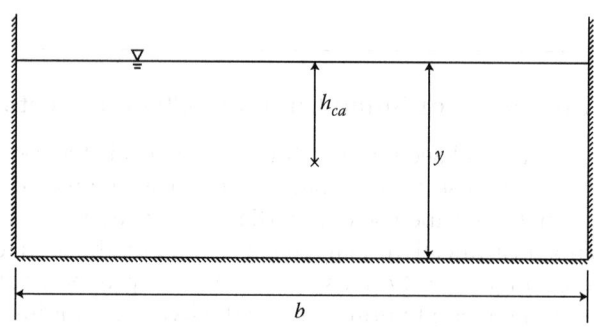

Cross-sectional area, $A = by$

Wetted perimeter, $P_w = 2y + b$

Hydraulic radius, $R_h = \dfrac{A}{P_w} = \dfrac{by}{2y + b}$

Height to the center of area, $h_{ca} = \dfrac{y}{2}$

Moment of area, $h_{ca}A = \dfrac{by^2}{2}$

FIGURE 9.23
Geometric properties of a rectangular channel section.

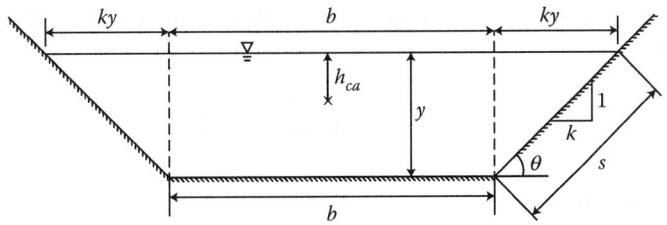

Cross-sectional area, $A = y\left(\dfrac{y}{\tan\theta} + b\right)$

Wetted perimeter, $P_w = \dfrac{2y}{\sin\theta} + b$

Hydraulic radius, $R_h = \dfrac{A}{P_w} = \dfrac{y\left(\dfrac{y}{\tan\theta} + b\right)}{\left(\dfrac{2y}{\sin\theta} + b\right)}$

Moment of area, $h_{ca}A = \dfrac{y^2}{6}\left(\dfrac{2y}{\tan\theta} + 3b\right)$

FIGURE 9.24
Geometric properties of a trapezoidal channel section.

where $k = 1/\tan\theta$ (θ is expressed in degrees); thus,

$$A = y\left(\frac{y}{\tan\theta} + b\right) \tag{9.153}$$

The wetted perimeter, P_w is given as follows:

$$P_w = 2s + b \tag{9.154}$$

where $s = y/\sin\theta$ (θ is expressed in degrees); thus,

$$P_w = \frac{2y}{\sin\theta} + b \tag{9.155}$$

The hydraulic radius, $R_h = A/P_w$ is given as follows:

$$R_h = \frac{A}{P_w} = \frac{y\left(\dfrac{y}{\tan\theta} + b\right)}{\left(\dfrac{2y}{\sin\theta} + b\right)} \tag{9.156}$$

The moment of area for the whole channel section, $h_{ca}A$ was determined above and is repeated as follows:

$$h_{ca}A = \frac{y^2}{6}\left(\frac{2y}{\tan\theta} + 3b\right) \tag{9.157}$$

where θ is expressed in degrees.

9.11.3 Geometric Properties of Triangular Channel Sections

The geometric properties of triangular channel sections required in the computation of typical open channel flow problems include the cross-sectional area, A; the wetted perimeter, P_w; the hydraulic radius, $R_h = A/P_w$; and the moment of area for the whole channel section, $h_{ca}A$. A triangular channel section is illustrated in Figure 9.25. The cross-sectional area is given as follows:

$$A = \frac{2y^2}{2} = y^2 \tag{9.158}$$

The wetted perimeter, P_w is given as follows:

$$P_w = 2y\sqrt{2} \tag{9.159}$$

The hydraulic radius, $R_h = /P_w$ is given as follows:

$$R_h = \frac{A}{P_w} = \frac{y^2}{2y\sqrt{2}} = \frac{y}{2\sqrt{2}} \tag{9.160}$$

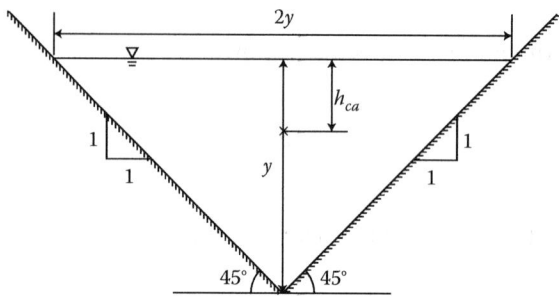

Cross-sectional area, $A = \dfrac{2y^2}{2} = y^2$

Wetted perimeter, $P_w = 2y\sqrt{2}$

Hydraulic radius, $R_h = \dfrac{A}{P_w} = \dfrac{y^2}{2y\sqrt{2}} = \dfrac{y}{2\sqrt{2}}$

Height to the center of area, $h_{ca} = \dfrac{y}{3}$

Moment of area, $h_{ca}A = \dfrac{y}{3}y^2 = \dfrac{y^3}{3}$

FIGURE 9.25
Geometric properties of a triangular channel section.

The moment of area for the whole channel section, $h_{ca}A$ is given as follows:

$$h_{ca}A = \frac{y}{3}y^2 = \frac{y^3}{3} \tag{9.161}$$

9.11.4 Geometric Properties of Partially Filled Circular Channel Sections

The geometric properties of partially filled circular channel sections required in the computation of typical open channel flow problems include the cross-sectional area, A; the wetted perimeter, P_w; and the hydraulic radius, $R_h = A/P_w$. A partially filled circular channel section is illustrated in Figure 9.26. The cross-sectional area is given as follows:

$$A = A_1 + A_2 + A_3 \tag{9.162}$$

$$A_1 = \frac{1}{2}(2ad) = ad = (r_o \sin \alpha)(r_o \cos \alpha) = (r_o \sin \theta)(-r_o \cos \theta) = -r_o^2 \sin \theta \cos \theta \tag{9.163}$$

$$A = -r_o^2 \sin \theta \cos \theta + \frac{\theta r_o^2}{2} + \frac{\theta r_o^2}{2} \tag{9.164}$$

$$A = r_o^2 (\theta - \sin \theta \cos \theta) \tag{9.165}$$

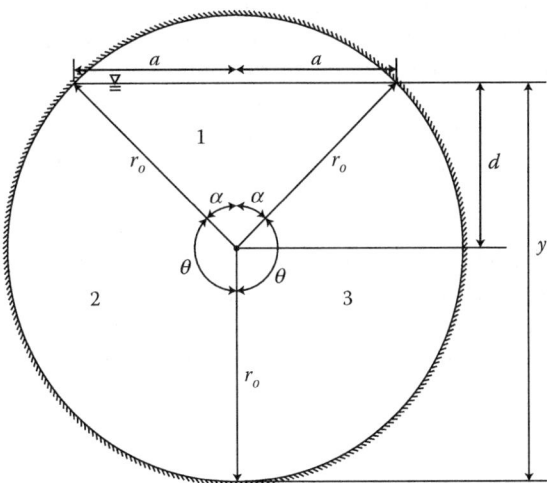

Cross-sectional area, $A = A_1 + A_2 + A_3 = r_o^2 (\theta - \sin \theta \cos \theta) = (y + r_o \cos \theta)^2 (\theta - \sin \theta \cos \theta)$

Wetted perimeter, $P_w = 2\theta r_o$

Hydraulic radius, $R_h = \dfrac{A}{P_w} = \dfrac{(y + r_o \cos \theta)^2 (\theta - \sin \theta \cos \theta)}{2\theta r_o}$

FIGURE 9.26
Geometric properties of a partially filled circular channel section (θ in radians).

However, it is of interest to express the cross-sectional area in terms of the depth of flow, $y = r_o + d$ as follows:

$$A = (y - d)^2 (\theta - \sin\theta \cos\theta) \tag{9.166}$$

$$A = (y + r_o \cos\theta)^2 (\theta - \sin\theta \cos\theta) \tag{9.167}$$

where θ is expressed in radians. The wetted perimeter, P_w is given as follows:

$$P_w = 2\theta r_o \tag{9.168}$$

where θ is expressed in radians. The hydraulic radius, $R_h = A/P_w$ is given as follows:

$$R_h = \frac{A}{P_w} = \frac{(y + r_o \cos\theta)^2 (\theta - \sin\theta \cos\theta)}{2\theta r_o} \tag{9.169}$$

where θ is expressed in radians.

9.12 Flow Depth and Reaction Force for Short Channel Transitions in Open Channel Flow: Ideal Flow

Because a gradual (vertical or horizontal) channel contraction/transition (not a "control"), an abrupt vertical or horizontal contraction (abrupt step upward or abrupt decrease in channel width) (a "control"), and a typical flow-measuring device ("controls") (such as sluice gates, weirs, overflow spillways, venturi flumes, and contracted openings) affect the flow over a short channel section, the major head loss due to flow resistance assumed to be negligible. As such, the Bernoulli equation (Eulerian/integral approach) is applied to determine either the upstream or downstream depth for the above listed channel transitions. One may note that the broad-crested weir is a special case of an abrupt vertical contraction, and the venturi flume a special case of an abrupt horizontal contraction. Furthermore, while there will be an unknown reaction force, R_x acting on the control volume due to the thrust of the flow on an open channel structure (a sluice gate, weir, overflow spillway, venturi flume, or a contracted opening, or any abrupt vertical or horizontal contraction), there is no corresponding reaction force, R_x acting on a gradual vertical or horizontal transition. As such, the integral form of the momentum equation is applied to solve for the unknown reaction force, R_x acting on the control volume that contains the open channel structure.

Because critical flow, which is defined by an analytical relationship between the flow depth and the discharge in the channel, occurs at the "controls" listed above, each may serve as a flow-measuring device; there is a minor head loss associated with the flow measurement. Furthermore, a gradual vertical or horizontal channel transition does not result in a "control," and thus, does not serve as a flow-measuring device.

9.12.1 Flow Depth for Gradual Channel Transitions: Not Controls

The Bernoulli equation (Eulerian/integral approach) is applied to determine either the upstream or downstream depth for a gradual (vertical or horizontal) channel transition and is given as follows:

$$\underbrace{\left(\underbrace{y_1 + \frac{v_1^2}{2g}}_{E_1} + z_1\right)}_{H_1} = \underbrace{\left(\underbrace{y_2 + \frac{v_2^2}{2g}}_{E_2} + z_2\right)}_{H_2} \tag{9.170}$$

Furthermore, as illustrated in a section below, the Bernoulli equation may also be applied between any two points in order to determine the unknown depth of flow at a gradual horizontal channel transition.

9.12.1.1 Flow Depth for a Gradual Upward Step

A rectangular channel section with a constant width, b and thus a constant unit discharge, q with a gradual vertical channel transition such as a gradual upward step, Δz along with the corresponding E-y plot is illustrated in Figure 9.27. Given that the channel bottom slope is classified as mild, the upstream uniform depth of flow, y_1 is subcritical. Due to a gradual upward step, the flow depth will decrease over the gradual upward step. One may note that because the gradual upward step does not result in a control, the flow does not become critical at the transition, and thus the downstream depth of flow does not become supercritical. Rather, the downstream depth of flow, y_2 is also subcritical. The unknown depth of flow (either y_1 or y_2) may be determined from the Bernoulli equation as follows:

$$y_1 + \frac{v_1^2}{2g} + z_1 = y_2 + \frac{v_2^2}{2g} + z_2 \tag{9.171}$$

Taking the channel bottom at the upstream point 1 as the datum, $z_1 = 0$, and thus, $z_2 = \Delta z$ yields the following:

$$y_1 + \frac{v_1^2}{2g} = y_2 + \frac{v_2^2}{2g} + \Delta z \tag{9.172}$$

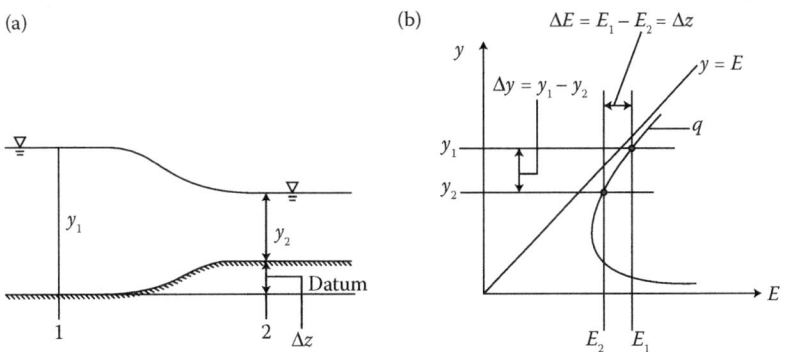

FIGURE 9.27
Decrease in flow depth due to gradual step upward for subcritical flow. (a) Elevation view, (b) E-y plot.

$$E_1 = E_2 + \Delta z \tag{9.173}$$

$$E_1 - E_2 = \Delta E = \Delta z \tag{9.174}$$

where the decrease in the specific energy, ΔE is a result of an increase in elevation, Δz. For instance, in the case where the upstream depth, y_1 and the upstream velocity, v_1 are known, the downstream depth, y_2 and the downstream velocity, v_2 are solved as follows:

$$E_1 = y_2 + \frac{v_2^2}{2g} + \Delta z \tag{9.175}$$

$$E_1 - \Delta z = y_2 + \frac{v_2^2}{2g} \tag{9.176}$$

$$E_1 - \Delta z = y_2 + \frac{q^2}{2gy_2^2} \tag{9.177}$$

Thus, for a given unit discharge, $q = v_1\, y_1 = v_2\, y_2$, the specific energy equation, E_2 is cubic in y_2. As such, there are two positive solutions and one negative solution for y_2, as illustrated by drawing a vertical line corresponding to E_2 in Figure 9.27b. While the negative solution is of no practical interest, the two positive solutions are. The larger of the two positive solutions is plotted on the upper subcritical limb, while the smaller of the two positive roots is plotted on the lower subcritical limb. However, because it was already established that the downstream depth, y_2 is subcritical, the desired solution to the above cubic specific energy equation, E_2 is the larger of the two positive roots. Using Mathcad to numerically solve for the unknown downstream subcritical depth, y_2, the critical depth of flow, y_c for the given unit discharge, q may be used as a lower limit to guide the guess value for y_2 as illustrated in the example below.

EXAMPLE PROBLEM 9.1

Water flows in a rectangular channel at a velocity of 1 m/sec and depth of 1.65 m. There is a gradual upward step, Δz of 0.15 m in the channel bed as illustrated in Figure EP 9.1. (a) Determine the flow regime (critical, subcritical, or supercritical) of the upstream depth, y_1. (b) Determine the flow regime of the downstream depth, y_2. (c) Determine the depth, y_2 and the velocity, v_2 of the water downstream of the step. (d) Determine E_1, E_2, and plot the E-y curve for the unit discharge, $q = v_1\, y_1 = v_2\, y_2$.

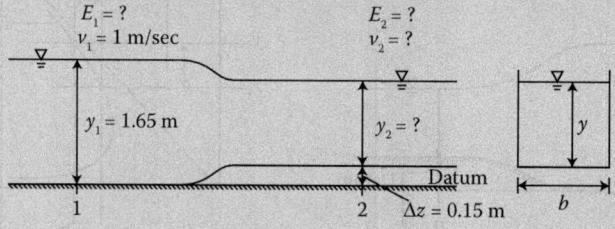

FIGURE EP 9.1
Water flows in a rectangular channel with gradual upward step, Δz in the channel bed.

Mathcad Solution

(a) In order to determine the flow regime of the upstream depth, y_1, one must first determine the critical depth of flow, y_c for the given unit discharge, q as follows:

$$v_1 := 1 \, \frac{m}{sec} \qquad y_1 := 1.65 \, m \qquad q := v_1 \cdot y_1 = 1.65 \, \frac{m^2}{sec}$$

$$g := 9.81 \, \frac{m}{sec^2} \qquad y_c := \sqrt[3]{\frac{q^2}{g}} = 0.652 \, m$$

Therefore, because $y_1 = 1.65 \, m$ is greater than $y_c = 0.652 \, m$, the depth of flow upstream of the step is subcritical.

(b) Because the gradual upward step does not result in a control, the flow does not become critical at the step (transition). Therefore, the flow regime of the downstream depth, y_2 is subcritical.

(c) The subcritical depth of the water downstream of the step, y_2 is computed by applying the Bernoulli equation as follows:

$$y_1 + \frac{v_1^2}{2g} + z_1 = y_2 + \frac{v_2^2}{2g} + z_2$$

Taking the channel bottom at the upstream point 1 as the datum, $z_1 = 0$, and thus, $z_2 = \Delta z$ yields the following:

$$y_1 + \frac{v_1^2}{2g} = y_2 + \frac{v_2^2}{2g} + \Delta z$$

Applying the unit discharge equation, $q = v_2 \, y_2$ yields the following:

$$y_1 + \frac{v_1^2}{2g} = y_2 + \frac{q^2}{2g y_2^2} + \Delta z$$

The subcritical downstream depth, y_2 is computed by assuming a guess value for y_2 that is greater than $y_c = 0.652 \, m$ as follows:

$$\Delta z := 0.15 \, m \qquad\qquad \text{Guess value:} \quad y_2 := 0.7 \, m$$

Given

$$y_1 + \frac{v_1^2}{2 \cdot g} = y_2 + \frac{q^2}{2 \cdot g \cdot y_2^2} + \Delta z$$

$$y_2 := \text{Find}(y_2) = 1.488 \, m$$

The downstream velocity, v_2 is computed by applying the unit discharge equation, $q = v_2\, y_2$ as follows:

$$v_2 := \frac{q}{y_2} = 1.109 \;\frac{m}{sec}$$

(d) The values for E_1 and E_2 are determined from the specific energy equation as follows:

$$E_1 := y_1 + \frac{v_1^2}{2 \cdot g} = 1.701 \; m \qquad\qquad E_2 := y_2 + \frac{v_2^2}{2 \cdot g} = 1.551 \; m$$

where:

$$E_1 := E_2 + \Delta z = 1.701 \; m$$

The E-y curve for the unit discharge, $q = v_1\, y_1 = v_2\, y_2 = 1.65 \; m^2/sec$ is plotted as follows:

$$E(y) := y + \frac{q_2}{2 \cdot g \cdot y^2} \qquad a \equiv 0.1 \; m \quad b \equiv 5 \; m \quad h \equiv 0.1 \; m \quad y := a,\, a + h..b$$

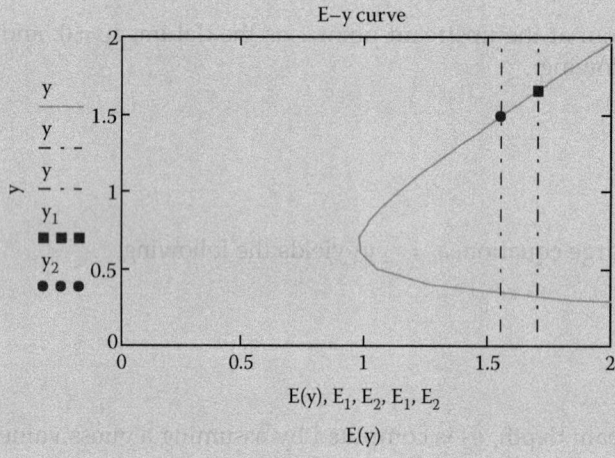

E−y curve

$E(y), E_1, E_2, E_1, E_2$

$E(y)$

9.12.1.2 Flow Depth for a Gradual Decrease in Channel Width

A rectangular channel section with a temporary gradual decrease in width, b and thus a gradual horizontal channel transition, along with the corresponding E-y plot, is illustrated in Figure 9.28. Given that the channel bottom slope is classified as mild, the upstream uniform depth of flow, y_1 is subcritical where the channel width, $b = b_1$ and thus $q = q_1$. Due to the gradual decrease in channel width at the channel transition, $b = b_t < b_1$, the unit discharge, $q = q_t > q_1$, will increase, and the flow depth, $y_t < y_1$ will decrease across the gradual decrease in channel width. The downstream depth of flow will resume to the upstream depth, $y_2 = y_1$ when the channel width resumes to the upstream channel

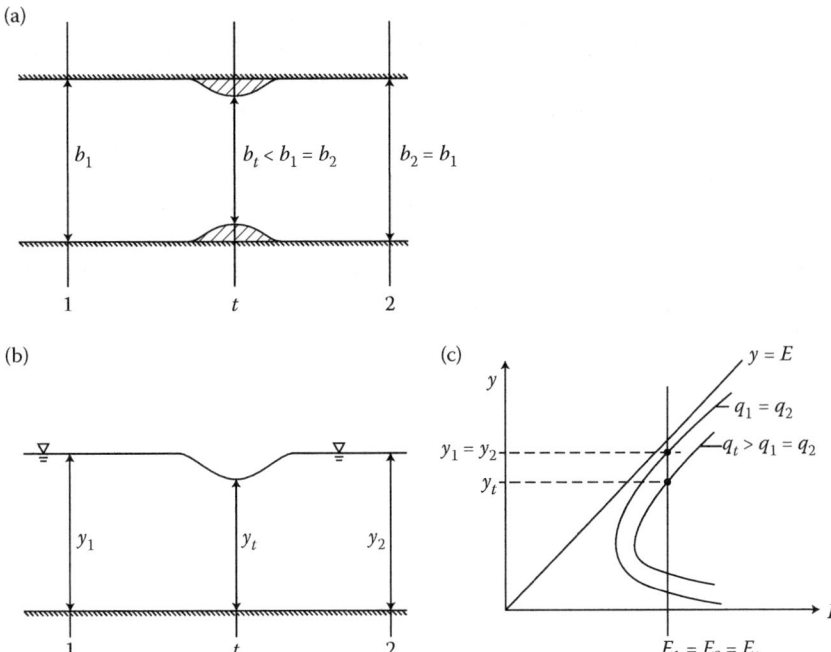

FIGURE 9.28
Temporary decrease in flow depth due to temporary gradual decrease in channel width for subcritical flow. (a) Plan view, (b) Elevation view, (c) E-y plot.

width, $b = b_2 = b_1$, and thus $q = q_2 = q_1$. One may note that because the gradual decrease in channel width does not result in a control, the flow at the transition remains subcritical; thus, the downstream depth of flow also remains subcritical. The unknown depth of flow (either y_1 or y_2) may be determined from the Bernoulli equation as follows:

$$y_1 + \frac{v_1^2}{2g} + z_1 = y_2 + \frac{v_2^2}{2g} + z_2 \tag{9.178}$$

Assuming that $z_1 = z_2$ yields the following:

$$y_1 + \frac{v_1^2}{2g} = y_2 + \frac{v_2^2}{2g} \tag{9.179}$$

$$E_1 = E_2 \tag{9.180}$$

For instance, in the case where the upstream depth, y_1 and the upstream velocity, v_1 are known, the downstream depth, y_2 and the downstream velocity, v_2 are solved as follows:

$$E_1 = y_2 + \frac{q_2^2}{2gy_2^2} \tag{9.181}$$

Thus, for a given unit discharge, $q_1 = v_1 y_1$, (where $q_1 = q_2 = v_2 y_2$) the specific energy equation, E_2 is cubic in y_2. As such, there are two positive solutions and one negative solution for

y_2, as illustrated by drawing a vertical line corresponding to $E_1 = E_2$ in Figure 9.28c. While the negative solution is of no practical interest, the two positive solutions are. The larger of the two positive solutions is plotted on the upper subcritical limb, while the smaller of the two positive solutions is plotted on the lower subcritical limb. However, because it was already established that the downstream depth, y_2 is subcritical, the desired solution to the above cubic specific energy equation, E_2 is the larger of the two positive roots, where $y_2 = y_1$. Using Mathcad to numerically solve for the unknown downstream subcritical depth, y_2, the critical depth of flow, y_c for the given unit discharge, q_2 may be used as a lower limit to guide the guess value for y_2 as illustrated in Example Problem 9.2 below.

Furthermore, at the gradual transition, the depth, y_t may be determined from the Bernoulli equation as follows:

$$y_1 + \frac{v_1^2}{2g} + z_1 = y_t + \frac{v_t^2}{2g} + z_t \tag{9.182}$$

Assuming that $z_1 = z_t$ yields the following:

$$y_1 + \frac{v_1^2}{2g} = y_t + \frac{v_t^2}{2g} \tag{9.183}$$

$$E_1 = E_t \tag{9.184}$$

Given that the upstream depth, y_1 and the upstream velocity, v_1 are known, the depth at the gradual transition, y_t and the unit discharge, q_t may solved as follows:

$$E_1 = y_t + \frac{q_t^2}{2g y_t^2} \tag{9.185}$$

where for a given constant discharge, $Q = b_t\, q_t = b_1 q_1$, and the specific energy equation, E_t is cubic in y_t. As such, there are two positive solutions and one negative solution for y_t, as illustrated by drawing a vertical line corresponding to $E_1 = E_t$ in Figure 9.28c. While the negative solution is of no practical interest, the two positive solutions are. The larger of the two positive solutions is plotted on the upper subcritical limb, while the smaller of the two positive solutions is plotted on the lower subcritical limb. However, because, it was already established that the depth at the transition, y_t is subcritical, the desired solution to the above cubic specific energy equation, E_t is the larger of the two positive roots. Using Mathcad to numerically solve for the unknown subcritical depth, y_t, the critical depth of flow, y_c for the given unit discharge, q_t may be used as a lower limit to guide the guess value for y_t, as illustrated in Example Problem 9.2 below.

EXAMPLE PROBLEM 9.2

Water flows in a rectangular channel of width 3 m at a velocity of 3 m/sec and depth of 3 m. There is a temporary gradual decrease in channel width to 2.55 m and thus a gradual horizontal channel transition, as illustrated in Figure EP 9.2. (a) Determine the flow regime of the upstream depth, y_1. (b) Determine the flow regime of the downstream depth, y_2. (c) Determine the depth, y_2 and the velocity, v_2 of the water downstream of the gradual transition. (d) Determine E_1 and E_2 and plot the E-y curve for the unit

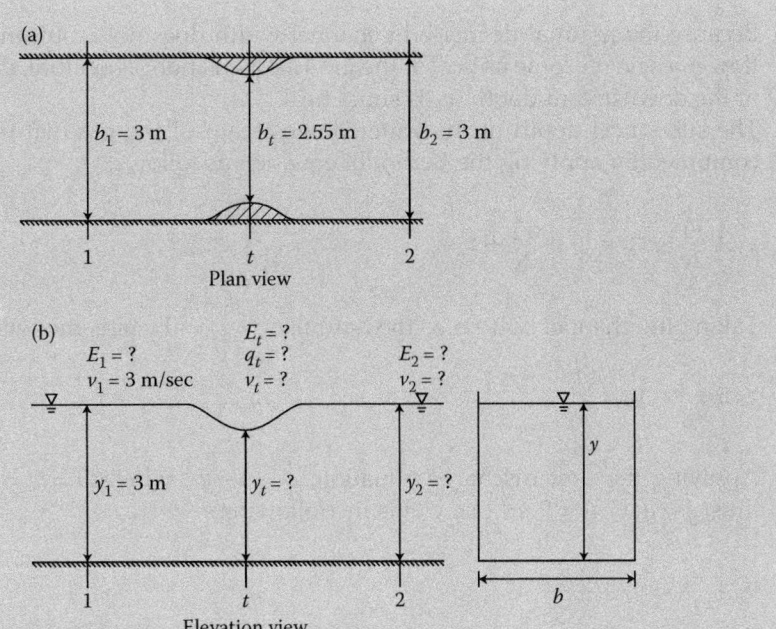

FIGURE EP 9.2
Water flows in a rectangular channel with temporary gradual decrease in channel width. (a) Plan view, (b) Elevation view.

discharge, $q = q_1 = q_2$, where $b = b_1 = b_2$. (e) Determine the depth, y_t; the velocity, v_t; and the unit discharge, q_t of the water at the gradual transition. (f) Determine E_t and plot the E-y curve for the unit discharge, $q = q_t > q_1 = q_2$, where $b = b_t < b_1 = b_2$.

Mathcad Solution

(a) In order to determine the flow regime of the upstream depth, y_1, one must first determine the critical depth of flow, y_{c1} for the upstream unit discharge, $q = q_1$ as follows:

$$v_1 := 3\ \frac{m}{sec} \qquad y_1 := 3\ m \qquad q_1 := v_1 \cdot y_1 = 9\ \frac{m^2}{sec}$$

$$g := 9.81\ \frac{m}{sec^2} \qquad y_{c1} := \sqrt[3]{\frac{q_1^2}{g}} = 2.021\ m$$

where:

$$b_1 := 3\ m \qquad Q := b_1 \cdot q_1 = 27\ \frac{m^3}{sec}$$

Therefore, because $y_1 = 3\ m$ is greater than $y_{c1} = 2.021\ m$, the depth of flow upstream of the gradual transition is subcritical.

(b) Because the gradual decrease in channel width does not result in a control, the flow does not become critical at the gradual transition. Therefore, the flow regime of the downstream depth, y_2 is subcritical.

(c) The subcritical depth of the water downstream of the gradual transition, y_2 is computed by applying the Bernoulli equation as follows:

$$y_1 + \frac{v_1^2}{2g} + z_1 = y_2 + \frac{v_2^2}{2g} + z_2$$

Taking the channel bottom as the datum, $z_1 = z_2 = 0$ yields the following:

$$y_1 + \frac{v_1^2}{2g} = y_2 + \frac{v_2^2}{2g}$$

Applying the unit discharge equation, $q_2 = v_2\, y_2$, where $b = b_1 = b_2 = 3$ m, and thus, $q = q_1 = q_2 = 9\ \mathrm{m^2/sec}$ yields the following:

$$y_1 + \frac{v_1^2}{2g} = y_2 + \frac{q_2^2}{2g y_2^2}$$

The subcritical downstream depth, y_2 is computed by assuming a guess value for y_2 that is greater than $y_{c1} = y_{c2} = 2.021$ m as follows:

$$q_2 := q_1 = 9\ \frac{\mathrm{m^2}}{\mathrm{sec}} \qquad\qquad \text{Guess value:}\quad y_2 := 2.2\ \mathrm{m}$$

Given

$$y_1 + \frac{v_1^2}{2 \cdot g} = y_2 + \frac{q_2^2}{2 \cdot g \cdot y_2^2}$$

$$y_2 := \mathrm{Find}(y_2) = 3\ \mathrm{m}$$

The downstream velocity, v_2 is computed by applying the unit discharge equation, $q_2 = v_2\, y_2$ as follows:

$$v_2 := \frac{q_2}{y_2} = 3\ \frac{\mathrm{m}}{\mathrm{sec}}$$

where:

$$b_2 := 3\ \mathrm{m} \qquad\qquad Q := b_2 \cdot q_2 = 27\ \frac{\mathrm{m^3}}{\mathrm{sec}}$$

(d) The values for E_1 and E_2 are determined from the specific energy equation as follows:

$$E_1 := y_1 + \frac{v_1^2}{2 \cdot g} = 3.459\ \mathrm{m} \qquad\qquad E_2 := y_2 + \frac{v_2^2}{2 \cdot g} = 3.459\ \mathrm{m}$$

where:

$E_1 := E_2 = 3.459\,m$

The E-y curve for the unit discharge, $q = q_1 = q_2 = 9\,m^2/sec$ is plotted as follows:

$$q := q_1 = 9\,\frac{m^2}{sec} \qquad\qquad E(y) := y + \frac{q^2}{2 \cdot g \cdot y^2}$$

$$a \equiv 0.1\,m \qquad b \equiv 5\,m \qquad h \equiv 0.1\,m \qquad y := a,\, a + h..b$$

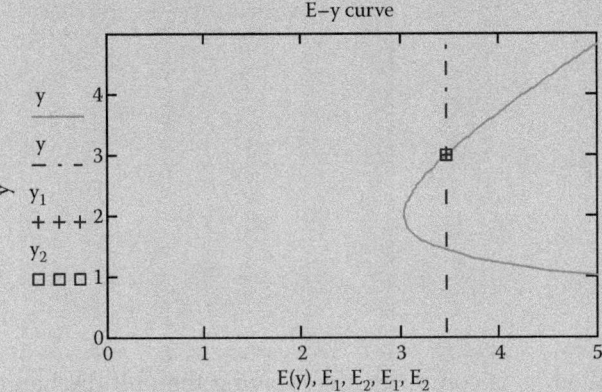

E–y curve

$E(y),\, E_1,\, E_2,\, E_1,\, E_2$

$E(y)$

(e) For a given constant discharge, $Q = b_t\, q_t = b_1 q_1 = b_2 q_2 = 27\,m^3/sec$, the unit discharge at the gradual transition, q_t and its corresponding critical depth of flow, y_{ct} are computed as follows:

$$b_t := 2.55\,m \qquad Q = 27\,\frac{m^3}{sec} \qquad q_t := \frac{Q}{b_t} = 10.588\,\frac{m^2}{sec}$$

$$g := 9.81\,\frac{m}{sec^2} \qquad y_{ct} := \sqrt[3]{\frac{q_t^2}{g}} = 2.252\,m$$

The subcritical depth of the water at the gradual transition, y_t is computed by applying the Bernoulli equation as follows:

$$y_1 + \frac{v_1^2}{2g} + z_1 = y_t + \frac{v_t^2}{2g} + z_t$$

Taking the channel bottom as the datum, $z_1 = z_t = 0$ yields the following:

$$y_1 + \frac{v_1^2}{2g} = y_t + \frac{v_t^2}{2g}$$

Applying the unit discharge equation, $q_t = v_t\, y_t$, where $b_t = 2.55$ m, yields the following:

$$y_1 + \frac{v_1^2}{2g} = y_t + \frac{q_t^2}{2gy_t^2}$$

The subcritical depth at the transition, y_t is computed by assuming a guess value for y_t that is greater than $y_{ct} = 2.252$ m as follows:

Guess value: $y_t := 2.3$ m

Given

$$y_1 + \frac{v_1^2}{2 \cdot g} = y_t + \frac{q_t^2}{2 \cdot g \cdot y_t^2}$$

$y_t := \text{Find}(y_t) = 2.637$ m

The velocity at the transition, v_t is computed by applying the unit discharge equation, $q_t = v_t\, y_t$ as follows:

$$v_t := \frac{q_t}{y_t} = 4.015 \ \frac{m}{sec}$$

(f) The value for E_t is determined from the specific energy equation as follows:

$$E_t := y_t + \frac{v_t^2}{2 \cdot g} = 3.459 \text{ m}$$

where $E_1 = E_2 = E_t = 3.459$ m

The E-y curve for the unit discharge, $q = q_t = 10.588$ m^2/sec (broken line) is plotted inside the E-y curve for the unit discharge, $q = q_1 = q_2 = 9$ m^2/sec (solid line) as follows:

$$q := \begin{pmatrix} q_1 \\ q_2 \\ q_t \end{pmatrix} = \begin{pmatrix} 9 \\ 9 \\ 10.588 \end{pmatrix} \frac{m^2}{sec} \qquad E(y, q) := y + \frac{q^2}{2 \cdot g \cdot y^2}$$

$a \equiv 0.1$ m $\qquad\qquad b \equiv 5$ m $\qquad h \equiv 0.1$ m $\qquad y := a,\, a + h..b$

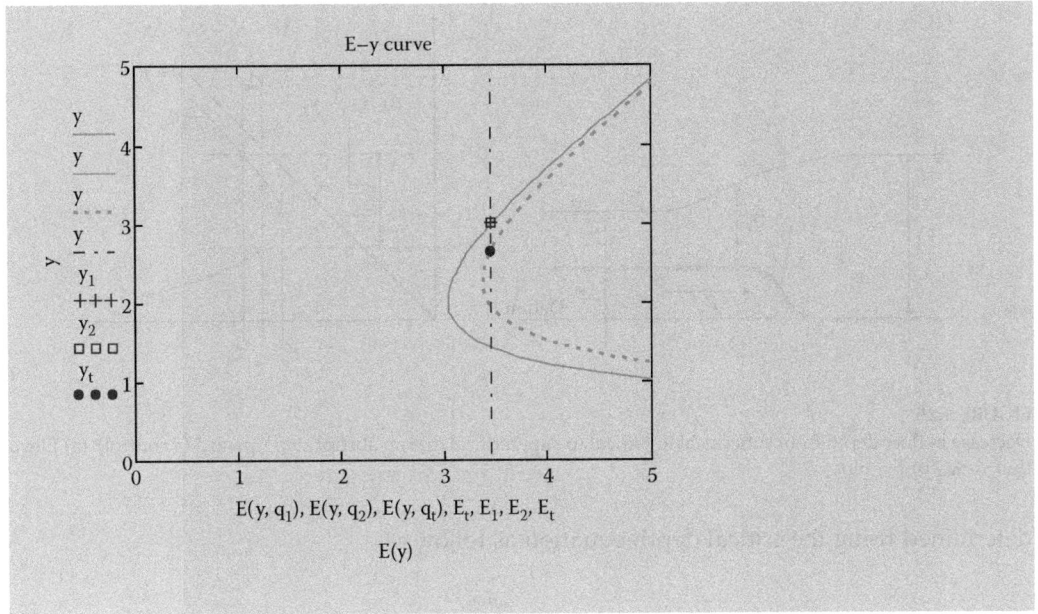

9.12.2 Flow Depth for Abrupt Channel Transitions: Controls

While the appropriate critical equation is used to determine the critical depth, y_c; the critical velocity, v_c; or the minimum specific energy, E_c at the control, the Bernoulli equation (Eulerian/integral approach) is applied to determine either the upstream or downstream depth for an abrupt (vertical or horizontal) channel transition and is given as follows:

$$\underbrace{\underbrace{\left(y_1 + \frac{v_1^2}{2g} + z_1\right)}_{E_1}}_{H_1} = \underbrace{\underbrace{\left(y_2 + \frac{v_2^2}{2g} + z_2\right)}_{E_2}}_{H_2} \qquad (9.186)$$

An abrupt vertical or horizontal channel transition is considered a control. An abrupt vertical channel contraction includes typical flow-measuring devices, such as sluice gates, weirs, and overflow spillways, and it includes a hydraulic drop. An abrupt horizontal channel contraction includes typical flow-measuring devices, such as venturi flumes and contracted openings.

9.12.2.1 Flow Depth for an Abrupt Upward Step

A rectangular channel section with a constant width, b and thus a constant unit discharge, q with an abrupt vertical channel transition such as an abrupt upward step, Δz, along with the corresponding E-y plot is illustrated in Figure 9.29. Given that the channel bottom slope is classified as mild, the upstream uniform depth of flow, y_1 is subcritical. The abrupt upward step at the channel transition serves as control, where the flow becomes critical, thus causing the downstream depth, y_2 to become supercritical. The critical depth, y_c at the control is

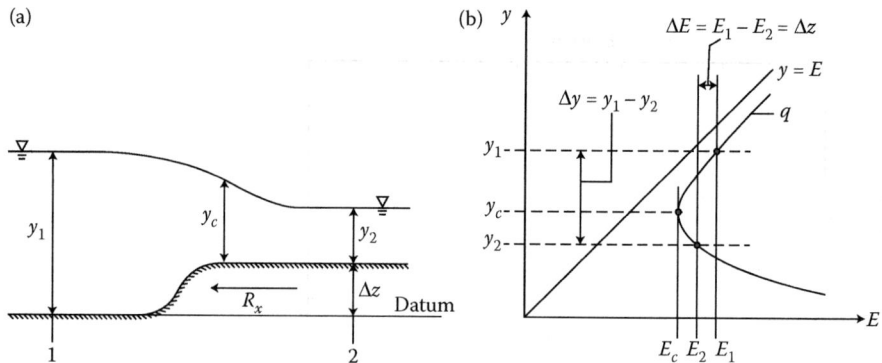

FIGURE 9.29
Decrease in flow depth from subcritical to critical to supercritical due an abrupt step upward (a control). (a) Elevation view. (b) E-y plot.

determined using the critical depth equation as follows:

$$y_c = \sqrt[3]{\frac{q^2}{g}} \tag{9.187}$$

which is plotted on the crest on the E-y plot for a given q, as illustrated in Figure 9.29b. The critical velocity, v_c and the minimum specific energy, E_c are computed as follows:

$$v_c = \sqrt{g y_c} \tag{9.188}$$

$$E_c = \frac{3}{2} y_c \tag{9.189}$$

The unknown depth of flow (either y_1 or y_2) may be determined from the Bernoulli equation as follows:

$$y_1 + \frac{v_1^2}{2g} + z_1 = y_2 + \frac{v_2^2}{2g} + z_2 \tag{9.190}$$

Taking the channel bottom at the upstream point 1 as the datum, $z_1 = 0$; thus, $z_2 = \Delta z$ yields the following:

$$y_1 + \frac{v_1^2}{2g} = y_2 + \frac{v_2^2}{2g} + \Delta z \tag{9.191}$$

$$E_1 = E_2 + \Delta z \tag{9.192}$$

$$E_1 - E_2 = \Delta E = \Delta z \tag{9.193}$$

where the decrease in the specific energy, ΔE is a result of an increase in elevation, Δz. For instance, in the case where the upstream depth, y_1 and the upstream velocity, v_1 are known, the downstream depth, y_2 and the downstream velocity, v_2 are solved as follows:

$$E_1 = y_2 + \frac{v_2^2}{2g} + \Delta z \tag{9.194}$$

$$E_1 - \Delta z = y_2 + \frac{v_2^2}{2g} \tag{9.195}$$

$$E_1 - \Delta z = y_2 + \frac{q^2}{2gy_2^2} \tag{9.196}$$

Thus, for a given unit discharge, $q = v_1\, y_1 = v_2\, y_2$, the specific energy equation, E_2 is cubic in y_2. As such, there are two positive and one negative solutions for y_2, as illustrated by drawing a vertical line corresponding to E_2 in Figure 9.29b. While the negative solution is of no practical interest, the two positive solutions are. The larger of the two positive solutions is plotted on the upper subcritical limb, while the smaller of the two positive roots is plotted on the lower subcritical limb. However, because it was already established that the downstream depth, y_2 is supercritical, the desired solution to the above cubic specific energy equation, E_2 is the smaller of the two positive roots. Using Mathcad to numerically solve for the unknown downstream supercritical depth, y_2, the critical depth of flow, y_c for the given unit discharge, q may be used as an upper limit to guide the guess value for y_2 as illustrated in Example Problem 9.3 below.

EXAMPLE PROBLEM 9.3
Water flows in a rectangular channel at a velocity of 1 m/sec and depth of 1.65 m. There is an abrupt upward step, Δz of 0.33 m in the channel bed as illustrated in Figure EP 9.3. (a) Determine the flow regime (critical, subcritical, or supercritical) of the upstream depth, y_1. (b) Determine the flow regime of the downstream depth, y_2. (c) Determine the critical velocity, v_c and the minimum specific energy, E_c at the control. (d) Determine the depth, y_2 and the velocity, v_2 of the water downstream of the step. (e) Determine E_1 and E_2 and plot the E-y curve for the unit discharge, $q = v_1\, y_1 = v_2\, y_2$.

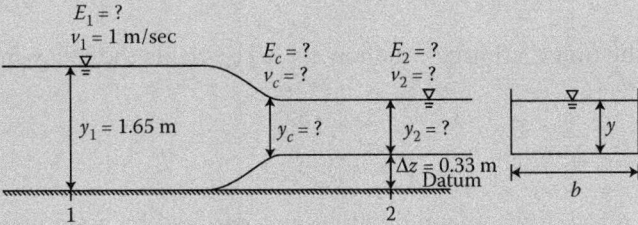

FIGURE EP 9.3
Water flows in a rectangular channel with abrupt upward step, Δz in the channel bed.

Mathcad Solution

(a) In order to determine the flow regime of the upstream depth, y_1, one must first determine the critical depth of flow, y_c for the given unit discharge, q as follows:

$$v_1 := 1 \frac{m}{\sec} \qquad y_1 := 1.65\,m \qquad q := v_1 \cdot y_1 = 1.65 \frac{m^2}{\sec}$$

$$g := 9.81 \frac{m}{\sec^2} \qquad y_c := \sqrt[3]{\frac{q^2}{g}} = 0.652\,m$$

Therefore, because $y_1 = 1.65\,m$ is greater than $y_c = 0.652\,m$, the depth of flow upstream of the step is subcritical.

(b) Because the abrupt upward step serves as a control, the flow becomes critical at the step (transition). Therefore, the flow regime of the downstream depth, y_2 is supercritical.

(c) The critical velocity, v_c at the control is computed as follows:

$$v_c := \sqrt{g \cdot y_c} = 2.53 \frac{m}{s}$$

The minimum specific energy, E_c at the control is computed as follows:

$$E_c := \frac{3}{2} y_c = 0.978\,m$$

(d) The supercritical depth of the water downstream of the step, y_2 is computed by applying the Bernoulli equation as follows:

$$y_1 + \frac{v_1^2}{2g} + z_1 = y_2 + \frac{v_2^2}{2g} + z_2$$

Taking the channel bottom at the upstream point 1 as the datum, $z_1 = 0$, and thus, $z_2 = \Delta z$ yields the following:

$$y_1 + \frac{v_1^2}{2g} = y_2 + \frac{v_2^2}{2g} + \Delta z$$

Applying the unit discharge equation, $q = v_2\, y_2$ yields the following:

$$y_1 + \frac{v_1^2}{2g} = y_2 + \frac{q^2}{2g y_2^2} + \Delta z$$

The supercritical downstream depth, y_2 is computed by assuming a guess value for y_2 that is less than $y_c = 0.652\,m$ as follows:

$$\Delta z := 0.33\,m \qquad\qquad \text{Guess value:} \quad y_2 := 0.5\,m$$

Given

$$y_1 + \frac{v_1^2}{2 \cdot g} = y_2 + \frac{q^2}{2 \cdot g \cdot y_2^2} + \Delta z$$

$$y_2 := \mathrm{Find}(y_2) = 0.373\,\mathrm{m}$$

The downstream velocity, v_2 is computed by applying the unit discharge equation, $q = v_2\, y_2$ as follows:

$$v_2 := \frac{q}{y_2} = 4.425\,\frac{\mathrm{m}}{\mathrm{sec}}$$

(e) The values for E_1 and E_2 are determined from the specific energy equation as follows:

$$E_1 := y_1 + \frac{v_1^2}{2 \cdot g} = 1.701\,\mathrm{m} \qquad\qquad E_2 := y_2 + \frac{v_2^2}{2 \cdot g} = 1.371\,\mathrm{m}$$

where:

$$E_1 := E_2 + \Delta z = 1.701\,\mathrm{m}$$

The E-y curve for the unit discharge, $q = v_1\, y_1 = v_2\, y_2 = 1.65\,\mathrm{m}^2/\mathrm{sec}$ is plotted as follows:

$$E(y) := y + \frac{q^2}{2 \cdot g \cdot y^2} \qquad a \equiv 0.1\,\mathrm{m} \quad b \equiv 5\,\mathrm{m} \quad h \equiv 0.1\,\mathrm{m} \quad y := a,\, a + h..b$$

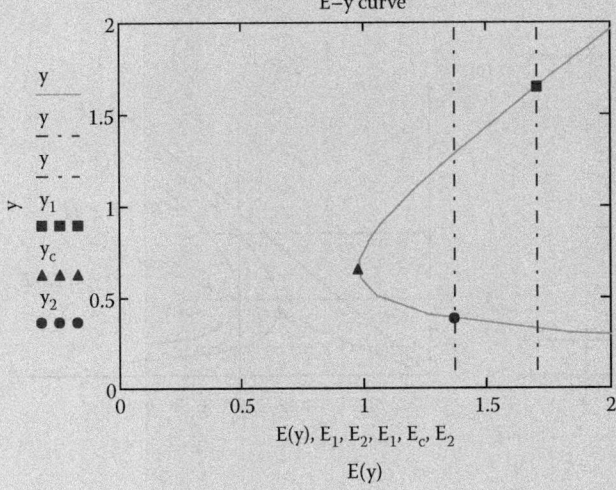

E–y curve

E(y), E_1, E_2, E_1, E_c, E_2

E(y)

9.12.2.2 *Flow Depth for an Abrupt Decrease in Channel Width*

A rectangular channel section with a temporary abrupt decrease in width, b and thus an abrupt horizontal channel transition, along with the corresponding E-y plot, is illustrated in Figure 9.30. Given that the channel bottom slope is classified as mild, the upstream uniform depth of flow, y_1 is subcritical where the channel width, $b = b_1$ and thus $q = q_1$. The abrupt decrease in channel width, $b = b_t \ll b_1$ at the channel transition serves as a control, where the flow becomes critical, thus causing the downstream depth, y_2 to become supercritical. The unit discharge, $q = q_t \gg q_1$, will increase, and the critical depth at the control, $y_t = y_c$ and is determined using the critical depth equation as follows:

$$y_c = \sqrt[3]{\frac{q_t^2}{g}} \tag{9.197}$$

where for a given constant discharge, $Q = b_t \, q_t = b_1 q_1$, and $y_t = y_c$ is plotted on the crest on the E-y plot for a given q_t, as illustrated in Figure 9.30c. The critical velocity, v_c and the minimum specific energy, $E_t = E_c$ are computed as follows:

$$v_c = \sqrt{g y_c} \tag{9.198}$$

$$E_c = \frac{3}{2} y_c \tag{9.199}$$

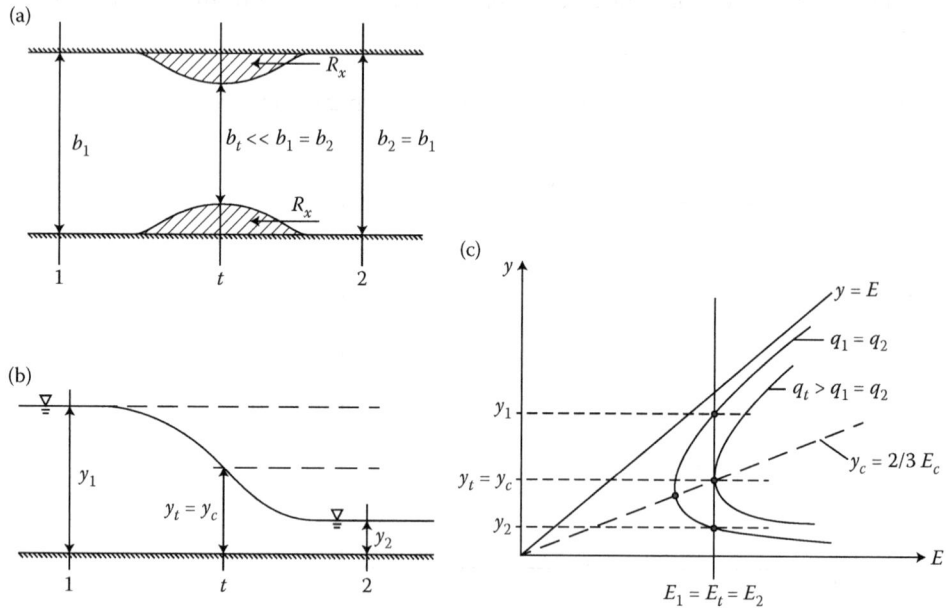

FIGURE 9.30
Decrease in flow depth from subcritical to critical to supercritical due an abrupt decrease in channel width (a control). (a) Plan view, (b) Elevation view, (c) E-y plot. (Adapted from Henderson, F. M., *Open Channel Flow*, Macmillan, New York, 1966, p. 33.)

Downstream of the control, the channel width resumes to the upstream channel width, $b = b_2 = b_1$; thus, $q = q_2 = q_1$. The unknown depth of flow (either y_1 or y_2) may be determined from the Bernoulli equation as follows:

$$y_1 + \frac{v_1^2}{2g} + z_1 = y_2 + \frac{v_2^2}{2g} + z_2 \tag{9.200}$$

Assuming that $z_1 = z_2$ yields the following:

$$y_1 + \frac{v_1^2}{2g} = y_2 + \frac{v_2^2}{2g} \tag{9.201}$$

$$E_1 = E_2 \tag{9.202}$$

For instance, in the case where the upstream depth, y_1 and the upstream velocity, v_1 are known, the downstream depth, y_2 and the downstream velocity, v_2 are solved as follows:

$$E_1 = y_2 + \frac{q_2^2}{2gy_2^2} \tag{9.203}$$

Thus, for a given unit discharge, $q_1 = v_1 y_1$ (where $q_1 = q_2 = v_2 y_2$), the specific energy equation, E_2 is cubic in y_2. As such, there are two positive solutions and one negative solution for y_2, as illustrated by drawing a vertical line corresponding to $E_1 = E_2$ in Figure 9.30c. While the negative solution is of no practical interest, the two positive solutions are. The larger of the two positive solutions is plotted on the upper subcritical limb, while the smaller of the two positive solutions is plotted on the lower subcritical limb. However, because it was already established that the downstream depth, y_2 is supercritical, the desired solution to the above cubic specific energy equation, E_2 is the smaller of the two positive roots. One may note that because $E_1 = E_2$, the upstream subcritical depth, y_1 and the downstream depth, y_2 are alternate depths of flow. Furthermore, one may note that because there is no change elevation between points 1, t, and 2, the specific energy does not change, and thus, $E_t = E_1 = E_2$. Using Mathcad to numerically solve for the unknown downstream subcritical depth, y_2, the critical depth of flow, $y_c = \sqrt[3]{q_2^2/g}$ for the given unit discharge, q_2 may be used as an upper limit to guide the guess value for y_2 as illustrated in Example Problem 9.4 below.

EXAMPLE PROBLEM 9.4

Water flows in a rectangular channel of width 3 m at a velocity of 3 m/sec and depth of 3 m. There is a temporary abrupt decrease in channel width to 2.462 m and thus an abrupt horizontal channel transition, as illustrated in Figure EP 9.4. (a) Determine the flow regime of the upstream depth, y_1. (b) Determine the flow regime of the downstream depth, y_2. (c) Determine the unit discharge, q_t; the critical depth, y_c; the critical velocity, v_c; and the minimum specific energy, $E_c = E_t$ for the unit discharge, q_t at the abrupt transition (control). (d) Determine the depth, y_2 and the velocity, v_2 of the water downstream of the abrupt transition. (e) Determine E_1, E_2, and plot the E-y curve for the unit discharge, $q = q_1 = q_2$, where $b = b_1 = b_2$. And, plot the E-y curve for the unit discharge, $q = q_t >> q_1 = q_2$, where $b = b_t << b_1 = b_2$.

(a)

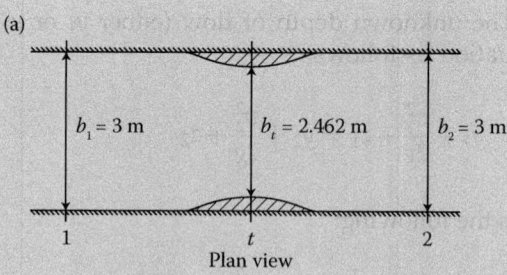

(b)

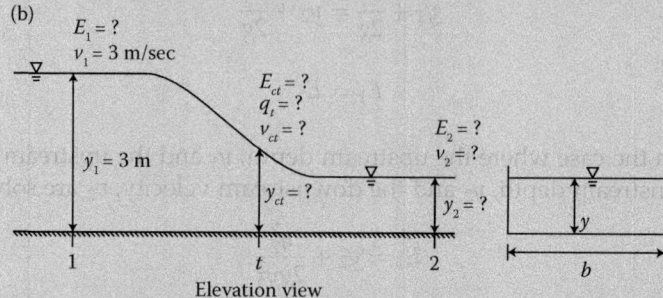

FIGURE EP 9.4
Water flows in a rectangular channel with temporary abrupt decrease in channel width. (a) Plan view, (b) Elevation view.

Mathcad Solution

(a) In order to determine the flow regime of the upstream depth, y_1, one must first determine the critical depth of flow, y_{c1} for the upstream unit discharge, $q = q_1$ as follows:

$$v_1 := 3 \, \frac{m}{sec} \qquad y_1 := 3 \, m \qquad q_1 := v_1 \cdot y_1 = 9 \, \frac{m^2}{sec}$$

$$g := 9.81 \, \frac{m}{sec^2} \qquad y_{c1} := \sqrt[3]{\frac{q_1^2}{g}} = 2.021 \, m$$

where:

$$b_1 := 3 \, m \qquad Q := b_1 \cdot q_1 = 27 \, \frac{m^2}{sec}$$

Therefore, because $y_1 = 3 \, m$ is greater than $y_{c1} = 2.021 \, m$, the depth of flow upstream of the abrupt transition is subcritical.

(b) Because the abrupt decrease in channel width serves as a control, the flow becomes critical at the abrupt transition. Therefore, the flow regime of the downstream depth, y_2 is supercritical.

(c) For a given constant discharge, $Q = b_t \, q_t = b_1 q_1 = b_2 q_2 = 27 \, m^3/sec$, the unit discharge, q_t at the abrupt transition (control) is computed as follows:

$$b_t := 2.462 \, m \qquad Q = 27 \, \frac{m^3}{sec} \qquad q_t := \frac{Q}{b_t} = 10.967 \, \frac{m^2}{sec}$$

The critical depth, y_c for the unit discharge, q_t at the abrupt transition (control), y_{ct} is computed as follows:

$$g := 9.81 \frac{m}{sec^2} \qquad y_{ct} := \sqrt[3]{\frac{q_t^2}{g}} = 2.306 \, m$$

The critical velocity, v_c for the unit discharge, q_t at the abrupt transition (control), v_{ct} is computed as follows:

$$v_{ct} := \sqrt{g \cdot y_{ct}} = 4.756 \frac{m}{s}$$

The minimum specific energy, $E_c = E_t$ for the unit discharge, q_t at the abrupt transition (control), E_{ct} is computed as follows:

$$E_{ct} := \frac{3}{2} y_{ct} = 3.459 \, m$$

(d) The supercritical depth of the water downstream of the abrupt transition, y_2 is computed by applying the Bernoulli equation as follows:

$$y_1 + \frac{v_1^2}{2g} + z_1 = y_2 + \frac{v_2^2}{2g} + z_2$$

Taking the channel bottom as the datum, $z_1 = z_2 = 0$ yields the following:

$$y_1 + \frac{v_1^2}{2g} = y_2 + \frac{v_2^2}{2g}$$

Applying the unit discharge equation, $q_2 = v_2 \, y_2$, where $b = b_1 = b_2 = 3 \, m$; thus, $q = q_1 = q_2 = 9 \, m^2/sec$ yields the following:

$$y_1 + \frac{v_1^2}{2g} = y_2 + \frac{q_2^2}{2g y_2^2}$$

The supercritical downstream depth, y_2 is computed by assuming a guess value for y_2 that is less than $y_{c1} = y_{c2} = 2.021$ m as follows:

$$q_2 := q_1 = 9 \frac{m^2}{sec} \qquad \text{Guess value:} \quad y_2 := 1.9 \, m$$

Given

$$y_1 + \frac{v_1^2}{2 \cdot g} = y_2 + \frac{q_2^2}{2 \cdot g \cdot y_2^2}$$

$$y_2 := Find(y_2) = 1.425 \, m$$

The downstream velocity, v_2 is computed by applying the unit discharge equation, $q_2 = v_2\, y_2$ as follows:

$$v_2 := \frac{q_2}{y_2} = 6.317\,\frac{m}{sec}$$

where:

$$b_2 := 3\,m \qquad Q := b_2 \cdot q_2 = 27\,\frac{m^3}{sec}$$

(e) The values for E_1 and E_2 are determined from the specific energy equation as follows:

$$E_1 := y_1 + \frac{v_1^2}{2g} = 3.459\,m \qquad\qquad E_2 := y_2 + \frac{v_2^2}{2g} = 3.459\,m$$

where $E_1 = E_2 = E_t = 3.459$ m. The E-y curve for the unit discharge, $q = q_t = 10.967$ m^2/sec (broken line) is plotted inside the E-y curve for the unit discharge, $q = q_1 = q_2 = 9$ m^2/sec (solid line) as follows:

$$q := \begin{pmatrix} q_1 \\ q_2 \\ q_t \end{pmatrix} = \begin{pmatrix} 9 \\ 9 \\ 10.967 \end{pmatrix}\,\frac{m^2}{sec} \qquad\qquad E(y, q) := y + \frac{q^2}{2 \cdot g \cdot y^2}$$

$$a \equiv 0.1\,m \qquad b \equiv 5\,m \qquad h \equiv 0.1\,m \qquad y := a,\, a + h .. b$$

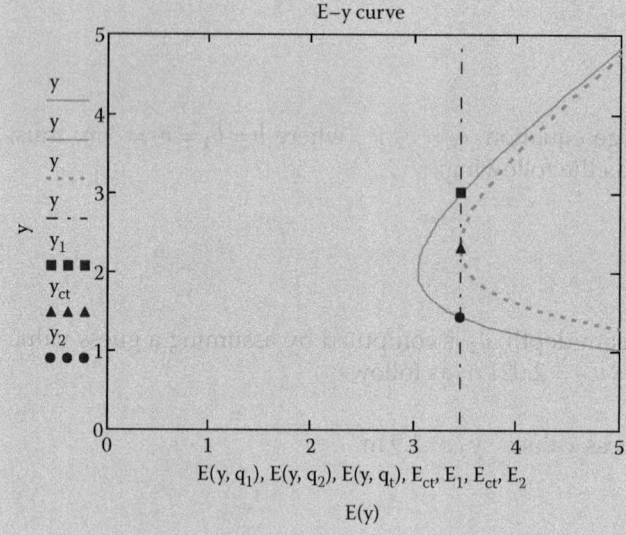

E–y curve

$E(y, q_1),\ E(y, q_2),\ E(y, q_t),\ E_{ct},\ E_1,\ E_{ct},\ E_2$

$E(y)$

9.12.2.3 Flow Depth for Typical Flow-Measuring Devices

Typical flow-measuring devices include sluice gates, weirs, overflow spillways, venturi flumes, and contracted openings. Critical flow occurs at these typical flow-measuring devices, and they are considered controls. Because the venturi flume and a contracted opening are a special case of an abrupt decrease in channel width, one may follow the procedure outlined above in Section 9.12.2.2 in order to determine the unknown flow depth (y_1 or y_2, and $y_t = y_c$). Furthermore, determination of the unknown flow depth for a sluice gate, a weir (either sharp-crested or broad-crested), and a spillway, illustrated in Figure 9.31, is similar regardless of the type of flow-measuring device, and is presented as follows.

Assuming a rectangular channel section with a constant width, b and thus a constant unit discharge, q allows the determination of the M-y and E-y plots, as illustrated for the respective flow-measuring devices such as sluice gate, a weir (either sharp-crested or broad-crested), and a spillway in Figure 9.31a and b, respectively. Given that the channel bottom slope is classified as mild, the upstream uniform depth of flow, y_1 is subcritical. The flow-measuring device serves as control, where the flow becomes critical, thus causing the downstream depth, y_2 to become supercritical. The critical depth, y_c at the control is determined using the critical depth equation as follows:

$$y_c = \sqrt[3]{\frac{q^2}{g}} \qquad (9.204)$$

which is plotted on the crest on the E-y plot for a given q, as illustrated in Figure 9.31b. The critical velocity, v_c and the minimum specific energy, E_c are computed as follows:

$$v_c = \sqrt{g y_c} \qquad (9.205)$$

$$E_c = \frac{3}{2} y_c \qquad (9.206)$$

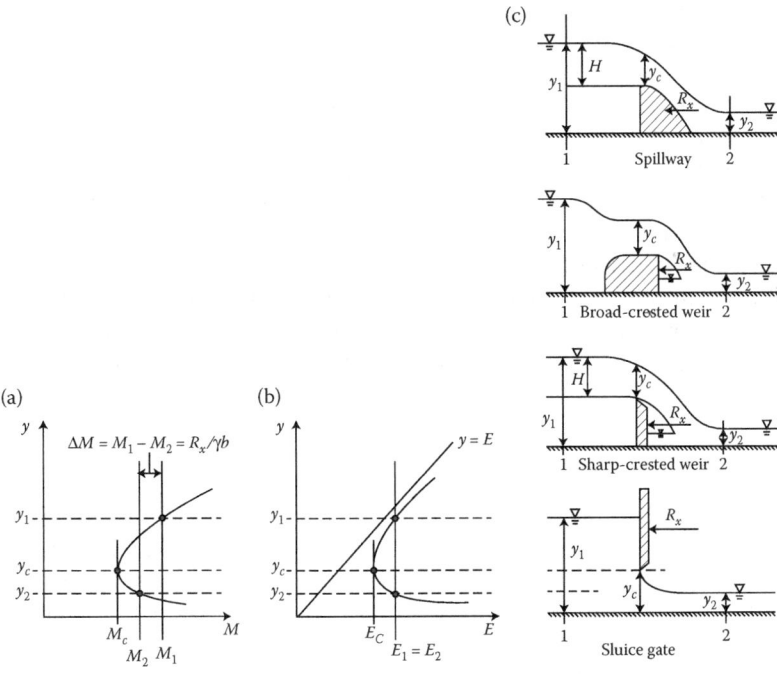

FIGURE 9.31
The M-y plot and the E-y plot for flow-measuring device. (a) M-y plot, (b) E-y plot, (c) Flow-measuring device.

The unknown depth of flow (either y_1 or y_2) may be determined from the Bernoulli equation as follows:

$$y_1 + \frac{v_1^2}{2g} + z_1 = y_2 + \frac{v_2^2}{2g} + z_2 \tag{9.207}$$

Assuming that $z_1 = z_2$ yields the following:

$$y_1 + \frac{v_1^2}{2g} = y_2 + \frac{v_2^2}{2g} \tag{9.208}$$

$$E_1 = E_2 \tag{9.209}$$

For instance, in the case where the upstream depth, y_1 and the upstream velocity, v_1 are known, the downstream depth, y_2 and the downstream velocity, v_2 are solved as follows:

$$E_1 = y_2 + \frac{v_2^2}{2g} \tag{9.210}$$

$$E_1 = y_2 + \frac{q^2}{2gy_2^2} \tag{9.211}$$

Thus, for a given unit discharge, $q = v_1\, y_1 = v_2\, y_2$, the specific energy equation, E_2 is cubic in y_2. As such, there are two positive solutions and one negative solution for y_2, as illustrated by drawing a vertical line corresponding to $E_1 = E_2$ in Figure 9.31b. While the negative solution is of no practical interest, the two positive solutions are. The larger of the two positive solutions is plotted on the upper subcritical limb, while the smaller of the two positive roots is plotted on the lower subcritical limb. However, because it was already established that the downstream depth, y_2 is supercritical, the desired solution to the above cubic specific energy equation, E_2 is the smaller of the two positive roots. One may note that because $E_1 = E_2$, the upstream subcritical depth, y_1 and the downstream depth, y_2 are alternate depths of flow. Using Mathcad to numerically solve for the unknown downstream supercritical depth, y_2, the critical depth of flow, y_c for the given unit discharge, q may be used as an upper limit to guide the guess value for y_2 as illustrated in Example Problem 9.5 below.

EXAMPLE PROBLEM 9.5

Water flows in a rectangular channel of width 3 m at a velocity of 2.5 m/sec and depth of 1.8 m upstream of a sluice gate as illustrated in Figure EP 9.5. (a) Determine the flow regime (critical, subcritical, or supercritical) of the upstream depth, y_1. (b) Determine the flow regime of the downstream depth, y_2. (c) Determine the critical velocity, v_c and the minimum specific energy, E_c at the control. (d) Determine the depth, y_2 and the velocity, v_2 of the water downstream of the sluice gate. (e) Determine E_1, E_2, and plot the E-y curve for the unit discharge, $q = v_1\, y_1 = v_2\, y_2$.

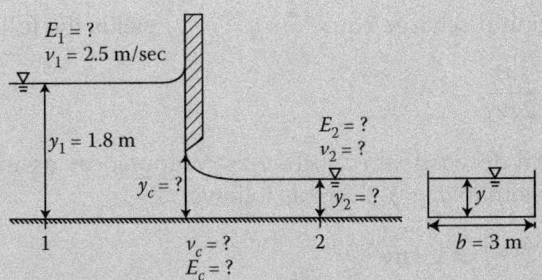

FIGURE EP 9.5
Water flows in a rectangular channel with a sluice gate inserted in the channel.

Mathcad Solution

(a) In order to determine the flow regime of the upstream depth, y_1, one must first determine the critical depth of flow, y_c for the given unit discharge, q as follows:

$$v_1 := 2.5 \, \frac{m}{sec} \qquad y_1 := 1.8 \, m \qquad q := v_1 \cdot y_1 = 4.5 \frac{m^2}{sec}$$

$$g := 9.81 \, \frac{m}{sec^2} \qquad y_c := \sqrt[3]{\frac{q^2}{g}} = 1.273 \, m$$

Therefore, because $y_1 = 1.8 \, m$ is greater than $y_c = 1.273 \, m$, the depth of flow upstream of the step is subcritical.

(b) Because the sluice gate serves as a control, the flow becomes critical at the gate (transition). Therefore, the flow regime of the downstream depth, y_2 is supercritical.

(c) The critical velocity, v_c at the control is computed as follows:

$$v_c := \sqrt{g \cdot y_c} = 3.534 \, \frac{m}{s}$$

The minimum specific energy, E_c at the control is computed as follows:

$$E_c := \frac{3}{2} y_c = 1.91 \, m$$

(d) The supercritical depth of the water downstream of the gate, y_2 is computed by applying the Bernoulli equation as follows:

$$y_1 + \frac{v_1^2}{2g} + z_1 = y_2 + \frac{v_2^2}{2g} + z_2$$

Taking the channel bottom as the datum, $z_1 = z_2 = 0$ yields the following:

$$y_1 + \frac{v_1^2}{2g} = y_2 + \frac{v_2^2}{2g}$$

Applying the unit discharge equation, $q = v_2\,y_2$ yields the following:

$$y_1 + \frac{v_1^2}{2g} = y_2 + \frac{q^2}{2gy_2^2}$$

The supercritical downstream depth, y_2 is computed by assuming a guess value for y_2 that is less than $y_c = 1.273\,\text{m}$ as follows:

Guess value: $y_2 := 1.2\,\text{m}$
Given

$$y_1 + \frac{v_1^2}{2 \cdot g} = y_2 + \frac{q^2}{2 \cdot g \cdot y_2^2}$$

$y_2 := \text{Find}(y_2) = 0.933\,\text{m}$

The downstream velocity, v_2 is computed by applying the unit discharge equation, $q = v_2\,y_2$ as follows:

$$v_2 := \frac{q}{y_2} = 4.823\,\frac{\text{m}}{\text{sec}}$$

(e) The values for E_1 and E_2 are determined from the specific energy equation as follows:

$$E_1 := y_1 + \frac{v_1^2}{2 \cdot g} = 2.119\,\text{m} \qquad\qquad E_2 := y_2 + \frac{v_2^2}{2 \cdot g} = 2.119\,\text{m}$$

where $E_1 = E_2 = 2.119\,\text{m}$, and the upstream subcritical depth, y_1 and the downstream supercritical depth, y_2 are alternate depths of flow. The E-y curve for the unit discharge, $q = v_1\,y_1 = v_2\,y_2 = 4.5\,\text{m}^2/\text{sec}$ is plotted as follows:

$$E(y) := y + \frac{q^2}{2 \cdot g\,y^2} \qquad a \equiv 0.1\,\text{m} \qquad b \equiv 5\,\text{m} \qquad h \equiv 0.1\,\text{m} \qquad y := a,\ a + h..b$$

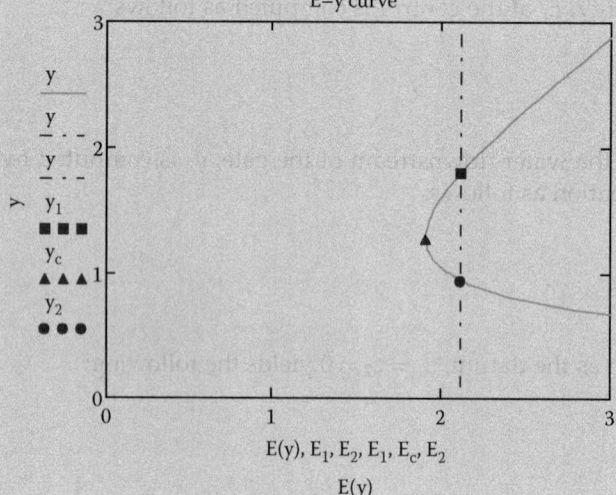

Assuming a nonrectangular channel section does not allow the definition of a unit discharge, q and thus the definition of the E-y plot or the M-y plot. Given that the channel bottom slope is classified as mild, the upstream uniform depth of flow, y_1 is subcritical. The flow-measuring device serves as control, where the flow becomes critical, and thus causes the downstream depth, y_2 to become supercritical. The critical depth, y_c at the control is determined using the critical depth equation for a nonrectangular channel as follows:

$$\frac{A_c}{B_c} = \frac{Q^2}{A_c^2}\frac{1}{g} \tag{9.212}$$

which is numerically solved for y_c using Mathcad. The critical velocity, v_c and the minimum specific energy, E_c are computed as follows:

$$v_c = \sqrt{g\frac{A_c}{B_c}} \tag{9.213}$$

$$E_c = \frac{3}{2}\frac{A_c}{B_c} \tag{9.214}$$

The unknown depth of flow (either y_1 or y_2) may be determined from the Bernoulli equation as follows:

$$y_1 + \frac{v_1^2}{2g} + z_1 = y_2 + \frac{v_2^2}{2g} + z_2 \tag{9.215}$$

Assuming that $z_1 = z_2$ yields the following:

$$y_1 + \frac{v_1^2}{2g} = y_2 + \frac{v_2^2}{2g} \tag{9.216}$$

$$E_1 = E_2 \tag{9.217}$$

For instance, in the case where the upstream depth, y_1 and the upstream velocity, v_1 are known, the downstream depth, y_2 and the downstream velocity, v_2 are solved as follows:

$$E_1 = y_2 + \frac{v_2^2}{2g} \tag{9.218}$$

$$E_1 = y_2 + \frac{Q^2}{2gA_2^2} \tag{9.219}$$

Thus, for a given discharge, $Q = v_1 A_1 = v_2 A_2$, the specific energy equation, E_2 is quintic in y_2 and where y_2 is solved for numerically using Mathcad. While the negative or nonreal solutions are of no practical interest, the two positive solutions are. The larger of the two positive solutions corresponds to the subcritical depth of flow, while the smaller of the two positive roots corresponds to the subcritical depth of flow. However, because it was already established that the downstream depth, y_2 is supercritical, the desired solution to the above quintic specific energy equation, E_2 is the smaller of the two positive roots. One may note that because $E_1 = E_2$, the upstream subcritical depth, y_1 and the downstream depth, y_2 are alternate depths of flow. Using Mathcad to numerically solve for the unknown downstream supercritical depth, y_2, the critical depth of flow, y_c for the given discharge, Q may be used as an upper limit to guide the guess value for y_2 as illustrated in Example Problem 9.6 below.

EXAMPLE PROBLEM 9.6

Water flows in a trapezoidal channel section as illustrated in Figure EP 9.6, where $b = 2$ m, and $\theta = 30°$. Upstream of the spillway, the velocity of the flow is 2 m/sec and the depth is 1.5 m. (a) Determine the flow regime (critical, subcritical, or supercritical) of the upstream depth, y_1. (b) Determine the flow regime of the downstream depth, y_2. (c) Determine the critical velocity, v_c and the minimum specific energy, E_c at the control. (d) Determine the depth, y_2 and the velocity, v_2 of the water downstream of the spillway. (e) Determine E_1 and E_2 for the discharge, $Q = v_1 A_1 = v_2 A_2$.

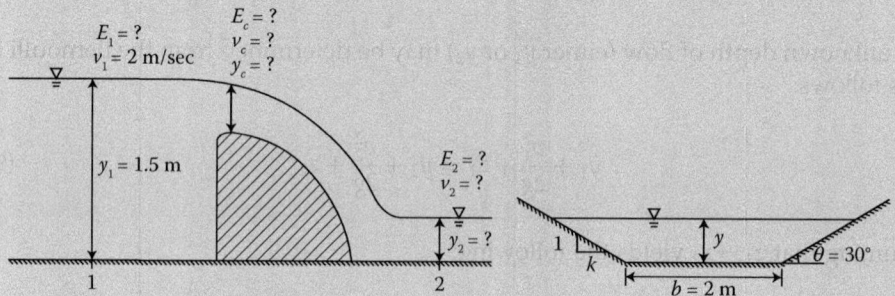

FIGURE EP 9.6
Water flows in a trapezoidal channel with a spillway inserted in the channel.

Mathcad Solution

(a) In order to determine the flow regime of the upstream depth, y_1, one must first determine the critical depth of flow, y_c for the given discharge, Q. It is important to note that the value for y_c does not equal to A_c/B_c, which is the generalization of the critical depth, y_c (as for the rectangular channel).

$$v_1 := 2\,\frac{m}{sec} \qquad y_1 := 1.5\,m \qquad \theta := 30\,deg \qquad b := 2\,m$$

$$k := \frac{1}{\tan(\theta)} \qquad B(y) := b + 2 \cdot k \cdot y \qquad A(y) := y \cdot (k \cdot y + b)$$

$$Q := v_1 \cdot A(y_1) = 13.794\,\frac{m^3}{sec} \qquad\qquad g := 9.81\,\frac{m}{sec^2}$$

Guess value: $y_c := 0.5\,m$

Given

$$\frac{A(y_c)}{B(y_c)} = \frac{Q^2}{A(y_c)^2} \frac{1}{g}$$

$$y_c := Find(y_c) = 1.206\,m$$

Just for comparison, the value for A_c/B_c is computed as follows:

$$\frac{A(y_c)}{B(y_c)} = 0.798\,m$$

Therefore, because $y_1 = 1.5\,m$ is greater than $y_c = 1.206\,m$, the depth of flow upstream of the spillway is subcritical.

(b) Because the spillway serves as a control, the flow becomes critical at the spillway (transition). Therefore, the flow regime of the downstream depth, y_2 is supercritical.

(c) The critical velocity, v_c at the control is computed as follows:

$$v_c := \sqrt{g\frac{A(y_c)}{B(y_c)}} = 2.798\,\frac{m}{s}$$

The minimum specific energy, E_c at the control is computed as follows:

$$E_c := \frac{3}{2}\frac{A(y_c)}{B(y_c)} = 1.197\,m$$

(d) The supercritical depth of the water downstream of the spillway, y_2 is computed by applying the Bernoulli equation as follows:

$$y_1 + \frac{v_1^2}{2g} + z_1 = y_2 + \frac{v_2^2}{2g} + z_2$$

Taking the channel bottom as the datum, $z_1 = z_2 = 0$ yields the following:

$$y_1 + \frac{v_1^2}{2g} = y_2 + \frac{v_2^2}{2g}$$

Applying the discharge equation, $Q = v_2\,A_2$ yields the following:

$$y_1 + \frac{v_1^2}{2g} = y_2 + \frac{Q^2}{2gA_2^2}$$

The supercritical downstream depth, y_2 is computed by assuming a guess value for y_2 that is less than $y_c = 1.206$ m as follows:

Guess value: $\qquad\qquad\qquad\qquad y_2 := 1.1\,m$

Given

$$y_1 + \frac{v_1^2}{2 \cdot g} = y_2 + \frac{Q^2}{2 \cdot g \cdot A(y_2)^2}$$

$$y_2 := \text{Find}(y_2) = 0.993 \text{ m}$$

The downstream velocity, v_2 is computed by applying the unit discharge equation, $Q = v_2 \, A_2$ as follows:

$$v_2 := \frac{Q}{A(y_2)} = 3.735 \, \frac{m}{s}$$

(e) The values for E_1 and E_2 are determined from the specific energy equation as follows:

$$E_1 := y_1 + \frac{v_1^2}{2 \cdot g} = 1.704 \text{ m} \qquad E_2 := y_2 + \frac{v_2^2}{2 \cdot g} = 1.704 \text{ m}$$

where $E_1 = E_2 = 1.704$ m, and the upstream subcritical depth, y_1 and the downstream supercritical depth, y_2 are alternate depths of flow.

9.12.3 Reaction Force on Open Channel Flow Structures/Controls (Abrupt Flow Transitions and Flow-Measuring Devices)

Once the critical depth, y_c at the control is determined from the critical equation and the unknown depth of flow (y_1 or y_2) has been determined from the Bernoulli equation, the reaction force, R_x acting on the control volume between points 1 and 2 due the thrust of flow on an open channel flow structure/control (abrupt flow transitions and flow-measuring devices) is solved using the integral form of the momentum equation (Eulerian/integral approach).

Assuming a rectangular channel section with a constant width, b and thus a constant unit discharge, q allows the determination of the M-y and E-y plots, as illustrated for the respective open channel flow structures/controls (abrupt flow transitions and flow-measuring devices) in Figures 9.29, 9.30, and 9.32 (abrupt flow transitions), and Figure 9.31 (flow-measuring devices). Furthermore, assuming that $z_1 = z_2$, and application of the Bernoulli equation for the control structure between points 1 and 2 yields that $E_1 = E_2$ and thus y_1 (subcritical) and y_2 (supercritical) are alternate depths of flow as illustrated on the E-y plot (see Figures 9.30 and 9.31). However, one may note that in the case where $z_1 \neq z_2$, as in the case of an abrupt upward step, application of the Bernoulli equation yields $\Delta E = \Delta z$, as illustrated on the E-y plot in Figure 9.29. The momentum equation between control face 1 and control face 2 is given as follows:

$$\frac{R_x}{\gamma b} = \underbrace{\left(\frac{q^2}{g y_1} + \frac{y_1^2}{2} \right)}_{M_1} - \underbrace{\left(\frac{q^2}{g y_2} + \frac{y_2^2}{2} \right)}_{M_2} \qquad (9.220)$$

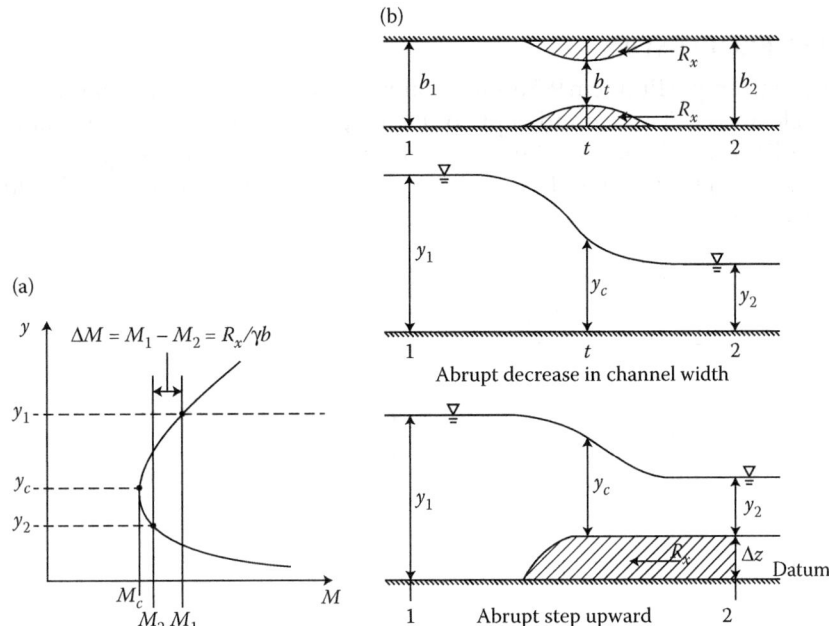

FIGURE 9.32
The *M-y* plot for abrupt channel transition. (a) *M-y* plot, (b) Abrupt channel transition.

Because the reaction force, R_x acting on the control volume between points 1 and 2 due to the thrust of flow on the open channel flow structure/control (abrupt flow transitions and flow-measuring devices) is not zero, the momentum function, M decreases between points 1 and 2 is given as follows:

$$M_1 \neq M_2 \tag{9.221}$$

where the decrease in the momentum function, ΔM is given as follows:

$$\Delta M = \frac{R_x}{\gamma b} \tag{9.222}$$

as illustrated on the *M-y* plot (see Figures 9.31 and 9.32), and the minimum momentum function, M_c occurs at the critical depth of flow, y_c is defined as follows:

$$M_c = \left(\frac{q^2}{g y_c} + \frac{y_c^2}{2} \right) \tag{9.223}$$

Finally, it is useful to define and apply the momentum function, over a short channel section when one flow depth is supercritical and one flow depth is subcritical, and are thus conjugate to one another. Such cases include an abrupt vertical or horizontal contraction (abrupt step upward or abrupt decrease in channel width), and a typical flow-measuring device (such as a sluice gate, weir, overflow spillway, venturi flume, and contracted opening).

EXAMPLE PROBLEM 9.7

Referring to Example Problem 9.5 above, water flows in a rectangular channel of width 3 m at a velocity of 2.5 m/sec and depth of 1.8 m upstream of a sluice gate as illustrated in Figure EP 9.7. (a) Determine the reaction force, R_x acting on the sluice gate due the thrust of flow. (b) Determine M_1, M_2, and M_c, and plot the M-y curve for the unit discharge, $q = v_1 y_1 = v_2 y_2$.

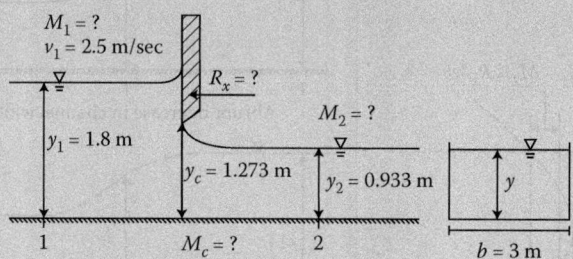

FIGURE EP 9.7
Water flows in a rectangular channel with a sluice gate inserted in the channel.

Mathcad Solution

(a) The reaction force, R_x acting on the sluice gate is computed by applying the momentum equation as follows:

$$\gamma := 9810 \frac{N}{m^3} \qquad g := 9.81 \frac{m}{sec^2} \qquad q := 4.5 \frac{m^2}{sec}$$

$$b := 3\,m \qquad y_1 := 1.8\,m \qquad y_2 := 0.933\,m$$

Guess value: $R_x := 1000\,N$

Given

$$\frac{R_x}{\gamma \cdot b} = \left(\frac{q^2}{g \cdot y_1} + \frac{y_1^2}{2}\right) - \left(\frac{q^2}{g \cdot y_2} + \frac{y_2^2}{2}\right)$$

$$R_x := Find(R_x) = 3.505 \times 10^3\,N$$

(b) The values for M_1 and M_2 are determined from the momentum function equation as follows:

$$M_1 := \frac{q^2}{g \cdot y_1} + \frac{y_1^2}{2} = 2.767\,m^2 \qquad\qquad M_2 := \frac{q^2}{g \cdot y_2} + \frac{y_2^2}{2} = 2.648\,m^2$$

where $M_1 - M_2 = R_x/\gamma b = 0.119\,m^2$, the upstream subcritical depth, y_1 and the downstream supercritical depth, y_2 are conjugate depths of flow, and the minimum momentum function, M_c which occurs at the critical depth of flow, $y_c = 1.273\,m$ is computed as follows:

$$y_c = 1.273\,m \qquad\qquad M_c := \frac{q^2}{g \cdot y_c} + \frac{y_c^2}{2} = 2.432\,m^2$$

The *M-y* curve for the unit discharge, $q = v_1 y_1 = v_2 y_2 = 4.5 \, \text{m}^2/\text{sec}$ is plotted as follows:

$$M(y) := \frac{q^2}{g \cdot y} + \frac{y^2}{2} \qquad a \equiv 0.1 \, \text{m} \qquad b \equiv 5 \, \text{m} \qquad h \equiv 0.1 \, \text{m} \qquad y := a, \, a + h..b$$

M–y curve

Assuming a nonrectangular channel section does not allow the definition of a unit discharge, q and thus the definition of the *E-y* plot or the *M-y* plot. Furthermore, assuming that $z_1 = z_2$, and application of the Bernoulli equation for the control structure between points 1 and 2 yields that $E_1 = E_2$ and thus y_1 (subcritical) and y_2 (supercritical) are alternate depths of flow. However, one may note that in the case where $z_1 \neq z_2$, as in the case of an abrupt upward step, application of the Bernoulli equation yields $\Delta E = \Delta z$. The momentum equation between control face 1 and control face 2 is given as follows:

$$\frac{R_x}{\gamma} = \underbrace{\left(\frac{Q^2}{gA_1} + h_{ca1}A_1 \right)}_{M_1} - \underbrace{\left(\frac{Q^2}{gA_2} + h_{ca2}A_2 \right)}_{M_2} \tag{9.224}$$

Because the reaction force, R_x acting on the control volume between points 1 and 2 due to the thrust of flow on the open channel flow structure is not zero, the momentum function, M decreases between points 1 and 2 is given as follows:

$$M_1 \neq M_2 \tag{9.225}$$

where the decrease in the momentum function, ΔM is given as follows:

$$\Delta M = \frac{R_x}{\gamma} \tag{9.226}$$

EXAMPLE PROBLEM 9.8

Referring to Example Problem 9.6 above, water flows in a trapezoidal channel section as illustrated in Figure EP 9.8, where $b = 2$ m, and $\theta = 30$ degrees. Upstream of the spillway, the velocity of the flow is 2 m/sec and the depth is 1.5 m. (a) Determine the reaction force, R_x acting on the spillway due the thrust of flow. (b) Determine M_1, M_2 for the discharge, $Q = v_1 A_1 = v_2 A_2$.

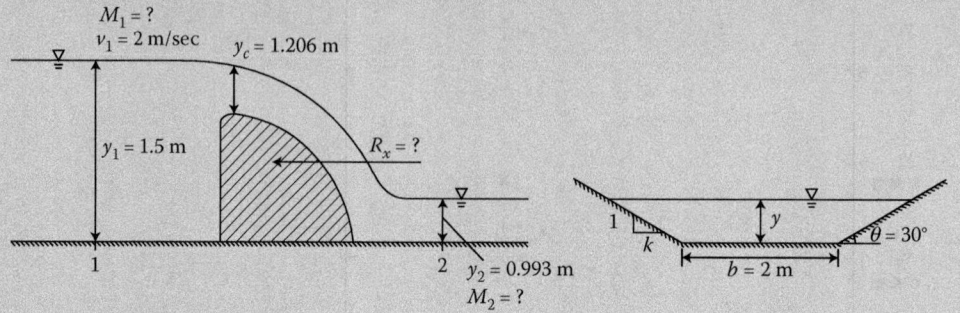

FIGURE EP 9.8
Water flows in a trapezoidal channel with a spillway inserted in the channel.

Mathcad Solution

(a) The reaction force, R_x acting on the spillway is computed by applying the momentum equation as follows:

$$\gamma := 9810 \, \frac{N}{m^3} \qquad g := 9.81 \, \frac{m}{sec^2} \qquad Q := 13.794 \, \frac{m^3}{sec}$$

$$\theta := 30 \text{ deg} \qquad k := \frac{1}{\tan(\theta)} \qquad b := 2 \, m$$

$$y_1 := 1.5 \, m \qquad y_2 := 0.993 \, m$$

$$B(y) := b + 2 \cdot k \cdot y \qquad A(y) := y \cdot (k \cdot y + b) \qquad \text{hcaA}(y) := \frac{y^2}{6}(2 \cdot k \cdot y + 3 \cdot b)$$

Guess value: $R_x := 1000 \, N$

Given

$$\frac{R_x}{\gamma} = \left(\frac{Q^2}{g \cdot A(y_1)} + \text{hcaA}(y_1) \right) - \left(\frac{Q^2}{g \cdot A(y_2)} + \text{hcaA}(y_2) \right)$$

$$R_x := \text{Find}(R_x) = 2.046 \times 10^3 \, N$$

(b) The values for M_1 and M_2 are determined from the momentum function equation as follows:

$$M_1 := \frac{Q^2}{g \cdot A(y_1)} + hcaA(y_1) = 7.011 \, \text{m}^3$$

$$M_2 := \frac{Q^2}{g \cdot A(y_2)} + hcaA(y_2) = 6.802 \, \text{m}^3$$

where $M_1 - M_2 = \dfrac{R_x}{\gamma} = 0.209 \, \text{m}^3$, and the upstream subcritical depth, y_1 and the downstream supercritical depth, y_2 are conjugate depths of flow.

9.13 Flow Depth and Major Head Loss for a Hydraulic Jump in Open Channel Flow

Although a critical depth of flow occurs at the hydraulic jump, it is not considered a "control" and thus does not serve as a "flow-measuring device"; rather, it serves an "energy dissipator," resulting in a major head loss. As discussed above in Section 9.10.1.1, numerous flow situations, such as the insertion of two controls in series in a channel, can result in the occurrence of a hydraulic jump. A hydraulic jump is a natural phenomenon that occurs when the open channel flow transitions from supercritical to subcritical flow. Specifically, when the fast-moving subcritical flow encounters the slow-moving subcritical flow, a hydraulic jump occurs at a critical depth of flow. The flow at the jump is characterized with a minimum specific energy and thus a maximum dissipation of energy. The resulting unknown major head loss, $h_{f,major}$ is due to a conversion of kinetic energy to heat and is modeled in the integral form of the energy equation. One may note that the head loss associated with a hydraulic jump is not due to frictional losses associated with the wall shear stress, τ_w (as in the case of flow resistance) but rather is due to intense agitation and turbulence and thus results in a high-energy loss. Therefore, because the high-energy head loss due to a hydraulic jump is not due to frictional losses associated with the wall shear stress, τ_w, there are no viscous forces to model in the integral form of the momentum equation. As such, the integral form of the momentum equation may be used to analytically solve for the unknown channel depth of flow (either upstream or downstream of the jump); thus, the derivation of the hydraulic jump equations (Eulerian/integral approach). Then, the integral form of the energy equation is used to solve for the unknown head loss due to the hydraulic jump (Eulerian/integral approach). Therefore, it is important to note that in addition to flow resistance, there is an additional source of major head loss in open channel flow due to the occurrence of a hydraulic jump. Furthermore, one may note that while the derivation of the hydraulic jump equations assumes a rectangular channel section, the corresponding solution approach to solve for the unknown channel depth of flow (either upstream or downstream of the jump) for non-rectangular channel sections is provided using a numerical solution approach that uses Mathcad in the solution procedure.

9.13.1 Critical Flow at the Hydraulic Jump

The insertion of two controls in series in a channel, as illustrated in Figure 9.17, can result in the occurrence of a hydraulic jump. A hydraulic jump is a natural phenomenon that occurs when the open channel flow transitions from supercritical to subcritical flow. Specifically, when the fast-moving subcritical flow encounters the slow-moving subcritical flow a hydraulic jump occurs at a critical depth of flow. The flow at the jump is characterized with a minimum specific energy and thus, a maximum dissipation of energy.

Assuming a rectangular channel section with a constant width, b and thus a constant unit discharge, q allows the determination of the M-y and E-y plots, as illustrated for a hydraulic jump in Figure 9.33. The uniform depth of flow upstream of the hydraulic jump, y_1 is supercritical. The hydraulic jump serves as an energy dissipator where the flow becomes critical. Furthermore, the uniform depth of flow downstream of the jump, y_2 is subcritical. The critical depth, y_c at the hydraulic jump is determined using the critical depth equation as follows:

$$y_c = \sqrt[3]{\frac{q^2}{g}} \tag{9.227}$$

which is plotted on the crest on the E-y plot for a given q, as illustrated in Figure 9.33b. The critical velocity, v_c and the minimum specific energy, E_c are computed as follows:

$$v_c = \sqrt{gy_c} \tag{9.228}$$

$$E_c = \frac{3}{2} y_c \tag{9.229}$$

Assuming a nonrectangular channel section does not allow the definition of a unit discharge, q and thus the definition of the E-y plot or the M-y plot. The uniform depth of flow upstream of the hydraulic jump, y_1 is supercritical. The hydraulic jump serves as an energy dissipator where the flow becomes critical. Furthermore, the uniform depth of flow downstream of the jump, y_2 is subcritical. The critical depth, y_c at the hydraulic jump is determined

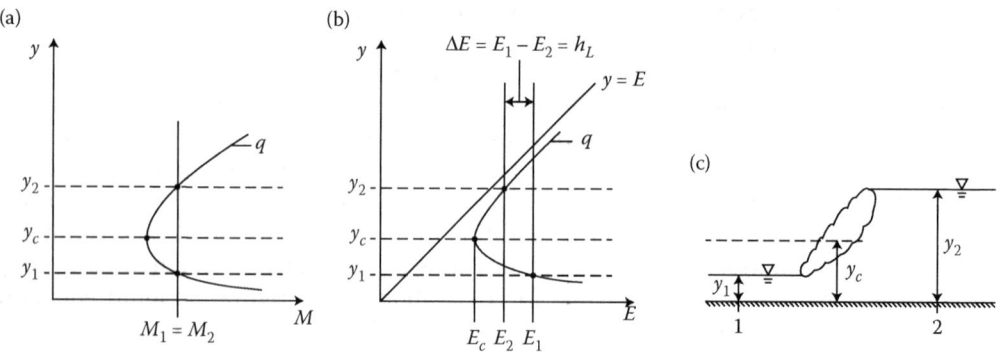

FIGURE 9.33
The M-y plot and the E-y plot for a hydraulic jump. (a) M-y plot, (b) E-y plot, (c) Hydraulic jump.

using the critical depth equation for a nonrectangular channel as follows:

$$\frac{A_c}{B_c} = \frac{Q^2}{A_c^2} \frac{1}{g} \tag{9.230}$$

which is numerically solved for y_c using Mathcad. The critical velocity, v_c and the minimum specific energy, E_c are computed as follows:

$$v_c = \sqrt{g \frac{A_c}{B_c}} \tag{9.231}$$

$$E_c = \frac{3}{2} \frac{A_c}{B_c} \tag{9.232}$$

9.13.2 Derivation of the Hydraulic Jump Equations: Rectangular Channel Sections

The derivation of the hydraulic jump equations assumes a rectangular channel section. Beginning with Equation 9.122 (the integral form of the momentum equation) the hydraulic jump equations, which are used to solve for the unknown channel depth of flow, either upstream or downstream of a simple hydraulic jump ($R_x = 0$) are derived as follows.

$$\frac{R_x}{\gamma b} = \underbrace{\left(\frac{q^2}{g y_1} + \frac{y_1^2}{2}\right)}_{M_1} - \underbrace{\left(\frac{q^2}{g y_2} + \frac{y_2^2}{2}\right)}_{M_2} \tag{9.233}$$

$$0 = \underbrace{\left(\frac{q^2}{g y_1} + \frac{y_1^2}{2}\right)}_{M_1} - \underbrace{\left(\frac{q^2}{g y_2} + \frac{y_2^2}{2}\right)}_{M_2} \tag{9.234}$$

Rearranging the momentum equation and grouping the similar terms for each control face (1 and 2) yields:

$$\left(\frac{q^2}{g y_1} - \frac{q^2}{g y_2}\right) = \left(\frac{y_2^2}{2} - \frac{y_1^2}{2}\right) \tag{9.235}$$

$$\frac{q^2}{g}\left(\frac{1}{y_1} - \frac{1}{y_2}\right) = \frac{1}{2}(y_2^2 - y_1^2) \tag{9.236}$$

$$\frac{q^2}{g}\left(\frac{y_2 - y_1}{y_1 y_2}\right) = \frac{1}{2}(y_2^2 - y_1^2) \tag{9.237}$$

$$\frac{q^2}{g}\left(\frac{1}{y_1 y_2}\right) = \frac{1}{2}\left(\frac{y_2^2 - y_1^2}{y_2 - y_1}\right) \tag{9.238}$$

Using the simplify command from the Mathcad Symbolic menu, the term in the right-hand side of the above equation can be simplified as follows:

$$\frac{y_2^2 - y_1^2}{y_2 - y_1} \text{ simplify} \rightarrow y_1 + y_2$$

thus, the momentum equation becomes as follows:

$$\frac{q^2}{g y_1 y_2} = \frac{1}{2}(y_1 + y_2) \tag{9.239}$$

At this point in the derivation of the hydraulic jump equations, it is necessary to state whether the upstream or the downstream flow condition are known. In practice, however, usually only the upstream conditions (y_1 and v_1) are known, while the downstream conditions (y_2 and v_2) are unknown. First, assuming the case for which the upstream conditions (y_1 and v_1) are known, one may derive the appropriate hydraulic jump equation in order to solve for the unknown downstream conditions (y_2 and v_2). Thus, substituting $q = v_1 y_1$ in Equation 9.239 above yields the following:

$$\frac{v_1^2 y_1^2}{g y_1 y_2} = \frac{1}{2}(y_1 + y_2) \tag{9.240}$$

$$\frac{v_1^2}{g} = \frac{1}{2}\frac{y_2}{y_1}(y_1 + y_2) \tag{9.241}$$

Next, dividing each side of the equation by y_1 in order to define the Froude number, F for the upstream section yields the following:

$$\frac{v_1^2}{g y_1} = F_1^2 = \frac{1}{2}\frac{y_2}{y_1}\frac{1}{y_1}(y_1 + y_2) \tag{9.242}$$

$$F_1^2 = \frac{1}{2}\frac{y_2}{y_1}\frac{1}{y_1}(y_1 + y_2) \tag{9.243}$$

$$F_1^2 = \frac{1}{2}\left(\frac{y_2}{y_1} + \frac{y_2^2}{y_1^2}\right) \tag{9.244}$$

$$F_1^2 = \frac{1}{2}\frac{y_2}{y_1}\left(\frac{y_2}{y_1} + 1\right) \tag{9.245}$$

which is known as the hydraulic jump equation for the case in which the upstream conditions (y_1 and v_1) are known. Note that the this equation is quadratic in the depth ratio, $r = y_2/y_1$, which may be solved using the Mathcad solve Symbolic solve block as follows:

Given

$$F_1{}^2 = \frac{1}{2} \cdot r \cdot (r+1)$$

$$\text{Find(r)} \rightarrow \left(\sqrt{2 \cdot F_1{}^2 + \frac{1}{4}} - \frac{1}{2} - \sqrt{2 \cdot F_1{}^2 + \frac{1}{4}} - \frac{1}{2} \right)$$

where the positive of the two roots is the physically possible depth ratio, $r = y_2/y_1$ and is given as follows:

$$\frac{y_2}{y_1} = -\frac{1}{2} + \frac{1}{2}\sqrt{1 + 8F_1{}^2} \tag{9.246}$$

$$\frac{y_2}{y_1} = \frac{1}{2}\left(-1 + \sqrt{1 + 8F_1{}^2}\right) \tag{9.247}$$

which presents the relationship between the upstream depths, y_1 and the downstream depth, y_2 (i.e., the conjugate depths) of a hydraulic jump. Alternatively, one may express the solution for the hydraulic jump equation as a function of the unknown downstream depth, y_2 as follows:

$$y_2 = \frac{y_1}{2}\left(-1 + \sqrt{1 + 8F_1{}^2}\right) \tag{9.248}$$

It is important to note that while the upstream depth y_1 (which is supercritical) is caused by a control upstream of it, the downstream depth, y_2 (which is subcritical) is caused by a control acting further downstream of it. Thus, only if the downstream control produces the required depth, y_2 will a hydraulic jump form; otherwise, it will not. Therefore, conjugate depths of flow (as presented in Equation 9.247) represent possible combinations of depths that could occur before and after the hydraulic jump. As such, if y_2 is conjugate to y_1, then the jump could occur anywhere in the channel between the upstream control and the downstream control, thus presenting an unstable location of the jump that causes it to move up and down the channel. Therefore, in order to stabilize the location of the jump, a special device known as a dentated sill is placed in between the two controls, which results in an "assisted" hydraulic jump.

In the case for which the downstream conditions (y_2 and v_2) are known, one may derive the appropriate hydraulic jump equation in order to solve for the unknown upstream conditions (y_1 and v_1). Thus, substituting $q = v_2\, y_2$ in Equation 9.239 above and following a similar procedure for the first case for which the upstream conditions (y_1 and v_1) are known, yields the solution for the hydraulic jump equation expressed as a function of the unknown upstream depth, y_1 as follows:

$$y_1 = \frac{y_2}{2}\left(-1 + \sqrt{1 + 8F_2^2}\right) \tag{9.249}$$

The momentum equation for the hydraulic jump between points 1 and 2 is given as follows:

$$\frac{R_x}{\gamma b} = \underbrace{\left(\frac{q^2}{gy_1} + \frac{y_1^2}{2}\right)}_{M_1} - \underbrace{\left(\frac{q^2}{gy_2} + \frac{y_2^2}{2}\right)}_{M_2} \tag{9.250}$$

Because, for a simple hydraulic jump, there is no reaction force, R_x acting on the control volume between points 1 and 2, the momentum function, M equation for the hydraulic jump between points 1 and 2 is given as follows:

$$M_1 = M_2 \tag{9.251}$$

One may note that y_1 (supercritical) and y_2 (subcritical) are known as conjugate depths of flow and are illustrated on the M-y plot. Conjugate depths of flow represent possible combinations of depths that could occur before and after the hydraulic jump. Furthermore, the minimum momentum function, M_c occurs at the critical depth of flow, y_c is defined as follows:

$$M_c = \left(\frac{q^2}{gy_c} + \frac{y_c^2}{2}\right) \tag{9.252}$$

9.13.3 Numerical Solution for a Hydraulic Jump: Nonrectangular Channel Sections

The corresponding solution approach to solve for the unknown channel depth of flow, either upstream or downstream of the hydraulic jump, for a nonrectangular channel section is provided using a numerical solution approach, which uses Mathcad in the solution procedure. Beginning with the integral form of the momentum equation for a nonrectangular channel section, and assuming a simple hydraulic jump ($R_x = 0$):

$$\frac{R_x}{\gamma} = \underbrace{\left(\frac{Q^2}{gA_1} + h_{ca1}A_1\right)}_{M_1} - \underbrace{\left(\frac{Q^2}{gA_2} + h_{ca2}A_2\right)}_{M_2} \tag{9.253}$$

$$0 = \underbrace{\left(\frac{Q^2}{gA_1} + h_{ca1}A_1\right)}_{M_1} - \underbrace{\left(\frac{Q^2}{gA_2} + h_{ca2}A_2\right)}_{M_2} \tag{9.254}$$

Assuming a common nonrectangular geometric channel section such as a trapezoidal or a triangular section, one may substitute the appropriate expressions for the cross-sectional area, A and the moment of area for the whole channel section, $h_{ca}A$, which are expressed as a function of the flow depth, y, into Equation 9.254 above. Furthermore, depending upon whether the upstream (y_1 and v_1) or the downstream (y_2 and v_2) flow condition is known, one may use a numerical Mathcad solve block to solve for the corresponding unknown flow depth as illustrated in the Example Problem 9.10 below.

9.13.4 Computation of the Major Head Loss due to a Hydraulic Jump

Once the integral form of the momentum equation is used to solve for the unknown depth of flow (regardless of the type of channel section, rectangular or nonrectangular), either upstream or downstream of the jump, the integral form of the energy equation is used to solve for the unknown major head loss due to the flow turbulence in the hydraulic jump as follows:

$$\underbrace{\left(y_1 + \frac{v_1{}^2}{2g} + z_1\right)}_{H_1} = \underbrace{\left(y_2 + \frac{v_2^2}{2g} + z_2\right)}_{H_2} + h_L \qquad (9.255)$$

Because $z_1 = z_2$, the specific energy equation for the hydraulic jump between points 1 and 2 is given as follows:

$$E_1 - h_L = E_2 \qquad (9.256)$$

Because the head loss, h_L acting within the control volume between points 1 and 1 due to the flow turbulence in the hydraulic jump is not zero, the specific energy, E decreases between points 1 and 1 and is given as follows:

$$E_1 \neq E_2 \qquad (9.257)$$

where the decrease in the specific energy, ΔE is given as follows:

$$\Delta E = h_L \qquad (9.258)$$

thus, y_1 (supercritical) and y_2 (subcritical), are not alternate depths of flow as illustrated on the E-y plot (assuming a rectangular channel section).

EXAMPLE PROBLEM 9.9

Referring to Example Problem 9.5, water flowing in a rectangular channel of width 3 m at a supercritical velocity of 4.823 m/sec and a supercritical depth of 0.933 m downstream of a sluice gate encounters a subcritical depth of flow produced by a second sluice gate located downstream of the first sluice gate. As such, a hydraulic jump occurs, as illustrated in Figure EP 9.9. (a) Determine the Froude number, F_1 for the supercritical depth, y_1 upstream of the hydraulic jump. (b) Determine the critical depth of flow, y_c critical velocity, v_c, and the minimum specific energy, E_c at the hydraulic jump. (c) Determine the subcritical depth, y_2 and the subcritical velocity, v_2 of the water downstream of the hydraulic jump. (d) Determine M_1, M_2, and M_c, and plot the M-y curve for the unit discharge, $q = v_1 \, y_1 = v_2 \, y_2$. (e) Determine E_1 and E_2 and plot the E-y curve for the unit discharge, $q = v_1 \, y_1 = v_2 \, y_2$. (f) Determine the head loss, h_L due to the flow turbulence in the hydraulic jump.

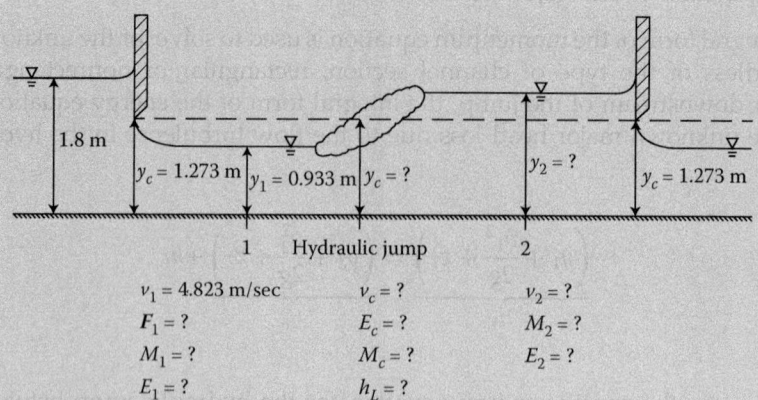

FIGURE EP 9.9
Water flows in a rectangular channel with two sluice gates inserted in the channel resulting in a hydraulic jump.

Mathcad Solution

(a) The Froude number, F_1 for the supercritical depth, y_1 upstream of the jump is computed as follows:

$$v_1 := 4.823 \frac{m}{sec} \qquad y_1 := 0.933 \, m \qquad F_1 := \frac{v_1}{\sqrt{g \cdot y_1}} = 1.594$$

(b) The critical depth of flow, y_c, critical velocity, v_c, and the minimum specific energy, E_c for the given unit discharge, q at the jump are determined as follows:

$$q := v_1 \cdot y_1 = 4.5 \frac{m^2}{sec} \qquad g := 9.81 \frac{m}{sec^2}$$

$$y_c := \sqrt[3]{\frac{q^2}{g}} = 1.273 \, m \qquad v_c := \sqrt{g \cdot y_c} = 3.534 \frac{m}{s} \qquad E_c := \frac{3}{2} y_c = 1.91 \, m$$

(c) Given that the upstream supercritical depth $y_1 = 0.933$ m and the upstream supercritical velocity $v_1 = 4.823$ m/sec are known, the unknown subcritical depth, y_2 of the water downstream of the hydraulic jump is computed by applying the appropriate hydraulic jump equation as follows:

$$y_2 := \frac{y_1}{2} \left(-1 + \sqrt{1 + 8F_1^2} \right) = 1.688 \, m$$

The downstream velocity, v_2 is computed by applying the unit discharge equation, $q = v_2 \, y_2$ as follows:

$$v_2 := \frac{q}{y_2} = 2.666 \frac{m}{sec}$$

(d) The values for M_1 and M_2 are determined from the momentum function equation as follows:

$$M_1 := \frac{q^2}{g \cdot y_1} + \frac{y_1^2}{2} = 2.648\,m^2 \qquad M_2 := \frac{q^2}{g \cdot y_2} + \frac{y_2^2}{2} = 2.648\,m^2$$

where $M_1 = M_2$, the upstream supercritical depth, y_1 and the downstream subcritical depth, y_2 are conjugate depths of flow, and the minimum momentum function, M_c which occurs at the critical depth of flow, $y_c = 1.273\,m$ is computed as follows:

$$y_c = 1.273\,m \qquad M_c := \frac{q^2}{g \cdot y_c} + \frac{y_c^2}{2} = 2.432\,m^2$$

The M-y curve for the unit discharge, $q = v_1\, y_1 = v_2\, y_2 = 4.5\,m^2/sec$ is plotted as follows:

$$M(y) := \frac{q^2}{g \cdot y} + \frac{y^2}{2} \qquad a \equiv 0.1\,m \qquad b \equiv 5\,m \qquad h \equiv 0.1\,m \qquad y := a,\, a + h..b$$

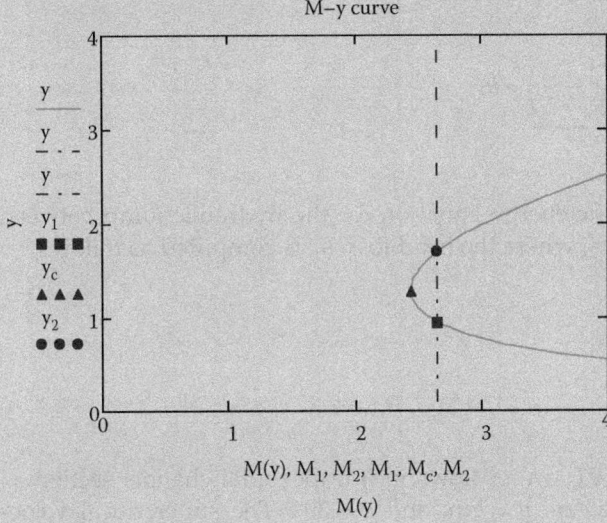

M–y curve

$M(y),\, M_1,\, M_2,\, M_1,\, M_c,\, M_2$

$M(y)$

(e) The values for E_1 and E_2 are determined from the specific energy equation as follows:

$$E_1 := y_1 + \frac{v_1^2}{2 \cdot g} = 2.119\,m \qquad\qquad E_2 := y_2 + \frac{v_2^2}{2 \cdot g} = 2.05\,m$$

where $E_1 - h_L = E_2$. The E-y curve for the unit discharge, $q = v_1\, y_1 = v_2\, y_2 = 4.5$ m^2/sec is plotted as follows:

$$E(y) := y + \frac{q^2}{2 \cdot g \cdot y^2} \qquad a \equiv 0.1\,m \qquad b \equiv 5\,m \qquad h \equiv 0.1\,m \qquad y := a,\, a + h..b$$

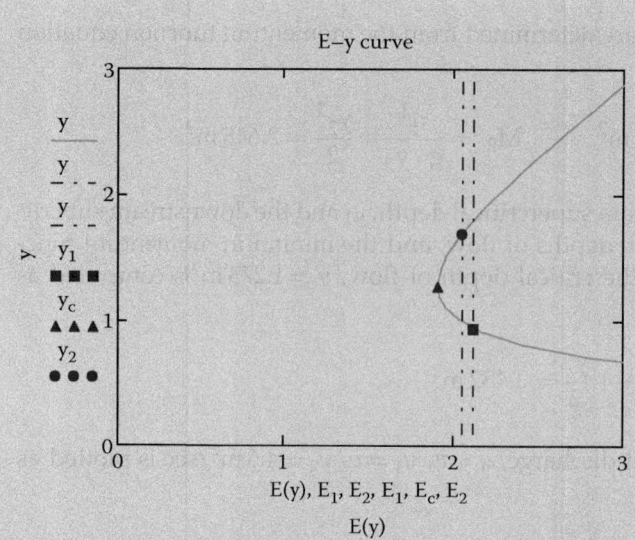

(f) The head loss, h_L due to the flow turbulence in the hydraulic jump is computed by applying the energy equation as follows:

$$\underbrace{\left(y_1 + \frac{v_1{}^2}{2g} + z_1 \right)}_{H_1} = \underbrace{\left(y_2 + \frac{v_2{}^2}{2g} + z_2 \right)}_{H_2} + h_L$$

Because $z_1 = z_2$, the specific energy equation for the hydraulic jump between points 1 and 2 is $E_1 - h_L = E_2$, where the head loss, h_L is computed as follows:

$$h_L := E_1 - E_2 = 0.068\,\text{m}$$

EXAMPLE PROBLEM 9.10

Referring to Example Problem 9.6, water flows in a trapezoidal channel section as illustrated in Figure EP 9.10, where $b = 2\,\text{m}$, and $\theta = 30°$. The supercritical velocity of the flow is 3.735 m/sec and the supercritical depth is 0.993 m downstream of the spillway encounters a subcritical depth of flow produced by a sluice gate located downstream of the spillway. As such, a hydraulic jump occurs, as illustrated in Figure EP 9.10. (a) Determine the Froude number, F_1 for the supercritical depth, y_1 upstream of the hydraulic jump. (b) Determine the critical depth of flow, y_c critical velocity, v_c and the minimum specific energy, E_c at the hydraulic jump. (c) Determine the subcritical depth, y_2 and the subcritical velocity, v_2 of the water downstream of the hydraulic jump. (d) Determine M_1, M_2 for the discharge, $Q = v_1\,A_1 = v_2\,A_2$. (e) Determine E_1 and E_2 for the discharge, $Q = v_1\,A_1 = v_2\,A_2$. (f) Determine the head loss, h_L due to the flow turbulence in the hydraulic jump.

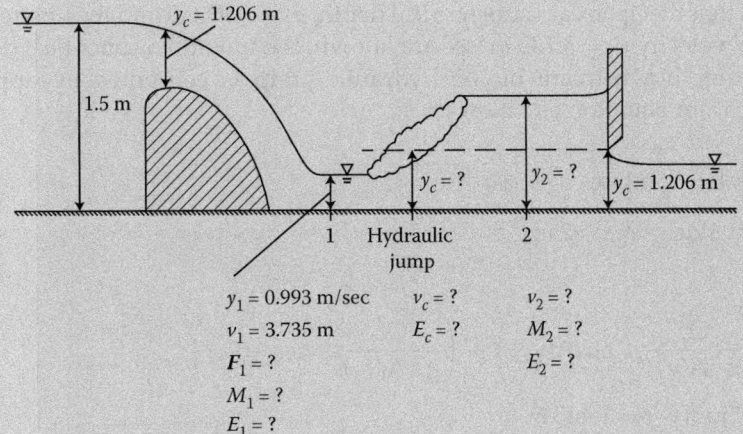

FIGURE EP 9.10

Water flows in a trapezoidal channel with a spillway and a sluice gate inserted in the channel resulting in a hydraulic jump.

Mathcad Solution

(a) The Froude number, F_1 for the supercritical depth, y_1 upstream of the hydraulic jump is computed as follows:

$$v_1 := 3.735 \frac{m}{sec} \qquad y_1 := 0.993\,m \qquad F_1 := \frac{v_1}{\sqrt{g \cdot y_1}} = 1.197$$

(b) The critical depth of flow, y_c for the given discharge, Q at the hydraulic jump is computed as follows:

$$\theta := 30\ deg \qquad b := 2\,m \qquad k := \frac{1}{\tan(\theta)} \qquad g := 9.81 \frac{m}{sec^2}$$

$$B(y) := b + 2\,k\,y \qquad A(y) := y \cdot (k \cdot y + b) \qquad Q := v_1\,A(y_1) = 13.797\,\frac{m^2}{sec}$$

Guess value: $\qquad y_c := 0.5\,m$

Given

$$\frac{A(y_c)}{B(y_c)} = \frac{Q^2}{A(y_c)^2}\frac{1}{g}$$

$$y_c := Find(y_c) = 1.206\,m$$

The critical velocity, v_c and the minimum specific energy, E_c for the given discharge, Q at the hydraulic jump are computed as follows:

$$v_c := \sqrt{g\,\frac{A(y_c)}{B(y_c)}} = 2.798\,\frac{m}{s} \qquad\qquad E_c := \frac{3}{2}\frac{A(y_c)}{B(y_c)} = 1.197\,m$$

(c) Given that the upstream supercritical depth, $y_1 = 0.993$ m and the upstream supercritical velocity $v_1 = 3.735$ m/sec are known, the unknown subcritical depth, y_2 of the water downstream of the hydraulic jump is computed by applying the momentum equation as follows:

$$\text{hcaA}(y) := \frac{y^2}{6}(2 \cdot k \cdot y + 3\,b)$$

Guess value: $y_2 := 2\,\text{m}$

Given

$$0 = \left(\frac{Q^2}{g \cdot A(y_1)} + \text{hcaA}(y_1)\right) - \left(\frac{Q^2}{g \cdot A(y_2)} + \text{hcaA}(y_2)\right)$$

$y_2 := \text{Find}(y_2) = 1.443\,\text{m}$

The downstream velocity, v_2 is computed by applying the unit discharge equation, $Q = v_2\,A_2$ as follows:

$$v_2 := \frac{Q}{A(y_2)} = 2.126\,\frac{\text{m}}{\text{s}}$$

(d) The values for M_1 and M_2 are determined from the momentum function equation as follows:

$$M_1 := \frac{Q^2}{g \cdot A(y_1)} + \text{hcaA}(y_1) = 6.804\,\text{m}^3$$

$$M_2 := \frac{Q^2}{g \cdot A(y_2)} + \text{hcaA}(y_2) = 6.804\,\text{m}^3$$

where $M_1 = M_2$, and the upstream supercritical depth, y_1 and the downstream subcritical depth, y_2 are conjugate depths of flow.

(e) The values for E_1 and E_2 are determined from the specific energy equation as follows:

$$E_1 := y_1 + \frac{v_1{}^2}{2 \cdot g} = 1.704\,\text{m} \qquad\qquad E_2 := y_2 + \frac{v_2{}^2}{2 \cdot g} = 1.673\,\text{m}$$

where $E_1 - h_L = E_2$.

(f) The head loss, h_L due to the flow turbulence in the hydraulic jump is computed by applying the energy equation as follows:

$$\underbrace{\left(y_1 + \frac{v_1{}^2}{2g} + z_1\right)}_{H_1} = \underbrace{\left(y_2 + \frac{v_2{}^2}{2g} + z_2\right)}_{H_2} + h_L$$

Because $z_1 = z_2$, the specific energy equation for the hydraulic jump between points 1 and 2 is $E_1 - h_L = E_2$, where the head loss, h_L is computed as follows:

$$h_L := E_1 - E_2 = 0.031\,\text{m}$$

9.14 Flow Depth and Major Head Loss in Uniform Open Channel Flow: Real Flow

Uniform flow will occur in the open channel in the absence of a channel or flow transition. Because the uniform flow creates a water surface profile and thus, affects the flow over a long channel section, the major head loss due to flow resistance is significant. The flow depth, y (or $S_o = S_f$, or v, or C [or n]) is solved using the resistance equation for uniform flow (Chezy or Manning's equation, where $C = R_h^{1/6}/n$, Equation 9.21) (Lagrangian/ differential approach) as follows (assuming the metric system) (see Section 9.6):

$$v = \underbrace{\frac{R_h^{1/6}}{n}}_{C} \sqrt{R_h S_f} = \frac{1}{n} R_h^{2/3} S_f^{1/2} \tag{9.259}$$

One may note that while the resistance equation for uniform flow may be analytically/ directly solved for $S_o = S_f$, or v, or C (or n), it is numerically solved for y using Mathcad. The major head loss, h_f is solved using major head loss equation (Equation 9.10) (Eulerian/ integral approach) as follows (assuming the metric system):

$$h_{f,maj} = \frac{\tau_w L}{\gamma R_h} = S_o L = S_f L = \frac{v^2 L}{C^2 R_h} = \frac{v^2}{\left(\frac{R_h^{1/6}}{n}\right)^2} \frac{L}{R_h} = \left(\frac{vn}{R_h^{2/3}}\right)^2 L \tag{9.260}$$

Because the resistance equation for uniform flow (Chezy or Manning's equation, expressed as the continuity equation) provides a relationship between the depth and discharge, uniform flow is considered a "control" in open channel flow and thus may serve as a flow-measuring device; there is a major head loss associated with the flow measurement. Thus, a less obvious control (and thus a flow measurement device) is due to the flow resistance because of the channel roughness that occurs in a uniform open channel flow, where the flow is not necessarily critical at the control. Depending upon the classification of the slope of the channel: mild, steep, or critical, the uniform flow will be subcritical, supercritical, or critical, respectively. The classification of the slope will depend on the channel roughness, the actual magnitude of the slope, and the discharge.

One may note that the critical equations are a special case of the resistance equation for uniform flow. In the case of a rectangular channel section, the critical depth and critical velocity equations are given, respectively, as follows:

$$y_c = \sqrt[3]{\frac{q^2}{g}} \tag{9.261}$$

$$v_c = \sqrt{g y_c} \tag{9.262}$$

In the case of a nonrectangular channel section, the critical depth and critical velocity equations are given, respectively, as follows:

$$\frac{A_c}{B_c} = \frac{Q^2}{A_c^2} \frac{1}{g} \tag{9.263}$$

which is numerically solved for y_c using Mathcad, and

$$v_c = \sqrt{g \frac{A_c}{B_c}}$$

(9.264)

EXAMPLE PROBLEM 9.11

Water flows at a discharge of 2.5 m³/sec in a rectangular channel that is 0.6 m wide with a channel bottom slope of 0.025 m/m, and a Manning's roughness coefficient of 0.015 m$^{-1/3}$ sec, as illustrated in Figure EP 9.11. (a) Determine the uniform depth of flow in the channel. (b) Determine the flow regime of the uniform flow in the channel, and the classification of the channel bottom slope. (c) Determine the head loss due to the channel resistance over a channel section length of 2 m. (d) Determine the critical channel bottom slope for the given discharge.

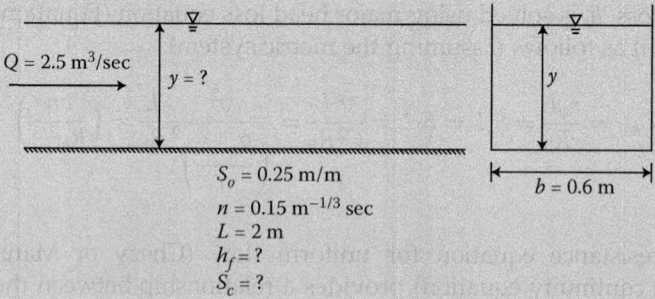

FIGURE EP 9.11
Water flows in a rectangular channel.

Mathcad Solution

(a) The uniform depth of flow, y in the channel is computed by applying the Manning's equation and the continuity equation, $Q = vA$ as follows:

$$Q := 2.5 \frac{m^3}{sec} \qquad b := 0.6 \, m \qquad S_o := 0.025 \frac{m}{m} \qquad n := 0.015 \, m^{\frac{-1}{3}} sec$$

$$A(y) := b \cdot y \qquad P_w(y) := (2 \cdot y + b) \qquad R_h(y) := \frac{A(y)}{P_w(y)}$$

Guess value: $y := 1 \, m$

Given

$$Q = \frac{1}{n} R_h(y)^{2/3} \cdot S_o^{1/2} \cdot A(y)$$

$$y := Find(y) = 1.044 \, m$$

(b) In order to determine the flow regime of the uniform flow in the channel, one must first determine the critical depth of flow, y_c for the given discharge, $Q = 2.5 \, \text{m}^3/\text{sec}$ as follows:

$$q := \frac{Q}{b} = 4.167 \, \frac{\text{m}^2}{\text{sec}} \qquad g := 9.81 \, \frac{\text{m}}{\text{sec}^2} \qquad y_c := \sqrt[3]{\frac{q^2}{g}} = 1.21 \, \text{m}$$

Therefore, because $y = 1.044 \, \text{m}$ is less than $y_c = 1.21 \, \text{m}$, the uniform depth of flow, y is supercritical. Furthermore, the channel bottom slope, $S_o = 0.025 \, \text{m/m}$ may be classified as a steep slope.

(c) The head loss, h_f due to the channel resistance over a channel section length, $L = 2$ m is computed by applying the major head loss equation as follows:

$$L := 2 \, \text{m} \qquad h_f := S_o \cdot L = 0.05 \, \text{m}$$

(d) The critical channel bottom slope, S_c for the given discharge, $Q = 2.5 \, \text{m}^3/\text{sec}$ is computed by applying the Manning's equation, and the continuity equation, $Q = vA$ for the critical depth of flow, $y_c = 1.21 \, \text{m}$ as follows:

$$Q = 2.5 \, \frac{\text{m}^3}{\text{sec}} \qquad y_c = 1.21 \, \text{m}$$

Guess value: $S_c := 0.01 \, \frac{\text{m}}{\text{m}}$

Given

$$Q = \frac{1}{n} \, R_h(y_c)^{2/3} \cdot S_c^{1/2} \cdot A(y_c)$$

$$S_c := \text{Find}(S_c) = 0.018 \, \frac{\text{m}}{\text{m}}$$

EXAMPLE PROBLEM 9.12

Water flows at a discharge of $2.8 \, \text{m}^3/\text{sec}$ in a trapezoidal channel section as illustrated in Figure EP 9.12. where $b = 6$ m. and $\theta = 30°$. with a channel bottom slope

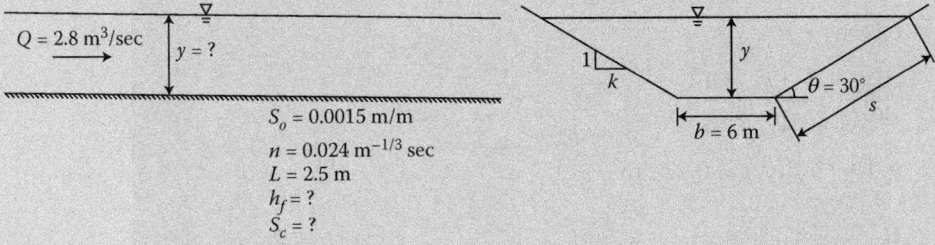

FIGURE EP 9.12
Water flows in a trapezoidal channel.

of 0.0015 m/m, and a Manning's roughness coefficient of $0.024\,\mathrm{m}^{-1/3}$ sec. (a) Determine the uniform depth of flow in the channel. (b) Determine the flow regime of the uniform flow in the channel, and the classification of the channel bottom slope. (c) Determine the head loss due to the channel resistance over a channel section length of 2.5 m. (d) Determine the critical channel bottom slope for the given discharge.

Mathcad Solution

(a) The uniform depth of flow, y in the channel is computed by applying the Manning's equation and the continuity equation, $Q = vA$ as follows:

$$Q := 2.8\,\frac{\mathrm{m}^3}{\mathrm{sec}} \qquad\qquad S_o := 0.0015\,\frac{\mathrm{m}}{\mathrm{m}} \qquad\qquad n := 0.024\,\mathrm{m}^{\frac{-1}{3}}\,\mathrm{sec}$$

$$b := 6\,\mathrm{m} \qquad\qquad \theta := 30\,\mathrm{deg} \qquad k := \frac{1}{\tan(\theta)} \qquad s(y) := \frac{y}{\sin(\theta)}$$

$$B(y) := b + 2 \cdot k \cdot y \qquad\qquad A(y) := y \cdot (k \cdot y + b)$$

$$P_w(y) := (2 \cdot s(y) + b) \qquad\qquad R_h(y) := \frac{A(y)}{P_w(y)}$$

Guess value: $y := 1\mathrm{m}$

Given

$$Q = \frac{1}{n}\,R_h(y)^{\frac{2}{3}} \cdot S_o^{\frac{1}{2}} \cdot A(y)$$

$$y := \mathrm{Find}(y) = 0.467\,\mathrm{m}$$

(b) In order to determine the flow regime of the uniform flow in the channel, one must first determine the critical depth of flow, y_c for the given discharge, $Q = 2.8\,\mathrm{m}^3/\mathrm{sec}$ as follows:

$$Q = 2.8\,\frac{\mathrm{m}^3}{\mathrm{sec}} \qquad\qquad g := 9.81\,\frac{\mathrm{m}}{\mathrm{sec}^2}$$

Guess value: $y_c := 1\,\mathrm{m}$

Given

$$\frac{A(y_c)}{B(y_c)} = \frac{Q^2}{A(y_c)^2} \cdot \frac{1}{g}$$

$$y_c := \mathrm{Find}(y_c) = 0.274\,\mathrm{m}$$

Therefore, because $y = 0.467\,\mathrm{m}$ is greater than $y_c = 0.274\,\mathrm{m}$, the uniform depth of flow, y is subcritical. Furthermore, the channel bottom slope, $S_o = 0.0015\,\mathrm{m/m}$ may be classified as a mild slope.

(c) The head loss, h_f due to the channel resistance over a channel section length, $L =$ 2.5 m is computed by applying the major head loss equation as follows:

$$L := 2.5 \text{ m} \qquad h_f := S_o \cdot L = 3.75 \times 10^{-3} \text{ m}$$

(d) The critical channel bottom slope, S_c for the given discharge, $Q = 2.8 \text{ m}^3/\text{sec}$ is computed by applying the Manning's equation and the continuity equation, $Q = vA$ for the critical depth of flow, $y_c = 0.274$ m as follows:

$$Q = 2.8 \frac{\text{m}^2}{\text{sec}} \qquad y_c = 0.274 \text{ m}$$

$$\text{Guess value:} \quad S_c := 0.002 \frac{\text{m}}{\text{m}}$$

Given

$$Q = \frac{1}{n} R_h(y_c)^{\frac{2}{3}} \cdot S_c^{\frac{1}{2}} \cdot A(y_c)$$

$$S_c := \text{Find}(S_c) = 9.164 \times 10^{-3} \frac{\text{m}}{\text{m}}$$

EXAMPLE PROBLEM 9.13

Water flows at a discharge of $4 \text{ m}^3/\text{sec}$ in a partially filled circular channel section as illustrated in Figure EP 9.13, where $r_o = 1$ m and $\theta = 2$ radians, with a channel bottom slope of 0.0013 m/m and a Manning's roughness coefficient of $0.025 \text{ m}^{-1/3}$ sec. (a) Determine the uniform depth of flow in the channel. (b) Determine the flow regime of the uniform flow in the channel and the classification of the channel bottom slope. (c) Determine the head loss due to the channel resistance over a channel section length of 5 m. (d) Determine the critical channel bottom slope for the given discharge.

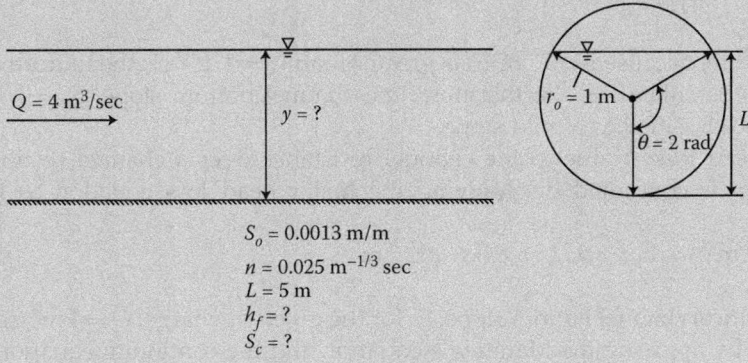

$$S_o = 0.0013 \text{ m/m}$$
$$n = 0.025 \text{ m}^{-1/3} \text{ sec}$$
$$L = 5 \text{ m}$$
$$h_f = ?$$
$$S_c = ?$$

FIGURE EP 9.13
Water flows in a partially filled circular channel.

Mathcad Solution

(a) The uniform depth of flow, y in the channel is computed by applying the Manning's equation and the continuity equation, $Q = vA$ as follows:

$$Q := 4 \frac{m^3}{sec} \qquad\qquad S_o := 0.0013 \frac{m}{m} \qquad\qquad n := 0.025 \, m^{\frac{-1}{3}} \, sec$$

$$r_o := 1 \, m \qquad \theta := 2 \, rad \qquad A(y) := (y + r_o \cdot \cos(\theta))^2 \cdot (\theta - \sin(\theta) \cdot \cos(\theta))$$

$$P_w(y) := 2 \, \theta \, r_o \qquad\qquad R_h(y) := \frac{A(y)}{P_w(y)}$$

Guess value: $y := 1 \, m$

Given

$$Q = \frac{1}{n} \, R_h(y)^{\frac{2}{3}} \, S_o^{\frac{1}{2}} \, A(y)$$

$$y := Find(y) = 1.578 \, m$$

(b) In order to determine the flow regime of the uniform flow in the channel, one must first determine the critical depth of flow, y_c for the given discharge, $Q = 4 \, m^3/sec$ as follows:

$$Q = 4 \frac{m^3}{sec} \qquad g := 9.81 \frac{m}{sec^2}$$

Guess value: $y_c := 1 \, m$

Given

$$\frac{A(y_c)}{B(y_c)} = \frac{Q^2}{A(y_c)^2} \frac{1}{g}$$

$$y_c := Find(y_c) = 1.467 \, m$$

Therefore, because $y = 1.587 \, m$ is greater than $y_c = 1.467 \, m$, the uniform depth of flow, y is subcritical. Furthermore, the channel bottom slope, $S_o = 0.0013 \, m/m$ may be classified as a mild slope.

(c) The head loss, h_f due to the channel resistance over a channel section length, $L = 5 \, m$ is computed by applying the major head loss equation as follows:

$$L := 5 \, m \qquad h_f := S_o \, L = 6.5 \times 10^{-3} \, m$$

(d) The critical channel bottom slope, S_c for the given discharge, $Q = 4 \, m^3/sec$ is computed by applying the Manning's equation and the continuity equation, $Q = vA$ for the critical depth of flow, $y_c = 1.467 \, m$ as follows:

$$Q = 4 \frac{m^3}{sec} \qquad\qquad y_c = 1.467 \, m$$

Guess value: $S_c := 0.002 \dfrac{m}{m}$

Given

$$Q = \frac{1}{n} R_h(y_c)^{\frac{2}{3}} S_c^{\frac{1}{2}} A(y_c)$$

$$S_c := \text{Find}(S_c) = 2.547 \times 10^{-3} \frac{m}{m}$$

EXAMPLE PROBLEM 9.14

Water flows at a discharge of $3.5 \, m^3/\text{sec}$ in a rectangular channel that is 0.7 m wide with a channel bottom slope of 0.03 m/m, and a uniform depth of flow of 1.5 m, as illustrated in Figure EP 9.14. (a) Determine the flow regime of the uniform flow in the channel, and the classification of the channel bottom slope. (b) Determine the Manning's roughness coefficient. (c) Determine the head loss due to the channel resistance over a channel section length of 4 m. (d) Determine the critical channel bottom slope for the given discharge.

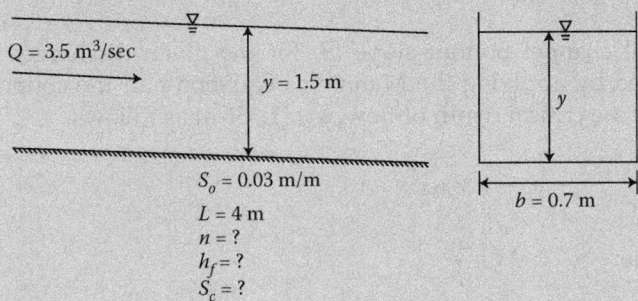

FIGURE EP 9.14
Water flows in a rectangular channel.

Mathcad Solution

(a) In order to determine the flow regime of the uniform flow in the channel, one must first determine the critical depth of flow, y_c for the given discharge, $Q = 3.5 \, m^3/\text{sec}$ as follows:

$$Q := 3.5 \frac{m^3}{\text{sec}} \qquad b := 0.7 \, m \qquad q := \frac{Q}{b} = 5 \frac{m^2}{\text{sec}}$$

$$g := 9.81 \frac{m}{\text{sec}^2} \qquad y_c := \sqrt[3]{\frac{q^2}{g}} = 1.366 \, m$$

Therefore, because $y = 1.5 \, m$ is greater than $y_c = 1.336 \, m$, the uniform depth of flow, y is subcritical. Furthermore, the channel bottom slope, $S_o = 0.03 \, m/m$ may be classified as a mild slope.

(b) The Manning's roughness coefficient, n is computed by applying the Manning's equation and the continuity equation, $Q = vA$ as follows:

$$Q = 3.5 \frac{m^3}{sec} \qquad S_o := 0.03 \frac{m}{m} \qquad y := 1.5 \, m$$

$$A(y) := by \qquad P_w(y) := (2\,y + b) \qquad R_h(y) := \frac{A(y)}{P_w(y)}$$

Guess value: $\qquad n := 0.01 \, m^{\frac{-1}{3}} sec$

Given

$$Q = \frac{1}{n} R_h(y)^{\frac{2}{3}} \cdot S_o^{\frac{1}{2}} \cdot A(y)$$

$$n := Find(n) = 0.022 m^{\frac{-1}{3}} \, sec$$

(c) The head loss, h_f due to the channel resistance over a channel section length, $L = 4$ m is computed by applying the major head loss equation as follows:

$$L := 4 \, m \qquad h_f := S_o \cdot L = 0.12 \, m$$

(d) The critical channel bottom slope, S_c for the given discharge, $Q = 3.5 \, m^3/sec$ is computed by applying the Manning's equation and the continuity equation, $Q = vA$ for the critical depth of flow, $y_c = 1.336$ m as follows:

$$Q = 3.5 \frac{m^3}{sec} \qquad y_c = 1.366 \, m$$

Guess value: $\quad S_c := 0.01 \frac{m}{m}$

Given

$$Q = \frac{1}{n} R_h(y_c)^{\frac{2}{3}} \cdot S_c^{\frac{1}{2}} \cdot A(y_c)$$

$$S_c := Find(S_c) = 0.037 \frac{m}{m}$$

9.15 Flow Depth and Major Head Loss in Nonuniform Open Channel Flow: Real Flow

Because "controls" such as a hydraulic drop, a break in the channel slope, or a free overfall (see Figure 9.34a, b and c, respectively) generate a nonuniform surface water profile and thus affect the flow over a long channel section, the major head loss due to flow resistance is significant. It is important to recall that in the case of the hydraulic drop, the flow is forced into subcritical flow by the banks of a lake, which is followed by a short crest (where $dz/dx = 0$,

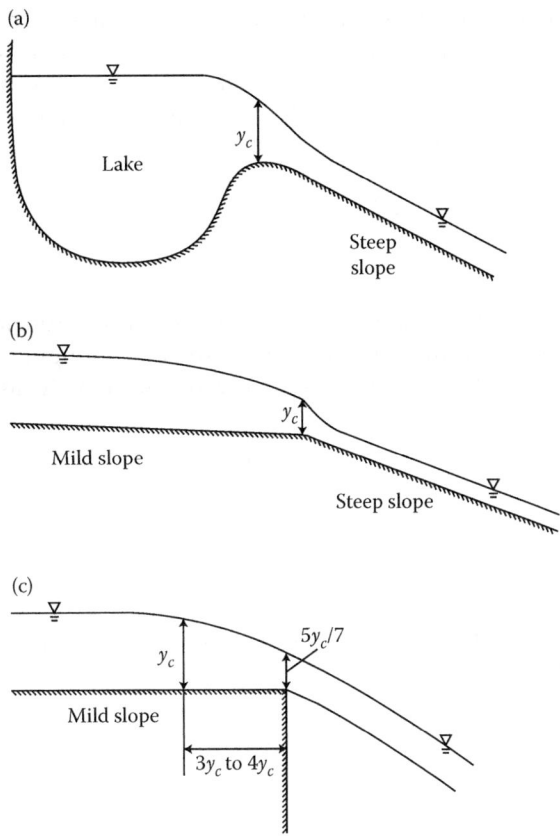

FIGURE 9.34
Controls that generate a nonuniform surface water profile over a long channel section. (a) Hydraulic drop, (b) Break in channel slope, (c) Free overfall. (Adapted from Henderson, F. M., *Open Channel Flow*, Macmillan, New York, 1966, p. 42, p. 107, p. 192.)

$dy/dx \neq 0$, and $F = 1$), then a channel with a steep slope where the flow is supercritical; the control provides a mechanism of release from subcritical to critical flow, followed by a steep channel bottom slope such that the channel resistance/roughness does not impose an effective restraint on the flow. In the case of the break in the channel slope, the flow is forced into subcritical flow (upstream depth of flow) by the channel roughness, then at the break in the channel slope, the flow becomes critical (where $S_o = S_f$, $dy/dx \neq 0$, and $F = 1$), followed by the steep channel, where the flow is supercritical (downstream depth of flow); the control provides a mechanism of release from subcritical to critical flow, followed by supercritical flow downstream of the control. In the case of the free overall, the flow is forced into subcritical flow (upstream depth of flow) by the channel roughness, then just upstream of the control, the flow becomes critical (where $S_o = S_f$, $dy/dx \neq 0$, and $F = 1$), followed by a falling jet; the control provides a mechanism of release from subcritical to critical flow, followed by a falling jet at the brink. Finally, it is important to recall that a steep channel slope (for which the uniform flow is supercritical) ending in a free overfall does not produce a draw-down curve (as in the case of a typical channel control) and thus does not serve as a channel control as illustrated in Figure 9.15.

The uniform depth of flow, y in the channel at a great distance upstream and/or downstream of the control, where the control ceases to affect the water surface profile, is determined by applying the resistance equation for uniform flow (Chezy or Manning's equation, where $C = (R_h^{1/6}/n)$ (Equation 9.21)) (Lagrangian/differential approach) as follows (assuming the metric system) (see Section 9.6):

$$v = \underbrace{\frac{R_h^{1/6}}{n}}_{C} \sqrt{R_h S_f} = \frac{1}{n} R_h^{2/3} S_f^{1/2} \tag{9.265}$$

Determination of the resulting nonuniform surface water profile begins by computation of the critical depth of flow, y_c at the control. In the case of a rectangular channel section, the critical depth and critical velocity equations are given, respectively, as follows:

$$y_c = \sqrt[3]{\frac{q^2}{g}} \tag{9.266}$$

$$v_c = \sqrt{g y_c} \tag{9.267}$$

In the case of a nonrectangular channel section, the critical depth and critical velocity equations are given, respectively, as follows:

$$\frac{A_c}{B_c} = \frac{Q^2}{A_c^2} \frac{1}{g} \tag{9.268}$$

which is numerically solved for y_c using Mathcad, and

$$v_c = \sqrt{g \frac{A_c}{B_c}} \tag{9.269}$$

The remaining nonuniform surface water profile is determined by applying the resistance equation for nonuniform flow derived above (see Equation 9.101) (Lagrangian/differential approach), $dy/dx = (S_o - S_f)/(1 - F^2)$ beginning just upstream and/or just downstream of the control and ending at the uniform depth of flow. It is important to note the calculation must actually begin at a point just upstream of the brink/control. The reason is that at the control, the flow is critical and thus the $F = 1$, for which the resistance equation would result in $dy/dx = \infty$, which would indicate a discontinuity in the water surface profile. The resistance equation for nonuniform flow is an ordinary differential equation that is numerically solved using Mathcad (because it is not, in general, explicitly solvable) in order to determine an unknown depth, y for a given point in the nonuniform open channel flow (proceeding in the direction(s) in which the control is being exercised, either upstream and/or downstream of the control). One may note that while there are numerous ordinary differential equation solvers in Mathcad, "rkfixed," which uses the fourth-order Runge-Kutta fixed-step method, is selected and applied in the examples below. The Chezy equation is used to solve for the friction slope, S_f as follows:

$$S_f = \frac{v^2}{C^2 R_h} \tag{9.270}$$

and the Froude number is given as follows:

$$F = \frac{v}{v_c} \tag{9.271}$$

The major head loss, h_f is solved using the major head loss equation (Equation 9.10) (Eulerian/integral approach) as follows:

$$h_f = \frac{\tau_w L}{\gamma R_h} = S_f L = \underbrace{\left[\frac{v^2}{C^2 R_h}\right]}_{S_f} L \tag{9.272}$$

One may note that a free overfall is a special case of the sharp-crested weir, where the height of the weir, $P = 0$.

Because critical flow, which is defined by an analytical relationship between the flow depth and the discharge in the channel, occurs at (or near) the following "controls": a break in the channel slope, a hydraulic drop, or a free overfall, (similar to the remaining controls that generate nonuniform flow, but affect the flow only over a short channel section), each may serve as a flow-measuring device; there is a small minor head loss associated with the flow measurement. It is important to recall from Section 9.9 that in the case of a hydraulic drop and a break in the channel slope, because there is no contraction of the flow and the flow pressure distribution is hydrostatic at the control, critical flow occurs at the control as illustrated in Figure 9.34a, b, respectively. As such, a hydraulic drop (a special case of an abrupt vertical contraction) and a break in the channel slope may serve as a "critical depth" meter. However, one may recall from Section 9.9 that in the case of the free overfall (a special case of a sharp-crested weir, $P = 0$), there is a contraction in the depth of flow (to a depth of approximately $5y_c/7$) and the flow pressure distribution is not hydrostatic at the control. As a result, the critical flow retreats just upstream of the brink where there is no contraction in the depth of flow and where the pressure distribution is hydrostatic as illustrated in Figure 9.34c. As such, the free overfall does not serve as a "critical depth" meter, where the critical depth actually occurs at a distance of $3y_c$ to $4\,y_c$ upstream of the brink. However, in the determination of the nonuniform surface water profile over a long channel section, critical flow is assumed at the brink for simplicity of calculation; in the case of a free overfall, because of the typical practical assumption of ideal flow made for flow-measuring devices in general (ideal flow meters), it is assumed that the critical flow occurs at the control/brink.

EXAMPLE PROBLEM 9.15

Water flows at a discharge of 30 m³/sec in a trapezoidal channel section as illustrated in Figure EP 9.15, where $b = 5$ m, and $\theta = 30°$, with a channel bottom slope of 0.001 m/m and a Manning's roughness coefficient of 0.026 m$^{-1/3}$ sec. The channel terminates in a free overfall (brink) as illustrated in Figure EP 9.15. (a) Determine the uniform depth of flow in the channel at a great distance upstream of the brink, where the control ceases to affect the water surface profile. (b) Determine the flow regime of the uniform flow in the channel and the classification of the channel bottom slope. (c) Determine and plot the resulting nonuniform surface water profile due to the drawdown. (d) Determine the head loss due to the channel resistance over a channel section length of 2500 m.

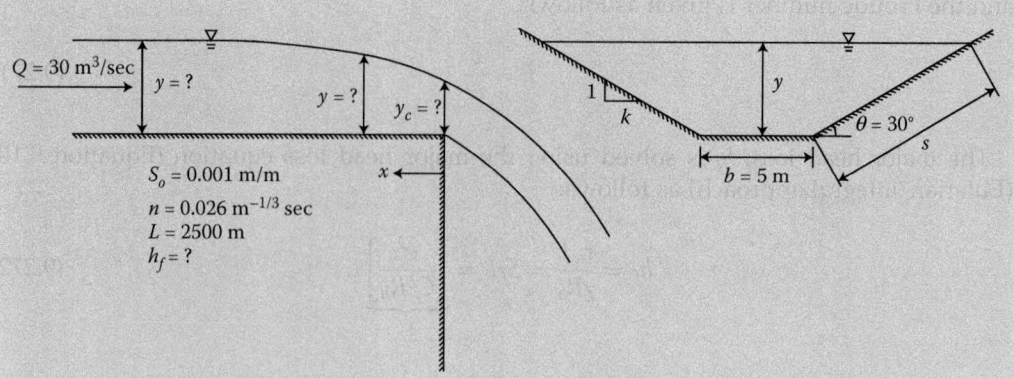

FIGURE EP 9.15
Water flows in a trapezoidal channel that terminates in a free overfall (brink).

Mathcad Solution

(a) The uniform depth of flow, y in the channel at a great distance upstream of the brink, where the control ceases to affect the water surface profile is computed by applying the Manning's equation and the continuity equation, $Q = vA$ as follows:

$$Q := 30 \; \frac{m^3}{sec} \qquad\qquad S_o := 0.001 \; \frac{m}{m} \qquad\qquad n := 0.026 \; m^{\frac{-1}{3}} \; sec$$

$$b := 5 \; m \qquad\qquad \theta := 30 \; deg \qquad\qquad k := \frac{1}{\tan(\theta)} \qquad s(y) := \frac{y}{\sin(\theta)}$$

$$B(y) := b + 2 \cdot k \cdot y \qquad A(y) := y \cdot (k \cdot y + b)$$

$$P_w(y) := (2 \cdot s(y) + b) \qquad R_h(y) := \frac{A(y)}{P_w(y)}$$

Guess value: $y := 1 \; m$

Given

$$Q = \frac{1}{n} \; R_h(y)^{\frac{2}{3}} \cdot S_o^{\frac{1}{2}} \cdot A(y)$$

$y := Find(y) = 2.216 \, m$

(b) In order to determine the flow regime of the uniform flow in the channel, one must first determine the critical depth of flow, y_c for the given discharge, $Q = 30 \; m^3/sec$ as follows:

$$Q = 30 \; \frac{m^3}{sec} \qquad g := 9.81 \; \frac{m}{sec^2}$$

Guess value: $y_c := 1 \; m$

Given

$$\frac{A(y_c)}{B(y_c)} = \frac{Q^2}{A(y_c)^2} \frac{1}{g}$$

$y_c := \text{Find}(y_c) = 1.315\,\text{m}$

Therefore, because $y = 2.216\,\text{m}$ is greater than $y_c = 1.315\,\text{m}$, the uniform depth of flow, y is subcritical. Furthermore, the channel bottom slope, $S_o = 0.001\,\text{m/m}$ may be classified as a mild slope.

(c) Determination of the resulting nonuniform surface water profile begins by computation of the critical depth of flow, y_c at the control/brink. The critical depth of flow, y_c was determined above by applying the critical depth equation, which yielded $y_c = 1.315\,\text{m}$. The remaining nonuniform surface water profile is determined by applying the resistance equation for nonuniform flow, $dy/dx = (S_o - S_f)/(1 - F^2)$ beginning just upstream of the control/brink and ending at the uniform depth of flow, $y = 2.216\,\text{m}$, which is at a great distance upstream of the brink/control. It is important to note the calculation must actually begin at a point just upstream of the brink/control. The reason is that at the brink/control, the flow is critical and thus the $F = 1$, for which the resistance equation would result in $dy/dx = \infty$, which would indicate a discontinuity in the water surface profile. Thus, the calculations will begin at $y = y_c + 0.001\,\text{m} = 1.316\,\text{m}$. The resistance equation for nonuniform flow is an ordinary differential equation that is numerically solved for y using an ordinary differential equation solver in Mathcad, "rkfixed," which uses the fourth-order Runge-Kutta fixed-step method as follows:

$$S_o := 0.001\,\frac{\text{m}}{\text{m}} \qquad n := 0.026\,\text{m}^{\frac{-1}{3}}\,\text{sec} \qquad Q := 30\,\frac{\text{m}^3}{\text{sec}}$$

$$y_c = 1.315\text{m} \qquad v_c := \sqrt{g\,\frac{A(y_c)}{B(y_c)}} = 3.135\,\frac{\text{m}}{\text{s}} \qquad v(y) := \frac{Q}{A(y)}$$

$$C(y) := \frac{R_h(y)^{\frac{1}{6}}}{n} \qquad S_f(y): \frac{v(y)^2}{C(y)^2\,R_h(y)} \qquad F(y) := \frac{v(y)}{v_c}$$

The first-order ordinary differential equation initial value problem is defined as follows:

$$\frac{d}{dx}y(x) = f(x, y) = \frac{S_o - S_f(y)}{1 - F(y)^2} \qquad y(x0) = y0 = y_c + 0.001$$

Enter the initial value problem specifics, noting that the derivative equation must be unit balanced (i.e., the arguments must have appropriate units attached):

$$f(x, y) := \frac{S_o - S_f(y \cdot m)}{1 - F(y \cdot m)^2}$$

Define the beginning of solution interval:

x0 := 0

y0 := 1.315 + 0.001 = 1.316

Enter the desired solution parameters, including the endpoint of solution interval, x1 and the number of solution values, N on [x0, x1]:

x1 := −2500

N := 10000

Define solver parameters, including the vector of initial solution values, *ic* and the derivative function, *D*, for which the second argument, *y* must be a vector of unknown function values:

ic := y0

D(x, y) := f(x, y)

Solution matrix, for which the independent variable values are *x* and the solution function values are *y*:

S := rkfixed(ic, x0, x1, N, D)

x := S$^{(0)}$

y := S$^{(1)}$

Values for *x* and *y* for the first 15 spatial points along the channel, beginning just upstream of the control/brink, for which $x = 0$ m and $y = 1.316$ m:

		0	1
S =	0	0	1.316
	1	-0.25	1.654
	2	-0.5	1.655
	3	-0.75	1.656
	4	-1	1.657
	5	-1.25	1.658
	6	-1.5	1.659
	7	-1.75	1.661
	8	-2	1.662
	9	-2.25	1.663
	10	-2.5	1.664
	11	-2.75	1.665
	12	-3	1.666
	13	-3.25	1.667
	14	-3.5	1.668
	15	-3.75	...

Values for *x* and *y* for the last 15 spatial points along the channel, ending at a great distance upstream of the control/brink, for which $x = -2500$ m and $y = 2.216$ m:

$$S = \begin{array}{|c|c|c|} \hline & 0 & 1 \\ \hline 9985 & -2.496 \cdot 10^3 & 2.216 \\ \hline 9986 & -2.497 \cdot 10^3 & 2.216 \\ \hline 9987 & -2.497 \cdot 10^3 & 2.216 \\ \hline 9988 & -2.497 \cdot 10^3 & 2.216 \\ \hline 9989 & -2.497 \cdot 10^3 & 2.216 \\ \hline 9990 & -2.498 \cdot 10^3 & 2.216 \\ \hline 9991 & -2.498 \cdot 10^3 & 2.216 \\ \hline 9992 & -2.498 \cdot 10^3 & 2.216 \\ \hline 9993 & -2.498 \cdot 10^3 & 2.216 \\ \hline 9994 & -2.498 \cdot 10^3 & 2.216 \\ \hline 9995 & -2.499 \cdot 10^3 & 2.216 \\ \hline 9996 & -2.499 \cdot 10^3 & 2.216 \\ \hline 9997 & -2.499 \cdot 10^3 & 2.216 \\ \hline 9998 & -2.499 \cdot 10^3 & 2.216 \\ \hline 9999 & -2.5 \cdot 10^3 & 2.216 \\ \hline 10000 & -2.5 \cdot 10^3 & ... \\ \hline \end{array}$$

The solution of the resistance equation was constrained to begin just upstream of the control/brink, for which $x = 0$ m and $y = 1.316$ m. However, because the critical depth of flow, $y_c = 1.315$ m actually occurs at the control/brink, the x-axis is redefined in order to reflect the occurrence of $y_c = 1.315$ m at $x = 0$ m:

$$y_c = 1.315 \text{ m} \qquad i := 0..N - 1 \qquad yy_{i+1} := y_i$$

$$y := yy \qquad\qquad y_0 := 1.315$$

Values for x and y for the first 15 spatial points along the channel, beginning at the control/brink, for which the x-axis has been redefined in order to reflect the occurrence of $y_c = 1.315$ m at $x = 0$ m:

$$x = \begin{array}{|c|c|} \hline & 0 \\ \hline 0 & 0 \\ \hline 1 & -0.25 \\ \hline 2 & -0.5 \\ \hline 3 & -0.75 \\ \hline 4 & -1 \\ \hline 5 & -1.25 \\ \hline 6 & -1.5 \\ \hline 7 & -1.75 \\ \hline 8 & -2 \\ \hline 9 & -2.25 \\ \hline 10 & -2.5 \\ \hline 11 & -2.75 \\ \hline 12 & -3 \\ \hline 13 & -3.25 \\ \hline 14 & -3.5 \\ \hline 15 & ... \\ \hline \end{array} \qquad y = \begin{array}{|c|c|} \hline & 0 \\ \hline 0 & 1.315 \\ \hline 1 & 1.316 \\ \hline 2 & 1.654 \\ \hline 3 & 1.655 \\ \hline 4 & 1.656 \\ \hline 5 & 1.657 \\ \hline 6 & 1.658 \\ \hline 7 & 1.659 \\ \hline 8 & 1.661 \\ \hline 9 & 1.662 \\ \hline 10 & 1.663 \\ \hline 11 & 1.664 \\ \hline 12 & 1.665 \\ \hline 13 & 1.666 \\ \hline 14 & 1.667 \\ \hline 15 & ... \\ \hline \end{array}$$

A plot of the resulting nonuniform surface water profile due to the drawdown at brink, which is known as a type $M2$ drawdown curve:

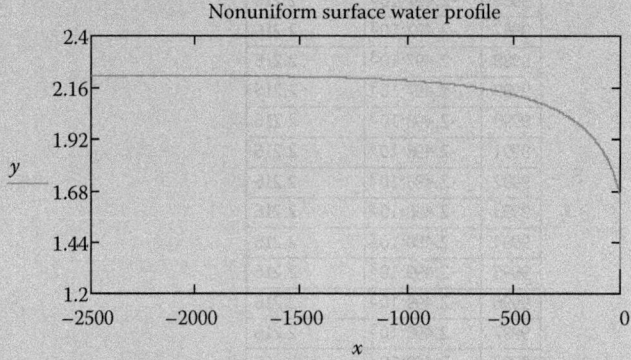

Nonuniform surface water profile

(d) The head loss, h_f due to the channel resistance over a channel section length of $L = 2500$ m is computed by applying the major head loss equation, beginning at the control/brink at $x = 0$ m to a great distance upstream of the control/brink at $x = -2500$ m and is given as follows:

$$x1 = -2.5 \times 10^3 \qquad N = 1 \times 10^4 \qquad \Delta x := \frac{|x1 \text{ m}|}{N} = 0.25 \text{ m}$$

$$S_f(y) := \frac{v(y)^2}{C(y)^2 R_h(y)} \qquad h_f(y) := S_f(y) \cdot \Delta x \qquad L := |x1 \text{ m}| = 2.5 \times 10^3 \text{m}$$

Values for y, S_f, and h_f, for the first 15 spatial points along the channel, beginning at the control/brink at $x = 0$ m are given as follows:

$y_i \cdot m =$	$S_f(y_i \cdot m) =$	$h_f(y_i \cdot m) =$
1.315 m	$7.289 \cdot 10^{-3}$	$1.822 \cdot 10^{-3}$ m
1.316	$7.268 \cdot 10^{-3}$	$1.817 \cdot 10^{-3}$
1.654	$3.093 \cdot 10^{-3}$	$7.732 \cdot 10^{-4}$
1.655	$3.085 \cdot 10^{-3}$	$7.712 \cdot 10^{-4}$
1.656	$3.077 \cdot 10^{-3}$	$7.692 \cdot 10^{-4}$
1.657	$3.069 \cdot 10^{-3}$	$7.673 \cdot 10^{-4}$
1.658	$3.061 \cdot 10^{-3}$	$7.653 \cdot 10^{-4}$
1.659	$3.053 \cdot 10^{-3}$	$7.634 \cdot 10^{-4}$
1.661	$3.046 \cdot 10^{-3}$	$7.614 \cdot 10^{-4}$
1.662	$3.038 \cdot 10^{-3}$	$7.595 \cdot 10^{-4}$
1.663	$3.031 \cdot 10^{-3}$	$7.577 \cdot 10^{-4}$
1.664	$3.023 \cdot 10^{-3}$	$7.558 \cdot 10^{-4}$
1.665	$3.016 \cdot 10^{-3}$	$7.539 \cdot 10^{-4}$
1.666	$3.008 \cdot 10^{-3}$	$7.521 \cdot 10^{-4}$
1.667	$3.001 \cdot 10^{-3}$	$7.503 \cdot 10^{-4}$
...	...	...

Values for y, S_f, and h_f, for the last 15 spatial points along the channel, ending at a great distance upstream of the control/brink at $x = -2500$ m are given as follows:

$y_i \cdot m =$	$S_f(y_i \cdot m) =$	$h_f(y_i \cdot m) =$
2.216 m	$1.001 \cdot 10^{-3}$	$2.503 \cdot 10^{-4}$ m
2.216	$1.001 \cdot 10^{-3}$	$2.503 \cdot 10^{-4}$
2.216	$1.001 \cdot 10^{-3}$	$2.503 \cdot 10^{-4}$
2.216	$1.001 \cdot 10^{-3}$	$2.503 \cdot 10^{-4}$
2.216	$1.001 \cdot 10^{-3}$	$2.503 \cdot 10^{-4}$
2.216	$1.001 \cdot 10^{-3}$	$2.503 \cdot 10^{-4}$
2.216	$1.001 \cdot 10^{-3}$	$2.503 \cdot 10^{-4}$
2.216	$1.001 \cdot 10^{-3}$	$2.503 \cdot 10^{-4}$
2.216	$1.001 \cdot 10^{-3}$	$2.503 \cdot 10^{-4}$
2.216	$1.001 \cdot 10^{-3}$	$2.503 \cdot 10^{-4}$
2.216	$1.001 \cdot 10^{-3}$	$2.503 \cdot 10^{-4}$
2.216	$1.001 \cdot 10^{-3}$	$2.503 \cdot 10^{-4}$
2.216	$1.001 \cdot 10^{-3}$	$2.503 \cdot 10^{-4}$
2.216	$1.001 \cdot 10^{-3}$	$2.503 \cdot 10^{-4}$
2.216	$1.001 \cdot 10^{-3}$	$2.503 \cdot 10^{-4}$
...	...	...

The total head loss, h_f due the channel resistance over a channel section length, $L = 2500$ m is given as follows:

$$\sum_i h_f(y_i\, m) = 2.861\, m$$

EXAMPLE PROBLEM 9.16

Water flows at a discharge of 0.6 m³/sec in a rectangular channel that is 1.6 m wide. There is a break in the channel slope, where the upstream channel section is characterized with a mild channel bottom slope of 0.0005 m/m and a length of 2000 m, and the downstream channel section is characterized with a steep channel bottom slope of 0.027 m/m and a length of 1500 m, as illustrated in Figure EP 9.16. The Manning's roughness coefficient is $0.013\, m^{-1/3}$ sec. (a) Determine the uniform depth of flow and flow regime in the channel at a great distance upstream of the break in the channel slope, where the control ceases to affect the water surface profile. (b) Determine the uniform depth of flow and flow regime in the channel at a great distance downstream of the break in the channel slope, where the control ceases to affect the water surface profile. (c) Determine and plot the resulting nonuniform surface water profile due to the break in the channel slope for the first channel section that is upstream of the control/break. (d) Determine the head loss due to the channel resistance over the first channel section with a length of 2000 m. (e) Determine and plot the resulting

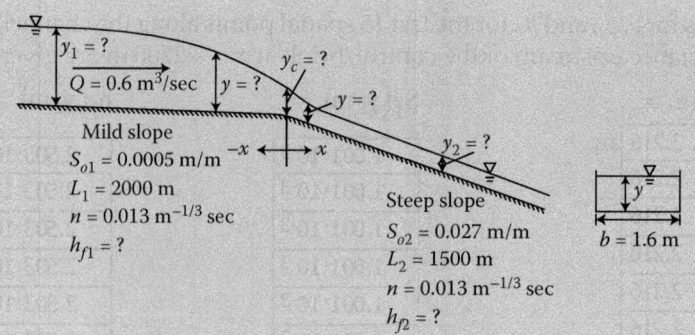

FIGURE EP 9.16
Water flows in a rectangular channel with a break in the channel slope.

nonuniform surface water profile due to the break in the channel slope for the second channel section that is downstream of the control/break. (f) Determine the head loss due to the channel resistance over the second channel section with a length of 1500 m. (g) Plot the resulting nonuniform surface water profile due to the break in the channel slope for the entire channel profile relative to the elevation of the channel bottom, which is defined as the hydraulic grade line, *HGL*, and determine the total head loss over the entire channel length of 3500 m.

Mathcad Solution

(a) The uniform depth of flow, y_1 in the channel at a great distance upstream of the break in the channel slope, where the control ceases to affect the water surface profile is computed by applying the Manning's equation and the continuity equation, $Q = vA$ as follows:

$$Q := 0.6 \, \frac{m^3}{sec} \qquad\qquad S_{o1} := 0.0005 \, \frac{m}{m} \qquad\qquad n := 0.013 \, m^{\frac{-1}{3}} \, sec$$

$$b := 1.6 \, m \qquad\qquad A(y) := b \cdot y \qquad\qquad P_w(y) := (2 \cdot y + b)$$

$$R_h(y) := \frac{A(y)}{P_w(y)}$$

Guess value: $y_1 := 1 \, m$

Given

$$Q = \frac{1}{n} \, R_h(y_1)^{\frac{2}{3}} \cdot S_{o1}^{\frac{1}{2}} \cdot A(y_1)$$

$$y_1 := Find(y)_1 = 0.485 \, m$$

In order to determine the flow regime of the uniform flow in the channel, y_1 one must first determine the critical depth of flow, y_c for the given discharge, $Q = 0.6 \, m^3/sec$ as follows:

$$Q = 0.6 \frac{m^3}{sec} \qquad\qquad b = 1.6\, m \qquad\qquad q := \frac{Q}{b} = 0.375 \frac{m^2}{sec}$$

$$g := 9.81 \frac{m}{sec^2} \qquad\qquad y_c := \sqrt[3]{\frac{q^2}{g}} = 0.243\, m$$

Therefore, because $y_1 = 0.485\, m$ is greater than $y_c = 0.243\, m$, the uniform depth of flow, y_1 is subcritical.

(b) The uniform depth of flow, y_2 in the channel at a great distance downstream of the break in the channel slope, where the control ceases to affect the water surface profile is computed by applying the Manning's equation and the continuity equation, $Q = vA$ as follows:

$$Q = 0.6 \frac{m^3}{sec} \qquad\qquad S_{o2} := 0.027 \frac{m}{m} \qquad\qquad n := 0.013\, m^{\frac{-1}{3}}\, sec$$

$$b := 1.6\, m$$

Guess value: $\qquad y_2 := 1\, m$

Given

$$Q = \frac{1}{n}\, R_h(y_2)^{\frac{2}{3}} \cdot S_{o2}^{\frac{1}{2}} \cdot A(y_2)$$

$$y_2 := Find(y_2) = 0.129\, m$$

Therefore, because $y_2 = 0.129\, m$ is less than $y_c = 0.243\, m$, the uniform depth of flow, y_2 is supercritical.

(c) Determination of the resulting nonuniform surface water profile for the first channel section that is upstream of the control/break begins by computation of the critical depth of flow, y_c at the control/break in channel slope. The critical depth of flow, y_c was determined above by applying the critical depth equation, which yielded $y_c = 0.243\, m$. The first part of the remaining nonuniform surface water profile is determined by applying the resistance equation for nonuniform flow, $dy/dx = (S_{o1} - S_f)/(1 - F^2)$ beginning just upstream of the control/break in channel slope and ending at the uniform depth of flow, $y_1 = 0.485\, m$, which is at a great distance upstream of the control/break in channel slope. Thus, the first set of calculations will begin at $y = y_c + 0.001\, m = 0.244\, m$. The resistance equation for nonuniform flow is an ordinary differential equation that is numerically solved for y using an ordinary differential equation solver in Mathcad, "rkfixed," which uses the fourth-order Runge-Kutta fixed-step method as follows:

$$S_{o1} := 0.0005 \frac{m}{m} \qquad\qquad n := 0.013\, m^{\frac{-1}{3}}\, sec \qquad\qquad Q := 0.6 \frac{m^3}{sec}$$

$$y_c := 0.243\, m \qquad\qquad v_c := \sqrt{g\, y_c} = 1.544 \frac{m}{s} \qquad\qquad v(y): \frac{Q}{A(y)}$$

$$C(y) := \frac{R_h(y)^{\frac{1}{6}}}{n} \qquad S_f(y) := \frac{v(y)^2}{C(y)^2 \, R_h(y)} \qquad F(y) := \frac{v(y)}{v_c}$$

The first-order ordinary differential equation initial value problem is defined as follows:

$$\frac{d}{dx} y(x) = f(x, y) = \frac{S_{o1} - S_f(y)}{1 - F(y)^2} \qquad\qquad y(x0) = y0 = y_c + 0.001$$

Enter the initial value problem specifics, noting that the derivative equation must be unit balanced (i.e., the arguments must have appropriate units attached):

$$f(x, y) := \frac{S_{o1} - S_f(y \cdot m)}{1 - F(y \cdot m)^2}$$

Define the beginning of the solution interval:

$x0 := 0$

$y0 := 0.243 + 0.001 = 0.244$

Enter the desired solution parameters, including the endpoint of the solution interval, $x1$ and the number of solution values, N on $[x0, x1]$:

$x1 := -2000$

$N := 10000$

Define solver parameters, including the vector of initial solution values, *ic* and the derivative function, *D* for which the second argument, *y* must be a vector of unknown function values:

$ic := y0$

$D(x, y) := f(x, y)$

Solution matrix, for which the independent variable values are x and the solution function values are y:

$S := \text{rkfixed}(ic, x0, x1, N, D)$

$x := S^{\langle 0 \rangle}$

$y := S^{\langle 1 \rangle}$

Values for x and y for the first 15 spatial points along the channel, beginning just upstream of the control, for which $x = 0$ m and $y = 0.244$ m:

$$S = \begin{array}{c|c|c}
 & 0 & 1 \\
\hline
0 & 0 & 0.244 \\
1 & -0.2 & 0.271 \\
2 & -0.4 & 0.273 \\
3 & -0.6 & 0.275 \\
4 & -0.8 & 0.277 \\
5 & -1 & 0.279 \\
6 & -1.2 & 0.28 \\
7 & -1.4 & 0.282 \\
8 & -1.6 & 0.283 \\
9 & -1.8 & 0.285 \\
10 & -2 & 0.286 \\
11 & -2.2 & 0.287 \\
12 & -2.4 & 0.289 \\
13 & -2.6 & 0.29 \\
14 & -2.8 & 0.291 \\
15 & -3 & \ldots
\end{array}$$

Values for x and y for the last 15 spatial points along the channel, ending at a great distance upstream of the control/break in channel slope, for which $x = -2000$ m and $y_1 = 0.485$ m:

$$S = \begin{array}{c|c|c}
 & 0 & 1 \\
\hline
9985 & -1.997 \cdot 10^3 & 0.485 \\
9986 & -1.997 \cdot 10^3 & 0.485 \\
9987 & -1.997 \cdot 10^3 & 0.485 \\
9988 & -1.998 \cdot 10^3 & 0.485 \\
9989 & -1.998 \cdot 10^3 & 0.485 \\
9990 & -1.998 \cdot 10^3 & 0.485 \\
9991 & -1.998 \cdot 10^3 & 0.485 \\
9992 & -1.998 \cdot 10^3 & 0.485 \\
9993 & -1.999 \cdot 10^3 & 0.485 \\
9994 & -1.999 \cdot 10^3 & 0.485 \\
9995 & -1.999 \cdot 10^3 & 0.485 \\
9996 & -1.999 \cdot 10^3 & 0.485 \\
9997 & -1.999 \cdot 10^3 & 0.485 \\
9998 & -2 \cdot 10^3 & 0.485 \\
9999 & -2 \cdot 10^3 & 0.485 \\
10000 & -2 \cdot 10^3 & \ldots
\end{array}$$

The solution of the resistance equation was constrained to begin just upstream of the control/break in channel slope, for which $x = 0$ m and $y = 0.244$ m. However, because the critical depth of flow, $y_c = 0.243$ m actually occurs at the control/break in channel slope, the x-axis is redefined in order to reflect the occurrence of $y_c =$ 0.243 m at $x = 0$ m:

$y_c = 0.243$ m $i := 0..N - 1$ $yy_{i+1} := y_i$

$y := yy$ $y_0 := 0.243$

Values for x and y for the first 15 spatial points along the channel, beginning at the control/break in channel slope, for which the x-axis has been redefined in order to reflect the occurrence of $y_c = 0.243$ m at $x = 0$ m:

	0
0	0
1	-0.2
2	-0.4
3	-0.6
4	-0.8
5	-1
6	-1.2
7	-1.4
8	-1.6
9	-1.8
10	-2
11	-2.2
12	-2.4
13	-2.6
14	-2.8
15	...

$x =$

	0
0	0.243
1	0.244
2	0.271
3	0.273
4	0.275
5	0.277
6	0.279
7	0.28
8	0.282
9	0.283
10	0.285
11	0.286
12	0.287
13	0.289
14	0.29
15	...

$y =$

A plot of the resulting nonuniform surface water profile due to the break in slope for the first channel section, which is known as a type *M2* drawdown curve:

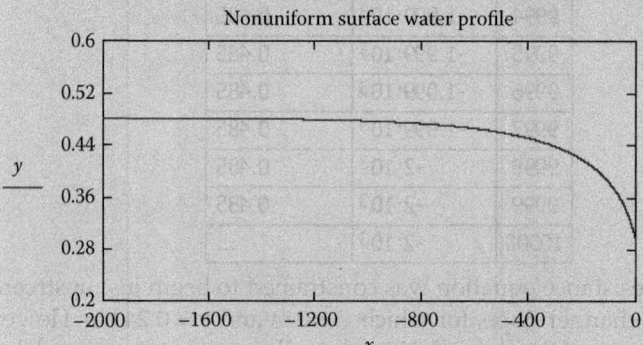

Store the x values in $X1$ and the y values in $Y1$ for the first channel section that is upstream of the control:

$X1 := x$

$Y1 := y$

(d) The head loss, h_{f1} due to the channel resistance over a channel section length of $L_1 = 2000$ m is computed by applying the major head loss equation, beginning at the control/break in channel slope at $x = 0$ m to a great distance upstream of the control/break in channel slope at $x = -2000$ m and is given as follows:

$$x1 = -2 \times 10^3 \qquad N = 1 \times 10^4 \qquad \Delta x := \frac{|x1 \text{ m}|}{N} = 0.2 \text{ m}$$

$$S_f(y) := \frac{v(y)^2}{C(y)^2 \cdot R_h(y)} \qquad h_f(y) := S_f(y) \cdot \Delta x \qquad L_1 := |x1 \cdot m| = 2 \times 10^3 \text{ m}$$

Values for y, S_f, and h_f, for the first 15 spatial points along the channel, beginning at the control/break in channel slope at $x = 0$ m are given as follows:

$y_i \cdot m =$	$S_f(y_i \cdot m) =$	$h_f(y_i \cdot m) =$
0.243 m	$3.78 \cdot 10^{-3}$	$7.561 \cdot 10^{-4}$ m
0.244	$3.734 \cdot 10^{-3}$	$7.467 \cdot 10^{-4}$
0.271	$2.721 \cdot 10^{-3}$	$5.443 \cdot 10^{-4}$
0.273	$2.658 \cdot 10^{-3}$	$5.315 \cdot 10^{-4}$
0.275	$2.601 \cdot 10^{-3}$	$5.201 \cdot 10^{-4}$
0.277	$2.549 \cdot 10^{-3}$	$5.099 \cdot 10^{-4}$
0.279	$2.503 \cdot 10^{-3}$	$5.006 \cdot 10^{-4}$
0.28	$2.46 \cdot 10^{-3}$	$4.92 \cdot 10^{-4}$
0.282	$2.42 \cdot 10^{-3}$	$4.84 \cdot 10^{-4}$
0.283	$2.383 \cdot 10^{-3}$	$4.766 \cdot 10^{-4}$
0.285	$2.349 \cdot 10^{-3}$	$4.697 \cdot 10^{-4}$
0.286	$2.316 \cdot 10^{-3}$	$4.632 \cdot 10^{-4}$
0.287	$2.286 \cdot 10^{-3}$	$4.571 \cdot 10^{-4}$
0.289	$2.257 \cdot 10^{-3}$	$4.514 \cdot 10^{-4}$
0.29	$2.229 \cdot 10^{-3}$	$4.459 \cdot 10^{-4}$
...	...	...

Values for y, S_f, and h_f, for the last 15 spatial points along the channel, ending at a great distance upstream of the control/break in channel slope at $x = -2000$ m are given as follows:

$y_i \cdot m =$	$S_f(y_i \cdot m) =$	$h_f(y_i \cdot m) =$
0.485 m		
0.485	$5.001 \cdot 10^{-4}$	$1 \cdot 10^{-4}$ m
0.485	$5.001 \cdot 10^{-4}$	$1 \cdot 10^{-4}$
0.485	$5.001 \cdot 10^{-4}$	$1 \cdot 10^{-4}$
0.485	$5.001 \cdot 10^{-4}$	$1 \cdot 10^{-4}$
0.485	$5.001 \cdot 10^{-4}$	$1 \cdot 10^{-4}$
0.485	$5.001 \cdot 10^{-4}$	$1 \cdot 10^{-4}$
0.485	$5.001 \cdot 10^{-4}$	$1 \cdot 10^{-4}$
0.485	$5.001 \cdot 10^{-4}$	$1 \cdot 10^{-4}$
0.485	$5.001 \cdot 10^{-4}$	$1 \cdot 10^{-4}$
0.485	$5.001 \cdot 10^{-4}$	$1 \cdot 10^{-4}$
0.485	$5.001 \cdot 10^{-4}$	$1 \cdot 10^{-4}$
0.485	$5.001 \cdot 10^{-4}$	$1 \cdot 10^{-4}$
0.485	$5.001 \cdot 10^{-4}$	$1 \cdot 10^{-4}$
0.485	$5.001 \cdot 10^{-4}$	$1 \cdot 10^{-4}$
0.485	$5.001 \cdot 10^{-4}$	$1 \cdot 10^{-4}$
...	...	...

The total head loss, h_{f1} due the channel resistance over a channel section length, $L_1 = 2000$ m is given as follows:

$$h_{f1} := \sum_i h_f(y_i \cdot m) = 1.119 m$$

(e) Determination of the resulting nonuniform surface water profile for the second channel section that is downstream of the control/break begins by computation of the critical depth of flow, y_c at the control/break in channel slope. The critical depth of flow, y_c was determined above by applying the critical depth equation, which yielded $y_c = 0.243$ m. The second part of the remaining nonuniform surface water profile is determined by applying the resistance equation for nonuniform flow, $dy/dx = (S_{o2} - S_f)/(1 - F^2)$ beginning just downstream of the control/break in channel slope and ending at the uniform depth of flow, $y_2 = 0.129$ m, which is at a great distance downstream of the control/break in channel slope. Thus, the second set of calculations will begin at $y = y_c - 0.001$ m $= 0.242$ m. The resistance equation for nonuniform flow is an ordinary differential equation that is numerically solved for y using an ordinary differential equation solver in Mathcad, "rkfixed," which uses the fourth-order Runge-Kutta fixed-step method as follows:

$$S_{o2} := 0.027 \ \frac{m}{m} \qquad n := 0.013 \ m^{\frac{-1}{3}} \ \sec \qquad Q := 0.6 \ \frac{m^3}{\sec}$$

$$y_c = 0.243 \text{ m} \qquad y_c := \sqrt{g \, y_c} = 1.544 \frac{\text{m}}{\text{s}} \qquad v(y) := \frac{Q}{A(y)}$$

$$C(y) := \frac{R_h(y)^{\frac{1}{6}}}{n} \qquad S_f(y) := \frac{v(y)^2}{C(y)^2 \, R_h(y)} \qquad F(y) := \frac{v(y)}{v_c}$$

The first-order ordinary differential equation initial value problem is defined as follows:

$$\frac{d}{dx} y(x) = f(x, y) = \frac{S_{o2} - S_f(y)}{1 - F(y)^2} \qquad\qquad y(x0) = y0 = y_c - 0.0001$$

Enter the initial value problem specifics, noting that the derivative equation must be unit balanced (i.e., the arguments must have appropriate units attached):

$$f(x, y) := \frac{S_{o2} - S_f(y \cdot m)}{1 - F(y \cdot m)^2}$$

Define the beginning of solution interval:

x0 := 0
y0 := 0.243 − 0.001 = 0.242

Enter the desired solution parameters, including the endpoint of solution interval, $x1$, and the number of solution values, N on $[x0, - x1]$:

x1 := 1500
N := 10000

Define solver parameters, including the vector of initial solution values, *ic*, and the derivative function, D, for which the second argument, y must be a vector of unknown function values:

ic := y0
D(x, y) := f(x, y)

Solution matrix, for which the independent variable values are x and the solution function values are y:

S := rkfixed(ic, x0, x1, N, D)

$x := S^{\langle 0 \rangle}$

$y := S^{\langle 1 \rangle}$

Values for x and y for the first 15 spatial points along the channel, beginning just downstream of the control, for which $x = 0$ m and $y = 0.242$ m:

$$S =$$

	0	1
0	0	0.242
1	0.15	0.192
2	0.3	0.188
3	0.45	0.184
4	0.6	0.18
5	0.75	0.177
6	0.9	0.175
7	1.05	0.172
8	1.2	0.17
9	1.35	0.168
10	1.5	0.166
11	1.65	0.164
12	1.8	0.162
13	1.95	0.16
14	2.1	0.159
15	2.25	...

Values for x and y for the last 15 spatial points along the channel, ending at a great distance downstream of the control/break in channel slope, for which $x = 1500$ m and $y_2 = 0.129$ m:

$$S =$$

	0	1
9985	$1.498 \cdot 10^3$	0.129
9986	$1.498 \cdot 10^3$	0.129
9987	$1.498 \cdot 10^3$	0.129
9988	$1.498 \cdot 10^3$	0.129
9989	$1.498 \cdot 10^3$	0.129
9990	$1.499 \cdot 10^3$	0.129
9991	$1.499 \cdot 10^3$	0.129
9992	$1.499 \cdot 10^3$	0.129
9993	$1.499 \cdot 10^3$	0.129
9994	$1.499 \cdot 10^3$	0.129
9995	$1.499 \cdot 10^3$	0.129
9996	$1.499 \cdot 10^3$	0.129
9997	$1.5 \cdot 10^3$	0.129
9998	$1.5 \cdot 10^3$	0.129
9999	$1.5 \cdot 10^3$	0.129
10000	$1.5 \cdot 10^3$	...

The solution of the resistance equation was constrained to begin just downstream of the control/break in channel slope, for which $x = 0$ m and $y = 0.242$ m.

However, because the critical depth of flow, $y_c = 0.243$ m actually occurs at the control/break in channel slope, the x-axis is redefined in order to reflect the occurrence of $y_c = 0.243$ m at $x = 0$ m:

$y_c = 0.243$ m $\qquad$ $i := 0..N - 1$ $\qquad$ $yy_{i+1} := y_i$

$y := yy$ $\qquad\qquad$ $y_0 := 0.243$

Values for x and y for the first 15 spatial points along the channel, beginning at the control/break in channel slope, for which the x-axis has been redefined in order to reflect the occurrence of $y_c = 0.243$ m at $x = 0$ m:

$x =$		0
	0	0
	1	0.15
	2	0.3
	3	0.45
	4	0.6
	5	0.75
	6	0.9
	7	1.05
	8	1.2
	9	1.35
	10	1.5
	11	1.65
	12	1.8
	13	1.95
	14	2.1
	15	...

$y =$		0
	0	0.243
	1	0.242
	2	0.192
	3	0.188
	4	0.184
	5	0.18
	6	0.177
	7	0.175
	8	0.172
	9	0.17
	10	0.168
	11	0.166
	12	0.164
	13	0.162
	14	0.16
	15	...

A plot of the resulting nonuniform surface water profile due to the break in slope for the second channel section, which is known as a type *S2* drawdown curve:

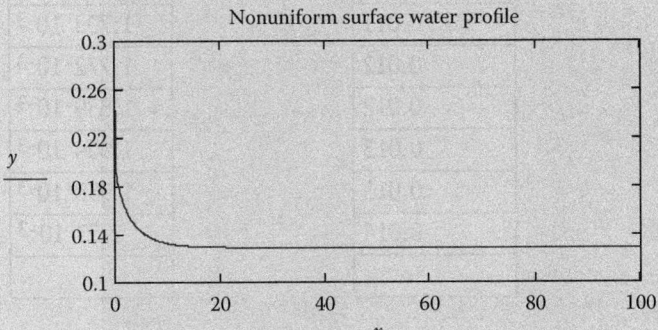

Nonuniform surface water profile

Store the x values in $X2$ and the y values in $Y2$ for the second channel section that is downstream of the control:

$$X2 := x$$

$$Y2 := y$$

(f) The head loss, h_{f2} due to the channel resistance over a channel section length of $L_2 = 1500$ m is computed by applying the major head loss equation, beginning at the control/break in channel slope at $x = 0$ m to a great distance downstream of the control/break in channel slope at $x = 1500$ m and is given as follows:

$$x1 = 1.5 \times 10^3 \qquad N = 1 \times 10^4 \qquad \Delta x := \frac{|x1\ m|}{N} = 0.15\ m$$

$$S_f(y) := \frac{v(y)^2}{C(y)^2\ R_h(y)} \qquad h_f(y) := S_f(y) \cdot \Delta x \qquad L_2 := |x1\ m| = 1.5 \times 10^3\ m$$

Values for y, S_f, and h_f, for the first 15 spatial points along the channel, beginning at the control/break in channel slope at $x = 0$ m are given as follows:

$y_i \cdot m =$	$S_f(y_i \cdot m) =$	$h_f(y_i \cdot m) =$
0.243 m	$3.78 \cdot 10^{-3}$	$5.67 \cdot 10^{-4}$ m
0.242	$3.828 \cdot 10^{-3}$	$5.742 \cdot 10^{-4}$
0.192	$7.72 \cdot 10^{-3}$	$1.158 \cdot 10^{-3}$
0.188	$8.302 \cdot 10^{-3}$	$1.245 \cdot 10^{-3}$
0.184	$8.856 \cdot 10^{-3}$	$1.328 \cdot 10^{-3}$
0.18	$9.388 \cdot 10^{-3}$	$1.408 \cdot 10^{-3}$
0.177	$9.902 \cdot 10^{-3}$	$1.485 \cdot 10^{-3}$
0.175	0.01	$1.56 \cdot 10^{-3}$
0.172	0.011	$1.632 \cdot 10^{-3}$
0.17	0.011	$1.703 \cdot 10^{-3}$
0.168	0.012	$1.772 \cdot 10^{-3}$
0.166	0.012	$1.839 \cdot 10^{-3}$
0.164	0.013	$1.904 \cdot 10^{-3}$
0.162	0.013	$1.968 \cdot 10^{-3}$
0.16	0.014	$2.03 \cdot 10^{-3}$
...	...	...

Values for y, S_f, and h_f, for the last 15 spatial points along the channel, ending at a great distance downstream of the control/break in channel slope at $x = 1500$ m are given as follows:

$y_i \cdot m =$	$S_f(y_i \cdot m) =$	$h_f(y_i \cdot m) =$
0.129 m	0.027	$4.05 \cdot 10^{-3}$ m
0.129	0.027	$4.05 \cdot 10^{-3}$
0.129	0.027	$4.05 \cdot 10^{-3}$
0.129	0.027	$4.05 \cdot 10^{-3}$
0.129	0.027	$4.05 \cdot 10^{-3}$
0.129	0.027	$4.05 \cdot 10^{-3}$
0.129	0.027	$4.05 \cdot 10^{-3}$
0.129	0.027	$4.05 \cdot 10^{-3}$
0.129	0.027	$4.05 \cdot 10^{-3}$
0.129	0.027	$4.05 \cdot 10^{-3}$
0.129	0.027	$4.05 \cdot 10^{-3}$
0.129	0.027	$4.05 \cdot 10^{-3}$
0.129	0.027	$4.05 \cdot 10^{-3}$
0.129	0.027	$4.05 \cdot 10^{-3}$
0.129	0.027	$4.05 \cdot 10^{-3}$
...	...	...

The total head loss, h_{f2} due the channel resistance over a channel section length, $L_2 = 1500$ m is given as follows:

$$h_{f2} := \sum_i h_f(y_i\, m) = 40.403 m$$

(g) Plot the resulting nonuniform surface water profile due to the break in the channel slope for the entire channel profile relative to the elevation of the channel bottom, which is defined as the hydraulic grade line, *HGL*, as follows. The x and y values for the first channel section that is upstream of the control were stored in X1 and Y1, while the x and y values for the second channel section that is downstream of the control were stored in X2 and Y2. In order to plot the entire surface water profile beginning at $x = -2000$ m and $y_1 = 0.485$ m, including $x = 0$ m and $y_c = 0.243$ m, and ending at $x = 1500$ m and $y_2 = 0.129$ m, the first step is to reverse the order of the values stored in the vectors X1 and Y1 for channel section 1 as follows:

X1 := reverse(X1)	
	0
0	$-2 \cdot 10^3$
1	$-2 \cdot 10^3$
2	$-2 \cdot 10^3$
3	$-1.999 \cdot 10^3$
4	$-1.999 \cdot 10^3$
5	$-1.999 \cdot 10^3$
6	$-1.999 \cdot 10^3$
7	$-1.999 \cdot 10^3$
8	$-1.998 \cdot 10^3$
9	$-1.998 \cdot 10^3$
10	$-1.998 \cdot 10^3$
11	$-1.998 \cdot 10^3$
12	$-1.998 \cdot 10^3$
13	$-1.997 \cdot 10^3$
14	$-1.997 \cdot 10^3$
15	...

X1 = (for the table above)

Y1 := reverse(Y1)	
	0
0	0.485
1	0.485
2	0.485
3	0.485
4	0.485
5	0.485
6	0.485
7	0.485
8	0.485
9	0.485
10	0.485
11	0.485
12	0.485
13	0.485
14	0.485
15	...

Y1 = (for the table above)

The second step is to store all the x values for the first and second channel sections in X and to store all the y values for the first and second channel section in Y as follows:

$$X := \text{stack}(X1, X2) \qquad Y := \text{stack}(Y1, Y2)$$

Values for X and Y for the first 15 spatial points along the channel, beginning at $x = -2000$ m and $y_1 = 0.485$ m are given as follows:

	0
0	$-2 \cdot 10^3$
1	$-2 \cdot 10^3$
2	$-2 \cdot 10^3$
3	$-1.999 \cdot 10^3$
4	$-1.999 \cdot 10^3$
5	$-1.999 \cdot 10^3$
6	$-1.999 \cdot 10^3$
7	$-1.999 \cdot 10^3$
8	$-1.998 \cdot 10^3$
9	$-1.998 \cdot 10^3$
10	$-1.998 \cdot 10^3$
11	$-1.998 \cdot 10^3$
12	$-1.998 \cdot 10^3$
13	$-1.997 \cdot 10^3$
14	$-1.997 \cdot 10^3$
15	...

X = (for the table above)

	0
0	0.485
1	0.485
2	0.485
3	0.485
4	0.485
5	0.485
6	0.485
7	0.485
8	0.485
9	0.485
10	0.485
11	0.485
12	0.485
13	0.485
14	0.485
15	...

Y = (for the table above)

Values for X and Y for the last 15 spatial points along the channel, ending at $x = 1500$ m and $y_2 = 0.129$ m are given as follows:

$$X = \begin{array}{|c|c|} \hline & 0 \\ \hline 19986 & 1.498 \cdot 10^3 \\ \hline 19987 & 1.498 \cdot 10^3 \\ \hline 19988 & 1.498 \cdot 10^3 \\ \hline 19989 & 1.498 \cdot 10^3 \\ \hline 19990 & 1.498 \cdot 10^3 \\ \hline 19991 & 1.499 \cdot 10^3 \\ \hline 19992 & 1.499 \cdot 10^3 \\ \hline 19993 & 1.499 \cdot 10^3 \\ \hline 19994 & 1.499 \cdot 10^3 \\ \hline 19995 & 1.499 \cdot 10^3 \\ \hline 19996 & 1.499 \cdot 10^3 \\ \hline 19997 & 1.499 \cdot 10^3 \\ \hline 19998 & 1.5 \cdot 10^3 \\ \hline 19999 & 1.5 \cdot 10^3 \\ \hline 20000 & 1.5 \cdot 10^3 \\ \hline 20001 & \ldots \\ \hline \end{array} \qquad Y = \begin{array}{|c|c|} \hline & 0 \\ \hline 19986 & 0.129 \\ \hline 19987 & 0.129 \\ \hline 19988 & 0.129 \\ \hline 19989 & 0.129 \\ \hline 19990 & 0.129 \\ \hline 19991 & 0.129 \\ \hline 19992 & 0.129 \\ \hline 19993 & 0.129 \\ \hline 19994 & 0.129 \\ \hline 19995 & 0.129 \\ \hline 19996 & 0.129 \\ \hline 19997 & 0.129 \\ \hline 19998 & 0.129 \\ \hline 19999 & 0.129 \\ \hline 20000 & 0.129 \\ \hline 20001 & \ldots \\ \hline \end{array}$$

The third step is to define the elevations $Z1$ and $Z2$ for each channel bottom slope, S_{o1} and S_{o2}, respectively, relative to the elevation at the break in the channel slope, which is defined as $Z0 = 0$ m as follows:

$$Z1 := -S_{o1}\, X1 \qquad Z2 := -S_{o2}\, X2$$

Values for $Z1$ for the last 15 spatial points along the first channel section, ending at the break in the channel slope where $Z1 = Z0 = 0$ m, and values for $Z2$ for the first 15 spatial points along the second channel section, beginning at the break in the channel slope where $Z2 = Z0 = 0$ m:

$$Z1 = \begin{array}{|c|c|} \hline & 0 \\ \hline 9985 & 1.5 \cdot 10^{-3} \\ \hline 9986 & 1.4 \cdot 10^{-3} \\ \hline 9987 & 1.3 \cdot 10^{-3} \\ \hline 9988 & 1.2 \cdot 10^{-3} \\ \hline 9989 & 1.1 \cdot 10^{-3} \\ \hline 9990 & 1 \cdot 10^{-3} \\ \hline 9991 & 9 \cdot 10^{-4} \\ \hline 9992 & 8 \cdot 10^{-4} \\ \hline 9993 & 7 \cdot 10^{-4} \\ \hline 9994 & 6 \cdot 10^{-4} \\ \hline 9995 & 5 \cdot 10^{-4} \\ \hline 9996 & 4 \cdot 10^{-4} \\ \hline 9997 & 3 \cdot 10^{-4} \\ \hline 9998 & 2 \cdot 10^{-4} \\ \hline 9999 & 1 \cdot 10^{-4} \\ \hline 10000 & \ldots \\ \hline \end{array} \qquad Z2 = \begin{array}{|c|c|} \hline & 0 \\ \hline 0 & 0 \\ \hline 1 & -4.05 \cdot 10^{-3} \\ \hline 2 & -8.1 \cdot 10^{-3} \\ \hline 3 & -0.012 \\ \hline 4 & -0.016 \\ \hline 5 & -0.02 \\ \hline 6 & -0.024 \\ \hline 7 & -0.028 \\ \hline 8 & -0.032 \\ \hline 9 & -0.036 \\ \hline 10 & -0.04 \\ \hline 11 & -0.045 \\ \hline 12 & -0.049 \\ \hline 13 & -0.053 \\ \hline 14 & -0.057 \\ \hline 15 & \ldots \\ \hline \end{array}$$

The fourth step is to store the Z1 values for the first channel section and Z2 for the second channel section in Z as follows:

$$Z := \text{stack}(Z1, Z2)$$

$$Z = \begin{array}{c|c} & 0 \\ \hline 0 & 1 \\ 1 & 1 \\ 2 & 1 \\ 3 & 1 \\ 4 & 1 \\ 5 & 1 \\ 6 & 0.999 \\ 7 & 0.999 \\ 8 & 0.999 \\ 9 & 0.999 \\ 10 & 0.999 \\ 11 & 0.999 \\ 12 & 0.999 \\ 13 & 0.999 \\ 14 & 0.999 \\ 15 & \cdots \end{array}$$

The fifth step is to define the *HGL* for the entire channel section as follows:

$$HGL := Y + Z$$

$$HGL = \begin{array}{c|c} & 0 \\ \hline 0 & 1.485 \\ 1 & 1.484 \\ 2 & 1.484 \\ 3 & 1.484 \\ 4 & 1.484 \\ 5 & 1.484 \\ 6 & 1.484 \\ 7 & 1.484 \\ 8 & 1.484 \\ 9 & 1.484 \\ 10 & 1.484 \\ 11 & 1.483 \\ 12 & 1.483 \\ 13 & 1.483 \\ 14 & 1.483 \\ 15 & \cdots \end{array}$$

The sixth step is to plot the *HGL* and the *Z* versus *X* for the entire channel length. One may note that in order to capture the variation in depth of flow for the entire channel length while accommodating the difference in the scale of variation in depth of flow for the two channel sections, only a snapshot of the entire length is actually plotted as follows:

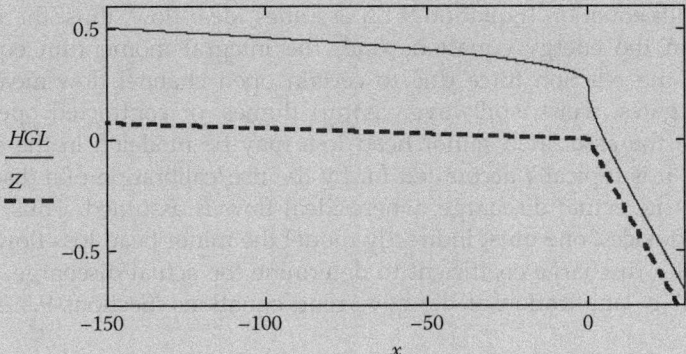

The total head loss over the entire channel length of 3500 m, $h_f = h_{f1} + h_{f2}$ is computed as follows:

$$h_f := h_{f1} + h_{f2} = 41.522 \, m$$

Thus, the total head loss over the entire channel length of $L = L_1 + L_2 = 3500 \, m$ is $h_f = h_{f1} + h_{f2} = 1.119 \, m + 41.522 \, m = 41.522 \, m$. It is interesting to note that although $L_1 = 2000 \, m$ is greater than $L_2 = 1500 \, m$, the resulting $h_{f1} = 1.119 \, m$ is significantly less than the resulting $h_{f2} = 41.522 \, m$. The reason for this is that the mild channel bottom slope, $S_{o1} = 0.0005 \, m/m$ results in a deep and slow-moving subcritical depth, decreasing from $y_1 = 0.485 \, m$ to $y_c = 0.243 \, m$. In comparison, the steep channel bottom slope $S_{o2} = 0.027 \, m/m$ results in a shallow and fast-moving supercritical depth, decreasing from $y_c = 0.243 \, m$ to $y_2 = 0.129 \, m$. As a result, the resulting friction slope, $S_f(y) = v(y)^2/C(y)^2 R_h(y)$ used in the computation of $h_{f1} = S_f(y) \, L_1$ is significantly less than the resulting friction slope, $S_f(y) = v(y)^2/C(y)^2 R_h(y)$ used in the computation of $h_{f2} = S_f(y) \, L_2$.

9.16 Actual Flowrate in Flow Measurement and Control Devices in Open Channel Flow: Ideal Flow Meters

Applications of the governing equations for ideal open channel flow include the analysis and design of an open channel flow-measuring device such as sluice gates, weirs, spillways, venturi flumes, and contracted openings. As such, the flow measurement devices in open channel flow are considered to be ideal flow meters. While the application of the continuity equation will always be necessary, the energy equation and the momentum equation play complementary roles; when one of the two equations of motion breaks down, the other may be used to find the additional unknown quantity. Furthermore, while the energy equation is useful only for a control volume, depending upon the general fluid flow problem, one may apply the continuity and momentum equations using either the integral or differential approach. However, for the specific case of ideal flow meters (addressed in Sections 9.16 and 9.17), while both the differential and integral continuity equations are required, only the integral momentum equation for a control volume is required. As such, for ideal flow meters, the Bernoulli equation (Equation 9.12) assumes ideal flow; thus, there is no head loss to evaluate in the energy equation. And, the integral momentum equation (Equation 9.13) models the reaction force due to certain open channel flow-measuring devices (such as sluice gates, weirs, spillways, venturi flumes, or contracted openings) (Section 9.12). Although the associated minor head loss may be modeled in the energy equation (Equation 9.1), it is typically accounted for by the use/calibration of a discharge coefficient to determine the actual discharge, where ideal flow is assumed. Thus, in the case of flow-measuring devices, one must indirectly model the minor head loss flow resistance term by the use of a discharge coefficient to determine the actual discharge, which requires a "subset level" of application of the governing equations (Sections 9.5.2.1 and 9.17 below).

The use/calibration of a discharge coefficient to determine the actual discharge is evaluated for certain open channel flow-measuring devices. There are numerous flow measurement devices for open channel flow, most of which can also serve as flow control devices. The most common types of flow measurement devices used in open channel flow include pitot-static tubes (not a control) (see Section 4.5.6.4 for open channel flow example problem), and typical flow-measuring devices (controls), which include sluice gates, weirs, overflow spillways, the Parshall flume (venturi flume), and contracted openings. Additional controls including a hydraulic drop, a break in channel bottom slope, a free overfall, and uniform flow may also serve as "flow-measuring devices." Because ideal flow is assumed where there is actually a minor head loss associated with the flow measurement for the pitot-static tube (see Section 4.5.6.4 for open channel flow example problem), and for the typical flow-measuring devices such as sluice gates, weirs, overflow spillways, the Parshall flume (venturi flume), and contracted openings, a discharge coefficient is typically evaluated for these flow-measuring devices in order to determine the actual discharge. The discharge coefficient accounts for the minor head loss due to the flow measurement, which is due to flow resistance, and, for some controls, flow contraction. However, because ideal flow is not assumed and there is a small minor head loss associated with the flow measurement for a hydraulic drop, and a break in the channel bottom slope, as well as a free overfall, a discharge coefficient is not typically evaluated for these flow-measuring devices. Additionally, except for the free overfall, there is no contraction of the flow for these flow measurement devices. Furthermore, the major head loss associated with the flow measurement for a uniform flow is accounted for in the resistance equation (resistance coefficient). The flow-measuring devices will vary depending on various factors including cost, space, size, accuracy, precision, sophistication, versatility, capacity, pressure drop, and the governing principle of operation.

In the case of certain flow-measuring devices (such as sluice gates, weirs, spillways, venturi flumes, or contracted openings), there are reaction forces to compute, which may solved by applying the integral form of the momentum equation (solved analytically, without the use of dimensional analysis) (see Section 9.12 and Chapter 5); note that the viscous forces have already been accounted for indirectly through the minor head loss term, $h_{f,minor}$ in the energy equation (or actually through the discharge coefficient, C_d, Section 9.17).

9.16.1 Flow-Measuring Devices in Open Channel Flow

The flow measurement devices used in open channel flow may be subdivided into three cat-egories: (1) pitot-static tubes (not a control); (2) typical flow-measuring devices (controls) such as sluice gates, weirs, overflow spillways, the Parshall flume (venturi flume), and con-tracted openings; and (3) additional controls such as a hydraulic drop, a break in the channel bottom slope, a free overfall, and uniform flow. While a "control" in the open channel can serve as a flow-measuring device, a flow-measuring device is not necessarily considered a "control." For instance, although a pitot-static tube is a common flow-measuring device, it is not a "control." Because the pitot-static tube and the typical flow-measuring devices affect the flow over a short channel section, the major loss due to the channel flow resistance is assumed to be negligible. Thus, these first two categories of flow measurement devices are considered to be "ideal flow" meters. However, because the additional controls in the third category of flow measurement devices (a hydraulic drop, a break in the channel bot-tom slope, a free overfall, and uniform flow) affect the flow over a long channel section, the major loss due to the channel flow resistance is assumed to be significant. Thus, they are not considered to be ideal flow meters. Rather (except for the free overfall), they are considered to be "critical depth" meters. Furthermore, the broad-crested weir, the venturi flume, and the contracted opening are also considered to be "critical depth" meters. A control is defined as a "critical depth" meter when there is no deviation from a hydrostatic pressure distribu-tion and there is no contraction in the depth of flow at the control, thus allowing the critical depth of flow, y_c to occur at the control.

9.16.1.1 Ideal Flow Meters

Ideal flow meters like pitot-static tubes, sluice gates, weirs, overflow spillways, the Parshall flume (venturi flume), and contracted openings affect the flow over a short channel section; thus, the major loss due to the channel flow resistance is assumed to be negligible. As such, in the evaluation of the actual discharge, which requires a "subset level" application of the governing equations (Sections 9.5.2.1, and 9.17 below), application of the Bernoulli equation (Equation 9.14) assumes no head loss in the determination of the ideal velocity. Further-more, application of the continuity equation yields the ideal flowrate, and application of the discharge coefficient yields the actual flowrate. The discharge coefficient accounts for the minor head loss due to the flow measurement that is due to flow resistance, and for some controls, flow contraction. In the case of a pitot tube, the fluid is momentarily brought to a stop, where the conversion of kinetic energy (dynamic pressure) in the flowing fluid into an ideal pressure rise (stagnation pressure-static pressure) results in the ideal pressure rise. However, in the case of the typical flow-measuring devices such as a sluice gate, weir, over-flow spillway, the Parshall flume (venturi flume), and contracted opening, the fluid is in motion through some type of constriction/obstruction, where the conversion of the pressure energy to kinetic energy results in an ideal pressure drop.

9.16.1.2 Critical Depth Meters

The additional controls in the third category of flow measurement devices (a hydraulic drop, a break in the channel bottom slope, a free overfall, and uniform flow) affect the flow over a long channel section, for which the major loss due to the channel flow resistance is assumed to be significant. As such, these are not treated as ideal flow meters. Rather, (except for the free overfall) they are considered to be "critical depth" meters. A control is defined as a "crit-ical depth" meter when there is no deviation from a hydrostatic pressure distribution, and there is no contraction in the depth of flow at the control, thus allowing the critical depth of flow, y_c to occur at the control. For the free overfall, experimental evidence indicates that the depth at the brink of the overfall is approximately $5y_c/7$. As such, the application of the analytical critical depth–discharge equation, which does not assume ideal flow, combined with a small minor head loss associated with the flow measurement (except for the uniform flow), and no contraction in the flow, does not require the evaluation of a discharge coefficient. Furthermore, for uniform flow, the major head loss associated with the flow measurement is accounted for in the resistance equation (resistance coefficient). Finally, although the broad-crested weir, the venturi flume, and the contracted opening are also considered to be "critical depth" meters, they are typically treated as "ideal flow" meters.

9.16.2 Controls Serving as Flow Measurement Devices in Open Channel Flow

Controls may serve as flow measurement devices and include both the second category of typical flow-measuring devices such as sluice gates, weirs, overflow spillways, the Parshall flume (venturi flume), and contracted openings, and the third category (additional controls: a hydraulic drop, a break in the channel bottom slope, a free overfall, and uniform flow) of flow measurement devices. These controls listed above may serve as a flow-measuring device, as each provides a relationship between the flow depth and discharge in the channel. As such, there are four types of controls. The first is an abrupt vertical channel contraction, which includes typical flow-measuring devices, such as sluice gates, weirs, and overflow spillways, and includes a hydraulic drop. The second is an abrupt horizontal channel contraction, which includes typical flow-measuring devices, such as venturi flumes and contracted openings. The third is a change in channel bottom slope, and includes a break in the channel slope and a free overfall. The fourth is uniform flow. One may note that a free overfall is a special case of the sharp-crested weir, where the height of the weir, $P = 0$.

9.16.2.1 Determination of the Depth–Discharge Relationship for Controls Serving as Flow-Measuring Devices

The controls listed above may serve as a flow-measuring device, as each provides a relationship between the flow depth and discharge in the channel. While the flow is critical at most controls (the flow is subcritical, supercritical, or critical for uniform flow as a control), the occurrence of critical flow does not necessarily provide a control. For instance, although the flow is critical at a hydraulic jump, it is not a control, and does not serve as a "flow-measuring device," but rather, it serves an "energy dissipator," resulting in major head loss. In the case of uniform flow, the resistance equation for uniform flow (Chezy or Manning's equation, expressed as the continuity equation, where $C = R_h^{1/6}/n$) provides a relationship between the depth and discharge as follows (assuming the metric system):

$$Q = vA = \underbrace{\frac{R_h^{1/6}}{n}}_{C} \sqrt{R_h S_o} A = \frac{1}{n} R_h^{2/3} S_o^{1/2} A \qquad (9.273)$$

for which the depth of flow, y is measured at the control and the discharge, Q is computed. Furthermore, one may note that the critical equations are a special case of the resistance equation for uniform flow. In the case of the remaining controls (sluice gates, weirs, overflow spillways, the Parshall flume (venturi flume), contracted openings, a hydraulic drop, a break in the channel bottom slope, and a free overfall), critical flow occurs at (or near) these controls in the open channel. The "critical flow" at (or near) the control fixes the relationship between the depth and the discharge in its neighborhood and provides a convenient theoretical basis for flowrate measurement. Specifically, the definite relationship between the depth and the discharge is the analytical expression for the critical depth of flow, y_c as a function of discharge, Q. For a rectangular channel, the critical depth–discharge relationship is given as follows:

$$y_c = \sqrt[3]{\frac{q^2}{g}} \tag{9.274}$$

for which the critical depth of flow, y_c is measured at the control and the unit discharge, q is computed. For a nonrectangular channel section, the critical depth–discharge relationship is given as follows:

$$\frac{A_c}{B_c} = \frac{Q^2}{A_c^2}\frac{1}{g} \tag{9.275}$$

for which the critical depth of flow, y_c is measured at the control and the discharge, Q is computed. As such, the setting of the particular control is guided by the analytical expression for y_c as a function of discharge, Q. The flow in the channel will be influenced by the control, as the control determines the depth–discharge relationship. Thus, for a certain setting of the control, there is a certain relationship between the upstream flow depth and the discharge, and likewise, there is a certain relationship between the downstream depth and the discharge. The discharge can be calculated once the depth is known. Thus, the control mechanism dictates what the flow depth must be for a given discharge, and what the discharge must be for a given flow depth, along the channel.

9.16.2.2 Deviations from the Assumptions for Critical Flow in Controls Serving as Flow-Measuring Devices

It is important to note that the critical depth of flow is derived based on the assumptions that (1) there is no contraction in the depth of flow and (2) the flow pressure distribution is hydrostatic. Due to the sharp-edged crest of the weir and the brink of the free overall, the fluid viscosity causes a contraction in the depth of flow, while gravity creates a pronounced curvature of the streamlines (forming a falling jet) in the vertical plane and thus a deviation from a hydrostatic pressure distribution at the sharp-edged crest and the brink (see Figures 9.3 and 9.4, respectively). Specifically, at the sharp-edged crest of the weir and at the brink of the overall, $\partial(p + \gamma z)/\partial z + \rho a_z = 0$, which is a strong deviation from a hydrostatic pressure distribution, with a mean pressure that is significantly less than hydrostatic (see Figures 9.3 and 9.4). Thus, while the critical flow condition does not disappear, it simply retreats upstream into the region where (1) there is no contraction in the depth of flow and (2) the pressure distribution is hydrostatic. As such, the occurrence of critical depth of flow, y_c occurs just upstream of the sharp-edged crest of the weir, and just upstream of the brink of the overfall. Experimental evidence has shown that the depth at the sharp-edged crest of the weir and at the brink of the overfall is approximately $5y_c/7$. Furthermore, in the case of a sharp-crested weir, $y_c \approx H$ (the head of the weir), which occurs just upstream of the sharp-edged crest of the weir. And in the case of the free overfall, the critical depth of flow, y_c occurs at a distance of approximately $3y_c$ to $4y_c$ upstream of the brink, depending on the value of the discharge.

Furthermore, it is important to note some differences and similarities between the various controls, regarding (1) contraction in the depth of flow and (2) the occurrence of hydrostatic pressure distribution at the control. Unlike the sharp-crested weir and the free overfall, for the remaining controls (a sluice gate, a spillway, a broad-crested weir, a venturi flume, a contracted opening, a break in the channel bottom slope, a hydraulic drop, and critical flow resulting from the flow resistance due to channel roughness in uniform flow), there is no deviation from a hydrostatic pressure distribution at the control. Furthermore, unlike a sharp-crested weir and a free overfall, in the cases of a broad-crested weir, a venturi flume, a contracted opening, a hydraulic drop, a break in the channel bottom slope, and critical flow due to flow resistance in uniform flow, there is no contraction in the depth of flow at the control, thus allowing the critical depth of flow, y_c to occur at the control; these are known as "critical depth" meters. However, similar to a sharp-crested weir and a free overfall, in the cases of a sluice gate and a spillway, the contraction in the depth of flow at the control causes the critical depth of flow, y_c to occur just upstream of the control.

9.16.2.3 Applied Depth–Discharge Relationship for Ideal versus Critical Flow Meters

As presented above in Section 9.16.2.1, the critical depth–discharge equation provides a convenient theoretical basis for flowrate measurement for the controls serving as flow-measuring devices. Additionally, for uniform flow, the resistance equation for uniform flow provides a convenient theoretical (and dimensional analysis) depth–discharge equation for uniform flow. Application of the critical depth–discharge equation is appropriate for the "critical depth" meters. However, deviations from the assumptions for critical flow for most of the remaining controls that are not considered to be "critical depth" meters, result in the definition of "ideal flow" meters. Thus, in the cases of a sharp-crested weir, a sluice gate, and a spillway (and a broad-crested weir, a venturi flume, a contracted opening), because they affect the flow over a short channel section and thus the major head loss due to the channel flow resistance is assumed to be negligible, these are considered to be "ideal flow" meters. As a result, the appropriate depth–discharge relationship for ideal flow meters is derived by applying the Bernoulli equation (Equation 9.14) in order to derive an expression for the ideal velocity, followed by applying the continuity equation in order to determine the ideal depth–discharge relationship. Additionally, because ideal flow was assumed, the application of a discharge coefficient accounts for the minor head loss due to flow resistance and for some controls, flow contraction (i.e., evaluation of the actual discharge requires a "subset level" application of the governing equations [Sections 9.5.2.1 and 9.17 below]). However, in the case of a free overfall, while it is not considered to be an ideal flow meter, it is not considered to be a critical depth meter either; however, using the experimental evidence that the depth at the brink is approximately $5y_c/7$, one may apply the critical depth–discharge equation accordingly.

9.16.3 Actual Flowrate for Ideal Flow Meters

In the case of the flow-measuring devices that include pitot-static tubes, sluice gates, weirs, overflow spillways, the Parshall flume (venturi flume), and contracted openings, they are considered to be "ideal flow" meters because they affect the flow over a short channel section and thus the major head loss due to the channel flow resistance is assumed to be negligible. For the case of a sharp-crested weir (there is also a deviation from a hydrostatic pressure distribution at the control), a sluice gate, and a spillway, because there is a contraction in the depth of flow at the control, y_c actually occurs just upstream of the control.

Therefore, the analytical critical depth–discharge relationship cannot be applied at the control (except for the ideal flow meters that are also considered to be critical flow meters: a broad-crested weir, a venturi flume, and a contracted opening). Therefore, the appropriate depth–discharge relationship for ideal flow meters is derived by applying the Bernoulli equation, the continuity equation, and a discharge coefficient (derived from the momentum equation and dimensional analysis). As such, application of the Bernoulli equation assumes no head loss in the determination of the ideal velocity. Furthermore, application of the continuity equation yields the ideal flowrate, and application of the discharge coefficient yields the actual flowrate. Because ideal flow was assumed, the discharge coefficient accounts for the minor head loss due to the flow measurement, which is due to flow resistance and for some controls, flow contraction (i.e., evaluation of the actual discharge requires a "subset level" application of the governing equations). (See Sections 9.5.2.1, and 9.17 below.)

The assumption of ideal flow and thus the application of the Bernoulli equation between any two points (upstream of the control, at the control, or downstream of the control) yields an expression for the ideal velocity as a function of an ideal pressure difference, Δy, which is directly measured. Furthermore, application of the continuity equation yields the ideal flowrate, and application of the discharge coefficient, $C_d = C_c C_v$ yields the actual flowrate. It is important to note that in the cases of a sharp-crested weir, a sluice gate, and a spillway, because there is a contraction in the depth of flow at the control, there is a need to apply a contraction coefficient, C_c to the computed discharge, and because ideal flow was assumed, there is a need to apply a velocity coefficient, C_v to the computed discharge (there is a minor head loss associated with the flow measurement) and thus the discharge coefficient, $C_d = C_c C_v$. However, in the cases of a broad-crested weir, a venturi flume, and a contracted opening (also considered to be "critical depth" meters, although they are typically treated as "ideal flow" meters), because there is no contraction in the depth of flow at the control, there is no need to apply a contraction coefficient, C_c to the computed discharge. Thus, in these cases, only a velocity coefficient, C_v, or where the discharge coefficient, $C_d = C_v$ is applied to the computed discharge in order to account for the minor head loss associated with the flow measurement.

9.16.4 Actual Flowrate for Critical Flow Meters

In the case of the flow-measuring devices that include a hydraulic drop, a break in the channel bottom slope, a free overfall, and uniform flow, they are not treated as ideal flow meters because they affect the flow over a long channel section, for which the major loss due to the channel flow resistance is assumed to be significant. Rather (except for the free overfall), they are considered to be "critical depth" meters. For the free overfall, experimental evidence indicates that the depth at the brink of the overfall is approximately $5y_c/7$. For the case of a hydraulic drop, a break in the channel bottom slope, and critical flow due to flow resistance in uniform flow (and a broad-crested weir, a venturi flume, and a contracted opening), the critical depth of flow, y_c actually occurs at the control because there is no deviation from a hydrostatic pressure distribution at the control, and there is no contraction in the depth of flow at the control. As such, these are called "critical depth" meters. Therefore, application of the critical depth–discharge equation can be applied at the controls for the "critical depth" meters. Thus, the relationship between the depth and the discharge is given by the analytical expression for the critical depth of flow, y_c as a function of discharge, Q, for which the critical depth of flow, y_c is measured at the control and the discharge, Q is computed. One may note that a break in the channel bottom slope constitutes an ideal critical depth meter, for which the depth of flow at the break is definitively critical. However, for the other critical depth meters, the critical depth of flow occurs in close vicinity to the control. Therefore, for a critical depth meter, while there is no contraction in flow, nor does the critical depth of flow retreat upstream of the control, the critical depth of flow cannot be exactly located at the control (except for the case of the break in the channel bottom slope).

The application of the analytical critical depth–discharge equation, which does not assume ideal flow, combined with a small minor head loss associated with the flow measurement (except for the uniform flow) and no contraction in the flow, does not require the evaluation of a discharge coefficient for critical depth meters. Furthermore, for uniform flow, the major head loss associated with the flow measurement is accounted for in the resistance equation (resistance coefficient). Finally, one may note that although the broad-crested weir, the venturi flume, and the contracted opening are also considered to be "critical depth" meters, the critical depth of flow cannot be exactly located at the control. As such, they are typically treated as "ideal flow" meters; application of a discharge coefficient accounts for the minor head loss due to the flow measurement, which is due to flow resistance (but not flow contraction), thus, $C_d = C_v$.

Finally, in the case of a free overfall, because it affects the flow over a long channel section, for which the major loss due to the channel flow resistance is assumed to be significant, it cannot be treated as an ideal flow meter. Additionally, because there is a contraction in the depth of flow at the control, and there is a deviation from a hydrostatic pressure distribution at the control, y_c actually occurs just upstream of the control. As a result, the location of the critical depth of flow, y_c does not coincide with the control, but rather is experimentally located at a distance of approximately $3y_c$ to $4y_c$ upstream of the brink, depending on the value of the discharge. Therefore, a free overfall is not considered to be a "critical depth" meter. As such, the analytical critical depth–discharge relationship cannot be applied at the control. However, the free overfall may serve as a flow-measuring device by using experimental evidence that the depth at the brink of the overfall is approximately $5y_c/7$. As a result, the analytical critical depth–discharge relationship can be applied to determine the resulting discharge accordingly. Furthermore, the application of the analytical critical depth–discharge equation, which does not assume ideal flow, combined with a small minor head loss associated with the flow measurement, and experimentally accounting for the contraction in the flow, does not require the evaluation of a discharge coefficient for a free overfall.

9.16.5 A Comparison between Ideal Flow Meters and Critical Depth Meters

The most common types of flow measurement devices used in open channel flow are considered to be "ideal flow" meters, and include pitot-static tubes (not a control), and typical flow-measuring devices (controls) like sluice gates, weirs, overflow spillways, the Parshall flume (venturi flume), and contracted openings. The additional controls that may also serve as flow-measuring devices are considered to be "critical depth" meters (except for the free overfall), and include a hydraulic drop, a break in the channel bottom slope, a free overfall, and uniform flow. Furthermore, although the broad-crested weir, the venturi flume, and the contracted opening are also considered to be "critical depth" meters, they are typically treated as "ideal flow" meters. It is of interest to compare the ideal flow meters to the critical depth meters and highlight some advantages and disadvantages of each type of meter. Furthermore, it is important to note that although the critical depth meters have certain advantages over ideal flow meters, the use of ideal flow meters are more practical than critical depth meters in open channel flow. As such, ideal flow meters are also known as "typical flow-measuring devices."

9.16.5.1 Advantages of Critical Depth Meters

There are two main advantages for the critical depth meters in comparison to ideal flow meters. The first advantage is that unlike most of the typical flow-measuring devices (ideal flow meters), a hydraulic drop, a break in the channel slope, a free overfall, a venturi flume, and a contracted opening (critical depth meters, except for the free overfall) may serve as flow-measuring devices without the existence of a dead water region upstream of the control where silt and debris can accumulate and significantly affect the depth–discharge relationship. The second advantage is that the critical depth meters apply the analytical critical depth–discharge equation, which does not assume ideal flow, combined with a small minor head loss associated with the flow measurement (except for the uniform flow), and no contraction in the flow and thus do not require the evaluation of a discharge coefficient. Furthermore, for uniform flow, the major head loss associated with the flow measurement is accounted for in the resistance equation (resistance coefficient).

9.16.5.2 Disadvantages of Critical Depth Meters

There are two main disadvantages for the critical depth meters in comparison to ideal flow meters. The first disadvantage is that although for a critical depth meter, while there is neither contraction in flow, nor does the critical depth of flow retreat upstream of the control, the critical depth of flow cannot be exactly located at the control (except for the case of the break in the channel bottom slope). A break in the channel bottom slope constitutes an ideal critical depth meter, for which the depth of flow at the break is definitively critical. However, for the other critical depth meters, the critical depth of flow occurs in the close vicinity of the control. Furthermore, it is interesting to note that although a break in the channel slope constitutes an ideal critical depth meter, in practical open channel flow situations, a long downstream supercritical slope is typically not available. As a result, a very steep and shorter downstream subcritical slope may be used. Thus, the case of a break in the channel slope approaches the case of a free overfall.

The second disadvantage deals with uniform flow as a control. A less obvious control is due to the flow resistance due to the channel roughness that occurs in a uniform open channel flow, where the flow may be subcritical, supercritical, or critical. The flow is not necessarily critical at the control. Depending upon the classification of the slope of the channel (mild, steep, or critical), the uniform flow will be subcritical, supercritical, or critical, respectively. The classification of the slope will depend on the channel roughness, the actual magnitude of the slope, and the discharge. As such, the setting of the slope is guided by the resistance equation. Thus, for a certain setting of the slope (mild, steep, or critical), there is a certain relationship between the uniform flow depth and the discharge, for a given flow regime: subcritical, supercritical, or critical, respectively. As such, a disadvantage would be in the attainment of critical slope in order to qualify the uniform flow as a critical depth meter.

9.16.5.3 Ideal Flow Meters: Typical Flow-Measuring Devices

Although critical depth meters have certain advantages over ideal flow meters, the use of ideal flow meters is more practical than critical depth meters in open channel flow. As such, ideal flow meters, which include sluice gates, weirs, overflow spillways, the Parshall flume (venturi flume), and contracted openings, are also known as "typical flow-measuring devices." The ideal depth meters assume ideal flow and apply the Bernoulli equation to determine the ideal velocity and thus require the application of a discharge coefficient in order to account for the minor head loss due to flow resistance and for some controls, flow contraction. Furthermore, although the broad-crested weir, the venturi flume, and the contracted opening are also considered to be critical depth meters, because the critical depth of flow cannot be exactly located at the control, they are typically treated as ideal flow meters.

9.17 Evaluation of the Actual Flowrate for Ideal Flow Meters in Open Channel Flow

Evaluation of the actual discharge flow resistance term requires a "subset level" application of the governing equations (see Chapters 6 and 7 and Section 9.5.2.1). Regardless of the type of open channel flow-measuring device, such as pitot-static tubes, sluice gates, weirs, overflow spillways, the Parshall flume (venturi flume), or contracted openings, the expression for the actual discharge in the differential continuity equation, which was given in Equation 9.16, is repeated as follows:

$$Q_a = \frac{\sqrt{2g\left(\Delta y - \frac{\tau_w L}{\gamma R_h}\right)}}{\sqrt{2g\Delta y}} \frac{A_a}{A_i} Q_i = C_d Q_i \tag{9.276}$$

where the discharge coefficient, C_d accounts for the minor head loss due to the flow measurement, which is due to the flow resistance, and, for some of the ideal flow meters, flow contraction. The assumption of ideal flow and thus the application of the Bernoulli equation between any two points yields an expression for the ideal velocity as a function of an ideal pressure difference, Δy, which is directly measured. Then, application of the differential continuity equation yields the ideal flowrate, Q_i. However, in real flow, the effects of viscosity, compressibility, and other real flow effects cause the actual velocity, v_a and actual area, A_a and thus the actual flowrate, Q_a to be less than the ideal velocity, v_i and ideal area, A_i and thus the ideal flowrate, Q_i, respectively. As a result, corrections for these real flow effects are made by applying a velocity or a discharge coefficient, respectively. The discharge coefficient represents both a velocity coefficient, which accounts for head loss due to friction, and a contraction coefficient, which accounts for the contraction of the streamlines of the fluid flow. As a result, the assumption of ideal flow is corrected by the use of a discharge coefficient, $C_d = C_c C_v$ (derived by application of the integral momentum supplemented by dimensional analysis), where application of the differential continuity equation yields the ideal flowrate. It is important to note that in the case of some ideal flow meters, because there is a contraction in the flow at the control, there is a need to apply a contraction coefficient, C_c to the computed discharge, and because ideal flow was assumed, there is a need to apply a velocity coefficient, C_v to the computed discharge (there is a minor head loss associated with the flow measurement) and thus the discharge coefficient, $C_d = C_c C_v$. However, in the case of a pitot-static tube (and some other ideal flow meters discussed below), because there is no contraction in the flow in the device there is no need to apply a contraction coefficient, C_c to the computed discharge. Thus, in these cases, only a velocity coefficient, C_v, or where the discharge coefficient, $C_d = C_v$ is applied to the computed discharge in order to account for the minor head loss associated with the flow measurement.

One may note that the evaluation of the actual flowrate for critical depth meters does not require the evaluation of a discharge coefficient, because the application of the analytical critical depth–discharge equation does not assume ideal flow. Furthermore, except for uniform flow, there is a small minor head loss associated with the flow measurement, and there is no contraction in the flow for critical depth meters. For uniform flow, the major head loss associated with the flow measurement is accounted for in the resistance equation (resistance coefficient). Finally, one may note that although the broad-crested weir, the venturi flume, and the contracted opening are also considered to be "critical depth" meters, because the critical depth of flow cannot be exactly located at the control, they are typically treated as ideal flow meters.

In the case of a pitot tube, the fluid is momentarily brought to a stop, where the conversion of kinetic energy (dynamic pressure) in the flowing fluid into an ideal pressure rise (stagnation pressure-static pressure) results in the ideal pressure rise. However, in the case of the typical flow-measuring devices like sluice gate, weir, overflow spillway, the Parshall flume (venturi flume), and contracted opening, the fluid is in motion through some type of constriction/obstruction, where the conversion of the pressure energy to kinetic energy results in an ideal pressure drop. Thus, the determination of the ideal velocity and thus the ideal discharge will vary depending upon whether there is an ideal pressure rise or pressure drop. As such, details of the evaluation of the actual flowrate for a pitot-static tube are presented in Section 8.8.2, while details of the evaluation of the actual flowrate a sluice gate, weir, overflow spillway, the Parshall flume (venturi flume), and contracted opening are presented in Sections 9.17.3 to 9.17.8.

Presented in the sections below is the detailed evaluation of the actual flowrate, Q_a for ideal flow meters in open channel flow, including evaluation of the ideal flowrate, Q_i, the discharge coefficient, C_d, the velocity coefficient, C_v and the contraction coefficient, C_c. The ideal flow meters include pitot-static tubes, sluice gates, weirs, overflow spillways, the Parshall flume (venturi flume), and contracted openings.

9.17.1 Evaluation of the Actual Flowrate for Ideal Flow Meters

Evaluation of the actual discharge flow resistance term requires a "subset level" application of the governing equations (see Chapters 6 and 7 and Section 9.5.2.1). The general approach for open channel flow-measuring devices is summarized below. Details for the actual flowrate for pitot-static tubes was presented in Section 8.8.2 for a pipe flow (also see section 4.5.6.4 for open channel flow example problem). And, details for the actual flowrate for the remaining ideal flow meters, which are also known as typical flow-measuring devices, that is sluice gates, weirs, overflow spillways, the Parshall flume (venturi flume), and contracted openings, are presented in Sections 9.17.3 to 9.17.8.

9.17.1.1 Application of the Bernoulli Equation

The flow measurement devices, which include sluice gates, weirs, overflow spillways, the Parshall flume (venturi flume), and contracted openings, are assumed to be ideal flow meters (or basically velocimeters). Therefore, the assumption of ideal flow and thus the application of the Bernoulli equation between any two points (upstream of the control, at the control, or downstream of the control), yields an expression for the ideal velocity, v_i as a function of an ideal pressure difference, Δy which is directly measured. For a given ideal flow meter, the critical flow, which occurs at (or near) the control fixes the relationship between the depth and the discharge in its neighborhood, and provides a convenient theoretical basis for flowrate measurement. As such, the setting of the particular control is guided by the analytical expression for y_c as a function of discharge, Q. The flow in the channel will be influenced by the control, as the control determines the depth–discharge relationship. Thus, for a certain setting of the control, there is a certain relationship between the subcritical upstream flow depth and the discharge, and, likewise, there is a certain relationship between the supercritical downstream depth and the discharge. Thus, the given control mechanism dictates what the flow depth must be for a given discharge, and what the discharge must be for a given flow depth, along the channel. Furthermore, the discharge can be calculated once the depth is known.

For the ideal flow meters that are not considered to be critical depth meters, which include a sluice gate, a sharp-crested weir (there is also a deviation from a hydrostatic pressure distribution at the control), and an overflow spillway, a contraction in the depth of flow at the control causes the critical depth of flow, y_c to occur just upstream of the control. Application of the Bernoulli equation between any two points (upstream of the control, at the control, or downstream of the control) yields an expression for the ideal velocity, v_i as a function of an ideal pressure difference, Δy, which is directly measured. As such, in the case of the sluice gate, the two points modeled in the Bernoulli equation are the subcritical upstream depth and the supercritical downstream depth, which are directly measured. However, in the case of the sharp-crested weir, whose theory also forms the basis for the design of an overflow spillway, the two points modeled in the Bernoulli equation are the subcritical upstream depth and the depth of flow at the control. Furthermore, because the critical flow does not occur at the control (but rather, just upstream of the control), the depth of flow at the control is analytically/empirically related/linked to the specific control parameters (such as the head of the weir or spillway, H), which are directly measured.

For the ideal flow meters that are also considered to be critical depth meters, which include a broad-crested weir, a venturi flume, and a contracted opening, because there is no contraction in the flow at the control, the critical depth of flow, y_c occurs in the close vicinity of the control. Application of the Bernoulli equation between any two points (upstream of the control, at the control, or downstream of the control) yields an expression for the ideal velocity, v_i as a function of an ideal pressure difference, Δy, which is directly measured. As such, for these critical depth meters, the two points modeled in the Bernoulli equation are the subcritical upstream depth and the depth of flow at the control. However, because the critical depth of flow, y_c cannot be exactly located at the control for these critical depth meters, it cannot be directly measured. Therefore, the critical depth of flow, which is assumed at the control, is analytically/empirically related/linked to the specific control parameters (such as in the head of the weir or flume, H), which are directly measured. Finally, one may note that this is probably why these critical depth meters are treated as ideal flow meters.

9.17.1.2 Application of the Continuity Equation, the Momentum Equation, and Dimensional Analysis

Then, application of the continuity equation is used to determine an expression for the ideal discharge, Q_i. However, in real flow, the effects of viscosity, compressibility, and other real flow effects cause the actual velocity, v_a and thus, the actual flowrate, Q_a to be less than the ideal velocity, v_i and the ideal flowrate, Q_i, respectively. As a result, corrections for these real flow effects are made by applying a velocity or a discharge coefficient, respectively. The discharge coefficient, C_d represents both a velocity coefficient, C_v which accounts for a minor head loss due to friction of the fluid flow, and a contraction coefficient, C_c which accounts for the contraction of the streamlines of the fluid flow. As a result, the assumption of ideal flow and the contraction of the streamlines is corrected by the use of a discharge coefficient, $C_d = C_c C_v$ which is derived using dimensional analysis, where the actual discharge is computed as $Q_a = C_d\, Q_i$, and where the discharge coefficient, C_d represents the flow resistance and flow contraction in some controls, and is empirically derived and evaluated.

9.17.1.3 Derivation of the Discharge Coefficient, C_d

The empirical derivation of the discharge coefficient, C_d which is defined as the product of the velocity coefficient, C_v and the contraction coefficient, C_c, was presented in Chapter 7, and is summarized as follows:

$$C_d = C_v C_c \tag{9.277}$$

where the discharge coefficient, C_d is defined from dimensional analysis as follows:

$$\frac{Q_a}{Q_i} = \frac{F_{dynpres}}{F_P} = \frac{v_a A_a}{v_i A_i} = \frac{v_a}{\sqrt{2g\Delta y}}\frac{A_a}{A_i} = \sqrt{\frac{E}{2}}\frac{A_a}{A_i} = \sqrt{\frac{F}{2}}\frac{A_a}{A_i} = C_v C_c = C_d = \phi\left(R, W, \frac{L_i}{L}\right) \tag{9.278}$$

where:

$$\sqrt{\frac{F}{2}} = \frac{v}{\sqrt{2gL}} = \frac{v}{\sqrt{2gy_c}} = \frac{v_a}{v_c} = \frac{v_a}{v_i} = C_v = \sqrt{\frac{E}{2}} \tag{9.279}$$

$$\frac{A_a}{A_i} = C_c \tag{9.280}$$

where A_a is the actual flow area; the ideal velocity, v_i equals the critical velocity, v_c; and the ideal area, A_i equals the critical area, A_c, and Q_i will vary for the various flow-measuring devices. The discharge coefficient, C_d represents both a velocity coefficient, C_v and a contraction coefficient, C_c; thus, the discharge coefficient, C_d is defined as the product of the velocity coefficient, C_v and the contraction coefficient, C_c as follows: $C_d = C_v C_c$. The velocity coefficient, $C_v = v_a/v_i = \sqrt{E/2} = v_a/v_c = \sqrt{F/2}$ accounts for minor head loss due to friction of the fluid flow. For ideal flow, the flow through the flow-measuring device (a constriction) is the conversion of flow energy (pressure energy) to kinetic energy. In real flow, the effects of viscosity, compressibility, and other real flow effects cause the actual velocity, v_a and thus the actual flowrate to be less than the ideal velocity, v_i and the ideal flowrate, respectively. The contraction coefficient, $C_c = A_a/A_i = A_a/A_c$ accounts for the contraction of the streamlines of the fluid flow (due to frictional pressure losses). As a result, corrections for these real flow effects are made by applying a velocity or a discharge coefficient, respectively. Although the contraction coefficient, C_c and the velocity coefficient, C_v which model the two nonideal flow effects, may be individually experimentally determined, their combined effect is usually modeled by the discharge coefficient, C_d which is determined experimentally and is given as follows: $C_d = C_v C_c$.

The drag coefficient, C_D (discharge coefficient, C_d) as modeled by the Euler number, $E =$ the Froude number, F is the dependent π term, as it models the dependent variable, Δy or S_f that cannot be derived analytically from theory, while the Reynolds number, R; the Weber number, W; and the geometry of the flow-measuring device, L_i/L (L represents the characteristic length of the flow meter, and L_i represents the geometry of the flow meter, where $i = 1, 2, 3, \ldots$) are the independent π terms. One may note that the drag coefficient, C_D (discharge coefficient, C_d) represents the flow resistance (and flow contraction in some controls), which causes an unknown pressure drop, Δy and an unknown S_f. As such, both the pressure force and the gravitational force will always play an important role; thus, the definition of the drag coefficient, C_D (discharge coefficient, C_d) is mainly a function of the Euler number, E and the Froude number, F, where $E = F$.

9.17.1.4 Evaluation of the Discharge Coefficient, C_d

Evaluation of the discharge coefficient, C_d for flow measurement devices in open channel flow is determined experimentally and presented in Sections 9.17.3 to 9.17.8 below on a particular flow measurement device (ideal flow meter). It is important to note that in the cases of a sharp-crested weir, a sluice gate, and a spillway, because there is a contraction in the depth of flow at the control, there is a need to apply a contraction coefficient, C_c to the computed discharge, and, because ideal flow was assumed, there is a need to apply a velocity coefficient, C_v to the computed discharge (there is a minor head loss associated with the flow measurement); thus, the discharge coefficient, $C_d = C_c C_v$. However, in the cases of a broad-crested weir, a venturi flume, and a contracted opening (also considered to be "critical depth" meters, although they are typically treated as "ideal flow" meters), because there is no contraction in the depth of flow at the control, there is no need to apply a contraction coefficient, C_c to the computed discharge. Thus, in these cases, only a velocity coefficient, C_v, or where the discharge coefficient, $C_d = C_v$ is applied to the computed discharge in order to account for the minor head loss associated with the flow measurement.

9.17.2 Sluice Gates, Weirs, Spillways, Venturi Flumes, and Contracted Openings in Open Channel Flow

An effective way to control and measure the flowrate in an open channel flow is to insert either an obstruction such as a sluice gate, a weir, or a spillway, or a constriction such as the Parshall flume (venturi flume), or a contracted opening. Sluice gates, weirs, spillways, and venturi flumes serve as "controls" by partially blocking (or in the case of a venturi flume or a contracted opening, serve as "controls" by constricting) the flow in the channel and thus allow a measurement of the flowrate in the open channel. The flow in an obstruction (sluice gates, weirs, and spillways) results in a two-dimensional flow, while the flow in a constriction (venturi flume and contracted openings) results in a one-dimensional flow. A sluice gate is an obstruction inserted in the channel that the fluid flows under, while a weir and a spillway are obstructions/overflow structures inserted in the channel that the fluid flows over. The venturi flume is a constriction inserted in the channel through which the fluid flows, while contracted openings occur a result of a bridge or a culvert in a stream.

It is interesting to note that the sluice gate and the weir for open channel flow are analogous to the orifice flow meter for pipe flow, while the venturi flume for open channel flow is analogous to a venturi flow meter for pipe flow. Depending upon the structural design, a weir may be either a sharp-crested weir or a broad-crested weir, as illustrated in Figure 9.35a and b, respectively. Furthermore, as illustrated below, the theory of a sharp-crested rectangular weir forms a basis for the analysis and design of a spillway, while the theory of a venturi flume forms a basis for the analysis and design of a contracted opening.

Evaluation of the actual discharge flow resistance term requires a "subset level" application of the governing equations (see Chapters 6 and 7 and Section 9.5.2.1), where evaluation

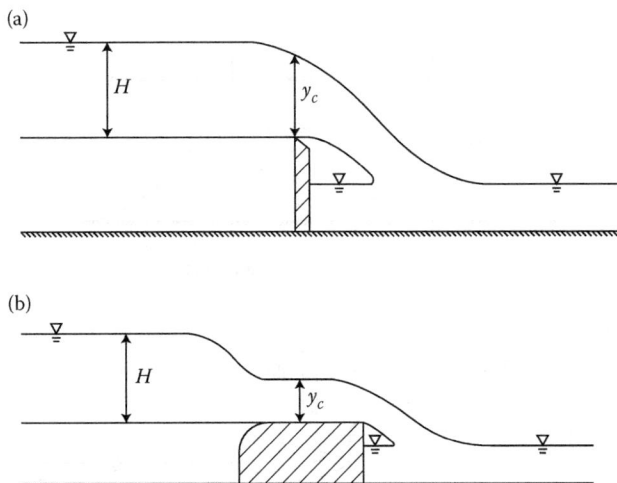

FIGURE 9.35
Weirs. (a) Sharp-crested weir, (b) Broad-crested weir.

of the actual flowrate and the respective discharge coefficient for a given flow measurement device (ideal flow meter) is presented in Sections 9.17.3 to 9.17.8 below. The ideal flow meters that are not considered to be critical depth meters include a sluice gate, a sharp-crested weir, and an overflow spillway. Furthermore, the ideal flow meters that are also considered to be critical depth meters include a broad-crested weir, a venturi flume, and a contracted opening.

9.17.3 Evaluation of the Actual Flowrate for a Sluice Gate

The sluice gate for open channel flow is analogous to the orifice flow meter for pipe flow. In particular, one may consider the sluice gate a special case of orifice flow. One may recall that for an orifice meter (see Figure 9.36a), a jet contraction occurs on both sides of the jet, and the pressure distribution in the vena contracta is not hydrostatic (it is dynamic), whereas for a sluice gate, as illustrated in Figure 9.36b, the jet contraction (vena contracta) occurs only on the top of the jet, and the pressure distribution in the vena contracta is hydrostatic (because y_1 is large compared to y_2). A sluice gate is a vertical, sharp-edged flat plate, which extends over the full width of the channel that is inserted in an open channel and is used to regulate and measure the flowrate in open channel flow. The flowrate, Q is a function of the upstream depth, y_1; the downstream depth, y_2; the gate opening or depth of flow at the gate, a; and the channel width, b. When the sluice gate is opened, the upstream subcritical flow at point 1 accelerates as it approaches the sluice gate (which is a control), reaching critical flow at the gate, and then accelerates further to supercritical flow downstream of the gate at point 2. One may note that the acceleration of the flow is due to the steady decline in the elevation of the free surface from point 1 to point 2; thus, the conversion of the elevation head at point 1 into a velocity head at point 2. It is interesting to note that the sluice gate serves as a "control," at which the critical depth of flow, y_c occurs, assuming ideal flow (thus, ideally, $a = y_c$). However, due to the sharp edge of the sluice gate, the fluid viscosity causes a contraction in the depth of flow both at the control and downstream of the control at point 2. Therefore, the contraction of flow causes (1) the critical depth of flow, y_c to occur just

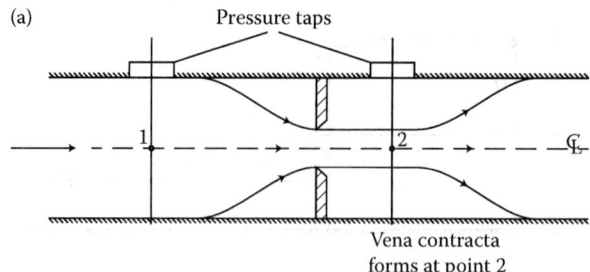

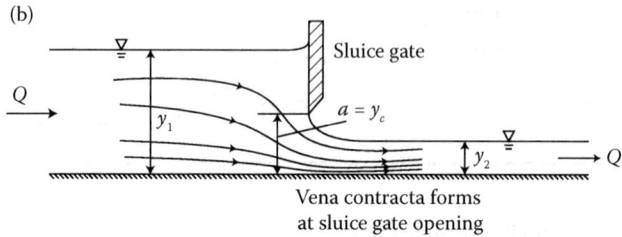

FIGURE 9.36
Sluice gate in open channel flow is analogous to orifice meter in pipe flow. (a) Orifice meter, (b) Sluice gate (with free outflow).

upstream of the sluice gate, and (2) the supercritical depth of flow at point 2, y_2 to be less than the gate opening or depth of flow at the gate, a by an unknown amount, as modeled by a contraction coefficient, $C_c = A_a/A_i = y_{2a}/a$. Finally, one may note that due to the contraction of flow at the sluice gate, this ideal flow meter is not considered to be a critical depth meter; thus, the discharge coefficient, C_d is a function of both a velocity coefficient, and a contraction coefficient, where $C_d = C_v C_c$.

The empirical derivation of the reduced/actual discharge, Q_a for a sluice gate requires a "subset level" application of the governing equations (see Chapters 6 and 7, and Section 9.5.2.1). Application of the Bernoulli equation yields an expression for the ideal velocity, v_i as a function of an ideal pressure difference, Δy which is directly measured. Assuming ideal flow, the Bernoulli equation between points 1 and 2 expressed in terms of pressure head is given as follows:

$$y_1 + z_1 + \frac{v_1{}^2}{2g} = y_2 + z_2 + \frac{v_2{}^2}{2g} \tag{9.281}$$

Assuming that $z_1 = z_2$ yields:

$$y_1 + \frac{v_1{}^2}{2g} = y_2 + \frac{v_2{}^2}{2g} \tag{9.282}$$

Assuming that the velocity profiles at the upstream point 1 and the downstream point 2 are uniform (ideal flow velocity profile), then the continuity equation between points 1 and 2 is given as:

$$Q = v_1 A_1 = v_2 A_2 \tag{9.283}$$

Applying the continuity equation and solving for the ideal velocity downstream of the restriction (sluice gate) at point 2, v_{2i} yields the following:

$$y_1 - y_2 = \frac{v_2^2}{2g} - \frac{v_2^2}{2g}\left(\frac{A_2}{A_1}\right)^2 = \frac{v_2^2}{2g}\left[1 - \left(\frac{A_2}{A_1}\right)^2\right] \tag{9.284}$$

$$v_{2i} = \frac{\sqrt{2g(y_1 - y_2)}}{\sqrt{1 - \left(\frac{A_2}{A_1}\right)^2}} = \frac{\sqrt{2g(y_1 - y_2)}}{\sqrt{1 - \left(\frac{by_2}{by_1}\right)^2}} = \frac{\sqrt{2g(y_1 - y_2)}}{\sqrt{1 - \left(\frac{y_2}{y_1}\right)^2}} \tag{9.285}$$

Next, application of the differential continuity equation yields an expression for the ideal (theoretical) flow rate, Q_i as follows:

$$Q_i = A_{2i}\frac{\sqrt{2g(y_1 - y_2)}}{\sqrt{1 - \left(\frac{y_2}{y_1}\right)^2}} \tag{9.286}$$

Finally, the reduced/actual discharge, Q_a is determined by application of the differential continuity equation and the momentum equation, which is supplemented by dimensional analysis in order to define the discharge coefficient; the discharge coefficient, C_d is applied to the ideal discharge, Q_i in order to define the actual discharge, Q_a. In real flow, the actual velocity, v_a will be less than the ideal velocity, v_i due to the frictional losses due to the fluid viscosity by an unknown amount, as modeled by a velocity coefficient, $C_v = v_a/v_i$. Furthermore, in real flow, due to the sharp edge of the sluice gate, the fluid viscosity causes a contraction in the depth of flow. As such, the supercritical depth of flow at point 2, y_2 will be less than the gate opening or depth of flow at the gate, a by an unknown amount, as modeled by a contraction coefficient, $C_c = A_a/A_i = y_{2a}/a$. Therefore, the actual discharge, Q_a will be less than the ideal discharge, Q_i and is experimentally determined by the use of a discharge coefficient. As a result, an experimentally determined discharge coefficient, $C_d = C_v C_c$ is applied to the ideal discharge, Q_i in order to determine the actual discharge, Q_a as follows:

$$Q_a = C_d Q_i = C_d A_{2i}\frac{\sqrt{2g(y_1 - y_2)}}{\sqrt{1 - \left(\frac{y_2}{y_1}\right)^2}} = C_c C_v A_{2i}\frac{\sqrt{2g(y_1 - y_2)}}{\sqrt{1 - \left(\frac{y_2}{y_1}\right)^2}} \tag{9.287}$$

which is analogous to the orifice discharge equation (see Equation 8.142), and where the effective head on the sluice gate is $(y_1 - y_2)$, and the discharge coefficient, C_d models two effects of real (nonideal) flow. The first effect is due to contraction in the flow as it flows under the sluice gate, where the area at point 2, A_2 is less than the area under the sluice gate, A_s by an unknown amount where:

$$C_c = \frac{A_a}{A_i} = \frac{A_{2a}}{A_{2i}} = \frac{A_{2a}}{A_s} = \frac{by_{2a}}{ba} = \frac{y_{2a}}{a} = \frac{y_2}{a} \tag{9.288}$$

where y_{2a} or y_2 is the actual flow depth at point 2, a is the height of the gate opening (which is the ideal flow depth at point 2), b is the width of the gate (and channel), and C_c is the contraction coefficient, which is < 1. The second effect is due to frictional pressure losses, which occur as a result of the turbulence and swirling motion near the downstream face of the sluice gate, yielding a head loss, which cannot be determined analytically. Instead, the frictional effects may be expressed as a velocity coefficient, C_v as follows:

$$C_v = \frac{v_{2a}}{v_{2i}}$$ (9.289)

where v_{2a} is the actual velocity at point 2, v_{2i} is the ideal velocity at point 2, and C_v is the velocity coefficient. Therefore, the actual discharge, Q_a is given as follows:

$$Q_a = C_d Q_i = C_d A_{2i} \frac{\sqrt{2g(y_1 - y_2)}}{\sqrt{1 - \left(\frac{y_2}{y_1}\right)^2}} = C_c C_v A_s \frac{\sqrt{2g(y_1 - y_2)}}{\sqrt{1 - \left(\frac{y_2}{y_1}\right)^2}} = C_c C_v ba \frac{\sqrt{2g(y_1 - y_2)}}{\sqrt{1 - \left(\frac{y_2}{y_1}\right)^2}}$$ (9.290)

where $A_{2i} = A_s = ba$.

It is important to note that although the contraction coefficient, C_c and the velocity coefficient, C_v which model the two nonideal flow effects, may be individually experimentally determined their combined effect is usually modeled by the discharge coefficient, C_d which is determined experimentally and is given as follows:

$$C_d = C_c C_v$$ (9.291)

where the value of the discharge coefficient, C_d is determined from dimensional analysis. One may note that the velocity coefficient, $C_v = v_a/v_i = \sqrt{E/2} = v_a/v_c = \sqrt{F/2}$, while the contraction coefficient, $C_c = A_a/A_i = y_2/a$. Furthermore, the discharge coefficient, $C_d = C_c C_v$ is also dependent on the boundary geometry of the sluice gate, $L_1/L = y_1/a$ and $L_2/L = y_2/a$. Therefore, the discharge coefficient, C_d is a function of both $C_c = A_a/A_i = L_2/a = y_2/a$ and $L_1/L = y_1/a$ as illustrated in Figure 9.37. For ideal flow the $C_d = 1$, however for actual flow $C_d < 1$. Experimental values for the discharge coefficient, C_d are presented in Figure 9.37, where typical values for a sluice gate with a free outflow range between 0.55 and 0.60. Additionally, one may note that for a sluice gate with a drowned outflow, as illustrated in Figure 9.38, the values for the discharge coefficient, C_d significantly drop and thus the actual flow rate decreases for similar upstream conditions. Finally, one may note that while the effective head on the sluice gate is $(y_1 - y_2)$, in the limit of $y_1 \gg y_2$, the sluice gate equation reduces to the following:

$$Q_a = C_d Q_i = C_d ba \sqrt{2g y_1}$$ (9.292)

where for such a limiting case, the kinetic energy of the fluid upstream of the sluice gate at point 1 is negligible and thus, yielding a downstream velocity, $v_2 = \sqrt{2g y_1}$.

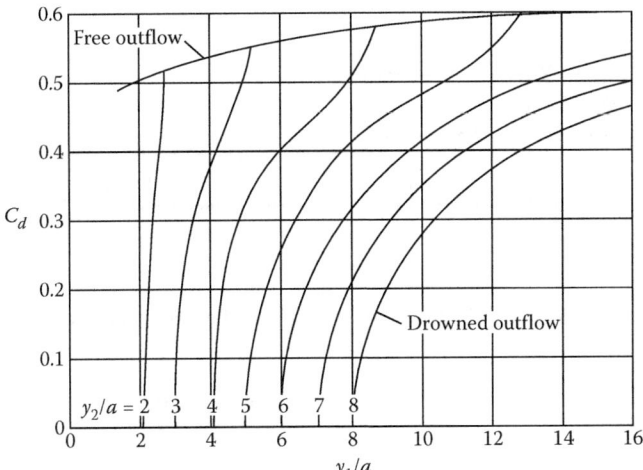

FIGURE 9.37
Discharge coefficient, C_d for a sluice gate is a function of contraction coefficient, $C_c = y_2/a$ and boundary geometry, $L_1/L = y_1/a$. (Adapted from Henry, H. R. et al., *Transactions of the American Society of Civil Engineers*, 115, 691, 1950.)

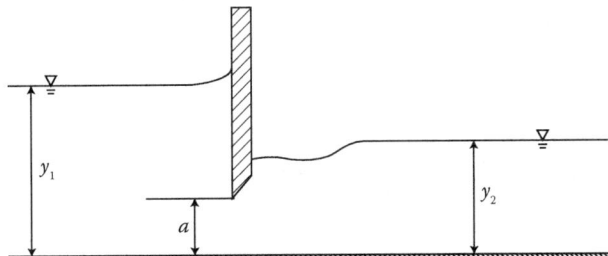

FIGURE 9.38
Sluice gate with drowned outflow.

EXAMPLE PROBLEM 9.17

Water flows in a rectangular channel of width 3 m and at a mild slope. A sluice gate is inserted in the channel as illustrated in Figure EP 9.17 in order to measure the flowrate. The opening of the gate is set at 1 m, the depth of flow upstream of the sluice gate is measured to be 2 m, and the depth of flow downstream of the sluice gate is measured to be 0.95 m. (a) Determine the ideal flowrate. (b) Determine the actual flowrate.

Mathcad Solution

(a) The opening of the gate, a is set high enough to result in a free overflow sluice gate as illustrated in Figure EP 9.17. The ideal flowrate, Q_i is determined from Equation 9.290 as follows:

$$b := 3\,m \qquad\qquad a := 1\,m \qquad\qquad y_1 := 2\,m \qquad\qquad y_2 := 0.95\,m$$

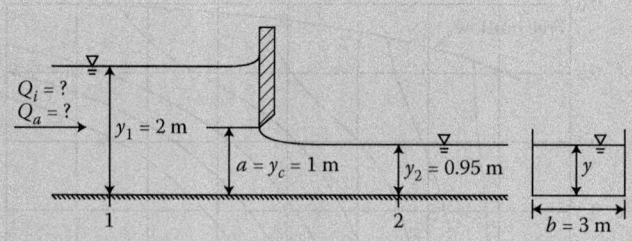

FIGURE EP 9.17
Water flows in a rectangular channel with a sluice gate inserted in the channel to measure the flowrate.

$$g := 9.81 \, \frac{m}{sec^2} \qquad\qquad A_{2i} := b \cdot a = 3 \, m^2$$

$$Q_i := A_{2i} \, \frac{\sqrt{2 \, g \, (y_1 - y_2)}}{\sqrt{1 - \left(\frac{y_2}{y_1}\right)^2}} = 15.474 \, \frac{m^3}{sec}$$

(b) The actual flowrate, Q_a is determined from Equation 9.290, $Q_a = C_d Q_i$, where the discharge coefficient, C_d for the free overflow sluice gate is a function of $L/L_i = y_1/a$, and is determined from Figure 9.37 as follows:

$$\frac{y_1}{a} = 2 \qquad\qquad C_d := 0.51$$

$$Q_a := C_d \cdot b \cdot a \cdot \frac{\sqrt{2 \cdot g \cdot (y_1 - y_2)}}{\sqrt{1 - \left(\frac{y_2}{y_1}\right)^2}} = 7.892 \, \frac{m^3}{sec}$$

9.17.4 Evaluation of the Actual Flowrate for a Sharp-Crested Weir

A sharp-crested weir is a vertical, sharp edged, notched flat plate, which extends over the full width of the channel that is inserted in an open channel and is used to regulate and measure the flowrate in open channel flow. The upstream fluid flows across the sharp edge of the weir and drops (contraction effect at the top of the nappe) by an amount of the draw down, Δh over the weir as the fluid starts its free overfall, and separates at the top edge of the weir (contraction effect at the bottom of the nappe), forming a nappe in the downstream section of the channel as illustrated in Figure 9.39, which also illustrates the head of the weir, H and the height of the weir, P. It may be noted that a free overfall is a special case of the sharp-crested weir (where the height of the weir, $P = 0$). One may note that the draw down, Δh over the weir causes the flow height over the weir to be considerably smaller than H. One may note that the sharp edge of the weir limits the opportunity for the development of a boundary layer; thus, the flow will be basically free from viscous effects and the resulting energy dissipation. The flowrate, Q is a result of the free overfall of the flow past the weir being clear from the weir (thus, formation of the nappe), and is a function of the head of the weir, H and the particular geometry of the notch in the weir. The particular shape

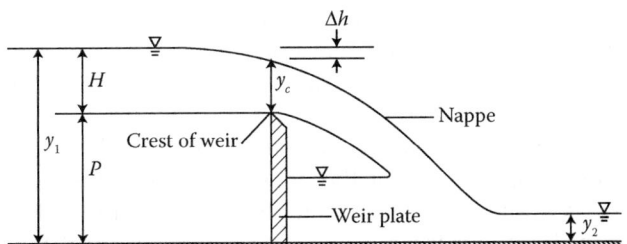

FIGURE 9.39
Geometry of a sharp-crested weir.

of the flow area/notch in the plane of the weir plate is used to specify the type of weir. Although the notch in the weir may be rectangular, triangular, trapezoidal, circular, parabolic, or any other geometric shape, the rectangular and the triangular are the most common shapes and are illustrated in Figure 9.40a and b, respectively, which also illustrates that the head of the weir, H is the height of the upstream fluid free surface above the bottom of the opening (notch) in the weir plate (i.e., the crest of the weir). When the weir is inserted in the open channel, the upstream subcritical flow at point 1 accelerates (as a result of pressure and gravitational forces) as it approaches the weir (which is a control), reaching critical flow at the crest of the weir, and then accelerates further to supercritical flow downstream of the weir at point 2. One may note that the acceleration of the flow is due to the steady decline in the elevation of the free surface from point 1 to point 2 and thus the conversion of the elevation head at point 1 into a velocity head at point 2. It is interesting to note that the weir serves as a "control," at which the critical depth of flow, y_c occurs, assuming ideal flow. However, due to the sharp-edged crest of the weir, the fluid viscosity causes a contraction in the depth of flow, while the gravity creates a pronounced curvature of the streamlines (forming a falling jet) in the vertical direction and thus, a deviation from a hydrostatic pressure distribution at the sharp-edged crest. As a result, this causes the actual critical depth of flow, y_c to occur just upstream of the crest of the weir, where $y_c \approx H$ (the head of the weir). Experimental evidence has shown that the depth at the sharp-edged crest of the weir is

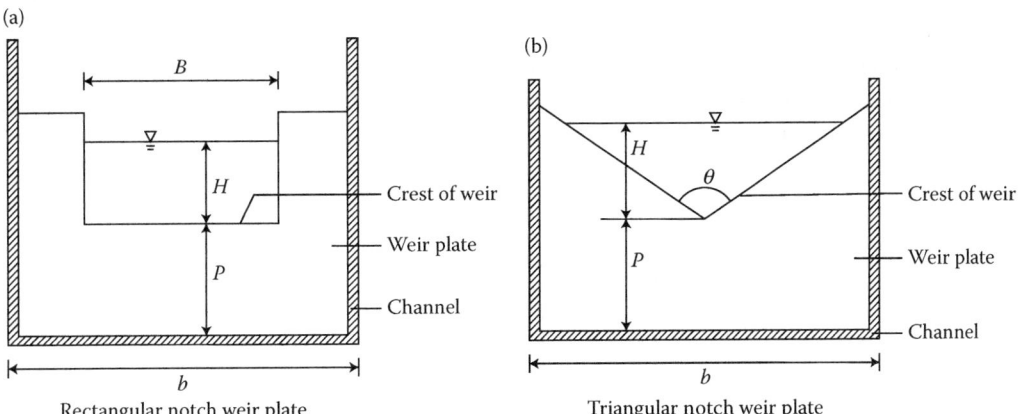

FIGURE 9.40
Two most common geometric shapes of notch in sharp-crested weir plate. (a) Rectangular notch weir plate, (b) Triangular notch weir plate.

approximately $5y_c/7$. Finally, one may note that due to the contraction of flow at the sharp-crested weir, this ideal flow meter is not considered to be a critical depth meter; thus, the discharge coefficient, C_d is a function of both a velocity coefficient, and a contraction coefficient, where $C_d = C_v C_c$.

The empirical derivation of the reduced/actual discharge, Q_a for a sharp-crested weir requires a "subset level" application of the governing equations (see Chapters 6 and 7 and Section 9.5.2.1). Assuming ideal flow implies that (1) the velocity profiles at the upstream point 1 and the downstream point 2 are uniform (ideal flow velocity profile); (2) there is no contraction in the fluid flow over the weir (the fluid flows horizontally over the weir crest) and thus the draw down, Δh over the weir as the fluid starts its free overfall is negligible compared to the head of the weir, H; thus, H is assumed to be the height of the upstream fluid free surface above the bottom of the opening (notch) in the weir plate; (3) the velocity profile within the nappe at the weir is nonuniform; (4) the pressure within the nappe is atmospheric; (5) frictional pressure losses due to the fluid viscosity as the fluid flows over the weir are negligible; and (6) the effects of surface tension is negligible. Thus, referring to Figure 9.41, the Bernoulli equation between points 1 (a point well upstream of the weir) and 3 (a point within the nappe above the weir) expressed in terms of pressure head is given as follows:

$$y_1 + z_1 + \frac{v_1{}^2}{2g} = y_3 + z_3 + \frac{v_3^2}{2g} \tag{9.293}$$

Assuming that $z_1 = z_3$ and setting $y = y_1 - y_3$ yields:

$$y_1 - y_3 = y = \frac{v_3^2}{2g} - \frac{v_1{}^2}{2g} \tag{9.294}$$

Assuming that y_1 is very large compared to y_3 (i.e., $P \gg H$), the kinetic energy upstream of the weir at point 1 is negligible compared to the kinetic energy at the weir at point 3. Then, solving for the ideal velocity at the weir at point 3, v_{3i} yields the following:

$$v_{3i} = \sqrt{2gy} \tag{9.295}$$

which is the ideal velocity flowing through the horizontal elemental strip of thickness, dy and elemental cross-sectional area, dA, where the definition of dA depends of the shape of the flow area/notch in the plane of the weir plate.

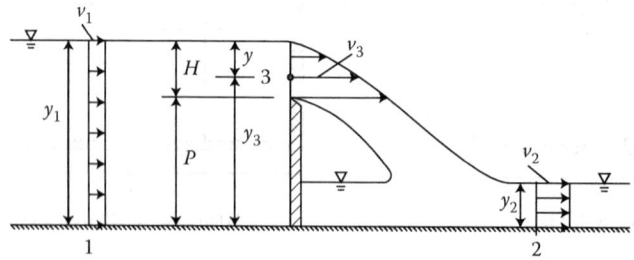

FIGURE 9.41
Assuming ideal flow for a sharp-crested weir.

9.17.4.1 Rectangular Sharp-Crested Weir

Assume a rectangular weir, $dA = B\,dy$, where B is the width of the rectangular weir opening/notch as illustrated in Figure 9.42. Therefore, from the differential continuity equation, the elemental ideal flowrate, dQ_i through the elemental area, dA is given as follows:

$$dQ_i = v_{3i}(dA) = \sqrt{2gy}(B\,dy) \tag{9.296}$$

And, integrating from $y = 0$ to $y = H$ yields the total ideal (theoretical) flow rate, Q_i over the rectangular weir as follows:

$$Q_i = \int dQ_i = \int_0^H B\sqrt{2g}\,y^{1/2}dy = B\sqrt{2g}\left[\frac{y^{3/2}}{\frac{3}{2}}\right]_0^H = \frac{2}{3}\sqrt{2g}BH^{3/2} \tag{9.297}$$

Finally, the reduced/actual discharge, Q_a is determined by application the differential continuity equation and the momentum equation, which is supplemented by dimensional analysis in order to define the discharge coefficient; the discharge coefficient, C_d is applied to the ideal discharge, Q_i in order to define the actual discharge, Q_a. In real flow, the actual velocity at point 3, v_3, will be less than the ideal velocity, v_{3i}, due to the frictional losses due to the fluid viscosity by an unknown amount, as modeled by a velocity coefficient, $C_v = v_a/v_i$. Furthermore, in real flow, due to the sharp crest of the weir, the fluid viscosity causes a contraction in the flow area (depth) above the weir, causing the draw down, Δh over the weir as the fluid starts its free overfall. Thus, the flow height over the weir, y is considerably smaller than H by an unknown amount, as modeled by a contraction coefficient, $C_c = A_a/A_i = y/H$. Therefore, the actual discharge, Q_a will be less than the ideal discharge, Q_i and is experimentally determined by the use of a discharge coefficient. As a result, an experimentally determined weir discharge coefficient, $C_d = C_v C_c$ is applied to the ideal discharge, Q_i in order to determine the actual discharge, Q_a as follows:

$$Q_a = C_d Q_i = C_d \frac{2}{3}\sqrt{2g}BH^{3/2} \tag{9.298}$$

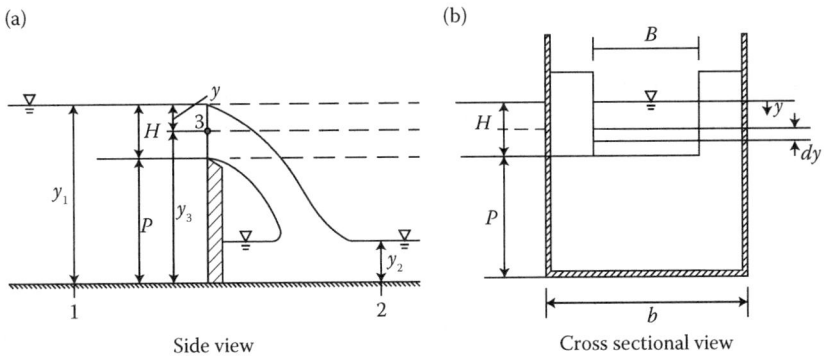

(a) Side view

(b) Cross sectional view

FIGURE 9.42
Assuming ideal flow for a rectangular sharp-crested weir. (a) Side view, (b) Cross-sectional view.

where the effective head on the weir is H, and the discharge the discharge coefficient, C_d models two effects of real (nonideal) flow. The first effect is due to contraction in the flow as it flows over the weir, where the flow height over the weir, y is considerably smaller than H by an unknown amount where:

$$C_c = \frac{A_a}{A_i} = \frac{y}{H} \tag{9.299}$$

where C_c is the contraction coefficient, which is < 1. The second effect is due to frictional pressure losses due to the fluid viscosity as the fluid flows over the weir, yielding a head loss, which cannot be determined analytically. Instead, the frictional effects may be expressed as a velocity coefficient, C_v as follows:

$$C_v = \frac{v_a}{v_i} \tag{9.300}$$

where v_a is the actual velocity over the weir, v_i is the ideal velocity over the weir, and C_v is the velocity coefficient. Therefore, the actual discharge, Q_a is given as follows:

$$Q_a = C_d Q_i = C_d \frac{2}{3}\sqrt{2g}BH^{3/2} = C_c C_v \frac{2}{3}\sqrt{2g}BH^{3/2} \tag{9.301}$$

It is important to note that although the contraction coefficient, C_c and the velocity coefficient, C_v which model the two nonideal flow effects, may be individually experimentally determined, their combined effect is usually modeled by the discharge coefficient, C_d which is determined experimentally and is given as follows:

$$C_d = C_c C_v \tag{9.302}$$

where the value of the discharge coefficient, C_d is determined from dimensional analysis. One may note that the velocity coefficient, $C_v = v_a/v_i = \sqrt{E/2} = v_a/v_c = \sqrt{F/2}$, while the contraction coefficient, $C_c = A_a/A_i = y/H$; thus, the discharge coefficient, C_d is a function of the head of the weir, $H \approx y_c$. Furthermore, the discharge coefficient, $C_d = C_c C_v$ is also dependent on the boundary geometry of the weir, $L_i/L = H/P$, the Reynolds number, R and the Weber number, W. For ideal flow the $C_d = 1$, however for actual flow $C_d < 1$. Experimental values for the rectangular weir discharge coefficient, C_d may be computed using the Rehbock formula assuming the English (BG) system (H and P in ft) as follows:

$$C_d = 0.605 + \frac{1}{305H} + 0.08\frac{H}{P} \tag{9.303}$$

The corresponding Rehbock formula for the SI units (H and P in m) is given as follows:

$$C_d = 0.605 + \frac{1}{1000H} + 0.08\frac{H}{P} \tag{9.304}$$

The Rehbock formula assumes a well-ventilated, sharp-crested rectangular weir with stilling devices upstream of the weir and yields an overall average value of the weir discharge coefficient, C_d of 0.62. The second term in the equation represents the effect of surface tension (capillarity), while the third term represents the velocity of approach, v_1 which is assumed to be uniform. The boundary geometry of the weir, $L_i/L = H/P$ is the most important parameter in determining the magnitude of the weir discharge coefficient, C_d, as this ratio has the greatest influence on the shape of the flow field (flow contraction and velocity profiles). The head of the weir, H is a direct measure of the Reynolds number, R and the Weber number, W as the effects of viscosity and surface tension are significant only when H is small. The effect of the Weber number, W is negligible except at low heads, where surface tension effects may be significant. And, the effect of the Reynolds number, R is small except at low heads, where viscous effects may be large. Since the flow over a weir typically involves water, the Reynolds number, R is usually high where the viscous effects are small. Furthermore, weirs serve as reliable flowrate measuring devices as long as the head, H is large enough so that the effects of viscosity and surface tension are negligible. When the head, H is large enough (0.08 to 2.0 ft; 0.025 to 0.60 m) compared to the values of the weir height, P (0.33 to 3.3 ft; 0.10 to 1.0 m), with a ratio H/P not greater than 1.0, the effects of viscosity and surface tension may be ignored and the weir discharge coefficient, C_d will increase with increasing head, H and decreasing weir height, P. Furthermore, the rectangular weir discharge coefficient, C_d will increase with decreasing values of the Reynolds number, R and the Weber number, W.

One may note that the main forces that govern the flow over a weir (rectangular, triangular, etc.) are the gravitational and inertia forces. The gravitational force accelerates the fluid from its free surface elevation upstream of the weir at point 1 to a larger velocity downstream of the weir at point 2. The gravitational force also causes the resulting draw down, Δh over the weir (contraction effect at the top of the nappe) and flow separation at the top edge of the weir (contraction effect at the top of the nappe). Furthermore, although the viscous and surface tension forces (important for a small value of H and thus low flowrates) are usually of secondary importance relative to the gravitational (and pressure) forces, their effects cannot totally be ignored. As such, the effects of all of these forces are modeled in the experimental determination of the weir discharge coefficient, C_d as presented in Equation 9.278. Thus, the weir discharge coefficient, C_d accounts for the nonideal flow conditions that were not included in the simplified assumptions made for ideal flow, with the most important difference between the ideal and real flow being the shape of the flow field (flow contraction and velocity profiles). Therefore, the weir discharge coefficient, C_d mainly represents a coefficient of contraction, C_c which models the extent of the contraction of the nappe in the real flow in comparison to the assumed no contraction in ideal flow. Furthermore, the effects of the remaining nonideal fluid effects (frictional and surface tension) are modeled by the weir discharge coefficient, C_d to the extent of how they affect the contraction of the flow.

EXAMPLE PROBLEM 9.18

Water flows in a rectangular channel of width 3 m and at a mild slope. A rectangular sharp-crested weir is inserted in the channel as illustrated in Figure EP 9.18 in order to measure the flowrate. The effective head on the weir is 1 m, the height of the weir is 1.5 m, and the width of the rectangular weir opening/notch is 2 m. (a) Determine the ideal flowrate. (b) Determine the actual flowrate.

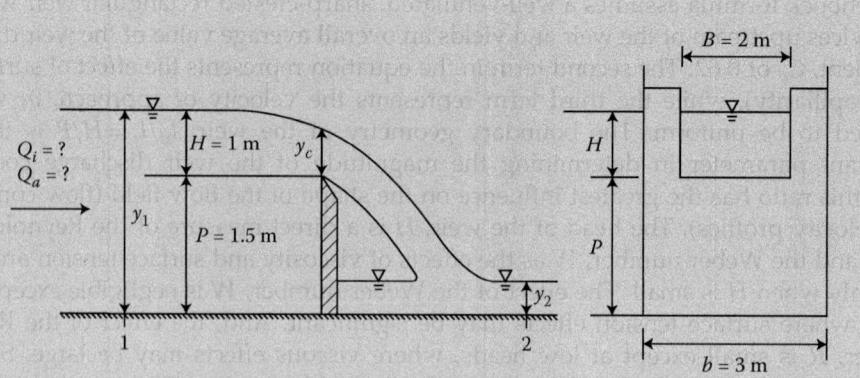

FIGURE EP 9.18
Water flows in a rectangular channel with a rectangular sharp-crested weir inserted in the channel to measure the flowrate.

Mathcad Solution

(a) The ideal flowrate, Q_i is determined from Equation 9.298 as follows:

$$H := 1\,m \qquad P := 1.5\,m \qquad B := 2\,m \qquad g := 9.81\,\frac{m}{sec^2}$$

$$Q_i := \frac{2}{3}\sqrt{2 \cdot g} \cdot B\,H^{\frac{3}{2}} = 5.906\,\frac{m^3}{sec}$$

(b) The actual flowrate, Q_a is determined from Equation 9.298 $Q_a = C_d Q_i$, where the discharge coefficient, C_d for the rectangular sharp-crested weir is determined from the Rehbock formula (Equation 9.304) as follows:

$$C_d := 0.605 + \frac{1\,m}{1000\,H} + 0.08\frac{H}{P} = 0.659$$

$$Q_a := C_d \frac{2}{3}\sqrt{2gB}H^{3/2} = 3.894\,\frac{m^3}{sec}$$

9.17.4.2 Triangular Sharp-Crested Weir

Assume a triangular weir, $dA = w\,dy$, where $w = 2\,(H-y)\tan(\theta/2)$ is the variable elemental width of the triangular weir opening/notch and θ is the angle of the V-notch as illustrated in Figure 9.43. Therefore, from the differential continuity equation, the elemental ideal flowrate, dQ_i through the elemental area, dA is given as follows:

$$dQ_i = v_{3i}(dA) = \sqrt{2gy}(w\,dy) \qquad (9.305)$$

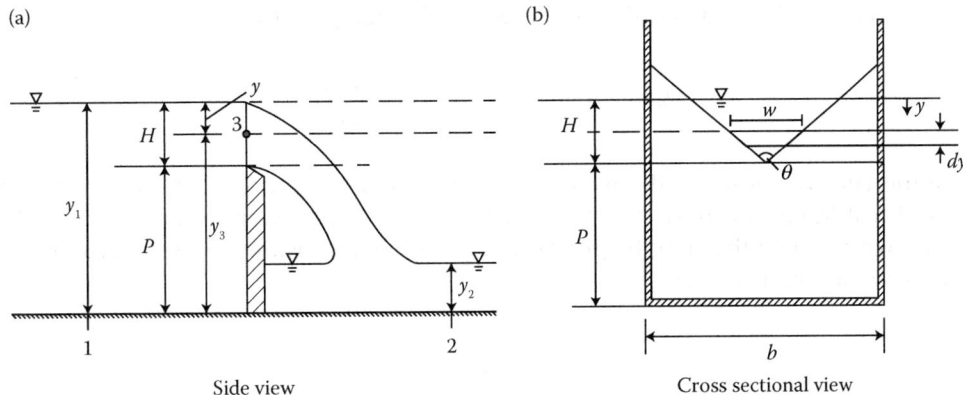

FIGURE 9.43
Assuming ideal flow for a triangular sharp-crested weir. (a) Side view, (b) Cross-sectional view.

And, integrating from $y = 0$ to $y = H$ yields the total ideal (theoretical) flow rate, Q_i over the triangular weir as follows:

$$Q_i = \int dQ_i = \int_0^H 2\sqrt{2g}\tan\left(\frac{\theta}{2}\right)(H - y)y^{1/2}dy \qquad (9.306)$$

Using Mathcad to integrate yields:

$$\int_0^H 2\sqrt{2g}\cdot\tan\left(\frac{\theta}{2}\right)\cdot(H - y)\cdot\sqrt{y}\,dy \rightarrow \frac{8}{15}\cdot H^{5/2}\cdot 2^{1/2}\cdot g^{1/2}\cdot\tan\left(\frac{1}{2}\cdot\theta\right)$$

Thus:

$$Q_i = \frac{8}{15}\sqrt{2g}\tan\left(\frac{\theta}{2}\right)H^{5/2} \qquad (9.307)$$

Finally, the reduced/actual discharge, Q_a is determined by application the differential continuity equation and the momentum equation, which is supplemented by dimensional analysis in order to define the discharge coefficient; the discharge coefficient, C_d is applied to the ideal discharge, Q_i in order to define the actual discharge, Q_a. In real flow, the actual velocity at point 3, v_3 will be less than the ideal velocity, v_{3i} because of the frictional losses due to the fluid viscosity by an unknown amount, as modeled by a velocity coefficient, $C_v = v_a/v_i$. Furthermore, in real flow, due to the sharp crest of the weir, the fluid viscosity causes a contraction in the flow area (depth) above the weir, causing the draw down, Δh over the weir as the fluid starts its free overfall. Thus, the flow height over the weir, y is considerably smaller than H by an unknown amount, as modeled by a contraction coefficient, $C_c = A_a/A_i = y/H$. Therefore, the actual discharge, Q_a will be less than the ideal discharge, Q_i, and is experimentally determined by the use of a discharge coefficient. As a result, an experimentally determined weir discharge coefficient, C_d is applied to the ideal discharge, Q_i in order to determine the actual discharge, Q_a as follows:

$$Q_a = C_d Q_i = C_d \frac{8}{15} \sqrt{2g} \tan\left(\frac{\theta}{2}\right) H^{5/2} \tag{9.308}$$

where the effective head on the weir is H, and the discharge coefficient, C_d models two effects of real (nonideal) flow. The first effect is due to contraction in the flow as it flows over the weir, where the flow height over the weir, y is considerably smaller than H by an unknown amount where:

$$C_c = \frac{A_a}{A_i} = \frac{y}{H} \tag{9.309}$$

where C_c is the contraction coefficient, which is < 1. The second effect is due to frictional pressure losses due to the fluid viscosity as the fluid flows over the weir, yielding a head loss, which cannot be determined analytically. Instead, the frictional effects may be expressed as a velocity coefficient, C_v as follows:

$$C_v = \frac{v_a}{v_i} \tag{9.310}$$

where v_a is the actual velocity over the weir, v_i is the ideal velocity over the weir, and C_v is the velocity coefficient. Therefore, the actual discharge, Q_a is given as follows:

$$Q_a = C_d Q_i = C_d \frac{8}{15} \sqrt{2g} \tan\left(\frac{\theta}{2}\right) H^{5/2} = C_c C_v \frac{8}{15} \sqrt{2g} \tan\left(\frac{\theta}{2}\right) H^{5/2} \tag{9.311}$$

It is important to note that although the contraction coefficient, C_c and the velocity coefficient, C_v (which model the two nonideal flow effects) may be individually experimentally determined, their combined effect is usually modeled by the discharge coefficient, C_d, which is determined experimentally and is given as follows:

$$C_d = C_c C_v \tag{9.312}$$

where the value of the discharge coefficient, C_d is determined from dimensional analysis. One may note that the velocity coefficient, $C_v = v_a/v_i = \sqrt{E/2} = v_a/v_c = \sqrt{F/2}$, while the contraction coefficient, $C_c = A_a/A_i = y/H$; thus, the discharge coefficient, C_d is a function of the head of the weir, $H \approx y_c$. Furthermore, the discharge coefficient, $C_d = C_c C_v$ is also dependent on the boundary geometry of the weir, $L_i/L = H/P$, the angle of the V-notch, θ; the Reynolds number, R; and the Weber number, W. For ideal flow $C_d = 1$; however, for actual flow, $C_d < 1$. Experimental values for the triangular weir discharge coefficient, C_d as a function of the head of the weir, H and the angle of the V-notch, θ, are presented in Figure 9.44, with an overall average value of 0.60. The triangular weir discharge coefficient, C_d typically increases with an increase in viscosity, surface tension, and the weir plate roughness. Furthermore, similar to the rectangular weir, the triangular weir discharge coefficient, C_d will increase with decreasing values of the Reynolds number, R and the Weber number, W.

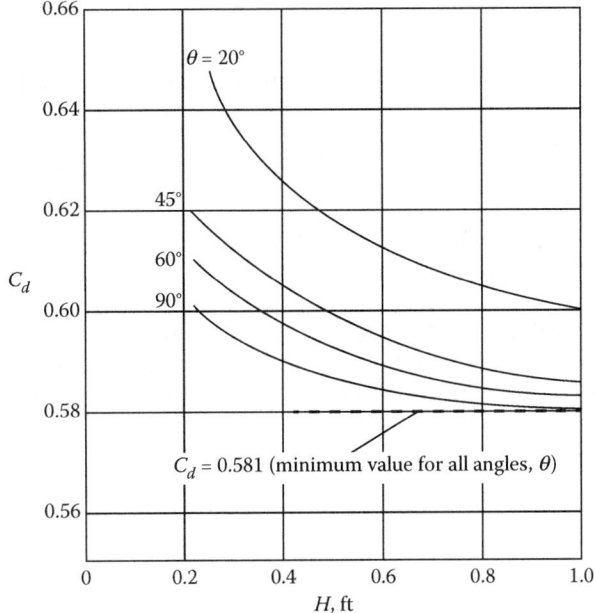

FIGURE 9.44
Discharge coefficient, C_d for a triangular sharp-crested weir is a function of head on weir, H and angle of the V-notch, θ. (From Lenz, A. T, *Transactions of the American Society of Civil Engineers*, 108, 766–767, 776–777, 778–779, 1943.)

EXAMPLE PROBLEM 9.19

Water flows in a rectangular channel of width 3 m and at a mild slope. A triangular sharp-crested weir is inserted in the channel as illustrated in Figure EP 9.19 in order to measure the flowrate. The effective head on the weir is 1 m, the height of the weir is 1.5 m, and the angle of the V-notch is 90°. (a) Determine the ideal flowrate. (b) Determine the actual flowrate.

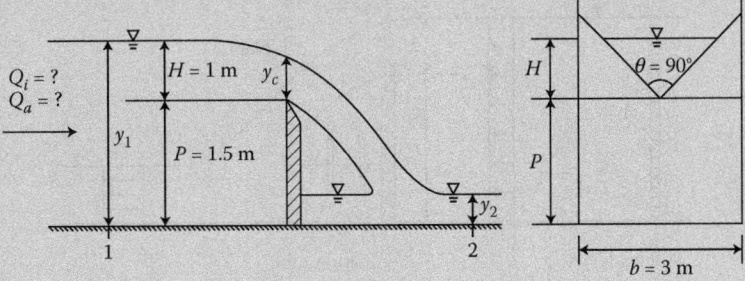

FIGURE EP 9.19
Water flows in a rectangular channel with a triangular sharp-crested weir inserted in the channel to measure the flowrate.

Mathcad Solution

(a) The ideal flowrate, Q_i is determined from Equation 9.308 as follows:

$$H := 1\,m \qquad P := 1.5\,m \qquad \theta := 90\,deg \qquad g := 9.81\,\frac{m}{sec^2}$$

$$Q_i := \frac{8}{15}\,\sqrt{2 \cdot g} \cdot \tan\left(\frac{\theta}{2}\right)\,H^{\frac{5}{2}} = 2.362\,\frac{m^3}{sec}$$

(b) The actual flowrate, Q_a is determined from Equation 9.308 $Q_a = C_d Q_i$, where the discharge coefficient, C_d for the triangular sharp-crested weir is a function of the head of the weir, H and the angle of the V-notch, θ, and is determined from Figure 9.44 as follows:

$$C_d := 0.58$$

$$Q_a := C_d \frac{8}{15}\,\sqrt{2 \cdot g} \cdot \tan\left(\frac{\theta}{2}\right) H^{5/2} = 1.37\,\frac{m^3}{sec}$$

9.17.4.3 A Comparison between a Rectangular and a Triangular Sharp-Crested Weir

One may note that the ideal flow assumption that the pressure within the nappe of a sharp-crested weir is atmospheric is typically more valid for the triangular weir than for the rectangular weir. Therefore, for rectangular weirs, it is sometimes necessary to install ventilation tubes in order to ensure atmospheric pressure in the nappe region (see Figure 9.45). Furthermore, the weir discharge coefficient, C_d accounts for this nonideal flow assumption as well. Additionally, it may be noted that the frictional resistance at the sidewalls of the channel will affect the flowrate, especially as the channel width, b becomes smaller. Furthermore, the fluid turbulence and the frictional resistance at both the sides and the bottom of the channel

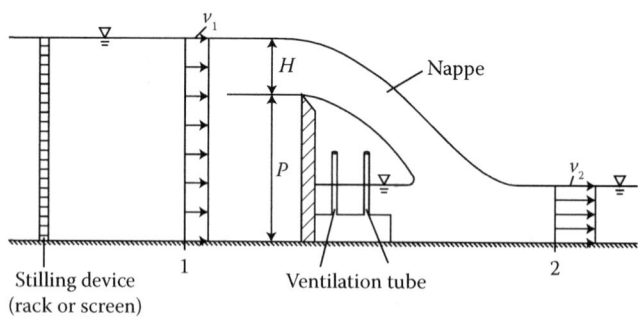

FIGURE 9.45
Ventilation tubes (ensure atmospheric pressure in nappe region) and stilling devices (ensure closer approximation to uniform velocity profile upstream of weir) validate ideal flow assumptions for rectangular sharp-crested weir. (Adapted from Vennard, J. K., and R. L., Street, *Elementary Fluid Mechanics*, 5th ed, John Wiley and Sons, New York, 1975, p. 567.)

section upstream of the weir also significantly contribute to the velocity profile. As a result, it becomes important to install stilling devices (such as racks and screens) (see Figure 9.45) sufficiently upstream of the weir in order to promote an even distribution of the fluid turbulence and thus a closer approximation to a uniform velocity profile.

It is interesting to note that the triangular weir is used to measure flowrates over a wider range of values than those using the rectangular weir. In particular, the triangular weir is more accurate for measuring small flowrates than the rectangular weir. This is because for small flowrates, Q the resulting head of the rectangular weir, H is very small in comparison to the resulting head of the triangular weir, H and thus would be difficult to measure accurately. As such, for small flowrates, the resulting head of the triangular weir, H is still reasonably large enough to accurately measure. Furthermore, the flowrate, Q is proportional only to $H^{3/2}$ for the rectangular weir, while the flowrate, Q is proportional to $H^{5/2}$ for the triangular weir.

9.17.5 Evaluation of the Actual Flowrate for a Spillway

A spillway is the overflow section of a dam that is designed to allow water to pass over its crest. An overflow spillway behaves in a similar manner to a weir, and is illustrated in Figure 9.46, which also illustrates the head of the spillway, H and the height of the spillway, P. A spillway that has a rounded-crest shape, similar to the underside of the nappe springing from a sharp-crested weir (see Figure 9.39), is called an ogee (S-shaped curve) spillway. The spillway is considered to be a special type of weir and may be used for flowrate measurement. However, because sharp-crested weirs consist of smooth, vertical, flat plates with sharp edges, they provide an easier and more reproducible approach for flowrate measurement.

The flowrate, Q is mainly a function of the head of the spillway. Similar to a weir, when the spillway is built in the open channel, the upstream subcritical flow at point 1 accelerates (as a result of pressure and gravitational forces) as it approaches the spillway (which is a control), reaching critical flow at the crest of the spillway, and then accelerates further to supercritical flow downstream of the spillway at point 2. It is interesting to note that the spillway serves as a "control," at which the critical depth of flow, y_c occurs, assuming ideal flow. However, due to the crest of the spillway, the fluid viscosity causes a contraction in the depth of flow because the flow is not ideal. As a result, the contraction of flow causes the actual critical depth of flow, y_c to occur just upstream of the crest of the spillway. Finally, one may note that due to the contraction of flow at the spillway, this ideal flow meter is not considered

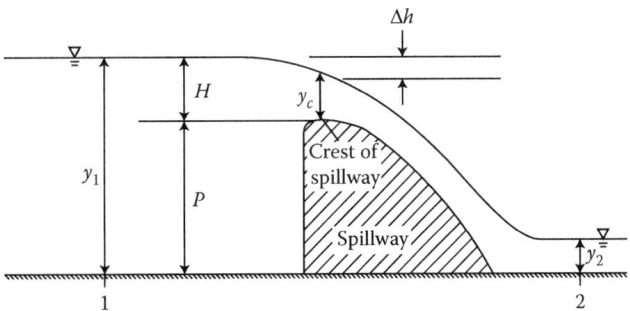

FIGURE 9.46
Geometry of a spillway.

to be a critical depth meter; thus, the discharge coefficient, C_d is a function of both a velocity coefficient, and a contraction coefficient, where $C_d = C_v C_c$.

The theory of a sharp-crested rectangular weir forms a basis for the analysis and design of a spillway. Therefore, the equation for the actual discharge, Q_a for a spillway is the same as that derived above for a sharp-crested weir and is repeated as follows:

$$Q_a = C_d Q_i = C_d \frac{2}{3} \sqrt{2g} B H^{3/2} \tag{9.313}$$

where H is the effective head on the spillway, B is the width of the crest of the spillway, and the discharge coefficient, C_d models two effects of real (nonideal) flow. The value of the spillway discharge coefficient, C_d may also be computed using the Rehbock formula given above for a sharp-crested rectangular weir (see Equations 9.303 and 9.304), with typical values ranging between 0.60 to 0.75. Furthermore, similar to the rectangular sharp-crested weir discharge coefficient, C_d as described by the Rehbock formula, the spillway discharge coefficient, C_d will increase with increasing head, H and decreasing weir height, P.

EXAMPLE PROBLEM 9.20

Water flows in a rectangular channel of width 3 m and at a mild slope. A spillway is inserted in the channel as illustrated in Figure EP 9.20 in order to measure the flowrate. The effective head on the spillway is 0.9 m, the height of the spillway is 1.2 m, and the width of the crest of the spillway is 3 m. (a) Determine the ideal flowrate. (b) Determine the actual flowrate.

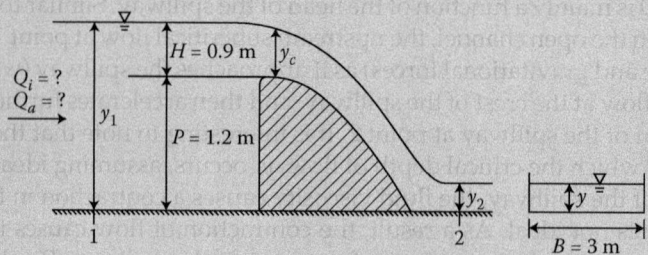

FIGURE EP 9.20
Water flows in a rectangular channel with a spillway inserted in the channel to measure the flowrate.

Mathcad Solution

(a) The ideal flowrate, Q_i is determined from Equation 9.313 as follows:

$$H := 0.9 \, m \qquad P := 1.2 \, m \qquad B := 3 \, m \qquad g := 9.81 \, \frac{m}{sec^2}$$

$$Q_i := \frac{2}{3} \sqrt{2 \cdot g} \cdot B \cdot H^{\frac{3}{2}} = 7.564 \, \frac{m^3}{sec}$$

(b) The actual flowrate, Q_a is determined from Equation 9.313 $Q_a = C_d Q_i$, where the discharge coefficient, C_d for the spillway is determined from the Rehbock formula (Equation 9.304) as follows:

$$C_d := 0.605 + \frac{1\,m}{1000\,H} + 0.008\frac{H}{P} = 0.666$$

$$Q_a := C_d \frac{2}{3}\sqrt{2 \cdot g} \cdot B \cdot H^{3/2} = 5.038 \frac{m^3}{\sec}$$

9.17.6 Evaluation of the Actual Flowrate for a Broad-Crested Weir

A broad-crested weir is a vertical rectangular structure in an open channel, as illustrated in Figure 9.47, which has a horizontal crest of length L_w and height, P; extends over the full width of the channel, b; and is used to regulate and measure the flowrate in open channel flow. One may note that the broad-crested weir is a special case of an abrupt vertical contraction. The flowrate, Q is mainly a function of the head of the weir, H. Similar to the sharp-crested weir, when the broad crested-weir is inserted in the open channel, the upstream subcritical flow at point 1 accelerates (as a result of pressure and gravitational forces) as it approaches the weir (which is a control), reaching critical flow over the broad crest of the weir, and then accelerates further to supercritical flow downstream of the weir at point 2. It is interesting to note that the broad-crested weir serves as a "control" at which the critical depth of flow, y_c occurs assuming ideal flow. However, unlike the sharp-crested weir and the spillway (discussed above), although the flow is not ideal, the broad crest allows the contraction of flow to occur just downstream of the crest and thus allows the critical depth of flow, y_c to occur at the crest. Specifically, because there is no contraction in the flow at the control, the critical depth of flow, y_c occurs in the close vicinity of the control. Therefore, the broad-crested weir, which is an ideal flow meter, is also considered to be a critical depth meter. Thus, because there is no contraction in the depth of low at the control, there is no need to apply the contraction coefficient, C_c to the computed discharge. As such, the discharge coefficient, C_d is only a function of a velocity coefficient, where $C_d = C_v$.

The empirical derivation of the reduced/actual discharge, Q_a for a broad-crested weir requires a "subset level" application of the governing equations (see Chapters 6 and 7 and Section 9.5.2.1). Assuming ideal flow implies the following: (1) the velocity profiles at

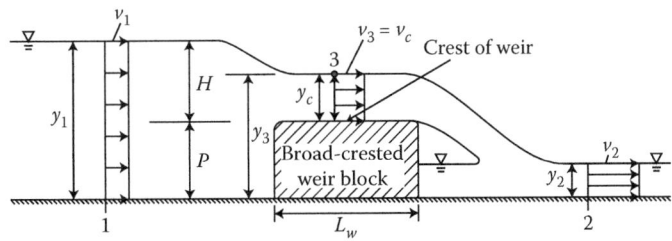

FIGURE 9.47
Geometry of a broad-crested weir.

the upstream point 1, the downstream point 2, and over the broad crest at point 3 are uniform (ideal flow velocity profile); (2) the flow over the broad crest is critical; (3) the pressure across the broad crest is hydrostatic; (4) frictional pressure losses due to the fluid viscosity as the fluid flows over the weir are negligible; and (5) the effects of surface tension are negligible. Thus, referring to Figure 9.47, the Bernoulli equation between points 1 (a point well upstream of the weir) and 3 (a point over the broad crest of the weir) expressed in terms of pressure head is given as follows:

$$y_1 + z_1 + \frac{v_1^2}{2g} = y_3 + z_3 + \frac{v_3^2}{2g} \tag{9.314}$$

Assuming that $z_1 = z_3$ and setting $y_1 = H + P$, $y_3 = y_c + P$, and $v_3 = v_c$ yields:

$$H + P + \frac{v_1^2}{2g} = y_c + P + \frac{v_c^2}{2g} \tag{9.315}$$

Assuming that y_1 is very large compared to y_3 (i.e., $P \gg H$), the kinetic energy upstream of the weir at point 1 is negligible compared to the kinetic energy at the weir at point 3. Then, solving for the ideal velocity at the weir at point 3, v_{3i} yields the following:

$$H = y_c + \frac{v_c^2}{2g} \tag{9.316}$$

Substituting the expression for the critical velocity, $v_c = \sqrt{gy_c}$ yields the following:

$$H = y_c + \frac{y_c}{2} = \frac{3y_c}{2} \tag{9.317}$$

Isolating the expression for the critical depth of flow, y_c yields:

$$y_c = \frac{2H}{3} \tag{9.318}$$

Next, application of the continuity equation yields an expression for the ideal (theoretical) flowrate. Assuming that the velocity profile at the point over the broad crest point 3 is uniform (ideal flow velocity profile), the ideal flowrate is computed as follows:

$$Q_i = v_{3i}A_3 = v_c A_c = v_c by_c = \sqrt{gy_c}by_c = b\sqrt{g}y_c^{3/2} = b\sqrt{g}\left(\frac{2}{3}\right)^{3/2} H^{3/2} \tag{9.319}$$

Finally, the reduced/actual discharge, Q_a is determined by application the differential continuity equation, and the momentum equation, which is supplemented by dimensional analysis in order to define the discharge coefficient; the discharge coefficient, C_d is applied to the ideal discharge, Q_i in order to define the actual discharge, Q_a. In real flow however, the actual velocity at point 3, v_3, will be less than the ideal velocity, v_{3i}, due to the frictional losses due to the fluid viscosity by an unknown amount, as modeled by a velocity coefficient, $C_v = v_a/v_i$. Therefore, the actual discharge, Q_a will be less than the ideal discharge, Q_i, and is experimentally determined by the use of a discharge coefficient. As a result, an experimentally determined weir discharge coefficient, $C_d = C_v$ is applied to the ideal discharge, Q_i in order to determine the actual discharge, Q_a as follows:

$$Q_a = C_d Q_i = C_d b \sqrt{g} \left(\frac{2}{3}\right)^{3/2} H^{3/2} = C_v b \sqrt{g} \left(\frac{2}{3}\right)^{3/2} H^{3/2} \tag{9.320}$$

where the discharge coefficient, C_d models only one effect of real (nonideal) flow. The one effect is due to frictional pressure losses due to the fluid viscosity as the fluid flows over the weir, yielding a head loss, which cannot be determined analytically. Instead, the frictional effects may be expressed as a velocity coefficient, C_v as follows:

$$C_v = \frac{v_{3a}}{v_{3i}} \tag{9.321}$$

where v_{3a} is the actual velocity at point 3, v_{3i} is the ideal velocity at point 3, and C_v is the velocity coefficient. The value of the discharge coefficient, $C_d = C_v$ is determined from dimensional analysis where the velocity coefficient, $C_v = v_a/v_i = \sqrt{E/2} = v_a/v_c = \sqrt{F/2}$. Furthermore, the discharge coefficient, $C_d = C_v$ is also dependent on the boundary geometry of the weir, $L_i/L = H/P$; the Reynolds number, R; and the Weber number, W. For ideal flow the $C_d = 1$, however for actual flow $C_d < 1$. Experimental values for the broad-crested weir discharge coefficient, C_d may be computed using an empirical equation by Chow (1959) assuming either the English (BG) or the SI system as follows:

$$C_d = \frac{0.65}{\sqrt{1 + \dfrac{H}{P}}} \tag{9.322}$$

where typical values for the broad-crested weir discharge coefficient, C_d are between 0.50 and 0.57. Similar to the rectangular sharp-crested weir discharge coefficient, C_d, as described by the Rehbock formula, the broad-crested weir discharge coefficient, C_d will increase with increasing head, H and decreasing weir height, P. In order to serve as reliable flowrate measuring devices, the broad-crested weirs are generally restricted to the range $0.08 < H/L_w < 0.50$. Furthermore, if $H/L_w < 0.08$ (the broad crest is too long), then viscous effects will be significant, and the flow over the broad crest will be supercritical; thus, critical flow would occur just upstream of the weir. If however, $H/L_w > 0.5$ (the broad crest is too short), then the flow may not be able to accelerate to critical velocity at the weir.

EXAMPLE PROBLEM 9.21

Water flows in a rectangular channel of width 3 m and at a mild slope. A broad-crested weir is inserted in the channel as illustrated in Figure EP 9.21, in order to measure the flowrate. The effective head on the weir is 1 m, the height of the weir is 1.5 m, the weir extends over the full width of the channel, and the horizontal crest of the weir is 2.5 m. (a) Determine the ideal flowrate. (b) Determine the actual flowrate.

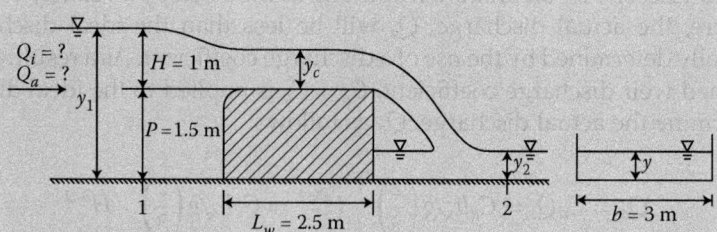

FIGURE EP 9.21
Water flows in a rectangular channel with a broad-crested weir inserted in the channel to measure the flowrate.

Mathcad Solution

(a) In order to ensure that a broad-crested weir serves as a reliable flowrate measuring device, check to see if $0.08 < H/L_w < 0.50$ as follows:

$$H := 1\,m \qquad\qquad P := 1.5\,m \qquad\qquad b := 3\,m \qquad\qquad L_w := 2.5\,m$$

$$g := 9.81\,\frac{m}{sec^2} \qquad\qquad \frac{H}{L_w} = 0.4$$

The ideal flowrate, Q_i is determined from Equation 9.320 as follows:

$$Q_i := b\sqrt{g}\left(\frac{2}{3}\right)^{3/2} H^{3/2} = 5.115\,\frac{m^3}{s}$$

(b) The actual flowrate, Q_a is determined from Equation 9.320 $Q_a = C_d Q_i$, where the discharge coefficient, C_d for the broad-crested weir is determined from the empirical equation by Chow (Equation 9.322) as follows:

$$C_d := \frac{0.65}{\sqrt{1+\dfrac{H}{P}}} = 0.503$$

$$Q_a := C_d \cdot b\sqrt{g}\left(\frac{2}{3}\right)^{3/2} H^{3/2} = 2.575\,\frac{m^3}{sec}$$

EXAMPLE PROBLEM 9.22

Water flows in a rectangular channel of width 3 m and at a mild slope. The actual flowrate in the channel ranges between $0.5\,m^3/sec$ and $4.5\,m^3/sec$. Refer to Example Problems 9.18, 9.19, and 9.21, which illustrated a rectangular sharp-crested weir, a triangular sharp-crested weir, and a broad-crested weir, respectively. In all three types of weirs, the height of the weir is 1.5 m, and the weir extends over the full width of the channel. For the rectangular sharp-crested weir, the width of the weir opening/notch is 2 m. For the triangular sharp-crested weir, the angle of the V-notch is 90°. And for the broad-crested weir, the horizontal crest of the weir is 2.5 m. (a) Plot the depth–discharge curve for the three different types of weirs. (b) Conduct a comparison between the three types of weirs for the given range of flowrates.

Mathcad Solution

(a) In order to plot the depth–discharge curve for the three different types of weirs, one must define the actual discharge equation for each type of weir. Note that the head on the weir, H and not the depth $(y = P + H)$ is actually plotted versus the actual discharge, Q_a as follows:

$$P := 1.5\,\text{m} \qquad g := 9.81\,\frac{\text{m}}{\text{sec}^2} \qquad b := 3\,\text{m}$$

For the rectangular sharp-crested weir:

$$B := 2\,\text{m} \qquad C_{drsc}(H) := 0.605 + \frac{1\,\text{m}}{1000\,H} + 0.08\,\frac{H}{P}$$

$$Q_{arsc}(H) := C_{drsc}(H)\,\frac{2}{3}\,\sqrt{2\,g} \cdot B\,H^{\frac{3}{2}}$$

Although the discharge coefficient, C_d for the triangular sharp-crested weir is a function of the head of the weir, H and the angle of the V-notch, θ, and is determined from Figure 9.44, for ease of application, assume a C_d of 0.58 as follows:

$$\theta := 90\,\text{deg} \qquad C_{dtsc}(H) := 0.58$$

$$Q_{atsc}(H) := C_{dtsc}(H)\,\frac{8}{15}\,\sqrt{2 \cdot g} \cdot \tan\left(\frac{\theta}{2}\right) H^{\frac{5}{2}}$$

In order to serve as reliable flowrate measuring devices, the broad-crested weirs are generally restricted to the range $0.08 < H/L_w < 0.5$ as follows:

$$L_w := 2.5\,\text{m} \qquad H_{min} := 0.08\,L_w = 0.2\,\text{m} \qquad H_{max} := 0.5\,L_w = 1.25\,\text{m}$$

$$C_{dbc}(H) := \frac{0.65}{\sqrt{1 + \dfrac{H}{P}}} \qquad Q_{abc}(H) := C_{dbc}(H) \cdot b\sqrt{g}\left(\frac{2}{3}\right)^{\frac{3}{2}} H^{\frac{3}{2}}$$

Plotting the depth–discharge curve:

$$Q_{amin} := 0.5 \cdot \frac{\text{m}^3}{\text{sec}} \qquad\qquad Q_{amax} := 4.5 \cdot \frac{\text{m}^3}{\text{sec}}$$

$$aa \equiv 0.2\,\text{m} \qquad\qquad bb \equiv 1.75\,\text{m} \qquad hh \equiv 0.1\,\text{m} \qquad H := aa, aa + hh..bb$$

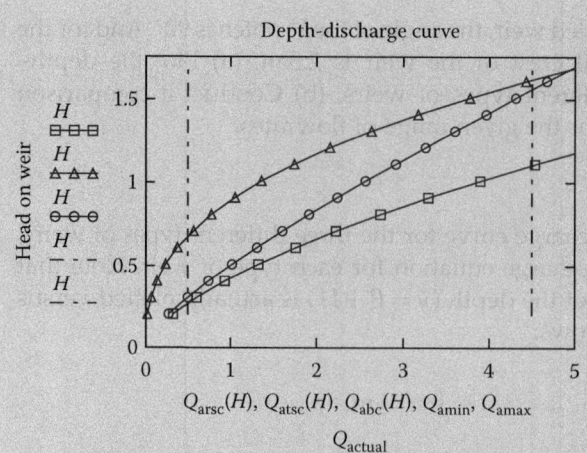

On the depth–discharge curve, the rectangular sharp-crested weir is illustrated by the rectangular symbol, the triangular sharp-crested weir is illustrated by the triangular symbol, and the broad-crested weir is illustrated by the circular symbol. The range for the actual flowrate in the channel, $Q_{amin} = 0.5 \, m^3/sec$ and $Q_{amax} = 4.5 \, m^3/sec$ is illustrated by the vertical broken lines on the depth–discharge curve.

(b) Two important points may be highlighted regarding a comparison between the three types of weirs. First, in order to serve as reliable flowrate measuring device, the broad-crested weir is generally restricted to the range $0.08 < H/L_w < 0.5$. Thus, if $H/L_w < 0.08$ (the broad crest is too long), then viscous effects will be significant, and the flow over the broad crest will be supercritical; thus, critical flow would occur just upstream of the weir. If, however, $H/L_w > 0.5$ (the broad crest is too short), then the flow may not be able to accelerate to critical velocity at the weir. Therefore, given that the horizontal crest of the weir, $L_w = 2.5 \, m$, $H_{min} = 0.2 \, m$, and $H_{max} = 1.25 \, m$. As such, the upper limit for the weir head, $H_{max} = 1.25 \, m$ for the broad-crested weir does not allow a reliable measurement of $Q_{amax} = 4.5 \, m^3/sec$. However, there is no such limitation imposed for the rectangular sharp-crested weir and the triangular sharp-crested weir. And second, in the cases of the broad-crested weir and the rectangular sharp-crested weir, the flowrate, Q is proportional only to $H^{3/2}$, while for the triangular sharp-crested weir, the flowrate, Q is proportional to $H^{5/2}$. As such, the depth–discharge curve plotted above illustrates that the triangular sharp-crested weir may be used to measure flowrates over a wider range of values than those using either the broad-crested weir or the rectangular sharp-crested weir. In particular, the triangular sharp-crested weir is more accurate for measuring small flowrates than the rectangular weir. This is because for small flowrates, such as $Q_{amin} = 0.5 \, m^3/sec$, the resulting head, H of the broad-crested or the rectangular sharp-crested weir is very small in comparison to the resulting head of the triangular sharp-crested weir, H and thus would be difficult to measure accurately. However, for small flowrates, the resulting head of the triangular sharp-crested weir, H is still reasonably large enough to accurately measure. Finally, this means that of the three weirs, the triangular sharp-crested weir would serve as the most reliable and most accurate flow-measuring device for the given open channel flow situation.

9.17.7 Evaluation of the Actual Flowrate for the Parshall Flume (Venturi Flume)

The venturi flume, as illustrated in Figure 9.48, is a section of an open channel with a gradually decreasing width followed by a gradually increasing width. As such, the venturi flume for open channel flow is analogous to a venturi flow meter for pipe flow, where the venturi flume is a special case of an abrupt horizontal contraction. It is important to note that although the decrease and subsequent increase in channel width is gradual, the constriction of width, b_c at the throat of the venturi flume is an abrupt/significant horizontal contraction in the channel walls, and thus serves as a "control," at which the critical depth of flow, y_c occurs. As explained above, if the contraction is not significant, then the constriction would not serve as a "control," and thus the critical depth of flow, y_c will not occur (rather, subcritical flow would be maintained).

The flowrate, Q is mainly a function of the head of the flume. Similar to the broad-crested weir, when the venturi flume is inserted in the open channel, the upstream subcritical flow at point 1 accelerates (as a result of pressure and gravitational forces) as it approaches the constriction (which is a control), reaching critical flow at the constriction, and then accelerates further to supercritical flow downstream of the constriction at point 2. It is interesting to note that the venturi flume serves as a "control," at which the critical depth of flow, y_c occurs, assuming ideal flow. However, unlike the sharp-crested weir and the spillway, although the flow is not ideal, the constriction does not cause a contraction in the flow and thus allows the critical depth of flow, y_c to occur at the constriction. Specifically, because there is no contraction in the flow at the control, the critical depth of flow, y_c occurs in the close vicinity of the control. Therefore, the venturi flume, which is an ideal flow meter, is also considered to be a critical depth meter or a critical depth flume. Thus, because there is no contraction in the depth of flow at the control, there is no need to apply the contraction coefficient, C_c to the computed discharge. As such, the discharge coefficient, C_d is only a function of a velocity coefficient, where $C_d = C_v$.

The theory of a broad-crested weir forms the basis for the analysis and design of a venturi flume. Therefore, the equation for the actual discharge, Q_a for a venturi is essentially the

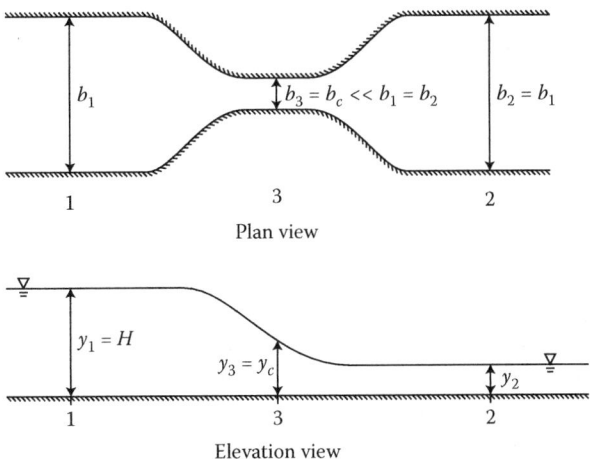

FIGURE 9.48
Geometry of a venturi flume. (a) Plan view, (b) Elevation view.

same as that derived above for a broad-crested weir and is derived as follows. The Bernoulli equation between points 1 (a point well upstream of the throat of the venturi flume) and 3 (a point at the throat of the venturi flume) expressed in terms of pressure head is given as follows:

$$y_1 + z_1 + \frac{v_1^2}{2g} = y_3 + z_3 + \frac{v_3^2}{2g} \tag{9.323}$$

Assuming that $z_1 = z_3$ and setting $y_1 = H$, $y_3 = y_c$, and $v_3 = v_c$ yields:

$$H + \frac{v_1^2}{2g} = y_c + \frac{v_c^2}{2g} \tag{9.324}$$

Assuming that y_1 is very large compared to y_3, the kinetic energy upstream of the throat at point 1 is negligible compared to the kinetic energy at the throat at point 3. Then, solving for the ideal velocity at the throat at point 3, v_{3i} yields the following:

$$H = y_c + \frac{v_c^2}{2g} \tag{9.325}$$

Substituting the expression for the critical velocity, $v_c = \sqrt{gy_c}$ yields the following:

$$H = y_c + \frac{y_c}{2} = \frac{3y_c}{2} \tag{9.326}$$

Isolating the expression for the critical depth of flow, y_c yields:

$$y_c = \frac{2H}{3} \tag{9.327}$$

Next, application of the differential continuity equation yields an expression for the ideal (theoretical) flowrate. Assuming that the velocity profile at the throat at point 3 is uniform (ideal flow velocity profile), the ideal flowrate is computed as follows:

$$Q_i = v_{3i}A_3 = v_cA_c = v_cb_cy_c = \sqrt{gy_c}b_cy_c = b_c\sqrt{g}y_c^{3/2} = b_c\sqrt{g}\left(\frac{2}{3}\right)^{3/2}H^{3/2} \tag{9.328}$$

where $b_c = b_3$.

Finally, the reduced/actual discharge, Q_a is determined by application the differential continuity equation, and the momentum equation, which is supplemented by dimensional analysis in order to define the discharge coefficient; the discharge coefficient, C_d is applied to the ideal discharge, Q_i in order to define the actual discharge, Q_a. In real flow, however, the actual velocity at point 3, v_3, will be less than the ideal velocity, v_{3i} due to the frictional losses due to the fluid viscosity by an unknown amount, as modeled by a velocity coefficient, $C_v = v_a/v_i$. Therefore, the actual discharge, Q_a will be less than the ideal discharge, Q_i and is experimentally determined by the use of a discharge coefficient. As a result, an experimentally determined venturi discharge coefficient, $C_d = C_v$ is applied to the ideal discharge, Q_i in order to determine the actual discharge, Q_a as follows:

$$Q_a = C_d Q_i = C_d b_c \sqrt{g} \left(\frac{2}{3}\right)^{3/2} H^{3/2} = C_v b_c \sqrt{g} \left(\frac{2}{3}\right)^{3/2} H^{3/2} \tag{9.329}$$

where the discharge coefficient, C_d models only one effect of real (nonideal) flow. The one effect is due to frictional pressure losses due to the fluid viscosity as the fluid flows through the flume, yielding a head loss, which cannot be determined analytically. Instead, the frictional effects may be expressed as a velocity coefficient, C_v as follows:

$$C_v = \frac{v_{3a}}{v_{3i}} \tag{9.330}$$

where v_{3a} is the actual velocity at point 3, v_{3i} is the ideal velocity at point 3, and C_v is the velocity coefficient. The value of the discharge coefficient, $C_d = C_v$ is determined from dimensional analysis where the velocity coefficient, $C_v = v_a/v_i = \sqrt{E/2} = v_a/v_c = \sqrt{F/2}$. Furthermore, the discharge coefficient, $C_d = C_v$ is also dependent on the boundary geometry of flume, $L_i/L = H/b_c$; the Reynolds number, R; and the Weber number, W. For ideal flow, $C_d = 1$; however, for actual flow, $C_d < 1$. Experimental values for the venture flume discharge coefficient, C_d range between 0.95 to 0.99 depending upon the exact geometry of the venturi flume.

It is interesting to note that although the use of a weir is probably the simplest method to measure the flowrate in an open channel, the resulting head loss for the sharp-crested weir (but not so much for the broad-crested weir) is relatively high, and if the water contains suspended particles, then sedimentation is deposited in the pool immediately upstream of the weir, which results in a gradual change in the discharge coefficient. Such difficulties may be overcome by the use of a venturi flume. The venturi flume is widely used in the measurement of irrigation water systems because a low head loss is required for its use, and the sediment is easily flushed through in case the water is silty. Thus, the advantages of venturi flumes is their ability to pass sediment-laden water without depositions and the small net change in the water level required between the entrance and the exit channels.

EXAMPLE PROBLEM 9.23

Water flows in a rectangular channel of width 3 m and at a mild slope. A venturi flume is inserted in the channel as illustrated in Figure EP 9.23 in order to measure the flowrate. The effective head on the flume is 1 m, and the constriction width is 0.5 m. Assume that the discharge coefficient, C_d for the venturi flume is 0.95. (a) Determine the ideal flowrate. (b) Determine the actual flowrate.

Mathcad Solution

(a) The ideal flowrate, Q_i is determined from Equation 9.329 as follows:

$$H := 1 \, m \qquad b_c := 0.5 \, m \qquad g := 9.81 \, \frac{m}{sec^2}$$

$$Q_i := b_c \sqrt{g} \left(\frac{2}{3}\right)^{3/2} H^{3/2} = 0.852 \, \frac{m^3}{sec}$$

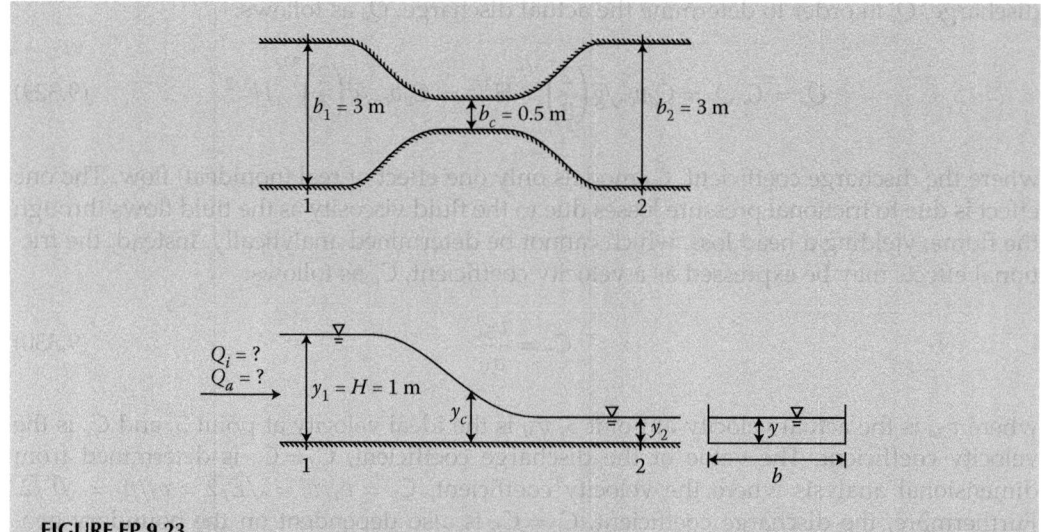

FIGURE EP 9.23
Water flows in a rectangular channel with a venturi flume inserted in the channel to measure the flowrate.

(b) The actual flowrate, Q_a is determined from Equation 9.329 $Q_a = C_d Q_i$, where the discharge coefficient, C_d for the venturi flume is assumed to be 0.95 as follows:

$$C_d := 0.95$$

$$Q_a := C_d \cdot b_c \sqrt{g} \left(\frac{2}{3}\right)^{3/2} H^{3/2} = 0.81 \frac{m^3}{sec}$$

9.17.8 Evaluation of the Actual Flowrate for a Contracted Opening

Flow though a contraction such as a bridge or a culvert in a stream can serve as a flow measurement device in open channel flow by applying the same principles used in a venturi flume/Parshall flume. Measurements are made while the water is flowing through the contracted opening. The cross-sectional opening of the water, which is flowing at a high velocity through the contracted opening, is measured. Furthermore, although there is a head loss from the upstream section to the contracted section where the area is measured, it is usually ignored because it is very small in comparison to the field measurement errors. Furthermore, similar to a venturi flume, a contracted opening, which is an ideal depth meter, is also considered to be a critical depth meter.

End-of-Chapter Problems

Problems with a "C" are conceptual problems. Problems with a "BG" are in English units. Problems with an "SI" are in metric units. Problems with a "BG/SI" are in both English and metric units. All "BG" and "SI" problems that require computations are solved using Mathcad.

Introduction

9.1C Explain the basic difference between pipe flow and open channel flow.

9.2C In the application of the governing equations (continuity, energy, and momentum) for open channel flow, it is important to make the distinction between real flow, ideal flow, and a hydraulic jump (see Table 4.1, and Section 9.5). Furthermore, in the application of the governing equations, although in some flow cases, one may ignore (or not need to model) the flow resistance, in most practical flow cases in general, one must account for the fluid viscosity and thus model the resulting shear stress and flow resistance. Explain when it is necessary to model the flow resistance and when it is not necessary.

9.3C In the application of the governing equations (continuity, energy, and momentum) for open channel flow, it is important to make the distinction between real flow, ideal flow, and a hydraulic jump (see Table 4.1 and Section 9.5). Furthermore, the application of the governing equations for real open channel flow and ideal open channel flow (flow-measuring devices) requires the modeling of flow resistance (frictional losses). The modeling of flow resistance requires a "subset level" application of the governing equations (Section 9.4). Explain how the flow resistance is modeled for real flow vs. ideal flow in the application of the governing equations.

9.4C Explain how the flow resistance is specifically modeled for real open channel flow vs. ideal open channel flow.

9.5C Briefly explain the role of transitions and controls in open channel flow.

9.6C The governing equations are tailored for open channel flow, yielding the definition of two topics that are unique to open channel flow. What are these two topics?

9.7C In the application of the governing equations (continuity, energy, and momentum) for open channel flow, it is important to make the distinction between real flow, ideal flow, and a hydraulic jump (see Table 4.1, and Section 9.5). List specific applications of the governing equations for real open channel flow, ideal open channel flow, and a hydraulic jump.

Application of the Eulerian (Integral) versus Lagrangian (Differential) Forms of the Governing Equations

9.8C Depending upon the flow problem, one may use either the Eulerian (integral) approach or the Lagrangian (differential) approach, or both, in the application of the three governing equations. Explain further.

9.9C In the case where the problem involves mass flow in and out of an open channel flow system, then the problem is modeled as a finite control volume. The integral form of the three governing equations are applied for the assumption of real flow, ideal flow, and hydraulic jump problems. Explain why and how one would apply the integral from of the three governing equations for the assumption of: (a) real open channel flow, (b) ideal open channel flow, and (c) hydraulic jump.

9.10C In the assumption of real flow, and for the specific case of a major head loss due to channel flow resistance, application of the differential form of the momentum equation may become necessary. List specific applications of the differential form of the momentum equation for real open channel flow.

The Occurrence of a Major Head Loss in Open Channel Flow

9.11C   Explain in detail the occurrence (and nonoccurrence) of a major head loss in open channel flow.

Modeling Flow Resistance in Open Channel Flow: A Subset Level Application of the Governing Equations

9.12C   The application of the governing equations for real open channel flow and ideal open channel flow (flow-measuring devices) (Section 9.5) requires the modeling of flow resistance (frictional losses). The modeling of flow resistance requires a "subset level" application of the governing equations. The distinction between ideal and real flow determines how the flow resistance is modeled in the "subset level" application of the governing equations. Explain how the flow resistance is evaluated in the application of the governing equations (continuity, energy, and momentum).

9.13C   The flow resistance in open channel flow (major head loss, and actual discharge) is modeled by a "subset level" application of the appropriate governing equations, as presented in Chapter 6 (and summarized in Sections 9.5.1.2 and 9.5.2.1). Explain what a "subset level" application of the appropriate governing equations implies, and explain when the use of dimensional analysis (Chapter 7) is needed in the derivation of the flow resistance equations.

Major Head Loss due to Flow Resistance in Real Open Channel Flow

9.14C   Applications of the governing equations for real open channel flow include the analysis and design of uniform (see Section 9.14; also see Example Problems 5.2, 5.11, and 5.12) and nonuniform (see Section 9.15) flow (generated by controls such as a break in channel slope, a hydraulic drop, or a free overfall). Furthermore, in the analysis and design of an open channel flow system, there may be channel friction (resistance) or controls (break in channel slope, hydraulic drop, or free overfall) in the open channel flow that will determine the significance of a major head loss and a nonuniform surface water profile. Explain the application of the governing equations (integral vs. differential) for the case of a straight channel section without flow-measuring devices.

Uniform versus Nonuniform Open Channel Flow

9.15C   The spatial variation in the cross-sectional area of the flow (or the existence of a channel or flow transition in an open channel, explained in Section 9.8) results in a spatial variation in the fluid velocity (convective acceleration) and thus spatially varied (nonuniform) flow. However, no spatial variation in the cross-sectional area of the flow (or the nonexistence of a channel or flow transition in an open channel) results in no spatial variation in the fluid velocity, and thus spatially uniform flow. Highlight the differences and similarities between the solutions of uniform vs. nonuniform flow problems.

Evaluation of the Major Head Loss: Poiseuille's Law for Laminar Flow and the Chezy Equation for Turbulent Flow

9.16C   Evaluation of the major head loss flow resistance term in the energy equation requires a "subset level" application of the governing equations (see Chapters 6 and 7 and Section 9.5.1.2). Depending upon the value for the Reynolds number, R real flow may be described as either laminar or turbulent. How is the major head loss due to channel friction modeled in the energy equation (Equation 9.1) evaluated for laminar vs. turbulent flow?

9.17C   Evaluation of the major head loss flow resistance term requires a "subset level" application of the governing equations (see Chapters 6 and 7 and Section 9.5.1.2). Explain when the major head loss in open channel flow may be determined analytically (i.e., without the need for dimensional analysis) vs. empirically (i.e., with the need for dimensional analysis).

9.18C   Discuss the Chezy coefficient, C. What is it a function of? Its range of values? Its units and dimensions? Its application in fluid flow? Compare it to the Darcy–Weisbach friction factor, f.

9.19C   Explain how the differential form of the Chezy equation, $S_f = v^2/C^2 R_h$ (Equation 9.19) is applied in turbulent flow.

9.20C   Derive the empirical expression for the major head loss, h_f due to channel friction for turbulent flow; assume the Chezy coefficient, C is used to model the flow resistance.

Turbulent Channel Flow Resistance Equations and Their Roughness Coefficients

9.21C   In addition to the Chezy equation, list two additional turbulent open channel flow resistance equations.

9.22C   In addition to Chezy coefficient, C alternate standard forms of the drag coefficient, C_D include empirical flow resistance coefficients including the Darcy–Weisbach friction factor, f (typically for pipe flow only) and the Manning's roughness coefficient, n. Explain the typical preferences of the various empirical flow resistance coefficients in their application in fluid flow situations.

Manning's Equation

9.23C   How are the Chezy coefficient, C and the Manning's roughness coefficient, n related?

9.24C   Beginning with the Chezy equation, $S_f = v^2/C^2 R_h$ (Equation 9.19), derive the Manning's equation (expressed as a function of the velocity, v). Furthermore, derive the Manning's head loss equation.

9.25C   Explain how the Manning's equation is applied in turbulent open channel flow.

Flow Type (State) and Flow Regime

9.26C   Describe the flow type (state) and the flow regime in open channel flow.

9.27C   Discuss the role and the significance of uniform flow in open channel flow.

9.28C   Define the various flow regimes for uniform open channel flow, and explain how the flow regime is dictated.

9.29C   Explain the transition of flow from uniform open channel flow to nonuniform open channel flow and the changes in the flow regime.

Transtions and Controls in Open Channel Flow

9.30C   Uniform flow will occur in the open channel in the absence of a channel or flow transition, whereas nonuniform flow will occur in the open channel as a result of a channel or flow transition. Define a channel transition and a flow transition, and state how each affects the flow.

9.31C Although channel bends (change in direction) produce only transient changes in the flow, most channel transitions (change in slope or cross section of the channel) produce a permanent change in the state of flow (from uniform to nonuniform). List some typical channel transitions and their impact on the flow.

9.32C Discuss the difference in the modeling of the major head loss for channel transitions that affect the flow over a short channel section vs. channel transitions that affect the flow over a long channel section.

The Definition of a Control in Open Channel Flow

9.33C Define a control in open channel flow.

9.34C A critical depth of flow occurs at the more obvious "controls," while a subcritical, supercritical, or critical depth of flow occurs at the less obvious "control." Explain this further.

9.35C Highlight the differences (in the relationship between the flow depth and channel discharge) and similarities (in modeling the major head loss due to flow resistance) between the roles of uniform vs. certain nonuniform flows as controls in open channel flow.

The Definition of a Flow-Measuring Device in Open Channel Flow

9.36C While a "control" in the open channel can serve as a flow-measuring device, a flow-measuring device is not necessarily considered a "control." For instance, although a pitot-static tube is a typical flow-measuring device, it is not a "control." Furthermore, while the flow is critical at most controls (the flow is subcritical, supercritical, or critical for uniform flow as a control), the occurrence of critical flow does not necessarily provide a control. For instance, although the flow is critical at a hydraulic jump, it is not a control. List the various flow-measuring devices in open channel flow.

Energy Concepts in Open Channel Flow

Specific Energy in Open Channel Flow

9.37C Define the total energy vs. the specific energy in open channel flow.

Specific Energy for Rectangular Channel Cross Sections

9.38C What does the assumption of a rectangular cross section in open channel flow allow for the continuity and specific energy equations?

9.39C The derivation of the analytical critical flow equations (expressions for y_c, v_c, and E_c) for a rectangular channel section considers the variation of E with y for a given constant value of unit discharge, q, at which the specific energy, E is at a minimum. Give the expressions for y_c, v_c, and E_c, and state any important information and applications for each equation.

Specific Energy for Nonrectangular Channel Cross Sections

9.40C The derivation of the analytical critical flow equations (expressions for y_c, v_c, and E_c) for a nonrectangular channel section considers the variation of E with y for a given constant value of discharge, Q, at which the specific energy, E is at a minimum. Give the expressions for y_c, v_c, and E_c for nonrectangular channel cross sections, and state any important information and applications for each equation.

Subcritical, Critical, and Supercritical Flow

9.41C What is the difference in depth between subcritical, critical, and supercritical flow, and what is the dimensionless number used to indicate the flow regime?

9.42C Give the expressions for the Froude number, F for rectangular channel sections and for nonrectangular channel sections.

Momentum Concepts in Open Channel Flow

9.43C Application of the momentum equation is useful using either an integral approach or a differential approach. Explain when it is appropriate to apply the differential approach vs. the integral approach in open channel flow.

The Momentum Function

9.44C Application of the momentum equation using the integral approach (between two points in a finite control volume) is appropriate for any channel or flow transition that affects the flow over a short channel section and thus assumes the flow resistance to be negligible. However, the momentum function in open channel flow is useful to define and apply over a short channel section when (1) the major head loss due to flow resistance is negligible, and (2) one flow depth is supercritical and one flow depth is subcritical, and are thus conjugate to one another. Give examples of such cases in open channel flow where it is useful to apply the momentum function.

9.45C Define a hydraulic jump, and give a flow situation that may result in a hydraulic jump.

9.46C Give the momentum function for rectangular channel sections.

9.47C Give the momentum function for nonrectangular channel sections.

Flow Depth and Reaction Force for Short Channel Transitions in Open Channel Flow: Ideal Flow

9.48C Because a gradual (vertical or horizontal) channel contraction/transition (not a "control"), an abrupt vertical or horizontal contraction (abrupt step upward or abrupt decrease in channel width) (a "control"), and a typical flow-measuring device ("controls") (such as sluice gates, weirs, overflow spillways, venturi flumes, and contracted openings) affect the flow over a short channel section, the major head loss due to flow resistance is assumed to be negligible. State how the upstream or downstream depth is determined, and state if there will be an unknown reaction force, R_x acting on the control volume due to the thrust of the flow on an open channel structure and how it is determined.

Flow Depth for Gradual Channel Transitions: Not Controls

Flow Depth for a Gradual Upward Step

9.49 SI Water flows in a rectangular channel at a velocity of 2 m/sec and depth of 2.11 m. There is a gradual upward step, Δz of 0.11 m in the channel bed as illustrated in

Figure EP 9.1. (a) Determine the flow regime (critical, subcritical, or supercritical) of the upstream depth, y_1. (b) Determine the flow regime of the downstream depth, y_2. (c) Determine the depth, y_2 and the velocity, v_2 of the water downstream of the step. (d) Determine E_1 and E_2 and plot the E-y curve for the unit discharge, $q = v_1\, y_1 = v_2\, y_2$.

Flow Depth for a Gradual Decrease in Channel Width

9.50SI   Water flows in a rectangular channel of width 4.6 m at a velocity of 2 m/sec and depth of 4 m. There is a temporary gradual decrease in channel width to 3.9 m and thus a gradual horizontal channel transition, as illustrated in Figure EP 9.2. (a) Determine the flow regime of the upstream depth, y_1. (b) Determine the flow regime of the downstream depth, y_2. (c) Determine the depth, y_2 and the velocity, v_2 of the water downstream of the gradual transition. (d) Determine E_1 and E_2 and plot the E-y curve for the unit discharge, $q = q_1 = q_2$, where $b = b_1 = b_2$. (e) Determine the depth, y_t, the velocity, v_t, and the unit discharge, q_t of the water at the gradual transition. (f) Determine E_t and plot the E-y curve for the unit discharge, $q = q_t > q_1 = q_2$, where $b = b_t < b_1 = b_2$.

Flow Depth for Abrupt Channel Transitions: Controls

Flow Depth for an Abrupt Upward Step

9.51SI   Water flows in a rectangular channel at a velocity of 1.5 m/sec and depth of 1.8 m. There is an abrupt upward step, Δz of 0.37 m in the channel bed as illustrated in Figure EP 9.3. (a) Determine the flow regime (critical, subcritical, or supercritical) of the upstream depth, y_1. (b) Determine the flow regime of the downstream depth, y_2. (c) Determine the critical velocity, v_c and the minimum specific energy, E_c at the control. (d) Determine the depth, y_2 and the velocity, v_2 of the water downstream of the step. (e) Determine E_1 and E_2 and plot the E-y curve for the unit discharge, $q = v_1\, y_1 = v_2\, y_2$.

Flow Depth for an Abrupt Decrease in Channel Width

9.52SI   Water flows in a rectangular channel of width 4 m at a velocity of 4 m/sec and depth of 4 m. There is a temporary abrupt decrease in channel width to 3.553 m and thus an abrupt horizontal channel transition, as illustrated in Figure EP 9.4. (a) Determine the flow regime of the upstream depth, y_1. (b) Determine the flow regime of the downstream depth, y_2. (c) Determine the unit discharge, q_t, the critical depth, y_c, the critical velocity, v_c, and the minimum specific energy, $E_c = E_t$ for the unit discharge, q_t at the abrupt transition (control). (d) Determine the depth, y_2 and the velocity, v_2 of the water downstream of the abrupt transition. (e) Determine E_1 and E_2 and plot the E-y curve for the unit discharge, $q = q_1 = q_2$, where $b = b_1 = b_2$. And, plot the E-y curve for the unit discharge, $q = q_t \gg q_1 = q_2$, where $b = b_t \ll b_1 = b_2$.

Flow Depth for Typical Flow-Measuring Devices

9.53SI   Water flows in a rectangular channel of width 5 m at a velocity of 3 m/sec and depth of 4 m upstream of a sluice gate as illustrated in Figure EP 9.5. (a) Determine the flow regime (critical, subcritical, or supercritical) of the upstream depth, y_1. (b) Determine the flow regime of the downstream depth, y_2. (c) Determine the critical velocity, v_c and the minimum specific energy, E_c at the control. (d) Determine the depth, y_2 and the velocity,

v_2 of the water downstream of the sluice gate. (e) Determine E_1 and E_2 and plot the E-y curve for the unit discharge, $q = v_1 \, y_1 = v_2 \, y_2$.

9.54SI Water flows in a trapezoidal channel section as illustrated in Figure EP 9.6, where $b = 2.7$ m, and $\theta = 32°$. Upstream of the spillway, the velocity of the flow is 2.4 m/sec and the depth is 1.8 m. (a) Determine the flow regime (critical, subcritical, or supercritical) of the upstream depth, y_1. (b) Determine the flow regime of the downstream depth, y_2. (c) Determine the critical velocity, v_c and the minimum specific energy, E_c at the control. (d) Determine the depth, y_2 and the velocity, v_2 of the water downstream of the spillway. (e) Determine E_1 and E_2 for the discharge, $Q = v_1 \, A_1 = v_2 \, A_2$.

Reaction Force on Open Channel Flow Structures/Controls (Abrupt Flow Transitions and Flow-Measuring Devices)

9.55SI Referring to ECP 9.53 (and its solution), water flows in a rectangular channel of width 5 m at a velocity of 3 m/sec and depth of 4 m upstream of a sluice gate as illustrated in Figure EP 9.7. (a) Determine the reaction force, R_x acting on the sluice gate due the thrust of flow. (b) Determine M_1, M_2, and M_c, and plot the M-y curve for the unit discharge, $q = v_1 \, y_1 = v_2 \, y_2$.

9.56SI Referring to ECP 9.54 (and its solution), water flows in a trapezoidal channel section as illustrated in Figure EP 9.8, where $b = 2.7$ m, and $\theta = 32°$. Upstream of the spillway, the velocity of the flow is 2.4 m/sec and the depth is 1.8 m. (a) Determine the reaction force, R_x acting on the spillway due the thrust of flow. (b) Determine M_1, M_2 for the discharge, $Q = v_1 \, A_1 = v_2 \, A_2$.

Flow Depth and Major Head Loss for a Hydraulic Jump in Open Channel Flow

9.57C In open channel flow, in addition to flow resistance, there is an additional source of major head loss in open channel flow due to the occurrence of a hydraulic jump. Explain how the unknown channel depth of flow (either upstream or downstream of the jump) is determined, and explain how the major head loss due to the jump is determined.

The Hydraulic Jump Equations: Rectangular Channel Sections

9.58C Give the hydraulic jump equations used to analytically solve for the unknown channel depth of flow (either upstream or downstream of the jump) derived from the integral form of the momentum equation.

9.59SI Referring to ECP 9.53 (and its solution), water flowing in a rectangular channel of width 5 m at a supercritical velocity and a supercritical depth downstream of a sluice gate encounters a subcritical depth of flow produced by a second sluice gate located downstream of the first sluice gate. As such, a hydraulic jump occurs, as illustrated in Figure EP 9.9. (a) Determine the Froude number, F_1 for the supercritical depth, y_1 upstream of the hydraulic jump. (b) Determine the critical depth of flow, y_c; critical velocity, v_c; and the minimum specific energy, E_c at the hydraulic jump. (c) Determine the subcritical depth, y_2 and the subcritical velocity, v_2 of the water downstream of the hydraulic jump. (d) Determine M_1, M_2, and M_c, and plot the M-y curve for the unit discharge, $q = v_1 \, y_1 = v_2 \, y_2$. (e) Determine E_1 and E_2 and plot the E-y curve for the unit discharge, $q = v_1 \, y_1 = v_2 \, y_2$. (f) Determine the head loss, h_L due to the flow turbulence in the hydraulic jump.

Numerical Solution for a Hydraulic Jump: Nonrectangular Channel Sections

9.60SI   Referring to ECP 9.54 (and its solution), water flows in a trapezoidal channel section as illustrated in Figure EP 9.10, where $b = 2.7$ m, and $\theta = 32°$. The supercritical velocity of the flow and the supercritical depth downstream of the spillway encounters a subcritical depth of flow produced by a sluice gate located downstream of the spillway. As such, a hydraulic jump occurs, as illustrated in Figure EP 9.10. (a) Determine the Froude number, F_1 for the supercritical depth, y_1 upstream of the hydraulic jump. (b) Determine the critical depth of flow, y_c; critical velocity, v_c; and the minimum specific energy, E_c at the hydraulic jump. (c) Determine the subcritical depth, y_2 and the subcritical velocity, v_2 of the water downstream of the hydraulic jump. (d) Determine M_1, M_2 for the discharge, $Q = v_1 A_1 = v_2 A_2$. (e) Determine E_1 and E_2 for the discharge, $Q = v_1 A_1 = v_2 A_2$. (f) Determine the head loss, h_L due to the flow turbulence in the hydraulic jump.

Flow Depth and Major Head Loss in Uniform Open Channel Flow: Real Flow

9.61C   Uniform flow will occur in the open channel in the absence of a channel or flow transition. Because the uniform flow creates a water surface profile and thus affects the flow over a long channel section, the major head loss due to flow resistance is significant. State how the flow depth, y (or $S_o = S_f$, or v, or C [or n]) is determined. Also, state how the major head loss, h_f is determined.

9.62SI   Water flows at a discharge of 6 m^3/sec in a rectangular channel that is 1.4 m wide with a channel bottom slope of 0.0029 m/m, and a Manning's roughness coefficient of 0.0018 m$^{-1/3}$ sec, as illustrated in Figure EP 9.11. (a) Determine the uniform depth of flow in the channel. (b) Determine the flow regime of the uniform flow in the channel and the classification of the channel bottom slope. (c) Determine the head loss due to the channel resistance over a channel section length of 3 m. (d) Determine the critical channel bottom slope for the given discharge.

9.63SI   Water flows at a discharge of 1.45 m^3/sec in a trapezoidal channel section as illustrated in Figure EP 9.12, where $b = 4$ m and $\theta = 25°$, with a channel bottom slope of 0.0017 m/m and a Manning's roughness coefficient of 0.026 m$^{-1/3}$ sec. (a) Determine the uniform depth of flow in the channel. (b) Determine the flow regime of the uniform flow in the channel, and the classification of the channel bottom slope. (c) Determine the head loss due to the channel resistance over a channel section length of 4 m. (d) Determine the critical channel bottom slope for the given discharge.

9.64SI   Water flows at a discharge of 9 m^3/sec in a partially filled circular channel section as illustrated in Figure EP 9.13, where $r_o = 1.9$ m and $\theta = 2$ radians, with a channel bottom slope of 0.0017 m/m and a Manning's roughness coefficient of 0.022 m$^{-1/3}$ sec. (a) Determine the uniform depth of flow in the channel. (b) Determine the flow regime of the uniform flow in the channel and the classification of the channel bottom slope. (c) Determine the head loss due to the channel resistance over a channel section length of 8 m. (d) Determine the critical channel bottom slope for the given discharge.

9.65SI   Water flows at a discharge of 2.3 m^3/sec in a rectangular channel that is 0.55 m wide with a channel bottom slope of 0.04 m/m and a uniform depth of flow of 3 m, as illustrated in Figure EP 9.14. (a) Determine the flow regime of the uniform flow in the channel and the classification of the channel bottom slope. (b) Determine the Manning's roughness coefficient. (c) Determine the head loss due to the channel resistance over a channel section length of 11 m. (d) Determine the critical channel bottom slope for the given discharge.

Flow Depth and Major Head Loss in Nonuniform Open Channel Flow: Real Flow

9.66C Because "controls" such as a hydraulic drop, a break in the channel slope, or a free overfall (see Figure 9.34a, b and c, respectively) generate a nonuniform surface water profile and thus, affect the flow over a long channel section, the major head loss due to flow resistance is significant. State how the flow depth, y (or $S_o = S_f$, or v, or C [or n]) is determined. Also, state how the major head loss, h_f is determined.

9.67SI Water flows at a discharge of $16\,\text{m}^3/\text{sec}$ in a trapezoidal channel section as illustrated in Figure EP 9.15, where $b = 4\,\text{m}$ and $\theta = 20°$, with a channel bottom slope of 0.003 m/m and a Manning's roughness coefficient of $0.029\,\text{m}^{-1/3}$ sec. The channel terminates in a free overfall (brink) as illustrated in Figure EP 9.15. (a) Determine the uniform depth of flow in the channel at a great distance upstream of the brink, where the control ceases to affect the water surface profile. (b) Determine the flow regime of the uniform flow in the channel and the classification of the channel bottom slope. (c) Determine and plot the resulting nonuniform surface water profile due to the drawdown. (d) Determine the head loss due to the channel resistance over a channel section length of 2500 m.

9.68SI Water flows at a discharge of $1.33\,\text{m}^3/\text{sec}$ in a rectangular channel that is 2.5 m wide. There is a break in the channel slope, where the upstream channel section is characterized with a mild channel bottom slope of 0.0006 m/m and a length of 2000 m, and the downstream channel section is characterized with a steep channel bottom slope of 0.029 m/m and a length of 1500 m, as illustrated in Figure EP 9.16. The Manning's roughness coefficient is $0.014\,\text{m}^{-1/3}/\text{sec}$. (a) Determine the uniform depth of flow and flow regime in the channel at a great distance upstream of the break in channel slope, where the control ceases to affect the water surface profile. (b) Determine the uniform depth of flow and flow regime in the channel at a great distance downstream of the break in channel slope, where the control ceases to affect the water surface profile. (c) Determine and plot the resulting nonuniform surface water profile due to the break in channel slope for the first channel section that is upstream of the control/break. (d) Determine the head loss due to the channel resistance over the first channel section with a length of 2000 m. (e) Determine and plot the resulting nonuniform surface water profile due to the break in channel slope for the second channel section that is downstream of the control/break. (f) Determine the head loss due to the channel resistance over the second channel section with a length of 1500 m. (g) Plot the resulting nonuniform surface water profile due to the break in channel slope for the entire channel profile relative to the elevation of the channel bottom, which is defined as the hydraulic grade line, *HGL*, and determine the total head loss over the entire channel length of 3500 m. NOTE: If instead of a break in channel slope, there was a hydraulic drop (see Figure 9.34a), then the solution to the hydraulic drop problem (steep slope) would be identical to the second section (steep) that is downstream of the control/break in channel slope.

Evaluation of the Actual Flowrate for Ideal Flow Meters in Open Channel Flow

Evaluation of the Actual Flowrate for a Sluice Gate

9.69SI Water flows in a rectangular channel of width 3 m and at a mild slope. A sluice gate is inserted in the channel as illustrated in Figure EP 9.17 in order to measure the flowrate. The opening of the gate is set at 1.273 m, the depth of flow upstream of the sluice gate is measured to be 1.8 m, and the depth of flow downstream of the sluice gate is measured to be 0.933 m. (a) Determine the ideal flowrate. (b) Determine the actual flowrate.

Evaluation of the Actual Flowrate for a Sharp-Crested Weir

9.70SI Water flows in a rectangular channel of width 4 m and at a mild slope. A rectangular sharp-crested weir is inserted in the channel as illustrated in Figure EP 9.18 in order to measure the flowrate. The effective head on the weir is 0.98 m, the height of the weir is 2.1 m, and the width of the rectangular weir opening/notch is 3 m. (a) Determine the ideal flowrate. (b) Determine the actual flowrate.

9.71SI Water flows in a rectangular channel of width 3.5 m and at a mild slope. A triangular sharp-crested weir is inserted in the channel as illustrated in Figure EP 9.19 in order to measure the flowrate. The effective head on the weir is 0.8 m, the height of the weir is 1.5 m, and the angle of the V-notch is 70°. (a) Determine the ideal flowrate. (b) Determine the actual flowrate.

Evaluation of the Actual Flowrate for a Spillway

9.72SI Water flows in a rectangular channel of width 4 m and at a mild slope. A spillway is inserted in the channel as illustrated in Figure EP 9.20 in order to measure the flowrate. The effective head on the spillway is 0.97 m, the height of the spillway is 1.6 m, and the width of the crest of the spillway is 4 m. (a) Determine the ideal flowrate. (b) Determine the actual flowrate.

Evaluation of the Actual Flowrate for a Broad-Crested Weir

9.73SI Water flows in a rectangular channel of width 2 m and at a mild slope. A broad-crested weir is inserted in the channel as illustrated in Figure EP 9.21 in order to measure the flowrate. The effective head on the weir is 1.35 m, the height of the weir is 2.4 m, the weir extends over the full width of the channel, and the horizontal crest of the weir is 2.8 m. (a) Determine the ideal flowrate. (b) Determine the actual flowrate.

Evaluation of the Actual Flowrate for the Parshall Flume (Venturi Flume)

9.74SI Water flows in a rectangular channel of width 3 m and at a mild slope. A venturi flume is inserted in the channel as illustrated in Figure EP 9.23 in order to measure the flowrate. The effective head on the flume is 1.97 m, and the constriction width is 0.66 m. Assume that the discharge coefficient, C_d for the venturi flume is 0.99. (a) Determine the ideal flowrate. (b) Determine the actual flowrate.

10

External Flow

10.1 Introduction

While the flow of fluid in both pipes (pressure/closed conduit flow) and open channels (gravity/open conduit flow) are contained in a conduit and thus are referred to as internal flows, the flow of fluid over a body that is immersed in the fluid is referred to as external flow. The results sought in a given flow analysis will depend upon whether the flow is internal or external. In an internal flow, energy or work is used to move/force the fluid through a conduit; thus, one is interested in the determination of the forces (reaction forces, pressure drops, flow depths), energy or head losses, and cavitation where energy is dissipated (see Chapters 3, 4, 5, 8, and 9). However, in an external flow (Chapter 10), energy or work is used to move a submerged body through the fluid, where the fluid is forced to flow over the surface of the body. Thus, one is interested in the determination of the flow pattern around the body; the lift force, F_L; the drag force, F_D (resistance to motion) on the body; and the patterns of viscous action in the fluid as it passes around the body. However, only when the body is moving at a high speed (and depending upon the shape of the body and the angle it makes with the direction of flow) does the lift force, F_L become dominant. One may note that while the continuity, energy, and momentum equations and the results of dimensional analysis are applied to internal flows, only the energy and momentum equations and the results of dimensional analysis are applied to external flows. Furthermore, the energy and momentum equations play complementary roles in the analysis of both internal and external flow; when one of the two equations of motion breaks down, the other may be used to find the additional unknown quantity. For external flow, the Eulerian (integral) approach is assumed in the application of the governing equations.

In the application of the governing equations (energy and momentum) for external flow, the assumption of ideal flow is made (see Table 4.1 and Section 10.4). The application of the governing equations for ideal external flow requires the modeling of flow resistance (frictional losses). The modeling of flow resistance requires a "subset level" application of the governing equations (Section 10.3). The distinction between ideal and real flow determines how the flow resistance is modeled in the "subset level" application of the governing equations. The assumption of "real" flow implies that the flow resistance is modeled in both the energy and the momentum equations (see Chapters 8 and 9). However, the assumption of "ideal" flow implies that the flow resistance is modeled only in the momentum equation (and thus, the subsequent assumption of "real" flow). The flow resistance due to friction in external flow and for velocity-measuring devices for external flow (modeled as a drag force, Section 10.5) is modeled by assuming "ideal" flow; although application of the Bernoulli equation does not model a head loss in the determination of ideal velocity, the associated minor head loss with the velocity measurement is accounted for by a drag coefficient (in the application of the momentum equation, where a subsequent assumption of "real" flow is made).

Applications of the governing equations for ideal external flow include the analysis and design of planes, cars, buildings, ships, submarines, and submerged pipes (Section 10.5). When a submerged body moves in a stationary homogenous real fluid, this fluid exerts pressure forces normal to the surface and shear forces parallel to the surface along the outer surface of the body because of the relative motion between the body and fluid (one may note that if the submerged body was not moving in the fluid, then only pressure forces would be exerted on the body's surface). The resultant of the combined pressure and shear (viscous) forces acting on the surface of the body is due to the pressure drop, Δp and has two components. Specifically, the component of the resultant pressure, F_p and shear, F_f (also referred to as F_V) forces that acts in the direction of the flow, s is called the drag force, F_D. The component of the resultant pressure and shear forces that acts normal to the direction of the flow is called the lift force, F_L and tends to move the submerged body in that direction. As noted above, only when the body is moving at a high speed (and depending upon the shape of the body and the angle it makes with the direction of flow) does the lift force, F_L become dominant.

10.1.1 Occurrence and Illustration of the Drag Force and the Lift Force

The drag force, F_D typically occurs on cars, airplanes, buildings, trees, submerged pipes, ships, and submarines, while the lift force, F_L typically occurs under the wings of airplanes and in the transport of precipitation and dust particles. Thus, consideration of both the drag and lift forces becomes important in the design of cars, airplanes, ships, submarines, and buildings, where the ultimate goal is to minimize the effect of the drag force while still generating the required lift for airplanes. Furthermore, in the design of the brakes of vehicles, the ultimate goal is to maximize the drag force, while in the design of cars, an additional goal would be to minimize the lift force.

Figure 10.1 illustrates that the fluid moving over a submerged stationary body exerts a drag force, F_D (in the same direction of moving fluid), and a lift force, F_L (normal to the direction of the flow). Alternatively, Figure 10.2 illustrates that a stationary fluid exerts a drag

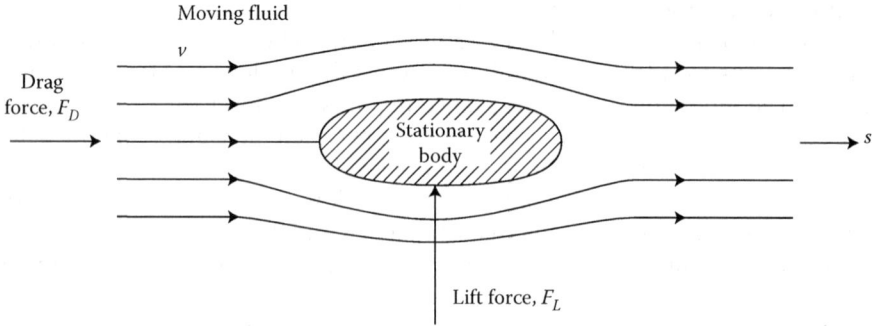

FIGURE 10.1
Fluid moving over a submerged stationary body exerts a drag force, F_D (in the same direction of moving fluid) and a lift force, F_L (normal to the direction of the flow).

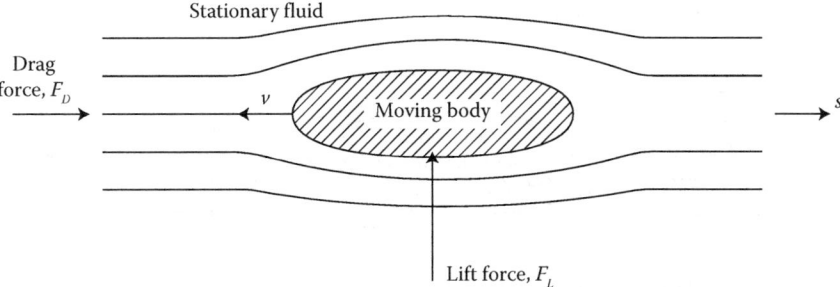

FIGURE 10.2
Stationary fluid exerts a drag force, F_D (in the opposite direction of moving body) and a lift force, F_L (normal to the direction of the flow) on a submerged moving body.

force, F_D on a moving body (in the opposite direction of the moving body) and a lift force, F_L (normal to the direction of the flow). These two cases are equivalent to each other, since what is significant is the relative motion between the fluid and the body. In either case, the motion is analyzed by fixing the coordinate system on the body.

One may note that, similarly to the pitot-static tube (see Chapters 4 and 6), there is a stagnation point on any stationary body that is placed into a flowing fluid. In the flow over a submerged body, some of the fluid flows over and under the body, while some of the fluid is brought to a stop at the nose of the body at the stagnation point, as illustrated in Figure 10.3 for a blunt body and in Figure 10.4 for a streamlined body. Furthermore, the dividing line is called the stagnation streamline and ends at the stagnation point on the submerged body, where the pressure is the stagnation pressure. The point of stagnation represents the conversion of all of the kinetic energy into an ideal pressure rise, where the fluid is momentarily brought to a stop. Therefore, at that stagnation point the drag force, F_D is composed only of the resultant pressure force, F_P that acts in the direction of the flow (there is no shear force, F_f component in the direction of the flow).

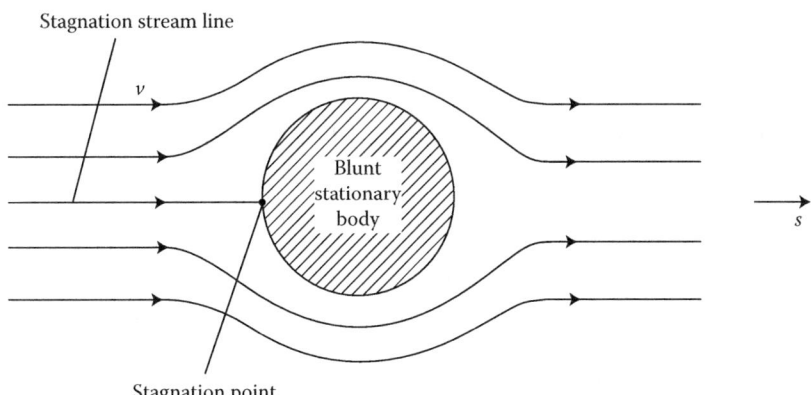

FIGURE 10.3
Stagnation point and stagnation streamline for a blunt stationary body placed into a flowing fluid.

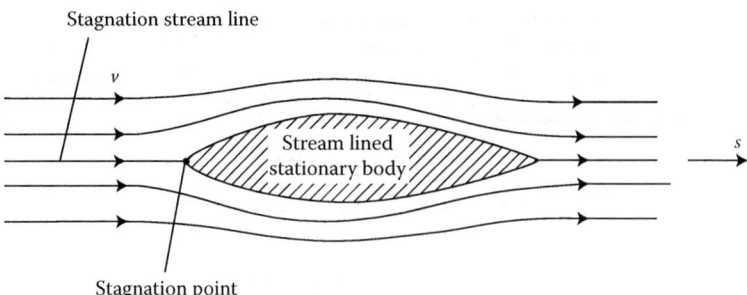

FIGURE 10.4
Stagnation point and stagnation streamline for a streamlined stationary body placed into a flowing fluid.

10.2 Application of the Eulerian (Integral) versus Lagrangian (Differential) Forms of the Governing Equations

While the continuity, energy, and momentum equations are applied to internal flows, only the energy and momentum equations are applied to external flows. In the case of internal flows (see Chapter 8 and 9), depending upon the flow problem, one may use either the Eulerian (integral) approach or the Lagrangian (differential) approach, or both, in the application of the three governing equations. Furthermore, in the case of internal flow, while the continuity and the momentum equations may be applied using either the integral (for a finite control volume) or the differential approach, application of the energy equation is useful only in the integral approach (for a finite control volume), as energy is work, which is defined over distance, L. However, in the case of external flow, the integral approach is assumed in the application of the energy equation, as well as the application of the momentum equation. Finally, the energy equation and the momentum equations (equations of motion) play complementary roles in the analysis of a given flow (internal or external) situation. Thus, when one of the two equations of motion breaks down, the other may be used to find the additional unknown quantity.

10.3 Modeling Flow Resistance in External Flow: A Subset Level Application of the Governing Equations

The application of the governing equations for ideal external flow and velocity-measuring devices (Section 10.4) requires the modeling of flow resistance (frictional losses). The modeling of flow resistance requires a "subset level" application of the governing equations. The distinction between ideal and real flow determines how the flow resistance is modeled in the "subset level" application of the governing equations. Specifically, the application of the integral form of the momentum equation for the case of external flow around an object given by Equation. 5.27 requires the evaluation of the drag force, $F_D = (F_P + F_f)_s$.

The flow resistance in external flow (drag force) is modeled by a "subset level" application of the appropriate governing equations, as presented in Chapter 6 (and summarized in

Section 10.4.1.1). A "subset level" application of the appropriate governing equations focuses only on the given element causing the flow resistance. Thus, assuming that flow resistance is to be accounted for, depending upon the specific flow situation, the flow type is assumed to be either a "real" flow or an "ideal" flow. The distinction between ideal and real flow determines how the flow resistance is modeled in the "subset level" application of the appropriate governing equations. The assumption of "real" flow implies that the flow resistance is modeled in both the energy and the momentum equations (see Chapters 8 and 9). However, the assumption of "ideal" flow implies that the flow resistance is modeled only in the momentum equation (and thus the subsequent assumption of "real" flow). The flow resistance due to friction in external flow and for velocity-measuring devices for external flow (modeled as a drag force; see Section 10.5) is modeled by assuming "ideal" flow; although application of the Bernoulli equation does not model a head loss in the determination of ideal velocity, the associated minor head loss with the velocity measurement is accounted for by a drag coefficient (in the application of the momentum equation, where a subsequent assumption of "real" flow is made). Furthermore, the use of dimensional analysis (Chapter 7) in supplement to a "subset level" application of the appropriate governing equations is needed to derive flow resistance equations when there is no theory to model the friction force (as in the case of turbulent external flow).

10.4 Application of the Governing Equations in External Flow

The governing equations applied in the analysis and design of external flow include the energy and momentum (supplemented with dimensional analysis) equations. The energy equation and the momentum equations (equations of motion) play complementary roles in the analysis of a given flow situation. Thus, when one of the two equations of motion breaks down, the other may be used to find the additional unknown quantity. Furthermore, in the application of the governing equations for external flow, the assumption of ideal flow is made (see Table 4.1). The application of the governing equations for ideal external flow requires the modeling of flow resistance (frictional losses). The modeling of flow resistance requires a "subset level" application of the governing equations (Sections 10.3 and 10.4.1.1)

10.4.1 Application of the Governing Equations for Ideal External Flow

When one of the two equations of motion breaks down, the other may be used to find the additional unknown. Assuming a finite control volume, for the assumption of incompressible, ideal, steady external flow, first, the integral energy equation (Bernoulli equation) is applied, which was given by Equation 4.46 and is repeated as follows:

$$\left(\frac{p_1}{\rho_1 g} + \frac{v_1^2}{2g} + z_1\right) = \left(\frac{p_2}{\rho_2 g} + \frac{v_2^2}{2g} + z_2\right) \tag{10.1}$$

which may be used to solve for an unknown "energy head" term (p/γ, z, or $v^2/2g$) for one of the two points in a finite control volume. Application of the Bernoulli equation assumes ideal flow; thus, there is no head loss term to evaluate in the energy equation. Therefore, for problems that affect the flow over a short external flow section, the major head loss due to flow resistance is assumed to be negligible, and one may apply the Bernoulli equation. These types of flow problems include velocity-measuring devices for external flow, which, however, have a minor loss associated with the velocity measurement. The associated minor head loss is typically accounted for by the use/calibration of a velocity/drag coefficient to determine the drag force, where ideal flow (Bernoulli equation) is assumed. Thus, in the case of velocity-measuring devices (e.g., pitot-static tubes), one must indirectly model the minor head loss flow resistance term by the use of a velocity/drag coefficient, which requires a "subset level" of application of the governing equations (Section 10.4.1.1). Then, application of the complementary integral momentum equation (Equation 5.27) is required only if there is an unknown reaction force acting on the object (see Example Problems 5.21, 10.12 and 10.15); note that the pressure and viscous forces are represented by the drag force, and the viscous forces have already been accounted for indirectly through the velocity/drag coefficient. As such, the integral form of the momentum equation is solved analytically, without the use of dimensional analysis.

10.4.1.1 Evaluation of the Drag Force

In the case of external flow around an object, the associated minor head loss is typically accounted for by the use/calibration of a drag coefficient to determine the drag force. Evaluation of the drag force requires a "subset level" application of the governing equations. A "subset level" application of the governing equations focuses only on the given element causing the flow resistance. The assumption of "ideal" flow implies that the flow resistance is modeled only in the momentum equation (and thus the subsequent assumption of "real" flow). The flow resistance equation for the drag force is derived in Chapters 6 and 7. In the derivation of the drag force for an external flow around an object, first, ideal flow is assumed and thus, the Bernoulli equation is applied as follows:

$$\left(\frac{p}{\gamma} + z + \frac{v^2}{2g}\right)_1 - \left(\frac{p}{\gamma} + z + \frac{v^2}{2g}\right)_2 = 0 \tag{10.2}$$

Assuming $z_1 = z_2$, this equation has one unknown, which is the ideal velocity of the external flow around an object. Thus, this equation yields an expression for the ideal velocity, $v_i = \sqrt{(2\Delta p)/\rho}$ as a function of an ideal pressure difference, Δp which is directly measured. Then, application of the momentum equation (supplemented by dimensional analysis) is used to determine an expression for the actual velocity, v_a and the drag force, F_D which was given in Equation 5.27 and is repeated as follows:

$$\sum F_s = (F_G + F_P + F_V + F_{other})_s = (\rho Q v_s)_2 - (\rho Q v_s)_1 \tag{10.3}$$

Thus, the resulting flow resistance equation for the drag force is given as follows:

$$F_D = (F_P + F_f)_s = (\Delta p A)_s = \left(\frac{\rho v^2}{2} A + \frac{\tau_w L}{R_h} A\right)_s = C_D \frac{1}{2} \rho v^2 A$$

where F_D is equal to the sum of the pressure and friction force in the direction of the flow, s, and the drag coefficient, C_D accounts for (indirectly models) the minor head loss associated with the velocity measurement, and the drag force.

The flow resistance (shear stress or drag force) in the external flow around an object is modeled as a resistance force/drag force, F_D in the integral form of the momentum equation.

The flow resistance causes an unknown pressure drop, Δp, which causes an unknown head loss, h_f, where the head loss is due to a conversion of kinetic energy to heat, which is modeled/displayed in the integral form of the energy equation. However, although the head loss, h_f causes the drag force, F_D, the head loss is not actually determined in the design of external flow around an object. The assumption of ideal flow and thus applying the Bernoulli equation to measure the ideal velocity of flow, $v_i = \sqrt{(2\Delta p)/\rho}$ by the use of a pitot-static tube (see Section 4.5.7.3 and Example Problem 4.18) yields an expression for the ideal velocity as a function of ideal pressure difference, Δp, which is directly measured. Therefore, the associated minor head loss with the velocity measurement is accounted for by the drag coefficient (in the application of the momentum equation). Thus, in the external flow around an object, the flow resistance is ultimately modeled as a drag force in the integral form of the momentum equation because the drag force is needed for the design of external flow around an object. The head loss, h_f causing the actual velocity,

$$v_a = \sqrt{\frac{2}{\rho}\left(\Delta p - \frac{\tau_w L}{R_h}\right)},$$

and the drag force,

$$F_D = (F_P + F_f)_s = (\Delta p A)_s = \left(\frac{\rho v^2}{2}A + \frac{\tau_w L}{R_h}A\right)_s$$

is caused by both pressure and friction forces, where the drag force, F_D is equal to the sum of the pressure and friction force in the direction of the flow, s. However, because the friction/viscous forces (due to shear stress, τ_w) due to the external flow cannot be theoretically modeled in the integral momentum equation, the actual pressure drop, Δp cannot be analytically determined; thus, the exact reduction in the velocity actual velocity, v_a, which is less than the ideal velocity, v_i cannot be theoretically determined. Furthermore, the exact component in the s-direction of the pressure and viscous forces cannot be theoretically determined. Therefore, one cannot derive an analytical expression for the drag force, F_D from the momentum equation. As a result, one must resort to dimensional analysis (supplements the momentum theory) in order to derive an expression for the drag force, F_D which involves the definition of a drag coefficient, C_D that represents the flow resistance. Drag forces on external flow are addressed in detail in Chapters 6 and 10 (see Section 10.5.1).

10.5 The Drag Force and the Lift Force in External Flow

Applications of the governing equations for ideal external flow include for instance the analysis and design of planes, cars, buildings, ships, and submerged pipes, which include a speed-measuring device such as a pitot-static tube, and the determination of a pressure difference in external flow from a velocity difference. The integral form of the governing equations (energy and momentum) for ideal flow are applied to a finite control volume in the analysis and design of external flows around a body. As such, for ideal external flow, the Bernoulli equation (Equation 10.1) assumes ideal flow; thus, there is no head loss to evaluate in the energy equation. And, the integral momentum equation (Equation 5.27) models the reaction force acting on the object (see Example Problems 5.21, 10.12, and 10.15). Although the associated minor head loss with the velocity-measuring devices for external flow may be modeled in the energy equation (Equation 10.1), it is typically accounted for by the use/ calibration of a velocity/drag coefficient to determine the drag force, where ideal flow is assumed. Thus, in the case of velocity-measuring devices (e.g., pitot-static tubes), one must indirectly model the minor head loss flow resistance term by the use of a velocity/drag coefficient to determine the drag force, which requires a "subset level" of application of the governing equations (Sections 10.4.1.1 and 10.5.1).

In the study of external flows, it is important to model the drag force, F_D. Furthermore, in the case where the flow of the fluid over the body is moving at a high speed (and depending upon the shape of the body and the angle it makes with the direction of flow), it is important to model the lift force, F_L as well as the drag force, F_D. It is important to note that the magnitude of both the drag force, F_D and the lift force, F_L are strongly dependent on the shape of the body. The head loss, h_f causing the actual velocity,

$$v_a = \sqrt{\frac{2}{\rho}\left(\Delta p - \frac{\tau_w L}{R_h}\right)}$$

and the drag force,

$$F_D = (F_P + F_f)_s = (\Delta p A)_s = \left(\frac{\rho v^2}{2}A + \frac{\tau_w L}{R_h}A\right)_s$$

are caused by both pressure and friction forces, where the drag force, F_D is equal to the sum of the pressure and friction force in the direction of the flow, s. The lift force, F_L is equal to the sum of the pressure and friction force normal to the direction of the flow, which tends to move the submerged body in that direction. However, because the friction/viscous forces (due to shear stress, τ_w) due to the external flow cannot be theoretically modeled in the integral momentum equation, the actual pressure drop, Δp cannot be analytically determined; thus, the exact reduction in the velocity actual velocity, v_a, which is less than the ideal velocity, v_i, cannot be theoretically determined. Furthermore, the exact component in the s-direction of the pressure and viscous forces cannot be theoretically determined. Therefore, one cannot derive an analytical expression for the drag force, F_D (or for the lift force, F_L) from the momentum equation. As a result, one must resort to dimensional analysis (supplements the momentum theory) in order to derive an expression for the drag force, F_D and the lift force, F_L, which involve the definition of a drag coefficient, C_D that represents the flow resistance, and a lift coefficient, C_L, respectively; C_D and C_L are empirically evaluated. Evaluation of the drag force equation along with the coefficient of drag, C_D for the external flow around an object is presented in Chapter 7, discussed in Chapter 6, and summarized in Section 10.4.1.1 above, as well as Section 10.5.1. Evaluation of the expression for the lift force, F_L and the lift coefficient, C_L may be accomplished in a similar manner, and is presented in Section 10.5.4. Furthermore, empirical evaluation of the coefficient of drag, C_D and the lift coefficient, C_L and application of the corresponding drag equation and lift equation are presented in Sections 10.5.1 and 10.5.4, respectively.

10.5.1 Evaluation of the Drag Force in External Flow

Evaluation of the drag force flow resistance term requires a "subset level" application of the governing equations (see Chapters 6 and 7 and Section 10.4.1.1). The expression for the drag force was given in Equation 10.4 and is repeated as follows:

$$F_D = (F_P + F_f)_s = (\Delta pA)_s = \left(\frac{\rho v^2}{2}A + \frac{\tau_w L}{R_h}A\right)_s = C_D \frac{1}{2}\rho v^2 A \tag{10.5}$$

where F_D is equal to the sum of the pressure and friction force in the direction of the flow, s and the drag coefficient, C_D accounts for (indirectly models) the minor head loss associated with the ideal velocity measurement, and the drag force. Furthermore, one can easily compute v_i used in Equation 10.5 by measuring Δp using a pitot-static tube. The assumption of ideal flow and thus the application of the Bernoulli equation between any two points yields an expression for the ideal velocity as a function of an ideal pressure difference, Δp, which is directly measured. However, in real flow, the effects of viscosity and other real flow effects cause the actual velocity, v_a to be less than the ideal velocity, v_i. Because the friction/viscous forces (due to shear stress, τ_w) due to the external flow cannot be theoretically modeled in the integral momentum equation, the actual pressure drop, Δp cannot be analytically determined; thus, the exact reduction in the velocity actual velocity, v_a which is less than the ideal velocity, v_i, cannot be theoretically determined. Furthermore, the exact component in the s-direction of the pressure and viscous forces cannot be theoretically determined. Therefore, one cannot derive an analytical expression for the drag force, F_D from the momentum equation. As a result, corrections for these real flow effects and the inability to theoretically determine the exact component in the s-direction of the pressure and viscous forces are made by applying a drag coefficient. Thus, the drag coefficient, C_D represents both a velocity coefficient, which accounts for head loss due to friction, and the inability to theoretically determine the exact component in the s-direction of the pressure and viscous forces. As a result, the assumption of ideal flow, and the inability to theoretically determine the exact component in the s-direction of the pressure and viscous forces, is corrected by the use of a drag coefficient, C_D.

10.5.1.1 Modeling the Drag Force in the Momentum Equation

The drag force, F_D on a submerged body has two components, a pressure drag (form drag), F_p and a friction drag (surface drag), F_f, where $F_D = (F_p + F_f)_s$. The pressure drag, F_p is due to the dynamic pressure (and is highly dependent on the form or the shape of the body) and acts normal to the surface of the body, while the friction drag, F_f is due to wall shear stress, τ_w and acts along the surface of the body. One may note that if the body is blunt in shape (spherical, for instance), then the pressure drag, F_p predominates the friction drag, F_f. Furthermore, if the body is more streamlined, then the friction drag, F_f predominates the pressure drag, F_p.

If a submerged body were moving in a stationary homogenous real fluid, then both pressure forces, F_p, and viscous forces, F_V (shear forces, F_f) would be exerted on the surface of the body as illustrated by the momentum equation as follows:

$$\sum F = \Delta pA = F_p + F_f \tag{10.6}$$

where the pressure drop is given as follows:

$$\underbrace{\Delta p}_{presdrop} = \underbrace{\frac{\rho v^2}{2}}_{dynpres} + \underbrace{\frac{\tau_w L}{R_h}}_{flowres} \tag{10.7}$$

Thus, the pressure drop, Δp is equal to the dynamic pressure, $\rho v^2/2$ plus an additional pressure drop, $\tau_w L/R_h$ due to the flow resistance, which makes the final pressure drop more than the ideal pressure drop. Thus, the pressure force is defined as:

$$F_P = \frac{\rho v^2}{2} A \qquad (10.8)$$

and the friction (viscous) force is defined as follows:

$$F_f = F_V = \frac{\tau_w L}{R_h} A \qquad (10.9)$$

Furthermore, the total force due to the pressure drop, Δp acting in the direction of the flow, s is called the drag force, F_D and is given as follows:

$$F_D = \sum F_s = (\Delta p A)_s = \underbrace{\left(\frac{\rho v^2}{2} A\right)}_{F_P}{}_s + \underbrace{\left(\frac{\tau_w L}{R_h} A\right)}_{F_f}{}_s \quad \text{(from theory)} \qquad (10.10a)$$

$$F_D = (F_P)_s + (F_f)_s \qquad (10.10b)$$

10.5.1.2 Derivation of the Drag Coefficient, C_D

Assume that the submerged body illustrated in Figure 10.2 is moving at a low speed in a stationary, homogeneous, viscous, compressible fluid, with a possible wave action at the free surface. As a result, the fluid would exert a drag force, F_D on the body, while the effect of the lift force, F_L would be negligible. For the given assumptions, other dominant forces to consider include the viscous, pressure, gravity, elastic, and inertia forces, and the relative surface roughness, ε/D. The derivation of the drag force equation along with the coefficient of drag, C_D for the external flow around an object was derived in Chapter 7, and is summarized as follows:

$$\frac{F_D}{\frac{\rho v^2}{2} A} = \frac{F_D}{F_{dynpres}} = \frac{F_P}{F_{dynpres}} = \frac{\Delta p A}{\frac{\rho v^2}{2} A} = \frac{\Delta p}{\frac{\rho v^2}{2}} = \frac{2}{E} = C_D = \phi\left(R, C, F, \frac{\varepsilon}{L}, \frac{L_i}{L}\right) \qquad (10.11)$$

where E = the Euler number, R = the Reynolds number, C = the Cauchy number, F = the Froude number, ε/L is the relative roughness, and L_i/L is the geometry. Thus, the resulting drag force, F_D derived from dimensional analysis is given as follows:

$$F_D = C_D \frac{1}{2} \rho v^2 A \qquad (10.12)$$

where A = the frontal area of the body (that is the area projected on a plane normal to the direction of flow). However, one may note that in the case where the body is slim (streamlined) and parallel to the direction of flow, then A = the planform area of the body (that is the area that is viewed looking at the body from above in a direction normal to the flow).

Furthermore, in the case of a flat plate or an airfoil that is parallel to the direction of flow, the planform area is basically the surface area that is parallel to the flow. Thus, in general, the drag force, F_D (and the lift force, F_L) on a submerged body depends on the density, ρ of the fluid, the velocity, v (of either the moving fluid or the moving body), and the size, A, the shape (blunt or streamlined), and the orientation with respect to the flow of the body.

10.5.2 Determination of the Drag Coefficient, C_D

One may note that because the drag coefficient, C_D represents the flow resistance, which causes an unknown pressure drop, Δp, the pressure force will always play an important role; thus, the definition of the drag coefficient, C_D is mainly a function of E. The viscous force will play an important role for streamlined bodies and in the case of low and intermediate velocities (laminar flow); thus, the definition of the drag coefficient, C_D will be dependent on R. For blunt bodies (unstreamlined bodies), the pressure force predominates the viscous force, so the definition of the drag coefficient, C_D will be mostly independent of R and mainly a function of E. Furthermore, for low and intermediate velocities (laminar flow), the drag coefficient, C_D is independent of the relative surface roughness, ε/L. In the case of high velocities (turbulent flow), the drag coefficient, C_D will be dependent on the relative surface roughness, less dependent on R, but dependent on the Mach number,

$$M = \sqrt{C} = \frac{v}{\sqrt{E_v/\rho}} = \frac{v}{c},$$

where C is the Cauchy number and c is the sonic velocity; thus, the elastic force (compressibility effects) will play an important role. The gravity force will play an important role if there is a wave action at the free surface; thus, the definition of the drag coefficient, C_D will be dependent on F; such examples include ships and open channel flow measuring devices/structures such as weirs and spillways. Evaluation of the drag coefficient, C_D is determined through experimentation, where typical values are presented and discussed in Section 10.5.3 as follows.

10.5.3 Evaluation of the Drag Coefficient, C_D

While the drag coefficient, C_D, in general, is a mainly a function of E, it will selectively vary with R, L_i/L, ε/L, M, and F depending mainly on the velocity of flow, the shape of the body, and whether there is a wave action at the free surface (see Table 10.1). Specifically, while the R divides the flow into three types: creeping, laminar, or turbulent, the L_i/L divides the shape of the body into two types: blunt or streamlined. The drag coefficient, C_D for creeping flow is dependent only on R for any shape body. Furthermore, for body shapes in general, the drag coefficient, C_D for laminar flow is dependent on R (especially for streamlined bodies, and less for blunt bodies) and L_i/L, but independent of ε/L, M, and F. And, for body shapes in general, the drag coefficient, C_D for turbulent flow is independent of R (except for streamlined bodies and round-shaped bodies, which are dependent on R), independent of F, but dependent on L_i/L, ε/L, and M. Finally, for body shapes in general, with a wave action at the free surface, the drag coefficient, C_D for laminar flow is dependent on R, L_i/L, and F, but independent of ε/L and M, while the drag coefficient, C_D for turbulent flow is

TABLE 10.1

The Variation of the Drag Coefficient, C_D for External Flow

$C_D = 2/E = f^n$ (Independent Pi Term)	Viscous Force $\equiv R$	Geometry $\equiv L_i/L$	Surface Roughness $\equiv \varepsilon/L$	Elastic Force $\equiv C = M^2$	Gravity Force $\equiv F$
Velocity of Flow and Shape of Body					
Creeping flow: any shape body, $C_D = 2/E = f^n$ (R)	yes	no	no	no	no
Laminar flow: any shape body (except round-shaped bodies)					
Streamlined bodies	yes	yes	no	no	no
Blunt bodies	*less*	yes	no	no	no
Turbulent flow: any shape body (except round-shaped bodies)					
Streamlined bodies	yes	yes	yes	yes	no
Blunt bodies	no	yes	yes	yes	no
Laminar flow: round-shaped bodies (circular cylinder or sphere)	yes	yes	no	no	no
Turbulent flow: round-shaped bodies (circular cylinder or sphere)	yes	yes	yes	yes	no
Laminar or turbulent flow with wave action at free surface: any shape body	yes	yes	no	no	yes

Note: For a blunt or streamlined body, the frontal area is used in the computation of the drag force. For a flat plate or an airfoil (very well streamlined body) that is parallel to the direction of flow, the planform area is used in the computation of the drag force.

dependent on R, L_i/L, ε/L, M, and F. For a detailed discussion on the evaluation of the drag coefficient, C_D, see Chapter 6.

10.5.3.1 The Role of the Velocity of Flow

The velocity of flow plays an important role in the evaluation of the drag coefficient, C_D. Furthermore, the velocity of flow is typically characterized by the Reynolds number, R. Depending upon the range of the Reynolds number, R, the flow is described as follows: (1) creeping flow for a low R, (2) laminar flow for a moderate R, or (3) turbulent flow for a high R. While creeping flow is typically defined at $R \leq 1$, regardless of the shape of the body, the range of R for laminar flow and turbulent flow will depend upon the shape of the body (further discussed in the sections below). While the velocity of flow and thus the magnitude of R, will determine the relative importance of the L_i/L, ε/L, and M in the evaluation of the drag coefficient, C_D, the occurrence of a wave action at the free surface will determine the relative importance of the F. Finally, assuming turbulent flow, the velocity of the flow will further characterize the flow by the Mach number, M. The flow is subsonic for $M < 1$, sonic for $M = 1$, supersonic for $M > 1$, and hypersonic for $M \gg 1$.

10.5.3.2 The Role of the Shape of the Body

The shape of the body plays an important role in the evaluation of the drag coefficient, C_D. Furthermore, the shape of the body is characterized by the geometry, L_i/L, which divides the

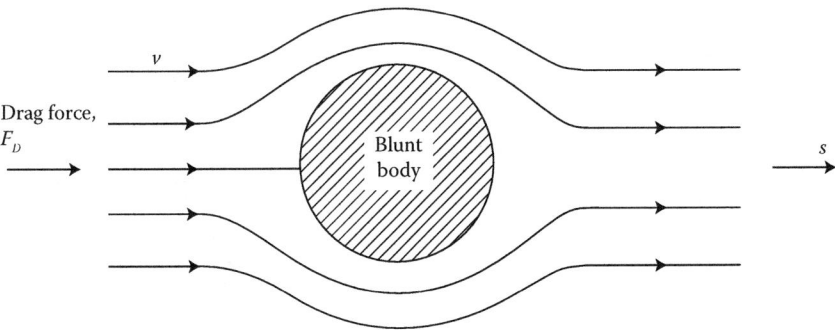

FIGURE 10.5
Shape of blunt body plays important role in evaluation of the drag coefficient, C_D and the drag force, F_D.

shape of the body into two types: blunt or streamlined. Thus, in general, the shape of the body may be characterized as either blunt (see Figure 10.5; other shapes may include people, structures, hot-air balloons, etc.) or streamlined (see Figure 10.6a, other shapes may include foils, airplanes, streamlined cars, etc.). Furthermore, the shape of the body and thus the type of external flow may be characterized as follows: (1) two-dimensional, (2) three-dimensional axisymmetric or (3) three-dimensional, as illustrated in Figures 10.7a–c, respectively. Two-dimensional bodies assume that the body of the object is infinitely long (or in practice, sufficiently long so that the end effects are insignificant) and of constant cross-sectional size and shape and the flow is normal to the body along the length of the body (for instance, a long circular rod). Three-dimensional axisymmetric bodies possess rotational symmetry about an axis in the direction of the flow (for instance, a circular disk). And, three-dimensional bodies may or may not possess a line or plane of symmetry, and thus, cannot be characterized as two-dimensional or three-dimensional axisymmetric flow (for instance, a structure). As such, the shape of the body influences the evaluation of the drag coefficient, as explained in the sections below.

The shape of the body has a significant effect on the evaluation of the drag coefficient, C_D due to the evaluation of the drag force, F_D. As defined above, the drag force, F_D on a submerged body is the component of the resultant pressure and shear forces that acts in the direction of flow. Thus, the drag force, F_D on a submerged body has two components,

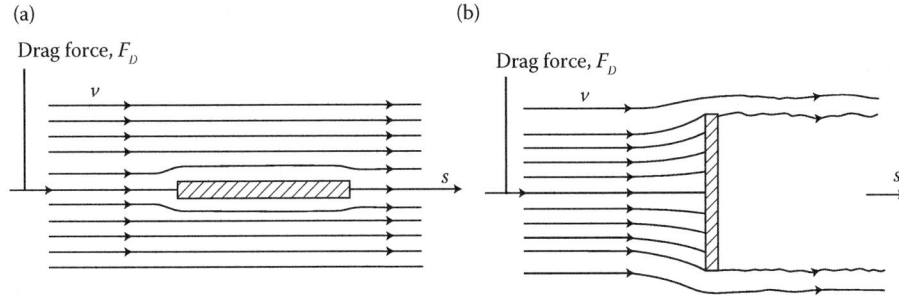

FIGURE 10.6
Shape and direction of streamlined body plays important role in evaluation of the drag coefficient, C_D and the drag force, F_D. (a) Thin flat plate parallel to direction of flow, (b) Thin flat normal to direction of flow.

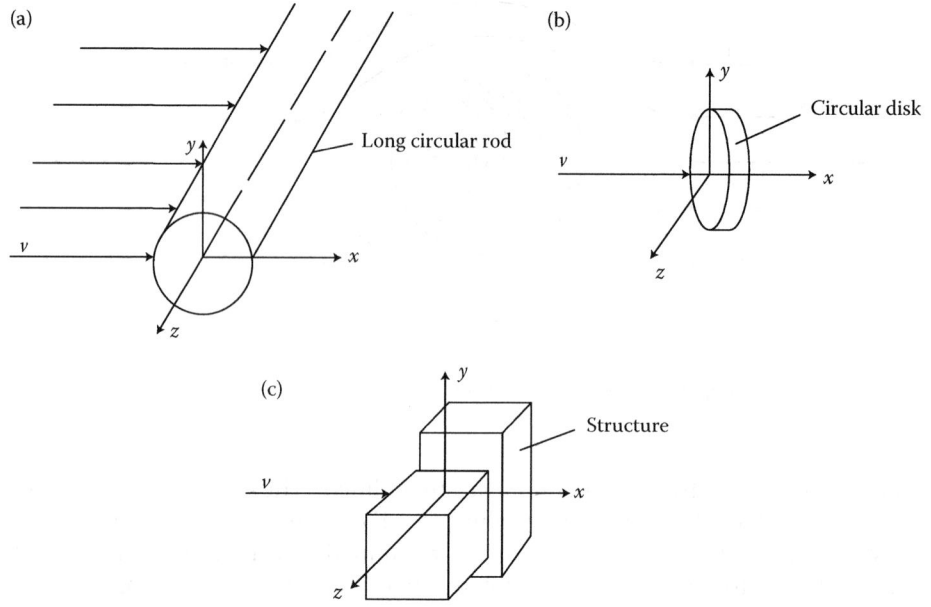

FIGURE 10.7
Shape of body characterizes type of flow. (a) Two-dimensional flow, (b) Three-dimensional axisymmetric flow, (c) Three-dimensional flow.

a pressure drag (form drag), F_p and a friction drag (surface drag), F_f. The pressure drag is due to pressure (and is highly dependent on the form or the shape of the body) and acts normal to the surface of the body, while the friction drag is due to wall shear stress, τ_w and acts along the surface of the body. One may note that if the body is blunt in shape (spherical, for instance), then the pressure drag predominates the friction drag. Furthermore, if the body is more streamlined, then the friction drag predominates the pressure drag.

The pressure drag, F_p is proportional to the *frontal area*, A and to the difference between the pressures acting on the front and the back of the submerged body. Figure 10.5 illustrates that the drag force, F_D on a blunt body is due mainly to pressure drag, F_p. Thus, the pressure drag, F_p is typically dominant for blunt bodies, while it is less dominant for streamlined bodies, and is at a minimum of zero for a thin flat surface that is parallel to the direction of flow (as illustrated in Figure 10.6a). The pressure drag, F_p is most dominant when the velocity of the fluid is too high for the fluid to follow the curvature of the body (as illustrated for the thin flat surface that is normal to the direction of flow in Figure 10.6b); therefore, the fluid will separate from the body at some point downstream of the body; this separation of the flow causes a very low pressure in the back of the body, creating a high pressure difference between the front and the back of the body and resulting in a significant pressure drag, F_p.

The friction drag, F_f is that component of the drag force, F_D that is due directly to the wall shear stress, τ_w acting in the direction of the flow; thus, its magnitude depends on the orientation of the body and the magnitude of the wall shear stress, τ_w. Because the friction drag, F_f is proportional to the *surface area*, A, bodies with a larger surface area experience a larger friction drag. Figure 10.6a illustrates that the friction drag, F_f is at a maximum (thus, the drag force, F_D is due entirely to friction drag, F_f) for a thin flat surface that is parallel to the direction of flow, where the friction drag equals to the total shear force on the surface of the body. Figure 10.6b illustrates that the friction drag, F_f is at a minimum of zero (thus, the drag force, F_D is due entirely to pressure drag, F_p) for a thin flat surface that is normal to the direction of flow.

When the drag force, F_D is broken down into its two components, pressure drag (form drag), F_p and a friction drag (surface drag), F_f, there are two separate drag coefficients, the pressure-drag coefficient, C_p and the friction-drag coefficient, C_f, where:

$$F_D = (F_p)_s + (F_f)_s \tag{10.13}$$

$$(F_p)_s = C_p \frac{1}{2} \rho v^2 A \tag{10.14}$$

$$(F_f)_s = C_f \frac{1}{2} \rho v^2 A \tag{10.15}$$

$$C_D = C_p + C_f \tag{10.16}$$

However, because it is usually difficult to determine the two drag coefficients (C_p and C_f) separately, and we are usually interested in the total drag force, F_D (rather than its two components, F_p and F_f), we are usually interested in evaluating the total drag coefficient, C_D.

10.5.3.3 Reducing the Total Drag Force by Optimally Streamlining the Body

In the design of cars, airplanes, ships, submarines, and buildings, the ultimate goal is to minimize the effect of the drag force, F_D. As noted above, the total drag force F_D has two components, a pressure drag, F_p and a friction drag, F_f. Furthermore, if the body is blunt in shape, then the pressure drag predominates, while if the body is more streamlined, then the friction drag predominates. Thus, in the optimization procedure to minimize the total drag force, F_D on the body, one must simultaneously consider the effects of both the pressure drag and the friction drag on the body. One must note that while streamlining a body will reduce flow separation and thus pressure drag by delaying boundary layer separation, it will increase the friction drag by increasing the surface area, A (the planform area in the case of a thin flat plate that is parallel to the direction of flow). Thus, the resulting drag force will be a function of which type of drag force (pressure or friction) dominates. Based on an optimization study conducted by Goldstein (1938), Figure 10.8 illustrates the variation of the friction, pressure, and total drag coefficients, C_f, C_p, C_D, respectively, of a streamlined strut with thickness-to-chord length (D to L) ratio for $R = 40,000$. It may be noted that for decreasing values of the D to L ratio (i.e., increase in streamlining), the friction drag increases, while the pressure drag decreases, where the minimum total drag occurs at $D/L = 0.25$ for the streamlined strut at $R = 40,000$, and where the planform area, $A = bL$ (as opposed to the frontal area, $A = bD$) is used to compute the drag force in the case of slim bodies such as airfoils and flat plates. Furthermore, as discussed below, the benefits of streamlining a blunt body will depend upon the velocity of the flow as well as the shape of the body.

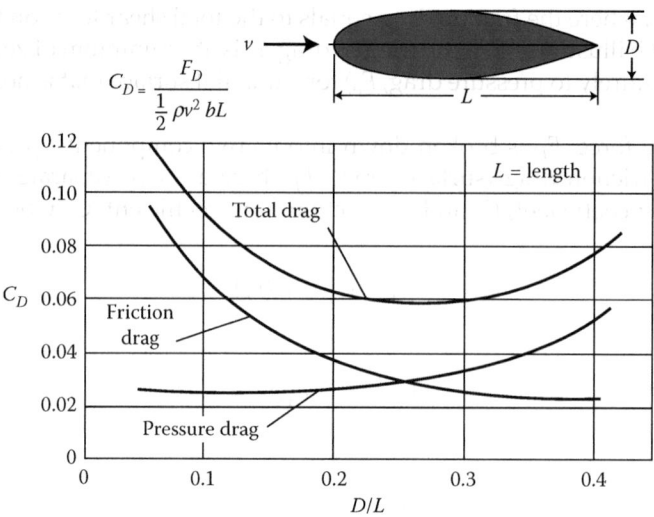

FIGURE 10.8

Variation of the friction, pressure, and total drag coefficients, C_f, C_p, C_D, respectively, of a streamlined strut with thickness-to-chord length (D to L) ratio for $R = 40,000$. (Adapted from Goldstein, S., *Modern Developments in Fluid Dynamics, Vols. I and II*, Clarendon Press, Oxford, 1938.)

EXAMPLE PROBLEM 10.1

The span of a strut is 1000 ft and the chord length is 0.85 ft, as illustrated in Figure EP 10.1. The velocity of air (standard atmosphere at sea level, $\rho = 0.0023768 \ \text{slugs/ft}^3$, $\mu = 0.37372 \times 10^{-6} \ \text{lb-sec/ft}^2$) flowing over the strut is 9 ft/sec. The drag coefficient for a streamlined strut varies with the thickness-to-chord length (D to L) ratio for a Reynolds number, $R = 40,000$, as illustrated in Figure 10.8. (a) Determine the effect of the least streamlined design thickness of the strut, $D/L = 0.4$ on the pressure, friction, and total drag. (b) Determine the effect of an optimally designed thickness of the strut $D/L = 0.25$ on the pressure, friction, and total drag. (c) Determine the effect of the most

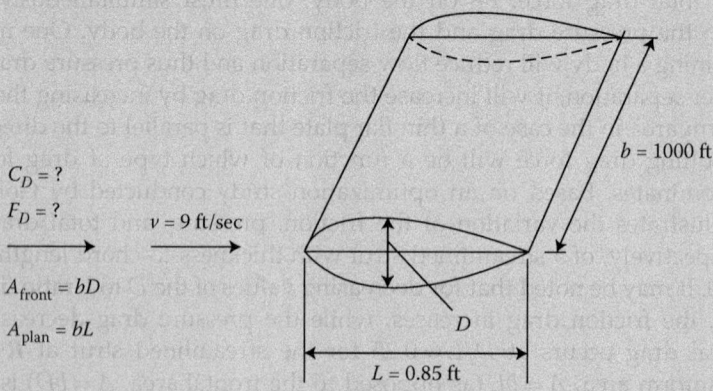

FIGURE EP 10.1
Air flows over a strut.

streamlined design thickness of the strut $D/L = 0.05$ on the pressure, friction, and total drag.

Mathcad Solution

(a)–(b) The thickness-to-chord length (D to L) ratio of the strut and the corresponding pressure, friction, and total drag coefficients, C_p, C_f, C_D, respectively, are determined from Figure 10.8. The Reynolds number, R is determined to ensure it is close enough to $R = 40,000$ (the chord length is assumed to represent the length characteristic of the strut). The frontal area is used to compute the pressure drag, while the planform area is used to compute both the friction drag and the total drag for the streamlined strut. Furthermore, the pressure drag, friction drag, and total drag force are determined by applying Equations 10.14, 10.15, and 10.12, respectively.

(a) The effect of the least streamlined design thickness of the strut, $D/L = 0.4$ on the pressure, friction, and total drag is determined as follows:

$$\text{slug} := 1 \, \text{lb} \frac{\text{sec}^2}{\text{ft}} \qquad b := 1000 \, \text{ft} \qquad L := 0.85 \, \text{ft} \qquad v := 9 \frac{\text{ft}}{\text{sec}}$$

$$\rho := 0.0023768 \frac{\text{slug}}{\text{ft}^3} \qquad\qquad \mu := 0.37372 \times 10^{-6} \, \text{lb} \frac{\text{sec}}{\text{ft}^2}$$

$$C_f := 0.025 \qquad\qquad C_p := 0.055 \qquad\qquad C_D := C_f + C_p = 0.08$$

$$\text{Guess value:} \qquad D := 1 \, \text{ft} \qquad A_{\text{front}} := 1 \, \text{ft}^2 \qquad\qquad A_{\text{plan}} := 1 \, \text{ft}^2$$

$$F_f := 1 \, \text{lb} \qquad F_p := 1 \, \text{lb} \qquad F_D := 2 \, \text{lb} \qquad R := 40000$$

Given

$$\frac{D}{L} = 0.4 \qquad A_{\text{front}} = b \cdot D \qquad A_{\text{plan}} = b \cdot L \qquad F_p = C_p \frac{1}{2} \rho \cdot v^2 \cdot A_{\text{front}}$$

$$F_f = C_f \frac{1}{2} \rho \cdot v^2 \cdot A_{\text{plan}} \qquad F_D = C_D \frac{1}{2} \rho v^2 \cdot A_{\text{plan}} \qquad R = \frac{\rho \cdot v \cdot L}{\mu}$$

$$\begin{pmatrix} D \\ A_{\text{plan}} \\ A_{\text{front}} \\ F_f \\ F_p \\ F_D \\ R \end{pmatrix} := \text{Find}(D, A_{\text{plan}}, A_{\text{front}}, F_f, F_p, F_D, R)$$

$$D = 0.34 \, \text{ft} \qquad A_{\text{plan}} = 850 \, \text{ft}^2 \qquad A_{\text{front}} = 340 \, \text{ft}^2$$

$$F_f = 2.046 \, \text{lb} \qquad F_p = 1.8 \, \text{lb} \qquad F_D = 6.546 \, \text{lb} \qquad R = 4.865 \times 10^4$$

However if the drag force is computed by applying Equation 10.13, a more accurate value is computed as follows:

$$F_D := F_f + F_p = 3.846 \, lb$$

Therefore, the least streamlined design thickness of the strut, $D/L = 0.4$ results in similar magnitudes (and thus contribution) for the pressure drag force and the friction drag force.

(b) The effect of an optimally designed thickness of the strut $D/L = 0.25$ on the pressure, friction, and total drag is determined as follows:

$$C_f := 0.03 \qquad\qquad C_p := 0.03 \qquad\qquad C_D := C_f + C_p = 0.06$$

Guess value: $D := 1 \, ft$ $A_{front} := 1 \, ft^2$ $A_{plan} := 1 \, ft^2$

$$F_f := 1 \, lb \qquad F_p := 1 \, lb \qquad F_D := 2 \, lb \qquad R := 40000$$

Given

$$\frac{D}{L} = 0.25 \qquad A_{front} = b \cdot D \qquad A_{plan} = b \cdot L \qquad F_p = C_p \frac{1}{2} \rho \cdot v^2 \cdot A_{front}$$

$$F_f = C_f \frac{1}{2} \rho \cdot v^2 \cdot A_{plan} \qquad\qquad F_D = C_D \frac{1}{2} \rho v^2 \cdot A_{plan} \qquad R = \frac{\rho \cdot v \cdot L}{\mu}$$

$$\begin{pmatrix} D \\ A_{plan} \\ A_{front} \\ F_f \\ F_p \\ F_D \\ R \end{pmatrix} := Find(D, A_{plan}, A_{front}, F_f, F_p, F_D, R)$$

$$D = 0.213 \, ft \qquad A_{plan} = 850 \, ft^2 \qquad A_{front} = 212.5 \, ft^2$$

$$F_f := 2.455 \, lb \qquad F_p := 0.614 \, lb \qquad F_D := 4.909 \, lb \qquad R = 4.865 \times 10^4$$

However if the drag force is computed by applying Equation 10.13, a more accurate value is computed as follows:

$$F_D := F_f + F_p = 3.068 \, lb$$

Therefore, although an optimally designed thickness of the strut $D/L = 0.25$ results in slight increase in the friction drag force, it results in a significant decrease in the pressure force and a decrease in the total drag force (minimum total drag force).

(c) The effect of the most streamlined design thickness of the strut $D/L = 0.05$ on the pressure, friction, and total drag is determined as follows:

$$C_f := 0.095 \qquad C_p := 0.025 \qquad C_D := C_f + C_p = 0.12$$

Guess value: $\quad D := 1 \text{ ft} \qquad A_{front} := 1 \text{ ft}^2 \qquad\qquad A_{plan} := 1 \text{ ft}^2$

$$F_f := 1 \text{ lb} \qquad F_p := 1 \text{ lb} \qquad F_D := 2 \text{ lb} \qquad R := 40000$$

Given

$$\frac{D}{L} = 0.05 \qquad A_{front} = b \cdot D \qquad A_{plan} = b \cdot L \qquad F_p = C_p \frac{1}{2} \rho \cdot v^2 \cdot A_{front}$$

$$F_f = C_f \frac{1}{2} \rho \cdot v^2 \cdot A_{plan} \qquad F_D = C_D \frac{1}{2} \rho \cdot v^2 \cdot A_{plan} \qquad R = \frac{\rho \cdot v \cdot L}{\mu}$$

$$\begin{pmatrix} D \\ A_{plan} \\ A_{front} \\ F_f \\ F_p \\ F_D \\ R \end{pmatrix} := \text{Find}(D, A_{plan}, A_{front}, F_f, F_p, F_D, R)$$

$$D = 0.043 \text{ ft} \qquad A_{plan} = 850 \text{ ft}^2 \qquad A_{front} = 42.5 \text{ ft}^2$$

$$F_f = 7.773 \text{ lb} \qquad F_p = 0.102 \text{ lb} \qquad F_D = 9.819 \text{ lb} \qquad R = 4.865 \times 10^4$$

However if the drag force is computed by applying Equation 10.13, a more accurate value is computed as follows:

$$F_D := F_f + F_p = 7.875 \text{ lb}$$

Therefore, while the most streamlined design thickness of the strut $D/L = 0.05$ results in the smallest pressure drag force, it results in the highest friction drag force and the highest total drag force. Thus, in conclusion, one may note that while optimally streamlining the strut from $D/L = 0.4$ to $D/L = 0.25$ reduces the respective pressure drag, friction drag, and total drag (reduces to the minimum total drag force), excessively streamlining the strut from $D/L = 0.25$ to $D/L = 0.05$ dramatically increases the friction drag and thus the total drag.

A study conducted by Blevins (1984) illustrated the effect of streamlining a long elliptical cylinder with different aspect ratios, L/D on the drag coefficient, C_D at $\mathbf{R} = 100,000$. In the definition of aspect ratio, L/D, L is the length of the ellipse in the direction of the flow, while D is the thickness of the ellipse as illustrated in Figure 10.9. The figure illustrates that when the aspect ratio, L/D is at a minimum value of 0, the cylinder becomes a flat plate normal to the direction of flow, and the drag coefficient, C_D is at maximum of about 1.9. As the aspect ratio, L/D increases, the drag coefficient, C_D decreases; for an aspect ratio, $L/D = 1$ (which would be a circular cylinder), the drag coefficient, C_D decreases to a value of about 1, and for an aspect ratio, $L/D \geq 4$, it reaches a minimum of about 0.2. Specifically, as the aspect ratio, L/D increases, the flat plate normal to the direction of flow ($C_D \cong 1.9$) becomes an ellipse normal to the direction of flow ($1 < C_D < 1.9$); then it becomes a circle normal to the direction of flow ($C_D \cong 1$); then it becomes an ellipse parallel to the direction of flow ($0.2 < C_D < 1$); then, finally, it becomes a flat plate parallel to the direction of flow ($C_D \cong 0.2$). It may be noted that as the aspect ratio, L/D increases (streamlining of the body and thus an increase in the friction drag), the decrease in the drag coefficient, C_D is due to the boundary layer remaining attached to the surface longer and a decrease in the pressure difference between the front and the back of the body (and thus a decrease in the pressure drag). Mathematically speaking, it may be noted that when the aspect ratio, $L/D \to \infty$ (decreasing D while holding L constant), the ellipse parallel to the direction of flow becomes a flat plate parallel to the direction of flow, and the drag coefficient, $C_D \to \infty$ because the frontal area, $A = bD \to 0$ as $D \to 0$ (where $b =$ the length of the cylinder), where A appears in the denominator in the definition of the drag coefficient, $C_D = F_D/(1/2)\rho v^2 bD$ and thus, mathematically, the drag force, F_D would drastically increase. However, practically speaking, because in the case a flat plate parallel to the direction of flow, the appropriate area, A is the planform area, $A = bL$ (rather than the frontal area, $A = bD$), the drag force, F_D actually

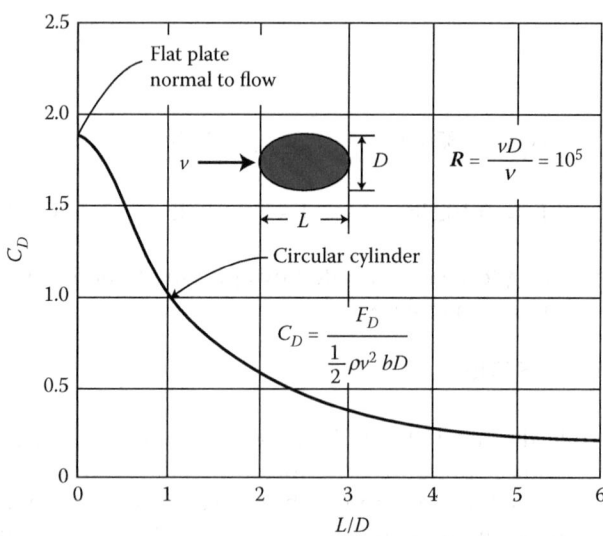

FIGURE 10.9
Variation of the drag coefficient, C_D for a long elliptical cylinder with aspect ratio, L/D for $\mathbf{R} = 100,000$. The frontal area, $A = bD$ is used for a flat plate normal to flow and blunt shapes, while the planform area, $A = bL$ is used for a flat plate parallel to flow and streamlined shapes. (Adapted from Blevins, R. D., *Applied Fluid Dynamics Handbook*, Von Nostrand Reinhold, New York, 1984, p. 347.)

decreases as the body becomes streamlined and flat. Furthermore, for flat bodies, the drag force, F_D is caused mostly by the friction drag, which is proportional to the surface area (i.e., the planform area, $A = bL$).

The decision to streamline a blunt body will depend upon the velocity of the flow as well as the shape of the body, as unnecessary streamlining may actually cause the total drag force to increase. It is important to note that that streamlining a blunt body moving at a high velocity (high R; for instance, $R = 100,000$, as assumed in Figure 10.9 above) not only reduces the total drag force (specifically, the pressure drag) on the body by reducing flow separation (as illustrated in Figure 10.9), but it also reduces the vibration and noise that typically results in such flow situations. However, if the blunt body is moving at low velocities (low R; for instance, creeping flows for which $R \leq 1$), where there is no significant flow separation, then it becomes unnecessary to streamline the body, as the drag force in low velocity situations is due mostly to friction drag; streamlining would simply increase the surface area, A and thus the total drag force.

EXAMPLE PROBLEM 10.2

The length of a long elliptical cylinder is 1500 ft and the thickness of the ellipse is 1 ft, as illustrated in Figure EP 10.2. The velocity of air (standard atmosphere at sea level, $\rho = 0.0023768$ slugs/ft^3, $\mu = 0.37372 \times 10^{-6}$ lb-sec/ft^2) flowing over the cylinder is 20 ft/sec. The drag coefficient for a long elliptical cylinder varies with the aspect ratio, L/D (L is the length of the ellipse in the direction of flow, and D is the thickness of the ellipse) for a Reynolds number, $R = 100,000$, as illustrated in Figure 10.9. (a) Determine the effect of the most blunt design aspect ratio, $L/D = 0$ (where the elliptical cylinder becomes a flat plate normal to the direction of flow) on the total drag. (b) Determine the effect of a blunt design aspect ratio, $L/D = 1$ (where the elliptical cylinder becomes a circular cylinder) on the total drag. (c) Determine the effect of a streamlined design aspect ratio, $L/D = 2$ (elliptical cylinder parallel to the direction of flow) on the total drag. (d) Determine the effect of the most streamlined design aspect ratio, $L/D \geq 4$ (where the elliptical cylinder becomes a flat plate parallel to the direction of flow) on the total drag.

Mathcad Solution

(a)–(d) The aspect ratio, L/D (L is the length of the ellipse in the direction of flow, and D is the thickness of the ellipse) of the elliptical cylinder and the corresponding total drag coefficient, C_D are determined from Figure 10.9. The Reynolds number, R is determined to ensure it is close enough to $R = 1,000,000$ (the thickness of the ellipse, D is assumed to represent the length characteristic of the elliptical cylinder). Furthermore, the total drag force is determined by applying Equation 10.12.

(a) The effect of the most blunt design aspect ratio, $L/D = 0$ (where the elliptical cylinder becomes a flat plate normal to the direction of flow) on the total drag is determined. The frontal area is used to compute the drag force for the flat plate normal to the direction of flow as follows:

$$\text{slug} := 1 \, \text{lb} \frac{\text{sec}^2}{\text{ft}} \qquad b := 1500 \, \text{ft} \qquad D := 1 \, \text{ft} \qquad v := 20 \frac{\text{ft}}{\text{sec}}$$

$$\rho := 0.0023768 \frac{\text{slug}}{\text{ft}^3} \qquad\qquad\qquad \mu := 0.37372 \times 10^{-6} \, \text{lb} \frac{\text{sec}}{\text{ft}^2}$$

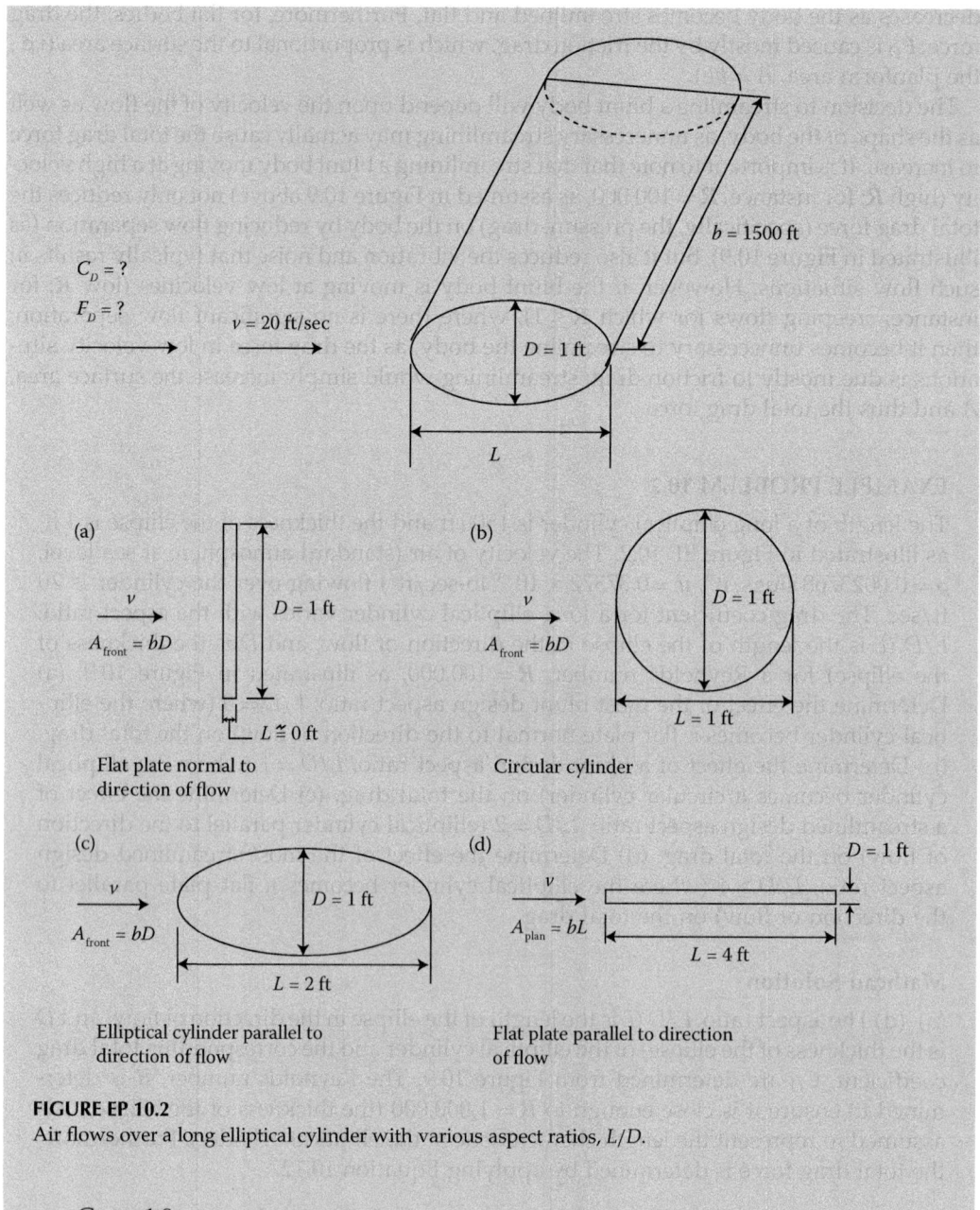

FIGURE EP 10.2
Air flows over a long elliptical cylinder with various aspect ratios, L/D.

$C_D := 1.9$

Guess value: $\qquad L := 1 \text{ ft} \qquad A_{front} := 1 \text{ ft}^2 \qquad F_D := 2 \text{ lb} \qquad R := 100000$

Given

$$\frac{L}{D} = 0 \qquad\qquad A_{front} = b \cdot D \qquad F_D = C_D \frac{1}{2} \rho \cdot v^2 \cdot A_{front} \qquad R = \frac{\rho \cdot v \cdot D}{\mu}$$

$$\begin{pmatrix} L \\ A_{front} \\ F_D \\ R \end{pmatrix} := \text{Find}(L, A_{front}, F_D, R)$$

$L = 0 \, \text{ft}$ $A_{front} = 1.5 \times 10^3 \, \text{ft}^2$ $F_D = 1.355 \times 10^3 \, \text{lb}$ $R = 1.272 \times 10^5$

(b) The effect of a blunt design aspect ratio, $L/D = 1$ (where the elliptical cylinder becomes a circular cylinder) on the total drag is determined. The frontal area is used to compute the drag force for the circular cylinder as follows:

$C_D := 1$

Guess value: $L := 1 \, \text{ft}$ $A_{front} := 1 \, \text{ft}^2$ $F_D := 2 \, \text{lb}$ $R := 100000$

Given

$\dfrac{L}{D} = 1$ $A_{front} = b \cdot D$ $F_D = C_D \dfrac{1}{2} \rho \cdot v^2 \cdot A_{front}$ $R = \dfrac{\rho \cdot v \cdot D}{\mu}$

$$\begin{pmatrix} L \\ A_{front} \\ F_D \\ R \end{pmatrix} := \text{Find}(L, A_{front}, F_D, R)$$

$L = 1 \, \text{ft}$ $A_{front} = 1.5 \times 10^3 \, \text{ft}^2$ $F_D = 713.04 \, \text{lb}$ $R = 1.272 \times 10^5$

(c) The effect of a streamlined design aspect ratio, $L/D = 2$ (elliptical cylinder parallel to the direction of flow) on the total drag is determined. The frontal area is used to compute the drag force for the elliptical cylinder as follows:

$C_D := 0.6$

Guess value: $L := 1 \, \text{ft}$ $A_{front} := 1 \, \text{ft}^2$ $F_D := 2 \, \text{lb}$ $R := 100000$

Given

$\dfrac{L}{D} = 2$ $A_{front} = b \cdot D$ $F_D = C_D \dfrac{1}{2} \rho \cdot v^2 \cdot A_{front}$ $R = \dfrac{\rho \cdot v \cdot D}{\mu}$

$$\begin{pmatrix} L \\ A_{front} \\ F_D \\ R \end{pmatrix} := \text{Find}(L, A_{front}, F_D, R)$$

$L = 2\,\text{ft}$ \qquad $A_{front} = 1.5 \times 10^3\,\text{ft}^2$ \qquad $F_D = 427.824\,\text{lb}$ \qquad $R = 1.272 \times 10^5$

(d) The effect of the most streamlined design aspect ratio, $L/D \geq 4$ (where the elliptical cylinder becomes a flat plate parallel to the direction of flow) on the total drag is determined. The planform area is used to compute the drag force for the flat plate parallel to the direction of flow as follows:

$C_D := 0.2$

Guess value: \qquad $L := 1\,\text{ft}$ \qquad\qquad $A_{plan} := 1\,\text{ft}^2$ \quad $F_D := 2\,\text{lb}$ \quad $R := 100000$

Given

$\dfrac{L}{D} = 4$ \qquad\qquad $A_{plan} = b \cdot L$ \qquad\qquad $F_D = C_D \dfrac{1}{2} \rho \cdot v^2 \cdot A_{plan}$ \qquad $R = \dfrac{\rho \cdot v \cdot D}{\mu}$

$$\begin{pmatrix} L \\ A_{plan} \\ F_D \\ R \end{pmatrix} := \text{Find}(L, A_{plan}, F_D, R)$$

$L = 4\,\text{ft}$ \qquad\qquad $A_{plan} = 6 \times 10^3\,\text{ft}^2$ \quad $F_D = 570.432\,\text{lb}$ \qquad\qquad $R = 1.272 \times 10^5$

Therefore, beginning with most blunt design aspect ratio, $L/D = 0$ (flat plate normal to the direction of flow), the drag force is the highest ($F_D = 1355\,\text{lb}$). Streamlining to a blunt design aspect ratio, $L/D = 1$ (circular cylinder) reduces the drag force ($F_D = 713.04\,\text{lb}$). Further streamlining to a design aspect ratio, $L/D = 2$ (elliptical cylinder) further reduces the drag force ($F_D = 427.824\,\text{lb}$). And even further streamlining to a design aspect ratio, $L/D \geq 4$ (flat plate parallel to the direction of flow) actually increases the drag force ($F_D = 570.432\,\text{lb}$), as the planform area (as opposed to the frontal area that is used in all of the other design aspect ratios) is used to compute the drag force.

Historically, efforts have been made to reduce the drag coefficients for vehicles. Because a significant amount of the aerodynamic drag on vehicles is due to the pressure drag, streamlining the shape of the car or truck will result in a significant reduction in the drag coefficient. Furthermore, a reduction in the drag force on a vehicle will improve its fuel efficiency. Table 10.2 illustrates the historical trend in the design of the body of vehicles in order to reduce the drag force. Additionally, Figure 10.10 further illustrates the historical trends in the reduction of vehicle drag coefficients as a result of streamlining. Table 10.3 illustrates typical frontal areas for several common classes of cars.

TABLE 10.2

The Effects of Historical Trend in Vehicle Design on the Drag Coefficient, C_D

Year of Vehicle	Body Shape	C_D Based on Frontal Area
1920		0.8
1930		0.46
1940–1945		0.54–0.58
1968–1969		0.36–0.38
1977		0.48
1990–1992		0.29–0.30
2000		0.29
2010		0.14
Tractor-trailer without deflector		0.75–0.95
Tractor-trailer with deflector		0.55–0.75

Source: Adapted from Finnemore, E. J., and J. B., Franzini, *Fluid Mechanics with Engineering Applications*, 10th ed, McGraw Hill, New York, 2002, p. 286; Crowe, C. T. et al., *Engineering Fluid Mechanics*, 8th ed, John Wiley and Sons, New York, 2005, p. 468; Blevins, R. D., *Applied Fluid Dynamics Handbook*, Von Nostrand Reinhold, New York, 1984, p. 348.

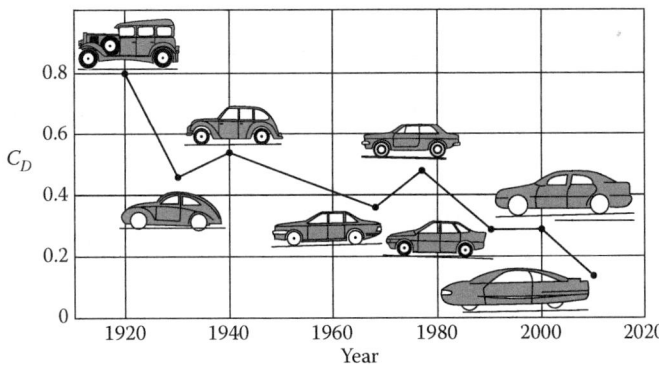

FIGURE 10.10

Historical trends in the reduction of vehicle drag coefficients, C_D as a result of streamlining (plot of data in Table 10.2). (Adapted from Finnemore, E. J., and J. B., Franzini, *Fluid Mechanics with Engineering Applications*, 10th ed, McGraw Hill, New York, 2002, p. 286; Crowe, C. T. et al., *Engineering Fluid Mechanics*, 8th ed, John Wiley and Sons, New York, 2005, p. 468; Blevins, R. D., *Applied Fluid Dynamics Handbook*, Von Nostrand Reinhold, New York, 1984, p. 348.)

TABLE 10.3

Typical Frontal Areas for the Four Common Classes of Automobiles

Class of Automobile	Frontal Area, A (ft^2)	Frontal Area, A (m^2)
Subcompact	19.4–21.5	1.8–2.0
Compact	20.2–23.7	1.9–2.2
Mid-size	21.5–25.8	2.0–2.4
Full-size	22.6–28.0	2.1–2.6

Source: Adapted from Esposito, A., *Fluid Mechanics with Applications*, Prentice-Hall, New Jersey, 1998, p. 598; Hucho, W-H., Ed., *Aerodynamics of Road Vehicles: From Fluid Mechanics to Vehicle Engineering*, Elsevier, Amsterdam, 2013, p. 37.

EXAMPLE PROBLEM 10.3

The drag coefficient for vehicles varies with the deign shape of the body, as illustrated in Table 10.2. Considering the four sedans (1920, 1940–1945, 1968–1969, and 1990–1992) illustrated in Figure EP 10.3, there is a historical trend in the reduction in the drag coefficient. Assume a given sedan moves at a speed of 55 mph in air (standard atmosphere at sea level, $\rho = 0.0023768$ slugs/ft^3, $\mu = 0.37372 \times 10^{-6}$ lb-sec/ft^2), and the average frontal area of a given sedan is 27 ft^2. (a) Determine the effect of the historical design and thus the drag coefficient on the magnitude of the drag force.

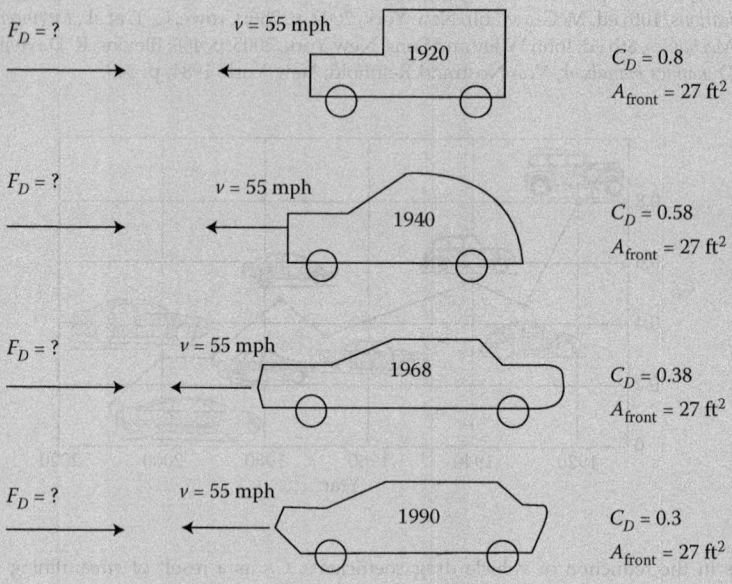

FIGURE EP 10.3
Four historical sedans move at a speed of 55 mph.

Mathcad Solution

(a) The effect of the historical design and thus the drag coefficient on the magnitude of the drag force is determined by comparing the four sedans. Furthermore, the drag force is determined by applying Equation 10.12 as follows:

$$\text{slug} := 1\,\text{lb}\frac{\sec^2}{\text{ft}} \qquad A_{front} := 27\,\text{ft}^2 \qquad v := 55\,\text{mph} = 80.667\frac{\text{ft}}{\sec}$$

$$\rho := 0.0023768\frac{\text{slug}}{\text{ft}^3} \qquad\qquad\qquad \mu := 0.37372 \times 10^{-6}\,\text{lb}\frac{\sec}{\text{ft}^2}$$

$$C_{D1920} := 0.8 \qquad C_{D1940} := 0.58 \qquad C_{D1968} := 0.38 \qquad C_{D1990} := 0.3$$

Guess value: $\qquad F_{D1920} := 2\,\text{lb} \qquad F_{D1940} := 2\,\text{lb} \qquad F_{D1968} := 2\,\text{lb}$

$$F_{D1990} := 2\,\text{lb}$$

Given

$$F_{D1920} = C_{D1920}\frac{1}{2}\rho \cdot v^2 \cdot A_{front} \qquad\qquad F_{D1940} = C_{D1940}\frac{1}{2}\rho \cdot v^2 \cdot A_{front}$$

$$F_{D1968} = C_{D1968}\frac{1}{2}\rho \cdot v^2 \cdot A_{front} \qquad\qquad F_{D1990} = C_{D1990}\frac{1}{2}\rho \cdot v^2 \cdot A_{front}$$

$$\begin{pmatrix} F_{D1920} \\ F_{D1940} \\ F_{D1968} \\ F_{D1990} \end{pmatrix} := \text{Find}(F_{D1920}, F_{D1940}, F_{D1968}, F_{D1990}) = \begin{pmatrix} 167.034 \\ 121.1 \\ 79.341 \\ 62.638 \end{pmatrix}\text{lb}$$

Therefore, due to the streamlining of the four sedans between 1920 and 1990, for a speed of 55 mph, the drag force has significantly been reduced from $F_D = 167.034$ lb to $F_D = 62.638$ lb, a reduction of nearly three fold, which contributes significantly to the vehicle's fuel efficiency.

10.5.3.4 The Occurrence of Flow Separation

In general, flow separation typically occurs in the cases of a blunt body moving at a high velocity ($R \geq 90$) and a streamlined body moving at a high velocity at a large angle of attack (such as an airfoil at an angle of attack of larger than $15°$). Furthermore, the location of separation depends on factors including R and the surface roughness. Finally, the exact location of flow separation is difficult to predict unless the body is characterized by sharp corners or abrupt changes in the shape of the body. For instance, Figure 10.11 illustrates how the location of the point of separation for the more streamlined body is delayed further downstream than the square-cornered body. Also, given are the respective values for the drag coefficient,

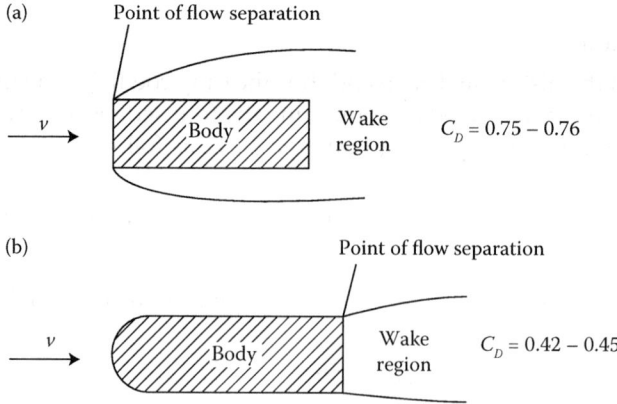

FIGURE 10.11
Location of flow separation and the drag coefficient, C_D is a function of the shape of body. (a) Square-cornered body, (b) Streamlined body. (Adapted from Schlichting, H., *Boundary Layer Theory*, 4th ed, McGraw Hill, New York, 1960, p. 34; Finnemore, E. J., and J. B., Franzini, *Fluid Mechanics with Engineering Applications*, 10th ed, McGraw Hill, New York, 2002, p. 378.)

C_D (0.42–0.45 for the streamlined body and 0.75–0.76 for the square-cornered body). Flow separation from the body results in a separated region between the body and the flow, thus creating a low-pressure region behind the body called the wake region. The wake comes to an end when the flow streams reattach.

EXAMPLE PROBLEM 10.4

The flow separation and thus the drag coefficient for vehicles varies with the deign shape of the body, as illustrated in Figure 10.11. Considering the two vans (blunt/square nose vs. streamlined/round nose) illustrated in Figure EP 10.4, there is a reduction in the drag coefficient for the streamlined van ($C_D = 0.75$ vs $C_D = 0.45$). Assume a given van moves at a speed of 65 mph in air (standard atmosphere at sea level, $\rho = 0.0023768$ slugs/ft^3, $\mu = 0.37372 \times 10^{-6}$ lb-sec/ft^2), and the average frontal area of a given van is 75 ft^2. (a) Determine the effect of the blunt versus streamlined design, and thus, the drag coefficient, on the magnitude of the drag force.

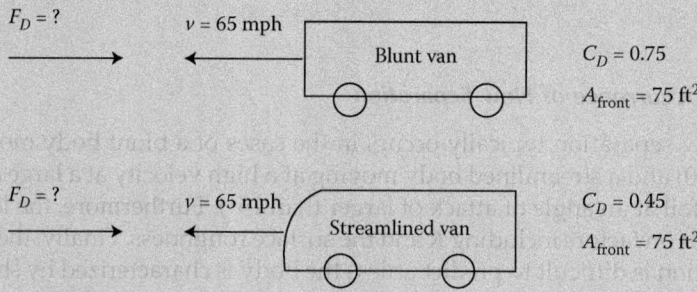

FIGURE EP 10.4
Blunt/square nose van and a streamlined/round nose van move at a speed of 65 mph.

Mathcad Solution

(a) The effect of the blunt versus streamlined design and thus the drag coefficient on the magnitude of the drag force is determined by comparing the two vans. Furthermore, the drag force is determined by applying Equation 10.12 as follows:

$$\text{slug} := 1\,\text{lb}\frac{\sec^2}{\text{ft}} \qquad A_{front} := 75\,\text{ft}^2 \qquad v := 65\,\text{mph} = 95.333\frac{\text{ft}}{\sec}$$

$$\rho := 0.0023768\frac{\text{slug}}{\text{ft}^3} \qquad\qquad \mu := 0.37372 \times 10^{-6}\,\text{lb}\frac{\sec}{\text{ft}^2}$$

$$C_{Dblunt} := 0.75 \qquad\qquad C_{Dstream1} := 0.45$$

Guess value: $\qquad F_{Dblunt} := 2\,\text{lb} \qquad F_{Dstream1} := 2\,\text{lb}$

Given

$$F_{Dblunt} = C_{Dblunt}\frac{1}{2}\rho\cdot v^2\cdot A_{front} \qquad\qquad F_{Dstream1} = C_{Dstream1}\frac{1}{2}\rho\cdot v^2\cdot A_{front}$$

$$\begin{pmatrix} F_{Dblunt} \\ F_{Dstream1} \end{pmatrix} := \text{Find}(F_{Dblunt}, F_{Dstream1}) = \begin{pmatrix} 607.54 \\ 364.524 \end{pmatrix}\text{lb}$$

Therefore, due to the streamlining of the van, for a speed of 65 mph, the drag force has significantly been reduced from $F_D = 607.54$ lb to $F_D = 364.524$ lb, a reduction of nearly 2-fold, which contributes significantly to the vehicle's fuel efficiency.

10.5.3.5 Reducing the Flow Separation (Pressure Drag)

In the design of the wings of airplanes, the ultimate goal is to minimize the effect of the drag force, F_D and maximize the effect of the lift force, F_L. For such a design, it is assumed that the body is both streamlined and is moving at a high velocity, in which case it is important to model the lift force, F_L as well as the drag force, F_D. As discussed above, streamlining a blunt body moving at a high velocity results in reducing the total drag force (specifically, the pressure drag) on the body by reducing flow separation. However, such a streamlined body moving at a high velocity with too large of an angle of attack subsequently increases the pressure drag force (and thus the total drag) by increasing the flow separation, and thus decreases the lift force. Therefore, in the design of the wings of airplanes, reducing the angle of attack (to less than 15°) of the streamlined wing will reduce the effect of the flow separation (and thus minimize the effect of the drag force, F_D), and will thus, in turn, maximize the effect of the lift force, F_L (see Section 10.5.4 for further discussion on lift and Example Problems 10.20 through 10.26 for examples on computation of the lift force).

10.5.3.6 The Importance of the Reynolds Number, R

The velocity of flow is typically characterized by the Reynolds number, R. Depending upon the range of the Reynolds number, R, the flow is described as follows: (1) creeping flow for a low R, (2) laminar flow for a moderate R, or (3) turbulent flow for a high R. While creeping flow is typically defined at $R \leq 1$, regardless of the shape of the body, the range of R for laminar flow and turbulent flow will depend upon the shape of the body (further discussed in the sections below). While the velocity of flow and thus the magnitude of R will determine the relative importance of the L_i/L, ε/L, and M in the evaluation of the drag coefficient, C_D, the occurrence of a wave action at the free surface will determine the relative importance of the F.

Although the drag coefficient, C_D, in general, is a mainly a function of E, it will selectively vary with R, L_i/L, ε/L, M, and F depending mainly on the velocity of flow, the shape of the body, and whether there is a wave action at the free surface (see Table 10.1). The drag coefficient, C_D for creeping flow (where $R \leq 1$ and there is no significant flow separation and thus the pressure drag is at a minimum) is dependent only on R for any shape body. Furthermore, for body shapes in general, the drag coefficient, C_D for laminar flow is dependent on R (especially for streamlined bodies, and less for blunt bodies) and L_i/L, but independent of ε/L, M, and F. And, for body shapes in general, the drag coefficient, C_D for turbulent flow is independent of R (except for streamlined bodies and round-shaped bodies, which are dependent on R), independent of F, but dependent on L_i/L, ε/L, and M. Finally, for body shapes in general, with a wave action at the free surface, the drag coefficient, C_D for laminar flow is dependent on R, L_i/L, and F, but independent of ε/L and M, while the drag coefficient, C_D for turbulent flow is dependent on R, L_i/L, ε/L, M, and F.

Furthermore, although the drag coefficient, C_D will vary with the R with varying degrees depending upon the range of the R (low [creeping; $R < 1$], moderate [laminar], or high [turbulent]) and the shape of the body, in general, the drag coefficient, C_D decreases with an increase in R. Figure 10.12 illustrates the variation of the drag coefficient, C_D with the R for two-dimensional bodies, while Figure 10.13 illustrates the variation of the drag coefficient, C_D with the R for three-dimensional axisymmetric bodies.

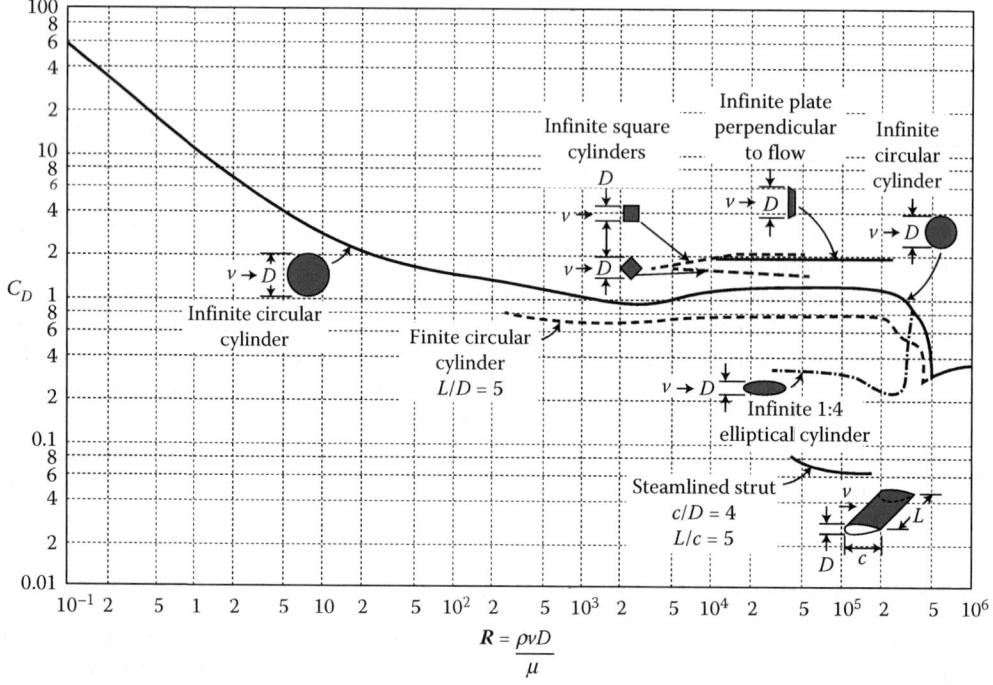

FIGURE 10.12

Variation of the drag coefficient, C_D with the R for two-dimensional bodies. (From Finnemore, E. J., and J. B., Franzini, *Fluid Mechanics with Engineering Applications*, 10th ed, McGraw Hill, New York, 2002, p. 383. With permission.)

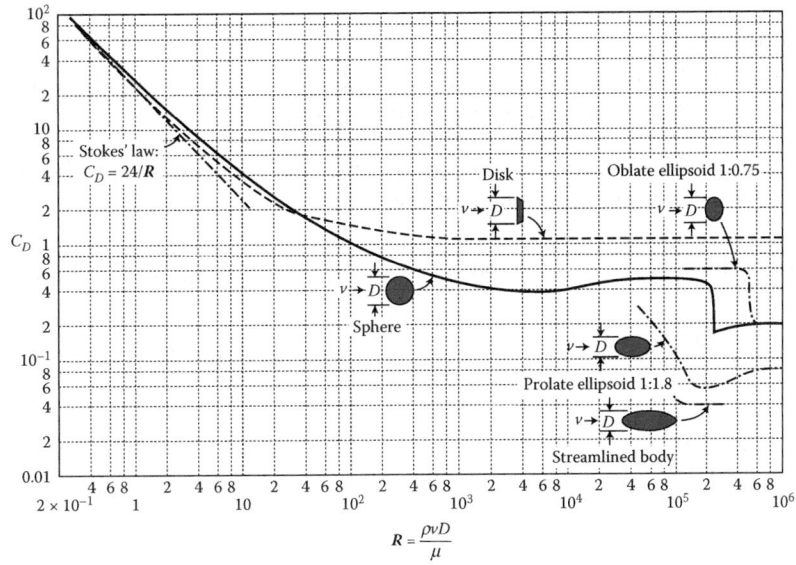

FIGURE 10.13
Variation of the drag coefficient, C_D with the R for three-dimensional axisymmetric bodies. (From Finnemore, E. J., and J. B., Franzini, *Fluid Mechanics with Engineering Applications*, 10th ed, McGraw Hill, New York, 2002, p. 376. With permission.)

EXAMPLE PROBLEM 10.5

Figure 10.12 illustrates the variation of the drag coefficient, C_D with the R for two-dimensional bodies. Consider a 2000-ft-long circular cylinder with a diameter of 3.5 ft. Assume water at 70°F flows over the cylinder, as illustrated in Figure EP 10.5. (a) Determine the drag coefficient and the drag force if the velocity of flow over the cylinder is 0.000155 ft/sec. (b) Determine the drag coefficient and the drag force if the velocity of flow over the cylinder is 3.1 ft/sec.

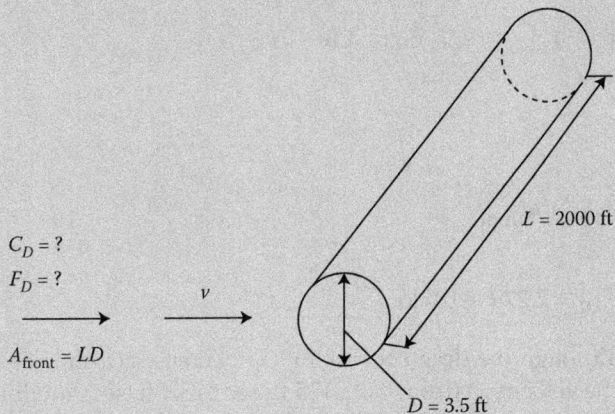

FIGURE EP 10.5
Water flows over a long circular cylinder.

Mathcad Solution

(a)–(b) The frontal area is used to compute the drag force for the circular cylinder, and the drag force is determined by applying Equation 10.12.

(a) The drag coefficient and the drag force for a velocity of flow over the cylinder of 0.000155 ft/sec is determined as follows:

$$\text{slug} := 1 \, \text{lb} \frac{\text{sec}^2}{\text{ft}} \qquad L := 2000 \, \text{ft} \qquad D := 3.5 \, \text{ft} \qquad A_{front} := L \, D = 7 \times 10^3 \, \text{ft}^2$$

$$\rho := 1.936 \frac{\text{slug}}{\text{ft}^3} \qquad\qquad\qquad \mu := 20.50 \times 10^{-6} \, \text{lb} \frac{\text{sec}}{\text{ft}^2}$$

$$v := 0.000155 \frac{\text{ft}}{\text{sec}} \qquad R = \frac{\rho \cdot v \cdot D}{\mu} = 51.233 \qquad C_D := 1.75$$

Guess value: $\qquad\qquad F_D := 1 \, \text{lb}$

Given

$$F_D = C_D \frac{1}{2} \rho \cdot v^2 \cdot A_{front}$$

$$F_D := \text{Find}(F_D) = 2.849 \times 10^{-4} \, \text{lb}$$

(b) The drag coefficient and the drag force for a velocity of flow over the cylinder of 3.1 ft/sec is determined as follows:

$$v := 3.1 \frac{\text{ft}}{\text{sec}} \qquad\qquad R := \frac{\rho \cdot v \cdot D}{\mu} = 1.025 \times 10^6 \qquad\qquad C_D := 0.35$$

Guess value: $\qquad\qquad F_D := 1 \, \text{lb}$

Given

$$F_D = C_D \frac{1}{2} \rho \cdot v^2 \cdot A_{front}$$

$$F_D := \text{Find}(F_D) = 2.279 \times 10^4 \, \text{lb}$$

Therefore, although the drag coefficient, C_D decreases (from 1.75 to 0.35) with an increase in the velocity, v (from 0.000155 ft/sec to 3.1 ft/sec) and thus an increase in the Reynolds number, R (from 51.233 to 1.025×10^6), the drag force, F_D increases (from 2.849×10^{-4} lb to 2.279×10^4 lb) with an increase in velocity, as F_D is directly proportional to v^2.

EXAMPLE PROBLEM 10.6

Figure 10.13 illustrates the variation of the drag coefficient, C_D with the R for three-dimensional axisymmetric bodies. Consider a disk with a diameter of 2 ft. Assume water at 70°F flows over the disk, as illustrated in Figure EP 10.6. (a) Determine the drag coefficient and the drag force if the velocity of flow over the disk is 0.00012 ft/sec. (b) Determine the drag coefficient and the drag force if the velocity of flow over the disk is 2.5 ft/sec.

FIGURE EP 10.6
Water flows over a disk.

Mathcad Solution

(a)–(b) The frontal area is used to compute the drag force for the disk, and the drag force is determined by applying Equation 10.12.

(a) The drag coefficient and the drag force for a velocity of flow over the disk of 0.00012 ft/sec is determined as follows:

$$\text{slug} := 1 \, \text{lb} \frac{\text{sec}^2}{\text{ft}} \qquad D := 2 \, \text{ft} \qquad A_{\text{front}} := \frac{\pi \cdot D^2}{4} = 3.142 \, \text{ft}^2$$

$$\rho := 1.936 \frac{\text{slug}}{\text{ft}^3} \qquad \qquad \mu := 20.50 \times 10^{-6} \, \text{lb} \frac{\text{sec}}{\text{ft}^2}$$

$$v := 0.00012 \frac{\text{ft}}{\text{sec}} \qquad R := \frac{\rho \cdot v \cdot D}{\mu} = 22.665 \qquad C_D := 2.1$$

Guess value: $\qquad F_D := 1 \, \text{lb}$

Given

$$F_D = C_D \frac{1}{2} \rho \cdot v^2 \cdot A_{\text{front}}$$

$$F_D := \text{Find}(F_D) = 9.196 \times 10^{-8} \, \text{lb}$$

(b) The drag coefficient and the drag force for a velocity of flow over the cylinder of 2.5 ft/sec is determined as follows:

$$v := 2.5 \frac{\text{ft}}{\text{sec}} \qquad R := \frac{\rho \cdot v \cdot D}{\mu} = 4.722 \times 10^5 \qquad C_D := 1.1$$

Guess value: $\qquad F_D := 1 \, \text{lb}$

Given

$$F_D = C_D \frac{1}{2} \rho \cdot v^2 \cdot A_{\text{front}}$$

$$F_D := \text{Find}(F_D) = 20.907 \text{ lb}$$

Therefore, although the drag coefficient, C_D decreases (from 2.1 to 1.1) with an increase in the velocity, v (from 0.00012 ft/sec to 2.5 ft/sec) and thus, an increase in the Reynolds number, R (from 22.665 to 4.722×10^5), the drag force, F_D increases (from 9.196×10^{-8} lb to 20.907 lb) with an increase in velocity, as F_D is directly proportional to v^2.

10.5.3.7 Creeping Flow ($R \leq 1$) for Any Shape Body, and Stokes Law for a Spherical Shaped Body

For creeping flows at $R \leq 1$, the shape of the body does not have a major role on the value of drag coefficient, C_D, which decreases with an increase in R. As such, for creeping flows at $R \leq 1$, where the pressure drag is at a minimum, the drag force is due mostly to friction drag; therefore, the drag coefficient, C_D is very much dependent on $R = F_I/F_v$, regardless of the shape of the body, whether blunt or streamlined (although one may note that the drag coefficient, C_D, in general, is a mainly a function of $E = F_I/F_P$). For creeping flows at $R \leq 1$, where the fluid smoothly wraps around the body, the drag coefficient, C_D will vary inversely with R. Figure 10.14 presents the drag coefficients, C_D for a sphere, a hemisphere, a circular disk that is normal to the direction of flow, and a circular disk that is parallel to the direction of flow for creeping flows at $R \leq 1$. This figure illustrates that the shape of the body does not have a major role on the value of drag coefficient, C_D (for creeping flows) and that the drag coefficient, C_D decreases with an increase in R. The drag coefficient, C_D for a sphere, in particular, for a creeping flow at $R \leq 1$ is given as follows: $C_D = 24/R$, which may be theoretically derived by applying Stokes law for a spherical shaped body as follows. For creeping flow at $R \leq 1$, the flow about the sphere is completely viscous, and the drag force, F_D acting on the spherical body is given by Stokes law as follows:

$$F_D = 3\pi \mu v D \tag{10.17}$$

Specifically, Stokes law illustrates that at $R \leq 1$, the drag force, F_D acting on a spherical body is directly proportional to the viscosity, the velocity of the fluid, and the diameter of the sphere. Thus, the drag coefficient, C_D for a sphere for a creeping flow at $R \leq 1$ may be theoretically derived as follows. Equating Stokes law for the drag force (Equation 10.17) to the drag force equation (Equation 10.12) and isolating the drag coefficient, C_D yields the drag coefficient, C_D for a sphere for a creeping flow at $R \leq 1$ as follows:

$$F_D = C_D \frac{1}{2} \rho v^2 A = C_D \frac{\rho v^2}{2} \frac{\pi D^2}{4} = 3\pi \mu v D \tag{10.18a}$$

$$C_D = \frac{24\mu}{\rho v D} = \frac{24}{R} \tag{10.18b}$$

Body	Drag coefficient, C_D
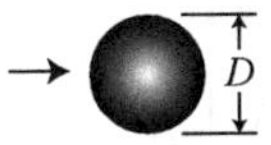 Sphere	$C_D = 24/R$
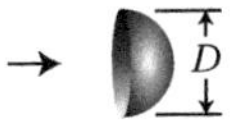 Hemisphere	$C_D = 22.2/R$
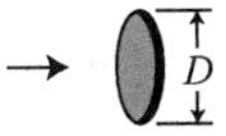 Circular disk normal to flow	$C_D = 20.4/R$
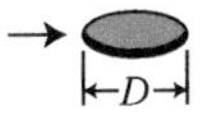 Circular disk parallel to flow	$C_D = 13.6/R$

FIGURE 10.14
Drag coefficient, C_D for creeping flows at $R \le 1$ for three-dimensional axisymmetric bodies ($R = \rho v D/\mu$ and $A = \pi D^2/4$). (Adapted from Happel, J., *Low Reynolds Number Hydrodynamics*, Prentice-Hall, New Jersey, 1965.)

where $A =$ frontal area of the sphere $= \pi D^2/4$ and $R = \rho v D/\mu$. Analogous to the theoretical friction factor for laminar flow, $f = 64/R$, which plots as a straight line (decreasing linearly with an increase in R, and independent of ε/D) on the Moody diagram (Figure 8.1), the drag coefficient, $C_D = 24/R$ for a sphere for a creeping flow at $R \le 1$, also plots as a straight line, decreasing linearly with an increase in R, and independent of L_i/L (i.e., regardless of the shape of the body), ε/L, M, and F, as illustrated in Figure 10.13.

Stokes law is typically applied in the study of transport of precipitation and dust particles, and suspended particles in water or air (see Example Problem 10.8a). Stokes law is also applied in the use of a falling sphere viscometer to determine the viscosity of a fluid (see Example Problem 10.8b).

EXAMPLE PROBLEM 10.7

Water at 70°F flows at a velocity of 0.0001 *ft/sec* over a hemisphere with a diameter of 0.1 *ft*, as illustrated in Figure EP 10.7. (a) Determine the drag coefficient and the drag force on the body.

Mathcad Solution

(a) The frontal area is used to compute the drag force for the hemisphere, and the drag force is determined by applying Equation 10.12.

$$C_D = \frac{22.2}{R}$$

$$F_D = ?$$

→ $v = 0.0001$ ft/sec

$$A_{front} = \frac{\pi D^2}{4}$$

$D = 0.1$ ft

FIGURE EP 10.7
Water flows over a hemisphere.

$$slug := 1 \, lb \frac{sec^2}{ft} \qquad\qquad D := 0.1 \, ft \qquad A_{front} := \frac{\pi \cdot D^2}{4} = 7.854 \times 10^{-3} \, ft^2$$

$$\rho := 1.936 \frac{slug}{ft^3} \qquad\qquad\qquad\qquad\qquad \mu := 20.50 \times 10^{-6} \, lb \frac{sec}{ft^2}$$

$$v := 0.0001 \frac{ft}{sec} \qquad\qquad R := \frac{\rho \cdot v \cdot D}{\mu} = 0.944$$

Because the Reynolds number, $R \leq 1$ (creeping flow), the drag coefficient, C_D varies inversely with R as presented in Figure 10.14. Thus, for a hemisphere, the drag coefficient, C_D is computed as follows:

$$C_D := \frac{22.2}{R} = 23.507$$

Guess value: $F_D := 1 \, lb$

Given

$$F_D = C_D \frac{1}{2} \rho \cdot v^2 \cdot A_{front}$$

$$F_D := Find(F_D) = 1.787 \times 10^{-9} \, lb$$

In general, regardless of the velocity of the flow, when a body is dropped into the atmosphere or a body of fluid (e.g., a falling sphere viscometer), because of its weight, the body will initially accelerate due to gravity. Acceleration occurs because the forces acting on the body in the direction of flow/fall are not balanced (see Figure 10.15a). However, due to the flow resistance acting on the body, there is a drag force acting in the opposite direction of the acceleration due to gravity (and velocity), as illustrated in Figure 10.15. Furthermore, because the drag force is directly promotional to v^2, as the velocity of the body increases (due to the gravitational acceleration), the drag force will also increase. As such, as the velocity increases, the drag force will also continue to increase, until the forces acting on the body in the direction of the flow/fall are

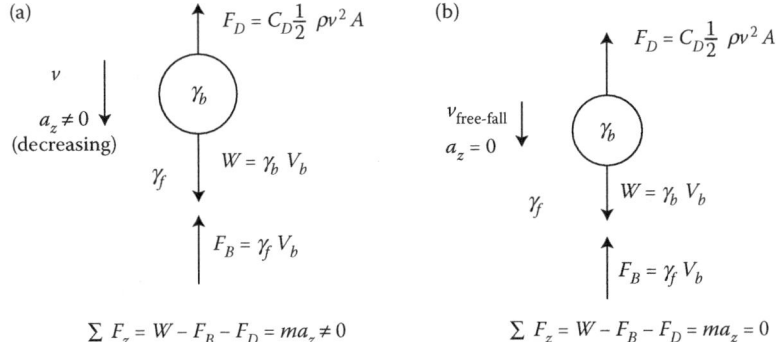

FIGURE 10.15
Forces acting on a body falling through a fluid. (a) Body is accelerating due to unbalanced forces, (b) Body is in free-fall (no acceleration) due to balanced forces.

balanced (Newton's second law of motion) (see Figure 10.15b), and the terminal (free-fall) velocity, v is reached as follows:

$$\sum_z F_z = W - F_B - F_D = ma_z = 0 \tag{10.19}$$

where the terminal velocity, v is the maximum velocity the falling body will reach during the fall, W is the weight of the body, F_B is the buoyant force (see Chapter 2 and Example Problems 10.8 and 10.13), and F_D is the drag force. Assuming that the fluid properties of the atmosphere or the body of fluid are constant, the terminal velocity, v remains a constant for the remainder of the fall.

EXAMPLE PROBLEM 10.8

(a) A dust particle with a diameter of 0.0012 in and a specific gravity of 3.6 falls in the air (standard atmosphere at sea level, $\rho = 0.0023768$ slugs/ft³, $\mu = 0.37372 \times 10^{-6}$ lb-sec/ft², $\gamma = 0.076472$ lb/ft³), as illustrated in Figure EP 10.8a, and settles to the ground after a windstorm. Determine the terminal (free-fall) velocity for the dust particle as it falls to the ground. (b) Mathematically illustrate how Stokes law is applied in the use of a falling sphere viscometer to determine the viscosity of a fluid.

Mathcad Solution

(a) In order to determine the drag force, Equation 10.19 (Newton's second law of motion) is applied in the vertical direction as follows:

$$\sum_z F_z = W - F_B - F_D = ma_z = 0$$

$$D_{dust} := 0.0012 \text{ in} = 1 \times 10^{-4} \text{ ft} \qquad V_{dust} := \frac{\pi \cdot D_{dust}^3}{6} = 5.236 \times 10^{-13} \text{ ft}^3$$

$$S_{dust} := 3.6 \qquad \gamma_w := 62.417 \frac{\text{lb}}{\text{ft}^3} \qquad \gamma_{dust} := S_{dust} \cdot \gamma_w = 224.701 \frac{\text{lb}}{\text{ft}^3}$$

$$W_{dust} := \gamma_{dust} \cdot V_{dust} = 1.177 \times 10^{-10} \text{ lb}$$

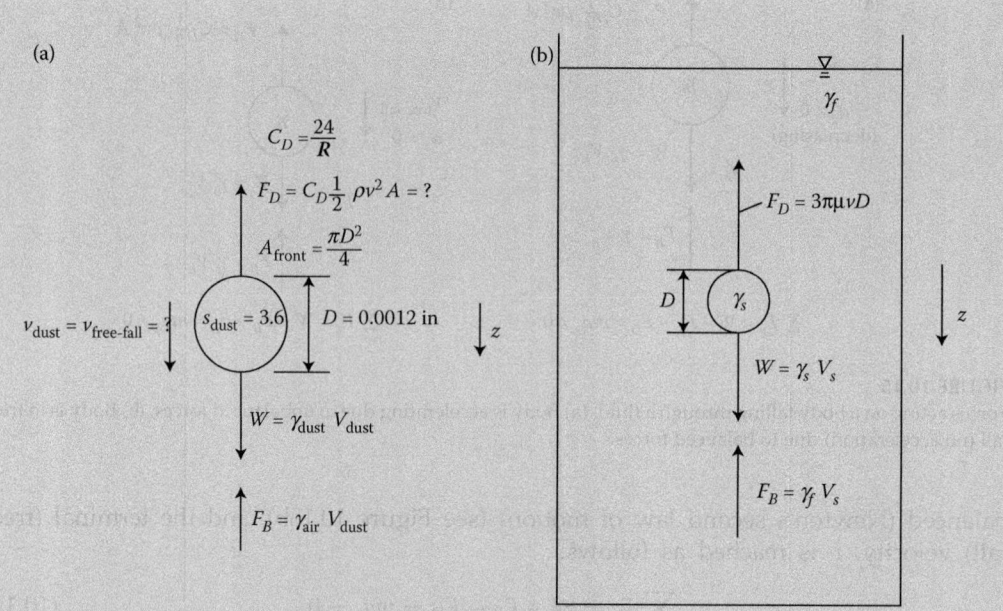

FIGURE EP 10.8
(a) Dust particle falls in the air at free-fall velocity before settling to the ground. (b) Falling sphere viscometer.

$$\gamma_{air} := 0.076472 \frac{lb}{ft^3} \qquad\qquad F_B := \gamma_{air} V_{dust} = 4.004 \times 10^{-14} \, lb$$

Guess value: $F_D := 1 \, lb$

Given

$$W_{dust} - F_B - F_D = 0$$

$$F_D := Find(F_D) = 1.176 \times 10^{-10} \, lb$$

In order to determine the terminal (free-fall) velocity for the dust particle as it falls to the ground, the drag force equation, Equation 10.12, is applied. Since the dust particle is small in diameter, assume creeping flow, $R \le 1$. Furthermore, since the dust particle is assumed to be a sphere, $C_D = 24/R$, this result for C_D from applying Stokes law is applied as follows:

$$slug := 1 \, lb \frac{sec^2}{ft} \qquad \rho_{air} := 0.0023768 \frac{slug}{ft^3} \qquad \mu_{air} := 0.37372 \times 10^{-6} \, lb \frac{sec}{ft^2}$$

$$A_{front} := \frac{\pi \, D_{dust}^2}{4} = 7.854 \times 10^{-9} \, ft^2$$

Guess value: $R := 0.1 \qquad C_D := 1 \qquad v_{dust} := 0.5 \frac{ft}{sec}$

Given

$$F_D = C_D \frac{1}{2} \rho_{air} \cdot v_{dust}^2 \cdot A_{front} \qquad R = \frac{\rho_{air} \cdot v_{dust} \cdot D_{dust}}{\mu_{air}} \qquad C_D = \frac{24}{R}$$

$$\begin{pmatrix} R \\ C_D \\ v_{dust} \end{pmatrix} := Find(R, C_D, v_{dust})$$

$$R = 0.212 \qquad C_D = 113.012 \qquad v_{dust} = 0.334 \frac{ft}{s}$$

Since $R = 0.212$, this confirms the assumption that $R \le 1$. Finally, it is important to note that when solving a numerical solve block (as opposed to an analytical solve block) in Mathcad, especially those problems that have more than one unknown to solve for, the solution procedure is highly sensitive to the guessed values. Thus, one may have to continue to adjust the guess values (for R, C_D, and v_{dust} above) until the numerical procedure converges to a solution. As such, application of the drag force equation and Newton's second law of motion serve as check, respectively, as follows:

$$F_D := C_D \frac{1}{2} \rho_{air} \cdot v_{dust}^2 \cdot A_{front} = 1.176 \times 10^{-10} \, lb$$

$$W_{dust} - F_B - F_D = 0 \, lb$$

(b) Stokes law is applied in the use of a falling sphere viscometer to determine the viscosity of a fluid as follows. Assume the fluid with an unknown viscosity, μ is placed in a very tall clear cylinder, as illustrated in Figure EP 10.8b. A small sphere (γ_s) with a given weight, $W = \gamma_s(\pi D^3/6)$ and a given diameter, D is dropped in the cylinder of fluid. In order to determine the unknown fluid (γ_f) viscosity, μ, first, Equation 10.19 (Newton's second law of motion) is applied in the vertical direction as follows:

$$\sum_z F_z = W - F_B - F_D = ma_z = 0$$

Then, assuming creeping flow at $R \le 1$, Stokes law (Equation 10.17) $F_D = 3\pi\mu v D$ is substituted in Equation 10.19 as follows

$$\sum_z F_z = \gamma_s \frac{\pi D^3}{6} - \gamma_f \frac{\pi D^3}{6} - 3\pi\mu v D = 0$$

Isolating the fluid viscosity, μ and simplifying the equation yields the following:

$$\mu = \frac{D^2(\gamma_s - \gamma_f)}{18v}$$

Furthermore, if the viscosity, μ of the fluid is known, then the expression for the terminal (free-fall) velocity is given as follows:

$$v = \frac{D^2(\gamma_s - \gamma_f)}{18\mu}$$

10.5.3.8 Laminar and Turbulent Flow for Any Shape Body except Round-Shaped Bodies

Although, in general, the drag coefficient, C_D decreases with an increase in R for moderate (laminar) to high (turbulent) values of R (see Figures 10.12 and 10.13), the shape of the body does play an important role in the value of the drag coefficient, C_D. For most geometric shapes (except for round-shaped bodies such as circular cylinders and spheres; these will be addressed in Section 10.5.3.9), for lower values of R, ($R < 10,000$), the drag coefficient, C_D is dependent on the R, with the drag coefficient, C_D generally decreasing with an increase in R, while it becomes independent of R at higher values of R, ($R > 10,000$), where the flow becomes fully turbulent. Figure 10.16 illustrates such a relationship between the drag coefficient, C_D and R for a disk-shaped body that is normal to the direction of flow. It may be noted that a similar relationship exists between the drag coefficient, C_D and R for most geometric shapes that are normal to the direction of flow. However, the relationship between the drag coefficient, C_D and R (and thus the dependency of C_D on R), for most geometric shapes that are parallel to the direction of flow and streamlined bodies, is more accurately illustrated by Figures 10.12 and 10.13 and Tables 10.4 through 10.6, which present the drag coefficient, C_D for numerous geometric shapes at various orientations to the direction of flow, for $R > 10,000$ (unless stated otherwise), for two-dimensional, three-dimensional axisymmetric, and three-dimensional bodies, respectively.

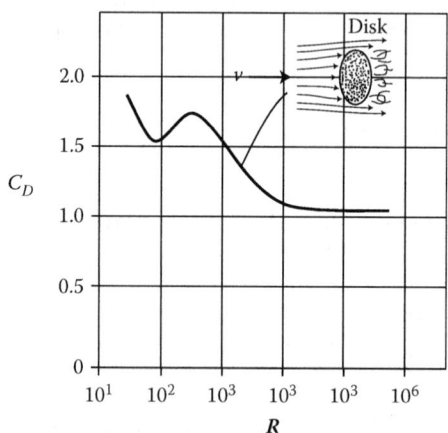

FIGURE 10.16

Variation of the drag coefficient, C_D with the R for a disk-shaped body that is normal to the direction of flow. (From Cengel, Y. A., and J. M., Cimbala, *Fluid Mechanics Fundamentals and Applications*, 3rd ed, McGraw Hill, New York, 2014, Copyright, p. 615. With permission.)

TABLE 10.4

The Drag Coefficient, C_D for Two-Dimensional Bodies at Various Orientations to the Direction of Flow, v for $R > 10,000$

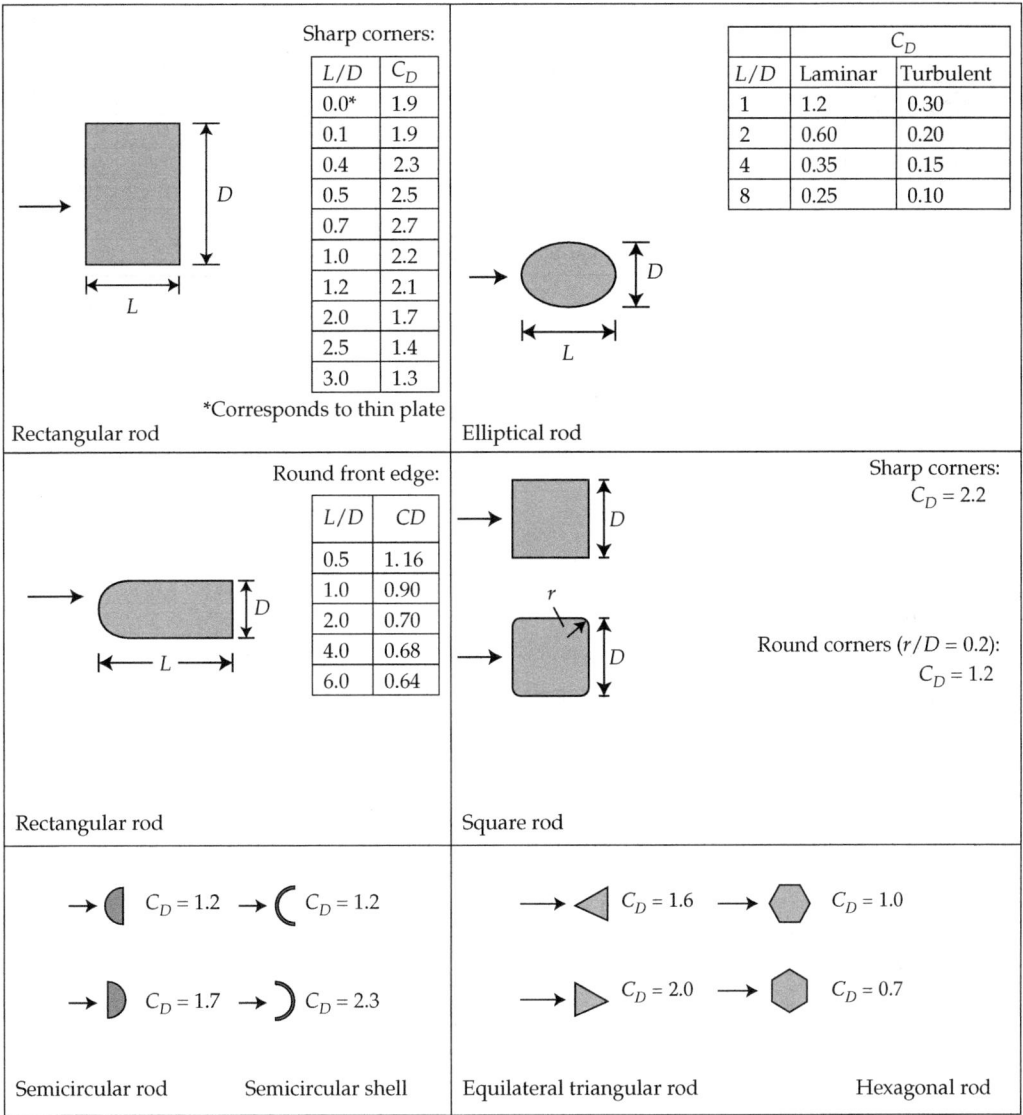

Rectangular rod (Sharp corners):

L/D	C_D
0.0*	1.9
0.1	1.9
0.4	2.3
0.5	2.5
0.7	2.7
1.0	2.2
1.2	2.1
2.0	1.7
2.5	1.4
3.0	1.3

*Corresponds to thin plate

Elliptical rod:

L/D	Laminar	Turbulent
1	1.2	0.30
2	0.60	0.20
4	0.35	0.15
8	0.25	0.10

Rectangular rod (Round front edge):

L/D	CD
0.5	1.16
1.0	0.90
2.0	0.70
4.0	0.68
6.0	0.64

Square rod:
Sharp corners: $C_D = 2.2$
Round corners ($r/D = 0.2$): $C_D = 1.2$

Semicircular rod: $C_D = 1.2$
Semicircular shell: $C_D = 1.2$, $C_D = 1.7$, $C_D = 2.3$

Equilateral triangular rod: $C_D = 1.6$, $C_D = 2.0$
Hexagonal rod: $C_D = 1.0$, $C_D = 0.7$

Source: White, F. M., *Fluid Mechanics*, 7th ed, McGraw Hill, New York, 2011, p. 489; Cengel, Y. A., and J. M., Cimbala, *Fluid Mechanics Fundamentals and Applications*, 3rd ed, McGraw Hill, New York, 2014, p. 619.
Note: C_D is based on the frontal area, $A = bD$, where $b =$ the length of the body in the direction normal to the page.

Furthermore, for laminar flows at $R < 10,000$, and in the case of streamlined bodies, the drag coefficient, C_D will be dependent on R, but independent of the relative surface roughness, ε/L, M, and F. Finally, for laminar flows at $R < 10,000$, and in the case of blunt bodies, the drag coefficient, C_D will be less dependent on R for blunt bodies (where it is mainly a function of E), but independent of the relative surface roughness, ε/L, M, and F. However,

TABLE 10.5

The Drag Coefficient, C_D for Three-Dimensional Axisymmetric Bodies at Given Orientation to the Direction of Flow, v for $R > 10,000$ Unless Otherwise Stated

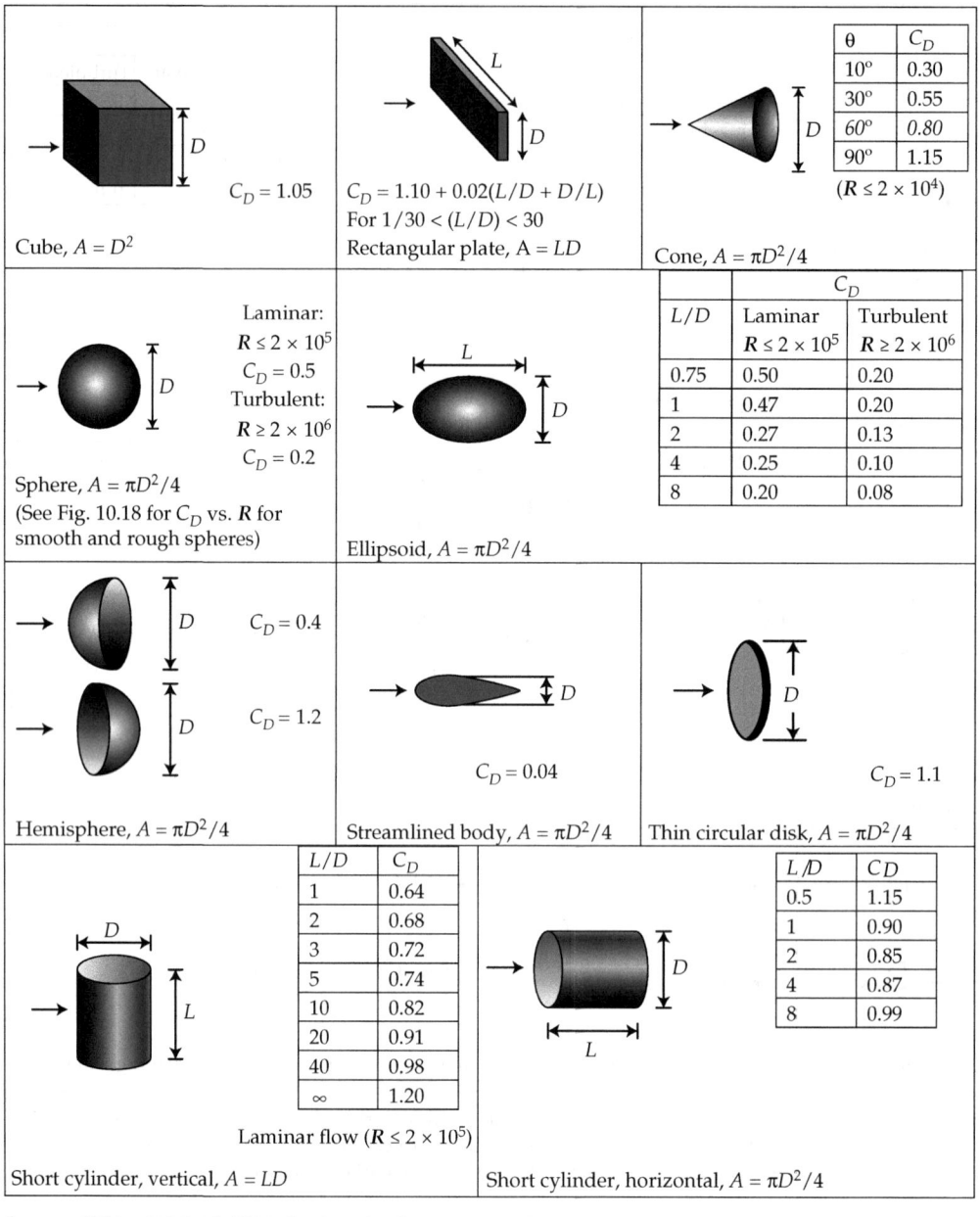

Cube, $A = D^2$: $C_D = 1.05$

Rectangular plate, A = LD:
$C_D = 1.10 + 0.02(L/D + D/L)$
For $1/30 < (L/D) < 30$

Cone, $A = \pi D^2/4$:

θ	C_D
10°	0.30
30°	0.55
60°	0.80
90°	1.15

$(R \le 2 \times 10^4)$

Sphere, $A = \pi D^2/4$
(See Fig. 10.18 for C_D vs. R for smooth and rough spheres)

Laminar:
$R \le 2 \times 10^5$
$C_D = 0.5$
Turbulent:
$R \ge 2 \times 10^6$
$C_D = 0.2$

Ellipsoid, $A = \pi D^2/4$

	C_D	
L/D	Laminar $R \le 2 \times 10^5$	Turbulent $R \ge 2 \times 10^6$
0.75	0.50	0.20
1	0.47	0.20
2	0.27	0.13
4	0.25	0.10
8	0.20	0.08

Hemisphere, $A = \pi D^2/4$: $C_D = 0.4$; $C_D = 1.2$

Streamlined body, $A = \pi D^2/4$: $C_D = 0.04$

Thin circular disk, $A = \pi D^2/4$: $C_D = 1.1$

Short cylinder, vertical, $A = LD$

L/D	C_D
1	0.64
2	0.68
3	0.72
5	0.74
10	0.82
20	0.91
40	0.98
∞	1.20

Laminar flow $(R \le 2 \times 10^5)$

Short cylinder, horizontal, $A = \pi D^2/4$

L/D	CD
0.5	1.15
1	0.90
2	0.85
4	0.87
8	0.99

Source: White, F. M., *Fluid Mechanics*, 7th ed, McGraw Hill, New York, 2011, p .491; Cengel, Y. A., and J. M., Cimbala, *Fluid Mechanics Fundamentals and Applications*, 3rd ed, McGraw Hill, New York, 2014, p. 620; Blevins, R. D., *Applied Fluid Dynamics Handbook*, Von Nostrand Reinhold, New York, 1984, pp. 332, 334.

Note: C_D is based on the frontal area, A.

TABLE 10.6

The Drag Coefficient, C_D for Three-Dimensional Bodies at Given Orientation to the Direction of Flow, v for $R > 10,000$

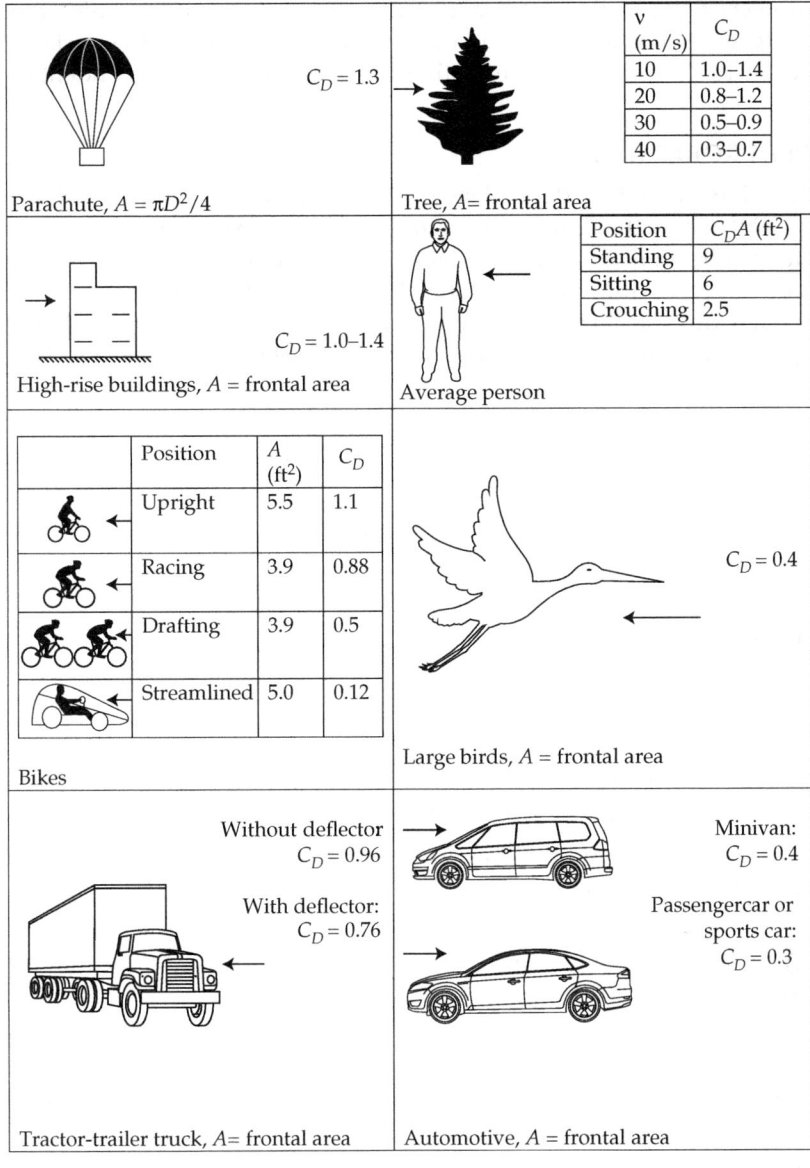

v (m/s)	C_D
10	1.0–1.4
20	0.8–1.2
30	0.5–0.9
40	0.3–0.7

Parachute, $A = \pi D^2/4$ $C_D = 1.3$

Tree, A = frontal area

High-rise buildings, A = frontal area $C_D = 1.0$–1.4

Position	$C_D A$ (ft^2)
Standing	9
Sitting	6
Crouching	2.5

Average person

Bikes

Position	A (ft^2)	C_D
Upright	5.5	1.1
Racing	3.9	0.88
Drafting	3.9	0.5
Streamlined	5.0	0.12

Large birds, A = frontal area $C_D = 0.4$

Tractor-trailer truck, A = frontal area

Without deflector $C_D = 0.96$

With deflector: $C_D = 0.76$

Automotive, A = frontal area

Minivan: $C_D = 0.4$

Passengercar or sports car: $C_D = 0.3$

Source: Adapted from White, F. M., *Fluid Mechanics*, 7th ed, McGraw Hill, New York, 2011, p. 491; Cengel, Y. A., and J. M., Cimbala, *Fluid Mechanics Fundamentals and Applications*, 3rd ed, McGraw Hill, New York, 2014, pp. 620–621; Munson, B. R. et al., *Fundamentals of Fluid Mechanics*, 3rd ed, John Wiley and Sons, New York, 1998, p. 613.

Note: C_D is based on the frontal area, A.

for fully turbulent flows at $R > 10{,}000$, and in the case of blunt bodies, although the drag coefficient, C_D is independent of R, it becomes dependent on the relative surface roughness, ε/L (see Section 10.5.3.11) and M (see Section 10.5.3.12), but still independent of F. And, finally for fully turbulent flows at $R > 10{,}000$, and in the case of streamlined bodies, the drag coefficient, C_D is dependent on R and also becomes dependent on the relative surface roughness, ε/L (see Section 10.5.3.11) and M (see Section 10.5.3.12), but still independent of F.

EXAMPLE PROBLEM 10.9

Figure 10.16 illustrates the independence of drag coefficient, C_D of R at higher values of R, $(R > 10{,}000)$, where the flow becomes fully turbulent, for a disk-shaped body that is normal to the direction of flow. Consider a disk with a diameter of 3 ft. Assume water at 70°F flows over the disk, as illustrated in Figure EP 10.9. (a) Determine the drag coefficient and the drag force if the velocity of flow over the disk is 0.3 ft/sec. (b) Determine the drag coefficient and the drag force if the velocity of flow over the disk is 2.3 ft/sec.

FIGURE EP 10.9
Water flows over a disk.

Mathcad Solution

(a)–(b) The frontal area is used to compute the drag force for the disk, and the drag force is determined by applying Equation 10.12.

(a) The drag coefficient and the drag force for a velocity of flow over the disk of 0.3 ft/sec is determined as follows:

$$\text{slug} := 1 \, \text{lb} \frac{\text{sec}^2}{\text{ft}} \qquad D := 3 \, \text{ft} \qquad A_{\text{front}} := \frac{\pi \cdot D^2}{4} = 7.069 \, \text{ft}^2$$

$$\rho := 1.936 \frac{\text{slug}}{\text{ft}^3} \qquad\qquad \mu := 20.50 \times 10^{-6} \, \text{lb} \frac{\text{sec}}{\text{ft}^2}$$

$$v := 0.3 \frac{\text{ft}}{\text{sec}} \qquad R = \frac{\rho \cdot v \cdot D}{\mu} = 8.5 \times 10^4 \qquad C_D := 1.15$$

Guess value: $F_D := 1 \, \text{lb}$

Given

$$F_D = C_D \frac{1}{2} \rho \cdot v^2 \cdot A_{\text{front}}$$

$$F_D := \text{Find}(F_D) = 0.708 \, \text{lb}$$

(b) The drag coefficient and the drag force for a velocity of flow over the cylinder of 2.3 ft/sec is determined as follows:

$$v := 2.3 \frac{ft}{sec} \qquad\qquad R = \frac{\rho \cdot v \cdot D}{\mu} = 6.516 \times 10^5 \qquad\qquad C_D := 1.15$$

Guess value: $\qquad\qquad F_D := 1\, lb$

Given

$$F_D = C_D \frac{1}{2} \rho \cdot v^2 \cdot A_{front}$$

$$F_D := Find(F_D) = 41.626\, lb$$

Therefore, although the drag coefficient, C_D remains a constant (1.15) with an increase in the velocity, v (from 0.3 ft/sec to 2.3 ft/sec) and thus an increase in the Reynolds number, R (from 8.5×10^4 to 6.516×10^5), the drag force, F_D increases (from 0.708 lb to 41.626 lb) with an increase in velocity, as F_D is directly proportional to v^2.

Although numerous experimental studies have been conducted to determine values for the drag coefficient, C_D, the reported drag coefficients are usually applicable only to flows at high values of R, ($R > 10,000$). As such, Tables 10.4 through 10.6 present the drag coefficient, C_D for numerous geometric shapes at various orientations to the direction of flow, for $R > 10,000$, for two-dimensional, three-dimensional axisymmetric and three-dimensional bodies, respectively. These three tables illustrate that the orientation of the body to the direction of flow plays a significant role on the value of the drag coefficient, C_D. Specifically, these tables illustrate for a high velocity ($R > 10,000$), the drag coefficient, C_D is mainly a function of the shape of the body, resulting from both the orientation of the body to the direction of flow and the smoothness of the edges of the body. An important point to note from these tables is that for blunt bodies with sharp corners (such as a rectangular block in Table 10.5, a flat plate normal to the flow in Table 10.5, or a long square rod in Table 10.4), flow separation occurs at the edges of the front and back surfaces, without any significant change in the type of flow (see Figure 10.11a). As a result, the drag coefficient in such cases is mostly independent of R, as the pressure drag predominates the friction drag. Furthermore, Table 10.4 illustrates that the drag coefficient, C_D of a long square rod may be reduced from 2.2 to 1.2 simply by smoothing (rounding) its corners. Table 10.4 illustrates that the drag coefficient, C_D decreases with an increase in R (from laminar to turbulent) for both a circular cylinder and an elliptical cylinder, while Table 10.5 illustrates that the drag coefficient, C_D decreases with an increase in R (from laminar to turbulent) for both a sphere and an ellipsoid. Table 10.4 illustrates that the drag coefficient, C_D decreases by streamlining a rectangular rod and an elliptical rod. Table 10.5 illustrates that the drag coefficient, C_D decreases by streamlining an ellipsoid, a short vertical cylinder, and a short horizontal cylinder. Table 10.6 illustrates that the drag coefficient, C_D decreases by streamlining a three-dimensional body.

EXAMPLE PROBLEM 10.10

Table 10.4 presents the drag coefficient, C_D for numerous geometric shapes at various orientations to the direction of flow, for $R > 10,000$, for two-dimensional bodies. Consider a 3000-ft-long equilateral triangular rod with a side of 5 ft. Assume water at 70°F flows at a velocity of 2.9 ft/sec over the rod, as illustrated in Figure EP 10.10. (a) Determine the drag coefficient and the drag force if the apex of the triangular rod is normal to the direction of flow as illustrated in Figure EP10.10a. (b) Determine the drag coefficient and the drag force if the base of the triangular rod is normal to the direction of flow as illustrated in Figure EP10.10b.

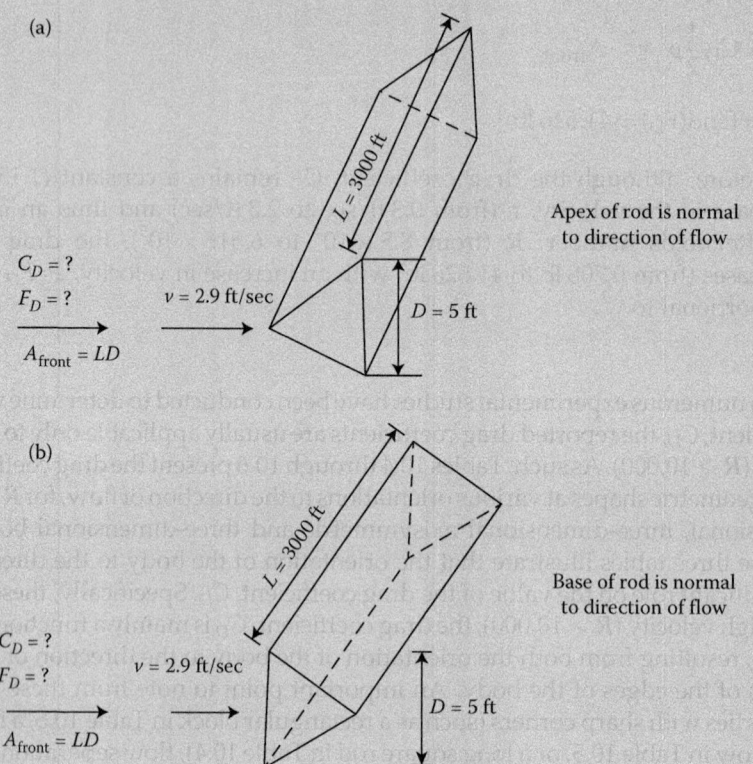

FIGURE EP 10.10
Water flows over an equilateral triangular rod.

Mathcad Solution

(a)–(b) The frontal area is used to compute the drag force for the equilateral triangular rod, and the drag force is determined by applying Equation 10.12.

(a) The drag coefficient and the drag force when the apex of the triangular rod is normal to the direction of flow is determined as follows:

$$\text{slug} := 1\,\text{lb}\frac{\text{sec}^2}{\text{ft}} \qquad L := 3000\,\text{ft} \qquad D := 5\,\text{ft} \qquad A_{front} := L \cdot D = 1.5 \times 10^4\,\text{ft}^2$$

$$\rho := 1.936\frac{\text{slug}}{\text{ft}^3} \qquad\qquad\qquad \mu := 20.50 \times 10^{-6}\,\text{lb}\frac{\text{sec}}{\text{ft}^2}$$

$$v := 2.9\frac{\text{ft}}{\text{sec}} \qquad R = \frac{\rho \cdot v \cdot D}{\mu} = 1.369 \times 10^6 \qquad C_D := 1.5$$

Guess value: $F_D := 1 \, \text{lb}$

Given

$$F_D = C_D \frac{1}{2} \rho \cdot v^2 \cdot A_{front}$$

$$F_D := \text{Find}(F_D) = 1.832 \times 10^5 \, \text{lb}$$

(b) The drag coefficient and the drag force when the base of the triangular rod is normal to the direction of flow is determined as follows:

$$v := 2.9 \frac{\text{ft}}{\text{sec}} \qquad R = \frac{\rho \cdot v \cdot D}{\mu} = 1.369 \times 10^6 \qquad C_D := 2.0$$

Guess value: $F_D := 1 \, \text{lb}$

Given

$$F_D = C_D \frac{1}{2} \rho \cdot v^2 \cdot A_{front}$$

$$F_D := \text{Find}(F_D) = 2.442 \times 10^5 \, \text{lb}$$

Therefore, the drag coefficient, C_D decreases (from 1.5 to 2.0) due to a change in the orientation of the triangular rod. Furthermore, assuming a constant velocity, v (2.9 ft/sec) results in a constant Reynolds number, R (1.369 × 10^6). And, the drag force, F_D increases (from 1.832×10^{-5} lb to 2.442×10^5 lb) due to a change in the orientation of the triangular rod and thus an increase in the drag coefficient, as F_D is directly proportional to C_D.

EXAMPLE PROBLEM 10.11

Table 10.5 presents the drag coefficient, C_D for numerous geometric shapes at various orientations to the direction of flow, for $R > 10,000$, three-dimensional axisymmetric bodies. Consider a hemisphere with a diameter of 3.3 ft. Assume water at 70°F flows at a velocity of 4.7 ft/sec over the hemisphere, as illustrated in Figure EP 10.11. (a) Determine the drag coefficient and the drag force if the curved surface of the hemisphere is normal to the direction of flow as illustrated in Figure EP 10.11a. (b) Determine the drag coefficient and the drag force if the flat surface of the hemisphere is normal to the direction of flow as illustrated in Figure EP 10.11b.

Mathcad Solution

(a)–(b) The frontal area is used to compute the drag force for the hemisphere, and the drag force is determined by applying Equation 10.12.

(a) $C_D = ?$

$F_D = ?$ $v = 4.7$ ft/sec

$A_{front} = \dfrac{\pi D^2}{4}$ $D = 3.3$ ft

(b) $C_D = ?$

$F_D = ?$ $v = 4.7$ ft/sec

$A_{front} = \dfrac{\pi D^2}{4}$ $D = 3.3$ ft

FIGURE EP 10.11
Water flows over a hemisphere.

(a) The drag coefficient and the drag force when the curved surface of the hemisphere is normal to the direction of flow is determined as follows:

$$\text{slug} := 1 \, lb \frac{sec^2}{ft} \qquad D := 3.3 \, ft \qquad A_{front} := \frac{\pi \cdot D^2}{4} = 8.553 \, ft^2$$

$$\rho := 1.939 \frac{slug}{ft^3} \qquad \mu := 20.50 \times 10^{-6} \, lb \frac{sec}{ft^2}$$

$$v := 4.7 \frac{ft}{sec} \qquad R = \frac{\rho \cdot v \cdot D}{\mu} = 1.465 \times 10^6 \qquad C_D := 0.4$$

Guess value: $F_D := 1 \, lb$

Given

$$F_D = C_D \frac{1}{2} \rho \cdot v^2 \cdot A_{front}$$

$$F_D := \text{Find}(F_D) = 73.156 \, lb$$

(b) The drag coefficient and the drag force when the flat surface of the hemisphere is normal to the direction of flow is determined as follows:

$$v := 4.7 \frac{ft}{sec} \qquad R = \frac{\rho \cdot v \cdot D}{\mu} = 1.465 \times 10^6 \qquad C_D := 1.2$$

Guess value: $F_D := 1 \, lb$

Given

$$F_D = C_D \frac{1}{2} \rho \cdot v^2 \cdot A_{front}$$

$$F_D := \text{Find}(F_D) = 219.467 \, lb$$

Therefore, the drag coefficient, C_D decreases (from 0.4 to 1.2) due to a change in the orientation of the hemisphere. Furthermore, assuming a constant velocity, v (4.7 ft/sec) results in a constant Reynolds number, R (1.465×10^6). And, the drag force, F_D increases (from 73.156 lb to 219.567 lb) due to a change in the orientation of the hemisphere and thus an increase in the drag coefficient, as F_D is directly proportional to C_D.

EXAMPLE PROBLEM 10.12

A rectangular street sign that is 2 ft wide and 0.5 ft high is secured to a 10-ft-long circular rod (post) with a diameter of 2 in that is secured into the ground, as illustrated in Figure EP 10.12. The sign is at sea level, and is designed to withstand a maximum wind speed of 55 mph (air at standard atmosphere at sea level, $\rho = 0.0023768$ slugs/ft³, $\mu = 0.37372 \times 10^{-6}$ lb-sec/ft²). (a) Determine the drag force on the street sign and pole. (b) Determine the reaction force acting on the sign and pole due to the thrust force of the maximum wind flow (anchoring force required to keep the sign and pole from detaching from the ground).

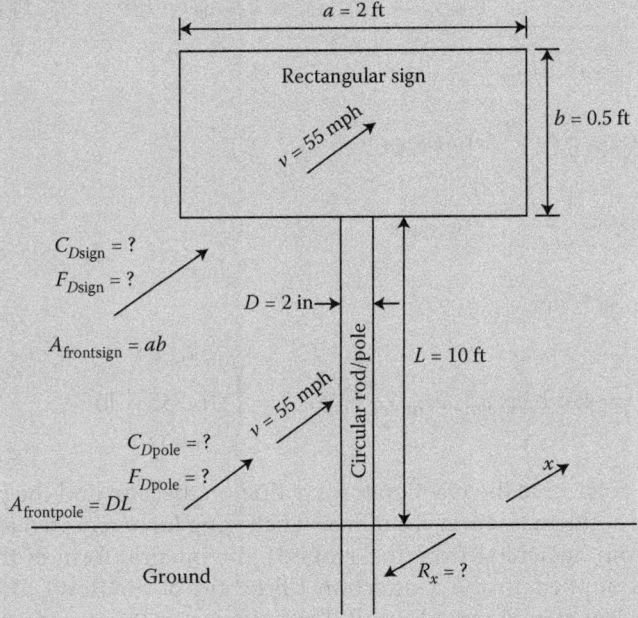

FIGURE EP 10.12
Air flows over a rectangular sign secured to the ground by a circular rod (post).

Mathcad Solution

(a) The frontal area is used to compute the drag force for both the sign and the pole, and the drag force on the street sign and pole is determined by applying Equation

10.12. The drag coefficient for the sign is given Table 10.5, and the drag coefficient for the pole is given in Figure 10.12.

$$slug := 1\,lb\,\frac{sec^2}{ft} \qquad\qquad v := 55\,mph = 80.667\,\frac{ft}{s}$$

$$\rho := 0.0023768\,\frac{slug}{ft^3} \qquad\qquad \mu := 0.37372 \times 10^{-6}\,lb\,\frac{sec}{ft^2}$$

$$a_{sign} := 2\,ft \qquad b_{sign} := 0.5\,ft \qquad A_{frontsign} := a_{sign} \cdot b_{sign} = 1\,ft^2$$

$$R_{sign} = \frac{\rho \cdot v \cdot b_{sign}}{\mu} = 2.565 \times 10^5 \qquad \frac{a_{sign}}{b_{sign}} = 4 \qquad C_{Dsign} := 1.17$$

$$D_{pole} := 2\,in = 0.167\,ft \qquad\qquad L_{pole} := 10\,ft$$

$$A_{frontpole} := D_{pole}\,L_{pole} = 1.667\,ft^2 \qquad R_{pole} = \frac{\rho \cdot v \cdot D_{pole}}{\mu} = 8.55 \times 10^4$$

$$C_{Dpole} := 1.3$$

Guess value $\qquad\qquad F_{Dsign} := 1\,lb \qquad F_{Dpole} := 1\,lb \qquad F_{Dtotal} := 2\,lb$

Given

$$F_{Dsign} = C_{Dsign}\frac{1}{2}\rho \cdot v^2 \cdot A_{frontsign}$$

$$F_{Dpole} = C_{Dpole}\frac{1}{2}\rho \cdot v^2 \cdot A_{frontpole}$$

$$F_{Dtotal} = F_{Dsign} + F_{Dpole}$$

$$\begin{pmatrix} F_{Dsign} \\ F_{Dpole} \\ F_{Dtotal} \end{pmatrix} := Find(F_{Dsign}, F_{Dpole}, F_{Dtotal}) = \begin{pmatrix} 9.048 \\ 16.755 \\ 25.803 \end{pmatrix} lb$$

(b) In order to determine the reaction force acting on the sign and the pole due to the thrust force of the maximum wind flow (anchoring force required to keep the sign and pole from detaching from the ground), the integral form of the momentum equation is applied in the x-direction (direction of air flow) for the sign and pole, as illustrated in Figure EP 10.12. The pressure and viscous forces in the direction of wind flow are represented by the drag force. Also, there is no gravitational force component along the x-axis because the wind flow (also negligible weight) is assumed to be horizontal.

$$\sum F_x = F_{Dtotal} + R_x = 0$$

Guess value: $\qquad R_x := 100\,lb$

Given

$F_{Dtotal} - R_x = 0$

$R_x := Find(R_x) = 25.803 \text{ lb}$

Thus, the anchoring force required to keep the sign and pole from detaching from the ground, R_x acts to left as assumed and illustrated in Figure EP 10.12.

EXAMPLE PROBLEM 10.13

Table 10.6 presents the drag coefficient, C_D for numerous geometric shapes at various orientations to the direction of flow, for $R > 10,000$, for three-dimensional bodies. A 5.5-ft-tall (assume a diameter of 2 ft) person weighing 150 lbs drops in the air (standard atmosphere at sea level, $\rho = 0.0023768 \text{ slugs/ft}^3$, $\mu = 0.37372 \times 10^{-6} \text{ lb-sec/ft}^2$, $\gamma = 0.076472 \text{ lb/ft}^3$) to the ground in a parachute with a diameter of 12 ft, weighing 15 lbs, and with ropes 10 ft high, as illustrated in Figure EP 10.13. (a) Determine the terminal (free-fall) velocity of the person in the parachute as he/she falls to the ground.

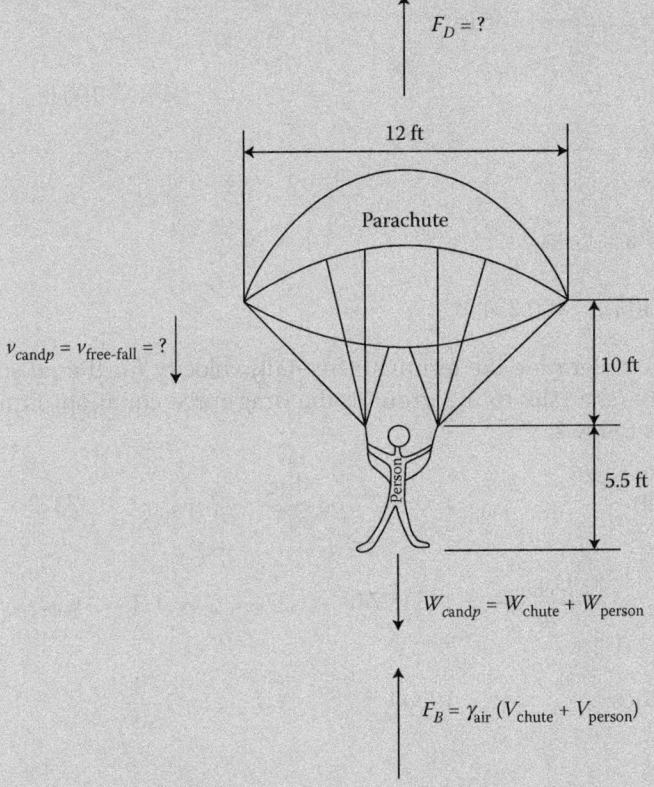

FIGURE EP 10.13
Person in a parachute falling in the air at free-fall velocity before hitting the ground.

Mathcad Solution

(a) In order to determine the drag force, Equation 10.19 (Newton's second law of motion) is applied in the vertical direction as follows:

$$\sum_z F_z = W_{candp} - F_B - F_D = ma_z = 0$$

$D_{chute} := 12 \text{ ft}$ $\qquad$ $h_{rope} := 10 \text{ ft}$ $\qquad$ $W_{chute} := 15 \text{ lb}$

$$V_{chute} := \frac{\pi \cdot D_{chute}^3}{12} + \frac{1}{3}\left(\frac{\pi \cdot D_{chute}^2 \cdot h_{rope}}{4}\right) = 829.38 \text{ ft}^3$$

$D_{person} := 2 \text{ ft}$ $\qquad$ $h_{person} := 5.5 \text{ ft}$ $\qquad$ $W_{person} := 150 \text{ lb}$

$$V_{person} := \frac{\pi \cdot D_{person}^2}{4} h_{person} = 17.279 \text{ ft}^3$$

$$W_{candp} := W_{chute} + W_{person} = 165 \text{ lb}$$

$$V_{candp} := V_{chute} + V_{person} = 846.659 \text{ ft}^3 \qquad\qquad \gamma_{air} := 0.076472 \frac{\text{lb}}{\text{ft}^3}$$

$$F_B := \gamma_{air} V_{candp} = 64.746 \text{ lb}$$

Guess value: $\qquad\qquad\qquad\qquad\qquad\qquad\qquad F_D := 100 \text{ lb}$

Given

$$W_{candp} - F_B - F_D = 0$$

$$F_D := \text{Find}(F_D) = 100.254 \text{ lb}$$

In order to determine the terminal (free-fall) velocity for the person in the parachute as he/she falls to the ground, the drag force equation, Equation 10.12, is applied as follows:

$$\text{slug} := 1 \text{ lb} \frac{\text{sec}^2}{\text{ft}} \qquad \rho_{air} := 0.0023768 \frac{\text{slug}}{\text{ft}^3} \qquad \mu_{air} := 0.37372 \times 10^{-6} \text{ lb} \frac{\text{sec}}{\text{ft}^2}$$

$$A_{frontchute} := \frac{\pi \cdot D_{chute}^2}{4} = 113.097 \text{ ft}^2 \qquad C_{Dchute} := 1.3 \quad AC_{Dperson} := 9 \text{ ft}^2$$

Guess values: $\qquad R := 10000 \qquad\qquad v_{candp} := 1\frac{\text{ft}}{\text{sec}}$

Given

$$F_D = \frac{1}{2}\rho_{air} v_{candp}^2 (C_{Dchute} \cdot A_{frontchute} + AC_{Dperson})$$

$$R = \frac{\rho_{air} \, v_{candp} \cdot D_{chute}}{\mu_{air}}$$

$$\begin{pmatrix} R \\ v_{candp} \end{pmatrix} := Find(R, v_{candp})$$

$$R = 1.775 \times 10^6 \qquad\qquad v_{candp} = 23.253 \frac{ft}{s}$$

Since $R = 1.775 \times 10^6$, this confirms the assumption that $R > 10{,}000$. Furthermore, application of the drag force equation and Newton's second law of motion serve as check, respectively, as follows:

$$F_D := \frac{1}{2} \cdot \rho_{air} \cdot v_{candp}^2 \cdot (C_{Dchute} \cdot A_{frontchute} + AC_{Dperson}) = 100.254 \, lb$$

$$W_{candp} - F_B - F_D = 0 \, lb$$

10.5.3.9 *Laminar and Turbulent Flow for Round-Shaped Bodies (Circular Cylinder or Sphere)*

For flows at moderate (laminar) to high (turbulent) values of R around/across round-shaped bodies such as circular cylinders and spheres, the drag coefficient, C_D, in general, decreases with an increase in R. Figure 10.17 illustrates that unlike Figure 10.16, which is for most other geometric shapes, the drag coefficient, C_D, in general, continues to

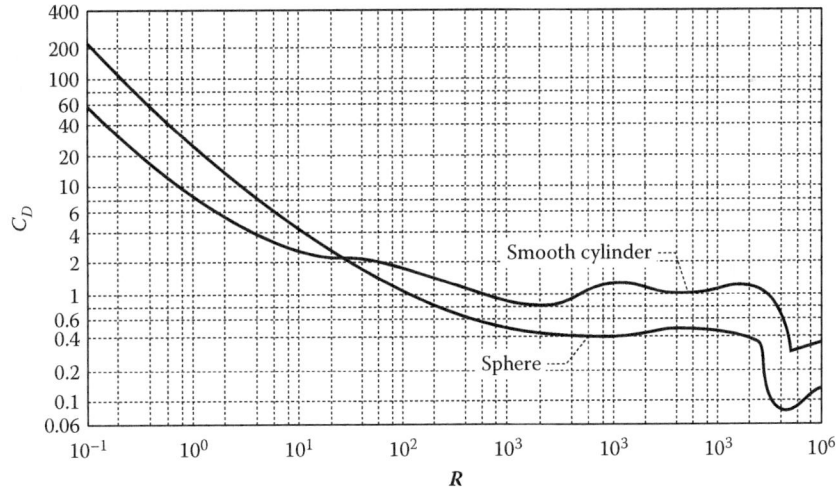

FIGURE 10.17
Variation of the drag coefficient, C_D with the R for smooth circular cylinders and spheres. (Adapted from Schlichting, H., *Boundary Layer Theory*, 7th ed, McGraw Hill, New York, 1979, p. 17.)

decrease with an increase in R for the full range of values for R, low (creeping; $R < 1$), moderate (laminar), and high (turbulent). For flow across circular cylinders and spheres, the critical value for R is 200,000; thus, the flow is defined as laminar for $R \leq 200,000$ and turbulent for $R \geq 200,000$, where $R = \rho v L / \mu = \rho v D / \mu$, where the characteristic length, L is the external diameter of the circular cylinder or sphere. The nature of the flow around a circular cylinder or sphere has a significant effect on the total drag coefficient, C_D, where both the friction drag and the pressure drag can be significant. For flows at low values of R ($R < 10$), the drag force is mostly due to friction drag. On the other hand, for flows at intermediate values of R ($5,000 < R < 10$), both effects of drag (friction and pressure) are significant. And, for flows at high values of R ($R > 5,000$), the drag force is mostly due to pressure drag (and thus a function of E). Figure 10.17 illustrates that for creeping flows at $R \leq 1$, where the fluid smoothly wraps around the circular cylinder and sphere, the drag coefficient, C_D will vary inversely with R. One may note that Figure 10.14 already illustrated that regardless of the shape of the body (blunt or streamlined; sphere, a hemisphere, a circular disk normal to the direction of flow, and a circular disk parallel to the direction of flow) for creeping flows at $R \leq 1$, C_D varies inversely with R, where Figure 10.17 illustrates this observation specifically for a circular cylinder and sphere. Furthermore, for creeping flows at $R \leq 1$, the fluid completely wraps around the cylinder or sphere (or other geometric shapes) and follows the body and thus there is no significant flow separation; for a sphere, there will be no flow separation. At higher velocities, while the fluid will still be attached to the cylinder or sphere on the frontal side, it will not remain attached to the top or the bottom of the body, as the flow is too fast to remain attached. Thus, flow separation occurs in the region behind the body at higher velocities, where a periodic vortex (vortex shedding) forms in the wake region. Specifically, at an $R \cong 10$, separation will begin to occur on the rear of the body, with vortex shedding taking place at an $R \cong 90$, and an increase in the region of separation with an increase in R up to an $R \cong 1000$. At an $R \cong 1000$, the total drag is mostly ($\cong 95\%$) due to pressure drag. For an R in the range of $10 < R < 1000$, although the drag coefficient, C_D continues to decrease with an increase in R, the drag force, F_D does not necessarily decrease, because the drag force is proportional to v^2, as $F_D = C_D(1/2 \rho v^2 A)$. Thus, the increase in velocity at higher values of R typically offsets the decrease in the drag coefficient, C_D. For an R in the moderate range of $1000 < R < 100,000$, the drag coefficient, C_D will remain relatively constant, which is characteristic behavior of such blunt bodies (circular cylinder and sphere); the flow that is attached to the body (boundary layer) is laminar, while the flow in the separated region past the body is highly turbulent with a wide turbulent wake region. Finally, in the high range of R, $100,000 < R < 1,000,000$ (typically at $R \cong 200,000$) the drag coefficient, C_D will suddenly drop in value due to turbulent flow in the boundary layer, which causes the flow separation point to occur further downstream of the rear of body and thus reduces both the wake and the magnitude of the pressure drag. It may be noted that, in contrast, in the case of streamlined bodies for which the boundary layer is turbulent, the drag coefficient, C_D increases mostly as a result of friction drag.

Finally, for laminar flows at $R \leq 200,000$, the drag coefficient, C_D will be independent of the relative surface roughness, ε/L, M, and F. However, for turbulent flows at $R \geq 200,000$, the drag coefficient, C_D becomes dependent on the relative surface roughness, ε/L (see Section 10.5.3.11) and M (see Section 10.5.3.12), but still independent of F. Furthermore, Tables 10.4, and 10.5 include the drag coefficient, C_D for laminar and turbulent flow of a circular cylinder and sphere, respectively.

EXAMPLE PROBLEM 10.14

Figure 10.17 illustrates that for flows at moderate (laminar) to high (turbulent) values of R around/across round-shaped bodies such as circular cylinders and spheres, the drag coefficient, C_D generally decreases with an increase in R. Consider a 5-mile-long oil pipeline with a diameter of 3 ft that is submerged in the ocean. Assume the ocean water is at 68°F flows over the pipeline, as illustrated in Figure EP 10.14. (a) Determine the drag coefficient and the drag force if the velocity of flow over the pipeline is 0.00025 ft/sec. (b) Determine the drag coefficient and the drag force if the velocity of flow over the pipeline is 2.5 ft/sec.

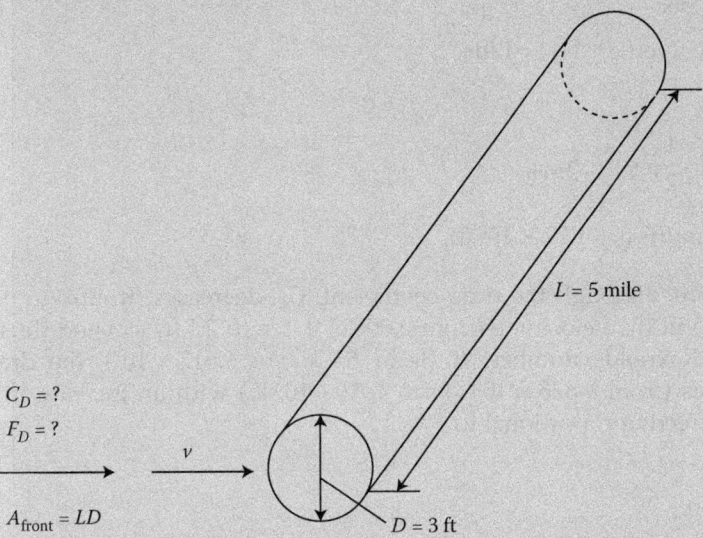

FIGURE EP 10.14
Ocean water flows over a circular pipeline.

Mathcad Solution

(a)–(b) The frontal area is used to compute the drag force for the circular pipeline, and the drag force is determined by applying Equation 10.12. The fluid properties for the ocean water (saltwater) are given in Table A.4 in Appendix A.

(a) The drag coefficient and the drag force for a velocity of flow over the cylinder of 0.00025 ft/sec is determined as follows:

$$slug := 1\,lb\,\frac{sec^2}{ft} \qquad L := 5\,mile = 2.64 \times 10^4\,ft \quad D := 3\,ft$$

$$A_{front} := L\,D = 7.92 \times 10^4\,ft^2 \quad \rho := 1.985\,\frac{slug}{ft^3} \qquad \mu := 22.50 \times 10^{-6}\,lb\,\frac{slug}{ft^2}$$

$$v := 0.00025\,\frac{ft}{sec} \qquad R := \frac{\rho \cdot v \cdot D}{\mu} = 66.167 \qquad C_D := 2$$

Guess value: $\qquad F_D := 1\,lb$

Given

$$F_D = C_D \frac{1}{2} \rho \cdot v^2 \cdot A_{front}$$

$$F_D := Find(F_D) = 9.826 \times 10^{-3} \, lb$$

(b) The drag coefficient and the drag force for a velocity of flow over the cylinder of 2.5 ft/sec are determined as follows:

$$v := 2.5 \frac{ft}{sec} \qquad R := \frac{\rho \cdot v \cdot D}{\mu} = 6.617 \times 10^5 \qquad C_D := 0.35$$

Guess value: $F_D := 1 \, lb$

Given

$$F_D = C_D \frac{1}{2} \rho \cdot v^2 \cdot A_{front}$$

$$F_D := Find(F_D) = 1.72 \times 10^5 \, lb$$

Therefore, although the drag coefficient, C_D decreases (from 2 to 0.35) with an increase in the velocity, v (from 0.00025 ft/sec to 2.5 ft/sec) and thus an increase in the Reynolds number, R (from 66.167 to 6.617×10^5), the drag force, F_D increases (from 9.826×10^{-3} lb to 1.72×10^5 lb) with an increase in velocity, as F_D is directly proportional to v^2.

EXAMPLE PROBLEM 10.15

Figure 10.17 illustrates that for flows at moderate (laminar) to high (turbulent) values of R around/across round-shaped bodies such as circular cylinders and spheres, the drag coefficient, C_D generally decreases with an increase in R. Consider a decorative spherical dome structure with a diameter of 4 ft that is secured into the ground, as illustrated in Figure EP 10.15. The dome is at sea level, and is designed to withstand a maximum wind speed of 30 mph (air at standard atmosphere at sea level, $\rho = 0.0023768$ slugs/ft^3, $\mu = 0.37372 \times 10^{-6}$ lb-sec/ft^2). (a) Determine the drag force on the dome.

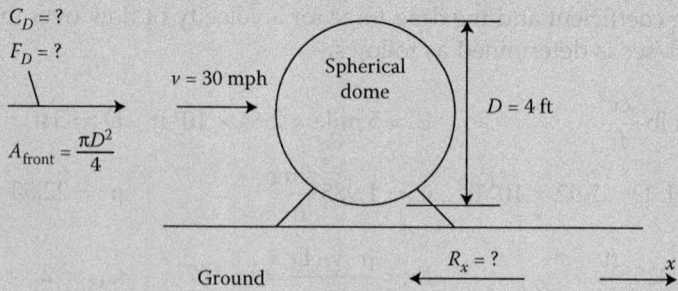

FIGURE EP 10.15
Air flows over a decorative spherical dome structure secured into the ground.

(b) Determine the reaction force acting on the dome due to the thrust force of the maximum wind flow (anchoring force required to keep dome from detaching from the ground).

Mathcad Solution

(a) The frontal area is used to compute the drag force for the spherical dome, and the drag force is determined by applying Equation. 10.12 as follows:

$$slug := 1\,lb\,\frac{sec^2}{ft} \qquad\qquad v := 30\,mph = 44\,\frac{ft}{s}$$

$$\rho := 0.0023768\,\frac{slug}{ft^3} \qquad\qquad \mu := 0.37372 \cdot 10^{-6}\,lb\,\frac{sec}{ft^2}$$

$$D := 4\,ft \qquad\qquad A_{front} := \frac{\pi \cdot D^2}{4} = 12.566\,ft^2$$

$$R := \frac{\rho \cdot v \cdot D}{\mu} = 1.119 \times 10^6 \qquad\qquad C_D := 0.14$$

Guess value: $\qquad\qquad F_D := 1\,lb$

Given

$$F_D = C_D \frac{1}{2}\rho \cdot v^2 \cdot A_{front}$$

$$F_D := Find(F_D) = 4.048\,lb$$

(b) In order to determine the reaction force acting on the dome due to the thrust force of the maximum wind flow (anchoring force required to keep dome from detaching from the ground), the integral form of the momentum equation is applied in the x-direction (direction of air flow) for the dome, as illustrated in Figure EP 10.15. The pressure and viscous forces in the direction of wind flow are represented by the drag force. And, there is no gravitational force component along the x-axis because the wind flow (also negligible weight) is assumed to be horizontal.

$$\sum F_x = F_D + R_x = 0$$

Guess value: $\qquad\qquad R_x := 10\,lb$

Given

$$F_D - R_x = 0$$

$$R_X := Find(R_X) = 4.048\,lb$$

Thus, the anchoring force required to keep the dome detaching from the ground, R_x acts to left as assumed and illustrated in Figure EP 10.15.

10.5.3.10 Laminar and Turbulent Flow with Wave Action at the Free Surface for Any Shape

For body shapes in general, with a wave action at the free surface, the drag coefficient, C_D for laminar or turbulent flow is dependent on R, L_i/L, and F (see Section 10.5.3.13), but independent of ε/L and M. For instance, a ship moving in a body of water may produce waves, which are caused by the product of the force and the velocity of the ship (where the energy production, or the power is force times velocity) and result in drag. While the nature of the waves produced by the ship depends on F (and thus the velocity of the ship) and the shape of the ship, the drag coefficient, C_D depends on both the F (which represents the wave-making effects and thus, the pressure effects) and the R (which represents the viscous effects). One may note that the wave and the viscous effects are often modeled separately, with the total drag being the sum of the two individual effects.

10.5.3.11 The Importance of the Relative Surface Roughness

The drag coefficient, C_D will vary with the relative surface roughness, ε/L, depending mainly on the shape of the body and velocity of flow. The drag coefficient, C_D will be independent of the relative surface roughness, ε/L in the case of low and intermediate flow velocities (laminar flow, which is dependent on R). However, the drag coefficient, C_D will be dependent on the relative surface roughness, ε/L in the case of high velocities (turbulent flow, which is dependent on M, but less dependent on R). For turbulent flow, the drag coefficient, C_D, in general (and especially for streamlined bodies), increases with an increase in the relative surface roughness, ε/L.

One may recall that for a streamlined body, the drag force, F_D is mostly due to the friction drag, F_f (due to the streamlined shape of the body) and thus the drag coefficient, C_D is mostly due to the friction-drag coefficient, C_f, which is highly dependent on the R (due to the streamlined shape of the body). For low values of R (laminar flow), the friction-drag coefficient, C_f will be independent of the relative surface roughness, ε/L, whereas for high values of R (turbulent flow), it will be highly dependent on the relative surface roughness, ε/L. Therefore, the friction-drag coefficient, C_f behaves similarly to the friction factor, f in pipe flow (see Chapter 8), where its value is a function of the flow regime (laminar or turbulent). Furthermore, the friction-drag coefficient, C_f is highly dependent on viscosity and thus, increases with increasing viscosity.

As noted above, for turbulent flow, the drag coefficient, C_D, generally increases with an increase in the relative surface roughness, ε/L. However, for blunt bodies (circular cylinder or sphere), an increase in the relative surface roughness, ε/L can actually cause a decrease in drag coefficient, C_D (and thus a decrease in drag force) for a specific range of R, as illustrated in Figure 10.18 for a sphere of diameter, D (where the relative surface roughness, $\varepsilon/L = \varepsilon/D$). Specifically, the roughening of sphere induces the boundary layer into turbulence at low values of R, causing the wake to become narrow and thus significantly reducing the pressure drag. Figure 10.18 illustrates that roughening the surface of a sphere may be used to reduce the drag for a very specific range of R. For instance, golf balls are roughened (dimpled) to induce turbulence at a lower R (the critical R for a roughened golf ball is $R \cong 40,000$) in order to take advantage of the sharp drop in the drag coefficient, C_D at the onset of turbulence in the boundary layer. Furthermore, the occurrence of turbulent flow at such a low R will reduce the drag coefficient, C_D in half; this translates into a longer distance of travel for the golf ball for a given hit.

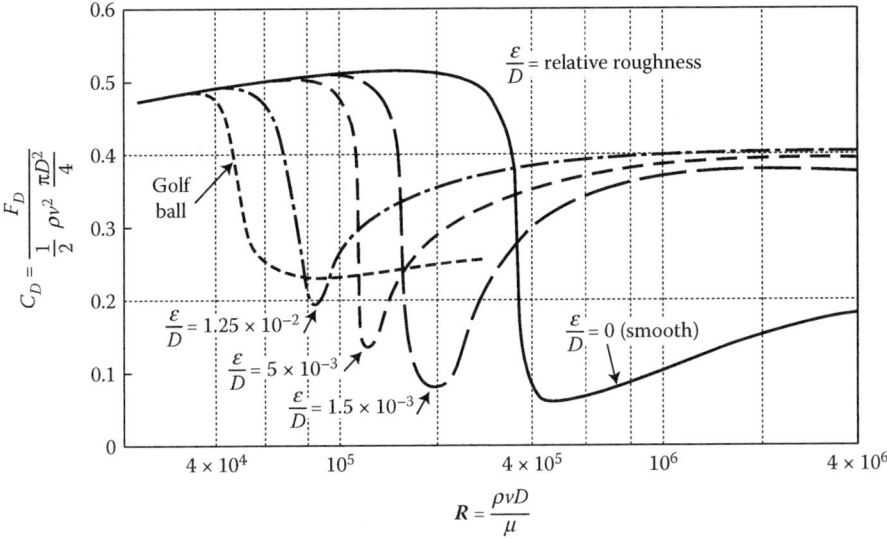

FIGURE 10.18
Variation of the drag the drag coefficient, C_D with the R and the relative surface roughness, ε/L for a sphere. (Adapted from Blevins, R. D., *Applied Fluid Dynamics Handbook*, Von Nostrand Reinhold, New York, 1984, p. 339.)

EXAMPLE PROBLEM 10.16

Figure 10.18 illustrates that for a specific range of R, for a sphere of diameter, D, an increase in the relative surface roughness, ε/L results in a decrease in drag coefficient, C_D. Consider a golf ball with a diameter or 1.68 in that weighs 1.62 oz. Assume the golf ball is hit and travels at a speed of 155 mph (air at standard atmosphere at sea level, $\rho = 0.0023768$ slugs/ft^3, $\mu = 0.37372 \times 10^{-6}$ lb-sec/ft^2), as illustrated in Figure EP 10.16. (a) Determine the drag force and the rate of deceleration of a standard (roughened/dimpled) golf ball. (b) Determine the drag force and the rate of deceleration of a smooth golf ball.

Mathcad Solution

(a)–(b) The frontal area is used to compute the drag force for the golf ball, and the drag force is determined by applying Equation 10.12. The rate of deceleration for the golf ball is determine by applying Newton's second law of motion in the x-direction. The pressure and viscous forces in the direction of velocity (the x-direction) are represented by the drag force. And there is no gravitational force component along the x-axis because the weight of the golf ball is assumed to be negligible; thus, $\Sigma F_x = F_D = ma_x$.

(a) The drag force and the rate of deceleration of a standard (roughened/dimpled) golf ball are determined as follows:

$$\text{slug} := 1\,\text{lb}\,\frac{\sec^2}{\text{ft}} \qquad\qquad v := 155\,\text{mph} = 227.333\,\frac{\text{ft}}{s}$$

$$\rho := 0.0023768\,\frac{\text{slug}}{\text{ft}^3} \qquad\qquad \mu := 0.37372 \times 10^{-6}\,\text{lb}\,\frac{\sec}{\text{ft}^2}$$

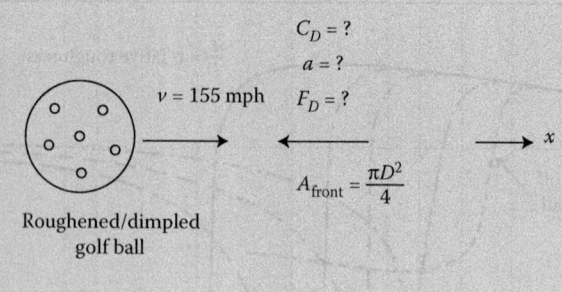

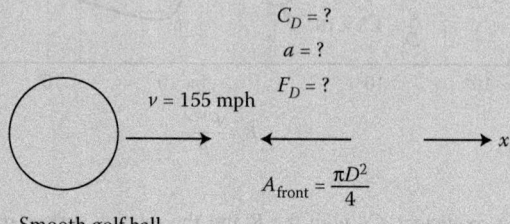

FIGURE EP 10.16
Roughened/dimpled golf ball and a smooth golf ball travel at a speed of 155 mph.

$D := 1.68 \text{ in} = 0.14 \text{ ft}$ $A_{front} := \dfrac{\pi \cdot D^2}{4} = 0.015 \text{ ft}^2$

$R := \dfrac{\rho \cdot v \cdot D}{\mu} = 2.024 \times 10^5$ $W := 1.62 \text{ oz} = 0.101 \text{ lb}$

$g := 32.174 \dfrac{\text{ft}}{\text{sec}^2}$ $m := \dfrac{W}{g} = 3.147 \times 10^{-3} \dfrac{\text{s}^2 \cdot \text{lb}}{\text{ft}}$

$C_{Drough} := 0.24$

Guess value: $F_{Drough} := 1 \text{ lb}$ $a_{rough} := 1 \dfrac{\text{ft}}{\text{sec}^2}$

Given

$F_{Drough} = C_{Drough} \dfrac{1}{2} \rho \cdot v^2 \cdot A_{front}$ $-F_{Drough} = -m \cdot a_{rough}$

$\begin{pmatrix} F_{Drough} \\ a_{rough} \end{pmatrix} := \text{Find}(F_{Drough}, a_{rough})$

$F_{Drough} = 0.227 \text{ lb}$ $a_{rough} = 72.103 \dfrac{\text{ft}}{\text{s}^2}$

(b) The drag force and the rate of deceleration of a smooth golf ball are determined as follows:

$C_{\text{Dsmooth}} := 0.51$

Guess value: $F_{\text{Dsmooth}} := 1\,\text{lb}$ $a_{\text{smooth}} := 1\,\dfrac{\text{ft}}{\text{sec}^2}$

Given

$F_{\text{Dsmooth}} = C_{\text{Dsmooth}}\dfrac{1}{2}\rho \cdot v^2 \cdot A_{\text{front}}$ $-F_{\text{Dsmooth}} = -m \cdot a_{\text{smooth}}$

$\begin{pmatrix} F_{\text{Dsmooth}} \\ a_{\text{smooth}} \end{pmatrix} := \text{Find}(F_{\text{Dsmooth}},\, a_{\text{smooth}})$

$F_{\text{Dsmooth}} = 0.482\,\text{lb}$ $a_{\text{smooth}} = 153.22\,\dfrac{\text{ft}}{\text{s}^2}$

Therefore, because for a given velocity (and thus a given R), the drag coefficient, C_D for the smooth golf ball is about 2 times that for the roughened golf ball, the corresponding drag force for the smooth golf ball is also about 2 times that for the roughened golf ball, as the drag force, F_D is directly proportional to the drag coefficient, C_D (see Equation 10.12). Furthermore, the rate of deceleration of the smooth golf ball is also about 2 times the rate of deceleration of the roughened golf ball, as the rate of deceleration is directly proportional to the drag force, F_D (or summation of forces, in general) (Newton's second law of motion) for a given mass.

10.5.3.12 The Importance of the Mach Number, M

The drag coefficient, C_D will vary with Mach number, $M = C^{0.5} = v/c$, depending mainly on the shape of the body and velocity of flow. The Mach number, M is defined as follows:

$$M = \sqrt{C} = \frac{v}{\sqrt{E_v/\rho}} = \frac{v}{c} \tag{10.20}$$

where C is the Cauchy number and c is the sonic velocity; thus, the elastic force (compressibility effects) will play an important role. The drag coefficient, C_D will be independent of M in the case of low and intermediate flow velocities (laminar flow, which is dependent on R). However, the drag coefficient, C_D will be dependent on M in the case of high velocities (turbulent flow, which is dependent on the relative surface roughness, ε/L, and less dependent on $R = \rho v L/\mu$), where compressibility effects become important. One may note that the introduction of the M effects, in addition to the R, in the evaluation of the drag coefficient, C_D complicates matters because their effects are often closely related, as both are directly proportional to the upstream velocity, v. Therefore, a change in the drag coefficient, C_D due to a change in velocity is due to both R and M. While the exact dependence of the drag coefficient, C_D on R and M is generally not very simple, the following simplification is typically made. For a low M, the drag coefficient, C_D is basically independent of M (compressibility effects are insignificant), while for a larger M, the drag coefficient, C_D may be highly dependent on M, and less dependent on R. Figure 10.19 illustrates such a relationship for two two-dimensional shapes (a blunt or square shape and a streamlined or elliptical shape) for subsonic flow ($M = v/c < 1$, where c is the sonic velocity; see Chapter 7). Specifically,

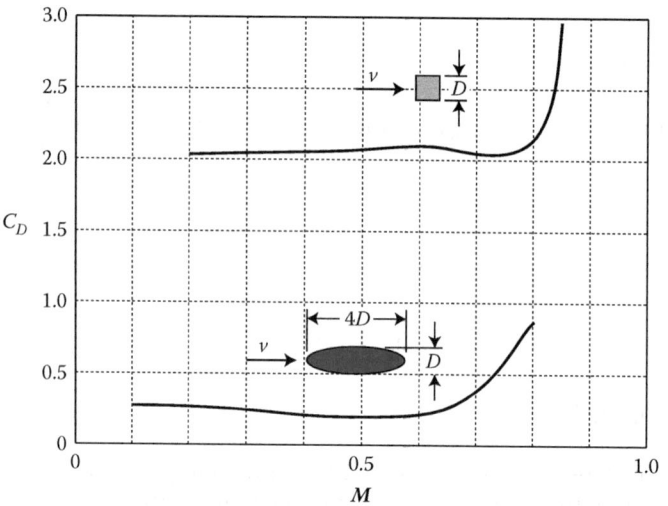

FIGURE 10.19

Variation of the drag the drag coefficient, C_D with the Mach number, M for two-dimensional shapes for subsonic flow. (Adapted from Blevins, R. D., *Applied Fluid Dynamics Handbook,* Von Nostrand Reinhold, New York, 1984, p. 341.)

Figure 10.19 illustrates that compressibility effects are insignificant if $M \leq 0.5$, while they are significant for $M > 0.5$. Figure 10.19 also illustrates that the drag coefficient, C_D for the blunt body is higher than that for the streamlined body, for the entire range of values for M ($0.1 \leq M \leq 0.9$).

EXAMPLE PROBLEM 10.17

Figure 10.19 illustrates the variation of the drag coefficient, C_D for blunt (square shape) and streamlined (elliptical) 2-D bodies, for subsonic flow ($M = v/c < 1$), where compressibility effects are insignificant if $M \leq 0.5$, while they are significant for $M > 0.5$. Consider a 1500-ft-long square-shaped cylinder with a side of 6 ft. Assume air at standard atmosphere at an altitude of 25,000 ft above sea level flows over the cylinder, as illustrated in Figure EP 10.17. (a) Determine the drag coefficient and the drag force if the velocity of flow over the cylinder is 400 ft/sec. (b) Determine the drag coefficient and the drag force if the velocity of flow over the cylinder is 800 ft/sec.

Mathcad Solution

(a)–(b) The frontal area is used to compute the drag force for the square cylinder, and the drag force is determined by applying Equation 10.12. The fluid properties for air at standard atmosphere at an altitude of 25,000 ft above sea level is given in Table A.1 in Appendix A.

(a) The drag coefficient and the drag force for a velocity of flow over the cylinder of 400 ft/sec are determined as follows:

$$\text{slug} := 1\,\text{lb}\frac{\text{sec}^2}{\text{ft}} \qquad\qquad L := 1500\,\text{ft} \qquad D := 6\,\text{ft} \qquad A_{\text{front}} := L \cdot D = 9 \times 10^3\,\text{ft}^2$$

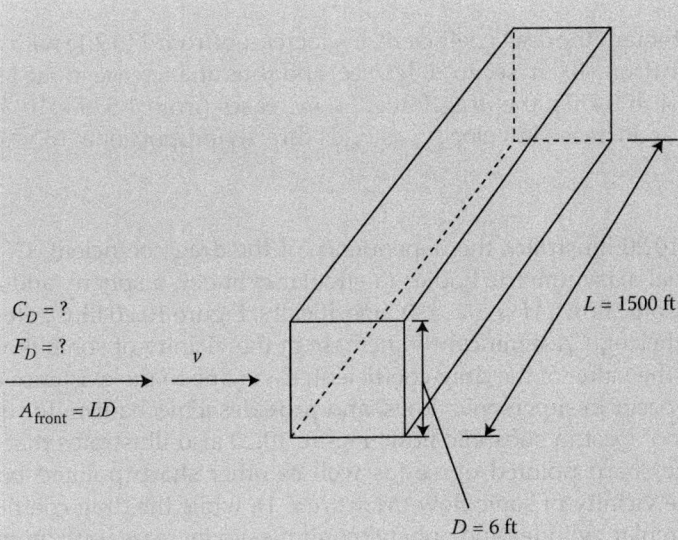

FIGURE EP 10.17
Air flows over a square-shaped cylinder.

$$\rho := 0.0010663 \frac{slug}{ft^3} \qquad c := 1016.11 \frac{ft}{sec} \qquad E_v := c^2 \cdot \rho = 1.101 \times 10^3 \frac{lb}{ft^2}$$

$$v := 400 \frac{ft}{sec} \qquad M := \frac{v}{\sqrt{\dfrac{E_v}{\rho}}} = 0.394 \qquad C_D := 2$$

Guess value: $\qquad F_D := 1 \, lb$

Given

$$F_D = C_D \frac{1}{2} \rho \cdot v^2 \cdot A_{front}$$

$$F_D := Find(F_D) = 1.535 \times 10^6 \, lb$$

(b) The drag coefficient and the drag force for a velocity of flow over the cylinder of 800 ft/sec are determined as follows:

$$v := 800 \frac{ft}{sec} \qquad M := \frac{v}{\sqrt{\dfrac{E_v}{\rho}}} = 0.787 \qquad C_D := 2.1$$

Guess value: $\qquad F_D := 1 \, lb$

Given

$$F_D = C_D \frac{1}{2} \rho \cdot v^2 \cdot A_{front}$$

$$F_D := Find(F_D) = 6.449 \times 10^6 \, lb$$

Therefore, although the drag coefficient, C_D increases (from 2 to 2.1) with an increase in the velocity, v (from 400 ft/sec to 800 ft/sec) and thus an increase in the Mach number, M (from 0.384 to 0.787), the drag force, F_D increases (from 1.535×10^{10} lb to 6.449×10^{11} lb) with an increase in velocity, as F_D is directly proportional to v^2.

Also, Figure 10.20 illustrates the dependence of the drag coefficient, C_D on M for three three-dimensional axisymmetric bodies (a circular cylinder, a sphere, and a sharp pointed ogive) for supersonic flow ($M = v/c > 1$). Specifically, Figure 10.20 illustrates that the values of the drag coefficient, C_D significantly increase in the vicinity of sonic flow ($M = v/c = 1$). Such a surge in the value of the drag coefficient, C_D is due to the existence of shock waves, which can only occur in supersonic flows, and provides a mechanism for the generation of drag that does not exist in subsonic flow. Figure 10.20 also illustrates that the drag coefficient, C_D for the sharp pointed ogive (as well as other sharp pointed bodies) reaches a maximum in the vicinity of sonic flow ($M = v/c = 1$), while the drag coefficient, C_D for the blunt bodies (circular cylinder and sphere) continues to increase with an increase in M for $M > 1$. Such an increase in the drag coefficient, C_D for the blunt bodies in supersonic flow is due to the nature of the shock wave structure and the flow separation associated with blunt bodies. As a result, the leading edges of subsonic aircraft wings are typically rounded and blunt, while those of supersonic aircraft are typically pointed and sharp.

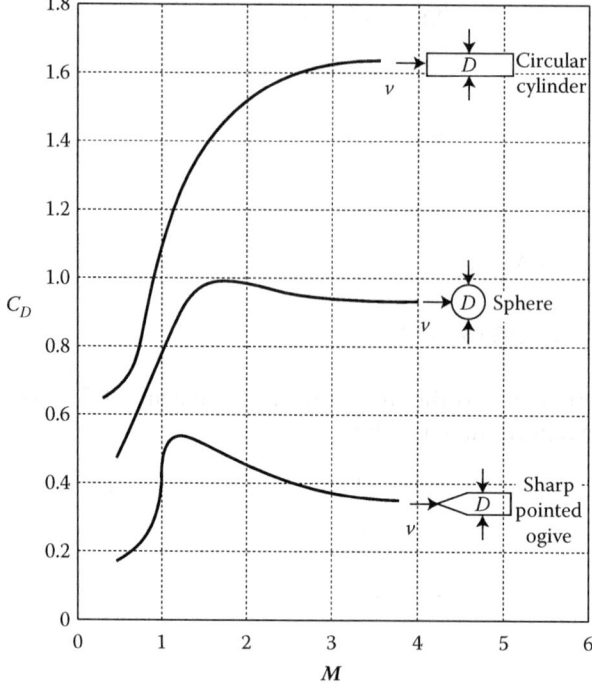

FIGURE 10.20
Variation of the drag the drag coefficient, C_D with the Mach number, M for three-dimensional axisymmetric bodies for supersonic flow. (Adapted from Vennard, J. K., and R. L., Street, *Elementary Fluid Mechanics*, 6th ed, John Wiley and Sons, New York, 1982, p. 648.)

EXAMPLE PROBLEM 10.18

Figure 10.20 illustrates the dependence of the drag coefficient, C_D on M for three three-dimensional axisymmetric bodies (a circular cylinder, a sphere, and a sharp pointed ogive) for supersonic flow ($M = v/c > 1$). Consider an ogive with a diameter of 3 in traveling in the air at standard atmosphere at an altitude of 30,000 ft above sea level, at a speed of 3800 ft/sec, as illustrated in Figure EP 10.18. (a) Determine the drag force acting on the ogive.

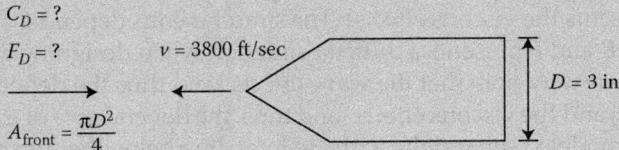

FIGURE EP 10.18
Ogive travelling at a speed of 3800 ft/sec.

Mathcad Solution

(a) The frontal area is used to compute the drag force for the ogive, and the drag force is determined by applying Equation 10.12. The speed of sound, $c = \sqrt{E_v/\rho}$ (along with the density) for air at standard atmosphere at an altitude of 30,000 ft above sea level is given in Table A.1 in Appendix A.

$$\text{slug} := 1\,\text{lb}\frac{\sec^2}{\text{ft}} \qquad D := 3\,\text{in} = 0.25\,\text{ft} \qquad A_{\text{front}} := \frac{\pi \cdot D^2}{4} = 0.049\,\text{ft}^2$$

$$v := 3800\frac{\text{ft}}{\sec} \qquad c := 994.85\frac{\text{ft}}{\sec} \qquad M := \frac{v}{c} = 3.82$$

$$C_D := 0.35 \qquad \rho := 0.00089065\frac{\text{slug}}{\text{ft}^3}$$

Guess value $F_D := 1\,\text{lb}$

Given

$$F_D = C_D \frac{1}{2}\rho \cdot v^2 \cdot A_{\text{front}}$$

$$F_D := \text{Find}(F_D) = 110.48\,\text{lb}$$

10.5.3.13 The Importance of the Froude Number, F

For body shapes in general, with a wave action at the free surface, the drag coefficient, C_D for laminar or turbulent flow is dependent on R, L_i/L, and F, but independent of ε/L and M. Thus, the drag coefficient, C_D will be dependent on F in the case of where there is a wave action at the top of the free surface. The drag coefficient, C_D will vary with $F = v/v_c$ (which is the ratio of the free-stream speed to a typical wave speed on the interface of two fluids, such as the surface of a body of water; ocean, lake, river) depending mainly on the shape of the body and velocity of flow. Such examples include ships and open channel flow measuring devices/structures such as weirs and spillways. For instance, a ship moving in the body of water may produce waves, which are caused by the product of the force and the velocity of the ship (where the energy production or the power is force times velocity), and results in drag. While the nature of the waves produced by the ship depends on F (and thus the velocity of the ship) and the shape of the ship, the drag coefficient, C_D depends on both the F (which represents the wave-making effects and thus the pressure effects) and the R (which represents the viscous effects). The simultaneous dependency of the drag coefficient, C_D on both R and F presents a difficult situation when designing a model of the prototype. However, one may note that the wave effects (and thus the dependence of the drag coefficient, C_D on F) and the viscous effects (and thus, the dependence of the drag coefficient, C_D on R) are often modeled separately, with the total drag being the sum of the two individual effects. Figure 10.12 illustrates the variation of the drag coefficient, C_D with the R for two-dimensional bodies, while Figure 10.13 illustrates the variation of the drag coefficient, C_D with the R for three-dimensional axisymmetric bodies. Additionally, Tables 10.4 through 10.6 present the drag coefficient, C_D for numerous geometric shapes at various orientations to the direction of flow, for $R > 10,000$, for two-dimensional, three-dimensional axisymmetric, and three-dimensional bodies, respectively. Figure 10.21 illustrates variation of the wave drag coefficient, C_D with both the F and the shape of the ship (a hull without a bow bulb/ streamlined body and a hull with a bow bulb/less streamlined body). It may be noted

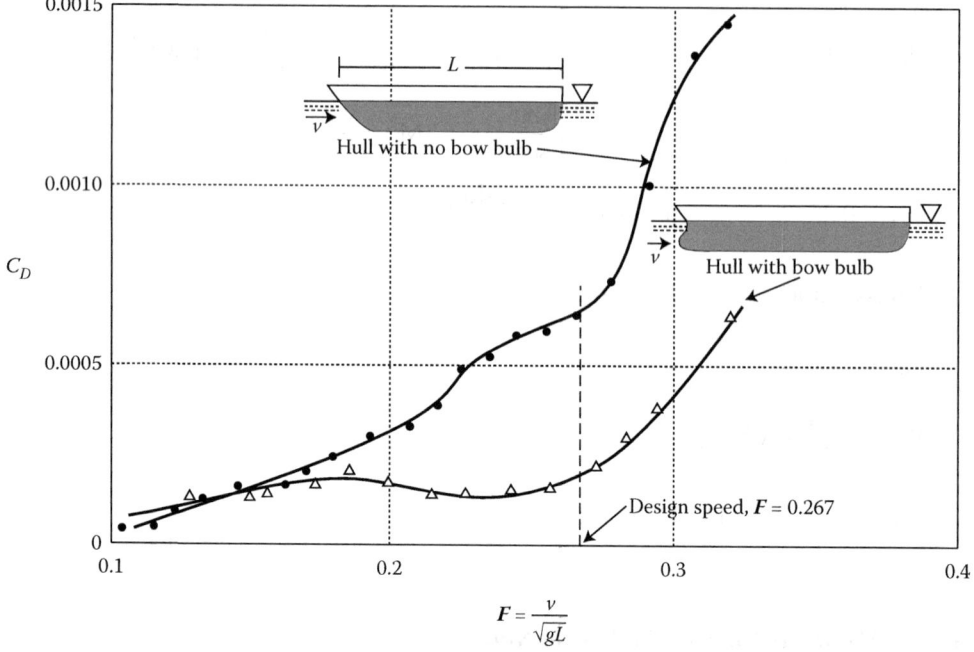

FIGURE 10.21
Variation of the drag the wave drag coefficient, C_D with the Froude number, F and the shape of the ship for three-dimensional bodies. (Adapted from Munson, B. R. et al., *Fundamentals of Fluid Mechanics*, 3rd ed, John Wiley and Sons, New York, 1998, p. 607; Inui, T., *Transactions of the Society of Naval Architects and Marine Engineers*, 70, 305, 326–327, 1962.)

that the "wavy" variation of the wave drag coefficient, C_D with the F (for either of the two body shapes) is due to the fact that the structure of the waves produced by the ship's hull is significantly a function of the velocity of the ship (and thus the F). Furthermore, Figure 10.21 illustrates that the wave and the corresponding wave drag that is mainly caused by the bow (front end of the ship) may be reduced by a reduction in streamlining of the body (i.e., adding a bow bulb to the hull). Finally, it is interesting to note that the more streamlined body (a hull without a bow bulb) actually results in a higher wave drag coefficient, C_D than the less streamlined body (a hull with a bow bulb) because it is the bow that induces the wave drag.

EXAMPLE PROBLEM 10.19

Figure 10.21 illustrates variation of the wave drag coefficient, C_D with both the F and the shape of the ship (a hull without a bow bulb/streamlined body and a hull with a bow bulb/less streamlined body). Consider a hull with a length of 50 ft that travels at a speed of 7.5 mph in the ocean, as illustrated in Figure EP 10.19. (a) Determine the drag force on a hull without a bow bulb/streamlined body. (b) Determine the drag force on a hull with a bow bulb/less streamlined body.

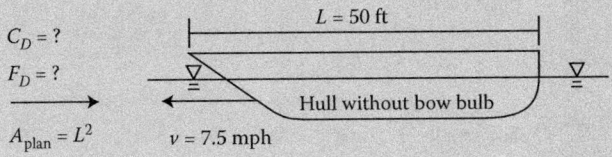

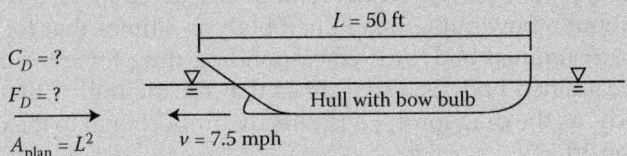

FIGURE EP 10.19
Hull without a bow bulb (streamlined body) and a hull with a bow bulb (less streamlined body) travel at a speed of 7.5 mph in the ocean.

Mathcad Solution

(a)–(b) The planform area is used to compute the drag force for the hull, and the drag force is determined by applying Equation 10.12. The fluid properties for the ocean water (saltwater) is given in Table A.4 in Appendix A.

(a) The drag force on a hull without a bow bulb/streamlined body is determined as follows:

$$\text{slug} := 1\,\text{lb}\frac{\sec^2}{\text{ft}} \qquad L := 50\,\text{ft} \qquad A_{\text{plan}} := L^2 = 2.5 \times 10^3\,\text{ft}^2$$

$$v := 7.5\,\text{mph} = 11\frac{\text{ft}}{\text{s}} \qquad g := 32.174\frac{\text{ft}}{\sec^2} \qquad F := \frac{v}{\sqrt{g \cdot L}} = 0.274 \qquad \rho := 1.985\frac{\text{slug}}{\text{ft}^3}$$

$C_D := 0.00075$

Guess value: $F_D := 1 \text{ lb}$

Given

$$F_D = C_D \frac{1}{2} \rho \cdot v^2 \cdot A_{plan}$$

$F_D := \text{Find}(F_D) = 225.173 \text{ lb}$

(b) The drag force on a hull with a bow bulb/less streamlined body is determined as follows:

$C_D := 0.00025$

Guess value: $F_D := 1 \text{ lb}$

Given

$$F_D = C_D \frac{1}{2} \rho \cdot v^2 \cdot A_{plan}$$

$F_D := \text{Find}(F_D) = 75.058 \text{ lb}$

Therefore, because for a given velocity (and thus, a given F) the drag coefficient, C_D for the hull without a bow bulb/streamlined body is 3 times that for the hull with a bow bulb/less streamlined body; the corresponding drag force for the hull without a bow bulb/streamlined body is also 3 times that for the hull with a bow bulb/less streamlined body, as the drag force, F_D is directly proportional to the drag coefficient, C_D (see Equation 10.12).

10.5.4 Evaluation of the Lift Force in External Flow

One may recall that the component of the resultant pressure and shear (viscous) forces that acts normal to the direction of the flow is called is called the lift force, F_L, which tends to move the submerged body in that direction. One may note that while some objects such as an airfoil are designed to generate lift, other objects, such as cars, are designed to reduce the generated lift. In the case where the submerged body illustrated in Figure 10.2 is moving at a high speed (and depending upon the shape of the body and the angle it makes with the direction of flow/angle of attack), then one must model the lift force, F_L as well as the drag force, F_D. Computation of the lift force, F_L is accomplished in similar manner to the computation of the drag force, F_D, and the equation is given as follows:

$$F_L = C_L \frac{1}{2} \rho v^2 A \tag{10.21}$$

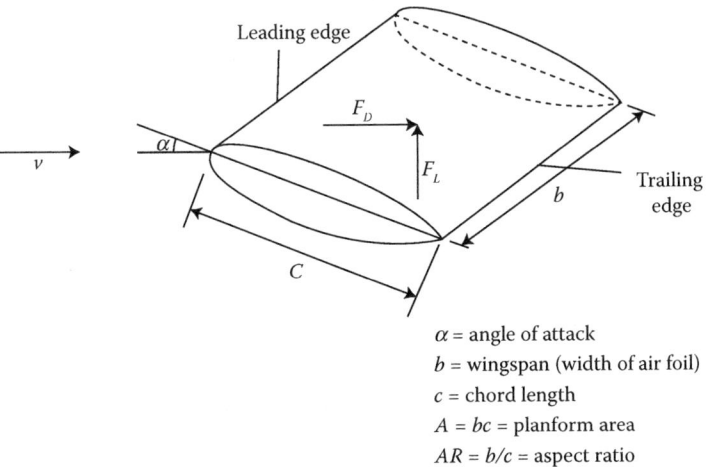

α = angle of attack
b = wingspan (width of air foil)
c = chord length
$A = bc$ = planform area
$AR = b/c$ = aspect ratio

FIGURE 10.22
Geometry of an airfoil.

where A = the planform area of the body (that is the area that is viewed looking at the body from above in a direction normal to the flow); in the case of a flat plate or an airfoil that is parallel to the direction of flow, the planform area is basically the surface area that is parallel to the flow; v is the upstream velocity of the fluid. Figure 10.22 illustrates that for an airfoil of width/span/wingspan, b and chord length, c (which is the length between the leading edge and the trailing edge), the planform area, $A = bc$. For an airplane, at takeoff and landing, and at a steady cruising speed at a constant altitude, the lift force, F_L must be equal to the total given weight of the aircraft, W. Thus, one may define the wing loading, which is the average lift per unit planform area of the wings, F_L/A. The lift coefficient, C_L is determined in a similar manner to the drag coefficient, C_D, as described in detail below. Furthermore, for an airplane, during a steady cruising speed, v_{cruise} at constant cruising altitude, and at a small angle of attack (that is, the weight component in the direction of flight is negligible), the net force in the direction of the flight are balanced (Newton's second law of motion); thus, the net force acting on the airplane in the direction of the flight is zero (see Figure 10.23) as follows:

$$\sum_s F_s = -F_T + F_D = ma_s = 0 \qquad (10.22)$$

and thus, the thrust force provided by the plane engines, F_T is equal to the drag force, F_D. The thrust force provided by the plane engines, F_T must be sufficient to overcome the wing drag force, F_D, where the power provided by the engines is defined as follows:

$$P_{engines} = F_T v_{cruise} \geq F_D v_{cruise} \qquad (10.23)$$

Most of the lift comes from the surface pressure distribution, as described by the Bernoulli equation. It may be noted that because most devices that generate lift typically operate in the large range of values for R (airfoils, spoilers, and fans), in which the viscous effects are confined to the boundary layers and the wake region, the wall shear stress, τ_w contributes very little to the lift. However, for bodies that operate in the very low range of values for R ($R < 1$)

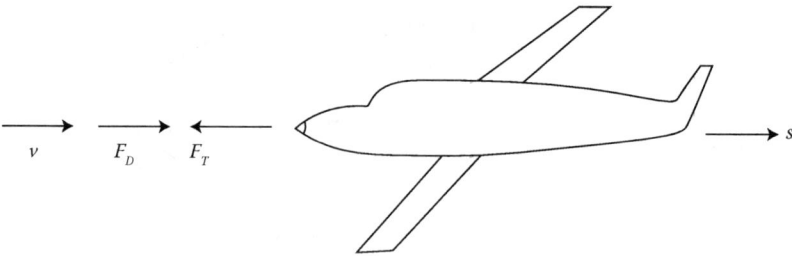

FIGURE 10.23
Forces acting on a plane in direction of flight, s.

(small insects and microscopic organisms), the viscous effects become important, and thus the contribution of the wall shear stress, τ_w to the lift may become as important as the contribution of the pressure distribution. It is of specific interest herein to examine devices that are specifically designed to generate lift, such as an airfoil as illustrated in Figure 10.22. Airfoils are designed to generate lift while keeping the drag at a minimum. Furthermore, because airfoils are streamlined and thus the wall shear is parallel to the surface of the body, the contribution of the viscous (shear) effects to the lift are typically negligible (unless the body is lightweight and flies at low velocities and thus at very low R). Additionally, it may be noted that at the typical high velocities (and thus at high R) of airfoils, the lift is mostly independent of the relative surface roughness, ε/L, as the surface roughness affects the wall shear stress, τ_w and not the pressure. As a result, in practice, the lift force, F_L is primarily due to the pressure distribution on the surface of the body; thus, the shape of the body plays a major role in the modeling of the lift force, F_L. In the design of airfoils, where the goal is to generate lift, the main objective is to minimize the average pressure at the upper surface of the body while maximizing it at the lower surface; the net result of the pressure distribution results in a net upward lift force, F_L that tends to lift the wing and thus the airplane's fuselage to which it is connected. This objective may be accomplished by applying the Bernoulli equation in order to identify the high-pressure and low-pressure regions on the body. Application of the Bernoulli equation indicates that pressure is low at points of high velocity of flow, while pressure is high at points of low velocity of flow (see Chapter 4). Assuming that the contribution of the viscous (shear) effects is negligible, the lift force, F_L may be determined by integrating the pressure distribution around the airfoil. Furthermore, for large values of R, the pressure distributions are typically directly proportional to the dynamic pressure, $\rho v^2/2$, with the viscous (shear) effects playing a secondary role.

A typical device that is designed to generate lift, such an airfoil, does so by generating a pressure distribution that is at a minimum at the top of the body and a maximum at the bottom of the body. The shape of the body and the angle of attack, α play a primary role in accomplishing this task. Although most devices that are designed to generate lift are nonsymmetrical in shape, symmetrical shapes at an appropriate angle of attack, α may also generate lift. Figure 10.24a illustrates that an angle of attack, $\alpha = 0°$ for the symmetrical-shaped airfoil will not produce the necessary pressure difference required to produce lift, while Figure 10.24b illustrates that an angle of attack, $\alpha = 0°$ for the nonsymmetrical-shaped airfoil will produce the necessary pressure difference required to produce lift. Furthermore, in addition to the shape of the body, an additional way to change both the lift and drag characteristics of an airfoil is to change the angle of attack, α, where the angle of attack, α is pitched up to increase lift.

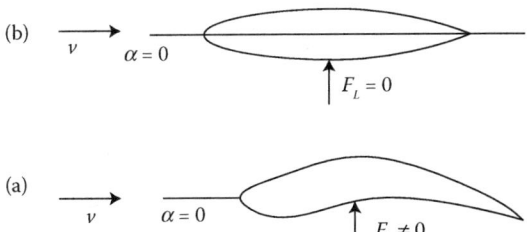

FIGURE 10.24
Shape of the body and the angle of attack, α play important role in generating the lift force. (a) Symmetrical airfoil, (b) Nonsymmetrical airfoil.

10.5.5 Evaluation of the Lift Coefficient

Similar to the drag coefficient, C_D, the value for the lift coefficient, C_L is mainly a function of E, and will primarily vary with the shape of the body, L_i/L and the angle of attack, α; however, in some cases it will also depend on R, the relative surface roughness, ε/L, M, and F. Thus,

$$C_L = \phi\left(R, C, F, \frac{\varepsilon}{L}, \frac{L_i}{L}, \alpha \right) \tag{10.24}$$

where $M = \sqrt{C} = v/\sqrt{E_v/\rho} = v/c$. The shape of the body has the most significant effect on the lift coefficient, C_L, followed by the angle of attack, α. While M plays an important role for relatively high-speed subsonic and supersonic flows ($M > 0.8$), R typically does not. Although the relative surface roughness, ε/L has a significant effect on the drag coefficient, C_D, it is relatively unimportant in determining the lift coefficient, C_L. Furthermore, F plays an important role if there is free surface present, such as in the case of an underwater wing that is used to support a high-speed hydrofoil surface ship.

The role of the shape of the body and the angle of attack in the evaluation of the lift coefficient, assuming an airfoil, are presented in the sections below. Specifically, the angle of attack is optimized in order to maximize lift while minimizing the drag. The shape of an airfoil may be optimized for takeoff and landing by the symmetry of the airfoil and the use of flaps. Furthermore, the shape of an airfoil may be optimized for cruising altitude through its aspect ratio.

10.5.5.1 The Role of the Shape of the Body and the Angle of Attack

Airfoils/wings of airplanes are designed (shaped and positioned/angled) specifically to generate lift with minimal drag. It is important to note that both the lift and drag (and thus, their respective coefficients) are strongly dependent on both the shape of the body and the angle of attack, α. One may recall (from discussion above) that for a high velocity ($R > 10,000$), the drag coefficient, C_D is mainly a function of the shape of the body (resulting from both the orientation of the body to the direction of flow and the smoothness of the edges of the body). Figure 10.25 illustrates how an airfoil is shaped and positioned/angled in order to generate sufficient lift during flight, meanwhile maintaining the drag at a minimum. With the goal of generating the maximum lift while producing a minimum drag for an airfoil, the main objective would be to simultaneously determine the optimum shape of the body and the optimum angle of attack, α to maintain during cruising as well as during

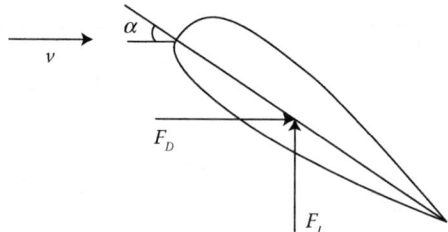

FIGURE 10.25
Shape and angle of attack, α of an airfoil are optimized to generate sufficient lift during flight while minimizing the drag.

takeoff and landing (in order to minimize the average pressure at the upper surface of the body while maximizing it at the lower surface; the net result of the pressure distribution results in a net upward lift force, F_L). Because the drag and lift are strongly dependent on the shape of the body, any effect that causes the shape to change will have a significant effect on both the drag and the lift. One may note that although the shape of the airfoil is mostly controlled by the design, weather conditions that lead to snow accumulation and ice formation may also contribute to a change in the shape of the wing; such an environmental change in the shape of the airfoil may cause a significant loss of lift, leading to a loss of altitude and potential crashing (or simply aborting on takeoff).

10.5.5.2 Optimizing the Shape of an Airfoil and the Angle of Attack

The ultimate goal in the design of an airfoil is to develop the optimum shape of the body and the optimum angle of attack, α in order to create an efficient (low-drag) airfoil with sufficient lift. Apart from the use of moveable slotted flaps to modify the shape of the airfoil (see Section 10.5.5.4), the symmetry of the airfoil plays an important role in achieving this ultimate goal. Figure 10.24b illustrates that a nonsymmetrical shaped airfoil at an angle of attack, $\alpha = 0°$ will produce the necessary pressure difference required to produce lift, while Figure 10.24a illustrates that a symmetrical-shaped airfoil at an angle of attack, $\alpha = 0°$ will not produce the necessary pressure difference required to produce lift. However, a symmetrical-shaped airfoil at an appropriate angle of attack, α (α is pitched up to increase lift) may also generate lift. Figure 10.26 illustrates how the lift coefficient, C_L varies as a function of the angle of attack, α for a symmetrical airfoil and a nonsymmetrical airfoil. It may be noted that regardless of the symmetry of the airfoil, the lift coefficient, C_L varies linearly (increasing severalfold by increasing α) with the angle of attack, α, reaching a maximum value at about $\alpha = 16°$, at which point it immediately begins to decline. At around an angle of attack, $\alpha = 16°$, $C_L = C_{L,max}$, where the velocity has reached v_{min}, and the airfoil begins to stall (explained below). The increase in the angle of attack, α to $\alpha = 16°$ causes flow separation and the formation of a wide wake region on the top surface of the airfoil. As discussed in Section 10.5.3.4 above, in the design of the wings of airplanes, reducing the angle of attack, α to less than 15° will reduce the effect of the flow separation (and thus minimize the effect of the drag), and this in turn will maximize the effect of the lift. Figure 10.26 also illustrates that the nonsymmetrical-shaped airfoil (with greater curvature at the top of the surface) at an angle of attack, $\alpha = 0°$ will produce the necessary pressure difference required to produce lift by yielding a nonzero lift coefficient, C_L, while the symmetrical-shaped airfoil at an angle of attack, $\alpha = 0°$ will not produce the necessary pressure difference required to

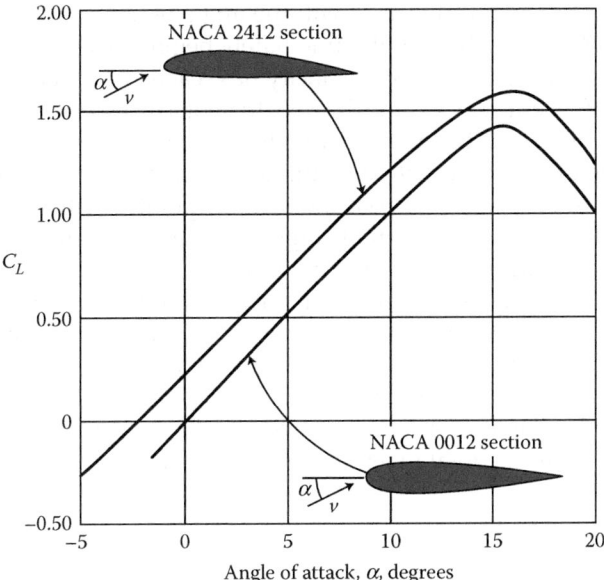

FIGURE 10.26
Variation of the lift coefficient, C_L with the angle of attack, α for a symmetrical airfoil and a nonsymmetrical airfoil. (Adapted from Abbott, I. H., and A. E., von Doenhoff, *Theory of Wing Sections, Including a Summary of Airfoil Data*, Dover, New York, 1959.)

produce lift by yielding a value of zero for the lift coefficient, C_L. Thus, aircraft with symmetrical wings must fly at a higher angle of attack, α in order to yield the same lift as the nonsymmetrical wings.

Because the drag force increases as the angle of attack is increased, it is important to determine how the drag coefficient, C_D varies with the angle of attack, α. Figure 10.27 illustrates both the lift coefficient, C_L versus the angle of attack, α and the drag coefficient, C_D versus the angle of attack, α for a typical cambered (nonsymmetrical) airfoil. Figure 10.27 illustrates

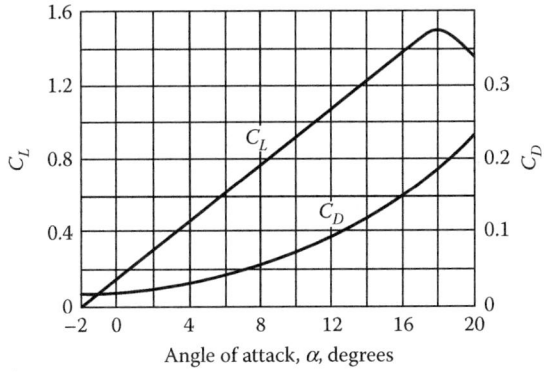

FIGURE 10.27
Variation of the lift coefficient, C_L and the drag coefficient, C_D with the angle of attack, α for a typical cambered (nonsymmetrical) airfoil. (Adapted from Blevins, R. D., *Applied Fluid Dynamics Handbook*, Von Nostrand Reinhold, New York, 1984, p. 357.)

that the lift coefficient, C_L varies linearly with the angle of attack, α, reaching a maximum value at about $\alpha = 18°$, at which point it immediately begins to decline. At around an angle of attack, $\alpha = 18°$, $C_L = C_{L,max}$, where the velocity has reached v_{min}, and the airfoil begins to stall. Similar to the lift coefficient, C_L, the drag coefficient, C_D also increases with the angle of attack, α. However, Figure 10.27 illustrates that the drag coefficient, C_D increases exponentially with the angle of attack, α, with a continuous increase in C_D, as the angle of attack increases from $-2°$ to $20°$, which is past the stalling point of $18°$. Therefore, in order to reduce the drag and thus conserve the fuel, large angles of attack should be used selectively for short periods of time (during takeoff and landing).

During takeoff and landing, the flight speed is at a minimum, v_{min}. As such, the lift force is also at a minimum, $F_{L,mim} = C_L(1/2)\rho v_{min}^2 A$, and must be equal to the total given weight of the aircraft, W. Furthermore, at the minimum flight velocity, v_{min} during takeoff and landing, the lift coefficient is a maximum, $C_{L,max} = (2F_{L,min}/\rho v_{min}^2 A)$. Thus, the minimum flight velocity v_{min} for both takeoff and landing is computed by assuming that the minimum lift force, F_{Lmin} is equal to the total given weight of the aircraft, W, and that $C_L = C_{L,max}$, as follows:

$$F_{L,mim} = W = C_{L,max} \frac{1}{2} \rho v_{min}^2 A \tag{10.25}$$

$$v_{min} = \sqrt{\frac{2W}{\rho C_{L,max} A}} \tag{10.26}$$

Because this minimum velocity, v_{min}, which is known as stall conditions, is a region of unstable operation of the aircraft, it must be avoided; as such, the Federal Aviation Administration (FAA) requires that the minimum safe speed for aircraft takeoff and landing be at least 1.2 times the stall speed (v_{min}). It is interesting to note that Equation 10.25 points out that the minimum allowable velocity for takeoff and landing (at least $1.2v_{min}$) is inversely proportional to the square root of the air density, ρ. As such, because the air density decreases with altitude, airports located at higher altitudes will require higher takeoff and landing velocities and thus will require longer runways. Furthermore, because the air density, ρ is inversely proportional to the air temperature, during hot summer days, where the air density is at its minimum, an even higher takeoff and landing velocity will be required, thus the further need for sufficiently long runways.

EXAMPLE PROBLEM 10.20

Figure 10.26 illustrates how the lift coefficient, C_L varies as a function of the angle of attack, α for a symmetrical airfoil and a nonsymmetrical airfoil. Consider an airplane that weighs 30,000 lb with nonsymmetrical wings (airfoils) with a span width of 25 ft and a chord length of 20 ft, as illustrated in Figure EP 10.20. Assume the airplane takes off and lands at sea level (air at standard atmosphere at sea level, $\rho = 0.0023768$ slugs/ft^3) at a maximum angle of attack. Also assume that the airplane cruises at an altitude of 35,000 ft (air at standard atmosphere at an altitude of 35,000 ft, $\rho = 0.0007319$ slugs/ft^3) at an angle of attack of 5°. (a) Determine the minimum takeoff and landing speed. (b) Determine the steady cruising speed at the cruising altitude.

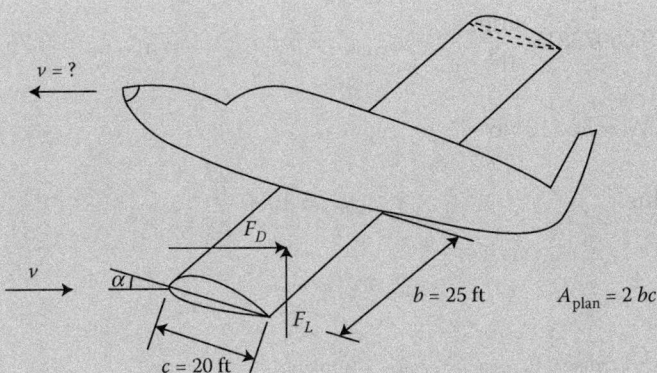

FIGURE EP 10.20
Plane with nonsymmetrical wings (airfoils) takes off, cruises, and lands.

Mathcad Solution

(a) The minimum takeoff and landing speed is determined by applying Equation 10.25. The lift force must equal the weight of the aircraft at takeoff and landing. Furthermore, the maximum angle of attack at takeoff and landing is read from Figure 10.26, where the lift coefficient is also a maximum.

$$\text{slug} := 1\,\text{lb}\,\frac{\sec^2}{\text{ft}} \qquad b := 25\,\text{ft} \qquad c := 20\,\text{ft} \qquad A_{plan} := 2 \cdot b \cdot c = 1 \times 10^3\,\text{ft}^2$$

$$\rho := 0.0023768\,\frac{\text{slug}}{\text{ft}^3} \qquad \alpha_{max} := 16\,\text{deg} \qquad C_{Lmax} := 1.6$$

$$W := 30000\,\text{lb} \qquad F_{Lmin} := W = 3 \times 10^4\,\text{lb}$$

Guess value: $\qquad v_{min} := 100\,\dfrac{\text{ft}}{\sec}$

Given

$$F_{Lmin} = C_{Lmax}\frac{1}{2}\rho \cdot v_{min}{}^2 \cdot A_{plan}$$

$$v_{min} := \text{Find}(v_{min}) = 125.609\,\frac{\text{ft}}{\text{s}} \qquad\qquad v_{min} = 85.642\,\text{mph}$$

which is the stall speed. However, the FAA requires that the minimum safe speed for aircraft takeoff and landing be at least 1.2 times the stall speed, v_{min} which is determined as follows:

$$v_{minsafe} := 1.2\,v_{min} = 150.73\,\frac{\text{ft}}{\text{s}} \qquad\qquad v_{minsafe} = 102.771\,\text{mph}$$

(b) The steady cruising speed at the cruising altitude is determined by applying Equation 10.21. The lift force must equal to the weight of the aircraft at steady cruising speed. Furthermore, the lift coefficient for the cruising angle of attack is read from Figure 10.26.

$$\rho_{cruise} := 0.00073819\,\frac{slug}{ft^3} \qquad \alpha_{cruise} := 5\ deg \qquad C_{Lcruise} := 0.75$$

$$F_{Lcruise} := W = 3 \times 10^4\ lb$$

Guess value: $\qquad\qquad\qquad v_{cruise} := 1000\,\dfrac{ft}{sec}$

Given

$$F_{Lcruise} = C_{Lcruise}\,\frac{1}{2}\,\rho_{cruise}\cdot v_{cruise}^{\ 2}\cdot A_{plan}$$

$$v_{cruise} := Find(v_{cruise}) = 329.201\,\frac{ft}{s} \qquad\qquad v_{cruise} = 224.455\ mph$$

EXAMPLE PROBLEM 10.21

Figure 10.27 illustrates both the lift coefficient, C_L versus the angle of attack, α, and the drag coefficient, C_D versus the angle of attack, α for a typical cambered (nonsymmetrical) airfoil. Consider an airplane that weighs 45,000 lb with cambered (nonsymmetrical) wings with a span width of 30 ft and a chord length of 25 ft, as illustrated in Figure EP 10.21. Assume that the airplane cruises at an altitude of 45,000 ft (air at standard atmosphere at an altitude of 45,000 ft, $\rho = 0.00046227$ slugs/ft^3) at a steady cruising speed of 250 mph. (a) Determine the angle of attack to cruise at a steady speed at cruising altitude. (b) Determine the drag force acting on the aircraft wings during steady cruising speed. (c) Determine the thrust power required by the plane engines in order to overcome the wing drag force.

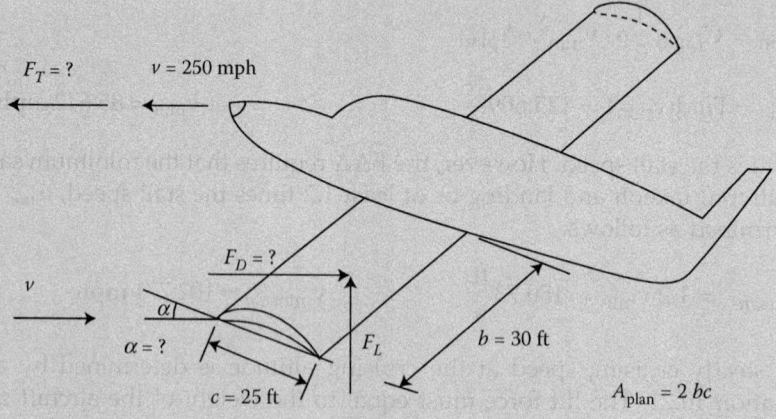

FIGURE EP 10.21
Plane with cambered nonsymmetrical wings (airfoils) cruises.

Mathcad Solution

(a) The lift coefficient is determined by applying Equation 10.21. The lift force must equal the weight of the aircraft at steady cruising speed.

$$\text{slug} := 1\,\text{lb}\,\frac{\sec^2}{\text{ft}} \qquad b := 30\,\text{ft} \qquad c := 25\,\text{ft} \qquad A_{\text{plan}} := 2 \cdot b \cdot c = 1.5 \times 10^3\,\text{ft}^2$$

$$\rho_{\text{cruise}} := 0.00046227\,\frac{\text{slug}}{\text{ft}^3} \qquad\qquad\qquad W := 45000\,\text{lb}$$

$$v_{\text{cruise}} := 250\,\text{mph} = 366.667\,\frac{\text{ft}}{\text{s}} \qquad\qquad F_{\text{Lcruise}} := W = 4.5 \times 10^4\,\text{lb}$$

Guess value: $\qquad\qquad\qquad\qquad C_{\text{Lcruise}} := 0.8$

Given

$$F_{\text{Lcruise}} = C_{\text{Lcruise}}\frac{1}{2}\rho_{\text{cruise}} \cdot v_{\text{cruise}}^2 \cdot A_{\text{plan}}$$

$$C_{\text{Lcruise}} = \text{Find}(C_{\text{Lcruise}}) = 0.965$$

Furthermore, the angle of attack to cruise at a steady speed at cruising altitude for the computed lift coefficient is determined from Figure 10.27 as follows:

$$\alpha := 10\,\text{deg}$$

(b) The drag force acting on the aircraft wings during steady cruising speed is determined by applying Equation 10.12. Furthermore, the drag coefficient is determined from Figure 10.27 for an angle of attack, $\alpha = 10°$.

$$C_{\text{Dcruise}} := 0.08$$

Guess value: $\qquad\qquad\qquad F_{\text{Dcruise}} := 100\,\text{lb}$

Given

$$F_{\text{Dcruise}} = C_{\text{Dcruise}}\frac{1}{2}\rho_{\text{cruise}} \cdot v_{\text{cruise}}^2 \cdot A_{\text{plan}}$$

$$F_{\text{Dcruise}} := \text{Find}(F_{\text{Dcruise}}) = 3.729 \times 10^3\,\text{lb}$$

(c) The thrust power required by the plane engines in order to overcome the wing drag force is determined by applying Equation 10.23 as follows:

$$F_{\text{Tcruise}} := F_{\text{Dcruise}} \qquad P_{\text{engines}} := F_{\text{Tcruise}} \cdot v_{\text{cruise}} = 1.367 \times 10^6\,\frac{\text{ft} \cdot \text{lb}}{\text{s}}$$

$$\text{hp} := 550\,\frac{\text{ft} \cdot \text{lb}}{\sec} \qquad P_{\text{engines}} = 2.486 \times 10^3\,\text{hp}$$

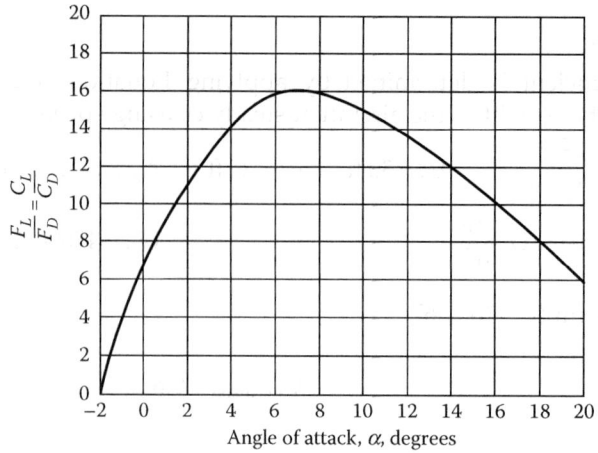

FIGURE 10.28
Variation of the lift-to-drag ratio, $F_L/F_D = C_L/C_D$ with the angle of attack, α for a typical cambered (nonsymmetrical) airfoil. (Adapted from Blevins, R. D., *Applied Fluid Dynamics Handbook*, Von Nostrand Reinhold, New York, 1984, p. 357.)

10.5.5.3 Optimizing the Performance of an Airfoil

Because it is desirable for airfoils to generate maximum lift while producing a minimum drag, a measure of performance for airfoils is the lift-to-drag ratio, $F_L/F_D = C_L/C_D$. As such, using Figure 10.27, which illustrates both the lift coefficient, C_L versus the angle of attack, α and the drag coefficient, C_D versus the angle of attack, α for a typical cambered (nonsymmetrical) airfoil, the lift-to-drag ratio, $F_L/F_D = C_L/C_D$ may be plotted versus different angles of attack, α as illustrated in Figure 10.28. Specifically, Figure 10.28 illustrates that the $F_L/F_D = C_L/C_D$ ratio increases with an increase in the angle of attack, α up to a value of about $F_L/F_D = C_L/C_D = 16$, at about $\alpha = 6.5°$, and then $F_L/F_D = C_L/C_D$ proceeds to decrease with an increase in the angle of attack, α. Thus, in general, regardless of the type of airfoil, the angle of attack, α that produces the maximum value of $F_L/F_D = C_L/C_D$ is the target operating point for an airfoil. Furthermore, as noted above, the minimum allowable velocity for take-off and landing is at least $1.2 v_{min}$.

EXAMPLE PROBLEM 10.22

Figure 10.28 illustrates the lift-to-drag ratio, $F_L/F_D = C_L/C_D$ versus different angles of attack, α for a typical cambered (nonsymmetrical) airfoil. Consider an airplane that weighs 50,000 lb with cambered (nonsymmetrical) wings with a span width of 40 ft and a chord length of 35 ft, as illustrated in Figure EP 10.22. Assume that the airplane cruises at an altitude of 45,000 ft (air at standard atmosphere at an altitude of 45,000 ft, $\rho = 0.00046227$ slugs/ft³) at a steady cruising speed of 300 mph. (a) Determine the lift force and the drag force acting on the aircraft wings during steady cruising speed at the target operating point for the airfoil.

Mathcad Solution

(a) The drag coefficient is determined by applying Equation 10.12. The lift force must equal the weight of the aircraft at steady cruising speed. Furthermore, the

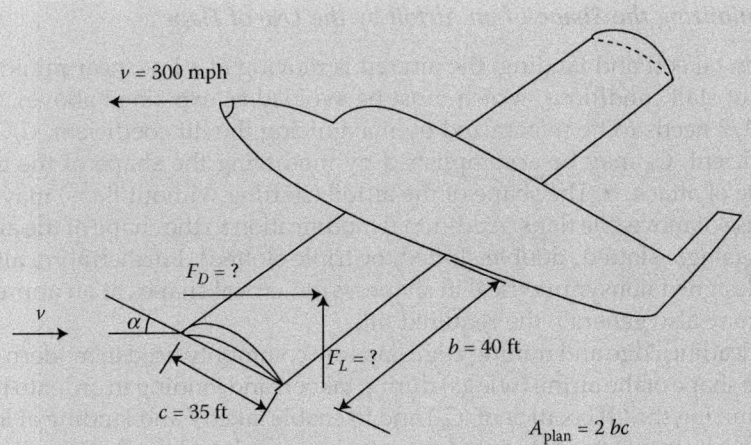

FIGURE EP 10.22
Plane with cambered nonsymmetrical wings (airfoils) cruises.

angle of attack, α that produces the maximum value of $F_L/F_D = C_L/C_D$, which is the target operating point for an airfoil, is determined from Figure 10.28.

$$\text{slug} := 1\,\text{lb}\frac{\text{sec}^2}{\text{ft}} \qquad b := 40\,\text{ft} \qquad c := 35\,\text{ft} \qquad A_{\text{plan}} := 2 \cdot b \cdot c = 2.8 \times 10^3\,\text{ft}^2$$

$$\rho_{\text{cruise}} := 0.00046227\frac{\text{slug}}{\text{ft}^3} \qquad\qquad\qquad W := 50000\,\text{lb}$$

$$v_{\text{cruise}} := 300\,\text{mph} = 440\frac{\text{ft}}{\text{s}} \qquad F_{\text{Lcruise}} := W = 5 \times 10^4\,\text{lb} \qquad \alpha := 6.5\,\text{deg}$$

Guess value: $\qquad C_{\text{Lcruise}} := 0.01 \qquad\qquad C_{\text{Dcruise}} := 0.01 \qquad F_{\text{Dcruise}} := 100\,\text{lb}$
Given

$$\frac{F_{\text{Lcruise}}}{F_{\text{Dcruise}}} = 16 \qquad\qquad\qquad \frac{C_{\text{Lcruise}}}{C_{\text{Dcruise}}} = 16$$

$$F_{\text{Dcruise}} = C_{\text{Dcruise}}\frac{1}{2}\rho_{\text{cruise}} \cdot v_{\text{cruise}}^2 \cdot A_{\text{plan}}$$

$$\begin{pmatrix} C_{\text{Lcruise}} \\ C_{\text{Dcruise}} \\ F_{\text{Dcruise}} \end{pmatrix} := \text{Find}(C_{\text{Lcruise}}, C_{\text{Dcruise}}, F_{\text{Dcruise}})$$

$$C_{\text{Lcruise}} = 0.399 \qquad\qquad C_{\text{Dcruise}} = 0.025 \qquad\qquad F_{\text{Dcruise}} = 3.125 \times 10^3\,\text{lb}$$

As a check, the lift force is computed by applying Equation 10.21 as follows:

$$F_{\text{Lcruise}} = C_{\text{Lcruise}}\frac{1}{2}\rho_{\text{cruise}} \cdot v_{\text{cruise}}^2 \cdot A_{\text{plan}} = 5 \times 10^4\,\text{lb}$$

10.5.5.4 *Optimizing the Shape of an Airfoil by the Use of Flaps*

Because upon takeoff and landing, the aircraft is moving at a low/near minimum velocity (v_{min} occurs at stall conditions, which must be avoided as explained above), the lift force, $F_L = C_L \, \rho v^2 A/2$ needs to be maximized by maximizing the lift coefficient, C_L. Maximizing the lift coefficient, C_L may be accomplished by modifying the shape of the airfoil (wing) and the angle of attack, α. The shape of the airfoil (starting without flaps) may be modified though the use of moveable flaps. Additional modification to the shape of the airfoil is to use slotted flaps (single-slotted, double-slotted, or triple-slotted). Furthermore, although most airfoils are designed nonsymmetrical in shape, symmetrical shapes at an appropriate angle of attack, α may also generate the required lift.

Moveable leading edge and trailing edge flaps are commonly used in modern large aircraft to change the shape of the airfoil (wings) during takeoff and landing in order to maximize the lift (by maximizing the lift coefficient, C_L) and to enable takeoff and landing at low speeds. It is interesting to note that the associated increase in drag (due to the flaps) during the takeoff and landing is not of any major concern because of the somewhat short time periods involved during takeoff and landing, as the flaps are retracted once the aircraft is at cruising altitude. Once the flaps are retracted, the wing resumes it normal shape characterized by an adequate lift coefficient, C_L (less than the lift coefficient, C_L during takeoff and landing) and a minimum drag coefficient C_D, which minimizes fuel consumption while cruising at a constant altitude. Furthermore, it is interesting to note that because the lift force, F_L is directly proportional to the square of the flow velocity ($F_L = C_L \, \rho v^2 A/2$), even a small lift coefficient, C_L can still generate a large lift force, F_L at large velocities (cruising aircraft speed).

The flapped airfoils (wings) that are used during takeoff and landing may be slotted (single-slotted, double-slotted, or triple-slotted) in order to prevent the separation of the boundary layer from the upper surface of the wings and the flaps and to increase the lift coefficient, C_L. Figure 10.29 illustrates how a flapped airfoil with a single-slot controls the boundary layer; the slot allows air to move from the high-pressure region under the wing into the low-pressure region on top of the wing, thus allowing the lift coefficient, C_L to

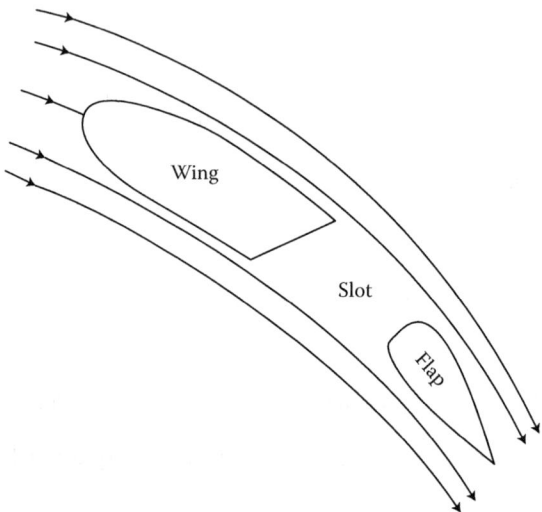

FIGURE 10.29
Flapped airfoil with a single-slot allows air to move from high-pressure region under wing into low-pressure region on top of wing, maximizing the lift coefficient, C_L.

approach its maximum value $C_L = C_{L,max}$ as the aircraft approaches its minimum velocity, v_{min} (stall conditions). Figure 10.30a illustrates the effects of flaps (no flap, single-slotted flap, and double-slotted flap) on the lift coefficient, C_L for a range of angles of attack, α. It is interesting to note that the maximum lift coefficient, $C_{L,max}$ (mostly occurring at the largest angle of attack, α in the experimented range of α) increases from 1.52 to 2.67 and then to 3.48 as the airfoil starts without flaps and progresses to the single-slotted flaps and then to the double-slotted flaps.

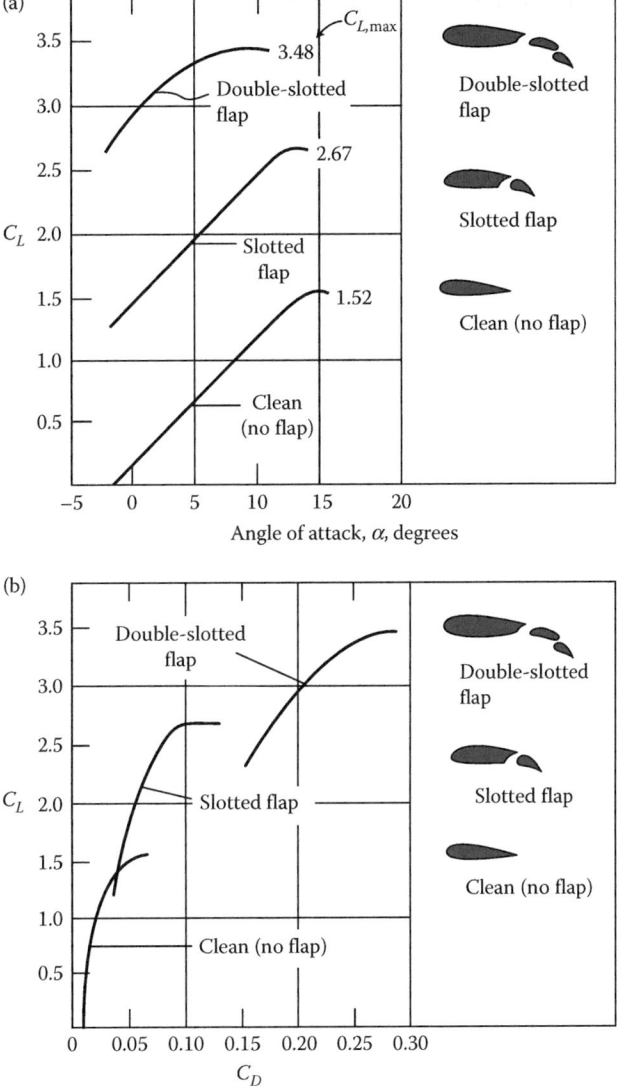

FIGURE 10.30
(a) Variation of the lift coefficient, C_L with the angle of attack, α for an airfoil with no flap, single-slotted flap, and double-slotted flap. (Adapted from Abbott, I. H., and A. E., von Doenhoff, *Theory of Wing Sections, including a Summary of Airfoil Data*, Dover, New York, 1959.) (b) The effects of flaps (no flap, single-slotted flap, and double-slotted flap) simultaneously on the lift coefficient, C_L and the drag coefficient, C_D for an airfoil. (Adapted from Abbott, I. H., and A. E., von Doenhoff, *Theory of Wing Sections, including a Summary of Airfoil Data*, Dover, New York, 1959.)

EXAMPLE PROBLEM 10.23

Figure 10.30a illustrates the effects of flaps (no flap, single-slotted flap, and double-slotted flap) on the lift coefficient, C_L for a range of angles of attack, α. Consider an airplane that weighs 30,000 lb with wings with a span width of 25 ft and a chord length of 20 ft, as illustrated in Figure EP 10.23. Assume the airplane takes off and lands at sea level (air at standard atmosphere at sea level, $\rho = 0.0023768$ slugs/ft^3) at an angle of attack of 10°. (a) Determine the minimum takeoff and landing speed for wings with no flaps. (b) Determine the minimum takeoff and landing speed for wings with slotted flaps.

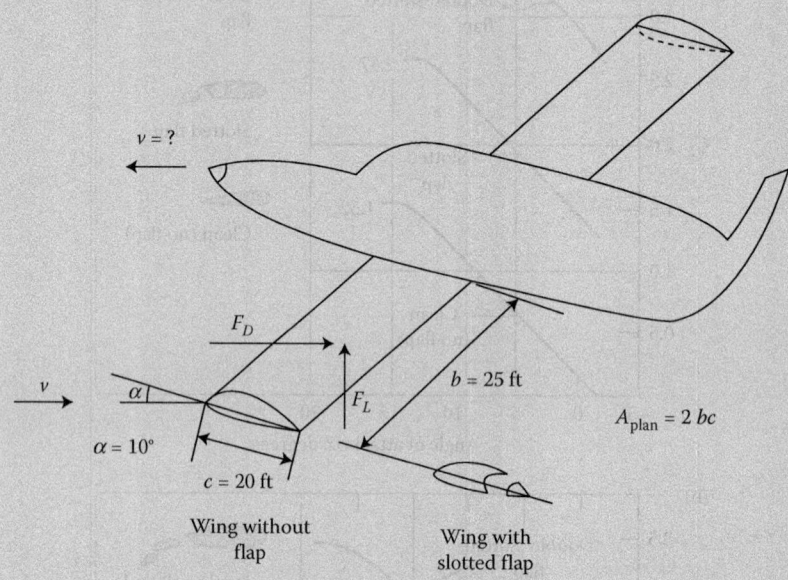

FIGURE EP 10.23
Plane with nonsymmetrical wings (airfoils) (with and without slotted flaps) takes off and lands.

Mathcad Solution

(a)–(b) The minimum takeoff and landing speed is determined by applying Equation 10.25. The lift force must equal the weight of the aircraft at takeoff and landing. Furthermore, the maximum lift coefficient at takeoff and landing is read from Figure 10.30a for the given angle of attack of 10°.

(a) The minimum takeoff and landing speed for wings with no flaps is determined as follows:

$$\text{slug} := 1 \,\text{lb} \frac{\text{sec}^2}{\text{ft}} \qquad b := 25 \,\text{ft} \qquad c := 20 \,\text{ft} \qquad A_{\text{plan}} := 2 \cdot b \cdot c = 1 \times 10^3 \,\text{ft}^2$$

$$\rho := 0.0023768 \frac{\text{slug}}{\text{ft}^3} \qquad \alpha := 10 \,\text{deg} \qquad W := 30000 \,\text{lb} \qquad F_{L\text{min}} := W = 3 \times 10^4 \,\text{lb}$$

$$C_{L\text{max}} := 1.25$$

Guess value: $v_{min} := 100 \dfrac{ft}{sec}$

Given

$$F_{Lmin} = C_{Lmax} \frac{1}{2} \rho \cdot v_{min}^2 \cdot A_{plan}$$

$v_{min} := Find(v_{min}) = 142.11 \dfrac{ft}{s}$ $v_{min} = 96.893 \, mph$

Furthermore, since this is not the stall speed, it does not have to be adjusted.

(b) The minimum takeoff and landing speed for wings with slotted flaps is determined as follows:

$C_{Lmax} := 2.5$

Guess value: $v_{min} := 100 \dfrac{ft}{sec}$

Given

$$F_{Lmin} = C_{Lmax} \frac{1}{2} \rho \cdot v_{min}^2 \cdot A_{plan}$$

$v_{min} := Find(v_{min}) = 100.487 \dfrac{ft}{s}$ $v_{min} = 68.514 \, mph$

Furthermore, since this is not the stall speed, it does not have to be adjusted (increased). And, finally, for the given angle of attack, α of 10°, the use of slotted flaps increases the lift coefficient, C_L from 1.25 to 2.5 and thus allows a decrease in the corresponding minimum takeoff and landing speed from 96.893 mph to 68.514 mph.

Furthermore, Figure 10.30b illustrates the effects of flaps (no flap, single-slotted flap, and double-slotted flap) simultaneously on the lift coefficient, C_L and the drag coefficient, C_D. This figure illustrates that the corresponding increase in the drag coefficient, C_D is even more significant than the increase in the lift coefficient, C_L; the maximum drag coefficient, C_D increases from about 0.06 for no flap to about 0.13 for a single-slotted flap, and to about 0.3 for a double-slotted flap. Thus, there is an undesirable increase in the drag coefficient, C_D that is accompanied by the desirable increase/maximization in the lift coefficient, C_L (resulting from both, addition of a flap/increase in slots in the flap and an increase in the angle of attack, α). As a result, the aircraft engines are required to work harder in order to provide the required thrust to overcome this drag; such a task may be accomplished by increasing the angle of attack, α and enlarging the wing area, $A = bc$ (this is done by extending the chord length, c, the length between the leading and the trailing edges of the wing) in order to maximize the lift coefficient, C_L and to maximize the lift force, F_L, respectively.

EXAMPLE PROBLEM 10.24

Figure 10.30b illustrates the effects of flaps (no flap, single-slotted flap, and double-slotted flap) simultaneously on the lift coefficient, C_L and the drag coefficient, C_D. Referring to Example Problem 10.23 above, consider an airplane that weighs 30,000 lb with wings with a span width of 25 ft and a chord length of 20 ft, as illustrated in Figure EP 10.24. Assume the airplane takes off and lands at sea level (air at standard atmosphere at sea level, $\rho = 0.0023768$ slugs/ft^3) at an angle of attack of 10°. (a) Determine the drag force during takeoff and landing for wings with no flaps. (b) Determine the drag force during takeoff and landing for wings with slotted flaps.

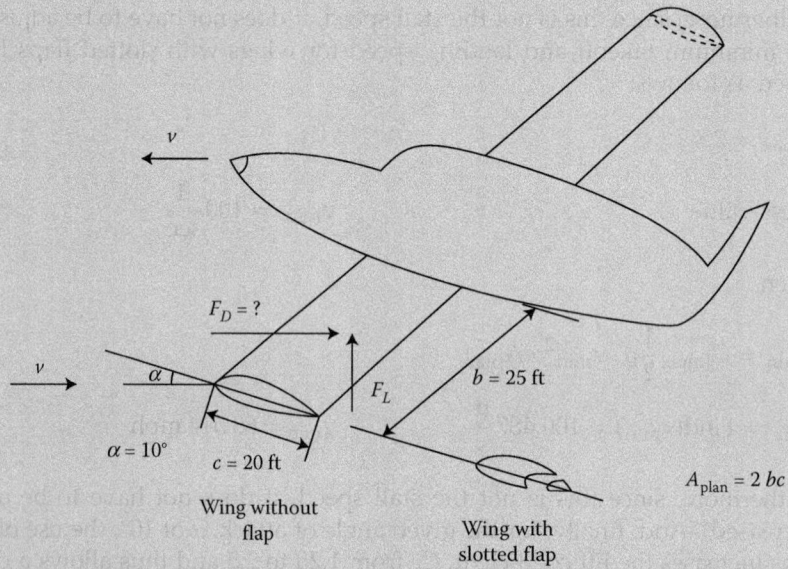

FIGURE EP 10.24
Plane with nonsymmetrical wings (airfoils) (with and without slotted flaps) takes off and lands.

Mathcad Solution

(a)–(b) The minimum takeoff and landing speed was determined in Example Problem 10.23 above. Furthermore, the maximum lift coefficient at takeoff and landing was read from Figure 10.30a for the given angle of attack of 10° in Example Problem 10.23 above. The lift force must equal the weight of the aircraft at takeoff and landing (see Example Problem 10.23 above). The drag force is determined by applying Equation 10.12, and the drag coefficient, C_D is determined from Figure 10.30b for the corresponding lift coefficient, C_L.

(a) The drag force during takeoff and landing for wings with no flaps is determined as follows:

$$\text{slug} := 1\,\text{lb}\frac{\sec^2}{\text{ft}} \qquad b := 25\,\text{ft} \qquad c := 20\,\text{ft} \qquad A_{plan} := 2 \cdot b \cdot c = 1 \times 10^3\,\text{ft}^2$$

$$\rho := 0.0023768\frac{\text{slug}}{\text{ft}^3} \qquad \alpha := 10\,\text{deg} \qquad W := 30000\,\text{lb} \qquad F_{Lmin} := W = 3 \times 10^4\,\text{lb}$$

$C_{Lmax} := 1.25$ $C_{Dmax} := 0.04$ $v_{min} := 96.893 \, mph = 142.11 \frac{ft}{s}$

Guess value: $F_{Dmin} := 100 \, lb$

Given

$$F_{Dmin} = C_{Dmax} \frac{1}{2} \rho \cdot v_{min}^2 \cdot A_{plan}$$

$F_{Dmin} = Find(F_{Dmin}) = 959.998 \, lb$

(b) The drag force during takeoff and landing for wings with slotted flaps is determined as follows:

$C_{Lmax} := 2.5$ $C_{Dmax} := 0.075$ $v_{min} := 68.514 \, mph = 100.487 \frac{ft}{s}$

Guess value: $F_{Dmin} := 100 \, lb$

Given

$$F_{Dmin} = C_{Dmax} \frac{1}{2} \rho \cdot v_{min}^2 \cdot A_{plan}$$

$F_{Dmin} = Find(F_{Dmin}) = 900.006 \, lb$

Therefore, although for the given angle of attack, α of $10°$, the use of slotted flaps increases the drag coefficient, C_D from 0.04 to 0.075, the corresponding drag force, F_D actually decreases from 959.998 lb to 900.006 lbs. The decrease in the drag force, F_D is due to a corresponding decrease in the minimum takeoff and landing speed from 96.893 mph to 68.514 mph, as the drag force is directly proportional to v_{min}^2.

10.5.5.5 Optimizing the Shape of an Airfoil by the Aspect Ratio

While modifying the shape of the airfoil, through the use of moveable flaps and the use of a symmetrical or a nonsymmetrical shape may be important for takeoff and landing of airplanes, modifying the shape of the airfoil through the aspect ratio of the airfoil is important during cruising altitude. The aspect ratio, AR is defined as the ratio of the square of the average span, b of an airfoil to the planform area, $A = bc$ as follows:

$$AR = \frac{b^2}{A} = \frac{b^2}{bc} = \frac{b}{c} \tag{10.27}$$

Specifically, the aspect ratio, AR is a measure of how narrow an airfoil is in the direction of flow. Figure 10.31a illustrates the variation of the lift coefficient, C_L as a function of angle of attack, α and the aspect ratio, AR. Figure 10.31b illustrates the variation of the drag

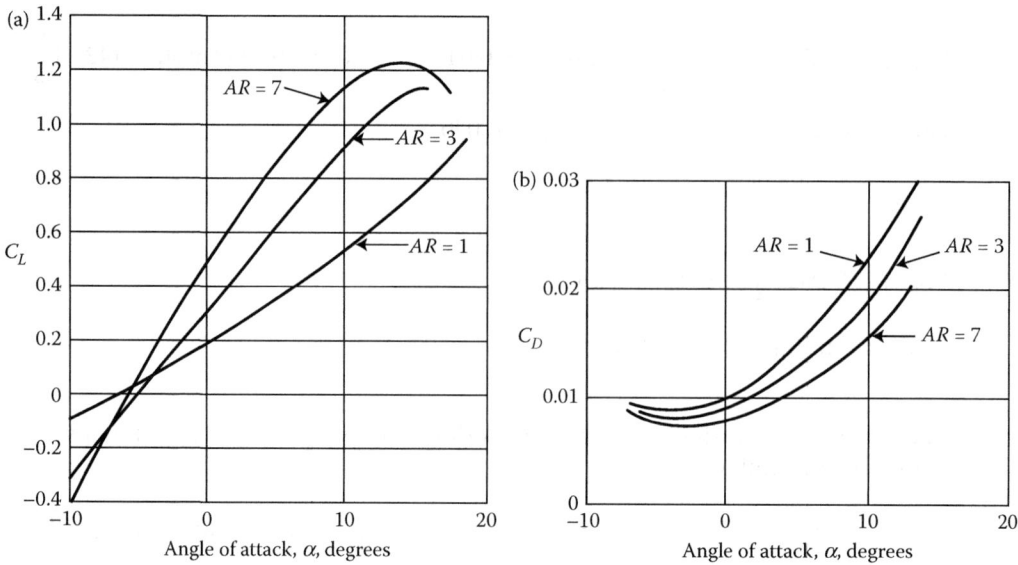

FIGURE 10.31
(a) The variation of the lift coefficient, C_L with the angle of attack, α and the aspect ratio, AR. (b) The variation of the drag coefficient, C_D with the angle of attack, α and the aspect ratio, AR. (Adapted from Prandtl, L., *National Advisory Committee for Aeronautics (NACA) Report 116,* Langley Memorial Aeronautical Laboratory, Langley Field, VA, 1921, p. 193.)

coefficient, C_D as a function of angle of attack, α and the aspect ratio, AR. In general, as the aspect ratio, AR increases, the lift coefficient, C_L increases while the drag coefficient, C_D decreases. Such an observation is due to the fact that a large aspect ratio, AR is representative of a long and narrow wing with a short tip length (see Figure 10.22), which yields smaller tip losses and smaller induced drag than a short and wide wing having the same planform area, $A = bc$. As a result, airfoils with large aspect ratios (such as those designed for large commercial aircraft) fly more efficiently (in terms of a reduced drag and an increased lift), although they are less maneuverable due to the increased moment of inertia that results from a larger distance from the wings to the center of the aircraft. Furthermore, airfoils with smaller aspect ratios (such as those designed for fighter jets) are less efficient (in terms of an increased drag and a reduced lift), while they are more maneuverable due to the proximity of the wings from the center of the aircraft (due to a reduction in the moment of inertia). The increased drag may be minimized by attaching devices called endplates or winglet at the tips of the wings, normal to the top surface.

EXAMPLE PROBLEM 10.25

Figure 10.31a illustrates the variation of the lift coefficient, C_L as a function of angle of attack, α and the aspect ratio, AR, while Figure 10.31b illustrates the variation of the drag coefficient, C_D as a function of angle of attack, α and the aspect ratio, AR. Consider an airplane with wings that have a planform area of 1600 ft², as illustrated in Figure EP 10.25. Assume that the airplane cruises at an altitude of 45,000 ft (air at standard atmosphere at an altitude of 45,000 ft, $\rho = 0.00046227$ slugs/ft³) at a steady cruising speed of 350 mph, with a cruising angle of attack of 5°. (a) Determine the lift force and the drag

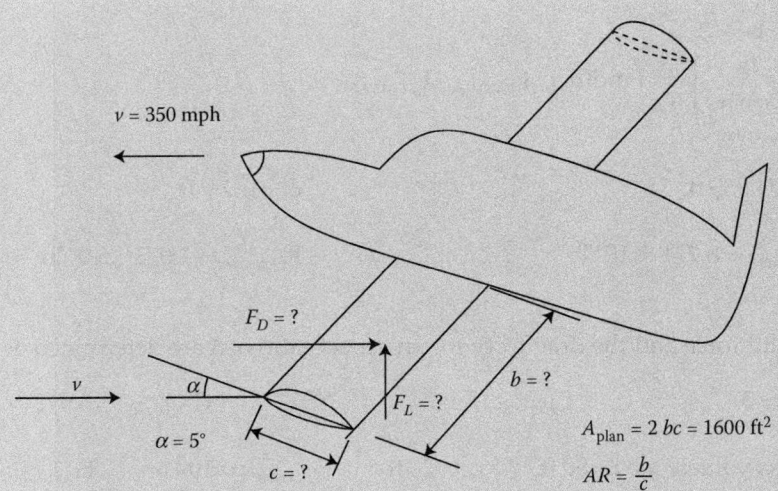

FIGURE EP 10.25
Plane cruises.

force for an aspect ratio of 7. (b) Determine the lift force and the drag force for an aspect ratio of 3.

Mathcad Solution

(a)–(b) The lift force is determined by applying Equation 10.21. Note, the lift force must equal the weight of the aircraft at steady cruising speed. The drag force acting on the aircraft wings during steady cruising speed is determined by applying Equation 10.12. Furthermore, the lift and drag coefficients are determined from Figures 10.31a and b, respectively, for the assumed aspect ratio, for an angle of attack, $\alpha = 5°$.

(a) The lift force and the drag force for an aspect ratio of 7 are determined as follows:

$$\text{slug} := 1\,\text{lb}\,\frac{\sec^2}{ft} \qquad A_{plan} := 1600\,ft^2 \qquad \alpha := 5\,\text{deg}$$

$$\rho_{cruise} := 0.00046227\,\frac{\text{slug}}{ft^3} \qquad v_{cruise} := 350\,\text{mph} = 513.333\,\frac{ft}{s}$$

$$AR := 7 \qquad C_{Lcruise} := 0.9 \qquad C_{Dcruise} := 0.011$$

Guess value: $b := 50\,ft$ $c := 20\,ft$ $F_{Dcruise} := 100\,lb$ $F_{Lcruise} := 100\,lb$

Given

$$AR = \frac{b}{c} \qquad A_{plan} = 2 \cdot b \cdot c$$

$$F_{Lcruise} = C_{Lcruise}\,\frac{1}{2}\,\rho_{cruise} \cdot v_{cruise}^2 \cdot A_{plan}$$

$$F_{Dcruise} = C_{Dcruise}\,\frac{1}{2}\,\rho_{cruise} \cdot v_{cruise}^2 \cdot A_{plan}$$

$$\begin{pmatrix} b \\ c \\ F_{Dcruise} \\ F_{Lcruise} \end{pmatrix} := Find(b, c, F_{Dcruise}, F_{cruise})$$

$b = 74.833\ ft$ $c = 10.69\ ft$

$F_{Lcruise} = 8.771 \times 10^4\ lb$ $F_{Dcruise} = 1.072 \times 10^3\ lb$

(b) The lift force and the drag force for an aspect ratio of 3 are determined as follows:

$AR := 3$ $C_{Lcruise} := 0.7$ $C_{Dcruise} := 0.012$

Guess value: $b := 50\ ft$ $c := 20\ ft$ $F_{Dcruise} := 100\ lb$ $F_{Lcruise} := 100\ lb$

Given

$AR = \dfrac{b}{c}$ $A_{plan} = 2 \cdot b \cdot c$

$F_{Lcruise} = C_{Lcruise} \dfrac{1}{2} \rho_{cruise} \cdot V_{cruise}^2 \cdot A_{plan}$

$F_{Dcruise} = C_{Dcruise} \dfrac{1}{2} \rho_{cruise} \cdot V_{cruise}^2 \cdot A_{plan}$

$$\begin{pmatrix} b \\ c \\ F_{Dcruise} \\ F_{Lcruise} \end{pmatrix} := Find(b, c, F_{Dcruise}, F_{Lcruise})$$

$b = 48.99\ ft$ $c = 16.33\ ft$

$F_{Lcruise} = 6.822 \times 10^4\ lb$ $F_{Dcruise} = 1.169 \times 10^3\ lb$

Therefore, for the given angle of attack, α of 5°, as the aspect ratio, AR decreases from 7 to 3, the lift coefficient, C_L decreases from 0.9 to 0.7, while the drag coefficient, C_D increases from 0.011 to 0.012. Furthermore, the corresponding lift force decreases from $8.771 \times 10^4\ lb$ to $6.822 \times 10^4\ lb$, while the drag force increases from $1.072 \times 10^3\ lb$ to $1.169 \times 10^3\ lb$, as the lift force and the drag force are directly proportional to the lift coefficient and the drag coefficient, respectively.

10.5.6 Estimating the Lift Force and Lift Coefficient for a Hot-Air Balloon

As stated above, in general, the component of the resultant pressure and shear (viscous) forces that acts normal to the direction of the flow is called the lift force, F_L, which tends to move the submerged body in that direction. Furthermore, the component of the resultant pressure and shear (viscous) forces that acts in the direction of the flow is called the drag force, F_D. However, for the case of an ascending hot-air balloon, both the lift force (acts upward) and the drag force (acts downward) act in the direction of flow, as illustrated in Figure EP 10.26. Assuming a constant velocity of ascent for the hot-air balloon, the application of Newton's second law of motion yields the following:

$$\sum_z F_z = F_L + F_B - W - F_D = ma_z = 0 \tag{10.28}$$

where W is the weight of the body and F_B is the buoyant force. Table 10.6 presented the drag coefficient, C_D for numerous geometric shapes at various orientations to the direction of flow, for $R > 10{,}000$, for three-dimensional bodies. Although Table 10.6 did not list a hot-air balloon, the drag coefficient, C_D for a hot-air balloon may be estimated by the drag coefficient, C_D for a parachute, where $C_D = 1.3$. Furthermore, because Section 10.5.5 focused on the evaluation of the lift coefficient, C_L for a airfoils, the lift coefficient for a hot-air balloon may be indirectly evaluated by the application of the equation for the lift force, F_L given by Equation 10.21. Such is illustrated in Example Problem 10.26.

EXAMPLE PROBLEM 10.26

A hot-air balloon with a diameter of 15 ft, height of 13 ft, weight of 200 lb, and cargo (4 ft × 4 ft × 4 ft) weighing 300 lb ascends at a constant velocity of 25 ft/sec in the atmosphere to a height of 10,000 ft above sea level (standard atmosphere at 10,000 ft above sea level, $\rho = 0.0017555$ slugs/ft^3, $\mu = 0.35343 \times 10^{-6}$ lb-sec/ft^2, $\gamma = 0.056424$ lb/ft^3), as illustrated in Figure EP 10.26. Assume the drag coefficient for the hot-air balloon, $C_D = 1.3$. (a) Determine the lift force. (b) Determine the lift coefficient.

Mathcad Solution

(a) In order to determine the lift force, Equation 10.28 (Newton's second law of motion) is applied in the vertical direction. The frontal area is used to compute the drag force for the hot-air balloon, and the drag force is determined by applying Equation 10.12 as follows:

$$\sum_z F_z = F_L + F_B - W - F_D = ma_z = 0$$

$$D_b := 15 \text{ ft} \qquad A_{frontb} := \frac{\pi \cdot D_b{}^2}{4} = 176.715 \text{ ft}^2 \qquad h_b := 13 \text{ ft} \qquad C_{Db} := 1.3$$

$$v := 25 \frac{\text{ft}}{\text{sec}} \qquad b_c := 4 \text{ ft} \qquad d_c := 4 \text{ ft} \qquad h_c := 4 \text{ ft} \qquad \gamma_{air} := 0.056424 \frac{\text{lb}}{\text{ft}^3}$$

$$V_{bc} := \frac{\pi \cdot D_b{}^3}{12} + \frac{1}{3}\left[\frac{\pi \cdot D_b{}^2 \,(h_b + h_c)}{4}\right] - \frac{1}{3}\left(\frac{\pi \cdot b_c{}^2 \, h_c}{4}\right) + b_c \cdot d_c \cdot h_c = 1.932 \times 10^3 \text{ ft}^3$$

$$W_b := 200 \text{ lb} \qquad\qquad W_c := 300 \text{ lb} \qquad\qquad W_{bc} := W_b + W_c = 500 \text{ lb}$$

Guess value: $\qquad F_B := 100 \text{ lb} \qquad F_D := 100 \text{ lb} \qquad F_L := 100 \text{ lb}$

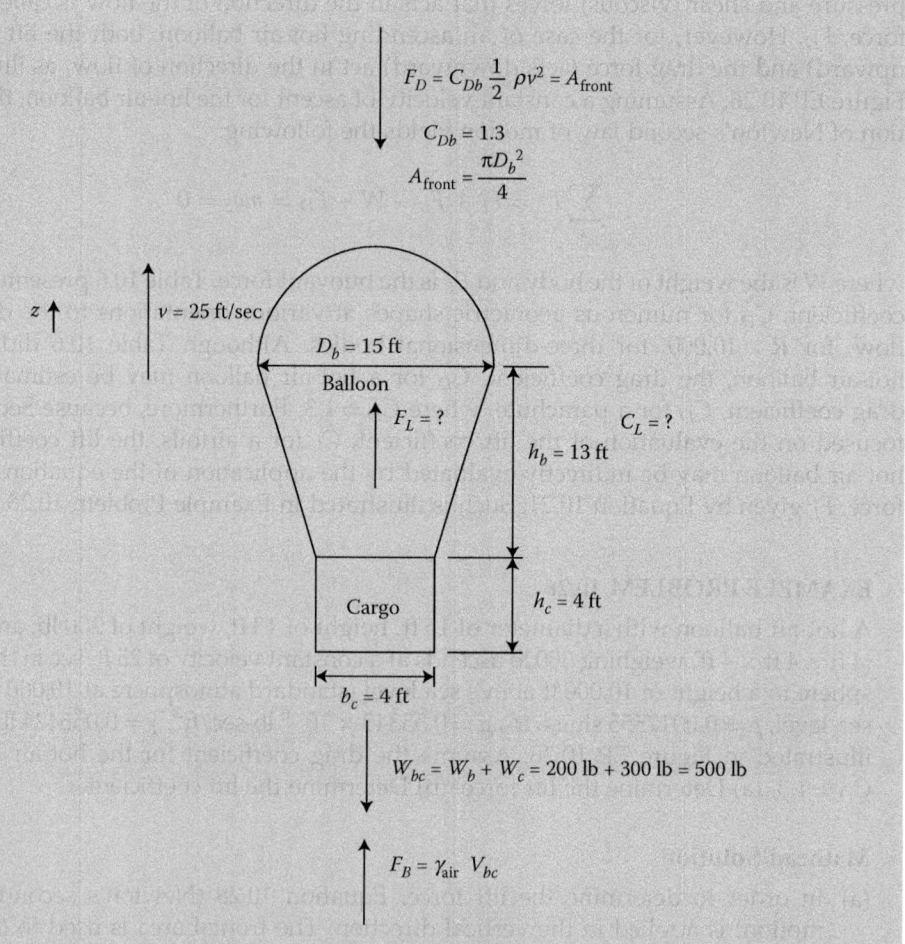

FIGURE EP 10.26
Hot-air balloon ascends in the atmosphere.

Given

$$F_L + F_B - W_{bc} - F_D = 0 \qquad\qquad F_B = \gamma_{air} \cdot V_{bc} \qquad F_D = C_{Db} \frac{1}{2} \rho_{air} \cdot v^2 \cdot A_{frontb}$$

$$\begin{pmatrix} F_B \\ F_D \\ F_L \end{pmatrix} := Find(F_B, F_D, F_L)$$

$$F_B = 109.022 \text{ lb} \qquad\qquad F_D = 170.631 \text{ lb} \qquad\qquad F_L = 561.609 \text{ lb}$$

(b) The planform area of the hot-air balloon is used to model the lift force for the hot-air balloon, and the lift coefficient is determined by applying the lift force equation, Equation 10.21, as follows:

$$slug := 1 \text{ lb} \frac{sec^2}{ft} \qquad\qquad \rho_{air} := 0.0017555 \frac{slug}{ft^3} \qquad\qquad \mu_{air} := 0.35343 \times 10^{-6} \text{ lb} \frac{sec}{ft^2}$$

$$\text{Aplanc} := \frac{\pi \cdot D_b{}^2}{4} = 176.715 \, \text{ft}^2$$

Guess value: $\qquad$ $R := 10000$ $\qquad$ $C_L := 1.0$

$$F_L = C_L \frac{1}{2} \rho_{air} \cdot v^2 \cdot \text{Aplanc} \qquad\qquad R = \frac{\rho_{air} \cdot v \cdot D_b}{\mu_{air}}$$

$$\begin{pmatrix} R \\ C_L \end{pmatrix} := \text{Find}(R, C_L)$$

$R = 1.863 \times 10^6$ $\qquad\qquad\qquad\qquad$ $C_L = 5.793$

Since $R = 1.863 \times 10^6$, this confirms the assumption that $R > 10,000$ (see Table 10.6). Furthermore, it is important to note that because such simplistic assumptions (for instance, the planform area of the hot-air balloon in the lift equation) have been made in the estimation and modeling of the lift force, F_L, the estimated value of the lift coefficient, C_L is a rough estimate at best.

End-of-Chapter Problems

Problems with a "C" are conceptual problems. Problems with a "BG" are in English units. Problems with an "SI" are in metric units. Problems with a "BG/SI" are in both English and metric units. All "BG" and "SI" problems that require computations are solved using Mathcad.

Introduction

10.1C Explain the difference between internal flow vs. external flow, and explain the difference in the results sought in a given flow analysis.

10.2C In external flow, one is interested in the determination of the flow pattern around the body; the lift force, F_L; the drag force, F_D (resistance to motion) on the body; and the patterns of viscous action in the fluid as it passes around the body. Explain when the lift force becomes dominant.

10.3C Explain what are the governing equations in external flow. Furthermore, are they applied using the differential approach, the integral approach, or both?

10.4C In the application of the governing equations (energy and momentum) for external flow, the assumption of ideal flow is made (see Table 4.1 and Section 10.4). Furthermore, the application of the governing equations for ideal external flow requires the modeling of flow resistance (frictional losses). The modeling of flow resistance requires a "subset level" application of the governing equations (Section 10.3). The distinction between ideal and real flow determines how the flow resistance is modeled in the "subset level" application of the governing equations. Explain the assumption of ideal flow and how the flow resistance is modeled.

10.5C Applications of the governing equations for ideal external flow include the analysis and design of planes, cars, buildings, ships, and submerged pipes, for instance (Section 10.5). Furthermore, in external flow, one is interested in the determination of the flow pattern around the body; the lift force, F_L; the drag force, F_D (resistance to motion) on the body; and the patterns of viscous action in the fluid as it passes around the body. Further explain the lift force, F_L and the drag force, F_D.

Modeling Flow Resistance in External Flow: A Subset Level Application of the Governing Equations

10.6C The application of the governing equations for ideal external flow and velocity-measuring devices (Section 10.4) requires the modeling of flow resistance (frictional losses). The modeling of flow resistance requires a "subset level" application of the governing equations. The distinction between ideal and real flow determines how the flow resistance is modeled in the "subset level" application of the governing equations. Explain how the flow resistance is evaluated in the application of the governing equations.

10.7C The flow resistance in external flow (drag force) is modeled by a "subset level" application of the appropriate governing equations, as presented in Chapter 6 (and summarized in Section 10.4.1.1). Explain what a "subset level" application of the appropriate governing equations implies, and explain when the use of dimensional analysis (Chapter 7) is needed in the derivation of the flow resistance equations.

The Drag Force and the Lift Force in External Flow

Evaluation of the Drag Force and the Drag Coefficient in External Flow

10.8C Evaluation of the drag force flow resistance term requires a "subset level" application of the governing equations (see Chapters 6 and 7 and Section 10.4.1.1). What is the expression for the drag force?

10.9C While the drag coefficient for external flow, C_D (in the expression for the drag force, $F_D = C_D(1/2)\rho v^2 A$) is a mainly a function of E, it will selectively vary with R, L_i/L, ε/L, M, and F (independent π terms) depending mainly on the velocity of flow, the shape of the body, and whether there is a wave action at the free surface. Explain the role of each independent π term in the evaluation of the drag coefficient, C_D.

10.10C While the drag coefficient for external flow, C_D (in the expression for the drag force, $F_D = C_D(1/2)\rho v^2 A$) is a mainly a function of E, it will selectively vary with R, L_i/L, ε/L, M, and F (independent π terms) depending mainly on the velocity of flow, the shape of the body, and whether there is a wave action at the free surface. Explain the role of the velocity of flow, and the shape of the body, in the evaluation of the drag coefficient, C_D.

10.11C In the design of cars, airplanes, ships, submarines, and buildings, the ultimate goal is to minimize the effect of the drag force, F_D. Explain how to minimize the effect of the drag force, F_D in design.

10.12BG The span of a strut is $900\,ft$ and the chord length is $0.4\,ft$, as illustrated in Figure EP 10.1. The velocity of air (standard atmosphere at sea level, $\rho = 0.0023768\,slugs/ft^3$, $\mu = 0.37372 \times 10^{-6}\,lb\text{-}sec/ft^2$) flowing over the strut is $13\,ft/sec$. The drag coefficient for a streamlined strut varies with the thickness-to-chord length (D to L) ratio for a Reynolds number, $R = 40{,}000$, as illustrated in Figure 10.8. (a) Determine the effect of the least streamlined design thickness of the strut, $D/L = 0.4$ on the pressure, friction, and total drag. (b) Determine the effect of an optimally designed thickness of the strut $D/L = 0.25$ on the pressure, friction, and total drag. (c) Determine the effect of the most streamlined design thickness of the strut $D/L = 0.05$ on the pressure, friction, and total drag.

10.13BG The length of a long elliptical cylinder is $1700\,ft$, and the thickness of the ellipse is $0.8\,ft$, as illustrated in Figure EP 10.2. The velocity of air (standard atmosphere at sea level, $\rho = 0.0023768\,slugs/ft^3$, $\mu = 0.37372 \times 10^{-6}\,lb\text{-}sec/ft^2$) flowing over the cylinder is $25\,ft/sec$. The drag coefficient for a long elliptical cylinder varies with the aspect ratio, L/D (L is the length of the ellipse in the direction of flow, and D is the thickness of the ellipse) for a Reynolds number, $R = 100{,}000$, as illustrated in Figure 10.9. (a) Determine the effect of the most blunt design aspect ratio, $L/D = 0$ (where the elliptical cylinder becomes a flat plate normal to the direction of flow) on the total drag. (b) Determine the effect of a blunt design aspect ratio, $L/D = 1$ (where the elliptical cylinder becomes a circular cylinder) on the total drag. (c) Determine the effect of a streamlined design aspect ratio, $L/D = 2$ (elliptical cylinder parallel to the direction of flow) on the total drag. (d) Determine the effect of the most streamlined design aspect ratio, $L/D \geq 4$ (where the elliptical cylinder becomes a flat plate parallel to the direction of flow) on the total drag.

10.14BG The drag coefficient for vehicles varies with the deign shape of the body, as illustrated in Table 10.2. Considering the four sedans (1920, 1940–1945, 1968–1969, and 1990–1992) illustrated in Figure EP 10.3, there is a historical trend in the reduction in the drag coefficient. Assume a given sedan moves at a speed of $75\,mph$ in air (standard atmosphere at sea level, $\rho = 0.0023768\,slugs/ft^3$, $\mu = 0.37372 \times 10^{-6}\,lb\text{-}sec/ft^2$), and the average frontal area of a given sedan is $29\,ft^2$. (a) Determine the effect of the historical design and thus the drag coefficient on the magnitude of the drag force.

10.15C Explain the occurrence of flow separation in external flow.

10.16BG The flow separation and thus the drag coefficient for vehicles vary with the deign shape of the body, as illustrated in Figure 10.11. Considering the two vans (blunt/square nose vs. streamlined/round nose) illustrated in Figure EP 10.4, there is a reduction in the drag coefficient for the streamlined van ($C_D = 0.75$ vs $C_D = 0.45$). Assume a given van moves at a speed of $45\,mph$ in air (standard atmosphere at sea level, $\rho = 0.0023768\,slugs/ft^3$, $\mu = 0.37372 \times 10^{-6}\,lb\text{-}sec/ft^2$), and the average frontal area of a given van is $85\,ft^2$. (a) Determine the effect of the blunt vs. streamlined design and thus the drag coefficient on the magnitude of the drag force.

10.17BG Figure 10.12 illustrates the variation of the drag coefficient, C_D with the R for two-dimensional bodies. Consider a 1000-ft-long circular cylinder with a diameter of $2.6\,ft$. Assume water at $70°F$ flows over the cylinder, as illustrated in Figure EP 10.5. (a) Determine the drag coefficient and the drag force if the velocity of flow over the cylinder is 0.000166 ft/sec. (b) Determine the drag coefficient and the drag force if the velocity of flow over the cylinder is $4.3\,ft/sec$.

10.18BG Figure 10.13 illustrates the variation of the drag coefficient, C_D with the R for three-dimensional axisymmetric, bodies. Consider a disk with a diameter of $3\,ft$. Assume water at $70°F$ flows over the disk, as illustrated in Figure EP 10.6. (a) Determine the drag coefficient and the drag force if the velocity of flow over the disk is $0.00015\,ft/sec$.

(b) Determine the drag coefficient and the drag force if the velocity of flow over the disk is 1.9 *ft/sec.*

10.19BG   Water at 70°F flows at a velocity of 0.0003 *ft/sec* over a hemisphere with a diameter of 0.2 *ft*, as illustrated in Figure EP 10.7. (a) Determine the drag coefficient and the drag force on the body.

10.20BG   (a) A dust particle with a diameter of 0.0015 in and a specific gravity of 4.3 falls in the air (standard atmosphere at sea level, $\rho = 0.0023768$ *slugs/ft³*, $\mu = 0.37372 \times 10^{-6}$ *lb-sec/ ft²*, $\gamma = 0.076472$ *lb/ft³*), as illustrated in Figure EP 10.8a, and settles to the ground after a windstorm. (a) Determine the terminal (free-fall) velocity for the dust particle as it falls to the ground. (b) Mathematically illustrate how Stokes law is applied in the use of a falling sphere viscometer to determine the viscosity of a fluid. Then, mathematically illustrate how Stokes law is applied to derive the expression for terminal (free-fall) velocity for the dust particle as it falls to the ground. (c) Compare your answers for parts (a) and (b).

10.21BG   Figure 10.16 illustrates the independence of drag coefficient, C_D of R at higher values of R, ($R > 10,000$), where the flow becomes fully turbulent, for a disk-shaped body that is normal to the direction of flow. Consider a disk with a diameter of 5 *ft*. Assume water at 70°F flows over the disk, as illustrated in Figure EP 10.9. (a) Determine the drag coefficient and the drag force if the velocity of flow over the disk is 0.45 *ft/sec*. (b) Determine the drag coefficient and the drag force if the velocity of flow over the disk is 3.7 *ft/sec*.

10.22BG   Table 10.4 presents the drag coefficient, C_D for numerous geometric shapes at various orientations to the direction of flow, for $R>10,000$, for two-dimensional bodies. Consider a 2500-*ft*-long equilateral triangular rod with a side of 4 *ft*. Assume water at 70°F flows at a velocity of 3.8 *ft/sec* over the rod, as illustrated in Figure EP 10.10. (a) Determine the drag coefficient and the drag force if the apex of the triangular rod is normal to the direction of flow as illustrated in Figure EP 10.10a. (b) Determine the drag coefficient and the drag force if the base of the triangular rod is normal to the direction of flow as illustrated in Figure EP10.10b.

10.23BG   Table 10.5 presents the drag coefficient, C_D for numerous geometric shapes at various orientations to the direction of flow, for $R > 10,000$, three-dimensional axisymmetric bodies. Consider a hemisphere with a diameter of 5.8 *ft*. Assume water at 70°F flows at a velocity of 6.66 *ft/sec* over the hemisphere, as illustrated in Figure EP 10.11. (a) Determine the drag coefficient and the drag force if the curved surface of the hemisphere is normal to the direction of flow as illustrated in Figure EP 10.11a. (b) Determine the drag coefficient and the drag force if the flat surface of the hemisphere is normal to the direction of flow as illustrated in Figure EP 10.11b.

10.24BG   A rectangular street sign that is 3 *ft* wide and 0.7 *ft* high is secured to a 12-*ft*-long circular rod (post) with a diameter of 2.3 in that is secured into the ground, as illustrated in Figure EP 10.12. The sign is at sea level, and is designed to withstand a maximum wind speed of 65 *mph* (air at standard atmosphere at sea level, $\rho = 0.0023768$ *slugs/ft³*, $\mu = 0.37372 \times 10^{-6}$ *lb-sec/ft²*). (a) Determine the drag force on the street sign and pole. (b) Determine the reaction force acting on the sign and pole due to the thrust force of the maximum wind flow (anchoring force required to keep sign and pole from detaching from the ground).

10.25BG   Table 10.6 presents the drag coefficient, C_D for numerous geometric shapes at various orientations to the direction of flow, for $R > 10,000$, for three-dimensional bodies. A 6.3-*ft*-tall (assume a diameter of 2 *ft*) person weighing 170 *lbs* drops in the air (standard atmosphere at sea level, $\rho = 0.0023768\ slugs/ft^3$, $\mu = 0.37372 \times 10^{-6}\ lb\text{-}sec/ft^2$, $\gamma = 0.076472$ lb/ft^3) to the ground in a parachute with a diameter of 13 ft, weighing 17 lbs, and with ropes 10 ft high, as illustrated in Figure EP 10.13. Determine the terminal (free-fall) velocity of the person in the parachute as he/she falls to the ground.

10.26BG Figure 10.17 illustrates that for flows at moderate (laminar) to high (turbulent) values of R around/across round-shaped bodies such as circular cylinders and spheres, the drag coefficient, C_D, in general, decreases with an increase in R. Consider an 8-mile-long oil pipeline with a diameter of 5 ft that is submerged in the ocean. Assume the ocean water is at 68°F flows over the pipeline, as illustrated in Figure EP 10.14. (a) Determine the drag coefficient and the drag force if the velocity of flow over the pipeline is 0.00038 ft/ sec. (b) Determine the drag coefficient and the drag force if the velocity of flow over the pipeline is 3.9 ft/sec.

10.27BG Figure 10.17 illustrates that for flows at moderate (laminar) to high (turbulent) values of R around/across round-shaped bodies such as circular cylinders and spheres, the drag coefficient, C_D, in general, decreases with an increase in R. Consider a decorative spherical dome structure with a diameter of 8 ft that is secured into the ground, as illustrated in Figure EP 10.15. The dome is at sea level, and is designed to withstand a maximum wind speed of 55 mph (air at standard atmosphere at sea level, $\rho = 0.0023768\ slugs/ft^3$, $\mu = 0.37372 \times 10^{-6}\ lb\text{-}sec/ft^2$). (a) Determine the drag force on the dome. (b) Determine the reaction force acting on the dome due to the thrust force of the maximum wind flow (anchoring force required to keep dome from detaching from the ground).

10.28BG Figure 10.18 illustrates that for a specific range of R, for a sphere of diameter, D, an increase in the relative surface roughness, ε/L results in a decrease in drag coefficient, C_D. Consider a golf ball with a diameter or 1.75 in that weighs 1.8 oz. Assume the golf ball is hit and travels at a speed of 165 mph (air at standard atmosphere at sea level, $\rho = 0.0023768$ $slugs/ft^3$, $\mu = 0.37372 \times 10^{-6}\ lb\text{-}sec/ft^2$), as illustrated in Figure EP 10.16. (a) Determine the drag force and the rate of deceleration of a standard (roughened/dimpled) golf ball. (b) Determine the drag force and the rate of deceleration of a smooth golf ball.

10.29BG Figure 10.19 illustrates the variation of the drag coefficient, C_D for blunt (square shape) and streamlined (elliptical) 2-D bodies, for subsonic flow ($M = v/c < 1$), where compressibility effects are insignificant if $M \le 0.5$, while they are significant for $M > 0.5$. Consider a 1800 ft long square shaped cylinder with a side of 7 ft. Assume air at standard atmosphere at an altitude of 25,000 ft above sea level flows over the cylinder, as illustrated in Figure EP 10.17. (a) Determine the drag coefficient and the drag force if the velocity of flow over the cylinder is 500 ft/sec. (b) Determine the drag coefficient and the drag force if the velocity of flow over the cylinder is 900 ft/sec.

10.30BG Figure 10.20 illustrates the dependence of the drag coefficient, C_D on M for three-dimensional axisymmetric bodies (a circular cylinder, a sphere, and a sharp pointed ogive) for supersonic flow ($M = v/c > 1$). Consider an ogive with a diameter of 4.5 in traveling in the air at standard atmosphere at an altitude of 30,000 ft above sea level, at a speed of 2400 ft/sec, as illustrated in Figure EP 10.18. Determine the drag force acting on the ogive.

10.31BG Figure 10.21 illustrates variation of the wave drag coefficient, C_D with both the F and the shape of the ship (a hull without a bow bulb/streamlined body and a hull with a bow bulb/less streamlined body). Consider a hull with a length of 75 ft that travels at a speed of 8.5 mph in the ocean, as illustrated in Figure EP 10.19. (a) Determine the drag force on a hull without a bow bulb/streamlined body. (b) Determine the drag force on a hull with a bow bulb/less streamlined body.

Evaluation of the Lift Force and the Lift Coefficient in External Flow

10.32C Evaluation of the lift force flow resistance term requires a "subset level" application of the governing equations (see Chapters 6 and 7). What is the expression for the lift force?

10.33C The lift coefficient, C_L is mainly a function of E, and will primarily vary with the shape of the body, L_i/L and the angle of attack, α, but in some cases, it will also depend on R, the relative surface roughness, ε/L, M, and F (independent π terms). Explain the role of each independent π term in the evaluation of the lift coefficient, C_L.

10.34C The lift coefficient, C_L is mainly a function of E, and will primarily vary with the shape of the body, L_i/L and the angle of attack, α, but in some cases, it will also depend on R, the relative surface roughness, ε/L, M, and F (independent π terms). Summarize the role of the shape of the body and the angle of attack in the evaluation of the lift coefficient, C_L, assuming an airfoil.

10.35BG Figure 10.26 illustrates how the lift coefficient, C_L varies as a function of the angle of attack, α for a symmetrical airfoil and a nonsymmetrical airfoil. Consider an airplane that weighs 40,000 lb with nonsymmetrical wings (airfoils) with a span width of 35 ft and a chord length of 22 ft, as illustrated in Figure EP 10.20. Assume the airplane takes off and lands at sea level (air at standard atmosphere at sea level, $\rho = 0.0023768$ $slugs/ft^3$) at a maximum angle of attack. Also assume that the airplane cruises at an altitude of 35,000 ft (air at standard atmosphere at an altitude of 35,000 ft, $\rho = 0.0007319$ $slugs/ft^3$) at an angle of attack of 7°. (a) Determine the minimum takeoff and landing speed. (b) Determine the steady cruising speed at the cruising altitude.

10.36BG Figure 10.27 illustrates both the lift coefficient, C_L vs. the angle of attack, α and the drag coefficient, C_D vs. the angle of attack, α for a typical cambered (nonsymmetrical) airfoil. Consider an airplane that weighs 50,000 lb with cambered (nonsymmetrical) wings with a span width of 40 ft and a chord length of 27 ft, as illustrated in Figure EP 10.21. Assume that the airplane cruises at an altitude of 45,000 ft (air at standard atmosphere at an altitude of 45,000 ft, $\rho = 0.00046227$ $slugs/ft^3$) at a steady cruising speed of 260 mph. (a) Determine the angle of attack to cruise at a steady speed at cruising altitude. (b) Determine the drag force acting on the aircraft wings during steady cruising speed. (c) Determine the thrust power required by the plane engines in order to overcome the wing drag force.

10.37BG Figure 10.28 illustrates the lift-to-drag ratio, $F_L/F_D = C_L/C_D$ vs. different angles of attack, α for a typical cambered (nonsymmetrical) airfoil. Consider an airplane that weighs 60,000 lb with cambered (nonsymmetrical) wings with a span width of 50 ft and a chord length of 39 ft, as illustrated in Figure EP 10.22. Assume that the airplane cruises at an altitude of 45,000 ft (air at standard atmosphere at an altitude of 45,000 ft, $\rho = 0.00046227$ $slugs/ft^3$) at a steady cruising speed of 335 mph. (a) Determine the lift force and the drag force acting on the aircraft wings during steady cruising speed, at the target operating point for the airfoil.

10.38BG Figure 10.30a illustrates the effects of flaps (no flap, single-slotted flap, and double-slotted flap) on the lift coefficient, C_L for a range of angles of attack, α. Consider an airplane that weighs 33,000 lb with wings with a span width of 29 ft and a chord length of 19 ft, as illustrated in Figure EP 10.23. Assume the airplane takes off and lands at sea level (air at standard atmosphere at sea level, $\rho = 0.0023768 \, slugs/ft^3$) at an angle of attack of 15°. (a) Determine the minimum takeoff and landing speed for wings with no flaps. (b) Determine the minimum takeoff and landing speed for wings with slotted flaps.

10.39BG Figure 10.30b illustrates the effects of flaps (no flap, single-slotted flap, and double-slotted flap) simultaneously on the lift coefficient, C_L and the drag coefficient, C_D. Referring to ECP 10.38, consider an airplane that weighs 33,000 lb with wings with a span width of 29 ft and a chord length of 19 ft, as illustrated in Figure EP 10.24. Assume the airplane takes off and lands at sea level (air at standard atmosphere at sea level, $\rho = 0.0023768 \, slugs/ft^3$) at an angle of attack of 15°. (a) Determine the drag force during takeoff and landing for wings with no flaps. (b) Determine the drag force during takeoff and landing for wings with slotted flaps.

10.40BG Figure 10.31a illustrates the variation of the lift coefficient, C_L as a function of angle of attack, α and the aspect ratio, AR, while Figure 10.31b illustrates the variation of the drag coefficient, C_D as a function of angle of attack, α and the aspect ratio, AR. Consider an airplane that has wings that have a planform area of 1,700 ft^2, as illustrated in Figure EP 10.25. Assume that the airplane cruises at an altitude of 45,000 ft (air at standard atmosphere at an altitude of 45,000 ft, $\rho = 0.00046227 \, slugs/ft^3$) at a steady cruising speed of 388 mph, with a cruising angle of attack of 10°. (a) Determine the lift force and the drag force for an aspect ratio of 3. (b) Determine the lift force and the drag force for an aspect ratio of 1.

10.41BG A hot-air balloon with a diameter of 16 ft, height of 15 ft, weight of 300 lb, and cargo (4 ft × 4 ft × 4 ft) weighing 400 lb ascends at a constant velocity of 35 ft/sec in the atmosphere to a height of 10,000 ft above sea level (standard atmosphere at 10,000 ft above sea level, $\rho = 0.0017555 \, slugs/ft^3$, $\mu = 0.35343 \times 10^{-6} \, lb\text{-}sec/ft^2$, $\gamma = 0.056424 \, lb/ft^3$), as illustrated in Figure EP 10.26. Assume the drag coefficient for the hot-air balloon, $C_D = 1.3$. (a) Determine the lift force. (b) Determine the lift coefficient.

11

Dynamic Similitude and Modeling

11.1 Introduction

The uncalibrated mathematical model resulting from the dimensional analysis procedure ($\pi_1 = \phi[nothing] = constant$ for a theoretical model, or $\pi_1 = \phi[\pi_2, \pi_3, ...]$ for an empirical model) (see Chapter 7) may be used to develop criteria/laws that govern dynamic similarity/similitude between two flow situations that are geometrically similar but different in size. Analogous to dimensional analysis, dynamic similitude (i.e., the laws governing dynamic similarity) is also an essential and invaluable tool in the deterministic mathematical modeling of fluid flow problems (or any problems other than fluid flow). Deterministic mathematical models are used in order to predict the performance of physical fluid flow situations. As such, in the deterministic mathematical modeling of physical fluid flow situations (both internal and external flow), while some fluid flow situations may be modeled by the application of pure theory (for e.g., creeping, laminar, sonic, or critical flow), some fluid flow situations (actually most, due to the assumption of real turbulent flow) are too complex for theoretical modeling, such as turbulent, subsonic, supersonic, hypersonic, subcritical, or supercritical flow. However, regardless of the complexity of the fluid flow situation, the analysis phase of mathematical modeling involves the formulation, calibration, and verification of the mathematical model, while the subsequent synthesis or design phase of mathematical modeling involves application of the mathematical model in order to predict the performance of the fluid flow situation. Analysis, by definition, is "to break apart" or "to separate into its fundamental constituents," where synthesis, by definition, is "to put together" or "to combine separate elements to form a whole." As such, dynamic similitude is an essential and invaluable tool used in both the analysis and synthesis phases of the deterministic mathematical modeling of fluid flow problems.

Regardless of the complexity of the fluid flow situation, dynamic similitude plays an important role in both the analysis and synthesis phases of the mathematical modeling of fluid flow problems (or any problem other than fluid flow). In the deterministic mathematical modeling of a fluid flow (or any) process, the model formulation step begins by applying the theories/principles (conservation of momentum, conservation of mass, conservation of energy, etc.) that govern the fluid flow situation (see Chapters 3 through 5). Then, the initial form of the model is determined by identifying and defining the physical forces and physical quantities that play an important role in the given flow situation. In the case where each physical force and physical quantity can be theoretically modeled, one may easily analytically/theoretically derive/formulate the mathematical model and apply the model without the need to calibrate or verify it (for e.g., see the definitions of Stokes Law, Poiseuille's law, sonic velocity, critical velocity, etc.). Additionally, if one wishes to experimentally formulate, calibrate, and verify the theoretical mathematical model, one may use the dimensional analysis procedure itself (Buckingham π theorem) (see Chapter 7) and the results of dimensional analysis (laws of dynamic similarity/similitude) (see Chapter 11, herein) to do so. However, in the case where there is at least one physical force or one physical quantity that cannot be theoretically modeled, one may not analytically/theoretically derive/formulate the mathematical model (as in the case of empirically modeling the flow resistance in real turbulent flow for variables such as major head loss, minor head loss, actual discharge, drag force, and the efficiency of a pump or turbine), and one needs to actually calibrate and verify the model before applying the model. In such a case, the dimensional analysis procedure itself (Buckingham π theorem) (see Chapter 7) is actually necessary in order to formulate (derive) the mathematical model while reducing/minimizing the number of variables required to be calibrated. Furthermore, the results of dimensional analysis (laws of dynamic similarity/similitude) (see Chapter 11) are necessary in order to design a geometrically scaled model (of convenient size and fluids, and of economic and time proportions) and allow the design of a systematic and well-guided approach to conduct and interpret the experimental investigations (greatly reducing the number of experiments) required to calibrate the mathematical model.

It is important to note that the topics of dimensional analysis and dynamic similitude are significantly related to one another as follows. While dimensional analysis provides an uncalibrated mathematical model of the fluid flow situation (Chapter 7), dynamic similitude provides a geometrically scaled physical model of the fluid flow situation (Chapter 11), which is used to calibrate the mathematical model. Specifically, dimensional analysis is used to formulate/derive an uncalibrated mathematical model ($\pi_1 = \phi[nothing] = constant$ for a theoretical model, or $\pi_1 = \phi[\pi_2, \pi_3, \ldots]$ for an empirical model) of the fluid flow situation. The uncalibrated mathematical model is used to develop laws of dynamic similitude, which are applied in order to design a geometrically scaled physical model (of convenient size and fluids, and of economic and time proportions) of the flow situation. The geometrically scaled physical model is used to calibrate the mathematical model (i.e., evaluate the π terms) by conducting experiments. The π terms in the uncalibrated mathematical model are used to design a systematic and well-guided approach to conduct and interpret the experimental investigations. Finally, the calibrated mathematical model may be applied to predict the performance of either the geometrically scaled physical model or the full-scale prototype of the fluid flow situation. Therefore, because the topics of dimensional analysis and dynamic similitude are significantly related to one another, many fluid mechanics textbooks present these two topics in the same chapter. However, in this textbook, while the chapter on dimensional analysis (Chapter 7, which provides an uncalibrated mathematical model) precedes the chapters on the applications of the calibrated mathematical models (Chapters 8 through 10), the chapter on dynamic similitude (Chapter 11, which designs geometrically scaled physical models of numerous fluid flow situations) is postponed until the last chapter. The reason is that in order for the reader to fully appreciate and understand how to design a geometrically scaled physical model of a given fluid flow situation, the author believes that it is important for the reader to first learn and understand how to apply the calibrated mathematical model of the fluid flow situation (the flow resistance equations), as presented in Chapters 8 through 10.

11.1.1 The Role of Dynamic Similitude in the Modeling of Fluid Flow

The important role that dynamic similitude plays in the deterministic mathematical modeling of fluid flow problems is highlighted in the following five steps conducted in the modeling process. First, the dimensional analysis procedure itself (Buckingham π theorem) (see

Chapter 7) is used to formulate a mathematical model of the fluid flow situation by arranging/reducing the two or more significant variables (identified from the governing equations) into one or more dimensionless coefficients/π terms ($\pi_1 = \phi[nothing] = constant$ for a theoretical model, or $\pi_1 = \phi[\pi_2, \pi_3, ...]$ for an empirical model), which represents the uncalibrated mathematical model. Second, this uncalibrated mathematical model is ideally calibrated (i.e., the π terms are evaluated) by conducting experiments using the full-scale physical prototype of the fluid flow situation. However, using the full-scale physical prototype to conduct experiments may present challenges such as inconvenient size (too large or too small), inconvenient fluids, economic constraints, time constraints, etc. Therefore, in order to solve this problem, one may instead design a geometrically scaled (either smaller or larger) physical model of the fluid flow situation. Fortunately, the uncalibrated mathematical model resulting from the dimensional analysis procedure ($\pi_1 = \phi[nothing] = constant$ for a theoretical model, or $\pi_1 = \phi[\pi_2, \pi_3, ...]$ for an empirical model) may be used to develop criteria/laws that govern dynamic similarity/similitude between two flow situations that are geometrically similar but different in size; the uncalibrated mathematical model is ready for the initial model application (synthesis) phase. These developed laws of dynamic similarity (similitude) may be applied in order to design a geometrically scaled model and thus allow the design of experiments using a model of convenient size and fluids and of economic and time proportions, that can predict the performance of the full-scale prototype (see Chapter 11). Therefore, the uncalibrated mathematical model is actually/practically calibrated (i.e., the π terms are evaluated) by conducting experiments using the geometrically scaled (either smaller or larger) physical model of the fluid flow situation. Third, the uncalibrated mathematical model is calibrated (i.e., the π terms are evaluated) by conducting experiments using the scaled physical model of the flow situation. Furthermore, the resulting one or more dimensionless coefficients/π terms may be used to design a systematic and well-guided approach to conducting and interpreting the experimental investigations, thus greatly reducing the number of experiments required to calibrate the resulting one or more dimensionless coefficients/π terms (note that while experimental investigations to calibrate the π terms were not conducted in the textbook, the historically documented calibrated π terms [drag coefficients, C_D] are presented and applied in Chapters 4, 5, and 8 through 10). Fourth (ideally), the calibrated mathematical model is verified using test data. And, fifth, the calibrated mathematical model is ready for the final model application (synthesis) phase; one can now use the calibrated mathematical model to predict the performance of either the geometrically scaled model or the full-scale prototype of the fluid flow situation (see applications of theoretical and empirical equations throughout the textbook).

Therefore, the dimensional analysis procedure itself and its results, including the laws of similitude, play an important role in the deterministic mathematical modeling of fluid flow problems. Specifically, the dimensional analysis procedure itself may be used to formulate both theoretical and empirical mathematical models. And, the results of the procedure, including the laws of similitude, may be used to design both the geometrically scaled models and the actual experiments using the scaled models that are used to calibrate the mathematical model. Thus, the results of the dimensional analysis procedure are used to develop the laws of similitude, which are developed and applied in Chapter 11 as follows. First, the uncalibrated mathematical model resulting from the dimensional analysis procedure ($\pi_1 = \phi[nothing] = constant$ for a theoretical model, or $\pi_1 = \phi[\pi_2, \pi_3, ...]$ for an empirical model) is used to develop criteria/laws that govern dynamic similarity/similitude between two flow situations that are geometrically similar but different in size; the uncalibrated mathematical model is used in the initial model application (synthesis) phase (see Sections 11.2 through 11.5). Second, these developed laws of dynamic similarity (similitude) are applied in order to design a geometrically scaled (either smaller or larger) physical model of the fluid flow situation and thus allow the design of experiments (to calibrate the mathematical model) using a model of convenient size and fluids, and of economic and time proportions, that can predict the performance of the full-scale prototype (see Sections 11.6 and 11.7). Third, the uncalibrated mathematical model ($\pi_1 = \phi[nothing] = constant$ for a theoretical model, or $\pi_1 = \phi[\pi_2, \pi_3, \ldots]$ for an empirical model) is calibrated (i.e., the π terms are evaluated) by conducting experiments using the geometrically scaled (either smaller or larger) physical model of the fluid flow prototype. The resulting one or more dimensionless coefficients/π terms may be used to design a systematic and well-guided approach to conducting, and interpreting the experimental investigations, thus greatly reducing the number of experiments required to calibrate the resulting one or more dimensionless coefficients/π terms. And, fourth, and finally, the calibrated mathematical model is ready for the final model application (synthesis) phase; one can now use the calibrated mathematical model to predict the performance of either the geometrically scaled model or the full-scale prototype of the fluid flow situation (see applications of theoretical and empirical equations throughout the textbook).

The relative importance of the role of the dimensional analysis procedure itself, and the results of the procedure, including the laws of similitude, in the modeling of fluid flow may be summarized as follows. The dimensional analysis procedure and its results, including the laws of similitude, may be used as an academic illustrative exercise, if one wishes, to experimentally formulate, calibrate, and verify a theoretical mathematical model. However, the dimensional analysis procedure and its results, including the laws of similitude, are actually necessary in order to formulate, calibrate, and verify the empirical modeling of flow resistance in internal and external flow and the efficiency of pumps and turbines. As such, the historically documented calibration results and applications of the respective flow resistance equations and the efficiency of pumps and turbines are presented in Chapters 4, 5, and 8 through 10.

11.1.2 The Role of Dynamic Similitude in the Empirical Modeling of Flow Resistance in Real Fluid Flow

A major application of the dimensional analysis procedure and its results, including the laws of similitude, in fluid flow is for problems that are too complex to be modeled by the application of pure theory. In particular, the modeling of flow resistance (due to turbulent flow) in internal and external real fluid flow is too complex for theoretical solution. Therefore, a very important application of the dimensional analysis procedure and its results in fluid flow, including the laws of similitude, is in the modeling of flow resistance in internal and external (real) flow. Specifically, the dimensional analysis procedure is applied in the derivation of a given flow resistance equation in internal and external flow (major head loss, minor head loss, actual discharge, and drag force). Furthermore, in the cases where the flow resistance is due to pipe or channel friction (major head loss), pipe component (minor head loss), and/or a flow-measuring device (minor head loss), the flow resistance contributes to the power losses and thus the determination of the efficiency of pumps and turbines. As such, dimensional analysis is also applied in the derivation of the efficiency of pumps and turbines. Then, the respective results of dimensional analysis (i.e., the laws of dynamic similarity/similitude) are necessary in order to design a geometrically scaled model (of convenient size and fluids and of economic and time proportions) and allow the design of a systematic and well-guided approach to conducting and interpreting the experimental investigations (greatly reducing the number of experiments) required to calibrate the mathematical model. Thus, once the flow resistance equations (major head loss, minor head loss, actual discharge, and drag force) and the efficiency of pumps and turbines are derived, calibrated, and verified, they may be directly applied in the appropriate flow situations.

11.1.2.1 Using Dynamic Similitude in the Empirical Modeling of Flow Resistance in Real Fluid Flow

Dimensional analysis and its results, including the laws of dynamic similitude, are used in the empirical modeling of flow resistance in real fluid flow. The important role that dynamic similitude plays in the empirical modeling of flow resistance in real fluid flow is highlighted in the following five steps conducted in the modeling process. First, as Chapter 7 demonstrated, the dimensional analysis procedure itself (Buckingham π theorem) is used to formulate empirical mathematical models, which supplement the momentum theory in the definition of the friction/viscous force and thus in the definition of the flow resistance equations. This step yielded the flow resistance equations and their corresponding drag coefficients, C_D and the empirical expressions for the efficiency of pumps and turbines, which are represented by the respective drag coefficients, C_D. Specifically, a given flow resistance equation, $\pi_1 = \phi(\pi_2, \pi_3, \ldots)$ represents the uncalibrated mathematical model; the dependent π term, $\pi_1 = C_D$ is a function of one or more independent π terms, $\pi_2, \pi_3, \ldots$. Second, this uncalibrated mathematical model (uncalibrated flow resistance equation) is ideally calibrated (i.e., the π terms are evaluated) by conducting experiments using the full-scale physical prototype of the fluid flow situation. However, using the full-scale physical prototype to conduct experiments may present challenges such as inconvenient size (too large or too small), inconvenient fluids, economic and time constraints, etc. Therefore, in order to solve this problem, one may instead design a geometrically scaled (either smaller or larger) physical model of the fluid flow situation. Fortunately, the uncalibrated mathematical model (flow resistance equation) resulting from the dimensional analysis procedure ($\pi_1 = \phi(\pi_2, \pi_3, \ldots)$) may be used to develop criteria/laws that govern dynamic similarity/similitude between two flow situations that are geometrically similar but different in size; at this point, the uncalibrated mathematical model is ready for the initial model application (synthesis) phase (see Sections 11.2 through 11.5). These developed laws of dynamic similarity (similitude) may be applied in order to design a geometrically scaled model and thus allow the design of experiments using a model of convenient size and fluids, and of economic and time proportions, that can predict the performance of the full-scale prototype (see Sections 11.6 and 11.7). Therefore, the uncalibrated mathematical model is actually calibrated (i.e., the π terms are evaluated) by conducting experiments using the geometrically scaled (either smaller or larger) physical model of the fluid flow situation. Third, the uncalibrated mathematical model (uncalibrated flow resistance equation) is calibrated (i.e., the π terms are evaluated) by conducting experiments using the scaled physical model of the flow situation. Furthermore, the resulting two or more dimensionless coefficients/π terms ($\pi_1 = \phi(\pi_2, \pi_3, \ldots)$) may be used to design a systematic and well-guided approach to conducting and interpreting the experimental investigations, thus greatly reducing the number of experiments required to calibrate the resulting two or more dimensionless coefficients/π terms. It is important to note that in this case, where the fluid flow situation is too complex to be modeled by the application of pure theory (e.g., turbulent, subsonic, supersonic, hypersonic, subcritical, or supercritical flow), the resulting dependent π term is a function of two or more independent π terms (i.e., $\pi_1 = \phi(\pi_2, \pi_3, \ldots)$; this represents an uncalibrated mathematical model. As such, calibration of the two or more π terms will require a well-guided limited number of experiments. The calibration results of the experiments may be graphically illustrated as follows. If $\pi_1 = \phi(\pi_2)$, then a plot of the dependent π term, π_1 versus the only one independent π term, π_2 will yield either a single linear or a single nonlinear plot. For instance, a single linear plot is illustrated on the Moody diagram in Figure 8.1 for laminar flow where the friction factor, $f = (64/R)$, where $\pi_1 = f$ and $\pi_2 = R$. A single nonlinear plot is illustrated in Figure 10.9 for the drag coefficient for external flow, C_D, where $\pi_1 = C_D$ and $\pi_2 = L/D$. And, another single nonlinear plot is illustrated in Figure 10.16 for the drag coefficient for external flow, $C_D = \phi(R)$, where $\pi_1 = C_D$ and $\pi_2 = R$. However, if $\pi_1 = \phi(\pi_2, \pi_3)$, then a plot of the dependent π term, π_1 versus two independent π terms, π_2 and π_3 will yield a family of lines or a family of curves. For instance, a family of curves is illustrated on the Moody diagram in Figure 8.1 for transitional and turbulent flow, where the friction factor, $f = \phi(R, \varepsilon/D)$ for various values of ε/D, where $\pi_1 = f$, $\pi_2 = R$, and $\pi_3 = \varepsilon/D$. And, another family of curves is illustrated in Figure 10.18 for the drag coefficient for external flow, $C_D = \phi(R, \varepsilon/L)$ for various values of ε/L, where $\pi_1 = C_D$, $\pi_2 = R$, and $\pi_3 = \varepsilon/L$. Fourth (ideally), the calibrated mathematical model is verified using test data. And, fifth and finally, the calibrated mathematical model is ready for the final model application (synthesis) phase; one can now use the calibrated empirical mathematical model to predict the performance of either the geometrically scaled model or the full-scale prototype of the fluid flow situation (see applications of the empirical flow resistance equations throughout the textbook).

11.1.3 Developing and Applying the Laws of Dynamic Similitude to Design Geometrically Scaled Physical Models of Real Fluid Flow

Although the laws of dynamic similitude may be developed and applied to any fluid flow situation, the main focus of this chapter will be to address real fluid flow situations that require the empirical modeling of flow resistance. The laws of dynamic similarity involve the definition of primary scale ratios and secondary/similitude scale ratios. The similitude scale ratios are needed in order to translate the model values (measurements) of the various physical quantities (geometric, kinematic, and dynamic) and physical properties of the model fluids into the corresponding prototype values. First, the primary scale ratios (see Section 11.2) and the secondary/similitude scale ratios (see Section 11.4) are defined. Then, the uncalibrated mathematical model formulated using dimensional analysis ($\pi_1 = \phi[nothing] = constant$ for a theoretical model, or $\pi_1 = \phi[\pi_2, \pi_3, \ldots]$ for an empirical model) is used to develop (and tailor for a given flow situation) the laws of dynamic similitude (see Sections 11.4 through 11.6). Finally, the laws of dynamic similitude are applied in order to design a geometrically scaled physical model (of convenient size and fluids and of economic and time proportions) of the flow situation (see Sections 11.6 and 11.7).

The development of the laws of dynamic similarity involves the following: (1) definition of primary scale ratios (see Section 11.2), (2) interpretation of the main π terms ratios (see Section 11.3), (3) definition of the laws governing dynamic similarity/similitude and determine secondary/similitude scale ratios (see Section 11.4), and (4) identification of the role and relative importance of the dynamic forces in dynamic similitude (see Section 11.5). The primary scale ratios model the geometric, kinematic, and dynamic similarities. The main π terms include the Euler number, E; the Froude number, F; the Reynolds number, R; the Cauchy number, C; and the Weber number, W. The laws governing dynamic similarity/similitude involve the definition of secondary/similitude scale ratios. And, the dynamic forces include the inertia force, F_I; the pressure force, F_P; the gravity, F_G; the viscous force, F_V; the elastic force, F_E; and the surface tension force, F_T.

Application of the laws of dynamic similitude involves the following: (1) development of guidelines in the application of the laws governing dynamic similarity (see Section 11.6), and (2) application of the laws governing dynamic similarity (see Section 11.7). The development of the guidelines in the application of the laws of dynamic similitude involves the definitions of true models versus distorted models. And, applications of the laws of dynamic similarity are illustrated for flow resistance equations and the efficiency of pumps and turbines. Application of the laws of similarity (similitude) allows the prediction of the performance of the "prototype" from experiments conducted with the "model." The secondary/similitude scale ratios are used to define the dynamic similarity requirements and thus establish a relationship between the model and its prototype, which enables the design of a model from its prototype, or a prototype from its model. Specifically, the secondary/similitude scale ratios are needed in order to translate the model values (measurements) of the various physical quantities (geometric, kinematic, and dynamic) and physical properties of the model fluids into the corresponding prototype values. Furthermore, the similitude scale ratios may be used to predict the behavior of the prototype to possible design changes using the model. The development of experiments of models of convenient size and fluids and of time and economic proportions is accomplished by the application of the laws that govern similarity/similitude. For instance, the laws that govern similarity allow the development of experiments with convenient fluids such as water or air, for instance, with the further application of the results to fluids that are less convenient to work with such as oil, hydrogen, or steam. Furthermore, significant experimental results may be achieved with the "model," which is a small-scale representation of the full-sized "prototype," at lower costs. Finally, different or the same fluids may be used for the model and the prototype, and the model may be either smaller or larger than the prototype.

11.2 Primary Scale Ratios

In the development of criteria/laws governing dynamic similarity/similitude, there are three types of similarity. The most basic is geometric similarity, followed by kinematic similarity, and, finally, dynamic similarity. A higher degree of similarity will also imply a lower degree of similarity. Furthermore, in order for a model to be similar to its prototype in the most complete sense (i.e., appropriate for practical design applications of fluid flow problems), dynamic similarity must be maintained between the model and its prototype. One may note that the three types of similarity are defined through primary scale ratios, which are: (1) the primary length scale ratio, (2) the primary velocity scale ratio, and (3) the primary force scale ratio. It is important to note that while the application of the primary force scale ratio in the design of a model or its prototype is sufficient in order to maintain dynamic similarity, the use of the other two primary scale ratios is not. Furthermore, application of the primary force scale ratio for a particular dominant force (in addition to the inertia force) (or the corresponding π term) in the design of a model or its prototype results in the derivation (and also implies the application) of secondary/similitude scale ratios for that dominant force/"force model" (discussed in a section below). As such, the three types of similarity for a particular dominant force are also defined through secondary/similitude scale ratios (see Tables 11.1 through 11.5). Furthermore, because the secondary/similitude velocity scale ratio, in particular, is directly derived from the primary force scale ratio, application of the former becomes sufficient in order to maintain dynamic similarity for a particular dominant force.

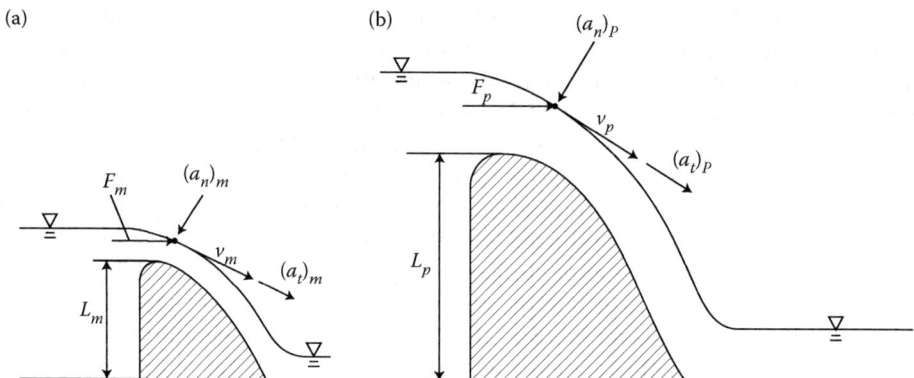

FIGURE 11.1
Geometric, kinematic, and dynamic similarities between (a) a model and (b) its prototype.

11.2.1 Geometric Similarity

Geometric similarity between a model and its prototype implies that the model and its prototype are identical in shape but differ only in size, where the model may be either smaller or larger than its prototype (see Figure 11.1). Geometric similarity is defined through the primary length scale ratio, L_r, which is given as follows:

$$L_r = \frac{L_p}{L_m} \tag{11.1}$$

where L_p = the linear dimensions of the prototype, and L_m = the linear dimensions of the model. A given value for the length scale ratio, L_r maintains a constant ratio of the linear dimensions of the prototype to the corresponding linear dimensions of the model at all points in the field of flow. For example, if the length scale ratio, $L_r = 20$, then every linear dimension in the prototype is 20 times the corresponding dimension on the model.

Geometric similarity implies that the ratio of all of the lengths of all corresponding points in the flow is the same. Furthermore, a given value of L_r also maintains a constant value for the primary area scale ratio, L_r^2, and the primary volume scale ratio, L_r^3. An additional term is also used to define geometric similarity, and is referred to as the model scale, or the model ratio, λ which is given as follows:

$$\lambda = \frac{1}{L_r} = \frac{L_m}{L_p} \tag{11.2}$$

11.2.2 Kinematic Similarity

Kinematic similarity between a model and its prototype implies that the model and its prototype are geometrically similar, and it implies that the ratio of all of the velocities and accelerations at all corresponding points in the flow are the same and have the same direction (see Figure 11.1). Kinematic similarity is defined through the primary velocity scale ratio, v_r, which is given as follows:

$$v_r = \frac{v_p}{v_m} \tag{11.3}$$

where v_p = the velocities of the prototype, and v_m = the velocities of the model. A given value for the velocity scale ratio, v_r maintains a constant ratio of the velocities of the proto-type to the corresponding velocities of the model at all points in the field of flow. From the definition of time, $T = L/v$, the primary time scale ratio, T_r is given as follows:

$$T_r = \frac{L_r}{v_r} \tag{11.4}$$

And, from the definition of acceleration, $a = L/T^2$, the primary acceleration scale ratio, a_r is given as follows:

$$a_r = \frac{L_r}{T_r^2} = \frac{v_r^2}{L_r} \tag{11.5}$$

Furthermore, it may be noted that the state of kinematic similarity can be maintained if and only if the state of dynamic similarity is maintained.

11.2.3 Dynamic Similarity

Dynamic similarity between a model and its prototype implies that the model and its pro-totype are geometrically similar and kinematically similar, and it implies that the ratio of all of the forces acting in a given flow situation are the same at all corresponding points in the field of flow (see Figure 11.1). Dynamic similarity is defined through the primary force scale ratio, F_r, which is given as follows:

$$F_r = \frac{F_p}{F_m} \tag{11.6}$$

where F_p = the forces of the prototype, and F_m = the forces of the model. A given value for the force scale ratio, F_r maintains a constant ratio of the forces of the prototype to the corresponding forces of the model at all points in the field of flow. As stated in Chapter 7, forces that may act in a flow situation include those due to pressure, F_P, gravity, F_G; viscos-ity, F_V; elasticity, F_E; surface tension, F_T; and the inertia force, F_I. As noted in Chapter 7, depending upon the fluid flow problem, while the inertia force will always be important, some of the above-listed forces may not play a significant role. Thus, assuming that all forces listed above are important, then dynamic similarity will be achieved if:

$$F_r = \frac{F_{P_p}}{F_{P_m}} = \frac{F_{G_p}}{F_{G_m}} = \frac{F_{V_p}}{F_{V_m}} = \frac{F_{E_p}}{F_{E_m}} = \frac{F_{T_p}}{F_{T_m}} = \frac{F_{I_p}}{F_{I_m}} \tag{11.7}$$

However, if one or more force is not important in a given flow situation, then it would not be necessary to satisfy its respective force scale ratio in the above equation in order to main-tain dynamic similarity.

In accordance with Newton's second law of motion, one may recall the importance of the role of the inertia force in maintaining steady-state flow. Thus, the necessity of satisfying, at a minimum, its force scale ratio in all dynamically similar steady-state flow situations may be expressed as follows:

$$F_r = \frac{F_{I_p}}{F_{I_m}} \tag{11.8}$$

where a given value for the force scale ratio, F_r maintains a constant ratio of the forces of the prototype to the corresponding forces of the model. Given that the inertia force will always be important, and given the importance of an additional specific force in the flow situation, a practical (and alternative) approach for expressing the force scale ratio, F_r that expresses the minimum requirement for maintaining both steady-state flow and dynamic similarity may be expressed as the ratio of the inertia force to the corresponding additional force as follows:

$$\text{Pressure Force:} \quad \left(\frac{F_I}{F_P}\right)_p = \left(\frac{F_I}{F_P}\right)_m = Euler\,Number = \mathbf{E} \tag{11.9}$$

$$\text{Gravity Force:} \quad \left(\frac{F_I}{F_G}\right)_p = \left(\frac{F_I}{F_G}\right)_m = Froude\,Number = \mathbf{F} \tag{11.10}$$

$$\text{Viscous Force:} \quad \left(\frac{F_I}{F_V}\right)_p = \left(\frac{F_I}{F_V}\right)_m = Reynolds\,Number = \mathbf{R} \tag{11.11}$$

$$\text{Elastic Force:} \quad \left(\frac{F_I}{F_E}\right)_p = \left(\frac{F_I}{F_E}\right)_m = Cauchy\,Number = \mathbf{C} \tag{11.12}$$

$$\text{Surface Tension Force:} \quad \left(\frac{F_I}{F_T}\right)_p = \left(\frac{F_I}{F_T}\right)_m = Weber\,Number = \mathbf{W} \tag{11.13}$$

where the ratio of the inertia force to each of the respective forces introduces the above-listed dimensionless numbers (main π terms), which were introduced in Chapter 7. Furthermore, a given value for the π term of importance (alternatively, a given value for the force scale ratio, F_r) maintains a constant ratio of the forces of the prototype to the corresponding forces of the model. For example, if in addition to the inertia force, the gravity force is also important, then a given value of the Froude number, $\mathbf{F}$ maintains both steady-state flow and dynamic similarity between the prototype and the model. Another example is: if in addition to the inertia force, the viscous force is also important, then a given value of the Reynolds number, $\mathbf{R}$ maintains both steady-state flow and dynamic similarity between the prototype and the model.

11.3 Interpretation of the Main π Terms

In the development of the laws of dynamic similarity, a practical alternative approach for expressing the primary force scale ratio, F_r that expresses the minimum requirement for maintaining both steady-state flow and dynamic similarity may be expressed as the ratio of the inertia force to the corresponding additional force. This yields five main π terms, which include the Euler number, $\mathbf{E}$; the Froude number, $\mathbf{F}$; the Reynolds number, $\mathbf{R}$; the

Cauchy number, C; and the Weber number, W. It is of interest to examine each π term and interpret the implication of its respective definition. It is important to realize that each of the five π terms represents a certain type of velocity ratio (actual to ideal).

11.3.1 The Euler Number

If in addition to the inertia force, F_I, the pressure force, F_P is also important in a fluid flow situation (such as in pressure flow or external flow, where pressure or a pressure difference is important), then in order to maintain steady-state flow and dynamic similarity, the Euler number, E governs the flow, and should remain a constant between the model and its prototype. The Euler number, E is defined as follows:

$$E = \frac{F_I}{F_P} = \frac{\rho v^2 L^2}{\Delta p L^2} = \frac{\rho v^2}{\Delta p} = \frac{v^2}{\frac{\Delta p}{\rho}} \tag{11.14}$$

where the Euler number, E expresses the relative importance of the fluid's inertia force, F_I compared to its pressure force, F_P. Alternatively, the Euler number, E may be expressed as the pressure (recovery) coefficient, C_p (especially in the case of a gradual pipe expansion) and is defined as follows:

$$\frac{2}{E} = C_p = \frac{\Delta p}{\frac{\rho v^2}{2}} = \frac{\Delta p_{actual}}{\Delta p_{ideal}} \tag{11.15}$$

For ideal flow, the flow through a contraction is the conversion of flow energy (pressure energy) to kinetic energy, while the flow through an expansion is the conversion of kinetic energy to flow energy (pressure energy). Furthermore, there is the possibility of cavitation (high velocity and thus low pressure) in the pipe flow, such as in the complex passages that may exist in valves. As such, Δp is defined as $(p_r - p_v)$, where p_r is some reference pressure and p_v is the vapor pressure. Therefore, the pressure coefficient, C_p becomes the dimensionless quantity known as the cavitation number, C_a as follows:

$$\frac{2}{E} = C_p = \frac{\Delta p}{\frac{\rho v^2}{2}} = \frac{p_r - p_v}{\frac{\rho v^2}{2}} = C_a \tag{11.16}$$

where both pressures must be absolute. Furthermore, the Euler number, E may be expressed as the velocity coefficient, C_v and is defined as follows:

$$\sqrt{E} = C_v = \frac{v}{\sqrt{\frac{\Delta p}{\rho}}} = \frac{v_a}{v_i} \tag{11.17}$$

which expresses the ratio of the actual velocity of the flow, v_a to the ideal velocity, v_i of the flow (ideal flow and thus no head loss due to friction). One may note that in the case of ideal flow, $C_v = 1$; however, in the case of real flow, $C_v < 1$, because of the head loss due to friction.

11.3.2 The Froude Number

If in addition to the inertia force, F_I, the gravity force, F_G is also important in a fluid flow situation (such as in open channel flow and external flow, where the flow has a free surface), then in order to maintain steady-state flow and dynamic similarity, the Froude number, F governs the flow, and should remain a constant between the model and its prototype. The Froude number, F is defined as follows:

$$F = \frac{F_I}{F_G} = \frac{\rho v^2 L^2}{\rho L^3 g} = \frac{v^2}{gL} \tag{11.18}$$

where the Froude number, F expresses the relative importance of the fluid's inertia force, F_I compared to its gravity force (or weight), F_G. Alternatively, the Froude number, F is expressed as follows:

$$\sqrt{F} = F = \frac{v}{\sqrt{gL}} = \frac{v_a}{v_i} = \frac{v_a}{v_c} \tag{11.19}$$

where the Froude number, F expresses the ratio of the actual velocity of the flow, v_a to the ideal velocity, v_i of the flow, which is equal to the critical velocity of flow, v_c. One may note that in the case of critical flow (ideal flow or at minimum specific energy), $F = 1$; however, in the case of subcritical flow (real flow at above the minimum specific energy), $F < 1$, and in the case of supercritical flow (real flow at above the minimum specific energy), $F > 1$.

11.3.3 The Reynolds Number

If in addition to the inertia force, F_I, the viscous force, F_V is also important in a fluid flow situation (such as in all types of fluid flow), then in order to maintain steady-state flow and dynamic similarity, the Reynolds number, R governs the flow, and should remain a constant between the model and its prototype. The Reynolds number, R is defined as follows:

$$R = \frac{F_I}{F_V} = \frac{\rho v^2 L^2}{\mu v L} = \frac{\rho v L}{\mu} \tag{11.20}$$

where the Reynolds number, R expresses the relative importance of the fluid's inertia force, F_I compared to its viscosity force, F_V. One may note that when both of these forces are important, the Reynolds number, R plays an important role in the flow situation. Alternatively, the Reynolds number, R may be expressed as follows:

$$R = \frac{v}{\frac{\mu}{\rho L}} = \frac{v_a}{v_i} = \frac{v_a}{v_{crp}} \tag{11.21}$$

where the Reynolds number, R expresses the ratio of the actual velocity of the flow, v_a to the ideal velocity, v_i of the flow, which is equal to the creeping velocity of flow, v_{crp}. One may note that in the case of creeping flow (ideal flow at a very low velocity for which the viscous forces are dominant and the inertia force is insignificant), $R \leq 1$. However, in the case of laminar flow (real flow at a small velocity), $R < 2000$, in the case of transitional flow (real flow at a moderate velocity), $2000 < R < 4000$, and in the case of turbulent flow (real flow at a high velocity), $R > 4000$. Furthermore, one may note that for laminar flow (small to moderate velocities), the viscous force, F_V is large enough to suppress the fluctuations in the flow due to the significance of the inertia force, F_I, which results in a small to moderate value for the Reynolds number, R. However, for turbulent flow (high velocities), the inertia force, F_I is large compared to the viscous force, F_V, which results in a large value for the Reynolds number, R.

11.3.4 The Cauchy Number

If in addition to the inertia force, F_I, the elastic force, F_E is also important in a fluid flow situation (such as in compressible flow), then in order to maintain steady-state flow and dynamic similarity, the Cauchy number, C governs the flow, and should remain a constant between the model and its prototype. The Cauchy number, C is defined as follows:

$$C = \frac{F_I}{F_E} = \frac{\rho v^2 L^2}{E_v L^2} = \frac{\rho v^2}{E_v} = \frac{v^2}{\dfrac{E_v}{\rho}} \tag{11.22}$$

where the Cauchy number, C expresses the relative importance of the fluid's inertia force, F_I compared to its elastic force, F_E. Alternatively, the Cauchy number, C may be expressed as the Mach number, M and is defined as follows:

$$\sqrt{C} = M = \frac{v}{\sqrt{\dfrac{E_v}{\rho}}} = \frac{v_a}{v_i} = \frac{v_a}{c} \tag{11.23}$$

where the Mach number, M expresses the ratio of the actual velocity of the flow, v_a (or the actual velocity of a body moving in a stationary fluid) to the ideal velocity, v_i of the flow, which is equal to the sonic velocity of flow, c. The sonic velocity, c is the speed of sound, or the celerity, which is the speed of an infinitesimally small pressure (sonic) wave in a compressible or elastic fluid. One may note that in the case of sonic flow (ideal flow at sonic velocity), $M = 1$, however, in the case of subsonic flow (real flow at below the sonic velocity), $M < 1$, and in the case of supersonic flow (real flow at above the sonic velocity), $M > 1$. Furthermore, for extremely high values of M, where $M \gg 1$, the flow is called hypersonic. It is important to note that when $M \leq 0.5$, the inertia force caused by the fluid motion is not sufficiently large to cause a significant change in the fluid density, ρ; thus, the compressibility of the fluid is negligible.

11.3.5 The Weber Number

If in addition to the inertia force, F_I, the surface tension force, F_T is also important in a fluid flow situation (such as in the formation of droplets, bubbles, or thin films; capillarity; sheet flows; or when the head on a weir is small), then in order to maintain steady-state flow and dynamic similarity, the Weber number, W governs the flow, and should remain a constant between the model and its prototype. The Weber number, W is defined as follows:

$$W = \frac{F_I}{F_T} = \frac{\rho v^2 L^2}{\sigma L} = \frac{\rho v^2 L}{\sigma} = \frac{v^2}{\dfrac{\sigma}{\rho L}} \tag{11.24}$$

where the Weber number, W expresses the relative importance of the fluid's inertia force, F_I compared to its surface tension force, F_T. Thus, when the surface tension force, F_T is relatively significant, then the Weber number, W is small: $W \leq 1$, and the droplet or bubble or thin film increases in size. However, when the surface tension force, F_T is relatively insignificant, then $W \gg 1$, and the inertia force overcomes the surface tension force and the droplet or bubble or thin film bursts/breaks up into smaller ones. In the case of a small head, H on a weir, one may alternatively express the Weber number, W as follows:

$$\sqrt{W} = W = \frac{v}{\sqrt{\dfrac{\sigma}{\rho L}}} = \frac{v_a}{v_i} = \frac{v_a}{v_c} \qquad (11.25)$$

where the Weber number, W expresses the ratio of the actual velocity of the flow, v_a to the ideal velocity, v_i of the flow, which is equal to the critical velocity flow, v_c. One may note that because the analysis of a weir assumes critical flow (ideal flow), the head, $H \approx y_c$. As such, the critical depth of flow, y_c serves as a minimum head on the weir, H. Therefore, regardless of the whether the head on the weir is small or large, for $H \approx y_c$, the Weber number, $W = (v_a/v_i) = (v_a/v_c) \leq 1$, where the effect of surface tension is significant only when the head, $H \approx y_c$ on the weir is small. Thus, the larger the head on the weir, H, the smaller the value of the Weber number, W (a lot smaller than 1) and thus the less significant the surface tension force. Finally, because the critical depth of flow, y_c serves as a minimum head on the weir, H (unlike the droplet, bubble, capillarity, or thin film flow), the Weber number, W will always be ≤ 1 in the case of a weir. It is important to note that in the case where the surface tension is negligible in the prototype, care should be taken in order that the surface tension does not become important in the model (smaller in scale than the prototype; this is usually accomplished by designing the model so that the head on the weir, H (or the depth in the channel) is no less than 1 or 2 inches).

11.3.6 Implications of the Definitions of the Main π Terms

The five main π terms $(E, F, R, C,$ and $W)$ represent a practical alternative approach for expressing the force scale ratio, F_r. Furthermore, one may note that the five main π terms may be expressed as the ratio of the actual velocity of flow, v_a to the ideal velocity of flow, v_i, which are called the secondary/similitude velocity scale ratios. Thus, the force scale ratio, F_r or the appropriate π term may be used to maintain both steady-state flow and dynamic similarity, with the requirement that each remain a constant between the model and its prototype. For instance, if in addition to the inertia force, the pressure force is also important, then either:

$$F_r = \frac{F_{P_p}}{F_{P_m}} = \frac{F_{I_p}}{F_{I_m}} = constant \qquad (11.26)$$

or:

$$\underbrace{\left[\left(\frac{F_I}{F_P}\right)_p\right]}_{E_p} = \underbrace{\left[\left(\frac{F_I}{F_P}\right)_m\right]}_{E_m} = Euler\ Number \qquad (11.27)$$

is required in order to maintain both steady-state flow and dynamic similarity between the model and its prototype. Furthermore, one may note that each of the five main π terms may be expressed as the ratio of the actual velocity of flow, v_a to the ideal velocity of flow, v_i (i.e., secondary/similitude velocity scale ratios) as follows:

$$\sqrt{E} = C_v = \frac{v}{\sqrt{\frac{\Delta p}{\rho}}} = \frac{v_a}{v_i} \tag{11.28}$$

$$\sqrt{F} = F = \frac{v}{\sqrt{gL}} = \frac{v_a}{v_i} = \frac{v_a}{v_c} \tag{11.29}$$

$$R = \frac{v}{\frac{\mu}{\rho L}} = \frac{v_a}{v_i} = \frac{v_a}{v_{crp}} \tag{11.30}$$

$$\sqrt{C} = M = \frac{v}{\sqrt{\frac{E_v}{\rho}}} = \frac{v_a}{v_i} = \frac{v_a}{c} \tag{11.31}$$

$$\sqrt{W} = W = \frac{v}{\sqrt{\frac{\sigma}{\rho L}}} = \frac{v_a}{v_i} = \frac{v_a}{v_c} \tag{11.32}$$

Thus, for instance, if in addition to the inertia force, the pressure force is also important, then the Euler number, E is required to remain a constant between the model and its prototype as follows:

$$\underbrace{\left[\frac{v}{\sqrt{\frac{\Delta p}{\rho}}}\right]_p}_{E_p} = \underbrace{\left[\frac{v}{\sqrt{\frac{\Delta p}{\rho}}}\right]_m}_{E_m} \tag{11.33}$$

One may recall that while the primary velocity scale ratio, $v_r = (v_p/v_m)$ is required to remain a constant between the model and its prototype in order to maintain kinematic similarity, the state of kinematic similarity can be maintained if and only if the state of dynamic similarity is maintained. As such, while the following is necessary:

$$v_r = \frac{v_p}{v_m} = constant \tag{11.34}$$

it is not sufficient in order to maintain kinematic similarity. Therefore, additionally, one must determine, in addition to the inertia force, the importance of any additional specific forces in the flow situation. Thus, for instance, if in addition to the inertia force, the pressure force is also important, then the Euler number, E is required to remain a constant between the model and its prototype as follows:

$$\underbrace{\left[\frac{v}{\sqrt{\frac{\Delta p}{\rho}}}\right]_p}_{E_p} = \underbrace{\left[\frac{v}{\sqrt{\frac{\Delta p}{\rho}}}\right]_m}_{E_m} \tag{11.35}$$

in order to maintain kinematic similarity as well as steady-state flow and dynamic similarity, where Equation 11.35 represents the secondary/similitude velocity scale ratio for the Euler number, E. Furthermore, one may recall that geometric similarity is implied by both kinematic and dynamic similarity. Thus, in summary, while the primary force scale ratio, F_r (or the main π terms, or the secondary/similitude velocity scale ratios) may be applied to maintain steady-state flow and dynamic similarity (which also implies kinematic and geometric similarity), the primary velocity scale ratio, v_r is not sufficient to maintain either kinematic or geometric similarity.

11.4 Laws Governing Dynamic Similirity: Secondary/Similitude Scale Ratios

In the development of the laws of dynamic similarity, the uncalibrated mathematical model formulated using dimensional analysis ($\pi_1 = \phi[\ nothing] = constant$ for a theoretical model, or $\pi_1 = \phi[\pi_2, \pi_3, ...]$ for an empirical model) is used to tailor the laws of dynamic similitude for a specific flow resistance model, depending upon the relative importance of a given dynamic force. Application of the primary force scale ratio, F_r for a particular dominant force (in addition to the inertia force) (or the corresponding π term) in the design of a model or its prototype results in the derivation (and also implies the application) of secondary/similitude scale ratios for that dominant force. As such, the three types of similarity for a particular dominant force are also defined through secondary/similitude scale ratios (see Tables 11.1 through 11.5). Furthermore, because the secondary/similitude velocity scale ratio, in particular, is directly derived from the primary force scale ratio (the dominant force and the inertia force define the secondary/similitude velocity scale ratio for a particular "force model," defined below), application of the secondary/similitude velocity scale ratio becomes sufficient in order to maintain dynamic similarity for a particular dominant force. As such, application of any secondary/similitude scale ratio that defines either the kinematic similarity (because they are defined in terms of the secondary/similitude velocity scale ratio) or the dynamic similarity (because they are defined in terms of the primary force scale ratio) (see Tables 11.1 through 11.5) is sufficient in order to maintain dynamic similarity for a particular dominant force.

Assuming that in addition to the inertia force, F_I a single force predominates a given flow situation, one may derive secondary/similitude scale ratios that represent a specific dominant force. The similitude scale ratios are used to establish a relationship between the model and its prototype and thus enable the design of a model from its prototype or a prototype from its model. If the predominant force is pressure, then one may derive similitude scale ratios for the "pressure model." If the predominant force is gravity, then one may derive similitude scale ratios for the "gravity model." If the predominant force is viscosity, then one may derive similitude scale ratios for the "viscosity model," etc. (as presented in Tables 11.1 through 11.5). It is important to note that one may derive a similitude scale ratio for all physical quantities (geometrics, kinematics, and dynamics) and all physical properties of fluids used with the model and the prototype. Furthermore, while the kinematic and the dynamic similitude scale ratios for the physical quantities will be unique for a specific type of "force model" (pressure model, gravity model, viscosity model, elastic model, and surface tension model), the geometric similitude scale ratios are not unique for a specific "force model". Additionally, the similitude scale ratios for the physical properties of the model and prototype fluids (ρ, s, γ, μ, v, σ, p_v, and E_v) will not be unique for a specific

TABLE 11.1

Similitude Scale Ratios for Physical Quantities for a Pressure Model

Physical Quantity	FLT System	MLT System	Primary Scale Ratios	Secondary/Similitude Scale Ratios for a Pressure Model
			$F_r = \dfrac{F_{P_p}}{F_{P_m}} = \dfrac{F_{I_p}}{F_{I_m}} = constant$	$\underbrace{\left[\left(\dfrac{\rho v^2}{\Delta p}\right)_p\right]}_{E_p} = \underbrace{\left[\left(\dfrac{\rho v^2}{\Delta p}\right)_m\right]}_{E_m}$
Geometrics				
Length, L	L	L	$L_r = \dfrac{L_p}{L_m}$	$L_r = \dfrac{L_p}{L_m}$
Area, A	L^2	L^2	$L_r^2 = \dfrac{L_p^2}{L_m^2}$	$L_r^2 = \dfrac{L_p^2}{L_m^2}$
Volume, V	L^3	L^3	$L_r^3 = \dfrac{L_p^3}{L_m^3}$	$L_r^3 = \dfrac{L_p^3}{L_m^3}$
Kinematics				
Time, T	T	T	$T_r = \dfrac{L_r}{v_r}$	$T_r = \dfrac{L_r}{v_r} = L_r \Delta p_r^{-1/2} \rho_r^{1/2}$
Velocity, v	LT^{-1}	LT^{-1}	$v_r = \dfrac{v_p}{v_m}$	$v_r = \dfrac{v_p}{v_m} = \dfrac{\left(\sqrt{\dfrac{\Delta p}{\rho}}\right)_p}{\left(\sqrt{\dfrac{\Delta p}{\rho}}\right)_m} = \Delta p_r^{1/2} \rho_r^{-1/2}$
Acceleration, a	LT^{-2}	LT^{-2}	$a_r = \dfrac{L_r}{T_r^2} = \dfrac{v_r^2}{L_r}$	$a_r = \dfrac{v_r^2}{L_r} = \Delta p_r \rho_r^{-1} L_r^{-1}$
Discharge, Q	$L^3 T^{-1}$	$L^3 T^{-1}$		$Q_r = v_r L_r^2 = \Delta p_r^{1/2} \rho_r^{-1/2} L_r^2$
Dynamics				
Mass, M	$FL^{-1}T^2$	M		$M_r = F_r a_r^{-1} = \rho_r L_r^3$
Force, F	F	MLT^{-2}	$F_r = \dfrac{F_{P_p}}{F_{P_m}} = \dfrac{F_{I_p}}{F_{I_m}}$	$F_r = \Delta p_r L_r^2 = \rho_r v_r^2 L_r^2$
Pressure, p	FL^{-2}	$ML^{-1}T^{-2}$		$p_r = F_r L_r^{-2} = \Delta p_r = \rho_r v_r^2$
Momentum, Mv or Impulse, FT	FT	MLT^{-1}		$F_r T_r = \rho_r^{1/2} L_r^3 \Delta p_r^{1/2}$
Energy, E or Work, W	FL	ML^2T^{-2}		$W_r = F_r L_r = \Delta p_r L_r^3$
Power, P	FLT^{-1}	ML^2T^{-3}		$P_r = W_r T_r^{-1} = \Delta p_r^{3/2} L_r^2 \rho_r^{-1/2}$

"force model" they will depend directly on the model and the prototype fluid. Finally, the acceleration due to gravity, g is assumed to be the same for the model and the prototype ($g_r = (g_p/g_m) = 1$). (Note: the secondary scale ratio, M_r is not unique to a specific model, although it is supposed to maintain dynamic similarity, as $M = F/a = FL^{-1}T^{-2}$.)

11.4.1 Similitude Scale Ratios for Physical Quantities

Beginning with the primary force scale ratio, F_r for a particular dominant force (in addition to the inertia force) or the corresponding π term, one may derive the similitude scale ratios for the physical quantities for the specific type of "force model." The geometric similitude

TABLE 11.2

Similitude Scale Ratios for Physical Quantities for a Gravity Model

Physical Quantity	FLT System	MLT System	Primary Scale Ratios	Secondary/Similitude Scale Ratios for a Gravity Model
			$F_r = \dfrac{F_{G_p}}{F_{G_m}} = \dfrac{F_{I_p}}{F_{I_m}} = constant$	$\underbrace{\left[\left(\dfrac{v}{\sqrt{gL}}\right)_p\right]}_{F_p} = \underbrace{\left[\left(\dfrac{v}{\sqrt{gL}}\right)_m\right]}_{F_m}$
Geometrics				
Length, L	L	L	$L_r = \dfrac{L_p}{L_m}$	$L_r = \dfrac{L_p}{L_m}$
Area, A	L^2	L^2	$L_r^2 = \dfrac{L_p^2}{L_m^2}$	$L_r^2 = \dfrac{L_p^2}{L_m^2}$
Volume, V	L^3	L^3	$L_r^3 = \dfrac{L_p^3}{L_m^3}$	$L_r^3 = \dfrac{L_p^3}{L_m^3}$
Kinematics				
Time, T	T	T	$T_r = \dfrac{L_r}{v_r}$	$T_r = \dfrac{L_r}{v_r} = L_r^{1/2}$
Velocity, v	LT^{-1}	LT^{-1}	$v_r = \dfrac{v_p}{v_m}$	$v_r = \dfrac{v_p}{v_m} = \dfrac{(\sqrt{gL})_p}{(\sqrt{gL})_m} = L_r^{1/2}$
Acceleration, a	LT^{-2}	LT^{-2}	$a_r = \dfrac{L_r}{T_r^2} = \dfrac{v_r^2}{L_r}$	$a_r = \dfrac{v_r^2}{L_r} = 1$
Discharge, Q	$L^3 T^{-1}$	$L^3 T^{-1}$		$Q_r = v_r L_r^2 = L_r^{5/2}$
Dynamics				
Mass, M	$FL^{-1}T^2$	M		$M_r = F_r a_r^{-1} = \rho_r L_r^3$
Force, F	F	MLT^{-2}	$F_r = \dfrac{F_{G_p}}{F_{G_m}} = \dfrac{F_{I_p}}{F_{I_m}}$	$F_r = \rho_r L_r^3 g_r = \rho_r v_r^2 L_r^2$
Pressure, p	FL^{-2}	$ML^{-1}T^{-2}$		$p_r = F_r L_r^{-2} = \rho_r L_r$
Momentum, Mv or Impulse, FT	FT	MLT^{-1}		$F_r T_r = \rho_r L_r^{7/2}$
Energy, E or Work, W	FL	$ML^2 T^{-2}$		$W_r = F_r L_r = \rho_r L_r^4$
Power, P	FLT^{-1}	$ML^2 T^{-3}$		$P_r = W_r T_r^{-1} = \rho_r L_r^{7/2}$

scale ratios include length, area, and volume. The kinematic similitude scale ratios include time, velocity, acceleration, and discharge. And, the dynamic similitude scale ratios include mass, force, pressure, momentum or impulse, energy or work, and power. As such, the derivation of the similitude scale ratios for the geometrics, kinematics, and dynamics for the pressure model, gravity model, viscosity model, elastic model, and surface tension model are presented in Tables 11.1 through 11.5.

11.4.2 Similitude Scale Ratios for Physical Properties of Fluids

The similitude scale ratios for the physical properties of the model and the prototype fluids (ρ, s, γ, μ, v, σ, p_v, and E_v) will not be unique for a specific "force model." Rather, they will

TABLE 11.3

Similitude Scale Ratios for Physical Quantities for a Viscosity Model

Physical Quantity	FLT System	MLT System	Primary Scale Ratios	Secondary/Similitude Scale Ratios for a Viscosity Model
			$F_r = \dfrac{F_{V_p}}{F_{V_m}} = \dfrac{F_{I_p}}{F_{I_m}} = constant$	$\underbrace{\left[\left(\dfrac{\rho v L}{\mu}\right)_p\right]}_{R_p} = \underbrace{\left[\left(\dfrac{\rho v L}{\mu}\right)_m\right]}_{R_m}$
Geometrics				
Length, L	L	L	$L_r = \dfrac{L_p}{L_m}$	$L_r = \dfrac{L_p}{L_m}$
Area, A	L^2	L^2	$L_r^2 = \dfrac{L_p^2}{L_m^2}$	$L_r^2 = \dfrac{L_p^2}{L_m^2}$
Volume, V	L^3	L^3	$L_r^3 = \dfrac{L_p^3}{L_m^3}$	$L_r^3 = \dfrac{L_p^3}{L_m^3}$
Kinematics				
Time, T	T	T	$T_r = \dfrac{L_r}{v_r}$	$T_r = \dfrac{L_r}{v_r} = L_r^2 \rho_r \mu_r^{-1}$
Velocity, v	LT^{-1}	LT^{-1}	$v_r = \dfrac{v_p}{v_m}$	$v_r = \dfrac{v_p}{v_m} = \dfrac{\left(\dfrac{\mu}{\rho L}\right)_p}{\left(\dfrac{\mu}{\rho L}\right)_m} = \mu_r \rho_r^{-1} L_r^{-1}$
Acceleration, a	LT^{-2}	LT^{-2}	$a_r = \dfrac{L_r}{T_r^2} = \dfrac{v_r^2}{L_r}$	$a_r = \dfrac{v_r^2}{L_r} = \mu_r^2 \rho_r^{-2} L_r^{-3}$
Discharge, Q	$L^3 T^{-1}$	$L^3 T^{-1}$		$Q_r = v_r L_r^2 = \mu_r \rho_r^{-1} L_r$
Dynamics				
Mass, M	$FL^{-1}T^2$	M		$M_r = F_r a_r^{-1} = \rho_r L_r^3$
Force, F	F	MLT^{-2}	$F_r = \dfrac{F_{V_p}}{F_{V_m}} = \dfrac{F_{I_p}}{F_{I_m}}$	$F_r = \mu_r v_r L_r = \mu_r^2 \rho_r^{-1} = \rho_r v_r^2 L_r^2$
Pressure, p	FL^{-2}	$ML^{-1}T^{-2}$		$p_r = F_r L_r^{-2} = \mu_r^2 \rho_r^{-1} L_r^{-2}$
Momentum, Mv or Impulse, FT	FT	MLT^{-1}		$F_r T_r = \mu_r L_r^2$
Energy, E or Work, W	FL	ML^2T^{-2}		$W_r = F_r L_r = \mu_r^2 L_r \rho_r^{-1}$
Power, P	FLT^{-1}	ML^2T^{-3}		$P_r = W_r T_r^{-1} = \mu_r^3 L_r^{-1} \rho_r^{-2}$

depend directly on the model and prototype fluid. As such, the similitude scale ratios for the physical properties of the model and prototype fluids are defined in Table 11.6.

11.5 The Role and the Relative Importance of the Dynamic Forces in Dynamic Similitude

In the development of the laws of dynamic similarity, the uncalibrated mathematical model formulated using dimensional analysis ($\pi_1 = \phi[$ *nothing*$] = constant$ for a theoretical model, or

TABLE 11.4

Similitude Scale Ratios for Physical Quantities for an Elastic Model

Physical Quantity	FLT System	MLT System	Primary Scale Ratios	Secondary/Similitude Scale Ratios for an Elastic Model
			$F_r = \dfrac{F_{E_p}}{F_{E_m}} = \dfrac{F_{I_p}}{F_{I_m}} = constant$	$\underbrace{\left[\left(\dfrac{\rho v^2}{E_v}\right)_p\right]}_{C_p} = \underbrace{\left[\left(\dfrac{\rho v^2}{E_v}\right)_m\right]}_{C_m}$
Geometrics				
Length, L	L	L	$L_r = \dfrac{L_p}{L_m}$	$L_r = \dfrac{L_p}{L_m}$
Area, A	L^2	L^2	$L_r^2 = \dfrac{L_p^2}{L_m^2}$	$L_r^2 = \dfrac{L_p^2}{L_m^2}$
Volume, V	L^3	L^3	$L_r^3 = \dfrac{L_p^3}{L_m^3}$	$L_r^3 = \dfrac{L_p^3}{L_m^3}$
Kinematics				
Time, T	T	T	$T_r = \dfrac{L_r}{v_r}$	$T_r = \dfrac{L_r}{v_r} = L_r E_{v_r}^{-1/2} \rho_r^{1/2}$
Velocity, v	LT^{-1}	LT^{-1}	$v_r = \dfrac{v_p}{v_m}$	$v_r = \dfrac{v_p}{v_m} = \dfrac{\left(\sqrt{\dfrac{E_v}{\rho}}\right)_p}{\left(\sqrt{\dfrac{E_v}{\rho}}\right)_m} = E_{v_r}^{1/2} \rho_r^{-1/2}$
Acceleration, a	LT^{-2}	LT^{-2}	$a_r = \dfrac{L_r}{T_r^2} = \dfrac{v_r^2}{L_r}$	$a_r = \dfrac{v_r^2}{L_r} = E_{v_r} \rho_r^{-1} L_r^{-1}$
Discharge, Q	$L^3 T^{-1}$	$L^3 T^{-1}$		$Q_r = v_r L_r^2 = E_{v_r}^{1/2} \rho_r^{-1/2} L_r^2$
Dynamics				
Mass, M	$FL^{-1}T^2$	M		$M_r = F_r a_r^{-1} = \rho_r L_r^3$
Force, F	F	MLT^{-2}	$F_r = \dfrac{F_{E_p}}{F_{E_m}} = \dfrac{F_{I_p}}{F_{I_m}}$	$F_r = E_{v_r} L_r^2 = \rho_r v_r^2 L_r^2$
Pressure, p	FL^{-2}	$ML^{-1}T^{-2}$		$p_r = F_r L_r^{-2} = E_{v_r}$
Momentum, Mv or Impulse, FT	FT	MLT^{-1}		$F_r T_r = E_{v_r}^{1/2} L_r^3 \rho_r^{1/2}$
Energy, E or Work, W	FL	ML^2T^{-2}		$W_r = F_r L_r = E_{v_r} L_r^3$
Power, P	FLT^{-1}	ML^2T^{-3}		$P_r = W_r T_r^{-1} = E_{v_r}^{3/2} L_r^2 \rho_r^{-1/2}$

$\pi_1 = \phi[\pi_2, \pi_3, \ldots]$ for an empirical model) is used to tailor the laws of dynamic similitude for a specific flow resistance model, depending upon the relative importance of a given dynamic force. Thus, in the development of the laws of dynamic similarity, it is important to summarize the role and the relative importance of the dynamic forces in dynamic similitude, which include the inertia force, F_I; the pressure force, F_P; the gravity, F_G; the viscous force, F_V; the elastic force, F_E; and the surface tension force, F_T. First, the inertia force, F_I will always be a dominant force in order to maintain steady-state flow, while its corresponding force ratio, $F_r = \left(F_{I_p}/F_{I_m}\right)$ will be important in order to maintain dynamic similarity. And, second, given the importance of an additional specific force in the flow situation (pressure, gravity, viscosity, elasticity, or surface tension), either its corresponding force ratio or alternatively, its corresponding π term (**E**, **F**, **R**, **C**, or **W**), respectively, will be important in

TABLE 11.5

Similitude Scale Ratios for Physical Quantities for a Surface Tension Model

Physical Quantity	FLT System	MLT System	Primary Scale Ratios	Secondary/Similitude Scale Ratios for a Surface Tension Model
			$F_r = \dfrac{F_{T_p}}{F_{T_m}} = \dfrac{F_{I_p}}{F_{I_m}} = constant$	$\underbrace{\left[\left(\dfrac{\rho v^2 L}{\sigma}\right)_p\right]}_{W_p} = \underbrace{\left[\left(\dfrac{\rho v^2 L}{\sigma}\right)_m\right]}_{W_m}$
Geometrics				
Length, L	L	L	$L_r = \dfrac{L_p}{L_m}$	$L_r = \dfrac{L_p}{L_m}$
Area, A	L^2	L^2	$L_r^2 = \dfrac{L_p^2}{L_m^2}$	$L_r^2 = \dfrac{L_p^2}{L_m^2}$
Volume, V	L^3	L^3	$L_r^3 = \dfrac{L_p^3}{L_m^3}$	$L_r^3 = \dfrac{L_p^3}{L_m^3}$
Kinematics				
Time, T	T	T	$T_r = \dfrac{L_r}{v_r}$	$T_r = \dfrac{L_r}{v_r} = L_r^{3/2}\sigma_r^{-1/2}\rho_r^{1/2}$
Velocity, v	LT^{-1}	LT^{-1}	$v_r = \dfrac{v_p}{v_m}$	$v_r = \dfrac{v_p}{v_m} = \dfrac{\left(\sqrt{\dfrac{\sigma}{\rho L}}\right)_p}{\left(\sqrt{\dfrac{\sigma}{\rho L}}\right)_m} = \sigma_r^{1/2}\rho_r^{-1/2}L_r^{-1/2}$
Acceleration, a	LT^{-2}	LT^{-2}	$a_r = \dfrac{L_r}{T_r^2} = \dfrac{v_r^2}{L_r}$	$a_r = \dfrac{v_r^2}{L_r} = \sigma_r\rho_r^{-1}L_r^{-2}$
Discharge, Q	$L^3 T^{-1}$	$L^3 T^{-1}$		$Q_r = v_r L_r^2 = \sigma_r^{1/2}\rho_r^{-1/2}L_r^{3/2}$
Dynamics				
Mass, M	$FL^{-1}T^2$	M		$M_r = F_r a_r^{-1} = \rho_r L_r^3$
Force, F	F	MLT^{-2}	$F_r = \dfrac{F_{T_p}}{F_{T_m}} = \dfrac{F_{I_p}}{F_{I_m}}$	$F_r = \sigma_r L_r = \rho_r v_r^2 L_r^2$
Pressure, p	FL^{-2}	$ML^{-1}T^{-2}$		$p_r = F_r L_r^{-2} = \sigma_r L_r^{-1}$
Momentum, Mv or Impulse, FT	FT	MLT^{-1}		$F_r T_r = \sigma_r^{1/2}L_r^{5/2}\rho_r^{1/2}$
Energy, E or Work, W	FL	ML^2T^{-2}		$W_r = F_r L_r = \sigma_r L_r^2$
Power, P	FLT^{-1}	ML^2T^{-3}		$P_r = W_r T_r^{-1} = \sigma_r^{3/2}L_r^{1/2}\rho_r^{-1/2}$

order to maintain dynamic similarity. Furthermore, it is important to distinguish between the role of the friction/viscous force for turbulent flow, F_f or $F_V = \tau A$ and the role of the friction/viscous force for laminar flow, $F_V = \tau A = \mu v L$ in dynamic similitude. While the friction/viscous force for turbulent flow, F_f or $F_V = \tau A$ models the flow resistance in real flow (both laminar and turbulent), the friction/viscous force for laminar flow, $F_V = \tau A = \mu v L$ models the relative importance of the fluid viscosity, μ.

One may recall from Chapter 7 that the friction/viscous force for turbulent flow, F_f or $F_V = \tau A$ will always be the first/main additional dominant force (in addition to the inertia force, F_I), which can be represented by a dominant physical quantity (F_D or h_f) that represents an energy loss or some form of flow resistance (and thus a flow resistance equation). In cases

TABLE 11.6

Similitude Scale Ratios for Physical Properties of the Model and Prototype Fluids

Physical Property	FLT System	MLT System	Secondary/Similitude Scale Ratios
Fluid Properties			
Density, ρ	$FL^{-4}T^2$	ML^{-3}	$\rho_r = \dfrac{\rho_p}{\rho_m}$
Specific Gravity, s	dimensionless	dimensionless	$s_r = \dfrac{s_p}{s_m}$
Specific Weight, γ	FL^{-3}	$ML^{-2}T^{-2}$	$\gamma_r = \dfrac{\gamma_p}{\gamma_m}$
Absolute Viscosity, μ	$FL^{-2}T$	$ML^{-1}T^{-1}$	$\mu_r = \dfrac{\mu_p}{\mu_m}$
Kinematic Viscosity, v	L^2T^{-1}	L^2T^{-1}	$v_r = \dfrac{v_p}{v_m}$
Surface Tension, σ	FL^{-1}	MT^{-2}	$\sigma_r = \dfrac{\sigma_p}{\sigma_m}$
Vapor Pressure, p_v	FL^{-2}	$ML^{-1}T^{-2}$	$p_{v_r} = \dfrac{p_{v_p}}{p_{v_m}}$
Bulk Modulus of Elasticity, E_v	FL^{-2}	$ML^{-1}T^{-2}$	$E_{v_r} = \dfrac{E_{v_p}}{E_{v_m}}$

where the flow resistance is due to pipe or channel friction (major head loss), pipe component (minor head loss), and/or a flow-measuring device (minor head loss), the flow resistance contributes to the power losses and thus the determination of the efficiency of pumps and turbines. Furthermore, one may also recall that in the dimensional analysis of fluid flow, although the first/main additional dominant force is the friction/viscous force for turbulent flow, $F_V = \tau A$, the type of flow will determine the cause of the friction/viscous force and thus the definition of the drag coefficient, C_D. As such, in external flow and in pipe flow, the first/main additional dominant friction/viscous force for turbulent flow, $F_V = \tau A$ is a result of the pressure force, $F_P = \Delta p A$. Therefore, in external and pipe flow, the drag coefficient, C_D will mainly be a function of the Euler number, $E = (F_I/F_P) = (\rho v^2/\Delta p)$, which is the dependent π term. However, in uniform open channel flow, the first/main additional dominant friction/viscous force for turbulent flow, $F_V = \tau A$ is a result of the gravity force, $F_G = Mg$. Therefore, in uniform open channel flow, the drag coefficient, C_D will mainly be a function of the Froude number, $F = (F_I/F_G) = (v^2/gL)$, which is the dependent π term. Finally, for flow-measuring devices in open channel flow, the first/main additional dominant friction/viscous force for turbulent flow, $F_V = \tau A$ is a result of both the pressure force, $F_P = \Delta p A = \gamma \Delta y A$ and the gravitational force, $F_G = Mg$. Therefore, for flow-measuring devices in open channel flow, the drag coefficient, C_D will mainly be a function of the Euler number, $E = (F_I/F_P) = (\rho v^2/\Delta p)$, which in this case is equal to the Froude number, $F = (F_I/F_G) = (v^2/gL)$, the dependent π term. The modeling of additional dominant forces or dominant physical quantities will introduce additional dimensionless numbers or independent π terms. For a given flow situation, the addition of the viscous force for laminar flow, $F_V = \tau A = \mu v L$; the elastic force, F_E; the surface tension force, F_T; and the gravity force, F_G, results in the following independent π terms: the Reynolds number, R; the Cauchy number, C; the Weber number, W; and the Froude number, F; respectively.

It is important to note that the single most common additional dominant force is the viscous force for laminar flow, $F_V = \tau A = \mu v L$, which results in the Reynolds number, R, an independent π term. In the case of external flow around an object, the drag coefficient, C_D (or the Euler number, E) used in the derivation of the drag force, F_D is dependent on the Reynolds number, R for streamlined bodies and in the case of low and intermediate velocities. In the case of internal pipe flow, the drag coefficient, C_D (or the Euler number, E) used in the derivation of the major head loss, h_f is dependent on the Reynolds number, R in the case of low and intermediate velocities (laminar flow). However, in the case of internal open channel flow, because the flow is mostly turbulent, the drag coefficient, C_D (or the Froude number, F) used in the derivation of the major head loss, h_f is virtually independent of the Reynolds number, R. Additionally, in the case of internal pipe flow with pipe devices, because the flow is mostly turbulent, the drag coefficient, C_D (or the Euler number, E) used in the derivation of the minor head loss, h_f is usually independent of the Reynolds number, R. In the case of flow measurement devices in pipe flow, the drag coefficient, C_D (or the Euler number, E) used in the derivation of the actual discharge, Q_a is dependent on the Reynolds number, R for low to moderate values, where the viscous effects are significant. In the case of flow measurement devices in open channel flow, the drag coefficient, C_D (or the Froude number, F) used in the derivation of the actual discharge, Q_a is dependent on the Reynolds number, R for low to moderate values in the case of a low head, H on the weir. Finally, in the case of the efficiency, η of pumps and turbines, the drag coefficient, $C_D = \eta$ (or the Euler number, E) is dependent on the Reynolds number, R for most flow situations.

11.6 Guidelines in the Application of the Laws Governing Dynamic Similarity

In the application of the laws of dynamic similarity, the uncalibrated mathematical model formulated using dimensional analysis ($\pi_1 = \phi(nothing) = constant$ for a theoretical model, or $\pi_1 = \phi(\pi_2, \pi_3, \ldots)$ for an empirical model) is used to tailor the laws of dynamic similitude for a specific flow resistance model, depending upon the relative importance of a given dynamic force. Thus, application of the laws governing dynamic similarity for real flow situations (flow resistance problems) involves: (1) the definition of the flow resistance prediction equation (and the prediction equation for any other physical quantity, such as the empirical expressions for the efficiency of pumps and turbines) between the model and its prototype, (2) the definition of the dynamic similarity requirements between the model and its prototype, (3) determination if the flow resistance prediction equation (or the prediction equation for any other physical quantity, such as the empirical expressions for the efficiency of pumps and turbines) results in a "true model" or a "distorted model," and (4) general guidelines in the application of "distorted models." Application of the laws of similarity (similitude) allows the prediction of the performance of the "prototype" from experiments conducted with the "model." The similitude scale ratios are used to define the dynamic similarity requirements and thus establish a relationship between the model and its prototype, which enables the design of a model from its prototype or a prototype from its model. Specifically, the similitude scale ratios are needed in order to translate the model values (measurements) of the various physical quantities (geometric, kinematic, and dynamic) and physical properties of the model fluids into the corresponding prototype values. Furthermore, the similitude scale ratios may be used to predict the behavior of the prototype to possible design changes using the model.

11.6.1 Definition of the Flow Resistance Prediction Equations (or Equations for Any Other Physical Quantity)

The definition of the flow resistance prediction equation (and the prediction equation for any other physical quantity, such as the empirical expressions for the efficiency of pumps and turbines) between the model and its prototype involves the application of the similitude scale ratios. The flow resistance prediction equation (and the prediction equation for any other physical quantity, such as the empirical expressions for the efficiency of pumps and turbines) is defined by equating the drag coefficient, C_D (dependent π term) for the model and the prototype. As such, a given flow resistance prediction equation (and the prediction equation for any other physical quantity, such as the empirical expressions for the efficiency of pumps and turbines) represents a dynamic similarity requirement. Dimensional analysis was applied in order to derive a given flow resistance equation and its corresponding drag coefficient, C_D (dependent π term), which is defined as a function of one or more independent π terms. Specifically, for the case of external and pipe flow, the drag coefficient, C_D as modeled by (is mainly a function of) the Euler number, E is the dependent π term. Thus, in the case of external and pipe flow, the similitude scale ratios for the "pressure model" are used to define the flow resistance prediction equation (and the prediction equation for any other physical quantity, such as the empirical expressions for the efficiency of pumps and turbines) between the model and its prototype. Additionally, for the case of open channel flow, the drag coefficient, C_D as modeled by (is mainly a function of) the Froude number, F is the dependent π term. Thus, in the case of open channel flow, the similitude scale ratios for the "gravity model" are used to define the flow resistance prediction equation (and the prediction equation for any other physical quantity, such as the empirical expressions for the efficiency of pumps and turbines) between the model and its prototype.

11.6.2 Definition of the Dynamic Similarity Requirements

The flow resistance prediction equation (and the prediction equation for any other physical quantity, such as the empirical expressions for the efficiency of pumps and turbines) is defined by equating the drag coefficient, C_D (dependent π term) for the model and the prototype. Depending upon the flow situation (external flow around an object, pipe flow, open channel flow, pipe flow with devices, pipe flow with a flow-measuring device, open channel flow with a flow-measuring device, flow through a pump or a turbine), the drag coefficient, C_D (dependent π term) will be a function of one or more of the following independent π terms: R, C, W, F, ε/L, and L_i/L. As such, the appropriate similitude scale ratios corresponding to the applicable independent π terms are used to define the dynamic similarity requirements (between the model and its prototype). Specifically, the criteria governing dynamic similarity requires that independent π terms remain a constant between the model and its prototype in order for the drag coefficient, C_D (dependent π term) to remain a constant between the model and its prototype. One may note that if there are no independent π terms (R, C, W, F, ε/L, or L_i/L) used in the definition of the dependent π term (either E or F, i.e., the drag coefficient, C_D), the drag coefficient, C_D is a constant and thus the flow resistance prediction equation (and the prediction equation for any other physical quantity, such as the empirical expressions for the efficiency of pumps and turbines) sufficiently defines the dynamic similarity requirements between the model and its prototype.

Because the similitude velocity scale ratio, in particular, is directly derived from the primary force scale ratio for a particular "force model," the similitude velocity scale ratio is unique for the particular "force model," and application of the similitude velocity scale ratio is sufficient in order to maintain dynamic similarity for a particular dominant force. Furthermore, application of any similitude scale ratio that defines either the kinematic similarity (because they are unique and are defined in terms of the similitude velocity scale ratio) or the dynamic similarity (because they are unique and are defined in terms of the primary force scale ratio) is sufficient in order to maintain dynamic similarity for a particular dominant force. However, one may note that, typically, it is the similitude velocity scale ratio that is applied in order to achieve and maintain steady-state and dynamic similarity between the model and its prototype. Therefore, in the case where more than one additional "force model" is required, simultaneous application of the respective similitude velocity scale ratios is sufficient in order to simultaneously keep the corresponding independent π terms a constant between the model and its prototype. This is required in order for the appropriate dependent π term (E, or F) to remain a constant between the model and its prototype.

11.6.3 "True Models" versus "Distorted Models"

In the application of the laws governing dynamic similarity, it is important to determine if the flow resistance prediction equation (and the prediction equation for any other physical quantity, such as the empirical expressions for the efficiency of pumps and turbines) results in a "true model" or a "distorted model." A "true model" is one in which the dynamic similarity requirements are met. A "distorted model" is one in which one or more dynamic similarity requirements are not met. It is important to note that although, in theory, it is always possible to impose that a given π term remain a constant between the model and its prototype, such may not always be possible in practice. As a result, one may note that "distorted models" for the flow resistance prediction equations (and the prediction equation for any other physical quantity, such as the empirical expressions for the efficiency of pumps and turbines) are typically due to the practical unavailability of convenient fluids; inability to maintain high speeds; difficulties in the geometric scaling including the surface roughness, ε; cost of model and model fluid; etc. Furthermore, one may note that one may successfully apply a "distorted model" for the flow resistance prediction equation (and the prediction equation for any other physical quantity, such as the empirical expressions for the efficiency of pumps and turbines) as long as the results are carefully interpreted. As such, the design and application of a "distorted model" requires a thorough understanding of the dynamic forces and the physical quantities involved in the flow situation.

Thus, depending upon the practical challenges imposed by the dynamic similarity requirements, the flow resistance prediction equation results (and the prediction equation for any other physical quantity, such as the empirical expressions for the efficiency of pumps and turbines) will result in either a "true model" or a "distorted model." In the case where only one "force model" is required (either the "pressure model" for external and pipe flow, or the "gravity model" for open channel flow), application of the specific flow resistance prediction equation (and the prediction equation for any other physical quantity, such as the empirical expressions for the efficiency of pumps and turbines) in order to impose that the corresponding dependent π term (E or F, respectively) and thus the corresponding drag coefficient, C_D remain a constant between the model and its prototype is typically straightforward (both in theory and in practice) and usually results in a "true model." Exceptions may include difficulties in the geometric scaling including the surface roughness, ε. Furthermore, in the case where one or more additional "force models" are required, if it is possible in practice for the corresponding independent π term to remain a constant (simultaneously in the case of more than one independent π terms) between the model and its prototype, then the dependent π term (E, or F) and thus the corresponding drag coefficient, C_D

will remain a constant between the model and its prototype; thus, the flow resistance prediction equation (and the prediction equation for any other physical quantity, such as the empirical expressions for the efficiency of pumps and turbines) usually results in a "true model." However, in the case where one or more additional "force models" are required, if it is not possible in practice for the corresponding independent π term to remain a constant (simultaneously in the case of more than one independent π terms) between the model and its prototype, then the dependent π term (E or F) and thus the corresponding drag coefficient, C_D will not remain a constant between the model and its prototype; thus, the flow resistance prediction equation (and the prediction equation for any other physical quantity, such as the empirical expressions for the efficiency of pumps and turbines) results in a "distorted model."

11.6.4 General Guidelines in the Application of "Distorted Models"

Because the occurrence of a "distorted model" for the flow resistance prediction equation (and the prediction equation for any other physical quantity, such as the empirical expressions for the efficiency of pumps and turbines, i.e., dependent π term) is common in modeling, it is important to identify the specific causes of distortion and to optimize the conditions for achieving dynamic similarity. Any deviation from the geometric, kinematic, or dynamic similarity between a model and its prototype for either the dependent or the independent π terms will result in a "distorted model." As such, a "distorted model" for the flow resistance prediction equation (and the prediction equation for any other physical quantity, such as the empirical expressions for the efficiency of pumps and turbines) may result from a deviation from the geometric, kinematic, or dynamic similarity existing (1) within the flow resistance prediction equation (and the prediction equation for any other physical quantity, such as the empirical expressions for the efficiency of pumps and turbines, i.e., dependent π term) itself, or (2) within one or more of the independent π terms. In the simplest case, if it is not possible in practice to maintain a constant L_r, geometric similarity cannot be achieved; thus, a "distorted model" for the flow resistance prediction equation (and the prediction equation for any other physical quantity, such as the empirical expressions for the efficiency of pumps and turbines) will result. Examples of such practical difficulties include the geometric scaling, which include the surface roughness, ε. Therefore, in a "true model" for the flow resistance prediction equation (and the prediction equation for any other physical quantity, such as the empirical expressions for the efficiency of pumps and turbines), the geometric similitude scale ratio, L_r (not unique for a given π term) will remain a constant in the model design regardless of the applicable dependent or independent π terms. In cases where it is not possible to achieve dynamic similarity (and thus kinematic and geometric similarity), a "distorted model" for the flow resistance prediction equation (and the prediction equation for any other physical quantity, such as the empirical expressions for the efficiency of pumps and turbines) will result.

When working with a "distorted model" for the flow resistance prediction equation (and the prediction equation for any other physical quantity, such as the empirical expressions for the efficiency of pumps and turbines), the goal is to optimize the conditions for achieving dynamic similarity in order for the "distorted model" to be as close as possible to a "true model." As such, the optimization procedure requires simultaneously achieving a set of objectives while imposing a set of practical limitations. Therefore, in order to optimize the conditions for achieving dynamic similarity, the objectives include that (1) the similitude length scale ratio, L_r remain a constant between the model and the prototype; (2) the similitude velocity scale ratio, v_r remain a constant between the model and the prototype; and

(3) the similitude force scale ratio, F_r (or the applicable dimensionless numbers: *E, F, R, C,* and *W*) remain a constant between the model and the prototype. Furthermore, the practical limitations imposed in model design include the following: (1) achieving an attainable model velocity, (2) minimizing the cost/scale of the model, (3) the given prototype fluid, and (4) minimizing the cost of the model fluid. Practical difficulties are encountered in the optimization procedure. For instance, achieving an attainable model velocity or minimizing the cost of the model fluid may require a large model scale/cost and/or relaxing the similarity requirement. Additionally, in order to accurately apply the similarity objectives, it is important that a similar flow type is maintained between the model and its prototype. For instance, if the flow in the prototype is turbulent, for which *R* does not play an important role in the similitude (no viscous effects), then the flow in the model should also be turbulent. Furthermore, for instance, if the flow in the prototype is subsonic, for which the *M* does not play an important role in the similitude (no compressibility/elastic effects), then the flow in the model should also be subsonic. Finally, for instance, if the flow in the prototype is not a thin layer of fluid, for which the *W* does not play an important role in the similitude (no surface tension effects), then the flow in the model should also not be a thin layer of fluid. However, if the flow in the prototype is laminar, supersonic, or a thin layer of fluid, for which *R, M,* or *C,* respectively, will play an important role in the similitude, then the flow in the model should also be laminar, supersonic, or a thin layer of film, respectively.

11.7 Application of the Laws Governing Dynamic Similarity for Flow Resistance Equations and Efficiency of Pumps and Turbines

In the application of the laws of dynamic similarity, the uncalibrated mathematical model formulated using dimensional analysis ($\pi_1 = \phi[nothing] = constant$ for a theoretical model, or $\pi_1 = \phi[\pi_2, \pi_3, ...]$ for an empirical model) is used to tailor the laws of dynamic similitude for a specific flow resistance model, depending upon the relative importance of a given dynamic force. One may recall that while one application of dimensional analysis is to supplement the momentum theory in the definition of the friction/viscous force (yielding the flow resistance equations and their corresponding drag coefficients, C_D and the empirical expressions for pump and turbine efficiencies, $\eta =$ the drag coefficients, C_D), an additional/further application is for the development of criteria governing dynamic similitude. The flow resistance equations (in the form of either F_D or h_f) include the drag force, the major head loss, the minor head loss, and the actual discharge. The ultimate goal is to design a model or a prototype that is guided through the use of dimensional analysis for dynamically similar situations. In such applications, the laws of similarity (similitude) allow the prediction of the performance of the "prototype" from experiments conducted with the "model." The laws of dynamic similarity involve the definition of primary scale ratios and secondary/similitude scale ratios. It is important to note the similitude scale ratios are needed in order to translate the model values (measurements) of the various physical quantities (geometric, kinematic, and dynamic) and physical properties of the model fluids into the corresponding prototype values.

Presented in the following sections are applications of the similitude scale ratios for each of the flow resistance predication equations and the pump and turbine efficiencies, with specific guidelines in the application of "distorted models." The specific guidelines in the application of "distorted models" (Section 11.7.1) are determined based upon the particular flow resistance equation and the empirical expressions for the efficiency of pumps and turbines, which are discussed in the following sections: (1) drag force in external flow (Section 11.7.2), (2) the major head loss in pipe flow (Section 11.7.3), (3) the major head loss in open channel flow (Section 11.7.4), (4) the minor head loss in pipe flow (Section 11.7.5), (5) the actual discharge in pipe flow (Section 11.7.6), (6) the actual discharge in open channel flow (Section 11.7.7), and (7) pump and turbine efficiencies (Section 11.7.8). Application of the similitude scale ratios for a given flow resistance predication equation (or pump or turbine efficiency) allows the prediction of the performance of the "prototype" from experiments conducted with the "model," and involves: (a) definition of the flow resistance prediction equation (and the prediction equation for any other physical quantity such as a pump or turbine efficiency) between the model and its prototype, (b) the definition of the dynamic similarity requirements between the model and its prototype, (c) example problems illustrating the application of the laws governing dynamic similarity for the flow resistance prediction equation (and the prediction equation for any other physical quantity such as a pump or turbine efficiency) between the model and its prototype, and (d) determination whether the flow resistance prediction equation (or pump or turbine efficiency) results in a "true model" or a "distorted model." It is important to note that the prediction of the performance of the "prototype" from experiments conducted with the "model" typically involve "actual measurements" (observed values) of variables (e.g., force, pressure, velocity, discharge, etc.) for either the prototype or the model or both. This type of data collection ("actual measurements"/observed values) would allow an on-site calibration of the appropriate drag coefficient, C_D for the given problem. However, in the example problems presented below, instead of assuming that "actual measurements" (observed values) of variables (e.g., force, pressure, velocity, discharge, etc.) for either the prototype or the model or both are made, these will be "predicted" (estimated) by applying the appropriate flow resistance prediction equation (or pump or turbine efficiency). As such, application of the appropriate flow resistance prediction equation (or pump or turbine efficiency) will apply the previously calibrated appropriate drag coefficient, C_D for the given problem. For instance, for an external flow problem, the previously calibrated appropriate drag coefficient, C_D presented in Chapter 10 would be used. And, for instance, for a pipe flow problem, the previously calibrated appropriate friction factor, $f = 8\, C_D$ presented in Chapter 8 would be used. Therefore, the example problems presented below do not demonstrate the actual data collection and thus the on-site calibration of the appropriate drag coefficient, C_D for the given problem. However, the example problems presented below provide the following: (1) an academic reinforcement of the application of the appropriate drag coefficient, C_D for the given problem by using previously calibrated (simulated) values; (2) determination (by simulation) whether the flow resistance prediction equation (or pump or turbine efficiency) would result in a "true model" or a "distorted model" for a given flow situation; and (3) application of the specific guidelines in order to accommodate a "distorted model" and thus guide in its application. Thus, the example problems presented below are more academic in nature rather than practical. As such, the example problems simulate the potential to develop a "true model" or a "distorted model" for various applications of a given flow resistance equation (or pump or turbine efficiency).

11.7.1 Specific Guidelines in the Application of "Distorted Models"

In the application of the laws governing dynamic similarity, it is important to determine if the flow resistance prediction equation (or pump or turbine efficiency) results in a "true model" or a "distorted model." A "true model" is one in which the dynamic similarity requirements are met. A "distorted model" is one in which one or more dynamic similarity

requirements are not met. For a given flow resistance equation (or pump or turbine efficiency), the dependent π term (i.e., the drag coefficient, C_D) will be a function of one or
more of the following independent π terms: R, C (or M), W, F, ε/L, and L_i/L. In the case where
the similarity requirements regarding the independent π term (s) are satisfied, then the dependent π term (i.e., the drag coefficient, C_D) will remain a constant between the model and its
prototype. Specifically, this implies that (1) the model fluid is practical and available, (2)
the model velocity (does not introduce any additional similarity requirements) and pressure
are attainable, (3) the model scale is reasonable, (4) and the model costs are reasonable. And
therefore, application of the given flow resistance prediction equation (or pump or turbine
efficiency) yields a "true model." However, in the case where the similarity requirements
regarding the independent π term(s) are not satisfied, then the dependent π term (i.e., the
drag coefficient, C_D) will not remain a constant between the model and its prototype. Specifically, this implies that (1) the model fluid is not practical or available, (2) the model velocity
(may introduce additional similarity requirements) and pressure are not attainable, (3) the
model scale is not reasonable, and/or (4) the model costs are not reasonable. And, therefore,
application of the given flow resistance prediction equation (or pump or turbine efficiency)
yields a "distorted model" for which the appropriate corrections measures are taken. Furthermore, the specific guidelines in the application of "distorted models" are determined based
upon the particular flow resistance equation (or pump or turbine efficiency).

In the case of the drag force in external flow, while the drag coefficient, C_D, in general, is a
mainly a function of E, it will selectively vary with R, L_i/L, ε/L, M, and F, depending mainly
on the velocity of flow, the shape of the body, and whether there is a wave action at the free
surface (see Table 11.7, which is the same as Table 10.1).

In the case of the major head loss in pipe flow, while the drag coefficient, C_D generally is
mainly a function of E, it will selectively vary with R and ε/L, depending mostly on the
velocity of flow (see Table 11.8).

In the case of the major head loss in open channel flow, while the drag coefficient, C_D generally is mainly a function of F, it will selectively vary with R and ε/L, depending mostly on
the velocity of flow (see Table 11.8).

In the case of the minor head loss due to pipe components in pipe flow, while the drag
coefficient, C_D generally is mainly a function of E, it will selectively vary with R, L_i/L,
and ε/L, depending mostly on the velocity of flow and the geometry of the pipe component
(see Table 11.8).

In the case of the actual discharge in pipe flow (flow-measuring device), while the drag
coefficient, C_D generally is mainly a function of E, it will selectively vary with R, and L_i/L,
depending mainly on the velocity of flow and the geometry of the flow-measuring device
(see Table 11.8).

In the case of the actual discharge in open channel flow (flow-measuring device), while the
drag coefficient, C_D generally is mainly a function of $E = F$, it will selectively vary with R,
L_i/L, and W, depending mainly on the velocity of flow, the type of flow-measuring device,
and the thickness of the fluid layer (see Table 11.8).

In the case of the pump and turbine efficiencies, η, while the drag coefficient, $C_D = \eta$ generally is mainly a function of E, it will (theoretically) selectively vary with R, ε/L, and C,
depending mostly on the velocity of flow and the type of fluid through the pump or turbine
(see Table 11.8). However, in practice, the drag coefficient, $C_D = \eta$ is assumed to be independent of C (assumption of incompressible flow), and will vary with R for laminar flow while
it will vary with ε/D for turbulent flow in the case where the prototype pump or turbine is
significantly larger than the model pump or turbine (or vice versa). Therefore, assuming
incompressible flow, for laminar flow the R must remain a constant between the model

TABLE 11.7

The Variation of the Drag Coefficient, C_D for External Flow

$C_D = 2/E = f''$ (Independent Π term)	Viscous Force $\equiv R$	Geometry $\equiv L_i/L$	Surface Roughness $\equiv \varepsilon/L$	Elastic Force $\equiv C = M^2$	Gravity Force $\equiv F$
Velocity of Flow and Shape of Body					
Creeping flow: any shape body, $C_D = 2/E = f''$ (R)	yes	no	no	no	no
Laminar flow: any shape body (except round-shaped bodies)					
• Laminar flow: streamlined bodies	yes	yes	no	no	no
• Laminar flow: blunt bodies	*less*	yes	no	no	no
Turbulent flow: any shape body (except round-shaped bodies)					
• Turbulent flow: streamlined bodies	yes	yes	yes	yes	no
• Turbulent flow: blunt bodies	no	yes	yes	yes	no
Laminar flow: round-shaped bodies (circular cylinder or sphere)	yes	yes	no	no	no
Turbulent flow: round-shaped bodies (circular cylinder or sphere)	yes	yes	yes	yes	no
Laminar or turbulent flow with wave action at free surface: any shape body	yes	yes	no	no	yes

Note: For a blunt or streamlined body, the frontal area is used in the computation of the drag force. For a flat plate or an airfoil (very well streamlined body) that is parallel to the direction of flow, the planform area is used in the computation of the drag force.

and its prototype (as a "true model" is sought), and for turbulent flow where the prototype pump or turbine is significantly larger than the model pump (or vice versa), the ε/D must remain a constant between the model and its prototype (as a "true model" is sought). Finally, it is important to note that for numerous practical pump or turbine flow problems, the drag coefficient, $C_D = \eta$ is assumed to be independent of all three independent π terms, R, C, and ε/D, which leads to the definition of the affinity laws (similarity rules) for the efficiency of homologous pumps and turbines; when dynamic similarity between the model and its prototype (or between two flow systems) is achieved, the model and its prototype (or the two flow systems) are said to be homologous (see Section 11.7.8).

In order to facilitate application of the specific guidelines in the application of "distorted models" for external and internal flow (pipe and open channel), the similarity requirements for a given type of external or internal flow are summarized in Table 11.9 (which also lists example problems presented in Sections 11.7.2 through 11.7.8), and the following sections below as follows: (1) Geometry, (2) Relative Roughness, (3) "Pressure Model," (4) "Viscosity Model," (5) "Elastic Model," (6) "Gravity Model," (7) "Surface Tension Model," (8) "Viscosity Model" and "Elastic Model," (9) "Viscosity Model" and "Gravity Model," (10) "Viscosity Model" and "Surface Tension" Model, and (11) "Gravity Model," "Viscosity Model," and "Surface Tension Model."

TABLE 11.8

The Variation of the Drag Coefficient, C_D for Internal Flow

Type and Velocity of Flow	$C_D = f''$ (Independent Π term)	Gravity Force $\equiv F$	Viscous Force $\equiv R$	Surface Roughness $\equiv \varepsilon/L$	Geometry $\equiv L_i/L$	Surface Tension Force $\equiv W$	Elastic Force $\equiv C = M^2$
Pipe flow							
• Laminar flow	$C_D = \dfrac{R_h}{L}\dfrac{1}{E}$	no	yes	no	no	no	no
• Turbulent flow		no	no	yes	no	no	no
• Transitional flow		no	yes	yes	no	no	no
Open channel flow: turbulent flow	$C_D = \dfrac{S_f}{2}\dfrac{1}{F}$	dep π term	no	yes	no	no	no
Pipe component: turbulent flow	$C_D = \dfrac{2}{E}$	no	no	yes	yes	no	no
Pipe flow-measuring device	$C_D = \sqrt{\dfrac{E}{2}\dfrac{A_a}{A_i}}$	no	yes	no	yes	no	no
Open channel flow-measuring device	$C_D = \sqrt{\dfrac{E}{2}\dfrac{A_a}{A_i}} = \sqrt{\dfrac{F}{2}\dfrac{A_a}{A_i}}$						
• Sluice gate or venturi meter		dep π term	no	no	yes	no	no
• Weir or spillway: large head		dep π term	no	no	yes	no	no
• Weir or spillway: small head		dep π term	yes	no	yes	yes	no
Efficiency of pumps	$C_D = \dfrac{1}{E}$						
• Laminar flow		no	yes	no	no	no	yes for compressible flow only
• Turbulent flow		no	no	yes	no	no	yes for compressible flow only
Efficiency of turbines	$C_D = E$						
• Laminar flow		no	yes	no	no	no	no
• Turbulent flow		no	no	yes	no	no	no

TABLE 11.9

Summary of Dynamic Similarity Requirements for External and Internal Flow

Example Problem	Similarity Requirement	Geometry	Relative Roughness	"Pressure Model"	"Viscosity Model"	"Elastic Model"	"Gravity Model"	"Viscosity Model" and "Elastic Model"	"Viscosity Model" and "Gravity Model"	"Viscosity Model," and "Surface Tension Model"
	$C_D = f''$ (Independent Π term)	L_i/L	ε/L	E	R	C (or M)	F	R and C (or M)	R and F	R and W
	External Flow									
11.1	Creeping flow				yes					
11.2	Laminar, streamlined	yes			yes					
11.3	Laminar, blunt	yes		**dep π term**						
11.5	Turbulent, streamlined	yes	yes					yes		
11.4	Turbulent, blunt	yes	yes			yes				
11.6	Laminar, round-shaped	yes	yes		yes			yes		
11.7	Turbulent, round-shaped	yes	yes							
11.8	Laminar,	yes								
11.9	turbulent, wave action, any shape								yes	
	Internal Flow									
11.10	Laminar, pipe				yes					
11.11	Turbulent, pipe		yes	**dep π term**						
11.12	Transitional, pipe		yes		yes					
11.13	Turbulent, open channel		yes				**dep π term**			

(Continued)

TABLE 11.9 (*Continued*)

Summary of Dynamic Similarity Requirements for External and Internal Flow

Example Problem	Similarity Requirement	Geometry	Relative Roughness	"Pressure Model"	"Viscosity Model"	"Elastic Model"	"Gravity Model"	"Viscosity Model" and "Elastic Model"	"Viscosity Model" and "Gravity Model"	"Viscosity Model," and "Surface Tension Model"
11.14	Turbulent pipe component	yes	yes	dep π term						
11.15	Pipe flow-measuring	yes			yes					
11.16	Sluice gate, venturi meter	yes					dep π term			
11.17	Weir, spillway, large head	yes					dep π term			
11.18	Weir, spillway, small head	yes					dep π term			yes
	Laminar, pump efficiency				yes					
	Turbulent, pump efficiency		yes	dep π term						
11.19	Pump efficiency (Affinity Laws)			dep π term						
	Laminar, turbine efficiency				yes					
	Turbulent, turbine efficiency		yes	dep π term						
11.20	Turbine efficiency (Affinity Laws)			dep π term						

It is important to note that while dynamic similarity requirements imply both kinematic and geometric similarity for a given flow type, Section 11.7.1.1 below on geometry similarity requirements outlines additional specific geometric similarity requirements (if any) regarding the shape of body for external flow, pipe components, pipe flow-measuring devices, and open channel flow-measuring devices for internal flow.

11.7.1.1 Geometry Similarity Requirements

The geometric similarity requirement that L_i/L remain a constant is modeled by the similitude length scale ratio, L_r as follows:

$$\left(\frac{L_i}{L}\right)_p = \left(\frac{L_i}{L}\right)_m \tag{11.36}$$

$$L_r = \frac{L_p}{L_m} = \frac{(L_i)_p}{(L_i)_m} = \frac{(L)_p}{(L)_m} \tag{11.37}$$

For an open channel flow-measuring device (weir or spillway with a head, H and height, P), the geometric similarity requirements are stated as follows. The geometric similarity requirement that H/P remain a constant is modeled by the similitude length scale ratio, L_r as follows:

$$\left(\frac{H}{P}\right)_p = \left(\frac{H}{P}\right)_m \tag{11.38}$$

$$L_r = \frac{L_p}{L_m} = \frac{(H)_p}{(H)_m} = \frac{(P)_p}{(P)_m} \tag{11.39}$$

11.7.1.2 Relative Roughness Similarity Requirements

The geometric similarity requirement that the relative roughness, ε/L remain a constant is modeled by the similitude length scale ratio, L_r as follows:

$$\left(\frac{\varepsilon}{L}\right)_p = \left(\frac{\varepsilon}{L}\right)_m \tag{11.40}$$

$$L_r = \frac{L_p}{L_m} = \frac{(\varepsilon)_p}{(\varepsilon)_m} = \frac{(L)_p}{(L)_m} \tag{11.41}$$

For a pipe with a diameter, D the geometric similarity requirements are stated as follows. The geometric similarity requirement that ε/D remain a constant is modeled by the similitude length scale ratio, L_r as follows:

$$\left(\frac{\varepsilon}{D}\right)_p = \left(\frac{\varepsilon}{D}\right)_m \tag{11.42}$$

$$L_r = \frac{L_p}{L_m} = \frac{(\varepsilon)_p}{(\varepsilon)_m} = \frac{(D)_p}{(D)_m} \tag{11.43}$$

Thus, in order to satisfy the similarity requirement for the model design as modeled by the relationship between the similitude length scale ratio, L_r; the similitude surface roughness scale ratio, ε_r; and the similitude pipe diameter scale ratio, D_r, one must consider the following. Because the similitude length scale ratio, L_r is typically much greater than one, then the required model surface roughness, ε_m needs to be smaller/smoother than the prototype surface roughness, ε_p, yet similar in pattern. Such a combined requirement is usually difficult to attain in practice. Furthermore, while this condition may be relaxed for laminar flow, where the ε/D does not play an important role, it is of importance for turbulent flow, for which ε/D is important.

For an open channel with a depth, y the geometric similarity requirements are stated as follows. The geometric similarity requirement that ε/y remain a constant is modeled by the similitude length scale ratio, L_r as follows:

$$\left(\frac{\varepsilon}{y}\right)_p = \left(\frac{\varepsilon}{y}\right)_m \tag{11.44}$$

$$L_r = \frac{L_p}{L_m} = \frac{(\varepsilon)_p}{(\varepsilon)_m} = \frac{(y)_p}{(y)_m} \tag{11.45}$$

Thus, in order to satisfy the similarity requirement for the model design as modeled by the relationship between the similitude length scale ratio, L_r; the similitude surface roughness scale ratio, ε_r; and the similitude channel depth scale ratio, y_r one must consider the following. Because the similitude length scale ratio, L_r is typically much greater than one, then the required model surface roughness, ε_m needs to be smaller/smoother than the prototype surface roughness, ε_p, yet similar in pattern. Such a combined requirement is usually difficult to attain in practice. Furthermore, while this condition may be relaxed for laminar flow, where the ε/y does not play an important role, it is of importance for turbulent flow, for which ε/y is important.

11.7.1.3 *"Pressure Model" Similarity Requirements*

The dynamic similarity requirement that E remain a constant is modeled by the similitude velocity scale ratio, v_r as follows:

$$v_r = \frac{v_p}{v_m} = \frac{\left(\sqrt{\frac{\Delta p}{\rho}}\right)_p}{\left(\sqrt{\frac{\Delta p}{\rho}}\right)_m} = \Delta p_r^{1/2} \rho_r^{-1/2} \tag{11.46}$$

or

$$v_r = \Delta p_r^{1/2} \rho_r^{-1/2} \tag{11.47}$$

Because the "pressure model" similarity requirements apply for both gas and liquid flow, fortunately, there are fluids (especially gases) that are available in order to satisfy the similarity requirement for the model design as modeled by the relationship between the similitude velocity scale ratio, v_r; the similitude pressure ratio, Δp_r; and the similitude density ratio, ρ_r. Thus, the resulting model would be "true" with respect to E.

11.7.1.4 *"Viscosity Model" Similarity Requirements*

The dynamic similarity requirement that R remain a constant is modeled by the similitude velocity scale ratio, v_r as follows:

$$v_r = \frac{v_p}{v_m} = \frac{\left(\frac{\mu}{\rho L}\right)_p}{\left(\frac{\mu}{\rho L}\right)_m} = \mu_r \rho_r^{-1} L_r^{-1} = \frac{\mu_r}{\rho_r} L_r^{-1} = v_r L_r^{-1} \tag{11.48}$$

or

$$v_r = v_r L_r^{-1} \tag{11.49}$$

where $v = (\mu/\rho)$ is the kinematic viscosity. Thus, in order to satisfy the similarity requirement for the model design as modeled by the relationship between the similitude velocity scale ratio, v_r; the similitude kinematic viscosity scale ratio, v_r; and the similitude length scale ratio, L_r, one may implement one of the following options.

One option would be to use the same fluid (liquid or gas) for the model and the prototype. This would result in the following relationship between v_r and L_r:

$$v_r = L_r^{-1} \tag{11.50}$$

$$\frac{v_p}{v_m} = \frac{L_m}{L_p} = \lambda \tag{11.51}$$

where λ is the model scale, which is typically less than one; thus, the required model velocity, v_m will be larger than the prototype velocity v_p. One may note that because the model scale, λ is usually much less than one, then the required model velocity, v_m may be much higher than the prototype velocity v_p. For instance, for a model scale, $\lambda = 1/20$ with a prototype velocity $v_p = 40\,mph$, the corresponding required model velocity, $v_m = 800\,mph$. First, the required model velocity, $v_m = 800\,mph$ results in turbulent flow, which would cause the relative roughness, ε/L and the elastic force, F_E (and thus C or M) to play an important role in the external flow around the object for the model, but not for the prototype (laminar flow). Furthermore, while the required model velocity, $v_m = 800\,mph$ may be too high to attain in practice for liquids, it would be attainable for gases.

A second option would be to use a different fluid (liquid or gas) for the model and the prototype, in order to reduce the required model velocity, v_m. This would maintain the similitude velocity scale ratio, v_r as follows:

$$v_r = v_r L_r^{-1} \tag{11.52}$$

$$\frac{v_p}{v_m} = \frac{v_p}{v_m} \frac{L_m}{L_p} \tag{11.53}$$

$$\frac{v_p}{v_m} = \frac{v_p}{v_m} \lambda \tag{11.54}$$

Because the model scale, λ is usually much less than one, the required model velocity, v_m will be much higher than the prototype velocity v_p, unless $(v_p/v_m) > 1$. One may note that although the absolute viscosity, μ and the density, ρ for a liquid such as water are larger than those corresponding to a gas such as air, the kinematic viscosity, $\nu = (\mu/\rho)$ for water is smaller than that for air by about 10 times. Therefore, in an effort to reduce the required model velocity, v_m, one may assume a prototype fluid of air and a model fluid of water. Furthermore, although such a model design would reduce the required model velocity, v_m, it may not reduce it significantly enough in order for it be practically attainable for a liquid.

A *third option* would be to increase the model scale, λ from the usually much less than one to a somewhat larger number less than one in order to reduce the required model velocity, v_m. This would maintain the similitude velocity scale ratio, v_r as follows:

$$\frac{v_p}{v_m} = \frac{v_p}{v_m}\lambda \tag{11.55}$$

Regardless of the prototype fluid, assuming air for the model fluid, a large wind tunnel may be used to model the prototype. Furthermore, while a relatively high speed may be attained for the model velocity, v_m, the design of such a large wind tunnel may run the risk of increased costs.

11.7.1.5 "Elastic Model" Similarity Requirements

The dynamic similarity requirement that C (or M) remain a constant is modeled by the similitude velocity scale ratio, v_r as follows:

$$v_r = \frac{v_p}{v_m} = \frac{\left(\sqrt{\dfrac{E_v}{\rho}}\right)_p}{\left(\sqrt{\dfrac{E_v}{\rho}}\right)_m} = E_{v_r}^{1/2}\rho_r^{-1/2} = c_r \tag{11.56}$$

or

$$v_r = c_r \tag{11.57}$$

where $c = \sqrt{E_v/\rho}$ is the sonic velocity. Because the "elastic model" similarity requirements typically apply for gas flow and not for liquid flow, fortunately, there are gases that are available in order to satisfy the similarity requirement for the model design, as modeled by the relationship between the similitude velocity scale ratio, v_r and the similitude sonic velocity scale ratio, c_r. Thus, the resulting model would be "true" with respect to C (or M).

11.7.1.6 "Gravity Model" Similarity Requirements

The dynamic similarity requirement that F remain a constant is modeled by the similitude velocity scale ratio, v_r as follows:

$$v_r = \frac{v_p}{v_m} = \frac{\left(\sqrt{gL}\right)_p}{\left(\sqrt{gL}\right)_m} = L_r^{1/2} \tag{11.58}$$

or

$$v_r = L_r^{1/2} = \frac{1}{\lambda^{1/2}} \tag{11.59}$$

where λ is the model scale. Because the "gravity model" similarity requirements apply for liquid flow, fortunately, there are fluids that are available in order to satisfy the similarity requirement for the model design as modeled by the relationship between the similitude velocity scale ratio, v_r and the model scale, λ. Thus, the resulting model would be "true" with respect to F.

11.7.1.7 "Surface Tension Model" Similarity Requirements

The dynamic similarity requirement that W remain a constant is modeled by the similitude velocity scale ratio, v_r as follows:

$$v_r = \frac{v_p}{v_m} = \frac{\left(\sqrt{\dfrac{\sigma}{\rho L}}\right)_p}{\left(\sqrt{\dfrac{\sigma}{\rho L}}\right)_m} = \sigma_r^{1/2}\rho_r^{-1/2}L_r^{-1/2} \tag{11.60}$$

or

$$v_r = \frac{\sigma_r^{1/2}}{\rho_r^{1/2}}\frac{1}{\lambda^{-1/2}} \tag{11.61}$$

where σ/ρ is the kinematic surface tension, and λ is the model scale. Because the "surface tension model" similarity requirements apply for liquid flow, fortunately, there are fluids that are available in order to satisfy the similarity requirement for the model design as modeled by the relationship between the similitude velocity scale ratio, v_r; the kinematic surface tension scale ratio, σ_r/ρ_r; and the model scale, λ. Thus, the resulting model would be "true" with respect to W.

11.7.1.8 "Viscosity Model" and "Elastic Model" Similarity Requirements

The dynamic similarity requirement that R and C (or M) simultaneously remain a constant is modeled by the simultaneous application of the respective similitude velocity scale ratios, v_r as follows:

$$v_r = \frac{v_p}{v_m} = \frac{\left(\dfrac{\mu}{\rho L}\right)_p}{\left(\dfrac{\mu}{\rho L}\right)_m} = \mu_r\rho_r^{-1}L_r^{-1} = \frac{\mu_r}{\rho_r}L_r^{-1} = v_r L_r^{-1} \tag{11.62}$$

$$v_r = \frac{v_p}{v_m} = \frac{\left(\sqrt{\dfrac{E_v}{\rho}}\right)_p}{\left(\sqrt{\dfrac{E_v}{\rho}}\right)_m} = E_{v_r}^{1/2}\rho_r^{-1/2} = c_r \tag{11.63}$$

Equating Equations 11.62 and 11.63 yields:

$$c_r = v_r L_r^{-1} \tag{11.64}$$

$$\frac{c_p}{c_m} = \frac{v_p}{v_m} \frac{L_m}{L_p} \tag{11.65}$$

$$\frac{c_p}{c_m} = \frac{v_p}{v_m} \lambda \tag{11.66}$$

where $c = \sqrt{E_v/\rho}$ is the sonic velocity and λ is the model scale. Thus, in order to satisfy the similarity requirement for the model design as modeled by the relationship between the similitude sonic velocity scale ratio, c_r; the similitude kinematic viscosity scale ratio, v_r; and the model scale, λ, one may implement one of the following options.

One option would be to use the same fluid for the model and the prototype. One may note that because of the high velocity assumed for the prototype, such a flow situation would typically assume a gas or air for both the model and the prototype fluid. This would result in the following:

$$\lambda = 1 \tag{11.67}$$

which implies that the model is the same size as the prototype. Because this is not a desirable outcome, another option is sought.

A *second option* would be to use two different fluids for the model and the prototype and recognize that a relation must exist between the model scale, the fluid viscosities, and the sonic speeds in the compressible fluids (gases) used in the model and the prototype in order to achieve dynamic similarity. Fortunately, unlike liquids, there are gases that are available that will allow both R and M to be satisfied simultaneously.

A *third option* would be to recognize that for a blunt body moving at a high velocity, the flow resistance (as modeled by C_D) is independent of R, but dependent on C or M. And, for a streamlined or round-shaped body (circular cylinder or sphere) moving at a high velocity (turbulent flow), the flow resistance (as modeled by C_D) is less dependent on R, while it is more dependent on C or M. As such, it becomes more important to satisfy the dynamic similarity requirement for C or M, while relaxing the requirement for R. Therefore, the resulting model would be "true" with respect to C or M, while it would be "distorted" with respect to R. Then, one may correct the results with experimental data for the unsatisfied R.

11.7.1.9 "Viscosity Model" and "Gravity Model" Similarity Requirements

The dynamic similarity requirement that R and F simultaneously remain a constant is modeled by the simultaneous application of the respective similitude velocity scale ratios, v_r as follows:

$$v_r = \frac{v_p}{v_m} = \frac{\left(\dfrac{\mu}{\rho L}\right)_p}{\left(\dfrac{\mu}{\rho L}\right)_m} = \mu_r \rho_r^{-1} L_r^{-1} = v_r L_r^{-1} \tag{11.68}$$

$$v_r = \frac{v_p}{v_m} = \frac{\left(\sqrt{gL}\right)_p}{\left(\sqrt{gL}\right)_m} = L_r^{1/2} \tag{11.69}$$

Equating Equations 11.68 and 11.69 yields:

$$v_r = L_r^{3/2} \tag{11.70}$$

where $v = (\mu/\rho)$ is the kinematic viscosity. Thus, in order to satisfy the similarity requirement for the model design as modeled by the relationship between the similitude kinematic viscosity scale ratio, v_r and the similitude length scale ratio, L_r, one may implement one of the following options.

One option would be to use fluids of different viscosities in the model and the prototype. Assuming that the prototype fluid is water and that the practical maximum value for L_r is about 4 or 5, a potential model fluid would be mercury. Therefore, while a model fluid with a proper viscosity may exist, thus yielding a "true model," it is usually impractical or impossible for application, especially for small model scales. The use of mercury becomes impractical because the model in this application is usually also relatively large; thus, the only practical model fluid would also be water.

A second option would be to use water in both the model and the prototype. This would result in the following:

$$\lambda = 1 \tag{11.71}$$

which implies that the model is the same size as the prototype. Because this is not a desirable outcome, another option is sought.

A third option would be to use water in both the model and the prototype, and to satisfy only the F similarity requirement, which results in different R in the model and the prototype. Then, one may correct the results with experimental data for the unsatisfied R.

Furthermore, it is important to note that there are many flow situations for which the operational value for R is large enough so that C_D is not dependent on R and thus the dynamic similarity requirement that R remain a constant between the model and the prototype may be relaxed.

A fourth option would be to use water in both the model and the prototype, and satisfy one of the two dimensionless numbers (R or F), then correct the results with experimental data depending on the other dimensionless number. One may note that the shear drag is a function of R, while the pressure drag is a function of F. Therefore, depending upon which type of drag (shear or pressure) predominates the flow situation, that will guide the determination regarding which of the two dimensionless numbers (R or F) should be satisfied. For instance, for an open channel flow of a highly viscous fluid, or for a flow with a relatively low R (where the effects of the R cannot be ignored), one may satisfy the F, then correct the results with experimental data based on the R. First, the F is satisfied as follows:

$$F_r = \frac{F_p}{F_m} = \frac{\left(\dfrac{v}{\sqrt{gL}}\right)_p}{\left(\dfrac{v}{\sqrt{gL}}\right)_m} = 1 \tag{11.72}$$

$$v_r = \frac{v_p}{v_m} = \frac{\left(\sqrt{gL}\right)_p}{\left(\sqrt{gL}\right)_m} = L_r^{1/2} \tag{11.73}$$

Second, for the assumption of water in both the model and the prototype, the viscosity ratio is evaluated as follows:

$$v_r = 1 \tag{11.74}$$

Thus, the resulting R_r is computed as follows:

$$R_r = \frac{R_p}{R_m} = \frac{\left(\dfrac{vL}{v}\right)_p}{\left(\dfrac{vL}{v}\right)_m} = \frac{v_r L_r}{v_r} \tag{11.75}$$

Substituting Equations 11.73 and 11.74 into Equation 11.75 yields the following:

$$R_r = \frac{v_r L_r}{v_r} = \frac{L_r^{1/2} L_r}{1} = L_r^{3/2} \tag{11.76}$$

Thus, the resulting R_r does not equal 1 (as it would if R was satisfied). Furthermore, assuming the geometric size of the model is smaller than the prototype (thus, $L_r > 1$), the resulting R_m is smaller than the resulting R_p. In practice, this is accommodated by approximate methods that are available for dealing with difficulties that arise from viscous effects (i.e., the effects of R). In the application of all of the approximate methods, one must be aware of the role that R plays and how it influences the drag coefficient, C_D. Although, $F_m = F_p$, $R_m \neq R_p$, and thus $C_{Dm} \neq C_{Dp}$, which is demonstrated by examining Figure 11.2. Figure 11.2 illustrates the typical variation of the drag coefficient, C_D with R for a blunt body (external flow)/rough surface (open channel flow), and a streamlined body (external flow)/smooth surface (open channel flow). Specifically, regardless of the shape of the body or the roughness of the channel, generally the drag coefficient, C_D decreases with an increase in R

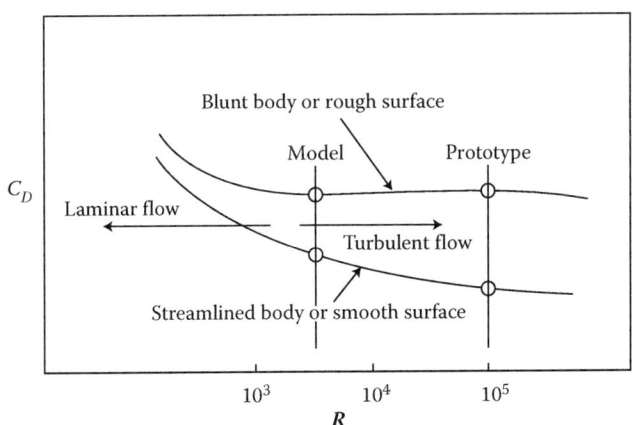

FIGURE 11.2
Typical variation of the drag coefficient, C_D with R for a blunt body (external flow)/rough surface (open channel flow), and a streamlined body (external flow)/smooth surface (open channel flow). (From Henderson, F. M., *Open Channel Flow*, 1st Ed, Pearson Education, Inc., New York, NY, Copyright 1966. Reprinted by permission.)

as the flow goes from laminar to turbulent, where a high value for the drag coefficient, C_D is due to the predominance of friction drag over pressure drag for laminar flow. In the case of a streamlined body (external flow)/smooth surface (open channel flow), the drag coefficient, C_D continues to decreases with an increase in R as the flow goes from laminar to turbulent. However, in the case of a blunt body (external flow)/rough surface (open channel flow), the drag coefficient, C_D becomes independent of R for turbulent flow. Thus, regardless of the shape of the body or the roughness of the channel, in the case where the resulting R_m is smaller than the resulting R_p, the corresponding drag coefficient, C_D in the model is higher than the corresponding drag coefficient, C_D in the prototype. Figure 11.2 assumes that shear drag predominates pressure drag and thus the drag coefficient, C_D is dependent on R. However, if pressure drag predominates friction drag (and thus the value of C_D is not as dependent on R, rather it is mostly a function of F), then the C_D in the model may be the same as in the prototype. This may occur when the R_m in the model is kept as high as possible in order to make the model flow fully turbulent, even though R_m may not be high enough to escape the effects of friction drag. A practical example where pressure drag predominates friction drag is the resistance of flow around a bridge pier. The drag coefficient, C_D is above 1, which indicates that pressure drag is a significant part of the total drag, so the R will be relatively unimportant. Furthermore, a practical example where friction drag predominates pressure drag is the resistance of flow on a long river channel, where the R will be important.

11.7.1.10 "Viscosity Model" and "Surface Tension Model" Similarity Requirements

The dynamic similarity requirement that R and W simultaneously remain a constant is modeled by the simultaneous application of the respective similitude velocity scale ratios, v_r as follows:

$$v_r = \frac{v_p}{v_m} = \frac{\left(\dfrac{\mu}{\rho L}\right)_p}{\left(\dfrac{\mu}{\rho L}\right)_m} = \mu_r \rho_r^{-1} L_r^{-1} = v_r L_r^{-1} \tag{11.77}$$

$$v_r = \frac{v_p}{v_m} = \frac{\left(\sqrt{\dfrac{\sigma}{\rho L}}\right)_p}{\left(\sqrt{\dfrac{\sigma}{\rho L}}\right)_m} = \sigma_r^{1/2} \rho_r^{-1/2} L_r^{-1/2} \tag{11.78}$$

Equating Equations 11.77 and 11.78 yields:

$$v_r = \frac{\sigma_r^{1/2}}{\rho_r^{1/2}} L_r^{1/2} \tag{11.79}$$

where $v = (\mu/\rho)$ is the kinematic viscosity and σ/ρ is the kinematic surface tension. Thus, in order to satisfy the similarity requirement for the model design as modeled by the relationship between the similitude kinematic viscosity scale ratio, v_r; the kinematic surface tension scale ratio, σ_r/ρ_r; and the similitude length scale ratio, L_r, one may implement one of the following options.

One option would be to use the same fluid for the model and the prototype, assuming the prototype fluid is water. This would result in the following:

$$\lambda = 1 \tag{11.80}$$

which implies that the model is the same size as the prototype. Because this is not a desirable outcome, another option is sought.

A second option would be to ignore the dependency of C_D on W and attempt to satisfy the dependency of C_D only on R and select one of the options listed in Section 11.7.1.4 above.

A third option would be to ignore the dependency of C_D on R (especially if the flow is turbulent) and attempt to satisfy the dependency of C_D only on W.

11.7.1.11 "Gravity Model," "Viscosity Model," and "Surface Tension Model" Similarity Requirements

The dynamic similarity requirement that F, R, and W simultaneously remain a constant is modeled by the simultaneous application of the respective similitude velocity scale ratios, v_r as follows. First, as presented in Section 11.7.1.8 above, the dynamic similarity requirement that R and F simultaneously remain a constant is modeled by the simultaneous application of the respective similitude velocity scale ratios, v_r as follows:

$$v_r = \frac{v_p}{v_m} = \frac{\left(\frac{\mu}{\rho L}\right)_p}{\left(\frac{\mu}{\rho L}\right)_m} = \mu_r \rho_r^{-1} L_r^{-1} = v_r L_r^{-1} \tag{11.81}$$

$$v_r = \frac{v_p}{v_m} = \frac{\left(\sqrt{gL}\right)_p}{\left(\sqrt{gL}\right)_m} = L_r^{1/2} \tag{11.82}$$

Equating Equations 11.81 and 11.82 yields:

$$v_r = L_r^{3/2} \tag{11.83}$$

where $v = (\mu/\rho)$ is the kinematic viscosity. The difficulties of simultaneously satisfying the dynamic requirements for R and F were discussed in Section 11.7.1.8 above. Then, to further add to the difficulties of dynamic requirements, the dynamic similarity requirement that W and F simultaneously remain a constant is modeled by the simultaneous application of the respective similitude velocity scale ratios, v_r as follows:

$$v_r = \frac{v_p}{v_m} = \frac{\left(\sqrt{\frac{\sigma}{\rho L}}\right)_p}{\left(\sqrt{\frac{\sigma}{\rho L}}\right)_m} = \sigma_r^{1/2} \rho_r^{-1/2} L_r^{-1/2} \tag{11.84}$$

$$v_r = \frac{v_p}{v_m} = \frac{\left(\sqrt{gL}\right)_p}{\left(\sqrt{gL}\right)_m} = L_r^{1/2} \tag{11.85}$$

Equating Equations 11.84 and 11.85 yields:

$$\frac{\sigma_r}{\rho_r} = L_r^2 \tag{11.86}$$

where σ/ρ is the kinematic surface tension. Finally, in order to satisfy the dynamic similarity requirement that R, W, and F simultaneously remain a constant is modeled by the simultaneous application of the respective similitude velocity scale ratios, v_r, which is accomplished by substituting Equation 11.83 into Equation 11.86, which yields:

$$v_r = \left(\frac{\sigma_r}{\rho_r}\right)^{3/4} \tag{11.87}$$

$$\frac{\mu_r}{\rho_r} = \left(\frac{\sigma_r}{\rho_r}\right)^{3/4} \tag{11.88}$$

Thus, in order to satisfy the similarity requirement for the model design as modeled by the relationship between the kinematic surface tension ratio, σ_r/ρ_r and the kinematic viscosity ratio, v_r, one may implement one of the following options.

One option would be to use the same fluid for the model and the prototype, assuming the prototype fluid is water. This would result in the following:

$$\lambda = 1 \tag{11.89}$$

which implies that the model is the same size as the prototype. Because this is not a desirable outcome, another option is sought.

A second option would be to ignore the dependency of C_D on W and attempt to satisfy the dependency of C_D only on F and R and select one of the options listed in Section 11.7.1.8 above.

A third option would be to recognize that for turbulent flow, the flow resistance (as modeled by C_D) is less dependent on W and R, while it is more dependent on F. As such, it becomes more important to satisfy the dynamic similarity requirement for F, while relaxing the requirement for W and R. Therefore, the resulting model would be "true" with respect to F, while it would be "distorted" with respect to W and R. Assuming that the prototype fluid is water, one may note that the because the model in this application is usually also relatively large, the only practical model fluid would also be water.

11.7.2 Application of the Similitude Scale Ratios for the Drag Force in External Flow

For the case of external flow around an object, dimensional analysis yielded the following uncalibrated mathematical model of the fluid flow situation:

$$\frac{F_D}{\frac{\rho v^2}{2} A} = \frac{F_D}{F_{dyn\,pres}} = \frac{F_P}{F_{dyn\,pres}} = \frac{\Delta p A}{\frac{\rho v^2}{2} A} = \frac{\Delta p}{\frac{\rho v^2}{2}} = \frac{2}{E} = C_D = \phi\left(R, C, F, \frac{\varepsilon}{L}, \frac{L_i}{L}\right) \tag{11.90}$$

where the resulting drag force, F_D was given as follows:

$$F_D = C_D \tfrac{1}{2} \rho v^2 A \qquad (11.91)$$

and where the drag coefficient, C_D as modeled by the Euler number, E is the dependent π term, as it models the dependent variable, Δp that cannot be derived analytically from theory, while the Reynolds number, R; the Cauchy number, C; the Froude number, F; the relative surface roughness, ε/L; and the geometry, L_i/L are the independent π terms. Recall that the Mach number, M is $M = \sqrt{C} = (v/\sqrt{E_v/\rho}) = \frac{v_a}{c}$, where c is the sonic velocity. One may note that because the drag coefficient, C_D represents the flow resistance, which causes an unknown pressure drop, Δp, the pressure force will always play an important role; thus, the definition of the drag coefficient, C_D is mainly a function of E.

Furthermore, because the drag coefficient, C_D is mainly a function of the Euler number, E, the inertia force, F_I and the pressure force, F_P will always be important (predominate) in the external flow around an object. Thus, the Euler number, E and the drag coefficient, $C_D = (2/E)$ (dependent π term) must remain a constant between the model and its prototype. However, depending upon the specifics of the external flow situation, the relative surface roughness, ε/L; the geometry, L_i/L; and the remaining forces, which include the viscous force, F_V; the elastic force, F_E; and the gravity force, F_G (and thus, their respective independent π terms, R, M, and F, respectively), may or may not play an important role in the definition of the drag coefficient, C_D. If in fact a given independent π term plays an important role in the definition of the drag coefficient, C_D then it must remain a constant between the model and its prototype (as a "true model" is sought).

Although the drag coefficient, C_D, in general, is a mainly a function of E, it will vary with R, L_i/L, ε/L, M, and F depending mainly on the velocity of flow, the shape of the body, and whether there is a wave action at the free surface (see Table 11.7, which is the same as Table 10.1). Specifically, while the R divides the flow into three types (creeping, laminar, or turbulent), the L_i/L divides the shape of the body into two types: blunt or streamlined. The drag coefficient, C_D for creeping flow is dependent only R for any shape body. Furthermore, for body shapes in general, the drag coefficient, C_D for laminar flow is dependent on R (especially for streamlined bodies, and less for blunt bodies) and L_i/L, but independent of ε/L, M, and F. And, for body shapes in general, the drag coefficient, C_D for turbulent flow is independent of R (except for streamlined bodies and round-shaped bodies, which are dependent on R), independent of F, but dependent on L_i/L, ε/L, and M. Finally, for body shapes in general, with a wave action at the free surface (such examples include ships and open channel flow-measuring devices/structures such as weirs and spillways), the drag coefficient, C_D for laminar flow is dependent on R, L_i/L, and F, but independent of ε/L and M, while the drag coefficient, C_D for turbulent flow is dependent on R, L_i/L, ε/L, M, and F.

11.7.2.1 Creeping Flow (R ≤ 1) for Any Shape Body

Although the drag coefficient, C_D will vary with the R with varying degrees depending upon the range of the R, low (creeping; $R \le 1$), moderate (laminar), or high (turbulent), and the shape of the body, in general, the drag coefficient, C_D decreases with an increase in R (see Figure 10.12 for two-dimensional bodies and Figure 10.13 for two-dimensional axisymmetric bodies). However, for creeping flows at $R \le 1$, the shape of the body (blunt or streamlined) does not have major role on the value of the drag coefficient, C_D, which decreases with an increase in R, as illustrated in Figure 10.14. Furthermore, the drag coefficient, C_D for creeping flow is dependent only on R and independent of L_i/L (i.e., regardless of the shape of the body), ε/L, M, and F.

In the case of creeping flow for any shape body, the flow resistance (as modeled by C_D) is only a function of R. As such, in addition to the inertia force, F_I and the pressure force, F_P, R is important in the external flow around an object. Therefore, because there is one independent π term, R used in the definition of the dependent π term, E and thus the drag coefficient, C_D, the appropriate similitude scale ratios corresponding to the independent π term are used to define the similarity requirements between the model and its prototype. Specifically, the criteria governing dynamic similarity requires the following:

$$\underbrace{\left[\left(\frac{\rho v L}{\mu}\right)_p\right]}_{R_p} = \underbrace{\left[\left(\frac{\rho v L}{\mu}\right)_m\right]}_{R_m} \qquad (11.92)$$

which states there is a dynamic similarity requirement that R remain a constant in order for the dependent π term (i.e., the drag coefficient, C_D) to remain a constant between the model and its prototype, where the flow resistance prediction equation is defined as follows:

$$\underbrace{\left[\frac{F_D}{\frac{1}{2}\rho v^2 A}\right]_p}_{C_{Dp}} = \underbrace{\left[\frac{F_D}{\frac{1}{2}\rho v^2 A}\right]_m}_{C_{Dm}} \qquad (11.93)$$

Furthermore, one may note that in the case where the similarity requirements regarding the independent π term, R ("viscosity model") are satisfied, then the dependent π term (i.e., the drag coefficient, C_D) will remain a constant between the model and its prototype. And, therefore, application of the flow resistance prediction equation yields a "true model." However, in the case where the similarity requirements regarding the independent π term, R are not satisfied, then the dependent π term (i.e., the drag coefficient, C_D) will not remain a constant between the model and its prototype. And, therefore, application of the flow resistance prediction equation yields a "distorted model" for which the appropriate correction measures are taken.

EXAMPLE PROBLEM 11.1

Water at 70°F flows over a prototype sphere with a diameter of 0.1 ft, as illustrated in Figure EP 11.1. An enlarged model of the smaller prototype is designed in order to study the flow characteristics of creeping flow over a sphere (Stokes law). The model fluid is oil ($s = 0.88$, $\mu = 150 \times 10^{-6}$ lb-sec/ft^2), the velocity of flow of the oil over the enlarged model sphere is 1.66×10^{-4} ft/sec, and the model scale, λ is 5. (a) Determine the drag force on the model sphere. (b) Determine the velocity of flow of the water over the prototype sphere in order to achieve dynamic similarity between the model and the prototype. (c) Determine the drag force on the prototype sphere in order to achieve dynamic similarity between the model and the prototype.

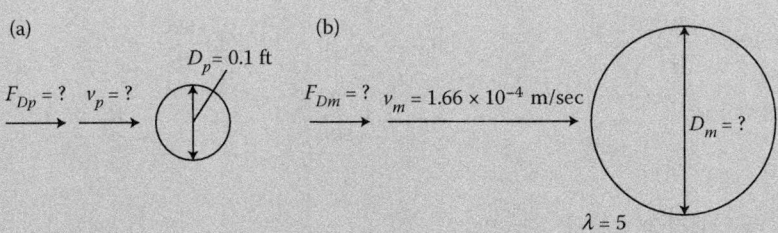

FIGURE EP 11.1
(a) Water flows over a prototype sphere. (b) Oil flows over an enlarged model sphere.

Mathcad Solution

(a) In order to determine the drag force on the model sphere, the drag force equation, Equation 11.91, is applied, where the frontal area of the sphere is used to compute the drag force. Because the Reynolds number, $R \leq 1$ (creeping flow) is assumed, the drag coefficient, C_D varies inversely with R as presented in Figure 10.14. Furthermore, in order to determine the diameter of the model sphere, the model scale, λ (inverse of the length ratio) is applied as follows:

$D_p := 0.1 \text{ ft}$ $\qquad\qquad\qquad \lambda := 5$

Guess value: $\qquad\qquad\qquad D_m := 0.3 \text{ ft}$

Given

$$\lambda = \frac{D_m}{D_p}$$

$D_m := \text{Find } (D_m) = 0.5 \text{ ft}$

$\text{slug} := 1 \text{ lb} \dfrac{\sec^2}{\text{ft}}$ $\qquad\qquad s_m := 0.88 \qquad\qquad \rho_w := 1.94 \dfrac{\text{slug}}{\text{ft}^3}$

$\rho_m := s_m \cdot \rho_w = 1.707 \dfrac{\text{slug}}{\text{ft}^3}$ $\qquad\qquad\qquad \mu_m := 150 \times 10^{-6} \text{ lb} \dfrac{\sec}{\text{ft}^2}$

$A_m := \dfrac{\pi \cdot D_m^2}{4} = 0.196 \text{ ft}^2$ $\qquad\qquad\qquad v_m := 1.66 \times 10^{-4} \dfrac{\text{ft}}{\sec}$

$R_m := \dfrac{\rho_m \cdot v_m \cdot D_m}{\mu_m} = 0.945$ $\qquad\qquad\qquad C_{Dm} := \dfrac{24}{R_m} = 25.406$

$F_{Dm} := C_{Dm} \cdot \dfrac{1}{2} \cdot \rho_m \cdot v_m^2 \cdot A_m = 1.173 \times 10^{-7} \text{ lb}$

(b) To determine the velocity of flow of the water over the prototype sphere in order to achieve dynamic similarity between the model and the prototype, for creeping flow, the R must remain a constant between the model and prototype as follows:

$$\underbrace{\left[\left(\frac{\rho v L}{\mu}\right)_p\right]}_{R_p} = \underbrace{\left[\left(\frac{\rho v L}{\mu}\right)_m\right]}_{R_m}$$

$$\rho_p := 1.936 \, \frac{\text{slug}}{\text{ft}^3} \qquad\qquad \mu_p := 20.50 \times 10^{-6} \, \text{lb} \, \frac{\text{sec}}{\text{ft}^2}$$

Guess value: $\qquad\qquad v_p := 1 \, \frac{\text{ft}}{\text{sec}} \qquad\qquad R_p := 1$

Given

$$R_p = \frac{\rho_p \cdot v_p \cdot D_p}{\mu_p}$$

$$R_p = R_m$$

$$\begin{pmatrix} v_p \\ R_p \end{pmatrix} := \text{Find} \, (v_p, R_p)$$

$$v_p = 1 \times 10^{-4} \, \frac{\text{ft}}{\text{s}} \qquad\qquad R_p = 0.945$$

(c) To determine the drag force on the prototype sphere in order to achieve dynamic similarity between the model and the prototype, for creeping flow, the drag coefficient, C_D must remain a constant between the model and the prototype (which is a direct result of maintaining a constant R between the model and the prototype) as follows:

$$\underbrace{\left[\frac{F_D}{\frac{1}{2}\rho v^2 A} \right]_p}_{C_{D_p}} = \underbrace{\left[\frac{F_D}{\frac{1}{2}\rho v^2 A} \right]_m}_{C_{D_m}}$$

Furthermore, the frontal area of the sphere is used to compute the drag force as follows:

$$A_p := \frac{\pi \cdot D_p^2}{4} = 7.854 \times 10^{-3} \, \text{ft}^2$$

Guess value: $\qquad\qquad F_{Dp} := 1 \, \text{lb} \qquad\qquad C_{Dp} := 1$

Given

$$C_{Dp} = \frac{F_{Dp}}{\frac{1}{2} \cdot \rho_p \cdot v_p^2 \cdot A_p}$$

$$C_{Dp} = C_{Dm}$$

$$\begin{pmatrix} F_{Dp} \\ C_{Dp} \end{pmatrix} := \text{Find} \, (F_{Dp}, C_{Dp})$$

$$F_{Dp} = 1.933 \times 10^{-9} \, \text{lb} \qquad\qquad C_{Dp} = 25.406$$

Therefore, although the similarity requirements regarding the independent π term, R ("viscosity model") are theoretically satisfied ($R_p = R_m = 0.945$), the dependent

π term (i.e., the drag coefficient, C_D) will actually/practically remain a constant between the model and its prototype $(C_{Dp} = 24/R_p = C_{Dm} = 24/R_m = 25.406)$ only if it is practical to maintain/attain the model velocity, drag force, fluid, scale, and cost.

11.7.2.2 Laminar Flow (R < 10,000) for Any Shape Body except Round-Shaped Bodies

Although the drag coefficient, C_D will vary with the R with varying degrees depending upon the range of the R, low (creeping; $R \leq 1$), moderate (laminar), or high (turbulent), and the shape of the body, in general, the drag coefficient, C_D decreases with an increase in R (see Figure 10.12 for two-dimensional bodies and Figure 10.13 for two-dimensional axisymmetric bodies). For most geometric shapes (except for round-shaped bodies such as circular cylinders and spheres; these will be addressed in Section 11.7.1.4 below), for laminar flows at $R < 10,000$, the drag coefficient, C_D is dependent on the R, with the drag coefficient, C_D generally decreasing with an increase in R, as illustrated in Figure 10.16. Although Figure 10.16 illustrates such a relationship between the drag coefficient, C_D and R for a disk-shaped body that is normal to the direction of flow, a similar relationship exists between the drag coefficient, C_D and R for most geometric shapes that are normal to the direction of flow and blunt bodies. However, the relationship between the drag coefficient, C_D and R (and thus the dependency of C_D on R) for most geometric shapes that are parallel to the direction of flow and streamlined bodies is more accurately illustrated by Figures 10.12 and 10.13 and Tables 10.4 through 10.6, which present the drag coefficient, C_D for numerous geometric shapes at various orientations to the direction of flow, for $R > 10,000$ (and for some shapes, for laminar flow, $R < 10,000$), for two-dimensional, two-dimensional axisymmetric, and three-dimensional bodies, respectively.

Furthermore, for laminar flows at $R < 10,000$, and in the case of streamlined bodies, the drag coefficient, C_D will be dependent on R, but independent of the relative surface roughness, ε/L, M, and F. Thus, in the case of laminar flow for streamlined bodies, the flow resistance (as modeled by C_D) is only a function of R and the geometry, L_i/L. As such, in addition to the inertia force, F_I and the pressure force, F_P, the viscous force, F_V and the geometry of the body, L_i/L are important in the external flow around an object. Therefore, because there are two independent π terms, R and L_i/L used in the definition of the dependent π term, E and thus the drag coefficient, C_D, the appropriate similitude scale ratios corresponding to the independent π terms are used to define the similarity requirements between the model and its prototype. Specifically, the criteria governing dynamic similarity require the following:

$$\underbrace{\left[\left(\frac{\rho v L}{\mu}\right)_p\right]}_{R_p} = \underbrace{\left[\left(\frac{\rho v L}{\mu}\right)_m\right]}_{R_m} \tag{11.94}$$

$$\left(\frac{L_i}{L}\right)_p = \left(\frac{L_i}{L}\right)_m \tag{11.95}$$

which states that in addition to the geometric similarity requirement for the geometry of the body, L_i/L, there is a dynamic similarity requirement that R remain a constant in order for the dependent π term (i.e., the drag coefficient, C_D) to remain a constant between the model and its prototype, where the flow resistance prediction equation is defined as follows:

$$\underbrace{\left[\frac{F_D}{\frac{1}{2}\rho v^2 A}\right]_p}_{C_{Dp}} = \underbrace{\left[\frac{F_D}{\frac{1}{2}\rho v^2 A}\right]_m}_{C_{Dm}} \tag{11.96}$$

Furthermore, one may note that in the case where the similarity requirements regarding the independent π terms, R ("viscosity model") and the geometry of the body, L_i/L are satisfied, then the dependent π term (i.e., the drag coefficient, C_D) will remain a constant between the model and its prototype. And, therefore, application of the flow resistance prediction equation yields a "true model." However, in the case where the similarity requirements regarding the independent π terms, R and the geometry of the body, L_i/L are not satisfied, then the dependent π term (i.e., the drag coefficient, C_D) will not remain a constant between the model and its prototype. And, therefore, application of the flow resistance prediction equation yields a "distorted model" for which the appropriate corrections measures are taken.

EXAMPLE PROBLEM 11.2

Water at 70°F flows over a prototype 1200-ft-long elliptical rod with the following geometry: an L of 8 ft and a D of 1 ft, as illustrated in Figure EP 11.2. A smaller model of the larger prototype is designed in order to study the flow characteristics of laminar flow over the streamlined body. The model fluid is glycerin ($s = 1.26, \mu = 31,200 \times 10^{-6}$ lb-sec/ft^2), the velocity of flow of the glycerin over the smaller model elliptical rod is 3 ft/sec, and the model scale, λ is 0.20. (a) Determine the drag force on the model elliptical rod. (b) Determine the velocity of flow of the water over the prototype elliptical rod in order to achieve dynamic similarity between the model and the prototype. (c) Determine the drag force on the prototype elliptical rod in order to achieve dynamic similarity between the model and the prototype.

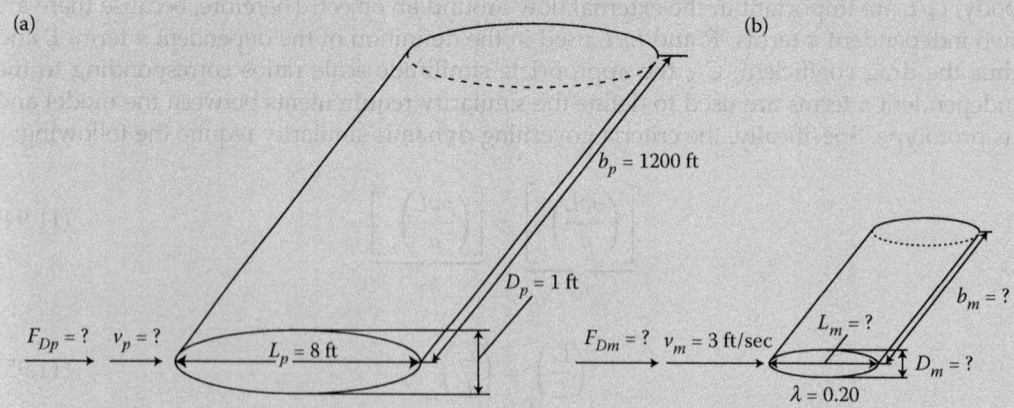

FIGURE EP 11.2
(a) Water flows over a prototype long elliptical rod. (b) Glycerin flows over a smaller model elliptical rod.

Mathcad Solution

(a) In order to determine the drag force on the model elliptical rod, the drag force equation, Equation 11.91, is applied, where the frontal area of the elliptical rod is used to compute the drag force. Because the Reynolds number, $R < 10,000$ (laminar flow) over a streamlined body is assumed, the drag coefficient, C_D decreases with an increase in R, as illustrated in Figure 10.12. However Table 10.4 for two-dimensional bodies presents an easier to read magnitude for the drag coefficient, C_D for the elliptical rod, assuming laminar flow. Furthermore, in order to determine the geometry L and D, and the length, b of the model elliptical rod, the model scale, λ (inverse of the length ratio) is applied as follows:

$D_p := 1 \, \text{ft}$ $\qquad\qquad$ $L_p := 8 \, \text{ft}$ $\quad$ $b_p := 1200 \, \text{ft}$ $\quad$ $\lambda := 0\,2$

Guess value: $\qquad\qquad\qquad$ $D_m := 0.1 \, \text{ft}$ $\quad$ $L_m := 0.8 \, \text{ft}$ $\quad$ $b_m := 120 \, \text{ft}$

Given

$$\lambda = \frac{D_m}{D_p} \qquad\qquad \lambda = \frac{L_m}{L_p} \qquad \lambda = \frac{b_m}{b_p}$$

$$\begin{pmatrix} D_m \\ L_m \\ b_m \end{pmatrix} := \text{Find}\,(D_m, L_m, b_m) = \begin{pmatrix} 0.2 \\ 1.6 \\ 240 \end{pmatrix} \text{ft}$$

$$\text{slug} := 1 \, \text{lb} \, \frac{\text{sec}^2}{\text{ft}} \qquad\qquad S_m := 1.26 \qquad \rho_w := 1.94 \, \frac{\text{slug}}{\text{ft}^3}$$

$$\rho_m := S_m \cdot \rho_w = 2.444 \, \frac{\text{slug}}{\text{ft}^3} \qquad\qquad \mu_m := 31{,}200 \times 10^{-6} \, \text{lb} \, \frac{\text{sec}}{\text{ft}^2}$$

$$A_m := b_m \cdot D_m = 48 \, \text{ft}^2 \qquad\qquad v_m := 3 \, \frac{\text{ft}}{\text{sec}}$$

$$R_m := \frac{\rho_m \cdot v_m \cdot D_m}{\mu_m} = 47.008 \qquad \frac{L_m}{D_m} = 8 \quad C_{DM} := 0.25$$

$$F_{DM} := C_{Dm} \cdot \frac{1}{2} \cdot \rho_m \cdot v_m^2 \cdot A_m = 131.998 \, \text{lb}$$

(b) To determine the velocity of flow of the water over the prototype elliptical rod in order to achieve dynamic similarity between the model and the prototype, for laminar flow over a streamlined body, the geometry L_i/L must remain a constant between the model and prototype as follows:

$$\left(\frac{L_i}{L}\right)_p = \left(\frac{L_i}{L}\right)_m$$

where the geometry is modeled as follows:

$$\frac{L_p}{D_p} = 8 \qquad\quad \frac{L_m}{D_m} = 8 \qquad\quad \frac{b_p}{D_p} = 1.2 \times 10^3 \qquad\quad \frac{b_m}{D_m} = 1.2 \times 10^3$$

Furthermore, the **R** must remain a constant between the model and prototype as follows:

$$\underbrace{\left[\left(\frac{\rho v L}{\mu}\right)_p\right]}_{R_p} = \underbrace{\left[\left(\frac{\rho v L}{\mu}\right)_m\right]}_{R_m}$$

$$\rho_p := 1.936 \, \frac{\text{slug}}{\text{ft}^3} \qquad\qquad \mu_p := 20.50 \times 10^{-6} \, \text{lb} \, \frac{\text{sec}}{\text{ft}^2}$$

Guess value: $\qquad\qquad v_p := 10 \, \frac{\text{ft}}{\text{sec}} \qquad\qquad R_p := 100$

Given

$$R_p = \frac{\rho_p \cdot v_p \cdot D_p}{\mu_p}$$

$$R_p = R_m$$

$$\begin{pmatrix} v_p \\ R_p \end{pmatrix} := \text{Find} \,(v_p, R_p)$$

$$v_p = 4.978 \times 10^{-4} \, \frac{\text{ft}}{\text{s}} \qquad\qquad\qquad R_p = 47.008$$

(c) To determine the drag force on the prototype elliptical rod in order to achieve dynamic similarity between the model and the prototype, for laminar flow over a streamlined body, the drag coefficient, C_D must remain a constant between the model and the prototype (which is a direct result of maintaining a constant **R**, and a constant L_i/L between the model and the prototype) as follows:

$$\underbrace{\left[\frac{F_D}{\frac{1}{2}\rho v^2 A}\right]_p}_{C_{Dp}} = \underbrace{\left[\frac{F_D}{\frac{1}{2}\rho v^2 A}\right]_m}_{C_{Dm}}$$

Furthermore, the frontal area of the elliptical rod is used to compute the drag force as follows:

$$A_p := b_p \cdot D_p = 1.2 \times 10^3 \, \text{ft}^2$$

Guess value: $\qquad\qquad F_{Dp} := 1 \, \text{lb} \qquad\qquad C_{Dp} := 1$

Given

$$C_{Dp} = \frac{F_{Dp}}{\frac{1}{2} \cdot \rho_p \cdot v_p^2 \cdot A_p}$$

$C_{Dp} = C_{Dm}$

$\begin{pmatrix} F_{Dp} \\ C_{Dp} \end{pmatrix} := \text{Find } (F_{Dp}, C_{Dp})$

$F_{Dp} = 7.195 \times 10^{-5} \, \text{lb}$ $\qquad\qquad\qquad\qquad C_{Dp} = 0.25$

Therefore, although the similarity requirements regarding the independent π term, L_i/L and the independent π term, R ("viscosity model") are theoretically satisfied ($R_p = R_m = 47.008$), the dependent π term (i.e., the drag coefficient, C_D) will actually/practically remain a constant between the model and its prototype ($C_{Dp} = C_{Dm} = 0.25$) only if it is practical to maintain/attain the model velocity, drag force, fluid, scale, and cost.

Finally, for laminar flows at $R < 10,000$, and in the case of blunt bodies, the drag coefficient, C_D will be less dependent on R for blunt bodies (where it is mainly a function of E), but independent of the relative surface roughness, ε/L, M, and F. Thus, in the case of laminar flow for blunt bodies, the flow resistance (as modeled by C_D) is only a function of the geometry, L_i/L. As such, in addition to the inertia force, F_I and the pressure force, F_P, the geometry, L_i/L is important in the external flow around an object. Therefore, because there is one independent π term, the geometry, L_i/L used in the definition of the dependent π term, E and thus the drag coefficient, C_D, the appropriate similitude scale ratios corresponding to the independent π term are used to define the similarity requirements between the model and its prototype. Specifically, the criteria governing dynamic similarity require the following:

$$\left(\frac{L_i}{L}\right)_p = \left(\frac{L_i}{L}\right)_m \tag{11.97}$$

which states that there is a geometric similarity requirement for the geometry of the body, L_i/L in order for the dependent π term (i.e., the drag coefficient, C_D) to remain a constant between the model and its prototype, where the flow resistance prediction equation is defined as follows:

$$\underbrace{\left[\frac{F_D}{\frac{1}{2}\rho v^2 A}\right]_p}_{C_{Dp}} = \underbrace{\left[\frac{F_D}{\frac{1}{2}\rho v^2 A}\right]_m}_{C_{Dm}} \tag{11.98}$$

Furthermore, one may note that in the case where the similarity requirements regarding the independent π term, L_i/L (geometric similarity) are satisfied, then the dependent π term (i.e., the drag coefficient, C_D) will remain a constant between the model and its prototype. And, therefore, application of the flow resistance prediction equation yields a "true model." However, in the case where the similarity requirements regarding the independent π term, L_i/L and the dependent π term, E ("pressure model") are not satisfied, then the dependent π term (i.e., the drag coefficient, C_D) will not remain a constant between the model and its prototype. And, therefore, application of the flow resistance prediction equation yields a "distorted model" for which the appropriate corrections measures are taken.

EXAMPLE PROBLEM 11.3

Oil ($s = 0.88, \mu = 150 \times 10^{-6}$ lb-sec/ft^2) flows at a velocity of 0.8 ft/sec over a prototype short cylinder with a diameter of 1 ft and height of 40 ft, as illustrated in Figure EP 11.3. A smaller model of the larger prototype is designed in order to study the flow characteristics of laminar flow over the blunt body. The model fluid is glycerin ($s = 1.26, \mu = 31{,}200 \times 10^{-6}$ lb-sec/ft^2) and the model scale, λ is 0.25. (a) Determine the drag force on the prototype short cylinder. (b) Determine the velocity of flow of the glycerin over the model short cylinder in order to achieve dynamic similarity between the model and the prototype. (c) Determine the drag force on the model short cylinder in order to achieve dynamic similarity between the model and the prototype.

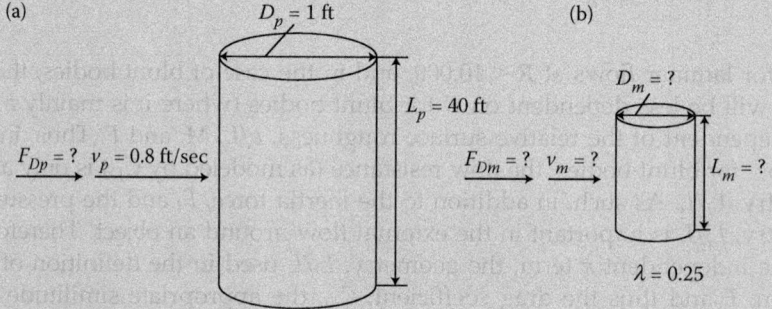

FIGURE EP 11.3
(a) Oil flows over a prototype short cylinder. (b) Glycerin flows over a smaller model short cylinder.

Mathcad Solution

(a) In order to determine the drag force on the prototype short cylinder, the drag force equation, Equation 11.91, is applied, where the frontal area of the short cylinder is used to compute the drag force. Because the Reynolds number, $R < 10{,}000$ (laminar flow) over a blunt body is assumed, the drag coefficient, C_D, in general, decreases with an increase in R, as illustrated in Figure 10.13. However, Table 10.5 for two-dimensional axisymmetric bodies presents an easier to read magnitude for the drag coefficient, C_D for the short cylinder, assuming laminar flow.

$$\text{slug} := 1 \, \text{lb} \, \frac{\text{sec}^2}{\text{ft}} \qquad s_p := 0.88 \qquad \rho_w := 1.94 \, \frac{\text{slug}}{\text{ft}^3}$$

$$\rho_p := s_p \cdot \rho_w = 1.707 \, \frac{\text{slug}}{\text{ft}^3} \qquad\qquad \mu_p := 150 \times 10^{-6} \, \text{lb} \, \frac{\text{sec}}{\text{ft}^2}$$

$$D_p := 1 \, \text{ft} \quad L_p := 40 \, \text{ft} \quad A_p := L_p \cdot D_p = 40 \, \text{ft}^2 \quad v_p := 0.8 \, \frac{\text{ft}}{\text{sec}}$$

$$R_p := \frac{\rho_p \cdot v_p \cdot D_p}{\mu_p} = 9.105 \times 10^3 \qquad\qquad C_{Dp} := 1.0$$

$$F_{Dp} := C_{Dp} \cdot \frac{1}{2} \cdot \rho_p \cdot v_p^2 \cdot A_p = 21.852 \, \text{lb}$$

(b)–(c) To determine the velocity of flow of the glycerin over the model short cylinder in order to achieve dynamic similarity between the model and the prototype, for laminar flow over a blunt body, and to determine the drag force on the model short cylinder in order to achieve dynamic similarity between the model and the prototype, for laminar flow over a blunt body, the geometry L_i/L must remain a constant between the model and prototype as follows:

$$\left(\frac{L_i}{L}\right)_p = \left(\frac{L_i}{L}\right)_m$$

Furthermore, in order to determine the diameter and the height of the model short cylinder, the model scale, λ (inverse of the length ratio) is applied as follows:

$\lambda := 0.25$

Guess value: $D_m := 1 \, \text{ft}$ $L_m := 1 \, \text{ft}$

Given

$$\lambda = \frac{D_m}{D_p} \qquad\qquad \lambda = \frac{L_m}{L_p}$$

$$\begin{pmatrix} D_m \\ L_m \end{pmatrix} := \text{Find} \, (D_m, L_m) = \begin{pmatrix} 0.25 \\ 10 \end{pmatrix} \text{ft}$$

And, where the geometry is modeled as follows:

$$\frac{L_p}{D_p} = 40 \qquad\qquad \frac{L_m}{D_m} = 40$$

However, because the drag coefficient, C_D will be less dependent on R for blunt bodies, R does not need to remain a constant between the model and the prototype.

(b)–(c) To determine the velocity of flow of the glycerin over the model short cylinder in order to achieve dynamic similarity between the model and the prototype, for laminar flow over a blunt body, and to determine the drag force on the model short cylinder in order to achieve dynamic similarity between the model and the prototype, for laminar flow over a blunt body, the drag coefficient, C_D must remain a constant between the model and the prototype (which is a direct result of maintaining a constant L_i/L between the model and the prototype and applying the "pressure model" similitude scale ratios, specifically the velocity ratio, v_r and the force ratio, F_r given in Table 11.1) as follows:

$$\underbrace{\left[\frac{F_D}{\frac{1}{2}\rho v^2 A}\right]_p}_{C_{D_p}} = \underbrace{\left[\frac{F_D}{\frac{1}{2}\rho v^2 A}\right]_m}_{C_{D_m}}$$

$$v_r = \frac{v_p}{v_m} = \frac{\left(\sqrt{\frac{\Delta p}{\rho}}\right)_p}{\left(\sqrt{\frac{\Delta p}{\rho}}\right)_m} = \Delta p_r^{\frac{1}{2}} \rho_r^{-\frac{1}{2}}$$

$$F_r = \Delta p_r L_r^2 = \rho_r v_r^2 L_r^2$$

Furthermore, the frontal area of the short cylinder is used to compute the drag force as follows:

$$s_m := 1.26 \qquad \rho_m := s_m \cdot \rho_w = 2.444 \frac{\text{slug}}{\text{ft}^3} \qquad \mu_m := 31{,}200 \times 10^{-6} \text{ lb} \frac{\text{sec}}{\text{ft}^2}$$

$$A_m := L_m \cdot D_m = 2.5 \text{ ft}^2 \qquad\qquad \Delta p_p := \frac{F_{D_p}}{A_p} = 0.546 \frac{\text{lb}}{\text{ft}^2}$$

Guess value: $\quad v_m := 0.1 \dfrac{\text{ft}}{\text{sec}} \qquad F_{Dm} := 1 \text{ lb} \quad \Delta p_m := 1 \dfrac{\text{lb}}{\text{ft}^2}$

$$C_{Dm} := 0.5$$

Given

$$C_{Dm} = \frac{F_{Dm}}{\frac{1}{2} \cdot \rho_m \cdot v_m^2 \cdot A_m} \qquad\qquad \frac{v_p}{\Delta p_p^{\frac{1}{2}} \cdot \rho_p^{\frac{-1}{2}}} = \frac{v_m}{\Delta p_m^{\frac{1}{2}} \cdot \rho_m^{\frac{-1}{2}}}$$

$$C_{Dm} = C_{Dp} \qquad\qquad \frac{F_{Dp}}{\Delta p_p \cdot A_p} = \frac{F_{Dm}}{\Delta p_m \cdot A_m}$$

$$\begin{pmatrix} v_m \\ F_{Dm} \\ \Delta p_m \\ C_{Dm} \end{pmatrix} := \text{Find}(v_m, F_{Dm}, \Delta p_m, C_{Dm})$$

$$v_m = 7.734 \frac{\text{ft}}{\text{s}} \qquad F_{Dm} = 182.769 \text{ lb} \qquad \Delta p_m = 73.108 \frac{\text{lb}}{\text{ft}^2} \qquad C_{Dm} = 1$$

Furthermore, the Euler number, E remains a constant between the model and the prototype as follows:

$$E_m := \frac{\rho_m \cdot v_m^2}{\Delta p_m} = 2 \qquad\qquad\qquad\qquad E_p := \frac{\rho_p \cdot v_p^2}{\Delta p_p} = 2$$

Therefore, although the similarity requirements regarding the independent π term, L_i/L and the dependent π term, E ("pressure model") are theoretically satisfied ($E_p = E_m = 2/C_D = 2$), the dependent π term (i.e., the drag coefficient, C_D) will actually/practically remain a constant between the model and its prototype ($C_{Dp} = C_{Dm} = 1.0$) only if it is practical to maintain/attain the model velocity, drag force, fluid, scale, and cost. Furthermore, because the drag coefficient, C_D will be less dependent on R for blunt bodies, R does not need to remain a constant between the model and the prototype as follows:

$$R_m := \frac{\rho_m \cdot v_m \cdot D_m}{\mu_m} = 151.484 \qquad\qquad R_p = 9.105 \times 10^3$$

11.7.2.3 Turbulent Flow (R > 10,000) for Any Shape Body except Round-Shaped Bodies

Although the drag coefficient, C_D will vary with the R with varying degrees depending upon the range of the R, low (creeping; $R \leq 1$), moderate (laminar), or high (turbulent), and the shape of the body, in general, the drag coefficient, C_D decreases with an increase in R (see Figure 10.12 for two-dimensional bodies and Figure 10.13 for two-dimensional axisymmetric bodies). For most geometric shapes (except for round-shaped bodies such as circular cylinders and spheres; these will be addressed in Section 11.7.1.4 below), for fully turbulent flows at $R > 10,000$, the drag coefficient, C_D becomes independent of R, as illustrated in Figure 10.16. Although Figure 10.16 illustrates such a relationship between the drag coefficient, C_D and R for a disk-shaped body that is normal to the direction of flow, a similar relationship exists between the drag coefficient, C_D and R for most geometric shapes that are normal to the direction of flow and blunt bodies. However, the relationship between the drag coefficient, C_D and R (and thus the dependency of C_D on R) for most geometric shapes that are parallel to the direction of flow and streamlined bodies is more accurately illustrated by Figures 10.12 and 10.13 and Tables 10.4 through 10.6, which present the drag coefficient, C_D for numerous geometric shapes at various orientations to the direction of flow, for $R > 10,000$ (and for some shapes, for laminar flow, $R < 10,000$), for two-dimensional, two-dimensional axisymmetric, and three-dimensional bodies, respectively.

Furthermore, for fully turbulent flows at $R > 10,000$, and in the case of blunt bodies, although the drag coefficient, C_D is independent of R, it becomes dependent on the relative surface roughness, ε/L and M, but still independent of F. Finally, for fully turbulent flows at $R > 10,000$, and in the case of streamlined bodies the drag coefficient, C_D is dependent on R and also becomes dependent on the relative surface roughness, ε/L and M, but still independent of F. The drag coefficient, C_D, generally (and especially for streamlined bodies) increases with an increase in relative surface roughness, ε/L. However, for blunt bodies (circular cylinder or sphere), an increase in relative surface roughness, ε/L can actually cause a decrease in the drag coefficient, C_D for a specific range of R, as illustrated in Figure 10.18 for a sphere of diameter, D (where the relative surface roughness, $\varepsilon/L = \varepsilon/D$).

The drag coefficient, C_D will be dependent on M for fully turbulent flows at $R > 10,000$ for blunt or streamlined bodies. Figure 10.19 illustrates such a relationship for two two-dimensional shapes (a blunt or square shape and a streamlined or elliptical shape) for subsonic flow ($M = v/c < 1$, where c is the sonic velocity; see Chapter 7). Specifically, Figure 10.19 illustrates that compressibility effects are insignificant if $M \leq 0.5$, while they are significant for $M > 0.5$. Figure 10.19 also illustrates that the drag coefficient, C_D for the blunt body is higher than that for the streamlined body for the entire range of values for M ($0.1 \leq M \leq 0.9$).

In the case of turbulent flow for blunt bodies (except for round-shaped bodies such as circular cylinders and spheres), the flow resistance (as modeled by C_D) is a function of C (or M), ε/L and the geometry, L_i/L. As such, in addition to the inertia force, F_I and the pressure force, F_P; the elastic force, F_E; the relative surface roughness, ε/L; and the geometry of the body, L_i/L are important in the external flow around an object. Therefore, because there are three independent π terms, C, ε/L, and L_i/L used in the definition of the dependent π term, E and thus the drag coefficient, C_D, the appropriate similitude scale ratios corresponding to the independent π terms are used to define the similarity requirements between the model and its prototype. Specifically, the criteria governing dynamic similarity requires the following:

$$\underbrace{\left[\left(\frac{\rho v^2}{E_v}\right)_p\right]}_{C_p} = \underbrace{\left[\left(\frac{\rho v^2}{E_v}\right)_m\right]}_{C_m} \tag{11.99}$$

$$\left(\frac{\varepsilon}{L}\right)_p = \left(\frac{\varepsilon}{L}\right)_m \tag{11.100}$$

$$\left(\frac{L_i}{L}\right)_p = \left(\frac{L_i}{L}\right)_m \tag{11.101}$$

which states that in addition to the geometric similarity requirement for the relative surface roughness, ε/L and the geometry of the body, L_i/L, there is a dynamic similarity requirement that C remain a constant in order for the dependent π term (i.e., the drag coefficient, C_D) to remain a constant between the model and its prototype, where the flow resistance prediction equation is defined as follows:

$$\underbrace{\left[\frac{F_D}{\frac{1}{2}\rho v^2 A}\right]_p}_{C_{Dp}} = \underbrace{\left[\frac{F_D}{\frac{1}{2}\rho v^2 A}\right]_m}_{C_{Dm}} \tag{11.102}$$

Furthermore, one may note that in the case where the similarity requirements regarding the independent π terms, C ("elastic model"), ε/L (geometric similarity), and L_i/L (geometric similarity) are satisfied, then the dependent π term (i.e., the drag coefficient, C_D) will remain a constant between the model and its prototype. And, therefore, application of the flow resistance prediction equation yields a "true model." However, in the case where the similarity requirements regarding the independent π terms C, ε/L, and L_i/L are not satisfied, then the dependent π term (i.e., the drag coefficient, C_D) will not remain a constant between the model and its prototype. And, therefore, application of the flow resistance prediction equation yields a "distorted model" for which the appropriate corrections measures are taken.

EXAMPLE PROBLEM 11.4

(Refer to Example Problem 10.17.) Air at standard atmosphere at an altitude of 25,000 ft above sea level flows at a velocity of 800 ft/sec over a prototype 1500-ft-long square shaped cylinder with a side of 6 ft, as illustrated in Figure EP 11.4. A smaller model of the larger prototype is designed in order to study the flow characteristics of turbulent flow over the blunt body. The model fluid is carbon dioxide at 68°F, and the model scale, λ is 0.20. (a) Determine the drag force on the prototype square cylinder. (b) Determine the velocity of flow of the carbon dioxide over the model square cylinder and the required pressure of the carbon dioxide in order to achieve dynamic similarity between the model and the prototype. (c) Determine the drag force on the model square cylinder in order to achieve dynamic similarity between the model and the prototype.

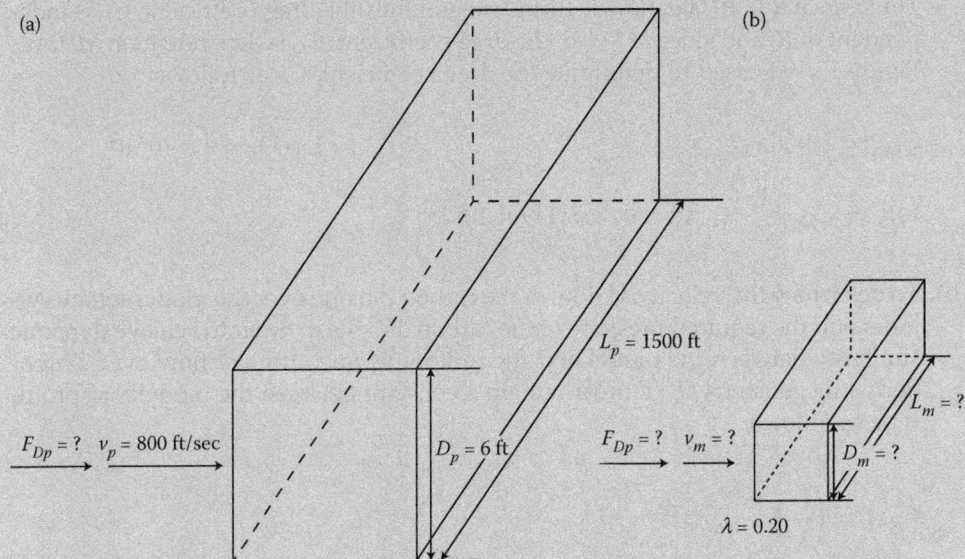

FIGURE EP 11.4
(a) Air flows over a prototype long square-shaped cylinder. (b) Carbon dioxide flows over a smaller model long square-shaped cylinder.

Mathcad Solution

(a) In order to determine the drag force on the prototype square cylinder, the drag force equation, Equation 11.91, is applied, where the frontal area of the square cylinder is used to compute the drag force. Because the Reynolds number, $R > 10,000$ (turbulent flow) over a blunt body is assumed, the drag coefficient, C_D becomes independent of R, as illustrated in Figure 10.12 for two-dimensional bodies. However, Table 10.4 for two-dimensional bodies presents an easier-to-read magnitude for the drag coefficient, C_D for the square cylinder, assuming turbulent flow. Furthermore, Figure 10.19 illustrates that for blunt (square-shaped) two-dimensional bodies, the drag coefficient, C_D is independent of M for $M \leq 0.5$, while the drag coefficient, C_D is dependent on M for $M > 0.5$. The fluid properties for air are given in Table A.1 in Appendix A. The R and $M = C^{0.5}$ are computed as follows:

$$\text{slug} := 1\,\text{lb}\,\frac{\sec^2}{\text{ft}} \qquad \rho_p := 0.0010663\,\frac{\text{slug}}{\text{ft}^3} \qquad \mu_p := 0.32166 \times 10^{-6}\,\text{lb}\,\frac{\sec}{\text{ft}^2}$$

$$c_p := 1016.11\,\frac{\text{ft}}{\sec} \qquad E_{vp} := c_p^2 \cdot \rho_p = 1.101 \times 10^3\,\frac{\text{lb}}{\text{ft}^2}$$

$$D_p := 6\,\text{ft} \qquad L_p := 1500\,\text{ft} \qquad v_p := 800\,\frac{\text{ft}}{\sec}$$

$$R_p := \frac{\rho_p \cdot v_p \cdot D_p}{\mu_p} = 1.591 \times 10^7 \quad C_p := \frac{\rho_p \cdot v_p^2}{E_{vp}} = 0.62 \quad M_p := \sqrt{C_p} = 0.787$$

Thus, since $R > 10{,}000$, flow is turbulent and thus the drag coefficient, C_D is independent of R, and since $M > 0.5$, the drag coefficient, C_D is dependent on M; thus, Figure 10.19 is used to determine the drag coefficient, C_D as follows:

$$C_{Dp} := 2.1 \qquad\qquad\qquad A_p := L_p \cdot D_p = 9 \times 10^3\,\text{ft}^2$$

$$F_{Dp} := C_{Dp} \cdot \frac{1}{2} \cdot \rho_p \cdot v_p^2 \cdot A_p = 6.449 \times 10^6\,\text{lb}$$

(b) To determine the velocity of flow of the carbon dioxide over the model square cylinder and the required pressure of the carbon dioxide in order to achieve dynamic similarity between the model and the prototype, for turbulent flow over a blunt body, the geometry, L_i/L must remain a constant between the model and prototype as follows:

$$\left(\frac{L_i}{L}\right)_p = \left(\frac{L_i}{L}\right)_m$$

Furthermore, in order to determine the geometry L and D of the model square cylinder, the model scale, λ (inverse of the length ratio) is applied as follows:

$$\lambda := 0.20$$

Guess value: $\qquad\qquad\qquad\qquad D_m := 1\,\text{ft} \qquad\qquad\qquad L_m := 1\,\text{ft}$

Given

$$\lambda = \frac{D_m}{D_p} \qquad\qquad\qquad\qquad \lambda = \frac{L_m}{L_p}$$

$$\begin{pmatrix} D_m \\ L_m \end{pmatrix} := \text{Find}(D_m, L_m) \begin{pmatrix} 1.2 \\ 300 \end{pmatrix}\,\text{ft}$$

And, the geometry is modeled as follows:

$$\frac{L_p}{D_p} = 250 \qquad\qquad\qquad\qquad\qquad\qquad\qquad \frac{L_m}{D_m} = 250$$

And, although the relative roughness, ε/L should remain a constant between the model and prototype as follows:

$$\left(\frac{\varepsilon}{L}\right)_p = \left(\frac{\varepsilon}{L}\right)_m$$

Figure 10.18 is for a sphere of diameter, D and not for a square shaped cylinder. As such, the dependence of the drag coefficient, C_D on the relative roughness, ε/L will not be modeled in the example problem herein. Furthermore, the C (or M) must remain a constant between the model and prototype as follows:

$$\underbrace{\left[\left(\frac{\rho v^2}{E_v}\right)_p\right]}_{C_p} = \underbrace{\left[\left(\frac{\rho v^2}{E_v}\right)_m\right]}_{C_m}$$

The fluid properties (gas constant and specific heat ratio, which are independent of the gas pressure and are only a function of the temperature of the gas) for carbon dioxide are given in Table A.5 in Appendix A. Furthermore, E_v for the carbon dioxide is determined by assuming isentropic conditions for the compression and expansion of the gas; thus, $E_v = kp$, where the pressure is determined by applying the ideal gas law, $p = \rho RT$ (see Chapter 1). The sonic velocity for the carbon dioxide, $c = \sqrt{E_v/\rho} = \sqrt{kRT}$, which is independent of the gas pressure and is only a function of the temperature of the gas.

$$T_m := 68°F = 527 \times 67°R \qquad Rgasc_m := 1123\frac{ft^2}{sec^2 \,°R} \qquad k_m := 1.28$$

Guess value:
$$v_m := 0.1\frac{ft}{sec} \qquad P_m := 1\frac{lb}{ft^2} \qquad E_{vm} := 1\frac{lb}{ft^2}$$

$$\rho_m := 0.00354\frac{slug}{ft^3} \qquad c_m := 1000\frac{ft}{sec} \qquad C_m := 0.5$$

Given

$$c_m = \frac{\rho_m \cdot v_m^2}{E_{vm}} \qquad P_m = \rho_m \cdot Rgasc_m \cdot T_m \qquad E_{vm} = k_m \cdot P_m$$

$$C_m = C_p \qquad c_m = \sqrt{\frac{E_{vm}}{\rho_m}} \qquad c_m = \sqrt{k_m \cdot Rgasc_m \cdot T_m}$$

$$\begin{pmatrix} v_m \\ P_m \\ E_{vm} \\ \rho_m \\ c_m \\ C_m \end{pmatrix} := Find\,(v_m,\, P_m,\, E_{vm},\, \rho_m,\, c_m,\, C_m)$$

$$v_m = 685.686\frac{ft}{s} \qquad P_m = 79.432\frac{lb}{ft^2} \qquad E_{vm} = 101.673\frac{lb}{ft^2}$$

$$\rho_m = 1.34 \times 10^{-4} \frac{\text{slug}}{\text{ft}^3} \qquad c_m = 870.916 \frac{\text{ft}}{\text{s}} \qquad C_m = 0.62$$

(c) To determine the drag force on the model square cylinder in order to achieve dynamic similarity between the model and the prototype for turbulent flow over a blunt body, the drag coefficient, C_D must remain a constant between the model and the prototype (which is a direct result of maintaining a constant C, and a constant L_i/L between the model and the prototype) as follows:

$$\underbrace{\left[\frac{F_D}{\frac{1}{2}\rho v^2 A}\right]_p}_{C_{Dp}} = \underbrace{\left[\frac{F_D}{\frac{1}{2}\rho v^2 A}\right]_m}_{C_{Dm}}$$

Furthermore, the frontal area of the square cylinder is used to compute the drag force as follows:

$$A_m := L_m \cdot D_m = 360 \text{ ft}^2$$

Guess value: $\qquad\qquad\qquad F_{Dm} := 1 \text{ lb} \qquad\qquad\qquad C_{Dm} := 1$

Given

$$C_{Dm} = \frac{F_{Dm}}{\frac{1}{2} \cdot \rho_m \cdot v_m^2 \cdot A_m}$$

$$C_{Dm} = C_{Dp}$$

$$\begin{pmatrix} F_{Dm} \\ C_{Dm} \end{pmatrix} := \text{Find} (F_{Dm}, C_{Dm})$$

$$F_{Dm} = 2.382 \times 10^4 \text{ lb} \qquad\qquad\qquad\qquad\qquad C_{Dm} = 2.1$$

Therefore, although the similarity requirements regarding the independent π term, L_i/L and the independent π term, C ("elastic model") are theoretically satisfied ($C_p = C_m = 0.62$), the dependent π term (i.e., the drag coefficient, C_D) will actually/practically remain a constant between the model and its prototype ($C_{Dp} = C_{Dm} = 2.1$) only if it is practical to maintain/attain the model velocity, drag force, fluid, scale, and cost. Furthermore, because the drag coefficient, C_D is independent of R for blunt bodies, R does not need to remain a constant between the model and the prototype as follows:

$$\mu_m := 0.310 \times 10^{-6} \text{ lb} \frac{\text{sec}}{\text{ft}^2}$$

$$R_m := \frac{\rho_m \cdot v_m \cdot D_m}{\mu_m} = 3.558 \times 10^5 \qquad\qquad\qquad R_p = 1.591 \times 10^7$$

One may ask: Is the model speed too high to be able to maintain for the model? If it is too high, then this is a "distorted model" that needs to be adjusted in order to achieve a "true model" or as close to it as possible. The answer is no, the speed of carbon dioxide under pressure is not too high for the model.

In the case of turbulent flow for streamlined bodies (except for round-shaped bodies such as circular cylinders and spheres), the flow resistance (as modeled by C_D) is a function of R, C (or M), ε/L, and the geometry. As such, in addition to the inertia force, F_I and the pressure force, F_P, the viscous force, F_V; the elastic force, F_E; the relative surface roughness, ε/L; and the geometry of the body, L_i/L are important in the external flow around an object. Therefore, because there are four independent π terms, R, C, ε/L, and L_i/L used in the definition of the dependent π term, E and thus the drag coefficient, C_D, the appropriate similitude scale ratios corresponding to the independent π terms are used to define the similarity requirements between the model and its prototype. Specifically, the criteria governing dynamic similarity require the following:

$$\underbrace{\left[\left(\frac{\rho v L}{\mu}\right)_p\right]}_{R_p} = \underbrace{\left[\left(\frac{\rho v L}{\mu}\right)_m\right]}_{R_m} \tag{11.103}$$

$$\underbrace{\left[\left(\frac{\rho v^2}{E_v}\right)_p\right]}_{C_p} = \underbrace{\left[\left(\frac{\rho v^2}{E_v}\right)_m\right]}_{C_m} \tag{11.104}$$

$$\left(\frac{\varepsilon}{L}\right)_p = \left(\frac{\varepsilon}{L}\right)_m \tag{11.105}$$

$$\left(\frac{L_i}{L}\right)_p = \left(\frac{L_i}{L}\right)_m \tag{11.106}$$

which states that in addition to the geometric similarity requirement for the relative surface roughness, ε/L and the geometry of the body, L_i/L, there is a dynamic similarity requirement that R and C simultaneously remain a constant, in order for the dependent π term (i.e., the drag coefficient, C_D) to remain a constant between the model and its prototype, where the flow resistance prediction equation is defined as follows:

$$\underbrace{\left[\frac{F_D}{\frac{1}{2}\rho v^2 A}\right]_p}_{C_{Dp}} = \underbrace{\left[\frac{F_D}{\frac{1}{2}\rho v^2 A}\right]_m}_{C_{Dm}} \tag{11.107}$$

Furthermore, one may note that in the case where the similarity requirements regarding the independent π terms, R ("viscous model"); C ("elastic model"); ε/L (geometric similarity); and L_i/L (geometric similarity) are satisfied, then the dependent π term (i.e., the drag coefficient, C_D) will remain a constant between the model and its prototype. And, therefore, application of the flow resistance prediction equation yields a "true model." However, in the case where the similarity requirements regarding the independent π terms, R, C, ε/L, and L_i/L are not satisfied, then the dependent π term (i.e., the drag coefficient, C_D) will not remain a constant between the model and its prototype. And, therefore, application of the flow resistance prediction equation yields a "distorted model" for which the appropriate corrections measures are taken.

EXAMPLE PROBLEM 11.5

Air at standard atmosphere at an altitude of 25,000 ft above sea level flows at a velocity of 700 ft/sec over a prototype 1600-ft-long elliptical cylinder with the following geometry: an L of 4 ft and a D of 1 ft, as illustrated in Figure EP 11.5. A smaller model of the larger prototype is designed in order to study the flow characteristics of turbulent flow over the streamlined body. The model fluid is carbon dioxide at 68°F, and the model scale, λ is 0.10. (a) Determine the drag force on the prototype elliptical cylinder. (b) Determine the velocity of flow of the carbon dioxide over the model elliptical cylinder and the required pressure of the carbon dioxide in order to achieve dynamic similarity between the model and the prototype. (c) Determine the drag force on the model elliptical cylinder in order to achieve dynamic similarity between the model and the prototype.

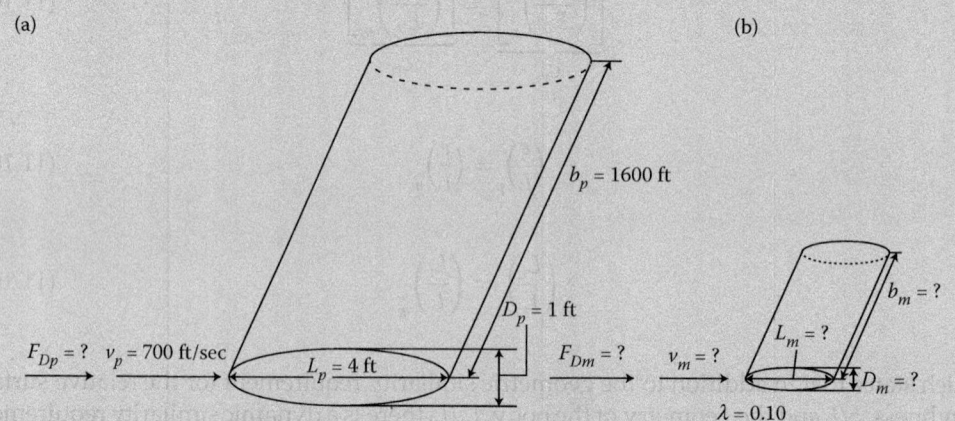

FIGURE EP 11.5
(a) Air flows over a prototype long elliptical cylinder. (b) Carbon dioxide flows over a smaller model long elliptical cylinder.

Mathcad Solution

(a) In order to determine the drag force on the prototype elliptical cylinder, the drag force equation, Equation 11.91, is applied, where the frontal area of the elliptical cylinder is used to compute the drag force. Because the Reynolds number, $R >$ 10,000 (turbulent flow) over a streamlined body is assumed, the drag coefficient, C_D is dependent on R, as illustrated in Figure 10.12 for two-dimensional bodies. However, Table 10.4 for two-dimensional bodies presents an easier-to-read magnitude for the drag coefficient, C_D for the elliptical cylinder, assuming turbulent flow. Furthermore, Figure 10.19 illustrates that for streamlined (elliptical-shaped) two-dimensional bodies, the drag coefficient, C_D is independent of M for $M \leq 0.5$, while the drag coefficient, C_D is dependent on M for $M > 0.5$. The fluid properties for air are given in Table A.1 in Appendix A. The R and $M = C^{0.5}$ are computed as follows:

$$\text{slug} := 1 \, \text{lb} \frac{\sec^2}{\text{ft}} \qquad \rho_p := 0.0010663 \frac{\text{slug}}{\text{ft}^3} \qquad \mu_p := 0.32166 \times 10^{-6} \, \text{lb} \frac{\sec}{\text{ft}^2}$$

$$c_p := 1016.11 \frac{\text{ft}}{\sec} \qquad E_{vp} := c_p^2 \cdot \rho_p = 1.101 \times 10^3 \frac{\text{lb}}{\text{ft}^2}$$

$$D_p := 1 \, \text{ft} \qquad L_p := 4 \, \text{ft} \qquad b_p := 1600 \, \text{ft} \qquad v_p := 700 \frac{\text{ft}}{\sec}$$

$$R_p := \frac{\rho_p \cdot v_p \cdot D_p}{\mu_p} = 2.32 \times 10^6 \quad C_p := \frac{\rho_p \cdot v_p^2}{E_{vp}} = 0.475 \quad M_p := \sqrt{C_p} = 0.689$$

Thus, since $R > 10,000$, flow is turbulent and thus, the drag coefficient, C_D is dependent of R for a streamlined body, Table 10.4 for two-dimensional bodies is used to determine the drag coefficient, C_D. Additionally, since $M > 0.5$, the drag coefficient, C_D is dependent on M for a streamlined body, Figure 10.19 is used to determine the drag coefficient, C_D. It is interesting to compare the two estimates for the drag coefficient, C_D as follows:

$$C_{DpR} := 0.15 \qquad\qquad C_{DpM} := 0.35$$

First, although the Reynolds number, R is important for streamlined bodies, the Mach number, M becomes important at high velocities (turbulent flow). And, second, Figure 10.19 yields a higher magnitude for the drag coefficient, C_D than Table 10.4 does. Therefore, the higher magnitude for the drag coefficient, C_D yielded by its dependence on M takes precedence over that yielded by its dependence on R as follows:

$$C_{Dp} := C_{DpM} = 0.35 \qquad\qquad A_p := b_p \cdot D_p = 1.6 \times 10^3 \, \text{ft}^2$$

$$F_{Dp} := C_{Dp} \cdot \frac{1}{2} \cdot \rho_p \cdot v_p^2 \cdot A_p = 1.463 \times 10^5 \, \text{lb}$$

(b) To determine the velocity of flow of the carbon dioxide over the model elliptical cylinder and the required pressure of the carbon dioxide in order to achieve dynamic similarity between the model and the prototype for turbulent flow over a streamlined body, the geometry, L_i/L must remain a constant between the model and prototype as follows:

$$\left(\frac{L_i}{L}\right)_p = \left(\frac{L_i}{L}\right)_m$$

Furthermore, in order to determine the geometry L and D, and the length, b of the model elliptical cylinder, the model scale, λ (inverse of the length ratio) is applied as follows:

$$\lambda := 1.10$$

Guess value $\qquad\qquad D_m := 1 \, \text{ft} \qquad\qquad L_m := 1 \, \text{ft} \qquad\qquad b_m := 100 \, \text{ft}$

Given

$$\lambda = \frac{D_m}{D_p} \qquad\qquad\qquad \lambda = \frac{L_m}{L_p} \qquad \lambda = \frac{b_m}{b_p}$$

$$\begin{pmatrix} D_m \\ L_m \\ b_m \end{pmatrix} := Find\,(D_m, L_m, b_m) = \begin{pmatrix} 0.1 \\ 0.4 \\ 160 \end{pmatrix} ft$$

And, the geometry is modeled as follows:

$$\frac{L_p}{D_p} = 4 \qquad\qquad \frac{L_m}{D_m} = 4 \qquad\qquad \frac{b_p}{D_p} = 1.6 \times 10^3 \qquad\qquad \frac{b_m}{D_m} = 1.6 \times 10^3$$

And, although the relative roughness, ε/L should remain a constant between the model and prototype as follows:

$$\left(\frac{\varepsilon}{L}\right)_p = \left(\frac{\varepsilon}{L}\right)_m$$

Figure 10.18 is for a sphere of diameter, D and not for an elliptical-shaped cylinder. As such, the dependence of the drag coefficient, C_D on the relative roughness, ε/L will not be modeled in the example problem herein. Furthermore, the R must remain a constant between the model and prototype as follows:

$$\underbrace{\left[\left(\frac{\rho v L}{\mu}\right)_p\right]}_{R_p} = \underbrace{\left[\left(\frac{\rho v L}{\mu}\right)_m\right]}_{R_m}$$

And, finally, the C (or M) must remain a constant between the model and prototype as follows:

$$\underbrace{\left[\left(\frac{\rho v^2}{E_v}\right)_p\right]}_{C_p} = \underbrace{\left[\left(\frac{\rho v^2}{E_v}\right)_m\right]}_{C_m}$$

The fluid properties (viscosity, gas constant, and specific heat ratio, which are independent of the gas pressure and are only a function of the temperature of the gas) for carbon dioxide are given in Table A.5 in Appendix A. Furthermore, E_v for the carbon dioxide is determined by assuming isentropic conditions for the compression and expansion of the gas; thus, $E_v = kp$, where the pressure is determined by applying the ideal gas law, $p = \rho RT$ (see Chapter 1). The sonic velocity for the carbon dioxide, $c = \sqrt{E_v/\rho} = \sqrt{kRT}$, which is independent of the gas pressure and is only a function of the temperature of the gas.

$$T_m := 68°F = 527.67°R \qquad\qquad\qquad \mu_m := 0.310 \times 10^{-6}\,lb\,\frac{sec}{ft^2}$$

$$Rgasc_m := 1123\,\frac{ft^2}{sec^2\,°R} \qquad\qquad\qquad k_m := 1.28$$

Guess value: $\qquad v_m := 0.1\,\frac{ft}{sec} \qquad P_m := 1\,\frac{lb}{ft^2} \qquad E_{vm} := 1\,\frac{lb}{ft^2}$

$$\rho_m := 0.00354\,\frac{slug}{ft^3} \qquad c_m := 1000\,\frac{ft}{sec} \qquad C_m := 0.5 \qquad R_m := 10{,}000$$

Given

$$C_m = \frac{\rho_m \cdot v_m^2}{E_{vm}} \qquad\qquad p_m = \rho_m \cdot Rgasc_m \cdot T_m \qquad\qquad E_{vm} = k_m \cdot p_m$$

$$C_m = C_p \qquad\qquad c_m = \sqrt{\frac{E_{vm}}{\rho_m}} \qquad\qquad c_m = \sqrt{k_m \cdot Rgasc_m \cdot T_m}$$

$$R_m = \frac{\rho_m \cdot v_m \cdot D_m}{\mu_m} \qquad\qquad R_m = R_p$$

$$\begin{pmatrix} v_m \\ p_m \\ E_{vm} \\ \rho_m \\ c_m \\ C_m \\ R_m \end{pmatrix} := \text{Find}\,(v_m, p_m, E_{vm}, \rho_m, c_m, C_m, R_m)$$

$$v_m = 599.975\,\frac{ft}{s} \qquad\qquad p_m = 7.105 \times 10^3\,\frac{lb}{ft^2} \qquad\qquad E_{vm} = 9.094 \times 10^3\,\frac{lb}{ft^2}$$

$$\rho_m = 0.012\,\frac{slug}{ft^3} \qquad\qquad c_m = 870.916\,\frac{ft}{s} \qquad\qquad C_m = 0.475 \qquad\qquad R_m = 2.32 \times 10^6$$

(c) To determine the drag force on the model elliptical cylinder in order to achieve dynamic similarity between the model and the prototype for turbulent flow over a streamlined body, the drag coefficient, C_D must remain a constant between the model and the prototype (which is a direct result of maintaining a constant R, a constant C, and a constant L_i/L between the model and the prototype) as follows:

$$\underbrace{\left[\frac{F_D}{\frac{1}{2}\rho v^2 A} \right]_p}_{C_{Dp}} = \underbrace{\left[\frac{F_D}{\frac{1}{2}\rho v^2 A} \right]_m}_{C_{Dm}}$$

Furthermore, the frontal area of the elliptical cylinder is used to compute the drag force as follows:

$$A_m := b_m \cdot D_m = 16\ ft^2$$

Guess value: $\qquad\qquad\qquad\qquad F_{Dm} := 1\ lb \qquad\qquad\qquad C_{Dm} := 1$

Given

$$C_{Dm} = \frac{F_{Dm}}{\frac{1}{2} \cdot \rho_m \cdot v_m^2 \cdot A_m}$$

$$C_{Dm} = C_{Dp}$$

$$\begin{pmatrix} F_{Dm} \\ C_{Dm} \end{pmatrix} := \text{Find} (F_{Dm}, C_{Dm})$$

$F_{Dm} = 1.208 \times 10^4 \text{ lb}$ $C_{Dm} = 0.35$

Therefore, although the similarity requirements regarding the independent π term, L_i/L; the independent π term, R ("viscosity model"); and the independent π term, C ("elastic model") are theoretically satisfied ($R_p = R_m = 2.32 \times 10^6$, $C_p = C_m = 0.475$), the dependent π term (i.e., the drag coefficient, C_D) will actually/practically remain a constant between the model and its prototype ($C_{Dp} = C_{Dm} = 0.35$) only of it is practical to maintain/attain the model velocity, drag force, fluid, scale, and cost.

11.7.2.4 Laminar and Turbulent Flow for Round-Shaped Bodies (Circular Cylinder or Sphere)

For flows at moderate (laminar) to high (turbulent) values of R around/across round-shaped bodies such as circular cylinders and spheres, the drag coefficient, C_D, in general, decreases with an increase in R. Figure 10.17 (Stokes law) illustrates that, unlike Figure 10.16, which is for most other geometric shapes, the drag coefficient, C_D, in general, continues to decrease with an increase in R for the full range of values for R; low (creeping; $R < 1$), moderate (laminar), and high (turbulent). For flow across circular cylinders and spheres, the critical value for R is 200,000; thus, the flow is defined as laminar for $R \leq 200{,}000$ and turbulent for $R \geq 200{,}000$, where $R = \rho v L / \mu = \rho v D / \mu$, where the characteristic length, L is the external diameter of the circular cylinder or sphere. Furthermore, Tables 10.4 and 10.5 include the drag coefficient, C_D for laminar and turbulent flow for a circular cylinder and sphere, respectively.

Furthermore, for laminar flows at $R \leq 200{,}000$, the drag coefficient, C_D will be dependent on R, but independent of the relative surface roughness, ε/L, C (or M), and F. Thus, in the case of laminar flow for round-shaped bodies such as circular cylinders and spheres, the flow resistance (as modeled by C_D) is only a function of R and the geometry, L_i/L. As such, in addition to the inertia force, F_I and the pressure force, F_P, the viscous force, F_V and the geometry of the body, L_i/L are important in the external flow around an object. Therefore, because there are two independent π terms, R and L_i/L, used in the definition of the dependent π term, E and thus the drag coefficient, C_D, the appropriate similitude scale ratios corresponding to the independent π terms are used to define the similarity requirements between the model and its prototype. Specifically, the criteria governing dynamic similarity require the following:

$$\underbrace{\left[\left(\frac{\rho v L}{\mu} \right)_p \right]}_{R_p} = \underbrace{\left[\left(\frac{\rho v L}{\mu} \right)_m \right]}_{R_m} \tag{11.108}$$

$$\left(\frac{L_i}{L} \right)_p = \left(\frac{L_i}{L} \right)_m \tag{11.109}$$

which states that in addition to the geometric similarity requirement for the geometry of the body, L_i/L, there is a dynamic similarity requirement that R remain a constant in order for the dependent π term (i.e., the drag coefficient, C_D) to remain a constant between the model and its prototype, where the flow resistance prediction equation is defined as follows:

$$\underbrace{\left[\frac{F_D}{\frac{1}{2}\rho v^2 A}\right]_p}_{C_{Dp}} = \underbrace{\left[\frac{F_D}{\frac{1}{2}\rho v^2 A}\right]_m}_{C_{Dm}} \tag{11.110}$$

Furthermore, one may note that in the case where the similarity requirements regarding the independent π terms, R ("viscosity model") and the geometry of the body, L_i/L are satisfied, then the dependent π term (i.e., the drag coefficient, C_D) will remain a constant between the model and its prototype. And, therefore, application of the flow resistance prediction equation yields a "true model." However, in the case where the similarity requirements regarding the independent π terms, R and the geometry of the body, L_i/L are not satisfied, then the dependent π term (i.e., the drag coefficient, C_D) will not remain a constant between the model and its prototype. And, therefore, application of the flow resistance prediction equation yields a "distorted model" for which the appropriate corrections measures are taken.

EXAMPLE PROBLEM 11.6

Water at 70°F flows over a prototype circular cylinder with a diameter of 6 ft and a length of 2000 ft, as illustrated in Figure EP 11.6. A smaller model of the larger prototype is designed in order to study the flow characteristics of laminar flow over the circular cylinder. The model fluid is crude oil at 68°F ($s = 0.86$, $\mu = 150 \times 10^{-6}$ lb-sec/ft^2), the velocity of flow of the oil over the smaller model circular cylinder is 2 ft/sec, and the model scale, λ is 0.25. (a) Determine the drag force on the model circular cylinder.

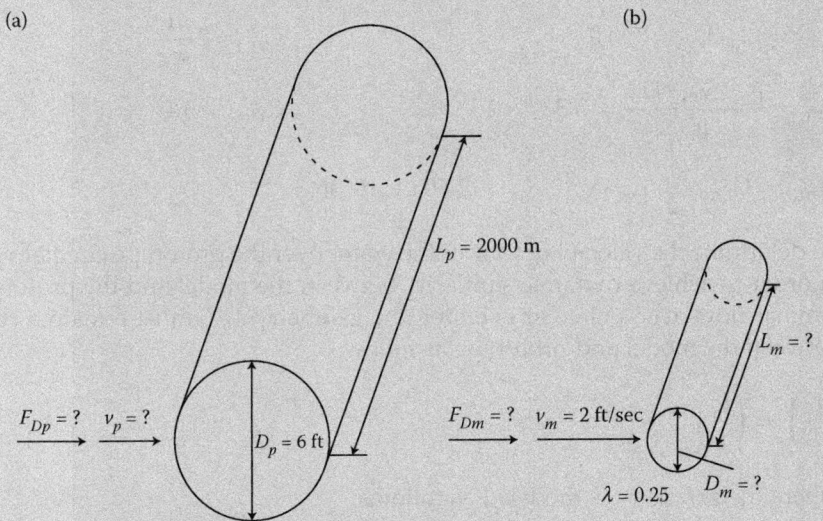

FIGURE EP 11.6
(a) Water flows over a prototype circular cylinder. (b) Crude oil flows over a smaller model circular cylinder.

(b) Determine the velocity of flow of the water over the prototype circular cylinder in order to achieve dynamic similarity between the model and the prototype. (c) Determine the drag force on the prototype circular cylinder in order to achieve dynamic similarity between the model and the prototype.

Mathcad Solution

(a) In order to determine the drag force on the model circular cylinder, the drag force equation, Equation 11.91, is applied, where the frontal area of the circular cylinder is used to compute the drag force. Because the Reynolds number, $R \leq 200,000$ (laminar flow) over a circular cylinder is assumed, the drag coefficient, C_D decreases with an increase in R, as illustrated in Figure 10.17 (Stokes law illustrated). Furthermore, in order to determine the geometry L and D of the model circular cylinder, the model scale, λ (inverse of the length ratio) is applied as follows:

$$D_p := 6 \text{ ft} \qquad\qquad L_p := 2000 \text{ ft} \qquad\qquad \lambda := 0.25$$

Guess value: $\qquad\qquad D_m := 0.1 \text{ ft} \qquad\qquad L_m := 0.8 \text{ ft}$

Given

$$\lambda = \frac{D_m}{D_p} \qquad\qquad\qquad \lambda = \frac{L_m}{L_p}$$

$$\begin{pmatrix} D_m \\ L_m \end{pmatrix} := \text{Find}(D_m, L_m) = \begin{pmatrix} 1.5 \\ 500 \end{pmatrix} \text{ ft}$$

$$\text{slug} := 1 \text{ lb} \frac{\text{sec}^2}{\text{ft}} \qquad\qquad S_m := 0.86 \qquad\qquad \rho_w := 1.94 \frac{\text{slug}}{\text{ft}^3}$$

$$\rho_m := S_m \cdot \rho_w = 1.668 \frac{\text{slug}}{\text{ft}^3} \qquad\qquad \mu_m := 150 \times 10^{-6} \text{ lb} \frac{\text{sec}}{\text{ft}^2}$$

$$A_m := L_m \cdot D_m = 750 \text{ ft}^2 \qquad\qquad v_m := 2 \frac{\text{ft}}{\text{sec}}$$

$$R_m := \frac{\rho_m \cdot v_m \cdot D_m}{\mu_m} = 3.337 \times 10^4 \qquad\qquad C_{Dm} := 1.0$$

$$F_{Dm} := C_{Dm} \cdot \frac{1}{2} \cdot \rho_m \cdot v_m^2 \cdot A_m = 2.503 \times 10^3 \text{ lb}$$

(b) To determine the velocity of flow of the water over the prototype circular cylinder in order to achieve dynamic similarity between the model and the prototype for laminar flow over a circular cylinder, the geometry L_i/L must remain a constant between the model and prototype as follows:

$$\left(\frac{L_i}{L}\right)_p = \left(\frac{L_i}{L}\right)_m$$

where the geometry is modeled as follows:

$$\frac{L_p}{D_p} = 333.333 \qquad\qquad\qquad \frac{L_m}{D_m} = 333.333$$

Furthermore, the **R** must remain a constant between the model and prototype as follows:

$$\underbrace{\left[\left(\frac{\rho v L}{\mu}\right)_p\right]}_{R_p} = \underbrace{\left[\left(\frac{\rho v L}{\mu}\right)_m\right]}_{R_m}$$

$$\rho_p := 1.936 \frac{\text{slug}}{\text{ft}^3} \qquad\qquad \mu_p := 20.50 \times 10^{-6} \text{ lb} \frac{\text{sec}}{\text{ft}^2}$$

Guess Value $\qquad\qquad v_p := 10 \frac{\text{ft}}{\text{sec}} \qquad\qquad R_p := 100$

Given

$$R_p = \frac{\rho_p \cdot v_p \cdot D_p}{\mu_p}$$

$$R_p = R_m$$

$$\begin{pmatrix} v_p \\ R_p \end{pmatrix} := \text{Find}\,(v_p, R_p)$$

$$v_p = 0.059 \frac{\text{ft}}{\text{s}} \qquad\qquad R_p = 3.337 \times 10^4$$

(c) To determine the drag force on the prototype circular cylinder in order to achieve dynamic similarity between the model and the prototype for laminar flow over a circular cylinder, the drag coefficient, C_D must remain a constant between the model and the prototype (which is a direct result of maintaining a constant **R** and a constant L_i/L between the model and the prototype) as follows:

$$\underbrace{\left[\frac{F_D}{\frac{1}{2}\rho v^2 A}\right]_p}_{C_{Dp}} = \underbrace{\left[\frac{F_D}{\frac{1}{2}\rho v^2 A}\right]_m}_{C_{Dm}}$$

Furthermore, the frontal area of the circular cylinder is used to compute the drag force as follows:

$$A_p := L_p \cdot D_p = 1.2 \times 10^4 \text{ ft}^2$$

Guess value: $\qquad\qquad F_{Dp} := 1 \text{ lb} \qquad\qquad C_{Dp} := 0.5$

Given

$$C_{Dp} = \frac{F_{DP}}{\frac{1}{2} \cdot \rho_p \cdot v_p^2 \cdot A_p}$$

$$C_{Dp} = C_{Dm}$$

$$\begin{pmatrix} F_{Dp} \\ C_{Dp} \end{pmatrix} := \text{Find} \left(F_{Dp}, C_{Dp} \right)$$

$F_{Dp} = 40.282 \, \text{lb}$ $\qquad\qquad\qquad\qquad\qquad\qquad\qquad\qquad$ $C_{Dp} = 1$

Therefore, although the similarity requirements regarding the independent π term, L_i/L and the independent π term, R ("viscosity model") are theoretically satisfied ($R_p = R_m = 3.337 \times 10^4$), the dependent π term (i.e., the drag coefficient, C_D) will actually/practically remain a constant between the model and its prototype ($C_{Dp} = C_{Dm} = 1.0$) only if it is practical to maintain/attain the model velocity, drag force, fluid, scale, and cost.

However, for turbulent flows at $R \geq 200{,}000$, the drag coefficient, C_D is dependent on R and becomes dependent on the relative surface roughness, ε/L and C (or M), but still independent of F. Specifically, an increase in the relative surface roughness, ε/L can actually cause a decrease in the drag coefficient, C_D for a specific range of R, as illustrated in Figure 10.18 for a sphere of diameter, D (where the relative surface roughness, $\varepsilon/L = \varepsilon/D$). Figure 10.20 illustrates the dependence of the drag coefficient, C_D on M for three two-dimensional axisymmetric bodies (a circular cylinder, a sphere, and a sharp pointed ogive) for supersonic flow ($M = v/c > 1$). Specifically, Figure 10.20 illustrates that the values of the drag coefficient, C_D significantly increase in the vicinity of sonic flow ($M = v/c = 1$). Such a surge in the value of the drag coefficient, C_D is due to the existence of shock waves, which can only occur in supersonic flows and provide a mechanism for the generation of drag that does not exist in subsonic flow. Figure 10.20 also illustrates that the drag coefficient, C_D for the sharp pointed ogive (as well as other sharp pointed bodies) reaches a maximum in the vicinity of sonic flow ($M = v/c = 1$), while the drag coefficient, C_D for the blunt bodies (circular cylinder and sphere) continues to increase with an increase in M for $M > 1$. Such an increase in the drag coefficient, C_D for the blunt bodies in supersonic flow is due to the nature of the shock wave structure and the flow separation associated with blunt bodies. As a result, the leading edges of subsonic aircraft wings are typically rounded and blunt, while those of supersonic aircraft are typically pointed and sharp.

In the case of turbulent flow for round-shaped bodies such as circular cylinders and spheres, the flow resistance (as modeled by C_D) is a function of R, C (or M), ε/L, and the geometry. As such, in addition to the inertia force, F_I and the pressure force, F_P, the viscous force, F_V; the elastic force, F_E; the relative surface roughness, ε/L; and the geometry of the body, L_i/L are important in the external flow around an object. Therefore, because there are four independent π terms, R, C, ε/L, and L_i/L, used in the definition of the dependent π term, E and thus the drag coefficient, C_D, the appropriate similitude scale ratios corresponding to the independent π terms are used to define the similarity requirements between the model and its prototype. Specifically, the criteria governing dynamic similarity requires the following:

$$\underbrace{\left[\left(\frac{\rho v L}{\mu} \right)_p \right]}_{R_p} = \underbrace{\left[\left(\frac{\rho v L}{\mu} \right)_m \right]}_{R_m} \qquad (11.111)$$

$$\underbrace{\left[\left(\frac{\rho v^2}{E_v}\right)_p\right]}_{C_p} = \underbrace{\left[\left(\frac{\rho v^2}{E_v}\right)_m\right]}_{C_m} \tag{11.112}$$

$$\left(\frac{\varepsilon}{L}\right)_p = \left(\frac{\varepsilon}{L}\right)_m \tag{11.113}$$

$$\left(\frac{L_i}{L}\right)_p = \left(\frac{L_i}{L}\right)_m \tag{11.114}$$

which states that in addition to the geometric similarity requirement for the relative surface roughness, ε/L and the geometry of the body, L_i/L, there is a dynamic similarity requirement that R and C simultaneously remain a constant in order for the dependent π term (i.e., the drag coefficient, C_D) to remain a constant between the model and its prototype, where the flow resistance prediction equation is defined as follows:

$$\underbrace{\left[\frac{F_D}{\frac{1}{2}\rho v^2 A}\right]_p}_{C_{Dp}} = \underbrace{\left[\frac{F_D}{\frac{1}{2}\rho v^2 A}\right]_m}_{C_{Dm}} \tag{11.115}$$

Furthermore, one may note that in the case where the similarity requirements regarding the independent π terms, R ("viscous model"), C ("elastic model"), ε/L (geometric similarity), and L_i/L (geometric similarity) are satisfied, then the dependent π term (i.e., the drag coefficient, C_D) will remain a constant between the model and its prototype. And, therefore, application of the flow resistance prediction equation yields a "true model." However, in the case where the similarity requirements regarding the independent π terms, R, C, ε/L, and L_i/L are not satisfied, then the dependent π term (i.e., the drag coefficient, C_D) will not remain a constant between the model and its prototype. And, therefore, application of the flow resistance prediction equation yields a "distorted model" for which the appropriate correction measures are taken.

EXAMPLE PROBLEM 11.7

Air at standard atmosphere at an altitude of 10,000 ft above sea level flows at a velocity of 1,100 ft/sec over a prototype sphere with a diameter of 2 ft and an absolute surface roughness of 0.025 ft, as illustrated in Figure EP 11.7. A smaller model of the larger prototype is designed in order to study the flow characteristics of turbulent flow over the sphere. The model fluid is carbon dioxide at 68°F, and the model scale, λ is 0.25. (a) Determine the drag force on the prototype sphere. (b) Determine the velocity of flow of the carbon dioxide over the model sphere and the required pressure of the carbon dioxide in order to achieve dynamic similarity between the model and the prototype. (c) Determine the drag force on the model sphere in order to achieve dynamic similarity between the model and the prototype.

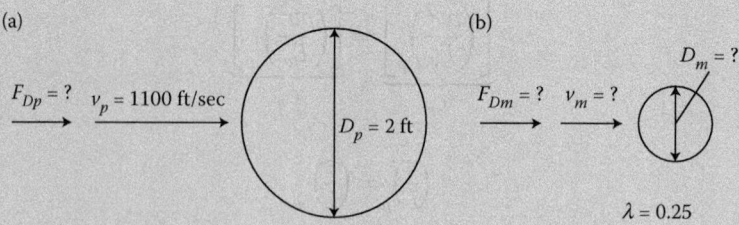

FIGURE EP 11.7
(a) Air flows over a prototype sphere. (b) Carbon dioxide flows over a smaller model sphere.

Mathcad Solution

(a) In order to determine the drag force on the prototype sphere, the drag force equation, Equation 11.91, is applied, where the frontal area of the sphere is used to compute the drag force. Because the Reynolds number, $R > 10{,}000$ (turbulent flow) over a sphere is assumed, the drag coefficient, C_D is dependent on R, and the relative surface roughness, ε/D as illustrated in Figure 10.18 for a sphere of diameter, D. Furthermore, Figure 10.20 illustrates the dependence of the drag coefficient, C_D on M for a sphere for supersonic flow ($M = v/c > 1$). The fluid properties for air are given in Table A.1 in Appendix A. The R and $M = C^{0.5}$ are computed as follows:

$$\text{slug} := 1\,\text{lb}\,\frac{\sec^2}{\text{ft}} \qquad \rho_p := 0.0017555\,\frac{\text{slug}}{\text{ft}^3} \qquad \mu_p := 0.35343 \times 10^{-6}\,\text{lb}\,\frac{\sec}{\text{ft}^2}$$

$$c_p := 1077.40\,\frac{\text{ft}}{\sec} \qquad E_{vp} := c_p^2 \cdot \rho_p = 2.038 \times 10^3\,\frac{\text{lb}}{\text{ft}^2}$$

$$D_p := 2.0\,\text{ft} \qquad \varepsilon_p := 0.025\,\text{ft} \qquad \frac{\varepsilon_p}{D_p} = 0.013 \qquad v_p := 1100\,\frac{\text{ft}}{\sec}$$

$$R_p := \frac{\rho_p \cdot v_p \cdot D_p}{\mu_p} = 1.093 \times 10^7 \qquad C_p := \frac{\rho_p \cdot v_p^2}{E_{vp}} = 1.042 \quad M_p := \sqrt{C_p} = 1.021$$

Thus, since $R > 10{,}000$, the flow is turbulent and thus the drag coefficient, C_D is dependent on R, and the relative surface roughness, ε/D for a sphere, Figure 10.18 is used to determine the drag coefficient, C_D. Additionally, since $M > 1$, the drag coefficient, C_D is dependent on M for a sphere body, Figure 10.20 is used to determine the drag coefficient, C_D. It is interesting to compare the two estimates for the drag coefficient, C_D as follows:

$$C_{DpR} := 0.4 \qquad\qquad C_{DpM} := 0.78$$

First, although the Reynolds number, R and the relative surface roughness, ε/D are important for a sphere, the Mach number, M becomes important at high velocities (turbulent flow). And, second, Figure 10.20 yields a higher magnitude for the drag coefficient, C_D than Figure 10.18 does. Therefore, the higher magnitude for the drag coefficient, C_D yielded by its dependence on M takes precedence over that yielded by its dependence on R and the relative surface roughness, ε/D as follows:

$$C_{Dp} := C_{DpM} = 0.78 \qquad\qquad A_p := \frac{\pi \cdot D_p^2}{4} = 3.142 \, \text{ft}^2$$

$$F_{Dp} := C_{Dp} \cdot \frac{1}{2} \cdot \rho_p \cdot v_p^2 \cdot A_p = 2.603 \times 10^3 \, \text{lb}$$

(b) To determine the velocity of flow of the carbon dioxide over the model sphere and the required pressure of the carbon dioxide in order to achieve dynamic similarity between the model and the prototype for turbulent flow over a sphere, the geometry, L_i/L must remain a constant between the model and prototype as follows:

$$\left(\frac{L_i}{L}\right)_p = \left(\frac{L_i}{L}\right)_m$$

And, the relative roughness, ε/L should remain a constant between the model and prototype as follows:

$$\left(\frac{\varepsilon}{L}\right)_p = \left(\frac{\varepsilon}{L}\right)_m$$

Furthermore, in order to determine the diameter and the absolute roughness of the model sphere, the model scale, λ (inverse of the length ratio) is applied.

$$\lambda := 0.25$$

Guess value: $\qquad\qquad D_m := 1 \, \text{ft} \qquad\qquad \varepsilon_m := 0.1 \, \text{ft}$

Given

$$\lambda = \frac{D_m}{D_p} \qquad\qquad\qquad \lambda = \frac{\varepsilon_m}{\varepsilon_p}$$

$$\begin{pmatrix} D_m \\ \varepsilon_m \end{pmatrix} := \text{Find} \, (D_m, \varepsilon_m) = \begin{pmatrix} 0.5 \\ 6.25 \times 10^{-3} \end{pmatrix} \text{ft}$$

And, the relative roughness similarity requirement is modeled as follows:

$$\frac{\varepsilon_p}{D_p} = 0.013 \qquad\qquad\qquad \frac{\varepsilon_m}{D_m} = 0.013$$

Furthermore, the R must remain a constant between the model and prototype as follows:

$$\underbrace{\left[\left(\frac{\rho v L}{\mu}\right)_p\right]}_{R_p} = \underbrace{\left[\left(\frac{\rho v L}{\mu}\right)_m\right]}_{R_m}$$

And, finally, the C (or M) must remain a constant between the model and prototype as follows:

$$\underbrace{\left[\left(\frac{\rho v^2}{E_v}\right)_p\right]}_{C_p} = \underbrace{\left[\left(\frac{\rho v^2}{E_v}\right)_m\right]}_{C_m}$$

The fluid properties (viscosity, gas constant, and specific heat ratio, which are independent of the gas pressure and are only a function of the temperature of the gas) for carbon dioxide are given in Table A.5 in Appendix A. Furthermore, E_v for the carbon dioxide is determined by assuming isentropic conditions for the compression and expansion of the gas; thus, $E_v = kp$, where the pressure is determined by applying the ideal gas law, $p = \rho RT$ (see Chapter 1). The sonic velocity for the carbon dioxide, $c = \sqrt{E_v/\rho} = \sqrt{kRT}$ which is independent of the gas pressure and is only a function of the temperature of the gas.

$T_m := 68°F = 527.67°R$ $\mu_m := 0.310 \times 10^{-6} \, lb\frac{sec}{ft^2}$

$Rgasc_m := 1123\frac{ft^2}{sec^2 \, °R}$ $k_m := 1.28$

Guess value: $v_m := 0.1\frac{ft}{sec}$ $p_m := 1\frac{lb}{ft^2}$ $E_{vm} := 1\frac{lb}{ft^2}$

$\rho_m := 0.001\frac{slug}{ft^3}$ $c_m := 1000\frac{ft}{sec}$ $C_m := 0.5$ $R_m := 10{,}000$

Given:

$C_m = \frac{\rho_m \cdot v_m^2}{E_{vm}}$ $p_m = \rho_m \cdot Rgasc_m \cdot T_m$ $E_{vm} = k_m \cdot p_m$

$C_m = C_p$ $c_m = \sqrt{\frac{E_{vm}}{\rho_m}}$ $c_m = \sqrt{k_m \cdot Rgasc_m \cdot T_m}$

$R_m = \frac{\rho_m \cdot v_m \cdot D_m}{\mu_m}$ $R_m = R_p$

$$\begin{pmatrix} v_m \\ p_m \\ E_{vm} \\ \rho_m \\ c_m \\ R_m \end{pmatrix} := Find\,(v_m,\, p_m,\, E_{vm},\, \rho_m,\, c_m,\, C_m,\, R_m)$$

$v_m = 889.184\frac{ft}{s}$ $p_m = 4.515 \times 10^3\frac{lb}{ft^2}$ $E_{vm} = 5.779 \times 10^3\frac{lb}{ft^2}$

$\rho_m = 7.619 \times 10^{-3}\frac{slug}{ft^3}$ $c_m = 870.916\frac{ft}{s}$ $C_m = 1.042$ $R_m = 1.093 \times 10^7$

(c) To determine the drag force on the model sphere in order to achieve dynamic similarity between the model and the prototype for turbulent flow over a sphere, the drag coefficient, C_D must remain a constant between the model and the prototype (which is a direct result of maintaining a constant R, a constant C, and a constant ε/D between the model and the prototype) as follows:

$$\underbrace{\left[\frac{F_D}{\frac{1}{2}\rho v^2 A} \right]_p}_{C_{Dp}} = \underbrace{\left[\frac{F_D}{\frac{1}{2}\rho v^2 A} \right]_m}_{C_{Dm}}$$

Furthermore, the frontal area of the sphere is used to compute the drag force as follows:

$$A_m := \frac{\pi \cdot D_m^2}{4} = 0.196 \, \text{ft}^2$$

Guess value: $\qquad\qquad F_{Dm} := 1 \, \text{lb} \qquad\qquad C_{Dm} := 1$

Given

$$C_{Dm} = \frac{F_{Dm}}{\frac{1}{2} \cdot \rho_m \cdot v_m^2 \cdot A_m}$$

$$C_{Dm} = C_{Dp}$$

$$\binom{F_{Dm}}{C_{Dm}} := \text{Find} \, (F_{Dm}, C_{Dm})$$

$F_{Dm} = 461.316 \, \text{lb} \qquad\qquad\qquad\qquad C_{Dm} = 0.78$

Therefore, although the similarity requirements regarding the independent π term, ε/D; the independent π term, R ("viscosity model"); and the independent π term, C ("elastic model") are theoretically satisfied ($R_p = R_m = 1.903 \times 10^7$, $C_p = C_m = 1.042$), the dependent π term (i.e., the drag coefficient, C_D) will actually/practically remain a constant between the model and its prototype ($C_{Dp} = C_{Dm} = 0.78$) only if it is practical to maintain/attain the model velocity, drag force, fluid, scale, and cost.

11.7.2.5 Laminar and Turbulent Flow with Wave Action at the Free Surface for Any Shape Body

For body shapes in general, with a wave action at the free surface, the drag coefficient, C_D for laminar or turbulent flow is dependent on R, L_i/L, and F, but independent of ε/L and C (or M). The drag coefficient, C_D will vary with $F = v/v_c$ (which is the ratio of the free-stream speed to a typical wave speed on the interface of two fluids, such as the surface of a body of water; e.g., ocean, lake, river), depending mainly on the velocity of flow and the shape of the body. Such examples include ships (discussed in this section) and open channel flow-measuring devices/structures such as weirs and spillways (discussed in Section 11.7.6). For instance, a ship moving in a body of water may produce waves, caused by the product of the force and the velocity of the ship (where the energy production, or the power, is force times velocity), resulting in drag. While the nature of the waves produced by the ship depends on F (and thus the velocity of the ship) and the shape of the ship, the drag coefficient, C_D depends on both the F (which represents the wave-making effects and thus the pressure effects) and the R (which represents the viscous effects). Figure 10.12 illustrates the variation of the drag coefficient, C_D with the R for two-dimensional bodies, while Figure 10.13 illustrates the variation of the drag coefficient, C_D with the R for three-dimensional axisymmetric bodies. Additionally, Tables 10.4 through 10.6 present the drag coefficient, C_D for numerous geometric shapes at various orientations to the direction of flow, for $R > 10,000$, for two-dimensional, two-dimensional axisymmetric, and three-dimensional bodies, respectively. Figure 10.21 illustrates variation of the wave drag coefficient, C_D with both the F and the shape of the ship (a hull without a bow bulb/ streamlined body and a hull with a bow bulb/less streamlined body). It may be noted that the "wavy" variation of the wave drag coefficient, C_D with the F (for either of the two body shapes) is due to the fact that the structure of the waves produced by the ship's hull is significantly a function of the velocity of the ship (and thus the F). Furthermore, Figure 10.21 illustrates that the wave and the corresponding wave drag that is mainly caused by the bow (front end of the ship) may be reduced by a reduction in streamlining of the body (i.e., adding a bow bulb to the hull). Finally, it is interesting to note that the more streamlined body (a hull without a bow bulb) actually results in a higher wave drag coefficient, C_D than the less streamlined body (a hull with a bow bulb) because it is the bow that induces the wave drag.

The simultaneous dependency of the drag coefficient, C_D on both R and F presents a difficult situation when designing a model of the prototype. There are two typical difficult situations that arise, and they are presented as follows. The first case is when a particular and practical model fluid is assumed and the velocity ratio is sought. As a result, it becomes impossible to simultaneously satisfy the dynamic requirements of R and F in the determination of the velocity ratio. The solution to this type of difficult situation is to note that that the wave effects (and thus the dependence of the drag coefficient, C_D on F) and the viscous effects (and thus the dependence of the drag coefficient, C_D on R) may be modeled separately, with the total drag being the sum of the two individual effects (see Example Problem 11.8 below). The second case is when both the model fluid and the velocity ratio are sought. While it is simple to simultaneously satisfy the dynamic requirements of R and F in the determination of both the model fluid and the velocity ratio, it is difficult to find a "practical" model fluid that meets the dynamic requirements of the fluid property (see Example Problem 11.9 below). The solution to this type of difficult situation is to realize that due to the wave action at the free surface, it becomes more important to satisfy the dynamic requirement for F. Although ignoring the dynamic requirement for R will result in a "distorted model" (which can be accommodated for by proper interpretation of the model results), the model will be less distorted and more of a "true model" when the R is large, which reduces the dependence of the drag coefficient, C_D on R.

In the case of laminar or turbulent flow with a wave action at the surface for any shape body, the flow resistance (as modeled by C_D) is only a function of R, F, and the geometry, L_i/L. One may note that drag force is a result of both the viscous effects developed along the hull of a ship (for instance) and the pressure effects developed due to the shape of the hull of the ship and the wave action at the surface. Thus, the shear drag is a function of R, while the pressure drag is a function of F. As such, in addition to the inertia force, F_I and the pressure force, F_P, the viscous force, F_V; the gravity force, F_G; and the geometry of the body, L_i/L are important in the external flow around an object. Therefore, because there are three independent π terms, R, F, and L_i/L, used in the definition of the dependent π term, E

and thus the drag coefficient, C_D, the appropriate similitude scale ratios corresponding to the independent π terms are used to define the similarity requirements between the model and its prototype. Specifically, the criteria governing dynamic similarity require the following:

$$\underbrace{\left[\left(\frac{\rho v L}{\mu}\right)_p\right]}_{R_p} = \underbrace{\left[\left(\frac{\rho v L}{\mu}\right)_m\right]}_{R_m} \tag{11.116}$$

$$\underbrace{\left[\left(\frac{v}{\sqrt{gL}}\right)_p\right]}_{F_p} = \underbrace{\left[\left(\frac{v}{\sqrt{gL}}\right)_m\right]}_{F_m} \tag{11.117}$$

$$\left(\frac{L_i}{L}\right)_p = \left(\frac{L_i}{L}\right)_m \tag{11.118}$$

which states that in addition to the geometric similarity requirement for L_i/L, there is a dynamic similarity requirement that R and F simultaneously remain a constant in order for the dependent π term (i.e., the drag coefficient, C_D) to remain a constant between the model and its prototype, where the flow resistance prediction equation is defined as follows:

$$\underbrace{\left[\frac{F_D}{\frac{1}{2}\rho v^2 A}\right]_p}_{C_{Dp}} = \underbrace{\left[\frac{F_D}{\frac{1}{2}\rho v^2 A}\right]_m}_{C_{Dm}} \tag{11.119}$$

Furthermore, one may note that in the case where the similarity requirements regarding the independent π terms R ("viscosity model"), F ("gravity model"), and L_i/L are satisfied, then the dependent π term (i.e., the drag coefficient, C_D) will remain a constant between the model and its prototype. And, therefore, application of the flow resistance prediction equation yields a "true model." However, in the case where the similarity requirements regarding the independent π terms, R, F, and L_i/L, are not satisfied, then the dependent π term (i.e., the drag coefficient, C_D) will not remain a constant between the model and its prototype. And, therefore, application of the flow resistance prediction equation yields a "distorted model" for which the appropriate corrections measures are taken.

EXAMPLE PROBLEM 11.8

Water at 50°F flows at a velocity of 9 ft/sec over a prototype 30-ft-long streamlined hull with a width of 30 ft and a depth of 4 ft, as illustrated in Figure EP 11.8. A smaller model of the larger prototype is designed in order to study the flow characteristics of the flow over the streamlined body with wave action at the free surface. The model fluid is water at 70°F, and the model scale, λ is 0.20. (a) Determine the drag force on the prototype streamlined hull. (b) Determine the velocity of flow of the water over the model streamlined hull in order to achieve dynamic similarity between the model and the prototype. (c) Determine the drag force on the model streamlined hull in order to achieve dynamic similarity between the model and the prototype.

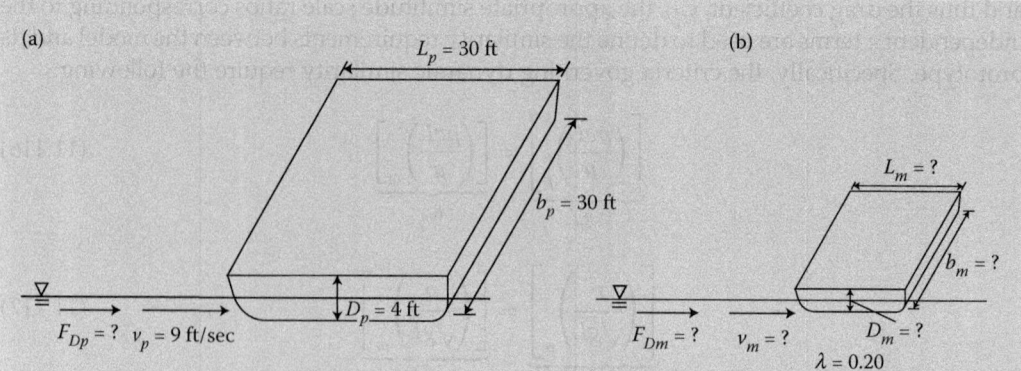

FIGURE EP 11.8
(a) Water flows over a prototype streamlined hull. (b) Water flows over a smaller model streamlined hull.

Mathcad Solution

(a) In order to determine the drag force on the prototype streamlined hull, the drag force equation, Equation 11.91, is applied. The drag coefficient, C_D is dependent on both R and F. Table 10.5 illustrates the variation of the drag coefficient, C_D with the R for two-dimensional axisymmetric bodies; note that a streamlined body is assumed to represent the streamlined hull, and the frontal area is assumed in the determination of the drag coefficient. Figure 10.21 illustrates the variation of the wave drag coefficient, C_D with the F for a streamlined hull; note that the planform area is assumed in the determination of the drag coefficient. The fluid properties for water are given in Table A.2 in Appendix A. Because in this example the model fluid is specified to be water, it will become impossible to simultaneously satisfy the dynamic requirements of R and F in the determination of the velocity ratio. Specifically, the "gravity" model requires that the velocity ratio be defined as follows:

$$v_r = \frac{v_p}{v_m} = \frac{(\sqrt{gL})_p}{(\sqrt{gL})_m} = L_r^{\frac{1}{2}}$$

However, the "viscosity model" requires that the velocity ratio be defined as follows:

$$v_r = \frac{v_p}{v_m} = \frac{\left(\frac{\mu}{\rho L}\right)_p}{\left(\frac{\mu}{\rho L}\right)_m} = \mu_r \rho_r^{-1} L_r^{-1} = \upsilon_r L_r^{-1}$$

Thus, equating the two velocity ratios yields the following dynamic requirement:

$$L_r^{\frac{1}{2}} = \upsilon_r L_r^{-1}$$

$$\upsilon_r = L_r^{\frac{3}{2}}$$

However, in fact $\upsilon_r \neq L_r^{\frac{3}{2}}$ as follows:

$$\text{slug} := 1\,\text{lb}\,\frac{\sec^2}{\text{ft}} \qquad \rho_p := 1.94\frac{\text{slug}}{\text{ft}^3} \qquad \mu_p := 27.35 \times 10^{-6}\,\text{lb}\,\frac{\sec}{\text{ft}^2}$$

$$\rho_m := 1.936\frac{\text{slug}}{\text{ft}^3} \qquad \mu_m := 20.50 \times 10^{-6}\,\text{lb}\,\frac{\sec}{\text{ft}^2} \qquad \lambda := 0.2 \qquad L_r := \frac{1}{\lambda} = 5$$

$$v_m := \frac{\mu_m}{\rho_m} = 1.059 \times 10^{-5}\,\frac{\text{ft}^2}{\sec} \qquad\qquad v_p := \frac{\mu_p}{\rho_p} = 1.41 \times 10^{-5}\,\frac{\text{ft}^2}{\sec}$$

$$v_r := \frac{v_p}{v_m} = 1.331 \qquad\qquad L_r^{\frac{2}{3}} = 11.18$$

which does not satisfy the dynamic requirement. The solution to this type of difficult situation is to note that that the wave effects (and thus the dependence of the drag coefficient, C_D on F) and the viscous effects (and thus the dependence of the drag coefficient, C_D on R) may be modeled separately, with the total drag being the sum of the two individual effects as follows:

$$g := 32.174\,\frac{\text{ft}}{\sec^2} \qquad L_p := 30\,\text{ft} \qquad b_p := 30\,\text{ft} \qquad D_p := 4\,\text{ft} \qquad v_p := 9\,\frac{\text{ft}}{\sec}$$

$$R_p := \frac{\rho_p \cdot v_p \cdot L_p}{\mu_p} = 1.915 \times 10^7 \qquad C_{DpR} := 0.04 \qquad A_{pfront} := b_p \cdot D_p = 120\,\text{ft}^2$$

$$F_{DpR} := C_{DpR} \cdot \frac{1}{2} \cdot \rho_p \cdot v_p^2 \cdot A_{pfront} = 377.136\,\text{lb}$$

$$F_p := \frac{v_p}{\sqrt{g \cdot L_p}} = 0.29 \qquad C_{DpF} := 0.00125 \qquad A_{pplan} := b_p \cdot L_p = 900\,\text{ft}^2$$

$$F_{DpF} := C_{DpF} \cdot \frac{1}{2} \cdot \rho_p \cdot v_p^2 \cdot A_{pplan} = 88.391\,\text{lb}$$

$$F_{Dp} := F_{DpR} + F_{DpF} = 465.527\,\text{lb}$$

(b) To determine the velocity of flow of the water over the model streamlined hull in order to achieve dynamic similarity between the model and the prototype for the flow over a streamlined body, the geometry, L_i/L must remain a constant between the model and prototype as follows:

$$\left(\frac{L_i}{L}\right)_p = \left(\frac{L_i}{L}\right)_m$$

Furthermore, in order to determine the length, L and width, b of the model streamlined hull, the model scale, λ (inverse of the length ratio) is applied as follows:

$$\lambda := 0.2$$

Guess value: $\qquad L_m := 10\,\text{ft} \qquad b_m := 5\,\text{ft} \qquad D_m := 1\,\text{ft}$

Given

$$\lambda = \frac{L_m}{L_p} \qquad\qquad \lambda = \frac{b_m}{b_p} \qquad\qquad \lambda = \frac{D_m}{D_p}$$

$$\begin{pmatrix} L_m \\ b_m \\ D_m \end{pmatrix} := \text{Find} (L_m, b_m, D_m) = \begin{pmatrix} 6 \\ 6 \\ 0.8 \end{pmatrix} \text{ft}$$

And, the geometry is modeled as follows:

$$\frac{b_p}{L_p} = 1 \qquad \frac{b_m}{L_m} = 1 \qquad \frac{D_p}{L_p} = 0.133 \qquad \frac{D_m}{L_m} = 0.133$$

Furthermore, the **R** must remain a constant between the model and prototype as follows:

$$\underbrace{\left[\left(\frac{\rho v L}{\mu}\right)_p\right]}_{R_p} = \underbrace{\left[\left(\frac{\rho v L}{\mu}\right)_m\right]}_{R_m}$$

Guess value: $v_{mR} := 100 \dfrac{\text{ft}}{\text{sec}}$ $R_m := 10{,}000$

Given

$$R_m = \frac{\rho_m \cdot v_{mR} \cdot L_m}{\mu_m}$$

$$R_m = R_p$$

$$\begin{pmatrix} v_{mR} \\ R_m \end{pmatrix} := \text{Find} (v_{mR}, R_m)$$

$$v_{mR} = 33.799 \frac{\text{ft}}{\text{s}} \qquad\qquad R_m = 1.915 \times 10^7$$

And, separately from **R**, the **F** must remain a constant between the model and prototype as follows:

$$\underbrace{\left[\left(\frac{v}{\sqrt{gL}}\right)_p\right]}_{F_p} = \underbrace{\left[\left(\frac{v}{\sqrt{gL}}\right)_m\right]}_{F_m}$$

Guess value: $v_{mF} := 100 \dfrac{\text{ft}}{\text{sec}}$ $F_m := 0.1$

Given

$$F_m = \frac{v_{mF}}{\sqrt{g \cdot L_m}}$$

$$F_m = F_P$$

$$\begin{pmatrix} v_{mF} \\ F_m \end{pmatrix} := \text{Find} (v_{mF}, F_m)$$

$$v_{mF} = 4.025 \frac{\text{ft}}{\text{s}} \qquad\qquad F_m = 0.29$$

It is interesting to note the two significantly different model velocities were yielded from satisfying the dynamic requirement for R, which yielded $v_{mR} = 33.799$ ft/sec, and satisfying the dynamic requirement for F, which yielded $v_{mF} = 4.025$ ft/sec.

(c) To determine the drag force on the model streamlined hull in order to achieve dynamic similarity between the model and the prototype for the flow over a streamlined body, the wave effects (and thus the dependence of the drag coefficient, C_D on F) and the viscous effects (and thus the dependence of the drag coefficient, C_D on R) may be modeled separately, with the total drag being the sum of the two individual effects. The drag coefficient, C_{DR} must remain a constant between the model and the prototype (which is a direct result of maintaining a constant R, and a constant L_i/L between the model and the prototype) as follows:

$$\underbrace{\left[\frac{F_{DR}}{\frac{1}{2}\rho v_R^2 A_{front}}\right]_p}_{C_{D_{pR}}} = \underbrace{\left[\frac{F_{DR}}{\frac{1}{2}\rho v_R^2 A_{front}}\right]_m}_{C_{D_{mR}}}$$

$A_{mfront} := b_m \cdot D_m = 4.8 \text{ ft}^2$

Guess value: $\qquad\qquad F_{DmR} := 1 \text{ lb} \qquad\qquad C_{DmR} := 1$

Given

$$C_{DmR} = \frac{F_{DmR}}{\frac{1}{2} \cdot \rho_m \cdot v_{mR}^2 \cdot A_m}$$

$C_{DmR} = C_{DpR}$

$$\begin{pmatrix} F_{DmR} \\ C_{DmR} \end{pmatrix} := \text{Find} \left(F_{DmR}, C_{DmR} \right)$$

$F_{DmR} = 8.685 \text{ lb} \qquad\qquad\qquad\qquad C_{DmR} = 0.04$

And, the drag coefficient, C_{DF} must remain a constant between the model and the prototype (which is a direct result of maintaining a constant F and a constant L_i/L between the model and the prototype) as follows:

$$\underbrace{\left[\frac{F_{DF}}{\frac{1}{2}\rho v_F^2 A_{plan}}\right]_p}_{C_{D_{pF}}} = \underbrace{\left[\frac{F_{DF}}{\frac{1}{2}\rho v_F^2 A_{plan}}\right]_m}_{C_{D_{mF}}}$$

$A_{mplan} := b_m \cdot L_m = 36 \text{ ft}^2$

Guess value: $\qquad\qquad F_{DmF} := 1 \text{ lb} \qquad\qquad C_{DmF} := 1$

Given

$$C_{DmF} = \frac{F_{DmF}}{\frac{1}{2} \cdot \rho_m \cdot v_{mF}^2 \cdot A_m}$$

$$C_{DmF} = C_{DpF}$$

$$\begin{pmatrix} F_{DmF} \\ C_{DmF} \end{pmatrix} := \text{Find}\,(F_{DmF},\, C_{DmF})$$

$F_{DmF} = 3.849 \times 10^{-3}\ \text{lb}$ $\qquad\qquad\qquad\qquad\qquad\qquad C_{DmF} = 1.25 \times 10^{-3}$

with the total drag being the sum of the two individual effects as follows:

$$F_{Dm} := F_{DmR} + F_{DmF} = 8.689\ \text{lb}$$

Therefore, although the similarity requirements regarding the independent π term, L_i/L, and, separately, the independent π term, R ("viscosity model") are theoretically satisfied ($R_p = R_m = 1.915 \times 10^7$), and the independent π term, F ("gravity model") is theoretically satisfied ($F_p = F_m = 0.29$), the dependent π term (i.e., the drag coefficient, C_D) will actually/practically remain a constant between the model and its prototype ($C_{DpR} = C_{DmR} = 0.04$, and $C_{DpF} = C_{DmF} = 0.00125$) only if it is practical to maintain/attain the model velocity, drag force, fluid, scale, and cost. Furthermore, because the dynamic requirements for R and F were satisfied separately (and not simultaneously) application of the drag force equation yields a "distorted model" as illustrated by this example.

EXAMPLE PROBLEM 11.9

(Refer to Example Problem 11.8.) Water at 50°F flows at a velocity of 9 ft/sec over a prototype 30-ft-long streamlined hull with a width of 30 ft and a depth of 4 ft, as illustrated in Figure EP 11.9. A smaller model of the larger prototype is designed in order to study

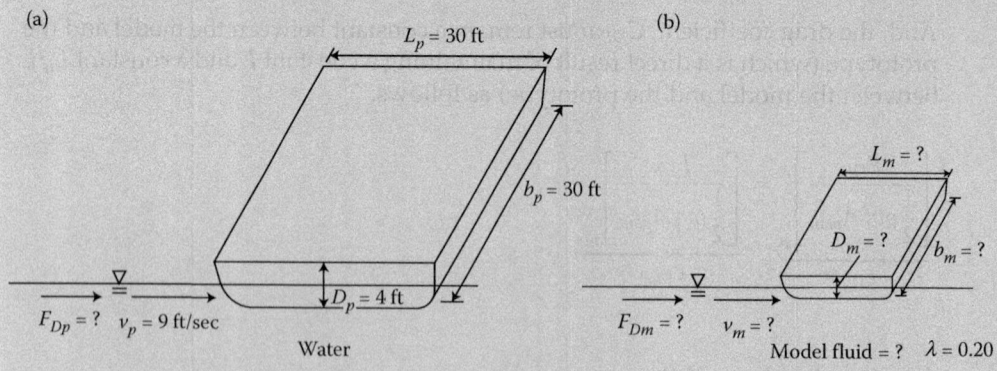

FIGURE EP 11.9
(a) Water flows over a prototype streamlined hull. (b) A model fluid flows over a smaller model streamlined hull.

the flow characteristics of the flow over the streamlined body with wave action at the free surface. The model scale, λ is 0.20. (a) Determine the drag force on the prototype streamlined hull. (b) Determine the model fluid and the velocity of flow of the model fluid over the model streamlined hull in order to achieve dynamic similarity between the model and the prototype. (c) Determine the drag force on the model streamlined hull in order to achieve dynamic similarity between the model and the prototype.

Mathcad Solution

(a) In order to determine the drag force on the prototype streamlined hull, the drag force equation, Equation 11.91, is applied. The drag coefficient, C_D is dependent on both R and F. Table 10.5 illustrates the variation of the drag coefficient, C_D with the R for two-dimensional axisymmetric bodies; note that a streamlined body is assumed to represent the streamlined hull, and the frontal area is assumed in the determination of the drag coefficient. Figure 10.21 illustrates the variation of the wave drag coefficient, C_D with the F for a streamlined hull; note that the planform area is assumed in the determination of the drag coefficient. The fluid properties for water are given in Table A.2 in Appendix A. Because in this example the model fluid is not specified, it is possible to simultaneously satisfy the dynamic requirements of R and F in the determination of the velocity ratio. However, typically it is difficult to find a practical fluid for the resulting model fluid property, as illustrated in (b) below.

$$\text{slug} := 1\,\text{lb}\,\frac{\sec^2}{\text{ft}} \qquad \rho_p := 1.94\,\frac{\text{slug}}{\text{ft}^3} \qquad \mu_p := 27.35 \times 10^{-6}\,\text{lb}\,\frac{\sec}{\text{ft}^2}$$

$$g := 32.174\,\frac{\text{ft}}{\sec^2} \qquad L_p := 30\,\text{ft} \qquad b_p := 30\,\text{ft} \qquad D_p := 4\,\text{ft} \qquad v_p := 9\,\frac{\text{ft}}{\sec}$$

$$R_p := \frac{\rho_p \cdot v_p \cdot L_p}{\mu_p} = 1.915 \times 10^7 \qquad C_{DpR} := 0.04 \qquad A_{\text{pfront}} := b_p \cdot D_p = 120\,\text{ft}^2$$

$$F_{DpR} := C_{DpR} \cdot \frac{1}{2} \cdot \rho_p \cdot v_p^2 \cdot A_{\text{pfront}} = 377.136\,\text{lb}$$

$$F_p := \frac{v_p}{\sqrt{g \cdot L_p}} = 0.29 \qquad C_{DpF} := 0.00125 \qquad A_{\text{pplan}} := b_p \cdot L_p = 900\,\text{ft}^2$$

$$F_{DpF} := C_{DpF} \cdot \frac{1}{2} \cdot \rho_p \cdot v_p^2 \cdot A_{\text{pplan}} = 88.391\,\text{lb}$$

$$F_{Dp} := F_{DpR} + F_{DpF} = 465.527\,\text{lb}$$

(b) To determine the model fluid and the velocity of flow of the model fluid over the model streamlined hull in order to achieve dynamic similarity between the model and the prototype for the flow over a streamlined body, the geometry, L_i/L must remain a constant between the model and prototype as follows:

$$\left(\frac{L_i}{L}\right)_p = \left(\frac{L_i}{L}\right)_m$$

Furthermore, in order to determine the length, L; width, b; and depth, D of the model streamlined hull, the model scale, λ (inverse of the length ratio) is applied as follows:

$$\lambda := 0.2$$

Guess value: $L_m := 10 \text{ ft}$ $b_m := 5 \text{ ft}$ $D_m := 1 \text{ ft}$

Given

$$\lambda = \frac{L_m}{L_p} \qquad\qquad \lambda = \frac{b_m}{b_p} \qquad\qquad \lambda = \frac{D_m}{D_p}$$

$$\begin{pmatrix} L_m \\ b_m \\ D_m \end{pmatrix} := \text{Find}(L_m, b_m, D_m) = \begin{pmatrix} 6 \\ 6 \\ 0.8 \end{pmatrix} \text{ ft}$$

And, the geometry is modeled as follows:

$$\frac{b_p}{L_p} = 1 \qquad\qquad \frac{b_m}{L_m} = 1 \qquad\qquad \frac{D_p}{L_p} = 0.133 \qquad\qquad \frac{D_m}{L_m} = 0.133$$

Furthermore, the R must remain a constant between the model and prototype as follows:

$$\underbrace{\left[\left(\frac{\rho v L}{\mu}\right)_p\right]}_{R_p} = \underbrace{\left[\left(\frac{\rho v L}{\mu}\right)_m\right]}_{R_m}$$

And, finally, the F must remain a constant between the model and prototype as follows:

$$\underbrace{\left[\left(\frac{v}{\sqrt{gL}}\right)_p\right]}_{F_p} = \underbrace{\left[\left(\frac{v}{\sqrt{gL}}\right)_m\right]}_{F_m}$$

Guess value: $v_m := 10 \dfrac{\text{ft}}{\text{sec}}$ $v_m := 1 \times 10^{-5} \dfrac{\text{ft}^2}{\text{sec}}$ $F_m := 0.1$ $R_m := 10{,}000$

$$\mu_m := 1 \times 10^{-5} \text{ lb} \frac{\text{sec}}{\text{ft}^2} \qquad\qquad \rho_m := 1 \frac{\text{slug}}{\text{ft}^3}$$

Given

$$F_m = \frac{v_m}{\sqrt{g \cdot L_m}} \qquad\qquad R_m = \frac{v_m \cdot L_m}{v_m} \qquad\qquad v_m = \frac{\mu_m}{\rho_m}$$

$$F_m = F_p \qquad\qquad\qquad R_m = R_p$$

$$\begin{pmatrix} v_m \\ v_m \\ \mu_m \\ \rho_m \\ F_m \\ R_m \end{pmatrix} := \text{Find}\,(v_m,\, v_m,\, \mu_m,\, \rho_m,\, F_m,\, R_m)$$

$$v_m = 4.025\,\frac{ft}{s} \qquad\qquad v_m = 1.261 \times 10^{-6}\,\frac{ft^2}{sec} \qquad\qquad \mu_m = 1.77 \times 10^{-6}\,\frac{lb\,s}{ft^2}$$

$$\rho_m = 1.403\,\frac{slug}{ft^3} \qquad\qquad F_m = 0.29 \qquad\qquad R_m = 1.915 \times 10^7$$

However, it is difficult to find a practical/convenient model fluid with a kinematic viscosity, $v = 1.261 \times 10^{-6}\,ft^2/sec$.

(c) To determine the drag force on the model streamlined hull in order to achieve dynamic similarity between the model and the prototype for the flow over a streamlined body, although the dependence of the drag coefficient, C_D on R and F are simultaneously achieved through the determination of a common model velocity, v_m and a common model fluid, the drag coefficient, C_{DR} assumes the frontal area, while the drag coefficient, C_{DF} assumes the planform area. Therefore, the total drag is the sum of the two types of drag.

The drag coefficient, C_{DR} must remain a constant between the model and the prototype (which is a direct result of maintaining a constant R and a constant L_i/L between the model and the prototype) as follows:

$$\underbrace{\left[\frac{F_{DR}}{\frac{1}{2}\rho v^2 A_{front}}\right]_p}_{C_{D_{pR}}} = \underbrace{\left[\frac{F_{DR}}{\frac{1}{2}\rho v^2 A_{front}}\right]_m}_{C_{D_{mR}}}$$

$$A_{mfront} := b_m \cdot D_m = 4.8\,ft^2$$

Guess value: $\qquad\qquad F_{DmR} := 1\,lb \qquad\qquad\qquad C_{DmR} := 1$

Given

$$C_{DmR} = \frac{F_{DmR}}{\frac{1}{2} \cdot \rho_m \cdot v_m^2 \cdot A_m}$$

$$C_{DmR} = C_{DpR}$$

$$\begin{pmatrix} F_{DmR} \\ C_{DmR} \end{pmatrix} := \text{Find}\,(F_{DmR},\, C_{DmR})$$

$$F_{DmR} = 0.089\,lb \qquad\qquad\qquad\qquad C_{DmR} = 0.04$$

And, the drag coefficient, C_{DF} must remain a constant between the model and the prototype (which is a direct result of maintaining a constant F and a constant L_i/L between the model and the prototype) as follows:

$$\underbrace{\left[\frac{F_{DF}}{\frac{1}{2}\rho v^2 A_{plan}}\right]_p}_{C_{D_{pF}}} = \underbrace{\left[\frac{F_{DF}}{\frac{1}{2}\rho v^2 A_{plan}}\right]_m}_{C_{D_{mF}}}$$

$A_{mplan} := b_m \cdot L_m = 36 \text{ ft}^2$

Guess value: $F_{DmF} := 1 \text{ lb}$ $C_{DmF} := 1$

Given

$$C_{DmF} = \frac{F_{DmF}}{\frac{1}{2} \cdot \rho_m \cdot v_m^2 \cdot A_m}$$

$C_{DmF} = C_{DpF}$

$\begin{pmatrix} F_{DmF} \\ C_{DmF} \end{pmatrix} := \text{Find}(F_{DmF}, C_{DmF})$

$F_{DmF} = 2.79 \times 10^{-3} \text{ lb}$ $C_{DmF} = 1.25 \times 10^{-3}$

with the total drag being the sum of the two types of drag as follows:

$F_{Dm} := F_{DmR} + F_{DmF} = 0.092 \text{ lb}$

Therefore, although the similarity requirements regarding the independent π term, L_i/L; the independent π term, R ("viscosity model"); and the independent π term, F ("gravity model") are theoretically satisfied ($R_p = R_m = 1.915 \times 10^6$, $F_p = F_m = 0.29$), the dependent π term (i.e., the drag coefficient, C_D) will actually/practically remain a constant between the model and its prototype ($C_{DpR} = C_{DmR} = 0.04$, and $C_{DpF} = C_{DmF} = 0.00125$) only if it is practical to maintain/attain the model velocity, drag force, fluid, scale, and cost. However, because it is difficult to find a practical/convenient model fluid with a kinematic viscosity, $v = 1.261 \times 10^{-6} \text{ ft}^2/\text{sec}$, the solution to this type of difficult situation is to realize that due to the wave action at the free surface, it becomes more important to satisfy the dynamic requirement for F. Although ignoring the dynamic requirement for R will result in a "distorted model" (which can be accommodated for by proper interpretation of the model results), the model will be less distorted and more of a "true model" when the R is large, which reduces the dependence of the drag coefficient, C_D on R. In this example, $R = 1.915 \times 10^6$, which is large; thus, not modeling the viscous effects may be insignificant. Thus, by not satisfying the dynamic requirement that the model fluid have a kinematic viscosity, $v = 1.261 \times 10^{-6} \text{ ft}^2/\text{sec}$ represents not satisfying the **R** dynamic requirement. Instead, typically, a practical/convenient model fluid such as water is used. It is interesting to compare this example to Example Problem 11.8 above. In Example Problem 11.8 above, the model fluid was assumed to be water, and the effects of **R** and **F** were modeled separately, which also resulted in a "distorted model." However, there were two significantly different model velocities that separately satisfied **R** and **F**, and the resulting drag force on the model was significantly higher than in this example problem. Which one is a less "distorted model"?

11.7.3 Application of the Similitude Scale Ratios for the Major Head Loss in Pipe Flow

For the case of internal pipe flow, dimensional analysis yielded the following uncalibrated mathematical model of the fluid flow situation:

$$\frac{\tau_w}{\rho v^2} = \frac{F_f}{F_I} = \frac{F_P}{F_I} = \frac{\Delta p R_h}{L}\frac{1}{\rho v^2} = \frac{R_h}{L}\frac{\Delta p}{\rho v^2} = \frac{R_h}{L}\frac{1}{E} = C_D = \phi\left(R, \frac{\varepsilon}{D}\right) \tag{11.120}$$

and,

$$\frac{F_f}{F_I} = f = 8C_D = 8\left[\frac{\tau_w}{\rho v^2}\right] = 8\left[\frac{D}{4L}\frac{\Delta p}{\rho v^2}\right] = \frac{D}{L}\frac{\Delta p}{\frac{\rho v^2}{2}} = \frac{D}{L}\frac{2}{E} = \phi\left(R, \frac{\varepsilon}{D}\right) \tag{11.121}$$

where the resulting major head loss, h_f was given as follows:

$$h_f = \frac{\tau_w L}{\gamma R_h} = \frac{\Delta p}{\gamma} = S_f L = C_D \rho v^2 \frac{L}{\gamma R_h} = \frac{v^2 L}{C^2 R_h} = f\frac{L}{D}\frac{v^2}{2g} = \left(\frac{vn}{R_h^{2/3}}\right)^2 L \tag{11.122}$$

where C is the Chezy coefficient and is defined as $C = \sqrt{g/C_D}$, $C = \sqrt{8g/f}$, and $C = \left(R_h^{1/6}/n\right)$, depending upon how the flow resistance is modeled, and where the drag coefficient, C_D as modeled by the Euler number, E is the dependent π term, as it models the dependent variable, Δp that cannot be derived analytically from theory, while the Reynolds number, R and the relative surface roughness, ε/D are the independent π terms. One may note that because the drag coefficient, C_D is mainly a function of the Euler number, E, the inertia force, F_I and the pressure force, F_P will always be important (predominate) in internal pipe flow. Thus, the Euler number, E and the drag coefficient, $C_D = (f/8) = (R_h/L)(1/E)$ must remain a constant between the model and its prototype. However, depending upon the specifics of the internal pipe flow situation, the relative surface roughness, ε/D and the remaining force, which includes the viscous force, F_V (and thus, its respective independent π term, R) may or may not play an important role in the definition of the drag coefficient, C_D.

11.7.3.1 Laminar Pipe Flow

In the case of laminar flow, the flow resistance (as modeled by C_D, C, f, or n) is only a function of R and independent of ε/D. As such, in addition to the inertia force, F_I and the pressure force, F_P, only the Reynolds number, R is important in pipe flow. Therefore, because there is one independent π term, R used in the definition of the dependent π term, E and thus the drag coefficient, C_D, the appropriate similitude scale ratio corresponding to the independent π term is used to define the similarity requirements between the model and its prototype. Specifically, the criteria governing dynamic similarity requires the following:

$$\underbrace{\left[\left(\frac{\rho v L}{\mu}\right)_p\right]}_{R_p} = \underbrace{\left[\left(\frac{\rho v L}{\mu}\right)_m\right]}_{R_m} \tag{11.123}$$

which states that there is a dynamic similarity requirement for the Reynolds number, R in order for the dependent π term (i.e., the drag coefficient, C_D) to remain a constant between the model and its prototype, where the flow resistance prediction equation is defined as follows:

$$\underbrace{\left[\frac{h_f}{\frac{\rho v^2 L}{\gamma R_h}}\right]_p}_{C_{D_p}} = \underbrace{\left[\frac{h_f}{\frac{\rho v^2 L}{\gamma R_h}}\right]_m}_{C_{D_m}} \tag{11.124}$$

$$\underbrace{\left[\frac{h_f}{\frac{v^2 L}{R_h}}\right]_p}_{\frac{1}{C_p^2}} = \underbrace{\left[\frac{h_f}{\frac{v^2 L}{R_h}}\right]_m}_{\frac{1}{C_m^2}} \tag{11.125}$$

$$\underbrace{\left[\frac{h_f}{\frac{v^2 L}{2gD}}\right]_p}_{f_p} = \underbrace{\left[\frac{h_f}{\frac{v^2 L}{2gD}}\right]_m}_{f_m} \tag{11.126}$$

$$\underbrace{\left[\frac{h_f}{\frac{v^2 L}{R_h^{4/3}}}\right]_p}_{n_p^2} = \underbrace{\left[\frac{h_f}{\frac{v^2 L}{R_h^{4/3}}}\right]_m}_{n_m^2} \tag{11.127}$$

Furthermore, one may note that in the case where the similarity requirements regarding the independent π term, R is satisfied, then the dependent π term (i.e., the drag coefficient, C_D) will remain a constant between the model and its prototype. And, therefore, application of the flow resistance prediction equation yields a "true model." However, in the case where the similarity requirements regarding the independent π term, R are not satisfied, then the dependent π term (i.e., the friction factor, f) will not remain a constant between the model and its prototype. And, therefore, application of the flow resistance prediction equation yields a "distorted model" for which the appropriate corrections measures are taken.

EXAMPLE PROBLEM 11.10

Crude oil at 68°F flows in a prototype 1500-ft-long circular pipe with a diameter of 2 ft, as illustrated in Figure EP 11.10. A smaller model of the larger prototype is designed in order to study the flow characteristics of laminar pipe flow. The model fluid is glycerin at 68°F, the velocity of glycerin in the smaller model pipe is 30 ft/sec, and the model scale, λ is 0.25. The flow resistance is modeled by the friction factor, $f = 8\,C_D$. (a) Determine the pressure drop (and head loss) in the flow of the glycerin in the model. (b) Determine the velocity flow of the crude oil in the prototype pipe flow in order to achieve dynamic similarity between the model and the prototype. (c) Determine the pressure drop (and head loss) in the flow of the crude oil in the prototype in order to achieve dynamic similarity between the model and the prototype.

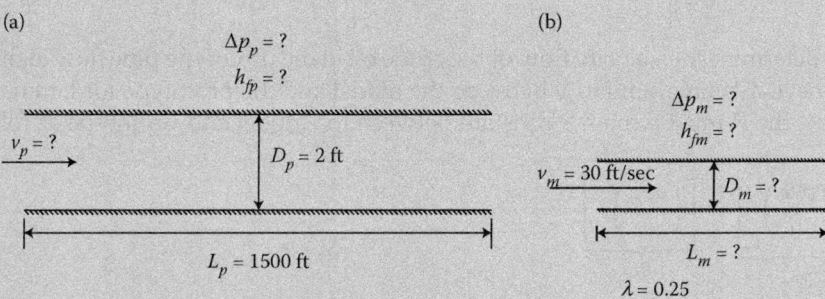

FIGURE EP 11.10
(a) Crude oil flows in a prototype pipe. (b) Glycerin flows in a smaller model pipe.

Mathcad Solution

(a) In order to determine the pressure drop (and head loss) in flow of the glycerin in the model, the major head loss equation, Equation 11.122, is applied as follows:

$$h_f = \frac{\Delta p}{\gamma} = f\frac{L}{D}\frac{v^2}{2g}$$

where the friction factor, f is used to model the flow resistance. Because the Reynolds number, $R < 2,000$ (laminar pipe flow) is assumed, the friction factor, f decreases with an increase in R, as illustrated in the Moody diagram in Figure 8.1, where $f = 64/R$. Furthermore, in order to determine the pipe length and diameter of the model pipe, the model scale, λ (inverse of the length ratio) is applied. The fluid properties for glycerin are given in Table A.4 in Appendix A.

$D_p := 2 \text{ ft}$ $L_P := 1500 \text{ ft}$ $\lambda := 0.25$

Guess value: $D_m := 0.1 \text{ ft}$ $L_m := 1 \text{ ft}$

Given

$$\lambda = \frac{D_m}{D_p} \qquad\qquad\qquad \lambda = \frac{L_m}{L_p}$$

$$\begin{pmatrix} D_m \\ L_m \end{pmatrix} := \text{Find}\,(D_m, L_m) = \begin{pmatrix} 0.5 \\ 375 \end{pmatrix} \text{ft}$$

$$\text{slug} := 1\,\text{lb}\frac{\sec^2}{\text{ft}} \qquad \rho_m := 2.44\frac{\text{slug}}{\text{ft}^3} \qquad \mu_m := 31{,}200 \times 10^{-6}\,\text{lb}\frac{\sec}{\text{ft}^2}$$

$$g := 32.174\frac{\text{ft}}{\sec^2} \qquad \gamma_m := \rho_m \cdot g = 78.505\frac{\text{lb}}{\text{ft}^3} \qquad v_m := 30\frac{\text{ft}}{\sec}$$

$$R_m := \frac{\rho_m \cdot v_m \cdot D_m}{\mu_m} = 1.173 \times 10^3 \qquad f_m := \frac{64}{R_m} = 0.055$$

$$h_{fm} := f_m \frac{L_m}{D_m}\frac{v_m^2}{2\,g} = 572.298\text{ ft} \qquad\qquad \Delta p_m := h_{fm} \cdot \gamma_m = 4.493 \times 10^4\frac{\text{lb}}{\text{ft}^2}$$

(b) To determine the velocity flow of the crude oil in the prototype pipe flow in order to achieve dynamic similarity between the model and the prototype for laminar pipe flow, the **R** must remain a constant between the model and prototype as follows:

$$\underbrace{\left[\left(\frac{\rho v L}{\mu}\right)_p\right]}_{R_p} = \underbrace{\left[\left(\frac{\rho v L}{\mu}\right)_m\right]}_{R_m}$$

The fluid properties for crude oil are given in Table A.4 in Appendix A.

$$\rho_p := 1.66\frac{\text{slug}}{\text{ft}^3} \qquad\qquad\qquad \mu_p := 105 \times 10^{-6}\,\text{lb}\frac{\sec}{\text{ft}^2}$$

Guess value $\qquad\qquad\qquad v_p := 10\frac{\text{ft}}{\sec} \qquad\qquad\qquad R_p := 1000$

Given

$$R_p = \frac{\rho_p \cdot v_p \cdot D_p}{\mu_p}$$

$$R_p = R_m$$

$$\begin{pmatrix} v_p \\ R_p \end{pmatrix} := \text{Find}\,(v_p, R_p)$$

$$v_p = 0.053\frac{\text{ft}}{\text{s}} \qquad\qquad\qquad\qquad R_p = 1.173 \times 10^3$$

(c) To determine the pressure drop (and head loss) in the flow of the crude oil in the prototype in order to achieve dynamic similarity between the model and the prototype for laminar pipe flow, the friction factor, f must remain a constant between the model and the prototype (which is a direct result of maintaining a constant **R** between the model and the prototype) as follows:

$$\left[\frac{h_f}{\frac{v^2 L}{2gD}}\right]_p = \left[\frac{h_f}{\frac{v^2 L}{2gD}}\right]_m$$

$$\underbrace{\qquad}_{f_p} \qquad \underbrace{\qquad}_{f_m}$$

$$\gamma_p := \rho_p \cdot g = 53.409 \frac{lb}{ft^3}$$

Guess value: $\quad h_{fp} := 1\,ft \qquad \Delta p_p := 1\frac{lb}{ft^2} \qquad f_p := 0.1$

Given

$$f_p = \frac{h_{fp}}{\left(\frac{v_p^2 \cdot L_P}{2 \cdot g \cdot D_P}\right)} \qquad\qquad \Delta p_p = h_{fp} \cdot \gamma_p$$

$$f_p = f_m$$

$$\begin{pmatrix} h_{fp} \\ \Delta p_p \\ f_p \end{pmatrix} := Find\,(h_{fp}, \Delta p_p, f_p)$$

$$h_{fp} = 1.786 \times 10^{-3}\,ft \qquad\qquad \Delta p_p = 0.095\frac{lb}{ft^2} \qquad f_b = 0.055$$

Therefore, although the similarity requirements regarding the independent π term, R ("viscosity model") are theoretically satisfied ($R_p = R_m = 1.173 \times 10^3$), the dependent π term (i.e., the friction factor, f) will actually/practically remain a constant between the model and its prototype ($f_p = f_m = 0.055$) only if it is practical to maintain/attain the model velocity, pressure, fluid, scale, and cost.

11.7.3.2 Completely Turbulent Pipe Flow (Rough Pipes)

In the case of a completely turbulent flow (rough pipes), the flow resistance (as modeled by C_D, C, f, or n) is only a function of ε/D and independent of R. As such, in addition to the inertia force, F_I and the pressure force, F_P, only the relative surface roughness, ε/D is important in pipe flow. Therefore, because there is one independent π term, ε/D used in the definition of the dependent π term, E and thus the drag coefficient, C_D, the appropriate similitude scale ratio corresponding to the independent π term is used to define the similarity requirements between the model and its prototype. Specifically, the criteria governing dynamic similarity requires the following:

$$\left(\frac{\varepsilon}{D}\right)_p = \left(\frac{\varepsilon}{D}\right)_m \tag{11.128}$$

which states that there is a geometric similarity requirement for the relative surface rough-ness, ε/D in order for the dependent π term (i.e., the drag coefficient, C_D) to remain a cons-tant between the model and its prototype, where the flow resistance prediction equation is defined as follows:

$$\underbrace{\left[\frac{h_f}{\dfrac{\rho v^2 L}{\gamma R_h}}\right]_p}_{C_{Dp}} = \underbrace{\left[\frac{h_f}{\dfrac{\rho v^2 L}{\gamma R_h}}\right]_m}_{C_{Dm}} \tag{11.129}$$

$$\underbrace{\left[\frac{h_f}{\dfrac{v^2 L}{R_h}}\right]_p}_{\dfrac{1}{C_p^2}} = \underbrace{\left[\frac{h_f}{\dfrac{v^2 L}{R_h}}\right]_m}_{\dfrac{1}{C_m^2}} \tag{11.130}$$

$$\underbrace{\left[\frac{h_f}{\dfrac{v^2 L}{2gD}}\right]_p}_{f_p} = \underbrace{\left[\frac{h_f}{\dfrac{v^2 L}{2gD}}\right]_m}_{f_m} \tag{11.131}$$

$$\underbrace{\left[\frac{h_f}{\dfrac{v^2 L}{R_h^{4/3}}}\right]_p}_{n_p^2} = \underbrace{\left[\frac{h_f}{\dfrac{v^2 L}{R_h^{4/3}}}\right]_m}_{n_m^2} \tag{11.132}$$

Furthermore, one may note that in the case where the similarity requirements regard-ing the independent π term, ε/D (geometric similarity) are satisfied, then the dependent π term (i.e., the drag coefficient, C_D) will remain a constant between the model and its pro-totype. And, therefore, application of the flow resistance prediction equation yields a "true model." However, in the case where the similarity requirements regarding the inde-pendent π term, ε/D and the dependent π term, E ("pressure model") are not satisfied, then the dependent π term (i.e., the friction factor, f) will not remain a constant between the model and its prototype. And, therefore, application of the flow resistance prediction equation yields a "distorted model" for which the appropriate corrections measures are taken.

EXAMPLE PROBLEM 11.11

Air at 68°F flows in a prototype 1300-ft-long circular pipe with a diameter of 3 ft and an absolute pipe roughness of 0.03 ft, as illustrated in Figure EP 11.11. A smaller model of the larger prototype is designed in order to study the flow characteristics of turbulent pipe flow. The model fluid is water at 70°F, the velocity of water in the smaller model pipe is 50 ft/sec, and the model scale, λ is 0.2. The flow resistance is modeled by the

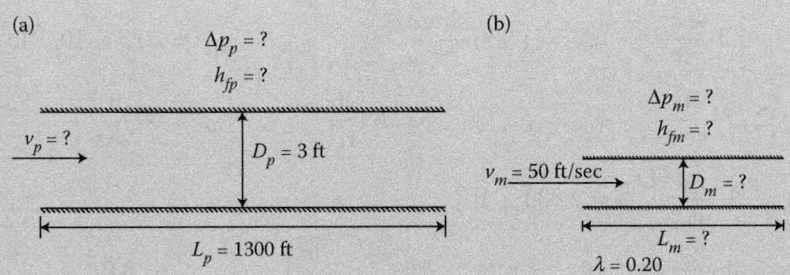

FIGURE EP 11.11
(a) Air flows in a prototype pipe. (b) Water flows in a smaller model pipe.

friction factor, $f = 8\,C_D$. (a) Determine the pressure drop (and head loss) in the flow of the water in the model. (b) Determine the velocity flow of the air in the prototype pipe flow in order to achieve dynamic similarity between the model and the prototype. (c) Determine the pressure drop (and head loss) in the flow of the air in the prototype in order to achieve dynamic similarity between the model and the prototype.

Mathcad Solution

(a) In order to determine the pressure drop (and head loss) in flow of the water in the model, the major head loss equation, Equation 11.122, is applied as follows:

$$h_f = \frac{\Delta p}{\gamma} = f\,\frac{L}{D}\frac{v^2}{2g}$$

where the friction factor, f is used to model the flow resistance. Because the Reynolds number, $R > 4{,}000$ (turbulent pipe flow) is assumed, the friction factor, f is only a function of ε/D and independent of R, as illustrated in by the Moody diagram in Figure 8.1. However the Colebrook equation, Equation 8.33, presents a mathematical representation of the Moody diagram in Figure 8.1. Furthermore, in order to determine the length, diameter, and the absolute pipe roughness of the model pipe, the model scale, λ (inverse of the length ratio) is applied. The fluid properties for water are given in Table A.2 in Appendix A.

$D_p := 3\ \text{ft}$ $\qquad$ $L_p := 1300\ \text{ft}$ $\qquad$ $\varepsilon_p := 0.03\ \text{ft}$ $\qquad$ $\lambda := 0.2$

Guess value: $\qquad$ $D_m := 0.1\ \text{ft}$ $\qquad$ $L_m := 1\ \text{ft}$ $\qquad$ $\varepsilon_m := 0.01\ \text{ft}$

Given

$$\lambda = \frac{D_m}{D_p} \qquad\qquad \lambda = \frac{L_m}{L_p} \qquad\qquad \lambda = \frac{\varepsilon_m}{\varepsilon_p}$$

$$\begin{pmatrix} D_m \\ L_m \\ \varepsilon_m \end{pmatrix} := \text{Find}\,(D_m,\,L_m,\,\varepsilon_m) = \begin{pmatrix} 0.6 \\ 260 \\ 6 \times 10^{-3} \end{pmatrix}\ \text{ft}$$

$$\text{slug} := 1\,\text{lb}\frac{\sec^2}{\text{ft}} \qquad \rho_m := 1.936\frac{\text{slug}}{\text{ft}^3} \qquad\qquad \mu_m := 20.5 \times 10^{-6}\,\text{lb}\frac{\sec}{\text{ft}^2}$$

$$g := 32.174\frac{\text{ft}}{\sec^2} \qquad \gamma_m := \rho_m \cdot g = 62.289\frac{\text{lb}}{\text{ft}^3} \qquad v_m := 50\frac{\text{ft}}{\sec}$$

$$R_m := \frac{\rho_m \cdot v_m \cdot D_m}{\mu_m} = 2.833 \times 10^6$$

Guess value: $\qquad h_{fm} := 1\,\text{ft} \qquad \Delta p_m := 1\frac{\text{lb}}{\text{ft}^2} \qquad f_m := 0.01$

Given

$$h_{fm} = f_m \frac{L_m}{D_m}\frac{v_m^2}{2\,g} \qquad\qquad \frac{1}{\sqrt{f_m}} = -2\log\left(\frac{\frac{\varepsilon_m}{D_m}}{3.7} + \frac{2.51}{R_m \cdot \sqrt{f_m}}\right)$$

$$\Delta p_m = h_{fm} \cdot \gamma_m$$

$$\begin{pmatrix} h_{fm} \\ \Delta p_m \\ f_m \end{pmatrix} := \text{Find}\,(h_{fm}, \Delta p_m, f_m)$$

$$h_{fm} = 638.493\,\text{ft} \qquad\qquad \Delta p_m = 3.977 \times 10^4\frac{\text{lb}}{\text{ft}^2} \qquad\qquad f_m = 0.038$$

(b)–(c) To determine the velocity flow of the air in the prototype pipe flow in order to achieve dynamic similarity between the model and the prototype for turbulent pipe flow and to determine the pressure drop (and head loss) in the flow of the air in the prototype in order to achieve dynamic similarity between the model and the prototype for turbulent pipe flow, the ε/D must remain a constant between the model and prototype as follows:

$$\left(\frac{\varepsilon}{D}\right)_p = \left(\frac{\varepsilon}{D}\right)_m$$

$$\frac{\varepsilon_p}{D_p} = 0.01 \qquad \frac{\varepsilon_m}{D_m} = 0.01$$

However, because the friction factor, f is independent of R for turbulent flow, R does not need to remain a constant between the model and the prototype.

(b)–(c) To determine the velocity flow of the air in the prototype pipe flow in order to achieve dynamic similarity between the model and the prototype for turbulent pipe flow and to determine the pressure drop (and head loss) in the flow of the air in the prototype in order to achieve dynamic similarity between the model and the prototype for turbulent pipe flow, the friction factor, f must remain a constant between the model and the prototype (which is a direct result of maintaining a constant ε/D between the model and the prototype, and applying the "pressure model" similitude scale ratio, specifically the velocity ratio, v_r given in Table 11.1) as follows:

$$\underbrace{\left[\dfrac{h_f}{\dfrac{v^2 L}{2gD}}\right]_p}_{f_p} = \underbrace{\left[\dfrac{h_f}{\dfrac{v^2 L}{2gD}}\right]_m}_{f_m}$$

$$v_r = \frac{v_p}{v_m} = \frac{\left(\sqrt{\dfrac{\Delta p}{\rho}}\right)_p}{\left(\sqrt{\dfrac{\Delta p}{\rho}}\right)_m} = \Delta p_r^{\frac{1}{2}} \rho_r^{\frac{-1}{2}}$$

The fluid properties for air are given in Table A.5 in Appendix A.

$$\rho_p := 0.00231 \frac{slug}{ft^3} \qquad \mu_p := 0.376 \times 10^{-6} \, lb \frac{sec}{ft^2} \qquad \gamma_p := \rho_p \cdot g = 0.074 \frac{lb}{ft^3}$$

Guess value: $\qquad v_p := 1 \dfrac{ft}{sec} \qquad h_{fp} := 1 \, ft \qquad \Delta p_p := 1 \dfrac{lb}{ft^2} \quad f_p := 0.01$

Given

$$f_p = \frac{h_{fp}}{\left(\dfrac{v_p^2 \cdot L_p}{2 \cdot g \cdot D_p}\right)} \qquad\qquad \frac{v_p}{\Delta p_p^{\frac{1}{2}} \cdot \rho_p^{\frac{-1}{2}}} = \frac{v_m}{\Delta p_m^{\frac{1}{2}} \cdot \rho_m^{\frac{-1}{2}}}$$

$$f_p = f_m \qquad\qquad \Delta p_p = h_{fp} \cdot \gamma_p$$

$$\begin{pmatrix} v_p \\ h_{fp} \\ \Delta p_p \\ f_p \end{pmatrix} := \text{Find}\,(v_p, h_{fp}, \Delta p_p, f_p)$$

$$v_p = 7.857 \frac{ft}{s} \qquad\qquad h_{fp} = 15.766 \, ft \qquad \Delta p_p = 1.172 \frac{lb}{ft^2} \qquad f_p = 0.038$$

Furthermore, the Euler number, E remains a constant between the model and the prototype as follows:

$$E_m := \frac{\rho_m \cdot v_m^2}{\Delta p_m} = 0.122 \qquad\qquad\qquad\qquad E_p := \frac{\rho_p \cdot v_p^2}{\Delta p_p} = 0.122$$

Therefore, although the similarity requirements regarding the independent π term, ε/D $((\varepsilon/D)_p = (\varepsilon/D)_m = 0.01)$ and the dependent π term, E ("pressure model") $(E_p = E_m = 0.122)$ are theoretically satisfied, the dependent π term (i.e., the friction factor, f) will actually/practically remain a constant between the model and its prototype $(f_p = f_m = 0.038)$ only if it is practical to attain/

maintain the model velocity, pressure, fluid, scale, and cost. Furthermore, because the friction factor, f is independent of R for turbulent flow, R does not need to remain a constant between the model and the prototype as follows:

$$R_m = 2.833 \times 10^6 \qquad\qquad R_p := \frac{\rho_p \cdot v_p \cdot D_p}{\mu_p} = 1.448 \times 10^5$$

11.7.3.3 Transitional Pipe Flow

In the case of transitional flow, the flow resistance (as modeled by C_D, C, f, or n) is a function of both ε/D and R. As such, in addition to the inertia force, F_I and the pressure force, F_P, the viscous force, F_V and the relative surface roughness, ε/D are important in pipe flow. Therefore, because there are two independent π terms, R and ε/D, used in the definition of the dependent π term, E and thus the drag coefficient, C_D, the appropriate similitude scale ratios corresponding to the independent π terms are used to define the similarity requirements between the model and its prototype. Specifically, the criteria governing dynamic similarity requires the following:

$$\underbrace{\left[\left(\frac{\rho v L}{\mu}\right)_p\right]}_{R_p} = \underbrace{\left[\left(\frac{\rho v L}{\mu}\right)_m\right]}_{R_m} \tag{11.133}$$

$$\left(\frac{\varepsilon}{D}\right)_p = \left(\frac{\varepsilon}{D}\right)_m \tag{11.134}$$

which states that in addition to the geometric similarity requirement for the relative surface roughness, ε/D, there is a dynamic similarity requirement that R remains a constant in order for the dependent π term (i.e., the drag coefficient, C_D) to remain a constant between the model and its prototype, where the flow resistance prediction equation is defined as follows:

$$\underbrace{\left[\frac{h_f}{\dfrac{\rho v^2 L}{\gamma R_h}}\right]_p}_{C_{Dp}} = \underbrace{\left[\frac{h_f}{\dfrac{\rho v^2 L}{\gamma R_h}}\right]_m}_{C_{Dm}} \tag{11.135}$$

$$\underbrace{\left[\frac{h_f}{\dfrac{v^2 L}{R_h}}\right]_p}_{\frac{1}{C_p^2}} = \underbrace{\left[\frac{h_f}{\dfrac{v^2 L}{R_h}}\right]_m}_{\frac{1}{C_m^2}} \tag{11.136}$$

$$\underbrace{\left[\frac{h_f}{\frac{v^2 L}{2gD}}\right]_p}_{f_p} = \underbrace{\left[\frac{h_f}{\frac{v^2 L}{2gD}}\right]_m}_{f_m} \tag{11.137}$$

$$\underbrace{\left[\frac{h_f}{\frac{v^2 L}{R_h^{4/3}}}\right]_p}_{n_p^2} = \underbrace{\left[\frac{h_f}{\frac{v^2 L}{R_h^{4/3}}}\right]_m}_{n_m^2} \tag{11.138}$$

Furthermore, one may note that in the case where the similarity requirements regarding the independent π terms, R ("viscosity model") and ε/D (geometric similarity) are satisfied, then the dependent π term (i.e., the drag coefficient, C_D) will remain a constant between the model and its prototype. And, therefore, application of the flow resistance prediction equation yields a "true model." However, in the case where the similarity requirements regarding the independent π terms, R and ε/D are not satisfied, then the dependent π term (i.e., the drag coefficient, C_D) will not remain a constant between the model and its prototype. And, therefore, application of the flow resistance prediction equation yields a "distorted model" for which the appropriate corrections measures are taken.

EXAMPLE PROBLEM 11.12

(Refer to Example Problem 11.11.) Air at 68°F flows in a prototype 1300-ft-long circular pipe with a diameter of 3 ft and an absolute pipe roughness of 0.03 ft, as illustrated in Figure EP 11.12. A smaller model of the larger prototype is designed in order to study the flow characteristics of transitional pipe flow. The model fluid is water at 70°F, the velocity of water in the smaller model pipe is 0.07 ft/sec, and the model scale, λ is 0.2. The flow resistance is modeled by the friction factor, $f = 8\,C_D$. (a) Determine the pressure drop (and head loss) in the flow of the water in the model. (b) Determine the velocity flow of the air in the prototype pipe flow in order to achieve dynamic similarity between the model and the prototype. (c) Determine the pressure drop (and head

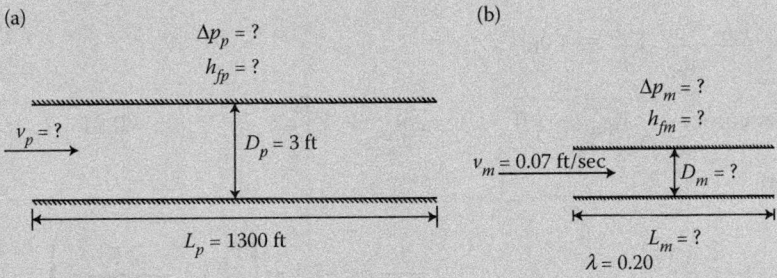

FIGURE EP 11.12
(a) Air flows in a prototype pipe. (b) Water flows in a smaller model pipe.

loss) in the flow of the air in the prototype in order to achieve dynamic similarity between the model and the prototype.

Mathcad Solution

(a) In order to determine the pressure drop (and head loss) in flow of the water in the model, the major head loss equation, Equation 11.122, is applied as follows:

$$
h_f = \frac{\Delta p}{\gamma} = f \frac{L}{D} \frac{v^2}{2g}
$$

where the friction factor, f is used to model the flow resistance. Because the Reynolds number, $2,000 < R < 4,000$ (transitional pipe flow) is assumed, the friction factor, f is a function of both ε/D and R, as illustrated by the Moody diagram in Figure 8.1. However the Colebrook equation, Equation 8.33, presents a mathematical representation of the Moody diagram in Figure 8.1. Furthermore, in order to determine the length, diameter, and the absolute pipe roughness of the model pipe, the model scale, λ (inverse of the length ratio) is applied. The fluid properties for water are given in Table A.2 in Appendix A.

$$D_p := 3 \text{ ft} \qquad L_p := 1300 \text{ ft} \qquad \varepsilon_p := 0.03 \text{ ft} \qquad \lambda := 0.2$$

Guess value: $\qquad D_m := 0.1 \text{ ft} \qquad L_m := 1 \text{ ft} \qquad \varepsilon_m := 0.01 \text{ ft}$

Given

$$\lambda = \frac{D_m}{D_p} \qquad\qquad \lambda = \frac{L_m}{L_p} \qquad\qquad \lambda = \frac{\varepsilon_m}{\varepsilon_p}$$

$$
\begin{pmatrix} D_m \\ L_m \\ \varepsilon_m \end{pmatrix} := \text{Find}(D_m, L_m, \varepsilon_m) = \begin{pmatrix} 0.6 \\ 260 \\ 6 \times 10^{-3} \end{pmatrix} \text{ ft}
$$

$$\text{slug} := 1 \text{ lb} \frac{\sec^2}{\text{ft}} \qquad \rho_m := 1.936 \frac{\text{slug}}{\text{ft}^3} \qquad\qquad \mu_m := 20.5 \times 10^{-6} \text{ lb} \frac{\sec}{\text{ft}^2}$$

$$g := 32.174 \frac{\text{ft}}{\sec^2} \qquad \gamma_m := \rho_m \cdot g = 62.289 \frac{\text{lb}}{\text{ft}^3} \qquad v_m := 0.07 \frac{\text{ft}}{\sec}$$

$$R_m := \frac{\rho_m \cdot v_m \cdot D_m}{\mu_m} = 3.966 \times 10^3$$

Guess value: $\qquad h_{fm} := 1 \text{ ft} \qquad \Delta p_m := 1 \frac{\text{lb}}{\text{ft}^2} \qquad f_m := 0.01$

Given

$$h_{fm} = f_m \frac{L_m}{D_m} \frac{v_m^2}{2 \, g} \qquad\qquad \frac{1}{\sqrt{f_m}} = -2 \log\left(\frac{\dfrac{\varepsilon_m}{D_m}}{3.7} + \frac{2.51}{R_m \cdot \sqrt{f_m}} \right)$$

$$\Delta p_m = h_{fm} \cdot \gamma_m$$

$$\begin{pmatrix} h_{fm} \\ \Delta p_m \\ f_m \end{pmatrix} := \text{Find} (h_{fm}, \Delta p_m, f_m)$$

$$h_{fm} = 1.622 \times 10^{-3} \, ft \qquad\qquad \Delta p_m = 0.101 \frac{lb}{ft^2} \qquad\qquad f_m = 0.049$$

(b) To determine the velocity flow of the air in the prototype pipe flow in order to achieve dynamic similarity between the model and the prototype for transitional pipe flow, the ε/D must remain a constant between the model and prototype as follows:

$$\left(\frac{\varepsilon}{D}\right)_p = \left(\frac{\varepsilon}{D}\right)_m$$

$$\frac{\varepsilon_p}{D_p} = 0.01 \qquad\qquad\qquad \frac{\varepsilon_m}{D_m} = 0.01$$

Furthermore, the R must remain a constant between the model and prototype as follows:

$$\underbrace{\left[\left(\frac{\rho v L}{\mu}\right)_p\right]}_{R_p} = \underbrace{\left[\left(\frac{\rho v L}{\mu}\right)_m\right]}_{R_m}$$

The fluid properties for air are given in Table A.5 in Appendix A.

$$\rho_p := 0.00231 \frac{slug}{ft^3} \qquad\qquad \mu_p := 0.376 \times 10^{-6} \, lb \frac{sec}{ft^2}$$

Guess value: $\qquad\qquad v_p := 1 \frac{ft}{sec} \qquad\qquad R_p := 3000$

Given

$$R_p = \frac{\rho_p \cdot v_p \cdot D_p}{\mu_p}$$

$$R_p = R_m$$

$$\begin{pmatrix} v_p \\ R_p \end{pmatrix} := \text{Find} (v_p, R_p)$$

$$v_p = 0.215 \frac{ft}{s} \qquad\qquad\qquad R_p = 3.966 \times 10^3$$

(c) To determine the pressure drop (and head loss) in the flow of the air in the prototype in order to achieve dynamic similarity between the model and the prototype for transitional pipe flow, the friction factor, f must remain a constant between the model and the prototype (which is a direct result of maintaining a constant ε/D

and a constant R between the model and the prototype) as follows:

$$\left[\frac{h_f}{\dfrac{v^2 L}{2gD}}\right]_p = \left[\frac{h_f}{\dfrac{v^2 L}{2gD}}\right]_m$$

$$\underbrace{}_{f_p} \qquad \underbrace{}_{f_m}$$

$$\gamma_p := \rho_p \cdot g = 0.074 \frac{lb}{ft^3}$$

Guess value: $h_{fp} := 1\ ft$ $\Delta p_p := 1 \frac{lb}{ft^2}$ $f_p := 0.1$

Given

$$f_p = \frac{h_{fp}}{\left(\dfrac{v_p^2 \cdot L_p}{2 \cdot g \cdot Dp}\right)} \qquad\qquad \Delta p_p = h_{fp} \cdot \gamma_p$$

$$f_p = f_m$$

$$\begin{pmatrix} h_{fp} \\ \Delta p_p \\ f_p \end{pmatrix} := Find\ (h_{fp},\ \Delta p_p,\ f_p)$$

$$h_{fp} = 0.015\ ft \qquad\qquad \Delta p_p = 1.139 \times 10^{-3} \frac{lb}{ft^2} \qquad\qquad f_p = 0.049$$

Therefore, although the similarity requirements regarding the independent π term, ε/D $((\varepsilon/D)_p = (\varepsilon/D)_m = 0.01)$ and the independent π term, R ("viscosity model") $(R_p = R_m = 3.966 \times 10^3)$ are theoretically satisfied, the dependent π term (i.e., the friction factor, f) will actually/practically remain a constant between the model and its prototype $(f_p = f_m = 0.049)$ only if it is practical to maintain/attain the model velocity, pressure, fluid, scale, and cost.

11.7.4 Application of the Similitude Scale Ratios for the Major Head Loss in Open Channel Flow

For the case of internal open channel flow, interpretation of the expression for the wall shear stress, τ_w derived from dimensional analysis for open channel flow yielded the following uncalibrated mathematical model of the fluid flow situation:

$$\frac{\tau_w}{\rho v^2} = \frac{F_f}{F_I} = \frac{\gamma R_h S_f}{\rho v^2} = \frac{\rho g y S_f}{2\rho v^2} = \frac{S_f}{2}\frac{gy}{v^2} = \frac{S_f}{2}\frac{1}{F} = C_D = \phi\left(R, \frac{\varepsilon}{y}\right) \qquad (11.139)$$

and,

$$\frac{F_f}{F_I} = f = 8C_D = 8\left[\frac{\tau_w}{\rho v^2}\right] = 8\left[\frac{S_f}{2}\frac{gy}{v^2}\right] = 4S_f\frac{gy}{v^2} = 4S_f\frac{1}{F} = \phi\left(R, \frac{\varepsilon}{y}\right) \qquad (11.140)$$

where the resulting major head loss, h_f was given as follows:

$$h_f = \frac{\tau_w L}{\gamma R_h} = S_f L = C_D\rho v^2\frac{L}{\gamma R_h} = \frac{v^2 L}{C^2 R_h} = f\frac{L}{(4R_h)}\frac{v^2}{2g} = \left(\frac{vn}{R_h^{2/3}}\right)^2 L \qquad (11.141)$$

where C is the Chezy coefficient and is defined as $C = \sqrt{g/C_D}$, $C = \sqrt{8g/f}$, and $C = \left(R_h^{1/6}/n\right)$, depending upon how the flow resistance is modeled, and where the drag coefficient, C_D as modeled by the Froude number, F is the dependent π term, as it models the dependent variable, S_f that cannot be derived analytically from theory, while the Reynolds number, R and the relative surface roughness, ε/y are the independent π terms. However, because the flow is mostly turbulent in open channels, the Reynolds number, R is usually so large that it has virtually no influence on the drag coefficient, C_D. Furthermore, one may recall that the Darcy–Weisbach friction factor, f is typically applied for pipe flow only. One may note that because the drag coefficient, C_D is mainly a function of the Froude number, F, the inertia force, F_I and the gravity force, F_G will always be important (predominate) in internal open flow. Thus, the Froude number, F and the drag coefficient, $C_D = (f/8) = (S_f/2)(1/F)$ must remain a constant between the model and its prototype. However, depending upon the specifics of the internal open channel flow situation, the relative surface roughness, ε/y and the remaining force, which includes the viscous force, F_V (and thus, their respective independent π terms) may or may not play an important role in the definition of the drag coefficient, C_D.

11.7.4.1 Turbulent Open Channel Flow

In the case of turbulent flow, the flow resistance (as modeled by C_D, C, f, or n) is only a function of ε/y and independent of R. As such, in addition to the inertia force, F_I and the gravity force, F_G, only the relative surface roughness, ε/y is important in open channel flow. Therefore, because there is one independent π term, ε/y used in the definition of the dependent π term, F and thus the drag coefficient, C_D, the appropriate similitude scale ratio corresponding to the independent π term is used to define the similarity requirements between the model and its prototype. Specifically, the criteria governing dynamic similarity requires the following:

$$\left(\frac{\varepsilon}{y}\right)_p = \left(\frac{\varepsilon}{y}\right)_m \qquad (11.142)$$

which states that there is a geometric similarity requirement for the relative surface roughness, ε/y in order for the dependent π term (i.e., the drag coefficient, C_D) to remain a constant between the model and its prototype, where the flow resistance prediction equation is defined as follows:

$$\underbrace{\left[\frac{h_f}{\frac{\rho v^2 L}{\gamma R_h}}\right]_p}_{C_{Dp}} = \underbrace{\left[\frac{h_f}{\frac{\rho v^2 L}{\gamma R_h}}\right]_m}_{C_{Dm}} \tag{11.143}$$

$$\underbrace{\left[\frac{h_f}{\frac{v^2 L}{R_h}}\right]_p}_{\frac{1}{C_p^2}} = \underbrace{\left[\frac{h_f}{\frac{v^2 L}{R_h}}\right]_m}_{\frac{1}{C_m^2}} \tag{11.144}$$

$$\underbrace{\left[\frac{h_f}{\frac{v^2 L}{2g(4R_h)}}\right]_p}_{f_p} = \underbrace{\left[\frac{h_f}{\frac{v^2 L}{2g(4R_h)}}\right]_m}_{f_m} \tag{11.145}$$

$$\underbrace{\left[\frac{h_f}{\frac{v^2 L}{R_h^{4/3}}}\right]_p}_{n_p^2} = \underbrace{\left[\frac{h_f}{\frac{v^2 L}{R_h^{4/3}}}\right]_m}_{n_m^2} \tag{11.146}$$

Furthermore, one may note that in the case where the similarity requirements regarding the independent π term, ε/y (geometric similarity) are satisfied, then the dependent π term (i.e., the drag coefficient, C_D) will remain a constant between the model and its prototype. And, therefore, application of the flow resistance prediction equation yields a "true model." However, in the case where the similarity requirements regarding the independent π term, ε/y and the dependent π term, F ("gravity model") are not satisfied, then the dependent π term (i.e., the Manning's roughness coefficient, n) will not remain a constant between the model and its prototype. And, therefore, application of the flow resistance prediction equation yields a "distorted model" for which the appropriate corrections measures are taken.

EXAMPLE PROBLEM 11.13

Water at 70°F flows in a prototype 1700-ft-long concrete rectangular open channel with a width of 5 ft, a uniform flow depth of 3 ft, and an absolute channel roughness of 0.01 ft, as illustrated in Figure EP 11.13. A smaller model of the larger prototype is designed in order to study the flow characteristics of turbulent open channel flow. The model fluid is also water at 70°F, the velocity of water in the smaller model channel is 60 ft/sec, and the model scale, λ is 0.25. The flow resistance is modeled by the Manning's roughness coefficient, $n = f''(C_D)$. (a) Determine the friction slope (and head loss)

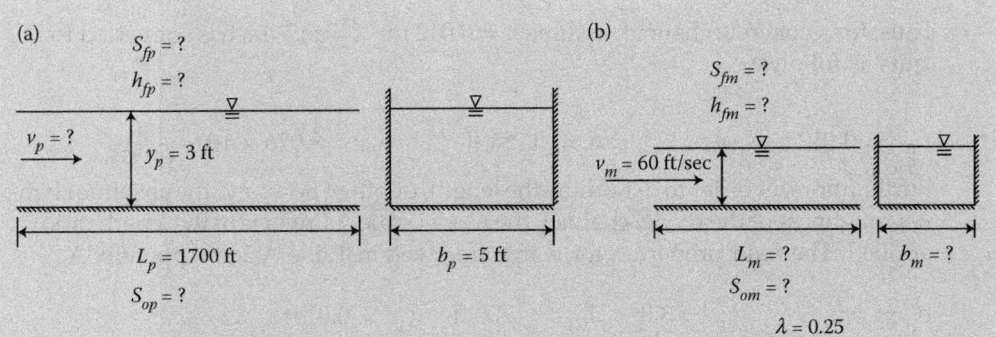

FIGURE EP 11.13
(a) Water flows in a prototype rectangular open channel. (b) Water flows in a smaller model rectangular open channel.

in the flow of the water in the model. (b) Determine the velocity flow of the water in the prototype open channel flow in order to achieve dynamic similarity between the model and the prototype. (c) Determine the friction slope (and head loss) in the flow of the water in the prototype in order to achieve dynamic similarity between the model and the prototype.

Mathcad Solution

(a) In order to determine the friction slope (and head loss) in flow of the water in the model, the major head loss equation, Equation 11.141, is applied as follows:

$$h_f = S_f L = \left(\frac{vn}{R_h^{2/3}} \right)^2 L$$

where the Manning's roughness coefficient, n is used to model the flow resistance. Empirical calibration of the Manning's roughness coefficient, n assumes turbulent flow, thus it is independent of R, and is only a function of ε/y. The absolute channel roughness, ε is indirectly modeled by the type of channel material, as illustrated in Table 8.6, which presents the Manning's roughness coefficient, n for various channel materials. It important to note that although the derivation/formulation of the Manning's equation assumes no specific units (SI or BG), the Manning's roughness coefficient, n has dimensions of $L^{-1/3} T$. Furthermore, because the Manning's roughness coefficient, n in Table 8.6 has been provided/calibrated in SI units $m^{-1/3}$ sec, it must be adjusted when using BG units. In order to convert the Manning's roughness coefficient, n with units of $m^{-1/3}$ sec to units of $ft^{-1/3}$ sec, note that 3.281 ft = 1 m; thus:

$$\frac{\left[\dfrac{3.281 \text{ ft}}{1 \text{ m}} \right]^{1/3}}{n[m^{-1/3} \text{ s}]} = \frac{1.486}{n[ft^{-1/3} \text{ sec}]}$$

Thus, for a concrete channel, assume $n = 0.012\,\mathrm{m}^{-1/3}\,\mathrm{sec}$, which is converted to BG units as follows:

$$n_m := 0.012\,\mathrm{m}^{\frac{-1}{3}}\,\mathrm{sec} \qquad m := 3.281\,\mathrm{ft} \qquad n_m = 8.076 \times 10^{-3}\,\frac{\mathrm{s}}{\mathrm{ft}^{0.333}}$$

Furthermore, in order to determine the length, depth of flow, and the absolute channel roughness of the model channel, the model scale, λ (inverse of the length ratio) is applied. The fluid properties for water are given in Table A.2 in Appendix A.

$$b_p := 5\,\mathrm{ft} \qquad y_p := 3\,\mathrm{ft} \qquad L_p := 1700\,\mathrm{ft} \qquad \varepsilon_p := 0.01\,\mathrm{ft} \qquad \lambda := 0.25$$

Guess value: $\quad b_m := 1\,\mathrm{ft} \qquad y_m := 1\,\mathrm{ft} \qquad L_m := 1\,\mathrm{ft} \qquad \varepsilon_m := 0.01\,\mathrm{ft}$

Given

$$\lambda = \frac{b_m}{b_p} \qquad\qquad \lambda = \frac{y_m}{y_p} \qquad\qquad \lambda = \frac{L_m}{L_p} \qquad\qquad \lambda = \frac{\varepsilon_m}{\varepsilon_p}$$

$$\begin{pmatrix} b_m \\ y_m \\ L_m \\ \varepsilon_m \end{pmatrix} := \mathrm{Find}(b_m, y_m, L_m, \varepsilon_m) = \begin{pmatrix} 1.25 \\ 0.75 \\ 425 \\ 2.5 \times 10^{-3} \end{pmatrix}\,\mathrm{ft}$$

$$\mathrm{slug} := 1\,\mathrm{lb}\,\frac{\mathrm{sec}^2}{\mathrm{ft}} \qquad\qquad \rho_m := 1.936\,\frac{\mathrm{slug}}{\mathrm{ft}^3} \qquad\qquad \mu_m := 20.5 \times 10^{-6}\,\mathrm{lb}\,\frac{\mathrm{sec}}{\mathrm{ft}^2}$$

$$v_m := 60\,\frac{\mathrm{ft}}{\mathrm{sec}} \qquad\qquad R_m := \frac{\rho_m \cdot v_m \cdot y_m}{\mu_m} = 4.25 \times 10^6$$

$$A_m := b_m \cdot y_m = 0.938\,\mathrm{ft}^2 \quad P_m := 2 \cdot y_m + b_m = 2.75\,\mathrm{ft} \quad R_{hm} := \frac{A_m}{P_m} = 0.341\,\mathrm{ft}$$

Guess value: $\qquad\qquad h_{fm} := 1\,\mathrm{ft} \qquad\qquad\qquad S_{fm} := 0.01\,\frac{\mathrm{ft}}{\mathrm{ft}}$

Given

$$h_{fm} = \left(\frac{v_m\,n_m}{R_{hm}^{\frac{2}{3}}}\right)^2 L_m \qquad\qquad\qquad S_{fm} = \frac{h_{fm}}{L_m}$$

$$\begin{pmatrix} h_{fm} \\ S_{fm} \end{pmatrix} := \mathrm{Find}(h_{fm}, S_{fm})$$

$$h_{fm} = 418.998\,\mathrm{ft} \qquad\qquad\qquad\qquad S_{fm} = 0.986\,\frac{\mathrm{ft}}{\mathrm{ft}}$$

It is important to note that for uniform flow, the channel bottom slope, S_o is equal to the friction slope, S_f. Furthermore, in order to satisfy the geometric similarity requirement when applying the model scale (and the dynamic similarity requirement), the prototype channel bottom slope, S_{op} must equal the model channel bottom slope, S_{om}; thus, the prototype friction slope, S_{fp}, must equal the model friction slope, S_{fm} (see (b), (c) below).

(b)–(c) To determine the velocity flow of the water in the prototype open channel flow in order to achieve dynamic similarity between the model and the prototype for turbulent open channel flow, and to determine the friction slope (and head loss) in the flow of the water in the prototype in order to achieve dynamic similarity between the model and the prototype, for turbulent open channel flow, the ε/y must remain a constant between the model and prototype as follows:

$$\left(\frac{\varepsilon}{y}\right)_p = \left(\frac{\varepsilon}{y}\right)_m$$

$$\frac{\varepsilon_p}{y_p} = 3.333 \times 10^{-3} \qquad \frac{\varepsilon_m}{y_m} = 3.333 \times 10^{-3}$$

However, because the Manning's roughness coefficient, n is independent of R, R does not need to remain a constant between the model and the prototype.

(b)–(c) To determine the velocity flow of the water in the prototype open channel flow in order to achieve dynamic similarity between the model and the prototype for turbulent open channel flow, and to determine the friction slope (and head loss) in the flow of the water in the prototype, in order to achieve dynamic similarity between the model and the prototype for turbulent open channel flow, the Manning's roughness coefficient, n must remain a constant between the model and the prototype (which is a direct result of maintaining a constant ε/y between the model and the prototype, and applying the "gravity model" similitude scale ratio; specifically the velocity ratio, v_r given in Table 11.2) as follows:

$$\underbrace{\left[\frac{h_f}{\frac{v^2 L}{R_h^{4/3}}}\right]_p}_{n_p^2} = \underbrace{\left[\frac{h_f}{\frac{v^2 L}{R_h^{4/3}}}\right]_m}_{n_m^2}$$

$$v_r = \frac{v_p}{v_m} = \frac{(\sqrt{gL})_p}{(\sqrt{gL})_m} = L_r^{\frac{1}{2}}$$

$$\rho_p := 1.936 \frac{slug}{ft^3} \qquad \mu_p := 20.5 \times 10^{-6}\, lb\, \frac{sec}{ft^2} \qquad g := 32.174 \frac{ft}{sec^2}$$

$$A_p := b_p \cdot y_p = 15\, ft^2 \qquad P_p := 2 \cdot y_p + b_p = 11\, ft \qquad R_{hp} := \frac{A_p}{P_p} = 1.364\, ft$$

Guess value: $\qquad v_p := 1 \frac{ft}{sec} \quad h_{fp} := 1\, ft \quad S_{fp} := 0.01 \frac{ft}{ft}$

$$n_p := 0.01\, ft^{\frac{-1}{3}}\, sec$$

Given

$$n_p^2 = \frac{h_{fp}}{\left(\dfrac{v_p^2 \cdot L_p}{R_{hp}^{\frac{4}{3}}}\right)} \qquad \frac{v_p}{\sqrt{g \cdot y_p}} = \frac{v_m}{\sqrt{g \cdot y_m}}$$

$$n_p = n_m \qquad\qquad S_{fp} = \frac{h_{fp}}{L_p} \qquad\qquad S_{fp} = S_{fm}$$

$$\begin{pmatrix} v_p \\ h_{fp} \\ S_{fp} \\ n_p \end{pmatrix} := \text{Find}\,(v_p, h_{fp}, S_{fp}, n_p)$$

$$v_p = 120\,\frac{ft}{s} \qquad h_{fp} = 1.676 \times 10^3 \, ft \qquad S_{fp} = 0.986\,\frac{ft}{ft}$$

$$n_p = 8.077 \times 10^{-3}\,\frac{s}{ft^{0.333}}$$

Furthermore, the Froude number, F remains a constant between the model and the prototype as follows:

$$F_m := \frac{v_m}{\sqrt{g \cdot y_m}} = 12.214 \qquad\qquad F_p := \frac{v_p}{\sqrt{g \cdot y_p}} = 12.214$$

Therefore, although the similarity requirements regarding the independent π term, ε/y $((\varepsilon/y)_p = (\varepsilon/y)_m = 3.333 \times 10^{-3})$, the dependent π term, F ("gravity model") $(F_p = F_m = 12.214)$, and the dependent π term, friction slope, S_f $(S_{fp} = S_{fm} = 0.986\,\text{ft/ft})$ are theoretically satisfied, the dependent π term (i.e., the friction factor, f) will actually/practically remain a constant between the model and its prototype $(n_p = n_m = 8.076 \times 10^{-3}\,\text{ft}^{-1/3}\,\text{sec})$ only if it is practical to maintain/attain the model velocity, slope, fluid, scale, and cost. Furthermore, because the Manning's roughness coefficient, n is independent of R, R does not need to remain a constant between the model and the prototype as follows:

$$R_m = 4.25 \times 10^6 \qquad\qquad R_p := \frac{\rho_p \cdot v_p \cdot y_p}{\mu_p} = 3.4 \times 10^7$$

11.7.5 Application of the Similitude Scale Ratios for the Minor Head Loss in Pipe Flow

For the case of internal pipe flow with pipe components/transitions or devices such as valves, fittings (tees, unions, elbows, and bends), entrances, exits, contractions, and expansions, dimensional analysis yielded the following uncalibrated mathematical model of the fluid flow situation:

$$\frac{\Delta p}{\dfrac{\rho v^2}{2}} = \frac{F_P}{F_I} = \frac{2}{E} = C_D = k = \phi\left(R,\,\frac{\varepsilon}{D},\,\frac{L_i}{L}\right) \tag{11.147}$$

where the resulting minor head loss, h_f was given as follows:

$$h_{f,\min} = \frac{\Delta p}{\gamma} = k\frac{v^2}{2g} \tag{11.148}$$

where the drag coefficient, C_D (or the minor loss coefficient, k) as modeled by the Euler number, E, is the dependent π term, as it models the dependent variable, Δp that cannot be derived analytically from theory, while the Reynolds number, R; the relative surface roughness, ε/D; and the geometry (design details, L_i/L) of the pipe device are the independent π terms. However, because the flow in pipes with devices is mostly turbulent, the Reynolds number, R is usually so large that the drag coefficient, C_D is usually independent of R.

One may note that because the drag coefficient, C_D is mainly a function of the Euler number, E, the inertia force, F_I and the pressure force, F_P will always be important (predominate) in internal pipe flow. Thus, the Euler number, E and the drag coefficient, $C_D = k = (2/E)$ must remain a constant between the model and its prototype. However, depending upon the specifics of the internal pipe flow with pipe components situation, the relative surface roughness, ε/D, the geometry of the pipe device, L_i/L and the remaining force, which includes the viscous force, F_V (and thus its respective independent π term, R) may or may not play an important role in the definition of the drag coefficient, C_D.

11.7.5.1 Turbulent Pipe Flow with Pipe Component

In the case of turbulent flow, the flow resistance (as modeled by C_D or k) is a function of ε/D, and the geometry of the pipe device, L_i/L, and is independent of R. As such, in addition to the inertia force, F_I and the pressure force, F_P, the relative surface roughness, ε/D and the geometry, L_i/L are important in a pipe flow device. Therefore, because there are two independent π terms, ε/D and L_i/L, used in the definition of the dependent π term, E and thus the drag coefficient, C_D, the appropriate similitude scale ratio corresponding to the independent π term is used to define the similarity requirements between the model and its prototype. Specifically, the criteria governing dynamic similarity require the following:

$$\left(\frac{\varepsilon}{D}\right)_p = \left(\frac{\varepsilon}{D}\right)_m \tag{11.149}$$

$$\left(\frac{L_i}{L}\right)_p = \left(\frac{L_i}{L}\right)_m \tag{11.150}$$

which states that there is a geometric similarity requirement for the relative surface roughness, ε/D and the geometry of the pipe device, L_i/L in order for the dependent π term (i.e., the drag coefficient, C_D) to remain a constant between the model and its prototype, where the flow resistance prediction equation is defined as follows:

$$\underbrace{\left[\frac{h_f}{\frac{v^2}{2g}}\right]_p}_{C_{Dp}=k_p} = \underbrace{\left[\frac{h_f}{\frac{v^2}{2g}}\right]_m}_{C_{Dm}=k_m} \tag{11.151}$$

Furthermore, one may note that in the case where the similarity requirements regarding the independent π terms, ε/D and L_i/L (geometric similarity) are satisfied, then the dependent π term (i.e., the drag coefficient, C_D) will remain a constant between the model and its prototype. And, therefore, application of the flow resistance prediction equation yields a "true model." However, in the case where the similarity requirements regarding the independent π terms ε/D and L_i/L are not satisfied, then the dependent π term (i.e., the minor loss coefficient, k) will not remain a constant between the model and its prototype. And, therefore, application of the flow resistance prediction equation yields a "distorted model" for which the appropriate corrections measures are taken.

EXAMPLE PROBLEM 11.14

Air at 68°F flows in a prototype circular pipe with a diameter of 4 ft and an absolute pipe roughness of 0.008 ft, which is fitted with a 90° bend with a round radius, r of 8 ft, as illustrated in Figure EP 11.14. A smaller model of the larger prototype is designed in order to study the flow characteristics of turbulent pipe flow with a pipe component. The model fluid is water at 70°F, the velocity of water in the smaller model pipe is 70 ft/sec, and the model scale, λ is 0.25. The flow resistance is modeled by the minor loss coefficient, $k = C_D$. (a) Determine the pressure drop (and head loss) in the flow of the water in the model. (b) Determine the velocity flow of the air in the prototype pipe flow in order to achieve dynamic similarity between the model and the prototype. (c) Determine the pressure drop (and head loss) in the flow of the air in the prototype in order to achieve dynamic similarity between the model and the prototype.

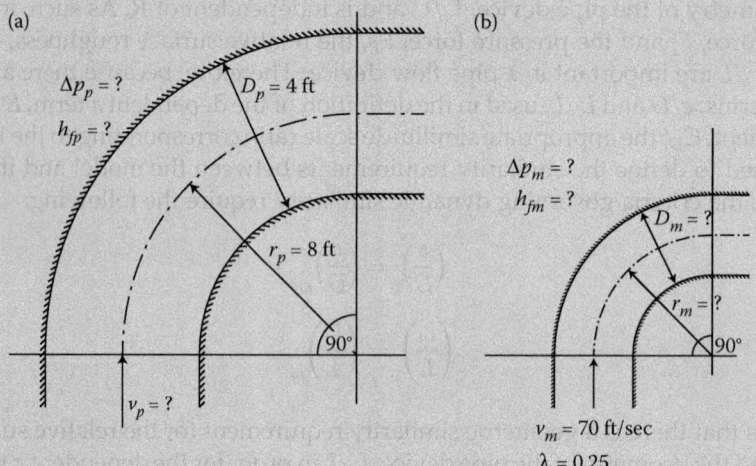

FIGURE EP 11.14
(a) Air flows in a prototype pipe fitted with a 90° bend. (b) Water flows in a smaller model pipe fitted with a 90° bend.

Mathcad Solution

(a) In order to determine the pressure drop (and head loss) in flow of the water in the model, the major head loss equation, Equation 11.148, is applied as follows:

$$h_{f,\,min} = \frac{\Delta p}{\gamma} = k\frac{v^2}{2g}$$

where the minor loss coefficient, k is used to model the flow resistance. Because the Reynolds number, $R > 4000$ (turbulent pipe flow with a pipe component) is assumed, the minor loss coefficient, k is only a function of ε/D and the geometry of the bend, and is independent of R, as illustrated in Figure 8.9. Furthermore, in order to determine the diameter and the absolute pipe roughness of the model pipe and the round radius of the $90°$ model bend, the model scale, λ (inverse of the length ratio) is applied. The fluid properties for water are given in Table A.2 in Appendix A.

$D_p := 4 \, \text{ft}$ $\qquad$ $r_p := 8 \, \text{ft}$ $\qquad$ $\varepsilon_p := 0.008 \, \text{ft}$ $\qquad$ $\lambda := 0.25$

Guess value: $\qquad$ $D_m := 0.1 \, \text{ft}$ $\qquad$ $r_m := 1 \, \text{ft}$ $\qquad$ $\varepsilon_m := 0.01 \, \text{ft}$

Given

$$\lambda = \frac{D_m}{D_p} \qquad\qquad \lambda = \frac{r_m}{r_p} \qquad\qquad \lambda = \frac{\varepsilon_m}{\varepsilon_p}$$

$$\begin{pmatrix} D_m \\ r_m \\ \varepsilon_m \end{pmatrix} := \text{Find}\,(D_m, r_m, \varepsilon_m) = \begin{pmatrix} 1 \\ 2 \\ 2 \times 10^{-3} \end{pmatrix} \text{ft}$$

Thus, the minor loss coefficient, k for the model is determined from Figure 8.9.

$$\frac{r_m}{D_m} = 2 \qquad\qquad \frac{\varepsilon_m}{D_m} = 2 \times 10^{-3} \qquad\qquad k_m := 0.35$$

$$\text{slug} := 1 \, \text{lb}\,\frac{\sec^2}{\text{ft}} \qquad \rho_m := 1.936\,\frac{\text{slug}}{\text{ft}^3} \qquad\qquad \mu_m := 20.5 \times 10^{-6}\,\text{lb}\,\frac{\sec}{\text{ft}^2}$$

$$g := 32.174\,\frac{\text{ft}}{\sec^2} \qquad \gamma_m := \rho_m \cdot g = 62.289\,\frac{\text{lb}}{\text{ft}^3} \qquad v_m := 70\,\frac{\text{ft}}{\sec}$$

$$R_m := \frac{\rho_m \cdot v_m \cdot D_m}{\mu_m} = 6.611 \times 10^6$$

Guess value: $\qquad$ $h_{fm} := 1 \, \text{ft}$ $\qquad\qquad$ $\Delta p_m := 1\,\frac{\text{lb}}{\text{ft}^2}$

Given

$$h_{fm} = k_m \frac{v_m^2}{2\,g}$$

$$\Delta p_m = h_{fm} \cdot \gamma_m$$

$$\begin{pmatrix} h_{fm} \\ \Delta p_m \end{pmatrix} := \text{Find}\,(h_{fm}, \Delta p_m)$$

$h_{fm} = 26.652 \, \text{ft}$ $\qquad$ $\Delta p_m = 1.66 \times 10^3\,\dfrac{\text{lb}}{\text{ft}^2}$

(b)–(c) To determine the velocity flow of the air in the prototype pipe flow in order to achieve dynamic similarity between the model and the prototype for turbulent pipe flow with a pipe component flow, and to determine the pressure drop (and head loss) in the flow of the air in the prototype in order to achieve dynamic similarity between the model and the prototype for turbulent pipe flow with a pipe component, the geometry, L_i/L must remain a constant between the model and prototype as follows:

$$\left(\frac{L_i}{L}\right)_p = \left(\frac{L_i}{L}\right)_m$$

$$\frac{r_p}{D_p} = 2 \qquad\qquad\qquad\qquad \frac{r_m}{D_m} = 2$$

And, the ε/D must remain a constant between the model and prototype as follows:

$$\left(\frac{\varepsilon}{D}\right)_p = \left(\frac{\varepsilon}{D}\right)_m$$

$$\frac{\varepsilon_p}{D_p} = 2 \times 10^{-3} \qquad\qquad\qquad\qquad \frac{\varepsilon_m}{D_m} = 2 \times 10^{-3}$$

However, because the minor loss coefficient, k is independent of R for turbulent flow with a pipe component, R does not need to remain a constant between the model and the prototype.

(b)–(c) To determine the velocity flow of the air in the prototype pipe flow in order to achieve dynamic similarity between the model and the prototype for turbulent pipe flow with a pipe component, and to determine the pressure drop (and head loss) in the flow of the air in the prototype in order to achieve dynamic similarity between the model and the prototype for turbulent pipe flow with a pipe component, the friction factor, f must remain a constant between the model and the prototype (which is a direct result of maintaining a constant ε/D and a constant L_i/L between the model and the prototype and applying the "pressure model" similitude scale ratio; specifically the velocity ratio, v_r given in Table 11.1) as follows:

$$\underbrace{\left[\frac{h_f}{\dfrac{v^2}{2g}}\right]_p}_{C_{D_p}=k_p} = \underbrace{\left[\frac{h_f}{\dfrac{v^2}{2g}}\right]_m}_{C_{D_m}=k_m}$$

$$v_r = \frac{v_p}{v_m} = \frac{\left(\sqrt{\dfrac{\Delta p}{\rho}}\right)_p}{\left(\sqrt{\dfrac{\Delta p}{\rho}}\right)_m} = \Delta p_r^{\frac{1}{2}} \rho_r^{\frac{-1}{2}}$$

The fluid properties for air are given in Table A.5 in Appendix A.

$$\rho_p := 0.00231 \frac{slug}{ft^3} \qquad \mu_p := 0.376 \times 10^{-6} \, lb \frac{sec}{ft^2} \qquad \gamma_p := \rho_p \cdot g = 0.074 \frac{lb}{ft^3}$$

Guess value: $\qquad\qquad v_p := 1 \frac{ft}{sec} \qquad h_{fp} := 1 \, ft \qquad \Delta p_p := 1 \frac{lb}{ft^2} \qquad k_p := 0.01$

Given

$$k_p = \frac{h_{fp}}{\left(\dfrac{v_p^2}{2 \cdot g}\right)} \qquad \frac{v_p}{\Delta p_p^{\frac{1}{2}} \cdot \rho_p^{\frac{-1}{2}}} = \frac{v_m}{\Delta p_m^{\frac{1}{2}} \cdot \rho_m^{\frac{-1}{2}}}$$

$$k_p = k_m \qquad \Delta p_p = h_{fp} \cdot \gamma_p$$

$$\begin{pmatrix} v_p \\ h_{fp} \\ \Delta p_p \\ k_p \end{pmatrix} := \text{Find}\,(v_p, h_{fp}, \Delta p_p, k_p)$$

$$v_p = 89.476\,\frac{ft}{s} \qquad h_{fp} = 43.546\,ft \qquad \Delta p_p = 3.236\,\frac{lb}{ft^2} \qquad k_p = 0.35$$

Furthermore, the Euler number, E remains a constant between the model and the prototype as follows:

$$E_m := \frac{\rho_m \cdot v_m^2}{\Delta p_m} = 5.714 \qquad\qquad E_p := \frac{\rho_p \cdot v_p^2}{\Delta p_p} = 5.714$$

Therefore, although the similarity requirements regarding the independent π term, ε/D $((\varepsilon/D)_p = (\varepsilon/D)_m = 0.002)$; the independent π term, L_i/L $(r_p/D_p = r_m/D_m = 2)$; and the dependent π term, E ("pressure model") $(E_p = E_m = 5.714)$ are theoretically satisfied, the dependent π term (i.e., the minor loss coefficient, k) will actually/practically remain a constant between the model and its prototype $(k_p = k_m = 0.35)$ only if it is practical to maintain/attain the model velocity, pressure, fluid, scale, and cost. Furthermore, because the minor loss coefficient, k is independent of R for turbulent flow with a pipe component, R does not need to remain a constant between the model and the prototype as follows:

$$R_m = 6.611 \times 10^6 \qquad\qquad R_p := \frac{\rho_p \cdot v_p \cdot D_p}{\mu_p} = 2.199 \times 10^6$$

11.7.6 Application of the Similitude Scale Ratios for the Actual Discharge in Pipe Flow

For the case of the actual discharge in pipe flow (flow-measuring devices such as orifice, nozzle, or venturi meters), dimensional analysis yielded the following uncalibrated mathematical model of the fluid flow situation:

$$\frac{Q_a}{Q_i} = \frac{F_I}{F_P} = \frac{v_a A_a}{v_i A_i} = \frac{v_a}{\sqrt{\dfrac{2\Delta p}{\rho}}}\,\frac{A_a}{A_i} = \sqrt{\frac{E}{2}}\,\frac{A_a}{A_i} = C_v C_c = C_d = \phi\left(R, \frac{L_i}{L}\right) \qquad (11.152)$$

where the resulting actual discharge, Q_a was given as follows:

$$Q_a = \underbrace{\left[\underbrace{\left(\frac{v_a}{\sqrt{\frac{2\Delta p}{\rho}}}\right)}_{C_v} \underbrace{\left(\frac{A_a}{A_i}\right)}_{C_c}\right]}_{C_d} Q_i \tag{11.153}$$

where the drag coefficient, C_D (or the discharge coefficient, C_d) as modeled by the Euler number, E is the dependent π term, as it models the dependent variable, Δp that cannot be derived analytically from theory, while the Reynolds number, R and the geometry, L_i/L of the flow-measuring device are the independent π terms. The definition of the drag coefficient, C_D (discharge coefficient, C_d) is mainly a function of E. The discharge coefficient, C_d is a function of geometry of the flow-measuring device, L_i/L; the Reynolds number, R; the location of the pressure taps; and the conditions of the upstream and downstream flow in the pipe. For low to moderate values of the Reynolds number, R where the viscous effects are significant, the experimentally determined discharge coefficient, C_d models both the frictional effects (C_v) and the contraction effects (C_c), whereas for relatively high values of the Reynolds number, R, where the viscous effects are insignificant, the experimentally determined discharge coefficient, C_d models mostly the contraction effects (C_c). Thus, on-site calibration is usually recommended in addition to the manufacturer's data. One may note that for real flow, the pressure drop, $\Delta p = p_1 - p_2$ is a function of the flowrate, Q; the density, ρ; the areas, A_1 and A_2; and the fluid viscosity, μ, which is modeled by the discharge coefficient, C_d.

One may note that because the drag coefficient, C_D is mainly a function of the Euler number, E, the inertia force, F_I and the pressure force, F_P will always be important (predominate) in a flow-measuring device for pipe flow. Thus, the Euler number, E and the drag coefficient, $C_D = C_d = \sqrt{(E/2)}(A_a/A_i)$ must remain a constant between the model and its prototype. However, depending upon the specifics of the flow-measuring device for pipe flow situation, the geometry of the flow-measuring device, and the remaining force, which includes the viscous force, F_V (and thus, its respective independent π term, R) may or may not play an important role in the definition of the drag coefficient, C_D.

11.7.6.1 Pipe Flow with a Flow-Measuring Device

In the typical case, the flow resistance (as modeled by C_D or C_d) is a function of both the geometry of the flow-measuring device, L_i/L and R. As such, in addition to the inertia force, F_I and the pressure force, F_P, the viscous force, F_V and the geometry L_i/L are important in a flow-measuring device in pipe flow. Therefore, because there are two independent π terms, R and L_i/L, used in the definition of the dependent π term, E and thus the drag coefficient, C_D, the appropriate similitude scale ratios corresponding to the independent π terms are used to define the similarity requirements between the model and its prototype. Specifically, the criteria governing dynamic similarity require the following:

$$\underbrace{\left[\left(\frac{\rho v L}{\mu}\right)_p\right]}_{R_p} = \underbrace{\left[\left(\frac{\rho v L}{\mu}\right)_m\right]}_{R_m} \tag{11.154}$$

$$\left(\frac{L_i}{L}\right)_p = \left(\frac{L_i}{L}\right)_m \tag{11.155}$$

which states that in addition to the geometric similarity requirement for the geometry, L_i/L, there is a dynamic similarity requirement that R remains a constant in order for the dependent π term (i.e., the drag coefficient, C_D) to remain a constant between the model and its prototype, where the flow resistance prediction equation is defined as follows:

$$\underbrace{\left[\frac{Q_a}{\left[\sqrt{\frac{2\Delta p}{\rho}}\right][A_i]}\right]_p}_{C_{Dp}=C_{dp}} = \underbrace{\left[\frac{Q_a}{\left[\sqrt{\frac{2\Delta p}{\rho}}\right][A_i]}\right]_m}_{C_{Dm}=C_{dm}} \tag{11.156}$$

Furthermore, one may note that in the case where the similarity requirements regarding the independent π terms, R ("viscosity model") and L_i/L (geometric similarity) are satisfied, then the dependent π term (i.e., the drag coefficient, C_D) will remain a constant between the model and its prototype. And, therefore, application of the flow resistance prediction equation yields a "true model." However, in the case where the similarity requirements regarding the independent π terms, R and L_i/L, are not satisfied, then the dependent π term (i.e., the drag coefficient, C_D) will not remain a constant between the model and its prototype. And, therefore, application of the flow resistance prediction equation yields a "distorted model" for which the appropriate corrections measures are taken.

EXAMPLE PROBLEM 11.15

Water at 70°F flows in a prototype circular pipe with a diameter of 6 ft and is fitted with an orifice meter with a diameter of 4.2 ft, as illustrated in Figure EP 11.15. A smaller model of the larger prototype is designed in order to study the flow characteristics of pipe flow in the flow-measuring device. The model fluid is crude oil at 68°F ($s = 0.86$, $\mu = 150 \times 10^{-6}$ lb-sec/ft^2), the ideal velocity (see Equation 8.140) of flow of the oil in the smaller model pipe fitted with an orifice is 2 ft/sec, and the model scale, λ is 0.25. (a) Determine the actual discharge in the model orifice meter. (b) Determine the ideal velocity of flow of the water in the prototype orifice meter in order to achieve dynamic similarity between the model and the prototype. (c) Determine the actual discharge in the prototype orifice meter in order to achieve dynamic similarity between the model and the prototype.

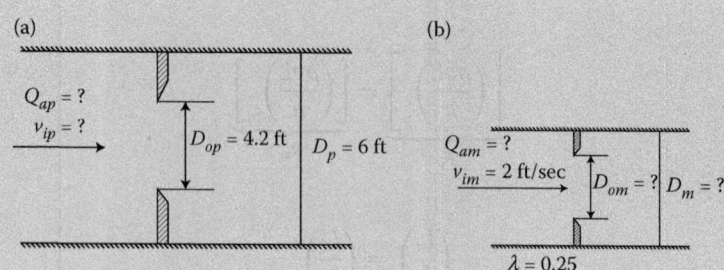

FIGURE EP 11.15

(a) Water flows in a prototype pipe fitted with an orifice meter. (b) Crude oil flows in a smaller model pipe fitted with an orifice meter.

Mathcad Solution

(a) In order to determine the actual discharge in the model orifice meter, the actual discharge equation, Equation 11.153, is applied as follows:

$$Q_a = \left[\underbrace{\left(\underbrace{\frac{v_a}{\sqrt{\frac{2\Delta p}{\rho}}}}_{C_v} \right) \underbrace{\left(\frac{A_a}{A_i} \right)}_{C_c}}_{C_d} \right] Q_i$$

where the C_D or C_d, or, in the case of an orifice meter, the orifice discharge coefficient, C_o, is used to model the flow resistance and is a function of both the geometry of the flow-measuring device, L_i/L and R, as illustrated in Figure 8.29. Furthermore, in order to determine the geometry D and D_o of the model pipe and orifice meter, the model scale, λ (inverse of the length ratio) is applied as follows:

$D_p := 6 \, ft$ $D_{op} := 4.2 \, ft$ $\lambda := 0.25$

Guess value: $D_m := 1 \, ft$ $D_{om} := 0.1 \, ft$

Given

$$\lambda = \frac{D_m}{D_p} \qquad\qquad \lambda = \frac{D_{om}}{D_{op}}$$

$$\begin{pmatrix} D_m \\ D_{om} \end{pmatrix} := \text{Find} \, (D_m, D_{om}) = \begin{pmatrix} 1.5 \\ 1.05 \end{pmatrix} ft$$

$slug := 1 \, lb \dfrac{sec^2}{ft}$ $s_m := 0.86$ $\rho_w := 1.94 \dfrac{slug}{ft^3}$

$\rho_m := s_m \cdot \rho_w = 1.668 \dfrac{slug}{ft^3}$ $\mu_m := 150 \times 10^{-6} \, lb \dfrac{sec}{ft^2}$

$A_{om} := \dfrac{\pi \cdot D_{om}^2}{4} = 0.866 \, ft^2$ $A_m := \dfrac{\pi \cdot D_m^2}{4} = 1.767 \, ft^2$ $v_{im} := 2 \dfrac{ft}{sec}$

$$R_m := \frac{\rho_m \cdot v_m \cdot D_m}{\mu_m} = 3.337 \times 10^4 \qquad \frac{D_{om}}{D_m} = 0.7 \qquad C_{om} := 0.63$$

$$Q_{im} := v_{im} \cdot A_{om} = 1.732 \frac{ft^3}{sec} \qquad Q_{am} := C_{om} \cdot Q_{im} = 1.091 \frac{ft^3}{sec}$$

(b) To determine the ideal velocity of flow of the water in the prototype orifice meter in order to achieve dynamic similarity between the model and the prototype for pipe flow in the flow-measuring device, the geometry L_i/L must remain a constant between the model and prototype as follows:

$$\left(\frac{L_i}{L}\right)_p = \left(\frac{L_i}{L}\right)_m$$

where the geometry is modeled as follows:

$$\frac{D_{op}}{D_p} = 0.7 \qquad\qquad \frac{D_{om}}{D_m} = 0.7$$

Furthermore, the **R** must remain a constant between the model and prototype as follows:

$$\underbrace{\left[\left(\frac{\rho v L}{\mu}\right)_p\right]}_{R_p} = \underbrace{\left[\left(\frac{\rho v L}{\mu}\right)_m\right]}_{R_m}$$

$$\rho_p := 1.936 \frac{slug}{ft^3} \qquad\qquad \mu_p := 20.50 \times 10^{-6} \, lb \frac{sec}{ft^2}$$

Guess value: $\qquad\qquad v_{ip} := 10 \frac{ft}{sec} \qquad\qquad R_p := 1000$

Given

$$R_p = \frac{\rho_p \cdot v_{ip} \cdot D_p}{\mu_p}$$

$$R_p = R_m$$

$$\binom{v_{ip}}{R_p} := Find\,(v_{ip}, R_p)$$

$$v_{ip} = 0.059 \frac{ft}{s} \qquad\qquad\qquad\qquad R_p = 3.337 \times 10^4$$

(c) To determine the actual discharge in the prototype orifice meter in order to achieve dynamic similarity between the model and the prototype for pipe flow in the flow-measuring device, the orifice discharge coefficient, C_o must remain a constant between the model and the prototype (which is a direct result of maintaining a constant **R** and a constant L_i/L between the model and the prototype)

as follows:

$$\underbrace{\left[\frac{Q_a}{\left[\sqrt{\frac{2\Delta p}{\rho}}\right][A_i]}\right]_p}_{C_{Dp}=C_{dp}} = \underbrace{\left[\frac{Q_a}{\left[\sqrt{\frac{2\Delta p}{\rho}}\right][A_i]}\right]_m}_{C_{Dm}=C_{dm}}$$

$$A_{op} := \frac{\pi \cdot D_{op}^2}{4} = 13.854\,\text{ft}^2 \qquad\qquad A_p := \frac{\pi \cdot D_p^2}{4} = 28.274\,\text{ft}^2$$

Guess value: $Q_{ap} := 1\dfrac{\text{ft}^3}{\text{sec}}$ $Q_{ip} := 1\dfrac{\text{ft}^3}{\text{sec}}$ $C_{op} := 0.5$

Given

$$C_{op} = \frac{Q_{ap}}{Q_{ip}} \qquad\qquad Q_{ip} = v_{ip} \cdot A_{op}$$

$$C_{op} = C_{om}$$

$$\begin{pmatrix} Q_{ap} \\ Q_{ip} \\ C_{op} \end{pmatrix} := \text{Find}\,(Q_{ap}, Q_{ip}, C_{op})$$

$$Q_{ap} = 0.514\frac{\text{ft}^3}{\text{sec}} \qquad\qquad Q_{ip} = 0.816\frac{\text{ft}^3}{\text{sec}} \qquad\qquad C_{op} = 0.63$$

Therefore, although the similarity requirements regarding the independent π term, L_i/L ($D_{op}/D_p = D_{om}/D_m = 0.7$) and the independent π term, R ("viscosity model") ($R_p = R_m = 3.337 \times 10^4$) are theoretically satisfied, the dependent π term (i.e., the orifice discharge coefficient, C_o) will actually/practically remain a constant between the model and its prototype ($C_{op} = C_{om} = 0.63$) only if it is practical to maintain/attain the model velocity, pressure, fluid, scale, and cost.

11.7.7 Application of the Similitude Scale Ratios for the Actual Discharge in Open Channel Flow

For the case of the actual discharge in open channel flow (i.e., flow-measuring devices such as sluice gates, weirs, spillways, venturi flumes, and contracted openings), dimensional analysis yielded the following uncalibrated mathematical model of the fluid flow situation:

$$\frac{Q_a}{Q_i} = \frac{F_{dynpres}}{F_P} = \frac{v_a A_a}{v_i A_i} = \frac{v_a}{\sqrt{2g\Delta y}}\frac{A_a}{A_i} = \sqrt{\frac{E}{2}}\frac{A_a}{A_i} = \sqrt{\frac{F}{2}}\frac{A_a}{A_i} = C_v C_c = C_d = \phi\left(R, W, \frac{L_i}{L}\right) \quad (11.157)$$

where the resulting actual discharge, Q_a was given as follows:

$$Q_a = \underbrace{\left[\underbrace{\left(\frac{v_a}{\sqrt{2g\Delta y}}\right)}_{C_v} \underbrace{\left(\frac{A_a}{A_i}\right)}_{C_c}\right]}_{C_d} Q_i \qquad (11.158)$$

where the drag coefficient, C_D (or the discharge coefficient, C_d) as modeled by the Euler number, $E =$ the Froude number, F is the dependent π term, as it models the dependent variable, Δy or S_f that cannot be derived analytically from theory, while the Reynolds number, R; the Weber number, W; and the geometry of the flow-measuring device, L_i/L are the independent π terms.

One may note that because the drag coefficient, C_D is mainly a function of the Euler number, E and the Froude number, F, where $E = F$, the inertia force, F_I and both the pressure force, F_P and the gravity force, F_G will always be important (predominate) in a flow-measuring device for open channel flow. Thus, the Euler number, E; the Froude number, F; and the drag coefficient, $C_D = C_d = \sqrt{(E/2)}(A_a/A_i) = \sqrt{(F/2)}(A_a/A_i)$ must remain a constant between the model and its prototype. However, depending upon the specifics of the flow-measuring device for open channel flow situation, the geometry of the flow-measuring device, L_i/L and the remaining forces, which includes the viscous force, F_V and the surface tension force, F_T (and thus, their respective independent π terms, R and W), may or may not play an important role in the definition of the drag coefficient, C_D.

11.7.7.1 Open Channel Flow with Sluice Gate or Venturi Meter

In the case of a sluice gate or a venturi flume, the flow resistance (as modeled by C_D or C_d) is a function of the geometry of the flow-measuring device. As such, in addition to the inertia force, F_I; the pressure force, F_P; and the gravity force, F_G, the geometry L/L_i is important in a flow-measuring device in open channel flow. Therefore, because there is one independent π term, L_i/L used in the definition of the dependent π term, $E = F$ and thus the drag coefficient, C_D, the appropriate similitude scale ratio corresponding to the independent π term is used to define the similarity requirements between the model and its prototype. Specifically, the criteria governing dynamic similarity requires the following:

$$\left(\frac{L_i}{L}\right)_p = \left(\frac{L_i}{L}\right)_m \qquad (11.159)$$

which states that there is a geometric similarity requirement for the geometry, L_i/L in order for the dependent π term (i.e., the drag coefficient, C_D) to remain a constant between the model and its prototype, where the flow resistance prediction equation is defined as follows:

$$\underbrace{\left[\frac{Q_a}{\sqrt{2g\Delta y}(A_i)}\right]_p}_{C_{Dp}=C_{dp}} = \underbrace{\left[\frac{Q_a}{\sqrt{2g\Delta y}(A_i)}\right]_m}_{C_{Dm}=C_{dm}} \qquad (11.160)$$

Furthermore, one may note that in the case where the similarity requirements regarding the independent π term, L_i/L (geometric similarity) are satisfied, then the dependent π term (i.e., the drag coefficient, C_D) will remain a constant between the model and its prototype. And, therefore, application of the flow resistance prediction equation yields a "true model." However, in the case where the similarity requirements regarding the independent π term, L/L_i and the dependent π term, F ("gravity model") are not satisfied, then the dependent π term (i.e., the discharge coefficient, C_d) will not remain a constant between the model and its prototype. And, therefore, application of the flow resistance prediction equation yields a "distorted model" for which the appropriate corrections measures are taken. For the following example, a sluice gate with a free outflow will be assumed, where the geometry, $L_1/L = y_1/a$, where L is represented by the opening of the sluice gate, a, and L_1 is represented by the upstream depth, y_1.

EXAMPLE PROBLEM 11.16

Water at 70°F flows in a prototype rectangular open channel with a width of 10 ft, and a sluice gate with a free outflow is inserted in the channel, as illustrated in Figure EP 11.16, in order to measure the flowrate. The opening of the gate is set at 3.28 ft, the depth of flow upstream of the sluice gate is 6.56 ft, and the depth of flow downstream of the sluice gate is 3.11 ft. A smaller model of the larger prototype is designed in order to study the flow characteristics of open channel flow in the flow-measuring device. The model fluid is also water at 70°F, the ideal downstream velocity (see Equation 9.285) of flow of the water, v_{2i} in the smaller model open channel fitted with a sluice gate is 8.46 ft/sec, and the model scale, λ is 0.25. (a) Determine the actual discharge in the model sluice gate. (b) Determine the ideal downstream velocity of flow of the water in the prototype sluice gate in order to achieve dynamic similarity between the model and the prototype. (c) Determine the actual discharge in the prototype sluice gate in order to achieve dynamic similarity between the model and the prototype.

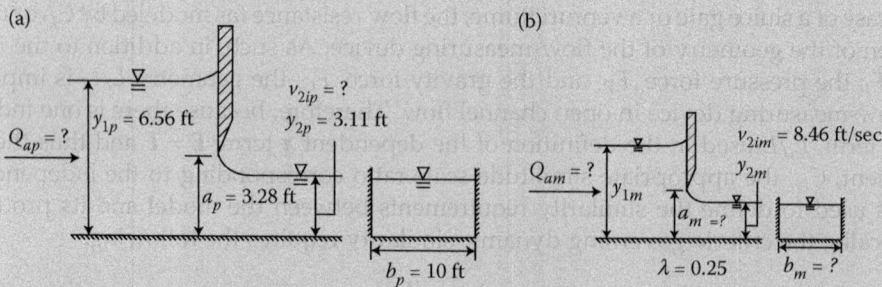

FIGURE EP 11.16
(a) Water flows in a prototype rectangular open channel with a sluice gate. (b) Water flows in a smaller model rectangular open channel with a sluice gate.

Mathcad Solution

(a) In order to determine the actual discharge in the model sluice gate, the actual discharge equation, Equation 11.158, is applied as follows:

$$Q_a = \left[\underbrace{\left(\frac{v_a}{\sqrt{2g\Delta y}} \right)}_{C_v} \underbrace{\left(\frac{A_a}{A_i} \right)}_{C_c} \right] Q_i$$
$$\underbrace{\hphantom{xxxxxxxxxxxxxx}}_{C_d}$$

where the C_D, or in the case of a sluice gate, the discharge coefficient, C_d, is used to model the flow resistance, and is a function of the geometry of the flow-measuring device, $L_1/L = y_1/a$, as illustrated in Figure 9.37. Furthermore, in order to determine the geometry y_1, y_2, and a of the model channel and sluice gate, the model scale, λ (inverse of the length ratio) is applied as follows:

$b_p := 10\,\text{ft}$ $y_{1p} := 6.56\,\text{ft}$ $y_{2p} := 3.11\,\text{ft}$ $a_p := 3.28\,\text{ft}$ $\lambda := 0.25$

Guess value: $b_m := 1\,\text{ft}$ $y_{1m} := 1\,\text{ft}$ $y_{2m} := 0.5\,\text{ft}$ $a_m := 0.5\,\text{ft}$

Given

$$\lambda = \frac{b_m}{b_p} \qquad\qquad \lambda = \frac{y_{1m}}{y_{1p}} \qquad\qquad \lambda = \frac{y_{2m}}{y_{2p}} \qquad\qquad \lambda = \frac{a_m}{a_p}$$

$$\begin{pmatrix} b_m \\ y_{1m} \\ y_{2m} \\ a_m \end{pmatrix} := \text{Find}\,(b_m, y_{1m}, y_{2m}, a_m) = \begin{pmatrix} 2.5 \\ 1.64 \\ 0.778 \\ 0.82 \end{pmatrix}\,\text{ft}$$

$$\text{slug} := 1\,\text{lb}\,\frac{\text{sec}^2}{\text{ft}} \qquad \rho_m := 1.936\,\frac{\text{slug}}{\text{ft}^3} \qquad \mu_m := 20.5 \times 10^{-16}\,\text{lb}\,\frac{\text{sec}}{\text{ft}^2}$$

$$v_{2im} := 8.46\,\frac{\text{ft}}{\text{sec}} \qquad\qquad R_m := \frac{\rho_m \cdot v_{2im} \cdot a_m}{\mu_m} = 6.551 \times 10^5$$

$$A_{2im} := b_m \cdot a_m = 2.05\,\text{ft}^2 \qquad\qquad Q_{im} := v_{2im} \cdot A_{2im} = 17.343\,\frac{\text{ft}^3}{\text{sec}}$$

$$\frac{y_{1m}}{a_m} = 2 \qquad C_{dm} := 0.51 \qquad\qquad Q_{am} := C_{dm} \cdot Q_{im} = 8.845\,\frac{\text{ft}^3}{\text{sec}}$$

(b)–(c) To determine the ideal downstream velocity of flow of the water in the prototype sluice gate in order to achieve dynamic similarity between the model and the prototype, and to determine the actual discharge in the prototype sluice gate in order to achieve dynamic similarity between the model and the prototype, the geometry L_i/L must remain a constant between the model and prototype as follows:

$$\left(\frac{L_i}{L}\right)_p = \left(\frac{L_i}{L}\right)_m$$

where the geometry is modeled as follows:

$$\frac{y_{1p}}{a_p} = 2 \qquad\qquad \frac{y_{1m}}{a_m} = 2 \qquad\qquad \frac{y_{2p}}{a_p} = 0.948 \qquad\qquad \frac{y_{2m}}{a_m} = 0.948$$

However, because the discharge coefficient, C_d is independent of R for a sluice gate, R does not need to remain a constant between the model and the prototype.

(b)–(c) To determine the ideal downstream velocity of flow of the water in the prototype sluice gate in order to achieve dynamic similarity between the model and the prototype, and to determine the actual discharge in the prototype sluice gate in order to achieve dynamic similarity between the model and the prototype, the discharge coefficient, C_d must remain a constant between the model and the prototype (which is a direct result of maintaining a constant L_i/L between the model and the prototype and applying the "gravity model" similitude scale ratio; specifically the velocity ratio, v_r given in Table 11.2) as follows:

$$\underbrace{\left[\frac{Q_a}{\sqrt{2g\Delta y(A_i)}}\right]_p}_{C_{Dp}=C_{dp}} = \underbrace{\left[\frac{Q_a}{\sqrt{2g\Delta y(A_i)}}\right]_m}_{C_{Dm}=C_{dm}}$$

$$v_r = \frac{v_p}{v_m} = \frac{\left(\sqrt{gL}\right)_p}{\left(\sqrt{gL}\right)_m} = L_r^{\frac{1}{2}}$$

$$\rho_p := 1.936\frac{slug}{ft^3} \qquad \mu_p := 20.5 \times 10^{-6}\,lb\frac{sec}{ft^2} \qquad g := 32.174\frac{ft}{sec^2}$$

$$A_{2ip} := b_p \cdot a_p = 32.8\,ft^2$$

Guess value: $\quad v_{2ip} := 1\frac{ft}{sec} \quad Q_{ap} := 1\frac{ft^3}{sec} \quad Q_{ip} := 1\frac{ft^3}{sec} \quad C_{dp} := 0.1$

Given

$$C_{dp} = \frac{Q_{ap}}{Q_{ip}} \qquad\qquad\qquad \frac{v_{2ip}}{\sqrt{g \cdot a_p}} = \frac{v_{2im}}{\sqrt{g \cdot a_m}}$$

$$C_{dp} = C_{dm} \qquad\qquad\qquad Q_{ip} = v_{2ip} \cdot A_{2ip}$$

$$\begin{pmatrix} v_{2ip} \\ Q_{ap} \\ Q_{ip} \\ C_{dp} \end{pmatrix} := Find\,(v_{2ip}, Q_{ap}, Q_{ip}, C_{dp})$$

$$v_{2ip} = 16.92\frac{ft}{s} \quad Q_{ap} = 283.038\frac{ft^3}{sec} \quad Q_{ip} = 554.976\frac{ft^3}{sec} \quad C_{dp} = 0.51$$

Furthermore, the Froude number, *F* remains a constant between the model and the prototype as follows:

$$F_m := \frac{v_{2im}}{\sqrt{g \cdot a_m}} = 1.647 \qquad\qquad F_p := \frac{v_{2ip}}{\sqrt{g \cdot a_p}} = 1.647$$

Therefore, although the similarity requirements regarding the independent π term, L_i/L ($y_{1p}/a_p = y_{1m}/a_m = 2$, and $y_{2p}/a_p = y_{2m}/a_m = 0.948$) and the dependent π term, *F* ("gravity model") ($F_p = F_m = 1.647$) are theoretically satisfied, the dependent π term (i.e., the discharge coefficient, C_d) will actually/practically remain a constant between the model and its prototype ($C_{dp} = C_{dm} = 0.51$) only if it is practical to maintain/attain the model velocity, flow depth, fluid, scale, and cost. Furthermore, because the discharge coefficient, C_d is independent of R for a sluice gate, R does not need to remain a constant between the model and the prototype as follows:

$$R_m = 6.551 \times 10^5 \qquad\qquad R_p := \frac{\rho_p \cdot v_{2ip} \cdot a_p}{\mu_p} = 5.241 \times 10^6$$

11.7.7.2 Open Channel Flow with Weir or Spillway with Large Head

In the case of a weir or a spillway with a large head, $H \approx y_c$ the flow resistance (as modeled by C_D or C_d) is a function of geometry of the weir or spillway, H/P (where P is the height of weir or spillway). The geometry, $L_i/L = H/P$, where L is represented by the P and L_i is represented by H. The boundary geometry of the weir, $(L_i/L) = (H/P)$, is the most important parameter in determining the magnitude of the weir/spillway discharge coefficient, C_d, as this ratio has the greatest influence on the shape of the flow field (flow contraction and velocity profiles). The head of the weir/spillway, H term is a direct measure of the Reynolds number, R, and the Weber number, W, as the effects of viscosity and surface tension are significant only when H is small. The effect of the Weber number, W is negligible except at low heads, where surface tension effects may be significant. And, the effect of the Reynolds number, R is small except at low heads, where viscous effects may be large. Since the flow over a weir/spillway typically involves water, the Reynolds number, R is usually high, where the viscous effects are small. Furthermore, weirs/spillways serve as reliable flowrate measuring devices as long as the head, H is large enough that the effects of viscosity and surface tension are negligible. As such, in addition to the inertia force, F_I; the pressure force, F_P; and the gravity force, F_G, H/P is important in a flow-measuring device in open channel flow. Therefore, because there is one independent π term, H/P used in the definition of the dependent π term, $E = F$ and thus the drag coefficient, C_D, the appropriate similitude scale ratio corresponding to the independent π term are used to define the similarity requirements between the model and its prototype. Specifically, the criteria governing dynamic similarity requires the following:

$$\left(\frac{H}{P}\right)_p = \left(\frac{H}{P}\right)_m \tag{11.161}$$

which states that there is a geometric similarity requirement for the geometry, H/P in order for the dependent π term (i.e., the drag coefficient, C_D) to remain a constant between the model and its prototype, where the flow resistance prediction equation is defined as follows:

$$\underbrace{\left[\frac{Q_a}{\sqrt{2g\,\Delta y(A_i)}}\right]_p}_{C_{Dp}=C_{dp}} = \underbrace{\left[\frac{Q_a}{\sqrt{2g\,\Delta y(A_i)}}\right]_m}_{C_{Dm}=C_{dm}} \tag{11.162}$$

Furthermore, one may note that in the case where the similarity requirements regarding the independent π term, H/P (geometric similarity) are satisfied, then the dependent π term (i.e., the drag coefficient, C_D) will remain a constant between the model and its prototype.

And, therefore, application of the flow resistance prediction equation yields a "true model." However, in the case where the similarity requirements regarding the independent π term, H/P and the dependent π term, F ("gravity model") are not satisfied, then the dependent π term (i.e., the drag coefficient, C_D) will not remain a constant between the model and its prototype. And, therefore, application of the flow resistance prediction equation yields a "distorted model" for which the appropriate corrections measures are taken.

EXAMPLE PROBLEM 11.17

Water at 70°F flows in a prototype rectangular open channel with a width of 10 ft, and a spillway with a large head is inserted in the channel, as illustrated in Figure EP 11.17. The head on the spillway is 2.95 ft, the height of the spillway is 3.9 ft, and the width of the crest of the spillway is 10 ft. A smaller model of the larger prototype is designed in order to study the flow characteristics of open channel flow in the flow-measuring device. The model fluid is also water at 70°F, the ideal discharge (see Equation 9.297) in the smaller model channel fitted with a spillway is 8.5 ft³/sec, and the model scale, λ is 0.25. (a) Determine the actual discharge in the model spillway. (b) Determine the ideal discharge in the prototype spillway in order to achieve dynamic similarity between the model and the prototype. (c) Determine the actual discharge in the prototype spillway in order to achieve dynamic similarity between the model and the prototype.

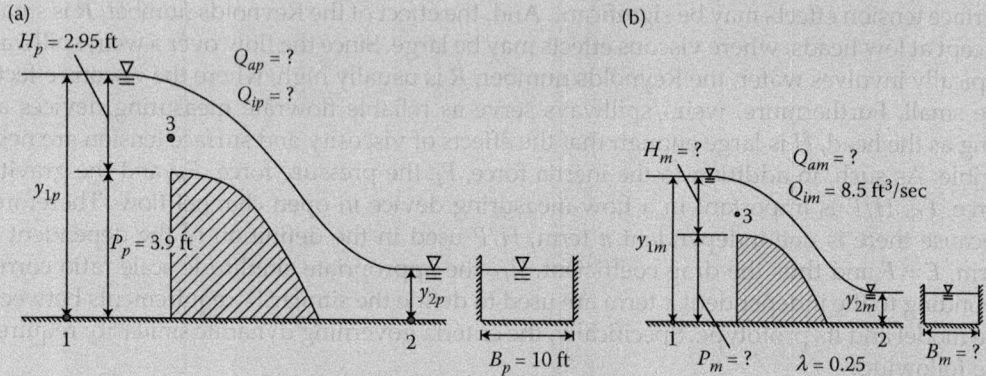

FIGURE EP 11.17
(a) Water flows in a prototype rectangular open channel with a spillway. (b) Water flows in a smaller model rectangular open channel with a spillway.

Mathcad Solution

(a) In order to determine the actual discharge in the model spillway, the actual discharge equation, Equation 11.158, is applied as follows:

$$
Q_a = \underbrace{\left[\underbrace{\left(\frac{v_a}{\sqrt{2g\Delta y}} \right)}_{C_v} \underbrace{\left(\frac{A_a}{A_i} \right)}_{C_c} \right]}_{C_d} Q_i
$$

where the C_D, or in the case of a spillway, the discharge coefficient, C_d, is used to model the flow resistance and is a function of the geometry of the flow-measuring device, $L_i/L = H/P$, as illustrated by the Rehbock formula (see Equation 9.303) as follows:

$$C_d = 0.605 + \frac{1}{305H} + 0.08\frac{H}{P}$$

Furthermore, in order to determine the geometry H, and P of the model channel and spillway, the model scale, λ (inverse of the length ratio) is applied as follows:

$B_p := 10\,\text{ft}$ $\qquad H_p := 2.95\,\text{ft}$ $\qquad P_p := 3.9\,\text{ft}$ $\qquad \lambda := 0.25$

Guess value: $\qquad B_m := 1\,\text{ft}$ $\qquad H_m := 1\,\text{ft}$ $\qquad P_m := 1\,\text{ft}$

Given

$$\lambda = \frac{B_m}{B_p} \qquad\qquad \lambda = \frac{H_m}{H_p} \qquad\qquad\qquad \lambda = \frac{P_m}{P_p}$$

$$\begin{pmatrix} B_m \\ H_m \\ P_m \end{pmatrix} := \text{Find}(B_m, H_m, P_m) = \begin{pmatrix} 2.5 \\ 0.738 \\ 0.975 \end{pmatrix}\text{ft}$$

$\text{slug} := 1\,\text{lb}\dfrac{\text{sec}^2}{\text{ft}}$ $\quad \rho_m := 1.936\dfrac{\text{slug}}{\text{ft}^3}$ $\qquad\qquad \mu_m := 20.5 \times 10^{-6}\,\text{lb}\dfrac{\text{sec}}{\text{ft}^2}$

$Q_{im} := 8.5\dfrac{\text{ft}^3}{\text{sec}}$ $\quad A_{3im} := B_m \cdot H_m = 1.844\,\text{ft}^2$ $\quad v_{3im} := \dfrac{Q_{im}}{A_{3im}} = 4.61\dfrac{\text{ft}}{\text{s}}$

$R_m := \dfrac{\rho_m \cdot v_{3im} \cdot H_m}{\mu_m} = 3.211 \times 10^5$

$C_{dm} := 0.605 + \dfrac{1\,\text{ft}}{305\,H_m} + 0.08\dfrac{H_m}{P_m} = 0.67$ $\qquad Q_{am} := C_{dm} \cdot Q_{im} = 5.695\dfrac{\text{ft}^3}{\text{sec}}$

(b)–(c) To determine the ideal discharge in the prototype spillway in order to achieve dynamic similarity between the model and the prototype, and to determine the actual discharge in the prototype spillway in order to achieve dynamic similarity between the model and the prototype, the geometry L_i/L must remain a constant between the model and prototype as follows:

$$\left(\frac{L_i}{L}\right)_p = \left(\frac{L_i}{L}\right)_m$$

where the geometry is modeled as follows:

$$\frac{H_p}{P_p} = 0.756 \qquad\qquad\qquad \frac{H_m}{P_m} = 0.756$$

However, because the discharge coefficient, C_d is independent of R for a spillway with a large head, R does not need to remain a constant between the model and the prototype.

(b)–(c) To determine the ideal discharge in the prototype spillway in order to achieve dynamic similarity between the model and the prototype, and to determine the actual discharge in the prototype spillway in order to achieve dynamic similarity between the model and the prototype, the discharge coefficient, C_d must remain a constant between the model and the prototype (which is a direct result of maintaining a constant L_i/L between the model and the prototype and applying the "gravity model" similitude scale ratio; specifically the velocity ratio, v_r given in Table 11.2) as follows:

$$\underbrace{\left[\frac{Q_a}{\sqrt{2g\Delta y}(A_i)}\right]_p}_{C_{Dp}=C_{dp}} = \underbrace{\left[\frac{Q_a}{\sqrt{2g\Delta y}(A_i)}\right]_m}_{C_{Dm}=C_{dm}}$$

$$v_r = \frac{v_p}{v_m} = \frac{\left(\sqrt{gL}\right)_p}{\left(\sqrt{gL}\right)_m} = L_r^{\frac{1}{2}}$$

$$\rho_p := 1.936 \frac{slug}{ft^3} \qquad \mu_p := 20.5 \times 10^{-6}\, lb\frac{sec}{ft^2} \qquad g := 32.174 \frac{ft}{sec^2}$$

$$A_{3ip} := B_p \cdot H_p = 29.5\, ft^2$$

Guess value: $\qquad v_{3ip} := 1 \dfrac{ft}{sec} \quad Q_{ap} := 1 \dfrac{ft^3}{sec} \quad Q_{ip} := 1 \dfrac{ft^3}{sec} \quad C_{dp} := 0.1$

Given

$$C_{dp} = \frac{Q_{ap}}{Q_{ip}} \qquad\qquad\qquad \frac{v_{3ip}}{\sqrt{g \cdot H_p}} = \frac{v_{3im}}{\sqrt{g \cdot H_m}}$$

$$C_{dp} = C_{dm} \qquad\qquad\qquad Q_{ip} = v_{3ip} \cdot A_{3ip}$$

$$\begin{pmatrix} v_{3ip} \\ Q_{ap} \\ Q_{ip} \\ C_{dp} \end{pmatrix} := Find\,(v_{3ip},\, Q_{ap},\, Q_{ip},\, C_{dp})$$

$$v_{3ip} = 9.22 \frac{ft}{s} \qquad Q_{ap} = 182.229 \frac{ft^3}{sec} \qquad Q_{ip} = 272 \frac{ft^3}{sec} \qquad C_{dp} = 0.67$$

Furthermore, the Froude number, F remains a constant between the model and the prototype as follows:

$$F_m := \frac{v_{3im}}{\sqrt{g \cdot H_m}} = 0.946 \qquad\qquad F_p := \frac{v_{3ip}}{\sqrt{g \cdot H_p}} = 0.946$$

Therefore, although the similarity requirements regarding the independent π term, L_i/L ($H_p/P_p = H_m/P_m = 0.756$) and the dependent π term, F ("gravity model") $(F_p = F_m = 0.946)$ are theoretically satisfied, the dependent π term (i.e., the discharge coefficient, C_d) will actually/practically remain a constant between the model and its prototype $(C_{dp} = C_{dm} = 0.67)$ only if it is practical to maintain/attain the model velocity, flow depth, fluid, scale, and cost. Furthermore, because the discharge coefficient, C_d is independent of R for a spillway with a large head, R does not need to remain a constant between the model and the prototype as follows:

$$R_m = 3.211 \times 10^5 \qquad R_p := \frac{\rho_p \cdot v_{3ip} \cdot H_p}{\mu_p} = 2.569 \times 10^6$$

11.7.7.3 Open Channel Flow with Weir or Spillway with Small Head

In the case of a weir or a spillway with a small head, $H \approx y_c$ the flow resistance (as modeled by C_D or C_d) is a function of geometry of the weir or spillway, H/P (where P is the height of weir), R, and W. The geometry, $L_i/L = H/P$, where L is represented by the P and L_i is represented by H. The boundary geometry of the weir, $(L_i/L) = (H/P)$ is the most important parameter in determining the magnitude of the weir/spillway discharge coefficient, C_d, as this ratio has the greatest influence on the shape of the flow field (flow contraction and velocity profiles). The head of the weir/spillway, H term is a direct measure of the Reynolds number, R and the Weber number, W, as the effects of viscosity and surface tension are significant only when H is small. The effect of the Weber number, W is negligible except at low heads, where surface tension effects may be significant. And, the effect of the Reynolds number, R is small except at low heads, where viscous effects may be large. Since the flow over a weir/spillway typically involves water, the Reynolds number, R is usually high, where the viscous effects are small. Furthermore, weirs/spillways serve as reliable flowrate-measuring devices as long as the head, H is large enough so that the effects of viscosity and surface tension are negligible. As such, in addition to the inertia force, F_I; the pressure force, F_P; and the gravity force, F_G, the viscous force, F_V; H/P; and the surface tension force, F_T are important in a flow-measuring device with a small head in open channel flow. Therefore, because there are three independent π terms, R, H/P, and W used in the definition of the dependent π term, $E = F$ and thus the drag coefficient, C_D, the appropriate similitude scale ratios corresponding to the independent π terms are used to define the similarity requirements between the model and its prototype. Specifically, the criteria governing dynamic similarity requires the following:

$$\underbrace{\left[\left(\frac{\rho v L}{\mu} \right)_p \right]}_{R_p} = \underbrace{\left[\left(\frac{\rho v L}{\mu} \right)_m \right]}_{R_m} \tag{11.163}$$

$$\underbrace{\left[\left(\frac{\rho v^2 L}{\sigma} \right)_p \right]}_{W_p} = \underbrace{\left[\left(\frac{\rho v^2 L}{\sigma} \right)_m \right]}_{W_m} \tag{11.164}$$

$$\left(\frac{H}{P} \right)_p = \left(\frac{H}{P} \right)_m \tag{11.165}$$

which states that in addition to the geometric similarity requirement for the geometry, H/P, there is a dynamic similarity requirement that R and W simultaneously remain a constant in order for the dependent π term (i.e., the drag coefficient, C_D) to remain a constant between the model and its prototype, where the flow resistance prediction equation is defined as follows:

$$\underbrace{\left[\frac{Q_a}{\sqrt{2g\Delta y}(A_i)}\right]_p}_{C_{Dp}=C_{dp}} = \underbrace{\left[\frac{Q_a}{\sqrt{2g\Delta y}(A_i)}\right]_m}_{C_{Dm}=C_{dm}} \qquad (11.166)$$

Furthermore, one may note that in the case where the similarity requirements regarding the independent π terms R ("viscosity model"), W ("surface tension model"), and H/P (geometric similarity) are satisfied, then the dependent π term (i.e., the drag coefficient, C_D) will remain a constant between the model and its prototype. And, therefore, application of the flow resistance prediction equation yields a "true model." However, in the case where the similarity requirements regarding the independent π terms R ("viscosity model"), W ("surface tension model"), and H/P (geometric similarity) are not satisfied, then the dependent π term (i.e., the drag coefficient, C_D) will not remain a constant between the model and its prototype. And, therefore, application of the flow resistance prediction equation yields a "distorted model" for which the appropriate corrections measures are taken.

EXAMPLE PROBLEM 11.18

Water at 70°F flows in a prototype rectangular open channel with a width of 10 ft, and a spillway with a small head is inserted in the channel, as illustrated in Figure EP 11.18. The head on the spillway is 0.95 ft, the height of the spillway is 3.9 ft, and the width of the crest of the spillway is 10 ft. A smaller model of the larger prototype is designed in order to study the flow characteristics of open channel flow in the flow-measuring device. The model fluid is also water at 70°F, the ideal discharge (see Equation 9.297) in the smaller model channel fitted with a spillway is 1.5 ft^3/sec, and the model scale, λ is 0.25. (a) Determine the actual discharge in the model spillway. (b) Determine

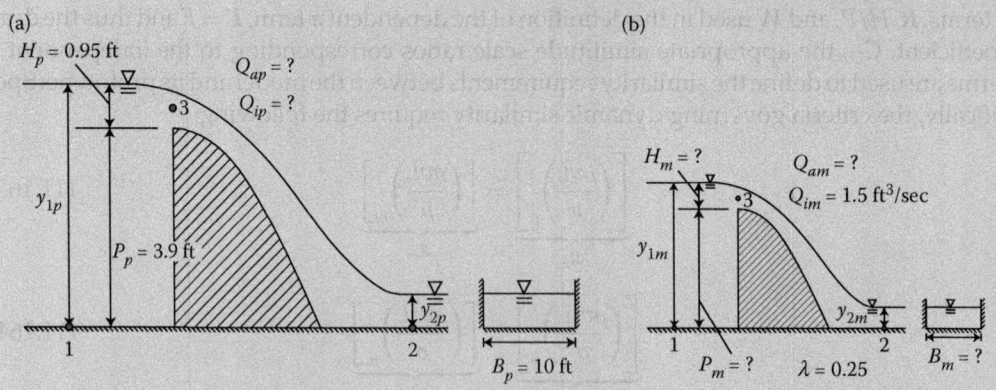

FIGURE EP 11.18
(a) Water flows in a prototype rectangular open channel with a spillway. (b) Water flows in a smaller model rectangular open channel with a spillway.

the ideal discharge in the prototype spillway in order to achieve dynamic similarity between the model and the prototype. (c) Determine the actual discharge in the prototype spillway in order to achieve dynamic similarity between the model and the prototype.

Mathcad Solution

(a) In order to determine the actual discharge in the model spillway, the actual discharge equation, Equation 11.158, is applied as follows:

$$Q_a = \underbrace{\left[\underbrace{\left(\frac{v_a}{\sqrt{2g\Delta y}} \right)}_{C_v} \underbrace{\left(\frac{A_a}{A_i} \right)}_{C_c} \right]}_{C_d} Q_i$$

where the C_D, or in the case of a spillway, the discharge coefficient, C_d, is used to model the flow resistance and is a function of the geometry of the flow-measuring device, $L_i/L = H/P$ (and R and W for a small head, H) as illustrated by the Rehbock formula (see Equation 9.303) as follows:

$$C_d = 0.605 + \frac{1}{305H} + 0.08\frac{H}{P}$$

where the head of the weir/spillway, H term is a direct measure of the Reynolds number, R and the Weber number, W, as the effects of viscosity and surface tension are significant only when H is small, as assumed in this example problem. Furthermore, in order to determine the geometry H, and P of the model channel and spillway, the model scale, λ (inverse of the length ratio) is applied as follows:

$$B_p := 10 \, \text{ft} \qquad H_p := 0.95 \, \text{ft} \qquad P_p = 3.9 \, \text{ft} \qquad \lambda := 0.25$$

Guess value: $\qquad B_m := 1 \, \text{ft} \qquad H_m := 1 \, \text{ft} \qquad P_m := 1 \, \text{ft}$

Given

$$\lambda = \frac{B_m}{B_p} \qquad\qquad \lambda = \frac{H_m}{H_p} \qquad\qquad \lambda = \frac{P_m}{P_p}$$

$$\begin{pmatrix} B_m \\ H_m \\ P_m \end{pmatrix} := \text{Find}(B_m, H_m, P_m) = \begin{pmatrix} 2.5 \\ 0.237 \\ 0.975 \end{pmatrix} \text{ft}$$

$$\text{slug} := 1 \, \text{lb} \frac{\text{sec}^2}{\text{ft}} \qquad \rho_m := 1.936 \frac{\text{slug}}{\text{ft}^3} \qquad \mu_m := 20.5 \times 10^{-6} \, \text{lb} \frac{\text{sec}}{\text{ft}^2}$$

$$\sigma_m := 0.00498 \frac{\text{lb}}{\text{ft}} \qquad Q_{im} := 1.5 \frac{\text{ft}^3}{\text{sec}} \qquad A_{3im} := B_m \cdot H_m = 0.594 \, \text{ft}^2$$

$$v_{3im} := \frac{Q_{im}}{A_{3im}} = 2.526 \frac{ft}{s}$$

$$R_m := \frac{\rho_m \cdot v_{3im} \cdot H_m}{\mu_m} = 5.666 \times 10^4 \qquad W_m = \frac{\rho_m \cdot v_{3im}^2 \cdot H_m}{\sigma_m} = 589.271$$

$$C_{dm} := 0.605 + \frac{1\,ft}{305 \cdot H_m} + 0.08 \frac{H_m}{P_m} = 0.638 \qquad Q_{am} := C_{dm} \cdot Q_{im} = 0.957 \frac{ft^3}{sec}$$

(b) To determine the ideal discharge in the prototype spillway in order to achieve dynamic similarity between the model and the prototype, the geometry L_i/L must remain a constant between the model and prototype as follows:

$$\left(\frac{L_i}{L}\right)_p = \left(\frac{L_i}{L}\right)_m$$

where the geometry is modeled as follows:

$$\frac{H_p}{P_p} = 0.244 \qquad\qquad \frac{H_m}{P_m} = 0.244$$

Furthermore, the **R** must remain a constant between the model and prototype as follows:

$$\underbrace{\left[\left(\frac{\rho v L}{\mu}\right)_p\right]}_{R_p} = \underbrace{\left[\left(\frac{\rho v L}{\mu}\right)_m\right]}_{R_m}$$

And, finally, the **W** must remain a constant between the model and prototype as follows:

$$\underbrace{\left[\left(\frac{\rho v^2 L}{\sigma}\right)_p\right]}_{W_p} = \underbrace{\left[\left(\frac{\rho v^2 L}{\sigma}\right)_m\right]}_{W_m}$$

Because in this example the model fluid is specified to be water, it will become impossible to simultaneously satisfy the dynamic requirements of **R** and **W** in the determination of the velocity ratio. Specifically, the "viscosity model" requires that the velocity ratio be defined as follows:

$$v_r = \frac{v_p}{v_m} = \frac{\left(\frac{\mu}{\rho L}\right)_p}{\left(\frac{\mu}{\rho L}\right)_m} = \mu_r \rho_r^{-1} L_r^{-1} = v_r L_r^{-1}$$

However, the "surface tension model" requires that the velocity ratio be defined as follows:

$$v_r = \frac{v_p}{v_m} = \frac{\left(\sqrt{\frac{\sigma}{\rho L}}\right)_p}{\left(\sqrt{\frac{\sigma}{\rho L}}\right)_m} = \sigma_r^{\frac{1}{2}} \rho_r^{\frac{-1}{2}} L_r^{\frac{-1}{2}}$$

Thus, equating the two velocity ratios yields the following dynamic requirement:

$$v_r = \frac{\sigma_r^{1/2}}{\rho_r^{1/2}} L_r^{1/2}$$

However, in fact $v_r \neq \left(\sigma_r^{1/2}/\rho_r^{1/2}\right)L_r^{1/2}$ as follows:

$$\rho_p := 1.936 \frac{slug}{ft^3} \qquad \mu_p := 20.5 \times 10^{-6} \, lb\frac{sec}{ft^2} \qquad \sigma_p := 0.00498 \frac{lb}{ft}$$

$$v_m := \frac{\mu_m}{\rho_m} = 1.059 \times 10^{-5} \frac{ft^2}{sec} \qquad\qquad v_p := \frac{\mu_p}{\rho_p} = 1.059 \times 10^{-5} \frac{ft^2}{sec}$$

$$L_r := \frac{1}{\lambda} = 4 \qquad \sigma_r := \frac{\sigma_p}{\sigma_m} = 1 \qquad\qquad \rho_r := \frac{\rho_p}{\rho_m} = 1$$

$$v_r := \frac{v_p}{v_m} = 1 \qquad \frac{\sigma_r^{\frac{1}{2}}}{\rho_r^{\frac{1}{2}}} \cdot L_r^{\frac{1}{2}} = 2$$

which does not satisfy the dynamic requirement. The solution to this type of difficult situation is to note that because the flow over a weir/spillway typically involves water, the Reynolds number, R is usually high where the viscous effects are small. Thus, it becomes more important to satisfy the dynamic requirement for W. Although ignoring the dynamic requirement for R will result in a "distorted model" (which can be accommodated for by proper interpretation of the model results), the model will be less distorted and more of a "true model" when the R is large, which reduces the dependence of the drag coefficient, C_d on R. In this example $R_m = 5.666 \times 10^4$, which is large; thus, not modeling the viscous effects may be insignificant.

$$A_{3ip} := B_P \cdot H_p = 9.5 \, ft^2$$

Guess value: $\qquad v_{3ip} := 1 \frac{ft}{sec} \qquad Q_{ip} := 1 \frac{ft^3}{sec} \qquad W_p := 500$

Given

$$W_p = \frac{\rho_p \cdot v_{3ip}^2 \cdot H_p}{\sigma_p}$$

$$W_p = W_m \qquad\qquad Q_{ip} = v_{3ip} \cdot A_{3ip}$$

$$\begin{pmatrix} v_{3ip} \\ Q_{ip} \\ W_p \end{pmatrix} := \text{Find}\,(v_{3ip},\, Q_{ip},\, W_p)$$

$$v_{3ip} = 1.263\,\frac{ft}{s} \qquad\qquad Q_{ip} = 12\,\frac{ft^3}{sec} \qquad\qquad W_p = 589.271$$

(c) To determine the actual discharge in the prototype spillway in order to achieve dynamic similarity between the model and the prototype, the discharge coefficient, C_d must remain a constant between the model and the prototype (which is a direct result of maintaining a constant W, and a constant L_i/L between the model and the prototype) as follows:

$$\underbrace{\left[\frac{Q_a}{\sqrt{2g\Delta y(A_i)}}\right]_p}_{C_{Dp}=C_{dp}} = \underbrace{\left[\frac{Q_a}{\sqrt{2g\Delta y(A_i)}}\right]_m}_{C_{Dm}=C_{dm}}$$

Guess value: $\qquad\qquad Q_{ap} := 1\,\dfrac{ft^3}{sec} \qquad\qquad C_{dp} := 0.1$

Given

$$C_{dp} = \frac{Q_{ap}}{Q_{ip}}$$

$$C_{dp} = C_{dm}$$

$$\begin{pmatrix} Q_{ap} \\ C_{dp} \end{pmatrix} := \text{Find}\,(Q_{ap},\, C_{dp})$$

$$Q_{ap} = 7.66\,\frac{ft^3}{sec} \qquad\qquad\qquad C_{dp} = 0.638$$

Therefore, although the similarity requirements regarding the independent π term, L_i/L ($H_p/P_p = H_m/P_m = 0.244$), and the independent π term, W ("surface tension model") ($W_p = W_m = 589.271$) are theoretically satisfied, the dependent π term (i.e., the discharge coefficient, C_d) will actually/practically remain a constant between the model and its prototype ($C_{dp} = C_{dm} = 0.638$) only if it is practical to maintain/attain the model velocity, flow depth, fluid, scale, and cost. And, thus, application of the actual discharge equation theoretically yields a "true model" only with respect to W. Furthermore, the similarity requirements regarding the independent π term, R ("viscosity model") ($R_p = 1.133 \times 10^5 \neq R_m = 5.666 \times 10^4$) are not satisfied and thus application of the actual discharge equation yields a "distorted model" with respect to R as illustrated by this example.

$$R_m = 5.666 \times 10^4 \qquad\qquad R_p := \frac{\rho_p \cdot v_{3ip} \cdot H_p}{\mu_p} = 1.133 \times 10^5$$

11.7.8 Application of the Similitude Scale Ratios for the Efficiency of Pumps and Turbines

Because a turbine is the exact opposite of a pump, an empirical derivation of the efficiency of a pump using dimensional analysis was actually conducted, and the empirical derivation of the efficiency of a turbine was interpreted from the results (Chapter 7). In the case of the pump and turbine efficiencies, η while the drag coefficient, $C_D = \eta$ in general is mainly a function of E, it will (theoretically) selectively vary with R, ε/L, and C, depending mostly on the velocity of flow and the type of fluid through the pump or turbine (see Table 11.8). However, in practice, the drag coefficient, $C_D = \eta$, assumed to be independent of C (assumption of incompressible flow), will vary with R for laminar flow, while it will vary with ε/D for turbulent flow in the case where the prototype pump or turbine is significantly larger than the model pump or turbine (or vice versa). Therefore, assuming incompressible flow, for laminar flow, the R must remain a constant between the model and its prototype (as a "true model" is sought), and for turbulent flow where the prototype pump or turbine is significantly larger than the model pump (or vice versa), the ε/D must remain a constant between the model and its prototype (as a "true model" is sought). Finally, it is important to note that for numerous practical pump or turbine flow problems, the drag coefficient, $C_D = \eta$ is assumed to be independent of all three independent π terms R, C, and ε/D, which leads to the definition of the affinity laws (similarity rules) for the efficiency of homologous pumps and turbines; when dynamic similarity between the model and its prototype (or between two flow systems) is achieved, the model and its prototype (or the two flow systems) are said to be homologous. In summary, in the case where the affinity laws are satisfied, then application of the pump or turbine design coefficients (C_H, C_Q, and C_P) yields a "true model." However, in the case where the affinity laws are not satisfied, then application of the pump or turbine design coefficients (C_H, C_Q, and C_P) yields a "distorted model" for which the appropriate corrections measures are taken. Specifically, in order to model the scaling effects between a model and its prototype (i.e., account for a "distorted model"), an empirical equation developed by Moody is applied. The empirical Moody efficiency correction equation for pumps and turbines implies that (models the fact that) the larger prototype pumps and turbines are typically more efficient than the smaller model pumps or turbines. The similitude scale ratios for pumps are addressed below in Sections 11.7.8.1 through 11.7.8.3, while the similitude scale ratios turbines are addressed in Sections 11.7.8.4 through 11.7.8.6.

11.7.8.1 Similitude Scale Ratios for the Efficiency of Pumps

Dimensional analysis yielded an empirical derivation of the efficiency of a pump (an uncalibrated mathematical model of the fluid flow situation) as follows:

$$\eta_{pump} = \frac{(P_p)_{out}}{(P_p)_{in}} = \frac{\gamma Q h_{pump}}{\omega T_{shaft,in}} = \frac{\gamma Q \frac{\Delta p}{\gamma}}{\omega T_{shaft,in}} = \frac{Q \Delta p}{\omega T_{shaft,in}} = \frac{v L^2 \Delta p}{\frac{vFL}{L}} = \frac{\Delta p L^2}{F} = \frac{F_P}{F_I} = \frac{\Delta p L^2}{\rho v^2 L^2}$$

$$= \frac{\Delta p}{\rho v^2} = \frac{1}{E} = C_D = \phi\left(C_H, C_Q, C_P, R, C, \frac{\varepsilon}{D}\right) \tag{11.167}$$

where the head added by the pump, $h_{pump} = (\Delta p/\gamma)$. The drag coefficient, C_D (pump efficiency, η_{pump}) as modeled by the Euler number, E is the dependent π term, as it models the dependent variable, Δp that cannot be derived analytically from theory, while the head coefficient, C_H; the capacity coefficient, C_Q; the power coefficient, C_P; the Reynolds number, R; the Cauchy number, C; and the relative blade surface roughness, ε/D are the independent π terms. One may note that the symbol, ϕ means that the drag coefficient, C_D (pump efficiency, η) is an unknown function of C_H, C_Q, C_P, R, C, and ε/D; thus, the drag coefficient, C_D (pump efficiency, η) is determined empirically from experimentation. Also, because the efficiency of a pump, η represents a head loss, h_f due to a pressure drop, Δp, the pressure force will always play an important role; thus, the definition of the drag coefficient, $C_D = \eta$ is mainly a function of E.

Based on the theoretical expression for the efficiency of a pump, $\eta_{pump} = ((P_p)_{out}/(P_p)_{in}) = (\gamma Q h_{pump}/\omega T_{shaft,in})$ (Equation 4.197), the empirical expression for the efficiency of a pump derived from dimensional analysis (Equation 11.167) can be expressed as a function of the pump design coefficients (C_H, C_Q, and C_P) as follows:

$$\eta_{pump} = \frac{(P_p)_{out}}{(P_p)_{in}} = \frac{\gamma Q h_{pump}}{\omega T_{shaft,in}} = \frac{(\rho Q)(g h_{pump})}{(\omega T_{shaft,in})} = \frac{(\rho \omega D^3 C_Q)(\omega^2 D^2 C_H)}{(\rho \omega^3 D^5 C_P)} = \frac{C_Q C_H}{C_P} = \frac{1}{E}$$
$$= C_D = \phi\left(R, C, \frac{\varepsilon}{D}\right) \tag{11.168}$$

Because the drag coefficient, C_D is mainly a function of the Euler number, E; the inertia force, F_I; and the pressure force, F_P will always be important (predominate) in the flow through a pump. Thus, the Euler number, E and the drag coefficient, $C_D = (1/E) = \eta$ (dependent π term) must remain a constant between the model and its prototype. Furthermore, because the efficiency of a pump, η_{pump} is now expressed as a known function of the pump design coefficients C_H, C_Q, and C_P, these pump design coefficients (C_H, C_Q, and C_P) must also remain a constant between the model and its prototype. However (theoretically), depending upon the specifics of the flow through a pump situation, the relative blade surface roughness, ε/D and the remaining forces, which include the viscous force, F_V and the elastic force, F_E (and thus their respective independent π terms, R and C, respectively), may or may not play an important role in the definition of the drag coefficient, C_D. If in fact a given independent π term plays an important role in the definition of the drag coefficient, C_D, then it must remain a constant between the model and its prototype (as a "true model" is sought).

Although the drag coefficient, $C_D = \eta_{pump} = (C_Q C_H/C_P)$ is mainly a function of E, it will vary with, R, C, and ε/D depending mainly on the velocity of the flow, the compressibility of the flow/fluid, and the degree of scaling between the model and the prototype (or vice versa). Specifically, while the C divides the flow into compressible or incompressible flow, for practical purposes, the flow is assumed to be incompressible; thus, the flow is independent of C. Furthermore, the R (and thus the velocity of the flow) divides the flow into laminar or turbulent. As such, based on the Moody diagram (see Figure 8.1), laminar flow is dependent on the R and independent of ε/D, while turbulent flow is independent of R and dependent on ε/D. However, for practical purposes, the effect of the relative blade surface roughness, ε/D on turbulent flow is usually not modeled unless the differences in absolute roughness, ε between the model and the prototype are large; this may occur if the degree of scaling between the model and the prototype (or vice versa) is very large, scaling from a significantly large pump to a significantly small pump (or vice versa). In the case where the prototype pump is significantly larger than the model pump ($L_r = L_p/L_m$ is typically much greater than one), then the required model surface roughness, ε_m needs to be smaller/smoother than the prototype surface roughness, ε_p, yet similar in pattern. However, such a combined requirement is usually difficult to attain in practice. In fact, the relative blade surface roughness, ε/D for the large prototype may be significantly smaller than the relative blade surface roughness, ε/D for the small model. Thus, in summary, the drag coefficient, $C_D = \eta_{pump} = (C_Q C_H / C_P)$, assumed to be independent of C (assumption of incompressible flow), will vary with R for laminar flow, while it will vary with ε/D for turbulent flow in the case where the prototype pump is significantly larger than the model pump (or vice versa). Therefore, assuming incompressible flow, for laminar flow, the R must remain a constant between the model and its prototype (as a "true model" is sought), and for turbulent flow where the prototype pump is significantly larger than the model pump (or vice versa), the ε/D must remain a constant between the model and its prototype (as a "true model" is sought). Finally, it is important to note that for numerous practical pump flow problems, the drag coefficient, $C_D = \eta_{pump} = (C_Q C_H / C_P)$ is assumed to be independent of all three independent π terms R, C, and ε/D in Equation 11.168, which leads to the definition of the affinity laws (similarity rules) for the efficiency of homologous pumps (and turbines); when dynamic similarity between the model and its prototype (or between two flow systems) is achieved, the model and its prototype (or the two flow systems) are said to be homologous.

11.7.8.2 Affinity Laws for the Efficiency of Homologous Pumps

The affinity laws (similarity rules) for the efficiency of homologous pumps (and turbines) assume that for numerous practical pump (and turbine) flow problems, the drag coefficient, $C_D = \eta_{pump} = (C_Q C_H / C_P)$ is independent of all three independent π terms R, C, and ε/D in Equation 11.168. Therefore, in order to achieve dynamic similarity between the model and its prototype (i.e., the model and the prototype are said to be homologous), the drag coefficient, $C_D = \eta_{pump} = (C_Q C_H / C_P)$ and thus the head coefficient, $C_H = (gh_{pump}/\omega^2 D^2)$; the capacity coefficient, $C_Q = (Q/\omega D^3)$; and the power coefficient, $C_P = (\omega T/\rho \omega^3 D^5)$ each must remain a constant between the model and its prototype as follows:

$$\underbrace{\left[\left(\eta_{pump}\right)_p\right]}_{\left(\frac{C_Q C_H}{C_P}\right)_p} = \underbrace{\left[\left(\eta_{pump}\right)_m\right]}_{\left(\frac{C_Q C_H}{C_P}\right)_m} \tag{11.169}$$

$$\underbrace{\left[\left(\frac{gh_{pump}}{\omega^2 D^2}\right)_p\right]}_{(C_H)_p} = \underbrace{\left[\left(\frac{gh_{pump}}{\omega^2 D^2}\right)_m\right]}_{(C_H)_m} \tag{11.170}$$

$$\underbrace{\left[\left(\frac{Q}{\omega D^3}\right)_p\right]}_{(C_Q)_p} = \underbrace{\left[\left(\frac{Q}{\omega D^3}\right)_m\right]}_{(C_Q)_m} \tag{11.171}$$

$$\underbrace{\left[\left(\frac{\omega T}{\rho \omega^3 D^5}\right)_p\right]}_{(C_P)_p} = \underbrace{\left[\left(\frac{\omega T}{\rho \omega^3 D^5}\right)_m\right]}_{(C_P)_m} \tag{11.172}$$

Equations 11.169 through 11.172 are known as the affinity laws (similarity rules) for the efficiency of homologous pumps (and turbines), and they represent the "pressure model" (i.e., $C_D = \eta_{pump} = (C_Q C_H / C_P) = (1/E)$ is independent of all three independent π terms R, C, and ε/D in Equation 11.168). This implies that the pump efficiency, η_{pump} (and thus the pump design coefficients C_H, C_Q, and C_P) (i.e., the affinity laws) sufficiently defines the dynamic similarity requirements between the model and its prototype. Thus, in addition to the inertia force, F_I, the pressure force, F_P is important in the design of pumps (and turbines). Therefore, there are no independent π terms used in the definition of the dependent π term, E and thus the drag coefficient, $C_D = (1/E) = \eta_{pump}$. As such, one may note that in the case where the similarity requirements regarding the dependent π term (i.e., $C_D = (1/E) = \eta_{pump}$ and thus, C_H, C_Q, and C_P) ("pressure model") between the model and its prototype (i.e., the affinity laws given in Equations 11.169 through 11.172) are satisfied, then application of the pump design coefficients yields a "true model." However, in the case where the similarity requirements regarding the dependent π term (i.e., $C_D = (1/E) = \eta_{pump}$ and thus C_H, C_Q, and C_P) ("pressure model") between the model and its prototype (i.e., the affinity laws given in Equations 11.169 through 11.172) ("pressure model") are not satisfied, then application of the pump design coefficients yields a "distorted model" for which the appropriate corrections measures are taken. Specifically, in order to model the scaling effects between a model and its prototype (i.e., account for a "distorted model"), an empirical equation developed by Moody is applied (see Section 11.7.8.3 below).

Because the efficiency of a pump, $\eta_{pump} = (C_Q C_H / C_P)$ is expressed as a function of the pump design coefficients C_H, C_Q, and C_P, the pump affinity laws (Equations 11.169 through 11.172) ("pressure model") are used in the design of pumps. Specifically, the affinity laws ("pressure model") may be used to scale up or scale down between a model and its prototype or between two flow systems. Typically, in the design of a pump, the pump design coefficients C_H, C_Q, and C_P are evaluated from test data (model pump), which are used to predict the performance of the prototype pump. However, it is important to note that, even though the application of the affinity laws (similarity rules) for the design of homologous pumps (and turbines) assumes that the pump efficiency, $\eta_{pump} = (C_Q C_H / C_P)$ is independent of all three independent π terms R, C, and ε/D in Equation 11.168 (i.e., assumes the "pressure model"), the assumption of a "pressure model" in itself may result in a "distorted model" that is due to the practical inability to maintain R or ε/D constant between a model and its prototype. Thus, it still remains important to address and account for (model) the effects of scaling (i.e., account for a "distorted model") between a model and its prototype resulting from the assumption of a "pressure model." Assuming a "pressure model," the dynamic similarity requirement that $C_D = (1/E) = \eta_{pump}$ (i.e., E) remain a constant is modeled by the similitude velocity scale ratio, v_r given by Equations 11.46 and 11.47 and repeated, respectively, as follows:

$$v_r = \frac{v_p}{v_m} = \frac{\left(\sqrt{\dfrac{\Delta p}{\rho}}\right)_p}{\left(\sqrt{\dfrac{\Delta p}{\rho}}\right)_m} = \Delta p_r^{1/2} \rho_r^{-1/2} \tag{11.173}$$

or

$$v_r = \Delta p_r^{1/2} \rho_r^{-1/2} \tag{11.174}$$

Because the "pressure model" similarity requirements apply for both gas and liquid flow, fortunately (technically, that is), there are fluids (especially gases) that are available in order to satisfy the similarity requirement for the model design as modeled by the relationship between the similitude velocity scale ratio, v_r (where $v_r = \omega_r D_r$); the similitude pressure ratio, Δp_r; and the similitude density ratio, ρ_r. Thus, the resulting model would be (technically, that is) "true" with respect to E. However, first, in the design of pumps (and turbines), the most practical and most common fluid is a liquid (namely, water), for which the required model velocity, v_m may be too high to attain in practice (especially for liquids). Specifically, the scale differences between the model and its prototype pump may cause the flow to transition from laminar to turbulent flow, where, for instance, the smaller model pump usually has a lower R than the larger prototype pump. Thus, the actual smaller model velocity, v_m may lead to laminar flow, which would require that R remain a constant between the model and its prototype. However, the fact is that in some cases it is not practically possible to keep R constant between a model and its prototype (especially for liquids), thus R may not even be a practical indicator of dynamic similarity between the model and its prototype. Second, in some cases, the prototype pump is significantly larger (or significantly smaller) than the model pump ($L_r = L_p/L_m$ is typically much greater than one), which would require that ε/D remain a constant between the model and its prototype. Thus, the required model surface roughness, ε_m needs to be smaller/smoother than the prototype surface roughness, ε_p, yet similar in pattern. However, such a combined (roughness and pattern) requirement is usually difficult to attain in practice. In fact, the relative blade surface roughness, ε/D for the large prototype may be significantly smaller than the relative blade surface roughness, ε/D for the small model. Therefore, the fact is that in most flow cases, it is not practically possible to keep ε/D constant between the model and its prototype, thus the ε/D may not even be a practical indicator of dynamic similarity between the model and its prototype. Thus, in conclusion, even though the affinity laws (the drag coefficient, $C_D = \eta_{pump} = (C_Q C_H/C_P)$ is independent of all three independent π terms R, C, and ε/D in Equation 11.168) ("pressure model") do not "directly" require that R and ε/D remain a constant between the model and its prototype, respectively, the "pressure model" itself does "indirectly" require that R and ε/D remain a constant between the model and its prototype; otherwise, a "distorted model" results. Thus, in summary, in the cases where there is the need but the practical inability to keep R or ε/D constant between a model and its prototype because, essentially, the "pressure model" scaling requirements are not satisfied, then application of the pump design coefficients will yield a "distorted model." If a "distorted model" results, appropriate correction measures are taken in order to model the scaling effects (i.e., account for a "distorted model") between the model and its prototype using an empirical equation developed by Moody (see Section 11.7.8.3 below).

11.7.8.3 Modeling of Scaling Effects for the Efficiency of Nonhomologous ("Distorted Models") Pumps Using the Moody Equation

The application of the affinity laws (similarity rules) for the design of homologous pumps (and turbines) provides a practical approach to scale up or scale down between a model and its prototype (or between two flow systems). In the case where the similarity requirements regarding the dependent π term (i.e., $C_D = (1/E) = \eta_{pump}$ and thus C_H, C_Q, and C_P) ("pressure model") between the model and its prototype (i.e., the affinity laws given in Equations 11.169 through 11.172) are satisfied, then application of the pump design coefficients yield a "true model" (i.e., homologous pumps). However, in the case where the similarity requirements regarding the dependent π term (i.e., $C_D = (1/E) = \eta_{pump}$ and thus C_H, C_Q, and C_P) ("pressure model") between the model and its prototype (i.e., the affinity laws given in Equations 11.169 through 11.172) ("pressure model") are not satisfied, then application of the pump design coefficients yield a "distorted model" (i.e., nonhomologous pumps) for which the appropriate corrections measures are taken. Specifically, in the cases where there is the need but the practical inability to keep R or ε/D constant between a model and its prototype, essentially, the "pressure model" scaling requirements (i.e., the affinity laws given in Equations 11.169 through 11.172) are not satisfied, then application of the pump design coefficients will yield a "distorted model" (i.e., nonhomologous pumps). If a "distorted model" results, appropriate correction measures are taken in order to model the scaling effects between the model and its prototype (i.e., account for a "distorted model") using an empirical equation developed by Moody. The empirical Moody efficiency correction equation for pumps (and turbines), which assumes geometrically similar pumps (and turbines), is given as follows:

$$\frac{1 - \eta_p}{1 - \eta_m} = \left(\frac{D_m}{D_p}\right)^{1/5} \tag{11.175}$$

The above empirical equation by Moody provides a measure of the scaling effects (i.e., accounts for a "distorted model") between the model and its prototype on the pump (or turbine) efficiency. Thus, Equation 11.175 may be used to account for the variation in efficiency, η between the model and its prototype for a "distorted model" (i.e., nonhomologous pumps or turbines). Specifically, in the design of a pump, while the pump design coefficients C_H, C_Q, and C_P that are evaluated from test data (model pump) may be used to predict the performance of the prototype pump, the pump efficiency computed from the test data (geometrically similar model pump) may be used to predict the efficiency of the prototype pump by applying Equation 11.175. Thus, the empirical Moody efficiency correction equation for pumps and turbines (Equation 11.175) implies that (models the fact that) the larger prototype pumps are typically more efficient than the smaller model pumps, and it is important to understand why.

Technically, in order to achieve dynamic similarity between the model and its prototype (i.e., the model and the prototype are said to homologous), the drag coefficient, $C_D = \eta_{pump} = (C_Q C_H/C_P)$ and thus the head coefficient, $C_H = (gh_{pump}/\omega^2 D^2)$; the capacity coefficient, $C_Q = (Q/\omega D^3)$; and the power coefficient, $C_P = (\omega T/\rho \omega^3 D^5)$, each must remain a constant between the model and its prototype as given by Equations 11.169 through 11.172 (the affinity laws or similarity rules for the efficiency of homologous pumps). However, although the model and its prototype may be geometrically similar (i.e., $L_r = (L_p/L_m) = (D_p/D_m) = constant$), that does not guarantee either kinematic similarly (i.e., $v_r = (v_p/v_m) = constant$, where $v_r = \omega_r \cdot D_r$) or dynamic similarity (i.e., $\eta_p = \eta_m$). Specifically, while it may be practically feasible to maintain geometric similarity between the model and its prototype (i.e., $L_r = (L_p/L_m) = (D_p/D_m) = constant$), it may not be practically feasible to maintain kinematic similarity— i.e., $(C_H)_p = (C_H)_m$, $(C_Q)_p = (C_Q)_m$, and $(C_P)_p = (C_P)_m$; therefore, it may not be practically feasible to achieve dynamic similarity, that is, $(\eta_{pump} = (C_Q C_H/C_P))_p = (\eta_{pump} = (C_Q C_H/C_P))_m$. In practice, the higher velocities that are attainable in the larger prototype pumps may not be attainable in the smaller model (laboratory) pumps. As explained above (see Section 11.7.8.2), in the design of pumps and turbines, the most practical and most common fluid is a liquid (namely, water), for which the required model velocity, v_m may be too high to attain in practice (especially for liquids). Thus, the smaller model pumps typically have smaller R than the larger prototype pumps, causing the flow to transition from laminar flow in the model (where the friction factor, f is high; see Moody diagram in Figure 8.1) to turbulent flow in the prototype (where the friction factor, f is low; see Moody diagram in Figure 8.1, and where the ε/D may be significantly smaller than in the model). Also, the larger prototype pump may have smaller gap thickness between the impeller blade tips (or turbine runner blade tips) and the pump or turbine housing, and smaller pump impeller blade thickness (or turbine runner blade thickness) both relative to the blade diameter in comparison to the smaller model pump or turbine. As a result, the frictional losses and fluid leakage for the larger prototype pump are significantly less than for the smaller model pump. Additionally, the smaller model pump that operates well at lower speeds may cavitate at higher speeds; thus, the larger prototype pumps are more efficient than the smaller model pumps; specifically, the smaller model pumps have smaller flow passages (increases chance of cavitation) than the larger prototype pumps. Thus, in summary, the empirical Moody efficiency correction equation for pumps and turbines (Equation 11.175) implies that (models the fact that) the larger prototype pumps are typically more efficient than the smaller model pumps.

EXAMPLE PROBLEM 11.19

A 3-hp prototype water pump is designed to deliver a total head of 140 ft at a discharge of 42 US gpm. The diameter of the prototype pump impeller blades is 8.5 in with an impeller rotational speed of 1750 rpm. A smaller model water pump of the prototype water pump is designed in order to study the flow characteristics of the flow through a pump. The model scale is 0.4. (a) Design a homologous model water pump. (b) Determine if the resulting designed model is a true model or a distorted model.

Mathcad Solution

(a) In order to design a homologous model water pump, the affinity laws (similarity rules) for the efficiency of homologous pumps (Equations 11.169 through 11.172) are applied as follows:
   But, first, the pump design coefficients, the hydraulic power output from the pump, the pump efficiency, and the Reynolds number are computed for the larger prototype water pump as follows:

$$\text{slug} := 1\,\text{lb}\frac{\sec^2}{\text{ft}} \qquad \rho_p := 1.936\frac{\text{slug}}{\text{ft}^3} \qquad g := 32.174\frac{\text{ft}}{\sec^2}$$

$$\gamma_p := \rho_p \cdot g := 62.289\frac{\text{lb}}{\text{ft}^3} \quad h_p := 550\frac{\text{ft}\cdot\text{lb}}{\sec} \qquad P_{pinp} := 3\cdot\text{hp} = 1.65\times10^3\frac{\text{ft}\cdot\text{lb}}{\text{s}}$$

$$h_{pumpp} := 140\,\text{ft} \qquad \text{USgal} := 0.133681\,\text{ft}^3 \quad Q_p := 42\frac{\text{USgal}}{\text{min}} = 0.094\frac{\text{ft}^3}{\sec}$$

$$D_p := 8.5\,\text{in} = 0.708\,\text{ft} \qquad\qquad \omega_p := 1750\,\text{rpm} = 183.26\frac{1}{\text{s}}$$

$$C_{Hp} := \frac{g\cdot h_{pumpp}}{\omega_p^2\cdot D_p^2} = 0.267 \qquad\qquad C_{Qp} := \frac{Q_p}{\omega_p\cdot D_p^3} = 1.437\times10^{-3}$$

$$T_p := \frac{P_{pinp}}{\omega_p} = 9.004\,\text{ftlb} \qquad\qquad C_{Pp} := \frac{\omega_p\cdot T_p}{\rho_p\cdot\omega_p^3\cdot D_p^5} = 7.766\times10^{-4}$$

$$\eta_{pumpp} := \frac{C_{Qp}\cdot C_{Hp}}{C_{Pp}} = 0.495 \qquad\qquad P_{poutp} := \gamma_p\cdot Q_p\cdot h_{pumpp} = 816.03\frac{\text{ft}\cdot\text{lb}}{\text{s}}$$

$$\eta_{pumpp} := \frac{P_{poutp}}{P_{pinp}} = 0.495 \qquad\qquad \mu_p := 21 \times 10^{-6} \frac{lb \cdot sec}{ft^2}$$

$$v_p := \omega_p \cdot D_p = 129.809 \frac{ft}{s} \qquad\qquad R_p := \frac{\rho_p \cdot v_p \cdot D_p}{\mu_p} = 8.477 \times 10^6$$

Then, in order to design a homologous smaller model water pump, the affinity laws (similarity rules) for the efficiency of homologous pumps (Equations 11.169 through 11.172) are applied as follows:

$$\lambda := 0.4 \qquad \rho_m := \rho_p = 1.936 \frac{slug}{ft^3} \qquad \gamma_m := \gamma_p = 62.289 \frac{lb}{ft^3}$$

Guess value: $D_m := 6\,in \quad h_{pumpm} := 50\,ft \quad \omega_m := 10{,}000\,rpm = 1.047 \times 10^3 \frac{1}{s}$

$$Q_m := 20 \frac{USgal}{min} = 0.045 \frac{ft^3}{sec} \qquad T_m := 2\,ft\,lb \qquad \eta_{pumpm} := 1$$

$$C_{Hm} := 1 \qquad\qquad C_{Qm} := 1 \qquad\qquad C_{Pm} := 1$$

Given

$$\lambda = \frac{D_m}{D_p} \qquad \frac{g \cdot h_{pumpp}}{\omega_p^2 \cdot D_p^2} = \frac{g \cdot h_{pumpm}}{\omega_m^2 \cdot D_m^2} \qquad \frac{Q_p}{\omega_p \cdot D_p^3} = \frac{Q_m}{\omega_m \cdot D_m^3}$$

$$\frac{\omega_p \cdot T_p}{\rho_p \cdot \omega_p^3 \cdot D_p^5} = \frac{\omega_m \cdot T_m}{\rho_m \cdot \omega_m^3 \cdot D_m^5} \qquad C_{Hm} = \frac{g \cdot h_{pumpm}}{\omega_m^2 \cdot D_m^2} \qquad C_{Qm} = \frac{Q_m}{\omega_m \cdot D_m^3}$$

$$C_{Pm} = \frac{\omega_m \cdot T_m}{\rho_m \cdot \omega_m^3 \cdot D_m^5} \qquad \eta_{pumpm} = \frac{C_{Qm} \cdot C_{Hm}}{C_{Pm}} \qquad \eta_{pumpp} = \eta_{pumpm}$$

$$\begin{pmatrix} D_m \\ h_{pumpm} \\ \omega_m \\ Q_m \\ T_m \\ \eta_{pumpm} \\ C_{Hm} \\ C_{Qm} \\ C_{Pm} \end{pmatrix} := Find\,(D_m,\,h_{pumpm},\,\omega_m,\,Q_m,\,T_m,\,\eta_{pumpm},\,C_{Hm},\,C_{Qm},\,C_{Pm})$$

$$D_m = 3.4\,in \qquad h_{pumpm} = 81.322\,ft \qquad \omega_m = 349.178 \frac{1}{s} \qquad \omega_m = 3.334 \times 10^3\,rpm$$

$$Q_m = 0.011 \frac{ft^3}{sec} \qquad Q_m = 5.122 \frac{USgal}{min} \qquad T_m = 0.335\,ft \cdot lb \qquad \eta_{pumpm} = 0.495$$

$$C_{Hm} = 0.267 \qquad\qquad C_{Qm} = 1.437 \times 10^{-3} \qquad\qquad C_{Pm} = 7.766 \times 10^{-4}$$

And, finally, the mechanical power input to the pump, the hydraulic power output from the pump, the pump efficiency, and the Reynolds number are computed for the smaller model water pump as follows:

$$P_{poutm} := \gamma_m \cdot Q_m \cdot h_{pumpm} = 57.803 \frac{ft \cdot lb}{s} \qquad P_{poutm} = 0.105 \, hp$$

$$P_{pinm} := \omega_m \cdot T_m = 116.876 \frac{ft \cdot lb}{s} \qquad P_{pinm} = 0.213 \, hp$$

$$\eta_{pumpm} := \frac{P_{poutm}}{P_{pinm}} = 0.495 \qquad \mu_m := 21 \times 10^{-6} \frac{lb \cdot sec}{ft^2}$$

$$v_m := \omega_m \cdot D_m = 98.934 \frac{ft}{s} \qquad R_m := \frac{\rho_m \cdot v_m \cdot D_m}{\mu_m} = 2.584 \times 10^6$$

(b) In order to determine if the resulting designed model is a true model or a distorted model, one must first determine if the application of the affinity laws (similarity rules) for the efficiency of homologous pumps (Equations 11.169 through 11.172) have yielded a practically feasible designed model or not. If the resulting designed smaller model water pump is practically feasible (design parameters: $D_m = 3.4$ in, $h_{pumpm} = 81.322$ ft, $\omega_m = 3334$ rpm, $Q_m = 0.011$ cfs $= 5.122$ US gpm, $T_m = 0.335$ ft-lb, $P_{pinm} = 0.213$ hp), then it is considered to be a true model (model and prototype are homologous pumps). However, if the resulting designed smaller model water pump is not practically feasible (at least one of the design parameters is not practically feasible), then it is considered to be a distorted model (model and prototype are nonhomologous pumps). Upon examination of the resulting model design parameters (design parameters: $D_m = 3.4$ in, $h_{pumpm} = 81.322$ ft, $\omega_m = 3334$ rpm, $Q_m = 0.011$ cfs $= 5.122$ US gpm, $T_m = 0.335$ ft-lb, $P_{pinm} = 0.213$ hp) and a comparison with some pump manufacture design data parameters, it appears that such a model water pump is indeed practically feasible. Furthermore, the $R_p = 8.477 \times 10^6$, while the $R_m = 2.584 \times 10^6$, which indicates that turbulent flow is maintained between the prototype pump and the model pump. However, in most laboratory situations, there will be some constraints imposed upon the model design; these are called "design conditions." For instance, a certain laboratory may impose a limitation on the available h_{pumpm} and/or a limitation on the available Q_m, or ω_m, etc. that it is capable of providing ("design conditions"). Furthermore, for a smaller model pump, it may not be practically feasible to attain high speeds, and/or cavitation may occur at higher speeds. Thus, when there are certain "design conditions" imposed due to practical limitations/constraints, the application of the affinity laws (similarity rules) for the efficiency of homologous pumps (Equations 11.169 through 11.172) will yield a distorted model (model and prototype are nonhomologous pumps). If a "distorted model" results, appropriate correction measures are taken in order to model the scaling effects (i.e., account for a "distorted model") between the model and its prototype, using the empirical equation developed by Moody (Equation 11.175) as follows:

Guess value: $\qquad \eta_{pumpm} := 0.1$

Given

$$\frac{1 - \eta_{pumpp}}{1 - \eta_{pumpm}} = \left(\frac{D_m}{D_p}\right)^{\frac{1}{5}}$$

$\eta_{pumpm} := \text{Find}\,(\eta_{pumpm}) = 0.393$

which indicates that the estimated pump efficiency of the designed smaller model water pump is only 0.393, rather than the designed pump efficiency of the model pump of 0.495; larger pumps are typically more efficient than smaller pumps.

11.7.8.4 Similitude Scale Ratios for the Efficiency of Turbines

The empirical expression of the efficiency of a turbine is interpreted by its definition (exact opposite of a pump) as follows:

$$\eta_{turbine} = \frac{(P_t)_{out}}{(P_t)_{in}} = \frac{\omega T_{shaft,out}}{\gamma Q h_{turbine}} = \frac{F_I}{F_P} = \frac{\rho v^2 L^2}{\Delta p L^2} = \frac{\rho v^2}{\Delta p} = E = C_D$$
$$= \phi\left(C_H, C_Q, C_P, R, C, \frac{\varepsilon}{D}\right) \tag{11.176}$$

Based on the theoretical expression for the efficiency of a turbine, $\eta_{turbine} = ((P_t)_{out}/(P_t)_{in}) = (\omega T_{shaft,out}/\gamma Q h_{turbine})$ (Equation 4.217), the empirical expression for the efficiency of a turbine derived from dimensional analysis (Equation 11.176) can be expressed as a function of the turbine design coefficients (C_H, C_Q, and C_P) as follows:

$$\eta_{turbine} = \frac{(P_t)_{out}}{(P_t)_{in}} = \frac{\omega T_{shaft,out}}{\gamma Q h_{turbine}} = \frac{(\omega T_{shaft,out})}{(\rho Q)(g h_{turbine})} = \frac{(\rho \omega^3 D^5 C_P)}{(\rho \omega D^3 C_Q)(\omega^2 D^2 C_H)} = \frac{C_P}{C_Q C_H} = E$$
$$= C_D = \phi\left(R, C, \frac{\varepsilon}{D}\right) \tag{11.177}$$

Because the drag coefficient, C_D is mainly a function of the Euler number, E; the inertia force, F_I; and the pressure force, F_P will always be important (predominate) in the flow through a turbine. Thus, the Euler number, E and the drag coefficient, $C_D = E = \eta$ (dependent π term) must remain a constant between the model and its prototype. Furthermore, because the efficiency of a turbine, $\eta_{turbine}$ is now expressed as a known function of the turbine design coefficients C_H, C_Q, and C_P, these turbine design coefficients (C_H, C_Q, and C_P) must also remain a constant between the model and its prototype. However (theoretically), depending upon the specifics of the flow through a turbine situation, the relative blade surface roughness, ε/D and the remaining forces, which include the viscous force, F_V and the elastic force, F_E (and thus their respective independent π terms, R and C, respectively) may or may not play an important role in the definition of the drag coefficient, C_D. If in fact a given independent π term plays an important role in the definition of the drag coefficient, C_D, then it must remain a constant between the model and its prototype (as a "true model" is sought).

Although the drag coefficient, $C_D = E = \eta$ is mainly a function of E, it will vary with R, C, and ε/D depending mainly on the velocity of the flow, the compressibility of the flow/fluid, and the degree of scaling between the model and the prototype (or vice versa) (see Section

11.7.8.1 on pumps above for a similar detailed discussion). In summary, the drag coefficient, $C_D = \eta_{turbine} = (C_P/C_QC_H)$ is assumed to be independent of C (assumption of incompressible flow) and will vary with R for laminar flow, while it will vary with ε/D for turbulent flow in the case where the prototype turbine is significantly larger than the model turbine (or vice versa). Therefore, assuming incompressible flow, for laminar flow, the R must remain a constant between the model and its prototype (as a "true model" is sought), and for turbulent flow, where the prototype turbine is significantly larger than the model turbine (or vice versa), the ε/D must remain a constant between the model and its prototype (as a "true model" is sought). Finally, it is important to note that for numerous practical turbine flow problems, the drag coefficient, $C_D = \eta_{turbine} = (C_P/C_QC_H)$ is assumed to be independent of all three independent π terms R, C, and ε/D in Equation 11.177, which leads to the definition of the affinity laws (similarity rules) for the efficiency of homologous turbines and pumps; when dynamic similarity between the model and its prototype (or between two flow systems) is achieved, the model and its prototype (or the two flow systems) are said to be homologous.

11.7.8.5 Affinity Laws for the Efficiency of Homologous Turbines

The affinity laws (similarity rules) for the efficiency of homologous turbines and pumps assume that for numerous practical turbine and pump flow problems, the drag coefficient, $C_D = \eta_{turbine} = (C_P/C_QC_H)$ is independent of all three independent π terms R, C, and ε/D in Equation 11.177. Therefore, in order to achieve dynamic similarity between the model and its prototype (i.e., the model and the prototype are said to be homologous), the drag coefficient, $C_D = \eta_{turbine} = (C_P/C_QC_H)$ and thus the head coefficient, $C_H = (gh_{turbine}/\omega^2D^2)$; the capacity coefficient, $C_Q = (Q/\omega D^3)$; and the power coefficient, $C_P = (\omega T/\rho\omega^3D^5)$ each must remain a constant between the model and its prototype as follows:

$$\underbrace{\left[(\eta_{turbine})_p\right]}_{\left(\frac{C_P}{C_QC_H}\right)_p} = \underbrace{\left[(\eta_{turbine})_m\right]}_{\left(\frac{C_P}{C_QC_H}\right)_m} \tag{11.178}$$

$$\underbrace{\left[\left(\frac{gh_{turbine}}{\omega^2D^2}\right)_p\right]}_{(C_H)_p} = \underbrace{\left[\left(\frac{gh_{turbine}}{\omega^2D^2}\right)_m\right]}_{(C_H)_m} \tag{11.179}$$

$$\underbrace{\left[\left(\frac{Q}{\omega D^3}\right)_p\right]}_{(C_Q)_p} = \underbrace{\left[\left(\frac{Q}{\omega D^3}\right)_m\right]}_{(C_Q)_m} \tag{11.180}$$

$$\underbrace{\left[\left(\frac{\omega T}{\rho\omega^3D^5}\right)_p\right]}_{(C_P)_p} = \underbrace{\left[\left(\frac{\omega T}{\rho\omega^3D^5}\right)_m\right]}_{(C_P)_m} \tag{11.181}$$

Equations 11.178 through 11.181 are known as the affinity laws (similarity rules) for the efficiency of homologous turbines and pumps, and they represent the "pressure model" (i.e., $C_D = \eta_{turbine} = (C_P/C_QC_H) = E$ is independent of all three independent π terms R, C, and ε/D in Equation 11.177). This implies that, the turbine efficiency, $\eta_{turbine}$ (and thus the turbine design coefficients C_H, C_Q, and C_P) (i.e., the affinity laws) sufficiently defines the dynamic similarity requirements between the model and its prototype. Thus, in addition to the inertia force, F_I, the pressure force, F_P is important in the design of turbines (and pumps). Therefore, there are no independent π terms used in the definition of the dependent π term, E and thus the drag coefficient, $C_D = E = \eta_{turbine}$. As such, one may note that in the case where the similarity requirements regarding the dependent π term (i.e., $C_D = E = \eta_{turbine}$ and thus C_H, C_Q, and C_P) ("pressure model") between the model and its prototype (i.e., the affinity laws given in Equations 11.178 through 11.181) are satisfied, then application of the turbine design coefficients yields a "true model." However, in the case where the similarity requirements regarding the dependent π term (i.e., $C_D = E = \eta_{turbine}$ and thus C_H, C_Q, and C_P) ("pressure model") between the model and its prototype (i.e., the affinity laws given in Equations 11.178 through 11.181) ("pressure model") are not satisfied, then application of the turbine design coefficients yields a "distorted model" for which the appropriate correction measures are taken. Specifically, in order to model the scaling effects between a model and its prototype (i.e., account for a "distorted model"), an empirical equation developed by Moody is applied (see Section 11.7.8.6 below).

Because the efficiency of a turbine, $\eta_{turbine} = (C_P/C_QC_H)$ is expressed as a function of the turbine design coefficients C_H, C_Q, and C_P, the turbine affinity laws (Equations 11.178 through 11.181) ("pressure model") are used in the design of turbines. Specifically, the affinity laws ("pressure model") may be used to scale up or scale down between a model and its prototype or between two flow systems. Typically, in the design of a turbine, the turbine design coefficients C_H, C_Q, and C_P are evaluated from test data (model turbine), which are used to predict the performance of the prototype turbine. However, it is important to note that, even though the application of the affinity laws (similarity rules) for the design of homologous turbines and pumps assumes that the turbine efficiency, $\eta_{turbine} = (C_P/C_QC_H)$ is independent of all three independent π terms R, C, and ε/D in Equation 11.177 (i.e., assumes the "pressure model"), the assumption of a "pressure model" in itself may result in a "distorted model" that is due to the practical inability to maintain R or ε/D constant between a model and its prototype. Thus, it still remains important to address and account for (model) the effects of scaling (i.e., account for a "distorted model") between a model and its prototype resulting from the assumption of a "pressure model" (see Section 11.7.8.2 on pumps above for a similar detailed discussion). In summary, in the cases where there is the need but the practical inability to keep R or ε/D constant between a model and its prototype (essentially, the "pressure model" scaling requirements are not satisfied), then application of the turbine design coefficients will yield a "distorted model." If a "distorted model" results, appropriate correction measures are taken in order to model the scaling effects (i.e., account for a "distorted model") between the model and its prototype using an empirical equation developed by Moody (see Section 11.7.8.6 below).

11.7.8.6 Modeling of Scaling Effects for the Efficiency of Nonhomologous ("Distorted Models") Turbines Using the Moody Equation

The application of the affinity laws (similarity rules) for the design of homologous turbines and pumps provides a practical approach to scale up or scale down between a model and its prototype (or between two flow systems). In the case where the similarity requirements regarding the dependent π term (i.e., $C_D = E = \eta_{turbine}$ and thus C_H, C_Q, and

C_P) ("pressure model") between the model and its prototype (i.e., the affinity laws given in Equations 11.178 through 11.181) are satisfied, then application of the turbine design coefficients yield a "true model" (i.e., homologous turbines). However, in the case where the similarity requirements regarding the dependent π term (i.e., $C_D = E = \eta_{turbine}$ and thus C_H, C_Q, and C_P) ("pressure model") between the model and its prototype (i.e., the affinity laws given in Equations 11.178 through 11.181) ("pressure model") are not satisfied, then application of the turbine design coefficients yield a "distorted model" (i.e., nonhomologous turbines) for which the appropriate correction measures are taken. Specifically, in the cases where there is the need but the practical inability to keep R or ε/D constant between a model and its prototype, essentially, the "pressure model" scaling requirements (i.e., the affinity laws given in Equations 11.178 through 11.181) are not satisfied, then application of the turbine design coefficients will yield a "distorted model" (i.e., nonhomologous turbines). If a "distorted model" (i.e., nonhomologous turbines) results, appropriate correction measures are taken in order to model the scaling effects between the model and its prototype (i.e., account for a "distorted model"), using an empirical equation developed by Moody. The empirical Moody efficiency correction equation for turbines (and pumps), which assumes geometrically similar turbines (and pumps), is given as follows:

$$\frac{1 - \eta_p}{1 - \eta_m} = \left(\frac{D_m}{D_p}\right)^{1/5} \tag{11.182}$$

The above empirical equation by Moody provides a measure of the scaling effects (i.e., accounts for a "distorted model") between the model and its prototype on the turbine (or pump) efficiency. Thus, Equation 11.182 may be used to account for the variation in efficiency, η between the model and its prototype for a "distorted model" (i.e., nonhomologous turbines or pumps). Specifically, in the design of a turbine, while the turbine design coefficients C_H, C_Q, and C_P that are evaluated from test data (model turbine) may be used to predict the performance of the prototype turbine, the turbine efficiency computed from the test data (geometrically similar model turbine) may be used to predict the efficiency of the prototype turbine by applying Equation 11.182. Thus, the empirical Moody efficiency correction equation for turbines and pumps (Equation 11.182) implies that (models the fact that) the larger prototype turbines are typically more efficient than the smaller model turbines. In order to understand why the empirical Moody efficiency correction equation for turbines and pumps (Equation 11.182) implies that (models the fact that) the larger prototype turbines are typically more efficient than the smaller model turbines, see Section 11.7.8.3 on pumps above for a similar detailed discussion.

EXAMPLE PROBLEM 11.20

A 0.25-hp prototype water impulse turbine extracts 13 ft of freely available head with a freely available discharge of 252 US gpm. The diameter of the prototype turbine runner blades is 9.5 in with a runner rotational speed of 800 rpm. A smaller model water turbine of the prototype water impulse is designed in order to study the flow characteristics of the flow through a turbine. The model scale is 0.5. (a) Design a homologous model water impulse turbine. (b) Determine if the resulting designed model is a true model or a distorted model.

Mathcad Solution

(a) In order to design a homologous model water impulse turbine, the affinity laws (similarity rules) for the efficiency of homologous turbines (Equations 11.178 through 11.181) are applied as follows:

But, first, the turbine design coefficients, the hydraulic power input to the turbine, the turbine efficiency, and the Reynolds number are computed for the larger prototype water impulse turbine as follows:

$$\text{slug} := 1\,\text{lb}\frac{\sec^2}{\text{ft}} \qquad \rho_p := 1.936\frac{\text{slug}}{\text{ft}^3} \qquad g := 32.174\frac{\text{ft}}{\sec^2}$$

$$\gamma_p := \rho_p \cdot g = 62.289\frac{\text{lb}}{\text{ft}^3} \quad \text{hp} := 550\frac{\text{ftlb}}{\sec} \qquad P_{toutp} := 0.25\,\text{hp} = 137.5\frac{\text{ft}\cdot\text{lb}}{s}$$

$$h_{turbinep} := 13\,\text{ft} \qquad \text{USgal} := 0.133681\,\text{ft}^3 \quad Q_p := 252\frac{\text{USgal}}{\text{min}} = 0.561\frac{\text{ft}^3}{\sec}$$

$$D_p := 9.5\,\text{in} = 0.792\,\text{ft} \qquad\qquad \omega_p := 800\,\text{rpm} = 83.776\frac{1}{s}$$

$$C_{Hp} := \frac{g\cdot h_{turbinep}}{\omega_p^2\cdot D_p^2} = 0.095 \qquad\qquad C_{Qp} := \frac{Q_p}{\omega_p\cdot D_p^3} = 0.014$$

$$T_p := \frac{P_{toutp}}{\omega_p} = 1.641\,\text{ft}\cdot\text{lb} \qquad C_{Pp} := \frac{\omega_p\cdot T_p}{\rho_p\cdot\omega_p^3\cdot D_p^5} = 3.884\times 10^{-4}$$

$$\eta_{turbinep} := \frac{C_{Pp}}{C_{Qp}\cdot C_{Hp}} = 0.302 \qquad P_{tinp} := \gamma_p\cdot Q_p\cdot h_{turbinep} = 454.645\frac{\text{ft}\cdot\text{lb}}{s}$$

$$\eta_{turbinep} := \frac{P_{toutp}}{P_{tinp}} = 0.302 \qquad\qquad \mu_p := 21\times 10^{-6}\frac{\text{lb}\cdot\sec}{\text{ft}^2}$$

$$v_p := \omega_p\cdot D_p = 66.323\frac{\text{ft}}{s} \qquad\qquad R_p := \frac{\rho_p\cdot v_p\cdot D_p}{\mu_p} = 4.84\times 10^6$$

Then, in order to design a homologous smaller model water turbine, the affinity laws (similarity rules) for the efficiency of homologous turbines (Equations 11.178 through 11.181) are applied as follows:

$$\lambda := 0.5 \qquad \rho_m := \rho_p = 1.936\frac{\text{slug}}{\text{ft}^3} \qquad\qquad \gamma_m := \gamma_p = 62.289\frac{\text{lb}}{\text{ft}^3}$$

Guess value: $D_m := 6\,\text{in}$ $h_{turbinem} := 10\,\text{ft}$ $\omega_m := 100\,\text{rpm} = 10.472\frac{-1}{s}$

$$Q_m := 20\frac{\text{USgal}}{\text{min}} = 0.045\frac{\text{ft}^3}{\sec} \qquad T_m := 2\cdot\text{ft}\cdot\text{lb} \quad \eta_{turbinem} := 1$$

$$C_{Hm} := 1 \qquad\qquad C_{Qm} := 1 \qquad\qquad C_{Pm} := 1$$

Given

$$\lambda = \frac{D_m}{D_p} \qquad \frac{g\cdot h_{turbinep}}{\omega_p^2\cdot D_p^2} = \frac{g\cdot h_{turbinem}}{\omega_m^2\cdot D_m^2} \qquad \frac{Q_p}{\omega_p\cdot D_p^3} = \frac{Q_m}{\omega_m\cdot D_m^3}$$

$$\frac{\omega_p \cdot T_p}{\rho_p \cdot \omega_p^3 \cdot D_p^5} = \frac{\omega_m \cdot T_m}{\rho_m \cdot \omega_m^3 \cdot D_m^5} \qquad C_{Hm} = \frac{g \cdot h_{turbinem}}{\omega_m^2 \cdot D_m^2} \qquad C_{Qm} = \frac{Q_m}{\omega_m \cdot D_m^3}$$

$$C_{Pm} = \frac{\omega_m \cdot T_m}{\rho_m \cdot \omega_m^3 \cdot D_m^5} \qquad \eta_{turbinem} = \frac{C_{Pp}}{C_{Qp} \cdot C_{Hp}} \qquad \eta_{turbinep} = \eta_{turbinem}$$

$$\begin{pmatrix} D_m \\ h_{turbinem} \\ \omega_m \\ Q_m \\ T_m \\ \eta_{turbinem} \\ C_{Hm} \\ C_{Qm} \\ C_{pm} \end{pmatrix} := \text{Find } (D_m, h_{utrbinem}, \omega_m, Q_m, T_m, \eta_{turbinem}, C_{Hm}, C_{Qm}, C_{Pm})$$

$D_m = 4.75 \text{ in}$ $\qquad\qquad h_{turbinem} = 0.241 \text{ ft}$

$\omega_m = 22.807 \dfrac{1}{s}$ $\qquad\qquad \omega_m = 217.793 \text{ rpm}$

$Q_m = 0.019 \dfrac{ft^3}{sec}$ $\qquad\qquad Q_m = 8.576 \dfrac{USgal}{min}$

$T_m = 3.801 \times 10^{-3} \text{ ft} \cdot \text{lb}$ $\qquad \eta_{turbinem} = 0.302$

$C_{Hm} = 0.095$ $\qquad\qquad C_{Qm} = 0.014$ $\qquad\qquad C_{Pm} = 3.884 \times 10^{-4}$

And, finally, the hydraulic power input to the turbine, the mechanical power output from the turbine, the turbine efficiency, and the Reynolds number are computed for the smaller model water turbine as follows:

$P_{tinm} := \gamma_m \cdot Q_m \cdot h_{turbinem} = 0.287 \dfrac{ft \cdot lb}{s}$ $\qquad P_{tinm} = 5.212 \times 10^{-4} \text{ hp}$

$P_{toutm} = \omega_m \cdot T_m = 0.08 \dfrac{ft \cdot lb}{s}$ $\qquad\qquad P_{toutm} = 1.576 \times 10^{-4} \text{ hp}$

$\eta_{turbinem} := \dfrac{P_{toutm}}{P_{tinm}} = 0.302$ $\qquad\qquad \mu_m := 21 \times 10^{-6} \dfrac{lb \cdot sec}{ft^2}$

$v_m := \omega_m \cdot D_m = 9.028 \dfrac{ft}{s}$ $\qquad\qquad R_m := \dfrac{\rho_m \cdot v_m \cdot D_m}{\mu_m} = 3.294 \times 10^5$

(b) In order to determine if the resulting designed model is a true model or a distorted model, one must first determine whether the application of the affinity laws (similarity rules) for the efficiency of homologous turbines (Equations 11.178 through 11.181) has yielded a practically feasible designed model. If the resulting designed smaller model water turbine is practically feasible (design parameters: $D_m = 4.75$

in, $h_{turbinem} = 0.241$ ft, $\omega_m = 217.793$ rpm, $Q_m = 0.019$ cfs $= 8.576$ US gpm, $T_m = 3.801 \times 10^{-3}$ ft-lb, $P_{toutm} = 1.576 \times 10^{-4}$ hp), then it is considered to be a true model (model and prototype are homologous turbines). However, if the resulting designed smaller model water turbine is not practically feasible (at least one of the design parameters is not practically feasible), then it is considered to be a distorted model (model and prototype are nonhomologous turbines). Upon examination of the resulting model design parameters (design parameters: $D_m = 4.75$ in, $h_{turbinem} = 0.241$ ft, $\omega_m = 217.793$ rpm, $Q_m = 0.019$ cfs $= 8.576$ US gpm, $T_m = 3.801 \times 10^{-3}$ ft-lb, $P_{toutm} = 1.576 \times 10^{-4}$ hp) and a comparison with some turbine manufacture design data parameters, it appears that such a model water turbine is indeed practically feasible. Furthermore, the $R_p = 4.84 \times 10^6$, while the $R_m = 3.294 \times 10^5$, which indicates that turbulent flow is maintained between the prototype turbine and the model turbine. However, in most laboratory situations, there will be some constraints imposed upon the model design; these are called "design conditions." For instance, a certain laboratory may impose a limitation on the available $h_{turbinem}$ and/or a limitation on the available Q_m or ω_m, etc. that it is capable of providing ("design conditions"). Furthermore, for a smaller model turbine, it may not be practically feasible to attain high speeds, and/or cavitation may occur at higher speeds. Thus, when there are certain "design conditions" imposed due to practical limitations/constraints, the application of the affinity laws (similarity rules) for the efficiency of homologous turbines (Equations 11.178 through 11.181) will yield a distorted model (model and prototype are nonhomologous turbines). If a "distorted model" results, appropriate correction measures are taken in order to model the scaling effects (i.e., account for a "distorted model") between the model and its prototype, using the empirical equation developed by Moody (Equation 11.182) as follows:

Guess value: $\qquad\qquad\qquad\qquad \eta_{turbinem} := 0.1$

Given

$$\frac{1 - \eta_{turbinep}}{1 - \eta_{turbinem}} = \left(\frac{D_m}{D_p}\right)^{\frac{1}{5}}$$

$\eta_{turbinem} := \text{Find}\,(\eta_{turbinem}) = 0.199$

which indicates that the estimated turbine efficiency of the designed smaller model water turbine is only 0.199, rather than the designed turbine efficiency of the model turbine of 0.302; larger turbines are typically more efficient than the smaller turbines.

End-of-Chapter Problems

Problems with a "C" are conceptual problems. Problems with a "BG" are in English units. Problems with an "SI" are in metric units. Problems with a "BG/SI" are in both English and metric units. All "BG" and "SI" problems that require computations are solved using Mathcad.

Introduction

11.1C Deterministic mathematical models are used in order to predict the performance of physical fluid flow situations. As such, in the deterministic mathematical modeling of physical fluid flow situations (both internal and external flow), while some fluid flow situations may be modeled by the application of pure theory (e.g., creeping flow, laminar flow, sonic flow, or critical flow), some (actually, most, due to the assumption of real turbulent flow) fluid flow situations are too complex for theoretical modeling (e.g., turbulent flow, subsonic [supersonic or hypersonic] flow, or subcritical [or supercritical] flow). However, regardless of the complexity of the fluid flow situation, the analysis phase of mathematical modeling involves the formulation, calibration, and verification of the mathematical model, while the subsequent synthesis or design phase of mathematical modeling involves application of the mathematical model in order to predict the performance of the fluid flow situation. Dimensional analysis and dynamic similitude (i.e., the laws governing dynamic similarity) are essential and invaluable tools in the deterministic mathematical modeling of fluid flow problems (or any problems other than fluid flow). Explain how the topics of dimensional analysis and dynamic similitude are significantly related to one another. Furthermore, explain the role of each (dimensional analysis and dynamic similitude) in the deterministic modeling of fluid flow problems.

11.2C In the deterministic mathematical modeling of physical fluid flow situations (both internal and external flow), while some fluid flow situations may be modeled by the application of pure theory (e.g., creeping flow, laminar flow, sonic flow, or critical flow), some (actually, most, due to the assumption of real turbulent flow) fluid flow situations are too complex for theoretical modeling (e.g., turbulent flow, subsonic [or supersonic or hypersonic] flow, or subcritical [or supercritical] flow). Explain the relative importance of the role of the dimensional analysis procedure itself and the results of the procedure, including the laws of similitude, in the theoretical vs. the empirical modeling of fluid flow.

The Role of Dynamic Similitude in the Empirical Modeling of Flow Resistance in Real Fluid Flow

11.3C A major application of the dimensional analysis procedure and its results, including the laws of similitude, in fluid flow is for problems that are too complex to be modeled by the application of pure theory. In particular, the modeling of flow resistance (due to turbulent flow) in internal and external real fluid flow is too complex for theoretical solution. List the flow resistance equations in fluid flow, which require the application of the dimensional analysis procedure and its results, including the laws of similitude. Furthermore, briefly explain the role of each (dimensional analysis and dynamic similitude) in the modeling of flow resistance.

Developing and Applying the Laws of Dynamic Similitude to Design Geometrically Scaled Physical Models of Real Fluid Flow

11.4C Briefly explain how the laws of dynamic similitude are developed and applied to any fluid flow situation, including real fluid flow situations that require the empirical modeling of flow resistance.

11.5C   Explain the steps involved in the development of the laws of dynamic similarity.

11.6C   What do the primary scale ratios model?

11.7C   What are the five main π terms?

11.8C   The laws governing dynamic similarity/similitude involve the definition of secondary/similitude scale ratios. What are the six sources of dynamic forces?

Primary Scale Ratios

11.9C   In the development of criteria/laws governing dynamic similarity/similitude, there are three types of similarity. List the three types of similarity, and explain how they are related to one another. Furthermore, what type of similarity is required in order for a model to be similar to its prototype in the most complete sense (i.e., appropriate for practical design applications of fluid flow problems)?

11.10C   How are three types of similarity defined? Furthermore, explain the relative importance of each in the design of a model or its prototype.

Geometric Similarity

11.11C   What does geometry similarity between a model and prototype imply? Give the expression for its definition. What is an alternative/additional term used to define geometric similarity?

Kinematic Similarity

11.12C   What does kinematic similarity between a model and prototype imply? Give the expression for its definition. Also, give the expressions for the primary time ratio, and the primary acceleration ratio.

Dynamic Similarity

11.13C   What does dynamic similarity between a model and prototype imply? Give the expression for its definition.

11.14C   Forces that may act in a flow situation include those due to pressure, F_P; gravity, F_G; viscosity, F_V; elasticity, F_E; surface tension, F_T; and the inertia force, F_I. As noted in Chapter 7, depending upon the fluid flow problem, while the inertia force will always be important, some of the above-listed forces may not play a significant role. Explain why the inertia force will always be important and how it is modeled in the dynamic similarity.

11.15C   Given that the inertia force will always be important, and given the importance of an additional specific force in the flow situation, a practical (and alternative to Equation 11.7) approach for expressing the force scale ratio, F_r that expresses the minimum requirement for maintaining both steady-state flow, and dynamic similarity may be expressed as the ratio of the inertia force to the corresponding additional force. Assuming that all other sources of forces are important in a fluid flow situation, give and name the resulting expressions (alternative to Equation 11.7) for the force scale ratio, F_r.

Interpretation of the Main Π Terms

11.16C In the development of the laws of dynamic similarity, a practical alternative approach (alternative to Equation 11.7) for expressing the primary force scale ratio, F_r that expresses the minimum requirement for maintaining both steady-state flow and dynamic similarity may be expressed as the ratio of the inertia force to the corresponding additional force. What five main π terms does this approach yield, and what common ratio (other than force) does each π term represent?

The Euler Number

11.17C Give the expression for the Euler number, E, and state how dynamic similarity is maintained when the pressure force is also (in addition to the inertia force in order to maintain steady-state flow) important in a fluid flow situation.

The Froude Number

11.18C Give the expression for the Froude number, F, and state how dynamic similarity is maintained when the gravity force is also (in addition to the inertia force in order to maintain steady-state flow) important in a fluid flow situation.

The Reynolds Number

11.19C Give the expression for the Reynolds number, R, and state how dynamic similarity is maintained when the viscous force is also (in addition to the inertia force in order to maintain steady-state flow) important in a fluid flow situation.

The Cauchy Number

11.20C Give the expression for the Cauchy number, C, and state how dynamic similarity is maintained when the elastic force is also (in addition to the inertia force in order to maintain steady-state flow) important in a fluid flow situation.

The Weber Number

11.21C Give the expression for the Weber number, W, and state how dynamic similarity is maintained when the surface tension force is also (in addition to the inertia force in order to maintain steady-state flow) important in a fluid flow situation.

Laws Governing Dynamic Similirity: Secondary/Similitude Scale Ratios

11.22C In the development of the laws of dynamic similarity, the uncalibrated mathematical model formulated using dimensional analysis, $\pi_1 = \phi(nothing) = constant$ for a theoretical model, or $\pi_1 = \phi(\pi_2, \pi_3, \ldots)$ for an empirical model, is used to tailor the laws of dynamic similitude for a specific flow resistance model, depending upon the relative importance of a given dynamic force. In the design of a model or its prototype, what does the application of the primary force scale ratio, F_r for a particular dominant force (in addition to the inertia force) (or the corresponding π term) result in the derivation of (and what does it imply?)

11.23C Assuming that in addition to the inertia force, F_I a single force predominates a given flow situation, one may derive secondary/similitude scale ratios that represent a specific dominant force. What are the similitude scale ratios used for?

11.24C   One may derive a similitude scale ratio for all physical quantities (geometrics, kinematics, and dynamics) and all physical properties of fluids used with the model and the prototype. Explain which of the similitude scale ratios are unique for a specific "force model" (pressure model, gravity model, viscosity model, elastic model, and surface tension model) and which are not.

Similitude Scale Ratios for Physical Quantities

11.25C   Beginning with the primary force scale ratio, F_r for a particular dominant force (in addition to the inertia force), or the corresponding π term, one may derive the similitude scale ratios for the physical quantities for the specific type of "force model." The similitude scale ratios are derived for what physical quantities?

Similitude Scale Ratios for Physical Properties of Fluids

11.26C   One may derive a similitude scale ratio for all physical properties of fluids used with the model and the prototype. The similitude scale ratios are derived for what typical fluid properties?

The Role and the Relative Importance of the Dynamic Forces in Dynamic Similitude

11.27C   In the development of the laws of dynamic similarity, the uncalibrated mathematical model formulated using dimensional analysis, $\pi_1 = \phi(nothing) = constant$ for a theoretical model, or $\pi_1 = \phi(\pi_2, \pi_3, \ldots)$ for an empirical model, is used to tailor the laws of dynamic similitude for a specific flow resistance model, depending upon the relative importance of a given dynamic force. Briefly summarize the role and the relative importance of the dynamic forces in dynamic similitude, which include the inertia force, F_I; the pressure force, F_P; the gravity, F_G; the viscous force, F_V; the elastic force, F_E; and the surface tension force, F_T.

11.28C   In the development of the laws of dynamic similarity, the uncalibrated mathematical model formulated using dimensional analysis, $\pi_1 = \phi(nothing) = constant$ for a theoretical model, or $\pi_1 = \phi(\pi_2, \pi_3, \ldots)$ for an empirical model, is used to tailor the laws of dynamic similitude for a specific flow resistance model, depending upon the relative importance of a given dynamic force. While the inertia force, F_I will always be a dominant force in order to maintain steady-state flow, given the importance of an additional specific force in the flow situation (pressure, gravity, viscosity, elasticity, or surface tension), either its corresponding force ratio or alternatively its corresponding π term (E, F, R, C, or W) will be important in order to maintain dynamic similarity. Explain why it is important to distinguish between the role of the friction/viscous force for turbulent flow, F_f or $F_V = \tau A$, and the role of the friction/viscous force for laminar flow, $F_V = \tau A = \mu v L$ in dynamic similitude.

Guidelines in the Application of the Laws Governing Dynamic Similarity

11.29C   In the application of the laws of dynamic similarity, the uncalibrated mathematical model formulated using dimensional analysis, $\pi_1 = \phi(nothing) = constant$ for a theoretical model, or $\pi_1 = \phi(\pi_2, \pi_3, \ldots)$ for an empirical model is used to tailor the laws of dynamic similitude for a specific flow resistance model, depending upon the relative importance of a given dynamic force. What does the application of the laws governing dynamic similarity for real flow situations (flow resistance problems) involve?

11.30C   What is the goal in the application of the laws of similarity (similitude), and, briefly, how is this goal accomplished?

11.31C In the application of the laws governing dynamic similarity for real flow situations (flow resistance problems), the first step is to define the flow resistance prediction equation (and the prediction equation for any other physical quantity, such as the empirical expressions for the efficiency of pumps and turbines) by equating the drag coefficient, C_D (dependent π term) for the model and the prototype. Explain the second step, which is to define the dynamic similarity requirements between the model and its prototype.

11.32C Explain the important role of the similitude velocity scale ratio in the application of the laws governing dynamic similarity for real flow situations (flow resistance problems). Furthermore, explain its important role when more than one additional "force model" is required to be satisfied.

11.33C In the application of the laws governing dynamic similarity for real flow situations (flow resistance problems), the first step is to define the flow resistance prediction equation by equating the drag coefficient, C_D (dependent π term) for the model and the prototype. The second step is to define the dynamic similarity requirements between the model and its prototype. Explain the third step, which is to determine if the flow resistance prediction results in a "true model" or a "distorted model." Furthermore, explain what causes the occurrence of a "distorted model." Finally, explain if one may successfully apply a "distorted model" for the flow resistance prediction equation.

11.34C In the application of the laws governing dynamic similarity for real flow situations (flow resistance problems), depending upon the practical challenges imposed by the dynamic similarity requirements, the flow resistance prediction equation results will result in either a "true model" or a "distorted model." Explain in detail the difference between a "true model" vs. a "distorted model."

11.35C In the application of the laws governing dynamic similarity for real flow situations (flow resistance problems), because the occurrence of a "distorted model" for the flow resistance prediction equation (dependent π term) is common in modeling, it is important to identify the specific causes of distortion, and to optimize the conditions for achieving dynamic similarity. What are some specific causes of model distortion?

11.36C When working with a "distorted model" for the flow resistance prediction equation (and the prediction equation for any other physical quantity, such as the empirical expressions for the efficiency of pumps and turbines), the goal is to optimize the conditions for achieving dynamic similarity in order for the "distorted model" to be as close as possible to a "true model." Explain how this goal is accomplished.

Application of the Laws Governing Dynamic Similarity for Flow Resistance Equations and Efficiency of Pumps and Turbines

Application of the Similitude Scale Ratios for the Drag Force in External Flow

11.37C For the case of external flow around an object, give the uncalibrated mathematical model of the fluid flow situation derived using dimensional analysis.

Creeping Flow (R ≤ 1) for Any Shape Body

11.38 BG Water at 70°F flows over a prototype sphere with a diameter of 0.6 ft, as illustrated in Figure EP 11.1. An enlarged model of the smaller prototype is designed in order to study the flow characteristics of creeping flow over a sphere (Stokes law). The model fluid is glycerin ($s = 1.258$, $\mu = 31{,}200 \times 10^{-6}$ lb-sec/ft^2), the velocity of flow of the glycerin over the enlarged model sphere is 2.35×10^{-4} ft/sec, and the model scale, λ is 4. (a) Determine the drag force on the model sphere. (b) Determine the velocity of flow of the water over the prototype sphere in order to achieve dynamic similarity between the model and the prototype. (c) Determine the drag force on the prototype sphere in order to achieve dynamic similarity between the model and the prototype.

Laminar Flow (R < 10,000) for Any Shape Body except Round-Shaped Bodies

11.39BG   Water at 90°F flows over a prototype 1400-ft-long elliptical rod with the following geometry: an L of 12 ft and a D of 0.41 ft, as illustrated in Figure EP 11.2. A smaller model of the larger prototype is designed in order to study the flow characteristics of laminar flow over the streamlined body. The model fluid is glycerin ($s = 1.26$, $\mu = 31{,}200 \times 10^{-6}$ lb-sec/ ft^2), the velocity of flow of the glycerin over the smaller model elliptical rod is 2 ft/sec, and the model scale, λ is 0.30. (a) Determine the drag force on the model elliptical rod. (b) Determine the velocity of flow of the water over the prototype elliptical rod in order to achieve dynamic similarity between the model and the prototype. (c) Determine the drag force on the prototype elliptical rod in order to achieve dynamic similarity between the model and the prototype.

11.40BG   Oil ($s = 0.88, \mu = 150 \times 10^{-6}$ lb-sec/ft^2) flows at a velocity of 1.1 ft/sec over a prototype short cylinder with a diameter of 5 ft and height of 50 ft, as illustrated in Figure EP 11.3. A smaller model of the larger prototype is designed in order to study the flow characteristics of laminar flow over the blunt body. The model fluid is glycerin ($s = 1.26$, $\mu = 31{,}200 \times 10^{-6}$ lb-sec/ft^2), and the model scale, λ is 0.5. (a) Determine the drag force on the prototype short cylinder. (b) Determine the velocity of flow of the glycerin over the model short cylinder in order to achieve dynamic similarity between the model and the prototype. (c) Determine the drag force on the model short cylinder in order to achieve dynamic similarity between the model and the prototype.

Turbulent Flow (R > 10,000) for Any Shape Body except Round-Shaped Bodies

11.41BG   (Refer to ECP 10.29.) Air at standard atmosphere at an altitude of 25,000 ft above sea level flows at a velocity of 700 ft/sec over a prototype 1800-ft-long square shaped cylinder with a side of 7 ft, as illustrated in Figure EP 11.4. A smaller model of the larger prototype is designed in order to study the flow characteristics of turbulent flow over the blunt body. The model fluid is oxygen at 68°F, and the model scale, λ is 0.30. (a) Determine the drag force on the prototype square cylinder. (b) Determine the velocity of flow of the oxygen over the model square cylinder and the required pressure of the oxygen in order to achieve dynamic similarity between the model and the prototype. (c) Determine the drag force on the model square cylinder in order to achieve dynamic similarity between the model and the prototype.

11.42BG   Air at standard atmosphere at an altitude of 25,000 ft above sea level flows at a velocity of 800 ft/sec over a prototype 1800-ft-long elliptical cylinder with the following geometry: an L of 2.5 ft and a D of 0.625 ft, as illustrated in Figure EP 11.5. A smaller model of the larger prototype is designed in order to study the flow characteristics of turbulent flow over the streamlined body. The model fluid is carbon dioxide at 68°F, and the model scale, λ is 0.20. (a) Determine the drag force on the prototype elliptical cylinder. (b) Determine the velocity of flow of the carbon dioxide over the model elliptical cylinder and the required pressure of the carbon dioxide in order to achieve dynamic similarity between the model and the prototype. (c) Determine the drag force on the model elliptical cylinder in order to achieve dynamic similarity between the model and the prototype.

Laminar and Turbulent Flow for Round-Shaped Bodies (Circular Cylinder or Sphere)

11.43BG Water at 70°F flows over a prototype circular cylinder with a diameter of 6 ft and a length of 2000 ft, as illustrated in Figure EP 11.6. A smaller model of the larger prototype is designed in order to study the flow characteristics of laminar flow over the circular cylinder. The model fluid is crude oil at 68°F ($s = 0.86$, $\mu = 150 \times 10^{-6}$ lb-sec/ft^2), the velocity of flow of the oil over the smaller model circular cylinder is 2 ft/sec, and the model scale, λ is 0.7. (a) Determine the drag force on the model circular cylinder. (b) Determine the velocity of flow of the water over the prototype circular cylinder in order to achieve dynamic similarity between the model and the prototype. (c) Determine the drag force on the prototype circular cylinder in order to achieve dynamic similarity between the model and the prototype.

11.44BG Air at standard atmosphere at an altitude of 10,000 ft above sea level flows at a velocity of 1300 ft/sec over a prototype sphere with a diameter of 2 ft and an absolute surface roughness of 0.025 ft, as illustrated in Figure EP 11.7. A smaller model of the larger prototype is designed in order to study the flow characteristics of turbulent flow over the sphere. The model fluid is helium at 68°F, and the model scale, λ is 0.33. (a) Determine the drag force on the prototype sphere. (b) Determine the velocity of flow of the helium over the model sphere and the required pressure of the helium in order to achieve dynamic similarity between the model and the prototype. (c) Determine the drag force on the model sphere in order to achieve dynamic similarity between the model and the prototype.

Laminar and Turbulent Flow with Wave Action at the Free Surface for Any Shape Body

11.45BG Water at 50°F flows at a velocity of 7 ft/sec over a prototype 40-ft-long streamlined hull with a width of 40 ft and a depth of 5 ft, as illustrated in Figure EP 11.8. A smaller model of the larger prototype is designed in order to study the flow characteristics of the flow over the streamlined body with wave action at the free surface. The model fluid is water at 70°F, and the model scale, λ is 0.25. (a) Determine the drag force on the prototype streamlined hull. (b) Determine the velocity of flow of the water over the model streamlined hull in order to achieve dynamic similarity between the model and the prototype. (c) Determine the drag force on the model streamlined hull in order to achieve dynamic similarity between the model and the prototype.

11.46BG (Refer to ECP 11.45.) Water at 50°F flows at a velocity of 7 ft/sec over a prototype 40-ft-long streamlined hull with a width of 40 ft and a depth of 5 ft, as illustrated in Figure EP 11.9. A smaller model of the larger prototype is designed in order to study the flow characteristics of the flow over the streamlined body with wave action at the free surface. The model scale, λ is 0.25. (a) Determine the drag force on the prototype streamlined hull. (b) Determine the model fluid and the velocity of flow of the model fluid over the model streamlined hull in order to achieve dynamic similarity between the model and the prototype. (c) Determine the drag force on the model streamlined hull in order to achieve dynamic similarity between the model and the prototype.

Application of the Similitude Scale Ratios for the Major Head Loss in Pipe Flow

11.47C For the case of internal pipe flow, give the uncalibrated mathematical model of the fluid flow situation derived using dimensional analysis.

Laminar Pipe Flow

11.48BG   SAE 30 oil at 68°F flows in a prototype 1800-ft-long circular pipe with a diameter of 3 ft, as illustrated in Figure EP 11.10. A smaller model of the larger prototype is designed in order to study the flow characteristics of laminar pipe flow. The model fluid is glycerin at 68°F, the velocity of glycerin in the smaller model pipe is 29 ft/sec, and the model scale, λ is 0.3. The flow resistance is modeled by the friction factor, $f = 8\,C_D$. (a) Determine the pressure drop (and head loss) in the flow of the glycerin in the model. (b) Determine the velocity flow of the SAE 30 oil in the prototype pipe flow in order to achieve dynamic similarity between the model and the prototype. (c) Determine the pressure drop (and head loss) in the flow of the SAE 30 oil in the prototype in order to achieve dynamic similarity between the model and the prototype.

Completely Turbulent Pipe Flow (Rough Pipes)

11.49BG   Water at 68°F flows in a prototype 2100-ft-long circular pipe with a diameter of 2 ft and an absolute pipe roughness of 0.04 ft, as illustrated in Figure EP 11.11. A smaller model of the larger prototype is designed in order to study the flow characteristics of turbulent pipe flow. The model fluid is water at 70°F, the velocity of water in the smaller model pipe is 60 ft/sec, and the model scale, λ is 0.4. The flow resistance is modeled by the friction factor, $f = 8\,C_D$. (a) Determine the pressure drop (and head loss) in the flow of the water in the model. (b) Determine the velocity flow of the water in the prototype pipe flow in order to achieve dynamic similarity between the model and the prototype. (c) Determine the pressure drop (and head loss) in the flow of the water in the prototype in order to achieve dynamic similarity between the model and the prototype.

Transitional Pipe Flow

11.50BG   (Refer to ECP 11.49.) Water at 68°F flows in a prototype 2100-ft-long circular pipe with a diameter of 2 ft and an absolute pipe roughness of 0.04 ft, as illustrated in Figure EP 11.12. A smaller model of the larger prototype is designed in order to study the flow characteristics of transitional pipe flow. The model fluid is water at 70°F, the velocity of water in the smaller model pipe is 0.04 ft/sec, and the model scale, λ is 0.4. The flow resistance is modeled by the friction factor, $f = 8\,C_D$. (a) Determine the pressure drop (and head loss) in the flow of the water in the model. (b) Determine the velocity flow of the water in the prototype pipe flow in order to achieve dynamic similarity between the model and the prototype. (c) Determine the pressure drop (and head loss) in the flow of the water in the prototype in order to achieve dynamic similarity between the model and the prototype.

Application of the Similitude Scale Ratios for the Major Head Loss in Open Channel Flow

11.51C   For the case of internal open channel flow, give the uncalibrated mathematical model of the fluid flow situation derived using dimensional analysis.

Turbulent Open Channel Flow

11.52BG   Water at 70°F flows in a prototype 2000-ft-long cement mortar surface rectangular open channel with a width of 6 ft, a uniform flow depth of 4 ft, and an absolute channel roughness of 0.02 ft, as illustrated in Figure EP 11.13. A smaller model of the larger prototype is designed in order to study the flow characteristics of turbulent open channel flow. The model fluid is also water at 70°F, the velocity of water in the smaller model channel is 75 ft/sec, and the model scale, λ is 0.2. The flow resistance is modeled by the Manning's roughness coefficient, $n = f'' (C_D)$. (a) Determine the friction slope (and head loss) in the flow of the water in the model. (b) Determine the velocity flow of the water in the prototype open channel flow in order to achieve dynamic similarity between the model and the prototype. (c) Determine the friction slope (and head loss) in the flow of the water in the prototype in order to achieve dynamic similarity between the model and the prototype.

Application of the Similitude Scale Ratios for the Minor Head Loss in Pipe Flow

11.53C For the case of internal pipe flow with pipe components (transitions or devices), give the uncalibrated mathematical model of the fluid flow situation derived using dimensional analysis.

Turbulent Pipe Flow with Pipe Component

11.54BG Air at 68°F flows in a prototype circular pipe with a diameter of 5 ft and an absolute pipe roughness of 0.005 ft fitted with a 90° bend with a round radius, r of 10 ft, as illustrated in Figure EP 11.14. A smaller model of the larger prototype is designed in order to study the flow characteristics of turbulent pipe flow with a pipe component. The model fluid is water at 70°F, the velocity of water in the smaller model pipe is 50 ft/sec, and the model scale, λ is 0.2. The flow resistance is modeled by the minor loss coefficient, $k = C_D$. (a) Determine the pressure drop (and head loss) in the flow of the water in the model. (b) Determine the velocity flow of the air in the prototype pipe flow in order to achieve dynamic similarity between the model and the prototype. (c) Determine the pressure drop (and head loss) in the flow of the air in the prototype in order to achieve dynamic similarity between the model and the prototype.

Application of the Similitude Scale Ratios for the Actual Discharge in Pipe Flow

11.55C For the case of the actual discharge in pipe flow (flow-measuring device), give the uncalibrated mathematical model of the fluid flow situation derived using dimensional analysis.

Pipe Flow with Flow-Measuring Device

11.56BG Water at 70°F flows in a prototype circular pipe with a diameter of 8 ft fitted with an orifice meter with a diameter of 3.2 ft, as illustrated in Figure EP 11.15. A smaller model of the larger prototype is designed in order to study the flow characteristics of pipe flow in the flow-measuring device. The model fluid is SAE 10 oil at 68°F ($s = 0.918$, $\mu = 1700 \times 10^{-6}$ lb-sec/ft^2), the ideal velocity (see Equation 8.140) of flow of the oil in the smaller model pipe fitted with an orifice is 4 ft/sec, and the model scale, λ is 0.3. (a) Determine the actual discharge in the model orifice meter. (b) Determine the ideal velocity of flow of the water in the prototype orifice meter in order to achieve dynamic similarity between the model and the prototype. (c) Determine the actual discharge in the prototype orifice meter in order to achieve dynamic similarity between the model and the prototype.

Application of the Similitude Scale Ratios for the Actual Discharge in Open Channel Flow

11.57C For the case of the actual discharge in open channel flow (flow-measuring device), give the uncalibrated mathematical model of the fluid flow situation derived using dimensional analysis.

Open Channel Flow with Sluice Gate or Venturi Meter

11.58BG   Water at 70°F flows in a prototype rectangular open channel with a width of 8 ft, and a sluice gate with a free outflow is inserted in the channel, as illustrated in Figure EP 11.16, in order to measure the flowrate. The opening of the gate is set at 2.584 ft, the depth of flow upstream of the sluice gate is 5.248 ft, and the depth of flow downstream of the sluice gate is 2.488 ft. A smaller model of the larger prototype is designed in order to study the flow characteristics of open channel flow in the flow-measuring device. The model fluid is also water at 70°F, the ideal downstream velocity (see Equation 9.285) of flow of the water, v_{2i} in the smaller model open channel fitted with a sluice gate is 6.768 ft/sec, and the model scale, λ is 0.2. (a) Determine the actual discharge in the model sluice gate. (b) Determine the ideal downstream velocity of flow of the water in the prototype sluice gate in order to achieve dynamic similarity between the model and the prototype. (c) Determine the actual discharge in the prototype sluice gate in order to achieve dynamic similarity between the model and the prototype.

Open Channel Flow with Weir or Spillway with Large Head

11.59BG   Water at 70°F flows in a prototype rectangular open channel with a width of 10 ft, and a spillway with a large head is inserted in the channel, as illustrated in Figure EP 11.17. The head on the spillway is 3.95 ft, the height of the spillway is 4.9 ft, and the width of the crest of the spillway is 10 ft. A smaller model of the larger prototype is designed in order to study the flow characteristics of open channel flow in the flow-measuring device. The model fluid is also water at 70°F, the ideal discharge (see Equation 9.297) in the smaller model channel fitted with a spillway is 9.5 ft³/sec, and the model scale, λ is 0.33. (a) Determine the actual discharge in the model spillway. (b) Determine the ideal discharge in the prototype spillway in order to achieve dynamic similarity between the model and the prototype. (c) Determine the actual discharge in the prototype spillway in order to achieve dynamic similarity between the model and the prototype.

Open Channel Flow with Weir or Spillway with Small Head

11.60BG   Water at 70°F flows in a prototype rectangular open channel with a width of 10 ft, and a spillway with a small head is inserted in the channel, as illustrated in Figure EP 11.18. The head on the spillway is 1.2 ft, the height of the spillway is 4.9 ft, and the width of the crest of the spillway is 10 ft. A smaller model of the larger prototype is designed in order to study the flow characteristics of open channel flow in the flow-measuring device. The model fluid is also water at 70°F, the ideal discharge (see Equation 9.297) in the smaller model channel fitted with a spillway is 2.5 ft³/sec, and the model scale, λ is 0.33. (a) Determine the actual discharge in the model spillway. (b) Determine the ideal discharge in the prototype spillway in order to achieve dynamic similarity between the model and the prototype. (c) Determine the actual discharge in the prototype spillway in order to achieve dynamic similarity between the model and the prototype.

Application of the Similitude Scale Ratios for the Efficiency of Pumps and Turbines

Similitude Scale Ratios for the Efficiency of Pumps

11.61C   For the efficiency of a pump, give the uncalibrated mathematical model of the fluid flow situation derived using dimensional analysis.

11.62BG   A 5-hp prototype water pump is designed to deliver a total head of 200 ft at a discharge of 66 US gpm. The diameter of the prototype pump impeller blades is 9.5 in with an impeller rotational speed of 1850 rpm. A smaller model water pump of the prototype water pump is designed in order to study the flow characteristics of the flow through a pump. The model scale is 0.5. (a) Design a homologous model water pump. (b) Determine if the resulting designed model is a true model or a distorted model.

Similitude Scale Ratios for the Efficiency of Turbines

11.63C   For the efficiency of a turbine, give the uncalibrated mathematical model of the fluid flow situation derived using dimensional analysis.

11.64BG   A 0.45-hp prototype water impulse turbine extracts 15 ft of freely available head with a freely available discharge of 275 US gpm. The diameter of the prototype turbine runner blades is 9.8 in with a runner rotational speed of 810 rpm. A smaller model water turbine of the prototype water impulse is designed in order to study the flow characteristics of the flow through a turbine. The model scale is 0.4. (a) Design a homologous model water impulse turbine. (b) Determine if the resulting designed model is a true model or a distorted model.

Appendix A: Physical Properties of Common Fluids

TABLE A.1

Physical Properties for the International Civil Aviation Organization (ICAO) Standard Atmosphere as a Function of Elevation above Sea Level

Elevation above Sea Level ft	Temperature (θ) °F	Absolute Pressure (p) psia	Density (ρ) slug/ft³	Specific Weight (γ) lb/ft³	Absolute (Dynamic) Viscosity (μ) 10^{-6} lb-sec/ft²	Kinematic Viscosity (ν) 10^{-3} ft²/sec	Speed of Sound (c) ft/sec	Acceleration due to Gravity (g) ft/sec²
0	59.000	14.69590	0.002376800	0.0764720	0.37372	0.15724	1116.45	32.174
5000	41.173	12.22830	0.002048100	0.0658640	0.36366	0.17756	1097.08	32.158
10,000	23.355	10.10830	0.001755500	0.0564240	0.35343	0.20133	1077.40	32.142
15,000	5.545	8.29700	0.001496100	0.0480680	0.34302	0.22928	1057.35	32.129
20,000	−12.255	6.75880	0.001267200	0.0406940	0.33244	0.26234	1036.94	32.113
25,000	−30.048	5.46070	0.001066300	0.0342240	0.32166	0.30167	1016.11	32.097
30,000	−47.832	4.37260	0.000890650	0.0285730	0.31069	0.34884	994.85	32.081
35,000	−65.607	3.46760	0.000738190	0.0236720	0.29952	0.40575	973.13	32.068
40,000	−69.700	2.73000	0.000587260	0.0188230	0.29691	0.50559	968.08	32.052
45,000	−69.700	2.14890	0.000462270	0.0148090	0.29691	0.64230	968.08	32.036
50,000	−69.700	1.69170	0.000363910	0.0116520	0.29691	0.81589	968.08	32.020
60,000	−69.700	1.04880	0.000225610	0.0072175	0.29691	1.31600	968.08	31.991
70,000	−67.425	0.65087	0.000139200	0.0044485	0.29836	2.14340	970.90	31.958
80,000	−61.976	0.40632	0.000085707	0.0027366	0.30182	3.52150	977.62	31.930
90,000	−56.535	0.25540	0.000053145	0.0016952	0.30525	5.74360	984.28	31.897
100,000	−51.099	0.16160	0.000033182	0.0010575	0.30865	9.30180	990.91	31.868

(Continued)

TABLE A.1 (*Continued*)

Physical Properties for the International Civil Aviation Organization (ICAO) Standard Atmosphere as a Function of Elevation above Sea Level

Elevation above Sea Level km	Temperature (θ) °C	Absolute Pressure (p) kPa abs	Density (ρ) kg/m^3	Specific Weight (γ) N/m^3	Absolute (Dynamic) Viscosity (μ) 10^{-6} N-sec/m^2	Kinematic Viscosity (v) 10^{-6} m2/sec	Speed of Sound (c) m/sec	Acceleration due to Gravity (g) m/sec^2
0	15.000	101.325	1.22500	12.0131	17.894	14.607	340.294	9.80665
1	8.501	89.876	1.11170	10.8987	17.579	15.813	336.430	9.80360
2	2.004	79.501	1.00660	9.8652	17.260	17.147	332.530	9.80050
3	−4.500	70.121	0.90925	8.9083	16.938	18.628	328.580	9.79740
4	−10.984	61.660	0.81935	8.0250	16.612	20.275	324.590	9.79430
5	−17.474	54.048	0.73643	7.2105	16.282	22.110	320.550	9.79120
6	−23.963	47.217	0.66011	6.4613	15.949	24.161	316.450	9.78820
8	−36.935	35.651	0.52579	5.1433	15.271	29.044	308.110	9.78200
10	−49.898	26.499	0.41351	4.0424	14.577	35.251	299.530	9.77590
12	−56.500	19.399	0.31194	3.0476	14.216	45.574	295.070	9.76970
14	−56.500	14.170	0.22786	2.2247	14.216	62.391	295.070	9.76360
16	−56.500	10.352	0.16647	1.6243	14.216	85.397	295.070	9.75750
18	−56.500	7.565	0.12165	1.1862	14.216	116.860	295.070	9.75130
20	−56.500	5.529	0.08891	0.8664	14.216	159.890	295.070	9.74520
25	−51.598	2.549	0.04008	0.3900	14.484	361.350	298.390	9.73000
30	−46.641	1.197	0.01841	0.1788	14.753	801.340	301.710	9.71470

Source: Adapted from Finnemore, E. J., and J. B. 2002. *Fluid Mechanics with Engineering Applications*, 10th ed. New York: McGraw Hill, 734.

TABLE A.2

Physical Properties for Water at Standard Sea-Level Atmospheric Pressure
as a Function of Temperature

Temperature (θ) $^\circ$F	Density (ρ) slug/ft^3	Specific Weight (γ) lb/ft^3	Absolute (Dynamic) Viscosity (μ) 10^{-6} lb-sec/ft^2	Kinematic Viscosity (ν) 10^{-6} ft^2/sec	Surface Tension (σ) lb/ft	Vapor Pressure (p_v) psia	Bulk Modulus of Elasticity (E_v) psi
32	1.940	62.42	37.46	19.31	0.00518	0.0885	293,000
40	1.940	62.43	32.29	16.64	0.00514	0.1220	294,000
50	1.940	62.41	27.35	14.10	0.00509	0.1780	305,000
60	1.938	62.37	23.59	12.17	0.00504	0.2560	311,000
70	1.936	62.30	20.50	10.59	0.00498	0.3630	320,000
80	1.934	62.22	17.99	9.30	0.00492	0.5070	322,000
90	1.931	62.11	15.95	8.26	0.00486	0.6980	323,000
100	1.927	62.00	14.24	7.39	0.00480	0.9490	327,000
110	1.923	61.86	12.84	6.67	0.00473	1.2750	331,000
120	1.918	61.71	11.68	6.09	0.00467	1.6920	333,000
130	1.913	61.55	10.69	5.58	0.00460	2.2200	334,000
140	1.908	61.38	9.81	5.14	0.00454	2.8900	330,000
150	1.902	61.20	9.05	4.76	0.00447	3.7200	328,000
160	1.896	61.00	8.38	4.42	0.00441	4.7400	326,000
170	1.890	60.80	7.80	4.13	0.00434	5.9900	322,000
180	1.883	60.58	7.26	3.85	0.00427	7.5100	318,000
190	1.876	60.36	6.78	3.62	0.00420	9.3400	313,000
200	1.868	60.12	6.37	3.41	0.00413	11.5200	308,000
212	1.860	59.83	5.93	3.19	0.00404	14.6900	300,000
$^\circ$C	kg/m^3	kN/m^3	N-sec/m^2	10^{-6} m^2/sec	N/m	kN/m^2 abs	10^6 kN/m^2
0	999.8	9.805	0.001781	1.785	0.0756	0.611	2.02
5	1000.0	9.807	0.001518	1.519	0.0749	0.872	2.06
10	999.7	9.804	0.001307	1.306	0.0742	1.230	2.10
15	999.1	9.798	0.001139	1.139	0.0735	1.710	2.14
20	998.2	9.789	0.001002	1.003	0.0728	2.340	2.18
25	997.0	9.777	0.000890	0.893	0.0720	3.170	2.22
30	995.7	9.765	0.000798	0.800	0.0712	4.240	2.25
40	992.2	9.731	0.000653	0.658	0.0696	7.380	2.28
50	988.0	9.690	0.000547	0.553	0.0679	12.330	2.29
60	983.2	9.642	0.000466	0.474	0.0662	19.920	2.28
70	977.8	9.589	0.000404	0.413	0.0644	31.160	2.25
80	971.8	9.530	0.000354	0.364	0.0626	47.340	2.20
90	965.3	9.467	0.000315	0.326	0.0608	70.100	2.14
100	958.4	9.399	0.000282	0.294	0.0589	101.330	2.07

Source: Adapted from Finnemore, E. J., and J. B. Franzini. 2002. *Fluid Mechanics with Engineering Applications*, 10th ed. New York: McGraw Hill, 732.

TABLE A.3

Physical Properties for Air at Standard Sea-Level Atmospheric Pressure as a Function of Temperature

Temperature (θ) °F	Density (ρ) slug/ft^3	Specific Weight (γ) lb/ft^3	Absolute (Dynamic) Viscosity (μ) 10^{-6} lb-sec/ft^2	Kinematic Viscosity (ν) 10^{-3} ft^2/sec
−40	0.002940	0.09460	0.312	0.106
−20	0.002807	0.09030	0.325	0.116
0	0.002684	0.08637	0.338	0.126
10	0.002627	0.08453	0.345	0.131
20	0.002572	0.08277	0.350	0.136
30	0.002520	0.08108	0.358	0.142
40	0.002470	0.07945	0.362	0.146
50	0.002421	0.07790	0.368	0.152
60	0.002374	0.07640	0.374	0.158
70	0.002330	0.07495	0.382	0.164
80	0.002286	0.07357	0.385	0.169
90	0.002245	0.07223	0.390	0.174
100	0.002205	0.07094	0.396	0.180
120	0.002129	0.06849	0.407	0.189
140	0.002058	0.06620	0.414	0.201
160	0.001991	0.06407	0.422	0.212
180	0.001929	0.06206	0.434	0.225
200	0.001871	0.06018	0.449	0.240
250	0.001739	0.05594	0.487	0.280
°C	kg/m^3	N/m^3	10^{-6} N-sec/m^2	10^{-6} m^2/sec
−40	1.515	14.86	14.9	9.8
−20	1.395	13.68	16.1	11.5
0	1.293	12.68	17.1	13.2
10	1.248	12.24	17.6	14.1
20	1.205	11.82	18.1	15.0
30	1.165	11.43	18.6	16.0
40	1.128	11.06	19.0	16.8
60	1.060	10.40	20.0	18.7
80	1.000	9.81	20.9	20.9
100	0.946	9.28	21.8	23.1
200	0.747	7.33	25.8	34.5

Source: Adapted from Finnemore, E. J., and J. B. Franzini. 2002. *Fluid Mechanics with Engineering Applications,* 10th ed. New York: McGraw Hill, 733.

TABLE A.4

Physical Properties for Some Common Liquids at Standard Sea-Level Atmospheric Pressure at Room Temperature (68°F or 20°C), in General

Liquid	Temperature (θ) °F	Density (ρ) slug/ft³	Specific Gravity[a] (s) –	Absolute (Dynamic) Viscosity (μ) 10^{-6} lb-sec/ft²	Surface Tension (σ) lb/ft	Vapor Pressure (p_v) psia	Bulk Modulus of Elasticity (E_v) psi	Specific Heat (c) ft-lb/(slug-°R) = ft²/(sec²-°R)
Benzene	68	1.700	0.876	14.370	0.0020	1.450000	150,000	10,290
Carbon tetra-chloride	68	3.080	1.588	20.350	0.0018	1.900000	160,000	5035
Crude oil	68	1.660	0.856	150.000	0.0020	–	–	–
Gasoline	68	1.320	0.680	6.100	–	8.000000	–	12,500
Glycerin	68	2.440	1.258	31200.000	0.0043	0.000002	630,000	14,270
Hydrogen	−430	0.143	0.074	0.435	0.0002	3.100000	–	12,000
Kerosene	68	1.570	0.809	40.000	0.0017	0.460000	–	12,000
Mercury	68	26.300	13.557	33.000	0.0320	0.000025	3,800,000	834
Oxygen	−320	2.340	1.206	5.800	0.0010	3.100000	–	~5760
SAE 10 oil	68	1.780	0.918	1700.000	0.0025	–	–	–
SAE 30 oil	68	1.780	0.918	9200.000	0.0024	–	–	–
Fresh water	68	1.936	0.998	21.000	0.0050	0.340000	318,000	25,000
Seawater	68	1.985	1.023	22.500	0.0050	0.340000	336,000	23,500

(Continued)

TABLE A.4 (Continued)

Physical Properties for Some Common Liquids at Standard Sea-Level Atmospheric Pressure at Room Temperature (68°F or 20°C), in General

Liquid	Temperature (θ) °C	Density (ρ) kg/m³	Specific Gravity[a] (s) –	Absolute (Dynamic) Viscosity (μ) 10^{-3} N-sec/m²	Surface Tension (σ) N/m	Vapor Pressure (p_v) kN/m² abs	Bulk Modulus of Elasticity (E_v) 10^6 N/m²	Specific Heat (c) N-m/(kg-°K) $= $ m²/(sec²-°K)
Benzene	20	876.0	0.876	0.650	0.0290	10.000000	1030	1720.0
Carbon tetra-chloride	20	1588.0	1.588	0.970	0.0260	13.100000	1100	842.0
Crude oil	20	856.0	0.856	7.200	0.0300	–	–	–
Gasoline	20	680.0	0.680	0.290	–	55.200000	–	2100.0
Glycerin	20	1258.0	1.258	1494.000	0.0630	0.000014	4344	2386.0
Hydrogen	−257	73.7	0.074	0.021	0.0029	21.400000	–	–
Kerosene	20	808.0	0.808	1.920	0.0250	3.200000	–	2000.0
Mercury	20	13,550.0	13.550	1.560	0.5100	0.000170	26200	139.4
Oxygen	−195	1206.0	1.206	0.278	0.0150	21.400000	–	~964.0
SAE 10 oil	20	918.0	0.918	82.000	0.0370	–	–	–
SAE 30 oil	20	918.0	0.918	440.000	0.0360	–	–	–
Fresh water	20	998.0	0.998	1.000	0.0730	2.340000	2171	4187.0
Seawater	20	1023.0	1.023	1.070	0.0730	2.340000	2300	3933.0

Source: Adapted from Finnemore, E. J., and J. B. Franzini. 2002. *Fluid Mechanics with Engineering Applications*, 10th ed. New York: McGraw Hill, 735.

[a] Relative to the reference fluid water at a standard temperature of 39.2°F and a standard sea-level atmospheric pressure of 14.696 psia, for which the density is 1.94 slug/ft³. Relative to the reference fluid water at a standard temperature of 4°C and a standard sea-level atmospheric pressure of 101.325 kPa abs, for which the density is 1000 kg/m³.

TABLE A.5

Physical Properties for Some Common Gases at Standard Sea-Level Atmospheric Pressure at Room Temperature (68°F or 20°C)

Gas at 68°F	Chemical Formula	Molar Mass (m) slug/slug-mol	Density (ρ) slug/ft³	Absolute (Dynamic) Viscosity (μ) 10^{-6} lb-sec/ft²	Gas Constant (R) ft-lb/(slug-°R) = ft²/(sec²·°R)	Specific Heat		Specific Heat Ratio, $k = c_p/c_v$
						c_p ft-lb/(slug-°R) = ft²/(sec²·°R)	c_v	
Air		28.960	0.002310	0.376	1715	6000	4285	1.40
Carbon dioxide	CO_2	44.010	0.003540	0.310	1123	5132	4009	1.28
Carbon monoxide	CO	28.010	0.002260	0.380	1778	6218	4440	1.40
Helium	He	4.003	0.000323	0.411	12,420	31,230	18,810	1.66
Hydrogen	H_2	2.016	0.000162	0.189	24,680	86,390	61,710	1.40
Methane	CH_4	16.040	0.001290	0.280	3100	13,400	10,300	1.30
Nitrogen	N_2	28.020	0.002260	0.368	1773	6210	4437	1.40
Oxygen	O_2	32.000	0.002580	0.418	1554	5437	3883	1.40
Water vapor	H_2O	18.020	0.001450	0.212	2760	11,110	8350	1.33
at 20°C		kg/kg-mol	kg/m³	10^{-6} N-sec/m²	N-m/(kg·°K) = m²/(sec²·°K)	N-m/(kg·°K) = m²/(sec²·°K)		
Air		28.960	1.2050	18.0	287	1003	716	1.40
Carbon dioxide	CO_2	44.010	1.8400	14.8	188	858	670	1.28
Carbon monoxide	CO	28.010	1.1600	18.2	297	1040	743	1.40
Helium	He	4.003	0.1660	19.7	2077	5220	3143	1.66
Hydrogen	H_2	2.016	0.0839	9.0	4120	14,450	10,330	1.40
Methane	CH_4	16.040	0.6680	13.4	520	2250	1730	1.30
Nitrogen	N_2	28.020	1.1600	17.6	297	1040	743	1.40
Oxygen	O_2	32.000	1.3300	20.0	260	909	649	1.40
Water vapor	H_2O	18.020	0.7470	10.1	462	1862	1400	1.33

Source: Adapted from Finnemore, E. J., and J. B. Franzini. 2002. *Fluid Mechanics with Engineering Applications*, 10th ed. New York: McGraw Hill, 736.

Appendix B: Geometric Properties of Common Shapes

TABLE B.1

The Area, A, the Center of Area (Centroid), y_{ca} and x_{ca}, the Centroidal Moment of Inertia, $I_{x\text{-}ca}$, and the Moment of Inertia about the y-Axis, I_y for Some Common Geometric Shapes

Shape	A	y_{ca}	x_{ca}	$I_{x\text{-}ca}$	I_y
Rectangle	wL	$\dfrac{L}{2}$		$\dfrac{wL^3}{12}$	$\dfrac{Lw^3}{12}$
Triangle	$\dfrac{BL}{2}$	$\dfrac{2L}{3}$		$\dfrac{BL^3}{36}$	
Circle	$\dfrac{\pi D^2}{4}$	$\dfrac{D}{2}$		$\dfrac{\pi D^4}{64}$	$\dfrac{\pi D^4}{64}$
Semicircle	$\dfrac{\pi D^2}{8}$	$\dfrac{2D}{3\pi}$		$\dfrac{\pi D^4}{128} - \dfrac{D^4}{18\pi}$	
Quarter of a circle	$\dfrac{\pi r^2}{4}$	$\dfrac{4r}{3\pi}$	$\dfrac{4r}{3\pi}$	$\dfrac{r^4(9\pi^2 - 64)}{144\pi}$	

(Continued)

TABLE B.1 (*Continued*)

The Area, A, the Center of Area (Centroid), y_{ca} and x_{ca}, the Centroidal Moment of Inertia, $I_{x\text{-}ca}$, and the Moment of Inertia about the y-Axis, I_y for Some Common Geometric Shapes

Shape	A	y_{ca}	x_{ca}	$I_{x\text{-}ca}$	I_y
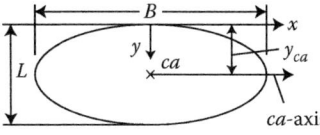 Ellipse	$\dfrac{\pi BL}{4}$	$\dfrac{L}{2}$		$\dfrac{\pi BL^3}{64}$	$\dfrac{\pi B^3 L}{16}$
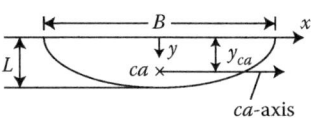 Semi ellipse	$\dfrac{\pi BL}{4}$	$\dfrac{4L}{3\pi}$		$\dfrac{\pi BL^3}{16} - \dfrac{4BL^3}{9\pi}$	
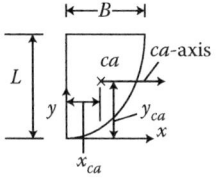 Half parabola	$\dfrac{2BL}{3}$	$\dfrac{3L}{5}$	$\dfrac{3B}{8}$	$\dfrac{2BL^3}{7} - \dfrac{18BL^3}{75}$	
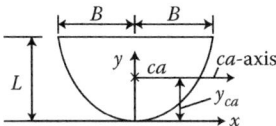 Full parabola	$2\left(\dfrac{2BL}{3}\right)$	$\dfrac{3L}{5}$		$2\left(\dfrac{2BL^3}{7} - \dfrac{18BL^3}{75}\right)$	

TABLE B.2

The Volume, V, and the Centroid, y_{ca} for Some Common Geometric Shapes

Shape	V	y_{ca}
Cylinder	$\dfrac{\pi D^2 L}{4}$	$\dfrac{L}{2}$
Cone	$\dfrac{1}{3}\left(\dfrac{\pi D^2 L}{4}\right)$	$\dfrac{3L}{4}$
Sphere	$\dfrac{\pi D^3}{6}$	$\dfrac{D}{2}$
Hemisphere	$\dfrac{\pi D^3}{12}$	$\dfrac{5r}{8}$
Semi ellipsoid	$\dfrac{2}{3}\left(\dfrac{\pi D^2 L}{4}\right)$	$\dfrac{5L}{8}$
Paraboloid	$\dfrac{1}{2}\left(\dfrac{\pi D^2 L}{4}\right)$	$\dfrac{2L}{3}$

References

1. Abbott, I. H. and A. E. von Doenhoff. 1959. *Theory of Wing Sections, Including a Summary of Airfoil Data*. New York: Dover.
2. ASHRAE. 1981. *ASHRAE Handbook of Fundamentals*. Atlanta: American Society of Heating, Refrigeration and Air-Conditioning Engineers, Inc. (ASHRAE).
3. Blevins, R. D. 1984. *Applied Fluid Dynamics Handbook*. New York: Von Nostrand Reinhold.
4. Bean, H. S. 1971. *Fluid Meters: Their Theory and Application, Report of ASME Research Committee on Fluid Meters*, 6th ed. New York: American Society of Mechanical Engineers.
5. Cengel, Y. A. and J. M. Cimbala. 2014. *Fluid Mechanics Fundamentals and Applications*, 3rd ed. New York: McGraw Hill.
6. Chow, V. T. 1959. *Open Channel Hydraulics*. New York: McGraw Hill.
7. Crowe, C. T., D. F. Elger, and J. A. Roberson. 2005. *Engineering Fluid Mechanics*, 8th ed. New York: John Wiley and Sons.
8. Esposito, A. 1998. *Fluid Mechanics with Applications*. Upper Saddle River, New Jersey: Prentice Hall.
9. Finnemore, E. J. and J. B. Franzini. 2002. *Fluid Mechanics with Engineering Applications*, 10th ed. New York: McGraw Hill.
10. Goldstein, S. 1938. *Modern Developments in Fluid Dynamics*, vols. I and II. Oxford: Clarendon Press. (Reprinted in paperback by Dover, New York, 1967.)
11. Happel, J. 1965. *Low Reynolds Number Hydrodynamics*. Upper Saddle River, New Jersey: Prentice Hall.
12. Henderson, F. M. 1966. *Open Channel Flow*. New York: Macmillan.
13. Henry, H. R. 1950. Discussion on "Diffusion of Submerged Jets," by M. L. Albertson, Y. B. Dai, R. A. Jensen, and H. Rouse. *Transactions of the American Society of Civil Engineers* 115: 687–695.
14. Hucho, W.-H. Ed. 2013. *Aerodynamics of Road Vehicles: From Fluid Mechanics to Vehicle Engineering*. Amsterdam: Elsevier.
15. Hwang, N. H. C. and R. J. Houghtalen. 1996. *Fundamentals of Hydraulic Engineering Systems*, 3rd ed. Upper Saddle River, New Jersey: Prentice Hall.
16. International Civil Aviation Organization (ICAO). 1993. *Manual of the ICAO Standard Atmosphere: Extended to 80 Kilometres (262 500 feet), Doc 7488-CD*, 3rd ed. Montreal: International Civil Aviation Organization.
17. International Organization for Standardization (ISO). 1980. *Measurement of Fluid Flow by Means of Orifice Plates, Nozzles, and Venturi Tubes Inserted in Circular Cross Section Conduits Running Full*, 1st ed. ISO 5167-1980 (E). Geneva: International Organization for Standardization.
18. Inui, T. 1962. Wave-making resistance of ships. *Transactions of the Society of Naval Architects and Marine Engineers* 70: 283–253.
19. Lenz, A. T. 1943. Viscosity and surface tension effects on V-notch weir coefficients. *Transactions of the American Society of Civil Engineers* 108:759–782.
20. Moody, L. F. 1944. Friction factors for pipe flow. *Transactions of the American Society of Mechanical Engineers* 66: 671–684.
21. Munson, B. R., D. F. Young, and T. H. Okiishi. 1998. *Fundamentals of Fluid Mechanics*, 3rd ed. New York: John Wiley and Sons.
22. Prandtl, L. 1921. Applications of modern hydrodynamics to aeronautics. *National Advisory Committee for Aeronautics (NACA)* Report 116. Langley Field, VA: Langley Memorial Aeronautical Laboratory.
23. Roberson, J. A. and C. T. Crowe. 1997. *Engineering Fluid Mechanics*, 6th ed. New York: John Wiley and Sons.
24. Schlichting, H. 1960. *Boundary Layer Theory*, 4th ed. New York: McGraw Hill.
25. Schlichting, H. 1979. *Boundary Layer Theory*, 7th ed. New York: McGraw Hill.

26. Streeter, V. L. Ed. 1961. *Handbook of Fluid Dynamics*. New York: McGraw Hill.
27. Vennard, J. K. and R. L. Street. 1975. *Elementary Fluid Mechanics*, 5th ed. New York: John Wiley and Sons.
28. Vennard, J. K. and R. L. Street. 1982. *Elementary Fluid Mechanics*, 6th ed. New York: John Wiley and Sons.
29. White, F. M. 1979. *Fluid Mechanics*. New York: McGraw Hill.
30. White, F. M. 2011. *Fluid Mechanics*, 7th ed. New York: McGraw Hill.
31. Williams, G. S., and A. Hazen. 1914. *Hydraulic Tables*, 2nd ed. New York: John Wiley and Sons.
32. http://www.infoniac.com/hi-tech/famous-scientists-their-inventions-and-discoveries.html
33. http://www.sciencekids.co.nz/pictures/scientists.html
34. https://www.google.com
35. http://www.gettyimages.com
36. https://pixabay.com
37. http://www.usbr.gov

Bibliography

Abbott, I. H. 1932. The drag of two streamline bodies as affected by protuberances and appendages. *National Advisory Committee for Aeronautics (NACA)* Report 451. Langley Field, Virginia: Langley Memorial Aeronautical Laboratory.

Abbott, I. H., A. E. von Doenhoff, and L. S. Stivers. 1945. Summary of airfoil data. *National Advisory Committee for Aeronautics (NACA)* Report 824. Langley Field, Virginia: Langley Memorial Aeronautical Laboratory.

Cengel, Y. A. and J. M. Cimbala. 2006. *Fluid Mechanics Fundamentals and Applications*. New York: McGraw Hill.

Cengel, Y. A. and J. M. Cimbala. 2010. *Fluid Mechanics Fundamentals and Applications*, 2nd ed. New York: McGraw Hill.

Colebrook, C. F. 1939. Turbulent flow in pipes with particular reference to the transition between the smooth and rough pipe laws. *Journal of the Institute of Civil Engineers London* 11: 133–156.

Daugherty, R. L., J. B. Franzini, and E. J. Finnemore. 1985. *Fluid Mechanics with Engineering Applications*, 8th ed. New York: McGraw Hill.

Durst, F. unknown date. *Fluid Mechanics Developments and Advancements in the 20th Century*. 4, D-91058 Erlangen, Germany: Institute of Fluid Mechanics, University of Erlangen-Nuremberg Cauerstr.

Eisner, F. 1930. Das Widerstandsproblem. *Proc. 3rd International Congress Applied Mechanics*.

El-Mongy, A. E., M. S. El-Bisy, and G. H. ElSaeed. 2001. *Fluid Mechanics. Supplement to Textbooks*. Cairo Egypt: Ain Shams University.

Gad-El-Hak, M. 1998. Fluid mechanics from the beginning to the third millennium. *International Journal for Engineering Ed.* 14: 177–185.

Giles. R. V., J. B. Evett, and C. Liu. 1994. *Fluid Mechanics and Hydraulics: Schaum's Outline Series*, 3rd ed. New York: McGraw Hill.

Granger, R. A. 1985. *Fluid Mechanics*. New York: Holt, Rinehart and Winston.

Hibbeler, R. C. 2015. *Fluid Mechanics*. New York: Pearson.

Hydraulic Institute. 1979. *Engineering Data Book*, 1st ed. Cleveland, Ohio Hydraulic Institute.

Larsen, R. W. 2004. *Introduction to Mathcad 11*. Upper Saddle River, New Jersey: Pearson Prentice Hall.

Liggett, J. A. 1994. *Fluid Mechanics*. New York: McGraw Hill.

Lindsey, W. F. 1938. Drag of cylinders of simple shapes. *National Advisory Committee for Aeronautics (NACA)* Report 619. Langley Field, VA: Langley Memorial Aeronautical Laboratory.

Mott, R. L. 2006. *Applied Fluid Mechanics*, 6th ed. Saddle River, New Jersey: Pearson Prentice Hall.

Munson, B. R., T. H. Okiishi, W. W. Huebsch, and A. P. Rothmayer. 2013. *Fundamentals of Fluid Mechanics*, 7th ed. New York: John Wiley and Sons.

Patterson, E. A. 2011. *Real Life Examples in Fluid Mechanics*. Department of Mechanical Engineering, Michigan State University.

Prandtl, L. 1923. *Ergebnisse der aerodynamischen Versuchsanstalt zu Gottingen*. Munich and Berlin: R. Oldenbourg.

Pritchard, P. J. and J. C. Leylegian. 2011. *Fox and McDonald's Introduction to Fluid Mechanics*, 8th ed. New York: John Wiley and Sons.

Sabersky, R. H., A. J. Acosta, E. G. Hauptmann, and E. M. Gates. 1999. *Fluid Flow: A First Course in Fluid Mechanics*, 4th ed. Saddle River, New Jersey: Prentice Hall.

Shames, I. H. 1992. *Mechanics of Fluids*, 3rd ed. New York: McGraw Hill.

Spiegel, M. R. 1968. *Mathematical Handbook of Formulas and Tables: Schaum's Outline Series in Mathematics*. New York: McGraw Hill.

Young, D. F., B. R. Munson, T. H. Okiishi, and W. W. Huebsch. 2011. *A Brief Introduction to Fluid Mechanics*, 5th ed. New York: John Wiley and Sons.

Zahm, A. F., R. H. Smith, and G. C. Hill. 1972. Point drag and total drag of navy struts No. 1 modified. *National Advisory Committee for Aeronautics (NACA) Report 137*. Langley Field, Virginia: Langley Memorial Aeronautical Laboratory.

Index

A

Absolute
 pipe roughness, 941
 pressure, 13, 69, 86, 126
 zero pressure scale, 126
Acceleration of system, 368
Acoustic velocity, *see* Sonic velocity
Actual area derivation, 721–722
Actual discharge
 derivation of, 720–721, 721–722
 equation for pipe flow, 722–723
 evaluation of, 719–720, 839–840
 exercises, 745–748
 in open channel flow, 764, 767, 768
 in pipe flow, 764, 766, 768
 similitude scale ratios for, 1515, 1520
Actual flowrate
 in control devices, 1253–1254
 for critical flow meters, 1259–1260
 evaluation, 1079–1080
 exercises, 1113
 for nozzle meter, 1093–1097
 for orifice meter, 1090–1093
 for pipe flow-measuring devices,
  1086–1090
 for pitot-static tube, 1081–1086
 for venturi meter, 1097–1100
Actual flowrate for ideal flow meters,
  1078–1079, 1258–1259, 1262–1263;
  see also Ideal flow meters; Open
  channel flow
 Bernoulli equation, 1263–1264
 continuity equation, 1264
 contracted openings, 1267
 contraction coefficient, 1264–1265
 discharge coefficient, 1264, 1265–1266
 evaluation for broad-crested weir,
  1285–1290
 evaluation for contracted opening, 1294
 evaluation for parshall flume, 1291–1294
 evaluation for sharp-crested weir, 1272–1274
 evaluation for sluice gate, 1267–1272
 evaluation for spillway, 1283–1285
 evaluation of discharge coefficient, 1266
 example problem, 1271–1272, 1277–1278,
  1281–1282, 1284–1285, 1287–1290,
  1293–1294

 exercises, 1303–1304
 sharp-crested weir, 1275–1278, 1278–1282,
  1282–1283
 sluice gates, 1266
 spillways, 1266
 velocity coefficient, 1264
 venturi flumes, 1266
 weirs, 1266, 1267
Actual velocity, 776
 derivation of, 661, 721–722
Adiabatic
 exponent, 86
 process, 456
Affinity laws (similarity rules), 1537
 pump efficiency, 1537–1539
 turbine efficiency, 1545–1547
Airfoil, 1372, 1374; *see also* Lift force in
  external flow
 attaching devices, 1390
 design, 1375
 example problem, 1378–1381, 1382–1383,
  1386, 1388, 1390–1392
 flapped, 1384
 geometry of, 1373
 with large aspect ratios, 1390
 lift coefficient variation with angle of attack,
  1377, 1385, 1390
 lift-to-drag ratio variation with angle of
  attack, 1382
 nonsymmetrical, 1375
 performance optimization,
  1382–1383
 shape optimization and angle of attack,
  1376–1381
 shape optimization by aspect ratio,
  1389–1392
 shape optimization by use of flaps,
  1384–1389
 with smaller aspect ratios, 1390
 symmetrical, 1375
Air properties, 1567
Alternate depths of flow, 1142
Altitude
 exercises, 96
 standard reference, 15
Archimedes principle, 230; *see also* Buoyancy
Aspect ratio, 1389
Atmospheric pressure, 64

B

Barometers, 129; *see also* Hydrostatic pressure
    measurement
  absolute atmospheric pressure, 130
  example problem, 131, 132–133
  exercises, 265
  mercury, 130
  torr, 132
Bends, 976
Bernoulli effect, 68, 391
Bernoulli equation, 375, 400, 577; *see also*
    Newton's second law of motion
  applications of, 379, 382, 720
  dynamic pressure modeling, 382–383
  energy grade line and hydraulic grade line,
    383–389
  exercises, 517, 541–542
  horizontal pipe flowing under pressure with
    piezometric tubes, 378
  hydraulic grade line and cavitation, 389–392
  for ideal external flow, 444–449
  ideal flow, 598, 614
  for ideal flow from tank, 423–442
  ideal open channel flow, 419–423
  for ideal pipe flow, 415–419
  pitot-static tube, 380–382, 650
  in terms of energy terms, 378
  in terms of pressure terms, 377
  Torricelli's theorem, 432–434
  upstream stagnation pressure, 382
  venturi meter, 653
  venturi tube, 383
BG, *see* British Gravitational
Boiling point, 68
Bourdon pressure gage, 129
Branching pipes, 1030–1031; *see also* Pipe systems
  cases for, 1033, 1045
  connected to reservoirs, 1031–1044
  connected to water supply source under
    pressure, 1044–1055
  example problem, 1033–1044, 1045–1055
  exercises, 1111
  governing equations for, 1032
British Gravitational (BG), 3, 6, 7; *see also* Units
  dimensions for, 3
  example problem, 10
  Rankine scale, 10
  units for, 9
British thermal unit (Btu), 81, 455
Broad-crested weir, 1176, 1267
  actual discharge, 1287
  evaluation of actual flowrate for, 1285–1290
  example problem, 1287–1290

  geometry of, 1285
  ideal flowrate, 1286
  velocity coefficient, 1287
Btu, *see* British thermal unit
Buckingham π theorem, 750, 753, 757
Bulk modulus of elasticity, 70, 71–75
  compression or expansion processes, 86
  determination for gases, 85–88
  example problem, 73–74, 86–87, 88
  exercises, 107–113
Buoyancy, 229, 230; *see also* Fluid statics; Buoyant
    body stability; Buoyant force
  applications, 239
  center of, 231
  example problem, 232–233, 234–235, 236–239,
    252–259
  exercises, 291–299
  free body diagram, 231
  neutral, 234
  of neutrally buoyant body, 233–235
  of partially submerged floating body, 235–239
  proportion above and below fluid line, 236
  of sinking body, 231–233
  types, 240
Buoyant body stability, 229, 240; *see also*
    Buoyancy; Fluid statics
  example problem, 252–259
  exercises, 291–299
  maximum angle of rotation, 246
  metacenter, 246
  metacentric height and moment, 247, 248–259
  metacentric height for top-heavy body,
    246–248
  neutrally buoyant body, 241, 242, 243
  rotational stability of floating body, 243–246
  rotational stability of suspended bodies,
    241–243
  stability of streamlined top-heavy floating
    body, 247
  in stable equilibrium, 244, 245, 246
  top-heavy floating body with negative
    metacentric height, 249
  in unstable equilibrium, 245
  vertical stability of floating body, 240–241
Buoyant force, 230–231, 239–240

C

Calorie, 455
Capillarity, 61; *see also* Surface tension
  example problem, 63
  meniscus, 62
  tension, 62

Capillary tubes, 60
Cauchy number, 92, 761, 1415; *see also* Dynamic
similitude
exercises, 1553
Cavitation, 64, 68–70, 389, 473, 503
Bernoulli effect, 391
cavitation bubbles, 69
example problem, 70
implosion, 70
of liquid due to pipe diameter reduction,
69, 392
number, 831, 1413
Center of buoyancy, 231; *see also* Buoyancy
Centistoke, 45
Channel section geometric properties,
1171–1172; *see also* Open channel flow
hydraulic radius, 1173, 1174
moment of area for whole channel
section, 1175
partially filled circular channel sections,
1175–1176
rectangular channel sections, 1172
trapezoidal channel sections, 1172–1174
triangular channel sections, 1174–1175
wetted perimeter, 1173, 1174
Channel transition, 1132
exercises, 1297
M-y plot for abrupt, 1205
Chezy coefficient, 690, 691, 708, 804–805,
1128–1129; *see also* Chezy equation;
Drag coefficient
evaluation, 690
exercises, 909
friction slope, 805
major head loss equation for turbulent flow,
799, 804–805
substitution in energy equation, 805–806
Chezy equation, 560, 668, 671, 686, 688, 796–797,
934, 1128–1129; *see also* Chezy
coefficient; Major head loss equation
for turbulent flow
application of, 690–691, 708–709
channel slope, 814
derivation, 690
exercises, 738–739
friction slope, 814
in pipe flow analysis and design, 935
substitution in energy equation, 691–692
Circular pipes, 699
Colebrook equation, 943
Compressibility, 70
exercises, 107–113
Mach number for fluids, 92

Compressible flow, 303
Condensation, 67
volume phenomena of, 68
Conjugate depths of flow, 1169
Conservation of energy principle, 452; *see also*
Energy equation; Thermodynamics,
first law of
Conservation of mass, 324–325; *see also*
Continuity equation
exercises, 351
fluid mechanics, 325
Conservation of momentum principle,
313, 365, 365–371; *see also* Newton's
second law of motion
in fluid mechanics, 366
Constant of integration, 682
Continuity equation, 301, 325–326, 555,
840–841
conservation of mass, 324
for control volume, 335–348
equations of motion, 301
example problem, 329–335, 339–348
exercises, 348
flow domain, 326
flow visualization, 304–307
fluid flow types, 302–304
fluid kinematics, 302
fluid motion, 307–313
for fluid system, 326–335
infinitesimal control volume, 326
laws governing fluid in motion, 313–324
one-dimensional, 328, 337
volume flowrate measurement, 348
Contracted openings, 1267
evaluation of actual flowrate for, 1294
Control, 1187, 1228, 1255; *see also* Critical depth
meters
generating nonuniform surface water
profile, 1229
in open channel, 1133, 1134, 1297
surface, 315
Control volume, 307, 1121; *see also* Energy
equation for control volume; Fluid
motion
continuity equation for, 335–348
energy equation for, 373–376
Eulerian point of view, 309–310
exercises, 355–362
fixed vs. moving, 311–312
fluid particle velocity, 310–311
Newton's second law of motion, 369–371
Convective acceleration, 369
Couette flow, 39

Creeping flow
   for any shape, 1447–1451
   example problem, 1448–1451
Critical depth meters, 1139, 1255–1256; *see also*
      Open channel flow
   actual flowrate, 1259–1260
   advantages and disadvantages, 1261
   comparison of ideal flow meters and, 1260
Critical flow, 1133, 1176; *see also* Open channel
      flow
   at controls by change in channel bottom
         slope, 1156–1159
   at controls by horizontal constriction,
         1153–1156
   at controls by vertical constriction, 1149–1153
   for nonrectangular channel sections,
         1146–1147
   occurrences at controls, 1148–1149
   for rectangular channel sections, 1141–1143
   steep channel slope, 1159
   at various controls, 1149

D

d'Alembert's paradox, 663
Dalton's law of partial pressure, 64
Darcy–Weisbach equation, 692, 806, 941; *see also*
      Flow resistance modeling in internal
      flow; Major head loss equation for
      turbulent flow
   application of, 815, 944–951
   example problem, 946–951
   exercises, 739, 741, 909–910
   head loss equation of, 694, 808–809, 942
   laminar and turbulent flow comparison
         using, 698–699, 700–702
   for noncircular pipe, 944
   problems for turbulent pipe flow, 946
   for turbulent flow, 701
Darcy–Weisbach friction coefficient, 692,
      941–942; *see also* Drag coefficient
   circular pipe flow, 806
   derivation of, 692–693, 806–808
   evaluation of, 693–694, 808, 809, 942–944
   example problem, 946–951
   exercises, 1105
   for laminar pipe flow, 809, 944
   Moody diagram, 942–943
Darcy–Weisbach friction factor, *see* Drag
      coefficient
Density, *see* Mass density
Depth–discharge curve, 1144

Dimensional analysis, 686, 688, 749; *see also*
      Dimensional analysis of fluid flow;
      Flow resistance modeling in external
      flow; Flow resistance modeling in
      internal flow; Flow resistance modeling
      in internal flow; Flow resistance
      modeling in pipe flow; Flow resistance
      modeling in pump and turbine
      efficiency
   Buckingham π theorem, 750
   drag coefficient evaluation, 757
   efficiency derivation for pumps and
         turbines, 757
   example problem, 895–901
   exercises, 901
   experimental formulation of theoretical
         equations, 893–901, 916–917
   in flow resistance, 752, 753–754
   flow resistance equation, 757
   flow resistance equation and drag coefficient
         derivation, 756
   flow resistance modeling, 752–753, 755–756
   flow types and, 755
   in fluid flow modeling, 750–752
   internal flow vs. external flow, 755
   laws of dynamic similarity, 750
   lift force, 755
   mathematically formulating equations, 893
   momentum theory with, 756
   π terms, 750
   procedure and results, 751
   single nonlinear plot, 754
Dimensional analysis of fluid flow, 757–758;
      see also Dimensional analysis
   actual discharge in open channel flow, 767
   actual discharge in pipe flow, 766
   application of, 774–775
   dimensional analysis theory, 757
   dimensionless numbers, 759–761
   dimensions and units in fluid flow, 760
   drag coefficient, 769–770
   drag force equation, 763
   exercises, 903
   flow resistance equations, 762–769
   forces acting on fluid element, 758–759
   major head loss, 764–765
   pi terms, 761–762
   pump (or turbine) efficiency, 770–774
   two-dimensional systems, 759
Dimensionless numbers, 759–761
Dimensions, 3
   for British Gravitational system, 3
   for density, 23

for energy, 81, 455
example problem, 5, 6
exercises, 94–95
for International System, 5
for number of moles of gas, 76
physical quantities in fluid flow, 4, 760
primary and secondary, 3
for specific weight, 28
Discharge coefficient, 239
derivation of, 721–722
Distorted models; *see also* Dynamic similitude
drag coefficient variation, 1432, 1433
dynamic similarity requirements, 1434–1435
elastic model similarity requirements, 1439
geometry similarity requirements, 1436
gravity model similarity requirements,
1439–1440, 1445
guidelines in application of, 1428–1436
pressure model similarity requirements, 1437
relative roughness similarity requirements,
1436–1437
surface tension model similarity
requirements, 1440, 1445
true models vs., 1427–1428
viscosity and elastic model similarity
requirements, 1440–1441
viscosity and gravity model similarity
requirements, 1441–1444
viscosity and surface tension model similarity
requirements, 1444–1445
viscosity model similarity requirements,
1438–1439, 1445
Drag coefficient, 645, 686, 762–769, 769–770,
816–820; *see also* Darcy–Weisbach
friction coefficient; Drag force in
external flow; Flow resistance
equations; Flow resistance modeling in
internal flow
for assumptions for open channel flow-
measuring devices, 861–866
derivation, 661, 713–714, 716, 778–786,
787–790, 1314–1315
determination, 1315
empirical evaluation of, 646
evaluation of, 1315–1316
example problem, 787–790
exercises, 728, 902–903, 904–905
inertia force, 784
Mach number, 785
for open channel flow-measuring devices,
854–861
for pipe flow measuring devices, 842–853
reduction trends in vehicle design, 1329

standard form of, 786, 799
substituting dynamic pressure force for
inertia force, 786
for 3D bodies, 1346, 1347
for 2D bodies, 1345
variation for external flow, 1316
variation with Froude number, 1370
variation with Mach number, 1366, 1368
variation with Reynolds number, 1334, 1335,
1344, 1357, 1363
Drag force, 776
flow resistance equation for, 778
inertia force, 785
Drag force equation
derivation guidelines, 763
derivation of, 661, 778–786, 787–790
exercises, 733–734
Drag force in external flow, 1306, 1311; *see also*
Drag coefficient; External flow
aspect ratio vs. drag coefficient, 1324
body shape, 1317
body shape and flow type, 1318
body shape role, 1316–1319
creeping flow, 1338–1344, 1447–1451
derived from dimensional analysis, 1314
evaluation of, 1312–1313
example problem, 1320–1323, 1325–1328,
1330–1331, 1332–1333, 1335–1344,
1348–1357, 1359–1361, 1363–1367,
1369, 1371–1372
exercises, 1396–1400
flow resistance equation, 1310
flow velocity role, 1316
fluid moving over submerged stationary
body, 1306
forces acting on body falling through
fluid, 1341
friction drag, 1318
friction-drag coefficient, 1319
friction force, 1314
frontal areas for automobiles, 1330
Froude number, 1369–1372
head loss, 1311, 1312
laminar and turbulent flow, 1344–1357
location of flow separation and drag
coefficient, 1332
Mach number, 1315, 1365–1369
modeling drag force in momentum equation,
1313–1314
occurrence of flow separation, 1331–1333
pressure-drag coefficient, 1319
pressure force, 1314
reducing flow separation, 1333

Drag force in external flow (*Continued*)
   reducing total drag force, 1319
   relative surface roughness, 1362–1365
   Reynolds number, 1333–1338
   similitude scale ratios for, 1446
   at stagnation point, 1307, 1308
   stationary fluid exerting, 1307
   Stokes law, 1338
   streamlined body, 1317
   vehicle drag coefficient reduction trends, 1329
   wake region, 1332
Dynamic forces
   acting on fluid element, 21, 758–759
   exercises, 96, 903
   and fluid properties, 22
   mathematical expressions, 758–759
Dynamic kinematic viscosity, 44; *see also*
      Viscosity
Dynamic pressure, 372, 650
   pitot-static tube, 651
Dynamic similitude, 1403, 1410, 1411–1412,
      1425, 1426–1427, 1429–1430; *see also*
      Distorted models; Similitude scale
      ratios
   applying laws of, 1408–1409
   Cauchy number, 1415
   dimensional analysis, 1406
   dynamic forces in, 1421–1425
   Euler number, 1413
   exercises, 1550, 1552
   flow resistance equations, 1429
   in flow resistance modeling, 1406–1408
   flow resistance prediction equation, 1426
   in fluid flow modeling, 1404–1406
   force scale ratio, 1412
   Froude number, 1414
   geometric similarity, 1410
   guidelines in application of distorted models,
      1428–1429
   implications of main π terms, 1416–1418
   interpretation of main π terms, 1412–1413
   kinematic similarity, 1410–1411
   laws governing, 1418–1419
   in mathematical modeling, 1403
   primary force scale ratio, 1411
   primary scale ratios, 1409
   ratio of inertia force to each of respective
      forces, 1412
   Reynolds number, 1414–1415
   similitude scale ratios, 1419–1421
   similitude velocity scale ratio, 1426–1427
   true models vs. distorted models, 1427–1428
   types of similarity, 1409

   uncalibrated mathematical model, 1403, 1405
   Weber number, 1415–1416

E

Effective roughness criterion, 705
EGL, *see* Energy grade line
Elastic force, 70
Elasticity, 70
   bulk modulus of, 70, 71–75
   exercises, 107–113
Elastic model
   similarity requirements, 1439
   similitude scale ratios, 1422, 1439
   and viscosity model similarity requirements,
      1440–1441
Elbows, 976
Electrical generator, 464
Energy, 371, 785
   calorie, 455
   dimensions and units for, 455
   internal, 453
   macroscopic, 453–454
   microscopic, 453
   thermal, 453
   types, 80
Energy equation, 1, 363; *see also*
      Thermodynamics, first law of;
      Newton's second law of motion
   conservation of energy, 452–457
   conservation of momentum, 365–371
   for control volume, 373–376, 458–460
   derivation of, 365
   exercises, 513
   fluid dynamics, 363–365
   for fluid system, 372–373, 457–458
   for hydraulic jump, 578
Energy equation for control volume, 373–374,
      458–460; *see also* Energy equation;
      Thermodynamics, first law of
   application, 376
   Bernoulli's equation, 375–376
   definitions, 374–375, 466–467
   electrical generator, 464
   exercises, 515–517
   expressed in energy head terms, 465–466
   expressed in power terms, 460–461
   generator efficiency, 464
   integral form of, 375
   mechanical energy loss, 459
   motor efficiency, 464
   practical assumptions, 375–376, 467
   pump and turbine losses, 461–463

pump efficiency, 464
pump–motor system, 461, 462, 464
shaft work, 461
total energy head, 374
turbine–generator system, 462, 463–465
Energy equation for hydraulic jump, 400
Energy grade line (EGL), 383, 930; *see also*
        Hydraulic grade line
    determination, 385, 387
    elevation head measurement, 388
    flow assumptions and, 384
    pressure drop, 388
    reduction due to friction head loss, 389, 390
    slope, 386
    stagnation pressure, 384
Energy transfer, 455
    adiabatic process, 456
    by heat, 455–456
    heat transfer, 455
    power, 456
    by work, 456–457
    work done by pressure forces on system, 457
Enthalpy, 79, 81, 84
    change in, 85
Entrance length, 676
Equation of state, 75
Equations of motion, 301, 392, 467, *see* Energy
        equation
Eulerian approach, 920–921, 1117
    exercises, 1102, 1295
Eulerian vs. Lagrangian forms of governing
        equations, 1116–1117
Euler number, 761, 1413; *see also* Dynamic
        similitude
    cavitation number, 1413
    exercises, 1553
    pressure coefficient, 1413
    velocity coefficient, 1413
Evaporation, 67
    of liquid in enclosed and vacuumed
        container, 67
    of liquid in open container, 67
    surface phenomena of, 67
    volume phenomena of, 68
External flow, 302–303, 647, 919, 1115, 1305;
        see also Drag force in external flow;
        Flow resistance modeling in external
        flow; Flow types; Lift force in external
        flow
    drag coefficient variation for, 1432
    drag force and lift force, 1306–1308, 1311
    dynamic similarity requirements for,
        1434–1435

Eulerian vs. Lagrangian forms of governing
        equations, 1308
    evaluation of drag force, 1310–1311
    exercises, 728, 1395
    flow resistance in, 659
    flow resistance modeling, 1308–1309
    governing equations in, 1309–1310
    velocity, 665

F

Flow
    domain, 326
    energy, 80, 454
    measurement devices, 718, 838, 1079
    transition, 1132, 1297
Flow depth, 1150, 1151; *see also* Open channel
        flow
    for abrupt channel transitions, 1187
    for abrupt decrease in channel width,
        1192–1196
    for abrupt upward step, 1187–1191
    alternate depths of, 1142
    conjugate depths, 1169
    critical depth, 1201
    critical flow at hydraulic jump, 1210–1211
    critical velocity, 1201
    decrease in, 1177, 1188, 1192
    example problem, 1178–1180, 1182–1187,
        1189–1190, 1193–1196, 1198–1200,
        1202–1204, 1206–1209, 1215–1220,
        1222–1228, 1231
    exercises, 1299–1301, 1302–1303
    Froude number for upstream section, 1212
    for gradual channel transitions, 1176–1177
    for gradual decrease in channel width,
        1180–1187
    for gradual upward step, 1177–1180
    hydraulic jump equations, 1211–1214
    major head loss, 1215–1220
    major head loss for hydraulic jump, 1209
    major head loss in nonuniform open channel
        flow, 1228–1253
    major head loss in uniform open channel
        flow, 1221–1228
    momentum function, 1205, 1207
    M-y plot for abrupt channel transition, 1205
    numerical solution for hydraulic jump, 1214
    and reaction force for short channel
        transitions, 1176
    reaction force on open channel flow control,
        1204–1209
    specific energy, minimum, 1197, 1201

Flow depth (*Continued*)
for typical flow-measuring devices, 1196–1204
unknown depth of flow, 1177, 1181
Flow field, 307
features, 304
fluid characteristics of, 310
Flow geometry, 304; *see also* Continuity equation
exercises, 349–350
path lines, 305
streamlines, 305–306
stream tubes, 306–307
Flow measuring devices, 1255; *see also* Open channel flow
assumption deviation in critical flow, 1257–1258
controls, 1256
critical depth meters, 1255–1256
depth–discharge relationship, 1256–1257, 1258
exercises, 1298
ideal flow meters, 1255
M-y plot and *E-y* plot for, 1197
for open channel flow, 650
pitot-static tube, 1135
Flow meters, *see* Flow—measuring devices
Flowrate, 329–335, 339–348
Flow regime, 1131–1132
exercises, 1297
occurrence of nonuniform flow and changes in, 1132
for uniform flow, 1131–1132
Flow resistance, 33, 348, 364, 443, 643, 645
drag coefficient, 645
exercises, 627
by friction, 404, 582, 644
friction/viscous force modeling, 645
in internal and external flow, 579
in laminar pipe flow, 704
modeling, 401–402, 578–580
by pipe components, 405, 582
prediction equation, 1426
pressure drop, 645
in turbulent pipe flow, 705
Flow resistance equations, 643, 762–769; *see also* Flow resistance modeling; Flow types; Laminar and turbulent internal flow; Major head loss equation for laminar flow; Major head loss equation for turbulent flow
for actual discharge, 413, 598, 599
application of, 646
derivation of, 645–646
for drag force, 442, 615, 778

exercises, 726–748, 902–903
for flow-measuring devices, 413
for major and minor head loss, 404
prediction equation, 1426
subset level application, 579
Flow resistance modeling, 643, 644–645, 646; *see also* Flow resistance equations
assumption of ideal flow, 644
exercises, 727–728, 748
in external flow, 643
in internal flow, 643, 645
as loss in turbine efficiency, 646, 725–726
subset level application of governing equations, 644–645
Flow resistance modeling in external flow, 659–660, 776; *see also* Dimensional analysis; Flow resistance equations
actual velocity, 661, 776
Bernoulli equation, 661, 777
d'Alembert's paradox, 663
dimensional analysis, 787–790
drag and drag coefficient derivation, 778–786, 787–790
drag coefficient, 661
drag force, 660, 664, 665, 776
drag force equation, 661
drag force evaluation, 660–661, 776–777
drag force in momentum equation, 662
example problem, 787–790, 791–795
exercises, 732, 907–909
external flow actual velocity, 665
friction force, 662, 663, 664
head loss, 660
ideal velocity, 661
lift and lift coefficient derivation, 791–795
lift coefficient, 791
lift force, 662, 665, 791
momentum equation and dimensional analysis, 661–662, 665–666, 777–778
pressure drop, 662, 663, 664
pressure force, 662, 663
submerged moving body in stationary ideal fluid, 663–664
submerged moving body in stationary real fluid, 664–665
submerged stationary body in stationary ideal fluid, 662–663
Flow resistance modeling in internal flow, 666–667, 718–719, 795, 838–839; *see also* Dimensional analysis; Flow resistance equations
actual area, 723
actual discharge, 724

actual discharge evaluation, 719–720, 839–840

Bernoulli equation, 720, 840

Chezy coefficient, 804–805

Chezy equation, 668, 671, 799, 804–805

Chezy equation substitution in energy equation, 805–806

continuity equation, 720–721, 840–841

Darcy–Weisbach equation, 806–809

dimensional analysis application, 800–804, 816–820

discharge coefficient, 239, 724

energy equation application, 668, 797

example problem, 849–853, 861–866

exercises, 734–736, 745–748, 909–914

flow resistance equation for major head loss, 798

friction slope, 670, 805

friction slope and wall shear stress, 798

governing equation, 669–671

head loss derivation, 797

ideal area, 723

integral momentum equation, 669, 722, 723, 842

major head loss equation for laminar and turbulent flow, 799–800

major head loss equation in open channel flow, 811–816

major head loss evaluation, 667–668, 796–797

Manning's equation, 809–811

momentum equation and dimensional analysis, 668–669, 721–724, 797–799, 841–842

pressure drop, 670

pressure drop and wall shear stress, 797

rectangular velocity distribution comparison, 724

reduced discharge and drag coefficient, 842

shear stress, 670

velocity profile comparison for ideal and real flows, 724–725

Flow resistance modeling in pipe flow, 709, 820; *see also* Dimensional analysis; Flow resistance equations

cavitation number, 831

drag coefficient, 829, 830

energy equation, 711–713, 715, 821–822

example problem, 834–838

exercises, 743–745, 911–912, 1102

governing equation, 921–922

major head loss, 709, 713, 717

minor head loss coefficient, 716–718, 824, 830, 832, 833

minor head loss equation, 714–715, 823–824

minor head loss evaluation, 710, 820–821

momentum equation and dimensional analysis, 713–714, 716, 822–823

pressure drop, 711, 824, 830

pressure rise, 712

Flow resistance modeling in pump and turbine efficiency, 866–867; *see also* Dimensional analysis

efficiency derivation, 867–878

efficiency evaluation, 867

efficiency for specific flow assumptions, 878–893

example problem, 878–893

exercises, 914–916

Flow types, 647; *see also* Flow regime; Flow resistance equations; Ideal flow; Real flow

exercises, 728

internal vs. external flow, 647

pipe vs. open channel flow, 647–648

real vs. ideal flow, 648–649

velocity distribution for open channel flow, 648

FLT, *see* Force-length time

Fluid, 21

air, 1567

bulk modulus of elasticity, 70

exercises, 97

gases, 1570

ICAO standard atmosphere, 1564–1565

liquids, 1568–1569

mass density, 22–25

properties of, 1, 15

specific gravity, 25–27

specific weight, 27

in static equilibrium, 2

surface tension, 55

vapor pressure, 64

viscosity, 22, 32

water, 1566

Fluid dynamics, 1, 301, 363–365; *see also* Energy equation

dimensional analysis, 1–2

exercises, 94, 348, 514

flow resistance, 364

internal flows, 2

Fluid flow

dimensionality of, 318

dimensions and units, 760

physical quantities of, 1

in pipes, 1115

Fluid flow types, 2–3, 302; *see also* Continuity
        equation
    compressible vs. incompressible flow, 303
    exercises, 349
    internal vs. external flow, 302–303
    laminar vs. turbulent flow, 303
    one-, two-, and three-dimensional flows, 304
    pressure vs. gravity flow, 303
    real vs. ideal flow, 303
    spatially varied vs. spatially uniform flow,
        303–304
    turbulent flow, 324
    unsteady vs. steady flow, 304
    velocity profile comparison, 323–324
Fluid kinematics, 1, 301, 302, 363, 555; *see also*
        Continuity equation
Fluid mechanics, 1, 301
    applications of, 718
    basic units of study, 301
    conservation of momentum principle in, 366
    exercises, 94
    fluid flow rate determination, 838
    governing equations, 301
Fluid motion, 307–308; *see also* Continuity
        equation
    control volume, 309–310
    deriving views from other, 312
    exercises, 350
    fixed vs. moving control volume, 311–312
    fluid particle velocity, 308–309
    fluid system, 308
    function of time, 312
    Lagrangian view vs. Eulerian view, 312–313
Fluid particle velocity, 308–309
Fluid properties
    exercises, 97
    relationship between dynamic forces and, 22
    standard atmospheric pressure for, 16
    variation in temperature for, 16
Fluid statics, 1, 115, 301; *see also* Buoyancy;
        Hydrostatic forces on submerged
        surfaces; Hydrostatic pressure
        measurement; Buoyant body stability
    exercises, 259
    fluid properties, 115
    hydrostatics, 116–125
    significant forces for, 115
Fluid system, 307; *see also* Fluid motion
    continuity equation for, 326–335
    energy equation for, 372–373, 457–458
    exercises, 351–355, 515, 620
    Lagrangian point of view, 308
    momentum equation, 556–558

Newton's second law of motion, 367–369
Force-length time (FLT), 759
Force scale ratio, 1412, 1416
Friction drag, 1318
Friction force, 662, 663
    submerged moving body, 663, 664
    submerged stationary body, 662
Friction slope; *see also* Major head loss equation
        for turbulent flow
    Chezy equation, 690–691
    empirical expression derivation for wall shear
        stress, 689
    in turbulent flow, 687–688
    and wall shear stress, 798
Friction/viscous force, 40, 53, 756
    modeling, 645
Froude number, 761, 762, 1147–1148, 1369–1372,
        1414; *see also* Dynamic similitude
    example problem, 1371–1372
    exercises, 1553

G

Gage pressure, 12
Gas
    constant, 76
    properties, 76, 1570
Generator efficiency, 464, 501
Geometric shape, 196, 1571
    area, 1572–1573
    center of area, 1572–1573
    centroid, 1574
    moment of inertia, 1572–1573
    volume, 1574
Geometric similarity, 1410; *see also* Dynamic
        similitude
    exercises, 1552
    requirements, 1436
    similitude length scale ratio, 1437
Gravity flow, 303
Gravity model
    similarity requirements, 1439–1440
    similitude scale ratios, 1420
    similitude velocity scale ratio, 1439
    viscosity model and surface tension model
        similarity requirements, 1445–1446
    and viscosity model similarity requirements,
        1441–1444

H

Hagan–Poiseuille law, *see* Poiseuille's law

Hazen–Williams equation, 696, 957, 958; *see also* Major head loss equation for turbulent flow
  application of, 959–962
  evaluation of roughness coefficient, 697
  example problem, 959–962
  exercises, 740–741, 1106
  head loss equation, 697
  problem types for turbulent pipe flow, 959
  for SI system, 958
Hazen–Williams roughness coefficient, 938, 957–958
Heat transfer, 455
HGL, *see* Hydraulic grade line
Hydraulic grade line (HGL), 383, 390, 391, 1137; *see also* Energy grade line
  and cavitation, 389–392
Hydraulic jump, 449, 752, 1132, 1163, 1209; *see also* Momentum equation; Newton's second law of motion
  applications of governing equations for, 450–451, 618–619, 1126
  critical flow at, 1210–1211
  energy and momentum equations for, 449–450, 618
  energy equation for, 400, 451–452, 578
  equation, 1212, 1213
  equations derivation, 1211–1214
  example problem, 451–452, 618–619
  exercises, 542, 641–642, 1301
  flow depth and major head loss for, 1209
  major head loss due to, 1215–1220
  M-y plot and E-y plot for, 1210
  numerical solution for, 1214
  use of controls in formation of, 1163–1164
Hydrostatic forces on submerged surfaces, 161–162; *see also* Fluid statics; Pressure prism variation
  center of pressure, 166
  distribution of, 166
  in enclosed and pressurized liquid, 184–196
  example problem, 168–169, 171, 172–173, 173–174, 175–176, 179–180, 183–184, 185–186, 188–189, 191–192, 195–196, 220–222
  exercises, 273
  in gas, 173–174
  horizontal component of force, 217–219
  hydrostatic pressure variation, 162
  in liquid open to atmosphere, 174–184
  magnitude and location of, 216–217
  in multilayered fluid, 223
  for plane surfaces, 165–166

resultant force acting in first fluid, 223–225
resultant force acting in multilayer fluid, 227–229
resultant force acting in second fluid, 225–227
resultant force magnitude, 166, 220–222
resultant force on plane surfaces, 166–173
uniform pressure prism, 163
vertical component of force, 219–220
Hydrostatic pressure, 117; *see also* Hydrostatics
  distribution, 123–124, 1136–1137, 1138, 1139
  equation, 116, 118–121
  head, 121–123
Hydrostatic pressure equation, 116, 118–121; *see also* Hydrostatics
  example problem, 121
  hydrostatic pressure vs. depth, 120
  Pascal's law, 118
  scalar ordinary differential equation of fluid statics, 119
Hydrostatic pressure head, 121; *see also* Hydrostatics
  example problem, 122–123
Hydrostatic pressure measurement, 126; *see also* Fluid statics; Manometers
  apparatus, 126, 129
  barometers, 129–133
  exercises, 263–272
  methods, 129
  pressure gages, 129
  pressure scales, 126–128
  pressure transducers, 129
  standard atmospheric pressure, 128
Hydrostatics, 116–117; *see also* Fluid statics; Hydrostatic pressure
  application of, 124–125
  exercises, 260–263
  forces acting on fluid in static equilibrium, 116
  free body diagram, 125
  Pascal's law, 116, 117, 118
Hypersonic flow, 92

I

ICAO, *see* International Civil Aviation Organization
Ideal discharge, derivation of, 720–721
Ideal external flow, 614; *see also* Momentum equation; Newton's second law of motion
  actual pressure drop, 443–444
  Bernoulli equation, 444–449, 614
  drag force evaluation, 442–444, 614–616

Ideal external flow (*Continued*)
   energy and momentum equations for, 442
   example problem, 447–448, 616–617
   exercises, 541–542, 640–641
   flow resistance equation for drag force, 615
   governing equations for, 444, 614, 616–617,
      1306, 1309–1310
   ideal pressure difference, 443
   ideal velocity, 444
   momentum equation, 443
   pitot-static tube, 445
   stagnation point in flow over submerged
      body, 449
   stagnation pressure, 446, 448
   stagnation streamline in flow over
      submerged body, 450
   subset level application, 442, 444, 614
Ideal flow, 303, 643, 649; *see also* Flow types
   Bernoulli equation, 650
   dynamic pressure, 650
   exercises, 729–730
   ignoring flow resistance in, 649–650
   modeling flow resistance in, 650–651
   pitot-static tube, 651–653
   real flow vs., 648–649
   static pressure, 650
   total pressure, 650
   velocity profiles for ideal and real flows,
      724–725
   venturi meter, 653–654
Ideal flow from tank; *see also* Ideal internal flow
   exercises, 637–640
   governing equations for, 605–614
Ideal flow meters, 414, 1116, 1255; *see also* Actual
      flowrate for ideal flow meters; Open
      channel flow; Venturi tube
   actual flowrate, 1078–1079, 1253–1254,
      1258–1259
   advantages of, 1261
   critical depth meters and, 1260
   disadvantages of, 1261
   flow-measuring devices, 1261–1262
Ideal fluid, 32
Ideal gas law, 75
   amount of matter, 76
   equation of state, 75
   example problem, 77–79
   gas constant, 76
   molecular weight, 76
Ideal internal flow; *see also* Ideal open channel
      flow; Momentum equation; Newton's
      second law of motion
   actual discharge evaluation, 413–414, 598–600

Bernoulli equation, 415, 419, 423, 598
   energy and momentum equations for, 412, 598
   example problem, 415–432, 434–442, 601–602,
      603–605, 605–611, 612–614
   exercises, 524, 634–640
   flow problems, 598
   flow resistance equation, 598, 599
   governing equations, 414–415, 600–601
   ideal flow from tank, 423, 605–614
   ideal pipe flow, 415–419, 601–602
   siphon, 439, 440
   Torricelli's theorem, 432–434
Ideal open channel flow, 419–423, 602–605,
      1123–1124; *see also* Ideal internal flow;
      Open channel flow
   application of governing equations, 602–605,
      1116, 1126
   evaluation of actual discharge, 1124–1125
   exercises, 636–637
   flow resistance equations for actual
      discharge, 1125
Ideal pipe flow, 927–928; *see also* Ideal internal
      flow; Pipe flow governing equations
   actual discharge evaluation, 928–929
   Bernoulli equation, 928
   exercises, 635–636, 1102
   governing equations, 601–602, 919, 929
Ideal plastic, 44
Ideal pressure difference, 443
Ideal specific heat
   at constant pressure, 82
   at constant volume, 82
   ratio, 79, 83
Ideal upstream velocity, 653
Ideal velocity, 840
   derivation of, 661, 720, 777
Implosion, 70
Impulse turbine, 486, 488; *see also* Turbine
   application of energy equation for, 489
Incompressible flow, 303
Integral momentum equation, 564–565, 722, 723,
      842; *see also* Momentum equation
   application of, 566–576
   example problem, 566–576
   exercises, 623–627
   fixed vs. moving control volume, 565–566
   velocity at given control face, 565
Internal energy, 80, 453
Internal flow, 2, 302–303, 643, 647, 795, 919, 1115;
      see also Flow types; Flow resistance
      modeling in internal flow
   dynamic similarity requirements for,
      1434–1435

exercises, 728
flow resistance in modeling, 643, 645
variation of drag coefficient for, 1433
International Civil Aviation Organization
      (ICAO), 14
  standard atmosphere, 1564–1565
International System (SI), 3, 6, 7; *see also* Units
  dimensions for, 5
  example problem, 12
  Kelvin scale, 11
  units for, 10
Inviscid, 32
  core, 679
Isentropic, 86, 87
Isothermal, 86

K

Kelvin scale, 11
Kinematic similarity, 1410–1411; *see also*
      Dynamic similitude
  exercises, 1552
  primary acceleration scale ratio, 1411
  primary time scale ratio, 1411
  primary velocity scale ratio, 1410
Kinematic viscosity, 44; *see also* Viscosity
Kinetic energy, 80

L

Lagrangian approach, 920, 921, 1118
  exercises, 1102, 1295
Laminar and turbulent flow, 698–699; *see also*
      Drag force in external flow; Major head
      loss equation for turbulent flow
  comparison using head loss equation,
    700–702
  evaluating Darcy–Weisbach friction
    coefficient, 700
  example problem, 1359–1361, 1471–1474,
    1475–1479, 1481–1491
  loss equation and Reynolds number, 699–700
  role of boundary roughness in, 705–706
  for round-shaped bodies, 1357–1361,
    1470–1479
  for shapes except round-shaped bodies,
    1344–1357
  velocity profile comparison, 704–705
  with wave action at free surface, 1362,
    1479–1491
Laminar and turbulent internal flow, 671–672;
     see also Flow resistance equations;
     Major head loss equation for laminar

flow; Major head loss equation for
    turbulent flow
  acceleration, 673
  developing vs. developed flow, 675–676,
    677, 678
  differential form of Newton's second law of
    motion, 672
  entrance length, 676
  exercises, 736
  extent of developing flow in pipe length,
    676–677
  forces acting on cylindrical fluid element in
    pipe flow, 673
  inviscid core, 679
  laminar vs. turbulent flow, 680–681
  magnitude of the pressure gradient, 680
  Newton's law of viscosity, 680
  pressure drop and wall shear stress variation
    with pipe length, 679–680
  shear stress, 674–675
  shear stress variation determination,
    672–675
  slope of total energy, 674
  velocity variation with pipe length, 678–679
  velocity variation with time, 675, 676
  viscous force, 673
Laminar flow, 36, 37, 303; *see also* Laminar and
    turbulent internal flow; Viscosity
  example problem, 41–43, 1452–1455,
    1456–1459
  flow resistance prediction equation,
    1452, 1455
  laminar Couette flow, 39
  major head loss equation derivation, 799
  Newton's law of viscosity for, 39–44
  pressure drop in, 685
  for shapes except round-bodies, 1451–1459
  shear stress distribution, 38–39
  shear stress of fluid, 39
  over solid boundary, 46
  velocity gradient, 40
Laminar pipe flow, 930–934, 1491–1495; *see also*
    Single pipes
  Darcy–Weisbach friction coefficient
    evaluation, 809
  Darcy–Weisbach friction factor and Reynolds
    number, 945
  differential form of Poiseuille's law, 931
  example problem, 932–934
  exercises, 1103
  Poiseuille's head loss equation, 932
  Poiseuille's law, 931, 944
  types of problems for, 932

Latitude
exercises, 96
standard reference, 15
Laws governing fluid in motion, 313–314; *see also*
Continuity equation
assumption of spatial dimensionality of fluid
flow, 318–319
assumptions for fluid flow type, 317–318
conservation of momentum, 313
exercises, 350–351
extensive and intensive fluid
properties, 314
governing laws assuming fluid system, 314
modeling of 1D flows, 320–323
modeling of 2D flows, 319–320
physical fluid parameters, 314
Reynolds transport theorem, 314–317
velocity profile comparison for various flow
types, 323–324
Laws of dynamic similarity, 750
Lift coefficient, 791–795
Lift force, 662, 755, 791
example problem, 791–795
Lift force in external flow, 1306, 1311; *see also*
External flow
airfoil performance optimization, 1382–1383
airfoil shape optimization, 1384–1392
evaluation of, 1372–1375
evaluation of lift coefficient, 1375
example problem, 1378–1381, 1382–1383,
1386, 1388, 1390–1392, 1393–1395
exercises, 1400–1401
forces acting on plane in direction of
flight, 1374
lift coefficient variation with angle of attack,
1377, 1385, 1390
lift force and lift coefficient for hot-air
balloon, 1392–1395
lift-to-drag ratio variation with angle of
attack, 1382
optimizing airfoil shape and angle of attack,
1376–1381
role of body shape and angle of attack,
1375–1376
stall conditions, 1378
stationary fluid exerting, 1307
wing loading, 1373
Linear variable differential transformer pressure
transducer (LVDT), 129
Liquid properties, 1568–1569
Liquids, 56
LVDT, *see* Linear variable differential
transformer pressure transducer

M

Mach number, 71, 762, 785, 1365–1369, 1415
example problem, 93, 1369
exercises, 113
for fluids, 92
Macroscopic energy, 453–454
Major head loss, 796; *see also* Single pipes
Chezy equation, 796–797
coefficient, 716–718, 824, 830, 832, 833
dimensional analysis, 764–765
equation, 714–715, 764, 765, 799–806,
811–816, 823–824
evaluation, 667–668, 710, 796–797, 820–821,
930, 1120–1122, 1128
exercises, 1557–1559
flow depth, 1215–1220, 1228–1253
by flow resistance, 1126–1131
flow resistance equation, 581, 798, 924, 1121
guidelines in derivation of, 764
for hydraulic jump, 1209, 1215–1220
in nonuniform flow, 1228–1253
in pipe flow, 709, 713, 717, 929
similitude scale ratios, 1491, 1504–1505
in uniform flow, 1221–1228
and wall shear stress, 706
Major head loss equation for laminar flow, 681,
799; *see also* Flow resistance equations
average velocity, 684
constant of integration, 682
exercises, 737
interpretation of Poiseuille's law, 685
Newton's law of viscosity, 682
parabolic velocity distribution for laminar
flow, 683
Poiseuille's law, 683–684
pressure drop, 681
shear stress, 682
steady and uniform laminar flow, 684
substituting Poiseuille's law into integral
form of energy equation, 684–685
velocity profile, 681–683
Major head loss equation for open channel flow,
811–813; *see also* Flow resistance
modeling in internal flow
channel slope, 814
Chezy equation, 814
Darcy–Weisbach equation, 815
dimensional analysis interpretation, 813–814
example problem, 816–820
exercises, 909–911
friction slope, 814
manning's equation for open channel flow,
815–816

Major head loss equation for turbulent flow,
686–687, 799–800; *see also* Flow
resistance equations; Flow resistance
modeling in internal flow
  boundary roughness role in laminar and
turbulent flow, 705–706
  Chezy coefficient, 690, 691, 805
  Chezy equation, 686, 688, 690–691,
708–709, 804
  Chezy equation into energy equation,
691–692, 805–806
  Darcy–Weisbach friction coefficient
evaluating, 700
  Darcy–Weisbach head loss equation and
Reynolds number, 699–700
  differential momentum equation, 687–688
  dimensional analysis, 686, 688, 800–804
  drag coefficient, 686, 689
  effective roughness criterion, 705
  exercises, 738
  flow resistance, 704, 705
  flow resistance coefficient, 698, 700
  friction forces, 686
  friction slope, 686, 687–688, 690, 708
  integral momentum equation, 687
  laminar and turbulent flow, 698–699,
700–702
  major head loss and wall shear stress, 706
  major head loss equation for open channel
flow, 706–707
  shear stress, 687
  surface roughness, 705
  velocity profiles, 702–705, 741–742
  wall shear stress, 686, 689
Manning's equation, 694, 695–696, 809, 810,
952–954; *see also* Flow resistance
modeling in internal flow; Major head
loss equation for turbulent flow
  application of, 954–956
  derivation and evaluation of roughness
coefficient, 695
  example problem, 955–956
  exercises, 739–740, 910, 1105–1106
  head loss equation, 696, 709, 811
  hydraulic radius, 695
  for open channel flow, 815–816, 1129–1131
  roughness coefficient, 809–810, 952–954,
1129–1131
  types of problems for turbulent pipe flow, 954
Manning's roughness coefficient, 809–810,
952–954, 1129–1131; *see also* Drag
coefficient
  empirical values for, 953

Manometers, 129, 133; *see also* Hydrostatic
pressure measurement
  differential, 150–161
  exercises, 265–272
  fluid, 139
  forces acting on column of fluid in static
equilibrium, 134
  manometry, 134–135
  open U-tube, 139–150
  piezometer columns/tubes, 135–139
  types of, 133
Manometers, differential, 150–152
  example problem, 153–154, 155–156, 157–158,
160–161
  multifluid, 152, 159–161
  single-fluid, 152–158
  with triple U-tube, 152
Mass density, 22
  dimensions and units for, 23
  example problem, 23, 24–25
  exercises, 97–98
  of fluid, 25
  specific volume, 24
  standard density for water, 24
Mass-length-time (MLT), 759
Mathcad, 939
Mechanical energy forms, 454
Mechanical pressure-measuring devices, 129
Mercury barometer, 130
Metacenter, 246
Metacentric height, 246; *see also* Buoyant body
stability
  average height body, 252
  computation of, 248–259
  length of, 247
  moments of inertia, 251
  rotational stability of top-heavy floating
body, 251
  for top-heavy floating body, 246–248
  top-heavy floating body with negative, 249
Microscopic energy, 453
Minor head loss, 820–821; *see also* Flow resistance
modeling in pipe flow
  alternative modeling in energy equation,
966–967
  associated, 980
  due to change in flow direction, 976
  coefficient, 824, 830–833
  dimensional analysis, 824–831
  energy equation, 821–822
  equation, 823–824
  evaluation, 963–964, 967–968
  example problem, 834–838, 972–974, 978–979

Minor head loss (*Continued*)
    exercises, 911–912, 1106–1108, 1559
    in fittings, 968–969, 974–979
    momentum equation and dimensional
        analysis, 822–823
    in pipe contractions, 989–990, 995–998
    in pipe entrances, 980–983
    in pipe exits, 983, 984
    in pipe expansions, 984–989, 990–994
    in pipe flow equation, 765
    pressure drop and drag coefficient,
        833–838
    similitude scale ratios for, 1510–1511
    substituting pressure drop into energy
        equation, 831
    in valves, 969–974
Minor head loss coefficient
    derivation of, 965
    evaluation of, 965–966
    for fittings, 969
    for 90° smooth bend, 977
    for pipe entrance, 983
    for valves, 968
MLT, *see* Mass-length-time
MLT system, *see* International System
Modeling of scaling effects
    for efficiency of nonhomologous pumps,
        1539–1544
    for efficiency of nonhomologous turbines,
        1546–1550
Model ratio, *see* Model scale
Model scale, 1410
Molecular weight, 76
Momentum equation, 1, 555; *see also* Hydraulic
        jump; Ideal external flow; Ideal internal
        flow; Newton's second law of motion;
        Real internal flow
    acceleration slope, 558
    Bernoulli equation, 577
    derivation of, 556
    differential form of, 558–564
    dimensionless slope, 557
    energy equation in complement to, 576–577
    exercises, 620
    fixed vs. moving control volume, 565–566
    for fluid system, 556–558
    friction slope term, 578
    governing equations, 577–578
    integral form of, 556, 564–576
    modeling flow resistance, 578–580
    slope of total energy, 558
    subset level application of governing
        equations, 579, 580

Momentum equation, differential form of,
        556–558, 584
    application of, 559–564
    Chezy equation, 560, 562, 564
    example problem, 560–561, 563–564
    exercises, 620–623
    friction slope, 560
    for nonuniform flow, 564
    for open channel flow, 562
    resistance equation, 564
Momentum equation for control volume, 564–565
    exercises, 623–627
    general statement, 565
    integral form of Newton's second law of
        motion, 564–565
Momentum function, 1162, 1165
    computation of moment of area, 1171
    conjugate depths of flow, 1169
    curve for rectangular channel sections,
        1166–1170
    exercises, 1299
    integral momentum equation for finite
        control volume, 1164
    M-y plot and *E-y* plot for sluice gate, 1167
    M-y plot for assisted hydraulic jump, 1169
    for nonrectangular channel sections,
        1170–1171
    for rectangular channel sections, 1164–1165
    use of controls in formation of hydraulic
        jump, 1163–1164
Moody diagram, 940
    Darcy–Weisbach friction coefficient, 942–943
Motor efficiency, 464, 482
Multifluid differential manometer
    with double U-tube, 152
    between two pipes, 159–161
Multi-fluid simple manometer, 139, 148–150
Multilayered fluid; *see also* Hydrostatic forces on
        submerged surfaces
    example problem, 227–229
    exercises, 291
    hydrostatic force for surfaces submerged
        in, 223
    resultant force acting in, 227–229

N

Newtonian fluids, 44
Newton's law of viscosity, 32, 680, 681; *see also*
        Viscosity
    example problem, 47–50, 51–53
    expression for shear stress for laminar flow,
        45–53

friction/viscous force, 40
laminar Couette flow, 39
for laminar flow, 39–44
Newton's second law of motion, 1, 19, 365–367,
672; *see also* Bernoulli equation; Energy
equation; Hydraulic jump; Ideal
external flow; Ideal internal flow;
Momentum equation; Real internal flow
acceleration of system, 19–21, 368
for control volume, 366, 369–371
convective acceleration, 369
differential form of, 367, 373, 672
energy equation based on, 371
energy equation for fluid system, 372–373
energy equation for hydraulic jump, 400
energy equation in complement to
momentum equation, 392
exercises, 96, 515
for fluid system, 367–369
governing equations, 392–401
integral form of, 564–565
laminar pipe flow, 682
modeling flow resistance, 401–402
1D Navier–Stokes equation, 372
sum of forces, 368
90-degree bends, 976–977
Noncircular pipes, 699
Non-Newtonian fluids, 44
Nonrectangular channel cross sections, 1145;
see also Open channel flow
derivation of critical flow for, 1146–1147
momentum function for, 1170–1171
numerical solution for hydraulic jump, 1214
Nonuniform flow; *see also* Open channel flow
and changes in flow regime, 1132
channel transition, 1132
exercises, 1296
flow depth and major head loss in, 1228–1253
Nonzero intercept, 44
No-slip condition, 46
Nozzle, 1088
Nozzle meter, 1093–1097
actual discharge, 1095
construction variations, 1094
evaluation of minor head loss due to,
1100–1101
example problem, 1096–1097
nozzle discharge coefficient, 1095

O

Obstruction flow meters, 1087
One-dimensional flow, 304, 318

modeling of, 320–323
Navier–Stokes equation, 372
uniform velocity profile, 321, 322
Open channel flow, 647–648, 1115; *see also* Actual
flowrate for ideal flow meter; Channel
section geometric properties; Critical
depth meters; Critical flow; Flow
depth; Flow regime; Flow types; Ideal
flow meters; Ideal open channel flow;
Momentum function; Nonrectangular
channel cross sections; Real open
channel flow; Rectangular channel
sections; Specific energy
actual discharge in, 764, 767, 768
actual flowrate in devices, 1253–1254
channel slope, 814
Chezy equation, 708–709
control in, 1133–1134
critical flow at controls, 1134–1135, 1148–1159
deviation from hydrostatic pressure
distribution, 1137
energy concepts in, 1135–1136
Eulerian approach, 1117
Eulerian vs. Lagrangian forms of governing
equations, 1116–1117
evaluation of actual flowrate for ideal flow
meters, 1262
exercises, 729, 742–743, 1294–1304
flow depth and major head loss, 1209,
1221–1228, 1228–1253
flow depth and reaction force, 1176
flow measurement devices, 650, 838, 1135
flow resistance modeling, 1118–1119
flow type and flow regime, 1131–1132
friction slope, 708, 814
Froude number, 1147–1148
geometric properties of channel sections,
1171–1176
governing equations, 1115, 1119–1126
hydrostatic pressure distribution, 1136–1137,
1138, 1139
Lagrangian approach, 1118
major head loss, 765, 1118, 1126–1131
major head loss and wall shear stress, 706
major head loss equation for, 706–707,
811–816
Manning's head loss equation, 709
measuring device, 403, 497, 1254, 1255
momentum concepts in, 1162
sharp-crested weir and free overfall,
1138–1139
similitude scale ratios for actual discharge in,
1520–1521

Open channel flow (*Continued*)
    specific energy in, 1139–1140
    structure, 1165
    subcritical and supercritical flow,
            1147–1148
    transitions and controls in, 1132–1133
    uniform flow at controls, 1159–1162
    velocity distribution for, 648
    with weir with large head, 1525–1529
    with weir with small head, 1529–1534
Open channel flow control, reaction force on,
        1204–1209
Open U-tube manometers, 139–140; *see also*
        Manometers
    example problem, 141–144, 145–147, 149–150
    manometer fluid, 139
    multifluid simple manometer, 139, 148–150
    single-fluid simple manometer, 139,
            140–147
Orifice, 1088
Orifice meter, 1090–1093
    minor head loss meter, 1100–1101

P

Partially filled circular channel sections,
        1175–1176
Partial pressure, 64
Pascal's law, 116, 117, 118; *see also* Hydrostatics
Path lines, 305, 308
Pelton wheel; *see also* Impulse turbine
Physical fluid parameters, 314
Piezometer, 136, 138
    columns, 135; *see also* Manometers
    example problem, 137, 138–139
Piezometric pressure, 372
Pipe components, 403
Pipe contractions
    EGL and HGL for sudden, 989
    example problem, 995–998
    minor losses in gradual, 995–998
Pipe devices, 655, 963
Pipe entrances, 980–983; *see also* Pipes with
        components
    associated minor losses, 980
    EGL and HGL for square-edged, 982
    and minor head loss coefficients, 981, 983
Pipe exits, 983
    EGL and HGL for reentrant, 985
    and minor head loss coefficients, 984
    minor losses, 983
Pipe expansions
    EGL and HGL for, 986, 991

    example problem, 992–994
    minor losses in sudden, 984–989
Pipe fittings, 971, 974; *see also* Pipes with
        components
    bends, 976–977
    example problem, 978–979
    tee, 974
    types of, 974
Pipe flow, 647–648, 919; *see also* Flow types;
        Pipe flow measurement and control
        devices; Pipe flow governing
        equations; Pipe systems; Single pipes
    actual discharge in, 764, 766
    Eulerian approach for pipe flow problems,
            920–921
    exercises, 729, 1101
    flow measurement devices, 838
    Lagrangian approach for pipe flow
            problems, 921
    major head loss, 764
    minor head loss, 765
    modeling flow resistance in, 921–922
Pipe flow governing equations, 919, 920, 922;
        see also Pipe flow
    differential approach for real pipe flow
            problems, 926–927
    evaluation of actual discharge, 928–929
    evaluation of head loss terms, 923–926
    evaluation of pump and turbine head
            terms, 926
    exercises, 1102
    for ideal pipe flow, 927–928, 929
    integral approach for real pipe flow
            problems, 923
    for real pipe flow, 923, 927
Pipe flow measurement and control devices,
        1078–1079; *see also* Pipe flow
    actual flowrate for ideal flow meters,
            1079–1080
    actual flowrate for nozzle meter, 1093–1097
    actual flowrate for orifice meter, 1090–1093
    actual flowrate for pipe flow-measuring
            devices, 1086–1090
    actual flowrate for pitot-static tube,
            1081–1086
    actual flowrate for venturi meter, 1097–1100
    Bernoulli equation, 1081, 1088
    discharge coefficient, 1080–1081
    example problem, 1084–1085, 1092–1093,
            1096–1097, 1099–1100
    exercises, 1113
    flow-measuring devices, 474, 492, 497,
            929, 1087

ideal flowrate, 1083
ideal pressure difference, 1085
minor head loss due to pipe flow-measuring
  devices, 1100–1101
obstruction flow meters, 1087
pitot-static tube, 1082, 1084, 1086
Pipe networks, 1063; *see also* Pipe systems
  Chezy equation, 1066
  continuity principle, 1063–1064
  energy principle, 1064–1067
  example problem, 1069–1078
  exercises, 1112
  governing equations for, 1067–1078
  Hazen–Williams equation, 1067
  head loss equation, 1065
  Manning equation, 1066
  momentum principle, 1067
  Poiseuille equation, 1066
Pipes with components, 962–963, 1001–1003;
    see also Pipe flow
  alternative modeling of minor head loss term,
    966–967
  control volume and EGL and HGL for pipe
    expansion, 986
  EGL and HGL for pipe expansion, 989, 991
  EGL and HGL for pipe system, 964
  example problem, 972–974, 978–979, 992–994,
    995–998, 998–1000
  exercises, 1106–1108
  minor head loss, 963–964, 967–968
  minor head loss coefficient, 965–966, 968
  minor losses in pipe entrances, 980–983
  minor losses in pipe exits, 983, 984
  minor losses in pipe expansions, 989–994
  minor losses in valves and fittings, 968–979
  minor loss in pipe contractions, 990–995
  pipe component categories, 963
  pipe fittings, 971
  pipe systems, 1001–1003
  pressure recovery coefficient, 965, 990
  pressure rise, 990
  with pump, 1004–1009
  reaction forces, 963, 998–1000
  static pressure, 965
Pipe systems, 1000; *see also* Branching pipes; Pipe
    flow; Pipe networks; Pipes with
    components
  example problem, 1001–1003, 1004–1009,
    1010–1016, 1018–1030, 1057–1063
  exercises, 1108
  pipes in loop, 1055–1063
  pipes in parallel, 1017–1030
  pipes in series, 1009–1016

pipes with pump, 1004–1009
single pipes, 1001
Pitot-static tube, 380, 445, 651, 1135; *see also*
    Bernoulli equation
  dynamic pressure, 651
  flow, 650
  for ideal velocity, 381
  to measure velocity, 652, 1082, 1084
  to parabolic velocity distribution, 1086
  pressure head, 651
  rise in pressure, 382, 653
  for upstream velocity, 653
π terms, 750, 761–762, 1412–1413, 1416–1418
  Cauchy number, 1415
  Euler number, 1413
  exercises, 903, 1553
  force scale ratio, 1416
  Froude number, 1414
  Reynolds number, 1414–1415
  Weber number, 1415–1416
Poiseuille's law, 681, 683–684, 931
  differential form of, 931
  exercises, 737, 1103
  head loss equation, 932
  interpretation of, 685
  properties of steady and uniform laminar
    flow, 684
  substituting into integral form of energy
    equation, 684–685
Potential energy, 80
Power, 456
Prandtl boundary layer theory, 678–679
Pressure
  exercises, 96
  flow, 303
  gages, 129
  recovery coefficient, 965
  standard reference, 15
  transducers, 129
Pressure coefficient, 1413
  common unit for, 117
Pressure drop, 388, 645
  derivation of, 713–714, 716
  expression for, 687
  for laminar and. turbulent flow, 685, 701
  pipe expansion, 656
  submerged moving body, 663, 664
  submerged stationary body, 662
  venturi meter, 653–654
  and wall shear stress, 797
Pressure force, 64
  submerged moving body, 663
  submerged stationary body, 662

Pressure model
    similarity requirements, 1437
    similitude scale ratios, 1419, 1437
Pressure prism; *see also* Hydrostatic forces on
        submerged surfaces
    trapezoidal, 170
    uniform, 163
    variation for surface submerged in enclosed
        pressurized liquid, 164–165
    variation for surface submerged in liquid,
        162–164
    variations for surface submerged
        in gas, 162
Pressure prism variation; *see also* Hydrostatic
        forces on submerged surfaces
    for surface submerged in gas, 162
    for surface submerged in liquid,
        162–164
    for surface submerged in liquid, enclosed,
        and pressurized, 164–165
    trapezoidal pressure prism, 170
Pressure scales, 12, 126; *see also* Hydrostatic
        pressure measurement
    absolute and gage, 127
    absolute pressure, 13, 126
    absolute zero pressure scale, 126
    example problem, 13
    exercises, 95
    gage pressure, 12
    positive and negative gage pressure, 128
    types of, 126
Primary
    acceleration scale ratio, 1411
    dimensions, 1
    force scale ratio, 1411
    time scale ratio, 1411
    velocity scale ratio, 1410
Pseudoplastic, 44
Pump efficiency, 464, 472, 481, 757, 772, 773–774
    example problem, 1541–1544
    exercises, 903, 905–907, 1560–1561
    similitude scale ratios for, 1535–1537
Pump–motor system, 461, 462, 464
Pumps
    affinity laws for efficiency of homologous,
        1537–1539
    empirical Moody efficiency correction
        equation, 1540
    example problem, 1541–1544
    scaling effects for efficiency of
        nonhomologous, 1539–1544
    similitude scale ratios for efficiency of,
        1535–1537

R

Rankine scale, 10
Reaction force on open channel flow control,
        1204–1209
Reaction turbine, 486, 487, 490; *see also* Turbine
    energy equation for, 489
    generator system analysis, 496
Real flow, 303, 643, 654, 755; *see also* Flow types;
        Laminar flow; Turbulent flow
    developing vs. developed flow, 659
    exercises, 729, 730–732
    vs. ideal flow, 648–649
    laminar vs. turbulent flow, 656–657
    modeling flow resistance in, 654–656
    pressure drop, 656
    shear stress, 655
    velocity profiles, 657–659, 724–725
    viscosity of, 649, 654
Real internal flow; *see also* Momentum equation;
        Newton's second law of motion;
        Reaction turbine; Real pipe flow;
        Thermodynamics, first law of
    cavitation, 503
    control volume, 581
    differential fluid element, 584–591
    dimensions and units for head terms, 470–471
    energy and momentum equations
        for, 403, 580
    energy equation for, 406–412, 468–470, 487
    evaluation of head loss terms, 403–405,
        580–583
    example problem, 406–412, 474–480, 482–486,
        492–496, 497–501, 504–507, 510–513,
        584–589, 590–591, 592–597
    exercises, 517–524, 543–554, 627–634
    finite control volume, 592–597
    flow resistance, 582
    flow resistance equation, 581
    fluid transport, 469
    generator efficiency, 491, 501
    governing equations, 406, 583–584
    impulse turbine, 486, 488, 489
    maximum allowable velocity, 473
    momentum equation, 584
    motor efficiency, 482
    power losses, 469–470
    with pump, 471–486
    pump efficiency, 472, 481
    shaft power output, 490
    with turbine, 486–513
    turbine efficiency, 501
    turbine–generator system efficiency, 491

Real open channel flow, 1120; *see also* Open
    channel flow
  Chezy equation and Chezy coefficient, 1122,
      1128–1129
  control volume, 1121
  differential approach for, 1122–1123
  evaluation of head loss term, 1120–1122, 1128
  evaluation of head terms, 1122
  exercises, 1296
  flow resistance equation for head loss, 1121,
      1126–1127
  governing equations, 1116, 1123
  integral approach, 1120
  Manning's equation and roughness
      coefficient, 1129–1131
  turbulent channel flow resistance
      equations, 1129
  uniform vs. nonuniform open
      channel flow, 1127
  unknown head loss, 1121
Real pipe flow; *see also* Pipe flow governing
    equations
  approach for problems, 923, 926–927
  Chezy equation, 925
  control volume, 924
  evaluation of head loss terms, 923–926
  evaluation of head terms, 926
  exercises, 1102
  flow resistance, 925
  flow resistance equation, 924, 925
  governing equations, 919, 923, 927
  head losses and reaction forces in, 962–963
  head losses in, 1000
Rectangular channel, 1140
Rectangular channel sections; *see also* Open
    channel flow
  critical flow for, 1141–1143
  depth–discharge curve for, 1143–1145
  exercises, 1301
  geometric properties of, 1172
  hydraulic jump equations, 1211–1214
  momentum function curve for, 1166–1170
  momentum function for, 1164–1165
  specific energy curve for, 1140–1141
  specific energy equation for, 1144
Reduced/actual discharge; *see also* Flow
    resistance modeling in internal flow
  equation, 722–724
  example problem, 861–866
  for open channel flow-measuring devices,
      854–861, 861–866
  for pipe flow measuring devices, 842–853
Rehbock formula, 1276–1277

Relative roughness
  similarity requirements, 1436–1437
  surface roughness, 1362–1365
Resistance equation, *see* Chezy equation
Reynolds number, 36, 761, 1414–1415; *see also*
    Dynamic similitude
  exercises, 1553
  for noncircular pipe, 944
Reynolds transport theorem, 314
  alternative statement of, 317
  control surface, 315
  control volume, 315, 316
  simplified version of, 315

S

Shaft power, *see* Shaft work
Shaft torque, *see* Shaft work
Shaft work, 461
Sharp-crested weir, 1267
  actual discharge, 1276, 1280
  discharge coefficient, 1276, 1281
  evaluation of actual flowrate for, 1272–1274
  example problem, 1277–1278, 1281–1282
  geometry of, 1273
  ideal flow assumptions, 1282
  ideal flow for, 1274, 1275, 1279
  rectangular, 1275–1278
  Rehbock formula, 1276–1277
  triangular, 1278–1283
  velocity coefficient, 1276
Shear stress
  derivation of, 713–714, 716
  friction slope and, 798
  pressure drop and, 797
SI, *see* International System
Similarity, 1409
Similarity rules, *see* Affinity laws
Similitude scale ratios, 1418, 1419–1421, 1437,
    1446–1447; *see also* Dynamic similitude
  creeping flow, 1447–1451
  for elastic model, 1422
  example problem, 1448–1451, 1452–1455,
      1456–1459, 1460–1464, 1466–1470,
      1471–1474, 1475–1479, 1481–1491
  exercises, 1553, 1555–1557
  flow resistance prediction equation,
      1452, 1455
  for gravity model, 1420
  laminar and turbulent flow for round-shaped
      bodies, 1470–1491
  laminar flow, 1451–1459
  for model and prototype fluids, 1424

Similitude scale ratios (*Continued*)
  for pressure model, 1419
  for surface tension model, 1423
  turbulent flow, 1459–1470
  for viscosity model, 1421
Similitude scale ratios for actual discharge
  example problem, 1517–1520, 1522–1525,
    1526–1529, 1530–1534
  exercises, 1559–1560
  in open channel flow, 1520–1521
  open channel flow with sluice gate,
    1521–1525
  open channel flow with weir, 1525–1534
  in pipe flow, 1515–1516
  pipe flow with flow-measuring device,
    1516–1520
Similitude scale ratios for efficiency, 1535
  affinity laws for homologous pumps,
    1537–1539
  affinity laws for homologous turbines,
    1545–1547
  example problem, 1541–1544, 1547–1550
  exercises, 1560–1561
  of pumps, 1535–1537
  scaling effects for nonhomologous pumps,
    1539–1544
  scaling effects for nonhomologous turbines,
    1546–1550
  of turbines, 1544–1545
Similitude scale ratios for major head loss in
    open channel flow, 1504–1505
  example problem, 1506–1510
  exercises, 1558–1559
  turbulent open channel flow, 1505–1510
Similitude scale ratios for major head loss in pipe
    flow, 1491
  example problem, 1493–1495, 1496–1500,
    1501–1504
  exercises, 1557–1558
  laminar pipe flow, 1491–1495
  transitional pipe flow, 1500–1504
  turbulent pipe flow, 1495–1500
Similitude scale ratios for minor head loss in
    pipe flow, 1510–1511
  example problem, 1512–1515
  exercises, 1559
  mathematical model of fluid flow situation,
    1510
  turbulent pipe flow with pipe component,
    1511–1515
Similitude velocity scale ratio, 1426–1427
  elastic model, 1439
  gravity model, 1439

  pressure model, 1437
  surface tension model, 1440
  viscosity model, 1438
Simple compressible system, 80, 454
Single-fluid differential manometer
  within single pipe, 152–154
  between two pipes, 154–158
Single-fluid simple manometer, 139, 140–147
Single pipes, 929; *see also* Pipe flow
  Colebrook equation, 943
  Darcy–Weisbach equation, 941–942, 942–951
  Darcy–Weisbach friction coefficient, 941–944
  differential form of Poiseuille's law, 931
  evaluation of major head loss term, 930
  example problem, 946–951, 955–956, 959–962
  exercises, 1103
  Hazen–Williams equation, 959–962
  Hazen–Williams roughness coefficient,
    957–958
  laminar pipe flow, 930–934
  Manning's equation, 952, 955, 956
  Manning's roughness coefficient, 952–954
  pipe systems, 1001
  Poiseuille's law, 931
  turbulent pipe flow, 934–937
  values for absolute pipe roughness, 941
Siphon, 439, 440
Sluice gates, 1266
  actual discharge, 1269, 1270
  discharge coefficient, 1270, 1271
  with drowned outflow, 1271
  evaluation of actual flowrate for, 1267–1272
  example problem, 1271–1272
  ideal flow rate, 1269
  in open channel flow, 1268
  velocity coefficient, 1270
Sonic velocity, 89, 92, 1415
  example problem, 89–90, 91
  exercises, 112
  wave of compression, 89
Spatially uniform flow, 303–304
Spatially varied flow, 303–304
Specific energy, 1140
  critical velocity, 1143
  curve for rectangular channel, 1140–1141
  depth–discharge curve, 1143–1145
  depths of flow, 1142, 1143
  equation for rectangular channel, 1144
  exercises, 1298–1299
  for nonrectangular channel, 1145
  nonrectangular channel flow, 1146–1147
  in open channel flow, 1139–1140
  for rectangular channel, 1140

rectangular channel flow, 1141–1143
subcritical flow, 1141
supercritical flow, 1141
total energy head, 1140
unit discharge, 1140, 1143
unit discharge equation, 1144
Specific gravity, 25–27
exercises, 97–98
Specific heat ratio, 71
Specific volume, 24
Specific weight, 27
dimensions and units for, 28
example problem, 29–30, 31–32
exercises, 97–98
for water, 30
Speed-measuring device, 444, 1311
Spillways, 1266
evaluation of flowrate for, 1283–1285
example problem, 1284–1285
geometry of, 1283
open channel flow, 1525–1529, 1529–1534
Stagnation pressure, 446
Stall conditions, 1378
Standard atmosphere, 13–15
Standard atmospheric pressure, 16, 128; *see also*
Hydrostatic pressure measurement
exercises, 96
for fluid properties, 16
specific weight of water at, 30
water density at, 24
Standard reference, 15
atmospheric pressure for fluid properties, 16
example problem, 16–18
exercises, 96
sea-level atmospheric pressure, temperature,
and gravity, 16
temperature assumed for reference fluid, 19
temperature impact on fluid properties, 16
Static pressure, 650, 965
Steady flow, 304
Streamlines, 305–306
Stream tubes, 306–307
Subcritical flow, 1147–1148
decrease in flow depth, 1150, 1177
increase in flow depth, 1151
Submerged curved surfaces; *see also* Hydrostatic
forces on submerged surfaces
example problem, 220–222
exercises, 286–291
horizontal component of force, 217–219
hydrostatic force, 216–217
resultant force, 220–222
vertical component of force, 219–220

Submerged nonrectangular planes, 196; *see also*
Hydrostatic forces on submerged
surfaces
example problem, 198–199, 199–200, 203–204,
207–208, 211–212, 215–216
exercises, 273–286
horizontal circular plane, 197–199
horizontal triangular plane, 199–200
sloping circular plane, 208–212
sloping triangular plane, 212–216
vertical, 200–201
vertical circular plane, 201–204
vertical triangular plane, 204–208
Submerged plane surfaces; *see also* Hydrostatic
forces on submerged surfaces
in enclosed and pressurized liquid, 184–196
exercises, 273–286
in gas, 173–174
in liquid open to atmosphere, 174–184
location of resultant hydrostatic force on,
169–173
magnitude of resultant hydrostatic force on,
166–169
Supercritical flow, 1132, 1147–1148
decrease in flow depth, 1152
increase in flow depth, 1151
Surface
friction, 657
phenomena, 68
roughness, 705
Surface tension, 55
capillarity, 61–63
capillary tubes, 60
contact angle, 61
droplet formation in atmosphere, 58–60
droplet formation on solid surface, 60, 61
example problem, 57–58, 59–60
exercises, 102–105
liquid membrane formation, 56
Surface tension model
gravity and viscosity model similarity
requirements, 1445–1446
similarity requirements, 1440
similitude scale ratios for, 1423
similitude velocity scale ratio, 1440
and viscosity model similarity requirements,
1444–1445

T

Tee, 974
Temperature
assumed for reference fluid, 19

Temperature (*Continued*)
  density for water at standard, 24
  exercises, 96
  sea-level temperature, 16
  specific weight for water at standard, 30
  standard reference, 15
  variation for fluid properties, 16
Thermal energy, 453, 455
Thermodynamics, first law of, 1, 79, 452–453;
          see also Energy equation; Energy
          equation for control volume; Real
          internal flow
  British thermal unit, 81
  dimensions and units for energy, 81
  dimensions and units for pump and turbine
          head terms, 470–471
  energy content, 80
  energy equation based on, 457
  energy equation, differential form of, 458
  energy equation for control volume, 458–467
  energy equation for fluid system, 457–458
  energy equation for real internal flow, 468
  energy equation in complement to
          momentum equation, 467–468
  energy transfer by heat, 455–456
  energy transfer by work, 456–457
  enthalpy, 79, 81, 84
  equations of motion, 467
  example problem, 83, 85
  exercises, 543
  flow energy, 80
  ideal specific heat, 82
  ideal specific heat ratio, 79, 83
  internal energy, 80, 82
  kinetic energy, 80
  metric unit of energy, 81
  potential energy, 80
  simple compressible system, 80
  specific heat, 84
  total energy, 80, 453–455
  types of energy, 80
Three-dimensional flows, 304
Torr, 132
Torricelli's theorem, 432–434
Total energy, 453–454; *see also* Thermodynamics,
          first law of
  dimensions and units for energy, 455
  flow energy, 454
  for fluid flow system, 454–455
  head, 374
  internal energy, 453
  macroscopic energy, 453–454
  microscopic energy, 453

per unit mass, 454
  simple compressible system, 454
  thermal energy, 453
Total pressure, 650
Transitional pipe flow, 1500–1504
Trapezoidal channel sections, 1172–1174
  hydraulic radius, 1173
  wetted perimeter, 1173
Trapezoidal pressure prism, 170; *see also*
          Pressure prism variation
Triangular channel sections, 1174
  hydraulic radius, 1174
  moment of area for, 1175
  wetted perimeter, 1174
True models, 1427–1428; *see also* Dynamic
          similitude
Turbine, 486, 1535; *see also* Real internal flow
  affinity laws for efficiency of, 1545–1547
  cavitation, 503
  efficiency, 463
  energy equation for pipe flow with, 487
  example problem, 504–507, 510–513,
          1547–1550
  exercises, 550–554, 1561
  generator system, 462, 464, 465, 491, 501
  impulse turbine, 486, 488, 489
  Moody efficiency correction equation, 1547
  reaction turbine, 486, 487, 489, 490, 496
  scaling effects for efficiency of, 1546–1550
  shaft power output from, 490
  similitude scale ratios for efficiency,
          1544–1545
Turbulence, 657
Turbulent flow, 36–37, 303, 324, 658; *see also*
          Major head loss equation for turbulent
          flow; Viscosity
  Darcy–Weisbach head loss equation for, 701
  developing vs. developed, 678
  example problem, 1460–1464, 1466–1470
  exercises, 738
  expression for shear stress for, 53–55
  fluid movement, 656–657
  integral form of energy equation, 686
  interpreting friction slope in, 687–688
  major head loss equation for, 799–806
  pressure drop equation for laminar and, 701
  pressure drop in, 685
  role of boundary roughness in laminar and
          turbulent flow, 705–706
  for round-shaped bodies, 1474
  for shape except round bodies, 1459–1470
  shear stress distribution for, 38–39
  similitude scale ratios, 1505–1510

surface roughness, 705
velocity profile, 702–705
Turbulent pipe flow, 934–937, 1495–1500; *see also*
    Single pipes
    Colebrook equation, 943
    drag coefficient, 937
    example problem, 936–937
    exercises, 1104
    flow resistance coefficients, 938–939
    flow resistance in, 705
    friction factor for, 942
    Hazen–Williams roughness coefficient,
        938, 939
    Mathcad, 939
    Moody diagram, 940
    problem types for, 935, 946, 954, 959
    resistance equations, 934, 939
    resistance equations and roughness
        coefficients, 937–938
Two-dimensional flow, 304, 759
    on airfoil, 320
    exercises, 903
    modeling of, 319–320
    nonuniform velocity profile, 319
    parabolic velocity profile, 321
    on weir of uniform cross section, 320

U

Uncalibrated mathematical model, 1403, 1405
Uniform flow, 1131; *see also* Open channel flow
    analysis of occurrence, 1159–1162
    exercises, 1296
    flow depth and major head loss in,
        1221–1228
    flow regimes for, 1132
    role and significance of, 1131–1132
    supercritical flow, 1132
Uniform pressure prism, 163
Unit discharge, 1143
    equation, 1144
Units, 3
    for British Gravitational system, 9–10
    for density, 23
    for energy, 81, 455
    example problem, 6, 7–9
    exercises, 94–95
    for International System, 10–12
    for number of moles of gas, 76
    physical quantities in fluid flow, 4, 760
    for specific weight, 28
    system of, 6, 7
Unsteady flow, 304

V

Valves; *see also* Pipes with components
    example problem, 972–974
    minor head loss in, 972
    minor losses in, 969–974
    types of, 970
Vaporization, 67
Vapor pressure, 64
    absolute pressure, 69
    atmospheric pressure, 64
    boiling point, 68
    cavitation, 68–70
    condensation, 67
    evaporation, 67
    example problem, 65–66
    exercises, 105–107
    of liquid, 67
    partial pressure, 64
    saturated, 67
    volume phenomena of evaporation,
        condensation, and boiling, 68
Variable viscosity, 44; *see also* Viscosity
Velocity
    coefficient, 1413
    measuring devices, 442, 1312
Venturi, 1088
Venturi flume, 1266
    actual discharge, 1293
    example problem, 1293–1294
    flowrate, 1291
    geometry of, 1291
    ideal flowrate, 1292
    velocity coefficient, 1293
Venturi meter, 653, 1097–1100
    actual discharge, 1098
    Bernoulli equation, 653
    discharge coefficient, 1098
    evaluation of minor head loss, 1100–1101
    example problem, 1099–1100
    ideal velocity, 653
    pressure drop, 653–654
Venturi tube, 383
Viscosity, 22, 32; *see also* Laminar flow; Newton's
    law of viscosity
    absolute, 40
    dynamic vs. kinematic, 44–45
    example problem, 33–36, 37, 45
    exercises, 98–102
    expression for shear stress for turbulent flow,
        53–55
    flow resistance, 33
    laminar pipe flow, 38

Viscosity (*Continued*)
  laminar vs. turbulent flow, 36–37
  shear stress and velocity gradient, 43
  shear stress distribution, 38–39
  variation with temperature, 33, 44
Viscosity model
  and elastic model similarity requirements,
      1440–1441
  gravity model, and surface tension model
      similarity requirements, 1445–1446
  and gravity model similarity requirements,
      1441–1444
  similarity requirements, 1438–1439
  similitude scale ratios f, 1421, 1438
  and surface tension model similarity
      requirements, 1444–1445
Volume flowrate measurement, 348; *see also*
      Continuity equation

example problem, 329–335, 339–348
Volume phenomena, 68

W

Wake region, 1332
Wall shear stress, 816–820
Water properties, 1566
Wave of compression, 89
Weber number, 761, 1415–1416; *see also*
              Dynamic similitude
  exercises, 1553
Weirs, 1266, 1267; *see also* Broad-crested weir;
              Sharp-crested weir
  open channel flow, 1525–1529, 1529–1534
Work, 785
  done by pressure forces on system, 457
  forms of, 456